HANDBOOK OF
HEAT TRANSFER

Other McGraw-Hill Handbooks of Interest

Avallone & Baumeister · MARKS' STANDARD HANDBOOK FOR MECHANICAL ENGINEERS
Bleier · FAN HANDBOOK
Brady et al. · MATERIALS HANDBOOK
Brink · HANDBOOK OF FLUID SEALING
Chironis & Sclater · MECHANISMS AND MECHANICAL DEVICES SOURCEBOOK
Czernik · GASKET HANDBOOK
Harris & Crede · SHOCK AND VIBRATION HANDBOOK
Hicks · HANDBOOK OF MECHANICAL ENGINEERING CALCULATIONS
Hicks · STANDARD HANDBOOK OF ENGINEERING CALCULATIONS
Lingaiah · MACHINE DESIGN DATA HANDBOOK
Parmley · STANDARD HANDBOOK OF FASTENING AND JOINING
Rothbart · MECHANICAL DESIGN HANDBOOK
Shigley & Mischke · STANDARD HANDBOOK OF MACHINE DESIGN
Suchy · DIE DESIGN HANDBOOK
Walsh · MCGRAW-HILL MACHINING AND METALWORKING HANDBOOK
Walsh · ELECTROMECHANICAL DESIGN HANDBOOK

HANDBOOK OF HEAT TRANSFER

Warren M. Rohsenow   Editor

Department of Mechanical Engineering
Massachusetts Institute of Technology

James P. Hartnett   Editor

Energy Resources Center
University of Illinois at Chicago

Young I. Cho   Editor

Department of Mechanical Engineering and Mechanics
Drexel University

Third Edition

MCGRAW-HILL

New York   San Francisco   Washington, D.C.   Auckland   Bogotá
Caracas   Lisbon   London   Madrid   Mexico City   Milan
Montreal   New Delhi   San Juan   Singapore
Sydney   Tokyo   Toronto

Library of Congress Cataloging-in-Publication Data

Handbook of heat transfer / editors, W.M. Rohsenow, J.P. Hartnett,
  Y.I. Cho. — 3rd ed.
     p.   cm.
  Includes bibliographical references and index.
  ISBN 0-07-053555-8 (alk. paper)
    1. Heat—Transmission—Handbooks, manuals, etc. 2. Mass transfer—
  Handbooks, manuals, etc. I. Rohsenow, W. M. (Warren M.) II. Hartnett,
J. P. (James P.) III. Cho, Y. I. (Young I.)
  QC320.4.H36   1998
  621.402'2—dc21

                                              97-51381
                                                 CIP

McGraw-Hill

*A Division of The **McGraw·Hill** Companies*

1 2 3 4 5 6 7 8 9 0 DOC/DOC 9 0 3 2 1 0 9 8

ISBN 0-07-053555-8 100136497/3

*The sponsoring editor for this book was Robert Esposito, the editing supervisor was
Stephanie S. Landis, and the production supervisor was Pamela A. Pelton. It was set in
Times Roman by North Market Street Graphics.*

Printed and bound by R. R. Donnelley & Sons Company.

This book is printed on acid-free paper.

CONTENTS

Chapter 4. Natural Convection   4.1

Chapter 5. Forced Convection, Internal Flow in Ducts   5.1

Chapter 6. Forced Convection, External Flows 6.1

Chapter 7. Radiation 7.1

Chapter 8. Microscale Transport Phenomena   **8.1**

Chapter 9. Heat Transfer in Porous Media   **9.1**

Chapter 10. Nonnewtonian Fluids

Chapter 11. Techniques to Enhance Heat Transfer

Chapter 12. Heat Pipes   **12.1**

Chapter 15. Boiling
15.1

Chapter 16. Measurement of Temperature and Heat Transfer
16.1

Chapter 17. Heat Exchangers   17.1

Chapter 18. Heat Transfer in Materials Processing

Index follows Chapter 18

CONTRIBUTORS

Bergles, Arthur E. *Department of Mechanical Engineering, Rensselaer Polytechnical Institute* (CHAP. 11, Techniques to Augment Heat Transfer), e-mail: abergles@aol.com

Bergman, Theodore L. *Department of Mechanical Engineering, University of Connecticut* (CHAP. 18, Heat Transfer in Materials Processing), e-mail: tberg@eng2.uconn.edu

Chauk, Shriniwas *Department of Chemical Engineering, Ohio State University* (CHAP. 13, Heat Transfer in Fluidized and Packed Beds)

Chen, Ping-Hai *Department of Mechanical Engineering, National Taiwan University, Taiwan, ROC* (CHAP. 16, Measurement of Temperature and Heat Transfer), e-mail: phchen@ccms.ntu.edu.tw

Chiang, Hwai Derg *Industrial Technology Research Institute, Taiwan, ROC* (CHAP. 16, Measurement of Temperature and Heat Transfer), e-mail: hdc@erl.itri.org.tw

Cho, Young I. *Department of Mechanical Engineering and Mechanics, Drexel University* (CHAP. 1, Basic Concepts of Heat Transfer; CHAP. 10, Nonnewtonian Fluids), e-mail: ycho@coe.drexel.edu

Dong, Z. F. *Department of Mechanical Engineering, Florida International University* (CHAP. 5, Forced Convection, Internal Flows), e-mail: zdong@agilis.com

Ebadian, M. A. *Hemispheric Center for Environmental Technology, Florida International University* (CHAP. 5, Forced Convection, Internal Flows), e-mail: ebadian@eng.fiu.edu

Fan, L. S. *Department of Chemical Engineering, Ohio State University* (CHAP. 13, Heat Transfer in Fluidized and Packed Beds), e-mail: fan@kcgl1.eng.ohio-state.edu

Goldstein, Richard J. *Department of Mechanical Engineering, University of Minnesota* (CHAP. 16, Measurement of Temperature and Heat Transfer), e-mail: goldstei@mailbox.mail.umn.edu

Hartnett, James P. *Energy Resources Center, University of Illinois, Chicago* (CHAP. 1, Basic Concepts of Heat Transfer; CHAP. 10, Nonnewtonian Fluids), e-mail: hartnett@uic.edu

Hewitt, Geoffrey F. *Department of Chemical Engineering and Chemical Technology, Imperial College of Science, Technology and Medicine, London, UK* (CHAP. 15, Boiling), e-mail: g.hewitt@ic.ac.uk

Hollands, K. G. T. *Department of Mechanical Engineering, University of Waterloo, Canada* (CHAP. 4, Natural Convection), e-mail: kholland@solar1.uwaterloo.ca

Howell, John R. *Department of Mechanical Engineering, University of Texas at Austin* (CHAP. 7, Radiation), e-mail: jhowell@mail.utexas.edu

Inouye, Mamoru *Ames Research Center—NASA (retired)* (CHAP. 6, Forced Convection, External Flows)

Irvine, Thomas F., Jr. *Department of Mechanical Engineering, State University of New York* (CHAP. 2, Thermophysical Properties), e-mail: tirvine@ccmail.sunysb.edu

Kaviany, Massoud *Department of Mechanics and Applied Mechanics Engineering, University of Michigan* (CHAP. 9, Heat Transfer in Porous Media), e-mail: kaviany@umich.edu

Majumdar, Arun *Department of Mechanical Engineering, University of California, Berkeley* (CHAP. 8, Microscale Heat Transfer), e-mail: majumdar@newton.berkeley.edu

Marto, Paul J. *Department of Mechanical Engineering, Naval Postgraduate School* (CHAP. 14, Condensation), e-mail: pmarto@nps.navy.mil

Mengüç, M. Pinar   *Department of Mechanical Engineering, University of Kentucky* (CHAP. 7, Radiation), e-mail: menguc@pop.engr.uky.edu

Peterson, G. P. Bud   *Department of Mechanical Engineering, Texas A&M University* (CHAP. 12, Heat Pipes), e-mail: gpp5386@teesmail.tamu.edu

Parikh, Pradip G.   *Boeing Commercial Airplane Group* (CHAP. 6, Forced Convection, External Flows), e-mail: pradip.g.parikh@boeing.com

Raithby, George D.   *Department of Mechanical Engineering, University of Waterloo, Canada* (CHAP. 4, Natural Convection), e-mail: graith@asc.on.ca

Rubesin, Morris W.   *Ames Research Center—NASA (retired)* (CHAP. 6, Forced Convection, External Flows), e-mail: mrubesin@aol.com

Sekulic, Dusan P.   *Department of Mechanical Engineering, University of Kentucky* (CHAP. 17, Heat Exchangers), e-mail: sekulicd@engr.uky.edu

Shah, Ramesh K.   *Delphi Harrison Thermal Systems, Lockport, NY* (CHAP. 17, Heat Exchangers), e-mail: rkshah@ibm.net

Viskanta, Raymond   *School of Mechanical Engineering, Purdue University* (CHAP. 18, Heat Transfer in Materials Processing), e-mail: viskanta@ecn.purdue.edu

Yovanovich, M. Michael   *Department of Mechanical and Electrical Engineering, University of Waterloo, Canada* (CHAP. 3, Conduction), e-mail: mmyov@mhtl.uwaterloo.ca

PREFACE

INTRODUCTION

Since the publication of the second edition of *Handbook of Heat Transfer,* there have been many new and exciting developments in the field, covering both fundamentals and applications. As the role of technology has grown, so too has the importance of heat transfer engineering. For example, in the industrial sector heat transfer concerns are critical to the design of practically every process. The same is true of such vitally important areas as energy production, conversion, and the expanding field of environmental controls. In the generation of electrical power, whether by nuclear fission or combustion of fossil fuels, innumerable problems remain to be solved. Similarly, further miniaturization of advanced computers is limited by the capability of removing the heat generated in the microprocessors. Heat transfer problems at the macro scale, as exemplified by global warming, also offer tremendous challenges. As technology advances, engineers are constantly confronted by the need to maximize or minimize heat transfer rates while at the same time maintaining system integrity. The upper and lower boundaries—system size, pressure, and temperature—are constantly expanding, confronting the heat transfer engineer with new design challenges.

In preparing this third edition, the goal of the editors was to provide, in a single volume, up-to-date information needed by practicing engineers to deal with heat transfer problems encountered in their daily work. This new edition of the handbook contains information essential for design engineers, consultants, research engineers, university professors, students, and technicians involved with heat transfer technology.

COVERAGE

The third edition of *Handbook of Heat Transfer* provides expanded treatment of the fundamental topics covered in earlier editions. More than half of the authors of these basic chapters on conduction, convection, radiation, condensation, and boiling are new, reflecting the fact that there are new leaders in the field. Those chapters in the second edition dealing with applications related to the so-called energy crisis (solar energy, energy storage, cooling towers, etc.) have been replaced by new chapters treating heat transfer problems encountered in materials processing, porous media, and micro scale systems. Sections on the following topics were retained and updated: thermophysical properties, heat transfer enhancement, heat exchangers, heat pipes, fluidized beds, nonnewtonian fluids, and measurement techniques.

UNITS

It is recognized at this time that the English Engineering System of units cannot be completely replaced by the International System (SI). Transition from the English system of units to SI will proceed at a rational pace to accommodate the needs of the profession, industry, and the public. The transition period will be long and complex, and duality of units probably will be demanded for at least one or two decades. Both SI and English units have been incorporated in this edition to the maximum extent possible, with the goal of making the handbook useful throughout the world. In general, numerical results, tables, figures, and equations in the handbook are given in both systems of units wherever presentation in dimensionless form is not given. In a few cases, some tables are presented in one system of units, mostly to save space, and conversion factors are printed at the end of such tables for the reader's convenience.

NOMENCLATURE

An attempt has been made by the editors to use a unified nomenclature throughout the handbook. Given the breadth of the technical coverage, some exceptions will be found. However, with few exceptions, one symbol has only one meaning within any given section. Each symbol is defined at the end of each section of the handbook. Both SI and English units are given for each symbol in the nomenclature lists.

INDEX

This edition provides a comprehensive alphabetical index designed to provide quick reference to information. Taken together with the Table of Contents, this index provides quick and easy access to any topic in the book.

ACKNOWLEDGMENTS

The editors acknowledge the outstanding performance of the contributing authors. Their cooperation on the contents and length of their manuscripts and in incorporating all of the previously mentioned specifications, coupled with the high quality of their work, has resulted in a handbook that we believe will fulfill the needs of the engineering community for many years to come. We also wish to thank the professional staff at McGraw-Hill Book Company, who were involved with the production of the handbook at various stages of the project, for their cooperation and continued support. The outstanding editorial work of Ms. Stephanie Landis of North Market Street Graphics is gratefully acknowledged.

The handbook is ultimately the responsibility of the editors. Care has been exercised to minimize errors, but it is impossible in a work of this magnitude to achieve an error-free publication. Accordingly, the editors would appreciate being informed of any errors so that these may be eliminated from subsequent printings. The editors would also appreciate suggestions from readers on possible improvements in the usefulness of the handbook so that these may be included in future editions.

W. M. Rohsenow
J. P. Hartnett
Y. I. Cho

HANDBOOK OF
HEAT TRANSFER

CHAPTER 1
BASIC CONCEPTS OF HEAT TRANSFER

Y. I. Cho
Drexel University

E. N. Ganic
University of Sarajevo

J. P. Hartnett
University of Illinois, Chicago

W. M. Rohsenow
Massachusetts Institute of Technology

HEAT TRANSFER MECHANISMS

Heat is defined as energy transferred by virtue of a temperature difference. It flows from regions of higher temperature to regions of lower temperature. It is customary to refer to different types of heat transfer mechanisms as *modes*. The basic modes of heat transfer are *conduction, radiation,* and *convection.*

Conduction

Conduction is the transfer of heat from one part of a body at a higher temperature to another part of the same body at a lower temperature, or from one body at a higher temperature to another body in physical contact with it at a lower temperature. The conduction process takes place at the molecular level and involves the transfer of energy from the more energetic molecules to those with a lower energy level. This can be easily visualized within gases, where we note that the average kinetic energy of molecules in the higher-temperature regions is greater than that of those in the lower-temperature regions. The more energetic molecules, being in constant and random motion, periodically collide with molecules of a lower energy level and exchange energy and momentum. In this manner there is a continuous transport of energy from the high-temperature regions to those of lower temperature. In liquids the molecules are more closely spaced than in gases, but the molecular energy exchange process is qualitatively similar to that in gases. In solids that are nonconductors of electricity (dielectrics), heat is conducted by lattice waves caused by atomic motion. In solids that are good conductors of electricity, this lattice vibration mechanism is only a small contribution to the energy transfer process, the principal contribution being that due to the motion of free electrons, which move in a similar way to molecules in a gas.

At the macroscopic level the heat flux (i.e., the heat transfer rate per unit area normal to the direction of heat flow) q'' is proportional to the temperature gradient:

$$q'' = -k \frac{dT}{dx} \tag{1.1}$$

where the proportionality constant k is a *transport* property known as the *thermal conductivity* and is a characteristic of the material. The minus sign is a consequence of the fact that heat is transferred in the direction of decreasing temperature. Equation 1.1 is the one-dimensional form of *Fourier's law* of heat conduction. Recognizing that the heat flux is a vector quantity, we can write a more general statement of Fourier's law (i.e., the conduction rate equation) as

$$\mathbf{q}'' = -k\, \nabla T \tag{1.2}$$

where ∇ is the three-dimensional del operator and T is the scalar temperature field. From Eq. 1.2 it is seen that the heat flux vector $\mathbf{q}''$ actually represents a current of heat (thermal energy) that flows in the direction of the steepest temperature gradient.

If we consider a one-dimensional heat flow along the x direction in the plane wall shown in Fig. 1.1a, direct application of Eq. 1.1 can be made, and then integration yields

$$q = \frac{kA}{\Delta x}(T_2 - T_1) \tag{1.3}$$

where the thermal conductivity is considered constant, Δx is the wall thickness, and T_1 and T_2 are the wall-face temperatures. Note that $q/A = q''$, where q is the heat transfer rate through an area A. Equation 1.3 can be written in the form

$$q = \frac{T_2 - T_1}{\Delta x / kA} = \frac{T_2 - T_1}{R_{th}} = \frac{\text{thermal potential difference}}{\text{thermal resistance}} \tag{1.4}$$

where $\Delta x/kA$ assumes the role of a *thermal resistance* R_{th}. The relation of Eq. 1.4 is quite like Ohm's law in electric circuit theory. The equivalent electric circuit for this case is shown in Fig. 1.1b.

The electrical analogy may be used to solve more complex problems involving both series and parallel resistances. Typical problems and their analogous electric circuits are given in many heat transfer textbooks [1–4].

In treating conduction problems it is often convenient to introduce another property that is related to the thermal conductivity, namely, the thermal diffusivity α,

$$\alpha = \frac{k}{\rho c} \tag{1.5}$$

where ρ is the density and c_p is the specific heat at constant pressure.

As mentioned above, heat transfer will occur whenever there exists a temperature difference in a medium. Similarly, whenever there exists a difference in the concentration or density of some chemical species in a mixture, mass transfer must occur. Hence, just as a temperature gradient constitutes the driving potential for heat transfer, the existence of a concentration gradient for some species in a mixture provides the driving potential for transport of that species. Therefore,

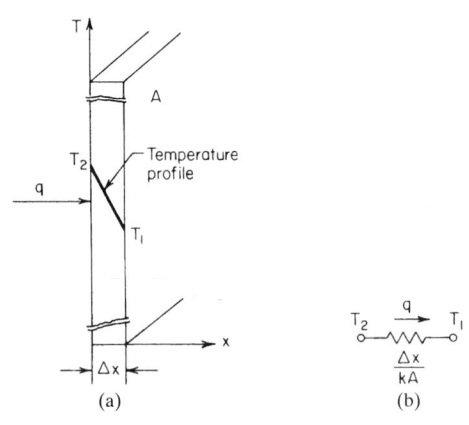

FIGURE 1.1 One-dimensional heat conduction through a plane wall (*a*) and electric analog (*b*).

the term *mass transfer* describes the relative motion of species in a mixture due to the presence of concentration gradients.

Since the same physical mechanism is associated with heat transfer by conduction (i.e., heat diffusion) and mass transfer by diffusion, the corresponding rate equations are of the same form. The rate equation for mass diffusion is known as *Fick's law,* and for a transfer of species 1 in a binary mixture it may be expressed as

$$j_1 = -\mathbf{D}\,\frac{dC_1}{dx} \tag{1.6}$$

where C_1 is a mass concentration of species 1 in units of mass per unit volume. This expression is analogous to Fourier's law (Eq. 1.1). Moreover, just as Fourier's law serves to define one important transport property, the thermal conductivity, Fick's law defines a second important transport property, namely the *binary diffusion coefficient* or mass diffusivity $\mathbf{D}$. The quantity j_1 [mass/(time × surface area)] is defined as the mass flux of species 1, i.e., the amount of species 1 that is transferred per unit time and per unit area perpendicular to the direction of transfer. In vector form Fick's law is given as

$$\mathbf{j}_1 = -\mathbf{D}\nabla C_1 \tag{1.7}$$

In general, the diffusion coefficient $\mathbf{D}$ for gases at low pressure is almost composition independent; it increases with temperature and varies inversely with pressure. Diffusion coefficients are markedly concentration dependent and generally increase with temperature.

Radiation

Radiation, or more correctly *thermal radiation,* is electromagnetic radiation emitted by a body by virtue of its temperature and at the expense of its internal energy. Thus thermal radiation is of the same nature as visible light, x rays, and radio waves, the difference between them being in their wavelengths and the source of generation. The eye is sensitive to electromagnetic radiation in the region from 0.39 to 0.78 μm; this is identified as the visible region of the spectrum. Radio waves have a wavelength of 1×10^3 to 2×10^{10} μm, and x rays have wavelengths of 1×10^{-5} to 2×10^{-2} μm, while the bulk of thermal radiation occurs in rays from approximately 0.1 to 100 μm. All heated solids and liquids, as well as some gases, emit thermal radiation. The transfer of energy by conduction requires the presence of a material medium, while radiation does not. In fact, radiation transfer occurs most efficiently in a vacuum. On the macroscopic level, the calculation of thermal radiation is based on the *Stefan-Boltzmann law,* which relates the energy flux emitted by an ideal radiator (or *blackbody*) to the fourth power of the absolute temperature:

$$e_b = \sigma T^4 \tag{1.8}$$

Here σ is the Stefan-Boltzmann constant, with a value of 5.669×10^{-8} W/(m²·K⁴), or 1.714×10^{-9} Btu/(h·ft²·°R⁴). Engineering surfaces in general do not perform as ideal radiators, and for real surfaces the above law is modified to read

$$e = \epsilon\sigma T^4 \tag{1.9}$$

The term ϵ is called the *emissivity* of the surface and has a value between 0 and 1. When two blackbodies exchange heat by radiation, the net heat exchange is then proportional to the difference in T^4. If the first body "sees" only body 2, then the net heat exchange from body 1 to body 2 is given by

$$q = \sigma A_1(T_1^4 - T_2^4) \tag{1.10}$$

When, because of the geometric arrangement, only a fraction of the energy leaving body 1 is intercepted by body 2,

$$q = \sigma A_1 F_{1-2}(T_1^4 - T_2^4) \qquad (1.11)$$

where F_{1-2} (usually called a *shape factor* or a *view factor*) is the fraction of energy leaving body 1 that is intercepted by body 2. If the bodies are not black, then the view factor F_{1-2} must be replaced by a new factor $\mathscr{F}_{1-2}$ which depends on the emissivity ϵ of the surfaces involved as well as the geometric view. Finally, if the bodies are separated by gases or liquids that impede the radiation of heat through them, a formulation of the heat exchange process becomes more involved (see Chap. 7).

Convection

Convection, sometimes identified as a separate mode of heat transfer, relates to the transfer of heat from a bounding surface to a fluid in motion, or to the heat transfer across a flow plane within the interior of the flowing fluid. If the fluid motion is induced by a pump, a blower, a fan, or some similar device, the process is called *forced convection*. If the fluid motion occurs as a result of the density difference produced by the temperature difference, the process is called *free* or *natural convection*.

Detailed inspection of the heat transfer process in these cases reveals that, although the bulk motion of the fluid gives rise to heat transfer, the basic heat transfer mechanism is *conduction*, i.e., the energy transfer is in the form of heat transfer by conduction within the moving fluid. More specifically, it is not *heat* that is being convected but *internal energy*.

However, there are convection processes for which there is, in addition, *latent heat exchange*. This latent heat exchange is generally associated with a phase change between the liquid and vapor states of the fluid. Two special cases are *boiling* and *condensation*.

Heat Transfer Coefficient.   In convective processes involving heat transfer from a boundary surface exposed to a relatively low-velocity fluid stream, it is convenient to introduce a heat transfer coefficient h, defined by Eq. 1.12, which is known as *Newton's law of cooling*:

$$q'' = h(T_w - T_f) \qquad (1.12)$$

Here T_w is the surface temperature and T_f is a characteristic fluid temperature.

For surfaces in unbounded convection, such as plates, tubes, bodies of revolution, etc., immersed in a large body of fluid, it is customary to define h in Eq. (1.12) with T_f as the temperature of the fluid far away from the surface, often identified as T_∞ (Fig. 1.2). For bounded convection, including such cases as fluids flowing in tubes or channels, across tubes in bundles, etc., T_f is usually taken as the enthalpy-mixed-mean temperature, customarily identified as T_m.

The heat transfer coefficient defined by Eq. 1.12 is sensitive to the geometry, to the physical properties of the fluid, and to the fluid velocity. However, there are some special situations in which h can depend on the temperature difference $\Delta T = T_w - T_f$. For example, if the surface is hot enough to boil a liquid surrounding it, h will typically vary as ΔT^2; or in the case of natural convection, h varies as some weak power of ΔT—typically as $\Delta T^{1/4}$ or $\Delta T^{1/3}$. It is important to note that Eq. 1.12 as a definition of h is valid in these cases too, although its usefulness may well be reduced.

As $q'' = q/A$, from Eq. 1.12 the thermal resistance in convection heat transfer is given by

$$R_{th} = \frac{1}{hA}$$

which is actually the resistance at a surface-to-fluid interface.

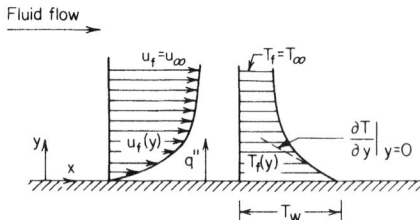

FIGURE 1.2   Velocity and temperature distributions in flow over a flat plate.

At the wall, the fluid velocity is zero, and the heat transfer takes place by conduction. Therefore, we may apply Fourier's law to the fluid at $y = 0$ (where y is the axis normal to the flow direction, Fig. 1.2):

$$q'' = -k \left. \frac{\partial T}{\partial y} \right|_{y=0} \tag{1.13}$$

where k is the thermal conductivity of fluid. By combining this equation with Newton's law of cooling (Eq. 1.12), we then obtain

$$h = \frac{q''}{T_w - T_f} = -\frac{k(\partial T/\partial y)|_{y=0}}{T_w - T_f} \tag{1.14}$$

so that we need to find the temperature gradient at the wall in order to evaluate the heat transfer coefficient.

Similar results may be obtained for *convective mass transfer*. If a fluid of species concentration $C_{1,\infty}$ flows over a surface at which the species concentration is maintained at some value $C_{1,w} \neq C_{1,\infty}$, transfer of the species by convection will occur. Species 1 is typically a vapor that is transferred into a gas stream by evaporation or sublimation at a liquid or solid surface, and we are interested in determining the rate at which this transfer occurs. As for the case of heat transfer, such a calculation may be based on the use of a convection coefficient [3, 5]. In particular we may relate the mass flux of species 1 to the product of a transfer coefficient and a concentration difference

$$j_1 = h_D(C_{1,w} - C_{1,\infty}) \tag{1.15}$$

Here h_D is the convection mass transfer coefficient and it has a dimension of L/t.

At the wall, $y = 0$, the fluid velocity is zero, and species transfer is due only to diffusion; hence

$$j_1 = -D \left. \frac{\partial C_1}{\partial y} \right|_{y=0} \tag{1.16}$$

Combining Eqs. 1.17 and 1.18, it follows that

$$h_D = -\frac{D(\partial C_1/\partial y)|_{y=0}}{C_{1,w} - C_{1,\infty}} \tag{1.17}$$

Therefore conditions that influence the surface concentration gradient $(\partial C_1/\partial y)|_{y=0}$ will also influence the convection mass transfer coefficient and the rate of species transfer across the fluid layer near the wall.

For convective processes involving high-velocity gas flows (high subsonic or supersonic flows), a more meaningful and useful definition of the heat transfer coefficient is given by

$$q'' = h(T_w - T_{aw}) \tag{1.18}$$

Here T_{aw}, commonly called the adiabatic wall temperature or the recovery temperature, is the equilibrium temperature the surface would attain in the absence of any heat transfer to or from the surface and in the absence of radiation exchange between the surroundings and the surface. In general the adiabatic wall temperature is dependent on the fluid properties and the properties of the bounding wall. Generally, the adiabatic wall temperature is reported in terms of a dimensionless recovery factor **r** defined as

$$T_{aw} = T_f + \mathbf{r} \frac{V^2}{2c_p} \tag{1.19}$$

The value of **r** for gases normally lies between 0.8 and 1.0. It can be seen that for low-velocity flows the recovery temperature is equal to the free-stream temperature T_f. In this case,

Eq. 1.15 reduces to Eq. 1.12. From this point of view, Eq. 1.18 can be taken as the generalized definition of the heat transfer coefficient.

Boundary Layer Concept.   The transfer of heat between a solid body and a liquid or gas flow is a problem whose consideration involves the science of fluid motion. On the physical motion of the fluid there is superimposed a flow of heat, and the two fields interact. In order to determine the temperature distribution and then the heat transfer coefficient (Eq. 1.14) it is necessary to combine the equations of motion with the energy conservation equation. However, a complete solution for the flow of a viscous fluid about a body poses considerable mathematical difficulty for all but the most simple flow geometries. A great practical breakthrough was made when Prandtl discovered that for most applications the influence of viscosity is confined to an extremely thin region very close to the body and that the remainder of the flow field could to a good approximation be treated as inviscid, i.e., could be calculated by the method of potential flow theory.

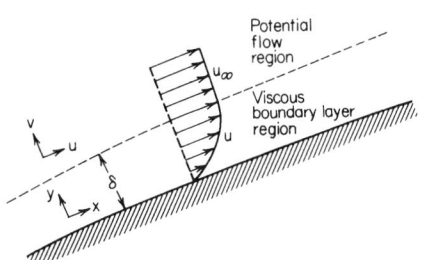

FIGURE 1.3   Boundary layer flow past an external surface.

The thin region near the body surface, which is known as the *boundary layer,* lends itself to relatively simple analysis by the very fact of its thinness relative to the dimensions of the body. A fundamental assumption of the boundary layer approximation is that the fluid immediately adjacent to the body surface is at rest relative to the body, an assumption that appears to be valid except for very low-pressure gases, when the mean free path of the gas molecules is large relative to the body [6]. Thus the *hydrodynamic* or *velocity boundary layer* δ may be defined as the region in which the fluid velocity changes from its free-stream, or potential flow, value to zero at the body surface (Fig. 1.3). In reality there is no precise "thickness" to a boundary layer defined in this manner, since the velocity asymptotically approaches the free-stream value. In practice we simply imply that the boundary layer thickness is the distance in which most of the velocity change takes place.

The viscous forces within the boundary layer region are described in terms of the shear stress τ between the fluid layers. If this stress is assumed to be proportional to the normal velocity gradient, we have the defining equation for viscosity

$$\tau = \mu \frac{du}{dy} \tag{1.20}$$

The constant of proportionality μ is called the *dynamic viscosity* (Pa·s), and Eq. 1.20 is sometimes referred to as Newton's law of shear [7] for a simple flow in which only the velocity component u exists. The ratio of the viscosity μ to the density ρ is known as the *kinematic viscosity* (m²/s) and is defined as

$$v = \frac{\mu}{\rho} \tag{1.21}$$

Flow inside a tube is a form of boundary layer problem in which, near the tube entrance, the boundary layer grows in much the same manner as over an external surface until its growth is stopped by symmetry at the centerline of the tube (Fig. 1.4). Thus the tube radius becomes the ultimate boundary layer thickness.

When there is heat transfer or mass transfer between the fluid and the surface, it is also found that in most practical applications the major temperature and concentration changes occur in a region very close to the surface. This gives rise to the concept of the *thermal boundary layer* δ_T and the *concentration boundary layer* δ_D. The influence of thermal conductivity k and mass diffusivity **D** is confined within these regions. Outside the boundary layer region the flow is essentially nonconducting and nondiffusing. The thermal (or concentration) boundary layer may be smaller than, larger than, or the same size as the velocity boundary layer. The development of the thermal boundary layer in the entrance region of a tube is shown in Fig. 1.5.

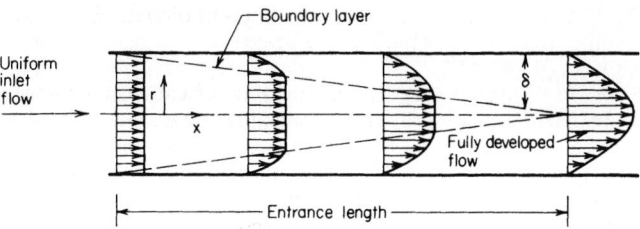

FIGURE 1.4 Velocity profile for laminar flow in a tube.

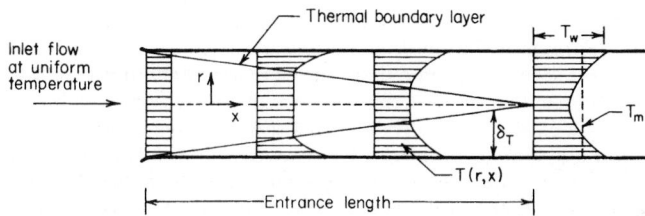

FIGURE 1.5 The development of temperature profile in the entrance region of a tube.

It is important to notice the similarity between Eqs. 1.1, 1.6, and 1.20. The heat conduction equation, Eq. 1.1, describes the transport of energy; the diffusion law, Eq. 1.6, describes the transport of mass; and the viscous shear equation, Eq. 1.20, describes the transport of momentum across fluid layers. We note also that the kinematic viscosity ν, the thermal diffusivity α, and the diffusion coefficient **D** all have the same dimensions L^2/t. As shown in Table 1.10, a dimensionless number can be formed from the ratio of any two of these quantities, which will give relative speeds at which momentum, energy, and mass diffuse through the medium.

Laminar and Turbulent Flows.   There are basically two different types of fluid motion, identified as *laminar* and *turbulent* flow. In previous sections we referred basically to laminar flow.

In the case of flow over a flat plate (Fig. 1.6), the flow near the leading edge is smooth and streamlined. Locally within the boundary layer the velocity is constant and invariant with time. The boundary layer thickness grows with increasing distance from the leading edge, and at some critical distance the inertial effects become sufficiently large compared to the viscous damping action that small disturbances in the flow begin to grow. As these disturbances

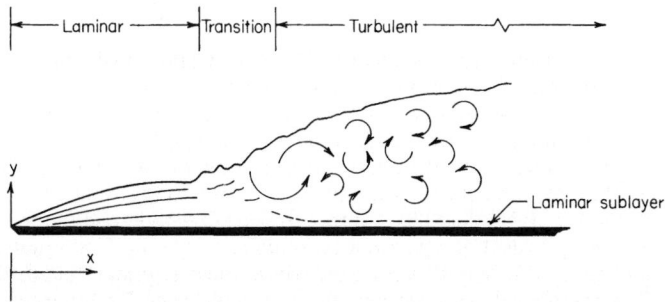

FIGURE 1.6 Laminar, transition, and turbulent boundary layer flow regimes in flow over a flat plate.

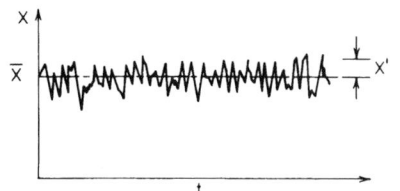

FIGURE 1.7 Property variation with time at some point in a turbulent boundary layer.

become amplified, the regularity of the viscous flow is disturbed and a transition from laminar to turbulent flow takes place. (However, there still must be a very thin laminar sublayer next to the wall, at least for a smooth plate.) These disturbances may originate from the free stream or may be induced by surface roughness.

In the turbulent flow region a very efficient mixing takes place, i.e., macroscopic chunks of fluid move across streamlines and transport energy and mass as well as momentum vigorously. The most essential feature of a turbulent flow is the fact that at a given point in it, the flow property X (e.g., velocity component, pressure, temperature, or a species concentration) is not constant with time but exhibits very irregular, high-frequency fluctuations (Fig. 1.7). At any instant, X may be represented as the sum of a *time-mean* value $\overline{X}$ and a fluctuating component X'. The average is taken over a time that is large compared with the period of typical fluctuation, and if $\overline{X}$ is independent of time, the time-mean flow is said to be *steady*.

The existence of turbulent flow can be advantageous in the sense of providing increased heat and mass transfer rates. However, the motion is extremely complicated and difficult to describe theoretically [3, 8]. In dealing with turbulent flow it is customary to speak of a *total* shear stress and *total* fluxes normal to the main flow direction (the main flow is in the x direction, and the y axis is normal to the flow direction), which are defined as

$$\tau_t = \mu \frac{\partial \overline{u}}{\partial y} - \rho \overline{u'v'} \tag{1.22}$$

$$q_t'' = -\left(k \frac{\partial \overline{T}}{\partial y} - \rho c_p \overline{v'T'}\right) \tag{1.23}$$

$$j_{1,t} = -\left(\mathbf{D} \frac{\partial \overline{C}_1}{\partial y} - \overline{v'C_1'}\right) \tag{1.24}$$

where the first term on the right side of Eqs. 1.22–1.24 is the contribution due to molecular diffusion and the second term is the contribution due to turbulent mixing. For example, $\overline{u'v'}$ is the time average of the product of u' and v'.

A simple conceptual model for turbulent flow deals with *eddies*, small portions of fluid in the boundary layer that move about for a short time before losing their identity [8]. The transport coefficient, which is defined as *eddy diffusivity for momentum transfer* ϵ_M, has the form

$$\epsilon_M \frac{\partial \overline{u}}{\partial y} = -\overline{u'v'} \tag{1.25}$$

Similarly, eddy diffusivities for heat and mass transfer, ϵ_H and ϵ_m, respectively, may be defined by the relations

$$\epsilon_H \frac{\partial \overline{T}}{\partial y} = -\overline{v'T'} \tag{1.26}$$

$$\epsilon_m \frac{\partial \overline{C}_1}{\partial y} = -\overline{v'C_1'} \tag{1.27}$$

Hence the total shear stress and total fluxes may be expressed, with the help of the relations of Eqs. 1.5 and 1.21, as

$$\tau_t = \rho(\nu + \epsilon_M) \frac{\partial \overline{u}}{\partial y} \tag{1.28}$$

$$q_t'' = -\rho c_p (\alpha + \epsilon_H) \frac{\partial \overline{T}}{\partial y} \tag{1.29}$$

$$j_{1,t} = -(\mathbf{D} + \epsilon_m) \frac{\partial \overline{C}_1}{\partial y} \tag{1.30}$$

In the region of a turbulent boundary layer far from the surface (the core region), the eddy diffusivities are much larger than the molecular diffusivities. The enhanced mixing associated with this condition has the effect of making velocity, temperature, and concentration profiles more uniform in the core. This behavior is shown in Fig. 1.8, which gives the measured velocity distributions for laminar and turbulent flow where the mass flow is the same in both cases [7]. It is evident from Fig. 1.8 that the velocity gradient at the surface, and therefore the surface shear stress, is much larger for turbulent flow than for laminar flow. Following the sameargument, the temperature or concentration gradient at the surface, and therefore the heat and mass transfer rates, are much larger for turbulent than for laminar flow. When the flow in the tube is turbulent, the mean velocity is about 83 percent of the center velocity. For laminar flow, the profile has a parabolic shape and the mean velocity is one-half the value at the center.

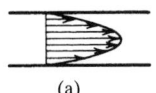 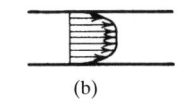

(a)   (b)

FIGURE 1.8 Velocity distribution in a tube: (a) laminar; (b) turbulent.

A fundamental problem in performing a turbulent flow analysis involves determining the eddy diffusivities as a function of the mean properties of the flow. Unlike the molecular diffusivities, which are strictly fluid properties, the eddy diffusivities depend strongly on the nature of the flow; they can vary from point to point in a boundary layer, and the specific variation can be determined only from experimental data.

For flow in circular tubes, the numerical value of the Reynolds number (defined in Table 1.10), based on mean velocity at which transition from laminar to turbulent flow occurs, was established as being approximately 2300, i.e.,

$$\mathrm{Re}_{cr} = \left(\frac{V_m D}{\nu} \right)_{cr} = 2300 \tag{1.31}$$

There exists, however, as demonstrated by numerous experiments [7], a lower value for Re_{cr} that is approximately at 2000. Below this value the flow remains laminar even in the presence of very strong disturbances. If the Reynolds number is greater than 10,000, the flow is considered to be fully turbulent. In the 2300 to 10,000 region, the flow is often described as transition flow. It is possible to shift these values by minimizing the disturbances in the inlet flow, but for general engineering application the numbers cited are representative.

For a flow over a flat plate, the transition to turbulent flow takes place at distance x, measured from the leading edge, as determined by

$$\mathrm{Re}_{cr} = \left(\frac{V_\infty x}{\nu} \right)_{cr} = 3.5 \times 10^5 \text{ to } 10^6 \tag{1.32}$$

but the values are dependent on the level of turbulence in the main stream. Here V_∞ is the free-stream velocity.

A particularly interesting phenomenon connected with transition in the boundary layer occurs with blunt bodies, e.g., spheres or circular cylinders. In the region of adverse pressure gradient (i.e., $\partial P/\partial x > 0$ in Fig. 1.9) the boundary layer separates from the surface. At this location the shear stress goes to zero, and beyond this point there is a reversal of flow in the vicinity of the wall, as shown in Fig. 1.9. In this

FIGURE 1.9 Velocity profile associated with separation on a circular cylinder in cross flow.

separation region, the analysis of the flow is very difficult and emphasis is placed on the use of experimental methods to determine heat and mass transfer.

Nonnewtonian Fluids.   In previous parts of this section we have mentioned only newtonian fluids. *Newtonian fluids* are those that have a linear relationship between the shear stress and the velocity gradient (or rate of strain), as in Eq. 1.20. The shear stress τ is equal to zero when du/dy equals zero. The viscosity, given by the ratio of shear stress to velocity gradient, is independent of the velocity gradient (or rate of strain), but may be a function of temperature and pressure.

Gases, and liquids such as water, usually exhibit newtonian behavior. However, many fluids, such as colloidal suspensions, polymeric solutions, paint, grease, blood, ketchup, slurry, etc., do not follow the linear shear stress–velocity gradient relation; these are called *nonnewtonian fluids.* Chapter 10 deals with the hydrodynamics and heat transfer of nonnewtonian fluids.

Combined Heat Transfer Mechanisms

In practice, heat transfer frequently occurs by two mechanisms in parallel. A typical example is shown in Fig. 1.10. In this case the heat conducted through the plate is removed from the plate surface by a combination of convection and radiation. An energy balance in this case gives

$$-k_sA\left.\frac{dT}{dy}\right|_w = hA(T_w - T_\infty) + \sigma A\epsilon(T_w^4 - T_a^4) \tag{1.33}$$

where T_a is the temperature of the surroundings, k_s is the thermal conductivity of the solid plate, and ϵ is the emissivity of the plate (i.e., in this special case $\mathscr{F}_{1-2} \equiv \epsilon$, as the area of the plate is much smaller than the area of the surroundings [3]). The plate and the surroundings are separated by a gas that has no effect on radiation.

There are many applications where radiation is combined with other modes of heat transfer, and the solution of such problems can often be simplified by using a thermal resistance R_{th} for radiation. The definition of R_{th} is similar to that of the thermal resistance for convection and conduction. If the heat transfer by radiation, for the example in Fig. 1.10, is written

$$q = \frac{T_w - T_a}{R_{th}} \tag{1.34}$$

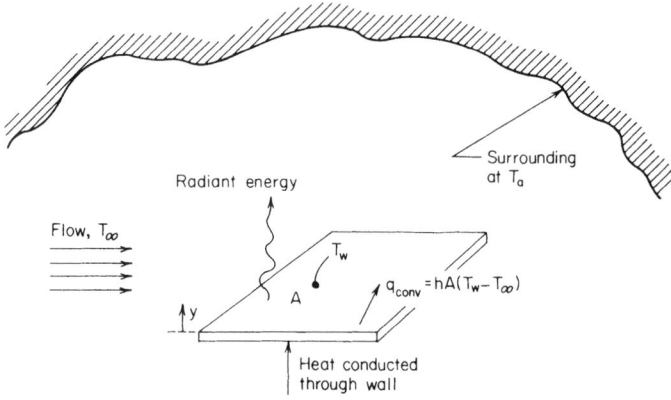

FIGURE 1.10   Combination of conduction, convection, and radiation heat transfer.

the resistance is given by

$$R_{\text{th}} = \frac{T_w - T_a}{\sigma A \epsilon (T_w^4 - T_a^4)} \tag{1.35}$$

Also, a heat transfer coefficient h_r can be defined for radiation:

$$h_r = \frac{1}{R_{\text{th}}A} = \frac{\sigma \epsilon (T_w^4 - T_a^4)}{T_w - T_a} = \sigma \epsilon (T_w + T_a)(T_w^2 + T_a^2) \tag{1.36}$$

Here we have linearized the radiation rate equation, making the heat rate proportional to a temperature difference rather than to the difference between two temperatures to the fourth power. Note that h_r depends strongly on temperature, while the temperature dependence of the convection heat transfer coefficient h is generally weak.

CONSERVATION EQUATIONS

Each time we try to solve a new problem related to momentum, heat, and mass transfer in a fluid, it is convenient to start with a set of equations based on basic laws of conservation for physical systems. These equations include:

1. The continuity equation (conservation of mass)
2. The equation of motion (conservation of momentum)
3. The energy equation (conservation of energy, or the first law of thermodynamics)
4. The conservation equation for species (conservation of species)

These equations are sometimes called the *equations of change,* inasmuch as they describe the change of velocity, temperature, and concentration with respect to time and position in the system.

The first three equations are sufficient for problems involving a pure fluid (a *pure* substance is a single substance characterized by an unvarying chemical structure). The fourth equation is added for a mixture of chemical species, i.e., when mass diffusion with or without chemical reactions is present.

- *The control volume.* When deriving the conservation equations it is necessary to select a control volume. The derivation can be performed for a volume element of any shape in a given coordinate system, although the most convenient shape is usually assumed for simplicity (e.g., a rectangular shape in a rectangular coordinate system). For illustration purposes, different coordinate systems are shown in Fig. 1.11. In selecting a control volume we

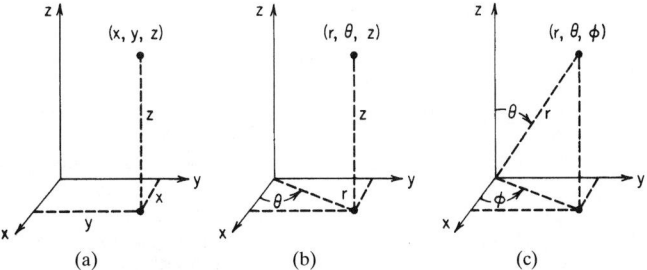

FIGURE 1.11 Coordinate systems: (*a*) rectangular, (*b*) cylindrical, (*c*) spherical.

have the option of using a volume fixed in space, in which case the fluid flows through the boundaries, or a volume containing a fixed mass of fluid and moving with the fluid. The former is known as the eulerian viewpoint and the latter is the lagrangian viewpoint. Both approaches yield equivalent results.

- *The partial time derivative ∂B/∂t.* The partial time derivative of $B(x, y, z, t)$, where B is any continuum property (e.g., density, velocity, temperature, concentration, etc.), represents the change of B with time at a fixed position in space. In other words, $\partial B/\partial t$ is the change of B with t as seen by a *stationary observer*.

- *Total time derivative dB/dt.* The total time derivative is related to the partial time derivative as follows:

$$\frac{dB}{dt} = \frac{\partial B}{\partial t} + \frac{dx}{dt}\frac{\partial B}{\partial x} + \frac{dy}{dt}\frac{\partial B}{\partial y} + \frac{dz}{dt}\frac{\partial B}{\partial z} \tag{1.37}$$

where dx/dt, dy/dt, and dz/dt are the components of the velocity of a moving observer. Therefore, dB/dt is the change of B with time as seen by the moving observer.

- *Substantial time derivative DB/Dt.* This derivative is a special kind of total time derivative where the velocity of the observer is just the same as the velocity of the stream, i.e., the observer drifts along with the current:

$$\frac{DB}{Dt} = \frac{\partial B}{\partial t} + u\frac{\partial B}{\partial x} + v\frac{\partial B}{\partial y} + w\frac{\partial B}{\partial z} \tag{1.38}$$

where u, v, and w are the components of the local fluid velocity $\mathbf{V}$. The substantial time derivative is also called the *derivative following the motion*. The sum of the last three terms on the right side of Eq. 1.38 is called the *convective contribution* because it represents the change in B due to translation.

The use of the operator D/Dt is always made when rearranging various conservation equations related to the volume element fixed in space to an element following the fluid motion. The operator D/Dt may also be expressed in vector form:

$$\frac{D}{Dt} = \frac{\partial}{\partial t} + (\mathbf{V} \cdot \nabla) \tag{1.39}$$

Mathematical operations involving ∇ are given in many textbooks. Applications of ∇ in various operations involving the conservation equations are given in Refs. 6 and 10. Table 1.1 gives the expressions for D/Dt in different coordinate systems.

TABLE 1.1   Substantial Derivative in Different Coordinate Systems

Rectangular coordinates (x, y, z):

$$\frac{D}{Dt} = \frac{\partial}{\partial t} + u\frac{\partial}{\partial x} + v\frac{\partial}{\partial y} + w\frac{\partial}{\partial z}$$

Cylindrical coordinates (r, θ, z):

$$\frac{D}{Dt} = \frac{\partial}{\partial t} + v_r\frac{\partial}{\partial r} + \frac{v_\theta}{r}\frac{\partial}{\partial \theta} + v_z\frac{\partial}{\partial z}$$

Spherical coordinates (r, θ, ϕ):

$$\frac{D}{Dt} = \frac{\partial}{\partial t} + v_r\frac{\partial}{\partial r} + \frac{v_\theta}{r}\frac{\partial}{\partial \theta} + \frac{v_\phi}{r \sin \theta}\frac{\partial}{\partial \phi}$$

The Equation of Continuity

For a volume element fixed in space,

$$\frac{\partial \rho}{\partial t} = -(\nabla \cdot \rho \mathbf{V})$$

$$\underbrace{\qquad}_{\substack{\text{net rate of} \\ \text{mass efflux per} \\ \text{unit volume}}}$$

(1.40)

The continuity equation in this form describes the rate of change of density at a fixed point in the fluid. By performing the indicated differentiation on the right side of Eq. 1.40 and collecting all derivatives of ρ on the left side, we obtain an equivalent form of the equation of continuity:

$$\frac{D\rho}{Dt} = -\rho(\nabla \cdot \mathbf{V})$$

(1.41)

The continuity equation in this form describes the rate of change of density as seen by an observer "floating along" with the fluid.

For a fluid of constant density (incompressible fluid), the equation of continuity becomes:

$$\nabla \cdot \mathbf{V} = 0$$

(1.42)

Table 1.2 gives the equation of continuity in different coordinate systems.

TABLE 1.2   Equation of Continuity in Different Coordinate Systems

Rectangular coordinates (x, y, z):

$$\frac{\partial \rho}{\partial t} + \frac{\partial}{\partial x}(\rho u) + \frac{\partial}{\partial y}(\rho v) + \frac{\partial}{\partial z}(\rho w) = 0$$

Cylindrical coordinates (r, θ, z):

$$\frac{\partial \rho}{\partial t} + \frac{1}{r}\frac{\partial}{\partial r}(\rho r v_r) + \frac{1}{r}\frac{\partial}{\partial \theta}(\rho v_\theta) + \frac{\partial}{\partial z}(\rho v_z) = 0$$

Spherical coordinates (r, θ, ϕ):

$$\frac{\partial \rho}{\partial t} + \frac{1}{r^2}\frac{\partial}{\partial r}(\rho r^2 v_r) + \frac{1}{r \sin \theta}\frac{\partial}{\partial \theta}(\rho v_\theta \sin \theta) + \frac{1}{r \sin \theta}\frac{\partial}{\partial \phi}(\rho v_\phi) = 0$$

Incompressible flow

Rectangular coordinates (x, y, z):

$$\frac{\partial u}{\partial x} + \frac{\partial v}{\partial y} + \frac{\partial w}{\partial z} = 0$$

Cylindrical coordinates (r, θ, z):

$$\frac{1}{r}\frac{\partial}{\partial r}(rv_r) + \frac{1}{r}\frac{\partial v_\theta}{\partial \theta} + \frac{\partial v_z}{\partial z} = 0$$

Spherical coordinates (r, θ, ϕ):

$$\frac{1}{r^2}\frac{\partial}{\partial r}(r^2 v_r) + \frac{1}{r \sin \theta}\frac{\partial}{\partial \theta}(v_\theta \sin \theta) + \frac{1}{r \sin \theta}\frac{\partial v_\phi}{\partial \phi} = 0$$

The Equation of Motion (Momentum Equation)

The momentum equation for a stationary volume element (i.e., a balance over a volume element fixed in space) with gravity as the only body force is given by

$$\frac{\partial \rho \mathbf{V}}{\partial t} = -(\nabla \cdot \rho \mathbf{V})\mathbf{V} - \nabla P + \nabla \cdot \tau + \rho \mathbf{g} \qquad (1.43)$$

rate of increase of momentum per unit volume	rate of momentum gain by convection per unit volume	pressure force on element per unit volume	rate of momentum gain by viscous transfer per unit volume	gravitational force on element per unit volume

Equation 1.43 may be rearranged, with the help of the equation of continuity, to give

$$\rho \frac{D\mathbf{V}}{Dt} = -\nabla P + \nabla \cdot \tau + \rho \mathbf{g} \qquad (1.44)$$

The last equation is a statement of Newton's second law of motion in the form *mass × acceleration = sum of forces.*

These two forms of the equation of motion (Eqs. 1.43 and 1.44), correspond to the two forms of the equation of continuity (Eqs. 1.40 and 1.41). As indicated, the only body force included in Eqs. 1.43 and 1.44 is gravity. In general, electromagnetic forces may also act on a fluid.

The scalar components of Eq. 1.44 are listed in Table 1.3 and the components of the stress tensor τ are given in Table 1.4.

For the flow of a newtonian fluid with varying density but constant viscosity μ, Eq. 1.44 becomes

$$\rho \frac{D\mathbf{V}}{Dt} = -\nabla P + \frac{1}{3}\mu\nabla(\nabla \cdot \mathbf{V}) + \mu\nabla^2\mathbf{V} + \rho \mathbf{g} \qquad (1.45)$$

If ρ and μ are constant, Eq. 1.44 may be simplified by means of the equation of continuity $(\nabla \cdot \mathbf{V} = 0)$ for a newtonian fluid to give

$$\rho \frac{D\mathbf{V}}{Dt} = -\nabla P + \mu\nabla^2\mathbf{V} + \rho \mathbf{g} \qquad (1.46)$$

This is the famous Navier-Stokes equation in vector form. The scalar components of Eq. 1.46 are given in Table 1.5.

For $\nabla \cdot \tau = 0$, Eq. 1.44 reduces to Euler's equation:

$$\rho \frac{D\mathbf{V}}{Dt} = -\nabla P + \rho \mathbf{g} \qquad (1.47)$$

which is applicable for describing flow systems in which viscous effects are relatively unimportant.

As mentioned before, there is a subset of flow problems, called *natural convection,* where the flow pattern is due to buoyant forces caused by temperature differences. Such buoyant forces are proportional to the coefficient of thermal expansion β, defined as:

$$\beta = -\frac{1}{\rho}\left(\frac{\partial \rho}{\partial T}\right)_P \qquad (1.48)$$

where T is absolute temperature. Using an approximation that applies to low fluid velocities and small temperature variations, it can be shown [9–11] that

$$\nabla P - \rho \mathbf{g} = \rho \beta \mathbf{g}(T - T_\infty) \qquad (1.49)$$

Then Eq. 1.44 becomes

$$\rho \frac{D\mathbf{V}}{Dt} = \nabla \cdot \tau - \rho \beta \mathbf{g}(T - T_\infty) \tag{1.50}$$

$$\underset{\substack{\text{buoyant force} \\ \text{on element per} \\ \text{unit volume}}}{}$$

The above equation of motion is used for setting up problems in natural convection when the ambient temperature T_∞ may be defined.

The Energy Equation

For a stationary volume element through which a pure fluid is flowing, the energy equation reads

$$\frac{\partial}{\partial t} \rho(\mathbf{u} + \tfrac{1}{2}V^2) = -\nabla \cdot \rho \mathbf{V}(\mathbf{u} + \tfrac{1}{2}V^2) - \nabla \cdot \mathbf{q}'' + \rho(\mathbf{V} \cdot \mathbf{g})$$

| rate of gain of energy per unit volume | rate of energy input per unit volume by convection | rate of energy input per unit volume by conduction | rate of work done on fluid per unit volume by gravitational forces |

$$- \nabla \cdot P\mathbf{V} + \nabla \cdot (\tau \cdot \mathbf{V}) + q''' \tag{1.51}$$

| rate of work done on fluid per unit volume by pressure forces | rate of work done on fluid per unit volume by viscous forces | rate of heat generation per unit volume ("source term") |

where **u** is the internal energy. The left side of this equation, which represents the rate of accumulation of internal and kinetic energy, does not include the potential energy of the fluid, since this form of energy is included in the work term on the right side. Equation 1.51 may be rearranged, with the aid of the equations of continuity and motion, to give [10, 19]

$$\rho \frac{D\mathbf{u}}{Dt} = -\nabla \cdot \mathbf{q}'' - P(\nabla \cdot \mathbf{V}) + \nabla \mathbf{V} : \tau + q''' \tag{1.52}$$

A summary of $\nabla \mathbf{V} : \tau$ in different coordinate systems is given in Table 1.6. For a newtonian fluid,

$$\nabla \mathbf{V} : \tau = \mu \Phi \tag{1.53}$$

and values of dissipation function Φ in different coordinate systems are given in Table 1.7. Components of the heat flux vector $\mathbf{q}'' = -k\nabla T$ are given in Table 1.8 for different coordinate systems.

Often it is more convenient to work with enthalpy rather than internal energy. Using the definition of enthalpy, $i = \mathbf{u} + P/\rho$, and the mass conservation equation, Eq. 1.41, Eq. 1.52 can be rearranged to give

$$\rho \frac{Di}{Dt} = \nabla \cdot k\nabla T + \frac{DP}{Dt} + \mu \Phi + q''' \tag{1.54}$$

TABLE 1.3   Equation of Motion in Terms of Viscous Stresses (Eq. 1.44)*

Rectangular coordinates (x, y, z)

x direction

$$\rho\left(\frac{\partial u}{\partial t} + u\frac{\partial u}{\partial x} + v\frac{\partial u}{\partial y} + w\frac{\partial u}{\partial z}\right) = -\frac{\partial P}{\partial x} + \left(\frac{\partial \tau_{xx}}{\partial x} + \frac{\partial \tau_{yx}}{\partial y} + \frac{\partial \tau_{zx}}{\partial z}\right) + \rho g_x$$

y direction

$$\rho\left(\frac{\partial v}{\partial t} + u\frac{\partial v}{\partial x} + v\frac{\partial v}{\partial y} + w\frac{\partial v}{\partial z}\right) = -\frac{\partial P}{\partial y} + \left(\frac{\partial \tau_{xy}}{\partial x} + \frac{\partial \tau_{yy}}{\partial y} + \frac{\partial \tau_{zy}}{\partial z}\right) + \rho g_y$$

z direction

$$\rho\left(\frac{\partial w}{\partial t} + u\frac{\partial w}{\partial x} + v\frac{\partial w}{\partial y} + w\frac{\partial w}{\partial z}\right) = -\frac{\partial P}{\partial z} + \left(\frac{\partial \tau_{xz}}{\partial x} + \frac{\partial \tau_{yz}}{\partial y} + \frac{\partial \tau_{zz}}{\partial z}\right) + \rho g_z$$

Cylindrical coordinates (r, θ, z)

r direction

$$\rho\left(\frac{\partial v_r}{\partial t} + v_r\frac{\partial v_r}{\partial r} + \frac{v_\theta}{r}\frac{\partial v_r}{\partial \theta} - \frac{v_\theta^2}{r} + v_z\frac{\partial v_r}{\partial z}\right) = -\frac{\partial P}{\partial r} + \left[\frac{1}{r}\frac{\partial}{\partial r}(r\tau_{rr}) + \frac{1}{r}\frac{\partial \tau_{\theta r}}{\partial \theta} - \frac{\tau_{\theta\theta}}{r} + \frac{\partial \tau_{zr}}{\partial z}\right] + \rho g_r$$

θ direction

$$\rho\left(\frac{\partial v_\theta}{\partial t} + v_r\frac{\partial v_\theta}{\partial r} + \frac{v_\theta}{r}\frac{\partial v_\theta}{\partial \theta} + \frac{v_r v_\theta}{r} + v_z\frac{\partial v_\theta}{\partial z}\right) = -\frac{1}{r}\frac{\partial P}{\partial \theta} + \left[\frac{1}{r^2}\frac{\partial}{\partial r}(r^2\tau_{r\theta}) + \frac{1}{r}\frac{\partial \tau_{\theta\theta}}{\partial \theta} + \frac{\partial \tau_{z\theta}}{\partial z}\right] + \rho g_\theta$$

z direction

$$\rho\left(\frac{\partial v_z}{\partial t} + v_r\frac{\partial v_z}{\partial r} + \frac{v_\theta}{r}\frac{\partial v_z}{\partial \theta} + v_z\frac{\partial v_z}{\partial z}\right) = -\frac{\partial P}{\partial z} + \left[\frac{1}{r}\frac{\partial}{\partial r}(r\tau_{rz}) + \frac{1}{r}\frac{\partial \tau_{\theta z}}{\partial \theta} + \frac{\partial \tau_{zz}}{\partial z}\right] + \rho g_z$$

Spherical coordinates (r, θ, ϕ)

r direction

$$\rho\left(\frac{\partial v_r}{\partial t} + v_r\frac{\partial v_r}{\partial r} + \frac{v_\theta}{r}\frac{\partial v_r}{\partial \theta} + \frac{v_\phi}{r\sin\theta}\frac{\partial v_r}{\partial \phi} - \frac{v_\theta^2 + v_\phi^2}{r}\right) = -\frac{\partial P}{\partial r}$$

$$+ \left[\frac{1}{r^2}\frac{\partial}{\partial r}(r^2\tau_{rr}) + \frac{1}{r\sin\theta}\frac{\partial}{\partial \theta}(\tau_{\theta r}\sin\theta) + \frac{1}{r\sin\theta}\frac{\partial \tau_{\phi r}}{\partial \phi} - \frac{\tau_{\theta\theta} + \tau_{\phi\phi}}{r}\right] + \rho g_r$$

θ direction

$$\rho\left(\frac{\partial v_\theta}{\partial t} + v_r\frac{\partial v_\theta}{\partial r} + \frac{v_\theta}{r}\frac{\partial v_\theta}{\partial \theta} + \frac{v_\phi}{r\sin\theta}\frac{\partial v_\theta}{\partial \phi} + \frac{v_r v_\theta}{r} - \frac{v_\phi^2\cot\theta}{r}\right) = -\frac{1}{r}\frac{\partial P}{\partial \theta}$$

$$+ \left[\frac{1}{r^2}\frac{\partial}{\partial r}(r^2\tau_{r\theta}) + \frac{1}{r\sin\theta}\frac{\partial}{\partial \theta}(\tau_{\theta\theta}\sin\theta) + \frac{1}{r\sin\theta}\frac{\partial \tau_{\phi\theta}}{\partial \phi} + \frac{\tau_{r\theta}}{r} - \frac{\tau_{\phi\phi}\cot\theta}{r}\right] + \rho g_\theta$$

ϕ direction

$$\rho\left(\frac{\partial v_\phi}{\partial t} + v_r\frac{\partial v_\phi}{\partial r} + \frac{v_\theta}{r}\frac{\partial v_\phi}{\partial \theta} + \frac{v_\phi}{r\sin\theta}\frac{\partial v_\phi}{\partial \phi} + \frac{v_\phi v_r}{r} + \frac{v_\theta v_\phi\cot\theta}{r}\right)$$

$$= -\frac{1}{r\sin\theta}\frac{\partial P}{\partial \phi} + \left[\frac{1}{r^2}\frac{\partial}{\partial r}(r^2\tau_{r\phi}) + \frac{1}{r}\frac{\partial \tau_{\theta\phi}}{\partial \theta} + \frac{1}{r\sin\theta}\frac{\partial \tau_{\phi\phi}}{\partial \phi} + \frac{\tau_{r\phi}}{r} + \frac{2\tau_{\theta\phi}\cot\theta}{r}\right] + \rho g_\phi$$

* Components of the stress tensor (τ) for newtonian fluids are given in Table 1.4. This equation may also be used for describing nonnewtonian flow. However, we need relations between the components of τ and the various velocity gradients; in other words, we have to replace the expressions given in Table 1.4 with other relations appropriate for the nonnewtonian fluid of interest. The expressions for τ for some nonnewtonian fluid models are given in Ref. 10. See also Chap. 10.

TABLE 1.4   Components of the Stress Tensor τ for Newtonian Fluids*

Rectangular coordinates (x, y, z)

$$\tau_{xx} = \mu\left[2\frac{\partial u}{\partial x} - \frac{2}{3}(\nabla \cdot \mathbf{V})\right] \qquad\qquad \tau_{xy} = \tau_{yx} = \mu\left[\frac{\partial u}{\partial y} + \frac{\partial v}{\partial x}\right]$$

$$\tau_{yy} = \mu\left[2\frac{\partial v}{\partial y} - \frac{2}{3}(\nabla \cdot \mathbf{V})\right] \qquad\qquad \tau_{yz} = \tau_{zy} = \mu\left[\frac{\partial v}{\partial z} + \frac{\partial w}{\partial y}\right]$$

$$\tau_{zz} = \mu\left[2\frac{\partial w}{\partial z} - \frac{2}{3}(\nabla \cdot \mathbf{V})\right] \qquad\qquad \tau_{zx} = \tau_{xz} = \mu\left[\frac{\partial w}{\partial x} + \frac{\partial u}{\partial z}\right]$$

$$(\nabla \cdot \mathbf{V}) = \frac{\partial u}{\partial x} + \frac{\partial v}{\partial y} + \frac{\partial w}{\partial z}$$

Cylindrical coordinates (r, θ, z)

$$\tau_{rr} = \mu\left[2\frac{\partial v_r}{\partial r} - \frac{2}{3}(\nabla \cdot \mathbf{V})\right]$$

$$\tau_{\theta\theta} = \mu\left[2\left(\frac{1}{r}\frac{\partial v_\theta}{\partial \theta} + \frac{v_r}{r}\right) - \frac{2}{3}(\nabla \cdot \mathbf{V})\right]$$

$$\tau_{zz} = \mu\left[2\frac{\partial v_z}{\partial z} - \frac{2}{3}(\nabla \cdot \mathbf{V})\right]$$

$$\tau_{r\theta} = \tau_{\theta r} = \mu\left[r\frac{\partial}{\partial r}\left(\frac{v_\theta}{r}\right) + \frac{1}{r}\frac{\partial v_r}{\partial \theta}\right]$$

$$\tau_{\theta z} = \tau_{z\theta} = \mu\left[\frac{\partial v_\theta}{\partial z} + \frac{1}{r}\frac{\partial v_z}{\partial \theta}\right]$$

$$\tau_{zr} = \tau_{rz} = \mu\left[\frac{\partial v_z}{\partial r} + \frac{\partial v_r}{\partial z}\right]$$

$$(\nabla \cdot \mathbf{V}) = \frac{1}{r}\frac{\partial}{\partial r}(rv_r) + \frac{1}{r}\frac{\partial v_\theta}{\partial \theta} + \frac{\partial v_z}{\partial z}$$

Spherical coordinates (r, θ, ϕ)

$$\tau_{rr} = \mu\left[2\frac{\partial v_r}{\partial r} - \frac{2}{3}(\nabla \cdot \mathbf{V})\right]$$

$$\tau_{\theta\theta} = \mu\left[2\left(\frac{1}{r}\frac{\partial v_\theta}{\partial \theta} + \frac{v_r}{r}\right) - \frac{2}{3}(\nabla \cdot \mathbf{V})\right]$$

$$\tau_{\phi\phi} = \mu\left[2\left(\frac{1}{r \sin\theta}\frac{\partial v_\phi}{\partial \phi} + \frac{v_r}{r} + \frac{v_\theta \cot\theta}{r}\right) - \frac{2}{3}(\nabla \cdot \mathbf{V})\right]$$

$$\tau_{r\theta} = \tau_{\theta r} = \mu\left[r\frac{\partial}{\partial r}\left(\frac{v_\theta}{r}\right) + \frac{1}{r}\frac{\partial v_r}{\partial \theta}\right]$$

$$\tau_{\theta\phi} = \tau_{\phi\theta} = \mu\left[\frac{\sin\theta}{r}\frac{\partial}{\partial \theta}\left(\frac{v_\phi}{\sin\theta}\right) + \frac{1}{r \sin\theta}\frac{\partial v_\theta}{\partial \phi}\right]$$

$$\tau_{\phi r} = \tau_{r\phi} = \mu\left[\frac{1}{r \sin\theta}\frac{\partial v_r}{\partial \phi} + r\frac{\partial}{\partial r}\left(\frac{v_\phi}{r}\right)\right]$$

$$(\nabla \cdot \mathbf{V}) = \frac{1}{r^2}\frac{\partial}{\partial r}(r^2 v_r) + \frac{1}{r \sin\theta}\frac{\partial}{\partial \theta}(v_\theta \sin\theta) + \frac{1}{r \sin\theta}\frac{\partial v_\phi}{\partial \phi}$$

---

* It should be noted that the sign convention adopted here for components of the stress tensor is consistent with that found in many fluid mechanics and heat transfer books; however, it is opposite to that found in some books on transport phenomena, e.g., Refs. 10, 11, and 14.

TABLE 1.5   Equation of Motion in Terms of Velocity Gradients for a Newtonian Fluid with Constant ρ and μ (Eq. 1.46)

Rectangular coordinates (x, y, z)

x direction

$$\rho\left(\frac{\partial u}{\partial t} + u\frac{\partial u}{\partial x} + v\frac{\partial u}{\partial y} + w\frac{\partial u}{\partial z}\right) = -\frac{\partial P}{\partial x} + \mu\left(\frac{\partial^2 u}{\partial x^2} + \frac{\partial^2 u}{\partial y^2} + \frac{\partial^2 u}{\partial z^2}\right) + \rho g_x$$

y direction

$$\rho\left(\frac{\partial v}{\partial t} + u\frac{\partial v}{\partial x} + v\frac{\partial v}{\partial y} + w\frac{\partial v}{\partial z}\right) = -\frac{\partial P}{\partial y} + \mu\left(\frac{\partial^2 v}{\partial x^2} + \frac{\partial^2 v}{\partial y^2} + \frac{\partial^2 v}{\partial z^2}\right) + \rho g_y$$

z direction

$$\rho\left(\frac{\partial w}{\partial t} + u\frac{\partial w}{\partial x} + v\frac{\partial w}{\partial y} + w\frac{\partial w}{\partial z}\right) = -\frac{\partial P}{\partial z} + \mu\left(\frac{\partial^2 w}{\partial x^2} + \frac{\partial^2 w}{\partial y^2} + \frac{\partial^2 w}{\partial z^2}\right) + \rho g_z$$

Cylindrical coordinates (r, θ, z)

r direction

$$\rho\left(\frac{\partial v_r}{\partial t} + v_r\frac{\partial v_r}{\partial r} + \frac{v_\theta}{r}\frac{\partial v_r}{\partial \theta} - \frac{v_\theta^2}{r} + v_z\frac{\partial v_r}{\partial z}\right) = -\frac{\partial P}{\partial r} + \mu\left[\frac{\partial}{\partial r}\left(\frac{1}{r}\frac{\partial}{\partial r}(rv_r)\right) + \frac{1}{r^2}\frac{\partial^2 v_r}{\partial \theta^2} - \frac{2}{r^2}\frac{\partial v_\theta}{\partial \theta} + \frac{\partial^2 v_r}{\partial z^2}\right] + \rho g_r$$

θ direction

$$\rho\left(\frac{\partial v_\theta}{\partial t} + v_r\frac{\partial v_\theta}{\partial r} + \frac{v_\theta}{r}\frac{\partial v_\theta}{\partial \theta} + \frac{v_r v_\theta}{r} + v_z\frac{\partial v_\theta}{\partial z}\right) = -\frac{1}{r}\frac{\partial P}{\partial \theta} + \mu\left[\frac{\partial}{\partial r}\left(\frac{1}{r}\frac{\partial}{\partial r}(rv_\theta)\right) + \frac{1}{r^2}\frac{\partial^2 v_\theta}{\partial \theta^2} + \frac{2}{r^2}\frac{\partial v_r}{\partial \theta} + \frac{\partial^2 v_\theta}{\partial z^2}\right] + \rho g_\theta$$

z direction

$$\rho\left(\frac{\partial v_z}{\partial t} + v_r\frac{\partial v_z}{\partial r} + \frac{v_\theta}{r}\frac{\partial v_z}{\partial \theta} + v_z\frac{\partial v_z}{\partial z}\right) = -\frac{\partial P}{\partial z} + \mu\left[\frac{1}{r}\frac{\partial}{\partial r}\left(r\frac{\partial v_z}{\partial r}\right) + \frac{1}{r^2}\frac{\partial^2 v_z}{\partial \theta^2} + \frac{\partial^2 v_z}{\partial z^2}\right] + \rho g_z$$

Spherical coordinates (r, θ, ϕ)*

r direction

$$\rho\left(\frac{\partial v_r}{\partial t} + v_r\frac{\partial v_r}{\partial r} + \frac{v_\theta}{r}\frac{\partial v_r}{\partial \theta} + \frac{v_\phi}{r\sin\theta}\frac{\partial v_r}{\partial \phi} - \frac{v_\theta^2 + v_\phi^2}{r}\right) = -\frac{\partial P}{\partial r} + \mu\left(\nabla^2 v_r - \frac{2v_r}{r^2} - \frac{2}{r^2}\frac{\partial v_\theta}{\partial \theta} - \frac{2v_\theta\cot\theta}{r^2} - \frac{2}{r^2\sin\theta}\frac{\partial v_\phi}{\partial \phi}\right) + \rho g_r$$

θ direction

$$\rho\left(\frac{\partial v_\theta}{\partial t} + v_r\frac{\partial v_\theta}{\partial r} + \frac{v_\theta}{r}\frac{\partial v_\theta}{\partial \theta} + \frac{v_\phi}{r\sin\theta}\frac{\partial v_\theta}{\partial \phi} + \frac{v_r v_\theta}{r} - \frac{v_\phi^2\cot\theta}{r}\right) = -\frac{1}{r}\frac{\partial P}{\partial \theta} + \mu\left(\nabla^2 v_\theta + \frac{2}{r^2}\frac{\partial v_r}{\partial \theta} - \frac{v_\theta}{r^2\sin^2\theta} - \frac{2\cos\theta}{r^2\sin^2\theta}\frac{\partial v_\phi}{\partial \phi}\right) + \rho g_\theta$$

ϕ direction

$$\rho\left(\frac{\partial v_\phi}{\partial t} + v_r\frac{\partial v_\phi}{\partial r} + \frac{v_\theta}{r}\frac{\partial v_\phi}{\partial \theta} + \frac{v_\phi}{r\sin\theta}\frac{\partial v_\phi}{\partial \phi} + \frac{v_\phi v_r}{r} + \frac{v_\theta v_\phi}{r}\cot\theta\right)$$

$$= -\frac{1}{r\sin\theta}\frac{\partial P}{\partial \phi} + \mu\left(\nabla^2 v_\phi - \frac{v_\phi}{r^2\sin^2\theta} + \frac{2}{r^2\sin\theta}\frac{\partial v_r}{\partial \phi} + \frac{2\cos\theta}{r^2\sin^2\theta}\frac{\partial v_\theta}{\partial \phi}\right) + \rho g_\phi$$

---

* For spherical coordinates the laplacian is

$$\nabla^2 = \frac{1}{r^2}\frac{\partial}{\partial r}\left(r^2\frac{\partial}{\partial r}\right) + \frac{1}{r^2\sin\theta}\frac{\partial}{\partial \theta}\left(\sin\theta\frac{\partial}{\partial \theta}\right) + \frac{1}{r^2\sin^2\theta}\left(\frac{\partial^2}{\partial \phi^2}\right)$$

TABLE 1.6 Summary of Dissipation Term $\nabla\mathbf{V}{:}\boldsymbol{\tau}$ in Different Coordinate Systems

Rectangular coordinates (x, y, z):

$$\nabla\mathbf{V}{:}\boldsymbol{\tau} = \tau_{xx}\left(\frac{\partial u}{\partial x}\right) + \tau_{yy}\left(\frac{\partial v}{\partial y}\right) + \tau_{zz}\left(\frac{\partial w}{\partial z}\right) + \tau_{xy}\left(\frac{\partial u}{\partial y} + \frac{\partial v}{\partial x}\right) + \tau_{yz}\left(\frac{\partial v}{\partial z} + \frac{\partial w}{\partial y}\right) + \tau_{zx}\left(\frac{\partial w}{\partial x} + \frac{\partial u}{\partial z}\right)$$

Cylindrical coordinates (r, θ, z):

$$\nabla\mathbf{V}{:}\boldsymbol{\tau} = \tau_{rr}\left(\frac{\partial v_r}{\partial r}\right) + \tau_{\theta\theta}\left(\frac{1}{r}\frac{\partial v_\theta}{\partial\theta} + \frac{v_r}{r}\right) + \tau_{zz}\left(\frac{\partial v_z}{\partial z}\right) + \tau_{r\theta}\left[r\frac{\partial}{\partial r}\left(\frac{v_\theta}{r}\right) + \frac{1}{r}\frac{\partial v_r}{\partial\theta}\right] + \tau_{\theta z}\left(\frac{1}{r}\frac{\partial v_z}{\partial\theta} + \frac{\partial v_\theta}{\partial z}\right) + \tau_{rz}\left(\frac{\partial v_z}{\partial r} + \frac{\partial v_r}{\partial z}\right)$$

Spherical coordinates (r, θ, ϕ):

$$\nabla\mathbf{V}{:}\boldsymbol{\tau} = \tau_{rr}\left(\frac{\partial v_r}{\partial r}\right) + \tau_{\theta\theta}\left(\frac{1}{r}\frac{\partial v_\theta}{\partial\theta} + \frac{v_r}{r}\right) + \tau_{\phi\phi}\left(\frac{1}{r\sin\theta}\frac{\partial v_\phi}{\partial\phi} + \frac{v_r}{r} + \frac{v_\theta\cot\theta}{r}\right) + \tau_{r\theta}\left(\frac{\partial v_\theta}{\partial r} + \frac{1}{r}\frac{\partial v_r}{\partial\theta} - \frac{v_\theta}{r}\right)$$

$$+ \tau_{r\phi}\left(\frac{\partial v_\phi}{\partial r} + \frac{1}{r\sin\theta}\frac{\partial v_r}{\partial\phi} - \frac{v_\phi}{r}\right) + \tau_{\theta\phi}\left(\frac{1}{r}\frac{\partial v_\phi}{\partial\theta} + \frac{1}{r\sin\theta}\frac{\partial v_\theta}{\partial\phi} - \frac{v_\phi\cot\theta}{r}\right)$$

TABLE 1.7 The Viscous Dissipation Function Φ

Rectangular coordinates (x, y, z):

$$\Phi = 2\left[\left(\frac{\partial u}{\partial x}\right)^2 + \left(\frac{\partial v}{\partial y}\right)^2 + \left(\frac{\partial w}{\partial z}\right)^2\right] + \left(\frac{\partial v}{\partial x} + \frac{\partial u}{\partial y}\right)^2 + \left(\frac{\partial w}{\partial y} + \frac{\partial v}{\partial z}\right)^2 + \left(\frac{\partial u}{\partial z} + \frac{\partial w}{\partial x}\right)^2 - \frac{2}{3}\left(\frac{\partial u}{\partial x} + \frac{\partial v}{\partial y} + \frac{\partial w}{\partial z}\right)^2$$

Cylindrical coordinates (r, θ, z):

$$\Phi = 2\left[\left(\frac{\partial v_r}{\partial r}\right)^2 + \left(\frac{1}{r}\frac{\partial v_\theta}{\partial\theta} + \frac{v_r}{r}\right)^2 + \left(\frac{\partial v_z}{\partial z}\right)^2\right] + \left[r\frac{\partial}{\partial r}\left(\frac{v_\theta}{r}\right) + \frac{1}{r}\frac{\partial v_r}{\partial\theta}\right]^2$$

$$+ \left[\frac{1}{r}\frac{\partial v_z}{\partial\theta} + \frac{\partial v_\theta}{\partial z}\right]^2 + \left(\frac{\partial v_r}{\partial z} + \frac{\partial v_z}{\partial r}\right)^2 - \frac{2}{3}\left[\frac{1}{r}\frac{\partial}{\partial r}(rv_r) + \frac{1}{r}\frac{\partial v_\theta}{\partial\theta} + \frac{\partial v_z}{\partial z}\right]^2$$

Spherical coordinates (r, θ, ϕ):

$$\Phi = 2\left[\left(\frac{\partial v_r}{\partial r}\right)^2 + \left(\frac{1}{r}\frac{\partial v_\theta}{\partial\theta} + \frac{v_r}{r}\right)^2 + \left(\frac{1}{r\sin\theta}\frac{\partial v_\phi}{\partial\phi} + \frac{v_r}{r} + \frac{v_\theta\cot\theta}{r}\right)^2\right]$$

$$+ \left[r\frac{\partial}{\partial r}\left(\frac{v_\theta}{r}\right) + \frac{1}{r}\frac{\partial v_r}{\partial\theta}\right]^2 + \left[\frac{\sin\theta}{r}\frac{\partial}{\partial\theta}\left(\frac{v_\phi}{\sin\theta}\right) + \frac{1}{r\sin\theta}\frac{\partial v_\theta}{\partial\phi}\right]^2 + \left[\frac{1}{r\sin\theta}\frac{\partial v_r}{\partial\phi} + r\frac{\partial}{\partial r}\left(\frac{v_\phi}{r}\right)\right]^2$$

$$- \frac{2}{3}\left[\frac{1}{r^2}\frac{\partial}{\partial r}(r^2 v_r) + \frac{1}{r\sin\theta}\frac{\partial}{\partial\theta}(v_\theta\sin\theta) + \frac{1}{r\sin\theta}\frac{\partial v_\phi}{\partial\phi}\right]^2$$

TABLE 1.8 Scalar Components of the Heat Flux Vector $\mathbf{q}''$

Rectangular (x, y, z)	Cylindrical (r, θ, z)	Spherical (r, θ, ϕ)
$q_x'' = -k\dfrac{\partial T}{\partial x}$	$q_r'' = -k\dfrac{\partial T}{\partial r}$	$q_r'' = -k\dfrac{\partial T}{\partial r}$
$q_y'' = -k\dfrac{\partial T}{\partial y}$	$q_\theta'' = -k\dfrac{1}{r}\dfrac{\partial T}{\partial\theta}$	$q_\theta'' = -k\dfrac{1}{r}\dfrac{\partial T}{\partial\theta}$
$q_z'' = -k\dfrac{\partial T}{\partial z}$	$q_z'' = -k\dfrac{\partial T}{\partial z}$	$q_\phi'' = -k\dfrac{1}{r\sin\theta}\dfrac{\partial T}{\partial\phi}$

For most engineering applications it is convenient to have the equation of thermal energy in terms of the fluid temperature and heat capacity rather than the internal energy or enthalpy. In general, for pure substances [11],

$$\frac{Di}{Dt} = \left(\frac{\partial i}{\partial P}\right)_T \frac{DP}{Dt} + \left(\frac{\partial i}{\partial T}\right)_P \frac{DT}{Dt} = \frac{1}{\rho}(1 - \beta T)\frac{DP}{Dt} + c_P \frac{DT}{Dt} \qquad (1.55)$$

where β is defined by Eq. 1.48. Substituting this into Eq. 1.54 we have the following general relation:

$$\rho c_p \frac{DT}{Dt} = \nabla \cdot k\nabla T + T\beta \frac{DP}{Dt} + \mu\Phi + q''' \qquad (1.56)$$

For an ideal gas, $\beta = 1/T$, and then

$$\rho c_p \frac{DT}{Dt} = \nabla \cdot k\nabla T + \frac{DP}{Dt} + \mu\Phi + q''' \qquad (1.57)$$

Note that c_p need not be constant.

We could have obtained Eq. 1.57 directly from Eq. 1.54 by noting that for an ideal gas, $di = c_p\,dT$ where c_p is constant and thus

$$\frac{Di}{Dt} = c_p \frac{DT}{Dt}$$

For an incompressible fluid with specific heat $c = c_p = c_v$ we go back to Eq. 1.52 ($d\mathbf{u} = c\,dT$) to obtain

$$\rho c \frac{DT}{Dt} = \nabla \cdot k\nabla T + \mu\Phi + q''' \qquad (1.58)$$

Equations 1.52, 1.54, and 1.56 can be easily written in terms of energy (heat) and momentum fluxes using relations for fluxes given in Tables 1.4, 1.6, and 1.8. The energy equation given by Eq. 1.58 (with $q''' = 0$ for simplicity) is given in Table 1.9 in different coordinate systems.

For *solids*, the density may usually be considered constant and we may set $\mathbf{V} = 0$, and Eq. 1.58 reduces to

$$\rho c \frac{\partial T}{\partial t} = \nabla \cdot k\nabla T + q''' \qquad (1.59)$$

which is the starting point for most problems in heat conduction.

The Energy Equation for a Mixture. The energy equations in the previous section are applicable for pure fluids. A thermal energy equation valid for a mixture of chemical species is required for situations involving simultaneous heat and mass transfer. For a pure fluid, conduction is the only diffusive mechanism of heat flow; hence Fourier's law is used, resulting in the term $\nabla \cdot k\nabla T$. More generally this term may be written $-\nabla\mathbf{q}''$, where $\mathbf{q}''$ is the diffusive heat flux, i.e., the heat flux relative to the mass average velocity. More specifically, for a mixture, $\mathbf{q}''$ is now made from three contributions: (1) ordinary conduction, described by Fourier's law, $-k\nabla T$, where k is the *mixture* thermal conductivity; (2) the contribution due to interdiffusion of species, given by $\sum_i \mathbf{j}_i i_i$; and (3) diffusional conduction (also called the diffusion-thermo effect or Dufour effect [6, 12]). The third contribution is of the second order and is usually negligible:

$$\mathbf{q}'' = -k\nabla T + \sum_i \mathbf{j}_i i_i \qquad (1.60)$$

TABLE 1.9   The Energy Equation* (for Newtonian Fluids of Constant ρ and k)

Rectangular coordinates (x, y, z):

$$\rho c_p \left(\frac{\partial T}{\partial t} + u \frac{\partial T}{\partial x} + v \frac{\partial T}{\partial y} + w \frac{\partial T}{\partial z} \right) = k \left(\frac{\partial^2 T}{\partial x^2} + \frac{\partial^2 T}{\partial y^2} + \frac{\partial^2 T}{\partial z^2} \right)$$

$$+ 2\mu \left\{ \left(\frac{\partial u}{\partial x} \right)^2 + \left(\frac{\partial v}{\partial y} \right)^2 + \left(\frac{\partial w}{\partial z} \right)^2 \right\} + \mu \left\{ \left(\frac{\partial u}{\partial y} + \frac{\partial v}{\partial x} \right)^2 + \left(\frac{\partial u}{\partial z} + \frac{\partial w}{\partial x} \right)^2 + \left(\frac{\partial v}{\partial z} + \frac{\partial w}{\partial y} \right)^2 \right\}$$

Cylindrical coordinates (r, θ, z):

$$\rho c_p \left(\frac{\partial T}{\partial t} + v_r \frac{\partial T}{\partial r} + \frac{v_\theta}{r} \frac{\partial T}{\partial \theta} + v_z \frac{\partial T}{\partial z} \right) = k \left[\frac{1}{r} \frac{\partial}{\partial r} \left(r \frac{\partial T}{\partial r} \right) + \frac{1}{r^2} \frac{\partial^2 T}{\partial \theta^2} + \frac{\partial^2 T}{\partial z^2} \right]$$

$$+ 2\mu \left\{ \left(\frac{\partial v_r}{\partial r} \right)^2 + \left[\frac{1}{r} \left(\frac{\partial v_\theta}{\partial \theta} + v_r \right) \right]^2 + \left(\frac{\partial v_z}{\partial z} \right)^2 \right\} + \mu \left\{ \left(\frac{\partial v_\theta}{\partial z} + \frac{1}{r} \frac{\partial v_z}{\partial \theta} \right)^2 + \left(\frac{\partial v_z}{\partial r} + \frac{\partial v_r}{\partial z} \right)^2 + \left[\frac{1}{r} \frac{\partial v_r}{\partial \theta} + r \frac{\partial}{\partial r} \left(\frac{v_\theta}{r} \right) \right]^2 \right\}$$

Spherical coordinates (r, θ, ϕ):

$$\rho c_p \left(\frac{\partial T}{\partial t} + v_r \frac{\partial T}{\partial r} + \frac{v_\theta}{r} \frac{\partial T}{\partial \theta} + \frac{v_\phi}{r \sin \theta} \frac{\partial T}{\partial \phi} \right) = k \left[\frac{1}{r^2} \frac{\partial}{\partial r} \left(r^2 \frac{\partial T}{\partial r} \right) + \frac{1}{r^2 \sin \theta} \frac{\partial}{\partial \theta} \left(\sin \theta \frac{\partial T}{\partial \theta} \right) + \frac{1}{r^2 \sin^2 \theta} \frac{\partial^2 T}{\partial \phi^2} \right]$$

$$+ 2\mu \left\{ \left(\frac{\partial v_r}{\partial r} \right)^2 + \left(\frac{1}{r} \frac{\partial v_\theta}{\partial \theta} + \frac{v_r}{r} \right)^2 + \left(\frac{1}{r \sin \theta} \frac{\partial v_\phi}{\partial \phi} + \frac{v_r}{r} + \frac{v_\theta \cot \theta}{r} \right)^2 \right\}$$

$$+ \mu \left\{ \left[r \frac{\partial}{\partial r} \left(\frac{v_\theta}{r} \right) + \frac{1}{r} \frac{\partial v_r}{\partial \theta} \right]^2 + \left[\frac{1}{r \sin \theta} \frac{\partial v_r}{\partial \phi} + r \frac{\partial}{\partial r} \left(\frac{v_\phi}{r} \right) \right]^2 + \left[\frac{\sin \theta}{r} \frac{\partial}{\partial \theta} \left(\frac{v_\phi}{\sin \theta} \right) + \frac{1}{r \sin \theta} \frac{\partial v_\theta}{\partial \phi} \right]^2 \right\}$$

* The terms contained in braces { } are associated with viscous dissipation and may usually be neglected except in systems with large velocity gradients.

Here $\mathbf{j}_i$ is a diffusive mass flux of species i, with units of mass/(area × time), as mentioned before. Substituting Eq. 1.60 in, for example, Eq. 1.54, we obtain the energy equation for a mixture:

$$\rho \frac{Di}{Dt} = \frac{DP}{Dt} + \nabla \cdot k \nabla T - \nabla \cdot \left(\sum_i \mathbf{j}_i i_i \right) + \mu \Phi + q''' \tag{1.61}$$

For a nonreacting mixture the term $\nabla \cdot (\sum_i \mathbf{j}_i i_i)$ is often of minor importance. But when endothermic or exothermic reactions occur, this term can play a dominant role. For reacting mixtures the species enthalpies

$$i_i = i_i^0 + \int_{T^0}^{T} c_{p,i} \, dT$$

must be written with a consistent set of heats of formation i_i^0 at T^0 [13].

The Conservation Equation for Species

For a stationary control volume, the conservation equation for species is

$$\underbrace{\frac{\partial C_i}{\partial t}}_{\substack{\text{rate of storage} \\ \text{of species } i \text{ per} \\ \text{unit volume}}} = \underbrace{-\nabla \cdot (C_i \mathbf{V})}_{\substack{\text{net rate of} \\ \text{convection of species} \\ i \text{ per unit volume}}} - \underbrace{\nabla \cdot \mathbf{j}_i}_{\substack{\text{net rate of diffusion} \\ \text{of species } i \text{ per} \\ \text{unit volume}}} + \underbrace{r_i'''}_{\substack{\text{production rate} \\ \text{of species } i \text{ per} \\ \text{unit volume}}} \tag{1.62}$$

Using the mass conservation equation, the above equation can be rearranged to obtain

$$\rho \frac{Dm_i}{Dt} = -\nabla \cdot \mathbf{j}_i + r_i''' \tag{1.63}$$

where m_i is mass fraction of species i, i.e., where $m_i = C_i/\rho$, where ρ is the density of the mixture, $\sum_i C_i = \rho$, and C_i is a partial density of species i (i.e., a mass concentration of species i).

The conservation equation for species can also be written in terms of mole concentration and mole fractions, as shown in Refs. 10, 12, and 13. The mole concentration of species i is $c_i = C_i/M_i$, where M_i is the molecular weight of the species. The mole fraction of species i is defined as $x_i = c_i/c$, where $c = \sum_i c_i$. As is obvious, $\sum_i m_i = 1$ and $\sum_i x_i = 1$.

Equations 1.62 and 1.63 written in different coordinate systems are given in Ref. 10.

Use of Conservation Equations to Set Up Problems

For a problem involving fluid flow and simultaneous heat and mass transfer, equations of continuity, momentum, energy, and chemical species (Eqs. 1.41, 1.44, 1.54, and 1.63) are a formidable set of partial differential equations. There are four *independent variables:* three space coordinates (say, x, y, z) and a time coordinate t.

If we consider a pure fluid, there are five equations: the continuity equation, three momentum equations, and the energy equation. The five accompanying *dependent variables* are pressure, three components of velocity, and temperature. Also, a thermodynamic equation of state serves to relate density to the pressure, temperature, and composition. (Notice that for natural convection flows the momentum and energy equations are coupled.)

For a mixture of n chemical species, there are n species conservation equations, but one is redundant, as the sum of mass fractions is equal to unity.

A complete mathematical statement of a problem requires specification of boundary and initial conditions. Boundary conditions are based on a physical statement or principle (for example: for viscous flow the component of velocity parallel to a stationary surface is zero at the wall; for an insulated wall the derivative of temperature normal to the wall is zero; etc.).

A general solution, even by numerical methods, of the full equations in the four independent variables is difficult to obtain. Fortunately, however, many problems of engineering interest are adequately described by simplified forms of the full conservation equations, and these forms can often be solved easily. The governing equations for simplified problems are obtained by deleting superfluous terms in the full conservation equations. This applies directly to laminar flows only. In the case of turbulent flows, some caution must be exercised. For example, on an average basis a flow may be two-dimensional and steady, but if it is unstable and as a result turbulent, fluctuations in the three components of velocity may be occurring with respect to time and the three spatial coordinates. Then the remarks about dropping terms apply only to the time-averaged equations [7, 12].

When simplifying the conservation equation given in a full form, we have to rely on physical intuition or on experimental evidence to judge which terms are negligibly small. Typical resulting classes of simplified problems are:

Constant transport properties

Constant density

Timewise steady flow (or quasi-steady flow)

Two-dimensional flow

One-dimensional flow

Fully developed flow (no dependence on the streamwise coordinate)

Stagnant fluid or rigid body

Terms may also be shown to be negligibly small by order-of-magnitude estimates [7, 12]. Some classes of flow that result are:

Creeping flows: inertia terms are negligible.

Forced flows: gravity forces are negligible.

Natural convection: gravity forces predominate.

Low-speed gas flows: viscous dissipation and compressibility terms are negligible.

Boundary-layer flows: streamwise diffusion terms are negligible.

DIMENSIONLESS GROUPS AND SIMILARITY IN HEAT TRANSFER

Modern engineering practice in the field of heat transfer is based on a combination of theoretical analysis and experimental data. Often the engineer is faced with the necessity of obtaining practical results in situations where, for various reasons, physical phenomena cannot be described mathematically or the differential equations describing the problem are too difficult to solve. An experimental program must be considered in such cases. However, in carrying the experimental program the engineer should know how to relate the experimental data (i.e., data obtained on the model under consideration) to the actual, usually larger, system (prototype). A determination of the relevant dimensionless parameters (groups) provides a powerful tool for that purpose.

The generation of such dimensionless groups in heat transfer (known generally as *dimensional analysis*) is basically done (1) by using differential equations and their boundary conditions (this method is sometimes called a *differential similarity*) and (2) by applying the dimensional analysis in the form of the Buckingham pi theorem.

The first method (differential similarity) is used when the governing equations and their boundary conditions describing the problem are known. The equations are first made dimensionless. For demonstration purposes, let us consider the relatively simple problem of a binary mixture with constant properties and density flowing at low speed, where body forces, heat source term, and chemical reactions are neglected. The conservation equations are, from Eqs. 1.42, 1.46, 1.58, and 1.63,

$$\text{Mass} \qquad\qquad \nabla \cdot \mathbf{V} = 0 \qquad\qquad\qquad (1.64)$$

$$\text{Momentum} \qquad\qquad \rho \frac{D\mathbf{V}}{Dt} = -\nabla P + \mu \nabla^2 \mathbf{V} \qquad\qquad (1.65)$$

$$\text{Thermal energy} \qquad\qquad \rho c \frac{DT}{Dt} = k\nabla^2 T + \mu \Phi \qquad\qquad (1.66)$$

$$\text{Species} \qquad\qquad \frac{Dm_1}{Dt} = \mathbf{D}\nabla^2 m_1 \qquad\qquad (1.67)$$

Using L and V as characteristic length and velocity, respectively, we define the dimensionless variables

$$x^* = \frac{x}{L} \qquad y^* = \frac{y}{L} \qquad z^* = \frac{z}{L} \qquad\qquad (1.68)$$

$$\mathbf{V}^* = \frac{\mathbf{V}}{V} \qquad\qquad (1.69)$$

$$t^* = \frac{t}{L/V} \tag{1.70}$$

$$P^* = \frac{P}{\rho V^2} \tag{1.71}$$

and also

$$T^* = \frac{T - T_w}{T_\infty - T_w} \tag{1.72}$$

$$m^* = \frac{m_1 - m_{1,w}}{m_{1,\infty} - m_{1,w}} \tag{1.73}$$

where the subscript ∞ refers to the external free-stream condition or some average condition and the subscript w refers to conditions adjacent to a bounding surface across which transfer of heat and mass occurs. If we introduce the dimensionless quantities (Eqs. 1.68–1.73), into Eqs. 1.64–1.67, we obtain, respectively,

$$\nabla^* \cdot \mathbf{V}^* = 0 \tag{1.74}$$

$$\frac{D\mathbf{V}^*}{Dt^*} = -\nabla^* P^* + \frac{1}{\text{Re}} \nabla^{*2}\mathbf{V}^* \tag{1.75}$$

$$\frac{DT^*}{Dt^*} = \frac{1}{\text{Re Pr}} \nabla^{*2}T^* + \frac{2\,\text{Ec}}{\text{Re}} \Phi^* \tag{1.76}$$

$$\frac{Dm^*}{Dt^*} = \frac{1}{\text{Re Sc}} \nabla^{*2}m^* \tag{1.77}$$

Obviously, the solutions of Eqs. 1.74–1.77 depend on the coefficients that appear in these equations. Solutions of Eqs. 1.74–1.77 are equally applicable to the model and prototype (where the model and prototype are geometrically similar systems of different linear dimensions in streams of different velocities, temperatures, and concentration), if the coefficients in these equations are the same for both model and prototype. These coefficients, Pr, Re, Sc, and Ec (called dimensionless parameters or similarity parameters), are defined in Table 1.10.

Focusing attention now on heat transfer, from Eq. 1.14, using the dimensionless quantities, the heat transfer coefficient is given as

$$h = \frac{k}{L} \left. \frac{\partial T^*}{\partial y^*} \right|_{y^*=0} \tag{1.78}$$

or, in dimensionless form,

$$\frac{hL}{k} = \left. \frac{\partial T^*}{\partial y^*} \right|_{y^*=0} = \text{Nu} \tag{1.79}$$

where the dimensionless group Nu is known as the Nusselt number. Since Nu is the dimensionless temperature gradient at the surface, according to Eq. 1.76 it must therefore depend on the dimensionless groups that appear in this equation; hence

$$\text{Nu} = f_1(\text{Re, Pr, Ec}) \tag{1.80}$$

For processes in which viscous dissipation and compressibility are negligible, which is the case in many industrial applications, we have

$$\text{Nu} = f_2(\text{Re, Pr}) \qquad \text{(forced convection)} \tag{1.81}$$

TABLE 1.10 Summary of the Chief Dimensionless Groups*

Group	Symbol	Definition	Physical significance (interpretation)	Main area of use
Biot number	Bi	$\dfrac{hL}{k_s}$	Ratio of internal thermal resistance of solid to fluid thermal resistance	Heat transfer between fluid and solid
Biot number[†] (mass transfer)	Bi_D	$\dfrac{h_D L}{D}$	Ratio of the internal species transfer resistance to the boundary layer species transfer resistance	Mass transfer between fluid and solid
Coefficient of friction (skin friction coefficient)	c_f	$\dfrac{\tau_w}{\rho V^2/2}$	Dimensionless surface shear stress	Flow resistance
Eckert number	Ec	$\dfrac{V_\infty^2}{c_p(T_w - T_\infty)}$	Kinetic energy of the flow relative to the boundary layer enthalpy difference	Forced convection (compressible flow)
Euler number	Eu	$\dfrac{\Delta P}{\rho V^2}$	Ratio of friction to velocity head	Fluid friction
Fourier number	Fo	$\dfrac{\alpha t}{L^2}$	Ratio of the heat conduction rate to the rate of thermal energy storage in a solid	Unsteady-state heat transfer
Fourier number (mass transfer)	Fo_D	$\dfrac{Dt}{L^2}$	Ratio of the species diffusion rate to the rate of species storage	Unsteady-state mass transfer
Froude number	Fr	$\dfrac{V^2}{gL}$	Ratio of inertial to gravitational force	Wave and surface behavior (mixed natural and forced convection)
Graetz number	Gz	$\text{Re Pr}\,\dfrac{D}{L} = \dfrac{\rho c_p V D^2}{kL}$	Ratio of the fluid stream thermal capacity to convective heat transfer	Forced convection
Grashof number	Gr	$\dfrac{g\beta\Delta T L^3}{v^2}$	Ratio of buoyancy to viscous forces	Natural convection
Colburn j factor (heat transfer)	j_H	$\text{St Pr}^{2/3}$	Dimensionless heat transfer coefficient	Forced convection (heat, mass, and momentum transfer analogy)
Colburn j factor (mass transfer)	j_D	$\text{St}_D\,\text{Sc}^{2/3}$	Dimensionless mass transfer coefficient	Forced convection (heat, mass, and momentum transfer analogy)
Jakob number	Ja	$\dfrac{\rho_l c_{pl}(T_w - T_{sat})}{\rho_g i_{lg}}$	Ratio of sensible heat absorbed by the liquid to the latent heat absorbed	Boiling

TABLE 1.10 Summary of the Chief Dimensionless Groups* (*Continued*)

Group	Symbol	Definition	Physical significance (interpretation)	Main area of use
Knudsen number	Kn	$\dfrac{\lambda}{L}$	Ratio of molecular mean free path to characteristic dimension	Low-pressure (low-density) gas flow
Lewis number	Le	$\dfrac{\alpha}{\mathbf{D}} = \dfrac{Sc}{Pr}$	Ratio of molecular thermal and mass diffusivities	Combined heat and mass transfer
Mach number	Ma	$\dfrac{V}{a}$	Ratio of the velocity of flow to the velocity of sound	Compressible flow
Nusselt number	Nu	$\dfrac{hL}{k}$	Basic dimensionless convective heat transfer coefficient (ratio of convection heat transfer to conduction in a fluid slab of thickness L)	Convective heat transfer
Péclet number	Pe	$Re\ Pr = \dfrac{\rho c_p VL}{k}$	Dimensionless independent heat transfer parameter (ratio of heat transfer by convection to conduction)	Forced convection
Péclet number (mass transfer)	Pe$_\mathbf{D}$	$Re\ Sc = \dfrac{VL}{\mathbf{D}}$	Dimensionless independent mass transfer coefficient (ratio of bulk mass transfer to diffusive mass transfer)	Mass transfer
Prandtl number	Pr	$\dfrac{\mu c_p}{k} = \dfrac{\nu}{\alpha}$	Ratio of molecular momentum and thermal diffusivities	Forced and natural convection
Rayleigh number	Ra	$Gr\ Pr = \dfrac{\rho g\beta\,\Delta T L^3}{\mu\alpha}$	Modified Grashof number (see interpretations for Gr and Pr)	Natural convection
Reynolds number	Re	$\dfrac{\rho VL}{\mu}$	Ratio of inertia to viscous forces	Forced convection; dynamic similarity
Schmidt number	Sc	$\dfrac{\nu}{\mathbf{D}}$	Ratio of molecular momentum and mass diffusivities	Mass transfer
Sherwood number	Sh	$\dfrac{h_\mathbf{D} L}{\mathbf{D}}$	Ratio of convection mass transfer to diffusion in a slab of thickness L	Convective mass transfer

Strouhal number[‡]	Sr	$\dfrac{Lf}{V}$	Ratio of the velocity of vibration Lf to the velocity of the fluid	Flow past tube (shedding of eddies)
Stanton number	St	$\dfrac{\mathrm{Nu}}{\mathrm{Re\,Pr}} = \dfrac{h}{\rho c_p V}$	Dimensionless heat transfer coefficient (ratio of heat transfer at the surface to that transported by fluid by its thermal capacity)	Forced convection
Stanton number (mass transfer)	St_D	$\dfrac{\mathrm{Sh}}{\mathrm{Re\,Sc}} = \dfrac{h_\mathrm{D}}{V}$	Dimensionless mass transfer coefficient	Convective mass transfer
Weber number	We	$\dfrac{\rho V^2 L}{\sigma}$	Ratio of inertia force to surface tension force	Droplet breakup; thin-film flow

* In these dimensionless groups L designates characteristic dimension (e.g., tube diameter, hydraulic diameter, length of the tube or plate, slab thickness, radius of a cylinder or sphere, droplet diameter, thin-film thickness, etc.). Physical properties are usually evaluated at mean temperature unless otherwise specified.

† Note: $\mathbf{D} = \mathbf{D}_{12}$ ($\mathbf{D}_{12}$ is also a commonly used symbol for binary diffusion coefficient; $\mathbf{D}_{ij}$ is the multicomponent diffusion coefficient). When species 1 is in very small concentration the symbol $\mathbf{D}_{1m}$ is occasionally used [12], representing an effective binary diffusion coefficient for species 1 diffusing through the mixture.

‡ In some engineering texts the symbol St is also used for this group.

In the case of buoyancy-induced flow, Eq. 1.65 should be replaced with the simplified version [16] of Eq. 1.50, and, following a similar procedure, we should obtain

$$\text{Nu} = f_3(\text{Gr, Pr}) \qquad \text{(natural convection)} \qquad (1.82)$$

where Gr is the Grashof number, defined in Table 1.10. Also, using the relation of Eq. 1.17 and dimensionless quantities,

$$h_{\mathbf{D}} = \frac{\mathbf{D}}{L} \left. \frac{\partial m^*}{\partial y^*} \right|_{y^*=0} \qquad (1.83)$$

or

$$h_{\mathbf{D}} \frac{L}{\mathbf{D}} = \left. \frac{\partial m^*}{\partial y^*} \right|_{y^*=0} = \text{Sh} \qquad (1.84)$$

This parameter, termed the Sherwood number, is equal to the dimensionless mass fraction (i.e., concentration) gradient at the surface, and it provides a measure of the convection mass transfer occurring at the surface. Following the same argument as before (but now for Eq. 1.77), we have

$$\text{Sh} = f_4(\text{Re, Sc}) \qquad \text{(forced convection, mass transfer)} \qquad (1.85)$$

The significance of expressions such as Eqs. 1.80–1.82 and 1.85 should be appreciated. For example, Eq. 1.81 states that convection heat transfer results, whether obtained theoretically or experimentally, can be represented in terms of three dimensionless groups instead of seven parameters (h, L, V, k, c_p, μ, and ρ). The convenience is evident. Once the form of the functional dependence of Eq. 1.81 is obtained for a particular surface geometry (e.g., from laboratory experiments on a small model), it is known to be universally applicable, i.e., it may be applied to different fluids, velocities, temperatures, and length scales, as long as the assumptions associated with the original equations are satisfied (e.g., negligible viscous dissipation and body forces). Note that the relations of Eqs. 1.80 and 1.85 are derived without actually solving the system of Eqs. 1.64–1.67. References 3, 7, 12, 15, 16, and 18 cover the above procedure in more detail and also include many different cases.

It is important to mention here that once the conservation equations are put in dimensionless form it is also convenient to make an order-of-magnitude assessment of all terms in the equations. Often a problem can be simplified by discovering that a term that would be very difficult to handle if large is in fact negligibly small [7, 12]. Even if the primary thrust of the investigation is experimental, making the equations dimensionless and estimating the orders of magnitude of the terms is good practice. It is usually not possible for an experimental test to include (simulate) all conditions exactly; a good engineer will focus on the most important conditions. The same applies to performing an order-of-magnitude analysis. For example, for boundary-layer flows, allowance is made for the fact that lengths transverse to the main flow scale with a much shorter length than those measured in the direction of main flow. References 7, 12, and 17 cover many examples of the order-of-magnitude analysis.

When the governing equations of a problem are unknown, an alternative approach of deriving dimensionless groups is based on use of dimensional analysis in the form of the Buckingham pi theorem [9, 11, 14, 16, 18]. The success of this method depends on our ability to select, largely from intuition, the parameters that influence the problem. For example, knowing in advance that the heat transfer coefficient in fully developed forced convection in a tube is a function of certain variables, that is, $h = f(V, \rho, \mu, c_p, k, D)$, we can use the Buckingham pi theorem to obtain Eq. 1.81, as shown in Ref. 11. However, this method is carried out without any consideration of the physical nature of the process in question, i.e., there is no way to ensure that all essential variables have been included. However, as shown above, starting with the differential form of the conservation equations we have derived the similarity parameters (dimensionless groups) in rigorous fashion.

In Table 1.10 those dimensionless groups that appear frequently in the heat and mass transfer literature have been listed. The list includes groups already mentioned above as well as those found in special fields of heat transfer. Note that, although similar in form, the Nusselt and Biot numbers differ in both definition and interpretation. The Nusselt number is defined in terms of thermal conductivity of the fluid; the Biot number is based on the solid thermal conductivity.

UNITS AND CONVERSION FACTORS

The dimensions that are used consistently in the field of heat transfer are length, mass, force, energy, temperature, and time. We should avoid using both force and mass dimensions in the same equation, since force is always expressible in dimensions of mass, length, and time, and vice versa. We do not make a practice of eliminating energy in terms of force times length, because the accounting of work and heat is practically always kept separate in heat transfer problems.

In this handbook both *SI* (the accepted abbreviation for *Système International d'Unités,* or International System of Units) and *English engineering units** are used simultaneously throughout. The base units for the English engineering units are given in the second column of Table 1.11. The unit of force in English units is the *pound force* (lb_f). However, the use of the *pound mass* (lb_m) and pound force in engineering work causes considerable confusion in the proper use of these two fundamentally different units.

TABLE 1.11 Conversion Factor g_c for the Common Unit Systems

Quantity	SI	English engineering*	cgs[†]	Metric engineering
Mass	kilogram, kg	pound mass, lb_m	gram, g	kilogram mass, kg
Length	meter, m	foot, ft	centimeter, cm	meter, m
Time	second, s	second, s, or hour, h	second, s	second, s
Force	newton, N	pound force, lb_f	dyne, dyn	kilogram force, kg_f
g_c	$1\ kg{\cdot}m/(N{\cdot}s^2)$[‡]	$32.174\ lb_m{\cdot}ft/(lb_f{\cdot}s^2)$	$1\ g{\cdot}cm/(dyn{\cdot}s^2)$	$9.80665\ kg{\cdot}m/(kg_f{\cdot}s^2)$

* In this system of units the temperature is given in degrees Fahrenheit (°F).
[†] Centimeter-gram-second: this system of units has been used mostly in scientific work.
[‡] Since $1\ kg{\cdot}m/s^2 = 1\ N$, then $g_c = 1$ in the SI system of units.

The two can be related as

$$1\ lb_f = \frac{1\ lb_m \times 32.174\ ft/s^2}{g_c}$$

whence

$$g_c = 32.174\ lb_m{\cdot}ft/(lb_f{\cdot}s^2)$$

Thus, g_c is merely a conversion factor and it should not be confused with the gravitational acceleration g. The numerical value of g_c is a constant depending only on the system of units involved and not on the value of the gravitational acceleration at a particular location. Values

---

* Also associated with this system of units are such names as *U.S. Customary Units, British engineering units, engineering units,* and *foot-pound-second system of units.* The name *English engineering units,* or, for short, *English units,* is selected in this handbook because it has been used by practicing engineers more frequently than the other names mentioned.

TABLE 1.12   SI Base and Supplementary Units

Quantity	Unit
Length	meter (m)
Mass	kilogram (kg)
Time	second (s)
Electric current	ampere (A)
Thermodynamic temperature	kelvin (K)
Amount of substance	mole (mol)
Luminous intensity	candela (cd)
Plane angle*	radian (rad)
Solid angle*	steradian (sr)

* Supplementary units.

of g_c corresponding to different systems of units found in engineering literature are given in Table 1.11.

The SI base units are summarized in Table 1.12. The SI units comprise a rigorously coherent form of the metric system, i.e., all remaining units may be derived from the base units using formulas that do not involve any numerical factors. For example, the unit of force is the newton (N); a 1-N force will accelerate a 1-kg mass at 1 m/s^2. Hence 1 N = 1 $kg \cdot m/s^2$. The unit of pressure is the N/m^2, often referred to as the pascal. In the SI system there is one unit of energy (thermal, mechanical, or electrical), the joule (J); 1 J = 1 N·m. The unit for energy rate, or power, is joules per second (J/s), where one J/s is equivalent to one watt (1 J/s = 1 W).

In the English system of units it is necessary to relate thermal and mechanical energy via the mechanical equivalent of heat J_c. Thus

$$J_c \times \text{thermal energy} = \text{mechanical energy}$$

The unit of heat in the English system is the British thermal unit (Btu). When the unit of mechanical energy is the pound-force-foot ($lb_f \cdot ft$), then

$$J_c = 778.16 \ lb_f \cdot ft/Btu$$

as 1 Btu = 778.16 $lb_f \cdot ft$. Happily, in the SI system the units of heat and work are identical and J_c is unity.

Since it is frequently necessary to work with extremely large or small numbers, a set of standard prefixes has been introduced to simplify matters (Table 1.13). Symbols and names for all units used in the handbook are given in Table 1.14. Conversion factors for commonly

TABLE 1.13   SI Prefixes (Decimal Multiples and Submultiples in SI Are Formed by Adding the Following Prefixes to the SI Unit)

Factor	Prefix	Symbol	Factor	Prefix	Symbol
10^{18}	exa-	E	10^{-1}	deci-	d
10^{15}	peta-	P	10^{-2}	centi-	c
10^{12}	tera-	T	10^{-3}	milli-	m
10^{9}	giga-	G	10^{-6}	micro-	μ
10^{6}	mega-	M	10^{-9}	nano-	n
10^{3}	kilo-	k	10^{-12}	pico-	p
10^{2}	hecto-	h	10^{-15}	femto-	f
10	deka-	da	10^{-18}	atto-	a

TABLE 1.14 Symbols and Names for Units Used in the Handbook

Symbol	Name	Symbol	Name
A	ampere	kg	kilogram mass
Btu	British thermal unit	kg_f	kilogram force
C	coulomb (= A·s)	lb_m	pound mass
°C	degree Celsius	lb_f	pound force
cal	calorie	m	meter
cm	centimeter	min	minute
deg	degree	mol	mole
dyn	dyne	N	newton
°F	degree Fahrenheit	Pa	Pascal (= N/m^2)
ft	foot	pdl	poundal
g	gram	°R	degree Rankine
H	henry (= V·s/A)	rad	radian (plane angle)
h	hour	s	second
hp	horsepower	sr	steradian (solid angle)
in	inch	T	tesla (= $V·s/m^2$)
J	joule (= N·m)	V	volt
K	kelvin (thermodynamic temperature)	W	watt (= J/s)

used quantities in heat transfer, from SI to English engineering units and vice versa, are given in Table 1.15.

Conversion factors for mass, density, pressure, energy, specific energy, specific heat, thermal conductivity, dynamic viscosity, and kinematic viscosity in different systems of units are also given in Chap. 2 (Tables 2.1–2.9).

TABLE 1.15 Conversion Factors for Commonly Used Quantities in Heat Transfer

Quantity	SI → English	English → SI*
Area	$1\ m^2 = 10.764\ ft^2$ $= 1550.0\ in^2$	$1\ ft^2 = 0.0929\ m^2$ $1\ in^2 = 6.452 \times 10^{-4}\ m^2$
Density	$1\ kg/m^3 = 0.06243\ lb_m/ft^3$	$1\ lb_m/ft^3 = 16.018\ kg/m^3$ $1\ slug/ft^3 = 515.379\ kg/m^3$
Energy[†]	$1\ J = 9.4787 \times 10^{-4}\ Btu$ $= 6.242 \times 10^{18}\ eV$	$1\ Btu = 1055.056\ J$ $1\ cal = 4.1868\ J$ $1\ lb_f·ft = 1.3558\ J$ $1\ hp·h = 2.685 \times 10^6\ J$
Energy per unit mass	$1\ J/kg = 4.2995 \times 10^{-4}\ Btu/lb_m$	$1\ Btu/lb_m = 2326\ J/kg$
Force	$1\ N = 0.22481\ lb_f$	$1\ lb_f = 4.448\ N$ $1\ pdl = 0.1382\ N$
Heat flux	$1\ W/m^2 = 0.3171\ Btu/(h·ft^2)$	$1\ Btu/(h·ft^2) = 3.1525\ W/m^2$ $1\ kcal/(h·m^2) = 1.163\ W/m^2$ $1\ cal/(s·cm^2) = 41.870 \times 10^3\ W/m^2$
Heat generation per unit volume	$1\ W/m^3 = 0.09665\ Btu/(h·ft^3)$	$1\ Btu/(h·ft^3) = 10.343\ W/m^3$
Heat transfer coefficient	$1\ W/(m^2·K) = 0.17612\ Btu/(h·ft^2·°F)$	$1\ Btu/(h·ft^2·°F) = 5.678\ W/(m^2·K)$ $1\ kcal/(h·m^2·°C) = 1.163\ W/(m^2·K)$ $1\ cal/(s·cm^2·°C) = 41.870 \times 10^3\ W/(m^2·K)$

TABLE 1.15 Conversion Factors for Commonly Used Quantities in Heat Transfer (*Continued*)

Quantity	SI → English	English → SI*
Heat transfer rate	1 W = 3.4123 Btu/h	1 Btu/h = 0.2931 W
Length	1 m = 3.2808 ft = 39.370 in	1 ft = 0.3048 m 1 in = 2.54 cm = 0.0254 m 1 yard = 0.9144 m 1 statute mile = 1609 m 1 mil = 0.001 in $= 2.54 \times 10^{-5}$ m 1 light-year = 9.46×10^{15} m 1 angstrom = 10^{-10} m 1 micron = 10^{-6} m
Mass	1 kg = 2.2046 lb_m	1 lb_m = 0.4536 kg 1 slug = 14.594 kg
Mass flow rate	1 kg/s = 7936.6 lb_m/h = 2.2046 lb_m/s	1 lb_m/h = 0.000126 kg/s 1 lb_m/s = 0.4536 kg/s
Power	1 W = 3.4123 Btu/h	1 Btu/h = 0.2931 W 1 Btu/s = 1055.1 W 1 $lb_f \cdot$ft/s = 1.3558 W 1 hp = 745.7 W
Pressure and stress[‡]	1 N/m^2 = 0.020886 lb_f/ft^2 $= 1.4504 \times 10^{-4}$ lb_f/in^2 $= 4.015 \times 10^{-3}$ in water $= 2.953 \times 10^{-4}$ in Hg	1 lb_f/ft^2 = 47.88 N/m^2 1 lb_f/in^2 = 6894.8 N/m^2 1 psi = 1 lb_f/in^2 = 6894.8 N/m^2 1 standard atmosphere = 1.0133×10^5 N/m^2 1 bar = 1×10^5 N/m^2
Specific heat	1 J/(kg·K) = 2.3886×10^{-4} Btu/($lb_m \cdot$°F)	1 Btu/($lb_m \cdot$°F) = 4187 J/(kg·K)
Surface tension	1 N/m = 0.06852 lb_f/ft	1 lb_f/ft = 14.594 N/m 1 dyn/cm = 1×10^{-3} N/m
Temperature	$T(K) = T(°C) + 273.15$ $= T(°R)/1.8$ $= [T(°F) + 459.67]/1.8$ $T(°C) = [T(°F) - 32]/1.8$	$T(°R) = 1.8T(K)$ $= T(°F) + 459.67$ $T(°F) = 1.8T(°C) + 32$ $= 1.8[T(K) - 273.15] + 32$
Temperature difference	1 K = 1°C = 1.8°R = 1.8°F	1°R = 1°F = 1 K/1.8 = 1°C/1.8
Thermal conductivity	1 W/(m·K) = 0.57782 Btu/(h·ft·°F)	1 Btu/(h·ft·°F) = 1.731 W/(m·K) 1 kcal/(h·m·°C) = 1.163 W/(m·K) 1 cal/(s·cm·°C) = 418.7 W/(m·K)
Thermal diffusivity	1 m^2/s = 10.7639 ft^2/s	1 ft^2/s = 0.0929 m^2/s 1 ft^2/h = 2.581×10^{-5} m^2/s
Thermal resistance	1 K/W = 0.52750 °F·h/Btu	1 °F·h/Btu = 1.8958 K/W
Velocity	1 m/s = 3.2808 ft/s	1 ft/s = 0.3048 m/s 1 knot = 0.5144 m/s
Viscosity (dynamic)[§]	1 $N \cdot s/m^2$ = 0.672 lb_m/(ft·s) $= 2.089 \times 10^{-2}$ $lb_f \cdot s/ft^2$	1 lb_m/(ft·s) = 1.4881 $N \cdot s/m^2$ 1 centipoise = 10^{-2} poise $= 1 \times 10^{-3}$ $N \cdot s/m^2$
Viscosity (kinematic)	1 m^2/s = 10.7639 ft^2/s	1 ft^2/s = 0.0929 m^2/s = 929 stoke 1 m^2/s = 10,000 stoke

TABLE 1.15 Conversion Factors for Commonly Used Quantities in Heat Transfer (*Continued*)

Quantity	SI → English	English → SI*
Volume	1 m^3 = 35.3134 ft^3	1 ft^3 = 0.02832 m^3
		1 in^3 = 1.6387 × 10^{-5} m^3
		1 gal (U.S. liq.) = 0.003785 m^3
		1 gal (U.K. liq.) = 0.004546 m^3
		1 m^3 = 1000 liter
		1 gal (U.S. liq.) = 4 quarts
		= 8 pints
		= 128 ounces
		1 quart = 0.946 × 10^{-3} m^3
Volume flow rate	1 m^3/s = 35.3134 ft^3/s	1 ft^3/h = 7.8658 × 10^{-6} m^3/s
	= 1.2713 × 10^5 ft^3/h	1 ft^3/s = 2.8317 × 10^{-2} m^3/s
		1 gal (U.S. liq.)/min = 6.309 ×
		10^{-5} m^3/s = 0.2271 m^3/hr

* Some units in this column belong to the cgs and mks metric systems.
† Definition of the units of energy based on thermal phenomena:

    1 Btu = energy required to raise 1 lb$_m$ of water 1°F at 68°F
    1 cal = energy required to raise 1 g of water 1°C at 20°C

‡ The SI unit for the quantity pressure is the pascal (Pa); 1 Pa = 1 N/m^2.
§ Also expressed in equivalent units of kg/(s·m).

NOMENCLATURE

Symbol, Definition, SI Units, English Units

A	heat transfer area: m^2, ft^2
a	acceleration: m/s^2, ft/s^2
a	speed of sound: m/s, ft/s
C	mass concentration of species: kg/m^3, lb$_m$/ft^3
c	specific heat: J/(kg·K), Btu/(lb$_m$·°F)
c_p	specific heat at constant pressure: J/(kg·K), Btu/(lb$_m$·°F)
c_v	specific heat at constant volume: J/(kg·K), Btu/(lb$_m$·°F)
D	tube inside diameter, diameter: m, ft
D	diffusion coefficient: m^2/s, ft^2/s
Ec	Eckert number (see Table 1.10)
e	emissive power: W/m^2, Btu/(h·ft^2)
e_b	blackbody emissive power: W/m^2, Btu/(h·ft^2)
F	force: N, lb$_f$
F_{1-2}	view factor (geometric shape factor for radiation from one blackbody to another)
$\mathscr{F}_{1-2}$	real body view factor (geometric shape and emissivity factor for radiation from one gray body to another)
f	frequency of vibration (see Table 1.10): s^{-1}
f_1, f_2, f_3, f_4	denotes function of Eqs. 1.80–1.82 and 1.85

Gr	Grashof number (see Table 1.10)
g	gravitational acceleration: m/s^2, ft/s^2
$\mathbf{g}$	gravitational acceleration (vector): m/s^2, ft/s^2
g_c	conversion factor (see Table 1.11): lb$_m$·ft/(lb$_f$·s^2)
h	heat transfer coefficient: W/(m^2·K), Btu/(h·ft^2·°F)
h_D	mass transfer coefficient: m/s, ft/s
i	enthalpy per unit mass: J/kg, Btu/lb$_m$
i_{lg}	latent heat of evaporation: J/kg, Btu/lb$_m$
i^0	heat of formation: J/kg, Btu/lb$_m$
j	mass diffusion flux of species: kg/(s·m^2), lb$_m$/(h·ft^2)
$\mathbf{j}$	mass diffusion flux of species (vector): kg/(s·m^2), lb$_m$/(h·ft^2)
k	thermal conductivity: W/(m·K), Btu/(h·ft·°F)
L	length: m, ft
M	mass: kg, lb$_m$
m	mass fraction of species (Eq. 1.63)
Nu	Nusselt number (see Table 1.10)
P	pressure: Pa (N/m^2), lb$_f$/ft^2
Pr	Prandtl number (see Table 1.10)
ΔP	pressure drop: Pa (N/m^2), lb$_f$/ft^2
q	heat transfer rate: W, Btu/h
$\mathbf{q}''$	heat flux (vector): W/m^2, Btu/(h·ft^2)
q''	heat flux: W/m^2, Btu/(h·ft^2)
q'''	volumetric heat generation: W/m^3, Btu/(h·ft^3)
R_{th}	thermal resistance: K/W, h·°F/Btu
Re	Reynolds number (see Table 1.10)
r	radial distance in cylindrical or spherical coordinate: m, ft
$\mathbf{r}$	recovery factor (Eq. 1.19)
r'''	volumetric generation rate of species: kg/(s·m^3), lb$_m$/(h·ft^3)
Sc	Schmidt number (see Table 1.10)
Sh	Sherwood number (see Table 1.10)
St	Stanton number (see Table 1.10)
T	temperature: °C, K, °F, °R
ΔT	temperature difference: °C, °F
t	time: s
u	velocity component in the axial direction (x direction) in rectangular coordinates: m/s, ft/s
$\mathbf{u}$	internal energy per unit mass: J/kg, Btu/lb$_m$
V	velocity: m/s, ft/s
$\mathbf{V}$	velocity (vector): m/s, ft/s
v	velocity component in the y direction in rectangular coordinates: m/s, ft/s
v_r	velocity component in the r direction: m/s, ft/s
v_z	velocity component in the z direction: m/s, ft/s

v_θ	velocity component in the θ direction: m/s, ft/s
v_ϕ	velocity component in the ϕ direction: m/s, ft/s
w	velocity component in the z direction in rectangular coordinates: m/s, ft/s
x	rectangular coordinate: m, ft
y	rectangular coordinate: m, ft
z	rectangular or cylindrical coordinate: m, ft

Greek

α	thermal diffusivity: m²/s, ft²/s
β	coefficient of thermal expansion: K^{-1}, $^\circ R^{-1}$
δ	hydrodynamic boundary layer thickness: m, ft
δ_D	concentration boundary layer thickness: m, ft
δ_T	thermal boundary layer thickness: m, ft
ϵ	emissivity
ϵ_H	eddy diffusivity of heat: m²/s, ft²/s
ϵ_M	eddy diffusivity of momentum: m²/s, ft²/s
ϵ_m	eddy diffusivity of mass: m²/s, ft²/s
θ	angle in cylindrical and spherical coordinates: rad, deg
λ	molecular mean free path: m, ft
μ	dynamic viscosity: Pa·s, lb_m/(s·ft)
ν	kinematic viscosity: m²/s, ft²/s
ρ	density: kg/m³, lb_m/ft³
σ	surface tension (see Table 1.10): N/m, lb_f/ft
σ	Stefan-Boltzmann radiation constant: W/(m²·K⁴), Btu/(h·ft²·$^\circ$R⁴)
τ	shear stress: N/m², lb_f/ft²
$\boldsymbol{\tau}$	shear stress tensor: N/m², lb_f/ft²
Φ	dissipation function (see Table 1.7): s⁻²
ϕ	angle in spherical coordinate system: rad, deg

Subscripts

a	surroundings
aw	adiabatic wall
cr	critical
f	fluid
g	gas (vapor)
i	species i
l	liquid
m	mean
r	radiation (Eq. 1.36)
s	solid
sat	saturation
t	total
w	wall

x	x component
y	y component
z	z component
θ	θ component
ϕ	ϕ component

Miscellaneous Subscripts

1	species 1 in binary mixture of 1 and 2
∞	free-stream condition

Superscripts

'	fluctuating component (for example, X' is the fluctuating component of X)
$-$	time average (for example, $\overline{X}$ is the time average of X)

Mathematical Operation Symbols

d/dx	derivative with respect to x: m^{-1}, ft^{-1}
$\partial/\partial t$	partial time derivative operator: s^{-1}
d/dt	total time derivative operator: s^{-1} (Eq. 1.37)
D/Dt	substantial time derivative operator: s^{-1} (Eq. 1.38)
∇	del operator (vector): m^{-1}, ft^{-1}
∇^2	laplacian operator: m^{-2}, ft^{-2}

REFERENCES

1. F. Kreith and W. Z. Black, *Basic Heat Transfer,* Harper & Row, New York, 1980.
2. J. P. Holman, *Heat Transfer,* 8th ed., McGraw-Hill, New York, 1997.
3. F. P. Incropera and D. P. DeWitt, *Fundamentals of Heat Transfer,* 4th ed., Wiley, New York, 1996.
4. M. N. Özisik, *Basic Heat Transfer,* McGraw-Hill, New York, 1977.
5. R. E. Treybal, *Mass-Transfer Operations,* 3d ed., McGraw-Hill, New York, 1980.
6. W. M. Kays and M. E. Crawford, *Convective Heat and Mass Transfer,* 2d ed., McGraw-Hill, New York, 1980.
7. H. Schlichting, *Boundary-Layer Theory,* 7th ed., McGraw-Hill, New York, 1979.
8. J. O. Hinze, *Turbulence,* 2d ed., McGraw-Hill, New York, 1975.
9. J. H. Lienhard, *A Heat Transfer Textbook,* Prentice-Hall, Englewood Cliffs, NJ, 1981.
10. R. B. Bird, W. E. Stewart, and E. N. Lightfoot, *Transport Phenomena,* Wiley, New York, 1960.
11. W. M. Rohsenow and H. Y. Choi, *Heat, Mass, and Momentum Transfer,* Prentice-Hall, Englewood Cliffs, NJ, 1961.
12. D. K. Edwards, V. E. Denny, and A. F. Mills, *Transfer Processes: An Introduction to Diffusion, Convection, and Radiation,* 2d ed., Hemisphere, Washington, DC, and McGraw-Hill, New York, 1979.
13. W. C. Reynolds and H. C. Perkins, *Engineering Thermodynamics,* 2d ed., McGraw-Hill, New York, 1977.
14. A. S. Foust, L. A. Wenzel, C. W. Clump, L. Mans, and L. B. Andersen, *Principles of Unit Operations,* 2d ed., Wiley, New York, 1980.
15. F. M. White, *Viscous Fluid Flow,* McGraw-Hill, New York, 1974.

16. B. Gebhart, *Heat Transfer,* 2d ed., McGraw-Hill, New York, 1971.

17. E. R. G. Eckert and R. M. Drake Jr., *Analysis of Heat and Mass Transfer,* McGraw-Hill, New York, 1972.

18. V. P. Isachenko, V. A. Osipova, and A. S. Sukomel, *Heat Transfer,* Mir Publishers, Moscow, 1977.

19. S. Whitaker, *Elementary Heat Transfer Analysis,* Pergamon, New York, 1976.

CHAPTER 2
THERMOPHYSICAL PROPERTIES

Thomas F. Irvine Jr.
State University of New York at Stony Brook

When organizing a chapter of thermophysical properties with limited space, some difficult decisions have to be made. Since this is a handbook for heat transfer practitioners, emphasis has been placed on transport rather than thermodynamic properties. The primary exception has been the inclusion of densities and isobaric specific heats, which are needed for the calculation of Prandtl numbers and thermal diffusivities.

In the spirit of today's computer usage, a number of gas properties are given in equation rather than tabular form. However, they are accompanied by skeleton tables to allow for program checks.

Because new refrigerants are being considered and used in technical applications, a number of transport and thermodynamic property tables are included for these substances.

Whenever possible, the properties in this chapter are divided into those for gases, liquids, and solids. There are unavoidable overlaps to this arrangement when the tables account for phase changes such as in the case of water.

CONVERSION FACTORS

TABLE 2.1 Conversion Factors for Units of Density

	kg/m^3	lb_m/ft^3	$lb_m/$(U.K. gal)	$lb_m/$(U.S. gal)	$slug/ft^3$	g/cm^3	t/m^3	U.K. ton/yd^3	U.S. ton/yd^3
kg/m^3	1	0.06243	0.01002	8.3454.−3	1.9403.−3	0.001	0.001	7.5248.−4	8.4278.−4
lb_m/ft^3	16.0185	1	0.16054	0.13368	0.03108	0.01602	0.01602	1.2054.−2	1.3500.−2
$lb_m/$(U.K. gal)	99.7763	6.22884	1	0.83268	0.19360	0.09976	0.09976	7.5080.−2	8.4090.−2
$lb_m/$(U.S. gal)	119.826	7.48052	1.20094	1	0.2325	0.11983	0.11983	9.0167.−2	1.0099.−1
$slug/ft^3$	515.38	32.1740	5.1653	4.3011	1	0.51538	0.51538	0.43435	0.43435
g/cm^3	1000	62.428	10.0224	8.34540	1.9403	1	1	0.75250	0.84280
t/m^3	1000	62.428	10.0224	8.34540	1.9403	1	1	0.75250	0.84280
U.K. ton/yd^3	1328.94	82.963	13.319	11.0905	2.5785	1.3289	1.3289	1	1.120
U.S. ton/yd^3	1186.5	74.075	11.892	9.9022	2.3023	1.1865	1.1865	0.89286	1

The notation 8.3454.−3 signifies 8.3454×10^{-3}.

TABLE 2.2 Conversion Factors for Units of Energy

	joule (J)	$ft \cdot lb_f$	cal_{th}	cal_{IT}	liter·atm	kJ	Btu	hp·h	kWh
joule (J)	1	0.73756	0.23901	0.23885	9.8690.−3	10^{-3}	9.4783.−4	3.7251.−7	2.7773.−7
$ft \cdot lb_f$	1.35582	1	0.32405	0.32384	1.33205.−2	1.3558.−3	1.2851.−3	5.0505.−7	3.7655.−7
cal_{th}	4.184	3.08596	1	0.99934	0.04129	4.184.−3	3.9657.−3	1.5586.−6	1.1620.−6
cal_{IT}	4.1868	3.08798	1.00066	1	0.04132	4.1868.−3	3.9683.−3	1.5596.−6	1.1628.−6
liter·atm	101.328	74.735	24.218	24.202	1	0.10325	9.6041.−2	3.7745.−5	2.8142.−5
kJ	1000	737.56	239.01	238.85	9.86896	1	0.94783	3.7251.−4	2.7773.−4
Btu	1055.05	778.16	252.16	252.00	10.4122	1.05505	1	3.9301.−4	2.9302.−4
hp·h	2.6845.+6	1.98.+6	641,617	641,197	26,494	2684.52	2544.5	1	0.74558
kWh	3.600.+6	2.6557.+6	860,564	8.6+5	35,534	3600	3412.8	1.34125	1
thermie	4.184.+6	3.087.+6	10^6	9.9934.+5	4.129.+3	4.184.+3	3.9657.+3	1.5586	1.1620

The notation 9.8690.−3, 4.184.+6 signifies 9.8690×10^{-3}, 4.184×10^6.

TABLE 2.3   Conversion Factors for Units of Mass

	g	lb_m	kg	slug	U.S. ton (short ton)	t (metric ton)	U.K. ton (long ton)
g	1	2.2046.–3	0.001	6.8522.–5	1.1023.–6	10^{-6}	9.8421.–7
lb_m	453.592	1	0.45359	0.031081	0.0005	4.5359.–4	4.4643.–4
kg	1000	2.20462	1	0.06852	1.1023.–3	0.001	9.8421.–4
slug	14,593.9	32.1740	14.5939	1	0.01609	0.01459	0.01436
U.S. ton (short ton)	907,185	2000	907.185	62.162	1	0.90719	0.89286
t (metric ton)	10^6	2204.62	1000	68.5218	1.10231	1	0.98421
U.K. ton (long ton)	1,016,047	2240	1016.05	69.621	1.12	1.01604	1

The notation 2.2046.–3 signifies 2.2046×10^{-3}.
Source: National Bureau of Standards Letter Circular 1071, 7 pp., 1976.

TABLE 2.4   Conversion Factors for Units of Pressure

	dyn/cm²*	N/m² = Pa	lb_f/ft^2	mmHg	in (H_2O)	in (Hg)	lb_f/in^2	kg/cm²	bar	atm
dyn/cm²	1	0.1	2.0886.–3	7.5006.–4	4.0148.–4	2.9530.–5	1.4504.–5	1.0197.–6	10^{-6}	9.8692.–7
N/m²	10	1	2.0886.–2	7.5006.–3	4.0148.–3	2.9530.–4	1.4504.–4	1.0197.–5	10^{-5}	9.8692.–6
lb_f/ft^2	478.79	47.879	1	0.35913	0.19221	1.4138.–2	6.9444.–3	4.8824.–4	4.7880.–4	4.7254.–4
mmHg	1333.22	133.32	2.7845	1	0.53526	0.03937	0.01934	1.3595.–3	1.3332.–3	1.3158.–3
in (H_2O)	2490.8	249.08	5.2023	1.8683	1	0.07355	0.03613	2.5399.–3	2.4908.–3	2.4585.–3
in (Hg)	33864	3386.4	70.727	25.400	13.596	1	0.49116	0.03453	0.03386	0.03342
lb_f/in^2	68,947	6894.7	144	51.715	27.680	2.03601	1	0.07031	0.06895	0.06805
kg/cm²	980,665	98,067	2048.2	735.57	393.71	28.959	14.223	1	0.98067	0.96784
bar	10^6	10^5	2088.5	750.06	401.47	29.530	14.504	1.01972	1	0.98692
atm	1,013,250	101,325	2116.2	760	406.79	29.921	14.696	1.03323	1.01325	1

* 1 dyn/cm² = 1 microbar.
The notation 2.0886.–3 signifies 2.0886×10^{-3}.

TABLE 2.5   Conversion Factors for Units of Specific Energy

	$ft \cdot lb_f/lb_m$	J/g	Btu/lb_m	cal/g
$ft \cdot lb_f/lb_m$	1	2.989.–3	1.285.–3	7.143.–4
J/g	334.54	1	0.4299	0.2388
Btu/lb_m	778.16	2.326	1	0.5556
cal/g	1400	4.184	1.8	1

TABLE 2.6   Conversion Factors for Units of Specific Energy per Degree

	$J/(g \cdot K)$	$Btu_{th}/(lb \cdot °F)$	$cal_{th}/(g \cdot °C)$	$Btu_{IT}/(lb_m \cdot °F)$	$cal_{IT}/(g \cdot °C)$
$J/(g \cdot K)$	1	0.23901	0.23901	0.23885	0.23885
$Btu_{th}/(lb_m \cdot °F)$	4.184	1	1	0.99933	0.99933
$cal_{th}/(g \cdot °C)$	4.184	1	1	0.99933	0.99933
$Btu_{IT}/(lb_m \cdot °F)$	4.1868	1.00067	1.00067	1	1
$cal_{IT}/(g \cdot °C)$	4.1868	1.00067	1.00067	1	1

TABLE 2.7 Conversion Factors for Units of Thermal Conductivity

	Btu·in/(h·ft²·°F)	W/(m·K)	kcal/(h·m·°C)	Btu/(h·ft·°F)	W/(cm·K)	cal/(s·cm·°C)	Btu·in/(s·ft²·°F)
Btu·in/(h·ft²·°F)	1	0.1441	0.1240	0.08333	1.441.–3	3.445.–4	2.777.–4
W/(m·K)	6.938	1	0.8604	0.5782	0.01	2.390.–3	1.926.–3
kcal/(h·m·°C)	8.064	1.162	1	0.6720	0.01162	2.778.–3	2.240.–3
Btu/(h·ft·°F)	12	1.730	1.488	1	0.01730	4.134.–3	3.333.–3
W/(cm·K)	694	100	86.04	57.82	1	0.2390	0.1926
cal/(s·cm·°C)	2903	418.4	360	241.9	4.184	1	0.8063
Btu·in/(s·ft²·°F)	3600	519.2	446.7	300	5.192	1.2402	1

The notation 1.441.–3 signifies 1.441×10^{-3}.

TABLE 2.8 Conversion Factors for Units of Dynamic Viscosity

	micropoise	lb$_m$/(ft·h)	centipoise	slug/(ft·h)	poise (P)	N·s/m²	Pa·s	lb$_m$/(s·ft)	lb$_f$·s/ft²
micropoise	1	2.4191.–4	10^{-4}	7.5188.–6	10^{-6}	10^{-7}	10^{-7}	6.7197.–8	2.0885.–9
lb$_m$/(ft·h)	4134	1	0.4134	3.1081.–2	4.1338.–3	4.1338.–4	4.1338.–4	2.7778.–4	8.6336.–6
centipoise	10^4	2.4191	1	7.5188.–2	0.01	0.001	0.001	6.7197.–4	2.0885.–5
slug/(ft·h)	1.3300.+5	32.174	13.300	1	0.1330	1.3300.–2	1.3300.–2	8.9372.–3	2.7778.–4
poise (P)	10^6	241.91	100	7.5188	1	0.1	0.1	6.7197.–2	2.0835.–3
N·s/m²	10^7	2419.1	1000	75.188	10	1	1	0.6720	2.0885.–2
Pa·s	10^7	2419.1	1000	75.188	10	1	1	0.6720	2.0885.–2
lb$_m$/(ft·s)	1.4882.+7	3600	1488.2	111.89	14.882	1.4882	1.4882	1	0.03108
lb$_f$·s/ft²	4.7880.+8	1.1583.+5	4.7880.+4	3600	478.80	47.880	47.880	32.174	1

1 lb$_m$/(ft·h) = 1 poundal·h/ft²; 1 P = 1 g/(cm·s).
The notation 2.4191.–4, 1.4882.+7 signifies 2.4191×10^{-4}, 1.4882×10^7.

TABLE 2.9 Conversion Factors for Units of Kinematic Viscosity

	ft²/h	stokes (St)	m²/h	ft²/s	m²/s
ft²/h	1	0.2581	0.0929	2.778.–4	2.581.–5
stokes (St)	3.8750	1	0.36	1.076.–3	10^{-4}
m²/h	10.7639	2.7778	1	2.990.–3	2.778.–4
ft²/s	3.600	929.03	334.45	1	0.09290
m²/s	38,750	10,000	3600	10.7639	1

The notation 2.581.–5 signifies 2.581×10^{-5}.
1 stoke = 1 cm²/s.

THERMOPHYSICAL PROPERTIES OF GASES

Table 2.10 treats the specific heats, dynamic viscosities, and thermal conductivities as functions of temperature only. To obtain the density of a gas, the perfect gas law may be used, i.e.,

$$P = \rho R T$$

From the specific heat and density and using other given properties, the thermal diffusivity and Prandtl number may be calculated.

For each gas, skeleton tables of the properties are given at several temperatures so that computer program checks can be made.

TABLE 2.10 Thermophysical Properties of Thirteen Common Gases Using Computer Equations

Air

| At/mol wt (kg/mol): 28.966
Gas constant (kJ/kg K): .287040
At/mol formula: (mixture) | Critical temperature (K): 132.6
Critical pressure (MPa): 3.77 |

$c_p = \sum [A(N)T^N]$	$k = \sum [C(N)T^N]$
A(0) = 0.103409E+1 A(1) = −0.2848870E-3 A(2) = 0.7816818E-6 A(3) = −0.4970786E-9 A(4) = 0.1077024E-12	Temperature range: $250 \leq T \leq 1050$ K Coefficients: C(0) = −2.276501E-3 C(4) = −1.066657E-13 C(1) = 1.2598485E-4 C(5) = 2.47663035E-17 C(2) = −1.4815235E-7 C(6) = 0.0 C(3) = 1.73550646E-10

$$\mu = \sum [B(N)T^N]$$

Temperature range: $250 \leq T < 600$ K	Temperature range: $600 \leq T \leq 1050$ K
Coefficients: B(0) = −9.8601E-1 B(4) = −5.7971299E-11 B(1) = 9.080125E-2 B(5) = 0.0 B(2) = −1.17635575E-4 B(6) = 0.0 B(3) = 1.2349703E-7	Coefficients: B(0) = 4.8856745 B(4) = −1.10398E-12 B(1) = 5.43232E-2 B(5) = 0.0 B(2) = −2.4261775E-5 B(6) = 0.0 B(3) = 7.9306E-9

Skeleton table

T (K)	c_p (kJ/kg K)	μ (Ns/m²) E6	k (W/m K) E3
300	1.0064	18.53	26.07
500	1.0317	26.82	39.48
1000	1.1415	41.77	67.21

Argon

| At/mol wt (kg/mol): 39.948
Gas constant (kJ/kg K): .208129
At/mol formula: Ar | Critical temperature (K): 150.8
Critical pressure (MPa): 4.87
Sat temp at one atmosphere (K): 87.5 |

$c_p = \sum [A(N)T^N]$	$k = \sum [C(N)T^N]$
Temperature range: $200 \leq T \leq 1600$ K Coefficients: A(0) = 0.52034 A(4) = 0.0 A(1) = 0.0 A(5) = 0.0 A(2) = 0.0 A(6) = 0.0 A(3) = 0.0	Temperature range: $200 \leq T \leq 1000$ K Coefficients: C(0) = −5.2839462E-4 C(4) = −3.22024235E-14 C(1) = 7.60706705E-5 C(5) = 1.17962552E-17 C(2) = −6.4749393E-8 C(6) = −1.86231745E-21 C(3) = 5.41874502E-11

$$\mu = \sum [B(N)T^N]$$

Temperature range: $200 \leq T < 540$ K	Temperature range: $540 \leq T \leq 1000$ K
Coefficients: B(0) = 1.22573 B(4) = 1.2939183E-9 B(1) = 5.9456964E-2 B(5) = −7.5027442E-13 B(2) = 1.897011E-4 B(6) = 0.0 B(3) = −8.171242E-7	Coefficients: B(0) = 4.03764 B(4) = −1.585569E-12 B(1) = 7.3665688E-2 B(5) = 0.0 B(2) = −3.3867E-5 B(6) = 0.0 B(3) = 1.127158E-8

Skeleton table

T (K)	c_p (kJ/kg K)	μ (Ns/m²) E6	k (W/m K) E3
300	0.5203	22.73	17.69
500	0.5203	33.66	26.42
1000	0.5203	53.52	42.71

Extracted from Ref. 4 with permission.
E-2 signifies ×10⁻², etc.

TABLE 2.10 Thermophysical Properties of Thirteen Common Gases Using Computer Equations (*Continued*)

n-Butane

At/mol wt (kg/mol): 58.124	Critical temperature (K): 408.1
Gas constant (kJ/kg K): .143044	Critical pressure (MPa): 3.65
At/mol formula: C_4H_{10}	Sat temp at one atmosphere (K): 261.5

$$c_p = \sum [A(N)T^N]$$

Temperature range: $280 \le T < 755$ K	Temperature range: $755 \le T \le 1080$ K
Coefficients:	Coefficients:
$A(0) = 2.3665134E\text{-}1$ $A(4) = 0.0$	$A(0) = 4.40126486$ $A(4) = 1.619382E\text{-}11$
$A(1) = 5.10573E\text{-}3$ $A(5) = 0.0$	$A(1) = -1.390866545E\text{-}2$ $A(5) = -2.966666E\text{-}15$
$A(2) = -4.16089E\text{-}7$ $A(6) = 0.0$	$A(2) = 3.471109E\text{-}5$ $A(6) = 0.0$
$A(3) = -1.1450804E\text{-}9$	$A(3) = -3.45278E\text{-}8$

$\mu = \sum [B(N)T^N]$	$k = \sum [C(N)T^N]$
Temperature range: $270 \le T \le 520$ K	Temperature range: $280 \le T \le 500$ K
Coefficients:	Coefficients:
$B(0) = -1.099487E\text{-}2$ $B(4) = 0.0$	$C(0) = 3.79912E\text{-}3$ $C(4) = 0.0$
$B(1) = 2.634504E\text{-}2$ $B(5) = 0.0$	$C(1) = -3.38011396E\text{-}5$ $C(5) = 0.0$
$B(2) = -3.54700854E\text{-}6$ $B(6) = 0.0$	$C(2) = 3.15886537E\text{-}7$ $C(6) = 0.0$
$B(3) = 0.0$	$C(3) = -2.25600514E\text{-}10$

Skeleton table

T (K)	c_p (kJ/kg K)	μ (Ns/m²) E6	k (W/m K) E3
300	1.700	7.573	16.00
500	2.542	12.27	37.67
1000	3.903	—	—

Carbon dioxide

At/mol wt (kg/mol): 44.01	Critical temperature (K): 304.1
Gas constant (kJ/kg K): .188919	Critical pressure (MPa): 7.38
At/mol formula: CO_2	Sat temp at one atmosphere (K): 194.7

$c_p = \sum [A(N)T^N]$	$\mu = \sum [B(N)T^N]$
Temperature range: $200 \le T \le 1000$ K	Temperature range: $200 \le T \le 1000$ K
Coefficients:	Coefficients:
$A(0) = 4.5386462E\text{-}1$ $A(4) = 2.862388E\text{-}12$	$B(0) = -8.095191E\text{-}1$ $B(4) = -1.47315277E\text{-}12$
$A(1) = 1.5334795E\text{-}3$ $A(5) = -1.6962E\text{-}15$	$B(1) = 6.0395329E\text{-}2$ $B(5) = 0.0$
$A(2) = -4.195556E\text{-}7$ $A(6) = 3.717285E\text{-}19$	$B(2) = -2.824853E\text{-}5$ $B(6) = 0.0$
$A(3) = -1.871946E\text{-}9$	$B(3) = 9.843776E\text{-}9$

$$k = \sum [C(N)T^N]$$

Temperature range: $200 \le T < 600$ K	Temperature range: $600 \le T \le 1000$ K
Coefficients:	Coefficients:
$C(0) = 2.971488E\text{-}3$ $C(4) = 2.68500151E\text{-}13$	$C(0) = 6.085375E\text{-}2$ $C(4) = 3.27864115E\text{-}13$
$C(1) = -1.33471677E\text{-}5$ $C(5) = 0.0$	$C(1) = -3.63680275E\text{-}4$ $C(5) = 0.0$
$C(2) = 3.14443715E\text{-}7$ $C(6) = 0.0$	$C(2) = 1.0134366E\text{-}6$ $C(6) = 0.0$
$C(3) = -4.75106178E\text{-}10$	$C(3) = -9.7042356E\text{-}10$

Skeleton table

T (K)	c_p (kJ/kg K)	μ (Ns/m²) E6	k (W/m K) E3
300	0.845	15.02	16.61
500	1.013	23.46	32.30
1000	1.234	39.71	68.05

Extracted from Ref. 4 with permission.
E-2 signifies ×10⁻², etc.

TABLE 2.10 Thermophysical Properties of Thirteen Common Gases Using Computer Equations (*Continued*)

Carbon monoxide

At/mol wt (kg/mol): 28.011	Critical temperature (K): 132.9
Gas constant (kJ/kg K): .296828	Critical pressure (MPa): 3.5
At/mol formula: CO	Sat temp at one atmosphere (K): 81.6

$$c_p = \sum [A(N)T^N]$$

Temperature range: $250 \le T \le 1050$ K

Coefficients:
A(0) = 1.020802	A(4) = −7.93722E-12
A(1) = 3.82075E-4	A(5) = 4.291972E-15
A(2) = −2.4945E-6	A(6) = −8.903274E-19
A(3) = 6.81145E-9	

$\mu = \sum [B(N)T^N]$	$k = \sum [C(N)T^N]$
Temperature range: $250 \le T \le 1050$ K	Temperature range: $250 \le T \le 1050$ K

Coefficients:
B(0) = −5.24575E-1	B(4) = −2.83747E-11	C(0) = −7.41704398E-4	C(4) = 3.65528473E-14
B(1) = 7.9606E-2	B(5) = 5.317831E-15	C(1) = 9.87435265E-5	C(5) = −1.2427179E-17
B(2) = −7.82295E-5	B(6) = 0.0	C(2) = −3.77511167E-8	C(6) = 0.0
B(3) = 6.2821488E-8		C(3) = −1.99334224E-11	

Skeleton table

T (K)	c_p (kJ/kg K)	μ (Ns/m²) E6	k (W/m K) E3
300	1.040	17.80	25.21
500	1.064	25.97	38.60
1000	1.184	40.62	64.44

Ethane

At/mol wt (kg/mol): 30.07	Critical temperature (K): 305.4
Gas constant (kJ/kg K): .276498	Critical pressure (MPa): 4.88
At/mol formula: C_2H_6	Sat temp at one atmosphere (K): 184.6

$$c_p = \sum [A(N)T^N]$$

Temperature range: $280 \le T < 755$ K	Temperature range: $755 \le T \le 1080$ K

Coefficients:
A(0) = 5.319795E-1	A(4) = 0.0	A(0) = 3.7183729	A(4) = 1.382794E-11
A(1) = 3.755877E-3	A(5) = 0.0	A(1) = −1.0891558E-2	A(5) = −2.52553E-15
A(2) = 1.789289E-6	A(6) = 0.0	A(2) = 2.95115E-5	A(6) = 0.0
A(3) = −2.13225E-9		A(3) = −2.95597E-8	

$\mu = \sum [B(N)T^N]$	$k = \sum [C(N)T^N]$
Temperature range: $200 \le T \le 1000$ K	Temperature range: $200 \le T \le 1000$ K

Coefficients:
B(0) = −5.107728E-1	B(4) = 0.0	C(0) = −3.83815197E-2	C(4) = −1.369896E-11
B(1) = 3.76582E-2	B(5) = 0.0	C(1) = 5.47282126E-4	C(5) = 1.05765043E-14
B(2) = −1.59412113E-5	B(6) = 0.0	C(2) = −2.80760648E-6	C(6) = −3.16347435E-18
B(3) = 3.906E-9		C(3) = 8.74854603E-9	

Skeleton table

T (K)	c_p (kJ/kg K)	μ (Ns/m²) E6	k (W/m K) E3
300	1.762	9.457	21.76
500	2.591	14.82	51.83
1000	4.081	25.11	163.9

Extracted from Ref. 4 with permission.
E-2 signifies ×10⁻², etc.

TABLE 2.10   Thermophysical Properties of Thirteen Common Gases Using Computer Equations (*Continued*)

Helium

At/mol wt (kg/mol): 4.003	Critical temperature (K): 5.189
Gas constant (kJ/kg K): 2.077022	Critical pressure (MPa): .23
At/mol formula: He	Sat temp at one atmosphere (K): 4.3

$$c_p = \sum [A(N)T^N]$$

Temperature range: $250 \le T \le 1050$ K

Coefficients:

A(0) = 5.1931	A(2) = 0.0	A(4) = 0.0	A(6) = 0.0
A(1) = 0.0	A(3) = 0.0	A(5) = 0.0	

$$\mu = \sum [B(N)T^N]$$

Temperature range: $250 \le T < 500$ K	Temperature range: $500 \le T \le 1050$ K

Coefficients:		Coefficients:	
B(0) = 3.9414E-1	B(4) = −2.4278655E-8	B(0) = 7.442412	B(4) = 0.0
B(1) = 1.7213335E-1	B(5) = 3.641644E-11	B(1) = 4.6649873E-2	B(5) = 0.0
B(2) = −1.38733E-3	B(6) = −2.14117E-14	B(2) = −1.0385665E-5	B(6) = 0.0
B(3) = 8.020045E-6		B(3) = 1.35269E-9	

$$k = \sum [C(N)T^N]$$

Temperature range: $250 \le T < 300$ K

Coefficients:

C(0) = 1.028793E-2	C(4) = −1.3477236E-11
C(1) = 8.51625139E-4	C(5) = 0.0
C(2) = −3.14258034E-6	C(6) = 0.0
C(3) = 1.02188556E-8	

Temperature range: $300 \le T < 500$ K	Temperature range: $500 \le T \le 1050$ K

Coefficients:		Coefficients:	
C(0) = −7.761491E-3	C(4) = 0.0	C(0) = −9.0656E-2	C(4) = −1.26457196E-13
C(1) = 8.66192033E-4	C(5) = 0.0	C(1) = 9.37593087E-4	C(5) = 0.0
C(2) = −1.5559338E-6	C(6) = 0.0	C(2) = −9.13347535E-7	C(6) = 0.0
C(3) = 1.40150565E-9		C(3) = 5.55037072E-10	

Skeleton table

T (K)	c_p (kJ/kg K)	μ (Ns/m^2) E6	k (W/m K) E3
300	5.193	19.94	149.7
500	5.193	28.17	211.5
1000	5.193	45.06	362.2

Extracted from Ref. 4 with permission.
E-2 signifies ×10^{-2}, etc.

TABLE 2.10 Thermophysical Properties of Thirteen Common Gases Using Computer Equations (*Continued*)

<table>
<tr><th colspan="2">Hydrogen</th></tr>
</table>

At/mol wt (kg/mol): 2.016
Gas constant (kJ/kg K): 4.124289
At/mol formula: H_2

Critical temperature (K): 33.3
Critical pressure (MPa): 1.3
Sat temp at one atmosphere (K): 20.4

$$c_p = \sum [A(N)T^N]$$

Temperature range: $250 \leq T < 425$ K

Coefficients:
A(0) = 5.0066253
A(1) = 1.01569422E-1
A(2) = -6.02891517E-4
A(3) = 2.7375894E-6

A(4) = -8.4758275E-9
A(5) = 1.43800374E-11
A(6) = -9.8072403E-15

Temperature range: $425 \leq T < 490$ K

Coefficients:
A(0) = 1.44947E+1   A(4) = 0.0
A(1) = 0.0   A(5) = 0.0
A(2) = 0.0   A(6) = 0.0
A(3) = 0.0

Temperature range: $490 \leq T \leq 1050$ K

Coefficients:
A(0) = 1.4920082E+1   A(4) = 0.0
A(1) = -1.996917584E-3   A(5) = 0.0
A(2) = 2.540615E-6   A(6) = 0.0
A(3) = -4.7588954E-10

$$\mu = \sum [B(N)T^N]$$

Temperature range: $250 \leq T < 500$ K

Coefficients:
B(0) = -1.35666E-1   B(4) = -5.23104E-9
B(1) = 6.84115878E-2   B(5) = 7.4490972E-12
B(2) = -3.928747E-4   B(6) = -4.250937E-15
B(3) = 1.8996E-6

Temperature range: $500 \leq T \leq 1050$ K

Coefficients:
B(0) = 2.72941   B(4) = -5.2889938E-13
B(1) = 2.3224377E-2   B(5) = 0.0
B(2) = -7.6287854E-6   B(6) = 0.0
B(3) = 2.92585E-9

$$k = \sum [C(N)T^N]$$

Temperature range: $250 \leq T < 500$ K

Coefficients:
C(0) = 2.009705E-2   C(4) = 5.52407932E-12
C(1) = 3.234622E-4   C(5) = 0.0
C(2) = 2.1637249E-6   C(6) = 0.0
C(3) = -6.49151204E-9

Temperature range: $500 \leq T \leq 1050$ K

Coefficients:
C(0) = 1.083105E-1   C(4) = 4.6468625E-14
C(1) = 2.21163789E-4   C(5) = 0.0
C(2) = 2.26380948E-7   C(6) = 0.0
C(3) = -1.74258636E-10

Skeleton table

T (K)	c_p (kJ/kg K)	μ (Ns/m^2) E6	k (W/m K) E3
300	14.27	8.949	181.3
500	14.50	12.72	256.6
1000	14.99	20.72	428.1

Extracted from Ref. 4 with permission.
E-2 signifies $\times 10^{-2}$, etc.

TABLE 2.10 Thermophysical Properties of Thirteen Common Gases Using Computer Equations (*Continued*)

Methane

At/mol wt (kg/mol): 16.043
Gas constant (kJ/kg K): .518251
At/mol formula: CH_4

Critical temperature (K): 190.5
Critical pressure (MPa): 4.6
Sat temp at one atmosphere (K): 111.5

$$c_p = \sum [A(N)T^N]$$

Temperature range: $280 \le T < 755$ K		Temperature range: $755 \le T \le 1080$ K	
Coefficients:		Coefficients:	
A(0) = 1.9165258	A(4) = 0.0	A(0) = 1.04356E+1	A(4) = 3.9030203E-11
A(1) = −1.09269E-3	A(5) = 0.0	A(1) = −4.2025284E-2	A(5) = −7.1345169E-15
A(2) = 8.696605E-6	A(6) = 0.0	A(2) = 8.849006E-5	A(6) = 0.0
A(3) = −5.2291144E-9		A(3) = −8.4304566E-8	

$$\mu = \sum [B(N)T^N]$$ $$k = \sum [C(N)T^N]$$

Temperature range: $200 \le T \le 1000$ K		Temperature range: $200 \le T \le 1000$ K	
Coefficients:		Coefficients:	
B(0) = 2.968267E-1	B(4) = 7.543269E-11	C(0) = −1.3401499E-2	C(4) = −9.1405505E-12
B(1) = 3.711201E-2	B(5) = −2.7237166E-14	C(1) = 3.6630706E-4	C(5) = 6.7896889E-15
B(2) = 1.218298E-5	B(6) = 0.0	C(2) = −1.82248608E-6	C(6) = −1.95048736E-18
B(3) = −7.02426E-8		C(3) = 5.93987998E-9	

Skeleton table

T (K)	c_p (kJ/kg K)	μ (Ns/m^2) E6	k (W/m K) E3
300	2.230	11.18	33.88
500	2.891	16.98	67.03
1000	4.491	27.54	169.0

Nitrogen

At/mol wt (kg/mol): 28.013
Gas constant (kJ/kg K): .296798
At/mol formula: N_2

Critical temperature (K): 126.2
Critical pressure (MPa): 3.4
Sat temp at one atmosphere (K): 77.3

$$c_p = \sum [A(N)T^N]$$

Temperature range: $280 \le T < 590$ K		Temperature range: $590 \le T \le 1080$ K	
Coefficients:		Coefficients:	
A(0) = 1.088047	A(4) = 0.0	A(0) = 1.4055077	A(4) = 2.08491259E-12
A(1) = −3.55968E-4	A(5) = 0.0	A(1) = −2.1894566E-3	A(5) = −3.7903033E-16
A(2) = 7.2907605E-7	A(6) = 0.0	A(2) = 4.7852898E-6	A(6) = 0.0
A(3) = −2.8861556E-10		A(3) = −4.540166E-9	

$$\mu = \sum [B(N)T^N]$$ $$k = \sum [C(N)T^N]$$

Temperature range: $250 \le T \le 1050$ K		Temperature range: $250 \le T \le 1050$ K	
Coefficients:		Coefficients:	
B(0) = 2.5465E-2	B(4) = −1.5622457E-11	C(0) = −1.5231785E-3	C(4) = −6.36537349E-14
B(1) = 7.5336535E-2	B(5) = 2.249666E-15	C(1) = 1.18879965E-4	C(5) = 1.47167023E-17
B(2) = −6.51566245E-5	B(6) = 0.0	C(2) = −1.2092845E-7	C(6) = 0.0
B(3) = 4.34945E-8		C(3) = 1.15567802E-10	

Skeleton table

T (K)	c_p (kJ/kg K)	μ (Ns/m^2) E6	k (W/m K) E3
300	1.039	17.82	25.90
500	1.056	25.94	38.61
1000	1.167	40.33	63.06

Extracted from Ref. 4 with permission.
E-2 signifies ×10^{-2}, etc.

TABLE 2.10 Thermophysical Properties of Thirteen Common Gases Using Computer Equations (*Continued*)

Oxygen

At/mol wt (kg/mol): 31.999	Critical temperature (K): 154.6
Gas constant (kJ/kg K): .259832	Critical pressure (MPa): 5.04
At/mol formula: O_2	Sat temp at one atmosphere (K): 90

$$c_p = \sum [A(N)T^N]$$

Temperature range: $250 \leq T < 590$ K		Temperature range: $590 \leq T \leq 1050$ K	
Coefficients:		Coefficients:	
$A(0) = 9.29247E\text{-}1$	$A(4) = 0.0$	$A(0) = 5.977293E\text{-}1$	$A(4) = -1.1772692E\text{-}13$
$A(1) = -3.220603E\text{-}4$	$A(5) = 0.0$	$A(1) = 1.183704E\text{-}3$	$A(5) = 0.0$
$A(2) = 1.166523E\text{-}6$	$A(6) = 0.0$	$A(2) = -1.156226E\text{-}6$	$A(6) = 0.0$
$A(3) = -7.1157865E\text{-}10$		$A(3) = 5.82171E\text{-}10$	

$$\mu = \sum [B(N)T^N]$$

Temperature range: $250 \leq T \leq 1050$ K

Coefficients:	
$B(0) = -3.97863E\text{-}1$	$B(4) = -1.690435E\text{-}11$
$B(1) = 8.7605894E\text{-}2$	$B(5) = 2.534147E\text{-}15$
$B(2) = -7.064124E\text{-}5$	$B(6) = 0.0$
$B(3) = 4.6287E\text{-}8$	

$$k = \sum [C(N)T^N]$$

Temperature range: $250 \leq T < 1000$ K		Temperature range: $1000 \leq T \leq 1050$ K	
Coefficients:		Coefficients:	
$C(0) = -7.6727798E\text{-}4$	$C(4) = 0.0$	$C(0) = -1.8654526E\text{-}1$	$C(4) = -7.84907953E\text{-}14$
$C(1) = 1.03560076E\text{-}4$	$C(5) = 0.0$	$C(1) = 7.05649428E\text{-}4$	$C(5) = 0.0$
$C(2) = -4.62034365E\text{-}8$	$C(6) = 0.0$	$C(2) = -7.71025034E\text{-}7$	$C(6) = 0.0$
$C(3) = 1.51980292E\text{-}11$		$C(3) = 4.02143777E\text{-}10$	

Skeleton table			
T (K)	c_p (kJ/kg K)	μ (Ns/m^2) E6	k (W/m K) E3
300	0.918	20.65	26.55
500	0.970	30.55	41.36
1000	1.090	48.48	71.79

Extracted from Ref. 4 with permission.
E-2 signifies $\times 10^{-2}$, etc.

TABLE 2.10 Thermophysical Properties of Thirteen Common Gases Using Computer Equations (*Continued*)

Propane

At/mol wt (kg/mol): 44.097
Gas constant (kJ/kg K): 0.207519
At/mol formula: C_3H_8

Critical temperature (K): 369.8
Critical pressure (MPa): 4.26
Sat temp at one atmosphere (K): 231.1

$$c_p = \sum [A(N)T^N]$$

Temperature range: $280 \leq T < 755$ K		Temperature range: $755 \leq T \leq 1080$ K	
Coefficients:		Coefficients:	
A(0) = 8.41607E-2	A(4) = 0.0	A(0) = 3.47456	A(4) = 1.2466175E-11
A(1) = 5.7701407E-3	A(5) = 0.0	A(1) = -9.4956207E-3	A(5) = -2.271073E-15
A(2) = -1.292127E-6	A(6) = 0.0	A(2) = 2.643558E-5	A(6) = 0.0
A(3) = -6.9945925E-10		A(3) = -2.6640384E-8	

$$\mu = \sum [B(N)T^N] \qquad\qquad k = \sum [C(N)T^N]$$

Temperature range: $270 \leq T \leq 600$ K		Temperature range: $270 \leq T \leq 500$ K	
Coefficients:		Coefficients:	
B(0) = -3.543711E-1	B(4) = 0.0	C(0) = -1.07682209E-2	C(4) = 0.0
B(1) = 3.080096E-2	B(5) = 0.0	C(1) = 8.38590352E-5	C(5) = 0.0
B(2) = -6.99723E-6	B(6) = 0.0	C(2) = 4.22059864E-8	C(6) = 0.0
B(3) = 0.0		C(3) = 0.0	

Skeleton table

T (K)	c_p (kJ/kg K)	μ (Ns/m²) E6	k (W/m K) E3
300	1.680	8.256	18.19
500	2.559	13.30	41.71
1000	3.969	—	—

Sulfur dioxide

At/mol wt (kg/mol): 64.063
Gas constant (kJ/kg K): .129784
At/mol formula: SO_2

Critical temperature (K): 430.7
Critical pressure (MPa): 7.88
Sat temp at one atmosphere (K): 268.4

$$c_p = \sum [A(N)T^N] \qquad\qquad \mu = \sum [B(N)T^N]$$

Temperature range: $300 \leq T \leq 1100$ K		Temperature range: $300 \leq T \leq 1100$ K	
Coefficients:		Coefficients:	
A(0) = 4.32805E-1	A(4) = 1.0409341E-12	B(0) = -1.141748	B(4) = 0.0
A(1) = 5.9994156E-4	A(5) = -2.5313735E-16	B(1) = 5.1281456E-2	B(5) = 0.0
A(2) = 4.593367E-7	A(6) = 0.0	B(2) = -1.3886282E-5	B(6) = 0.0
A(3) = -1.433024E-9		B(3) = 2.15266E-9	

$$k = \sum [C(N)T^N]$$

Temperature range: $300 \leq T \leq 900$ K

Coefficients:	
C(0) = -1.86270694E-2	C(4) = -7.53585825E-12
C(1) = 3.19110134E-4	C(5) = 5.48078289E-15
C(2) = -1.73644245E-6	C(6) = -1.56355469E-18
C(3) = 5.09847985E-9	

Skeleton table

T (K)	c_p (kJ/kg K)	μ (Ns/m²) E6	k (W/m K) E3
300	0.623	13.05	9.623
500	0.726	21.30	19.98
900	0.834	35.33	39.98

Extracted from Ref. 4 with permission.
E-2 signifies ×10⁻², etc.

TABLE 2.11   Compressibility Factors

		Compressibility factor Z of air*												
		Pressure, bar												
T (K)	1	5	10	20	40	60	80	100	150	200	250	300	400	500
75	0.0052	0.0260	0.0519	0.1036	0.2063	0.3082	0.4094	0.5099	0.7581	1.0025	—	—	—	—
80	—	0.0250	0.0499	0.0995	0.1981	0.2958	0.3927	0.4887	0.7258	0.9588	1.1931	1.4139	—	—
90	0.9764	0.0236	0.0471	0.0940	0.1866	0.2781	0.3686	0.4581	0.6779	0.8929	1.1098	1.3110	1.7161	2.1105
100	0.9797	0.8872	0.0453	0.0900	0.1782	0.2635	0.3498	0.4337	0.6386	0.8377	1.0395	1.2227	1.5937	1.9536
120	0.9880	0.9373	0.8660	0.6730	0.1778	0.2557	0.3371	0.4132	0.5964	0.7720	0.9530	1.1076	1.5091	1.7366
140	0.9927	0.9614	0.9205	0.8297	0.5856	0.3313	0.3737	0.4340	0.5909	0.7699	0.9114	1.0393	1.3202	1.5903
160	0.9951	0.9748	0.9489	0.8954	0.7803	0.6603	0.5696	0.5489	0.6340	0.7564	0.8840	1.0105	1.2585	1.4970
180	0.9967	0.9832	0.9660	0.9314	0.8625	0.7977	0.7432	0.7084	0.7180	0.7986	0.9000	1.0068	1.2232	1.4361
200	0.9978	0.9886	0.9767	0.9539	0.9100	0.8701	0.8374	0.8142	0.8061	0.8549	0.9311	1.0185	1.2054	1.3944
250	0.9992	0.9957	0.9911	0.9822	0.9671	0.9549	0.9463	0.9411	0.9450	0.9713	1.0152	1.0702	1.1990	1.3392
300	0.9999	0.9987	0.9974	0.9950	0.9917	0.9901	0.9903	0.9930	1.0074	1.0326	1.0669	1.1089	1.2073	1.3163
350	1.0000	1.0002	1.0004	1.0014	1.0038	1.0075	1.0121	1.0183	1.0377	1.0635	1.0947	1.1303	1.2116	1.3015
400	1.0002	1.0012	1.0025	1.0046	1.0100	1.0159	1.0229	1.0312	1.0533	1.0795	1.1087	1.1411	1.2117	1.2890
450	1.0003	1.0016	1.0034	1.0063	1.0133	1.0210	1.0287	1.0374	1.0614	1.0913	1.1183	1.1463	1.2090	1.2778
500	1.0003	1.0020	1.0034	1.0074	1.0151	1.0234	1.0323	1.0410	1.0650	1.0913	1.1183	1.1463	1.2051	1.2667
600	1.0004	1.0022	1.0039	1.0081	1.0164	1.0253	1.0340	1.0434	1.0678	1.0920	1.1172	1.1427	1.1947	1.2475
800	1.0004	1.0020	1.0038	1.0077	1.0157	1.0240	1.0321	1.0408	1.0621	1.0844	1.1061	1.1283	1.1720	1.2150
1000	1.0004	1.0018	1.0037	1.0068	1.0142	1.0215	1.0290	1.0365	1.0556	1.0744	1.0948	1.1131	1.1515	1.1889

		Compressibility factor Z of argon†										
		Pressure, bar										
T (K)	Sat. liquid	Sat. vapor	1	50	100	150	200	250	300	400	500
85	0.0031	0.9706	0.0040	—	—	—	—	—	—	—	—
90	0.0052	0.9579	0.9684	0.1919	0.3801	0.5648	0.7467	0.9260	—	—	—
95	0.0080	0.9415	0.9731	0.1859	0.3675	0.5456	0.7205	0.8928	1.0625	1.3959	—
100	0.0119	0.9220	0.9773	0.1807	0.3567	0.5288	0.6975	0.8634	1.0267	1.3470	1.6932
120	0.0418	0.8112	0.9866	0.1683	0.3280	0.4818	0.6311	0.7770	0.9197	1.1981	1.4978
140	0.1153	0.6144	0.9915	0.1737	0.3230	0.4636	0.5985	0.7294	0.8568	1.1040	1.3699
160	—	—	0.9943	0.6161	0.3610	0.4766	0.5954	0.7122	0.8265	1.0478	1.2866
180	—	—	0.9962	0.7754	0.5432	0.5405	0.6246	0.7014	0.8200	1.0165	1.2321
200	—	—	0.9972	0.8509	0.7121	0.6540	0.6870	0.7555	0.8360	1.0051	1.1982
250	—	—	0.9988	0.9374	0.8877	0.8602	0.8591	0.8812	0.9208	1.0263	1.1713
300	—	—	0.9995	0.9730	0.9552	0.9482	0.9533	0.9694	0.9950	1.0673	1.1786
350	—	—	0.9998	0.9911	0.9880	0.9915	0.9987	1.0179	1.0399	1.0971	1.1902
400	—	—	1.0001	1.0006	1.0056	1.0148	1.0280	1.0450	1.0656	1.1157	1.1976
450	—	—	1.0001	1.0063	1.0154	1.0276	1.0427	1.0602	1.0804	1.1258	1.2002
500	—	—	1.0002	1.0090	1.0205	1.0342	1.0501	1.0678	1.0874	1.1301	1.1997
600	—	—	1.0003	1.0118	1.0250	1.0394	1.0553	1.0723	1.0904	1.1291	1.1933
700	—	—	1.0003	1.0128	1.0261	1.0399	1.0551	1.0709	1.0874	1.1224	1.1821
800	—	—	1.0003	1.0126	1.0258	1.0396	1.0532	1.0678	1.0830	1.1147	1.1707
900	—	—	1.0003	1.0122	1.0250	1.0378	1.0506	1.0643	1.0782	1.1068	1.1596
1000	—	—	1.0002	1.0119	1.0239	1.0364	1.0484	1.0608	1.0736	1.0999	1.1497

Note: See page 2.15 for footnotes.

TABLE 2.11 Compressibility Factors (*Continued*)

Compressibility factor Z of carbon dioxide[‡]

	Pressure, bar											
T (°C)	1	5	10	20	40	60	80	100	200	300	400	500
0	0.9933	0.9658	0.9294	0.8496	—	—	—	—	—	—	—	—
50	0.9964	0.9805	0.9607	0.9195	0.8300	0.7264	0.5981	0.4239	—	—	—	—
100	0.9977	0.9883	0.9764	0.9524	0.9034	0.8533	0.8022	0.7514	0.5891	0.6420	—	—
150	0.9985	0.9927	0.9853	0.9705	0.9416	0.9131	0.8854	0.8590	0.7651	0.7623	0.8235	0.9098
200	0.9991	0.9953	0.9908	0.9818	0.9640	0.9473	0.9313	0.9170	0.8649	0.8619	0.8995	0.9621
250	0.9994	0.9971	0.9943	0.9886	0.9783	0.9684	0.9593	0.9511	0.9253	0.9294	0.9508	1.0096
300	0.9996	0.9982	0.9967	0.9936	0.9875	0.9822	0.9773	0.9733	0.9640	0.9746	1.0030	1.0464
350	0.9998	0.9991	0.9983	0.9964	0.9938	0.9914	0.9896	0.9882	0.9895	1.0053	1.0340	1.0734
400	0.9999	0.9997	0.9994	0.9989	0.9982	0.9979	0.9979	0.9984	1.0073	1.0266	1.0559	1.0928
450	1.0000	1.0000	1.0003	1.0005	1.0013	1.0023	1.0038	1.0056	1.0170	1.0412	1.0709	1.1067
500	1.0000	1.0004	1.0008	1.0015	1.0035	1.0056	1.0079	1.0107	1.0282	1.0522	1.0820	1.1165
600	1.0000	1.0007	1.0013	1.0030	1.0062	1.0093	1.0129	1.0168	1.0386	1.0648	1.0948	1.1277
700	1.0003	1.0010	1.0017	1.0036	1.0073	1.0161	1.0155	1.0198	1.0436	1.0707	1.1000	1.1318
800	1.0002	1.0009	1.0019	1.0040	1.0082	1.0122	1.0168	1.0212	1.0458	1.0731	1.1016	1.1324
900	1.0002	1.0009	1.0020	1.0041	1.0083	1.0128	1.0171	1.0221	1.0463	1.0726	1.1012	1.1303
1000	1.0002	1.0009	1.0021	1.0042	1.0084	1.0128	1.0172	1.0218	1.0460	1.0725	1.0725	1.1274

Compressibility factor Z of methane[§]

	Pressure, bar											
T (K)	1	5	10	20	40	60	80	100	200	300	400	500
100	0.0044	0.0219	0.0437	0.0874	0.1741	0.2604	0.3459	0.4313	0.8498	1.2585	1.6579	2.0492
150	0.9856	0.9243	0.8333	0.0708	0.1401	0.2078	0.2748	0.3405	0.6573	0.9602	1.2519	1.5359
200	0.9937	0.9682	0.9350	0.8629	0.6858	0.3755	0.3218	0.3657	0.6148	0.8564	1.0894	1.3145
250	0.9972	0.9841	0.9678	0.9356	0.8694	0.8035	0.7403	0.6889	0.6953	0.8593	1.0383	1.2172
300	0.9982	0.9915	0.9828	0.9663	0.9342	0.9042	0.8773	0.8548	0.8280	0.9140	1.0417	1.1812
350	0.9988	0.9954	0.9905	0.9821	0.9657	0.9513	0.9390	0.9293	0.9226	0.9775	1.0678	1.1751
400	0.9995	0.9976	0.9957	0.9908	0.9833	0.9771	0.9721	0.9691	0.9783	1.0258	1.0968	1.1821
450	0.9999	0.9996	0.9991	0.9965	0.9941	0.9923	0.9917	0.9922	1.0128	1.0577	1.1195	1.1916
500	1.0000	1.0000	1.0000	1.0003	1.0009	1.0021	1.0043	1.0068	1.0335	1.0780	1.1347	1.1990
600	1.0002	1.0010	1.0021	1.0040	1.0083	1.0128	1.0175	1.0227	1.0555	1.0989	1.1495	1.2049
700	1.0003	1.0014	1.0028	1.0061	1.0116	1.0177	1.0237	1.0298	1.0646	1.1056	1.1522	1.2023
800	1.0003	1.0017	1.0034	1.0068	1.0130	1.0198	1.0264	1.0331	1.0680	1.1071	1.1500	1.1956
900	1.0004	1.0018	1.0036	1.0071	1.0137	1.0206	1.0274	1.0340	1.0680	1.1056	1.1457	1.1878
1000	1.0004	1.0014	1.0036	1.0072	1.0142	1.0208	1.0275	1.0342	1.0678	1.1033	1.1400	1.1790

TABLE 2.11 Compressibility Factors (*Continued*)

Compressibility factor Z of nitrogen[¶]

					Pressure, bar							
T (K)	1	5	10	20	40	60	80	100	200	300	400	500
70	0.0057	0.0287	0.0573	0.1143	0.2277	0.3400	0.4516	0.5623	1.1044	1.6308	Solid	Solid
80	0.9593	0.0264	0.0528	0.1053	0.2093	0.3122	0.4140	0.5148	1.0061	1.4797	1.9396	2.3879
90	0.9722	0.0251	0.0500	0.0996	0.1975	0.2938	0.3888	0.4826	0.9362	1.3700	1.7890	2.1962
100	0.9798	0.8910	0.0487	0.0966	0.1905	0.2823	0.3720	0.4605	0.8840	1.2852	1.6707	2.0441
120	0.9883	0.9397	0.8732	0.7059	0.1975	0.2822	0.3641	0.4438	0.8188	1.1684	1.5015	1.8223
140	0.9927	0.9635	0.9253	0.8433	0.6376	0.4251	0.4278	0.4799	0.7942	1.0996	1.3920	1.6726
160	0.9952	0.9766	0.9529	0.9042	0.8031	0.7017	0.6304	0.6134	0.8107	1.0708	1.3275	1.5762
180	0.9967	0.9846	0.9690	0.9381	0.8782	0.8125	0.7784	0.7530	0.8550	1.0669	1.2893	1.5105
200	0.9978	0.9897	0.9791	0.9592	0.9212	0.8882	0.8621	0.8455	0.9067	1.0760	1.2683	1.4631
250	0.9992	0.9960	0.9924	0.9857	0.9741	0.9655	0.9604	0.9589	1.0048	1.1143	1.2501	1.3962
300	0.9998	0.9990	0.9983	0.9971	0.9964	0.9973	1.0000	1.0052	1.0559	1.1422	1.2480	1.3629
350	1.0001	1.0007	1.0011	1.0029	1.0069	1.0125	1.0189	1.0271	1.0810	1.1560	1.2445	1.3405
400	1.0002	1.0011	1.0024	1.0057	1.0125	1.0199	1.0283	1.0377	1.0926	1.1609	1.2382	1.3216
450	1.0003	1.0018	1.0033	1.0073	1.0153	1.0238	1.0332	1.0430	1.0973	1.1606	1.2303	1.3043
500	1.0004	1.0020	1.0040	1.0081	1.0167	1.0257	1.0350	1.0451	1.0984	1.1575	1.2213	1.2881
600	1.0004	1.0021	1.0040	1.0084	1.0173	1.0263	1.0355	1.0450	1.0951	1.1540	1.2028	1.2657
800	1.0004	1.0017	1.0036	1.0074	1.0157	1.0237	1.0320	1.0402	1.0832	1.1264	1.1701	1.2140
1000	1.0003	1.0015	1.0034	1.0067	1.0136	1.0205	1.0275	1.0347	1.0714	1.1078	1.1449	1.1814

Compressibility factor Z of oxygen**

					Pressure, bar							
T (K)	1	5	10	20	40	60	80	100	200	300	400	500
75	0.0043	0.0213	0.0425	0.0849	0.1693	0.2533	0.3368	0.4200	0.8301	1.2322	1.6278	2.0175
80	0.0041	0.0203	0.0406	0.0811	0.1616	0.2418	0.3214	0.4007	0.7912	1.1738	1.5495	1.9196
90	0.0038	0.0188	0.0376	0.0750	0.1494	0.2233	0.2966	0.3696	0.7281	1.0780	1.4211	1.7580
100	0.9757	0.0177	0.0354	0.0705	0.1404	0.2096	0.2783	0.3464	0.6798	1.0040	1.3206	1.6309
120	0.9855	0.9246	0.8367	0.0660	0.1302	0.1935	0.2558	0.3173	0.6148	0.8999	1.1762	1.4456
140	0.9911	0.9535	0.9034	0.7852	0.1334	0.1940	0.2527	0.3099	0.5815	0.8374	1.0832	1.3214
160	0.9939	0.9697	0.9379	0.8689	0.6991	0.3725	0.2969	0.3378	0.5766	0.8058	1.0249	1.2364
180	0.9960	0.9793	0.9579	0.9134	0.8167	0.7696	0.5954	0.5106	0.6043	0.8025	0.9990	1.1888
200	0.9970	0.9853	0.9705	0.9399	0.8768	0.8140	0.7534	0.6997	0.6720	0.8204	0.9907	1.1623
250	0.9987	0.9938	0.9870	0.9736	0.9477	0.9237	0.9030	0.8858	0.8563	0.9172	1.0222	1.1431
300	0.9994	0.9968	0.9941	0.9884	0.9771	0.9676	0.9597	0.9542	0.9560	0.9972	1.0689	1.1572
350	0.9998	0.9990	0.9979	0.9961	0.9919	0.9890	0.9870	0.9870	1.0049	1.0451	1.1023	1.1722
400	1.0000	1.0000	1.0000	1.0000	1.0003	1.0011	1.0022	1.0045	1.0305	1.0718	1.1227	1.1816
450	1.0002	1.0007	1.0015	1.0024	1.0048	1.0074	1.0106	1.0152	1.0445	1.0859	1.1334	1.1859
500	1.0002	1.0011	1.0022	1.0038	1.0075	1.0115	1.0161	1.0207	1.0523	1.0927	1.1380	1.1866
600	1.0003	1.0014	1.0024	1.0052	1.0102	1.0153	1.0207	1.0266	1.0582	1.0961	1.1374	1.1803
800	1.0003	1.0014	1.0026	1.0055	1.0109	1.0164	1.0219	1.0271	1.0565	1.0888	1.1231	1.1582
1000	1.0003	1.0013	1.0026	1.0053	1.0101	1.0149	1.0198	1.0253	1.0507	1.0783	1.1072	1.1369

TABLE 2.11 Compressibility Factors (*Continued*)

<table>
<tr><td colspan="14" align="center">Compressibility factor Z of propylene[††]</td></tr>
<tr><td></td><td colspan="13" align="center">Pressure, bar</td></tr>
<tr><td>T (K)</td><td>1</td><td>5</td><td>10</td><td>20</td><td>40</td><td>60</td><td>80</td><td>100</td><td>200</td><td>400</td><td>600</td><td>800</td><td>1000</td></tr>
<tr><td>200</td><td>0.004</td><td>0.008</td><td>0.039</td><td>0.079</td><td>0.157</td><td>0.236</td><td>—</td><td>—</td><td>—</td><td>—</td><td>—</td><td>—</td><td>—</td></tr>
<tr><td>250</td><td>0.975</td><td>0.018</td><td>0.035</td><td>0.070</td><td>0.139</td><td>0.207</td><td>—</td><td>—</td><td>—</td><td>—</td><td>—</td><td>—</td><td>—</td></tr>
<tr><td>300</td><td>0.986</td><td>0.927</td><td>0.840</td><td>0.067</td><td>0.132</td><td>0.195</td><td>—</td><td>—</td><td>—</td><td>—</td><td>—</td><td>—</td><td>—</td></tr>
<tr><td>350</td><td>0.992</td><td>0.957</td><td>0.909</td><td>0.623</td><td>0.148</td><td>0.207</td><td>—</td><td>—</td><td>—</td><td>—</td><td>—</td><td>—</td><td>—</td></tr>
<tr><td>400</td><td>0.995</td><td>0.972</td><td>0.943</td><td>0.881</td><td>0.715</td><td>0.563</td><td>0.405</td><td>0.399</td><td>0.611</td><td>1.058</td><td>1.478</td><td>1.878</td><td>2.265</td></tr>
<tr><td>450</td><td>0.996</td><td>0.979</td><td>0.962</td><td>0.922</td><td>0.829</td><td>0.759</td><td>0.678</td><td>0.616</td><td>0.667</td><td>1.044</td><td>1.420</td><td>1.781</td><td>2.129</td></tr>
</table>

<table>
<tr><td colspan="12" align="center">Compressibility factor Z of water substance</td></tr>
<tr><td></td><td colspan="11" align="center">Pressure, bar</td></tr>
<tr><td>T (K)</td><td>1</td><td>5</td><td>10</td><td>15</td><td>20</td><td>25</td><td>30</td><td>40</td><td>50</td><td>60</td><td>80</td></tr>
<tr><td>400</td><td>0.990</td><td>0.003</td><td>0.006</td><td>0.009</td><td>0.012</td><td>0.014</td><td>0.017</td><td>0.023</td><td>0.029</td><td>0.035</td><td>0.046</td></tr>
<tr><td>450</td><td>0.993</td><td>0.003</td><td>0.006</td><td>0.009</td><td>0.012</td><td>0.014</td><td>0.016</td><td>0.022</td><td>0.027</td><td>0.033</td><td>0.043</td></tr>
<tr><td>500</td><td>0.996</td><td>0.980</td><td>0.958</td><td>0.930</td><td>0.901</td><td>0.878</td><td>0.016</td><td>0.021</td><td>0.026</td><td>0.031</td><td>0.042</td></tr>
<tr><td>550</td><td>0.997</td><td>0.985</td><td>0.969</td><td>0.956</td><td>0.939</td><td>0.922</td><td>0.904</td><td>0.865</td><td>0.822</td><td>0.773</td><td>0.042</td></tr>
<tr><td>600</td><td>0.998</td><td>0.990</td><td>0.979</td><td>0.970</td><td>0.961</td><td>0.948</td><td>0.935</td><td>0.910</td><td>0.885</td><td>0.858</td><td>0.798</td></tr>
<tr><td>650</td><td>0.999</td><td>0.992</td><td>0.984</td><td>0.977</td><td>0.968</td><td>0.959</td><td>0.958</td><td>0.937</td><td>0.919</td><td>0.902</td><td>0.864</td></tr>
<tr><td>700</td><td>1.000</td><td>0.994</td><td>0.988</td><td>0.984</td><td>0.976</td><td>0.967</td><td>0.966</td><td>0.952</td><td>0.941</td><td>0.929</td><td>0.900</td></tr>
<tr><td>750</td><td>1.000</td><td>0.996</td><td>0.991</td><td>0.988</td><td>0.981</td><td>0.975</td><td>0.971</td><td>0.961</td><td>0.955</td><td>0.945</td><td>0.927</td></tr>
<tr><td>800</td><td>1.000</td><td>0.997</td><td>0.993</td><td>0.991</td><td>0.985</td><td>0.982</td><td>0.976</td><td>0.970</td><td>0.966</td><td>0.957</td><td>0.945</td></tr>
<tr><td>850</td><td>1.000</td><td>0.997</td><td>0.995</td><td>0.992</td><td>0.989</td><td>0.984</td><td>0.981</td><td>0.977</td><td>0.973</td><td>0.967</td><td>0.957</td></tr>
<tr><td>900</td><td>1.000</td><td>0.998</td><td>0.997</td><td>0.993</td><td>0.992</td><td>0.989</td><td>0.986</td><td>0.982</td><td>0.979</td><td>0.974</td><td>0.965</td></tr>
<tr><td>950</td><td>1.000</td><td>0.998</td><td>0.997</td><td>0.994</td><td>0.994</td><td>0.993</td><td>0.991</td><td>0.985</td><td>0.983</td><td>0.980</td><td>0.973</td></tr>
<tr><td>1000</td><td>1.000</td><td>0.999</td><td>0.998</td><td>0.995</td><td>0.995</td><td>0.994</td><td>0.993</td><td>0.990</td><td>0.987</td><td>0.985</td><td>0.978</td></tr>
<tr><td>1200</td><td>1.000</td><td>1.000</td><td>0.999</td><td>0.998</td><td>0.998</td><td>0.997</td><td>0.997</td><td>0.995</td><td>0.994</td><td>0.994</td><td>0.992</td></tr>
<tr><td>1400</td><td>1.000</td><td>1.000</td><td>1.000</td><td>1.000</td><td>1.000</td><td>1.000</td><td>1.000</td><td>0.999</td><td>0.998</td><td>0.998</td><td>0.998</td></tr>
<tr><td>1600</td><td>1.000</td><td>1.000</td><td>1.000</td><td>1.000</td><td>1.000</td><td>1.000</td><td>1.000</td><td>1.000</td><td>1.000</td><td>1.000</td><td>1.000</td></tr>
<tr><td>1800</td><td>1.001</td><td>1.001</td><td>1.001</td><td>1.000</td><td>1.000</td><td>1.000</td><td>1.000</td><td>1.000</td><td>1.000</td><td>1.001</td><td>1.002</td></tr>
<tr><td>2000</td><td>1.003</td><td>1.002</td><td>1.002</td><td>1.002</td><td>1.002</td><td>1.002</td><td>1.002</td><td>1.002</td><td>1.002</td><td>1.003</td><td>1.003</td></tr>
</table>

* Calculated from values of pressure, volume (or density), and temperature in A. A. Vasserman, Y. Z. Kazavchinskii, and V. A. Rabinovich, *Thermophysical Properties of Air and Air Components,* Nauka, Moscow, 1966, and NBS-NSF Trans. TT 70-50095, 1971; and A. A. Vasserman and V. A. Rabinovich, *Thermophysical Properties of Liquid Air and Its Components,* Moscow, 1968, and NBS-NSF Trans. 69-55092, 1970.

[†] Calculated from *P-v-T* values tabulated in A. A. Vasserman and V. A. Rabinovich, *Thermophysical Properties of Liquid Air and Its Components,* Israeli Program for Scientific Translations TT 69-55092, 235 pp., 1970; A. A. Vasserman, Y. Z. Kazavchinskii, and V. A. Rabinovich, *Thermophysical Properties of Air and Air Components,* IPST TT 70-50095, 383 pp., 1971.

[‡] Calculated from density-pressure-temperature data in Vukalovitch and Altunin, *Thermophysical Properties of Carbon Dioxide,* Atomizdat, Moscow, 1965, and Collet's, London, 1968, trans.

[§] Computed from pressure-volume-temperature tables in Zagoruchenko and Zhuravlev, *Thermophysical Properties of Gaseous and Liquid Methane,* Moscow, 1969, and NBS-NSF TT 70-50097, 1970 translation.

[¶] Computed from tables in A. A. Vasserman, Y. Z. Kazavchinskii, and V. A. Rabinovich, *Thermophysical Properties of Air and Air Components,* Nauka, Moscow, 1966, NBS-NSF Trans. TT 70-50095, 1971.

[**] Computed from tables in A. A. Vasserman, Y. Z. Kazavchinskii, and V. A. Rabinovich, *Thermophysical Properties of Air and Air Components,* Nauka, Moscow, 1966, and NBS-NSF Trans. TT 70-50095, 1971.

[††] Calculated from *P-v-T* tables of D. M. Vashchenko, Y. F. Voinov, et al., Standartov, Moscow, Monograph 8, 1971; NBS IR 75-763, NTIS COM-75-11276, 203 pp., 1972; republished 1975.

TABLE 2.12 Isobaric Specific Heats to High Temperatures

T (K)	Ar	CCl$_2$F$_2$	CH$_4$	CH$_3$OH	CO	CO$_2$	H$_2$	H$_2$O	He	N$_2$	NH$_3$	NO	N$_2$O	O$_2$	SO$_2$	Air*	T (K)
100	2.500	4.780	4.000	4.323	3.501	3.512	—	4.006	2.500	3.500	4.003	3.886	3.530	3.501	4.032	3.5824	100
200	2.500	7.021	4.026	4.830	3.501	3.881	—	4.010	2.500	3.501	4.058	3.659	4.043	3.503	4.375	3.5062	200
300	2.500	8.721	4.295	5.531	3.505	4.460	—	4.040	2.500	3.503	4.281	3.590	4.655	3.534	4.803	3.5059	300
400	2.500	9.900	4.871	6.530	3.529	4.952	—	4.120	2.500	3.518	4.622	3.602	5.134	3.621	5.229	3.5333	400
500	2.500	10.706	5.574	7.563	3.583	5.346	3.520	4.236	2.500	3.558	5.000	3.667	5.515	3.739	5.600	3.5882	500
600	2.500	11.258	6.282	8.502	3.661	5.669	3.527	4.368	2.500	3.621	5.376	3.758	5.828	3.860	5.897	3.6626	600
700	2.500	11.644	6.951	9.327	3.749	5.938	3.540	4.508	2.500	3.699	5.738	3.853	6.088	3.967	6.127	3.7455	700
800	2.500	11.920	7.569	10.051	3.837	6.163	3.562	4.656	2.500	3.781	6.084	3.942	6.305	4.057	6.304	3.828	800
900	2.500	12.122	8.131	10.686	3.918	6.351	3.593	4.808	2.500	3.860	6.413	4.021	6.486	4.132	6.441	3.906	900
1000	2.500	12.274	8.635	11.245	3.991	6.509	3.632	4.962	2.500	3.932	6.722	4.089	6.638	4.194	6.550	3.979	1000
1100	2.500	12.391	9.084	11.735	4.054	6.643	3.677	5.114	2.500	3.998	7.010	4.147	6.765	4.246	6.636	4.046	1100
1200	2.500	12.482	9.482	12.165	4.110	6.756	3.726	5.262	2.500	4.056	7.275	4.197	6.872	4.290	6.707	4.109	1200
1300	2.500	12.555	9.832	12.543	4.158	6.852	3.777	5.404	2.500	4.107	7.517	4.239	6.962	4.328	6.765	4.171	1300
1400	2.500	12.613	10.140	12.875	4.199	6.934	3.829	5.538	2.500	4.151	7.737	4.275	7.040	4.363	6.814	4.230	1400
1500	2.500	12.661	10.410	13.167	4.235	7.004	3.880	5.663	2.500	4.190	7.935	4.306	7.107	4.395	6.855	4.289	1500
1600	2.500	12.700	10.649	13.424	4.266	7.065	3.931	5.780	2.500	4.224	8.113	4.333	7.164	4.426	6.891	4.352	1600
1700	2.500	12.734	10.859	13.650	4.294	7.118	3.979	5.887	2.500	4.254	8.274	4.356	7.215	4.455	6.922	4.418	1700
1800	2.500	12.762	11.044	13.851	4.318	7.164	4.026	5.987	2.500	4.281	8.419	4.377	7.260	4.483	6.950	4.487	1800
1900	2.500	12.785	11.208	14.029	4.339	7.205	4.070	6.079	2.500	4.304	8.549	4.395	7.299	4.511	6.975	4.566	1900
2000	2.500	12.806	11.354	14.187	4.358	7.242	4.112	6.164	2.500	4.325	8.667	4.411	7.335	4.539	6.997	4.662	2000
2100	2.500	12.823	11.483	14.328	4.375	7.274	4.152	6.242	2.500	4.344	8.773	4.425	7.367	4.567	7.017	4.781	2100
2200	2.500	12.839	11.599	14.454	4.390	7.303	4.189	6.314	2.500	4.360	8.869	4.438	7.395	4.594	7.036	4.947	2200
2300	2.500	12.852	11.703	14.567	4.404	7.329	4.224	6.381	2.500	4.375	8.956	4.450	7.422	4.621	7.053	5.179	2300
2400	2.500	12.864	11.796	14.668	4.416	7.353	4.257	6.443	2.500	4.389	9.035	4.461	7.446	4.647	7.069	5.484	2400
2500	2.500	12.875	11.880	14.760	4.427	7.375	4.288	6.500	2.500	4.401	9.107	4.471	7.468	4.673	7.084	5.882	2500
2600	2.500	12.884	11.955	14.843	4.437	7.395	4.318	6.553	2.500	4.413	9.172	4.480	7.488	4.699	7.099	6.40	2600
2700	2.500	12.892	12.024	14.918	4.447	7.413	4.346	6.603	2.500	4.423	9.232	4.489	7.508	4.724	7.112	7.06	2700
2800	2.500	12.900	12.086	14.987	4.456	7.430	4.372	6.649	2.500	4.433	9.287	4.497	7.526	4.748	7.125	7.87	2800
2900	2.500	12.906	12.143	15.049	4.464	7.445	4.397	6.692	2.500	4.442	9.338	4.504	7.542	4.771	7.137	8.86	2900
3000	2.500	12.913	12.194	15.106	4.471	7.460	4.421	6.733	2.500	4.450	9.384	4.511	7.558	4.794	7.149	9.96	3000
3100	2.500	12.918	12.242	15.158	4.478	7.474	4.444	6.771	2.500	4.457	9.427	4.518	7.573	4.816	7.160	—	3100
3200	2.500	12.923	12.285	15.206	4.485	7.486	4.465	6.807	2.500	4.464	9.467	4.524	7.588	4.837	7.171	—	3200
3300	2.500	12.928	12.325	15.250	4.491	7.499	4.486	6.841	2.500	4.471	9.504	4.530	7.601	4.858	7.182	—	3300
3400	2.500	12.932	12.361	15.290	4.497	7.510	4.505	6.873	2.500	4.477	9.538	4.535	7.614	4.877	7.192	—	3400
3500	2.500	12.936	12.395	15.327	4.502	7.521	4.524	6.903	2.500	4.483	9.570	4.541	7.627	4.896	7.202	—	3500
3600	2.500	12.939	12.427	15.362	4.508	7.531	4.542	6.932	2.500	4.489	9.600	4.546	7.639	4.913	7.212	—	3600
3700	2.500	12.942	12.455	15.394	4.513	7.541	4.559	6.960	2.500	4.494	9.628	4.551	7.651	4.930	7.222	—	3700
3800	2.500	12.945	12.482	15.424	4.517	7.550	4.576	6.986	2.500	4.499	9.654	4.556	7.662	4.946	7.231	—	3800
3900	2.500	12.948	12.507	15.451	4.522	7.559	4.592	7.011	2.500	4.504	9.678	4.560	7.673	4.961	7.240	—	3900
4000	2.500	12.951	12.530	15.477	4.526	7.568	4.608	7.035	2.500	4.508	9.701	4.565	7.683	4.976	7.250	—	4000
4100	2.500	12.953	12.552	15.501	4.531	7.576	4.623	7.058	2.500	4.513	9.723	4.569	7.694	4.989	7.259	—	4100
4200	2.500	12.955	12.572	15.523	4.535	7.584	4.637	7.080	2.500	4.517	9.743	4.573	7.704	5.002	7.267	—	4200
4300	2.500	12.957	12.591	15.544	4.538	7.592	4.651	7.102	2.500	4.521	9.763	4.577	7.714	5.015	7.276	—	4300
4400	2.500	12.959	12.609	15.564	4.542	7.599	4.665	7.122	2.500	4.525	9.781	4.581	7.723	5.026	7.285	—	4400
4500	2.500	12.961	12.625	15.582	4.546	7.606	4.678	7.142	2.500	4.528	9.798	4.585	7.733	5.037	7.293	—	4500
4600	2.500	12.963	12.641	15.599	4.549	7.614	4.691	7.161	2.500	4.532	9.815	4.589	7.742	5.048	7.302	—	4600
4700	2.500	12.964	12.655	15.616	4.553	7.620	4.704	7.180	2.500	4.535	9.831	4.593	7.751	5.058	7.310	—	4700
4800	2.500	12.966	12.669	15.631	4.556	7.627	4.717	7.198	2.500	4.539	9.845	4.596	7.760	5.068	7.319	—	4800
4900	2.500	12.967	12.682	15.645	4.559	7.634	4.729	7.216	2.500	4.542	9.860	4.600	7.769	5.078	7.327	—	4900
5000	2.500	12.968	12.694	15.659	4.563	7.640	4.740	7.233	2.500	4.545	9.873	4.604	7.778	5.087	7.335	—	5000

All table values are for the dimensionless ratio c_p/R, where R is the gas constant. To obtain values of c_p, multiply the tabular values by the appropriate gas constant. Thus, for specific heats in units of kJ(kg mol)(K), multiply by 8.31434; for specific heats in Btu/(lb mol)(°R), multiply by 1.986, etc.

* Data for air from "Tables of Thermal Properties of Gases," U.S. Department of Commerce, National Bureau of Standards, Circular 564, 1955.

Source: R. A. Svehla, "Estimated Viscosities and Thermal Conductivities at High Temperatures," *NASA Tech. Rep.* R-132, 1962.

TABLE 2.13 Thermophysical Properties of Selected Gases

Substance	Data	Property (at low pressure)	T (°C) -150 / T (K) 123.15	-100 / 173.15	-50 / 223.15	0 / 273.15	25 / 298.15	100 / 373.15	200 / 473.15	300 / 573.15	400 / 673.15	600 / 873.15	800 / 1 073.15	1 000 / 1 273.15	1 200 / 1 473.15
Acetone	Chemical formula: C_3H_6O Molecular weight: 58.08 Normal density (at 0°C, 101.3 kPa): 2.59 kg/m³ Boiling point: 56.1°C Critical temperature: 235.0°C Critical pressure: 4.761 MPa	Specific heat capacity c_{pg} (kJ/kg K)	S	S	L	L	L	1.557	1.838	2.093	2.311	2.659	2.906	3.098	3.006
		Thermal conductivity λ_g [(W/m²)/(K/m)]	S	S	S	L	L	0.018	(0.027)	(0.038)	(0.051)	(0.076)	—	—	—
		Dynamic viscosity η_g (10⁻⁵ Ns/m²)	S	S	L	L	L	0.931	1.21	1.46	1.72	2.20	2.64	3.05	3.42
Acetylene	Chemical formula: C_2H_2 Molecular weight: 26.04 Normal density (at 0°C, 101.3 kPa): 1.17 kg/m³ Boiling point: −83.95°C Critical temperature: 35.55°C Critical pressure: 6.24 MPa	Specific heat capacity c_{pg} (kJ/kg K)	S	S	1.503	1.616	1.687	1.871	2.047	2.177	2.286	2.462	2.613	2.734	2.834
		Thermal conductivity λ_g [(W/m²)/(K/m)]	S	S	0.013	0.018	0.021	0.030	0.042	0.053	0.066	0.087	0.107	0.125	0.143
		Dynamic viscosity η_g (10⁻⁵ Ns/m²)	S	S	0.785	0.960	1.04	1.28	1.55	1.83	2.08	2.53	2.93	3.30	3.65
Ammonia	Chemical formula: NH_3 Molecular weight: 17.03 Normal density (at 0°C, 101.3 kPa): 0.76 kg/m³ Boiling point: −33.4°C Critical temperature: 132.4°C Critical pressure: 11.29 MPa	Specific heat capacity c_{pg} (kJ/kg K)	S	S	L	2.056	2.093	2.219	2.366	2.516	2.663	2.805	3.538	4.099	4.509
		Thermal conductivity λ_g [(W/m²)/(K/m)]	S	S	L	0.022	0.024	0.033	0.047	0.067	0.088	0.109	0.209	0.304	0.388
		Dynamic viscosity η_g (10⁻⁵ Ns/m²)	S	S	L	0.930	1.00	1.28	1.65	1.99	2.34	2.67	4.16	5.40	6.49
Benzene	Chemical formula: C_6H_6 Molecular weight: 78.11 Normal density (at 0°C, 101.3 kPa): 3.49 kg/m³ Boiling point: 80.1°C Critical temperature: 289.45°C Critical pressure: 4.924 MPa	Specific heat capacity c_{pg} (kJ/kg K)	S	S	S	L	L	1.336	1.679	1.959	2.186	2.525	2.767	2.943	3.077
		Thermal conductivity λ_g [(W/m²)/(K/m)]	S	S	S	L	L	0.020	0.030	(0.036)	(0.047)	(0.070)	(0.092)	(0.112)	(0.130)
		Dynamic viscosity η_g (10⁻⁵ Ns/m²)	S	S	S	L	L	0.951	1.20	1.45	(1.65)	(2.10)	(2.53)	(2.95)	(3.35)

S, solid; L, liquid; values in parentheses are estimated values.

TABLE 2.13 Thermophysical Properties of Selected Gases (*Continued*)

Substance	Data	Property (at low pressure)	−150 / 123.15	−100 / 173.15	−50 / 223.15	0 / 273.15	25 / 298.15	100 / 373.15	200 / 473.15	300 / 573.15	400 / 673.15	600 / 873.15	800 / 1073.15	1000 / 1273.15	1200 / 1473.15
Bromine	Chemical formula: Br$_2$ Molecular weight: 159.81 Normal density (at 0°C, 101.3 kPa): 7.13 kg/m^3 Boiling point: 58.75°C Critical temperature: 310.85°C Critical pressure: 10.34 MPa	Specific heat capacity c_{pg} (kJ/kg K)	S	S	S	L	L	0.227	0.229	0.230	0.231	0.232	0.234	0.235	0.237
		Thermal conductivity λ_g [(W/m^2)/(K/m)]	S	S	S	L	L	0.006	0.007	0.009	0.011	0.013	0.021	0.026	0.032
		Dynamic viscosity η_g (10^{-5} Ns/m^2)	S	S	S	L	L	1.88	2.37	2.92	3.40	3.87	5.98	7.73	9.25
Carbon Tetra-chloride	Chemical formula: CCL$_4$ Molecular weight: 153.82 Normal density (at 0°C, 101.3 kPa): 6.87 kg/m^3 Boiling point: 76.7°C Critical temperature: 283.2°C Critical pressure: 4.56 MPa	Specific heat capacity c_{pg} (kJ/kg K)	S	S	S	L	L	0.586	0.624	0.645	0.657	0.670	0.691	0.696	0.699
		Thermal conductivity λ_g [(W/m^2)/(K/m)]	S	S	S	L	L	0.009	0.012	0.015	0.019	0.021	0.032	0.041	0.049
		Dynamic viscosity η_g (10^{-5} Ns/m^2)	S	S	S	L	L	1.23	1.53	1.83	2.12	2.38	3.45	4.35	5.15
Chlorine	Chemical formula: Cl$_2$ Molecular weight: 70.91 Normal density (at 0°C, 101.3 kPa): 3.16 kg/m^3 Boiling point: −34.04°C Critical temperature: 144.0°C Critical pressure: 7.710 MPa	Specific heat capacity c_{pg} (kJ/kg K)	S	L	L	0.473	0.477	0.494	0.507	0.515	0.523	0.528	0.536	0.544	0.548
		Thermal conductivity λ_g [(W/m^2)/(K/m)]	S	L	L	0.008	0.009	0.012	0.015	0.018	0.021	0.024	0.035	0.045	0.054
		Dynamic viscosity η_g (10^{-5} Ns/m^2)	S	L	L	1.23	1.34	1.68	2.10	2.50	2.86	3.22	4.68	5.90	6.99
Ethanol	Chemical formula: C$_2$H$_6$O Molecular weight: 46.07 Normal density (at 0°C, 101.3 kPa): 2.06 kg/m^3 Boiling point: 78.31°C Critical temperature: 243.1°C Critical pressure: 6.39 MPa	Specific heat capacity c_{pg} (kJ/kg K)	S	L	L	L	L	1.825	2.114	2.370	2.596	2.964	3.245	3.458	3.622
		Thermal conductivity λ_g [(W/m^2)/(K/m)]	S	L	L	L	L	0.023	0.039	0.047	(0.059)	(0.079)	—	—	—
		Dynamic viscosity η_g (10^{-5} Ns/m^2)	S	L	L	L	L	1.09	1.38	1.65	1.88	2.36	2.78	3.17	3.52

Column headers shown as T (°C) / T (K).

Substance	Data	Property (at low pressure)	T (°C)												
			T (K)												
			-150	-100	-50	0	25	100	200	300	400	500	1 000	1 500	2 000
			123.15	173.15	223.15	273.15	298.15	373.15	473.15	573.15	673.15	773.15	1 273.15	1 773.15	2 273.15
Ethylene	Chemical formula: C_2H_4 Molecular weight: 28.05 Normal density (at 0°C, 101.3 kPa): 1.26 kg/m³ Boiling point: −103.72°C Critical temperature: 9.50°C Critical pressure: 5.06 MPa	Specific heat capacity c_{pg} (kJ/kg K)	L	1.654	1.319	1.461	1.553	1.830	2.177	2.479	2.738	3.157	3.475	3.722	3.910
		Thermal conductivity λ_g [(W/m²)/(K/m)]	L	0.009	0.013	0.017	0.021	0.031	0.044	0.060	0.075	0.106	0.136	0.162	0.188
		Dynamic viscosity η_g (10^{-5} Ns/m²)	L	0.592	0.770	0.939	1.02	1.27	1.55	1.78	2.01	2.44	2.79	3.07	3.45
Ethylene Glycol	Chemical formula: $C_2H_6O_2$ Molecular weight: 62.07 Normal density (at 0°C, 101.3 kPa): 2.77 kg/m³ Boiling point: 197.25°C Critical temperature: 371.85°C Critical pressure: 7.7 MPa	Specific heat capacity c_{pg} (kJ/kg K)	S	S	S	L	L	L	(1.826)	(2.057)	(2.260)	(2.590)	—	—	—
		Thermal conductivity λ_g [(W/m²)/(K/m)]	S	S	S	L	L	L	(0.029)	(0.040)	(0.052)	(0.076)	—	—	—
		Dynamic viscosity η_g (10^{-5} Ns/m²)	S	S	S	L	L	L	(1.31)	(1.59)	(1.86)	(2.35)	—	—	—
Fluorine	Chemical formula: F_2 Molecular weight: 38.00 Normal density (at 0°C, 101.3 kPa): 1.70 kg/m³ Boiling point: −187.95°C Critical temperature: −129.15°C Critical pressure: 5.32 MPa	Specific heat capacity c_{pg} (kJ/kg K)	0.766	0.755	0.795	0.816	0.825	0.862	0.904	0.921	0.938	0.950	0.988	1.001	1.009
		Thermal conductivity λ_g [(W/m²)/(K/m)]	0.010	0.015	0.020	0.024	0.027	0.033	0.040	0.047	0.053	0.060	0.091	0.115	0.137
		Dynamic viscosity η_g (10^{-5} Ns/m²)	0.890	1.25	1.56	2.09	2.42	2.79	3.30	3.90	4.37	4.81	7.67	10.3	12.5
Glycerol	Chemical formula: $C_3H_8O_3$ Molecular weight: 92.09 Normal density (at 0°C, 101.3 kPa): 4.11 kg/m³ Boiling point: 289.85°C Critical temperature: 452.85°C Critical pressure: 6.69 MPa	Specific heat capacity c_{pg} (kJ/kg K)	S	S	S	S	L	L	L	(2.15)	(2.29)	(2.53)	—	—	—
		Thermal conductivity λ_g [(W/m²)/(K/m)]	S	S	S	S	L	L	L	(0.030)	(0.040)	(0.062)	—	—	—
		Dynamic viscosity η_g (10^{-5} Ns/m²)	S	S	S	S	L	L	L	(1.42)	(1.66)	(2.16)	—	—	—

S, solid; L, liquid; values in parentheses are estimated values.

2.19

TABLE 2.13 Thermophysical Properties of Selected Gases (*Continued*)

Substance	Data	Property (at low pressure)	-150 / 123.15	-100 / 173.15	-50 / 223.15	0 / 273.15	25 / 298.15	100 / 373.15	200 / 473.15	300 / 573.15	400 / 673.15	500 / 773.15	1000 / 1273.15	1500 / 1773.15	2000 / 2273.15
Heptane	Chemical formula: C_7H_{16} Molecular weight: 100.20 Normal density (at 0°C, 101.3 kPa): 4.47 kg/m³ Boiling point: 98.45°C Critical temperature: 267.46°C Critical pressure: 2.736 MPa	Specific heat capacity c_{pg} (kJ/kg K)	S	S	L	L	L	2.026	2.437	2.793	3.070	3.571	3.936	4.212	4.417
		Thermal conductivity λ_g [(W/m²)/(K/m)]	S	S	L	L	L	0.017	0.029	0.041	0.054	0.080	0.104	0.124	(0.142)
		Dynamic viscosity η_g (10^{-5} Ns/m²)	S	S	L	L	L	0.76	0.95	1.14	1.32	1.65	1.97	2.26	(2.55)
Hexane	Chemical formula: C_6H_{14} Molecular weight: 86.18 Normal density (at 0°C, 101.3 kPa): 3.85 kg/m³ Boiling point: 68.73°C Critical temperature: 234.29°C Critical pressure: 3.031 MPa	Specific heat capacity c_{pg} (kJ/kg K)	S	S	L	L	L	2.026	2.441	2.801	3.120	3.583	3.957	4.237	4.446
		Thermal conductivity λ_g [(W/m²)/(K/m)]	S	S	L	L	L	0.019	0.030	0.043	0.056	0.084	0.109	0.132	(0.152)
		Dynamic viscosity η_g (10^{-5} Ns/m²)	S	S	L	L	L	0.822	1.04	1.23	1.48	1.90	2.12	2.40	2.66
Methanol	Chemical formula: CH_4O Molecular weight: 32.04 Normal density (at 0°C, 101.3 kPa): 1.43 kg/m³ Boiling point: 64.7°C Critical temperature: 240°C Critical pressure: 7.95 MPa	Specific heat capacity c_{pg} (kJ/kg K)	S	S	L	L	L	1.595	1.823	2.064	2.273	2.629	3.01	3.23	3.40
		Thermal conductivity λ_g [(W/m²)/(K/m)]	S	S	L	L	L	0.026	0.045	0.055	0.071	0.104	0.136	0.167	0.197
		Dynamic viscosity η_g (10^{-5} Ns/m²)	S	S	L	L	L	1.22	1.56	1.89	2.20	2.79	3.33	3.82	4.28

Substance	Data	Property (at low pressure)	-150 / 123.15	-100 / 173.15	-50 / 223.15	0 / 273.15	25 / 298.15	100 / 373.15	200 / 473.15	300 / 573.15	400 / 673.15	600 / 873.15	800 / 1073.15	1000 / 1273.15	1200 / 1473.15
Ketene	Chemical formula: C_2H_2O Molecular weight: 42.04 Normal density (at 0°C, 101.3 kPa): 1.88 kg/m³ Boiling point: −41.15°C Critical temperature: 106.85°C Critical pressure: 6.48 MPa	Specific heat capacity c_{pg} (kJ/kg K)	S	L	L	1.093	1.143	1.290	1.461	1.599	1.717	1.905	2.043	2.148	2.227
		Thermal conductivity λ_g [(W/m²)/(K/m)]	S	L	L	(0.015)	(0.017)	(0.024)	(0.034)	(0.045)	(0.055)	(0.070)	—	—	—
		Dynamic viscosity η_g (10^{-5} Ns/m²)	S	L	L	(1.05)	(1.15)	(1.43)	(1.78)	(2.10)	(2.40)	(2.94)	—	—	—

Substance / Properties	Property													
Krypton — Chemical formula: Kr; Molecular weight: 83.80; Normal density (at 0°C, 101.3 kPa): 3.74 kg/m^3; Boiling point: −153.35°C; Critical temperature: −63.755°C; Critical pressure: 5.502 MPa	Specific heat capacity c_{pg} (kJ/kg K)	0.247	0.247	0.247	0.247	0.247	0.247	0.247	0.247	0.247	0.247	0.247	0.247	0.247
	Thermal conductivity λ_g [(W/m^2)/(K/m)]	0.004	0.006	0.007	0.009	0.010	0.012	0.014	0.016	0.018	0.021	0.030	0.035	0.041
	Dynamic viscosity η_g (10^{-5} Ns/m^2)	1.05	1.49	1.91	2.33	2.52	3.06	3.74	4.38	4.91	5.39	7.55	9.39	11.02
Nitric Oxide — Chemical formula: NO; Molecular weight: 30.01; Normal density (at 0°C, 101.3 kPa): 1.34 kg/m^3; Boiling point: −151.75°C; Critical temperature: −93.15°C; Critical pressure: 6.48 MPa	Specific heat capacity c_{pg} (kJ/kg K)	0.971	0.971	0.971	0.971	0.971	0.980	1.005	1.030	1.059	1.089	1.176	1.218	1.239
	Thermal conductivity λ_g [(W/m^2)/(K/m)]	0.013	0.018	0.021	0.024	0.026	0.031	0.038	0.046	0.053	0.059	0.088	0.113	0.135
	Dynamic viscosity η_g (10^{-5} Ns/m^2)	0.85	1.21	1.49	1.79	1.92	2.27	2.68	3.12	3.47	3.85	5.29	6.55	7.72
Nitrogen Dioxide — Chemical formula: NO_2; Molecular weight: 46.01; Normal density (at 0°C, 101.3 kPa): 2.05 kg/m^3; Boiling point: 21.1°C; Critical temperature: 158.2°C; Critical pressure: 10.13 MPa	Specific heat capacity c_{pg} (kJ/kg K)	S	S	S	L	0.808	0.858	0.929	0.984	1.034	1.080	1.193	1.256	1.281
	Thermal conductivity λ_g [(W/m^2)/(K/m)]	S	S	S	L	1.18	0.065	0.033	0.040	0.047	0.055	0.085	—	—
	Dynamic viscosity η_g (10^{-5} Ns/m^2)	S	S	S	L	(1.49)	1.84	2.26	2.65	2.99	3.32	4.55	—	—
Neon — Chemical formula: Ne; Molecular weight: 20.18; Normal density (at 0°C, 101.3 kPa): 0.90 kg/m^3; Boiling point: −246.06°C; Critical temperature: −228.75°C; Critical pressure: 2.654 MPa	Specific heat capacity c_{pg} (kJ/kg K)	1.030	1.030	1.030	1.030	1.030	1.030	1.030	1.030	1.030	1.030	1.030	1.030	1.030
	Thermal conductivity λ_g [(W/m^2)/(K/m)]	0.027	0.034	0.041	0.046	0.049	0.057	0.067	0.077	0.087	0.097	0.132	0.154	0.180
	Dynamic viscosity η_g (10^{-5} Ns/m^2)	1.67	2.14	2.58	2.99	3.12	3.65	4.26	4.89	5.32	5.81	7.81	9.95	11.68
Pentane — Chemical formula: C_5H_{12}; Molecular weight: 72.15; Normal density (at 0°C, 101.3 kPa): 3.22 kg/m^3; Boiling point: 36.05°C; Critical temperature: 196.45°C; Critical pressure: 3.369 MPa	Specific heat capacity c_{pg} (kJ/kg K)	S	L	L	L	L	2.026	2.445	2.809	3.115	3.613	3.990	4.275	4.488
	Thermal conductivity λ_g [(W/m^2)/(K/m)]	S	L	L	L	L	0.021	0.034	0.047	0.061	0.090	0.117	0.142	(0.162)
	Dynamic viscosity η_g (10^{-5} Ns/m^2)	S	L	L	L	L	0.860	1.09	1.29	1.49	1.85	2.17	2.46	2.74

S, solid; L, liquid; values in parentheses are estimated values.

TABLE 2.13 Thermophysical Properties of Selected Gases (*Continued*)

										T (°C)					
			-150	-100	-50	0	25	100	200	300	400	600	800	1000	1200
										T (K)					
Substance	Data	Property (at low pressure)	123.15	173.15	223.15	273.15	298.15	373.15	473.15	573.15	673.15	873.15	1073.15	1273.15	1473.15
Propylene	Chemical formula: C_3H_6 Molecular weight: 42.08 Normal density (at 0°C, 101.3 kPa): 1.90 kg/m³ Boiling point: −47.7°C Critical temperature: 91.65°C Critical pressure: 4.61 MPa	Specific heat capacity c_{pg} (kJ/kg K)	L	L	L	1.424	1.520	1.800	2.160	2.479	2.755	3.203	3.542	3.802	3.998
		Thermal conductivity λ_g [(W/m²)/(K/m)]	L	L	L	0.014	0.017	0.026	0.039	0.054	0.069	0.099	0.127	0.155	0.180
		Dynamic viscosity η_g (10^{-5} Ns/m²)	L	L	L	0.780	0.860	1.07	1.34	1.59	1.82	2.23	2.62	2.97	3.29
Toluene	Chemical formula: C_7H_8 Molecular weight: 92.14 Normal density (at 0°C, 101.3 kPa): 4.11 kg/m³ Boiling point: 110.63°C Critical temperature: 320.85°C Critical pressure: 4.05 MPa	Specific heat capacity c_{pg} (kJ/kg K)	S	S	L	L	L	L	1.758	2.047	2.286	2.650	2.914	3.102	3.245
		Thermal conductivity λ_g [(W/m²)/(K/m)]	S	S	L	L	L	L	0.032	0.042	(0.052)	(0.072)	(0.092)	(0.112)	(0.130)
		Dynamic viscosity η_g (10^{-5} Ns/m²)	S	S	L	L	L	L	1.12	1.33	1.545	1.95	(2.33)	(2.68)	(3.01)
Xenon	Chemical formula: Xe Molecular weight: 131.30 Normal density (at 0°C, 101.3 kPa): 5.86 kg/m³ Boiling point: −108.15°C Critical temperature: 16.55°C Critical pressure: 5.822 MPa	Specific heat capacity c_{pg} (kJ/kg K)	S	0.159	0.159	0.159	0.159	0.159	0.159	0.159	0.159	0.159	0.159	0.159	0.159
		Thermal conductivity λ_g [(W/m²)/(K/m)]	S	0.003	0.004	0.005	0.006	0.007	0.008	0.010	0.012	0.013	0.018	0.022	0.026
		Dynamic viscosity η_g (10^{-5} Ns/m²)	S	1.39	1.78	2.11	2.29	2.83	3.50	4.15	4.73	5.24	7.38	9.22	1.084

S, solid; L, liquid; values in parentheses are estimated values.
Source: Ref. 5 with permission.

TABLE 2.14 Fickian Diffusion Coefficients $[(m^2/s) \times 10^{-4}]$ at Atmospheric Pressure

T (K)	D_{ij}	T (K)	D_{ij}	T (K)	D_{ij}	T (K)	D_{ij}
Air–carbon dioxide [20]		*Carbon dioxide–argon [20]*		*Water–carbon dioxide [4]*		*Neon-argon [15]*	
276.2	0.1420	276.2	0.1326	307.5	0.202	273.0	0.276
317.2	0.1772	317.2	0.1652	328.6	0.211	288.0	0.300
Ammonia-helium [23]				352.4	0.245	303.0	0.327
274.2	0.668	*Nitrogen-nitrogen [7]*				318.0	0.357
308.2	0.783	77.5	0.0168	*Water-helium [4]*			
331.1	0.881	194.5	0.104	307.2	0.902	*Neon-neon [7]*	
		273	0.185	328.5	1.011	77.5	0.0492
Ammonia-neon [23]		298	0.212	352.5	1.121	194.5	0.255
274.2	0.298	353	0.287			273	0.452
308.4	0.378			*Water-hydrogen [3]*		298	0.516
333.1	0.419	*Nitrogen-xenon [17]*		293.1	0.850	353	0.703
		242.2	0.0854	322.7	1.012		
Ammonia-xenon [23]		274.6	0.1070	365.6	1.24	*Neon-xenon [14]*	
274.2	0.114	303.45	0.1301	365.6	1.26	273.0	0.186
308.4	0.145	334.2	0.1549	372.5	1.28	288.0	0.202
333.1	0.173					303.0	0.221
		Oxygen-argon [24]		*Hydrogen (trace)-*		318.0	0.244
Argon-argon [7]		293.2	0.200	*oxygen [2]*			
77.5	0.0134	*Oxygen-argon [16]*		300	0.820	*Nitrogen-argon [17]*	
90	0.0180	243.2	0.135	400	1.40	244.2	0.1348
194.5	0.0830	274.7	0.168	500	2.10	274.6	0.1689
273	0.156	304.5	0.202	600	2.89	303.55	0.1999
295	0.178	334.0	0.239	700	3.81	334.7	0.2433
353	0.249			800	4.74		
		Oxygen-helium [16]		900	5.74	*Nitrogen-helium [17]*	
Argon-argon [12]		244.2	0.536			243.2	0.477
273	0.156	274.0	0.640	*Hydrogen-neon [10]*		275.0	0.596
293	0.175	304.4	0.761	242.2	0.792	303.55	0.719
303	0.186	334.0	0.912	274.2	0.974	332.5	0.811
318	0.204			303.2	1.150		
		Oxygen-oxygen [7]		341.2	1.405	*Helium-nitrogen*	
Argon-helium [11]		77.5	0.0153			*(20% N_2) [27]*	
287.9	0.697	194.5	0.104	*Hydrogen-xenon [10]*		190	0.305
354.0	0.979	273	0.187	242.2	0.410	298	0.712
418.0	1.398	298	0.232	274.2	0.508	300	0.738
		353	0.301	303.9	0.612	305	0.747
Argon-helium [12]				341.2	0.751	310	0.740
273.0	0.640	*Oxygen-water [4]*				320	0.812
288.0	0.701	307.9	0.282	*Methane-methane [7]*		330	0.857
303.0	0.760	328.8	0.318	90	0.0266	340	0.881
318.0	0.825	352.2	0.352	194.5	0.0992	350	0.946
				273	0.206	360	0.967
Argon-xenon [12]		*Oxygen-xenon [16]*		298	0.240	370	1.035
273.0	0.0943	242.2	0.084	353	0.318	380	1.051
288.0	0.102	274.75	0.100			390	1.107
303.0	0.114	303.55	0.126	*Methane-methane [21]*		400	1.157
318.0	0.128	333.6	0.149	298.2	0.235		
				353.6	0.315	*Helium-nitrogen*	
Argon-xenon [13]		*Water-air [3]*		382.6	0.360	*(50% N_2) [27]*	
194.7	0.0508	289.9	0.244			190	0.310
273.2	0.0962	365.6	0.357	*Methane-water [4]*		298	0.725
329.9	0.1366	372.5	0.377	307.5	0.292	300	0.751
378.0	0.1759			328.6	0.331	305	0.758
		Water–carbon dioxide [3]		352.1	0.356		
Carbon dioxide–argon [25]		296.1	0.164				
293	0.139	365.6	0.249				
		372.6	0.259				

Note: See page 2.25 for footnotes and references.

TABLE 2.14   Fickian Diffusion Coefficients [(m²/s) × 10⁻⁴] at Atmospheric Pressure (*Continued*)

T (K)	D_{ij}	T (K)	D_{ij}	T (K)	D_{ij}	T (K)	D_{ij}
Helium-nitrogen (50% N₂)		*Hydrogen (trace)-argon [9]*		*Helium-nitrogen*		*Carbon dioxide–oxygen*	
[27] (Continued)		295	0.83	*(trace) [18]*		*(trace) [2]*	
310	0.759	448	1.76	298	0.687	300	0.160
320	0.827	628	3.21	323	0.766	400	0.270
330	0.879	806	4.86	353	0.893	500	0.400
340	0.899	958	6.81	383	1.077	600	0.565
350	0.966	1069	8.10	413	1.200	700	0.740
360	0.985			443	1.289	800	0.928
370	1.058	*Helium-argon (trace) [18]*		473	1.569	900	1.14
380	1.068	413	1.237	498	1.650	1000	1.39
390	1.144	443	1.401				
400	1.180	473	1.612	*Helium (trace)-*		*Carbon monoxide–*	
		498	1.728	*nitrogen [1]*		*carbon monoxide [22]*	
Helium-nitrogen (100% N₂				300	0.743	194.7	0.109
extrapolated) [27]		*Helium (trace)-argon [8]*		400	1.21	273.2	0.190
190	0.317	300	0.76	500	1.76	319.6	0.247
298	0.740	400	1.26	600	2.40	373.0	0.323
300	0.766	500	1.86	700	3.11		
305	0.774	600	2.56	800	3.90	*Carbon monoxide–*	
310	0.775	700	3.35	900	4.76	*nitrogen [22]*	
320	0.845	800	4.23	1000	5.69	194.7	0.105
330	0.902	900	5.20	1200	7.74	273.2	0.186
340	0.921	1000	6.25			319.6	0.242
350	0.989	1100	7.38	*Carbon dioxide–nitrogen*		373.0	0.318
360	1.013			*(trace) [1]*			
370	1.086	*Helium–carbon dioxide [20]*		300	0.177	*Carbon monoxide*	
380	1.094	276.2	0.5312	400	0.300	*(trace)–oxygen [2]*	
390	1.168	317.2	0.6607	500	0.445	300	0.212
400	1.210	346.2	0.7646	600	0.610	400	0.376
				700	0.798	500	0.552
Helium-oxygen		*Helium–carbon dioxide*		800	0.998	600	0.746
(trace) [18]		*(trace) [18]*		900	1.22	700	0.961
298	0.729	298	0.612	1000	1.47	800	1.22
323	0.809	323	0.678	1100	1.70		
353	0.987	353	0.800			*Helium-air [20]*	
383	1.120	583	0.884	*Carbon dioxide–*		276.2	0.6242
413	1.245	413	1.040	*nitrogen [26]*		317.2	0.7652
443	1.420	443	1.133	295	0.159	346.2	0.9019
473	1.595	473	1.279	1156	1.78		
498	1.683	498	1.414	1158	1.92	*Helium-argon [20]*	
				1286	2.34	276.2	0.6460
Helium-xenon [12]		*Helium–methyl alcohol*		1333	2.26	317.2	0.7968
273.0	0.501	*(trace) [18]*		1426	2.55	346.2	0.9244
288.0	0.550	423	1.032	1430	2.72		
303.0	0.604	443	1.135	1469	2.85	*Helium-argon*	
318.0	0.655	463	1.218	1490	2.92	*(trace) [18]*	
		483	1.335	1653	3.32	298	0.729
Hydrogen-argon [10]		503	1.389			323	0.809
242.2	0.562	523	1.475	*Carbon dioxide–*		353	0.978
274.2	0.698			*nitrous oxide [19]*		383	1.122
303.9	0.830	*Helium-neon [14]*		194.8	0.0531		
341.2	1.010	273.0	0.906	273.2	0.0996	*Carbon dioxide–*	
		288.0	0.986	312.8	0.1280	*argon [26]*	
Hydrogen-argon [11]		303.0	1.065	362.6	0.1683	295	0.139
287.9	0.828	318.0	1.158			1181	1.88
354.2	1.111					1207	1.88
418.0	1.714					1315	2.38

TABLE 2.14   Fickian Diffusion Coefficients $[(m^2/s) \times 10^{-4}]$ at Atmospheric Pressure (*Continued*)

T (K)	D_{ij}	T (K)	D_{ij}	T (K)	D_{ij}	T (K)	D_{ij}
Carbon dioxide–		*Carbon dioxide–*		*Carbon dioxide–*		*Carbon dioxide–*	
argon [26] (Continued)		*carbon dioxide [19]*		*carbon dioxide [6]*		*nitrogen [24]*	
1368	2.59	194.8	0.0516	296	0.109	289	0.158
1383	2.13	273.2	0.0970	298	0.109		
1427	2.53	312.8	0.1248	1180	1.73	*Water-hydrogen [4]*	
1445	2.66	362.6	0.1644		1.84	307.3	1.020
1495	2.65			1218	2.04	328.6	1.121
1503	2.84	*Carbon dioxide–*		1330	2.38	352.7	1.200
1538	3.08	*carbon dioxide [5]*		1445	2.80		
1676	3.21	233	0.0662		2.86	*Water-nitrogen [4]*	
		253	0.0794	1450	2.56	307.6	0.256
Carbon dioxide–		274	0.0925	1487	2.88	328.6	0.303
carbon dioxide [7]		293	0.1087	1490	2.98	352.2	0.359
194.7	0.0500	313	0.1239	1520	2.78		
273	0.0907	333	0.1395	1576	3.12	*Xenon-xenon [13]*	
298	0.113	363	0.1613	1580	3.33	194.7	0.0257
353	0.153	393	0.1876	1665	3.29	273.2	0.0480
		423	0.2164	1680	3.50	293.0	0.0443
		453	0.2477			300.5	0.0576
		483	0.2892			329.9	0.0684
						378.0	0.0900

All the D_{ij} values are in $(m^2/s) \times 10^{-4}$. For example, at 276.2 K the interdiffusion coefficient for the air–carbon dioxide mixture is 1.420×10^{-5} m²/s.

For an extensive review with formula fits but no data tables, see Marrero and Mason, *J. Phys. Chem. Ref. Data,* **1**:3–118 (1972). Interpolation from a graph of log D_{ij} versus log T is often simple.

References for Fickian interdiffusion coefficients

1. R. E. Walker and A. A. Westenberg, "Molecular Diffusion Studies in Gases at High Temperatures. II. Interpretation of Results on the He-N₂ and CO₂-N₂ Systems," *J. Chem. Phys.,* **29**:1147, 1958.

2. R. E. Walker and A. A. Westenberg, "Molecular Diffusion Studies in Gases at High Temperatures. IV. Results and Interpretation of the CO₂-O₂, CH₄-O₂, H₂-O₂, CO-O₂ and H₂O-O₂ Systems," *J. Chem. Phys.,* **32**:436, 1960.

3. M. Trautz and W. Müller, "Die Reibung, Wärmeleitung und Diffusion in Gasmischungen. XXXIII. Die Korrektion der bisher mit der Verdampfungsmethode gemessenen Diffusionskonstanten," *Ann. Physik,* **22**:333, 1935.

4. F. A. Schwertz and J. E. Brow, "Diffusivity of Water Vapor in Some Common Gases," *J. Chem. Phys.,* **19**:640, 1951.

5. K. Schäfer and P. Reinhard, "Zwischenmolekulare Kräfte und die Temperaturabhängigkeit der Selbstdiffusion von CO₂," *Z. Naturforsch,* **18**:187, 1963.

6. G. Ember, J. R. Ferron, and K. Wohl, "Self-Diffusion Coefficients of Carbon Dioxide at 1180°–1680°K," *J. Chem. Phys.,* **37**:891, 1962.

7. E. B. Winn, "The Temperature Dependence of the Self-Diffusion Coefficients of Argon, Neon, Nitrogen, Oxygen, Carbon Dioxide, and Methane," *Phys. Rev.,* **80**:1024, 1950.

8. R. E. Walker and A. A. Westenberg, "Molecular Diffusion Studies in Gases at High Temperature. III. Results and Interpretation of the He-A System," *J. Chem. Phys.,* **31**:319, 1959.

9. A. A. Westenberg and G. Frazier, "Molecular Diffusion Studies in Gases at High Temperature. V. Results for the H₂-Ar System," *J. Chem. Phys.,* **36**:3499, 1962.

10. R. Paul and I. B. Srivastava, "Mutual Diffusion of the Gas Pairs H₂-Ne, H₂-Ar, and H₂-Xe at Different Temperatures," *J. Chem. Phys.,* **35**:1621, 1961.

11. R. A. Strehlow, "The Temperature Dependence of the Mutual Diffusion Coefficient for Four Gaseous Systems," *J. Chem. Phys.,* **21**:2101, 1953.

12. K. P. Srivastava, "Mutual Diffusion of Binary Mixtures of Helium, Argon, and Xenon at Different Temperatures," *Physica,* **25**:571, 1959.

13. I. Amdur and T. F. Schatzki, "Diffusion Coefficients of the Systems Xe-Xe and A-Xe," *J. Chem. Phys.,* **27**:1049, 1957.

14. K. P. Srivastava and A. K. Barua, "The Temperature Dependence of Interdiffusion Coefficient for Some Pairs of Rare Gases," *Indian J. Phys.,* **33**:229, 1959.

15. B. N. Srivastava and K. P. Srivastava, "Mutual Diffusion of Pairs of Rare Gases at Different Temperatures," *J. Chem. Phys.,* **30**:984, 1959.

16. R. Paul and I. B. Srivastava, "Studies on Binary Diffusion of the Gas Pairs O₂-A, O₂-Xe, and O₂-He," *Indian J. Phys.,* **35**:465, 1961.

17. R. Paul and I. B. Srivastava, "Studies on the Binary Diffusion of the Gas Pairs N₂-A, N₂-Xe, and N₂-He," *Indian J. Phys.,* **35**:523, 1961.

18. S. L. Seager, L. R. Geertson, and J. C. Giddings, "Temperature Dependence of Gas and Vapor Diffusion Coefficients," *J. Chem. Eng. Data,* **8**:168, 1963.

19. I. Amdur, J. W. Irvine, Jr., E. A. Mason, and J. Ross, "Diffusion Coefficients of the Systems CO₂-CO₂ and CO₂-N₂O," *J. Chem. Phys.,* **20**:436, 1952.

20. J. N. Holsen and M. R. Strunk, "Binary Diffusion Coefficients in Nonpolar Gases," *Ind. Eng. Chem. Fund.,* **3**:143, 1964.

21. C. R. Mueller and R. W. Cahill, "Mass Spectrometric Measurement of Diffusion Coefficients," *J. Chem. Phys.,* **40**:651, 1964.

22. I. Amdur and L. M. Shuler, "Diffusion Coefficients of the Systems CO-CO and CO-N₂," *J. Chem. Phys.,* **38**:188, 1963.

23. I. B. Srivastava, "Mutual Diffusion of Binary Mixtures of Ammonia with He, Ne and Xe," *Indian J. Phys.,* **36**:193, 1962.

24. L. E. Boardman and N. E. Wild, "The Diffusion of Pairs of Gases with Molecules of Equal Mass," *Proc. Royal Soc.* **A162**:511, 1937.

25. L. Waldmann, "Die Temperaturerscheinungen bei der Diffusion in ruhenden Gasen und ihre messtechnische Anwendung," *Z. Phys.,* **124**:2, 1947.

26. T. A. Pakurar and J. R. Ferron, "Measurement and Prediction of Diffusivities to 1700°K in Binary Systems Containing Carbon Dioxide," *Univ. of Delaware Tech. Rept.* DEL-14-P, 1964.

27. J.-W. Yang, "A New Method of Measuring the Mass Diffusion Coefficient and Thermal Diffusion Factor in a Binary Gas System," doctoral dissertation, Univ. of Minnesota, 1966.

28. R. E. Walker, L. Monchick, A. A. Westenberg, and S. Favin, "High Temperature Gaseous Diffusion Experiments and Intermolecular Potential Energy Functions," *Planet. Space Sci.,* **3**:221, 1961.

THERMOPHYSICAL PROPERTIES OF LIQUIDS

TABLE 2.15 Thermophysical Properties of Saturated Water and Steam

T (°C)	Liquid				Steam			
	$\eta \cdot 10^7$	$\lambda \cdot 10^3$	c_p	Pr	$\eta \cdot 10^7$	$\lambda \cdot 10^3$	c_p	Pr
0	17 525	569	4.217	12.99	80.4	17.6	1.864	0.85
10	12 992	586	4.193	9.30	84.5	18.2	1.868	0.87
20	10 015	602	4.182	6.96	88.5	18.8	1.874	0.88
30	7 970	617	4.179	5.40	92.6	19.4	1.883	0.90
40	6 513	630	4.179	4.32	96.6	20.1	1.894	0.91
50	5 440	643	4.181	3.54	100	20.9	1.907	0.92
60	4 630	653	4.185	2.97	105	21.6	1.924	0.94
70	4 005	662	4.190	2.54	109	22.3	1.944	0.95
80	3 510	669	4.197	2.20	113	23.1	1.969	0.96
90	3 113	675	4.205	1.94	117	23.9	1.999	0.98
100	2 790	680	4.216	1.73	121	24.8	2.034	0.99
110	2 522	683	4.229	1.56	124	25.8	2.075	1.00
120	2 300	685	4.245	1.43	128	26.7	2.124	1.02
130	2 110	687	4.263	1.31	132	27.8	2.180	1.04
140	1 950	687	4.285	1.22	135	28.8	2.245	1.05
150	1 810	686	4.310	1.14	139	30.0	2.320	1.08
160	1 690	684	4.339	1.07	142	31.3	2.406	1.09
170	1 585	681	4.371	1.02	146	32.6	2.504	1.12
180	1 493	676	4.408	0.97	149	34.1	2.615	1.14
190	1 412	671	4.449	0.94	153	35.7	2.741	1.17
200	1 338	664	4.497	0.91	156	37.5	2.883	1.20
210	1 273	657	4.551	0.88	160	39.4	3.043	1.24
220	1 215	648	4.614	0.86	163	41.5	3.223	1.27
230	1 162	639	4.686	0.85	167	43.9	3.426	1.30
240	1 114	629	4.770	0.85	171	46.5	3.656	1.34
250	1 070	617	4.869	0.84	174	49.5	3.918	1.38
260	1 030	604	4.985	0.85	178	52.8	4.221	1.42
270	994	589	5.13	0.86	182	56.6	4.574	1.47
280	961	573	5.30	0.89	187	60.9	4.996	1.53
290	930	557	5.51	0.92	193	66.0	5.51	1.61
300	901	540	5.77	0.96	198	71.9	6.14	1.69
310	865	522	6.12	1.01	205	79.1	6.96	1.80
320	830	503	6.59	1.09	214	87.8	8.05	1.96
330	790	482	7.25	1.19	225	98.9	9.59	2.18
340	748	460	8.27	1.34	238	113	11.92	2.51
350	700	435	10.08	1.62	256	130	15.95	3.14
360	644	401	14.99	2.41	282	150	26.79	5.04
370	564	338	53.9	8.99	335	183	112.9	20.66

Viscosity η (N·s/m^2), thermal conductivity λ (W/m·deg), heat capacity c_p (kJ/kg·deg), Prandtl number Pr.
Source: Ref. 2 with permission.

TABLE 2.16 Thermophysical Properties of Water and Steam at Various Temperatures and Pressures

Pressure, bar		300	350	400	450	500	550	600	650	700	800	900	1000
								T (K)					
1	μ	8.57.−4	3.70.−4	1.32.−5	1.52.−5	1.73.−5	1.94.−5	2.15.−5	2.36.−5	2.57.−5	2.98.−5	3.39.−5	3.78.−5
	c_p	4.18	4.19	1.99	1.97	1.98	2.00	2.02	2.06	2.09	2.15	2.22	2.29
	k	0.614	0.668	0.0268	0.0311	0.0358	0.0410	0.0464	0.0521	0.0581	0.0710	0.0843	0.0981
	Pr	5.81	2.32	0.980	0.967	0.955	0.945	0.936	0.928	0.920	0.906	0.891	0.881
5	μ	8.57.−4	3.70.−4	2.17.−4	1.49.−5	1.72.−5	1.93.−5	2.15.−5	2.36.−5	2.57.−5	2.98.−5	3.39.−5	3.78.−5
	c_p	4.18	4.19	4.26	2.21	2.10	2.07	2.07	2.08	2.11	2.16	2.23	2.29
	k	0.614	0.668	0.689	0.0335	0.0369	0.0416	0.0469	0.0526	0.0585	0.0713	0.0846	0.0984
	Pr	5.82	2.32	1.34	0.983	0.973	0.959	0.947	0.937	0.925	0.907	0.892	0.881
10	μ	8.57.−4	3.70.−4	2.17.−4	1.51.−4	1.71.−5	1.93.−5	2.15.−5	2.37.−5	2.58.−5	2.99.−5	3.39.−5	3.78.−5
	c_p	4.18	4.19	4.25	4.39	2.29	2.17	2.13	2.13	2.13	2.18	2.24	2.30
	k	0.615	0.668	0.689	0.677	0.0380	0.0423	0.0474	0.0530	0.0590	0.0717	0.0851	0.0988
	Pr	5.82	2.32	1.34	0.981	1.028	0.988	0.963	0.949	0.931	0.908	0.892	0.881
20	μ	8.56.−4	3.71.−4	2.18.−4	1.51.−4	1.68.−5	1.92.−5	2.15.−5	2.38.−5	2.59.−5	3.00.−5	3.40.−5	3.79.−5
	c_p	4.17	4.19	4.25	4.39	2.84	2.41	2.26	2.22	2.19	2.21	2.26	2.32
	k	0.616	0.669	0.689	0.679	0.0402	0.0435	0.0485	0.0539	0.0599	0.0726	0.0859	0.0996
	Pr	5.80	2.32	1.34	0.979	1.19	1.063	0.999	0.977	0.946	0.912	0.893	0.881
40	μ	8.55.−4	3.71.−4	2.18.−4	1.52.−4	1.19.−4	1.89.−5	2.15.−5	2.40.−5	2.61.−5	3.02.−5	3.42.−5	3.80.−5
	c_p	4.17	4.19	4.25	4.38	4.65	3.18	2.60	2.42	2.32	2.28	2.30	2.34
	k	0.617	0.671	0.690	0.680	0.644	0.0488	0.0516	0.0564	0.0620	0.0744	0.0877	0.101
	Pr	5.78	2.31	1.34	0.977	0.862	1.23	1.08	1.031	0.975	0.924	0.895	0.881
60	μ	8.54.−4	3.72.−4	2.19.−4	1.53.−4	1.20.−4	9.84.−5	2.14.−5	2.43.−5	2.63.−5	3.04.−5	3.43.−5	3.82.−5
	c_p	4.16	4.18	4.24	4.37	4.63	5.26	3.11	2.68	2.47	2.35	2.34	2.37
	k	0.619	0.672	0.692	0.682	0.646	0.579	0.0561	0.0594	0.0645	0.0764	0.0895	0.103
	Pr	5.74	2.31	1.34	0.976	0.859	0.893	1.19	1.095	1.008	0.934	0.899	0.879
80	μ	8.53.−4	3.72.−4	2.19.−4	1.53.−4	1.20.−4	9.92.−5	2.14.−5	2.46.−5	2.66.−5	3.06.−5	3.45.−5	3.83.−5
	c_p	4.16	4.18	4.24	4.36	4.62	5.21	3.88	3.00	2.65	2.43	2.39	2.40
	k	0.620	0.674	0.693	0.684	0.648	0.583	0.0628	0.0631	0.0672	0.0785	0.0914	0.105
	Pr	5.72	2.31	1.34	0.975	0.856	0.886	1.33	1.17	1.046	0.946	0.902	0.877
100	μ	8.52.−4	3.73.−4	2.20.−4	1.53.−4	1.21.−4	9.98.−5	2.14.−5	2.49.−5	2.69.−5	3.08.−5	3.47.−5	3.85.−5
	c_p	4.15	4.17	4.23	4.35	4.60	5.17	5.22	3.42	2.85	2.52	2.44	2.44
	k	0.622	0.675	0.694	0.685	0.651	0.588	0.0730	0.0679	0.0704	0.0807	0.0934	0.107
	Pr	5.69	2.31	1.34	0.975	0.853	0.879	1.74	1.25	1.088	0.960	0.905	0.876

The notation 8.57.−4 signifies 8.57×10^{-4}.

TABLE 2.17   Isobaric Specific Heat for Water and Steam at Various Temperatures and Pressures

T (°C)	Pressure, bar							
	0.1	1	10	20	40	60	80	100
0	4.218	4.217	4.212	4.207	4.196	4.186	4.176	4.165
50	1.929	4.181	4.179	4.176	4.172	4.167	4.163	4.158
100	1.910	2.038	4.214	4.211	4.207	4.202	4.198	4.194
120	1.913	2.007	4.243	4.240	4.235	4.230	4.226	4.221
140	1.918	1.984	4.283	4.280	4.275	4.269	4.263	4.258
160	1.926	1.977	4.337	4.334	4.327	4.320	4.313	4.307
180	1.933	1.974	2.613	4.403	4.395	4.386	4.378	4.370
200	1.944	1.975	2.433	4.494	4.483	4.472	4.461	4.450
220	1.954	1.979	2.316	2.939	4.601	4.586	4.571	4.557
240	1.964	1.985	2.242	2.674	4.763	4.741	4.720	4.700
260	1.976	1.993	2.194	2.505	3.582	4.964	4.932	4.902
280	1.987	2.001	2.163	2.395	3.116	4.514	5.25	5.20
300	1.999	2.010	2.141	2.321	2.834	3.679	5.31	5.70
320	2.011	2.021	2.126	2.268	2.649	3.217	4.118	5.79
340	2.024	2.032	2.122	2.239	2.536	2.943	3.526	4.412
350	2.030	2.038	2.125	2.235	2.504	2.861	3.350	4.043
360	2.037	2.044	2.127	2.231	2.478	2.793	3.216	3.769
365	2.040	2.048	2.128	2.227	2.462	2.759	3.134	3.655
370	2.043	2.050	2.128	2.222	2.446	2.725	3.072	3.546
375	2.046	2.053	2.127	2.218	2.428	2.690	3.018	3.446
380	2.049	2.056	2.127	2.212	2.412	2.657	2.964	3.356
385	2.052	2.059	2.126	2.207	2.396	2.627	2.913	3.274
390	2.056	2.061	2.125	2.202	2.381	2.600	2.867	3.201
395	2.059	2.065	2.125	2.200	2.369	2.575	2.826	3.137
400	2.062	2.068	2.126	2.197	2.358	2.553	2.789	3.078
405	2.066	2.071	2.127	2.195	2.349	2.534	2.756	3.025
410	2.069	2.074	2.128	2.193	2.340	2.517	2.727	2.979
415	2.072	2.077	2.129	2.192	2.334	2.501	2.700	2.936
420	2.076	2.080	2.131	2.192	2.327	2.487	2.675	2.898
425	2.079	2.083	2.132	2.190	2.321	2.474	2.653	2.863
430	2.082	2.086	2.134	2.190	2.316	2.462	2.632	2.830
440	2.089	2.093	2.138	2.190	2.307	2.441	2.596	2.773
450	2.095	2.099	2.141	2.191	2.300	2.424	2.565	2.726
460	2.102	2.106	2.146	2.192	2.294	2.409	2.538	2.684
480	2.116	2.119	2.154	2.196	2.286	2.385	2.496	2.618
500	2.129	2.132	2.164	2.201	2.281	2.368	2.464	2.569
520	2.142	2.146	2.175	2.208	2.280	2.357	3.441	2.531
540	2.156	2.159	2.185	2.216	2.280	2.349	2.423	2.502
560	2.170	2.173	2.197	2.226	2.285	2.349	2.416	2.487
580	2.184	2.187	2.208	2.233	2.285	2.342	2.401	2.465
600	2.198	2.200	2.219	2.240	2.287	2.336	2.389	2.445
620	2.212	2.213	2.230	2.250	2.291	2.334	2.381	2.431
640	2.226	2.227	2.243	2.260	2.298	2.337	2.379	2.423
660	2.240	2.241	2.256	2.272	2.307	2.343	2.381	2.421
680	2.254	2.255	2.270	2.286	2.317	2.352	2.388	2.424
700	2.268	2.270	2.283	2.299	2.330	2.362	2.398	2.429
800	2.339	2.341	2.352	2.364	2.389	2.414	2.440	2.465

TABLE 2.17 Isobaric Specific Heat for Water and Steam at Various Temperatures and Pressures
(*Continued*)

T (°C)	Pressure, bar							
	150	175	200	210	220	225	230	240
0	4.141	4.129	4.117	4.113	4.108	4.106	4.103	4.099
50	4.148	4.142	4.137	4.135	4.133	4.132	4.131	4.129
100	4.183	4.178	4.173	4.171	4.169	4.168	4.167	4.165
120	4.209	4.204	4.198	4.196	4.194	4.193	4.192	4.189
140	4.245	4.238	4.232	4.229	4.227	4.226	4.224	4.222
160	4.291	4.283	4.276	4.273	4.270	4.268	4.267	4.264
180	4.350	4.340	4.331	4.328	4.324	4.322	4.320	4.317
200	4.425	4.413	4.402	4.397	4.393	4.390	4.388	4.384
220	4.523	4.508	4.492	4.486	4.481	4.478	4.475	4.469
240	4.653	4.632	4.611	4.603	4.595	4.591	4.588	4.580
260	4.832	4.801	4.772	4.760	4.749	4.744	4.738	4.728
280	5.09	5.04	4.997	4.979	4.963	4.955	4.947	4.931
300	5.50	5.41	5.33	5.31	5.28	5.26	5.25	5.23
320	6.23	6.05	5.89	5.84	5.79	5.76	5.74	5.69
340	8.14	7.45	7.01	6.87	6.74	6.68	6.63	6.53
350	8.68	9.27	9.10	7.81	7.56	7.45	7.35	7.17
360	6.86	12.57	11.37	10.18	9.40	9.10	8.84	8.41
365	6.15	9.84	19.72	13.77	11.62	10.94	10.40	9.58
370	5.69	8.36	18.38	75.67	18.38	15.56	13.84	11.79
375	5.33	7.40	12.71	19.03	52.7	81.49	29.52	17.44
380	5.02	6.68	10.19	13.14	19.19	25.71	40.95	68.4
385	4.750	6.13	8.68	10.49	13.38	15.62	18.88	33.4
390	4.520	5.68	7.65	8.90	10.68	11.88	13.42	18.21
395	4.325	5.32	6.90	7.83	9.06	9.84	10.77	13.29
400	4.155	5.02	6.33	7.06	7.97	8.53	9.16	10.76
405	4.007	4.770	5.87	6.46	7.18	7.60	8.06	9.20
410	3.879	4.556	5.50	5.99	6.57	6.90	7.26	8.12
415	3.764	4.371	5.19	5.61	6.09	6.36	6.65	7.32
420	3.664	4.211	4.933	5.29	5.70	5.92	6.16	6.71
425	3.573	4.069	4.711	5.02	5.37	5.56	5.77	6.22
430	3.491	4.945	4.520	4.795	5.10	5.26	5.44	5.83
440	3.350	3.734	4.205	4.424	4.664	4.791	4.927	5.22
450	3.235	3.564	3.959	4.139	4.333	4.435	4.544	4.77
460	3.138	3.424	3.761	3.912	4.074	4.159	4.247	4.43
480	2.986	3.210	3.465	3.576	3.695	3.756	3.819	3.95
500	2.875	3.056	3.257	3.343	3.434	3.481	3.529	3.63
520	2.791	2.940	3.104	3.174	3.247	3.284	3.322	3.40
540	2.726	2.852	2.989	3.046	3.106	3.136	3.167	3.23
560	2.683	2.791	2.906	2.954	3.003	3.028	3.054	3.10
580	2.638	2.733	2.833	2.875	2.918	2.939	2.961	3.01
600	2.598	2.682	2.770	2.807	2.844	2.863	2.882	2.92
620	2.566	2.640	2.717	2.709	2.781	2.798	2.814	2.85
640	2.542	2.607	2.675	2.703	2.731	2.746	2.760	2.79
660	2.528	2.585	2.644	2.669	2.694	2.707	2.719	2.75
680	2.520	2.572	2.625	2.646	2.669	2.680	2.691	2.71
700	2.518	2.565	2.613	2.632	2.652	2.662	2.672	2.69
800	2.531	2.564	2.598	2.611	2.625	2.632	2.639	2.65

TABLE 2.17   Isobaric Specific Heat for Water and Steam at Various Temperatures and Pressures (*Continued*)

T (°C)	Pressure, bar							
	250	270	300	400	500	600	800	1000
0	4.095	4.086	4.073	4.032	3.993	3.956	3.882	3.800
50	4.127	4.123	4.117	4.098	4.080	4.064	4.035	4.010
100	4.163	4.159	1.153	4.135	4.117	4.100	4.068	4.039
120	4.187	4.183	4.177	4.156	4.137	4.119	4.085	4.054
140	4.220	4.215	4.208	4.185	4.163	4.143	4.105	4.071
160	4.261	4.255	4.247	4.220	4.196	4.172	4.130	4.092
180	4.313	4.306	4.296	4.265	4.235	4.208	4.159	4.116
200	4.379	4.371	4.358	4.319	4.284	4.252	4.195	4.145
220	4.464	4.452	4.437	4.388	4.344	4.305	4.237	4.180
240	4.572	4.558	4.537	4.474	4.419	4.371	4.290	4.223
260	4.717	4.697	4.669	4.584	4.514	4.453	4.354	4.276
280	4.916	4.886	4.845	4.728	4.633	4.555	4.432	4.340
300	5.20	5.16	5.09	4.920	4.788	4.683	4.524	4.411
320	5.65	5.57	5.46	5.19	4.996	4.848	4.633	4.485
340	6.43	6.27	6.07	5.60	5.30	5.08	4.766	4.552
350	7.02	6.76	6.45	5.81	5.45	5.20	4.871	4.663
360	8.07	7.56	7.03	6.10	5.64	5.34	4.954	4.719
365	8.99	8.18	7.43	6.27	5.73	5.40	4.987	4.737
370	10.56	9.12	7.98	6.48	5.84	5.47	5.03	4.764
375	13.76	10.67	8.76	6.70	5.96	5.56	5.08	4.802
380	23.37	13.51	9.90	6.97	6.10	5.65	5.14	4.843
385	73.1	20.07	11.68	7.30	6.26	5.75	5.20	4.884
390	28.04	38.02	14.60	7.71	6.43	5.84	5.25	4.919
395	17.31	33.71	19.68	8.19	6.61	5.94	5.30	4.949
400	13.02	21.11	25.71	8.78	6.81	6.05	5.34	4.974
405	10.67	15.32	24.85	9.47	7.04	6.16	5.38	4.996
410	9.17	12.22	19.59	10.25	7.29	6.27	5.42	5.02
415	8.12	10.30	15.45	11.12	7.57	6.40	5.46	5.04
420	7.35	8.99	12.70	12.00	7.87	6.54	5.51	5.06
425	6.74	8.04	10.83	12.73	8.18	6.69	5.56	5.08
430	6.26	7.32	9.49	13.13	8.50	6.84	5.61	5.10
440	5.54	6.28	7.73	12.54	9.08	7.17	5.72	5.15
450	5.02	5.58	6.62	10.89	9.48	7.47	5.84	5.20
460	4.631	5.08	5.87	9.28	9.52	7.71	5.97	5.26
480	4.089	4.389	4.902	7.08	8.55	7.87	6.19	5.40
500	3.731	3.951	4.316	5.81	7.20	7.48	6.31	5.51
520	3.481	3.650	3.926	5.02	6.13	6.76	6.28	5.58
540	3.295	3.431	3.650	4.487	5.37	6.03	6.10	5.56
560	3.158	3.268	3.442	4.095	4.796	5.38	5.75	5.43
580	3.051	3.144	3.290	3.823	4.387	5.890	5.39	5.28
600	2.960	3.040	3.165	3.614	4.082	4.510	5.03	5.08
620	2.882	2.952	3.060	3.446	3.845	4.216	4.724	4.871
640	2.819	2.880	2.974	3.308	3.654	3.981	4.465	4.669
660	2.771	2.824	2.906	3.197	3.500	3.791	4.249	4.485
680	2.736	2.783	2.855	3.110	3.376	3.637	4.068	4.322
700	2.713	2.755	2.819	3.044	3.279	3.513	3.916	4.178
800	2.666	2.694	2.736	2.879	3.024	3.168	3.441	3.669

Source: Ref. 2 with permission.

TABLE 2.18 Dynamic Viscosity $[\eta \cdot 10^7 \ (\text{N·s/m}^2)]$ of Water and Steam at Various Temperatures and Pressures

T (°C)	\multicolumn{9}{c}{Pressure, bar}
	1	20	40	60	80	100	150	200	210
0	17,525	17,514	17,502	17,491	17,480	17,468	17,439	17,411	17,405
10	12,992	12,986	12,980	12,975	12,969	12,963	12,948	12,934	12,931
20	10,015	10,013	10,010	10,008	10,005	10,003	9,997	9,991	9,990
30	7,971	7,970	7,970	7,970	7,970	7,969	7,968	7,968	7,968
40	6,513	6,514	6,515	6,516	6,517	6,519	6,521	6,524	6,525
50	5,441	5,443	5,445	5,447	5,449	5,451	5,456	5,461	5,462
60	4,630	4,633	4,636	6,638	4,641	4,644	4,650	4,657	4,658
70	4,004	4,007	4,010	4,013	4,016	4,019	4,027	4,036	4,038
80	3,509	3,513	3,516	3,520	3,523	3,527	3,535	3,544	3,546
90	3,113	3,116	3,120	3,124	3,128	3,131	3,141	3,150	3,152
100	121	2,793	2,797	2,801	2,805	2,809	2,819	2,828	2,830
110	125	2,526	2,530	2,534	2,538	2,542	2,552	2,563	2,565
120	129	2,303	2,307	2,311	2,315	2,319	2,330	2,340	2,342
130	133	2,114	2,118	2,123	2,127	2,131	2,142	2,152	2,154
140	137	1,953	1,957	1,962	1,966	1,970	1,981	1,992	1,994
150	141	1,814	1,818	1,823	1,827	1,832	1,843	1,854	1,856
160	146	1,693	1,698	1,702	1,707	1,711	1,722	1,734	1,736
170	150	1,588	1,592	1,597	1,601	1,606	1,617	1,628	1,631
180	154	1,495	1,500	1,504	1,509	1,513	1,525	1,536	1,538
190	158	1,413	1,417	1,422	1,426	1,431	1,442	1,454	1,456
200	162	1,339	1,343	1,348	1,353	1,358	1,369	1,381	1,383
210	166	1,275	1,278	1,282	1,287	1,292	1,303	1,315	1,317
220	170	164	1,218	1,223	1,228	1,232	1,244	1,256	1,258
230	174	169	1,164	1,169	1,174	1,179	1,190	1,202	1,204
240	178	174	1,115	1,120	1,125	1,129	1,141	1,153	1,156
250	182	179	1,070	1,075	1,080	1,084	1,096	1,108	1,111
260	186	183	180	1,033	1,039	1,043	1,055	1,067	1,069
270	190	188	185	995	1,000	1,005	1,017	1,029	1,031
280	194	193	191	189	964	969	981	993	996
290	198	197	196	194	931	936	948	960	963
300	202	202	201	200	199	904	917	929	932
310	207	206	206	206	206	866	881	895	898
320	211	211	211	212	212	213	843	859	862
330	215	216	216	218	219	221	800	820	824
340	219	220	222	224	226	229	749	777	782
350	223	225	227	229	232	236	248	727	734
360	227	229	231	234	237	241	255	661	673
370	231	233	236	239	243	246	259	298	335
380	235	238	240	243	246	250	263	288	297
390	239	242	244	247	250	254	266	286	292
400	243	246	248	251	254	258	268	286	290
410	247	250	252	255	258	261	272	287	291
420	251	254	256	259	262	265	275	288	292
430	255	258	260	263	266	269	278	290	294
440	260	262	264	267	269	272	281	293	296
450	264	266	268	270	273	276	285	296	298
460	268	270	272	274	277	280	288	298	301
470	272	274	276	278	281	284	292	301	304
480	276	278	280	282	285	288	295	304	307
490	280	282	284	286	289	291	299	308	310

TABLE 2.18 Dynamic Viscosity [$\eta \cdot 10^7$ (N·s/m^2)] of Water and Steam at Various Temperatures and Pressures (*Continued*)

T (°C)	\multicolumn{9}{c}{Pressure, bar}
	1	20	40	60	80	100	150	200	210
500	284	286	288	290	293	295	302	311	313
520	292	294	296	298	301	303	310	318	320
540	300	302	304	306	308	311	317	324	326
560	308	310	312	314	316	319	325	332	333
580	316	318	320	322	324	326	332	339	340
600	325	326	328	330	332	334	340	346	347
620	333	334	336	338	340	342	348	353	355
640	341	342	344	346	348	350	355	361	362
660	349	351	352	354	356	358	363	368	370
680	357	359	360	362	364	366	371	376	377
700	365	367	368	370	372	374	378	384	385

T (°C)	220	230	240	250	300	400	500	600	800
0	17,399	17,394	17,388	17,382	17,353	17,296	17,239	17,182	17,067
10	12,928	12,925	12,922	12,919	12,905	12,875	12,846	12,817	12,759
20	9,988	9,987	9,986	9,985	9,979	9,967	9,954	9,942	9,918
30	7,967	7,967	7,967	7,967	7,966	7,965	7,963	7,962	7,959
40	6,225	6,526	6,526	6,527	6,529	6,535	6,540	6,546	6,557
50	5,463	5,464	5,465	5,466	5,471	5,481	5,491	5,502	5,522
60	4,660	4,661	4,662	4,664	4,670	4,684	4,697	4,711	4,737
70	4,038	4,040	4,041	4,043	4,051	4,066	4,082	4,098	4,129
80	3,548	3,549	3,551	3,553	3,561	3,579	3,596	3,614	3,648
90	3,154	3,155	3,157	3,159	3,168	3,187	3,206	3,224	3,261
100	2,832	2,834	2,836	2,838	2,848	2,867	2,887	2,906	2,945
110	2,567	2,569	2,571	2,573	2,583	2,603	2,623	2,644	2,684
120	2,344	2,347	2,349	2,351	2,361	2,382	2,403	2,424	2,465
130	2,157	2,159	2,161	2,163	2,174	2,195	2,216	2,237	2,280
140	1,996	1,998	2,000	2,003	2,013	2,035	2,057	2,078	2,122
150	1,858	1,860	1,862	1,865	1,876	1,898	1,920	1,941	1,985
160	1,738	1,740	1,742	1,745	1,756	1,778	1,800	1,822	1,867
170	1,633	1,635	1,637	1,640	1,651	1,674	1,696	1,718	1,763
180	1,540	1,543	1,545	1,547	1,559	1,581	1,604	1,627	1,672
190	1,458	1,461	1,463	1,465	1,477	1,500	1,523	1,546	1,591
200	1,385	1,388	1,390	1,392	1,404	1,427	1,450	1,473	1,519
210	1,320	1,322	1,324	1,327	1,338	1,362	1,385	1,408	1,455
220	1,261	1,263	1,265	1,268	1,279	1,303	1,326	1,350	1,397
230	1,207	1,209	1,212	1,214	1,226	1,249	1,273	1,297	1,344
240	1,158	1,160	1,163	1,165	1,177	1,201	1,225	1,248	1,296
250	1,113	1,116	1,118	1,120	1,132	1,156	1,180	1,204	1,252
260	1,072	1,074	1,077	1,079	1,091	1,115	1,140	1,164	1,212
270	1,034	1,036	1,038	1,041	1,053	1,077	1,102	1,126	1,175
280	998	1,001	1,003	1,006	1,018	1,042	1,067	1,091	1,140
290	965	968	970	972	985	1,009	1,034	1,059	1,108
300	934	937	939	941	954	978	1,004	1,028	1,078
310	901	904	906	909	922	948	972	997	1,045
320	865	868	871	874	888	915	940	964	1,012
330	827	831	834	837	853	881	908	932	980
340	786	790	794	798	817	848	876	901	949
350	740	745	751	756	779	815	845	871	920
360	683	692	700	707	738	781	814	842	891
370	596	617	633	646	692	746	784	813	864

TABLE 2.18 Dynamic Viscosity [$\eta \cdot 10^7$ (N·s/m²)] of Water and Steam at Various Temperatures and Pressures (*Continued*)

T (°C)	Pressure, bar								
	220	230	240	250	300	400	500	600	800
380	311	340	468	537	630	703	748	783	840
390	300	310	324	348	561	667	721	759	817
400	296	303	311	321	458	627	692	735	797
410	295	300	306	313	380	580	660	710	777
420	296	300	304	310	352	529	626	683	758
430	297	300	304	309	340	479	591	656	737
440	299	302	305	309	334	438	555	628	716
450	301	304	307	310	331	411	521	599	695
460	303	306	309	312	330	394	495	572	674
470	306	308	311	314	330	383	466	546	654
480	309	311	313	316	331	376	446	522	633
490	312	314	316	318	332	371	432	502	614
500	315	317	319	321	334	369	421	485	596
520	321	323	325	327	338	367	408	460	563
540	328	330	331	333	343	368	402	444	537
560	335	336	338	340	348	370	399	435	516
580	342	343	345	346	354	374	399	430	502
600	349	350	352	353	361	379	401	428	491
620	356	357	359	360	367	384	404	428	484
640	363	365	366	367	374	389	408	429	480
660	371	372	373	374	381	395	412	432	477
680	378	379	380	382	388	401	418	435	477
700	386	387	388	389	395	408	422	439	478

Source: Ref. 2 with permission.

TABLE 2.19 Thermal Conductivity [$\lambda \cdot 10^3$ (W/m·deg)] of Water and Steam at Various Temperatures and Pressures

T (°C)	Pressure, bar							
	1	20	40	60	80	100	150	200
0	569	570	572	574	575	577	581	585
10	588	589	590	592	594	595	599	603
20	603	605	607	608	610	612	616	620
30	617	620	622	623	625	627	631	634
40	630	633	635	637	638	640	644	648
50	643	645	647	648	650	651	655	659
60	653	655	657	658	660	661	665	669
70	662	664	665	667	668	670	674	677
80	669	671	673	674	676	677	681	684
90	675	677	679	680	682	683	687	690
100	24.5	682	684	685	686	688	691	694
110	25.2	686	687	688	690	691	694	698
120	26.0	688	689	691	692	693	697	700
130	26.9	689	690	692	693	694	698	701
140	27.7	689	690	692	693	694	698	701
150	28.6	688	689	690	692	693	696	700
160	29.5	685	687	688	690	691	694	698
170	30.4	682	683	685	686	688	691	695

TABLE 2.19   Thermal Conductivity [$\lambda \cdot 10^3$ (W/m·deg)] of Water and Steam at Various Temperatures and Pressures (*Continued*)

T (°C)				Pressure, bar				
	1	20	40	60	80	100	150	200
180	31.3	677	679	680	682	683	687	691
190	32.2	672	673	675	677	678	682	686
200	33.1	665	667	668	670	672	676	681
210	34.1	657	659	661	663	665	670	674
220	35.1	40.0	650	652	654	656	662	667
230	36.1	40.3	640	643	645	647	653	658
240	37.1	40.8	629	632	634	637	643	649
250	38.1	41.4	616	619	622	625	632	639
260	39.1	42.1	48.9	606	609	612	620	628
270	40.1	42.9	48.7	590	594	598	607	616
280	41.2	43.8	48.8	58.1	578	582	593	602
290	42.3	44.7	49.1	56.8	560	565	577	587
300	43.3	45.7	49.6	56.1	66.9	545	559	571
310	44.4	46.7	50.3	55.8	64.7	523	539	553
320	45.5	47.7	51.0	55.9	63.3	75.2	516	532
330	46.7	48.8	51.8	56.2	62.5	72.0	491	509
340	47.8	49.9	52.7	56.7	62.1	69.9	462	483
350	49.0	51.0	53.7	57.3	62.1	68.8	104	454
360	50.1	52.1	54.7	58.0	62.3	68.1	94.8	420
370	51.3	53.2	55.7	58.8	62.7	67.8	89.3	163
380	52.5	54.4	56.7	59.7	63.3	67.8	85.9	129
390	53.6	55.5	57.8	60.6	64.0	68.1	83.6	115
400	54.8	56.7	58.9	61.6	64.7	68.6	82.2	107
410	56.0	57.9	60.1	62.6	65.6	69.1	81.2	102
420	57.3	59.1	61.2	63.7	66.5	69.8	80.8	98.3
430	58.5	60.3	62.4	64.8	67.5	70.6	80.6	95.7
440	59.7	61.5	63.6	65.9	68.5	71.4	80.6	94.1
450	61.0	62.8	64.8	67.0	69.5	72.4	81.0	93.3
460	62.2	64.0	66.0	68.2	70.6	73.3	81.5	92.4
470	63.5	65.3	67.2	69.4	71.7	74.3	82.0	92.1
480	64.8	66.5	68.5	70.6	72.9	75.4	82.7	92.1
490	66.0	67.8	69.7	71.8	74.0	76.5	83.5	92.2
500	67.3	69.1	71.0	73.0	75.2	77.6	84.3	92.6
520	69.9	71.7	73.5	75.5	77.6	79.9	86.2	93.7
540	72.5	74.3	76.1	78.1	80.1	82.3	88.2	95.2
560	75.2	76.9	78.7	80.6	82.7	84.7	90.4	96.9
580	77.8	79.6	81.4	83.3	85.2	87.3	92.7	98.8
600	80.5	82.3	84.1	85.9	87.8	89.8	95.1	101
620	83.2	85.0	86.7	88.6	90.5	92.4	97.6	103
640	85.9	87.7	89.5	91.3	93.2	95.1	100	105
660	88.7	90.4	92.2	94.0	95.8	97.7	103	108
680	91.4	93.1	94.9	96.7	98.5	100	105	110
700	94.2	95.9	97.7	99.5	101	103	108	113

T (°C)	210	220	230	240	250	300	400	500
0	586	586	587	588	589	592	599	606
10	604	605	606	606	607	611	617	624
20	620	621	622	623	623	627	634	640
30	635	636	637	637	638	642	648	654
40	648	649	650	650	651	654	661	666
50	660	660	661	662	662	666	672	678
60	670	670	671	672	672	676	682	687
70	678	679	679	680	681	684	690	695

TABLE 2.19 Thermal Conductivity [$\lambda \cdot 10^3$ (W/m·deg)] of Water and Steam at Various Temperatures and Pressures (*Continued*)

T (°C)	Pressure, bar							
	210	220	230	240	250	300	400	500
80	685	686	686	687	688	691	697	702
90	691	691	692	693	693	696	702	708
100	695	696	696	697	698	701	707	713
110	698	699	700	700	701	704	710	716
120	700	701	702	702	703	706	712	718
130	702	702	703	703	704	707	714	720
140	701	702	703	703	704	707	714	720
150	700	701	702	702	703	706	713	720
160	698	699	700	700	701	705	711	718
170	696	696	697	698	698	702	709	716
180	692	692	693	694	695	698	706	713
190	687	688	688	689	690	694	702	709
200	681	682	683	684	685	689	697	704
210	675	676	677	678	678	683	691	699
220	668	669	670	671	672	676	685	693
230	660	661	662	663	664	669	678	686
240	650	652	653	654	655	660	670	679
250	640	642	643	644	646	651	662	671
260	630	631	632	634	635	642	653	663
270	617	619	621	622	624	631	643	653
280	604	606	608	609	611	619	633	643
290	590	592	594	595	597	606	622	633
300	573	576	578	580	582	592	609	622
310	555	558	561	563	566	577	596	610
320	535	538	541	544	547	560	582	597
330	513	516	520	523	526	541	566	583
340	488	491	495	499	503	520	548	568
350	458	463	467	472	476	496	529	552
360	<u>425</u>	430	435	440	445	468	504	537
370	206	<u>392</u>	385	396	406	437	479	514
380	147	170	185	269	322	398	453	490
390	126	140	150	165	188	338	423	465
400	115	124	134	144	156	262	388	439
410	108	114	124	132	141	206	348	411
420	103	108	116	123	130	177	307	382
430	99.8	104	109	116	122	160	271	352
440	97.6	101	105	110	116	148	241	323
450	96.0	99.2	103	106	111	139	217	297
460	95.0	97.9	101	104	108	131	198	274
470	94.5	97.0	99.7	103	106	125	184	253
480	94.2	96.5	99.0	102	104	120	172	236
490	94.2	96.4	98.7	101	103	118	163	220
500	94.4	96.4	98.5	101	103	116	155	207
520	95.3	97.1	98.9	101	103	113	142	186
540	96.6	98.2	99.8	102	103	112	136	170
560	98.3	99.7	101	103	104	112	133	159
580	100	101	103	104	106	113	131	153
600	102	103	105	106	107	114	130	149
620	104	105	107	108	109	116	130	147
640	106	108	109	110	111	117	131	147
660	109	110	111	112	113	119	132	146
680	111	112	113	115	116	121	133	147
700	114	115	116	117	118	124	135	148

Source: Ref. 2 with permission.

TABLE 2.20   Surface Tension [σ (dynes/cm)] of Water in Air

T (°C)	σ	T (°C)	σ	T (°C)	σ	T (°C)	σ
0	75.50	130	52.90	260	23.73	362	1.53
10	74.40	140	50.79	270	21.33	363	1.37
20	72.88	150	48.68	280	18.94	364	1.22
30	71.20	160	46.51	290	16.60	365	1.07
40	69.48	170	44.38	300	14.29	366	0.93
50	67.77	180	42.19	310	12.04	367	0.79
60	66.07	190	40.00	320	9.84	368	0.66
70	64.36	200	37.77	330	7.69	369	0.54
80	62.69	210	35.51	340	5.61	370	0.42
90	60.79	220	33.21	350	3.64	371	0.31
100	58.91	230	30.88	355	2.71	372	0.20
110	56.97	240	28.52	360	1.85	373	0.10
120	54.96	250	26.13	361	1.68	374.15	0

Source: Ref. 2 with permission.

TABLE 2.21   Surface Tension (N/m) of Various Liquids

Substance	T (K)										
	250	260	270	280	290	300	320	340	360	380	400
Acetone	0.0292	0.0280	0.0267	0.0254	0.0241	0.0229	0.0203	0.0178	0.016	0.014	0.012
Benzene	—	—	0.0321	0.0307	0.0293	0.0279	0.0253	0.0228	0.0204	0.0180	0.0156
Bromine	0.047	0.046	0.045	0.044	0.0425	0.041	0.038	0.035	0.032	0.030	0.027
Butane	0.0176	0.0164	0.0152	0.0140	0.0128	0.0116	0.0092	0.0069	0.0049	0.0031	0.0016
Chlorine	0.0243	0.0227	0.0212	0.0197	0.0182	0.0167	0.0137	0.0107	0.0079	0.0051	0.0037
Decane	0.0278	0.0269	0.0260	0.0251	0.0241	0.0233	0.0215	0.0196	0.0178	0.0161	0.0145
Diphenyl	—	—	—	—	—	0.0416	0.0388	0.0362	0.0338	0.0316	0.0295
Ethane	0.0061	0.0049	0.0037	0.0026	0.0015	0.0007	—	—	—	—	—
Ethanol	—	—	0.0247	0.0239	0.0231	0.0222	0.0204	0.0186	0.0167	0.0148	0.0126
Ethylene	0.0033	0.0020	0.0009	0.0002	—	—	—	—	—	—	—
Heptane	0.0242	0.0233	0.0224	0.0214	0.0204	0.0194	0.0175	0.0156	0.0137	0.0118	0.0100
Hexane	0.0230	0.0219	0.0207	0.0198	0.0187	0.0176	0.0154	0.0134	0.0116	0.0096	0.0077
Methanol	0.0266	0.0257	0.0248	0.0238	0.0229	0.0221	0.0204	0.0187	0.0169	0.0150	0.0129
Nonane	0.0270	0.0261	0.0251	0.0242	0.0232	0.0223	0.0204	0.0186	0.0167	0.0148	0.0129
Octane	0.0256	0.0247	0.0237	0.0228	0.0219	0.0210	0.0191	0.0173	0.0155	0.0138	0.0123
Pentane	0.0210	0.0198	0.0186	0.0175	0.0164	0.0153	0.0131	0.0108	0.0088	0.0069	0.0053
Propane	0.0128	0.0114	0.0101	0.0088	0.0076	0.0064	0.0043	0.0025	0.0007	—	—
Propanol	0.0274	0.0266	0.0258	0.0249	0.0241	0.0232	0.0214	0.0198	0.0182	0.0168	0.0155
Propylene	0.0132	0.0119	0.0105	0.0090	0.0077	0.0064	0.0041	0.0022	0.0005	—	—
R 12	0.0147	0.0134	0.0121	0.0108	0.0095	0.0082	0.0057	0.0034	—	—	—
Toluene	0.0345	0.0330	0.0315	0.0301	0.0288	0.0275	0.0251	0.0227	0.0205	0.0185	0.0165
Water	—	—	—	0.0747	0.0733	0.0717	0.0685	0.0651	0.0615	0.0576	0.0536

TABLE 2.22   Isobaric Expansion Coefficient of Water (β) at one bar

T (°C)	$\beta \times 10^4$ (1/K)	T (°C)	$\beta \times 10^4$ (1/K)	T (°C)	$\beta \times 10^4$ (1/K)	T (°C)	$\beta \times 10^4$ (1/K)
10	0.883	35	3.47	60	5.22	85	6.69
15	1.51	40	3.86	65	5.53	90	6.96
20	2.08	45	4.23	70	5.82	95	7.22
25	2.59	50	4.57	75	6.12	99.63	7.46
30	3.05	55	4.90	80	6.40		

Calculated from data in Ref. 7.

TABLE 2.23 Heat Capacity of Seawater (kJ/kg K) at Various Temperatures and Salinities

T (°C)									Salinity, g/kg							
	150	140	130	120	110	100	90	80	70	60	50	40	30	20	10	0
0	3.516	3.543	3.573	3.606	3.641	3.679	3.720	3.763	3.809	3.858	3.910	3.964	4.021	4.081	4.143	4.209
10	3.518	3.547	3.579	3.612	3.648	3.686	3.727	3.770	3.815	3.863	3.913	3.965	4.020	4.077	4.136	4.198
20	3.521	3.552	3.584	3.619	3.656	3.694	3.735	3.777	3.822	3.868	3.917	3.967	4.020	4.074	4.131	4.189
30	3.525	3.557	3.591	3.626	3.663	3.702	3.743	3.785	3.829	3.874	3.922	3.971	4.021	4.074	4.128	4.184
40	3.529	3.562	3.597	3.633	3.671	3.710	3.751	3.793	3.836	3.881	3.927	3.975	4.024	4.075	4.127	4.180
50	3.533	3.568	3.604	3.641	3.679	3.719	3.759	3.801	3.844	3.888	3.934	3.981	4.029	4.078	4.128	4.180
60	3.538	3.574	3.611	3.649	3.687	3.727	3.768	3.810	3.853	3.896	3.941	3.987	4.034	4.082	4.131	4.181
70	3.544	3.581	3.618	3.657	3.696	3.736	3.777	3.819	3.861	3.905	3.950	3.995	4.041	4.088	4.137	4.186
80	3.551	3.588	3.626	3.665	3.704	3.745	3.786	3.828	3.871	3.914	3.959	4.004	4.050	4.096	4.144	4.192
90	3.558	3.595	3.634	3.673	3.713	3.754	3.795	3.837	3.880	3.924	3.968	4.014	4.059	4.106	4.154	4.202
100	3.565	3.603	3.642	3.682	3.722	3.763	3.805	3.847	3.891	3.934	3.979	4.025	4.071	4.118	4.165	4.213
110	3.573	3.612	3.651	3.690	3.731	3.772	3.815	3.857	3.901	3.946	3.991	4.037	4.083	4.131	4.179	4.228
120	3.582	3.620	3.659	3.700	3.740	3.782	3.825	3.868	3.912	3.957	4.003	4.050	4.097	4.146	4.195	4.245
130	3.591	3.629	3.669	3.709	3.750	3.792	3.835	3.879	3.924	3.970	4.016	4.064	4.113	4.162	4.213	4.264
140	3.601	3.639	3.678	3.718	3.760	3.802	3.845	3.890	3.936	3.982	4.030	4.079	4.129	4.181	4.233	4.286
150	3.611	3.649	3.688	3.728	3.769	3.812	3.856	3.902	3.948	3.996	4.045	4.096	4.148	4.201	4.255	4.311
160	3.622	3.659	3.698	3.738	3.780	3.823	3.867	3.913	3.961	4.010	4.061	4.113	4.167	4.222	4.279	4.338
170	3.634	3.670	3.708	3.748	3.790	3.833	3.878	3.926	3.974	4.025	4.078	4.132	4.188	4.246	4.306	4.367
180	3.646	3.681	3.719	3.758	3.800	3.844	3.890	3.938	3.988	4.041	4.095	4.152	4.210	4.271	4.334	4.399

T (°C)					Salinity, g/kg						
	40	39	38	37	36	35	34	33	32	31	30
0	3.964	3.970	3.975	3.981	3.987	3.992	3.998	4.004	4.010	4.015	4.021
10	3.965	3.971	3.976	3.981	3.987	3.992	3.998	4.003	4.009	4.014	4.020
20	3.967	3.973	3.978	3.983	3.988	3.993	3.999	4.004	4.009	4.015	4.020
30	3.971	3.976	3.981	3.986	3.991	3.996	4.001	4.006	4.011	4.016	4.021
40	3.975	3.980	3.985	3.990	3.995	4.000	4.004	4.009	4.014	4.019	4.024
50	3.981	3.985	3.990	3.995	4.000	4.004	4.009	4.014	4.019	4.024	4.029
60	3.987	3.992	3.997	4.001	4.006	4.011	4.015	4.020	4.025	4.029	4.034
70	3.995	4.000	4.004	4.009	4.013	4.018	4.023	4.027	4.032	4.037	4.041
80	4.004	4.008	4.013	4.017	4.022	4.027	4.031	4.036	4.040	4.045	4.050
90	4.014	4.018	4.023	4.027	4.032	4.036	4.041	4.046	4.050	4.055	4.059
100	4.025	4.029	4.034	4.038	4.043	4.048	4.052	4.057	4.061	4.066	4.071
110	4.037	4.041	4.046	4.051	4.055	4.060	4.065	4.069	4.074	4.079	4.083
120	4.050	4.054	4.059	4.064	4.069	4.073	4.078	4.083	4.088	4.092	4.097
130	4.064	4.069	4.074	4.078	4.083	4.088	4.093	4.098	4.103	4.108	4.113
140	4.079	4.084	4.089	4.094	4.099	4.104	4.109	4.114	4.119	4.124	4.129
150	4.096	4.101	4.106	4.111	4.116	4.121	4.127	4.132	4.137	4.142	4.148
160	4.113	4.119	4.124	4.129	4.135	4.140	4.145	4.151	4.156	4.162	4.167
170	4.132	4.137	4.143	4.149	4.154	4.160	4.165	4.171	4.177	4.182	4.188
180	4.152	4.157	4.163	4.169	4.175	4.181	4.187	4.192	4.198	4.204	4.120

Source: Ref. 3 with permission.

TABLE 2.24 Dynamic Viscosity of Seawater (10^{-3} Ns/m^2) at Various Temperatures and Salinities

T (°C)	Salinity, g/kg 0	10	20	30	40	50	60	70	80	90	100	110	120	130	140	150
0	1.775	1.802	1.831	1.861	1.893	1.928	1.965	2.005	2.049	2.096	2.147	2.202	2.261	2.326	2.395	2.470
10	1.304	1.327	1.350	1.375	1.401	1.429	1.459	1.491	1.526	1.563	1.603	1.646	1.693	1.743	1.797	1.855
20	1.002	1.021	1.041	1.061	1.083	1.106	1.131	1.157	1.185	1.216	1.248	1.283	1.321	1.361	1.404	1.451
30	0.797	0.814	0.830	0.848	0.866	0.886	0.906	0.929	0.952	0.977	1.004	1.033	1.064	1.098	1.133	1.171
40	0.653	0.667	0.681	0.696	0.712	0.729	0.747	0.765	0.786	0.807	0.830	0.854	0.880	0.908	0.938	0.970
50	0.546	0.559	0.571	0.585	0.599	0.613	0.629	0.645	0.662	0.681	0.700	0.721	0.744	0.768	0.793	0.821
60	0.466	0.477	0.488	0.500	0.512	0.525	0.539	0.553	0.568	0.584	0.602	0.620	0.639	0.660	0.682	0.706
70	0.404	0.414	0.424	0.434	0.445	0.457	0.469	0.481	0.495	0.509	0.524	0.540	0.558	0.576	0.595	0.616
80	0.355	0.364	0.373	0.382	0.392	0.402	0.413	0.424	0.436	0.449	0.463	0.477	0.492	0.508	0.525	0.544
90	0.315	0.323	0.331	0.340	0.349	0.358	0.368	0.378	0.389	0.400	0.412	0.425	0.439	0.453	0.469	0.485
100	0.282	0.290	0.297	0.305	0.313	0.322	0.331	0.340	0.350	0.360	0.371	0.383	0.395	0.408	0.422	0.436
110	0.255	0.262	0.269	0.276	0.284	0.291	0.300	0.308	0.317	0.326	0.336	0.347	0.358	0.370	0.382	0.395
120	0.232	0.239	0.245	0.252	0.259	0.266	0.273	0.281	0.289	0.298	0.307	0.317	0.327	0.337	0.349	0.361
130	0.213	0.219	0.225	0.231	0.237	0.244	0.251	0.258	0.266	0.273	0.282	0.291	0.300	0.310	0.320	0.331
140	0.196	0.201	0.207	0.213	0.219	0.225	0.231	0.238	0.245	0.252	0.260	0.268	0.277	0.286	0.295	0.305
150	0.181	0.187	0.192	0.197	0.203	0.208	0.214	0.221	0.227	0.234	0.241	0.249	0.256	0.265	0.273	0.283
160	0.169	0.173	0.178	0.183	0.189	0.194	0.200	0.205	0.211	0.218	0.224	0.231	0.239	0.246	0.254	0.263
170	0.157	0.162	0.167	0.171	0.176	0.181	0.186	0.192	0.198	0.203	0.210	0.216	0.223	0.230	0.237	0.245
180	0.147	0.152	0.156	0.161	0.165	0.170	0.175	0.180	0.185	0.191	0.196	0.202	0.209	0.215	0.222	0.230

T (°C)	Salinity, g/kg 30	31	32	33	34	35	36	37	38	39	40
0	1.861	1.864	1.867	1.871	1.874	1.877	1.880	1.883	1.887	1.890	1.893
10	1.375	1.377	1.380	1.382	1.385	1.388	1.390	1.393	1.396	1.398	1.401
20	1.061	1.063	1.065	1.068	1.070	1.072	1.074	1.076	1.078	1.081	1.083
30	0.848	0.850	0.851	0.853	0.855	0.857	0.859	0.861	0.862	0.864	0.866
40	0.696	0.698	0.699	0.701	0.702	0.704	0.706	0.707	0.709	0.710	0.712
50	0.585	0.586	0.587	0.589	0.590	0.592	0.593	0.594	0.596	0.597	0.599
60	0.500	0.501	0.503	0.504	0.505	0.506	0.507	0.509	0.510	0.511	0.512
70	0.434	0.435	0.437	0.438	0.439	0.440	0.441	0.442	0.443	0.444	0.445
80	0.382	0.383	0.384	0.385	0.386	0.387	0.388	0.389	0.390	0.391	0.392
90	0.340	0.341	0.342	0.343	0.343	0.344	0.345	0.346	0.347	0.348	0.349
100	0.305	0.306	0.307	0.308	0.308	0.309	0.310	0.311	0.312	0.312	0.313
110	0.276	0.277	0.278	0.278	0.279	0.280	0.281	0.281	0.282	0.283	0.284
120	0.252	0.252	0.253	0.254	0.254	0.255	0.256	0.257	0.257	0.258	0.259
130	0.231	0.231	0.232	0.233	0.233	0.234	0.235	0.235	0.236	0.237	0.237
140	0.213	0.213	0.214	0.215	0.215	0.216	0.216	0.217	0.218	0.218	0.219
150	0.197	0.198	0.198	0.199	0.199	0.200	0.200	0.201	0.202	0.202	0.203
160	0.183	0.184	0.184	0.185	0.186	0.186	0.187	0.187	0.188	0.188	0.189
170	0.171	0.172	0.172	0.173	0.173	0.174	0.174	0.175	0.175	0.176	0.176
180	0.161	0.161	0.161	0.162	0.162	0.163	0.163	0.164	0.164	0.165	0.165

Source: Ref. 3 with permission.

TABLE 2.25 Thermal Conductivity of Seawater (mW/m K) at Various Temperatures and Salinities

T (°C)	0	10	20	30	35*	40	50	60	70	80	90	100	110	120	130	140	150
						Salinity, g/kg											
0	572	570	569	567	566	565	563	562	560	558	556	554	552	550	548	546	544
10	589	587	586	584	584	583	581	580	578	577	575	573	571	570	568	566	564
20	604	603	602	600	600	599	598	597	595	594	592	591	589	588	586	585	583
30	618	617	616	615	614	614	613	612	611	609	608	607	606	604	603	602	600
40	630	629	629	628	628	627	626	626	625	624	623	622	621	620	618	617	616
50	641	641	640	640	639	639	639	638	637	637	636	635	634	633	632	631	630
60	651	651	650	650	650	650	649	649	649	648	648	647	646	646	645	644	644
70	659	659	659	659	659	659	659	659	658	658	658	658	657	657	656	656	655
80	666	666	667	667	667	667	667	667	667	667	667	667	667	666	666	666	666
90	672	672	673	673	673	674	674	674	674	675	675	675	675	675	675	675	675
100	676	677	678	678	679	679	680	680	681	681	681	682	682	682	682	682	683
110	680	681	682	683	683	683	684	685	685	686	687	687	688	688	688	689	689
120	682	683	684	685	686	686	687	688	689	690	691	691	692	693	693	694	694
130	683	685	686	687	688	688	690	691	692	693	694	695	695	696	697	698	699
140	684	685	687	688	689	689	691	692	693	694	696	697	698	699	700	701	702
150	683	684	686	688	688	689	691	692	694	695	696	698	699	700	701	702	703
160	681	683	684	686	687	688	690	691	693	694	696	697	699	700	701	703	704
170	678	680	682	684	685	686	687	689	691	693	694	696	698	699	701	702	704
180	674	676	678	680	681	682	684	686	686	690	692	694	695	697	699	700	702

* "Normal" seawater.
Source: Ref. 3 with permission.

TABLE 2.26 Prandtl Number of Seawater at Various Temperatures and Salinities

T (°C)	0	10	20	30	35*	40	50	60	70	80	90	100	110	120	130	140	150
						Salinity, g/kg											
0	13.1	13.1	13.1	13.2	13.2	13.3	13.4	13.5	13.6	13.8	14.0	14.3	14.5	14.8	15.2	15.5	16.0
10	9.29	9.35	9.39	9.46	9.49	9.53	9.62	9.72	9.84	9.97	10.1	10.3	10.5	10.7	11.0	11.2	11.6
20	6.95	6.99	7.04	7.11	7.13	7.17	7.24	7.33	7.43	7.53	7.67	7.80	7.96	8.13	8.32	8.52	8.76
30	5.40	5.45	5.49	5.54	5.58	5.60	5.67	5.74	5.82	5.92	6.01	6.12	6.24	6.39	6.54	6.69	6.88
40	4.33	4.38	4.41	4.46	4.48	4.51	4.57	4.63	4.70	4.78	4.86	4.95	5.05	5.16	5.28	5.42	5.56
50	3.56	3.60	3.64	3.68	3.71	3.73	3.77	3.83	3.89	3.95	4.02	4.10	4.18	4.28	4.38	4.48	4.60
60	2.99	3.03	3.06	3.10	3.12	3.14	3.19	3.24	3.28	3.34	3.40	3.47	3.54	3.61	3.69	3.78	3.88
70	2.57	2.60	2.63	2.66	2.68	2.70	2.74	2.78	2.82	2.87	2.92	2.98	3.04	3.11	3.18	3.25	3.33
80	2.23	2.26	2.29	2.32	2.34	2.35	2.39	2.42	2.46	2.50	2.55	2.60	2.65	2.71	2.77	2.83	2.90
90	1.97	2.00	2.02	2.05	2.06	2.08	2.11	2.14	2.18	2.21	2.25	2.29	2.34	2.39	2.44	2.50	2.56
100	1.75	1.78	1.80	1.83	1.84	1.86	1.88	1.92	1.94	1.98	2.01	2.05	2.09	2.13	2.18	2.23	2.28
110	1.59	1.61	1.63	1.65	1.66	1.68	1.70	1.73	1.75	1.78	1.81	1.84	1.88	1.92	1.96	2.00	2.05
120	1.44	1.47	1.49	1.51	1.51	1.53	1.55	1.57	1.60	1.62	1.65	1.68	1.71	1.75	1.78	1.82	1.86
130	1.33	1.35	1.37	1.38	1.39	1.40	1.42	1.44	1.46	1.49	1.51	1.54	1.57	1.60	1.63	1.66	1.70
140	1.23	1.24	1.26	1.28	1.29	1.30	1.31	1.33	1.35	1.37	1.39	1.42	1.44	1.47	1.50	1.53	1.56
150	1.14	1.16	1.18	1.19	1.20	1.21	1.22	1.24	1.26	1.27	1.30	1.32	1.34	1.36	1.39	1.42	1.45
160	1.08	1.08	1.10	1.11	1.12	1.13	1.14	1.16	1.17	1.19	1.21	1.23	1.25	1.28	1.30	1.32	1.35
170	1.01	1.03	1.04	1.05	1.06	1.06	1.07	1.09	1.10	1.12	1.13	1.16	1.17	1.20	1.22	1.24	1.26
180	0.959	0.975	0.983	0.997	1.00	1.00	1.02	1.03	1.04	1.06	1.07	1.09	1.10	1.13	1.14	1.17	1.19

* "Normal" seawater.
Source: Ref. 3 with permission.

TABLE 2.27 Density of Seawater (kg/m³) at Various Temperatures and Salinities

Salinity, g/kg

T (°C)	40	39	38	37	36	35*	34	33	32	31	30
0	1,032.0	1,031.2	1,030.4	1,029.6	1,028.9	1,028.1	1,027.3	1,026.5	1,025.7	1,024.9	1,024.2
10	1,030.8	1,030.0	1,029.3	1,028.5	1,027.7	1,027.0	1,026.2	1,025.4	1,024.7	1,023.9	1,023.2
20	1,028.3	1,027.5	1,026.8	1,026.0	1,025.3	1,024.5	1,023.8	1,023.0	1,022.3	1,021.5	1,020.8
30	1,025.1	1,024.4	1,023.6	1,022.9	1,022.1	1,021.4	1,020.6	1,019.9	1,019.1	1,018.4	1,017.6
40	1,021.4	1,020.6	1,019.9	1,019.1	1,018.4	1,017.7	1,016.9	1,016.2	1,015.4	1,014.7	1,013.9
50	1,017.1	1,016.3	1,015.6	1,014.8	1,014.1	1,013.4	1,012.6	1,011.9	1,011.2	1,010.4	1,009.7
60	1,012.2	1,011.5	1,010.8	1,010.0	1,009.3	1,008.6	1,007.8	1,007.1	1,006.3	1,005.6	1,004.9
70	1,006.9	1,006.2	1,005.4	1,004.7	1,003.9	1,003.2	1,002.5	1,001.7	1,001.0	1,000.3	999.5
80	1,001.1	1,000.3	999.6	998.8	998.1	997.4	996.6	995.9	995.2	994.4	993.7
90	994.7	994.0	993.3	992.5	991.8	991.1	990.3	989.6	988.8	988.1	987.4
100	988.0	987.2	986.5	985.8	985.0	984.3	983.5	982.8	982.1	981.3	980.6
110	980.8	980.0	979.3	978.6	977.8	977.1	976.3	975.6	974.8	974.1	973.3
120	973.2	972.4	971.7	970.9	970.2	969.4	968.7	967.9	967.2	966.4	965.7
130	965.2	964.4	963.7	962.9	962.1	961.4	960.6	959.9	959.1	958.4	957.6
140	956.8	956.0	955.3	954.5	953.7	953.0	952.2	951.4	950.7	949.9	949.1
150	948.1	947.3	946.5	945.7	945.0	944.2	943.4	942.6	941.8	941.1	940.3
160	939.0	938.2	937.4	936.6	935.8	935.1	934.3	933.5	932.7	931.9	931.1
170	929.6	928.8	928.0	927.2	926.4	925.6	924.8	924.0	923.2	922.4	921.6
180	919.9	919.1	918.3	917.5	916.7	915.8	915.0	914.2	913.4	912.6	911.7

T (°C)	150	140	130	120	110	100	90	80	70	60	50	40	30	20	10
0	1,120.4	1,112.5	1,104.4	1,096.2	1,088.0	1,079.7	1,071.6	1,063.5	1,055.5	1,047.6	1,039.8	1,032.0	1,024.2	1,016.2	1,008.1
10	1,118.0	1,110.1	1,102.0	1,093.9	1,085.7	1,077.6	1,669.6	1,061.6	1,053.8	1,046.0	1,038.4	1,030.2	1,023.2	1,015.5	1,007.7
20	1,114.2	1,106.2	1,098.2	1,090.3	1,082.4	1,074.5	1,066.7	1,058.9	1,051.2	1,043.5	1,035.9	1,028.3	1,020.8	1,013.3	1,005.8
30	1,109.9	1,102.0	1,094.1	1,086.3	1,078.5	1,070.8	1,063.1	1,055.4	1,047.8	1,040.2	1,032.6	1,025.1	1,017.6	1,010.2	1,002.8
40	1,105.2	1,097.4	1,089.6	1,081.9	1,074.2	1,066.6	1,059.0	1,051.4	1,043.8	1,036.3	1,028.8	1,021.4	1,013.9	1,006.6	999.2
50	1,100.1	1,092.4	1,084.8	1,077.1	1,069.5	1,062.0	1,054.4	1,046.9	1,039.4	1,031.9	1,024.5	1,017.1	1,009.7	1,002.3	995.0
60	1,094.8	1,087.1	1,079.5	1,072.0	1,064.4	1,056.9	1,049.4	1,041.9	1,034.5	1,027.0	1,019.6	1,012.2	1,004.9	997.5	990.2
70	1,089.1	1,081.5	1,074.0	1,066.4	1,058.9	1,051.4	1,043.9	1,036.5	1,029.1	1,021.7	1,014.3	1,006.9	999.5	992.2	984.9
80	1,083.1	1,075.6	1,068.0	1,060.5	1,053.0	1,045.5	1,038.1	1,030.6	1,023.2	1,015.8	1,008.4	1,001.1	993.7	986.4	979.0
90	1,076.8	1,069.3	1,061.8	1,054.3	1,046.8	1,039.3	1,031.8	1,024.4	1,017.0	1,009.5	1,002.1	994.7	987.4	980.0	972.7
100	1,070.3	1,062.7	1,055.2	1,047.7	1,040.2	1,032.7	1,025.2	1,017.7	1,010.3	1,002.9	995.4	988.0	980.6	973.2	965.8
110	1,063.4	1,055.9	1,048.3	1,040.8	1,033.2	1,025.7	1,018.2	1,010.7	1,003.2	995.7	988.0	980.8	973.3	965.9	958.5
120	1,056.3	1,048.7	1,041.2	1,033.6	1,026.0	1,018.4	1,010.9	1,003.3	995.8	988.2	980.7	973.2	965.7	958.2	950.7
130	1,049.0	1,041.3	1,033.7	1,026.1	1,018.5	1,010.8	1,003.2	995.6	988.0	980.4	972.8	965.2	957.6	950.0	942.4
140	1,041.4	1,033.7	1,026.0	1,018.3	1,010.6	1,002.9	995.2	987.6	979.9	972.2	964.5	956.8	949.1	941.4	933.8
150	1,033.6	1,025.8	1,018.0	1,010.3	1,002.5	994.8	987.0	979.2	971.4	963.7	955.9	948.1	940.3	932.5	924.7
160	1,025.5	1,017.7	1,009.9	1,002.0	994.2	986.3	978.5	970.6	962.7	954.8	946.9	939.0	931.1	923.2	915.2
170	1,017.2	1,009.3	1,001.4	993.5	985.6	977.6	969.7	961.7	953.7	945.7	937.7	929.6	921.6	913.5	905.4
180	1,008.7	1,000.8	992.8	984.8	976.8	968.7	960.7	952.6	944.4	936.3	928.1	919.9	911.7	903.5	895.3

* "Normal" seawater.

Source: Ref. 2, with permission.

TABLE 2.28 Thermophysical Properties of Selected Liquids at Temperatures Below Their Boiling Points

Substance	Data	Property	T (°C) −150	−100	−75	−50	−25	0	20	50	100	150	200	250	300
			T (K) 123.15	173.15	198.15	223.15	248.15	273.15	293.15	323.15	373.15	423.15	473.15	523.15	573.15
Acetone	Chemical formula: C_3H_6O Molecular weight: 58.08 Melting point: −93.2°C Boiling point: 56.1°C Critical temperature: 235°C Critical pressure: 4.761 MPa	Density ρ_f (kg/m³)	S	S	893	868	840	812	791	756	V	V	V	V	V
		Specific heat capacity c_{pl} (kJ/kg K)	S	S	2.010	2.039	2.072	2.102	2.156	2.252	V	V	V	V	V
		Thermal conductivity λ_f [(W/m²)/(K/m)]	S	S	0.179	0.175	0.170	0.165	0.160	0.154	V	V	V	V	V
		Dynamic viscosity η_{lf} (10^{-5} Ns/m²)	S	S	134.1	82.0	56.0	39.8	32.5	24.9	V	V	V	V	V
Acetylene	Chemical formula: C_2H_2 Molecular weight: 26.04 Melting point: −80.75°C Boiling point: −83.95°C Critical temperature: 35.55°C Critical pressure: 6.24 MPa	Density ρ_f (kg/m³)	S	S	612	V	V	V	V	V	V	V	V	V	V
		Specific heat capacity c_{pl} (kJ/kg K)	S	S	(3.1)	V	V	V	V	V	V	V	V	V	V
		Thermal conductivity λ_f [(W/m²)/(K/m)]	S	S	(0.54)	V	V	V	V	V	V	V	V	V	V
		Dynamic viscosity η_{lf} (10^{-5} Ns/m²)	S	S	(16)	V	V	V	V	V	V	V	V	V	V
Benzene	Chemical formula: C_6H_6 Molecular weight: 78.11 Melting point: 5.55°C Boiling point: 80.11°C Critical temperature: 289.45°C Critical pressure: 4.924 MPa	Density ρ_f (kg/m³)	S	S	S	S	S	S	879	847	V	V	V	V	V
		Specific heat capacity c_{pl} (kJ/kg K)	S	S	S	S	S	S	1.729	1.821	V	V	V	V	V
		Thermal conductivity λ_f [(W/m²)/(K/m)]	S	S	S	S	S	S	0.144	0.134	V	V	V	V	V
		Dynamic viscosity η_{lf} (10^{-5} Ns/m²)	S	S	S	S	S	S	64.9	43.6	V	V	V	V	V
Dowtherm A	Chemical formula: Mixture $(C_6H_5)_2O$ (73.5%); $(C_6H_5)_2$ (26.5%) Molecular weight: 166 Melting point: 12°C Boiling point: 257.1°C Critical temperature: 497°C Critical pressure: 3.134 MPa	Density ρ_f (kg/m³)	S	S	S	S	S	S	1,060	1,036	995	951	906	858	V
		Specific heat capacity c_{pl} (kJ/kg K)	S	S	S	S	S	S	1.574	1.660	1.800	1.947	2.087	2.219	V
		Thermal conductivity λ_f [(W/m²)/(K/m)]	S	S	S	S	S	S	0.141	0.137	0.132	0.125	0.119	0.113	V
		Dynamic viscosity η_{lf} (10^{-5} Ns/m²)	S	S	S	S	S	S	380	215	100	58	39	28	—
Dowtherm J	Chemical formula: $C_{10}H_{14}$ Molecular weight: 134 Melting point: — Boiling point: 181°C Critical temperature: 383°C Critical pressure: 2.837 MPa	Density ρ_f (kg/m³)	S	S	S	917	897	888	872	842	801	754	V	V	V
		Specific heat capacity c_{pl} (kJ/kg K)	S	S	S	1.650	1.713	1.772	1.830	1.924	2.093	2.278	V	V	V
		Thermal conductivity λ_f [(W/m²)/(K/m)]	S	S	S	0.137	0.135	0.134	0.133	0.130	0.126	0.122	V	V	V
		Dynamic viscosity η_{lf} (10^{-5} Ns/m²)	S	S	S	410	225	140	90	62	36	22	V	V	V

S, solid; V, vapor; values in parentheses are estimated values.

TABLE 2.28 Thermophysical Properties of Selected Liquids at Temperatures Below Their Boiling Points (*Continued*)

Substance	Data	Property	−150	−100	−75	−50	−25	0	20	50	100	150	200	250	300
		T (°C)	123.15	173.15	198.15	223.15	248.15	273.15	293.15	323.15	373.15	423.15	473.15	523.15	573.15
		T (K)													
Ethanol	Chemical formula: C_2H_6O Molecular weight: 46.07 Melting point: −114.5°C Boiling point: 78.3°C Critical temperature: 243.10°C Critical pressure: 6.39 MPa	Density ρ_l (kg/m³)	S	892	870	850	825	806	789	763	V	V	V	V	V
		Specific heat capacity c_{pl} (kJ/kg K)	S	1.901	1.947	2.014	2.093	2.232	2.395	2.801	V	V	V	V	V
		Thermal conductivity λ_l [(W/m²)/(K/m)]	S	0.197	0.193	0.188	0.183	0.177	0.173	0.165	V	V	V	V	V
		Dynamic viscosity η_l (10⁻⁵ Ns/m²)	S	4,701	1,526	640	324.1	1768.6	120.1	70.1	V	V	V	V	V
Ethylene	Chemical formula: C_2H_4 Molecular weight: 28.05 Melting point: −169.15°C Boiling point: −103.72°C Critical temperature: 9.5°C Critical pressure: 5.06 MPa	Density ρ_l (kg/m³)	630	V	V	V	V	V	V	V	V	V	V	V	V
		Specific heat capacity c_{pl} (kJ/kg K)	2.433	V	V	V	V	V	V	V	V	V	V	V	V
		Thermal conductivity λ_l [(W/m²)/(K/m)]	0.242	V	V	V	V	V	V	V	V	V	V	V	V
		Dynamic viscosity η_l (10⁻⁵ Ns/m²)	41.0	V	V	V	V	V	V	V	V	V	V	V	V
Ethylene Glycol	Chemical formula: $C_2H_6O_2$ Molecular weight: 62.07 Melting point: −12.95°C Boiling point: 197.25°C Critical temperature: 371.85C Critical pressure: 7.7 MPa	Density ρ_l (kg/m³)	S	S	S	S	S	1,128	1,115	1091	1,055	1,016	V	V	V
		Specific heat capacity c_{pl} (kJ/kg K)	S	S	S	S	S	2.261	2.357	2.500	2.847	(2.94)	V	V	V
		Thermal conductivity λ_l [(W/m²)/(K/m)]	S	S	S	S	S	0.254	0.256	0.260	0.265	(0.252)	V	V	V
		Dynamic viscosity η_l (10⁻⁵ Ns/m²)	S	S	S	S	S	5,701	2,041	707	202	85.9	V	V	V
Glycerol	Chemical formula: $C_3H_8O_3$ Molecular weight: 92.09 Melting point: 17.85°C Boiling point: 290°C Critical temperature: 452.85°C Critical pressure: 6.69 MPa	Density ρ_l (kg/m³)	S	S	S	S	S	S	1,260	1,242	1,209	1,154	1,090	(1007)	V
		Specific heat capacity c_{pl} (kJ/kg K)	S	S	S	S	S	S	2.366	2.512	2.805	3.06	3.34	(3.74)	V
		Thermal conductivity λ_l [(W/m²)/(K/m)]	S	S	S	S	S	S	0.286	0.290	0.297	0.300	0.295	0.282	V
		Dynamic viscosity η_l (10⁻⁵ Ns/m²)	S	S	S	S	S	S	149900	(18000)	1300	170	22.0	(3.0)	V
Heptane	Chemical formula: C_7H_{16} Molecular weight: 100.20 Melting point: −90.55°C Boiling point: 98.45°C Critical temperature: 267.46°C Critical pressure: 2.736 MPa	Density ρ_l (kg/m³)	S	S	761	741	721	701	684	658	V	V	V	V	V
		Specific heat capacity c_{pl} (kJ/kg K)	S	S	2.104	2.035	2.081	2.144	2.198	2.307	V	V	V	V	V
		Thermal conductivity λ_l [(W/m²)/(K/m)]	S	S	0.156	0.148	0.139	0.131	0.124	0.114	V	V	V	V	V
		Dynamic viscosity η_l (10⁻⁵ Ns/m²)	S	S	129.0	96.6	72.5	52.6	41.3	30.2	V	V	V	V	V

Thermophysical Properties

Hexane
Chemical formula: C_6H_{14}
Molecular weight: 86.18
Melting point: −95.32°C
Boiling point: 68.73°C
Critical temperature: 234.29°C
Critical pressure: 3.031 MPa

Density ρ (kg/m³)	S	S	S	742	721	700	678	659	631	V	V	V	V	V	V
Specific heat capacity c_{pl} (kJ/kg K)	S	S	S	1.993	2.035	2.093	2.165	2.227	(2.37)	V	V	V	V	V	V
Thermal conductivity λ_l [(W/m²)/(K/m)]	S	S	S	0.156	0.146	0.137	0.127	0.120	0.110	V	V	V	V	V	V
Dynamic viscosity η_{lf} (10⁻⁵ Ns/m²)	S	S	S	92.0	68.5	51.5	38.3	30.8	22.9	V	V	V	V	V	V

Ketene
Chemical formula: C_2H_2O
Molecular weight: 42.04
Melting point: −135.15°C
Boiling point: −41.15°C
Critical temperature: 106.85°C
Critical pressure: 6.48 MPa

Density ρ (kg/m³)	S	S	(1080)	(1030)	(979)	V	V	V	V	V	V	V	V	V	V
Specific heat capacity c_{pl} (kJ/kg K)	S	S	(1.79)	(1.92)	(2.02)	V	V	V	V	V	V	V	V	V	V
Thermal conductivity λ_l [(W/m²)/(K/m)]	S	S	(0.267)	(0.250)	(0.233)	V	V	V	V	V	V	V	V	V	V
Dynamic viscosity η_{lf} (10⁻⁵ Ns/m²)	S	S	—	(110)	—	V	V	V	V	V	V	V	V	V	V

Naphthalene
Chemical formula: $C_{10}H_8$
Molecular weight: 128.17
Melting point: 80.35°C
Boiling point: 217.95°C
Critical temperature: 475.25°C
Critical pressure: 4.05 MPa

Density ρ (kg/m³)	S	S	S	S	S	S	S	S	S	963	922	(878)	V	V	V
Specific heat capacity c_{pl} (kJ/kg K)	S	S	S	S	S	S	S	S	S	1.805	1.993	2.139	V	V	V
Thermal conductivity λ_l [(W/m²)/(K/m)]	S	S	S	S	S	S	S	S	S	0.137	0.130	0.123	V	V	V
Dynamic viscosity η_{lf} (10⁻⁵ Ns/m²)	S	S	S	S	S	S	S	S	S	77.4	52.0	37.5	V	V	V

Nitrogen Dioxide
Chemical formula: NO_2
Molecular weight: 46.01
Melting point: −11.25°C
Boiling point: 21.15°C
Critical temperature: 158.25°C
Critical pressure: 1.013 MPa

Density ρ (kg/m³)	S	S	S	S	S	S	1,494	1,446	V	V	V	V	V	V	V
Specific heat capacity c_{pl} (kJ/kg K)	S	S	S	S	S	S	1.505	1.535	V	V	V	V	V	V	V
Thermal conductivity λ_l [(W/m²)/(K/m)]	S	S	S	S	S	S	0.140	0.130	V	V	V	V	V	V	V
Dynamic viscosity η_{lf} (10⁻⁵ Ns/m²)	S	S	S	S	S	S	49.4	4.21	V	V	V	V	V	V	V

Pentane
Chemical formula: C_5H_{12}
Molecular weight: 72.15
Melting point: −129.75°C
Boiling point: 36.05°C
Critical temperature: 196.45°C
Critical pressure: 3.369 MPa

Density ρ (kg/m³)	S	S	737	715	693	670	646	626	V	V	V	V	V	V	V
Specific heat capacity c_{pl} (kJ/kg K)	S	S	1.972	2.001	2.060	2.123	2.206	2.273	V	V	V	V	V	V	V
Thermal conductivity λ_l [(W/m²)/(K/m)]	S	S	0.155	0.151	0.148	0.144	0.140	0.136	V	V	V	V	V	V	V
Dynamic viscosity η_{lf} (10⁻⁵ Ns/m²)	S	S	125.0	66.0	48.4	36.4	27.7	22.7	V	V	V	V	V	V	V

Propylene
Chemical formula: C_3H_6
Molecular weight: 42.08
Melting point: −185.25°C
Boiling point: −47.7°C
Critical temperature: 91.65°C
Critical pressure: 4.61 MPa

Density ρ (kg/m³)	729	671	641	612	V	V	V	V	V	V	V	V	V	V	V
Specific heat capacity c_{pl} (kJ/kg K)	2.098	2.085	2.123	2.177	V	V	V	V	V	V	V	V	V	V	V
Thermal conductivity λ_l [(W/m²)/(K/m)]	0.217	0.179	0.160	0.145	V	V	V	V	V	V	V	V	V	V	V
Dynamic viscosity η_{lf} (10⁻⁵ Ns/m²)	129.1	37.0	26.5	19.2	V	V	V	V	V	V	V	V	V	V	V

S, solid; V, vapor; values in parentheses are estimated values.

2.43

TABLE 2.28 Thermophysical Properties of Selected Liquids at Temperatures Below Their Boiling Points (*Continued*)

Toluene — Chemical formula: C_7H_8; Molecular weight: 92.14; Melting point: −94.99°C; Boiling point: 110.63°C; Critical temperature: 320.85°C; Critical pressure: 4.05 MPa

Property													
T (°C)	−150	−100	−75	−50	−25	0	20	50	100	150	200	250	300
T (K)	123.15	173.15	198.15	223.15	248.15	273.15	293.15	323.15	373.15	423.15	473.15	523.15	573.15
Density ρ_l (kg/m³)	S	S	955	932	908	885	867	839	793	V	V	V	V
Specific heat capacity c_{pl} (kJ/kg K)	S	S	1.465	1.507	1.553	1.612	1.717	1.800	1.968	V	V	V	V
Thermal conductivity λ_l [(W/m²)/(K/m)]	S	S	0.156	0.152	0.148	0.144	0.141	0.136	0.128	V	V	V	V
Dynamic viscosity η_l (10^{-5} Ns/m²)	S	S	500	212	117.0	77.3	58.6	41.9	26.9	V	V	V	V

Ammonia — Chemical formula: NH_3; Molecular weight: 17.03; Melting point: −77.7°C; Boiling point: −33.41°C; Critical temperature: 132.4°C; Critical pressure: 11.29 MPa

Bromine — Chemical formula: Br_2; Molecular weight: 159.81; Melting point: −8.25°C; Boiling point: 58.75°C; Critical temperature: 310.85°C; Critical pressure: 10.3 MPa

Carbon Tetrachloride — Chemical formula: CCl_4; Molecular weight: 153.82; Melting point: −22.9°C; Boiling point: 76.7°C; Critical temperature: 283.21°C; Critical pressure: 4.56 MPa

Substance	Property													
	T (°C)	−200	−180	−160	−140	−120	−100	−50	0	20	50	100	150	200
	T (K)	73.15	93.15	113.15	133.15	153.15	173.15	223.15	273.15	293.15	323.15	373.15	423.15	473.15
Ammonia	Density ρ_l (kg/m³)	S	S	S	S	S	S	695	V	V	V	V	V	V
	Specific heat capacity c_{pl} (kJ/kg K)	S	S	S	S	S	S	4.45	V	V	V	V	V	V
	Thermal conductivity λ_l [(W/m²)/(K/m)]	S	S	S	S	S	S	0.547	V	V	V	V	V	V
	Dynamic viscosity η_l (10^{-5} Ns/m²)	S	S	S	S	S	S	31.7	V	V	V	V	V	V
Bromine	Density ρ_l (kg/m³)	S	S	S	S	S	S	S	3,208	3,140	(3040)	V	V	V
	Specific heat capacity c_{pl} (kJ/kg K)	S	S	S	S	S	S	S	0.448	0.452	0.456	V	V	V
	Thermal conductivity λ_l [(W/m²)/(K/m)]	S	S	S	S	S	S	S	(0.129)	0.124	0.117	V	V	V
	Dynamic viscosity η_l (10^{-5} Ns/m²)	S	S	S	S	S	S	S	124	99.6	76.2	V	V	V
Carbon Tetrachloride	Density ρ_l (kg/m³)	S	S	S	S	S	S	S	1,633	1,594	1,534	V	V	V
	Specific heat capacity c_{pl} (kJ/kg K)	S	S	S	S	S	S	S	0.842	0.850	0.862	V	V	V
	Thermal conductivity λ_l [(W/m²)/(K/m)]	S	S	S	S	S	S	S	0.107	0.106	0.105	V	V	V
	Dynamic viscosity η_l (10^{-5} Ns/m²)	S	S	S	S	S	S	S	134.9	96.1	65.4	V	V	V

Chlorine

Chemical formula: Cl_2
Molecular weight: 70.91
Melting point: −100.50°C
Boiling point: −34.04°C
Critical temperature: 144.0°C
Critical pressure: 7.71 MPa

Property																		
Density ρ_f (kg/m³)	S	S	S	S	S	S	1,717	1,598	V	V	V	V	V	V	V	V	V	V
Specific heat capacity c_{pl} (kJ/kg K)	S	S	S	S	S	S	0.883	0.892	V	V	V	V	V	V	V	V	V	V
Thermal conductivity λ_f [(W/m²)/(K/m)]	S	S	S	S	S	S	0.198	0.186	V	V	V	V	V	V	V	V	V	V
Dynamic viscosity η_{lf} (10^{-5} Ns/m²)	S	S	S	S	S	S	104.0	55.4	V	V	V	V	V	V	V	V	V	V

Fluorine

Chemical formula: F_2
Molecular weight: 38.00
Melting point: −220.15°C
Boiling point: −187.95°C
Critical temperature: −129.15°C
Critical pressure: 5.32 MPa

Property																		
Density ρ_f (kg/m³)	1,140	V	V	V	V	V	V	V	V	V	V	V	V	V	V	V	V	V
Specific heat capacity c_{pl} (kJ/kg K)	1.51	V	V	V	V	V	V	V	V	V	V	V	V	V	V	V	V	V
Thermal conductivity λ_f [(W/m²)/(K/m)]	(0.155)	V	V	V	V	V	V	V	V	V	V	V	V	V	V	V	V	V
Dynamic viscosity η_{lf} (10^{-5} Ns/m²)	34.9	V	V	V	V	V	V	V	V	V	V	V	V	V	V	V	V	V

S, solid; V, vapor; values in parentheses are estimated values.
Source: Ref. 5 with permission.

TABLE 2.29   Thermophysical Properties of Liquid Metals

Composition	Melting point (K)	T (K)	ρ (kg/m³)	c_p (kJ/kg·K)	$\nu \cdot 10^7$ (m²/s)	k (W/m·K)	$\alpha \cdot 10^5$ (m²/s)	Pr
Bismuth	544	589	10,011	0.1444	1.617	16.4	0.138	0.0142
		811	9,739	0.1545	1.133	15.6	1.035	0.0110
		1033	9,467	0.1645	0.8343	15.6	1.001	0.0083
Lead	600	644	10,540	0.159	2.276	16.1	1.084	0.024
		755	10,412	0.155	1.849	15.6	1.223	0.017
		977	10,140	—	1.347	14.9	—	—
Potassium	337	422	807.3	0.80	4.608	45.0	6.99	0.0066
		700	741.7	0.75	2.397	39.5	7.07	0.0034
		977	674.4	0.75	1.905	33.1	6.55	0.0029
Sodium	371	366	929.1	1.38	7.516	86.2	6.71	0.011
		644	860.2	1.30	3.270	72.3	6.48	0.0051
		977	778.5	1.26	2.285	59.7	6.12	0.0037
NaK (45%/55%)	292	366	887.4	1.130	6.522	25.6	2.552	0.026
		644	821.7	1.055	2.871	27.5	3.17	0.0091
		977	740.1	1.043	2.174	28.9	3.74	0.0058
NaK (22%/78%)	262	366	849.0	0.946	5.797	24.4	3.05	0.019
		672	775.3	0.879	2.666	26.7	3.92	0.0068
		1033	690.4	0.883	2.118	—	—	—
PbBi (44.5%/55.5%)	398	422	10,524	0.147	—	9.05	0.586	—
		644	10,236	0.147	1.496	11.86	0.790	0.189
		922	9,835	—	1.171	—	—	—

Adapted from *Liquid Materials Handbook,* 23rd ed., the Atomic Energy Commission, Department of the Navy, Washington, DC, 1952.

THERMOPHYSICAL PROPERTIES OF SOLIDS

TABLE 2.30   Density of Selected Elements (kg/m³)

T (K)	Symbol								
	Al	Sb*	Ba	Be*	Bi*	Cd*	Ca	Ce	Cs
50	2736	6734	3650	1863	9880	8830	1572		1962
100	2732	6726	3640	1862	9870	8800	1568		1944
150	2726	6716	3630	1861	9850	8760	1563		1926
200	2719	6706	3620	1860	9830	8720	1559		1907
250	2710	6695	3610	1858	9810	8680	1554		1887
300	2701	6685	3600	1855	9790	8640	1550	6860	<u>1866</u>
400	2681	6662	3580	1848	9750	8560	1539	6850	1781
500	2661	6638	3555	1840	<u>9710</u>	<u>8470</u>	1528	6840	1723
600	2639	6615	3530	1831		8010	1517	6820	1666
800	<u>2591</u>	<u>6569</u>		1812		7805		6790	1552
1000	2365	6431	<u> </u>	1790		7590	<u> </u>	<u>6760</u>	1438
1200	2305	6307		1768					1311
1400	2255	6170		<u>1744</u>					1182
1600									
1800									
2000									

* Polycrystalline form tabulated. Above the horizontal line the condensed phase is solid; below, it is liquid.
† Hysteresis effect present.

TABLE 2.30 Density of Selected Elements (kg/m³) (*Continued*)

T (K)	Cr	Cu	Co	Dy*	Er	Eu*	Gd*	Ga	Ge	Au	Hf
										Symbol	
50	7160	9019	8925	8578	9120		7966		5363	19,490	13,350
100	7155	9009	8919	8579	9105		7960		5358	19,460	13,340
150	7150	8992	8905	8581	9090		7954		5353	19,420	13,330
200	7145	8973	8892	8580	9080		7949		5348	19,380	13,320
250	7140	8951	8876	8567	9070		†		5344	19,340	13,310
300	7135	8930	8860	8554	9060	5240	†	5910	5340	19,300	13,300
400	7120	8884	8823	8530	9030	5190	†	6010	5330	19,210	13,275
500	7110	8837	8784	8507	9000	5155	7926	5946	5320	19,130	13,250
600	7080	8787	8744	8484	8970	5127	7907	5880	5310	19,040	13,220
800	7040	8686	8642	8431	8910		7866	5770	5290	18,860	13,170
1000	7000	8568	8561	8377	8840	____	7818	5650	5265	18,660	13,110
1200	6945	8458	8475		8740		7754	5540	5240	18,440	13,050
1400	6890	7920					____	5420		17,230	
1600	6830	7750								16,950	
1800	6760	7600	7630								
2000	6700	7460	7410								____

	Ho	In*	Ir	Fe	La*	Pb	Li	Lu*	Mg	Mo
50	8820	7460	22,600	7910	6203	11,570	547	9830	1765	10,260
100	8815	7430	22,580	7900	6200	11,520	546	9840	1762	10,260
150	8810	7400	22,560	7890	6196	11,470	543	9840	1757	10,250
200	8800	7370	22,540	7880	6193	11,430	541	9850	1752	10,250
250	8790	7340	22,520	7870	6190	11,380	537	9840	1746	10,250
300	8780	7310	22,500	7860	6187	11,330	533	9830	1740	10,240
400	8755	7230	22,450	7830	6180	11,230	526	9800	1736	10,220
500	8730	6980	22,410	7800	6160	11,130	492	9770	1731	10,210
600	8700	6810	22,360	7760	6170	11,010	482	9740	1726	10,190
800	8650		22,250	7690	6140	10,430	462	9660	1715	10,160
1000	8600		22,140	7650	6160	10,190	442	9580	1517	10,120
1200			22,030	7620		9,940	442	9500	1409	10,080
1400			21,920	7520			402			10,040
1600	____		21,790	7420			381			10,000
1800			21,660	7320			361	____		9,950
2000			21,510	7030			341			9,900

	Ni	Nb	Os	Pd	Pt	Pu	K	Pa*
50	8960	8610	22,550	12,110	21,570	20,270	905	
100	8960	8600	22,540	12,100	21,550	20,170	898	
150	8940	8590	22,520	12,090	21,530	20,080	890	
200	8930	8580	22,510	12,070	21,500	19,990	882	
250	8910	8570	22,490	12,050	21,470	19,860	873	
300	8900	8570	22,480	12,030	21,450	19,730	863	15,370
400	8860	8550	22,450	11,980	21,380	17,720	814	15,320
500	8820	8530	22,420	11,940	21,330	17,920	790	15,280
600	8780	8510	22,390	11,890	21,270	15,300	767	15,230
800	8690	8470	22,320	11,790	21,140	16,370	720	15,150

* Polycrystalline form tabulated. Above the horizontal line the condensed phase is solid; below, it is liquid.
† Hysteresis effect present.

TABLE 2.30   Density of Selected Elements (kg/m³) (*Continued*)

T (K)	Ni	Nb	Os	Pd	Pt	Pu	K	Pa*
							Symbol	
1000	8610	8430	22,250	11,680	21,010		672	15,050
1200	8510	8380		11,570	20,870		623	14,910
1400	8410	8340			20,720		574	
1600	8320	8290			20,570		527	____
1800	7690	8250		____	20,400			
2000	7450	8200	____		20,220			

	Re*	Rh	Rb	Sc*	Ag	Na	Sr	Ta
50	21,100	12,490			10,620	1014	2655	16,500
100	21,070	12,480			10,600	1007	2638	16,490
150	21,040	12,470			10,575	999	2632	16,480
200	21,020	12,460			10,550	990	2621	16,460
250	21,010	12,445			10,520	980	2618	16,450
300	21,000	12,430	____	3000	10,490	970	2615	16,440
400	20,960	12,400	14,320	2990	10,430	921		16,410
500	20,920	12,360	13,860	2980	10,360	897		16,370
600	20,880	12,330	13,400	2970	10,300	874		16,340
800	20,800	12,250	12,340	2950	10,160	826		16,270
1000	20,710	12,170	11,560	2930	10,010	779	____	16,200
1200	20,630	12,080	10,640	2910	9,850	731		16,130
1400	20,540	11,980	9,720		9,170	683		16,060
1600	20,450	11,880			8,980	638		15,980
1800	20,350			____				15,910
2000	20,250	____						15,820

	Tl	Th	Tm*	Sn	Ti	W	U*	V	Yb	Y*	Zn*	Zr*
50	12,080	11,745	9370		4530	19,320	19,240	6080		4500	7280	6540
100	12,040	11,740	9360		4510	19,310	19,210	6074		4490	7260	6535
150	12,000	11,745	9350		4515	19,300	19,170	6068		4485	7230	6530
200	11,950	11,750	9340		4520	19,290	19,140	6062		4480	7200	6525
250	11,900	11,735	9330		4515	19,280	19,100	6056		4475	7170	6520
300	11,850	11,720	9320	7280	4510	19,270	19,070	6050	7020	4470	7135	6515
400	11,730	11,680	9280	____	4490	19,240	18,980	6030	6960	4450	7070	6510
500	11,500	11,630	9250	6900	4480	19,220	18,890	6010	6900	4440	7000	6490
600	11,250	11,590	9210	6900	4470	19,190	18,790	6000	6850	4420	6935	6480
800	10,960	11,500	9150	6760	4440	19,130	18,550	5960	6720	4390	6430	6450
1000		11,400	9080	6620	4410	19,080	18,110	5920	6590	4360	6260	6420
1200		11,300		6480	4380	19,020	17,760	5880		4320		6410
1400				6340	4350	18,950	17,530	5830				6380
1600					4320	18,890		5780			____	6340
1800			____		____	18,830		5730				6300
2000					4110	18,760		____				6260

* Polycrystalline form tabulated. Above the horizontal line the condensed phase is solid; below, it is liquid.
† Hysteresis effect present.

TABLE 2.31 Heat Capacity of Selected Elements (kJ/kg K)

Values are tabulated against temperature T (K).

Symbol	10	15	20	25	30	40	50	60	80	100	150	200	250	300	400	500	600	800	1000	1200
Al	0.0014	0.0040	0.0089	0.0175	0.0315	0.0775	0.142	0.214	0.357	0.481	0.683	0.797	0.859	0.902	0.949	0.997	1.042	1.134	0.921	0.921
Sb	0.0021	0.0069	0.0260	0.0402	0.0546	0.0832	0.103	0.135	0.160	0.169	0.191	0.200	0.205	0.209	0.213	0.219	0.225	0.237	0.258	0.258
Ba	—	—	—	—	—	—	—	—	—	—	—	—	—	0.192	0.202	0.213	0.222	0.247	0.209	0.229
Be	0.0003	0.0009	0.0014	0.0028	0.0042	—	0.0186	—	—	0.195	0.610	1.109	1.537	1.840	2.191	2.442	2.605	2.823	3.018	3.217
Bi	0.0104	0.0240	0.0340	0.0487	0.0579	0.0729	0.0855	0.092	0.102	0.109	0.117	0.120	0.121	0.122	0.123	—	0.142	0.136	0.131	—
Cd	0.0082	0.0233	0.0462	0.0636	0.0860	0.118	0.145	0.159	0.183	0.198	0.213	0.221	0.227	0.231	0.242	0.252	—	—	—	—
Ca	0.0042	0.0157	0.0396	0.0647	0.0930	0.194	0.271	0.340	0.427	0.486	0.573	0.617	0.640	0.656	0.685	0.729	0.763	0.843	0.991	0.772
Ce	0.0314	0.0340	0.0526	0.0735	0.0920	—	0.0926	—	—	0.193	0.200	0.206	0.209	0.212	0.218	0.230	0.242	0.266	0.290	—
Cs	0.0831	0.1231	0.1470	0.1599	0.1687	—	0.1826	—	—	0.1939	0.202	0.208	0.218	—	0.240	0.232	0.224	0.217	0.231	0.248
Cr	0.0008	0.0012	0.0021	0.0045	0.0077	0.0107	0.038	0.059	0.127	0.190	0.317	0.382	0.424	0.450	0.501	0.537	0.565	0.611	0.653	0.692
Co	0.0012	0.0026	0.0048	0.0106	0.0171	0.0404	0.070	0.110	0.184	0.234	0.329	0.376	0.406	0.426	0.451	0.484	0.509	0.543	0.631	0.651
Cu	0.0009	0.0027	0.0076	0.0158	0.0270	0.059	0.099	0.137	0.203	0.254	0.323	0.357	0.377	0.386	0.396	0.406	0.431	0.448	0.446	0.480
Dy	0.0046	0.0154	0.0345	0.0566	0.0783	—	0.142	—	—	0.214	0.280	0.179	0.173	0.168	0.170	0.176	0.181	0.190	0.198	0.205
Er	0.0118	0.0400	0.1256	0.0933	0.1151	—	0.170	—	—	0.147	0.155	0.162	0.165	0.168	0.172	0.176	0.179	0.187	0.194	0.200
Eu	0.0256	0.0573	0.0655	0.0911	—	—	0.068	—	—	—	—	0.137	—	0.176	0.182	0.187	0.193	0.204	0.215	0.200
Gd	0.0048	0.0122	0.0282	0.0471	0.0649	—	0.1235	—	—	0.184	0.208	0.230	0.265	0.231	0.186	0.191	0.195	0.204	0.213	—
Ga	0.0035	0.0150	0.0322	0.0504	0.0714	0.110	0.154	0.177	0.216	0.266	0.316	0.341	0.359	0.377	—	—	—	—	—	—
Ge	0.0008	0.0044	0.0129	0.0236	0.0363	0.0619	0.0860	0.108	0.153	0.192	0.257	0.286	0.305	0.323	0.343	0.355	0.364	0.377	0.390	0.396
Au	0.0026	0.0074	0.0163	0.0263	0.0370	0.0569	0.072	0.084	0.100	0.109	0.119	0.124	0.127	0.129	0.131	0.133	0.136	0.141	0.147	0.153
Hf	0.0009	0.0038	0.0096	0.0180	0.0281	—	—	—	—	0.115	0.131	0.137	0.141	0.143	0.146	0.149	0.151	0.157	0.163	0.169
Ho	0.0162	0.0398	0.0580	0.0756	0.0931	0.140	0.149	0.176	0.193	0.214	0.161	0.161	0.163	0.165	0.170	0.174	0.178	0.187	0.195	—
In	0.0155	0.0367	0.0608	0.0857	0.108	—	0.159	—	—	—	0.220	0.224	0.227	0.233	0.252	—	—	—	—	—
Ir	0.0003	0.0008	0.0021	0.0048	0.0094	—	0.0381	—	—	0.0903	0.113	0.122	0.128	0.131	0.133	0.137	0.140	0.146	0.152	—
Fe	0.0013	0.0026	0.0039	0.0075	0.0124	0.0276	0.054	0.086	0.154	0.216	0.324	0.384	0.422	0.450	0.491	0.524	0.555	0.692	1.034	—
La	0.0078	0.0241	0.0446	0.0663	0.0750	0.113	0.133	0.145	0.161	0.170	0.182	—	—	0.200	0.205	0.210	0.215	0.224	0.234	—
Pb	0.0135	0.0351	0.0531	0.0678	0.0796	0.0944	0.103	0.108	0.114	0.118	0.122	0.125	0.127	0.129	0.132	0.137	0.142	—	—	—
Li	0.0090	0.0259	0.0573	0.1025	0.1688	—	0.549	—	—	1.923	2.701	3.105	3.377	3.54	3.76	4.34	4.26	4.17	4.15	4.14
Lu	0.0029	0.0096	0.0210	0.0349	0.0483	—	0.091	—	—	0.129	0.141	0.147	0.151	0.154	0.158	0.161	0.165	0.172	0.179	—
Mg	0.0017	0.0066	0.0148	0.0310	0.0568	0.138	0.243	0.336	0.513	0.648	0.842	0.929	0.985	1.005	1.082	1.131	1.177	1.263	—	—
Mn (α)	0.0031	0.0052	0.0091	0.0145	0.0251	0.046	0.088	0.127	0.213	0.268	0.365	0.420	0.454	0.481	0.510	0.551	0.581	0.635	0.688	—
Hg	0.0225	0.0359	0.0515	0.0633	0.0737	0.0895	0.0993	0.107	0.116	0.121	0.129	0.136	0.141	0.139	0.136	0.135	0.135	0.104	—	—
Mo	0.0005	0.0013	0.0029	0.0058	0.0096	0.0236	0.0410	0.0610	0.105	0.140	0.196	0.223	0.241	0.248	0.261	0.268	0.274	0.280	0.292	—

TABLE 2.31 Heat Capacity of Selected Elements (kJ/kg K) (Continued)

Symbol	T (K)																			
	10	15	20	25	30	40	50	60	80	100	150	200	250	300	400	500	600	800	1000	1200
Nd	0.0365	0.0519	0.0711	0.0827	0.0983	0.120	0.150	0.160	0.178	0.185	0.196	—	—	—	0.225	0.240	0.255	0.287	0.318	—
Ni	0.0018	0.0031	0.0058	0.0100	0.0166	0.0380	0.068	0.103	0.173	0.232	0.329	0.383	0.416	0.444	0.490	0.540	0.590	0.530	0.556	0.582
Nb	0.0022	0.0054	0.0173	0.0210	0.0350	0.0680	0.099	0.127	0.173	0.202	0.238	0.254	0.263	0.268	0.272	0.277	0.281	0.290	0.298	0.307
Os	—	—	—	—	—	—	—	—	—	—	—	—	—	0.131	0.133	0.135	0.137	0.141	0.145	0.148
Pd	0.0021	0.0047	0.0091	0.0161	0.0259	0.0509	0.077	0.101	0.141	0.168	0.208	0.228	0.238	0.245	0.250	0.255	0.261	0.271	0.282	0.293
Pt	0.0011	0.0034	0.0077	0.0139	0.0211	0.0382	0.054	0.069	0.088	0.101	0.118	0.127	0.132	0.134	0.136	0.138	0.140	0.146	0.152	0.158
Pu	—	—	—	—	—	—	—	—	—	—	0.096	0.111	0.124	0.132	—	—	—	—	—	—
K	0.0847	0.1444	0.1875	0.2198	0.2399	—	—	—	—	—	0.672	0.694	0.718	0.768	0.805	0.785	0.771	0.761	0.792	0.846
Pr	0.0294	0.0600	0.0944	0.1290	0.1505	—	0.184	—	—	0.186	0.191	0.193	0.195	0.197	0.201	0.211	0.220	0.240	0.258	—
Pa	—	—	—	—	—	—	—	—	—	—	—	—	—	0.126	0.131	0.137	0.143	0.153	0.165	—
Re	—	—	0.0034	0.0072	0.0121	—	0.043	—	—	0.097	0.120	0.130	0.137	0.138	0.139	0.142	0.145	0.151	0.156	—
Rh	0.0007	0.0014	0.0027	0.0056	0.0106	0.0266	0.0489	0.072	0.114	0.147	0.195	0.220	0.234	0.246	0.257	0.265	0.274	0.290	0.307	—
Rb	0.0847	0.1444	0.1875	0.2198	0.2399	—	0.2741	—	—	0.299	0.310	0.321	0.335	0.365	0.367	—	—	—	—	—
Ru	0.0004	0.0009	0.0017	0.0035	0.0070	—	0.0368	—	—	0.134	0.187	0.215	0.229	0.238	0.242	0.248	0.255	0.267	0.279	—
Sc	0.0035	0.0081	0.0158	0.0270	0.0437	—	0.1433	—	—	0.365	0.470	0.520	0.548	0.564	0.570	0.580	0.589	0.610	0.630	—
Ag	0.0019	0.0066	0.0159	0.0291	0.0443	0.0778	0.108	0.133	0.166	0.187	0.213	0.225	0.232	0.236	0.240	0.245	0.251	0.264	0.276	0.291
Na	—	—	—	—	—	—	—	—	—	—	—	—	—	—	1.37	1.33	1.30	1.26	1.26	1.29
Sr	—	—	—	—	—	—	—	—	—	—	—	—	—	0.301	0.318	0.334	0.349	0.382	0.454	0.353
Ta	0.0012	0.0036	0.0082	0.0152	0.0237	0.0421	0.0590	0.075	0.095	0.108	0.125	0.132	0.137	0.141	0.145	0.148	0.149	0.152	0.160	—
Te	0.0069	0.0203	0.0354	0.0508	0.0737	0.0922	0.116	0.132	0.155	0.169	0.186	0.193	0.197	0.201	0.206	0.211	0.216	0.225	—	—
Tl	0.0166	0.0326	0.0491	0.0651	0.0778	0.0920	0.103	0.108	0.116	0.120	0.124	0.126	0.128	0.130	0.136	0.143	—	—	—	—
Th	0.0029	0.0094	0.0200	0.0325	0.0433	—	0.073	—	—	0.099	0.108	0.112	0.115	0.118	0.124	0.129	0.134	0.145	0.156	0.167
Tm	0.0116	0.0327	0.0629	0.0973	0.1305	—	0.226	—	—	0.150	0.154	0.157	0.158	0.160	0.163	0.167	0.171	0.178	0.186	—
Sn	0.0078	0.0226	0.0400	0.0582	0.0760	0.108	0.130	0.149	0.173	0.189	0.206	0.214	0.220	0.222	0.245	0.267	0.257	0.257	0.257	—
Ti	0.0013	0.0033	0.0069	0.0140	0.0248	0.0516	0.094	0.144	0.227	0.295	0.406	0.464	0.501	0.525	0.555	0.578	0.597	0.627	0.652	—
W	0.0002	0.0007	0.0019	0.0042	0.0078	0.0184	0.0332	0.048	0.072	0.089	0.113	0.125	0.131	0.135	0.137	0.139	0.140	0.144	0.148	—
U	0.0015	0.0055	0.0128	0.0230	0.0339	—	0.0659	—	—	0.094	0.103	0.109	0.114	0.117	0.124	0.134	0.145	0.174	0.178	—
V	0.0023	0.0043	0.0072	0.0107	0.0189	0.0420	0.0730	0.115	0.190	0.257	0.379	0.434	0.462	0.483	0.512	0.528	0.540	0.563	0.598	—
Yb	0.0085	0.0254	0.0457	0.0653	0.0808	—	0.116	—	—	0.139	0.145	0.149	0.151	0.155	0.160	0.171	0.172	0.178	0.185	0.213
Y	0.0026	0.0089	0.0212	0.0329	0.0593	—	0.137	—	—	0.233	0.265	0.282	0.292	0.298	0.305	0.313	0.321	0.338	0.354	0.372
Zn	0.0025	0.0109	0.0269	0.0493	0.0760	0.123	0.170	0.205	0.258	0.295	0.345	0.366	0.380	0.389	0.404	0.419	0.435	0.479	0.479	—
Zr	0.0014	0.0046	0.0119	0.0220	0.0344	0.0941	0.101	0.108	0.116	0.120	0.124	0.126	0.128	0.130	0.136	0.143	0.153	0.153	0.153	—

TABLE 2.32 Thermal Conductivity and Density of Selected Elements

Substance	Chemical formula	T (°C)	T (°K)	Density ρ (kg/m³)	Thermal conductivity λ (W/m K)
Aluminum, 99.75%	Al	−190	83.15		255.860
		0	273.15	2,700	229.111
		200	473.15		229.111
		300	573.15		222.133
		800	1073.15		125.604
99%		−100	173.15	—	209.340
		0	273.15		209.340
		100	373.15		207.014
		300	573.15		222.133
Antimony, very pure	Sb	−190	83.15		20.934
		−100	173.15		19.190
		0	273.15		17.678
		100	373.15	6,690	16.282
		300	573.15		15.817
		500	773.15		18.608
Beryllium, 99.5%	Be	−250	23.15		94.203
		−100	173.15		125.604
		0	273.15	1,850	160.494
		100	373.15		190.732
		200	473.15		215.155
Bismuth	Bi	−190	83.15		25.586
		−100	173.15		12.095
		0	273.15	9,800	8.374
		100	373.15		7.211
		200	473.15		7.211
Cadmium, pure	Cd	−190	83.15		104.670
		−100	173.15		96.529
		0	273.15	8,620	93.040
		100	373.15		91.877
		200	473.15		91.296
		300	573.15		87.807
Cobalt, 97.1%	Co	20	293.15	≈8,900	69.780
Copper, pure 99.9–98%	Cu	−180	93.15		464.037
		−100	173.15		407.050
		0	273.15	8,930	386.116
		100	373.15		379.138
		200	473.15		373.323
		400	673.15		364.019
		600	873.15		353.552
Commercial		20	293.15	8,300	372.160
Electrolytic, pure		−180	93.15		488.460
		0	273.15	8,900	395.420
		100	373.15		391.931
		300	573.15		381.464
		800	1073.15		367.508
Gold 99.999%	Au	−190	83.15		327.966
		0	273.15	19,290	310.521
		100	373.15		310.521
		300	573.15		304.706
99.98%		0	273.15		294.239
		100	373.15		294.239

TABLE 2.32 Thermal Conductivity and Density of Selected Elements (*Continued*)

Substance	Chemical formula	T (°C)	T (°K)	Density ρ (kg/m³)	Thermal conductivity λ (W/m K)
Iridium, pure	Ir	0	273.15	22,420	59.313
		100	373.15		56.987
Iron (Armc)	Fe	20	293.15	7,850	73.169
99.92%		100	373.15		67.454
		200	473.15		61.639
		400	673.15		48.846
		600	873.15		38.379
		800	1073.15		29.075
Cast, 1% Ni		20	293.15	7,280	50.009
		100	373.15		49.428
		300	573.15		46.520
		500	773.15		37.216
Cast, 3% C		20	293.15	7,280	55.824 . . . 63.965
Steel, 99.2%		0	273.15	7,800	45.357
Fe, 0.2% C		100	373.15		45.357
		300	573.15		43.031
		500	773.15		37.216
		800	1073.15		30.238
Wrought, pure		0	273.15	7,800	59.313
		100	373.15		56.987
		200	473.15		52.335
		400	673.15		44.194
		600	873.15		37.216
		800	1073.15		29.075
Lead, pure	Pb	−250	23.15		48.846
		−200	73.15		40.705
		−100	173.15		36.867
		0	273.15		35.123
		20	293.15	11,340	34.774
		100	373.15		33.378
		300	573.15		29.773
		500	773.15		16.747
Lithium, pure	Li	0	273.15	530	70.943
		100	373.15		70.943
Magnesium, pure	Mg	−190	83.15		186.080
		0	273.15	1,740	172.124
		200	473.15		162.820
99.6%		0	273.15	≈1,740	144.212
		100	373.15		139.560
		300	573.15		131.419
		500	773.15		131.419
Manganese	Mn	0	273.15	7,300	50.242
Mercury	Hg	−190	83.15		48.846
		−100	173.15		36.053
		−50	223.15		27.912
	(Liquid)	0	273.15	13,595	8.141

TABLE 2.32 Thermal Conductivity and Density of Selected Elements (*Continued*)

Substance	Chemical formula	T (°C)	T (°K)	Density ρ (kg/m³)	Thermal conductivity λ (W/m K)
Molybdenum 99.84%	Mo	−180	93.15		174.450
		−100	173.15		138.397
		0	273.15	10,200	137.234
		100	373.15		137.234
		1000	1273.15		98.855
Nickel 99.94%	Ni	−180	93.15		110.485
		0	273.15	8,800	93.040
		100	373.15		82.573
		200	473.15		73.269
		300	573.15		63.965
		400	673.15		59.313
		500	773.15		61.639
99.2%		0	273.15		67.454
		100	373.15		62.802
		200	473.15	—	58.150
		400	673.15		52.335
		600	873.15		56.987
		800	1073.15		62.802
97 to 99%		−100	173.15		55.824
		0	273.15		58.150
		100	373.15		56.987
		200	473.15	—	54.661
		400	673.15		48.846
		600	873.15		53.498
		800	1073.15		58.150
Palladium, pure	Pd	−190	83.15		76.758
		0	273.15	—	68.617
		100	373.15		73.269
Platinum, pure	Pt	−190	83.15		77.921
		0	273.15	21,400	70.013
		100	373.15		71.408
		300	573.15		75.595
		500	773.15		79.084
		800	1073.15		86.062
		1000	1273.15		89.551
Potassium, pure	K	0	273.15	860	136.071
		100	373.15		118.626
Rhodium, pure	Rh	−190	83.15		212.829
		0	273.15	12,500	88.388
		100	373.15		80.247
Silver > 99.98%	Ag	−190	83.15		425.658
		0	273.15	10,500	418.680
		100	373.15		416.354
		300	573.15		407.050
99.9%		−100	173.15		419.843
		0	273.15	10,500	410.539
		100	373.15		391.931
		300	573.15		361.693
		500	773.15		362.856

TABLE 2.32 Thermal Conductivity and Density of Selected Elements (*Continued*)

Substance	Chemical formula	T (°C)	T (°K)	Density ρ (kg/m³)	Thermal conductivity λ (W/m K)
Sodium, pure	Na	−100	173.15		154.679
		0	273.15	970	100.018
		50	323.15		93.040
		100	373.15		83.736
Tantalum	Ta	0	273.15	16,650	54.661
		100	373.15		54.080
		1000	1273.15		63.965
		1400	1673.15		72.106
		1800	2073.15		82.573
Thallium, pure	Tl	−190	83.15		62.802
		0	273.15	11,840	51.172
		100	373.15		41.868
Tin, pure	Sn	−150	123.15		79.084
		−100	173.15		74.432
		0	273.15	7,300	66.058
		100	373.15		59.313
		200	473.15		56.987
Wolfram	W	−190	83.15		217.481
		0	273.15	19,300	166.309
		100	373.15		151.190
		500	773.15		119.789
		1000	1273.15		98.855
		1500	1773.15		113.974
		2000	2273.15		136.071
		2400	2673.15		146.538
Zinc, pure	Zn	−100	173.15		115.137
		0	273.15	7,130	112.811
		100	373.15		109.904
		200	473.15		105.833
		300	573.15		101.181

Source: Ref. 2 with permission.

TABLE 2.33 Thermal Diffusivity of Selected Elements (m²/s)

T (K)	Element											
	Al	Sb	Be	Cd	Ca	Ce	Cs	Cr	Co	Cu	Dy	Er
10	9.90	0.0339		0.0148			4.2–4	0.072	0.025	2.30	2.5–4	6.6–5
15	2.40	0.00735		0.0020			2.5–4	0.061	0.016	0.70	1.0–4	2.3–5
20	0.50	0.00177		6.1–4			1.9–4	0.038	0.011	0.16	4.7–5	6.8–6
25	0.15	6.8–4		3.5–4			1.7–4	0.018	0.005	0.047	3.1–5	1.0–5
30	0.06	4.1–4		2.2–4			1.5–4	0.010	0.003	0.018	2.4–5	8.4–6
40	0.012	2.0–4		1.5–5			1.3–4	3.7–3	1.1–3	4.0–3	1.5–5	6.8–6
50	3.3–3	1.2–4		1.1–4			1.26–4	7.5–4	4.8–4	1.4–3	1.1–5	6.0–6
60	1.4–3	8.8–5		8.6–5			1.20–4	5.9–4	2.6–4	6.9–4	9.0–6	5.9–6
80	4.4–4	5.6–5		6.8–5			1.12–4	2.0–4	1.2–4	3.1–4	6.5–6	7.4–6
100	2.3–4	4.2–5	3.58–3	6.3–5			1.06–4	1.2–4	7.7–5	2.2–4	5.5–6	9.1–6
150	1.3–4	2.8–5	4.0–4	5.6–5			9.7–5	5.7–5	4.7–5	1.9–4	3.8–6	9.7–6
200	1.1–4	2.3–5	1.5–4	5.2–5			9.3–5	4.1–5	3.7–5	1.3–4	6.3–6	9.9–6
250	1.1–4	2.0–5	8.3–5	5.0–5	2.1–4		8.8–5	3.3–5	3.1–5	1.2–4	7.0–6	9.9–6
300	9.7–5	1.7–5	5.9–5	4.9–5	2.0–4	8.0–6	8.0–5	2.9–5	2.7–5	1.2–4	7.4–6	9.4–6
400	9.4–5	1.5–5	4.0–5	4.6–5	1.8–4	9.0–6	4.8–5	2.6–5	2.2–5	1.1–4	7.5–6	9.0–6
500	8.9–5	1.3–5	3.1–5	4.3–5	1.6–4	1.0–5	5.1–5	2.3–5	1.8–5	1.1–4	7.7–6	8.9–6
600	8.4–5	1.2–5	2.6–5	1.8–5	1.5–4	1.0–5	5.5–5	2.0–5	1.5–5	1.0–4	7.9–6	8.9–6
800	7.4–5	1.1–5	2.1–5	2.0–5	—	1.1–5	5.8–5	1.7–5	1.2–5	9.0–5	8.6–6	9.0–6
1000	6.6–5		1.7–5	2.4–5		1.1–5	5.3–5	1.4–5	1.0–5	9.0–5	9.2–6	9.1–6
1200	6.1–5		1.4–5				4.6–5	1.3–5	9.0–6	8.0–5		
1400			1.2–5				3.9–5	1.2–5			—	—
1600								1.1–5				
1800												
2000												

T (K)	Element											
	Gd	Ge	Au	Hf	Ho	In	Ir	Fe	La	Pb	Li	Lu
10	8.2–3	0.46	0.060	8.0–4	9.8–5	4.3–3	0.183	0.133		1.1–3	0.124	7.1–4
15	3.5–3	0.072	0.015	3.0–4	4.0–5	9.1–4	0.091	0.075		2.0–4	0.052	2.5–4
20	1.4–3	0.021	0.005	1.4–4	2.9–5	4.0–4	0.046	0.043		9.3–5	0.023	1.2–4
25	7.5–4	0.010	0.002	9.0–5	2.2–5	2.3–4	0.016	0.025	3.4–5	6.4–5	0.011	0.7–4
30	4.8–4	0.006	0.001	6.0–5	1.9–5	1.6–4	0.007	0.013	2.6–5	5.1–5	0.006	0.5–4
40	2.7–4	2.4–3	4.5–4	3.8–5	1.5–5	1.1–4		3.2–3	1.4–5	3.9–5	2.3–3	3.0–5
50	1.6–4	1.3–3	3.0–4	2.8–5	1.2–5	8.9–5	5.6–4	1.2–3	1.1–5	3.5–5	0.8–3	2.4–5
60	1.4–4	8.2–4	2.3–4	2.3–5	1.1–5	7.8–5		4.9–4	1.0–5	3.3–5	0.5–3	2.0–5
80	1.0–4	4.0–4	1.8–4	2.1–5	0.9–5	7.0–5		1.6–4	0.9–5	3.1–5	0.2–3	1.7–5

Above the solid line a substance is solid; below it, it is liquid.
The notation 5.4.–3 signifies 5.4 × 10⁻³.

TABLE 2.33 Thermal Diffusivity of Selected Elements (m²/s) (*Continued*)

T (K)	Element											
	Gd	Ge	Au	Hf	Ho	In	Ir	Fe	La	Pb	Li	Lu
100	9.0–5	2.2–4	1.5–4	1.7–5	0.7–5	6.2–5	8.4–5	8.2–5	1.0–5	2.9–5	0.1–3	1.5–5
150	7.0–5	1.0–4	1.40–4	1.4–5	0.9–5	5.8–5	6.3–5	4.1–5	1.0–5	2.7–5	6.5–5	1.3–5
200	6.0–5	6.3–5	1.34–4	1.3–5	1.0–5	5.4–5	5.6–5	3.1–5	1.0–5	2.6–5	5.4–5	1.2–5
250		4.6–5	1.31–4	1.3–5	1.1–5	5.1–5	5.2–5	2.6–5	1.1–5	2.5–5	4.8–5	1.1–5
300		3.5–5	1.27–4	1.2–5	1.1–5	4.8–5	5.0–5	2.2–5	1.1–5	2.4–5	4.5–5	1.1–5
400		2.4–5	1.23–4	1.2–5	1.2–5	4.1–5	4.8–5	1.8–5	1.2–5	2.3–5	4.5–5	1.1–5
500		1.8–5	1.19–4	1.1–5	1.2–5	2.2–5	4.6–5	1.5–5	1.3–5	2.2–5	2.1–5	
600		1.4–5	1.15–4	1.1–5		2.4–5	4.4–5	1.3–5	1.4–5	2.0–5	2.3–5	
800		1.0–5	1.07–4	1.0–5			4.1–5	1.1–5	1.5–5	1.3–5	2.8–5	
1000		0.9–5	9.8–5	1.0–5			3.5–5	1.0–5	1.6–5	1.5–5	3.3–5	
1200		0.8–5	9.0–5	9.0–5			3.3–5			1.7–5	3.7–5	
1400			4.1–5				3.1–5					
1600			4.4–5				3.0–5					
2000												

T (K)	Mg	Mo	Ni	Nb	Pd	Pt	K	Pu	Rh	Rb	Ag	Na
10	0.395	0.0292	0.163	1.6–2	0.0428	0.0517			0.357		0.835	
15	0.116	0.0167	0.079	6.7–3	0.0166	0.0114			0.224		0.140	
20	0.050	0.0095	0.033	2.6–3	0.0054	0.0029			0.115		0.031	
25	0.023	0.0057	0.014	1.0–3	0.0021	0.0010			0.044		0.009	
30	0.010	0.0037	0.006	4.6–4	0.0009	0.0005			0.017		0.004	
40	2.6–3	1.4–3	1.7–3	1.6–4	2.8–4	1.6–4			5.1–3		1.3–3	
50	9.2–4	7.1–4	6.2–4	1.3–4	1.3–4	9.2–5	2.3–4		9.3–4		6.6–4	
60	4.9–4	4.0–4	3.1–4	6.0–5	8.2–5	6.3–5	2.1–4		4.1–4		4.5–4	
80	2.1–4	2.0–4	1.3–4	3.9–5	4.8–5	4.3–5	1.9–4		1.8–4		2.8–4	
100	1.5–4	1.3–4	8.0–5	3.2–5	3.8–5	3.6–5	1.8–4	2.1–6	1.0–4		2.3–4	
150	1.1–4	7.4–5	4.2–5	2.6–5	2.9–5	2.9–5		2.1–6	6.5–5		1.9–4	
200	1.0–4	6.3–5	3.1–5	2.4–5	2.6–5	2.7–5		2.2–6	5.6–5		1.8–4	
250	9.1–5	5.7–5	2.6–5	2.4–5	2.5–5	2.5–5		2.3–6	5.2–5		1.8–4	
300	8.9–5	5.4–5	2.3–5	2.3–5	2.4–5	2.5–5		2.6–6	4.9–5		1.7–4	
400	8.2–5	5.1–5	1.9–5	2.4–5	2.5–5	2.5–5	7.3–5		4.6–5	6.1–5	1.7–4	6.9–5
500	7.7–5	4.8–5	1.5–5	2.4–5	2.5–5	2.5–5	7.3–5		4.3–5	5.9–5	1.7–4	6.8–5
600	7.3–5	4.5–5	1.3–5	2.4–5	2.6–5	2.5–5	7.2–5		4.0–5	5.8–5	1.6–4	6.7–5
800	6.7–5	4.2–5	1.4–5	2.5–5	2.8–5	2.5–5	6.7–5		3.6–5	5.5–5	1.5–4	6.4–5

Continuation band (high-temperature rows, printed above the main table):

T	Ta	Te	Th	Sn	Ti	W	U	V	Y	Zn	Zr
1000	4.3–5	3.8–5	1.5–5	2.6–5	2.9–5	2.5–5	6.0–5		3.2–5	1.4–4	6.0–5
1200		3.5–5	1.5–5	2.6–5	3.0–5	2.5–5	4.7–5		3.0–5	1.3–4	5.3–5
1400		3.3–5	1.6–5	2.7–5		2.6–5			2.8–5		4.7–5
1600		3.1–5	1.6–5	2.7–5		2.6–5			2.6–5		4.1–5
1800		3.0–5		2.8–5		2.7–5					
2000		2.8–5		2.8–5	—	2.8–5					

Main table:

T	Ta	Te	Th	Sn	Ti	W	U	V	Y	Zn	Zr
10	5.4–3	8.0–4			2.4–3	1.140	3.4–4	1.0–3			1.1–2
15	2.4–3	2.4–4			1.3–3	0.350	1.3–4	7.7–4		2.5–2	3.8–3
20	1.1–3	1.4–4			8.0–4	0.105	4.5–5	5.7–4		4.6–3	1.4–3
25	5.4–4	9.1–5			4.7–4	0.039	3.8–5	4.6–4		1.7–3	6.3–4
30	2.9–4	7.2–5			3.1–4	0.013	2.7–5	3.0–4		7.0–4	3.3–4
40	1.3–4	5.8–5			1.6–4	2.5–3	2.0–5	1.5–4		3.1–4	9.6–5
50	7.7–5	5.0–5			9.6–5	7.5–4	1.5–5	9.2–5	2.4–5	1.7–4	7.4–5
60	5.5–5	4.6–5			6.3–5	3.5–4	1.4–5	5.9–5	2.0–5	1.0–4	5.9–5
80	4.0–5	4.1–5			3.8–5	2.3–4	1.3–5	3.4–5	1.8–5	7.0–5	5.0–5
100	3.3–5	3.9–5	5.2–5		2.7–5	1.6–4	1.2–5	2.3–5	1.5–5	5.5–5	4.3–5
150	2.8–5	3.5–5	4.5–5				1.2–5	1.4–5	1.4–5	5.1–5	
200	2.6–5	3.3–5	4.1–5				1.2–5	1.2–5	1.3–5	4.7–5	
250	2.5–5	3.1–5	4.0–5				1.2–5	1.1–5	1.3–5	4.3–5	
300	2.4–5	3.0–5	3.9–5				1.2–5	1.1–5	1.3–5	4.1–5	
400	2.4–5	2.8–5	3.8–5				1.3–5	1.0–5	1.3–5	3.9–5	
500	2.4–5	2.6–5	3.7–5	1.8–4			1.3–5	1.0–5	1.4–5	3.7–5	
600	2.4–5		3.6–5	2.1–4			1.3–5	1.0–5	1.4–5	3.4–5	
800	2.4–5		3.4–5	2.4–4			1.3–5	1.1–5	1.5–5	1.8–5	
1000	2.3–5		3.2–5	2.7–4			1.3–5	1.1–5	1.5–5	2.2–5	
1200	2.3–5		3.1–5				1.4–5	1.1–5	1.6–5		
1400	2.3–5										
1600	2.4–5										
1800											
2000											

Above the solid line a substance is solid; below it, it is liquid.
The notation 5.4.–3 signifies 5.4×10^{-3}.

TABLE 2.34   Density and Thermal Conductivity of Alloys

Alloy	Composition (%)	T (°C)	T (K)	Density ρ (kg/m³)	Thermal conductivity λ (W/m K)
Aluminum alloys	96 Al, 1.8 Cu, 0.9 Fe, 0.9 Cr, 0.4 Si	20	293.15	—	104.670
Aluminum bronze	95 Cu, 5Al	20	293.15	7800	82.573
Aluminum magnesium	92 Al, 8 Mg	−180	93.15		75.595
		−100	173.15		84.899
		0	273.15		102.344
		20	293.15	≈2600	105.833
		100	373.15		123.278
		200	473.15		147.701
Alusil	80 Al, 20 Si	−180	93.15		122.115
		−100	173.15		141.886
		0	273.15		158.168
		20	293.15	≈2650	160.494
		100	373.15		168.635
		200	473.15		174.450
Bismuth-antimony	80 Bi, 20 Sb	0	273.15	—	6.606
		100	373.15		8.618
	50 Bi, 50 Sb	0	273.15	—	8.327
		100	373.15		9.374
	30 Bi, 70 Sb	0	273.15	—	9.653
		100	373.15		11.660
Brass	90 Cu, 10 Zn	−100	173.15		88.388
		0	273.15	≈8600	102.344
		100	373.15		117.463
		200	473.15		133.745
		300	573.15		148.864
		400	673.15		166.309
		500	773.15		180.265
		600	873.15		195.384
	70 Cu, 30 Zn	0	273.15	≈8600	105.833
		100	373.15		109.322
		200	473.15		110.485
		300	573.15		113.974
		400	673.15		116.300
		500	773.15		119.789
		600	873.15		120.952
	66 Cu, 33 Zn	0	273.15	≈8600	100.018
		100	373.15		106.996
		200	473.15		112.811
		300	573.15		120.952
		400	673.15		127.930
		500	773.15		134.908
		600	873.15		151.190
	60 Cu, 40 Zn	0	273.15	≈8600	105.833
		100	373.15		119.789
		200	473.15		137.234
		300	573.15		152.353
		400	673.15		168.635
		500	773.15		186.080
		600	873.15		200.036

TABLE 2.34 Density and Thermal Conductivity of Alloys (*Continued*)

Alloy	Composition (%)	T (°C)	T (K)	Density ρ (kg/m³)	Thermal conductivity λ (W/m K)
Brass	61.5 Cu, 38.5 Zn	20	293.15		79.084
		100	373.15		88.388
Bronze	90 Cu, 10 Sn	20	293.15	8766	41.868
	75 Cu, 25 Sn	20	293.15	≈8900	25.586
	88 Cu, 10 Sn, 2 Zn	20	293.15	≈8800	47.683
	84 Cu, 6 Sn, 9 Zn, 1 Pb	20	293.15	—	58.150
	86 Cu, 7 Zn, 6.4 Sn	20	293.15	≈8600	60.476
		100	373.15		70.943
Chrome-nickel steel	0.8 Cr, 3.5 Ni, 0.4 C	20	293.15	8100 . . .	34.890
		100	373.15	8700	36.053
		200	473.15		37.216
		400	673.15		37.216
		600	873.15		31.401
	Cr . . . Ni	20	293.15	7900	13.956
		200	473.15		17.445
		500	773.15		20.934
	17 . . . 19 Cr, 8 Ni, 0.1 . . . 0.2 C	20	293.15	8100 . . .	14.538
		100	373.15	9000	15.701
		200	473.15		16.864
		300	573.15		18.608
		500	773.15		20.934
	10 Cr, 34 Ni	20	293.15		12.212
		100	373.15		13.375
		200	473.15	—	15.119
		300	573.15		16.282
		500	773.15		19.190
	15 Cr, 27 Ni, 3 W, 0.5 C	20	293.15		11.281
		100	373.15		12.793
		200	473.15	—	13.956
		300	573.15		15.119
		500	773.15		18.608
	15 Cr, 13 Ni, 2 W, 0.5 C	20	293.15		11.630
		100	373.15		11.630
		200	473.15	—	11.630
		300	573.15		12.212
		500	773.15		12.793
		800	1073.15		16.282
Chrome steel	0.8 Cr, 0.2 C	100	373.15	≈7850	39.542
		200	473.15		37.216
		400	673.15		31.401
		600	873.15		26.749
	5 Cr, 0.5 Mn, 0.1 C	20	293.15	8100 . . .	37.216
		100	373.15	9000	31.635
		200	473.15		31.053
		500	773.15		33.727
	15 Cr, 0.1 C	20	293.15	8100 . . .	25.586
		500	773.15	9000	25.586

TABLE 2.34 Density and Thermal Conductivity of Alloys (*Continued*)

Alloy	Composition (%)	T (°C)	T (K)	Density ρ (kg/m³)	Thermal conductivity λ (W/m K)
Chrome steel	14 Cr, 0.3 C	20	293.15	8100 . . .	24.423
		100	373.15	9000	25.005
		200	473.15		25.586
		300	573.15		25.586
		500	773.15		25.586
	16 Cr, 0.9 C	100	373.15	8100 . . .	23.842
		200	473.15	9000	23.260
		300	573.15		23.260
		500	773.15		23.260
		800	1073.15		23.260
	26 Cr, 0.1 C	20	293.15	8100 . . .	19.771
		100	373.15	9000	20.934
		200	473.15		22.097
		300	573.15		22.911
		500	773.15		24.423
Cobalt steel	5 . . . 10 Co	20	293.15	≈7800	40.705
Constantin	60 Cu, 40 Ni	−100	173.15		20.934
		0	273.15		22.213
		20	293.15	8800	22.679
		100	373.15		25.586
Copper alloys	92 Al, 8 Cu	−180	93.15		89.551
		−100	173.15		109.322
		0	273.15		127.930
		20	293.15	≈2800	131.419
		100	373.15		143.049
		200	473.15		152.353
Copper-manganese	70 Cu, 30 Mn	20	293.15	≈7800	12.793
Copper-nickel	90 Cu, 10 Ni	20	293.15	≈8800	58.150
		100	373.15		75.595
	80 Cu, 20 Ni	20	293.15	≈8500	33.727
		100	373.15		40.705
	40 Cu, 60 Ni	20	293.15	≈8400	22.097
		100	373.15		25.586
	18 Cu, 82 Ni	20	293.15		25.586
		100	393.15		25.586
Duralumin	94 . . . 96 Al, 3 . . . 5 Cu, 0.5 Mg	−180	93.15		90.714
		−100	173.15		125.604
		0	273.15		159.331
		20	293.15	≈2800	165.146
		100	373.15		181.428
		200	473.15		194.221
Electron alloy	93 Mg, 4 Zn, 0.5 Cu	20	293.15	1800	116.300
German alloy	88 Al, 10 Zn, 2 Cu	0	273.15	2900	143.049
		20	293.15		145.375
		100	373.15		154.679
Gold-copper alloy	88 Au, 12 Cu	0	273.15	—	55.824
		100	373.15		67.454
	27 Au, 73 Cu	0	273.15	—	90.714
		100	373.15		113.974

TABLE 2.34 Density and Thermal Conductivity of Alloys (*Continued*)

Alloy	Composition (%)	T (°C)	T (K)	Density ρ (kg/m³)	Thermal conductivity λ (W/m K)
Invar	35 Ni, 65 Fe	20	293.15	8130	11.049
Lautal	95 Al, 4.5 . . . 5.5 Cu, 0.3 Si	20	293.15	—	139.560
Magnesium-aluminum	92 Mg, 8 Al	−180	93.15		41.868
		−100	173.15		50.009
		0	273.15	≈1800	60.476
		20	293.15		61.639
		100	373.15		69.780
		200	473.15		79.084
	2.5 Al	20	293.15	—	85.597
	4.2 Al	20	293.15	—	69.082
	6.2 Al	20	293.15	—	55.591
	10.3 Al	20	293.15	—	43.496
Magnesium-aluminum-silicone	88 Mg, 10 Al, 2 Si	−180	93.15		30.238
		−100	173.15		40.705
		0	273.15	≈1850	55.824
		20	293.15		58.150
		100	373.15		68.617
		200	473.15		75.595
Magnesium-copper	92 Mg, 8 Cu	−180	93.15		88.388
		−100	173.15		106.996
		0	273.15	≈2400	124.441
		20	293.15		125.604
		100	373.15		130.256
		200	473.15		132.582
	93.7 Mg, 6.3 Cu	20	293.15		131.419
Manganese-nickel steel	12 Mn, 3 Ni, 0.75 C	20	293.15		13.956
		100	373.15		14.770
		200	473.15	—	16.282
		300	573.15		17.445
		500	773.15		19.771
Manganese steel	1.6 Mn, 0.5 C	20	293.15	≈7850	40.705
		100	373.15		40.705
		300	573.15		37.216
		500	773.15		34.890
	2 Mn	20	293.15	≈7850	32.564
	5 Mn	20	293.15	≈7850	18.608
Manganine	84 Cu, 4 Ni, 12 Mn	−100	173.15		16.282
		0	273.15	8400	20.934
		20	293.15		21.864
		100	373.15		26.400
Monel	29 Cu, 67 Ni, 2 Fe	20	293.15	8710	22.097
		100	373.15		24.423
		200	473.15		27.563
		300	573.15		30.238
		400	673.15		33.727

TABLE 2.34   Density and Thermal Conductivity of Alloys (*Continued*)

Alloy	Composition (%)	T (°C)	T (K)	Density ρ (kg/m³)	Thermal conductivity λ (W/m K)
New silver	62 Cu, 15 Ni, 22 Zn	−150	123.15	8433	17.678
		−100	173.15		19.170
		+20	293.15		25.005
		100	373.15		31.401
		200	473.15		39.542
		300	573.15		45.357
		400	673.15		48.846
Nickel alloy	70 Ni, 28 Cu, 2 Fe	20	293.15	≈8200	34.890
Nickel-chrome	90 Ni, 10 Cr	0	273.15	≈8220	17.096
		20	293.15		17.445
		100	373.15		18.957
		200	473.15		20.934
		300	573.15		22.795
		400	673.15		24.656
	80 Ni, 20 Cr	0	273.15	≈8200	12.212
		20	293.15		12.560
		100	373.15		13.840
		200	473.15		15.584
		300	573.15		17.212
		400	673.15		18.957
		600	873.15		22.562
Nickel-chrome steel	61 Ni, 15 Cr, 20 Fe, 4 Mn	20	293.15	≈8190	11.630
		100	373.15		11.863
		200	473.15		12.212
		300	573.15		12.444
		400	673.15		12.677
		600	873.15		13.142
		800	1073.15		13.956
	61 Ni, 16 Cr, 23 Fe	0	273.15	≈8190	11.863
		20	293.15		12.095
		100	373.15		13.258
		200	473.15		14.654
		300	573.15		16.049
		400	673.15		17.445
	70 Ni, 18 Cr, 12 Fe	20	293.15	—	11.514
	62 Ni, 12 Cr, 26 Fe	20	293.15	≈8100	13.491
Nickel-silver	—	0	273.15	—	29.308
		100	373.15		37.216
Nickel steel	5 Ni	20	293.15	8130	34.890
	10 Ni	20	293.15		27.912
	15 Ni	20	293.15		22.097
	20 Ni	20	293.15		18.608
	25 Ni	20	293.15		15.119
	30 Ni	20	293.15		12.212
	35 Ni	20	293.15		11.049
	40 Ni	20	293.15		11.049
	50 Ni	20	293.15		14.538
	60 Ni	20	293.15		19.190
	70 Ni	20	293.15		25.586
	80 Ni	20	293.15		32.564

TABLE 2.34 Density and Thermal Conductivity of Alloys (*Continued*)

Alloy	Composition (%)	T (°C)	T (K)	Density ρ (kg/m³)	Thermal conductivity λ (W/m K)
Nickel steel	30 Ni, 1 Mn, 0.25 C	20	293.15	8190	12.095
		100	373.15		13.607
	36 Ni, 0.8 Mn	20	293.15	—	12.095
	1.4 Ni, 0.5 Cr, 0.3 C	20	293.15	≈7850	45.357
		100	373.15		44.194
		300	573.15		40.705
		500	773.15		37.216
Phosphor bronze	92.8 Cu, 5 Sn, 2 Zn, 0.15 P	20	293.15	≈8766	79.084
	91.7 Cu, 8 Sn, 0.3 P	20	293.15	8800	45.357
		100	373.15		52.335
		200	473.15		61.639
	87.8 Cu, 10 Sn, 2 Zn, 0.2 P	20	293.15	—	41.868
	87.2 Cu, 12.4 Sn, 0.4 P	20	293.15	8700	36.053
Piston alloy, cast	91.5 Al, 4.6 Cu, 1.8 Ni, 1.5 Mg	0	273.15	≈2800	143.049
		20	293.15		144.212
		100	373.15		151.190
		200	473.15		158.168
	84 Al, 12 Si, 1.2 Cu, 1 Ni	0	273.15	≈2800	134.908
		20	293.15		134.908
		100	373.15		137.234
		200	473.15		139.560
Platinum-iridium	90 Pt, 10 Ir	0	273.15	—	30.936
		100	373.15		31.401
Platinum-rhodium	90 Pt, 10 Rh	0	273.15	—	30.238
		100	373.15		30.587
Rose's metal	50 Bi, 25 Pb, 25 Sn	20	293.15	—	16.282
Silumin	86 ... 89 Al, 11 ... 14 Si	0	273.15	2600	159.331
		20	293.15		161.657
		100	373.15		170.961
Steel	0.1 C	0	273.15	7850	59.313
		100	373.15		52.335
		200	473.15		52.335
		300	573.15		46.520
		400	673.15		44.194
		600	873.15		37.216
		900	1173.15		33.727
	0.2 C	20	293.15	7850	50.009
	0.6 C	20	293.15	7850	46.520
—Bessemer	0.52 C, 0.34 Si	20	293.15	7850	40.240
Tungsten steel	1 W, 0.6 Cr, 0.3 C	20	293.15	7900	39.542
		100	373.15		38.379
		300	573.15		36.053
		500	773.15		33.727
V 1 A steel	—	20	293.15	—	20.934
V 2 A steel	—	20	293.15	7860	15.119
Wood's metal	48 Bi, 26 Pb, 13 Sn, 13 Cd	20	293.15	—	12.793

Source: Ref. 1 with permission.

TABLE 2.35   Thermophysical Properties of Miscellaneous Materials

Description/composition	Typical properties at 300 K		
	Density ρ (kg/m^3)	Thermal conductivity k (W/m·K)	Specific heat c_p (J/kg·K)
Structural building materials			
Building boards			
Asbestos-cement board	1920	0.58	—
Gypsum or plaster board	800	0.17	—
Plywood	545	0.12	1215
Sheathing, regular density	290	0.055	1300
Acoustic tile	290	0.058	1340
Hardboard, siding	640	0.094	1170
Hardboard, high density	1010	0.15	1380
Particle board, low density	590	0.078	1300
Particle board, high density	1000	0.170	1300
Woods			
Hardwoods (oak, maple)	720	0.16	1255
Softwoods (fir, pine)	510	0.12	1380
Masonry materials			
Cement mortar	1860	0.72	780
Brick, common	1920	0.72	835
Brick, face	2083	1.3	—
Clay tile, hollow			
1 cell deep, 10 cm thick	—	0.52	—
3 cells deep, 30 cm thick	—	0.69	—
Concrete block, 3 oval cores			
Sand/gravel, 20 cm thick	—	1.0	—
Cinder aggregate, 20 cm thick	—	0.67	—
Concrete block, rectangular core			
2 core, 20 cm thick, 16 kg	—	1.1	—
Same with filled cores	—	0.60	—
Plastering materials			
Cement plaster, sand aggregate	1860	0.72	—
Gypsum plaster, sand aggregate	1680	0.22	1085
Gypsum plaster, vermiculite aggregate	720	0.25	—

TABLE 2.35 Thermophysical Properties of Miscellaneous Materials (*Continued*)

Description/composition	Typical properties at 300 K		
	Density ρ (kg/m³)	Thermal conductivity k (W/m·K)	Specific heat c_p (J/kg·K)
Insulating materials and systems			
Blanket and batt			
Glass fiber, paper faced	16	0.046	—
	28	0.038	—
	40	0.035	—
Glass fiber, coated; duct liner	32	0.038	835
Board and slab			
Cellular glass	145	0.058	1000
Glass fiber, organic bonded	105	0.036	795
Polystyrene, expanded			
Extruded (R-12)	55	0.027	1210
Molded beads	16	0.040	1210
Mineral fiberboard; roofing material	265	0.049	—
Wood, shredded/cemented	350	0.087	1590
Cork	120	0.039	1800
Loose fill			
Cork, granulated	160	0.045	—
Diatomaceous silica, coarse powder	350	0.069	—
	400	0.091	—
Diatomaceous silica, fine powder	200	0.052	—
	275	0.061	—
Glass fiber, poured or blown	16	0.043	835
Vermiculite, flakes	80	0.068	835
	160	0.063	1000
Formed/foamed in place			
Mineral wool granules with asbestos/inorganic binders, sprayed	190	0.046	—
Polyvinyl acetate cork mastic, sprayed or troweled	—	0.100	—
Urethane, two-part mixture; rigid foam	70	0.026	1045
Reflective			
Aluminum foil separating fluffy glass mats; 10–12 layers; evacuated; for cryogenic application (150 K)	40	0.00016	—
Aluminum foil and glass paper laminate; 75–150 layers; evacuated; for cryogenic application (150 K)	120	0.000017	—
Typical silica powder, evacuated	160	0.0017	—

TABLE 2.35 Thermophysical Properties of Miscellaneous Materials (*Continued*)

Description/composition	Maximum service T (K)	Typical density (kg/m³)	Typical thermal conductivity k (W/m·K) at various temperatures													
			200 K	215 K	230 K	240 K	255 K	270 K	285 K	300 K	310 K	365 K	420 K	530 K	645 K	750 K
Industrial insulation																
Blankets																
Blanket, mineral fiber, metal																
reinforced	920	96–192	—	—	—	—	—	—	—	—	0.038	0.046	0.056	0.078	—	—
	815	40–96	—	—	—	—	—	—	—	—	0.035	0.045	0.058	0.088	—	—
Blanket, mineral fiber, glass;	450	10	—	—	—	—	0.038	0.040	0.043	0.048	0.052	0.076	—	—	—	—
fine fiber, organic bonded		12	—	—	—	—	0.036	0.039	0.042	0.046	0.049	0.069	—	—	—	—
		16	—	—	—	0.033	0.035	0.036	0.039	0.042	0.046	0.062	—	—	—	—
		24	—	—	—	0.030	0.032	0.033	0.036	0.039	0.040	0.053	—	—	—	—
		32	—	—	—	0.029	0.030	0.032	0.033	0.036	0.038	0.048	—	—	—	—
		48	—	—	—	0.027	0.029	0.030	0.032	0.033	0.035	0.045	—	—	—	—
Blanket, alumina-silica fiber	1530	48	—	—	—	—	—	—	—	—	—	—	—	0.071	0.105	0.150
		64	—	—	—	—	—	—	—	—	—	—	—	0.059	0.087	0.125
		96	—	—	—	—	—	—	—	—	—	—	—	0.052	0.076	0.100
		128	—	—	—	—	—	—	—	—	—	—	—	0.049	0.068	0.091
Felt, semirigid; organic bonded	480	50–125	0.023	0.025	—	0.027	—	0.035	0.036	0.038	0.039	0.051	0.063	—	—	—
	730	50	—	—	0.026	—	0.029	0.030	0.032	0.033	0.035	0.051	0.079	—	—	—
Felt, laminated; no binder	920	120	—	—	—	—	—	—	—	—	—	—	0.051	0.065	0.087	—
Blocks, boards, and pipe insulations																
Asbestos paper, laminated and corrugated																
4-ply	420	190	—	—	—	—	—	—	—	0.078	0.082	0.098	—	—	—	—
6-ply	420	255	—	—	—	—	—	—	—	0.071	0.074	0.085	—	—	—	—
8-ply	420	300	—	—	—	—	—	—	—	0.068	0.071	0.082	—	—	—	—
Magnesia, 85%	590	185	—	—	—	—	—	—	—	—	0.051	0.055	0.061	—	—	—
Calcium silicate	920	190	—	—	—	—	—	—	—	—	0.055	0.059	0.063	0.075	0.089	0.104
Cellular glass	700	145	—	—	0.046	0.048	0.051	0.052	0.055	0.058	0.062	0.069	0.079	—	—	—
Diatomaceous silica	1145	345	—	—	—	—	—	—	—	—	—	—	—	0.092	0.098	0.104
	1310	385	—	—	—	—	—	—	—	—	—	—	—	0.101	0.100	0.115
Polystyrene, rigid																
Extruded (R-12)	350	56	0.023	0.023	0.022	0.023	0.023	0.025	0.026	0.027	0.029	—	—	—	—	—
Extruded (R-12)	350	35	0.023	0.023	0.025	0.023	0.025	0.026	0.027	0.029	—	—	—	—	—	—
Molded beads	350	16	0.026	0.029	0.030	0.033	0.035	0.036	0.038	0.040	—	—	—	—	—	—
Rubber, rigid foamed	340	70	—	—	—	—	—	0.029	0.030	0.032	0.033	—	—	—	—	—
Insulating cement																
Mineral fiber (rock, slag or glass)																
With clay binder	1255	430	—	—	—	—	—	—	—	—	0.071	0.079	0.088	0.105	0.123	—
With hydraulic setting binder	922	560	—	—	—	—	—	—	—	—	0.108	0.115	0.123	0.137	—	—
Loose fill																
Cellulose, wood, or paper pulp	—	45	—	—	—	—	—	—	0.038	0.039	0.042	—	—	—	—	—
Perlite, expanded	—	105	0.036	0.039	0.042	0.043	0.046	0.049	0.051	0.053	0.056	—	—	—	—	—
Vermiculite, expanded	—	122	—	—	—	0.058	0.061	0.063	0.065	0.068	0.071	—	—	—	—	—
	—	80	—	—	—	0.051	0.055	0.058	0.061	0.063	0.066	—	—	—	—	—

TABLE 2.35 Thermophysical Properties of Miscellaneous Materials (*Continued*)

Description/composition	T (K)	Density ρ (kg/m³)	Thermal conductivity k (W/m·K)	Specific heat c_p (J/kg·K)
Other materials				
Asphalt	300	2115	0.062	920
Bakelite	300	1300	1.4	1465
Brick, refractory				
Carborundum	872	—	18.5	—
	1672	—	11.0	—
Chrome brick	473	3010	2.3	835
	823	—	2.5	—
	1173	—	2.0	—
Diatomaceous silica, fired	478	—	0.25	—
	1145	—	0.30	—
Fire clay, burnt 1600 K	773	2050	1.0	960
	1073	—	1.1	—
	1373	—	1.1	—
Fire clay, burnt 1725 K	773	2325	1.3	960
	1073	—	1.4	—
	1373	—	1.4	—
Fire clay brick	478	2645	1.0	960
	922	—	1.5	—
	1478	—	1.8	—
Magnesite	478	—	3.8	1130
	922	—	2.8	—
	1478	—	1.9	—
Clay	300	1460	1.3	880
Coal, anthracite	300	1350	0.26	1260
Concrete (stone mix)	300	2300	1.4	880
Cotton	300	80	0.06	1300
Foodstuffs				
Banana (75.7% water content)	300	980	0.481	3350
Apple, red (75% water content)	300	840	0.513	3600
Cake batter	300	720	0.223	—
Cake, fully done	300	280	0.121	—
Chicken meat, white	198	—	1.60	—
(74.4% water content)	233	—	1.49	—
	253	—	1.35	—
	263	—	1.20	—
	273	—	0.476	—
	283	—	0.480	—
	293	—	0.489	—
Glass				
Plate (soda lime)	300	2500	1.4	750
Pyrex	300	2225	1.4	835
Ice	273	920	0.188	2040
	253	—	0.203	1945
Leather (sole)	300	998	0.013	—
Paper	300	930	0.011	1340
Paraffin	300	900	0.020	2890

TABLE 2.35   Thermophysical Properties of Miscellaneous Materials (*Continued*)

Description/composition	T (K)	Density ρ (kg/m^3)	Thermal conductivity k (W/m·K)	Specific heat c_p (J/kg·K)
Other materials (continued)				
Rock				
Granite, Barre	300	2630	2.79	775
Limestone, Salem	300	2320	2.15	810
Marble, Halston	300	2680	2.80	830
Quartzite, Sioux	300	2640	5.38	1105
Sandstone, Berea	300	2150	2.90	745
Rubber, vulcanized				
Soft	300	1100	0.012	2010
Hard	300	1190	0.013	—
Sand	300	1515	0.027	800
Soil	300	2050	0.52	1840
Snow	273	110	0.049	—
		500	0.190	—
Teflon	300	2200	0.35	—
	400	—	0.45	—
Tissue, human				
Skin	300	—	0.37	—
Fat layer (adipose)	300	—	0.2	—
Muscle	300	—	0.41	—
Wood, cross grain				
Balsa	300	140	0.055	—
Cypress	300	465	0.097	—
Fir	300	415	0.11	2720
Oak	300	545	0.17	2385
Yellow pine	300	640	0.15	2805
White pine	300	435	0.11	—
Wood, radial				
Oak	300	545	0.19	2385
Fir	300	420	0.14	2720

Source: Ref. 6 with permission.

THERMOPHYSICAL PROPERTIES OF SATURATED REFRIGERANTS

TABLE 2.36 Saturation Properties for Refrigerant 22

T_s (°C)	P_s (MPa)	ρ (kg/m^3)	C_ρ (kJ/kg K)	μ (Pas) × 10^6	κ (mW/m K)	σ (mN/m)
−140	—	1675.3 L	—	—	—	35.70
		— V	0.445	—	—	
−120	0.00023	1624.0 L	—	—	—	32.00
		0.01571 V	0.470	—	—	
−100	0.00200	1571.7 L	—	—	—	28.37
		0.12051 V	0.497	—	—	
−80	0.01035	1518.3 L	1.070	—	—	24.83
		0.56129 V	0.527	—	—	
−60	0.03747	1463.6 L	1.076	—	123.1	21.39
		1.86102 V	0.563	—	5.61	
−40*	0.10132	1409.1 L	1.092	—	114.1	18.18
		4.7046 V	0.606	—	6.93	
−20	0.24529	1346.8 L	1.125	260.1	104.8	—
		10.797 V	0.667	—	8.27	
0.00	0.49811	1281.8 L	1.171	210.1	96.2	—
		21.263 V	0.744	11.80	9.50	
20	0.91041	1210.0 L	1.238	169.1	87.8	—
		38.565 V	0.849	—	10.71	
40	1.5341	1128.4 L	1.338	136.3	79.8	—
		66.357 V	1.009	—	11.90	
60	2.4274	1030.5 L	1.528	—	—	—
		111.73 V	1.307	—	—	
80	3.6627	894.8 L	2.176	—	—	—
		195.69 V	2.268	—	—	
96.14†	4.9900	523.8 L	—	—	—	0.00
		523.8 V	—	—	—	

* Boiling point.
† Critical point.
L, liquid; V, vapor.
Extracted from Ref. 8 with permission.

TABLE 2.37 Saturation Properties for Refrigerant 123

T_s (°C)	P_s (MPa)	ρ (kg/m³)	C_ρ (kJ/kg K)	μ (Pas) × 10⁶	κ (mW/m K)	σ (mN/m)
−107.15*	0.0000	1770.9 L	0.9287	—	—	—
		0.00047 V	0.4737	—	—	—
−100	0.00001	1754.5 L	0.9259	—	—	—
		0.00123 V	0.4863	—	—	—
−80	0.00013	1709.5 L	0.9325	—	—	—
		0.01195 V	0.5202	—	—	—
−60	0.00081	1665.0 L	0.9319	—	—	—
		0.06977 V	0.5529	—	—	—
−40	0.00358	1619.9 L	0.9480	—	—	23.19
		0.28314 V	0.5850	—	—	
−20	0.01200	1573.7 L	0.9681	735.33	89.320	20.65
		0.87999 V	0.6174	9.085	8.051	
0	0.03265	1526.0 L	0.9902	564.55	83.816	18.18
		2.2417 V	0.6508	9.838	9.089	
20	0.07561	1476.5 L	1.0135	442.57	78.512	15.77
		4.9169 V	0.6861	10.562	10.163	
27.46†	0.10000	1457.5 L	1.0226	405.86	76.581	14.88
		6.3917 V	0.6999	10.825	10.576	
40	0.15447	1424.7 L	1.0384	352.37	73.388	13.42
		9.6296 V	0.7242	11.259	11.291	
60	0.28589	1369.9 L	1.0662	233.84	68.417	11.15
		17.310 V	0.7667	11.939	12.496	
80	0.48909	1311.2 L	1.0996	230.53	63.563	8.97
		29.188 V	0.8162	12.625	13.807	
100	0.78554	1246.9 L	1.1432	188.08	58.769	6.88
		46.996 V	0.8779	13.370	15.260	
120	1.1989	1174.3 L	1.2072	153.35	—	4.91
		73.471 V	0.9643	14.289	—	
140	1.7562	1088.2 L	1.3177	123.81	—	3.08
		113.71 V	1.1106	15.646	—	
160	2.4901	975.66 L	1.5835	—	—	1.44
		180.24 V	1.4728	—	—	
180	3.4505	765.88 L	4.5494	—	—	0.14
		341.95 V	5.6622	—	—	
183.68‡	3.6618	550.00 L	—	—	—	0.00
		550.00 V	—	—	—	

* Triple point.
† Normal boiling point.
‡ Critical point.
L, liquid; V, vapor.
Extracted from Ref. 9 with permission.

TABLE 2.38 Saturation Properties for Refrigerant 134a

T_s (°C)	P_s (MPa)	ρ (kg/m^3)	C_p (kJ/kg K)	μ (Pas) $\times 10^6$	κ (mW/m K)	σ (mN/m)
−103.30*	0.00039	1591.1 L	1.1838	2186.6	—	28.15
		0.02817 V	0.5853	6.63	—	
−100	0.0056	1582.3 L	1.1842	1958.2	—	27.56
		0.03969 V	0.5932	6.76	—	
−80	0.00367	1529.0 L	1.1981	1109.9	—	24.11
		0.23429 V	0.6416	7.57	—	
−60	0.01591	1474.3 L	1.2230	715.4	121.1	20.81
		0.92676 V	0.6923	8.38	—	
−40	0.05121	1417.7 L	1.2546	502.2	111.9	17.66
		2.7695 V	0.7490	9.20	8.19	
−26.08†	0.10133	1376.6 L	1.2805	363.1	105.1	15.54
		5.2566 V	0.7941	9.90	9.55	
−20	0.13273	1358.2 L	1.2930	337.2	102.4	14.51
		6.7845 V	0.8158	10.16	10.11	
0	0.2928	1294.7 L	1.3410	265.3	93.67	11.56
		14.428 V	0.8972	11.02	11.96	
20	0.5717	1225.3 L	1.4048	208.7	84.78	8.76
		27.780 V	1.0006	11.91	13.93	
40	1.0165	1146.7 L	1.4984	162.7	75.69	6.13
		50.085 V	1.1445	12.89	16.19	
60	1.6817	1052.8 L	1.6601	124.1	66.36	3.72
		87.379 V	1.3868	14.15	19.14	
80	2.6332	928.24 L	2.0648	89.69	57.15	1.60
		115.07 V	2.0122	16.31	24.0	
101.06‡	4.0592	511.94 L	—	—	—	0.0
		511.94 V	—	—	—	

* Triple point.
† Boiling point.
‡ Critical point.
Extracted from Ref. 10 with permission.

TABLE 2.39 Saturation Properties for Refrigerant 502 (Azeotrope of R22 and R115)

T_s (°C)	P_s (MPa)	ρ (kg/m³)	C_ρ (kJ/kg K)	μ (Pas) × 10⁶	κ (mW/m K)	σ (mN/m)
−70	0.02757	1557.6 L	1.024	543.6	—	—
		1.8501 V	—	—		
−60	0.04872	1527.2 L	1.042	469.7	97.9	17.41
		3.1417 V	0.574	—	—	
−45.42*	0.10132	1481.5 L	1.071	383.9	92.1	15.16
		6.2181 V	0.600	—	—	
−40	0.12964	1464.0 L	1.082	358.1	90.0	14.35
		7.8315 V	0.609	—	7.11	
−20	0.29101	1396.4 L	1.128	282.6	82.4	11.42
		16.818 V	0.649	—	8.47	
0	0.57313	1322.5 L	1.178	229.2	74.8	8.64
		32.425 V	0.709	11.69	9.80	
20	1.0197	1239.4 L	1.234	—	67.1	—
		58.038 V	0.804	12.84	11.21	
40	1.6770	1140.7 L	1.295	—	—	—
		99.502 V	0.949	13.99	12.81	
60	2.6014	1010.5 L	—	—	—	—
		171.23 V	—	—	—	
82.2†	4.075	561 L	—	—	—	—
		561 V	—	—	—	

* Boiling point.
† Critical point.
Extracted from Ref. 8 with permission.

TABLE 2.40 Saturation Properties for Ammonia

T_{sat} (K)	239.75	250	270	290	310	330	350	370	390	400
p_{sat} (kPa)	101.3	165.4	381.9	775.3	1424.9	2422	3870	5891	8606	10,280
ρ_ℓ, kg/m³	682	669	643	615	584	551	512	466	400	344
ρ_g, kg/m³	0.86	1.41	3.09	6.08	11.0	18.9	31.5	52.6	93.3	137
h_ℓ, kJ/kg	808.0	854.0	945.7	1039.6	1135.7	1235.7	1341.9	1457.5	1591.4	1675.3
h_g, kJ/kg	2176	2192	2219	2240	2251	2255	2251	2202	2099	1982
$\Delta h_{g,\ell}$, kJ/kg	1368	1338	1273	1200	1115	1019	899	744	508	307
$c_{p,\ell}$, kJ/(kg K)	4.472	4.513	4.585	4.649	4.857	5.066	5.401	5.861	7.74	
$c_{p,g}$, kJ/(kg K)	2.12	2.32	2.69	3.04	3.44	3.90	4.62	6.21	8.07	
$\eta\ell$, µNs/m²	285	246	190	152	125	105	88.5	70.2	50.7	39.5
η_g, µNs/m²	9.25	9.59	10.30	11.05	11.86	12.74	13.75	15.06	17.15	19.5
λ_ℓ (mW/m²)/(K/m)	614	592	569	501	456	411	365	320	275	252
λ_g (mW/m²)/(K/m)	18.8	19.8	22.7	25.2	28.9	34.3	39.5	50.4	69.2	79.4
Pr_ℓ	2.06	1.88	1.58	1.39	1.36	1.32	1.34	1.41	1.43	
Pr_g	1.04	1.11	1.17	1.25	1.31	1.34	1.49	1.70	1.86	
σ, mN/m	33.9	31.5	26.9	22.4	18.0	13.7	9.60	5.74	2.21	0.68
$\beta_{e,\ell}$, kK⁻¹	1.90	1.98	2.22	2.63	3.18	4.01	5.50	8.75	19.7	29.2

Source: Ref. 3 with permission.

ACKNOWLEDGMENT

The author gratefully acknowledges the use of a number of thermophysical property tables from previous editions of *Handbook of Heat Transfer*. These include Table 12 from the first edition, prepared by Professor Warren Ibele of the University of Minnesota, and Tables 11, 12, 14, 16, 21, 29, 30, and 33, prepared by Professor Peter Liley of Purdue University.

NOMENCLATURE

Symbol, Definition, SI Units, English Units

c_p	specific heat at constant pressure: kJ/(kg·K), Btu/(lb$_m$·°F)
c_{pf}	specific heat at constant pressure of saturated liquid: kJ/(kg·K), Btu/(lb$_m$·°F)
c_v	specific heat at constant volume: kJ/(kg·K), Btu/(lb$_m$·°F)
D_{ij}	diffusion coefficient: m^2/s, ft^2/s
g	gravitational acceleration: m/s^2, ft/s^2
k	thermal conductivity: W/(m·K), Btu/(h·ft·°F)
k_f	thermal conductivity of saturated liquid: W/(m·K), Btu/(h·ft·°F)
M	molecular weight: kg/(kilogram-mole), lb$_m$/(pound-mole)
P	pressure: bar, lb$_f$/in^2 (psi)
Pr	Prandtl number, $\mu c_p/k$, dimensionless
R	gas constant: kJ/(kg·K), Btu/(lb$_m$·°R)
T	temperature: K, °R, °C
v	specific volume: m^3/kg, ft^3/lb$_m$
Z	compressibility factor, Pv/RT, dimensionless

Greek Symbols

α	thermal diffusivity: m^2/s, ft^2/s
β	coefficient of volumetric thermal expansion: K^{-1}, °R^{-1}
λ or κ	thermal conductivity: W/mK, Btu/h·ft·°F
η or μ	dynamic viscosity: Pa·s, lb$_m$/(h·ft)
ν	kinematic viscosity: m^2/s, ft^2/s
ρ	density: kg/m^3, lb$_m$/ft^3
σ	surface tension: N/m, lb$_f$/ft

REFERENCES

1. K. Raznjavić, *Handbook of Thermodynamic Tables*, 2d ed., 392 pp., Begell House, New York, ISBN 1-56700-046-0, 1996.

2. N. B. Vargaftik, Y. K. Vinogradov, and V. S. Yargan, *Handbook of Physical Properties of Liquids and Gases*, 1370 pp., Begell House, New York, ISBN 1-56700-063-0, 1996.

3. C. F. Beaton and G. F. Hewitt, *Physical Property Data for the Design Engineer*, 394 pp., Hemisphere Publishing, New York, ISBN 0-89116-739-0, 1989.

4. T. F. Irvine Jr. and P. Liley, *Steam and Gas Tables with Computer Equations,* Academic Press, San Diego, ISBN 0-12-374080-0, 1984.

5. G. F. Hewitt, ed., *Handbook of Heat Exchanger Design,* Begell House, New York, ISBN 1-56700-000-2, 1992.

6. F. P. Incropera and D. P. De Witt, *Fundamentals of Heat and Mass Transfer,* 3d ed., Wiley, New York, ISBN 0-471042711-X, 1990.

7. L. Hoar, J. S. Gallagher, and G. S. Kell, *NBS/NRC Steam Tables,* Hemisphere Publishing, New York, ISBN 0-89116-354-9, 1984.

8. American Society of Heating and Air Conditioning Engineers, *1993 ASHRAE Handbook, Fundamentals, SI Edition,* ISBN 0-910110-97-2, 1993.

9. International Institute of Refrigeration, *Thermodynamic and Physical Properties, R123,* Paris, ISBN 2-903633-70-3, 1995.

10. International Institute of Refrigeration, *Extended Thermophysical Properties, R134a,* Paris, ISBN 2-903633-73-8, 1995.

SELECTED ADDITIONAL SOURCES OF THERMOPHYSICAL PROPERTIES

1. D. S. Viswanath and G. Natarajan, *Data Book on the Viscosity of Liquids,* 990 pp., Hemisphere Publishing, New York, ISBN 0-89116-778-1, 1989.

2. Y. S. Touloukian, R. W. Powell, C. Y. Ho, and P. G. Klemens, *Thermophysical Properties of Matter,* vol. 1, *Thermal Conductivity, Metallic Elements and Alloys,* 1469 pp., IFI/Plenum, New York, SBN 306-67021-6, 1970.

3. Y. S. Touloukian, R. W. Powell, C. Y. Ho, and P. G. Klemens, *Thermophysical Properties of Matter,* vol. 2, *Thermal Conductivity, Nonmetallic Solids,* 1172 pp., IFI/Plenum, New York, SBN 306-67022-4, 1970.

4. Y. S. Touloukian, P. E. Liley, and S. C. Saxena, *Thermophysical Properties of Matter,* vol. 3, *Thermal Conductivity of Nonmetallic Liquids and Gases,* 531 pp., IFI/Plenum, New York, SBN 306-67023-2, 1970.

5. Y. S. Touloukian, S. C. Saxena, and P. Hestermans, *Thermophysical Properties of Matter,* vol. 11, *Viscosity,* 643 pp., IFI/Plenum, New York, ISBN 0-306-67031-3, 1975.

6. B. Platzer, A. Polt, and G. Maurer, *Thermophysical Properties of Refrigerants,* Springer-Verlag, Berlin, 1990.

7. J. T. R. Watson, *Viscosity of Gases in Metric Units,* National Engineering Laboratory, HSMO, Edinburgh, 1972.

8. R. P. Danner and T. E. Daubert, *Physical and Thermodynamic Properties of Pure Chemicals,* DIPPR, Hemisphere Publishing, New York, 1989.

9. *Warmeatlas,* VDI-Verlag GMB H, Dusseldorf, 1984.

10. A. L. Harvath, *Physical Properties of Inorganic Compounds SI Units,* Crane, Russak & Co., New York, 1975.

11. C. L. Yaws, *Physical Properties,* McGraw-Hill, New York, 1972.

CHAPTER 3
CONDUCTION AND THERMAL CONTACT RESISTANCES (CONDUCTANCES)

M. M. Yovanovich
University of Waterloo

INTRODUCTION

When steady-state conduction occurs within and outside solids, or between two contacting solids, it is frequently handled by means of conduction shape factors and thermal contact conductances (or contact resistances), respectively. This chapter covers the basic equations, definitions, and relationships that define shape factors and the thermal contact, gap, and joint conductances for conforming, rough surfaces, and nonconforming, smooth surfaces.

Shape factors for two- and three-dimensional systems are presented. General expressions formulated in orthogonal curvilinear coordinates are developed. The general expression is used to develop numerous general expressions in several important coordinate systems such as (1) circular, elliptical, and bicylinder coordinates and (2) spheroidal coordinates (spherical, oblate spheroidal, and prolate spheroidal). The integral form of the shape factor for an ellipsoid is presented and then used to obtain analytical expressions and numerical values for the shape factors of several isothermal geometries (spheres, oblate and prolate spheroids, circular and elliptical disks). It is demonstrated that the dimensionless shape factor is a weak function of the geometry (shape and aspect ratio) provided that the square root of the total *active* surface area is selected as the characteristic body length. A general dimensionless expression is proposed for accurate estimations of shape factors of three-dimensional bodies such as cuboids. Shape factor expressions are presented for two-dimensional systems bounded by isothermal coaxial (1) regular polygons, (2) internal circles and outer regular polygons, and (3) internal regular polygons and outer circles. A method is given for estimating the shape factors of systems bounded by two isothermal cubes and other combinations of internal and external geometries. The shape factor results of this chapter are used in the chapter on natural convection to model heat transfer from isothermal bodies of arbitrary shape.

Transient conduction within solids and into full and half-spaces is presented for a wide range of two- and three-dimensional geometries.

Steady-state and transient constriction (spreading) resistances for a range of geometries for isothermal and isoflux boundary conditions are given. Analytical solutions for half-spaces and heat flux tubes and channels are reported.

Elastoconstriction resistance and gap and joint resistances for line and point contacts are presented. Contact conductances of conforming rough surfaces that undergo (1) elastic, (2) plastic, and (3) elastoplastic deformation are reported. The gap conductance integral is presented. The overall joint conductance is considered.

Analytical solutions and correlation equations are presented rather than graphic results. The availability of many computer algebra systems such as *Macsyma, MathCad, Maple, MATLAB,* and *Mathematica,* as well as spreadsheets such as *Excel* and *Quattro Pro* that provide symbolic, numerical, and plotting capabilities, makes the analytical solutions amenable to quick, accurate computations. All equations and correlations reported in this chapter have been verified in *Maple* worksheets and *Mathematica* notebooks. These worksheets and notebooks will be available on my home page on the Internet. Some spreadsheet solutions will also be developed and made available on the Internet.*

BASIC EQUATIONS, DEFINITIONS, AND RELATIONSHIPS

Shape factors of isothermal, three-dimensional convex bodies having complex shapes and small to large aspect ratios are of considerable interest for applications in the nuclear, aerospace, microelectronic, and telecommunication industries. The shape factor S also has applications in such diverse areas as antenna design, electron optics, electrostatics, fluid mechanics, and plasma dynamics [27].

In electrostatics, for example, the capacitance C is the total charge Q_e required to raise the potential ϕ_e of an isolated body to the electrical potential V_e, and the relationship between them is (e.g., Greenspan [27], Morse and Feshbach [68, 69], Smythe [98], and Stratton [111])

$$C = \frac{Q_e}{V_e} = \iint_A -\epsilon \frac{\partial \phi_e}{\partial n} \, dA$$

where ϵ is the permittivity of the surrounding space, ϕ_e is the nondimensional electric potential, n is the outward-directed normal on the surface, and A is the total surface area of the body.

Mathematicians prefer to deal with the capacity C^* of a body, which they [81, 113] define as

$$C^* = \frac{1}{4\pi} \iint_A -\frac{\partial \phi_e}{\partial n} \, dA$$

Shape Factor, Thermal Resistance, and Diffusion Length. The shape factor S, the thermal resistance R, and the thermal diffusion length Δ are three useful and related thermal parameters. They are defined by the following relationships:

$$S = \frac{1}{kR} = \frac{A}{\Delta} = \frac{Q}{k(T_0 - T_\infty)} = \iint_A -\frac{\partial \phi}{\partial n} \, dA \tag{3.1}$$

where k is the thermal conductivity, T_0 is the temperature of the isothermal body, T_∞ is the temperature of points remote from the body, and ϕ is the dimensionless temperature $(T(\vec{r}) - T_\infty)/(T_0 - T_\infty)$.

The relationships between the shape factor S, the capacitance C, and the capacity C^* are

$$S = \frac{C}{\epsilon} = 4\pi C^* \tag{3.2}$$

The three parameters have units of length.

Analytical solutions are available for a small number of geometries such as the family of geometries related to the ellipsoid (e.g., sphere, oblate and prolate spheroids, elliptical and circular disks). Precise numerical values of S for other axisymmetric convex bodies have been obtained by various numerical methods such as that proposed by Greenspan [27] and that proposed by Wang and Yovanovich [123].

---

* The Internet address is mmyovemhtl.uwaterloo.ca.

Chow and Yovanovich [15] showed, by analytical and numerical methods, that the capacitance is a slowly changing function of the conductor shape and aspect ratio provided the total area of the conductor is held constant.

Wang and Yovanovich [123] showed that the dimensionless shape factor

$$S^*_{\sqrt{A}} = \frac{S\mathcal{L}}{A} = \frac{\mathcal{L}}{A} \iint_A \frac{-\partial\phi}{\partial n}\, dA \tag{3.3}$$

where the characteristic scale length $\mathcal{L}$ was chosen to be $\sqrt{A}$ as recommended by Yovanovich [133], Yovanovich and Burde [134], and Yovanovich [144–146], when applied to a range of axisymmetric, convex bodies is a weak function of the body shape and its aspect ratios.

This chapter reports and demonstrates, through inclusion of additional accurate numerical results of Greenspan [27] for complex body shapes such as cubes, ellipsoids, and circular and elliptical toroids, a lens that is formed by the intersection of two spheres such that $S^*_{\sqrt{A}}$ is a relatively weak function of the body shape and its aspect ratios.

This chapter also introduces the geometric length Δ, called the diffusion length, and shows that this physical length scale is closely related to the square root of the total body surface area when the body is convex.

The dimensionless geometric parameter $\sqrt{A}/\Delta$ is proposed as an alternate parameter for determination of shape factors of complex convex bodies.

Shape Factors

Formulation of the Problem in General Coordinates.   Consider the steady flow of heat Q from an isothermal surface A_1 at temperature T_1 through a homogeneous medium of thermal conductivity k to a second isothermal surface A_2 at temperature $T_2(T_1 > T_2)$. The stationary temperature field depends on the geometry of the isothermal boundary surfaces. When these isothermal surfaces can be made coincident with a coordinate surface by a judicious choice of coordinates, then the temperature field will be one-dimensional in that coordinate system. In other words, heat conduction occurs across two surfaces of an orthogonal curvilinear parallelepiped (Fig. 3.1a), and the remaining four coordinate surfaces are adiabatic.

Let the general coordinates u_1, u_2, u_3 be so chosen that $T = T(u_1)$ and, therefore, $\partial T/\partial u_2 = \partial T/\partial u_3 = 0$. Under these conditions, the heat flux vector will have one component in the u_1 direction:

$$q_1 = -k(dT/ds) = -k(dT/\sqrt{g_1}\, du_1)$$

where $\sqrt{g_1}$ is the metric coefficient in the u_1 direction. The metric coefficients are defined by the general line element ds expressed in terms of the differentials of arc lengths on the coordinate lines [67]

$$(ds)^2 = g_1(du_1)^2 + g_2(du_2)^2 + g_3(du_3)^2$$

The product terms such as $du_i\, du_j$ $(i \neq j)$ do not appear because of the orthogonality property of the chosen coordinate system. These metric coefficients can also be generated by means of the following formula [67]:

$$g_i \equiv (\partial x/\partial u_i)^2 + (\partial y/\partial u_i)^2 + (\partial z/\partial u_i)^2 \qquad i = 1, 2, 3$$

provided that the Cartesian coordinates x, y, z can be expressed in terms of the new coordinates u_1, u_2, u_3 by the equations

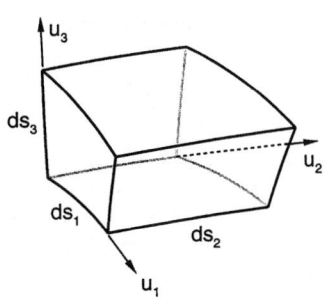

FIGURE 3.1a   Orthogonal curvilinear parallelepiped.

$$x = x(u_1, u_2, u_3), \qquad y = y(u_1, u_2, u_3), \qquad z = z(u_1, u_2, u_3)$$

The elemental coordinate surface located at u_1, which is orthogonal to the u_1 direction, is therefore

$$dA_1 = ds_2 \, ds_3 = \sqrt{g_2 g_3} \, du_2 \, du_3$$

and the heat flow per unit time through this surface into the volume element is, according to Fourier's law of conduction,

$$Q_1 = -k \, dA_1 (dT/ds_1) = -k(\sqrt{g/g_1})(dT/du_1) \, du_2 \, du_3$$

where $g \equiv g_1 g_2 g_3$ [67]. The heat flow rate out of the volume element is

$$Q_1 + (dQ_1/ds_1) \, ds_1 = Q_1 + (dQ_1/du_1) \, du_1$$

neglecting the higher-order terms of the Taylor series expansion of Q_1 about u_1.

The net rate of heat conduction out of the volume element in the u_1 direction is

$$(d/du_1)[k(\sqrt{g/g_1})(dT/du_1)] \, du_1 \, du_2 \, du_3$$

For steady-state conditions with no heat sources within the volume element, the Laplace equation in general coordinates is obtained by dividing by the elemental volume $\sqrt{g} \, du_1 \, du_2 \, du_3$ and equating to zero. Therefore,

$$(1/\sqrt{g})(d/du_1)[k(\sqrt{g/g_1})(dT/du_1)] = 0$$

The above equation is the governing differential equation and it is nonlinear when k is a function of temperature.

The isothermal temperature boundary conditions are

$$u_1 = a, \qquad T = T_1$$
$$u_1 = b, \qquad T = T_2$$

The above equation can be reduced to a linear differential equation by the introduction of a new temperature θ related to the temperature T by the Kirchhoff transformation [4, 11]:

$$\theta = (1/k_0) \int_0^{T_0} k \, dT$$

where k_0 denotes the value of the thermal conductivity at some convenient reference temperature, say $T = 0$. It follows that

$$d\theta/du_1 = (k/k_0)(dT/du_1)$$

and, therefore, we have

$$(d/du_1)[(\sqrt{g/g_1})(d\theta/du_1)] = 0$$

after multiplying through by $\sqrt{g}/k_0$. The boundary conditions become

$$u_1 = a, \qquad \theta = \theta_1 = (1/k_0) \int_0^{T_1} k \, dT$$

$$u_1 = b, \qquad \theta = \theta_2 = (1/k_0) \int_0^{T_2} k \, dT$$

The solution of the linear equation is

$$\theta = C_1 \int_0^{u_1} (g_1/\sqrt{g}) \, du_1 + C_2$$

where the constants C_1 and C_2 are obtained from

$$C_1 = -(\theta_1 - \theta_2)\Big/\int_a^b (g_1/\sqrt{g})\,du_1$$

and

$$C_2 = \theta_1 - C_1 \int_0^a (g_1/\sqrt{g})\,du_1$$

Temperature Distribution in Orthogonal Curvilinear Coordinates.   The temperature distribution in orthogonal curvilinear coordinates is

$$\frac{\theta_1 - \theta}{\theta_1 - \theta_2} = \frac{\displaystyle\int_a^{u_1} (g_1/\sqrt{g})\,du_1}{\displaystyle\int_a^b (g_1/\sqrt{g})\,du_1}, \qquad a < u_1 < b \tag{3.4}$$

The local heat flux in the u_1 direction is

$$q_1 = \frac{-k_0}{\sqrt{g_1}}\frac{d\theta}{du_1} = \frac{k_0(\theta_1 - \theta_2)}{\sqrt{g_2 g_3}\displaystyle\int_a^b (g_1/\sqrt{g})\,du_1} \tag{3.5}$$

The heat flow rate through the elemental surface dA_1 is

$$q_1 dA_1 = \frac{k_0(\theta_1 - \theta_2)}{\displaystyle\int_a^b (g_1/\sqrt{g})\,du_1}\,du_2\,du_3 \tag{3.6}$$

The total heat flow rate through the thermal system can be obtained by integration between the appropriate limits. Therefore,

$$Q = k_0(\theta_1 - \theta_2)\int_{u_3}\int_{u_2}\frac{du_2\,du_3}{\displaystyle\int_a^b (g_1/\sqrt{g})\,du_1} \tag{3.7}$$

An examination of the above equation shows that

$$k_0(\theta_1 - \theta_2) = \int_{T_2}^{T_1} k\,dT = k_a(T_1 - T_2) \tag{3.8}$$

where k_a, the average value of the thermal conductivity, is given by

$$k_a = k_0[1 + (\beta/2)(T_1 + T_2)] \tag{3.9}$$

if $k = k_0(1 + \beta T)$.

Shape Factor and Thermal Resistance in Orthogonal Curvilinear Coordinates.   The definition of thermal resistance of a system (total temperature drop across the system divided by the total heat flow rate) yields the following general expression for the thermal resistance R and the conduction shape factor S:

$$S = (Rk_a)^{-1} = \int_{u_3}\int_{u_2}\frac{du_2\,du_3}{\displaystyle\int_a^b (g_1/\sqrt{g})\,du_1} \tag{3.10}$$

The right side of the previous equation has units of length and depends on the geometry of the body only. There are several very important and useful coordinate systems that can be used to solve many seemingly complex conduction problems. Since each coordinate system

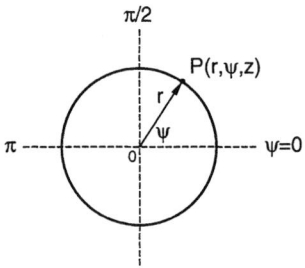

FIGURE 3.1b   Circular cylinder coordinates.

has three principal directions, there are three sets of shape factors corresponding to each of these directions. The conduction shape factors for several coordinate systems are given in the following section. This section by no means represents the total number of coordinate systems that are amenable to this type of analysis. It does, however, contain the most frequently used coordinate systems.

General Expressions for Conduction Shape Factors
Circular Cylinder Coordinates (r, ψ, z): Fig. 3.1b

r direction. Let $u_1 = r$; therefore, $u_2 = \psi$, $u_3 = z$, and $g_1/\sqrt{g} = 1/r$. If

$$a < r < b$$
$$0 < \psi < \beta \qquad \beta_{max} = 2\pi$$
$$0 < z < L$$

then

$$S^{-1} = Rk_a = \frac{\ln (b/a)}{\beta L} \tag{3.11}$$

ψ direction. Let $u_1 = \psi$; therefore, $u_2 = z$, $u_3 = r$, and $g_1/\sqrt{g} = r$. Limits of integration are given just previously.

$$S^{-1} = Rk_a = \frac{\beta}{L \ln (b/a)} \tag{3.12}$$

z direction. Let $u_1 = z$; therefore, $u_2 = r$, $u_3 = \psi$, and $g_1/\sqrt{g} = 1/r$. Limits of integration were given previously.

$$S^{-1} = Rk_a = \frac{2L}{\beta(b^2 - a^2)} \tag{3.13}$$

Spherical Coordinates (r, θ, ψ): Fig. 3.1c

$$(ds)^2 = (dr)^2 + r^2(d\theta)^2 + r^2 \sin^2 \theta(d\psi)^2$$

$$g_r = 1, \qquad g_\theta = r^2, \qquad g_\psi = r^2 \sin^2 \theta, \qquad \sqrt{g} = r^2 \sin \theta$$

r direction. Let $u_1 = r$; therefore, $u_2 = \theta$, $u_3 = \psi$, and $g_1/\sqrt{g} = 1/(r^2 \sin \theta)$. If

$$a < r < b$$
$$\beta_1 < \theta < \beta_2 \qquad \beta_{min} = 0 \qquad \text{and} \qquad \beta_{max} = \pi$$
$$0 < \psi < \gamma \qquad \gamma_{max} = 2\pi$$

then

$$S^{-1} = Rk_a = \frac{[(1/a) - (1/b)]}{\gamma(\cos \beta_1 - \cos \beta_2)} \tag{3.14}$$

θ direction. Let $u_1 = \theta$; therefore, $u_2 = \psi$, $u_3 = r$, and $g_1/\sqrt{g} = 1/(\sin \theta)$. Limits of integration are given just previously.

$$S^{-1} = Rk_a = \frac{\ln [\tan (\beta_2/2)] - \ln [\tan (\beta_1/2)]}{\gamma(b - a)} \tag{3.15}$$

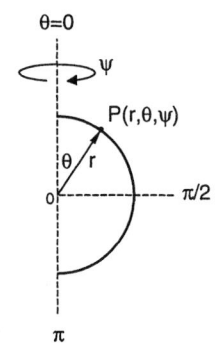

FIGURE 3.1c   Spherical coordinates.

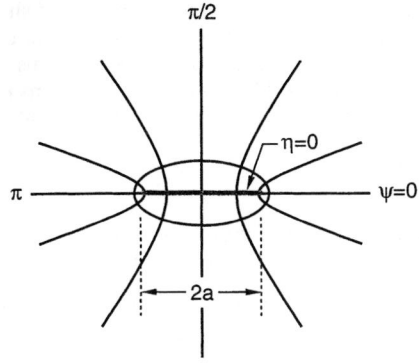

FIGURE 3.1d Elliptical cylinder coordinates.

ψ *direction.* Let $u_1 = \psi$; therefore, $u_2 = r$, $u_3 = \theta$, and $g_1/\sqrt{g} = \sin \theta$. Limits of integration were given previously.

$$S^{-1} = Rk_a = \frac{\gamma}{(b-a)\{\ln [\tan (\beta_2/2)] - \ln [\tan (\beta_1/2)]\}} \qquad (3.16)$$

Elliptical Cylinder Coordinates (η, ψ, z): *Fig. 3.1d*

$$(ds)^2 = a^2(\cosh^2 \eta - \cos^2 \psi)[(d\eta)^2 + (d\psi)^2] + (dz)^2$$

$$g_\eta = g_\psi = a^2(\cosh^2 \eta - \cos^2 \psi), \qquad g_z = 1,$$

$$\sqrt{g} = a^2(\cosh^2 \eta - \cos^2 \psi)$$

η *direction.* Let $u_1 = \eta$; therefore, $u_2 = \psi$, $u_3 = z$, and $g_1/\sqrt{g} = 1$. If

$$\begin{array}{lll} \eta_1 < \eta < \eta_2 & \eta_{min} = 0, & \eta_{max} = \infty \\ 0 < \psi < \beta & \beta_{max} = 2\pi & \\ 0 < z < L & & \end{array}$$

then

$$S^{-1} = Rk_a = \frac{(\eta_2 - \eta_1)}{L\beta} \qquad (3.17)$$

where

$$\eta_1 = \frac{1}{2} \ln \left[\frac{b_1 + c_1}{b_1 - c_1} \right], \qquad \eta_2 = \frac{1}{2} \ln \left[\frac{b_2 + c_2}{b_2 - c_2} \right]$$

and

$$a = \sqrt{b_1^2 - c_1^2} = \sqrt{b_2^2 - c_2^2}$$

ψ *direction.* Let $u_1 = \psi$; therefore, $u_2 = z$, $u_3 = \eta$, and $g_1/\sqrt{g} = 1$. Limits of integration were given previously.

$$S^{-1} = Rk_a = \frac{\beta}{L(\eta_2 - \eta_1)} \qquad (3.18)$$

z *direction.* Let $u_1 = z$; therefore, $u_2 = r$, $u_3 = \psi$, and

$$\frac{g_1}{\sqrt{g}} = \frac{1}{a^2(\cosh^2 \eta - \cos^2 \psi)}$$

Limits of integration were given previously.

$$S^{-1} = Rk_a = \frac{L}{a \displaystyle\int_{\eta_1}^{\beta} \int_{\eta_1}^{\eta_2} (\cosh^2 \eta - \cos^2 \psi) \, d\eta \, d\psi} \qquad (3.19)$$

Bicylinder Coordinates (η, ψ, z): *Fig. 3.1e*

$$(ds)^2 = \frac{a^2}{(\cosh \eta - \cos \psi)^2} [(d\eta)^2 + (d\psi)^2] + (dz)^2$$

$$g_\eta = g_\psi = \frac{a^2}{(\cosh \eta - \cos \psi)^2}, \qquad g_z = 1, \qquad \sqrt{g} = \frac{a^2}{(\cosh \eta - \cos \psi)^2}$$

η *direction.* Let $u_1 = \eta$; therefore, $u_2 = \psi$, $u_3 = z$, and $g_1/\sqrt{g} = 1$. If

$$\begin{array}{lll} \eta_1 > \eta > \eta_2 & \eta_{min} = -\infty, & \eta_{max} = +\infty \\ 0 < \psi < \beta & \beta_{max} = 2\pi & \\ 0 < z < L & & \end{array}$$

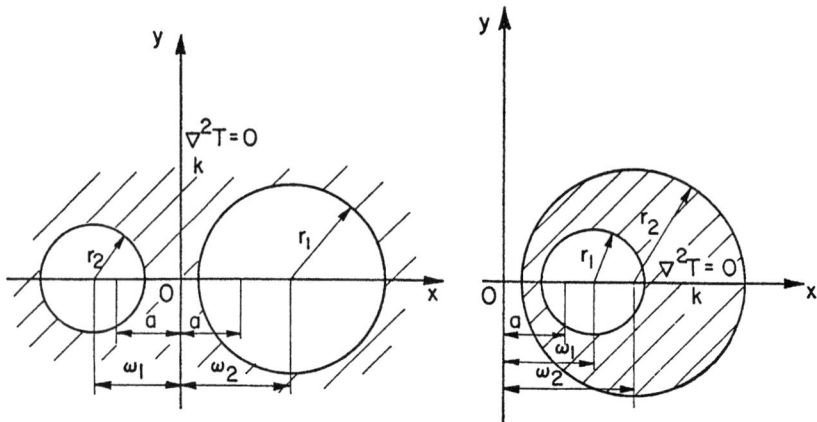

FIGURE 3.1e   Bicylinder coordinates.

then
$$S^{-1} = Rk_a = \frac{(\eta_2 - \eta_1)}{L\beta} \tag{3.20}$$

where
$$|\eta_1| = \sinh^{-1}\sqrt{\left(\frac{w_1}{r_1}\right)^2 - 1} \quad \text{and} \quad |\eta_2| = \sinh^{-1}\sqrt{\left(\frac{w_2}{r_2}\right)^2 - 1}$$

ψ *direction.* Let $u_1 = \psi$; therefore, $u_2 = z$, $u_3 = \eta$, and $g_1/\sqrt{g} = 1$. Limits of integration were given previously.

$$S^{-1} = Rk_a = \frac{\beta}{L(\eta_2 - \eta_1)} \tag{3.21}$$

z *direction.* Let $u_1 = z$; therefore, $u_2 = \eta$, $u_3 = \psi$, and

$$\frac{g_1}{\sqrt{g}} = \frac{(\cosh \eta - \cos \psi)^2}{a^2}$$

Limits of integration were given previously.

$$S^{-1} = Rk_a = \frac{L}{a^2 \int_0^\beta \int_{\eta_2}^{\eta_1} [d\eta \, d\psi/(\cosh \eta - \cos \psi)^2]} \tag{3.22}$$

where
$$a = \sqrt{w_1^2 - r_1^2} = \sqrt{w_2^2 - r_2^2}$$

Oblate Spheroidal Coordinates (η, θ, ψ)*: Fig. 3.1f*

$$(ds)^2 = a^2(\cosh^2 \eta - \sin^2 \theta)[(d\eta)^2 + (d\theta)^2] + a^2 \cosh^2 \eta \sin^2 \theta (d\psi)^2$$

$$g_\eta = g_\theta = a^2(\cosh^2 \eta - \sin^2 \theta)$$

$$g_\psi = a^2 \cosh^2 \eta \sin^2 \theta$$

$$\sqrt{g} = a^3(\cosh^2 \eta - \sin^2 \theta) \cosh \eta \sin \theta$$

η *direction.* Let $u_1 = \eta$; therefore, $u_2 = \theta$, $u_3 = \psi$, and

$$\frac{g_1}{\sqrt{g}} = \frac{1}{a \cosh \eta \sin \theta}$$

If
$$\begin{array}{lll} \eta_1 < \eta < \eta_2 & \eta_{min} = 0, & \eta_{max} = \infty \\ \beta_1 < \theta < \beta_2 & \beta_{min} = 0, & \beta_{max} = \pi \\ 0 < \psi < \gamma & \gamma_{max} = 2\pi \end{array}$$

then
$$S^{-1} = Rk_a = \frac{\tan^{-1}(\sinh \eta_2) - \tan^{-1}(\sinh \eta_1)}{a\gamma[\cos \beta_1 - \cos \beta_2]} \tag{3.23}$$

where
$$a = \sqrt{b_1^2 - c_1^2} = \sqrt{b_2^2 - c_2^2}$$

and
$$\eta_1 = \tanh^{-1}(c_1/b_1), \qquad \eta_2 = \tanh^{-1}(c_2/b_2)$$

θ *direction.* Let $u_1 = \theta$; therefore, $u_2 = \psi$, $u_3 = \eta$, and

$$\frac{g_1}{\sqrt{g}} = \frac{1}{a \cosh \eta \sin \theta}$$

Limits of integration were given previously:

$$S^{-1} = Rk_a = \frac{\ln[\tan(\beta_2/2)] - \ln[\tan(\beta_1/2)]}{a\gamma(\sinh \eta_2 - \sinh \eta_1)} \tag{3.24}$$

ψ *direction.* Let $u_1 = \psi$; therefore, $u_2 = \eta$, $u_3 = \theta$, and

$$\frac{g_1}{\sqrt{g}} = \frac{\cosh \eta \sin \theta}{a(\cosh^2 \eta - \sin^2 \theta)}$$

Limits of integration were given previously.

$$S^{-1} = Rk_a = \frac{\gamma}{a \displaystyle\int_{\eta_1}^{\eta_2} \int_{\beta_1}^{\beta_2} [(\cosh^2 \eta - \sin^2 \theta)/\cosh \eta \sin \theta]\, d\theta\, d\eta} \tag{3.25}$$

Prolate Spheroidal Coordinates (η, θ, ψ): *Fig. 3.1g*

$$(ds)^2 = a^2(\sinh^2 \eta + \sin^2 \theta)[(d\eta)^2 + (d\theta)^2] + a^2 \sinh^2 \eta \sin^2 \theta (d\psi)^2$$

$$g_\eta = g_\theta = a^2(\sinh^2 \eta + \sin^2 \theta)$$

$$g_\psi = a^2 \sinh^2 \eta \sin^2 \theta$$

$$\sqrt{g} = a^3(\sinh^2 \eta + \sin^2 \theta) \sinh \eta \sin \theta$$

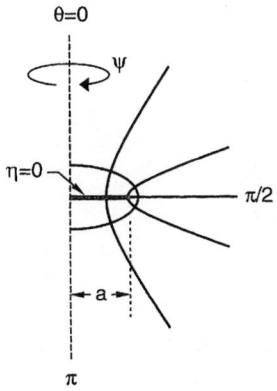

FIGURE 3.1f Oblate spheroidal coordinates.

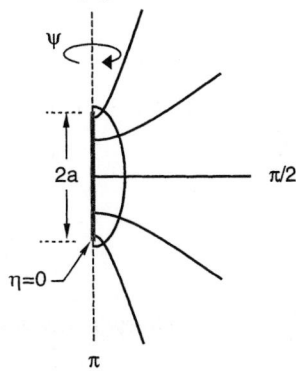

FIGURE 3.1g Prolate spheroidal coordinates.

η *direction.* Let $u_1 = \eta$; therefore, $u_2 = \theta$, $u_3 = \psi$, and

$$\frac{g_1}{\sqrt{g}} = \frac{1}{a \sinh \eta \sin \theta}$$

If

$$\begin{array}{lll} \eta_1 < \eta < \eta_2 & \eta_{min} = 0, & \eta_{max} = \infty \\ \beta_1 < \theta < \beta_2 & \beta_{min} = 0, & \beta_{max} = \pi \\ 0 < \psi < \gamma & \gamma_{max} = 2\pi \end{array}$$

then

$$S^{-1} = Rk_a = \frac{\ln [\tanh (\eta_2/2)] - \ln [\tanh (\eta_1/2)]}{a\gamma(\cos \beta_1 - \cos \beta_2)} \tag{3.26}$$

where

$$a = \sqrt{b_1^2 - c_1^2} = \sqrt{b_2^2 - c_2^2}$$

and

$$\eta_1 = \frac{1}{2} \ln \left[\frac{b_1 + c_1}{b_1 - c_1} \right], \qquad \eta_2 = \frac{1}{2} \ln \left[\frac{b_2 + c_2}{b_2 - c_2} \right]$$

θ *direction.* Let $u_1 = \theta$; therefore, $u_2 = \psi$, $u_3 = \eta$, and

$$\frac{g_1}{\sqrt{g}} = \frac{1}{a \sinh \eta \sin \theta}$$

Limits of integration were given previously.

$$S^{-1} = Rk_a = \frac{\ln [\tan (\beta_2/2)] - \ln [\tan (\beta_1/2)]}{a\gamma(\cosh \eta_2 - \cosh \eta_1)} \tag{3.27}$$

ψ *direction.* Let $u_1 = \psi$; therefore, $u_2 = \eta$, $u_3 = \theta$, and

$$\frac{g_1}{\sqrt{g}} = \frac{\sinh \eta \sin \theta}{a(\sinh^2 \eta + \sin^2 \theta)}$$

Limits of integration were given previously.

$$S^{-1} = Rk_a = \frac{\gamma}{a \displaystyle\int_{\eta_1}^{\eta_2} \int_{\beta_1}^{\beta_2} [(\sinh^2 \eta - \sin^2 \theta)/\sinh \eta \sin \theta] \, d\theta \, d\eta} \tag{3.28}$$

The basic relations given above for several curvilinear coordinates can be used to obtain expressions for the shape factor for many problems of interest to thermal analysts. Several typical two- and three-dimensional examples are presented in Fig. 3.2. The material in the following section provides shape factors for three-dimensional isothermal bodies in full space.

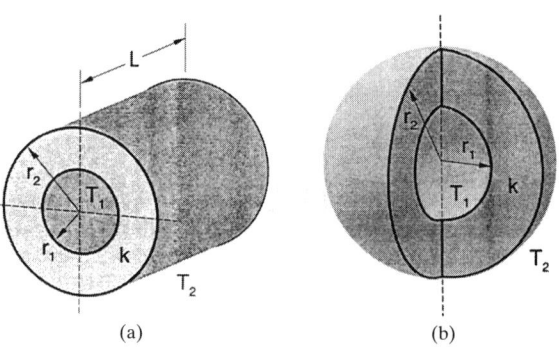

(a)   (b)

FIGURE 3.2   (*a*) circular pipe wall; (*b*) single spherical shell.

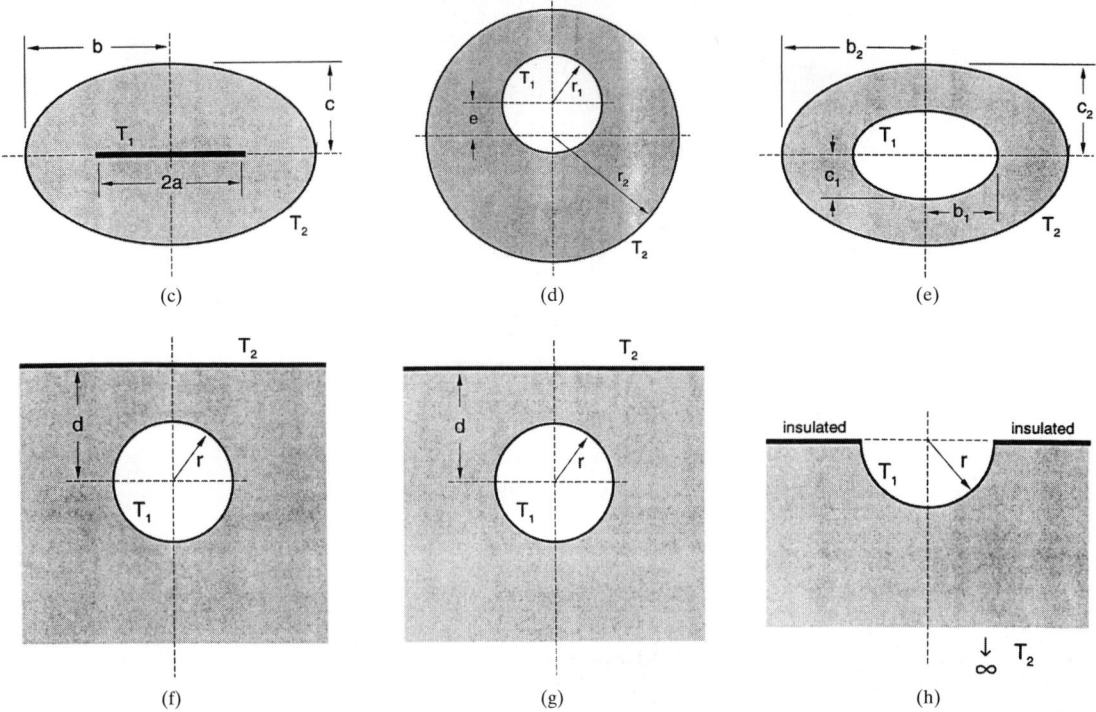

FIGURE 3.2 (*Continued*)   (*c*) strip in elliptical cylinder; (*d*) pipe in eccentric circular cylinder; (*e*) confocal elliptical cylinders; (*f*) sphere embedded in a half-space; (*g*) circular pipe embedded in a half-space; (*h*) hemispherical cavity in a half-space.

Shape Factors for Ellipsoids: Integral Form for Numerical Calculations

The capacity and/or the capacitance of an isopotential ellipsoid are presented in several texts and handbooks such as those by Flugge [22], Jeans [40], Kellogg [43], Mason and Weaver [62], Morse and Feshbach [68, 69], Smythe [98], and Stratton [111]. The results presented in these texts are used to develop expressions for the shape factors of several bodies: spheres, oblate and prolate spheroids (see Fig. 3.3), circular and elliptical disks, and ellipsoids. The shape factor for the ellipsoid is general; it reduces to the shape factor for the other bodies.

The capacity of an isopotential ellipsoid having semiaxes $a \geq b \geq c$ was given in the integral form [113]:

$$\frac{1}{C^*} = \frac{1}{2} \int_0^\infty \frac{dv}{\sqrt{(a^2 + v)(b^2 + v)(c^2 + v)}} \tag{3.29}$$

where v is a dummy variable. This expression is used to develop the expression for the dimensionless shape factor of isothermal ellipsoids. Since $S = 4\pi C^*$, one can set the space variable $v = a^2 t$, where t is a dimensionless variable. Next we normalize the two smaller axes b, c of the ellipsoid with respect to the largest semiaxis a such that $\beta = b/a$ and $\gamma = c/a$. This leads to the following dimensionless integral [150]:

$$\frac{8\pi a}{S} = I(\beta, \gamma) = \int_0^\infty \frac{dt}{\sqrt{(1 + t)(\beta^2 + t)(\gamma^2 + t)}}, \qquad 0 \leq \gamma \leq \beta \leq 1 \tag{3.30}$$

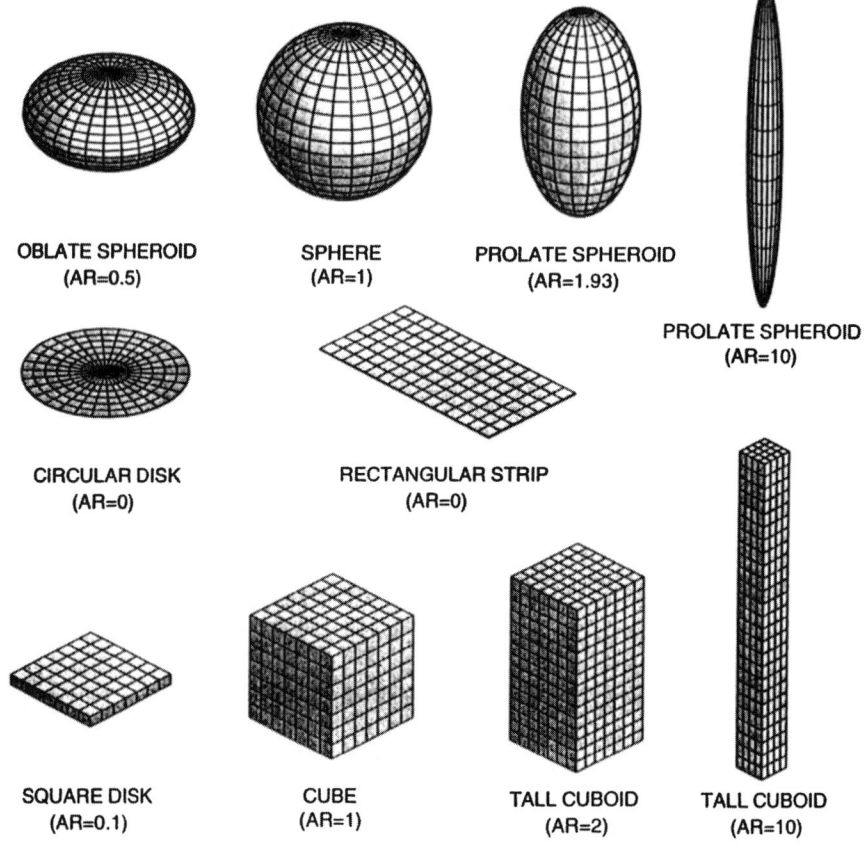

OBLATE SPHEROID
(AR=0.5)

SPHERE
(AR=1)

PROLATE SPHEROID
(AR=1.93)

PROLATE SPHEROID
(AR=10)

CIRCULAR DISK
(AR=0)

RECTANGULAR STRIP
(AR=0)

SQUARE DISK
(AR=0.1)

CUBE
(AR=1)

TALL CUBOID
(AR=2)

TALL CUBOID
(AR=10)

FIGURE 3.3   Three-dimensional bodies.

The ellipsoidal integral can be expressed in terms of the incomplete elliptical integral of the first kind $F(\kappa, \phi)$ [10, 60]:

$$I(\beta, \gamma) = \frac{2}{\sqrt{1 - \gamma^2}}\, F\left(\sin^{-1}\sqrt{1 - \gamma^2},\ \sqrt{\frac{1 - \beta^2}{1 - \gamma^2}}\right) \tag{3.31}$$

where κ and ϕ are the modulus and amplitude angle, respectively. Computer algebra systems can be used to evaluate the above special function accurately and quickly.

The ellipsoidal integral reduces to several special cases, which are presented next.

Sphere. $a = b = c$; $\beta = \gamma = 1$.

$$I(1, 1) = \int_0^\infty \frac{dt}{(1 + t)^{3/2}} = 2 \tag{3.32}$$

which gives $S = 4\pi a$, a well-known result.

Circular disk. $a = b$, $c = 0$; $\beta = 1$, $\gamma = 0$.

$$I(1, 0) = \int_0^\infty \frac{dt}{(1 + t)\sqrt{t}} = \pi \tag{3.33}$$

which gives $S = 8a$, also a well-known result.

Elliptical disk. $a \geq b$, $c = 0$; $0 < \beta \leq 1$, $\gamma = 0$.

$$I(\beta, 0) = \int_0^\infty \frac{dt}{\sqrt{(1+t)(\beta^2+t)t}} = 2K(\sqrt{1-\beta^2}) \tag{3.34}$$

where $K(\kappa)$ is the complete elliptical integral of the first kind of modulus $\kappa = \sqrt{1-\beta^2}$. There are several methods available to compute accurately the complete elliptical integral [1]. Here are two simple approximations:

$$K(\sqrt{1-\beta^2}) \approx \frac{2\pi}{(1+\sqrt{\beta})^2}, \qquad 0.2 < \beta \leq 1 \tag{3.35}$$

and

$$K(\sqrt{1-\beta^2}) \approx \ln\left(\frac{4}{\beta}\right), \qquad 0 < \beta < 0.2 \tag{3.36}$$

Oblate spheroid. $a = b > c$; $\beta = 1$, $0 \geq \gamma < 1$.

$$I(1, \gamma) = \int_0^\infty \frac{dt}{(1+t)\sqrt{\gamma^2+t}} = \frac{2}{\sqrt{1-\gamma^2}} \cos^{-1} \gamma \tag{3.37}$$

Prolate spheroid. $a > b = c$; $\beta = \gamma \leq 1$.

$$I(\gamma, \gamma) = \int_0^\infty \frac{dt}{(\gamma^2+t)\sqrt{1+t}} = \frac{1}{\sqrt{1-\gamma^2}} \ln\left[\frac{1+\sqrt{1-\gamma^2}}{1-\sqrt{1-\gamma^2}}\right] \tag{3.38}$$

The above results correspond to an important family of axisymmetric, convex geometries. The results presented in dimensional form or in nondimensional form as given above do not reveal an important property presented by this family of geometries and other geometries when the appropriate physical characteristic scale length is used for the nondimensionalization.

The numerical values S_a^* for oblate spheroids ($\beta = 1$, $0 \leq \gamma \leq 1$), prolate spheroids ($\beta = \gamma$, $1 \geq \gamma \geq 10$), and elliptical disks ($0 \leq \beta \leq 1$, $\gamma = 0$) are presented in Tables 3.1–3.3.

TABLE 3.1   Shape Factors and Diffusion Lengths for Oblate Spheroids

$\dfrac{a}{c}$	S_a^*	$\dfrac{\sqrt{A}}{\Delta}$	$\dfrac{a}{c}$	S_a^*	$\dfrac{\sqrt{A}}{\Delta}$
1	12.5664	3.54491	8	8.62546	3.36841
2	10.3923	3.52903	9	8.55700	3.35413
3	9.62476	3.49392	10	8.50206	3.34194
4	9.23085	3.45939	10^2	8.05085	3.21098
5	8.99090	3.42994	10^3	8.00509	3.19356
6	8.82932	3.40553	10^4	8.00051	3.19174
7	8.71308	3.38530			

TABLE 3.2   Shape Factors and Diffusion Lengths for Prolate Spheroids

$\dfrac{a}{c}$	S_a^*	$\dfrac{\sqrt{A}}{\Delta}$	$\dfrac{a}{c}$	S_a^*	$\dfrac{\sqrt{A}}{\Delta}$
1	12.5664	3.54491	6	5.00047	3.87533
2	8.26359	3.56613	7	4.72205	3.95878
3	6.72115	3.62769	8	4.50319	4.04005
4	5.89664	3.70638	9	4.32539	4.11883
5	5.37092	3.79053	10	4.17723	4.19508

TABLE 3.3 Shape Factors and Diffusion Lengths for Elliptical Disks

$\dfrac{a}{b}$	S_a^*	$\dfrac{\sqrt{A}}{\Delta}$	$\dfrac{a}{b}$	S_a^*	$\dfrac{\sqrt{A}}{\Delta}$
1	8.00000	3.19154	6	3.93511	3.84541
2	5.82716	3.28763	7	3.75763	3.96618
3	4.96964	3.43397	8	3.61576	4.07995
4	4.48606	3.57936	9	3.49888	4.18755
5	4.16641	3.71670	10	3.40033	4.28974

The range of dimensionless shape factor for the oblate spheroids is $8 \le S_a^* \le 4\pi$. The highest and lowest values correspond to the sphere and the circular disk, respectively. The radii of the disk and sphere are set to one unit.

The dimensionless shape factor range for the prolate spheroids is approximately $4.177 \le S_a^* \le 4\pi$ for the aspect ratio range $1 \le a/b \le 10$, and the major axis is set to $2a = 2$.

The dimensionless shape factor range for the elliptical disks is approximately $3.4 \le S_a^* \le 8$. The highest value corresponds to a circular disk of unit radius, and the lowest value corresponds to an elliptical disk with a 10 to 1 aspect ratio.

The range of all values of S_a^* presented in the three tables is quite large. The ratio of the largest and smallest values is approximately 3.7. These values correspond to the sphere and the elliptical disks of large aspect.

In the next section the range of the dimensionless shape factors will be reduced significantly by the introduction of a scale length based on the square root of the total surface area.

Surface Area of Ellipsoids. The expression for the total surface area of an ellipsoid is written as [150]:

$$\frac{A}{2\pi a^2} = \gamma^2 + \frac{\beta}{\sin\phi}\left[\gamma^2 F(\phi, \kappa) + (1 - \gamma^2)E(\phi, \kappa)\right] = \mathcal{A}(\beta, \gamma) \tag{3.39}$$

with
$$\phi = \cos^{-1}\gamma \quad \text{and} \quad \kappa = \left(\frac{1 - (\gamma/\beta)^2}{1 - \gamma^2}\right)^{1/2}$$

The total surface area related to the semimajor axis is a function of the two aspect ratios β and γ. The special functions $F(\phi, \kappa)$ and $E(\phi, \kappa)$ are incomplete elliptical integrals of the first and second kind, respectively. They depend on the amplitude angle ϕ and the modulus κ. These special functions can be computed quickly and accurately by means of computer algebra systems such as *Mathematica* [153]. Their properties are given in Abramowitz and Stegun [1].

The relationship between the square root of the total surface area and the semimajor axis is [150]:

$$\frac{\sqrt{A}}{a} = \sqrt{2\pi\mathcal{A}(\beta, \gamma)} \tag{3.40}$$

Dimensionless Shape Factor and Diffusion Length of Ellipsoids. The dimensionless shape factor $S_{\sqrt{A}}^*$ and the proposed dimensionless diffusion length $\sqrt{A}/\Delta$ for isothermal ellipsoids can be obtained from the shape factor integral $I(\beta, \gamma)$ and the relationship $\sqrt{A}/a$ given previously. The equation is

$$S_{\sqrt{A}}^* = \frac{\sqrt{A}}{\Delta} = \frac{4\sqrt{2\pi}}{I(\beta, \gamma)\sqrt{\mathcal{A}(\beta, \gamma)}} \tag{3.41}$$

The functions that appear in this expression were computed quickly and accurately using *Mathematica*. The numerical values for oblate spheroids are presented in the third column of

Table 3.1. The range of values has been significantly reduced. The ratio of the values for the sphere $a/c = 1$ and the circular disk $a/c \to \infty$ has been reduced from 1.57 to 1.11.

The numerical values for prolate spheroids are presented in the third column of Table 3.2. Here the reduction in the range is much greater. The ratio of the values for the sphere $a/b = 1$ and the long prolate spheroid $a/b = 10$ has been reduced from 3.0 to 1.18.

The numerical values for elliptical disks are presented in the third column of Table 3.3. Here, also, we observe that the reduction in the range is much greater. The ratio of the values for the circular disk $a/b = 1$ and the long elliptical disk $a/b = 10$ has been reduced from 2.35 to 1.34.

There is another benefit when $\mathcal{L} = \sqrt{A}$ is used as the characteristic body length. The differences between the values for the elliptical disks and the prolate spheroids are greatly reduced, becoming negligible for large aspect ratios. The largest difference of approximately 11 percent occurs when the aspect ratio is 1, i.e., when a sphere and a circular disk are compared.

This shows that elliptical disks (zero-thickness bodies) and prolate spheroids that have identical total surface areas and similar aspect ratios possess shape factors that are close in magnitude.

This important finding is demonstrated further in the subsequent sections, where a wide range of body shapes is considered.

Raithby and Hollands, in the chapter on natural convection, have developed correlation equations for external convection from isothermal bodies. In the correlation equations, the conduction Nusselt number is based on the shape factors developed in these sections.

Shape Factors for Three-Dimensional Bodies in Unbounded Domains

The shape factors for steady conduction within two- and three-dimensional systems that are bounded by isothermal surfaces are available. Dimensionless shape factors for several three-dimensional bodies are presented next. The results are based on analytical and numerical techniques.

Circular Toroid. The circular toroid is characterized by the ring diameter d and the toroid diameter D (Fig. 3.4a). The analytical solution [94] for the shape factor is written as an infinite series in which each term consists of toroidal or ring functions [1]:

$$S^*_{\sqrt{A}} = \frac{\sqrt{A}}{\Delta} = \frac{4}{\pi}\sqrt{\xi - 1/\xi}\left\{\frac{Q_{-1/2}(\xi)}{P_{-1/2}(\xi)} + 2\sum_{n=1}^{\infty}\frac{Q_{n-1/2}(\xi)}{P_{n-1/2}(\xi)}\right\} \qquad (3.42)$$

with $\xi = D/d > 1$. The special functions are accurately computed using *Mathematica*. The series converges very slowly for $D/d \to 1$, which corresponds to toroids with small inner diameters. When $D/d = 1$, $S^*_{\sqrt{A}} = 3.482761$, which is approximately 1.8 percent smaller than the value for the sphere. In the narrow range: $1 \le D/d \le 2$, $S^*_{\sqrt{A}} = 3.449$ to within 1 percent.

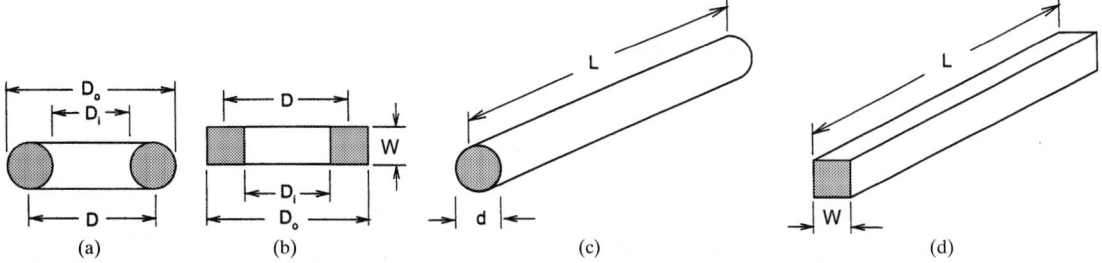

FIGURE 3.4   (*a*) Circular toroid; (*b*) square toroid; (*c*) finite circular cylinder; (*d*) finite square cylinder.

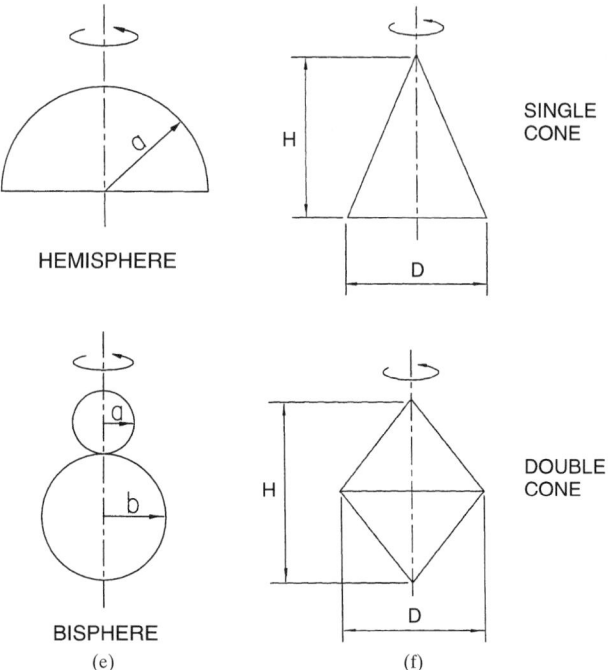

FIGURE 3.4 (*Continued*) (*e*) bisphere and hemisphere; (*f*) single and double cones.

For $D/d > 5$, the dimensionless shape factor for the toroid approaches the asymptote:

$$S^*_{\sqrt{A}} = \frac{\sqrt{A}}{\Delta} = 2\pi\, \frac{\sqrt{D/d}}{\ln{(8D/d)}} \tag{3.43}$$

For the practical range $2 \le D/d \le 10$, the shape factor is approximated with a maximum error of about 0.7 percent by multiplying the asymptotic expression by the empirical correlation equation:

$$C_{CT} = \frac{81}{80} + \frac{e^{-D/d}}{\sqrt{4.5}}$$

Square Toroid. The accurate numerical values of shape factors for square toroids, which are characterized by the inner and outer diameters D_i and D_o, respectively (Fig. 3.4*b*), were reported by Wang [122] for a wide range of the diameter ratio D_i/D_o. The dimensionless results were found to be in close agreement with the analytical results for the equivalent circular toroid defined by

$$\frac{D}{d} = \frac{\pi}{4}\, \frac{1 + D_i/D_o}{1 - D_i/D_o}, \qquad \frac{D_i}{D_o} > 0.1 \tag{3.44}$$

The above relationship was obtained by the application of two geometric rules: (1) set the surface area and (2) set the mean perimeter of the equivalent circular toroid equal to the surface area and the mean perimeter of the square toroid. When the circular toroid asymptotic expression given above is multiplied by the empirical correlation equation

$$C_{ST} = \frac{161}{160} + \frac{e^{-D/d}}{\sqrt{18}}$$

with D/d defined by the previous equation, the numerical values and the predicted values differ by less than 0.8 percent.

Finite Circular Cylinder.   The dimensionless shape factor for an isothermal right circular cylinder of length L and diameter d (Fig. 3.4c) was obtained from the analytical solution for the capacitance [96, 97]. Using the square root of the total surface area, the result is recast as

$$S^*_{\sqrt{A}} = \frac{\sqrt{A}}{\Delta} = \frac{3.1915 + 2.7726(L/d)^{0.76}}{\sqrt{1 + 2L/d}}, \qquad 0 \le L/d \le 8 \qquad (3.45)$$

The dimensionless shape factor for the right circular cylinder is in very close agreement with the values for the oblate spheroid in the range $0 \le L/d \le 1$ and with the values for the prolate spheroid in the range $1 \le L/d \le 8$. The difference when $L/d = 1$ is less than 1 percent. This shows that the results for the sphere and a finite circular cylinder of unit aspect ratio are very close. The simple expression obtained from the Smythe solution can be used to estimate the shape factors of circular disks, oblate spheroids, and prolate spheroids in the range $0 \le L/d \le 8$. For $L/d > 8$, the prolate spheroid asymptotic result can be used to provide accurate results for long circular cylinders and other equivalent bodies.

Finite Square Cylinder.   The dimensionless shape factors for finite square cylinders of length L and side dimension W (Fig. 3.4d) can be calculated using the finite circular expression by means of the equivalent aspect ratio

$$\frac{L}{d} = \frac{1}{2^{1/4}} \frac{L}{W} \qquad (3.46)$$

which was obtained by the following rule: take the geometric mean of the aspect ratios obtained by (1) inscribing and (2) circumscribing the square cylinder with circular cylinders. This procedure produces results that are in close agreement with reported numerical results in the range $0 \le L/W \le 10$.

Cube, Bisphere, and Hemisphere.   The dimensionless shape factors are as follows: for the cube (Fig. 3.3), $S^*_{\sqrt{A}} = 3.391$, for the bisphere (Fig. 3.4e), $S^*_{\sqrt{A}} = 3.4749$, and for the hemisphere (Fig. 3.4e), $S^*_{\sqrt{A}} = 3.4601$. These numerical and analytical results are approximately 4.5 percent, 2 percent, and 2.5 percent smaller than the value for the sphere.

Single and Double Cones.   The single and double cones (Fig. 3.4f) are characterized by the height dimension H and the maximum diameter D. The dimensionless shape factors were obtained by means of an accurate numerical technique [122].

Single cone.   The single cone numerical results for the shape factor are predicted by the following two correlation equations.

for $0.001 \le x = H/D \le 1$:

$$S^*_{\sqrt{A}} = 3.19399 + 0.629823x - 0.933731x^2 + 0.862597x^3 - 0.312459x^4 \qquad (3.47)$$

for $1 \le x = H/D \le 8$:

$$S^*_{\sqrt{A}} = 3.280967 + 1.61022(x/10) - 0.047366(x/10)^2 - 0.30067(x/10)^3 + 2.99117 \times 10^{-3}(x/10)^4$$

$$(3.48)$$

Double Cone. The double cone numerical data for the shape factors are correlated by the following two polynomials.

for $0.001 \leq x = H/D \leq 1$:

$$S^*_{\sqrt{A}} = 3.194264 + 0.626604x - 0.477791x^2 + 0.0751056x^3 + 0.0531827x^4 \qquad (3.49)$$

and for $1 \leq x = H/D \leq 10$:

$$S^*_{\sqrt{A}} = 3.41318 + 0.419048(x/10) + 2.02734(x/10)^2 - 2.23961(x/10)^3 + 0.80661(x/10)^4 \quad (3.50)$$

Circular and Rectangular Annulus. The dimensionless shape factors of isothermal circular and rectangular annuli are presented next.

Circular annulus. The circular annulus has inner and outer radii a and b, respectively. The two capacitance analytical solutions of Smythe [95] are recast into the following two expressions, which relate the dimensionless shape factor to the radii ratio $\epsilon = a/b$:

$$S^*_{\sqrt{A}} = \frac{\sqrt{A}}{\Delta} = \pi\sqrt{2\pi}\,\sqrt{\frac{1+\epsilon}{1-\epsilon}}\cdot\frac{1}{\ln 16 + \ln[(1+\epsilon)/(1-\epsilon)]} \qquad (3.51)$$

which is restricted to the range $1.000 < 1/\epsilon < 1.1$; and

$$S^*_{\sqrt{A}} = \frac{\sqrt{A}}{\Delta} = \frac{8}{\pi}\sqrt{\frac{2}{\pi}}\,\frac{1}{\sqrt{1-\epsilon^2}}[\cos^{-1}\epsilon + \sqrt{1-\epsilon^2}\tanh^{-1}\epsilon][1 + 0.0143\epsilon^{-1}\tan^3(1.28\epsilon)] \quad (3.52)$$

which is valid in the range $0 < \epsilon < 1/1.1$.

Rectangular annulus. The rectangular annulus is characterized by its outer length L and outer width W. The width $\mathcal{W}$ of the annular area is uniform. The interior open region has dimensions $L - 2\mathcal{W}$ by $W - 2\mathcal{W}$.

The dimensionless shape factor for the isothermal rectangular annulus is derived from the correlation equation of Schneider [89], who obtained accurate numerical values of the thermal constriction resistance of doubly connected rectangular contact areas by means of the boundary integral equation method:

$$S^*_{\sqrt{A}} = \frac{\sqrt{A}}{\Delta} = \frac{\sqrt{2}}{[C_1(\mathcal{W}/2W)^{C_2} + C_3]}\cdot\left[\frac{\mathcal{W}}{W} + \frac{L}{W} - 2\left(\frac{\mathcal{W}}{W}\right)^2\right]^{-1/2} \qquad (3.53)$$

with the recommended correlation coefficients

$$C_1 = -0.00232\frac{L}{W} + 0.03128$$

$$C_2 = 0.2927\left(\frac{L}{W} + 0.7463\right)^{-1} + 0.4316$$

$$C_3 = 0.6786\left(\frac{L}{W} + 0.8145\right)^{-1} + 0.0346$$

The correlation equation is restricted to the ranges $1 \leq L/W \leq 4$ and $0.01 \leq \mathcal{W}/W \leq 0.5$, with a maximum error of 1.45 percent between the correlation predictions and the numerical values. Selected values of the dimensionless diffusion length $\sqrt{A}/\Delta$ are given in Table 3.4.

The dimensionless diffusion length is a weak function of the two parameters L/W and $\mathcal{W}/W$ over a practical range of these parameters. The values are close to the circular and elliptical disk values reported earlier.

TABLE 3.4 Dimensionless Diffusion Length for Rectangular Annulus

L/W	$\mathcal{W}/W$	$\sqrt{A}/\Delta$	L/W	$\mathcal{W}/W$	$\sqrt{A}/\Delta$
1	0.1	3.2918	3	0.1	3.7001
	0.3	3.1978		0.3	3.6106
	0.5	3.3575		0.5	3.6370
2	0.1	3.4893	4	0.1	3.8774
	0.3	3.4044		0.3	3.7829
	0.5	3.4681		0.5	3.7884

Rectangular Plate.   The dimensionless shape factor for isothermal rectangular plates of length L and width W, where $L \geq W$, is calculated by means of the following two semianalytical expressions:

$$S^*_{\sqrt{A}} = 0.8 \frac{(1 + \sqrt{L/W})^2}{\sqrt{L/W}}, \qquad 1 \leq L/W \leq 5 \tag{3.54}$$

$$S^*_{\sqrt{A}} = \frac{\sqrt{8\pi L/W}}{\ln(4L/W)}, \qquad 5 < L/W < \infty \tag{3.55}$$

The maximum difference between the predicted values and the numerical values for the range $1 \leq L/W \leq 4$ is less than 1 percent.

Cuboid.   The dimensionless shape factor of cuboids with side dimensions $L_1 \geq L_2 \geq L_3$ is approximated by the expression [150]

$$S^*_{\sqrt{A}} = [S^*_{\sqrt{A}}]_{\text{rect}} \frac{(1 + 0.8688(L_1/D_{\text{GM}})^{0.76})}{\sqrt{1 + 2L_1/D_{\text{GM}}}} \tag{3.56}$$

where the equivalent circular cylinder diameter is based on the cuboid lengths L_2 and L_3:

$$D_{\text{GM}} = \sqrt{\frac{2}{\pi}(L_2 + L_3)\sqrt{L_2^2 + L_3^2}}$$

The values for $[S^*_{\sqrt{A}}]_{\text{rect}}$ are obtained from the expressions for the rectangular plates given above. This parameter depends on the cuboid lengths L_2 and L_3. The proposed expression for cuboids predicts the dimensionless shape factor to within ±5 percent.

Three-Dimensional Bodies With Layers: Langmuir Method

Shape factors for three-dimensional systems such as regions bounded by isothermal concentric spheres or concentric cubes; inner sphere and outer cube; and inner cube and outer sphere are presented in this section. The systems fall into two categories: (1) uniform thickness layers, and (2) nonuniform thickness layers. Warrington et al. [124] reported in graphic form numerical results for the cube-in-cube, sphere-in-cube, and cube-in-sphere systems. The numerical results for the shape factors were normalized with respect to the classical sphere-in-sphere solution and plotted against the ratio D_i/D_o, where D_i and D_o are the inner and outer equivalent diameters, respectively. Hassani and Hollands [33] proposed an approximate method for calculating shape factors for a region of uniform thickness surrounding an isothermal convex body of arbitrary shape. The proposed method is based on the asymptotic results corresponding to very thin layers where the shape factor is given by

$$S_0(\Delta) = \frac{A_i}{\Delta} \tag{3.57}$$

where A_i and Δ are the surface area of the inner body and the layer thickness, respectively, and the result corresponding to *infinitely* thick layers

$$S_\infty = C_\infty \sqrt{A_i} \tag{3.58}$$

where C_∞ is a constant that is close to the value $2\sqrt{\pi}$ for many body shapes [15] as shown in the sections on transient one-dimensional conduction in half-spaces and external transient conduction from long cylinders. Hassani and Hollands [33] set $C_\infty = 3.51$ for all bodies and proposed the following equation, which accurately interpolates between the two asymptotic solutions:

$$S^n(\Delta) = [S_0^n + S_\infty^n]^{1/n} \tag{3.59}$$

where n is a constant that is a function of the body shape. By trial and error, Hassani and Hollands found the following empirical formula:

$$n = \left[1.26 - \frac{(2 - \sqrt{A_i}/L_S)}{9\sqrt{1.0 - 4.79V_i^{2/3}/A_i}}, 1.0 \right]_{max} \tag{3.60}$$

where V_i and A_i are the volume and surface area of the inner body, and L_S is the longest straight line passing through the inner body. They gave the following rule: $Y = [x_1, x_2]_{max}$ means that $Y = x_1$ if $x_1 > x_2$ and $Y = x_2$ if $x_2 > x_1$. The shape parameter n was found to lie in the range $1 \le n \le 1.2$ for a very wide range of body shapes. The results obtained through this method show agreement to within about 5 percent with those obtained from numerical or existing analytical techniques.

Shape Factors for Two-Dimensional Systems

The shape factors for two-dimensional systems are available for (1) long cylinders bounded by homologous, regular polygons having N sides (Fig. 3.5), (2) long cylinders bounded internally by circles and bounded externally by regular polygons (Fig. 3.6), and (3) long cylinders bounded internally by regular polygons and bounded externally by circles (Fig. 3.7). In all three cases, as the number of sides N of the regular polygon becomes large ($N > 10$), the shape factor approaches the shape factor for the system bounded by two coaxial circular cylinders.

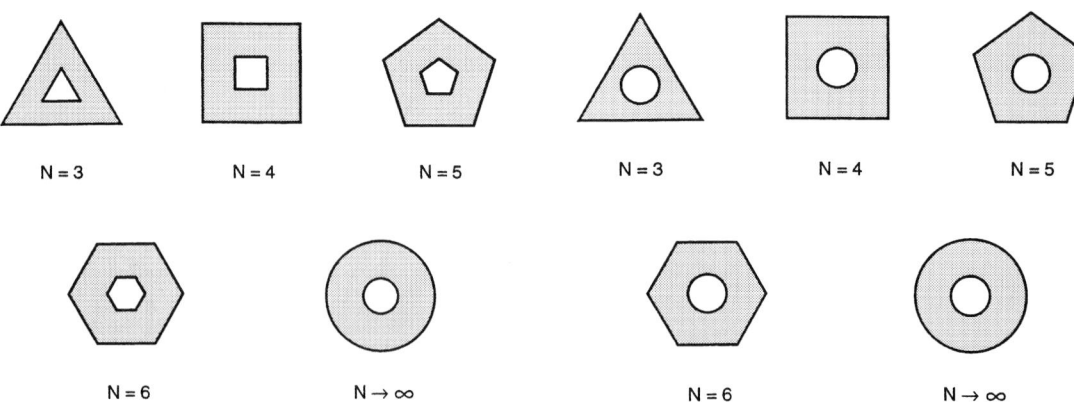

$N = 3$ $N = 4$ $N = 5$ $N = 3$ $N = 4$ $N = 5$

$N = 6$ $N \to \infty$ $N = 6$ $N \to \infty$

FIGURE 3.5   Regions bounded by regular polygons.

FIGURE 3.6   Regions bounded by inner circles and outer regular polygons.

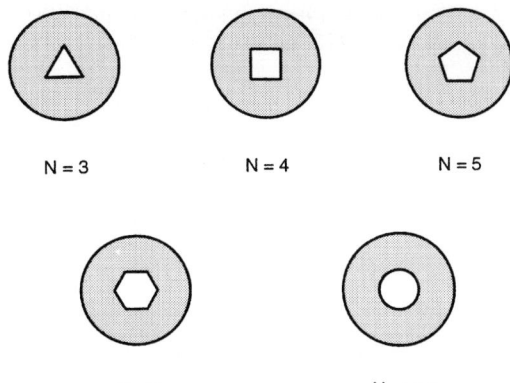

FIGURE 3.7   Regions bounded by inner regular polygons and outer circles.

Regular Polygon Inside a Regular Polygon.   The shape factor of two-dimensional regions bounded internally and externally by isothermal regular polygons of sides $N \geq 3$ (Fig. 3.5) is obtained by means of the general expression

$$\frac{S}{L} = \frac{4\pi}{\ln\left[1 + \dfrac{(D/d)^2 - 1}{(N/\pi)\tan(\pi/N)}\right]}, \qquad N \geq 3 \tag{3.61}$$

where d and D are the diameters of the inscribed circles of the inner and outer polygons, respectively. For the region bounded by two squares ($N = 4$), the general expression reduces to

$$\frac{S}{L} = \frac{4\pi}{\ln\left[1 + (\pi/4)[(D/d)^2 - 1]\right]} \tag{3.62}$$

The square/square problem has a complex analytical solution that requires the numerical computation of complete elliptical integrals of the first kind (see Ref. 8). For a large range of the parameter d/D, the analytical solution provides an accurate asymptotic expression:

$$\frac{S}{L} = 4\frac{[1 + (d/D)]}{[1 - (d/D)]} - \frac{8}{\pi}\ln 2, \qquad 0.3 \leq d/D \leq 1 \tag{3.63}$$

Two correlation equations based on electrical measurements were reported by Smith et al. [93]:

$$\frac{S}{L} = \frac{2\pi}{0.93\ln[0.947(D/d)]}, \qquad \frac{D}{d} > 1.4 \tag{3.64}$$

and

$$\frac{S}{L} = \frac{2\pi}{0.785\ln(D/d)}, \qquad \frac{D}{d} < 1.4 \tag{3.65}$$

Circle Inside a Regular Polygon.   Several expressions have been developed for this system (Fig. 3.6). Two relationships that give results to within a fraction of 1 percent are given. The first is [52]

$$\frac{S}{L} = \frac{2\pi}{\ln[A_s\beta]} \tag{3.66}$$

where $\beta = b/a > 1$ is the ratio of the radius of the inner circle to the radius of the inscribed circle. The parameter A_s is obtained by means of a numerical integration of the integral

$$A_s = \left[\int_0^1 (1 + u^N)^{-2/N} \, du \right]^{-1}, \qquad N \geq 3 \tag{3.67}$$

Laura and Susemihl [52] gave several values of the parameter A_s for several values of N. The alternate relationship [91] is

$$\frac{S}{L} = \frac{4N}{A\sqrt{A^2 + 1}} \tan^{-1} \left[\frac{\sqrt{A^2 + 1}}{A} \tan \frac{\pi}{N} \right], \qquad N \geq 3 \tag{3.68}$$

where the parameter $A = \sqrt{2 \ln \beta}$. The second relationship does not require a numerical integration.

Regular Polygon Inside a Circle. Numerous analytical, numerical, and experimental studies have produced results for shape factors and resistances for regions bounded internally by isothermal regular polygons of N sides where $N \geq 3$ and externally by an isothermal circle (Fig. 3.7). Lewis [55] gave the following general analytical result

$$\frac{S}{L} = \frac{2\pi}{\ln [A_N(D/d)]} \tag{3.69}$$

where the coefficients are given by

$$A_N = [(\sqrt{N} - \sqrt{N - 2})]^{2/N}(N - 2)^{1/N}, \qquad N \geq 3 \tag{3.70}$$

The above relationship is limited to the range $0 < d/D < \cos \pi/N$, where d and D are the diameters of the inscribed circle and the outer circle, respectively. The relationship gives accurate values of S for small values of d/D and for all values of N. The accuracy decreases for values of $d/D \rightarrow \cos \pi/N$ for small values of N. Ramachandra Murthy and Ramachandran [82] obtained two empirical correlation equations for regions bounded by squares and hexagons. Their correlation equations were developed from electrical measurements and have been shown to be in good agreement with the above relationship for a limited range of values of d/D.

Polygons With Layers. Hassani et al. [34] presented a procedure for obtaining a close upper bound for shape factors for a uniform thickness two-dimensional layer on cylinders having cross sections of the following forms: an equilateral triangle, a square, a rhombus, and a rectangle (Fig. 3.8). The shape factor per unit length of the inner cylinder is obtained from

$$\frac{S}{L} = \frac{2\pi}{\ln [1 + (2\pi B/P_i)]} \tag{3.71}$$

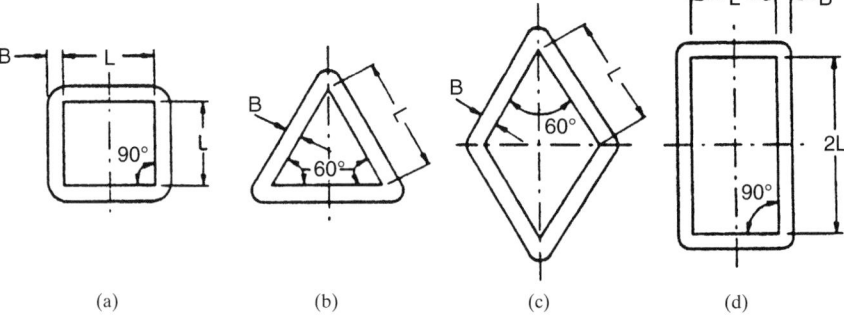

(a)   (b)   (c)   (d)

FIGURE 3.8 Polygons with uniform thickness layers.

where B is the layer thickness and P_i is the perimeter of the inner boundary. The accuracy of the above relationship was verified by comparison of the predicted values against numerical values for the layer thickness–to–side dimension range $0.10 \leq B/L \leq 3.00$. The proposed relationship overpredicts the shape factor by approximately 1–3 percent.

TRANSIENT CONDUCTION

Introduction

Transient conduction internal and external to various bodies subjected to the boundary conditions of the (1) first kind (Dirichlet), (2) second kind (Neumann), and (3) third kind (Robin) are presented in this section. Analytical solutions are presented in the form of series or integrals. Since these analytical solutions can be computed quickly and accurately using computer algebra systems, the solutions are not presented in graphic form.

Internal Transient Conduction

Internal one-dimensional transient conduction within infinite plates, infinite circular cylinders, and spheres is the subject of this section. The dimensionless temperature $\phi = \theta/\theta_i$ is a function of three dimensionless parameters: (1) dimensionless position $\zeta = x/\mathscr{L}$, (2) dimensionless time $\text{Fo} = \alpha t/\mathscr{L}^2$, and (3) the Biot number $\text{Bi} = h\mathscr{L}/k$, which depends on the convective boundary condition. The characteristic length $\mathscr{L}$ is the half-thickness L of the plate and the radius a of the cylinder or the sphere. The thermophysical properties k, α, the thermal conductivity and the thermal diffusivity, are constant.

The basic solutions for the plate and the cylinder can be used to obtain solutions within rectangular plates, cuboids, and finite circular cylinders. The equations and the initial and boundary conditions are well known [4, 11, 23, 28, 29, 38, 49, 56, 80, 87]. The solutions presented below follow the recent review of Yovanovich [151]. The Heisler [36] cooling charts for dimensionless temperature are obtained from the series solution:

$$\phi = \sum_{n=1}^{\infty} A_n \exp(-\delta_n^2 \, \text{Fo}) S_n(\delta_n \zeta) \tag{3.72}$$

with the temperature Fourier coefficients A_n for the plate, cylinder, and sphere, respectively, given in Table 3.5. The spatial functions for the three basic geometries are given in Table 3.6. The eigenvalues δ_n are the positive roots of the characteristic equations found in Table 3.6 where Bi, the physical parameter, ranges between 0 and ∞.

TABLE 3.5   Fourier Coefficients for Temperature and Heat Loss

Geometry	A_n	B_n
Plate	$\dfrac{2 \sin \delta_n}{\delta_n + \sin \delta_n \cos \delta_n}$	$\dfrac{2\text{Bi}^2}{\delta_n^2(\text{Bi}^2 + \text{Bi} + \delta_n^2)}$
Cylinder	$\dfrac{2 J_1(\delta_n)}{\delta_n[J_0^2(\delta_n) + J_1^2(\delta_n)]}$	$\dfrac{4\text{Bi}^2}{\delta_n^2(\delta_n^2 + \text{Bi}^2)}$
Sphere	$(-1)^{n+1} \dfrac{2\text{Bi}\,[\delta_n^2 + (\text{Bi} - 1)^2]^{1/2}}{(\delta_n^2 + \text{Bi}^2 - \text{Bi})}$	$\dfrac{6\text{Bi}^2}{\delta_n^2(\delta_n^2 + \text{Bi}^2 - \text{Bi})}$

TABLE 3.6   Space Functions and Characteristic Equations

Geometry	S_n	Characteristic equation
Plate	$\cos(\delta_n\zeta)$	$x\sin x = \text{Bi}\cos x$
Cylinder	$J_0(\delta_n\zeta)$	$xJ_1(x) = \text{Bi}\,J_0(x)$
Sphere	$\dfrac{\sin(\delta_n\zeta)}{(\delta_n\zeta)}$	$(1-\text{Bi})\sin x = x\cos x$

The Grober charts [29] for the heat loss fraction Q/Q_i, where $Q_i = \rho c_p V\theta_i$ is the initial total internal energy, are obtained from the series solution:

$$\frac{Q}{Q_i} = 1 - \sum_{n=1}^{\infty} B_n \exp(-\delta_n^2\,\text{Fo}) \tag{3.73}$$

The Fourier coefficients B_n are given in Table 3.5 for the three geometries. These coefficients depend on the Biot number.

Lumped Capacitance Model

When the Biot number is sufficiently small (Bi < 0.2), the series solutions converge to the first term for all values of Fo > 0. The values of the Fourier coefficients A_1 and B_1 approach 1, and the dimensionless temperature and the heat loss fraction approach the general lumped capacitance solutions

$$\phi = e^{-(hS/\rho c_p V)t}$$

and

$$\frac{Q}{Q_i} = 1 - e^{-(hS/\rho c_p V)t}$$

where S and V are the total active surface area and the volume. The lumped capacitance solutions for the three geometries are given in Table 3.7.

TABLE 3.7   Lumped Capacitance Solutions Bi < 0.2

Geometry	ϕ	Q/Q_i
Plate	$e^{-\text{Bi Fo}}$	$1 - e^{-\text{Bi Fo}}$
Cylinder	$e^{-2\text{Bi Fo}}$	$1 - e^{-2\text{Bi Fo}}$
Sphere	$e^{-3\text{Bi Fo}}$	$1 - e^{-3\text{Bi Fo}}$

Heisler and Grober Charts—Single-Term Approximations

The Heisler [36] cooling charts and the Grober [29] heat loss fraction charts for the three geometries can be calculated accurately by the single-term approximations [28, 56]

$$\frac{\theta}{\theta_i} = A_1 \exp(-\delta_1^2\,\text{Fo})S_1(\delta_1\zeta)$$

and

$$\frac{Q}{Q_i} = 1 - B_1 \exp(-\delta_1^2\,\text{Fo})$$

TABLE 3.8 Asymptotic Values of First Eigenvalues,
Correlation Parameter n, and Critical Fourier Number

Geometry	Bi $\to 0$	Bi $\to \infty$	n	Fo_c
Plate	$\delta_1 \to \sqrt{Bi}$	$\delta_1 \to \pi/2$	2.139	0.24
Cylinder	$\delta_1 \to \sqrt{2Bi}$	$\delta_1 \to 2.4048255$	2.238	0.21
Sphere	$\delta_1 \to \sqrt{3Bi}$	$\delta_1 \to \pi$	2.314	0.18

for all values of the Biot number, provided Fo $\geq Fo_c$. The critical values of the Fourier number for the three geometries are given in Table 3.8. The first eigenvalue can be computed accurately by means of the correlation [151]

$$\delta_1 = \frac{\delta_{1,\infty}}{[1 + (\delta_{1,\infty}/\delta_{1,0})^n]^{1/n}}$$

which is valid for all values of Bi. The asymptotic values of the first eigenvalues corresponding to very small and very large values of Bi are given in Table 3.8. The correlation parameter n is also given in Table 3.8. The correlation equation predicts values of δ_1 that differ from the exact values by less than 0.4 percent.

For Fo < Fo_c, additional terms in the series solutions must be included. It is therefore necessary to use numerical methods to compute the higher-order eigenvalues δ_n that lie in the intervals $n\pi < \delta_n < (n + 1/2)\pi$ for the plate and $(n - 1)\pi < \delta_n < n\pi$ for the cylinder and the sphere. Computer algebra systems are very effective in computing the eigenvalues.

Chen and Kuo [12] have presented approximate solutions of θ/θ_i and Q/Q_i for the plates and long cylinders. The accuracy of these solutions is acceptable for engineering calculations.

Multidimensional Systems

The basic solutions for the infinite plates and infinitely long cylinders can be used to obtain solutions for multidimensional systems such as long rectangular plates, cuboids, and finite circular cylinders with end cooling. The texts on conduction heat transfer [4, 11, 23, 29, 38, 49, 56, 87] should be consulted for the proofs of the method and other examples.

Langston [51] showed how to obtain the heat loss from multidimensional systems using the one-dimensional solutions given above. Two-dimensional systems such as long rectangular plates and finite circular cylinders are characterized by two Biot numbers, two Fourier numbers, and two dimensionless position parameters. Three-dimensional systems such as cuboids are characterized by three sets of values of Bi, Fo, and ζ corresponding to the three cartesian coordinates. When the two or three sets of Fourier numbers are greater than the critical values given in Table 3.8, then the first-term approximate solutions discussed above can be used to develop composite solutions. Yovanovich [151] has discussed the application of the basic solutions to long rectangular plates, cuboids, and finite-length circular cylinders.

Dimensionless Temperature and Heat Loss Fraction for Finite-Length Cylinders.   The finite-length circular cylinder of radius R and length $2X$, shown in Fig. 3.9, has constant properties and is cooled through the sides and the two ends by uniform film coefficients h_r and h_x, respectively. The sys-

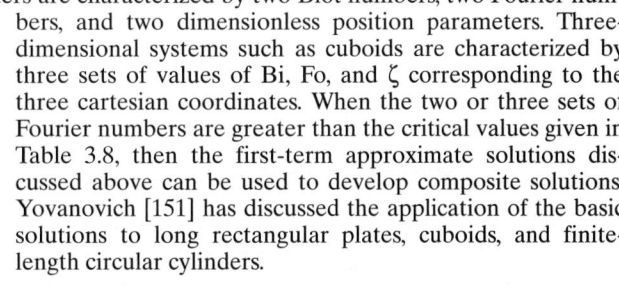

FIGURE 3.9   Circular cylinder of finite length.

tem is characterized by four physical parameters: $Bi_x = h_x X/k$, $Bi_c = h_r R/k$, $Fo_x = \alpha t/X^2$, and $Fo_r = \alpha t/R^2$. The dimensionless temperature within the cylinder is obtained from the product solution:

$$\phi_{xr} = \phi_x \phi_r$$

where ϕ_x and ϕ_r are the solutions corresponding to the x and r coordinates, respectively. According to Langston [51], the heat loss fraction can be obtained from

$$\left(\frac{Q}{Q_i}\right)_{xr} = \left(\frac{Q}{Q_i}\right)_x + \left(\frac{Q}{Q_i}\right)_r - \left(\frac{Q}{Q_i}\right)_x \cdot \left(\frac{Q}{Q_i}\right)_r$$

where $Q_i = \rho c_p 2\pi X R^2 \theta_i$. The subscripts x and r denote solutions corresponding to the x and r coordinates, respectively.

Transient One-Dimensional Conduction in Half-Spaces

The analytical solutions for transient one-dimensional conduction in half-spaces $x \geq 0$ are well known and appear in most heat transfer texts. The solutions are given here for completeness and to review important characteristics of the solutions.

Equation and Initial and Boundary Conditions.   The diffusion equation and the initial and boundary conditions are presented first, followed by the solutions with some important relationships.

$$\frac{\partial^2 \theta}{\partial x^2} = \frac{1}{\alpha} \frac{\partial \theta}{\partial t}, \qquad t > 0, \qquad x > 0 \tag{3.74}$$

where $\theta = T(x, t) - T_i$ is the instantaneous temperature rise within the half-space. The initial condition is

$$\theta = 0, \qquad t = 0, \qquad x > 0 \tag{3.75}$$

and the boundary condition at remote points in the half-space is

$$\theta \rightarrow 0, \qquad t > 0, \qquad x \rightarrow \infty \tag{3.76}$$

There are three options for the boundary condition at $x = 0$.

Dirichlet Condition

$$\theta = T_0 - T_i, \qquad t > 0 \tag{3.77}$$

where T_0 is the fixed temperature on the surface.

Neumann Condition

$$\frac{\partial \theta}{\partial x} = -\frac{q_0}{k}, \qquad t > 0 \tag{3.78}$$

where q_0 is the constant heat flux imposed on the surface.

Robin Condition

$$\frac{\partial \theta}{\partial x} = -\frac{h}{k}(\theta_f - \theta), \qquad t > 0 \tag{3.79}$$

where h is the constant film or contact conductance that connects the surface to the heat source and $\theta_f = T_f - T_i$ is the constant temperature difference between the heat source temperature and the initial temperature. The solutions have been obtained by several analytical methods. Introducing the dimensionless parameters $\phi = (T(x, t) - T_i)/(T_0 - T_i)$, $\eta = x/(2\sqrt{\alpha t})$, and $Bi = (h/k)\sqrt{\alpha t}$, the three solutions are given below.

Dirichlet Solution

$$\phi = \text{erfc}\,(\eta), \qquad \eta > 0 \tag{3.80}$$

which gives the instantaneous and time-average surface fluxes

$$q_0(t) = \frac{1}{\sqrt{\pi}}\,\frac{k(T_0 - T_i)}{\sqrt{\alpha t}} \tag{3.81}$$

and

$$\overline{q}_0(t) = \frac{2}{\sqrt{\pi}}\,\frac{k(T_0 - T_i)}{\sqrt{\alpha t}} \tag{3.82}$$

The time average value of any function $f(t)$ is defined as $\overline{f}(t) = (1/t)\int_0^t f(t)\,dt$.

Neumann Solution

$$\frac{k[T(x, t) - T_i]}{2q_0\sqrt{\alpha t}} = \frac{1}{\sqrt{\pi}}\,\exp(-\eta^2) - \eta\,\text{erfc}\,(\eta) \tag{3.83}$$

which gives the following relationships for the instantaneous and time-average surface temperatures

$$\frac{k[T(0, t) - T_i]}{2q_0\sqrt{\alpha t}} = \frac{1}{\sqrt{\pi}} \tag{3.84}$$

and

$$\frac{k[\overline{T}(0) - T_i]}{2q_0\sqrt{\alpha t}} = \frac{2}{3\sqrt{\pi}} \tag{3.85}$$

Robin Solution

$$\frac{T(x, t) - T_i}{T_f - T_i} = \text{erfc}\,(\eta) - \exp(2\eta\,Bi + Bi^2)\,\text{erfc}\,(\eta + Bi) \tag{3.86}$$

which yields the following two relationships for the instantaneous surface temperature and the surface heat flux:

$$\frac{T(0, t) - T_i}{T_f - T_i} = 1 - \exp(Bi^2)\,\text{erfc}\,(Bi) \tag{3.87}$$

and

$$\frac{q_0(t)\sqrt{\alpha t}}{k(T_f - T_i)} = Bi\,\exp(Bi^2)\,\text{erfc}\,(Bi) \tag{3.88}$$

For large values of the parameter $Bi \geq 100$, the Robin solution approaches the Dirichlet solution.

The three one-dimensional solutions presented above give important *short*-time results that appear in other solutions such as the external transient three-dimensional conduction from isothermal bodies of arbitrary shape into large regions. These solutions are presented in the next section.

External Transient Conduction From Long Cylinders

Introduction.   Transient one-dimensional conduction external to long circular cylinders is considered in this section. The conduction equation, the boundary and initial conditions, and the solutions for the Dirichlet and Neumann conditions are presented. The conduction equation for the instantaneous temperature rise $\theta(r, t) - T_i$ in the region external to a long circular cylinder of radius a is

$$\frac{\partial^2 \theta}{\partial r^2} + \frac{1}{r}\frac{\partial \theta}{\partial r} = \frac{1}{\alpha}\frac{\partial \theta}{\partial t}, \qquad t > 0, \qquad r > a \tag{3.89}$$

The initial condition is

$$\theta = 0, \qquad t = 0$$

and the boundary condition at remote points in the full space is

$$\theta \to 0, \qquad r \to \infty$$

Two types of boundary conditions at the cylinder boundary $r = a$ will be considered: (1) Dirichlet and (2) Neumann.

Dirichlet Condition

$$\theta = T_0 - T_i, \qquad t > 0$$

where T_0 is the fixed surface temperature.

Neumann Condition

$$\frac{\partial \theta}{\partial r} = -\frac{q_0}{k}, \qquad t > 0$$

where q_0 is the constant heat flux on the cylinder surface.

The solutions for the two boundary conditions are reported in Carslaw and Jaeger [11]. The solutions were obtained by means of the Laplace transform method. The solutions are given as infinite integrals and the integrand consists of Bessel functions of the first and second kinds of order zero.

Dirichlet Solution

$$\frac{\theta}{\theta_i} = 1 + \frac{2}{\pi}\int_0^\infty e^{-\text{Fo}\,\beta^2}\frac{J_0(\beta r/a)Y_0(\beta) - Y_0(\beta r/a)J_0(\beta)}{J_0^2(\beta) + Y_0^2(\beta)}\frac{d\beta}{\beta} \tag{3.90}$$

with $\text{Fo} \equiv \alpha t/a^2 > 0$. The integral can be evaluated accurately and easily for all dimensionless times using computer algebra systems.

Instantaneous Surface Heat Flux.   The instantaneous surface flux, defined as $q(t) = -k\partial\theta/\partial r$ at $r = a$, is given by the integral

$$\frac{aq(t)}{k\theta_i} = \frac{4}{\pi^2}\int_0^\infty e^{-\text{Fo}\,\beta^2}\frac{d\beta}{[J_0^2(\beta) + Y_0^2(\beta)]\beta} \tag{3.91}$$

Carslaw and Jaeger [11] presented short-time expressions for the instantaneous temperature rise and the surface heat flux.

Short-Time Temperature Rise

$$\frac{\theta}{\theta_i} = \frac{1}{\sqrt{\rho}}\,\text{erfc}\left[\frac{\rho - 1}{2\sqrt{\text{Fo}}}\right] + \frac{(\rho - 1)\sqrt{\text{Fo}}}{4\rho^{3/2}}\,\text{ierfc}\left[\frac{\rho - 1}{2\sqrt{\text{Fo}}}\right] + \frac{(9 - 2\rho - 7\rho^2)}{32\rho^{5/2}}\,\text{i}^2\text{erfc}\left[\frac{\rho - 1}{2\sqrt{\text{Fo}}}\right]$$

with $\rho \equiv r/a \geq 1$. The special functions ierfc (x) and i^2erfc (x) are integrals of the complementary error function and are defined in Carslaw and Jaeger [11].

Short-Time Surface Heat Flux.   The instantaneous surface heat flux is given by

$$\frac{aq(t)}{k\theta_i} = \frac{1}{\sqrt{\pi}\sqrt{Fo}} + \frac{1}{2} - \frac{1}{4}\left(\frac{Fo}{\pi}\right)^{1/2} + \frac{1}{8}\,Fo$$

The first term corresponds to the half-space solution when the dimensionless time is very small, i.e., $Fo < 10^{-3}$.

Neumann Solution.   The instantaneous temperature rise for arbitrary dimensionless time $Fo > 0$ at arbitrary radius $r/a \geq 1$ is given by the integral solution:

$$\frac{k\theta}{q_0} = -\frac{2}{\pi}\int_0^\infty (1 - e^{-Fo\,\beta^2})\,\frac{J_0(\beta r/a)Y_1(\beta) - Y_0(\beta r/a)J_1(\beta)}{J_1^2(\beta) + Y_1^2(\beta)}\,\frac{d\beta}{\beta^2}$$

Carslaw and Jaeger [11] presented an approximate short-time solution for arbitrary radius:

$$\frac{k\theta}{q_0} = \frac{2}{\sqrt{\rho}}\,\sqrt{Fo}\left\{ \mathrm{ierfc}\left[\frac{\rho - 1}{2\sqrt{Fo}}\right] - \frac{(3\rho + 1)}{4\rho}\,i^2\mathrm{erfc}\left[\frac{\rho - 1}{2\sqrt{Fo}}\right]\right\}$$

with $\rho = r/a \geq 1$ and $Fo = \alpha t/a^2$. The instantaneous surface temperature rise θ_0 for short times is given by

$$\frac{k\theta_0}{q_0} = \frac{2}{\sqrt{\pi}}\,\sqrt{Fo} - \frac{1}{2}$$

Transient External Conduction From Spheres

Introduction.   Solutions of transient conduction from a sphere of radius a into an isotropic space whose properties are constant and whose initial temperature T_i is constant are considered here. The dimensionless equation is

$$\frac{\partial^2\phi}{\partial\rho^2} + \frac{2}{\rho}\frac{\partial\phi}{\partial\rho} = \frac{\partial\phi}{\partial\,Fo}, \qquad Fo > 0, \qquad \rho > 1 \tag{3.92}$$

where ϕ is the dimensionless temperature, $\rho = r/a$, and $Fo = \alpha t/a^2$ is the dimensionless time defined with respect to the sphere radius. Solutions are available in Carslaw and Jaeger [11] for three boundary conditions: (1) the Dirichlet condition where $T(a, t) = T_0$, (2) the Neumann condition where $\partial T(a, t)/\partial r = -q_0/k$, and (3) the Robin condition where $\partial T(a, t)/\partial r = -(h/k)[T_f - T(a, t)]$. The thermophysical parameters T_0, T_f, q_0, and h are constants. These boundary conditions in dimensionless form are $\phi = 1$, $\partial\phi/\partial\rho = -1$, and $\partial\phi/\partial\rho = -Bi(1 - \phi)$ with $Bi = ha/k$ for the three boundary conditions, respectively. The three definitions of dimensionless temperature are presented below.

The three dimensionless solutions [11] are:

Dirichlet Solution

$$\phi = \frac{1}{\rho}\,\mathrm{erfc}\left(\frac{\rho - 1}{2\sqrt{Fo}}\right) \tag{3.93}$$

where $\phi = (T(r, t) - T_i)/(T_0 - T_i)$.

Neumann Solution

$$\phi = \frac{1}{\rho} \operatorname{erfc}\left(\frac{\rho - 1}{2\sqrt{Fo}}\right) - \frac{1}{\rho} \exp(\rho - 1 + Fo) \operatorname{erfc}\left(\frac{\rho - 1}{2\sqrt{Fo}} + \sqrt{Fo}\right) \tag{3.94}$$

where $\phi = k(T(r, t) - T_i)/(aq_0)$.

Robin Solution

$$\phi = \frac{Bi}{Bi + 1} \frac{1}{\rho} \operatorname{erfc}\left(\frac{\rho - 1}{2\sqrt{Fo}}\right)$$

$$-\frac{Bi}{Bi + 1} \frac{1}{\rho} \exp[(Bi + 1)(\rho - 1) + (Bi + 1)^2 \, Fo] \operatorname{erfc}\left[\frac{\rho - 1}{2\sqrt{Fo}} + (Bi + 1)\sqrt{Fo}\right] \tag{3.95}$$

where $\phi = (T(r, t) - T_i)/(T_f - T_i)$.

Instantaneous Surface Temperature and Heat Flux. The previous solutions give the following important results for the instantaneous surface temperature and surface heat flux.

Dirichlet Condition. The instantaneous surface heat flux is given by

$$\frac{aq(a, t)}{k(T_0 - T_i)} = 1 + \frac{1}{\sqrt{\pi}\sqrt{Fo}} \tag{3.96}$$

Neumann Condition. The instantaneous surface temperature is given by

$$\frac{k(T(a, t) - T_i)}{aq_0} = 1 - e^{Fo} \operatorname{erfc}(\sqrt{Fo}) \tag{3.97}$$

Robin Condition. The Robin solution given above yields expressions for the instantaneous surface temperature and the instantaneous surface heat flux. They are as follows:

$$\frac{(T(a, t) - T_i)}{(T_f - T_i)} = \frac{Bi}{Bi + 1} \{1 - e^{(Bi + 1)^2 \, Fo} \operatorname{erfc}[(Bi + 1)\sqrt{Fo}]\} \tag{3.98}$$

and

$$\frac{aq(a, t)}{k(T_f - T_i)} = \frac{Bi}{Bi + 1} \{1 + e^{(Bi + 1)^2 \, Fo} \operatorname{erfc}[(Bi + 1)\sqrt{Fo}]\} \tag{3.99}$$

Instantaneous Thermal Resistance

Resistance Definition. The instantaneous thermal resistance for the three boundary conditions is defined as $R = (T(a, t) - T_i)/Q$ where $Q = q(a, t)4\pi a^2$. The results given above yield the following expressions.

Dirichlet Condition Resistance

$$4\pi ka R_D = \frac{1}{[1 + (1/\sqrt{\pi}\sqrt{Fo})]} \tag{3.100}$$

Neumann Condition Resistance

$$4\pi ka R_N = 1 - e^{Fo} \operatorname{erfc}(\sqrt{Fo}) \tag{3.101}$$

Robin Condition Resistance

$$4\pi ka R_R = \frac{1 - e^{z^2}\,\mathrm{erfc}\,(z)}{1 + e^{z^2}\,\mathrm{erfc}\,(z)} \tag{3.102}$$

where $z = (\mathrm{Bi} + 1)\sqrt{\mathrm{Fo}}$.

The three previous expressions approach the steady-state result $4\pi ka R = 1$ for large dimensionless time.

The previous expression for the Robin condition can be calculated by the following rational approximation with a maximum error of less than 1.2 percent:

$$4\pi ka R_R = \frac{1 - a_1 s - a_2 s^2 - a_3 s^3}{1 + a_1 s + a_2 s^2 + a_3 s^3} \tag{3.103}$$

where $s = 1/(1 + pz)$ and the coefficients are $a_1 = 0.3480242$, $a_2 = -0.0958798$, $a_3 = 0.7478556$, $p = 0.47047$.

The three solutions corresponding to the three boundary conditions can be used to obtain approximate solutions for other convex bodies, such as a cube, for which there are no analytical solutions available. The dimensionless parameters Bi and Fo are defined with respect to the equivalent sphere radius, which is obtained by setting the surface area of the sphere equal to the surface area of the given body, i.e., $a = \sqrt{A/(4\pi)}$. This will be considered in the following section, which covers transient external conduction from isothermal convex bodies.

Transient External Conduction From Isothermal Convex Bodies

External transient conduction from an isothermal convex body into a surrounding space has been solved numerically (Yovanovich et al. [149]) for several axisymmetric bodies: circular disks, oblate and prolate spheroids, and cuboids such as square disks, cubes, and tall square cuboids (Fig. 3.10). The sphere has a complete analytical solution [11] that is applicable for all dimensionless times $\mathrm{Fo}_{\sqrt{A}} = \alpha t/A$. The dimensionless instantaneous heat transfer rate is $Q^*_{\sqrt{A}} = Q\sqrt{A}/(kA\theta_0)$, where k is the thermal conductivity of the surrounding space, A is the total area of the convex body, and $\theta_0 = T_0 - T_i$ is the temperature excess of the body relative to the initial temperature of the surrounding space. The analytical solution for the sphere is given by

$$Q^*_{\sqrt{A}} = 2\sqrt{\pi} + \frac{1}{\sqrt{\pi}\,\sqrt{\mathrm{Fo}_{\sqrt{A}}}} \tag{3.104}$$

which consists of the linear superposition of the steady-state solution (dimensionless shape factor) and the small-time solution (half-space solution). This observation was used to propose a simple approximate solution for all body shapes of the form

$$Q^*_{\sqrt{A}} = S^*_{\sqrt{A}} + \frac{1}{\sqrt{\pi}\,\sqrt{\mathrm{Fo}_{\sqrt{A}}}} \tag{3.105}$$

where $S^*_{\sqrt{A}}$ is the dimensionless shape factor for isothermal convex bodies in full space. This parameter is a relatively weak function of shape and aspect ratio; its values lie in the range $3.192 \le S^*_{\sqrt{A}} \le 4.195$ for the wide range of bodies examined (Fig. 3.10). The simple model predicts values with RMS and maximum differences between the predicted and numerical values in the ranges 0.40–6.31 percent and 0.98–11.52 percent, respectively, as shown in Table 3.9. The maximum differences lie in the intermediate range of dimensionless time: $10^{-3} < \mathrm{Fo}_{\sqrt{A}} < 10^{-1}$.

A more accurate model based on the method of Churchill and Usagi [13] was proposed for all bodies:

$$Q^*_{\sqrt{A}} = \left[(S^*_{\sqrt{A}})^n + \left(\frac{1}{\sqrt{\pi}\,\sqrt{\mathrm{Fo}_{\sqrt{A}}}} \right)^n \right]^{1/n} \tag{3.106}$$

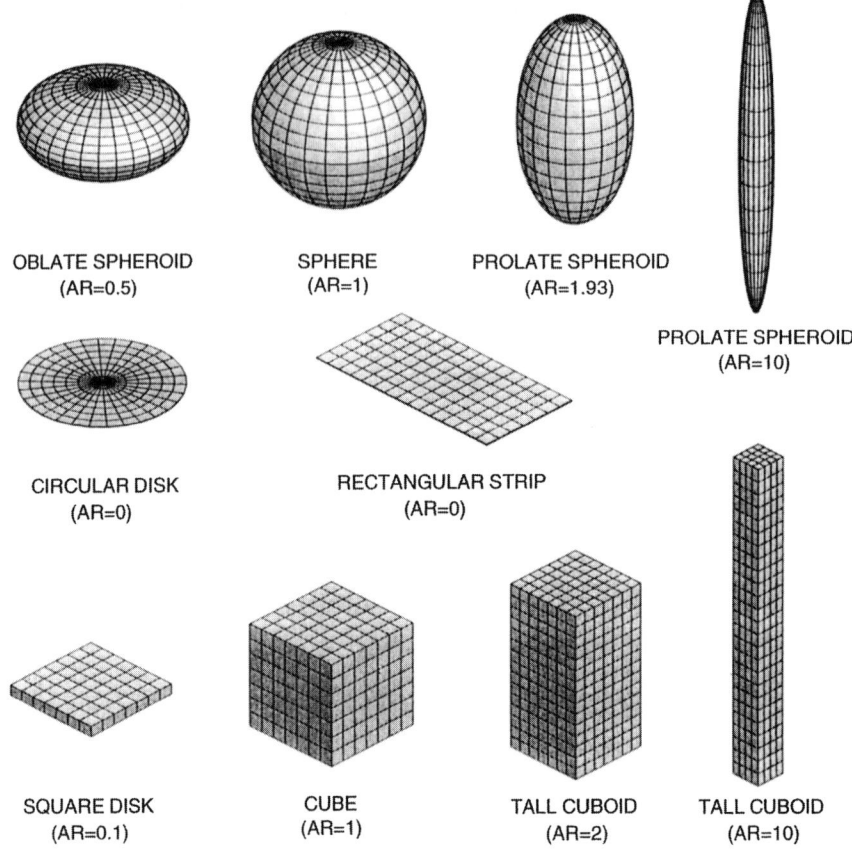

OBLATE SPHEROID
(AR=0.5)

SPHERE
(AR=1)

PROLATE SPHEROID
(AR=1.93)

PROLATE SPHEROID
(AR=10)

CIRCULAR DISK
(AR=0)

RECTANGULAR STRIP
(AR=0)

SQUARE DISK
(AR=0.1)

CUBE
(AR=1)

TALL CUBOID
(AR=2)

TALL CUBOID
(AR=10)

FIGURE 3.10 Various convex bodies.

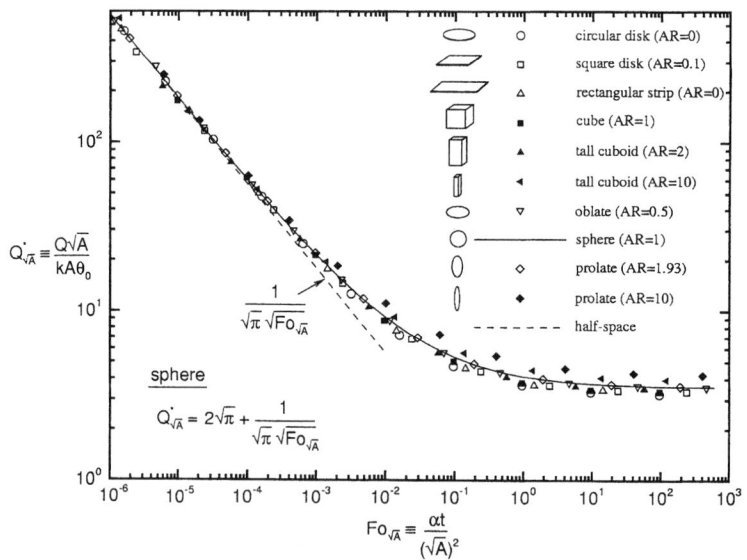

circular disk (AR=0)
square disk (AR=0.1)
rectangular strip (AR=0)
cube (AR=1)
tall cuboid (AR=2)
tall cuboid (AR=10)
oblate (AR=0.5)
sphere (AR=1)
prolate (AR=1.93)
prolate (AR=10)
half-space

$$Q^{*}_{\sqrt{A}} \equiv \frac{Q\sqrt{A}}{kA\theta_0}$$

$$\frac{1}{\sqrt{\pi}\,\sqrt{Fo_{\sqrt{A}}}}$$

sphere

$$Q^{*}_{\sqrt{A}} = 2\sqrt{\pi} + \frac{1}{\sqrt{\pi}\,\sqrt{Fo_{\sqrt{A}}}}$$

$$Fo_{\sqrt{A}} \equiv \frac{\alpha t}{(\sqrt{A})^2}$$

FIGURE 3.11 Comparison of numerical results and proposed model.

TABLE 3.9 Comparison of Superposition Model and Numerical Results

Body	$S^*_{\sqrt{A}}$	Max. % diff.	RMS % diff.
Circular disk ($AR = 0$)	3.192	7.05	3.36
Rectangular strip ($AR = 0$)	3.303	4.99	2.94
Square disk ($AR = 0.1$)	3.343	2.60	1.91
Cube ($AR = 1$)	3.388	3.78	2.52
Tall cuboid ($AR = 2$)	3.406	2.77	1.87
Oblate spheroid ($AR = 0.5$)	3.529	0.98	0.40
Prolate spheroid ($AR = 1.93$)	3.564	1.65	0.73
Tall cuboid ($AR = 10$)	3.945	3.15	1.63
Prolate spheroid ($AR = 10$)	4.195	11.52	6.31

The values of the parameter n that reduce the RMS and maximum differences between the model and the numerical data were found by trial and error to lie in the range $0.87 \leq n \leq 1.10$ for RMS and maximum differences less than 1.34 percent and 2.06 percent, respectively. The largest values of n were required for the thin bodies and the smallest values were required for the tallest bodies. For bodies having aspect ratios close to unity, the values of n were found to lie close to unity like the sphere. The values of n, and the corresponding values of the maximum and RMS percent differences for the various bodies are given in Table 3.10. The numerical data and the model predictions are shown in Fig. 3.11.

Table 3.11 gives recommended values of n for axisymmetric bodies (spheroids) and cuboids for various aspect ratios.

TABLE 3.10 Comparison of Blended Model and Numerical Results

Body	n	Max. % diff.	RMS % diff.
Circular disk ($AR = 0$)	1.10	1.83	0.80
Rectangular strip ($AR = 0$)	1.07	1.28	0.62
Square disk ($AR = 0.1$)	1.05	1.44	0.67
Cube ($AR = 1$)	1.05	1.95	1.04
Tall cuboid ($AR = 2$)	1.03	1.65	0.81
Oblate spheroid ($AR = 0.5$)	0.994	0.82	0.36
Prolate spheroid ($AR = 1.93$)	0.99	1.59	0.53
Tall cuboid ($AR = 10$)	0.96	2.08	1.13
Prolate spheroid ($AR = 10$)	0.87	2.06	1.34

TABLE 3.11 Blending Parameter and Recommendations

Body shape	Aspect ratio	n
Spheroids		
Thin disks	$AR \approx 0$	1.1
Oblates and prolates	$0.5 < AR < 2$	1.0
Tall prolates	$AR \gg 2$	0.9
Cuboids		
Thin disks	$AR \approx 0$	1.07
Disks and cubes	$0.1 < AR < 1$	1.05
Tall cuboids	$AR \approx 2$	1.03
Square cylinders	$AR \gg 2$	0.96

SPREADING (CONSTRICTION) RESISTANCE

Introduction

Spreading (constriction) resistance is an important thermal parameter that depends on several factors such as (1) geometry (singly or doubly connected areas, shape, aspect ratio), (2) domain (half-space, flux tube), (3) boundary condition (Dirchlet, Neumann, Robin), and (4) time (steady-state, transient). The results are presented in the form of infinite series and integrals that can be computed quickly and accurately by means of computer algebra systems. Accurate correlation equations are also provided.

Definitions of Spreading Resistance

Half-Space Spreading Resistance. When conduction occurs in a region whose dimensions are two or more orders of magnitude larger than the largest dimension of the source area (Fig. 3.12), the spreading resistance is defined as the difference between the heat source temperature and the heat sink temperature divided by the total heat transfer rate from the heat source. If the flux over the heat source area is uniform, the source temperature may be based on the area-average source temperature or the centroid temperature, which is the maximum temperature or close to it. If the heat sink temperature is spatially variable, then the area-average temperature is used in the definition. Thus

$$R_s = \frac{\overline{T}_{\text{source}} - \overline{T}_{\text{sink}}}{Q} \tag{3.107}$$

Flux Tube or Channel Spreading Resistance. When conduction occurs in a confined region such as a finite or an infinitely long flux tube or flux channel (Fig. 3.13), then the one-dimensional conduction resistance and other resistances must be accounted for. The definition proposed by Mikic and Rohsenow [65] can be used to define the spreading resistance. It is as follows:

$$R_s = \frac{\overline{T}_{\text{source}} - \overline{T}_{\text{contact plane}}}{Q} \tag{3.108}$$

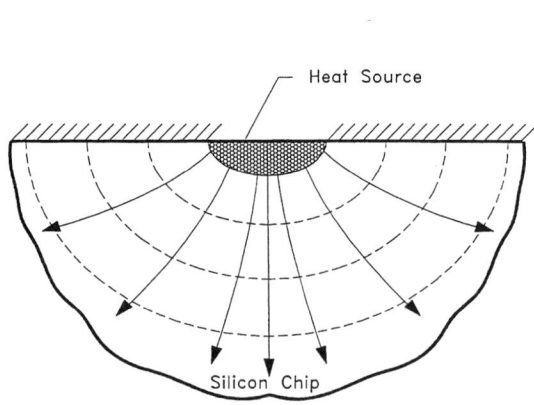

FIGURE 3.12 Spreading in a half-space.

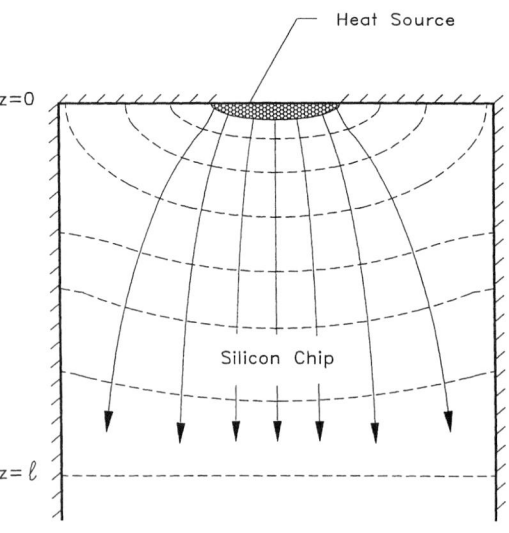

FIGURE 3.13 Spreading in a flux tube or channel.

Dimensionless Spreading Resistance.   Whichever definition is used, the dimensionless spreading resistance is generally defined as $R_s^* = k\mathscr{L}R_s$, where k is the thermal conductivity of the region and $\mathscr{L}$ is some length scale related to the contact area. It will be shown for arbitrary, singly connected contact areas that $\mathscr{L} = \sqrt{A}$, where A is the active area of the heat source and $\mathscr{L}$ is the appropriate length scale.

Spreading Resistance of Isoflux Arbitrary Areas on Half-Space

Circular, Rectangular, and Square Areas.   The discussion of spreading resistance in isotropic half-spaces begins with the circular area, which has analytical solutions for the isothermal and isoflux boundary conditions, and the rectangular contact, which has an isoflux solution. The solutions are reported in Carslaw and Jaeger [11]. From the circular contact solutions, one finds that the spreading resistance (1) for the isothermal condition is $R_s = 1/(4ka)$, where a is the contact radius, (2) for the isoflux condition is $R_s = 8/(3\pi^2 ka)$ based on the area-average temperature, and (3) for the isoflux condition is $R_s = 1/(\pi ka)$ based on the centroid temperature.

This geometry establishes the effect of boundary condition and the choice of source temperature used in the definition of the spreading resistance. The spreading resistances are related as follows:

$$kaR_s(\text{centroid}) > kaR_s(\text{area average}) > kaR_s(\text{isothermal}) \tag{3.109}$$

The temperature ratios are $T_0/\overline{T} = 1.1781$ and $\overline{T}/T(\text{isothermal}) = 1.0807$. These relationships are approximately valid for other geometries.

The centroid and area-average temperatures of the rectangular contact area of length $2a$ and width $2b$ with $a \geq b$ are given in Carslaw and Jaeger [11]. The centroid temperature is

$$T_0 = \frac{2}{\pi} \frac{qa}{k} \left[\sinh^{-1} \frac{1}{\epsilon} + \epsilon \sinh^{-1} \epsilon \right] \tag{3.110}$$

where the aspect ratio parameter is $\epsilon = a/b \geq 1$. The area-average temperature is

$$\overline{T} = \frac{2}{\pi} \frac{qa}{k} \left\{ \sinh^{-1} \frac{1}{\epsilon} + \frac{1}{\epsilon} \sinh^{-1} \epsilon + \frac{\epsilon}{3} \left[1 + \frac{1}{\epsilon^3} - \left(1 + \frac{1}{\epsilon^2} \right)^{3/2} \right] \right\} \tag{3.111}$$

The two expressions for the rectangular area give the results $kaR_s = 0.2806$ and $kaR_s = 0.2366$ for the spreading resistance of a square contact area based on the centroid and area-average temperatures, respectively. The ratio of centroid temperature to area-average temperature is $T_0/\overline{T} = 1.1857$. The differences between the results for the circle and those for the square are very small when the spreading resistances are based on $\mathscr{L} = \sqrt{A}$. The dimensionless spreading resistances become $k\sqrt{A}R_0 = 0.5611$ and $k\sqrt{A}R_0 = 0.5642$ for the square and circle, respectively, and $k\sqrt{A}R = 0.4728$ and $k\sqrt{A}R = 0.4787$ for the square and circle, respectively, where $\overline{R}$ is based on the area average temperature.

Regular Polygonal Areas.   The equilateral triangle, square, pentagon, and hexagon and the circle are members of the regular polygonal contact family as shown in Fig. 3.14. The dimensionless spreading resistance based on the centroid temperature is obtained from the integral [132].

$$k\sqrt{A}R_0 = \frac{1}{\pi} \sqrt{\frac{N}{\tan(\pi/N)}} \ln \left[\frac{1 + \sin(\pi/N)}{\cos(\pi/N)} \right], \qquad N \geq 3 \tag{3.112}$$

The above expression gives $k\sqrt{A}R_0 = 0.5516$ for the triangle ($N = 3$), which is approximately 2.3 percent smaller than the value for the circle where ($N = \infty$). The corresponding value for the area-average basis was reported by Yovanovich and Burde [134] to be $k\sqrt{A}\,\overline{R} = 0.4600$, which is approximately 4 percent smaller than the value for the circle.

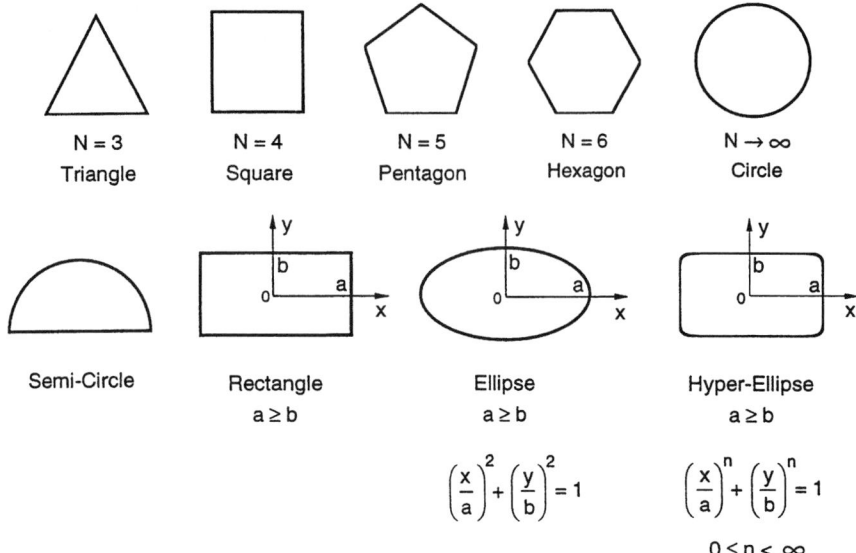

FIGURE 3.14   Singly connected planar areas.

Hyperellipse Contact Areas.   The family of contact areas defined by the hyperellipse $(x/a)^n + (y/b)^n = 1$ with $b \leq a$ is shown in Fig. 3.14. Yovanovich et al. [137] used a numerical technique to determine the spreading resistances for several values of the shape parameter $n = 0.5, 1, 2, 4, \infty$ and for a range of the aspect ratio $0.1 \leq \alpha = b/a \leq 1$.

The centroid temperature–based spreading resistance is obtained from the integral [137]

$$k\sqrt{A}R_0 = \frac{1}{\pi}\sqrt{\alpha}\sqrt{\frac{n}{B}}\int_0^{\pi/2}\frac{d\omega}{[\sin^n\omega + \alpha^n\cos^n\omega]^{1/n}} \tag{3.113}$$

where $\alpha = b/a$ is the aspect ratio parameter and B is the beta function [1], which depends on the shape parameter n:

$$B = B\left(1 + \frac{1}{n}, \frac{1}{n}\right)$$

Nonsymmetric Contact Areas.   Yovanovich and Burde [134] reported centroidal and area-average spreading resistances of several nonsymmetric contact areas such as semicircles, isosceles triangles of different aspect ratios, and squares with corners removed. For these contact areas Yovanovich and Burde reported that all numerical results lie in the ranges $0.4424 \leq k\sqrt{A}\,\overline{R} \leq 0.4733$ and $0.5197 \leq k\sqrt{A}R_0 \leq 0.5614$, and that $\overline{T}/T_0 = 0.84 \pm 1.7$ percent.

Circular Annular Contact Areas on Half-Space

Isoflux Annulus.   Yovanovich and Schneider [135] reported the temperature distribution within a circular annular contact $a \leq r \leq b$ subject to a uniform flux q:

$$T(r) = \frac{2}{\pi}\frac{qb}{k}\left\{E\left(\frac{r}{b}\right) - \left(\frac{r}{b}\right)E\left(\frac{a}{r}\right) + \left(\frac{r}{b}\right)\left[1 - \left(\frac{a}{r}\right)^2\right]K\left(\frac{a}{r}\right)\right\} \tag{3.114}$$

where k is the thermal conductivity of the half-space. The special functions that appear in the preceding expression, $K(x)$ and $E(x)$, are complete elliptical integrals of the first and second kinds, respectively [1, 10]. The heat sink temperature was set to zero for convenience. The thermal spreading resistance is based on the area-average temperature:

$$\overline{T} = \frac{1}{\pi(b^2 - a^2)} \int_a^b T(r) 2\pi r\, dr$$

The dimensionless thermal spreading resistance with $\epsilon = a/b < 1$ is given by

$$4kbR_s = \frac{32}{3\pi^2} \frac{1}{(1-\epsilon^2)^2} [1 + \epsilon^3 - (1 + \epsilon^2)E(\epsilon) + (1 - \epsilon^2)K(\epsilon)] \tag{3.115}$$

The above result reduces to the case of a circular contact area $\epsilon = 0$ on a half-space where $4kbR_s = 32/(3\pi^2)$.

Isothermal Annulus.   Smythe [95] obtained the solution for the capacitance of a thin circular annulus. The solutions were recast [135] into the following dimensionless thermal spreading resistance expressions:

$$4kbR_s = \frac{4}{\pi^2} \left[\frac{\ln 16 + \ln [(1 + \epsilon)/(1 - \epsilon)]}{(1 + \epsilon)} \right] \tag{3.116}$$

for $1.000 < 1/\epsilon < 1.10$, and

$$4kbR_s = \frac{\pi/2}{[\cos^{-1} \epsilon + \sqrt{1 - \epsilon^2} \tanh^{-1} \epsilon][1 + 0.0143\epsilon^{-1} \tan^3 (1.28\epsilon)]} \tag{3.117}$$

for $1.1 < 1/\epsilon < \infty$.

Doubly Connected Isoflux Contact Areas on Half-Space

The numerical data of Martin [154] for the dimensionless thermal spreading resistance $k\sqrt{A_c}\, R_s$ for three doubly connected regular polygons (Fig. 3.15)—equilateral triangle, square, and circle—are functions of $\epsilon = \sqrt{A_i/A_o}$, where A_i and A_o are the inner and outer projected areas of the polygons (Fig. 3.15). The contact area is $A_c = A_o - A_i$. Accurate correlation equations with a maximum relative error of 0.6 percent were given for the range $0 \le \epsilon \le 0.995$:

$$k\sqrt{A_c}\, R_s = a_0 \left[1 - \left(\frac{\epsilon}{a_1} \right)^{a_2} \right]^{a_3} \tag{3.118}$$

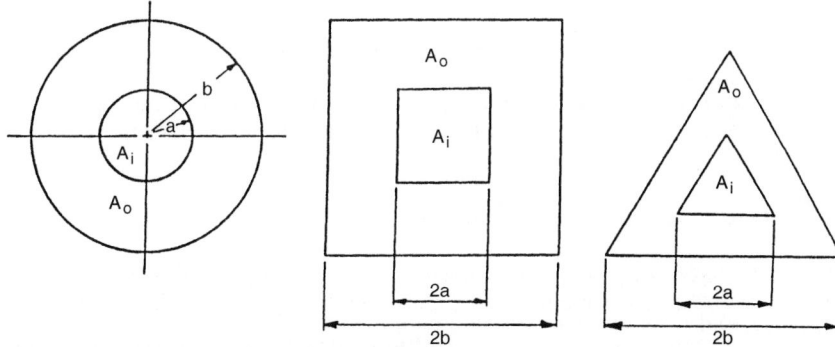

FIGURE 3.15   Doubly connected regular polygonal areas.

and for the range $0.995 \leq \epsilon \leq 0.9999$:

$$kP_o R_s = a_5 \ln \left[\frac{a_4}{(1/\epsilon) - 1} \right] \tag{3.119}$$

where P_o is the outer perimeter of the polygons, and the correlation coefficients a_0 through a_5 are given in Table 3.12.

TABLE 3.12 Coefficients for Doubly Connected Polygons

	Circle	Square	Triangle
a_0	0.4789	0.4732	0.4602
a_1	0.99957	0.99980	1.00010
a_2	1.5056	1.5150	1.5101
a_3	0.35931	0.37302	0.38637
a_4	39.66	68.59	115.91
a_5	0.31604	0.31538	0.31529

The correlation coefficient a_0 represents the dimensionless spreading resistance of the full contact area, in agreement with results presented previously. Since the results for the square and the circle are very close for all values of the parameter ϵ, the correlation equations for the square or the circle may be used for other doubly connected regular polygons such as pentagons, hexagons, etc.

Effect of Contact Conductance on Spreading Resistance

Martin et al. [61] used a numerical technique to determine the effect of a uniform contact conductance h on the spreading resistance of square and circular contact areas. The dimensionless spreading resistance values were correlated with an accuracy of 0.1 percent by the following expression:

$$k\sqrt{A}R_s = c_1 - c_2 \tanh (c_3 \ln \text{Bi} - c_4), \qquad 0 \leq \text{Bi} < \infty \tag{3.120}$$

with $\text{Bi} = h\sqrt{A}/k$. The correlation coefficients c_1 through c_4 are presented in Table 3.13.

TABLE 3.13 Coefficients for Square and Circle

	Circle	Square
c_1	0.46159	0.45733
c_2	0.017499	0.016463
c_3	0.43900	0.47035
c_4	1.1624	1.1311

When $\text{Bi} \leq 0.1$, the values predicted by the expression approach the values corresponding to the isoflux boundary condition; and when $\text{Bi} \geq 100$, the predicted values are within 0.1 percent of the values obtained for the isothermal boundary condition. The transition from the isoflux values to the isothermal values occurs in the range $0.1 \leq \text{Bi} \leq 100$.

Spreading Resistance in Flux Tubes and Channels

General Expression for Circular Contact Area With Arbitrary Flux on Circular Flux Tube.
The general expression for the dimensionless spreading (constriction) resistance $4kaR_s$ for a circular contact subjected to an arbitrary axisymmetric flux distribution $f(u)$ [131–133] is obtained from the series

$$4kaR_s = \frac{(8/\pi)}{\int_0^1 uf(u)\,du} \sum_{n=1}^{\infty} \frac{J_1(\delta_n\epsilon)}{\delta_n^2 J_0^2(\delta_n)} \int_0^1 uf(u)J_0(\delta_n\epsilon u)\,du \tag{3.121}$$

where δ_n are the roots of $J_1(\cdot) = 0$ and $\epsilon = a/b$ is the relative size of the contact area.

General Expression for Flux Distributions of the Form $(1 - u^2)^\mu$.   Yovanovich [131–133] reported the following general solution for axisymmetric flux distributions of the form $f(u) = (1 - u^2)^\mu$, where the parameter μ accounts for the shape of the flux distribution. The above general expression reduces to the following general expression:

$$4kaR_s = \frac{16}{\pi}(\mu + 1)2^\mu \Gamma(\mu + 1)\frac{1}{\epsilon}\sum_{n=1}^{\infty}\frac{J_1(\delta_n\epsilon)J_{\mu+1}(\delta_n\epsilon)}{\delta_n^3 J_0^2(\delta_n)(\delta_n\epsilon)^\mu} \tag{3.122}$$

where $\Gamma(x)$ is the gamma function [1] and $J_\nu(x)$ is the Bessel function of arbitrary order ν.

The above general expression can be used to obtain specific solutions for various values of the flux distribution parameter μ. Three particular solutions are considered next.

Equivalent Isothermal Contact Area: $\mu = -\frac{1}{2}$.   The isothermal contact area requires the solution of a difficult mathematical problem that has received much attention from numerous researchers [24, 42, 65, 70, 71, 84, 132, 133].

Mikic and Rohsenow [65] proposed the use of the flux distribution corresponding to $\mu = -\frac{1}{2}$ to approximate an isothermal contact area for small relative contact areas: $0 < \epsilon < 0.5$. The general expression becomes

$$4kaR_s = \frac{8}{\pi}\frac{1}{\epsilon}\sum_{n=1}^{\infty}\frac{J_1(\delta_n\epsilon)\sin(\delta_n\epsilon)}{\delta_n^3 J_0^2(\delta_n)} \tag{3.123}$$

An accurate correlation equation of this series solution is given later.

Isoflux Contact Area: $\mu = 0$.   The preceding general solution with $\mu = 0$ yields the isoflux solution reported by Mikic and Rohsenow [65]:

$$4kaR_s = \frac{16}{\pi}\frac{1}{\epsilon}\sum_{n=1}^{\infty}\frac{J_1^2(\delta_n\epsilon)}{\delta_n^3 J_0^2(\delta_n)} \tag{3.124}$$

An accurate correlation equation of this series solution is given later.

Parabolic Flux Distribution: $\mu = \frac{1}{2}$.   Yovanovich [131–133] reported the solution for the parabolic flux distribution corresponding to $\mu = \frac{1}{2}$.

$$4kaR_s = \frac{24}{\pi}\frac{1}{\epsilon}\sum_{n=1}^{\infty}\frac{J_1(\delta_n\epsilon)\sin(\delta_n\epsilon)}{\delta_n^3 J_0^2(\delta_n)}\left\{\frac{1}{(\delta_n\epsilon)^2} - \frac{1}{\delta_n\epsilon\tan(\delta_n\epsilon)}\right\} \tag{3.125}$$

Effect of Flux Distribution on Circular Contact Area on Half-Space

The three series solutions just given converge very slowly as $\epsilon \to 0$, which corresponds to the case of a circular contact area on a half-space. The corresponding half-space results are given by Strong et al. [112]: $4kaR_s (\mu = -\frac{1}{2}) = 1$, $4kaR_s (\mu = 0) = 32/(3\pi^2)$, $4kaR_s (\mu = \frac{1}{2}) = 1.1252$.

Simple Correlation Equations of Spreading Resistance for Circular Contact Area

Yovanovich [131–133] recommended the following simple correlations for the three flux distributions:

$$4kaR_s = a_1(1 - a_2\epsilon) \tag{3.126}$$

in the range $0 < \epsilon \leq 0.1$ with a maximum error of 0.1 percent and

$$4kaR_s = a_1(1 - \epsilon)^{a_3} \tag{3.127}$$

in the range $0 < \epsilon \leq 0.3$ with a maximum error of 1 percent. The correlation coefficients for the three flux distributions are given in Table 3.14.

TABLE 3.14 Correlation Coefficients for $\mu = -\frac{1}{2}, 0, \frac{1}{2}$

μ	$-\frac{1}{2}$	0	$\frac{1}{2}$
a_1	1	1.0808	1.1252
a_2	1.4197	1.4111	1.4098
a_3	1.50	1.35	1.30

Accurate Correlation Equations for Various Combinations of Contact Area, Flux Tubes, and Boundary Condition

General Accurate Correlation Equation. Solutions are also available for various combinations of contact areas and flux tubes such as circle/circle and circle/square for the uniform flux, true isothermal, and equivalent isothermal boundary conditions [71].

Numerical results were correlated using the polynomial

$$4kaR_s = C_0 + C_1\epsilon + C_3\epsilon^3 + C_5\epsilon^5 + C_7\epsilon^7 \tag{3.128}$$

The dimensionless spreading (constriction) resistance coefficient C_0 is the half-space value, and the correlation coefficients C_1 through C_7 are given in Table 3.15.

General Approximate Correlation Equation for Applications. For microelectronic applications, an accurate engineering approximation [74] that is valid for a circular contact on a circular or square flux tube or a square contact on a square flux tube is

$$k\sqrt{A_c}R_s = 0.475 - 0.62\epsilon + 0.13\epsilon^3 \tag{3.129}$$

where $\epsilon = \sqrt{A_c/A_t}$, and A_c, A_t are the contact and flux tube areas, respectively. The maximum error with respect to the exact solution is less than 2 percent for $0 \leq \epsilon \leq 0.5$ and less than 4 percent for $0 \leq \epsilon \leq 0.7$.

TABLE 3.15 Coefficients for Correlations of Dimensionless Spreading Resistance $4kaR_s$

Flux tube geometry and contact boundary condition	C_0	C_1	C_3	C_5	C_7
Circle/circle, uniform flux	1.08076	−1.41042	0.26604	−0.00016	0.058266
Circle/circle, true isothermal	1.00000	−1.40978	0.34406	0.04305	0.02271
Circle/square, uniform flux	1.08076	−1.24110	0.18210	0.00825	0.038916
Circle/square, equivalent isothermal flux	1.00000	−1.24142	0.20988	0.02715	0.02768

Square Contact Area on Square Flux Tube. Mikic and Rohsenow [65] reported the solution for an isoflux square contact area on the end of a semi-infinite square flux tube. The solution was recast [74] to give the dimensionless spreading resistance:

$$k\sqrt{A_c}R_s = \frac{2}{\pi^3\epsilon}\left[\sum_{m=1}^{\infty}\frac{\sin^2(m\pi\epsilon)}{m^3} + \frac{1}{\pi^2\epsilon^2}\sum_{m=1}^{\infty}\sum_{n=1}^{\infty}\frac{\sin^2(m\pi\epsilon)\sin^2(n\pi\epsilon)}{m^2n^2\sqrt{m^2+n^2}}\right] \quad (3.130)$$

Circular Contact Area on Square Flux Tube. Sadhal [155] reported the general solution for an isoflux or equivalent isothermal elliptical contact area on a rectangular flux tube. His general solution gives the dimensionless spreading resistance for an isoflux square contact area on a circular flux tube that has the form

$$k\sqrt{A_c}R_s = \frac{2}{\pi^3\epsilon}\left[\sum_{n=1}^{\infty}\frac{J_1^2(2n\sqrt{\pi}\epsilon)}{n^3} + \sum_{m=1}^{\infty}\sum_{n=1}^{\infty}\frac{J_1^2(2\sqrt{\pi}\epsilon\sqrt{m^2+n^2})}{(m^2+n^2)^{3/2}}\right] \quad (3.131)$$

with the relative size parameter defined as $\epsilon = \sqrt{A_c/A_t} = (\sqrt{\pi}/2)a/b$. The dimensionless isoflux spreading resistance $k\sqrt{A_c}R_s$ has the half-space values 0.47890 and 0.47320 for the circular and square contact areas, respectively, as $\epsilon \to 0$. Negus et al. [74] reported accurate correlations for the circle/circle, circle/square, and square/square combinations corresponding to the preceding series solutions given by Eq. 3.128.

General Spreading Resistance Expression for Circular Annular Area on Circular Flux Tube

General Expression for Arbitrary Flux. The spreading resistance of a circular annulus of inner and outer radii a and b, respectively, on one end of a semi-infinite circular flux tube of radius c and thermal conductivity k is considered here.

The general expression for the dimensionless spreading resistance $4kbR_s$ for a circular annular contact subjected to an arbitrary axisymmetric flux distribution $f(u)$ is given by [131]

$$4kbR_s = \frac{(8/\pi)}{(1-\epsilon^2)\int_\epsilon^1 uf(u)\,du}\sum_{n=1}^{\infty}\frac{J_1(\beta\delta_n)\{1-\epsilon[J_1(\beta\epsilon\delta_n)]/[J_1(\beta\delta_n)]\}}{\delta_n^2 J_0^2(\delta_n)}\int_\epsilon^1 uf(u)J_0(\beta\delta_n u)\,du \quad (3.132)$$

Spreading Resistance of Isoflux Annular Contact on Circular Flux Tube. The previous general expression reduces for the isoflux case that corresponds to $f(u) = 1$ to the following series [128, 131–133]:

$$4kbR_s = \frac{16}{\pi\beta(1-\epsilon^2)^2}\sum_{n=1}^{\infty}\frac{J_1^2(\beta\delta_n)}{\delta_n^3 J_0^2(\delta_n)}\left[1-\epsilon\frac{J_1(\beta\epsilon\delta_n)}{J_1(\beta\delta_n)}\right]^2 \quad (3.133)$$

with the parameters $0 < \beta = b/c < 1$ and $0 \leq \epsilon < 1$. The eigenvalues δ_n are the roots of $J_1(\cdot) = 0$. This solution reduces to the solution for the isoflux circular contact area on the end of a circular flux tube when $\beta = 0$. For small values of β and $0 < \epsilon < 1$ one can use the closed-form solution reported previously for a circular annular contact on a half-space.

Spreading Resistance Within Two-Dimensional Channels

General Expression for Arbitrary Flux. The steady-state spreading resistance due to conduction through a strip of width $2a$ on one end of an infinitely long two-dimensional channel of width $2b$ and thermal conductivity k (Fig. 3.16) has solutions reported by Smythe [98], Mikic and Rohsenow [65], Veziroglu and Chandra [119], and Sexl and Burkhard [90]. Solu-

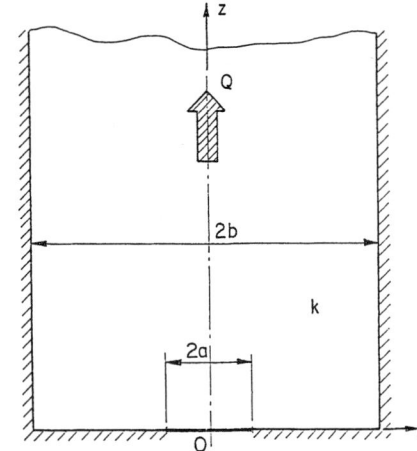

FIGURE 3.16 Two-dimensional flux channel.

tions have been obtained for the isothermal strip, the isoflux strip, and a general solution developed for arbitrary flux distribution over the strip. Yovanovich [130] reported the following general solution:

$$kR_s = \frac{1}{\pi^2} \frac{1}{\int_0^1 f(u)\, du} \left(\frac{1}{\epsilon}\right) \sum_{n=1}^{\infty} \frac{\sin{(n\pi\epsilon)}}{n^2} \int_0^1 f(u)\, \cos{(n\pi\epsilon u)}\, du$$

(3.134)

where $f(u)$ with $0 \le u = x/a < 1$ represents the arbitrary flux distribution over the contact strip, and $\epsilon = a/b < 1$ is the relative size of the contact strip.

General Expression for Flux Distribution of the Form $(1 - u^2)^\mu$. Yovanovich [130] chose the general flux function $f(u) = (1 - u^2)^\mu$ with parameter μ, which gives (1) the isoflux contact when $\mu = 0$, (2) the equivalent isothermal strip when $\mu = -\frac{1}{2}$, and (3) the parabolic flux distribution when $\mu = \frac{1}{2}$ to develop another general solution:

$$kR_s = \frac{1}{\pi^2} \Gamma\left(\mu + \frac{3}{2}\right)\left(\frac{1}{\epsilon}\right) \sum_{n=1}^{\infty} \frac{\sin{(n\pi\epsilon)}}{n^2} \left[\frac{2}{n\pi\epsilon}\right]^{\mu + (1/2)} J_{\mu + (1/2)}(n\pi\epsilon)$$

(3.135)

Setting $\mu = -\frac{1}{2}$ and $\mu = 0$ in the above general solution gives the two solutions reported by Mikic and Rohsenow [65].

Equivalent Isothermal Contact Area: $\mu = -\frac{1}{2}$. For $\mu = -\frac{1}{2}$

$$kR_s = \frac{1}{\pi^2} \left(\frac{1}{\epsilon}\right) \sum_{n=1}^{\infty} \frac{\sin{(n\pi\epsilon)}}{n^2} J_0(n\pi\epsilon)$$

(3.136)

Isoflux Contact Area: $\mu = 0$. For $\mu = 0$

$$kR_s = \frac{1}{\pi^3} \left(\frac{1}{\epsilon}\right)^2 \sum_{n=1}^{\infty} \frac{\sin^2{(n\pi\epsilon)}}{n^3}$$

(3.137)

Parabolic Flux Distribution: $\mu = \frac{1}{2}$. The parabolic flux distribution $\mu = \frac{1}{2}$ gives

$$kR_s = \frac{2}{\pi^3} \left(\frac{1}{\epsilon}\right)^2 \sum_{n=1}^{\infty} \frac{\sin{(n\pi\epsilon)}}{n^3} J_1(n\pi\epsilon)$$

(3.138)

True Isothermal Contact Area. Veziroglu and Chandra [119] used a conformal mapping technique to obtain the closed-form solution for the true isothermal strip:

$$kR_s = \frac{1}{\pi} \ln\left\{\frac{1}{\sin{[(\pi/2)\epsilon]}}\right\}$$

(3.139)

The true isothermal strip solution and the equivalent isothermal flux solution predict values of the dimensionless spreading resistance that are in close agreement provided $\epsilon < 0.4$. The parabolic flux distribution gives the greatest values of the spreading resistance, followed by the isoflux values, which are greater than the values for the isothermal strip.

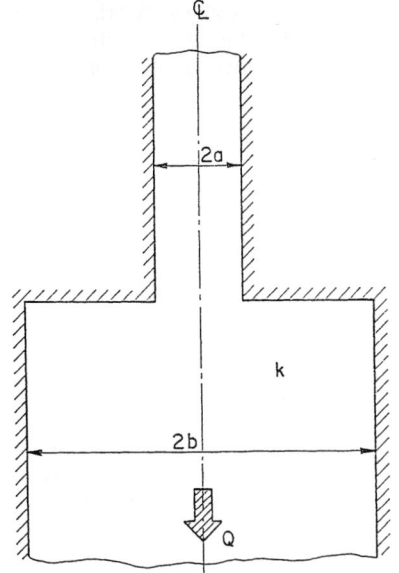

FIGURE 3.17   Two-dimensional flux channel with area change.

Effect of Area Change.   Smythe [98] reported the analytical solution for the spreading resistance for the case of two coaxial channels of widths $2a$ and $2b$ where $a < b$ are in perfect contact of width $2a$ (Fig. 3.17):

$$kR_s = \frac{1}{2\pi} \left\{ \left(\epsilon + \frac{1}{\epsilon} \right) \ln \left[\frac{1+\epsilon}{1-\epsilon} \right] + 2 \ln \left[\frac{1-\epsilon^2}{4\epsilon} \right] \right\} \quad (3.140)$$

Effect of Single and Multiple Layers (Coatings) on Spreading Resistance

The influence of single and multiple layers of different thermal conductivities and thicknesses is of great interest. The solutions given below can be used to determine the effect of oxide layers and coatings on the spreading resistance. Solutions are available for semi-infinite flux tubes and finite systems such as circular disks.

Spreading Resistance Within Compound Disks With Conductance.   The compound disk is shown in Fig. 3.18. The disk consists of two isotropic materials of thicknesses t_1, t_2 and thermal conductivities k_1, k_2 that are in perfect contact. The radius of the compound disk is denoted b and its thickness is denoted $t = t_1 + t_2$. The lateral boundary $r = b$ is adiabatic; the face at $z = t$ is either cooled by a fluid through the film conductance h or in contact with a heat sink through a contact conductance h. In either case h is assumed to be uniform. The face at $z = 0$ consists of the heat source area of radius a, and the remainder of that face $a < r \leq b$ is adiabatic. The boundary condition over the contact area can be modeled as (1) uniform heat flux or (2) isothermal. The complete solution for these two boundary conditions has been reported by Yovanovich et al. [138]. The general solution for the dimensionless constriction parameter $4k_1aR_c$ depends on several dimensionless geometric and thermophysical parameters: $\tau = t/b$, $\tau_1 = t_1/b$, $\tau_2 = t_2/b$, $\epsilon = a/b$, $\kappa = k_1/k_2$, $\mathrm{Bi} = hb/k_2$, μ. The parameter μ describes

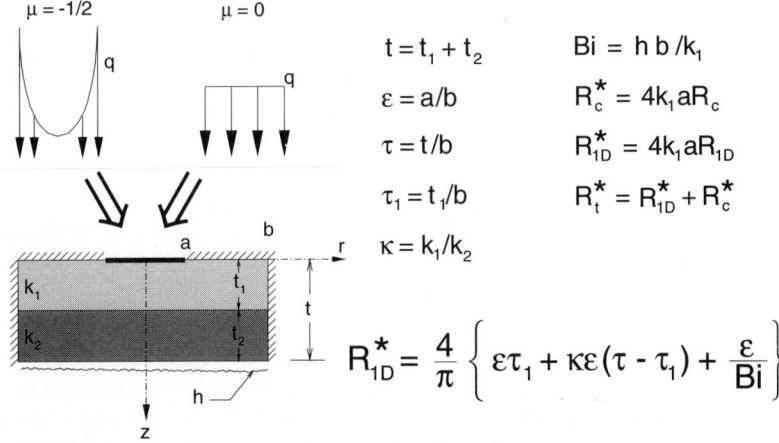

FIGURE 3.18   Compound circular disk with conductance.

the heat flux distribution over the contact area. When $\mu = 0$, the heat flux is uniform, and when $\mu = -\frac{1}{2}$, this heat flux distribution is called the equivalent isothermal distribution because it produces an *almost* isothermal contact area provided $a/b < 0.6$.

The general solution is given as

$$4k_1 a R_s = \frac{8(\mu + 1)}{\pi \epsilon} \sum_{n=1}^{\infty} A_n(n, \epsilon) B_n(n, \tau, \tau_1) \frac{J_1(\delta_n \epsilon)}{\delta_n \epsilon} \tag{3.141}$$

The coefficients A_n are functions of the heat flux parameter μ. They become, for $\mu = -\frac{1}{2}$:

$$A_n = \frac{-2\epsilon \sin \delta_n \epsilon}{\delta_n^2 J_0^2(\delta_n)}$$

and for $\mu = 0$:

$$A_n = \frac{-2\epsilon J_1(\delta_n \epsilon)}{\delta_n^2 J_0^2(\delta_n)}$$

The function B_n is defined as

$$B_n = \frac{\phi_n \tanh (\delta_n \tau_1) - \phi_n}{1 - \phi_n} \tag{3.142}$$

and the two functions that appear in the above relationship are defined as

$$\phi_n = \frac{\kappa - 1}{\kappa} \cosh (\delta_n \tau_1)[\cosh (\delta_n \tau_1) - \phi_n \sinh (\delta_n \tau_1)] \tag{3.143}$$

and

$$\phi_n = \frac{\delta_n + \text{Bi} \tanh (\delta_n \tau)}{\delta_n \tanh (\delta_n \tau) + \text{Bi}} \tag{3.144}$$

The eigenvalues δ_n are the positive roots of $J_1(\cdot) = 0$.

Characteristics of ϕ_n. This function accounts for the effects of the parameters δ_n, τ, and Bi. For limiting values of the parameter Bi it reduces to

$$\phi_n = \tanh (\delta_n \tau) \qquad \text{Bi} \to \infty$$

and

$$\phi_n = \coth (\delta_n \tau) \qquad \text{Bi} \to 0$$

For all $0 < \text{Bi} < \infty$ and for all values $\tau > 0.72$, $\tanh (\delta_n \tau) = 1$ for all $n \geq 1$. Therefore $\phi_n = 1$ for $n \geq 1$.

Characteristics of B_n. When $\tau_1 > 0.72$, $\tanh (\delta_n \tau) = 1$, $\phi_n = 1$ for all $0 < \text{Bi} < \infty$, therefore $B_n = 1$ for $n \geq 1$. These characteristics lead to the previously discussed flux tube solutions.

The general solution for the compound disk can be used to obtain the spreading resistances for the several cases shown in Figs. 3.19 and 3.20.

Spreading Resistance Within Isotropic Finite Disks With Conductance. The dimensionless constriction resistance for isotropic ($\kappa = 1$) finite disks ($\tau_1 < 0.72$) with negligible thermal resistance at the heat sink interface ($\text{Bi} = \infty$) is given by the following solutions.

For $\mu = -\frac{1}{2}$:

$$4ka R_s = \frac{8}{\pi \epsilon} \sum_{n=1}^{\infty} \frac{J_1(\delta_n \epsilon) \sin (\delta_n \epsilon)}{\delta_n^3 J_0^2(\delta_n)} \tanh (\delta_n \tau) \tag{3.145}$$

For $\mu = 0$:

$$4ka R_s = \frac{16}{\pi \epsilon} \sum_{n=1}^{\infty} \frac{J_1^2(\delta_n \epsilon)}{\delta_n^3 J_0^2(\delta_n)} \tanh (\delta_n \tau) \tag{3.146}$$

If the external resistance is negligible ($\text{Bi} \to \infty$), the temperature at the lower face of the disk is assumed to be isothermal. The solutions for isoflux $\mu = 0$ heat source and isothermal base temperature were given by Kennedy [42] for (1) the centroid temperature and (2) the area-average contact area temperature.

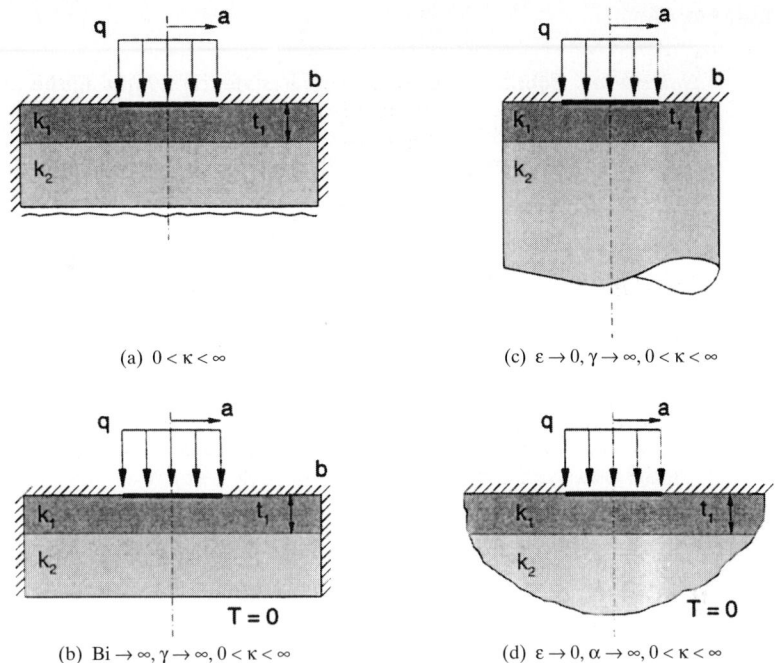

(a) $0 < \kappa < \infty$

(c) $\varepsilon \to 0, \gamma \to \infty, 0 < \kappa < \infty$

(b) $\mathrm{Bi} \to \infty, \gamma \to \infty, 0 < \kappa < \infty$

(d) $\varepsilon \to 0, \alpha \to \infty, 0 < \kappa < \infty$

FIGURE 3.19 Special cases of the compound disk with $\kappa \neq 1$.

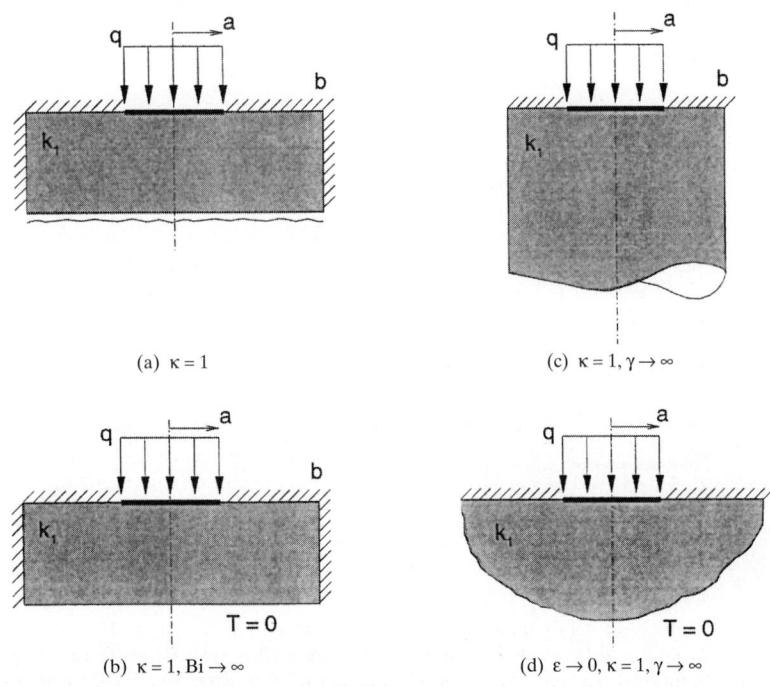

(a) $\kappa = 1$

(c) $\kappa = 1, \gamma \to \infty$

(b) $\kappa = 1, \mathrm{Bi} \to \infty$

(d) $\varepsilon \to 0, \kappa = 1, \gamma \to \infty$

FIGURE 3.20 Special cases of the compound disk with $\kappa = 1$.

Correlation Equation for Spreading Resistance Within Finite Disk With Conductance.
The solution for the isoflux boundary condition and with external thermal resistance was
recently reexamined by Song et al. [156] and Lee et al. [157]. These researchers nondimensionalized the constriction resistance based on the centroid and area-average temperatures
using the square root of the contact area as recommended by Chow and Yovanovich [15] and
Yovanovich [132, 137, 144–146, 150], and compared the analytical results against the numerical results reported by Nelson and Sayers [158] over the full range of the independent parameters Bi, ϵ, and τ. Nelson and Sayers [158] also chose the square root of the contact area to
report their numerical results. The analytical and numerical results were reported to be in
excellent agreement.

Lee et al. [157] recommended a simple closed-form expression for the dimensionless constriction resistance based on the area-average and centroid temperatures. They defined the
dimensionless spreading resistance parameter as $\psi = \sqrt{\pi}kaR_c$ and recommended the following approximations.

For the area-average temperature: $\psi_{ave} = \frac{1}{2}(1 - \epsilon)^{3/2}\phi_c$ (3.147)

For the centroid temperature: $\psi_{max} = \frac{1}{\sqrt{\pi}}(1 - \epsilon)\varphi_c$ (3.148)

with $\varphi_c = \dfrac{\text{Bi} \tanh (\delta_c\tau) + \delta_c}{\text{Bi} + \delta_c \tanh (\delta_c\tau)}$

and $\delta_c = \pi + \dfrac{1}{\sqrt{\pi\epsilon}}$

The above approximations are within ±10 percent of the analytical results [156, 157] and
the numerical results [158]. They do not, however, indicate where the maximum errors
occur.

Circular Contact Area on Single Layer (Coating) on Half-Space

Equivalent Isothermal Circular Contact.   Dryden [16] obtained the solution for an equivalent isothermal circular contact of radius a that is in perfect contact with an isotropic layer of
thermal conductivity k_1 and thickness t_1 that is also in perfect thermal contact with a substrate
of thermal conductivity k_2. This system is shown in Fig. 3.21.
Dryden based his solution on the axisymmetric flux distribution:

$$q(r) = \frac{Q}{2\pi a(a^2 - r^2)^{1/2}}$$

where Q is the heat transfer rate through the contact area.
The spreading resistance, based on the area-average temperature, is obtained from the integral:

$$R_s = \frac{1}{\pi k_1 a} \int_0^\infty \left[\frac{\lambda_2 \exp(\zeta t_1/a) + \lambda_1 \exp(-\zeta t_1/a)}{\lambda_2 \exp(\zeta t_1/a) - \lambda_1 \exp(-\zeta t_1/a)} \right] \frac{J_1(\zeta) \sin \zeta}{\zeta^2} d\zeta$$

(3.149)

with $\lambda_1 = (1 - k_2/k_1)/2$ and $\lambda_2 = (1 + k_2/k_1)/2$. The parameter ζ
is a dummy variable of integration. The constriction resis-
tance depends on the thermal conductivity ratio k_1/k_2 and

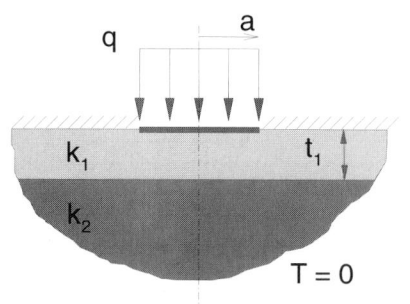

FIGURE 3.21   Circular contact area on single layer
(coating) on half-space.

the relative layer thickness t_1/a. Dryden gave simple asymptotes for the constrictions for thin layers ($t_1/a \leq 0.1$) and thick layers ($t_1/a \geq 10$). These asymptotes were also presented as dimensionless spreading resistances defined as $4k_2aR_s$. They are as follows.

Thin-Layer Asymptote

$$(4k_2aR_s)_{\text{thin}} = 1 + \left(\frac{4}{\pi}\right)\left(\frac{t_1}{a}\right)\left[\frac{k_2}{k_1} - \frac{k_1}{k_2}\right] \tag{3.150}$$

Thick-Layer Asymptote

$$(4k_2aR_s)_{\text{thick}} = \frac{k_2}{k_1} - \left(\frac{2}{\pi}\right)\left(\frac{a}{t_1}\right)\left(\frac{k_2}{k_1}\right) \ln\left(\frac{2}{1 + k_1/k_2}\right) \tag{3.151}$$

These asymptotes provide results that are within 1 percent of the full solution for relative layer thickness $t_1/a < 0.5$ and $t_1/a > 2$.

The dimensionless constriction is based on the substrate thermal conductivity k_2. The preceding general solution is valid for conductive layers where $k_1/k_2 > 1$ as well as for resistive layers where $k_1/k_2 < 1$. The infinite integral can be evaluated numerically by means of computer algebra systems, which provide accurate results.

Isoflux Circular Contact.   Hui and Tan [39] gave the solution for a circular contact of radius a that is subjected to a uniform and steady heat flux q and that is in perfect contact with an isotropic layer of thermal conductivity k_1 and thickness t_1 that is also in perfect thermal contact with a substrate of thermal conductivity k_2. This system is shown in Fig. 3.21.

Hui and Tan [39] reported the following integral solution for the dimensionless spreading resistance:

$$4k_2aR_s = \frac{32}{3\pi^2}\left(\frac{k_2}{k_1}\right)^2 + \frac{8}{\pi}\left[1 - \left(\frac{k_2}{k_1}\right)^2\right]\int_0^\infty \frac{J_1^2(\zeta)\,d\zeta}{[1 + (k_1/k_2)\tanh(\zeta t_1/a)]\zeta^2} \tag{3.152}$$

which depends on the thermal conductivity ratio k_1/k_2 and the relative layer thickness t_1/a. The dimensionless constriction is based on the substrate thermal conductivity k_2. The above general solution is valid for conductive layers where $k_1/k_2 > 1$ as well as for resistive layers where $k_1/k_2 < 1$.

Circular Contact Area on Multiple Layers on Circular Flux Tube

The effect of single and multiple isotropic layers or coatings on the end of a circular flux tube has been determined by Antonetti [2] and Sridhar et al. [107]. The heat enters the end of the circular flux tube of radius b and thermal conductivity k_3 through a coaxial, circular contact area that is in perfect thermal contact with an isotropic layer of thermal conductivity k_1 and thickness t_1. This layer is in perfect contact with a second layer of thermal conductivity k_2 and thickness t_2 that is in perfect contact with the flux tube having thermal conductivity k_3 (Fig. 3.22). The lateral boundary of the flux tube is adiabatic and the contact plane outside the contact area is also adiabatic. The boundary condition over the contact area may be (1) isoflux or (2) isothermal. The dimensionless constriction resistance $\psi_{2\text{ layers}} = 4k_3aR_c$ is defined with respect to the thermal conductivity of the flux

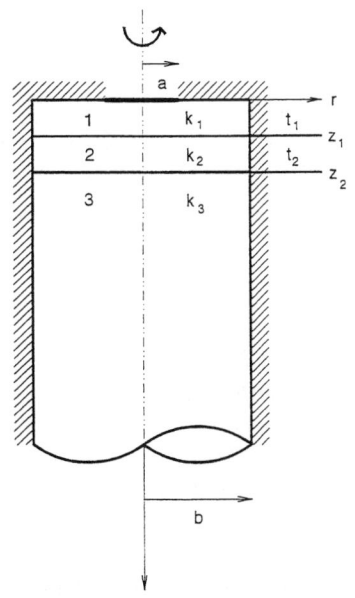

FIGURE 3.22 Flux tube with two layers.

tube, which is often referred to as the substrate. This constriction resistance depends on several dimensionless parameters: relative contact size $\epsilon = a/b$ where $0 < \epsilon < 1$; two conductivity ratios, $\kappa_{21} = k_2/k_1$ and $\kappa_{32} = k_3/k_2$; two relative layer thicknesses, $\tau_1 = t_1/a$ and $\tau_2 = t_2/a$; and the boundary condition over the contact area. The solution for two layers is given as

$$\psi_{2\,\text{layers}} = \frac{16}{\pi} \frac{1}{\epsilon} \sum_{n=1}^{\infty} \phi_{n,\epsilon} \kappa_{21} \kappa_{32} \frac{\vartheta^+}{\vartheta^-} \tag{3.153}$$

where

$$\phi_{n,\epsilon} = \frac{J_1^2(\delta_n \epsilon)}{\delta_n^3 J_0^2(\delta_n)} \rho_{n,\epsilon} \tag{3.154}$$

and the boundary condition parameter is defined as

$$\rho_{n,\epsilon} = \begin{cases} \dfrac{\sin(\delta_n \epsilon)}{2 J_1(\delta_n \epsilon)} & \text{isothermal contact} \\ 1 & \text{isoflux contact} \end{cases} \tag{3.155}$$

The thermal conductivity ratios are defined above. The layer parameters ϑ^+ and ϑ^- come from the following general function:

$$\begin{aligned} \vartheta^{\pm} = {} & (1 + \kappa_{21})(1 + \kappa_{32}) \pm (1 - \kappa_{21})(1 + \kappa_{32}) \exp(-2\delta_n \epsilon \tau_1) \\ & + (1 - \kappa_{21})(1 - \kappa_{32}) \exp(-2\delta_n \epsilon \tau_2) \pm (1 + \kappa_{21})(1 + \kappa_{32}) \exp(-2\delta_n \epsilon(\tau_1 + \tau_2)) \end{aligned}$$

The eigenvalues δ_n that appear in the solution are the roots of $J_1(\cdot) = 0$.

The two-layer solution can be used to obtain the solution for a single layer of thermal conductivity k_1 and thickness t_1 in perfect contact with a flux tube of thermal conductivity k_2. In this case the dimensionless constriction resistance $\psi_{1\,\text{layer}}$ depends on the relative contact size ϵ, the conductivity ratio κ_{21}, and the relative layer thickness τ_1:

$$\psi_{1\,\text{layer}} = \frac{16}{\pi} \frac{1}{\epsilon} \sum_{n=1}^{\infty} \phi_{n,\epsilon} \kappa_{21} \frac{\vartheta^+}{\vartheta^-} \tag{3.156}$$

and the general layer function reduces to

$$\vartheta^{\pm} = 2[(1 + \kappa_{21})(1 - \kappa_{21})] \exp(-2\delta_n \epsilon \tau_1)$$

Transient Spreading Resistance

Introduction.   Transient spreading resistance occurs during startup and is important in certain micro-electronic systems. The spreading resistance can be defined with respect to the area-average temperature as a single point temperature such as the centroid. Solutions have been reported for isoflux contact areas on half-spaces, circular contact areas on circular flux tubes, and strips on channels.

Spreading Resistance of Isoflux Circular Contact Area on Half-Space.   Beck [6] reported the following integral solution for a circular area of radius a that is subjected to a uniform and constant flux q for $t > 0$:

$$4kaR_s = \frac{8}{\pi} \int_0^{\infty} J_1^2(\zeta) \, \text{erf}\,(\zeta \sqrt{\text{Fo}}) \frac{d\zeta}{\zeta^2} \tag{3.157}$$

where the dimensionless time is defined as $\text{Fo} = \alpha t/a^2$. The spreading resistance is based on the area average temperature. Steady state is obtained when $\text{Fo} \to \infty$ and the solution goes to $4kaR_s = 32/(3\pi^2)$.

Beck gave approximate solutions for short and long times. For short times (Fo < 0.6),

$$4kaR_s = \frac{8}{\pi}\left[\sqrt{\frac{\text{Fo}}{\pi}} - \frac{\text{Fo}}{\pi} + \frac{\text{Fo}^2}{8\pi} + \frac{\text{Fo}^3}{32\pi} + \frac{15\text{Fo}^4}{512\pi}\right] \tag{3.158}$$

and for long times (Fo ≥ 0.6),

$$4kaR_s = \frac{32}{3\pi^2} - \frac{2}{\pi^{3/2}\sqrt{\text{Fo}}}\left[1 - \frac{1}{3(4\text{Fo})} + \frac{1}{6(4\text{Fo})^2} - \frac{1}{12(4\text{Fo})^3}\right] \tag{3.159}$$

The maximum errors of about 0.18 percent and 0.07 percent occur at Fo = 0.6 for the short- and long-time expressions, respectively.

Transient Spreading Resistance of Isoflux Hyperellipse Contact Area on Half-Space

The hyperellipse is defined by $(x/a)^n + (y/b)^n = 1$ with $b \le a$, where n is the shape parameter and a and b are the axes along the x and y axes, respectively, as shown in Fig. 3.14. The hyperellipse reduces to many special cases by setting the values of n and the aspect ratio parameter $\gamma = b/a$, which lies in the range $0 \le \gamma \le 1$. Therefore the solution developed for the hyperellipse can be used to obtain solutions for many other geometries, such as ellipses and circles, rectangles and squares, diamondlike geometries, etc. The transient dimensionless centroid constriction resistance $k\sqrt{A}R_0$ where $R_0 = T_0/Q$ is given by the double-integral solution [152]:

$$k\sqrt{A}R_0 = \frac{2}{\pi\sqrt{A}}\int_0^{\pi/2}\int_0^{r_0} \text{erfc}\left(\frac{r}{2\sqrt{A}\sqrt{\text{Fo}_{\sqrt{A}}}}\right) dr\, d\omega \tag{3.160}$$

where the area of the hyperellipse is given by $A = (4\gamma/n)B(1 + 1/n, 1/n)$ and $B(x, y)$ is the beta function. The dimensionless time is defined as $\text{Fo}_{\sqrt{A}} = \alpha t/A$. The upper limit of the radius is given by $r_0 = \gamma/[(\sin \omega)^n + \gamma^n(\cos \omega)^n]^{1/n}$ and the aspect ratio parameter $\gamma = b/a$. The preceding solution has the following characteristics: (1) for small dimensionless times, $\text{Fo}_{\sqrt{A}} \le 4 \times 10^{-2}$ and $k\sqrt{A}R_0 = (2/\sqrt{\pi})\sqrt{\text{Fo}_{\sqrt{A}}}$ for all values of n and γ; (2) for long dimensionless times $\text{Fo}_{\sqrt{A}} \ge 10^3$. The results are within 1 percent of the steady-state values, which are given by the single integral:

$$k\sqrt{A}R_0 = \frac{2\gamma}{\pi\sqrt{A}}\int_0^{\pi/2}\frac{d\omega}{[(\sin \omega)^n + \gamma^n(\cos \omega)^n]^{1/n}} \tag{3.161}$$

which depends on the aspect ratio γ and the shape parameter n. The dimensionless spreading resistance depends on the three parameters $\text{Fo}_{\sqrt{A}}$, γ, and n in the transition region: $4 \times 10^{-2} \le \text{Fo}_{\sqrt{A}} \le 10^3$ in some complicated manner that can be deduced from the solution for the circular area. For this axisymmetric shape we put $\gamma = 1$, $n = 2$ into the hyperellipse double integral, which yields the following closed-form result valid for all dimensionless time [152]:

$$k\sqrt{A}R_0 = \sqrt{\text{Fo}_{\sqrt{A}}}\left[\frac{1}{\sqrt{\pi}} - \frac{1}{\sqrt{\pi}}\exp(-1/(4\pi\,\text{Fo}_{\sqrt{A}})) + \frac{1}{2\sqrt{\pi}\sqrt{\text{Fo}_{\sqrt{A}}}}\,\text{erfc}\left(\frac{1}{2\sqrt{\pi}\sqrt{\text{Fo}_{\sqrt{A}}}}\right)\right] \tag{3.162}$$

where the dimensionless time for the circle of radius a is $\text{Fo}_{\sqrt{A}} = \alpha t/(\pi a^2)$.

Transient Spreading Resistance of Isoflux Regular Polygonal Contact Area on Half-Space

For regular polygons having sides $N \geq 3$ as depicted in Fig. 3.14, the area is $A = Nr_i^2 \tan \pi/N$, where r_i is the radius of the inscribed circle. The dimensionless constriction resistance based on the centroid temperature $k\sqrt{A}R_0$ is given by the following double integral [152]:

$$k\sqrt{A}R_0 = 2\sqrt{\frac{N}{\tan(\pi/N)}} \int_0^{\pi/N} \int_0^{1/\cos\omega} \mathrm{erfc}\left(\frac{r}{2\sqrt{N\tan(\pi/N)}\,\sqrt{\mathrm{Fo}_{\sqrt{A}}}}\right) dr\,d\omega \qquad (3.163)$$

where the polygonal area is expressed in terms of the number of sides N and for convenience the inscribed radius has been set to unity. This double-integral solution has identical characteristics to those of the double-integral solution given above for the hyperellipse, i.e., for small dimensionless time, $\mathrm{Fo}_{\sqrt{A}} \leq 4 \times 10^{-2}$ and $k\sqrt{A}R_0 = (2/\sqrt{\pi})\sqrt{\mathrm{Fo}_{\sqrt{A}}}$ for all polygons $N \geq 3$; for long dimensionless time, $\mathrm{Fo}_{\sqrt{A}} \geq 10^3$. The results are within 1 percent of the steady-state values given by the following closed-form expression [152]:

$$k\sqrt{A}R_0 = \frac{1}{\pi}\sqrt{\frac{N}{\tan(\pi/N)}}\ln\left[\frac{1+\sin(\pi/N)}{\cos(\pi/N)}\right] \qquad (3.164)$$

The dimensionless spreading resistance $k\sqrt{A}R_0$ depends on the parameters $\mathrm{Fo}_{\sqrt{A}}$ and N in the transition region $4 \times 10^{-2} \leq \mathrm{Fo}_{\sqrt{A}} \leq 10^3$ in some complicated manner that, as described above, can be deduced from the solution for the circular area. The steady-state solution yields ψ_0 values of 0.5617, 0.5611, and 0.5642 for the equilateral triangle ($N = 3$), the square ($N = 4$), and the circle ($N \to \infty$). The difference between the values for the triangle and the circle is approximately 2.2 percent, whereas the difference between the values for the square and the circle is less than 0.6 percent. The following procedure is proposed for computation of the centroid-based transient spreading resistance for the range $4 \times 10^{-2} \leq \mathrm{Fo}_{\sqrt{A}} \leq 10^6$. The closed-form solution for the circle is the basis of the proposed method. For any planar, singly connected contact area subjected to a uniform heat flux take, $\psi_0 = k\sqrt{A}R_0$.

$$\frac{\psi_0}{\psi_0(\mathrm{Fo}_{\sqrt{A}} \to \infty)} = 2\sqrt{\mathrm{Fo}_{\sqrt{A}}}\left[1 - \exp(-1/(4\pi\,\mathrm{Fo}_{\sqrt{A}})) + \frac{1}{2\sqrt{\mathrm{Fo}_{\sqrt{A}}}}\mathrm{erfc}\left(\frac{1}{2\sqrt{\pi}\sqrt{\mathrm{Fo}_{\sqrt{A}}}}\right)\right]$$

$$(3.165)$$

The right side of the above equation can be considered to be a *universal* dimensionless time function that accounts for the transition from small times to near steady state. The proposed procedure should provide quite accurate results for any planar, singly connected area. A simpler expression based on the Greene approximation [25] of the complementary error function is proposed [152]:

$$\frac{\psi_0}{\psi_0(\mathrm{Fo}_{\sqrt{A}} \to \infty)} = \frac{1}{z\sqrt{\pi}}\left[1 - e^{-z^2} + a_1\sqrt{\pi}z e^{-a_2(z+a_3)^2}\right] \qquad (3.166)$$

where $z = 1/(2\sqrt{\pi}\sqrt{\mathrm{Fo}_{\sqrt{A}}})$ and the three correlation coefficients are $a_1 = 1.5577$, $a_2 = 0.7182$, $a_3 = 0.7856$. This approximation provides values of ψ_0 with maximum errors of less than 0.5 percent for $\mathrm{Fo}_{\sqrt{A}} \geq 4 \times 10^{-2}$.

Transient Spreading Resistance Within Semi-Infinite Flux Tubes and Channels

Isoflux Circular Contact Area on Circular Flux Tube. Turyk and Yovanovich [118] reported the analytical solutions for transient spreading resistance within semi-infinite circular flux tubes and two-dimensional channels. The circular contact and the rectangular strip are subjected to uniform and constant heat fluxes. The dimensionless spreading resistance for the flux tube is given by the series solution

$$4kaR_s = \frac{16}{\pi} \frac{1}{\epsilon} \sum_{n=1}^{\infty} \frac{J_1^2(\delta_n \epsilon) \; \mathrm{erf} \, (\delta_n \epsilon \sqrt{\mathrm{Fo}})}{\delta_n^3 J_0^2(\delta_n)} \tag{3.167}$$

where $\epsilon = a/b < 1$, Fo $= \alpha t/a^2$, and δ_n are the roots of $J_1(x) = 0$. The series solution approaches the steady-state solution presented in an earlier section when the dimensionless time satisfies the criterion Fo $\geq 1/\epsilon^2$ or when the real time satisfies the criterion $t \geq a^2/(\alpha\epsilon^2)$.

Isoflux Strip on Two-Dimensional Channel. The dimensionless spreading resistance within a two-dimensional channel of width $2b$ and thermal conductivity k was reported as

$$kR_s = \frac{1}{\pi^3 \epsilon} \sum_{m=1}^{\infty} \frac{\sin^2 (m\pi\epsilon) \; \mathrm{erf} \, (m\pi\epsilon\sqrt{\mathrm{Fo}})}{m^3} \tag{3.168}$$

where $\epsilon = a/b < 1$ is the relative size of the contact strip, and the dimensionless time is defined as Fo $= \alpha t/a^2$. There is no half-space solution for the two-dimensional channel. The transient solution is within 1 percent of the steady-state solution when the dimensionless time satisfies the criterion Fo $\geq 1.46/\epsilon^2$.

CONTACT, GAP, AND JOINT RESISTANCES AND CONTACT CONDUCTANCES

Point and Line Contact Models

The thermal resistance models for steady-state conduction through contact areas and gaps formed when nonconforming smooth surfaces are placed in contact are based on the Hertz elastic deformation model [37, 41, 117, 121] and the thermal spreading (constriction) results presented previously. The general elastoconstriction models for point and line contacts are reviewed by Yovanovich [143]. In the general case the contact area is elliptical and its dimensions are much smaller than the dimensions of the contacting bodies. The gap that is formed is a function of the shape of the contacting bodies, and in general the local gap thickness is described by complex integrals and special functions called elliptical integrals [8, 10]. Two important special cases are considered in the following sections: sphere-flat and circular cylinder-flat contacts. The review of Yovanovich [143] can be consulted for the general case.

Elastoconstriction Resistance of Sphere-Flat Contacts. The contact resistance of the sphere-flat contact shown in Fig. 3.23 is discussed in this section. The thermal conductivities of the sphere and flux tube are k_1 and k_2, respectively. The total contact resistance is the sum of the constriction resistance in the sphere and the spreading resistance within the flux tube. The contact radius a is much smaller than the sphere diameter D and the tube diameter. Assuming isothermal contact area, the general elastoconstriction resistance model [143] becomes:

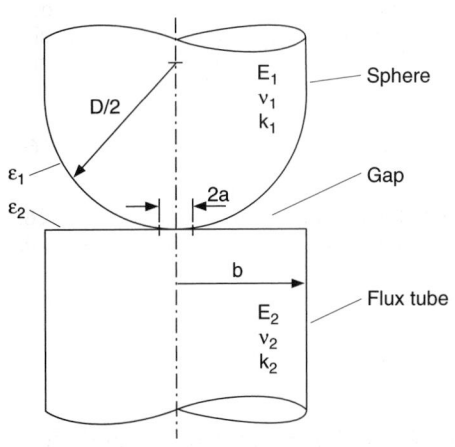

FIGURE 3.23 Sphere-flat contact with gap.

$$R_c = \frac{1}{2k_s a} \tag{3.169}$$

where $k_s = 2k_1 k_2/(k_1 + k_2)$ is the harmonic mean thermal conductivity of the contact, and the contact radius is obtained from the Hertz elastic model [117]:

$$\frac{2a}{D} = \left[\frac{3F\Delta}{D^2}\right]^{1/3} \tag{3.170}$$

where F is the mechanical load at the contact, and the physical parameter is defined as

$$\Delta = \frac{1}{2}\left[\frac{(1 - v_1^2)}{E_1} + \frac{(1 - v_2^2)}{E_2}\right] \tag{3.171}$$

where v_1 and v_2 are the Poisson's ratio and E_1 and E_2 are the elastic modulus of the sphere and flat contacts, respectively.

Elastoconstriction Resistance of Cylinder-Flat Contacts. The thermal contact resistance model for the contact formed by a smooth circular cylinder of diameter D and thermal conductivity k_1, and a smooth flat of thermal conductivity k_2, was reported by McGee et al. [63] to be

$$k_s R_c = \frac{1}{2\pi}\frac{k_s}{k_1}\ln\left(\frac{\pi}{F^*}\right) - \frac{k_s}{2k_2} - \frac{1}{2\pi}\frac{k_s}{k_2}\ln(4\pi F^*) \tag{3.172}$$

where k_s is the harmonic mean thermal conductivity and Δ is the Hertz physical parameter defined above. The dimensionless mechanical load is defined as $F^* = F\Delta/(DL)$, where F is the total load at the contact strip and L is the length of the cylinder and the flat.

Gap Resistance Model of Sphere-Flat Contacts. The general elastogap resistance model for point contacts [143] reduces for the sphere-flat contact to

$$\frac{1}{R_g} = \left(\frac{D}{L}\right)k_{g,0}I_{g,p} \tag{3.173}$$

where $L = D/(2a)$ is the relative contact size defined previously. The gap integral for point contacts proposed by Kitscha and Yovanovich [46] is defined as

$$I_{g,p} = \int_1^L \frac{2x \tan^{-1}\sqrt{x^2 - 1}}{(2\delta/D) + (2M/D)}\,dx \tag{3.174}$$

The local gap thickness δ is obtained from

$$\frac{2\delta}{D} = 1 - \left(1 - \left(\frac{x}{L}\right)^2\right)^{1/2} + \frac{1}{\pi L^2}\left[(2 - x^2)\sin^{-1}\left(\frac{1}{x}\right) + \sqrt{x^2 - 1}\right] - \frac{1}{L^2} \tag{3.175}$$

where $L = D/(2a)$, $x = r/a$, and $1 \le x \le L$. The gap gas rarefaction parameter is defined as

$$M = \alpha\beta\Lambda \tag{3.176}$$

with $\alpha = (2 - \alpha_1)/\alpha_1 + (2 - \alpha_2)/\alpha_2$, where α_1 and α_2 are the accommodation coefficients at the gas-solid interfaces, respectively. The gas parameter $\beta = (2\gamma)/[(\gamma + 1)/\text{Pr}]$, where $\gamma = C_p/C_v$ and Pr is the Prandtl number. The mean free path Λ of the gas molecules is given in terms of $\Lambda_{g,0}$, the mean free path at some reference gas temperature T_0 and reference gas pressure P_0, as follows:

$$\Lambda = \Lambda_{g,0}\frac{T_g}{T_{g,0}}\frac{P_{g,0}}{P_g}$$

Gap Resistance Model of Cylinder-Flat Contacts.   The general elastogap resistance model for line contacts [143] reduces for the circular cylinder-flat contact to

$$\frac{1}{R_g} = \frac{4aL_{\text{cyl}}}{D} k_{g,0} I_{g,l} \tag{3.177}$$

where D and L are the diameter and length of the cylinder, and a is the half-width of the contact strip. The gap integral for line contacts [63] is

$$I_{g,l} = \frac{2}{\pi} \int_1^L \frac{\cosh^{-1}(\zeta)}{(2\delta/D) + (2M/D)} \, d\zeta \tag{3.178}$$

The local gap thickness for line contacts [121] is

$$\frac{2\delta}{D} = \left(1 - \frac{1}{L^2}\right)^{1/2} - \left(1 - \frac{\zeta^2}{L^2}\right)^{1/2} + \frac{1}{2L}[\zeta(\zeta^2 - 1)^{1/2} - \cosh^{-1}(\zeta) - \zeta^2 + 1] \tag{3.179}$$

where $L = D/(2a)$, $\zeta = x/L$ and $1 \leq x \leq L$. The dimensionless contact strip width is obtained from the Hertz theory [117]

$$\frac{2a}{D} = 4\sqrt{\frac{F\Delta}{\pi D L_{\text{cyl}}}} \tag{3.180}$$

where F is the total load at the contact and Δ is the physical parameter defined earlier.

Radiative Resistance Model of Sphere-Flat Contacts.   The radiative resistance of a gap formed by two bodies in elastic contact, such as a sphere-flat or cylinder-flat contact, respectively, is complex because it depends on the geometry of the gap—the surface emissivities of the boundaries, which includes the side walls that form the enclosure.

Kitscha and Yovanovich [46] and Kitscha [47] proposed the following radiative resistance model for a sphere-disk contact with bounding side walls. All surfaces were assumed to be gray with constant emissivity values ϵ_1, ϵ_2, and ϵ_3 for the sphere, disk, and side walls, respectively. The sphere and disk were assumed to be isothermal at temperatures T_1 and T_2, respectively. The following expression was proposed:

$$R_r = \frac{1}{A_2 \overline{F}_{12} 4\sigma T_m^3} \tag{3.181}$$

where A_2 is the surface area of the disk, σ is the Stefan-Boltzmann constant, and $T_m = (T_1 + T_2)/2$ is the mean absolute temperature of the contact. The radiative parameter $\overline{F}_{12}$ is defined as

$$\frac{1}{\overline{F}_{12}} = \frac{1 - \epsilon_1}{2\epsilon_1} + \frac{1 - \epsilon_2}{\epsilon_2} + 1.104 \tag{3.182}$$

Joint Resistance Model of Sphere-Flat Contacts.   The joint thermal resistance of a contact formed by elastic bodies, such as a sphere-flat contact is obtained from the model proposed by Kitscha and Yovanovich [46] and Kitscha [47]:

$$\frac{1}{R_j} = \frac{1}{R_c} + \frac{1}{R_g} + \frac{1}{R_r} \tag{3.183}$$

which is a function of the constriction resistance R_c, the gap resistance R_g, and the radiative resistance R_r, which are in parallel. The accuracy of the proposed model was verified by numerous experiments.

Experimental Verification of Elastoconstriction and Elastogap Models.   Experimental data have been obtained for the elastoconstriction resistance of point contacts [47] and line contacts [63] for a range of sphere and cylinder diameters, material properties, and mechanical loads. Data were obtained for the verification of the elastogap model for the point contact [46] and line contact [63]. The elastogap models have been tested with air, argon, helium, and nitrogen as the gap fluid at gas pressures from 10^{-6} torr to atmospheric pressure.

Some representative test data for the elastoconstriction and elastogap resistances compared with the theoretical values are given in the following sections.

Sphere-Flat Test Results.   Kitscha [47] performed experiments on steady heat conduction through 25.4- and 50.8-mm sphere-flat contacts in an air and argon environment at pressures between 10^{-5} torr and atmospheric pressure. He obtained vacuum data for the 25.4-mm-diameter smooth sphere in contact with a polished flat having a surface roughness of approximately 0.13 μm RMS. The mechanical load ranged from 16 to 46 N. The mean contact temperature ranged between 321 and 316 K. The harmonic mean thermal conductivity of the sphere-flat contact was found to be 51.5 W/mK. The emissivities of the sphere and flat were estimated to be $\epsilon_1 = 0.2$ and $\epsilon_2 = 0.8$, respectively.

The contact, gap, radiative, and joint resistances were nondimensionalized as $R^* = Dk_s R$.

The dimensionless radiative resistance for the sphere-flat contact given above becomes

$$R_r^* = \frac{3.82 \times 10^{10}}{T_m^3}$$

The dimensionless constriction resistance is $R_c^* = L$ and the dimensionless joint resistance in a vacuum is

$$\frac{1}{R_j^*} = \frac{1}{L} + \frac{1}{R_r^*} \tag{3.184}$$

The model predictions and the vacuum experimental results are compared in Table 3.16.

TABLE 3.16   Dimensionless Load, Constriction, Radiative, and Joint Resistances [47]

N, newtons	L, D/2a	T_m, K	R_r^*	R_j^*	R_j^* test
16.0	115.1	321	1155	104.7	107.0
22.2	103.2	321	1155	94.7	99.4
55.6	76.0	321	1155	71.3	70.9
87.2	65.4	320	1164	61.9	61.9
195.7	50.0	319	1177	48.0	48.8
266.9	45.1	318	1188	43.4	42.6
467.0	37.4	316	1211	36.4	35.4

The radiative resistance was approximately 10 times the constriction resistance at the lightest load and 30 times at the highest load. The largest difference between the theory and experiments is approximately −4.7 percent, within the probable experimental error. These and other vacuum tests [47] verified the accuracy of the elastoconstriction and the radiation models.

The elastogap model for a point contact was verified by Kitscha [47] and Ogniewicz [159]. For air, the gas parameter M depends on T_m, P_g as follows:

$$M = 1.373 \times 10^{-4} \frac{T_m}{DP_g} \tag{3.185}$$

where D is in cm, T_m in K and P_g in mmHg. The numerical value M is based on air properties at $T_{g,0} = 288$K and $P_{g,0} = 760$ mmHg.

TABLE 3.17 Elastogap Resistance Theory and Air Data [47]: $D = 25.4$ mm, $D/(2a) = 115.1$

T_m, K	P_g, mm Hg	R_g^*	R_r^*	R_j^* theory	R_j^* test
309	400.0	76.9	1293	44.5	46.8
310	100.0	87.4	1280	47.8	49.6
311	40.0	97.1	1268	50.6	52.3
316	4.4	137.2	1209	59.5	59.0
318	1.8	167.2	1186	64.5	65.7
321	0.6	227.9	1153	71.7	73.1
322	0.5	246.6	1143	73.4	74.3
325	0.2	345.4	1111	80.1	80.3
321	Vacuum	∞	1155	104.7	107.0

The elastogap model and the experimental results are compared over a range of gas pressures in Table 3.17. Although tests were conducted at smaller values of the dimensionless parameter L over a range of gas pressures, sphere diameters, and gases, the results given in Table 3.17 are representative of the other data and they also correspond to the case that challenges the validity of the proposed elastoconstriction and elastogap models. First note that the radiative resistance is approximately 10 times the constriction resistance. Second, observe that the gap resistance is approximately ⅔ of the constriction resistance at the highest gas pressure, approximately equal to the constriction resistance at a gas pressure between 4 and 40 mmHg, and finally 3 times the constriction resistance at $P_g = 0.2$ mmHg. The agreement between the theory and the tests is very good to excellent.

The largest difference occurs at the highest gas pressures, where the theory predicts lower joint resistances by approximately 5 percent. The agreement between theory and experiment improves with decreasing gas pressure.

It can also be seen in Table 3.17 that the air within the sphere-flat gap significantly decreases the joint resistance when compared with the vacuum result.

Thermal Contact, Gap, and Joint Conductance Models

Thermal contact, gap, and joint conductance models developed by many researchers over the past five decades are reviewed and summarized in several articles [20, 23, 50, 58, 143, 147, 148] and in the recent text of Madhusudana [59]. The models are, in general, based on the assumption that the contacting surfaces are conforming (or flat) and that the surface asperities have particular height and asperity slope distributions [26, 116, 125]. The models assume either plastic or elastic deformation of the contacting asperities, and require the thermal spreading (constriction) resistance results presented above.

Plastic Contact Conductance Model of Cooper, Mikic, and Yovanovich.   The thermal contact conductance models are based on three fundamental models: (1) the metrology model (surface roughness and asperity slope), (2) the contact mechanics model (deformation of the softer contacting surface asperities), and (3) the thermal constriction (spreading) resistance model for the microcontact areas. Cooper et al. [14] presented the contact conductance model, which is based on the Gaussian distribution of the asperity heights and slopes, the plastic deformation of the contacting asperities, and the constriction resistance, which is based on the isothermal circular contact area on a circular flux tube result. The development of the dimensionless contact conductance model for conforming rough surfaces has been presented in several publications [14, 65, 139, 143, 147, 148]. The theoretical dimensionless contact conductance has the form

$$C_c = \frac{\sigma}{m} \frac{h_c}{k_s} = \frac{1}{2\sqrt{2\pi}} \frac{\exp(-x^2)}{(1-\epsilon)^{1.5}} \tag{3.186}$$

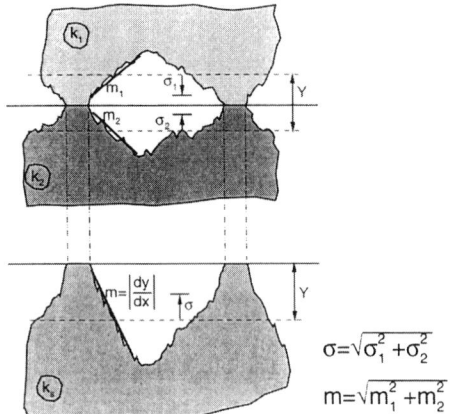

$$\sigma = \sqrt{\sigma_1^2 + \sigma_2^2}$$

$$m = \sqrt{m_1^2 + m_2^2}$$

FIGURE 3.24 Conforming rough surface geometric parameters.

with $x = \mathrm{erfc}^{-1}(2P/H_c)$ and $\epsilon = \sqrt{P/H_c}$, where P is the contact pressure and H_c is the flow pressure at the plastically deformed surface asperities. The surface parameters are $\sigma = \sqrt{\sigma_1^2 + \sigma_2^2}$ and $m = \sqrt{m_1^2 + m_2^2}$, where σ_1 and σ_2 represent the RMS surface roughness of the two contacting surfaces and m_1 and m_2 represent the absolute mean slopes of the surface asperities of the contacting surfaces (Fig. 3.24). The interface effective thermal conductivity is defined as $k_s = 2k_1k_2/(k_1 + k_2)$. The metrology model also gives the following geometric relationships [139–141].

Relative Real Contact Area.

$$\epsilon^2 \equiv \frac{A_r}{A_a} = \frac{1}{2}\,\mathrm{erfc}\,(x)$$

where A_r is the total real contact area and A_a is the corresponding total apparent area.

Contact Spot Density.

$$n = \frac{1}{16}\left(\frac{m}{\sigma}\right)^2 \frac{\exp(-2x^2)}{\mathrm{erfc}\,(x)}$$

Mean Contact Spot Radius.

$$a = \sqrt{\frac{8}{\pi}}\left(\frac{\sigma}{m}\right)\exp(x^2)\,\mathrm{erfc}\,(x)$$

where $x = (1/\sqrt{2})(Y/\sigma)$ and Y is the mean plane separation (Fig. 3.24). The relative mean plane separation is obtained from

$$\frac{Y}{\sigma} = \sqrt{2}\,\mathrm{erfc}^{-1}\left(\frac{2P}{H_c}\right)$$

which is approximated by [139–141]

$$\frac{Y}{\sigma} = 1.184\left[-\ln\left(3.132\,\frac{P}{H_c}\right)\right]^{0.547}$$

which is valid in the ranges $4.75 \geq Y/\sigma \geq 2.0$ and $10^{-6} \leq P/H_c \leq 2 \times 10^{-2}$, and it has a maximum error of approximately 1 percent. The relative mean plane separation appears in the gap conductance model.

Yovanovich [139] proposed the correlation equation for the contact conductance model:

$$C_c = 1.25\left(\frac{P}{H_c}\right)^{0.95} \qquad (3.187)$$

which is valid for the wide range $10^{-6} \leq P/H_c \leq 2 \times 10^{-2}$ and has a maximum error of approximately 1.5 percent.

Yovanovich et al. [140] and Hegazy [35] recognized the importance of the effect of work-hardened layers on the deformation of the contacting asperities. They proposed an iterative procedure to obtain the appropriate value of the contact microhardness H_c from Vickers microhardness H_v measurements [41, 114]. Song and Yovanovich [102] reported the explicit expression for calculating the relative contact pressure:

$$\frac{P}{H_c} = \left[\frac{0.9272P}{c_1(1.62\sigma/m)^{c_2}}\right]^{1/(1+0.071c_2)} \qquad (3.188)$$

where the correlation coefficients c_1 and c_2 are obtained from Vickers microhardness measurements. Sridhar and Yovanovich [108] developed correlation equations for the Vickers coefficients:

$$\frac{c_1}{3178} = [4.0 - 5.77H_B^* + 4.0(H_B^*)^2 - 0.61(H_B^*)^3]$$

and

$$c_2 = -0.370 + 0.442\left(\frac{H_B}{c_1}\right)$$

where H_B is the Brinell hardness [41, 114] and $H_B^* = H_B/3178$. The correlation equations are valid for the Brinell hardness range of 1300–7600 MPa. The above correlation equations were developed for a range of metals: Ni200, SS304, Zr alloys, Ti alloys, and tool steel. Sridhar and Yovanovich [108] also reported a correlation equation that relates the Brinell hardness number to the Rockwell C hardness number:

$$BHN = 43.7 + 10.92HRC - \frac{HRC^2}{5.18} + \frac{HRC^3}{340.26}$$

for the range $20 \leq HRC \leq 65$. It has been demonstrated that the above plastic contact conductance model predicts accurate values of h_c for a range of surface roughness σ/m, a range of metals (SS304, Ni200, Zr alloys, etc.), and a range of relative contact pressure P/H_c [2, 35, 105, 106, 109].

Plastic Contact Conductance Model of Greenwood and Williamson.   Sridhar and Yovanovich [109] developed correlation equations for the contact conductance of conforming rough surfaces based on the Greenwood and Williamson [26] surface model using the plastic deformation model described above. The dimensionless contact conductance correlation is

$$C_c \equiv \frac{\sigma}{m}\frac{h_c}{k_s} = 0.91\alpha^{0.31}\left\{\frac{0.9272P}{c_1[(2.47/\alpha^{0.269})(\sigma/m)]^{c_2}}\right\}^{(0.971\alpha^{1/251.93})/(1+0.038c_2)} \tag{3.189}$$

which is valid in the relative contact pressure range $10^{-5} \leq P/H_c \leq 10^{-2}$. The surface parameter α is the bandwidth parameter, which depends on the variance of surface heights, the variance of surface slopes, and the variance of the second derivative of surface heights. The correlation is valid for the range $5 \leq \alpha \leq 100$. σ and m are the surface roughness and the surface slope as defined previously. The parameters c_1 and c_2 are the Vickers microhardness correlation coefficients, which can be determined by means of the relationships just presented.

Sridhar and Yovanovich [109] found that the Greenwood and Williamson (GW) model [26] and the Cooper, Mikic, and Yovanovich (CMY) model [14] are in very good agreement when $\alpha = 5$. For all values of $\alpha > 5$, the GW model predicts values of C_c that are greater than those predicted by the CMY model.

Elastic Contact Conductance Models of Mikic and Greenwood and Williamson.   Sridhar and Yovanovich [106] reviewed the elastic contact models proposed by Greenwood and Williamson [26] and Mikic [66] and compared the correlation equation with data obtained for five different metals. The models were developed for conforming rough surfaces; they differ in the description of the surface metrology and the contact mechanics. The thermal model developed by Cooper et al. [14] was used. The details of the development of the models and the correlation equations are reviewed by Sridhar and Yovanovich [106]. The correlation equation derived from the Mikic [66] surface and asperity contact models is

$$C_c \equiv \frac{\sigma}{m}\frac{h_c}{k_s} = 1.54\left(\frac{\sqrt{2}P}{mE'}\right)^{0.94} \tag{3.190}$$

where σ and m are the contact surface roughness and slope. The equivalent modulus is defined as $E' = [(1 - v_1^2)/E_1 + (1 - v_2^2)/E_2]^{-1}$ where v_1 and v_2 are the Poisson's ratio, and E_1 and E_2 are the elastic modulus of the contacting asperities.

The correlation equation derived from the Greenwood and Williamson [26] surface and asperity contact models is

$$C_c \equiv \frac{\sigma}{m} \frac{h_c}{k_s} = (1.18 + 0.161 \ln \alpha)\left(\frac{\sqrt{2}P}{mE'}\right)^{0.922\alpha^{1/205.54}} \tag{3.191}$$

where α is the bandwidth parameter discussed above.

The two elastic contact correlation equations [106] were developed for the ranges $10^{-5} \leq \sqrt{2}P/mE' \leq 10^{-2}$ and $5 \leq \alpha \leq 100$. The two correlation equations predict values of C_c that are in close agreement when $\alpha = 5$ and for the largest value of α the difference between the predicted values is approximately 35 percent.

The correlation equation developed from the Mikic [66] models is simpler; therefore it is recommended for predicting contact conductance for elastic contacts.

Elastoplastic Contact Model and Relationships.   Sridhar and Yovanovich [106] developed an elastoplastic contact conductance model that is summarized below in terms of the geometric parameters (1) A_r/A_a, the real to apparent area ratio; (2) n, the contact spot density; (3) a, the mean contact spot radius; and (4) λ, which is the dimensionless mean plane separation:

$$\frac{A_r}{A_a} = \frac{f_{ep}}{2} \, \text{erfc}\,(\lambda/\sqrt{2})$$

$$n = \frac{1}{16}\left(\frac{m}{\sigma}\right)^2 \frac{\exp(-\lambda^2)}{\text{erfc}\,(\lambda/\sqrt{2})}$$

$$a = \sqrt{\frac{8}{\pi}} \cdot \sqrt{f_{ep}} \cdot \frac{\sigma}{m} \exp(\lambda^2/2)\,\text{erfc}\,(\lambda/\sqrt{2})$$

$$h_c = \frac{k_s}{2\sqrt{2\pi}} \frac{m}{\sigma} \frac{\sqrt{f_{ep}} \cdot \exp(-\lambda^2/2)}{[1 - \sqrt{(f_{ep}/2)}\,\text{erfc}\,(\lambda/\sqrt{2})]^{1.5}}$$

$$\lambda \equiv \frac{Y}{\sigma} = \sqrt{2}\,\text{erfc}^{-1}\left(\frac{1}{f_{ep}} \cdot \frac{2P}{H_{ep}}\right)$$

The elastoplastic parameter f_{ep} is a function of the dimensionless contact strain ϵ_c^*, which depends on the amount of work hardening. This physical parameter lies in the range $0.5 \leq f_{ep} \leq 1.0$. The smallest and largest values correspond to zero and infinitely large contact strain, respectively. The dimensionless contact strain is defined as

$$\epsilon_c^* = 1.67\left(\frac{mE'}{S_f}\right)$$

where S_f is the material yield or flow stress [41], which is a complex physical parameter that must be determined by experiment for each metal. The elastoplastic microhardness H_{ep} can be determined by means of an iterative procedure that requires the following relationship:

$$H_{ep} = \frac{2.76S_f}{[1 + (6.5/\epsilon_c^*)^2]^{1/2}} \tag{3.192}$$

The proposed elastoplastic contact conductance model moves smoothly between the elastic contact model of Mikic [66] and the plastic contact conductance model of Cooper, Mikic, and

Yovanovich [14], which was modified by Yovanovich and co-workers [102, 139, 140] to include the effect of work-hardened layers on the deformation of the contacting asperities.

Gap Conductance Model and Integral

The gap conductance model for conforming rough surfaces was developed, modified, and verified by Yovanovich and co-workers [35, 73, 100–104]. The gap contact model is based on surfaces having Gaussian height distributions. It also accounts for the mechanical deformation of the contacting surface asperities. The development of the gap conductance model appears in several papers [139, 143, 147].

The gap conductance model is expressed in terms of an integral:

$$h_g = \frac{k_g}{\sigma} \frac{1}{\sqrt{2\pi}} \int_0^\infty \frac{\exp[-(Y/\sigma - u)^2/2]}{u + M/\sigma}\, du \qquad (3.193)$$

where k_g is the thermal conductivity of the gas trapped in the gap, σ is the effective surface roughness of the interface, and $u = t/\sigma$ is the dimensionless local gap thickness. The integral depends on two parameters: Y/σ, which is the mean plane separation, and M/σ, which is the relative gas rarefaction parameter defined and discussed previously. The gas rarefaction parameter $M = \alpha\beta\Lambda$ has also been defined and discussed previously.

Accommodation Coefficients.   The accommodation coefficient α accounts for the efficiency of gas-surface energy exchange. There is a large body of research dealing with experimental and theoretical aspects of α for different gases in contact with metallic surfaces under various surface conditions and temperatures [32, 92, 120, 126]. Song and Yovanovich [100] and Song et al. [101, 104] examined the gap conductance models available in the literature, the experimental data, and the models for the accommodation coefficients. Song and Yovanovich [100] developed a correlation for the accommodation coefficient for "engineering" surfaces (i.e., surfaces with adsorbed layers of gases and oxides). They proposed a correlation based on experimental results of numerous investigators for monatomic gases. The relation was extended by the introduction of a "monatomic equivalent molecular weight" to diatomic and polyatomic gases. The correlation equation has the form

$$\alpha = \exp(C_0 T)[M_g/(C_1 + M_g) + \{1 - \exp(C_0 T)\}\{2.4\mu/(1 + \mu)^2\}]$$

where $C_0 = -0.57$
$T = (T_s - T_0)/T_0$
$M_g = M_g$ for monatomic gases
$\quad\; = 1.4\, M_g$ for diatomic and polyatomic gases
$C_1 = 6.8$, units of M_g [g/mole]
$\mu = M_g/M_s$

where T_s and T_0 are the absolute temperatures of the surface and the gas, and M_g and M_s are the molecular weights of the gas and the solid, respectively. The agreement between the predictions according to this correlation and the published data for diatomic and polyatomic gases was within ±25 percent.

Wesley and Yovanovich [127] compared the predictions of the proposed gap conductance model and experimental measurements of gaseous gap conductance between the fuel and clad of a nuclear fuel rod. The agreement was very good and the model was recommended for fuel pin analysis codes.

Gap Conductance Correlation Equations.   Although the gap integral can be computed accurately and easily by means of computer algebra systems, Negus and Yovanovich [73] proposed the following correlation equations for the gap integral:

$$I_g = \frac{f_g}{(Y/\sigma) + (M/\sigma)}$$

with $f_g = 1.063 + 0.0471(4 - Y/\sigma)^{1.68}[\ln(\sigma/M)]^{0.84}$ for $2 \le Y/\sigma \le 4$ and $0.01 \le M/\sigma \le 1$

and $f_g = 1 + 0.06(\sigma/M)^{0.8}$ for $2 \le Y/\sigma \le 4$ and $1 \le M/\sigma \le \infty$

The correlations have a maximum error of approximately 2 percent.

ACKNOWLEDGMENTS

The author is grateful to the Natural Science and Engineering Research Council for its continued financial support of the research for the preparation of this chapter. Thanks to P. Teertstra, J. R. Culham, Y. S. Muzychka and M. Stevanovic for help with the figures and other assistance.

NOMENCLATURE

Symbol, Definition, SI Units

A, A_c, A_t	area, contact, and flux tube areas: m^2
A_a, A_r	apparent and real contact areas, $A_r/A_a < 1$: m^2
A_s, A_N	shape factor parameters
A_n, B_n	Fourier coefficients for temperature and heat loss
AR	aspect ratio of bodies
$\mathcal{A}(\beta, \gamma)$	dimensionless ellipsoid area function $= A/(2\pi a^2)$
a	semimajor axis of ellipse, hyperellipse, rectangle: m
a	point contact radius, line contact half-width: m
a, b, c	semiaxes of ellipsoid: m
a, b	radii of bisphere: m
a_1, a_2, a_3, a_4, a_5	correlation coefficients
$B(x, y)$	beta function
B	layer thickness
Bi	Biot number $= h\mathcal{L}/k$ (typical)
$\text{Bi}_{\sqrt{A}}$	Biot number based on square root of area
BHN	Brinell hardness number
C	capacitance: coulomb/volt
C^*	capacity: m
C_c	dimensionless contact conductance $= \sigma h_c/(m k_s)$
C_p, C_v	specific heat at constant pressure and volume: J/(kg·°C)
C_0, C_1, C_3, C_5, C_7	correlation coefficients
C_{CT}, C_{ST}	correlation coefficients for circular and square toroids
C_1, C_2, C_3	correlation coefficients
c	specific heat: J/(kg·°C)
c_1, c_2	Vickers microhardness correlation coefficients

D	diameter of cylinders, spheres, and toroids: m
D_i, D_o	inner and outer diameters of circular and square toroids
D_{GM}	geometric mean diameter: m
d	diameter of cylinders and toroids: m
E	modulus of elasticity: N/m^2
$E(\kappa)$	complete elliptic integral of the second kind
$E(\phi, \kappa)$	incomplete elliptic integral of second kind
E'	equivalent modulus $= [(1 - v_1^2)/E_1 + (1 - v_2^2)/E_2]^{-1}$
erf, erfc	error and complementary error functions
erfc^{-1}	inverse complementary error function
F	contact load: N
$F*$	dimensionless line contact load $= F\Delta/(DL)$
$F(\phi, \kappa)$	incomplete elliptic integral of the first kind
Fo	Fourier number $= \alpha t/\mathscr{L}^2$ (typical)
Fo$_{\sqrt{A}}$	Fourier number based on square root of area
Fo$_c$	critical value of Fourier number
$\overline{F}_{12}$	radiative parameter for point contact
f_{ep}	elastoplastic contact parameter
f_g	gap conductance correlation equation
$g_i, \sqrt{g}$	metric coefficients, jacobian
H	height of single and double cones
HRC	Rockwell C hardness number
H_B	Brinell hardness
H_c	surface contact microhardness: MPa
H_{ep}	elastoplastic microhardness: MPa
H_V	Vickers microhardness: MPa
h, h_c, h_g, h_j	heat transfer coefficient and contact, gap and joint conductances: W/(m$^2\cdot$K)
$I(\beta, \gamma)$	shape factor integral
I_n	nth-order modified Bessel function of first kind
I_g	gap conductance integral $= f_g/(Y/\sigma + M/\sigma)$
$I_{g,p}$	point contact gap integral
$I_{g,l}$	line contact gap integral
i^nerfc	integrals of complementary error function
J_n	nth-order Bessel function of first kind
$K(\kappa)$	complete elliptic integral of first kind of modulus κ
K_n	nth-order modified Bessel function of first kind
k	thermal conductivity: W/(m$\cdot$K)
k_1, k_2	layer and substrate conductivities: W/(m$\cdot$K)
k_a	conductivity at average temperature: W/(m$\cdot$K)
k_0	conductivity at reference temperature: W/(m$\cdot$K)
k_s	harmonic mean conductivity $= 2k_1k_2/(k_1 + k_2)$: W/(m$\cdot$K)
L	length of circular and square cylinders and rectangular plate: m

L	dimensionless point contact parameter: $D/(2a)$
L_{cyl}	cylinder length: m
L_1, L_2, L_3	side dimensions of cuboid: m
$\mathscr{L}$	arbitrary length scale: m
M	gap gas parameter $= \alpha\beta\Lambda$: m
M_g, M_s	gas and solid molecular weights: g/mole
m	effective surface asperity slope $= \sqrt{m_1^2 + m_2^2}$
m_1, m_2	surface mean asperity slope of contacting rough surfaces
N	number sides of regular polygons
n	outward directed normal, shape parameter
n	interpolation parameter
n	contact spot density: m^{-2}
O	area
P, P_i, P_0	perimeter, inner, and outer perimeters: m
P	contact pressure: MPa
P_g	gas pressure: kPa
$P_{g,0}$	reference gas pressure: kPa
Pr	Prandtl number $= \mu c/k$
$P_{n-1/2}(\zeta)$	toroidal function
p	correlation coefficient
Q	heat flow rate: W
$Q_{\mathscr{L}}^*$	dimensionless heat flow rate $= (Q\mathscr{L})/(kA\theta_0)$ (typical)
Q_i	initial internal energy $= \rho c V\theta_i$ (typical): J
Q_e	total electrical charge
$Q_{n-1/2}(\zeta)$	toroidal function
q	heat flux: W/m^2
q_0	surface heat flux: W/m^2
R	thermal resistance, spreading resistance $= (\overline{T}_{source} - \overline{T}_{sink})/Q$ (typical): K/W
R_0	spreading resistanced based on centroid temperature: K/W
R_c, R_s	constriction and spreading resistances: K/W
R_n	gap resistance: K/W
R_j	joint or overall resistance $= (1/R_c + 1/R_g + 1/R_r)^{-1}$: K/W
R_r	radiation resistance: K/W
R^*	dimensionless resistance $= k\mathscr{L}R$ (typical)
R_s^*	dimensionless spreading resistance $= 4k\mathscr{L}R_s$ (typical)
R^*	dimensionless contact, gap, or radiation resistance $= k_s DR$ (typical)
R_c^*	dimensionless point contact resistance $= L$
R_g^*	dimensionless gap resistance
R_r^*	dimensionless radiative resistance $= (3.82 \times 10^{10})/T_m^3$
R_j^*	dimensionless joint resistance $= (1/L + 1/R_r^*)^{-1}$
r	radius, polar, and spherical coordinates: m
$\overrightarrow{r}$	radius vector: m

r_0	radius for hyperellipse: m
S	shape factor: m
$S^*_{\sqrt{A}}$	dimensionless shape factor
S_f	material yield or flow stress: MPa
S_0, S_∞	shape factors for thin and thick conduction layers: m
$S^*_{\mathscr{L}}$	dimensionless shape factor $= (S\mathscr{L})/A$
S	active surface area: m^2
s	dimensionless parameter $= 1/(1 + pz)$
s	dummy variable
$T, \overline{T}$	temperature, area average temperature: K
T_f	fluid temperature: K
T_i	initial solid temperature: K
T_m	mean temperature $= (T_1 + T_2)/2$: K
$T_{g,0}$	reference gas temperature: K
t	time: s
t	variable gap thickness: m
t_1, t_2	coating or layer thicknesses: m
u	dimensionless position variable
u	dimensionless variable gap thickness $= t/\sigma$
u_i	curvilinear coordinates
V	volume: m^3
V_e	electrical potential: V
W	width of annular area: m
$\mathscr{W}$	width of rectangular plate: m
x, y, z	cartesian coordinates: m
x	dimensionless parameter for single and double cones
x	dimensionless mean plane separation $= (1/\sqrt{2})(Y/\sigma)$
Y	mean plane separation $= \sigma\sqrt{2}\ \mathrm{erfc}^{-1}(2P/H_c)$: m
Y_n	nth-order Bessel function of second kind
z	dimensionless parameter $= (\mathrm{Bi} + 1)\sqrt{\mathrm{Fo}}$
z	dimensionless time parameter $= 1/(2\sqrt{\pi}\sqrt{\mathrm{Fo}_{\sqrt{A}}})$

Greek Symbols

α	thermal diffusivity $= k/(\rho c)$: m^2/s
α	accommodation coefficient $= (2 - \alpha_1)/\alpha_1 + (2 - \alpha_2)/\alpha_2$
α_1, α_2	accommodation coefficients for gap
α	bandwith surface roughness parameter $5 \le \alpha \le 100$
β	gas parameter: $= (2\gamma)/[(\gamma + 1)/\mathrm{Pr}]$
β	dimensionless parameter in ellipsoidal integral $= b/a$
β	thermal conductivity temperature coefficient
β_1, β_2	integration limits
β	dimensionless parameter in ellipsoidal integral $= b/a$

β	dimensionless parameter in ellipsoidal integral $= b/a$
$\Gamma(x)$	gamma function
γ	dimensionless parameter in ellipsoidal integral $= c/a$
γ	specific heat ratio $= C_p/C_v$
γ	aspect ratio parameter $= b/a \leq 1$
Δ	diffusion thickness $= A/S$: m
Δ	elastic parameter $= [(1 - v_1^2)/E_1 + (1 - v_2^2)/E_2]/2$: $(\text{MPa})^{-1}$
δ_n	nth eigenvalue
δ_1	first eigenvalue
$\delta_{1,0}$	first eigenvalue for Bi $\rightarrow 0$
$\delta_{1,\infty}$	first eigenvalue for Bi $\rightarrow \infty$
δ_c	approximate first eigenvalue $= \pi + 1/(\sqrt{\pi}\epsilon)$
ϵ	permittivity of space: farad
ϵ	relative size $= a/b$ (typical)
ϵ_c	contact strain
ϵ_c^*	dimensionless contact strain $= 1.67(mE'/S_f)$
ζ	arguments of toroidal functions $= D/d$
ζ	dimensionless position $= x/L$ (typical)
ζ	dummy variable
η	elliptic cylinder, bicylinder, oblate and prolate spheroidal coordinate
η_1, η_2	integration limits
$\eta_{\min}, \eta_{\max}$	minimum and maximum integration limits
θ	steady or transient temperature excess $= [T(\vec{r}, t) - T_\infty]$ (typical): K
θ_f	maximum system temperature excess $= T_f - T_i$ (typical): K
θ_i	initial temperature excess $= T_f - T_i$ (typical): K
θ	Kirchhoff temperature $= 1/k_0 \int_0^{T_0} k \, dT$ (typical): K
θ	spherical, oblate, and prolate spheroidal coordinate
$\vartheta^{\pm}$	multiple layer thickness and conductivity parameter
κ	modulus of complete and incomplete elliptic integrals of first and second kind
κ	thermal conductivity ratio $= k_1/k_2$ (typical)
κ_{21}	conductivity ratio $= k_2/k_1$
κ_{32}	conductivity ratio $= k_3/k_2$
Λ	molecular mean free path: m
$\Lambda_{g,0}$	molecular mean free path at reference conditions: m
λ	relative gap thickness $= Y/\sigma$
λ_1	conductivity ratio parameter $= (1 - k_2/k_1)/2$
λ_2	conductivity ratio parameter $= (1 + k_2/k_1)/2$
μ	heat flux distribution parameter $q^* = (1 - u^2)^{\mu}$
μ	order of Bessel function of the first kind $J_{\mu + (1/2)}(n\pi\epsilon)$
μ	ratio of molecular weights $= M_g/M_s$
v_1, v_2	Poisson's ratios

v	elliptic cylinder, bicylinder, oblate and prolate spheroidal coordinate
v	dummy variable in ellipsoidal integral
ρ	mass density: kg/m^3
ρ	dimensionless position in cylindrical and spherical coordinates $= r/a$ (typical)
$\rho_{n,\epsilon}$	heat flux parameter $= 1$, isothermal; $= \sin(\delta_n\epsilon)/[2J_1(\delta_n\epsilon)]$, isoflux
σ_1, σ_2	RMS surface rough of contacting surfaces: m
σ	effective RMS surface rough of interface: m
τ	relative layer thickness $= t/a$ (typical)
υ	
ϕ	dimensionless temperature $= [T(\vec{r}) - T_\infty]/(T_0 - T_\infty)$
ϕ_e	nondimensional electrical potential
ϕ_x, ϕ_r	dimensionless temperatures for plate and cylinder
ϕ_c	approximation function for compound cylinder
ϕ_n	compound cylinder function
φ_n	compound cylinder function
ψ	dimensionless spreading resistance $= 4k\mathcal{L}R$ (typical)
ψ	dimensionless spreading resistance $= \sqrt{\pi}kaR_c$
ψ_{ave}	dimensionless spreading resistance $= (1-\epsilon)^{3/2}k\phi_c/2$
ψ_{max}	dimensionless spreading resistance $= (1-\epsilon)k\phi_c/\sqrt{\pi}$
ψ_0	dimensionless centroid based spreading resistance $= k\sqrt{A}R_0$
ω	angle

Subscripts

$\sqrt{A}$	scale length is square root of the area
a	apparent
ave	based on average temperature
c	contact, constriction, critical
cyl	cylinder
contact plane	apparent
CT, ST	circular and square toroids
e	electrical
ep	elastoplastic
f	fluid
f	flow stress
D	Dirichlet condition solution
GM	geometric mean
g	gap, gas
$g,0$	at reference gas temperature and pressure
g,l	gap for line contact
g,p	gap for point contact
i	initial value
i, o	inner and outer

i	inner body volume and area
j	joint
L	longest straight line through inner body
l	line contact
m	mean value
max	based on centroid temperature
min	based on area average temperature
N	Neumann condition solution
n	nth
p	point contact
p	constant pressure condition
R	Robin condition solution
r	radiative
rect	rectangular area
s	spreading
s	based on harmonic mean value
source	based on source temperature
sink	based on sink temperature
thick	thick layer
thin	thin layer
v	constant volume condition
r, ψ, z	cylindrical coordinates
r, ψ, θ	spherical coordinates
η, ψ, z	elliptical cylinder coordinates
η, ψ, z	bicylinder coordinates
η, θ, ψ	oblate and spheroidal coordinates
0	zero thickness limit
0	based on centroid temperature
0, 1	zeroeth order, first order
$0, \infty$	value on the surface and at infinity
∞	infinite thickness limit
1, 0	first eigenvalue value at zero Biot number limit
$1, \infty$	first eigenvalue value at infinite Biot number limit
1, 2	solids 1 and 2; surfaces 1 and 2
1, 2, 3	cuboid side dimensions
12	net radiative transfer
$1D$	one-dimensional conduction

Superscripts

*	dimensionless value
'	effective value of parameter
$\pm$	parameter with positive or negative terms

Overscores

–	mean value

REFERENCES

1. M. Abramowitz and I. A. Stegun, *Handbook of Mathematical Functions,* Dover, New York, 1965.
2. V. W. Antonetti, "On the Use of Metallic Coatings to Enhance Thermal Contact Conductance," PhD thesis, University of Waterloo, Waterloo, Ontario, Canada, 1983.
3. V. W. Antonetti and M. M. Yovanovich, "Enhancement of Thermal Contact Conductance By Metallic Coatings: Theory and Experiments," *Journal of Heat Transfer* (107): 513–519, 1985.
4. V. S. Arpaci, *Conduction Heat Transfer,* Addison-Wesley, Reading, MA, 1966.
5. M. J. Balcerzak and S. Raynor, "Steady State Temperature Distribution and Heat Flow in Prismatic Bars with Isothermal Boundary Conditions," *Int. J. Heat Mass Transfer,* Vol. 3, pp. 113–125, 1961.
6. J. V. Beck, "Average Transient Temperature Within A Body Heated By A Disk Heat Source," in *Heat Transfer, Thermal Control, and Heat Pipes, Progress in Aeronautics and Astronautics,* Vol. 70, pp. 3–24, AIAA, New York, 1979.
7. J. H. Blackwell, "Transient Heat Flow from a Thin Circular Disk Small-Time Solution," *Journal of the Australian Mathematical Society,* Vol. XIV, pp. 433–442, 1972.
8. F. Bowman, *Introduction to Elliptic Integrals with Applications,* Dover Publications, New York, 1961.
9. A. W. Bush, R. D. Gibson, and T. R. Thomas, "The Elastic Contact of a Rough Surface," *Wear,* Vol. 35, pp. 87–111, 1975.
10. P. F. Byrd and M. D. Friedman, *Handbook of Elliptic Integrals for Engineers and Scientists,* 2d ed., Springer-Verlag, New York, 1971.
11. H. S. Carslaw and J. C. Jaeger, *Conduction of Heat in Solids,* 2d ed., Oxford Press, London, 1959.
12. R. Y. Chen and T. L. Kuo, "Closed Form Solutions for Constant Temperature Heating in Solids," *Mech. Eng. News* (16/1): 20, 1979.
13. S. W. Churchill and R. Usagi, "A General Expression for the Correlation of Rates of Transfer and Other Phenomena," *AIChE J.,* Vol. 18, pp. 1121–1132, 1972.
14. M. G. Cooper, B. B. Mikic, and M. M. Yovanovich, "Thermal Contact Conductance," *Int. J. Heat Mass Transfer,* Vol. 12, pp. 279–300, 1969.
15. Y. L. Chow and M. M. Yovanovich, "The Shape Factor of the Capacitance of a Conductor," *J. Appl. Phys.* (53/12): 8470–8475, 1982.
16. J. R. Dryden, "The Effect of a Surface Coating on the Constriction Resistance of a Spot on an Infinite Half-Plane," *ASME Journal of Heat Transfer,* Vol. 105, pp. 408–410, 1983.
17. J. R. Dryden, M. M. Yovanovich, and A. S. Deakin, "The Effect of Coatings on the Steady-State and Short Time Constriction Resistance for an Arbitrary Axisymmetric Flux," *ASME Journal of Heat Transfer* (107): 33–38, 1985.
18. E. R. G. Eckert and R. M. Drake, *Analysis of Heat and Mass Transfer,* McGraw-Hill, New York, 1972.
19. N. J. Fisher and M. M. Yovanovich, "Thermal Constriction Resistance of Sphere/Layered Flat Contacts: Theory and Experiments," *ASME Journal of Heat Transfer,* Vol. 111, pp. 249–256, 1989.
20. L. S. Fletcher, "Recent Developments in Contact Conductance Heat Transfer," *Trans. ASME, J. Heat Transfer,* Vol. 110, pp. 1059–1070, 1988.
21. L. S. Fletcher, "A Review of Thermal Enhancement Techniques for Electronic Systems," *IEEE Trans. Components, Hybrids, Manufacturing Technology* (13/4): 1012–1021, 1990.
22. S. Flugge, *Handbuch der Physic,* Vol. XVI, pp. 145–147, Springer, Berlin, 1958.
23. B. Gebhart, *Heat Conduction and Mass Diffusion,* McGraw-Hill, New York, 1993.
24. R. D. Gibson, "The Contact Resistance for a Semi-Infinite Cylinder in a Vacuum," *Appl. Energy,* Vol. 2, pp. 57–65, 1976.
25. P. R. Greene, "A Useful Approximation to the Error Function: Applications to Mass, Momentum, and Energy Transport in Shear Layers," *J. Fluids Engineering,* Vol. 111, pp. 224–226, 1989.
26. J. A. Greenwood and J. B. Williamson, "Contact of Nominally Flat Surfaces," *Proc. Roy. Soc. Lond. A*(295): 300–319, 1966.
27. D. Greenspan, "Resolution of Classical Capacity Problems by Means of a Digital Computer," *Canadian Journal of Physics,* Vol. 44, pp. 2605–2614, 1966.

28. U. Grigull and H. Sandner, *Heat Conduction,* Springer-Verlag, New York, 1984.

29. H. Grober, S. Erk, and U. Grigull, *Fundamentals of Heat Transfer,* McGraw-Hill, New York, 1961.

30. E. Hahne and U. Grigull, "A Shape Factor Scheme for Point Source Configurations," *Int. J. Heat Mass Transfer,* Vol. 17, pp. 267–272, 1974.

31. E. Hahne and U. Grigull, "Formfaktor und Formwiderstand der Stationaren Mehrdimensionalen Warmeleitung," *Int. J. Heat Mass Transfer,* Vol. 18, pp. 751–766, 1975.

32. J. P. Hartnett, "A Survey of Thermal Accommodation Coefficients," *Rarefied Gas Dynamics,* L. Talbot ed., pp. 1–28, Academic Press, New York, 1961.

33. A. V. Hassani and K. G. T. Hollands, "Conduction Shape Factor for a Region of Uniform Thickness Surrounding a Three-Dimensional Body of Arbitrary Shape," *Journal of Heat Transfer,* Vol. 112, pp. 492–495, 1990.

34. A. V. Hassani, K. G. T. Hollands, and G. D. Raithby, "A Close Upper Bound for the Conduction Shape Factor of a Uniform Thickness, 2D Layer," *Int. J. Heat Mass Transfer* (36/12): pp. 3155–3158, 1993.

35. A. A. Hegazy, "Thermal Joint Conductance of Conforming Rough Surfaces," PhD thesis, University of Waterloo, Waterloo, Ontario, Canada, 1985.

36. M. P. Heisler, "Temperature Charts for Induction Heating and Constant-Temperature," *ASME,* Vol. 69, pp. 227–236, 1947.

37. H. R. Hertz, *Miscellaneous Papers,* English Translation, Macmillan and Co., London, 1896.

38. J. M. Hill and J. N. Dewynne, *Heat Conduction,* Blackwell Scientific Publications, Palo Alto, CA, 1987.

39. P. Hui and H. S. Tan, "Temperature Distributions in a Heat Dissipation System Using a Cylindrical Diamond Heat Spreader on a Copper Sink," *J. Appl. Phys.* (75/2): 748–757, 1994.

40. J. Jeans, *The Mathematical Theory of Electricity and Magnetism,* pp. 244–249, Cambridge University Press, Cambridge, UK, 1963.

41. K. L. Johnson, *Contact Mechanics,* Cambridge University Press, Cambridge, UK, 1985.

42. D. P. Kennedy, "Spreading Resistance in Cylindrical Semiconductor Devices," *Journal of Applied Physics* (31/8): 1490–1497, 1960.

43. O. D. Kellogg, *Foundations of Potential Theory,* pp. 188–189, Dover Publications, New York, 1953.

44. N. R. Keltner, "Transient Heat Flow in Half-Space Due to an Isothermal Disk on the Surface," *Journal of Heat Transfer* (95): 412–414, 1973.

45. N. R. Keltner, B. L. Bainbridge, and J. V. Beck, "Rectangular Heat Source on a Semi-Infinite Solid— An Analysis for a Thin Film Flux Calibration," 84-HT-46, 1984.

46. W. W. Kitscha and M. M. Yovanovich, "Experimental Investigation on the Overall Thermal Resistance of Sphere-Flat Contacts," in *Heat Transfer With Thermal Control Applications, AIAA Prog. in Astronautics and Aeronautics,* Vol. 39, pp. 93–110, MIT Press, Cambridge, MA, 1975.

47. W. W. Kitscha, "Thermal Resistance of Sphere-Flat Contacts," MASc thesis, Department of Mechanical Engineering, University of Waterloo, Waterloo, Ontario, Canada, 1982.

48. A. D. Kraus and A. Bar-Cohen, *Thermal Analysis and Control of Electronic Equipment,* pp. 199–214, McGraw-Hill, New York, 1983.

49. S. S. Kutateladze, *Fundamentals of Heat Transfer,* Edward Arnold Ltd., London, 1963.

50. M. A. Lambert and L. S. Fletcher, "A Review of Models for Thermal Contact Conductance of Metals," *AIAA-96-0239, 34th Aerospace Sciences Meeting and Exhibit,* Reno, NV, January 15–18, 1996.

51. L. S. Langston, "Heat Transfer from Multidimensional Objects Using One-Dimensional Solutions for Heat Loss," *Int. J. Heat Mass Transfer,* Vol. 25, pp. 149–150, 1982.

52. P. A. A. Laura and E. A. Susemihl, "Determination of Heat Flow Shape Factors for Hollow, Regular Polygonal Prisms," *Nuclear Engineering and Design,* Vol. 25, pp. 409–412, 1973.

53. P. A. A. Laura and G. Sanchez Sarmiento, "Heat Flow Shape Factors for Circular Rods with Regular Polygonal Concentric Inner Bore," *Nuclear Engineering and Design* (47): 227–229, 1978.

54. P. A. A. Laura and G. Sanchez Sarmiento, "Analytical Determination of Heat Flow Shape Factors for Composite, Prismatic Bars of Doubly-Connected Cross Section," *Nuclear Engineering and Design,* Vol. 50, pp. 397–385, 1978.

55. G. K. Lewis, "Shape Factors in Conduction Heat Flow for Circular Bars and Slabs with Internal Geometries," *Int. J. Heat Mass Transfer,* Vol. 11, pp. 985–992, 1968.

56. A. V. Luikov, *Analytical Heat Diffusion Theory,* Academic Press, New York, 1968.

57. C. V. Madhusudana, "The Effect of Interface Fluid on Thermal Contact Conductance," *Int. J. Heat Mass Transfer,* Vol. 18, pp. 989–991, 1975.

58. C. V. Madhusudana and L. S. Fletcher, "Contact Heat Transfer—The Last Decade," *AIAA Journal* (24/3): 510–523, 1986.

59. C. V. Madhusudana, *Thermal Contact Conductance,* Springer, New York, 1996.

60. W. Magnus, F. Oberhettinger, and R. P. Soni, *Formulas and Theorems for Special Functions of Mathematical Physics,* Springer-Verlag, New York, 1966.

61. K. A. Martin, M. M. Yovanovich, and Y. L. Chow, "Method of Moments Formulation of Thermal Constriction Resistance of Arbitrary Contacts," *AIAA-84-1745, AIAA 19th Thermophysics Conference,* Snowmass, CO, June 25–28, 1984.

62. M. Mason and W. Weaver, *The Electromagnetic Field,* pp. 130–131, Dover Publications, New York, 1929.

63. G. R. McGee, M. H. Schankula, and M. M. Yovanovich, "Thermal Resistance of Cylinder-Flat Contacts: Theoretical Analysis and Experimental Verification of a Line-Contact Model," *Nuclear Engineering and Design,* Vol. 86, pp. 369–381, 1985.

64. T. H. McWaid and E. Marschall, "Applications of the Modified Greenwood and Williamson Contact Model for Prediction of Thermal Contact Resistance," *Wear,* Vol. 152, pp. 263–277, 1992.

65. B. B. Mikic and W. M. Rohsenow, *Thermal Contact Resistance,* Mechanical Engineering Report No. DSR 74542-41, MIT, Cambridge, MA, 1966.

66. B. B. Mikic, "Thermal Contact Conductance; Theoretical Considerations," *Int. J. Heat Mass Transfer,* Vol. 17, pp. 205–214, 1974.

67. P. Moon and D. E. Spencer, *Field Theory Handbook,* Springer-Verlag, Berlin, 1971.

68. P. M. Morse and H. Feshbach, *Methods of Theoretical Physics,* part I, McGraw-Hill, New York, 1953.

69. P. M. Morse and H. Feshbach, *Methods of Theoretical Physics,* part II, p. 1308, McGraw-Hill, New York, 1953.

70. K. J. Negus and M. M. Yovanovich, "Constriction Resistance of Circular Flux Tubes With Mixed Boundary Conditions by Linear Superposition of Neumann Solutions," ASME 84-HT-84, 1984.

71. K. J. Negus and M. M. Yovanovich, "Application of the Method of Optimized Images to Steady Three-Dimensional Conduction Problems," ASME 84-WA/HT-110, 1984.

72. K. J. Negus, M. M. Yovanovich, and J. C. Thompson, "Thermal Constriction Resistance of Circular Contacts on Coated Surfaces: Effect of Contact Boundary Condition," *AIAA-85-1014, AIAA 20th Thermophysics Conference,* Williamsburg, VA, June 19–21, 1985.

73. K. J. Negus and M. M. Yovanovich, "Correlation of Gap Conductance Integral for Conforming Rough Surfaces," *Journal of Thermophysics and Heat Transfer* (2): 279–281, 1988.

74. K. J. Negus, M. M. Yovanovich, and J. V. Beck, "On the Nondimensionalization of Constriction Resistance for Semi-Infinite Heat Flux Tubes," *Journal of Heat Transfer* (111): 804–807, 1989.

75. K. J. Negus and M. M. Yovanovich, "Transient Temperature Rise At Surfaces Due To Arbitrary Contacts On Half-Spaces," *CSME Trans.* (13, 1/2): 1–9, 1989.

76. K. M. Nho and M. M. Yovanovich, "Effect of Oxide Layers on Measured and Theoretical Contact Conductances in Finned-Tube Heat Exchangers," in *Compact Heat Exchangers, A Festschrift for A.L. London,* Hemisphere, New York, pp. 397–420, 1990.

77. E. J. Normington and J. H. Blackwell, "Transient Heat Flow From Constant Temperature Spheroids and the Thin Circular Disk," *Quarterly Journal of Mechanics and Applied Mathematics,* Vol. 17, pp. 65–72, 1964.

78. E. J. Normington and J. H. Blackwell, "Transient Heat Flow From a Thin Circular Disk-Small Time Solution," *Journal of the Australian Mathematical Society,* Vol. 14, pp. 433–442, 1972.

79. R. A. Onions and J. F. Archard, "The Contact of Surfaces Having a Random Structure," *J. Phys., D: Appl. Phys.,* Vol. 6, pp. 289–304, 1973.

80. M. N. Ozisik, *Heat Conduction,* Wiley, New York, 1980.

81. G. Polya and G. Szego, *Isoperimetric Inequalities in Mathematical Physics,* Princeton University Press, Princeton, NJ, 1951.

82. M. L. Ramachandra Murthy and A. Ramachandran, "Shape Factors in Conduction Heat Transfer," *British Chemical Engineering Design* (12/5): 730–731, 1967.

83. G. D. Raithby and K. G. T. Hollands, "A General Method of Obtaining Approximate Solutions to Laminar and Turbulent Free Convection Problems," in *Advances in Heat Transfer,* Vol. 11, pp. 265–315, Academic Press, 1975.

84. L. C. Roess, *Theory of Spreading Conductance,* appendix A, Beacon Laboratories of Texas Company, Beacon, NY (unpublished report), 1950.

85. R. S. Sayles and T. R. Thomas, "Thermal Conductance of a Rough Elastic Contact," *Applied Energy,* Vol. 2, pp. 249–267, 1976.

86. M. H. Schankula, D. W. Patterson, and M. M. Yovanovich, "The Effect of Oxide Films on the Thermal Resistance Between Contacting Zirconium Alloys," *Materials in Nuclear Energy, Proceedings of Conference of the American Society for Metals,* Huntsville, Ontario, Canada, pp. 106–111, 1983.

87. P. J. Schneider, *Conduction Heat Transfer,* Addison-Wesley, Reading, MA, 1957.

88. P. J. Schneider, *Temperature Response Charts,* Wiley, New York, 1963.

89. G. E. Schneider, "Thermal Resistance Due to Arbitrary Dirichlet Contacts on a Half-Space," *Progress in Astronautics and Aeronautics* (65): pp. 103–119, 1978.

90. R. U. Sexl and D. G. Burkhard, "An Exact Solution for Thermal Conduction Through a Two-Dimensional Eccentric Constriction," *Progress in Astronautics and Aeronautics* (21): 617–620, New York, 1969.

91. L. M. Simeza and M. M. Yovanovich, "Shape Factors for Hollow Prismatic Cylinders Bounded by Isothermal Inner Circles and Outer Regular Polygons," *Int. J. Heat Mass Transfer* (30): 812–816, 1987.

92. Y. G. Semyonov, S. E. Borisov, and P. E. Suetin, "Investigation of Heat Transfer in Rarefied Gases Over a Wide Range of Knudsen Numbers," *Int. J. Heat Mass Transfer,* Vol. 27, pp. 1789–1799, 1984.

93. J. C. Smith, J. E. Lind, and D. S. Lermond, "Shape Factors for Conductive Heat Flow," *AIChE Journal* (3/3): 330–331, 1958.

94. T. Smith, G. E. Schneider, and M. M. Yovanovich, "Numerical Study of Conduction Heat Transfer from Toroidal Surfaces Into an Infinite Domain," *AIAA-92-2941, AIAA 27th Thermophysics Conference,* Nashville, TN, 1992.

95. W. R. Smythe, "The Capacitance of a Circular Annulus," *American Journal of Physics* (22/8): 1499–1501, 1951.

96. W. R. Smythe, "Charged Right Circular Cylinder," *Journal of Applied Physics* (27/8): 917–920, 1956.

97. W. R. Smythe, "Charged Right Circular Cylinder," *Journal of Applied Physics* (33/10): 2966–2967, 1962.

98. W. R. Smythe, *Static and Dynamic Electricity,* 3d ed., McGraw-Hill, New York, 1968.

99. I. N. Sneddon, *Mixed Boundary Value Problems in Potential Theory,* North-Holland, Amsterdam, 1966.

100. S. Song and M. M. Yovanovich, "Correlation of Thermal Accommodation Coefficients for Engineering Surfaces," *ASME HTD,* Vol. 69, pp. 107–116, 1987.

101. S. Song, M. M. Yovanovich, and K. Nho, "Thermal Gap Conductance: Effects of Gas Pressure and Mechanical Load," *J. Thermophysics and Heat Transfer* (6/1): 62–68, 1992.

102. S. Song and M. M. Yovanovich, "Relative Contact Pressure: Dependence on Surface Roughness and Vickers Microhardness," *AIAA Journal of Thermophysics and Heat Transfer* (2/1): 43–47, 1988.

103. S. Song, "Analytical and Experimental Study of Heat Transfer Through Gas Layers of Contact Interfaces," PhD thesis, University of Waterloo, Waterloo, Ontario, Canada, 1988.

104. S. Song, M. M. Yovanovich, and F. O. Goodman, "Thermal Gap Conductance of Conforming Rough Surfaces in Contact," *Trans. ASME J. Heat Transfer* (115): 533–540, 1993.

105. M. R. Sridhar, "Elastoplastic Contact Models for Sphere-Flat and Conforming Rough Surface Applications," PhD thesis, University of Waterloo, Waterloo, Ontario, Canada, 1994.

106. M. R. Sridhar and M. M. Yovanovich, "Review of Elastic and Plastic Contact Conductance Models: Comparison with Experiment," *J. Thermophysics and Heat Transfer* (8/4): 633–640, 1994.

107. M. R. Sridhar and M. M. Yovanovich, "Thermal Contact Conductance of Tool Steel and Comparison with Model," *Int. J. of Heat Mass Transfer* (39/4): 831–839, 1996.

108. M. R. Sridhar and M. M. Yovanovich, "Empirical Methods to Predict Vickers Microhardness," *Wear,* Vol. 193, pp. 91–98, 1996.

109. M. R. Sridhar and M. M. Yovanovich, "Elastoplastic Contact Model for Isotropic Conforming Rough Surfaces and Comparison with Experiments," *ASME J. of Heat Transfer* (118/1): 3–9, 1996.

110. M. R. Sridhar and M. M. Yovanovich, "Elastoplastic Constriction Resistance Model for Sphere-Flat Contacts," *ASME J. of Heat Transfer* (118/1): 202–205, 1996.

111. J. A. Stratton, *Electromagnetic Theory,* pp. 207–211, McGraw-Hill, New York, 1941.

112. A. Strong, G. Schneider, and M. M. Yovanovich, "Thermal Constriction Resistance of a Disc with Arbitrary Heat Flux—Finite Difference Solution in Oblate Spheroidal Coordinates," *AIAA-74-690, AIAA/ASME 1974 Thermophysics and Heat Transfer Conference,* Boston, MA, July 15–17, 1974.

113. G. Szego, "On the Capacity of a Condenser," *American Mathematical Society, 2* (51): 325–352, 1945.

114. D. Tabor, *The Hardness of Metals,* Oxford University Press, London, 1951.

115. L. B. Thomas, *Rarefied Gas Dynamics,* Academic Press, New York, 1967.

116. T. R. Thomas, *Rough Surfaces,* Longman Group Limited, London, 1982.

117. S. P. Timoshenko and J. N. Goodier, *Theory of Elasticity,* 3d ed., McGraw-Hill, New York, 1970.

118. P. J. Turyk and M. M. Yovanovich, "Transient Constriction Resistance for Elemental Flux Channels Heated by Uniform Heat Sources," 84-HT-52, 1984.

119. T. N. Veziroglu and S. Chandra, "Thermal Conductance of Two-Dimensional Constrictions," *Progress in Astronautics and Aeronautics* (21): 617–620, New York, 1969.

120. H. Y. Wachman, "The Thermal Accommodation Coefficient: A Critical Survey," *ARS J,* Vol. 32, pp. 2–12, 1962.

121. J. A. Walowit and J. N. Anno, *Modern Developments in Lubrication Mechanics,* Applied Science Publishers, Barking, UK, 1975.

122. C. S. Wang, "Surface Element Method Based on Ring Sources and Line Sources for Accurate Calculation of Shape Factors for Arbitrary Isothermal Axisymmetric Surfaces and Some Two Dimensional Problems," MASc thesis, Department of Mechanical Engineering, University of Waterloo, Waterloo, Ontario, Canada, 1993.

123. C. S. Wang and M. M. Yovanovich, "Ring Source Method for Calculation of Shape Factors of Complex Axisymmetric Bodies," *AIAA 27th Thermophysics Conference,* Nashville, TN, July 6–8, 1994.

124. R. O. Warrington Jr., R. E. Rowe, and R. L. Mussulman, "Steady Conduction in Three-Dimensional Shells," *Journal of Heat Transfer* (104): 393–394, 1982.

125. D. J. Whitehouse and J. F. Archard, "The Properties of Random Surfaces of Significance in Their Contact," *Proc. Roy. Soc. Lond. A* (316): 97–121, 1970.

126. M. L. Wiedmann and P. R. Trumpler, "Thermal Accommodation Coefficients," *Trans. ASME,* Vol. 68, pp. 57–64, 1946.

127. D. A. Wesley and M. M. Yovanovich, "A New Gaseous Gap Conductance Relationship," *Nuclear Technology,* Vol. 72, pp. 70–74, 1986.

128. F. C. Yip, "Thermal Contact Constriction Resistance," PhD thesis, Department of Mechanical Engineering, University of Calgary, Calgary, Alberta, Canada, 1969.

129. F. C. Yip, "Effect of Oxide Films on Thermal Contact Resistance," in *Progress in Astronautics and Aeronautics, Heat Transfer With Thermal Control,* M. M. Yovanovich ed., Vol. 39, pp. 45–64, MIT Press, Cambridge, MA, 1974.

130. M. M. Yovanovich, "A General Expression for Predicting Conduction Shape Factors," in *Thermophysics and Spacecraft Thermal Control, AIAA Progress in Astronautics and Aeronautics,* Vol. 35, pp. 265–291, MIT Press, Cambridge, MA, 1974.

131. M. M. Yovanovich, "General Thermal Constriction Resistance Parameter for Annular Contacts on Circular Flux Tubes," *AIAA Journal* (14/6): 822–824, 1976.

132. M. M. Yovanovich, "General Expressions for Constriction Resistances of Arbitrary Flux Distributions," in *Radiative Transfer and Thermal Control, AIAA Progress in Astronautics and Aeronautics,* Vol. 49, pp. 381–396, New York, 1976.

133. M. M. Yovanovich, "Thermal Constriction of Contacts on a Half-Space: Integral Formulation," in *Radiative Transfer and Thermal Control,* Vol. 49, pp. 397–418, AIAA, New York, 1976.

134. M. M. Yovanovich and S. S. Burde, "Centroidal and Area Average Resistances of Nonsymmetric, Singly Connected Contacts," *AIAA Journal* (15/10): 1523–1525, 1977.

135. M. M. Yovanovich and G. E. Schneider, "Thermal Constriction Resistance Due to a Circular Annular Contact," *AIAA Progress in Astronautics and Aeronautics,* Vol. 56, pp. 141–154, 1977.

136. M. M. Yovanovich, "General Conduction Resistance for Spheroids, Cavities, Disks, Spheroidal and Cylindrical Shells," *AIAA 77-742, AIAA 12th Thermophysics Conference,* Albuquerque, New Mexico, June 27–29, 1977.

137. M. M. Yovanovich, S. S. Burde, and J. C. Thompson, "Thermal Constriction Resistance of Arbitrary Planar Contacts With Constant Flux," *Thermophysics of Spacecraft And Outer Planet Entry Probes, AIAA Progress in Astronautics and Aeronautics,* Vol. 56, pp. 127–139, 1977.

138. M. M. Yovanovich, C. H. Tien, and G. E. Schneider, "General Solution of Constriction Resistance Within a Compound Disk," *Heat Transfer, Thermal Control, and Heat Pipes, AIAA Progress in Astronautics and Aeronautics,* Vol. 70, pp. 47–62, 1980.

139. M. M. Yovanovich, "Thermal Contact Correlations," in *Progress in Astronautics and Aeronautics, Spacecraft Radiative Transfer and Temperature Control,* Thomas E. Horton ed., Vol. 83, New York, 1982.

140. M. M. Yovanovich, A. A. Hegazy, and J. DeVaal, "Surface Hardness Distribution Effects Upon Contact, Gap and Joint Conductances," *AIAA-82-0887, AIAA/ASME 3rd Joint Thermophysics, Fluids, Plasma and Heat Transfer Conference,* St. Louis, MO, June 7–11, 1982.

141. M. M. Yovanovich, J. DeVaal, and A. A. Hegazy, "A Statistical Model to Predict Thermal Gap Conductance Between Conforming Rough Surfaces," *AIAA-82-0888, AIAA/ASME Third Joint Thermophysics, Fluids, Plasma and Heat Transfer Conference,* St. Louis, MO, June 1982.

142. M. M. Yovanovich, K. J. Negus, and J. C. Thompson, "Transient Temperature Rise of Arbitrary Contacts with Uniform Flux by Surface Element Methods," *AIAA-84-0397, AIAA 22nd Aerospace Sciences Meeting,* Reno, NV, January 9–12, 1984.

143. M. M. Yovanovich, "Recent Developments in Thermal Contact, Gap and Joint Conductance Theories and Experiments," *Eighth Int. Heat Transfer Conference,* San Francisco, CA, Vol. 1, pp. 35–45, 1986.

144. M. M. Yovanovich, "New Nusselt and Sherwood Numbers for Arbitrary Isopotential Geometries at Near Zero Peclet and Rayleigh Numbers," *AIAA-87-1643, AIAA 22nd Thermophysics Conference,* Honolulu, HI, 1987.

145. M. M. Yovanovich, "Natural Convection from Isothermal Spheroids in the Conductive to Laminar Flow Regimes," *AIAA-87-1587, AIAA 22nd Thermophysics Conference,* Honolulu, HI, 1987.

146. M. M. Yovanovich, "On the Effect of Shape, Aspect Ratio and Orientation Upon Natural Convection from Isothermal Bodies of Complex Shape," *ASME HTD* (82): 121–129, 1987.

147. M. M. Yovanovich and V. W. Antonetti, "Application of Thermal Contact Resistance Theory to Electronic Packages," in *Advances in Thermal Modeling of Electronic Components and Systems,* A. Bar-Cohen and A. D. Kraus eds., Vol. 1, Chap. 2, pp. 79–128, Hemisphere Publishing, New York, 1988.

148. M. M. Yovanovich, "Theory and Applications of Constriction and Spreading Resistance Concepts For Microelectronic Thermal Management," in *Cooling Techniques For Computers,* W. Aung ed., pp. 277–332, Hemisphere Publishing Corp., New York, 1991.

149. M. M. Yovanovich, P. Teertstra, and J. R. Culham, "Modeling Transient Conduction From Isothermal Convex Bodies of Arbitrary Shape," *Journal of Thermophysics and Heat Transfer* (9/3): 385–390, 1995.

150. M. M. Yovanovich, "Dimensionless Shape Factors and Diffusion Lengths of Three-Dimensional Bodies," *ASME/JSME Thermal Engineering Joint Conference* (1): 103–114, 1995.

151. M. M. Yovanovich, "Simple Explicit Expressions for Calculation of the Heisler-Grober Charts," *AIAA-96-3968, 1996 National Heat Transfer Conference,* Houston, TX, August 3–6, 1996.

152. M. M. Yovanovich, "Transient Spreading Resistance of Arbitrary Isoflux Contact Areas: Development of a Universal Time Function," *AIAA-97-2458, AIAA 32nd Thermophysics Conference,* Atlanta, GA, June 23–25, 1997.

153. *Mathematica,* Wolfram Research Inc., Champaign, IL, 1996.

154. K. A. Martin, "Thermal Constriction Resistance of Arbitrary Contacts With the Boundary Condition of the Third Kind," M.A.Sc. Thesis, Department of Mechanical Engineering, University of Waterloo, Waterloo, Ontario, Canada, 1980.

155. S. S. Sadhal, "Exact Solutions for the Steady and Unsteady Diffusion Problems for a Rectangular Prism: Cases of Complex Neumann Conditions," *ASME 84-HT-83, 22nd Heat Transfer Conference,* Niagara Falls, NY, Aug. 6–8, 1984.

156. S. Song, S. Lee, and V. Au, "Closed-Form Equation for Thermal Constriction/Spreading Resistances With Variable Resistance Boundary Condition," *Proc. IEPS Conference,* Atlanta, GA, pp. 111–121, 1994.

157. S. Lee, S. Song, V. Au, and K. P. Moran, "Constriction/Spreading Resistance Model for Electronics Packaging," *Proc. 4th ASME/JSME Thermal Engineering Joint Conference,* Maui, HI, pp. 199–206, March 19–24, 1995.

158. D. J. Nelson and W. A. Sayers, "A Comparison of Two-Dimensional Planar, Axisymmetric and Three-Dimensional Spreading Resistances," *Proc. 8th IEEE SEMI-THERM Symposium on Semiconductor, Thermal Measurements, and Management,* Austin, TX, pp. 62–68, February, 1992.

159. Y. Ogniewicz, "Conduction in Basic Cells of Packed Beds," M.A.Sc. Thesis, Department of Mechanical Engineering, University of Waterloo, Waterloo, Ontario, Canada, 1975.

CHAPTER 4
NATURAL CONVECTION

G. D. Raithby and K. G. T. Hollands
University of Waterloo

INTRODUCTION

Natural convection is the motion that results from the interaction of gravity with density differences within a fluid. The differences may result from gradients in temperature, concentration, or composition. This chapter deals with the heat transfer associated with natural convection driven by temperature gradients in a newtonian fluid.

The first section presents some fundamental ideas that are frequently referred to in the remainder of the chapter. The next three sections deal with the major topics in natural convection. The first of these addresses problems of heat exchange between a body and an extensive quiescent ambient fluid, such as that depicted in Fig. 4.1a. Open cavity problems, such as natural convection in fin arrays or through cooling slots (Fig. 4.1b), are considered next. The last major section deals with natural convection in enclosures, such as in the annulus between cylinders (Fig. 4.1c). The remaining sections present results for special topics including transient convection, natural convection with internal heat generation, mixed convection, and natural convection in porous media.

In response to the main needs of engineers, this chapter focuses on the average (rather than the local) heat transfer, and it provides correlation equations in preference to tabulated data. The advantage is that even complex equations are easy to program into computer codes used for design.

The forms of the correlation equations are based, whenever possible, on general principles and an approximate solution method, both briefly discussed in the section on basics. Confidence in the use of the equations will be enhanced by reading that section. Where accuracy is important, the underlying assumptions and experimental validation should be understood, and where accuracy is critical, no equation should be used beyond its range of experimental validation.

BASICS

Equations of Motion and Their Simplification

In this section the full equations of motion for the external problem sketched in Fig. 4.1a are simplified by using approximations appropriate to natural convection, and the resulting equations are nondimensionalized to bring to light the important dimensionless groups. Although

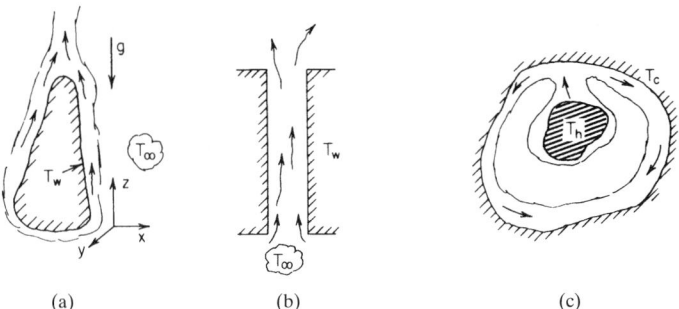

FIGURE 4.1 Natural convection on external surfaces (a), through open cavities (b), and in enclosures (c).

this development is written around the external problem, similar developments may be made for the open cavity and internal problems (Fig. 4.1b and c).

Full Equations of Motion. The complete equations of motion in cartesian coordinates and for a newtonian fluid are presented in Chap. 1. The coordinate z will be used to represent the upward vertical direction as shown in Fig. 1a.

Simplified Equations of Motion. The local pressure in the fluid can (for convenience) be broken into three terms:

$$P \equiv P_{\text{ref}} - \int_{z_{\text{ref}}}^{z} g\rho_\infty(z)\, dz + P_d \tag{4.1}$$

The first term is the pressure at some reference point ($x_{\text{ref}}, y_{\text{ref}}, z_{\text{ref}}$) in the ambient fluid far from the body; the second term is the hydrostatic pressure of this ambient fluid; and the third term, P_d, is the pressure component associated with dynamics of the flow. Substituting Eq. 4.1 for P into the full equations of motion, we find that P_d simply replaces P in the x and y momentum equations. But in the z momentum equation there arises the additional term $\rho_\infty(z)g$, representing the hydrostatic pressure gradient force. The local gravitational body force ρg also appears in the z momentum equation, and the imbalance between these two forces, represented by the difference $\rho_\infty(z)g - \rho g$, is the driving force of natural convection. By introducing $\beta = -(1/\rho)(\partial\rho/\partial T)_P$ and $\gamma = 1/\rho(\partial\rho/\partial P)_T$, this imbalance, called the buoyancy force, can be expressed by

$$(\rho_\infty(z) - \rho)g = \rho g \left[\exp\left(-\int_T^{T_\infty(z)} \beta\, dT + \int_P^{P_\infty(z)} \gamma\, dP \right) - 1 \right] \tag{4.2}$$

where T and P (like ρ) are local values at x, y, z, and $T_\infty(z)$ and $P_\infty(z)$, like $\rho_\infty(z)$, are values in the ambient fluid at the same z.

Equation 4.2 may be simplified. Because the difference between $P(x, y, z)$ and $P_\infty(z)$ is small, the term involving γ may be dropped. Further simplifications are different for gases and liquids. For gases (assumed ideal), $\beta = 1/T$ and Eq. 4.2 reduces to

$$(\rho_\infty(z) - \rho)g = \rho g\, \frac{1}{T_\infty}\, (T - T_\infty) \approx \rho_0 g \beta_\infty (T - T_\infty) \tag{4.3}$$

where $\beta_\infty = 1/\overline{T}_\infty$ and ρ_0 is evaluated at a mean film temperature T_f: $T_f = 0.5(\overline{T}_w + \overline{T}_\infty)$. [$\overline{T}_w$ and $\overline{T}_\infty$ are T_w and $T_\infty(z)$ averaged over the vertical height of the body.] The replacement of ρ by

ρ_0 in the third term in Eq. 4.3 has assumed that changes in density are small compared to density itself. For liquids Eq. 4.2 is approximated as follows:

$$(\rho_\infty(z) - \rho)g \approx \rho g\{\exp[-\beta_0(T_\infty - T)] - 1\} \approx \rho g\beta_0(T - T_\infty) \approx \rho_0 g\beta_0(T - T_\infty) \qquad (4.4)$$

where both β_0 and ρ_0 are evaluated at T_f. Thus, except for the reference temperature at which β is evaluated, the final expression for the buoyancy force is the same for liquids and gases. An extra degree of approximation was, however, required in Eq. 4.4 as compared to Eq. 4.3, and the temperature range over which Eq. 4.4 is valid will be considerably less than that for Eq. 4.3. For water near 4°C (40°F), the range of validity is so small that, for practical problems, Eq. 4.4 must be replaced by a nonlinear relation between ρ and T [110].

Thus the appropriate equations of motion for natural convection [107] are produced by introducing Eq. 4.1 for P into the full equations of motion and then introducing Eq. 4.3 or 4.4 for the term $(\rho_\infty - \rho)g$. These equations are then further simplified by taking each of the properties ρ, μ, c_p, and k as constant at their respective values, ρ_0, μ_0, c_{p0}, and k_0, evaluated at T_f.

The resulting equations are nondimensionalized by substituting $x = x^*L$, $y = y^*L$, $z = z^*L$, $T = \overline{T}_\infty + \theta\Delta T$, $u = v_0 u^*$, $v = v_0 v^*$, $w = v_0 w^*$, $\underline{P_d = \frac{1}{2}\rho_0 v_0^2 P_d^*}$, $\underline{\text{and } t = (L/v_0)t^*}$, where L is a characteristic dimension of the body, $\Delta T = T_w - T_\infty$, and $v_0 = \sqrt{g\beta_1 \Delta TL/(1 + v_0/\alpha_0)}$, with $\beta_1 = \beta_\infty$ for gases and β_0 for liquids. With the assumption that $T_\infty(z)$ and T_w are constant, the resulting equations for mass, z momentum, and energy become

$$\frac{\partial u^*}{\partial x^*} + \frac{\partial u^*}{\partial y^*} + \frac{\partial w^*}{\partial z^*} = 0 \qquad (4.5)$$

$$\begin{array}{cccc} \text{I} & \text{II} & \text{III} & \text{IV} \end{array}$$

$$\frac{Dw^*}{Dt^*} = -\frac{\partial P_d^*}{\partial z^*} + \sqrt{\frac{\text{Pr}\,(1 + \text{Pr})}{\text{Ra}}}\,\nabla^{*2}w^* + (1 + \text{Pr})\theta \qquad (4.6)$$

$$\begin{array}{ccccc} \text{I} & \text{II} & \text{III} & \text{IV} & \text{V} \end{array}$$

$$\frac{D\theta}{Dt^*} = \sqrt{\frac{1 + \text{Pr}}{\text{Ra Pr}}}\,\nabla^{*2}\theta - \frac{\beta TgL}{c_{p0}\Delta T} + \beta_1 T\,\text{Ec}\,\frac{DP_d^*}{Dt^*} + \frac{\text{Ec}}{\text{Re}}\left[\left(\frac{\partial w^*}{\partial z^*}\right)^2 + \cdots\right] \qquad (4.7)$$

where, with $\alpha_0 = k_0/\rho_0 c_{p0}$ and $v_0 = \mu_0/\rho_0$, Ra (the Rayleigh number) $= g\beta_1 \Delta TL^3/v_0\alpha_0$, Pr (the Prandtl number) $= v_0/\alpha_0$, Re (the Reynolds number) $= Lv_0/v_0$, and Ec (the Eckert number) $= v_0^2/(c_{p0}\Delta T)$. See the end of the nomenclature list for the definition of the operators D/Dt^* and ∇^{*2}. The nondimensional x momentum equation is identical to Eq. 4.6 except that u^* replaces w^*, $\partial P_d^*/\partial x^*$ replaces $\partial P_d^*/\partial z^*$, and the term in θ does not occur; an analogous description applies to the nondimensional y momentum equation. The subscripts 0 and 1 on the properties will, for simplicity, be dropped in the rest of this chapter.

Terms III–V in Eq. 4.7 often have only minor influence and can usually be dropped. Term III reflects the fact that the pressure of the ambient fluid decreases with elevation so that, if the fluid moves upward, it does work on expanding at the expense of internal energy and this produces a cooling effect; since $\beta Tg/c_p$ (the adiabatic lapse rate) is usually small (0.01°C/m or 0.006°F/ft for gases) compared to $\Delta T/L$, this term can be safely dropped. Term IV represents a similar effect, this time due to dynamic pressure changes, and can likewise be neglected. Term V represents viscous heating and has a negligible effect.

With terms III–V in Eq. 4.7 deleted, Eqs. 4.5–4.7, together with the x and y momentum equations, constitute the "simplified equations of motion" appropriate to natural convection problems. For constant T_w and T_∞, the boundary conditions on these equations are $\theta = 1$ and $u^* = v^* = w^* = 0$ on the body and $\theta = 0$ far from the body. *Steady-state laminar* solutions to these equations are those that are obtained after setting the time partials (i.e., terms containing partial derivatives with respect to t^*) in the equations equal to zero. *Steady-state turbulent* solutions are those that are obtained after setting time partials of *mean* quantities equal to zero but that incorporate the effect of the unsteady (turbulent) motion.

Important Dimensionless Groups.   The average Nusselt number for steady-state heat transfer from the body shown in Fig. 4.1a depends on the dimensionless groups that arise in the nondimensionalized equations of motion and their boundary conditions [78]. With T_w and T_∞ constant, the only dimensionless groups that appear in the boundary conditions are those associated with the body shape. Provided the simplified equations (Eqs. 4.5–4.7) are valid, the only other dimensionless groups are the Rayleigh and Prandtl numbers. Thus, for a given body shape,

$$\text{Nu} = \frac{qL}{A\Delta Tk} = f(\text{Ra}, \text{Pr}) \tag{4.8}$$

where A is the surface area transferring the total heat rate q.

The role of the Prandtl number and its relative importance can be surmised by inspecting Eqs. 4.6 and 4.7. For Pr >> 1, the buoyancy force that drives the motion (term IV in Eq. 4.6) can be seen to be balanced almost exclusively by the viscous force (term III). With this approximation, and because, for Pr >> 1, Pr disappears from Eq. 4.7, Ra is the only remaining dimensionless group in the simplified equations. For Pr << 1 the viscous force is unimportant, and the buoyancy force must be balanced by the dynamic pressure and inertial terms; in this case the product Ra Pr remains as the only dimensionless group. It is therefore concluded that

$$\text{Nu} = f(\text{Ra}) \qquad \text{for Pr} >> 1 \tag{4.9a}$$

$$\text{Nu} = f(\text{Ra Pr}) \qquad \text{for Pr} << 1 \tag{4.9b}$$

and the Prandtl number governs the relative importance of viscous and inertial forces in the problem.

Validation of the Simplified Equations.   Steady-state laminar solutions to the simplified equations for flow around a long horizontal circular cylinder immersed in air with both T_∞ and T_w constant have been obtained by Kuehn and Goldstein [165] and Fujii et al. [101], using numerical methods. Their results, in terms of the Nusselt number, are plotted in Fig. 4.2,

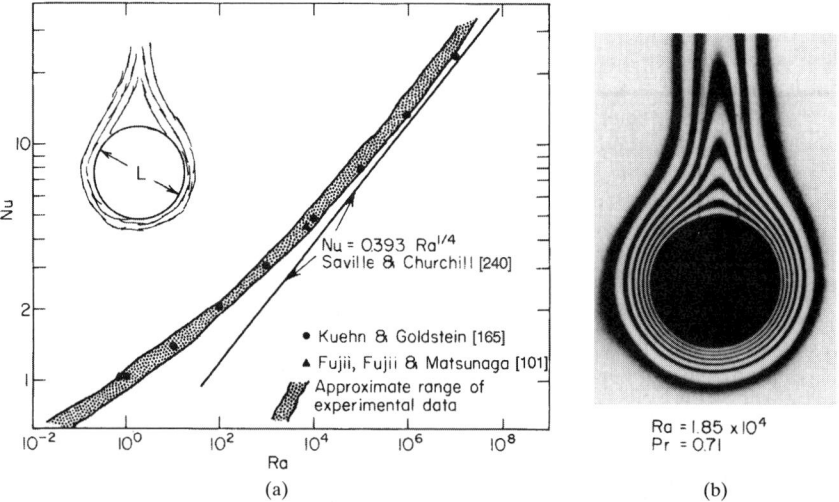

FIGURE 4.2   Measurements and predictions for a circular cylinder (a). (b) shows isotherms for natural convection around a horizontal cylinder. (*Courtesy of* ERG Eckert.)

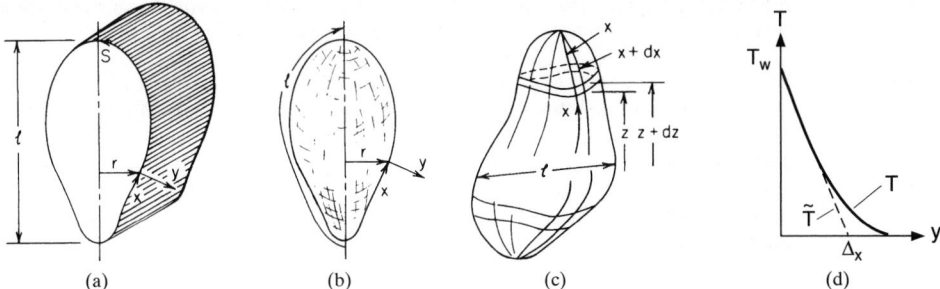

FIGURE 4.3 Surface coordinate systems for two-dimensional (a), axisymmetric (b), and three-dimensional (c) natural convection flows; (d) is a schematic of the temperature distribution near the boundary.

together with the band in which the experimental data for this problem lie. The agreement is good; the differences for Ra $\geq 10^6$ are believed to be due to turbulence, which is not included in their steady-state laminar solution. This comparison lends support to the approximations inherent in the simplified equations of motion.

Thin-Layer Approximation. Laminar analyses often make the further approximation that the boundary layer is so thin that when the simplified equations of motion are rewritten in terms of local surface coordinates, i.e., in terms of the x and y of Fig. 4.3a, several terms normally associated with curvature effects can be dropped. The Nusselt number equation, based on solutions to such laminar "thin-layer" equation sets, always takes the form

$$Nu = c\, Ra^{1/4} \qquad (4.10)$$

where c is independent of Ra but does depend on geometry and Prandtl number. For example, $c = 0.393$ for the horizontal circular cylinder in air (Pr ≈ 0.71) [240]. Plotted in Fig. 4.2, Eq. 4.10 with this value of c falls consistently below the experimental data; also (since no single power law can fit the data) its functional form is incorrect. At high Ra it approaches Kuehn and Goldstein's laminar solution to the complete simplified equations of motion, but, as already discussed, the presence of turbulence has rendered laminar solution invalid in this range.

The thin-layer approximation fails because natural convective boundary layers are not thin. From the interferometric fringes in Fig. 4.2b (which are essentially isotherms), the thermal boundary layer around a circular cylinder is seen to be nearly 30 percent of the cylinder diameter. For such thick boundary layers, curvature effects are important. Despite this failure, thin-layer solutions provide an important foundation for the development of correlation equations, as explained in the section on heat transfer correlation method.

Problem Classification

Classification of Flow. The flow near bodies like that in Fig. 4.1a can be classified into one of three types: two-dimensional (2D), axisymmetric, and three-dimensional (3D). The flow is 2D if the body is invariant in cross-sectional shape along a long horizontal axis (e.g., a long horizontal circular cylinder). An axisymmetric flow takes place near a body (Fig. 4.3b) whose shape can be generated by revolving a body contour about a vertical line; for example, a sphere is generated by rotating a semicircle. If the body meets neither the 2D or axisymmetric requirements, its flow is classified as 3D; this class includes the flow around 2D and axisymmetric bodies whose axes have been tilted.

In addition to this geometric classification, the flow can be classified as fully laminar (i.e., laminar over the entire body), fully turbulent (turbulent over the entire body), or laminar and turbulent (i.e., laminar over one portion and turbulent over the other). Laminar flow is confined to a boundary layer, but turbulent flow may be either attached or detached, as discussed in the section on turbulent Nusselt number.

Thermal Boundary Conditions.   The simplest thermal boundary conditions are T_w and T_∞, both specified constants. If T_∞ varies, it is assumed to be a function only of the elevation z; also, dT_∞/dz is always positive, since, with β positive, a situation where dT_∞/dz is negative is always unstable and cannot be maintained in an extensive fluid. Either T_w or q'' may be specified on the body surface. The difference between the temperature at a given point on the body and T_∞ at the same elevation is denoted ΔT; $\overline{\Delta T}$ is the area-weighted average value of ΔT over the surface, and ΔT_0 is an arbitrary reference temperature difference, usually set equal to $\overline{\Delta T}$. If ΔT is positive over the lower part of the body and negative over the upper part, the surface flows over each part are in opposite directions, and these two flows meet and detach from the body near the line on the surface where $\Delta T = 0$.

Direction of Heat Transfer (Cooling versus Heating).   The relationship of Nu to Ra and Pr for a heated body is precisely the same as when the body is inverted and cooled, except that $T_w - T_\infty$ is replaced by $T_\infty - T_w$. This principle implies, for example, that the relationship for the heated triangular cylinder in Fig. 4.4a is identical to that for the inverted cooled triangular cylinder in Fig. 4.4b. In this chapter it is normally assumed that the surface is heated, and the results for a cooled surface can be inferred from results for a corresponding heated surface by using the above principle.

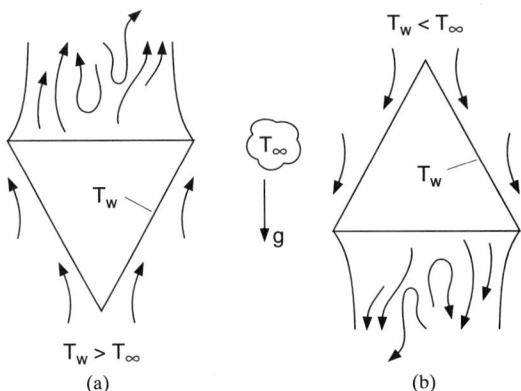

FIGURE 4.4   The Nusselt number for a heated body (a) is the same if the same body is cooled and inverted (b).

Definition of Surface Angle ζ.   Let $\hat{\mathbf{b}}$ be the unit vector in the direction of the buoyancy force as follows. For $T_w > T_\infty$, $\hat{\mathbf{b}}$ is directed vertically upward as shown in Fig. 4.5a and c, and for $T_w < T_\infty$, $\hat{\mathbf{b}}$ is directed vertically downward (Fig. 4.5b and d). The unit outward normal from the surface is $\hat{\mathbf{n}}$, and the angle between $\hat{\mathbf{n}}$ and $\hat{\mathbf{b}}$ is defined as the surface angle ζ. For $0° \le \zeta < 90°$, as in Fig. 4.5c and d, the buoyancy force has components along and *away from* the surface. For $90° < \zeta \le 180°$, Fig. 4.5a and b, the buoyancy force is directed along and *into* the surface. Clearly, ζ varies locally over a curved surface.

Heat Transfer Correlation Method

Overview.   The heat transfer correlations given in subsequent sections draw on a certain paradigm, which is outlined in this section. Strictly, the reader need not be aware of this

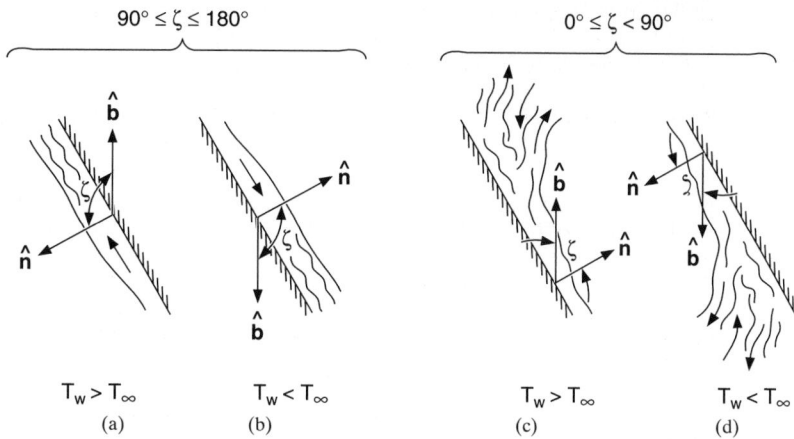

$90° \leq \zeta \leq 180°$ $0° \leq \zeta < 90°$

$T_w > T_\infty$ $T_w < T_\infty$ $T_w > T_\infty$ $T_w < T_\infty$

(a)   (b)   (c)   (d)

FIGURE 4.5   Definition of surface angle ζ for heated and cooled surfaces. For $90° \leq \zeta \leq 180°$ the flow remains attached, while it may detach if $0° \leq \zeta < 90°$.

paradigm to use the equations. On the other hand, knowing something of it will assist the reader in interpreting the results and in extending the results to shapes not mentioned. Briefly, the idea is to obtain an equation for the laminar Nusselt number Nu_ℓ that would be valid for the hypothetical situation where the flow is always fully laminar over the entire surface, to obtain an equation for the turbulent Nusselt number, Nu_t, that would be valid for the hypothetical situation where the flow is always fully turbulent at all surface locations, and to combine the equations for Nu_ℓ and Nu_t to obtain an expression for Nu that is valid for all cases: wholly laminar, wholly turbulent, or partly laminar and partly turbulent.

The equation for the laminar Nusselt number Nu_ℓ is obtained in a two-step procedure. In the first step, not only is the flow idealized as everywhere laminar, but the boundary layer is treated as thin. There results from this idealization the equation for the laminar thin-layer Nusselt number Nu^T. As already explained, natural convection boundary layers are generally not thin, so the second step is to correct Nu^T to account for thick boundary layers. This correction uses the method of Langmuir [175]. The corrected Nusselt number is the laminar Nusselt number Nu_ℓ.

Conduction Nusselt Number Nu_{COND}.   It is the aim of the method to give expressions for Nu (defined by Eq. 4.8, where L is any convenient length scale) that cover the entire range in Ra from zero to infinity, so one needs to consider the Nusselt number in the limit where $Ra \rightarrow 0$. For $Ra \rightarrow 0$, there is no fluid motion, and (with the possible exception of radiation) heat transfer is by conduction only. The Nusselt number that applies in this limit is therefore called the conduction Nusselt number Nu_{COND}. Nu_{COND} is zero for 2D flows and nonzero for 3D flows. For a sphere of radius D, for example, $Nu_{COND} = 2L/D$. Except for simple body shapes—for which values of Nu_{COND} may be available in the literature (see, for example, Ref. 292)—a detailed conduction analysis may be required to determine Nu_{COND}. On the other hand, the work of Chow and Yovanovich [296] and, later, of Yovanovich [292] has shown that the quantity $(\sqrt{A}/L) \, Nu_{COND}$ is highly insensitive to the body shape, ranging from 3.391 to 3.609 over a very wide range of shapes, including cubes, lenses, ellipsoids, and toroids; for an ellipsoid, $(\sqrt{A}/L)$ Nu_{COND} ranges from 3.55 to 3.19 as the ratio of major to minor axis ranges from 1 to 10^4. In the absence of a conduction analysis for a particular body shape, the Yovanovich method provides a quick method for estimating Nu_{COND}.

Laminar Thin-Layer Nusselt Number Nu^T.   Raithby and Hollands [223] derived the following equation (similar to Eq. 4.10) as a close approximation to the full solution of the thin-layer equations:

$$\mathrm{Nu}^T = G\overline{C}_\ell \, \mathrm{Ra}^{1/4} \tag{4.11}$$

where

$$\mathrm{Ra} = \frac{g\beta\Delta T_0 L^3}{\nu\alpha} \tag{4.12}$$

and L is any convenient reference length (the dependence on this length cancels out of the equations). G is a ratio of integrals (both of which can often be evaluated in closed form) that depend on the shape of the body and on thermal boundary conditions, and $\overline{C}_\ell$ is an approximately universal function of Prandtl number Pr given by Churchill and Usagi [54]

$$\overline{C}_\ell = \frac{0.671}{[1 + (0.492/\mathrm{Pr})^{9/16}]^{4/9}} \tag{4.13}$$

$\overline{C}_\ell$ is tabulated in Table 4.1. For the special case of a heated body with constant T_w and T_∞, the G ratio is given for both 2D and axisymmetric flows by

$$G = \frac{L^{1/4}}{\int_0^S r^i \, dx} \left[\int_0^S r^{4i/3} \sin^{1/3} \zeta \, dx \right]^{3/4} \tag{4.14}$$

where index $i = 0$ if the flow is 2D and $i = 1$ if the flow is axisymmetric (see the section on classification of flow). The body coordinates r and x in Eq. (4.14) are defined in Fig. 4.3, x being the distance measured along the surface of the body from the lowest point on the body, and S is the value of x at the highest point on the body. The angle ζ in the integral is the surface angle explained previously and in Fig. 4.5. Raithby and Hollands [225–227] and (for the special case of $\mathrm{Pr} \to \infty$) Stewart [264] have given more general expressions covering the cases of 3D flows and nonuniform T_w and T_∞. Hassani and Hollands [126, 127] gave the following approximate equation for G for 3D flows: $G = (L_f L \mathcal{P}^2 / A^2)^{1/4}$, where L_f is the vertical distance from the lowest point on the body to the highest, and $\mathcal{P}$ is the vertically averaged perimeter of the body defined by

$$\mathcal{P} = \frac{1}{L_f} \int_0^{L_f} \mathcal{P}(z) \, dz \tag{4.15}$$

where $\mathcal{P}(z)$ is the local perimeter of the body at elevation z above the lowest point on the body. [A horizontal plane at height z will intersect the surface of the body at a flat curve; $\mathcal{P}(z)$ is the circumference of this curve.] For 2D flow, this equation for G reduces to $G = (4L_f L / P_i^2)^{1/4}$, where P_i is the perimeter of the cylinder's cross section.

To further characterize the laminar thin-layer heat transfer, it is helpful to introduce the idea of the local and average conduction layer thicknesses, Δ_x and $\overline{\Delta}$, respectively: Δ_x is defined in Fig. 4.3d as the local distance from the wall at which the linear $\tilde{T}$ profile intersects T_∞, it being understood that the $\tilde{T}$ profile is a hypothetical linear temperature profile that has the same value at the surface and the same slope at the surface as the true temperature profile. Stated differently, Δ_x is the local thickness of stationary fluid that would offer the same ther-

TABLE 4.1   Prandtl Number Dependence of Various Coefficients

	Mercury	Gases			Water			Oils	
Pr	0.022	0.71	1.0	2.0	4.0	6.0	50	100	2000
$\overline{C}_\ell$	0.287	0.515	0.534	0.568	0.595	0.608	0.650	0.656	0.668
C_t^V	0.055*	0.103	0.106	0.112	0.113	0.113	0.097	0.091	0.064
C_t^U	0.14*	0.140	0.140	0.140	0.140	0.141	0.143	0.145	0.149
$\overline{H}_\ell$	0.397	0.624	0.641	0.671	0.694	0.704	0.738	0.744	0.754

* No experimental validation available.

mal resistance as the thin boundary layer. The average conduction layer thickness $\overline{\Delta}$ is an area-weighted harmonic-mean average of Δ_x taken over the whole body surface. One can show that $\overline{\Delta}$ is related to the thin-layer average Nusselt number by

$$\overline{\Delta} = \frac{L}{\mathrm{Nu}^T} \qquad (4.16)$$

Fully Laminar Nusselt Number $\mathrm{Nu}_{l^{\bullet}}$. As the next step in the correlation method, the thin-layer solution is corrected to account for thick-layer effects [175, 223]. The body is surrounded by a uniform layer of stationary fluid of thickness $\overline{\Delta}$, and outside that thickness the fluid temperature is taken to be T_∞. The heat transfer that would occur across this layer is determined by a conduction analysis and converted to a Nusselt number, and this Nusselt number is Nu_l.

For example, to determine Nu_l for the case where the body is a very long horizontal isothermal circular cylinder of diameter D, the relevant heat transfer would then be that by heat conduction across a cylindrical annulus of inner diameter D, inner temperature T_w, outer diameter $D + 2\overline{\Delta}$, and outer temperature T_∞ (assumed constant). Calculating this heat transfer by standard methods, substituting Eq. 4.16, and converting to a Nusselt number yields

$$\mathrm{Nu}_\ell = \frac{C_1}{\ln\,(1 + C_1/\mathrm{Nu}^T)}; \qquad C_1 = \frac{2\pi L}{P_i} \qquad (4.17)$$

where the reference length L is equated to the diameter D, and P_i is the perimeter of the cylinder. The work of Hassani et al. [129] on conduction across layers of uniform thickness on very long cylinders of various cross section, has shown, in effect, that Eq. 4.17 applies to triangular, rectangular, and trapezoidal cross sections as well, and by implication to cylinders of almost any convex cross section.

As a second example, which also will be extended to a more general case, consider the case where the body is an isothermal sphere and, again, T_∞ is uniform. The relevant conduction heat transfer in this case is that between two concentric spheres of diameters D and $D + 2\overline{\Delta}$, respectively. Solving this elementary problem in heat conduction, recasting in terms of the Nusselt number, substituting from Eq. 4.16, and then recognizing that the $2L/D$ is the conduction Nusselt number $\mathrm{Nu}_{\mathrm{COND}}$, one obtains

$$\mathrm{Nu}_\ell = \mathrm{Nu}_{\mathrm{COND}} + \mathrm{Nu}^T \qquad (4.18)$$

The work of Hassani and Hollands [128] on conduction across layers of uniform thickness applied to 3D bodies of various shape has shown that Eq. 4.18 applies approximately to other 3D body shapes as well. (Note that the actual $\mathrm{Nu}_{\mathrm{COND}}$ for the body at hand must be used.) The work of Hassani and Hollands also shows that slightly better results for such nonspherical bodies would be obtained if Eq. 4.18 is modified as follows:

$$\mathrm{Nu}_\ell = ((\mathrm{Nu}_{\mathrm{COND}})^n + (\mathrm{Nu}^T)^n)^{1/n} \qquad (4.19)$$

where n is a parameter best determined by fitting to experimental data. If data are not available, $n \approx 1.07$ can be used as a rough approximation; alternatively, a more accurate value can be obtained by using a formula proposed by Hassani and Hollands [128].

A special case arises when the length of a horizontal cylinder greatly exceeds its diameter. If the cylinder is treated as infinite, Eq. 4.17 would yield the result $\mathrm{Nu}_\ell \to 0$ as $\mathrm{Ra} \to 0$. Nu_ℓ should, however, approach $\mathrm{Nu}_{\mathrm{COND}}$ as $\mathrm{Ra} \to 0$, where $\mathrm{Nu}_{\mathrm{COND}}$ is small but nonzero because the cylinder length is not truly infinite. As a rough approximation for this case:

$$\mathrm{Nu}_\ell = \left[\frac{C_1}{\ln\,(1 + C_1/\mathrm{Nu}^T)}, \mathrm{Nu}_{\mathrm{COND}}\right]_{\mathrm{max}}; \qquad C_1 = \frac{2\pi L}{P_i} \qquad (4.20)$$

where $[x,\,y]_{\mathrm{max}}$ requires that the maximum of x and y is to be taken.

Turbulent Nusselt Number Nu$_t$.   This section presents a model for the Nusselt number–Rayleigh number relation applying in the limiting case where the flow is turbulent at all locations on the surface, i.e., in the asymptote Ra $\rightarrow \infty$.

Model Description.   The chief characteristic of turbulent heat transfer is that (for a given Prandtl number and Rayleigh number) the heat transfer at a point on the surface depends only on the local surface angle ζ (Fig. 4.5), and is independent of how far the point is from the leading edge. It follows that

$$\text{Nu}_{t,x} = C_t(\zeta)\,\text{Ra}_x^{1/3} \tag{4.21}$$

where Nu$_{t,x}$ is the local Nusselt number and the function $C_t(\zeta)$ depends only on the Prandtl number. (The appearance of x in this equation is an artifact of the nomenclature; when Nu$_{t,x}$ and Ra$_x$ are replaced by their definitions, the x on each side of the equation cancels out.) If one integrates this local Nusselt number over the entire body surface to get the total heat transfer, one obtains

$$\text{Nu}_t = \overline{C}_t\,\text{Ra}^{1/3} \tag{4.22}$$

where

$$\overline{C}_t = \frac{1}{A}\int_A C_t(\zeta)\left(\frac{T_w - T_\infty}{\Delta T_0}\right)^{4/3} dA \tag{4.23}$$

Equation for C$_t$.   A number of experiments at different Prandtl numbers (mostly on tilted plates) have been carried out that permit the function $C_t(\zeta)$ to be modeled. Observation has also revealed that there are two patterns of turbulent flow: detached and attached. Attached flow, where the flow sticks to the body contour, is best exemplified by the flow on a vertical plate, and the $C_t(\zeta)$, $= C_t (90°)$ applying in this situation is denoted C_t^V. Detached flow, where turbulent eddies rise away from the heated surface, is best exemplified by the flow on a horizontal upward-facing (heated) plate, and the $C_t(\zeta) = C_t(0°)$ applying in this situation is denoted C_t^U. The first step in establishing the $C_t(\zeta)$ function has been to model how C_t^U and C_t^V depend on the Prandtl number; the equations for these quantities (justified later) that have been found to best fit currently available data are

$$C_t^V = \frac{0.13\,\text{Pr}^{0.22}}{(1 + 0.61\,\text{Pr}^{0.81})^{0.42}} \tag{4.24}$$

and

$$C_t^U = 0.14\left(\frac{1 + 0.0107\,\text{Pr}}{1 + 0.01\,\text{Pr}}\right) \tag{4.25}$$

C_t^V and C_t^U are tabulated in Table 4.1.

For other angles, we can make use of the fact that the attached flow on an inclined plate would look just like the corresponding flow on the vertical plate if the gravity on the vertical plate were changed from g to $g \sin \zeta$: in other words, the flow is driven by the component of the buoyancy force that is directed along the surface. Since (from Eq. 4.22) Nu$_x \propto$ Ra$_x^{1/3}$, it follows that, in attached flow,

$$C_t = C_t^V \sin^{1/3} \zeta \tag{4.26}$$

In detached flow, the mixing of the boundary layer fluid and the ambient fluid is driven by the component of gravity normal to the surface. Since this mixing dominates the heat exchange, the heat transfer from a tilted upward-facing plate can be obtained from the equation for a horizontal plate by replacing g by $g \cos \zeta$. For the flow to remain detached, however, $g \cos \zeta$ must remain positive. Since Nu$_x \propto$ Ra$_x^{1/3}$, it follows directly that, in detached flow,

$$C_t = C_t^U [\cos \zeta, 0]_{\max}^{1/3} \tag{4.27}$$

Equation 4.26 applies for attached flow and Eq. 4.27 for detached flow. Experience has indicated that the flow pattern actually observed at a given location on the body will be the one that maximizes the local heat transfer. From this it follows that

$$C_t = [C_t^U [\cos \zeta, 0]_{\max}^{1/3}, C_t^V \sin^{1/3} \zeta]_{\max} \qquad (4.28)$$

Equation 4.28 for C_t is substituted into Eq. 4.23 to obtain $\overline{C}_t$. This requires an integration over the surface. The values of $\overline{C}_t$ in the equations provided in this chapter were obtained in this way.

For geometries not covered in this chapter, it may be more convenient to use the following approximate equation for $\overline{C}_t$ [128], which applies for 0.7 Pr < 2000:

$$\overline{C}_t = 0.0972 - (0.0157 + 0.462C_t^V) \frac{A_h}{A} + (0.615C_t^V - 0.0548 - 6 \times 10^{-6} \text{ Pr}) \frac{L_f \mathcal{P}}{A} \qquad (4.29)$$

A_h is the area of any horizontal downward-facing heated parts (or upward-facing horizontal cooled parts) of the body's surface.

The recommended equation for $C_t(\zeta)$ is provisional because of the lack of data and the disagreement among different sets of measurements. Since the recommended $C_t(\zeta)$ affects many of the correlations in this chapter, the user of these correlations should understand the experimental foundation for Eq. 4.24 for C_t^V and Eq. 4.25 for C_t^U. The following paragraphs provide the necessary background.

Discussion of C_t^V Equation.   Equation 4.24 for C_t^V has been forced to pass through 0.103 for gases (Pr = 0.71), and through the value of 0.064 measured by Lloyd et al. [187] for Pr = 2000. There is no experimental confirmation of Eq. 4.24 for Pr < 0.7. The data for gases (Pr ≈ 0.7) are tightly clustered around 0.103, provided $\Delta T/T \ll 1$. For water, the measurements of Fujii and Fujii [99] yield $C_t^V \approx 0.13$, compared to 0.11 by Vliet and Liu [275]; the equation passes near the lower of these values. The equation agrees with values obtained from the data of Fujii and Fujii [99] for oils (Pr = 20–200) if all properties are evaluated at a reference temperature of $T_e = 0.75T_w + 0.25T_\infty$. For Pr = 2000, the value obtained from the data of Moran and Lloyd [197] falls about 15 percent below the benchmark value of 0.064 obtained by Lloyd et al. [187] using the same technique.

The form of Eq. 4.24 is surprising since, based on Eq. 4.9a and b, one would expect C_t^V to increase monotonically with increasing Pr and approach the Pr → ∞ asymptote from below. It is therefore not clear why the high-quality data of Lloyd et al. [187] at Pr ≈ 2000 give a C_t^V value less than that for water and air. The apparent corroboration by Fujii and Fujii [99] for oils is also highly uncertain because of the extreme sensitivity to the reference temperature used. Using $T_f = 0.5T_w + 0.5T_\infty$ as the reference temperature, instead of T_e, raises the value of C_t^V by about 20 percent, to almost the same value as for water.

The C_t^V values for gases given by Eq. 4.24 assume that for $\Delta T/T_\infty \ll 1$, or $T_w/T_\infty \to 1$. If this condition is not met for a vertical plate, Eq. 4.21 should be rewritten as

$$\text{Nu}_{t,x} = C_t^V f\left(\frac{T_w}{T_\infty}\right) \text{Ra}_x^{1/3} \qquad (4.30)$$

Measurements by Pirovano et al. [220, 221], Siebers et al. [251], and Clausing et al. [61, 63] suggest quite different values of f. Until further data become available, the recommended equation for f in Eq. 4.30 is

$$f\left(\frac{T_w}{T_\infty}\right) = 1.0 + 0.078\left(\frac{T_w}{T_\infty} - 1\right) \qquad (4.31)$$

The equations provided in this chapter assume that $\Delta T/T_\infty \to 0$. For large ΔT, the equation for C_t^V should be replaced by fC_t^V, where f is given by Eq. 4.31. It is concluded that there is considerable uncertainty in the recommended equation for C_t^V for Pr > 1 and for large T_w/T_∞. This is perhaps the most important unresolved fundamental issue in the equations used to estimate natural convection heat transfer.

Discussion of C_t^U Equation.   Equation 4.25 for C_t^U has been forced to agree with the value 0.15 measured by Lloyd and Moran [297] at Pr ≈ 2000, and forced to pass through $C_t^U = 0.14$ for air (Pr = 0.71) and water (Pr ≈ 6). The latter values are deduced from measurements of heat transfer across horizontal fluid layers (see the section on natural convection within enclosures) that, for $Ra \rightarrow \infty$, are correlated by $Nu = c\,Ra^{1/3}$; in this limit the relation between c and C_t^U is $C_t^U = c2^{4/3}$. $C_t^U = 0.14$ for air and water agrees to within experimental scatter with measurements of heat transfer from horizontal upward-facing plates obtained by Yousef et al. [290], Bovy and Woelk [23], Clausing and Berton [62], Grober et al. [119], Hassan and Mohamed [123], Fishenden and Saunders [98], Al-Arabi and El-Riedy [4], Weiss [282], and Fujii and Imura [103]. The single data point for mercury ($Pr = 0.024$), deduced from the horizontal layer measurements of Globe and Dropkin [114], is $C_t^U = 0.13$, which compares well with 0.14 from Eq. 4.25.

Blending of Laminar and Turbulent Nusselt Numbers.   The previous two sections provided heat transfer equations for the cases where there is laminar heat transfer from the entire body, (Nu_ℓ), and turbulent heat transfer from the entire body, (Nu_t). To obtain a fit to heat transfer data over the entire range of Ra, the blending equation of Churchill and Usagi [54] is used:

$$Nu = (Nu_\ell^m + Nu_t^m)^{1/m} \qquad (4.32)$$

The appropriate value of m, which generally lies in the range from 4 to 20, is chosen so as to give the best fit to the experimental data. Where no data for the body shape at hand are available, the value for a known body of similar shape is tentatively recommended; Tables 4.2 (found on page 4.24) and 4.3 (found on pages 4.26–4.31) give values of m for a range of shapes. Minor modifications to this prescription are sometimes implemented in order to more accurately represent the dependence of heat transfer on Ra through the transition.

The aim of the correlation method has been to give an equation for Nu (Ra) that covers the entire range of Ra from zero to infinity. On the other hand, the Rayleigh number range of interest in the particular problem at hand may be low enough to ensure that laminar heat transfer dominates. In this case it may be preferable to simplify the equation by assuming that the turbulent term in Eq. 4.32 does not contribute appreciably, and take Nu as being equal to Nu_ℓ. This is allowable if, at the highest Rayleigh number of interest, the difference between the calculated values of Nu and Nu_ℓ is acceptably small.

Although certain of the steps in the development of the Nu (Ra) function given in this section have assumed that the body is convex, in practice, a minor degree of concavity has not been found to compromise its usefulness. For example, it has been applied to long cylinders with an "apple core" cross section and to 3D bodies in the shape of an apple core. A test is to examine the degree to which the conduction layer on one part of the surface overlaps with that on another part; if the degree of overlap is modest, the method should be useful. An extreme case of concavity is the case of a body with a vertical hole passing through it. Provided the conduction layer thickness $\bar{\Delta}$ (Eq. 4.16) is thinner than, say, about ⅓ the hole radius, the method should continue to work satisfactorily. If not, the hole should be treated separately as an open cavity, using methods explained in the section on open cavity problems.

EXTERNAL NATURAL CONVECTION

Equations are presented in this section for evaluating the heat transfer by natural convection from the external surfaces of bodies of various shapes. The correlation equations are of the form described in the section on the heat transfer correlation method, and the orientation of the surface is given by the surface angle ζ defined in Fig. 4.4. Supporting experimental evidence for each such equation set is outlined after each equation tabulation. The correlations are in terms of Nu, Ra, and Pr, parameters that involve physical properties, a length scale, and a reference temperature difference. Rules for the evaluation of property values are provided in the nomenclature, and the relevant length scale and reference temperature difference are provided in a separate definition sketch for each problem.

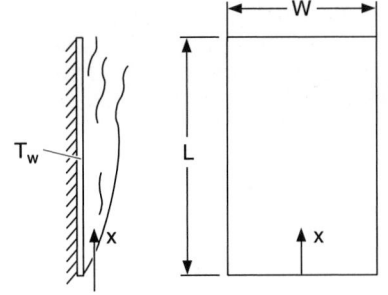

$$Nu = \frac{\overline{q}''L}{\Delta T k} \quad Nu_x = \frac{q_x''x}{\Delta T k} \quad Ra = \frac{g\beta\overline{\Delta T}L^3}{\nu\alpha} \quad Ra_x = \frac{g\beta\Delta T x^3}{\nu\alpha}$$

$$Ra^* = \frac{g\beta q''L^4}{\nu\alpha k} = Ra\ Nu \quad Ra_x^* = \frac{g\beta q''x^4}{\nu\alpha k} = Ra_x\ Nu_x$$

FIGURE 4.6 Definition sketch for natural convection on a vertical plate with uniform wall temperature or uniform heat flux.

Flat Plates

Vertical Flat Plate with Uniform T_w and T_∞, $\zeta = 90°$

Correlation. In terms of quantities defined in Fig. 4.6 and the nomenclature, the total heat transfer from a wide ($W \gg L$) vertical isothermal plate can be estimated from the following equations:

$$Nu^T = \overline{C}_\ell\ Ra^{1/4} \tag{4.33a}$$

$$Nu_\ell = \frac{2.0}{\ln(1 + 2.0/Nu^T)} \tag{4.33b}$$

$$Nu_t = C_t^V\ Ra^{1/3}/(1 + 1.4 \times 10^9\ Pr/Ra) \tag{4.33c}$$

$$Nu = [(Nu_\ell)^m + (Nu_t)^m]^{1/m} \quad m = 6 \tag{4.33d}$$

For large $\Delta T/T$, C_t^V should be replaced by $C_t^V f$, where f is given by Eq. 4.31.

Comparison With Data. Figure 4.7 compares Eq. 4.33 to four sets of measurements for gases for small $\Delta T/T$. The range of experimental data is $10^{-1} < Ra < 10^{12}$, and the RMS deviation of the data from the equation is 5 percent. The data of Saunders [239] have been reevaluated using more accurate property values. The data of Pirovano et al. [220, 221] are for their lowest T_w/T_∞ ratio, and only the laminar data of Clausing and Kempka [63] are shown. Equation 4.33 also closely fits the data for other fluids compiled by Churchill [56].

Outstanding Issues. There are no data at low Ra for Pr < 0.7. There is also uncertainty surrounding the expression for C_t^V (Eq. 4.24) for both Pr $\ll$ 1 and Pr $\gg$ 1. The transition in Eq. 4.33 is assumed to depend exclusively on Ra/Pr; in fact, the transition appears to also depend on the vertical stratification within the ambient medium and on the size of ΔT.

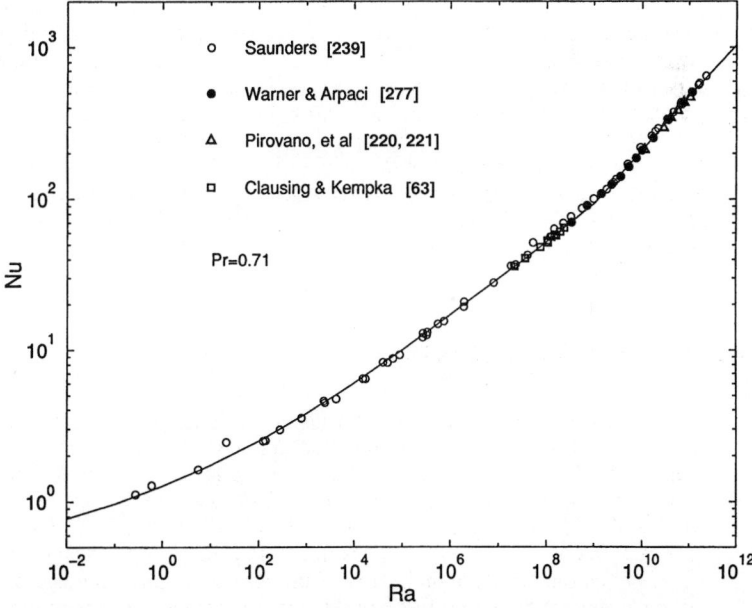

FIGURE 4.7 Comparison of Eq. 4.33 with data for an isothermal, vertical flat plate in air.

Vertical Plate With Uniform Surface Heat Flux and Constant T_∞.   When the surface heat flux, q''_w, is uniform and known, values of Nu_x and Nu are used respectively to compute the local temperature difference ΔT and the average temperature difference $\overline{\Delta T}$. Parameters related to this problem are defined in Fig. 4.6 and in the nomenclature.

Calculation of Local ΔT.   The local ΔT can be found from the value of Nu_x calculated from the following equation for $Ra^* \gtrsim 0.1$.

$$Nu_x^T = H_\ell (Ra_x^*)^{1/5} \qquad H_\ell = \left(\frac{Pr}{4 + 9\sqrt{Pr} + 10\,Pr}\right)^{1/5} \tag{4.34a}$$

$$Nu_{\ell,x} = 0.4/\ln\,(1 + 0.4/Nu_x^T) \tag{4.34b}$$

$$Nu_{t,x} = \frac{(C_t^V)^{3/4}(Ra_x^*)^{1/4}}{1 + (C_2\,Pr/Ra_x^*)^3} \qquad \begin{matrix} 10^{12} < C_2 < 2 \times 10^{13} \\ C_2 = 7 \times 10^{12} \quad \text{(nominal)} \end{matrix} \tag{4.34c}$$

$$Nu_x = ((Nu_{\ell,x})^m + (Nu_{t,x})^m)^{1/m} \qquad m = 3.0 \tag{4.34d}$$

The value of C_2 determines the transition between laminar and turbulent heat transfer. A nominal value is $C_2 = 7 \times 10^{12}$.

Comparison With Data.   Equation 4.34 is in excellent agreement with the data of Humphreys and Welty [147] and Chang and Akins [40] for $Pr = 0.023$ (mercury), but the data lie only in the laminar regime. There is also good agreement with measurements of Goldstein and Eckert [115], Vliet and Liu [275], and Qureshi and Gebhart [222] for water, although the observed transition depends on the level of heat flux. By choosing an appropriate value for C_2 for each q'', Fig. 4.8 shows that Eq. 4.34 can be made to fit each data set. For a nominal value

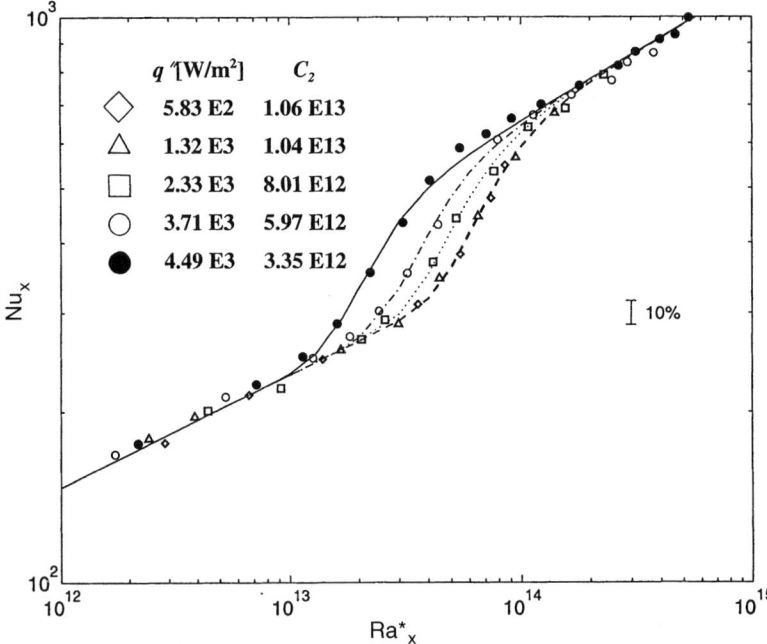

FIGURE 4.8   Comparison of local heat transfer measurements of Qureshi and Gebhart [222] with Eq. 4.34 for a uniform heat flux, vertical flat plate in water.

by $C_2 = 7 \times 10^{12}$, Eq. 4.34 gives good agreement with measurements for oils, Pr ≈ 60 and Pr ≈ 140, except in the transition region. Fujii and Fujii [99] have shown that strong vertical temperature gradient in the ambient fluid also affects the value of Nu_x in the transition regime.

Calculation of $\overline{\Delta T}$. In some cases involving uniform q'', it is sufficient to know the average temperature difference, defined as

$$\overline{\Delta T} = \frac{1}{A} \int_A (T_w - T_\infty) \, dA \tag{4.35}$$

This is obtained from the average Nusselt number Nu defined in Fig. 4.6. A rough estimate of Nu can be obtained for Ra ≥ 10^5 by using the equations in the section on vertical flat plates with uniform T_w and T_∞, $\zeta = 90°$, with ΔT replaced by $\overline{\Delta T}$. For higher accuracy, and for convenience since Ra* is used in place of Ra, use the following equation set:

$$Nu^T = \overline{H}_\ell \, Ra^{*1/5} \qquad \overline{H}_\ell = \frac{6}{5} \left(\frac{Pr}{4 + 9\sqrt{Pr} + 10\,Pr} \right)^{1/5} \tag{4.36a}$$

$$Nu_\ell = \frac{1.0}{\ln (1 + 1.0/Nu^T)} \tag{4.36b}$$

$$Nu_t = (C_t^V)^{3/4} \, Ra^{*1/4} \Big/ \left(1 + \left(\frac{C_2 \, Pr}{Ra^*} \right)^{0.4} \right) \qquad C_2 \approx 7 \times 10^{12} \tag{4.36c}$$

$$Nu = (Nu_\ell^m + Nu_t^m)^{1/m} \qquad m = 6.0 \tag{4.36d}$$

$\overline{H}_\ell$ is tabulated in Table 4.1.

Vertical Plates of Various Planforms. There is a class of vertical plates, such as shown in Fig. 4.9, for which any vertical line on the plate intersects the edge only twice. When such plates are isothermal, the thin-layer laminar Nusselt number Nu^T can be calculated [225] by dividing the surface into strips of width Δx and length $S(x)$, applying Eq. 4.33a to compute the heat transfer from each strip, adding these to obtain the total (thin-layer) heat transfer, and using this heat transfer to calculate Nu^T. The result can be expressed as

$$Nu^T = G\overline{C}_\ell \, Ra^{1/4} \qquad G = \frac{\ell^{1/4} \int_0^W S^{3/4} \, dx}{A} \tag{4.37a}$$

where ℓ is any characteristic dimension of the surface. This also results from the application of Eq. 4.14. Figure 4.9 shows examples of the constants G for three body shapes.

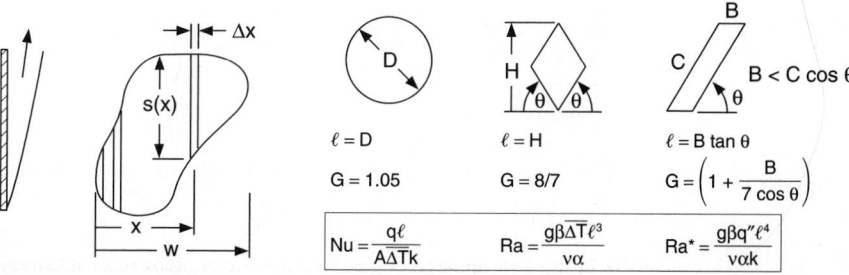

FIGURE 4.9 Definition sketch for natural convection from plates of various planform.

For laminar heat transfer, the correlations for such disks have been given by Eq. 4.18 [227]:

$$\text{Nu}_\ell = \text{Nu}_{\text{COND}} + \text{Nu}^T \tag{4.37b}$$

Yovanovich and Jafarpur [294] have shown that, for the length scales in Fig. 4.9,

$$\text{Nu}_{\text{COND}} = \frac{\ell}{\sqrt{A}}\, \text{Nu}_{\sqrt{A},\text{COND}} \tag{4.37c}$$

If the plate is heated on one side only, but is immersed in a full space $\text{Nu}_{\sqrt{A},\text{COND}} \approx 3.55$, if the plate is heated on one side and immersed in a half-space (i.e., the heated plate is embedded in an infinite adiabatic surface) $\text{Nu}_{\sqrt{A},\text{COND}} \approx 2.26$, and if the plate is heated on both sides and immersed in a full space $\text{Nu}_{\sqrt{A},\text{COND}} \approx 3.19$. In all cases A is the surface area that is active. For a general equation for Nu: substitute Eqs. 4.37a and c into Eq. 4.37b to obtain the equation for Nu_ℓ, use Eq. 4.33 for Nu_t, and substitute these equations for Nu_ℓ and Nu_t into Eq. 4.33 to obtain the equation for Nu.

The isothermal surface correlation just provided can also be used to estimate Nu for a uniform flux boundary by replacing ΔT by $\overline{\Delta T}$. For higher accuracy, use Eq. 4.36a to find the thin-layer $\overline{\Delta T}$ for each strip, find $\overline{\Delta T}$ for the entire surface by area weighting the $\overline{\Delta T}$ for each strip (Eq. 4.35), and calculate Nu^T from the area weighted $\overline{\Delta T}$. This results in

$$\text{Nu}^T = C_2 \bar{H}_\ell\, \text{Ra}^{*1/5} \qquad C_2 = \frac{A\ell^{1/5}}{\displaystyle\int_0^w S^{6/5}\, dx} \tag{4.38}$$

where ℓ is again any characteristic dimension of the surface, and terms are defined in Fig. 4.9. To find Nu, substitute Eq. 4.38 for Nu^T and Eq. 4.37c for Nu_{COND} into Eq. 4.37b to obtain an expression for Nu_ℓ. Use this Nu_ℓ and Eq. 4.36c for Nu_t in Eq. 4.36d to obtain Nu.

Horizontal Heated Upward-Facing Plates ($\zeta = 0°$) With Uniform T_w and T_∞

Correlation. For horizontal isothermal plates of various planforms with unrestricted inflow at the edges as shown in Fig. 4.10, the heat transfer is correlated for $1 < \text{Ra} < 10^{10}$ by the equation:

$$\text{Nu}^T = 0.835 \overline{C}_\ell\, \text{Ra}^{1/4} \tag{4.39a}$$

$$\text{Nu}_\ell = \frac{1.4}{\ln(1 + 1.4/\text{Nu}^T)} \tag{4.39b}$$

$$\text{Nu}_t = C_t^U\, \text{Ra}^{1/3} \tag{4.39c}$$

$$\text{Nu} = ((\text{Nu}_\ell)^m + (\text{Nu}_t)^m)^{1/m} \qquad m = 10 \tag{4.39d}$$

Use of the length scale L^* [116], defined in Fig. 4.10c, is intended to remove explicit dependence on the planform from the correlation.

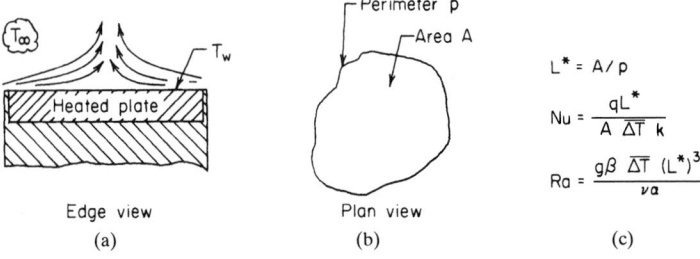

FIGURE 4.10 Definition sketch for natural convection on a horizontal upward-facing plate of arbitrary planform. Only the top heated surface of area A is heated.

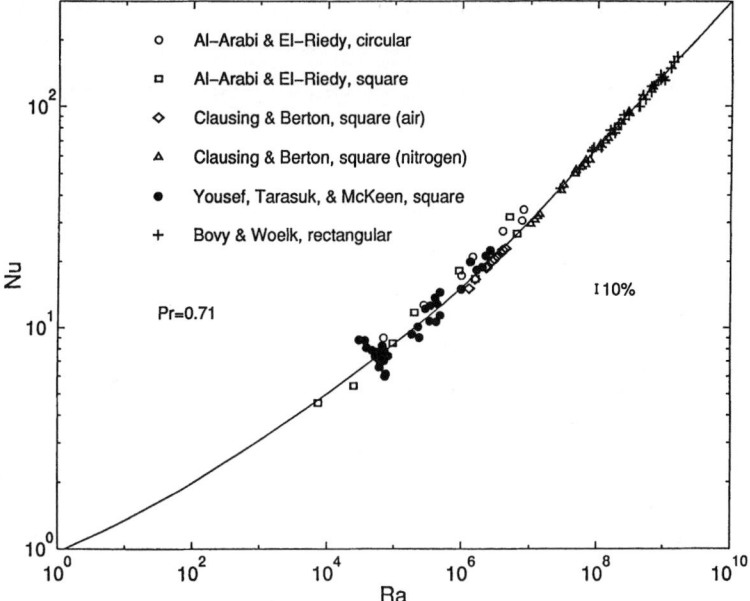

FIGURE 4.11   Comparison of Eq. 4.39 to data for upward-facing heated plates of various planform in air.

Comparison With Data.   Figure 4.11 shows that Eq. (39) is in good agreement with measurements for gases (Pr ≈ 0.7). The Clausing and Berton [62] data shown in the figure have been extrapolated to $T_w/T_\infty = 1$ using their correlation. The scatter in the data of Yousef et al. [290] is due to temporal changes in the heat transfer. Excellent agreement was also found with the data of Goldstein et al. [116] for 1.9 < Pr < 2.5 for a variety of shapes, but the data of Sahraoui et al. [237] for a disk and flat annular ring, 1 < Ra < 10^3, fall below this equation. For Pr > 100, Eq. 4.39 is in excellent agreement with the measurements of Lloyd and Moran [297] for 10^7 < Ra < 10^9, and agrees, within the experimental scatter, with the measurements of Lewandowski et al. [179] for 10^2 < Ra < 10^4.

Horizontal Heated Upward-Facing Plates (ζ = 0°) With Uniform Heat Flux
Correlation.   The nomenclature for this problem is also given in Fig. 4.10, where $\overline{\Delta T}$ is the surface average temperature difference. Equation 4.39 should also be used for this case, where the calculated Nu value provides the average temperature difference $\overline{\Delta T}$.

Comparison With Data.   Equation 4.39 agrees to within about 10 percent with the data of Fujii and Imura [103] and Kitamura and Kimura [161], for heat transfer in water (Pr ≈ 6), for 10^4 < Ra < 10^{11}. Both these experiments were performed using effectively infinite strips of finite width.

Horizontal Isothermal Heated Downward-Facing Plates (ζ = 180°)
Correlation.   Definitions and a typical flow pattern for this problem are shown in Fig. 4.12. The heat transfer relations given here assume that the downward-facing surface is substantially all heated; if the heated surface is set into a larger surface, the heat transfer will be reduced. Since the buoyancy force is mainly into the surface, laminar flow prevails up to very high Rayleigh numbers. The following equation can be used for 10^3 < Ra ≤ 10^{10}:

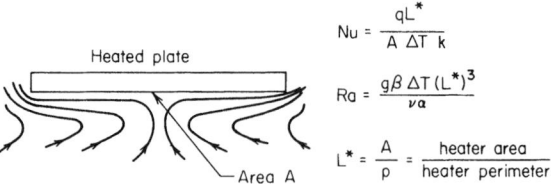

FIGURE 4.12 Definition sketch for natural convection on a downward-facing plate. Only the bottom surface of area A is heated.

$$\mathrm{Nu}^T = \frac{0.527}{(1 + (1.9/\mathrm{Pr})^{9/10})^{2/9}}\,\mathrm{Ra}^{1/5} \qquad (4.40a)$$

$$\mathrm{Nu}_\ell = \frac{2.5}{\ln(1 + 2.5/\mathrm{Nu}^T)} \qquad (4.40b)$$

The coefficient in Eq. 4.40a was obtained by fitting the results of the integral analysis of Fujii et al. [102].

Comparison With Data. Measurements for isothermal plates in air are compared to Eq. 4.40 in Fig. 4.13 for rectangular plates. Data lie within about ±20 percent of the correlation. Measurements have also been done using water, but only with a uniform heat flux boundary condition. For water, the data of Fujii and Imura [103] for a simulated 2D strip lie about 30 percent below Eq. 4.40, but the data of Birkebak and Abdulkadir [20] lie about 3 percent

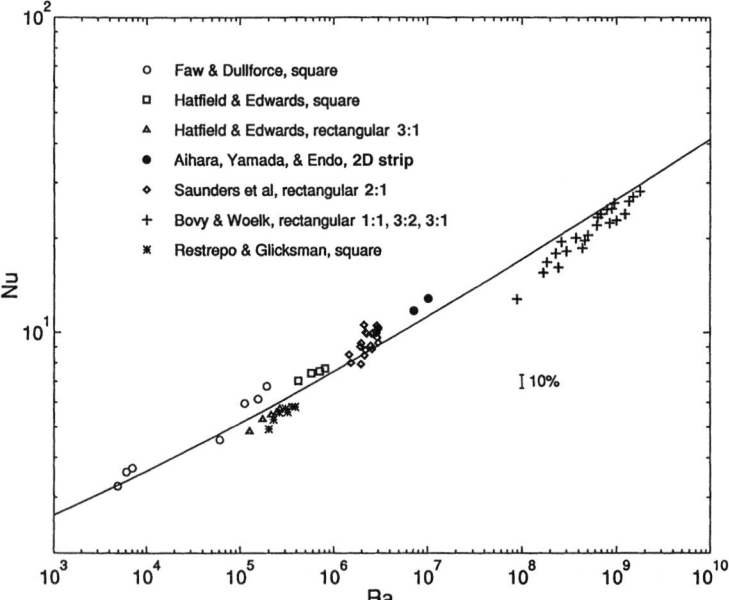

FIGURE 4.13 Comparison of Eq. 4.40 to data for downward-facing heated plates of various planform in air.

above the correlation. The importance of the heat transfer from the outer edge of the plate is believed to be the cause of such large discrepancies.

Plates at Arbitrary Angle of Tilt.   The previous sections provide equations from which to compute the total heat transfer from vertical plates ($\zeta = 90°$), horizontal upward-facing plates ($\zeta = 0°$), and horizontal downward-facing surfaces ($\zeta = 180°$). These equations are the basis for obtaining the heat transfer from tilted plates.

For a wide isothermal plate at any angle of tilt, first compute the heat transfer from Eq. 4.33 (vertical plate) but with g replaced by $g \sin \zeta$, then compute the heat transfer from Eq. 4.40 (downward-facing plate) with g replaced by $g(0, -\cos \zeta)_{max}$, then compute the heat transfer from Eq. 4.39 (upward-facing plate) with g replaced by $g(0, \cos \zeta)_{max}$, and take the maximum of the three heat transfer rates. It is important to take maximum heat transfer rather than the maximum Nusselt number, because the Nusselt numbers are based on different length scales.

For plates with small aspect ratio, such as shown in Fig. 4.9, follow the same procedure except use Eq. 4.37b in place of Eq. 4.33b.

Vertical Isothermal Plate in Stably Stratified Ambient

Correlation.   For an isothermal vertical plate (see Fig. 4.6) in an ambient fluid whose temperature T_∞ increases linearly with height x, the heat transfer depends on the stratification parameter **S**, defined by

$$S = \frac{L}{\overline{\Delta T}} \frac{dT_\infty}{dx} \qquad (4.41)$$

where the mean temperature difference $\overline{\Delta T}$ is also the value of $T_w - T_\infty$ at the mid-height of the surface. For $0 \le S < 2$, $T_w - T_\infty$ is positive over the entire plate; for $S = 2$, $T_w = T_\infty$ at $x = L$; and for $S \to \infty$ the plate temperature is lower than T_∞ over the top half of the plate and greater than T_∞ over the bottom half.

From laminar thin-layer analysis [41, 225, 226] the value of Nu^T, corrected for stratification effects, is

$$Nu^T = (1 + S/a)^b \overline{C}_\ell \, Ra^{1/4} \qquad S \le 2 \qquad (4.42a)$$

$$= cS^{1/4} \overline{C}_\ell \, Ra^{1/4} \qquad S \ge 2 \qquad (4.42b)$$

For gases: $a = 1$, $b = 0.38$, and $c = 1.28$; for water ($Pr \approx 6$): $a = 2$, $b = 0.5$, and $c = 1.19$. For turbulent flow everywhere on the plate and for $S \le 2$, Nu_t is given by

$$Nu_t \approx \frac{3}{7S} \left[\left(1 + \frac{S}{2}\right)^{7/3} - \left(1 - \frac{S}{2}\right)^{7/3} \right] \frac{C_t^V \, Ra^{1/3}}{(1 + 1.4 \times 10^9 \, Pr/Ra)}; \qquad S \le 2 \qquad (4.43)$$

To estimate Nu at intermediate Ra for $S \le 2$, substitute Eq. 4.42 for Nu^T into Eq. 4.33b to obtain Nu_ℓ and use this Nu_ℓ and Nu_t from Eq. 4.43 in Eq. 4.33d to find the Nusselt number.

Confirmation of Procedure.   This procedure leads to good agreement with the data of Chen and Eichhorn [41] for water, for $2 \times 10^6 < Ra < 3 \times 10^7$, but it is unconfirmed outside this range. The accuracy of Eq. 4.43 is particularly questionable through transition since the transition to turbulence depends on S.

Cylinders

Long Vertical Cylinders, Circular or Noncircular

Correlation.   The objective is to calculate the heat transfer from the lateral surface of the long vertical cylinder or wire, where heat transfer from the ends is ignored. See Fig. 4.14 for relevant nomenclature.

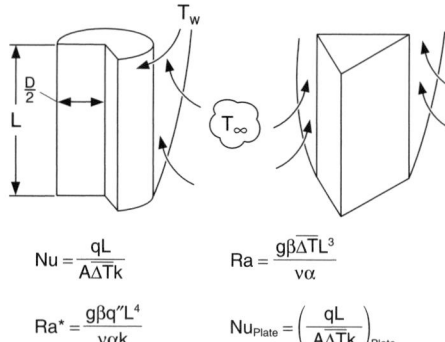

$$Nu = \frac{qL}{A\Delta Tk} \qquad Ra = \frac{g\beta\overline{\Delta T}L^3}{\nu\alpha}$$

$$Ra^* = \frac{g\beta q''L^4}{\nu\alpha k} \qquad Nu_{Plate} = \left(\frac{qL}{A\overline{\Delta T}k}\right)_{Plate}$$

FIGURE 4.14   Definition sketch for natural convection from long vertical cylinders.

For a vertical cylinder of length L and diameter D, first calculate the Nusselt number Nu^T and Nu_ℓ for a vertical flat plate of height L, using Eq. 4.33 if the cylinder is isothermal and Eq. 4.36 if it has uniform heat flux. These Nusselt numbers are based on the length L of the plate (see Fig. 4.6) and are renamed Nu^T_{Plate} and $Nu_{\ell,Plate}$ here. The laminar Nusselt number for the vertical cylinder (defined in Fig. 4.14), Nu_ℓ, is then calculated from

$$Nu_\ell = \frac{\zeta}{\ln(1+\zeta)} Nu_{\ell,Plate}; \qquad \zeta = \frac{1.8L/D}{Nu^T_{Plate}} \qquad (4.44)$$

For an isothermal boundary, calculate Nu_t from Eq. 4.33c and substitute this Nu_t and Eq. 4.44 for Nu_ℓ into Eq. 4.33d to obtain Nu. For an isoflux boundary, calculate Nu_t from Eq. 4.36c, calculate Nu_ℓ from Eq. 4.44, and find Nu by substituting these values of Nu_t and Nu_ℓ into Eq. 4.36d.

For a cylinder of noncircular convex cross section, such as the triangular cylinder shown in Fig. 4.14, follow the same procedure but replace D in Eq. 4.44 by $D_{\text{eff}} = P_i/\pi$, where P_i is the perimeter of the cylinder.

Discussion.   This procedure is compared to available measurements in Fig. 4.15. The predictions fall about 10 percent higher than the measurements of Nagendra et al. [203] for a uniform flux into water. There is excellent agreement with the measurements of Li and Tarasuk [180] for $14 < L/D < 21$ for a uniform wall temperature in air, and predictions fall slightly below the measurements of Al-Arabi and Salman [6], also for air. There is also good agreement with the data of Kyte et al. [172] for a vertical wire of extremely small diameter ($D = 0.08$ mm, $L/D = 5430$) in air where the transverse curvature correction in Eq. 4.44 is very large (the measured Nu values in Fig. 4.15 fall below the predictions at the low end of the Nu range because of rarefied gas effects).

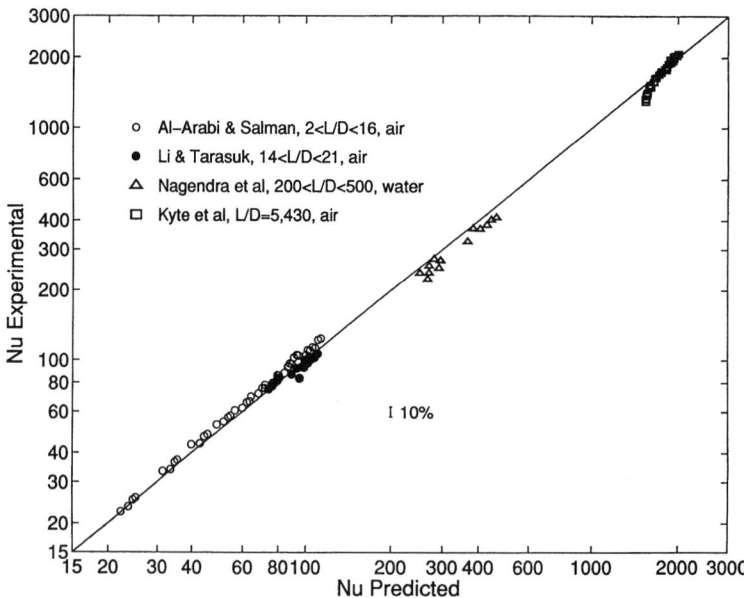

FIGURE 4.15   Comparison of recommended procedure for calculating heat transfer to data for vertical cylinders in air and water.

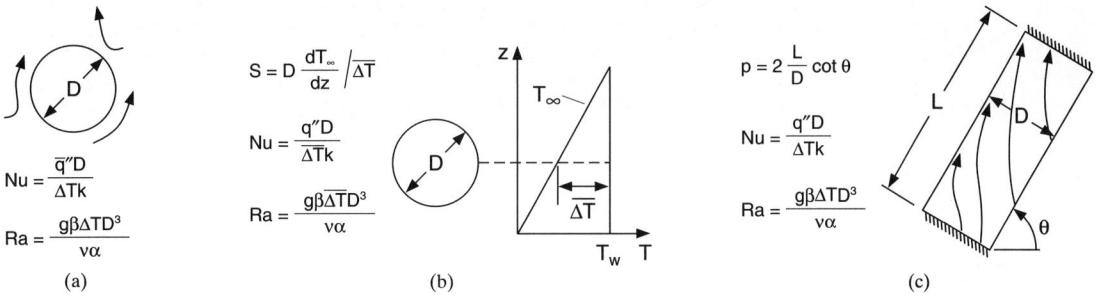

FIGURE 4.16 Definition sketch for long horizontal cylinder in an isothermal medium (*a*), in a stratified medium (*b*), and for a tilted cylinder (*c*).

The present recommendation leads to good agreement with existing data, but none of these data lie in the turbulent range, and the tested range of Pr and L/D is also limited.

A correlation for short vertical cylinders, including heat transfer from the ends, is provided in the section on bodies with small aspect ratios.

Long Horizontal Circular Cylinders

Correlation. For a long isothermal horizontal circular cylinder in an isothermal environment, and using the nomenclature in Fig. 4.16, Nu is obtained from the following equations:

$$\text{Nu}^T = 0.772\overline{C}_\ell\,\text{Ra}^{1/4} \tag{4.45a}$$

$$\text{Nu}_\ell = \frac{2f}{\ln(1 + 2f/\text{Nu}^T)} \qquad f = 1 - \frac{0.13}{(\text{Nu}^T)^{0.16}} \tag{4.45b}$$

$$\text{Nu}_t = \overline{C}_t\,\text{Ra}^{1/3} \tag{4.45c}$$

$$\text{Nu} = [(\text{Nu}_\ell)^m + (\text{Nu}_t)^m]^{1/m} \qquad m \approx 10 \tag{4.45d}$$

$\overline{C}_t$ can be found from Table 4.2 (elliptical cylinders, $C/L = 1.0$). If $\text{Ra} \geq 10^{-4}$, the expression for f in Eq. 4.45b can be replaced by $f = 0.8$.

Discussion. Equation 4.45 follows the recommendation of Raithby and Hollands [227] except that the value of m has been improved. There is a large body of data for this problem, because the shape is so frequently encountered in practice. Five sets of data, thought to be especially reliable, are compared to Eq. 4.45 in Fig. 4.17 for air. The RMS deviation from the equation is less than 5 percent for $10^{-10} < \text{Ra} < 10^7$. Similar agreement is found with the data of Chen and Eichhorn [42] for water and with the data of Schütz [300] for $\text{Pr} \approx 2000$. High Rayleigh number data are still needed to confirm the value of $\overline{C}_t$ in Eq. 4.45c and Table 4.2.

For a uniform heat flux condition, apply Eq. 4.45 to obtain the average temperature difference $\overline{\Delta T}$.

Stratified Medium. For a long horizontal circular cylinder in a thermally stratified environment in which the temperature increases linearly with height (see Fig. 4.16b for nomenclature), $\overline{\Delta T}$ is the temperature difference at the mid-height of the cylinder. First calculate the laminar isothermal Nusselt number Nu_ℓ from Eq. 4.45 with $\Delta T = \overline{\Delta T}$ and rename it $\text{Nu}_{\ell,\text{iso}}$; the corresponding calculated heat flow is q_{iso}. The laminar Nusselt number Nu_ℓ corrected to account for the stratification is then estimated from

$$\frac{q}{q_{\text{iso}}} = \frac{\text{Nu}_\ell}{\text{Nu}_{\ell,\text{iso}}} = \left(1 + \frac{S}{a}\right)^b \qquad S \leq 2 \tag{4.46a}$$

$$= cS^{1/4} \qquad S \geq 2 \tag{4.46b}$$

For gases: $a = 1$, $b = 0.39$, and $c = 1.29$. For water ($\text{Pr} \approx 6$): $a = 2$, $b = 0.53$, and $c = 1.21$.

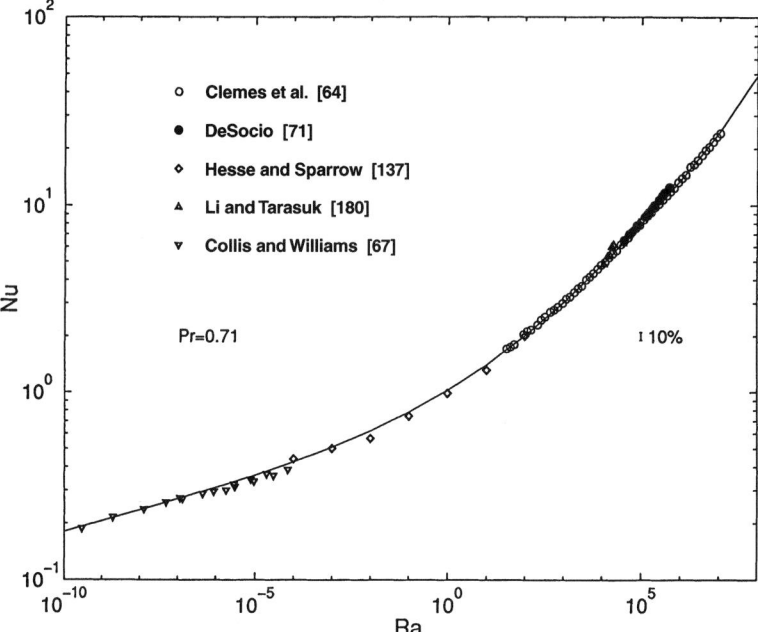

FIGURE 4.17 Comparison of Eq. 4.45 to data for long, horizontal, isothermal cylinders in air.

Discussion of Stratified Medium.   Equation 4.46 was obtained by fitting the analytical results of Chen and Eichhorn [42]. There is agreement with their measurements of Nu/Nu_{iso} to within about 10 percent. All their data were obtained for $2 \times 10^5 < Ra < 3 \times 10^7$. Use of Eq. 4.46 is not recommended for Ra much outside this range, because turbulent heat transfer will not be properly accounted for at larger Ra and because thick boundary layer effects may be poorly approximated for smaller Ra.

Long Inclined Circular Cylinder, Laminar Flow.   See Fig. 4.16c for nomenclature. The cylinder axis is inclined from the horizontal at an angle θ. The heat transfer from the ends is ignored, which will be valid for insulated ends or for $L/D \geq 5$. The analysis of Raithby and Hollands [225, 226] led to the following correlation:

$$Nu^T = \left\{ 0.772 + \frac{0.228}{1 + 0.676p^{1.23}} \right\} \left(\cos \theta + \frac{D}{L} \sin \theta \right)^{1/4} \overline{C}_\ell \, Ra^{1/4} \qquad (4.47a)$$

$$Nu_\ell = \frac{1.8}{\ln (1 + 1.8/Nu^T)} \qquad p = 2\frac{L}{D} \cot \theta \qquad (4.47b)$$

Equation 4.47a can be shown to reduce exactly to Eqs. 4.33a and 4.45a for the vertical ($\theta = 90°$) and horizontal ($\theta = 0°$) orientation, respectively.

Discussion.   Equation 4.47 agrees with the data of Oosthuizen [213] and Li and Tarasuk [180] to within RMS and maximum errors of 6 and 12 percent, respectively. The data range is $2.0 \times 10^4 \leq Ra \leq 2.5 \times 10^5$, $Pr = 0.71$, $8 < LD < 21$, and $0° \leq \theta \leq 90°$, and the fluid is air. Figure 4.18 presents a comparison for one case, and shows that the heat transfer falls as the cylinder is rotated from the horizontal ($\theta = 0°$) to the vertical ($\theta = 90°$) orientation.

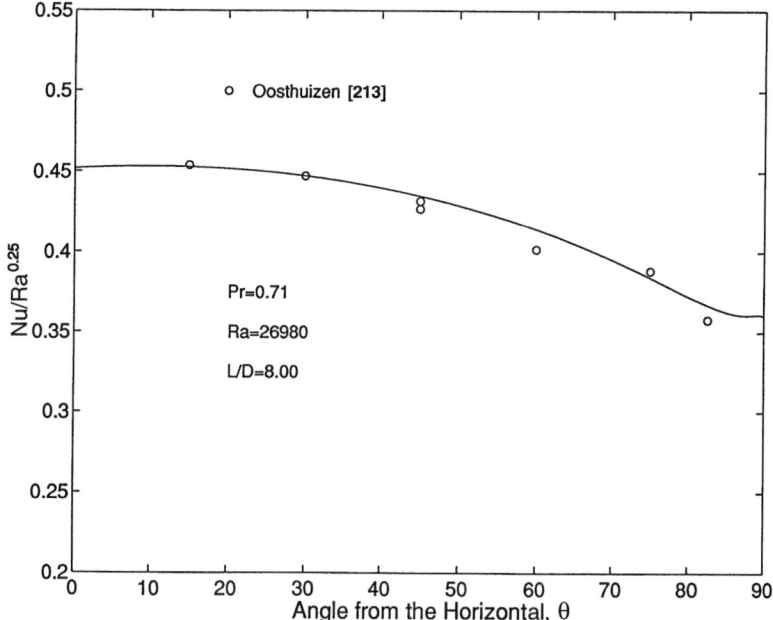

FIGURE 4.18 Dependence of heat transfer on the inclination angle of a cylinder (defined in Fig. 4.16c): comparison of Eq. 4.47 with the data of Oosthuizen [213].

Long Horizontal Noncircular Cylinders

Correlation for Specific Shapes.   For the four shapes shown in Table 4.2, the heat transfer is calculated from the following equation:

$$Nu^T = G\overline{C}_\ell\, Ra^{1/4} \tag{4.48a}$$

$$Nu_\ell = \frac{C_2}{\ln\,(1 + C_2/Nu^T)} \tag{4.48b}$$

$$Nu_t = \overline{C}_t\, Ra^{1/3} \tag{4.48c}$$

$$Nu = [(Nu_\ell)^m + (Nu_t)^m]^{1/m} \tag{4.48d}$$

The constants G, C_2, $\overline{C}_t$, and m are tabulated in Table 4.2, which also gives the definitions of Nu and Ra for each cylinder shape in addition to the equation for the perimeter P that is required to find the surface area.

Discussion of Correlation for Specific Shapes.   The equation for elliptical cylinders fits the approximate analysis of Raithby and Hollands [224] but has not been verified by experiment except in the limiting cases of a vertical plate ($C/L = 0$) and a circular cylinder ($C/L = 1.0$). The vertical plate predictions by Eq. 4.48 are slightly different than those based on the more accurate specialized equations given in the section on external natural convection in flat plates.

Equation 4.48 agrees with the measurements by Clemes et al. [64] for square cylinders at angles of rotation of 0°, 15°, 30°, and 45° in air (Pr = 0.71) for $10^1 < Ra < 10^7$ to an RMS deviation of less than 3 percent and a maximum error of 8 percent. Similar comparisons for the semicircular cylinder at $\theta = 0°$, 45°, 90°, –45°, and –90° give RMS and maximum deviations from the measurements of Clemes et al. [64] of 2.5 and 5 percent, respectively. For the slotted cylinder, the RMS and maximum deviations were 2 and 5 percent, respectively.

TABLE 4.2 Constants for Use in Eq. 4.48 for Long Horizontal Cylinders in an Isothermal Environment

Geometry and geometric parameters	C/L / θ	f_1	G	C_2	$\bar{C}_t$ Pr = 0.71	6.0	100	2000	m	Comments
Isothermal elliptical cylinder $\mathrm{Nu} = \bar{h}L/k$, $\mathrm{Ra} = \dfrac{g\beta\Delta TL^3}{\nu\alpha}$, $P = Lf_1$										Fit to analysis of Raithby and Hollands [224]
	1.0	3.142	0.772	1.6	0.103	0.109	0.097	0.088	10	
	0.8	2.836	0.819	1.7	0.103	0.109	0.096	0.087	10	See separate sections for
	0.6	2.553	0.870	1.8	0.103	0.110	0.095	0.084	10	circular cylinder (C/L = 1.0)
	0.4	2.301	0.924	1.9	0.103	0.111	0.094	0.080	10	and vertical plate (C/L = 0.0)
	0.2	2.101	0.973	2.0	0.103	0.112	0.092	0.073	10	
	0.1	2.032	0.991	2.0	0.103	0.112	0.091	0.068	10	
	0.0	2.000	1.000	2.0	0.103	0.113	0.091	0.064	10	
Square cylinder $\mathrm{Nu} = \bar{h}L/k$, $\mathrm{Ra} = \dfrac{g\beta\Delta TL^3}{\nu\alpha}$, $P = 4L$	θ									Fits data of Clemes et al. [64] for air to RMS deviation of <3%
	0°		0.735	1.3	0.087	0.091	0.080	0.070	4.5	
	15°		0.720	1.3	0.102	0.108	0.094	0.087	4.5	
	30°		0.786	1.3	0.106	0.110	0.101	0.094	4.5	
	45°		0.797	1.3	0.108	0.113	0.103	0.095	4.5	$10^3 < \mathrm{Ra} < 10^9$
	60°		same as θ = 30°							
	75°		same as θ = 15°							
Semicircular cylinder $\mathrm{Nu} = \bar{h}D/k$, $\mathrm{Ra} = \dfrac{g\beta\Delta TD^3}{\nu\alpha}$, $P = 2.57D$	θ									Fits data of Clemes et al. [64] for air with RMS deviation <3%
	0°		0.900	2.7	0.103	0.111	0.095	0.079	3.5	
	45°		0.891	2.7	0.102	0.106	0.097	0.091	4.5	
	90°		0.808	2.7	0.077	0.081	0.072	0.065	15	
	-45°		0.874	2.7	0.102	0.106	0.097	0.091	4.5	$10 < \mathrm{Ra} < 10^8$
	-90°		0.776	2.7	0.097	0.098	0.095	0.099	4.0	
Slotted cylinder $\mathrm{Nu} = \bar{h}L/k$, $\mathrm{Ra} = \dfrac{g\beta\Delta TL^3}{\nu\alpha}$, $P = 3.63D$	θ									Fits data of Clemes et al. [64] for air with RMS deviation <2%
	0°		0.700	2.1	0.098	0.104	0.093	0.083	6.0	
	30°		0.700	2.1	0.104	0.108	0.098	0.091	6.0	
	60°		0.700	2.1	0.105	0.110	0.098	0.090	6.0	
	90°		0.700	2.1	0.095	0.100	0.089	0.080	6.0	$10 < \mathrm{Ra} < 10^7$

4.24

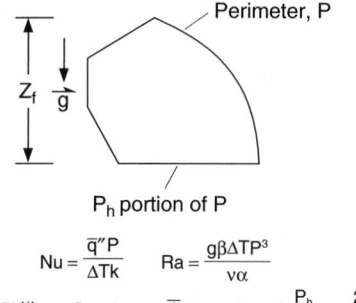

$$\text{Nu} = \frac{\overline{q}''P}{\Delta Tk} \qquad \text{Ra} = \frac{g\beta\Delta TP^3}{\nu\alpha}$$

$$G = (4Z_f/P)^{1/4} \qquad C_2 = 2\pi \qquad \overline{C}_t = a - (a-b)\frac{P_h}{P} + e\frac{2Z_f}{P}$$

$$a = 0.0972 \qquad b = 0.0815 - 0.462 C_t^V$$

$$e = 0.615 C_t^V - 0.0548 - 6 \times 10^{-6} \text{Pr}\, C_t^V$$

FIGURE 4.19 Definition sketch for long horizontal convex cylinder of arbitrary shape. The constants G, C_2, and $\overline{C}_t$ are for use in Eq. 4.48.

There appears to be no verification of Eq. 4.48 for these shapes for fluids other than air, and for Rayleigh numbers (defined in Table 4.2) outside the range $10^1 < \text{Ra} < 10^8$. The turbulent asymptote of the equations has therefore not been verified for any fluid.

General Correlation for Convex Horizontal Cylinders. For cylinders of arbitrary convex shape, such as shown in Fig. 4.19, Eq. 4.48 can be used to find the Nusselt number, provided the coefficients G, C_2, and $\overline{C}_t$ are known. General expressions for these coefficients, based on the recommendations of Clemes et al. [64] and Hassani [125], are given in Fig. 4.19. Note that Nu and Ra in Fig. 4.19 are based on the perimeter of the cylinder, P. Z_f is the height of the cylinder, C_t^V is given by Eq. 4.24, and P_h/P is the fraction of the perimeter that lies in the horizontal plane and faces downward for a heated cylinder (it is the fraction that is horizontal and upward-facing for a cooled cylinder). This correlation gives results in close agreement with Eq. 4.48 (using the constants in Table 4.2 and allowing for the different definitions of Nu and Ra). There is also agreement to within about 12 percent with the data of Nakamura and Asako [204] for cylinders of modified square and triangular cross section in the range $3 \leq \text{Pr} \leq 8$.

Bodies with Small Aspect Ratios

Correlations for 3D and Axisymmetric Flows. For bodies of "small aspect ratio," there is no single dimension that greatly exceeds all other dimensions. Examples are shown in Table 4.3a. For these bodies, the correlation equation has the form

$$\text{Nu}^T = G\overline{C}_\ell\, \text{Ra}^{1/4} \tag{4.49a}$$

$$\text{Nu}_\ell = [(\text{Nu}_{\text{COND}})^n + (\text{Nu}^T)^n]^{1/n} \tag{4.49b}$$

$$\text{Nu}_t = \overline{C}_t\, \text{Ra}^{1/3} \tag{4.49d}$$

$$\text{Nu} = [\text{Nu}_\ell^m + \text{Nu}_t^m]^{1/m} \tag{4.49c}$$

The length scales on which Nu and Ra are based, and values of the constants, are provided in Table 4.3a. The basis for these relations was discussed in the section on heat transfer correlation method.

In the rightmost column in Table 4.3a, NA is an abbreviation for *none available,* meaning that the correlation is based entirely on the approximate method [227]. Values of $n = 1.07$, recommended by Hassani and Hollands [126, 127], and $m = 10$ are used for these cases. When data are available, the values of G have been adjusted to provide a best fit of the data. Except for the thin oblate spheroid, $C/D = 0.1$, the value G was never adjusted by more than a few percent from the value predicted by the approximate method. $\overline{C}_t$ was never adjusted, mainly because so few of the available data fall in the turbulent regime. Table 4.3b provides references to all data used, and the RMS error (E_{RMS}) and maximum error (E_{MAX}) from Eq. 4.49 fit using the constants in Table 4.3a. The range of Ra and Pr covered by the data is also noted.

The Nusselt number in the Ra $\to$ 0 limit is the conduction Nusselt number Nu_{COND}. This was calculated in all cases using the recommendations of Yovanovich et al. [291–294].

Table 4.3b shows that Eq. 4.49 fits the data very closely. The range of data is, however, limited.

For body shapes not treated in Table 4.3, the general correlation of Hassani and Hollands [126, 127] is recommended.

Correlations for Spheres in a Thermally Stratified Medium. Consider the case of an isothermal sphere in a thermally stratified medium with constant vertical temperature gradient dT_∞/dz and with a temperature difference at the mid-height of the sphere of $\overline{\Delta T}$. A Nusselt number Nu_{iso} is first calculated for an isothermal sphere in an isothermal environment

TABLE 4.3a Constants for Use in Eq. 4.49 for Objects with Small Aspect Ratios (Axisymmetric and 3D), Isothermal Body, and Constant T_∞

Sphere

$$Nu = \frac{\overline{h}L}{k} \qquad Ra = \frac{g\beta\Delta T L^3}{\nu\alpha} \qquad A_s = \pi L^2$$

	Nu_{COND}	G	n	$\overline{C}_t$ Pr = 0.71	Pr = 6.0	Pr = 100	Pr = 2000	m	Data
	2.00	0.878	1.0	0.104	0.111	0.098	0.086	15	See Table 4.3b.

Prolate spheroid

$$Nu = \frac{\overline{h}L}{k} \qquad Ra = \frac{g\beta\Delta T L^3}{\nu\alpha} \qquad A_s = \pi L c\, f_1/2$$

C/L	f_1	Nu_{COND}	G	n	Pr = 0.71	Pr = 6.0	Pr = 100	Pr = 2000	m	Data
0.1	1.578	8.43	1.012	1.07	0.103	0.112	0.091	0.066	10	NA
0.2	1.598	5.35	1.005	1.07	0.103	0.112	0.091	0.069	10	NA
0.4	1.665	3.51	0.980	1.07	0.103	0.111	0.093	0.076	10	NA
0.5	1.709	3.08	0.964	1.07	0.103	0.111	0.094	0.078	10	See Table 4.3b.
0.6	1.759	2.76	0.948	1.07	0.103	0.111	0.095	0.080	10	NA
0.8	1.873	2.31	0.913	1.07	0.104	0.111	0.097	0.083	10	NA
1.0	2.000	See sphere								

Oblate spheroid

$$Nu = \frac{\overline{h}D}{k} \qquad Ra = \frac{g\beta\Delta T D^3}{\nu\alpha} \qquad A_s = \pi D^2 f_1/2$$

C/D	f_1	Nu_{COND}	G	n	Pr = 0.71	Pr = 6.0	Pr = 100	Pr = 2000	m	Data
0.1	1.030	2.63	0.713	1.04	0.094	0.097	0.094	0.090	10	NA
0.2	1.094	2.62	0.745	1.07	0.100	0.103	0.098	0.092	10	NA
0.4	1.274	2.48	0.845	1.07	0.103	0.108	0.100	0.092	10	NA
0.5	1.380	2.40	0.866	1.07	0.104	0.109	0.101	0.091	10	See Table 4.3b.
0.6	1.494	2.31	0.877	1.07	0.104	0.110	0.100	0.090	10	NA
0.8	1.739	2.15	0.884	1.07	0.104	0.110	0.099	0.088	10	NA
1.0	2.000	See sphere								

Short vertical circular cylinder—square ends

$$Nu = \frac{\overline{h}D}{k} \qquad Ra = \frac{g\beta\Delta T D^3}{\nu\alpha} \qquad A_s = \frac{\pi D^2}{2}\left(1 + \frac{2L}{D}\right)$$

L/D	Nu_{COND}	G	n	Pr = 0.71	Pr = 6.0	Pr = 100	Pr = 2000	m	Data
0.0	2.55	0.670	1.07	0.070	0.070	0.073	0.075	10	NA
0.1	2.44	0.730	1.06	0.076	0.077	0.076	0.073	15	See Table 4.3b.
0.5	1.93	0.792	1.07	0.087	0.091	0.082	0.069	10	NA
1.0	1.59	0.839	1.11	0.092	0.098	0.085	0.068	10	See Table 4.3b.
2.0	1.26	0.733	1.07	0.096	1.04	0.087	0.066	10	NA
4.0	0.99	0.657	1.07	0.099	1.08	0.089	0.065	10	NA

Short horizontal cylinder—square ends

$$Nu = \frac{\bar{h}D}{k}$$

$$Ra = \frac{g\beta\Delta TD^3}{v\alpha}$$

$$A_s = \frac{\pi D^2}{2}\left(1 + \frac{2L}{D}\right)$$

L/D									
0.0	2.55	1.052	1.07	0.103	0.113	0.091	0.064	10	NA
0.1	2.44	1.005	1.11	0.103	0.112	0.092	0.068	10	See Table 4.3b.
0.5	1.93	0.912	1.07	0.103	0.111	0.094	0.076	10	NA
1.0	1.59	0.889	1.11	0.103	0.110	0.095	0.080	8	See Table 4.3b.
2.0	1.26	0.828	1.07	0.103	0.110	0.096	0.083	10	NA
4.0	0.99	0.803	1.07	0.103	0.109	0.096	0.085	10	NA

Short tilted cylinder—flat ends, L/D = 1

$$Nu = \frac{\bar{h}D}{k}$$

$$Ra = \frac{g\beta\Delta TD^3}{v\alpha}$$

$$A_s = \frac{\pi D^2}{2}\left(1 + \frac{2L}{D}\right)$$

$\theta = 0°$	See short vertical circular cylinder								
$\theta = 45°$	1.59	0.897	1.18	0.106	0.112	0.101	0.089	11	See Table 4.3b.
$\theta = 90°$	See short horizontal circular cylinder								

Vertical cylinder with spherical end caps

$$Nu = \frac{\bar{h}D}{k}$$

$$Ra = \frac{g\beta\Delta TD^3}{v\alpha}$$

$$A_s = \pi DL$$

L/D									
1.0	See sphere								
2.0	1.42	0.815	1.03	0.104	0.112	0.095	0.075	12	See Table 4.3b.
3.0	1.18	0.742	1.07	0.103	0.112	0.093	0.071	10	NA
4.0	1.05	0.695	1.07	0.103	0.113	0.093	0.070	10	NA
5.0	0.96	0.659	1.07	0.103	0.113	0.092	0.068	10	NA

Horizontal cylinder with spherical end caps

$$Nu = \frac{\bar{h}D}{k}$$

$$Ra = \frac{g\beta\Delta TD^3}{v\alpha}$$

$$A_s = \pi DL$$

L/D									
1.0	See sphere								
2.0	1.42	0.873	1.1	0.104	0.110	0.098	0.087	9.5	See Table 4.3b.
3.0	1.18	0.807	1.07	0.103	0.110	0.097	0.087	10	NA
4.0	1.05	0.799	1.07	0.103	0.110	0.097	0.087	10	NA
5.0	0.96	0.793	1.07	0.103	0.109	0.097	0.088	10	NA

4.27

TABLE 4.3a Constants for Use in Eq. 4.49 for Objects with Small Aspect Ratios (Axisymmetric and 3D), Isothermal Body, and Constant T_∞ (*Continued*)

Geometry and geometric parameters	Nu_{COND}	G	n	$\bar{C}_t$ Pr = 0.71	6.0	100	2000	m	Data
Bisphere									
$Nu = \bar{h}D/k$									
$Ra = \dfrac{g\beta\Delta T D^3}{\nu\alpha}$	1.39	0.745	1.02	0.104	0.111	0.098	0.086	15	See Table 4.3b.
$A_s = 2\pi D^2$									
Short vertical square cylinder									
$Nu = \bar{h}D/k$ L/D									
0.0	2.26	0.670	1.07	0.070	0.070	0.072	0.075	10	NA
0.1	2.16	0.730	1.0	0.076	0.076	0.076	0.073	15	See Table 4.3b.
0.5	1.68	0.792	1.07	0.087	0.092	0.082	0.070	10	NA
$Ra = \dfrac{g\beta\Delta T D^3}{\nu\alpha}$ 1.0	See cube face up								
2.0	1.08	0.731	1.07	0.096	0.104	0.087	0.066	10	NA
$A_s = 2D^2\left(1 + \dfrac{2L}{D}\right)$ 4.0	0.83	0.656	1.07	0.092	0.108	0.081	0.059	10	NA
Cube—corner up									
$Nu = \bar{h}D/k$									
$Ra = \dfrac{g\beta\Delta T D^3}{\nu\alpha}$	1.38	0.825	1.1	0.103	0.113	0.091	0.064	5.0	See Table 4.3b.
$A_s = 6D^2$									
Cube—edge up									
$Nu = \bar{h}D/k$									
$Ra = \dfrac{g\beta\Delta T D^3}{\nu\alpha}$	1.38	0.815	1.1	0.106	0.113	0.094	0.085	6.0	See Table 4.3b.
$A_s = 6D^2$									

Cube—face up

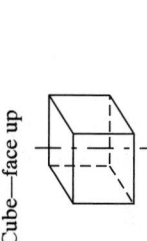

$$Nu = \bar{h}D/k$$
$$Ra = \frac{g\beta\Delta TD^3}{\nu\alpha}$$
$$A_s = 6D^2$$

1.38	0.758	1.05	0.092	0.099	0.085	0.068	4.5	See Table 4.3b.

Short square cylinder—edge up

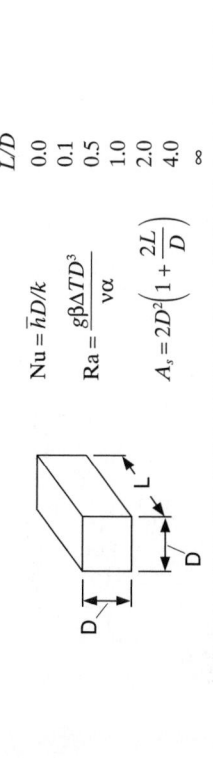

$$Nu = \bar{h}D/k$$
$$Ra = \frac{g\beta\Delta TD^3}{\nu\alpha}$$
$$A_s = 2D^2\left(1 + \frac{2L}{D}\right)$$

L/D									
0.0	2.26	1.048	1.07	0.103	0.113	0.091	0.064	10	NA
0.1	2.16	1.036	1.2	0.104	0.113	0.093	0.069	15	NA
0.5	1.68	0.910	1.07	0.106	0.113	0.097	0.080	10	See Table 4.3b.
1.0	See cube edge up								
2.0	1.08	0.826	1.07	0.107	0.113	0.101	0.089	10	NA
4.0	0.83	0.802	1.07	0.107	0.113	0.102	0.090	10	NA
∞	See Table 4.2								

Short square cylinder—face up

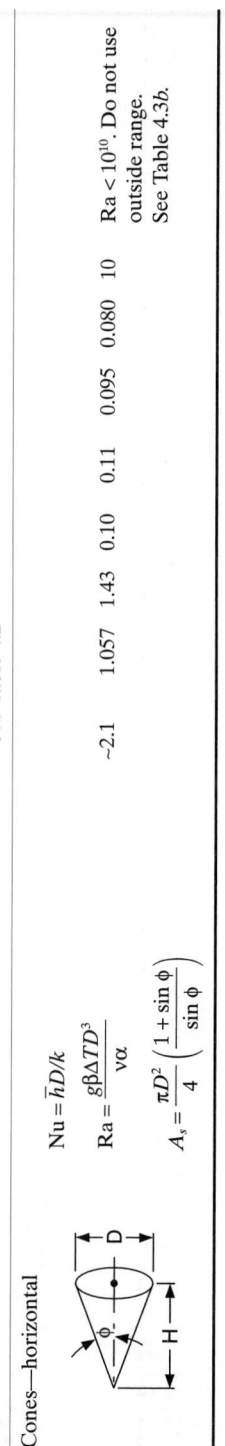

$$Nu = \bar{h}D/k$$
$$Ra = \frac{g\beta\Delta TD^3}{\nu\alpha}$$
$$A_s = 2D^2\left(1 + \frac{2L}{D}\right)$$

L/D									
0.0	2.26	1.000	1.07	0.103	0.113	0.091	0.064	10	NA
0.1	2.16	1.010	1.2	0.100	0.110	0.089	0.065	10	NA
0.5	1.68	0.836	1.07	0.095	0.102	0.086	0.067	10	See Table 4.3b.
1.0	See cube face up								
2.0	1.08	0.735	1.07	0.090	0.096	0.084	0.068	10	NA
4.0	0.83	0.704	1.07	0.083	0.094	0.083	0.069	10	NA
∞	See Table 4.2								

Cones—horizontal

$$Nu = \bar{h}D/k$$
$$Ra = \frac{g\beta\Delta TD^3}{\nu\alpha}$$
$$A_s = \frac{\pi D^2}{4}\left(\frac{1+\sin\phi}{\sin\phi}\right)$$

~2.1	1.057	1.43	0.10	0.11	0.095	0.080	10	Ra < 10^10. Do not use outside range. See Table 4.3b.

TABLE 4.3a Constants for Use in Eq. 4.49 for Objects with Small Aspect Ratios (Axisymmetric and 3D), Isothermal Body, and Constant T_∞ (*Continued*)

Geometry and geometric parameters		$\mathrm{Nu_{COND}}$	G	n	$\overline{C}_l$ Pr =				m	Data
					0.71	6.0	100	2000		
L corner $\mathrm{Nu} = \dfrac{\overline{h}L}{k}$ $\mathrm{Ra} = \dfrac{g\beta\Delta T L^3}{\nu\alpha}$ $A_s = 2LW$	$W/L \to \infty$	0.0	0.0	1.0	—	0.154	—	—	1.0	$10^6 < \mathrm{Ra} < 10^9$, Pr = 6. Do not use outside range. See Table 4.3b.
V corner $\mathrm{Nu} = \dfrac{\overline{h}L}{k}$ $\mathrm{Ra} = \dfrac{g\beta\Delta T L^3}{\nu\alpha}$ $A_s = 2WL$	$W/L \to \infty$	0.0	0.938	1.0	—	0.0	—	—	1.0	$10^6 < \mathrm{Ra} \le 5.6 \times 10^7$, Pr = 6.
	$W/L \to \infty$		0.0	1.0	—	0.132	—	—	1.0	$5.6 \times 10^7 < \mathrm{Ra} < 2 \times 10^9$, Pr = 6. Do not use outside range. See Table 4.3b.

TABLE 4.3b Measurements for the Shapes in Table 4.3a. E_{RMS} and E_{MAX} Are the RMS and Maximum Differences of Eq. 4.49 From the Data

Shape	Reference	Ra range	Pr	E_{RMS} (%)	E_{MAX} (%)
Sphere	38, 39	10^1–10^8	0.71	0.7	4.1
	242	10^9–10^{12}	~6	10.7	18.5
	301	10^8–10^{10}	2000	5.2	9.5
	280	10^7–10^{11}	2000	9.7	18.9
Prolate spheroid					
$C/L = 0.52$	124, 126, 127	10^1–10^8	0.71	1.8	4.7
	229	10^3–10^7	0.71	6.0	9.3
Oblate spheroid					
$C/L = 0.5$	124, 126, 127	10^1–10^7	0.71	3.3	6.4
$C/L = 0.1$	229	10^3–10^6	0.71	7.2	11.9
	124, 126, 127	10^1–10^8	0.71	3.3	7.4
Short vertical cylinder					
$L/D = 0.1$	124, 126, 127	10^1–10^8	0.71	3.1	10.2
$LD = 1.0$	255	10^1–10^7	0.71	1.4	5.1
		10^4–10^5	0.71	5.0	5.9
Short horizontal cylinder					
$L/D = 0.1$	124, 126, 127	10–10^7	0.71	1.3	3.0
$L/D = 1.0$	124, 126, 127	10–10^7	0.71	1.0	2.9
$0.069 \leq L/D \leq 0.155$	301	10^2–10^5	0.71	~3.0	~10.0
Short inclined cylinder					
$L/D = 1.0, \theta = 45°$	124, 126, 127	10–10^7	0.71	2.4	9.9
Vertical cylinder with spherical end caps					
$L/D = 2.0$	124, 126, 127	1–10^7	0.71	2.0	4.9
Horizontal cylinder with spherical end caps					
$L/D = 2.0$	124, 126, 127	1–10^7	0.71		
Bisphere					
	124, 126, 127	1–10^7	0.71	1.9	5.6
	280	10^8–10^9	2000	?	32%
Short vertical square cylinder					
$L/D = 0.1$	124, 126, 127	10–10^8	0.71	2.7	4.9
Cube—corner up					
	38, 39	1–10^7	0.71	1.5	4.3
	262	10^5–10^7	0.71	0.9	1.8
	262	10^5–10^7	6.0	2.9	5.6
	280	10^8–10^{11}	2000		
Cube—edge up	38, 39	1–10^7	0.71	2.1	7.3
	262	10^3–10^6	0.71	1.9	3.2
	262	10^5–10^7	~60	2.5	3.7
	280	10^8–10^{10}	2000	7.9	19
Cube—face up	38, 39	10^2–10^7	0.71	0.9	2.6
	262	10^3–10^6	0.71	0.9	1.5
	262	10^5–10^7	6.0	5.3	9.7
	286	10^6–10^{10}	2000	7.8	15.1
	280	10^6–10^9	2000	8.4	14.2
Short square cylinder—edge up					
$L/D = 0.1$	124, 126, 127	10^2–10^7	0.71	1.7	4.2
$L/D = 1.0$	see Cube—edge up				
Short square cylinder—face up					
$L/D = 0.1$	124, 126, 127	10–10^8	0.71	1.6	3.7
$L/D = 1.0$	see Cube—face up				
Horizontal cones					
$3.5 < \phi < 11.5$	212	10^5–10^6	0.71	3.1	10.6
	280	10^6–10^{11}	2000	10.1	33.9
L corner	234	10^6–10^9	6.0	4.2	
V corner	234	10^6–10^9	6.0	3.4	4.0

NA, none available.

(use Eq. 4.49) with the constants in Table 4.3a and with a temperature difference of $\overline{\Delta T}$. The corresponding total heat flow is q_{iso}. The actual heat flow q is corrected to account for the stratification as follows:

$$\frac{q}{q_{iso}} = \frac{\text{Nu}}{\text{Nu}_{iso}} = \left(1 + \frac{S}{a}\right)^b \qquad S \le 2 \tag{4.50a}$$

$$= cS^{1/4} \qquad S \ge 2 \tag{4.50b}$$

For gases: $a = 1$, $b = 0.36$, and $c = 1.25$. For water ($\text{Pr} \approx 6$): $a = 2.0$, $b = 0.47$, and $c = 1.17$.

Equation 4.50 was obtained by fitting the analytical results of Chen and Eichhorn [42]. They also obtained measurements for $0 < S < 3.5$, $\text{Pr} \approx 6$, and $10^6 < \text{Ra} < 10^8$, and Eq. 4.50 agrees with these data to within about 10 percent. Extrapolation beyond the range of the experimental data is not recommended.

OPEN CAVITY PROBLEMS

In open cavity problems, buoyancy generated by heat exchange with the enclosure walls drives flow through the cavity (Fig. 4.20a). Either the wall temperature or the heat flux can be specified on the cavity walls, and cavities may take a variety of forms (Fig. 4.20). The fluid temperature far from the cavity is assumed constant at T_∞. The cooling of electronic equipment and the augmentation of heat transfer using finned surfaces are two important areas where open cavity problems arise.

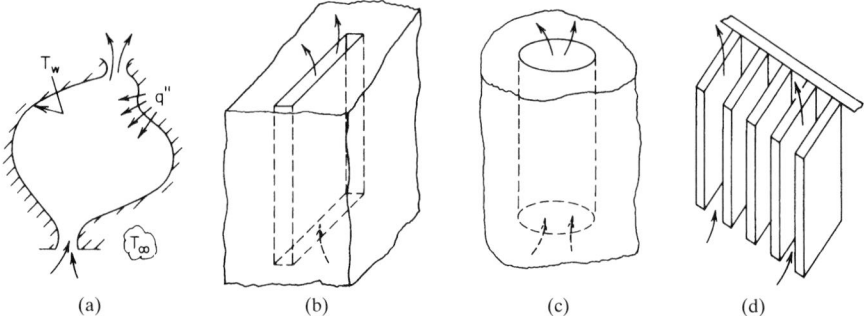

(a)   (b)   (c)   (d)

FIGURE 4.20   Various configurations in which "open cavity" natural convection occurs.

In the equations presented in this section, the characteristic temperature difference appearing in the Nusselt and Rayleigh numbers is $(\overline{T}_w - T_\infty)$, where $\overline{T}_w$ is the average wall temperature of the cavity. Properties are to be evaluated at the film temperature $T_f = 0.5(\overline{T}_w + T_\infty)$ unless otherwise specified.

Cooling Channels

Cooling channels of the type depicted in Fig. 4.20b and c will first be discussed. For channels that are very long relative to the spacing of the vertical surfaces, the flow and heat transfer become fully developed (i.e., velocity and temperature profiles become invariant with distance along the channel) and are described by simple equations. For short channels or widely spaced vertical surfaces, a boundary layer regime is observed in which the boundary layers on the vertical walls remain well separated. In the latter case the heat transfer relation is similar in form to that for a vertical plate but the heat transfer is usually found to be slightly higher. This augmentation results from the flow induced through the cavity by the chimney effect.

Parallel Isothermal Plates.   For parallel isothermal plates of either equal or different temperatures (see Fig. 4.21), Aung [11] has shown that the Nusselt number in the fully developed regime is given by

$$\text{Nu}_{fd} = \frac{4T^{*2} + 7T^* + 4}{90(1 + T^*)^2}\,\text{Ra} \approx \frac{\text{Ra}}{24} \qquad \text{Ra} \le 10 \qquad (4.51)$$

Both plate temperatures are assumed in the analysis to be equal to or greater than T_∞, and q in the definition of Nu in Fig. 4.21 is the total heat delivered to the fluid from both plates; the remaining symbols are defined in the figure.

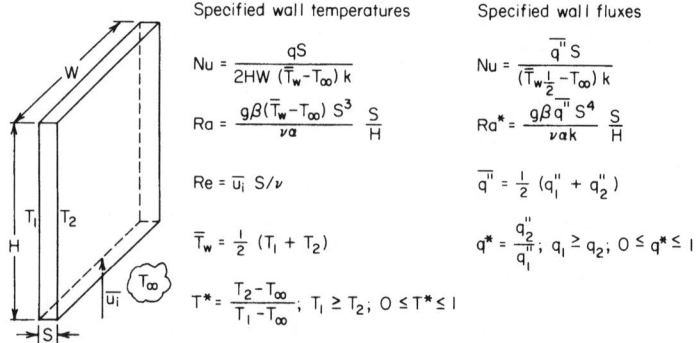

FIGURE 4.21   Geometry and nomenclature for natural convection heat transfer from a wide ($W \gg S$) rectangular cooling slot with heat flux or temperature-specified conditions on the walls.

For Ra > 10, the observed values of Nu depart from the value of Nu_{fd} given in Eq. 4.51 and beyond Ra $\approx 10^3$ follow a relation of the form

$$\text{Nu} = c\overline{C}_\ell\,\text{Ra}^{1/4} \qquad (4.52)$$

where $\overline{C}_\ell$ is given by Eq. 4.13 and Table 4.1. This is called the laminar boundary layer regime. For air, measurements of Elenbaas [88, 89] and Aung et al. [12] indicate that $c \approx 1.20$, which gives 20 percent higher heat transfer than a vertical isolated plate. The analysis of Aung et al. [12] and Bodoia and Osterle [21] yielded a value of $c \approx 1.32$. The measurements of Novotny [210] for Pr = 6 and the analysis of Miyatake and Fujii [196] for Pr = 10 suggest that c approaches 1.0 at higher Pr.

For all Ra up to Ra $\approx 10^5$, the following heat transfer equation is recommended:

$$\text{Nu} = [(\text{Nu}_{fd})^m + (c\overline{C}_\ell\,\text{Ra}^{1/4})^m]^{1/m} \qquad m = -1.9 \qquad (4.53)$$

While property values are normally evaluated at $0.5(\overline{T}_w + T_\infty)$, for large temperature differences and small Ra better agreement between Eq. 4.53 and measurements is obtained (e.g., Ref. 178) by evaluating the properties at $\overline{T}_w$.

The heat transfer per unit surface area, and the heat transfer coefficient, are relatively insensitive to the plate spacing until the spacing is reduced to the point at which the thermal boundary layers begin to interfere. If the objective is to transfer the maximum heat from a given volume of height H, adding more channels proportionately increases the heat transfer by adding surface area, until the boundary layers begin to interfere. The addition of still more channels decreases the spacing of each channel, so that the heat transfer coefficient falls, but the total heat transfer continues to increase because the product of heat transfer coefficient and surface area, or equivalently the heat transfer per unit cross-sectional area of the channel q/A_x, increases. As the plate spacing decreases still further, q/A_x passes through a maximum and then falls. Levy et al. [178] and Raithby and Hollands [223] have shown that the boundary layers begin to interfere at $400 < Ra < 800$. From Eq. 4.53, q/A_x can be shown to pass through a maximum at

$$\mathrm{Ra} = \mathrm{Ra}_{max} = (24c\overline{C}_\ell 2^{-(1/m)})^{4/3} \approx 60 \tag{4.54}$$

Given Ra_{max}, the spacing S can be found that corresponds to this maximum q/A_x.

In a related problem, Sparrow and Prakash [261] have shown that the heat transfer from a series of parallel-plate channels can be substantially increased, for the same surface area, by breaking the continuous vertical plates into a staggered array of discrete vertical plates.

Uniform Heat Flux Parallel Plates. If the heat fluxes are specified as q_1'' and q_2'', respectively, on the surfaces of the vertical plates (where $q_1'' > q_2''$ and both q_1'' and q_2'' are positive, denoting heat transfer from the plate to the fluid), the Nusselt number for the fully developed regime for air is given by [11]

$$\mathrm{Nu}_{fd} = 0.29(\mathrm{Ra}^*)^{1/2} \qquad \mathrm{Ra}^* \leqslant 5 \tag{4.55}$$

where the terms are defined in Fig. 4.21. The temperature difference that appears in Nu is, in this case, $\overline{T}_{w(1/2)} - T_\infty$, where $\overline{T}_{w(1/2)}$ is the average wall temperature at the channel mid-height.

In the laminar boundary layer regime, the Nusselt number relation is of the form of that for a vertical flat plate

$$\mathrm{Nu} = c\overline{H}_\ell(\mathrm{Ra}^*)^{1/5} \qquad 10^2 \leqslant \mathrm{Ra}^* \leqslant 10^4 \tag{4.56}$$

where $\overline{H}_\ell$ is given by Eq. 4.36a and Table 4.1 and c is given in the following text. Equation 4.56 may apply well past the upper limit indicated. The following equation satisfactorily fits the results of Aung et al. [12] obtained for $\mathrm{Ra}^* < 10^4$:

$$\mathrm{Nu} = [(\mathrm{Nu}_{fd})^m + (c\overline{H}_\ell(\mathrm{Ra}^*)^{1/5})^m]^{1/m} \qquad m = -3.5 \tag{4.57}$$

The Nu value calculated from Eq. 4.57 provides the average temperature at the mid-height of the channel. For air, the analysis suggests $c = 1.15$, while the data of Sobel et al. [254] yield $c = 1.07$; the latter value is recommended. These values may be compared to $c = 1.00$ for an isolated vertical plate in air (i.e., Eq. 4.36a).

Bar-Cohen and Rohsenow [14] provide a relation, similar in form to Eq. 4.57, for the local Nusselt number, from which the maximum plate temperature can be calculated.

Extensive measurements for mercury ($\mathrm{Pr} \approx 0.022$) have been reported by Colwell and Welty [68] and Humphreys and Welty [147]. The data at this Prandtl number do not support Eq. 4.57, and the original references should be consulted.

Sobel et al. [254] showed through measurements that the heat transfer from an array of uniform heat flux parallel plates could be substantially increased by interrupting and staggering the plate surfaces.

Isothermal Circular Channels. Measurements and analyses by Elenbaas [88, 89] and Dyer [77] show that the heat transfer from isothermal cylindrical cooling channels (Fig. 4.20c) can be represented by

$$\mathrm{Nu} = \left[\left(\frac{\mathrm{Ra}}{16}\right)^m + (c\overline{C}_\ell \,\mathrm{Ra}^{1/4})^m\right]^{1/m} \qquad m = -1.03 \tag{4.58}$$

The nomenclature is defined in Fig. 4.22. The experiments for $\mathrm{Pr} = 0.71$ (air) indicate $c \approx 1.17$, while analysis gives 1.22. The value of Ra for the maximum heat transfer per unit of cross-

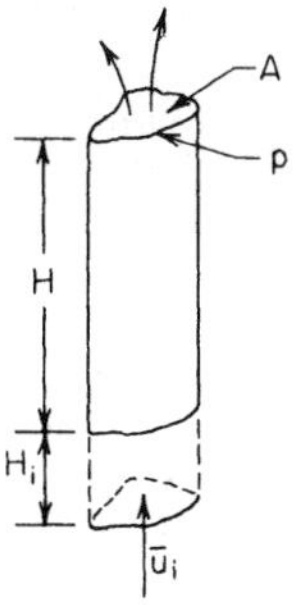

$$r = 2A/p$$

Specified wall temperature   Specified wall flux

$$\mathrm{Nu} = \frac{qr}{pH\,(T_w - T_\infty)\,k} \qquad \mathrm{Nu} = \frac{q''r}{(\overline{T}_w - T_\infty)\,k}$$

$$\mathrm{Ra} = \frac{g\beta\,(T_w - T_\infty)\,r^3}{\nu\alpha}\,\frac{r}{H} \qquad \mathrm{Ra}^* = \frac{g\beta q''r^4}{\nu\alpha k}\,\frac{r}{H}$$

$$\mathrm{Re} = \frac{u_i r}{\nu}$$

FIGURE 4.22   Natural convection through a cylindrical cooling channel.

sectional area of the tube (for a given height H and temperature difference) is given by Eq. 4.54 with 24 replaced by 16; this yields $\mathrm{Ra}_{max} \approx 50$ for air.

Dyer [77] also presents results for air for the case when an unheated entry length H_i is added to the tube as shown in Fig. 4.22. For fully developed flow, the 16 in Eqs. 4.54 and 4.58 is, in this case, replaced by $16(1 + H_i/H)$. The heat flow is also reduced in the boundary layer regime by increasing H_i, and if H_i becomes sufficiently large, the throughflow is reduced and a large portion of the wall cooling is provided by a downward flow through the central portion of the top of the tube and a return flow upward along the tube walls (thermosiphon exchange).

Uniform Flux Cylindrical Channels.   The heat flow results from the analysis of Dyer [76] for air and for vertical circular cooling channels with uniform heat flux at the boundary can be represented closely by

$$\mathrm{Nu} = \left\{ \left(\sqrt{\frac{\mathrm{Ra}^*}{8}} \right)^m + (0.67(\mathrm{Ra}^*)^{1/5})^m \right\}^{1/m} \qquad 0.1 < \mathrm{Ra}^* < 10^5 \qquad (4.59)$$

where $m = -1.7$, $\mathrm{Pr} = 0.71$ (air), and the remaining symbols are defined in Fig. 4.22. From the definition of Nu, the Nu value obtained from Eq. 4.59 provides the average wall temperature. Data from experiments in the range $5 < \mathrm{Ra}^* \leq 5 \times 10^3$ agree well with Eq. 4.59.

Isothermal Channels of Other Shapes.   For isothermal cooling channels of other shapes,

$$\mathrm{Nu} = \left[\left(\frac{\mathrm{Ra}}{f\,\mathrm{Re}} \right)^m + (c\overline{C}_\ell\,\mathrm{Ra}^{1.4})^m \right]^{1/m} \qquad \mathrm{Ra} \lesssim 10^4 \qquad m \approx -1.5 \qquad (4.60)$$

where the nomenclature is defined in Fig. 4.22, $f\,\mathrm{Re}$ is the friction factor–Reynolds number product in Table 4.4, and c has a value of about 1.20 for air and should decrease toward 1.0 with increasing Pr. Equation 4.54 with 24 replaced by $f\,\mathrm{Re}$ yields Ra_{max}, the Rayleigh number for maximum heat transfer per unit of cross-sectional area for a channel of given length H and given $T_w - T_\infty$. The relationship of the fully developed Nusselt number to $f\,\mathrm{Re}$ was originally pointed out by Elenbaas [88, 89].

TABLE 4.4   Values of $f\,\mathrm{Re}$ for Internal Flow in Ducts of Various Shapes

Duct shape			60°	b	2b, b	4b, b		C, B, C/B = 0.9
$f\,\mathrm{Re}$	24	16	13.3	14.225	15.55	18.70	15.05	18.23

Extended Surfaces

Heat transfer from each of the extended surfaces (fins), shown in Fig. 4.23, is now discussed. The prediction of the heat transfer requires a solution for complex 3D motion, so few analyses are as yet available. Experimental data have been obtained exclusively using air.

In many practical applications there is a significant temperature drop between the base of the fin and its tip, and this affects both the natural convection flow and heat transfer. Very little information is available on the coupling between fin conduction and fluid convection, so that attention is restricted to isothermal fins. As a first approximation, the heat transfer coefficient for isothermal fins can be used for the case where the fins are not isothermal, because there is such a weak dependence on the temperature difference. Property values are to be evaluated at $0.5(T_w + T_\infty)$ unless otherwise indicated.

Rectangular Isothermal Fins on Vertical Surfaces.   Vertical rectangular fins, such as shown in Fig. 4.23a, are often used as heat sinks. If $W/S \geq 5$, Aihara [1] has shown that the heat transfer coefficient is essentially the same as for the parallel-plate channel (see the section on parallel isothermal plates). Also, as $W/S \to 0$, the heat transfer should approach that for a vertical flat plate. Van De Pol and Tierney [270] proposed the following modification to the Elenbaas equation [88, 89] to fit the data of Welling and Wooldridge [283] in the range $0.6 < \mathrm{Ra} < 100$, $\mathrm{Pr} = 0.71$, $0.33 < W/S < 4.0$, and $42 < H/S < 10.6$:

$$\mathrm{Nu} = \frac{\mathrm{Ra}}{\Psi} \left\{ 1 - \exp\left[-\Psi\left(\frac{0.5}{\mathrm{Ra}} \right)^{3/4} \right] \right\} \tag{4.61a}$$

where

$$\Psi = \frac{24(1 - 0.483 e^{-0.17/\alpha^*})}{\{(1 + \alpha^*/2)[1 + (1 - e^{-0.83\alpha^*})(9.14\sqrt{\alpha^*} e^{\bar{s}} - 0.61)]\}^3} \tag{4.61b}$$

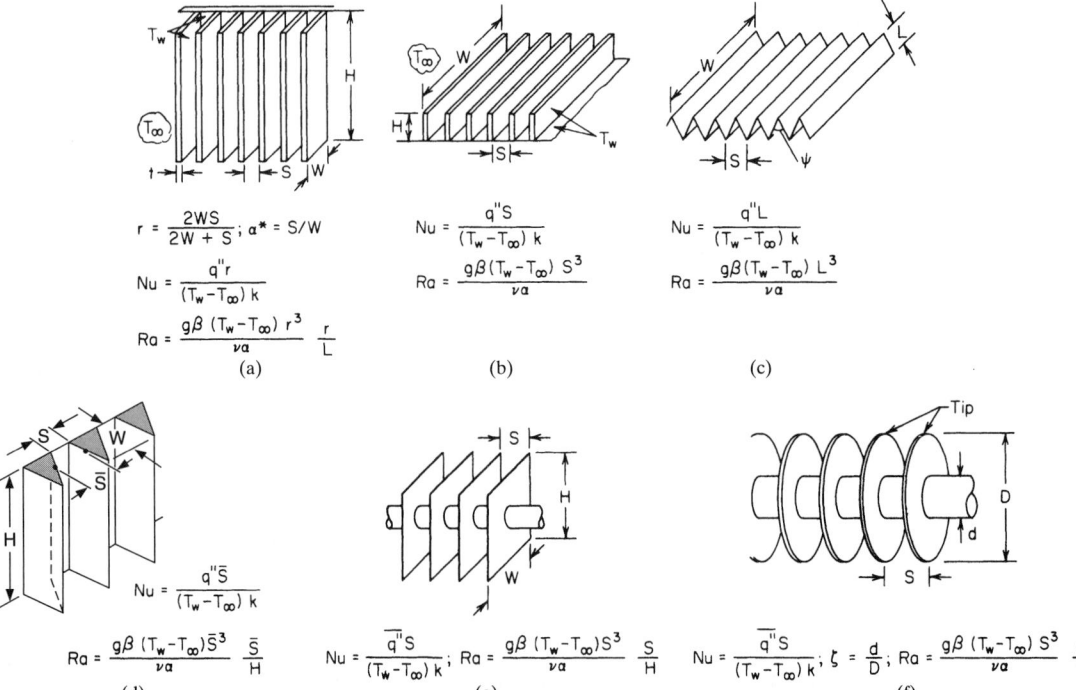

FIGURE 4.23   Flow configurations and nomenclature for various open cavity problems.

and where $\overline{S}$ is dimensionless and equal to $-4.65S$ (for S in cm) or $-11.8S$ (for S in inches). Other nomenclature is defined in Fig. 4.23a.

For a given base plate area there are two fin spacings, S_1 and S_2, of particular interest. If the fin spacing is decreased, starting from a large value, the heat transfer coefficient remains relatively constant until the spacing S_1 (which corresponds to $Ra = Ra_1$) is reached, at which it begins to fall rapidly because of fin interference. As the spacing is decreased within a specified volume, more fins are added to the base plate, thereby increasing the total surface area for heat transfer. Since total heat flow is proportional to the product of heat transfer coefficient and surface area, decreasing the spacing below S_1 still improves the total heat transfer until the spacing S_2 (which corresponds to $Ra = Ra_{max}$) is reached; below S_2, the total heat transfer falls. For long fins ($\alpha^* \ll 1$), S_1 and S_2 will fall to roughly the same values as for parallel-plate channels: $Ra_1 \approx 600$, $Ra_{max} \approx 60$. For short fins, these spacings can be established from Eq. 4.61.

Rectangular Isothermal Fins on Horizontal Surfaces.   The heat transfer from rectangular finned surfaces such as shown in Fig. 4.23b (upward-facing for $T_w > T_\infty$ or downward-facing for $T_w < T_\infty$) has been measured by Jones and Smith [150], Starner and McManus [263], and Harahap and McManus [120]. For a given fin width, $W = 0.254$ m (0.833 ft), Jones and Smith were able to correlate their measured heat transfer to within about ±25 percent on an Nu-Ra plot. The following equation closely represents this correlation over the data range $2 \times 10^2 < Ra < 6 \times 10^5$, $Pr = 0.71$, $0.026 \le H/W \le 0.19$, and $0.016 \le S/W \le 0.20$:

$$Nu = \left[\left(\frac{Ra}{1500} \right)^m + (0.081\ Ra^{0.39})^m \right]^{1/m} \qquad m = -2 \qquad (4.62)$$

This simple equation ignores the effect of the geometric parameters H/S and H/W. While H/S does not appear to play a strong role, H/W is known to have significant effect.

A parametric study using Eq. 4.62 shows that, for a given base area and temperature difference, the curve of total heat transfer versus fin spacing displays a sharp maximum for high fins (large H) and a less well-defined peak for short fins. This results from the rapid increase in total surface area with decreasing fin spacing (i.e., as fins are added to the base plate) for high fins. Because the total heat transfer falls off very sharply for spacings below the optimum, a conservative design would use spacings larger than the optimum S_2 (defined in the section on rectangular isothermal fins on vertical surfaces) calculated from Eq. 4.62.

Horizontal Corrugated Surfaces.   A heated horizontal isothermal corrugated surface, such as that shown in Fig. 4.23c, can also be considered a finned surface. The heat transfer is from the top surface only. Al-Arabi and El-Refaee [3] have measured heat transfer rates to air ($Pr = 0.71$) over the range $1.8 \times 10^4 < Ra < 1.4 \times 10^7$ (the nomenclature is defined in Fig. 4.23c), and have provided the following correlations:

$$Nu = \left(\frac{0.46}{\sin{(\Psi/2)}} - 0.32 \right) Ra^m \qquad \text{for } 1.8 \times 10^4 < Ra < Ra_c \qquad (4.63)$$

where $\qquad Ra_c = (15.8 - 14.0 \sin{(\Psi/2)}) \times 10^5$

$\qquad m = 0.148 \sin{(\Psi/2)} + 0.187$

and $\qquad Nu = \left(0.090 + \frac{0.054}{\sin{(\Psi/2)}} \right) Ra^{1/3} \qquad \text{for } Ra_c < Ra < 1.4 \times 10^7 \qquad (4.64)$

The measurements and the above correlation were intended to represent the case where W is asymptotically large. The dependence of the heat transfer on W remains to be determined.

Vertical Triangular Fins.   For the vertical fin array in Fig. 4.23d, $\overline{S}$ is the fin spacing measured at the mid-height of the fin, so that $\overline{S}W$ is the cross-sectional area of the flow channel formed by the sides of adjacent fins, the base of width S, and the vertical plane passing through the fin tips. Nu and Ra are defined in the figure. The correlation of Karagiozis et al. [152] is

$$\text{Nu} = \text{Nu}_{\text{COND}} + \overline{C}_\ell \, \text{Ra}^{1/4} \left[1 + \left(\frac{3.26}{\text{Ra}^{0.21}} \right)^3 \right]^{-(1/3)} + \delta \, \text{Nu} \qquad (4.65a)$$

$$\delta \, \text{Nu} = [(0.147 \, \text{Ra}^{0.39} - 0.158 \, \text{Ra}^{0.46}), \, 0]_{\max} \qquad (4.65b)$$

where Nu_{COND} is the (conduction) Nusselt number for the entire fin array in the limit as $\text{Ra} \rightarrow 0$. A conservative design (i.e., the heat transfer is underestimated) results if $\text{Nu}_{\text{COND}} = 0$ is used.

Otherwise, to estimate Nu_{COND}, suppose that triangular fins are mounted on a rectangular base plate of dimension L_1 and H, where the base plate entirely covers the vertical surface on which it is mounted. The surface and base plate therefore have area $A_b = L_1 \times H$. Suppose further that the fin height is small compared to the base plate dimensions (i.e., $W \ll L_1$ and $W \ll H$), and that there are no adjacent cool surfaces to which heat can be directly conducted. The lower bound on Nu_{COND} (see Refs. 293 and 294) can be found from

$$\text{Nu}_{\text{COND}} = \frac{\overline{S}(1 + \sqrt{r})^2}{\sqrt{\pi A_b r}} \qquad r \leq 5 \qquad (4.66a)$$

$$= \overline{S} \frac{\sqrt{4\pi r}}{\sqrt{A_b} \, \ln{(4r)}} \qquad r \geq 5 \qquad (4.66b)$$

where $r = (L_1/H, \, H/L_1)_{\max}$. The upper bound for Nu_{COND} is roughly 1.57 times the value calculated from Eq. 4.66; this factor is obtained by extrapolating the recommendation of Yovanovich and Jafarpur [293], intended for $r \approx 1$. In the absence of better information, use the average of these two values.

If the base plate of the fins is attached to a vertical isothermal surface whose dimensions are much larger than both H and L_1, Nu_{COND} will approach zero. If, however, there is a nearby cool surface to which heat can be transferred, Nu_{COND} can be larger than the upper bound just described; this effect becomes important when the nearest distance to the cool surface is less than $\overline{S}/(\text{Nu} - \text{Nu}_{\text{COND}})$ and where $(\text{Nu} - \text{Nu}_{\text{COND}})$ is obtained from Eq. 4.65.

Figure 4.24 compares Eq. 4.65, shown by the solid line, to the data of Karagiozis et al. [152, 153]. The data approach the asymptote for laminar heat transfer from a vertical flat plate (dotted line) at high Ra. The dashed line shows Eq. 4.65a with no correction (i.e., $\delta \, \text{Nu} = 0$); this correction is seen to be significant only at low Ra.

Finned surfaces are often installed on a vertical surface with the fin tips running horizontally (instead of vertically). Karagiozis [153] has shown that this reduces the heat transfer by up to a factor of 2, and the use of this orientation is not recommended.

Square Isothermal Fins on a Horizontal Tube.   Square fins attached to a horizontal tube, as shown in Fig. 4.23e, are commonly used in heat exchangers. The experimental data of Elenbaas [88, 89] for square plates, without the tube, covered the range $0.2 < \text{Ra} < 4 \times 10^4$, for $\text{Pr} = 0.71$. These data are closely correlated by

$$\text{Nu} = \left[\left(\frac{\text{Ra}^{0.89}}{18} \right)^m + (0.62 \, \text{Ra}^{1/4})^m \right]^{1/m} \qquad m = -2.7 \qquad (4.67)$$

Recent heat transfer measurements by Sparrow and Bahrami [256, 257] lie about a factor of 10 higher than those of Elenbaas near the lower end of the Ra range. Tsubouchi and Masuda [269] recommended that the heat transfer from square fins be calculated from their equations for circular fins using an equivalent diameter $D = 1.23H$. Their procedure for circular fins is outlined in the following section.

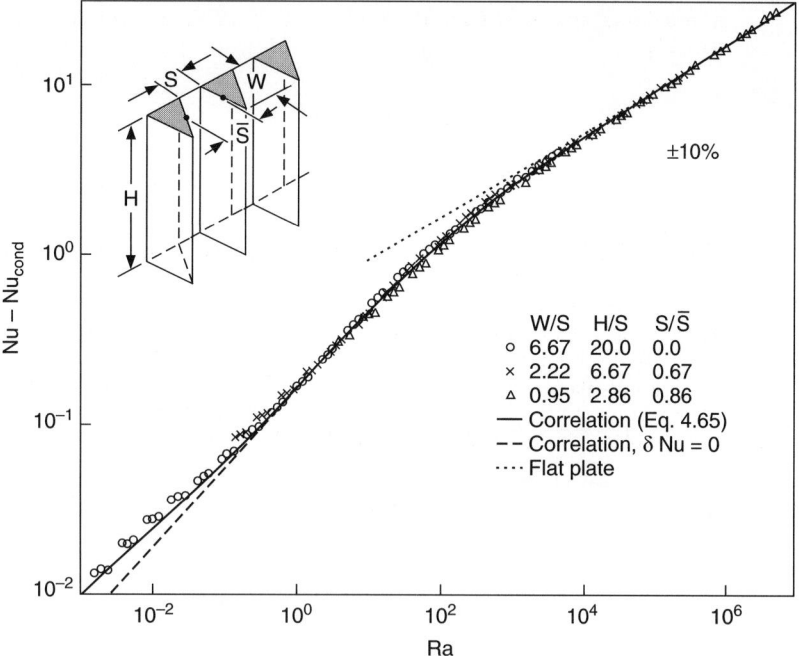

FIGURE 4.24 Comparison of Eq. 4.65 with data of Karagiozis et al. [152] for vertical isothermal triangular fins.

Circular Isothermal Fins on a Horizontal Tube.   Tsubouchi and Masuda [269] measured the heat transfer by natural convection in air from circular fins attached to circular tubes, as in the configuration shown in Fig. 4.23*f.* Correlations for the heat transfer from the tips of the fins (see the figure for definition), and from the cylinder plus vertical fin surfaces, were reported separately.

The average heat transfer from the tips was correlated by

$$\text{Nu} = c\,\text{Ra}^b \tag{4.68}$$

where Nu and Ra are defined in Fig. 4.23*f.* Data were obtained for $2 \leq \text{Ra} \leq 10^4$, and c and b are listed in Table 4.5 for various values of fin-to-cylinder diameter D/d. For $1.36 < D/d < 3.73$, the following approximations can be used: $b = 0.29$, $c = 0.44 + 0.12D/d$.

The heat transfer from the lateral fin surfaces together with the supporting cylinder were correlated [269] for high fins, $1.67 < D/d < \infty$, by

$$\text{Nu} = \frac{\text{Ra}}{12\pi}\left\{2 - \exp\left[-\left(\frac{C_1}{\text{Ra}}\right)^{3/4}\right] - \exp\left[-\beta\left(\frac{C_1}{\text{Ra}}\right)^{3/4}\right]\right\} \tag{4.69a}$$

TABLE 4.5 Values of c and b Calculating the Heat Loss From Fin Tips (Eq. 4.68)

	D/d					
	3.73	3.00	2.45	1.82	1.36	1.14
c	0.9	0.8	0.66	0.66	0.62	0.59
b	0.29	0.29	0.29	0.29	0.29	0.27

where
$$\beta = 0.17\zeta + e^{-4.8\zeta}; \qquad \zeta = d/D$$

$$C_1 = \left[\frac{23.7 - 1.1(1 + 152\zeta^2)^{1/2}}{1 + \beta} \right]^{4/3} \qquad (4.69b)$$

Properties are based on the wall temperature. This equation is in excellent agreement with Tsubouchi and Masuda's data over the measurement range $3 < \mathrm{Ra} < 10^4$.

For shorter fins, $1.67 \geq D/d \geq 1.0$, Eq. 4.69a is replaced by

$$\mathrm{Nu} = C_0 \, \mathrm{Ra}_0^p \left\{ 1 - \exp\left[-\left(\frac{C_1}{\mathrm{Ra}_0} \right)^{C_2} \right] \right\}^{C_3} \qquad (4.70a)$$

where
$$C_0 = -0.15 + 0.3\zeta + 0.32\zeta^{16} \qquad C_1 = -180 + 480\zeta - 1.4\zeta^{-8}$$

$$C_2 = 0.04 + 0.9\zeta \qquad\qquad C_3 = 1.3(1 - \zeta) + 0.0017\zeta^{-12} \qquad (4.70b)$$

$$p = \tfrac{1}{4} + C_2 C_3 \qquad\qquad \mathrm{Ra}_0 = \mathrm{Ra}/\zeta = \mathrm{Ra}\ D/d$$

Properties are again evaluated at the wall temperature.

Edwards and Chaddock [81] correlated their data for heat transfer from the entire surface, including the tip, for $D/d = 1.94$, $5 < \mathrm{Ra} < 10^4$, by

$$\mathrm{Nu} = 0.125 \, \mathrm{Ra}^{0.55} \left[1 - \exp\left(-\frac{137}{\mathrm{Ra}} \right) \right]^{0.294} \qquad (4.71)$$

where properties are evaluated at $T_\infty + 0.62(T_w + T_\infty)$. The measurements of Jones and Nwizu [151] fall slightly below Eq. 4.71. The fact that these equations do not have the expected fully developed behavior ($\mathrm{Nu} \propto \mathrm{Ra}$) as $\mathrm{Ra} \to 0$ has been attributed [269] to tip effects.

NATURAL CONVECTION WITHIN ENCLOSURES

Introduction

Enclosure problems (Fig. 4.1c) arise when a solid surface completely envelops a cavity containing a fluid and, possibly, interior solids. This section is concerned with heat transfer by natural convection within such enclosures. Problems without interior solids include the heat transfer between the various surfaces of a rectangular cavity or a cylindrical cavity. These problems, along with problems with interior solids including heat transfer between concentric or eccentric cylinders and spheres and enclosures with partitions, are discussed in the following sections. Property values (including β) in this section are to be taken at $T_m = (T_h + T_c)/2$.

Geometry and List of Parameters for Cavities Without Interior Solids

The problem of natural convection in a cavity without interior solids is exemplified by the two situations sketched in Fig. 4.25. In both situations, the fluid-filled cavity is bounded by two isothermal parallel "plates" that are inclined at angle θ from horizontal, spaced at distance L, and held at different temperatures. The temperature T_h is assumed to be larger than T_c, so cavities with $\theta = 0°$ are described as horizontal with heating from below, those with $\theta = 90°$ are described as vertical with heating from the side, and those with $\theta = 180°$ are described as horizontal with heating from above.

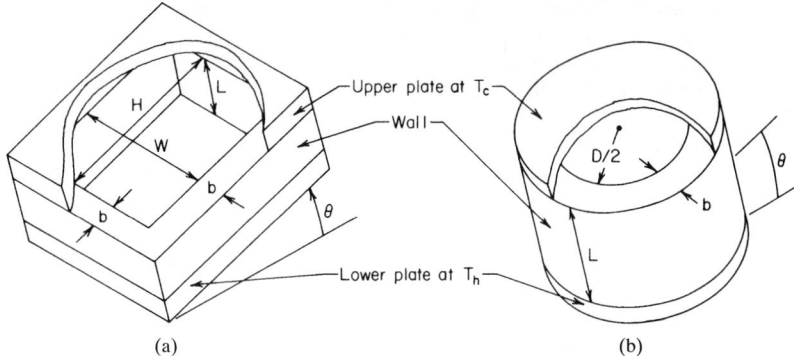

FIGURE 4.25 The (a) rectangular parallelepiped and (b) truncated circular cylinder cavities. Angle θ is measured from the horizontal.

In Fig. 4.25a the cavity is a rectangular parallelepiped, in Fig. 4.25b a truncated circular cylinder; the area over which the fluid contacts each plate is $A_f = HW$ for the first, and $A_f = \pi D^2/4$ for the second. The "wall" (as distinct from the plates) bounding the fluid on the sides is of uniform thickness b and thermal conductivity k_w. The upper and lower plates extend over the wall as shown, and each therefore contacts the wall over an area $A_w = (2b + H)(2b + W) - HW$ for the first shape, and $A_w = (\pi/4)[(D + 2b)^2 - D^2]$ for the second. The thermal boundary condition on the outside faces of the wall (i.e., those faces not bounded by either the fluid or the plates) is adiabatic. The fluid is assumed to be either completely transparent or completely opaque to thermal radiation. Usually liquids are opaque and gases transparent. When the fluid is transparent, radiant exchange can affect the free convection by altering the wall temperature distribution: hemispherical emissivities ϵ_h, ϵ_c, and ϵ_w are assigned to the surfaces of the hot plate, the cold plate, and the walls, respectively.

Two different Nusselt numbers can be assigned for both cavity problems. The first is defined by

$$\text{Nu} = 1 + \frac{(q_f - q_{f0})L}{k(T_h - T_c)A_f} \tag{4.72}$$

and the second by

$$\text{Nu} = 1 + \frac{[q_f + q_r + q_w - (q_{f0} + q_{r0} + q_{w0})]L}{k(T_h - T_c)A_f} \tag{4.73}$$

where q_f is the nonradiative heat transfer rate from the inner area A_f of the hot plate, q_r is the net radiant heat transfer from that same area into the cavity, and q_w is the heat transfer conducted from the area A_w of the hot plate and into the wall. The heat transfers q_{f0}, q_{r0}, and q_{w0} are the respective values of q_f, q_r, and q_w when the fluid is completely stationary and therefore behaves thermally like a solid—a situation reached in the limit Ra → 0. For fluids that are radiantly opaque,

$$q_r = q_{r0} = 0 \qquad q_{f0} = \frac{kA_f(T_h - T_c)}{L} \qquad q_{w0} = \frac{k_w A_w(T_h - T_c)}{L}$$

If the fluid is transparent, $q_{r0} + q_{f0} + q_{w0}$ must be determined from a combined radiative-conductive analysis—see, for example, Hollands et al. [143]. Such an analysis is beyond the scope of this chapter, whose function is to report the additional heat transfer associated with free convective motion. This motion usually alters the temperature distribution in the wall from that which exists when the fluid is stationary. In so doing, it alters not only q_f but also q_w and q_r; the alteration in q_r and q_w is not incorporated into Nu as defined by Eq. 4.72, but it is incorporated into Nu as defined by Eq. 4.73. Thus the latter (Eq. 4.73) is to be preferred, and it will therefore be the meaning of Nu used in this section.

A dimensional analysis reveals that, in the most general case,

$$\text{Nu} = \text{Nu}\left(\text{Ra, Pr, } \theta, \frac{H}{L}, \frac{W}{H}, \frac{b}{L}, \frac{k_w}{k}, \frac{\sigma T_m^3 L}{k}, \epsilon_w, \epsilon_h, \epsilon_c, \frac{T_h}{T_c}\right) \tag{4.74}$$

for the rectangular parallelepiped cavity, and that for the circular cylinder cavity,

$$\text{Nu} = \text{Nu}\left(\text{Ra, Pr, } \theta, \frac{D}{L}, \frac{b}{L}, \frac{k_w}{k}, \frac{\sigma T_m^3 L}{k}, \epsilon_w, \epsilon_h, \epsilon_c, \frac{T_h}{T_c}\right) \tag{4.75}$$

The Rayleigh number is based on dimension L:

$$\text{Ra} = \frac{g\beta(T_h - T_c)L^3}{\nu\alpha} \tag{4.76}$$

Under certain (and probably most) conditions, the parameter lists given by Eqs. 4.74 and 4.75 can be considerably shortened. Table 4.6 lists some of the more common conditions and the shortening each permits. Of particular interest are the adiabatic wall and the perfectly conducting wall. The latter imposes a linear temperature rise from the cold plate to the hot plate, regardless of the convective strength.

TABLE 4.6   Conditions Under Which the List of Parameters in Eq. 4.74 or 4.75 Can Be Shortened

Entry	Condition	Changes in parameter list permitted	Name of condition or comment
1.	Fluid is opaque	Can drop $\sigma T_m^3 L/k$, ϵ_w, ϵ_h, ϵ_c, and T_h/T_c	Opaque field
2.	Fluid is opaque and $k_w/k \ll 1$ or $b=0$	Can drop $\sigma T_m^3 L/k$, ϵ_w, ϵ_h, ϵ_c, T_h/T_c, k_w/k, and b/L	Adiabatic walls
3.	$k_w/k \gg 1$	Can drop $\sigma T_m^3 L/k$, ϵ_w, ϵ_h, ϵ_c, T_h/T_c, k_w/k, and b/L	Perfectly conducting walls
4.	$b/D \leq 0.05$	b/L and k_w/k, and b/L can be dropped, but the single group $k_w b/kD$ must be added	Finlike walls on a cylinder
5.	$b/H \leq 0.05$ and $b/W \leq 0.05$	b/L and k_w/k, and b/L can be dropped, but the single group $k_w b/(kH)$ must be added	Finlike walls on a rectangular parallelepiped
6.	Walls behave like fins and $k_w/(kL) > 25$ Nu	Can drop $\sigma T_m^3 L/k$, ϵ_w, ϵ_h, ϵ_c, T_h/T_c, and $k_w b/kD$	Perfectly conducting walls (obtained with finlike walls)
7.	$k_w/k < \text{Nu}/5$ and $b/L > 0.75$	Can drop b/L	Very thick walls (which behave like $b = \infty$)
8.	$T_h/T_c \approx 1$	Can drop T_h/T_c	Radiation effects can be linearized
9.	$H/L \gtrsim 10$ and $W/L \gtrsim 10$	Can drop k_w/k, b/L, $\sigma T_m^3 L/k$, ϵ_w, ϵ_h, ϵ_c, and T_h/T_c	Extensive plates (for which heat transfer between walls and fluid is unimportant)
10.	$D/L \gtrsim 40$, or $W/L \gtrsim 10$ and $H/L \gtrsim 40$	Can drop k_w/k, b/L, $\sigma T_m^3 L/k$, ϵ_w, ϵ_h, ϵ_c, T_h/T_c, and either D/L or W/L and H/L	Very extensive plates (for which extent of walls is unimportant)

The Conduction Layer Model

The concept of surrounding the surfaces by a layer of stationary fluid, called the conduction layer, is useful for the present enclosure problem as well as for the external and open cavity problems. Unless the conduction layer thickness is greater than the cavity dimensions, a central region is produced (Fig. 4.26a and b), which experience has shown takes up a nearly uniform temperature; this region can therefore be modeled as isothermal. Once the thicknesses of the conduction layers have been specified, finding the heat transfer and the temperature T_{cr} of this central region is a relatively straightforward heat conduction problem.

The conduction layer thickness Δ on each individual surface of the enclosure may be calculated using the equation $\Delta = X/\mathrm{Nu}_X$, where X is the characteristic dimension used for that surface and Nu_X is the Nusselt number on that surface calculated using the methods previously discussed; that is, in calculating the Nusselt number for a particular surface, one assumes that the surface is immersed in an infinite fluid of uniform temperature T_{cr}. Since T_{cr} depends on the conduction layer thickness, the method will, in fact, require some iteration to find T_{cr}; an initial guess for T_{cr} is required. For the side walls (as opposed to the plates), we take for the surface temperature the average of T_h and T_c. Once calculated, the appropriate conduction layers are applied to all the surfaces, where they are all treated as solids of conductivity equal to the fluid conductivity. The remaining core fluid is treated as material of infinite conductiv-

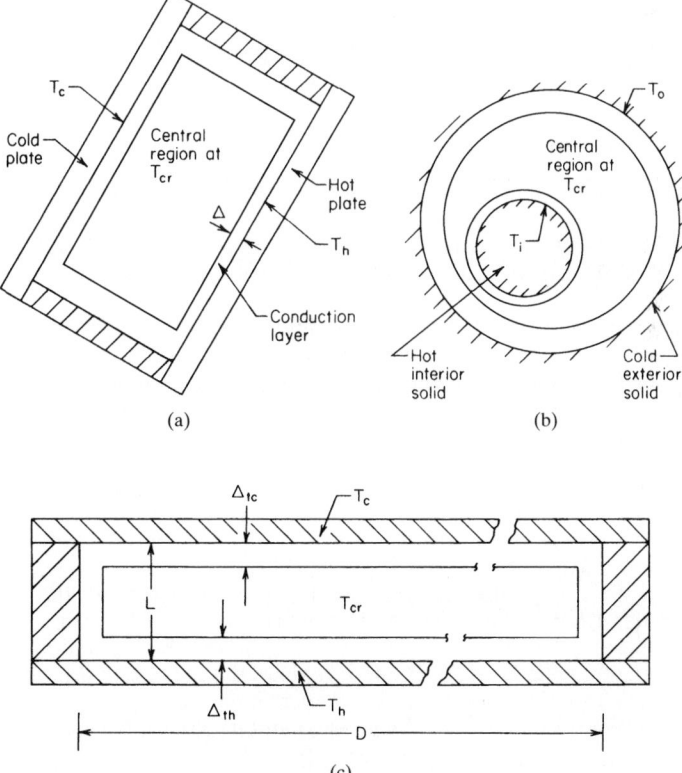

FIGURE 4.26   The conduction layer model: (*a*) conduction layer growth on plates and wall for a cavity without interior solids; (*b*) similar growth for a cavity with an interior solid; (*c*) conduction layer model applied to a horizontal cavity having $\theta = 0$ and $D \gg L$.

ity, and from a simple conduction solution, the resultant heat transfer on each plate can be calculated, as required. (Also calculated from the conduction analysis is a new value of T_{cr}, to be compared to the previously assumed value in the iterative scheme.) If the conduction layers overlap to the degree that the central region disappears, the heat transfer may simply be equated to that for pure conduction across the fluid.

These ideas can be illustrated by considering a cylindrical cavity with $D \gg L$ and $\theta = 0$ (Fig. 4.26c). The conduction layer thicknesses on the plates are found to be $\Delta_h = [\nu\alpha/\{g\beta(T_h - T_{cr})\}]^{1/3}/C_t^U$ on the hot plate and $\Delta_c = [\nu\alpha/\{g\beta(T_{cr} - T_c)\}]^{1/3}/C_t^U$ on the cold plate. By symmetry, $T_{cr} = (T_h + T_c)/2$, so that each conduction layer is in fact of equal thickness $\Delta = 2^{1/3}[\nu\alpha/\{g\beta(T_h - T_c)\}]^{1/3}/C_t^U$. Since the central core offers no thermal resistance and since $D \gg L$, the combined unit area thermal resistance of the fluid is $2\Delta/k$, so that $q'' = k(T_h - T_c)/2\Delta$ and Nu is found to be given by Nu $= C_t^U 2^{-4/3}$ Ra$^{1/3}$. As Ra is decreased, the conduction layers thicken until at some particular value of Ra they touch in the middle of the layer (that is, until $2\Delta = L$) and Nu becomes unity. According to the model, Nu remains at unity for smaller values of Ra. Thus the conduction layer model prediction for this problem is

$$\text{Nu} = [1, C_t^U 2^{-4/3} \text{ Ra}^{1/3}]_{\max} \tag{4.77}$$

This prediction will be compared with measurement in the next section.

As yet, the conduction layer approach has only been tested quantitatively on those problems in which the influence of the side walls is unimportant. Even for these problems the model has met with only mixed success in closely predicting the heat transfer. However, it does predict the correct trends and the correct asymptotes, it is useful in correlating experimental data, and does afford a simple physical understanding to problems that, when viewed from a different perspective, often appear very complex. The practitioner may find it useful for problems for which there is insufficient information from other sources.

Horizontal Rectangular Parallelepiped and Circular Cylinder Cavities

Cavities Extensive in the Horizontal Direction (H $\gg$ L and W $\gg$ L, or D $\gg$ L). This section deals with situations covered by entry 10 in Table 4.6, with the additional proviso that either $\theta = 0$ or $\theta = 180°$.

When $\theta = 180°$, the hot, light fluid lies above the cold, heavy fluid, so the stationary fluid layer (in which there is no fluid motion) is inherently stable, and Nu = 1 for all Ra. (In terms of the conduction layer model, for $\theta = 180°$ both the conduction layers are infinite, so the conduction layers always overlap, and Nu = 1.)

In the $\theta = 0°$ orientation, hot, light fluid lies below the cold, heavy fluid, so the stationary fluid layer is inherently unstable. Despite this inherent instability, the fluid remains stationary provided Ra is less than a "critical Rayleigh number" denoted by Ra$_c$. The value of Ra$_c$ for this particular geometry is 1708. For Ra > Ra$_c$, the instability leads to a steady-state convective motion, the form and strength of which depends on both Ra and Pr. For Ra only slightly greater than Ra$_c$, it consists of steady rolls of order L in size, but as Ra is further increased, more complex flow patterns are observed, and eventually the flow becomes unsteady. At very high Ra it becomes fully turbulent. The heat transfer characteristics reflect the existence of these various flow regimes: for Ra < Ra$_c$ the fluid is stationary, so Nu is unity; the cellular motion initiated at Ra$_c$ produces a sharp rise in Nu with Ra, which ultimately becomes asymptotic to the relation Nu $\propto$ Ra$^{1/3}$ at very large Ra.

For $\theta = 0°$, the recommended equation [140] for Nu is:

$$\text{Nu} = 1 + \left[1 - \frac{1708}{\text{Ra}}\right]^{\cdot}\left[k_1 + 2\left(\frac{\text{Ra}^{1/3}}{k_2}\right)^{1 - \ln(\text{Ra}^{1/3}/k_2)}\right] + \left[\left(\frac{\text{Ra}}{5803}\right)^{1/3} - 1\right]^{\cdot} \tag{4.78}$$

where square brackets with dots indicate that only positive values of the argument are to be taken, i.e.,

$$[X]^{\cdot} = \left(\frac{|X| + X}{2}\right) \tag{4.79}$$

X being any quantity. Values of the parameters k_1 and k_2, both functions of Pr, are tabulated in Table 4.7 for several values of Pr. This table also cites the experiments from which the values were inferred, and gives the range in Ra over which Eq. 4.78 has been tested for each Pr. The following equations fit the dependence of k_1 and k_2 on Pr exhibited in Table 4.7.

$$k_1 = \frac{1.44}{1 + 0.018/Pr + 0.00136/Pr^2} \tag{4.80}$$

$$k_2 = 75 \exp(1.5\, Pr^{-1/2}) \tag{4.81}$$

The form of Eq. 4.80 resulted from an approximate analysis [118], but the constants in the denominator have been chosen to fit the values given in Table 4.7 for Pr = 0.7 and Pr = 0.024, and they are therefore based on limited data. Caution is advised in using this equation when Pr < 0.7, except when Pr ≈ 0.024. Also, for some values of Pr, the narrow range of Ra over which Eq. 4.78 has been tested should be noted. The data of Kek and Müller's [158] recent experiments using liquid sodium with Pr = 0.0058 are fit reasonably well by Eqs. 4.78–4.81, but the fit is improved considerably if k_1 is set equal to 0.087 and the power on (Ra/5830) is changed from ⅓ to ¼.

TABLE 4.7 Values of k_1 and k_2 to Be Used in Eq. 4.78

Pr (approximate)	k_1	k_2	Range of Ra tested	Reference
0.02	0.35	>200	$10^3 \le Ra \le 10^8$	233
0.7	1.40	>400	$10^3 \le Ra \le 10^{11}$	See 142 for list.
6	1.44	140	$10^3 \le Ra \le 2 \times 10^5$	See 142 for list; also 117
34	1.44	100	$10^3 \le Ra \le 10^5$	243
100	1.44	~85	$10^3 \le Ra \le 3 \times 10^6$	243
200	1.44	85	$10^3 \le Ra \le 5 \times 10^5$	233
3000	1.44	~75	$10^3 \le Ra \le 3 \times 10^4$	243

Equation 4.78 with k_1 and k_2 having values appropriate to water at moderate temperatures (Pr ≈ 6) is plotted in Fig. 4.27, together with relevant data for water. Figure 4.28 shows a plot of Eq. 4.78 for various values of Pr, covering only those ranges in Ra at which the equation has been tested. Also plotted are the predictions of the conduction layer model given by Eq. 4.77. This model is seen to be correct only in the limit of small Ra (Ra < 1708) and large Ra

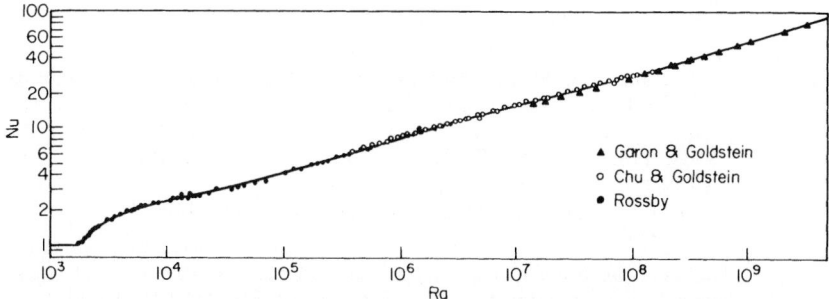

FIGURE 4.27 Comparison of Eq. 4.78 (solid curve) and the data of Garon and Goldstein [104], Chu and Goldstein [53], and Rossby [233] for water (Pr ≈ 6).

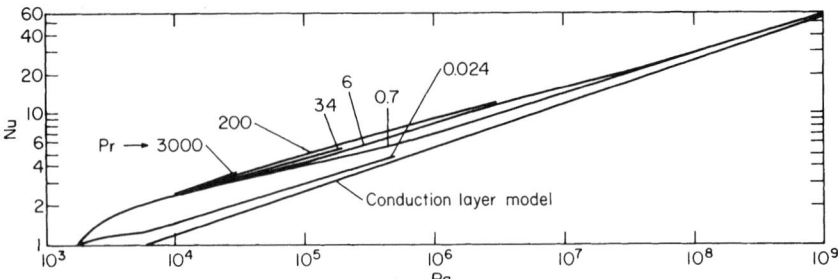

FIGURE 4.28 Plot of Eq. 4.78 for various values of Prandtl numbers, describing the heat transfer across a horizontal cavity with $D/L \gtrsim 10$ and heating from below; also shown is the heat transfer predicted by the conduction layer model (Eq. 4.77).

($Ra \gtrsim 10^8$). The maximum error, which occurs near $Ra = 5830$, varies from 20 percent for $Pr = 0.024$ to 50 percent for $Pr \geq 0.7$.

Critical Rayleigh Numbers for Horizontal Cavities Restricted in the Horizontal Direction.
A critical Rayleigh number Ra_c governs the initiation of convective motion in the horizontal cavity that is restricted in the horizontal direction, just as in the extensive cavity discussed in the previous section. In restricted cavities, Ra_c depends on the geometric parameters describing the cavity and on the thermal properties of the wall, but not on the Prandtl number. For $\theta = 180°$, $Ra_c = \infty$, as for horizontally extensive cavities. For $\theta = 0°$, Ra_c is bounded between two values, Ra_{cp} and Ra_{ci}. Evaluated by Catton [31–33], and Buell and Catton [27], these bounds are tabulated in Table 4.8 for rectangular parallelepiped cavities and in Table 4.9 for circular cylinder cavities. The greater of the two, Ra_{cp}, applies to the perfectly conducting wall case (Table 4.6, entry 3 or 5); and the lesser, Ra_{ci}, applies to the adiabatic wall case (see Table 4.6, entry 2; note that because of radiation, a wall with $k_w = 0$ is not necessarily adiabatic). Interpolation and extrapolation in Tables 4.8 and 4.9 may be assisted by knowledge of certain asymptotes: for the circular cylinder,

$$\text{as} \quad \frac{D}{L} \to \infty, \quad Ra_{cp} \to 1708 \quad \text{and} \quad Ra_{ci} \to 1708$$

and [216, 289]

$$\text{as} \quad \frac{D}{L} \to 0, \quad Ra_{cp}\left(\frac{D}{L}\right)^4 \to 3456 \quad \text{and} \quad Ra_{ci}\left(\frac{D}{L}\right)^4 \to 1086.4$$

For the rectangular parallelepiped,

$$\text{as} \quad \frac{H}{L} \text{ and } \frac{W}{L} \to \infty, \quad Ra_{cp} \to 1708 \quad \text{and} \quad Ra_{ci} \to 1708$$

For the rectangular parallelepiped having $W/H = \infty$ [285],

$$\text{as} \quad \frac{H}{L} \to 0, \quad Ra_{cp}\left(\frac{H}{L}\right)^4 \to 97.4 \quad \text{and} \quad Ra_{ci}\left(\frac{H}{L}\right)^2 \to 473.7$$

The asymptotic relations for the cylinder suggest that for the square-section ($H/W = 1$) rectangular parallelepiped, $Ra_{cp}(H/L)^4$ and $Ra_{ci}(H/L)^4$ should approach constant values as $H/L \to 0$. Extrapolating the values in Table 4.8 gives the following estimates for these asymptotes:

TABLE 4.8 Critical Rayleigh Numbers Ra_{cp} and Ra_{ci} for Horizontal Rectangular Parallelepiped Cavities Having a Perfectly Conducting Wall (Ra_{cp}) or an Adiabatic Wall (Ra_{ci}) [31–33]

| | H/L | | | | | | | | | |
| | 0.125 | | 0.25 | | 0.5 | | 1.00 | | 2.00 | |
W/L	Ra_{ci}	Ra_{cp}	Ra_{ci}	Ra_{cp}	Ra_{ci}	Ra_{cp}	Ra_{ci}	Ra_{cp}	Ra_{ci}	Ra_{cp}
0.125	3,011,718	9,802,960								
0.25	333,013	1,554,480	203,163	638,754						
0.50	70,040	606,001	28,452	115,596	17,307	48,178				
1.00	37,689	469,377	11,962	64,271	5,262	14,615	3446	6974		
1.50	39,798		12,540		6,341		3270			
2.00	36,262	444,995	11,020	53,529	4,524	11,374	2789	5138	2276	3774
2.50	37,058		11,251		4,567		2754		2222	
3.00	35,875	444,363	10,757	50,816	4,330	9,831	2622	3906	2121	2754
3.50	36,209		10,858		4,355		2609		2098	
4.00	35,664	457,007	10,635	50,136	4,245		2552	3634	2057	2531
4.50	35,794		10,666		4,261	9,312	2545		2044	
5.00	35,486	473,725	10,544	50,088	4,186		2502	3446	2009	2360
5.50	35,556		10,571		4,196	9,099	2498		2001	
6.00	35,380	494,741	10,499	50,410	4,158		2480	3558	1989	2286
6.50	35,451		10,518		4,165	8,980	2447		1984	
12.00	35,193		10,426		4,118		2453		1967	

| | H/L | | | | | | | | | |
| | 3.00 | | 4.00 | | 5.00 | | 6.00 | | 12.00 | |
	Ra_{ci}	Ra_{cp}	Ra_{ci}	Ra_{cp}	Ra_{ci}	Ra_{cp}	Ra_{ci}	Ra_{cp}	Ra_{ci}	Ra_{cp}
3.00	2004	2557								
3.50	1978									
4.00	1941	2337	1894	2270						
4.50	1927		1878							
5.00	1897	2174	1852	2111		2082				
5.50	1888		1842							
6.00	1879	2101	1833	2037	1810	2008	1797	1992		
6.50	1871		1826		1803		1789		1741	
12.00	1855		1808		1783		1768			

TABLE 4.9 Critical Rayleigh Numbers Ra_{cp} and Ra_{ci} at Different Values of D/L for Horizontal Circulary Cavities Having a Perfectly Conducting Wall (Ra_{cp}) or an Adiabatic Wall (Ra_{ci}) [27]

	0.4	0.5	0.7	1.0	1.4	2	3	4	∞
Ra_{cp}	151,200	66,600	21,300	8010	4350	2540	2010	1880	1708
Ra_{ci}	51,800	23,800	8,420	3770	2650	2260	1900	1830	1708

$$\text{as } \frac{H}{L} \to 0, \quad \text{Ra}_{cp}\left(\frac{H}{L}\right)^4 \to 2350 \quad \text{and} \quad \text{Ra}_{ci}\left(\frac{H}{L}\right)^4 \to 710$$

For the limit H/L equal to infinity, Daniels and Ong [70] have obtained the following values for Ra_{cp}: 8955 at $W/L = 0.5$; 2944 at $W/L = 1$; 1870 at $W/L = 2$; and 1719 at $W/L = 4$.

For opaque fluids contained in rectangular parallelepiped cavities with finlike walls (see Table 4.6, entry 5), Catton [32, 33] has calculated Ra_c as a function of L/H, L/W, and the wall thermal admittance C_{ar}, defined by

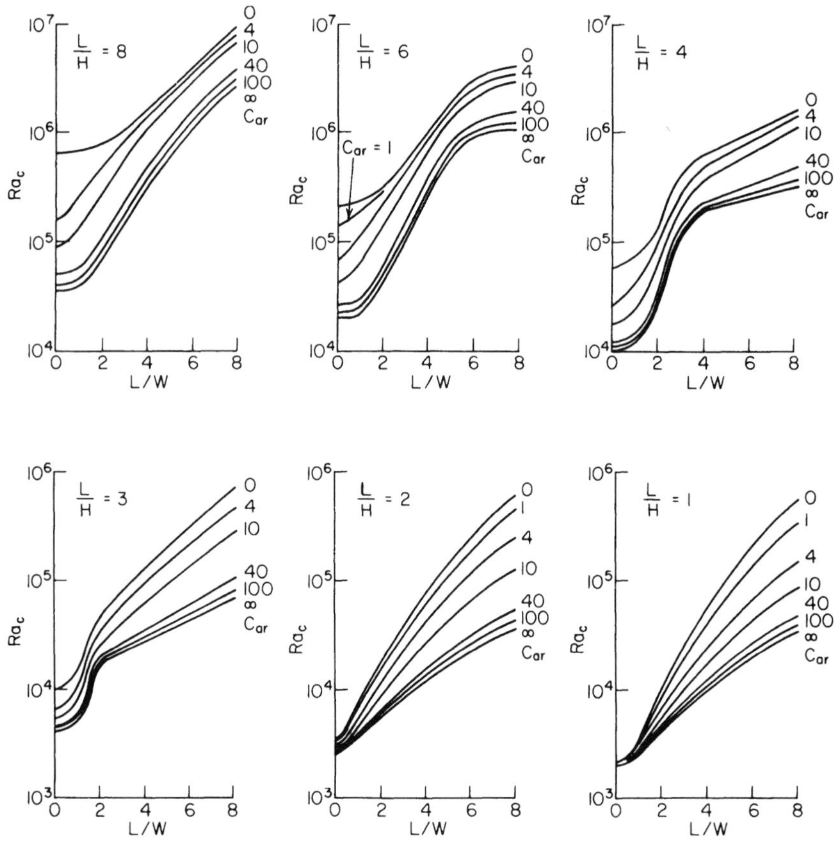

FIGURE 4.29   Catton's [32, 33] plots of Ra_c as a function of L/H, L/W, and C_{ar} for horizontal rectangular parallelepiped cavities with finlike walls of arbitrary conductivity.

$$C_{ar} = \frac{kL}{k_w b} \qquad (4.82)$$

The results are plotted in Fig. 4.29. Note that $Ra_c \to Ra_{ci}$ as $C_{ar} \to \infty$, and $Ra_c \to Ra_{cp}$ as $C_{ar} \to 0$. (Values of Ra_c for cases when opposing walls have different admittances are also available [32, 33].) Figure 4.30 gives similar plots taken from the data of Buell and Catton [27] for the circular cylinder with a finlike wall (Table 4.6, entry 4), for which the wall thermal admittance C_{ac} is defined by

$$C_{ac} = \frac{kD}{2k_{wt}^* b} \qquad (4.83)$$

where (for the finlike wall and an opaque fluid) $k_{wt}^* = k_w$.

No exact solutions for Ra_c are available when the walls are not finlike and L is finite. But for circular cylinder cavities with thick walls and for $L/D \to \infty$, Ostroumov [216] showed that Fig. 4.30 is valid, provided k_{wt}^* in Eq. 4.83 is equated to k_{wt}, where k_{wt} (the equivalent finlike wall conductivity of a very thick wall material—see Table 4.6, entry 7) is defined by

$$k_{wt} = \frac{k_w D}{2b} \frac{(D + 2b)^2 - D^2}{(D + 2b)^2 + D^2} \qquad (4.84)$$

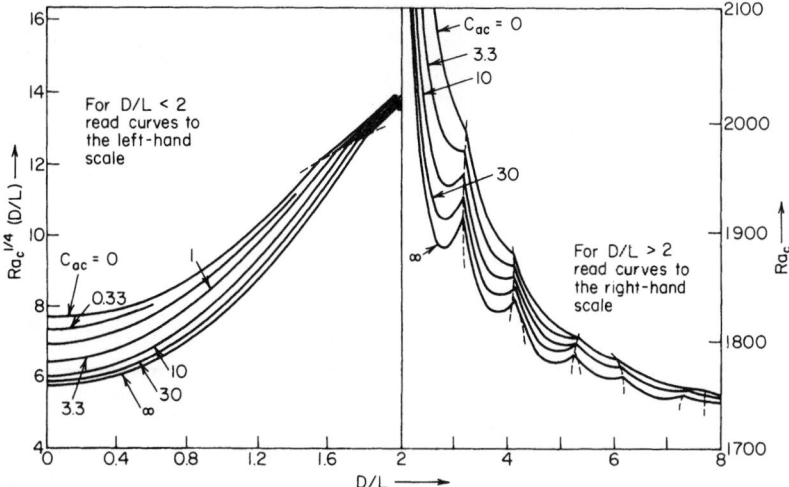

FIGURE 4.30 Buell's [26] plots of Ra_c as a function of D/L and C_{ac} for horizontal cylindrical cavities with finlike walls of arbitrary conductivity.

The effect of the radiant exchange in a cavity containing a transparent fluid is similar to that of an increase in wall conductivity. For circular cylinders in which $T_h/T_c \approx 1$, Edwards and Sun [83] showed that for large L/D the results of Fig. 4.30 apply to radiantly transparent fluids if k_{wt}^* in Eq. 4.83 is given by

$$k_{wt}^* = k_{wt} + \frac{8\sigma T_m^3 D \epsilon_w}{4 - \epsilon_w} \tag{4.85}$$

where $T_m = (T_h + T_c)/2$. Thus the effect of radiation is to raise the apparent conductivity of the walls and thereby raise Ra_c. Experiments by Hollands [141] and Cane et al. [29] confirmed that, although technically correct only in the limit $L/D \to \infty$, Eq. 4.85 can be applied with little error for L/D as small as unity. They also showed that Eq. 4.85 can be applied to square-celled ($H/W = 1$) rectangular parallelepiped cavities with finlike walls if D is set equal to H; that is, be redefining C_{ar} for transparent fluids contained in square cavities as

$$C_{ar} = \frac{kL}{b[k_w + 8\sigma T_m^3 H \epsilon_w/(4 - \epsilon_w)]} \tag{4.86}$$

the points on Fig. 4.29 pertaining to $H/W = 1$ are made valid for both transparent and opaque fluids. For rectangular parallelepiped gas-filled cavities in which $W \gg H$ and $T_h/T_c \approx 1$, Edwards and Sun [82] and Sun [265] showed that for large L/H the results of Fig. 4.29 pertaining to $W \gg H$ are valid, provided C_{ar} is redefined as

$$C_{ar} = \frac{kL}{bk_w}\left(1 + \frac{\sigma T_m^3 L^2 \epsilon (1 - S_k)}{\{8b[1 - (1 - \epsilon)S_k]\}}\right)^{-1} \tag{4.87}$$

where, for $L/H > 2$, S_k is given by [253]:

$$S_k = 1.0102 - 1.4388\frac{H}{L} - 9.4653\left(\frac{H}{L}\right)^2 + 31.44\left(\frac{H}{L}\right)^3 - 27.515\left(\frac{H}{L}\right)^4 \tag{4.88}$$

This method has been found [253] to predict Ra_c with reasonable accuracy for L/H at least as small as 3.

Heat Transfer Across Horizontal Cavities Restricted in the Horizontal Direction. The recommended equation [140] for the heat transfer across horizontally nonextensive cavities is

$$\text{Nu} = 1 + \left[1 - \frac{\text{Ra}_c}{\text{Ra}}\right]^{\bullet}\left[k_1 + 2\left(\frac{\text{Ra}^{1/3}}{k_2}\right)^{1 - \ln(\text{Ra}^{1/3}/k_2)}\right]$$

$$+ \left[\left(\frac{\text{Ra}}{5830}\right)^{1/3} - 1\right]^{\bullet}\left(1 - \exp\left\{-0.95\left[\left(\frac{\text{Ra}}{\text{Ra}_c}\right)^{1/3} - 1\right]^{\bullet}\right\}\right) \quad (4.89)$$

where Ra_c is the critical Rayleigh number appropriate to the particular cavity, calculated by methods outlined in the previous section, and k_1 and k_2 are as given by Eqs. 4.80 and 4.81. This largely empirical equation has been tested against experimental data for gases [29, 141, 253], and liquids of various Prandtl numbers [36, 265, 266] (but not liquid metals) using circular cylinder cavities (as approximated by hexagons) with $0.2 \leq D/L \leq 5$, and rectangular parallelepiped cavities with $1 \leq H/L \leq 10$ and planforms ranging from square ($H/W = 1$) to long ($W/H \gg 1$). For circular cylinder, and for rectangular parallelepiped, cavities with $W = H$, Eq. 4.89 generally agrees with measurements to within 10 percent, but for rectangular parallelepiped cavities with $W \gg H$, differences of up to 25 percent occur. The equation from Smart et al. [253]

$$\text{Nu} = 1 + 0.131\,\text{Ra}^{1/3}\left(1 - \exp\left\{-0.18\left[\left(\frac{\text{Ra}}{\text{Ra}_c}\right)^{0.513} - 1\right]^{\bullet}\right\}\right) \quad (4.90)$$

fits the data for $W \gg H$ better than Eq. 4.89, but in contrast to Eq. 4.89, it does not have the proper asymptote, as $\text{Ra} \to \infty$. Equation 4.90 agrees well with data for $\text{Ra} < 100\,\text{Ra}_c$ and $L/H = 3, 5$, and 10. It is not recommended for $\text{Ra} > 100\,\text{Ra}_c$.

Figure 4.31 shows a plot of Eq. 4.89 for a circular cylinder cavity with perfectly conducting walls and various values of D/L. As is clear from the graph, the Nusselt number rises very steeply with Ra after initiation of convection, and very rapidly approaches the value of Nu for the horizontally extensive cavity. This behavior is consistent with the conduction layer model: at high Ra, the conduction layers on the walls at the sides are so thin that they have no effect on the heat transfer; at sufficiently low Ra, they are so thick that they overlap (even though those on the horizontal plates do not), so that their presence governs the condition for a stationary fluid.

Heat Transfer in Vertical Rectangular Parallelepiped Cavities: θ = 90°

Cavities with H/L ≥ 5 and W/L ≳ 5. In contrast to the horizontal cavity, for which there is flow only when $\text{Ra} > \text{Ra}_c$, the vertical cavity experiences flow for any finite Ra. At small Ra,

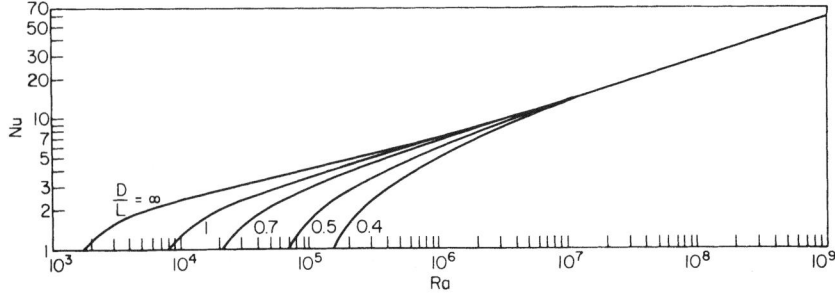

FIGURE 4.31 Relation between Nu and Ra for horizontal cylindrical cavities with perfectly conducting walls, for various values of D/L, as given by Eq. 4.89 with Pr = 6.0.

however, the velocities are small and essentially parallel to the plates, so that they contribute little to the heat transfer, and for all practical purposes, Nu = 1. These conditions constitute the conduction regime. The development of the flow as Ra increases beyond this regime depends on H/L.

If $H/L \gtrsim 40$, the conduction regime becomes unstable at a critical Rayleigh number Ra_c, which is plotted as a function of Pr in Fig. 4.32 (from Ref. 162). Increases in Ra past Ra_c lead through a turbulent transition regime and finally into a fully developed turbulent boundary layer regime characterized by turbulent boundary layers on each plate and a well-mixed core between them in which there is a vertical temperature gradient of about $0.36(T_h - T_c)/H$ [170]. If $H/L \lesssim 40$, the flow enters a laminar boundary layer regime before becoming unstable and entering the turbulent transition regime. This laminar boundary layer regime [112] is characterized by laminar boundary layers on each plate with an essentially stationary core between them: this core is nearly isothermal in the horizontal direction, but it has a positive gradient in the vertical direction of approximately $0.5(T_h - T_c)/H$. The stability analyses [17] of the laminar boundary layer regime predict higher critical Rayleigh numbers than for the conduction regime plotted in Fig. 4.32. In summary, for $H/L \gtrsim 40$, the regimes encountered as Ra increases are first conduction, then turbulent transition, and then turbulent boundary layer; for $H/L \lesssim 40$ they are conduction, then laminar boundary layer, then turbulent boundary layer.

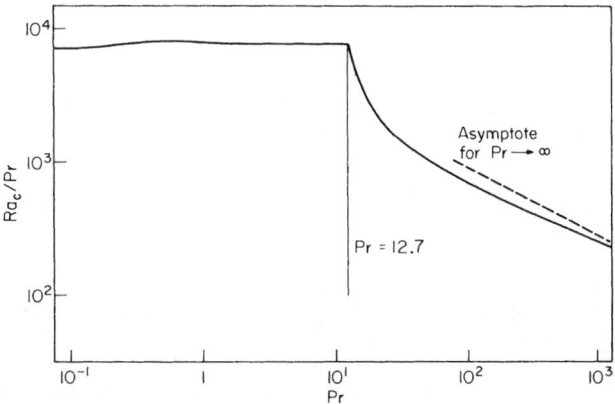

FIGURE 4.32   Korpela's [162] plot of Ra_c governing the stability of the conduction regime in a vertical rectangular parallelepiped cavity with $W/L \gtrsim 5$ and $H/L \gtrsim 40$. For Pr < 12.7, the instability leads to stationary, horizontal axis rolls; for Pr > 12.7, it leads to unsteady, vertically traveling waves.

The recommended correlation equations for the Nusselt number relation are based on experimental data. For Pr $\approx$ 0.7 (gases), and $H/L > 40$, the equation of Shewen et al. [250] is recommended:

$$\mathrm{Nu} = \left[1 + \left[\frac{0.0665\,\mathrm{Ra}^{1/3}}{1 + (9000/\mathrm{Ra})^{1.4}} \right]^2 \right]^{1/2} \tag{4.91}$$

This equation has been validated for Ra < 10^6 and 40 < H/L < 110. For 5 < H/L < 40, the equation of ElSherbiny et al. [84, 85] equates Nu to the maximum of three Nusselt numbers as follows:

$$\mathrm{Nu} = [\mathrm{Nu}_{ct}, \mathrm{Nu}_l, \mathrm{Nu}_t]_{\max} \tag{4.92}$$

where
$$\mathrm{Nu}_{ct} = \left[1 + \left\{ \frac{0.104\,\mathrm{Ra}^{0.293}}{1 + (6310/\mathrm{Ra})^{1.36}} \right\}^3 \right]^{1/3} \tag{4.93}$$

$$\mathrm{Nu}_l = 0.242 \left(\frac{\mathrm{Ra}\, L}{H} \right)^{0.273} \tag{4.94}$$

$$\mathrm{Nu}_t = 0.0605\, \mathrm{Ra}^{1/3} \tag{4.95}$$

Nu_{ct} applies to the conduction and the turbulent transition regime, Nu_l to the laminar boundary layer regime, and Nu_t to the turbulent boundary layer regime. Equation 4.92 fits the data with a maximum deviation of about 10 percent and mean deviation of about 4 percent. It compares well with computer simulations [177]. A more accurate but also more complex set of equations is also available in [84, 85]. Equation 4.92 has been validated up to Ra $(H/L)^3 =$ 1.5×10^{10} for $H/L = 5, 20,$ and 40. Equation 4.92 is based on data for perfectly conducting walls, but for $H/L \gtrsim 10$ the effect of wall properties is not expected to be important (see Table 4.6, entry 9).

For fluids with Pr $\gtrsim 4$, the recommended equations are based on the proposals of Seki et al. [248]. For Ra $(H/L)^3 < 4 \times 10^{12}$,

$$\mathrm{Nu} = \left[1,\, 0.36\, \mathrm{Pr}^{0.051} \left(\frac{L}{H} \right)^{0.36} \mathrm{Ra}^{0.25},\, 0.084\, \mathrm{Pr}^{0.051} \left(\frac{L}{H} \right)^{0.1} \mathrm{Ra}^{0.3} \right]_{\max} \tag{4.96}$$

and for Ra $(H/L)^3 > 4 \times 10^{12}$,

$$\mathrm{Nu} = 0.039\, \mathrm{Ra}^{1/3} \tag{4.97}$$

These equations have been tested for values of H/L ranging from 5 to 47.5. The middle term in Eq. 4.96 has been tested for $3 < \mathrm{Pr} < 40{,}000$, and the last term for $3 \lesssim \mathrm{Pr} \lesssim 200$. Equation 4.97 has been tested only for $\mathrm{Pr} \approx 5$, and may underpredict measurements by as much as 20 percent. For $5 \le H/L < 10$, the equations should be most accurate for adiabatic walls. For $H/L \gtrsim 10$, the plates are extensive (Table 4.6, entry 9), and the wall thermal properties are not important.

Vertical Cavities ($\theta = 90°$) with L/H ≥ 2 and W/L $\gtrsim 5$.   Except in an end region immediately adjacent to the two vertical plates, the flow in a cavity with $L \gg H$ is everywhere parallel to the horizontal walls, with hot fluid in the upper half of the cavity streaming toward the cold plate and cold fluid in the lower half streaming toward the hot plate (only at very high Rayleigh numbers, where turbulent eddies of a scale smaller than H are possible, will this simple flow pattern break down). The plates at temperatures T_h and T_c deflect the streams into boundary layers on each vertical surface. The predictions of Bejan and Tien [16] for *adiabatic walls* are correlated to within 8 percent by their equation

$$\mathrm{Nu} = 1 + \left\{ \left[\gamma_1\, \mathrm{Ra}^2 \left(\frac{H}{L} \right)^8 \right]^m + \left[\gamma_2\, \mathrm{Ra}^{1/5} \left(\frac{L}{H} \right)^{2/5} \right]^m \right\}^{1/m} \tag{4.98}$$

in which $m = -0.386$, $\gamma_1 = 2.756 \times 10^{-6}$, and $\gamma_2 = 0.623$. The analysis is for laminar flow and hence this equation is not recommended for large Ra $(H/L)^3$. Because of the dominance of the walls in this problem, departures from the adiabatic wall conditions can be expected to have a marked effect on Nu.

Rectangular cavities of practical interest are very often not isolated cells but rather members of a multicellular array, such as that sketched in Fig. 4.33. When $\theta = 0$, the central plane of each partition forms an adiabatic plane of symmetry, so that each cell behaves like an isolated cell (of the type defined in the section on geometry and parameters for cavities without interior solids) having wall thickness b equal to one-half the partition thickness. When $\theta \ne 0$, there is usually heat transfer between cells, the magnitude of which is established by the coupling parameter $k_w L/kb$.

Smart et al. [253] carried out an experimental study on multicellular arrays with air as the fluid and $\theta = 90°$ and found that for Ra at least as high as 10^7, Eq. 4.98 fit their data provided

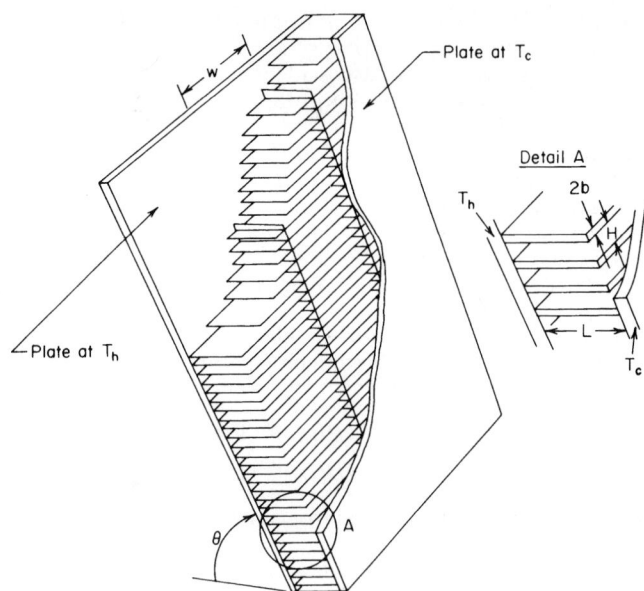

FIGURE 4.33 Sketch of a multicellular array in which many rectangular parallelepiped cavities such as sketched in Fig. 4.25 may be contained.

the values of γ_1 and γ_2 were slightly altered. The altered values of γ_1 and γ_2, tabulated in Table 4.10, depended on the conductive and radiative wall properties, as noted in the table. Whether the changes in γ_1 and γ_2 were attributable to the multicellular array effect, the radiative effect, or both, cannot be resolved from the data.

Vertical Cavities with 0.5 < H/L < 5 and W/L $\gtrsim$ 5. In this intermediate range of H/L, the low-to-moderate Rayleigh number flow consists of a two-dimensional roll. The problem, particularly with $L/H = 1$, has been the subject of many numerical studies, and indeed for the adiabatic wall case with Pr = 0.7, it has formed the basis of a "benchmark problem" [72] for computational fluid dynamic (CFD) codes (even though it is virtually impossible to duplicate this situation in the real world because real fluids with Pr = 0.7 can never be properly insulated). For both the perfectly conducting and the adiabatic boundary conditions, Table 4.11 gives a tabulation of Nu as a function of Ra for values of H/L of 0.5, 1, 2, and 5, as calculated by Catton et al. [35] (and reported by Catton [34]) for very large Pr, by Wong and Raithby [284] and Raithby and Wong [230] for Pr = 0.7, and by Le Quéré [176]. The effect of Pr over

TABLE 4.10 Values of γ_1 and γ_2 to Be Used in Eq. 4.98 for Air-Filled Cavities in Multicellular Arrays [253]

L/H	ϵ_w	ϵ_h	ϵ_c	$kL/k_w b$	$\gamma_1 \times 10^6$	γ_2
3	0.13	0.065	0.065	100	1.274	0.415
5	0.13	0.065	0.065	166	1.324	0.474
5	0.9	0.065	0.065	42	0.970	0.594
5	0.9	0.9	0.065	42	1.524	0.430
5	0.9	0.9	0.9	42	4.76	0.511
10	0.13	0.065	0.065	332	3.952	0.502

See Fig. 4.33 for the meaning of b.

the range $0.7 < \text{Pr} < \infty$ is seen to be quite modest. But for $\text{Pr} < 0.7$, the effect of Pr has been found to be stronger: the effect of low Prandtl number on cavities with adiabatic walls and $H/L = 1$ was investigated, using a CFD code, by Lage and Bejan [173]. They found that at $\text{Ra} = 1 \times 10^5$, Nu went from 4.9 at $\text{Pr} = 1$, to 3.35 at $\text{Pr} = 0.1$, and to 2.77 at $\text{Pr} = 0.01$. These workers also give a criterion to establish whether the flow is laminar or turbulent. Extension to Table 4.11 to higher Ra appears only to have been made for the "benchmark" configuration (i.e., adiabatic walls, $\text{Pr} = 0.7$). Thus Kuyper et al. [171] correlated their CFD results by the equations

$$\text{Nu} = 0.171 \, \text{Ra}^{0.282} \quad \text{for} \quad 10^4 < \text{Ra} < 10^8 \tag{4.99a}$$

$$\text{Nu} = 0.050 \, \text{Ra}^{0.341} \quad \text{for} \quad 10^8 < \text{Ra} < 10^{12} \tag{4.99b}$$

Equation 4.99b fits predictions obtained using a turbulence model. Hsieh and Wang [146] correlated their high Rayleigh number experimental results on adiabatic-walled cavities by the equations

$$\text{Nu} = 0.321 \, \text{Ra}^{0.241} \, (H/L)^{-0.095} \, \text{Pr}^{0.053} \quad \text{for} \quad 10^6 < \text{Ra} < 1.4 \times 10^7 \tag{4.100}$$

and $$\text{Nu} = 0.133 \, \text{Ra}^{0.301} \, (H/L)^{-0.095} \, \text{Pr}^{0.053} \quad \text{for} \quad \text{Ra} > 1.4 \times 10^7 \leqslant \text{Ra} \leqslant 2 \times 10^9 \tag{4.101}$$

This equation pair was derived from data covering the range $0.7 < \text{Pr} < 464$ and $3 < H/L < 5$.

Heat Transfer in Vertical Cavities With W/L $\gtrsim$ 5: An Overview.   Figure 4.34 presents a compilation of some of the data of the previous three sections in terms of Nu versus H/L, with Ra as a parameter, for adiabatic walls with $\epsilon_w = 0$ and $\text{Pr} = 0.7$. The figure shows a peak that moves to lower values of H/L as Ra is increased.

TABLE 4.11   Tabulation of Numerically Computed Nusselt Number for Vertical ($\theta = 90°$) Rectangular Parallelepiped Cavities Having $W/L \gtrsim 5$ and $0.5 < H/L < 5$.

			H/L		
	0.5	1		2	5
Prandtl number	∞	0.7	∞	0.7	0.7
Reference	35	176, 230	35	230	230
Perfectly conducting walls					
$\text{Ra} = 10^3$	1.00	1.05	1.05	1.11	1.05
3×10^3	1.01	1.25	—	1.42	1.28
10^4	1.07	1.75	1.77	1.97	1.81
3×10^4	1.48	2.41	2.50	2.61	2.45
10^5	2.51	3.40	3.66	3.53	3.30
3×10^5	3.64	4.47	—	—	—
Adiabatic walls					
$\text{Ra} = 10^3$	1.00	1.12	1.12	1.19	1.09
3×10^3	1.05	1.50	—	1.64	1.39
10^4	1.30	2.24	2.24	2.34	2.00
3×10^4	2.18	3.14	3.16	3.12	2.72
10^5	3.82	4.51	4.52	4.26	3.68
10^6	—	8.83	—	—	—
10^7	—	16.52	—	—	—
10^8	—	30.22	—	—	—

See Fig. 4.25 for meaning of symbols.

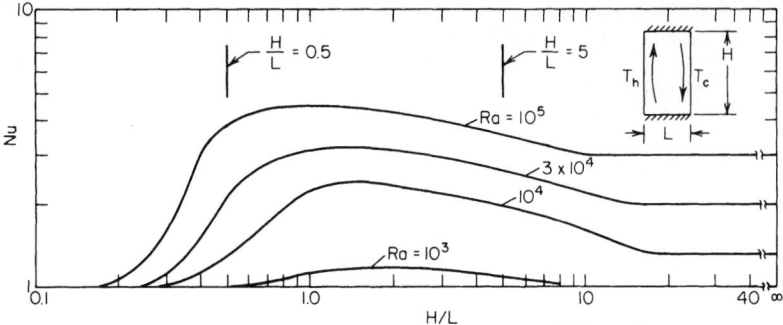

FIGURE 4.34   Nu as a function of H/L for various values of Ra for a vertical rectangular parallelepiped cavity with $W/L \gtrsim 5$, Pr = 0.7, and $\epsilon_w = 0$ (no radiation effects). For $H/L < 0.5$, the plot is based on Eq. 4.98; for $0.5 \le H/L \le 5$, it is based on Table 4.11; for $H/L > 5$, it is based on Eq. 4.92.

Effect of* W/L *on the Heat Transfer in Vertical Cavities.   For $L/H = 4$, and for cavities with adiabatic walls, Arnold et al. [9] showed experimentally that Nu is unaffected when W/H is decreased from essentially infinity to 2. On the other hand, Edwards et al. [80] demonstrated that when W/H is further reduced to unity, a very substantial reduction in Nu results. For cavities with $2 \le L/H \le 10$ within a multicellular array (Fig. 4.33), the experimental data of Cane et al. [29] with $W/H = 1$ show a constant reduction factor in $\text{Nu} - 1$ of about 2.2 when compared to the $W/H \to \infty$ data of Smart et al. [253]. That is, for all other factors (including Ra) constant,

$$\frac{(\text{Nu} - 1)_{W/H=1}}{(\text{Nu} - 1)_{W/H \to \infty}} = 0.45 \pm 0.05 \qquad (4.102)$$

The reduction relation given by Eq. 4.102 has been tested for $\text{Ra} \le 10^6$ with $L/H = 2, 3, 4$, and 5, and for $\text{Ra} \le 3 \times 10^6$ for $L/H = 10$.

Very little information exists on the effect of W on Nu at moderate to high values of H/L. Edwards et al. [80] indicate that when $H/L \ge 0.5$, provided $W/L \ge 2$, Nu is insensitive to W. A somewhat similar insensitivity to W/L was found in the CFD study of Fusegi et al. [298] for adiabatic-walled cavities; they found that changing W/L from ∞ to 1 decreased Nu by only 8.3 percent at most. At higher Ra an even smaller decrease in Nu would be expected. More work is required to sort out the effect of W/L.

Heat Transfer in Inclined Rectangular Cavities

Angular Scaling.   Depending on the inclination θ, flow in an inclined cavity with $W/H > 8$ can resemble that in the corresponding horizontal cavity or that in the corresponding vertical cavity; it rarely combines the characteristics of both. Consequently, with few exceptions, the Nusselt number in the inclined cavity can be determined, to a reasonable approximation, from either the vertical or the horizontal Nusselt number relation, by means of simple angular scaling laws.

In this section Nu_H (Ra) will refer to the Nusselt number–Rayleigh number relation for a horizontal cavity (as determined by methods given in the section on natural convection in these cavities) having the same values for all the other relevant dimensionless groups as the inclined cavity at hand. Similarly, Nu_V (Ra) will refer to the Nu (Ra) relation for the corresponding cavity at $\theta = 90°$ (as determined by methods given in the section on heat transfer in vertical rectangular parallelepiped cavities), while Nu_θ (Ra) will be the sought Nu (Ra) relation at the angle θ.

The scaling laws are found to be slightly dependent on the Prandtl number; the laws will first be reported for $Pr \gtrsim 4$ (nonmetallic liquids), then for $Pr \approx 0.7$ (gases).

Cavities with Pr ≳ 4 and W/H ≳ 8. For $90° < \theta < 180°$, i.e., cavities heated from above, the scaling law suggested by Arnold et al. [7],

$$Nu_\theta \, (Ra) = 1 + (Nu_V \, (Ra) - 1) \sin \theta \tag{4.103}$$

has been experimentally validated by Arnold et al. [8] for cavities with $H/L = 1, 3, 6,$ and 12. For $0 < \theta < 90°$ (heating from below), two scaling laws are particularly useful: the horizontal scaling law of Clever [65],

$$Nu_\theta \, (Ra) = Nu_H \, (Ra \cos \theta) \tag{4.104}$$

and the vertical scaling law,

$$Nu_\theta \, (Ra) = Nu_V \, (Ra \sin \theta) \tag{4.105}$$

for $H/L \geq 6$, the maximum of the values of Nu_θ given by each is recommended, i.e.,

$$Nu_\theta \, (Ra) = [Nu_H \, (Ra \cos \theta), \, Nu_V \, (Ra \sin \theta)]_{max} \tag{4.106}$$

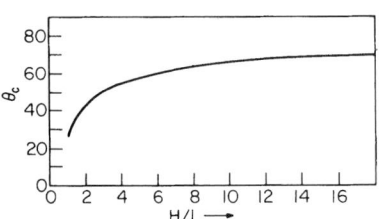

FIGURE 4.35 Plot of the crossover angle θ_c governing the transition from horizontal-like flow to vertical-like flow in an inclined cavity containing nonmetallic liquids. Adapted from Arnold et al. [8]. θ_c is in degrees.

Equation 4.106 yields a single value of the crossover angle θ_c, defined so that for $\theta < \theta_c$ the horizontal scaling law applies and for $\theta > \theta_c$ the vertical scaling law applies. Angle θ_c is obtained by equating $Nu_H \, (Ra \cos \theta)$ to $Nu_V \, (Ra \sin \theta)$ and solving for θ. This angle (which also locates a minimum in Nu_θ when Nu_θ is plotted against θ with Ra held constant) is plotted as a function of H/L in Fig. 4.35, as given by Arnold et al. [8]. Equation 4.106, with a different but very similar vertical scaling law, was validated by Arnold et al. [8] for $H/L = 6$ and 12 and for $10^4 \leq Ra \leq 10^6$.

For $1 < H/L \leq 6$, the variation of Nu_θ over the range $0 < \theta < 90°$ is found to be quite modest, and the horizontal scaling law breaks down. For this regime in H/L the following relations, similar to those proposed by Catton [34], are recommended:

$$Nu_\theta \, (Ra) = Nu_V \, (Ra \sin \theta) \qquad \theta_c < \theta < 90° \tag{4.107}$$

$$Nu_\theta \, (Ra) = Nu_H \, (Ra) \left[\frac{Nu_V \, (Ra \sin \theta_c)}{Nu_H \, (Ra)} \right]^{\theta/\theta_c} \qquad 0 < \theta < \theta_c \tag{4.108}$$

where θ_c is obtained from Fig. 4.35. Equations 4.107 and 4.108 are in agreement with the data of Arnold et al. [8] for $H/L = 1$ and 3. Agreement with the data and theory of Ozoe et al. [217] is not so satisfactory, but the differences are mostly within 10 percent. For $1 \leq H/L \leq 6$, and for $10^4 < Ra < 5 \times 10^5$, the $Nu_V \, (Ra)$ relation can be approximated by a ¼-power law so that for this range Eq. 4.107 can be approximated by [35]

$$Nu_\theta \, (Ra) = Nu_V \, (Ra) \sin^{1/4} \theta \tag{4.109}$$

For $H/L \leq 0.25$, Eq. 4.106 is again recommended; when $Nu_H \, (Ra \cos \theta)$ is equated to $Nu_V \, (Ra \sin \theta)$ to find the crossover angle θ_c, this angle is generally found to be greater than 60°. Equation 4.106 agrees within about 10 percent with the data of Arnold et al. [9] for $H/L = 0.25$, and Edwards et al. [80] for $H/L = 0.25$ and $H/L = 0.14$. No data on which to base scaling laws seem to be available for $0.25 \leq H/L \leq 1$.

Cavities with Pr ≈ 0.7 and W/H ≳ 8. For $H/L > 5$ and $0 \le \theta \le 60°$, direct application of the horizontal scaling law (Eq. 4.104) introduces significant errors when $Pr \approx 0.7$ and $Ra \approx 10^4$ [144]; these errors have been shown [66, 235] to result from a secondary instability that appears at a Rayleigh number only slightly greater than that for the primary instability discussed in the section on horizontal rectangular parallelepiped and circular cylinder cavities. A modified scaling relation, taken from Hollands et al. [144], is recommended for $0 \le \theta \le 60°$:

$$\text{Nu}_\theta \, (\text{Ra}) = 1 + 1.44 \left[1 - \frac{1708}{\text{Ra}\cos\theta} \right]^{\bullet} \left[1 - \frac{1708\,(\sin 1.8\theta)^{1.6}}{\text{Ra}\cos\theta} \right] + \left[\left(\frac{\text{Ra}\cos\theta}{5830} \right)^{1/3} - 1 \right]^{\bullet} \quad (4.110)$$

This equation agrees well with data up to $Ra \approx 10^5$; for $Ra = 10^6$ it underestimates the measurements of ElSherbiny et al. [84, 85] by 10 percent. (See Eq. 4.79 for meaning of dots.)

For $\theta = 60°$ the recommended relation, taken from ElSherbiny et al. [84, 85], is

$$\text{Nu}_{60} \, (\text{Ra}) = [\text{Nu}_{60}^1, \text{Nu}_{60}^2]_{\max} \quad (4.111)$$

where
$$\text{Nu}_{60}^1 = \left[1 + \frac{(0.0936\,\text{Ra}^{0.314})^7}{1 + G} \right]^{1/7} \qquad G = \frac{0.5}{[1 + (\text{Ra}/3165)^{20.6}]^{0.1}} \quad (4.112)$$

and
$$\text{Nu}_{60}^2 = \left(0.1044 + 0.1750\,\frac{L}{H} \right) \text{Ra}^{0.283} \quad (4.113)$$

For $60° \le \theta \le 90°$, linear interpolation between the 60° and 90° relations is recommended:

$$\text{Nu}_\theta \, (\text{Ra}) = \frac{90° - \theta}{30°} \, \text{Nu}_{60} \, (\text{Ra}) + \frac{\theta - 60°}{30°} \, \text{Nu}_V \, (\text{Ra}) \quad (4.114)$$

The tested range of validity of Eqs. 4.111–4.114 is the same as that for Eq. 4.92.

For $0.5 \le H/L \le 5$ and $45° \le \theta \le 90°$, Meyer et al. [191] recommend a relation that can be fitted by

$$\text{Nu}_\theta \, (\text{Ra}) = 1.06 \, \text{Nu}_V \, (\text{Ra}) \qquad\qquad 45° \le \theta \le 60° \quad (4.115)$$

$$\text{Nu}_\theta \, (\text{Ra}) = \left[1 + \frac{0.06(90° - \theta)}{30°} \right] \text{Nu}_V \, (\text{Ra}) \qquad 60° \le \theta \le 90° \quad (4.116)$$

For $3 \le L/H \le 10$ and $0 \le \theta \le 75°$, Smart et al. [253] found that the horizontal scaling law (Eq. 4.99) is valid. For ranges of θ not covered in this section, the relevant scaling laws for $Pr \gtrsim 4$ are tentatively recommended (see the preceding section).

Effect of W/H on Nu for Inclined Cavities. The data of Edwards et al. [80] indicate that for $L/H = 4$, horizontal scaling (which had been found valid for $0 \le \theta \le 60°$ when $W/H = 24$ and 8) becomes invalid for $W/H = 4$. The data of Cane et al. [29], for which $W/H = 1$, agree well with the following relation:

$$\text{Nu} = 1 + 1.15[\text{Nu}_V \, (\text{Ra}) - 1] \cos (\theta - 60°) \left(\frac{\text{Ra}\,H^4}{2840L^4} \right)^{1.64(1 - \sin\theta)} \quad (4.117)$$

provided $\theta \ge 30°$ and $Ra\,H^4/L^4 \le 6000$. Equation 4.117 was validated for $L/H = 2, 3, 4$, and 5. Plotted experimental data for a number of combinations of W/H, H/W, and θ are given by Edwards et al. [80].

Heat Transfer in Enclosures with Interior Solids at Prescribed Temperature

Region between Concentric or Eccentric Cylinders.   The geometry and dimensions are as shown in cross section in Fig. 4.36a, the two cylinders being assumed to have parallel axes. The dimension E represents the perpendicular distance from the axis of the inner cylinder to the axis of the outer cylinder. Thus, for concentric cylinders, $E = 0$. Each cylinder is taken to be isothermal but at a different temperature.

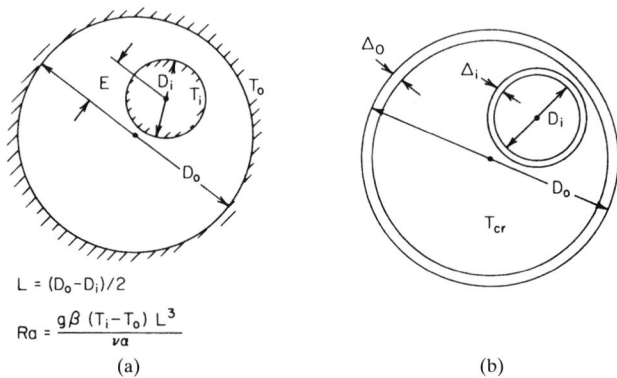

$$L = (D_o - D_i)/2$$

$$Ra = \frac{g\beta\,(T_i - T_o)\,L^3}{\nu a}$$

(a)   (b)

FIGURE 4.36   Sketch of concentric and eccentric cylinder and sphere problems.

The stages of flow development with increasing Ra parallel those in the vertical rectangular parallelepiped cavity. For small Ra, the fluid flow (which is present for any finite Ra) is sufficiently feeble that the heat transfer is, for all practical purposes, by conduction only. As Ra is increased, a laminar boundary layer regime is established wherein the flow is largely restricted to boundary layers on each of the cylinders. The central region in the regime is stably stratified and almost stationary: it generally contains one or two large, slowly rotating eddies on each side of the cavity, but the details of this flow structure have little effect on the heat transfer. At high Ra the boundary layers can be expected to become turbulent.

The Nusselt number for this problem is defined by

$$Nu = \frac{q'\ln\,(D_o/D_i)}{2\pi(T_i - T_o)k} \tag{4.118}$$

where q' is the heat transfer by conduction and convection from the inner cylinder to the outer one per unit axial length of cylinder. The temperatures T_i and T_o correspond to the inside diameter D_i and the outside diameter D_o. By Nu_{COND} we mean the Nusselt number that applies when the fluid is stationary and so there is no convection. The definition for Nu prescribed by Eq. 4.118 makes $Nu_{COND} = 1$ when the cylinders are concentric ($E = 0$), and

$$Nu_{COND} = \frac{\ln\,(D_o/D_i)}{\cosh^{-1}\,[(D_o^2 + D_i^2 - 4E^2)/2D_oD_i]} \tag{4.119}$$

when $0 \le E < L$.

Figure 4.36b shows conduction layers applied to each cylinder. Both conduction layer thicknesses are shown as being much less than the spacing L so that they do not touch—this situation will always occur if Ra is large enough to make $\Delta_o + \Delta_i < L - E$. Since, according to the conduction layer model, the central region is isothermal, this model predicts that, provided the conduction layers do not touch, Nu will be independent of E. In fact, this is what is observed; for cylinders with vertical axes and for Ra = 4.8×10^4, D_o/D_i = 2.5, and Pr = 0.7, Kuehn and Goldstein [164] found that Nu changed by no more than 10 percent when E was varied from 0 to $\frac{2}{3}L$, regardless of the direction of displacement of this inner cylinder. (Calculations indicate that for this Ra, $\Delta_i \approx \Delta_o \approx L/4$, so a slight overlap of conduction layers would have occurred at $E = \frac{2}{3}L$.)

Cylinders With Horizontal Axes. The conduction layer model has been shown to accurately predict the heat transfer for horizontal cylinders [164, 223], but because of the need to iteratively solve for the central region temperature it does not yield an explicit expression for Nu. However, by making additional approximations Raithby and Hollands [223] were able to derive an explicit relation for the heat transfer when the (assumed laminar) conduction layers do not overlap:

$$\mathrm{Nu}_l = 0.603\overline{C}_l \frac{\ln (D_o/D_i) \, \mathrm{Ra}^{1/4}}{[(L/D_i)^{3/5} + (L/D_o)^{3/5}]^{5/4}} \qquad (4.120)$$

For $E = 0$, the conduction layers just touch when Ra falls to the value at which Eq. 4.120 predicts $\mathrm{Nu}_l = \mathrm{Nu}_{\mathrm{COND}} = 1$; for still smaller values of Ra, Nu remains at $\mathrm{Nu}_{\mathrm{COND}}$, so that for the conduction *and* laminar flow regimes

$$\mathrm{Nu} = [\mathrm{Nu}_{\mathrm{COND}}, \mathrm{Nu}_l]_{\mathrm{max}} \qquad (4.121)$$

where Nu_l is given by Eq. 4.120 and $\mathrm{Nu}_{\mathrm{COND}}$ is equal to unity. For $E \neq 0$, Eq. 4.121 can still be used, with Eq. 4.119 being used for $\mathrm{Nu}_{\mathrm{COND}}$; however, since the conduction layers do not touch uniformly as Ra decreases toward the conduction regime, some error must be expected around $\mathrm{Nu}_l \approx \mathrm{Nu}_{\mathrm{COND}}$. Well removed from this region, the equation should be accurate.

Equation 4.121 was tested against earlier (i.e. prior to 1975) data with $E = 0$ over the ranges $2 \times 10^2 \leq \mathrm{Ra} \leq 8 \times 10^7$, $0.7 \leq \mathrm{Pr} \leq 6000$, and $1.15 \leq D_o/D_i \leq 8$; agreement was almost invariably within 10 percent. The more recent data of Kuehn and Goldstein [164] at $D_o/D_i = 2.6$ and at Pr = 0.7, which covered the range $2 \times 10^2 \leq \mathrm{Ra} \leq 8 \times 10^7$, agree with Eq. 4.116 with a mean deviation of 2.1 percent and a maximum error, occurring when Nu ≈ 1, of 8 percent. The equation also agrees very well with the numerical solutions of Farouk and Güçeri [96], which were for $D_o/D_i = 2.6$ and ranged in Ra from 10^3 to 10^7. For Ra > 10^8, turbulence effects neglected in the derivation of Eq. 4.115 may become important, and the nonexplicit conduction layer model, expanded on by Kuehn and Goldstein [163, 164], should be adopted. Also, the equation may not be accurate at very low Prandtl numbers. An equation with a wider Prandtl number range has been recommended by Hessami et al. [136] on the basis of their own data (which covered the ranges 0.023 < Pr < 10,000 and 1.15 < D_o/D_i < 11.4) as well as those of some other workers. Their equation is

$$\mathrm{Nu} = 0.265[\ln (D_o/D_i)][\mathrm{Ra} \, (D_i/L)^3)(1 - D_i/D_o)/(1 + 0.952/\mathrm{Pr})]^{1/4} \qquad (4.122)$$

Cylinders With Vertical Axes. In the case of vertical cylinders, the height H of the cylinders also plays a role, and so do the two flat end faces, which are assumed to be adiabatic surfaces. Several studies have been done on this problem; Kumar and Kalam [169] have provided a summary of much of that work. For laminar flow and outside the conduction regime, they recommend the equation

$$\mathrm{Nu}_l = 0.09 \, \mathrm{Ra}^{0.278} \, (D_o/D_i)^{0.34 + 0.329D_i/D_o}(L/H)^{0.122}(D_i/L) \ln (D_o/D_i) \qquad (4.123)$$

Once Nu_l has been calculated from this equation, Eq. 4.121 should then be applied. The range of validity of this approach should extend at least over the following ranges: Ra < 10^6, 2 < (D_o/D_i) < 15, 1 < H/L < 10.

Region Between Concentric and Eccentric Spheres. The geometry, dimensions, and Rayleigh number definition are as sketched in Fig. 4.36a, the centers of the (isothermal) spheres being distance E apart. The flow regimes and heat transfer relations closely parallel those for the circular cylinder. The Nusselt number is defined by

$$\text{Nu} = \frac{qL}{\pi D_i D_o (T_i - T_o) k} \tag{4.124}$$

which makes the Nu_{COND} (i.e., the Nu for a stationary fluid) equal to unity when $E = 0$. Nu_{COND} for $0 < E < L$ is found by conduction analysis to be [M. M. Yovanovich, personal communication]

$$\text{Nu}_{\text{COND}} = \frac{D_o - D_i}{\Phi(\eta_i) D_o - \Phi(\eta_o) D_i} \tag{4.125}$$

$$\Phi(\eta) = \left[1 + \sum_{n=1}^{\infty} \frac{\sinh \eta}{\sinh (n+1)\eta} \right]^{-1} \tag{4.126}$$

$$\eta_i = \cosh^{-1} \sqrt{\frac{D_o^2 - D_i^2 - 4E^2}{4 D_i E}}, \qquad \eta_o = \cosh^{-1} \left(\frac{2E}{D_o} + \frac{D_i}{D_o} \sqrt{\frac{D_o^2 - D_i^2 - 4E^2}{4 D_i E}} \right) \tag{4.127}$$

When $0.3 \leq \eta \leq 1.2$, Eq. 4.126 can be approximated to within 1 percent by $\Phi(\eta) = 0.659 \eta^{0.42}$; when $\eta \geq 1.2$, it can be similarly approximated by $\Phi(\eta) = (2 \cosh \eta - 1)/(2 \cosh \eta)$. As was the case for cylinders, Nu is expected to be independent of E when Ra is large enough to make $\Delta_i + \Delta_o < L - E$. Using a modified conduction layer method, Raithby and Hollands [223] obtained an explicit relation for the Nusselt number Nu: namely Eq. 4.121 with Nu_{COND} given by Eqs. 4.125–4.127 (or given by unity if $E = 0$) and Nu_l given by

$$\text{Nu}_l = 1.16 \overline{C}_l \left(\frac{L}{D_i} \right)^{1/4} \frac{\text{Ra}^{1/4}}{[(D_i/D_o)^{3/5} + (D_o/D_i)^{4/5}]^{5/4}} \tag{4.128}$$

This equation was shown to closely fit the $E = 0$ data of Scanlan et al. [241], which covered the ranges $1.3 \times 10^3 \leq \text{Ra} \leq 6 \times 10^8$, $5 \leq \text{Pr} \leq 4000$, and $1.25 \leq D_o/D_i \leq 2.5$. The measurements of Weber et al. [281] for eccentric spheres, where the displacement from the concentric position is vertically up or down, showed that for downward displacement E had little effect on Nu for $0 \leq E/L \leq 0.75$, but for upward displacement with $0.25 \leq E/L \leq 0.75$, the Nusselt number was observed to be about 10 percent higher than that given by Eq. 4.128. The same reservations as discussed for the cylinder when $\text{Nu}_{\text{COND}} \approx \text{Nu}_l$ apply here.

Other 3D Enclosures With Interior Solids. Warrington and Powe [278] showed that so far as the heat transfer is concerned, cubes and stubby cylinders behave similarly to equivalent spheres of the same volume. This appears to be the case for both the inner and outer body shape. So Eqs. 4.121, 4.124, and 4.128 appear to be applicable to other inner and outer body shapes as well, it being understood that $D_o = (6 V_o/\pi)^{1/3}$ and $D_i = (6 V_i/\pi)^{1/3}$, where V_o and V_i are the inner and outer body volumes, respectively. Sparrow and Charmichi [258], using stubby cylinders for the inner and outer body shapes, confirmed the conduction layer model prediction that the heat transfer is independent of eccentricity E when Ra (based on inner cylinder diameter) is greater than about 1500.

Partitioned Enclosures

Classification of Partitions. Partitions are relatively thin, solid walls mounted inside the enclosure, as for example in Fig. 4.33. The partitions are "passive" in the sense that neither their temperature nor heat flux is prescribed. Depending on their extent and orientation, such partitions can have profound effect on the heat transfer. Partitions can be classified in various ways. *Complete partitions* run continuously from one side to another; *partial partitions* have breaks, or *windows.* Referring to Fig. 4.25, *parallel partitions* run parallel to the hot and cold plates; *perpendicular partitions* run perpendicular to these plates, or parallel to the side walls as shown in Fig. 4.33. Partitions can also be either *single* (one partition only) or *multiple* (more than one partition).

Complete, Parallel Partitions.   We first consider a single parallel partition. Such a partition will divide the enclosure into two *subenclosures.* Usually it is a fair approximation to treat each subenclosure as a regular enclosure of the type discussed in the previous parts of this section. Such a strategy requires the determination of the partition's temperature, and this can generally be found by trial and error through an energy balance on the partition itself. (If the fluid is a gas it is very important to include radiation in the heat balance—for a sample of such calculations, see Hollands and Wright [145].) The problem with such a strategy is that it inherently assumes that the partition is isothermal. Under certain conditions, for example at very high Rayleigh numbers or in the triple-paned vertical windows [287], the assumption would appear to introduce little or essentially zero error. On the other hand, it is known to introduce substantial error in at least one case: the horizontal layer at near-critical conditions.

Thus consider the rectangular parallelepiped of Fig. 4.25a with $\theta = 0$ and with a thin partition running parallel to the plates and extending the whole H by W distance between the walls. It is further assumed that the cavity is extensive in the horizontal direction, i.e., $W \gg L$ and $H \gg L$, and (for the moment) that the fluid is opaque to thermal radiation. Catton and Lienhard [37] have treated the problem of predicting the critical Rayleigh number in such a situation, for arbitrary thickness L_B and conductivity k_B of the partition and for arbitrary spacings L_1 and L_2 between the lower plate and the partition and between the upper plate and the partition, respectively. (Note that $L = L_1 + L_2 + L_B$.) For $L_1 = L_2$ they found that the critical Rayleigh number for each layer varied with k/k_B, from 1708 at $k/k_B = 0$ to very close to 1296 at large values of k/k_B. (The isothermal partition model described in the previous paragraph would predict a critical Rayleigh number of 1708.) Keeping to the restriction that $L_1 = L_2$, and for thin partitions (implying in this case that $L_B/L < 0.2$), their predicted critical Rayleigh number Ra_c for each layer was found to vary with the group $k(L_1 + L_2)/(k_B L_B)$ as follows: for $k(L_1 + L_2)/(k_B L_B) = 1$, $\text{Ra}_c \approx 1580$; for $k(L_1 + L_2)/(k_B L_B) = 3$, $\text{Ra}_c \approx 1480$; for $k(L_1 + L_2)/(k_B L_B) = 10$, $\text{Ra}_c \approx 1375$; and for $k(L_1 + L_2)/(k_B L_B) = 100$, $\text{Ra}_c \approx 1305$. When $L_1 \neq L_2$, they found that the critical Rayleigh number for the thicker fluid layer is always bounded between 1708 and 1296. Lienhard [182] extended these results to layers of fluids (like gases) that are transparent to thermal radiation, where the partition is opaque. In the case where $L_1 = L_2$, the effect of radiation is to raise the upper-layer Ra_c, applying for the case of the limit of large k/k_B, from the value of 1296 that applied for no radiation to a value intermediate between 1296 and 1708, given by

$$\text{Ra}_c = 1502 - 206 \tanh\left(\ln\left[1.636\left(\frac{T_c}{T_h}\right)^{0.5331}\left(\frac{k}{4\sigma T_h^4 L_1}\right)^{0.46511}\right]\right) \tag{4.129}$$

Once convection starts in either of the two layers, it drives a fluid motion in the other. Using a certain model (the Landau model), Lienhard and Catton [299] predicted the heat transfer in the Rayleigh number range slightly greater than critical. Use of Eq. 4.78 for both layers, with the Ra_c the one relevant to thicker layer, is also tentatively recommended.

Like the single-partition enclosure, the multipartitioned enclosure may be treated by analyzing each subenclosure separately, assuming isothermal partitions, and then making energy balances on the individual partitions to determine the partition temperatures (e.g., see Hollands and Wright [145]). But, as for the single-partition case, substantial errors can arise around the critical condition at $\theta = 0$. Finding the critical Rayleigh numbers for each layer has been treated by Hieber [139] and Lienhard [181] for some special cases. For the case where the spacing between partitions is constant, the Ra_c at the limit of diminishingly small k_B is found to be only 720 for the inner layers and 1296 for the layers that are next to the hot and cold plates. These should be compared to an expected value of 1708 on the basis of the isothermal partition. For further details the interested reader is referred to these papers.

Complete Perpendicular Partitions.   For this case, where the partition(s) are perpendicular to the plates (as illustrated in Fig. 4.33), two or more subenclosures are formed, the partitions themselves constituting at least part of the side walls of these subenclosures. In the conceptual process of forming these subenclosures, the partitions' thickness should be split down the middle, with half of the partition being inside each of the adjacent subenclosures. With certain reservations, the heat transfer across these subenclosures can then be determined using the methods for unpartitioned enclosures described in previous sections. The reservations concern the boundary conditions applying on the outside faces of the side walls. In the section on geometry and parameters for cavities without interior solids (which refers to Fig. 4.25), these outside faces were prescribed as being adiabatic surfaces. Thus, strictly speaking, for the methods of previous sections to apply, the central faces of the split partitions should be treatable as adiabatic. Under certain conditions—for example, because of symmetry—this may be realistic, although because of the possibility of corotating cells in adjacent subenclosures, even then the adiabatic assumption may not be strictly correct. On the other hand, the errors associated with the nonadiabatic conditions should be acceptably small. If the thermal analyst is not satisfied with the uncertainties associated with this approximation, more refined analyses do exist (e.g., see Asako et al. [10]) and experimental data are available (e.g., see Cane et al. [29], Smart et al. [253]), but these will be only for certain limited methods of partitioning that may or may not match those of interest to the analyst.

Partial Parallel Partitions.   In a number of studies researchers have investigated the rectangular parallelepiped cavity of Fig. 4.25a, at $\theta = 90°$, in which a partial partition is inserted parallel to and midway between the plates. In most of the experimental work, the Rayleigh number was on the order of 10^{10} (so the conduction boundary layers on the vertical walls are very much smaller than the cavity dimensions) and the Prandtl number was close to 6. However, when expressed in terms of the equations below, the results should have wider applicability. In the early studies, the partition projected some vertical distance, say H_p, either down from the ceiling or up from the floor, and it ran the entire width W of the cavity. For the situation in which the partition runs up from the floor, Lin and Bejan [186] found that the part of the cold side of the cavity that is below the top of the partition in elevation contains only a very weak circulation cell that hardly contributes to the heat transfer, whereas a strong circulatory cell runs—in a single boundary layer—up the hot plate, across the ceiling to the cold plate, down the cold plate to the elevation of the top of the partition, across the cavity to the top of the partition, down the partition on the hot side, across the floor, and back to the hot plate. Unless the opening is very small, this cell carries almost the entire heat transfer observed, which means that only the upper part of the cold plate receives significant heat transfer. When the partition projects down from the ceiling, an analogous pair of cells is produced. On the basis of this model Lin and Bejan developed a quantitative heat transfer model that fit their data and those of other workers. After some reworking, their model can be expressed as

$$\mathrm{Nu} = \mathrm{Nu}_{np} - (\mathrm{Nu}_{np} - \mathrm{Nu}_{fp})\left(1 - \frac{1.5H^{-3/4}}{(H_0 + H_{\mathrm{MIN}})^{-3/4} + 0.5(H_0 + H_{\mathrm{MAX}})^{-3/4}}\right)\left(1 - \frac{1.5H^{-3/4}}{H_{\mathrm{MIN}}^{-3/4} + 0.5H_{\mathrm{MAX}}^{-3/4}}\right)^{-1}$$

(4.130)

where $H_0 = H - H_p$ is the height of the "window" left in the partition, $H_{\mathrm{MAX}} = H_p$, and for the moment $H_{\mathrm{MIN}} = 0$. To apply this equation, one must first evaluate the Nusselt number Nu with the partition fully removed; this is the Nu_{np} (or Nu for no partition) in Eq. 4.130. Next one evaluates the Nu with the full partition in place, i.e., with $H_p = H$; this is the Nu_{fp} (or Nu for full partition) in Eq. 4.130. Both Nu_{np} and Nu_{np} can be evaluated using the methods of previous sections. Equation 4.130 provides an interpolation scheme to be used between these two limiting values for Nu. The work of Nansteel and Greif [205] showed that moving the partition—so that it is no longer central but is closer to either the hot plate or the cold plate—has only a minor effect on the heat transfer provided the movement is $H/4$ or less.

An extension can be made to Lin and Bejan's analysis to allow for there being both an upper and a lower partition, the upper partition extending down, say, distance H_U from the ceiling, and the lower partition extending upward, say, distance H_L from the floor, leaving an opening of height $H_0 = H - H_L - H_U$. The result of such an analysis is again expressed by Eq. 4.130, provided one now interprets H_{MAX} as the maximum of H_U and H_L and H_{MIN} as the minimum of H_U and H_L.

It may be that the opening in the partition does not extend the full width W of the cavity shown in Fig. 4.25. Say it only extends over a width W_0, leaving widths W_L and W_R on the left and right, respectively. The flow in the parts of the cavity on the right and left of the opening should be very similar to the flow in the corresponding complete cavity with the full partition, while the flows in the part having the opening should be very similar to the flow in the corresponding complete cavity with the opening extending the full width. Thus it is recommended that one determine the Nusselt number from a weighted average of the Nu_{fp} (giving it weight $(W_L + W_R)/W$) and the Nu calculated from Eq. 4.130 (giving it weight W_0/W). Karki et al. [154] have numerically investigated such enclosure in the low Rayleigh number range: $10^4 < \mathrm{Ra} < 10^7$.

TRANSIENT NATURAL CONVECTION

External Transient Convection

Overview. Suppose a body, such as that shown in Fig. 4.37a, is initially in equilibrium with its surroundings at T_∞, but at time $t = 0$ its surface temperature is changed impulsively to T_w. The surface heat flux rises to infinity (theoretically) and then falls off quickly with increasing t as shown in Fig. 4.37b. Since there is a delay in initiating the fluid motion, the heat flow into the fluid is initially by conduction, and the surface average heat flux $\overline{q''}$ follows the conduction curve in Fig. 4.37b to the departure time t_D. At a later time $(t > t_\infty)$, steady-state convection is achieved. These two regimes are called the conduction regime $(0 \le t \le t_D)$ and the steady-state regime $(t \ge t_\infty)$. The transition regime lies in the range $t_D < t < t_\infty$.

At time t_i, the heat flow by conduction matches the steady-state convective heat transfer from the body. If conditions are met to initiate convection before t_i (this will depend on Pr and Ra), the heat flow falls monotonically in the transition regime, as along path A in Fig. 4.37b. Otherwise convection will not be initiated until $t_D > t_i$, so the heat flow will have fallen below the steady-state value and must therefore recover from the undershoot in the transition regime as shown by path B.

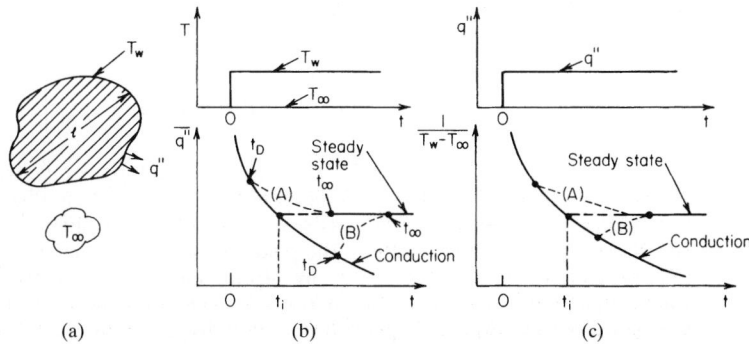

FIGURE 4.37 Transient response of heat flow from a surface (a) following a step change in wall temperature (b), and of the wall temperature following a suddenly applied heat flux (c).

The correct order of magnitude of the time constant for the transient response of the fluid is given by t_i. For a flat plate (or for any body shape up to the time that the penetration distance of the heat conduction from the boundary is small compared to the radius of curvature of the body), the conduction heat transfer is given by [244]

$$q'' = \frac{k(T_w - T_\infty)}{\sqrt{\pi \alpha t}} \tag{4.131}$$

The value of t_i is found by equating this value of q'' to the steady-state average heat transfer q''_∞. For example, for a vertical flat plate 0.1 m (0.32 ft) long that is subjected to an impulsive 10°C (16°F) temperature change and is immersed in various fluids at 20°C (68°F), the values of t_i are as follows: for air, $t_i = 0.5$ s; for water, $t_i = 5$ s; for oil (Pr = 10^4), $t_i = 80$ s; for mercury, $t_i = 2$ s.

An imposed temporal change of T_w, other than the step change, is also of interest. In fact, because the heat capacity of any body or wall is finite, a step change could never be achieved. If the time constant associated with the prescribed change in T_w is much larger than t_i, the heat transfer at each instant in the transient can be accurately calculated from the steady-state natural convection equation; this is called a quasi-static transient. From the above estimates of t_i it will be appreciated that the quasi-static approximation will usually be valid for gases.

The analogous problem of response to a step change in surface heat flux is depicted in Fig. 4.37c. The inverse of $T_w - T_\infty$ (proportional to the heat transfer coefficient) again either may fall monotonically to its steady-state value or may undershoot; the possibility of an undershoot, which corresponds to a temperature overshoot, may have serious ramifications in practical problems and has provided much of the incentive for studying the external transient problem. As before, a step change in surface heat flux could never be achieved in practice because of finite surface heat capacitance. If the time constant for changes in surface heat flux at the surface is large compared to t_i, either because of body heat capacitance or because heating is only gradually applied, the quasi-static approximation will again be accurate.

Vertical Surfaces.   Analyses and measurements related to transient heat transfer on vertical surfaces have been reviewed by Ede [79]. For negligible solid heat capacitance, results of analyses are discussed for step changes in wall temperature and in surface heat flux, and for periodic specified temperature or flux, as well as for other prescribed variations. Most recent analyses have confirmed the estimate of Siegel [252] that the local time for departure from the conduction regime at distance position x from the leading of a vertical flat plate after a step change in wall temperature is

$$t_D = 1.8(1.5 + \text{Pr})^{1/2} \left\{ \frac{x}{[g\beta(T_w - T_\infty)]} \right\}^{1/2} \tag{4.132a}$$

The steady-state regime is attained in time t_∞, where

$$t_\infty = 5.24(0.952 + \text{Pr})^{1/2} \left\{ \frac{x}{[g\beta(T_w - T_\infty)]} \right\}^{1/2} \tag{4.132b}$$

Gebhart and Adams [105, 106, 108] have predicted the average wall temperature response of a vertical flat plate following a sudden application of internal heating to the wall (applied by switching on a current). The plate was assumed to be isothermal in any horizontal plane, and the average (over the length of the plate) wall temperature was reported. The predictions and measurements for air and water are reproduced in Fig. 4.38. In Fig. 4.38, ψ, τ^*, and $\mathbf{Q}$ are dimensionless average surface temperature, time, and thermal capacitance defined in Fig. 4.38 by Gebhart and Adams. In these relations $\mathbf{b}$ and M are Prandtl-number-dependent parameters given in Table 4.12, $\overline{T}_w$ is the average wall temperature (the unknown), $\mathbf{P}''$ in the definition of Ra* is the rate of internal generation of energy per unit heat transfer surface area A, and C''_w in the definition of $\mathbf{Q}$ is the total heat capacitance of the wall per unit of heat

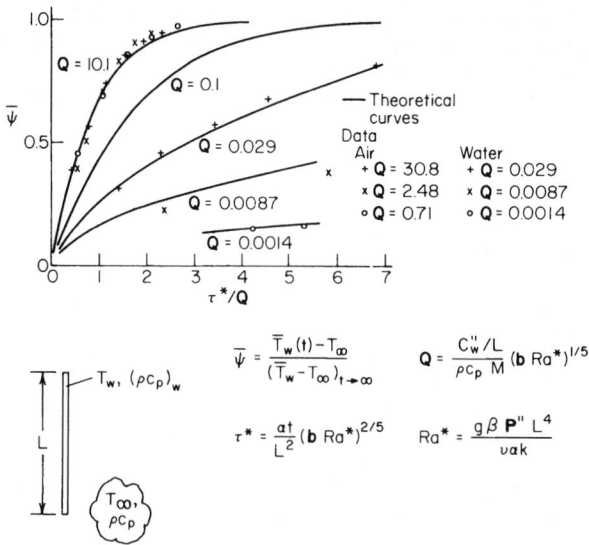

FIGURE 4.38 Temperature response of a vertical plate to a step heat input including the effects of thermal capacitance (from Gebhart and Adams [108]).

transfer surface area. All the data collapse into a single curve for $\mathbf{Q} \geq 1$ because the quasi-static approximation applies in this range of $\mathbf{Q}$.

Horizontal Wires. Attention is restricted to horizontal cylinders (see Fig. 4.39 for nomenclature) that are nearly isothermal on any plane perpendicular to the axis of the cylinder (this will occur if $\bar{h}D/k_w \leq 0.1$, where $\bar{h}$ is the average heat transfer coefficient on the outside surface and k_w is the thermal conductivity of the wire). If the wire is suddenly heated by switching on a current, its temperature response in the conduction regime for a given heat capacity ratio $(\rho c_p)_w / \rho c_p$ can be found [30, 218]. The trends are schematically illustrated in Fig. 4.39 for different values of Ra*.

The end of the conduction regime at time τ_D (see Fig. 4.39 for definition of τ) is associated with the top-heavy instability (see the section on horizontal rectangular parallelepiped and circular cylinder cavities) that occurs along the separation line on the top of the cylinder [272]. The dimensionless time τ_D for air, water, and alcohol was adequately correlated by

$$\tau_D = 80.2 (\text{Ra}^*)^{-2/3} \tag{4.133}$$

where P'', which appears in the definition of Ra* in Fig. 4.39, is the internal generation rate per unit surface area.

In the transition regime ($\tau_D < \tau < \tau_\infty$), if the conduction heat transfer at τ_D exceeds the steady-state heat transfer for high Ra*, Nu falls monotonically as shown in Fig. 4.39. For small

TABLE 4.12 Constants for the Gebhart Model (Fig. 4.38)

	Pr					
	0.01	0.72	1.0	10	100	1000
M	1.88	1.79	1.79	1.77	1.77	1.76
$\mathbf{b} \times 10^4$	1.41	40.2	71.2	87.4	137	118

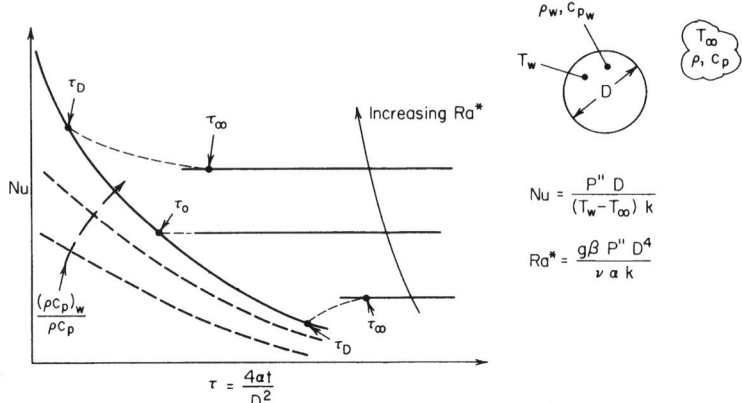

FIGURE 4.39 Temporal response of the Nusselt number to suddenly applied internal heating in a horizontal wire.

Ra* the heat transfer at τ_D falls below the steady-state value, causing the wall temperature to overshoot its steady-state value. The value of Ra* at which the steady-state heat transfer is the same as the heat transfer at the departure time τ_D is defined as Ra_0^*, and the corresponding τ_D is denoted in the figure as τ_0; in this case the transition regime is virtually nonexistent.

Because the existence of an overshoot in the wall temperature is of crucial concern, it is desirable to know both Ra_0^* and τ_0. The value of Ra_0^* can be calculated from the following equation, which fits the prediction of Parsons and Mulligan [218] for intermediate values of Pr (air, water, alcohol):

$$Ra_0^* = [20, 20c^{3/2}]_{\min} \qquad c = \frac{2\rho c_p}{(\rho c_p)_w} \tag{4.134}$$

where the minimum of the two quantities in brackets is to be used. Subscript w refers to the wire, and ρc_p is the volumetric heat capacitance of the fluid. τ_0 is the value of τ_D evaluated from Eq. 4.133 at $Ra = Ra_0^*$.

Parsons and Mulligan [218] have correlated the time to reach steady state, τ_∞, by the equation

$$(\tau_\infty - \tau_0) = 2.5|\tau_D - \tau_0| \tag{4.135}$$

Internal Transient Convection

Overview. This section deals with the transient heating or cooling of a fluid in an enclosure, such as a tank. Applications include storage of cryogenic fluids and transients in startup or shutdown of operating systems. Literature reviews by Clark [59] and Hess and Miller [135] are available.

Response to Step Change in Wall Temperature. Following a sudden change in wall temperature there is an initial transient in which conduction dominates, as described in the previous section. This heat transfer depends on the shape of the enclosure but is given roughly by Eq. 4.131 at small enough time.

Following the conduction regime there is a transition period, during which convection becomes established, to a quasi-steady regime in which the bulk temperature of the fluid in the enclosure gradually approaches the wall temperature. It is heat transfer in this quasi-steady regime that is the subject of this section.

Spheres.   Schmidt [242] measured heat transfer to the fluid inside a sphere following what was essentially a step change in wall temperature. Fluids tested were water and three alcohols: methyl, ethyl, and butyl. For $2 \times 10^8 < \mathrm{Ra} < 5 \times 10^{11}$ the Nusselt number was correlated by

$$\mathrm{Nu} = \frac{q''D}{(T_w - \overline{T})k} = 0.098\,\mathrm{Ra}^{0.345} \qquad \mathrm{Ra} = \frac{g\beta(T_w - \overline{T})D^3}{\nu\alpha} \tag{4.136}$$

where T_w is the wall temperature and $\overline{T}$ is the average temperature of the fluid within the sphere. Properties are to be evaluated at $\overline{T}$. This equation (with $\overline{T}$ replaced by T_∞) agrees to within 18 percent with Eq. 4.49, with constants in Table 4.3a, for steady turbulent heat transfer from the external surface of a sphere. No corresponding data seem to be available for the laminar flow regime.

Cylinders.   Evans and Stefany [92] measured the heat transfer to fluids in short cylindrical enclosures and, with length-to-diameter ratios $0.75 < L/D < 2.0$, following an abrupt increase or decrease in wall temperature. Measurements were obtained for both horizontally and vertically oriented cans, $7 < \mathrm{Pr} < 7000$ and $6 \times 10^5 < \mathrm{Ra} < 6 \times 10^9$, and all the data in the quasi-steady regime were correlated to within about 10 percent by

$$\mathrm{Nu} = \frac{q''L}{(T_w - \overline{T})k} = 0.55\,\mathrm{Ra}^{1/4} \qquad \mathrm{Ra} = \frac{g\beta(T_w - T_i)L^3}{\nu\alpha} \tag{4.137}$$

In this equation $\overline{T}$ is the mean fluid temperature at any time, T_w is the wall temperature, T_i is the mean temperature of the quiescent fluid before the transient is initiated, and L is the length of the cylinder. All physical properties are evaluated at T_w. The length scales L and D were not sufficiently different to affect the correlation much, and L was arbitrarily chosen. The heat transfer coefficient $q''/(T_w - \overline{T})$ remained constant throughout the quasi-steady period. Hiddink et al. [138] found that Eq. 4.137 also correlated data for heating only the bottom and side walls of vertical cylinders provided the cylinder diameter replaced L, and a coefficient of 0.52 rather than 0.55 was used. Their experiments were performed for length-to-diameter ratios of 0.25 to 2.0 and for $5 \le \mathrm{Pr} \le 83{,}000$.

Hauf and Grigull [133–135] precisely measured the natural convection heat transfer inside a tube following a step change in the temperature of a fluid in forced convection over the outside of the tube. In this case the heat transfer coefficient on the outer surface is constant throughout the transient, and the heat capacity of the wall plays an important role. Cheng et al. [50] have studied conditions leading to the formation of ice inside horizontal tubes (without throughflow), also with uniform heat transfer coefficient between the outside boundary and a cold environment.

Uniform Flux or Linear Rise in Boundary Temperature.   If there is a uniform heat flux to the fluid in the enclosure, all temperatures increase linearly with time in the quasi-steady regime, but velocities and spatial gradients become virtually independent of time. A similar quasi-steady regime also exists when the boundary temperature is isothermal but increases linearly with time.

The equations governing the fluid motion and heat transfer in these quasi-steady regimes are (if the property values and boundary conditions are the same) identical to those for steady-state convection in the same geometry with a uniform internal generation of energy [73]. The heat transfer equations from one situation can therefore be readily transferred to the other by replacing the constant $\rho c_p\,\partial T/\partial t$ by the internal generation rate q''' (in $\mathrm{W/m^3}$ or $\mathrm{Btu/h \cdot ft^3}$).

Heat transfer to water in a long cylindrical tube subjected to a linearly increasing wall temperature was measured by Deaver and Eckert [73]. Their data in the laminar quasi-steady regime can be closely represented by

$$\mathrm{Nu} = 8 + 3 \Big/ \ln\left(\frac{\mathrm{Nu}^T + 2\epsilon}{((\mathrm{Nu}^T - 3),\,\epsilon)_{max}}\right) \tag{4.138}$$

where Nu^T is given by Eq. 4.45a and where diameter and the difference between the wall temperature and the mean fluid temperature $(T_w - \overline{T})$ are the length scale and temperature difference used in Nu and Ra, and $\epsilon \approx 10^{-30}$. The first term in Eq. 4.138 represents the conduction solution, while the second term is the laminar Nusselt number for a cylinder corrected for curvature effects. For any selected rate of increase of temperature with time, an energy balance gives the heat that must be added through the boundaries, and Eq. 4.138 gives the corresponding wall-to-fluid temperature difference.

A number of studies [50, 91, 247] have discussed the thermal development of the flow prior to the quasi-steady state under uniform flux conditions. Quasi-steady results for a uniform flux applied to the side wall of a short vertical cylinder are described by Hess and Miller [135].

NATURAL CONVECTION WITH INTERNAL GENERATION

Internal Problems

Background.   Natural convection driven by internal heat sources is of interest in geophysics, and the heat transfer associated with such motion is important in the design of tanks in which fermentation or other chemical reactions occur and in the safety analysis of nuclear reactors where a core meltdown is postulated. The last of these applications has led to the intensive study of internally generating horizontal fluid layers.

The connection between internal generation problems and the quasi-steady transient problems described in the previous section permits data obtained in one area to be applied in the other.

Horizontal Fluid Layers.   A uniform volumetric heat production q''' in a horizontal layer bounded above by an isothermal surface and on the sides and bottom by adiabatic surfaces is depicted in Fig. 4.40. For a stationary fluid, the Nusselt number defined in the figure is Nu = 2, and the temperature difference used to construct the Rayleigh number is $T_0 - T_1 = q'''L^2/2k$. As Ra increases from zero, the layer remains stable and heat flow is by conduction until a critical Rayleigh number of 1386 is reached [167]. Thereafter convection promotes a monotonic increase in Nu with Ra. For water $(2.5 \leq Pr \leq 7)$, and for $Ra \leq 10^{12}$, the heat transfer data of Kulacki et al. [166–168] are accurately represented by

$$Nu = [2.0, 0.389\ Ra^{0.228}]_{max} \qquad (4.139)$$

The lateral extent of the layer and the boundary shape appear to play a small role in this problem until the smallest horizontal dimension of the fluid cavity is equal to or smaller than the depth of the layer [174].

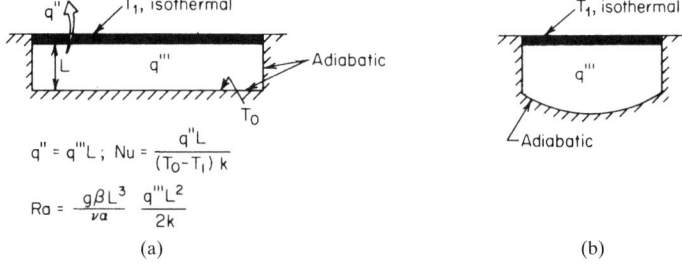

FIGURE 4.40   Geometry and nomenclature for some natural problems where motion is driven by internal heat generation within the fluid.

The transient thermal response of the horizontal layer after turning on or shutting off of the internal generation has been studied extensively also by Kulacki et al. [166–168] and Keylani and Kulacki [159].

When both bottom and top surfaces are maintained at constant temperatures and there is internal generation, there is a superposition of the horizontal layer problem discussed in the section on natural convection within enclosures and the internal generation problem previously described. These are characterized by the "external" Rayleigh number defined in the section on natural convection within enclosures and the "internal" Rayleigh number defined in Fig. 4.40a. The dependence of the layer stability on these parameters has been discussed by Ning et al. [208]. The heat transfer at the top and bottom surfaces has been estimated for these conditions by Baker et al. [13], Suo-Anttila and Catton [276], and Cheung [51].

Other Enclosure Geometries.   The steady and transient heat transfer in internally generating fluid layers bounded above by an isothermal flat surface, below by an adiabatic spherical segment, and on the sides by an adiabatic cylinder, as shown in Fig. 4.40b, has been measured by Min and Kulacki [192, 193]. Kee et al. [157] present numerical heat transfer predictions for a heat-generating fluid within an isothermal sphere. Murgatroyd and Watson [202] and Watson [279] have examined the corresponding problem for closed vertical cylinders, and Bergholz [18] has presented an approximate analysis for a rectangular enclosure. The quasi-steady data of Deaver and Eckert [73] for convection in a long cylinder can be converted to the corresponding internal generation problem, as already described.

Convective heat transfer across two immiscible stably stratified layers, bounded above by an isothermal plate and below by an adiabatic plate, and with the bottom layer heated internally, has been measured by Nguyen and Kulacki [206]. A similar problem was earlier studied by Schramm and Reineke [245].

CONVECTION IN POROUS MEDIA

A porous medium consists of a packed bed of solid particles in which the fluid in the pores between particles is free to move. The superficial fluid velocity $\vec{V}$ is defined as the volumetric flow rate of the fluid per unit of cross-sectional area normal to the motion. It is the imbalance between the pressure gradient ($\vec{\nabla}P$) and the hydrostatic pressure gradient ($\rho\vec{g}$) that drives the fluid motion. The relation that includes both viscous and inertial effects is the Forscheimer equation [47]

$$-\vec{\nabla}P' \equiv -(\vec{\nabla}P - \rho\vec{g}) = \frac{\mu}{K}\,\vec{V} + \frac{\rho\chi}{K}\,|\vec{V}|\,\vec{V} \qquad (4.140)$$

where ρ and μ are the fluid properties, K is the permeability of the medium, and χ is the Forscheimer coefficient. The coefficients K and χ are approximated by [90]

$$K = \frac{d^2\phi^3}{150(1-\phi)^2} \qquad \chi = \frac{1.75d}{150(1-\phi)} \qquad (4.141)$$

where d is the average particle diameter and ϕ is the medium porosity, or volume fraction occupied by the fluid. For $Re_p/(1-\phi) \leq 10$, where $Re_p = \rho d\,|\vec{V}|/\mu$ is the particle Reynolds number, the Forscheimer term, the term with coefficient χ, in Eq. 4.140 can be dropped, and the resulting linear relation between velocity and pressure gradient is Darcy's equation. The relations in the chapter are restricted to Darcy flow.

Properties and Dimensionless Groups

Heat transfer in a porous medium depends on the thermal properties of both the fluid and the solid. For porous media k_m is the thermal conductivity of the medium (fluid and solid) in the absence of no fluid motion. Relations here are from Kaviany [155]. If the heat transfer through the solid and fluid are assumed to act in parallel,

$$k_m = \phi k_f + (1 - \phi)k_s \qquad (4.142)$$

whereas in series

$$\frac{1}{k_m} = \frac{\phi}{k_f} + \frac{(1 - \phi)}{k_s} \qquad (4.143)$$

The true value of k_m will lie somewhere above that given by Eq. 4.143 and below that from Eq. 4.142. Provided k_f is not very much greater than k_s, k_m can be approximated by

$$k_m = k_s^{1-\phi} k_f^{\phi} \qquad (4.144)$$

Properties α_m and σ are the thermal diffusivity and heat capacity ratio of the medium defined by

$$\alpha_m = \frac{k_m}{\phi(\rho C_p)_f + (1 - \phi)(\rho C_p)_s} \qquad \sigma = \frac{\phi(\rho C_p)_f + (1 - \phi)(\rho C_p)_s}{(\rho C_p)_f} \qquad (4.145)$$

where subscripts f and s refer to the fluid and solid, respectively. Typical values of porosity are given in Table 4.13.

TABLE 4.13   Values of Porosity for Porous Media
From Scheidegger (Data selected from Table 2.1 [155]
and Table 15.1.1 [302]

Fiberglass	0.88–0.93
Silica grains	0.65
Black slate powder	0.57–0.66
Leather	0.56–0.65
Catalyst (Fisher-Tropsch, granules only)	0.45
Silica powder	0.37–0.49
Spherical beads	
Simple cubic packing	0.476
Body-centered packing	0.32
Face-centered packing	0.26
Well shaken	0.36–0.43
Cigarette filters	0.17–0.49
Brick	0.12–0.34
Hot compacted copper powder	0.09–0.34
Concrete	0.02–0.07
Coal	0.02–0.12
Granular crushed rock	0.44–0.45
Soil	0.43–0.54
Sand	0.37–0.50
Sedimentary rock	
Sandstone	0.1–0.3
Limestone	0.06–0.2
Chalk	0.29
Chet	0.038
Conglomerate	0.17
Dolomite	0.04–0.28
Shale	0.05–0.21
Siltstone	0.097

The Darcy-modified Rayleigh number, based on characteristic dimension L, is defined as follows

$$\tilde{Ra} = Ra \times Da = \frac{\rho g \beta \Delta T L^3}{\mu \alpha_m} \times \frac{K}{L^2} = \frac{\rho g \beta \Delta T K L}{\mu \alpha_m} \qquad (4.146)$$

where $Da = K/L^2$ is the Darcy number. The heat transfer rate can be recovered from the Nusselt number, defined in Figs. 4.41 and 4.42 for the problems considered in this section.

External Heat Transfer Correlations

Vertical Flat Plate (Fig. 4.41a).   Based on a similarity solution [49] for Darcian flow, for an isothermal plate,

$$Nu = 0.89 \, \tilde{Ra}^{1/2} \qquad (4.147)$$

This relation fits the data of Kaviany and Mittal [156] over the range of their data, $1 < \tilde{Ra} < 10^2$, for polyurethane foams with $\phi \approx 0.98$ and 0.4 to 4 pores per millimeter.

Horizontal Upward-Facing Plate (Fig. 4.41b).   Based on the similarity solution of Cheng and Chang [48], the average heat transfer from an isothermal plate in Darcian flow is given by

$$Nu = 1.26 \, \tilde{Ra}^{1/2} \qquad (4.148)$$

There appears to be no experimental validation.

Horizontal Cylinder (Fig. 4.41c).   The boundary layer solution of Cheng, as reported by Fand et al. [94], yields

$$Nu = 0.565 \, \tilde{Ra}^{1/2} \qquad (4.149)$$

This is found to correlate their data to within about 30 percent for $0.5 < \tilde{Ra} < 10^2$. Departures from Eq. 4.149 in the low Ra region occur for oils (high Pr) because the boundary layer approximations become invalid, and in the high Ra region because the flow is no longer in the Darcy regime. Fand et al. [94] give correlations that fit their data more precisely, but the generality of these equations is unknown.

Spheres (Fig. 4.41c).   Cheng [46] reports the similarity solution for heat transfer from an isothermal sphere to be

$$Nu = 0.362 \, \tilde{Ra}_D^{1/2} \qquad (4.150)$$

There appears to be no experimental validation of this relation.

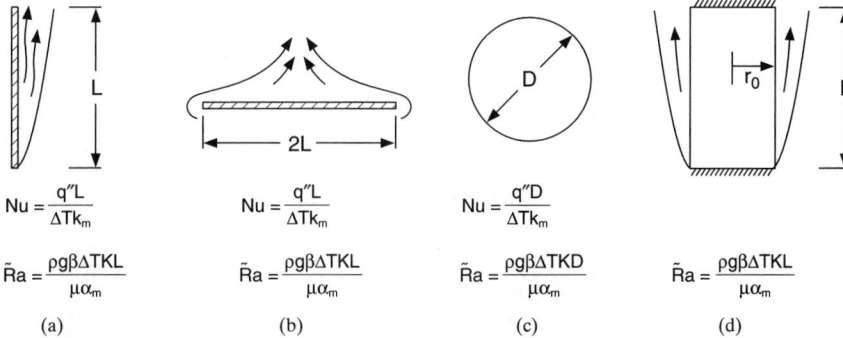

$$Nu = \frac{q''L}{\Delta T k_m} \qquad Nu = \frac{q''L}{\Delta T k_m} \qquad Nu = \frac{q''D}{\Delta T k_m}$$

$$\tilde{Ra} = \frac{\rho g \beta \Delta T K L}{\mu \alpha_m} \qquad \tilde{Ra} = \frac{\rho g \beta \Delta T K L}{\mu \alpha_m} \qquad \tilde{Ra} = \frac{\rho g \beta \Delta T K D}{\mu \alpha_m} \qquad \tilde{Ra} = \frac{\rho g \beta \Delta T K L}{\mu \alpha_m}$$

(a)   (b)   (c)   (d)

FIGURE 4.41   Definition sketch for vertical plate (a), long horizontal strip (b), long horizontal circular cylinder (c), and short vertical cylinder with insulated ends (d), all in an isothermal porous medium.

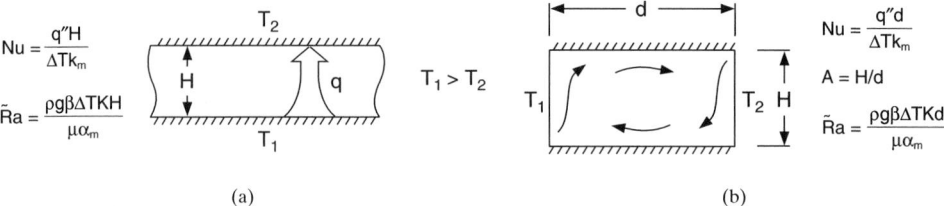

FIGURE 4.42 Definition sketch for heat transfer between extensive horizontal plates (*a*) and in long cavity with isothermal side walls and insulated top and bottom. The medium is porous.

Vertical Cylinder With Adiabatic Ends (Fig. 4.41d). The heat transfer q_c from the curved boundary of the vertical cylinder in Fig. 4.41*d* is related to the heat transfer from a vertical plate (q_p) of the same height and surface area by [194]

$$\frac{q_c}{q_p} = 1 + 0.26\zeta_L \qquad \zeta_L = \frac{2L}{r_0 \tilde{\text{Ra}}^{1/2}} \tag{4.151}$$

q_c/q_p increases as Ra falls, similar to that for a continuous fluid medium. There is apparently no experimental validation of Eq. 4.151.

Internal Heat Transfer Correlations

Horizontal Layer, Heated from Below (Fig. 4.42a). For an extensive horizontal layer heated from below, the Nusselt number is approximately given by [87]

$$\text{Nu} = \left[1, \frac{\tilde{\text{Ra}}}{40}\right]_{\max} \tag{4.152}$$

There is a very large scatter of experimental data from this relation [46] that has been attributed to the effect of Prandtl number and the ratio of bead diameter to the depth of the porous layer being too large for the analysis to hold.

Rectangular Cavity, Heated From the Side (Fig. 4.42). The rectangular cavity of width *w* and height *H* in Fig. 4.42*b* has side boundaries held at different temperatures, T_1 and T_2, and is insulated on the top and bottom. The cavity is considered to be extensive in the third dimension. Bejan [15] compiled heat transfer calculations for aspect ratios $0.1 < A < 30$, and this is reproduced in Fig. 4.43.

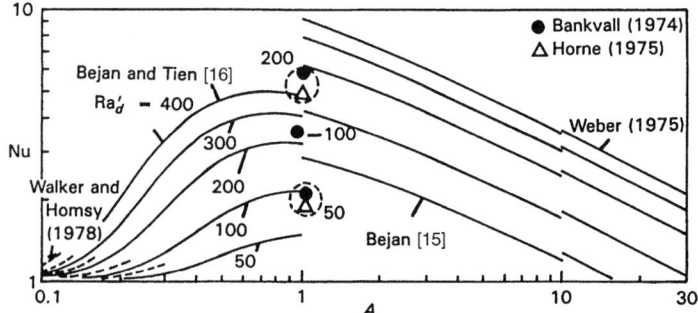

FIGURE 4.43 Summary of heat transfer calculation for a cavity filled with a fluid-saturated porous medium, from Bejan [15]. (*Reprinted by permission of Pergamon Press.*)

MIXED CONVECTION

Mixed convection occurs when both natural convection and forced convection play significant roles in the transfer of heat. In applications it is important to first establish whether satisfactory predictions will result by ignoring either one, or if the combined effects must be considered. Guidelines for delineating the forced, natural, and mixed convection regimes are reported for external and internal flows in the sections that follow. Some design equations and graphs for heat transfer in the mixed convection regime are also given. Attention is focused on laminar flows, in which mixed convection effects are most frequently important.

External Flow

Introduction.   For the problem depicted in Fig. 4.44, the heat transfer by "pure" forced convection would increase monotonically with Reynolds number along the curve shown. The heat transfer by pure natural convection from the same surface for various Ra is denoted by the horizontal lines in the figure. If Re is slowly increased from zero in the real problem, the measured values of Nu would at first follow the natural convection curve, since the superimposed forced convection velocities are too feeble to affect the heat transfer. If the forced convection "assists" the natural convection, the Nu curve in Fig. 4.44 will break upward along path A at larger Re and approach the pure forced convection curve from above. If the flows are "opposed," Nu passes through a minimum along path B in Fig. 4.44 and approaches the forced convection curve from below. Mixed convection occurs when the heat transfer is significantly different from that for either pure natural convection or pure forced convection.

From the Nusselt number for pure natural convection, Nu_N, and that for pure forced convection, Nu_F, a rough estimate of the actual Nusselt number for a given problem is

$$\mathrm{Nu} = [\mathrm{Nu}_N, \mathrm{Nu}_F]_{\max} \tag{4.153}$$

That is, the maximum of the two Nusselt numbers is used. The error in this equation is often less than 25 percent, with the maximum deviation near the intersection point of the curves, denoted by the dot in Fig. 4.44.

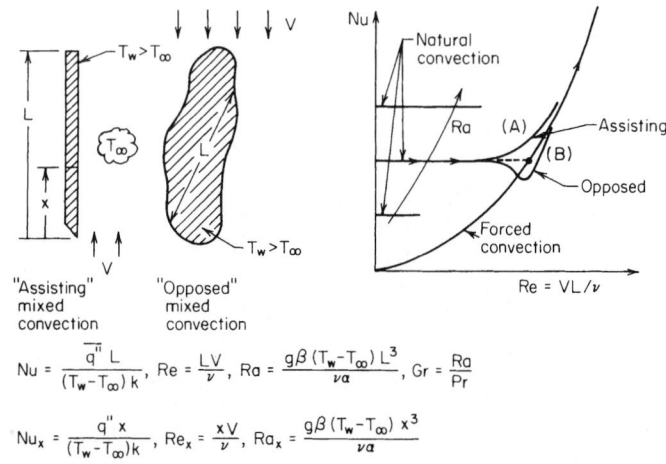

$$\mathrm{Nu} = \frac{\overline{q''}\,L}{(T_w - T_\infty)\,k}, \quad \mathrm{Re} = \frac{LV}{\nu}, \quad \mathrm{Ra} = \frac{g\beta\,(T_w - T_\infty)\,L^3}{\nu\alpha}, \quad \mathrm{Gr} = \frac{\mathrm{Ra}}{\mathrm{Pr}}$$

$$\mathrm{Nu}_x = \frac{q''\,x}{(T_w - T_\infty)\,k}, \quad \mathrm{Re}_x = \frac{xV}{\nu}, \quad \mathrm{Ra}_x = \frac{g\beta\,(T_w - T_\infty)\,x^3}{\nu\alpha}$$

FIGURE 4.44   Definition of terms relating to mixed convection in opposed and assisting flows.

The intersection points of the pure natural convection and pure forced convection equation also provide valuable information on the conditions for which forced and natural convection are equally important. For example, for laminar flow along the heated isothermal vertical plate in Fig. 4.6 if Eq. 4.33a for Nu_N is equated to the forced convection Nusselt number given by

$$\mathrm{Nu}_F = 0.664\mathrm{Re}^{1/2}\,\mathrm{Pr}^{1/3} \tag{4.154}$$

This yields the following relation between Gr_i and Re_i^2:

$$\mathrm{Gr}_i = \left(\frac{0.664\mathrm{Pr}^{1/12}}{\overline{C}_l}\right)^4 \mathrm{Re}_i^2 \tag{4.155}$$

where i refers to the intersection point. For a 0.30-m (1-ft) vertical plate at 60°C (140°F) immersed in air at 20°C (68°F), the superimposed forced convection velocity that satisfies Eq. 4.155 is 0.25 m/s (0.82 ft/s); under the same conditions but in water, this velocity is 0.084 m/s (0.28 ft/s).

The intersection relation (see, for example, Eq. 4.155) also permits one to roughly estimate when mixed convection should be considered. If, for a given Reynolds number, the value of Rayleigh number Ra for a laminar problem greatly exceeds the intersection Rayleigh number Ra_i, forced convection effects can be ignored. If Ra is much less than Ra_i, natural convection effects may be ignored. The change from the natural convection regime to the forced convection regime occurs over a smaller range of Ra, for a given Re, for turbulent flow than for laminar.

Vertical Plates.   If Eq. 4.33a (Eq. 4.33b should be used if Ra ≤ 10⁴ is of interest) and Eq. 4.33c are used for laminar and turbulent natural convection, respectively, and if Eq. 4.154 and $\mathrm{Nu}_F = (0.037\mathrm{Re}^{0.8} - 871.3)\,\mathrm{Pr}^{1/3}$ are used, respectively, for laminar and turbulent forced convection, then Eq. 4.155 gives the laminar-laminar (i.e., laminar forced convection and laminar natural convection) intersection, while the laminar-turbulent intersection is given by

$$\mathrm{Gr}_i = \frac{0.664}{C_t^V}\,\mathrm{Re}_i^{3/2} \tag{4.156}$$

and the turbulent-turbulent intersection by

$$\mathrm{Gr}_i = \left[\frac{0.037 - 871.3/\mathrm{Re}_i^{0.8}}{C_t^V}\right]^3 \mathrm{Re}_i^{2.4} \tag{4.157}$$

These are plotted in Fig. 4.45 as a solid line. The dashed curves are drawn at $\mathrm{Ra} = 10\,\mathrm{Ra}_i$ and $\mathrm{Ra} = 0.1\,\mathrm{Ra}_i$ as estimates of the bounds on the mixed convection regime.

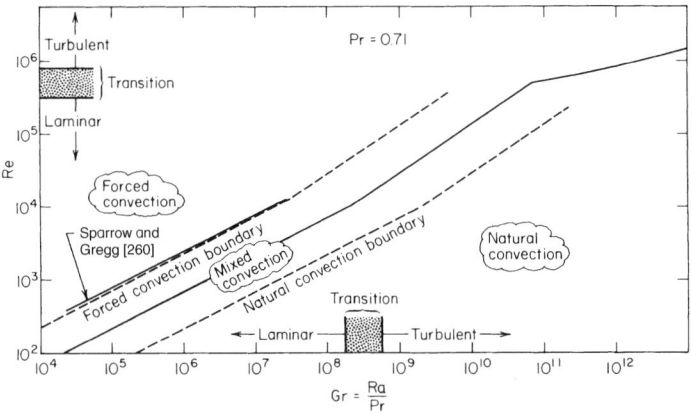

FIGURE 4.45   Regimes of forced, mixed, and natural convection for flow along a vertical plate.

Sparrow and Gregg [260] established by a perturbation analysis that the forced convection Nusselt number was altered by less than 5 percent by either an assisting or opposing natural convection if

$$|Gr| \leq 0.225Re^2 \qquad (4.158)$$

for $0.01 \leq Pr \leq 10$. The curve representing this equation in Fig. 4.45 agrees closely with the estimate given by the dashed lines of the forced convection boundary of the mixed convection regime.

The mean heat transfer in the mixed convection regime for assisting flows has been correlated by Churchill [55] using the equation

$$Nu = [(Nu_F)^m + (Nu_N)^m]^{1/m} \qquad (4.159)$$

$m = 3$ was found to best fit the results of laminar boundary layer analyses. The same equation was found to apply to both isothermal and uniform heat flux plates. Oosthuizen and Bassey [214] also correlated their data with Eq. 4.159, but with $m = 4$; the Reynolds and Grashof numbers in their experiments were below the ranges where boundary layer theory holds, and they also provide measurements (but no correlation) for opposed mixed convection. Churchill [55] reviews other available data.

The authors of most experimental studies and analyses have focused their attention on the local values defined in Fig. 4.44. Churchill [55] fitted the local Nu_x values for assisting flow using Eq. 4.159 with $m = 3$ and with each average value on the right side of the equation replaced by its local value counterpart. Mucoglu and Chen [201] have solved the inclined flat plate problem for uniform wall temperature and heat flux and have presented local heat transfer results for mixed convection.

Horizontal Flow.   For laminar flow over the upper surface of a horizontal heated plate (or over the bottom surface of a cooled plate), the center of the mixed convection regime can again be estimated by equating the forced convection Nusselt number from Eq. 4.154 to that for natural convection from Eq. 4.39c (for detached turbulent convection). This results in

$$Gr_i = \left(\frac{0.664}{C_t^U}\right)^3 Re_i^{3/2} \approx 107Re_i^{3/2} \qquad (4.160)$$

where the terms are defined in Fig. 4.46. As a rough approximation, one would expect that for $Gr \lesssim 11Re^{3/2}$ forced convection dominates, and for $Gr \gtrsim 1100Re^{3/2}$ natural convection dominates.

The details of the flow in the mixed convection regime have been clarified by Gilpin et al. [113]. After an initial development of the laminar forced convection boundary layer, rolls with axes aligned with the flow appear at the location marked *Onset* in Fig. 46. These persist until the end of the transition regime, marked *Breakup,* after which the motion appears as fully detached turbulent natural convection flow.

The experiments for water, $7 \leq Pr \leq 10$, revealed that the rolls first become visible at the x location at which

$$Gr\left(\frac{x}{L}\right)^{3/2} = K \, Re^{3/2} \qquad (4.161)$$

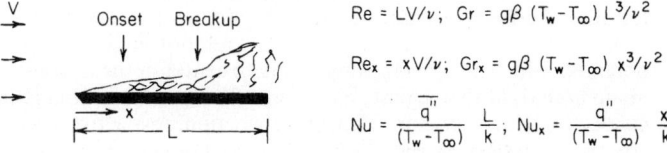

FIGURE 4.46   Mixed convection on a heated horizontal plate.

where $K \approx 100$. The stability analysis for longitudinal rolls by Wu and Cheng [288] is in good agreement with these observations and also predicts a Prandtl number dependence of K, but Chen and Mucoglu [44] have questioned the validity of these predictions. Despite the uncertainties that still exist, there is general agreement that buoyancy effects can greatly decrease the value of Re_x at which the first instability appears.

To obtain an average heat transfer equation, the flow is modeled as purely forced convection up to x_c and purely natural convection for $x > x_c$; from the experiments [148] for water, x_c is given roughly by Eq. 4.161 with $K \approx 155$. Integrating the local heat transfer relations results in the following expression for average heat transfer:

$$\text{Nu} = \sqrt{\zeta} 0.664 \text{Re}^{1/2} \text{Pr}^{1/3} + (1 - \zeta) C_t^U \text{Ra}^{1/3} \tag{4.162a}$$

$$\zeta = \left[\frac{29\text{Re}}{\text{Gr}^{2/3}}, 1 \right]_{\min} \qquad \text{Ra} = \text{Gr Pr} \tag{4.162b}$$

where the minimum of the two quantities in brackets is to be used in Eq. 4.162b. The coefficient 29 in Eq. 4.162b will likely be somewhat dependent on Prandtl number, but data are not available to resolve this dependence.

For laminar flow above a cooled surface or below a heated surface, the presence of buoyancy forces stabilizes the flow (inhibits transition) and tends to diminish the heat transfer. The analysis of Chen et al. [43] predicts that natural convection will alter the local convective heat transfer by less than 5 percent if $|\text{Gr}|(x/L)^{3/2}/\text{Re}^{5/2} \leq 0.03$ for Pr = 0.7. Robertson et al. [232] show that for $|\text{Gr}|/\text{Re}^{5/2} > 0.8$ and Pr = 0.7, buoyancy may inhibit the flow so strongly that a separation bubble may form over the surface.

In turbulent flow, stable stratification significantly damps turbulence and reduces heat transfer in the vertical direction.

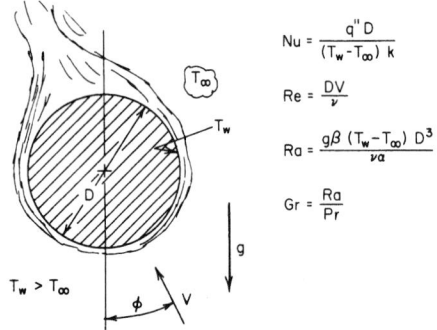

$$\text{Nu} = \frac{q'' D}{(T_w - T_\infty) k}$$

$$\text{Re} = \frac{DV}{\nu}$$

$$\text{Ra} = \frac{g\beta (T_w - T_\infty) D^3}{\nu \alpha}$$

$$\text{Gr} = \frac{\text{Ra}}{\text{Pr}}$$

$T_w > T_\infty$

FIGURE 4.47   Perpendicular flow across a horizontal circular cylinder in mixed convection.

Horizontal Cylinders.   For a heated horizontal cylinder in perpendicular cross flow, the angle of the approaching stream, ϕ in Fig. 4.47, greatly affects the heat flow in the mixed convection regime. For $\phi = 0$ the forced flow assists the natural convection and the dependence of the average Nusselt number on Re resembles path A in Fig. 4.44. For $\phi = 90°$ there is a sharper transition from natural to forced convection than when $\phi = 0$, while for opposed flow ($\phi = 180°$) there is a minimum as shown by path B in Fig. 4.44. For a cooled cylinder the same description applies except that ϕ is measured from the vertical axis extending upward from the cylinder.

Equating the Nusselt numbers for pure natural convection and pure forced convection provides a good estimate of the Ra-Re curve along which mixed convection effects are most important, as already discussed. After a careful study of available data, Morgan [198] proposed the following equation for forced convection heat transfer from a cylinder for cross flow in a low-turbulence airstream:

$$\text{Nu}_F = a \, \text{Re}^n \tag{4.163}$$

where a and n are given in Table 4.14. This equation can also be used as a first approximation for other fluids if the right side is multiplied by the factor $(\text{Pr}/0.71)^{1/3}$. If Eq. 4.163 is equated to Eq. 4.45 for Nu_N, the Re_i versus Gr_i relation indicated by the solid curve in Fig. 4.48 is obtained, which denotes the approximate center of the mixed convection regime. The approximate bounds of this regime, based on a 5 percent deviation in heat transfer from pure forced convection and from pure natural convection, respectively, have been estimated by Morgan [198] to lie in the shaded bands.

TABLE 4.14 Constants for Forced Convection Over a Circular Cylinder (Eq. 4.163)

	Re range						
	10^{-4} to 4×10^{-3}	4×10^{-3} to 9×10^{-2}	9×10^{-2} to 1.0	1.0 to 35	35 to 5×10^{3}	5×10^{3} to 5×10^{4}	5×10^{4} to 2×10^{5}
a	0.437	0.565	0.800	0.795	0.583	0.148	0.0208
n	0.0895	0.136	0.280	0.384	0.471	0.633	0.814

A procedure for calculating the heat transfer in the mixed convection regime for the problem has also been proposed by Morgan [198] on the basis of work of Börner [22] and Hatton et al. [131]. For a given Ra and Re, the value of Nu_N is computed from Eq. 4.45. For the given Re, the constants a and n are chosen from Table 4.14. The value of Re_i is then found from Eq. 4.163 with $\mathrm{Nu}_F = \mathrm{Nu}_N$; that is,

$$\mathrm{Re}_i = \left[\frac{\mathrm{Nu}_N}{a(\mathrm{Pr}/0.71)^{1/3}} \right]^{1/n} \tag{4.164}$$

An effective Reynolds number $\mathrm{Re}_{\mathrm{eff}}$ is then calculated from

$$\mathrm{Re}_{\mathrm{eff}} = [(\mathrm{Re}_i + \mathrm{Re} \cos \phi)^2 + (\mathrm{Re} \sin \phi)^2]^{1/2} \tag{4.165}$$

and Nu is computed by insertion of $\mathrm{Re}_{\mathrm{eff}}$ into Eq. 4.163 to obtain

$$\mathrm{Nu} = a\, \mathrm{Re}_{\mathrm{eff}}^n \left(\frac{\mathrm{Pr}}{0.71} \right)^{1/3}$$

It will be seen that if $\mathrm{Re}_i \gg \mathrm{Re}$, the natural convection result is recovered, while if $\mathrm{Re}_i \ll \mathrm{Re}$, $\mathrm{Nu} \approx \mathrm{Nu}_F$. Morgan showed that this calculation procedure gave good agreement with experiments for air and water for $\phi = 90°$. For assisting flow ($\phi = 0°$) the agreement was poorer, and it was still poorer for opposed flow ($\phi = 180°$).

Gebhart et al. [109, 111] have provided accurate heat transfer measurements spanning the mixed convection zone for extremely fine wires for $\mathrm{Pr} = 0.7$, 6.3, and 63, and have also provided equations for the bounds of the mixed convection zone.

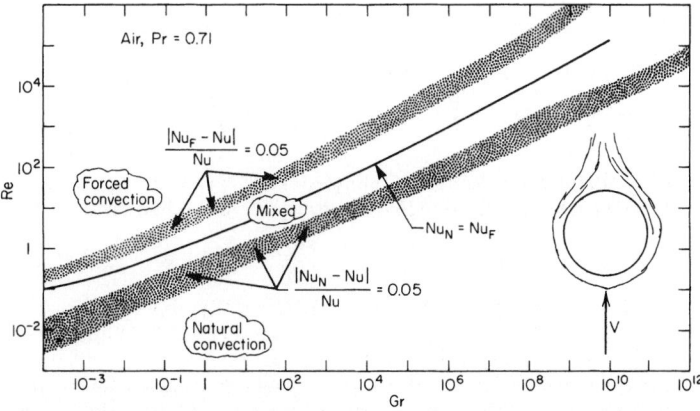

FIGURE 4.48 Regimes of forced, mixed, and natural convection for assisting flow over a horizontal circular cylinder.

Other Shapes.   Churchill [55] proposed that Eq. (4.159) with $m = 3$ be used also for assisting mixed convection flow around other surface shapes, such as spheres and cylinders. The appropriate expressions for Nu_F and Nu_N for the body shape of interest must be used.

Internal Flows

Horizontal Tubes

Uniform Heat Flux.   For laminar flow in a horizontal tube where uniform heat flux is applied at the outer boundary of the tube, the bulk temperature T_b, increases linearly in the axial direction. To maintain the heat flow to the fluid, the wall temperature must remain higher than the fluid temperature, and under these conditions a fully developed natural convection motion becomes established in which velocity and temperature gradients become independent of the axial location. Because the fully developed Nusselt number for laminar pure forced convection is small ($Nu_F \to 4.36$), the buoyancy-induced mixing motion can greatly enhance the heat transfer.

Marcos and Bergles [189] correlated their data for fully developed heat transfer for water and ethylene glycol in glass and metal tubes by the equation

$$Nu_{fd} = \left\{ 4.36^2 + \left[0.055 \left(\frac{Ra \, Pr^{0.35}}{P_w^{1/4}} \right)^{0.4} \right]^2 \right\}^{1/2} \tag{4.166}$$

where the nomenclature is defined in Fig. 4.49. The P_w term accounts for the redistribution by circumferential conduction of the uniform heat flux on the outside surface of the tube before entering the fluid. Fluid properties are to be evaluated at the mean film temperature, $0.5(\overline{T}_{wi} + T_b)$. The data are in the range $3 \times 10^4 < Ra < 10^6$, $4 < Pr < 175$, $2 < P_w < 66$.

The heat transfer immediately downstream of the location where heating begins will be dominated by forced convection and will depend on the velocity profile. For a parabolic inlet profile, the forced convection Nusselt number can be approximated by [249]:

$$Nu_F = 1.30 \left(\frac{Re \, Pr}{x/D} \right)^{1/3} \qquad \frac{x/D}{Re \, Pr} \leq 0.01 \tag{4.167}$$

For $Ra \leq 5 \times 10^5$ the transition from forced (Eq. 4.167) to natural convection (Eq. 4.166) is very sharp [249] but it becomes more rounded with increasing Ra. To find the total heat trans-

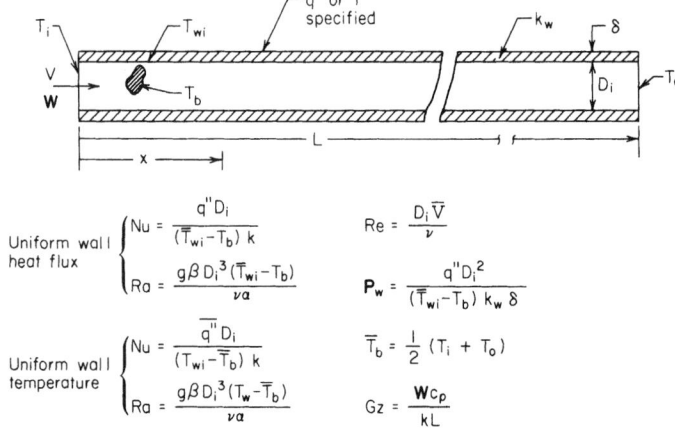

FIGURE 4.49   Geometry and definition of terms for mixed convection inside a horizontal tube.

fer over some length of tube, numerically integrate from the inlet of the tube using, at each location, the maximum value of local heat transfer given by Eqs. 4.166 and 4.167, and using the energy balance at each step to update the bulk temperature T_b.

As a result of natural convection, the fully developed condition is reached much farther upstream than for pure forced convection. For example, for Ra $\approx 10^6$, Nu_{fd} is reached at roughly $x/D = 0.001\mathrm{Re}\,\mathrm{Pr}$ as opposed to $x/D \approx 0.06\mathrm{Re}\,\mathrm{Pr}$ for pure forced convection (i.e., 1 diameter as opposed to 60 diameters for Re Pr $= 1000$).

When circular tubes subjected to a uniform heat flux are tilted upward, the Nusselt number has been shown to monotonically decrease with increasing angle for air [236].

Isothermal Wall.   Natural convection also affects the laminar thermal development in a tube with an isothermal wall. In this case the temperature differences in the fluid near the tube inlet initiate a natural convection motion, but as the fluid temperature approaches the wall temperature far downstream, the motion slows and the fully developed Nusselt number ($\mathrm{Nu}_F = 3.66$) is approached.

On the basis of data available up to 1964, Metais and Eckert [190] established the forced convection boundary of the mixed convection regime, and their results are presented in Fig. 4.50. The line was drawn where natural convection was thought to alter the heat transfer from that for pure forced convection by 10 percent. Figure 4.49 defines the nomenclature for this problem.

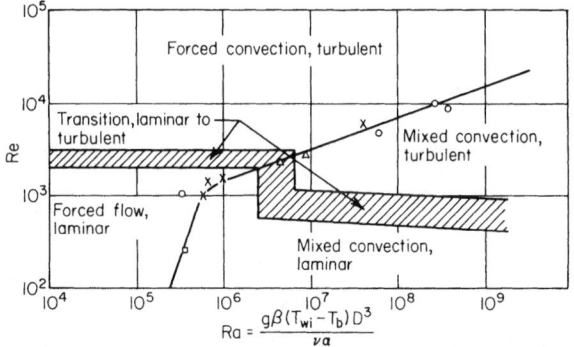

FIGURE 4.50   Regimes of forced and mixed convection for flow through horizontal tubes with uniform wall temperature, for $10^{-2} <$ (Pr D/L) < 1. From Metais and Eckert [190].

Heat transfer relations in the mixed convection regime have been proposed by several investigators, but the equation proposed by Depew and August [74] appears to be most successful. This relation agrees with their measurements for water (Pr $= 6.5$), ethyl alcohol (Pr ≈ 15), and a glycerol-water mixture (Pr $= 375$), as well as with the data from other authors, to about ± 40 percent:

$$\mathrm{Nu} = 1.75\left(\frac{\mu_b}{\mu_w}\right)^{1/4}[\mathrm{Gz} + 0.12(\mathrm{Gz}\,\mathrm{Ra}^{1/3}\,\mathrm{Pr}^{0.03})^{0.88}]^{1/3} \qquad (4.168)$$

where b and w refer to average bulk and wall conditions, Nu, Gz, and Ra are defined in Fig. 4.49, and $\overline{T}_b$ is obtained by arithmetically averaging the inlet and outlet temperature. All physical properties are to be evaluated at $\overline{T}_b$. The data on which this equation was derived lie in the following ranges: $28 < L/D < 193$, $10 < \mathrm{Gz} < 456$, and $30 < \mathrm{Gr} < 2 \times 10^7$.

Vertical Tubes.   The flow regime chart for vertical tube flow shown in Fig. 4.51 was prepared by Metais and Eckert [190] for either a uniform-heat-flux or uniform-wall-temperature boundary condition. The two boundaries of the mixed convection are defined in such a way

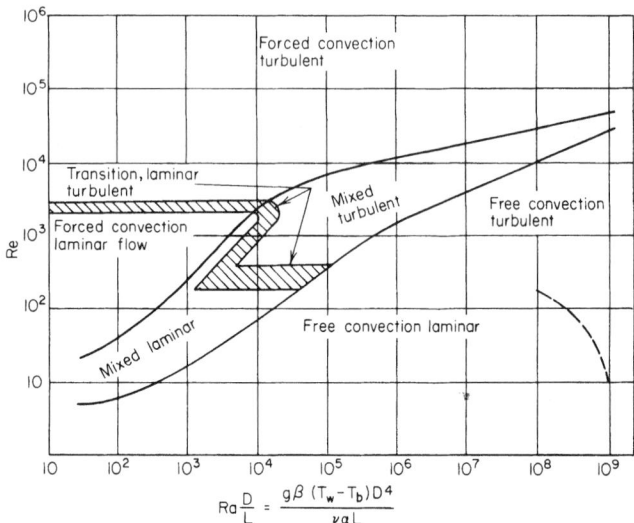

FIGURE 4.51   Regimes of natural, forced, and mixed convection for flow through vertical tubes with uniform wall temperature of heat flux, for $10^{-2} < (\text{Pr } D/L) < 1$. From Metais and Eckert [190].

that the Nusselt number at these locations does not deviate by more than 10 percent from the values for pure natural convection and pure forced convection.

ACKNOWLEDGMENTS

The authors are grateful to the Natural Science and Engineering Research Council of Canada for financial support of the research required to prepare this chapter. Thanks to Andrew Woronko and Skye Legon for digitizing and analyzing data, and to Anita Fonn for typing the manuscript.

NOMENCLATURE

Symbol, Definition, SI Units, English Units

A — heat transfer surface area of body: m^2, ft^2

A_f — area over which fluid contacts each plate in enclosure problem (Fig. 4.25): m^2, ft^2

A_h — flat horizontal heated area that faces downward (or cooled horizontal area that faces upward on a cold body): m^2, ft^2

A_w — area over which wall contacts each plate in enclosure problem (Fig. 4.25): m^2, ft^2

a — proportionality constant in Morgan's law (Eq. 4.163) for forced convection heat transfer from cylinders (see Table 4.14)

a — constant in stratified medium (Eqs. 4.42, 4.46, and 4.50)

b — wall thickness (Fig. 4.25): m, ft

b	constant in stratified medium (Eqs. 4.42, 4.46, and 4.50)
b	function of Pr given in Table 4.12 and used in Fig. 4.38
$\hat{\mathbf{b}}$	unit vector in direction of buoyancy force (Fig. 4.5)
C	minor axis of elliptical cylinder or spheroid (Tables 4.2 and 4.3a): m, ft
C_1	constant (Eq. 4.20)
C_1	constant, defined separately for different problems
C_2	functions of C/L for elliptic cylinders and spheroids, tabulated in Tables 4.2 and 4.3a
C_2	constant (e.g., Eqs. 4.34c and 4.38)
C_{ac}	wall admittance parameter for circular cylinder enclosures, given by Eq. 4.83.
C_{ar}	wall admittance parameter for rectangular parallelepiped enclosures, given by Eq. 4.82 for opaque fluids and (approximately) by Eq. 4.86 for transparent fluids
$\overline{C}_l$	function of Pr (Eq. 4.13 and Table 4.1)
C_t	function of Pr and ζ, given by Eq. 4.28
$\overline{C}_t$	average value of C_t over a body, defined by Eq. 4.23 (see Tables 4.2 and 4.3 for $\overline{C}_t$ for various body shapes)
C_t^U	function of Pr given by Eq. 4.25; see Table 4.1
C_t^V	function of Pr given by Eq. 4.24; see Table 4.1
c	constant, defined separately for each problem
c	constant in stratified medium equations (Eqs. 4.42, 4.46, and 4.50)
c_p, c_{p0}	specific heat of fluid evaluated at T_f: J/(kg·K), Btu/(lb$_{\mathrm{m}}$·°F)
D	diameter of sphere, cylinder, or disk: m, ft
Da	Darcy number (Eq. 4.146)
D_i	for a pair of eccentric or concentric spheres or cylinders, the diameter of the inner one (Fig. 4.36): m, ft
D_o	for a pair of eccentric or concentric spheres or cylinders, the diameter of the outer one (Fig. 4.36): m, ft
d	tube diameter (Fig. 4.23): m, ft
d	average particle diameter in porous medium: m, ft
E	perpendicular distance between axes of eccentric cylinders, or distance between centers of eccentric spheres (Fig. 4.36): m, ft
Ec	Eckert number, $v_0^2/(c_p \, \Delta T)$
f Re	friction factor–Reynolds number product for forced flow in a duct (Table 4.4)
f	quantity defined by Eq. 4.45b
G	geometry-dependent constant (Eqs. 4.11, 4.14, and 4.37a and Tables 4.2 and 4.3)
Gr	Grashof number, Ra/Pr
Gr$_i$	Grashof number at which $\mathrm{Nu}_F = \mathrm{Nu}_N$
Gz	Graetz number, defined in Fig. 4.49
g	acceleration of gravity: m/s^2, ft/s^2
H	height, defined by Figs. 4.21–4.23 and 4.25: m, ft
H_L	function of Prandtl number (Eq. 4.34a)
H_L	height of partition extending up from the floor of a cavity: m, ft

H_U	height of partition hanging down from the ceiling of a cavity: m, ft
$\overline{H}_L$	function of Prandtl number (Eq. 4.36a, Table 4.1)
H_p	partition height: m, ft
H_{MIN}, H_{MAX}	minimum and maximum, respectively, of H_U and H_L: m, ft
i	flow index: $i = 0$ for 2D flow problems and $i = 1$ for axisymmetric flow problems
k	thermal conductivity of fluid, evaluated at T_f for external and open cavity problems and at T_m for enclosure problems (unless otherwise specifically directed): W/(m·K), Btu/h·ft·°F
k_1, k_2	constants in Hollands equation (Eq. 4.78); given by Eqs. 4.80–4.81 and Table 4.7
k_m	effective thermal conductivity of porous medium: W/mK, Btu/hr ft·°F
k_f	thermal conductivity of fluid in a porous medium: W/mK, Btu/hr ft·°F
k_s	thermal conductivity of solid in a porous medium: W/mK, Btu/hr ft·°F
k_w	thermal conductivity of wall: W/(m·K), Btu/h·ft·°F
k_{wt}	equivalent enclosure wall thermal conductivity, given by Eq. 4.84: W/(m·K), Btu/h·ft·°F
k_{wt}^*	special wall conductivity, equal to k_w for enclosures with finlike walls and opaque fluids, and given by Eq. 4.85 (approximately) for enclosures with thick walls and transparent fluids: W/(m·K), Btu/h·ft·°F
L	characteristic length, chosen separately for each problem: m, ft
L	spacing between isothermal surfaces (Figs. 4.25 and 4.26), or length of a plate or a cylinder (Figs. 4.6 and 4.14): m, ft
L_f	vertical distance between lowest and highest points on a body: m, ft
L^*	characteristic length for a horizontal plate, equal to A/p (Fig. 4.10): m, ft
l	characteristic dimension of body or enclosure: m, ft
m	exponent used in the Churchill-Usagi fit (Eqs. 4.32, 4.34d, 4.36d, etc.)
$\hat{n}$	unit vector normal to the surface (Fig. 4.5)
Nu	Nusselt number, usually given by $qL/(A \, \Delta Tk)$, but see separate defining sketch for each problem
Nu_{COND}	Nusselt number when fluid is stationary and transfer is by conduction only
Nu_F	Nusselt number for pure forced convection
Nu_{fd}	Nusselt number for fully developed flow
Nu_H (Ra)	Nu-Ra relation which results when a given inclined cavity is rotated to the horizontal position
Nu_{iso}	Nusselt number that is obtained for constant T_w and T_∞ with $T_w - T_\infty = \Delta T_m$
Nu_{np}	Nusselt number for a cavity with no partition
Nu_{fp}	Nusselt number for a cavity with full interior partition
Nu_l	average Nusselt number taken over entire body, assuming laminar flow prevails over entire body
Nu_N	Nusselt number for pure natural convection
Nu_t	average Nusselt number taken over entire body, assuming turbulent heat transfer prevails over whole body
$Nu_{t,x}$	local Nusselt number for turbulent heat transfer at location x, $q''x/\Delta Tk$

Nu_V (Ra)	Nu-Ra relation that results when a given inclined cavity is rotated to the vertical position
Nu_x^t	local Nusselt number for a turbulent boundary layer at point x
Nu^T	average "thin-layer-solution" Nusselt number for laminar flow
Nu_θ (Ra)	Nu-Ra relation for a cavity inclined at angle θ
n	Churchill usage constant used to determine Nu_ℓ (Eq. 4.19, Table 4.3)
n	exponent in Eq. 4.163 for forced convection heat transfer over cylinders (Table 4.14)
P	pressure: Pa, $\mathrm{lb_f/ft^2}$
P_d	pressure component associated with dynamics of flow (Eq. 4.1): Pa, $\mathrm{lb_f/ft^2}$
P_i	perimeter of a cylinder at the intersection of its surface with the plane normal to its axis: m, ft
P_{ref}	reference pressure, equal to pressure at $z = z_{\mathrm{ref}}$ far from solid surface: Pa, $\mathrm{lb_f/ft^2}$
P_∞	pressure at height z and far from solid surface or wall: Pa, $\mathrm{lb_f/ft^2}$
$\mathbf{P}_w$	"circumferential heat flux" dimensionless group for internal mixed convection problems (Fig. 4.49)
P''	rate of internal generation of energy in a wall per unit of heat transfer surface area of the wall: $\mathrm{W/m^2}$, $\mathrm{Btu/(h \cdot ft^2)}$
Pr	Prandtl number, equal to ν/α, evaluated at T_f (or T_m for enclosure problems) unless otherwise specifically directed
p	perimeter of plate (Fig. 4.10b): m, ft
p	tilt parameter for long tilted cylinder (Eq. 4.47b)
$\mathcal{P}(z)$	perimeter of body along the intersection with a horizontal plane at elevation z: m, ft
$\mathcal{P}$	average body perimeter, defined by Eq. 4.15
$\mathbf{Q}$	dimensionless thermal capacity rate (Fig. 4.38)
q	heat transfer rate, total heat delivered to fluid moving through an open cavity: W, Btu/h
q^*	heat flux ratio for rectangular open cavity, equal to q_2/q_1 (Fig. 4.21)
q'	heat flow per unit length of surface: W/m, Btu (h·ft)
q''	heat flow per unit area of surface: $\mathrm{W/m^2}$, Btu (h·ft^2)
$\overline{q''}$	average heat flux over surface, equals q/A: $\mathrm{W/m^2}$, Btu (h·ft^2)
q'''	rate of internal generation of heat within the fluid: $\mathrm{W/m^3}$, Btu (h·ft^3)
q_f	heat transfer from hot plate to fluid in enclosure problem: W, Btu/h
q_{f0}	value of q_f when fluid is stationary: W, Btu/h
q_r	radiative heat transfer from hot plate in enclosure problem: W, Btu/h
q_{r0}	value of q_r when fluid is stationary: W, Btu/h
q_w	heat transfer over A_w from plate to wall in enclosure problem (see Fig. 4.25 for definition of plate and wall): W, Btu/h
q_{w0}	value of q_w when fluid is stationary: W, Btu/h
q_x''	local heat flux at location x: $\mathrm{W/m^2}$, $\mathrm{Btu/(h \cdot ft^2)}$
Δq	heat flow through area ΔA: W, Btu/h
Ra	Rayleigh number in terms of the reference temperature difference ΔT_0, usually given by $g\beta\Delta T_0 L^3/\nu\alpha$ (but see separate definition sketch for each problem discussed)

Ra_c	critical Rayleigh number governing the initiation of small eddy convective motion in a fluid
Ra_{ci}, Ra_{cp}	critical Rayleigh numbers for cavities with adiabatic and perfectly conducting walls, respectively
Ra_x	local value of Ra based on local temperature difference, given by $g\beta\Delta Tx^3/\nu\alpha$
Ra_{max}	value of Ra at which the heat transfer per unit volume of vertical channels is maximum for a given channel height H
$\tilde{R}a$	Rayleigh number for porous medium (Eq. 4.146)
Ra^*	Rayleigh number in terms of surface flux q'', usually given by $g\beta\overline{q''}L^4/\nu\alpha k$ (but see separate definition sketch for each problem)
Ra_x^*	local value of Ra^* defined in separate definition sketch for each problem (e.g., Fig. 4.6)
Re	Reynolds number, usually equal to Lv_0/ν (but see separate definition sketch for each problem)
Re_{fd}	Reynolds number when flow and heat transfer are fully developed
Re_i	Reynolds number at which $Nu_F = Nu_N$
Re_p	Reynolds number based on particle diameter in a porous medium
r	radius of axisymmetric body in horizontal plane measured from vertical axis characteristic dimension (Fig. 4.3b): m, ft
S	spacing between fins or width of open cavity (Fig. 4.21 or 4.23): m, ft
$\overline{S}$	average spacing between vertical triangular fins (Fig. 4.23d)
S	stratification number, given by Eq. 4.41, with $L = D$ for cylinders, spheres
$S(\chi)$	vertical height of plate at location χ (Fig. 4.9): m, ft
T	local fluid temperature: K, °F
$\overline{T}$	mean fluid temperature: K, °F
T^*	wall temperature ratio for open rectangular cavity (Fig. 4.21)
$\overline{T}_b$	average of inlet and outlet bulk fluid temperatures for internal tube flow: K, °F
T_c	temperature of cold plate of enclosure (Figs. 4.25 and 4.26): K, °F
T_{cr}	temperature of the central region in a cavity (Fig. 4.26): K, °F
T_f	film temperature $(T_\infty + T_w)/2$, unless specifically given another meaning in section of interest: K, °F
T_h	temperature of hot plate in enclosure (Figs. 4.25 and 4.26): K, °F
T_i	initial temperature of quiescent fluid before transient is initiated: K, °F
T_m	$(T_h + T_c)/2$: K, °F
T_w	temperature of solid surface of body or wall: K, °F
$\overline{T}_w$	mean value of T_w taken over surface of body: K, °F
T_{wi}	temperature of inside of tube wall (Fig. 4.49): K, °F
$\overline{T}_{w(1/2)}$	mean wall temperature halfway up the rectangular open cavity (Fig. 4.21): K, °F
T_∞	fluid temperature far from solid surface or wall at a given elevation z: K, °F
$\overline{T}_\infty$	mean value of T_∞ taken over vertical height of body
t	time: s
t_D	time period over which the fluid near a wall subjected to a step change in temperature or heat flux behaves like a stationary fluid (Figs. 4.37 and 4.39): s

t_i	time after the initiation of a transient at which the purely conductive heat transfer solution matches the steady-state convective heat transfer (Figs. 4.37 and 4.39): s
t_0	value of t_D when Nu at t_D is equal to Nu at t_∞ (Fig. 4.39): s
t_∞	time after the initiation of a transient at which the heat transfer at the wall is essentially the steady-state value (Fig. 4.37): s
t^*	dimensionless time, tv_0/L
u	fluid velocity in the x direction: m/s, ft/s
$V, \overline{V}$	fluid velocity and mean fluid velocity in tube: m/s, ft/s
v	fluid velocity in the y direction: m/s, ft/s
v_0	reference velocity, $\sqrt{g\beta\Delta TL(1+\mathrm{Pr})}$: m/s, ft/s
W	width of plate (Fig. 4.6), of cavity (Figs. 4.25 and 4.33), or of open cavity (Fig. 4.23): m, ft
w	fluid velocity in the z direction: m/s, ft/s
x	spatial coordinate (Fig. 4.1); distance measured along a surface streamline starting from the beginning of the streamline (Fig. 4.3): m, ft
x^*	x/L
x_f	value of x at trailing edge of body: m, ft
y	spatial coordinate (Fig. 4.1); coordinate measured normal to surface of body and into the fluid (Fig. 4.3): m, ft
y^*	y/L
z	spatial coordinate (Fig. 4.1): m, ft; elevation of a point above reference level (Fig. 4.3c): m, ft
z_f	elevation of end of streamline passing through a particular point on a body
z_i	elevation of start of streamline passing through a particular point on a body: m, ft
z_{ref}	z at reference level: m, ft
z^*	z/L

Greek Symbols

α	thermal diffusivity evaluated at T_f (or at T_m for enclosure problems) unless specifically otherwise directed: m²/s, ft²/s
α_m	thermal diffusivity at porous medium (Eq. 4.145): m²/s, ft²/s
α^*	aspect ratio, S/W in Fig. 4.23a
β	coefficient of thermal expansion; evaluate β at T_f for liquids and T_∞ for gases, except for enclosure problems, where β is evaluated at T_m (unless otherwise specifically directed): K⁻¹, °R⁻¹
β_0	β evaluated at T_f: K⁻¹, °R⁻¹
β_1	equals β_0 for liquids and β_∞ for gases
β_∞	β evaluated at $\overline{T}_\infty$: K⁻¹, °R⁻¹
γ	pressure coefficient of expansion: Pa⁻¹, ft²/lb$_f$
γ_1, γ_2	constants in the Bejan-Tien correlation equation (Eq. 4.98)
Δ_c	conduction layer thickness on the cold plate (Fig. 4.26): m, ft
Δ_{tx}	local turbulent conduction layer thickness on cold plate: m, ft
Δ_h	conduction layer thickness on hot plate (Fig. 4.26): m, ft

Δ_x	local conduction layer thickness for a laminar or turbulent boundary layer; $\Delta_x = k(T_w - T_\infty)/q''$: m, ft
ΔT	temperature difference $|T_w - T_\infty|$, $|T_h - T_c|$, $|T_i - T_0|$, and so on: K or °C, °F
$\overline{\Delta}$	average conduction layer thickness around body (Eq. 4.16)
$\overline{\Delta T}$	area-weighted average value of $|T_w - T_\infty|$ over the surface (or part of the surface) of body (Eq. 4.35): K, °F
ΔT_m	ΔT evaluated at the mid-height of a plate, cylinder, or sphere
ΔT_0	reference temperature difference, defined separately for each problem: K, °F
Δx	segment of length in x direction: m, ft
δ	thickness of tube wall (Fig. 4.49): m, ft
ϵ_c	emissivity of cold plate in enclosure problem (Fig. 4.25)
ϵ_h	emissivity of hot plate in enclosure problem (Fig. 4.25)
ϵ_w	emissivity of wall in enclosure problem (Fig. 4.25)
ζ	d/D (Fig. 4.23f)
ζ	local surface angle (Fig. 4.5)
ζ	parameter that accounts for transverse curvature of vertical cylinder (Eq. 4.44)
θ	angle of inclination from horizontal (Figs. 4.9, 4.16, and 4.25): rad, deg
θ_c	for enclosure problem, the crossover value of θ defined so that for $\theta < \theta_c$, Eq. 4.107 applies and for $\theta_c \le \theta \le 90°$, Eq. 4.108 applies: rad, deg
θ	dimensionless temperature, $(T - T_\infty)/(T_w - T_\infty)$
μ	dynamic viscosity, evaluated at T_f (or T_m for enclosure problems) unless specifically directed otherwise: Pa·s, lb$_m$/(s·ft)
μ_b	viscosity of fluid evaluated at bulk temperature: Pa·s, lb$_m$/(s·ft)
ν	kinematic viscosity, evaluated at T_f (or T_m for enclosure problems) unless specifically directed otherwise: m²/s, ft²/s
ρ	local density of fluid: kg/m³, lb$_m$/ft³
ρ_0	reference density, evaluated at film temperature T_f and P_{ref}: kg/m³, lb$_m$/ft³
ρ_∞	density of fluid at $T = T_\infty$, $P = P_{ref}$: kg/m³, lb$_m$/ft³
$(\rho c_p)_f$	heat capacity per unit volume for fluid: kJ/(m³·K), Btu/(ft³·°F)
$(\rho c_p)_w$	heat capacity per unit volume for wall adjacent to fluid: kJ/(m³·K), Btu/(ft³·°F)
σ	Stefan-Boltzman constant: W/m²K⁴, Btu/(h·ft²·°R⁴)
σ	heat capacitance ratio for a porous medium (Eq. 4.145)
$\tau, \tau_D, \tau_0, \tau_\infty$	dimensionless time, $4\alpha t/D^2$, $4\alpha t_D/D^2$, $4\alpha t_0/D^2$, $4\alpha t_\infty/D^2$, respectively
ϕ	volume fraction occupied by fluid for a porous medium
Ψ	function of α^*, given by Eq. 4.61b
ψ	angle of opening of V-corrugated surface (Fig. 4.23c): rad, deg
χ	Forscheimer coefficient (Eq. 4.141)

Special Brackets

$[\ ]^\bullet$	indicates that only positive values of quantities in brackets are to be taken: $[X]^\bullet = (|X| + X)/2$
$[x_1, x_2, \ldots, x_n]_{min}$	indicates that the minimum member of set $\{x_i\}$ is to be taken
$[x_1, x_2, \ldots, x_n]_{max}$	indicates that the maximum member of the set $\{x_i\}$ is to be taken

Operators

$$\frac{DX}{Dt^*} = \frac{\partial X}{\partial t^*} + u^* \frac{\partial X}{\partial x^*} + v^* \frac{\partial X}{\partial y^*} + w^* \frac{\partial X}{\partial z^*}$$

$$\nabla^{*2} X = \frac{\partial^2 X}{\partial x^{*2}} + \frac{\partial^2 X}{\partial y^{*2}} + \frac{\partial^2 X}{\partial z^{*2}}$$

REFERENCES

1. T. Aihara, "Natural Convective Heat Transfer in Vertical Parallel Fins of Rectangular Profile," *Jpn. Soc. Mech. Eng.* (34): 915–926, 1968.

2. T. Aihara, Y. Yamada, and S. Endo, "Free Convection Along a Downward-Facing Surface of a Heated Horizontal Plate," *Int. J. Heat Mass Transfer* (15): 2535–2549, 1972.

3. M. Al-Arabi and M. M. El-Rafaee, "Heat Transfer by Natural Convection From Corrugated Plates to Air," *Int. J. Heat Mass Transfer* (21): 357–359, 1978.

4. M. Al-Arabi and M. K. El-Riedy, "Natural Convection Heat Transfer From Isothermal Horizontal Plates of Different Shapes," *Int. J. Heat Mass Transfer* (19): 1399–1404, 1976.

5. M. Al-Arabi and B. Sakr, "Natural Convection Heat Transfer From Inclined Isothermal Plates," *Int. J. Heat Mass Transfer* (31): 559–566, 1988.

6. M. Al-Arabi and Y. K. Salman, "Laminar Natural Convection Heat Transfer from an Inclined Cylinder," *Int. J. Heat Mass Transfer* (23): 45–51, 1980.

7. J. N. Arnold, P. N. Bonaparte, I. Catton, and D. K. Edwards, "Experimental Investigation of Natural Convection in a Finite Rectangular Region Inclined at Various Angles from 0° to 180°," *Proc. 1974 Heat Transfer Fluid Mech. Inst., Corvallis, Ore.,* Stanford University Press, Stanford, CA, pp. 321–329, 1974.

8. J. N. Arnold, I. Catton, and D. K. Edwards, "Experimental Investigation of Natural Convection in Inclined Rectangular Regions of Differing Aspect Ratios," *J. Heat Transfer* (98): 67–71, 1976.

9. J. N. Arnold, D. K. Edwards, and I. Catton, "Effect of Tilt and Horizontal Aspect Ratio on Natural Convection in Rectangular Honeycombs," *J. Heat Transfer* (99): 120–122, 1977.

10. Y. Asako, H. Nakamura, and M. Faghri, "Three-Dimensional Laminar Natural Convection in a Vertical Air Slot With Hexagonal Honeycomb Core," *J. Heat Transfer* (112): 130–136, 1990.

11. W. Aung, "Fully Developed Laminar Free Convection Between Vertical Plates Heated Asymmetrically," *Int. J. Heat Mass Transfer* (15): 1577–1580, 1972.

12. W. Aung, L. S. Fletcher, and V. Sernas, "Developing Laminar Free Convection Between Vertical Flat Plates With Asymmetric Heating," *Int. J. Heat Mass Transfer* (15): 2293–2308, 1972.

13. L. Baker, R. E. Faw, and F. A. Kulacki, "Post-Accident Heat Removal, Part I, Heat Transfer Within an Internally Heated Nonboiling Liquid Layer," *Nucl. Sci. Eng.* (61): 222–230, 1976.

14. A. Bar-Cohen and W. M. Rohsenow, "Thermally Optimum Spacing of Vertical, Natural Convection Cooled, Parallel Plates," *J. Heat Transfer* (106): 116–123, 1984.

15. A. Bejan, "On the Boundary Layer Regime in a Vertical Enclosure Filled with a Porous Medium," *Letters Heat Mass Transfer* (6): 93–102, 1979.

16. A. Bejan and C. L. Tien, "Laminar Natural Convection Heat Transfer in a Horizontal Cavity with Different End Temperatures," *J. Heat Transfer* (100): 641–647, 1978.

17. R. F. Bergholz, "Instability of Steady Natural Convection in a Vertical Fluid Layer," *J. Fluid Mech.* (84/4): 743–768, 1978.

18. R. F. Bergholz, "Natural Convection of a Heat Generating Fluid in a Closed Cavity," *J. Heat Transfer* (102): 242–247, 1980.

19. A. E. Bergles and R. R. Simonds, "Combined Forced and Free Convection for Laminar Flow in Horizontal Tubes with Uniform Heat Flux," *Int. J. Heat Mass Transfer* (14): 1989–2000, 1971.

20. R. C. Birkebak and A. Abdulkadir, "Heat Transfer by Natural Convection From the Lower Side of Finite Horizontal, Heated Surface," *Proc. Int. Heat Trans. Conf.,* Elsevier Publishing, Amsterdam, paper NC 2.2, 1970.

21. J. R. Bodoia and J. F. Osterle, "The Development of Free Convection Between Heated Vertical Plates," *J. Heat Transfer* (84): 40–44, 1962.

22. H. Börner, "Über den Wärme- und Stoffübertragung an umspültan Einzelköpern bei Uberlagerung von freier und erzungener Strömung," *VDI Forschungsh.,* 512, 1965.

23. A. J. Bovy, and G. Woelk, "Untersuchungen zur freier Konvektion an ebenen Wänden," *Z. Wärme Stoffübertrag.* (4): 105–112, 1971.

24. W. H. Braun, S. Ostrach, and J. E. Heighway, "Free-Convection Similarity Flows About Two-Dimensional and Axisymmetric Bodies With Closed Lower Ends," *Int. J. Heat Mass Transfer* (2): 121–135, 1961.

25. A. Brown, "Relative Importance of Viscous Dissipation and Pressure Stress Effects on Laminar Free Convection," *Trans. Inst. Chem. E.* (56): 77–80, 1978.

26. J. C. Buell, "The Effect of Rotation and Wall Conductance on the Stability of a Fully Enclosed Fluid Heated From Below," MScE Thesis, University of California, Los Angeles, School of Engineering and Applied Science, Los Angeles, CA, 1981.

27. J. C. Buell and I. Catton, "The Effect of Wall Conduction on the Stability of a Fluid in a Right Circular Cylinder Heated From Below," *J. Heat Transfer* (105): 255–260, 1983.

28. H. K. Calkin, "Axisymmetric Free Convection Boundary-Layer Flow Past Slender Bodies." *Int. J. Heat Mass Transfer* (11): 1141–1153, 1968.

29. R. L. D. Cane, K. G. T. Hollands, G. D. Raithby, and T. E. Unny, "Free Convection Heat Transfer Across Inclined Honeycomb Panels." *J. Heat Transfer* (99/1): 86–91, 1977.

30. H. S. Carslaw, and J. C. Jaeger, *Conduction of Heat in Solids,* 2d ed., pp. 342–345, Oxford University Press, Oxford, UK, 1959.

31. I. Catton, "Convection in a Closed Rectangular Region: The Onset of Motion," *J. Heat Transfer* (92): 186–187, 1970.

32. I. Catton, "Effect of Wall Conduction on the Stability of a Fluid in a Rectangular Region Heated From Below," *J. Heat Mass Transfer* (94/4): 446–452, 1972.

33. I. Catton, "The Effect of Insulating Vertical Walls on the Onset of Motion in a Fluid Heated From Below," *Int. J. Heat Mass Transfer* (15): 665–672, 1972.

34. I. Catton, "Natural Convection in Enclosures," *Proc. 6th Int. Heat Transfer Conf.,* Hemisphere Publishing, Washington, DC, vol. 6, pp. 13–31, 1978.

35. I. Catton, P. S. Ayyaswamy, and R. M. Clever, "Natural Convection Flow in a Finite Rectangular Slot," *Int. J. Heat Mass Transfer* (17): 173–184, 1974.

36. I. Catton and D. K. Edwards, "Effect of Side Walls on Natural Convection Between Horizontal Plates Heated From Below," *J. Heat Transfer* (89): 295–299, 1967.

37. I. Catton and J. H. Lienhard V, "Thermal Stability of Two Fluid Layers Separated by a Solid Interlayer of Finite Thickness and Thermal Conductivity," *J. Heat Transfer* (106): 605–612, 1984.

38. M. J. Chamberlain, "Free Convection Heat Transfer from a Sphere, Cube and Vertically Aligned Bi-Sphere," MASc Thesis, University of Waterloo, Waterloo, Ontario, Canada, 1983.

39. M. J. Chamberlain, K. G. T. Hollands, and G. D. Raithby, "Experiments and Theory on Natural Convection Heat Transfer From Bodies of Complex Shape," *J. Heat Transfer* (107): 624–629, 1985.

40. B. H. Chang and R. G. Akins, "An Experimental Investigation of Natural Convection in Mercury at Low Grashof Numbers," *Int J. Heat Mass Transfer* (15): 513–525, 1972.

41. C. C. Chen and R. Eichhorn, "Natural Convection From a Vertical Surface to a Thermally Stratified Fluid," *J. Heat Transfer* (98): 446–451, 1976.

42. C. C. Chen and R. Eichhorn, "Natural Convection From Simple Bodies Immersed in Thermally Stratified Fluids," *Rep. UKY TR 105-ME14-77,* College of Engineering, University of Kentucky, Lexington, KY, October 1977.

43. T. S. Chen, E. M. Sparrow, and A. Mucoglu, "Mixed Convection in Boundary Layer on a Horizontal Plate," *J. Heat Transfer* (99): 66–71, 1977.

44. T. C. Chen and A. Mucoglu, "Wave Instability of Mixed Convection Flow Over a Horizontal Flat Plate," *Int. J. Heat Mass Transfer* (22): 185–196, 1979.

45. P. S. Chen, W. R. Wilcox, and P. J. Schlichta, "Free Convection About a Regular Crystal Growing From a Solution," *Int. J. Heat Mass Transfer* (22): 1669–1679, 1979.

46. P. Cheng, "Geothermal Heat Transfer," in *Handbook of Heat Transfer Applications,* W. M. Rohsenow, J. P. Hartnett, and E. N. Ganiç eds., Chap. 11, McGraw-Hill, New York, 1985.

47. P. Cheng, "Natural Convection Porous Medium: External Flows," in *Natural Convection: Fundamentals and Applications,* S. Kakac, W. Aung, and R. Viskanta eds., pp. 475–513, Hemisphere Publishing, Washington, DC, Springer-Verlag, Berlin, 1985.

48. P. Cheng and I.-D. Chang, "Buoyancy Induced Flows in a Porous Medium Adjacent to Impermeable Horizontal Surfaces," *Int. J. Heat Mass Transfer* (19): 1267–1272, 1976.

49. P. Cheng and W. J. Minkowycz, "Free Convection About a Vertical Flat Plate in a Porous Medium With Application to Heat Transfer from a Dike," *J. Geophysical Research* (82): 2040–2044, 1977.

50. K. C. Cheng, M. Takeuchi, and R. R. Gilpin, "Transient Natural Convection in Horizontal Water Pipes With Maximum Density Effect and Supercooling," *Num. Heat Transfer* (1): 101–115, 1978.

51. F. B. Cheung, "Correlation Equation for Turbulent Thermal Convection in a Horizontal Fluid Layer Heated Internally and From Below," *J. Heat Transfer* (100): 416–422, 1978.

52. F. B. Cheung, S. W. Shiah, D. H. Cho, and L. Baker Jr., "Turbulent Natural Convection in a Horizontal Layer of Small-Prandtl-Number Fluid," *ASME J. Heat Transfer* (113): 919–925, 1991.

53. T. Y. Chu and R. J. Goldstein, "Turbulent Convection in a Horizontal Layer of Water," *J. Fluid Mech.* (60/1): 141–159, 1973.

54. S. W. Churchill and R. Usagi, "A General Expression for the Correlation of Rates of Transfer and Other Phenomena," *AIChE J.* (18): 1121–1128, 1972.

55. S. W. Churchill, "A Comprehensive Correlating Equation for Laminar Assisting Forced and Free Convection," *AIChE J.* (23): 10–16, 1977.

56. S. W. Churchill, *Free Convection Around Immersed Bodies Heat Exchanger Design Handbook,* E. U. Schlünder, editor-in-chief, Hemisphere Publishing, New York, 1983.

57. S. W. Churchill and H. H. S. Chu, "Correlating Equations for Laminar and Turbulent Free Convection From a Horizontal Cylinder," *Int. J. Heat Mass Transfer* (18): 1049–1053, 1975.

58. S. W. Churchill and H. Ozoe, "A Correlation of Laminar Free Convection From a Vertical Plate," *J. Heat Transfer* (95): 540–541, 1973.

59. J. A. Clark, "A Review of Pressurization, Stratification and Interfacial Phenomena," *Adv. Cryogen. Eng.* (8): 259, 1964.

60. A. M. Clausing, "Simple Functional Representations for the Thermophysical Properties of Gases at Standard Pressure," *ASHRAE Transactions* (88/2): 1982.

61. A. M. Clausing, "Natural Convection Correlations for Vertical Surfaces Including Influences of Variable Property Values," *J. Heat Transfer* (105): 138–143, 1983.

62. A. M. Clausing and J. J. Berton, "An Experimental Investigation of Natural Convection From an Isothermal Horizontal Plate," *J. Heat Transfer* (111): 904–908, 1989.

63. A. M. Clausing and S. N. Kempka, "The Influences of Property Variations on Natural Convection From Vertical Surfaces," *J. Heat Transfer* (103): 609–612, 1981.

64. S. B. Clemes, K. G. T. Hollands, and A. P. Brunger, "Natural Convection Heat Transfer From Long Horizontal Isothermal Cylinders," *J. Heat Transfer* (116): 96–104, 1994.

65. R. M. Clever, "Finite Amplitude Longitudinal Convection Rolls in Inclined Layers," *J. Heat Transfer* (95): 407–408, 1973.

66. R. M. Clever and F. H. Busse, "Instabilities of Longitudinal Convection Rolls in an Inclined Layer," *J. Fluid Mech.* (81): 107–127, 1977.

67. D. C. Collis and M. J. Williams, "Free Convection of Heat From Fin Wires," Aeronautical Research Laboratory, Note 140, Melbourne, Australia, 1954.

68. R. G. Colwell and J. R. Welty, "An Experimental Investigation of Natural Convection With a Low Prandtl Number Fluid in a Vertical Channel With Uniform Wall Heat Flux," *J. Heat Transfer* (96): 448–454, 1974.

69. M. Combarnous, "Natural Convection in Porous Media and Geothermal Systems," *Proc. 6th Int. Heat Transfer Conf.,* vol. 6, pp. 45–59, Hemisphere Publishing, Washington, DC, 1978.

70. P. G. Daniels and C. F. Ong, "Linear Stability of Convection in a Rigid Channel Uniformly Heated From Below," *Int. J. Heat Mass Transfer* (33/1): 55–60, 1990.

71. L. M. De Socio, "Laminar Free Convection Around Horizontal Circular Cylinders," *Int. J. Heat Mass Transfer* (26): 1669–1677, 1983.

72. G. de Vahl Davis, "Natural Convection of Air in a Square Cavity: A Bench Mark Numerical Solution," *Int. J. Numerical Methods Fluids* (3): 249–264, 1983.

73. F. K. Deaver and E. R. G. Eckert, "An Interferometric Investigation of Convective Heat Transfer in a Horizontal Cylinder With Wall Temperature Increasing at a Uniform Rate," *Proc. 4th Int. Heat Transfer Conf.,* vol. 4, paper NC 1.1, Elsevier Publishing, Amsterdam, 1970.

74. C. A. Depew and S. E. August, "Heat Transfer Due to Combined Free and Forced Convection in a Horizontal and Isothermal Tube," *J. Heat Transfer* (93): 380–384, 1971.

75. V. K. Dhir and J. H. Lienhard, "Laminar Film Condensation on Plane and Axisymmetric Bodies on Non-Uniform Gravity," *J. Heat Transfer* (93): 97–100, 1971.

76. J. R. Dyer, "The Development of Laminar Natural Convection Flow in a Vertical Uniform Heat Flux Duct," *Int. J. Heat Mass Transfer* (18): 1455–1465, 1975.

77. J. R. Dyer, "Natural Convective Flow Through a Vertical Duct With Restricted Entry," *Int. J. Heat Mass Transfer* (211): 1341–1354, 1978.

78. E. R. G. Eckert and Robert M. Drake, *Analysis of Heat and Mass Transfer,* McGraw-Hill, New York, 1972.

79. A. J. Ede, "Advances in Free Convection," in *Advances in Heat Transfer,* T. F. Irvine and J. P. Hartnett eds., vol. 4, pp. 1–64, Academic, New York, 1968.

80. D. K. Edwards, I. Catton, J. N. Arnold, et al., "Studies of Heat Transfer in Thermally Driven Flows," *Rep. No. UCLA-Eng-7828,* University of California, Los Angeles, CA, May 1978.

81. J. A. Edwards and J. B. Chaddock, "An Experimental Investigation of the Radiation and Free-Convection Heat Transfer From a Cylindrical Disk Extended Surface," *Trans. ASHRAE* (69): 313–322, 1963.

82. D. K. Edwards and W. M. Sun, "Prediction of the Onset of Natural Convection in Rectangular Honeycomb Structures," *Int. Solar Energy Soc. Conf.,* Paper No. 7/62, Melbourne, Australia, 1970.

83. D. K. Edwards and W. M. Sun, "Effect of Wall Radiation on Thermal Instability in a Vertical Cylinder," *Int. J. Heat Mass Transfer* (14): 15–18, 1971.

84. S. M. ElSherbiny, G. D. Raithby, and K. G. T. Hollands, "Heat Transfer by Natural Convection Across Vertical and Inclined Cavities," *J. Heat Transfer* (104): 96–102, 1982.

85. S. M. ElSherbiny, K. G. T. Hollands, and G. D. Raithby, "Effect of Thermal Boundary Conditions on Natural Convection in Vertical and Inclined Air Layers," *J. Heat Transfer* (104): 515–520, 1982.

86. M. K. El-Riedy, "Analogy Between Heat and Mass Transfer by Natural Convection From Air to Horizontal Tubes," *Int. J. Heat Mass Transfer* (24): 365–369, 1981.

87. J. W. Elder, "Steady Natural Convection in a Porous Medium Heated From Below," *J. Fluid Mech.* (27): 29–48, 1967.

88. W. Elenbaas, "Heat Dissipation of Parallel Plates by Free Convection," *Physica* (IX/1): 2–28, 1942.

89. W. Elenbaas, "The Dissipation of Heat by Free Convection: The Inner Surface of Vertical Tubes of Different Shapes of Cross-Section," *Physica* (IX/8): 865–874, 1942.

90. S. Ergun, "Fluid Flow Through Packed Columns," *Chem. Eng. Prog.* (48): 89–94, 1952.

91. L. B. Evans, R. C. Reid, and E. M. Drake, "Transient Natural Convection in a Vertical Cylinder," *AIChE J.* (14): 251–259, 1968.

92. L. B. Evans and N. E. Stefany, "An Experimental Study of Transient Heat Transfer in Cylindrical Enclosures," *Chem. Eng. Prog. Symp. Ser.* (62/64): 209–215, 1960.

93. T. Z. Fahidy, "On the Flat Plate Approximation to Laminar Convection From the Outer Surface of a Vertical Cylinder," *Int. J. Heat Mass Transfer* (17): 159–160, 1974.

94. M. R. Fand, T. E. Steinberger, and P. Cheng, "Natural Convection Heat Transfer From a Horizontal Cylinder in a Porous Medium," *Int. J. Heat Transfer* (29/1): 119–133, 1986.

95. M. R. Fand, E. W. Morris, and M. Lum, "Natural Convection Heat Transfer From Horizontal Cylinders to Air, Water and Silicone Oils for Rayleigh Numbers Between 3×10^2 and 2×10^7," *Int. J. Heat Mass Transfer* (20): 1173–1184, 1977.

96. B. Farouk and S. I. Güçeri, "Laminar and Turbulent Natural Convection in the Annulus Between Horizontal Concentric Cylinders," *J. Heat Transfer* (104): 631–636, 1982.

97. R. E. Faw and T. A. Dullforce, "Holographic Interferometry Measurement of Convective Heat Transport Beneath a Heated Horizontal Plate in Air," *Int. J. Heat Mass Transfer* (24/5): 859–869, 1981.

98. M. Fishenden and O. A. Saunders, *An Introduction to Heat Transfer,* pp. 89–99, Clarendon Press, Oxford, UK, 1957.

99. T. Fujii and M. Fujii, "Experiments on Natural Convection Heat Transfer From the Outer Surface of a Vertical Cylinder to Liquids," *Int. J. Heat Mass Transfer* (13): 753–787, 1970.

100. T. Fujii and M. Fujii, "The Dependence of Local Nusselt Number on Prandtl Number in the Case of Free Convection Along a Vertical Surface With Uniform Heat Flux," *Int. J. Heat Mass Transfer* (19): 121–122, 1976.

101. T. Fujii, M. Fujii, and T. Matsunaga, "A Numerical Analysis of Laminar Free Convection Around an Isothermal Horizontal Circular Cylinder," *Num. Heat Transfer* (2): 329–344, 1979.

102. T. Fujii, H. Honda, and I. Morioka, "A Theoretical Study of Natural Convection Heat Transfer From Downward-Facing Horizontal Surfaces With Uniform Heat Flux," *Int. J. Heat Mass Transfer* (16): 611–627, 1973.

103. T. Fujii and H. Imura, "Natural Convection Heat Transfer From a Plate With Arbitrary Inclination," *Int. J. Heat Mass Transfer* (15): 755–767, 1972.

104. A. M. Garon and R. J. Goldstein, "Velocity and Heat Transfer Measurements in Thermal Convection," *Phys. Fluids* (16/11): 1818–1925, 1973.

105. B. Gebhart, "Transient Natural Convection From Vertical Elements," *J. Heat Transfer* (83): 61–70, 1961.

106. B. Gebhart, "Natural Convection Cooling Transients," *Int. J. Heat Mass Transfer* (7): 479–483, 1964.

107. B. Gebhart, *Heat Transfer,* McGraw-Hill, New York, 1971.

108. B. Gebhart and D. E. Adams, "Measurements of Transient Natural Convection on Flat Vertical Surfaces," *J. Heat Transfer* (85): 25–28, 1963.

109. B. Gebhart, T. Audunson, and L. Pera, "Forced, Mixed, and Natural Convection From Long Horizontal Wires, Experiments at Various Prandtl Numbers," *Proc. 4th Int. Heat Transfer Conf.,* Elsevier Publishing, vol. 4, paper No. 3.2, Amsterdam, 1970.

110. B. Gebhart and J. C. Mollendorf, "A New Density Relation for Pure and Saline Water," *Deep Sea Research* (244): 831–848, 1977.

111. B. Gebhart and L. Pera, "Mixed Convection From Long Horizontal Cylinders," *J. Fluid Mech.* (45): 49–64, 1970.

112. A. E. Gill, "The Boundary Layer Regime for Free Convection in a Rectangular Cavity," *J. Fluid Mech.* (26/3): 515–536, 1966.

113. R. R. Gilpin, H. Imura, and K. C. Cheng, "Experiments on the Onset of Longitudinal Vortices in Horizontal Blasius Flow Heated From Below," *J. Heat Transfer* (100): 71–77, 1978.

114. S. Globe and D. Dropkin, "Natural Convection Heat Transfer in Liquid Confined by Two Horizontal Plates and Heated From Below," *ASME J. Heat Transfer* (81): 24–30, 1959.

115. R. J. Goldstein and E. R. G. Eckert, "The Steady and Transient Free Convection Boundary Layer on a Uniformly Heated Vertical Plate," *Int. J. Heat Mass Transfer* (1): 208–218, 1960.

116. R. J. Goldstein, E. M. Sparrow, and D. C. Jones, "Natural Convection Mass Transfer Adjacent to Horizontal Plates," *Int. J. Heat Mass Transfer* (16): 1025–1035, 1973.

117. R. J. Goldstein and S. Tokuda, "Heat Transfer by Thermal Convection at High Rayleigh Numbers," *Int. J. Heat Mass Transfer* (23): 738–740, 1980.

118. D. O. Gough, E. A. Spregel, and J. Toomre, "Modal Equations for Cellular Convection," *J. Fluid Mech.* (68/4): 695–719, 1975.

119. H. Grober, S. Erk, and U. Grigull, *Grundgesetze der Wärmeübertragung,* Springer-Verlag, 1961.

120. F. Harahap and H. N. McManus, "Natural Convection Heat Transfer From Horizontal Rectangular Fin Arrays," *J. Heat Transfer* (89): 32–38, 1967.

121. G. M. Harpole and I. Catton, "Laminar Natural Convection About Downward Facing Heated Blunt Bodies to Liquid Metals," *J. Heat Transfer* (98): 208–212, 1976.

122. M. M. Hasan and R. Eichhorn, "Local Non-Similarity Solution for Free Convection Flow and Heat Transfer From an Inclined Plate," *J. Heat Transfer* (101): 642–647, 1979.

123. K. E. Hassan and S. A. Mohamed, "Natural Convection From Isothermal Flat Surfaces," *Int. J. Heat Mass Transfer* (13): 1873–1886, 1970.

124. A. Hassani, "An Investigation of Free Convection Heat Transfer From Bodies of Arbitrary Shape," PhD thesis, Department of Mechanical Engineering, University of Waterloo, Waterloo, Ontario, Canada, 1987.

125. A. V. Hassani, "Natural Convection Heat Transfer From Cylinders of Arbitrary Cross Section," *J. Heat Transfer* (114): 768–773, 1992.

126. A. V. Hassani and K. G. T. Hollands, "On Natural Convection Heat Transfer From Three-Dimensional Bodies of Arbitrary Shape," *J. Heat Transfer* (111): 363–371, 1989.

127. A. V. Hassani and K. G. T. Hollands, "Prandtl Number Effect on Natural Convection on External Natural Convection Heat Transfer From Irregular Three-Dimensional Bodies," *Int. J. Heat Mass Transfer* (32/11): 2075–2080, 1989.

128. A. V. Hassani and K. G. T. Hollands, "Conduction Shape Factor for a Region of Uniform Thickness Surrounding a Three-Dimensional Body of Arbitrary Shape," *J. Heat Transfer* (112): 492–495, 1990.

129. A. V. Hassani, K. G. T. Hollands, and G. D. Raithby, "A Close Upper Bound for the Conduction Shape Factor of a Uniform Thickness, 2D Layer," *Int. J. Heat Mass Transfer* (36/12): 3155–3158, 1993.

130. D. W. Hatfield and D. K. Edwards, "Edge and Aspect Ratio Effects on Natural Convection From the Horizontal Heated Plate Facing Downwards," *Int. J. Heat Mass Transfer* (24/6): 1019–1024, 1981.

131. A. P. Hatton, D. D. James, and H. W. Swire, "Combined Forced and Natural Convection With Low-Speed Air Flow Over Horizontal Cylinders," *J. Fluid Mech.* (42): 17–31, 1970.

132. W. Hauf and U. Grigull, "Instationärer Wärmeübergang durch freie Konvektion in horizontalen zylindrischen Behältern," *Proc. 41st Int. Heat Transfer Conf.*, vol. 4, paper NC 1.3, Elsevier Publishing, Amsterdam, 1970.

133. W. Hauf and U. Grigull, "Instationärer Warmeübergang in Horizontalen, Zylindrischen Behältern," *Z. Wärme Stoffübertrag.* (8): 57–68, 1975.

134. W. Hauf and U. Grigull, "Wärmeübergangsmessungen am Horizontalen, Zylindrischen Behaltern. Massgebliche Parameter," *Z. Wärme Stoffübertrag.* (9): 21–28, 1976.

135. C. F. Hess and C. W. Miller, "Natural Convection in a Vertical Cylinder Subject to Constant Heat Flux," *Int. J. Heat Mass Transfer* (22): 421–430, 1979.

136. M. A. Hessami, A. Pollard, R. D. Rowe, and D. W. Ruth, "A Study of Free Convective Heat Transfer in a Horizontal Annulus With a Large Radii Ratio," *J. Heat Transfer* (107): 603–610, 1985.

137. G. Hesse and E. M. Sparrow, "Low Rayleigh Number Natural Convection Heat Transfer From High-Temperature Wires to Gases," *Int. J. Heat Mass Transfer* (17): 796–798, 1974.

138. J. Hiddink, J. Schenk, and S. Bruin, "Natural Convection Heating of Liquids in Closed Containers," *Appl. Sci. Res.* (32): 217–237, 1976.

139. C. A. Hieber, "Multilayer Rayleigh-Benard Instability via Shooting Method," *J. Heat Transfer* (109): 538–540, 1987.

140. K. G. T. Hollands, "Multi-Prandtl Number Correlation Equations for Natural Convection in Layers and Enclosures," *Int. J. Heat Mass Transfer* (27/3): 466–468, 1984.

141. K. G. T. Hollands, "Natural Convection in Horizontal Thin-Walled Honeycomb Panels," *J. Heat Transfer* (95/4): 439–444, 1973.

142. K. G. T. Hollands, G. D. Raithby, and L. Konicek, "Correlation Equations for Free Convection Heat Transfer in Horizontal Layers of Air and Water," *Int. J. Heat Mass Transfer* (18): 879–884, 1975.

143. K. G. T. Hollands, G. D. Raithby, R. B. Russell, and R. G. Wilkinson, "Coupled Radiative and Conductive Heat Transfer Across Honeycomb Panels and Through Single Cells," *Int. J. Heat Mass Transfer* (27/11): 2119–2131, 1984.

144. K. G. T. Hollands, T. E. Unny, G. D. Raithby, and L. Konicek, "Free Convection Heat Transfer Across Inclined Air Layers," *J. Heat Transfer* (98): 189–193, 1976.

145. K. G. T. Hollands and J. L. Wright, "Heat Loss Coefficients and Effective * Products for Flat-Plate Collectors With Diathermanous Covers," *Solar Energy* (30/3): 211–216, 1983.

146. S.-S. Hsieh and C.-Y. Wang, "Experimental Study of Three-Dimensional Natural Convection in Enclosures With Different Working Fluids," *Int. J. Heat Mass Transfer* (37/17): 2687–2698, 1994.

147. W. W. Humphreys and J. R. Welty, "Natural Convection With Mercury in a Uniformly Heated Vertical Channel During Unstable Laminar and Transitional Flow," *AIChE Journal* (21/2): 268–274, 1975.

148. H. Imura, R. R. Gilpin, and K. C. Cheng, "An Experimental Investigation of Heat Transfer and Buoyancy Induced Transition From Laminar Forced Convection to Turbulent Free Convection Over a Horizontal Isothermally Heated Plate," *J. Heat Transfer* (100): 429–434, 1978.

149. F. P. Incropera and D. P. De Witt, *Fundamentals of Heat and Mass Transfer,* Wiley, New York, 1990.

150. C. D. Jones and L. F. Smith, "Optimum Arrangement of Rectangular Fins on Horizontal Surfaces for Free Convection Heat Transfer," *J. Heat Transfer* (92): 6–10, 1970.

151. C. D. Jones and E. I. Nwizu, "Optimum Spacing of Circular Fins on Horizontal Tubes for Natural Convection Heat Transfer," *ASHRAE Symp. Bull. DV-69-3,* 11–15, 1969.

152. A. Karagiozis, G. D. Raithby, and K. G. T. Hollands, "Natural Convection Heat Transfer From Arrays of Isothermal Triangular Fins to Air," *J. Heat Transfer* (116): 105–111, 1994.

153. A. Karagiozis, "An Investigation of Laminar Free Convection Heat Transfer From Isothermal Finned Surfaces," PhD thesis, University of Waterloo, Waterloo, Ontario, Canada, 1991.

154. K. C. Karki, P. S. Sathyamurthy, and S. V. Patankar, "Natural Convection in a Partitioned Cubic Enclosure," *Int. J. Heat Mass Transfer* (114): 410–417, 1992.

155. M. Kaviany, "Principles of Heat Transfer in Porous Media," *Mechanical Engineering Series,* Springer-Verlag, New York, 1991.

156. M. Kaviany and M. Mittal, "Natural Convection Heat Transfer From a Vertical Plate to High Permeability Porous Media: An Experiment and an Approximate Solution," *Int. J. Heat Mass Transfer* (30/5): 967–977, 1987.

157. R. J. Kee, C. S. Landram, and J. C. Miles, "Natural Convection of a Heat Generating Fluid Within Closed Vertical Cylinders and Spheres," *J. Heat Transfer* (98): 55–61, 1976.

158. V. Kek and U. Müller, "Low Prandtl Number Convection in Layers Heated From Below," *Int. J. Heat Mass Transfer* (36/11): 2795–2804, 1993.

159. M. Keylani and F. A. Kulacki, "An Experimental Study of Turbulent Thermal Convection With Time-Dependent Volumetric Energy Sources," *U.S. Nuclear Regulatory Rep. NUREG-RF710968, NRC-3,* 1979.

160. W. J. King, "The Basic Laws and Data of Heat Transmission," *Mech. Eng.* (54): 347–353, 1932.

161. K. Kitamura and F. Kimura, "Heat Transfer and Fluid Flow of Natural Convection Adjacent to Upward-Facing Horizontal Plates," *Int. J. Heat Mass Transfer* (38/17): 3149–3159, 1995.

162. S. E. Korpela, D. Gozum, and C. B. Baxi, "On the Stability of the Conduction Regime of Natural Convection in a Vertical Slot," *Int. J. Heat Mass Transfer* (16): 1683–1690, 1973.

163. T. H. Kuehn and R. J. Goldstein, "Correlating Equations for Natural Convection Heat Transfer Between Horizontal Circular Cylinders," *Int. J. Heat Mass Transfer* (19): 1127–1134, 1976.

164. T. H. Kuehn and R. J. Goldstein, "An Experimental Study of Natural Convection Heat Transfer in Concentric and Eccentric Cylindrical Annuli," *J. Heat Transfer* (100): 635–640, 1978.

165. T. H. Kuehn and R. J. Goldstein, "Numerical Solution to the Navier-Stokes Equations for Laminar Natural Convection About a Horizontal Isothermal Circular Cylinder," *Int. J. Heat Mass Transfer* (23): 971–979, 1980.

166. F. A. Kulacki and A. A. Emara, "Steady and Transient Thermal Convection in a Fluid Layer With Volumetric Energy Sources," *J. Fluid Mech.* (83): 375–395, 1977.

167. F. A. Kulacki and R. J. Goldstein, "Hydrodynamic Instability in Fluid Layers With Uniform Volumetric Energy Sources," *Appl. Sci. Res.* (31): 81–109, 1975.

168. F. A. Kulacki and M. E. Nagle, "Natural Convection in a Horizontal Fluid Layer With Volumetric Energy Sources," *J. Heat Transfer* (91): 204–211, 1975.

169. F. R. Kumar and A. Kalam, "Laminar Thermal Convection Between Vertical Coaxial Isothermal Cylinders," *Int. J. Heat Mass Transfer* (34/2): 513–524, 1991.

170. S. S. Kutateladze, V. P. Ivakin, A. G. Kirdyashkin, and A. N. Kekalov, "Thermal Free Convection in a Liquid in a Vertical Slot Under Turbulent Flow Conditions," *Heat Transfer Sov. Res.* (10/5): 118–125, 1978.

171. R. A. Kuyper, T. H. Van Der Meer, C. J. Hoogendoorn, and R. A. W. M. Henkes, "Numerical Study of Laminar and Turbulent Natural Convection in an Inclined Square Cavity," *Int. J. Heat Mass Transfer* (36/11): 2899–2911, 1993.

172. J. R. Kyte, A. J. Madden, and E. L. Piret, "Natural Convection Heat Transfer at Reduced Pressure," *Chemical Engineering Progress* (49/12): 653–662, 1953.

173. J. L. Lage and A. Bejan, "The Ra-Pr Domain of Laminar Natural Convection in an Enclosure Heated From the Side," *Numerical Heat Transfer Part A* (19): 21–41, 1991.

174. N. K. Lambha, S. A. Korpela, and F. A. Kulacki, "Thermal Convection in a Cylindrical Cavity With Volumetric Energy Generation," *Proc. 6th Int. Heat Transfer Conf.,* vol. 2, pp. 311–316, Hemisphere Publishing, Washington, DC, 1978.

175. I. Langmuir, "Convection and Conduction of Heat in Gases," *Phys. Rev.* (34): 401–422, 1912.

176. P. Le Quéré, "Accurate Solutions to the Square Thermally Driven Cavity at High Rayleigh Number," *Computers Fluids* (20/1): 29–41, 1991.

177. Y. Lee and S. A. Korpela, "Multicellular Natural Convection in a Vertical Slot," *J. Fluid Mechanics* (126): 91–121, 1983.

178. E. K. Levy, P. A. Eichen, W. R. Cintani, and R. R. Shaw, "Optimum Plate Spacings for Laminar Natural Convection Heat Transfer From Parallel Vertical Isothermal Flat Plates: Experimental Verification," *J. Heat Transfer* (97): 474–476, 1975.

179. W. M. Lewandowski, P. Kubski, and H. Bieszk, "Heat Transfer From Polygonal Horizontal Isothermal Surfaces," *Int. J. Heat Mass Transfer* (37/5): 855–864, 1994.

180. J. Li and J. D. Tarasuk, "Local Free Convection Around Inclined Cylinders in Air: An Interferometric Study," *Experimental Thermal and Fluid Science* (5): 235–242, 1992.

181. J. H. Lienhard, "An Improved Approach to Conductive Boundary Conditions for the Rayleigh-Bénard Instability," *J. Heat Transfer* (109): 378–387, 1987.

182. J. H. Lienhard, "Thermal Radiation in Rayleigh-Bernard Instability," *J. Heat Transfer* (112/1): 100–109, 1990.

183. J. H. Lienhard and I. Catton, "Heat Transfer Across a Two-Fluid-Layer Region," *J. Heat Transfer* (108): 198–205, 1989.

184. J. H. Lienhard V, J. R. Lloyd, and W. R. Moran, "Natural Convection Adjacent to Horizontal Surface of Various Planforms," *J. Heat Transfer* (96): 443–447, 1974.

185. F. N. Lin and B. T. Chao, "Laminar Free Convection Over Two-Dimensional and Axisymmetric Bodies of Arbitrary Contour," *J. Heat Transfer* (96): 435–442, 1974.

186. N. N. Lin and A. Bejan, "Natural Convection in a Partly Divided Enclosure," *Int. J. Heat Mass Transfer* (26/12): 1867–1878, 1983.

187. J. R. Lloyd, E. M. Sparrow, and E. R. G. Eckert, "Laminar, Transition and Turbulent Natural Convection Adjacent to Inclined and Vertical Surfaces," *Int. J. Heat Mass Transfer* (15): 457–473, 1972.

188. S. Makai and T. Okazaki, "Heat Transfer From a Horizontal Wire at Small Reynolds and Grashof Numbers, II," *Int. J. Heat Mass Transfer* (18): 397–413, 1975.

189. S. M. Marcos and A. E. Bergles, "Experimental Investigation of Combined Forced and Free Laminar Convection in Horizontal Tubes," *J. Heat Transfer* (97): 212–219, 1975.

190. B. Metais and E. R. G. Eckert, "Forced, Mixed, and Free Convection Regimes," *J. Heat Transfer* (86): 295–296, 1964.

191. B. A. Meyer, J. W. Mitchell, and M. M. ElWakel, "Natural Convection Heat Transfer in Moderate Aspect Ratio Enclosures," *J. Heat Transfer* (101): 655–659, 1979.

192. J. H. Min and F. A. Kulacki, "Transient Natural Convection in a Single-Phase Heat Generating Pool Bounded From Below by a Segment of a Sphere," *Nucl. Eng. Des.* (54): 267–278, 1979.

193. J. H. Min and F. A. Kulacki, "An Experimental Study of Thermal Convection With Volumetric Energy Sources in a Fluid Layer Bounded From Below by a Segment of Sphere," *Proc. 6th Int. Heat Transfer Conf.,* vol. 5, pp. 155–160, Hemisphere Publishing, Washington, DC, 1978.

194. W. J. Minkowycz and P. Cheng, "Free Convection About a Vertical Cylinder in a Porous Medium," *Int. J. Heat Mass Transfer* (19): 805–813, 1976.

195. W. J. Minkowycz and E. M. Sparrow, "Local Nonsimilar Solutions for Natural Convection on a Vertical Cylinder," *J. Heat Transfer* (96): 178–183, 1974.

196. O. Miyatake and T. Fujii, "Free Convection Heat Transfer Between Vertical Parallel Plates—One Plate Isothermally Heated and the Other Thermally Insulated," *Heat Transfer Jpn. Res.* (1): 30–38, 1972.

197. W. R. Moran and J. R. Lloyd, "Natural Convection Mass Transfer Adjacent to Vertical and Downward-Facing Inclined Surfaces," *J. Heat Transfer* (97): 472–474, 1975.

198. V. T. Morgan, "The Overall Convective Heat Transfer From Smooth Circular Cylinders," in *Advances in Heat Transfer,* T. F. Irvine and J. P. Hartnett eds., vol. 11, pp. 199–264, Academic, New York, 1975.

199. V. T. Morgan, "The Heat Transfer From Base Stranded Conductors by Natural and Forced Convection in Air," *Int. J. Heat Mass Transfer* (16): 2023–2034, 1973.

201. A. Mucoglu and T. S. Chen, "Mixed Convection on Inclined Surfaces," *J. Heat Transfer* (101): 422–426, 1979.

202. W. Murgatroyd and A. Watson, "An Experimental Investigation of the Natural Convection of a Heat Generating Fluid Within a Closed Vertical Cylinder," *J. Mech. Eng. Sci.* (12): 354–363, 1970.

203. H. R. Nagendra, M. D. Tirunarayan, and A. Ramachandran, "Laminar Free Convection From Vertical Cylinders With Uniform Heat Flux," *J. Heat Transfer* (92): 191–194, 1970.

204. H. Nakamura and Y. Asako, "Laminar Free Convection From a Horizontal Cylinder With Uniform Cross-Section of Arbitrary Shape," *Bull. JSME* (21/153): 471–478, 1978.

205. M. W. Nansteel and R. Greif, "An Investigation of Natural Convection in Enclosures With Two- and Three-Dimensional Partitions," *Int. J. Heat Mass Transfer* (27/4): 561–571, 1984.

206. A. T. Nguyen and F. A. Kulacki, "Convection in Multi-Layer Systems," in *Mechanisms of Continental Drift and Plate Tectonics,* P. A. Davies and S. K. Runcorn eds., pp. 259–266, Academic, New York, 1980.

207. D. A. Nield and A. Bejan, *Convection in Porous Media,* Springer-Verlag, New York, 1992.

208. K. S. Ning, R. E. Faw, and T. W. Lester, "Hydrodynamic Stability in Horizontal Fluid Layers With Uniform Volumetric Energy Sources," *J. Heat Transfer* (100): 729–730, 1978.

209. K. Noto and R. Matsumoto, "Turbulent Heat Transfer by Natural Convection Along an Isothermal Vertical Plate Surface," *J. Heat Transfer* (97): 621–624, 1975.

210. J. L. Novotny, "Laminar Free Convection Between Finite Vertical Parallel Plates," in *Progress in Heat and Mass Transfer,* T. F. Irvine Jr. ed., vol. 2, pp. 13–22, Pergamon Press, New York, 1968.

211. P. H. Oosthuizen and J. T. Paul, "An Experimental Study of the Free-Convective Heat Transfer From Horizontal Non-Circular Cylinders, Fundamentals of Natural Convection/Electronic Equipment," *ASME HTD* (32): 91–97, 1984.

212. P. H. Oosthuizen, "Free Convection Heat Transfer From Horizontal Cones," *J. Heat Transfer* (95): 409–410, 1973.

213. P. H. Oosthuizen, "Experimental Study of Free Convection Heat Transfer From Inclined Cylinders," *J. Heat Transfer* (98): 672–674, 1976.

214. P. H. Oosthuizen and M. Bassey, "An Experimental Study of Combined Forced- and Free-Convective Heat Transfer From Flat Plates to Air at Low Reynolds Numbers," *J. Heat Transfer* (95): 120–121, 1973.

215. P. H. Oosthuizen and E. Donaldson, "Free Convective Heat Transfer From Vertical Cones," *J. Heat Transfer* (94): 330–331, 1972.

216. G. A. Ostroumov, "Free Convection Under Conditions of the Internal Problem," *NACA TM 1407* 29–37, 1958.

217. H. Ozoe, H. Sayami, and S. W. Churchill, "Natural Convection in an Inclined Rectangular Channel at Various Aspect Ratios and Angles—Experimental Measurements," *Int. J. Heat Mass Transfer* (18): 1425–1431, 1975.

218. J. R. Parsons and J. C. Mulligan, "Transient Free Convection From a Suddenly Heated Horizontal Wire," *J. Heat Transfer* (100): 423–428, 1978.

219. J.-L. Peube and D. Blay, "Convection Naturelle Laminaire Tridimensionnelle Autour de Surfaces," *Int. J. Heat Mass Transfer* (21): 1125–1131, 1978.

220. A. Pirovano, S. Viannay, and M. Jannot, "Convection Naturelie en Régime Turbulent le Long d'une Plaque Plane Verticale," *Rep. EUR 4489f,* Commission of the European Communities, 1970.

221. A. Pirovano, S. Viannay, and M. Jannot, "Convection Naturelle en Régime Turbulent le Long d'une Plaque Plane Verticale," *Proc. 4th Int. Heat Transfer Conf.,* Elsevier Publishing, Amsterdam, paper NC 1.8, 1970.

222. Z. H. Qureshi and B. Gebhart, "Transition and Transport in a Buoyancy Driven Flow in Water Adjacent to a Vertical Uniform Flux Surface," *Int. J. Heat Mass Transfer* (21): 1467–1479, 1978.

223. G. D. Raithby and K. G. T. Hollands, "A General Method of Obtaining Approximate Solutions to Laminar and Turbulent Free Convection Problems," in *Advances in Heat Transfer,* T. F. Irvine and J. P. Hartnett eds., vol. 11, pp. 266–315, Academic, New York, 1975.

224. G. D. Raithby and K. G. T. Hollands, "Laminar and Turbulent Free Convection From Elliptic Cylinders, With a Vertical Plate and Horizontal Circular Cylinder as Special Cases," *J. Heat Transfer* (98): 72–80, 1976.

225. G. D. Raithby and K. G. T. Hollands, "Analysis of Heat Transfer by Natural Convection (or Film Condensation) for Three-Dimensional Flows," *Proc. 6th Int. Heat Transfer Conf.,* Hemisphere Publishing, Washington, DC, vol. 2, pp. 187–192, 1978.

226. G. D. Raithby and K. G. T. Hollands, "Heat Transfer by Natural Convection Between a Vertical Surface and a Stably Stratified Fluid," *J. Heat Transfer* (100): 378–381, 1978.

227. G. D. Raithby and K. G. T. Hollands, "Natural Convection," in *Handbook of Heat Transfer Fundamentals,* W. M. Rohsenow, J. P. Hartnett, and E. N. Ganiç eds., chap. 6, McGraw-Hill, New York, 1985.

228. G. D. Raithby, K. G. T. Hollands, and T. E. Unny, "Analyses of Heat Transfer by Natural Convection Across Vertical Fluid Layers," *J. Heat Transfer* (99): 287–293, 1977.

229. G. D. Raithby, A. Pollard, K. G. T. Hollands, and M. M. Yovanovich, "Free Convection Heat Transfer From Spheroids," *J. Heat Transfer* (98): 452–458, 1976.

230. G. D. Raithby and H. H. Wong, "Heat Transfer by Natural Convection Across Vertical Air Layers," *Num. Heat Transfer* (4): 447–457, 1981.

231. F. Restrepo and L. R. Glicksman, "The Effect of Edge Conditions on Natural Convection From a Horizontal Plate," *Int. J. Heat Mass Transfer* (17): 135–142, 1974.

232. G. E. Robertson, J. H. Seinfeld, and L. G. Leal, "Combined Forced and Free Convection Flow Past a Horizontal Plate," *AIChE J.* (19): 998–1008, 1972.

233. H. T. Rossby, "A Study of Benard Convection With and Without Rotation," *J. Fluid Mech.* (36/2): 309–335, 1969.

234. R. Ruiz and E. M. Sparrow, "Natural Convection in V-Shaped and L-Shaped Corners," *Int. J. Heat Mass Transfer* (30/12): 2539–2548, 1987.

235. D. W. Ruth, K. G. T. Hollands, and G. D. Raithby, "On Free Convection Experiments in Inclined Air Layers Heated From Below," *J. Fluid Mech.* (96/3): 461–479, 1980.

236. J. A. Sabbagh, A. Aziz, A. S. El-Ariny, and G. Hamad, "Combined Free and Forced Convection in Inclined Circular Tubes," *J. Heat Transfer* (98): 322–324, 1976.

237. M. Sahraoui, M. Kaviany, and H. Marshall, "Natural Convection From Horizontal Disks and Rings," *J. Heat Transfer* (112): 110–116, 1990.

238. O. A. Saunders, M. Fishenden, and H. D. Mansion, "Some Measurements of Convection by an Optical Method," *Engineering* (May): 483–485, 1935.

239. O. A. Saunders, "The Effect of Pressure on Natural Convection in Air," *Proc. Royal Soc., Ser. A* (157): 278–291, 1936.

240. D. A. Saville and S. W. Churchill, "Laminar Free Convection in Boundary Layers Near Horizontal Cylinders and Vertical Axisymmetric Bodies," *J. Fluid Mech.* (29): 391–399, 1967.

241. J. A. Scanlan, E. H. Bishop, and R. E. Powe, "Natural Convection Heat Transfer Between Concentric Spheres," *Int. J. Heat Mass Transfer* (13): 1857–1872, 1970.

242. E. Schmidt, "Versuche zum Wärmeübergang bei natürlicher Konvektion," *Chem. Eng. Tech.* (28): 175–180, 1956.

243. E. Schmidt and P. L. Silveston, "Natural Convection in Horizontal Liquid Layers," *Chem. Eng. Prog. Symp. Ser.* (55/29): 163–169, 1959.

244. P. J. Schneider, *Conduction Heat Transfer,* p. 242, Addison-Wesley, Reading, MA, 1957.

245. R. Schramm and H. H. Reineke, "Natural Convection in a Horizontal Layer of Two Different Fluids With Internal Heat Sources," *Proc. 6th Int. Heat Transfer Conf.* (2): 299–304, 1978.

246. G. Scats, "Natural Convection Mass Transfer Measurements on Spheres and Horizontal Cylinders by Electrochemical Method," *Int. J. Heat Mass Transfer* (6): 873–879, 1963.

247. R. G. Schwind and G. C. Vliet, "Observations and Interpretations of Natural Convection and Stratification in Vessels," *Proc. Heat Transfer Fluid Mech. Inst.* (51): 52–68, 1964.

248. N. Seki, S. Fukusako, and H. Inaba, "Heat Transfer of Natural Convection in a Rectangular Cavity With Vertical Walls of Different Temperatures," *Bull. JSME* (21/152): 246–253, 1978.

249. J. R. Sellars, M. Tribus, and T. S. Klein, "Heat Transfer to Laminar Flow in a Round Tube or Flat Conduit—The Graetz Problem Extended," *Trans. ASME* (78): 441–448, 1956.

250. E. Shewen, K. G. T. Hollands, and G. D. Raithby, "Heat Transfer By Natural Convection Across a Vertical Air Cavity of Large Aspect Ratio," *J. Heat Transfer* (118): 993–995, 1996.

251. D. L. Siebers, R. F. Moffatt, and R. G. Schwind, "Experimental, Variable Properties Natural Convection From a Large, Vertical, Flat Surface," *J. Heat Transfer* (107): 124–132, 1985.

252. R. Siegel, "Transient Free Convection from a Vertical Flat Plate," *Trans. ASME* (80): 347–359, 1958.

253. D. R. Smart, K. G. T. Hollands, and G. D. Raithby, "Free Convection Heat Transfer Across Rectangular-Celled Diathermous Honeycombs." *J. Heat Transfer* (102/1): 75–80, 1980.

254. N. Sobel, F. Landis, and W. K. Mueller, "Natural Convection Heat Transfer in Short Vertical Channels Including the Effects of Stagger," *Proc. Int. Heat Transfer Conf.,* American Institute of Chemical Engineers, New York, pp. 121–125, 1966.

255. E. M. Sparrow and M. A. Ansari, "A Refutation of King's Rule for Multi-Dimensional External Natural Convection," *Int. J. Heat Mass Transfer* (36/9): 1357–1364, 1983.

256. E. M. Sparrow and P. A. Bahrami, "Experiments on Natural Convection From Vertical Parallel Plates With Either Open or Closed Edges," *J. Heat Transfer* (102): 221–227, 1980.

257. E. M. Sparrow and P. A. Bahrami, "Experiments on Natural Convection Heat Transfer on the Fins of a Finned Horizontal Tube," *Int. J. Heat Mass Transfer* (23): 1555–1560, 1980.

258. E. M. Sparrow and M. Charmichi, "Natural Convection Experiments in an Enclosure Between Eccentric or Concentric Vertical Cylinders of Different Height and Diameter," *Int. J. Heat Mass Transfer* (26/1): 133–143, 1983.

259. E. M. Sparrow and J. L. Gregg, "Laminar Free Convection Heat Transfer From the Outer Surface of a Vertical Circular Cylinder," *Trans. ASME* (78): 1823–1829, 1956.

260. E. M. Sparrow and J. L. Gregg, "Buoyancy Effects in Forced-Convection Flow and Heat Transfer," *J. Appl. Mech.* (26): 133–134, 1959.

261. E. M. Sparrow and C. Prakash, "Enhancement of Natural Convection Heat Transfer by a Staggered Array of Discrete Vertical Plates," *J. Heat Transfer* (102): 215–220, 1980.

262. E. M. Sparrow and A. J. Stretton, "Natural Convection From Variously Oriented Cubes and From Other Bodies of Unity Aspect Ratio," *Int. J. Heat Mass Transfer* (28/4): 741–752, 1985.

263. K. E. Starner and H. N. McManus, "An Experimental Investigation of Free Convection Heat Transfer From Rectangular Fin Arrays," *J. Heat Transfer* (85): 273–278, 1963.

264. W. E. Stewart, "Asymptotic Calculation of Free Convection in Laminar Three-Dimensional Systems," *Int. J. Heat Mass Transfer* (14): 1013–1031, 1971.

265. W. M. Sun, "Effect of Arbitrary Wall Conduction and Radiation on Free Convection in a Cylinder," PhD thesis, University of California School of Engineering and Applied Science, Los Angeles, CA, 1970.

266. W. M. Sun and D. K. Edwards, "Natural Convection in Cells With Finite Conducting Side Walls Heated From Below," *Proc. 4th Int. Heat Transfer Conf.,* paper NC 2.3, Elsevier Publishing, Amsterdam, 1970.

267. A. J. Suo-Anttila and I. Catton, "The Effect of a Stabilizing Temperature Gradient on Heat Transfer from a Molten Fuel Layer With Volumetric Heating," *J. Heat Transfer* (97): 544–548, 1976.

268. A. Suwono, "Laminar Free Convection Boundary-Layer in Three-Dimensional Systems," *Int. J. Heat Mass Transfer* (23): 53–61, 1980.

269. T. Tsubouchi and H. Masuda, "Natural Convection Heat Transfer From Horizontal Cylinders With Circular Fins," *Proc. 6th Int. Heat Transfer Conf.,* Elsevier Publishing, Amsterdam, paper NC 1.10, 1970.

270. D. W. Van De Pol and J. K. Tierney, "Free Convection Nusselt Number for Vertical U-Shaped Channels," *J. Heat Transfer* (95): 542–543, 1973.

271. N. B. Vargaftik, *Tables on the Thermophysical Properties of Liquids and Gases,* 2d ed., pp. 587–596, Hemisphere, New York, 1975.

272. C. M. Vest and M. L. Lawson, "Onset of Convection Near a Suddenly Heated Horizontal Wire," *Int. J. Heat Mass Transfer* (15): 1281–1283, 1972.

273. G. C. Vliet and C. K. Liu, "An Experimental Study of Turbulent Natural Convection Boundary Layers," *J. Heat Transfer* (91): 517–531, 1969.

274. G. C. Vliet and D. C. Ross, "Turbulent Natural Convection on Upward and Downward Facing Inclined Constant Heat Flux Surfaces," *J. Heat Transfer* (97): 549–555, 1975.

275. G. C. Vliet and C. K. Liu, "An Experimental Study of Turbulent Natural Convection Boundary Layers," *J. Heat Transfer* (92): 517–531, 1969.

276. I. C. Walton, "Second-Order Effects in Free Convection," *J. Fluid Mech.* (62): 793–809, 1974.

277. C. Y. Warner and V. S. Arpaci, "An Experimental Investigation of Turbulent Natural Convection in Air at Low Pressure on a Vertical Heated, Flat Plate," *Int. J. Heat Mass Transfer* (11): 397–406, 1968.

278. R. O. Warrington Jr. and R. E. Powe, "The Transfer of Heat by Natural Convection Between Bodies and Their Enclosures," *Int. J. Heat Mass Transfer* (28/2): 319–330, 1985.

279. A. Watson, "Natural Convection of a Heat Generating Fluid in a Closed Vertical Cylinder. An Examination of Theoretical Predictions," *J. Mech. Eng. Sci.* (13): 151–156, 1971.

280. M. E. Weber, P. Astrauskas, and S. Petsalis, "Natural Convection Mass Transfer to Nonspherical Objects at High Rayleigh Number," *Canadian J. Chem. Eng.* (62): 68–72, 1984.

281. N. Weber, R. E. Powe, E. H. Bishop, and J. A. Scanlan, "Heat Transfer by Natural Convection Between Vertically Eccentric Spheres," *J. Heat Transfer* (95): 47–52, 1973.

282. H. P. Weiss, "Verfahren zur Berechnung des Wärmeübergangs und der Konvektionsströmung Über Einer Horizontalen, Beheizten Platte," *Z. Angew. Math. Phys.* (28): 409–417, 1977.

283. J. R. Welling and C. B. Wooldridge, "Free Convection Heat Transfer Coefficients From Rectangular Vertical Fins," *J. Heat Transfer* (87): 439–444, 1965.

284. H. H. Wong and G. D. Raithby, "Improved Finite Difference Methods Based on a Critical Evaluation of the Approximation Errors," *Num. Heat Transfer* (2): 139–163, 1979.

285. R. A. Wooding, "Instability of a Viscous Liquid of Variable Density in a Vertical Hele-Shaw Cell," *J. Fluid Mech.* (7): 501–515, 1960.

286. D. H. Worthington, M. A. Patrick, and A. A. Wragg, "Effect of Shape on Natural Convection Heat and Mass Transfer at Horizontally Oriented Cuboids," *Chem. Eng. Res. Des.* (65): 131–138, 1987.

287. J. L. Wright and J. F. Sullivan, "Simulation and Measurement of Windows with Low Emissivity Coatings Used in Conjunction With Teflon Inner Glazings," *ISES Solar World Congress,* Hamburg, Germany, W. H. Blass and F. Pfisterer eds., vol. 4, pp. 3136–3140, Pergamon Press, Oxford, UK, September 1987.

288. R. S. Wu and K. C. Cheng, "Thermal Instability of Blasius Flow Along Horizontal Plates," *Int. J. Heat Mass Transfer* (19): 907–913, 1976.

289. C. S. Yih, "Thermal Stability of Viscous Flows," *Q. Appl. Math.* (17): 25–42, 1959.

290. W. W. Yousef, J. D. Tarasuk, and W. J. McKeen, "Free Convection Heat Transfer From Upward-Facing Isothermal Horizontal Surfaces," *J. Heat Transfer* (104): 493–500, 1982.

291. M. M. Yovanovich, "New Nusselt and Sherwood Numbers for Arbitrary Isopotential Geometries at Near Zero Peclet and Rayleigh Numbers," *AIAA 22nd Thermophysics Conference,* p. 1643, AIAA, 1987.

292. M. M. Yovanovich, "Dimensionless Shape Factors and Diffusion Lengths of Three-Dimensional Bodies," *ASME/JSME Thermal Engineering Conference,* Lahaina, HI, vol. 1, pp. 103–114, American Society of Mechanical Engineers, March 19–24, 1995.

293. M. M. Yovanovich and K. Jafarpur, "Bounds on Laminar Natural Convection From Isothermal Disks and Finite Plates of Arbitrary Shape From All Orientations and Prandtl Numbers," *ASME HTD* (264): 93–110, 1993.

294. M. M. Yovanovich and K. Jafarpur, "Models of Laminar Natural Convection From Vertical and Horizontal Isothermal Cuboids for All Prandtl Numbers and All Rayleigh Numbers Below 10^{11}," *ASME HTD* (264): 111–126, 1993.

295. A. A. Wragg and R. P. Loomba, "Free Convection Flow Patterns at Horizontal Surfaces With Ionic Mass Transfer," *Int. J. Heat Mass Transfer* (13): 439–442, 1970.

296. Y. L. Chow and M. M. Yovanovich, "The Shape Factor of the Capacitance of a Conductor," *J. Applied Physics* (53/12): 8470–8475, 1982.

297. J. R. Lloyd and W. R. Moran, "Natural Convection Adjacent to Horizontal Surface of Various Platforms," *J. Heat Transfer* (96): 443–447, 1974.

298. T. Fusegi, J. M. Hyun, K. Kuwahara, and B. Farouk, "A Numerical Study of Three-Dimensional Natural Convection in a Differentially Heated Cubical Enclosure," *Int. J. Heat Mass Transfer* (34): 1543–1557, 1991.

299. V. Lienhard and I. Catton, "Heat Transfer Across a Two-Fluid Layer Region," *J. Heat Transfer* (108): 198–205, 1986.

300. G. Schütz, "Natural Convection Mass Transfer Measurements on Spheres and Horizontal Cylinders by Electrochemical Method," *Int. J. Heat Mass Transfer* (6): 873–879, 1963.

301. C. J. Kobus and G. L. Wedekind, "An Experimental Investigation Into Forced, Natural and Combined Forced and Natural Convective Heat Transfer From Stationary Circular Disks," *Int. J. Heat Mass Transfer* (38/18): 3329–3339, 1995.

302. B. Gebhart, Y. Jaluria, R. Mahajan, and R. Sammakia, *Bouyancy-Induced Flows and Transport,* Hemisphere, 1998.

CHAPTER 5

FORCED CONVECTION, INTERNAL FLOW IN DUCTS

M. A. Ebadian and Z. F. Dong
Florida International University

INTRODUCTION

Scope of the Chapter

This chapter deals with internal flow and heat transfer in ducts such as circular pipes, rectangular pipes, and other pipes with irregular cross sections. The scope of the chapter is restricted to the study of the steady, incompressible flow of newtonian fluids. The effects of natural convection, phase change, mass transfer, and chemical reactions have been ignored. This chapter is organized according to duct geometry. Hydrodynamics and heat transfer characteristics will be presented for each duct in terms of mathematical expressions, tables, and graphs in the corresponding sections. To the authors' knowledge, the most accurate and updated correlations and data for the friction factor and Nusselt number are provided for use in practical calculations.

Characteristics of Laminar Flow in Ducts

As a result of the development of the hydrodynamic and thermal boundary layers, four types of laminar flows occur in ducts, namely, fully developed, hydrodynamically developing, thermally developing (hydrodynamically developed and thermally developing), and simultaneously developing (hydrodynamically and thermally developing). In this chapter, the term *fully developed flow* refers to fluid flow in which both the velocity profile and temperature profile are fully developed (i.e., hydrodynamically and thermally developed flow). In such cases, the velocity profile and dimensionless temperature profile are constant along the flow direction. The friction factor and Nusselt number are also constant.

Hydrodynamically developing flow is isothermal fluid flow in which the velocity profile varies in the flow direction. Fluid flow from the entrance of the duct to the location at which the fully developed velocity profile forms is referred to as hydrodynamically developing flow. The distance over which the velocity distribution changes and the hydrodynamic boundary layer develops is referred to as the *hydrodynamic entrance length*. The friction factor in the hydrodynamic entrance is a function of the axial location.

The term *thermally developing flow* refers to fluid flow in which the temperature profile is developing and the velocity profile has already developed (i.e., the velocity distribution is invariant with axial length, and the nondimensional temperature profile changes with axial length). In other words, the hydrodynamic boundary layer is already developed while the thermal boundary is developing. This kind of flow is alternately termed *thermal entrance flow.* The distance over which the nondimensional temperature distribution changes or the thermal boundary layer develops is termed *thermal entrance length,* corresponding to hydrodynamic entrance length in hydrodynamically developing flow. The Nusselt number for thermally developing flow changes with axial location.

Simultaneously developing flow is fluid flow in which both the velocity and the temperature profiles are developing. The hydrodynamic and thermal boundary layers are developing in the entrance region of the duct. Both the friction factor and Nusselt number vary in the flow direction. Detailed descriptions of fully developed, hydrodynamically developing, thermally developing, and simultaneously developing flows can be found in Shah and London [1] and Shah and Bhatti [2].

Characteristics of Turbulent Flow in Ducts

In turbulent flow, the fluid particles do not travel in a well-ordered pattern. These particles possess velocities with macroscopic fluctuations at any point in the flow field. Even in steady turbulent flow, the local velocity components transverse to the main flow direction change in magnitude with respect to time. Instantaneous velocity consists of time-average velocity and its fluctuating component. When heat transfer is involved in turbulent flow, the instantaneous temperature is composed of the time-average temperature and its fluctuating components.

Similar to laminar flow in ducts, turbulent flow can be divided into four types—fully developed, hydrodynamically developing, thermally developing, and simultaneously developing. Nevertheless, the hydrodynamic and thermal entrance lengths in turbulent duct flow are characteristically much shorter than their corresponding lengths in laminar duct flow. Consequently, the results of fully developed turbulent flow and heat transfer are frequently used in design calculations without reference to the hydrodynamic and thermal entrance regions. However, caution must be taken in using the fully developed results for low Prandtl number fluids such as liquid metals inasmuch as entrance effects are very important for these fluids. Table 5.1 illustrates the relationships between the types of flow, boundary layers, velocity and temperature distributions, the friction factor, and the Nusselt number.

TABLE 5.1 Terminology for Flow Types

Flow type	Hydrodynamic boundary layer	Velocity distribution in the flow direction	Friction factor	Thermal boundary layer	Dimensionless temperature distribution in the flow	Nusselt number
Fully developed flow	Developed	Invariant	Constant	Developed	Invariant	Constant
Hydrodynamically developing flow	Developing	Variant	Variant	—	—	—
Thermally developing flow	Developed	Invariant	Constant	Developing	Variant	Variant
Simultaneously developing flow	Developing	Variant	Variant	Developing	Variant	Variant

Hydraulic Diameter

For fluid flow and heat transfer inside a duct, various dimensionless parameters are used. In these parameters, a characteristic length of the duct is commonly involved. The hydraulic diameter D_h of the duct serves this purpose. It is defined as follows:

$$D_h = 4A_c/P \tag{5.1}$$

where A_c is the flow cross-sectional area and P is the wetted perimeter of the duct.

For a circular duct, the hydraulic diameter is equal to its physical diameter. For a noncircular duct, it is convenient to use the hydraulic diameter to substitute for the characteristic physical dimension. However, for ducts with very sharp corners (e.g., triangular and cusped ducts), the use of the hydraulic diameter results in unacceptably large errors in the turbulent flow friction and heat transfer coefficients determined from the circular duct correlation. Other dimensions have been proposed as substitutes for hydraulic diameter. These equivalent diameters, provided for specific ducts only, will be presented elsewhere in this chapter.

Fluid Flow Parameters

One of the flow parameters most commonly used in practice is the friction factor, also referred to as the *Fanning friction factor f,* which is defined as follows:

$$f = \frac{\tau_w}{\rho u_m^2/2} \tag{5.2}$$

where τ_w is wall shear stress, u_m is the mean velocity, and ρ is the density of fluid.

The Reynolds number Re, the parameter that represents the status of the flow, is defined as

$$\mathrm{Re} = \frac{u_m D_h}{v} \tag{5.3}$$

The hydrodynamic entrance length L_{hy} is defined as the axial distance required to attain 99 percent of the ultimate fully developed maximum velocity when the entering flow is uniform. The dimensionless hydrodynamic entrance length is expressed as $L_{hy}^+ = L_{hy}/D_h\,\mathrm{Re}$.

In this hydrodynamic entrance region, the apparent friction factor f_{app} is employed to incorporate the combined effects of wall shear and the change in momentum flow rate due to the developing velocity profile. Based on the total axial pressure drop from the duct inlet ($x = 0$) to the point of interest, the apparent friction factor is defined as follows:

$$\Delta p^* = \frac{p_0 - p}{\rho u_m^2/2} = f_{\mathrm{app}} \frac{2x}{D_h} \tag{5.4}$$

The incremental pressure drop number $K(x)$ in the hydrodynamic entrance region is expressed as

$$K(x) = (f_{\mathrm{app}} - f_{fd}) \frac{2x}{D_h} \tag{5.5}$$

where f_{fd} is the friction factor for fully developed laminar flow. $K(x)$ is sometimes referred to as the incremental pressure defect. It increases from a value of zero at $x = 0$ to a constant value $K(\infty)$ in the hydrodynamically developed region at $x > L_{hy}$.

The relationship between the friction factor, axial pressure drop, and incremental pressure drop number is the following:

$$\Delta p^* = (f_{app}\,\text{Re})(4x^+) = K(x) + (f\,\text{Re})(4x^+) \tag{5.6}$$

where x^+ is the dimensionless axial length, defined as $x/(D_h\,\text{Re})$.

Heat Transfer Parameters

The most useful parameters for heat transfer are the fluid bulk mean temperature and the heat transfer coefficient. The fluid bulk mean temperature T_m, also known as the mixing cup or flow average temperature, is defined as

$$T_m = \frac{1}{A_c u_m} \int_{A_c} uT\,dA_c \tag{5.7}$$

The circumferentially averaged but axial local heat transfer coefficient h_x is defined by

$$q_x'' = h_x(T_{w,m} - T_m) \tag{5.8}$$

where $T_{w,m}$ is the wall mean temperature and T_m is the fluid bulk mean temperature given by Eq. 5.7. In Eq. 5.8, the heat flux q_x'' and the temperature difference $T_{w,m} - T_m$ are vector quantities. The direction in which the heat is transferred is from the wall to the fluid, and the temperature consistently drops from the wall to the fluid.

The average circumferential and axial heat transfer coefficient can be obtained by means of the following expression:

$$h_m = \frac{1}{x}\int_0^x h_x\,dx \tag{5.9}$$

Correspondingly, the circumferentially averaged but axially local Nusselt number is defined as

$$\text{Nu}_x = \frac{h_x D_h}{k} \tag{5.10}$$

The mean Nusselt number based on h_m in the thermal entrance region is defined as

$$\text{Nu}_m = \frac{1}{x}\int_0^x \text{Nu}_x\,dx = \frac{h_m D_h}{k} \tag{5.11}$$

Two dimensionless axial distances will be used in this chapter. The term x^+, which denotes hydrodynamically developing flow, is defined as

$$x^+ = \frac{x/D_h}{\text{Re}} \tag{5.12}$$

The term x^*, denoting thermally and simultaneously developing flows, is expressed as:

$$x^* = \frac{x/D_h}{\text{Pe}} = \frac{x/D_h}{\text{Re Pr}} \tag{5.13}$$

where Pe is the Péclet number, defined as Re Pr. The relationship between x^+ and x^* is simply given by

$$x^* = x^+/\text{Pr} \tag{5.14}$$

Corresponding to the hydrodynamic entrance length, the thermal entrance length L_{th} is defined as the axial distance needed to achieve a value of the local Nusselt number Nu_x, which is 1.05 times the fully developed Nusselt number value. The dimensionless thermal entrance length is expressed as $L_{th}^+ = L_{th}/(D_h\,Pe)$.

Thermal Boundary Conditions

The following thermal boundary conditions are the most important and frequently encountered in practical use:

1. Uniform wall temperature, denoted by ⓣ. This boundary condition is present when the duct has a constant wall temperature in both the circumferential and axial directions. Uniform wall temperature has frequent application in condensers, evaporators, and automotive radiators with high flow rates.

2. Convective boundary condition, denoted by ⓣ₃. The convection between the duct wall and the environment is taken into consideration in the ⓣ₃ boundary condition. The duct wall temperature is considered to be uniform in the axial direction. The practical applications are the same as those in the ⓣ boundary condition, except for the case of finite thermal resistance in the wall.

3. Radiative boundary condition, denoted by ⓣ₄. This boundary condition involves radiative heat transfer from the duct to the environment. The wall heat flux is proportional to the fourth power of the absolute wall temperature, and the environment temperature is uniform in the axial direction. This boundary condition can be found in high-temperature systems such as space radiators, liquid-metal exchangers, and heat exchangers involving heat-radiating gases.

4. Uniform wall heat flux axially, but uniform wall temperature circumferentially, denoted by Ⓗ₁. This boundary condition is found in electric resistance heating, nuclear heating, and counterflow heat exchangers, in which two fluids have nearly the same fluid capacity rates and the wall is highly conductive.

5. Uniform wall heat flux axially and circumferentially, denoted by Ⓗ₂. This boundary condition has applications that are similar to those listed for the Ⓗ₁ boundary condition except that the thermal conductivity of the wall material is low and the wall thickness is uniform.

6. Conductive boundary condition, denoted by Ⓗ₄. This boundary condition refers to the axial uniform wall heat flux and finite heat condition along the wall circumference. The same applications can be found as those for the Ⓗ₁ boundary condition except for the existence of heat conduction in the circumferential direction.

7. Exponential wall heat flux, denoted by Ⓗ₅. This boundary condition represents a duct with circumferentially constant wall temperature and exponentially varying wall heat flux along the axial direction. Exponential wall heat flux can be seen in parallel and counterflow heat exchangers with an appropriate value of m.

The preceding boundary conditions are applicable to both singly connected and doubly connected ducts. Detailed descriptions of the various boundary conditions are available in Shah and London [1] and Shah and Bhatti [2].

CIRCULAR DUCTS

In industry, circular ducts are widely used in various applications. Fluid and heat transfer inside circular ducts have been analyzed in great detail. A discussion of laminar and turbulent flows and heat transfer in circular ducts is presented in the following sections.

Laminar Flow

For a circular duct with a diameter of $2a$, the characteristics of laminar flow and heat transfer for four kinds of flows, namely, fully developed, hydrodynamically developing, thermally developing, and simultaneously developing, are outlined in the following sections.

Fully Developed Flow

Velocity Profile and Friction Factor.   The velocity profile of fully developed laminar flow of a constant-property fluid in a circular duct with an origin at the duct axis is given by the Hagen-Poiseuille parabolic profile, as follows:

$$\frac{u}{u_m} = 2\left[1 - \left(\frac{r}{a}\right)^2\right] \tag{5.15}$$

where u_m can be obtained by the following equation:

$$u_m = -\frac{1}{8\mu}\left(\frac{dp}{dx}\right)a^2 \tag{5.16}$$

The product of the friction factor and the Reynolds number for fully developed flow in a circular duct is found to be constant. This is obtained from Eq. 5.15 as:

$$f\,\mathrm{Re} = 16 \tag{5.17}$$

Heat Transfer on Walls With Uniform Temperature.   For this boundary condition, denoted as Ⓣ, temperature distribution in a circular duct for fully developed laminar flow in the absence of flow work, thermal energy sources, and fluid axial conduction has been solved by Bhatti [3] and presented by Shah and Bhatti [2], as follows:

$$\frac{T_w - T}{T_w - T_m} = \sum_{n=0}^{\infty} C_{2n}\left(\frac{r}{a}\right)^{2n} \tag{5.18}$$

where the coefficients C_{2n} are given by

$$C_0 = 1 \qquad C_2 = -\frac{\lambda_0^2}{2^2} = -1.828397$$

$$C_{2n} = \frac{\lambda_0^2}{(2n)^2}\,(C_{2n-4} - C_{2n-2}) \qquad \lambda_0 = 2.7043644199 \tag{5.19}$$

The bulk mean temperature of the fluid can be obtained by:

$$\frac{T_w - T_m}{T_w - T_e} = 0.819048\,\exp(-2\lambda_0^2 x^+) \tag{5.20}$$

It should be noted that Eq. 5.20 is valid for $x^* > 0.0335$ [3].

The Nusselt number corresponding to Eq. 5.18 is as follows:

$$\mathrm{Nu_T} = \frac{\lambda_0}{2} = 3.657 \tag{5.21}$$

The Péclet number is introduced to consider the effect of fluid axial conduction. It has been found that the axial conduction of fluid can be ignored when the Péclet number is greater than 10. For a small Péclet number, the following formulas by Michelsen and Villadsen [4] are recommended:

$$\mathrm{Nu_T} = \begin{cases} 4.1807(1 - 0.0439\mathrm{Pe} + \cdots) & \mathrm{Pe} < 1.5 \\ 3.6568(1 + 1.227/\mathrm{Pe}^2 + \cdots) & \mathrm{Pe} > 5 \end{cases} \tag{5.22}$$

The Brinkmann number $\mathrm{Br} = (\mu u_m^2)/[k(T_{w,m} - T_e)]$ is introduced to account for the influence of viscous dissipation, such as heating or cooling of the fluid due to internal friction in high-velocity flow, highly viscous fluid, or in cases in which viscous dissipation cannot be ignored. When viscous dissipation is considered, the asymptotic Nusselt number in a very long pipe, found by Ou and Cheng [5], is 9.6 and independent of the Brinkmann number.

Heat Transfer on Walls With Uniform Heat Flux. For circular ducts with symmetrical heating, the same heat transfer results for fully developed flow and developing flow are obtained for boundary conditions Ⓗ through Ⓗ④. Therefore, the uniform wall heat flux boundary conditions are simply designated as the Ⓗ boundary condition. Shah and Bhatti [2] have derived the temperature distribution and Nusselt number by recasting the results reported by Tyagi [6] for heat transfer in circular ducts. These follow:

$$\frac{T_w - T}{T_w - T_m} = 6\left[1 - \left(\frac{r}{a}\right)^2\right]\left[\frac{(12 + \gamma) - (4 + \gamma)(r/a)^2}{44 + 3\gamma}\right] \tag{5.23}$$

$$\mathrm{Nu_H} = \frac{q_w'' D_h}{k(T_w - T_m)} = \frac{192}{44 + 3\gamma} \qquad \gamma = S^* + 64\,\mathrm{Br}' \tag{5.24}$$

where S^* is the dimensionless thermal energy source number ($S^* = SD_h/q_w''$) and Br' is the dimensionless Brinkmann number, defined as $\mathrm{Br}' = \mu u_m^2/q_w'' D_h$, for the uniform wall heat flux boundary condition. For the case of negligible viscous dissipation and no thermal energy sources ($\gamma = 0$), the Nusselt number can be obtained from Eq. 5.24:

$$\mathrm{Nu_H} = 4.364 \tag{5.25}$$

Eq. 5.24 can also be applied in the case of finite axial fluid conduction.

Heat Transfer on Convection Duct Walls. For this boundary condition, denoted as Ⓣ③, the wall temperature is considered to be constant in the axial direction, and the duct has convection with the environment. An external heat transfer coefficient is incorporated to represent this case. The dimensionless Biot number, defined as $\mathrm{Bi} = h_e D_h/k_w$, reflects the effect of the wall thermal resistance, induced by external convection.

For cases in which the external heat transfer coefficient h_e is constant, Hickman [7] developed the following formula to calculate the Nusselt number:

$$\mathrm{Nu_{T3}} = \frac{4.3636 + \mathrm{Bi}}{1 + 0.2682\mathrm{Bi}} \tag{5.26}$$

From the preceding equation, the uniform wall heat flux and uniform wall temperature can be obtained when $\mathrm{Bi} = 0$ and $\mathrm{Bi} = \infty$, respectively.

Heat Transfer on Radiative Duct Walls. Heat transfer on radiative duct walls is of the Ⓣ④ boundary condition type. Kadaner et al. [8] obtained the following equation for the fully developed Nusselt number under the Ⓣ④ boundary condition:

$$\mathrm{Nu_{T4}} = \frac{8.728 + 3.66\mathrm{Sk}\,(T_a/T_e)^3}{2 + \mathrm{Sk}\,(T_a/T_e)^3} \tag{5.27}$$

where Sk is the dimensionless Stark number, defined as $\mathrm{Sk} = \varepsilon_w \sigma T_e^3 D_h/k$, and T_a and T_e are the absolute temperatures of the external environment and the internal fluid at the location of the impingement of the radiation flux, respectively. In Eq. 5.27, when Sk = 0 and Sk = ∞, Nu_{T4} reduces to Nu_H and Nu_T.

Heat Transfer on the Walls With Exponentially Varying Heat Flux.   Exponentially varying wall heat flux is delineated by the Ⓗ⑤ boundary condition and represented by $q''_w = q''_0 \exp(mx^*)$, where the exponent m can have both positive and negative values corresponding to the exponential growth or decay of the wall heat flux.

Shah and London [1] have obtained the Nusselt number for $-51.36 \leq m \leq 100$ as the following correlation with a maximum error of 3 percent:

$$Nu_{H5} = 4.3573 + 0.0424m - 2.8368 \times 10^{-8}m^2 + 3.6250 \times 10^{-6}m^3$$
$$- 7.6497 \times 10^{-8}m^4 + 9.1222 \times 10^{-4}m^5 - 3.8446 \times 10^{-12}m^6 \quad (5.28)$$

When $m = 0$ and $m = -14.627$, the Ⓗ⑤ boundary condition corresponds to uniform wall heat flux and uniform wall temperature, respectively. Recently, Piva [9] obtained the fully developed Nusselt number for the Ⓗ⑤ boundary condition based on confluent hypergeometric functions. The Nusselt number for very large m can be found in Piva [9].

Hydrodynamically Developing Flow.   Solutions for three different flow conditions are provided as follows.

Solutions for Very Large Reynolds Number Flows.   For very large Reynolds number flow, boundary layer theory simplifications are involved in the solutions. The numerical solution found by Hornbeck [10] is the most accurate of the various solutions reviewed by Shah and London [1]. The dimensionless axial velocity and pressure drop obtained by Hornbeck [10] are presented in Table 5.2.

Based on these results, the hydrodynamic entrance length was found to be:

$$L_{hy}^+ = 0.0565 \qquad K(\infty) = 1.28 \qquad (5.29)$$

It should be noted, however, that the solutions for very large Reynolds numbers are inaccurate in the duct inlet. For practical computations, the following correlation proposed by Shah

TABLE 5.2   Axial Velocity and Pressure Drop in the Hydrodynamic Entrance Region of a Circular Duct [10]

x^+	Axial velocity u/u_m											Dimensionless pressure drop Δp^*
	$r/a = 0$	0.1	0.2	0.3	0.4	0.5	0.6	0.7	0.8	0.9	1.0	
0.00000	1.0000	1.0000	1.0000	1.0000	1.0000	1.0000	1.0000	1.0000	1.0000	1.0000	0	0.0000
0.00050	1.1503	1.1503	1.1503	1.1503	1.1503	1.1503	1.1502	1.1485	1.1293	0.8434	0	0.3220
0.00125	1.2269	1.2269	1.2269	1.2269	1.2268	1.2264	1.2230	1.2016	1.0950	0.6893	0	0.5034
0.00250	1.3126	1.3126	1.3125	1.3124	1.3115	1.3068	1.2867	1.2144	1.0098	0.5908	0	0.7204
0.00375	1.3782	1.3781	1.3779	1.3770	1.3733	1.3596	1.3160	1.2000	0.9511	0.5417	0	0.8960
0.00500	1.4332	1.4331	1.4234	1.4299	1.4214	1.3959	1.3292	1.1814	0.9107	0.5102	0	1.0506
0.00750	1.5239	1.5232	1.5204	1.5120	1.4902	1.4395	1.3349	1.1476	0.8585	0.4720	0	1.3212
0.01000	1.5977	1.5960	1.5893	1.5727	1.5358	1.4623	1.3308	1.1218	0.8261	0.4496	0	1.5610
0.01250	1.6595	1.6562	1.6448	1.6188	1.5675	1.4751	1.3245	1.1023	0.8040	0.4346	0	1.7822
0.01750	1.7555	1.7488	1.7269	1.6831	1.6073	1.4874	1.3125	1.0757	0.7756	0.4159	0	2.1900
0.02250	1.8240	1.8142	1.7829	1.7244	1.6306	1.4927	1.3034	1.0588	0.7584	0.4047	0	2.5692
0.03000	1.8920	1.8785	1.8366	1.7626	1.6509	1.4962	1.2943	1.0433	0.7429	0.3947	0	3.1064
0.04000	1.9431	1.9266	1.8763	1.7901	1.6650	1.4981	1.2875	1.0321	0.7319	0.3877	0	3.7894
0.05000	1.9698	1.9517	1.8969	1.8042	1.6721	1.4990	1.2840	1.0264	0.7263	0.3840	0	4.4520
0.06250	1.9863	1.9872	1.9095	1.8128	1.6764	1.4996	1.2818	1.0229	0.7229	0.3718	0	5.2688
∞	2.0000	1.9800	1.9200	1.8200	1.6800	1.5000	1.2800	1.0200	0.7200	0.3800	0	—

and London [1] can be used to calculate the dimensionless axial pressure drop in the hydro-dynamic entrance region:

$$\Delta p^* = 13.74(x^+)^{1/2} + \frac{1.25 + 64x^+ - 13.74(x^+)^{1/2}}{1 + 0.00021(x^+)^{-2}} \tag{5.30}$$

Solutions for the Flow with Re < 400. It has been found that the effects of axial momentum diffusion and radial pressure variation are significant only in the duct inlet of $x^+ < 0.005$. Chen [11] obtained the dimensionless hydrodynamic entrance length L_{hy}^+ and the fully developed incremental pressure drop number $K(\infty)$, which are given by

$$L_{hy}^+ = 0.056 + \frac{0.60}{\text{Re} \ (1 + 0.035\text{Re})} \tag{5.31}$$

$$K(\infty) = 1.20 + \frac{38}{\text{Re}} \tag{5.32}$$

For $x^+ > 0.005$, the solutions proposed by Hornbeck [10] are quite satisfactory.

Solutions for Small Reynolds Number (Re → 0) Flows. For small Reynolds number flows, such as creeping flow, in which viscous forces completely overwhelm the inertia forces, the hydrodynamic entrance length L_{hy}^+ has been found to approach the value of 0.60 as Re → 0 with the uniform flow at the inlet of the circular duct [1].

The following expression, proposed by Weissberg [12] and verified experimentally by Linehan and Hirsch [13], can be used to compute pressure drop in the entrance region of a circular duct with a very small Reynolds number flow:

$$\Delta p^* = 64\left(x^+ + \frac{3\pi}{16\text{Re}}\right) \tag{5.33}$$

Thermally Developing Flow

Heat Transfer on Walls With Uniform Wall Temperature. Heat transfer in a duct with uniform wall temperature is known as the Graetz or Graetz-Nusselt problem. In this case, a fluid with a fully developed velocity distribution (Eq. 5.15) and a uniform temperature flows into the entrance, and the fluid axial conduction, viscous dissipation flow work, and energy resources are negligible. Graetz [14] and Nusselt [15] solved this problem as follows:

$$\theta = \frac{T_w - T}{T_w - T_e} = \sum_{n=0}^{\infty} C_n R_n\left(\frac{r}{a}\right) \exp(-2\lambda_n^2 x^*) \tag{5.34}$$

$$\theta_m = \frac{T_w - T_m}{T_w - T_e} = 8 \sum_{n=0}^{\infty} \frac{G_n}{\lambda_n^2} \exp(-2\lambda_n^2 x^*) \tag{5.35}$$

$$\text{Nu}_{x,\text{T}} = \frac{\sum\limits_{n=0}^{\infty} G_n \exp(-2\lambda_n^2 x^*)}{2 \sum\limits_{n=0}^{\infty} (G_n/\lambda_n^2) \exp(-2\lambda_n^2 x^*)} \tag{5.36}$$

$$\text{Nu}_{m,\text{T}} = -\frac{\ln \theta_m}{4x^*} \tag{5.37}$$

where λ_n, $R_n(r/a)$, and C_n denote the eigenvalues, eigenfunctions, and constants, respectively, and $G_n = -(C_n/2)R_n'(1)$, where $R_n'(1)$ is the derivative of $R_n(r/a)$ evaluated at $r/a = 1$.

The eigenvalues and constants in Eqs. 5.34–5.36 are listed in Table 5.3, while those for $R_n(r/a)$ are given in Table 5.4. When n is greater than 10, the following correlations are used to calculate the λ_n and G_n [16]:

$$\lambda_n = \lambda + S_1\lambda^{-4/3} + S_2\lambda^{-8/3} + S_3\lambda^{-10/3} + S_4\lambda^{-11/3} + o(\lambda^{-14/3}) \tag{5.38}$$

$$G_n = \frac{C}{\lambda^{1/3}}\left[1 + \frac{L_1}{\lambda^{4/3}} + \frac{L_2}{\lambda^{6/3}} + \frac{L_3}{\lambda^{7/3}} + \frac{L_4}{\lambda^{10/3}} + \frac{L_5}{\lambda^{11/3}} + o(\lambda^{-4})\right] \tag{5.39}$$

where
$$\lambda = 4n + 8/3 \qquad n = 0, 1, 2, \ldots \tag{5.40}$$

$$S_1 = 0.159152288 \qquad S_2 = 0.0114856354 \qquad S_3 = -0.224731440$$
$$S_4 = -0.033772601 \qquad C = 1.012787288 \tag{5.41}$$

and
$$L_1 = 0.144335160 \qquad L_2 = 0.115555556 \qquad L_3 = -0.21220305$$
$$L_4 = -0.187130142 \qquad L_5 = 0.0918850832 \tag{5.42}$$

TABLE 5.3 Eigenvalues and Constants for Eqs. 5.34–5.36 [17]

n	λ_n	C_n	G_n
0	2.70436	1.47644	0.74877
1	6.67903	-0.80612	0.54383
2	10.67338	0.58876	0.46286
3	14.67108	-0.47585	0.41542
4	18.66987	0.40502	0.38292
5	22.66914	-0.35576	0.35869
6	26.66866	0.31917	0.33962
7	30.66832	-0.29073	0.32406
8	34.66807	0.26789	0.31101
9	38.66788	-0.24906	0.29984
10	42.66773	0.23323	0.29012

TABLE 5.4 Eigenfunctions, $R_n(r/a)$, for Eq. 5.34 [17,18]

n	$r/a = 0$	0.1	0.2	0.3	0.4	0.5	0.6	0.7	0.8	0.9	1.0
0	1.0	0.98184	0.92889	0.84547	0.73809	0.61460	0.48310	0.35101	0.22426	0.10674	0.0
1	1.0	0.89181	0.60470	0.23386	-0.10959	-0.34214	-0.43218	-0.39763	-0.28449	-0.14113	0.0
2	1.0	0.73545	0.15247	-0.31521	-0.39208	0.14234	0.16968	0.33149	0.30272	0.16262	0.0
3	1.0	0.53108	-0.23303	-0.35914	0.06793	0.31507	0.11417	-0.19604	-0.29224	-0.17762	0.0
4	1.0	0.30229	-0.40260	0.00054	0.29907	-0.07973	-0.25523	0.03610	0.25918	-0.18817	0.0
5	1.0	0.07488	-0.32121	0.28982	-0.04766	-0.20532	0.19750	0.10372	-0.20893	-0.19522	0.0
6	1.0	0.12642	-0.07613	0.20122	-0.25168	0.19395	-0.01391	-0.18883	0.14716	0.19927	0.0
7	1.0	-0.28107	0.17716	-0.10751	0.03452	0.05514	-0.15368	0.20290	-0.07985	-0.20068	0.0
8	1.0	-0.37523	0.29974	-0.25305	0.22174	-0.20502	0.19303	-0.15099	0.01298	0.19967	0.0
9	1.0	-0.40326	0.23915	-0.08558	-0.02486	0.08126	-0.09176	0.05652	0.04787	-0.19645	0.0
10	1.0	-0.36817	0.04829	0.16645	-0.20058	0.13289	-0.06474	0.04681	-0.09797	0.19120	0.0
11	1.0	-0.28088	-0.15310	0.19847	0.01714	-0.15931	0.16099	-0.12577	0.13375	-0.18409	0.0
12	1.0	-0.15836	-0.24999	-0.00845	0.18456	-0.01927	-0.13393	0.15742	-0.15311	0.17527	0.0
13	1.0	-0.02118	-0.19545	-0.18955	-0.01074	0.15967	0.01258	-0.13539	0.15549	-0.16491	0.0
14	1.0	0.10953	-0.03182	-0.13083	-0.17183	-0.08560	0.10927	0.07069	-0.14189	0.15319	0.0

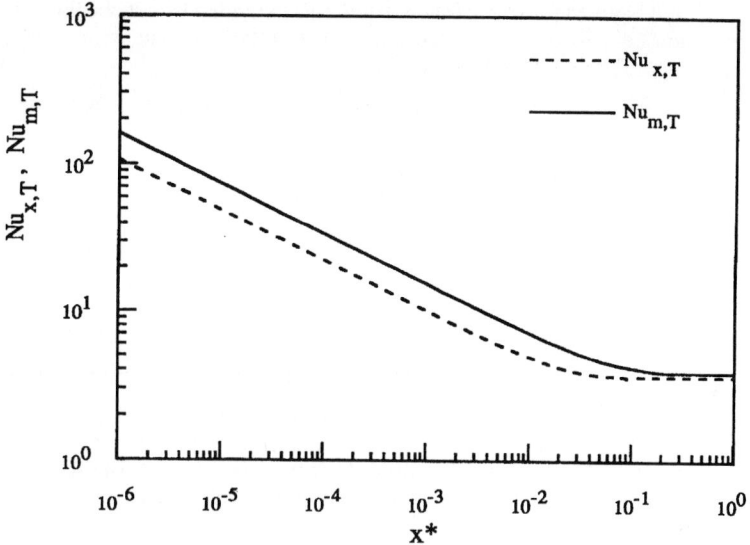

FIGURE 5.1   Local and mean Nusselt numbers $\mathrm{Nu}_{x,\mathrm{T}}$ and $\mathrm{Nu}_{m,\mathrm{T}}$ for thermal developing flow in a circular duct.

The local Nusselt number and mean Nusselt number computed from Eqs. 5.36 and 5.37 are shown in Fig. 5.1. The data corresponding to this figure can be found in Shah and London [1].

The thermal entrance length for thermally developing flow in circular ducts can be obtained using the following expression:

$$L_{th,\mathrm{T}}^{*} = 0.0335 \tag{5.43}$$

It should be noted that the solutions derived from Eqs. 5.34 through 5.36 are inaccurate when $x^{*} < 10^{-4}$. Therefore, Lévêque's [19] asymptotic solution is introduced. The local and mean Nusselt numbers can be computed from the following formulas [1]:

$$\mathrm{Nu}_{x,\mathrm{T}} = \begin{cases} 1.077x^{*-1/3} - 0.7 & \text{for } x^{*} \leq 0.01 \\ 3.657 + 6.874(10^{3}x^{*})^{-0.488}e^{-57.2x^{*}} & \text{for } x^{*} > 0.01 \end{cases} \tag{5.44}$$

$$\mathrm{Nu}_{m,\mathrm{T}} = \begin{cases} 1.615x^{*-1/3} - 0.7 & \text{for } x^{*} \leq 0.005 \\ 1.615x^{*-1/3} - 0.2 & \text{for } 0.005 < x^{*} < 0.03 \\ 3.657 + (0.0499/x^{*}) & \text{for } x^{*} \geq 0.03 \end{cases} \tag{5.45}$$

Shah and Bhatti [2] and Hausen [20] have obtained the following correlations for the local and mean Nusselt numbers for the entire x^{+}.

$$\mathrm{Nu}_{x,\mathrm{T}} = 3.66 + \frac{0.0018}{x^{*1/3}(0.04 + x^{*2/3})^{2}} \tag{5.46}$$

$$\mathrm{Nu}_{m,\mathrm{T}} = 3.66 + \frac{0.0668}{x^{*1/3}(0.04 + x^{*2/3})} \tag{5.47}$$

The effects of fluid axial conduction on the Graetz solution have been reviewed extensively by Shah and London [1]. Furthermore, Laohakul et al. [21], Ebadian and Zhang [22], and Nguyen [23] have investigated this extended Graetz problem with axial heat conduction. For Pe < 50, $Nu_{m,T}$ can be calculated with the following expression [4]:

$$Nu_{m,T} = \begin{cases} 3.6508\left(1 + \dfrac{1.227}{Pe^2} + \cdots\right) & \text{for Pe} > 5 \\ 4.1807(1 - 0.0439Pe + \cdots) & \text{for Pe} < 1.5 \end{cases} \quad (5.48)$$

It has been confirmed that the effect of fluid axial conduction can be neglected for Pe > 50 [24]. However, the thermal entrance length $L_{th,T}^+$ varies with the Péclet number. Nguyen [23] expresses $L_{th,T}^+$ as follows:

$$L_{th,T}^* = \begin{cases} -0.003079 + 0.4663/Pe & \text{for } 1 \le Pe \le 5 \\ 0.02020 + 0.3550/Pe & \text{for } 5 \le Pe \le 20 \\ 0.03258 + 0.1295/Pe & \text{for } 20 \le Pe \le 1000 \end{cases} \quad (5.49)$$

Other extended Graetz problems in which the effect of viscous dissipation, inlet velocity, and temperature profiles are considered are reviewed in detail by Shah and London [1].

Heat Transfer on Walls With Uniform Heat Flux.   The temperature profile and the local and mean Nusselt numbers for thermally developing flow in a circular duct with uniform wall heat flux are provided by Siegel et al. [25] as follows:

$$\Theta = \frac{T - T_e}{q_w''(D_h/k)} = 4x^* + \frac{1}{2}\left(\frac{r}{a}\right)^2 - \frac{1}{8}\left(\frac{r}{a}\right)^4 - \frac{7}{48} + \frac{1}{2}\sum_{n=1}^{\infty} C_n R_n\left(\frac{r}{a}\right)\exp(-2\beta_n^2 x^*) \quad (5.50)$$

$$\Theta_m = \frac{T_m - T_e}{q_w''(D_h/k)} = 4x^* \quad (5.51)$$

$$Nu_{x,H} = \left(\frac{11}{48} + \frac{1}{2}\sum_{n=1}^{\infty} C_n R_n(1)\exp(-2\beta_n^2 x^*)\right)^{-1} \quad (5.52)$$

$$Nu_{m,H} = \left(\frac{11}{48} + \frac{1}{2}\sum_{n=1}^{\infty} C_n R_n(1)\frac{1 - \exp(-2\beta_n^2 x^*)}{2\beta_n^2 x^*}\right)^{-1} \quad (5.53)$$

where β_n, $R_n(r/a)$, and C_n are eigenvalues, eigenfunctions, and constants, respectively.

Hsu [26] extended the work conducted by Siegel et al. [25] and reported the first 20 values for β_n^2, $R_n(1)$, and C_n. These are listed in Table 5.5. In addition, Hsu [26] presented approximate formulas for higher eigenvalues and constants. The following are of particular interest:

$$\beta_n = 4n + \tfrac{4}{3} \quad (5.54)$$

$$R_n(1) = (-1)^n 0.774759003\beta_n^{-1/3} \quad (5.55)$$

$$C_n = (-1)^n 3.099036005\beta_n^{-4/3} \quad (5.56)$$

The local and mean Nusselt numbers for the $\oplus$ boundary condition are displayed in Fig. 5.2. For the inlet of the circular duct, the local and mean Nusselt numbers can be computed by

$$Nu_{x,H} = \begin{cases} 1.302x^{*-1/3} - 1 & \text{for } x^* \le 0.00005 \\ 1.302x^{*-1/3} - 0.5 & \text{for } 0.00005 \le x^* \le 0.0015 \\ 4.364 + 8.68(10^3 x^*)^{-0.506}e^{-41x^*} & \text{for } x^* \ge 0.0015 \end{cases} \quad (5.57)$$

$$Nu_{m,H} = \begin{cases} 1.953x^{*-1/3} - 1 & \text{for } x^* \le 0.03 \\ 4.364 + 0.0722/x^* & \text{for } x^* > 0.03 \end{cases} \quad (5.58)$$

TABLE 5.5 Eigenvalues and Constants for Eqs. 5.50–5.53 [26]

n	β_n^2	$R_n(1)$	C_n
1	25.679611	−0.49251658	0.40348318
2	83.861753	0.39550848	−0.17510993
3	174.16674	−0.34587367	0.10559168
4	296.53630	0.31404646	−0.073282370
5	450.94720	−0.29125144	0.055036482
6	637.38735	0.27380693	−0.043484355
7	855.849532	−0.25985296	0.035595085
8	1106.329035	0.24833186	−0.029908452
9	1388.822594	−0.23859024	0.025640098
10	1703.3278521	0.23019903	−0.022333685
11	2049.843045	−0.22286280	0.019706916
12	2438.366825	0.21637034	−0.017576456
13	2838.898142	−0.21056596	0.015818436
14	3281.436173	0.20533190	−0.014346369
15	3755.980271	−0.20057716	0.013098171
16	4262.529926	0.19623013	−0.012028202
17	4801.084747	−0.19223350	0.011102223
18	5371.644444	0.18854081	−0.010294071
19	5974.208812	−0.18511389	0.0095834495
20	6608.777727	0.18192104	−0.0089543767

The thermal entrance length for thermally developing flow under the uniform wall heat flux boundary condition is equal to the following:

$$L_{th,H}^* = 0.0430 \tag{5.59}$$

The effects of fluid axial conduction on the thermal entrance problem with uniform wall heat flux are negligible for Pe > 10 when $x^* \geq 0.005$ [24]. However, the thermal entrance length $L_{th,H}^*$ obtained by Nguyen [23] is expressed in terms of Pe:

$$L_{th,H}^* = \begin{cases} -0.000518 + 0.4686/\text{Pe} & \text{for } 1 \leq \text{Pe} \leq 5 \\ 0.03263 + 0.3090/\text{Pe} & \text{for } 5 \leq \text{Pe} \leq 20 \\ 0.04217 + 0.1309/\text{Pe} & \text{for } 20 \leq \text{Pe} \leq 1000 \end{cases} \tag{5.60}$$

The effects of viscous dissipation on the thermal entrance problem with the uniform wall heat flux boundary condition can be found in Brinkman [27], Tyagi [6], Ou and Cheng [28], and Basu and Roy [29]. Other effects, such as inlet temperature, internal heat source, and wall heat flux variation, are reviewed by Shah and London [1] in detail.

Heat Transfer on the Walls With Exponential Heat Flux.   Heat transfer on walls with exponential wall heat flux is denoted as the Ⓗⓔ boundary condition. According to the analysis by Siegel et al. [25], the local Nusselt number for a circular duct with exponential variation of the wall heat flux, as represented by $q_w'' = q_0'' \exp(mx^*)$, can be determined using the following formula:

$$\text{Nu}_{x,H5} = \left(\sum_{n=1}^{\infty} \frac{-C_n R_n(1)\beta_n^2}{m + 2\beta_n^2} \{1 - \exp[-(m + 2\beta_n^2)x^*]\} \right)^{-1} \tag{5.61}$$

The constants C_n, $R_n(1)$, and β_n^2 in Eq. 5.61 can be obtained from Table 5.5 and Eqs. 5.54 through 5.55.

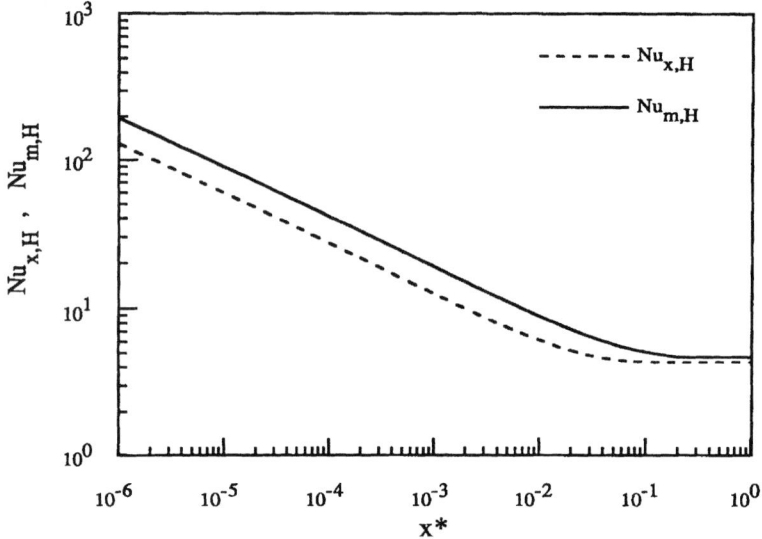

FIGURE 5.2 Local and mean Nusselt numbers $Nu_{x,H}$ and $Nu_{m,H}$ for thermal developing flow in a circular duct.

Heat Transfer on Walls With External Convection. Figure 5.3 presents the results obtained by Hsu [30] for the thermal entrance problem with the convective duct wall boundary condition Ⓣ₃ without consideration of viscous dissipation, fluid axial conduction, flow work, or internal heat sources. As limiting cases of the Ⓣ₃ boundary condition, the curves corresponding to Bi = 0 and Bi = ∞ are identical to $Nu_{x,H}$ and $Nu_{x,T}$, respectively. Significant viscous dissipation effects have been found by Lin et al. [31] for larger Bi values.

Heat Transfer on Walls With Radiation. The local Nusselt numbers normalized with respect to $Nu_{x,H}$ have been obtained by Kadaner et al. [8] for thermally developing flow with the radioactive duct wall boundary condition Ⓣ₄. This is expressed as:

$$\frac{Nu_{x,T4}}{Nu_{x,H}} = 0.94 - \frac{0.0061 - 0.0053 \ln x^+}{1 + 0.0242 \ln x^+} \ln\left(\frac{Sk}{2}\right) \tag{5.62}$$

Equation 5.62 is valid in the ranges $0.001 < x^+ < 0.2$ and $0.2 < Sk < 100$ for zero ambient temperature. It should be noted that Nusselt numbers $Nu_{x,T4}$ with $Sk = 0$ and $Sk = \infty$ are identical to $Nu_{x,H}$ and $Nu_{x,T}$, respectively.

Simultaneously Developing Flow. Simultaneously developing flow usually occurs when the fluid exhibits a moderate Prandtl number. In such a flow, the velocity and the temperature profiles develop simultaneously along the flow direction. Therefore, the heat transfer rate strongly depends on the Prandtl number of the fluid and the thermal boundary condition.

Heat Transfer on Walls With Uniform Temperature. The local and mean Nusselt numbers $Nu_{x,T}$ and $Nu_{m,T}$ are shown in Figs. 5.4 and 5.5, respectively. The data corresponding to these figures can be found in Jensen [32] and Shah and London [1]. It can be observed that for fluids with a large Prandtl number (Pr → ∞), the solution for simultaneously developing flow corresponds to the solution for thermally developing flow because the velocity profile develops before the temperature profile begins developing. However, for a fluid with a very small Prandtl number (Pr = 0), the temperature profile develops much more quickly than the velocity profile, while the velocity remains uniform. This is termed *slug flow*. The local and mean Nusselt numbers for slug flow (Pr = 0) and for Pr → ∞ are shown in Figs. 5.4 and 5.5, respectively.

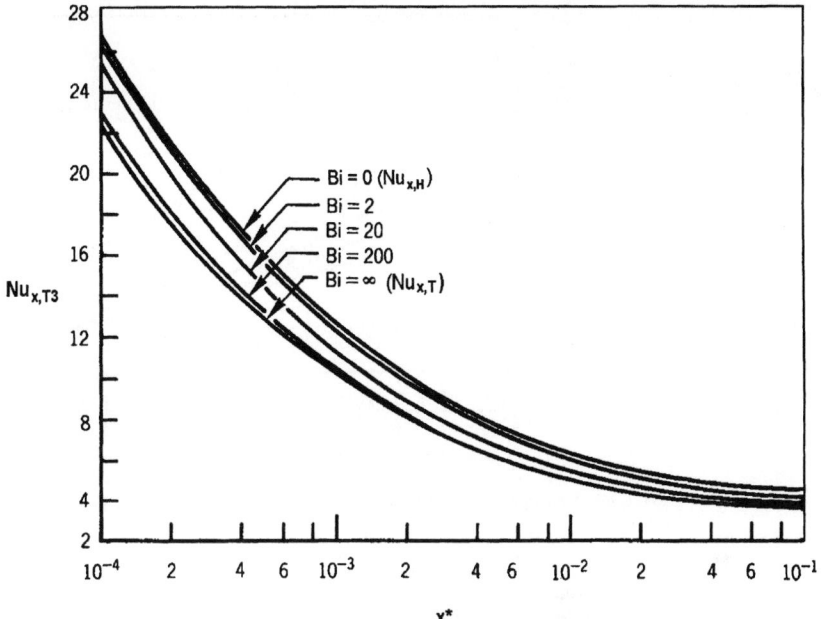

FIGURE 5.3   Local Nusselt number $Nu_{x,T3}$ for thermally developing flow in a circular duct [30].

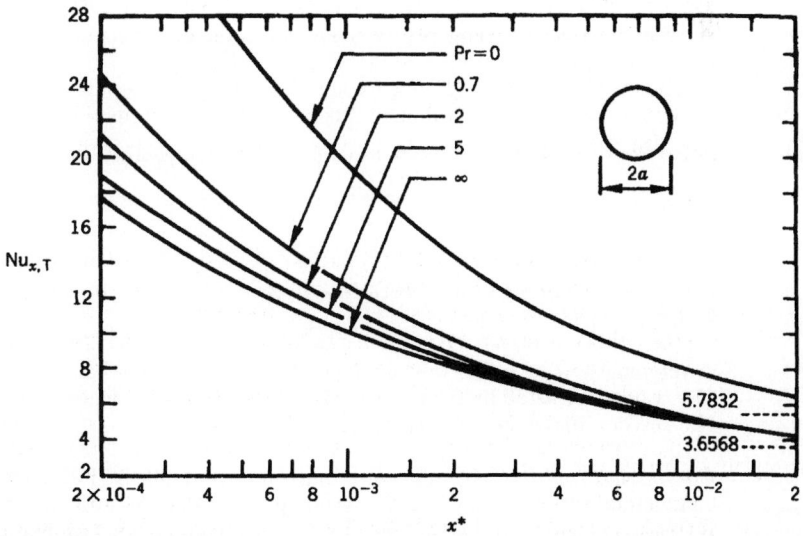

FIGURE 5.4   Local Nusselt number $Nu_{x,T}$ for simultaneously developing flow in a circular duct [1].

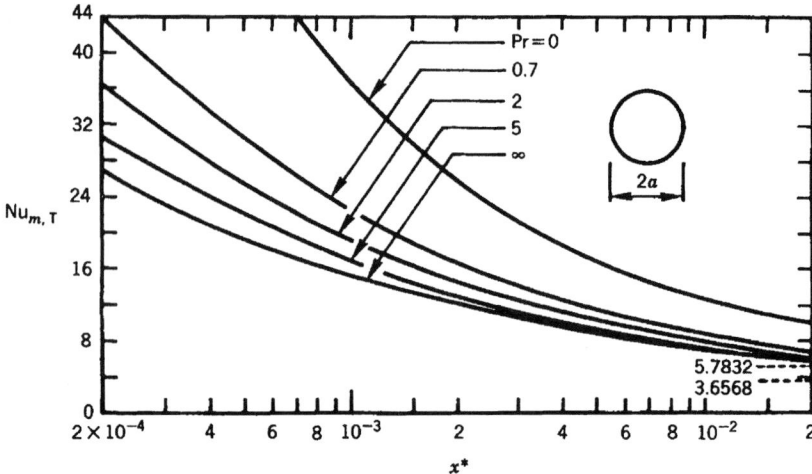

FIGURE 5.5 Mean Nusselt number $Nu_{m,T}$ for simultaneously developing flow in a circular duct [1].

The thermal entrance lengths for simultaneously developing flow with the thermal boundary condition of uniform wall temperature provided by Shah and London [1] are as follows:

$$L_{th,T}^* = \begin{cases} 0.028 & \text{for } Pr = 0 \\ 0.037 & \text{for } Pr = 0.7 \\ 0.033 & \text{for } Pr = \infty \end{cases} \tag{5.63}$$

Heat Transfer on the Walls With Uniform Heat Flux. The solutions for simultaneously developing flow in circular ducts with uniform wall heat flux Ⓗ are reviewed by Shah and London [1]. Recently, a new integral or boundary layer solution has been obtained by Al-Ali and Selim [33] for the same problem. However, the most accurate results for the local Nusselt numbers [1] are presented in Table 5.6.

The thermal entrance lengths for simultaneously developing flow with the thermal boundary condition of uniform wall heat flux [1, 34] can be obtained by

$$L_{th,T}^* = \begin{cases} 0.042 & \text{for } Pr = 0 \\ 0.053 & \text{for } Pr = 0.7 \\ 0.043 & \text{for } Pr = \infty \end{cases} \tag{5.64}$$

The axial diffusions of heat and momentum were considered by Pagliarini [35] for simultaneously developing flow, whereas the viscous dissipation effect has been taken into account by Barletta [36].

Heat Transfer With the Convective Boundary Condition. The solution for simultaneously developing flow with the convective boundary condition Ⓣ₃ has been obtained by Javeri [37]. The results are listed in Table 5.6. It should be noted that when $Bi = \infty$, $Nu_{x,T3}$ is identical to $Nu_{x,T}$. When $Bi = 0$, $Nu_{x,T3}$ is the same as $Nu_{x,H}$.

Conjugate Problem. To this point, uniform wall thickness has been assumed, and no heat conduction in the wall has been involved, meaning that the wall has infinite heat conductivity. If heat conduction in the wall is considered, forced convection and conduction in the wall must be analyzed simultaneously. The solution for this combined problem, referred to as a *conjugate problem,* entails several additional parameters. An extensive review has been per-

TABLE 5.6   Local Nusselt Number $Nu_{x,T3}$ for Simultaneously Developing Flow in a Circular Duct [1]

				$Nu_{x,T3}$			
Bi	x^*	$Pr = \infty$	$Pr = 10$	$Pr = 1.0$	$Pr = 0.7$	$Pr = 0.1$	$Pr = 0$
200	0.00025	15.87	18.27	24.78	26.24	34.64	43.38
	0.00050	12.98	14.04	17.05	17.81	22.93	30.17
	0.00125	9.621	9.837	11.64	12.04	14.54	18.78
	0.0025	7.558	7.501	8.597	8.874	10.64	13.71
	0.0050	6.061	5.977	6.514	6.691	7.914	10.23
	0.0125	4.672	4.677	4.776	4.872	5.649	7.385
	0.025	4.026	4.027	4.037	4.090	4.666	6.259
	0.050	3.722	3.726	3.733	3.740	4.176	5.869
	0.125	3.669	3.669	3.669	3.669	3.892	5.832
	0.25	3.666	3.666	3.667	3.667	3.750	5.832
20	0.00025	18.12	22.52	29.69	31.26	40.59	50.35
	0.00050	14.31	16.41	20.75	21.72	28.14	37.00
	0.00125	10.40	11.02	13.43	13.97	17.49	23.70
	0.0025	8.154	8.276	9.738	10.10	12.47	17.00
	0.0050	6.469	6.471	7.217	7.441	9.043	12.36
	0.0125	4.924	4.927	5.126	5.247	6.204	8.518
	0.025	4.193	4.194	4.244	4.307	4.974	6.942
	0.050	3.838	3.838	3.853	3.866	4.348	6.310
	0.125	3.763	3.763	3.764	3.764	4.008	6.224
	0.25	3.763	3.763	3.763	3.763	3.855	6.224
10	0.00025	18.68	22.99	30.56	32.40	43.45	53.19
	0.00050	14.75	16.81	21.58	22.89	30.11	38.89
	0.00125	10.77	11.39	14.09	14.76	18.72	25.32
	0.0025	8.468	8.590	10.26	10.70	13.34	18.35
	0.0050	6.705	6.715	7.608	7.872	9.651	13.41
	0.0125	5.110	5.113	5.368	5.504	6.576	9.255
	0.025	4.329	4.331	4.407	4.479	5.223	7.473
	0.050	3.935	3.937	3.965	3.974	4.495	6.684
	0.125	3.845	3.846	3.846	3.846	4.105	6.547
	0.25	3.844	3.844	3.844	3.844	3.941	6.547
4	0.00025	20.55	26.09	34.99	34.96	49.72	54.41
	0.00050	15.82	18.69	23.87	24.62	33.26	39.32
	0.00125	11.43	12.36	15.22	15.74	20.13	25.68
	0.0025	8.938	9.287	11.02	11.43	14.32	19.02
	0.0050	7.112	7.021	8.159	8.429	10.39	14.26
	0.0125	5.389	5.391	5.760	5.884	7.118	10.09
	0.025	4.574	4.578	4.684	4.756	5.621	8.181
	0.050	4.120	4.129	4.165	4.171	4.759	7.289
	0.125	4.001	4.003	4.003	4.003	4.285	7.088
	0.25	4.000	4.000	4.000	4.000	4.105	7.087

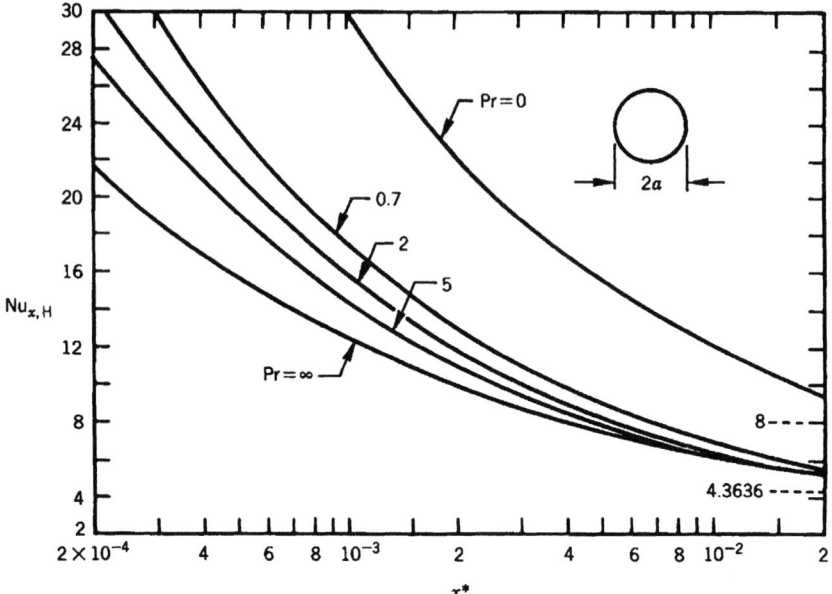

FIGURE 5.6 Local Nusselt number $Nu_{x,H}$ for simultaneously developing flow in a circular duct [1].

formed by Barozzi and Pagliarini [38] for fully developed flow. The reader can also refer to Kuo and Lin [39] and Pagliarini [40] for developing flow.

Turbulent Flow

In this section, turbulent flow and heat transfer in a circular duct with a diameter of $2a$ is discussed for fully developed and developing flow.

Critical Reynolds Number.   The Reynolds number, defined as $u_m D_h / v$, is widely adopted to identify flow status such as laminar, turbulent, and transition flows. A great number of experimental investigations have been performed to ascertain the critical Reynolds number at which laminar flow transits to turbulent flow. It has been found that the transition from laminar flow to fully developed turbulent flow occurs in the range of $2300 \leq Re < 10^4$ for circular ducts [41]. Correspondingly, flow in this region is termed *transition flow*. More conservatively, the lower end of the critical Reynolds number is set at 2100 in most applications.

Generally, the duct inlet configuration and surface roughness have significant effects on the value of Re_{crit}. Other factors, such as noise, vibration, and flow pulsation, affect Re_{crit} as well. Caution should be taken to choose Re_{crit} for the particular application. On the other hand, flow and heat transfer characteristics are difficult to predict in transition flow. The reader is encouraged to consult the literature for the cases not mentioned in this chapter.

Fully Developed Flow.   In this section, the characteristics of the fully developed turbulent flow and heat transfer are presented for both a smooth and a rough circular duct with a diameter of $2a$.

The Power Law Velocity Distribution.   The solution for the power law velocity distribution is introduced in Prandtl [42] in the following form:

$$\frac{u}{u_{max}} = \left(\frac{y}{a}\right)^{1/n} \qquad \frac{u}{u_{max}} = \frac{2n^2}{(n+1)(2n+1)} \tag{5.65}$$

where $y = a - r$ represents the radial distance measured from the wall. The exponent n varies with the Reynolds number. The values of n are listed in Table 5.7; these were obtained from the measurements by Nikuradse [43].

TABLE 5.7   Constants in Eqs. 5.65, 5.71, and 5.72

Re	n	m	C
4000	6	3.5	0.1064
2.3×10^4	6.6	3.8	0.0880
1.1×10^5	7	4	0.0804
1.1×10^6	8.8	4.9	0.0490
2.0×10^6	10	5.5	0.0363
3.2×10^6	10	5.5	0.0366

Universal Velocity-Defect Law.   Figure 5.7 shows the velocity profile computed from Eq. 5.65 together with Nikuradse's [43] measurement data. It can be observed that the velocity profile becomes flatter over most of the duct section, and the exponent $1/n$ of the power law of the velocity distribution, Eq. 5.65, decreases as Re increases. This observation led to the derivation of another form for velocity distribution, the universal velocity-defect law. The for-

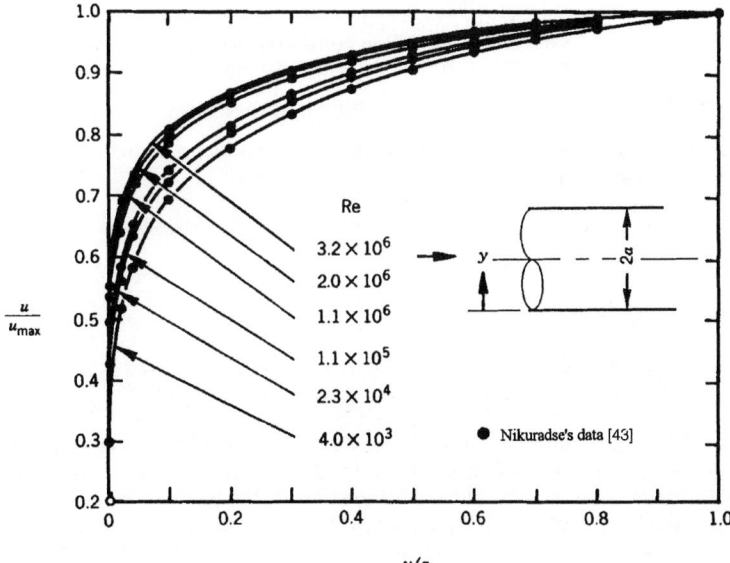

FIGURE 5.7   Power law distribution for fully developed turbulent flow in a smooth circular duct [45].

mula, which follows, was discovered by Prandtl [44] and can be used in high Reynolds number flow, applicable in the turbulent core:

$$\frac{u_{\max} - u}{u_t} = 2.5 \ln \frac{a}{y} \tag{5.66}$$

where

$$u_t = \sqrt{\frac{\tau_w}{\rho}} = u_m \sqrt{\frac{f}{2}} \tag{5.67}$$

denotes friction velocity.

Von Kármán [46] derived the following form for the universal velocity-defect law:

$$\frac{u_{\max} - u}{u_t} = -2.5 \left[\ln \left(1 - \sqrt{1 - \frac{y}{a}} \right) + \sqrt{1 - \frac{y}{a}} \right] \tag{5.68}$$

Wang [47] proposed a third form for the universal velocity-defect law. It follows:

$$\frac{u_{\max} - u}{u_t} = 2.5 \left[\ln \frac{1 + \sqrt{1 - y/a}}{1 - \sqrt{1 - y/a}} - 2 \tan^{-1} \sqrt{1 - \frac{y}{a}} - 0.572 \ln \frac{2.53 - y/a + 1.75\sqrt{1 - y/a}}{2.53 - y/a - 1.75\sqrt{1 - y/a}} \right]$$

$$+ 1.43 \tan^{-1} \frac{1.75\sqrt{1 - y/a}}{0.53 + y/a} \tag{5.69}$$

Darcy's [48] experimental measurements led to the following formula for the velocity-defect law:

$$\frac{u_{\max} - u}{u_t} = 5.08 \left(1 - \frac{y}{a} \right)^{3/2} \tag{5.70}$$

Figure 5.8 displays the velocity distributions in terms of the velocity defect obtained from Eqs. 5.66, 5.68, 5.69, and 5.70. When compared to the experimental data presented by Nikuradse [43], it can be observed that the Eq. 5.69 is in overall best accord with the data; however, it is too complicated to be used. Equation 5.70 agrees well with the data except near the wall, $y/a < 0.25$.

Friction Factor.   From the power law velocity distribution of Eq. 5.65, the friction factor can be expressed as:

$$f = \frac{C}{\mathrm{Re}^{1/m}} \tag{5.71}$$

where C is an experimentally determined constant and m is related to the n in Eq. 5.66 as follows:

$$m = \frac{n + 1}{2} \tag{5.72}$$

The constants n, m, and C have been determined based on Nikuradse's [43] measurements, as is shown in Table 5.7.

Several friction factor correlations for fully developed turbulent flow in smooth, circular ducts are listed in Table 5.8. According to Bhatti and Shah [45], these formulas were derived from highly accurate experimental data for a certain Reynolds number range.

The Prandtl-Kármán-Nikuradse (PKN) correlation is based on the universal velocity-defect law with the coefficients slightly modified to fit the highly accurate experimental data reported by Nikuradse [43], which is known to be the most accurate. This correlation is also referred to as the universal law of friction. However, since the PKN formula gives f values

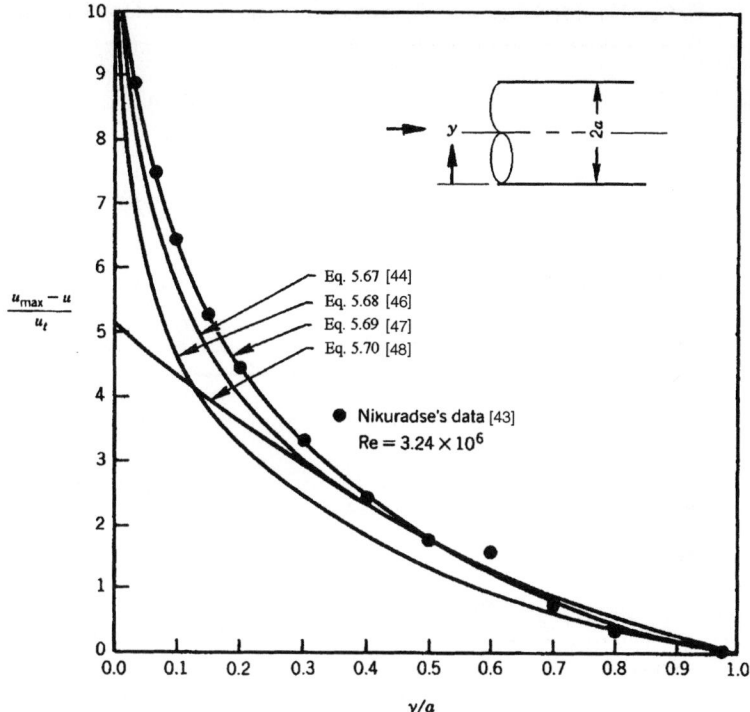

FIGURE 5.8   Universal velocity-defect law distribution for fully developed turbulent flow in a smooth circular duct [45].

implicitly, the explicit formulas by Colebrook [54] and Techo et al. [56], which are close approximations to the PKN formula, may be used.

Velocity Distribution and the Friction Factor for Rough Circular Ducts.   Fully developed velocity distribution in a completely rough circular duct has been expressed by Schlichting [57] as follows:

$$u^+ = 2.5 \ln \frac{y^+}{Re_\varepsilon} + 8.5 \tag{5.73}$$

where

$$u^+ = \frac{u}{u_t} = \frac{u}{\sqrt{\tau_w/\rho}} \; ; \qquad y^+ = \frac{yu_t}{v} = \frac{y\sqrt{\tau_w/\rho}}{v} \tag{5.74}$$

and Re_ε is the roughness Reynolds number, defined as $Re_\varepsilon = \varepsilon u_t/v$. The term ε denotes the surface-roughness element height. The value $Re_\varepsilon < 5$ corresponds to the hydraulically smooth regime; $5 \leq Re_\varepsilon \geq 70$ corresponds to the transition from the hydraulically smooth to the completely rough regime; and $Re > 70$ corresponds to the completely rough regime. Furthermore, $y^+ < 5$ corresponds to the laminar sublayer region, whereas $y^+ > 70$ is the fully turbulent region and $5 < y^+ < 70$ is the transition region.

The friction factor correlations for fully developed turbulent flow in a rough circular duct are summarized in Table 5.9. The friction factor for turbulent flow in an artificially roughed circular duct can be found in Rao [59].

Moody's [58] plot, shown in Fig. 5.9, gives the friction factor for laminar and turbulent flow in both smooth and rough circular ducts. Relative roughness ε/D_h is used as a parameter for

TABLE 5.8   Fully Developed Turbulent Flow Friction Factor in Smooth, Circular Ducts [45]

Investigators	Correlation	Recommended Re range	Remarks
Blasius [49]	$f = 0.0791\ \mathrm{Re}^{-0.25}$	4×10^3 to 10^5	Within +2.6 and −1.3% of PKN (see the following)
McAdams [50]	$f = 0.46\ \mathrm{Re}^{-0.2}$ $f = 0.036\ \mathrm{Re}^{-0.1818}$	3×10^4 to 10^6 4×10^4 to 10^7	Within +2.6 and −0.4% of PKN
Bhatti and Shah [45]	$f = 0.0366\ \mathrm{Re}^{-0.1818}$	4×10^4 to 10^7	Within +2.4 and −3% of PKN
Nikuradse [43]	$f = 0.0008 + 0.0553\ \mathrm{Re}^{-0.237}$	10^5 to 10^7	Within −2% of PKN
Drew et al. [51]	$f = 0.00128 + 0.1143\ \mathrm{Re}^{-0.311}$ $f = 0.0014 + 0.125\ \mathrm{Re}^{-0.32}$	4×10^3 to 5×10^6 4×10^3 to 10^7	Within +3% of PKN
Bhatti and Shah [45]	$f = 0.00128 + 0.1143\ \mathrm{Re}^{-0.311}$	4×10^3 to 10^7	Within +1.2 and −2% of PKN
Prandtl [52] von Kármán [53] Nikuradse [43]	$\dfrac{1}{\sqrt{f}} = 1.7272\ \ln\ (\mathrm{Re}\ \sqrt{f}) - 0.3946$	4×10^3 to 10^7	Classical correlation, here called PKN, has a theoretical basis and is valid for arbitrarily large Re. Its predictions agree with extensive experimental measurements within ±2%.
Colebrook [54]	$\dfrac{1}{\sqrt{f}} = 1.5635\ \ln \left(\dfrac{\mathrm{Re}}{7} \right)$	4×10^3 to 10^7	Mathematical approximation to PKN, yielding numerical values within ±1% of PKN.
Filonenko [55]	$\dfrac{1}{\sqrt{f}} = 1.58\ \ln\ \mathrm{Re} - 3.28$	10^4 to 10^7	Within ±1.8% of PKN
Techo et al. [56]	$\dfrac{1}{\sqrt{f}} = 1.7372\ \ln\ \dfrac{\mathrm{Re}}{1.964\ \ln\ \mathrm{Re} - 3.8215}$	10^4 to 10^7	Explicit form of PKN;

turbulent curves. The broken line demarcating fully turbulent flow and transition flow, obtained by Moody [58], is as follows:

$$\sqrt{f} = \frac{100}{\mathrm{Re}\ (\varepsilon/D_h)} \tag{5.75}$$

It should be noted that the horizontal portions of the curves to the right of the broken line are represented by Nikuradse's [60] correlation, which is presented in Table 5.9. The downward-sloping line for the smooth turbulent flow is represented by the PKN correlation shown in Table 5.8. The downward-sloping line for laminar flow is represented by Eq. 5.17. Relative roughness ε can be obtained from Table 5.10 for a variety of commercial pipes.

Heat Transfer in Smooth Circular Ducts.   For gases and liquids (Pr > 0.5), very little difference exists between the Nusselt number for uniform wall temperature and the Nusselt number for uniform wall heat flux in smooth circular ducts. However, for Pr < 0.1, there is a difference between $\mathrm{Nu_T}$ and $\mathrm{Nu_H}$. Table 5.11 presents the fully developed turbulent flow Nusselt number in a smooth circular duct for Pr > 0.5. The correlation proposed by Gnielinski [69] is recommended for Pr > 0.5, as are those suggested by Bhatti and Shah [45]. In this table, the f in the equation is calculated using the Prandtl [52]-von Kármán [53]-Nikuradse [43]; Colebrook [54]; Filonenko [55]; or Techo et al. [56] correlations shown in Table 5.8.

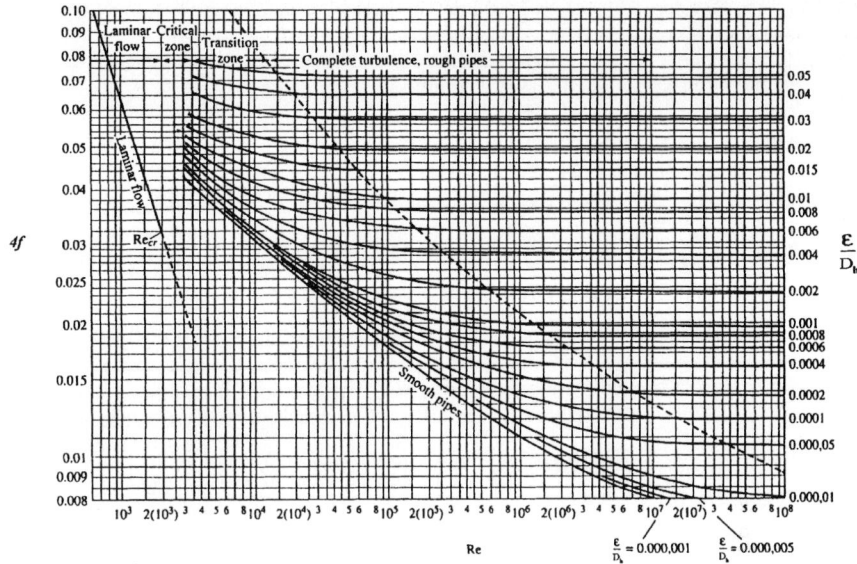

FIGURE 5.9 Moody's [58] friction factor diagram for fully developed flow in a rough circular duct [45].

For liquid metal (Pr < 0.1), the most accurate correlations for Nu_T and Nu_H are those put forth by Notter and Sleicher [80]:

$$Nu_T = 4.8 + 0.0156 Re^{0.85} Pr^{0.93} \tag{5.76}$$

$$Nu_H = 6.3 + 0.0167 Re^{0.85} Pr^{0.93} \tag{5.77}$$

These equations are valid for $0.004 < Pr < 0.1$ and $10^4 < Re < 10^6$.

Heat Transfer in Rough Circular Ducts. The Nusselt number for a complete, rough flow regime in a circular duct is given in Table 5.12. The term f in this table denotes the friction factor for fully rough flow. It is given by the Nikuradse [60] correlation shown in Table 5.9. The recommended equations for practical calculations are those correlations by Bhatti and Shah [45] shown in Table 5.12.

Artificially roughed circular ducts are also often used to enhance heat transfer. The Nusselt numbers for artificially roughed ducts have been reviewed by Rao [59].

Hydrodynamically Developing Flow. An analytical, close-form solution for hydrodynamically developing flow in rough circular ducts has been obtained by Zhiqing [87]. The velocity distribution in the hydrodynamic entrance region is given as

$$\frac{u}{u_{max}} = \begin{cases} (y/\delta)^{1/7} & \text{for } 0 \le y \le \delta \\ 1 & \text{for } \delta \le y \le a \end{cases} \tag{5.78}$$

$$\frac{u}{u_{max}} = 1 - \frac{1}{4}\left(\frac{\delta}{a}\right) + \frac{1}{15}\left(\frac{\delta}{a}\right)^2 \tag{5.79}$$

where δ is the hydrodynamic boundary layer thickness, which varies with axial coordinate x in accordance with the following relation:

TABLE 5.9 Fully Developed Turbulent Flow Friction Factor Correlations for a Rough Circular Duct [48] (a = tube radius)

Investigators	Correlations	Remarks
von Kármán [46]	$\dfrac{1}{\sqrt{f}} = 3.36 - 1.763 \ln \dfrac{\varepsilon}{a}$	This explicit theoretical formula is applicable for $\mathrm{Re}_\varepsilon > 70$.
Nikuradse [60]	$\dfrac{1}{\sqrt{f}} = 3.48 - 1.737 \ln \dfrac{\varepsilon}{a}$	This experimentally derived formula renders very nearly the same results as the von Kármán [46] formula.
Colebrook [54]	$\dfrac{1}{\sqrt{f}} = 3.48 - 1.7372 \ln \left(\dfrac{\varepsilon}{a} + \dfrac{9.35}{\mathrm{Re}\sqrt{f}} \right)$	This implicit formula is applicable for $5 \leq \mathrm{Re}_\varepsilon \leq 70$, spanning the transition, hydraulically smooth, and completely rough flow regimes.
Moody [58]	$f = 1.375 \times 10^{-3} \left[1 + 21.544 \left(\dfrac{\varepsilon}{a} + \dfrac{100}{\mathrm{Re}} \right)^{1/3} \right]$	Shows a maximum derivation of -15.78% from the Colebrook-White equation for $4000 \leq \mathrm{Re} \leq 10^8$ and $2 \times 10^{-8} \leq \varepsilon/a \leq 0.1$.
Wood [61]	$f = 0.08 \left(\dfrac{\varepsilon}{a} \right)^{0.225} + 0.265 \left(\dfrac{\varepsilon}{a} \right) + 66.69 \left(\dfrac{\varepsilon}{a} \right)^{0.4} \mathrm{Re}^{-n}$ where $n = 1.778 \left(\dfrac{\varepsilon}{a} \right)^{0.134}$	Applicable only for $\varepsilon/a \geq 2 \times 10^{-5}$; shows a maximum deviation of 6.16% from the Colebrook-White equation for $4000 \leq \mathrm{Re} \leq 10^8$ and $2 \times 10^{-8} \leq \varepsilon/a \leq 0.1$.
Swamee and Jain [62]	$\dfrac{1}{\sqrt{f}} = 3.4769 - 1.7372 \ln \left[\dfrac{\varepsilon}{a} + \dfrac{42.48}{\mathrm{Re}^{0.9}} \right]$	Shows a maximum deviation of 3.19% from the Colebrook-White equation for $4000 \leq \mathrm{Re} \leq 10^8$ and $2 \times 10^{-8} \leq \varepsilon/a \leq 0.1$.
Churchill [63]	$f = 2 \left[\left(\dfrac{8}{\mathrm{Re}} \right)^{12} + \dfrac{1}{(A_1 + B_1)^{3/2}} \right]^{1/12}$ where $A_1 = \left\{ 2.2088 + 2.457 \ln \left[\dfrac{\varepsilon}{a} + \dfrac{42.683}{\mathrm{Re}^{0.9}} \right] \right\}^{16}$ $B_1 = \left(\dfrac{37.530}{\mathrm{Re}} \right)^{16}$	Unlike other equations in the table, this equation applies to all three flow regimes—laminar, transition, and turbulent. Its predictions for laminar flow are in agreement with $f = 16/\mathrm{Re}$. The predictions for transition flow are subject to some uncertainty. However, the predictions for turbulent flow are comparable with those rendered by the preceding equation.
Chen [64]	$\dfrac{1}{\sqrt{f}} = 3.48 - 1.7372 \ln \left[\dfrac{\varepsilon}{a} - \dfrac{16.2446}{\mathrm{Re}} \ln A_2 \right]$ where $A_2 = \dfrac{(\varepsilon/a)^{1.1098}}{6.0983} + \left(\dfrac{7.149}{\mathrm{Re}} \right)^{0.8981}$	This explicit equation is consistently in good agreement with the Colebrook-White equation for $4000 \leq \mathrm{Re} \leq 10^8$ and $2 \times 10^{-8} \leq \varepsilon/a \leq 0.1$, the maximum deviation being -0.39%.
Round [65]	$\dfrac{1}{\sqrt{f}} = 4.2146 - 1.5635 \ln \left[\dfrac{\varepsilon}{a} + \dfrac{96.2963}{\mathrm{Re}} \right]$	Comparable with Moody's equation.
Zigrang and Sylvester [66]	$\dfrac{1}{\sqrt{f}} = 3.4769 - 1.7372 \ln \left[\dfrac{\varepsilon}{a} - \dfrac{16.1332}{\mathrm{Re}} \ln A_3 \right]$ where $A_3 = \dfrac{\varepsilon/a}{7.4} - 2.1802 \ln \left[\dfrac{\varepsilon}{a} - 16.1332 \ln A_3 \right]$	Shows a maximum deviation of $+0.96\%$ from the Colebrook-White equation for $4000 \leq \mathrm{Re} \leq 10^8$ and $2 \times 10^{-8} \leq \varepsilon/a \leq 0.1$
Zigrang and Sylvester [66]	$\dfrac{1}{\sqrt{f}} = 3.4769 - 1.7372 \ln \left[\dfrac{\varepsilon}{a} - 16.1332 \ln A_4 \right]$ where $A_4 = \dfrac{\varepsilon/a}{7.4} - 2.1802 \ln \left[\dfrac{\varepsilon/a}{7.4} + \dfrac{13}{\mathrm{Re}} \right]$	Predictions not substantially different from those of the preceding equation.

TABLE 5.9 Fully Developed Turbulent Flow Friction Factor Correlations for a Rough Circular Duct [48] (*Continued*)

Investigators	Correlations	Remarks
Haaland [67]	$\dfrac{1}{\sqrt{f}} = 3.4735 - 1.5635 \ln\left[\left(\dfrac{\varepsilon}{a}\right)^{1.11} + \dfrac{63.6350}{\mathrm{Re}}\right]$	Shows a maximum deviation of +1.21% from the Colebrook-White equation for $4000 \le \mathrm{Re} \le 10^8$ and $2 \times 10^{-8} \le \varepsilon/a \le 0.1$.
Serghides [68]	$\dfrac{1}{\sqrt{f}} = A_5 - \dfrac{(A_5 - B_2)^2}{A_5 - 2B + C_1}$ where $A_5 = -0.8686 \ln\left(\dfrac{\varepsilon/a}{7.4} + \dfrac{12}{\mathrm{Re}}\right)$ $B_2 = -0.86868 \ln\left(\dfrac{\varepsilon/a}{7.4}\right) + \dfrac{2.5A_5}{\mathrm{Re}}$ $C_1 = -0.8686 \ln\left(\dfrac{\varepsilon/a}{7.4} + \dfrac{2.51B_2}{\mathrm{Re}}\right)$	Shows a maximum deviation of +0.14% from the Colebrook-White equation for $4000 \le \mathrm{Re} \le 10^8$ and $2 \times 10^{-8} \le \varepsilon/a \le 0.1$.
Serghides [68]	$\dfrac{1}{\sqrt{f}} = 4.781 - \dfrac{(A_5 - 4.781)^2}{(4.781 - 2A_5 - B_2)}$	Shows a maximum deviation of −0.45% from the Colebrook-White equation for $4000 \le \mathrm{Re} \le 10^8$ and $2 \times 10^{-8} \le \varepsilon/a \le 0.1$.

$$\frac{x/D_h}{\mathrm{Re}^{1/4}} = 1.4039\left(\frac{\delta}{a}\right)^{5/4}\left[1 + 0.1577\left(\frac{\delta}{a}\right) - 0.1793\left(\frac{\delta}{a}\right)^2 - 0.0168\left(\frac{\delta}{a}\right)^3 + 0.0064\left(\frac{\delta}{a}\right)^4\right] \quad (5.80)$$

The axial pressure drop Δp^*, the incremental pressure drop number $K(x)$, and the apparent Fanning friction factor f_{app} are given as follows [87]:

$$\Delta p^* = \left(\frac{u_{\max}}{u_m}\right)^2 - 1 \quad (5.81)$$

$$K(x) = \Delta p^* - 0.316\,\frac{x/D_h}{\mathrm{Re}^{1/4}} \quad (5.82)$$

$$f_{\mathrm{app}}\,\mathrm{Re}^{1/4} = \frac{\Delta p^*}{4x/(D_h\,\mathrm{Re}^{0.25})} \quad (5.83)$$

TABLE 5.10 Average Roughness of Commercial Pipes

Material (new)	Roughness ε (mm)
Riveted steel	0.9–9
Reinforced concrete	0.3–3
Wood	0.18–0.9
Cast iron	0.26
Galvanized steel	0.15
Asphalted cast iron	0.12
Bitumen-coated steel	0.12
Structural and forged steel	0.045
Drawn tubing	0.0015
Glass	Smooth

TABLE 5.11   Fully Developed Turbulent Flow Nusselt Numbers in a Smooth, Circular Duct for Gases and Liquids (Pr > 0.5) [48]

Investigators	Correlations	Application range
Dittus and Boelter [70]	$\text{Nu} = \begin{cases} 0.024\ \text{Re}^{0.8}\ \text{Pr}^{0.4} & \text{for heating} \\ 0.026\ \text{Re}^{0.8}\ \text{Pr}^{0.3} & \text{for cooling} \end{cases}$	$0.7 \le \text{Pr} \le 120$ and $2500 \le \text{Re} \le 1.24 \times 10^5$, $L/d > 60$
Colburn [71]	$\text{Nu} = (f/2)\ \text{Re}\ \text{Pr}^{1/3}$ $\text{Nu} = 0.023\ \text{Re}^{0.8}\ \text{Pr}^{1/3}$	$0.5 \le \text{Pr} \le 3$ and $10^4 \le \text{Re} \le 10^5$
von Kármán [72]	$\text{Nu} = \dfrac{(f/2)\ \text{Re}\ \text{Pr}}{1 + 5(f/2)^{1/2}\left[\text{Pr} - 1 + \ln\left(\dfrac{5\ \text{Pr} + 1}{6}\right)\right]}$	$0.5 \le \text{Pr} \le 10$ and $10^4 \le \text{Re} \le 5 \times 10^6$
Prandtl [52]	$\text{Nu} = \dfrac{(f/2)\ \text{Re}\ \text{Pr}}{1 + 8.7(f/2)^{1/2}(\text{Pr} - 1)}$	$0.5 \le \text{Pr} \le 5$ and $10^4 \le \text{Re} \le 5 \times 10^6$
Drexel and McAdams [73]	$\text{Nu} = 0.021\ \text{Re}^{0.8}\ \text{Pr}^{0.4}$	$\text{Pr} \le 0.7$ and $10^4 \le \text{Re} \le 5 \times 10^6$
Friend and Metzner [74]	$\text{Nu} = \dfrac{(f/2)\ \text{Re}\ \text{Pr}}{1.2 + 11.8(f/2)^{1/2}(\text{Pr} - 1)\ \text{Pr}^{-1/3}}$	$50 < \text{Pr} < 600$ and $5 \times 10^4 \le \text{Re} \le 5 \times 10^6$
Petukhov, Kirillov, and Popov [75]	$\text{Nu} = \dfrac{(f/2)\ \text{Re}\ \text{Pr}}{C + 12.7(f/2)^{1/2}(\text{Pr}^{2/3} - 1)}$ where $C = 1.07 + 900/\text{Re} - [0.63/(1 + 10\ \text{Pr})]$	$0.5 < \text{Pr} < 10^6$ and $4000 \le \text{Re} \le 5 \times 10^6$
Hausen [76]	$\text{Nu} = 0.037(\text{Re}^{0.75} - 180)\ \text{Pr}^{0.42}\ [1 + (x/D)^{-2/3}]$	$0.7 \le \text{Pr} \le 3$ and $10^4 \le \text{Re} \le 10^5$
Webb [77]	$\text{Nu} = \dfrac{(f/2)\ \text{Re}\ \text{Pr}}{1.07 + 9(f/2)^{1/2}(\text{Pr} - 1)\ \text{Pr}^{1/4}}$	$0.5 \le \text{Pr} \le 100$ and $10^4 \le \text{Re} \le 5 \times 10^6$
Gnielinski [69]	$\text{Nu} = \dfrac{(f/2)(\text{Re} - 1000)\ \text{Pr}}{1 + 12.7(f/2)^{1/2}(\text{Pr}^{2/3} - 1)}$	$0.5 \le \text{Pr} \le 2000$ and $2300 \le \text{Re} \le 5 \times 10^6$,
	$\text{Nu} = 0.0214(\text{Re}^{0.8} - 100)\ \text{Pr}^{0.4}$	$0.5 \le \text{Pr} \le 1.5$ and $10^4 \le \text{Re} \le 5 \times 10^6$
	$\text{Nu} = 0.012(\text{Re}^{0.87} - 280)\ \text{Pr}^{0.4}$	$1.5 \le \text{Pr} \le 500$ and $3 \times 10^3 \le \text{Re} \le 10^6$
Sieder and Tate [78]	$\text{Nu} = 0.027\ \text{Re}^{4/5}\ \text{Pr}^{1/3}\left(\dfrac{\mu}{\mu_w}\right)^{0.14}$	$0.7 < \text{Pr} < 16{,}700$ and $\text{Re} > 10^4$
Sandall et al. [79]	$\text{Nu} = \dfrac{\sqrt{f/2}\ \text{Re}\ \text{Pr}}{12.48\ \text{Pr}^{2/3} - 7.853\ \text{Pr}^{1/3} + 3.613 \ln \text{Pr} + 5.8 + C}$ where $C = 2.78 \ln \left(\sqrt{f/2}\ \text{Re}/45\right)$	$0.5 \le \text{Pr} < 2000$ and $10^4 \le \text{Re} \le 5 \times 10^6$

The hydrodynamic entrance length L_{hy}/D_h can be calculated by the following equation [87]:

$$\frac{L_{hy}}{D_h} = 1.3590\ \text{Re}^{1/4} \tag{5.84}$$

The results from Eq. 5.84 agree fairly well with experimental data [88].

Thermally Developing Flow.   Numerous investigators [80, 89–94] have carried out the investigation of turbulent thermally developing flow in a smooth circular duct with uniform wall temperature and uniform wall heat flux boundary conditions. It has been found that the dimensionless temperature and the Nusselt number for thermally developing turbulent flow have the same formats as those for laminar thermally developing flow (i.e., Eqs. 5.34–5.37 and Eqs. 5.50–5.53). The only differences are the eigenvalues and constants in the equations.

TABLE 5.12 Nusselt Numbers for Fully Developed Turbulent Flow in the Fully Rough Flow Regime of a Circular Duct [45]

Investigators	Correlations	Remarks
Martinelli [81]	$\mathrm{Nu} = \dfrac{\mathrm{Re}\,\mathrm{Pr}\,\sqrt{f/2}}{5[\mathrm{Pr} + \ln(1 + 5\,\mathrm{Pr}) + 0.5\ln(\mathrm{Re}\,\sqrt{f/2}/60)]}$	This equation differs from that derived by Martinelli [81] for a smooth duct by the omission of the temperature ratio $(T_w - T_c)/(T_w - T_m)$.
Nunner [82]	$\mathrm{Nu} = \dfrac{\mathrm{Re}\,\mathrm{Pr}\,(f/2)}{1 + 1.5\,\mathrm{Re}^{-1/8}\,\mathrm{Pr}^{-1/6}\,[\mathrm{Pr}\,(f/f_s) - 1]}$	This correlation is valid for $\mathrm{Pr} \approx 0.7$; it does not give satisfactory results for $\mathrm{Pr} > 1$.
Dipprey and Sabersky [83]	$\mathrm{Nu} = \dfrac{\mathrm{Re}\,\mathrm{Pr}\,(f/2)}{1 + \sqrt{f/2}[5.19\,\mathrm{Re}_\varepsilon^{0.2}\,\mathrm{Pr}^{0.44} - 8.48]}$	This correlation is valid for $0.0024 \le \varepsilon/D_h \le 0.049$, $1.2 \le \mathrm{Pr} \le 5.94$, and $1.4 \times 10^4 \le \mathrm{Re} \le 5 \times 10^5$.
Gowen and Smith [84]	$\mathrm{Nu} = \dfrac{\mathrm{Re}\,\mathrm{Pr}\,\sqrt{f/2}}{4.5 + [0.155(\mathrm{Re}\,\sqrt{f/2})^{0.54} + \sqrt{2/f}]\sqrt{\mathrm{Pr}}}$	This correlation is valid for $0.0021 \le \varepsilon/D_h \le 0.095$, $0.7 \le \mathrm{Pr} \le 14.3$, and $10^4 \le \mathrm{Re} \le 5 \times 10^4$.
Kawase and Ulbrecht [85]	$\mathrm{Nu} = 0.0523\,\mathrm{Re}\,\sqrt{\mathrm{Pr}}\,\sqrt{f}$	The predictions of this correlation are somewhat lower than those of the following correlation.
Kawase and De [86]	$\mathrm{Nu} = 0.0471\,\mathrm{Re}\,\sqrt{\mathrm{Pr}}\,\sqrt{f}$ $(1.11 + 0.44\,\mathrm{Pr}^{-1/3} - 0.7\,\mathrm{Pr}^{-1/6})$	The predictions of this correlation are in reasonable agreement with the experimental data for $0.0024 \le \varepsilon/D_h \le 0.165$, $5.1 < \mathrm{Pr} < 390$, and $5000 < \mathrm{Re} < 5 \times 10^5$.
Bhatti and Shah [45]	$\mathrm{Nu} = \dfrac{(\mathrm{Re}\,\mathrm{Pr}\,(f/2))}{1 + \sqrt{f/2}(4.5\,\mathrm{Re}_\varepsilon^{0.2}\,\mathrm{Pr}^{0.5} - 8.48)}$	This correlation is valid for $0.5 < \mathrm{Pr} < 10$, $0.002 < 0.002 < \varepsilon/D_h < 0.05$, and $\mathrm{Re} > 10^4$. Its predictions are within $\pm5\%$ of the available measurements.
Bhatti and Shah [45]	$\mathrm{Nu} = \dfrac{(\mathrm{Re} - 1000)\,\mathrm{Pr}\,(f/2)}{1 + \sqrt{f/2}[(17.42 - 13.77\,\mathrm{Pr}_t^{0.8})\,\mathrm{Re}_\varepsilon^{0.5} - 8.48]}$	This correlation is valid for $0.5 < \mathrm{Pr} < 5000$, $0.001 < 0.002 < \varepsilon/D_h < 0.05$, and $\mathrm{Re} > 2300$. Its predictions are within $\pm15\%$ of the available measurements.

Bhatti and Shah [45] listed these eigenvalues and constants. On the other hand, the local Nusselt numbers for uniform wall temperature and uniform wall heat flux $\mathrm{Nu}_{x,T}$ and $\mathrm{Nu}_{x,H}$ in turbulent developing flow are nearly identical for $\mathrm{Pr} > 0.2$. Therefore, the subscripts T or H are dropped in the equations in this section.

The mean Nusselt number Nu_m for thermally developing flow with uniform wall temperature or uniform wall heat flux conditions can be calculated using Al-Arabi's [95] correlation:

$$\frac{\mathrm{Nu}_m}{\mathrm{Nu}_\infty} = 1 + \frac{C}{x/D_h} \qquad (5.85)$$

where Nu_∞ stands for the fully developed Nusselt number Nu_T or Nu_H and

$$C = \frac{(x/D_h)^{0.1}}{\mathrm{Pr}^{1/6}}\left(0.68 + \frac{3000}{\mathrm{Re}^{0.81}}\right) \qquad (5.86)$$

However, the thermal entrance lengths for uniform wall temperature and uniform wall heat flux are much different. These are shown in Figs. 5.10 and 5.11, respectively.

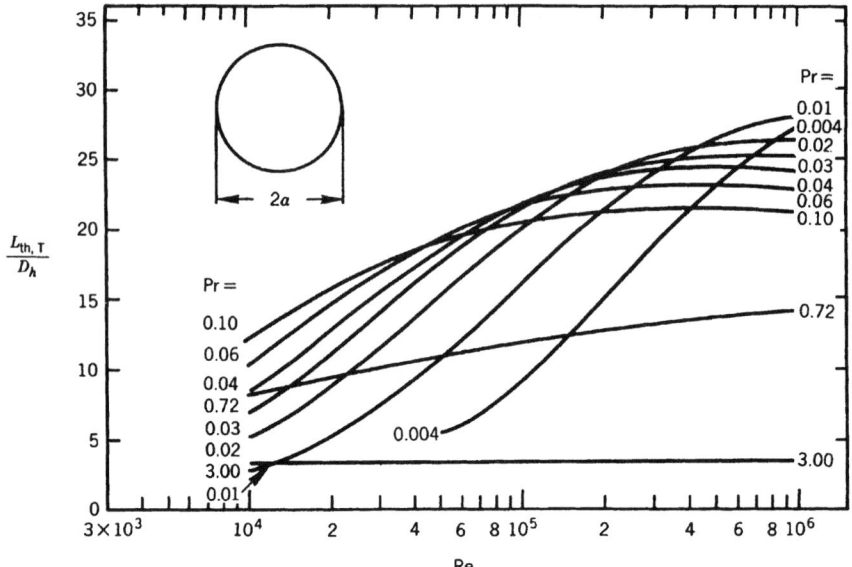

FIGURE 5.10   Thermal entrance lengths for the turbulent Graetz problem for a smooth circular duct [80].

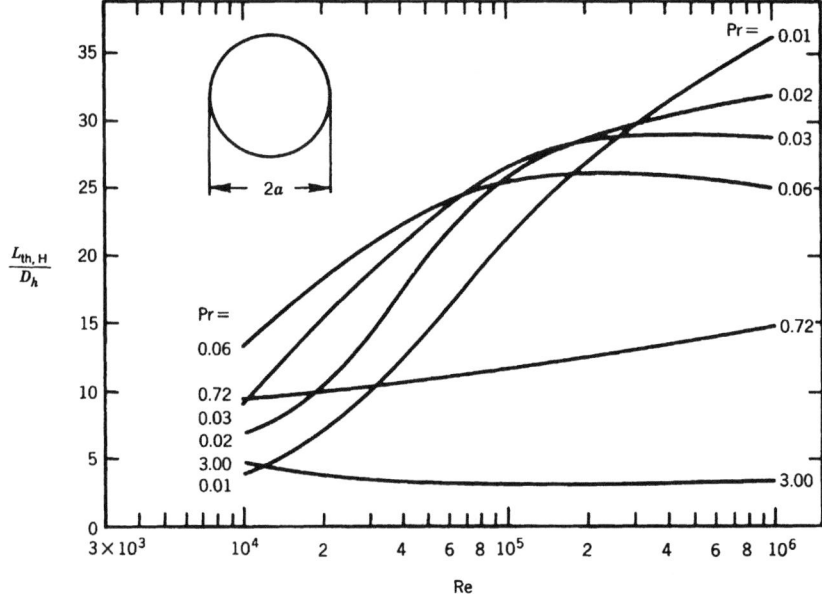

FIGURE 5.11   Thermal entrance lengths for the turbulent thermal entrance problem for a smooth circular duct with uniform wall heat flux [80].

For liquid metals ($Pr < 0.03$), when $x/D_h > 2$ and $Pe > 500$, the local and mean Nusselt numbers for the uniform wall temperature boundary condition have been proposed by Chen and Chiou [96] as follows:

$$\frac{Nu_{x,T}}{Nu_T} = 1 + \frac{2.4}{x/D_h} - \frac{1}{(x/D_h)^2} \tag{5.87}$$

$$\frac{Nu_{m,T}}{Nu_T} = 1 + \frac{7}{x/D_h} + \frac{2.8}{x/D_h} \ln\left(\frac{x/D_h}{10}\right) \tag{5.88}$$

$$Nu_T = 4.5 + 0.0156 Re^{0.85} Pr^{0.86} \tag{5.89}$$

When the uniform heat flux condition is applied to the duct wall, Eqs. 5.87 and 5.88 are still used, but with $Nu_{x,H}$, $Nu_{m,H}$, and Nu_H replacing $Nu_{x,T}$, $Nu_{m,T}$, and Nu_T, respectively, while Nu_H can be obtained from:

$$Nu_H = 5.6 + 0.0165 Re^{0.85} Pr^{0.86} \tag{5.90}$$

The thermal entrance length $L_{th,H}$ for liquid metals has been found by Genin et al. [97]:

$$\frac{L_{th,H}}{D_h} = \frac{0.04 Pe}{1 + 0.002 Pe} \tag{5.91}$$

Simultaneously Developing Flow.   The local Nusselt numbers obtained theoretically by Deissler [92] for simultaneously developing velocity and temperature fields in a smooth circular duct subject to uniform wall temperature and the uniform heat flux for $Pr = 0.73$ are plotted in Fig. 5.12. It can be seen from this figure that the Nusselt numbers for two different thermal boundary conditions are identical for $x/D_h > 8$.

It is worth noting that the duct entrance configuration affects simultaneously developing flow [98, 99]. The local Nusselt number is different for each duct entrance configuration. For practice usage, Bhatti and Shah [45] suggest the following formula for the calculation of the mean Nusselt number:

$$\frac{Nu_m}{Nu_\infty} = 1 + \frac{C}{(x/D_h)^n} \tag{5.92}$$

where Nu_∞ denotes the fully developed Nusselt number Nu_H or Nu_T. The terms C and n have been determined from the $Nu_{m,H}$ measurement given by Mills [99] for air ($Pr = 0.7$). Table 5.13 lists the resulting C and n of Eq. 5.92 for each configuration. Equation 5.92 may be used in the case of the uniform wall temperature and uniform wall heat flux boundary conditions.

For liquid metals ($Pr < 0.03$), Chen and Chiou [95] have obtained the correlations for simultaneously developing flow in a smooth circular duct with a uniform velocity profile at the inlet. These follow:

$$\frac{Nu_x}{Nu_\infty} = 0.88 + \frac{2.4}{x/D_h} - \frac{1.25}{(x/D_h)^2} - A \tag{5.93}$$

$$\frac{Nu_x}{Nu_\infty} = 1 + \frac{5}{x/D_h} + \frac{1.86}{x/D_h} \ln\left(\frac{x/D_h}{10}\right) - B \tag{5.94}$$

where for the uniform wall temperature boundary condition,

$$A = \frac{40 - x/D_h}{190}, \qquad B = 0.09 \tag{5.95}$$

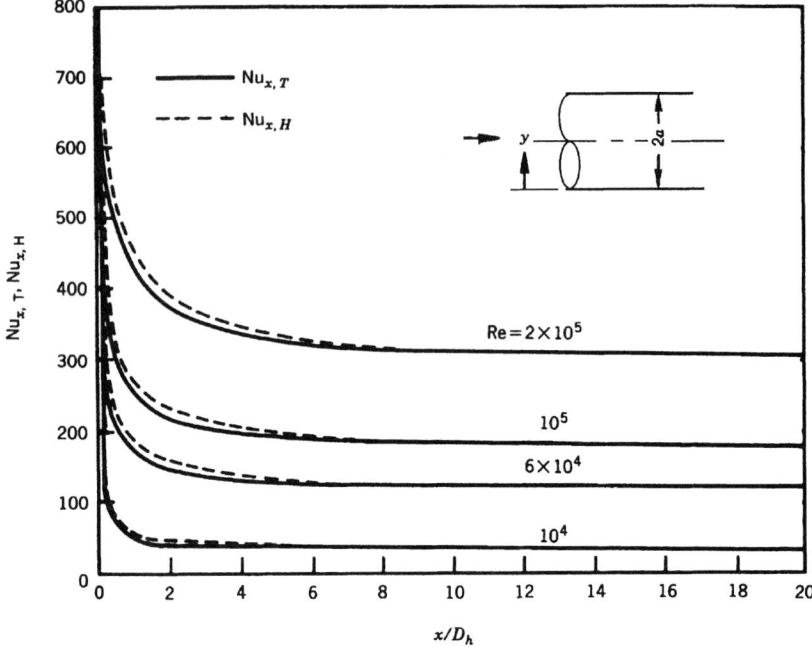

FIGURE 5.12   Local Nusselt numbers $Nu_{x,T}$ and $Nu_{x,H}$ for simultaneously developing turbulent flow in a smooth circular duct for Pr = 0.73 [92].

and for the uniform wall heat flux boundary condition,

$$A = B = 0 \tag{5.96}$$

The Nu_x in Eqs. 5.93 and 5.94 denotes the local Nusselt number $Nu_{x,T}$ or $Nu_{x,H}$, and Nu_∞ represents the fully developed Nu_T or Nu_H. Equations 5.93 and 5.94 are valid for $2 \le x/D_h < 35$ and Pe > 500.

Transition Flow

As seen in the previous section, flow is considered to be laminar when Re < 2300 and turbulent when Re > 10^4. Transition flow occurs in the range of 2300 < Re < 10^4. Few correlations or formulas for computing the friction factor and heat transfer coefficient in transition flow are available. In this section, the formula developed by Bhatti and Shah [45] is presented to compute the friction factor. It follows:

$$f = A + \frac{B}{Re^{1/m}} \tag{5.97}$$

Equation 5.97 is applicable to the laminar, transition, and turbulent flow regions. For laminar flow (Re < 2100), $A = 0$, $B = 16$, and $m = 1$. For transition flow, 2100 < Re $\le$ 4000, $A = 0.0054$, $B = 2.3 \times 10^{-8}$, and $m = -2/3$. For turbulent flow, (Re > 4000) $A = 1.28 \times 10^{-3}$, $B = 0.1143$, and $m = 3.2154$. Blasius's [49] formula (see Table 5.8) is also applicable for calculating the friction factor in the range of $4000 \le Re \le 10^5$.

TABLE 5.13 Nusselt Number Ratios for a Smooth Circular Duct with Various Entrance Configurations for Pr = 0.7 [2]

Entrance configurations	Schematics	C	n
Long calming section		0.9756	0.760
Square entrance		2.4254	0.676
180° Round bend		0.9759	0.700
90° Round bend		1.0517	0.629
90° Elbow		2.0152	0.614

Heat transfer results for transition flow are rather uncertain due to the fact that so many parameters are needed to characterize heat-affected flow. In the range of $0 < \mathrm{Pr} < \infty$ and $2100 \le \mathrm{Re} \le 10^6$, Churchill [100] recommends that the following equations be used to calculate the Nusselt number for the laminar, transition, and turbulent regimes:

$$\mathrm{Nu}^{10} = \mathrm{Nu}_l^{10} + \left\{ \frac{\exp[(2200 - \mathrm{Re})/365]}{\mathrm{Nu}_l^2} + \frac{1}{\mathrm{Nu}_t^2} \right\}^{-5} \tag{5.98}$$

$$\mathrm{Nu}_l = \begin{cases} 3.657 & \text{for the uniform wall temperature boundary condition} \\ 4.364 & \text{for the uniform wall heat flux boundary condition} \end{cases} \tag{5.99}$$

$$\mathrm{Nu}_t = \mathrm{Nu}_0 + \frac{0.079(f/2)^{1/2}\,\mathrm{Re}\,\mathrm{Pr}}{(1 + \mathrm{Pr}^{4/5})^{5/6}} \tag{5.100}$$

$$\mathrm{Nu}_0 = \begin{cases} 4.8 & \text{for the uniform wall temperature boundary condition} \\ 6.3 & \text{for the uniform wall heat flux boundary condition} \end{cases} \tag{5.101}$$

where superscript 10 indicates transition region and Nu_0 denotes an overall Nusselt number associated with the convection boundary condition.

Kaupas et al. [101] experimentally investigated heat transfer in transition gas flow in a circular duct at high heat flux in the range of $2 \times 10^3 \le \mathrm{Re} \le 3 \times 10^4$.

More research is needed to determine reliable friction factors and Nusselt numbers for transition flow.

CONCENTRIC ANNULAR DUCTS

Concentric annular ducts are a common and important geometry for fluid flow and heat transfer devices. The double pipe heat exchanger is a simple example. In this device, one fluid flows through an inside pipe, while the other flows through the concentric annular passages. The friction factor and the heat transfer coefficient are essential for the design of such heat transfer devices.

Four Fundamental Thermal Boundary Conditions

As shown in Fig. 5.13, there are two walls in concentric annular ducts. Either or both of them can be involved in heat transfer to a flowing fluid in the annulus. Four fundamental thermal boundary conditions, which follow, can be used to define any other desired boundary condition. Correspondingly, the solutions for these four fundamental boundary conditions can be adopted to obtain the solutions for other boundary conditions using superposition techniques.

The four fundamental thermal boundary conditions are as follows:

First kind.   Uniform temperature (different from the entering fluid temperature) at one wall; the other wall at the uniform entering fluid temperature

Second kind.   Uniform wall heat flux at one wall; the other wall insulated (i.e., adiabatic with zero heat flux)

Third kind.   Uniform temperature (different from the entering fluid temperature) at one wall; the other wall insulated

Fourth kind.   Uniform wall heat flux at one wall; the other wall maintained at the entering fluid temperature.

The previously mentioned boundary conditions can be applied in both laminar and turbulent flow.

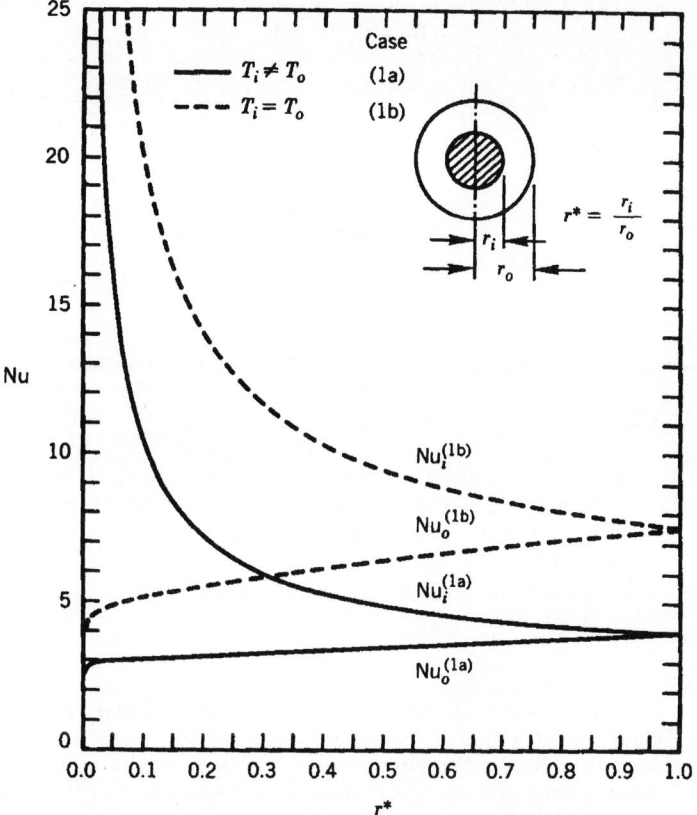

FIGURE 5.13   Fully developed Nusselt numbers for uniform temperatures at both walls in concentric annular ducts [1].

Laminar Flow

In this section, the characteristics of laminar flow and heat transfer in concentric annular ducts are presented, and the effect of eccentricity is discussed.

Fully Developed Flow.   Velocity distribution, the friction factor, and heat transfer for fully developed laminar flow in concentric annular ducts are described sequentially.

Velocity Distribution and the Friction Factor.   For a concentric annular duct with inner radius r_i and outer radius r_o, the velocity distribution and friction factor for fully developed flow in a concentric annular duct are as follows [1]:

$$u = -\frac{1}{4\mu}\left(\frac{dp}{dx}\right)r_o^2\left[1 - \left(\frac{r}{r_o}\right) + 2r_m^{*2}\ln\left(\frac{r}{r_o}\right)\right] \tag{5.102}$$

$$u_m = -\frac{1}{8\mu}\left(\frac{dp}{dx}\right)r_o^2(1 + r^{*2} - 2r_m^{*2}) \tag{5.103}$$

$$\frac{u_{\max}}{u_m} = \frac{2(1 - r_m^{*2} + 2r_m^{*2}\ln r_m^*)}{1 + r^{*2} - 2r_m^{*2}} \tag{5.104}$$

$$f_i \operatorname{Re} = -\frac{1}{\mu}\left(\frac{dp}{dx}\right)\frac{D_h}{u_m}\left(\frac{r_m^2 - r_i^2}{r_i}\right), \qquad D_h = 2(r_o - r_i) \tag{5.105}$$

$$f_o \operatorname{Re} = -\frac{1}{\mu}\left(\frac{dp}{dx}\right)\frac{D_h}{u_m}\left(\frac{r_o^2 - r_m^2}{r_o}\right) \tag{5.106}$$

$$f \operatorname{Re} = \frac{16(1-r^*)^2}{1 + r^{*2} - 2r_m^{*2}} \tag{5.107}$$

where r_m in the preceding equations is the radius where the maximum velocity achieves $[(\partial u/\partial r) = 0]$, and r_m^* is its dimensionless form, which is defined as

$$r_m^* = \frac{r_m}{r_o} = \left(\frac{1-r^{*2}}{2\ln(1/r^*)}\right)^{1/2} \tag{5.108}$$

The terms f_i and f_o represent the friction factor at the inner and the outer walls, respectively. The circumferentially averaged friction factor is related to f_i and f_o as follows:

$$f = \frac{f_i r_i + f_o r_o}{r_i + r_o} \tag{5.109}$$

Natarajan and Lakshmanan [102] present a simple equation for $f \operatorname{Re}$ that is easy to use. This equation agrees with the values calculated from Eq. 5.107 within ±2 percent for $r^* \geq 0.005$. It follows:

$$f \operatorname{Re} = 24 r^{*0.035} \tag{5.110}$$

Heat Transfer. Fundamental solutions for boundary conditions of the first, second, and third kinds for fully developed flow in concentric annular ducts are given in Table 5.14. The nomenclature used in describing the corresponding solutions can best be explained with reference to the specific heat transfer parameters $\theta_{lj}^{(k)}$ and $\theta_{mj}^{(k)}$, which are the dimensionless duct wall and fluid bulk mean temperature, respectively. The superscript k denotes the type of the fundamental solution according to the four types of boundary conditions described in the section entitled "Four Fundamental Thermal Boundary Conditions." Thus, $k = 1, 2, 3,$ or 4. The subscript l in $\theta_{lj}^{(k)}$ refers to the particular wall at which the temperature is evaluated; $l = i$ or o when the temperature is evaluated at the inner or the outer wall. The subscript j in $\theta_{lj}^{(k)}$ refers

TABLE 5.14 Fundamental Solutions of the First, Second, and Third Kinds of Boundary Conditions for Fully Developed Flow in Concentric Annular Ducts [1]

r^*	$\Phi_{ii}^{(1)}$	$\mathrm{Nu}_{ii}^{(1)}$	$\mathrm{Nu}_{oo}^{(1)}$	$\theta_{io}^{(2)} - \theta_{mo}^{(2)}$	$\mathrm{Nu}_{ii}^{(2)}$	$\mathrm{Nu}_{oo}^{(2)}$	$\mathrm{Nu}_{ii}^{(3)}$	$\mathrm{Nu}_{oo}^{(3)}$
0	∞	∞	2.66667	−0.145833	∞	4.36364	∞	3.6568
0.01	42.99515	50.45396	2.90834	−0.130725	54.01669	4.69234	—	—
0.02	25.05098	30.17942	2.94836	−0.127945	32.70512	4.73424	32.337	3.9934
0.04	14.91204	18.61387	2.99928	−0.124122	20.50925	4.77803	—	—
0.05	12.68471	16.05843	3.01887	−0.122568	17.81128	4.79098	17.460	4.0565
0.08	9.10628	11.94251	3.06751	−0.118559	13.46806	4.80270	—	—
0.10	7.81730	10.45870	3.09528	−0.116214	11.90578	4.83421	11.560	4.1135
0.15	5.97397	8.34163	3.15708	−0.110999	9.68703	4.86026	—	—
0.25	4.32809	6.47139	3.26700	−0.102207	7.75347	4.90475	7.3708	4.2321
0.40	3.27407	5.30511	3.42077	−0.091495	6.58330	4.97917	—	—
0.50	2.88539	4.88896	3.52035	−0.085513	6.18102	5.03653	5.7382	4.4293
0.80	2.24071	4.23035	3.81134	−0.071409	5.57849	5.23654	—	—
1.00	2.00000	4.00000	4.00000	−0.064286	5.38462	5.38462	4.8608	4.8608

to the wall at which $T \neq T_e$ or $q_w'' \neq 0$ (i.e., the active wall that participates in the heat transfer with flowing fluid). The role of k and j in the $\theta_{mj}^{(k)}$ is the same as in $\theta_{ij}^{(k)}$, while m represents fluid bulk mean temperature.

The other heat transfer results related to the four fundamental solutions can be obtained using the following equations in conjunction with Table 5.14 [1]:

$$\Phi_{oo}^{(1)} = -\Phi_{oi}^{(1)} = r^*\Phi_{ii}^{(1)} = -r^*\Phi_{io}^{(1)} \tag{5.111}$$

$$\theta_{mi}^{(1)} = 1 - \theta_{mo}^{(1)} \tag{5.112}$$

$$Nu_{io}^{(1)} = Nu_{ii}^{(1)} = \frac{\Phi_{ii}^{(1)}}{1 - \theta_{mi}^{(1)}} = \frac{\Phi_{ii}^{(1)}}{\theta_{mo}^{(1)}} \tag{5.113}$$

$$Nu_{io}^{(1)} = Nu_{oo}^{(1)} = \frac{\Phi_{ii}^{(1)}}{1 - \theta_{mo}^{(1)}} = \frac{r^*\Phi_{ii}^{(1)}}{\theta_{mi}^{(1)}} \tag{5.114}$$

$$\theta_{oi}^{(2)} - \theta_{mi}^{(2)} = r^*[\theta_{io}^{(2)} - \theta_{mo}^{(2)}] \tag{5.115}$$

$$Nu_{ii}^{(2)} = \frac{1}{\theta_{ii}^{(2)} - \theta_{mi}^{(2)}} \tag{5.116}$$

$$Nu_{oo}^{(2)} = \frac{1}{\theta_{oo}^{(2)} - \theta_{mo}^{(2)}} \tag{5.117}$$

$$Nu_{oi}^{(2)} = Nu_{io}^{(2)} = 0 \tag{5.118}$$

$$\Phi_{ii}^{(3)} = \Phi_{oo}^{(3)} = 0 \tag{5.119}$$

$$\theta_{oi}^{(3)} = \theta_{io}^{(3)} = \theta_{mi}^{(3)} = \theta_{mo}^{(3)} = 1 \tag{5.120}$$

$$\Phi_{oi}^{(4)} = \frac{1}{\Phi_{io}^{(4)}} = -r^* \tag{5.121}$$

$$\theta_{ii}^{(4)} = r^*\theta_{oo}^{(4)} = \frac{1}{Nu_{ii}^{(1)}} + \frac{r^*}{Nu_{oo}^{(1)}} \tag{5.122}$$

$$Nu_{oo}^{(4)} = Nu_{io}^{(4)} = \frac{1}{\theta_{ii}^{(4)} - \theta_{mi}^{(4)}} = \frac{1}{r^*\theta_{mo}^{(4)}} \tag{5.123}$$

$$Nu_{oo}^{(4)} = Nu_{oi}^{(4)} = \frac{1}{\theta_{oo}^{(4)} - \theta_{mo}^{(4)}} = \frac{r^*}{\theta_{mi}^{(4)}} = Nu_{oo}^{(1)} \tag{5.124}$$

The direct use of these four fundamental solutions is rare in engineering applications. The solutions for practical problems must be developed. The following examples should be of great interest with respect to the application of these fundamental solutions.

Uniform Temperature at Both Walls. When $T_i \neq T_o$, the problem is designated as 1a, and the fully developed Nusselt numbers at the two walls are designated as $Nu_i^{(1a)}$ and $Nu_o^{(1a)}$. When $T_i = T_o$, the problem is designated as 1b, and the fully developed Nusselt numbers at the two walls are designated as $Nu_i^{(1b)}$ and $Nu_o^{(1b)}$. These are presented in Fig. 5.13. Tabulated values for these and the subsequent solutions are available in Shah and London [1].

A circumferentially averaged Nusselt number in the case of $T_i = T_o$, designated as Nu_T, can be obtained from $Nu_i^{(1b)}$ and $Nu_o^{(1b)}$ by means of the following relation [1]:

$$Nu_T = \frac{Nu_o^{(1b)} + r^* Nu_i^{(1b)}}{1 + r^*} \tag{5.125}$$

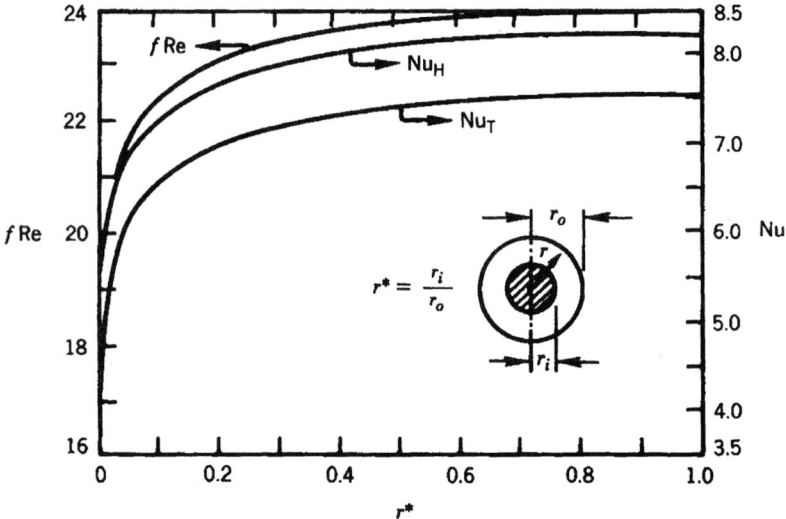

FIGURE 5.14 Fully developed friction factor and Nusselt numbers for concentric annular ducts [2].

The Nu_T values for Eq. 5.125 are presented in Fig. 5.14, in which Nu_H, described later, and f Re, calculated from Eq. 5.107, are also displayed.

Uniform Heat Fluxes at Both Walls.   When $q_i'' = q_o''$, the problem is designated as 2*a*, and the fully developed Nusselt numbers at the two walls are designated as $Nu_i^{(2a)}$ and $Nu_o^{(2a)}$. When $q_i'' \neq q_o''$, the problem is designated as 2*b* and the fully developed Nusselt numbers at the two walls are denoted $Nu_i^{(2b)}$ and $Nu_o^{(2b)}$. These Nusselt numbers are shown in Fig. 5.15. In the case of $q_i'' = q_o''$, the circumferentially averaged Nusselt number Nu_H can be obtained from $Nu_i^{(2b)}$ and $Nu_o^{(2b)}$ via Eq. 5.125 by replacing subscript T with H and superscript 1 with 2. The Nu_H obtained in this way is shown in Fig. 5.14.

It should be noted that the heat flux is positive if the heat transfer occurs from the wall to the fluid. Therefore, a negative Nusselt number like those shown in Fig. 5.15 signifies that heat transfer has taken place in the opposite direction (i.e., from the fluid to the wall). In both aforementioned cases, $T_w - T_m$ is considered positive. Therefore, the infinite Nusselt numbers in Figs. 5.13 and 5.15 indicate that $T_w = T_m$, not infinite heat flux.

Uniform Temperature at One Wall and Uniform Heat Flux at the Other.   The subscripts 1 and 2 refer to either the inside or the outside wall. When $T_1 \neq T_2$, the problem is known as 4*a*, and when $T_1 = T_2$ it is known as 4*b*. It has been shown by Shah and London [1] that

$$Nu_i^{(4a)} = Nu_i^{(4b)} = Nu_i^{(1a)} \tag{5.126}$$

$$Nu_o^{(4a)} = Nu_o^{(4b)} = Nu_o^{(1a)} \tag{5.127}$$

Hydrodynamically Developing Flow.   Shah and London [1] summarize the solutions for the hydrodynamic development of laminar flow in concentric annuli. The apparent friction factor in the hydrodynamic entrance region, derived by Shah [103], is expressed as:

$$f_{app} \, Re = 3.44(x^+)^{-.05} + \frac{K(\infty)/(4x^+) + f \, Re - 3.44(x^+)^{-0.5}}{1 + C(x^+)^{-2}} \tag{5.128}$$

The values of $K(\infty)$, f Re, and C in Eq. 5.128 are given in Table 5.15. A very good agreement, within ±3 percent, with the various analytical predictions has been achieved using Eq. 5.128.

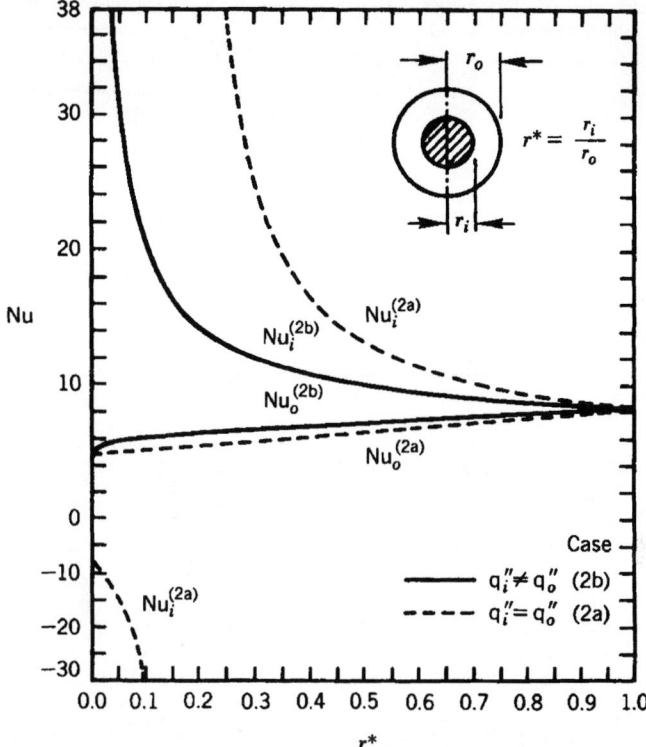

FIGURE 5.15 Fully developed Nusselt numbers for uniform axial heat fluxes at both walls in concentric annular ducts [1].

In addition, the hydrodynamic entrance length L_{hy}^+, recommended by Shah and London [1], is given in Table 5.15.

Thermally Developing Flow. The solutions for thermally developing flow in concentric annular ducts under each of the four fundamental thermal boundary conditions are tabulated in Tables 5.16, 5.17, 5.18, and 5.19. These results have been taken from Shah and London [1]. Additional quantities can be determined from the correlations listed at the bottom of each table using the data presented.

TABLE 5.15 Hydrodynamically Developing Flow Parameters and Constants to Use in Conjunction with Eq. 5.128 for Concentric Annular Ducts [103]

r^*	L_{hy}^+	$K(\infty)$	$f\,\mathrm{Re}$	C
0	0.0541	1.250	16.000	0.000212
0.05	0.0206	0.830	21.567	0.000050
0.10	0.0175	0.784	22.343	0.000043
0.50	0.0116	0.688	23.813	0.000032
0.75	0.0109	0.678	23.967	0.000030
1.0	0.0108	0.674	24.000	0.000029

TABLE 5.16 Fundamental Solutions of the First Kind for Thermally Developing Flow in Concentric Annular Ducts (compiled from Shah and London [1])

r^*	x^*	$\Phi_{x,ii}^{(1)}$	$\theta_{x,mi}^{(1)}$	$Nu_{x,ii}^{(1)}$	$Nu_{x,oi}^{(1)}$	$\Phi_{x,oo}^{(1)}$	$\theta_{x,mo}^{(1)}$	$Nu_{x,oo}^{(1)}$	$Nu_{x,io}^{(1)}$
0.02	0.00001	—	—	—	—	51.081	0.00303	51.236	—
	0.00005	—	—	—	—	29.350	0.00876	29.609	—
	0.0001	78.5	0.0011	78.5	—	23.033	0.01380	23.355	—
	0.0005	57.5	0.0031	57.7	—	12.934	0.03930	13.463	—
	0.001	50.87	0.00519	51.14	—	9.993	0.06134	10.646	—
	0.005	39.28	0.01874	40.03	—	5.272	0.16848	6.340	—
	0.01	35.475	0.03328	36.697	0.043	3.881	0.25664	5.220	0.567
	0.05	28.381	0.11294	31.994	2.133	1.537	0.60540	3.896	19.962
	0.1	26.124	0.15146	30.787	2.748	0.835	0.75734	3.440	27.500
	0.5	25.051	0.16993	30.179	2.948	0.501	0.83006	2.948	30.179
	∞	25.051	0.16993	30.179	2.948	0.501	0.83006	2.948	30.179
0.05	0.00001	—	—	—	—	51.627	0.00297	51.781	—
	0.00005	—	—	—	—	29.676	0.00860	29.934	—
	0.0001	52.0	0.0014	52.1	—	23.296	0.01355	23.616	—
	0.0005	35.4	0.0045	35.6	—	13.095	0.03862	13.621	—
	0.001	30.43	0.00759	30.67	—	10.125	0.06031	10.774	—
	0.005	22.03	0.02652	22.63	—	5.356	0.16591	6.422	—
	0.01	19.397	0.04606	20.334	0.054	3.951	0.25247	5.286	0.166
	0.05	14.671	0.14681	17.195	2.241	1.590	0.59338	3.911	11.085
	0.1	13.269	0.19140	16.409	2.841	0.915	0.73191	3.413	14.856
	0.5	12.685	0.21009	16.058	3.019	0.634	0.78991	3.019	16.058
	∞	12.685	0.21009	16.058	3.019	0.634	0.78991	3.019	16.058
0.10	0.00001	80.290	0.00043	80.324	—	52.186	0.00287	52.336	—
	0.00005	49.632	0.00129	49.696	—	30.019	0.00830	30.270	—
	0.0001	40.682	0.00210	40.767	—	23.576	0.01308	23.888	—
	0.0005	26.249	0.00662	26.424	—	13.275	0.03732	13.789	—
	0.001	21.949	0.01094	22.192	—	10.276	0.05832	10.912	—
	0.005	13.833	0.03542	14.341	—	5.461	0.16087	6.509	—
	0.01	12.918	0.06131	13.762	0.064	4.044	0.24530	5.359	0.155
	0.05	9.227	0.18382	11.305	2.343	1.670	0.57485	3.927	7.491
	0.1	8.199	0.23388	10.702	2.933	1.022	0.70058	3.413	9.792
	0.5	7.817	0.25256	10.459	3.095	0.782	0.74744	3.095	10.459
	∞	7.817	0.25256	10.459	3.095	0.782	0.74744	3.095	10.459
0.25	0.00001	66.502	0.00079	66.555	—	53.276	0.00257	53.414	—
	0.00005	39.733	0.00234	39.827	—	30.710	0.00746	30.940	—
	0.0001	31.947	0.00375	32.067	—	24.150	0.01176	24.438	—
	0.0005	19.482	0.01130	19.704	—	13.665	0.03368	14.141	—
	0.001	15.843	0.01826	16.138	—	10.613	0.05273	11.204	—
	0.005	9.975	0.05639	10.571	—	5.717	0.14658	6.699	—
	0.01	8.236	0.09229	9.073	0.083	4.277	0.22473	5.517	0.130
	0.05	5.315	0.25334	7.118	2.526	1.885	0.52830	3.996	4.844
	0.1	4.567	0.31231	6.641	3.120	1.276	0.63474	3.494	6.141
	0.5	4.328	0.33120	6.471	3.267	1.082	0.66880	3.267	6.471
	∞	4.328	0.33120	6.471	3.267	1.082	0.66880	3.267	6.471

TABLE 5.16 Fundamental Solutions of the First Kind for Thermally Developing Flow in Concentric Annular Ducts (compiled from Shah and London [1]) (*Continued*)

r^*	x^*	$\Phi_{x,ii}^{(1)}$	$\theta_{x,mi}^{(1)}$	$\mathrm{Nu}_{x,ii}^{(1)}$	$\mathrm{Nu}_{x,oi}^{(1)}$	$\Phi_{x,oo}^{(1)}$	$\theta_{x,mo}^{(1)}$	$\mathrm{Nu}_{x,oo}^{(1)}$	$\mathrm{Nu}_{x,io}^{(1)}$
0.50	0.00001	60.470	0.00121	60.543	—	54.613	0.00220	54.733	—
	0.00005	35.541	0.00354	35.668	—	31.583	0.00637	31.785	—
	0.0001	28.295	0.00563	28.455	—	24.889	0.01007	25.142	—
	0.0005	16.711	0.01658	16.993	—	14.190	0.02897	14.614	—
	0.001	13.339	0.02642	13.701	—	11.077	0.04549	11.605	—
	0.005	7.930	0.07824	8.603	—	6.086	0.12797	6.979	—
	0.01	6.341	0.12488	7.246	0.092	4.622	0.19773	5.761	0.116
	0.05	3.713	0.32233	5.480	2.744	2.204	0.47154	4.171	3.751
	0.1	3.073	0.38995	5.037	3.374	1.615	0.56325	3.698	4.671
	0.5	2.885	0.40982	4.889	3.520	1.443	0.59018	3.520	4.889
	∞	2.885	0.40982	4.889	3.520	1.443	0.59018	3.520	4.889
1.0	0.00001	56.804	0.00171	56.901	—	56.804	0.00171	56.901	—
	0.00005	33.046	0.00498	33.211	—	33.046	0.00498	33.211	—
	0.0001	26.141	0.00788	26.349	—	26.141	0.00788	26.349	—
	0.0005	15.106	0.02288	15.459	—	15.106	0.02288	15.459	—
	0.001	11.895	0.03613	12.341	—	11.895	0.03613	12.341	—
	0.005	6.750	0.10371	7.531	—	6.750	0.10371	7.531	—
	0.01	5.235	0.16249	6.251	0.064	5.235	0.16249	6.251	0.064
	0.05	2.762	0.39926	4.597	3.112	2.762	0.39926	4.597	3.112
	0.1	2.168	0.47770	4.151	3.835	2.168	0.47770	4.151	3.835
	0.5	2.000	0.50000	4.000	4.000	2.000	0.50000	4.000	4.000
	∞	2.000	0.50000	4.000	4.000	2.000	0.50000	4.000	4.000

$$\Phi_{x,oi}^{(1)} = -\theta_{x,mi}^{(1)}\,\mathrm{Nu}_{x,oi} \qquad \theta_{x,ii}^{(1)} = 1 \qquad \theta_{x,oi}^{(1)} = 0$$
$$\Phi_{x,oi}^{(1)} = -\theta_{x,mo}^{(1)}\,\mathrm{Nu}_{x,io}^{(1)} \qquad \theta_{x,oo}^{(1)} = 1 \qquad \theta_{x,oi}^{(1)} = 0$$

The thermal entrance lengths for thermally developing flow with these four fundamental thermal boundary conditions are given in Table 5.20.

Using the four fundamental solutions presented in Tables 5.16–5.19, thermally developing flow with thermal boundary conditions different from the fundamental conditions presented in the section entitled "Four Fundamental Thermal Boundary Conditions" can be obtained by the superposition method. Three examples are detailed in the following sections.

Uniform Temperatures at Both Walls. As mentioned in the section "Fully Developed Flow," when $T_i \neq T_o$, the problem is designated as 1a, and its solution is expressed in terms of the following equations [1]:

$$T_{x,m}^{(1a)} = T_e + (T_i - T_e)\theta_{x,mi}^{(1)} + (T_o - T_e)\theta_{x,mo}^{(1)} \tag{5.129}$$

$$q_{x,i}^{\prime\prime(1a)} = \frac{k}{D_h}\left[(T_i - T_e)\Phi_{x,ii}^{(1)} + (T_o - T_e)\Phi_{x,io}^{(1)}\right] \tag{5.130}$$

$$q_{x,o}^{\prime\prime(1a)} = \frac{k}{D_h}\left[(T_i - T_e)\Phi_{x,oi}^{(1)} + (T_o - T_e)\Phi_{x,oo}^{(1)}\right] \tag{5.131}$$

$$\mathrm{Nu}_{x,i}^{1a} = \frac{(T_i - T_e)\Phi_{x,ii}^{(1)} + (T_o - T_e)\Phi_{x,mo}^{(1)}}{(T_o - T_e)(1 - \theta_{x,mo}^{(1)}) - (T_i - T_e)\theta_{x,mi}} \tag{5.132}$$

$$\mathrm{Nu}_{x,o}^{1a} = \frac{(T_o - T_e)\Phi_{x,oo}^{(1)} + (T_i - T_e)\Phi_{x,mo}^{(1)}}{(T_o - T_e)(1 - \theta_{x,mo}^{(1)}) - (T_i - T_e)\theta_{x,mi}} \tag{5.133}$$

TABLE 5.17 Fundamental Solutions of the Second Kind for Thermally Developing Flow in Concentric Annular Ducts (compiled from Shah and London [1])

r^*	x^*	$\theta^{(2)}_{x,ii}$	$\theta^{(2)}_{x,oi}$	$Nu^{(2)}_{x,ii}$	$\theta^{(2)}_{x,oo}$	$\theta^{(2)}_{x,io}$	$Nu^{(2)}_{x,oo}$
0.02	0.00005	—	—	—	0.027853	—	36.157
	0.0001	0.0115	—	86.9	0.035376	—	28.584
	0.0005	0.0164	—	61.2	0.062242	—	16.589
	0.001	0.01859	—	54.01	0.079885	—	13.164
	0.005	0.02429	—	41.85	0.146217	—	7.898
	0.01	0.027036	0.000002	38.093	0.193018	0.000097	6.502
	0.05	0.034110	0.001567	33.126	0.401928	0.078339	4.585
	0.1	0.038397	0.005296	32.729	0.603071	0.264808	4.741
	0.5	0.069792	0.036657	32.705	2.172011	1.832839	4.734
	1.0	0.109008	0.075872	32.705	4.132796	3.793624	4.734
	∞	∞	∞	32.705	∞	∞	4.734
0.05	0.00005	—	—	—	0.027554	—	36.544
	0.0001	0.01725	—	58.0	0.034990	—	28.894
	0.0005	0.02602	—	38.6	0.061520	—	16.774
	0.001	0.03034	—	33.17	0.078920	—	13.314
	0.005	0.04227	—	24.20	0.144201	—	7.990
	0.01	0.04837	0.000005	21.521	0.190125	0.000108	6.578
	0.05	0.064801	0.003872	18.091	0.393912	0.077433	4.916
	0.1	0.075142	0.012947	17.827	0.589334	0.258931	4.799
	0.5	0.151832	0.089110	17.811	2.113444	1.782195	4.792
	1.0	0.246620	0.184348	17.811	4.018208	3.686958	4.792
	∞	∞	∞	17.811	∞	∞	4.792
0.10	0.00005	0.017254	—	58.019	0.027252	—	36.940
	0.0001	0.021194	—	47.265	0.034596	—	29.212
	0.0005	0.033486	—	30.027	0.060756	—	16.976
	0.001	0.04043	—	24.96	0.077880	—	13.469
	0.005	0.06024	—	17.12	0.141892	—	8.083
	0.01	0.070738	0.000012	14.903	0.186700	0.000116	6.652
	0.05	0.100634	0.007642	12.128	0.383391	0.074621	4.961
	0.1	0.120267	0.024794	11.918	0.571090	0.247944	4.841
	0.5	0.265811	0.170197	11.906	2.052041	1.701967	4.834
	1.0	0.447629	0.352015	11.906	3.843224	3.520150	4.834
	∞	∞	∞	11.906	∞	∞	4.834
0.25	0.00005	0.021101	—	47.480	0.026676	—	37.713
	0.0001	0.026327	—	38.099	0.033832	—	29.840
	0.0005	0.043591	—	23.153	0.059215	—	17.357
	0.001	0.053885	—	18.838	0.075739	—	13.786
	0.005	0.086819	—	12.075	0.136845	—	8.275
	0.01	0.105856	0.000030	10.219	0.179005	0.000118	6.803
	0.05	0.165971	0.016492	7.938	0.358324	0.065968	4.042
	0.1	0.208792	0.054572	7.764	0.523546	0.218290	4.913
	0.5	0.528974	0.374448	7.753	1.803883	1.497792	4.905
	1.0	0.928974	0.774448	7.753	3.403883	3.097793	4.905
	∞	∞	∞	7.753	∞	∞	4.905
0.50	0.00005	0.023334	—	42.960	0.025994	—	38.668
	0.0001	0.029345	—	34.233	0.032919	—	30.626
	0.0005	0.049805	—	20.351	0.057314	—	17.864
	0.001	0.062474	—	16.356	0.073054	—	14.207
	0.005	0.105411	—	10.127	0.130327	—	8.548
	0.01	0.131911	0.000052	8.433	0.168984	0.000105	7.027
	0.05	0.224075	0.024263	6.353	0.326157	0.054896	5.186

TABLE 5.17 Fundamental Solutions of the Second Kind for Thermally Developing Flow in Concentric Annular Ducts (compiled from Shah and London [1])

r^*	x^*	$\theta^{(2)}_{x,ii}$	$\theta^{(2)}_{x,oi}$	$\mathrm{Nu}^{(2)}_{x,ii}$	$\theta^{(2)}_{x,oo}$	$\theta^{(2)}_{x,io}$	$\mathrm{Nu}^{(2)}_{x,oo}$
	0.1	0.294840	0.090599	6.192	0.464851	0.181605	5.046
	0.5	0.828452	0.623910	6.181	1.531881	1.247819	5.037
	1.0	1.495118	1.290576	6.181	2.865214	2.581152	5.037
	∞	∞	∞	6.181	∞	∞	5.037
1.0	0.00005	0.024940	—	40.257	0.024940	—	40.257
	0.0001	0.031498	—	31.950	0.031498	—	31.950
	0.0005	0.054322	—	18.754	0.054322	—	18.754
	0.001	0.068821	—	14.965	0.068821	—	14.965
	0.005	0.120121	—	9.081	0.120121	—	9.081
	0.01	0.153517	0.000080	7.490	0.153517	0.000080	7.490
	0.05	0.280307	0.041118	5.546	0.280307	0.000853	6.773
	0.1	0.385362	0.136006	5.395	0.385362	0.136066	5.395
	0.5	1.185714	0.935714	5.385	1.185714	0.935714	5.385
	∞	∞	∞	5.385	∞	∞	5.385

$$\theta^{(2)}_{x,mi} = \left(\frac{4r^*}{1+r^*}\right)x^* \qquad \Phi^{(2)}_{x,ii} = 1 \qquad \Phi^{(2)}_{x,oi} = 0$$

$$\theta^{(2)}_{x,mo} = \left(\frac{4r^*}{1+r^*}\right)x^* \qquad \Phi^{(2)}_{x,oo} = 1 \qquad \Phi^{(2)}_{x,oi} = 0$$

When $T_i = T_o$, the problem is designated as 1b. The solutions for this problem can be obtained from Eqs. 5.129 through 5.133 with $T_o = T_i$. The circumferentially averaged Nusselt number for this problem can be determined with the following expression:

$$\mathrm{Nu}_{x,T} = \frac{\mathrm{Nu}^{(1b)}_{x,o} + r^* \mathrm{Nu}^{(1b)}_{x,i}}{1 + r^*} \tag{5.134}$$

Uniform Heat Fluxes at Both Walls. As is mentioned in the section "Fully Developed Flow," when $q''_i \neq q''_o$, the problem is designated as 2a. The solution to this problem is expressed in terms of the following equations [1]:

$$T^{(2a)}_{x,i} = T_e + \frac{D_h}{k}\left[q''_i \theta^{(2)}_{x,ii} + q''_o \theta^{(2)}_{x,io}\right] \tag{5.135}$$

$$T^{(2a)}_{x,o} = T_e + \frac{D_h}{k}\left[q''_i \theta^{(2)}_{x,oi} + q''_o \theta^{(2)}_{x,oo}\right] \tag{5.136}$$

$$T^{(2a)}_{x,m} = T_e + \frac{D_h}{k}\left[q''_i \theta^{(2)}_{x,mi} + q''_o \theta^{(2)}_{x,mo}\right] \tag{5.137}$$

$$\mathrm{Nu}^{(2a)}_{x,i} = \frac{q''_i}{q''_i[\theta^{(2)}_{x,oo} - \theta^{(2)}_{x,mi}] - q''_o[\theta^{(2)}_{x,mo} - \theta^{(2)}_{x,io}]} \tag{5.138}$$

$$\mathrm{Nu}^{(2a)}_{x,o} = \frac{q''_o}{q''_o[\theta^{(2)}_{x,oo} - \theta^{(2)}_{x,mo}] - q''_i[\theta^{(2)}_{x,mi} - \theta^{(2)}_{x,oi}]} \tag{5.139}$$

TABLE 5.18   Fundamental Solutions of the Third Kind for Thermally Developing Flow in Concentric Annular Ducts (Shah and London [1])

r^*	x^*	$\Phi^{(3)}_{x,ii}$	$\theta^{(3)}_{x,oi}$	$\theta^{(3)}_{x,mi}$	$Nu^{(3)}_{x,ii}$	$\Phi^{(3)}_{x,oo}$	$\theta^{(3)}_{x,io}$	$\theta^{(3)}_{x,mo}$	$Nu^{(3)}_{x,oo}$
0.02	0.01	35.394	0.00012	0.0331	36.775	3.8810	0.00118	0.25616	5.217
	0.05	28.207	0.05729	0.13407	32.574	1.5241	0.34276	0.62191	4.031
	0.1	24.666	0.16699	0.23740	32.345	0.68894	0.69722	0.82751	3.994
	0.5	8.9413	0.69793	0.72350	32.337	0.00131	0.99942	0.99967	3.993
	1.0	2.5157	0.91501	0.92220	32.337	0.00000	1.00000	1.00000	3.993
	∞	0	1	1	32.337	0	1	1	3.993
0.05	0.01	19.405	0.00020	0.04562	20.332	3.9517	0.00140	0.25254	5.287
	0.05	14.605	0.07801	0.16980	17.592	1.5705	0.34310	0.61631	4.093
	0.1	12.273	0.21677	0.29722	17.464	0.71766	0.69444	0.82311	4.057
	0.5	3.2443	0.79290	0.81419	17.460	0.00148	0.99937	0.99963	4.057
	1.0	0.6151	0.96074	0.96477	17.460	0.00000	1.00000	1.00000	4.057
	∞	0	1	1	17.460	0	1	1	4.057
0.10	0.01	12.920	0.00032	0.061221	13.762	4.0442	0.00153	0.24535	5.359
	0.05	9.1601	0.10177	0.21362	11.648	1.6427	0.33688	0.60416	4.150
	0.1	7.3655	0.26918	0.36295	11.562	0.76967	0.68400	0.81292	4.114
	0.5	1.3705	0.86398	0.88144	11.560	0.00194	0.99920	0.99953	4.114
	1.0	0.1675	0.98337	0.98551	11.560	0.00000	1.00000	1.00000	4.114
	∞	0	1	1	11.560	0	1	1	4.114
0.25	0.01	8.2382	0.00058	0.09229	9.076	4.2773	0.00159	0.22480	5.518
	0.05	5.2488	0.14493	0.29346	7.429	1.8451	0.31296	0.56785	4.269
	0.1	3.8763	0.36106	0.47417	7.372	0.92804	0.64888	0.78074	4.233
	0.5	0.3664	0.93959	0.95028	7.371	0.00412	0.99844	0.99903	4.232
	1.0	0.0192	0.99683	0.99739	7.371	0.00000	1.00000	1.00000	4.232
	∞	0	1	1	7.371	0	1	1	4.232
0.50	0.01	6.3404	0.00087	0.1250	7.246	4.6214	0.00147	0.19789	5.762
	0.05	3.6492	0.18808	0.3692	5.785	2.1535	0.27967	0.51803	4.468
	0.1	2.4675	0.44405	0.5700	5.739	1.1814	0.59887	0.73330	4.430
	0.5	0.1156	0.97394	0.9798	5.738	0.01048	0.99644	0.99763	4.429
	1.0	0.0025	0.99943	0.9996	5.738	0.00003	0.99999	0.99999	4.429
	∞	0	1	1	5.738	0	1	1	4.429
1.0	0.01	5.2424	0.00119	0.16254	6.260	5.2421	0.00119	0.16254	6.260
	0.05	2.7028	0.23561	0.44867	4.902	2.7028	0.23561	0.44867	4.902
	0.1	1.6468	0.52780	0.66124	4.861	1.6468	0.52780	0.66124	4.861
	0.5	0.0337	0.99033	0.99306	4.861	0.0337	0.99033	0.99306	4.861
	∞	0	1	1	4.861	0	1	1	4.861

$$\Phi^{(3)}_{x,io} = 0 \qquad\qquad \theta^{(3)}_{x,ii} = 1 \qquad\qquad Nu^{(3)}_{x,oi} = 0$$

$$\Phi^{(3)}_{x,oi} = 0 \qquad\qquad \theta^{(3)}_{x,oo} = 1 \qquad\qquad Nu^{(3)}_{x,io} = 0$$

TABLE 5.19   Fundamental Solutions of the Fourth Kind for Thermally Developing Flow in Concentric Annular Ducts (Shah and London [1])

r^*	x^*	$\theta_{x,mi}^{(4)}$	$Nu_{x,ii}^{(4)}$	$Nu_{x,oi}^{(4)}$	$\theta_{x,mo}^{(4)}$	$Nu_{x,oo}^{(4)}$	$Nu_{x,io}^{(4)}$
0.02	0.01	0.0007837	38.093	0.030	0.039215	6.809	—
	0.05	0.0034689	32.689	1.976	0.196078	4.938	15.128
	0.1	0.0052665	31.260	2.648	0.37264	4.529	22.761
	0.5	0.0067801	30.181	2.948	1.1911	3.375	29.338
	1.0	0.0067830	30.179	2.948	1.5257	3.057	29.995
	∞	0.0067834	30.179	2.948	1.6568	2.948	30.179
0.05	0.01	0.0019050	21.522	0.037	0.037508	6.552	
	0.05	0.0084154	17.776	2.038	0.18544	4.807	8.355
	0.1	0.012796	16.798	2.713	0.34771	4.379	12.442
	0.5	0.016555	16.060	3.018	1.0081	3.289	15.728
	1.0	0.016563	16.059	3.019	1.2005	3.067	16.006
	∞	0.016563	16.058	3.019	1.2455	3.019	16.058
0.10	0.01	0.0036361	14.902	0.042	0.036254	6.646	0.034
	0.05	0.016093	11.864	2.093	0.17599	4.836	5.739
	0.1	0.024625	11.072	2.778	0.32377	4.350	8.309
	0.5	0.032288	10.460	3.095	0.83845	3.271	10.304
	1.0	0.032307	10.459	3.095	0.94176	3.116	10.442
	∞	0.032307	10.459	3.095	0.95614	3.095	10.459
0.25	0.01	0.0079989	10.225	0.050	0.031985	6.802	0.066
	0.05	0.035729	7.710	2.190	0.15253	4.880	3.800
	0.1	0.055780	7.040	2.908	0.27133	4.321	5.322
	0.5	0.076431	6.474	3.266	0.58532	3.344	6.421
	1.0	0.076523	6.471	3.267	0.61639	3.271	6.649
	∞	0.076523	6.471	3.267	0.61810	3.267	6.471
0.50	0.01	0.013332	8.433	0.055	0.026664	7.026	0.064
	0.05	0.060343	6.137	2.317	0.12508	4.998	3.007
	0.1	0.096747	5.503	3.095	0.21530	4.400	4.125
	0.5	0.14163	4.894	3.518	0.40000	3.554	4.870
	1.0	0.14203	4.889	3.520	0.40889	3.521	4.889
	∞	0.14203	4.889	3.520	0.40908	3.520	4.889
1.0	0.01	0.02009	7.495	0.254	0.02009	7.495	0.254
	0.05	0.09211	5.341	2.559	0.09211	5.341	2.559
	0.1	0.15285	4.723	3.453	0.15285	4.723	3.453
	0.5	0.24801	4.013	3.993	0.24801	4.013	3.993
	1.0	0.24998	4.000	4.000	0.24998	4.000	4.000
	∞	0.25000	4.000	4.000	0.25000	4.000	4.000

$$\Phi_{x,oi}^{(4)} = -\theta_{x,mi}^{(4)}\, Nu_{x,oi}^{(4)} \qquad \Phi_{x,ii}^{(4)} = 1 \qquad \theta_{x,oi}^{(4)} = 0$$

$$\Phi_{x,io}^{(4)} = -\theta_{x,mo}^{(4)}\, Nu_{x,io}^{(4)} \qquad \Phi_{x,oo}^{(4)} = 1 \qquad \theta_{x,io}^{(4)} = 0$$

TABLE 5.20 Thermal Entrance Lengths for Thermally Developing Flows in Concentric Annular Ducts (Shah and London [1])

r^*	$L_{th,i}^{*(1)}$	$L_{th,o}^{*(1)}$	$L_{th,i}^{*(2)}$	$L_{th,o}^{*(2)}$	$L_{th,i}^{*(3)}$	$L_{th,o}^{*(3)}$	$L_{th,i}^{*(4)}$	$L_{th,o}^{*(4)}$
0.02	0.05840	0.1650	0.02699	0.03901	0.02252	0.03001	0.07962	0.04241
0.05	0.06488	0.1458	0.03043	0.03886	0.02429	0.02970	0.09493	0.6638
0.10	0.06953	0.1311	0.03334	0.03911	0.02558	0.02960	0.1309	0.5284
0.25	0.07621	0.1126	0.03726	0.04006	0.02720	0.02964	0.1721	0.2875
0.50	0.08237	0.1003	0.03975	0.04090	0.02829	0.02956	0.1721	0.2875
1.00	0.09023	0.09023	0.04101	0.04101	0.02913	0.02913	0.2201	0.2201

Uniform Temperature at One Wall and Uniform Heat Flux at the Other.   In this case, the subscripts 1 and 2 refer to either the inside or the outside wall. The thermal entrance length solution to this problem is expressible in terms of the following equations [1]:

$$T_{x,2} = T_e + (T_1 - T_e)\theta_{x,21}^{(3)} + \frac{D_h}{k}\, q_2'' \theta_{x,22}^{(4)} \tag{5.140}$$

$$T_{x,m} = T_e + (T_1 - T_e)\theta_{x,m1}^{(3)} + \frac{D_h}{k}\, q_2'' \theta_{x,m2}^{(4)} \tag{5.141}$$

$$q_{w,1}'' = \frac{k}{D_h}\, (T_1 - T_e)\Phi_{x,11}^{(3)} + q_2'' \Phi_{x,12}^{(4)} \tag{5.142}$$

$$\mathrm{Nu}_{x,1} = \frac{(T_1 - T_e)\theta_{x,11}^{(3)}(q_2'' D_h/k)\Phi_{x,12}^{(4)}}{(T_1 - T_e)(1 - \theta_{x,m1}^{(3)})(q_2'' D_h/k)\theta_{x,m2}^{(4)}} \tag{5.143}$$

$$\mathrm{Nu}_{x,2} = \frac{1}{(\theta_{x,22}^{(4)} - \theta_{x,m2}^{(4)}) + [(T_1 - T_e)k/q_2'' D_h](\theta_{x,21}^{(3)}\theta_{x,m1}^{(3)})} \tag{5.144}$$

In Eqs. 5.129 through 5.144, the coefficients θ and Φ in various combinations of subscripts and superscripts can be found in Tables 5.16 through 5.19.

Simultaneously Developing Flow.   For the four fundamental thermal boundary conditions, the solutions to simultaneously developing velocity and temperature fields in concentric annuli with $r^* = 0.1, 0.25, 0.50$, and 1.0 and $\mathrm{Pr} = 0.01, 0.7$, and 10 have been obtained by Kakaç and Yücel [104]. Presented in Tables 5.21 to 5.23 are the results for $\mathrm{Pr} = 0.7$. The results for $\mathrm{Pr} = 0.01$ and $\mathrm{Pr} = 10$ have also been tabulated in Kakaç and Yücel [104].

Unlike thermally developing flow, the superposition method cannot be applied directly to the simultaneously developing flow because of the dependence of the velocity profile on the axial locations. However, certain influence coefficients are introduced to determine the local Nusselt number for simultaneous developing flow in concentric annuli with thermal boundary conditions that are different from the four fundamental conditions; the influence coefficients θ_1^* through θ_{12}^*, determined by Kakaç and Yücel [104] are listed in Tables 5.24 and 5.25.

Several examples of the application of the influence coefficients and fundamental solutions are detailed in the following paragraphs.

The fundamental solution of the first kind, presented in Table 5.21, is only valid if one of the duct walls is at the same temperature as the entering fluid. When the duct walls are maintained at uniform and equal or unequal temperatures T_i and T_o, the local Nusselt numbers $\mathrm{Nu}_{x,i}$ and $\mathrm{Nu}_{x,o}$ at the two walls can be determined from the following [1]:

$$\frac{\mathrm{Nu}_{x,i}}{\mathrm{Nu}_{x,ii}^{(1)}} = \frac{1 - [(T_o - T_e)/(T_i - T_e)]\theta_1^*}{1 - [(T_o - T_e)/(T_i - T_e)]\theta_2^*} \tag{5.145}$$

TABLE 5.21 Fundamental Solution of the First Kind for Simultaneously Developing Flow in Concentric Annular Ducts for Pr = 0.7 [104]

r^*	x^*	$Nu^{(1)}_{x,ii}$	$Nu^{(1)}_{x,oi}$	$Nu^{(1)}_{x,oo}$	$Nu^{(1)}_{x,io}$	r^*	x^*	$Nu^{(1)}_{x,ii}$	$Nu^{(1)}_{x,oi}$	$Nu^{(1)}_{x,oo}$	$Nu^{(1)}_{x,io}$
0.10	0.00005	68.030	—	57.450	—	0.50	0.00005	61.930	—	56.310	—
	0.0001	46.990	—	36.860	—		0.0001	39.820	—	36.560	—
	0.0005	26.960	—	17.480	—		0.0005	19.170	—	17.700	—
	0.001	22.020	—	12.910	—		0.0010	14.600	—	13.110	—
	0.005	15.030	—	6.690	—		0.0025	10.510	—	9.060	—
	0.01	13.330	0.0649	5.310	0.2473		0.01	7.085	0.1066	5.640	0.1485
	0.05	11.162	2.3436	3.856	7.6064		0.05	5.465	2.7629	4.108	3.8121
	0.1	10.567	2.9307	3.353	9.7125		0.1	5.030	3.3749	3.658	4.6824
	∞	10.450	3.0970	3.095	10.4603		∞	4.892	3.5228	3.518	4.8881
0.25	0.00005	63.500	—	56.990	—	1.00	0.00005	58.850	—	58.850	—
	0.0001	41.700	—	36.690	—		0.0001	37.615	—	37.615	—
	0.0005	21.310	—	17.620	—		0.0005	18.140	—	18.140	—
	0.0010	16.660	—	13.000	—		0.001	13.440	—	13.440	—
	0.0025	12.460	—	8.930	—		0.005	7.484	—	7.484	—
	0.01	8.870	0.0843	5.400	0.1772		0.01	6.126	0.1152	6.126	0.1152
	0.05	7.099	2.5334	3.859	4.9398		0.5	4.576	3.1321	4.576	3.1321
	0.1	6.626	3.1222	3.394	6.1506		0.1	4.140	3.8392	4.140	3.8392
	∞	6.471	3.2669	3.267	6.4713		∞	4.000	4.0000	4.000	4.0000

TABLE 5.22 Fundamental Solution of the Second and Third Kinds for Simultaneously Developing Flow in Concentric Annular Ducts for Pr = 0.7 [104]

r^*	x^*	$Nu^{(2)}_{x,ii}$	$Nu^{(2)}_{x,oo}$	$Nu^{(3)}_{x,ii}$	$Nu^{(3)}_{x,oo}$	r^*	x^*	$Nu^{(2)}_{x,ii}$	$Nu^{(2)}_{x,oo}$	$Nu^{(3)}_{x,ii}$	$Nu^{(3)}_{x,oo}$
0.10	0.00005	91.410	82.510	68.030	57.450	0.50	0.00005	83.340	81.370	61.930	56.310
	0.0001	64.670	55.520	46.990	36.860		0.0001	58.640	54.870	39.820	36.560
	0.0005	33.240	24.300	26.960	17.480		0.0005	25.900	24.490	19.170	17.700
	0.001	26.350	17.660	22.020	12.910		0.0010	19.240	17.860	14.600	13.110
	0.005	16.890	9.014	15.030	6.690		0.0025	13.395	12.090	10.510	9.060
	0.01	14.630	7.044	13.330	5.313		0.10	8.500	7.250	7.085	5.639
	0.05	12.043	4.969	11.500	4.099		0.05	6.351	5.188	5.777	4.405
	0.1	11.840	4.841	11.416	4.045		0.1	6.190	5.044	5.734	4.378
	∞	11.900	4.834	11.560	4.113		∞	6.181	5.036	5.738	4.429
0.25	0.00005	87.590	82.050	63.500	56.990	1.00	0.00005	83.620	83.620	58.850	58.850
	0.0001	60.170	55.240	41.700	36.690		0.0001	56.220	56.220	37.615	37.615
	0.0005	27.870	24.370	21.310	17.620		0.0005	24.880	24.880	18.140	18.140
	0.0010	21.160	17.720	16.660	13.000		0.01	18.270	18.270	13.440	13.440
	0.0025	15.220	11.940	12.460	8.930		0.005	9.601	9.601	7.484	7.484
	0.01	10.190	7.100	8.870	5.397		0.01	7.631	7.631	6.126	6.126
	0.05	7.931	5.046	7.412	4.142		0.05	5.542	5.542	4.890	4.890
	0.1	7.759	4.915	7.357	4.084		0.1	5.387	5.387	4.847	4.847
	∞	7.735	4.904	7.370	4.232		∞	5.384	5.384	4.860	4.860

TABLE 5.23 Fundamental Solution of the Fourth Kind for Simultaneously Developing Flow in Concentric Annular Ducts for Pr = 0.7 [104]

r^*	x^*	$Nu^{(4)}_{x,ii}$	$Nu^{(4)}_{x,oi}$	$Nu^{(4)}_{x,oo}$	$Nu^{(4)}_{x,io}$	r^*	x^*	$Nu^{(4)}_{x,ii}$	$Nu^{(4)}_{x,oi}$	$Nu^{(4)}_{x,oo}$	$Nu^{(4)}_{x,io}$
0.10	0.00005	91.410	—	82.510	—	0.50	0.00005	86.340	—	81.370	—
	0.0001	64.670	—	55.520	—		0.0001	58.640	—	54.870	—
	0.005	33.240	—	34.300	—		0.005	25.900	—	24.490	—
	0.001	26.350	—	17.660	—		0.0010	19.240	—	17.860	—
	0.0025	16.890	—	9.014	—		0.0025	13.395	—	12.090	—
	0.01	14.626	—	7.044	—		0.01	8.497	0.0752	7.249	0.0752
	0.05	11.780	−2.0910	4.854	5.6765		0.05	6.136	2.3150	5.000	3.0098
	0.1	10.997	−2.7783	4.352	8.2064		0.1	5.502	3.0995	4.399	4.1245
	∞	10.450	−3.960	3.111	10.4592		∞	4.890	3.5211	3.518	4.8912
0.25	0.00005	87.590	—	87.5900	—	1.00	0.00005	83.620	—	83.620	—
	0.001	60.170	—	60.170	—		0.0001	56.220	—	56.220	—
	0.005	27.870	—	24.370	—		0.0005	24.880	—	24.880	—
	0.0010	21.160	—	17.720	—		0.001	18.270	—	18.270	—
	0.0025	15.220	—	11.940	—		0.0025	9.601	—	9.601	—
	0.01	10.190	—	7.100	0.0939		0.01	7.631	0.0503	7.631	0.0503
	0.05	7.703	2.197	4.884	3.7974		0.05	5.339	2.5453	5.339	2.5453
	0.1	7.034	2.9006	4.321	5.3139		0.1	4.719	3.4571	4.719	3.4571
	∞	6.471	3.2671	3.267	6.4714		∞	4.000	4.0000	4.000	4.0000

$$\frac{Nu_{x,o}}{Nu^{(1)}_{x,oo}} = \frac{1 - [(T_i - T_e)/(T_o - T_e)]\theta_3^*}{1 - [(T_i - T_e)/(T_o - T_e)]\theta_4^*} \tag{5.146}$$

where $Nu^{(1)}_{x,ii}$ and $Nu^{(1)}_{x,oo}$ are available in Table 5.21 and θ_1^*, θ_2^*, θ_3^*, and θ_4^* are listed in Table 5.24.

The fundamental solution of the second kind presented in Table 5.22 is valid in the case of one duct wall's being adiabatic, that is, either $q_i'' = 0$ or $q_o'' = 0$. However, for both duct walls subjected to uniform and equal or unequal wall fluxes q_i'' and q_o'', the local Nusselt numbers $Nu_{x,i}$ and $Nu_{x,o}$ at the two walls can be determined from the following expression [1]:

$$\frac{Nu_{x,i}}{Nu^{(2)}_{x,ii}} = \frac{1}{1 - (q_o''/q_i'')\theta_5^*} \tag{5.147}$$

$$\frac{Nu_{x,o}}{Nu^{(2)}_{x,oo}} = \frac{1}{1 - (q_i''/q_o'')\theta_6^*} \tag{5.148}$$

where $Nu^{(2)}_{x,ii}$ and $Nu^{(1)}_{x,oo}$ are taken from Table 5.22 and θ_1^*, θ_2^*, θ_3^*, and θ_4^* are listed in Table 5.24.

The third example is for the case of uniform heat flux at the outer wall and uniform temperature on the inner wall, that is, $q_w'' = q_o''$ at $r = r_o$ and $T_w = T_i$ at $r = r_i$. The local Nusselt numbers at the two walls are determined from the fundamental solutions of the third and fourth kinds from Tables 5.22 and 5.23 and the influence coefficients from Table 5.25, which are given as

$$\frac{Nu_{x,i}}{Nu^{(3)}_{x,ii}} = \frac{1 - [(q_o''D_h/k)/(T_i - T_e)]\theta_7^*}{1 - [(T_i - T_e)/(T_o - T_e)]\theta_8^*} \tag{5.149}$$

$$\frac{Nu_{x,o}}{Nu^{(4)}_{x,oo}} = \frac{1}{1 - [(T_i - T_e)/(q_o''D_h/k)]\theta_{12}^*} \tag{5.150}$$

If $T_w = T_o$ at $r = r_o$ and $q_w'' = q_i''$ at $r = r_i$, the local Nusselt numbers $Nu_{x,i}$ and $Nu_{x,o}$ at the two walls are given by

TABLE 5.24 Influence Coefficients from Fundamental Solutions of the Third and Fourth Kinds for Simultaneously Developing Flow in Concentric Annular Ducts for Pr = 0.7 [104]

r^*	x^*	θ_1^*	θ_2^*	θ_3^*	θ_4^*	θ_5^*	θ_6^*
0.10	0.00005	—	0.0191	—	0.0022	0.0160	0.0015
	0.0001	—	0.0273	—	0.0033	0.228	0.0020
	0.0005	—	0.0595	—	0.0086	0.0587	0.0044
	0.001	—	0.0891	—	0.0136	0.0944	0.0064
	0.005	—	0.1996	—	0.0460	0.3060	0.0165
	0.01	0.0054	0.2930	0.0011	0.0849	0.5289	0.0256
	0.05	0.4885	0.7167	0.2698	0.4434	1.2855	0.0538
	0.1	0.8442	0.9184	0.6908	0.7892	1.3705	0.0565
	∞	1.0000	1.0000	1.0000	1.0000	1.3835	0.0562
0.22	0.00005	—	0.0169	—	0.0044	0.0136	0.0031
	0.0001	—	0.0241	—	0.0066	0.0187	0.0043
	0.0005	—	0.0530	—	0.0159	0.0433	0.0096
	0.0010	—	0.0762	—	0.0242	0.0667	0.0140
	0.0025	—	0.1246	—	0.0440	0.1209	0.0238
	0.01	0.0056	0.2744	0.0020	0.1263	0.3242	0.0565
	0.05	0.4996	0.7182	0.3588	0.5474	0.7443	0.1189
	0.1	0.8596	0.9262	0.7909	0.8604	0.7897	0.1249
	∞	1.0000	1.0000	1.0000	1.0000	0.7932	0.1250
0.50	0.00005	—	0.0142	—	0.072	0.0113	0.0051
	0.0001	—	0.0202	—	0.0106	0.0153	0.0070
	0.0005	—	0.0449	—	0.0245	0.0336	0.0160
	0.0010	—	0.0652	—	0.0365	0.0506	0.0235
	0.0025	—	0.1081	—	0.0637	0.0887	0.0400
	0.01	0.0051	0.2481	0.0030	0.1674	0.2252	0.0960
	0.05	0.4922	0.7058	0.4922	0.7058	0.4979	0.2037
	0.1	0.8624	0.9265	0.8280	0.8971	0.5270	0.2147
	∞	1.0000	1.0000	1.0000	1.0000	0.5288	0.2160
1.00	0.00005	—	0.0087	—	0.0087	0.0064	0.0064
	0.0001	—	0.0138	—	0.0138	0.0095	0.0095
	0.0005	—	0.0339	—	0.0339	0.0234	0.0234
	0.001	—	0.0503	—	0.0503	0.0354	0.0354
	0.005	—	0.1321	—	0.1321	0.0951	0.0951
	0.01	0.0041	0.2101	0.0041	0.2101	0.1512	0.1512
	0.05	0.4635	0.4745	0.4635	0.4745	0.3257	0.3257
	0.1	0.8512	0.9165	0.8512	0.9165	0.3427	0.3247
	∞	1.0000	1.0000	1.0000	1.0000	0.3460	0.3460

$$\frac{\mathrm{Nu}_{x,i}}{\mathrm{Nu}_{x,ii}^{(4)}} = \frac{1}{1 - [(T_o - T_e)/(q_i'' D_h/k)]\theta_{11}^*} \tag{5.151}$$

$$\frac{\mathrm{Nu}_{x,o}}{\mathrm{Nu}_{x,oo}^{(3)}} = \frac{1 - [(q_i'' D_h/k)/(T_o - T_e)]\theta_{11}^*}{1 - [(q_i'' D_h/k)/(T_o - T_e)]\theta_{10}^*} \tag{5.152}$$

where $\mathrm{Nu}_{x,oo}^{(3)}$ and $\mathrm{Nu}_{x,ii}^{(4)}$ are found in Tables 5.22 and 5.23. The terms θ_9^*, θ_{10}^*, and θ_{11}^* are listed in Table 5.25.

It should be noted that the fundamental solution of the third kind, which is presented in Table 5.22, is restricted in the case of uniform temperature (different from T_e) at one wall with

TABLE 5.25 Influence Coefficients from Fundamental Solutions of the Third and Fourth Kinds for Simultaneously Developing Flow in Concentric Annular Ducts for Pr = 0.7 [104]

r^*	x^*	θ_7^*	θ_8^*	θ_9^*	θ_{10}^*	θ_{11}^*	θ_{12}^*
0.10	0.00005	—	0.0002	—	—	1.7416	0.1754
	0.0001	—	0.0004	—	0.0001	1.7578	0.1782
	0.0005	—	0.0018	—	0.0002	1.9608	0.1956
	0.001	—	0.0037	—	0.0004	2.2146	0.2198
	0.005	—	0.0188	—	0.0023	3.2452	0.3348
	0.01	0.0003	0.0387	0.0001	0.0051	3.9806	0.4318
	0.05	0.1102	0.2233	0.0217	0.0425	3.0630	0.5460
	0.1	0.6352	0.5080	0.0944	0.1374	1.3722	0.4086
	∞	∞	∞	∞	∞	0.0000	0.0000
0.25	0.00005	—	0.0002	—	0.0001	1.4746	0.3572
	0.0001	—	0.0003	—	0.0001	1.4424	0.3562
	0.0005	—	0.0016	—	0.0004	1.4536	0.3668
	0.0010	—	0.0033	—	0.0009	1.5770	0.3966
	0.0025	—	0.0083	—	0.0023	1.8228	0.4630
	0.01	0.0003	0.0353	0.0001	0.0106	2.5066	0.6700
	0.05	0.1105	0.2159	0.0448	0.0849	1.9258	0.7256
	0.1	0.3732	0.5168	0.1860	0.2612	0.9059	0.4882
	∞	∞	∞	∞	∞	0.0000	0.0000
0.50	0.00005	—	0.0002	—	0.0001	1.2192	0.5752
	0.0001	—	0.0003	—	0.0002	1.1744	0.5700
	0.0005	—	0.0014	—	0.0014	1.1354	0.5746
	0.0010	—	0.0027	—	0.0027	1.2106	0.6104
	0.0025	—	0.0070	—	0.0070	1.3650	0.6920
	0.01	0.0003	0.0307	0.0002	0.0170	1.8152	0.9452
	0.05	0.1036	0.1991	0.0813	0.1456	1.4446	0.9036
	0.1	0.3620	0.5034	0.2612	0.4486	0.7276	0.5518
	∞	∞	∞	∞	∞	0.0000	0.0000
1.00	0.00005	—	0.0001	—	0.0001	0.7224	0.7224
	0.0001	—	0.0002	—	0.0002	0.7670	0.7670
	0.0005	—	0.0010	—	0.0010	0.8152	0.8152
	0.001	—	0.0021	—	0.0021	0.8750	0.8750
	0.005	—	0.0112	—	0.0112	1.1202	1.1202
	0.01	0.0003	0.0241	0.0003	0.0241	1.3146	1.3146
	0.05	0.0881	0.1683	0.0881	0.1683	1.1274	1.1274
	0.1	0.3243	0.4548	0.3243	0.4548	0.6208	0.6208
	∞	∞	∞	∞	∞	0.0000	0.0000

the other adiabatic. The fundamental solution of the fourth kind (Table 5.23) is valid when one wall is at uniform temperature ($\neq T_e$) and the other wall has a uniform temperature equal to T_e. The reader should therefore exercise caution when using Tables 5.21–5.25.

Effects of Eccentricity. In practice, a perfect concentric annular duct cannot be achieved because of manufacturer tolerances, installation, and so forth. Therefore, eccentric annular ducts are frequently encountered. The velocity profile for fully developed flow in an eccentric annulus has been analyzed by Piercy et al. [105]. Based on Piercy's solution, Shah and London [1] have derived the friction factor formula, as follows:

$$f\,Re = 16(1 - r^{*2})(1 - r^*)^2 \times \left(1 - r^{*4} + Z - 8e^{*2}(1 - r^*)^2 S^2 \sum_{n=1}^{\infty} \frac{n\,\exp[-n(\alpha + \beta)]}{\sinh\,[n(\beta - \alpha)]}\right) \quad (5.153)$$

where

$$S = \frac{1 - r^*}{2e^*} (1 - e^{*2})^{1/2} \left[\left(\frac{1 + r^*}{1 - r^*} \right)^2 - e^{*2} \right]^{1/2}$$ (5.154)

$$\alpha = \sinh^{-1} S \qquad \beta = \sinh^{-1} (S/r^*)$$

$$Z = \frac{4e^{*2}(1 - r^*)^2}{\alpha - \beta} S^2$$

where $r^* = r_i/r_o$ and $e^* = \varepsilon/(r_o - r_i)$. The term ε denotes the distance between the centers of the two circular walls. Equation 5.153 is valid for $0 < e^* < 1$ and $0 \le r^* < 1$.

For $e^* = 1$ and $0 \le r^* < 1$, the following equation obtained by Tiedt [106] can be applied:

$$f\,\mathrm{Re} = \frac{16(1 - r^{*2})(1 - r^*)^2}{1 - r^{*4} - 4r^{*2}\psi'[1/(1 - r^*)]}$$ (5.155)

where ψ' is the so-called trigamma function with the argument $1/(1 - r^*)$. This is given by

$$\psi'\left(\frac{1}{1 - r^*} \right) = \sum_{n=0}^{\infty} \left(\frac{1}{n + [1/(1 - r^*)]} \right)^2$$ (5.156)

From the results obtained by Becker [107] for $0 \le e^* \le 1$ and $r^* \to 1$, Tiedt [106] also demonstrated that

$$f\,\mathrm{Re} = \frac{24}{1 + 1.5e^{*2}}$$ (5.157)

It should be noted that when $e^* = 0$, the eccentric annular duct is reduced to a concentric annular duct.

Cheng and Hwang [108] analyzed the heat transfer problem in eccentric annular ducts. The Nusselt numbers for fully developed flow in eccentric annular ducts with the Ⓗ1 and Ⓗ2 thermal boundary conditions are given in Table 5.26. For eccentric annular ducts with boundary conditions different from the four described in the section entitled "Four Fundamental

TABLE 5.26 Nusselt Number Nu_{H1} and Nu_{H2} for Fully Developed Laminar Flow in Eccentric Annular Ducts [1, 108]

r^*	$e^* = 0$	0.01	0.05	0.1	0.2	0.4	0.6	0.8	0.9	0.95	0.99
					Nu_{H1}						
0.25	7.804	7.800	—	7.419	6.524	4.761	3.735	3.203	3.038	—	2.925
0.50	8.117	8.111	—	7.608	6.473	4.393	3.247	2.644	2.446	—	2.305
0.75	8.214	8.208	—	7.659	6.432	4.227	3.024	2.384	2.171	—	2.016
0.90	8.232	8.226	—	7.667	6.422	4.192	2.975	2.324	2.106	—	1.947
					Nu_{H2}						
0.02	—	—	6.347	6.224	5.826	4.913	4.537	4.157	4.149	4.160	—
0.05	—	—	6.777	6.582	5.956	4.701	4.289	3.849	3.827	3.833	—
0.10	—	—	7.139	6.815	5.854	4.228	3.518	3.334	3.322	3.341	—
0.20	—	—	7.430	6.793	5.191	3.191	2.517	2.360	2.361	2.370	—
0.30	—	—	7.452	6.395	4.241	2.244	1.694	1.571	1.571	1.575	—
0.40	—	—	7.267	5.688	3.198	1.484	1.080	0.990	0.988	0.989	—
0.50	—	—	6.839	4.691	2.206	0.918	0.650	0.590	0.586	0.586	—
0.60	—	—	6.064	3.463	1.364	0.522	0.362	0.326	0.323	0.322	—
0.70	—	—	4.776	2.154	0.725	0.262	0.179	0.160	0.158	0.157	—
0.80	—	—	3.273	1.002	0.299	0.104	0.070	0.062	0.061	0.061	—
0.90	—	—	0.887	0.246	0.068	0.023	0.015	0.013	0.013	0.013	—
0.95	—	—	0.229	0.059	0.016	0.005	0.003	0.003	0.003	0.003	—

Thermal Boundary Conditions," caution must be taken in using the superposition technique. The reader is strongly recommended to consult the literature.

Turbulent Flow

Presented in this section are the friction factor and Nusselt number for turbulent flow and heat transfer in concentric annular ducts. The effects of eccentricity on the friction factor and Nusselt number are also discussed.

Critical Reynolds Number.   For concentric annular ducts, the critical Reynolds number at which turbulent flow occurs varies with the radius ratio. Hanks [109] has determined the lower limit of Re_{crit} for concentric annular ducts from a theoretical perspective for the case of a uniform flow at the duct inlet. This is shown in Fig. 5.16. The critical Reynolds number is within ±3 percent of the selected measurements for air and water [109].

Fully Developed Flow.   Knudsen and Katz [110] obtained the following velocity distributions for fully developed turbulent flow in a smooth concentric annular duct in terms of wall coordinates u^+ and y^+:

$$u_o^+ = 3.0 + 2.6492 \ln y_o^+ \qquad \text{for } r_m \leq r \leq r_o \tag{5.158}$$

$$u_i^+ = 6.2 + 1.9109 \ln y_i^+ \qquad \text{for } r_i \leq r \leq r_m \tag{5.159}$$

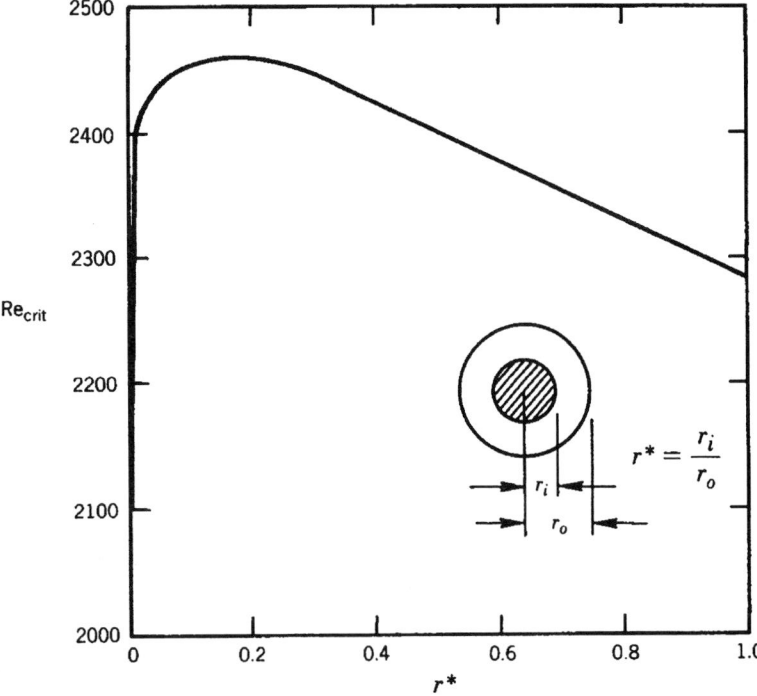

FIGURE 5.16   Lower limits of the critical Reynolds numbers for concentric annular ducts with uniform velocity at the inlet [109].

where
$$u_o^+ = \frac{u}{u_{t,o}} \qquad u_i^+ = \frac{u}{u_{t,i}}$$

$$y_o^+ = \frac{u_{t,o}(r_o - r)}{\nu} \qquad y_i^+ = \frac{u_{t,i}(r - r_i)}{\nu} \qquad (5.160)$$

and
$$u_{t,o} = \sqrt{\frac{\tau_o}{\rho}} \qquad u_{t,i} = \sqrt{\frac{\tau_i}{\rho}}$$

The radius of maximum velocity r_m^* in Eqs. 5.158 and 5.159 can be determined by the formula obtained by Kays and Leung [111]. It follows:

$$r_m^* = \frac{r_m}{r_o} = r^{*0.343}(1 + r^{*0.657} - r^*) \qquad (5.161)$$

A critical review of the extensive friction factor data has been made by Jones and Leung [112]. The researchers recommend that the fully developed friction factor formulas for smooth circular ducts given in Table 5.8 be used for calculating the friction factor for concentric annular ducts by replacing $2a$ with the laminar equivalent diameter D_l for concentric annular ducts. The term D_l is defined by

$$D_l = D_h \frac{1 + r^{*2} + (1 - r^{2*})/\ln r^*}{(1 - r^*)^2} \qquad (5.162)$$

where $r^* = r_i/r_o$; $D_h = 2(r_o - r_i)$ and r_i and r_o are the radii of the inner and outer tubes, respectively.

The fully developed Nusselt numbers Nu_o and Nu_i at the outer and inner walls of a smooth concentric annular duct can be determined from the following relations for uniform wall heat fluxes q_o'' and q_i'' at the outer and inner walls:

$$\text{Nu}_o = \frac{h_o D_h}{k} = \frac{\text{Nu}_{oo}}{1 - (q_i''/q_o'')\theta_o^*} \qquad (5.163)$$

$$\text{Nu}_i = \frac{h_i D_h}{k} = \frac{\text{Nu}_{ii}}{1 - (q_o''/q_i'')\theta_i^*} \qquad (5.164)$$

where
$$q_o'' = h_o(T_o - T_m), \qquad q_i'' = h_i(T_i - T_m) \qquad (5.165)$$

The terms T_o and T_i denote the duct wall temperatures at the outer and inner walls. The temperature difference $T_o - T_i$ is given by

$$T_o - T_i = \frac{D_h}{k}\left[q_o''\left(\frac{1}{\text{Nu}_{oo}} + \frac{\theta_i^*}{\text{Nu}_{ii}}\right) - q_i''\left(\frac{1}{\text{Nu}_{ii}} + \frac{\theta_o^*}{\text{Nu}_{oo}}\right)\right] \qquad (5.166)$$

The Nusselt numbers Nu_{oo} and Nu_{ii}, as well as the influence coefficients θ_o^* and θ_i^* in Eqs. 5.163, 5.164, and 5.166 are provided by Kays and Leung [111]. These are given in Table 5.27 for wide ranges of Re and Pr and for $r^* = 0.1, 0.2, 0.5$, and 0.8.

For $r^* = 1$, the concentric annular duct is reduced to a parallel plate duct. The applicable results are given in Table 5.28, the simple Nu being used for the Nusselt number at the heated wall.

It should be noted that for laminar flow (Re < 2300) in parallel plate ducts, Nu is equal to 5.385 and θ^* is equal to 0.346 for all values of Pr.

Dwyer [113] has developed semiempirical equations for liquid metal flow (Pr < 0.03) in a concentric annular duct $(0 < r^* \le 1)$ with one wall subjected to uniform heat flux and the other

TABLE 5.27 Nusselt Numbers and Influence Coefficients for Fully Developed Turbulent Flow in a Concentric Annular Duct with Uniform Heat Flux at One Wall and the Other Wall Insulated [111]

$r^* = 0.10$	Heating from outer wall with inner wall insulated									
	Re = 10^4		3×10^4		10^5		3×10^5		10^6	
Pr	Nu_{oo}	θ^*	Nu_{oo}	θ^*	Nu_{oo}	θ^*	Nu_{oo}	θ^*	Nu_{oo}	θ^*
0	6.00	0.077	6.12	0.079	6.32	0.081	6.50	0.084	6.68	0.085
0.001	6.00	0.077	6.12	0.079	6.40	0.082	6.60	0.082	7.20	0.082
0.003	6.00	0.077	6.24	0.081	6.55	0.083	7.34	0.082	10.8	0.071
0.01	6.13	0.076	6.50	0.081	7.80	0.077	12.1	0.067	26.4	0.052
0.03	6.45	0.076	7.95	0.075	13.7	0.065	28.2	0.051	71.8	0.036
0.5	24.8	0.039	53.4	0.032	134	0.028	320.0	0.025	860.0	0.022
0.7	29.8	0.032	66.0	0.028	167	0.024	409.0	0.022	1100.0	0.020
1.0	36.5	0.026	81.8	0.023	212	0.021	520.0	0.019	1430.0	0.017
3	61.5	0.013	147.0	0.013	395	0.012	1000.0	0.012	2830.0	0.011
10	99.2	0.006	246.0	0.006	685	0.006	1780.0	0.006	5200.0	0.006
30	143.0	0.003	360.0	0.003	1030	0.003	2720.0	0.003	8030.0	0.003
100	205.0	0.002	525.0	0.002	1500	0.002	4030.0	0.002	12,100	0.002
1000	378.0		980.0		2850		7600.0		23,000	

$r^* = 0.10$	Heating from inner wall with outer wall insulated									
	Re = 10^4		3×10^4		10^5		3×10^5		10^6	
Pr	Nu_{ii}	θ_i^*	Nu_{ii}	θ_i^*	Nu_{ii}	θ_i^*	Nu_{ii}	θ_i^*	Nu_{ii}	θ_i^*
0	11.5	1.475	11.5	1.502	11.5	1.500	11.5	1.460	11.6	1.477
0.001	11.5	1.475	11.5	1.502	11.5	1.480	11.7	1.462	12.3	1.410
0.003	11.5	1.475	11.5	1.475	11.7	1.473	12.6	1.391	17.0	1.124
0.01	11.8	1.472	11.8	1.442	13.5	1.323	19.4	1.090	39.0	0.760
0.03	12.5	1.472	14.1	1.330	21.8	1.027	42.0	0.760	103.0	0.526
0.5	40.8	0.632	81.0	0.486	191.0	0.394	443.0	0.339	1160.0	0.294
0.7	48.5	0.512	98.0	0.407	235.0	0.338	550.0	0.292	1510.0	0.269
1.0	58.5	0.412	120.0	0.338	292.0	0.286	700.0	0.256	1910.0	0.232
3	93.5	0.202	206.0	0.175	535.0	0.162	1300.0	0.152	3720.0	0.148
10	140.0	0.089	328.0	0.081	890.0	0.078	2300.0	0.078	6700.0	0.077
30	195.0	0.041	478.0	0.039	1320.0	0.038	3470.0	0.038	10,300.0	0.040
100	272.0	0.017	673.0	0.015	1910.0	0.015	5030.0	0.016	15,200.0	0.018
1000	486.0	0.004	1240.0	0.003	3600.0	0.003	9600.0	0.004	28,700.0	0.004

$r^* = 0.2$	Heating from outer wall with inner wall insulated									
	Re = 10^4		3×10^4		10^5		3×10^5		10^6	
Pr	Nu_{oo}	θ_o^*	Nu_{oo}	θ_o^*	Nu_{oo}	θ_o^*	Nu_{oo}	θ_o^*	Nu_{oo}	θ_o^*
0	5.83	0.140	5.92	0.145	6.10	0.151	6.16	0.152	6.35	0.157
0.001	5.83	0.140	5.92	0.144	6.10	0.151	6.30	0.154	6.92	0.153
0.003	5.83	0.140	6.00	0.146	6.22	0.150	6.90	0.150	10.2	0.136
0.01	5.95	0.140	6.20	0.146	7.40	0.144	11.4	0.131	24.6	0.102
0.03	6.22	0.140	7.55	0.140	12.7	0.125	26.3	0.098	80.0	0.074
0.5	22.5	0.071	51.5	0.064	130.0	0.055	310.0	0.049	823.0	0.044
0.7	29.4	0.063	64.3	0.055	165.0	0.049	397.0	0.044	1070.0	0.040
1.0	35.5	0.051	80.0	0.046	206.0	0.042	504.0	0.039	1390.0	0.035
3	60.0	0.026	145.0	0.026	390.0	0.024	980.0	0.024	2760.0	0.023
10	98.0	0.013	243.0	0.013	680.0	0.012	1750.0	0.012	4980.0	0.012
30	142.0	0.004	360.0	0.006	1030.0	0.006	2700.0	0.006	7850.0	0.006
100	205.0	0.003	520.0	0.003	1500.0	0.003	4000.0	0.003	12,000.0	0.003
1000	380.0	0.001	980.0	0.001	2830.0	0.001	7500.0	0.001	22,500.0	0.001

TABLE 5.27 Nusselt Numbers and Influence Coefficients for Fully Developed Turbulent Flow in a Concentric Annular Duct with Uniform Heat Flux at One Wall and the Other Wall Insulated [111] (*Continued*)

$r^* = 0.2$	Heating from inner wall with outer wall insulated									
	$Re = 10^4$		3×10^4		10^5		3×10^5		10^6	
Pr	Nu_{ii}	θ_i^*	Nu_{ii}	θ_i^*	Nu_{ii}	θ_i^*	Nu_{ii}	θ_i^*	Nu_{ii}	θ_i^*
0	8.40	1.009	8.30	1.028	8.30	1.020	8.30	1.038	8.30	1.020
0.001	8.40	1.009	8.40	1.040	8.30	1.020	8.40	1.014	8.90	0.976
0.003	8.40	1.009	8.40	1.027	8.50	1.025	9.05	0.980	12.5	0.834
0.01	8.50	1.000	8.60	1.018	9.70	0.944	14.0	0.796	33.6	0.748
0.03	9.00	1.012	10.1	0.943	15.8	0.771	31.7	0.600	81.0	0.374
0.5	31.2	0.520	64.0	0.398	157.0	0.333	370.0	0.295	980.0	0.262
0.7	38.6	0.412	79.8	0.338	196.0	0.286	473.0	0.260	1270.0	0.235
1.0	46.8	0.339	99.0	0.284	247.0	0.248	600.0	0.229	1640.0	0.209
3.0	77.4	0.172	175.0	0.151	465.0	0.143	1150.0	0.137	3250.0	0.135
10.0	120.0	0.120	290.0	0.074	800.0	0.072	2050.0	0.073	6000.0	0.077
30.0	172.0	0.036	428.0	0.034	1210.0	0.035	3150.0	0.036	9300.0	0.038
100.0	243.0	0.014	617.0	0.014	1760.0	0.015	4630.0	0.016	13,800.0	0.016
1000.0	448.0	0.004	1400.0	0.002	3280.0	0.002	8800.0	0.004	26,000.0	0.003

$r^* = 0.5$	Heating from outer wall with inner wall insulated									
	$Re = 10^4$		3×10^4		10^5		3×10^5		10^6	
Pr	Nu_{oo}	θ_o^*	Nu_{oo}	θ_o^*	Nu_{oo}	θ_o^*	Nu_{oo}	θ_o^*	Nu_{oo}	θ_o^*
0	5.66	0.281	5.78	0.294	5.80	0.296	5.83	0.302	5.95	0.310
0.001	5.66	0.281	5.78	0.294	5.80	0.296	5.92	0.302	6.40	0.304
0.003	5.66	0.281	5.78	0.294	5.85	0.294	6.45	0.301	9.00	0.278
0.01	5.73	0.281	5.88	0.289	6.80	0.289	10.3	0.264	22.6	0.217
0.03	6.03	0.279	7.05	0.284	11.6	0.258	24.4	0.214	64.0	0.163
0.5	2.6	0.162	49.8	0.142	125.0	0.123	298.0	0.111	795.0	0.098
0.7	28.3	0.137	62.0	0.119	158.0	0.107	380.0	0.097	1040.0	0.090
1.0	34.8	0.111	78.0	0.101	200.0	0.092	490.0	0.085	1340.0	0.078
3.0	60.5	0.059	144.0	0.058	384.0	0.055	960.0	0.054	2730.0	0.052
10.0	100.0	0.028	246.0	0.028	680.0	0.028	1750.0	0.028	5030.0	0.028
30.0	143.0	0.013	365.0	0.013	1030.0	0.014	2700.0	0.014	8000.0	0.015
100.0	207.0	0.006	530.0	0.006	1500.0	0.006	4000.0	0.006	12,000.0	0.006
1000.0	387.0	0.001	990.0	0.001	2830.0	0.001	7600.0	0.001	23,000.0	0.001

$r^* = 0.5$	Heating from inner wall with outer wall insulated									
	$Re = 10^4$		3×10^4		10^5		3×10^5		10^6	
Pr	Nu_{ii}	θ_i^*	Nu_{ii}	θ_i^*	Nu_{ii}	θ_i^*	Nu_{ii}	θ_i^*	Nu_{ii}	θ_i^*
0	6.28	0.620	6.30	0.632	6.30	0.651	6.30	0.659	6.30	0.654
0.001	6.28	0.620	6.30	0.632	6.30	0.651	6.40	0.659	6.75	0.644
0.003	6.28	0.620	6.30	0.632	6.40	0.656	6.85	0.637	9.40	0.585
0.01	6.37	0.622	6.45	0.636	7.30	0.623	10.8	0.540	23.2	0.427
0.03	6.75	0.627	7.53	0.598	12.0	0.533	24.8	0.430	35.5	0.333
0.5	24.6	0.343	52.0	0.292	130.0	0.253	310.0	0.299	835.0	0.208
0.7	30.9	0.300	66.0	0.258	166.0	0.225	400.0	0.206	1080.0	0.185
1.0	38.2	0.247	83.5	0.218	212.0	0.208	520.0	0.183	1420.0	0.170
3.0	66.8	0.129	152.0	0.121	402.0	0.115	1010.0	0.114	2870.0	0.111
10.0	106.0	0.059	260.0	0.059	715.0	0.059	1850.0	0.059	5400.0	0.061
30.0	153.0	0.028	386.0	0.027	1080.0	0.028	2850.0	0.031	8400.0	0.032
100.0	220.0	0.006	558.0	0.006	1600.0	0.006	4250.0	0.007	12,600.0	0.007
1000.0	408.0	0.002	1040.0	0.002	3000.0	0.002	8000.0	0.002	24,000.0	0.002

TABLE 5.27 Nusselt Numbers and Influence Coefficients for Fully Developed Turbulent Flow in a Concentric Annular Duct with Uniform Heat Flux at One Wall and the Other Wall Insulated [111] (*Continued*)

$r^* = 0.8$	Heating from outer wall with inner wall insulated									
	$Re = 10^4$		3×10^4		10^5		3×10^5		10^6	
Pr	Nu_{oo}	θ_o^*	Nu_{oo}	θ_o^*	Nu_{oo}	θ_o^*	Nu_{oo}	θ_o^*	Nu_{oo}	θ_o^*
0	5.65	0.379	5.70	0.386	5.75	0.398	5.80	0.407	5.85	0.409
0.001	5.65	0.379	5.70	0.386	5.75	0.398	5.88	0.406	6.25	0.407
0.003	5.65	0.379	5.70	0.386	5.84	0.397	6.35	0.407	8.80	0.374
0.01	5.75	0.381	5.85	0.386	6.72	0.390	9.95	0.361	21.0	0.286
0.03	6.10	0.388	6.90	0.380	11.1	0.339	23.2	0.290	62.0	0.216
0.5	22.4	0.225	48.0	0.191	121.0	0.169	292.0	0.153	790.0	0.136
0.7	28.0	0.192	61.0	0.166	156.0	0.150	378.0	0.136	1020.0	0.122
1.0	34.8	0.159	76.5	0.141	197.0	0.129	483.0	0.120	1330.0	0.111
3.0	61.3	0.083	142.0	0.079	382.0	0.078	960.0	0.076	2730.0	0.073
10.0	100.0	0.039	243.0	0.039	670.0	0.039	1740.0	0.040	5050.0	0.040
30.0	146.0	0.019	365.0	0.019	1040.0	0.020	2720.0	0.021	8000.0	0.022
100.0	209.0	0.008	533.0	0.008	1500.0	0.009	4000.0	0.009	12,000.0	0.010
1000.0	385.0	0.002	1000.0	0.002	2870.0	0.002	7720.0	0.002	23,000.0	0.002

$r^* = 0.8$	Heating from inner wall with outer wall insulated									
	$Re = 10^4$		3×10^4		10^5		3×10^5		10^6	
Pr	Nu_{ii}	θ_i^*	Nu_{ii}	θ_i^*	Nu_{ii}	θ_i^*	Nu_{ii}	θ_i^*	Nu_{ii}	θ_i^*
0	5.87	0.489	5.90	0.505	5.92	0.515	5.95	0.525	5.97	0.528
0.001	5.87	0.489	5.90	0.505	5.92	0.515	6.00	0.518	6.33	0.516
0.003	5.87	0.489	5.90	0.505	6.03	0.485	6.40	0.504	8.80	0.468
0.01	5.95	0.485	6.07	0.506	6.80	0.493	10.0	0.452	21.7	0.382
0.03	6.20	0.478	7.05	0.485	11.4	0.445	23.0	0.357	61.0	0.276
0.5	22.9	0.268	49.5	0.250	123.0	0.214	296.0	0.193	800.0	0.174
0.7	28.5	0.244	62.3	0.212	157.0	0.186	384.0	0.172	1050.0	0.160
1.0	35.5	0.200	78.3	0.181	202.0	0.166	492.0	0.154	1350.0	0.140
3.0	63.0	0.108	145.0	0.102	386.0	0.097	973.0	0.096	2750.0	0.093
10.0	102.0	0.051	248.0	0.051	693.0	0.052	1790.0	0.051	5150.0	0.051
30.0	147.0	0.027	370.0	0.027	1050.0	0.027	2750.0	0.029	8100.0	0.030
100.0	215.0	0.010	540.0	0.010	1540.0	0.010	4050.0	0.011	12,100.0	0.012
1000.0	393.0	0.002	1000.0	0.002	2890.0	0.002	7700.0	0.002	23,000.0	0.002

TABLE 5.28 Nusselt Numbers and Influence Coefficients for Fully Developed Turbulent Flow in a Smooth Concentric Annular Duct With $r^* = 1$ (Parallel Plates Duct With Uniform Heat Flux at One Wall and the Other Wall Insulated* [111]

	$Re = 10^4$		3×10^4		10^5		3×10^5		10^6	
Pr	Nu	θ^*	Nu	θ^*	Nu	θ^*	Nu	θ^*	Nu	θ^*
0.0	5.70	0.428	5.78	0.445	5.80	0.456	5.80	0.460	5.80	0.468
0.001	5.70	0.428	5.78	0.445	5.80	0.456	5.88	0.460	6.23	0.460
0.003	5.70	0.428	5.80	0.445	5.90	0.450	6.32	0.450	8.62	0.422
0.01	5.80	0.428	5.92	0.455	6.70	0.440	9.80	0.407	21.5	0.333
0.03	6.10	0.428	6.90	0.428	11.0	0.390	23.0	0.330	61.2	0.255
0.5	22.5	0.256	47.8	0.222	120.0	0.193	290.0	0.174	780.0	0.157
0.7	27.8	0.220	61.2	0.192	155.0	0.170	378.0	0.156	1030.0	0.142
1.0	35.0	0.182	76.8	0.162	197.0	0.148	486.0	0.138	1340.0	0.128
3.0	60.8	0.095	142.0	0.092	380.0	0.089	966.0	0.087	2700.0	0.084
10.0	101.0	0.045	214.0	0.045	680.0	0.045	1760.0	0.045	5080.0	0.046
30.0	147.0	0.021	367.0	0.022	1030.0	0.022	2720.0	0.023	8000.0	0.024
100.0	210.0	0.009	214.0	0.009	1520.0	0.010	4030.0	0.010	12,000.0	0.011
1000.0	390.0	0.002	997.0	0.002	2880.0	0.002	7650.0	0.002	23,000.0	0.002

wall insulated. For the case of the outer wall's being heated, the semiempirical equations are as follows:

$$\text{Nu}_{oo} = A_o + B_o(\beta \, \text{Pe})^{n_o} \tag{5.167}$$

where
$$A_o = 5.26 + \frac{0.05}{r^*} \tag{5.168}$$

$$B_o = 0.01848 + \frac{0.003154}{r^*} - \frac{0.0001333}{r^{*2}} \tag{5.169}$$

$$n_o = 0.78 - \frac{0.01333}{r^*} + \frac{0.000833}{r^{*2}} \tag{5.170}$$

$$\beta = 1 - \frac{1.82}{\text{Pr} \, (\varepsilon_m/\nu)_{\text{max}}^{1.4}} \tag{5.171}$$

where $(\varepsilon_m/\nu)_{\text{max}}$ can be calculated from the relation

$$\left(\frac{\varepsilon_m}{\nu}\right)_{\text{max}} = \frac{1}{2}\left(\frac{\varepsilon_m}{\nu}\right)_{\text{max},c} \tag{5.172}$$

An expression for $(\varepsilon_m/\nu)_{\text{max},c}$ applicable to a circular duct ($r^* = 0$) was developed by Bhatti and Shah [45]. It is given by

$$\left(\frac{\varepsilon_m}{\nu}\right)_{\text{max},c} = 0.037 \text{Re} \, \sqrt{f} \tag{5.173}$$

In Eq. 5.173, the friction factor f can be calculated from the explicit formula given by Techo et al. [56], which is shown in Table 5.8.

For a concentric annular duct with the inner wall heated, the semiempirical equations developed by Dwyer [113] are applicable:

$$\text{Nu}_{ii} = A_i + B_i(\beta \, \text{Pe})^{n_i} \tag{5.174}$$

where
$$A_i = 4.63 + \frac{0.686}{r^*} \tag{5.175}$$

$$B_i = 0.02154 - \frac{0.000043}{r^*} \tag{5.176}$$

$$n_i = 0.752 + \frac{0.01657}{r^*} - \frac{0.000883}{r^{*2}} \tag{5.177}$$

The values of β for this case can also be calculated from Eqs. 5.171 to 5.173. Both Eqs. 5.166 and 5.173 are valid for Pe values above the critical values. For Pr = 0.005, 0.01, 0.02, and 0.03, the critical Pe values are 270, 300, 330, and 345, respectively. For liquid metals, only the heat transfer mode for Pe < Pe_{crit} is molecular conduction.

Hydrodynamically Developing Flow.   Hydrodynamically developing turbulent flow in concentric annular ducts has been investigated by Rothfus et al. [114], Olson and Sparrow [115], and Okiishi and Serouy [116]. The measured apparent friction factors at the inner wall of two concentric annuli ($r^* = 0.3367$ and $r^* = 0.5618$) with a square entrance are shown in Fig. 5.17 ($r^* = 0.5618$), where f_i is the fully developed friction factor at the inner wall. The values of f_i equal 0.01, 0.008, and 0.0066 for Re = 6000, 1.5×10^4, and 3×10^4, respectively [114].

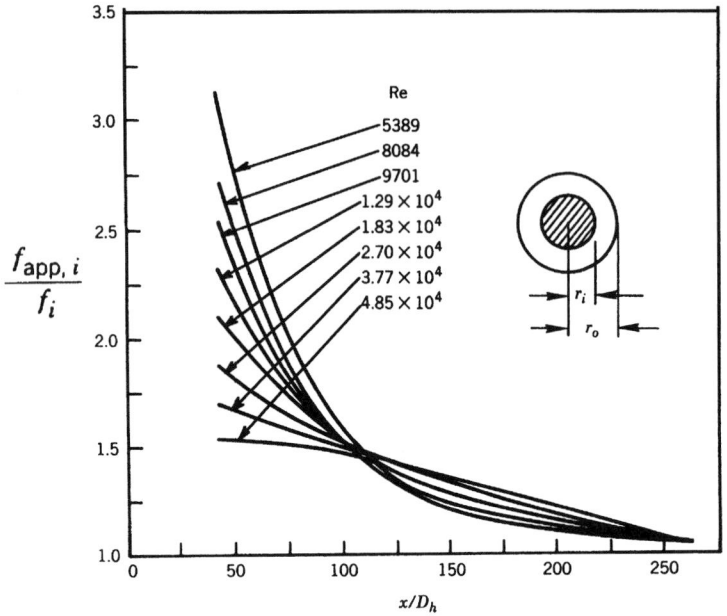

FIGURE 5.17   Normalized apparent friction factors for turbulent flow in the hydro-dynamic entrance region of a smooth concentric annular duct ($r^* = 0.5168$) [114].

Having determined $f_{app,i}$ from Fig. 5.17, the apparent friction factor $f_{app,o}$ at the outer wall can be determined from

$$\frac{f_{app,o}}{f_{app,i}} = \frac{r^*(1 - r_m^{*2})}{r_m^{*2} - r^{*2}} \tag{5.178}$$

where r_m^* is given by

$$r_m^* = r^{*0.343}(1 + r^{*0.657} - r^*) \tag{5.179}$$

Having identified both $f_{app,o}$ and $f_{app,i}$, the circumferentially averaged friction factor can be calculated as follows:

$$f_{app} = \frac{f_{app,o}r_o + f_{app,i}r_i}{r_o + r_i} \tag{5.180}$$

Thermally Developing Flow.   Kays and Leung [111] present experimental results for thermally developing turbulent flow in four concentric annular ducts, $r^* = 0.192, 0.255, 0.376$, and 0.500, with the boundary condition of one wall at uniform heat flux and the other insulated, that is, the fundamental solution of the second kind. In accordance with this solution, the local Nusselt numbers $\text{Nu}_{x,o}$ and $\text{Nu}_{x,i}$ at the outer and inner walls are expressed as

$$\text{Nu}_{x,o} = \frac{\text{Nu}_{x,oo}}{1 - \theta_{x,o}^* q_i''/q_o''}, \qquad \text{Nu}_{x,i} = \frac{1}{1 - \theta_{x,ii} - \theta_{x,mi}} \tag{5.181}$$

where q_o'' and q_i'' are the uniform heat fluxes at the outer and inner walls. Both q_o'' and q_i'' are positive whenever heat is added to the fluid and negative whenever heat is transferred out of the fluid. The Nusselt numbers $Nu_{x,oo}$ and $Nu_{x,ii}$ and the influence coefficients $\theta_{x,o}^*$ and $\theta_{x,i}^*$ are given by:

$$Nu_{x,oo} = \frac{1}{\theta_{x,oo} - \theta_{x,mo}} \qquad Nu_{x,ii} = \frac{1}{\theta_{x,ii} - \theta_{x,mi}} \qquad (5.182)$$

$$\theta_{x,o}^* = \frac{\theta_{x,mi} - \theta_{x,mo}}{\theta_{x,oo} - \theta_{x,mo}} \qquad \theta_{x,i}^* = \frac{\theta_{x,mo} - \theta_{x,io}}{\theta_{x,ii} - \theta_{x,mi}} \qquad (5.183)$$

and
$$\theta_{x,mo} = \frac{4(x/D_h)}{Re\ Pr\ (1 + r^*)} \qquad \theta_{x,mi} = \frac{4r^*(x/D_h)}{Re\ Pr\ (1 + r^*)} \qquad (5.184)$$

The nondimensional temperatures $\theta_{x,oo}$, $\theta_{x,ii}$, $\theta_{x,oi}$, and $\theta_{x,io}$ for $r^* = 0.192$ and 0.5 are presented in Fig. 5.18 as an example. Additional graphical results for $r^* = 0.192$, 0.255, and 0.376 are available in Kays and Leung [111].

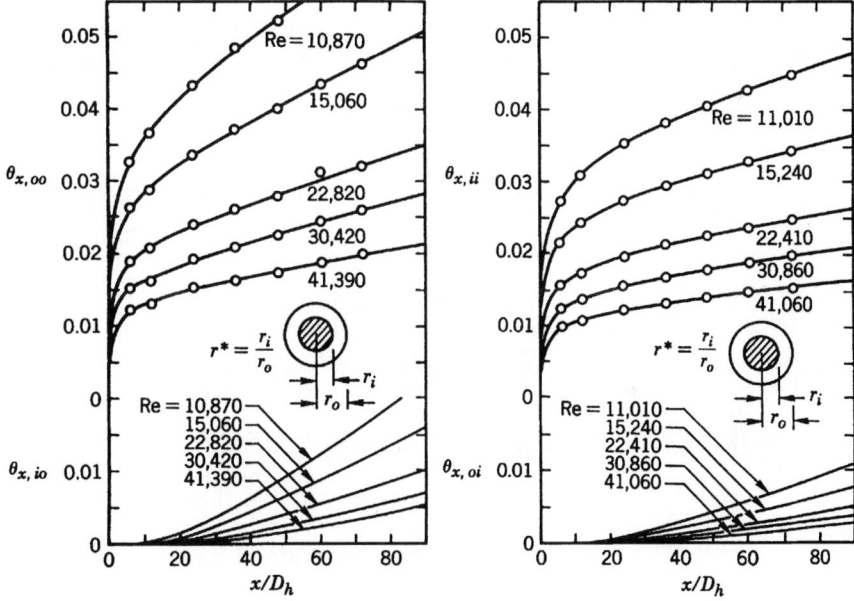

FIGURE 5.18 $\theta_{x,oo}$, $\theta_{x,io}$, $\theta_{x,ii}$, $\theta_{x,oi}$ for use with Eqs. 5.182 and 5.183 for thermally developing flow in a smooth concentric annular duct with $r^* = 0.5$ and $Pr = 0.7$ [111].

The preceding solution is restricted to a fluid with $Pr = 0.7$, $10^4 < Re < 1.61 \times 10^5$, and $0.192 < r^* < 0.5$. Cross plotting and interpolation can be employed to increase the application range of the results in terms of Re and r^*. For $Pr = 0.01$ and $Pr = 1000$, an eigenvalue solution to the fundamental problem of the second kind for four concentric annular ducts ($r^* = 0.02$, 0.1067, 0.1778, and 0.3422) can be found in Quarmby and Anand [117].

Simultaneously Developing Flow.   Little information is available on simultaneously developing turbulent flow in concentric annular ducts. However, the theoretical and experimental studies by Roberts and Barrow [118] indicate that the Nusselt numbers for simultaneously developing flow are not significantly different from those for thermally developing flow.

Effects of Eccentricity.   Jonsson and Sparrow [119] have conducted a careful experimental investigation of fully developed turbulent flow in smooth, eccentric annular ducts. The researchers have provided the velocity measurements graphically in terms of the wall coordinate u^+ as well as the velocity-defect representation. From their results, the circumferentially averaged fully developed friction factor is correlated by a power-law relationship of the following type:

$$f = \frac{C}{\text{Re}^n} \tag{5.185}$$

where C is a strong function of e^*, a relatively weak function of r^*, and independent of the Reynolds number, which is given in Fig. 5.19. A single value, $n = 0.18$, has been suggested by Jonsson and Sparrow [119] for all r^*, e^*, and Re. More details regarding the friction factors f_i and f_o for each of the two surfaces are also available [120]. Other investigations of fully developed turbulent flow in eccentric annular ducts have been conducted by Lee and Barrow [121], Deissler and Tayler [122], Yu and Dwyer [123], and Ricker et al. [124].

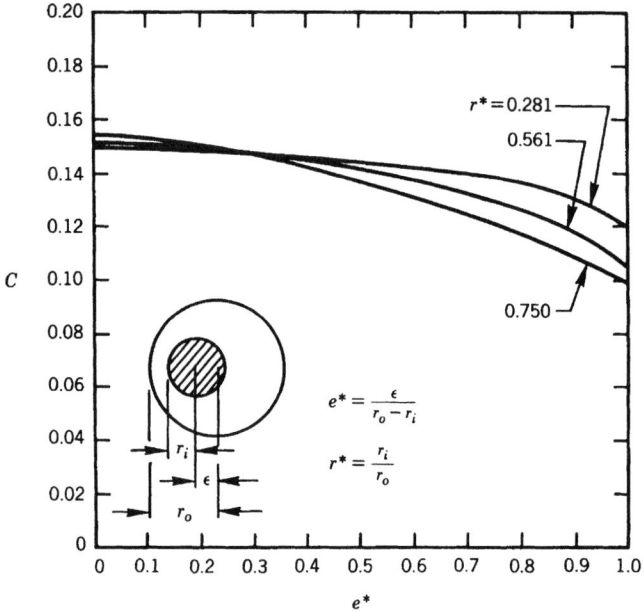

FIGURE 5.19   Empirical constant C in Eq. 5.185 [119].

The effects of the eccentricity on turbulent heat transfer in eccentric annular ducts have been investigated by Judd and Wade [125], Leung et al. [126], Lee and Barrow [121], and Yu and Dwyer [123] for the boundary condition of a uniform wall heat flux on the inner or outer surfaces while the other wall is insulated. The results were obtained under specific conditions. Further details can be found in the previously mentioned references.

Few investigations have been conducted on hydrodynamically developing flow in eccentric annular ducts. Jonsson [120] has obtained experimental information on the pressure gradient in hydrodynamically developing flow and provided the hydrodynamic lengths L_{hy}/D_h for $1.8 \times 10^4 < \text{Re} < 1.8 \times 10^5$. These are presented in Table 5.29.

TABLE 5.29 Turbulent Flow Hydrodynamic
Entrance Lengths for Smooth, Eccentric
Annular Ducts [120]

	L_{hy}/D_h			
r^*	$e^* = 0$	0.5	0.9	1.0
0.281	29	32	38	38
0.561	26	38	59	78
0.750	28	50	69	91

Few results that can be used in practice are available for thermally developing flow and simultaneously developing flow in eccentric annuli. According to the discussion in Bhatti and Shah [45], the Nusselt numbers may be estimated from the corresponding results for concentric annuli ($e^* = 0$).

PARALLEL PLATE DUCTS

Parallel plate ducts, also referred to as *flat ducts* or *parallel plates,* possess the simplest duct geometry. This is also the limiting geometry for the family of rectangular ducts and concentric annular ducts. For most cases, the friction factor and Nusselt number for parallel plate ducts are the maximum values for the friction factor and the Nusselt number for rectangular ducts and concentric annular ducts.

Laminar Flow

Laminar flow and heat transfer in parallel plate ducts are described in this section. The friction factor and Nusselt number are given for practical calculations.

Fully Developed Flow. For a parallel plate duct with hydraulic diameter $D_h = 4b$ (b being the half-distance between the plates) and the origin at the duct axis, the velocity distribution and friction factor are given by the following expression:

$$\frac{u}{u_m} = \frac{3}{2}\left[1 - \left(\frac{y}{b}\right)^2\right]$$ (5.186)

$$u_m = -\frac{1}{3\mu}\left(\frac{dp}{dx}\right)b^2, \qquad f\,\mathrm{Re} = 24$$ (5.187)

Similar to the four fundamental thermal boundary conditions for concentric annuli, the four kinds of fundamental conditions for parallel plate ducts are shown in Fig. 5.20. The fully developed Nusselt numbers for the four boundary conditions follow [1]:

First kind:	$\mathrm{Nu}_1 = \mathrm{Nu}_2 = 4$	(5.188)
Second kind:	$\mathrm{Nu}_1 = 0 \qquad \mathrm{Nu}_2 = 5.385$	(5.189)
Third kind:	$\mathrm{Nu}_1 = 0 \qquad \mathrm{Nu}_2 = 4.861$	(5.190)
Fourth kind:	$\mathrm{Nu}_1 = \mathrm{Nu}_2 = 4$	(5.191)

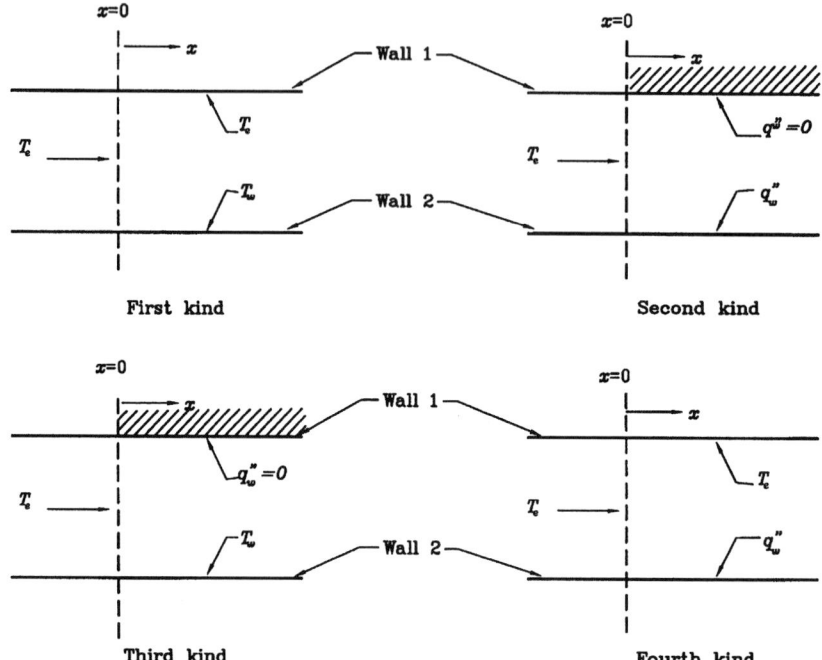

FIGURE 5.20   Four fundamental boundary conditions for a parallel plate duct [2].

Examples of the application of these fundamental solutions to obtain the fully developed Nusselt number for a duct with three different boundary conditions follow. The Nusselt numbers are defined as

$$\text{Nu}_j = \frac{q''_{wj} D_h}{k(T_j - T_m)} \tag{5.192}$$

where j denotes wall 1 or 2, and T_j is the temperature of the jth wall.

Uniform Temperature at Each Wall.   When the temperatures on two walls are equal, $T_{w1} = T_{w2}$, then $\text{Nu}_1 = \text{Nu}_2 = \text{Nu}_T$. The value of Nu_T is given by Shah and London [1] as follows:

$$\text{Nu}_T = 7.541 \tag{5.193}$$

When the temperatures on two walls are different, $T_{w1} \neq T_{w2}$, then $\text{Nu}_1 = \text{Nu}_2 = 4$, as shown in Eq. 5.188.

When the effect of viscous dissipation is considered, the following formulas developed by Cheng and Wu [127] for the case $T_{w1} > T_{w2}$ are used to compute the Nusselt numbers:

$$\text{Nu}_1 = \frac{4(1 - 6\text{Br})}{1 - 48/35\text{Br}} \qquad \text{Nu}_2 = \frac{4(1 + 6\text{Br})}{1 + 48/35\text{Br}} \tag{5.194}$$

When the effects of viscous dissipation and flow work are considered together for the case of $T_{w1} = T_{w2}$, Ou and Cheng [128] have shown that for fully developed flow, $\text{Nu}_T = 0$ and the dimensionless temperature distribution is as follows:

$$\frac{T_w - T}{T_w - T_e} = \frac{9}{8} \text{Br} \left[1 - \left(\frac{y}{b}\right)^2 \right]^2, \qquad \frac{T_w - T_m}{T_w - T_e} = \frac{27}{35} \text{Br} \tag{5.195}$$

Taking the fluid axial condition into account, Pahor and Strand [129] and Grosjean et al. [130] have obtained the following asymptotic formulas for the Nusselt number in the case of $T_{w1} = T_{w2}$:

$$\mathrm{Nu_T} = \begin{cases} 7.540(1 + 3.79/\mathrm{Pe}^2 + \cdots) & \text{for Pe} \gg 1 \\ 8.11742(1 - 0.030859\mathrm{Pe} + 0.0069436\mathrm{Pe}^2 - \cdots) & \text{for Pe} \ll 1 \end{cases} \quad (5.196)$$

Uniform Heat Flux at Each Wall. When the heat fluxes on the two walls of parallel plate ducts are equal, $q''_{w1} = q''_{w2}$, the temperature distribution of fully developed laminar flow is given by:

$$\frac{T_w - T_{\mathrm{H}}}{T_w - T_{m,\mathrm{H}}} = \frac{35}{136}\left[5 - 6\left(\frac{y}{b}\right)^2 + \left(\frac{y}{b}\right)^4\right], \qquad \frac{T_w - T_{m,\mathrm{H}}}{q''_w D_h/k} = \frac{17}{140} \quad (5.197)$$

and the Nusselt number is obtained as:

$$\mathrm{Nu_1} = \mathrm{Nu_2} = \mathrm{Nu_H} = 8.235 \quad (5.198)$$

When $q''_{w1} \neq q''_{w2}$, then

$$\mathrm{Nu_1} = \frac{140}{26 - 9(q''_{w2}/q''_{w1})}, \qquad \mathrm{Nu_2} = \frac{140}{26 - 9(q''_{w1}/q''_{w2})} \quad (5.199)$$

To consider the effect of internal energy generation and viscous dissipation, the following formula obtained by Tyagi [6] is recommended:

$$\mathrm{Nu_H} = \frac{140}{17 + \tfrac{3}{4}S^* + 108\mathrm{Br}'} \quad (5.200)$$

Uniform Temperature at One Wall and Uniform Heat Flux at the Other. When the two walls of a parallel plate duct are subject to a thermal boundary condition such as uniform temperature at one wall and uniform heat flux at the other, the Nusselt numbers for fully developed laminar flow for $q''_w = 0$ and $q''_w \neq 0$ are determined to be:

$$\mathrm{Nu_1} = 4.8608 \qquad \mathrm{Nu_2} = 0 \qquad \text{for } q''_w = 0 \quad (5.201)$$

$$\mathrm{Nu_1} = \mathrm{Nu_2} = 4 \qquad \text{for } q''_w \neq 0 \quad (5.202)$$

The Convective Boundary Condition ⑬. The fully developed Nusselt number with the convective boundary condition at both walls can be computed from Hickman's [7] analysis:

$$\mathrm{Nu_{T3}} = \frac{4620 + 561\mathrm{Bi}}{561 + 74\mathrm{Bi}} \quad (5.203)$$

The Exponential Wall Heat Flux Boundary Condition ⑮. When both walls of parallel plate duct are subjected to the exponential heat flux of $q''_w = q''_0 \exp(mx^*)$, the fully developed Nusselt number can be obtained as follows [2]:

$$\mathrm{Nu_{H5}} = 8.24 + 2.1611 \times 10^{-3}m - 4.4397 \times 10^{-5}m^2 + 1.2856 \times 10^{-7}m^3 - 2.7035 \times 10^{-10}m^4 \quad (5.204)$$

When $m = -30.16$, then $\mathrm{Nu_{H5}} = \mathrm{Nu_T}$, and when $m = 0$, then $\mathrm{Nu_{H5}} = \mathrm{Nu_H}$.

Hydrodynamically Developing Flow. For hydrodynamically developing flow in parallel plate ducts, Shah and London [1] obtained the apparent friction factor f_{app} and Chen [11] has obtained L_{hy}, $K(\infty)$ as the function of x^+ and Re, respectively, as shown in the following equations:

$$f_{\mathrm{app}}\,\mathrm{Re} = \frac{3.44}{(x^+)^{1/2}} + \frac{24 + 0.674/(4x^+) - 3.44/(x^+)^{1/2}}{1 + 0.000029(x^+)^{-2}} \quad (5.205)$$

$$\frac{L_{hy}}{D_h} = 0.011\,\mathrm{Re} + \frac{0.315}{1 + 0.0175\,\mathrm{Re}} \tag{5.206}$$

$$K(\infty) = 0.64 + \frac{38}{\mathrm{Re}} \tag{5.207}$$

Thermally Developing Flow. The results for thermally developing flow in parallel plate ducts are presented for the following practical thermal boundary conditions of interest.

Equal and Uniform Temperatures on Both Walls. The local and mean Nusselt numbers for parallel plate ducts with equal and uniform temperatures on both walls can be computed from Nusselt's [131] solution, which is displayed in Fig. 5.21. The tabulated values for Fig. 5.21 are available in Shah and London [1].

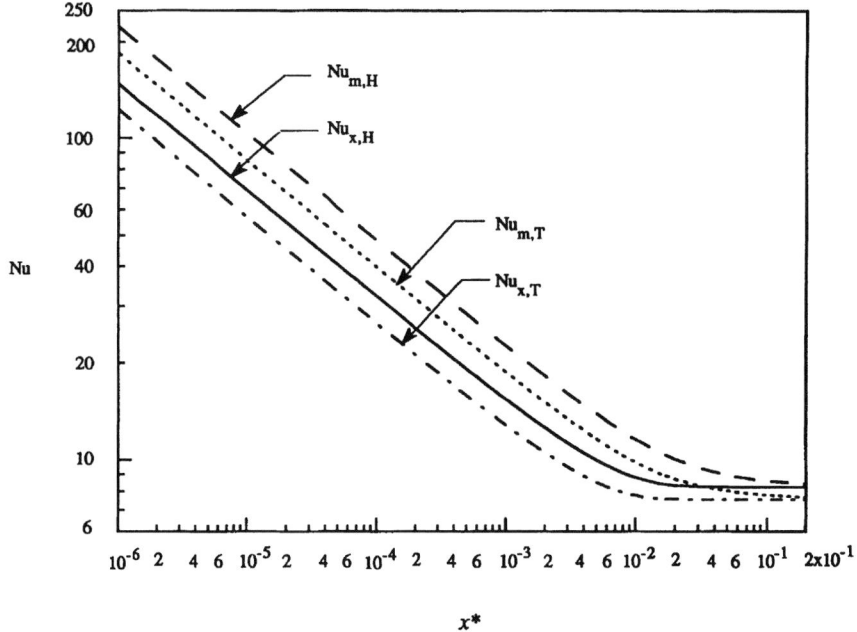

FIGURE 5.21 Local and mean Nusselt numbers in the thermal entrance region of a parallel plate duct with the ⓉT and ⒽH boundary conditions [1].

From these results, the dimensionless thermal entrance length can be determined as follows:

$$L_{th,T}^{*} = 0.00797 \tag{5.208}$$

It is also suggested that the following set of empirical equations proposed by Shah and London [1] be used for the practical calculation of the local Nusselt number:

$$\mathrm{Nu}_{x,T} = \begin{cases} 1.233x^{*-1/3} + 0.4 & \text{for } x^* < 0.001 \\ 7.541 + 6.874(10^3 x^*)^{-0.488}e^{-245x^*} & \text{for } x^* > 0.001 \end{cases} \tag{5.209}$$

$$\mathrm{Nu}_{m,T} = \begin{cases} 1.849x^{*-1/3} & \text{for } x^* \le 0.0005 \\ 1.849x^{*-1/3} + 0.6 & \text{for } 0.0005 < x^* \le 0.006 \\ 7.541 + 0.0235/x^* & \text{for } x^* > 0.006 \end{cases} \tag{5.210}$$

Uniform and Equal Heat Flux at Both Walls.   Thermally developing flow in a parallel plate duct with uniform and equal heat flux at both walls has been investigated by Cess and Shaffer [132] and Sparrow et al. [133] in terms of a series format for the local and mean Nusselt numbers. The dimensionless thermal entrance length for this problem has been found by Shah and London [1] to be as follows:

$$L_{th,H} = 0.01154 \tag{5.211}$$

The local and mean Nusselt numbers are also displayed in Fig. 5.21.

The following set of equations are recommended for practical computations without loss of accuracy [1]:

$$\mathrm{Nu}_{x,H} = \begin{cases} 1.490x^{*-1/3} & \text{for } x^* \leq 0.0002 \\ 1.490x^{*-1/3} - 0.4 & \text{for } 0.0002 < x^* \leq 0.001 \\ 8.235 + 8.68(10^3x^*)^{-0.506}e^{-164x^*} & \text{for } x^* \geq 0.001 \end{cases} \tag{5.212}$$

$$\mathrm{Nu}_{m,H} = \begin{cases} 2.236x^{*-1/3} & \text{for } x^* \leq 0.001 \\ 2.236x^{*-1/3} + 0.9 & \text{for } 0.001 < x^* \leq 0.01 \\ 8.235 + 0.0364/x^* & \text{for } x^* \geq 0.01 \end{cases} \tag{5.213}$$

It has been concluded that except in the neighborhood of the duct inlet ($x^* < 10^{-2}$), the effect of the fluid axial conduction is negligible for Pe > 50 [134, 135].

Convective Boundary Condition at Both Walls or One Wall.   The solutions for the convective boundary condition on both walls or one wall are reviewed in Shah and London [1], where more detailed descriptions are available.

Simultaneously Developing Flow.   The results for simultaneously developing flow in parallel plate ducts are provided for the following thermal boundary conditions.

Equal and Uniform Temperatures at Both Walls.   For simultaneously developing flow in a parallel plate duct with fluids of 0.1 < Pr < 1000, the following equations are recommended for the computation of the local and mean Nusselt numbers [2, 136, 137]:

$$\mathrm{Nu}_{x,T} = 7.55 + \frac{0.024x^{*-1.14}[0.0179\mathrm{Pr}^{0.17}\,x^{*-0.64} - 0.14]}{[1 + 0.0358\mathrm{Pr}^{0.17}\,x^{*-0.64}]^2} \tag{5.214}$$

$$\mathrm{Nu}_{m,T} = 7.55 + \frac{0.024x^{*-1.14}}{1 + 0.0358\mathrm{Pr}^{0.17}\,x^{*-0.64}} \tag{5.215}$$

The thermal entrance length $L_{th,T}^*$ has been found to be 0.0064 for 0.01 < Pr < 10,000 [138, 139]. A detailed description can be found in Shah and London [1] for the solutions for Pr = ∞ and Pr = 0.

When one duct wall is insulated and the other is at a uniform temperature, the local and mean Nusselt numbers for simultaneously developing flow have been obtained for fluids of $0.1 \leq \mathrm{Pr} \leq 10$. These follow [1]:

$$\mathrm{Nu}_{x,T} = 4.86 + \frac{0.0606x^{*-1.2}[0.0455\mathrm{Pr}^{0.17}\,x^{*-0.7} - 0.2]}{[1 + 0.0909\mathrm{Pr}^{0.17}\,x^{*-0.7}]^2} \tag{5.216}$$

$$\mathrm{Nu}_{m,T} = 4.86 + \frac{0.0606x^{*-1.2}}{1 + 0.0909\mathrm{Pr}^{0.17}\,x^{*-0.7}} \tag{5.217}$$

Uniform and Equal Heat Flux at Both Walls.   The local Nusselt number for heat transfer of laminar flow in a parallel plate duct with uniform and equal heat flux at both walls is displayed in Fig. 5.22 for different Prandtl numbers, Pr = 0 [34] and Pr = 0.01, 0.7, 1, 10, and ∞ [136]. The thermal entrance lengths obtained from the results presented in this figure are 0.016, 0.030, 0.017, 0.014, 0.012, and 0.0115, for Pr = 0, 0.01, 0.7, 1, 10, and ∞, respectively.

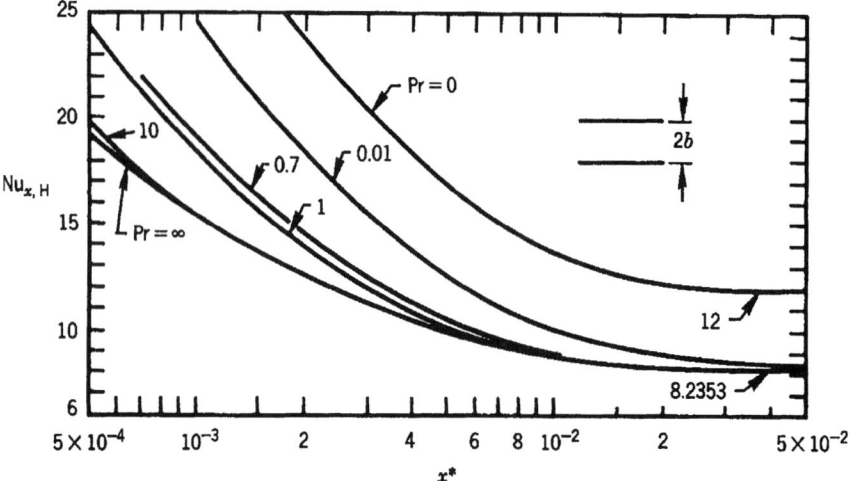

FIGURE 5.22 Local and mean Nusselt numbers for simultaneously developing flow in a parallel plate duct with the ⊞ boundary condition [34, 136].

The local Nusselt number is displayed in Fig. 5.23 for Pr = 0.0, 0.01, 0.7, 10, and ∞ when one wall of the parallel plate duct is insulated and the other wall is subjected to uniform heat flux heating [140]. Included in Fig. 5.23 are the results for Pr = ∞, obtained from the concentric annular duct corresponding to $r^* = 1$. The local and mean Nusselt numbers for Pr = 0 were obtained by Bhatti [34].

In addition, Nguyen [141] has obtained the apparent friction factor and Nusselt numbers for low Reynolds number simultaneously developing flow in a parallel plate duct with a con-

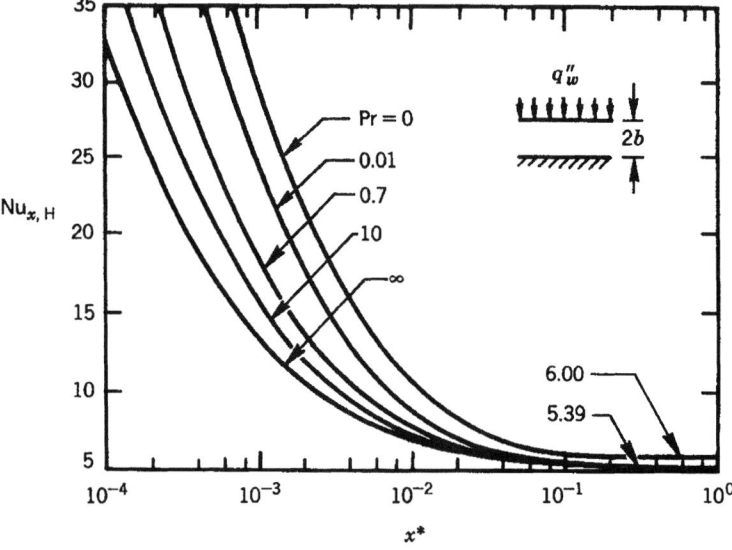

FIGURE 5.23 Local and mean Nusselt numbers for simultaneously developing flow in a flat duct with uniform heat flux at one wall and the other wall insulated [34, 140].

stant temperature and constant wall heat flux thermal boundary conditions. Results are presented for $Pr = 0.7$ and Reynolds numbers between 1 and 20. The results for $Pr = 0.2, 0.7, 2, 7, 10,$ and 100 as well as for the Reynolds numbers between 40 and 200 have been numerically determined by Nguyen and Maclaine-Cross [142].

Turbulent Flow

The characteristics of turbulent flow and heat transfer in parallel plate ducts are discussed in this section. The friction factor for transition flow is also addressed.

Transition Flow.   The lower limit of the critical Reynolds number Re_{crit} for a parallel plate duct is reported to be between 2200 and 3400, depending on the entrance configurations and disturbance sources [143]. The following friction factor formula developed by Hrycak and Andrushkiw [144] is recommended for transition flow in the range of $2200 \leq Re \leq 4000$:

$$f = -2.56 \times 10^{-3} + 4.085 \times 10^{-6}\,Re - 5.5 \times 10^{-10}\,Re^2 \qquad (5.218)$$

The mean Nusselt number in the thermal entrance region of a parallel plate duct with uniform wall temperature at both walls in the range of $2300 < Re < 6000$ is given by Hausen [20] as follows:

$$Nu_{m,T} = 0.116(Re^{2/3} - 160)\,Pr^{1/3}\left[1 + \left(\frac{x}{D_h}\right)^{-2/3}\right] \qquad (5.219)$$

Fully Developed Flow.   Beavers et al. [145] obtained the following friction factor for fully developed turbulent flow in a parallel plate duct for $5000 < Re < 1.2 \times 10^6$ from very accurate experimental data:

$$f = \frac{0.1268}{Re^{0.3}} \qquad (5.220)$$

For $1.2 \times 10^4 < Re < 1.2 \times 10^6$, Dean [146] has developed the following equation based on a comprehensive survey of the available data:

$$f = \frac{0.0868}{Re^{1/4}} \qquad (5.221)$$

Comparisons of precision using Eqs. 5.220 and 5.221 and Blasius's formula (Table 5.8) in which the diameter of circular duct $2a$ is replaced by hydraulic diameter $4b$, b being the half-space between two plates, have been conducted by Bhatti and Shah [45]. In the range of $5000 \leq Re \leq 3 \times 10^4$, Eq. 5.220 is recommended; otherwise, Eq. 5.221 should be used to obtain the friction factor for fully developed turbulent flow in a parallel plate duct. However, use of the hydraulic diameter to substitute for the circular duct diameter in the Blasius equation is reasonable for the prediction of the fraction factor [45].

Kays and Leung [111] analyzed turbulent heat transfer in a parallel flat plate duct for arbitrarily prescribed heat flux on the two duct walls. The fully developed Nusselt number Nu_H can be obtained from the following expression:

$$Nu_H = \frac{Nu}{1 - \gamma\theta^*} \qquad (5.222)$$

where γ is the ratio of the prescribed heat fluxes on the two duct walls. The Nusselt number Nu and the influence coefficient θ^* in Eq. 5.222 are given in Table 5.28 for different Re and Pr numbers. It should be noted that $\gamma = 0$ signifies that one wall is heated and the other is insulated; $\gamma = 1$ indicates that uniform heat fluxes of equal magnitudes are applied to both walls;

and $\gamma = -1$ refers to heat transfer into one wall and out of the other wall, while the absolute values of the heat fluxes at both walls are the same.

Bhatti and Shah [45] and Sparrow and Lin [133] have performed a comparison of Nusselt numbers predicted using Eq. 5.222 or other equations for parallel plate ducts and the Nusselt number calculated using the equation for circular ducts replacing $2a$ with the hydraulic diameter of the parallel plate duct. It was concluded that the Nusselt number for parallel plate ducts can be determined using the circular duct correlations.

Analogous to circular ducts, the fully developed turbulent Nusselt numbers for uniform wall temperature and uniform wall heat flux boundary conditions in parallel plate ducts are nearly identical for $Pr > 0.7$ and $Re > 10^5$. This is also true for the Nusselt number of turbulent thermally developing flow in a parallel plate duct [147].

For liquid metal, when one wall of the parallel plate duct is heated and the other is adiabatic, the following empirical equation is recommended for $Pr < 0.03$ by Duchatelle and Vautrey [148]:

$$Nu_H = 5.14 + 0.0127\, Pe^{0.8} \tag{5.223}$$

Fully developed fluid flow and heat transfer results for rough parallel plate ducts can be predicted using the results for rough circular ducts with the use of hydraulic diameter [45].

Hydrodynamically Developing Flow. Hydrodynamically developing flow in smooth parallel plate ducts with uniform velocity at the duct inlet has been analyzed by Deissler [92] by means of an integral method. The apparent friction factors f_{app} in the hydrodynamic entrance are presented in Fig. 5.24.

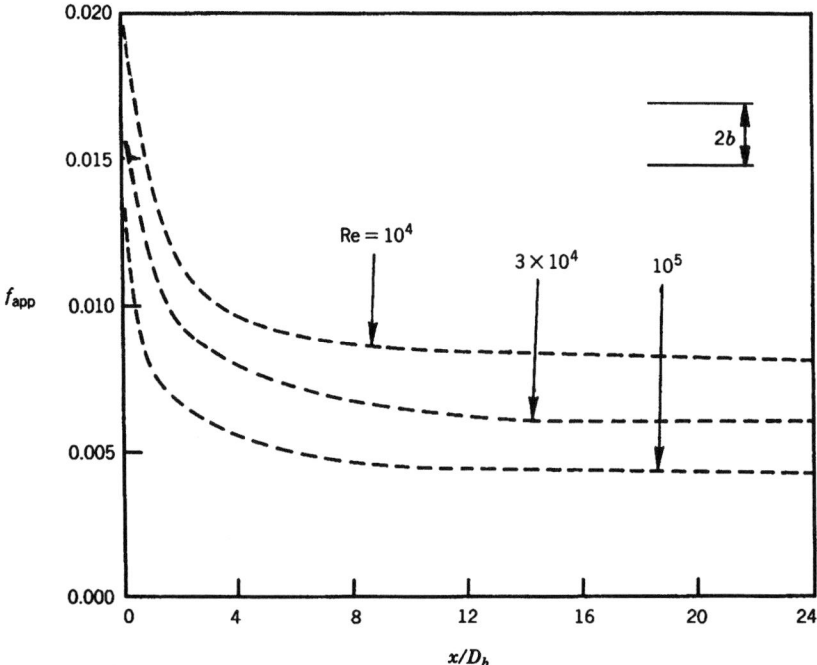

FIGURE 5.24 Turbulent flow apparent friction factors in the hydrodynamic entrance region of a parallel plate duct with uniform inlet velocity [45].

Thermally Developing and Simultaneously Developing Flow.   Thermally developing turbulent flow in a parallel plate duct with uniform and equal temperatures at both walls has been solved by Sakakibara and Endo [149] and by Shibani and Özisik [150]. A discussion of the solution can be found in Bhatti and Shah [45]. Hatton and Quarmby [151] and Sakakibara and Endo [149] have obtained the solution for a thermally developing turbulent flow problem in a flat duct with one wall at uniform temperature and the outer wall insulated (i.e., the fundamental solution of the third kind). Hatton and Quarmby [151], Hatton et al. [152], and Sakakibara [153] have analyzed thermally developing turbulent flow in a flat duct with uniform heat flux at one wall and the other insulated (i.e., the fundamental solution of the second kind). Özisik et al. [154] have obtained the solution of the thermal entry region heat transfer of turbulent developing flow in a parallel plate duct with uniform wall temperature. Detailed discussions of these solutions can be found in the previously mentioned references.

Few investigations have been conducted for simultaneously developing flow in parallel plate ducts. Therefore, no correlations are provided for practical usage.

RECTANGULAR DUCTS

Rectangular ducts are also often used in the design of heat transfer devices such as compact heat exchangers. Unlike circular and parallel plate ducts, two-dimensional analysis is required to obtain the friction factors and Nusselt numbers for rectangular ducts.

Laminar Flow

In this section, the friction factors and Nusselt numbers for fully developed, hydrodynamically developing, thermally developing, and simultaneously developing laminar flows in rectangular ducts are presented.

Fully Developed Flow

Velocity Distribution and the Friction Factor.   Marco and Han [155] have obtained the fully developed velocity distribution in a rectangular duct with cross-sectional dimensions $2a$ and $2b$. It follows:

$$u = -\frac{16}{\pi^3}\left(\frac{dp}{dx}\right)\frac{a^2}{\mu}\sum_{n=1,3\ldots}^{\infty}\frac{(-1)^{(n-1)/2}}{n^3}\left(1 - \frac{\cosh(n\pi y/2a)}{\cosh(n\pi b/2a)}\right)\cos\left(\frac{n\pi z}{2a}\right) \qquad (5.224)$$

where the pressure gradient dp/dx is related to u_m as follows:

$$u_m = -\frac{1}{3}\left(\frac{dp}{dx}\right)\frac{a^2}{\mu}\left[1 - \frac{192}{\pi^5}\left(\frac{a}{b}\right)\sum_{n=1,3\ldots}^{\infty}\frac{1}{n^5}\tanh\left(\frac{n\pi b}{2a}\right)\right] \qquad (5.225)$$

The origin of the Cartesian coordinate is at the center of the rectangular duct. To avoid computational complexity, the following simple approximations have been suggested [156]:

$$\frac{u}{u_{\max}} = \left[1 - \left(\frac{y}{b}\right)^n\right]\left[1 - \left(\frac{z}{a}\right)^m\right] \qquad (5.226)$$

$$\frac{u_{\max}}{u_m} = \left(\frac{m+1}{m}\right)\left(\frac{n+1}{n}\right) \qquad (5.227)$$

Natarajan and Lakshmanan [157] have provided the relation for the values of m and n:

$$m = 1.7 + 0.5\alpha^{*-1.4} \qquad (5.228)$$

$$n = \begin{cases} 2 & \text{for } \alpha^* \le \frac{1}{3} \\ 2 + 0.3(\alpha^* - \frac{1}{3}) & \text{for } \alpha^* \ge \frac{1}{3} \end{cases} \tag{5.229}$$

where $\alpha^* = b/a$ is the aspect ratio. The exact expression for the fully developed friction factor is

$$f\,\text{Re} = \frac{24}{\left(1 + \dfrac{1}{\alpha^*}\right)^2 \left(1 - \dfrac{192}{\pi^5 \alpha^*} \displaystyle\sum_{n=1,3,\dots}^{\infty} \dfrac{\tanh\,(n\pi\alpha^*/2)}{n^5}\right)} \tag{5.230}$$

However, for ease in practical calculations, the following empirical equation suggested by Shah and London [1] is used to approximate Eq. 5.230:

$$f\,\text{Re} = 24(1 - 1.3553\alpha^* + 1.9467\alpha^{*2} - 1.7012\alpha^{*3} + 0.9564\alpha^{*4} - 0.2537\alpha^{*5}) \tag{5.231}$$

Heat Transfer. The fully developed Nusselt numbers Nu_T for the case of the uniform temperature at four walls are approximated by the following formula [1]:

$$\text{Nu}_T = 7.541(1 - 2.610\alpha^* + 4.970\alpha^{*2} - 5.119\alpha^{*3} + 2.702\alpha^{*4} + 0.548\alpha^{*5}) \tag{5.232}$$

For rectangular ducts with uniform temperature at one or more walls, the Nusselt numbers, available in Shah and London [1], are displayed in Table 5.30.

For rectangular ducts with a uniform wall heat flux at four walls under the ⊕ boundary condition, the fully developed Nusselt numbers Nu_{H1} can be computed with the following formula [1]:

$$\text{Nu}_{H1} = 8.235(1 - 2.0421\alpha^* + 3.0853\alpha^{*2} - 2.4765\alpha^{*3} + 1.0578\alpha^{*4} - 0.1861\alpha^{*5}) \tag{5.233}$$

For rectangular ducts with one or more walls subjected to the ⊕ boundary condition with the other wall insulated, the fully developed Nusselt numbers Nu_{H1} are displayed in Table 5.30.

TABLE 5.30 Nusselt Number for Fully Developed Laminar Flow in Rectangular Ducts With One Wall or More Walls Heating

α^*	Nu_T	Nu_{H1}	Nu_T	Nu_{H1}	Nu_T	Nu_{H1}	Nu_T	Nu_{H1}	Nu_T	Nu_{H1}
0.0	7.541	8.235	7.541	8.235	7.541	8.235	0	0	4.861	5.385
0.1	5.858	6.785	6.095	6.939	6.399	7.248	0.457	0.538	3.823	4.410
0.2	4.803	5.738	5.195	6.072	5.703	6.561	0.833	0.964	3.330	3.914
0.3	4.114	4.990	4.579	5.393	5.224	5.997	1.148	1.312	2.996	3.538
0.4	3.670	4.472	4.153	4.885	4.884	5.555	1.416	1.604	2.768	3.279
0.5	3.383	4.123	3.842	4.505	4.619	5.203	1.647	1.854	2.613	3.104
0.6	3.198	3.895	—	—	—	—	—	—	2.509	2.987
0.7	3.083	3.750	3.408	3.991	4.192	4.662	2.023	2.263	2.442	2.911
0.8	3.014	3.664	—	—	—	—	—	—	2.401	2.866
0.9	2.980	3.620	—	—	—	—	—	—	2.381	2.843
1.0	2.970	3.608	3.018	3.556	3.703	4.094	2.437	2.712	2.375	2.836
2.0	3.383	4.123	2.602	3.146	2.657	2.947	3.185	3.539	2.613	2.911
2.5	3.670	4.472	2.603	3.169	2.333	2.598	3.390	3.777	2.768	3.279
5.0	4.803	5.738	2.982	3.636	1.467	1.664	3.909	4.411	3.330	3.914
10.0	5.858	6.785	3.590	4.252	0.843	0.975	4.270	4.851	3.823	4.410
∞	7.541	8.235	4.861	5.385	0	0	4.861	5.385	4.861	5.385

TABLE 5.31 Fully Developed f Re, Nu_T, Nu_{H1}, and Nu_{H2} for Laminar Flow in Rectangular Ducts With All Four Walls Transferring Heat [1]

α^*	f Re	Nu_T	Nu_{H1}	Nu_{H2}	α^*	f Re	Nu_T	Nu_{H1}	Nu_{H2}
1.000	14.227	2.970	3.60795	3.091	0.250	18.233	4.439	5.33106	2.94
0.900	14.261	2.980	3.62045	—	0.200	19.071	4.803	5.73769	2.93
1/1.2	14.328	—	3.64531	—	1/6	19.702	5.137	6.04946	2.93
0.800	14.378	3.014	3.66382	—	1/7	20.193	—	6.29404	2.94
0.750	14.476	—	3.70052	—	0.125	20.585	5.597	6.49033	2.94
1/1.4	14.565	3.077	3.73419	—	1/9	20.904	—	6.65107	2.94
0.700	14.605	3.083	3.74961	—	1/10	21.169	5.858	6.78495	2.95
2/3	14.701	3.117	3.79033	—	1/12	21.583	—	6.99507	—
0.600	14.980	3.198	3.89456	—	1/15	22.019	—	7.21683	—
0.500	15.548	3.383	4.12330	3.02	1/20	22.477	—	7.45083	—
0.400	16.368	3.670	4.47185	—	1/50	23.363	—	7.90589	—
1/3	17.090	3.956	4.79480	2.97	0	24.000	7.541	8.23529	8.235
0.300	17.512	4.114	4.98989	—					

For rectangular ducts with four walls heated under the Ⓗ2 thermal boundary condition, the fully developed Nu_{H2} can be determined from Table 5.31 [1]. The f Re, Nu_T, and Nu_{H1} are also given in Table 5.31 for convenience of usage.

Hydrodynamically Developing Flow.   Shah and London [1] have reviewed and compared several analytical and experimental investigations of hydrodynamically developing flow in rectangular ducts. They concluded that the numerical results reported by Curr et al. [158] and the analytical results reported by Tachibana and Iemoto [159] best fit the experimental data. The apparent friction factors obtained by Curr et al. [158] are shown in Fig. 5.25.

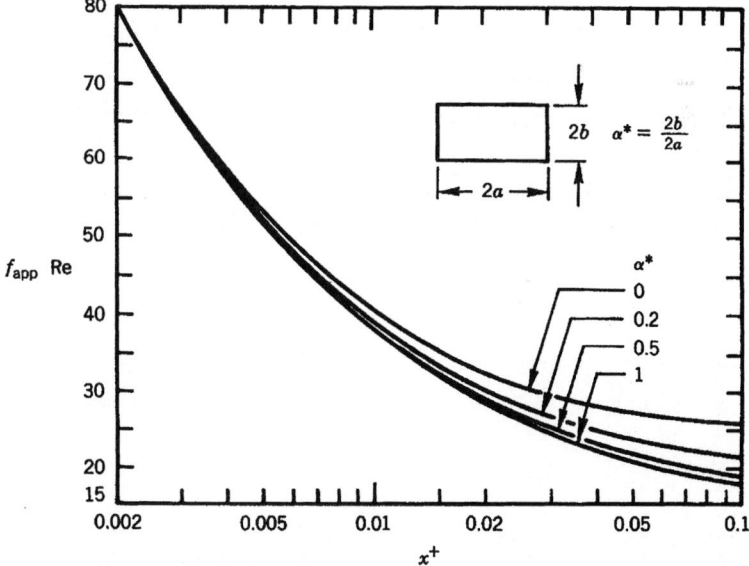

FIGURE 5.25   Apparent Fanning friction factors for hydrodynamically developing flow in rectangular ducts [158].

Thermally Developing Flow.   Wibulswas [160] and Aparecido and Cotta [161] have solved the thermal entrance problem for rectangular ducts with the thermal boundary condition of uniform temperature and uniform heat flux at four walls. However, the effects of viscous dissipation, fluid axial conduction, and thermal energy sources in the fluid are neglected in their analyses. The local and mean Nusselt numbers $Nu_{x,T}$, $Nu_{m,T}$, and $Nu_{x,H1}$ and $Nu_{m,H1}$ obtained by Wibulswas [160] are presented in Tables 5.32 and 5.33.

For square ducts, $\alpha^* = 1$, the Nusselt numbers for the Ⓣ, Ⓗ①, and Ⓗ② thermal boundary conditions have been obtained by Chandrupatla and Sastri [162]. As recommended by Shah and London [1], the results obtained by Chandrupatla and Sastri [162], shown in Table 5.34, are more accurate than those presented by Wibulswas [160].

The thermal entrance lengths for rectangular ducts in the Ⓣ boundary condition $L^*_{th,T}$ are determined to be 0.008, 0.054, 0.049, and 0.041 for $\alpha^* = 0, 0.25, 0.5$, and 1, respectively [1]. The thermal entrance lengths in the Ⓗ① boundary condition $L^*_{th,H1}$ are found to be 0.0115, 0.042,

TABLE 5.32   Local and Mean Nusselt Numbers in the Thermal Entrance Region of Rectangular Ducts With the Ⓣ Boundary Condition [160]

$\dfrac{1}{x^*}$	$Nu_{x,T}$						$Nu_{m,T}$					
	$\alpha^* = 1.0$	0.5	⅓	0.25	0.2	⅙	1.0	0.5	⅓	0.25	0.2	⅙
0	2.65	3.39	3.96	4.51	4.92	5.22	2.65	3.39	3.96	4.51	4.82	5.22
10	2.86	3.43	4.02	4.53	4.94	5.25	3.50	3.95	4.54	5.00	5.36	5.66
20	3.08	3.54	4.17	4.65	5.05	5.34	4.03	4.46	5.00	5.44	5.77	6.04
30	3.24	3.70	4.29	4.76	5.13	5.41	4.47	4.86	5.39	5.81	6.13	6.37
40	3.43	3.85	4.42	4.87	5.22	5.48	4.85	5.24	5.74	6.16	6.45	6.70
60	3.78	4.16	4.67	5.08	5.40	5.64	5.50	5.85	6.35	6.73	7.03	7.26
80	4.10	4.46	4.94	5.32	5.62	5.86	6.03	6.37	6.89	7.24	7.53	7.77
100	4.35	4.72	5.17	5.55	5.83	6.07	6.46	6.84	7.33	7.71	7.99	8.17
120	4.62	4.92	5.42	5.77	6.06	6.27	6.86	7.24	7.74	8.13	8.39	8.63
140	4.85	5.15	5.62	5.98	6.26	6.47	7.22	7.62	8.11	8.50	8.77	9.00
160	5.03	5.34	5.80	6.18	6.45	6.66	7.56	7.97	8.45	8.86	9.14	9.35
180	5.24	5.54	5.99	6.37	6.63	6.86	7.87	8.29	8.77	9.17	9.46	9.67
200	5.41	5.72	6.17	6.57	6.80	7.02	8.15	8.58	9.07	9.47	9.79	10.01

TABLE 5.33   Local and Mean Nusselt Numbers in the Thermal Entrance Region of Rectangular Ducts With the Ⓗ① Boundary Condition [160]

$\dfrac{1}{x^*}$	$Nu_{x,H1}$				$Nu_{m,H1}$			
	$\alpha^* = 1.0$	0.5	⅓	0.25	1.0	0.5	⅓	0.25
0	3.60	4.11	4.77	5.35	3.60	4.11	4.77	5.35
10	3.71	4.22	4.85	5.45	4.48	4.94	5.45	6.03
20	3.90	4.38	5.00	5.62	5.19	5.60	6.06	6.57
30	4.18	4.61	5.17	5.77	5.76	6.16	6.60	7.07
40	4.45	4.84	5.39	5.87	6.24	6.64	7.09	7.51
60	4.91	5.28	5.82	6.26	7.02	7.45	7.85	8.25
80	5.33	5.70	6.21	6.63	7.66	8.10	8.48	8.87
100	5.69	6.05	6.58	7.00	8.22	8.66	9.02	9.39
120	6.02	6.37	6.92	7.32	8.69	9.13	9.52	9.83
140	6.32	6.68	7.22	7.63	9.09	9.57	9.93	10.24
160	6.60	6.96	7.50	7.92	9.50	9.96	10.31	10.61
180	6.86	7.23	7.76	8.18	9.85	10.31	10.67	10.92
200	7.10	7.46	8.02	8.44	10.18	10.64	10.97	11.23

TABLE 5.34 Local and Mean Nusselt Numbers in the Thermal Entrance Region of a Square Duct With the Ⓣ, Ⓗ, and Ⓗ Boundary Conditions [162]

$\dfrac{1}{x^*}$	$Nu_{x,T}$	$Nu_{m,T}$	$Nu_{x,H1}$	$Nu_{m,H1}$	$Nu_{x,H2}$	$Nu_{m,H2}$
0	2.975	2.975	3.612	3.612	3.095	3.095
10	2.976	3.514	3.686	4.549	3.160	3.915
20	3.074	4.024	3.907	5.301	3.359	4.602
25	3.157	4.253	4.048	5.633	3.471	4.898
40	3.432	4.841	4.465	6.476	3.846	5.656
50	3.611	5.173	4.720	6.949	4.067	6.083
80	4.084	5.989	5.387	8.111	4.654	7.138
100	4.357	6.435	5.769	8.747	4.993	7.719
133.3	4.755	7.068	6.331	9.653	5.492	8.551
160	—	—	6.730	10.279	5.848	9.128
200	5.412	8.084	7.269	11.103	6.330	9.891

0.048, 0.057, and 0.066 for $\alpha^* = 0$, 0.25, ⅓, 0.5, and 1, respectively [1]. Thermally developing flow in rectangular ducts with one wall or more insulated is reviewed in Shah and London [1].

Simultaneously Developing Flow. Table 5.35 presents the results for simultaneously developing flow in rectangular ducts; these were obtained by Wibulswas [160] for the Ⓣ and Ⓗ boundary conditions for air (Pr = 0.72). Transverse velocity is neglected in this analysis. However, Chandrupatla and Sastri [163] include transverse velocity in their analysis for a square duct with the Ⓗ boundary condition.

The $Nu_{x,H1}$ and $Nu_{m,H1}$ obtained by Chandrupatla and Sastri [163] are illustrated in Table 5.36. It should be noted that in Table 5.36, Pr = 0 corresponds to slug flow, whereas Pr = ∞ corresponds to hydrodynamically developed flow.

TABLE 5.35 Local and Mean Nusselt Numbers for Simultaneously Developing Flow in Rectangular Ducts With the Ⓣ and Ⓗ Boundary Conditions [160]

$\dfrac{1}{x^*}$	$Nu_{x,H1}$				$Nu_{m,H1}$				$Nu_{m,T}$				
	$\alpha^* = 1.0$	0.5	⅓	0.25	1.0	0.5	⅓	0.25	1.0	0.5	⅓	0.25	⅙
5	—	—	—	—	4.60	5.00	5.58	6.06	—	—	—	—	—
10	4.18	4.60	5.18	5.66	5.43	5.77	6.27	6.65	3.75	4.20	4.67	5.11	5.72
20	4.66	5.05	5.50	5.92	6.60	6.94	7.31	7.58	4.39	4.79	5.17	5.56	6.13
30	5.07	5.40	5.82	6.17	7.52	7.83	8.13	8.37	4.88	5.23	5.60	5.93	6.47
40	5.47	5.75	6.13	6.43	8.25	8.54	8.85	9.07	5.28	5.61	5.96	6.27	6.78
50	5.83	6.09	6.44	6.70	8.90	9.17	9.48	9.70	5.63	5.95	6.28	6.61	7.07
60	6.14	6.42	6.74	7.00	9.49	9.77	10.07	10.32	5.95	6.27	6.60	6.90	7.35
80	6.80	7.02	7.32	7.55	10.53	10.73	11.13	11.35	6.57	6.88	7.17	7.47	7.90
100	7.38	7.59	7.86	8.08	11.43	11.70	12.00	12.23	7.10	7.42	7.70	7.98	8.38
120	7.90	8.11	8.37	8.58	12.19	12.48	12.78	13.03	7.61	7.91	8.18	8.48	8.85
140	8.38	8.61	8.84	9.05	12.87	13.15	13.47	13.73	8.06	8.37	8.66	8.93	9.28
160	8.84	9.05	9.38	9.59	13.50	13.79	14.10	14.48	8.50	8.80	9.10	9.36	9.72
180	9.28	9.47	9.70	9.87	14.05	14.35	14.70	14.95	8.91	9.20	9.50	9.77	10.12
200	9.69	9.88	10.06	10.24	14.55	14.88	15.21	15.49	9.30	9.60	9.91	10.18	10.51
220	—	—	—	—	15.03	15.36	15.83	16.02	9.70	10.00	10.30	10.58	10.90

TABLE 5.36 Local and Mean Nusselt Numbers for Simultaneously Developing Flow in a Square Duct ($\alpha^* = 1$) With the Ⓗ Boundary Condition [163]

x^*	$\mathrm{Nu}_{x,\mathrm{H1}}$					$\mathrm{Nu}_{m,\mathrm{H1}}$				
	Pr = 0.0	0.1	1.0	10.0	∞	0.0	0.1	1.0	10.0	∞
0.005	14.653	11.659	8.373	7.329	7.269	21.986	17.823	13.390	11.200	11.103
0.0075	12.545	9.597	7.122	6.381	6.331	19.095	15.391	11.489	9.737	9.653
0.01	11.297	8.391	6.379	5.716	5.769	17.290	13.781	10.297	8.823	8.747
0.0125	10.459	7.615	5.877	5.480	5.387	16.003	12.620	9.461	8.181	8.111
0.02	9.031	6.353	5.011	4.759	4.720	13.622	10.475	7.934	7.010	6.949
0.025	8.500	5.883	4.683	4.502	4.465	12.647	9.601	7.315	6.533	6.476
0.04	7.675	5.108	4.152	4.080	4.048	10.913	8.043	6.214	5.682	5.633
0.05	7.415	4.826	3.973	3.939	3.907	10.237	7.426	5.782	5.347	5.301
0.1	7.051	4.243	3.687	3.686	3.686	8.701	5.948	4.783	4.580	4.549
∞	7.013	3.612	3.612	3.612	3.612	7.013	3.612	3.612	3.612	3.612

Turbulent Flow

Entrance configuration is the key factor affecting flow transition in rectangular ducts. The lower limit of the critical Reynolds numbers $\mathrm{Re}_{\mathrm{crit}}$ along with entrance configuration has been investigated by Davies and White [164], Allen and Grunberg [165], Cornish [166], and Hartnett et al. [167]. The lower limits of the critical Reynolds numbers for a smooth rectangular duct with two entrance configurations are given in Table 5.37.

For most engineering calculations of friction factors and Nusselt numbers for fully developed flow in rectangular ducts, it is sufficiently accurate to use the circular duct correlations by replacing the circular duct diameter $2a$ with the hydraulic diameter $D_h = 4ab/(a + b)$ or with D_l, defined by the following equations to consider the shape effect [168]:

$$D_l = \frac{2}{3} D_h (1 + \alpha^*)^2 \left(1 - \frac{192\alpha^*}{\pi^5} \sum_{n=0}^{\infty} \frac{1}{(2n+1)^5} \tanh \frac{(2n+1)\pi\alpha^*}{2} \right) \qquad (5.234)$$

An approximate expression for D_l is:

$$D_l = \frac{2}{3} D_h + \frac{11}{24} \alpha^*(2 - \alpha^*) \qquad (5.235)$$

which yields D_l values within ±2 percent of those given by Eq. 5.234.

TABLE 5.37 Lower Limits of the Critical Reynolds Numbers for Smooth Rectangular Ducts

Entrance configurations	Schematics	Aspect ratio α^*	Critical Reynolds number $\mathrm{Re}_{\mathrm{crit}}$
Smooth entrance		0	3400
		0.1	4400
		0.2	7000
		0.3333	6000
		1.0	4300
Abrupt entrance		0	3100
		0.1	2920
		0.2	2500
		0.2555	2400
		0.3425	2360
		1.0	2200

The fully developed friction factor and heat transfer coefficients for turbulent flow in an asymmetrically heated rectangular duct have been reported by Rao [59]. In this investigation, the experimental region of the Reynolds number was from 10^4 to 5×10^4.

For fully developed Nusselt numbers for the turbulent flow of liquid metals in rectangular ducts, a simple correlation has been derived for the Ⓣ and Ⓗ boundary conditions [169]. This correlation follows:

$$Nu = \tfrac{2}{3}Nu_{slug} + 0.015Pe^{0.8} \qquad (5.236)$$

where Nu_{slug} is the Nusselt number corresponding to slug flow (Pr = 0) through rectangular ducts, which is given in Fig. 5.26 as a function of α^* for rectangular ducts under the Ⓣ and Ⓗ boundary conditions.

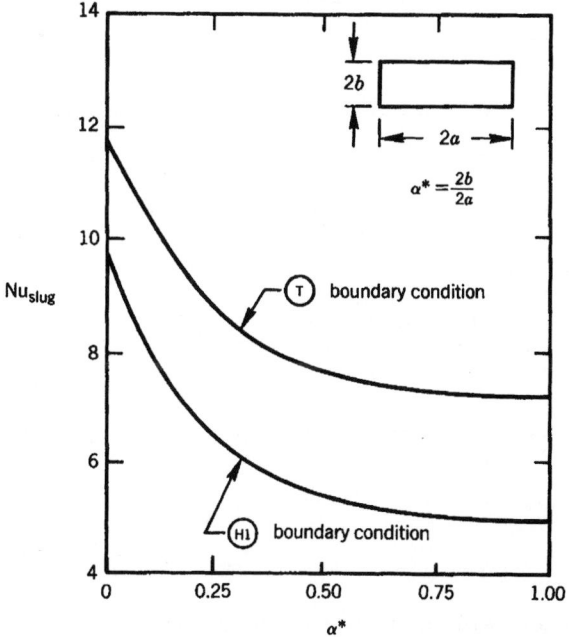

FIGURE 5.26   Slug flow Nusselt numbers for rectangular ducts [169].

TRIANGULAR DUCTS

The flow and heat transfer characteristics of triangular ducts, as shown in Fig. 5.27, are explained in this section. The coordinates shown in Fig. 5.27 are used in the presentation of the results.

Laminar Flow

In this section, the laminar flow and heat transfer characteristics are explained for different triangular ducts.

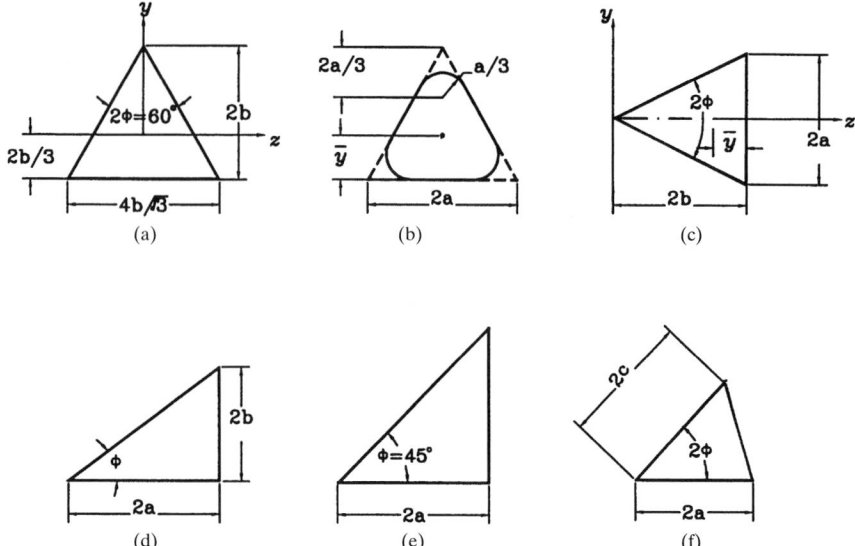

FIGURE 5.27 Triangular ducts: (a) equilateral; (b) equilateral with rounded corners; (c) isosceles; (d) and (e) right; and (f) arbitrary.

Fully Developed Flow

Equilateral Triangular Ducts. For equilateral triangle ducts as shown in Fig. 5.27a, the fully developed laminar flow velocity profile and friction factor have been obtained by Marco and Han [155]:

$$\frac{u}{u_m} = \frac{15}{8}\left[\left(\frac{y}{b}\right)^3 - 3\left(\frac{y}{b}\right)\left(\frac{z}{b}\right)^2 - 2\left(\frac{y}{b}\right)^2 - 2\left(\frac{z}{b}\right)^2 + \frac{32}{37}\right] \tag{5.237}$$

$$u_m = -\frac{b^2}{15\mu}\left(\frac{dp}{dx}\right), \qquad f\,\mathrm{Re} = \frac{40}{3} = 13.333 \tag{5.238}$$

When the three walls of the equilateral triangles are subjected to the uniform wall temperature boundary condition ⓣ, the fully developed Nusselt number Nu_T is equal to 2.49 [170]. However, when the uniform wall heat flux with the uniform circumferential wall temperature boundary condition ⓗ is applied, the Nusselt number Nu_{H1} is determined by the following equation [6]:

$$\mathrm{Nu}_{H1} = \frac{^{28}\!/_{9}}{1 + \tfrac{1}{12}S^* + {}^{40}\!/_{11}\mathrm{Br}'} \tag{5.239}$$

The internal heat source and viscous dissipation effects are considered in Eq. 5.239.

The Nusselt number for uniform wall heat flux in both the axial and circumferential directions under the ⓗ₂ boundary condition Nu_{H2} is found to be 1.892 [171].

Since sharp triangular ducts are rarely seen in practical use, triangular ducts with rounded corners, such as that presented in Fig. 5.27b, have been investigated by Shah [172]. His results are presented in Table 5.38, in which the y and $y_{\max}$ refer to the distances measured from the duct base to the centroid and to the point of maximum fluid velocity, respectively.

TABLE 5.38   Fully Developed Laminar Flow and Heat Transfer Characteristics of Equilateral Triangular Ducts With Rounded Corners [172]

	No rounded corners	One rounded corner	Two rounded corners	Three rounded corners
$D_h/2a$	0.57735	0.59745	0.62115	0.64953
$y/2a$	0.28868	0.26778	0.30957	0.28868
$y_{max}/2a$	0.28868	0.28627	0.29117	0.28868
u_{max}/u_m	2.222	2.172	2.115	2.064
$K(\infty)$	1.818	1.698	1.567	1.441
L_{hy}^+	0.0398	0.0659	0.0319	0.0284
$f\,Re$	13.333	14.057	14.899	15.993
Nu_{H1}	3.111	3.401	3.756	4.205
Nu_{H2}	1.892	2.196	2.715	3.780

Isosceles Triangular Ducts.   For isosceles triangular ducts like those shown in Fig. 5.27c, the velocity distribution and friction factors for fully developed laminar flow are expressed by the following set of equations suggested by Migay [173]:

$$u = -\frac{1}{2\mu}\left(\frac{dp}{dx}\right)\frac{y^2 - z^2\tan^2\phi}{1 - \tan^2\phi}\left[\left(\frac{z}{2b}\right)^{B-2} - 1\right] \tag{5.240}$$

$$u_m = -\frac{2b^2}{3\mu}\left(\frac{dp}{dx}\right)\frac{(B-2)\tan^2\phi}{(B+2)(1 - \tan^2\phi)} \tag{5.241}$$

$$f\,Re = \frac{12(B+2)(1 - \tan^2\phi)}{(B-2)[\tan\phi + (1+\tan^2\phi)^{1/2}]^2} \tag{5.242}$$

$$B = [4 + \tfrac{5}{2}(\cot^2\phi - 1)]^{1/2} \tag{5.243}$$

Apparently, when $2\phi = 90°$, $f\,Re$ is indeterminate from Eq. 5.242. Migay [173] obtained another relation for $2\phi = 90°$, as follows:

$$f\,Re = \frac{12(B+2)(1 - 3\tan^2\phi)}{\{\tfrac{3}{2}\tan\phi[4\tan^2\phi + \tfrac{5}{2}(1-\tan^2\phi)]^{-1/2} - 2\}[\tan\phi + (1+\tan^2\phi)^{1/2}]^2} \tag{5.244}$$

The remaining flow and heat transfer characteristics, represented by $K(\infty)$, L_{hy}^+, Nu_T, Nu_{H1}, and Nu_{H2}, together with $f\,Re$, are given in Table 5.39 [1].

The fully developed Nusselt numbers Nu_T and Nu_{H1} for laminar flow in isosceles triangular ducts with one wall or more heated are given in Table 5.40.

Right Triangular Ducts.   For right triangular ducts, shown in Fig. 5.27d and e, the fully laminar developed flow and heat transfer characteristics $f\,Re$, $K(\infty)$, Nu_{H1}, and Nu_{H2} are given in Fig. 5.28 [2]. The data for this figure were taken from Haji-Sheikh et al. [170], Sparrow and Haji-Sheikh [174], and Iqbal et al. [175].

Arbitrary Triangular Ducts.   For triangular ducts with arbitrary angles such as that shown in Fig. 5.27f, the fully developed friction factors and Nusselt numbers are presented for fully developed laminar flow in Figs. 5.29 and 5.30 [2].

Thermally and Simultaneously Developing Flows.   Hydrodynamically developing laminar flow in triangular ducts has been solved by different investigators as is reviewed by Shah and London [1]. Wibulswas [160] obtained a numerical solution for the problem of simultaneously

TABLE 5.39 Flow and Heat Transfer Characteristics for Fully Developed Laminar Flow in Isosceles Triangular Ducts [1]

$2b/2a$	2ϕ	$K(\infty)$	L_{hy}^+	$f\,Re$	Nu_T	Nu_{H1}	Nu_{H2}
∞	0	2.971	0.1048	12.000	0.943	2.059	0
8.000	7.15	2.521	0.0648	12.352	1.46	2.348	0.039
5.715	10.00	2.409	0.0590	12.474	1.61	2.446	0.080
4.000	14.25	2.271	0.0533	12.636	1.81	2.575	0.173
2.836	20.00	2.128	0.0484	12.822	2.00	2.722	0.366
2.000	28.07	1.991	0.0443	13.026	2.22	2.880	0.747
1.866	30.00	1.966	0.0436	13.065	2.26	2.910	0.851
1.500	36.87	1.898	0.0418	13.181	2.36	2.998	1.22
1.374	40.00	1.876	0.0412	13.222	2.39	3.029	1.38
1.072	50.00	1.831	0.0401	13.307	2.45	3.092	1.76
1.000	53.13	1.824	0.0399	13.321	2.46	3.102	1.82
0.866	60.00	1.818	0.0398	13.333	2.47	3.111	1.89
0.750	67.38	1.824	0.0399	13.321	2.45	3.102	1.84
0.714	70.00	1.829	0.0400	13.311	2.45	3.095	1.80
0.596	80.00	1.860	0.0408	13.248	2.40	3.050	1.59
0.500	90.00	1.907	0.0421	13.153	2.34	2.982	1.34
0.289	120.00	2.165	0.0490	12.744	2.00	2.680	0.62
0.250	126.87	2.235	0.0515	12.622	1.90	2.603	0.490
0.134	150.00	2.543	0.0644	12.226	1.50	2.325	0.156
0.125	151.93	2.574	0.0659	12.196	1.47	2.302	0.136
0	180.00	2.971	0.1048	12.000	0.943	2.059	0

TABLE 5.40 Fully Developed Nu_T and Nu_{H1} for Heat Transfer in Isosceles Triangular Ducts With One Wall or More Walls Heated [1]

$\dfrac{2b}{2a}$	ϕ degrees	Nu_T				Nu_{H1}			
∞	0	1.885	0.000	1.215	1.215	2.059	0	1.346	1.346
5.000	5.71	—	0.822	1.416	1.312	2.465	1.003	1.824	1.739
2.500	11.31	2.058	1.268	1.849	1.573	2.683	1.515	2.274	1.946
1.667	16.70	2.227	1.525	2.099	1.724	2.796	1.807	2.541	2.074
1.250	21.80	2.312	1.675	2.237	1.802	2.845	1.978	2.695	2.141
1.000	26.56	2.344	1.758	2.301	1.831	2.849	2.076	2.773	2.161
0.833	30.96	—	—	2.319	1.822	—	—	2.801	2.146
0.714	34.99	2.311	1.812	2.306	1.787	2.778	2.146	2.792	2.107
0.625	38.66	—	—	2.274	1.735	—	—	2.774	2.053
0.556	41.99	—	—	2.232	1.673	—	—	2.738	1.989
0.500	45.00	2.162	1.765	2.183	1.606	2.594	2.111	2.696	1.921
0.450	48.01	—	—	2.127	1.529	—	—	2.646	1.843
0.400	51.34	—	—	2.055	1.433	—	—	2.583	1.746
0.350	55.01	1.923	1.633	1.968	1.315	2.332	1.991	2.505	1.628
0.300	59.04	—	—	1.861	1.173	—	—	2.412	1.486
0.250	63.43	1.671	1.471	1.733	1.004	2.073	1.843	2.301	1.316
0.200	68.20	1.512	1.361	1.581	0.805	1.917	1.746	2.174	1.114
0.150	73.30	1.330	1.229	1.401	0.578	1.748	1.635	2.032	0.874
0.100	78.69	1.126	1.071	1.182	0.332	1.576	1.515	1.881	0.587
0.050	84.29	0.895	0.878	0.893	0.106	1.418	1.398	1.737	0.244
0	90.00	0.6076	0.608	—	—	1.346	1.346	—	—

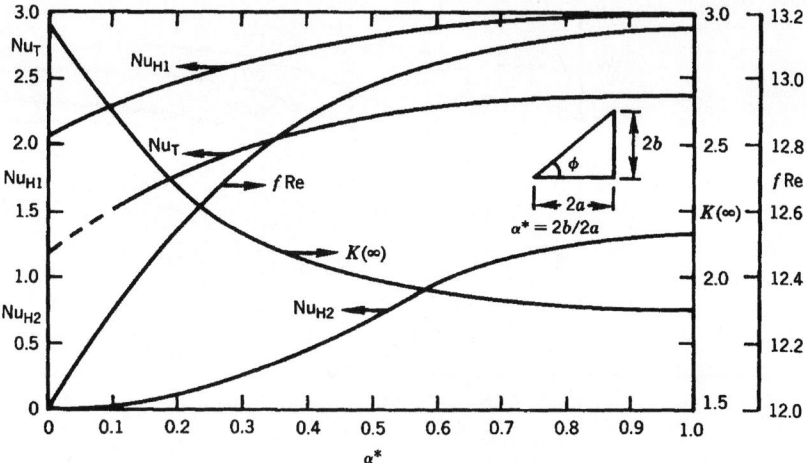

FIGURE 5.28 Fully developed f Re and Nu for right triangular ducts [2].

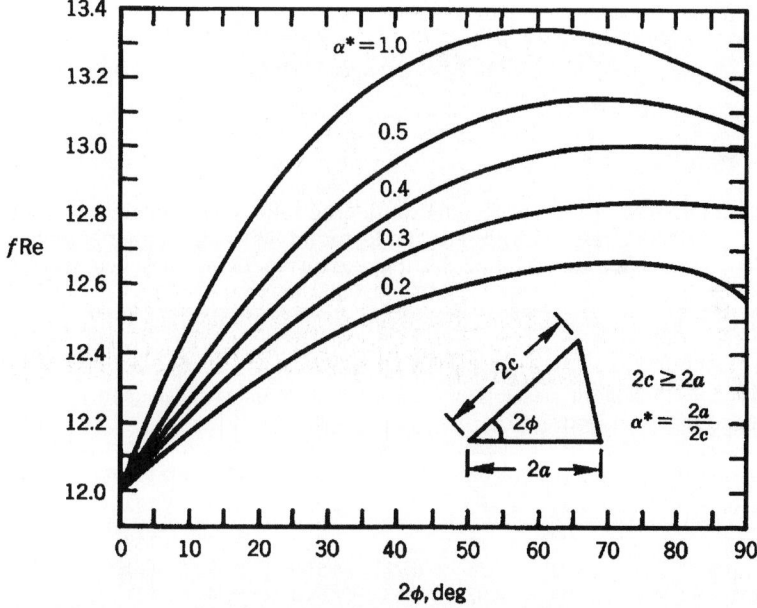

FIGURE 5.29 Fully developed friction factors for arbitrary triangular ducts [2].

developing laminar flow for equilateral triangular and right-angled isosceles triangular ducts for Pr = 0, 0.72, and ∞. His results for uniform wall temperature and axial uniform wall heat flux with circumferential uniform wall temperature boundary conditions are presented in Tables 5.41 and 5.42. Since Pr = ∞ implies that the flow is hydrodynamically developed, the results for Pr = ∞ can be applied to all fluids in thermally developing laminar flow.

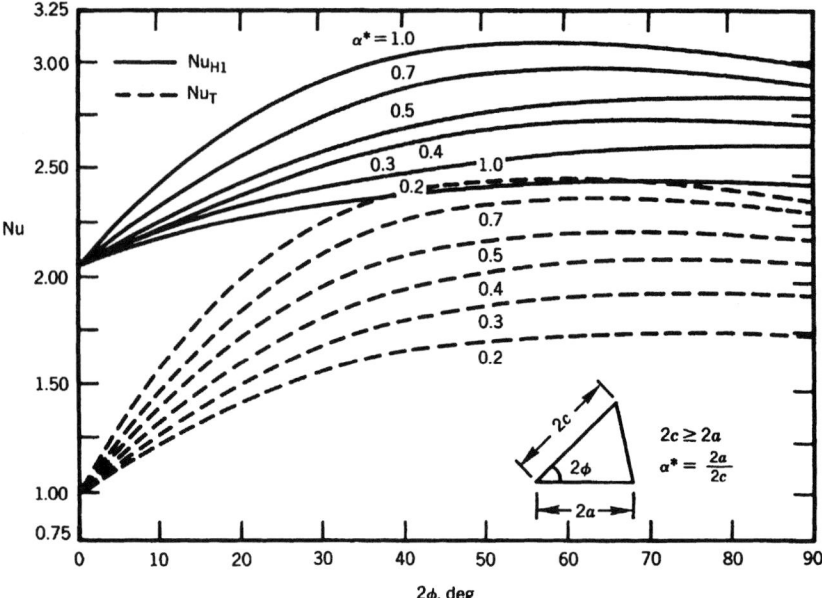

FIGURE 5.30   Fully developed Nusselt numbers for arbitrary triangular ducts [2].

Turbulent Flow

The lower limit of Re_{crit} is considered to be approximately 2000 in triangular ducts [45]. No reliable results for the friction factor and Nusselt number are available for transition flow in triangular ducts. In this section, the turbulent flow and heat transfer characteristics for equilateral, isosceles, and right triangular ducts are presented.

TABLE 5.41   Local and Mean Nusselt Numbers for Thermally and Simultaneously Developing Laminar Flows and Equilateral Triangular Ducts [160]

$\frac{1}{x^*}$	$Nu_{x,T}$			$Nu_{m,T}$			$Nu_{x,H1}$			$Nu_{m,H1}$		
	$Pr = \infty$	0.72	0	∞	0.72	0	∞	0.72	0	∞	0.72	0
10	2.57	2.80	3.27	3.10	3.52	4.65	3.27	3.58	4.34	4.02	4.76	6.67
20	2.73	3.11	3.93	3.66	4.27	5.79	3.48	4.01	5.35	4.76	5.87	8.04
30	2.90	3.40	4.46	4.07	4.88	6.64	3.74	4.41	6.14	5.32	6.80	9.08
40	3.08	3.67	4.89	4.43	5.35	7.32	4.00	4.80	6.77	5.82	7.57	9.96
50	3.26	3.93	5.25	4.75	5.73	7.89	4.26	5.13	7.27	6.25	8.20	10.65
60	3.44	4.15	5.56	5.02	6.08	8.36	4.49	5.43	7.66	6.63	8.75	11.27
80	3.73	4.50	6.10	5.49	6.68	9.23	4.85	6.03	8.26	7.27	9.73	12.35
100	4.00	4.76	6.60	5.93	7.21	9.98	5.20	6.56	8.81	7.87	10.60	13.15
120	4.24	4.98	7.03	6.29	7.68	10.59	5.50	7.04	9.30	8.38	11.38	13.82
140	4.47	5.20	7.47	6.61	8.09	11.14	5.77	7.50	9.74	8.84	12.05	14.46
160	4.67	5.40	7.88	6.92	8.50	11.66	6.01	7.93	10.17	9.25	12.68	15.02
180	4.85	5.60	8.20	7.18	8.88	12.10	6.22	8.33	10.53	9.63	13.27	15.50
200	5.03	5.80	8.54	7.42	9.21	12.50	6.45	8.71	10.87	10.02	13.80	16.00

TABLE 5.42 Local and Mean Nusselt Numbers for Thermally and Simultaneously Developing Laminar Flows in Right-Angled Isosceles Triangular Ducts [160]

$\dfrac{1}{x^*}$	$Nu_{x,T}$			$Nu_{m,T}$			$Nu_{x,H1}$			$Nu_{m,H1}$		
	$Pr = \infty$	0.72	0	∞	0.72	0	∞	0.72	0	∞	0.72	0
10	2.40	2.52	3.75	2.87	3.12	4.81	3.29	4.00	5.31	4.22	5.36	6.86
20	2.53	2.76	4.41	3.33	3.73	5.85	3.58	4.73	6.27	4.98	6.51	7.97
30	2.70	2.98	4.82	3.70	4.20	6.48	3.84	5.23	6.85	5.50	7.32	8.68
40	2.90	3.18	5.17	4.01	4.58	6.97	4.07	5.63	7.23	5.91	7.95	9.20
50	3.05	3.37	5.48	4.28	4.90	7.38	4.28	5.97	7.55	6.25	8.50	9.67
60	3.20	3.54	5.77	4.52	4.17	7.73	4.47	6.30	7.85	6.57	8.99	10.07
80	3.50	3.85	6.30	4.91	5.69	8.31	4.84	6.92	8.37	7.14	9.80	10.75
100	3.77	4.15	6.75	5.23	6.10	8.80	5.17	7.45	8.85	7.60	10.42	11.32
120	4.01	4.43	7.13	5.52	6.50	9.18	5.46	7.95	9.22	8.03	10.90	11.77
140	4.21	4.70	7.51	5.78	6.82	9.47	5.71	8.39	9.58	8.40	11.31	12.14
160	4.40	4.96	7.84	6.00	7.10	9.70	5.95	8.80	9.90	8.73	11.67	12.47
180	4.57	5.22	8.10	6.17	7.33	9.94	6.16	9.14	10.17	9.04	12.00	12.75
200	4.74	5.49	8.38	6.33	7.57	10.13	6.36	8.50	10.43	9.33	12.29	13.04

Fully Developed Flow

Equilateral Triangular Ducts.   The friction factor for an equilateral triangular duct has been measured by Altemani and Sparrow [176]. Their data are fitted by the following equation in the region of $4000 < \mathrm{Re} < 8 \times 10^4$:

$$f = \frac{0.0425}{\mathrm{Re}^{0.2}} \tag{5.245}$$

These researchers have also obtained the fully developed Nusselt numbers for air ($Pr = 0.7$) in the range of $4000 < \mathrm{Re} < 8 \times 10^4$ in an equilateral triangular duct with the Ⓗ boundary condition on two walls and the third wall insulated as follows:

$$Nu_{H1} = 0.019 \mathrm{Re}^{0.781} \tag{5.246}$$

Isosceles Triangular Ducts.   Bhatti and Shah [45] recommended that the friction factor for fully developed turbulent flow in isosceles triangular ducts can be determined using different correlations. For $0 < 2\phi < 60°$, the circular duct correlations in Table 5.8 can be used with D_h replaced by D_g, as defined by Bandopadhayay and Ambrose [177]:

$$D_g = \frac{D_h}{2\pi} \left[3 \ln \cot \frac{\theta}{2} - 2 \ln \tan \frac{\phi}{2} - \ln \tan \frac{\phi}{2} \right] \tag{5.247}$$

where $\theta = (90° - \phi)/2$. For $2\phi = 60°$, the circular duct correlations in Table 5.8 should be used with D_h replaced by D_l, which is equal to $\sqrt{3}a$. For $60° < 2\phi < 90°$, the use of circular duct correlations with D_h is probably accurate enough. For $2\phi = 90°$, the circular duct correlations in Table 5.8 can be used. For $2\phi > 90°$, no definite recommendation can be made at this moment.

Right Triangular Ducts.   The fully developed turbulent friction factor for two right-angled triangular ducts and an equilateral triangular duct are shown in Fig. 5.31 [45]. The data are from Nikuradse [178] and Schiller [179]. Also provided in this figure are the correlations for computing the friction factor for each triangular duct.

Thermally Developing Flow.

Altemani and Sparrow [176] have conducted experimental measurements of the thermally developing flow of air ($Pr = 0.7$) in an equilateral triangular duct with the Ⓗ boundary condition on two walls and the third wall insulated. The local Nusselt numbers $Nu_{x,H1}$ and the thermal entrance lengths from their results are given in Figs. 5.32 and 5.33, respectively.

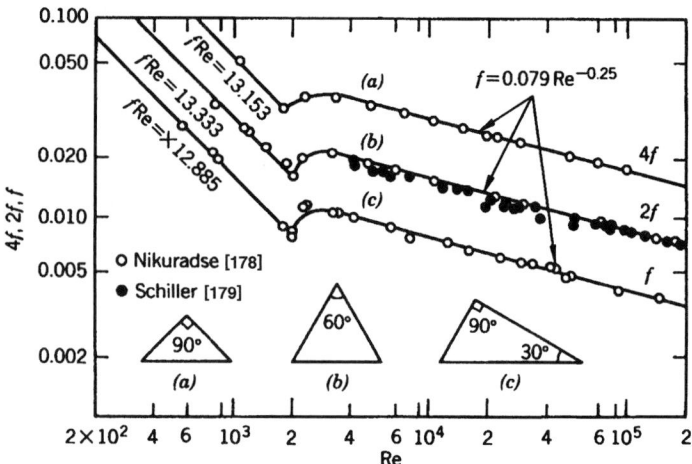

FIGURE 5.31 Fully developed friction factor for turbulent flow in smooth right-angled and equilateral triangular ducts [45].

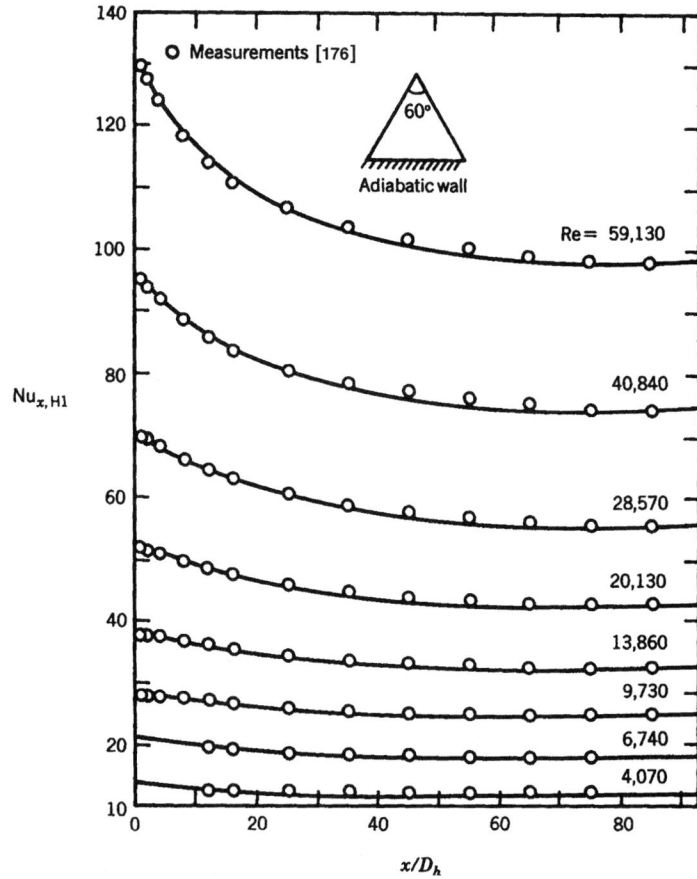

FIGURE 5.32 Local Nusselt numbers $Nu_{x,H1}$ for thermally developing and hydrodynamically developing turbulent airflow ($Pr = 0.7$) in a smooth equilateral triangular duct [176].

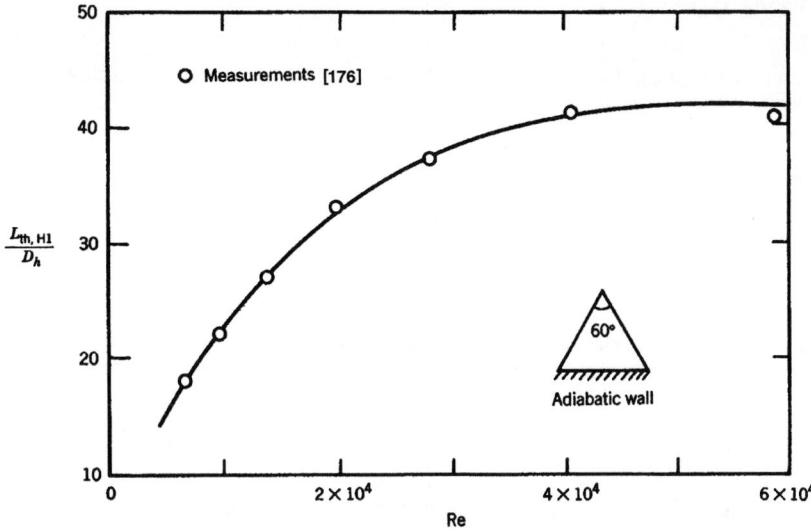

FIGURE 5.33 Thermal entrance lengths for thermally developing and hydrodynamically developing turbulent airflow (Pr = 0.7) in a smooth equilateral triangular duct [176].

For equilateral triangular ducts having rounded corners with a ratio of the corner radius of curvature to the hydraulic diameter of 0.15, Campbell and Perkins [180] have measured the local friction factor and heat transfer coefficients with the ⒽⒽ boundary condition on all three walls over the range $6000 < \text{Re} < 4 \times 10^4$. The results are reported in terms of the hydrodynamically developed flow friction factor in the thermal entrance region with the local wall (T_w) to fluid bulk mean (T_m) temperature ratio in the range $1.1 < T_w/T_m < 2.11$, $6000 < \text{Re} < 4 \times 10^4$, and $7.45 < x/D_h < 72$. These data were correlated by

$$\frac{f}{f_{\text{iso}}} = \left(\frac{T_w}{T_m}\right)^{-0.40 + (x/D_h)^{-0.67}} \tag{5.248}$$

where f_{iso} denotes the friction factor for isothermal flow, which can be calculated from either the Blasius formula or the PKN formula presented in Table 5.8. In these calculations, kinematic viscosity ν entering $\text{Re} = u_m D_h/\nu$ must be evaluated at the duct wall temperature T_w.

The following correlation has been obtained by Campbell and Perkins [180] from their measurements for local Nusselt numbers:

$$\text{Nu}_{x,\text{H1}} = 0.021 \text{Re}^{0.8} \, \text{Pr}^{0.4} \left(\frac{T_w}{T_m}\right)^{0.7} \Phi \tag{5.249}$$

For $6 < x/D_h \leq 50$, the correction factor Φ is given by

$$\Phi = \left[1 + \left(\frac{x}{D_h}\right)^{-0.7} \left(\frac{T_w}{T_m}\right)^{0.7}\right] \tag{5.250}$$

and for $x/D_h > 50$, $\Phi = 1$. Equation 5.249 is valid for $6 < x/D_h < 123$ and $1.10 < T_w/T_m < 2.11$. The fluid properties ν, α, and k entering Eq. 5.249 must be taken at the fluid bulk mean temperature T_m.

ELLIPTICAL DUCTS

Elliptical ducts can be thought as a family of ducts, including geometries ranging from lenticular to circular ducts. The major and minor axes lengths are represented by $2a$ and $2b$, respectively, in this section. The origin of the coordinate is the intersection of the major and minor axes.

Laminar Flow

Presented in this section are the friction factors and Nusselt numbers for laminar flow in elliptical ducts.

Fully Developed Flow. The velocity distribution for fully developed laminar flow in elliptic ducts with major axis $2a$ and minor axis $2b$ is given by Shah and London [1] as follows:

$$\frac{u}{u_m} = 2\left[1 - \left(\frac{z}{a}\right)^2 - \left(\frac{y}{b}\right)^2\right] \tag{5.251}$$

$$u_m = -\frac{1}{4\mu}\left(\frac{dp}{dx}\right)\left(\frac{b^2}{1+\alpha^{*2}}\right), \qquad f\,\text{Re} = 2(1+\alpha^{*2})\left(\frac{\pi}{E(m)}\right)^2 \tag{5.252}$$

In the preceding equations, $m = 1 + \alpha^{*2}$, and $E(m)$ is the complete elliptic integral of the second kind. The hydraulic diameter of the elliptical duct is $D_h = \pi b/E(m)$, and the cross-sectional area is $A_c = \pi ab$.

The expression for hydrodynamic entrance length L_{hy}^+ developed by Bhatti [181] is as follows:

$$L_{hy}^+ = \frac{0.5132}{1+\alpha^{*2}}\left(\frac{E(m)}{\pi}\right)^2 \tag{5.253}$$

The fully developed incremental pressure drop number $K(\infty)$ for elliptic ducts has been found to be independent of the duct aspect ratio $\alpha^* = 2b/2a$ [187]. The value of $K(\infty)$ is recommended to be 1.26 for practical calculations [2].

The fully developed Nusselt numbers Nu_T, Nu_{H1}, Nu_{H2} are displayed in Fig. 5.34. The values of the data used in this figure are derived from the results by Tyagi [6], Tao [182], Iqbal et al. [175], and Dunwoody [183].

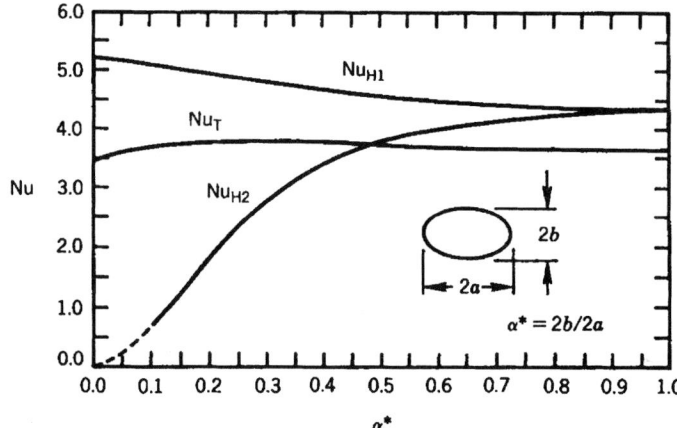

FIGURE 5.34   Fully developed Nusselt numbers for elliptical ducts [2].

Hydrodynamically Developing Flow.   Bhatti [181] has analyzed hydrodynamically developing flow in elliptic ducts. The apparent friction factors and incremental pressure drop numbers can be expressed as:

$$f_{app} \, Re = \frac{\Delta p^*}{4x^+}, \qquad \Delta p^* = \frac{2(1-\eta)(1+3\eta) - (1+\eta)^2 \ln \eta^3}{3(1+\eta)^2} \qquad (5.254)$$

$$K(x) = \frac{(3\eta^3 + 9\eta^2 + 21\eta + 7)(1-\eta)}{6(1+\eta)^2} \qquad (5.255)$$

where η is a boundary layer parameter that refers to the fraction of the duct cross section carrying inviscid flow. The term η is determined from the following equation implicitly as a function of x^+:

$$16(1 + \alpha^{*2}) \left(\frac{\pi}{E(m)} \right)^2 x^+ = \eta^2 - 1 - \ln \eta^2 \qquad (5.256)$$

Thermally Developing Flow.   The local Nusselt number for thermally developing flow in elliptic ducts with uniform wall temperature $Nu_{x,T}$ was obtained by Dunwoody [183] in terms of a double infinite series. These results are considered the most accurate as $x^* > 0.005$. Dunwoody's formula for calculating the mean Nusselt number is as follows:

$$Nu_{m,T} = \frac{\lambda}{4} \left(1 + \frac{C}{x^*} \right) \qquad (5.257)$$

The λ and C values for Eq. 5.257 are given in Table 5.43.

TABLE 5.43   The Constants λ and C for Eq. 5.257 [183]

α^*	λ	C
0.0625	14.59	0.0578
0.125	14.90	0.0388
0.25	15.17	0.0239
0.5	14.97	0.0158
0.8	14.67	0.0138

For $x^* < 0.005$, the following formula obtained by Richardson [184] is recommended:

$$Nu_{m,T} = \frac{3}{\Gamma(\tfrac{4}{3})(9x^+)^{1/3}} \left(1 + \frac{(1-\alpha^*)^2 + (1-\alpha^*)^3}{36} \right) \qquad (5.258)$$

For elliptic ducts subjected to the Ⓗ thermal boundary condition, Someswara et al. [185] have solved thermally developing flow. The mean Nusselt number $Nu_{m,H1}$ can be computed using the following expression:

$$Nu_{m,H1} = \frac{2.61F}{x^{*1/3}} \qquad (5.259)$$

where factor F is a function of α^* and can be calculated by the following [2]:

$$F = 1.2089 - 0.795\alpha^* - 4.3011\alpha^{*2} + 23.8465\alpha^{*3} - 44.7053\alpha^{*4} + 37.0874\alpha^{*5} - 11.4809\alpha^{*6}$$
$$(5.260)$$

Equation 5.260 is accurate within ±3 percent error to the original data given by Someswara et al. [185].

Turbulent Flow

The friction factors for fully developed turbulent flow have been measured by Barrow and Roberts [186] in elliptical ducts with $\alpha^* = 0.316$ and 0.415 in the range of $10^3 < \text{Re} < 3.10^5$ and by Cain and Duffy [187] in elliptical ducts with $\alpha^* = 0.5$ and 0.667 in the range $2 \times 10^4 < \text{Re} < 1.3 \times 10^5$. Based on the data presented by Barrow and Roberts [186] and Cain and Duffy [187], Bhatti and Shah [45] have derived the following correlation to calculate the friction factor:

$$\frac{f}{f_c} = 0.4443 + 2.2168\alpha^* - 2.0431\alpha^{*2} + 0.3821\alpha^{*3} \qquad (5.261)$$

where f_c is the friction factor for a smooth circular duct ($\alpha^* = 1$), which can be obtained from the Blasius equation in Table 5.8. It should be noted that Eq. 5.261 is valid for $0.136 < \alpha^* < 1$.

Heat transfer in fully developed turbulent flow in elliptical ducts has been determined in several investigations. A comparison of the different results has been presented by Bhatti and Shah [45]. It was concluded that the Gnielinski correlation for circular ducts can confidently be used to calculate the fully developed Nusselt number for elliptical ducts for fluids of $\text{Pr} \geq 0.5$.

For liquid metals, the fully developed Nusselt numbers can be determined using Eq. 5.236 for elliptical ducts with the Ⓗ boundary condition. The values of Nu_{slug} required in Eq. 5.236 are given in Fig. 5.35.

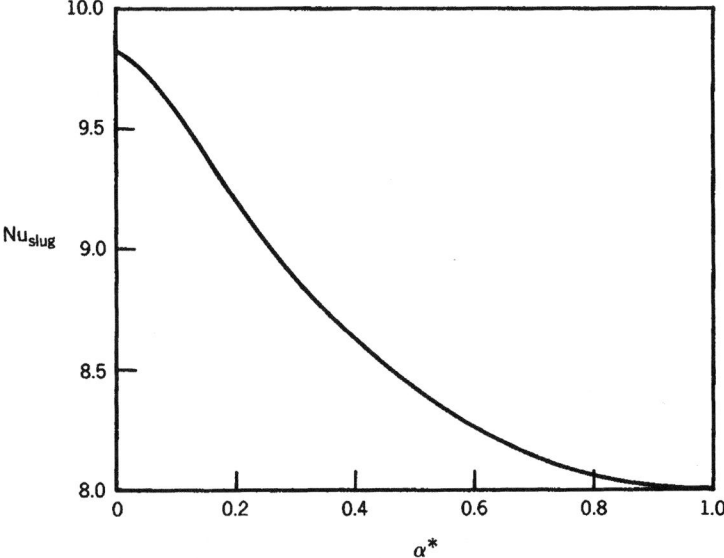

FIGURE 5.35   Slug flow Nusselt numbers for elliptical ducts with the Ⓗ boundary condition [169].

CURVED DUCTS AND HELICOIDAL PIPES

The most prominent characteristic of flow in curved ducts and helicoidal pipes is the secondary flow induced by centrifugal force due to the curvature of the pipe. Consequently, the friction factor is higher in curved pipes than that in straight pipes for the same Reynolds number. The pitch of the helicoidal pipe also has an effect on the flow. As a result, the heat transfer rate is higher in curved or helicoidal pipes than in straight pipes. Therefore, curved or helicoidal pipes are widely used in engineering applications.

Spiral coils are curved ducts with varying curvature. The friction factor and heat transfer rate for spiral coils are also included in this section. In addition to the dimensionless parameters used in straight pipes, the following parameters are particularly useful in the case of curved ducts or helicoidal pipes: the Dean number De; the helical number He, and the effective radius of curvature R_c. These are defined as follows:

$$\text{De} = \text{Re} \left(\frac{a}{R} \right)^{1/2} \tag{5.262}$$

$$\text{He} = \text{Re} \left(\frac{a}{R_c} \right)^{1/2} = \text{De} \left[1 + \left(\frac{b}{2\pi R} \right)^2 \right]^{1/2} \tag{5.263}$$

$$R_c = R \left[1 + \left(\frac{b}{2\pi R} \right) \right] \tag{5.264}$$

where a denotes the radius of a circular pipe, b represents the coil pitch, and R is the radius of the coil.

In this section, emphasis will be given to the correlations used for calculating the friction factors and Nusselt numbers for laminar and turbulent flows in curved ducts, helicoidal pipes, and spiral ducts. These will be presented as the ratio of the friction factor in curved ducts to the friction factor in straight ducts f_c/f_s and the ratio of the Nusselt number in curved ducts to the Nusselt number in straight ducts Nu_c/Nu_s, in most cases. The subscript c represents curved ducts or helicoidal pipes, while the subscript s denotes straight pipes of the same shape.

Fully Developed Laminar Flow

Dean [188, 189] first studied the velocity profile of flow in helicoidal pipes using perturbation analysis. His result is valid only for De < 20, where the velocity distribution is almost identical to that found in straight ducts. Mori and Nakayama [190] have obtained the solution for De > 100 for a coil with $R > a$. Their results are in agreement with the experimental data reported by Mori and Nakayama [190] and Adler [191] and numerical simulations [192].

The friction factors for fully developed flow in helicoidal pipe proposed by Srinivasan et al. [193] in the range of 7 < R/a < 104 follow:

$$\frac{f_c}{f_s} = \begin{cases} 1 & \text{for De} < 30 \\ 0.419\text{De}^{0.275} & \text{for } 30 < \text{De} < 300 \\ 0.1125\text{De}^{0.5} & \text{for De} > 300 \end{cases} \tag{5.265}$$

However, after reviewing the available experimental data and theoretical predications, Manlapaz and Churchill [194] recommended the following correlations, which contain both the Dean number De and radius ratio of R/a:

$$\frac{f_c}{f_s} = \left[\left(1.0 - \frac{0.18}{[1 + (35/\text{De})^2]^{0.5}} \right)^m + \left(1.0 + \frac{a/R}{3} \right)^2 \left(\frac{\text{De}}{88.33} \right) \right]^{0.5} \tag{5.266}$$

where $m = 2$ for De < 20; $m = 1$ for 20 < De < 40; and $m = 0$ for De > 40. It can be observed that Eq. 5.265 does not include the parameter R/a and will not be used for all the range of R/a. After a comparison of Eq. 5.266 with experimental data, Shah and Joshi [195] suggested that Eq. 5.265 be used for the coils with R/a < 3 and that either Eq. 5.265 or Eq. 5.266 be used for coils with R/a > 3.

The friction factors for spiral coils can be obtained using the following correlation [193], which is valid for $500 < \text{Re} \, (b/a)^{0.5} < 20{,}000$ and $7.3 < b/a < 15.5$.

$$f_c = \frac{0.62(n_2^{0.7} - n_1^{0.7})^2}{\text{Re}^{0.6} \, (b/a)^{0.3}} \tag{5.267}$$

where n_1 and n_2 represent the numbers of turns from the origin to the start and the end of a spiral.

The critical Reynolds number, which is used to identify the transition from laminar to turbulent flow in curved or helicoidal pipes, has been recommended for design purposes by Srinivasan et al. [193]:

$$\text{Re}_{\text{crit}} = 2100\left[1 + 12\left(\frac{R}{a}\right)^{-0.5}\right] \tag{5.268}$$

However, for spiral pipes, no single Re_{crit} exists because of varying curvature. The minimum value of Re_{crit} can be obtained using $R_{\max}$ to replace R in Eq. 5.268, and the maximum value of Re_{crit} can be determined using $R_{\min}$ to replace the R in Eq. 5.268.

The fully developed Nusselt numbers for laminar flow in helicoidal pipes subjected to the uniform wall temperature have been obtained theoretically and experimentally by Mori and Nakayama [196], Tarbell and Samuels [197], Dravid et al. [198], Akiyama and Cheng [199], and Kalb and Seader [200]. A comparison of these results has been made using the Manlapaz-Churchill [201] correlation.

In Fig. 5.36, experimental and theoretical results [196–200] are compared with the following Manlapaz-Churchill [201] correlation based on a regression analysis of the available data:

$$\text{Nu}_T = \left[\left(3.657 + \frac{4.343}{x_1}\right)^3 + 1.158\left(\frac{\text{De}}{x_2}\right)^{3/2}\right]^{1/3} \tag{5.269}$$

where

$$x_3 = \left(1.0 + \frac{957}{\text{De}^2 \, \text{Pr}}\right)^2, \qquad x_2 = 1.0 + \frac{0.477}{\text{Pr}} \tag{5.270}$$

It can be seen that the prediction calculated from Eq. 5.269 agrees well with the experimental data in most cases, except for $\text{Pr} = 0.01$ and 0.1 at intermediate He values.

The fully developed Nusselt numbers for spiral coils with uniform wall temperature are suggested by Kubair and Kuldor [202, 203], as follows:

$$\text{Nu}_T = \left(1.98 + 1.8\,\frac{a}{R_{\text{ave}}}\right) \text{Gz}^{0.7} \tag{5.271}$$

which is valid in the ranges of $9 \le \text{Gz} \le 1000$, $80 < \text{Re} < 6000$, and $20 < \text{Pr} < 100$. Although this equation can be used to obtain the Nusselt number in the thermal entrance region, the fully developed Nusselt number may be calculated by substituting $\text{Gz} = 20$.

For helicoidal pipes with the H1 thermal boundary condition, the Nusselt number has been developed by Manlapaz and Churchill [201]:

$$\text{Nu}_{\text{H1}} = \left[\left(4.364 + \frac{4.636}{x_3}\right)^3 + 1.816\left(\frac{\text{De}}{x_4}\right)^{3/2}\right]^{1/3} \tag{5.272}$$

where

$$x_3 = \left(1.0 + \frac{1342}{\text{De}^2 \, \text{Pr}}\right)^2 \qquad x_4 = 1.0 + \frac{1.15}{\text{Pr}} \tag{5.273}$$

Figure 5.37 compares Eq. 5.272 with some of the available theoretical predictions [204, 205] and experimental Nusselt number data [190, 198]. The figure indicates a fairly good agreement between the correlation and most of the available data.

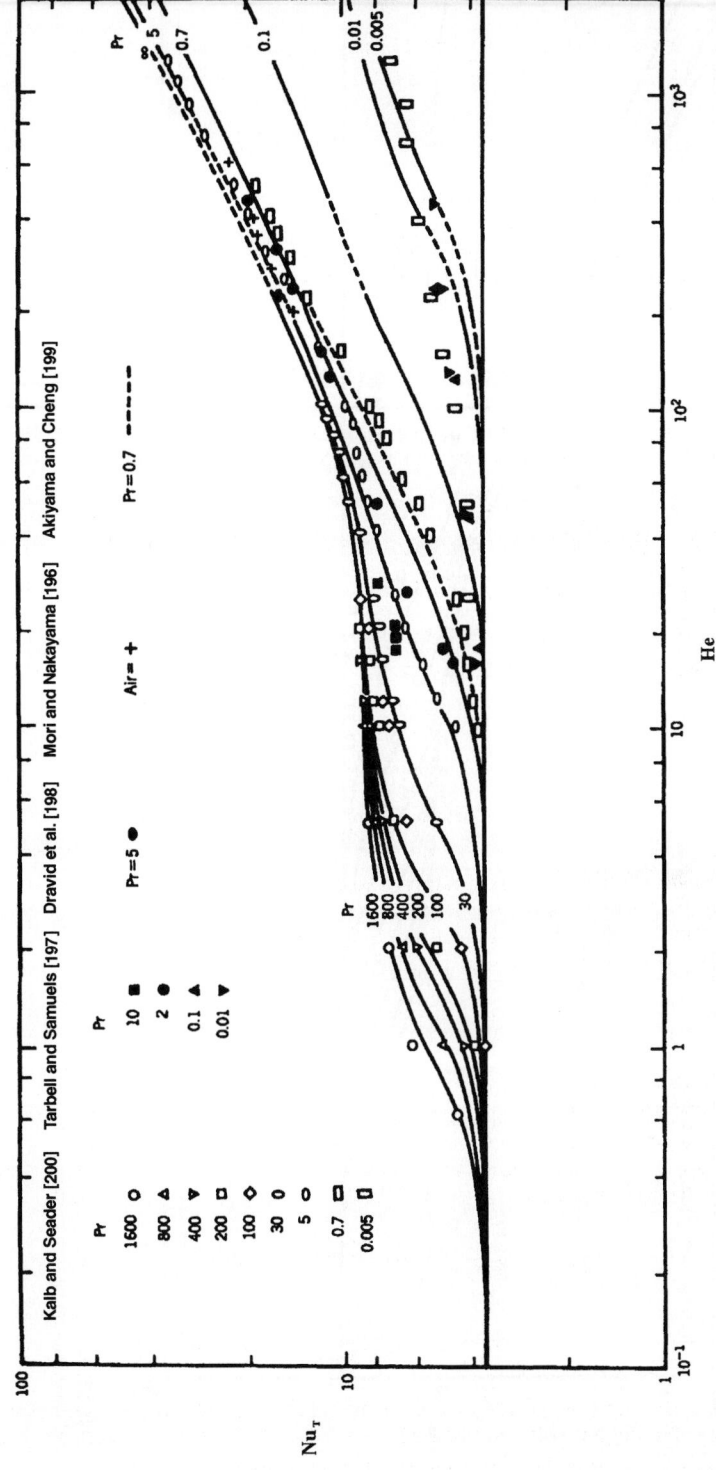

FIGURE 5.36 A comparison of the recommended design correlation (Eq. 5.268, solid line) with theoretical and experimental Nusselt numbers for the ⓣ boundary condition [201].

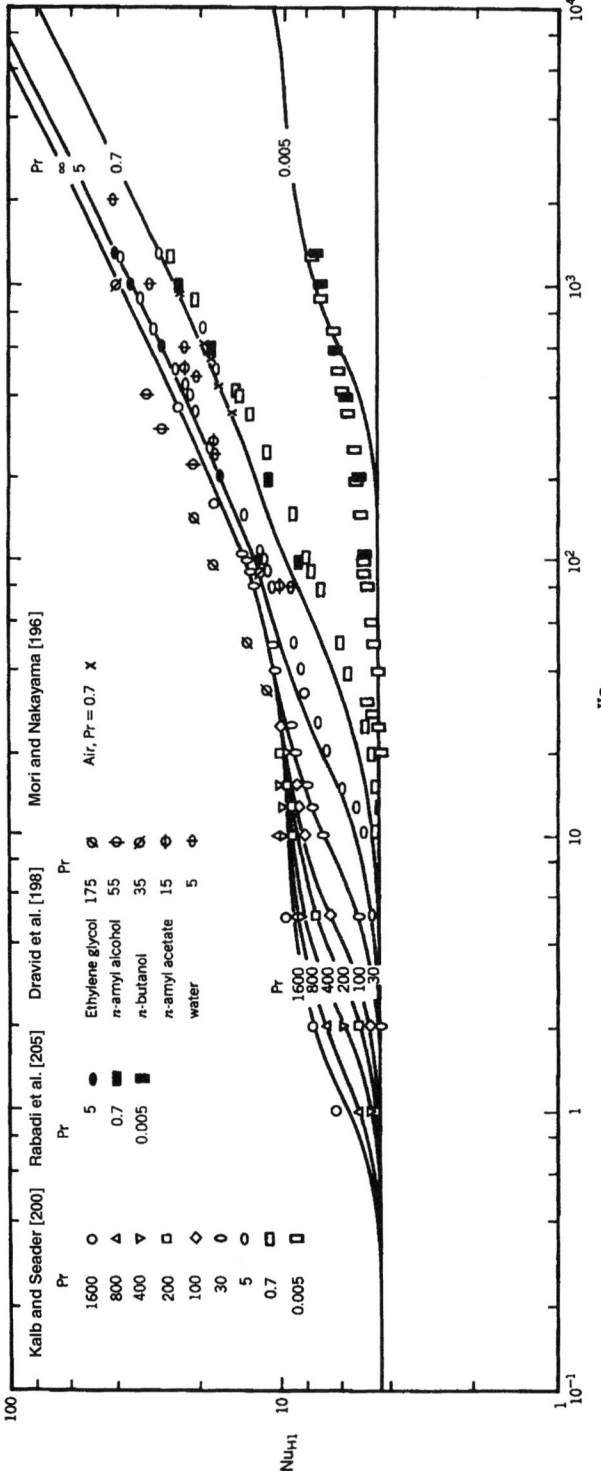

FIGURE 5.37 A comparison of the recommended design correlation (Eq. 5.271, solid line) with theoretical and experimental Nusselt numbers for the Ⓗ boundary condition [201].

TABLE 5.44 Numerically Calculated Nu_{H2} for Helical Coils of Circular Cross Sections [201]

					$Nu_{p,H2}$			
R/a	b/R	Re	De	He	Pr = 0.1	0.3162	1.0	10.0
5.0	0.0	9.196	4.113	4.113	4.642	4.639	4.633	4.620
	0.5	9.197	4.113	4.100	4.462	4.639	4.633	4.620
	1.0	9.194	4.112	4.061	4.462	4.640	4.634	4.621
5.0	0.0	46.70	20.88	20.88	4.769	4.759	4.936	8.447
	0.5	47.72	20.89	20.83	4.768	4.758	4.934	8.438
	1.0	46.79	20.93	20.67	4.765	4.755	4.929	8.414
10.0	0.0	392.6	124.14	124.14	5.604	7.541		
	0.5	393.0	124.29	123.90	5.602	7.535		
	1.0	394.4	124.72	123.17	5.596	7.518		
5.0	0.0	402.5	180.01	180.01	6.058	9.312		
	0.5	403.1	180.28	179.71	6.078	9.307		
	1.0	404.9	181.07	178.82	6.071	9.292		
10.0	0.0	1008	318.8	318.8	7.120	14.30		
	0.5	1009	319.1	318.1	7.114	14.27		
	1.0	1013	320.5	316.5	7.103	14.23		
5.0	0.0	1043	466.6	466.6	9.680			
	0.5	1045	467.4	465.9	9.600			
	1.0	1051	469.8	464.0	9.588			

The fully developed Nusselt numbers Nu_{H2} for helicoidal pipes have been obtained numerically by Manlapaz and Churchill [201]. Their results are listed in Table 5.44.

In Table 5.44, it can be seen that the pitch of the helicoidal coil has almost no influence on the Nusselt number. However, the studies by Yang et al. [206, 207] have shown a positive effect of the pitch on the Nusselt number when Pr > 1. In addition, the experiments conducted by Austen and Soliman [208] indicated that the Nusselt number for the laminar flow of water (3 < Pr < 6) in the uniformly heated helicoidal pipe is in good agreement with the prediction from Manlapaz and Churchill [201].

To consider the effect of variable viscosity, the viscosity ratio $(\mu_m/\mu_w)^{0.14}$ is applied. The use of Eqs. 5.269 and 5.272 with their right sides multiplied by $(\mu_m/\mu_w)^{0.14}$ is recommended. The density change of fluids leads to natural convection; consequently, heat transfer is normally enhanced.

An experimental correlation has been obtained by Abul-Hamayel and Bell [209] to account for the density and viscosity variations in helicoidal pipe. Experiments with water, ethylene glycol, and n-butyl alcohol in a helicoidal pipe with the Ⓗ boundary condition were conducted. The following equation was derived from the measurement data:

$$Nu_{H1} = \left[4.36 + 2.84\left(\frac{Gr}{Re^2}\right)^{3.94}\right]\left[1 + 0.9348\left(\frac{Gr}{Re^2}\right)^{2.78} x_5\right]\left[1 + 0.0276\,De^{0.75}\,Pr^{0.197}\left(\frac{u_m}{u_w}\right)^{0.14}\right]$$

(5.274)

where

$$x_5 = \exp\left(-\frac{1.33\,Gr'}{De^2}\right)$$

(5.275)

This correlation is valid for 92 < Re < 5500, 2.2 < Pr < 101 and 760 < Gr' < 10^6. It reduces to the straight circular duct forced convection Nusselt number value of 4.36 upon neglecting the coil effect (De $\to$ 0). Equation 5.274 is recommended for those fluids whose densities are strongly dependent on temperature.

Developing Laminar Flow

Hydrodynamically developing laminar flow, thermally developing laminar flow, and simultaneously developing laminar flow in helical coils are still under investigation [210–212]. Accurate formulas for engineering applications are limited. However, the entrance region of a helical coil is approximately 20 to 50 percent shorter than that of a straight tube. For most engineering applications, when De $\geq$ 200, the design can be based on fully developed values without significant errors [195].

Turbulent Flow in Coils With Circular Cross Sections

The research conducted by Hogg [213] has indicated that turbulent flow entrance length in coils with circular cross sections is much shorter than that for laminar flow. Turbulent flow can become fully developed within the first half-turn of the coil. Therefore, most of the turbulent flow and heat transfer analyses concentrate on the fully developed region.

Ito [214] has proposed the following correlation to calculate the friction factor for turbulent flow in helicoidal coils:

$$f_c \left(\frac{R}{a}\right)^{0.5} = 0.00725 + 0.076 \left[\text{Re} \left(\frac{R}{a}\right)^{-2} \right]^{-0.25} \qquad \text{for } 0.034 < \text{Re} \left(\frac{R}{a}\right)^{-2} < 300 \qquad (5.276)$$

However, Srinivasan et al. [193] has obtained another formula for the turbulent friction factor, as follows:

$$f_c \left(\frac{R}{a}\right)^{0.5} = 0.084 \left[\text{Re} \left(\frac{R}{a}\right)^{-2} \right]^{-0.2} \qquad \text{for Re} \left(\frac{R}{a}\right)^{-2} < 700 \quad \text{and} \quad 7 < \frac{R}{a} < 104 \qquad (5.277)$$

Either Eq. 5.276 or Eq. 5.277 can be used for design purposes since they are very similar and agree quite well (within ±10 percent) with the experimental data for air [215] and water [216] and with the numerical predictions by Patankar et al. [217].

The friction factor for spiral coils can be obtained using the following experimental correlation [193]:

$$f_c = \frac{0.0074(n_2^{0.9} - n_1^{0.9})^{1.5}}{[\text{Re} \, (b/a)^{0.5}]^{0.2}} \qquad (5.278)$$

Equation 5.278 is valid for 40,000 < Re $(b/a)^{0.5}$ < 150,000 and 7.3 < b/a < 15.5.

Since the turbulent Nusselt numbers are independent of the thermal boundary condition for Pr $\geq$ 0.7, the Nusselt numbers that appear in this section will not be specified with thermal boundary conditions.

The following correlation, developed by Schmidt [218] to calculate the turbulent Nusselt number, is suggested for $2 \times 10^4 < \text{Re} < 1.5 \times 10^5$ and $5 < R/a < 84$:

$$\frac{\text{Nu}_c}{\text{Nu}_s} = 1.0 + 3.6 \left[1 - \left(\frac{a}{R}\right) \right] \left(\frac{a}{R}\right)^{0.8} \qquad (5.279)$$

Equation 5.279 was obtained using air and water flow in coils. For low Reynolds numbers, Pratt's [219] correlation is recommended:

$$\frac{\text{Nu}_c}{\text{Nu}_s} = 1 + 3.4 \frac{a}{R} \qquad \text{for } 1.5 \times 10^3 < \text{Re} < 2 \times 10^4 \qquad (5.280)$$

This correlation was obtained from experiments using water and isopropyl alcohol.

When the variable thermal properties of the fluid are considered, Orlov and Tselishchev [220] recommend the following correlation:

$$\frac{Nu_c}{Nu_s} = \left(1 + 3.54\,\frac{a}{R}\right)\left(\frac{Pr_m}{Pr_w}\right)^{0.025} \quad \text{for } \frac{R}{a} > 6 \tag{5.281}$$

The pitch effect in helicoidal circular pipe has been considered in the investigation conducted by Yang and Ebadian [221]. These researchers have concluded that the effect of pitch is minimum on heat transfer.

Fully Developed Laminar Flow in Curved, Square, and Rectangular Ducts

The following formulas are suggested by Shah and Joshi [195] to compute the friction factor for fully developed laminar flow in curved square ducts:

$$\frac{(f\,Re)_c}{(f\,Re)_s} = 0.1520De^{0.5}\,(1.0 - 0.216De^{0.5} + 0.473De^{-1} + 111.6De^{-1.5} - 256.1De^{-2})$$

$$\text{for } De < 100 \tag{5.282}$$

$$\frac{(f\,Re)_c}{(f\,Re)_s} = 0.2576De^{0.39} \quad \text{for } 100 < De < 1500 \tag{5.283}$$

$$\frac{(f\,Re)_c}{(f\,Re)_s} = 0.1115De^{0.5} \quad \text{for } De > 1500 \tag{5.284}$$

The preceding three equations were obtained through the comparison of theoretical investigations [222–224] and experimental measurement [225]. The influence of the pitch of coil on the friction factor has been found to be negligible [226, 227]. Friction factors for curved rectangular ducts are provided by Cheng et al. [222], as follows:

$$\frac{f_c}{f_s} = C_0\,De^{*0.5}\,(1.0 + C_1\,De^{-1/2} + C_2\,De^{-1} + C_3\,De^{-3/2} + C_4\,De^{-2}) \tag{5.285}$$

Equation 5.284 is valid for De ≤ 700. C_0, C_1, C_2, C_3, and C_4 in Eq. 5.284 are constants given in Table 5.45 and De is defined as Re $(D_h/R)^{1/2}$.

The following correlation, obtained by Cheng et al. [228], is recommended for curved square ducts:

$$Nu_{H2} = Nu_T = 0.152 + 0.627(1.414De)^{0.5}\,Pr^{0.25} \tag{5.286}$$

Equation 5.286 is valid for $0.7 \le Pr \le 5$ and $20 \le De \le 705$. This correlation agrees quite well with the experimental data for air for the Ⓗ boundary condition [224]. It also represents the Nusselt number for the Ⓣ boundary condition [229] quite well.

TABLE 5.45   Constants for Eq. 5.285 [222]

α^*	C_0	C_1	C_2	C_3	C_4
0.5	0.0974	4.366	−13.56	131.8	−182.6
1.0	0.1278	−0.257	0.699	187.7	−512.2
2.0	0.2736	−24.79	325.2	−1591.0	2728.0
5.0	0.0805	−5.217	104.4	−202.8	0.0

It should be noted again that the effect of the wall thermal boundary condition on the Nusselt number for coils is not significant for the fluids with $Pr \geq 0.7$. Equation 5.286 can also be used for the Ⓣ, Ⓗ①, and Ⓗ② thermal boundary conditions. Furthermore, the appropriate correlation for circular cross section coiled tubes can be adopted with the substitution of the appropriate hydraulic diameter for $2a$ to calculate the Nusselt number when the parameters are out of the application range as is the case in Eq. 5.286.

Fully Developed Turbulent Flow in Curved Rectangular and Square Ducts

For curved rectangular ducts as well as square ducts, when $Re^* < 8000$, the fully developed friction factors can be computed from the following correlation obtained by Butuzov et al. [230] and Kadambi [231]:

$$\frac{f_c}{f_s} = 0.435 \times 10^{-3} \, Re^{*0.96} \left(\frac{R}{d^*} \right)^{0.22} \tag{5.287}$$

where d^* is the short-side length of the rectangular duct and Re^* is defined as $u_m d^*/\nu$. The term f_s refers to the friction factor in a straight duct with the same aspect ratio as that of curved coil. For $Re^* \geq 8000$, Eq. 5.276 or Eq. 5.277 for circular ducts can be used with a replaced by $0.5D_h$, where D_h is the hydraulic diameter of the rectangular duct.

The Nusselt numbers for turbulent flow in curved rectangular ducts have been studied by Butuzov et al. [230] and Kadambi [231]. The correlation suggested by Butuzov et al. [230] is as follows:

$$\frac{Nu_c}{Nu_s} = 0.117 \times 10^{-2} \, Re^{*0.93} \left(\frac{R}{d^*} \right)^{0.24} \tag{5.288}$$

This correlation is valid for $450 \leq Re^* (R/d^*)^{0.5} \leq 7500$ and $25 \leq R/d^* \leq 164$. The term Nu_s in Eq. 5.288 is fully developed Nusselt number for a straight duct.

Laminar Flow in Coiled Annular Ducts

Xin et al. [232] experimentally investigated the laminar flow and turbulent flow in coiled annular ducts. The pressure drop was measured for air and water flows. Based on these experimental measurements, the friction factor data can be correlated for laminar and turbulent flow as follows:

$$f = 0.02985 + 75.89 \left[0.5 - a \tan \left(\frac{De - 39.88}{77.56} \right) \middle/ \pi \right] \left(\frac{d_o - d_i}{D} \right)^{1.45} \tag{5.289}$$

where D is coil diameter. This equation is valid in the region of $De = 35\text{–}20{,}000$, $d_o/d_i = 1.61\text{–}1.67$, and $D/(d_o - d_i) = 21\text{–}32$.

For the heat transfer in laminar flow in coiled annular ducts, Garimella et al. [233] experimentally obtained the following correlation to calculate the heat transfer coefficient:

$$Nu = 0.027 De^{0.94} \, Pr^{0.69} \left(\frac{d_o - d_i}{D} \right)^{0.01} \tag{5.290}$$

This equation indeed shows that the Dean number represents the heat transfer in laminar flow; the coil ratio $(d_o - d_i)/D$ is another factor to affect the heat transfer.

Laminar Flow in Curved Ducts With Elliptic Cross Sections

Dong and Ebadian [234] numerically obtained the friction factor for laminar flow in curved elliptic ducts. The friction factor ratio f_c/f_s is represented by the following expression:

$$\frac{f_c}{f_s} = 1 + 0.0031\alpha^{*3}\,\mathrm{De}^{1.07} \tag{5.291}$$

where f_s is the friction factor for straight elliptic ducts and α^* is the ratio of the minor axis to the major axis of the elliptic duct.

In subsequent research [235], thermally developing flow in curved elliptic ducts is analyzed for different α^* and Prandtl numbers. The local Nusselt numbers along the flow direction are shown in graph form, and the asymptotic values of the Nusselt numbers have been obtained, as is shown in Table 5.46. In a related study, the effects of buoyancy on laminar flow in curved elliptic ducts are discussed by Dong and Ebadian [236].

TABLE 5.46 The Asymptotic Values of the Nusselt Number for Curved Elliptic Ducts [235]

α^*	R/D_h	Re	De	Pr 0.1	0.7	5	50
0.2	4	849.16	424.6	9.70	19.22	26.65	52.79
	10	105.15	33.3	3.81	4.18	7.48	11.68
	100	1977.6	197.8	6.31	11.55	16.51	37.79
0.5	4	1271.3	635.7	8.92	23.23	35.93	75.51
	10	1058.0	334.6	6.73	15.20	23.10	54.58
	100	1514.1	151.4	5.07	9.38	13.83	32.84
0.8	4	881.7	440.8	6.57	18.07	28.99	64.91
	10	1336.4	422.6	6.33	16.76	27.58	61.62
	100	118.6	11.9	3.68	3.75	4.62	9.46

LONGITUDINAL FLOW BETWEEN CYLINDERS

Longitudinal flow between cylinders is encountered in the fuel elements of nuclear power reactors, shell-and-tube heat exchangers, boilers, and condensers, among other applications. A *cylinder* is considered to be a long circular pipe or rod. The flow and heat transfer characteristics between the cylinders are dependent on their arrangement (e.g., triangular array, square array, etc.) as well as the Reynolds number. In this section, the fully developed friction factor and Nusselt number for longitudinal flow between the cylinders in triangular and square arrays are introduced. For longitudinal flow in other channels formed by the cylinders and the walls, the reader is encouraged to refer to Shah and London [1] and Rehme [237].

Laminar Flow

The friction factor and Nusselt number for longitudinal laminar flow between a triangular array and a square array are discussed in this section.

Triangular Array.   A triangular array is shown in Fig. 5.38. The fluid longitudinally flows in the virtual channel formed by the triangular array. The friction factor for fully developed laminar flow in this configuration has been proposed by Rehme [237] as follows:

$$f\,\mathrm{Re} = \begin{cases} 5.1777(P/D - 1)^{0.404} & \text{for } 1.02 \le P/D < 1.12 & (5.292) \\ 36.713(P/D - 1)^{0.24} & \text{for } 1.12 \le P/D < 1.6 & (5.293) \\ 36.947(P/D - 1)^{0.372} & \text{for } 1.6 \le P/D < 2.0 & (5.294) \\ \dfrac{16(r_*^2 - 1)^3}{4r_*^2 \ln r_* - 3r_*^4 + 4r_*^2 - 1} & \text{for } P/D > 2.1 & (5.295) \end{cases}$$

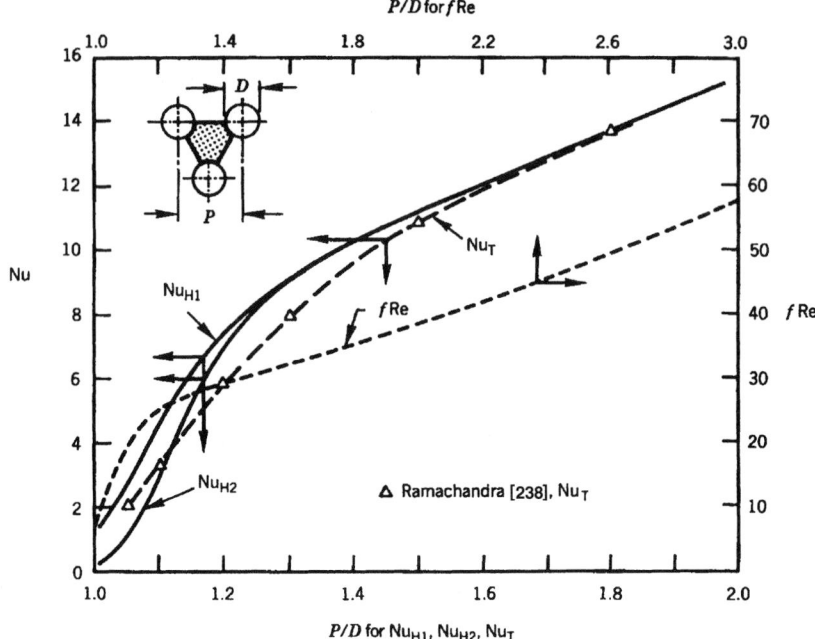

FIGURE 5.38 Fully developed $f\,\mathrm{Re}$ and Nusselt numbers for longitudinal laminar flow between cylinders in a triangular array [237].

where
$$r_* = \frac{P}{D}\sqrt{\frac{2\sqrt{3}}{\pi}} \approx 1.05\,\frac{P}{D} \tag{5.296}$$

Equations 5.292 through 5.295 were obtained as a result of comparison with numerous investigations such as those by Rosenberg [239], Sparrow and Loeffler [240], Axford [241, 242], Shih [243], Rehme [244, 245], Johannsen [246], Malák et al. [247], Ramachandra [238], Mikhailov [248], Subbotin et al. [249], Dwyer and Berry [250], Rehme [251], and Cheng and Todreas [252]. The $f\,\mathrm{Re}$ calculated from Eqs. 5.292 through 5.294 is shown in Fig. 5.38.

The fully developed Nusselt number for longitudinal flow in a triangular array with uniform cylinder temperature has been analyzed by Ramachandra [238] and is shown in Fig. 5.38. The fully developed Nusselt numbers for the Ⓗ1 and Ⓗ2 boundary conditions have been studied by Sparrow et al. [253], Dwyer and Berry [250], Hsu [254], and Ramachandra [238]. The differences for Nu_{H1} and Nu_{H2} reported by these investigators are small (1 percent). The fully developed Nu_{H1} and Nu_{H2} are shown in Fig. 5.38.

Miyatake and Iwashita [255] conducted a numerical analysis to determine the characteristics of developing laminar flow between a triangular array of cylinders with a uniform wall temperature and various ratios of pitch to diameter (P/D). The relationships between the local Nusselt number $\mathrm{Nu}_{x,T}$ and local Graetz number Gz_x and between the logarithmic mean Nusselt number $\mathrm{Nu}_{lm,T}$ and Graetz number Gz were obtained as follows:

for $P/D = 1.0$–1.1:
$$\mathrm{Nu}_{x,T} = 9.26(1 + 0.0022\,\mathrm{Gz}_x^{1.46})^{1/4} \tag{5.297}$$

$$\mathrm{Nu}_{lm,T} = 9.26(1 + 0.0179\,\mathrm{Gz}^{1.46})^{1/4} \tag{5.298}$$

for $P/D = 1.1$–4.0
$$\mathrm{Nu}_{x,T} = (a^2 + b^2\,\mathrm{Gz}_x^{2/3})^{1/2} \tag{5.299}$$

$$\mathrm{Nu}_{lm,T} = [a^2 + (3b/2)^2\,\mathrm{Gz}^{2/3}]^{1/2} \tag{5.300}$$

where
$$a = \frac{8.92[1 + 2.82(P/D - 1)]}{1 + 6.86(P/D - 1)^{5/3}} \tag{5.301}$$

$$b = \frac{2.34[1 + 24(P/D - 1)]}{[1 + 36.5(P/D - 1)^{5/4}][2\sqrt{3}(P/D - 1)^2 - \pi]^{1/3}} \tag{5.302}$$

$$\mathrm{Gz}_x = \dot{m}c_p/kx \tag{5.303}$$

$$\mathrm{Gz} = \dot{m}c_p/kL \tag{5.304}$$

$$\mathrm{Nu}_{x,T} = h_x D/k \tag{5.305}$$

$$\mathrm{Nu}_{lm,T} = h_m D/k = \frac{\dot{m}c_p(T_b - T_0)_{x=L}}{\pi L \Delta T_{lm}} \tag{5.306}$$

$$\Delta T_{lm} = \frac{(T_w - T_0)(T_w - T_b)_{x=L}}{\ln\,[(T_w - T_0)/(T_w - T_b)_{x=L}]} \tag{5.307}$$

In Eqs. 5.306 and 5.307, T_0, T_w, and T_b are the inlet, wall, and fluid bulk temperatures, respectively; L is the length of the cylinder. It is noted that the fully developed Nusselt number can be calculated using $\mathrm{Gz} \to 0$ ($L \to \infty$) in the corresponding equations.

In the study of Miyatake and Iwashita [256], the relationship of local Nusselt number and Graetz number is formulated for developing longitudinal flow between a triangular array of cylinders with a uniform heat flux and various pitch-to-diameter ratios. For $P/D = 1.01$–1.1:

$$\mathrm{Nu}_{x,H2} = \frac{b\,\mathrm{Gz}_x^{1/3} - a}{1 + 451\mathrm{Gz}_x^{-[15.4P/D - 14.937]}} + 1 \qquad \text{when } \mathrm{Gz}_x \geq (a/b)^3 \tag{5.308}$$

$$\mathrm{Nu}_{x,H2} = a; \qquad \text{when } \mathrm{Gz}_x < (a/b)^3 \tag{5.309}$$

For $P/D = 1.1$–4.0:
$$\mathrm{Nu}_{x,H2} = (a^2 + b^2\,\mathrm{Gz}_x^{2/3})^{1/2} \tag{5.310}$$

where
$$a = \frac{3.1(P/D - 1)^{0.1} + 324(P/D - 1)^{1.6}}{1 + 69.5(P/D - 1)^{2.4}}$$

$$b = \frac{1.536[1 + 8.24(P/D - 1)^{0.39}]}{[2\sqrt{3}(P/D - 1)^2 - \pi]^{1/3}[1 + 6.37(P/D - 1)^{0.73}]} \tag{5.311}$$

Square Array.   A square array is displayed in Fig. 5.39. The fully developed friction factor for longitudinal flow in such a virtual channel has been investigated by Sparrow and Loeffler [240], Shih [243], Rehme [244, 245], Malák et al. [247], Meyder [257], Kim [258], Ramachandra [238], and Ohnemus [259]. The f Re is given in Fig. 5.39. It can be approximated by the following equation [237]:

$$f\,\mathrm{Re} = 40.70\left(\frac{P}{D} - 1\right)^{0.435} \tag{5.312}$$

Equation 5.312 is valid in the range of $1.05 \leq P/D \leq 2.0$.

The fully developed Nusselt numbers for the Ⓗ① and Ⓗ② boundary conditions in square arrays have been analyzed by Kim [258], Ramachandra [238], Ohnemus [259], and Chen et al. [260]. The fully developed Nu_{H1} and Nu_{H2} are shown in Fig. 5.39.

Miyatake and Iwashita [255] also investigated the developing longitudinal laminar flow between a square array of cylinders with uniform wall temperature. The local and logarithmic Nusselt number can be obtained using the following correlations:

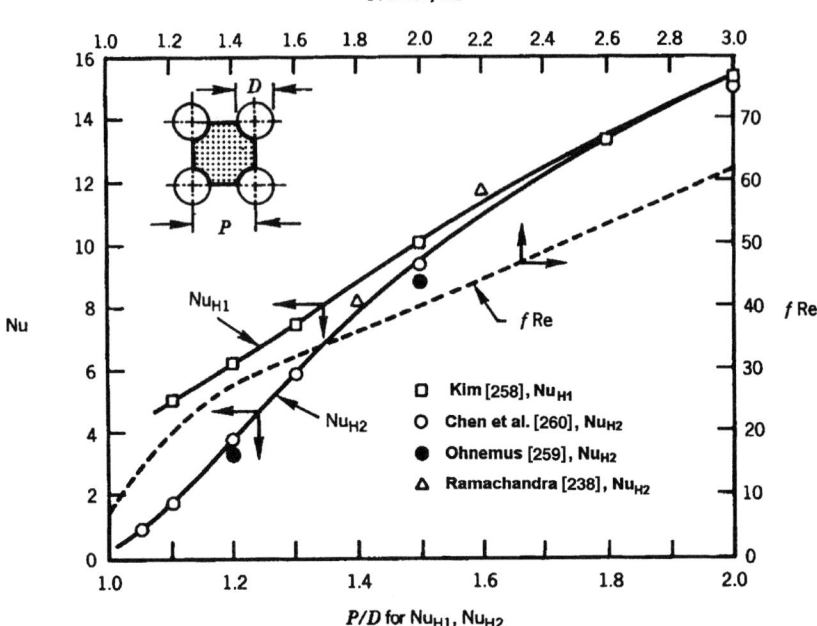

FIGURE 5.39 Fully developed f Re and Nusselt numbers for longitudinal laminar flow in a square array [237].

For $P/D = 1.0$–1.2:

$$\mathrm{Nu}_{x,\mathrm{T}} = 4.08(1 + 0.0058\mathrm{Gz}_x^{1.46})^{1/4} \tag{5.313}$$

$$\mathrm{Nu}_{lm,\mathrm{T}} = 4.08(1 + 0.0349\mathrm{Gz}_x^{1.46})^{1/4} \tag{5.314}$$

For $P/D = 1.2$–4.0, the same equations as Eqs. 5.299 and 5.300 are used, but the a and b are different:

$$a = \frac{4.00[1 + 0.509(P/D - 1)]}{1 + 0.765(P/D - 1)^{5/3}} \tag{5.315}$$

$$b = \frac{1.69[1 + 9.1(P/D - 1)]}{[1 + 10.8(P/D - 1)^{5/4}][4(P/D - 1)^2 - 2]^{1/3}} \tag{5.316}$$

For developing longitudinal laminar flow between a square array of cylinders with a uniform wall heat flux, the local Nusselt number correlations were made by Miyatake and Iwashita [256] as follows:

For $P/D = 1.01$–1.2:

$$\mathrm{Nu}_{x,\mathrm{H2}} = \frac{b\,\mathrm{Gz}_x^{1/3} - a}{1 + 94\mathrm{Gz}_x^{-[7.66P/D - 7.379]}} + a \qquad \text{when } \mathrm{Gz}_x \ge (a/b)^3 \tag{5.317}$$

$$\mathrm{Nu}_{x,\mathrm{H2}} = a \qquad \text{when } \mathrm{Gz}_x < (a/b)^3 \tag{5.318}$$

For $P/D = 1.2$–4.0:

$$\mathrm{Nu}_{x,\mathrm{H2}} = (a^2 + b^2\,\mathrm{Gz}_x^{2/3})^{1/2} \tag{5.319}$$

where

$$a = \frac{3.6(P/D - 1)^{0.2} + 32.2(P/D - 1)^{1.5}}{1 + 9.1(P/D - 1)^{2.2}} \qquad (5.320)$$

$$b = \frac{1.224[1 + 4.40(P/D - 1)^{0.39}]}{[4(P/D - 1)^2 - \pi]^{1/3}[1 + 2.66(P/D - 1)^{0.73}]} \qquad (5.321)$$

Fully Developed Turbulent Flow

Fully developed turbulent flow and heat transfer in triangular and square arrays have been analyzed by Deissler and Taylor [261, 262]. The friction factors for longitudinal flow between the cylinders in a triangular and a square array are given in Fig. 5.40. Correspondingly, the Nusselt numbers, in terms of the Stanton number, defined as Nu/(Re Pr), are given in Fig. 5.41, where the cylinders are considered to be uniformly heated.

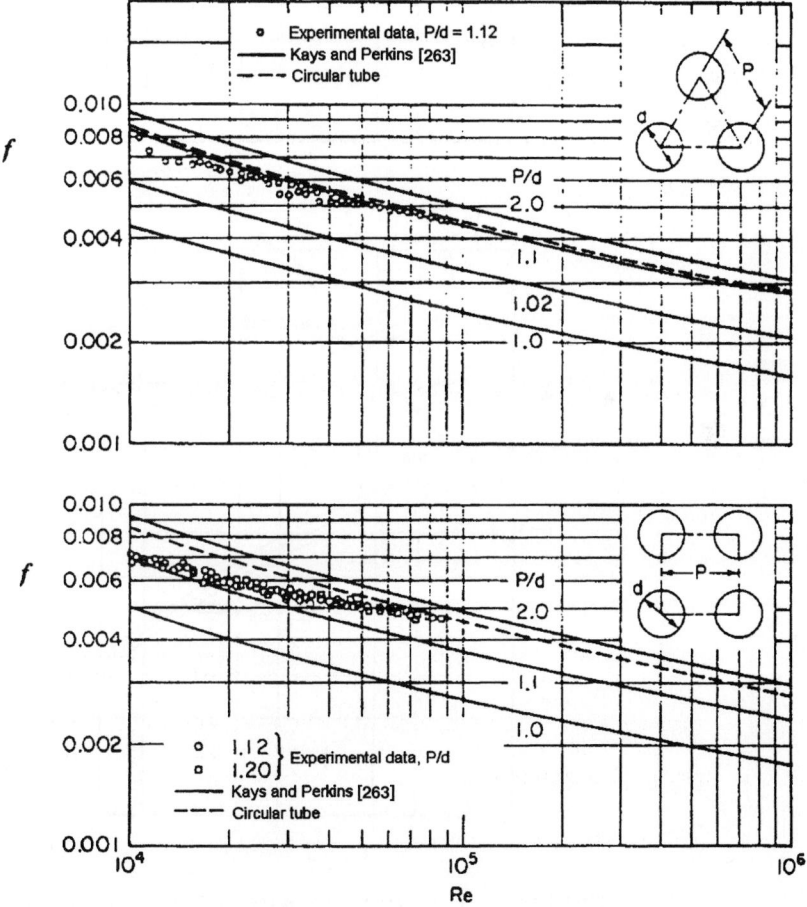

FIGURE 5.40   Fully developed friction factor for longitudinal turbulent flow between a triangular and a rectangular array [261].

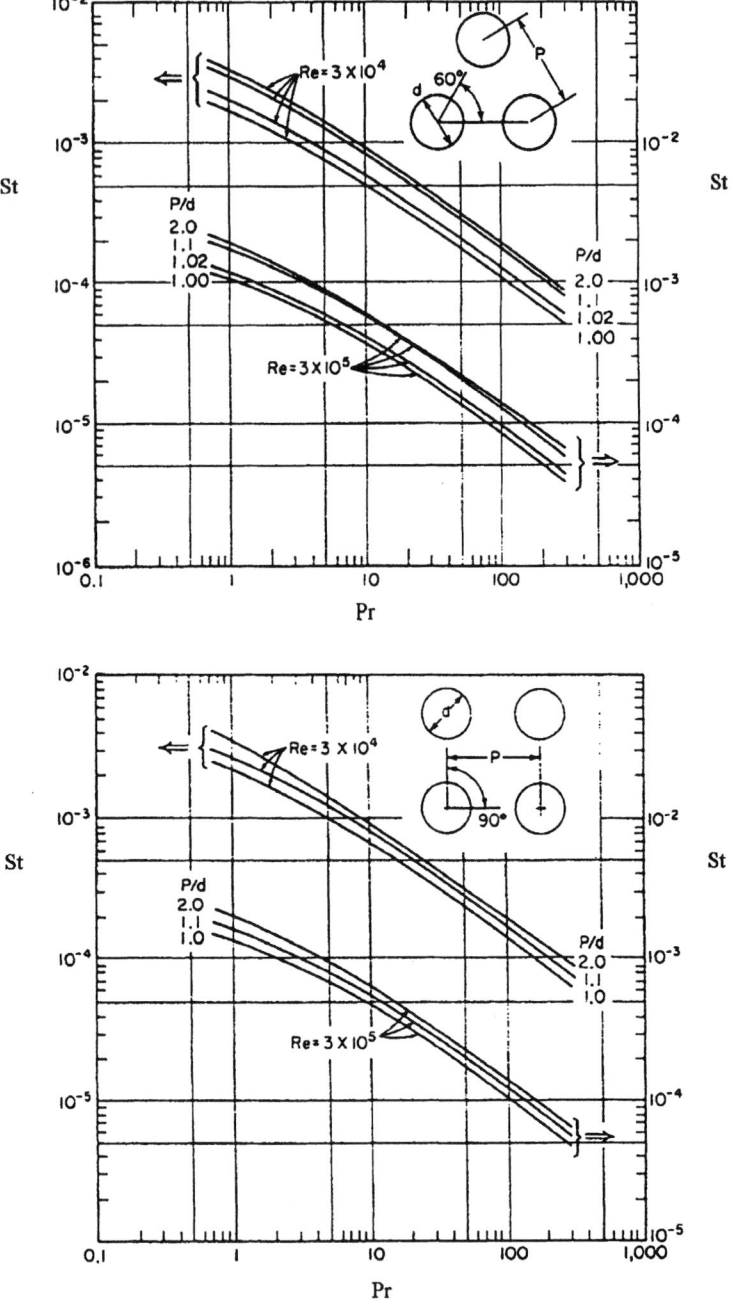

FIGURE 5.41   Fully developed heat transfer characteristics for longitudinal turbulent flow between a triangular array and a square array [261].

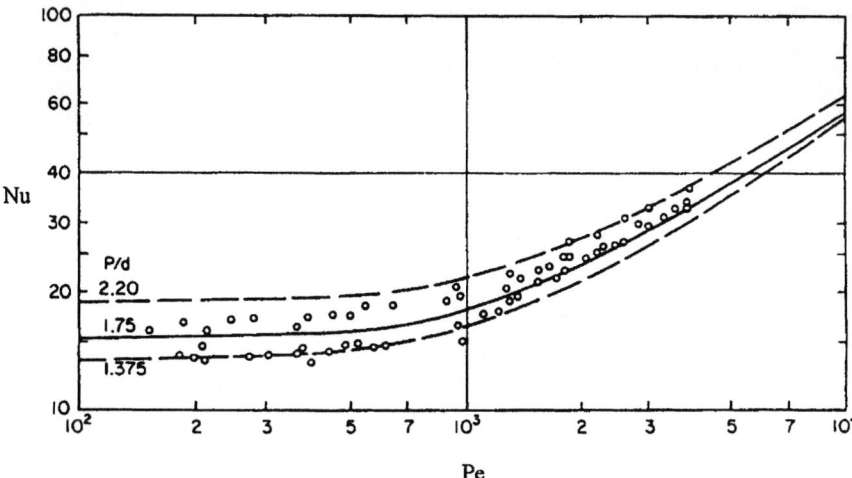

FIGURE 5.42   Nusselt numbers for fully developed longitudinal flow between cylinders in a triangular array [263].

Maresca and Dwyer [264] have analyzed the heat transfer of liquid metal flow in a triangular array with uniform longitudinal heat flux. The Nusselt number resulting from their analysis is given in Fig. 5.42.

INTERNALLY FINNED TUBES

Internally finned tubes are ducts with internal longitudinal fins. These tubes are widely used in compact heat exchangers. The friction factor–Reynolds number product and the Nusselt number for such internally finned tubes, designated as $(f\,\mathrm{Re})_d$ and $\mathrm{Nu}_{bc,d}$, respectively, are computed from the following definitions:

$$f_d = \Delta p^* \left(\frac{D_{h,\mathrm{finless}}}{4L} \right) = \left(\frac{\rho \Delta p}{2LW^2} \right) (A_c^2 D_h)_{\mathrm{finless}} \tag{5.322}$$

$$\mathrm{Re}_d = \frac{G D_{h,\mathrm{finless}}}{\mu} = \left(\frac{u_m}{\mu} \right) \left(\frac{D_h}{A_c} \right)_{\mathrm{finless}} \tag{5.323}$$

$$\mathrm{Nu}_{bc,d} = \frac{q'' D_{h,\mathrm{finless}}}{k(t_w - t_m)} = \frac{q'}{4k(t_w - t_m)} \left(\frac{D_h^2}{A_c} \right)_{\mathrm{finless}} \tag{5.324}$$

where $D_{h,\mathrm{finless}}$ is the hydraulic diameter corresponding to finless ducts. Based on actual geometry, the $D_{h,\mathrm{finned}}$ is used in the $f\,\mathrm{Re}$ and Nu_{bc}. The relationships between $(f\,\mathrm{Re})_d$, $\mathrm{Nu}_{bc,d}$, and $f\,\mathrm{Re}$, Nu_{bc} are given in the following expressions:

$$(f\,\mathrm{Re})_d = (f\,\mathrm{Re}) \left(\frac{D_{h,\mathrm{finless}}}{D_{h,\mathrm{finned}}} \right)^2 \left(\frac{A_{c,\mathrm{finless}}}{A_{c,\mathrm{finned}}} \right) \tag{5.325}$$

$$\mathrm{Nu}_{bc,d} = \mathrm{Nu}_{bc} \left(\frac{D_{h,\mathrm{finless}}}{D_{h,\mathrm{finned}}} \right)^2 \left(\frac{A_{c,\mathrm{finned}}}{A_{c,\mathrm{finless}}} \right) \tag{5.326}$$

Circular Ducts With Thin Longitudinal Fins

Hu and Chang [265] have obtained the friction factors and Nusselt numbers for fully developed laminar flow through a circular duct having longitudinal rectangular fins equally spaced along the wall. The fin's efficiency was treated as 100 percent, while its thickness was treated as zero. The fully developed $(f\,\mathrm{Re})_d$ and $\mathrm{Nu}_{H2,d}$ for laminar flow in a circular duct with longitudinal fins are given in Table 5.47, in which l^* and n are relative fin length and the number of fins, respectively.

Prakash and Liu [266] have numerically analyzed laminar flow and heat transfer in the entrance region of an internally finned circular duct. In this study, the fully developed $f\,\mathrm{Re}$ is compared with those reported by Hu and Chang [265] and Masliyah and Nandakumar [267]. The incremental pressure drop $K(\infty)$ and hydrodynamic entrance length L_{hy}^+ together with $f\,\mathrm{Re}$ are given in Table 5.48, in which the term n refers to the number of fins, while l^* denotes the relative length of the fins.

TABLE 5.47   The Fully Developed $(f\,\mathrm{Re})_d$ and $\mathrm{Nu}_{H2,d}$ for Laminar Flow in a Circular Duct With Longitudinal Fins [1]

					$(f\,\mathrm{Re})_d$			
n	$l^* = 0.2$	0.4	0.6	0.7	0.79	0.795	0.8	0.9
2	17.28	20.83	27.42	31.89	35.68	35.98	36.64	40.54
8	21.22	42.87	101.10	139.55	161.03	162.03	164.84	172.70
12	—	—	—	—	—	286.66	—	—
16	25.99	69.57	219.54	348.86	434.40	439.37	448.43	481.12
20	—	—	—	—	607.72	616.52	632.11	—
22	—	—	—	—	701.75	712.76	732.60	—
24	—	—	—	—	—	813.67	838.23	—
28	—	—	—	—	—	1025.6	1062.7	—
32	30.43	91.65	372.37	773.69	1221.0	1251.6	1298.7	1546.8

					$\mathrm{Nu}_{H2,d}$			
n	$l^* = 0.2$	0.4	0.6	0.7	0.79	0.795	0.8	0.9
2	4.25	4.32	4.88	5.38	6.11	6.16	6.23	6.93
8	4.27	4.67	8.66	16.79	29.49	30.10	30.65	27.26
12	—	—	—	—	—	53.65	—	—
16	4.12	4.04	7.29	21.65	72.66	73.48	71.06	31.85
20	—	—	—	—	81.89	83.60	80.41	—
22	—	—	—	—	84.11	86.82	84.02	---
24	—	—	—	—	—	85.00	83.70	—
28	—	—	—	—	—	75.32	78.06	—
32	3.84	3.39	4.10	8.62	55.76	62.43	67.05	25.15

TABLE 5.48   Flow Characteristics for the Entrance Problem in an Internally Finned Circular Duct

n	8			16			24		
l^*	$(f\,\mathrm{Re})$	$K(\infty)$	L_{hy}^+	$(f\,\mathrm{Re})$	$K(\infty)$	L_{hy}^+	$(f\,\mathrm{Re})$	$K(\infty)$	L_{hy}^+
0	15.96	1.25	0.0415	—	—	—	—	—	—
0.3	27.88	2.44	0.0443	39.18	4.11	0.0438	46.00	5.40	0.0417
0.6	97.37	2.85	0.0320	208.1	10.7	0.0540	293.0	23.5	0.0622
1	171.8	1.58	0.00524	477.4	1.79	0.00235	933.8	1.93	0.00136

TABLE 5.49 Heat Transfer Characteristics for Fully Developed Flow in a Finned Circular Duct

n	8				16				24			
l^*	$\mathrm{Nu}_{T,d}$	$L^+_{T,d}$	$\mathrm{Nu}_{H1,d}$	$L^+_{H1,d}$	$\mathrm{Nu}_{T,d}$	$L^+_{T,d}$	$\mathrm{Nu}_{H1,d}$	$L^+_{H1,d}$	$\mathrm{Nu}_{T,d}$	$L^+_{T,d}$	$\mathrm{Nu}_{H1,d}$	$L^+_{H1,d}$
0	3.658	0.0421	4.371	0.0571	—	—	—	—	—	—	—	—
0.3	4.110	0.0392	5.245	0.0658	3.993	0.0301	5.107	0.0653	3.859	0.0258	4.830	0.0618
0.6	8.779	0.00549	17.26	0.0848	6.545	0.0411	13.32	0.121	5.313	0.0116	9.154	0.116
1.0	33.25	0.00574	41.58	0.00774	80.55	0.00336	106.5	0.00379	143.7	0.00246	200.0	0.00227

The fully developed Nusselt numbers for the thermal boundary conditions of uniform wall temperature and axial uniform wall heat flux with circumferential uniform temperature obtained by Prakash and Liu [266] are given in Table 5.49, along with the corresponding thermal entrance lengths. The term n in Table 5.49 denotes the number of fins, whereas l^* represents the relative length of the fins.

Square Ducts With Thin Longitudinal Fins

Gangal and Aggarwala [268] have analytically obtained the $f\,\mathrm{Re}$ and Nu_{H1} for fully developed flow in a square duct with four equal internal fins, as that shown in Fig. 5.43. The fins were treated as having zero thickness and 100 percent efficiency. The results of $f\,\mathrm{Re}$ and $\mathrm{Nu}_{H1,d}$ for fully developed flow are provided in Table 5.50.

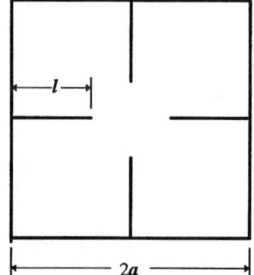

FIGURE 5.43 A square duct with four equal longitudinal thin fins.

TABLE 5.50 Longitudinal Four Thin Fins Within a Square Duct: $(f\,\mathrm{Re})_d$ and $\mathrm{Nu}_{H1,d}$ for Fully Developed Laminar Flow [268]

l^*	$(f\,\mathrm{Re})_d$	$\mathrm{Nu}_{H1,d}$
0	14.261	3.609
0.125	15.285	3.721
0.250	18.281	4.160
0.375	23.630	5.172
0.500	31.877	7.309
0.625	42.527	11.096
0.750	52.341	14.025
1	56.919	14.431

Rectangular Ducts With Longitudinal Thin Fins from Opposite Walls

The fully developed $(f\,\mathrm{Re})_d$ and $\mathrm{Nu}_{H1,d}$ for rectangular ducts with two fins and four fins on opposite walls have beenobtained by Aggarwala and Gangal [269] and Gangal [270]. These are shown in Fig. 5.44.

Circular Ducts With Longitudinal Triangular Fins

Nandakumar and Masliyah [271] and Masliyah and Nandakumar [267] have analyzed fully developed laminar flow in a circular duct with equally spaced triangular fins, as shown in the inset in Fig. 5.45. The flow area and wetted perimeter for this type of duct are given by

$$A_{c,\text{finned}} = \pi a^2 - n[a^2\phi - a(a-l)\sin\phi] \qquad (5.327)$$

$$P_{\text{finned}} = 2\pi a + 2nl' - 2n\phi a \qquad (5.328)$$

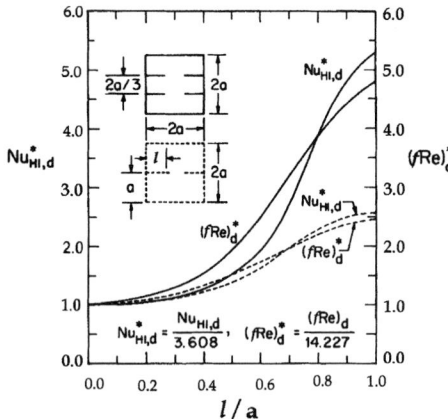

FIGURE 5.44 Friction factor and Nusselt number for fully developed laminar flow in rectangular ducts with longitudinal thin fins from opposite walls [1].

where

$$l' = [a^2 + (a - l)^2 - 2a(a - l) \cos \phi]^{1/2} \quad (5.329)$$

The hydraulic diameter can be calculated from $D_{h,\text{finned}} = 4(A_c/P)_{\text{finned}}$. The results of $(f\,\text{Re})_d$ and $\text{Nu}_{\text{H1},d}$ are provided and Figs. 5.45 and 5.46, in which the case of $2\phi = 0°$ represents longitudinal fins of zero thickness.

Circular Ducts With Twisted Tape

The enhancement of heat transfer inside a circular duct is often achieved by inserting a thin, metal tape in such a way that the tape is twisted about its longitudinal axis, as indicated in Fig. 5.47. Swirl flow is created in this manner. The width of the tape is usually the same as the internal diameter of the duct. The tape twist ratio X_L is defined as H/d. When X_L approaches infinity, the circular duct with the twisted tape becomes two semicircular straight ducts separated by the tape.

Manglik and Bergles [272, 273] made an extensive review on the study of laminar and turbulent flow in circular ducts with inserted tape. For laminar flow, the dimensional swirl parameter Sw was incorporated in the correlation of friction factor. This parameter considers the thickness of inserted tape δ,

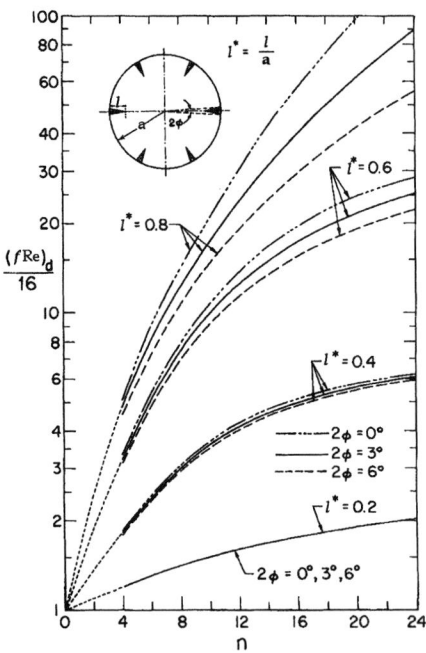

FIGURE 5.45 Friction factors for fully developed laminar flow in a circular duct with longitudinal triangular fins [1].

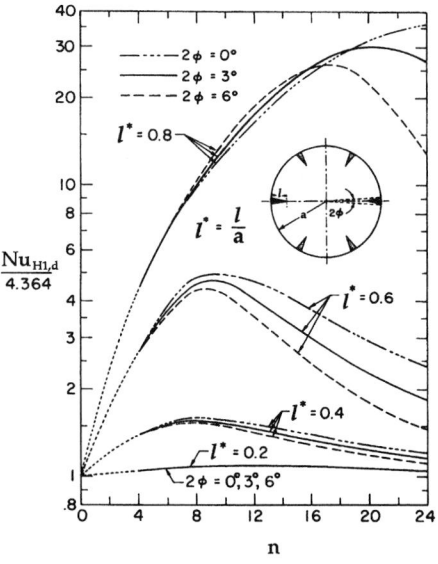

FIGURE 5.46 Nusselt numbers for fully developed laminar flow in a circular duct with longitudinal triangular fins [1].

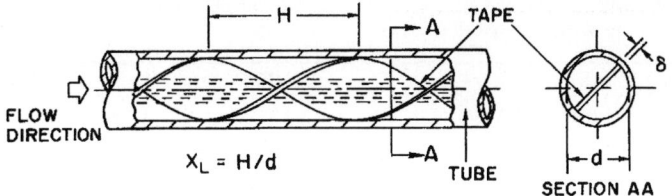

FIGURE 5.47 A circular tube with a twisted tape inserted [1].

the twist ratio X_L, and the helicoidally twisting flow velocity. The definition of dimensional swirl parameter Sw is given as:

$$Sw = \mathrm{Re}_{sw} \Big/ \sqrt{X_L} \left(\frac{\pi}{\pi - 4\delta/d} \right) \left(1 + \frac{\pi^2}{4X_L^2} \right) \tag{5.330}$$

For most applications, the following equation is recommended for the calculation of friction factor in laminar flow in circular ducts with inserted tape:

$$f\,\mathrm{Re} = 15.767\pi \left(\pi + 2 - \frac{2\delta}{d} \right)^2 \left(\pi - 4\frac{\delta}{d} \right)^{-3} \left(1 + \frac{\pi}{2X_L} \right) (1 + 10^{-6}Sw^{2.55})^{1/6} \tag{5.331}$$

where $f\,\mathrm{Re}$ is based on the empty tube diameter d. The above equation can be applied in the range $0 < (\delta/d) < 0.1$ and $300 \le Sw \le 1400$.

The mean Nusselt number for the laminar flow in isothermal circular ducts with inserted tape can be obtained from the following equation suggested by Manglik and Bergles [272]:

$$\mathrm{Nu}_m = 4.612\{[(1 + 0.0951\mathrm{Gz}^{0.894})^{2.5} + 6.413 \times 10^{-9}(Sw \cdot \mathrm{Pr}^{0.391})^{3.853}]^{0.2}$$
$$+ 2.132 \times 10^{-14}(\mathrm{Re}_{ax} \cdot \mathrm{Ra})^{2.23}\}^{0.1} \left(\frac{\mu_m}{\mu_w} \right)^{0.14} \tag{5.332}$$

Where Gz, Re_{ax}, and Ra are the Graetz number, the Reynolds number based on axial velocity, and the Rayleigh number, respectively. Their definitions are expressed as follows:

$$\mathrm{Gz} = \dot{m}c_p/KL \tag{5.333}$$

$$\mathrm{Re}_{ax} = \frac{\dot{m}\mu}{\pi d/4 - \delta\lambda} \tag{5.334}$$

$$\mathrm{Ra} = \frac{\rho g d^3 \beta \Delta Tw}{\mu a} \tag{5.335}$$

For the turbulent flow in circular duct with inserted tape, it was proposed by Manglik and Bergles [273] that the friction factor can be calculated by the following equation:

$$f = \frac{0.0791}{\mathrm{Re}^{0.25}} \left(\frac{\pi}{\pi - 4\delta/d} \right)^{1.75} \left(1 + \frac{2.752}{X_L^{1.29}} \right) \tag{5.336}$$

It was found that the flow rates with $\mathrm{Re} \ge 10^4$ can be considered as fully developed turbulent flow. Therefore, the above equation is a more generalized correlation that covers a broad database of available empirical data for turbulent flow [273].

TABLE 5.51 The Fully Developed $(f\,Re)_d$ and Nu_d Values for Forced Convection of Laminar Flow in a Semicircular Duct With Internal Fins [274]

n	Fin length (l^*)									
	0.1	0.2	0.3	0.4	0.5	0.6	0.7	0.8	0.9	1.0
	$(f\,Re)_d$									
1	43.307	46.129	50.369	55.717	61.674	67.462	72.148	75.047	76.175	76.314
3	—	53.345	—	87.943	—	140.762	—	170.601	—	173.382
5	46.836	60.205	84.306	122.503	175.639	234.651	280.714	303.208	309.012	309.420
8	49.207	69.103	105.634	167.660	265.071	391.297	506.379	568.948	585.854	587.054
11	—	75.927	—	201.532	—	547.204	—	910.726	—	953.762
17	—	84.656	—	243.454	—	815.772	—	1798.80	—	1959.23
	$Nu_{H1,d}$									
1	6.806	7.196	7.895	8.896	10.086	11.159	11.765	11.839	11.839	11.821
3	—	7.531	—	12.171	—	24.816	—	25.843	—	25.180
5	6.878	7.668	9.383	13.104	21.892	38.581	47.001	44.127	42.667	42.558
8	6.904	7.627	8.954	11.760	19.148	41.329	77.458	79.918	76.329	76.021
11	—	7.467	—	10.253	—	32.084	—	124.440	—	118.145
17	—	7.176	—	8.808	—	19.041	—	211.243	—	226.977

A generalized correlation of mean Nusselt number for turbulent heat transfer in an isothermal circular duct with inserted tape was developed by Manglik and Bergles [273] based on the experimental data. It is expressed as:

$$Nu = 0.023(1 + 0.769/X_L)\,Re^{0.8}\,Pr^{0.4}\left(\frac{\pi}{\pi - 4\delta/d}\right)^{0.8}\left(\frac{\pi + 2 + 2\delta/d}{\pi - 4\delta/d}\right)^{0.2}\phi \qquad (5.337)$$

where

$$\phi = \left(\frac{\mu_m}{\mu_w}\right)^n \text{ or } \left(\frac{T_m}{T_w}\right)^m \qquad (5.338)$$

$n = 0.18$ for liquid heating and 0.30 for liquid cooling; $m = 0.45$ for gas heating and 0.15 for gas cooling.

Semicircular Ducts With Internal Fins

Dong and Ebadian [274] have used a very fine grid to perform a numerical analysis of fully developed laminar flow in a semicircular duct with internal longitudinal fins. The fins are considered to have zero thickness, and the number of fins n and relative fin length $l^* = l/a$ are taken into account. The ⒣ thermal boundary condition is applied. Their results are given in Table 5.51.

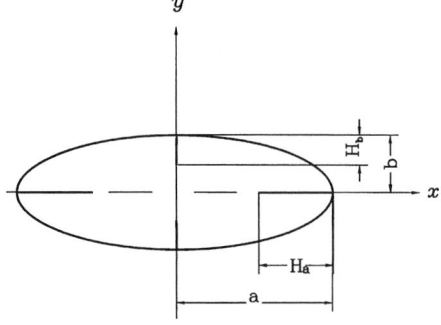

FIGURE 5.48 An elliptical duct with internal fins.

Elliptical Ducts With Internal Longitudinal Fins

An elliptical duct with four internal longitudinal fins mounted on the major and minor axes, as shown in Fig. 5.48, has been analyzed by Dong and Ebadian [275] for fully developed laminar flow and heat transfer. In this analysis, the fins are considered to have zero thickness. The ⒣ thermal boundary condition is applied to the duct wall, and l^* is defined as a ratio of $H_a/a = H_b/b$. The friction factors and Nusselt numbers for fully developed laminar flow are given in Table 5.52.

TABLE 5.52 Friction Factors and Nusselt Numbers for Fully
Developed Flow in an Elliptical Duct With Internal Fins [275]

α^*	l^*	$(f\,\mathrm{Re})_d$	$\mathrm{Nu}_{\mathrm{H1},d}$	$(f\,\mathrm{Re})$	$\mathrm{Nu}_{\mathrm{H1}}$
	0.0	72.20	3.78	72.20	3.78
0.25	0.5	108.61	3.78	45.36	1.58
	0.8	270.63	12.51	69.34	3.20
	0.9	301.73	16.04	71.53	3.80
	1.0	313.62	45.85	67.26	3.40
0.5	0.0	67.26	3.75	67.26	3.75
	0.5	133.39	4.97	50.02	1.86
	0.9	297.78	16.31	68.78	3.76
	1.0	309.20	16.11	61.71	3.21
0.8	0.0	64.33	3.67	64.33	3.67
	0.5	135.53	5.36	50.72	2.00
	0.9	391.06	16.19	67.34	3.74
	1.0	303.38	15.96	58.91	3.10
1.0	0.0	63.99	3.67	63.99	3.67
	0.5	140.57	5.60	51.46	2.05
	0.9	294.18	16.14	65.89	3.61
	1.0	301.82	15.96	58.40	3.09

OTHER SINGLY CONNECTED DUCTS

The fluid flow and heat transfer characteristics for 14 types
of singly connected ducts are described in this section.

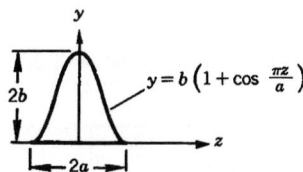

FIGURE 5.49 A sine duct.

$$y = b\left(1 + \cos\frac{\pi z}{a}\right)$$

Sine Ducts

A sine duct with associated coordinates is shown in Fig. 5.49.
The characteristics of fully developed laminar flow and heat
transfer in such a duct are given in Table 5.53. These results
are based on the analysis by Shah [172].

TABLE 5.53 Fully Developed Fluid Flow and Heat Transfer Characteristics
of Sine Ducts [172]

b/a	$K(\infty)$	L_{hy}^+	$f\,\mathrm{Re}$	Nu_T	$\mathrm{Nu}_{\mathrm{H1}}$	$\mathrm{Nu}_{\mathrm{H2}}$
∞	3.218	0.1701	15.303	0.739	2.521	0
2	1.884	0.0403	14.553	—	3.311	0.95
$3/2$	1.806	0.0394	14.022	2.60	3.267	1.37
1	1.744	0.0400	13.023	2.45	3.102	1.555
$\sqrt{3}/2$	1.739	0.0408	12.630	—	3.014	1.47
$3/4$	1.744	0.0419	12.234	2.33	2.916	1.34
$1/2$	1.810	0.0464	11.207	2.12	2.617	0.90
$1/4$	2.013	0.0553	10.123	1.80	2.213	0.33
$1/8$	2.173	0.0612	9.746	—	2.017	0.095
0	2.271	0.0648	9.600	1.178	1.920	0

Trapezoidal Ducts

A trapezoidal duct is displayed in the inset of Fig. 5.50. Fully developed laminar flow and the heat transfer characteristics of trapezoidal ducts have been analyzed by Shah [172]. The fully developed f Re, $\mathrm{Nu_{H1}}$, and $\mathrm{Nu_{H2}}$ are given in Figs. 5.50 and 5.51. Farhanieh and Sunden [276] numerically investigated the laminar flow and heat transfer in the entrance region of trapezoidal ducts. The fully developed values of f Re, $K(\infty)$, and Nu were in accordance with the results from Shah [172].

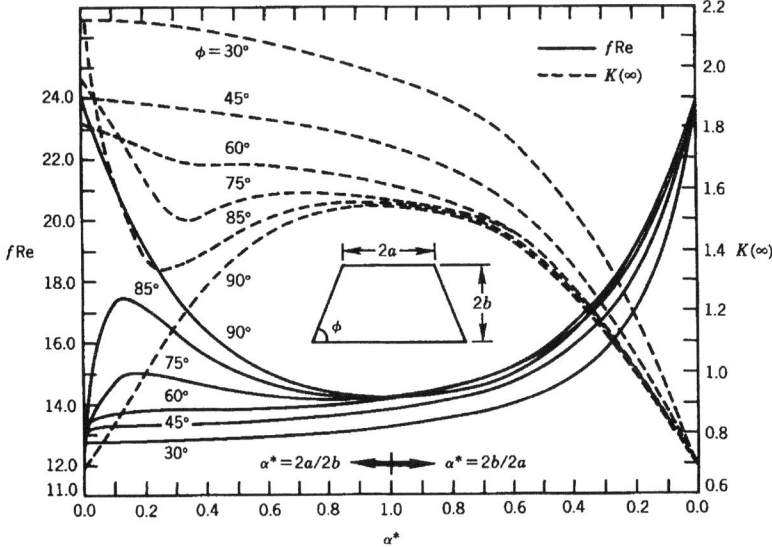

FIGURE 5.50   Fully developed f Re and $K(\infty)$ for laminar flow in a trapezoidal duct [172].

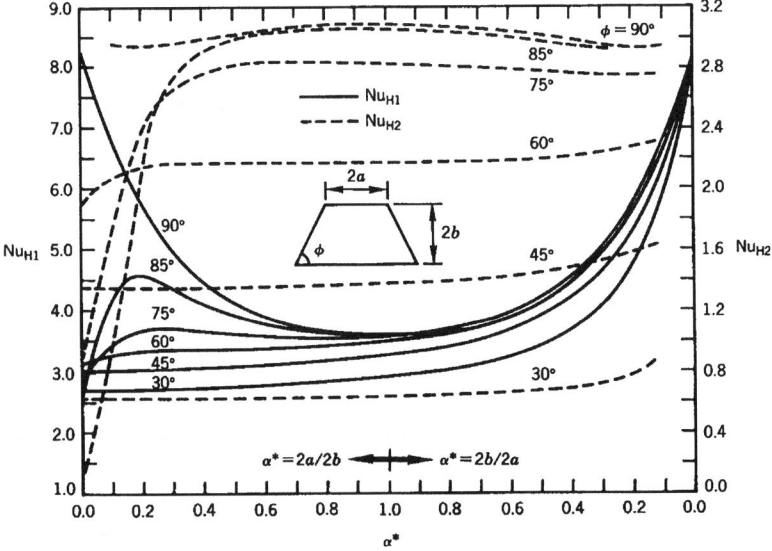

FIGURE 5.51   Fully developed Nusselt numbers for laminar flow in a trapezoidal pipe [172].

Chiranjivi and Rao [277] experimentally obtained a correlation for laminar and turbulent flow in trapezoidal ducts with one side heated, which is expressed as:

$$\text{Nu} = a \, \text{Re}^b \, \text{Pr}^{0.52} \left(\frac{\mu_w}{\mu} \right) \qquad (5.339)$$

where $a = 6.27$ and $b = 0.14$ for laminar flow, and $a = 0.79$ and $b = 0.4$ for turbulent flow of Reynolds number from 3000 to 15,000.

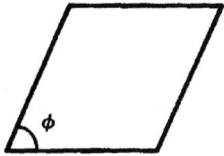

FIGURE 5.52   A rhombic duct.

Rhombic Ducts

A rhombic duct is depicted in Fig. 5.52. The fully developed flow and heat transfer characteristics of rhombic ducts obtained by Shah [172] are shown in Table 5.54.

Quadrilateral Ducts

A quadrilateral duct is shown schematically in Fig. 5.53. Nakamura et al. [278] analyzed fully developed laminar flow and heat transfer in arbitrary polygonal ducts. Their results are presented in Table 5.55.

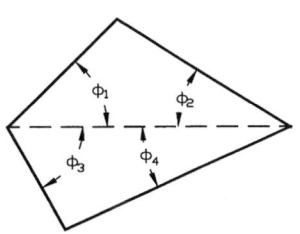

FIGURE 5.53   A schematic drawing of a quadrilateral duct.

Regular Polygonal Ducts

Fully developed flow and heat transfer in a regular polygonal duct with n equal sides, each subtending an angle of $360°/n$ at the duct center, have been reviewed by Shah and London [1]. The f Re and Nu are given in Table 5.56.

TABLE 5.54   Fully Developed Laminar Flow and Heat Transfer Characteristics of Rhombic Ducts [172]

ϕ	$K(\infty)$	L_{hy}^+	f Re	Nu_{H1}	Nu_{H2}	ϕ	$K(\infty)$	L_{hy}^+	f Re	Nu_{H1}	Nu_{H2}
0	2.971	0.1048	12.000	2.059	0	50	1.778	0.0380	13.542	3.188	1.62
10	2.693	0.0732	12.073	2.216	0.070	60	1.673	0.0353	13.830	3.367	2.16
20	2.384	0.0570	12.416	2.457	0.279	70	1.603	0.0336	14.046	3.500	2.64
30	2.120	0.0477	12.803	2.722	0.624	80	1.564	0.0327	14.181	3.581	2.97
40	1.925	0.0419	13.193	2.969	1.09	90	1.551	0.0324	14.227	3.608	3.09
45	1.850	0.0397	13.381	3.080	1.34						

TABLE 5.55   Fully Developed Friction Factors, Incremental Pressure Drop Numbers, and Nusselt Numbers for Some Quadrilateral Ducts [278]

ϕ_1 (deg)	ϕ_2 (deg)	ϕ_3 (deg)	ϕ_4 (deg)	f Re	$K(\infty)$	Nu_{H1}	Nu_{H2}
60	70	45	32.23	14.16	1.654	3.45	2.80
50	60	30	21.67	14.36	1.612	3.55	2.90
60	30	45	71.57	14.69	1.522	3.72	3.05
60	30	60	79.11	14.01	1.707	3.35	2.68

TABLE 5.56   Fully Developed Laminar Flow
Characteristics of Regular Polygonal Ducts [1]

n	$f\,\text{Re}$	Nu_{H1}	Nu_{H2}	Nu_T
3	13.333	3.111	1.892	2.47
4	14.227	3.608	3.091	2.976
5	14.737	3.859	3.605	—
6	15.054	4.002	3.862	—
7	15.31	4.102	4.009	—
8	15.412	4.153	4.100	—
9	15.52	4.196	4.159	—
10	15.60	4.227	4.201	—
20	15.88	4.329	4.328	—
∞	16.000	4.364	4.364	3.657

For practical calculations, Schenkel [279] proposed the following formula to compute the $f\,\text{Re}$ in regular polygonal ducts:

$$f\,\text{Re} = 16\left(\frac{n^2}{0.44 + n^2}\right)^4 \tag{5.340}$$

The values of the predictions from Eq. 5.337 agree well (within ±1 percent) with the tabulated values in Table 5.56.

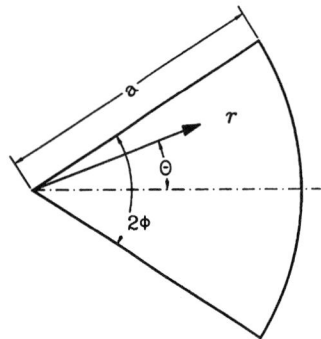

FIGURE 5.54   A circular sector duct.

Circular Sector Ducts

A schematic drawing of a circular sector duct is presented in Fig. 5.54. The fully developed $f\,\text{Re}$ and Nu for circular sector ducts have been obtained by Eckert and Irvine [280], Sparrow and Haji-Sheikh [174], Hu and Chang [265], and Ben-Ali et al. [281]. The results are summarized in Table 5.57.

Soliman et al. [282] numerically analyzed the problem of laminar flow development in circular sector ducts. Their $f_{app}\,\text{Re}$ and L_{hy}^+ results for $2\phi = 11.25$, 22.5, 45, and 90° are presented in Table 5.58. Furthermore, the hydrodynamic entrance lengths are 0.235, 0.144, 0.108, and 0.0786 for hydrodynamically developing flow in circular sector ducts with $2\phi = 11.25$, 22.5, 45, and 90°, respectively.

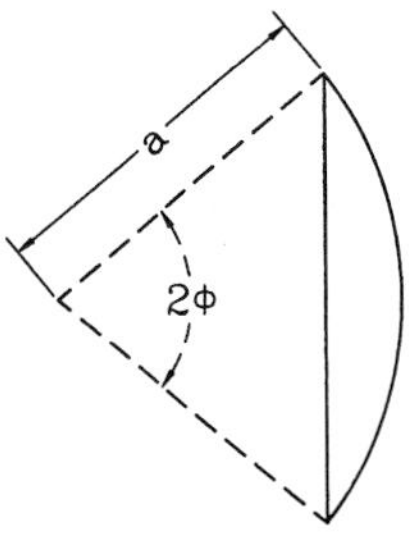

Circular Segment Ducts

A circular segment duct is depicted in Fig. 5.55. The fully developed flow and heat transfer characteristics obtained by Sparrow and Haji-Sheikh [283] are given in Table 5.59.

Hong and Bergles [284] have analyzed the thermal entrance solution of heat transfer for a circular segment duct with $2\phi = 180°$ (i.e., a semicircular duct). Two kinds of thermal boundary conditions are used: (1) a constant wall heat flux along the axial flow direction with a constant wall temperature along the duct circumference, and (2) a constant wall heat flux along the axial flow direction and a constant wall temperature along the semicircular arc, with zero heat

FIGURE 5.55   A circular segment duct.

TABLE 5.57   Fully Developed f Re, $K(\infty)$, and Nu for Laminar Flow in a Circular Sector Duct

2ϕ	f Re	$K(\infty)$	Nu_{H1}	Nu_{H2}	Nu_T	2ϕ	f Re	$K(\infty)$	Nu_{H1}	Nu_{H2}	Nu_T
0	12.000	2.971	2.059	—		80	14.592	1.530	3.671	—	—
5	12.33	—	2.245	0.018	1.423	90	14.79	—	3.730	2.984	3.060
8	12.411	2.480	2.384	—	—	100	14.929	1.504	3.806	—	—
10	12.504	—	—	0.081	1.692	120	15.200	1.488	3.906	2.898	3.191
15	12.728	2.235	2.619	0.195	1.901	150	15.54	—	3.999	2.995	3.268
20	12.98	—	2.742	0.354	2.073	160	15.611	1.468	4.04	—	—
30	13.310	1.855	3.005	0.838	2.341	180	15.767	1.463	4.089	2.923	—
36	13.510	—	—	1.174	—	210	15.98	—	4.127	2.871	3.347
40	13.635	—	—	1.400	2.543	240	16.15	—	4.171	2.821	3.370
45	13.782	1.657	3.27	1.667	—	270	16.29	—	4.208	2.781	3.389
50	13.95	—	3.337	1.990	2.70	300	16.42	—	4.244	2.749	3.407
60	14.171	1.580	3.479	2.421	2.822	330	16.54	—	4.280	2.723	3.427
72	14.435	—	—	2.608	—	350	16.62	—	4.304	2.708	3.443

TABLE 5.58   Flow Parameters for Hydrodynamically Developing Flow in Circular Sector Ducts [282]

$\dfrac{x^+}{L_{hy}^+}$	$2\phi = 11.25°$		$22.5°$		$45°$		$90°$	
	f_{app} Re	$K(x)$	f_{app} Re	$K(x)$	f_{app} Re	$K(x)$	f_{app} Re	$K(x)$
0.001	109.3	0.207	147.9	0.177	180.4	0.171	226.0	0.154
0.003	66.09	0.335	81.62	0.281	98.27	0.266	115.7	0.241
0.006	48.27	0.456	60.18	0.377	63.69	0.352	78.29	0.314
0.010	39.52	0.538	48.30	0.469	53.91	0.432	60.26	0.380
0.020	28.93	0.758	35.64	0.628	39.59	0.567	43.93	0.492
0.030	25.09	0.887	30.20	0.739	33.47	0.662	37.37	0.571
0.040	22.70	0.991	27.02	0.830	29.89	0.738	33.50	0.635
0.050	21.18	1.076	24.92	0.901	27.51	0.801	30.88	0.689
0.070	19.90	1.211	22.20	1.019	24.41	0.903	27.44	0.778
0.100	17.34	1.364	19.91	1.153	21.78	1.021	24.40	0.881
0.150	15.82	1.540	17.81	1.313	19.40	1.165	21.58	1.007
0.200	14.95	1.664	16.65	1.429	18.05	1.269	19.95	1.100
0.250	14.47	1.756	15.89	1.517	17.09	1.350	18.87	1.171
0.300	14.07	1.828	16.35	1.588	16.54	1.412	18.07	1.229
0.350	13.93	1.885	14.95	1.646	16.06	1.464	17.51	1.275
0.400	13.59	1.930	14.65	1.695	15.66	1.510	17.05	1.315
0.450	13.51	1.970	14.41	1.735	15.34	1.546	16.70	1.347
0.500	13.35	1.997	14.22	1.770	15.11	1.577	16.41	1.375
0.600	13.19	2.051	13.95	1.825	14.79	1.625	15.96	1.418
0.700	13.03	2.091	13.76	1.868	14.55	1.661	15.67	1.451
0.800	12.95	2.118	13.63	1.902	14.39	1.689	15.46	1.475
0.900	12.90	2.139	13.52	1.930	14.23	1.710	15.31	1.494
1.000	12.87	2.156	13.47	1.951	14.15	1.728	15.19	1.509

TABLE 5.59   The f Re, $K(\infty)$, Nu_{H1}, and Nu_{H2} for Fully Developed Laminar Flow in Circular Segment Ducts [283]

2ϕ	f Re	$K(\infty)$	Nu_{H1}	Nu_{H2}	2ϕ	f Re	$K(\infty)$	Nu_{H1}	Nu_{H2}
0	15.555	1.740	3.580	0	120	15.690	1.571	3.894	1.608
10	15.558	1.739	3.608	0.013	180	15.767	1.463	4.089	2.923
20	15.560	1.734	3.616	0.052	240	15.840	1.385	4.228	3.882
40	15.575	1.715	3.648	0.207	300	15.915	1.341	4.328	4.296
60	15.598	1.686	3.696	1.456	360	16.000	1.333	4.364	4.364
80	15.627	1.650	3.756	0.785					

TABLE 5.60   Local Nusselt Numbers in the Thermal Entrance Region of a Semicircular Duct [284]

x^*	$\mathrm{Nu}_{x,H1}$ BC1	BC2	x^*	$\mathrm{Nu}_{x,H2}$ BC1	BC2
0.000458	17.71	17.43	0.0279	4.767	4.339
0.000954	13.72	13.41	0.0351	4.562	4.037
0.00149	11.80	11.37	0.0442	4.429	3.830
0.00208	10.55	10.08	0.0552	4.276	3.686
0.00271	9.605	9.141	0.0686	4.217	3.543
0.00375	7.475	8.127	0.0849	4.156	3.425
0.00493	7.723	7.375	0.105	4.124	3.330
0.00627	7.137	6.788	0.130	4.118	3.265
0.00777	6.556	6.312	0.159	4.108	3.208
0.00946	6.300	5.912	0.196	—	3.171
0.0128	5.821	5.368	0.241	—	3.161
0.0168	5.396	4.935	0.261	—	3.160
0.0217	5.077	4.579	∞	4.089	3.160

flux along the diameter. The local Nusselt numbers for these two boundary conditions are presented in Table 5.60. The terms BC1 and BC2 in Table 5.60 refer to the previously mentioned first and second boundary conditions.

Annular Sector Ducts

An annular sector duct is displayed in the inset of Fig. 5.56. Shah and London [1] have calculated the $f\,\mathrm{Re}$ value to a high degree of accuracy using the analytical solution proposed by Sparrow et al. [285]. The results of $f\,\mathrm{Re}$ are presented in Fig. 5.56.

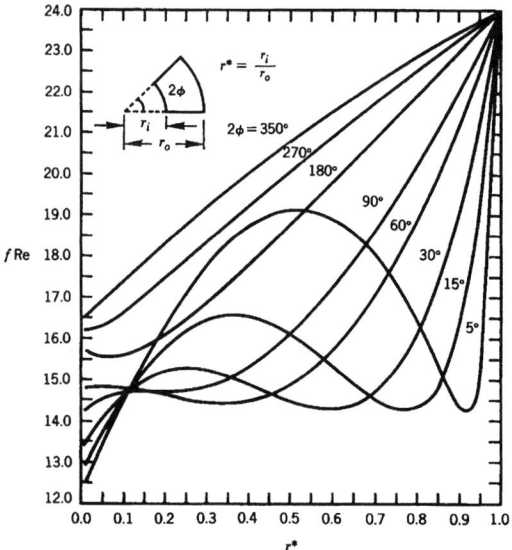

FIGURE 5.56   Fully developed friction factor for laminar flow in an annular sector duct [1].

Schenkel [279] has developed the following approximate equation for f Re in an annular secular duct:

$$f\,\text{Re} = \frac{24}{\left[1 - \dfrac{0.63}{\phi}\left(\dfrac{1-r^*}{1+r^*}\right)\right]\left[1 + \dfrac{1}{\phi}\left(\dfrac{1-r^*}{1+r^*}\right)\right]^2} \qquad (5.341)$$

This equation is valid for $\phi \geq \phi_{\min}(r^*)$. The values of $\phi_{\min}$ for $r^* = 0, 0.1, 0.2, 0.3, 0.4, 0.5, 0.6, 0.7, 0.8,$ and 0.9 are $60°, 50°, 42°, 35°, 28.5°, 22.5°, 17.5°, 13°, 8.5°,$ and $4°$, respectively. For $\phi \geq \phi_{\min}(r^*)$, the predictions of Eq. 5.341 are in excellent accord with the results presented in Fig. 5.56.

Ben-Ali et al. [281] and Soliman [286] have investigated fully developed flow and heat transfer in annular sector ducts with the Ⓣ, Ⓗ①, and Ⓗ② boundary conditions. The Nusselt numbers obtained by those investigators can be found in Table 5.61.

Simultaneously developing flow in annular sector ducts for air (Pr = 0.7) has been analyzed by Renzoni and Prakash [287]. In their analysis, the outer curved wall is treated as adiabatic, and the Ⓗ① boundary condition is imposed on the inner curved wall as well as on the two straight walls of the sector. The fully developed friction factors, incremental pressure drop numbers, hydrodynamic entrance lengths, and thermal entrance lengths are presented in Table 5.62. The term L_{hy}^+ used in Table 5.62 is defined as the dimensionless axial distance at which f_{app} Re $= 1.05f$ Re. The fully developed Nusselt numbers are represented by Nu_{fd} in order not to confuse the reader since the thermal boundary condition applied in Renzoni and Prakash [287] is different from those defined in the section.

Stadium-Shaped Ducts

A stadium-shaped duct and a modified stadium-shaped duct are displayed in the insets of Fig. 5.57. Zarling [288] has obtained the f Re and Nu_{H1} for fully developed laminar flow in stadium-shaped ducts. Cheng and Jamil [289] have determined the f Re and Nu for the mod-

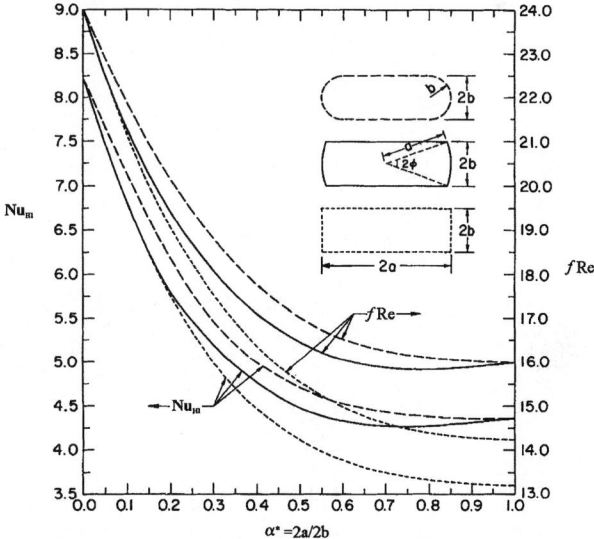

FIGURE 5.57 Fully developed f Re and Nu_{H1} for stadium-shaped and modified stadium-shaped ducts [1].

TABLE 5.61 Fully Developed Nusselt Numbers for Annular Sector Duct [281, 286]

$r^* =$		0.1	0.2	0.3	0.4	0.5	0.6	0.7	0.8	0.9
$2\phi = 5°$	Nu_{H1}	2.723	3.254	3.850	4.460	4.976	5.243	5.101	4.466	3.648
	Nu_{H2}	0.0326	0.0618	0.1253	0.2756	0.6420	1.426	2.419	2.922	3.083
	Nu_T	1.707	2.045	2.440	2.886	3.355	3.773	3.962	3.634	3.005
$2\phi = 10°$	Nu_{H1}	2.896	3.387	3.893	4.326	4.554	4.471	4.098	3.675	3.889
	Nu_{H2}	0.1427	0.2746	0.5298	1.045	1.849	2.603	3.015	3.077	3.049
	Nu_T	1.996	2.341	2.717	3.096	3.405	3.523	3.337	3.026	3.197
$2\phi = 15°$	Nu_{H1}	3.041	3.488	3.903	4.177	4.207	3.997	3.709	3.641	4.462
	Nu_{H2}	0.3145	0.6250	1.135	1.881	2.575	2.964	3.111	3.085	2.987
	Nu_T	2.208	2.541	2.876	3.166	3.316	3.249	3.051	3.000	3.662
$2\phi = 20°$	Nu_{H1}	3.163	3.564	3.888	4.031	3.951	3.747	3.616	3.834	4.979
	Nu_{H2}	0.6253	1.083	1.760	2.456	2.887	3.051	3.129	3.056	2.947
	Nu_T	2.374	2.684	2.971	3.163	3.194	3.077	2.980	3.153	4.009
$2\phi = 30°$	Nu_{H1}	3.354	3.656	3.821	3.802	3.686	3.627	3.785	4.371	5.726
	Nu_{H2}	1.348	2.000	2.602	2.946	3.065	3.088	3.062	2.993	2.912
	Nu_T	2.615	2.868	3.045	3.090	3.029	2.988	3.113	3.582	4.782
$2\phi = 40°$	Nu_{H1}	3.490	3.697	3.746	3.683	3.640	3.743	4.106	4.875	6.208
	Nu_{H2}	2.041	2.603	2.942	3.065	3.086	3.066	3.069	2.953	2.898
	Nu_T	2.780	2.966	3.046	3.025	2.997	3.079	3.366	4.008	5.250
$2\phi = 50°$	Nu_{H1}	3.589	3.711	3.690	3.658	3.712	3.948	4.446	5.290	6.537
	Nu_{H2}	2.531	2.901	3.053	3.085	3.070	3.035	3.040	2.930	2.889
	Nu_T	2.893	3.015	3.032	3.010	3.053	3.241	3.642	4.337	5.573
$2\phi = 60°$	Nu_{H1}	3.660	3.715	3.685	3.697	3.844	4.183	4.763	5.624	6.775
	Nu_{H2}	2.821	3.025	3.079	3.072	3.046	3.005	2.957	2.916	2.883
	Nu_T	2.973	3.038	3.029	3.041	3.157	3.426	3.910	4.686	5.812
$2\phi = 90°$	Nu_{H1}	3.782	3.761	3.815	4.006	4.351	4.855	5.510	6.299	7.208
	Nu_{H2}	3.060	3.060	3.037	3.009	2.979	2.954	2.916	2.896	2.868
	Nu_T	3.097	3.086	3.128	3.280	3.562	3.987	4.576	5.342	6.238
$2\phi = 120°$	Nu_{H1}	3.863	3.893	4.072	4.392	4.836	5.383	6.010	6.700	7.443
	Nu_{H2}	3.028	3.003	2.982	2.960	2.937	2.916	2.900	2.887	2.869
	Nu_T	3.157	3.180	3.326	3.590	3.966	4.457	5.055	5.744	6.488
$2\phi = 150°$	Nu_{H1}	3.956	4.079	4.360	4.757	5.239	5.778	6.357	6.964	7.590
	Nu_{H2}	2.967	2.953	2.941	2.927	2.913	2.901	2.892	2.876	2.874
	Nu_T	3.222	3.322	3.553	3.892	4.322	4.828	5.396	5.997	6.646
$2\phi = 180°$	Nu_{H1}	4.067	4.284	4.639	5.077	5.565	6.080	6.611	7.150	7.691
	Nu_{H2}	2.915	2.916	2.912	2.905	2.899	2.891	2.885	2.868	2.881
	Nu_T	3.301	3.480	3.782	4.169	4.621	5.118	5.651	6.175	6.775
$2\phi = 210°$	Nu_{H1}	4.192	4.490	4.894	5.351	5.835	6.334	6.845	7.368	7.884
	Nu_{H2}	2.876	2.888	2.892	2.891	2.890	2.886	2.880	2.868	2.887
	Nu_T	3.392	3.641	3.999	4.415	4.874	5.352	5.843	6.323	6.834
$2\phi = 240°$	Nu_{H1}	4.322	4.685	5.120	5.580	6.045	6.505	6.956	7.394	7.820
	Nu_{H2}	2.847	2.868	2.878	2.882	2.882	2.881	2.874	2.869	2.894
	Nu_T	3.491	3.807	4.198	4.631	5.080	5.539	5.989	6.437	6.894
$2\phi = 270°$	Nu_{H1}	4.453	4.867	5.320	5.777	6.225	6.660	7.078	7.479	7.864
	Nu_{H2}	2.825	2.853	2.867	2.875	2.879	2.879	2.867	2.871	2.900
	Nu_T	3.594	3.961	4.379	4.818	5.263	5.700	6.099	6.528	6.941
$2\phi = 300°$	Nu_{H1}	4.581	5.034	5.496	5.946	6.377	6.788	7.178	7.548	7.899
	Nu_{H2}	2.808	2.842	2.860	2.870	2.875	2.875	2.866	2.874	2.905
	Nu_T	3.699	4.107	4.542	4.982	5.413	5.825	6.205	6.601	6.979
$2\phi = 330°$	Nu_{H1}	4.704	5.186	5.652	6.093	6.510	6.904	7.279	7.638	7.976
	Nu_{H2}	2.794	2.833	2.853	2.865	2.871	2.869	2.866	2.877	2.909
	Nu_T	3.798	4.241	4.688	5.124	5.540	5.926	6.295	6.664	7.010
$2\phi = 350°$	Nu_{H1}	4.782	5.279	5.745	6.179	6.583	6.960	7.311	7.638	7.945
	Nu_{H2}	2.787	2.827	2.849	2.861	2.870	2.869	2.867	2.878	2.911
	Nu_T	3.864	4.327	4.776	5.208	5.620	5.993	6.347	6.699	7.705

TABLE 5.62 Fully Developed Fluid Flow and Heat Transfer Characteristics of Annular Sector Ducts with Adiabatic OuterCurved Wall [287]

r^*	$f\,\mathrm{Re}$	$K(\infty)$	L_{hy}^+	Nu_{fd}	L_{th}^*
\multicolumn{6}{c}{$2\phi = 15°$}					
0.2	15.65	1.77	0.0775	3.433	0.1530
0.5	16.01	1.32	0.0500	4.372	0.0924
0.8	14.21	1.42	0.0529	3.340	0.0898
\multicolumn{6}{c}{$2\phi = 22.5°$}					
0.2	15.35	1.64	0.0703	3.493	0.1320
0.5	14.90	1.37	0.0516	3.933	0.0838
0.8	15.03	1.33	0.0476	3.113	0.1090
\multicolumn{6}{c}{$2\phi = 45°$}					
0.2	14.73	1.46	0.0574	3.461	0.1070
0.5	14.29	1.42	0.0529	3.235	0.0972
0.8	17.58	1.07	0.0303	3.327	0.1230

ified stadium-shaped ducts. The $f\,\mathrm{Re}$ and Nu_{H1} for fully developed laminar flow in rectangular ducts are also included for the purpose of comparison.

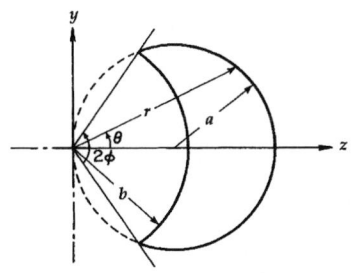

FIGURE 5.58 A moon-shaped duct.

Moon-Shaped Ducts

A moon-shaped duct is depicted in Fig. 5.58. Shah and London [1] have determined the fully developed $f\,\mathrm{Re}$ and the velocity profile for moon-shaped ducts. These follow:

$$u = \frac{c_1}{4}(r^2 - b^2)\left(1 - \frac{2a\cos\theta}{r}\right) \tag{5.342}$$

$$f\,\mathrm{Re} = -\frac{c_1 D_h^2}{2u_m} \tag{5.343}$$

where

$$D_h = 2a\left[\frac{(2 - \alpha^{*2})\phi + \sin 2\phi}{(2 + \alpha^*)\phi}\right] \tag{5.344}$$

$$u_m = \frac{c_1 a^2}{4}\frac{(\tfrac{1}{2}\alpha^{*4} + 2\alpha^{*2} - 1)\phi - \tfrac{8}{3}\alpha^{*3}\sin\phi + (\alpha^{*2} - \tfrac{2}{3})\sin 2\phi - \tfrac{1}{12}\sin 4\phi}{(2 - \alpha^{*2})\phi + \sin 2\phi} \tag{5.345}$$

In the preceding equations, $\alpha^* = b/a$ and $c_1 = \mu\,dp/dx$.

Corrugated Ducts

Three corrugated ducts are schematically shown in the insets of Fig. 5.59. Hu and Chang [265] have analyzed the $f\,\mathrm{Re}$ for fully developed laminar flow in circumferentially corrugated circular ducts with n sinusoidal corrugations over the circumference as shown in Fig. 5.59, inset a, for $e^* = \varepsilon/a = 0.06$. The perimeter and hydraulic diameter of these ducts must be evaluated numerically. However, their free flow area A_c is given by $A_c = \pi a^2(1 + 0.5\varepsilon^2)$.

The $f\,\mathrm{Re}$, Nu_{H1}, and Nu_{H2} values determined by Hu and Chang [265] for circumferentially corrugated circular ducts with sinusoidal corrugations are presented in Table 5.63 as functions

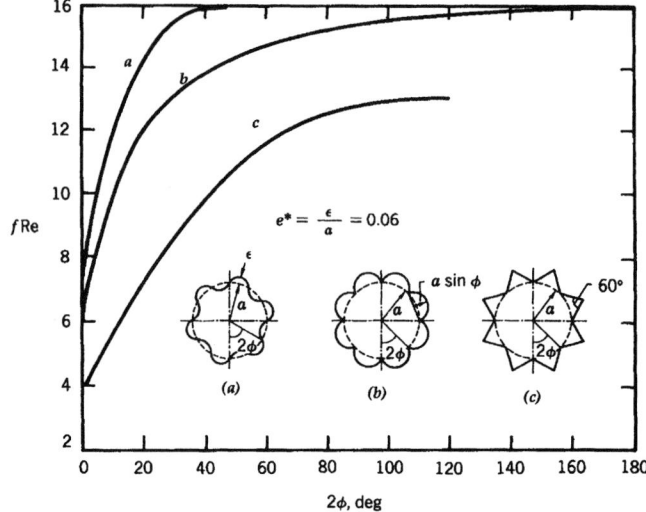

FIGURE 5.59   Fully developed friction factors for circumferentially corrugated circular ducts [2].

of n and e^*, which are defined in Fig. 5.59. Angle 2ϕ in Fig. 5.59 is related to n simply as $2\phi = 360°/n$.

Schenkel [279] has determined the fully developed friction factors for circular ducts with semicircular corrugations, as that shown in Fig. 5.59, inset b. For this kind of duct,

$$A_c = \pi a^2 \frac{\sin \phi}{\phi} \left[\frac{\pi}{2} \sin \phi + \cos \phi \right], \qquad P = \pi^2 a \frac{\sin \phi}{\phi} \qquad (5.346)$$

TABLE 5.63   Fully Developed Friction Factors and Nusselt Numbers for Circumferentially Corrugated Circular Ducts With Sinusoidal Corrugations [1]

n	e^*	$f\,\mathrm{Re}$	$\mathrm{Nu_{H1}}$	$\mathrm{Nu_{H2}}$	$D_h/2a$
8	0.02	15.990	4.356	4.357	0.9986
	0.04	15.962	4.334	4.335	0.9944
	0.06	15.915	4.297	4.299	0.9874
	0.08	15.850	4.244	4.246	0.9776
	0.10	15.765	4.176	4.177	0.9650
	0.12	15.678	4.090	4.089	0.9501
12	0.02	15.952	4.340	4.340	0.9966
	0.04	15.806	4.267	4.267	0.9863
	0.06	15.559	4.142	4.140	0.9689
	0.08	15.200	3.962	3.956	0.9439
	0.10	14.711	3.723	3.709	0.9107
16	0.02	15.887	4.316	4.316	0.9938
	0.04	15.542	4.168	4.167	0.9747
	0.06	14.943	3.912	3.906	0.9418
	0.08	14.051	3.540	3.527	0.8934
24	0.02	15.679	4.245	4.245	0.9856
	0.04	14.671	3.875	3.870	0.9402
	0.06	12.872	3.231	3.219	0.8583

The radius of the semicircular corrugation is $a \sin \phi$. The f Re values for ducts with semicircular corrugations can be determined using the following expression given by Schenkel [279]:

$$f\,\mathrm{Re} = 6.4537 + 0.8350\phi - 3.6909 \times 10^{-2}\phi^2 + 8.6674 \times 10^{-4}\phi^3 - 1.0588 \times 10^{-5}\phi^4$$
$$+ 6.2094 \times 10^{-8}\phi^5 - 1.3261 \times 10^{-4}\phi^6 \quad (5.347)$$

where ϕ is in degrees. Equation 5.347 is valid for $0 \le 2\phi \le 180°$. When $2\phi = 180$, this geometry reduces to a circular duct. The prediction of f Re = 16 was obtained from Eq. 5.347 for circular ducts.

Schenkel [279] has also determined the fully developed friction factors for laminar flow in circular ducts having triangular corrugations with an angle of 60°, as shown in Fig. 5.59, inset *c*. For this type of duct, the cross section of the fluid flow area A_c and wetted perimeter P can be calculated as follows:

$$A_c = \pi a^2 \frac{\cos\phi + \sqrt{3}\sin\phi}{\phi}, \qquad P = 4\pi a \frac{\sin\phi}{\phi} \qquad (5.348)$$

The f Re values for ducts with triangular corrugations can be obtained with the following expression [279]:

$$f\,\mathrm{Re} = 3.8952 + 0.3692\phi - 3.2483 \times 10^{-3}\phi^2 - 3.3187 \times 10^{-5}\phi^3 + 4.5962 \times 10^{-7}\phi^4 \qquad (5.349)$$

where ϕ is in degrees. Equation 5.349 is valid for $0 \le 2\phi \le 120°$.

A comparison of f Re for these three types of corrugated ducts with $e^* = 0.06$ is displayed in Fig. 5.59.

Parallel Plate Ducts With Spanwise Periodic Corrugations at One Wall

Two types of corrugations (triangular and rectangular) in parallel plate ducts are displayed in the insets of Figs. 5.60 and 5.61, respectively. Sparrow and Charmchi [290] have obtained the solutions for fully developed laminar flow in these ducts. The flow in the duct is considered to be perpendicular to the plane of the paper. Both ducts are assumed to be infinite in the span-

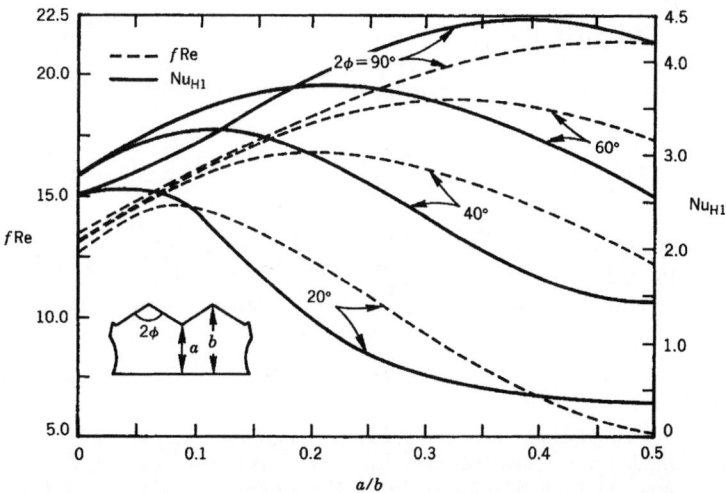

FIGURE 5.60 Fully developed friction factors and Nusselt numbers for flat ducts with spanwise-periodic triangular corrugations at one wall [290].

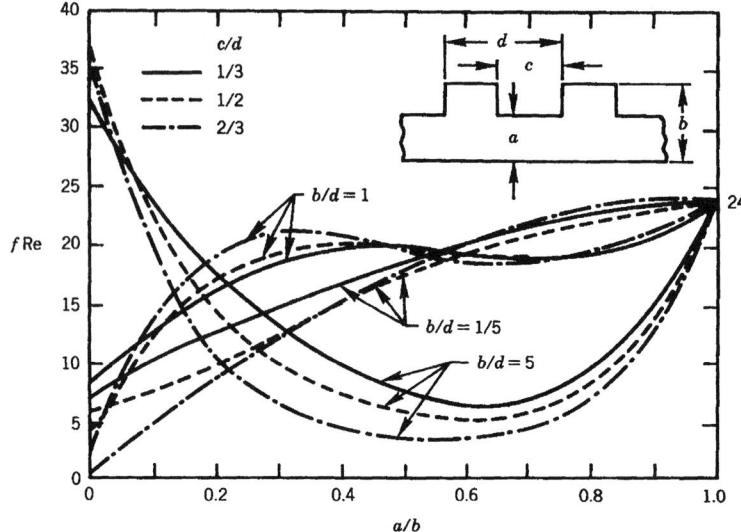

FIGURE 5.61 Fully developed friction factors and Nusselt numbers for flat ducts with spanwise-periodic rectangular corrugations at one wall [291].

wise direction; therefore, the end effects due to the short bounding walls are neglected. The corrugated wall is subjected to the Ⓗ thermal boundary condition, while the flat wall is considered to be adiabatic.

The cross-sectional area and perimeter of a flat duct with spanwise triangular corrugation can be found by:

$$A_c = n(b^2 - a^2)\tan\phi, \qquad P = 2n(b-a)\frac{1+\sin\phi}{1+\cos\phi} \qquad (5.350)$$

where n represents the number of triangular corrugations and 2ϕ is the angle of the top vertex of the triangle.

The fully developed f Re and $\mathrm{Nu_{H1}}$ values obtained by Sparrow and Charmchi [290] are shown in Fig. 5.60, which is taken from Shah and Bhatti [2]. If $a/b = 0$, the duct with triangular corrugations reduces to an array of isosceles triangles. The f Re and $\mathrm{Nu_{H1}}$ values from Fig. 5.60 agree well with the values obtained from the corresponding figures in the section concerning triangular ducts.

Fully developed laminar flow and heat transfer in a parallel plate duct with spanwise-periodic rectangular corrugations at one wall have been investigated by Sparrow and Chukaev [291]. The end effect is also ignored in their analysis. The fully developed f Re is shown in Fig. 5.61, which is based on the results reported by Sparrow and Chukaev [291] and the extension by Shah and Bhatti [2]. The heat transfer characteristics for the three pairs of geometric parameters can be found in Sparrow and Chukaev [291].

Cusped Ducts

A *cusped duct,* also referred to as a *star-shaped duct,* such as the one shown in Fig. 5.62, is made up of concave circular arcs. The fully developed f Re, $\mathrm{Nu_{H1}}$, and $\mathrm{Nu_T}$, in laminar flow are given in Table 5.64, in which n is the number of the concave circular arcs in the cusped ducts. The values for f Re, $\mathrm{Nu_T}$, and $\mathrm{Nu_{H1}}$ are taken from Shah and London [1], Dong et al. [292], and

TABLE 5.64 Fully Developed $f\,\text{Re}$, Nu_T, and Nu_{H1} for Laminar Flow in Cusped Ducts

n	$f\,\text{Re}$	Nu_T	Nu_{H1}
3	6.503	0.92	—
4	6.606	1.09	1.352
5	6.634	1.23	—
6	6.639	—	—
8	6.629	—	—

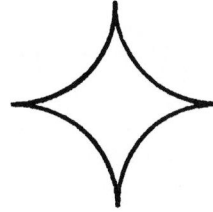

FIGURE 5.62 A cusped duct with four concave walls.

Dong and Ebadian [293]. An analysis of thermally developing laminar flow in cusped ducts can be found in Dong et al. [292].

Cardioid Ducts

A cardioid duct is shown in Fig. 5.63. Fully developed laminar flow and heat transfer under the Ⓗ① boundary condition have been analyzed by Tyagi [294]. The $f\,\text{Re}$ and Nu_{H1} values derived from this analysis are 5.675 and 4.208, respectively. The Nusselt number for the Ⓗ② thermal boundary condition was found to be 4.097 [1].

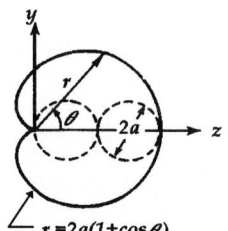

FIGURE 5.63 A cardioid duct.

Unusual Singly Connected Ducts

For the fully developed friction factors for laminar flow in unusual singly connected ducts, interested readers are encouraged to consult Shah and Bhatti [2].

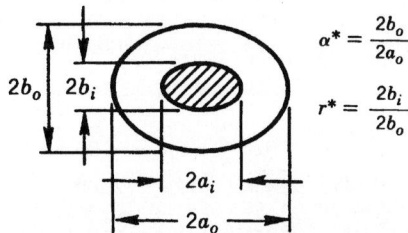

$$\alpha^* = \frac{2b_o}{2a_o}$$

$$r^* = \frac{2b_i}{2b_o}$$

FIGURE 5.64 A confocal elliptical duct.

OTHER DOUBLY CONNECTED DUCTS

Fully developed laminar flow and heat transfer in several doubly connected ducts are discussed in the following sections.

Confocal Elliptical Ducts

A confocal elliptical duct is shown in Fig. 5.64. According to the analysis by Topakoglu and Arnas [295], the friction factor for fully developed laminar flow in confocal elliptical ducts can be computed by

$$f\,\text{Re} = \frac{256 A_c^3}{\pi I_{oo} P^2 (a_o + b_o)^4} \tag{5.351}$$

where
$$I_{oo} = \frac{1}{4}\,(1 - \omega^4)\left(1 + \frac{m^8}{\omega^4}\right) - 2m^4\,\frac{1 - \omega^2}{1 + \omega^2} + \frac{1}{4\ln\omega}\,(1 - \omega^2)^2\left(1 - \frac{m^4}{\omega^2}\right)^2 \tag{5.352}$$

$$\frac{A_c}{(a_o + b_o)^2} = \frac{\pi}{4}(1 - \omega^2)\left(1 + \frac{m^4}{\omega^2}\right) \tag{5.353}$$

$$\frac{P}{a_o + b_o} = 2\left[(1 + m^2)E_1 + \left(1 + \frac{m^2}{\omega^2}\right)\omega E_\omega\right] \tag{5.354}$$

$$\omega = \left(\frac{a_i + b_i}{a_o + b_o}\right) = \frac{\alpha^* r^* + [1 - \alpha^{*2}(1 - r^*)^2]^{1/2}}{1 + \alpha^*} \tag{5.355}$$

$$m = \left(\frac{1 - \alpha^*}{1 + \alpha^*}\right)^{1/2}, \qquad \alpha^* = \frac{b_o}{a_o}, \qquad r^* = \frac{b_i}{b_o} \tag{5.356}$$

E_1 and E_ω are the complete elliptical integrals of the second kind. These are evaluated using the arguments $1 - b_o^2/a_o^2$ and $1 - b_i^2/a_i^2$, respectively. In addition, b_i/a_i is related to ω and m by means of the following:

$$\frac{b_i}{a_i} = \frac{1 - (m^2/\omega^2)}{1 + (m^2/\omega^2)} \tag{5.357}$$

The fully developed Nusselt numbers Nu_{H1} determined from the analysis of Topakoglu and Arnas [295], together with the f Re calculated from Eq. 5.351, are displayed in Table 5.65.

Regular Polygonal Ducts With Centered Circular Cores

The product of fully developed friction factor and Reynolds number in laminar flow f Re obtained by Ratkowsky and Epstein [296] for polygonal ducts with centered circular cores (see the inset in Fig. 5.65) are shown in Fig. 5.65. The fully developed Nu_{H1} obtained by Cheng and Jamil [297] are given in Fig. 5.66. It can be observed that as $n \to \infty$, the value of f Re approaches 6.222 for $\alpha^* = 1$ (annular duct); f Re approaches 16 for $\alpha^* = 0$ (circular duct).

Circular Ducts With Centered Regular Polygonal Cores

The product of fully developed friction factors and the Reynolds number f Re obtained by Hagen and Ratkowsky [298] for laminar flow in circular ducts with centered regular polygo-

TABLE 5.65   The f Re and Nu_{H1} for fully Developed Laminar Flow in Confocal Elliptical Ducts [1]

	$\alpha^* = 0.2$		0.4		0.6		0.8		0.9		0.95	
r^*	f Re	Nu_{H1}	f Re	Nu_{H1}	f Re	Nu_{H1}	f Re	Nu_{H1}	f Re	Nu_{H1}	f Re	Nu_{H1}
0.02	19.419	5.1237	19.468	5.1231	20.291	5.4782	21.766	6.5083	22.436	7.1933	22.620	7.4100
0.1	19.452	5.1252	19.622	5.1395	20.622	5.5534	22.388	6.7384	23.151	7.5273	23.366	7.7940
0.2	19.478	5.1230	19.759	5.1479	20.965	5.6162	22.750	6.8973	23.454	7.6945	23.643	7.9574
0.3	19.495	5.1185	19.871	5.1541	21.201	5.6770	22.974	7.0218	23.610	7.7961	23.777	8.0427
0.4	19.507	5.1130	19.973	5.1626	21.404	5.7441	23.135	7.1325	23.708	7.8696	23.855	8.0955
0.5	19.516	5.1072	20.072	5.1751	21.585	5.7441	23.257	7.1325	23.773	7.8696	23.903	8.0955
0.6	19.525	5.1016	20.171	5.1922	21.749	5.8966	23.257	7.3224	23.819	7.9699	23.934	8.1546
0.7	19.534	5.0965	20.268	5.2137	21.896	5.9779	23.429	7.4023	23.851	8.0046	23.953	8.1711
0.8	19.544	5.0921	20.365	5.2390	22.029	6.0597	23.490	7.4724	23.874	8.0320	23.966	8.1825
0.9	19.555	5.0885	20.460	5.2676	22.148	6.1404	23.539	7.5336	23.890	8.0536	23.973	8.1901
0.95	19.561	5.0788	20.506	5.2836	22.203	6.1801	23.560	7.5606	23.896	8.0621	23.975	8.1928
0.98	19.565	—	20.534	—	22.234	—	23.572	—	23.900	—	23.976	—

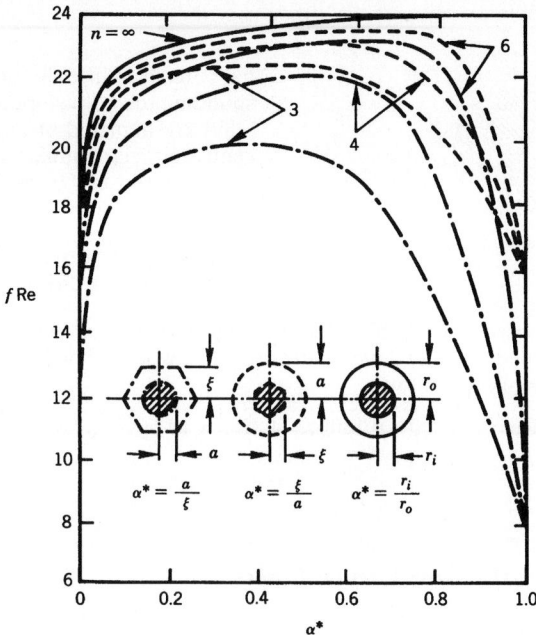

FIGURE 5.65 Fully developed friction factors for regular polygonal ducts with centered circular cores and circular ducts with centered rectangular polygonal cores [2].

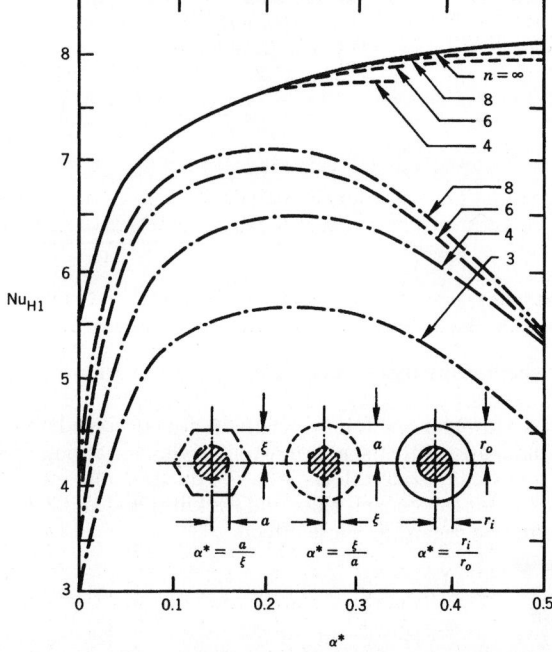

FIGURE 5.66 Fully developed Nusselt numbers for regular polygonal ducts with centered circular cores and circular ducts with centered rectangular polygonal cores [2].

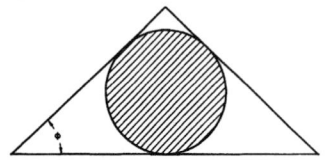

FIGURE 5.67   An isosceles triangular duct with an inscribed circular core.

nal cores (see inset in Fig. 5.65) is shown in Fig. 5.65. Corresponding fully developed Nu_{H1} obtained by Cheng and Jamil [289] are depicted in Fig. 5.66. The f Re and Nu_{H1} for concentric circular annular ducts are shown in Figs. 5.65 and 5.66 for the purpose of comparison.

Isosceles Triangular Ducts With Inscribed Circular Cores

An isosceles triangular duct with an inscribed circular core is shown in Fig. 5.67. The f Re obtained for fully developed laminar flow in such a duct by Bowen [299] can be expressed in terms of ϕ, as follows:

$$f\,\mathrm{Re} = 12.0000 - 0.1605\phi + 4.2883 \times 10^{-3}\phi^2 - 1.0566 \times 10^{-4}\phi^3$$
$$+ 1.6251 \times 10^{-6}\phi^4 - 1.04821 \times 10^{-8}\phi^5 \quad (5.358)$$

where ϕ is in degrees.

Elliptical Ducts With Centered Circular Cores

For elliptical ducts with centered circular cores, fully developed laminar flow has been analyzed by Shivakumar [300]. The f Re values are given in Table 5.66, in which α^* denotes the ratio of the length of the minor axis to the length of the major axis of the ellipse and r^* is the ratio of the diameter of the circular core to the length of the minor axis.

TABLE 5.66   Fully Developed Friction Factors for Elliptical Ducts With Centered Circular Cores [300]

		f Re		
α^*	$r^* = 0.5$	0.6	0.7	0.95
0.5	19.321	—	—	—
0.7	21.694	—	19.402	—
0.9	23.519	23.435	23.159	16.816

CONCLUDING REMARKS

This chapter discusses forced convection in various ducts. The formulas, correlations, equations, tables, and figures included in this chapter are given for the purpose of practical calculations. However, the following effects are not considered: a detailed discussion of heat source and dissipation effects, non-newtonian fluids, varying thermal property effects, porous wall ducts, unsteady-state effects, rotating ducts, combined radiation, and convection. The interested reader can consult Kays and Perkins [263] and Kakaç, Shah, and Aung [301] for further information regarding these effects.

NOMENCLATURE

A_c	flow cross-sectional area, m²
a	radius of a circular duct, m; half-length of major axis of an elliptic duct, m; half-length of the width of a rectangular duct, m

Bi	Biot number $= h_e D_h/k$
Br	Brinkmann number for the ① boundary condition, $= \mu u_m^2/k(T_{w,m} - T_e)$
Br′	Brinkmann number for the ⑪ boundary condition, $= \mu u_m^2/q''D_h$
b	half-spacing of a parallel plate duct, m; coil spacing, m; half length of minor axis of an elliptic duct, m; half length of height of a rectangular duct, m
C	constant
C_p	specific heat of the fluid at constant pressure, J/(kg·K)
D	diameter of a circular cylinder, m
De	Dean number $= \mathrm{Re}\ \sqrt{a/R}$
De*	modified Dean number $= \mathrm{Re}\ \sqrt{D_h/R}$
D_g	general length
D_h	hydraulic diameter of the duct $= 4A_c/P$, m
D_l	laminar equivalent diameter, m
$E(m)$	complete elliptic integral of the second kind with argument m, which is defined by Eq. 2.252
e^*	eccentricity of the eccentric annular duct $= \varepsilon/(r_o - r_i)$; amplitude of the circular duct with sinusoidal corrugation $= \varepsilon/a$
F	a multiplicative factor entering various expressions
f	circumferentially averaged fully developed friction factor $= \tau_w/(\rho u_m^2/2)$
f_{app}	apparent Fanning friction factor $= \Delta p^*/(2x/D_h)$
f_c	friction factor for curved ducts $= \tau_w/(\rho u_m^2/2)$
f_s	friction factor for straight ducts
$G_n(\ )$	eigenfunctions
Gr	Grashof number $= \beta g a^3 \Delta T/\nu^2$
Gr′	modified Grashof number $= \beta g a^4 q''/k\nu$
Gz	Graetz number $= mcp/kL = p/(4D_h x^*)$
⑪	uniform wall heat flux boundary conditions
⑪1	thermal boundary condition referring to uniform axial wall heat flux with uniform peripheral wall temperature
⑪2	thermal boundary condition referring to axially and circumferentially uniform wall heat flux
⑪4	conductive thermal boundary condition
⑪5	thermal boundary condition referring to exponential wall heat flux
He	helical coil number
H_a	length of the fin on the major axis in an elliptic duct, m
H_b	length of the fin on the minor axis in an elliptic duct, m
h	convective heat transfer coefficient, W/(m²·K)
h_e	convective heat transfer coefficient for the duct exterior, W/(m²·K)
$J_i(\ )$	Bessel functions of the first kind and orders 0 or 1 corresponding to $i = 0$ or 1
K	wall conductivity parameter $= ks/k_w \delta_w$
$K(x)$	incremental pressure drop number, defined by Eq. 5.5
k	thermal conductivity, W/(m·K)
l	length of fin, m
l^*	relative length of fins $= l/a$

L	length of the duct, m
L_{hy}	hydrodynamic entrance length, m
L_{hy}^+	dimensionless hydrodynamic entrance length $= L_{hy}/D_h\,\mathrm{Re}$
L_{th}	thermal entrance length, m
L_{th}^*	dimensionless thermal entrance length $= L_{th}/D_h\,\mathrm{Pe}$
m	argument of complete integral of the second kind; an exponent; a constant
n	number of sides of a duct; a dimensionless constant
n_1	number of coil turns at the beginning of the spiral $= L/(2\pi bN) - N/2$
n_2	number of coil turns at the end of the spiral $= L/(2\pi bN) + N/2$
N	number of spiral coil turns $= n_2 - n_1$
Nu	Nusselt number $= hD_h/k$
Nu_{bc}	circumferentially averaged mean Nusselt number in the fully developed region for a given boundary condition
$\mathrm{Nu}_{x,bc}$	circumferentially averaged but axially local Nusselt number
$\mathrm{Nu}_{m,bc}$	mean Nusselt number for the thermal entrance region for the specified thermal boundary condition
$\mathrm{Nu}_{x,lj}^{(k)}$	local Nusselt number for a doubly connected duct $= \Phi_{lj}^{(k)}/(\theta_{lj}^{(k)} - \theta_{mj}^{(k)})$
$\mathrm{Nu}_{x,i}^{(k)}, \mathrm{Nu}_{x,o}^{(k)}$	local Nusselt number at inner and outer walls of a concentric or eccentric annular duct
Nu_o	overall Nusselt number associated with the Ⓣ③ boundary condition
$\mathrm{Nu}_{p,bc}$	peripherally local Nusselt number in the fully developed region for a given boundary condition
P	wetted perimeter of the duct, m; distance between circular cylinders, m
Pe	Péclet number $= u_m D_h/\alpha = \mathrm{Re}\,\mathrm{Pr}$
Pr	Prandtl number $= \nu/\alpha$
p	fluid static pressure, Pa
Δp	fluid static pressure drop in the flow direction between two cross sections of interest, Pa
Δp^*	dimensionless fluid static pressure drop $= \Delta p/(\rho u_m^2/2)$
q_w''	wall heat flux, heat transfer rate per unit heat transfer area of the duct (average value with respect to perimeter), W/m^2
q	heat transfer rate, W
q'	heat transfer rate per unit length, w/m
q''	heat flux, w/m^2
Re	Reynolds number $= u_m D_h/\nu$
$\mathrm{Re}_{\mathrm{crit}}$	critical Reynolds number
Re_ε	roughness Reynolds number
$R_n(\ )$	eigenfunctions
r	radial coordinate in the cylindrical coordinate system, m
r_i, r_o	inner and outer tube radii of a concentric annular duct, m
r^*	aspect ratio r_i/r_o of a concentric or eccentric annular duct
r_m^*	radius in annular ducts where maximum velocity is achieved, m
R	radius of curvature, m

S	dimensionless parameter for eccentric annular duct; thermal energy source function, rate of thermal energy generated per unit volume of the fluid, W/m^3
s	duct dimension, m
S^*	thermal energy source number, $= SD_h^2/k(T_{w,m} - T_e)$ for $\textcircled{T}$ boundary condition; $= SD_h/q''$ for $\textcircled{H}$ boundary conditions
Sk	Stark number $= \varepsilon_w \sigma T^3 D_h/k$
T	fluid temperature, K
T_a	ambient fluid temperature, K,
T_m	fluid bulk mean temperature, K
T_w	wall temperature at the inside duct periphery, K
$T_{w,m}$	circumferentially averaged wall temperature, K
$T_{w,\max}^*$	dimensionless maximum wall temperature
$T_{w,\min}^*$	dimensionless minimum wall temperature
$\textcircled{T}$	uniform wall temperature boundary condition
$\textcircled{T3}$	convection boundary condition
$\textcircled{T4}$	radiative boundary condition
u	fluid velocity, fluid axial velocity in x direction, m/s
u_m	fluid mean axial velocity, m/s
$u_{\max}$	fluid maximum axial velocity for fully developed flow, m/s
u_t	turbulent friction or shear velocity $= \sqrt{\tau_w/\rho}$, m/s
u^+	wall coordinate $= u/u_t$, dimensionless
v	fluid velocity component in y or r direction, m/s
w	fluid velocity component in the z or θ direction, m/s
x	axial (streamwise) coordinate in the Cartesian or cylindrical coordinate system, m
x^+	dimensionless axial coordinate for the hydrodynamic entrance region, $= x/D_h$ Re
x^*	dimensionless axial coordinate for the thermal entrance region, $= x/D_h$ Pe
X_L	twist ratio
y	Cartesian coordinate across the flow cross section, m; distance measured from the duct wall, m
y^+	wall coordinate $= yu_t/v$
$\bar{y}$	distance of the centroid of the duct cross section measured from the base, m
$y_{\max}$	normal distance from the base to the point where umax occurs in the duct cross section, m
z	Cartesian coordinates across the flow cross section, m; distance from the apex of a triangle, m

Greek Symbols

α	fluid thermal diffusivity $= k/\rho c_p$, m^2/s
α^*	aspect ratio of a rectangular channel $= 2b/2a$; ratio of the minor axis to the major axis of an elliptic duct, $2b/2a$
β	coefficient of thermal expansion, 1/K
β_n	eigenvalues

$\Gamma(\ )$	gamma function
γ	dimensionless parameter defined by Eq. 5.24; ratio of heat fluxes at two walls of a parallel plate duct
δ	hydrodynamic boundary layer thickness, m; thickness of a twisted tape, m
δ_w	duct wall thickness, m
ε	distance between centers of two circles of an eccentric annular duct, m; amplitude of a circular duct with sinusoidal corrugations, m; roughness of duct wall, m
ε_w	emissivity of the duct wall material; eddy diffusivity, m²/s
Θ	dimensionless fluid temperature for the boundary condition of axially constant wall heat flux, $= (T - T_e)/q_w'' D_h/k$
θ_m	dimensionless fluid bulk mean temperature $= (T_m - T_w)/(T_e - T_w)$
$\theta_j^{(k)}$	dimensionless fluid temperature for a doubly connected duct, defined in Shah and London [1]
$\theta_{lj}^{(k)}$	dimensionless circumferentially averaged wall temperature ($l = i$ for the inner wall, $l = o$ for outer wall) for the fundamental boundary condition of kind k when the inner or outer wall ($j = i$ or o) is heated or cooled; dimensionless fluid bulk mean temperature if $l = m$
$\theta_{mj}^{(k)}$	fluid bulk mean temperature for the fundamental boundary condition of kind k when the inner wall ($j = i$) or outer wall ($j = o$) is heated or cooled
θ_i^*	influence coefficients derived from the fundamental solutions of the second kind, $= (\theta_{mo}^{(2)} - \theta_{io}^{(2)})/(\theta_{ii}^{(2)} - \theta_{mi}^{(2)})$
θ_o^*	influence coefficients derived from the fundamental solutions of the second kind, $= (\theta_{mi}^{(2)} - \theta_{oi}^{(2)})/(\theta_{oo}^{(2)} - \theta_{mo}^{(2)})$
μ	fluid dynamic viscosity coefficient, Pa·s
ν	fluid kinematic viscosity coefficient $= \mu/\rho$, m²/s
ρ	fluid density, kg/m³
σ	Stefan-Boltzmann constant $= 5.6697 \times 10^{-8}$ W/(m²·K⁴)
τ	wall shear stress, Pa
$\Phi_j^{(k)}$	dimensionless heat flux at a point in the flow field for the jth wall of a doubly connected duct, defined in Ref. 1
$\Phi_{lj}^{(k)}$	dimensionless wall heat flow defined in a manner similar to $\theta_{lj}^{(k)} = q_l'' D_h / k(T_j - T_e)$ for $k = 1, 3$; $= q''/q_{lj}''$ for $k = 2, 4$
$\Phi_{m,T}$	dimensionless mean wall heat flux for boundary condition of axially constant wall temperature, $= q'' D_h/k(T_w - T_e)$
$\Phi_{x,T}$	dimensionless local wall heat flux for boundary condition of axially constant wall temperature, $= q_x'' D_h/k(T_w - T_e)$
$\Phi_i^{(k)}, \Phi_o^{(k)}$	dimensionless heat flux at a point in the flow field for the inner or outer wall of a concentric or eccentric annular duct
ϕ	apex angle or half-apex angle of a duct; angle of tube curvature
Φ	coefficient, defined by Eq. 5.250

Subscripts

bc	thermal boundary condition
c	center, centroid, or curved
d	finned duct
e	initial value at the entrance of the duct or where the heat transfer begins

f	fluid
fd	fully developed flow
finless	finless duct
H	Ⓗ boundary condition
H1	Ⓗ¹ boundary condition
H2	Ⓗ² boundary condition
H4	Ⓗ⁴ boundary condition
H5	Ⓗ⁵ boundary condition
hy	hydrodynamic
i	inner surface of a doubly connected duct
in	inlet
j	heated wall of a doubly connected duct, $= i$ or o
l	laminar flow
m	mean
max	maximum
min	minimum
o	outer surface of a doubly connected duct
p	peripheral value
s	smooth, straight duct
slug	slug flow
T	Ⓣ boundary condition
T3	Ⓣ³ boundary condition
T4	Ⓣ⁴ boundary conditions
t	turbulent
th	thermal
x	an arbitrary section along the duct length; a local value as opposed to a mean value; axial
w	wall or fluid at the wall
∞	fully developed value at $x = \infty$

REFERENCES

1. R. K. Shah, and A. L. London, "Laminar Flow Forced Convection in Ducts," *Supplement 1 to Advances in Heat Transfer,* eds. T. F. Irvine and J. P. Hartnett, Academic Press, New York, 1978.

2. R. K. Shah, and M. S. Bhatti, "Laminar Convection Heat Transfer in Ducts," *Handbook of Single-Phase Convective Heat Transfer,* eds. S. Kakaç, R. K. Shah, and W. Aung, Wiley-Interscience, John Wiley & Sons, New York, 1987.

3. M. S. Bhatti, "Fully Developed Temperature Distribution in a Circular Tube with Uniform Wall Temperature," unpublished paper, Owens-Corning Fiberglass Corporation, Granville, Ohio, 1985.

4. M. L. Michelsen, and J. Villadsen, "The Graetz Problem with Axial Heat Conduction," *Int. J. Heat Mass Transfer,* (17): 1391–1402, 1974.

5. J. W. Ou, and K. C. Cheng, "Viscous Dissipation Effects on Thermal Entrance Heat Transfer in Laminar and Turbulent Pipe Flows with Uniform Wall Temperature," *AIAA,* paper no. 74-743 or *ASME* paper no. 74-HT-50, 1974.

6. V. P. Tyagi, "Laminar Forced Convection of a Dissipative Fluid in a Channel," *J. Heat Transfer,* (88): 161–169, 1966.

7. H. J. Hickman, "An Asymptotic Study of the Nusselt-Graetz Problem, Part 1: Large *x* Behavior," *J. Heat Transfer,* (96): 354–358, 1974.

8. Y. S. Kadaner, Y. P. Rassadkin, and E. L. Spektor, "Heat Transfer in Laminar Liquid Flow through a Pipe Cooled by Radiation," *Heat Transfer-Sov. Res.,* (3/5): 182–188, 1971.

9. S. Piva, "An Analytical Approach to Fully Developed Heating of Laminar Flows in Circular Pipes," *Int. Comm. Heat Mass Transfer,* (22/6): 815–824, 1995.

10. R. W. Hornbeck, "Laminar Flow in the Entrance Region of a Pipe," *Appl. Sci. Res.,* (A13): 224–232, 1964.

11. R. Y. Chen, "Flow in the Entrance Region at Low Reynolds Numbers," *J. Fluids Eng.,* (95): 153–158, 1973.

12. H. L. Weissberg, "End Correction for Slow Viscous Flow through Long Tubes," *Phys. Fluids,* (5): 1033–1036, 1962.

13. J. H. Linehan, and S. R. Hirsch, "Entrance Correction for Creeping Flow in Short Tubes," *J. Fluids Eng.,* (99): 778–779, 1977.

14. L. Graetz, "Über die Wärmeleitungs Fähigkeit von Flüssigkeiten (On the Thermal Conductivity of Liquids)," part 1, *Ann Phys. Chem.,* (18): 79–94, 1883; part 2, *Ann. Phys. Chem.,* (25): 337–357, 1885.

15. W. Nusselt, "Die Abhängigkeit der Wärmübergangszahl von der Rohrlänge (The Dependence of the Heat-Transfer Coefficient on the Tube Length)", *VDIZ,* (54): 1154–1158, 1910.

16. J. Newman, "The Graetz Problem," *The Fundamental Principles of Current Distribution and Mass Transfer in Electrochemical Cells,* ed. A. J. Bard, vol. 6, Dekker, New York, pp. 187–352, 1973.

17. G. M. Brown, "Heat or Mass Transfer in a Fluid in Laminar Flow in a Circular or Flat Conduit," *AIChE J.,* (6): 179–183, 1960.

18. B. K. Larkin, "High-Order Eigenfunctions of the Graetz Problem," *AIChE J.,* (7): 530, 1961.

19. A. Lévêque, "Les Lois de la Transmission de Chaleur par Convection," *Ann. Mines, Mem.,* ser. 12, (13): 201–299, 305–362, 381–415, 1928.

20. H. Hausen, "Dartellung des Wärmeüberganges in Rohren durch verallgemeinerte Potenzbeziehyngen," *VDIZ., Suppl.* "Verfahrenstechnik," (4): 91–98, 1943.

21. C. Laohakul, C. Y. Chan, K. Y. Look, and C. W. Tan, "On Approximate Solutions of the Graetz Problem with Axial Conduction," *Int. J. Heat Mass Transfer,* (28): 541–545, 1985.

22. M. A. Ebadian, and H. Y. Zhang, "An Exact Solution of Extended Graetz Problem with Axial Heat Conduction," *Int. J. Heat Mass Transfer,* (32): 1709–1717, 1989.

23. T. V. Nguyen, "Laminar Heat Transfer for Thermally Developing Flow in Ducts," *Int. J. Heat Mass Transfer,* (35): 1733–1741, 1992.

24. D. K. Hennecke, "Heat Transfer by Hagen-Poiseuille Flow in the Thermal Development Region with Axial Conduction," *Wärme-Stoffübertrag.,* (1): 177–184, 1968.

25. R. Siegel, E. M. Sparrow, and T. M. Hallman, "Steady Laminar Heat Transfer in a Circular Tube with Prescribed Wall Heat Flux," *Appl. Sci. Res.,* (A7): 386–392, 1958.

26. C. J. Hsu, "Heat Transfer in a Round Tube with Sinusoidal Wall Heat Flux Distribution," *AIChE J.,* (11): 690–695, 1965.

27. H. C. Brinkmann, "Heat Effects in Capillary Flow," *Appl., Sci. Res.,* (A2): 120–124, 1951.

28. J. W. Ou, and K. C. Cheng, "Viscous Dissipation Effects on Thermal Entrance Region Heat Transfer in Pipes with Uniform Wall Heat Flux," *Appl. Sci. Res.,* (28): 289–301, 1973.

29. T. Basu, and D. N. Roy, "Laminar Heat Transfer in a Tube with Viscous Dissipation," *Int. J. Heat Mass Transfer,* (28): 699–701, 1985.

30. C. J. Hsu, "Exact Solution to Entry-Region Laminar Heat Transfer with Axial Conduction and Boundary Conditions of the Third Kind," *Chem. Eng. Sci.,* (23): 457–468, 1968.

31. T. F. Lin, K. H. Hawks, and W. Leidenfrost, "Analysis of Viscous Dissipation Effect on Thermal Entrance Heat Transfer in Laminar Pipe Flows with Convective Boundary Conditions," *Wärme-und Stoffübertragung,* (17): 97–105, 1983.

32. M. K. Jensen, "Simultaneously Developing Laminar Flow in an Isothermal Circular Tube," *Int. Comm. Heat Mass Transfer,* (16): 811–820, 1989.

33. H. H. Al-Ali, and M. S. Selim, "Analysis of Laminar Flow Forced Convection Heat Transfer with Uniform Heating in the Entrance Region of a Circular Tube," *Can. J. Chem. Eng.,* (70): 1101–1107, 1992.

34. M. S. Bhatti, "Limiting Laminar Heat Transfer in Circular and Flat Ducts by Analogy with Transient Heat Conduction Problems," unpublished paper, Owens-Corning Fiberglass Corporation, Granville, Ohio, 1985.

35. G. Pagliarini, "Steady Laminar Heat Transfer in the Entry Region of Circular Tubes with Axial Diffusion of Heat and Momentum," *Int. J. Heat Mass Transfer,* (32/6): 1037–1052, 1989.

36. A. Barletta, "On Forced Convection in a Circular Duct with Slug Flow and Viscous Dissipation," *Int. Comm. Heat Mass Transfer,* (23): 69–78, 1996.

37. V. Javeri, "Simultaneous Development of the Laminar Velocity and Temperature Fields in a Circular Duct for the Temperature Boundary Condition of the Third Kind," *Int. J. Heat Mass Transfer,* (19): 943–949, 1976.

38. G. S. Barozzi, and G. Pagliarini, "A Method to Solve Conjugate Heat Transfer Problems: The Case of Fully Developed Laminar Flow in a Pipe," *J. Heat Transfer,* (107): 77–83, 1985.

39. J. C. Kuo, and T. F. Lin, "Steady Conjugate Heat Transfer in Fully Developed Laminar Pipe Flows," *J. of Thermophysics and Heat Transfer,* (2/3): 281–283, 1988.

40. G. Pagliarini, "Conjugate Heat Transfer for Simultaneously Developing Laminar Flow in a Circular Tube," *J. of Heat Transfer,* (113): 763–766, 1991.

41. W. Pfenniger, "Experiments with Laminar Flow in the Inlet Length of a Tube at High Reynolds Numbers With and Without Boundary Layer Suction," technical report, Northrop Aircraft Inc., Hawthorne, California, 1952.

42. L. Prandtl, "Über den Reibungswiderstand Strömender Luft," *Ergebnisse der Aerodynamics Versuchstalt zu Göttingen,* edition 3, pp. 1–5, 1927.

43. J. Nikuradse, "Gesetzmässigkeiten der turbulenten Strömung in glatten Rohren," *Forsch. Arb. Ing. Wes.,* (356), 1932; English translation, *NASA TT F-10,* 359, 1966.

44. L. Prandtl, "Neuere Ergebnisse der Turbulenzforschung," *VDIZ.,* (77/5): 105–114, 1933; English translation *NACA TM 7320,* 1933.

45. M. S. Bhatti, and R. K. Shah, "Turbulent and Transition Flow Convective Heat Transfer in Ducts," *Handbook of Single-Phase Convective Heat Transfer,* eds. S. Kakaç, R. K. Shah, and W. Aung, Wiley-Interscience, New York, 1987.

46. T. von Kármán, "Mechanische Ähnlichkeit und Turbulenz," *Nachr. Ges. Wiss Göttingen, Math. Phys. Klasse,* (5): 58–76, 1930; English translation, *NACA TM 611,* 1931.

47. C. Wang, "On the Velocity Distribution of Turbulent Flow in Pipes and Channels of Constant Cross Section," *J. Appl. Mech.,* (68): A85–A90, 1946.

48. H. Darcy, "Recherches Expérimentales Relatives Aux Mouvements de l'Eau Dans Tuyauz," *Mem. Prés. Acad. des Sci. Inst. Frances,* (15): 141, 1858.

49. H. Blasius, "Das Ähnlichkeitsgesetz bei Reibungsvorgängen in Flüssigkeiten," *Forschg. Arb. Ing.-Wes.,* (131): Berlin, 1913.

50. W. H. McAdams, *Heat Transmission,* 3d ed., McGraw-Hill, New York, 1954.

51. T. B. Drew, E. C. Koo, and W. H. McAdams, "The Friction Factor for Clean Round Pipes," *Trans. AIChE,* (28): 56–72, 1932.

52. L. Prandtl, *Führrer durch die Stömungslehre,* Vieweg, Braunschweig, p. 359, 1944.

53. T. von Kármán, "Turbulence and Skin Friction," *J. Aerosp. Sci.,* (7): 1–20, 1934.

54. C. F. Colebrook, "Turbulent Flow in Pipes with Particular Reference to the Transition Region between the Smooth and Rough Pipes Laws," *J. Inst. Civil Eng.,* (11): 133–156, 1939.

55. G. K. Filonenko, "Hydraulic Resistance in Pipes (in Russia), *Teplonergetika,* (1/4): 40–44, 1954.

56. R. Techo, R. R. Tickner, and R. E. James, "An Accurate Equation for the Computation of the Friction Factor for Smooth Pipes from the Reynolds Number," *J. Appl. Mech.* (32): 443, 1965.

57. H. Schlichting, *Boundary Layer Theory,* 7th ed., McGraw-Hill, New York, 1979.

58. L. F. Moody, "Friction Factors for Pipe Flow," *Trans. ASME,* (66): 671–684, 1944.

59. M. R. Rao, "Forced Convection Heat Transfer and Fluid Friction in Fully Developed Turbulent Flow in Smooth and Rough Tubes," *Indian ASI,* pp. 3.8–3.23, 1989.

60. J. Nikuradse, "Strömungsgesetze in rauhen Rohren," *Forsch. Arb. Ing.-Wes.,* (361), 1933; English translation, *NACA TM 1292.*

61. D. J. Wood, "An Explicit Friction Factor Relationship," *Civ. Eng.,* (36): 60–61, 1966.

62. P. K. Swamee, and A. K. Jain, "Explicit Equation for Pipe-Flow Problems," *J. Hydraulic Div. ASCE,* (102): 657–664, 1976.

63. S. W. Churchill, "Friction Factor Equation Spans All Fluid Flow Regimes," *Chem. Eng.,* 91–92, 1977.

64. N. H. Chen, "An Explicit Equation for Friction Factor in Pipe," *Ind. Eng. Chem. Fund.,* (18): 296–297, 1979.

65. G. F. Round, "An Explicit Approximation for the Friction Factor–Reynolds Number Relation for Rough and Smooth Pipes," *Can. J. Chem. Eng.,* (58): 122–123, 1980.

66. D. J. Zigrang, and N. D. Sylvester, "Explicit Approximations to the Solution of Colebrook's Friction Factor Equation," *AIChE J.,* (28): 514–515, 1982.

67. S. E. Haaland, "Simple and Explicit Formulas for the Friction Factor in Turbulent Pipe Flow," *J. Fluids Eng.,* (105): 89–90, 1983.

68. T. K. Serghides, "Estimate Friction Factor Accurately," *Chem. Eng.,* (91): 63–64, 1984.

69. V. Gnielinski, "New Equations for Heat and Mass Transfer in Turbulent Pipe and Channel Flow," *Int. Chem. Eng.,* (16): 359–368, 1976.

70. P. W. Dittus, and L. M. K. Boelter, "Heat Transfer in Automobile Radiators of the Tubular Type," *Univ. Calif. Pub. Eng.,* (2/13): 443–461, 1930; reprinted in *Int. Comm. Heat Mass Transfer,* (12): 3–22, 1985.

71. A. P. Colburn, "A Method of Correlating Forced Convection Heat Transfer Data and a Comparison With Fluid Friction," *Trans. AIChE,* (19): 174–210, 1933; reprinted in *Int. J. Heat Mass Transfer,* (7): 1359–1384, 1964.

72. T. von Kármán, "The Analogy Between Fluid Friction and Heat Transfer," *Trans. ASME,* (61): 705–710, 1939.

73. R. E. Drexel, and W. H. McAdams, "Heat Transfer Coefficients for Air Flowing in Round Tubes, and Around Finned Cylinders," *NACA ARR No. 4F28;* also *Wartime Report W-108,* 1945.

74. W. L. Friend, and A. B. Metzner, "Turbulent Heat Transfer inside Tubes and the Analogy among Heat, Mass, and Momentum Transfer," *AIChE J.,* (4): 393–402, 1958.

75. B. S. Petukhov, and V. V. Kirillov, "The Problem of Heat Exchange in the Turbulent Flow of Liquids in Tubes," (in Russian) *Teploenergetika,* (4/4): 63–68, 1958; see also B. S. Petukhov and V. N. Popov, "Theoretical Calculation of Heat Exchange in Turbulent Flow in Tubes of an Incompressible Fluid with Variable Physical Properties," *High Temp.,* (1/1): 69–83, 1963.

76. H. Hausen, "Neue Gleichungen für die Wärmeübertragung bei freier oder erzwungener Stromung," *Allg. Warmetchn.,* (9): 75–79, 1959.

77. R. L. Webb, "A Critical Evaluation of Analytical Solutions and Reynolds Analogy Equations for Turbulent Heat and Mass Transfer in Short Tubes," *Wärme-und Stóffübertrag.,* (4): 197–204, 1971.

78. E. N. Sieder, and G. E. Tate, "Heat Transfer and Pressure Drop of Liquids in Tubes," *Ind. Eng. Chem,* (28): 1429–1436, 1936.

79. O. C. Sandall, O. T. Hanna, and P. R. Mazet, "A New Theoretical Formula for Turbulent Heat and Mass Transfer with Gases or Liquids in Tube Flow," *Can. J. Chem. Eng.,* (58): 443–447, 1980.

80. R. H. Notter, and C. A. Sleicher, "A Solution to the Turbulent Graetz Problem III. Fully Developed and Entry Region Heat Transfer Rates," *Chem. Eng. Sci.,* (27): 2073–2093, 1972.

81. R. C. Martinelli, "Heat Transfer to Molten Metals," *Trans. ASME,* (69): 947–959, 1947.

82. W. Nunner, "Wärmeübergang und Druckabfall in Rauhen Rohren," *VDI-Forschungsheft 445,* ser. B, (22): 5–39, 1956; English translation, (786), *Atomic Energy Research Establishment,* Harwell, United Kingdom.

83. D. F. Dipprey, and R. H. Sabersky, "Heat and Momentum Transfer in Smooth and Rough Tubes at Various Prandtl Numbers," *Int. J. Heat Mass Transfer,* (6): 329–353, 1963.

84. R. A. Gowen, and J. W. Smith, "Turbulent Heat Transfer from Smooth and Rough Surfaces," *Int. J. Heat Mass Transfer,* (11): 1657–1673, 1968.

85. Y. Kawase, and J. J. Ulbrecht, "Turbulent Heat and Mass Transfer in Dilute Polymer Solutions," *Chem. Eng. Sci.,* (37): 1039–1046, 1982.

86. Y. Kawase, and A. De, "Turbulent Heat and Mass Transfer in Newtonian and Dilute Polymer Solutions Flowing through Rough Tubes," *Int. J. Heat Mass Transfer,* (27): 140–142, 1984.

87. W. Zhiqing, "Study on Correction Coefficients of Laminar and Turbulent Entrance Region Effect in Round Pipe," *Appl. Math. Mech.,* (3/3): 433–446, 1982.

88. A. R. Barbin, and J. B. Jones, "Turbulent Flow in the Inlet Region of a Smooth Pipe," *J. Basic Eng.,* (85): 29–34, 1963.

89. H. Latzko, "Der Wärmeübergang and einen turbulenten Flüssigkeitsoder Gasstrom," *Z. Angew. Math. Mech.,* (1/4): 268–290, 1921; English translation, *NACA TM 1068,* 1944.

90. C. A. Sleicher, and M. Tribus, "Heat Transfer in a Pipe with Turbulent Flow and Arbitrary Wall-Temperature Distribution," *1956 Heat Transfer and Fluid Mechanics Institute,* Stanford University Press, Stanford, pp. 59–78, 1956.

91. H. L. Becker, "Heat Transfer in Turbulent Tube Flow," *Appl. Sci. Res.,* ser. A, (6): 147–191, 1956.

92. R. G. Deissler, "Analysis of Turbulent Heat Transfer and Flow in the Entrance Regions of Smooth Passages," *NACA TN 3016,* 1953.

93. E. M. Sparrow, T. M. Hallman, and R. Siegel, "Turbulent Heat Transfer in the Thermal Entrance Region of a Pipe with Uniform Heat Flux," *Appl. Sci. Res.,* ser. A, (7): 37–52, 1957.

94. J. A. Malina, and E. M. Sparrow, "Variable-Property, Constant-Property, and Entrance Region Heat Transfer Results for Turbulent Flow of Water and Oil in a Circular Tube," *Chem. Eng. Sci.,* (19): 953–962, 1964.

95. M. Al-Arabi, "Turbulent Heat Transfer in the Entrance Region of a Tube," *Heat Transfer Eng.,* (3): 76–83, 1982.

96. C. J. Chen, and J. S. Chiou, "Laminar and Turbulent Heat Transfer in the Pipe Entrance Region for Liquid Metals," *Int. J. Heat Mass Transfer,* (24): 1179–1189, 1981.

97. L. G. Genin, E. V. Kudryavtseva, Y. A. Pakhotin, and V. G. Svindov, "Temperature Fields and Heat Transfer for a Turbulent Flow of Liquid Metal on an Initial Thermal Section," *Teplofiz. Vysokikh Temp.,* (16/6): 1243–1249, 1978.

98. L. M. K. Boelter, G. Young, and H. W. Iverson, "An Investigation of Aircraft Heaters XXVII—Distribution of Heat Transfer Rate in the Entrance Section of a Circular Tube," *NACA TN 1451,* 1948.

99. A. F. Mills, "Experimental Investigation of Turbulent Heat Transfer in the Entrance Region of a Circular Conduit," *J. Mech. Eng. Sci.,* (4): 63–77, 1962.

100. S. W. Churchill, "Comprehensive Correlating Equations for Heat, Mass and Momentum Transfer in Fully Developed Flow in Smooth Tubes," *Ind. Eng. Chem. Found.,* (16/1): 109–116, 1977.

101. V. E. Kaupas, P. S. Poskas, and J. V. Vilemas, "Heat Transfer to a Transition-Range Gas Flow in a Pipe at High Heat Fluxes (2. Heat Transfer in Laminar to Turbulent Flow Transition)," *Heat Transfer-Sov. Res.,* (21/3): 340–351, 1989.

102. N. M. Natarajan, and S. M. Lakshmanan, "Laminar Flow through Annuli: Analytical Method of Pressure Drop," *Indian Chem. Eng.,* (5/3): 50–53, 1973.

103. R. K. Shah, "A Correlation for Laminar Hydrodynamic Entry Length Solutions for Circular and Noncircular Ducts," *J. Fluids Eng.,* (100): 177–179, 1978.

104. S. Kakaç, and O. Yücel, *Laminar Flow Heat Transfer in an Annulus with Simultaneous Development of Velocity and Temperature Fields,* Technical and Scientific Council of Turkey, TUBITAK, ISITEK No. 19, Ankara, Turkey, 1974.

105. N. A. V. Piercy, M. S. Hooper, and H. F. Winny, "Viscous Flow through Pipes with Cores," London Edinburgh Dublin, *Philos. Mag. J. Sci.,* (15): 647–676, 1933.

106. W. Tiedt, "Berechnung des laniinaren und turbulenten Reibungswiderstandes konzentrischer und exzentrischer Ringspalte," part I, *Chem.-Ztg. Chem. Appar.,* (90): 813–821, 1966: part II, *Chem.-Ztg. Chem. Appar.,* (91): 17–25, 1967.

107. E. Becker, "Strömungsvorgänge in ringförmigen Spalten und ihre Beziehung zum Poiseuilleschen Gesetz," *Forsch. Geb. Ingenieurwes., VDI,* (48): 1907.

108. K. C. Cheng, and G. J. Hwang, "Laminar Forced Convection in Eccentric Annuli," *AIChE J.,* (14): 510–512, 1968.

109. R. W. Hanks, "The Laminar-Turbulent Transition for Flow in Pipes, Concentric Annuli, and Parallel Plates," *AIChE J.,* (9): 45–48, 1963.

110. J. G. Knudsen, and D. L. Katz, *Fluid Dynamics and Heat Transfer,* Robert E. Kneger, Huntingdon, N.Y., pp. 191–193, 1979.

111. W. M. Kays, and E. Y. Leung, "Heat Transfer in Annular Passages: Hydrodynamically Developed Turbulent Flow with Arbitrarily Prescribed Heat Flux," *Int. J. Heat Mass Transfer,* (6): 537–557, 1963.

112. O. C. Jones Jr., and J. C. M. Leung, "An Improvement in the Calculation of Turbulent Friction in Smooth Concentric Annuli," *J. Fluids Eng.,* (103): 615–623, 1981.

113. O. E. Dwyer, "Eddy Transport in Liquid Metal Heat Transfer," *AIChE J.,* (9): 261–268, 1963.

114. R. R. Rothfus, C. C. Monrad, K. G. Sikchi, and W. J. Heideger, "Isothermal Skin Friction in Flow through Annular Sections," *Ind. Eng. Chem.,* (47): 913–918, 1955.

115. R. M. Olson, and E. M. Sparrow, "Measurements of Turbulent Flow Development in Tubes and Annuli with Square or Rounded Entrances," *AIChE J.,* (9): 766–770, 1963.

116. T. H. Okiishi, and G. K. Serovy, "An Experimental Study of the Turbulent Flow Boundary Layer Development in Smooth Annuli," *J. Basic Eng.,* (89): 823–836, 1967.

117. A. Quarmby, and R. K. Anand, "Turbulent Heat Transfer in the Thermal Entrance Region of Concentric Annuli with Uniform Wall Heat Flux," *Int. J. Heat Mass Transfer,* (13): 395–411, 1970.

118. A. Roberts and H. Barrow, "Turbulent Heat Transfer in the Thermal Entrance Region of an Internally Heated Annulus," *Proc. Inst. Mech. Engrs.,* (182/3H): 268–276, 1967.

119. V. K. Jonsson, and E. M. Sparrow, "Experiments on Turbulent Flow Phenomena in Eccentric Annular Ducts," *J. Fluid Mech.,* (25): 65–68, 1966.

120. V. K. Jonsson, "Experimental Studies of Turbulent Flow Phenomena in Eccentric Annuli," Ph.D. thesis, University of Minnesota, Minneapolis, 1965.

121. Y. Lee, and H. Barrow, "Turbulent Flow and Heat Transfer in Concentric and Eccentric Annuli," *Proc. Thermodyn. and Fluid Mech. Convention,* Institute of Mechanical Engineers, London, paper no. 12, 1964.

122. R. G. Deissler, and M. F. Taylor, "Analysis of Fully Developed Turbulent Heat Transfer in an Annulus of Various Eccentricities," *NACA TN 3451,* 1955.

123. W. S. Yu, and O. E. Dwyer, "Heat Transfer to Liquid Metals Flowing Turbulently in Eccentric Annuli—I," *Nucl. Sci. Eng.,* (24): 105–117, 1966.

124. W. Ricker, J. H. T. Wade, and W. Wilson, "On the Velocity Fields in Eccentric Annuli," *ASME,* paper 68-WA/FE-35, 1968.

125. R. L. Judd, and J. H. T. Wade, "Forced Convection Heat Transfer in Eccentric Annular Passages," Heat Transfer and Fluid Mechanics Institute, Stanford University, Stanford, pp. 272–288, 1963.

126. E. Y. Leung, W. M. Kays, and W. C. Reynolds, "Heat Transfer with Turbulent Flow in Concentric and Eccentric Annuli With Constant and Variable Heat Flux," *TR AHT-4, Mech. Eng. Dept.,* Stanford University, Stanford, 1962.

127. K. C. Cheng, and R. S. Wu, "Viscous Dissipation Effects on Convective Instability and Heat Transfer in Plane Poiseuille Flow Heated from Below," *Appl. Sci. Res.,* (32), 327–346, 1976.

128. J. W. Ou, and K. C. Cheng, "Effects of Pressure Work and Viscous Dissipation on Graetz Problem for Gas Flow in Parallel-Plate Channels," *Wärme-Stoffübertrag.,* (6): 191–198, 1973.

129. S. Pahor, and J. Strand, "A Note on Heat Transfer in Laminar Flow through a Gap," *Appl. Sci Res.,* (A10): 81–84, 1961.

130. C. C. Grosjean, S. Pahor, and J. Strand, "Heat Transfer in Laminar Flow through a Gap," *Appl. Sci. Res.,* (A11): 292–294, 1963.

131. W. Nusselt, "Der Wärmeaustausch am Berieselungskühler," *VDIZ.,* (67): 206–210, 1923.

132. R. D. Cess, and E. C. Shaffer, "Heat Transfer to Laminar Flow between Parallel Plates with a Prescribed Wall Heat Flux," *Appl. Sci. Res.,* (A8): 339–344, 1959.

133. E. M. Sparrow, and S. H. Lin, "Turbulent Heat Transfer in a Parallel-Plate Channel," *Int. J. Heat Mass Transfer,* (6): 248–249, 1963.

134. A. S. Jones, "Two-Dimensional Adiabatic Forced Convection at Low Péclet Number," *Appl. Sci. Res.,* (125): 337–348, 1972.

135. C. J. Hsu, "An Exact Analysis of Low Péclet Number Thermal Entry Region Heat Transfer in Transversely Nonuniform Velocity Fields," *AIChE J.,* (17): 732–740, 1971.

136. C. L. Hwang, and L. T. Fan, "Finite Difference Analysis of Forced Convection Heat Transfer in Entrance Region of a Flat Rectangular Duct," *Appl. Sci. Res.,* (A13): 401–422, 1964.

137. K. Stephan, "Wärmeübergand und druckabfall bei nicht ausgebildeter Laminarströmung in Rohren und in ebenen Spalten," *Chem-Ing-Tech.,* (31): 773–778, 1959.

138. M. S. Bhatti, and C. W. Savery, "Heat Transfer in the Entrance Region of a Straight Channel; Laminar Flow With Uniform Wall Temperature," *J. Heat Transfer,* (100): 539–542, 1978.

139. R. Das, and A. K. Mohanty, "Forced Convection Heat Transfer in the Entrance Region of a Parallel Plate Channel," *Int. J. Heat Mass Transfer,* (26): 1403–1405, 1983.

140. H. S. Heaton, W. C. Reynolds, and W. M. Kays, "Heat Transfer in Annular Passages: Simultaneous Development of Velocity and Temperature Fields in Laminar Flow," *Int. J. Heat Mass Transfer,* (7): 763–781, 1964.

141. T. V. Nguyen, "Low Reynolds Number Simultaneously Developing Flows in the Entrance Region of Parallel Plates," *Int. J. Heat Mass Transfer,* (34): 1219–1225, 1991.

142. T. V. Nguyen, and I. L. MacLaine-Cross, "Simultaneously Developing Laminar Flow, Forced Convection in the Entrance Region of Parallel Plates," *J. Heat Transfer,* (113): 837–842, 1991.

143. G. S. Beavers, E. M. Sparrow, and R. A. Magnuson, "Experiments on the Breakdown of Laminar Flow in a Parallel-Plate Channel," *Int. J. Heat Mass Transfer,* (13): 809–815, 1970.

144. P. Hrycak, and R. Andrushkiw, "Calculation of Critical Reynolds Number in Round Pipes and Infinite Channels and Heat Transfer in Transition Regions," *Heat Transfer 1974,* (II): 183–187, 1974.

145. G. S. Beavers, E. M. Sparrow, and J. R. Lloyd, "Low Reynolds Number Flow in Large Aspect Ratio Rectangular Ducts," *J. Basic Eng.,* (93): 296–299, 1971.

146. R. B. Dean, "Reynolds Number Dependence of Skin Friction and Other Bulk Flow Variables in Two-Dimensional Rectangular Duct Flow," *J. Fluids Eng.,* (100): 215–223, 1978.

147. S. Kakaç, and S. Paykoc, "Analysis of Turbulent Forced Convection Heat Transfer between Parallel Plates," *J. Pure Appl. Sci.,* (1/1): 27–47, 1968.

148. L. Duchatelle, and L. Vautrey, "Determination des Coefficients de Convection d'un Alliage NaK en Ecoulement Turbulent Entre Plaques Planes Paralleles," *Int. J. Heat Mass Transfer,* (7): 1017–1031, 1964.

149. M. Sakakibara, and K. Endo, "Analysis of Heat Transfer for Turbulent Flow between Parallel Plates," *Int. Chem. Eng.,* (18): 728–733, 1976.

150. A. A. Shibani, and M. N. Özisik, "A Solution to Heat Transfer in Turbulent Flow between Parallel Plates," *Int. J. Heat Mass Transfer,* (20): 565–573, 1977.

151. A. P. Hatton, and A. Quarmby, "The Effect of Axially Varying and Unsymmetrical Boundary Conditions on Heat Transfer with Turbulent Flow between Parallel Plates," *Int. J. Heat Mass Transfer,* (6): 903–914, 1963.

152. A. P. Hatton, A. Quarmby, and I. Grundy, "Further Calculations on the Heat Transfer with Turbulent Flow between Parallel Plates," *Int. J. Heat Mass Transfer,* (7): 817–823, 1964.

153. M. Sakakibara, "Analysis of Heat Transfer in the Entrance Region with Fully Developed Turbulent Flow between Parallel Plates—The Case of Uniform Wall Heat Flux," *Mem. Fac. of Eng. Fukui Univ.,* (30/2): 107–120, 1982.

154. M. N. Özisik, R. M. Cotta, and W. S. Kim, "Heat Transfer in Turbulent Forced Convection between Parallel-Plates," *Can. J. Chem. Eng.,* (67): 771–776, 1989.

155. S. M. Marco, and L. S. Han, "A Note on Limiting Laminar Nusselt Number in Ducts with Constant Temperature Gradient by Analogy to Thin-Plate Theory," *Trans. ASME,* (77): 625–630, 1955.

156. H. F. P. Purday, *Streamline Flow,* Constable, London, 1949; same as *An Introduction to the Mechanics of Viscous Flow,* Dover, New York, 1949.

157. N. M. Natarajan, and S. M. Lakshmanan, "Laminar Flow in Rectangular Ducts: Prediction of Velocity Profiles and Friction Factor," *Indian J. Technol.,* (10): 435–438, 1972.

158. R. M. Curr, D. Sharma, and D. G. Tatchell, "Numerical Prediction of Some Three-Dimensional Boundary Layers in Ducts," *Comput. Methods Appl. Mech. Eng.,* (1): 143–158, 1972.

159. M. Tachibana, and Y. Iemoto, "Steady Laminar Flow in the Inlet Region of Rectangular Ducts," *Bull. JSME,* (24/193): 1151–1158, 1981.

160. P. Wibulswas, "Laminar Flow Heat Transfer in Non-Circular Ducts," Ph.D. thesis, London University, London, 1966.

161. J. B. Aparecido, and R. M. Cotta, "Thermally Developing Laminar Flow inside Rectangular Ducts," *Int. J. Heat Mass Transfer,* (33): 341–347, 1990.

162. A. R. Chandrupatla, and V. M. K. Sastri, "Laminar Forced Convection Heat Transfer of a Non-Newtonian Fluid in a Square Duct," *Int. J. Heat Mass Transfer,* (20): 1315–1324, 1977.

163. A. R. Chandrupatla, and V. M. K. Sastri, "Laminar Flow and Heat Transfer to a Non-Newtonian Fluid in an Entrance Region of a Square Duct with Prescribed Constant Axial Wall Heat Flux," *Numer. Heat Transfer,* (1): 243–254, 1978.

164. S. J. Davies, and C. M. White, "An Experimental Study of the Flow of Water in Pipes of Rectangular Section," *Proc. Roy. Soc.,* (A119): 92–107, 1928.

165. J. Allen, and N. D. Grunberg, "The Resistance to the Flow of Water along Smooth Rectangular Passages and the Effect of a Slight Convergence or Divergence of the Boundaries," *Philos. Mag., Ser.* (7): 490–502, 1937.

166. R. J. Cornish, "Flow in a Pipe of Rectangular Cross Section," *Proc. Roy. Soc. London,* (A120): 691–700, 1928.

167. J. P. Hartnett, and C. Y. Koh, and S. T. McComas, "A Comparison of Predicted and Measured Friction Factors for Turbulent Flow through Rectangular Ducts," *J. Heat Transfer,* (84): 82–88, 1962.

168. O. C. Jones Jr., "An Improvement in the Calculation of Turbulent Friction in Rectangular Ducts," *J. Fluids Eng.,* (98): 173–181, 1976.

169. J. P. Hartnet, and T. F. Irvine Jr., "Nusselt Values for Estimating Liquid Metal Heat Transfer in Noncircular Ducts," *AIChE J.,* (3): 313–317, 1957.

170. A. Haji-Sheikh, M. Mashena, and M. J. Haji-Sheikh, "Heat Transfer Coefficient in Ducts with Constant Wall Temperature," *J. Heat Transfer,* (105): 878–883, 1983.

171. K. C. Cheng, "Laminar Forced Convection in Regular Polygonal Ducts with Uniform Peripheral Heat Flux," *J. Heat Transfer,* (91): 156–157, 1969.

172. R. K. Shah, "Laminar Flow Friction and Forced Convection Heat Transfer in Ducts of Arbitrary Geometry," *Int. J. Heat Mass Transfer,* (18): 849–862, 1975.

173. V. K. Migay, "Hydraulic Resistance of Triangular Channels in Laminar Flow (in Russia)," *Izv. Vyssh. Uchebn. Zared. Energ.,* (6/5): 122–124, 1963.

174. E. M. Sparrow, and A. Haji-Sheikh, "Laminar Heat Transfer and Pressure Drop in Isosceles Triangular, Right Triangular, and Circular Sector Ducts," *J. Heat Transfer,* (87): 426–427, 1965.

175. M. Iqbal, A. K. Khatry, and B. D. Aggarwala, "On the Second Fundamental Problem of Combined Free and Forced Convection through Vertical Non-Circular Ducts," *Appl. Sci. Res.,* (26): 183–208, 1972.

176. C. A. C. Altemani, and E. M. Sparrow, "Turbulent Heat Transfer and Fluid Flow in an Unsymmetrically Heated Triangular Duct," *J. Heat Transfer,* (102): 590–597, 1980.

177. P. C. Bandopadhayay, and C. M. Ambrose, "A Generalized Length Dimension for Noncircular Ducts," *Lett. Heat Mass Transfer,* (7): 323–328, 1980.

178. J. Nikuradse, "Untersuchungen uber Turbulent Stromung in nicht kreisformigen Rohren," *Ing.-Arch.,* (1): 306–332, 1930.

179. L. Schiller, "Über den Strömungswiderstand von Rohren Verschiedenen Querschnitts und Rauhigkeitsgrades," *Z. Angew. Malh. Mech.,* (3): 2–13, 1923.

180. D. A. Campbell, and H. C. Perkins, "Variable Property Turbulent Heat and Momentum Transfer for Air in a Vertical Rounded Corner Triangular Duct," *Int. J. Heat Mass Transfer,* (11): 1003–1012, 1968.

181. M. S. Bhatti, "Laminar Flow in the Entrance Region of Elliptical Ducts," *J. Fluids Eng.,* (105): 290–296, 1983.

182. L. N. Tao, "On Some Laminar Forced-Convection Problems," *J. Heat Transfer,* (83): 466–472, 1961.

183. N. T. Dunwoody, "Thermal Results for Forced Convection through Elliptical Ducts," *J. Appl. Mech.,* (29): 165–170, 1962.

184. S. M. Richardson, "Leveque Solution for Flow in an Elliptical Duct," *Letters in Heat and Mass Transfer,* (7): 353–362, 1980.

185. R. P. Someswara, N. C. Ramacharyulu, and V. V. G. Krishnamurty, "Laminar Forced Convection in Elliptical Ducts," *Appl. Sci. Res.,* (21): 185–193, 1969.

186. H. Barrow, and A. Roberts, "Flow and Heat Transfer in Elliptic Ducts," *Heat Transfer 1970,* paper no. FC 4.1, Versailles, 1970.

187. D. Cain, and J. Duffy, "An Experimental Investigation on Turbulent Flow in Elliptical Ducts," *Int. J. Mech. Sci.,* (13): 451–459, 1971.

188. W. R. Dean, "Note on the Motion of a Fluid in a Curved Pipe," *Philos. Mag.,* ser. 7, (4): 208–223, 1927.

189. W. R. Dean, "The Streamline Motion of Fluid in a Curved Pipe," *Philos. Mag.,* ser. 7, (5/30): 673–695, 1928.

190. Y. Mori, and W. Nakayama, "Study on Forced Convective Heat Transfer in Curved Pipes (1st Report, Laminar Region)," *Int. J. Heat Mass Transfer,* (8): 67–82, 1965.

191. M. Adler, "Flow in a Curved Tube," *Z. Angew. Math. Mech.,* (14): 257–265, 1934.

192. S. V. Patankar, V. S. Pratap, and D. B. Spalding, "Prediction of Laminar Flow and Heat Transfer in Helically Coiled Pipes," *J. Fluid Mech.,* (62/3): 539–551, 1974.

193. P. S. Srinivasan, S. S. Nandapurkar, and S. S. Holland, "Friction Factors for Coils," *Trans. Inst. Chem. Eng.,* (48): T156–T161, 1970.

194. R. L. Manlapaz, and S. W. Churchill, "Fully Developed Laminar Flow in a Helically Coiled Tube of Finite Pitch," *Chem. Eng. Commun.,* (7): 57–78, 1980.

195. R. K. Shah, and S. D. Joshi, "Convective Heat Transfer in Curved Ducts," *Handbook of Single Phase Convective Heat Transfer,* eds. S. Kakaç, R. K. Shah, and W. Aung, Wiley Interscience, John Wiley & Sons, New York, 1987.

196. Y. Mori, and W. Nakayama, "Study on Forced Convective Heat Transfer in Curved Pipes (3rd Report, Theoretical Analysis under the Condition of Uniform Wall Temperature and Practical Formulae)," *Int. J. Heat Mass Transfer,* (10): 681–695, 1967.

197. J. M. Tarbell, and M. R. Samuels, "Momentum and Heat Transfer in Helical Coils," *Chem. Eng., J.—Lausanne (Netherlands),* (5): 117–127, 1973.

198. N. A. Dravid, K. A. Smith, E. W. Merrill, and P. L. T. Brian, "Effect of Secondary Fluid Motion on Laminar Flow Heat Transfer in Helically Coiled Tubes," *AIChE J.,* (17): 1114–1122, 1971.

199. M. Akiyama, and K. C. Cheng, "Laminar Forced Convection Heat Transfer in Curved Pipes with Uniform Wall Temperature," *Int. J. Heat Mass Transfer,* (15): 1426–1431, 1972.

200. C. E. Kalb, and J. D. Seader, "Fully Developed Viscous-Flow Heat Transfer in Curved Circular Tubes with Uniform Wall Temperature," *AIChE J.,* (20): 340–346, 1974.

201. R. L. Manlapaz, and S. W. Churchill, "Fully Developed Laminar Convection from a Helical Coil," *Chem. Eng. Commun.,* (9): 185–200, 1981.

202. V. Kubair, and N. R. Kuldor, "Heat Transfer to Newtonian Fluids in Spiral Coils at Constant Tube Wall Temperature in Laminar Flow," *Indian Journal Tech.,* (3): 144–146, 1965.

203. V. Kubair, and N. R. Kuldor, "Heat Transfer to Newtonian Fluids in Coiled Pipes in Laminar Flow," *Int. J. Heat Mass Transfer,* (9): 63–75, 1966.

204. C. E. Kalb, and J. D. Seader, "Heat Mass Transfer Phenomena for Viscous Flow in Curved Circular Tubes," *Int. J. Heat Mass Transfer,* (15): 801–817, 1972.

205. N. J. Rabadi, J. C. F. Chow, and H. A. Simon, "An Efficient Numerical Procedure for the Solution of Laminar Flow and Heat Transfer in Coiled Tubes," *Numer. Heat Transfer,* (2): 279–289, 1979.

206. G. Yang, Z. F. Dong, and M. A. Ebadian, "Convective Heat Transfer in a Helicoidal Pipe Heat Exchanger," *J. Heat Transfer,* (115): 796–800, 1993.

207. G. Yang, Z. F. Dong, and M. A. Ebadian, "Laminar Forced Convection in a Helicoidal Pipe with Finite Pitch," *Int. J. Heat Mass Transfer,* (38): 853–862, 1995.

208. D. S. Austen, and H. M. Soliman, "Laminar Flow and Heat Transfer in Helically Coiled Tubes with Substantial Pitch," *Experimental Thermal and Fluid Science,* (1): 183–194, 1988.

209. M. A. Abul-Hamayel, and K. J. Bell, "Heat Transfer in Helically Coiled Tubes with Laminar Flow," *ASME paper,* no. 79-WA/HT-11, 1979.

210. C. X. Lin, P. Zhang, and M. A. Ebadian, "Laminar Forced Convection in the Entrance Region of Helicoidal Pipes," *Int. J. Heat and Mass Transfer.* In press.

211. Z. F. Dong, and M. A. Ebadian, "Computer Simulation of Laminar and Turbulent Flow in Helicoidal Pipes," in *Computer Simulations in Compact Heat Exchanges,* B. Sunden and M. Faghri eds., Computational Mechanics Publications, Southampton, UK. In press.

212. S. Liu, and J. H. Masliyah, "Developing Convective Heat Transfer in Helicoidal Pipes with Finite Pitch," *Int. J. Heat and Fluid Flow,* (15/1): 66–74, 1994.

213. G. W. Hogg, "The Effect of Secondary Flow on Point Heat Transfer Coefficients for Turbulent Flow inside Curved Tubes," Ph.D. thesis, University of Idaho, Moscow, ID, 1968.

214. H. Ito, "Friction Factors for Turbulent Flow in Curved Pipes," *J. Basic Eng.,* (81): 123–134, 1959.

215. B. E. Boyce, J. G. Coiller, and J. Levy, "Hold Up and Pressure Drop Measurements in the Two Phase Flow of Air Water Mixtures in Helical Coils," *Co-current Gas Liquid Fluid,* Plenum Press, London, pp. 203–231, 1969.

216. G. F. C. Rogers, and Y. R. Mayhew, "Heat Transfer and Pressure Loss in Helically Coiled Tubes with Turbulent Flow," *Int. J. of Heat Mass Transfer,* (7): 1207–1216, 1964.

217. S. V. Patankar, V. S. Pratap, and D. B. Spalding, "Prediction of Turbulent Flow in Curved Pipes," *J. Fluid Mech.,* (67/3): 583–595, 1975.

218. E. F. Schmidt, "Wärmeübergang und Druckverlust in Rohrschlangen," *Chem. Ing. Tech.,* (39): 781–789, 1967.

219. N. H. Pratt, "The Heat Transfer in a Reaction Tank Cooled by Means of a Coil," *Trans. Inst. Chem. Eng.,* (25): 163–180, 1947.

220. V. K. Orlov, and P. A. Tselishchev, "Heat Exchange in a Spiral Coil with Turbulent Flow of Water," *Thermal Eng.,* (translated from *Teploenergetika*), (11/12): 97–99, 1964.

221. G. Yang, and M. A. Ebadian, "Turbulent Forced Convection in a Helicoidal Pipe with Substantial Pitch," *Int. J. Heat Mass Transfer,* (39): 2015–2022, 1996.

222. K. C. Cheng, R. C. Lin, and J. W. Ou, "Fully Developed Laminar Flow in Curved Rectangular Channels," *Journal of Fluids Eng.,* (98): 41–48, 1976.

223. K. C. Cheng, and M. Akiyama, "Laminar Forced Convection Heat Transfer in a Curved Rectangular Channel," *Int. J. Heat Mass Transfer,* (13): 471–490, 1970.

224. Y. Mori, Y. Uchida, and T. Ukon, "Forced Convective Heat Transfer in a Curved Channel with a Square Cross Section," *Int. J. Heat Mass Transfer,* (14): 1787–1805, 1976.

225. J. A. Baylis, "Experiments on Laminar Flow in Curved Channels of Square Cross Section," *J. Fluid Mech.,* (48/3): 417–422, 1971.

226. B. Joseph, E. P. Smith, and R. J. Adler, "Numerical Treatment of Laminar Flow in a Helically Coiled Tube of Square Cross Section: Part 1—Stationary Helically Coiled Tubes," *AIChE J.,* (21): 965–979, 1975.

227. J. H. Masliyah, and K. Nanadakumar, "Fully Developed Laminar Flow in a Helical Tube of Finite Pitch," *Chem. Eng. Commun.,* (29): 125–138, 1984.

228. K. C. Cheng, R. C. Lin, and J. W. Ou, "Graetz Problem in Curved Square Channels," *J. of Heat Transfer,* (97): 244–248, 1975.

229. K. C. Cheng, R. C. Lin, and J. W. Ou, "Graetz Problem in Curved Rectangular Channels with Convective Boundary Condition—The Effect of Secondary Flow on Liquid Solidification-Free Zone," *Int. J. Heat Mass Transfer,* (18): 996–999, 1975.

230. A. Butuzov, M. K. Bezrodnyy, and M. M. Pustovit, "Hydraulic Resistance and Heat Transfer in Forced Flow in Rectangular Coiled Tubes," *Heat Transfer—Sov. Res.,* (7/4): 84–88, 1975.

231. V. Kadambi, "Heat Transfer and Pressure Drop in a Helically Coiled Rectangular Duct," *ASME* paper no. 83-WA/HT-1, 1983.

232. R. C. Xin, A. Awwad, Z. F. Dong, and M. A. Ebadian, "An Experimental Study of Single-Phase and Two-phase Flow Pressure Drop in Annular Helicoidal Pipes," *Int. J. of Heat and Fluid Flow,* in press.

233. S. Garimella, D. E. Richards, and R. N. Christensen, "Experimental Investigation of Heat Transfer in Coiled Annular Ducts," *J. of Heat Transfer,* (110): 329–336, 1988.

234 Z. F. Dong, and M. A. Ebadian, "Numerical Analysis of Laminar Flow in Curved Elliptic Ducts," *J. Fluids Eng.,* (113): 555–562, 1991.

235. Z. F. Dong, and M. A. Ebadian, "Thermal Developing Flow in a Curved Duct of Elliptic Cross Section," *Numer. Heat Transfer,* part A, (24): 197–212, 1993.

236. Z. F. Dong, and M. A. Ebadian, "Effects of Buoyancy on Laminar Flow in Curved Elliptic Duct," *J. Heat Transfer,* (114): 936–943, 1992.

237. K. Rehme, "Convective Heat Transfer over Rod Bundles," *Handbook of Single Phase Convective Heat Transfer,* eds. S. Kakaç, R. K. Shah, and W. Aung, Wiley-Interscience, John Wiley & Sons, New York, 1987.

238. V. Ramachandra, "The Numerical Prediction of Flow and Heat Transfer in Rod Bundle Geometries," Ph.D. thesis, Imperial College of Science and Technology, London, 1979.

239. H. Rosenberg, "Numerical Solution of the Velocity Profile in Axial Laminar Flow through a Bank of Touching Rods in a Triangular Array," *Trans. Am. Nucl. Soc.,* (1): 55–57, 1958.

240. E. M. Sparrow, and A. L. Loeffler Jr., "Longitudinal Laminar Flow between Cylinders Arranged in Regular Array," *AIChE J.,* (5): 325–330, 1959.

241. R. A. Axford, "Two-Dimensional Multiregion Analysis of Temperature Fields and Heat Fluxes in Tube Bundles with Internal Solid Nuclear Heat Sources," LA-3167, *Los Alamos Scientific Laboratory,* Los Alamos, New Mexico, 1964.

242. R. A. Axford, "Two-Dimensional Multiregional Analysis of Temperature Fields in Reactor Tube Bundles," *Nucl. Eng. Design,* (6): 25–42, 1967.

243. F. S. Shih, "Laminar Flow in Axisymmetric Conduits by a Rational Approach," *Can. J. Chem. Eng.,* (45): 285–294, 1967.

244. K. Rehme, "Laminarströmung in Stabbündeln," *Chemie-Ingenieur-Technik,* (43): 962–966, 1971.

245. K. Rehme, "Laminarströmung in Stabbündeln," *Reaktortagung 1971,* Deutsches Atomforum, Bonn, Germany, pp. 130–133, 1971.

246. K. Johannsen, "Longitudinal Flow over Tube Bundles," *Low Reynolds Number Flow Heat Exchangers,* eds. S. Kakaç, R. K. Shah, and A. E. Bergles, Hemisphere, New York, pp. 229–273, 1983.

247. J. Malák, J. Hejna, and J. Schmid, "Pressure Losses and Heat Transfer in Non-Circular Channels with Hydraulically Smooth Walls," *Int. J. Heal Mass Transfer,* (18): 139–149, 1975.

248. M. D. Mikhailov, "Finite Element Analysis of Turbulent Heat Transfer in Rod Bundles," *Turbulent Forced Convection in Channels and Bundles,* eds. S. Kakaç, and D. B. Spalding, (1): 250–277, 1979.

249. V. I. Subbotin, M. K. Ibragimov et al., *Hydrodynamics and Heat Transfer in Nuclear Power Systems,* Atomizdat, Moscow, 1975.

250. O. E. Dwyer, and H. C. Berry, "Laminar Flow Heat Transfer for In-Line Flow through Unbaffled Rod Bundles," *Nucl. Sci. Eng.,* (42): 81–88, 1970.

251. K. Rehme, "Simple Method of Predicting Friction Factors of Turbulent Flow in Non-Circular Channels," *Int. J. Heat Mass Transfer,* (10): 933–950, 1973.

252. S. K. Cheng and N. E. Todreas, "Hydrodynamic Models and Correlations for Bare and Wire-Wrapped Hexagonal Rod Bundles—Bundle Friction Factors, Subchannel Friction Factors and Mixing Parameters," *Nuclear Engineering and Design,* (92): 227–251, 1986.

253. E. M. Sparrow, A. L. Loeffler Jr., and H. A. Hubbard, "Heat Transfer to Longitudinal Laminar Flow between Cylinders," *Trans. ASME, J. Heat Transfer,* (83): 415–422, 1961.

254. C. J. Hsu, "Laminar and Slug Flow Heat Transfer Characteristics of Fuel Rods Adjacent to Fuel Subassembly Walls," *Nucl. Sci. Eng.,* (49): 398–404, 1972.

255. O. Miyatake, and H. Iwashita, "Laminar-Flow Heat Transfer to a Fluid Flowing Axially Between Cylinders with a Uniform Surface Temperature," *Int. J. Heat Mass Transfer,* (33): 417–425, 1990.

256. O. Miyatake, and H. Iwashita, "Laminar-Flow Heat Transfer to a Fluid Flowing Axially Between Cylinders with a Uniform Wall Heat Flux," *Int. J. Heat Mass Transfer,* (34): 322–327, 1991.

257. R. Meyder, "Solving the Conservation Equations in Fuel Rod Bundles Exposed to Parallel Flow by Means of Curvilinear-Orthogonal Coordinates," *J. Comp. Physics,* (17): 53–67, 1975.

258. J. H. Kim, "Heat Transfer in Longitudinal Laminar Flow along Circular Cylinders in Square Array," *Fluid Flow and Heat Transfer over Rod or Tube Bundles,* eds. S. C. Yao and P. A. Pfund, *ASME,* New York, pp. 155–161, 1979.

259. J. Ohnemus, *Wärmeubergang und Druck verlust in einem Zenbtralkanal enies Stabbnüdels in quadratischer Anordnung,* Diplomarbeit, Inst. Für Neutrönenphysik und Reaktortechnik, Kernforschungszentrum Karlsruhe, 1982.

260. B. C-J. Chen, T. H. Chien, W. T. Sha, and J. H. Kim, "Solution of Flow in an Infinite Square Array of Circular Tubes by Using Boundary Fitted Coordinate Systems," *Numerical Grid Generation,* ed. J. F. Thompson, Elsevier, New York, pp. 619–632, 1982.

261. R. G. Deissler, and M. F. Taylor, "Analysis of Turbulent Flow and Heat Transfer in Noncircular Passages," *NASA TR-31,* 1959.

262. R. G. Deissler, and M. F. Taylor, "Analysis of Axial Turbulent Flow and Heat Transfer through Banks of Rods or Tubes," *Reactor Heat Transfer Conf.,* New York, TID 75299, part 1, pp. 416–461, 1956.

263. W. M. Kays, and H. C. Perkins, "Forced Convection, Internal Flow in Ducts," *Handbook of Heat Transfer,* McGraw-Hill, New York, 1985.

264. M. W. Maresca, and O. E. Dwyer, "Heat Transfer to Mercury Flowing in Line through a Bundle of Circular Rods," *J. Heat Transfer,* (86): 180–185, 1964.

265. M. H. Hu, and Y. P. Chang, "Optimization of Finned Tubes for Heat Transfer in Laminar Flow," *J. Heat Transfer,* (95): 332–338, 1973; for numerical results, see M. H. Hu, "Flow and Thermal Analysis for Mechanically Enhanced Heat Transfer Tubes," Ph.D. thesis, State University of New York at Buffalo, 1973.

266. C. Prakash, and Y. Liu, "Analysis of Laminar Flow and Heat Transfer in the Entrance Region of an Internally Finned Circular Duct," *J. Heat Transfer,* (107): 84–91, 1985.

267. J. H. Masliyah, and K. Nandakumar, "Heat Transfer in Internally Finned Tubes," *J. Heat Transfer,* (98): 257–261, 1976.

268. M. K. Gangal, and B. D. Aggarwala, "Combined Free and Forced Convection in Laminar Internally Finned Square Ducts," *Z. Angew. Math. Phys.,* (28): 85–96, 1977.

269. B. D. Aggarwala, and M. K. Gangal, "Heat Transfer in Rectangular Ducts with Fins from Opposite Walls," *Z. Angew. Math. Phys.,* (56): 253–266, 1976.

270. M. K. Gangal, "Some Problems in Channel Flow," Ph.D. thesis, University of Calgary, Calgary, 1974.

271. K. Nandakumar, and J. H. Masliyah, "Fully Developed Viscous Flow in Internally Finned Tubes," *Chem. Eng. J.,* (10): 113–120, 1975.

272. R. M. Manglik, and A. E. Bergles, "Heat Transfer and Pressure Drop Correlations for Twisted-Tape Inserts in Isothermal Tubes: Part I—Laminar Flows," *J. Heat Transfer,* (115): 881–889, 1993.

273. R. M. Manglik, and A. E. Bergles, "Heat Transfer and Pressure Drop Correlations for Twisted-Tape Inserts in Isothermal Tubes: Part II—Transition and Turbulent Flows," *J. Heat Transfer,* (115): 890–896, 1993.

274. Z. F. Dong, and M. A. Ebadian, "Analysis of Combined Natural and Forced Convection in Vertical Semicircular Ducts with Radial Internal Fins," *Numer. Heat Transfer,* part A, (27): 359–372, 1995.

275. Z. F. Dong, and M. A. Ebadian, "A Numerical Analysis of Thermally Developing Flow in an Elliptic Duct with Fins," *Int. J. Heat Fluid Flow,* (12): 166–172, 1991.

276. B. Farhanieh, and B. Sunden, "Three Dimensional Laminar Flow and Heat Transfer in the Entrance Region of Trapezoidal Ducts," *Int. J. Numerical Methods in Fluids,* (13): 537–556, 1991.

277. C. Chiranjivi, and P. S. Rao, "Laminar and Turbulent Convection Heat Transfer in a Symmetric Trapezoidal Channel," *Indian Journal of Technology,* (9): 416–420, 1971.

278. N. Nakamura, S. Hiraoka, and I. Yamada, "Flow and Heat Transfer of Laminar Forced Convection in Arbitrary Polygonal Ducts," *Heat Transfer-Jpn. Res.,* (2/4): 56–63, 1974.

279. G. Schenkel, *Laminar Durchströmte Profilkanäle; Ersatzradien und Widerstansbeiwerte,* Fortschritt-Beriche der VDI Zeitschriften, Reihe: Stromungstechnik, vol. 7, no. 62, 1981.

280. E. R. G. Eckert, and T. F. Irvine Jr., "Flow in Corners of Passages with Noncircular Cross Sections," *Trans. ASME,* (78): 709–718, 1956.

281. T. M. Ben-Ali, H. M. Soliman, and E. K. Zariffeh, "Further Results for Laminar Heat Transfer in Annular Section and Circular Sector Ducts," *J. Heat Transfer,* (111): 1090–1093, 1989.

282. H. M. Soliman, A. A. Menis, and A. C. Trupp, "Laminar Flow in the Entrance Region of Circular Sector Ducts," *J. Appl. Mech.,* (49): 640–642, 1982.

283. E. M. Sparrow, and A. Haji-Sheikh, "Flow and Heat Transfer in Ducts of Arbitrary Shape with Arbitrary Thermal Boundary Conditions," *J. Heat Transfer,* (88): 351–358, 1966. Discussion by C. F. Neville, *J. Heat Transfer,* (91): 588–589, 1969.

284. S. W. Hong, and A. E. Bergles, "Augmentation of Laminar Flow Heat Transfer in Tubes by Means of Twisted-Tape Inserts," tech. rep. HTL-5, ISU-EMI-Ames 75011, *Eng. Res. Inst.,* Iowa State University, Ames, 1974.

285. E. M. Sparrow, T. S. Chen, and V. K. Jonsson, "Laminar Flow and Pressure Drop in Internally Finned Annular Ducts," *Int. J. Heat Mass Transfer,* (7): 583–585, 1964.

286. H. M. Soliman, "Laminar Heat Transfer in Annular Sector Ducts," *J. Heat Transfer,* (109): 247–249, 1987.

287. P. Renzoni, and C. Prakash, "Analysis of Laminar Flow and Heat Transfer in the Entrance Region of an Internally Finned Concentric Circular Annular Ducts," *J. Heat Transfer,* (109): 532–538, 1987.

288. J. P. Zarling, "Application of Schwarz-Neumann Technique to Fully Developed Laminar Heat Transfer in Noncircular Ducts," *J. Heat Transfer,* (99): 332–335, 1977.

289. K. C. Cheng, and M. Jamil, "Laminar Flow and Heat Transfer in Circular Ducts with Diametrically Opposite Flat Sides and Ducts of Multiply Connected Cross Sections," *Can. J. Chem. Eng.,* (48): 333–334, 1970.

290. E. M. Sparrow, and M. Charmchi, "Heat Transfer and Fluid Flow Characteristics of Spanwise-Periodic Corrugated Ducts," *Int. J. Heat Mass Transfer,* (23): 471–481, 1980.

291. E. M. Sparrow, and A. Chukaev, "Forced-Convection Heat Transfer in a Duct Having Rectangular Protuberances," *Numer. Heat Transfer,* (3): 149–167, 1980.

292. Z. F. Dong, M. A. Ebadian, and A. Campo, "Numerical Analysis of Convective Heat Transfer in the Entrance Region of Cusped Ducts," *Numer. Heat Transfer,* part A, (20): 459–472, 1991.

293. Z. F. Dong, and M. A. Ebadian, "Mixed Convection in the Cusped Duct," *J. Heat Transfer,* (116): 250–253, 1994.

294. V. P. Tyagi, "A General Non-Circular Duct Convective Heat-Transfer Problem for Liquids and Gases," *Int. J. Heat Mass Transfer,* (9): 1321–1340, 1966.

295. H. C. Topakoglu, and O. A. Arnas, "Convective Heat Transfer for Steady Laminar Flow between Two Confocal Elliptical Pipes with Longitudinal Uniform Wall Temperature Gradient," *Int. J. Heat Mass Transfer,* (17): 1487–1498, 1974.

296. D. A. Ratkowsky, and N. Epstein, "Laminar Flow in Regular Polygonal Ducts with Circular Centered Cores," *Can. J. Chem. Eng.,* (46): 22–26, 1968.

297. K. C. Cheng, and M. Jamil, "Laminar Flow and Heat Transfer in Ducts of Multiply Connected Cross Sections," *ASME,* paper no. 67-HT-6, 1967.

298. S. L. Hagen, and D. A. Ratkowsky, "Laminar Flow in Cylindrical Ducts Having Regular Polygonal Shaped Cores," *Can. J. Chem. Eng.,* (46): 387–388, 1968.

299. B. D. Bowen, "Laminar Flow in Unusual-Shaped Ducts," B.S. thesis, University of British Columbia, Vancouver, 1967.

300. P. N. Shivakumar, "Viscous Flow in Pipes Whose Cross Sections are Doubly Connected Regions," *Appi. Sci. Res.,* (27): 355–365, 1973.

301. S. Kakaç, R. K. Shah, and W. Aung, *Handbook of Single-Phase Convective Heat Transfer,* Wiley-Interscience, John Wiley & Sons, New York, 1987.

CHAPTER 6
FORCED CONVECTION, EXTERNAL FLOWS

M. W. Rubesin
Retired, Ames Research Center—NASA

M. Inouye
Retired, Ames Research Center—NASA

P. G. Parikh
Boeing Commercial Airplane Group

INTRODUCTION

In the current era of large electronic computers, many complex problems in convection are being solved precisely by numerical solution of equations expressing basic principles. Keen insight into the fine points of such problems, however, requires extensive parametric studies that consume computer time, and therefore the numerical approach is usually applied only to a few examples. Further, these numerical programs may occupy so much computer storage space that their use as subroutines within generalized systems studies becomes impractical. Thus there still exists a need for general formulas and data correlations that can be used in preliminary design, in systems studies where convection is only one of many inputs, in creative design where inventiveness is based on understanding the influences of the variables of a problem, and in verifying computer codes for convective heat transfer to complex bodies. This chapter provides many of these tools for the case of forced convection over simple bodies.

Specifically, theoretical equations and correlations of data are presented for evaluating the local rate of heat transfer between the surface of a body and an encompassing fluid at different temperatures and in relative motion. *Forced convection* requires either that the fluid be pumped past the body, as for a model in a wind tunnel, or the body be propelled through the fluid, as an aircraft in the atmosphere. The methods presented apply equally to either situation when velocities are expressed relative to the body. Gravity forces are usually negligible under these conditions. Further, the contents of this chapter are confined to those conditions where the fluid behaves as a continuum.

The evaluation of forced convection to bodies has become a major problem in many aspects of modern technology. A few examples of applications include the following: thermally de-icing aircraft surfaces; turbine blade cooling; furnace tube bundles; and protecting high-performance aircraft, missile nose cones, and reentry bodies from intense aerodynamic heating. Formulas for evaluating convective heat transfer rates are generally established through a combination of theoretical analysis and experimentation. Analysis is almost universally based on *boundary layer theory*—the mathematical solution of conservation equations of individual species, overall mass, momentum, and energy that are applicable to the thin region of fluid adjacent to the surfaces of bodies where the effects of shear, heat conduction, and species diffusion are controlling. Separated flows are not considered here. The experimentation involves the measurement of solid and fluid temperatures and of heat flux in a multitude of ways.

DEFINITION OF TERMS

The local convective heat flux from a point on a body is often expressed through Newton's law of cooling, generalized as

$$q''_w = \rho_e u_e \, \text{St} \, (i_w - i_{e,\text{eff}}) \tag{6.1}$$

Enthalpy is used as the measure of the thermal driving potential to broaden the application of Eq. 6.1 to thermally perfect gases with temperature-dependent specific heats where

$$i = \int c_p(T) \, dT \tag{6.2}$$

With bodies having constant surface temperatures, at low speeds

$$i_{e,\text{eff}} = i_e = c_p T_e \tag{6.3}$$

At high speeds where frictional heating takes place,

$$i_{e,\text{eff}} = i_e + r(0)(u_e^2/2) = i_{aw} \tag{6.4}$$

where $r(0)$ is the recovery factor, having approximate values for air of 0.85 and 0.9 for laminar and turbulent flow, respectively.

For bodies with nonuniform surface temperature distributions, $i_{e,\text{eff}}$ depends not only on the conditions at the boundary layer edge but also on the distribution of the surface temperature upstream of the location being considered.

For constant fluid properties, Eq. 6.1 correlates both theoretical and experimental results for a wide range of flow and temperature conditions through the single parameter Stanton number St. As will be seen in subsequent sections, this correlation is even useful when St is dependent on i_e or $i_{e,\text{eff}}$ and the equation is nonlinear.

The Reynolds analogy, defined as the ratio of the Stanton number to the local skin friction coefficient $\text{St}/(c_f/2)$ is a function of the Prandtl number and is extremely useful for estimating heat transfer. Pressure drop can be used to predict heat transfer in pipes, and the skin friction can be used to predict Stanton number for external flows.

When mass transfer of a foreign gas occurs at a surface, an equation similar to Eq. 6.1 is employed to define the local mass flux of the species i of a binary mixture as

$$j_{iw} = \rho_e u_e c_{mi}(K_{iw} - K_{ie}) \tag{6.5}$$

The Reynolds analogy can be extended to express $c_{mi}/(c_f/2)$ as a function of the Lewis number.

TWO-DIMENSIONAL LAMINAR BOUNDARY LAYER

Uniform Free-Stream Conditions

The most studied configuration for forced convection has been the "flat plate," a surface at constant pressure with a sharp leading edge. The simplicity of this configuration so facilitates the solution of the boundary layer equations, even for a variety of surface boundary conditions, that the bulk of heuristic theoretical boundary layer research is identified with the flat plate. These results are useful because much that is learned can be extended to more realistic body shapes using computer codes and applied directly to platelike surfaces (e.g., supersonic aircraft wings or fins having wedge cross sections and attached shock waves).

Uniform Surface Temperature

Governing Differential Equations. For a fluid as general as a gas in chemical equilibrium, the boundary layer equations for laminar flow over a flat plate are:

$$\frac{\partial}{\partial x}(\rho u) + \frac{\partial}{\partial y}(\rho v) = 0 \tag{6.6}$$

$$\rho u \frac{\partial u}{\partial x} + \rho v \frac{\partial u}{\partial y} = \frac{\partial}{\partial y}\left(\mu \frac{\partial u}{\partial y}\right) \tag{6.7}$$

$$\rho u \frac{\partial I}{\partial x} + \rho v \frac{\partial I}{\partial y} = \frac{\partial}{\partial y}\left[\frac{\mu}{Pr_T}\frac{\partial I}{\partial y} + \mu\left(1 - \frac{1}{Pr_T}\right)\frac{\partial(u^2/2)}{\partial y}\right] \tag{6.8}$$

with the boundary conditions

$$\left.\begin{array}{llll}
x = 0 & y > 0; & u = u_e & I = I_e \\
x > 0 & y \to \infty; & u \to u_e & I \to I_e \\
& y = 0; & u = 0 & v = 0 \\[2mm]
\multicolumn{4}{c}{I = i_w = \text{constant} \quad\text{or}\quad \dfrac{\partial i}{\partial y} = 0 \quad\text{for}\quad I = i_{aw}}
\end{array}\right\} \tag{6.9}$$

The leading edge of the plate is located at $x = 0$. The surface boundary conditions at $y = 0$ reflect the assumed conditions of zero mass transfer, a prescribed uniform temperature including the case of zero heat flux and an implied condition of a smooth surface.

A stream function ψ defined as

$$\rho u = \rho_e \frac{\partial \psi}{\partial y} \qquad \rho v = -\rho_e \frac{\partial \psi}{\partial x} \tag{6.10}$$

immediately satisfies Eq. 6.6.

Fluid With Constant Properties. When the density and viscosity in Eqs. 6.6 and 6.7 are constant, the velocity field is independent of the temperature field. Blasius [1] collapsed the partial differential equations (Eqs. 6.6 and 6.7) to a single ordinary differential equation by transforming the coordinate system from x and y to ζ and η, defined as

$$\zeta = x \qquad \eta = \sqrt{yu_e/\nu_e x} \tag{6.11}$$

The stream function defined by Eq. (6.10) is expressed as

$$\psi = m(\zeta)f(\eta)$$

where $m(\zeta) = \sqrt{u_e \nu_e \zeta}$

and the velocity components are expressed in terms of the similarity variable η

$$\frac{u}{u_e} = f'(\eta) \qquad \frac{v}{u_e} = \frac{1}{2}\frac{[\eta f'(\eta) - f(\eta)]}{\sqrt{Re_{x_e}}} \tag{6.12}$$

where $f(\eta)$ is the solution of the ordinary differential equation

$$f''' + \tfrac{1}{2}ff'' = 0 \tag{6.13}$$

with the boundary conditions

$$\begin{aligned} \eta = 0 \qquad & f = 0, f' = 0 \\ \eta \to \infty \qquad & f' \to 1 \end{aligned} \tag{6.14}$$

The u velocity profile is shown in Fig. 6.1. The velocity ratio reaches a value of 0.99 at a boundary layer thickness of

$$\frac{\delta}{x} = \frac{5}{\sqrt{\mathrm{Re}_{x_e}}} \tag{6.15}$$

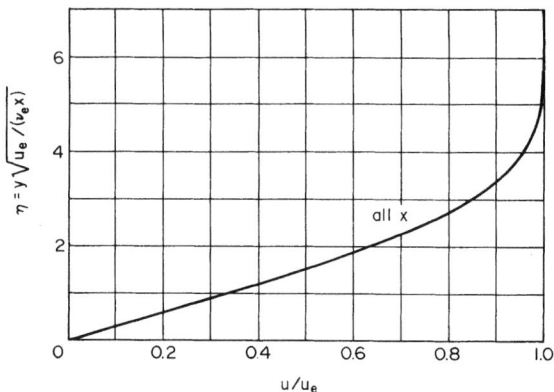

FIGURE 6.1   Similar velocity profile in the laminar boundary layer on a flat plate—constant fluid properties.

The local skin friction coefficient is

$$\frac{c_f}{2} = \frac{f''(0)}{\sqrt{\rho_e u_e x/\mu_e}} = \frac{0.332}{\sqrt{\mathrm{Re}_{x_e}}} \tag{6.16}$$

The average skin friction coefficient for the length of the plate up to x is defined as

$$\frac{\bar{c}_f}{2} = \frac{1}{x}\int_0^x \frac{c_f}{2}\,dx = \frac{0.664}{\sqrt{\mathrm{Re}_{x_e}}} \tag{6.17}$$

Equation 6.17 indicates that on a flat plate the average skin friction coefficient is equal to twice the local skin friction coefficient at the trailing edge.

Experimental verification of the Blasius theory has been hindered by the difficulty in reproducing the ideal flat plate boundary conditions in the laboratory. Whenever uniform pressure was attained and the effects of a real leading edge were accounted for, however, it was found that the preceding calculated results were always verified to within the accuracy of the experiment.

Pohlhausen [2] utilized the Blasius coordinate system and velocity distribution to evaluate the convective heating processes within the constant-property boundary layer on a flat plate. He solved two problems:

1. The convective heat transfer rate to a plate with uniform surface temperature for fluid speeds sufficiently low to make viscous dissipation negligible
2. The temperature attained by an insulated plate (zero surface heat transfer) when exposed to a high-speed stream where viscous dissipation is important

The latter is the *plate thermometer* or *adiabatic wall problem*. Eckert and Drewitz [3] showed that the general problem of heat transfer to a uniform-surface-temperature plate in constant-property high-speed flow is merely the superposition of the two Pohlhausen solutions.

For a uniform-surface-temperature plate in a low-speed flow ($u^2 \ll 2c_pT$), Eq. 6.8 simplifies to

$$\rho u \frac{\partial T}{\partial x} + \rho v \frac{\partial T}{\partial y} = \frac{\mu}{\text{Pr}} \frac{\partial^2 T}{\partial y^2} \tag{6.18}$$

with the boundary conditions

$$
\begin{aligned}
x = 0 \quad & y > 0; && T = T_e \\
x > 0 \quad & y = 0; && T = T_w \\
& y \to \infty; && T \to T_e
\end{aligned}
$$

When Eq. 6.18 is transformed to the independent variables (Eq. 6.11) and the new normalized dependent variable

$$Y_0(\eta) = \frac{T - T_e}{T_w - T_e} \tag{6.19}$$

is introduced, the ordinary homogeneous differential equation that results is

$$Y_0'' + \tfrac{1}{2}\text{Pr}\, f Y_0' = 0 \tag{6.20}$$

where f is the Blasius stream function. The transformed boundary conditions are

$$
\left.
\begin{aligned}
\eta = 0 \quad & Y_0 = 1 \\
\eta \to \infty \quad & Y_0 \to 0
\end{aligned}
\right\} \tag{6.21}
$$

Solutions for $\text{Pr} = 0.5$ and 1.0 are shown in Fig. 6.2 as solid curves. The abscissa of this figure is the thermal boundary layer thickness parameter η_H, consisting of the Blasius boundary layer similarity parameter multiplied by $\text{Pr}^{1/3}$. The close agreement of the two solid curves suggests for Pr near unity that the thermal boundary layer thickness where $Y_0 = 0.01$ is inversely proportional to approximately $\text{Pr}^{1/3}$ or

$$\frac{\delta_T}{x} = \frac{5\text{Pr}^{-1/3}}{\sqrt{\text{Re}_{x_e}}} = \frac{\delta}{x}\,\text{Pr}^{-1/3} \tag{6.22}$$

Thus, fluids with Pr less than unity have thermal boundary layers that are thick relative to their flow boundary layers. Conversely, fluids with Pr greater than unity have relatively thin thermal boundary layers.

This latter condition suggests a particularly simple solution of Eq. 6.20 for very large Pr [4] because the temperature variations occur where the velocity distribution is still linear in η (see Fig. 6.1 for $\eta < 2.0$). The linear velocity condition in Eq. 6.20 permits expressing Y_0 explicitly in terms of η_H. The solution for this case of large Pr is shown as a dashed line in Fig. 6.2 and agrees quite well with the calculations based on the more exact velocity distributions for Pr near unity. This agreement indicates that Eq. 6.22 is applicable over a large range of Pr from values characteristic of gases to those for heavy oils.

The local Stanton number found by Pohlhausen is represented very well by

$$\text{St} = \frac{0.332\text{Pr}^{-2/3}}{\sqrt{\text{Re}_{x_e}}} \tag{6.23}$$

a form consistent with the parameter η_H.

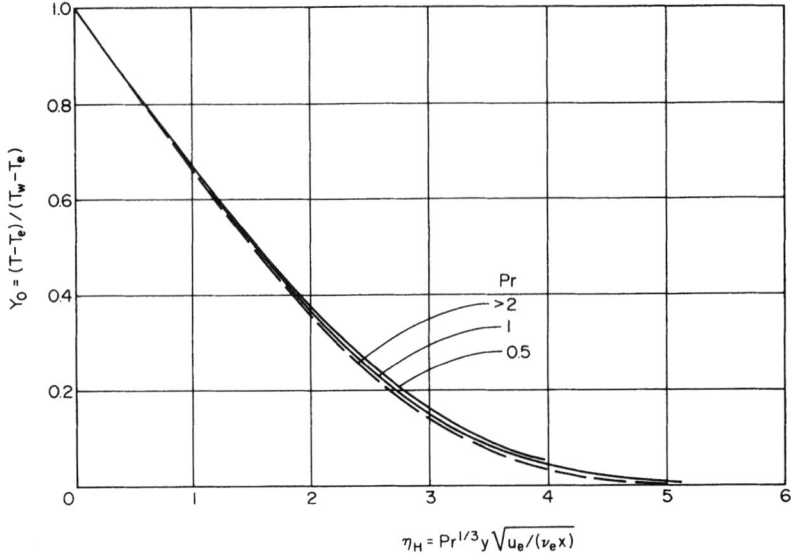

FIGURE 6.2   Temperature distributions in the laminar boundary layer on a flat plate at uniform temperature—constant property, low-speed flow.

The modified Reynolds analogy from Eqs. 6.16 and 6.23 is

$$\text{St} = \frac{c_f}{2} \, \text{Pr}^{-2/3} \tag{6.24}$$

The excellent agreement of this formula with the precise numerical results of Pohlhausen [2] over a large range of Pr is shown graphically in Fig. 6.4 (solid curves are the numerical results). The dashed line, labeled 1.02 Pr$^{-2/3}$, results from the analysis employing a linear velocity distribution throughout the boundary layer [5]. Equation 6.24 has been shown to be consistent with experimental results through a successive series of data correlations dating back to Colburn [6].

The average Stanton number up to station x is

$$\overline{\text{St}} = \frac{1}{x} \int_0^x \text{St} \, dx = 2\text{St} \tag{6.25}$$

For an insulated plate in high-speed flow with constant properties, Eq. 6.8 combined with Eq. 6.7 reduces to

$$\rho u \frac{\partial T}{\partial x} + \rho v \frac{\partial T}{\partial y} = \frac{\mu}{\text{Pr}} \frac{\partial^2 T}{\partial y^2} + \frac{\mu}{c_p} \left(\frac{\partial u}{\partial y} \right)^2 \tag{6.26}$$

with boundary conditions

$$\left. \begin{aligned} x = 0 \quad & y > 0; \quad && T = T_e \\ x > 0 \quad & y \to \infty; \quad && T \to T_e \\ & y = 0; \quad && \frac{\partial T}{\partial y} = 0 \end{aligned} \right\} \tag{6.27}$$

When the independent variables are transformed to the Blasius variables and a new dependent variable

$$\mathbf{r}(\eta) = \frac{T - T_e}{u_e^2/2c_p} \tag{6.28}$$

is introduced into Eq. 6.26, the ordinary inhomogeneous equation that results is

$$\mathbf{r}'' + \tfrac{1}{2}\mathrm{Pr}\, f\, \mathbf{r}' = -2\mathrm{Pr}\, f''^2 \tag{6.29}$$

with boundary conditions

$$\left.\begin{array}{ll} \eta = 0 & \mathbf{r}' = 0 \\ \eta \to \infty & \mathbf{r} \to 0 \end{array}\right\} \tag{6.30}$$

The solutions of this problem are indicated in Fig. 6.3. The temperature distributions shown are based on calculations employing exact velocity distributions for Pr near unity and a linear velocity distribution for very large Pr. These temperatures have been normalized by the temperature rise at the surface, and the abscissa is the η_H utilized in Fig. 6.2.

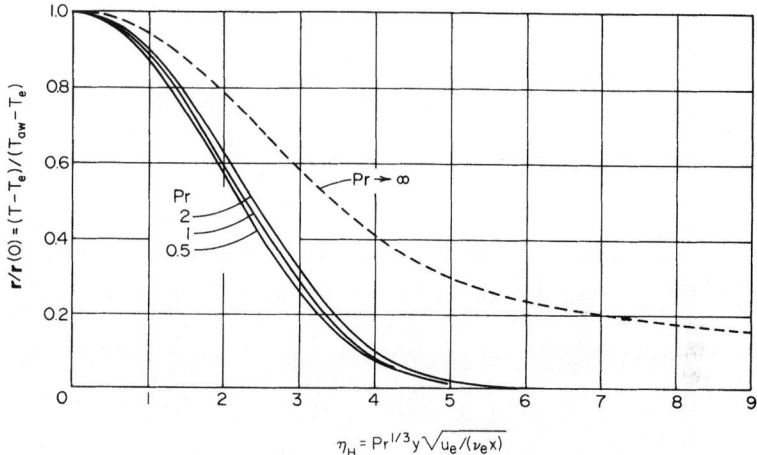

FIGURE 6.3 Temperature profiles in the laminar boundary layer on an insulated plate—constant-property, high-speed flow.

Figure 6.3 shows less correlation for different Pr than was exhibited for the uniform-surface-temperature case in Fig. 6.2. The implication here is that the thermal boundary layer produced by viscous dissipation grows at a rate different from $\mathrm{Pr}^{-1/3}$ for all but the very large values of Pr. For Pr near unity, the growth factor is closer to $\mathrm{Pr}^{-0.28}$.

The adiabatic wall temperature (recovery temperature) is given by

$$T_{aw} = T_e + \mathbf{r}(0)\,\frac{u_e^2}{2c_p} \tag{6.31}$$

Figure 6.4 shows the dependence of the recovery factor $\mathbf{r}(0)$ on Pr as given by Refs. 2 and 5 (solid line). In the region $0.5 < \mathrm{Pr} < 2$, the formula

$$\mathbf{r}(0) = \mathrm{Pr}^{1/2} \tag{6.32}$$

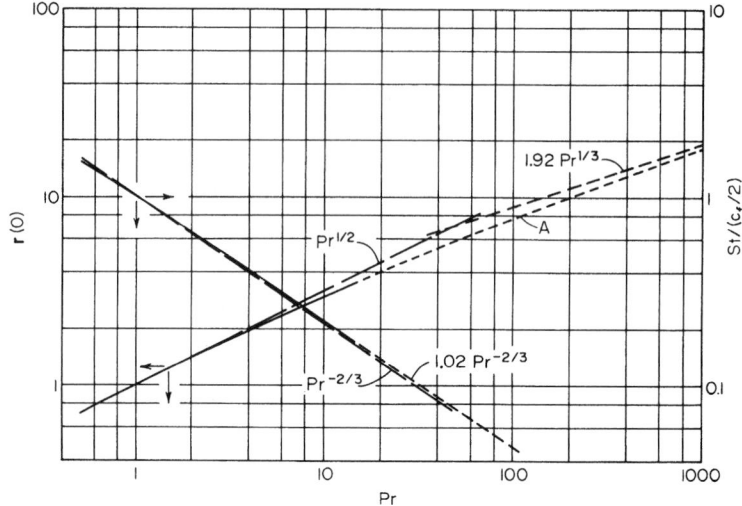

FIGURE 6.4 Influence of Prandtl number on the recovery factor and modified Reynolds analogy for a laminar boundary layer on a flat plate.

represents the calculated values to within 1 percent. For Pr = 7, Eq. 6.32 yields results high by 5.4 percent. The dashed line labeled A in Fig. 6.4 represents an extrapolation of the exact numerical results for Pr < 15 to approach asymptotically the limiting value

$$\mathbf{r}(0) = 1.92 \mathrm{Pr}^{1/3} \tag{6.33}$$

resulting explicitly when a linear velocity distribution exists throughout the thermal boundary layer.

For a uniform-surface-temperature plate in high-speed flow, the temperature distribution within the boundary layer is expressed by a superposition of the two Pohlhausen solutions [3]. This is permissible because the energy equation (Eq. 6.8) with constant properties is linear in temperature. The general solution of the energy equation is the sum of the general solution of the homogeneous equation (Eq. 6.18) and a particular solution of the inhomogeneous equation (Eq. 6.26):

$$T - T_e = (T_w - T_{aw})Y_0(\eta) + \frac{u_e^2}{2c_p}\,\mathbf{r}(\eta) \tag{6.34}$$

$Y_0(\eta)$ and $r(\eta)$ (Eqs. 6.19 and 6.28) are indicated in Figs. 6.2 and 6.3, respectively. The local heat flux is expressed as

$$q_w'' = P_e/\rho_e u_e c_p \ \mathrm{St} \ (T_w - T_{aw}) \tag{6.35}$$

The appropriate Stanton number is again represented by Eq. 6.23, T_{aw} is given by Eq. 6.31, and $\mathbf{r}(0)$ is given by Eq. 6.32 or Fig. 6.4.

Liquids With Variable Viscosity. When the temperature difference between a liquid and a surface becomes significant, it is necessary to consider the temperature dependence of the viscosity across the boundary layer. Calculations of convective heating were made [7] for a liquid whose viscosity varies as

$$\frac{\mu}{\mu_w} = \left(\frac{T_w + T_c}{T + T_c}\right)^b$$

where b and T_c are constants. The boundary layer equations are solved through a transformation of independent variables identical in form with Eq. 6.11, but with the kinematic viscosity ν evaluated at the surface temperature. The resulting transformed momentum equation is

$$\left(\frac{\mu}{\mu_w} f''\right)' + \frac{1}{2} ff'' = 0$$

and the energy equation, where viscous dissipation has been neglected, is identical with Eq. 6.20, but with Pr evaluated at the wall temperature. In Ref. 7 the form of the solution requires a choice of the constant b ($b = 3$ in most of the examples) but avoids the necessity of predetermining an explicit value of the constant T_c. The skin friction and heat transfer are expressed directly in terms of the viscosity ratio across the boundary layer μ_w/μ_e and the Prandtl number at the surface.

Figure 6.5 shows the velocity distributions in a boundary layer of a liquid with $Pr_w = 100$ (e.g., sulfuric acid at room temperature). For this Prandtl number, the thermal boundary layer penetration into the liquid is much less than the flow boundary layer, and the regions where viscosity variations occur are confined close to the surface. The curve corresponding to $\mu_w/\mu_e = 1$ is the Blasius solution (see Fig. 6.1). The curve labeled $\mu_w/\mu_e = 0.23$ corresponds to a heated surface where the low viscosity near the surface requires steeper velocity gradients to maintain a continuity of shear with the outer portion of the boundary layer. The heated free-stream cases reveal the opposite effects. In general, the outer portions of the flow boundary layers act similarly to the velocity distribution of the Blasius case except for their being displaced in or out by the effects that have taken place in the thermal boundary layer.

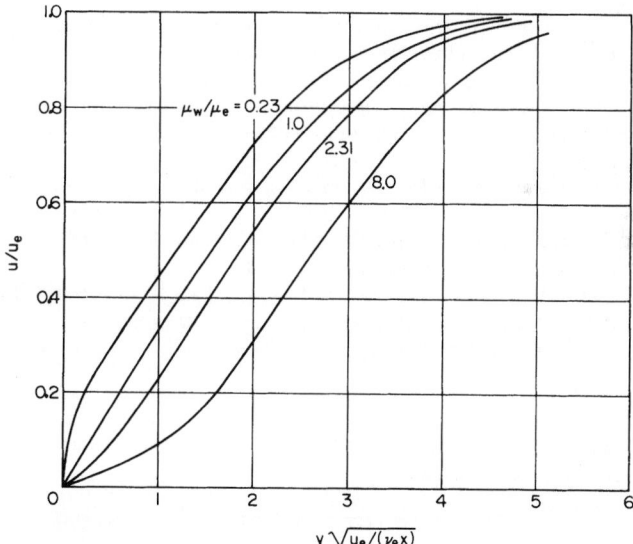

FIGURE 6.5   Velocity profiles in the laminar boundary layer of a liquid, $Pr_w = 100$ [7].

The temperature profiles for different Pr_w and μ_w/μ_e are indicated in Fig. 6.6. Note that the curves for $Pr_w = 10$ apply equally well for greater Prandtl numbers because of the use of the thermal boundary layer thickness parameter as the abscissa (see Fig. 6.2).

The effects of the viscosity variation across the boundary layer on the surface shear stress and heat flux are shown in Fig. 6.7. The shear stress is normalized by the value obtained from the Bla-

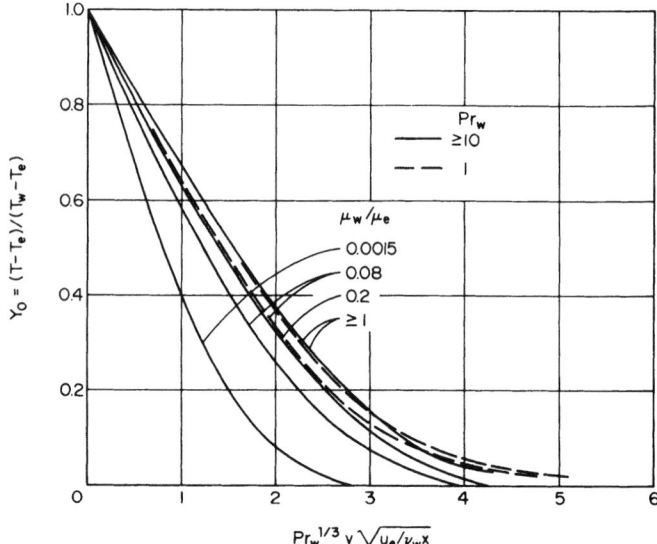

FIGURE 6.6   Temperature profiles in the laminar boundary layer of a liquid [7].

sius solution with the same free-stream properties. The heat flux is normalized by the Pohlhausen value with the viscosity and Prandtl number evaluated at the wall temperature. Note that at the higher Prandtl numbers the wall shear becomes less dependent on the fluid properties.

Ideal Gases at High Temperatures.   The speeds of modern military aircraft, missiles, or reentry bodies are so high that the resulting recovery temperatures are several times to orders of

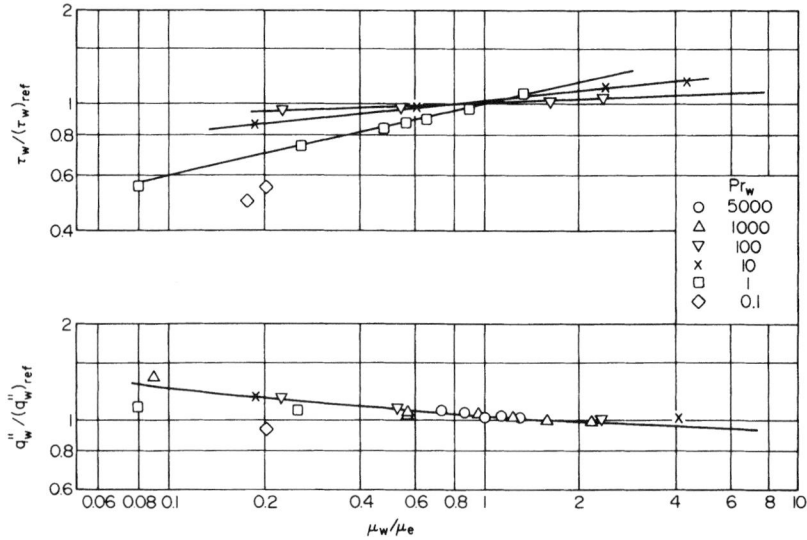

FIGURE 6.7   Effects of viscosity changes across a laminar liquid boundary layer on surface shear and heat flux—reference shear from Blasius solution with free-stream properties, reference heat flux from Pohlhausen solution with wall properties [7].

magnitude larger than the ambient atmospheric temperatures. Under these conditions, the behavior of atmospheric gases within the boundary layer changes from that of an ideal gas to that of a real gas, including the physical effects of rotational and vibrational excitation, dissociation, and even ionization. The real-gas behavior is so complex that numerical analysis is the only means of introducing it into boundary layer problems. Because the cases of simpler gas behavior are directly applicable to many aircraft problems and are guides for correlating the numerical results of the real-gas computations, examples of ideal-gas solutions will be presented first.

Equations 6.6–6.8 are converted to a convenient form through the Howarth-Dorodnitsyn transformation of the independent variables [8] from x, y to ζ, η as follows:

$$\zeta = u_e \mu_r \rho_r x$$

$$\eta = \sqrt{\frac{u_e}{\nu_e x}} \int_0^y \frac{\rho}{\rho_r} \, dy \tag{6.36}$$

The subscript r represents a reference condition for the properties, usually the conditions at the edge of the boundary layer or at the surface.

The transformed momentum and energy equations for a uniform-temperature flat plate are

$$(C_r f'')' + \tfrac{1}{2} ff'' = 0 \tag{6.37}$$

$$\left(\frac{C_r \bar{I}'}{\mathrm{Pr}}\right)' + \frac{1}{2} f\bar{I}' + \frac{u_e^2}{I_e}\left[C_r\left(1 - \frac{1}{\mathrm{Pr}}\right)f'f''\right]' = 0 \tag{6.38}$$

where

$$f' = \frac{u}{u_e} \qquad \bar{I} = \frac{I}{I_e} \qquad C_r = \frac{\mu\rho}{\mu_r\rho_r} \qquad \frac{\partial}{\partial\eta} = (\ )' \tag{6.39}$$

with the boundary conditions

$$\eta = 0 \quad f = f' = 0 \quad \bar{I} = \frac{i_w}{I_e}$$

$$\eta \to \infty \quad f' \to 1, \quad \bar{I} \to 1$$

The fluid properties are introduced into these equations only through C_r and Pr, which involve combinations of individual physical properties. Certain values of C_r and Pr permit simplifications that lead to useful general relationships.

Prandtl Number Equal to Unity. If Pr = 1, considerable simplification results. Equation 6.38 acquires a form identical with Eq. 6.37, $\bar{I}$ being analogous to f'. A solution of the energy equation, therefore, is directly expressible in terms of the velocity distribution as

$$\bar{I} = \left(1 - \frac{i_w}{I_e}\right)f' + \frac{i_w}{I_e} \tag{6.40}$$

after the boundary conditions at the surface and boundary layer edge have been introduced. When the wall enthalpy equals the total enthalpy, Eq. 6.40 indicates that $\bar{I} = 1$, i.e., the total enthalpy is constant throughout the boundary layer. For other wall enthalpies, the local total enthalpy is linearly dependent on the local velocity. The corresponding static enthalpy distribution is

$$\frac{i}{i_e} = 1 + \frac{i_w - I_e}{i_e}(1 - f') + \frac{u_e^2}{2i_e}(1 - f'^2) \tag{6.41}$$

The Reynolds analogy is given by

$$\mathrm{St} = \frac{c_f}{2} \tag{6.42}$$

with the requirement that the recovery enthalpy be

$$i_{aw} = I_e \qquad \text{or} \qquad \mathbf{r}(0) = 1$$

Note that $c_f/2$ in Eq. 6.42 differs in magnitude from that given by Eq. 6.16 because of the departure from the Blasius solution by the existence of $C_r(\eta)$ in Eq. 6.37.

Viscosity-Density Product Equal to a Constant.   If $C_r = \bar{C}$ and Pr are constant, a natural transformation suggests itself [9], where η is replaced by

$$\eta_c = \frac{\eta}{\sqrt{\bar{C}}} \tag{6.43}$$

and

$$F'(\eta_c) = f'(\eta) \tag{6.44}$$

$$\mathbf{I}(\eta_c) = \bar{I}(\eta) \tag{6.45}$$

Equations 6.37 and 6.38 become

$$F''' + \tfrac{1}{2}FF'' = 0 \tag{6.46}$$

$$\mathbf{I}'' + \tfrac{1}{2}\text{Pr } F\mathbf{I}' + \frac{u_e^2}{I_e}(\text{Pr} - 1)(F'F'')' = 0 \tag{6.47}$$

with boundary conditions

$$\begin{aligned} \eta_c &= 0 & F = F' &= 0, & \mathbf{I} &= \bar{I}_w \\ \eta_c &\to \infty & F' &\to 1, & \mathbf{I} &\to 1 \end{aligned}$$

The assumption of constant C_r, therefore, permits separation of the momentum equation from its dependence on the energy equation and results in an energy equation that is linear in $\mathbf{I}$ so that general solutions can be obtained from a superposition of individual solutions.

Equation 6.46 with its boundary conditions is the Blasius problem again. The energy equation is satisfied by

$$\frac{i}{i_e} = \frac{i_w - i_{aw}}{i_e} Y_0(\eta_c) + \frac{u_e^2}{2i_e} \mathbf{r}(\eta_c) + 1 \tag{6.48}$$

where Y_0 and $\mathbf{r}$ are obtainable from Figs. 6.2 and 6.3 when η_H is adjusted according to Eq. 6.43. The recovery factor $\mathbf{r}(0)$ is independent of $\bar{C}$ and therefore is identical to the constant-property value given in Fig. 6.4 and Eq. 6.32.

$$\frac{c_f}{2} = \frac{0.332}{\sqrt{\text{Re}_{x_e}}} \frac{\mu_w \rho_w / \mu_e \rho_e}{\sqrt{\bar{C}}} = \left(\frac{c_f}{2}\right)_i \frac{\mu_w \rho_w / \mu_e \rho_e}{\sqrt{\bar{C}}} \tag{6.49}$$

The term $(c_f/2)_i$ represents the skin friction coefficient corresponding to a constant-property boundary layer at the same local length Reynolds number. Similarly, the Stanton number is given by

$$\text{St} = \frac{0.332}{\sqrt{\text{Re}_{x_e}}} \frac{\mu_w \rho_w / \mu_e \rho_e}{\sqrt{\bar{C}}} \text{Pr}^{-2/3} = \text{St}_i \frac{\mu_w \rho_w / \mu_e \rho_e}{\sqrt{\bar{C}}} \tag{6.50}$$

where St_i is the corresponding constant-property Stanton number.

For a gas that satisfies the perfect gas equation of state and whose viscosity obeys the equation

$$\frac{\mu}{\mu_e} = \frac{T}{T_e} \tag{6.51}$$

where T_e is the reference condition, the constant $\overline{C}$ becomes

$$\overline{C} = C_e = \frac{\mu\rho}{\mu_e\rho_e} = \frac{\mu_w\rho_w}{\mu_e\rho_e} \equiv 1 \qquad (6.52)$$

Thus, Eqs. 6.49 and 6.50 indicate that the skin friction coefficient and Stanton number remain equal to their constant-property values. In terms of these dimensionless transfer coefficients, the effects of the linear dependence of viscosity on temperature just cancel those of the perfect gas variation of the density. It should be noted, however, that the density variation itself still affects the boundary layer thickness.

For a constant-temperature plate, Chapman and Rubesin [9] modified Eq. 6.51

$$\frac{\mu}{\mu_e} = C_{ew} \frac{T}{T_e} \qquad (6.53)$$

in order to approximate better the actual viscosity distribution near the surface. Equation 6.53 with the perfect gas equation of state yields

$$\overline{C} = C_{ew} = \frac{\mu_w\rho_w}{\mu_e\rho_e}$$

and Eqs. 6.49 and 6.50 become

$$\frac{c_f}{2} = \left(\frac{c_f}{2}\right)_i \sqrt{\frac{\mu_w\rho_w}{\mu_e\rho_e}} \qquad (6.54)$$

$$\mathrm{St} = \frac{c_f}{2} \mathrm{Pr}^{-2/3} \qquad (6.55)$$

$$\mathrm{St} = \mathrm{St}_i \sqrt{\frac{\mu_w\rho_w}{\mu_e\rho_e}} \qquad (6.56)$$

The above skin friction relationship was deduced intuitively many years earlier by von Kármán [10], who assumed local wall properties would control the skin friction law when property variations occur. Thus, Eq. 6.54 is equivalent to

$$\left(\frac{c_f}{2}\right)_w = \frac{0.332}{\sqrt{\rho_w u_e x/\mu_w}}$$

where

$$\tau_w = \rho_w u_e^2 \left(\frac{c_f}{2}\right)_w$$

Similarly, since Pr has been assumed constant in Eq. 6.47, the modified Reynolds analogy also applies under these conditions with

$$\mathrm{St}_w = \left(\frac{c_f}{2}\right)_w \mathrm{Pr}^{-2/3}$$

where

$$q_w'' = \mathrm{St}_w\, \rho_w u_e (i_w - i_{aw})$$

Sutherland Law Viscosity.   Crocco [5] solved equations equivalent to Eqs. 6.37 and 6.38 utilizing the rather accurate Sutherland viscosity relationship and a constant value of Pr other than unity. For a gas that satisfies the ideal equation of state, the quantity C_r, referred to free-stream conditions, becomes

$$C_r = \frac{\mu\rho}{\mu_e\rho_e} = \sqrt{\frac{T}{T_e}} \left(\frac{1+\theta_0}{T/T_e + \theta_0}\right) \qquad (6.57)$$

where $\theta_0 = T_{sc}/T_e$ and T_{sc} is the Sutherland constant corresponding to the specific gas. Values of θ_0 are indicated in Table 6.1 for a variety of gases and boundary layer edge temperatures characteristic of those occurring in the stratosphere, under room conditions, and in products of combustion. Crocco obtained numerical results for $\theta_0 = 0$, $\frac{1}{3}$, 1, and 3. Because enthalpy is employed as the thermodynamic dependent variable in Eq. 6.38 to account for specific heat variations, it is necessary to express C_e in terms of enthalpy rather than temperature as in Eq. 6.57. This is no problem when attention is confined to a specific gas where the enthalpy and temperature are uniquely related at specified pressures. Crocco, however, chose to avoid this approach because it would confine his results to specific gases and thermodynamic conditions. To retain generality, he made the assumption that the temperature ratio in Eq. 6.57 can be replaced by the enthalpy ratio i/i_e without introducing serious errors.

TABLE 6.1   Prandtl Number and Sutherland Constant for Gases [5]

Gas	Pr, $T = 230$ K	Sutherland constant T_{sc}, K	$\theta_0 = T_{sc}/T_e$ T_e 218 K	T_e 300 K	T_e 3000 K
H_2	0.717	90	0.413	0.300	0.030
CO	0.765	104	0.477	0.347	0.035
N_2	0.739	112	0.514	0.373	0.037
Air	0.725	116	0.532	0.387	0.039
O_2	0.731	131	0.601	0.437	0.044
CO_2	0.805	266	1.220	0.887	0.089
H_2O	1.08	673	3.09	2.24	0.224

A major result of Crocco's numerical solutions was the discovery that the functional dependence of the local enthalpy on the local velocity is independent of the particular law of viscosity employed. Thus, $i(f')$ found for the simplified case of $C_e \equiv 1$ applies for all values of θ. The conclusions deduced from this discovery are that the modified Reynolds analogy of Eq. 6.24 or Fig. 6.4 and the recovery factor expression of Eq. 6.32 or Fig. 6.4 apply to all gases, regardless of their viscosity laws, as long as Prandtl number is constant. Another significant consequence of this discovery is that it simplifies the solution of Eqs. 6.37 and 6.38 by avoiding either a simultaneous solution of two differential equations or a sequential iteration process. The simpler process uses Eq. 6.48 and the Blasius solution to relate i and f'. Then C_e is evaluated in terms of f' through Eq. 6.57 with T/T_e replaced by i/i_e and with the proper θ_0, and Eq. 6.37 is solved to yield the final velocity distribution. The local enthalpy distribution in terms of the local velocity is given by

$$\frac{i}{i_e} = 1 + \frac{i_w - i_{aw}}{i_e} Y_0(f') + \frac{u_e^2}{2i_e} \mathbf{r}(0) \frac{\mathbf{r}(f')}{\mathbf{r}(0)} \tag{6.58}$$

where the enthalpy profile functions Y_0 and $\mathbf{r}/\mathbf{r}(0)$ are plotted in Figs. 6.8 and 6.9 for several Prandtl numbers. Many authors have argued that Eq. 6.41 can be modified to account for Prandtl number deviations from unity by replacing I_e by i_{aw} and multiplying the last term by $\mathbf{r}(0)$. A comparison with Eq. 6.58 reveals that this suggestion is equivalent to retaining Y_0 and $\mathbf{r}/\mathbf{r}(0)$ characteristic of Pr $= 1$ for all Prandtl numbers. Reference to Figs. 6.8 and 6.9 indicates the errors introduced by this procedure. For example, on a plate at constant temperature, the local temperature at $f' = u/u_e = 0.5$ is 10 percent higher at Pr $= 0.725$ from Fig. 6.8 than would be given by the aforementioned rule.

Correlation equations that fit these results within 0.015 of the ordinates are

$$Y_0 = 1 - \text{Pr}^{1/2} \frac{u}{u_e} - (1 - \text{Pr}^{1/3}) \left(\frac{u}{u_e}\right)^{6.3\text{Pr}^{-1/2}} \tag{6.59}$$

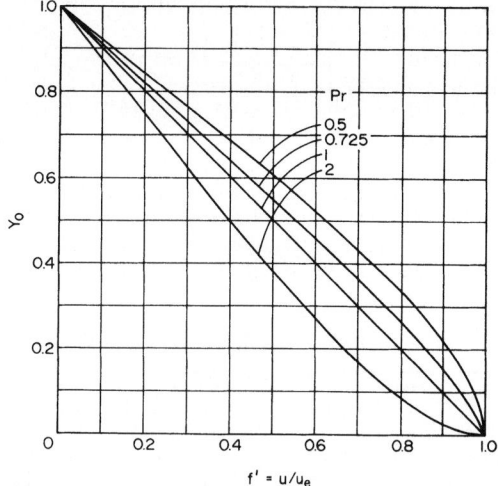

FIGURE 6.8 Laminar boundary layer enthalpy profile function on a uniform-temperature flat plate, $C_e = 1$ [5].

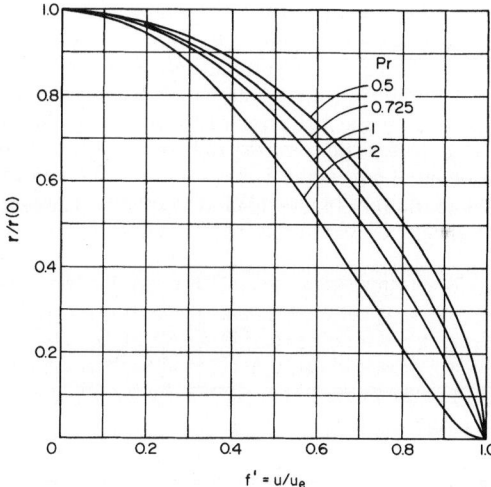

FIGURE 6.9 Laminar boundary layer enthalpy profile function on an insulated flat plate, $C_e = 1$ [5].

and
$$\frac{\mathbf{r}}{\mathbf{r}(0)} = 1 - \mathrm{Pr}^{1/2}\left(\frac{u}{u_e}\right)^2 - (1 - \mathrm{Pr}^{1/2})\left(\frac{u}{u_e}\right)^{7.3\mathrm{Pr}^{-0.38}} \tag{6.60}$$

An example of the skin friction results obtained by Crocco for the case of $\theta_0 = 3$ is shown in Fig. 6.10. The individual curves represent values for constant surface temperatures where $i_w/i_e = 0.25, 0.50, 1.0, 1.5$, and 2.0, and for an insulated plate.

For slender aircraft flying in the stratosphere, the temperature at the edge of the boundary layer is 218 K (−67.6°F) and the value of θ_0 for air based on the Sutherland constant is 0.505. Van Driest [11] repeated Crocco's analysis for these conditions and Pr = 0.75. Graphs of the

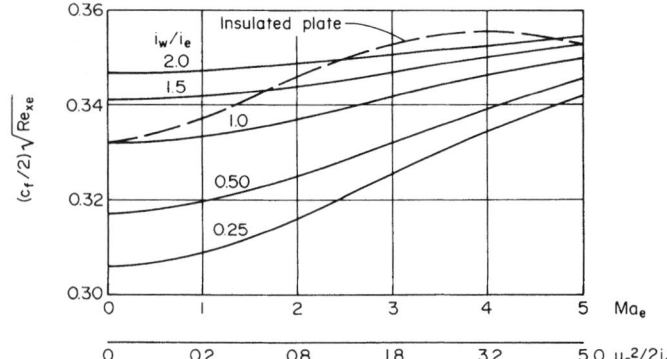

FIGURE 6.10 Local laminar boundary layer skin friction coefficient on a flat plate at uniform temperature, $Pr = 0.725$ and $\theta_0 = 3$ [5].

local velocity, temperature, and Mach number profiles for an extensive range of conditions are presented in Ref. 11. The local skin friction coefficient is indicated in Fig. 6.11 for Ma_e up to 20, for wall temperatures $0.25 \leq i_w/i_e \leq 6$, and for an insulated plate. Two examples of the solutions based on a constant value of $C_e = C_{ew}$ are indicated for comparison with the insulated plate curve (circled points). Note that the use of wall properties underestimates the skin friction here by about 5 percent.

Air as a Real Gas in Chemical Equilibrium. At reentry speeds, the high enthalpies introduce Prandtl number variations and the nonideal effects of dissociation and ionization in the behavior of equilibrium air. Several studies [12–17] have determined the effects of these property variations on the behavior of the laminar boundary layer for successively increasing speeds. A characteristic common to these theories because of the complexity of the behavior of air at elevated enthalpies is the reliance on completely numerical computation of a relatively limited number of examples. The results, however, are not markedly different from the

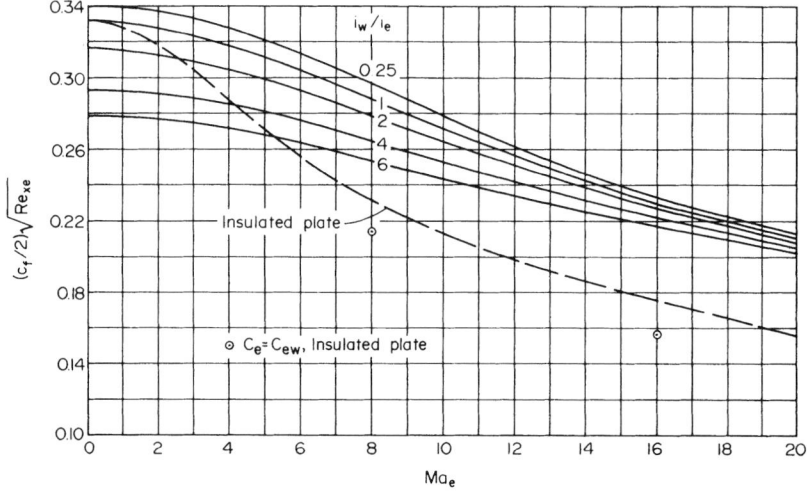

FIGURE 6.11 Local skin friction coefficient for air flowing in a laminar boundary layer on a flat plate, $Pr = 0.75$, $\theta_0 = 0.505$ [11].

ideal-gas cases. The variations in Pr_T cause the recovery factor to be dependent on the surface temperature and Mach number at the edge of the boundary layer. When this relatively small effect is taken into account, the skin friction and heat transfer coefficients exceed the van Driest results of the previous section by less than 15 percent for enthalpies characteristic of flight speeds less than 25,000 ft/s (7620 m/s) and wall temperatures below the sublimation temperature of carbon (6000°R or 3333 K). When ionization takes place, however, a marked increase occurs in both the skin friction and heat transfer.

Errors are inherent in the above solutions because of the uncertainties in the transport properties of air at very high temperatures. The theories of Refs. 15 and 17 employed total properties from different sources, while Ref. 16 accounted for equilibrium air by using frozen properties and Le = 1 in the diffusional heat flux contribution. A comparison of skin friction and heat transfer coefficients reveals differences of less than 10 percent between the results of Refs. 15 and 16 and only a few percent between the results of Refs. 16 and 17. Thus, prior to ionization the errors in convective heating predictions caused by property uncertainties are rather small. With the onset of ionization, large errors may have been introduced because of the large uncertainty of the thermal conductivity of ionized air as influenced primarily by the charge-transfer cross section of atomic nitrogen. Hence, the marked increase in heat transfer rate with the presence of ionization [17] can only be considered qualitatively correct.

A technique for correlating the results for the convective heating behavior of any ideal gas and real air is given in the following section.

Reference Enthalpy Method. The behavior of the skin friction coefficient indicated in Figs. 6.10 and 6.11 can be correlated to a very good approximation by the modified incompressible formula

$$\frac{c_f}{2} = \frac{0.332}{\sqrt{\text{Re}_{x_e}}} \sqrt{\frac{\rho'\mu'}{\rho_e\mu_e}} \qquad (6.61)$$

where the properties designated with the prime are evaluated at a reference enthalpy i' and the boundary layer edge pressure. This correlation technique was expressed originally in terms of a reference temperature T' [12, 18–20] and later as a reference enthalpy i' to account for variations in specific heat [14, 21].

A convenient form of the reference enthalpy is

$$\frac{i'}{i_e} = \mathbf{a}* + \mathbf{b}* \frac{i_w}{i_e} + \mathbf{c}* \frac{i_{aw}}{i_e}$$

On evaluation of the coefficients $\mathbf{a}*$, $\mathbf{b}*$, and $\mathbf{c}*$ based on the van Driest values of skin friction in Fig. 6.11, it is found that they differ from those given in the earlier references. Therefore, the following formula is adopted:

$$\frac{i'}{i_e} = 0.32 + 0.50 \frac{i_w}{i_e} + 0.18 \frac{i_{aw}}{i_e} \qquad (6.62)$$

A convenient form of the skin friction coefficient compatible with the Crocco and van Driest formulations is

$$\frac{c_f}{2} = \frac{0.332}{\sqrt{\text{Re}_{x_e}}} \sqrt{\left(\frac{i'}{i_e}\right)^{1/2} \frac{1+\theta_0}{i'/i_e + \theta_0}} \qquad (6.63)$$

Note that Eq. 6.62 differs slightly from the popular Eckert relationship

$$\frac{i'}{i_e} = 0.28 + 0.50 \frac{i_w}{i_e} + 0.22 \frac{i_{aw}}{i_e} \qquad (6.64)$$

The relative accuracies of these formulations are indicated in Table 6.2 for the skin friction coefficients shown in Figs. 6.10 and 6.11.

TABLE 6.2 Comparison of Skin Friction Coefficients
From Reference Enthalpy Methods

Source	% RMS error [11] Air	% RMS error [5] $\theta_0 = 0$	$\theta_0 = \frac{1}{3}$	$\theta_0 = 1$	$\theta_0 = 3$
Eq. 6.64	2.0	2.3	1.1	0.65	0.83
Eq. 6.62	0.64	1.2	0.30	0.50	0.64

The figures of merit indicated in this table represent the RMS errors of the formulas at the matrix of points $Ma_e = 0$ and 5, $i_w/i_e = 0.25, 2.0$, and i_{aw}/i_e in Fig. 6.10 for $\theta_0 = 3$ and for similar points for the other values of θ_0; and at $Ma_e = 0, 5,$ and 10, $i_w/i_e = 0.25, 6,$ and i_{aw}/i_e in Fig. 6.11 for $\theta_0 = 0.505$.

From Table 6.2 for air, it can be seen that Eq. 6.62 gives some improvement in comparison with the van Driest results over the older formulas. At the high speeds where air behaves as a real gas, Wilson [15] shows that equations equivalent to Eqs. 6.61 and 6.62 yield skin friction coefficients that agree with those found from numerical integrations of the boundary layer equations to within 5 percent for total enthalpies corresponding to free-stream speeds up to 25,000 ft/s (7620 m/s). Similar close agreement is achieved between the use of the Eckert reference enthalpy and results of Cohen [16].

For real air, the total Prandtl number varies in an oscillating manner with enthalpy distribution across the boundary layer. In view of this behavior, it would not be expected a priori that evaluation of Pr_T at the reference enthalpy would be appropriate for evaluating the recovery factor from Eq. 6.32 or the modified Reynolds analogy from Eq. 6.24. Comparison with the numerical results of Refs. 12, 13, and 15, however, reveals that this interpretation of the reference enthalpy technique yields results for the recovery factor correct to within 2.5 percent, and for the Reynolds analogy correct to within 5 percent. Note that Wilson [15] suggests the use of Pr_{T_w} evaluated at wall enthalpy in the Reynolds analogy. Comparison of this method with the use of Pr_T' evaluated at the reference enthalpy for surface temperatures below the sublimation temperature of carbon reveals little difference. Because Pr_T' rather than Pr_w or Pr_e yields better agreement with the modified Reynolds analogy of van Driest [13], the consistent use of Pr_T' in both the recovery factor and Reynolds analogy as suggested by Eckert is still appropriate. Further, when the assumption that $Le = 1$ in Ref. 16 is interpreted as equivalent to the assumption that $Pr_F = Pr_T$, the use of Pr_T' based on the reference enthalpy for the recovery factor and Reynolds analogy is again borne out by these independent calculations to the accuracies quoted previously.

With the onset of ionization, the reference enthalpy technique yields results that begin to depart from the exact calculations, and recourse to the latter [16, 17] is recommended for accurate predictions.

Nonuniform Surface Temperature. The previous section was devoted to uniform-temperature plates. In practice, however, this ideal condition seldom occurs, and it is necessary to account for the effects of surface temperature variations along the plate on the local and average convective heat transfer rates. This is required especially in the regions directly downstream of surface temperature discontinuities, e.g., at seams between dissimilar structural elements in poor thermal contact. These effects cannot be accounted for by merely utilizing heat transfer coefficients corresponding to a uniform surface temperature coupled with the local enthalpy or temperature potentials. Such an approach not only leads to serious errors in magnitude of the local heat flux, but can yield the wrong direction, i.e., whether the heat flow is into or out of the surface.

It has been shown that, for property variations for which superposition of solutions is permitted, a series of solutions corresponding to a step in surface temperature can be utilized to represent an arbitrary surface temperature [22]. This approach is identical with the Duhamel method used in heat conduction problems to satisfy time-dependent boundary conditions with a series of solutions involving sudden changes in surface boundary conditions [23]. The resulting convective heat flux distribution expressed in terms of the surface enthalpy distribution is

$$q_w''(x) = \rho_e u_e \, \text{St} \, (x, 0) \left\{ [i_w(0) - i_{aw}] + \int_0^x \frac{\text{St} \, (x, s)}{\text{St} \, (x, 0)} \frac{d[i_w(s) - i_{aw}]}{ds} \, ds \right.$$

$$\left. + \sum_{j=1}^k \frac{\text{St} \, (x, s_j)}{\text{St} \, (x, 0)} [i_w(s_j^+) - i_w(s_j^-)] \right\} \quad (6.65)$$

Here, St (x, s) represents the local Stanton number on a plate at a uniform temperature for $x > s$ preceded by an unheated portion of length s, and St $(x, 0)$ is the Stanton number on a plate with a uniform temperature over its entire length. The first term in parentheses in the enthalpy potential arises from the difference between the leading-edge enthalpy of the plate and the recovery enthalpy. The integral term accounts for the portions where continuous surface enthalpy variations occur. The last term sums over a k number of discontinuous jumps in surface enthalpy that may occur downstream of the leading edge. The terms $i_w(s_j^-)$ and $i_w(s_j^+)$ represent the surface enthalpy just upstream and downstream of s_j where the jth jump in enthalpy occurs.

The effect of a stepwise discontinuity in surface temperature on a flat plate can be expressed as

$$\frac{\text{St} \, (x, s)}{\text{St} \, (x, 0)} = \left[1 - \left(\frac{s}{x} \right)^{3/4} \right]^{-1/3} \quad (6.66)$$

This closed-form equation was obtained through similarity solutions of the energy equation by investigators who assumed that the velocity profile in the boundary layer is linear in η_c for the case of constant C_e and Pr_T or in y for the case of constant fluid properties [24, 25]. Note that the right side does not contain terms involving the fluid properties, a direct consequence of C_e and Pr_T being assumed constant throughout the boundary layer. Again, an intuitive approach to include property variations is to use the local surface enthalpy in the reference enthalpy technique for evaluating St $(x, 0)$. The stepwise discontinuous-surface-temperature solutions are used solely to define the functional form of the enthalpy potential appropriate to an arbitrary surface temperature. A plot of Eq. 6.66 is given in Fig. 6.12 ($\beta_p = 0$ for a flat plate).

The preceding theory has been verified by several experiments. For example, in Ref. 26, local heat transfer rates were measured in the presence of ramplike temperature distributions that began downstream of the leading edge (see inset in Fig. 6.13). The data shown in Fig. 6.13 agree with the theory (solid line) to within ±10 percent, the estimated accuracy of the data. The dot-dashed line in the figure represents the use of a local temperature potential in estimating the heat flux and yields large errors for this particular form of the surface temperature distribution. Had the leading edge been raised above the recovery temperature, the error in neglecting the variable-surface-temperature effects would have diminished. In general, if continuous variations in the surface temperature or enthalpy are large compared to the overall driving potential, the variable-surface-temperature methods must be utilized. For discontinuous surface temperatures, much smaller variations are important.

Stepwise and Arbitrary Heat Flux Distribution. It is often necessary to evaluate the surface temperatures resulting from a prescribed heat flux distribution. The superposition of solutions yields the surface enthalpy distribution as

$$i_w(x) - i_{aw} = \frac{0.207}{\rho_e u_e \, \text{St} \, (x, 0) x} \int_0^x \frac{q_w''(x')}{[1 - (x'/x)^{3/4}]^{2/3}} \, dx' \quad (6.67)$$

Surface With Mass Transfer. An effective method of alleviating the intense convective heating of surfaces exposed to very-high-enthalpy streams is by means of mass transfer cooling systems. The coolant is introduced, usually in gaseous form, into the hot boundary

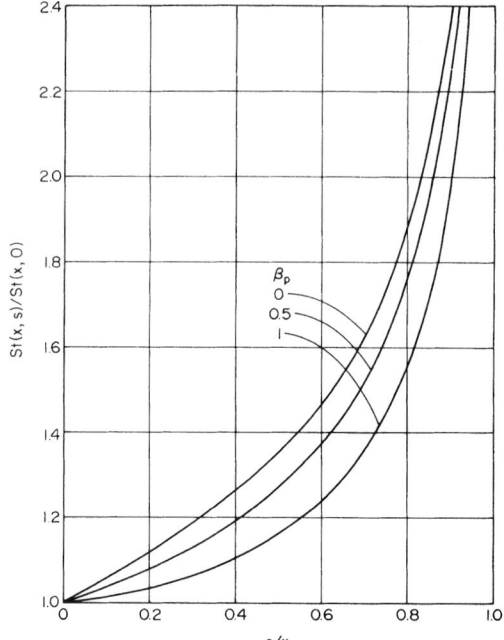

FIGURE 6.12 Effect of a stepwise surface temperature discontinuity on the local Stanton number—Eq. 6.66 for a flat plate ($\beta_p = 0$) and Eq. 6.124 for a surface with a pressure gradient ($\beta_p > 0$)

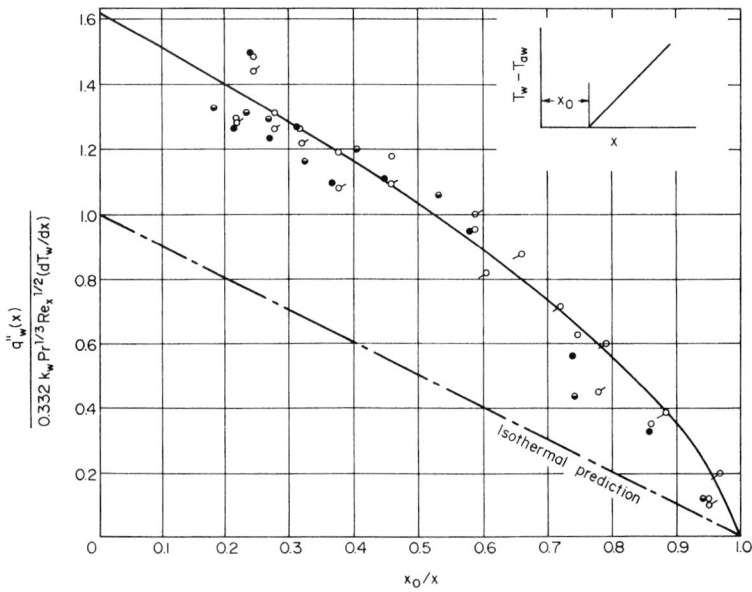

FIGURE 6.13 Comparison of data and theory for a flat plate with a delayed ramp surface temperature distribution [26].

layer through the surface being protected. Mass transfer cooling is particularly applicable to leading edges of wings, fins, and nose tips of aircraft; turbine blades; and reentry capsules and missile nose cones. The types of systems include transpiration, ablation, and film cooling.

A *transpiration cooling system* is characterized by a porous surface material that remains intact while the coolant is being pumped through the pores toward the hot boundary layer. The coolant may be a gas or a liquid that changes phase within the porous surface or after it emerges from the surface. This system operates best when pore sizes are so small that the coolant leaves at the surface temperature of the porous solid. These systems are complicated in that they require coolant storage vessels, pumps, controls, distribution ducts, and filters to avoid pore clogging. It is also difficult to fabricate strong, aerodynamically smooth porous surfaces. Another drawback of these systems is that they are unstable because a clogged pore resulting from local overheating seals off the flow of coolant and causes local failure. The advantages of transpiration cooling systems are their versatility in the choice of coolant and coolant distribution, the reusability of the system, and the retention of intact aerodynamically contoured surfaces.

An *ablation cooling system* is one where the gas entering the boundary layer has been generated by the thermal destruction of a sacrificial solid thermally protecting an underlying structure. The simplest ablation mechanism is the sublimation of a homogeneous material. More complex ablation involves the thermal degradation of composite materials such as reinforced plastics where a heat-absorbing pyrolysis occurs below the surface. The gaseous products of pyrolysis pass through a carbonaceous char, gaining additional sensible heat or chemical heat through endothermic reactions, and then pass into the boundary layer to absorb additional heat. Ablation cooling has the enormous advantage of being self-controlled and requiring no active elements. The disadvantages are that surface dimensions are usually altered, part of the char is eroded mechanically by shear forces rather than through heat-absorbing phase change, and often the heavy gaseous products are not as effective in blocking the incoming convective heat as light gases. Furthermore, ablation systems are generally not reusable. For short-time applications, dimensional stability has been achieved in ablation systems by employing porous refractory metal surfaces that have been impregnated with lower-melting-temperature metals that absorb heat by melting, vaporizing, and transpiring through the porous refractory matrix.

A *film cooling system* is one where a surface is protected by a film of coolant introduced into the boundary layer from either a finite length of porous surface or a slot upstream. A liquid can be used as the coolant to absorb heat by vaporization as it is drawn along the surface by the main stream gas. A severe limitation on such systems is the requirement that gravity or inertial forces act in a direction that will keep the liquid film stable and against the surface. In addition, a film cooling system requires all the active elements of a transpiration cooling system. Its main advantage over the latter is in the simpler construction of limited areas of porous surfaces or slots and in localized ducting.

The boundary layer behavior over a continuously transpiration-cooled surface and an ablation-cooled surface is generally the same, differing only as a result of the specific chemical identity of the coolant. The effects of a porous surface when the pore size is below some threshold dimension that is a small fraction of the local boundary layer thickness, and of the flow of liquid char over ablating surfaces, do not appreciably alter the behavior of the boundary layer and can be neglected in design considerations. Thus, boundary layer theory with continuous mass injection is applicable to both types of cooling systems. Further, results of experiments involving transpiration systems can be utilized in the prediction of the behavior of ablation systems. In film cooling systems, because of the discontinuities formed by slots or limited porous regions, the boundary-layer profiles at various stations along the surface are dissimilar so that prediction methods are quite complex and rely on experimental data or rather complicated numerical analyses.

Uniform Surface Temperature, Air as Coolant.   When the boundary layer and coolant gases are the same, the equations controlling boundary layer behavior are Eqs. 6.6–6.8. The mass injection at the surface simply alters the boundary conditions (Eq. 6.9) at the wall to be

$$x > 0,\ y = 0 \qquad u = 0,\ v = v_w(x)$$

$$I = \text{constant} = i_w \text{ or } i_{aw},\ \text{i.e.,} \qquad \frac{\partial i}{\partial y} = 0$$

where

$$f(0) = -\frac{2\rho_w v_w}{\rho_e u_e} \sqrt{\frac{\rho_e u_e x}{\mu_e}} \qquad (6.68)$$

As boundary layer similarity requires $f(0)$ to be independent of x, $\rho_w v_w$ must be proportional to $x^{-1/2}$. A simple heat balance on an element of a porous surface with a transpired gas, or of a subliming surface, reveals that this distribution of gaseous injection is uniquely compatible with the requirements of a constant surface temperature. Thus, $\rho_w = $ constant, and $v_w \sim x^{-1/2}$. This mass injection distribution has practical importance because the porous surface can operate at its maximum allowable temperature everywhere, thereby minimizing the coolant required.

The boundary layer equations, together with these boundary conditions, were solved in a series of similarity theories such as those described in the section on the two-dimensional laminar boundary layer, beginning with solutions for constant properties and progressing to ideal gases with variable properties. A rather complete bibliography of these theories and corroborating experiments is given in Ref. 27. The assumption of constant $C_e = \mu\rho/\mu_e\rho_e$ proved equally useful in this problem as with the impervious plate in extending constant-property solutions to high Mach number cases where air still behaves as an ideal gas. The results shown here are predominantly from Refs. 28–30.

It is found in these analyses that blowing, i.e., negative values of $f(0)$, thickens both the flow and thermal boundary layers. In addition, the velocity profiles take on an S shape characteristic of boundary layers approaching separation (see Fig. 6.19). Separation, $(\partial u/\partial y)_w = 0$, occurs when $f(0) = -1.238$ [28]. These S-shaped profiles are less stable, thereby decreasing transition Reynolds numbers with increased blowing rates. Near the surface, blowing reduces the velocity and temperature gradients and the corresponding shear and heat transfer rates, respectively. The heat transfer does not drop as rapidly as the shear and, consequently, Reynolds analogy becomes dependent on the blowing rate. The recovery factor, $\mathbf{r}(0)$, also depends on the blowing rate when Pr does not equal unity. For the case of a slender body where $\mathbf{r}(0)u_e^2/2 \gg I_w - I_e$, the reduction in the heat flux q_w is dependent on the product of the reduction in Stanton number and the recovery factor. This is shown in Fig. 6.14 by the line labeled Air-Air.

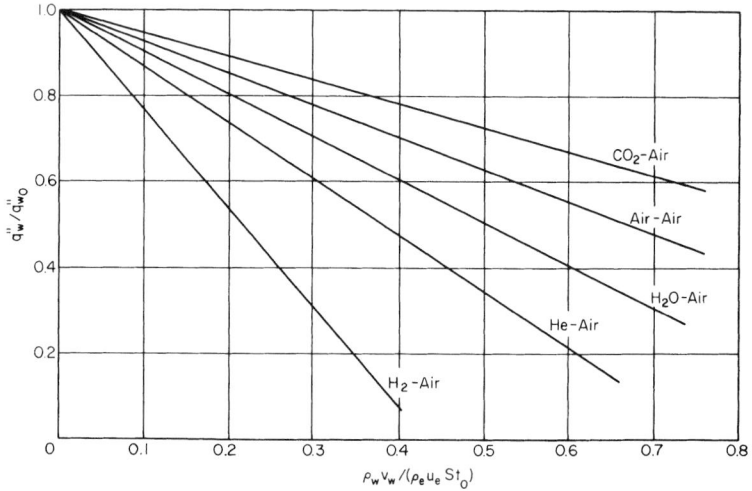

FIGURE 6.14 Reduction of the local heat flux by surface mass transfer of a foreign gas, $T_w - T_e < \mathbf{r}(0)u_e^2/2c_{p_1}$ [31].

Uniform Surface Temperature, Foreign Gas as Coolant. The effectiveness of air injection in reducing convective heat flux stimulated investigations into the use of other coolants. With the introduction of a foreign species into the boundary layer, the boundary layer equations reduce to the continuity equation (Eq. 6.6), the momentum equation (Eq. 6.7), the energy equation

$$\rho u \frac{\partial I}{\partial x} + \rho v \frac{\partial I}{\partial y} = \frac{\partial}{\partial y} \left\{ \frac{\mu}{\mathrm{Pr}_F} \left[\frac{\partial I}{\partial y} - (i_1 - i_2)(1 - \mathrm{Le}) \frac{\partial K_1}{\partial y} \right] + \left[\mu \left(1 - \frac{1}{\mathrm{Pr}_F} \right) \frac{\partial (u^2/2)}{\partial y} \right] \right\} \quad (6.69)$$

and the diffusion equation

$$\rho u \frac{\partial K_1}{\partial x} + \rho v \frac{\partial K_1}{\partial y} = \frac{\partial}{\partial y} \left(\frac{\mu \, \mathrm{Le}}{\mathrm{Pr}_F} \frac{\partial K_1}{\partial y} \right) \quad (6.70)$$

Here, $\mathrm{Pr}_F/\mathrm{Le} = \mathrm{Sc}$ is the Schmidt number, K_1 is the coolant mass concentration, and i_1 and i_2 are the coolant and air enthalpies, respectively. These equations are the bases for the bulk of the analyses involving foreign gas transpiration.

The boundary conditions for Eqs. 6.69 and 6.70 are

$$\begin{aligned} x = 0, \; y \geq 0 & \quad K_1 = 0 \\ x > 0, \; y \to \infty & \quad K_1 \to 0 \\ y = 0 & \quad K_1 = K_{1w} \end{aligned}$$

The value of K_{1w} is dependent on the blowing rate, and some hypothesis is required to establish this dependency. Most authors have assumed a zero net mass transfer of air into the surface. This requires that the air carried away from the surface by the mass motion normal to the surface be balanced by the diffusion of air toward the surface. Since $K_1 = 1 - K_2$ and $j_1 = -j_2$, this balance of air motion can be expressed directly in terms of the coolant properties as

$$\rho_w v_w = -\frac{\rho_w \mathcal{D}_{12}}{1 - K_{1w}} \left(\frac{\partial K_1}{\partial y} \right)_w = \frac{j_{1w}}{1 - K_{1w}} \quad (6.71)$$

Equation 6.71 is the required relationship between the blowing rate of the coolant and its concentration at the wall. When the diffusion flux is expressed in terms of the mass transfer coefficient, the concentration of the coolant at the wall is given by

$$K_{1w} = \frac{1}{1 + (\rho_e u_e / \rho_w v_w) c_{m_1}} \quad (6.72)$$

Note that the mass transfer coefficient is defined differently in Ref. 31.

Although the total heat flux at the surface in a binary gas is composed of the sum of a conduction term and a diffusion term, the results of analyses are expressed solely in terms of the heat conduction term. The reason is that this term is equal to the heat gained by the coolant in passing from its reservoir to the surface in contact with the boundary layer. This simple heat balance is

$$k_w \left(\frac{\partial T}{\partial y} \right)_w = \rho_w v_w (i_{1w} - i_{1c}) \quad (6.73)$$

where the subscript c represents the initial coolant condition in the reservoir. Further, the recovery factor is defined in terms of the surface temperature that results in zero heat conduction at the surface for a given mass transfer rate. This adiabatic condition is achieved in experiments by setting $T_{1c} = T_w$ for a given mass flow rate so that the coolant gains no heat. In terms of the Stanton number, the heat balance indicated by Eq. 6.73 can be expressed as

$$\frac{q_w''}{q_{w0}''} = \frac{\rho_w v_w}{\rho_e u_e \, \mathrm{St}_0} \frac{i_{1w} - i_{1c}}{i_{2aw_0} - i_{2w}} \quad (6.74)$$

The subscript 0 in this equation signifies zero mass transfer at the surface ($f(0) = 0$). The subscript 2 indicates that the enthalpy potential contributing to the heat flux by conduction alone is dependent only on the specific heat of the air and not that of the coolant.

The effect of mass transfer on heat flux from slender bodies is shown in Fig. 6.14 for a variety of coolant gases. The corresponding mass transfer coefficients are shown in Fig. 6.15.

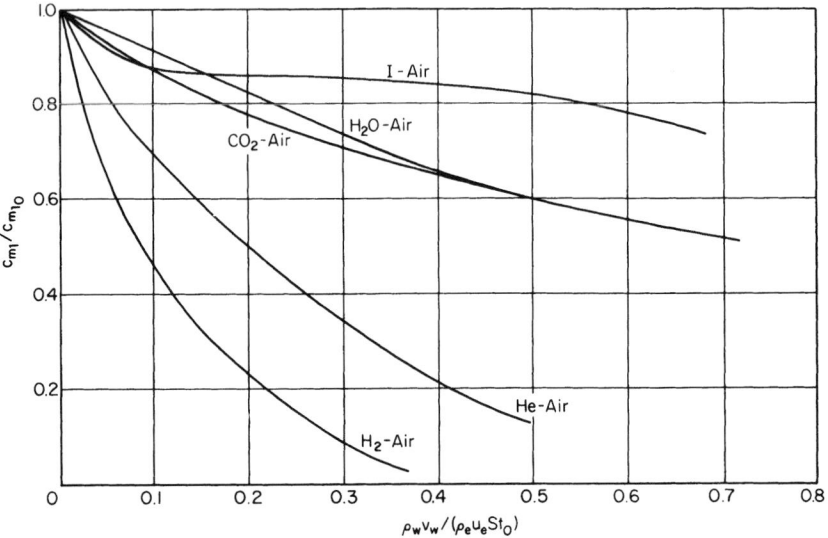

FIGURE 6.15   Reduction of the local mass transfer coefficient by surface mass transfer of a foreign gas [31]. (Reprinted by permission of Pergamon Press.)

An important result from these figures is that lighter gases are more effective in reducing the transport coefficients and surface heat flux. For a range of calculations including Mach numbers as high as 12 and surface temperatures from free-stream (392°R; 218 K) to recovery temperatures, the maximum departure from these correlation lines of individual solutions is ±15 percent for q_w'' and ±25 percent for c_{m_i}. The bulk of the discrete numerical results lie within about one-third of these bandwidths. The maximum spread of results is obtained with the mass transfer of hydrogen, and the spread is smaller with helium and much smaller with the heavier gases. The differences in results from different calculations with the light gases are due primarily to the use of different gas properties [32].

Figures 6.14 and 6.15 are particularly useful for obtaining the mass transfer rate required in a transpiration cooling system. Usually the following quantities are specified: the coolant and its reservoir conditions, the porous surface material and its maximum allowable surface temperature, and the inviscid flow conditions outside the boundary layer. For cases where the difference between the surface temperature and the boundary layer edge temperature is small compared to the temperature rise by frictional heating, Fig. 6.14 can be used directly. For the specified conditions, the factor $(i_{1w} - i_{1c})/(i_{2aw_0} - i_{2w})$ can be readily established from the thermodynamic properties of the coolant and air and $i_{2aw_0} = i_{aw}$. This factor is used as the slope of a straight line that represents Eq. 6.74 and passes through the origin of Fig. 6.14. The intercept of this line with the appropriate heat blockage curve on the figure is the operating condition required to yield the specified surface temperature. The abscissa of this point yields the required local mass transfer rate.

The behavior of a coolant composed of mixtures of He, Ar, Xe, and N_2 injected into a nitrogen boundary layer has been analyzed in Ref. 33. Examination of the results reveals that at a given mass transfer rate the skin friction coefficient is correlated for different coolants within ±5 percent using the mean molecular weight of the coolant. Thus, for a coolant composed of a mixture of n gases, the average molecular weight of the coolant at reservoir conditions defined as

$$M_{1_{\text{avg}}} = \frac{1}{\sum_i^n (K_{ic}/M_i)} \tag{6.75}$$

can be used to interpolate between the curves of Figs. 6.14 and 6.15.

Uniform Surface Injection. Although a mass transfer distribution yielding a uniform surface temperature is most efficient, it is much easier to construct a porous surface with a uniform mass transfer distribution. Libby and Chen [34] have considered the effects of uniform foreign gas injection on the temperature distribution of a porous flat plate. For these conditions, however, boundary layer similarity does not hold. Libby and Chen extended the work of Iglisch [35] and Lew and Fanucci [36], where direct numerical solutions of the partial differential equations were employed. An example of the nonuniform surface enthalpy and coolant concentrations resulting from these calculations is shown in Fig. 6.16.

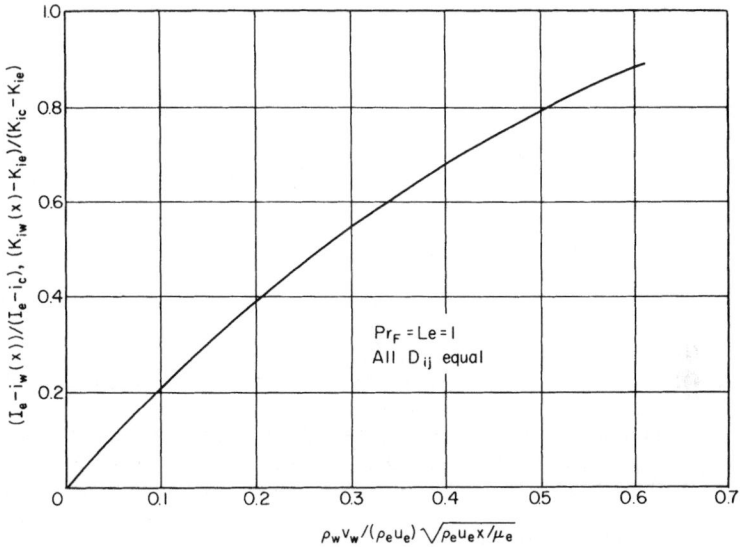

FIGURE 6.16 Surface temperature and coolant concentration distribution along a plate with uniform foreign gas injection, Le = Pr = C_e = 1, all binary diffusion coefficients equal [34]. (Reprinted by permission from *The Physics of Fluids.*)

Film Cooling With Upstream Transpiration. Film cooling systems provide protection to a surface exposed to high-enthalpy streams by injecting a coolant into the hot boundary layer upstream of the surface. Injection can take place through local porous sections or slots at various angles to the surface. The coolant may be the same gas as in the boundary layer, a foreign gas, or a liquid. In the upstream porous section, more coolant is provided than is required to maintain safe surface temperatures. The excess coolant is entrained in the boundary layer close to the surface and is carried downstream, providing an insulating layer between the hot free-stream gas and the surface. This layer is dissipated by laminar mixing while flowing with the boundary layer in such a way that protection is afforded over a limited distance. Because of the discontinuous nature of the surface injection, the boundary layer downstream of the discontinuity is far from similar. Numerical solutions of higher-order integral equations [37] or of finite difference forms of the partial differential equations [38] have been used for evaluating the local convection to the surface.

Examples of thermal protection offered by an upstream transpiration system are indicated in Fig. 6.17 (see Ref. 37). The figure shows the temperature rise that occurs on an insulated surface downstream of a transpiration-cooled section for two amounts of blowing. Also indicated are the corresponding surface temperatures for the case where the upstream section was cooled internally, i.e., $f(0) = 0$. The quantity T_{wL} is the upstream surface temperature in either case. The differences between the curves labeled $f(0) = -0.5$ and $f(0) = -1$ and the $f(0)$ curve show how the presence of the transpired coolant within the downstream boundary layer distorts the temperature profiles so as to afford greater downstream protection depending on the amount of blowing.

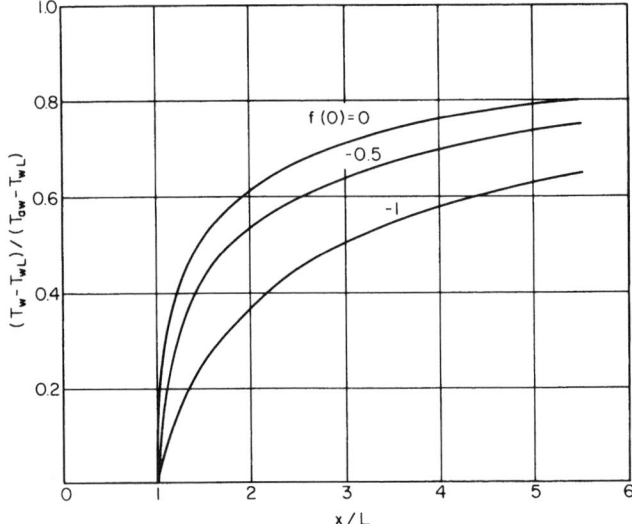

FIGURE 6.17   Effect of upstream transpiration cooling on the temperature distribution of the impervious portion of a flat plate.

Cone in Supersonic Flow.   The preceding solutions for a flat plate may be applied to a cone in supersonic flow through the Mangler transformation [39], which in its most general form relates the boundary layer flow over an arbitrary axisymmetric body to an equivalent flow over a two-dimensional body. This transformation is contained in Eq. 6.89, which results in transformed axisymmetric momentum and energy equations equivalent to the two-dimensional equations (Eqs. 6.95 and 6.97). Hence, solutions of these equations are applicable to either a two-dimensional or an axisymmetric flow, the differences being contained solely in the coordinate transformations.

For the case of an arbitrary pressure distribution, it is just as convenient to solve the axisymmetric problem directly. When the solution for the equivalent two-dimensional flow already exists, however, as for a flat plate, then the results for the corresponding axisymmetric problem can be obtained by direct comparison. This correspondence exists for a cone in supersonic flow when the surface pressure is constant. Solutions of Eqs. 6.95 and 6.97 for a flat plate expressed in terms of ζ and η may then be applied to a cone. Illustrative examples are presented in the following subsections.

Uniform Surface Temperature.   Transformations (see Eq. 6.89) for a flat plate ($\mathbf{k} = 0$) or a cone ($\mathbf{k} = 1$) become

$$
\left.
\begin{aligned}
\zeta &= \frac{\rho_w \mu_w u_e}{\mu_r^2} \left(\frac{\sin \theta_c}{\mathbf{L}} \right)^{2\mathbf{k}} \frac{x^{2\mathbf{k}+1}}{2\mathbf{k}+1} \\[2mm]
\eta &= \sqrt{\frac{(2\mathbf{k}+1)u_e}{2\rho_w \mu_w x}} \int_0^y \rho \, dy
\end{aligned}
\right\}
\tag{6.76}
$$

with ρ_w, μ_w, and u_e equal to constants, $\bar{\zeta} = 0$, and $r = x \sin \theta_c$. The transformed momentum and energy equations (Eqs. 6.95 and 6.97) are essentially the same as Eqs. 6.37 and 6.38 for a flat plate. For the same wall and boundary layer edge conditions, then, the solutions for $f(\eta)$ and $I(\eta)$ are the same for a flat plate or a cone. These results may be expressed in terms of physical variables as

$$
y = \sqrt{\frac{2\rho_w \mu_w x}{(2\mathbf{k}+1)u_e}} \int_0^\eta \frac{1}{\rho} \, d\eta
\tag{6.77}
$$

$$
\tau_w = \mu_w \left(\frac{\partial u}{\partial y} \right)_w = \rho_e u_e^2 \sqrt{\frac{(2\mathbf{k}+1)\mu_e}{2\rho_e u_e x}} \frac{\mu_w \rho_w}{\mu_e \rho_e} f''(0)
\tag{6.78}
$$

$$
q_w'' = -k_w \left(\frac{\partial T}{\partial y} \right)_w = -\frac{k_w I_e}{\mu_w c_{p_w}} \sqrt{\frac{(2\mathbf{k}+1)\mu_e}{2\rho_e u_e x}} \frac{\mu_w \rho_w}{\mu_e \rho_e} \bar{I}'(0)
\tag{6.79}
$$

For a given value of x and the same flow properties, the boundary layer on a cone ($\mathbf{k} = 1$) is thinner by a factor of $1/\sqrt{3}$, and the surface shear stress and heat transfer are larger by a factor of $\sqrt{3}$. The local skin friction and heat transfer coefficients are related similarly:

$$
\frac{(c_f)_{\text{cone}}}{(c_f)_{\text{flat plate}}} = \frac{\text{St}_{\text{cone}}}{\text{St}_{\text{flat plate}}} = \sqrt{3}
\tag{6.80}
$$

The local and average coefficients for a cone are related as follows:

$$
\left(\frac{\bar{c}_f}{c_f} \right)_{\text{cone}} = \left(\frac{\overline{\text{St}}}{\text{St}} \right)_{\text{cone}} = \frac{4}{3}
\tag{6.81}
$$

These relationships may be used to obtain cone flow results from the flat-plate results of the section on uniform free-stream conditions. Real-gas solutions for air obtained in this manner are given in Ref. 17.

Nonuniform Surface Temperature.   Transformations (Eq. 6.76) are applicable to flows with nonuniform surface temperature provided a linear viscosity law is assumed ($\mu\rho$ = constant). The flat-plate results given previously for constant Pr_T may be applied to a cone with the requirement that the surface boundary conditions be the same in terms of ζ. For a flat plate, $\zeta \sim x$, and for a cone, $\zeta \sim x^3$. Therefore, the flat-plate results must be modified in such a way that lengths in the x direction are replaced by x^3 to obtain the cone results. For example, Eq. 6.66, which expresses the effect of a stepwise surface temperature for a flat plate, becomes for a cone

$$
\left[\frac{\text{St}(x, s)}{\text{St}(x, 0)} \right]_{\text{cone}} = \left[1 - \left(\frac{s}{x} \right)^{9/4} \right]^{-1/3}
\tag{6.82}
$$

Similar expressions may be derived for the effect of an arbitrary heat flux distribution.

Mass Transfer, Uniform Surface Temperature.   The uniform-surface-temperature results above may be extended to include mass transfer. Similarity requires that $f(0)$ be constant, which determines the blowing distribution along the surface. The normal velocity component from Eq. 6.99 is

$$\frac{v_w}{u_e} = -\sqrt{\frac{(2\mathbf{k}+1)\mu_w}{2\rho_w u_e x}}\, f(0) \tag{6.83}$$

For a cone, $v_w \sim x^{-1/2}$ as for a flat plate, but is larger by the factor $\sqrt{3}$ for given values of x and $f(0)$ (thus, the blowing parameter $\rho_w v_w / \rho_e u_e\, St_0$ remains unchanged). With this difference, the flat-plate results may be applied to a cone according to the equations above.

For a nonsimilar blowing distribution, for example, v_w = constant, Eq. 6.83 is not applicable. Solutions to this problem may be found in Ref. 40.

Surface With Streamwise Pressure Gradient

Gas With Uniform Elemental Composition in Chemical Equilibrium.   Except for a few configurations such as a flat plate and wedges or cones in supersonic flow, the pressure varies over the body surface as determined by inviscid flow theory. The influence of pressure gradients on forced convection in laminar boundary layers is presented in this section. Axisymmetric bodies at zero angle of attack and yawed cylinders of infinite length will be treated together to illustrate their close relationship (see Fig. 6.18). Because boundary layer theory requires negligible pressure gradients across the boundary layer, the techniques presented here apply only to those bodies whose local surface radius of curvature (r_n in Fig. 6.18) is large compared with the boundary layer thickness, thereby minimizing centrifugal forces.

Governing Differential Equations.   In the absence of foreign gas injection, the boundary layer can be considered to have uniform elemental composition and be in chemical equilibrium, and is governed by the following equations:

$$\frac{\partial}{\partial x}(\rho u r^{\mathbf{k}}) + \frac{\partial}{\partial y}(\rho v r^{\mathbf{k}}) = 0 \tag{6.84}$$

$$\rho u \frac{\partial u}{\partial x} + \rho v \frac{\partial u}{\partial y} = -\frac{dP}{dx} + \frac{\partial}{\partial y}\left(\mu \frac{\partial u}{\partial y}\right) \tag{6.85}$$

$$\rho u \frac{\partial w}{\partial x} + \rho v \frac{\partial w}{\partial y} = \frac{\partial}{\partial y}\left(\mu \frac{\partial w}{\partial y}\right) \qquad \text{(for } \mathbf{k} = 0 \text{ only)} \tag{6.86}$$

$$\rho u \frac{\partial I}{\partial x} + \rho v \frac{\partial I}{\partial y} = \frac{\partial}{\partial y}\left[\frac{\mu}{Pr_T}\frac{\partial I}{\partial y} + \mu\left(1 - \frac{1}{Pr_T}\right)\frac{\partial}{\partial y}\left(\frac{u^2 + w^2}{2}\right)\right] \tag{6.87}$$

where $\mathbf{k} = 1$ for the axisymmetric body and $\mathbf{k} = 0$ for the yawed cylinder.

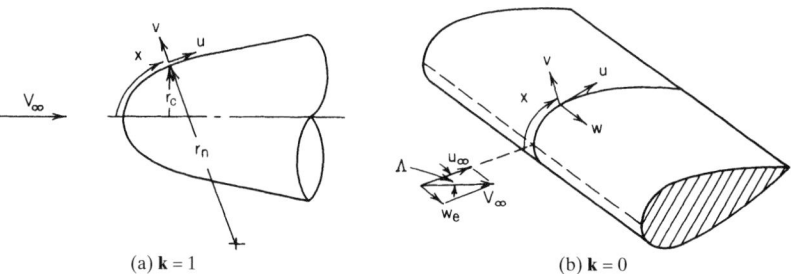

(a) $\mathbf{k} = 1$ (b) $\mathbf{k} = 0$

FIGURE 6.18   Sketch of coordinates employed for related two-dimensional flows. (*a*) axisymmetric body; (*b*) yawed cylinder of infinite length.

The boundary conditions are

$$
\begin{aligned}
x = 0,\, y > 0 \quad & u = u_e(0) \\
& I = I_e \\
& w = w_e = V_\infty \sin \Lambda \quad \text{for} \quad k = 0 \\
x > 0,\, y \to \infty \quad & u \to u_e(x) \\
& I \to I_e \\
& w \to w_e = V_\infty \sin \Lambda \quad \text{for} \quad k = 0 \\
y = 0 \quad & u = w = 0 \\
& v = 0 \text{ for an impervious surface or } v_w(x) \text{ with transpiration} \\
& I = i_w(x) \quad \text{for} \quad q''_w \neq 0 \quad \text{or} \quad \frac{\partial I}{\partial y} = 0 \quad \text{for} \quad I = i_{aw}(x)
\end{aligned}
\tag{6.88}
$$

The yaw angle Λ is the complement of the angle between the free-stream direction and the cylinder axis (see Fig. 6.18).

Transformation of Variables and Equations. The extensions of transformations Eq. 6.36 to include the effects of pressure gradients are

$$
\zeta = \frac{1}{\mu_r^2} \int_0^x \rho_w \mu_w u_e \left(\frac{r}{L} \right)^{2k} dx = \zeta(x)
$$

$$
\eta = \frac{u_e (r/L)^k}{\mu_r \sqrt{2(\zeta - \overline{\zeta})}} \int_0^y \rho \, dy
\tag{6.89}
$$

where μ_r is a reference viscosity and L is a reference length introduced to make ζ and η dimensionless [16, 41]. The function $\overline{\zeta}(\zeta)$ is yet to be determined and is a key element in the extension of similarity solutions to flows where the inviscid boundary conditions do not permit boundary layer similarity. (Note the change in symbols employed here from those of Refs. 16 and 41.)

The dependent variables are defined as

$$
f_\eta = \frac{u}{u_e(x)} \qquad \overline{I} = \frac{I}{I_e} \qquad \overline{w} = \frac{w}{w_e} \qquad \text{(for } k = 0 \text{ only)}
\tag{6.90}
$$

Additional parameters that enter the equations are

$$
\bar{t} = \frac{i}{I_e} = \overline{I} - (1 - \bar{t}_s)\overline{w}^2 - (\bar{t}_s - \bar{t}_e)f_\eta^2
\tag{6.91}
$$

where

$$
\bar{t}_s = 1 - \frac{w_e^2}{2I_e}
\tag{6.92}
$$

and

$$
\bar{t}_e = 1 - \frac{u_e^2 + w_e^2}{2I_e}
\tag{6.93}
$$

The pressure gradient parameter is defined as

$$
\beta_p = \frac{2(\zeta - \overline{\zeta})}{u_e} \frac{\bar{t}_s}{\bar{t}_e} \frac{du_e}{d\zeta}
\tag{6.94}
$$

Both $\bar{t}_e$ and β_p are functions of x through their dependence on $u_e(x)$.

The transformed partial differential equations of momentum and energy are

$$(C_w f_{\eta\eta})_\eta + \left(1 - \frac{d\overline{\zeta}}{d\zeta}\right) f f_{\eta\eta} = \beta_p\left[f_\eta^2 - \frac{\overline{I}}{\overline{I}_s} + \frac{1 - \overline{I}_s}{\overline{I}_s}\,\overline{w}^2 - \frac{\overline{I}_e}{\overline{I}_s}\left(\frac{\rho_e}{\rho} - \frac{\overline{I}}{\overline{I}_e}\right)\right]$$
$$+ 2(\zeta - \overline{\zeta})(f_\eta f_{\eta\zeta} - f_\zeta f_{\eta\eta}) \quad (6.95)$$

$$(C_w \overline{w}_\eta)_\eta + \left(1 - \frac{d\overline{\zeta}}{d\zeta}\right) f \overline{w}_\eta = 2(\zeta - \overline{\zeta})(f_\eta \overline{w}_\zeta - f_\zeta \overline{w}_\eta) \quad (6.96)$$

$$\left(\frac{C_w}{\mathrm{Pr}_T}\,\overline{I}_\eta\right)_\eta + \left(1 - \frac{d\overline{\zeta}}{d\zeta}\right) f \overline{I}_\eta = \left\{C_w\left(\frac{1}{\mathrm{Pr}_T} - 1\right)[(\overline{I}_s - \overline{I}_e)(f_\eta^2)_\eta + (1 - \overline{I}_s)(\overline{w}^2)_\eta]\right\}_\eta$$
$$+ 2(\zeta - \overline{\zeta})(f_\eta \overline{I}_\zeta - f_\zeta \overline{I}_\eta) \quad (6.97)$$

The boundary conditions for zero mass transfer are

$$\eta = 0 \qquad f(\zeta, 0) = f_\eta(\zeta, 0) = \overline{w}(\zeta, 0) = 0$$
$$\overline{I}(\zeta, 0) = \overline{I}_w(\zeta) = \overline{i}_w(\zeta) \quad \text{for } q_w'' \neq 0 \Bigg\}$$
$$\text{and} \qquad \overline{I}_\eta(\zeta, 0) = 0 \quad \text{for } q_w'' = 0 \qquad (6.98)$$
$$\eta \to \infty \qquad f_\eta(\zeta, \infty) = \overline{w}(\zeta, \infty) = \overline{I}(\zeta, \infty) \to 1$$

With mass transfer, $f(\zeta, 0)$ depends on $v_w(x)$ as follows:

$$\frac{v_w(x)}{u_e(x)} = -\frac{\mu_w}{\mu_r}\left(\frac{r}{\mathbf{L}}\right)^k\left[\frac{f(\zeta, 0)}{\sqrt{2\zeta}} + \sqrt{2\zeta}\,f_\zeta(\zeta, 0)\right] \quad (6.99)$$

Similar Solutions

Similarity Criteria and Reduced Equations.   The partial differential equations (Eqs. 6.95–6.97) are not amenable to solution except by numerical methods utilizing high-speed computers. Considerable simplifications can result, as in the case of the flat plate, if these equations are reduced to ordinary differential equations through the similarity concept where the dependent variables f, $\overline{w}$, and $\overline{I}$ are assumed functions of η alone. Equations 6.95–6.97 become, for $\overline{\zeta} = $ constant

$$(C_w f'')' + f f'' = \beta_p\left[f'^2 - \frac{\overline{I}}{\overline{I}_s} + \frac{1 - \overline{I}_s}{\overline{I}_s}\,\overline{w}^2 - \frac{\overline{I}_e}{\overline{I}_s}\left(\frac{\rho_e}{\rho} - \frac{\overline{I}}{\overline{I}_e}\right)\right] \quad (6.100)$$

$$(C_w \overline{w}')' + f \overline{w}' = 0 \quad (6.101)$$

$$\left(\frac{C_w}{\mathrm{Pr}_T}\,\overline{I}'\right)' + f\overline{I}' = \left\{C_w\left(\frac{1}{\mathrm{Pr}_T} - 1\right)[(\overline{I}_s - \overline{I}_e)(f'^2)' + (1 - \overline{I}_s)(\overline{w}^2)']\right\}' \quad (6.102)$$

Consistent with the similarity assumption, none of the terms that appear in these equations or in the related boundary conditions can be dependent on x or ζ; that is, β_p, $\overline{I}_e$, and $\overline{I}_w$, as well as $\overline{\zeta}$, must be constant. The parameter $\overline{I}_e$, defined by Eq. 6.93, however, violates this requirement when u_e varies with x. The similarity assumption is also violated by the terms that contain $\overline{I}_e$ explicitly in Eqs. 6.100 and 6.102 and by the gas properties C_w, Pr_T, ρ_e/ρ, which can be expressed in terms of $\overline{I}$, which in turn depends on $\overline{I}_e$ through Eq. 6.91. Consequently, exact similar solutions are not possible under general stagnation region flow conditions.

Exact similar solutions are possible for stagnation regions where $u_e \approx 0$ and $\overline{I}_e$ is a constant and equal to unity for an axisymmetric body and to $\overline{I}_s$ for a yawed cylinder. The terms involving $\overline{I}_e$ drop out of Eqs. 6.100–6.102, and similarity occurs for constant i_w and β_p.

For similar flows, the pressure gradient parameter expressed as follows must be constant:

$$\beta_p = \frac{2}{u_e} \frac{\bar{\imath}_s}{\bar{\imath}_e} \frac{du_e}{dx} \frac{\int_0^x \rho_w \mu_w u_e \left(\frac{r}{L}\right)^{2k} dx}{\rho_w \mu_w u_e \left(\frac{r}{L}\right)^{2k}} \tag{6.103}$$

In a stagnation region, the fluid properties are nearly constant, and $u_e \sim x$; also $r \approx x$. Thus, $\beta_p = \frac{1}{2}$ for an axisymmetric body, and $\beta_p = 1$ for a yawed cylinder.

The skin friction coefficient and Stanton number are defined under the conditions of similarity on axisymmetric and two-dimensional bodies as follows. The components of the shear stress in the x_i direction are given by

$$\tau_{wi} = \sqrt{\frac{\mu_w \rho_w \bar{\imath}_s u_e^2}{\beta_p \bar{\imath}_e} \frac{du_e}{dx} \frac{\partial}{\partial \eta} \left(\frac{u_i}{u_{ie}}\right)\bigg|_w} \tag{6.104}$$

where the subscript $i = 1$ or 3 represents the x or z direction, respectively. The skin friction coefficient is defined as

$$\frac{c_{f_n}}{2} = \frac{\tau_{vn}}{\rho_w u_e u_{ie}} \tag{6.105}$$

with

$$\frac{c_{f_1}}{2} \sqrt{\frac{\rho_w u_e x}{\mu_w}} = \sqrt{\frac{\bar{\imath}_s}{\beta_p \bar{\imath}_e} \frac{x}{u_e} \frac{du_e}{dx}} f_w'' \tag{6.106}$$

$$\frac{c_{f_3}}{2} \sqrt{\frac{\rho_w u_e x}{\mu_w}} = \sqrt{\frac{\bar{\imath}_s}{\beta_p \bar{\imath}_e} \frac{x}{u_e} \frac{du_e}{dx}} \overline{w}_w' \tag{6.107}$$

Alternative forms of these equations that are sometimes more convenient are

$$\frac{c_{f_1}}{2} \sqrt{\frac{\rho_w u_e x_{\text{eff}}}{\mu_w}} = \frac{1}{\sqrt{2}} f_w'' \tag{6.108}$$

$$\frac{c_{f_3}}{2} \sqrt{\frac{\rho_w u_e x_{\text{eff}}}{\mu_w}} = \frac{1}{\sqrt{2}} \overline{w}_w' \tag{6.109}$$

where

$$x_{\text{eff}} = \frac{\int_0^x \rho_w \mu_w u_e \left(\frac{r}{L}\right)^{2k} dx}{\rho_w \mu_w u_e \left(\frac{r}{L}\right)^{2k}} \tag{6.110}$$

The corresponding Stanton number expressions for a surface at constant temperature are

$$\text{St} \sqrt{\frac{\rho_w u_e x}{\mu_w}} = \sqrt{\frac{\bar{\imath}_s}{\beta_p \bar{\imath}_e} \frac{x}{u_e} \frac{du_e}{dx}} \frac{1}{\text{Pr}_{\bar{T}w}} \frac{\bar{I}_w'}{\bar{I}_w - \bar{I}_{aw}} \tag{6.111}$$

$$\text{St} \sqrt{\frac{\rho_w u_e x_{\text{eff}}}{\mu_w}} = \frac{1}{\sqrt{2}} \frac{1}{\text{Pr}_{\bar{T}w}} \frac{\bar{I}_w'}{\bar{I}_w - \bar{I}_{aw}} \tag{6.112}$$

Uniform Surface Temperature, Ideal Gas With Viscosity-Density Product and Prandtl Number Equal to Unity. For the case of an ideal gas with $C_w = \text{Pr} = 1$, similarity is possible away from the stagnation region of a body. Equations 6.100 and 6.102 for an axisymmetric body or a cylinder normal to the free stream reduce to

$$f''' + ff'' = \beta_p(f'^2 - \bar{I})$$ (6.113)

$$\bar{I}'' + f\bar{I}' = 0$$ (6.114)

when u_e satisfies

$$\frac{u_e}{\sqrt{\bar{t}_e}} = A\zeta^{\beta_p/2}$$ (6.115)

These equations are equivalent to those solved in Ref. 42 for a uniform surface temperature. Examples of the extensive solutions in Refs. 16 and 42 are presented in this section.

Figure 6.19 shows the velocity distribution in the boundary layer of an axisymmetric body or an unswept cylinder for a cold wall ($\bar{I}_w = 0$). (Note that the value of η used here is a factor of $\sqrt{2}$ smaller than the one employed in sections on the flat plate.) An accelerating free stream ($\beta_p > 0$, u_e increasing) reduces the flow boundary layer thickness and increases the velocity gradient rather uniformly through the boundary layer. A decelerating free stream ($\beta_p < 0$) thickens the flow boundary layer rather severely and causes the velocity distribution to acquire an S shape. Eventually, the boundary layer will separate; that is, $f''_w = 0$. For the boundary conditions of Fig. 6.19, two solutions are possible for negative values of β_p near separation. It is argued in Ref. 42 that true similarity with negative β_p cannot occur physically because Eq. 6.115 would require $u_e \rightarrow \infty$ as $\zeta \rightarrow 0$. Thus, similar solutions with negative β_p can only be approached after a period of nonsimilar flow, and depending on the conditions, one or the other of the similar solutions for a given β_p can be attained. In Ref. 42, experimental evidence from Ref. 43 for turbulent flow is cited for the possible existence of double-valued flow-field behavior. The velocity profiles shown are characteristic of those for either a cooled surface or one at the total enthalpy of the fluid. For a heated surface and $\beta_p > 0$, it is possible for the velocity ratio f' in the outer portion of the boundary layer to attain a value greater than unity before approaching unity at the outer edge. The physical reason for this is that the acceleration of lower-density fluid by the favorable pressure gradient exceeds the retardation by the viscous forces.

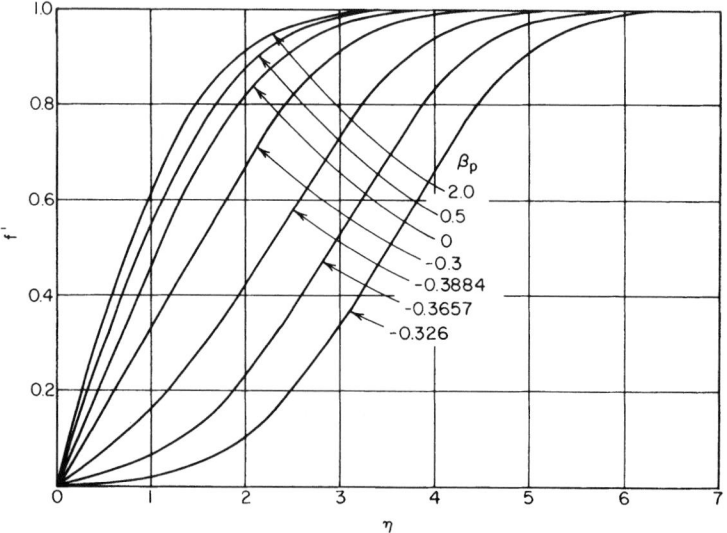

FIGURE 6.19   Similar velocity distributions for body with surface pressure gradient β_p defined by Eq. 6.103, $\bar{I}_w = 0$, $\bar{t}_s = 1$ [42].

The total enthalpy distribution in terms of the velocity is shown in Fig. 6.20 for $\bar{I}_w = 0$ and $\bar{i}_s = 1$. A pressure gradient can cause significant departures from the Crocco relationship (Eq. 6.40, represented by the straight line labeled $\beta_p = 0$). The latter is often used for approximate calculations even when pressure gradients are present.

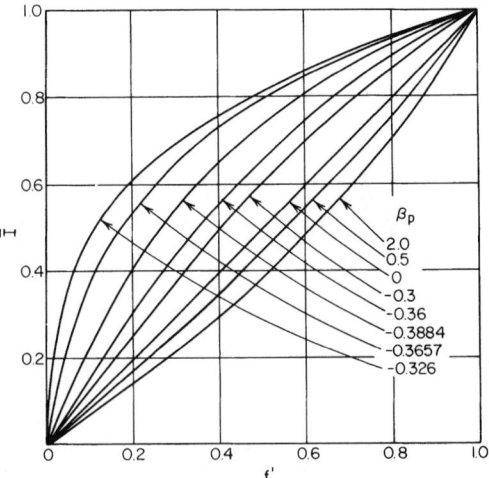

FIGURE 6.20   Enthalpy and velocity relationship within similar boundary layers on a body with pressure gradient β_p defined by Eq. 6.103, $\bar{I}_w = 0$, $\bar{i}_s = 1$ [42].

Figure 6.21 shows the wall shear parameter f_w'' required to evaluate the local skin friction coefficients by Eq. 6.106 or 6.108. These curves apply for the case where $\bar{i}_s = 1$. The double-valued nature of f_w'' for a cooled surface ($\bar{I}_w = 0$) for β_p near separation ($f_w'' = 0$) is evident. Generally, f_w'' is more sensitive to variations in β_p for a hot surface. In fact, for cold wall conditions ($\bar{I}_w = 0$), the variation of f_w'' with β_p for $\beta_p > 0$ is quite modest. Also, a cooled surface tends to retard separation; that is, $f_w'' = 0$ at a smaller value of β_p.

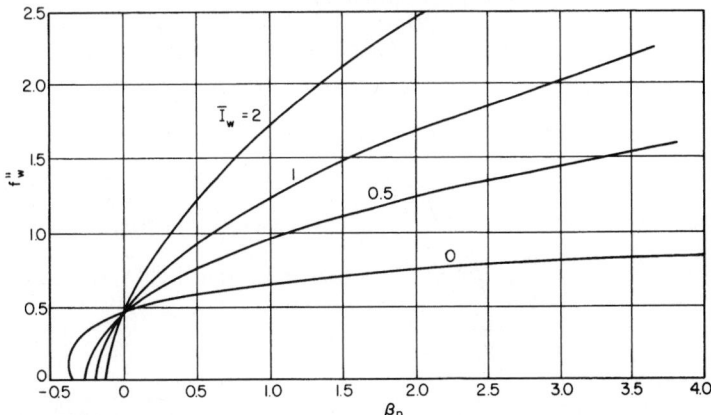

FIGURE 6.21   Effect of pressure gradient on the skin friction parameter f_w'' for various wall temperature levels, $\bar{i}_s = 1$ [42].

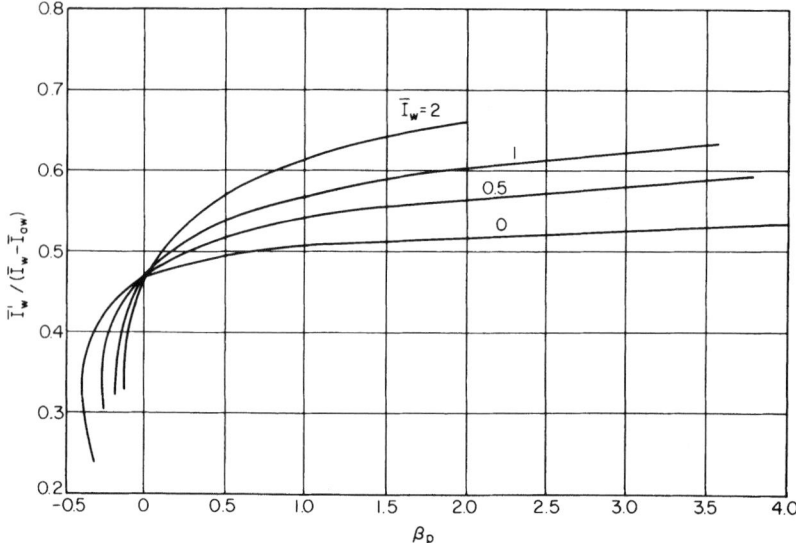

FIGURE 6.22   Effect of pressure gradient on the heat transfer parameter $\bar{I}'_w$ for various wall temperature levels, $\bar{t}_s = 1$, Pr = 1 [42].

The heat transfer parameter required in Eq. 6.111 or 6.112 to calculate Stanton number is shown in Fig. 6.22 for $\bar{t}_s = 1$. It should be noted that because of the similarity of Eqs. 6.101 and 6.102 for Pr = 1,

$$\overline{w}'_w = \frac{\bar{I}'_w}{\bar{I}_w - \bar{I}_{aw}} \tag{6.116}$$

Hence, the ordinate in Fig. 6.22 can also be used in conjunction with Eq. 6.107 or 6.109 to calculate the cross flow skin friction coefficient for cases of very small yaw angles ($\bar{t}_s \approx 1$). Note that $\bar{I}_{aw}$ is equal to unity because the solution of Eq. 6.102 with Pr = 1 and an insulated surface is $\bar{I} \equiv 1$. Although the trends exhibited in Figs. 6.21 and 6.22 are generally similar, it must be cautioned that such large variations in the Reynolds analogy factor occur that the latter is no longer a useful concept. The heat transfer parameter for a cooled surface shows a rather small variation with β_p for $\beta_p \geq \frac{1}{2}$, a fact first utilized in Ref. 44 to obtain relatively simple expressions for the local heat flux to blunt bodies in hypersonic flow.

Cylinder Normal to the Free Stream, Fluid With Constant Properties. For constant fluid properties, Eqs. 6.100 and 6.102 reduce to

$$f''' + ff'' = \beta_p(f'^2 - 1) \tag{6.117}$$

$$\bar{I}'' + \text{Pr}\, f\bar{I}' = 2\text{Pr}\,\zeta f'\bar{I}_\zeta \tag{6.118}$$

For a cylinder normal to the free stream, the inviscid velocity distribution is given by

$$u_e = Ax^m \tag{6.119}$$

with

$$\beta_p = \frac{2m}{1+m} \tag{6.120}$$

The term on the right side of Eq. 6.118 has been added to account for a nonuniform surface temperature.

Similar solutions for Prandtl numbers other than unity may be obtained from Eqs. 6.117 and 6.118 or their equivalent. A major simplification is the independence of the momentum equation (Eq. 6.117), from the energy equation (Eq. 6.118), which makes f independent of $\bar{I}$. Also, the linear form of the energy equation in $\bar{I}$ permits handling arbitrary surface temperature distributions as in the case of the flat plate. (See the section on the two-dimensional laminar boundary layer.)

Solutions of the momentum equation (Eq. 6.117) [45] yield velocity distributions generally similar to those of Fig. 6.19, and the skin friction parameter f_w'' shown by the line labeled 1 in Fig. 6.21. The skin friction coefficient is given by

$$\frac{c_f}{2} \sqrt{\frac{\rho_e u_e x}{\mu_e}} = \frac{f_w''}{\sqrt{2 - \beta_p}} \tag{6.121}$$

For a uniform surface temperature, solutions of the energy equation (Eq. 6.118) [46, 47] can be expressed as

$$\text{St}(x, 0) \sqrt{\frac{\rho_e u_e x}{\mu_e}} = \frac{\text{Pr}^{-a}}{\sqrt{2 - \beta_p}} \left(\frac{\bar{I}_w'}{\bar{I}_w - \bar{I}_{aw}} \right)_{\text{Pr} = 1} \tag{6.122}$$

where the heat transfer parameter in parentheses on the right is given by the line labeled 1 in Fig. 6.22. The exponent **a** for the Prandtl number is given in Table 6.3 as a function of β_p.

TABLE 6.3   Exponent of Pr in Eq. 6.122

β_p	−0.199	0	1	1.6
a	0.746	0.673	0.645	0.633

In Ref. 46, an ingenious set of transformations is employed to evaluate the recovery factor away from the stagnation line. The results for $\beta_p = 1$ show a significant departure (≈ -10 percent for Pr = 0.7) from $\mathbf{r}(0) = \text{Pr}^{1/2}$. These values, however, do not agree with calculations performed in Ref. 49. Perhaps the discrepancy is due to the evaluation of $\mathbf{r}(0)$ in Ref. 46 by taking the derivative of a function. Slight errors in the function itself could easily account for a 10 percent error in the derivative. For accuracies of $\mathbf{r}(0)$ within a few percent [48], it is recommended that

$$\mathbf{r}(0) = \text{Pr}^{1/2} \tag{6.123}$$

be employed for all β_p.

For a stepwise and arbitrary surface temperature distribution, the local heat flux distribution is given by Eq. 6.65. The term St $(x, 0)$ represents the Stanton number for the same flow conditions but with a uniform surface temperature. Equation 6.65 was derived formally with the assumption that i_{aw} is uniform along the surface; however, small continuous variations in i_{aw} or T_{aw} are permissible. The value of i_{aw} at $x = 0$ is used in the first term on the right side of Eq. 6.65, and the local value of i_{aw} is employed within the integral. Although Eq. 6.65 appears to be reciprocal in variations of i_w and i_{aw}, this is not the case. Changes in the same direction of both i_w and i_{aw} do not necessarily cancel, because a change in i_{aw} takes effect gradually downstream.

The kernel function St (x, s)/[St $(x, 0)$] in Eq. 6.65 represents the behavior of the heat transfer coefficient after a jump in wall temperature at $x = s$. This function was obtained in Refs. 24 and 25 by solving the energy equation with the assumption of a linear velocity profile and is given by

$$\frac{\text{St}(x, s)}{\text{St}(x, 0)} = \left[1 - \left(\frac{s}{x} \right)^{a^*} \right]^{-b^*} \tag{6.124}$$

$$a^* = \frac{3}{2(2 - \beta_p)} \qquad b^* = \frac{1}{3} \tag{6.125}$$

The analogous function for a flat plate is given by Eq. 6.66. These results were improved upon in Ref. 49, where a power-law velocity profile was assumed, $u/u_e = (y/y_e)^{d^*}$, with d^* found to best fit Hartree's calculations [45] as listed in Table 6.4. The form of the kernel function (Eq. 6.124) is the same, but the exponents are changed to

$$a^* = \frac{2 + d^*}{(1 + d^*)(2 - \beta_p)} \qquad b^* = \frac{1}{2 + d^*} \tag{6.126}$$

TABLE 6.4   Exponent d^* for Velocity Function, $u/u_e = (y/y_e)^{d^*}$

m	0	1/9	1/3	1	4
β_p	0	0.2	0.5	1	1.6
d^*	0.88	0.86	0.80	0.76	0.66

Use of these values to calculate the heat flux distribution from Eqs. 6.65 and 6.124 yields excellent agreement with the exact solutions of Refs. 9, 47, and 50. Hence, for $m \geq 0$, values of a^* and b^* given by Eq. 6.126 are preferred to those given by Eq. 6.125. For $m = 0$, there is little difference between the kernel functions (Eq. 6.124) based on the two different sets of exponents. Values of the kernel function Eq. 6.124 are shown in Fig. 6.12 for $\beta_p = 0.5$ and 1. The effect of the upstream temperature jump decays more rapidly with increasing β_p. As $\beta_p \to 2$, the kernel function becomes unity for all $s/x < 1$, as is seen directly from the functional form of Eqs. 6.124 and 6.126.

Axisymmetric Stagnation Point, High-Speed Flow.   The axisymmetric stagnation point has received attention from many investigators because of its importance in the assessment of the convective heat load of missile nose cones and atmospheric entry vehicles. The speeds involved in these applications produce stagnation enthalpies where real-gas behavior must be considered in the evaluation of the forced convection. Because of the very complex behavior of the physical properties of real gases, a characteristic common to all the studies is the reliance on numerical solutions followed by correlations of the results in terms of parameters involving the fluid properties. In addition to air, other gases have been treated because of the current interest in the exploration of the planetary neighbors of Earth. The contents of this section are confined to gases in chemical equilibrium with uniform elemental composition and to flows where boundary layer similarity occurs—namely, the immediate vicinity of the stagnation point—and where there is a uniform surface temperature. The effects of surface mass transfer of the same gas as exists in the free stream are also included.

Equations 6.100 and 6.102 with $t_s = t_e = 1$, $t = I$ have been solved for real air in Ref. 16 and in Refs. 51–55, with the latter references utilizing the concept of total properties k_T, c_{pT}, Pr_T. The air properties of Ref. 56 were employed in all the studies except that of Ref. 55, which employed properties evaluated in Refs. 57 and 58, where careful consideration was given to the effect of dominant resonant charge exchange cross sections in establishing the thermal conductivity of ionized nitrogen.

For speeds under 30,000 ft/s (9144 m/s), which represent relatively moderate entry conditions into the earth's atmosphere without appreciable ionization, the numerical results of Ref. 16 for 10^{-4} atm $< P_{st} < 10^2$ atm and $540°R$ (300 K) $< T_w < 3100°R$ (1722 K) are correlated by

$$\frac{Nu_w}{Pr_{Tw}^{0.4} \, Re_w^{0.5}} = 0.767 \left(\frac{\mu_e \rho_e}{\mu_w \rho_w} \right)^{0.43} \tag{6.127}$$

where $\quad \text{Nu}_w = \dfrac{q_w'' c_{pw} \mathbf{L}}{(i_w - I_e) k_w}$

$$\text{Re}_w = \dfrac{(du_e/dx)\mathbf{L}^2 \rho_w}{\mu_w}$$

At the stagnation point (where $u_e = 0$) the recovery enthalpy is identical to the stagnation enthalpy even for $\text{Pr}_T \neq 1$. Thus, from the definition of Nu_w and Re_w, the local heat flux at the stagnation point in airflow for the speed range up to 30,000 ft/s (9144 m/s) can be expressed as

$$q_w'' = \left(\frac{\text{Nu}_w}{\sqrt{\text{Re}_w}} \right) \frac{I_e(\bar{I}_w - 1)}{\text{Pr}_{\bar{Tw}}} \sqrt{\mu_w \rho_w} \sqrt{\frac{du_e}{dx}} \qquad (6.128)$$

or from Eq. 6.127 as

$$q_w'' = 0.767 \text{Pr}_{\bar{Tw}}^{-0.6} (\mu_e \rho_e)^{0.43} (\mu_w \rho_w)^{0.07} \sqrt{\frac{du_e}{dx}} \, I_e(\bar{I}_w - 1) \qquad (6.129)$$

Here, a negative value of q_w'' represents heat flux toward the body.

At speeds greater than 10,000 ft/s (3048 m/s), where $I_e \approx V_\infty^2/2$, Eq. 6.129 can be represented in much simpler form when a relatively cold surface temperature (below dissociation temperature) is assumed:

$$q_w'' \sqrt{\frac{r_n}{P_{st}}} = 121 \left(\sqrt{\frac{T_w}{492}} \frac{1.38}{T_w/492 + 0.38} \right)^{0.07} \sqrt{\frac{r_n}{V_\infty} \frac{du_e}{dx}} \, \bar{U}^{2.21}(\bar{I}_w - 1) \qquad (6.130a)$$

For the heat flux expressed in Btu/(s·ft^2), the dimensionless velocity is expressed as $\bar{U} = V_\infty/$ (10,000 ft/s), T_w is expressed in °R, and P_{st} is expressed in atm. In SI units, Eq. 6.130a becomes

$$q_w'' \sqrt{\frac{r_n}{P_{st}}} = 2382 \left(\sqrt{\frac{T_w}{273}} \frac{1.38}{T_w/273 + 0.38} \right)^{0.07} \sqrt{\frac{r_n}{V_\infty} \frac{du_e}{dx}} \, \bar{U}^{2.21}(\bar{I}_w - 1) \qquad (6.130b)$$

where $\bar{U} = V_\infty/(3048 \text{ m/s})$, q_w'' is in W/m^2, T_w is expressed in K, and P_{st} is in N/m^2. Note that

$$\frac{r_n}{V_\infty} = \frac{1.54}{V_\infty} \sqrt{\frac{P_{st} - P_\infty}{\rho_{st}}} \qquad (6.131)$$

For stagnation-point heating in gases other than air, correlations similar to Eq. 6.128 were obtained in Ref. 53 for speeds up to 30,000 ft/s (9144 m/s). The correlation is of the form

$$\frac{\text{Nu}_w}{\sqrt{\text{Re}_w}} = A^* \left(\frac{\mu_e \rho_e}{\mu_w \rho_w} \right)^{B^*} \qquad (6.132)$$

with coefficients A^* and B^* given in Table 6.5 for the various gases considered. There is rather close agreement between Eqs. 6.127 and 6.132 for air. The heat flux is obtained from Eq. 6.128.

TABLE 6.5   Coefficients for Eq. 6.132 [53]

Gas	A^*	B^*
Air	0.718	0.475
N_2	0.645	0.398
H_2	0.675	0.358
CO_2	0.649	0.332
A	0.515	0.110

For speeds above 30,000 ft/s (9144 m/s), where the total enthalpy reaches values where ionization significantly lowers the viscosity μ_e, the correlation for air (Eq. 6.127) from Ref. 16 begins to break down. Similar behavior was observed on pointed cones in Ref. 17. Similarly, it was found for argon in Ref. 53 that only the solutions for the lowest wall temperatures were correlated well by the property parameter $\mu_e\rho_e/\mu_w\rho_w$. Thus, extrapolation of the Nusselt number relations beyond the range of $\mu_e\rho_e/\mu_w\rho_w$ actually used in the correlations could yield significant errors. Two alternate approaches avoid this problem. In Ref. 52 the intermediate Nusselt number correlation was bypassed, and a correlation for air was achieved directly in terms of the heat flux *for speeds up to 50,000 ft/s (15,200 m/s)* in English units as follows:

$$q_w''\sqrt{\frac{r_n}{P_{st}}} = 119\sqrt{\frac{r_n}{V_\infty}\frac{du_e}{dx}}\ \overline{U}^{2.19}(\overline{I}_w - 1) \qquad (6.133)$$

(In SI units, the coefficient is 2342.) It will be noted that this equation is quite similar to Eq. 6.130. The implication of this similarity is that the large variations in μ_e associated with ionization, which is not present in the range of velocities for which Eq. 6.130 is valid, can be ignored in the evaluation of heat flux. In fact, it is systematically demonstrated in Ref. 53 that the surface heat flux is quite insensitive to variations of the gas properties at the boundary layer edge and is controlled instead by the gradients of these properties near the surface. Apparently, correlations such as Eq. 6.127 or 6.132 result because in the speed range where they are applicable, the physical properties vary monotonically through the boundary layer, and their derivatives in the inner portion of the boundary layer are related to the ratio of properties across the boundary layer. For a variety of gases, the resulting heat flux expression, using strong shock relationships, is given by

$$q_w''\sqrt{\frac{r_n}{P_{st}}} = F_0\sqrt{\frac{r_n}{V_\infty}\frac{du_e}{dx}}\ \overline{U}^{2.2}(\overline{I}_w - 1) \qquad (6.134)$$

where the units are the same as in Eqs. 6.130a and 6.130b, and F_0 is given in Table 6.6.

From the value of F_0 for air in Table 6.6 and the form of Eq. 6.134, it is apparent that the correlation of Ref. 53 yields results essentially identical to those of Eq. 6.133 taken from Ref. 52 and to Eq. 6.130 derived by directly extrapolating the lower–speed range equation of Ref. 16. Thus, it is recommended that Eq. 6.134 with the coefficients of Table 6.6 be utilized to predict stagnation-point heat flux at speeds greater than 10,000 ft/s (3048 m/s). At lower speeds, Eqs. 6.128 and 6.132 with the coefficients of Table 6.5 are appropriate. A comparison of these techniques with existing stagnation-point measurements is shown in Fig. 6.23. It should be noted that convection predominated over radiation in these measurements despite speeds up to 50,000 ft/s (15,200 m/s) because of the small nose radii for the models. For body dimensions exceeding a few feet, shock layer radiation begins to compete with convection at speeds of about 25,000 ft/s (7620 m/s) and becomes the predominant heating mechanism at higher speeds.

TABLE 6.6 Coefficient F_0

Gas (% volume)	$\dfrac{\text{Btu}}{\text{s}\cdot\text{ft}^2}\left(\dfrac{\text{ft}}{\text{atm}}\right)^{1/2}$	$\dfrac{\text{W}}{\text{m}^2}\left(\dfrac{\text{m}}{\text{N/m}^2}\right)^{1/2}$
Air	121	2382
N_2	121	2382
CO_2	141	2775
A	165	3248
91% N_2–10% CO_2	120	2362
50% N_2–50% CO_2	134	2638
40% N_2–10% CO_2–50% A	144	2834
65% CO_2–35% A	142	2795

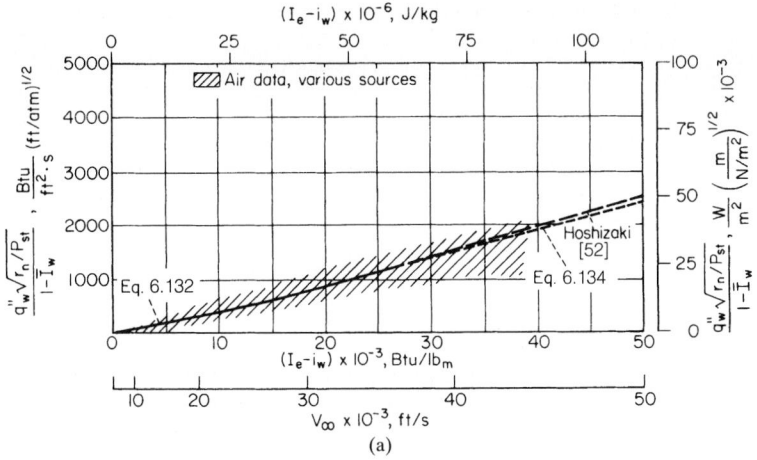

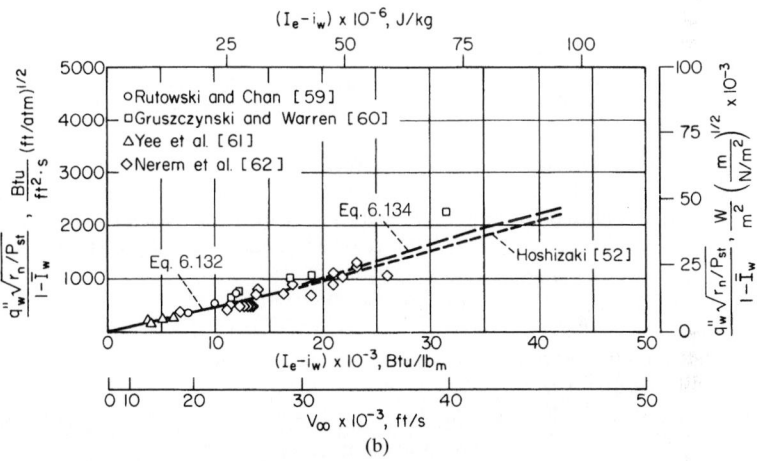

FIGURE 6.23   Comparison of heat transfer rates predicted by Eq. 6.134 and the coefficients of Table 6.6 with data [53]. (*a*) air; (*b*) carbon dioxide.

Stagnation Line on a Cylinder in High-Speed Flow.   The stagnation line on a uniform-temperature cylinder of infinite length with its axis either normal to the free stream or swept back at angle Λ is characterized by boundary layer similarity solutions with $\beta_p = 1$. The solution in Ref. 16 of Eqs. 6.100–6.102 for $\beta_p = 1$ yields the correlation

$$\frac{\mathrm{Nu}_w}{\mathrm{Pr}_{Tw}^{0.4}\,\mathrm{Re}_w^{0.5}} = 0.57\left(\frac{\mu_e\rho_e}{\mu_w\rho_w}\right)^{0.45} \tag{6.135}$$

for $i_w/I_e < \bar{\imath}_s$ (see Eq. 6.92 for definition of $\bar{\imath}_s$) and $V_\infty \cos \Lambda < 29{,}000$ ft/s (8840 m/s). Although Eq. 6.135 was established from calculations that employed real-air properties, the resulting coefficient and exponents are consistent with those for low-speed calculations for either constant properties or ideal gases. In terms of heat flux, Eq. 6.135 becomes

$$q_w'' = 0.57\mathrm{Pr}_{Tw}^{-0.6}\,(\mu_e\rho_e)^{0.45}(\mu_w\rho_w)^{0.05}\sqrt{\frac{du_e}{dx}}\,(i_w - I_{aw}) \tag{6.136}$$

where the recovery enthalpy I_{aw} is given by

$$I_{aw} = I_e - (1 - \Pr_{Tw}^{0.5}) \frac{w_e^2}{2} = I_e - (1 - \Pr_{Tw}^{0.5}) \frac{V_\infty^2 \sin^2 \Lambda}{2} \tag{6.137}$$

and a negative q_w'' represents heat flow into the body. An alternate form of Eq. 6.136, valid for $V_\infty > 10{,}000$ ft/s (3048 m/s) and having the same form as Eq. 6.130 for an axisymmetric stagnation point in airflow, is

$$q_w'' \sqrt{\frac{r_n}{P_{st}}} = 87.3 \left[\left(\frac{T_w}{492}\right)^{1/2} \frac{1.38}{T_w/492 + 0.38} \right]^{0.05} \sqrt{\frac{r_n}{V_\infty} \frac{du_e}{dx}} \, \overline{U}^{2.2}(\overline{I}_w + 0.15 \sin^2 \Lambda - 1) \tag{6.138a}$$

with the following units: q_w'', Btu/(s·ft²); r_n, ft; P_{st}, atm; T_w, °R; V_∞, ft/s; and $\overline{U} = V_\infty/(10{,}000$ ft/s). In SI units, this equation is

$$q_w'' \sqrt{\frac{r_n}{P_{st}}} = 1718 \left[\left(\frac{T_w}{273}\right)^{1/2} \frac{1.38}{(T_w/273) + 0.38} \right]^{0.05} \sqrt{\frac{r_n}{V_\infty} \frac{du_e}{dx}} \, \overline{U}^{2.2}(\overline{I}_w + 0.15 \sin^2 \Lambda - 1) \tag{6.138b}$$

with q_w'', W/m²; r_n, m; P_{st}, N/m²; T_w, K; V_∞, m/s; and $\overline{U} = V_\infty/(3048$ m/s).

The velocity gradient in Eq. 6.138 is obtained from Eq. 6.131, but with 1.54 changed to 1.43. In these equations, P_{st} and ρ_{st} are the inviscid flow conditions on the stagnation line of the swept cylinder. For an ideal gas in hypersonic flow, the inviscid flow relationships are particularly simple, and Eq. 6.138 shows that the heat flux is reduced with sweep by approximately $\cos^{3/2} \Lambda$. Equation 6.138 may be extended to gases other than air by setting the quantity in brackets equal to unity and replacing the coefficient 87.3 by $0.72F_0$, where F_0 is given in Table 6.6.

Mass Transfer in Stagnation Region. As on a flat plate, surface mass transfer is an effective means for alleviating convective heating in the stagnation regions of axisymmetric bodies and cylinders. The effect of surface mass transfer of a gas with the same elemental composition as the free stream will be treated initially. Consistent with the similarity requirements following Eqs. 6.100–6.102, the surface mass transfer rate is given by

$$\rho_w v_w = -f(0) \sqrt{\frac{\mu_w \rho_w \bar{\imath}_s}{\beta_p \bar{\imath}_e} \frac{du_e}{dx}} \tag{6.139}$$

In the vicinity of the stagnation region and with a uniform surface temperature, the terms under the radical sign are constants. Thus, boundary layer similarity, that is, $f(0)$ being independent of x, requires a uniform mass injection rate along the surface rather than one varying as $x^{-1/2}$ as on a flat plate. A convenient correlation parameter, as in the case of the flat plate, is

$$B_m = \frac{\rho_w v_w}{\rho_e \mu_e \, \text{St}_0} = \frac{\rho_w v_w (i_w - I_{aw})}{q_{w0}''} = -\frac{f(0) \, \Pr_{Tw}}{\sqrt{\beta_p}(\text{Nu}_w/\sqrt{\text{Re}_w})_0} \tag{6.140}$$

where the subscript 0 denotes zero surface mass transfer.

The effect of surface mass transfer in reducing the Stanton number is indicated in Fig. 6.24. Note that for the stagnation point on a body of revolution or an unswept cylinder, the recovery enthalpy is equal to the total enthalpy. Hence, $q_w'' \sim$ St so that Fig. 6.24 indicates the reduction of the heat flux as well as the Stanton number. The line representing the axisymmetric stagnation point correlates all the gases listed in Table 6.5 to within a few percent [54]. Although the curves for the cylinder are based on air calculations, the correlation for the various gases for the axisymmetric stagnation point implies that the cylinder results can be applied to other gases when the injected and free-stream gas are the same. Note that in the coordinate system of Fig. 6.24, the effect of sweep is quite small.

Along the stagnation line of a swept cylinder, the recovery enthalpy is less than the stagnation enthalpy. Thus, the effect of surface mass transfer on the recovery factor, as shown in Fig. 6.25, should be considered in establishing the proper driving potential for the

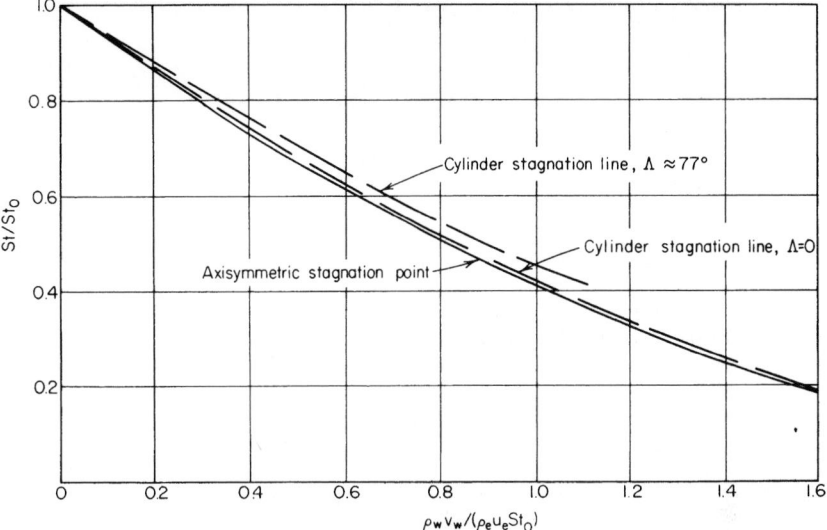

FIGURE 6.24 The reduction of Stanton number in stagnation regions by surface mass transfer.

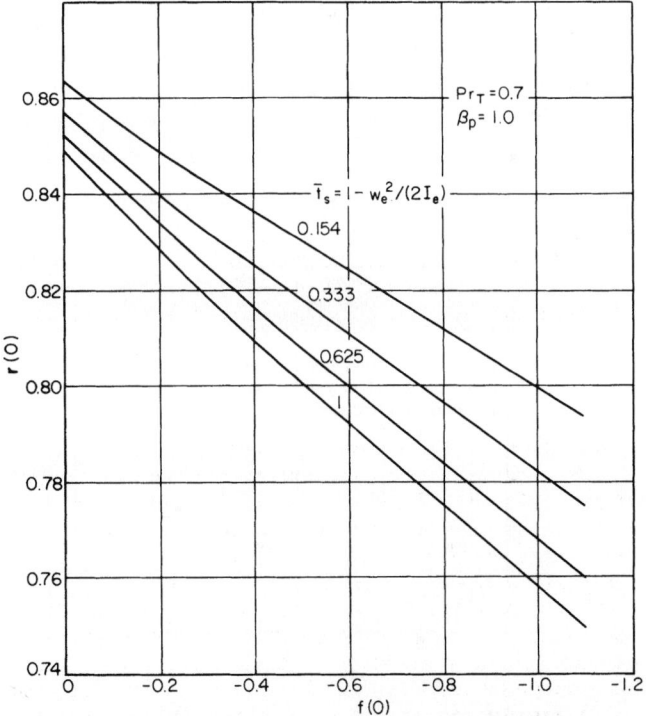

FIGURE 6.25 The effect of surface mass transfer on the recovery factor on the stagnation line of a yawed cylinder of infinite length [48].

heat flux [48]. The heat flux for a swept cylinder normalized by its value with zero mass transfer is

$$\frac{q_w''}{q_{w_0}''} = \frac{\text{St}}{\text{St}_0} \left[\frac{1 - (1 - \mathbf{r}(0)) \sin^2 \Lambda - \bar{I}_w}{1 - (1 - \mathbf{r}_0(0)) \sin^2 \Lambda - \bar{I}_w} \right] \tag{6.141}$$

Because the term containing the recovery factor depends on $\sin^2 \Lambda$, modification of the recovery factor to account for surface mass transfer need be made only for large sweep angles. The graphic procedure for using these figures in establishing the required mass flow rate to yield a prescribed surface temperature was described on page 6.24.

The evaluation of the effectiveness of a transpiration cooling system utilizing a foreign gas is quite formidable and requires the use of complex computer codes.

Heat Transfer Over a Single Cylinder and Arrays of Cylinders in Low-Speed Cross Flow.
A boundary layer subjected to a significant static pressure increase in the streamwise direction thickens considerably and may eventually separate from the surface of the body. For steady separation, the velocity gradient normal to the surface vanishes at the separation point, and downstream the flow direction is reversed. Although the skin friction vanishes, significant amounts of heat transfer can occur at the point of separation. Separation is commonly encountered in cross flow over blunt bodies such as circular cylinders. The flow pattern downstream of separation is very complex and is often accompanied by unsteadiness and vortex shedding. Theoretical treatment of such flows is still in an early stage of development [63], and heat transfer predictions must rely on experimental data. The following sections summarize the experimental data available on heat transfer rates from single cylinders and arrays of cylinders in cross flow.

Single Cylinder.   The classic experiments on the average heat transfer rates from a cylinder in cross flow were performed by Hilpert [64] for a wide range of Reynolds numbers in air. Hilpert's results are shown in Fig. 6.26 as average Nusselt number versus the cross flow Reynolds number.

In Ref. 66, Morgan made an extensive review of more recent heat transfer data obtained on a cylinder in cross flow. He found that the average Nusselt number could be correlated as

$$\overline{\text{Nu}}_d = (A + B \, \text{Re}_d^{n^*}) \, \text{Pr}^{m^*} \tag{6.142}$$

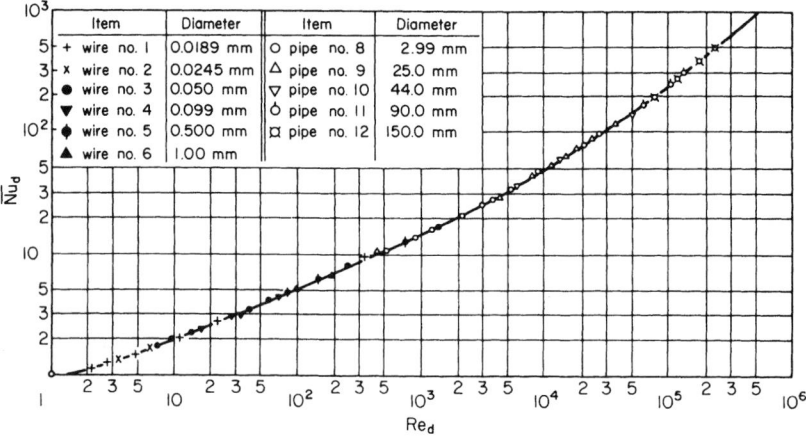

FIGURE 6.26   Nusselt number for average heat transfer from circular cylinder in cross flow of air [64]. (Reprinted from Ref. 65 by permission of McGraw-Hill.)

where A, B, n^*, and m^* are constants. The value of the exponent m^* found in various experiments lies between 0.3 and 0.4; Zukauskas [67] recommends a value of 0.37. The data were obtained mostly in air. Depending on the Reynolds number, the scatter in Nusselt numbers calculated from Eq. 6.142, using constants from various investigations, ranges from 10 percent to 29 percent. Other types of correlations were proposed and tested by Morgan and found to be less accurate than the above. Morgan attributed the scatter in the heat transfer data to three factors: (1) aspect ratio (length/diameter) of the cylinder, (2) free-stream turbulence level, and (3) wind tunnel blockage effects.

Corrections by Morgan [66] for the combined effects of free-stream turbulence and wind tunnel blockage on Nusselt numbers measured in air are shown in Fig. 6.27. Here, d/d_T is the ratio of the cylinder diameter to the wind tunnel height or diameter, $\delta \overline{Nu}_d$ is the increase in the Nusselt number over Hilpert's measurements (Fig. 6.26), and Tu is the intensity of the longitudinal turbulent fluctuations in the free stream. In addition to these corrections, Morgan also proposed the Nusselt number correlations shown in Table 6.7 that are applicable to an extremely wide range of Reynolds numbers in air. At the higher Reynolds numbers, these Nusselt numbers are quite consistent with those of Fig. 6.26.

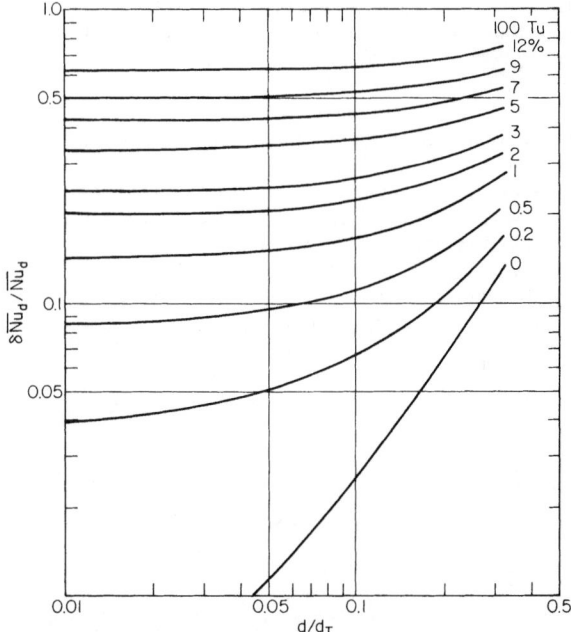

FIGURE 6.27 Correction factors to Nusselt numbers for combined effects of wind tunnel blockage and free-stream turbulence [66]. (Reprinted by permission of Academic Press.)

Local heat transfer rates from the surface of a cylinder in cross flow in air were measured by Schmidt and Wenner [68] and are shown in Fig. 6.28. The local Nusselt number is based on the local heat transfer coefficient and the cylinder diameter. Note that for subcritical Reynolds numbers ($Re_d < 170,000$), the local Nusselt number decreases initially along the surface from the forward stagnation point to a minimum at the separation point and subsequently reaches high values again in the separated portion of the flow on the back surface. For

TABLE 6.7 Correlation of Cross-Flow Forced Convection From Cylinders in Air [66]

	$\overline{Nu}_d = D_2\,Re_d^{n_1}$		
Re_d			
From	To	D_2	n_1
10^{-4}	4×10^{-3}	0.437	0.0895
4×10^{-3}	9×10^{-2}	0.565	0.136
9×10^{-2}	1	0.800	0.280
1	35	0.795	0.384
35	5×10^3	0.583	0.471
5×10^3	5×10^4	0.148	0.633
5×10^4	2×10^5	0.0208	0.814

Reynolds numbers above the critical value, transition from laminar to turbulent flow in the upstream attached boundary layer causes a dramatic increase in the local heat transfer, followed by a sharp decrease in the separated flow region.

The local heat transfer distribution is extremely sensitive to the free-stream turbulence intensity according to the measurements of Kestin and Maeder [69]; this is reflected in the correction terms shown in Fig. 6.27 for the average Nusselt number.

Arrays of Cylinders. The heat transfer behavior of a tube in a bank differs considerably from that of a single tube immersed in a flow of infinite extent. The presence of adjacent tubes in an array and the turbulence and unsteadiness generated by upstream tubes generally tend to increase the overall heat transfer from a particular tube. After the flow has passed through several rows of tubes, however, the heat transfer from individual tubes becomes independent of their location and just a function of the Reynolds number with a parametric dependence on the array geometry. Average and local heat transfer data for tube banks have been summarized by Zukauskas [67].

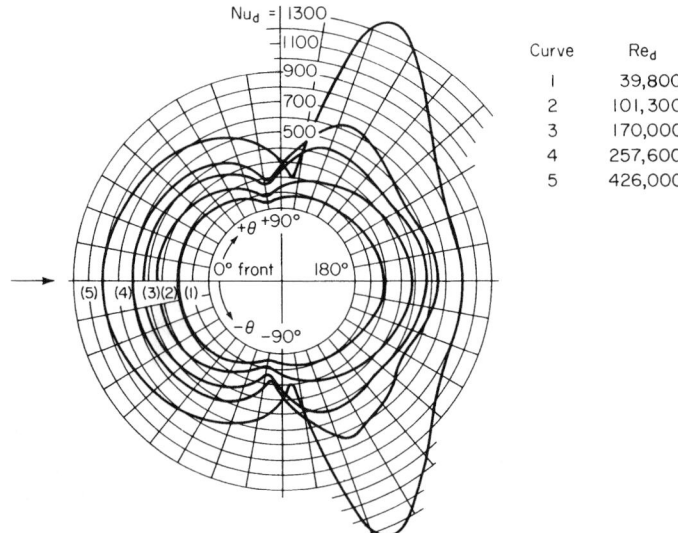

Curve	Re_d
1	39,800
2	101,300
3	170,000
4	257,600
5	426,000

FIGURE 6.28 Distribution of local heat transfer on the surface of a circular cylinder in cross flow in air [66]. (Reprinted from Ref. 65 by permission of McGraw-Hill.)

The overall heat transfer data for a tube in an infinite array are correlated by Kays and London [70] as:

$$\overline{St} = c_h \, Pr^{-2/3} \, Re_d^{-0.4} \tag{6.143}$$

or

$$\overline{Nu}_d = c_h \, Pr^{-1/3} \, Re_d^{0.6} \tag{6.144}$$

The Reynolds number in this correlation is based on the flow velocity at the minimum area section normal to the flow direction.

A typical correlation for c_h is shown in Fig. 6.29 for the case of a staggered array. For effects of array geometry, Refs. 67 and 70 should be consulted.

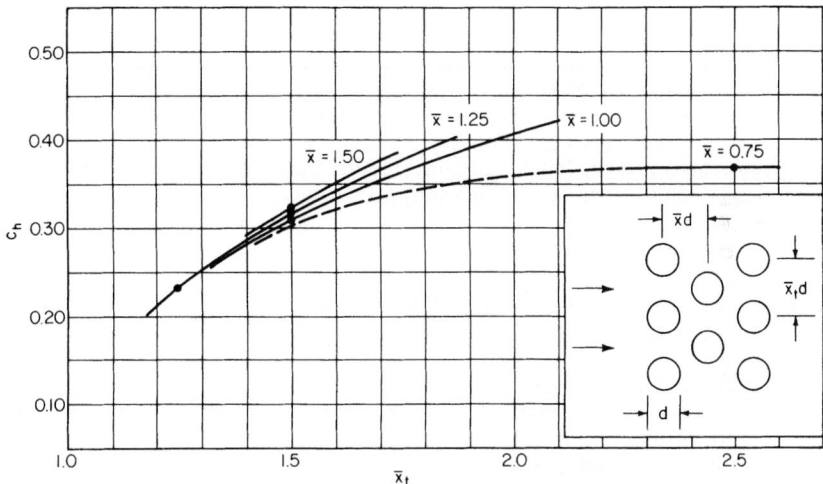

FIGURE 6.29   Correlation of coefficient in Eq. 6.143 or 6.144 for an infinite bank of staggered array, $300 < Re_d < 15{,}000$ [70]. (Reprinted by permission of McGraw-Hill.)

For a tube located near the front of the bank, the overall or average heat transfer is lower than that predicted by correlations for infinite arrays. The necessary correction factor as a function of row number in an array as presented by Kays and London [70] is shown in Fig. 6.30.

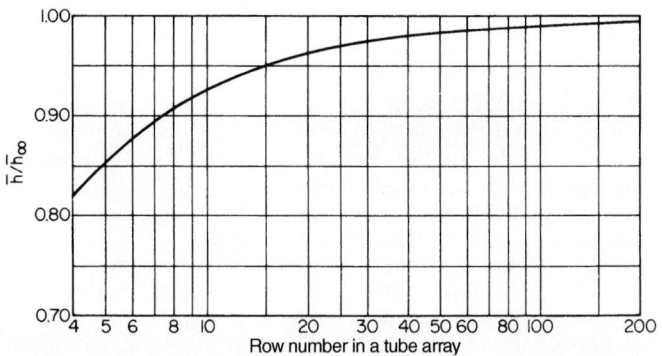

FIGURE 6.30   Correlation factor to account for row-to-row variation in heat transfer from a tube in a staggered or in-line bank [70]. (Reprinted by permission of McGraw-Hill.)

TWO-DIMENSIONAL TURBULENT BOUNDARY LAYER

Turbulence Transport Mechanisms and Modeling

In turbulent flows, the transport of momentum, heat, and/or individual species within gradients of velocity, temperature, and concentration is caused predominantly by the chaotic motion of elements of fluid (eddies). This mixing process transports properties much more effectively than the molecular processes identified with viscosity, thermal conductivity, and diffusion. A rather complete description of these processes is given in Refs. 71–73.

Currently there exist computers with sufficient storage capacity and speed to allow computation of these time-dependent motions for rather simple flows with finite difference meshes sufficiently fine to resolve the larger eddies of the motion. Even with such computations, however, it is necessary to model the effects of the eddies that are too small to be resolved. It is believed that since the transport of properties is governed by the larger eddies, the modeling process is less critical in these computations than where the entire turbulence is modeled. These "turbulence simulations" are still too costly for routine engineering computations and are used primarily to study the "physics" of particular turbulent flows. In fact, the results provide much more information than an engineer may ever want or need.

In engineering computations, the turbulent transport of properties is usually treated in a statistical manner, where computations are concerned with the *mean* velocities, temperatures, and/or concentrations. This statistical approach, however, masks many of the actual physical processes in the dynamic flow field, which must be recovered by the modeling at some level of the turbulence statistics. This modeling was originally guided by the results from experiments, but currently this guidance can rely on "simulations" as well.

The statistical turbulence models generally employed at present are based on time-averaging, at a single point in space, the instantaneous dynamic equations representing the conservation of mass, momentum components, energy, and species concentrations. These equations, in their most general form, apply to compressible, viscous, heat-conducting, and diffusing fluids. The statistical representation is initiated by expressing the dependent variables as the sum of a mean and fluctuating quantity, e.g.,

$$u = \bar{u} + u'$$
$$\rho = \bar{\rho} + \rho'$$
$$K_i = \bar{K}_i + K'_i$$

Substitution of such a decomposition of the dependent variables into the basic conservation equations is then followed by time-averaging the equations according to the following definition:

$$\bar{f} = \lim_{t \to \infty} \frac{1}{t} \int_0^t f(t') \, dt'$$

It is seen that this definition of mean quantities eliminates terms that are linear in the fluctuating quantities. Moments of the fluctuating quantities that are retained in a boundary layer are $\rho'v'$, $\overline{(\rho v)'u'}$, $\overline{(\rho v)'w'}$, $\overline{(\rho v)'i'}$, and $\overline{(\rho v)'K'_i}$, which represent the turbulent flux of mass, momentum, heat, and species concentration in the direction normal to a surface. These quantities are added to their molecular counterparts. Details of these derivations can be found in several sources, e.g., Refs. 65, 71, 74, and 75.

The evaluation of these statistical second moments is the goal of turbulence models. These models fall into two categories. First are models in which the turbulent fluxes are expressed in the same functional form as their laminar counterparts, but in which the molecular properties of viscosity, thermal conductivity, and diffusion coefficient are supplemented by corresponding eddy viscosities, conductivities, and diffusivities. The primary distinction is the recognition that the eddy coefficients are properties of the turbulent flow field, not the physical properties of the fluid. The second category of turbulence models includes those that express the turbulent moments in terms of partial differential field equations. When this is done, new moments (some of higher order, others involving pressure fluctuations, and still others involving space derivatives of the fluctuating quantities) appear in these equations. The number of moments grows faster than the number of the additional moment equations; thus, the set of equations cannot be closed. References 73 and 76 demonstrate this closure problem in detail.

The classical turbulence models express the eddy viscosity algebraically in terms of a turbulence scale and intensity that are related, respectively, to the characteristic length dimensions of the flow field and the local mean velocity gradients. This implies an equilibrium between the local turbulence and the mean motion. This requirement of equilibrium has been relaxed in some eddy viscosity models where the intensity and scale of turbulence used to evaluate the eddy viscosity are expressed by partial differential equations for the turbulence kinetic energy and its dissipation rate. This latter class of models is presented in detail, for example, in Refs. 76 and 77.

A 1982 conference held at Stanford University [63] was devoted to assessing the merits of existing turbulence models in the prediction of the mean velocity fields for both simple and complex turbulent flows. The flow fields employed as standards for comparison were selected on the basis of their being well-documented experimentally. It was found that the field models of turbulence, the second-order closure of the Reynolds stress equations or the two-equation eddy viscosity models, while having a broader range of application than particular algebraic eddy viscosity models, did not show dramatic improvement in accuracy over the simpler models for flow situations similar to those experiments on which the simpler models were based. In view of these observations, and the analytical advantages of the simpler models in the analysis of convection, classical algebraic eddy viscosity models will be used to represent turbulent transport in this chapter.

The turbulent flux vector for the local shear stress is given by

$$\tau_t = \bar{\rho}\epsilon_M \frac{\partial \bar{u}}{\partial y} = -\overline{(\rho v)' u'} \tag{6.145}$$

and the heat flux by

$$q_t'' = -\bar{\rho} c_p \epsilon_H \frac{\partial \bar{T}}{\partial y} = \overline{(\rho v)' i'} \tag{6.146}$$

The quantity ϵ_M is called the eddy diffusivity for momentum, and ϵ_H is the eddy diffusivity for heat. They are related through the turbulent Prandtl number $\mathrm{Pr}_t = \epsilon_M/\epsilon_H$.

Although these eddy diffusivities act in the same manner as the kinematic viscosity and thermal diffusivity in laminar flow, the critical difference is that the eddy diffusivities are not properties of the fluid but are dependent largely on the dynamic behavior of the fluid motion. In this section the fluid dynamic bases for evaluating these eddy diffusivities are given. They will then be used in a variety of convective heating situations to yield formulas useful in engineering computations.

Mean Velocity Characteristics for Constant Fluid Properties. Mean velocity distributions measured in turbulent boundary layers when the fluid properties are uniform (low speeds and small temperature differences between the free stream and surface) are described in the reviews of Refs. 78 and 79. At a given station, the turbulent boundary layer is composed of two regions with velocity profiles described by the "law of the wall" and the "law of the wake" after Coles [80, 81].

Wall Region. The region near the wall possesses a universal velocity profile when the data are correlated in terms of the coordinates $u^+ = u/u^*$ and $y^+ = u^* y/v$. The quantity $u^* = \sqrt{\tau_w/\rho}$ is called the friction velocity and is the appropriate characteristic velocity in this region. The corresponding characteristic length is v/u^*.

The near-wall region is composed of three layers as shown in Fig. 6.31. The layer immediately adjacent to the surface ($y^+ \lesssim 5$) is called the laminar sublayer where, because of the presence of the surface, the turbulence has been damped into a fluctuating laminar flow. In this layer, viscosity predominates over the eddy viscosity, and the velocity distribution may be approximated by

$$u^+ = y^+ \tag{6.147}$$

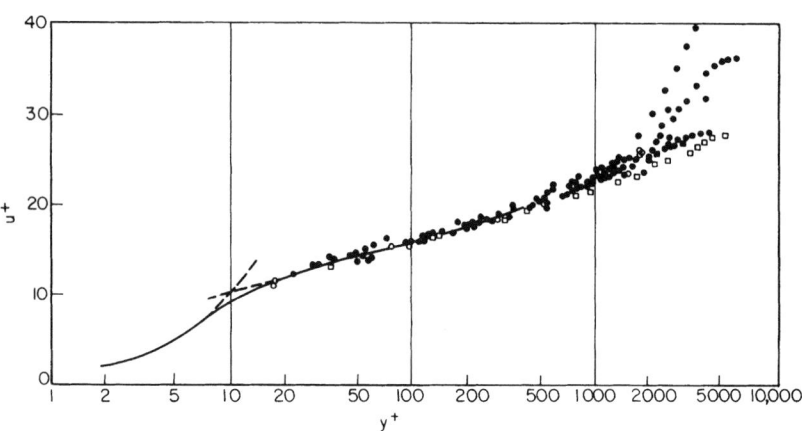

FIGURE 6.31 Universal velocity profile for an incompressible turbulent boundary layer near the surface ("law of the wall") [79]. (Reprinted by permission of Academic Press.)

For $y^+ > 50$, the turbulent processes completely dominate the local shear, and the resulting correlation can be represented by

$$u^+ = \frac{1}{\kappa} \ln y^+ + \mathbf{B} \tag{6.148}$$

The parameter κ is called the von Kármán constant, and the value that fits most of the data is 0.41. The corresponding value of $\mathbf{B}$ is 5.0. Intermediate between these layers is the buffer layer, where both shear mechanisms are important. The essential feature of this data correlation is that the wall shear completely controls the turbulent boundary layer velocity distribution in the vicinity of the wall. So dominant is the effect of the wall shear that even when pressure gradients along the surface are present, the velocity distributions near the surface are essentially coincident with the data obtained on plates with uniform surface pressure [82]. Within this region for a flat plate, the local shear stress remains within about 10 percent of the surface shear stress. It is noted that this shear variation is often ignored in turbulent boundary layer theory.

Wake Region. The experimental data toward the outer edge of the boundary layer do not correlate in a plot of $u^+(y^+)$. Correlation of these data can be achieved by utilizing the boundary layer thickness δ as the characteristic dimension and expressing the velocity as a decrement relative to its value at the boundary layer edge. Such a "wake" correlation is shown in Fig. 6.32 for a plate with uniform pressure. Although δ is rather arbitrary because of the asymptotic approach of the velocity to its free-stream value, no serious error results in these correlations if a consistent definition such as $u(\delta) = 0.995u_e$ is adopted. The velocity distribution for the combined wall and wake region of a fully developed turbulent boundary layer was formulated by Coles [81] as

$$u^+ = \frac{1}{\kappa} \ln y^+ + \mathbf{B} + \frac{\tilde{\pi}}{\kappa} \, \mathbf{w}\!\left(\frac{y}{\delta}\right) \tag{6.149}$$

where $\mathbf{w}(y/\delta)$ is the wake function indicated in Fig. 6.33. This function is approximated quite well by $1 - \cos{(\pi y/\delta)}$. Equation 6.149 applies to equilibrium boundary layers, as defined by Clauser [79], where $(\delta^*/\tau_w)\,dP/dx$ is constant along the surface. Under these conditions and at large Reynolds numbers, the parameter $\tilde{\pi}$ in Eq. 6.149 is independent of position.

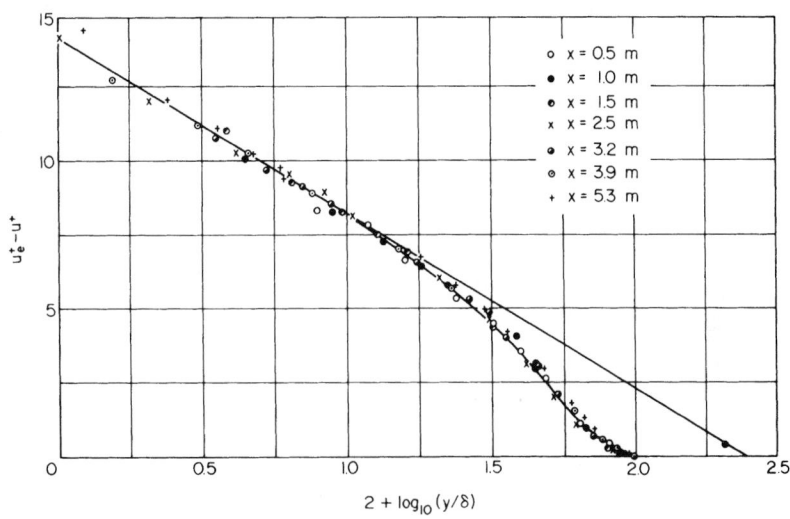

FIGURE 6.32 Velocity decrement for an incompressible turbulent boundary layer away from the surface ("law of the wake") [83].

Coles [84] used Eq. 6.149, with $\kappa = 0.41$ and $\mathbf{B} = 5.0$, to evaluate $c_f/2$, δ, and $\tilde{\pi}$ from a large set of equilibrium boundary layer data. In this evaluation, Coles excluded data for $y^+ < 50$, where Eq. 6.149 does not apply, and data for values of y near the boundary layer edge, where Eq. 6.149 provides poor values of the slope du^+/dy^+. For flat plates with uniform free-stream velocities, Coles found agreement with the data when $\tilde{\pi} = 0.62$. On the other hand, if experimental skin friction data obtained at very high Reynolds numbers are used to define $\tilde{\pi}$, a value of 0.55 is favored. The use of these different values of $\tilde{\pi}$ produces differences of only a few percent in the skin friction and boundary layer integral parameters, such as the displacement thickness

$$\delta^* = \int_0^\delta \left(1 - \frac{\rho u}{\rho_e u_e}\right) dy \tag{6.150}$$

and the momentum thickness

$$\theta = \int_0^\delta \frac{\rho u}{\rho_e u_e}\left(1 - \frac{\rho u}{\rho_e u_e}\right) dy \tag{6.151}$$

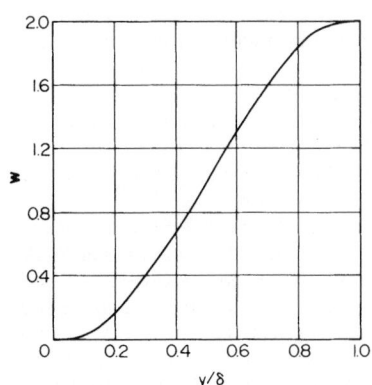

FIGURE 6.33 Coles' "law of the wake" function in Eq. 6.149 [81].

The value of $\tilde{\pi}$ recommended in regions of zero or positive pressure gradients is

$$\tilde{\pi} = 0.55 + 0.47 \frac{\delta^*}{\tau_w} \frac{dP}{dx} \tag{6.152}$$

In regions of negative pressure gradients, $\tilde{\pi}$ can be estimated by doubling the constant 0.47 in Eq. 6.152.

The parameter $\tilde{\pi}$ is influenced by low Reynolds numbers even when the boundary layer is fully turbulent. In zero-pressure-gradient flow, $\tilde{\pi}$ can be represented approximately by

$$\tilde{\pi} = 0.6 \left(\frac{\mathrm{Re}_\theta - 470}{\mathrm{Re}_\theta} \right)^{0.85} \tag{6.153}$$

for $\mathrm{Re}_\theta \leq 5000$; $\tilde{\pi} = 0.55$ for $\mathrm{Re}_\theta > 5000$.

If Eq. 6.149 is used throughout the boundary layer to define the integral parameters, the displacement thickness becomes

$$\frac{\delta^*}{\delta} = \frac{1 + \tilde{\pi}}{\kappa} \sqrt{\frac{c_f}{2}} \tag{6.154}$$

and the shape factor is

$$H = \frac{\delta^*}{\theta} = \left(1 - \frac{2 + 3.179\tilde{\pi} + 1.5\tilde{\pi}^2}{1 + \tilde{\pi}} \frac{\sqrt{c_f/2}}{\kappa} \right)^{-1} \tag{6.155}$$

The approximations introduced in the integrations leading to these equations, that is, utilizing Eq. 6.149 throughout the sublayer and buffer layer and ignoring the nonzero value of $\partial u^+/\partial y^+$ near the outer edge of the boundary layer, are valid at Reynolds numbers of practical interest.

Equation (6.149) also leads directly to the skin friction expression

$$\frac{1}{\sqrt{c_f/2}} = \frac{1}{\kappa} \ln \mathrm{Re}_{\delta^*} + \frac{1}{\kappa} \ln \frac{\kappa}{1 + \tilde{\pi}} + \mathbf{B} + \frac{2\tilde{\pi}}{\kappa} \tag{6.156}$$

For $2000 < \mathrm{Re}_\theta < 50{,}000$ and $0.55 < \tilde{\pi} < 2.0$ and with $\kappa = 0.41$ and $\mathbf{B} = 5.0$, Eqs. 6.155 and 6.156 yield skin friction coefficients within ± 10 percent of the well-established Ludwieg-Tillman correlation equation [82]

$$\frac{c_f}{2} = 0.123 \times 10^{-0.678H} \mathrm{Re}_\theta^{-0.268} \tag{6.157}$$

Given the momentum thickness at some location on a body with streamwise pressure gradients, the evaluation of the local skin friction coefficient from Eq. 6.157 requires knowledge of the shape factor H. Alternatively, the same input information can be used to evaluate both the skin friction coefficient and the shape factor by solving Eqs. 6.152–6.156 iteratively.

Clauser [79] made two observations regarding turbulent boundary layer velocity profiles that have direct influence on convective heating analyses. First, he noted that the region characterized by the "law of the wall" equilibrates very quickly after it is disturbed by some external force. Thus, this region is nearly in equilibrium with its local surface boundary conditions. In contrast, the region characterized by the "law of the wake" possesses a long memory of the upstream events. Clauser also found that the eddy diffusivity in the wake region is essentially independent of the distance from the surface y and is related to the local boundary layer thickness, which reflects the growth of the boundary layer from the leading edge. The resulting expression for the eddy diffusivity in the wake region on a flat plate or a surface with an equilibrium turbulent boundary layer $((\delta^*/\tau_w)(dP/dx) = \text{constant})$ is

$$\frac{\overline{\epsilon}}{\nu} = 0.018 \frac{u_e \delta^*}{\nu} \tag{6.158}$$

where $\overline{\epsilon} = \nu + \epsilon_M$.

Flat Plate With Zero Pressure Gradient.   With values of $\kappa = 0.41$, $\mathbf{B} = 5.0$, and $\tilde{\pi} = 0.55$, Eqs. 6.154 and 6.155 reduce to

$$\frac{\delta^*}{\delta} = 3.8 \sqrt{\frac{c_f}{2}} \tag{6.159}$$

and

$$H = \frac{\delta^*}{\theta} = \left(1 - 6.6 \sqrt{\frac{c_f}{2}}\right)^{-1} \tag{6.160}$$

The skin friction in the range $5000 < R_\theta < 50{,}000$ can be expressed as

$$\sqrt{\frac{2}{c_f}} = 2.44 \ln \mathrm{Re}_{\delta^*} + 4.4 \tag{6.161}$$

A convenient relationship expressed explicitly in terms of the momentum-thickness Reynolds number that agrees with Eqs. 6.160 and 6.161 within a few percent in the range $5000 < \mathrm{Re}_\theta < 50{,}000$ is given by

$$\frac{c_f}{2} = \frac{1}{(2.43 \ln \mathrm{Re}_{\theta_e} + 5.0)^2} \tag{6.162}$$

This form of equation originated with von Kármán; however, the coefficients appearing here are slightly different because they are based on more recent data.

It is often convenient to evaluate the local skin friction coefficient with the following simple equations:

$$\frac{c_f}{2} = \frac{0.0128}{\mathrm{Re}_{\theta_e}^{1/4}} \qquad \mathrm{Re}_{\theta_e} < 4000 \tag{6.163}$$

$$\frac{c_f}{2} = \frac{0.0065}{\mathrm{Re}_{\theta_e}^{1/6}} \qquad \mathrm{Re}_{\theta_e} > 4000 \tag{6.164}$$

after Blasius and Falkner [85], respectively. Equations 6.162–6.164 are compared with high–Reynolds number data [86] in Fig. 6.34.

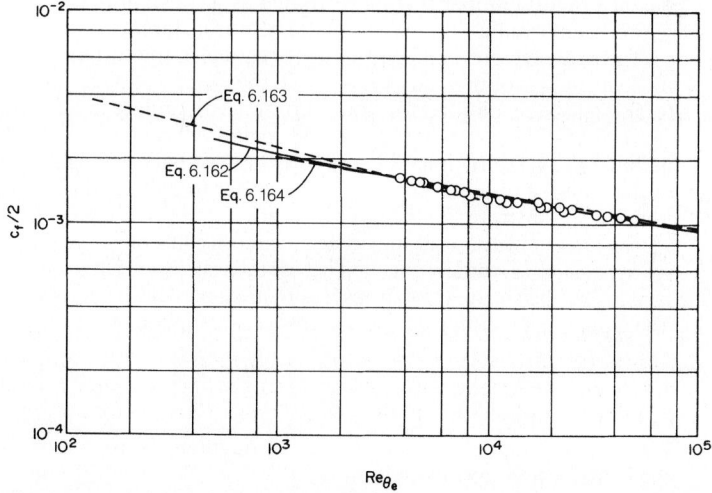

FIGURE 6.34 Comparison of local skin friction coefficients predicted by Eqs. 6.162–6.164 with the experimental data of Ref. 86.

On an impervious flat plate with uniform pressure, the momentum integral equation is

$$\frac{\tau_w}{\rho_e u_e^2} = \frac{d\theta}{dx} \tag{6.165}$$

This equation leads directly to an expression for the local length Reynolds number,

$$\mathrm{Re}_{x_e} = \int_0^{\mathrm{Re}_{\theta_e}} \frac{d\,\mathrm{Re}_{\theta_e}}{c_f/2} \tag{6.166}$$

If the laminar and transitional zones are short, the turbulent boundary layer can be assumed to begin at the leading edge of the plate. In this case, for surface length Reynolds number of a few million, Eq. 6.163 can be employed in Eq. 6.166 to establish the following simple relationship:

$$\frac{c_f}{2} = \frac{0.0296}{\mathrm{Re}_{x_e}^{1/5}} \tag{6.167}$$

For length Reynolds numbers in the tens of millions, the upstream portion of the boundary layer is not significant, and Eq. 6.164 may be used in Eq. 6.166 all the way from $\mathrm{Re}_{\theta_e} = 0$ to yield

$$\frac{c_f}{2} = \frac{0.0131}{\mathrm{Re}_{x_e}^{1/7}} \tag{6.168}$$

A more general equation proposed by Schultz-Grunow [83] that covers the full Reynolds number range is given by

$$\frac{c_f}{2} = \frac{0.185}{(\log \mathrm{Re}_{x_e})^{2.584}} \tag{6.169}$$

Eddy Diffusivity Models.   The mean velocity data described in the previous section provide the bases for evaluating the eddy diffusivity for momentum (eddy viscosity) in heat transfer analyses of turbulent boundary layers. These analyses also require values of the turbulent Prandtl number for use with the eddy viscosity to define the eddy diffusivity of heat. The turbulent Prandtl number is usually treated as a constant that is determined from comparisons of predicted results with experimental heat transfer data.

The presence of "wall" and "wake" regions in the turbulent boundary layer is reflected in the distribution of the eddy viscosity. In the outer wake region, Clauser's empirical form (Eq. 6.158), has been adopted for finite-difference boundary layer computations, although Ref. 87 suggests that the constant in Eq. 6.158 be altered to 0.0168. In analyses, however, the need for defining the wake region eddy viscosity has not been critical, largely because of the nearness of the value of the turbulent Prandtl number to unity and the use of the Crocco transformation, as demonstrated in the next section.

Evaluation of the wall region eddy viscosity has been facilitated by the fact that the shear stress remains essentially constant across the wall layer. On a flat plate with uniform free-stream conditions, the shear stress drops less than 10 percent from its wall value across the wall layer. The equation governing low-speed flow on a flat plate in the absence of a pressure gradient is

$$u \frac{\partial u}{\partial x} + v \frac{\partial u}{\partial y} = \frac{\partial}{\partial y}\left[(\nu + \epsilon_M)\frac{\partial u}{\partial y}\right] \tag{6.170}$$

If the shear is equal to the wall shear throughout the region of interest, Eq. 6.170 reduces to the Couette flow approximation

$$(\nu + \epsilon_M)\frac{du}{dy} = \frac{\tau_w}{\rho} \tag{6.171}$$

Expressed in near-wall coordinates, $u^+ = u/u^*$ and $y^+ = u^*y/\nu$, Eq. 6.171 becomes

$$\frac{\epsilon_M}{\nu} = \frac{1}{du^+/dy^+} - 1 \tag{6.172}$$

Equation 6.172 directly relates the mean velocity data to the eddy viscosity.

From Eqs. 6.147 and 6.172, it is seen that the sublayer is physically identified with zero eddy viscosity or no turbulent transport. Beyond $y^+ = 50$, Eqs. 6.148 and 6.172 indicate

$$\frac{\epsilon_M}{\nu} = \kappa y^+ - 1 \tag{6.173}$$

For $5 \le y^+ \le 30$, von Kármán [88] fitted the buffer layer velocity data with a semi-logarithmic relationship, $u^+ = 5[1 + \ln (y^+/5)]$, that provided

$$\frac{\epsilon_M}{\nu} = \frac{y^+}{5} - 1 \tag{6.174}$$

Spalding [89] defined a simple expression for the eddy viscosity applicable to the entire law of the wall region as

$$\frac{\epsilon_M}{\nu} = \frac{\kappa}{\mathbf{E}}\left[e^{\kappa u^+} - 1 - \kappa u^+ - \frac{(\kappa u^+)^2}{2!} - \frac{(\kappa u^+)^3}{3!} \right] \tag{6.175}$$

where $\kappa = 0.4$ and $9 < \mathbf{E} < 10$. For consistency with Eq. 6.148, it is necessary to set $\kappa = 0.41$ and $\mathbf{E} = 7.8$. Another form of ϵ_M/ν, recommended by van Driest [90] and currently popular in finite difference calculations of turbulent boundary layers, is

$$\frac{\epsilon_M}{\nu} = l^{+2}\left| \frac{du^+}{dy^+} \right| \tag{6.176}$$

with

$$l^+ = \kappa y^+ \left[1 - \exp\left(-\frac{y^+}{A^+} \right) \right] \tag{6.177}$$

When $\kappa = 0.41$, it is required that $A^+ = 24.7$ for consistency with Eq. 6.148. Finite difference calculations favor $A^+ = 26$ on a flat plate, and Refs. 87 and 91 give corrections to A^+ required in regions with pressure gradients and when mass transfer occurs at the surface. With a pressure gradient, the A^+ in a constant-property fluid becomes, according to Ref. 87,

$$A^+ = \frac{26}{[1 - 11.8(\nu u_e/u^{*3})(du_e/dx)]^{0.5}} \tag{6.178}$$

This modification is rather small, e.g., even where $\bar{\pi}$ has risen to 10, A^+ is decreased by only 6 percent. This behavior is consistent with the observed insensitivity of the law of the wall to pressure gradients.

The behavior of the alternate forms of ϵ_M/ν in the near-wall region of a turbulent boundary layer is shown in Fig. 6.35. The classical Prandtl-Taylor model assumes a sudden change from laminar flow ($\epsilon_M/\nu = 0$) to fully turbulent flow (Eq. 6.173) at $y^+ = 10.8$. The von Kármán model [88] allows for the buffer region and interposes Eq. 6.174 between these two regions. The continuous models depart from the fully laminar conditions of the sublayer around $y^+ = 5$ and asymptotically approach limiting values represented by Eq. 6.173. In finite difference calculations, ϵ_M/ν is allowed to increase until it reaches the value given by Eq. 6.158 and then is either kept constant at this value or diminished by an intermittency factor found experimentally by Klebanoff [92].

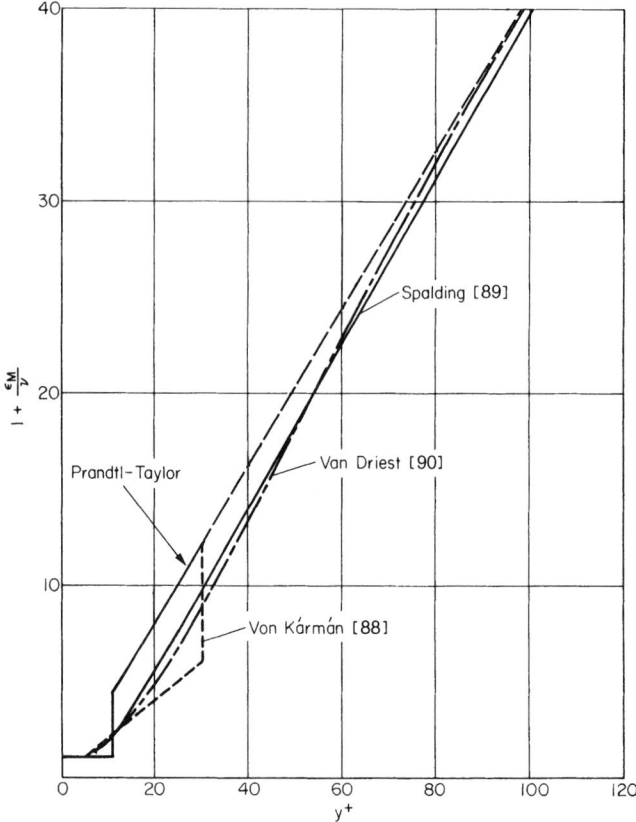

FIGURE 6.35   Near-wall eddy viscosity models.

Uniform Free-Stream Conditions

Uniform Surface Temperature

Low-Speed Flow.   Heat transfer is best found from the Reynolds analogy, the relationship between heat transfer and skin friction evaluated through analyses utilizing the empirical velocity distributions cited earlier. Knowledge of the flow field, which is independent of the temperature field when the fluid properties are constant, can be used directly to define the temperature field for a variety of thermal conditions and to evaluate the resulting convective heat transfer rates. For low-speed, constant-property flow, the energy equation is

$$\rho c_p u \, \frac{\partial T}{\partial x} + \rho c_p v \, \frac{\partial T}{\partial y} = \frac{\partial}{\partial y} \left[\left(\frac{\mu}{\mathrm{Pr}_T} + \frac{\rho \epsilon_M}{\mathrm{Pr}_t} \right) c_p \, \frac{\partial T}{\partial y} \right] \tag{6.179}$$

When this equation (Eq. 6.170) and the continuity equation are transformed from the x, y coordinate system to one with x and u as independent variables (the Crocco transformation), a formal solution results for the temperature field expressed in terms of the velocity field. This expression for the temperature, developed by van Driest [93], is

$$\frac{T}{T_e} = \frac{T_w}{T_e} - \left(\frac{T_w}{T_e} - 1 \right) \frac{\mathbf{S}(u/u_e)}{\mathbf{S}(1)} \tag{6.180}$$

where
$$\mathbf{S}(u/u_e) = \int_0^{u/u_e} \mathrm{Pr}_M \exp\left(-\int_{\tau_w}^{\tau} \frac{1-\mathrm{Pr}_M}{\tau} d\tau\right) d\left(\frac{u}{u_e}\right) \qquad (6.181)$$

and
$$\mathrm{Pr}_M = \frac{1 + \epsilon_M/\nu}{1/\mathrm{Pr}_T + (1/\mathrm{Pr}_t)(\epsilon_M/\nu)} \qquad (6.182)$$

It follows directly that
$$\frac{\mathrm{St}}{c_f/2} = \frac{1}{\mathbf{S}(1)} \qquad (6.183)$$

In terms of wall layer coordinates, the temperature distribution can be expressed from Eqs. 6.180–6.183 as

$$t^+ = \frac{\sqrt{c_f/2}}{\mathrm{St}} \frac{T_w - T}{T_w - T_e} = \int_0^{u^+} \mathrm{Pr}_M \exp\left(-\int_{\tau_w}^{\tau} \frac{1-\mathrm{Pr}_M}{\tau} d\tau\right) du^+ \qquad (6.184)$$

Reference [78] gives an excellent review of the hypotheses employed by various analysts to empirically define the integrand and the intermediate limits of integration in Eq. 6.181 for the different regions within the turbulent boundary layer.

The most widely used formula resulting from integrating Eq. 6.181 was due to von Kármán [88]. In effect, he eliminated the exponential term in Eq. 6.181 by setting the shear stress equal to its wall value near the surface and utilizing $\mathrm{Pr}_t = 1$ in the fully turbulent region. This truncated form of $\mathbf{S}(u/u_e)$ was integrated in the sublayer $(0 \leq y^+ \leq 5)$ with $\mathrm{Pr}_M = \mathrm{Pr}_T$, in the buffer layer $(5 \leq y^+ \leq 30)$ with Pr_M evaluated using the ϵ_M/ν indicated in Fig. 6.35, and in the outer portion of the boundary layer with $\mathrm{Pr}_M = \mathrm{Pr}_t = 1.0$. Deissler [94] extended this analysis by replacing the discontinuous formulations of ϵ_M/ν in the sublayer and buffer layer with $\epsilon_M/\nu = n_D^2 u^+ y^+$ where n_D is an empirical constant. When this value of ϵ_M/ν reaches that given by Eq. 6.173, the latter is used through the rest of the boundary layer. Deissler retained the assumption that $\mathrm{Pr}_t = 1$ in this fully turbulent region.

Because the wake region encompasses more of the boundary layer as the Reynolds number is increased, later analyses attempted to account for the wake region in different ways. Van Driest [93] permitted the shear in the boundary layer to approach zero at the outer edge. He also introduced nonunity turbulent Prandtl numbers to permit his recovery factor expression for air to agree with data, consistent with the findings of an earlier analysis [95]. The van Driest expression is

$$\frac{c_f/2}{\mathrm{St}} = \mathrm{Pr}_t\left(1 + 5\sqrt{\frac{c_f}{2}}\left\{\frac{1-\mathrm{Pr}_t}{5\kappa}\left[\frac{\pi^2}{6} + \frac{3}{2}(1-\mathrm{Pr}_t)\right] + \left(\frac{\mathrm{Pr}_T}{\mathrm{Pr}_t} - 1\right) + \ln\left[1 + \frac{5}{6}\left(\frac{\mathrm{Pr}_T}{\mathrm{Pr}_t} - 1\right)\right]\right\}\right) \qquad (6.185)$$

where $0.7 \leq \mathrm{Pr}_t \leq 1$ and $\kappa = 0.4$. Equation 6.185 reduces to the von Kármán analogy for $\mathrm{Pr}_t = 1$.

Spalding [89] abandoned the Couette flow hypothesis and solved Eq. 6.179 through the use of the von Mises transformation obtaining

$$\frac{\partial T}{\partial x^+} = \frac{1}{\epsilon_M^+ u^+} \frac{\partial}{\partial u^+}\left(\frac{\epsilon_H^+}{\epsilon_M^+} \frac{\partial T}{\partial u^+}\right) \qquad (6.186)$$

where
$$x^+ = \int_0^{\mathrm{Re}_{x_e}} \sqrt{\frac{c_f}{2}} \, d\,\mathrm{Re}_{x_e}$$

$$\epsilon_M^+ = 1 + \frac{\epsilon_M}{\nu} \qquad (6.187)$$

$$\epsilon_H^+ = \frac{1}{\mathrm{Pr}_T} + \frac{1}{\mathrm{Pr}_t}\frac{\epsilon_M}{\nu}$$

Spalding used Eq. 6.175 for ϵ_M/ν throughout the boundary layer.

Ferrari [96] also solved the von Mises form of the energy equation, but with the velocity distributions in the inner and outer portions of the boundary layer represented by polynomials of a velocity potential function. This analysis is applicable to gases in that it accounts for compressibility, but the solutions may be in error for high Prandtl numbers because of the rather approximate fit to the law of the wall.

Spalding [97] examined the results of other investigators who followed his approach and deduced that Pr_t must differ from unity to yield agreement with experimental data that show a Reynolds number dependence of the Reynolds analogy factor. The requirement of a nonunity turbulent Prandtl number for gases seems to be independent of the details of the analyses used to define the Reynolds analogy factor. From velocity distributions in gases, Spalding suggested $Pr_t = 0.9$. He showed, further, how previous solutions can be adjusted to accommodate nonunity Pr_t and found that the numerical results can be fitted with the interpolation formula

$$\frac{c_f/2}{St} = Pr_t \sqrt{\frac{c_f}{2}} \left\{ 6.76 \left(\frac{x^+}{Pr_t} \right)^{1/9} + 11.57 \left[\left(\frac{Pr_T}{Pr_t} \right)^{3/4} - 1 \right] \right\} \qquad (6.188)$$

Later, Spalding [98] extended his analysis to include the wake region of the boundary layer. The numerical results for heat transfer rates to a plate agree with those given by Eq. 6.188 to a few percent. This would be expected because at high Prandtl numbers the thermal boundary layer is confined to the inner portion of the flow boundary layer and is insensitive to the flow characteristics in the outer portion of the flow boundary layer.

The numerical results of the various Reynolds analogy factors are compared in Fig. 6.36 for laminar Prandtl numbers ranging from those characteristic of gases to those of oils and for $Re_{x_e} = 10^7$. Results for very low laminar Prandtl numbers, characteristic of liquid metals, are not shown because the assumptions for the velocity distributions in the various analyses are

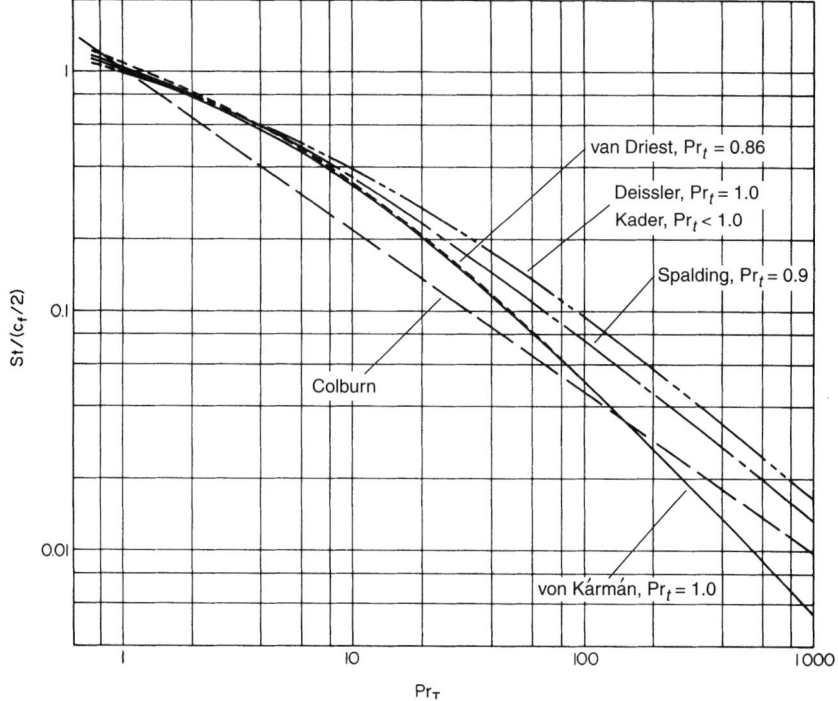

FIGURE 6.36 Reynolds analogy factor for incompressible turbulent boundary layer.

not valid when the thermal boundary layer is much thicker than the flow boundary layer. A comparison of the von Kármán analogy with the van Driest results indicates that use of $Pr_t = 0.89$ rather than $Pr_t = 1$ affects the results only for laminar Prandtl numbers less than about 10 and causes the Reynolds analogy factor to exceed unity when $Pr_T = 1$. These observations are also evident from the form of Eq. 6.188. The Deissler and Spalding results, although crossing in the vicinity of $Pr_T = 1$, both approach an asymptotic limit proportional to $Pr_T^{-3/4}$ for very large Prandtl numbers. Most significant is the departure of all the analyses from the Colburn analogy $Pr_T^{-2/3}$. An explanation of this can be deduced from the Deissler analysis [94], where the Reynolds analogy is also applied to the case of pipe flow through the calculation of bulk-mean properties. Here the analytical results agree reasonably well with the Colburn analogy. Thus, a deductive extension of the behavior of turbulent pipe flow data and laminar boundary layer theory on a flat plate, both of which yield a Reynolds analogy factor of $Pr_T^{-2/3}$, to the turbulent boundary layer on a flat plate yields erroneous results, particularly for $Pr_T \leq 10$.

If Eq. 6.184, restricted to a constant value of shear, is used to evaluate the temperature profile in a heated boundary layer, there results

$$t^+ = \frac{Pr_t}{\kappa} \ln y^+ + 5Pr_t \ln \left(1 + 5 \frac{Pr_T}{Pr_t}\right) + 5Pr_T - 8.5Pr_t \tag{6.189}$$

This formula agrees very well with the near-wall temperature profile data of Blackwell et al. [99] obtained in air if $Pr_T = 0.7$ and $Pr_t = 0.88$. It should be noted that good agreement with the data was achieved here without consideration of near-wall turbulent Prandtl number variations observed by some investigators [100].

Kader [101] presented an empirical correlation formula for boundary layer temperature profiles that were demonstrated to fit experimental data in fluids with Prandtl numbers in the range from 0.7 to 60. The formula is

$$t^+ = Pr_T \, y^+ \exp(-\overline{\Gamma}) + \left\{2.12 \ln \left[(1 + y^+) \frac{2.5(2 - y/\delta)}{1 + 4(1 - y/\delta)^2}\right] + \overline{\beta}(Pr_T)\right\} \exp\left(-\frac{1}{\overline{\Gamma}}\right) \tag{6.190}$$

where

$$\overline{\beta}(Pr_T) = (3.85Pr_T^{1/3} - 1.3)^2 + 2.12 \ln Pr_T \tag{6.191}$$

and

$$\overline{\Gamma} = \frac{10^{-2}(Pr_T \, y^+)^4}{1 + 5Pr_T^3 \, y^+} \tag{6.192}$$

The Reynolds analogy factor corresponding to these relationships can be written as

$$\frac{St}{c_f/2} = \frac{1}{\sqrt{c_f/2}\left[2.12 \ln \left(2.5Pr_T \, Re_{\delta*} \frac{\kappa}{1 + \tilde{\pi}}\right) + (3.85Pr_T^{1/3} - 1.3)^2\right]} \tag{6.193}$$

When Eq. 6.193 is evaluated on a flat plate at $Re_x = 10^7$, it essentially coincides with the line attributed to the Deissler theory shown in Fig. 6.36 for Pr_T up to 100. It is recommended that Eq. 6.193 be used for Prandtl numbers considerably greater than unity.

For Prandtl numbers slightly less than unity (gases), it is quite obvious that the Colburn analogy is at variance with the analytical results. For $Pr_T = 0.7$, excellent low-speed data [102] indicate values of $2St/c_f \approx Pr_T^{-0.4} \approx 1.15$. Other data in air at higher speeds [103], though less accurate and sometimes containing disturbances such as small pressure gradients, favor the value $2St/c_f = 1.2$. Since these data were obtained over a small Prandtl number range, it is difficult to ascertain empirically the influence of Pr_t and the functional dependence on Pr_T. Accordingly, it is recommended that the Spalding and van Driest results, represented by Eqs. 6.188 and 6.185, be utilized for $Pr_T \approx 1$ in view of their general agreement with the data cited. These conclusions are supported further by the agreement of Eq. 6.188 with the data of Ref. 99.

High-Speed Flow. Frictional dissipation occurs in the turbulent boundary layer for high-speed flow because of combined viscous and turbulent shear mechanisms. To account for variations in the specific heat of the fluid, it is best to express the heat potential as enthalpy rather than temperature. It was shown by van Driest [93] that under high-speed conditions, Eq. 6.180 is replaced by

$$\frac{i}{i_e} = \frac{i_w}{i_e} - \left(\frac{i_w}{i_e} - 1\right)\frac{\mathbf{S}(u/u_e)}{\mathbf{S}(1)} + \frac{u_e^2}{i_e}\left[\frac{\mathbf{S}(u/u_e)}{\mathbf{S}(1)}\overline{R}(1) - \overline{R}\left(\frac{u}{u_e}\right)\right] \tag{6.194}$$

The function $\mathbf{S}(u/u_e)$ is still represented by Eq. 6.181. The function $\overline{R}(u/u_e)$ is

$$\overline{R}\left(\frac{u}{u_e}\right) = \int_0^{u/u_e} \mathrm{Pr}_M \exp\left(-\int_{\tau(0)}^{\tau}\frac{1-\mathrm{Pr}_M}{\tau}\,d\tau\right)\left\{\int_0^{u/u_e}\exp\left[\int_{\tau(0)}^{\tau}\frac{1-\mathrm{Pr}_M}{\tau}\,d\tau\right]d\left(\frac{u}{u_e}\right)\right\}d\left(\frac{u}{u_e}\right) \tag{6.195}$$

As with a laminar boundary layer, the local surface heat flux can be written as

$$q_w'' = \rho_e u_e\,\mathrm{St}\,(i_w - i_{aw}) \tag{6.196}$$

where

$$i_{aw} = i_e + \mathbf{r}(0)\frac{u_e^2}{2} \tag{6.197}$$

It is found from Eq. 6.194 that the recovery factor is

$$\mathbf{r}(0) = 2\overline{R}(1) \tag{6.198a}$$

When Eq. 6.195 is solved under the same assumptions employed in Eq. 6.181 [93, 95, 96], it is found that the recovery factor for air ($\mathrm{Pr}_T = 0.7$, $\mathrm{Pr}_t = 1$) experiences a marked Reynolds number dependence, contrary to a rather profuse accumulation of data that yield values of $\mathbf{r}(0)$ between 0.87 and 0.89. Thus, solutions of Eq. 6.198a, rather than yielding values of $\mathbf{r}(0)$, have been used to estimate the turbulent Prandtl number necessary to eliminate the dependence of $\mathbf{r}(0)$ on Re_{x_e}. It is in this manner that van Driest found $\mathrm{Pr}_t = 0.89$ for use in Eq. 6.183. Values of $\mathbf{r}(0)$ within a percent of most existing data in air can be obtained by using

$$\mathbf{r}(0) = \mathrm{Pr}_T^{1/3} \tag{6.198b}$$

Liquids With Variable Viscosity. Deissler [94] considered a fluid where the viscosity is temperature dependent but all the other fluid properties remain constant. Solutions for $\mathrm{Pr}_T = 10$ and

$$\frac{\mu}{\mu_w} = \left(\frac{T}{T_w}\right)^{-4} \tag{6.199}$$

are shown in Figs. 6.37 and 6.38. The ordinate in Fig. 6.38 is defined as

$$t^+ = \frac{\sqrt{c_f/2}}{\mathrm{St}}\frac{T_w - T}{T_w - T_e}$$

and the parameter

$$\beta_T = \frac{\mathrm{St}}{\sqrt{c_f/2}}\frac{T_w - T_e}{T_w} = \frac{1}{t_e^+}\frac{T_w - T_e}{T_w} \tag{6.200}$$

It is noted in Fig. 6.37 that the effects of variable viscosity cause the velocity distributions outside the buffer layer to displace while remaining essentially parallel, so that

$$\frac{u}{\sqrt{\tau_w/\rho}} = A\ln\left(\frac{\sqrt{\tau_w/\rho}\,\rho y}{\mu_w}\right) + B(\beta_T) \tag{6.201}$$

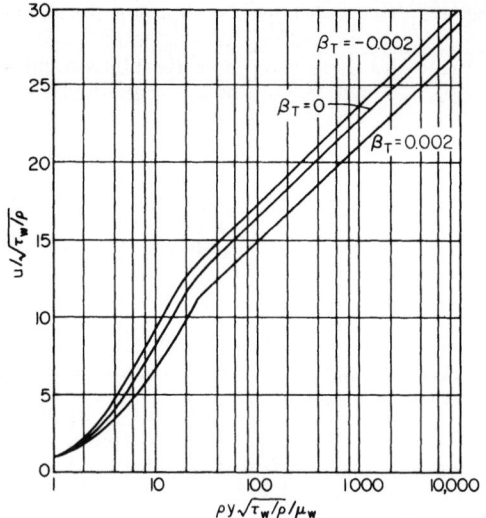

FIGURE 6.37   Universal velocity profiles for a liquid with $\mu/\mu_w = (T/T_w)^{-4}$; $\beta_T = (St/\sqrt{c_f/2})[(T_w - T_e)/T_w]$ [94].

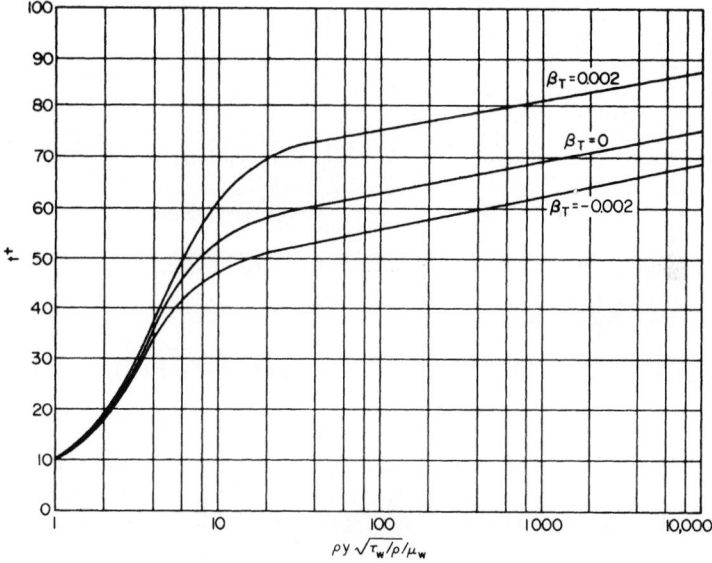

FIGURE 6.38   Universal temperature profiles for a liquid with velocity profiles as in Fig. 6.37.

where $B(\beta_T)$ is shown in Fig. 6.39. The parallel curves lead to the conclusion that $c_f/2$ can be computed from $c_f(\text{Re}_{x_e})$ relationships, e.g., Eq. 6.167 or 6.168, if the Reynolds number employed is replaced by

$$\text{Re}_{\text{eff}} \approx \frac{\rho_e u_e x}{\mu_w}\, e^{B(\beta_T) - B(0)} \tag{6.202}$$

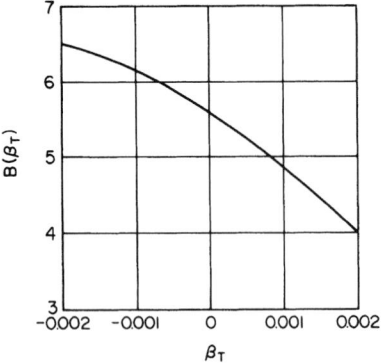

FIGURE 6.39 Displacement function of Eq. 6.201.

A simple iterative procedure for evaluating the local heat transfer rate is begun by assuming a value of β_T. The effective Reynolds number is evaluated from Fig. 6.39 and Eq. 6.202. Equation 6.167 or 6.168 yields $c_f/2$. The ordinate in Fig. 6.37 set equal to $\sqrt{2/c_f}$ yields the limiting value at the edge of the boundary layer of $\rho\delta\sqrt{\tau_w/\rho}/\mu_w$ that is then used in Fig. 6.38 to obtain a value of t_e^+. This value of t_e^+ and the wall and boundary layer edge temperatures are substituted in Eq. 6.200 to yield the next approximation for β_T. The procedure is repeated until convergence on a value of β_T is achieved. The local heat flux is then found from

$$q_w'' = \rho c_p u_e T_w \beta_T \sqrt{\frac{c_f}{2}} \tag{6.203}$$

Ideal Gases at High Temperature. Three fundamentally different approaches have been applied to the treatment of the turbulent boundary layer with variable fluid properties; all are restricted to air behaving as an ideal, calorically perfect gas. First, the Couette flow solutions have been extended to permit variations in viscosity and density. Second, mathematical transformations, analogous to Eq. 6.36 for a laminar boundary layer, have been used to transform the variable-property turbulent boundary layer differential equations into constant-property equations in order to provide a direct link between the low-speed boundary layer and its high-speed counterpart. Third, empirical correlations have been found that directly relate the variable-property results to incompressible skin friction and Stanton number relationships. Examples of the latter are reference temperature or enthalpy methods analogous to those used for the laminar boundary layer, and the method of Spalding and Chi [104].

Generalized Coordinate Transformation. The coordinate transformation rules for extending constant-property skin friction and Stanton number formulas to account for property variations as performed by Spalding and Chi [104] can be expressed functionally as

$$\tilde{c}_f = f_1(\tilde{R}e_\theta) \tag{6.204}$$

and

$$\tilde{c}_f = f_2(\tilde{R}e_x) \tag{6.205}$$

the generalized formulas are

$$c_f F_c = f_1(F_{R\theta} \, Re_{\theta_e}) \tag{6.206}$$

$$c_f F_c = f_2(F_{R_x} \, Re_{x_e}) \tag{6.207}$$

In general, the coordinate transformation factors F_c, $F_{R\theta}$, F_{R_x} are functions of Ma_e, T_w/T_e, Re_{x_e}, and T_e. (If the viscosity is expressed as a power-law function of temperature, the dependence on T_e can be eliminated.) From the von Kármán momentum integral equation for a flat plate,

$$\frac{c_f}{2} = \frac{d \, Re_\theta}{d \, Re_{x_e}} \tag{6.208}$$

it follows that

$$F_{R_x} = \frac{F_{R\theta}}{F_c} \tag{6.209}$$

The expressions for F_c and $F_{R\theta}$ given by the various methods are presented in Table 6.8. Values of these functions obtained from the van Driest theory [105] are shown in Figs. 6.40 and 6.41 for $T_e = 400$ R (222 K), $0 < Ma_e < 10$, and $T_w/T_e = 1, 2, 3, 5, T_{aw}/T_e$. It is noted that only the Coles method allows F_c and $F_{R\theta}$ to depend on the Reynolds number (through the parameter $\tilde{c}_f$).

TABLE 6.8 Expressions for F_c and $F_{R\theta}$ for Air as an Ideal Gas

Method	F_c	$F_{R\theta}$	Supplemental formulas and notes
Eckert [20]	ρ_e/ρ'	μ_e/μ'	$\dfrac{T'}{T_e} = 0.28 + 0.50\,\dfrac{T_w}{T_e} + 0.22\,\dfrac{T_{aw}}{T_e}$
Sommer and Short [107]	ρ_e/ρ'	μ_e/μ'	$\dfrac{T'}{T_e} = 0.36 + 0.45\,\dfrac{T_w}{T_e} + 0.19\,\dfrac{T_{aw}}{T_e}$
van Driest (von Kármán mixing length) [105]	$\dfrac{T_{aw}/T_e - 1}{(\sin^{-1}\alpha + \sin^{-1}\beta)^2}$	μ_e/μ_w	$\alpha = \dfrac{T_{aw}/T_e + T_w/T_e - 2}{\sqrt{(T_{aw}/T_e - T_w/T_e)^2 + 4(T_w/T_e)(T_{aw}/T_e - 1)}}$ $\beta = \dfrac{T_{aw}/T_e - T_w/T_e}{\sqrt{(T_{aw}/T_e - T_w/T_e)^2 + 4(T_w/T_e)(T_{aw}/T_e - 1)}}$
Coles [106]	$\dfrac{\mu_e\rho_e}{\mu_w\rho_w}\dfrac{\mu_{ss}}{\mu_e}$	μ_e/μ_{ss}	$\dfrac{T_{ss}}{T_e} = \dfrac{T_w}{T_e} + 17.2\sqrt{\dfrac{\bar{c}_f}{2}}\left(\dfrac{T_{aw}}{T_e} - \dfrac{T_w}{T_e}\right) - 305\,\dfrac{\bar{c}_f}{2}\left(\dfrac{T_{aw}}{T_e} - 1\right)$
Spalding and Chi [104]	$\dfrac{T_{aw}/T_e - 1}{(\sin^{-1}\alpha + \sin^{-1}\beta)^2}$	$\dfrac{(T_{aw}/T_e)^{0.772}}{(T_w/T_e)^{1.474}}$	α, β, same as van Driest $\mu \sim T^{0.76}$ assumed in empirical exponent of $F_{R\theta}$

The recommended design method for Mach numbers up to supersonic ($Ma_e < 4$) and adiabatic wall conditions ($T_w = T_{aw}$) is arbitrary since all the methods yield essentially the same results. The design methods for hypersonic Mach numbers ($Ma_e > 4$) and cold wall conditions ($T_w < T_{aw}$) should be based on conservatism. Comparison of the available skin friction data reveals differences of ±20 percent on surfaces near adiabatic wall temperature and as much as a factor of 2 for highly cooled walls ($T_w \sim 0.2T_{aw}$). If only the most recent skin friction data for

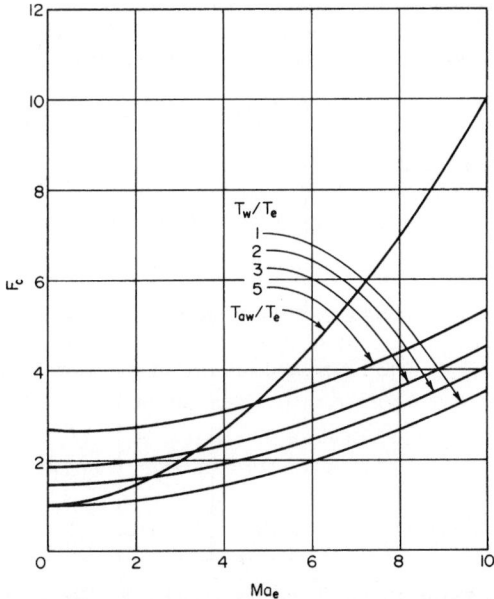

FIGURE 6.40 Compressible turbulent boundary layer transformation parameter for skin friction coefficient, $r(0) = 0.9$, $T_e = 400$ R (222 K).

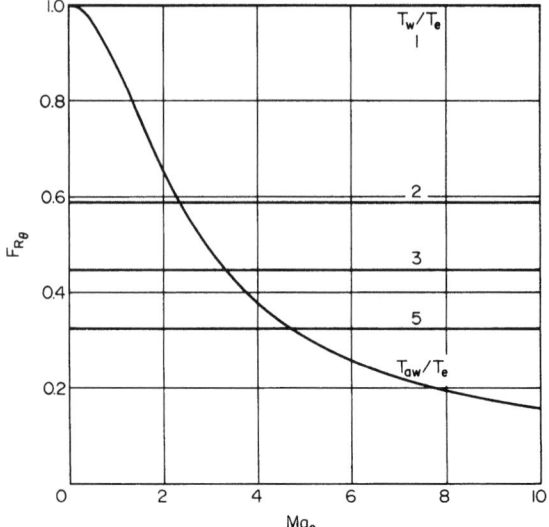

FIGURE 6.41 Compressible turbulent boundary layer transformation parameter for momentum-thickness Reynolds number, $r(0) = 0.9$, $T_e = 400$ R (222 K).

$0.3 < T_w/T_{aw} < 1.0$ are considered, particularly those where Re_{θ_e} was obtained from boundary-layer surveys, a recent evaluation [108] favors the Coles and van Driest methods. The Sommer and Short method and the Spalding and Chi method may underpredict the skin friction as much as 30 percent. For very cold walls ($T_w < 0.3T_{aw}$), none of the methods predict the effect of wall temperature ratio in the available skin friction data. The van Driest skin friction predictions are shown in Fig. 6.42 for $T_e = 400$ R (222 K), $0 < Ma_e < 10$, and $T_w/T_e = 1, 2, 3, 5, T_{aw}/T_e$.

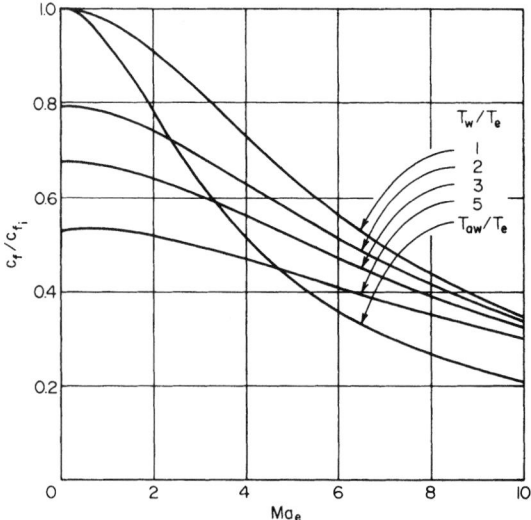

FIGURE 6.42 Local skin friction coefficient for a compressible turbulent boundary layer on a flat plate, $r(0) = 0.9$, $T_e = 400$ R (222 K).

The heat transfer is obtained by using a Reynolds analogy factor in combination with a skin friction prediction. For Mach numbers up to supersonic and near-adiabatic wall conditions, the recommended Reynolds analogy factor is the same as for low-speed flow, namely, 1.16. Any of the previously discussed skin friction theories may be used. For hypersonic Mach numbers and cold wall conditions, there are indications that the Reynolds analogy factor is more nearly equal to unity [108]. Use of this value and either the van Driest or the Coles theory for skin friction results in heat transfer predictions within ±10 percent of the data for $5 < \mathrm{Ma}_e < 7.5$ and $0.1 < T_w/T_{aw} < 0.6$ [108]. Alternatively, to similar accuracy one can use the usually accepted Reynolds analogy factor of 1.16 and the Spalding and Chi skin friction theory.

Nonuniform Surface Temperature.   Nonuniform surface temperatures affect the convective heat transfer in a turbulent boundary layer similarly as in a laminar case except that the turbulent boundary layer responds in shorter downstream distances. The heat transfer to surfaces with arbitrary temperature variations is obtained by superposition of solutions for convective heating to a uniform-temperature surface preceded by a surface at the recovery temperature of the fluid (Eq. 6.65). For the superposition to be valid, it is necessary that the energy equation be linear in T or i, which imposes restrictions on the types of fluid property variations that are permitted. In the turbulent boundary layer, it is generally required that the fluid properties remain constant; however, under the assumption that boundary layer velocity distributions are expressible in terms of the local stream function rather than y for ideal gases, the energy equation is also linear in T [96].

The effect of a stepwise discontinuity in surface temperature on a flat plate is expressible as

$$\frac{\mathrm{St}\,(x, s)}{\mathrm{St}\,(x, 0)} = \left[1 - \left(\frac{s}{x} \right)^{9/10} \right]^{-1/9} \tag{6.210}$$

For fluids other than gases, use of the Spalding relationship (Eq. 6.188) results in

$$\frac{\mathrm{St}\,(x, s)}{\mathrm{St}\,(x, 0)} = \frac{\dfrac{5.62\mathrm{Re}_{x_e}^{1/10}}{\mathrm{Pr}_t^{1/9}} + 11.57\left[\left(\dfrac{\mathrm{Pr}}{\mathrm{Pr}_t} \right)^{3/4} - 1 \right]}{\dfrac{5.62\mathrm{Re}_{x_e}^{1/10}}{\mathrm{Pr}_t^{1/9}}\left[1 - \left(\dfrac{s}{x} \right)^{9/10} \right]^{1/9} + 11.57\left[\left(\dfrac{\mathrm{Pr}}{\mathrm{Pr}_t} \right)^{3/4} - 1 \right]} \tag{6.211}$$

A short distance downstream from the step in surface temperature, for $\mathrm{Pr} = 0.7$ and $\mathrm{Pr}_t = 0.9$, the second term in both the numerator and denominator of Eq. 6.211 becomes comparatively small, and Eq. 6.211 reduces to Eq. 6.210. For very high Prandtl numbers, the Stanton number behind a step in surface temperature is essentially the same as on a uniform-temperature surface.

The intuitive approach for including variable fluid properties is to evaluate the local St $(x, 0)$ from the variable-property techniques of the previous section, employing the local surface temperature or enthalpy as a parameter. Equations 6.210 and 6.211 are used primarily to define the form of the enthalpy potential appropriate to an arbitrary surface temperature. (See Eq. 6.65).

Stepwise and Arbitrary Heat Flux Distribution.   A corollary of the problem treated in the previous section is the problem of finding the surface temperature distribution (or enthalpy distribution) resulting from a prescribed heat flux distribution. The superposition of solutions based on Eq. 6.210 yields

$$i_w(x) - i_{aw} = \frac{0.0979}{\rho_e u_e\,\mathrm{St}\,(x, 0)} \int_0^1 \frac{q_w''(s/x)}{[1 - (s/x)^{9/10}]^{8/9}}\, d\!\left(\frac{s}{x} \right) \tag{6.212}$$

Surface With Mass Transfer.   The advantages of mass transfer cooling systems in certain applications were discussed in the section on two-dimensional laminar boundary layers beginning on page 6.19. Reference 109 should be consulted for a complete description of the performance of different forms of mass transfer cooling systems. This section highlights the evaluation of turbulent boundary layer behavior in transpiration cooling systems.

Uniform Fluid Properties.   Analyses of turbulent boundary layers experiencing surface transpiration employ a hierarchy of increasingly complex models of the turbulent transport mechanisms. Most of the analyses, supported by complementary experiments, have emphasized the transpiration of air into low-speed airstreams [110–112]. Under these conditions, the fluid properties in the boundary layer are essentially constant, and the turbulent boundary layer can be described mathematically with Eqs. 6.170 and 6.179. In addition, when small quantities of a foreign species are introduced into the boundary layer for diagnostic purposes or by evaporation, the local foreign species concentration in the absence of thermal diffusion is given by

$$\rho u \frac{\partial K_1}{\partial x} + \rho v \frac{\partial K_1}{\partial y} = \frac{\partial}{\partial y} \left[\left(\frac{\mu}{Sc} + \frac{\rho \epsilon_M}{Sc_t} \right) \frac{\partial K_1}{\partial y} \right] \tag{6.213}$$

where the laminar and turbulent Schmidt numbers are $Sc = \nu/\mathcal{D}$ and $Sc_t = \epsilon_M/\epsilon_D$, respectively. The boundary conditions at the surface are $v = v_w(x)$ and $T = T_w(x)$, where one of these is specified and the other is determined from the heat balance represented by Eq. 6.74.

The simpler analyses reduce the boundary layer equations to ordinary differential equations, with the distance normal to the surface as the independent variable. This results from the assumption of Couette flow, where changes of the dependent variables in the streamwise direction may be neglected. The continuity and momentum equations then become

$$\frac{dv}{dy} = 0 \tag{6.214}$$

$$v \frac{du}{dy} = \frac{d}{dy} \left[(\nu + \epsilon_M) \frac{du}{dy} \right] \tag{6.215}$$

Solved simultaneously, these equations yield

$$v = v_w \tag{6.216}$$

and
$$\frac{c_f}{2} = \frac{\ln(1 + B_f)}{B_f} \frac{1}{u_e} \left(\int_0^\delta \frac{dy}{\nu + \epsilon_M} \right)^{-1} \tag{6.217}$$

where
$$B_f = \frac{v_w}{u_e} \frac{1}{c_f/2} \tag{6.218}$$

The simplest analyses, e.g., Ref. 110, assume that the value of the integral in Eq. 6.217 is independent of the amount of surface mass transfer at a given location or Re_x. This implies that mass transfer affects the boundary layer thickness and the eddy viscosity equally, thus compensating for the change in the limit of integration with a proportionate change in the reciprocal of the integrand. The resulting simple expression

$$\left. \frac{c_f}{c_{f0}} \right|_{Re_x} = \frac{\ln(1 + B_f)}{B_f} \tag{6.219}$$

predicts experimental data surprisingly well [113].

Sometimes it is more convenient to scale the transpiration rate to c_{f0}, the skin friction coefficient in the absence of surface mass transfer. With

$$b_f = \frac{v_w}{u_e} \frac{1}{c_{f0}/2} \tag{6.220}$$

Eq. 6.219 becomes

$$\left.\frac{c_f}{c_{f0}}\right|_{\mathrm{Re}_x} = \frac{b_f}{\exp(b_f) - 1} \tag{6.221}$$

The quantity c_{f0} in Eqs. 6.219 and 6.221 is found from the equations given in the section on uniform free-stream conditions. For example, at moderate Reynolds numbers on a flat plate, the skin friction coefficient is given by

$$\frac{c_f}{2} = 0.0296 \mathrm{Re}_x^{-0.2} \frac{\ln(1 + B_f)}{B_f} \tag{6.222}$$

Expressions based on Re_θ are found through the use of the integral momentum equation for a flat plate,

$$\frac{d\,\mathrm{Re}_\theta}{d\,\mathrm{Re}_x} = \frac{c_f}{2}(1 + B_f) \tag{6.223}$$

and yield

$$\frac{c_f}{2} = 0.0129 \left[\frac{\ln(1 + B_f)}{B_f} \right]^{1.25} (1 + B_f)^{0.25}\, \mathrm{Re}_\theta^{0.25} \tag{6.224}$$

A Couette flow analysis of the energy equation leads to an expression similar to Eq. 6.219 for the reduction in Stanton number due to blowing:

$$\left.\frac{\mathrm{St}}{\mathrm{St}_0}\right|_{\mathrm{Re}_x} = \frac{\ln(1 + B_h)}{B_h} \tag{6.225}$$

where

$$B_h = \frac{v_w}{u_e\,\mathrm{St}} \tag{6.226}$$

At moderate Reynolds numbers and with an empirical Reynolds analogy factor independent of blowing, the Stanton number on a flat plate can be expressed as

$$\mathrm{St}\,\mathrm{Pr}^{0.4} = 0.0296 \mathrm{Re}_x^{-0.2} \frac{\ln(1 + B_h)}{B_h} \tag{6.227}$$

The energy integral equation reduces for low-speed flow over a flat plate to

$$\mathrm{St} + \frac{v_w}{u_e} = \frac{d\,\mathrm{Re}_\Gamma}{d\,\mathrm{Re}_x} \tag{6.228a}$$

where

$$\Gamma = \int_0^{\delta_T} \frac{\rho u}{\rho_c u_e}\,dy = \left(\frac{I - I_e}{i_w - I_e}\right)dy \tag{6.228b}$$

Equation 6.227 can then be expressed as

$$\mathrm{St}\,\mathrm{Pr}^{0.4} = 0.0129 \mathrm{Re}_\Gamma - 0.25 \left[\frac{\ln(1 + B_h)}{B_h} \right]^{1.25} (1 + B_h)^{0.25} \tag{6.229}$$

Figure 6.43 shows Eqs. 6.224 and 6.229, based on local boundary layer thicknesses, to be applicable locally to boundary layers experiencing acceleration, deceleration, and moderate variations in blowing along the surface.

While the analysis described above is useful to define the surface parameters c_f, St, and, by inference, c_{mi}, more detailed analysis is required to define the velocity, temperature, and species concentration profiles in the boundary layer (see Refs. [112] and [113]).

Variable Fluid Properties. The fluid properties within a boundary layer may vary because of large temperature differences caused by frictional dissipation at high Mach numbers or because the coolant introduced through the surface possesses properties quite different from those of the main stream.

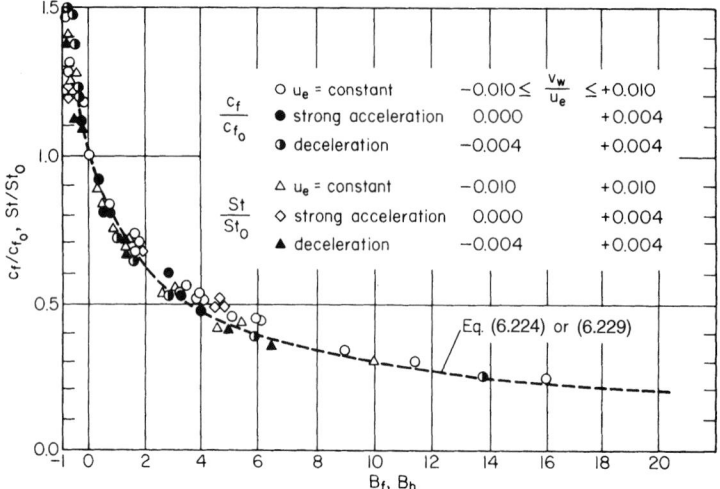

FIGURE 6.43   Effect of transpiration at the surface on skin friction and heat transfer [113]. (Reprinted from Ref. 74 by permission of McGraw-Hill.)

Early theories for transpiration of air into air [114, 115] were based on the Couette flow approximation. Reference 114 extended the Reynolds analogy to include mass transfer by defining a two-part boundary layer consisting of a laminar sublayer and a fully turbulent core. Here, $l^+ = 0$ in the sublayer ($y^+ < y_a^+$), and $l^+ = 0.4y^+$ and $\mu = 0$ in the fully turbulent region. The density was permitted to vary with temperature. The effect of foreign gas injection in a low-speed boundary layer was studied in Ref. 116, and all these theories were improved upon in Ref. 117.

The aforementioned theories show very small effects of Mach number on the normalized skin friction and heat transfer data c_f/c_{f0} and St/St_0, respectively. Some experiments [118] support these results, while others [119, 120, 121] show surface mass transfer to be less effective in reducing skin friction and heat transfer as Mach number increases. In view of these differences, it is necessary to rely on the aggregate of data presented in Ref. 109 to assess the uncertainties involved in a particular approach.

For preliminary design of ablation systems, in particular, it has been found convenient to use a figure such as Fig. 6.44, which represents a composite of experimental results with air injection [122, 123–126]. This figure reflects a conservative choice of a fixed recovery factor equal to 0.9 even in the presence of light gas injection, and introduces the effects of foreign gas injection by empirically adjusting the blowing parameter. The modified parameter is $(\rho_w v_w / \rho_e u_e \, St_0)(M_2/M_1)^{a_m}$, where the subscripts 1 and 2 refer to the coolant and air, respectively, and $a_m = 0.35$ when $M_1 > M_2$, $a_m = 0.6$ when $M_1 < M_2$. The graphic technique for using this figure to establish $\rho_w v_w$ is to determine the intersection of the curves in Fig. 6.44 with a straight line drawn from the origin with slope

$$\frac{i_{1w} - i_{1c}}{i_{2aw} - i_{2w}} \left(\frac{M_1}{M_2} \right)^{a_m}$$

In an ablating system, the change in the enthalpy of the coolant, $i_{1w} - i_{1c}$, includes the effective heat of ablation, the heat absorbed by phase change, and the chemical processes that take place in the char, if present. In evaporating systems, $i_{1w} - i_{1c}$ includes the heat of vaporization.

Surface Roughness.   Up to this point, the turbulent boundary layer has been assumed to form on a surface that is aerodynamically smooth, namely, a surface whose roughness elements are small compared with the thickness of the viscous sublayer. As many surfaces in practical appli-

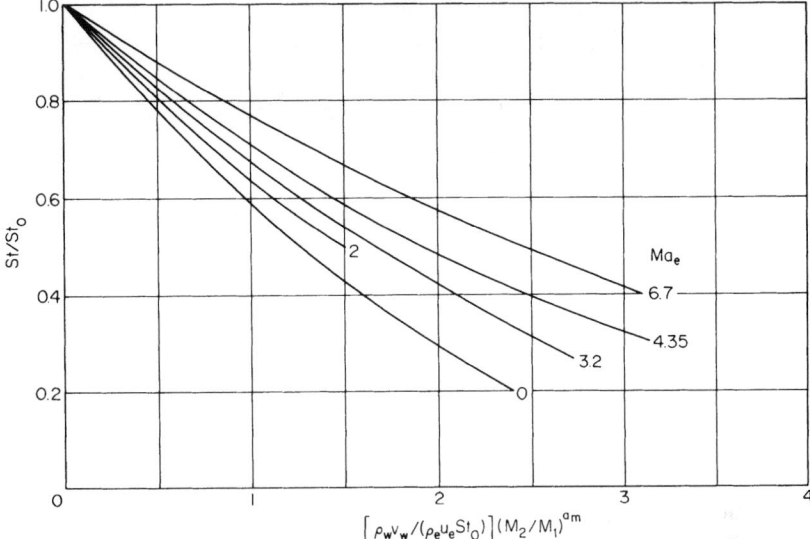

FIGURE 6.44 Compressibility effects on the reduction of the Stanton number by surface mass transfer on bodies with zero axial pressure gradient and including effects of foreign gas injection.

cations are not aerodynamically smooth, the effects of surface roughness must be accounted for in describing the hydrodynamic and thermal behavior of turbulent boundary layers.

Skin Friction.   The earliest investigation of the effects of surface roughness was conducted by Nikuradse [127], who determined the friction factors for flow through artificially roughened pipes. The pipes were coated with various sizes of sand grains in dense arrays. Nikuradse found that the friction factor could be correlated with two parameters, Re_d and k_s/d, where k_s is the size of the sand grains. For a given roughness, the friction factor becomes independent of the Reynolds number when the latter is sufficiently large.

Similar behavior is observed in a turbulent boundary layer over a rough flat plate, where Prandtl and Schlichting [128] showed that the important parameter is k_s/x. The local skin friction coefficient is shown in Fig. 6.45 as a function of Re_x with x/k_s as a parameter.

The lowest curve represents the skin friction for a smooth flat plate (Eq. 6.167). The other curves apply for rough plates with u_ek_s/ν and x/k_s as parameters. The region above the dashed line is defined to be "fully rough." At a point on a moderately rough surface where x and x/k_s are fixed and the Reynolds number is low, the skin friction coefficient is the same as on a smooth surface. An increase in velocity or Reynolds number causes the skin friction coefficient at that point to rise from its smooth-surface value. When the Reynolds number reaches a critical high value, the skin friction coefficient increases asymptotically to a constant value. The dashed line defines the critical Reynolds numbers, above which fully rough conditions exist. With uniform roughness and fixed unit Reynolds number, the skin friction coefficient along the plate behaves as indicated by the curves generally parallel to, but higher than, the smooth-plate curve.

Schlichting [65] gives the following correlations for local and average skin friction coefficients in the fully rough regime:

$$c_f = \left(2.87 + 1.58 \ln \frac{x}{k_s} \right)^{-2.5} \tag{6.230}$$

$$\bar{c}_f = \left(1.89 + 1.62 \ln \frac{x}{k_s} \right)^{-2.5} \tag{6.231}$$

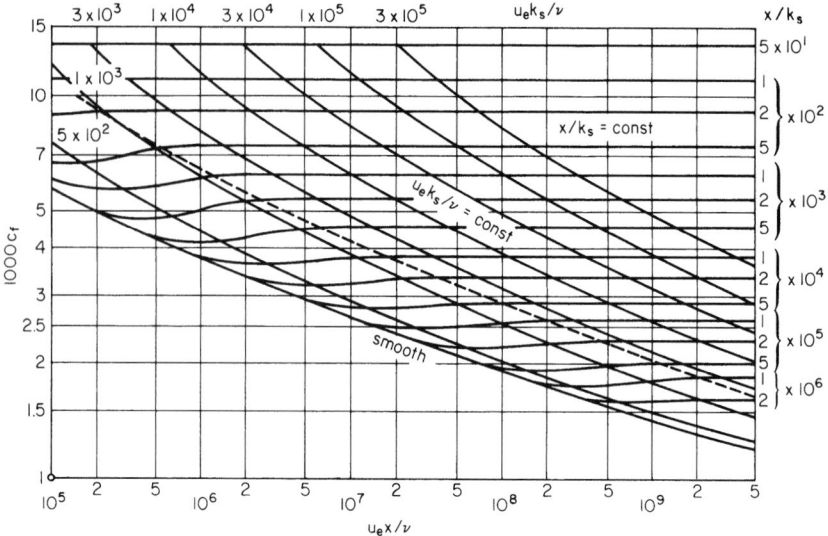

FIGURE 6.45   Variation of local skin friction coefficient along a sand-roughened flat plate [65]. (Reprinted by permission of McGraw-Hill.)

Effects of surface roughness are also evident in the boundary layer mean velocity profiles shown in Fig. 6.46. The profiles still exhibit a near-wall logarithmic behavior, but with a dependence on the roughness Reynolds number $k^+ = k_s u^*/\nu$. The law of the wall for a rough surface may be written as

$$u^+ = \frac{1}{\kappa} \ln y^+ + 5.0 - \Delta u^+ \tag{6.232}$$

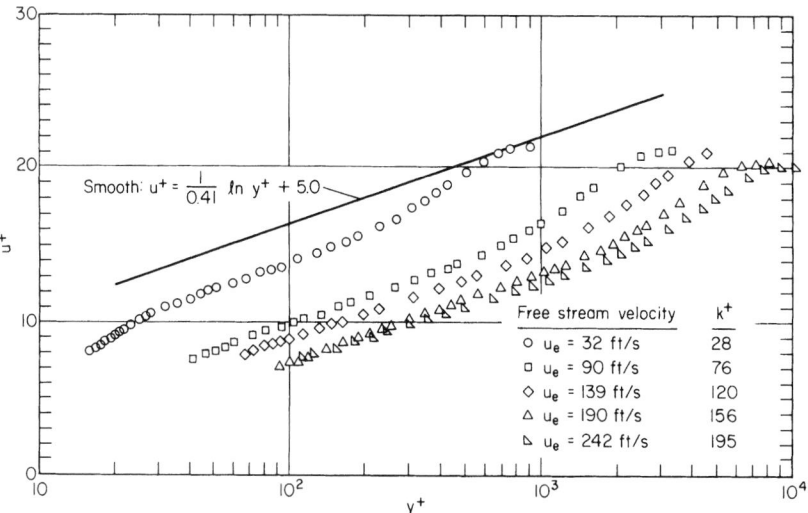

FIGURE 6.46   Velocity profiles in wall coordinates on a rough plate with u_e constant [129]. (Reprinted by permission of the authors.)

where Δu^+ is a function of k^+. This functional relationship is shown in Fig. 6.47 for various roughness geometries. For a uniform particle distribution, Δu^+ approaches zero for $k^+ < 5.0$. This value of k^+ defines an aerodynamically smooth surface consistent with Nikuradse's definition. The surface is transitionally rough for $5 < k^+ < 70$ and fully rough for $k^+ > 70$. For surfaces having a wide distribution of particle size, Δu^+ remains nonzero even for small k^+ based on average particle size. For this geometry, even a small number of large particles disturbs significantly the near-wall flow. For larger values of k^+, Δu^+ becomes proportional to $\ln k^+$ with the proportionality constant nearly equal to $1/\kappa$. The law of the wall then becomes

$$u^+ = \frac{1}{\kappa} \ln \frac{y}{k_s} + \mathbf{B}* \qquad (6.233)$$

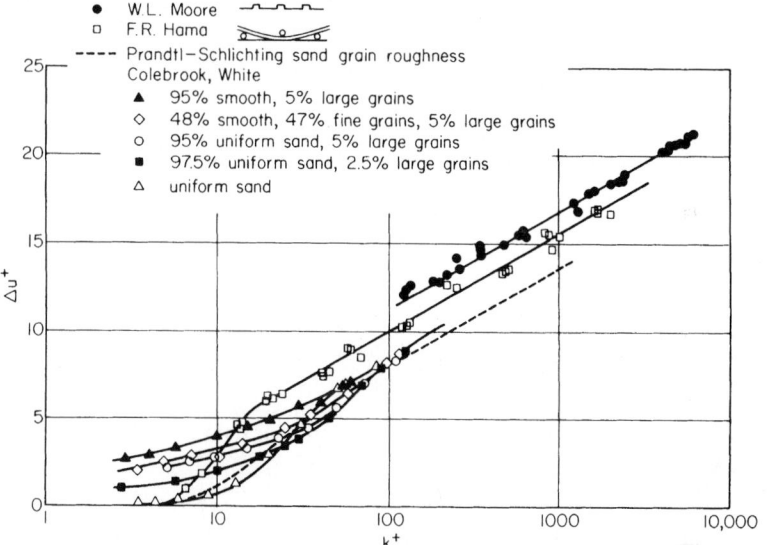

FIGURE 6.47 Effect of roughness size and type on universal velocity profiles in a turbulent boundary layer over a rough flat plate [130]. (Reprinted by permission of Hemisphere Publishing.)

where $\mathbf{B}*$ is a function of the surface geometry but independent of k^+ provided $k^+ > 70$. The behavior of the velocity profile (Eq. 6.233) and the constant value of the skin friction coefficient in the fully rough regime suggest that the integral parameters also attain unique values. Based on the data of Ref. 131, Kays and Crawford [74] recommend the following correlation for the skin friction coefficient on a fully rough flat plate:

$$\frac{c_f}{2} = \frac{0.168}{[\ln (864 \, \theta/k_s)]^2} \qquad (6.234)$$

where θ is the momentum thickness.

Heat Transfer.   The Stanton number over a rough surface behaves similarly to the skin friction coefficient; at sufficiently high roughness Reynolds numbers k^+, the Stanton number becomes independent of the free-stream velocity. At a given Re_x or Re_δ, roughness causes an increase in local Stanton number over the smooth-plate value. These effects are shown in Fig. 6.48 for five values of the free-stream velocity. The geometry of the rough surface used in these experiments was the densest array of spheres of radius r as shown in Fig.

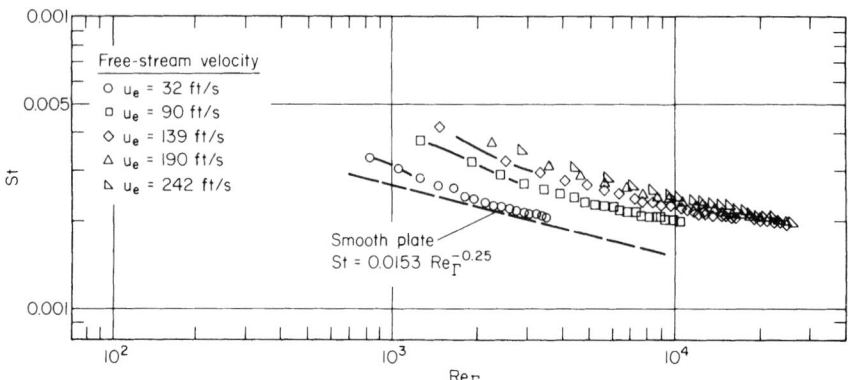

FIGURE 6.48   Rough-surface Stanton number versus energy-thickness Reynolds number [129]. (Reprinted by permission of the authors.)

20.24 of Schlichting [65]. For a smooth plate, data for all five velocities collapse on a single correlation:

$$St = 0.0129 Re_\Gamma^{-0.25} Pr^{-0.4}$$

$$\approx 0.0153 Re_\Gamma^{-0.25} \qquad \text{for air at room temperature} \tag{6.235}$$

The effect of roughness is seen as an increase in the local Stanton number with increasing u_e at a fixed value of the Reynolds number based on energy thickness $Re_{\Gamma E}$. The roughness Reynolds number range corresponding to the five values of u_e is also shown in Fig. 6.48. The same data plotted as St against Γ_E/r in Fig. 6.49, however, follow a single correlation showing no dependence on u_e. The Stanton number data of Ref. 131 in the fully rough regime on the same test surface can be correlated as

$$St = 0.00317\left(\frac{\Gamma}{r}\right)^{-0.175} \tag{6.236}$$

in the range $1.5 < \Gamma/r < 10.0$.

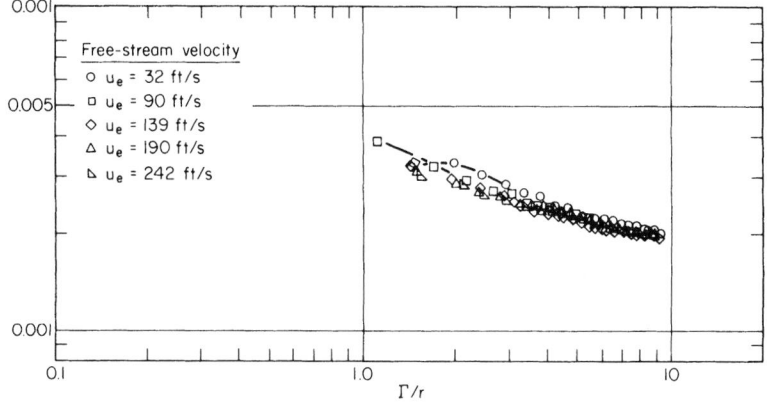

FIGURE 6.49   Rough-surface Stanton number versus normalized energy thickness for a flat-plate boundary layer [129]. (Reprinted by permission of the authors.)

Streamwise Curvature.   Streamwise surface curvature, e.g., on a highly cambered turbine blade, has a significant effect on the local rate of convection. Convex surfaces tend to reduce convection rates from those on a flat surface experiencing boundary layers with the same thickness and edge conditions, whereas concave surfaces tend to increase convection rates. In laminar boundary layers, these effects can be evaluated by transforming the cartesian coordinates of the analysis to an orthogonal set with x representing the distance along the curved surface and y locally normal to the surface. The principal change in the governing equations is that

$$\frac{\partial(\ )}{\partial y} \text{ is replaced by } \frac{1}{1 + y/r_c} \cdot \frac{\partial(\ )}{\partial y},$$

where r_c is the local streamwise radius of curvature of the surface. Here, it is assumed that the boundary layer is very thin compared with r_c, which is positive for a convex surface and negative for a concave one. At sufficiently small values of δ/r_c, the instabilities inherent in flow over a concave surface do not generate Taylor-Görtler vortices, and the change in convection, or skin friction, for positive or negative r_c is similar but of opposite sign. For boundary layers that are somewhat thicker, the next level of approximation requires additional terms identified with second-order boundary layer theory; see Ref. 132, where it is shown that the skin friction coefficient for a laminar boundary layer on a curved surface is related to its flat-plate counterpart by

$$\frac{c_f}{c_{f\text{flat plate}}} = 1 - \frac{0.87\delta}{r_c} \tag{6.237}$$

When Taylor-Görtler vortices develop, the boundary layer on a concave surface possesses an additional mixing mechanism and is capable of transferring heat and momentum at a greater rate than suggested by Eq. 6.237.

The early heat transfer experiments by Thomann [133] demonstrated that a turbulent boundary layer behaves qualitatively the same as a laminar boundary layer, but that the magnitude of the curvature effect is about an order of magnitude greater. Bradshaw's careful review [134] of this topic identified the relatively large effect of the streamwise curvature in a turbulent boundary layer as not due to the geometric effects on the mean motion but due to extra rates of strain on the turbulence production. In this view, as the surface curvature changes, the local turbulence and mean motion are out of equilibrium, which implies a breakdown of the eddy viscosity concept. Others [135, 136] have retained the eddy viscosity approach by modifying it empirically to alter its magnitude—a new mixing length form—and allowing for nonequilibrium of the mean motion and the turbulence through an empirical lag equation. This approach employs finite difference calculations of the boundary layer equations in partial differential form. It has been shown [137] that finite difference computations of the mean conservation equations that utilize the transport equations for the components of Reynolds stress lead to solutions for skin friction that represent curvature effects without model modifications but merely through geometric changes appropriate to thin layers.

Free-Stream Turbulence and Unsteadiness.   It was shown in Fig. 6.27 that the free-stream turbulence level significantly affects local and overall heat transfer from single cylinders in cross flow. This is caused primarily by early transition of the laminar boundary layer on the forward portion of the cylinder and subsequently by delayed separation of the turbulent boundary layer from the surface of the cylinder. Free-stream turbulence and unsteadiness also affect, to varying degrees, the heat transfer behavior of a turbulent boundary layer in the absence of transition-point shift and separation.

A summary of the available experimental data on the effects of free-stream turbulence on heat transfer was presented by Kestin [138]. The experiments of Refs. 139–141 showed that for free-stream turbulence intensities ranging from 0.75 percent to about 4 percent, the skin friction and heat transfer coefficients at a fixed Re_x remained practically unchanged.

A later study by Hancock [142] revealed a significant dependence of the boundary layer momentum thickness on the free-stream turbulence level. Consequently, significant effects were observed when skin friction data were compared at the same momentum-thickness

Reynolds number Re_θ. The correlated data are shown in Fig. 6.50, where the increase in the skin friction is plotted versus a parameter that accounts for both the intensity and the scale of the free-stream turbulence. The effect of free-stream turbulence is primarily in the outer region of the boundary layer, where the law of the wake is modified, and, hence, the boundary layer integral thicknesses are modified.

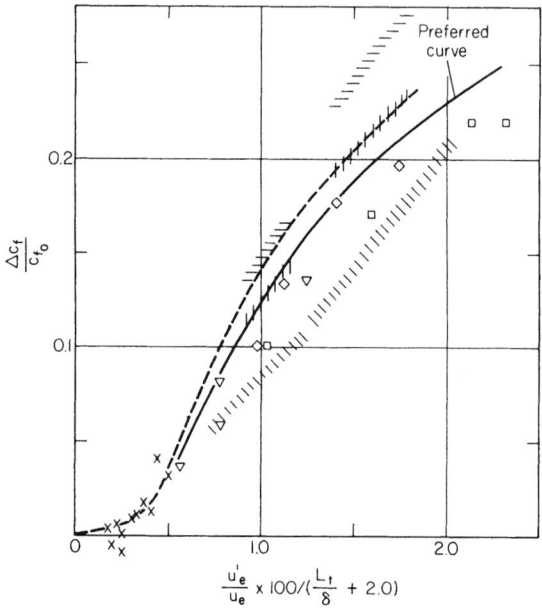

FIGURE 6.50   Effect of free-stream turbulence on skin friction for a flat-plate turbulent boundary layer [142]. (All details about different data points are given in Ref. 142.) (Reprinted by permission of the authors.)

The problem of turbulent boundary layers with an oscillatory free stream has received considerable attention recently. Such flows are encountered with turbine blades, reciprocating cylinder walls, and helicopter rotor blades. The experiments of Refs. 143–146 have shown that even at amplitudes as large as 40 percent of the mean and frequencies ranging from quasi-steady to twice the bursting frequency, approximately $\frac{1}{6}\, u/\delta$, the mean velocity and turbulence intensity profiles in the boundary layer remain unaffected and are indeed the same as those measured with free-stream velocity distributions held steady at the mean value. This shows that there is apparently no energy transfer between the imposed organized oscillations and the random turbulent fluctuations in the boundary layer. This being the case, the behavior of unsteady turbulent boundary layers can be predicted satisfactorily using turbulence models developed for steady flows [145].

TRANSITIONAL BOUNDARY LAYERS

Transitional Boundary Layers for Uniform Free-Stream Velocity

Because the transition zone from a laminar to a turbulent boundary layer often covers a major portion of the exposed surface of a body, it is necessary to be able to predict the rapidly changing convection rate in this zone. The position of the onset of turbulence and the extent of the transition zone for a specific configuration depend on many factors such as the scale and spectral content of the free-stream turbulence and sound field, the free stream Mach number, and the surface characteristics of smoothness, waviness, temperature, compliance, and mass transfer. To date, there is no universal correlation of these factors that will permit the prediction of the position and extent of the transition zone. What is presented here is a technique for predicting transitional boundary layer convection on a plate, given the position of the transition region. If the transition zone is not well known, one design approach is to arbitrarily assign a series of positions of the onset of turbulence and to set the length of the transition zone equal to the length of the fully laminar boundary layer. The sensitivity of the final design to changes in the position of transition must be determined; a high degree of sensitivity suggests the need for careful experimentation with prototype models.

The contents of this section are an extension of the work of Ref. 147 to include the effects of variable fluid properties. The ideas employed are based on the observations of Ref. 148 that on a flat plate the distribution of β, the fraction of time a surface point is covered by a fully turbulent boundary layer, is closely approximated by a Gaussian integral curve throughout the transition zone, i.e.,

$$\beta(x) = \int_0^x P(s_t) \frac{ds_t}{\overline{\sigma}} \tag{6.238}$$

where $\overline{\sigma}$ is the standard deviation of the transition location about its mean $\overline{s}_t$, and $P(s_t)$ is the probability that transition initiates between s_t and $s_t + ds_t$:

$$P(s_t) = \frac{1}{\sqrt{2\pi}} \exp\left[-\frac{1}{2}\left(\frac{s_t - \overline{s}_t}{\overline{\sigma}}\right)^2\right] \tag{6.239}$$

Use of Eq. 6.239 requires that the transition take place sufficiently downstream so that the boundary layer is fully laminar at all times near the leading edge ($\overline{s}_t > 2\sigma$).

If the term $q''_{wt}(x, s_t)$ is defined as the heat flux at point x in a turbulent boundary layer with instantaneous transition from laminar to turbulent flow at s_t, and $q''_{wl}(x)$ is defined as the heat flux at point x with a laminar boundary layer beginning at the leading edge, the heat flux caused by the intermittency of turbulence in the transition zone is then

$$q''_w(x) = [1 - \beta(x)]q''_{wl}(x) + \int_0^x P(s_t)q''_{wt}(x, s_t) \frac{ds_t}{\overline{\sigma}} \tag{6.240}$$

The first term is the product of the laminar heat flux and the fraction of time the boundary layer is laminar at x. The second term accounts for both the fraction of time the boundary layer is turbulent and the effect of the moving transition location. The term $q''_{wt}(x, s_t)$ is sufficiently complex mathematically that Eq. 6.240 is normally solved by numerical integration. If it is assumed that the energy thickness remains unchanged as the laminar boundary layer changes instantaneously into a turbulent boundary layer, then

$$\frac{q''_{wt}(x, s_t)}{q''_{wt}(x, 0)} = \left\{1 - \frac{s_t}{x}\left[1 - \frac{36.9}{\text{Pr}^{1/3}}\left(\frac{z_l}{z_t}\right)^{5/4} \frac{F_c F_{R\theta}^{1/4}}{\left(\frac{\mu'\rho'}{\mu_e\rho_e}\right)^{5/8}\left(\frac{\rho_e u_e s_t}{\mu_e}\right)^{3/8}}\right]\right\}^{-1/5} \tag{6.241}$$

where

$$\frac{\overline{z}_l}{\overline{z}_t} = \frac{i_e + \text{Pr}^{1/2}(u_e^2/2) - i_w}{i_e + \text{Pr}^{1/3}(u_e^2/2) - i_w} \tag{6.242}$$

The form of Eq. 6.241 applies for $\text{Re}_{x_e} < 4 \times 10^6$, where the Blasius skin friction equation (Eq. 6.16) is reasonably accurate and $2\ \text{St}/c_f = \text{Pr}^{-2/3}$ and $\text{Pr}^{-0.4}$ for laminar and turbulent flow, respectively. It also uses the laminar reference enthalpy approach to define $\mu'\rho'/\mu_e\rho_e$ (see the section on uniform free-stream conditions) and uses the turbulent boundary layer transformations F_c and $F_{R\theta}$, which are assumed insensitive to Reynolds number variations (see the section on ideal gases at high temperature). Thus, given Ma_e, T_e, T_w, $\bar{s}_t$, σ, Eq. 6.240 provides the distribution of heat flux in the prescribed transition zone by techniques consistent with those for the fully laminar and fully turbulent boundary layers given previously.

COMPLEX CONFIGURATIONS

The material presented earlier was confined to steady-state flows over simply shaped bodies such as flat plates, with and without pressure gradients in the streamwise direction, or stagnation regions on blunt bodies. The simplicity of these flow configurations allows reduction of the problems to the solution of steady-state ordinary differential equations. The evaluation of convective heat transfer to more complex three-dimensional configurations, characteristic of real aerodynamic vehicles, involves the solution of partial differential equations. Even when the latter are confined to steady-state problems, they require extensive use of computers in the solution of finite difference or finite element formulations. Nonsteady flows further complicate the problems by introducing another dimension, namely, time.

Recent years have shown considerable progress in the development of methods for solving these more complex problems. Larger and faster computers have become more accessible and solution algorithms are more efficient. Complex flow fields undergoing chemical reactions between many species are being performed routinely. For simpler configurations, time-dependent calculations of the dynamic behavior of chaotic turbulent flows have been performed to provide numerical experiments with much more detail than can be provided by physical experiments.

With regard to accuracy, laminar flows can be solved accurately provided care is given to specifying a computational mesh that can resolve shock waves and/or the regions very near surfaces. The introduction of surface mass transfer or nonuniform streamwise surface temperatures is straightforward. With transitional or turbulent boundary layers, however, the state of the art is less satisfactory. A truly predictive method of computing the flow over practical three-dimensional shapes, where the laminar, transitional, and turbulent boundary portions of the flow have been computed in a time-dependent manner and with mesh spacing that resolves all the significant eddies, is not currently available and is likely to be too expensive for general engineering use far into the foreseeable future. Practical mainstream studies of turbulent flows have been confined to steady-state computations dealing only with mean motions. Experimental guidance has been used to define the transition regions and the turbulent regions have been computed with a hierarchy of turbulence models ranging from the simple algebraic eddy viscosity models presented here to second-order models where the concept of the eddy viscosity can be dropped and the Reynolds stresses themselves are evaluated with field equations.

Second-order turbulence models depend on partial differential equations describing measures of the turbulence, i.e., the kinetic energy, the dissipation rate, and/or the individual Reynolds stresses. Thus the number of field equations defining a flow field is increased significantly. Two-equation models that define the eddy viscosity in terms of the turbulence kinetic energy and the dissipation rate (or a dissipation rate per unit kinetic energy) add two partial differential equations to the system and represent a 20 to 40 percent increase in computer storage needs; in addition, these models often reduce convergence rates because of their stiffness. Reynolds stress computations more than double computational costs for similar reasons. Despite their complexity and cost, the second-order models become advantageous for flows involving steep pressure gradients, separated boundary layers, or surface curvature. The models also often lack universality, working very well for one or more flows but then failing when applied to some other type of flow. As a consequence, codes with particular turbulence models must be verified by comparing their results with data from physical or numerical experiments for similar flow fields. Only when an algorithm, a mesh field, and a turbulence model have been verified for a particular shape and set of flow conditions can the technique be applied with confidence to other generally similar configurations and free-stream flow conditions. In their early stages of development, these codes should be applied to flat plates and stagnation regions and compared to the time-tested methods shown in the previous sections as a first step in the code verification process. Also, because computations with these codes are costly, they are used sparingly. For example, in the computation of the total heat transferred to a body on a trajectory, standard practice is to employ these codes at specific, but widely separated, intervals and then to interpolate for the times between these solutions with guidance for the influence of various parameters from formulas such as those presented earlier.

NOMENCLATURE

Symbol, Definition, SI Units, English Units

A	constant, Eqs. 6.119, 6.142, and 6.201
A^*	coefficient, Eq. 6.132, Table 6.5
A^+	van Driest wall damping parameter, Eq. 6.178
a^*	coefficient, Eq. 6.124
$\mathbf{a}^*$	constant in reference enthalpy method
$\mathbf{a}$	exponent, Eq. 6.122, Table 6.3
a_m	molecular weight ratio exponent, Fig. 6.44
B	constant, Eq. 6.142
$B(\beta_T)$	function, Eq. 6.201, Fig. 6.39
B^*	coefficient, Eq. 6.132, Table 6.5
$\mathbf{B}$	coefficient in law of the wall, Eq. 6.148
$\mathbf{B}^*$	surface geometry function, Eq. 6.233
B_f	blowing function, Eq. 6.218
B_h	blowing function, Eq. 6.225
B_m	blowing function, compressible flow, Eq. 6.140
b	exponent in viscosity law for liquids
b^*	coefficient, Eq. 6.124
$\mathbf{b}^*$	constant in reference enthalpy equation
b_f	blowing function, Eq. 6.220
$\overline{C}$	constant value of C_r, Eq. 6.43
C_e	normalized viscosity-density product referred to boundary layer edge conditions
C_{ew}	Chapman-Rubesin constant, Eq. 6.53
C_r	normalized viscosity-density product referred to conditions at r, Eq. 6.39
$\mathbf{c}^*$	constant in reference enthalpy equation
c_f	local skin friction coefficient
$\overline{c}_f$	average skin friction coefficient
c_{fw}	local skin friction coefficient in terms of wall properties, Eq. 6.105
c_h	coefficient, Eqs. 6.143 and 6.144
c_{mi}	mass transfer coefficient for species i

c_p	specific heat at constant pressure: J/(kg·K), Btu/(lb$_m$·°F)
c_v	specific heat at constant volume: J/(kg·K), Btu/(lb$_m$·°F)
D_2	cross flow forced convection coefficient, Table 6.7
$\mathcal{D}$	diffusion coefficient: m²/s, ft²/s
$\mathcal{D}_{ij}$	binary diffusion coefficient: m²/s, ft²/s
d	body or tube diameter: m, ft
d^*	exponent in power-law velocity profile, Table 6.4
d_T	wind tunnel diameter or height: m, ft
$\mathbf{E}$	coefficient for turbulent boundary layer, Eq. 6.175
Ec	Eckert number, $V^2/2i$
F	dimensionless boundary layer stream function, Eq. 6.41
F_c	turbulent boundary layer transformation function, $\tilde{c}_f/c_f$, Table 6.8
F_{R_x}	turbulent boundary layer transformation function, $\tilde{Re}_{x_e}/Re_{x_e}$, Eq. 6.209
$F_{R\theta}$	turbulent boundary layer transformation function, $\tilde{Re}_{\theta_e}/Re_{\theta_e}$, Table 6.8
F_0	function, Table 6.6: (W/m²)[m/(N·m²)]$^{1/2}$, [Btu/(s·ft²)](ft/atm)$^{1/2}$, Eq. 6.134
f	Blasius stream function, Eq. 6.12
$f(x)$	denotes function of x
H	form factor in two-dimensional flow, δ^*/θ
h	local heat transfer coefficient: W/(m²·K), Btu/(h·ft²·°F)
$\bar{h}$	average heat transfer coefficient for row of tubes: W/(m²·K), Btu/(h·ft²·°F)
$\bar{h}_\infty$	average heat transfer coefficient for infinite array of tubes: W/(m²·K), Btu/(h·ft²·°F)
I	total enthalpy per unit mass: J/kg, Btu/lb$_m$, $I = i + (u^2 + v^2 + w^2)/2$
$\bar{I}$	normalized total enthalpy, Eq. 6.39
$\mathbf{I}$	normalized total enthalpy, Eq. 6.45
i	enthalpy per unit mass: J/kg, Btu/lb$_m$
j	diffusion flux: kg/(m²·s), lb$_m$/(ft²·s)
K_i	mass concentration of species i
k	thermal conductivity: W/(m·K), Btu/(s·ft·°R)
k^+	roughness Reynolds number, wall layer coordinates
$\mathbf{k}$	index, Eq. 6.76
k_s	roughness size: m, ft
L	length of porous section in film cooling: m, ft
$\mathbf{L}$	reference length, e.g., Eq. 6.76 and Eq. 6.89: m, ft
L_t	free-stream turbulence length scale, Fig. 6.50: m, ft
Le	Lewis number, $\rho\mathcal{D}c_p/k$, or ϵ_D/ϵ_H
l	mixing length: m, ft
l^+	normalized mixing length, lu^*/ν
M	molecular weight: kg/kmol, lb$_m$/lb$_m$-mol
M_c	mean molecular weight of coolant: kg/kmol, lb$_m$/lb$_m$-mol
Ma	Mach number
m	exponent, Eq. 6.119
m^*	constant, Eq. 6.142

Nu	Nusselt number, $St\, Re_L\, Pr$
$\overline{Nu}$	mean or average Nusselt number, $\overline{St}\, Re_L\, Pr$
n	number of species
$n*$	constant, Eq. 6.142
n_D	Deissler empirical constant
n_1	exponent, Table 6.7
P	pressure: N/m^2, lb_f/ft^2 (atm)
$P(\ )$	probability function of transition, Eq. 6.239
Pr	Prandtl number, $\mu c_p/k$, single gas or liquid
Pr_F	frozen Prandtl number, gas properties weighted over all species present
Pr_M	effective Prandtl number, Eq. 6.182
Pr_T	Prandtl number based on total properties for gas in chemical equilibrium
Pr_t	turbulent Prandtl number, ϵ_M/ϵ_H
q''	heat flux: W/m^2, $Btu/(s\cdot ft^2)$
$\overline{q''}$	average heat flux (over space): W/m^2, $Btu/(s\cdot ft^2)$
$\overline{R}$	enthalpy profile function, Eq. 6.195
R	universal gas constant: $m^2/(s^2\cdot K)$ or $J\cdot kmol^{-1}\cdot K^{-1}$, $ft^2/(s^2\cdot{}^\circ R)$
Re_d	Reynolds number based on diameter d
Re_{eff}	effective Reynolds number, Eq. 6.202
Re_L	Reynolds number based on length L; $L = x$, θ, $\delta*$, δ, or Γ
r	radius of spherical roughness: m, ft
r	dimensionless temperature variable, Eq. 6.28
r(0)	recovery factor
r_c	streamwise radius of curvature: m, ft
r_n	nose radius: m, ft
S	enthalpy profile function, Eq. 6.181
St	local Stanton number, $h/\rho c_p u_e$
$\overline{St}$	average Stanton number, $\overline{h}/\rho c_p u_e$
St (x, s)	local Stanton number on a plate with a temperature jump at $x = s$
s	distance to location of surface temperature discontinuity: m, ft
s_j^+	downstream of surface temperature jump: m, ft
s_j^-	upstream of surface temperature jump: m, ft
s_t	distance from leading edge to transition from laminar to turbulent flow: m, ft
T	temperature: K, $^\circ R$
T_c	constant in viscosity law for liquids: K, $^\circ R$
T_{sc}	Sutherland constant: K
Tu	intensity of turbulence, u'/v_e
T_{wL}	porous section temperature
t	time: s
$\bar{t}$	normalized local enthalpy, Eq. 6.91
t^+	normalized temperature parameter, Eq. 6.184
$\bar{t}_e$	normalized free-stream enthalpy, Eq. 6.93

$\bar{\imath}_s$	normalized free-stream enthalpy, Eq. 6.92
U_∞	velocity component normal to leading edge of yawed cylinder, Fig. 6.18
$\overline{U}$	normalized free-stream velocity: $V_\infty/(10{,}000 \text{ ft/s})$, $V_\infty/(3048 \text{ m/s})$
u	velocity component in x direction: m/s, ft/s
u^*	friction velocity, $\sqrt{\tau_w/\rho}$: m/s, ft/s
u^+	normalized velocity, u/u^*
Δu^+	shift in normalized velocity due to roughness
u	internal energy per unit mass: J/kg, Btu/lb$_\text{m}$
u_e	free-stream velocity component in the x direction: m/s, ft/s
u_e'	free-stream RMS fluctuation of velocity: m/s, ft/s
u_i	velocity component in the x_i direction: m/s, ft/s
V_∞	free-stream velocity: m/s, ft/s
v	velocity component in the y direction: m/s, ft/s
v_0^+	velocity component in the y direction at surface normalized by u^*
w	velocity component in the z direction: m/s, ft/s
$\overline{w}$	normalized cross flow velocity, Eq. 6.90
w	Coles wake function, Fig. 6.33
x	rectangular coordinate: m, ft
$\bar{x}$	dimensionless tube spacing in the free-stream direction
x_eff	effective distance, laminar flow, Eq. 6.110
x_i	generalized orthogonal coordinate: m, ft
$\bar{x}_t$	dimensionless tube spacing normal to the free-stream direction
x^+	normalized x_1 distance, Eq. 6.187
$Y_0(\eta)$	dimensionless temperature function, Eq. 6.19
y	rectangular coordinate, normal to surface: m, ft
y^+	normalized distance from wall, u^*y/ν

Greek Symbols

α	van Driest function, Table 6.8
β	fraction of time a surface point is covered by a fully turbulent boundary layer, Eq. 6.238; van Driest function, Table 6.8
$\overline{\beta}$	Prandtl number function, Eq. 6.191
β_p	pressure gradient parameter, Eq. 6.94
β_T	normalized temperature parameter, Eq. 6.200
$\overline{\Gamma}$	function, Eq. 6.192
Γ	energy thickness, Eq. 6.228b: m, ft
Δu^+	roughness effect on velocity profile
δ	boundary layer thickness: m, ft; or incremental value
δ^*	displacement thickness, two-dimensional flow, Eq. 6.150: m, ft
δ_T	thermal boundary layer thickness: m, ft
$\overline{\epsilon}$	sum of molecular kinematic and eddy viscosities
ϵ_{D_i}	turbulent eddy diffusivity for species i: m^2/s, ft^2/s

ϵ_H	thermal eddy diffusivity: m^2/s, ft^2/s
ϵ_H^+	normalized thermal eddy diffusivity, Eq. 6.187
ϵ_M	momentum eddy diffusivity; momentum eddy diffusivity in the x direction: m^2/s, ft^2/s
ϵ_M^+	normalized momentum eddy diffusivity, Eq. 6.187
ζ	transformed boundary layer variable, Eqs. 6.11, 6.36 and 6.89
$\bar{\zeta}$	function, Eq. 6.89
η	transformed boundary layer variable, Eqs. 6.11, 6.36, and 6.89
η_c	modified transformed boundary layer variable, Eq. 6.43
η_H	thermal boundary layer thickness parameter, Figs. 6.2 and 6.3
θ	momentum thickness, two-dimensional flow, Eq. 6.151: m, ft
θ_c	cone angle: rad, deg
θ_0	ratio of Sutherland constant to edge temperature
κ	von Kármán mixing length constant, Eq. 6.148
Λ	sweepback angle: rad, deg
μ	dynamic viscosity: kg/(m·s), lb$_m$/(ft·s)
ν	kinematic viscosity: m^2/s, ft^2/s
$\tilde{\pi}$	parameter in Coles wake velocity distribution, Eq. 6.149
ρ	density: kg/m^3, lb$_m$/ft^3
$\bar{\sigma}$	standard deviation of transition location about its mean
τ	shear stress between fluid layers: N/m^2, lb$_f$/ft^2
τ_w	shear stress at wall: N/m^2, lb$_f$/ft^2
ψ	stream function: m^2/s, ft^2/s, Eq. 6.10

Subscripts

aw	adiabatic wall conditions
c	initial coolant condition in reservoir
d	diameter
e	evaluated at the boundary layer edge
eff	effective value
F	frozen properties
i	component i of a mixture
i	constant-property case
i	defining orthogonal direction, Eq. 6.104
j	component j of a mixture
l	laminar
r, ref	reference condition
st	stagnation point on body
T	total properties
t	turbulent
w	wall
0	zero mass transfer

1	component 1 (coolant) of mixture
2	component 2 (air) of mixture
∞	free-stream conditions

Superscripts

$-$	mean value
$'$	random fluctuating value; dummy variable, reference property, ordinary derivative
$*$	properties to be evaluated at reference enthalpy or temperature condition
$\sim$	constant-property case

REFERENCES

1. H. Blasius, "The Boundary Layer in Fluids With Little Friction," *NACA Tech. Mem. 1256,* 1950, transl. of "Grenzschichten in Flüssigkeiten mit kleiner Reibung," *Z. Math. Phys.* (56): 1–37, 1908.

2. E. Pohlhausen, "Der Wärmeaustausch zwischen festen Körpern und Flüssigkeiten mit kleiner Reibung und kleiner Wärmeleitung," *Z. angew. Math. Mech.* (1): 115–121, 1921.

3. E. R. G. Eckert and O. Drewitz, "The Heat Transfer to a Plate in Flow at High Speed," *NACA Tech. Mem. 1045,* 1943, transl. of "Der Wärmeübergang an eine mit grosser Geschwindigkeit längs angeströmte Platte," *Forschung auf dem Gebiete des Ingenieurwesens* (11): 116–124, 1940.

4. G. W. Morgan and W. H. Warner, "On Heat Transfer in Laminar Boundary Layers at High Prandtl Numbers," *J. Aeronaut. Sci.* (23): 937–948, 1956.

5. L. Crocco, "Lo strato limite laminare nei gas," *Monografie Sci. di Aeronaut. 3,* Rome, 1946, transl. as *North American Aviation Aerophys. Lab. Rep. APL/NAA/CF-1038,* 1948.

6. A. P. Colburn, "A Method for Correlating Forced Convection Heat Transfer Data and a Comparison With Fluid Friction," *Trans. Am. Inst. Chem. Eng.* (29): 174–210, 1933.

7. R. A. Seban, "Laminar Boundary Layer of a Liquid With Variable Viscosity," in *Heat Transfer Thermodynamics and Education, Boelter Anniversary Volume,* H. A. Johnson ed., pp. 319–329, McGraw-Hill, New York, 1964.

8. L. Howarth, "Concerning the Effect of Compressibility in Laminary Boundary Layers and Their Separation," *Proc. Roy Soc. London* (194): 16–42, 1948.

9. D. R. Chapman and M. W. Rubesin, "Temperature and Velocity Profiles in the Compressible Laminar Boundary Layer With Arbitrary Distribution of Surface Temperature," *J. Aeronaut. Sci.* (16): 547–565, 1949.

10. T. von Kármán, "The Problem of Resistance in Compressible Fluids," *Atti del Convegno della Fondazione Alessandro Volta 1935,* 223–326, 1936.

11. E. R. van Driest, "Investigation of Laminar Boundary Layer in Compressible Fluids Using the Crocco Method," *NACA Tech. Note 2597,* 1952.

12. G. B. W. Young and E. Janssen, "The Compressible Boundary Layer," *J. Aeronaut. Sci.* (19): 229–236, 288, 1952.

13. E. R. van Driest, "The Laminar Boundary Layer With Variable Fluid Properties," *North American Aviation Rep. AL-1866,* Los Angeles, 1954.

14. M. F. Romig and F. J. Dore, "Solutions of the Compressible Laminar Boundary Layer Including the Case of a Dissociated Free Stream," *Convair Rep. ZA-7-012,* San Diego, 1954.

15. R. E. Wilson, "Real-Gas Laminar-Boundary-Layer Skin Friction and Heat Transfer," *J. Aerosp. Sci.* (29): 640–647, 1962.

16. N. B. Cohen, "Boundary-Layer Similar Solutions and Correlation Equations for Laminar Heat Transfer Distribution in Equilibrium Air at Velocities up to 41,000 Feet per Second," *NASA Tech. Rep. R-118,* 1961.

17. G. T. Chapman, "Theoretical Laminar Convective Heat Transfer and Boundary Layer Characteristics on Cones at Speeds of 24 km/sec," *NASA Tech. Note D-2463,* 1964.

18. M. W. Rubesin and H. A. Johnson, "A Summary of Skin Friction and Heat Transfer Solutions of the Laminar Boundary Layer on a Flat Plate," *Proc. 1948 Heat Transfer Fluid Mech. Inst.;* also *Trans. ASME* (71): 383–388, 1949.

19. A. D. Young, "Boundary Layers," in *Modern Developments in Fluid Dynamics: High Speed Flow,* L. Howarth ed., vol. 1, chap. 10, p. 422, Oxford University Press, New York, 1953.

20. E. R. G. Eckert, "Survey on Heat Transfer at High Speeds," *Wright Air Development Center Tech. Rep. 54-70,* Dayton, 1954.

21. E. R. G. Eckert, "Engineering Relations for Heat Transfer and Friction in High Velocity Laminar and Turbulent Boundary Layer Flow Over Surfaces With Constant Pressure and Temperature," *Trans. ASME* (78): 1273–1283, 1956.

22. M. W. Rubesin, "The Effect of an Arbitrary Surface Temperature Variation along a Flat Plate on the Convective Heat Transfer in an Incompressible Turbulent Boundary Layer," *NACA Tech. Note 2345,* 1951.

23. H. S. Carslaw and J. C. Jaeger, *Conduction of Heat in Solids,* 2d ed., Oxford University Press, New York, 1959.

24. R. Bond, "Heat Transfer to a Laminar Boundary Layer With Nonuniform Free Stream Velocity and Nonuniform Wall Temperature," *Univ. California Inst. Eng. Res. Rep.,* University of California, Berkeley, CA, 1950.

25. M. J. Lighthill, "Contributions to the Theory of Heat Transfer Through a Laminar Boundary Layer," *Proc. Roy. Sci. London* (202): 359–377, 1950.

26. R. Eichhorn, E. R. G. Eckert, and A. D. Anderson, "An Experimental Study of the Effects of Nonuniform Wall Temperature on Heat Transfer in Laminar and Turbulent Axisymmetric Flow Along a Cylinder," *Wright Air Development Center Tech. Rep.* 58-33, Dayton, 1958.

27. B. M. Leadon, "The Status of Heat Transfer Control by Mass Transfer for Permanent Structures," in *Aerodynamically Heated Structures,* P. E. Glaser ed., p. 171, Prentice-Hall, Englewood Cliffs, NJ, 1962.

28. H. W. Emmons and D. Leigh, "Tabulation of the Blasius Function With Blowing and Suction," *Harvard Univ. Combust. Aerodynamics Lab. Tech. Rep. 9,* Harvard University, Cambridge, MA, 1953.

29. J. P. Hartnett and E. R. G. Eckert, "Mass Transfer Cooling in the Laminar Boundary Layer With Constant Fluid Properties," *Trans. ASME* (79): 247–254, 1957.

30. G. M. Low, "The Compressible Laminar Boundary Layer with Fluid Injection," *NACA Tech. Note 3404,* 1955.

31. J. F. Gross, J. P. Hartnett, D. J. Masson, and C. Gazley Jr., "A Review of Binary Laminar Boundary Layer Characteristics," *Int. J. Heat Mass Transfer* (3): 198–221, 1961.

32. E. R. G. Eckert, A. A. Hayday, and W. J. Minkowycz, "Heat Transfer, Temperature Recovery and Skin Friction on a Flat Plate With Hydrogen Release into a Laminar Boundary Layer," *Int. J. Heat Mass Transfer* (4): 17–29, 1961.

33. P. A. Libby and P. Sepri, "Laminar Boundary Layer With Complex Composition," *Phys. Fluids* (10): 2138–2146, 1967.

34. P. A. Libby and K. Chen, "Laminar Boundary Layer with Uniform Injection," *Phys. Fluids* (8): 568–574, 1965.

35. R. Iglisch, "Exact Calculations of Laminar Boundary Layers in Longitudinal Flow Over a Flat Plate With Homogeneous Suction," *NACA Tech. Mem. 1205,* 1949, transl. of "Exakte Berechnung der laminaren Grenzschicht an der längsangeströmten ebenen Platte mit homogener Absaugung," *Schriften der Deutschen Akademie der Luftfahrtforschung,* Band 8 B, Heft 1, 1944.

36. H. G. Lew and J. B. Fanucci, "On the Laminar Compressible Boundary Layer Over a Flat Plate With Suction or Injection," *J. Aeronaut. Sci.* (22): 589–597, 1955.

37. A. J. Pallone, "Nonsimilar Solutions of the Compressible Laminar Boundary Layer Equations With Applications to the Upstream Transpiration Cooling Problem," *J. Aerosp. Sci.* (28): 449–456, 492, 1961.

38. J. T. Howe, "Some Finite Difference Solutions of the Laminar Compressible Boundary Layer Showing Effects of Upstream Transpiration Cooling," *NASA Mem. 2-26-59A,* 1959.

39. W. Mangler, "Zusammenhang zwischen ebenen und rotationssymmetrischen Grenzschichten in kompressiblen Flüssigkeiten," *Z. angew. Math. Mech.* (28): 97–103, 1948.

40. P. A. Libby, "Laminar Boundary Layer on a Cone With Uniform Injection," *Phys. Fluids* (8): 2216–2218, 1965.

41. I. E. Beckwith and N. B. Cohen, "Application of Similar Solutions to Calculations of Laminar Heat Transfer on Bodies With Yaw and Large Gradient in High Speed Flow," *NASA Tech. Note D-625,* 1961.

42. C. B. Cohen and E. Reshotko, "Similar Solutions for the Compressible Laminar Boundary Layer With Heat Transfer and Pressure Gradient," *NACA Rep. 1293,* 1956.

43. F. H. Clauser, "Turbulent Boundary Layers in Adverse Pressure Gradients," *J. Aeronaut. Sci.* (21): 91–108, 1954.

44. L. Lees, "Laminar Heat Transfer Over Blunt-Nosed Bodies at Hypersonic Flight Speeds," *Jet Propulsion* (26): 259–269, 1956.

45. D. R. Hartree, "On an Equation Occurring in Falkner and Skan's Approximate Treatment of the Equation of the Boundary Layer," *Proc. Cambridge Philosoph. Soc.* (33): 223–239, 1937.

46. A. N. Tifford, "The Thermodynamics of the Laminar Boundary Layer of a Heated Body in a High-Speed Gas Flow Field," *J. Aeronaut. Sci.* (12): 241–251, 1945.

47. S. Levy, "Heat Transfer to Constant-Property Laminar Boundary Layer Flows With Power-Function Free-Stream Velocity and Wall-Temperature Variation," *J. Aeronaut. Sci.* (19): 341–348, 1952.

48. I. E. Beckwith, "Similar Solutions for the Compressible Boundary Layer on a Yawed Cylinder With Transpiration Cooling," *NASA Tech. Rep. R-42,* 1959.

49. D. R. Davies and D. E. Bourne, "On the Calculation of Heat and Mass Transfer in Laminar and Turbulent Boundary Layers, I: The Laminar Case," *Q. J. Mech. Appl. Math.* (9): 457–466, 1956.

50. A. N. Tifford and S. T. Chu, "Heat Transfer in Laminar Boundary Layers Subject to Surface Pressure and Temperature Distributions," *Proc. 2d Midwestern Conf. Fluid Mech.,* p. 363, 1952.

51. A. Pallone and W. Van Tassell, "Stagnation Point Heat Transfer for Air in the Ionization Regime," *ARS J.* (32): 436–437, 1962.

52. H. Hoshizaki, "Heat Transfer in Planetary Atmospheres at Super-Satellite Speeds," *ARS J.* (32): 1544–1551, 1962.

53. J. G. Marvin and G. S. Deiwert, "Convective Heat Transfer in Planetary Gases," *NASA Tech. Rep. R-224,* 1965.

54. J. G. Marvin and R. B. Pope, "Laminar Convective Heating and Ablation in the Mars Atmosphere," *AIAA J.* (5): 240–248, 1967.

55. P. DeRienzo and A. J. Pallone, "Convective Stagnation-Point Heating for Re-entry Speeds up to 70,000 fps Including Effects of Large Blowing Rates," *AIAA J.* (5): 193–200, 1967.

56. C. F. Hansen, "Approximations for the Thermodynamic and Transport Properties of High-Temperature Air," *NASA Tech. Rep. R-50,* 1959.

57. S. Bennett, J. M. Yos, C. F. Knopp, J. Morris, and W. L. Bade, "Theoretical and Experimental Studies of High-Temperature Gas Transport Properties," *AVCO Corp. RAD-TR-65-7,* 1965.

58. W. F. Ahtye, "A Critical Evaluation of Methods for Calculating Transport Coefficients of Partially Ionized Gas," *NASA Tech. Mem. X-54,* 1964.

59. R. W. Rutowski and K. K. Chan, "Shock Tube Experiments Simulating Entry Into Planetary Atmospheres," *Lockheed Missiles and Space Co. LMSD 288139,* vol. 1, part 2, 1960.

60. J. S. Gruszczynski and W. R. Warren, "Measurements of Hypervelocity Stagnation Point Heat Transfer in Simulated Planetary Atmospheres," *General Electric Space Sci. Lab. R63SD29,* 1963.

61. L. Yee, H. E. Bailey, and H. T. Woodward, "Ballistic Range Measurements of Stagnation-Point Heat Transfer in Air and Carbon Dioxide at Velocities up to 18,000 feet per second," *NASA Tech. Note D-777,* 1961.

62. R. M. Nerem, C. J. Morgan, and B. C. Graber, "Hypervelocity Stagnation Point Heat Transfer in a Carbon Dioxide Atmosphere," *AIAA J.* (1): 2173–2175, 1963.

63. S. J. Kline, *Proceedings of the AFOSR-HTTM-Stanford Conference on Complex Turbulent Flows,* Stanford University Department of Mechanical Engineering, Stanford, CA, 1982.

64. R. Hilpert, "Wärmeabgabe von geheizten Drähten und Rohren in Luftstrom," *Forsch. Ingenieurwes.* (4): 215–224, 1933.

65. H. Schlichting, *Boundary Layer Theory,* 6th ed., McGraw-Hill, New York, 1968.

66. V. T. Morgan, "The Overall Convective Heat Transfer From Smooth Circular Cylinders," in *Advances in Heat Transfer,* T. F. Irvine Jr., and J. P. Hartnett eds., vol. 11, pp. 199–264, Academic, New York, 1975.

67. A. Zukauskas, "Heat Transfer From Tubes in Cross Flow," in *Advances in Heat Transfer,* J. P. Hartnett and T. F. Irvine Jr. eds., vol. 8, pp. 93–160, Academic, New York, 1972.

68. E. Schmidt and K. Wenner, "Wärmeabgabe über den Umfang eines angeblasenen geheizten Zylinders," *Forsch. Ingenieurwes.* (12): 65–73, 1941.

69. J. Kestin and P. F. Maeder, "Influence of Turbulence on Transfer of Heat from Cylinders," *NACA TN 4018,* 1954.

70. W. M. Kays and A. L. London, *Compact Heat Exchangers,* 2d ed., McGraw-Hill, New York, 1964.

71. J. O. Hinze, *Turbulence: An Introduction to Its Mechanism and Theory,* 2d ed., McGraw-Hill, New York, 1975.

72. P. Bradshaw, ed., *Turbulence,* Springer-Verlag, New York, 1976.

73. H. Tennekes and J. L. Lumley, *A First Course in Turbulence,* MIT Press, Cambridge, MA, 1972.

74. W. M. Kays and M. E. Crawford, *Convective Heat and Mass Transfer,* 2d ed., McGraw-Hill, New York, 1980.

75. E. R. van Driest, "Turbulent Boundary Layer in Compressible Fluids," *J. Aeronaut. Sci.* (18): 145–160, 216, 1951.

76. B. E. Launder and D. B. Spalding, *Mathematical Models of Turbulence,* Academic, New York, 1972.

77. W. Rodi, "Turbulence Models and Their Applications in Hydraulics," International Association for Hydraulic Research State-of-the-Art Paper, Delft, the Netherlands, June 1980.

78. J. Kestin and P. D. Richardson, "Heat Transfer Across Turbulent Incompressible Boundary Layers," *Int. J. Heat Mass Transfer* (6): 147–189, 1963.

79. F. H. Clauser, "The Turbulent Boundary Layer," in *Advances in Applied Mechanics,* H. L. Dryden et al. eds., vol. 4, Academic, New York, 1956.

80. D. Coles, "The Law of the Wall in Turbulent Shear Flow," in *Sonderdruck aus 50 Jahre Grenzschichtforschung,* H. Goertler and W. Tollmien eds., Friedrich Vieweg & Sohn, Brunswick, Germany, 1955.

81. D. Coles, "The Law of the Wake in Turbulent Boundary Layer," *J. Fluid Mech.* (1): 191–226, 1956.

82. H. Ludwieg and W. Tillman, "Investigation of the Wall Shearing Stress in Turbulent Boundary Layers," *NACA Tech. Mem. 1285,* 1959, transl. of "Untersuchungen über die Wandschubspannung in turbulenten Reibungsschichten," *Ing. Arch.* (17): 288–299, 1949.

83. F. Schultz-Grunow, "New Frictional Resistance Law for Smooth Plates," *NACA Tech. Mem. 986,* 1941, transl. of "Neues Widerstandsgesetz für glatte Platten," *Luftfahrforsch.* (17): 239–246, 1940.

84. D. Coles, "The Young Person's Guide to the Data," in *Proceedings of the Computation of Turbulent Boundary Layers—1968, AFOSR-IFP-Stanford Conf. 1968,* D. Coles and E. A. Hirst eds., Stanford University, Stanford, CA, 1968.

85. G. B. Schubauer and C. M. Tchen, "Turbulent Flow," in *Turbulent Flows and Heat Transfer,* C. C. Lin ed., sec. B, pp. 119–122, Princeton University Press, Princeton, NJ, 1959.

86. D. W. Smith and J. H. Walker, "Skin-Friction Measurements in Incompressible Flow," *NASA Tech. Rep. R-26,* 1959.

87. T. Cebeci and A. M. O. Smith, *Analysis of Turbulent Boundary Layers,* Academic, New York, 1974.

88. T. von Kármán, "The Analogy Between Fluid Friction and Heat Transfer," *Trans. ASME* (61): 705–710, 1939.

89. D. B. Spalding, "Heat Transfer to a Turbulent Stream From a Surface With a Stepwise Discontinuity in Wall Temperature, International Developments in Heat Transfer," in *Conf. Int. Dev. Heat Transfer,* part 2, pp. 439–446, ASME, New York, 1961.

90. E. R. van Driest, "On Turbulent Flow Near a Wall," *J. Aeronaut. Sci.* (23): 1007–1011, 1956.

91. W. M. Kays, R. J. Moffat, and W. H. Thielbahr, "Heat Transfer to the Highly Accelerated Turbulent Boundary Layer With and Without Mass Addition," *J. Heat Transfer* (92): 499–505, 1970.

92. P. S. Klebanoff, "Characteristics of Turbulence in a Boundary Layer With Zero Pressure Gradient," *NACA Tech. Note 3178,* 1954.

93. E. R. van Driest, "The Turbulent Boundary Layer With Variable Prandtl Number," in *Sonderdruck aus 50 Jahre Grenzschichtforschung,"* H. Goertler and W. Tollmien eds., Friedrich Vieweg & Sohn, Brunswick, Germany, 1955.

94. R. G. Deissler, "Analysis of Turbulent Heat Transfer, Mass Transfer, and Friction in Smooth Tubes at High Prandtl and Schmidt Numbers," *NACA Rep. 1210,* 1954.

95. M. W. Rubesin, "A Modified Reynolds Analogy for the Compressible Turbulent Boundary Layer on a Flat Plate," *NACA Tech. Note 2917,* 1953.

96. C. Ferrari, "Effect of Prandtl Number on the Heat Transfer Properties of a Turbulent Boundary Layer When the Temperature Distribution Along the Wall Is Arbitrarily Assigned," *Z. angew. Math. Mech.* (36): 116–135, 1956.

97. D. B. Spalding, "Contribution to the Theory of Heat Transfer Across a Turbulent Boundary Layer," *Int. J. Heat Mass Transfer* (7): 743–761, 1964.

98. D. B. Spalding, "A Unified Theory of Friction, Heat Transfer, and Mass Transfer in the Turbulent Boundary Layer and Wall Jet," *Aeronaut. Res. Council (England) ARC-CP-829,* 1965.

99. B. F. Blackwell, W. M. Kays, and R. J. Moffat, "The Turbulent Boundary Layer on a Porous Plate: An Experimental Study of the Heat Transfer Behavior With Adverse Pressure Gradients," *Stanford Univ. Dept. Mech. Eng. Rep. HMT-16,* Stanford University, Stanford, CA, August 1972.

100. R. A. Antonia, "Behaviour of the Turbulent Prandtl Number Near the Wall," *Int. J. Heat Mass Transfer* (23): 906–908, 1980.

101. B. A. Kader, "Temperature and Concentration Profiles in Fully Turbulent Boundary Layers," *Int. J. Heat Mass Transfer* (24): 1541–1544, 1981.

102. W. C. Reynolds, W. M. Kays, and S. J. Kline, "Heat Transfer in the Turbulent Incompressible Boundary Layer, I—Constant Wall Temperature," *NASA Mem. 12-1-58W,* 1958.

103. A. Seiff, "Examination of the Existing Data on the Heat Transfer of Turbulent Boundary Layers at Supersonic Speeds From the Point of View of Reynolds Analogy," *NACA Tech. Note 3284,* 1954.

104. D. B. Spalding and S. W. Chi, "The Drag of a Compressible Turbulent Boundary Layer on a Smooth Flat Plate With and Without Heat Transfer," *J. Fluid Mech.* (18): 117–143, 1964.

105. E. R. van Driest, "The Problem of Aerodynamic Heating," *Aeronaut. Eng. Rev.* (15): 26–41, 1956.

106. D. Coles, "The Turbulent Boundary Layer in a Compressible Fluid," *Rand Corp. Rep. R-403-PR,* Santa Monica, CA, 1962.

107. S. C. Sommer and B. J. Short, "Free-Flight Measurements of Turbulent Boundary Layer Skin Friction in the Presence of Severe Aerodynamic Heating at Mach Numbers From 2.8 to 7.0," *NACA Tech. Note 3391,* 1955.

108. E. J. Hopkins, M. W. Rubesin, M. Inouye, E. R. Keener, G. G. Mateer, and T. E. Polek, "Summary and Correlation of Skin-Friction and Heat-Transfer Data for a Hypersonic Turbulent Boundary Layer on Simple Shapes," *NASA Tech. Note D-5089,* 1969.

109. J. P. Hartnett, "Mass Transfer Cooling," in *Handbook of Heat Transfer Applications,* W. M. Rohsenow, J. P. Hartnett, and E. N. Ganić eds., chap. 1, McGraw-Hill, New York, 1985.

110. H. S. Mickley, R. C. Ross, A. L. Squyers, and W. E. Stewart, "Heat, Mass, and Momentum Transfer for Flow Over a Flat Plate With Blowing or Suction," *NACA TN 3208,* 1954.

111. W. M. Kays, "Heat Transfer to the Transpired Turbulent Boundary Layer," *Stanford Univ. Dept. Mech. Eng. Rep. HMT-14,* Stanford University, Stanford, CA, June 1971.

112. R. M. Kendall, M. W. Rubesin, T. J. Dahm, and M. R. Mendenhall, "Mass, Momentum, and Heat Transfer Within a Turbulent Boundary Layer With Foreign Gas Mass Transfer at the Surface, Pt. 1: Constant Fluid Properties," *Itek Corp. Vidya Div. Rept. 111,* 1964.

113. W. M. Kays and R. J. Moffat, "The Behavior of Transpired Turbulent Boundary Layers," *Stanford Univ. Dept. Mech. Eng. Rep. HMT-20,* Stanford University, Stanford, CA, 1975.

114. M. W. Rubesin, "An Analytical Estimation of the Effect of Transpiration Cooling on the Heat Transfer and Skin Friction Characteristic of a Compressible Turbulent Boundary Layer," *NACA Tech. Note 3341,* 1954.

115. W. H. Dorrance and F. J. Dore, "The Effect of Mass Transfer on the Compressible Turbulent Boundary Layer Skin Friction and Heat Transfer," *J. Aeronaut. Sci.* (21): 404–410, 1954.

116. M. W. Rubesin and C. C. Pappas, "An Analysis of the Turbulent Boundary Layer Characteristics on a Flat Plate With Distributed Light Gas Injection," *NACA Tech. Note 4149,* 1958.

117. E. L. Knuth and H. Dershin, "Use of Reference States in Predicting Transport Rates in High-Speed Turbulent Flows With Mass Transfer," *Int. J. Heat Mass Transfer* (6): 999–1018, 1963.

118. R. L. P. Voisinet, "Influence of Roughness and Blowing on Compressible Turbulent Boundary Layer Flow," *Naval Surface Weapons Center TR 79-153,* Silver Spring, MD, June 1979.

119. C. C. Pappas and A. F. Okuno, "Measurements of Skin Friction of the Compressible Turbulent Boundary Layer on a Cone with Foreign Gas Injection," *J. Aerosp. Sci.* (27): 321–333, 1960.

120. H. S. Mickley and R. S. Davis, "Momentum Transfer for Flow Over a Flat Plate With Blowing," *NACA Tech. Note 4017,* 1957.

121. T. Tendeland and A. F. Okuno, "The Effect of Fluid Injection on the Compressible Turbulent Boundary Layer—The Effect on Skin Friction of Air Injected Into the Boundary Layer of a Cone at $M = 2.7$," *NACA Res. Mem. A56DO5,* 1956.

122. R. J. Moffat and W. M. Kays, "The Turbulent Boundary Layer on a Porous Plate: Experimental Heat Transfer With Uniform Blowing and Suction," *Stanford Univ. Dept. Mech. Eng. Rep. HMT-1,* Stanford University, Stanford, CA, 1967.

123. C. C. Pappas and A. F. Okuno, "Measurements of Heat Transfer and Recovery Factor of a Compressible Turbulent Boundary Layer on a Sharp Cone With Foreign Gas Injection," *NASA Tech. Note D-2230,* 1964.

124. E. R. Bartle and B. M. Leadon, "The Effectiveness as a Universal Measure of Mass Transfer Cooling for a Turbulent Boundary Layer," *Proc. 1962 Heat Transfer and Fluid Mechanics Inst.,* Stanford University Press, Stanford, CA, pp. 27–41, 1962.

125. J. E. Danberg, "Characteristics of the Turbulent Boundary Layer With Heat and Mass Transfer at Mach Number 6.7," *Proc. 5th U.S. Navy Symp. Aeroballistics,* U.S. Naval Ordnance Lab., 1961.

126. C. J. Scott, G. E. Anderson, and D. R. Elgin, "Laminar, Transitional and Turbulent Mass Transfer Cooling Experiments at Mach Numbers From 3 to 5," *Univ. Minnesota Inst. Tech. Res. Rep. 162,* 1959.

127. J. Nikuradse, "Laws of Flow in Rough Pipes," *NACA Tech. Mem. 1292,* November 1950, transl. of "Strömungsgesetze in rauhen Rohren," *VDI Forschungsheft,* No. 361, 1933.

128. L. Prandtl and H. Schlichting, *Das Widerstandsgesetz rauher Platten,* Werft, Reederei, Hafen 1–4, 1934.

129. J. M. Healzer, R. J. Moffat, and W. M. Kays, "The Turbulent Boundary Layer on a Rough Porous Plate: Experimental Heat Transfer With Uniform Blowing," *Stanford Univ. Dept. Mech. Eng. Rep. HMT-18,* Stanford University, Stanford, CA, 1974.

130. T. Cebeci and P. Bradshaw, *Momentum Transfer in Boundary Layers,* Hemisphere, Washington, DC, 1977.

131. M. M. Pimenta, R. J. Moffat, and W. M. Kays, "The Turbulent Boundary Layer: An Experimental Study of the Transport of Momentum and Heat With the Effect of Roughness," *Stanford Univ. Dept. Mech. Eng. Rep. HMT-21,* Stanford University, Stanford, CA, 1975.

132. M. D. Van Dyke, "Higher-Order Boundary-Layer Theory," in *Annual Reviews of Fluid Mechanics,* pp. 265–292, Palo Alto, CA, 1969.

133. H. Thomann, "Effect of Streamwise Curvature on Heat Transfer in a Turbulent Boundary Layer," *J. Fluid Mech.* (33/2): 383–392, 1968.

134. P. Bradshaw, "Effects of Streamline Curvature on Turbulent Flow," *AGARDograph No. 169,* 1973.

135. R. E. Mayle, M. F. Blair, and F. C. Kopper, "Turbulent Boundary Layer Heat Transfer on Curved Surfaces," *J. Heat Transfer* (101): 521–525, August 1979.

136. S. A. Eide and J. P. Johnston, "Prediction of the Effects of Longitudinal Wall Curvature and System Rotation on Turbulent Boundary Layers," *Stanford Univ. Dept. Mech. Eng. Rep. PD-19,* Stanford University, Stanford, CA, November 1974.

137. D. C. Wilcox and M. W. Rubesin, "Progress in Turbulence Modeling for Complex Flow Fields Including Effects of Compressibility," *NASA Tech. Paper 1517,* April 1980.

138. J. Kestin, "Effect of Free-Stream Turbulence on Heat Transfer Rates," in *Advances in Heat Transfer,* T. F. Irvine Jr., and J. P. Hartnett, eds., vol. 3, pp. 1–32, Academic, New York, 1966.

139. A. Edwards and B. N. Furber, "The Influence of Free-Stream Turbulence on Heat Transfer by Convection From an Isolated Region of a Plane Surface in Parallel Air Flow," *Proc. Inst. Mech. Eng.* (170): 941, 1956.

140. W. C. Reynolds, W. M. Kays, and S. J. Kline, "Heat Transfer in the Turbulent Incompressible Boundary Layer, IV—Effect of Location of Transition and Prediction of Heat Transfer in a Known Transition Region," *NASA Mem. 12-4-58W,* 1958.

141. J. Kestin, P. F. Maeder, and H. E. Wang, "Influence of Turbulence on the Transfer of Heat From Plates With and Without a Pressure Gradient," *Int. J. Heat Mass Transfer* (3): 133–154, 1961.

142. P. E. Hancock, "The Effect of Free-Stream Turbulence on Turbulent Boundary Layers," PhD thesis, Imperial College, London, 1980.

143. S. K. F. Karlsson, "An Unsteady Turbulent Boundary Layer," *J. Fluid Mech.* (5/2): 622–636, 1959.

144. M. H. Patel, "On Turbulent Boundary Layers in Oscillatory Flow," *Proc. Royal Soc. London A* (353): 121–144, 1977.

145. J. Coustiex, R. Houdeville, and M. Raynaud, "Oscillating Turbulent Boundary Layer With a Strong Mean Pressure Gradient," *Proc. 2d Symp. Turbulent Shear Flows,* London, pp. 6.12–6.17, 1979.

146. P. G. Parikh, W. C. Reynolds, R. Jayaraman, and L. Carr, "Dynamics of an Unsteady Turbulent Boundary Layer," *Proc. 3d Symp. Turbulent Shear Flows,* Davis, CA, pp. 8.35–8.40, 1981.

147. E. R. van Driest, "Turbulent Boundary Layer on a Cone in a Supersonic Flow at Zero Angle of Attack," *J. Aeronaut. Sci.* (19): 55–57, 72, 1952.

148. G. B. Schubauer and P. S. Klebanoff, "Contribution on the Mechanics of Boundary Layer Transition," *NACA Rep. 1289,* 1956.

CHAPTER 7
RADIATION

John R. Howell
The University of Texas at Austin

M. Pinar Mengüç
University of Kentucky

INTRODUCTION

The field of radiative transfer is undergoing major advances in the capability to analyze complex problems. A well-developed theoretical base exists that can be applied to the solution of most (but not all) engineering problems. Major increases in computational speed and capacity have opened the way to solutions of problems that include complex geometries, spectral effects, and inhomogeneous properties. The need for such a capability is driven by applications that include high-temperature manufacturing processes and materials processing, improved efficiency, and more accurate design methods necessary for energy conversion devices, the use of new materials, hypersonic flow analysis, and others.

Some areas of radiative transfer analysis that had seen sparse research for some time have undergone a renaissance. This is true for radiative transfer among surfaces with no participating medium. The need to provide fast and accurate computer visualization for use in data depiction, virtual reality, real-time animation, and other areas requiring accurate modeling of radiative transfer has revived research on surface radiative transfer algorithms. These same needs have brought forth new research on surface property measurement and modeling so that accurate spectral and directional effects can be included in computer visualization.

Other areas of radiative transfer have been driven by increased capability of analysis due to the great strides in computer capability. Just a few years ago, two-dimensional problems of radiative transfer in enclosures with a participating medium were at the edge of computational capability. Now, these are routine, and many three-dimensional cases have been analyzed. Because of the need in applications such as utility steam generator design to analyze three-dimensional geometries with up to tens of thousands of surface and volume computational elements, much research is now focused on further increases in computational speed. Massively parallel computers may well provide the required computational capability for such problems.

In this chapter, we present the fundamentals of radiative transfer analysis. We begin with a review of the properties of the ideal radiating body, the *blackbody*, by describing radiative exchange among ideal (black) surfaces and then extending the analysis to surfaces with real radiative properties. We further extend the analysis to the case of a medium between the bounding surfaces that can absorb, emit, and scatter radiation. We provide information on the radiative properties of such participating media and conclude with a discussion of methods for treating combined-mode heat transfer problems in which radiative transfer is important. We present the bounds of applicability of the theory based on current state-of-the-art and available data.

Additional information can be found in the recent texts by Siegel and Howell [1], Brewster [2], and Modest [3].

Radiation Intensity and Flux

Any substance at a finite temperature emits electromagnetic energy in discrete energy quanta called *photons*. The energy of each photon is equal to $h\nu = hc/\lambda$, where h is the Planck constant, c is the speed of light, and ν and λ are the frequency and wavelength of the emitted energy, respectively. As it will be outlined below, with decreasing wavelength, or increasing frequency, the energy associated with photons increases.

Thermal radiation is associated with a temperature range of approximately 30 to 30,000 K and wavelength range of 0.1 to 100 μm. For most practical purposes, however, we are interested in a wavelength range of 0.4 (near-UV) to about 15 μm (near-IR). The photons whose energies correspond to this wavelength/temperature range are capable of changing the discrete vibrational, rotational, and electronic energy states of atoms and molecules of the material on which they are incident, and this, in turn, changes the internal energy and the corresponding temperature of the material. As a result of this, energy is transferred from a hot object to a colder one via thermal radiation as opposed to transfer by phonons when the objects are in contact.

Radiative heat transfer from one small volume or surface element to another is determined by accounting for the energies of photons of all wavelengths, emitted in all directions over a certain time interval. Depending on the location of each element and its orientation with respect to others, the amount of radiant energy exchange between elements will vary. In order to determine the contribution of each element to the radiation balance, we introduce a fundamental and mathematically convenient quantity termed *radiation intensity*.

By definition, the fractional radiant energy de_λ propagating through (or originating from) an infinitesimally small area dA_n in the direction $\hat{\Omega}(\theta, \phi)$, confined within an infinitesimally small solid angle $d\Omega$ around $\hat{\Omega}(\theta, \phi)$, within a wavelength interval $d\lambda$ around the wavelength of λ, and within a time interval of dt is called the *radiation intensity*:

$$I_\lambda(\hat{\Omega}) = I_\lambda(\theta, \phi) = \lim_{(dA, d\Omega, d\lambda, dt) \to 0} \frac{de_\lambda(\theta, \phi)}{dA_n \, d\Omega \, d\lambda \, dt} \tag{7.1}$$

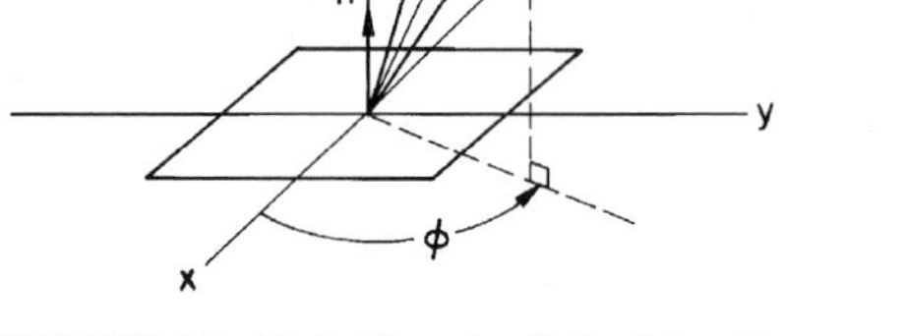

FIGURE 7.1 Definition of radiative intensity.

The direction of propagation $\hat{\Omega}$ is defined in spherical coordinates in terms of the zenith angle θ and azimuthal angle ϕ (see Fig. 7.1). dA_n is the area normal to the direction of propagation $\hat{\Omega}$ and is equal to $dA \cos \theta$, with θ being the zenith angle (i.e., the angle between the surface normal $\hat{n}$ and $\hat{\Omega}$).

The solid angle is defined as the ratio of infinitesimal area normal to $\hat{\Omega}$, to the square of the distance between two infinitesimal surface elements exchanging radiation (see Fig. 7.2). Note that radiation intensity may vary as a function of location $\vec{r} = \vec{r}(x, y, z)$, direction $\hat{\Omega} = \hat{\Omega}(\theta, \phi)$, time t, and wavelength λ; therefore, it is a function of seven independent parameters. For most calculations of interest, the transient nature of radiation intensity is not critical. Recently, however, with the advances in femto- and picosecond pulsed lasers, the transient radiative transfer applications have started becoming important (see, e.g., Ref. 4); nevertheless, we will not cover these advances in this chapter.

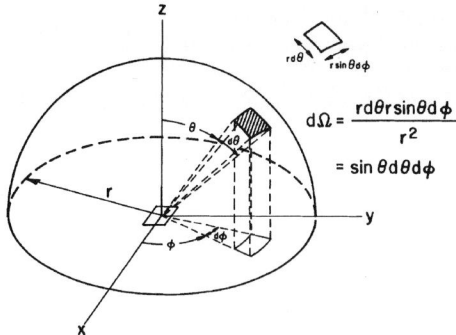

FIGURE 7.2 Definition of solid angle.

For heat transfer predictions, the radiative heat flux through a surface (with normal $\hat{n}$) is required, which is determined by integrating the radiation intensity incident from all directions (i.e., 4π steradians) as

$$\vec{q}_\lambda(\vec{r}) = \int_{\Omega = 4\pi} I_\lambda(\vec{r}, \hat{\Omega})\hat{n} \cdot \hat{\Omega} \, d\Omega \qquad (7.2)$$

where $\hat{n}$ is the unit vector normal to the surface (see Fig. 7.1). Note that by replacing $\hat{\Omega}$ with $\hat{n} \cdot \hat{\Omega}$, which is equivalent to the direction cosine for a given surface, the flux across the surface is obtained. Radiative flux on an opaque (i.e., nontransmitting) surface element is obtained by evaluating the integral over a single hemisphere (i.e., $\Omega = 2\pi$ steradians). The total radiative flux is determined by integrating $\vec{q}_\lambda(\vec{r})$ over the entire wavelength spectrum.

Blackbody Radiation

The blackbody is the standard against which the behavior of all real radiating materials is gauged. It has well-defined characteristics that are firmly based in theory and experiment. Here, these characteristics will be outlined, as understanding them is paramount to conceptualizing the radiative transfer phenomenon.

General Definitions and Characteristics. The most important attributes of a blackbody can be listed as:

- A blackbody is defined as a surface or volume that absorbs all incident radiation. This includes radiation at every wavelength and from every direction.
- The blackbody is the best possible emitter of radiation at every wavelength and in every direction.
- Radiation emitted by a blackbody increases monotonically at every wavelength with absolute temperature.
- Radiation within an isothermal enclosure with blackbody boundaries is isotropic; that is, uniform in all directions.

With such qualities, the blackbody is seen to be a convenient standard for comparing the properties of real materials. All real materials will reflect some incident radiant energy and are thus not perfect absorbers. Because they do not absorb as much as the ideal blackbody, they must emit less than an ideal blackbody to remain in thermal equilibrium with their surroundings. A real surface thus emits less than the blackbody (again, at every wavelength and in every direction).

It is possible to construct a nearly ideal blackbody by the artifice of defining the surface over the small entrance to a deep cavity as the blackbody. Little of the radiation crossing this fictitious "black" surface reflects back through the cavity opening, especially if the internal cavity surfaces are made as nonreflecting as possible and are oriented so as not to face the cavity opening directly. If the interior cavity surfaces are maintained at a uniform constant temperature and the cavity opening is small, then the radiation within the cavity is isotropic and the energy leaving the cavity opening will be quite close to that of a blackbody at the cavity temperature. These observations are the basis for producing experimental blackbodies used in making comparisons of radiation from real material surfaces for property measurements. To analyze radiative heat transfer, quantitative measures of the blackbody characteristics described in this section must be provided.

Blackbody Intensity, Emissive Power.   It can be shown that the intensity leaving a black surface is independent of θ and φ; that is, the blackbody emitted intensity is isotropic [1]. This fact provides another convenient benchmark for comparing the behavior of real surfaces.

For a blackbody, the spectral intensity is given by the *Planck distribution of blackbody intensity* [5]:

$$I_{\lambda b} = \frac{2C_1}{n^2\lambda^5(e^{C_2/n\lambda T} - 1)} \tag{7.3}$$

where T is the absolute temperature (in K) and C_1 and C_2 are constants with values

$$C_1 = hc_o^2 = 0.59552 \cdot 10^8 \text{ W } \mu\text{m}^4/\text{m}^2$$

$$C_2 = hc_o/k = 14{,}388 \text{ } \mu\text{mK} \tag{7.4}$$

with
$$h = 6.626075 \cdot 10^{-34} \text{ Js (Planck constant)} \tag{7.5}$$

$$k = 1.380658 \cdot 10^{-23} \text{ J/K (Boltzmann constant)} \tag{7.6}$$

$$c_o = 2.99792458 \cdot 10^8 \text{ m/s (speed of light in vacuum)} \tag{7.7}$$

and n is the index of refraction of the medium. To determine the energy leaving a black surface in all directions, $I_{\lambda b}$ is integrated over the hemisphere of solid angles $d\Omega = \sin\theta \, d\theta \, d\phi$ to give the *spectral emissive power of a blackbody* $e_{\lambda b}$:

$$e_{\lambda b} = \int_{\theta=0}^{\pi}\int_{\phi=0}^{2\pi} I_{\lambda b} \cos\theta \sin\theta \, d\theta \, d\phi = \frac{2\pi C_1}{n^2\lambda^5(e^{C_2/n\lambda T} - 1)} = \pi I_{\lambda b} \tag{7.8}$$

Figure 7.3 depicts the blackbody function for different temperatures, including the solar temperature of 5762 K. Equation 7.8 can be simplified by dividing by $n^3 T^5$ to give

$$\frac{e_{\lambda b}}{n^3 T^5} = \frac{2\pi C_1}{(n\lambda T)^5(e^{C_2/n\lambda T} - 1)} \tag{7.9}$$

This relation is plotted in Fig. 7.4.

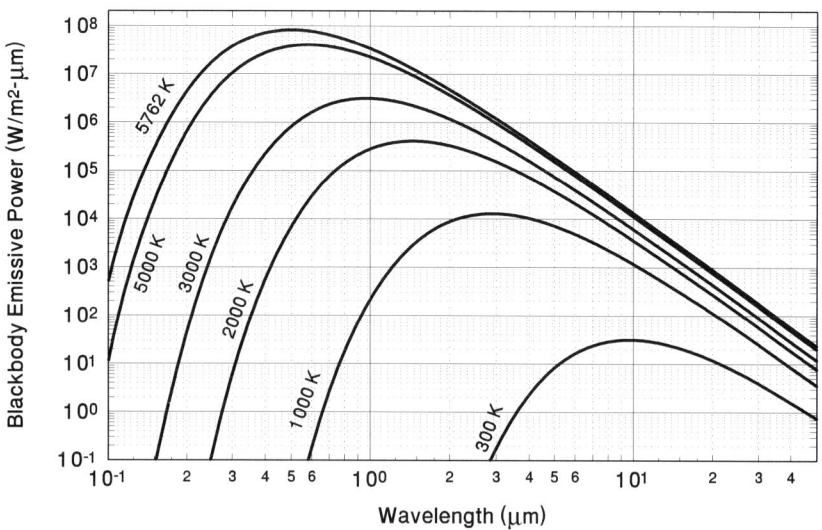

FIGURE 7.3   The Planck distribution of blackbody emissive power.

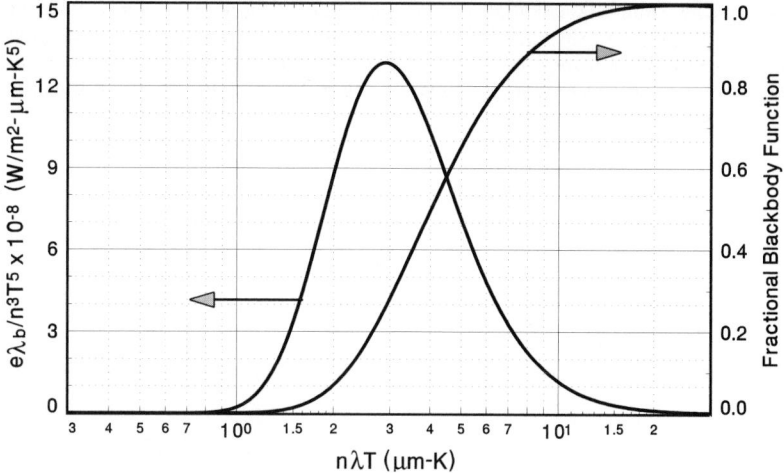

FIGURE 7.4   Normalized and fractional blackbody emission distributions.

Note that, with decreasing temperature, the area under the curve (i.e., the *total* energy emitted by the blackbody) decreases. The maximum value of the curve shown in Fig. 7.4 can be found mathematically by taking the partial derivative of Eq. 7.9 with respect to wavelength and setting the result equal to zero. The maximum of the curve occurs at a fixed value of

$$C_3 = (n\lambda T)_{max} = 2897.8 \ \mu mK \tag{7.10}$$

This relation is known as *Wien's displacement law,* and it provides a convenient means of determining where the wavelength of peak radiated energy occurs for a given temperature. For example, the solar radiation spectrum has a peak at about $\lambda = 0.5 \ \mu m$, as $T_{sun} = 5762$ K. On the other hand, an object at room temperature ($T_{room} = 300$ K) would emit maximum energy at about $\lambda \approx 10 \ \mu m$. It is obvious that, as absolute temperature increases, the wavelength at which the maximum emissive power occurs becomes smaller.

Values of $e_{\lambda b}/n^3 T^5$ are tabulated in many sources (usually for $n = 1$; see Refs. 1, 2, 3, and 6) and easily computed from Eq. 7.9. If $\exp(C_2/n\lambda T) \gg 1$, the Planck blackbody distribution can be simplified to

$$e_{\lambda b} = \frac{2\pi C_1}{n^2 \lambda^5} \ e^{-C_2/n\lambda T} \tag{7.11}$$

This expression, known as *Wien's distribution of blackbody radiation,* is accurate over a wide range of wavelengths and temperatures.

Equation 7.9 can be integrated with respect to wavelength to obtain the total blackbody emissive power (i.e., the rate of energy emitted by a blackbody into all directions at all wavelengths):

$$e_b = \int_{\lambda=0}^{\infty} e_{\lambda b} \ d\lambda = n^2 \sigma T^4 \tag{7.12}$$

Equation 7.12 is known as the *Stefan-Boltzmann equation,* and σ is the Stefan-Boltzmann constant, which has the value

$$\sigma = 2C_1 \pi^5 / 15 C_2^4 = 2\pi^5 k^4 / (15 h^3 c_o^2) = 5.6705 \cdot 10^{-8} \ W/(m^2 K^4) \tag{7.13}$$

The value of σ calculated from the accepted values of Planck's constant h, the Boltzmann constant k, and the speed of light in a vacuum c_o agrees with experimentally determined values within experimental error of 1.4 parts in 10^4 [7].

Equation 7.12 shows that the rate of energy emitted by a blackbody increases in proportion to the absolute temperature to the fourth power, so that radiation will generally be the dominating heat transfer mode at high absolute temperatures.

For many practical applications, the blackbody emission is needed within finite, relatively narrow wavelength intervals, rather than the entire spectrum. To calculate the blackbody energy between wavelengths λ_1 and λ_2, we write

$$\int_{\lambda_1}^{\lambda_2} e_{\lambda b}\, d\lambda = \frac{C_1}{C_2^4} \int_{\xi_1}^{\xi_2} n^2 T^4 \frac{\xi^3\, d\xi}{e^\xi - 1} \tag{7.14}$$

$$= \frac{C_1}{C_2^4} \left[\int_0^{\xi_2} n^2 T^4 \frac{\xi^3\, d\xi}{e^\xi - 1} - \int_0^{\xi_1} n^2 T^4 \frac{\xi^3\, d\xi}{e^\xi - 1} \right] \tag{7.15}$$

$$= (F_{0 - n\lambda_2 T} - F_{0 - n\lambda_1 T}) E_b(T) \tag{7.16}$$

where $\xi = C_2/n\lambda T$. Here, $F_{0-n\lambda T}$ is the fraction of the energy in the blackbody distribution that lies below some given $n\lambda T$ value and is called the fractional blackbody function of the first kind. This fraction is defined as

$$F_{0-n\lambda T} = \frac{\int_0^{\lambda T} (e_{\lambda b}/T^5)\, d(\lambda T)}{\int_0^\infty (e_{\lambda b}/T^5)\, d(\lambda T)} = \frac{\int_0^\lambda e_{\lambda b}\, d\lambda}{\int_0^\infty e_{\lambda b}\, d\lambda} \tag{7.17}$$

where the F fraction can be evaluated numerically and is listed in most textbooks. An analytical formulation for this integral was suggested [8], resulting in

$$F_{0-n\lambda T} = \frac{15}{\pi^4} \sum_{j=1}^\infty \left[\frac{e^{-j\xi}}{j} \left(\xi^3 + \frac{3\xi^2}{j} + \frac{6\xi}{j^2} + \frac{6}{j^3} \right) \right] \tag{7.18}$$

This relation converges very quickly, usually within three terms of the series except at very large values of ξ. Figure 7.4 also depicts the fractional blackbody function versus $n\lambda T$ product.

Nonblack Surfaces and Materials

No real materials act as blackbodies, so measures of their deviation from blackbody behavior are used to define the spectral, directional, and temperature dependence of real surfaces relative to those of a blackbody.

The notation used is to denote a spectrally dependent property with a λ subscript as before, and to denote a directionally dependent property with a prime ($'$). These symbols are omitted when the property in question has been averaged over one or both of the dependencies.

Emissivity. The ability of a surface to emit radiation in comparison with the ideal emission by a blackbody is defined as the emissivity of the surface. The emissivity can be defined on a spectral, directional, or total basis.

*Directional Spectral Emissivity**

$$\varepsilon_\lambda' = \frac{I_\lambda}{I_{\lambda b}} \tag{7.19}$$

---

* The prime denotes the directional quantity. Note that the radiation intensity I is directional by definition.

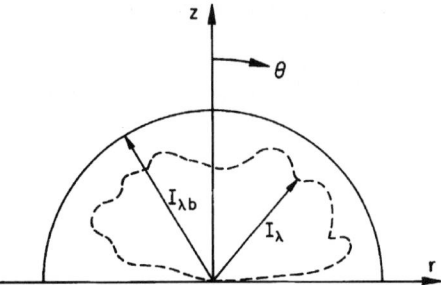

FIGURE 7.5 Schematic of angular intensity distributions leaving an ideal blackbody ($I_{\lambda b}$) and a real surface (I_λ).

Figure 7.5 depicts the angular distribution of $I_{\lambda b}$ and I_λ schematically for an ideal blackbody and a typical real surface. Note that only the azimuthal angle (θ) dependence is shown for the sake of clarity. Integrating the emitted energy over all wavelengths at a particular direction results in the directional total emissivity.

Directional Total Emissivity

$$\varepsilon' = \frac{I}{I_b} = \frac{\pi \int_{\lambda=0}^{\infty} I_\lambda \, d\lambda}{\sigma T^4} = \frac{\pi \int_{\lambda=0}^{\infty} \varepsilon'_\lambda I_{\lambda b} \, d\lambda}{\sigma T^4} = \frac{\pi I}{\sigma T^4} \qquad (7.20)$$

Integrating the energy emitted over all directions at a particular wavelength gives hemispherical-spectral emissivity.

Hemispherical-Spectral Emissivity

$$\varepsilon_\lambda = \frac{e_\lambda}{e_{\lambda b}} = \frac{1}{\pi} \int_{\phi=0}^{2\pi} \int_{\theta=0}^{\pi/2} \varepsilon'_\lambda \cos\theta \sin\theta \, d\theta \, d\phi \qquad (7.21)$$

Integrating the emitted energy over both wavelength and direction and comparing with the similar integrated quantity for a blackbody yields hemispherical total emissivity.

Hemispherical Total Emissivity

$$\varepsilon = \frac{e}{e_b} = \frac{\int_{\lambda=0}^{\infty} \int_{\phi=0}^{2\pi} \int_{\theta=0}^{\pi/2} \varepsilon'_\lambda I_{\lambda b} \cos\theta \sin\theta \, d\theta \, d\phi \, d\lambda}{\sigma T^4}$$

$$= \frac{\pi \int_{\lambda=0}^{\infty} \varepsilon_\lambda I_{\lambda b} \, d\lambda}{\sigma T^4} = \frac{\int_{\lambda=0}^{\infty} \varepsilon_\lambda e_{\lambda b} \, d\lambda}{\sigma T^4} \qquad (7.22)$$

The various integrated emissivities allow calculation using data for the detailed emissivities.

Absorptivity. The absorptivity is the property that defines the fraction of the incident energy that is absorbed by a surface. This property may also be dependent upon the direction and wavelength of the incident radiation. These properties will usually depend on the temperature of the absorbing surface (T_s); however, notation to indicate this fact is omitted.

Directional Spectral Absorptivity

$$\alpha'_\lambda = \frac{d^3 q_{\lambda,a}}{d^3 q_{\lambda,i}} = \frac{d^3 q_{\lambda,a}}{I_{\lambda,i} \cos\theta \, d\Omega \, d\lambda} \qquad (7.23)$$

where the superscript 3 indicates that the flux q is a dependent function of three variables: wavelength λ, direction (θ, ϕ), and location $r(x, y, z)$. Integrating the absorbed energy over all wavelengths at a particular direction of incidence results in directional total absorptivity.

Directional Total Absorptivity

$$\alpha' = \frac{d^2 q_a}{d^2 q_i} = \frac{\pi \int_{\lambda=0}^{\infty} \alpha'_\lambda I_{\lambda,i} \, d\lambda}{\int_{\lambda=0}^{\infty} I_{\lambda,i} \, d\lambda} \qquad (7.24)$$

By integrating Eq. 7.23 over all incident directions, the energy absorbed from all directions at a particular wavelength is obtained as hemispherical-spectral absorptivity.

Hemispherical-Spectral Absorptivity

$$\alpha_\lambda = \frac{d^2 q_{\lambda,a}}{d^2 q_{\lambda,i}} = \frac{\int_{\phi=0}^{2\pi} \int_{\theta=0}^{\pi/2} \alpha'_\lambda I_{\lambda,i} \cos\theta \sin\theta \, d\theta \, d\phi}{\int_{\phi=0}^{2\pi} \int_{\theta=0}^{\pi/2} I_{\lambda,i} \cos\theta \sin\theta \, d\theta \, d\phi} \tag{7.25}$$

Finally, integrating the absorbed energy over both wavelength and direction and comparing with the integrated incident energy gives hemispherical total absorptivity.

Hemispherical Total Absorptivity

$$\alpha = \frac{dq_a}{dq_i} = \frac{\int_{\lambda=0}^{\infty} \int_{\phi=0}^{2\pi} \int_{\theta=0}^{\pi/2} \alpha'_\lambda I_{\lambda,i} \cos\theta \sin\theta \, d\theta \, d\phi \, d\lambda}{\int_{\lambda=0}^{\infty} \int_{\phi=0}^{2\pi} \int_{\theta=0}^{\pi/2} I_{\lambda,i} \cos\theta \sin\theta \, d\theta \, d\phi \, d\lambda} = \frac{\int_{\phi=0}^{2\pi} \int_{\theta=0}^{\pi/2} \alpha' I_i \cos\theta \sin\theta \, d\theta \, d\phi}{\int_{\phi=0}^{2\pi} \int_{\theta=0}^{\pi/2} I_i \cos\theta \sin\theta \, d\theta \, d\phi} \tag{7.26}$$

Kirchhoff's Law.   Through an energy balance at thermodynamic equilibrium, it can be shown that the directional-spectral emissivity is always equal to the directional-spectral absorptivity of a surface, or

$$\alpha'_\lambda = \varepsilon'_\lambda \tag{7.27}$$

This relation is known as Kirchhoff's law. Equation 7.27 may be substituted into the various relationships for the integrated emissivity or absorptivity. However, it does not follow that such quantities as directional total, hemispherical-spectral, or hemispherical total emissivity and absorptivity are necessarily equal. In fact, the integrated properties are only equal if certain restrictions are met. These are given in Table 7.1.

TABLE 7.1   Requirements for Application of Kirchhoff's Law

Relation	Restrictions
I: $\alpha'_\lambda = \varepsilon'_\lambda$	None
II: $\alpha_\lambda = \varepsilon_\lambda$	Incident radiation has equal intensity from all angles, or $\alpha'_\lambda = \varepsilon'_\lambda$ are independent of angle
III: $\alpha' = \varepsilon'$	Incident radiation has a spectral distribution proportional to that of a blackbody at the temperature of the surface, or $\alpha'_\lambda = \varepsilon'_\lambda$ are independent of wavelength
IV: $\alpha = \varepsilon$	One restriction each from II and III above

Reflectivity.   Reflectivity is the property of a surface that defines the fraction of incident energy that is reflected by the surface. This property depends not only on the wavelength and directional characteristics, but it must also describe the directional distribution of the reflected radiation. It therefore has more independent variables than the properties discussed so far. Integrated properties are defined by integration over incident angle, reflected angle, wavelength, and combinations of these. Many of the reflectivities are rarely used in practice; the most useful ones are defined here. Note that the same notation is used as for absorptivity and emissivity, except that a double prime indicates a reflectivity that depends on both direction of incidence and direction of reflection.

The most fundamental reflectivity defines the intensity reflected into a particular direction resulting from energy incident from a given direction. At a particular wavelength, this is as follows.

Bidirectional Spectral Reflectivity

$$\rho''_\lambda = \frac{I_{\lambda,r}}{I_{\lambda,i} \cos\theta_i \sin\theta_i \, d\theta_i \, d\phi_i} \tag{7.28}$$

where double prime ($''$) indicates the directional nature of both incident and reflected radiant energy.

Integrating ρ''_λ over all angles of reflection gives:

Directional-Hemispherical Spectral Reflectivity

$$\rho'_{\lambda i} = \frac{\int_{\phi_r=0}^{2\pi} \int_{\theta_r=0}^{\pi/2} I_{\lambda,r} \cos\theta_r \sin\theta_r \, d\theta_r \, d\phi_r}{I_{\lambda,i} \cos\theta_i \sin\theta_i \, d\theta_i \, d\phi_i} = \int_{\phi_r=0}^{2\pi} \int_{\theta_r=0}^{\pi/2} \rho''_\lambda \cos\theta_r \sin\theta_r \, d\theta_r \, d\phi_r \qquad (7.29)$$

Because of the reciprocity of the bidirectional reflectivity ρ''_λ, the hemispherical-directional spectral reflectivity for isotropic incident intensity $\rho'_{\lambda r}$ is equal to the directional-hemispherical reflectivity $\rho'_{\lambda i}$. Here, a single prime ($'$) is used to denote the directional nature of incident radiation.

The fraction of the incident energy from all directions that is reflected into all directions at a particular wavelength is written as hemispherical spectral reflectivity.

Hemispherical Spectral Reflectivity

$$\rho_\lambda = \frac{\int_{\phi_i=0}^{2\pi} \int_{\theta_i=0}^{\pi/2} \rho'_{\lambda,i} I'_{\lambda,i} \cos\theta_i \sin\theta_i \, d\theta_i \, d\phi_i}{\int_{\phi_i=0}^{2\pi} \int_{\theta_i=0}^{\pi/2} I'_{\lambda,i} \cos\theta_i \sin\theta_i \, d\theta_i \, d\phi_i} \qquad (7.30)$$

By integrating over all wavelengths, the reflectivities defined above become total values and are given as follows.

Bidirectional Total Reflectivity

$$\rho'' = \frac{\int_{\lambda=0}^{\infty} I_{\lambda,r} \, d\lambda}{\cos\theta_i \sin\theta_i \, d\theta_i \, d\phi_i \int_{\lambda=0}^{\infty} I_{\lambda,r} \, d\lambda} = \frac{\int_{\lambda=0}^{\infty} \rho''_\lambda I_{\lambda,i} \, d\lambda}{\int_{\lambda=0}^{\infty} I_{\lambda,i} \, d\lambda} \qquad (7.31)$$

Directional-Hemispherical Total Reflectivity

$$\rho'_i = \frac{\int_{\lambda=0}^{\infty} \rho_{\lambda,i} I_{\lambda,i} \, d\lambda}{\int_{\lambda=0}^{\infty} I_{\lambda,i} \, d\lambda} \qquad (7.32)$$

Hemispherical Total Reflectivity

$$\rho = \frac{\int_{\lambda=0}^{\infty} \int_{\phi_i=0}^{2\pi} \int_{\theta_i=0}^{\pi/2} \rho'_{\lambda,i} I'_{\lambda,i} \cos\theta_i \sin\theta_i \, d\theta_i \, d\phi_i \, d\lambda}{\int_{\lambda=0}^{\infty} \int_{\phi_i=0}^{2\pi} \int_{\theta_i=0}^{\pi/2} I'_{\lambda,i} \cos\theta_i \sin\theta_i \, d\theta_i \, d\phi_i \, d\lambda} = \frac{\int_{\phi_i=0}^{2\pi} \int_{\theta_i=0}^{\pi/2} \rho'_i I'_i \cos\theta_i \sin\theta_i \, d\theta_i \, d\phi_i}{\int_{\phi_i=0}^{2\pi} \int_{\theta_i=0}^{\pi/2} I'_i \cos\theta_i \sin\theta_i \, d\theta_i \, d\phi_i} \qquad (7.33)$$

Relations among Surface Properties.   Because incident energy onto an opaque surface is either reflected or absorbed by the surface, it follows that one unit of radiant energy incident on an opaque surface at a given wavelength and from a given direction will have a fraction absorbed and the remainder reflected, or

$$1 = \alpha'_\lambda + \rho'_\lambda \qquad (7.34)$$

Similarly, for radiation at all wavelengths, it follows that

$$1 = \alpha' + \rho' \qquad (7.35)$$

Total radiation onto a surface at all wavelengths and from all directions follows a similar conservation relation,

$$1 = \alpha + \rho \tag{7.36}$$

Kirchhoff's law (Eq. 7.27) may be used to replace the absorptivity in these relations with emissivity if the restrictions of Table 7.1 are observed. Thus, data on one of the radiative properties can often be used to generate the others, although care must be used to avoid violating the restrictions of Table 7.1.

If the surface is not opaque but semitransparent, then Eqs. 7.34–7.36 need to include transmissivity τ and be replaced by the following equations:

$$1 = \alpha'_\lambda + \rho'_\lambda + \tau'_\lambda \tag{7.37}$$
$$1 = \alpha' + \rho' + \tau' \tag{7.38}$$
$$1 = \alpha + \rho + \tau \tag{7.39}$$

where τ'_λ, τ', and τ are defined similar to the corresponding α equations (see Eqs. 7.23, 7.24, and 7.25).

Values for Surface Properties. In solving radiative transfer problems involving surfaces, property values must be available. Radiative surface properties given in the literature show wide variations, because the values are altered greatly by the presence of surface contaminants (oxidation, fingerprints, etc.), the presence of surface texture or roughness (machining marks, grain structure), and tailored surface characteristics such as coatings, films, or geometric structures specifically designed to affect spectral or directional characteristics. Thus, tabulated values serve at best as rough guidelines to the properties of engineering surfaces. Tabulated property values for selected substances are given in the first two tables in App. A (Tables A.7.1 and A.7.2).

The available data for radiative properties of metallic and nonmetallic surfaces can be found in a number of data bases [9, 10]; they are too extensive to be included in this chapter. Also, the reader is referred to a number of recent handbooks and reviews for more detailed data [11–17].

In the absence of tabulated or measured properties for a given surface, various options are available. The behavior of a surface can be computed based on fundamental theories, such as Maxwell's electromagnetic wave theory; the surface characteristics can be assumed based on extrapolation from the behavior of similar surfaces; a model of the surface behavior can be constructed based on simplified assumed surface characteristics; or greatly simplified characteristics can be assumed to be accurate enough for use. In the third table in App. A (Table A.7.3), the spectral complex index of refraction data for a number of metals are listed (from Ref. 16), which can be used to determine the surface absorption, reflection, and transmission characteristics as discussed below.

Computed Properties. For perfect dielectric materials and for highly conducting materials, electromagnetic wave theory can be used to predict radiative properties [1, 18]. These predictions are based on the assumption that the surface is *optically smooth;* that is, the surface roughness is small compared with the wavelength of the radiation (otherwise, the problem may be numerically intractable). In addition, the composition of the material must be well known so that necessary properties such as the simple refractive index n and the attenuation coefficient k that comprise the complex refractive index $m = n - ik$ can be assigned (Fig. 7.6); this implies that the surface is not contaminated by oxide films or other impurities. If these restrictions are met, then properties can be computed with reasonable accuracy from the following relations, which apply for radiation from medium 1, assumed to be non- or weakly absorbing (i.e., $n_1 \approx 1$ and $k_1 \approx 0$), onto the surface of absorbing/reflecting surface of medium 2:

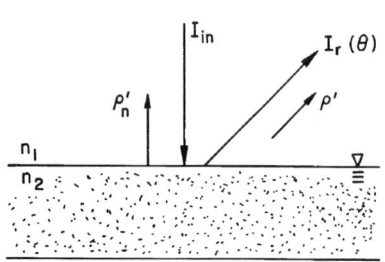

FIGURE 7.6 Definition of radiative properties at the interfaces.

- Dielectrics (insulating materials, with $n = n_2/n_1$ and $k_1 = k_2 = 0$)

  —Normal-hemispherical total or spectral reflectivity

  $$\rho_n' = \frac{(n-1)^2}{(n+1)^2} \tag{7.40}$$

  —Directional-hemispherical total or spectral reflectivity

  $$\rho' = \frac{1}{2}\left[\left(\frac{n^2\cos\theta - [n^2 - \sin^2\theta]^{1/2}}{n^2\cos\theta + [n^2 - \sin^2\theta]^{1/2}}\right)^2 + \left(\frac{[n^2 - \sin^2\theta]^{1/2} - \cos\theta}{[n^2 - \sin^2\theta]^{1/2} + \cos\theta}\right)^2\right] \tag{7.41}$$

  —Normal total emissivity

  $$\varepsilon_n' = \frac{4n}{(n+1)^2} \tag{7.42}$$

  —Hemispherical total emissivity

  $$\varepsilon = \frac{1}{2} - \frac{(3n+1)(n-1)}{6(n+1)^2} - \frac{n^2(n^2-1)^2}{(n^2+1)^3}\ln\left(\frac{n-1}{n+1}\right) + \frac{2n^3(n^2+2n-1)}{(n^2+1)(n^4-1)} - \frac{8n^4(n^4+1)}{(n^2+1)(n^4-1)^2}\ln n \tag{7.43}$$

- Metals (electromagnetically attenuating materials; for incidence on a surface through most gases, $n_1 = 1$ and $k_1 = 0$):

  —Normal hemispherical reflectivity

  $$\rho_n' = \frac{(n_2 - n_1)^2 + (k_2 - k_1)^2}{(n_2 + n_1)^2 + (k_2 + k_1)^2} = \frac{(n_2 - 1)^2 + (k_2)^2}{(n_2 + 1)^2 + (k_2)^2} \tag{7.44}$$

  —Normal spectral emissivity ($n_1 = 1$ and $k_1 = 0$),

  $$\varepsilon_{\lambda,n}' = 36.5\left(\frac{r_e}{\lambda}\right)^{1/2} - 464\frac{r_e}{\lambda} \tag{7.45}$$

  where r_e is the electrical resistivity of the metal in ohm − cm and λ is in μm

  —Normal total emissivity ($n_1 = 1$ and $k_1 = 0$)

  $$\varepsilon' = 0.578(r_e T)^{1/2} - 0.178 r_e T + 0.0584(r_e T)^{3/2} \tag{7.46}$$

  —Hemispherical total emissivity ($n_1 = 1$ and $k_1 = 0$)

  $$\varepsilon = 0.766(r_e T)^{1/2} - [0.309 - 0.0889\ln(r_e T)]r_e T - 0.0175(r_e T)^{3/2} \tag{7.47}$$

The relations of Eqs. 7.40–7.47 as well as Kirchhoff's law (with the restrictions of Table 7.1) may be used to find other properties from the electromagnetic theory relations.

The relations given for both dielectrics and metals are for unpolarized incident radiation; if polarization is important, then more detailed analysis must be used (see Refs. 1, 3, 18, 19). Also, the refractive index and absorption index may show spectral dependence, in which case the computed radiative properties will also be spectral in nature.

Property Approximations.   Because of the lack of accurate radiative properties in many situations, it is common practice to invoke certain approximations for the property behavior. The most common assumptions are that the surface properties are independent of wavelength (a gray surface), independent of direction (a diffuse surface), that the surface behaves as an ideal mirror (a specular surface), or that the surface is black. The assumption of a gray-diffuse surface is the most commonly invoked. For a surface that is truly both gray and diffuse, Kirchhoff's law applies for all of the property sets; that is, $\alpha = \varepsilon$, and the computation of radiative exchange among surfaces and in enclosures is considerably simplified. However, for surfaces with directional or spectral property variations, these assumptions can introduce considerable error into the energy exchange calculations, particularly when radiation is the dominant heat transfer mode.

RADIATIVE EXCHANGE: ENCLOSURES CONTAINING A NONPARTICIPATING MEDIUM

In this section, we deal with energy exchange among surfaces and in enclosures when no medium is present between the surfaces that affects the transfer; that is, no scattering, absorption, or emission occurs within the medium. Such effects are covered in the next major section of this chapter.

Black Surfaces

The radiative exchange between two black surfaces depends only on the absolute temperatures of the surfaces and their shapes and relative positions. If the fraction of blackbody energy leaving area element dA_j and incident on area element dA_k is defined as dF_{dj-dk}, then the energy emitted in wavelength interval $d\lambda$ around λ by black element j and incident on element k is $d^3q_{\lambda,dj-dk}$ and given by (see Fig. 7.7)

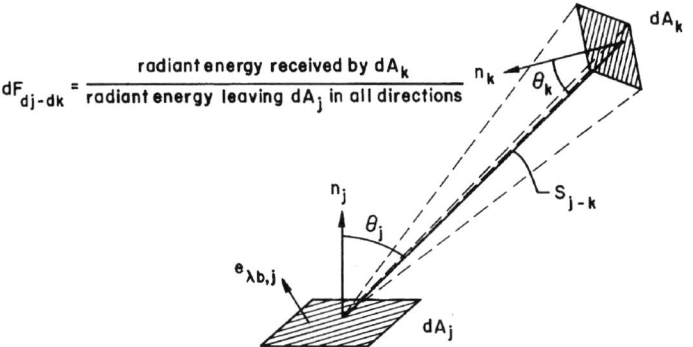

$$dF_{dj-dk} = \frac{\text{radiant energy received by } dA_k}{\text{radiant energy leaving } dA_j \text{ in all directions}}$$

FIGURE 7.7   Nomenclature and schematic for radiative exchange between two surfaces.

$$d^3q_{\lambda,dj \to dk} = e_{\lambda b,j}\, dA_j\, dF_{dj-dk}\, d\lambda \qquad (7.48)$$

where, again, the superscript 3 means that the flux is a function of three independent parameters, wavelength, direction, and spatial coordinate. The dF_{dj-dk} is called the configuration factor, shape factor, view factor, or angle factor. It is independent of wavelength because for black surfaces, the directional distribution of emission from a surface does not depend on wavelength but is diffuse at every wavelength.

Equation 7.48 gives the energy leaving surface j that is incident upon (and therefore absorbed by) black surface k. The energy leaving k and absorbed by j is given by a similar equation:

$$d^3q_{\lambda,dk \to dj} = e_{\lambda b,k}\, dA_k\, dF_{dk-dj}\, d\lambda \qquad (7.49)$$

The net exchange between the two surface elements $d^3q_{\lambda,dj-dk}$ (instead of $d^3q_{\lambda,dj \rightarrow dk}$) is then

$$d^3q_{\lambda,dj-dk} = d^3q_{\lambda,dj \rightarrow dk} - d^3q_{\lambda,dk \rightarrow dj} = e_{\lambda b,j} dA_j dF_{dj-dk} d\lambda - e_{\lambda b,k} dA_k dF_{dk-dj} d\lambda \qquad (7.50)$$

If surfaces j and k are at the same temperature, then the net energy exchange must be zero, and it follows from Eq. 7.50 that

$$dA_k dF_{dk-dj} = dA_j dF_{dj-dk} \qquad (7.51)$$

Substituting this reciprocity relation into Eq. 7.50 results in

$$d^3q_{\lambda,dj-dk} = (e_{\lambda b,j} - e_{\lambda b,k}) dA_j dF_{dj-dk} d\lambda \qquad (7.52)$$

Integrating over all wavelengths gives

$$d^2q_{dj-dk} = (e_{b,j} - e_{b,k}) dA_j dF_{dj-dk} = \sigma(T_j^4 - T_k^4) dA_j dF_{dj-dk} \qquad (7.53)$$

If surface k is finite in extent and dA_k is an element of that surface, then Eq. 7.53 can be integrated over the entire area of surface k to find the net energy exchange of surface element j with surface k:

$$dq_{dj-k} = \sigma \int_{A_k} [T_j^4 - T_k^4] \, dF_{dj-dk} \, dA_j \qquad (7.54)$$

Isothermal Surfaces.   If surface k is isothermal, Eq. 7.54 can be easily integrated to obtain

$$dq_{dj-k} = \sigma[T_j^4 - T_k^4] F_{dj-k} dA_j \qquad (7.55)$$

where

$$F_{dj-k} = \int_{A_k} dF_{dj-dk} \qquad (7.56)$$

If surface j is also isothermal, then Eq. 7.55 can be integrated to give

$$dq_{j-k} = \sigma[T_j^4 - T_k^4] \int_{A_j} dF_{dj-k} \, dA_j = \sigma[T_j^4 - T_k^4] F_{j-k} A_j \qquad (7.57)$$

where

$$F_{j-k} = \frac{1}{A_j} \int_{A_j} F_{dj-k} \, dA_j \qquad (7.58)$$

By writing the net radiative energy flux on surface k (rather than j), it is easily found that

$$A_j dF_{j-dk} = dA_k F_{dk-j}$$

$$A_k dF_{k-dj} = dA_j F_{dj-k}$$

$$A_k F_{k-j} = A_j F_{j-k} \qquad (7.59)$$

These *reciprocity* relations will be used in later sections.

If there are N surfaces forming an enclosure, then the net radiative energy flux on surface j is given by summing the contributions from all surfaces forming the interior of the enclosure:

$$dq_j = \sigma \sum_{k=1}^{N} (T_j^4 - T_k^4) dF_{dj-k} dA_j \qquad (7.60)$$

A total of N temperatures and radiative energy fluxes must be known to solve for the others. If the temperatures of all of the N surfaces are available, then the radiative fluxes at all surfaces are easily computed from Eq. 7.60. If radiative energy fluxes are given at M of the N surfaces, and temperatures are given at the others, then the set of M linear equations must be solved for the unknown temperatures. The methods given in the section "Exchange among Gray Diffuse Surfaces" can be used.

Nonisothermal Surfaces.   If the temperature of a surface varies, determination of the radiative energy flux or temperature profile on all surfaces of an enclosure requires the solution of integral equations (see Eq. 7.54). For the general case of all surfaces having varying temperatures, the relation for an element on surface j becomes

$$dq(r_j)_{dj-k} = \sigma \int_{A_k} [T_j^4(r_j) - T_k^4(r_k)] \, dF_{dj-dk} \, dA_j \qquad (7.61)$$

and, summing over all surfaces in the enclosure,

$$dq(r_j)_{dj} = \sigma \sum_{k=1}^{N} \int_{A_k} [T_j^4(r_j) - T_k^4(r_k)] \, dF_{dj-dk} \, dA_j \qquad (7.62)$$

Configuration Factors.   To solve for the radiative transfer among surfaces using the previous black-surface equations, expressions for the configuration factors must be available. Many factors for common geometries have been derived and presented in the literature. Compilations are given in Siegel and Howell [1] (42 factors and references to 234 factors available in the literature); Brewster [2] (13 factors); Modest [3] (51 factors); and Howell [20] (over 278 factors with an annotated bibliography). Some useful factors are illustrated in App. B.

Configuration Factor Algebra.   When the configuration factor F_{A-B} between two surfaces is known, the reciprocity relation (Eq. 7.59) can be used to find F_{B-A}. Other relations can also be developed that allow simple calculations of new factors from known factors.

If surface B can be subdivided into N nonoverlapping surfaces that completely cover surface B, then

$$F_{A-B} = F_{A-1} + F_{A-2} + F_{A-3} + \cdots + F_{A-N} = \sum_{n=1}^{N} F_{A-n} \qquad (7.63)$$

because all energy fractions from surface A to parts of surface B must equal the fraction of the total energy leaving A that is incident on all of B.

Suppose that surface 1 is completely enclosed by a set of M surfaces. In that case, all energy leaving surface A must strike some other surface forming the enclosure. In terms of configuration factors:

$$F_{1-1} + F_{1-2} + F_{1-3} + \cdots + F_{1-M} = \sum_{m=1}^{M} F_{1-m} = 1 \qquad (7.64)$$

Note the term F_{1-1} must be included in the summation if surface A is concave to account for the fraction of energy leaving surface A which is incident on itself.

The reciprocity relations plus Eqs. 7.63 and 7.64 form the basis of what is called *configuration factor algebra.* Using these relations, new factors can be computed from a set of known factors; sometimes, factors can be generated from the algebra alone. The procedure is best illustrated by example.

Consider two right isosceles triangles that are joined along a short side, as shown in Fig. 7.8. The triangles are perpendicular to one another. To find F_{1-2}, note that an enclosure can be formed by first joining the free corners of the triangles by a line of length l as shown in Fig. 7.9. This forms a corner cavity with the third congruent triangle. The enclosure is completed by placing an equilateral triangle of side l

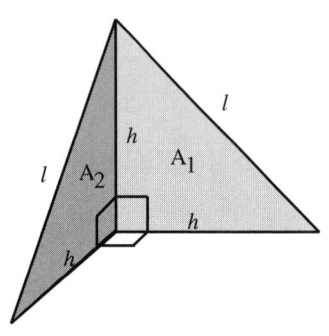

FIGURE 7.8 Perpendicular right isosceles triangles joined along their short sides.

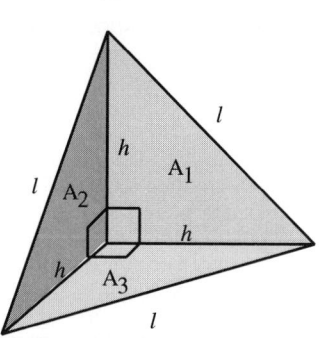

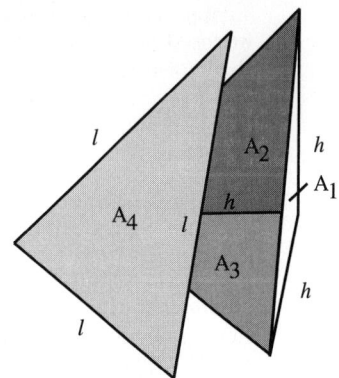

FIGURE 7.9   Construction of corner cavity by addition of line connecting free corners of triangle.

FIGURE 7.10   Completion of enclosure by addition of equilateral triangle, surface 4.

(and area A_4) over the cavity formed by the three isosceles triangles, which have equal areas A_1, A_2, and A_3. This is shown in Fig. 7.10.

Now, apply configuration factor algebra. Eq. 7.64 gives

$$F_{1-1} + F_{1-2} + F_{1-3} + F_{1-4} = 1 \qquad (7.65)$$

Because surface 1 is planar, $F_{1-1} = 0$. By symmetry, $F_{1-2} = F_{1-3}$. Thus, Eq. 7.65 reduces to

$$2F_{1-2} + F_{1-4} = 1 \qquad (7.66)$$

For surface 4 of the enclosure, Eq. 7.64 plus the use of symmetry gives

$$F_{4-1} + F_{4-2} + F_{4-3} + F_{4-4} = 3F_{4-1} = 1 \qquad (7.67)$$

Using reciprocity, Eq. 7.59 results in

$$F_{1-4} = \frac{A_4}{A_1} F_{4-1} = \frac{A_4}{3A_1} \qquad (7.68)$$

Substituting into Eq. 7.66 results in

$$F_{1-2} = \frac{1 - F_{1-4}}{2} = \frac{1 - (A_4/3A_1)}{2} \qquad (7.69)$$

Using geometry, $A_1 = h^2/2$ and $A_4 = \sqrt{3}l^2/4 = \sqrt{3}h^2/2$, giving

$$F_{1-2} = \frac{1 - (1/\sqrt{3})}{2} \approx 0.21132 \qquad (7.70)$$

This is the desired answer. The other configuration factors can be determined in a similar manner.

Siegel and Howell [1] note on p. 220 that, for an N-surfaced enclosure of all planar or convex surfaces (i.e., $F_{i-i} = 0$ for all i), $N(N-3)/2$ factors must be found from a catalog of factors or by calculation. The remaining factors can then be determined by configuration factor algebra. If M of the surfaces ($M < N$) are concave (i.e., have $F_{i-i} \geq 0$), then $[N(N-3)/2] + M$ factors must be known before configuration factor algebra can determine the remaining factors. The presence of symmetry may further reduce the number of factors that must be known before the rest can be determined.

When the values of certain factors are known only approximately, then the constraints imposed on the factors by reciprocity and conservation in an enclosure can be used to refine the known values. Methods for this purpose have been proposed in Refs. 21–24.

Exchange Among Gray Diffuse Surfaces

When an enclosure can be assumed to be made up of a set of black or gray diffuse surfaces, well-developed techniques are available for determining the exchange among the surfaces.

In general, one boundary condition (either surface temperature or heat flux) must be specified for every surface of the enclosure. It is possible by more advanced techniques to obtain solutions for the case when some surfaces have both conditions prescribed and others have neither condition specified. This situation causes the solution methods described here to fail, as the equation set is then ill-conditioned. So-called *inverse solution methods* must be invoked. Methods of handling such a problem are given for black-surfaced enclosures in Ref. 25 and for enclosures with black or gray diffuse surfaces in Refs. 26, 27, and 28.

Use of Configuration Factors.   For diffuse nonblack surfaces, the radiation leaving the surface is made up of both diffusely emitted and diffusely reflected energy. Thus, the directional distribution of all radiation is diffuse. It follows that we can continue to use the concept of configuration factors to determine the fraction of the leaving radiation that strikes another surface. Thus, all of the information generated about configuration factors for black surfaces also applies to the more general case of diffusely reflecting and emitting surfaces. If the surfaces are diffuse and have wavelength-dependent properties, the methods to be outlined here will apply for energy exchange at each wavelength; a final summation or integration over all wavelengths will then be necessary to compute total radiant exchange.

The outgoing radiation energy flux from a given location on surface k, $q_{o,k}$ is made up of the emitted and reflected flux from that surface, or

$$q_{o,k} = \varepsilon_k \sigma T_k^4 + \rho_k q_{i,k} \qquad (7.71)$$

where all quantities are evaluated at a particular location on surface k. The quantity $q_{i,k}$ is the radiation flux incident at the given location from all other surfaces in the enclosure, including surface k itself, if it is concave. The quantity $q_{o,k}$ is often called the *radiosity* of the surface, and $q_{i,k}$, the *irradiance*. Note that, contrary to the practice in most of heat transfer, these energy fluxes carry a *directionality*—the radiosity is the portion of the radiant energy flux with the component *away* from the surface, while the irradiance is the portion directed *toward* the surface.

The *net* radiative heat flux *leaving* surface $k (q_k)$ is the difference between the radiosity and the irradiance, or

$$q_k = q_{o,k} - q_{i,k} \qquad (7.72)$$

This flux corresponds to the usual concept used in heat transfer, as the net energy flux is taken as positive if in the direction parallel to the surface normal of the position on k.

The final equation for energy transfer quantifies the irradiance as the sum of the radiant energies reaching a location on surface k from all other areas on the enclosure surface. This relation can have various forms depending on the degree of approximation used in the analysis.

Circuit Analogy.   Simple problems in radiant transfer can be diagrammed and formulated in analogy with a simple electrical circuit [29]. This is done by observing that Eq. 7.57 for black surfaces is in the form of Ohm's Law if dq_{j-k} is analogous to the current and $(T_j^4 - T_k^4)$ is analogous to the driving potential difference. In this case, the corresponding resistance is $1/(\sigma F_{j-k} A_j)$. The analogy can also be extended to gray diffuse surfaces. In making this analogy, all of the usual assumptions are still present: all surfaces are gray and diffuse, and the radiosity leaving each surface is uniform over the surface (implying that the temperature, heat flux, and incident radiation are uniform over each surface). The circuit analogy is not a useful approach when more than about four surfaces are treated, and it will not be elaborated here. For more details, see Refs. 1 and 3.

Net Radiation Method.   A more powerful method for describing radiative transfer is the net radiation method. In this method, radiative energy balances are constructed for each surface, and the resulting set of equations is then solved. (Some equations as written may fail in the limit of black surfaces and must be slightly modified starting from the original relations.)

Uniform Surface Radiosity.   Here, first, we limit consideration to gray diffuse surfaces with uniform radiosity. In that case, Eqs. 7.71 and 7.72 apply. Because of the assumption of gray diffuse surfaces, Eq. 7.71 can be rewritten as

$$q_{o,k} = \varepsilon_k \sigma T_k^4 + (1 - \varepsilon_k) q_{i,k} \tag{7.73}$$

An additional equation for the radiative energy incident on surface k is

$$q_{i,k} A_k = \sum_{j=1}^{N} q_{o,j} A_j F_{j-k} = A_k \sum q_{o,j} F_{k-j} \tag{7.74}$$

The three equations, 7.72–7.74, can be written for each of the N surfaces in the enclosure. If either T_k or q_k is prescribed for each surface, this results in $3N$ equations in $3N$ unknowns: the unknowns being $q_{i,k}$, $q_{o,k}$, and either q_k or T_k values for each surface.

For m of the N surfaces in the enclosure having specified temperatures and the remaining $N - m$ surfaces having specified heat flux, $q_{i,k}$ can be eliminated to give the useful forms

$$\sum_{j=1}^{N} [\delta_{kj} - (1 - \varepsilon_k) F_{k-j}] q_{o,j} = \varepsilon_k \sigma T_k^4 \qquad 1 \le k \le m \tag{7.75}$$

and

$$\sum_{j=1}^{N} (\delta_{kj} - F_{k-j}) q_{o,j} = q_k \qquad m + 1 \le k \le N \tag{7.76}$$

where δ_{kj} is the Kronecker delta function, defined such that if $k = j$, $\delta_{kj} = 1$ and if $k \ne j$, $\delta_{kj} = 0$. This results in a set of N equations in N unknowns that can be solved directly for the $q_{o,k}$ of each surface. Once $q_{o,k}$ is known for each surface, the unknown q_k or T_k for each surface is found from the relation

$$q_k = \frac{\varepsilon_k}{1 - \varepsilon_k} (\sigma T_k^4 - q_{o,k}) \tag{7.77}$$

This approach is useful when values of the radiosity $q_{o,k}$ are needed, for example, to predict or interpret the readings observed by radiometric devices such as pyrometers.

If values of the radiosity are not needed, an alternative method is to eliminate both $q_{o,k}$ and $q_{i,k}$ from Eqs. 7.75, 7.76, and 7.77. The result can be put in the form

$$\sum_{j=1}^{N} \left(\frac{\delta_{kj}}{\varepsilon_j} - F_{k-j} \frac{1 - \varepsilon_j}{\varepsilon_j} \right) q_j = \sum_{j=1}^{N} (\delta_{kj} - F_{k-j}) \sigma T_j^4 \tag{7.78}$$

Surfaces with Nonuniform Radiosity.   If the radiosity across a given surface does not meet the assumption of uniformity, then the surface may be subdivided into subsurfaces, each of which approximates the condition of uniformity. In the limit, this reduces to relations in the form of integral equations. In this case, the net radiation method can be extended. Note that Eqs. 7.72 and 7.73 still apply to every position on surface k, but Eq. 7.74 must be modified to remove the assumption of uniform radiosity. The third equation for the net radiation method is the relation for incident radiation onto a particular location on surface k from all other surfaces, each of which can have a variable radiosity. The resulting relations are

$$q_{o,k}(\bar{r}_k) = \varepsilon_k(\bar{r}_k)\sigma T_k^4(\bar{r}_k) + [1 - \varepsilon_k(\bar{r}_k)]q_{i,k}(\bar{r}_k) \tag{7.79}$$

$$q_k(\bar{r}_k) = q_{o,k}(\bar{r}_k) - q_{i,k}(\bar{r}_k) \tag{7.80}$$

$$q_{i,k}(\bar{r}_k) = \sum_{j=1}^{N} \int_{A_j} q_{o,j}(\bar{r}_j)\, dF_{dk-dj}(\bar{r}_k, \bar{r}_j) \tag{7.81}$$

These can be combined to eliminate $q_{i,k}$ as for the uniform radiosity case to provide equations for the radiosity of the surfaces 1 to m (with known temperature distributions) and surfaces $m + 1$ to N (with known heat flux distributions) as

$$q_{o,k}(\bar{r}_k) - [1 - \varepsilon_k(\bar{r}_k)] \sum_{j=1}^{N} \int_{A_j} q_{o,j}(\bar{r}_j)\, dF_{dk-dj}(\bar{r}_k, \bar{r}_j) = \varepsilon_k(\bar{r}_k)\sigma T_k^4(\bar{r}_k) \qquad 1 \le k \le m \tag{7.82}$$

$$q_{o,k}(\bar{r}_k) - \sum_{j=1}^{N} \int_{A_j} q_{o,j}(\bar{r}_j)\, dF_{dk-dj}(\bar{r}_k, \bar{r}_j) = q_k(\bar{r}_k) \qquad m + 1 \le k \le N \tag{7.83}$$

The relation between radiosity and the unknown local net radiative flux or temperature is found from Eqs. 7.84 and 7.85:

$$q_k(\bar{r}_k) = \frac{\varepsilon_k(\bar{r}_k)}{1 - \varepsilon_k(\bar{r}_k)} [\sigma T_k^4(\bar{r}_k) - q_{o,k}(\bar{r}_k)] \qquad 1 \le k \le m \tag{7.84}$$

$$\sigma T_k^4(\bar{r}_k) = \frac{1 - \varepsilon_k(\bar{r}_k)}{\varepsilon_k(\bar{r}_k)} q_k(\bar{r}_k) + q_{o,k}(\bar{r}_k) \qquad m + 1 \le k \le N \tag{7.85}$$

An alternative form is found by eliminating both $q_{i,k}$ and $q_{o,k}$ from Eqs. 7.79 through 7.81 to obtain

$$\frac{q_k(\bar{r}_k)}{\varepsilon_k} - \sum_{j=1}^{N} \frac{1 - \varepsilon_j(\bar{r}_j)}{\varepsilon_j(\bar{r}_j)} \int_{A_j} q_j(\bar{r}_j)\, dF_{dk-dj}(\bar{r}_k, \bar{r}_j) = \sigma T_k^4(\bar{r}_k) - \sum_{j=1}^{N} \int_{A_j} \sigma T_j^4(\bar{r}_j)\, dF_{dk-dj}(\bar{r}_k, \bar{r}_j) \tag{7.86}$$

Solution Techniques

Surfaces with Uniform Radiosity.   For the case of gray diffuse surfaces with uniform radiosities, the solution of either Eq. 7.82 or 7.83 for $q_{o,k}$ or Eq. 7.86 for the unknown temperatures and heat fluxes requires the solution of a set of simultaneous equations. These are of the form

$$[a_{kj}][x_j] = [C_k] \tag{7.87}$$

For determining $x_j = q_{o,j}$ using Eqs. 7.82 and 7.83, the matrix coefficients are given by

$$\begin{aligned} a_{kj} &= \delta_{kj} - (1 - \varepsilon_k)F_{k-j} & 1 \le k \le m \\ a_{kj} &= \delta_{kj} - F_{k-j} & m + 1 \le k \le N \end{aligned} \tag{7.88}$$

and the right-hand-side vector elements are given by

$$\begin{aligned} C_k &= \varepsilon_k \sigma T_k^4 & 1 \le k \le m \\ C_k &= q_k & m + 1 \le k \le N \end{aligned} \tag{7.89}$$

For finding the unknown q_k or T_k using Eq. 7.86, the values of the unknowns can be rearranged so that the unknown m values of q_k are given by $x_j = q_j$ for $1 \le k \le m$ and $x_j = T_j$ for $m + 1 \le k \le N$. The matrix coefficients in this case are given for row k by

$$a_{kj} = \frac{\delta_{jk}}{\varepsilon_j} - F_{k-j}\frac{1 - \varepsilon_j}{\varepsilon_j} \qquad 1 \le k \le m$$

$$a_{kj} = \sigma(\delta_{jk} - F_{k-j}) \qquad m + 1 \le k \le N \tag{7.90}$$

and the elements for the $[C_k]$ vector are given for all k by

$$C_k = \sum_{j=1}^{m} (\delta_{kj} - F_{k-j})\sigma T_j^4 \qquad\qquad 1 \le j \le m \qquad\qquad (7.91)$$

$$C_k = \sum_{j=m+1}^{N} \left(\frac{\delta_{kj}}{\varepsilon_j} - F_{k-j}\frac{1-\varepsilon_j}{\varepsilon_j}\right) q_j \qquad m+1 \le j \le N \qquad (7.92)$$

Equation 7.87 for either solution method is now solved using standard matrix inversion routines to find the inverse of the matrix $[a_{kj}]$, denoted by $[a_{kj}]^{-1}$, and then using standard matrix multiplication routines to solve for the unknown elements of the $[x_j]$ vector by

$$[x_j] = [a_{kj}]^{-1}[a_{kj}][x_j] = [a_{kj}]^{-1}[C_k] \qquad\qquad (7.93)$$

Solutions for nonuniform radiosity relations: analytical solution of radiative transfer is possible for only the simplest cases. In general, a numerical solution is necessary.

RADIATIVE EXCHANGE WITH A PARTICIPATING MEDIUM

Fundamentals and Definitions

Absorption, Extinction, and Scattering Coefficients. If a beam of radiation of intensity $I_\lambda(\vec{r}, \hat{\Omega})$ propagates within a participating media along the direction $\hat{\Omega}$, it will lose its energy due to absorption and scattering. In mathematical terms, the changes in radiation intensity due to the absorption and scattering of the radiation beam within a small length element ds along the direction $\hat{\Omega}$ are proportional to the local intensity, and are expressed as

$$\text{absorption:} \quad -\kappa_\lambda I_\lambda(\hat{\Omega})ds \qquad\qquad (7.94)$$

and

$$\text{scattering:} \quad -\sigma_\lambda I_\lambda(\hat{\Omega})ds \qquad\qquad (7.95)$$

Here, κ_λ and σ_λ are the spectral absorption and scattering coefficients, respectively. The attenuation of radiation intensity by the medium is proportional to both of these coefficients. The spectral attenuation (or extinction) coefficient is expressed by the sum of the absorption and scattering coefficients, as

$$\text{extinction coefficient:} \quad \beta_\lambda = \kappa_\lambda + \sigma_\lambda \qquad\qquad (7.96)$$

The single scattering albedo is defined as the fraction of energy attenuated due to scattering:

$$\text{single scattering albedo:} \quad \omega_\lambda = \sigma_\lambda/\beta_\lambda \qquad\qquad (7.97)$$

Scattering Phase Function. Energy absorbed is converted primarily to thermal energy, whereas scattered radiant energy is redistributed in the medium. The directional distribution of scattered energy is expressed by the phase function, $\Phi_\lambda(\hat{\Omega}', \hat{\Omega})$, which represents the fraction of energy incident in direction $\hat{\Omega}'$ and scattered into direction $\hat{\Omega}$. The angle between the directions of the incident beam ($\hat{\Omega}'$) and the scattered beam ($\hat{\Omega}$) is called as the scattering angle Θ. It is related to the azimuthal and zenith angles via the following equation:

$$\cos\Theta = \cos\theta\cos\theta' + \sin\theta\sin\theta'\cos(\phi - \phi') \qquad\qquad (7.98)$$

Figure 7.11 shows the relation between the incident and scattered intensities schematically.

The phase function is normalized so that the sum of scattered light in all directions adds up to 100 percent:

$$\frac{1}{4\pi}\int_{\Omega=4\pi}\Phi_\lambda(\hat{\Omega}', \hat{\Omega})\,d\Omega' = \frac{1}{4\pi}\int_{\phi=0}^{2\pi}\int_{\theta=0}^{\pi}\Phi_\lambda(\theta', \phi'; \theta, \phi)\sin\theta\,d\theta\,d\phi = 1 \qquad (7.99)$$

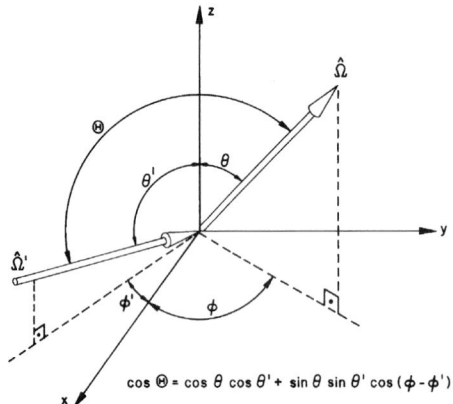

$$\cos \Theta = \cos \theta \cos \theta' + \sin \theta \sin \theta' \cos (\phi - \phi')$$

FIGURE 7.11 Relation between the incident and scattered intensities and the definition of scattering angle.

Examples of phase functions for spherical particles of different sizes are depicted in Fig. 7.12 as computed using electromagnetic wave theory (so-called Mie scattering). In Fig. 7.12(*a*), the real part of the index of refraction is 1.8, and the imaginary part is 0.8. The size parameter $x = \pi D/\lambda$, which is the scaling parameter relating the particle diameter D and the radiation wavelength λ, is changing from 0.01 to 10. It is obvious that, with increasing x, the particles become more forward-scattering. In Fig. 7.12(*b*), the size parameter x is kept constant, and the index of refraction is varied. For this particular size parameter, the effects of n and k on the phase function are not large. For nonhomogeneous spheres, especially if there is a size distribution, the small variation on the complex index of refraction does not affect the phase function significantly. However, for particle characterization studies based on laser/light diagnostic techniques, more accurate information about the spectral complex index of refraction data is needed.

For numerical computations, an explicit form of the phase function is required. A Legendre polynomial expansion of the phase function is the most common form. For an azimuthally symmetric medium (i.e., one independent of azimuthal angle ϕ), the phase function is written as:

$$\Phi_\lambda(\mu', \mu) = \sum_{i=0}^{N} a_i P_i(\mu') P_i(\mu) \tag{7.100}$$

where $a_0 = 1$, P_i are the ith order Legendre polynomials, and $\mu = \cos \theta$ and $\mu' = \cos \theta'$ are the direction cosines for the directions of the scattered and the incident beams, respectively.

Isotropic scattering indicates that the radiant energy incident on a volume element is uniformly distributed to all directions. For an isotropically scattering medium, all a_i coefficients of the phase function are zero, except a_0. If only a_0 and the first coefficient a_1 are considered, then one obtains the linearly anisotropic phase function, which means that the phase function is a linear function of $\cos \Theta$ (or, in the case of an azimuthally symmetric medium, a linear function of $\mu = \cos \theta$).

If the size parameter is small ($x \leq 0.05$), then the phase function can be approximated by the Rayleigh phase function, which reads as

$$\Phi_\lambda = \frac{3}{4} (1 + \cos \Theta) \tag{7.101}$$

On the other hand, for particles with diameters much larger than the wavelength of radiation (that is, for large-size parameters), several terms of the series expansion need to be considered to account for the strong forward-scattering peak. This yields a significant increase in computational effort. To avoid this, it is possible to use a δ-Eddington phase function [30, 31, 32], which is given as:

$$\Phi_\lambda(\theta', \phi'; \theta, \phi) = 2f\delta(1 - \cos \Theta) + (1 - f)(1 + 3g \cos \Theta) \tag{7.102}$$

where δ represents the Dirac delta function, Θ is the scattering angle, and the f and g parameters are related to the coefficients of the Legendre polynomial expansion of the phase function as

$$g = (a_1 - a_2)/(1 - a_2), \quad f = a_2 \tag{7.103}$$

This approximation divides the directional distribution of scattered energy into two components: one highly peaked component in the forward direction, and an isotropic distribution for all other directions. Improvement of this phase function is possible if the second term on the right-hand side of Eq. 7.102 is comprised of more than a single term. Crosbie and David-

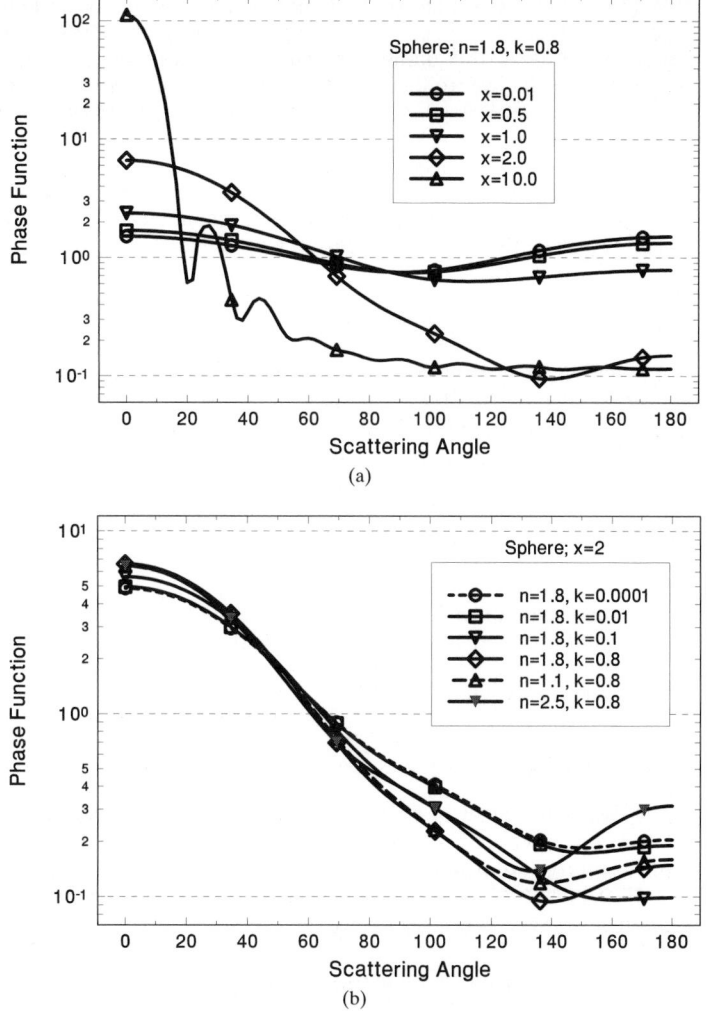

FIGURE 7.12 Scattering phase functions: (*a*) For different size parameters $x = \pi D/\lambda$; (*b*) for a fixed size parameter of $x = 2$ and different complex index of refraction values.

son [31] showed that, if a full phase function is expressed as a sum of a Dirac-delta term and series expansion, the number of terms required in the series of the new expression is usually an order of magnitude smaller than that required for the original phase function.

Lambert-Beer Law. The radiation intensity of a beam propagating through an absorbing slab is reduced exponentially, as a function of slab thickness y and slab absorption coefficient κ_λ:

$$I_\lambda(y, \hat{\Omega}) = I_{0,\lambda}(y, \hat{\Omega}) \exp(-\kappa_\lambda y) = I_{0,\lambda}(y, \hat{\Omega}) \exp(-\tau_\lambda) \qquad (7.104)$$

where $I_{0,\lambda}(y, \hat{\Omega})$ is the intensity of the incident beam at the slab boundary, and $\tau_\lambda = \kappa_\lambda y$ is the optical thickness. According to the Lambert law, the intensity exiting the slab would be the same if the absorption coefficient is doubled and the slab thickness is reduced by one half.

The Beer law is similar to the Lambert law with the exception that the absorption coefficient is expressed as the product of a unit absorption coefficient and the concentration of particles. In general, this definition is more fundamental and appropriate for application to dispersed media, where the concentration can be directly measured. If there are particles in the medium, depending on their size with respect to the wavelength of the incident radiation, they scatter as well as absorb the incident radiation. With increasing concentration, the multiple scattering effect becomes significant, and the Beer law deviates from the experimental measurements, especially if the size of the particles is comparable to the wavelength of radiation. Under these conditions, the complete radiative transfer equation should be solved. For solid materials, the Beer and Lambert laws are identical.

Radiative Transfer Equation.   Although the theoretical basis for radiative transfer is well established for most problems of engineering interest, there are problems that cannot be accurately handled. The radiative transfer equation (RTE) describes the spatial dependence of radiation on absorption, scattering, and emission by a surrounding medium. This equation is the foundation for almost all radiative transfer analysis, and all of the standard techniques assume that the equation is valid. However, some common engineering systems do not conform to the assumptions buried in this equation.

The RTE is a simplified form of the complete Maxwell equations describing the propagation of an electromagnetic wave in an attenuating medium. The simplified RTE does not include the effects of polarization of the radiation or the influence of nearby particles on the radiation scattered or absorbed by other particles (dependent scattering or absorption). For example, if polarization effects are present (as they are when reflections occur at off-normal incidence from polished surfaces or in reflections from embedded interfaces), then the analyst should revert to complete solution of the Maxwell equations, which is a formidable task in complex geometries! Delineating the bounds of applicability of the radiative transfer equation is an area of active research.

The RTE is based on the conservation of radiant energy* along a direction $\hat{\Omega}$ in a small absorbing, emitting, and scattering volume element dV. Intuitively, it is written by following the graphical representation shown in Fig. 7.13 as:

(Change in energy of a pencil of radiation along a direction $\hat{\Omega}$ and pathlength of ds within a volume element dV) = +① (Contribution of emitted radiant energy by dV along $\hat{\Omega}$) − ② (Energy absorbed by dV along $\hat{\Omega}$) + ③ (Contribution of energy emitted/scattered from all volume/surface elements in the medium which is incident on dV and scattered in the direction $\hat{\Omega}$) − ④ (Energy scattered out of the direction $\hat{\Omega}$ by dV)

---

* Note that *conservation of radiant energy* should not be confused with *conservation of energy;* any *loss* in radiant energy within a control volume via the absorption mechanism is a *gain* for the overall energy balance and vice versa.

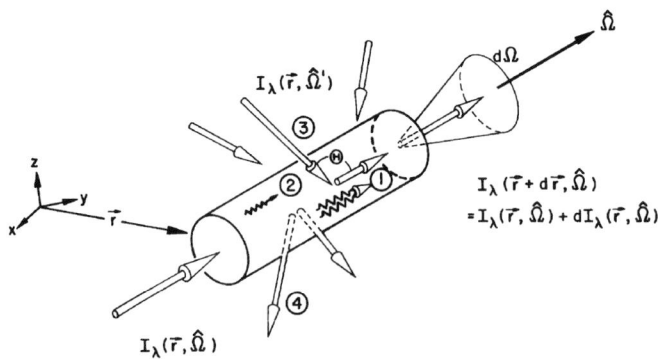

FIGURE 7.13   Conservation of radiant energy principle.

where circled numbers correspond to the physical phenomena depicted in the figure. After some manipulation, the radiative energy balance is expressed in terms of intensity as:

$$\frac{1}{c}\frac{\partial I_\lambda}{\partial t} + \vec{\nabla}\cdot\hat{\Omega}I_\lambda + \beta_\lambda I_\lambda = \kappa_\lambda I_{b\lambda}(T) + \frac{\sigma_\lambda}{4\pi}\int_{\Omega'=4\pi}\Phi_\lambda(\hat{\Omega}',\hat{\Omega})I_\lambda(\hat{\Omega}')\,d\Omega' \qquad (7.105)$$

where c is the speed of light. For most practical applications, the transient behavior of the radiation intensity is negligibly small and the first term is omitted. The $\vec{\nabla}\cdot\hat{\Omega}$ operator is to denote the incremental change in intensity along the direction $\hat{\Omega}$ and is expressed as a function of the coordinate system chosen for a specific geometry. For one-dimensional media, $\vec{\nabla}\cdot\hat{\Omega} = d/ds$, where s is the coordinate along the propagation direction of the beam. Specific expressions of $\vec{\nabla}\cdot\hat{\Omega}$ for other geometries are available in the literature [33, 34]. Note that the radiation originating from every volume/surface element in the medium directed to the control volume may be rescattered in the direction of interest; the integral term at the right-hand side of Eq. 7.105 is to account for this contribution to the radiation intensity in the direction $\hat{\Omega}$.

If the medium is neither emitting (cold) nor scattering, the RTE, as given by Eq. 7.105, is simplified to:

$$\vec{\nabla}\cdot\hat{\Omega}I_\lambda = -\kappa_\lambda I_\lambda \qquad (7.106)$$

Integration of this equation along the line of sight gives the Lambert law, Eq. 7.104.

Boundary Conditions for the RTE. The solution of the radiative transfer equation in a given geometry is subject to boundary conditions, which give the radiation intensity distribution on the boundaries. The boundary intensity is comprised of two components: (1) contribution due to emission at the boundary surfaces and (2) contribution due to diffuse and specular reflection of radiation intensity incident on the boundaries. The radiation incident on the boundary is due to intensity emitted from all volume and surface elements in the medium. In mathematical terms, the general boundary condition on any surface element is written as [1, 6]:

$$I_\lambda^+(y,\hat{\Omega}) = \varepsilon_\lambda I_{b\lambda}(T) + \frac{\rho_\lambda^d}{\pi}\int_{\Omega'=2\pi}I_\lambda^-(y,\hat{\Omega}')\hat{\Omega}\cdot\hat{n}\,d\Omega' + \rho_\lambda^s I_\lambda^-(y,-\hat{\Omega}) \qquad (7.107)$$

where ε_λ, ρ_λ^d and ρ_λ^s are the diffuse emissivity and the diffuse and specular reflectivity of the surface, respectively. T is the surface temperature, and $-\hat{\Omega}$ is to denote the direction of specular (mirrorlike) reflection. In most radiative transfer applications, boundaries are assumed to emit and reflect diffusely only.

Conservation of Radiant Energy Equation. The radiative transfer equation represents the change in radiation intensity for a single beam in the direction $\hat{\Omega}$ along an infinitesimal path length ds within a small control volume dV in the medium. As discussed before, the net change in radiation intensity along a given direction is the difference between emission and in-scattering gains and absorption and out-scattering losses. Absorption loss and emission gain directly affect the temperature of volume element dV. The scattered radiant energy, however, has no direct impact on the energy balance of dV, although the scattered photons may be absorbed by the medium later and ultimately influence the energy balance. The control volume temperature is affected by the absorption/emission of radiation incident on dV from all possible directions. The net contribution of radiative transfer to the energy balance of dV can be determined by integrating the RTE over the entire solid angle range of $\Omega = 4\pi$, for all possible wavelengths, which yields

$$\vec{\nabla}\cdot\vec{q} = \int_0^\infty \kappa_\lambda[4\pi I_{b\lambda}(T) - G_\lambda]\,d\lambda \qquad (7.108)$$

where the incident radiation G_λ is defined as

$$G_\lambda = \int_{\Omega = 4\pi} I_\lambda \, d\Omega \tag{7.109}$$

Equation 7.108 describes the conservation of radiative transfer in a unit volume (includes the radiative energy incident from *all* directions), and $\vec{\nabla} \cdot \vec{q}$ is the radiative flux divergence. The scattering coefficient does not appear in Eq. 7.108, because scattering does not directly contribute to the local energy balance in the medium. However, scattering of thermal radiation has an indirect effect via the G_λ term. Also note that the sign of $\vec{\nabla} \cdot \vec{q}$ is a function of the difference between total emission $\int_0^\infty 4\pi\kappa_\lambda I_{\lambda b} \, d\lambda$ and total absorption of incident radiation $\int_0^\infty \kappa_\lambda G_\lambda \, d\lambda$. The difference in the spectral emission and absorption, however, may be positive in certain bands of the wavelength spectrum and negative in others, and the distribution of spectral absorption coefficient and radiation intensity in the medium must be known to calculate $\vec{\nabla} \cdot \vec{q}$ accurately.

If $\vec{\nabla} \cdot \vec{q}$ is positive, then the control volume dV is emitting more energy than it is absorbing; therefore it is radiatively cooling. If $\vec{\nabla} \cdot \vec{q}$ is negative, then the medium is heated due to the absorption of radiation incident on it. In other words, the $\vec{\nabla} \cdot \vec{q}$ term determines the effect of radiation on the total energy balance of dV, and the negative of it appears as a radiation source term in the conservation of total energy equation. The intimate coupling between the radiative transfer equation and the total energy equation is obvious.

Solution Techniques for the RTE

The radiative transfer equation (RTE) is an integro-differential equation; it is difficult to develop a closed-form solution to it in general multidimensional and nonhomogeneous media. After introducing a number of approximations, however, reasonably accurate models of the RTE can be obtained. In all models, the objective is to solve the RTE, or a modified form of it, in terms of radiation intensity or its moments (such as flux) and then calculate the distribution of the divergence of radiative flux $\vec{\nabla} \cdot \vec{q}$ everywhere in the medium. In this section, we will discuss the approximate models of the RTE which can be extended to multidimensional geometries.

The dependent variable in the RTE is the radiation intensity, which is, as stated earlier, a function of seven independent variables: three spatial coordinates (x, y, and z, in Cartesian coordinates, or r, z, and θ_r, in cylindrical coordinates), two angular coordinates (θ, ϕ), wavelength (λ) or frequency (ν) of radiation, and time (t). Although the intensity is also a function of local temperature, local absorption and scattering coefficients, and scattering phase function, they are not written explicitly, but implied through the $\vec{r}$-dependence. For most practical problems, a transient analysis of radiative transfer is not required. (Transient dependence enters through the coupling with the energy equation.) The wavelength dependency of radiation intensity, on the other hand, is extremely important and needs special consideration. Given the fact that the spectral variation of intensity arises primarily due to wavelength-dependent radiative properties, we can delay the discussion of this effect until the next section. It is sufficient to say that, if a radiative transfer model is developed for monochromatic (single-wavelength) radiation, then the same model can be used in sequence at all wavelengths of significance, and then the total effect is determined.

In most radiative transfer models, the objective is to reduce the number of geometric independent parameters to a minimum. The simplest model can be obtained if the spatial and angular dependency of radiation intensity can be neglected. In this case, it is assumed that the gas inside an enclosure (for example, a combustion system) has uniform distribution and constant temperature. Then the radiative heat transfer from a source (e.g., flame) to sink (e.g., refractories and combustion gases) can be calculated from a single expression called the total radiation exchange area [35]. Although this expression is a complicated function of geometry and gas and surface radiative properties, it needs to be calculated only once. This approach is known as a stirred-vessel model, and its usefulness lies in its simplicity and capability of estimating global furnace performance [35]. However, the spatial distribution of radiative flux cannot be predicted using this approach, and, because of this, it should be used with caution.

More accurate radiative transfer models can be obtained by separating the angular variation of radiation intensity from its spatial variation. Next, the angular distribution is simplified further by considering only a finite number of directions found either deterministically or statistically, and, in each of these angular subdivisions, the intensity is assumed to be invariant with angle. In theory, if the number of subdivisions increases, the accuracy of the model also increases; however, both the numerical difficulty and computational cost of the solution escalate. Brief discussions of several different radiative transfer models developed over the years using this type of independent variable separation technique are given in the following subsections.

Zonal Method.   The zonal method is one of the most widely employed models for calculating the radiative transfer in combustion chambers [35]. The basic idea behind this approach is similar to that of radiation exchange between finite size surfaces of an enclosure containing a nonparticipating medium (see the previous section). In its most simple version, the walls of an enclosure are divided into many uniform-property, uniform-temperature surface elements. This division determines the spatial distribution of the radiation field (intensity). The net radiation exchange between these elements is determined using a radiosity-irradiation approach along with appropriate radiation configuration factors where radiosities at all surface elements are determined simultaneously. The radiation configuration factor between arbitrary surfaces i and j is determined by considering the fraction of radiative energy leaving surface i that arrives at surface j and is a purely geometric parameter (see Fig. 7.7) [1]. Each surface element sees others within a hemispherical solid angle. The relative orientation of all i surfaces with respect to the j surface divides the hemispherical solid angle to many small discrete solid angles. This discretization implicitly represents the angular subdivision of the radiation field for the j surface within the medium.

For an absorbing, but nonemitting (cold) medium, the amount of radiation energy emanating from surface i and reaching surface j is reduced because of the attenuation of the beam by the medium along the line of sight joining the two surfaces. This energy loss can be accounted for in the model easily. Here, the entire volume is divided into many elements, each presumed to have uniform and constant properties. Then, the interaction of each volume and surface element is written explicitly and cast into a matrix form, and the required radiosities are determined computationally.

For an emitting medium, however, a more fundamental modification of the method is needed. In this case, the emission contribution of each small-volume element to the radiative flux distribution on every surface and volume element is to be determined.

For nonparticipating media, either Poljak's (see Chap. 8 of Ref. 1) or Gebhart's (see App. D of Ref. 1) formulation can be followed to develop the required equations to determine the radiative flux distribution on the walls. If the medium is absorbing and emitting, then Hottel's zonal method can be followed (Chap. 11 of Ref. 35). Originally, the zonal method was developed for nonscattering media. Later, Noble [36] extended the methodology to include scattering by the medium. Smith and his coworkers presented a numerical study in which they showed the feasibility of including isotropic scattering by the volume elements [37]. Byun and Smith extended this formulation to include anisotropically scattering media, although their formulation was only for a one-dimensional geometry [38].

An alternative zonal method, which includes the isotropic scattering effects, was given by Larsen and Howell [22, 39]. Another variation of the zonal method was proposed by Naraghi and Chung [40], who developed an explicit matrix formulation. Later Naraghi et al. [41] presented a continuous exchange method for the solution of the RTE. The exchange factor definitions given in these works are different from Hottel's or Larsen's formulations. A modified form of this method was also suggested by Naraghi et al. [42, 43]. They called it the *discrete exchange method,* where the integral equations of the continuous exchange method are discretized using a numerical integration scheme. The method yields accurate results in both two- and three-dimensional enclosures, especially if a Gaussian quadrature integration is performed.

Differential and Moment Methods.   The radiative transfer equation can be approximated using the moments of radiation intensity instead of intensity itself. A moment is defined as the integral of intensity multiplied by a power of a direction cosine over a predetermined solid angle division. For example,

$$I_n(\vec{r}) = \int_{4\pi} I(\vec{r}, \hat{\Omega}) \mu^n \, d\Omega \qquad (7.110)$$

is the nth moment of intensity in the z-direction, and $\mu = \cos\theta$ is the corresponding direction cosine. The first moment of intensity ($n = 1$) is the radiative flux in x, y, and z directions for Cartesian direction cosines, which are defined, respectively, as:

$$\xi = \sin\theta\cos\phi \qquad \eta = \sin\theta\sin\phi \qquad \mu = \cos\theta \qquad (7.111)$$

Here, θ and ϕ are the spherical polar components, used to define the angular dependency of direction of propagation of radiation intensity.

A classical moment formulation of radiative transfer is obtained by multiplying the entire RTE by a single direction cosine or a multiplicative combination of different direction cosines, and then by integrating over a solid angle range. This procedure eliminates the integral in the RTE (the last term on the right-hand side of Eq. 7.105), and yields a series of equations in terms of different orders of moments. Using a closure condition, which is an approximation between the highest moment used and the lower-order ones, the number of unknowns is equated to that of available equations. As one may guess, if the order of the highest moment being considered is large, then this approximation for the closure condition yields a smaller error, and the predictions of the approximate radiation model approaches those obtained from exact models.

In order to take advantage of moment methods, the radiation intensity is expressed as a series of products of angular and spatial functions. If the angular dependence is expressed using a simple power series, the *moment method* (*MM*) is obtained; if spherical harmonics are employed to express the intensity, then the method is called the *spherical harmonics* (*SH*) *approximation.* In principle, the first-order moment and the first-order spherical harmonics approximations are identical to each other, as well as to the lowest order *discrete ordinates (DO) approximation* [34].

The spherical harmonics approximation to the RTE in multidimensional geometries has been formulated and solved by several different researchers. Liou and Ou [44] gave the formulation for the first-order P_1 approximation for modeling cuboidal clouds; Bayazitoglu and Higenyi [45] presented analyses for both P_1 and P_3 approximations for a nonscattering medium in Cartesian, cylindrical, and spherical geometries; Higenyi and Bayazitoglu [46] and Ahluwalia and Im [47] included isotropic scattering in the formulation for a cylindrical medium; Ratzel and Howell [48] developed the solution for the P_1 and P_3 approximations in two-dimensional rectangular enclosures, and they accounted for isotropic scattering. Mengüç and Viskanta [49, 50] developed the general equations for the solution of both P_1 and P_3 approximations in absorbing, emitting, and anisotropically scattering cylindrical and three-dimensional rectangular enclosures.

The first-order spherical harmonics (P_1) approximation is one of the most simple RTE models, as it can be cast into a single second-order differential equation. In general, it does not yield very accurate results, and the error in radiative flux predictions can be as large as 50 percent for low optical thicknesses. It can, however, be modeled with little effort, and, therefore, its use is strongly recommended if the alternative is not to account for the radiation effects in a comprehensive heat transfer model or to use a simple zero-dimensional stirred-vessel method.

The P_3-approximation is more difficult to solve in multidimensional geometries. It yields improved predictions over those obtained from the P_1-approximation, and its accuracy is comparable to the similar-order discrete ordinates approximation (S_4), which is computationally more efficient. For one-dimensional systems, it is possible to develop higher-order P_N approximations [51, 52]; however, for multidimensional geometries, even the P_5-approximation is extremely complicated.

Instead of using higher-order spherical harmonics approximations, reasonably accurate RTE models can be developed by modifying the relatively simple P_1 approximation. Modest [53] presented such a modified differential approximation based on the formulation given by Olfe [54]. He divided the intensity into two components to separate the medium contribution from the wall contribution and then assumed that the medium contribution can be accounted for, roughly, using the P_1 approximation. The results show significant improvement over the classical P_1 approximation and have a possible use for more complicated geometries. However, if the phase function is highly forward- or backward-scattering, or if there are large temperature gradients in the medium, this method may not yield accurate solutions.

Another improvement of the P_1 approximation is obtained by defining moments not in the entire solid angle range but within predetermined solid angle divisions [55, 56]. In this approach, the number of resulting moment equations is proportional to the number of solid angle divisions used. For planar systems, the 4π solid angle domain is divided into two hemispheres, and the double P_1 (DP_1) approximation is obtained. For a two-dimensional rectangular geometry, a similar derivation yields a quadruple P_1 (QP_1) approximation. Both approximations yield better predictions than the P_3 approximation.

DP_1 Model for Planar Geometries.   Here, the formulation of a relatively simple model for plane parallel media is presented. This hybrid model was introduced and evaluated by Mengüç and Iyer [55]. It is a much better alternative to regular flux approximations or moment methods, and it is simpler to use than more accurate F_N method or higher-order DO and SH approximations. It can be considered as a hybrid between the first-order SH (P_1) method and a two-flux approximation.

Let us assume that, in a plane-parallel and azimuthally symmetric medium, the intensities in positive and negative z-directions are given as (see Fig. 7.14):

$$M^+(\tau, \mu) = A_0 + A_1\mu \qquad (7.112)$$

$$M^-(\tau, \mu) = B_0 + B_1\mu \qquad (7.113)$$

The zeroth and the first-order moments of intensity can be obtained by multiplying the M^+ and M^- by 1 and μ and integrating over the corresponding hemispheres. For M^+, this yields:

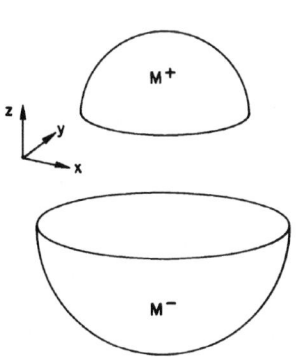

FIGURE 7.14   Representation of angular division of intensity for DP_1 approximation [55].

$$M_0^+ = \int_{\phi=0}^{2\pi} \int_{\mu=0}^{1} M^+ \, d\Omega = 2\pi A_0 + \pi A_1 \qquad (7.114)$$

$$M_1^+ = \int_{\phi=0}^{2\pi} \int_{\mu=0}^{1} \mu M^+ \, d\Omega = \pi A_0 + (2\pi/3)A_1 \qquad (7.115)$$

Equations 7.114 and 7.115 can be solved for the A_0 and A_1 coefficients and substituted back into Eq. 7.112 to obtain

$$M^+(\tau, \mu) = \frac{1}{\pi} [(2M_0^+ - 3M_1^+) + (-3M_0^+ + 6M_1^+)\mu] \qquad (7.116)$$

The second order moment of intensity is obtained by multiplying Eq. 7.116 by μ^2 and integrating over the upper hemisphere:

$$M_{11}^+ = \int_{\phi=0}^{2\pi} \int_{\mu=0}^{1} \mu^2 M^+ \, d\Omega = -(1/6)M_0^+ + M_1^+ \tag{7.117}$$

Following the same procedure, the intensity and second-order moment for the lower hemisphere are found as:

$$M^-(\tau, \mu) = \frac{1}{\pi} \left[(2M_0^- + 3M_1^-) + (3M_0^- + 6M_1^-)\mu \right] \tag{7.118}$$

$$M_{11}^- = \int_{\phi=0}^{2\pi} \int_{\mu=-1}^{0} \mu^2 M^- \, d\Omega = -(1/6)M_0^- - M_1^- \tag{7.119}$$

If the phase function is approximated as a linearly anisotropic one (there is no justification for the use of a more elaborate one since the method itself will not pick the resolution), such as:

$$\phi(\mu, \mu') = 1 + a_1 \mu \mu' \tag{7.120}$$

then, the RTE for each domain becomes

$$\frac{\mu}{\tau_0} \frac{\partial M^+}{\partial \tau} + M^+ = (1-\omega)I_b(T) + \frac{\omega}{4\pi} \left[(M_0^+ + M_0^-) + a_1 \mu (M_1^+ + M_1^-) \right] \tag{7.121}$$

$$\frac{\mu}{\tau_0} \frac{\partial M^-}{\partial \tau} + M^- = (1-\omega)I_b(T) + \frac{\omega}{4\pi} \left[(M_0^+ + M_0^-) + a_1 \mu (M_1^+ + M_1^-) \right] \tag{7.122}$$

As it is performed for regular P_1 approximations, these equations are multiplied by 1 and μ and integrated over corresponding hemispheres to obtain the four first-order differential equations:

$$\frac{1}{\tau_0} \frac{\partial M_1^+}{\partial \tau} = 2\pi(1-\omega)I_b + \left(\frac{\omega}{2} - 1 \right)M_0^+ + \frac{\omega}{2} M_0^- + \frac{\omega a_1}{4} M_1^+ + \frac{\omega a_1}{4} M_1^- \tag{7.123}$$

$$\frac{1}{\tau_0} \frac{\partial M_0^+}{\partial \tau} = 6\pi(1-\omega)I_b + \left(\frac{3\omega}{2} - 6 \right)M_0^+ + \frac{3\omega}{2} M_0^- + \left(\frac{\omega a_1}{2} + 6 \right)M_1^+ + \frac{\omega a_1}{2} M_1^- \tag{7.124}$$

$$\frac{1}{\tau_0} \frac{\partial M_1^-}{\partial \tau} = 2\pi(1-\omega)I_b + \frac{\omega}{2} M_0^+ + \left(\frac{\omega}{2} - 1 \right)M_0^- - \frac{\omega a_1}{4} M_1^+ - \frac{\omega a_1}{4} M_1^- \tag{7.125}$$

$$\frac{1}{\tau_0} \frac{\partial M_0^-}{\partial \tau} = -6\pi(1-\omega)I_b - \frac{3\omega}{2} M_0^+ + \left(6 - \frac{3\omega}{2} \right)M_0^- + \frac{\omega a_1}{2} M_1^+ + \left(6 + \frac{\omega a_1}{2} \right)M_1^- \tag{7.126}$$

For the boundary conditions, Marshak's boundary relations are imposed. This means that the moments of the intensity at each of the boundaries are determined from:

$$\int_{\Omega_j} I(\tau, \mu) l \, d\Omega_j = \int_{\Omega_j} I^+ l \, d\Omega_j \qquad (j = 1, 2) \tag{7.127}$$

where l is the direction cosine, and the I^+ for a diffusely emitting and reflecting surface is given as (see Eq. 7.108):

$$I^+ = \varepsilon_w I_b(T_w) + (\rho_1^d/\pi) \int_{\Omega'=2\pi} I^-(\tau, \mu') l \, d\Omega' \tag{7.128}$$

or, for each surface:

$$I_1^+ = \varepsilon I_b(T_1) - (\rho_1^d/\pi)M_1^- \qquad I_2^+ = \varepsilon_2 I_b(T_2) + (\rho_2^d/\pi)M_1^+ \tag{7.129}$$

After substituting Eqs. 7.129 into Eq. 7.127 and carrying out the integration for $l = 1$ and $l = \mu$, we obtain the necessary boundary conditions:

$$M_0^+ = 2\pi I_1^+, \quad M_1^+ = \pi I_1^+, \quad M_0^- = 2\pi I_2^+, \quad M_1^- = -\pi I_2^+ \tag{7.130}$$

The radiative flux at any given point in the medium is found from

$$q = \int_{4\pi} I\mu \, d\Omega = \int_{\phi=0}^{2\pi} \int_{\mu=-1}^{1} I\mu \, d\Omega = M_1^+ + M_1^- \tag{7.131}$$

Figure 7.15 shows the accuracy of this approach compared to the more accurate P_9 approximation. Note that for the P_9, a nine-term phase function was used, and it can be considered as exact (see Ref. 55 for comparisons). Additional calculations showed that, even for nonuniform media, the DP_1 approximation yields accurate results [55].

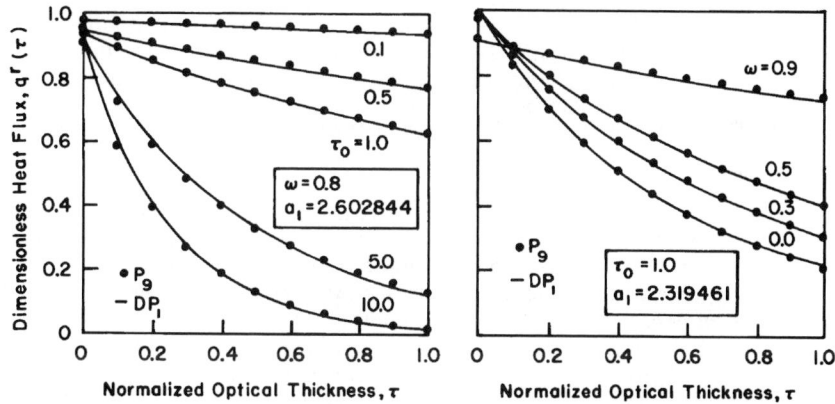

FIGURE 7.15   Comparison of DP_1 approximation predictions with those of the exact P_9 approximation [55].

Multiflux and Discrete Ordinates Approximations.   In order to solve the radiative transfer equations, the spatial and angular variations of intensity are usually expressed in two different functional forms. The entire angular domain is divided into several subdomains, and the radiation intensity in each is considered to be uniform. Then the intensities in each angular domain are related and solved for as functions of position. In principle, this is the same approach followed in the zonal method; however, in deriving the flux methods, the angular subdomains are obtained by following a mathematical reasoning rather than using the relative orientation of the different surfaces with respect to each other. These angular divisions are fixed and do not change as a function of spatial coordinates.

The simplest multiflux model can be developed for a one-dimensional planar medium or a one-dimensional axisymmetric cylindrical geometry by dividing the entire solid angle range ($4\pi \, sr$) into two components: one in the forward and the other in the backward hemisphere. It is assumed that the incident radiation and flux are proportional to each other in each of these hemispheres. Then, the question is finding a closure condition, or, in other words, a proportionality factor between flux and integrated intensity in each half-sphere. This choice yields different two-flux methods, such as Schuster-Hamaker, Schuster-Schwarzchild or modified two-flux models [1, 19, 51].

In multidimensional geometries, the number of angular subdomains needs to be increased. A natural extension of the two-flux method in two- and three-dimensional systems would be four- and six-flux methods, respectively. However, these straightforward extensions to multidimensional systems do not yield acceptable accuracy in radiative transfer predictions. Extensive lists of various multiflux approximations proposed in the past are available in the literature (see, e.g., Refs. 1, 3, and 34). Because the intuitive flux approximations are inferior to the mathematically rigorous multiflux models, they will not be discussed here.

The discrete ordinates (DO) approximation is also a multiflux model. The discrete ordinates approximation was originally suggested by Chandrasekhar [19] for astrophysical applications, and a detailed derivation of the related equations was discussed by several researchers for application to neutron transport problems [33, 57–61]. During the last two decades, the method has been applied to various heat transfer problems [62–81].

In DO models, a Gaussian or Lobatto quadrature scheme is used in the integration of in-scattered radiation, and, for a three-dimensional system, the entire solid angle range is divided into $N(N + 2)$ angular subdomains, where N is determined by the order of quadrature scheme used. Such a DO approximation is referred to as the S_N approximation [61]. For axisymmetric systems, the number of angular subdivisions required is reduced by one-half. The intensity is assumed to remain uniform within each of these angular domains whose extent is obtained from the weights of the quadrature scheme employed. For one-dimensional planar media, the simplest DO approximation is identical to the modified two-flux method. Higher-order DO approximations can easily be used for a plane-parallel medium if the exact solution to the RTE is needed [82].

The DO methods are better than the well-known multiflux methods, which usually cannot couple angular divisions adequately. The DO methods avoid this by discretizing the angular domain such that no direction lies on a coordinate direction. In general, the DO approximation yields acceptable results with an increasing number of directions considered. However, if there are strong local sinks and sources or discontinuous boundary conditions, then the ray effects begin to affect the accuracy of the results [61]. Even with increasing order of the S_N approximation, these ray effects show an oscillatory behavior on the boundary fluxes. Therefore, care should be exercised in interpreting the S_N method predictions especially near flame zones or boundaries with discontinuities in temperature or heat flux. In the following section, we will give the formulation of the DO approximation for cylindrical geometries.

Fiveland and Jessee [77] and Krebs et al. [79] have recently discussed a modified formulation of the DO approximation. They adapted the even-parity formulation originally discussed by Lewis and Miller [61]. They showed that this formulation yields parabolic equations; therefore, a boundary value problem is to be solved instead of an initial value problem. Given that several solution techniques are available for the solution of governing equations (particularly in the field of computational fluid dynamics [CFD]), a solution to the RTE can be obtained with little difficulty. This formulation does not necessarily produce more accurate results, and the ray effects still persist.

Also, an adaptive mesh refinement (AMR) algorithm was developed by Jessee et al. [83]. The results showed that the AMR algorithm did not necessarily produce any improvement over single-grid algorithms. Recently, the accuracy of the discrete ordinates approximation in three-dimensional rectangular enclosures were evaluated carefully [77, 84, 85]. Selçuk and Kayakol [85] have observed that the standard S_N quadrature scheme as well as the ones suggested by Fiveland [65] yield more accurate results than the Gaussian quadrature scheme.

Tan [86, 87] and Hsu et al. [88, 89, 90] have proposed and explored the YIX method. This technique uses the discrete ordinate formulation for angular discretization but precalculates and stores the kernel functions for the distance integration along each ordinate direction. The resulting computational grid is not uniform, but it makes each unequal distance increment along a given ordinate direction contribute an approximately equal amount to the flux at the origin element. This approach allows the path integrations along each ordinate direction to be replaced with fast summations.

The YIX method requires precomputation of the kernel functions but makes the actual radiative transfer computation very efficient. For problems with constant absorption and scattering coefficient, or for problems with known fixed distributions of these properties, the YIX method is an improvement over the standard DO method. If, however, the properties are temperature-dependent, then the kernel functions will change at every iteration, and much of the benefit of the method is lost.

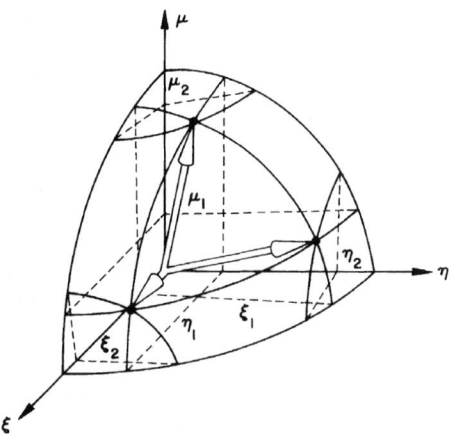

S_2 and S_4 *Models for Cylindrical Geometries.* For the solution of the RTE in cylindrical media, the formulations for the S_2 and S_4 discrete ordinates (DO) approximations (based on Ref. 69) will be presented here. Note that, in a more recent study, Jendoubi et al. [75] used a similar DO approximation in cylindrical geometry and evaluated the effect of anisotropic phase function on the accuracy of the model.

As discussed before, the idea in the DO approximation is to solve the RTE (Eq. 7.105) in a finite number of directions. If these directions are chosen using the values of the Gaussian quadrature points, then the number of directions considered (N_m) will be related to the order N of Gaussian quadratures as $N_m = N(N + 2)$. Here, N is also the order of the DO approximation S_N. Each direction chosen is uniquely determined by the Gaussian quadratures and corresponds to a point on a surface of a unit sphere (see Fig. 7.16). It is assumed that the solution is valid for an angular range, which corresponds to an area w_m on the surface of the unit sphere. The w_m values are nothing but the weights corresponding to the Gaussian quadratures. They satisfy the requirement:

FIGURE 7.16 Representation of the division of spherical domain by a Gaussian quadrature scheme.

$$\sum_m w_m = 4\pi \qquad (7.132)$$

Using this nomenclature, the RTE in each of the Gaussian quadrature directions is written as

$$\frac{\xi_m}{r}\frac{\partial(rI_m)}{\partial r} - \frac{1}{r}\frac{\partial(\eta_m I_m)}{\partial \phi} + \mu_m\frac{\partial I_m}{\partial z} = -(\kappa + \sigma)I_m + \kappa I_b + \frac{\sigma}{4\pi}\int_{4\pi} \Phi(\Omega, \Omega')I_m\, d\Omega' \qquad (7.133)$$

where, the direction cosines ξ_m, η_m, and μ_m are written in terms of zenith and azimuthal angles, θ and ϕ (see Fig. 7.1).

Considering diffusely emitting and reflecting cylindrical boundaries, the corresponding boundary conditions are:

$$\text{at } r = R: \quad I_m = \varepsilon_w I_{bw} + (1 - \varepsilon_w)\frac{q_r^+}{\pi}; \quad \xi_m < 0$$

$$\text{at } r = 0: \quad I_m = I'_m; \quad \xi'_m = \xi_m$$

$$\text{at } z = 0: \quad I_m = \varepsilon_w I_{bw} + (1 - \varepsilon_w)\frac{q_z^-}{\pi}; \quad \mu_m > 0$$

$$\text{at } z = L: \quad I_m = \varepsilon_w I_{bw} + (1 - \varepsilon_w)\frac{q_z^+}{\pi}; \quad \mu_m < 0 \qquad (7.134)$$

Radiative fluxes in $\pm r$ and $\pm z$ directions become (see Eq. 7.2):

$$q_r^+ = \sum_m w_m\xi_m I_m; \quad \xi_m > 0$$

$$q_r^- = \sum_m w_m|\xi_m|I_m; \quad \xi_m < 0$$

$$q_z^+ = \sum_m w_m \mu_m I_m; \quad \mu_m > 0$$

$$q_z^- = \sum_m w_m |\mu_m| I_m; \quad \mu_m < 0 \tag{7.135}$$

In this formulation, the angular derivative term (i.e., the second term in Eq. 7.133) is the most difficult to evaluate. Since the original work of Carlson and Lathrop [33], a direct differencing procedure has been used to simplify this term, such as:

$$\frac{1}{r} \frac{\delta(\eta_m I_m)}{\delta \phi} = \frac{1}{r} \frac{\alpha_{m+1/2} I_{m+1/2} - \alpha_{m-1/2} I_{m-1/2}}{w_m} \tag{7.136}$$

As mentioned by Jamaluddin and Smith [69], the $m \pm 1/2$ directions correspond to the angular range of w_m. An explicit relation can be obtained assuming a uniform radiation field in the corresponding angular domain:

$$\alpha_{m+1/2} - \alpha_{m-1/2} = w_m \xi_m \tag{7.137}$$

If Eq. 7.133 is multiplied by $2\pi r dr dz$ and integrated over the ring-shaped volume element, a difference formulation of the RTE is obtained:

$$\xi_m(A_{i+1} I_{i+1} - A_i I_i) - (A_{i+1} - A_i)[(\alpha_{m+1/2} I_{m+1/2} - \alpha_{m-1/2} I_{m-1/2})/w_m] + \mu_m(B_{j+1} I_{j+1} - B_j I_j)$$

$$= -(\kappa + \sigma) V_p I_m + \kappa V_p I_b + \frac{\sigma V_p}{4\pi} \sum_m \Phi(\Omega, \Omega') w_m I_m \tag{7.138}$$

Here, let us assume that the phase function is a linearly-anisotropic one, expressed as:

$$\Phi(\Omega, \Omega') = 1 + a_0 \cos \phi = 1 + a_0(\xi_m \xi_m' + \mu_m \mu_m') \tag{7.139}$$

Using a central-differencing scheme, the intensities can be related to I_m

$$I_i + I_{i+1} = I_{m+1/2} + I_{m-1/2} = I_j + I_{j+1} = 2 I_m \tag{7.140}$$

where I_m is the intensity at the center of the volume element:

$$I_m = \frac{\xi_m A I_i + \beta_m I_{m-1/2} + \mu_m B I_j + V_p(\kappa I_b + \sigma I_s)}{\xi_m A + \beta_m + \mu_m B + V_p(\kappa + \sigma)} \tag{7.141}$$

with

$$A = A_i + A_{i+1}$$

$$B = B_j + B_{j+1}$$

$$\beta_m = -(\alpha_{m+1/2} + \alpha_{m-1/2})(A_{i+1} - A_i)/w_m$$

$$I_s = \sum_m w_m I_m \left[\frac{1 + a_o(\xi_m \xi_m' + \mu_m \mu_m')}{4\pi} \right]$$

Equation 7.141 is written for positive ξ_m and μ_m values. For other combinations, similar equations need to be developed. Note that V_p is the volume of the computational cell.

For the solution of the governing equations, an iterative scheme is followed [69]. After determining the intensity at a cell center (see Eq. 7.141), the intensity downstream of the surface element can be determined via extrapolation using Eq. 7.140. The central differencing used in Eq. 7.140 may result in negative intensities, particularly if the change in the radiative field is steep. A numerical solution to this problem was recommended by Truelove [67], where a mixture of central and upward differencing is used:

$$I_{i+1} = (1 + f) I_m - f I_i \tag{7.142}$$

where $0 \leq f \leq 1$ and $\xi_m, \mu_m > 0$. If f is 1.0, it is central differencing, and if $f = 0.0$, it is upwind differencing, and the intensity is always positive.

For a scattering medium, the first solution excludes the in- and out-scattering terms; they are included in further iterations. Tables 7.2 and 7.3 list the quadratures used for the S_2 and S_4 approximations.

TABLE 7.2   The Ordinates and Weights for the S_2 Approximation in Axisymmetric Cylindrical Enclosures

SN	Ordinates			Weights
Point (m)	μ	η	ξ	w
1	−0.5	0.7071	−0.5	π
2	0.5	0.7071	−0.5	π
3	−0.5	0.7071	0.5	π
4	0.5	0.7071	0.5	π

TABLE 7.3   The Ordinates and Weights for the S_4 Approximation in Axisymmetric Cylindrical Enclosures

SN	Ordinates			Weights
Point (m)	μ	η	ξ	w
1	−0.2959	0.2959	−0.9082	$\pi/3$
2	0.2959	0.2959	−0.9082	$\pi/3$
3	−0.9082	0.2959	−0.2959	$\pi/3$
4	−0.2959	0.9082	−0.2959	$\pi/3$
5	0.2959	0.9082	−0.2959	$\pi/3$
6	0.9082	0.2959	−0.2959	$\pi/3$
7	−0.9082	0.2959	0.2959	$\pi/3$
8	−0.2959	0.9082	0.2959	$\pi/3$
9	0.2959	0.9082	0.2959	$\pi/3$
10	0.9082	0.2959	0.2959	$\pi/3$
11	−0.2959	0.2959	0.9082	$\pi/3$
12	0.2959	0.2959	0.9082	$\pi/3$

Figure 7.17 depicts the comparison of S_2 and S_4 DO approximation predictions [69] with Monte Carlo calculations [91] in an absorbing, emitting, and scattering medium. Overall, the accuracy of the agreement is acceptable, for a wide range of optical thicknesses and single scattering albedo values considered.

Note that the extension of the DO approximation to three-dimensional rectangular enclosures was attempted by a number of researchers (see, e.g., Refs. 65 and 85). Even though the formulation of the three-dimensional model will not be given here, its governing equations can be derived with little difficulty. As shown in Refs. 65 and 85, the original S_N quadratures yield accurate results in three-dimensional solutions; they are listed in Table 7.4 for S_2, S_4, S_6, and S_8 approximations.

Statistical Models.   By the Lambert-Beer law, the radiative transfer equation is derived from a macroscopic point of view using the principle of the conservation of radiative energy. In order to solve the RTE using any of the available methods, certain mathematical assumptions are introduced, and the physics of the problem are retained. If the assumptions are correct, then the RTE model yields physically acceptable results.

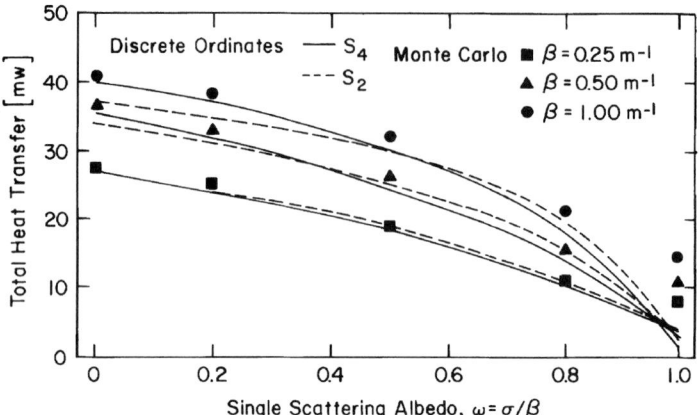

FIGURE 7.17   Comparison of S_2 and S_4 predictions with the Monte Carlo results in cylindrical geometry (adapted from Ref. 69).

On the other hand, it is possible to determine the radiative energy balance everywhere in the medium by considering a large number of photons, which originate from each volume/ surface element, propagate in all directions, and are absorbed and scattered. The radiative energy gain or loss in each element is calculated by considering the effect of an infinitely large number of photons, as each photon affects the total energy balance. In doing so, the direction and the path length of each photon bundle are determined statistically. This procedure, called

TABLE 7.4   Ordinates and Weights for the S_N Approximations in Rectangular Geometries (First Quadrant Values)

SN Approximation	Point (m)	Ordinates			Weights
		μ	ξ	η	w
S2	1	0.5773503	0.5773503	0.5773503	1.5707963
S4	1	0.2958759	0.2958759	0.9082483	0.5235987
	2	0.9082483	0.2958759	0.2958759	0.5235987
	3	0.2958759	0.9082483	0.2958759	0.5235987
S6	1	0.1838670	0.1838670	0.9656013	0.1609517
	2	0.6950514	0.1838670	0.6950514	0.3626469
	3	0.9656013	0.1838670	0.1838670	0.1609517
	4	0.1838670	0.6950514	0.6950514	0.3626469
	5	0.6950514	0.6950514	0.1838670	0.3626469
	6	0.1838670	0.9656013	0.1838670	0.1609517
S8	1	0.1422555	0.1422555	0.9795543	0.1412359
	2	0.5773503	0.1422555	0.8040087	0.0992284
	3	0.8040087	0.1422555	0.5773503	0.0992284
	4	0.9795543	0.1422555	0.1422555	0.1712359
	5	0.1422555	0.5773503	0.8040087	0.0992284
	6	0.5773503	0.5773503	0.5773503	0.4617179
	7	0.8040087	0.5773503	0.1422555	0.0992284
	8	0.1422555	0.8040087	0.5773503	0.0992284
	9	0.5773503	0.8040087	0.1422555	0.0992284
	10	0.1422555	0.9795543	0.1422555	0.1712359

a statistical or Monte-Carlo approach, is straightforward to implement, although, computationally, it may be expensive.

The Monte Carlo method seeks to replace mathematical descriptions of radiative transfer such as those described in the subsections on the zonal method and the differential and moment method with a simulation of the physical processes that are occurring. The method simulates radiative transfer by following the histories of small amounts of radiative energy, often called energy *bundles*. The bundles are followed throughout the series of events such as emission, reflection, scattering, and absorption that occur over the bundle lifetime. By following a sufficiently large number of bundles, the radiative flux or flux divergence distributions can be found over surfaces or within volumes of participating media.

The Monte Carlo method results in statistically averaged results. The accuracy of the results depends upon the number of sample bundles chosen for study in the same way that experimental accuracy depends upon the number of replications of the experiment. This dependence on the number of samples is the greatest drawback of the method, as it requires a tradeoff between computer time and statistical accuracy of the results.

Brief Outline of the Monte Carlo Method.   Monte Carlo is based on determining the probability of a particular event in the history of an energy bundle. Whenever such an event occurs based on the physics of the problem, a choice is made such that, after many such events, the choices have the correct distribution. This is done by relating the variable being chosen to a random number R, where $0 \leq R \leq 1$, through a cumulative distribution function, or CDF. For example, the direction of emission from a diffuse surface is determined by the two angles (θ, ϕ). For a particular energy bundle being emitted from the surface, these angles are determined from Table 7.5 to be

$$\sin \theta = R^{1/2} \tag{7.143}$$

$$\phi = 2\pi R \tag{7.144}$$

A separate random number must be chosen for each of these relations, so that they will be independent of one another. Each event will have a different CDF. Derivations of the statistical relations necessary to determine the CDF relations needed to implement the Monte Carlo technique are given in Refs. 1, 2, and 3 and are not repeated here; however, the resulting relations for implementing the method are given in Table 7.5.

These relations are based on the assumption that wavelength and directional dependence are independent of one another; for example, the directional distribution of emitted energy is independent of the wavelength of emission. When this is not the case, somewhat more complex relations are necessary [1, 92]. For use in a Monte Carlo calculation, we need a relation of the form $\mu = F(R)$ rather than $R = G(\mu)$. These inversions can be generated from the integral forms shown in the table by substituting the required properties, carrying out the integrations, and then curve-fitting the results. Haji-Sheikh [93] has provided such a relation for $\lambda = F(R)$ for black or gray surfaces that is reproduced in Siegel and Howell [1].

Examples.   To illustrate the Monte Carlo method, two radiative transfer examples are presented.

First, take the case of determining the radiative transfer between two infinite parallel gray diffuse directly opposed plates separated by a distance D and of width W, with known temperatures T_1 and T_2 and equal emissivities ε (Fig. 7.18). No attenuating medium is present between the surfaces.

To begin the model, consider radiant energy that is initiated by emission from surface 1. The first choice in the history of emission will be the position x along plate 1, at which point emission occurs. This can be chosen at random as $x = RW$, or it can be chosen in sequence such that for sample number n of the N bundles, we let $x = (n/N)W$. Either method will provide uniform emission along the constant temperature plate, but the latter method avoids repetitive

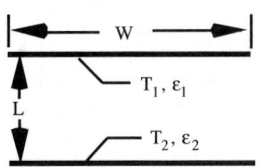

FIGURE 7.18 Geometry for the Monte Carlo example problem.

TABLE 7.5   Monte Carlo Relations for Surface-Surface and Participating Media Radiative Exchange

Phenomenon	Variable	Relation
Emission or reflection from diffuse gray or black surface	Cone angle θ Circumferential angle ϕ Wavelength λ	$\sin\theta = R^{1/2}$ $\phi = 2\pi R$ $F_{0-\lambda} = R$
Emission from gray θ-directional surface	Cone angle θ	$R = \dfrac{2\int_0^\theta \varepsilon(\theta^*)\sin\theta^*\cos\theta^*\,d\theta^*}{\varepsilon}$
Emission from nongray θ-directional surface	Cone angle θ	$R = \dfrac{2\pi\int_0^\theta \int_0^\infty \varepsilon(\lambda,\theta^*)i_{\lambda b}(\lambda)\sin\theta^*\cos\theta^*\,d\lambda\,d\theta^*}{\varepsilon\sigma T^4}$
Emission from diffuse nongray surface	Wavelength λ	$R = \dfrac{\pi\int_0^\lambda \varepsilon(\lambda^*)i_{\lambda b}(\lambda^*)\,d\lambda^*}{\varepsilon\sigma T^4}$
Emission from nongray θ-directional surface	Wavelength λ	$R = \dfrac{2\pi\int_0^\lambda \int_0^{\pi/2} \varepsilon(\lambda^*,\theta)i_{\lambda b}(\lambda^*)\sin\theta\cos\theta\,d\theta\,d\lambda^*}{\varepsilon\sigma T^4}$
Emission from a volume element	Cone angle θ Circumferential angle ϕ Wavelength λ Wavelength λ	$\cos\theta = 1 - 2R$ $\phi = 2\pi R$ Gray medium: $F_{0-\lambda} = R$ Nongray medium: $R = \dfrac{\int_0^\lambda a_\lambda(\lambda^*)i_{\lambda b}(\lambda^*)\,d\lambda^*}{\int_0^\infty a_\lambda(\lambda^*)i_{\lambda b}(\lambda^*)\,d\lambda^*}$
Attenuation by medium	Path length s	Uniform properties: $s = -\dfrac{1}{K_\lambda}\ln R$ Nonuniform properties: $\ln R = -\displaystyle\int_0^s K_\lambda(s^*)\,ds^*$
Isotropic scattering	Cone angle θ Circumferential angle ϕ	$\cos\theta = 1 - 2R$ $\phi = 2\pi R$
Anisotropic scattering	Cone angle θ Circumferential angle ϕ	$R = \dfrac{1}{2}\displaystyle\int_0^\theta \Phi(\theta^*)\sin\theta^*\,d\theta^*$ $\phi = 2\pi R$

calls of the random number generator and is usually both faster and introduces less statistical variation.

The rate of energy emission per unit area can be broken into N bundles of energy per unit time and area, each carrying an amount of energy w. Therefore,

$$e = wN_1 = \varepsilon\sigma T_1^4 \tag{7.145}$$

The programmer can choose the value of N to provide the desired degree of accuracy in the final solution.

The direction of emission at point x is chosen using Eqs. 7.143 and 7.144. Now we determine whether the bundle intersects the opposing plate by determining the x-position at which the bundle intersects the plane containing plate 2; in other words, $0 \le x_2 \le W$, where

$$x_2 = x_1 + L\tan\theta\cos\phi \tag{7.146}$$

If the bundle misses plate 2, then we choose another bundle; if it strikes plate 2, then we know the position where it strikes, x_2, from Eq. 7.146. At this position, we need to see how much energy is absorbed. Two common strategies are employed for this. In the first, we choose a new random number and see whether $R \geq \alpha = \varepsilon$. If the answer is yes, then the bundle is completely reflected from surface 2. If the answer is no, then the entire energy of the bundle is absorbed on surface 2 at x_2. In the second strategy, we simply absorb an amount of energy $w\alpha$ at x_2 and reflect a bundle with remaining energy $(1 - \alpha)w$. In this strategy, the bundle energy is diminished at each reflection, and the process is usually ended by depositing all remaining energy when the bundle energy has been reduced to some small fraction of the original value.

The first of these strategies requires somewhat less bookkeeping but has larger statistical uncertainty. The second method is probably more efficient for most cases, where efficiency is the product of CPU time and solution variance.

If the bundle is reflected from diffuse surface 2, then angles of reflection are chosen by Eqs. 7.143 and 7.144, and trigonometry is again applied to determine whether the reflected bundle strikes surface 1, and, if so, at what location. This procedure of reflections is continued until the bundle misses the opposite surface, or it is either absorbed or its energy is reduced below the given threshold value. If some or all of the energy is absorbed, the amount absorbed is tallied at the position of incidence.

To complete the solution, bundles must be emitted from surface 2. If they have the same energy per bundle as those emitted from surface 1, then we must emit $N_2 = \varepsilon \sigma T_2^4/w = (\varepsilon_2 T_2^4/\varepsilon_1 T_1^4)N_1$ bundles from surface 2. The histories follow the same sequence of emission and reflection as before, except that the bundle histories are initiated at surface 2.

A flow chart for this process is shown in Fig. 7.19. In this case, the plates are assumed to be black, and the environment is at $T_e = 0$. The S_k and S_n are counters used to tally the number of

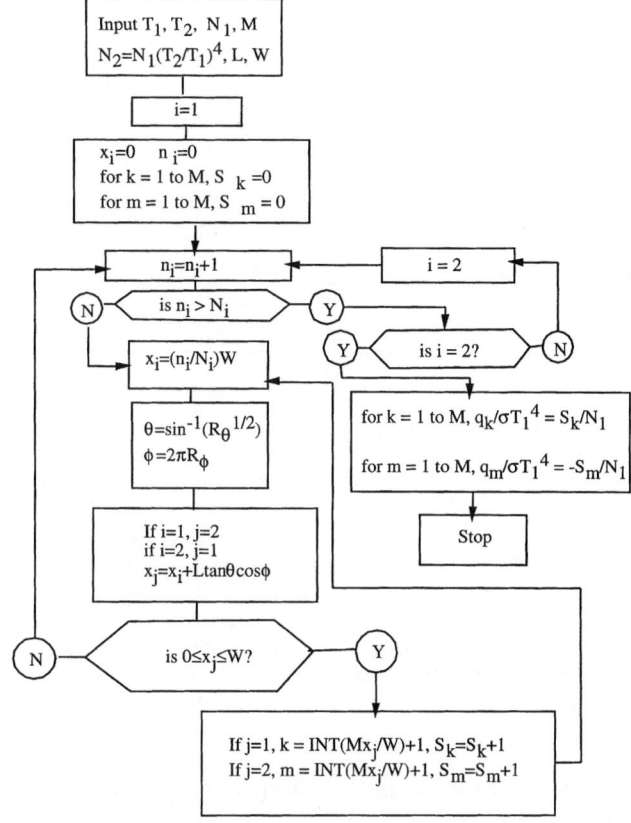

FIGURE 7.19   Flowchart for the Monte Carlo solution.

absorptions at positions on surfaces 1 and 2, respectively, and M is the number of Δx increments on the surfaces.

For a second example, consider a gray absorbing medium with uniform properties contained between black infinite parallel plates at different temperatures. Find the heat transfer between the plates and the temperature distribution in the medium, assuming radiative equilibrium (i.e., no sources or sinks in the medium).

Energy bundles are followed through their histories after emission from each surface until absorption at a boundary. Because of the assumption of radiative equilibrium, any bundles absorbed within a medium volume element must be balanced by an emission from that element; this is simulated by simply reemitting an absorbed bundle in a new direction and continuing the history until final absorption at a boundary. The medium temperature distribution is computed by equating the emission from the element to the absorption. The flow chart for this case is shown in Fig. 7.20.

Table 7.6 gives the results for the net radiative flux between the plates at an optical thickness of 1.0 for various numbers of energy bundles, compared with the exact solution of this problem, which is 0.5532. The solution was programmed according to the flow chart in Fig. 7.20.

Discussion of Monte Carlo methods for radiative transfer calculations have been given in Refs. 61, 93, and 94. It was used to model radiative heat transfer in large scale furnaces by, among others, Taniguchi and Funazu [95], Xu [96], Gupta et al. [91], and Richter [97]. In general, Monte Carlo techniques do yield reliable results if certain precautions are taken and sufficient numbers of photon bundles are utilized and biasing techniques are applied [61]. As mentioned by Howell [94], one of the main drawbacks of a Monte Carlo technique is the grid

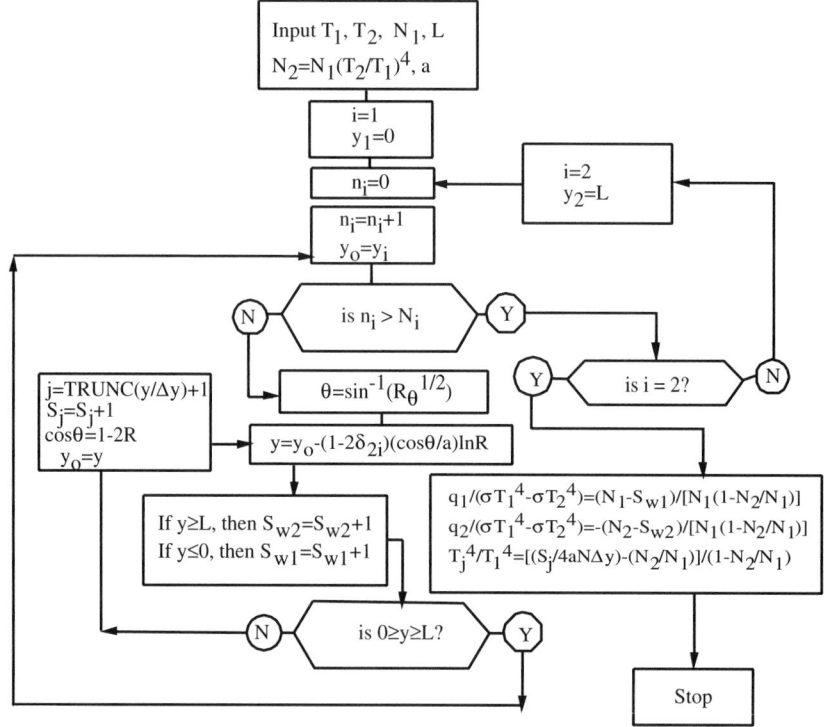

FIGURE 7.20 Monte Carlo flowchart for radiative exchange between opposed black plates of finite width.

TABLE 7.6 Results of Monte Carlo Example Problem; Normalized Radiative Flux at Surface 1 for Various Sample Sizes

N_1, bundles emitted by surface 1	Radiative flux at surface 1, $q_1/\sigma T_1^4$
100	0.592
500	0.542
1000	0.553
5000	0.561
10,000	0.567
50,000	0.5564

incompatibility with the conservation of mass and energy equations. This problem, however, is likely to be eliminated with increasing computational power of computers. The main advantage of the Monte Carlo techniques is its possible application to any complex geometry.

Recent work on the Monte Carlo technique includes that by Farmer and Howell [98–101], who investigated the optimal strategies for implementing the Monte Carlo technique on both serial and parallel computers. For up to 32 parallel processors, the CPU time on the parallel machines was very nearly inversely proportional to the number of processors, indicating that the method is very well suited to parallel machines.

Numerical Models.   With the advancement of computers, it is theoretically possible to solve any differential or integral equation using an appropriate numerical discretization scheme. The challenge here is to devise a scheme such that the physics of the problem are retained. All the radiative transfer models developed in the past attempt to do this; however, they also take advantage of different analytical techniques to reduce the complexity of equations and the required computational power as much as possible. In numerical solution techniques, the computational requirements may be decreased by using certain numerical simplifications and/or by choosing coarser grids. In general, it is safe to say that, with the availability of supercomputers, finer grids compatible with the grid scheme used for the solution of the other conservation equations can be used with little difficulty and computational penalty. Therefore, numerical techniques are likely to have wide acceptance in the future to model the radiative heat transfer in complicated geometries, especially for nonscattering media (see Refs. 102 and 103).

Raithby and Chui [103] suggested a finite volume formulation of the RTE in one-dimensional planar and two-dimensional rectangular enclosures. The model is capable of handling scattering by the medium and yields very good agreement with the exact solutions. In principle, it is very similar to the discrete ordinates approximation, the main difference being the selection of the angular discretization to be used. In DO approximations, this selection is analytically determined, whereas Raithby and Chui use an arbitrary criterion for the selection. Therefore, a trial-and-error approach is required to determine the optimum angular discretization.

The finite element techniques have been used to model radiative heat transfer in multidimensional enclosures. Razzaque et al. [104, 105] considered the combined conduction-radiation problem in an absorbing/emitting medium. Chung and Kim [106] and Sökmen and Razzaque [107] allowed for isotropic scattering in the medium. Tan [108] introduced a product integration method, which is similar to the finite element technique. This approach requires much less computational time than the other finite element solution techniques. Burns et al. [109, 110] formulated a detailed finite element radiative transfer code that uses grids generated by commercial FEM grid generators. The code computes the local radiative flux divergence based on a given temperature distribution in the medium; the computed divergence can then be used in the energy equation in a commercial FEM code for treating combined-mode problems. Note that the finite element approach was also used for the solution of the reduced equations of the DO approximations [73, 74, 76].

Hybrid Techniques. Almost all the methods discussed before have certain disadvantages. It is sometimes possible to combine the features of two or more methods to develop a more efficient technique to model radiative heat transfer in furnaces. For example, if the furnace geometry is very complicated, the zone method cannot be used effectively, as it is quite difficult to determine all the required exchange factors. Here, a Monte Carlo technique can be adopted, as done by Vercammen and Froment [111], to calculate the exchange factors between volume and surface elements, and then the radiative heat transfer between each element is calculated using the zonal method. Edwards [112] also suggested a similar approach where he obtained the exchange factors using a Monte Carlo technique and calculated the radiative flux distribution using a radiosity-irradiosity approach. With this hybrid approach, exchange factors in most complicated geometries can be calculated with little difficulty. Also, a significant amount of time is saved by using the zone method instead of a Monte Carlo technique to determine the radiative heat flux distribution in the medium. In addition to that, possible statistical errors due to a Monte Carlo technique are avoided.

The basic principles of Monte Carlo techniques have been used by Lockwood and Shah [113] in developing the so-called *discrete transfer method.* Instead of choosing the direction of the intensities originating (due to emission, scattering, or reflection) from each volume/surface element randomly, they suggested a deterministic approach. Although the method was shown to be accurate and computationally efficient for nonscattering media, it did not yield accurate results if scattering particles were present. Recently, Selçuk and Kayakol [114] evaluated this approach and outlined the problems related to its implementation. They noted that, for relatively simple, homogeneous, and nonscattering media, this approach yielded about 10 percent error for the radiative source term near the corners, even if 64 rays were considered. Additionally, it required three times more CPU time than the S_4 approximation to converge in three-dimensional rectangular enclosures [85].

An alternative approach similar to the discrete transfer method was reported by Taniguchi et al. [115, 116, 117]. They showed that, for nonscattering media, the method yields very good results with significant computational time savings over standard Monte Carlo techniques.

Richter [97] suggested a similar semistochastic approach, where he developed a solution scheme based on the principles of a Monte Carlo technique, yet the directions of photons emitted by each volume/surface element were predetermined. He applied this approach to several large-scale furnaces and showed that, if the scattering is not accounted for, the model yields reasonable results even if as few as 10,000 photons are considered.

Another hybrid approach is to combine the diffusion approximation for optically thick media with the Monte Carlo technique. The diffusion method can be applied in geometric or spectral regions that are optically thick with good accuracy; Monte Carlo is used in geometric or spectral regions with intermediate or thin optical depth and near boundaries. Farmer and Howell [98, 99] have implemented two forms of such a hybrid, finding good accuracy and greatly reduced computer time over conventional Monte Carlo.

Strategies for Choosing a Radiative Transfer Model. One of the most important decisions an engineer has to make regarding the modeling of radiative heat transfer in large-scale furnaces is the choice of an appropriate radiative transfer model. Considering the fact that the solution of the RTE requires information about the temperature and spatial distribution of radiative properties in the medium, the radiative transfer model is to be coupled with the models for flow, chemical kinetics, turbulence, and so on. Therefore, the design engineer must choose a model that will be compatible with the solution techniques for the other governing equations. The model should also be reliable and able to predict accurately the radiative flux and the divergence of radiative flux distributions in the medium. In addition, the model should be computationally efficient.

It is not always necessary or desirable to choose the most accurate radiative transfer model available. If the accuracy of the radiative property data used in predictions is not as good as the accuracy of the model itself, it is difficult to justify the extra computational effort required by a refined model. A simpler approximate model may be more appropriate.

It is important to ask the following questions before a specific model is chosen. (1) Is the medium geometry simple? (2) Are there steep temperature and species concentration distributions in the medium? (3) Are there anisotropically scattering particles? (4) If there are, what kind of scattering phase function approximations can be used for them? Having some approximate answers to these questions will help to expedite the selection process.

For the solution of the radiative transfer equation, there are several different models available in the literature, as summarized in the foregoing sections. None of these models, however, can be used on a universal basis and applied to all different types of practical problems. It is up to the researcher to decide which model should be used for what type of application. In order to make such a choice, he or she should know the advantages, disadvantages, range of applicability, and versatility of each model. In this section, we will attempt to introduce some simple guidelines to make the selection procedure less time-consuming. For this purpose, the advantages and disadvantages of different models are listed in Table 7.7 to help the reader in choosing an appropriate model.

As discussed previously, the radiative transfer equation is written in terms of radiation intensity, which is a function of seven independent parameters. The RTE is developed phenomenologically and is a mathematical expression of a physical model (i.e., the conservation of the radiative energy). It is a complicated integro-differential equation. There is no available analytical solution to the RTE in its general form. In order to solve it, physical and mathematical approximations are to be introduced individually or in tandem.

We can consider possible approximations under three different types of broad categories: (1) simplification of the spectral nature of properties; (2) angular discretization of the intensity field; and (3) spatial discretization of the medium (for all parameters).

It is important to realize that the solution of the radiative transfer equation is required only to obtain the divergence of the radiative flux vector that is a total quantity (i.e., inte-

TABLE 7.7 Comparison of Radiative Transfer Models

	Angular resolution	Spatial resolution	Spectral resolution	Scattering medium 2D	3D
Flux methods					
Multi flux approaches	••	••••	••••	••	••
Discrete transfer method	••	••••	••••	••	••
Discrete ordinates approx.	•••	••••	••••	•••	•••
YIX DO	•••	••••	••••	•••	•••
Even parity/odd parity DO	•••	••••	••••	•••	•••
Moment methods					
Moment method	•	••••	••••	•	•
Spherical harmonics approx.	••	•••	••••	•••	•••
double/quadruple SHA	••	••••	••••	•••	•••
Zone methods	••	•••	••	••	••
Monte Carlo techniques	••••	•••	•••	••••	••••
Numerical approaches					
Finite difference techniques	••	••••	•••	••	••
Finite element techniques	••	••••	•••	••	••

Potential accuracy: •••• Very good; ••• Good; •• Acceptable; • Not good. (Ratings are subjective and given for applications to multidimensional complex problems; they are likely to change with the increasing availability of faster computers).

grated over the entire wavelength range spectrum). Since eventually we need to obtain a spectrally integrated quantity, one natural way to simplify the problem is by using spectrally averaged radiative properties. The averaging over the wavelength spectrum can be performed over predefined narrow ranges or over the entire wavelength spectrum.

The radiative flux itself is an integrated quantity over a predetermined hemispherical angular domain. After solving the radiative transfer equation in terms of intensity in all directions, the radiative flux is calculated. If a radiative heat transfer model based on fluxes or similar integrated quantities is developed, then the required mathematical complexity will be greatly reduced. The disadvantage of this approach is the eventual suppression of directional variations of intensity. In other words, if there are localized sinks or sources that contribute to radiative transfer, their effect cannot be predicted accurately by such an approximate model. If smaller angular divisions are employed, then the accuracy is increased.

Finally, the medium is to be discretized spatially to perform the numerical calculations. It is preferable to employ the same discretization for radiative transfer calculations as for flow and other scalar field calculations. This is a very time-consuming approach if, for example, the zonal method is used. The multiflux approximations are more useful for this type of discretization.

After this brief background, we can start discussing a logical procedure for selecting a radiative transfer model. If the physical system being considered is not large and has relatively simple geometry, and if a high degree of accuracy is required, then a very narrow discretization of angular radiation field can be chosen, and computations can be carried out by considering the RTE over each of these directions. If the medium is not scattering, then the integral term in Eq. 7.105 vanishes, and the problem becomes relatively straightforward and significantly simpler. This approach is likely to yield a highly accurate solution for radiation intensity.

To be able to take advantage of such a detailed solution scheme, one must model the radiative property variations very accurately, and the effect of temperature on these properties should be considered. Once the divergence of the radiative flux is determined, it enters into the energy equation as a source term, which affects the temperature profile in the medium directly and the concentration distributions indirectly. Since the medium properties are dependent on the temperature and concentration distributions, the radiative properties need to be updated before each iteration in a comprehensive model.

If the medium is scattering, and if the scattering phase function is not mathematically simple, then direct simulation of the radiative transfer equation becomes extremely prohibitive. There are basically two exceptions to this case. If the scattering particles are highly forward-scattering (i.e., most of the energy scattered is in the same direction as the incident beam), then the phase function may be modeled using a δ-Eddington approximation. After introducing appropriate scaling laws, the scattering is modeled with isotropic or linearly anisotropic phase functions. In this case, the problem is still tractable, although not simple. This complication is greatly reduced if the optical thickness is small (i.e., the maximum value of the integral of path-length extinction coefficient product in the medium is less than 0.1 so that a single scattering assumption can be made).

The second exception is if we are interested in the propagation of a collimated light source (i.e., a laser). In this case, since only one incident direction is to be considered, the problem can be modeled by direct simulation, even for multiple scattering media up to intermediate optical thicknesses ($\tau \simeq 1$). In general, the direct simulation of the radiative transfer equation is to be chosen if a fundamental understanding of radiation-combustion or radiation-turbulence interactions is required.

If the level of accuracy required is not so high, then a Monte Carlo (MC) simulation of the radiative transfer equation can be considered. The MC technique can be effectively used for complex geometries, and it is possible to account for spectral property variations in detail. Its main drawback is the requirement for extensive computational time, which, probably, will not be an important issue in the near future.

For comprehensive modeling of combustion systems, a relatively less detailed multiflux or moment method is preferable to the statistical MC approach. For example, the S_4 discrete ordinates approximation has been successfully used by several researchers for both cylindrical and rectangular systems. The method allows the user to account for sufficient detail in spectral properties of combustion products and reasonably accurate discretization of the angular radiation field, and it is compatible with the finite difference schemes used for flow and energy equations. The S_6 approximation yields slightly better results, and its advantage is more visible if there are large temperature gradients and scattering particles in the medium. Of course, its use can be justified only if the properties are known with good accuracy. It is computationally more costly than the S_4 approximation. Because of this, an innovative strategy is to be devised to avoid computational difficulties. For example, the S_2 or S_4 approximation can be used for the initial calculations of the radiative flux distribution, and S_6 can be used only for the final iterations. This approach is likely to yield more accurate predictions with little computational penalty. For systems with uniform or fixed distributions of properties, the YIX modification of the S_N method should be considered [87–90].

On the other hand, the use of even higher-order DO approximations is not warranted given the current level of the accuracy and availability of radiative properties. In multidimensional systems, moment, spherical harmonics, and hybrid multiflux approximations usually do not yield results with the accuracy or efficiency of the DO approximations.

For nonscattering media, the zone method yields accurate results; however, it may not be directly compatible with the flow and energy equations. This problem has been solved by considering two separate finite difference schemes: one for radiation calculations and the other for flow and energy equations [118]. Another alternative is to solve the RTE using a finite difference or finite element scheme. This approach will guarantee the compatibility of the equations [103, 109]. Also worth consideration is the semistochastic model suggested by Richter [97], which is very fast for nonscattering media, and can easily be extended to account for scattering.

Solutions to Benchmark Problems

The choice of modeling method, particularly when a participating medium is present, is not yet a clear one. Each method discussed above has drawbacks as well as positive attributes. In an effort to clarify the choice of methods, the American Society of Mechanical Engineers has sponsored a series of workshops in the form of technical sessions and discussions. A particular series of problems was proposed by the workshop organizers, Professor Timothy Tong of Colorado State University and Dr. Russ Skocypec of Sandia, Albuquerque. Contributors were asked to provide solutions to the benchmark problems at the first workshop. The problems were simple geometries in one, two, and three dimensions. A participating medium was specified with given spectrally dependent anisotropic scattering properties and a given model for spectral band absorptance. (Edward's exponential wideband approximation for a given mole fraction of CO_2 in nitrogen was specified.) A temperature distribution within the medium was given, and the bounding surfaces were specified as cold and black. Researchers were asked to apply their favorite method of solution and provide the workshop with numerical values of boundary heat flux distributions and the divergence of the radiative flux at various locations within the medium.

Solutions were presented based on a generalized zonal method, three Monte Carlo solutions, a modified discrete ordinates method (the YIX method), and two specialized approaches. The conclusions noted by Tong and Skocypec [119] are that the boundary heat flux values were in better agreement than were the flux divergence values, but even the boundary fluxes varied by as much as 40 percent among investigators for two-dimensional geometries and as much as a factor of 2 for the three-dimensional geometries. The poor agreement found in this exercise has led to continued dialogue among all of those concerned. Solutions have been reexamined and modified among many of the researchers, and the conclusions at this time are that the major methods provide good agreement (within a few percent) for two- and three-dimensional geometries for the case of a gray medium with and without anisotropic scattering, but there remain considerable differences in the solutions when spectral effects are included. It is not clear as yet why this is the case.

Recently, benchmarking attempts have been extended to more complicated L-shaped configurations [90, 120, 121, 122]. Hsu and Tan [90] have also presented comparisons between different approaches and suggested that the ray effects could be minimized by increasing the order of angular quadrature used.

There are not many exact solutions to the RTE in multidimensional enclosures. The few exceptions include those by Cheng [123] and Crosbie and Lee [124] for inhomogeneous media and Selçuk [125] and Selçuk and Tahiroglu [126] for homogeneous systems, although they are all related to "simple" geometries. Benchmarking studies against the "real" data, as those obtained from careful experiments, are still lacking!

RADIATIVE PROPERTIES FOR PARTICIPATING MEDIA

The medium that interacts with radiation may contain particles and gases which absorb and scatter the radiant energy. In combustion chambers, for example, soot, char, fly-ash, coal particles and spray droplets affect the propagation of radiant energy. Among various gases, carbon dioxide and water vapor are the major participants to radiative transfer, both in combustion chambers and in the atmosphere.

In this section, we will give a short review for the radiative properties of gases and then present some easy-to-use formulations.

Radiative Properties of Gases

If an electromagnetic wave is incident on a gas cloud, it interacts with the individual molecules. This interaction can be considered as a radiative transition and results in a change of energy level in each molecule. If the molecule absorbs the energy from the EM wave, there may be a transition between the nondissociated molecular states (bound-bound transition), between the nondissociated and dissociated states (bound-free transition), or within the dissociated states. For most radiative heat transfer applications, the energy level of interest is such that, when the EM wave is incident on a molecule, it results in a bound-bound transition, if any. According to quantum mechanics, this means that, for an EM wave being absorbed by a molecule, it has to have just the right amount of energy to raise the molecular energy state to higher levels.

It is also possible to consider this transition as the absorption of a photon (energy quantum). The energy of a photon is expressed as $E = h\nu$, where h is the Planck constant and ν is the frequency of the wave, which is related to the wavelength via the speed of light: $c = c_o/n = \nu\lambda = \nu\lambda_o/n = \nu/\eta$. Here, subscript o is used to denote the vacuum, and n is the real part of the complex index of refraction of the medium. It is obvious that a molecule can absorb the energy in discrete amounts of $h\nu$ or hc/λ, which results in a bound-bound transition. This suggests that, if a beam of radiation spanning a wavelength interval is incident on a gas cloud, it will lose its energy at certain wavelengths but will not be affected at others.

For most gas molecules, there are several hundreds, even thousands, of possible molecular energy states. Therefore, one expects to find a large number of wavelengths at which the molecules absorb the incident energy, which makes the prediction of gas absorption a very difficult problem. Additionally, each of these absorption wavelengths can be broadened because of pressure and temperature as well as the uncertainty principle of Heisenberg. It is obvious that exact consideration of all these active frequencies/wavelengths may not be desirable for most engineering applications, although it will be within the realm of possibility with increasing availability of high-speed, large-memory computers.

Even though most of the discussion given above refers only to gas absorption, it can be readily extended to emission. It should be understood, however, that the emission is related to the gas temperature, whereas absorption depends on both gas temperature and the temperature of the source.

Line Radiation.   As briefly mentioned under ideal conditions above, a molecule absorbs or emits radiation at a fixed wavelength. Then, the change in the energy of the beam incident along path s within the gas cloud is written as (Eq. 7.105 with scattering neglected):

$$\frac{dI_\eta}{ds} = \kappa_\eta(I_{b\eta} - I_\eta) \tag{7.147}$$

where I_η is the spectral radiation intensity and η $(=1/\lambda)$ is the wavenumber (usually expressed in units of cm^{-1}). The absorption coefficient κ_v is equal to the emission coefficient based on Kirchhoff's law.

The absorption and emission by gas molecules are realized not at a single frequency but over a very narrow band of frequencies. The reason for this is mainly the change in the energy level of molecular states due to molecular collisions, temperature, pressure, or relative motion of molecules with respect to the beam of radiation. The result is the broadening of absorption/emission lines; the most well known are the collision, natural line, and Doppler broadenings. The shape of these narrow spectral lines is exponential, peaked at the center, with rapid decay away from the center frequency. The corresponding line width is in the order of $d\eta = 0.05$ cm^{-1}. Here, $\eta = 1/\lambda$ is the wavenumber corresponding to the wavelength λ; the unit cm^{-1} is read as "wavenumbers." For example, at wavelength of $\lambda = 1$ μm, $\eta = 10,000$ cm^{-1}, and for $d\eta = 0.05$ cm^{-1}, $d\lambda = -d\eta/\eta^2 = 0.001$ μm.

The shape and width of each absorption line are functions of temperature and pressure of the medium. The Lorenz profile is usually used to define the shape of these lines at moderate temperatures under local thermodynamic equilibrium conditions. Although other profiles, such as Doppler or Stark profiles, can also be used to define the line shapes, the Lorenz profile, which adequately describes collision-broadened lines, is more appropriate for most applications, including those in combustion systems. For more details, the reader is referred to Siegel and Howell [1] or Modest [3].

Narrowband Models.   Given that detailed spectral calculations with wavelength resolution on the order of 10^{-3} μm is neither computationally efficient nor justifiable for many engineering applications, it is better to develop more affordable models. A close look at the broadened absorption lines depicts that several of them are positioned very close to each other and may overlap, especially for the vibration-rotation transitions of diatomic and polyatomic molecules. The absorption coefficients of individual lines can be added to find the absorption coefficient of the narrowband:

$$\kappa_\eta = \sum_j \kappa_{\eta j} \tag{7.148}$$

Two of the best-known models used for this purpose are the Elsasser and the statistical Goody models, both of which employ the Lorenz profile for description of individual line shapes. These models give very accurate predictions over a bandwidth of approximately 50 cm^{-1}, which is considered narrow for most practical purposes. (At $\lambda = 1$ μm, this bandwidth translates to about $\Delta\lambda = 0.05$ μm.) Because of this, the model is called the *narrowband model*. Although this technique is significantly simpler than the line-by-line models, it still requires an extensive database about the species considered and significant computational effort. Such a detailed model can be considered useful only if the species concentration distribution is known very accurately, which is usually not the case.

Detailed discussions of narrowband models were given by Tien [127], Ludwig et al. [128], Edwards [129], and Tiwari [130]. The discussion of these approaches can also be found in standard texts [1, 2, 3].

Wideband Models.   The narrowband models introduce significant simplification over the line-by-line calculations; however, the accurate predictions depend not only on the rigor of the model but also on the accuracy of input data, such as the local temperature, the temperature profile, the partial pressures of the radiating gases, and so on. In most practical systems, these data are not available with good accuracy. This suggests that even simpler approximations may be more appropriate for calculation of gas properties in practical systems.

Water vapor, carbon dioxide, and carbon monoxide are the most important contributors to nonluminous radiation in combustion systems. They are the byproducts of any hydrocarbon combustion and absorb and emit radiation selectively only at certain wavelengths. Although there are other gases such as NO_x and SO_2 present in the combustion products, their partial pressures are very small, and their contributions can be safely neglected in radiative transfer predictions. Among these, SO_2 is more important, as it has usually an order of magnitude higher concentration in large-scale furnaces than that of NO. The emissivity of SO_2 is about the same order of magnitude as the H_2O, and, therefore, it is possible to account for the SO_2 contribution easily by adding its partial pressure to that of H_2O and considering only the CO_2 and H_2O contributions in the calculations [131].

For combustion gases such as water vapor, carbon dioxide, and carbon monoxide, the number of wavelength ranges important for thermal radiation calculations is not excessive. Figure 7.21 shows the important absorption bands of the CO_2 and H_2O gases at two different temperatures [132]. For H_2O, there are four important widebands centered at wavelengths of 1.38, 1.84, 2.7, and 6.3 μm. For CO_2, there are bands at 2.7 and 4.3 μm, and, for CO, there is one band centered at 4.4 μm. The width of each of these so-called widebands is an adjustable parameter that is determined by assuming the absorption and emission of radiation in the wideband is equal to the effective absorption and emission of several narrowbands present within the band. Therefore, the shape of the band chosen determines the width.

Edwards and his coworkers first developed basic concepts of wideband models (see Ref. 129). There are different variations of the model, such as the box or exponential wideband models. The underlying idea in these approaches is to represent the gas absorption over a relatively wide spectral band (about 0.1 to 0.5 μm) with a simple function. If this function is an exponential, the resulting model is called as the *exponential wideband model* [129, 133], which is the most well-known and successful of all different models. It is possible to simplify this even further by assuming absorption remains constant over a prescribed wavelength interval. This yields the box model, which is not as good as the former one, as expected.

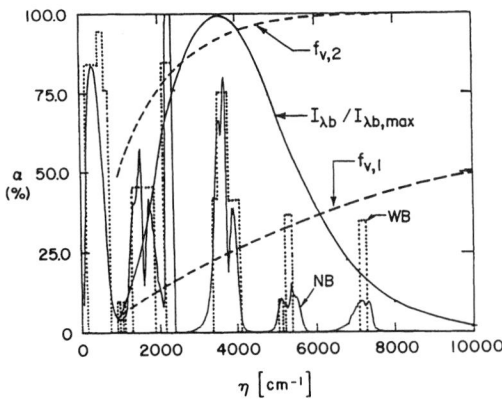
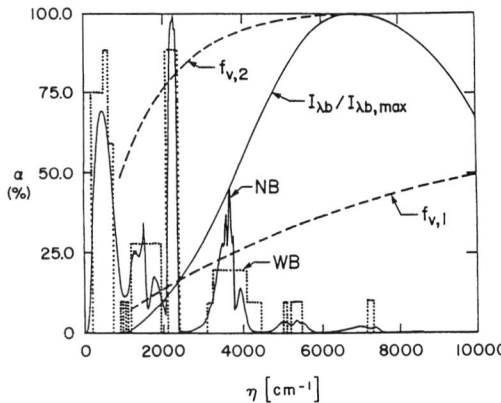

FIGURE 7.21   The important absorption/emission bands of combustion gases at different temperatures [132].

In this section, we will outline the formulation of the Edwards-Menard model, as it can be readily used for different applications.

Formulation of the Exponential Wideband Models.   In exponential wideband models, the band strength parameter, or an integrated absorption coefficient of a given wideband, is defined as:

$$\alpha \equiv \int_0^\infty \kappa_\eta \, d\eta \tag{7.149}$$

where κ_η is the spectral absorption coefficient, based on either line-by-line calculations or narrowband models. Note that the integration is evaluated from 0 to ∞, even though the band itself is finite in size. The Edwards-Menard model assumes an exponentially decaying function for the smoothed absorption coefficient (sometimes written as $\tilde{\kappa}_\eta = (S/d)_\eta$, where S and d are parameters defined similar to those used for individual absorption lines).

The exponential wideband model assumes that for a symmetric band, $\tilde{\kappa}_\eta$ is given as:

$$\tilde{\kappa}_\eta = \frac{\alpha}{\omega} e^{-f(\eta)/\omega} = \frac{\alpha}{\omega} e^{-2|\eta_c - \eta|/\omega} \tag{7.150}$$

where α is defined by Eq. (149) and ω is the so-called bandwidth parameter, which is the width of the band at $1/e$ of the maximum intensity. Most of the absorption bands of the gases of engineering interest are symmetric, and the corresponding η_c wavenumbers are listed in Table 7.8 (adapted from Ref. 3). If the bands are not symmetric, the $f(\eta)$ function is slightly modified to represent a cut band shape, which simulates the real profile more accurately. Then, either an upper limit η_u or lower limit η_l is used, and $f(\eta)$ is defined as $(\eta_u - \eta)$ or $(\eta - \eta_l)$. Note that, for only a few bands (η_u or η_l), values are reported in Table 7.9. (Even though these correlations were developed more than three decades ago, they are still the most convenient and accurate of all wideband model expressions, see Refs. 1 and 3 for details.)

The total band absorption is defined as

$$A \equiv \int_{band} \varepsilon_\eta \, d\eta = \int_0^\infty (1 - e^{-\kappa_\eta x}) \, d\eta \tag{7.151}$$

with x being the path length. The normalized total band absorptance is:

$$A^* = A/\omega = A^*(\alpha, \beta, \tau_0) \tag{7.152}$$

Here, $\tau_0 = \kappa_{\eta_c} x$ is the optical thickness at the band center. Table 7.8 lists all the exponential wideband model correlations as derived by Edwards and Menard for an isothermal gas. They are calculated using the equations given below:

$$\alpha(T) = \alpha_0 \frac{\Psi^*(T)}{\Psi^*(T_0)} = \alpha_0 \frac{\left[1 - \exp\left(-\sum_{k=1}^m u_k \delta_k\right)\right]\Psi(T)}{\left[1 - \exp\left(-\sum_{k=1}^m u_{0,k} \delta_k\right)\right]\Psi(T_0)} \tag{7.153}$$

$$\beta(t) = \gamma P_e = \gamma_0 \sqrt{\frac{T_0}{T}} \frac{\Phi(T)}{\Phi(T_0)} P_e \tag{7.154}$$

$$\omega(T) = \omega_0 \sqrt{T/T_0} \tag{7.155}$$

where
$$\Psi(T) = \frac{\prod_{k=1}^m \sum_{v_k = v_{0,k}}^\infty [(v_k + g_k + \delta_k - 1)!/(g_k - 1)!v_k!]e^{-u_k v_k}}{\prod_{k=1}^m \sum_{v_k = 0}^\infty [(v_k + g_k - 1)!/(g_k - 1)!v_k!]e^{-u_k v_k}} \tag{7.156}$$

TABLE 7.8   Exponential Wide Band Correlations for an Isothermal Gas [133]

$\beta \le 1$	$0 \le \tau_0 \le \beta$	$A^* = \tau_0$	Linear regime
	$\beta \le \tau_0 \le 1/\beta$	$A^* = 2\sqrt{\tau_0 \beta} - \beta$	Square root regime
	$1/\beta \le \tau_0 \le \infty$	$A^* = \ln(\tau_0 \beta) + 2 - \beta$	Logarithmic regime
$\beta \ge 1$	$0 \le \tau_0 \le 1$	$A^* = \tau_0$	Linear regime
	$1 \le \tau_0 \le \infty$	$A^* = \ln(\tau_0) + 1$	Logarithmic regime

TABLE 7.9   Wideband Correlation Parameters for Important Gases (Adapted from Ref. 3)

Band location		Pressure parameters		Correlation parameters		
λ (µm)	η_c (cm^{-1})	(δ_k)	n	b [cm^{-1}/(g/m^2)]	α_0	γ_0 (cm^{-1})	ω_0

λ (µm)	η_c (cm^{-1})	(δ_k)	n	b [cm^{-1}/(g/m^2)]	α_0	γ_0 (cm^{-1})	ω_0
H$_2$O	$m = 3$, $\eta_1 = 3652$ cm^{-1}, $\eta_2 = 1595$ cm^{-1}, $\eta_3 = 3756$ cm^{-1}, $g_k = (1,1,1)$						
71 µm*	140 cm^{-1}	(0,0,0)	1	B§ + 0.5	44.205	0.14311	69.3
6.3 µm	1600 cm^{-1}	(0,1,0)	1	B + 0.5	41.2	0.09427	56.4
2.7 µm	3760 cm^{-1}	(0,2,0)	1	B + 0.5	0.2	0.13219†‡	60.0†
		(1,0,0)	1	B + 0.5	2.3	0.13219†‡	60.0†
		(0,0,1)	1	B + 0.5	23.4	0.13219†‡	60.0†
1.87 µm	5350 cm^{-1}	(0,1,1)	1	B + 1.5	3.0	0.08169	43.1*
1.38 µm	7250 cm^{-1}	(1,0,1)	1	B + 1.5	2.5	0.11628	32.0
CO$_2$	$m = 3$, $\eta_1 = 1351$ cm^{-1}, $\eta_2 = 666$ cm^{-1}, $\eta_3 = 2396$ cm^{-1}, $g_k = (1,2,1)$						
15 µm	667 cm^{-1}	(0,1,0)	0.7	1.3	19.0	0.06157	12.7
10.4 µm	960 cm^{-1}	(−1,0,1)	0.8	1.3	2.47×10^{-9}	0.04017	13.4
9.4 µm	1060 cm^{-1}	(0,−2,1)	0.8	1.3	2.48×10^{-9}	0.11888	10.1
4.3 µm	2410 cm^{-1} ($= \eta_u$)	(0,0,1)	0.8	1.3	110.0	0.24723	11.2
2.7 µm	3660 cm^{-1}	(1,0,1)	0.65	1.3	4.0	0.13341	23.5
2.0 µm	5200 cm^{-1}	(2,0,1)	0.65	1.3	0.060	0.39305	34.5
CO	$m = 1$, $\eta_1 = 2143$ cm^{-1}, $g_1 = 1$						
4.7 µm	2143 cm^{-1}	(1)	0.8	1.1	20.9	0.07506	25.5
2.35 µm	4260 cm^{-1}	(2)	0.8	1.0	0.14	0.16758	20.0
CH$_4$	$m = 4$, $\eta_1 = 2914$ cm^{-1}, $\eta_2 = 1526$ cm^{-1}, $\eta_3 = 3020$ cm^{-1}, $\eta_4 = 1306$ cm^{-1}, $g_k = (1,2,3,3)$						
7.7 µm	1310 cm^{-1}	(0,0,0,1)	0.8	1.3	28.0	0.08698	21.0
3.3 µm	3020 cm^{-1}	(0,0,1,0)	0.8	1.3	46.0	0.06973	56.0
2.4 µm	4220 cm^{-1}	(1,0,0,1)	0.8	1.3	2.9	0.35429	60.0
1.7 µm	5861 cm^{-1}	(1,1,0,1)	0.8	1.3	0.42	0.68598	45.0
NO	$m = 1$, $\eta_1 = 1876$ cm^{-1}, $g_1 = 1$						
5.3 µm	1876 cm^{-1}	(1)	0.65	1.0	9.0	0.18050	20.0
SO$_2$	$m = 3$, $\eta_1 = 1151$ cm^{-1}, $\eta_2 = 519$ cm^{-1}, $\eta_3 = 1361$ cm^{-1}, $g_k = (1,1,1)$						
19.3 µm	519 cm^{-1}	(0,1,0)	0.7	1.28	4.22	0.05291	33.1
8.7 µm	1151 cm^{-1}	(1,0,0)	0.7	1.28	3.67	0.05952	24.8
7.3 µm	1361 cm^{-1}	(0,0,1)	0.65	1.28	29.97	0.49299	8.8
4.3 µm	2350 cm^{-1}	(2,0,0)	0.6	1.28	0.423	0.47513	16.5
4.0 µm	2512 cm^{-1}	(1,0,1)	0.6	1.28	0.346	0.58937	10.9

* For the rotational band $\alpha = \alpha_0 \exp(-9\sqrt{T_0/T})$, $\gamma = \gamma_0 \sqrt{T_0/T}$.
† Combination of three bands, all but weak (0,2,0) band are fundamental bands, $\alpha_0 = 25.9$ cm^{-1}.
‡ Line overlap for overlapping bands from Eqs. (7.163, 7.164).
§ B $= 8.6\sqrt{T_0/T}$.

$$\Phi(T) = \frac{\left[\prod_{k=1}^{m} \sum_{v_k=v_{0,k}}^{\infty} ([(v_k + g_k + \delta_k - 1)!/(g_k - 1)!v_k!]e^{-u_k v_k})^{1/2}\right]^2}{\prod_{k=1}^{m} \sum_{v_k=v_{0,k}}^{\infty} [(v_k + g_k + \delta_k - 1)!/(g_k - 1)!v_k!]e^{-u_k v_k}} \qquad (7.157)$$

and

$$u_k = hc\eta_k/kT, \quad u_{0,k} = hc\eta_k/kT_0 \qquad (7.158)$$

where $T_0 = 100$ K.

$$v_{0,k} = \begin{pmatrix} 0 & \text{for } \delta_k \geq 0 \\ |\delta_k| & \text{for } \delta_k \leq 0 \end{pmatrix} \qquad (7.159)$$

$$P_e = \left[\frac{p}{p_0}\left(1 + (b-1)\frac{p_a}{p}\right)\right]^n, \quad (p_0 = 1 \text{ atm}) \qquad (7.160)$$

The parameters required for these calculations are listed in Table 7.9 for six common gases. Here, v_k is the vibrational quantum number, δ_k is the change in v_k during transition, and g_k is the statistical weight for the corresponding transition. Most of the absorption bands of the gases given in Table 7.9 have nonnegative δ_k values, and, for most, $v_{0,k}$ is zero. Under these conditions, the denominator of Eq. 7.157 for Φ, which is identical to the numerator of Eq. 7.156 for Ψ, becomes:

$$\sum_{v_k=0}^{\infty} \frac{(v_k + g_k + \delta_k - 1)!}{(g_k - 1)!} e^{-u_k v_k} = \frac{(g_k + \delta_k - 1)!}{(g_k - 1)!}(1 - e^{-u_k})^{-g_k - \delta_k} \qquad (7.161)$$

If the values for $v_{0,k}$ are not zero, corresponding values should be subtracted from the above expression [3].

The reference values $\Psi(T_0) = \Psi_0$ and $\Phi(T_0) = \Phi_0$ are obtained at a very low temperature of $T_0 = 100$ K:

$$\Psi_0 \approx \prod_{k=1}^{m} \frac{(g_k + \delta_k - 1)!}{(g_k - 1)!}, \quad \Phi_0 \approx 1 \qquad (7.162)$$

Note that the Ψ and Φ expressions given in Eqs. 7.156 and 7.157 have infinite-summation terms. Recently, Lallemant and Weber [134] suggested a simplified formulation based on polynomial expression of the Φ functions, which reduces the required computational time significantly.

Sometimes, the absorption bands of different species overlap. For these cases, the band parameters are calculated as:

$$\alpha = \sum_{j=1}^{J} \alpha_j \qquad (7.163)$$

$$\beta = \frac{1}{\alpha}\left[\sum_{j=1}^{J} \sqrt{\alpha_j \beta_j}\right]^2 \qquad (7.164)$$

where J is the number of overlapping bands.

Felske and Tien [135] presented a relatively simple correlation for the band absorption A^* as:

$$A^* = 2E_1\left(\sqrt{\frac{\tau_0 \beta}{1 + \beta/\tau_0}}\right) + E_1\left(\frac{1}{2}\sqrt{\frac{\tau_0/\beta}{1 + \beta/\tau_0}}\right) - E_1\left(\frac{1 + 2\beta}{2}\sqrt{\frac{\tau_0/\beta}{1 + \beta/\tau_0}}\right)$$

$$+ \ln\left(\frac{\tau_0}{(1 + \beta/\tau_0)(1 + 2\beta)}\right) + 2\gamma_E \quad (7.165)$$

where E_1 is the exponential integral function and $\gamma_E = 0.57221\ldots$ is the Euler constant [136].

Nonisothermal Gases.   The spectral emissivity for a nonisothermal path within a gas cloud is expressed as:

$$\varepsilon_\eta = 1 - \exp\left(-\int_0^X \kappa_\eta \, dX\right) \tag{7.166}$$

Using this expression, the total band absorptance can be determined:

$$A = \int_0^\infty \varepsilon_\eta \, d\eta = \int_0^\infty \left[1 - \exp\left(-\int_0^X \kappa_\eta \, dX\right)\right] d\eta \tag{7.167}$$

If a simple wideband model is going to be used, the path-averaged values of α, β, and ω are needed. The correlations for these values were listed in Refs. 129 and 137. If a weak line limit is considered, the following simple relations yield acceptable accuracy in predictions [137]:

$$\tilde{\alpha} \equiv \frac{1}{X}\int_0^X \int_0^\infty \kappa_\eta \, d\eta \, dX = \frac{1}{X}\int_0^X \alpha \, dX \tag{7.168}$$

$$\tilde{\omega} \equiv \frac{1}{\tilde{\alpha}X}\int_0^X \omega\alpha \, dX \tag{7.169}$$

$$\tilde{\beta} \equiv \frac{1}{\tilde{\omega}\tilde{\alpha}X}\int_0^X \beta\omega\alpha \, dX \tag{7.170}$$

Total Emissivity and Absorptivity Models.   For many engineering applications, spectrally integrated gas emission and absorption are needed. The total emissivity is defined as the ratio of the total radiation emitted from a path of length X to the maximum possible emission:

$$\varepsilon \equiv \frac{\int_0^\infty I_{b\eta}\varepsilon_\eta \, d\eta}{\int_0^\infty I_{b\eta} \, d\eta} = \frac{\int_0^\infty I_{b\eta}(1 - e^{-\kappa_\eta X}) \, d\eta}{\int_0^\infty I_{b\eta} \, d\eta}$$

$$= \sum_{i=1}^N \left(\frac{\pi I_{b\eta 0}}{\sigma T^4}\right)_i \int_{\Delta\eta_{\text{band}}} (1 - e^{\kappa_\eta X}) \, d\eta = \sum_{i=1}^N \left(\frac{\pi I_{b\eta 0}}{\sigma T^4}\right)_i A_i \tag{7.171}$$

The last expression can be easily calculated using the data supplied before. In the derivation of this expression, there are two implicit assumptions: (1) the spectral wideband is still very narrow, such that the Planck blackbody emission does not vary over the spectral interval of the band; and (2) there are no overlapping bands. Even though the first assumption is easy to justify, the second is not. For example, at about $\lambda = 2.7$ μm, both H_2O and CO_2 have strong absorption bands, which contribute significantly to their emission. For completely overlapped bands, the total emissivity is expressed as:

$$\varepsilon_{a+b} = \varepsilon_a + \varepsilon_b - \varepsilon_a\varepsilon_b \tag{7.172}$$

If the bands overlap only partially, the calculations become more cumbersome and will not be given here (see Refs. 1 and 3 for further discussion).

Most textbooks present the total emissivity and absorptivity in charts, an approach first introduced by Hottel [35]. These charts are based on the following expressions:

$$\varepsilon = \varepsilon(p_aL, p, T_g) \tag{7.173}$$

$$\alpha = \alpha(p_aL, p, T_g, T_s) \approx \left(\frac{T_g}{T_s}\right)^{1/2}\varepsilon\left(p_aL\,\frac{T_g}{T_s}, p, T_s\right) \tag{7.174}$$

where T_g is the gas temperature and T_s is the temperature of the source (for absorption calculations).

Figures 7.22 and 7.23 depict the gas emissivity ε for H_2O and CO_2, the most common combustion gases. These values are valid for individual gases and mixtures with a total pressure of 1 atm. If the pressure is higher and/or if both H_2O and CO_2 are present, then the total emissivity is to be corrected according to the relation:

$$\varepsilon_{total} = C_{H_2O}\varepsilon_{H_2O} + C_{CO_2}\varepsilon_{CO_2} - \Delta\varepsilon \qquad (7.175)$$

Figure 7.24 shows the correction factor $\Delta\varepsilon$ for the overlap of CO_2 and H_2O bands for different pressures and path lengths.

Absorption Coefficient Based Gas Properties. In the gas models discussed above, the final product is the band absorptance and total emissivity or absorptivity. For most radiative transfer models, however, the absorption coefficient is the desired input. For this purpose, either polynomial expressions or the weighted-sum-of-gray-gases (WSGG) models are developed, which are usually based on the pseudo absorption coefficient concept [35]:

$$\varepsilon = \sum_{i=0}^{I} b_{\varepsilon,i}\{1 - e^{K_i p_j L}\} \qquad (7.176)$$

where $b_{\varepsilon,i}$ and K_i are the parameters used to fit this polynomial to radiative properties predicted by the narrow or wideband models, and they are usually functions of pressure and temperature. Here L is the physical length of the gas cloud and p_j is the partial pressure of the jth species. It is obvious that K_i, by definition, is similar to the absorption coefficient. This suggests that the RTE can be solved N times for different values of the K_i using any standard formulation, such as the DO, SH, or MC techniques. Note that if $i = 0$, $b_{\varepsilon,i}$ is also set to zero to account for the transparent window of the wavelength range.

Polynomial gas property models [139, 140, 141] as well as the weighted-sum-of-gray-gas models [142–155] have been studied extensively. All of these models are for predicting the gas radiation along a homogeneous line of sight (i.e., constant pressure and temperature).

One of the most elaborate of these approaches is that developed by Denison and Webb [147–153]. Instead of fitting the parameters of the WSGG model to the total emissivity data (as done for all other models), they used the high resolution transmission molecular absorption database of Rothman et al. [156]. By doing so, they replaced the spectral integration over wavenumber by a quadrature over an absorption cross section. Their approach is also known as the correlated k-distribution method.

Detailed comparison of different WSGG models are available in the literature. Spectral calculations in a medium with homogeneous absorption coefficient profile are given by Modest [146]. Song [157] compared the WSGG model predictions against wideband model results in a planar medium. Denison and Webb [147, 148] evaluated their model against the detailed line-by-line calculations for water-vapor in planar media. Comparisons of WSGG against a statistical narrowband model were given for both planar and axisymmetric combustion chambers by Soufiani and Djavdan [158]. Pierrot et al. [159] presented detailed comparisons for a planar medium containing an H_2O-N_2 mixture. Their results indicated that WSGG models are very time-efficient yet may yield inaccurate predictions. The most significant weakness of these models is realized if there is a large difference between the temperatures of the radiation source and the absorbing gases. Also, if particle and wall properties considered are not gray, the model becomes quite cumbersome. Several different WSGG models need to be used at different wavelength intervals over the spectrum, as the model properties lose their universality [159].

Tang and Brewster [160] used the correlated k-distribution method proposed in [161] to account for nongray gas contribution when absorbing, emitting, and anisotropically scattering particles are present. They obtained very good accuracy, and showed that this approach is compatible with the discrete ordinates and Monte Carlo techniques. Marin and Buckius [162] also used a correlated-k distribution method based on wideband formulation for absorbing, emitting, and scattering media and reported reasonable accuracy.

In a recent study, Lallemant and Weber [134] compared the performance of various WSGG models with their simplified exponential wideband model. These results indicate that,

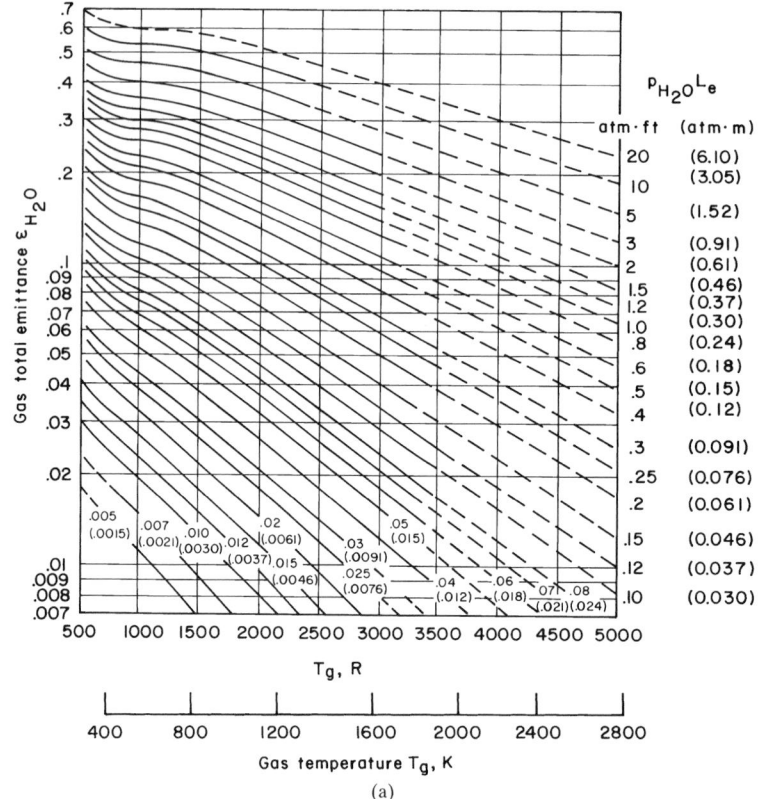

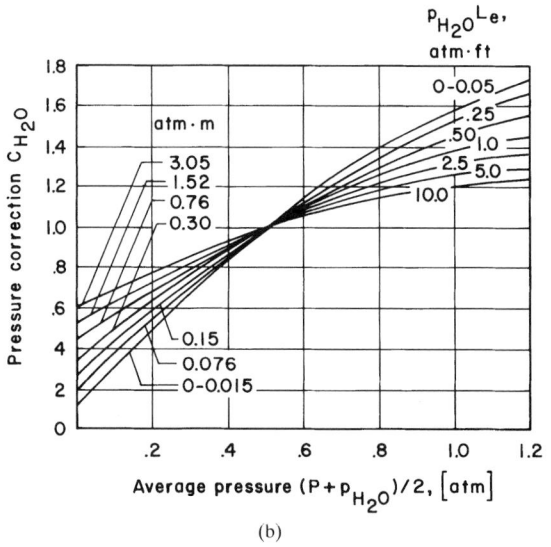

FIGURE 7.22   (a) Total emissivity ε_{H_2O} for water vapor as a function of T_{gas} and $p_{H_2O}L_e$ product; (b) the corresponding correction factor C_{H_2O} for different total pressures.

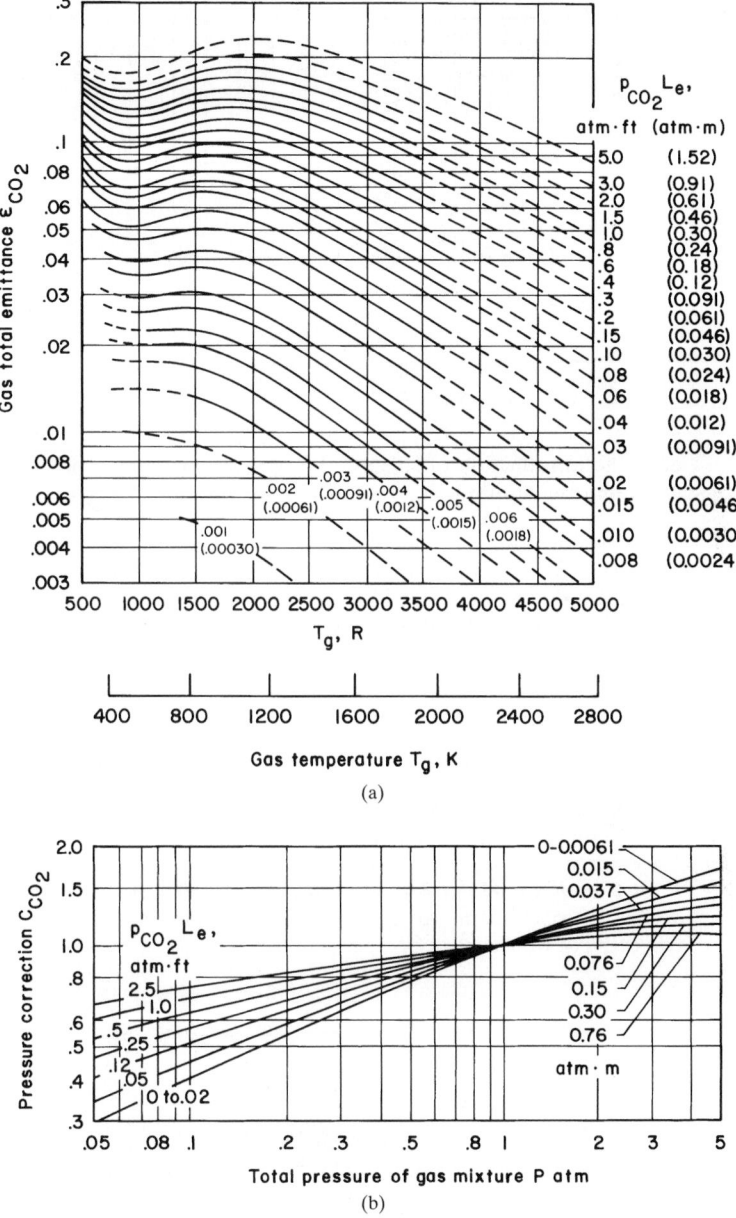

FIGURE 7.23 Total emissivity ε_{CO_2} for carbon dioxide as a function of T_{gas} and $p_{CO_2}L_e$ product; (b) the corresponding correction factor C_{CO_2} for different total pressures.

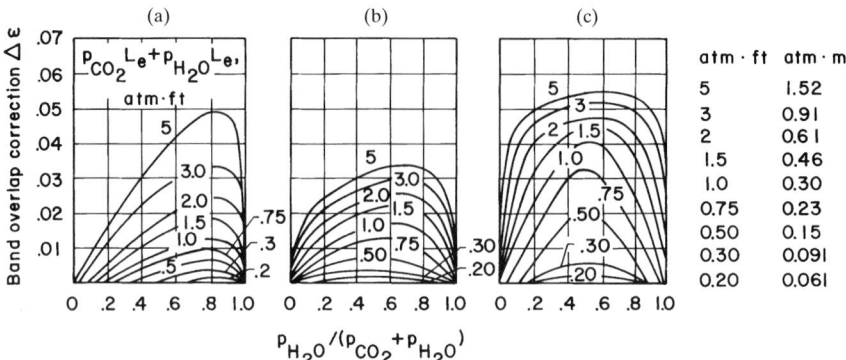

FIGURE 7.24 $\Delta\varepsilon$ band overlap correction factors for water vapor and carbon dioxide total emissivities (see Eq. 7.175). (a) $T_{gas} = 400$ K (720 R); (b) $T_{gas} = 810$ K (1460 R); (c) $T_{gas} \geq 1200$ K (2160 R).

after the introduction of suggested modifications, the model given in Ref. 134 yields more accurate results, although it requires slightly more CPU time.

For most combustion chamber applications, pressure, temperature, and concentration along the line of sight of radiation do change. Under these circumstances, one should exercise care in using total emissivity and absorptivity models. For inhomogeneous paths, either the equivalent line model [163, 164] or total transmittance nonhomogeneous method [165] can be used. These models are quite complicated and computer-intensive and can therefore be efficiently used only for simple systems (such as the interpretation of results obtained from optical diagnostic techniques) rather than for multidimensional systems. Also, they can be coupled with particle properties, as suggested by Grosshandler and Modak [166].

Denison and Webb [150, 151] applied the spectral line WSGG (SLW) model to inhomogeneous paths. They reported that, compared to line-by-line calculations, the SLW model was off about 15 to 20 percent; yet, the computational savings were significant. Another simple and effective model was developed by Parthasarathy et al. [154] and applied to different media with both uniform and nonuniform temperature profiles.

Planck and Rosseland Mean Absorption Coefficients.   Two of the most common *mean absorption* coefficient expressions are the Planck and Rosseland coefficients, κ_P and κ_R, respectively, which are defined as [1, 3, 34]:

$$\kappa_P = \frac{\displaystyle\int_0^\infty I_{b\eta}\kappa_\eta \, d\eta}{\displaystyle\int_0^\infty I_{b\eta} \, d\eta} = \frac{\pi}{\sigma T^4}\int_0^\infty I_{b\eta}\kappa_\eta \, d\eta = \sum_{i=1}^N \left(\frac{\pi I_{b\eta 0}}{\sigma T^4}\right)_i \alpha_i \tag{7.177}$$

$$\frac{1}{\kappa_R} = \frac{\displaystyle\int_0^\infty (1/\kappa_\eta)(dI_{b\eta}/dT) \, d\eta}{\displaystyle\int_0^\infty (dI_{b\eta}/dT) \, d\eta} = \frac{\pi}{4\sigma T^3}\int_0^\infty \frac{1}{\kappa_\eta}\frac{dI_{b\eta}}{dT} \, d\eta \tag{7.178}$$

Note that these coefficients are obtained over the entire wavelength range; therefore, the spectral details are lost. For practical applications where the spectral structure of radiation field is important, the use of these mean coefficients is not recommended. The Planck coefficient yields acceptable results if the medium optical thickness is small (less than unity), whereas the Rosseland coefficient is more useful at large optical thicknesses (i.e., larger than five over the entire spectrum).

Radiative Properties of Particulates

In many engineering applications, the radiative properties of particles are much more important than those of combustion gases because particles absorb, emit, and scatter radiation continuously across the entire wavelength spectrum. By contrast, combustion gases participate radiatively only in narrow bands centered around discrete wavelengths. The radiative properties required for typical radiative transfer calculations are absorption and scattering coefficients and scattering phase function. These properties are dependent on the partial pressures and chemical composition of combustion gases, material and physical structure of the particles, particle size and volume fraction distributions in the medium, and the wavelength of radiation.

The material structure of the particulate matter determines its complex index of refraction, which is considered to be the most fundamental property. The real part of the complex refractive index is the ratio of the speed of light in vacuum to that within the particle for light at normal incidence. In this case, the imaginary part, which is also termed the attenuation, extinction, or absorption index, is directly related to the rate of attenuation of radiation with depth within the material. For other than normal incidence, the relations between the complex index of refraction, speed of light, and attenuation within the particle are complicated and require rigorous solution of the electromagnetic (EM) wave equations (i.e., Maxwell's equations) within the medium of interest with appropriate boundary conditions.

As one may expect, if the shape of the particle is simple (i.e., mathematically tractable), then the solution to the problem can be obtained relatively easily. An analytical solution to the propagation of electromagnetic waves in spherical particles was first given by Lorenz in 1890 and Mie in 1903 (see Ref. 167 for historical discussion and Refs. 167 and 168 for the detailed formulations). The theory, which is widely referred to as the Lorenz-Mie theory, is used to determine how the incident wave is absorbed and scattered by a homogeneous spherical obstacle, provided that the diameter of the sphere, the wavelength of the incident electromagnetic wave, and the complex index of refraction of particle and surroundings are available. For spherical particles, there is a natural scaling factor:

$$x = \pi D/\lambda \qquad (7.179)$$

which is called the *size parameter*. Two particles of different sizes but with the same refractive index absorb and scatter the EM wave similarly, if they have the same size parameter (i.e., if the ratio of their diameter to the wavelength of the incident radiation is the same).

The Lorenz-Mie theory yields information about the absorption and scattering of an incident EM wave by an isolated single spherical particle. It predicts the rate at which the energy is absorbed and scattered by the particle as a function of its complex index of refraction. The ratio of this rate to the incident irradiance gives the absorption and scattering cross sections, C_{abs} and C_{sca}, respectively. These cross sections have the units of area, and indicate the size of an imaginary surface which extracts the same amount of energy from the incident beam. It is preferable to express these parameters in dimensionless form. If we consider the ratio of the absorption/scattering cross section to the geometric area cross section of a particle, we obtain the efficiency factors:

$$Q_{abs} = \frac{C_{abs}}{\pi D^2/4}; \quad Q_{sca} = \frac{C_{sca}}{\pi D^2/4} \qquad (7.180)$$

and

$$Q_{ext} = Q_{abs} + Q_{sca} \qquad (7.181)$$

The extinction efficiency factor Q_{ext} approaches a constant asymptotic value of 2 with increasing size parameter. It means that a large particle attenuates (absorbs and scatters) twice as much energy from an incident beam as characterized by its own geometric area. The additional attenuation is because of diffraction of the EM waves by the particle. This phenomenon is known as the *extinction paradox*. Note that, at the large particle limit, the values of Q_{abs} and Q_{sca} are not necessarily equal to 1 (i.e., one half of the value of Q_{ext}); rather, their asymptotic limits are dependent on their complex index of refraction.

The volumetric absorption, scattering, and extinction coefficients of polydispersions are related to efficiency factors as

$$\eta_\lambda = \int_0^\infty Q_{\eta,\lambda} \frac{\pi D^2}{4} \, dN(D) \tag{7.182}$$

where η_λ is either κ_λ, σ_λ, or β_λ, $Q_{\eta,\lambda}$ is the corresponding spectral efficiency factor Q_{abs}, Q_{sca}, or Q_{ext}; and $dN(D)$ is the particle size distribution (number of particles of diameter D per unit volume).

Wiscombe and Mugnai [169] showed that differences exist in the scattering properties of irregularly shaped particles when compared to the Lorenz-Mie theory calculations. For irregularly shaped particles, the oscillations in the efficiency factors versus size parameter are damped out, and more side-scattering and less back-scattering are observed. The agreement with the Lorenz-Mie theory calculations becomes worse for radiative properties as the size parameter increases past $x = \pi D/\lambda = 5$. It is stated that the spherical shape is a singularity in nature rather than a norm, and using a single effective diameter may yield erroneous results in radiative property predictions [170].

Analytical expressions similar to those for spherical particles have been derived for infinite-length cylinders in perpendicular incidence as well as in oblique incidence, for elliptic cylinders, and for spheroids (see Refs. 168 and 169). With increasing complexity of the shape of the particle, even with as little change as from sphere to spheroid, the analytical solution to the problem becomes formidable. Then, the use of numerical solution techniques may be preferable to analytical techniques.

The numerical techniques used to determine the particle absorption and scattering characteristics can be divided into two subgroups [169]: (1) integral equation methods and (2) differential equation approaches. Integral equation methods use Green's functions in the formulation of the vector in the Helmholtz equations, and the boundary conditions between the obstacle and the surrounding medium are accounted for directly in the formulation. The extended boundary condition method (EBCM) which is also known as the T-matrix method is the most successful integral equation method [169, 170, 171]; a recent detailed review is available in the literature [171]. Differential equation methods, on the other hand, were developed employing finite element methods and can be extended to inhomogeneous obstacles. They are simpler in concept compared to integral equation methods; however, they require significantly more computational time.

Another method was suggested by Purcell and Pennypacker [172]. They replaced the scattering obstacle by a cubic array of point dipoles spaced no farther than $\lambda/4\pi$ apart and considered mutual dipole-dipole interactions exactly. This technique provides a highly general prediction of scattering cross sections of irregularly shaped and nonhomogeneous particles. Later extensions of this approach were given by Draine [173]. Currently, the only drawback of this technique appears to be the maximum size parameter that can be considered; because of convergence problems, results obtained are not reliable beyond $\pi D/\lambda \geq 4$. A recent review of these techniques has been given by Manickavasagam and Mengüç [174].

Although most particles in combustion chambers are nonspherical, because of large uncertainties in particle shape and material structure (which affects the complex index of refraction), the use of more sophisticated models to determine particle radiative properties is not always warranted. There are special applications where the exact shape and composition of the particle may be very critical to the understanding of the physical phenomena. For example, it is important to know the soot agglomerate structure to determine the soot formation mechanism in flames using laser/light diagnostics. Also important is the effect of pores and material inhomogeneity in pulverized coal particles on the radiation-combustion interaction. In other words, particle shape and structure are very critical in understanding the microscale phenomena being observed in flames; however, for predicting the average radiative properties for radiative heat transfer calculations, the use of simpler models may be sufficient.

Simple engineering models for coal/char particles were suggested by Buckius and Hwang [175], Viskanta et al. [176], and Mengüç and Viskanta [177]. In Ref. 175, the coal/char radiative properties were calculated from the Lorenz-Mie theory for different size distributions and complex index of refraction values, and then a curve-fitting technique was employed to obtain simple engineering equations. It was shown that these "empirical" equations are valid for a wide variety of size distributions. The same conclusion was also drawn in Ref. 177, where analytical expressions were developed starting from anomalous diffraction theory. More recently, Im and Ahluwalia [178] and Kim et al. [179] presented a series of approximations for pulverized-coal and particulate-gas mixtures. Any of these models can be easily incorporated into complicated global computer algorithms.

Skocypec and Buckius [180, 181] presented an analytical formulation to obtain the radiation heat transfer from a mixture of combustion gases and scattering particles. They considered band models for the gases and accounted for the absorption and scattering by particles. They developed charts similar to Hottel charts for combustion gases. The results presented can be used to obtain the average radiative properties if the particle loading information is not known accurately. (See also Refs. 182–184 for a discussion on the limits of this formulation.)

In the following sections, we will present simple techniques to calculate the optical/radiative properties of soot, fly ash, and coal/char particles. Also, the necessary physical parameters will be summarized.

Soot Particles/Agglomerates. Soot is one of the most important contributors to radiative transfer in combustion chambers. It is formed during the combustion of almost all hydrocarbons, and in combustion chambers it exists almost everywhere. Unlike combustion gases such as H_2O or CO_2, which absorb and emit only at certain wavelengths, soot participates in radiative transfer at all wavelengths. Therefore, the radiant energy emitted from a sooty flame is significantly more than, for example, a clean-burning natural gas flame. Although this can be considered as an advantage, having unburned carbon particles emitted in the exhaust is a major drawback. In order to predict the contribution of soot to overall radiative transfer phenomena in combustion chambers, one must know the soot shape, size, size distribution, and optical properties.

In a combustion chamber, soot volume fraction or number densities are usually not known. It is generally accepted that the primary soot particles are spherical in shape and about 20–60 nm in diameter (see the review by Charalampopoulos [185]). However, depending on the flow and combustion characteristics of the system, they agglomerate to form irregularly shaped large particles. The shape can be clusters of spheres or cylindrical long tails attached to burning coal particles.

If the agglomeration is not considered, calculation of required radiative properties of soot particles will be straightforward. Since the size of an individual soot sphere is much smaller than the wavelength of radiation, the Rayleigh limit (for small $x = \pi D/\lambda$) to the Lorenz-Mie theory can be used. Then, the soot absorption and scattering efficiency factors are given as

$$Q_{abs,Rayleigh} = 12x\,\frac{N_1}{M_1} = x\left\{\frac{24n_\lambda k_\lambda}{(n_\lambda^2 - k_\lambda^2 + 2)^2 + 4n_\lambda^2 k_\lambda^2}\right\} \tag{7.183}$$

$$Q_{sca,Rayleigh} = \frac{8}{3}x^4\left(1 - 3\,\frac{M_2}{M_1}\right) \tag{7.184}$$

where
$$M_1 = |2 + m^2|^2 = N_1^2 + (2 + N_2)^2 \qquad M_2 = 1 + 2N_2$$
$$N_1 = 2nk \equiv -I(m^2) \qquad N_2 = n^2 - k^2 \equiv R(m^2) \tag{7.185}$$

Selamet and Arpaci [186, 187] have investigated the accuracy of the Rayleigh approximation for soot particles and proposed a simple extension based on the Penndorf approximation. According to their study, for larger particles, the extinction efficiency should be modified to:

$$Q_{ext} = Q_{abs,Rayleigh} + 2x^3 \left[N_1 \left(\frac{1}{15} + \frac{5}{3} \frac{1}{M_4} + \frac{6}{5} \frac{M_5}{M_1^2} \right) + \frac{4}{3} \frac{M_6}{M_1^2} x \right] \qquad (7.186)$$

where
$$M_4 = 4N_1^2 + (3 + 2N_2)^2 \qquad M_5 = 4(N_2 - 5) + 7N_3$$

$$M_6 = (N_2 + N_3 - 2)^2 - 9N_1^2 \qquad N_3 = (n^2 + k^2)^2 = N_1^2 + N_2^2 \qquad (7.187)$$

These equations can be used up to size parameter of $x = 0.8$ for soot index of refraction range of $1.5 \le n \le 4.0$ and $0.5 \le k \le 3.0$ with acceptable accuracy [187].

The soot absorption coefficient is calculated from Eq. 7.182, which can be simplified further for a monodisperse particle cloud as:

$$\kappa_\lambda = Q_{abs} N A_c = \frac{3 Q_{abs} f_v}{2D} \qquad (7.188)$$

where A_c is the particle cross sectional area and f_v is the volume fraction of soot particles in the medium.

For modeling soot radiation, a very simple engineering equation was suggested by Hottel and Sarofim [35] in the form of:

$$\kappa_\lambda = K f_v / \lambda \qquad (7.189)$$

where K is a constant, which depends on the complex index of refraction. Hottel and Sarofim used a value of 7 for K in their calculations. Later, Siegel [188] suggested different values for different fuels: 6.3 for oil flames, 4.9 and 4.0 for propane and acetylene flames, and a range between 3.7 to 7.5 for coal flames. Modak [140] has incorporated this expression into a computer code to calculate the radiant emission from a mixture of soot and H_2O and CO_2 gases. Note that this approximation implies that the soot particles are small compared to the wavelength of radiation, and scattering is not important.

The size parameter $x = \pi D/\lambda$ is in the order of 10^{-1} if λ is about 1 or 2 μm. Therefore, Q_{sca}, which is proportional to x^4, is small. Because of this, scattering of radiation by *individual* soot particles is safely neglected for radiation heat transfer calculations. This conclusion, however, is only applicable to small soot spheres; if these small particles agglomerate, the importance of scattering increases. If the agglomerates are treated in calculations as if they are small spheres, then the radiative energy gain and loss cannot be predicted accurately. This, in turn, affects the predictions of temperature and species distribution, devolatilization, chemical kinetics, and soot formation.

There have been several attempts in the literature to model soot agglomerates as homogeneous solid particles. These approaches yield acceptable results for calculation of soot absorption coefficients, even though they are not exact. This is because the soot volume fraction is required along with absorption cross section to calculate the absorption coefficient (see Eq. 7.183), and the uncertainty in the value of local soot volume fraction is usually larger than that for the cross sections. For example, soot agglomerates were simulated as prolate and oblate spheroids and as infinite-length cylinders [189–192].

For a cloud of randomly oriented small spheroids, the spectral absorption ($\simeq$ extinction) coefficient is given as (Lee and Tien [189]):

$$\kappa_\lambda = \frac{4\pi n_\lambda k_\lambda}{3\lambda} f_v \sum_{j=1,2} \frac{j}{[1 + (n_\lambda^2 - k_\lambda^2 - 1)P_j]^2 + (2n_\lambda k_\lambda P_j)^2} \qquad (7.190)$$

where P_1 and P_2 are the depolarization factors, and $P_1 = 1 - 2P_2$. For prolate spheroids:

$$P_1 = \frac{1 - e^2}{e^2} \left[\frac{1}{2e} \ln \left(\frac{1 + e}{1 - e} \right) - 1 \right] \qquad (7.191)$$

and for oblate spheroids:

$$P_1 = \frac{1}{e^2}\left[1 - \left(\frac{1-e^2}{e^2}\right)^{1/2} \sin^{-1} e \right] \qquad (7.192)$$

where e is the eccentricity factor, defined as $D_{\text{minor}}/D_{\text{major}}$.

For cylinders, it is difficult to obtain a similar expression analytically. However, an approximate expression for the absorption efficiency factor was given as [189]:

$$Q_{\text{abs}} = x\gamma\,\frac{\pi}{4}\,n_\lambda k_\lambda \left\{ 1 + \frac{4}{(n_\lambda^2 - k_\lambda^2 + 1)^2 + 4n_\lambda^2 k_\lambda^2} \right\} \qquad (7.193)$$

where γ is a constant. Lee and Tien [189] recommended $\gamma = 1.1$ for infrared wavelengths. Charalampopoulos and Hahn [192] suggested a functional form for γ, as

$$\gamma = a + by + cy^2 \qquad (7.194)$$

where $y = x/|m|$, and $|m|$ is the modulus of the complex index of refraction. They recommended the following values for the coefficients: $a = 1.005563$, $b = 5.313441$, $c = -8.011187$. This expression yields very good agreement with the exact results over the wavelength spectrum of 0.4–15 μm and temperature range of 600–2000 K.

For cylindrical-shaped soot agglomerates, Mackowski et al. [190] presented more detailed expressions for both absorption and scattering efficiency factors, and showed that the expressions were reliable from the visible wavelength range up to 5 μm. They also applied a hybrid sphere/cylinder-shaped soot agglomerate model to interpret the experimental results obtained for the hydrocarbon flames, and showed that a larger fraction of soot agglomerates behave like cylinders [190, 191].

Determining the scattering characteristics of an agglomerate is more complicated than that for a homogeneous but irregularly shaped particle. For the latter, a solution is obtained by matching the amplitude and phase of the electromagnetic waves inside and outside of the particle, which is mathematically difficult but straightforward. For agglomerates, however, the interaction between individual spheres is to be considered. This can be done by superposing the electric field components of scattered electromagnetic waves from each individual sphere.

Coherent scattering of radiation by soot agglomerates has been studied by various researchers. The first model developed to examine irregularly shaped particles and agglomerates in a comprehensive way, including the retardation effects, is the discrete-dipole approximation (DDA) of Purcell and Pennypacker [172], who used it to study interstellar dust. The DDA is a general and flexible method for computing the approximate extinction and scattering cross sections of particles of arbitrary shape. The particle is modeled by an array of N polarizable elements (called dipoles) in vacuum. In their original study, Purcell and Pennypacker presented solutions for particles with $N < 256$. The reader is referred to Ref. 174 for a recent review of agglomerate approaches. The following is a brief survey to highlight some of the milestones.

Draine [173] presented a comprehensive treatment of the problem and discussed the validity of different approximations. A detailed review of DDA for scattering calculations and its relation to the other methods was presented by Draine and Flatau [193]. Vaglieco et al. [194] applied this approach to investigate the spectral characteristics of diesel soot. Ivezic and Mengüç [195] used the DDA to study dependent/independent scattering by agglomerates. A different approach to predict scattering and absorption coefficients of multiple-sphere clusters was presented by Mackowski [196] and Fuller [197], who considered superposition of radiative interactions among spheres. Analytical expressions were derived for orientation-averaged scattering and absorption coefficients.

For agglomerates of small spheres, a solution technique developed by Jones [198, 199] has received considerable attention [191, 200–205]. In this approach, the primary soot particles are assumed to be in the Rayleigh limit. The accuracy of the model decreases significantly as the individual soot spheres become larger compared to the wavelength of radiation, i.e., when they cannot be considered as Rayleigh spheres. Note that, even for primary soot particles as small as 50 nm in diameter, the Rayleigh theory yields errors at visible wavelengths. In another model proposed by Iskander et al. [206], an irregularly shaped particle was divided into a number of cubical cells, and, inside each cell, the electromagnetic field was assumed to be uniform. Then a control volume approach was employed to reduce the governing equations to a set of linear algebraic equations.

Ku and Shim [204] used these formulations [172, 198, 199, 206] to model soot agglomerates with different shapes. They compared the extinction coefficient, single scattering albedo, and the phase function of agglomerates determined from these three approaches. They concluded that, among all available techniques, the approach given by Iskander et al. [206] is the most reliable. Their results show that the scattering coefficient and phase function of soot agglomerates are strongly dependent on the overall size parameter and the number of individual spheres that make the agglomerate.

Recent studies have shown that the morphology of soot agglomerates can be represented as fractals (see Refs. 174, 185, 207–214). The fractal-like structure of the agglomerates obeys the relation $N = k_f [R_g/d_p]^{D_f}$. Here, N is the number of primary spheres in the agglomerate, R_g is the radius of gyration of the agglomerate, d_p is the diameter of the primary sphere, k_f is the prefactor to be determined, and D_f is the fractal dimension of the agglomerate. For most flame soot, D_f is around 1.75, and $k_f = 8.0$ [213, 214]. A major consensus of the available studies is that the fractal geometry is better suited for understanding of soot morphology and for accurately interpreting the data obtained from optical diagnostic techniques.

Complex Index of Refraction of Soot.   Soot refractive index has been measured by several researchers. The experimental techniques used can be broadly categorized as in situ and ex situ techniques. In the former, the measurements are performed nonintrusively in a flame environment. The necessary information is retrieved either from spectral transmission data or both the transmission and scattering information, as in Refs. 215–224. The ex situ measurements involve the reflection/transmission of incident spectral radiation on planar pellets of soot, and the optical properties are determined using the Fresnel relations [225]. An alternative ex situ technique was used by Janzen [226], who dispersed the soot particles in a KBr matrix and used transmission measurements to extract the required optical properties.

The Fresnel equations give a relation between transmission and reflection of radiation incident on a plane-parallel homogeneous layer and the real and imaginary parts of the complex refractive index, $m_\lambda = n_\lambda - ik_\lambda$. Therefore, if a homogeneous thin waffle of soot particles can be prepared and experiments are performed at several different wavelengths, corresponding complex index of refraction data can be obtained. There are, however, two main drawbacks of this technique. First of all, it is difficult to claim that the optical properties of the compressed waffles would be identical to those of individual particles. The waffle samples prepared for the experiments are usually not homogeneous, and the physical and molecular structure of particles are altered during the sample preparation, which requires compression and sometimes heating. It is almost impossible to prepare smooth surfaces for the samples, even after careful polishing. For application of the Fresnel equations, it is very important to have surface roughness below a fraction of the wavelength of incident radiation. Otherwise, surface scattering becomes dominant over reflection, which yields physically incorrect refractive index data.

In spite of these problems, the Fresnel reflection-transmission technique is relatively easy to use, and, because of that, it has been employed by various researchers to determine the coal refractive index (see Refs. 227 and 228 for reviews).

In Fig. 7.25, the spectral variations of the real and the imaginary part of the complex index of refraction of soot as reported by a number of different researchers are depicted (based on dispersion relations reported in Ref. 228).

One of the most important parameters required to account for the soot contribution to radiative energy balance is the soot volume fraction. This information is to be obtained from rigorous soot formation models or should be measured experimentally. Unless this information is available, the error induced due to the uncertainty in soot agglomerate size, shape, and

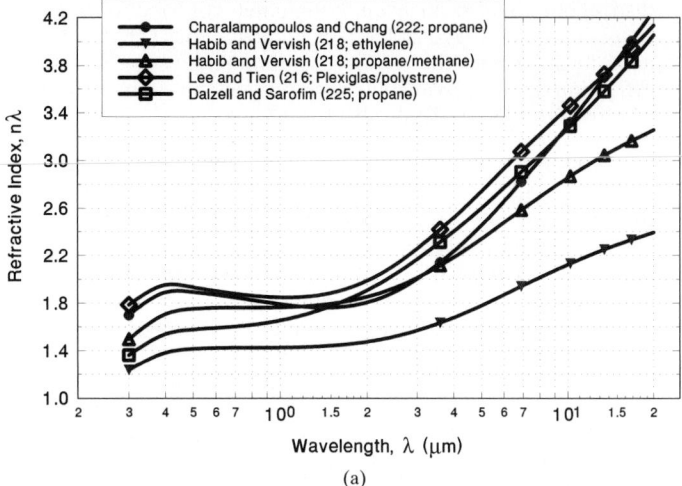

(a)

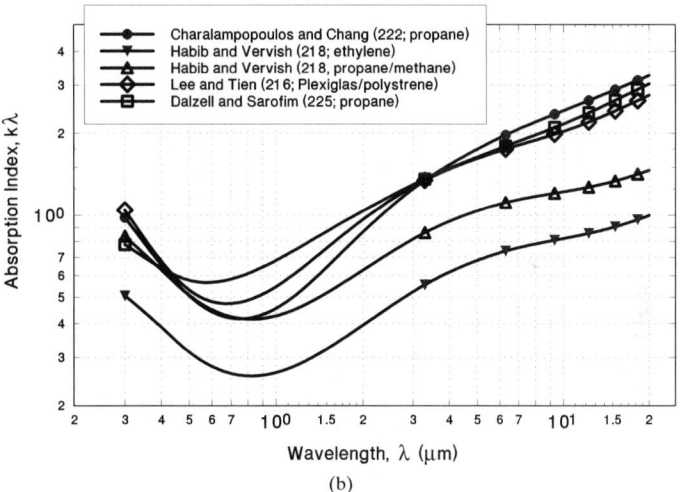

(b)

FIGURE 7.25   Spectral refractive index data for soot particles as determined by different researchers: (*a*) Refractive index, n_λ; (*b*) absorption index (k_λ).

complex index of refraction will always be secondary in calculating radiative heat transfer in furnaces. Unfortunately, prediction of soot volume fraction distribution in combustion systems (or even in small flames) is complicated, as it requires the equations for fluid flow, energy transfer, turbulence, chemical kinetics, and radiative transfer to be solved simultaneously. A number of approaches have recently appeared for modeling soot formation, which were reviewed by Fletcher et al. [228].

Coal and Char Particles.   Coal and char particles affect the absorption/scattering characteristics of particle-laden flames. It is important to have the absorption and scattering coefficients of these inhomogeneous, irregularly shaped particles to predict the temperature distribution in furnaces, even though their contribution is mostly restricted to the flame zone. If coal particles are considered as homogeneous spheres, then the Lorenz-Mie theory can be used to determine the required parameters, provided that the spectral complex index of refraction data are available. Note that, even if there were agreement on the complex index of refraction data of coals, this would not necessarily mean that radiative properties can be determined exactly from the theory. This is because pulverized-coal and char particles have irregular shapes that change during the combustion process. In a combustion environment, it is very difficult to classify the particles into a single, unique shape. The simple shape assumptions are usually made to reduce the computational effort.

The real part n_λ of the refractive index for all coals, as reported by all researchers, is between 1.5 and 2.1 within the spectral range of 1 to 20 μm. The imaginary part k_λ, however, shows a larger variation (and higher uncertainty), as the reported values are between 0.01 and 1.2. For anthracite coal, for example, Blokh [230] reported k_λ values close to 0.9 at near infrared wavelengths; within the visible spectrum, however, the values are closer to 0.1. For bituminous coals, k_λ is constant at about 0.3 up to 6 μm and then increases linearly with the wavelength. These n_λ and k_λ values are plotted in Fig. 7.26.

Brewster and Kunitomo [231] attempted to obtain coal refractive indices with a different approach. Following an earlier work by Janzen [226], they suspended coal particles in a KBr matrix and measured spectral transmission from the samples. Assuming particles are spherical and knowing the size distribution and volume fraction of coal in the sample, they predicted the extinction efficiency factor Q_{ext}. Using a dispersion equation curve fitting for Q_{ext}, they determined the complex index of refraction of different coal samples. They reported values for the absorption index k_λ that was an order of magnitude smaller than earlier studies.

Although there is no doubt that the particle extinction method is superior to classical reflection/transmission techniques for determining the effective complex index of refraction of coal, it is more suitable for spherical particles rather than irregularly shaped particles. For coal and char particles, it is very likely that material and shape nonhomogeneity will affect the results obtained in experiments. Therefore, the particle extinction technique can be used to obtain *average* index of refraction data for coal particles.

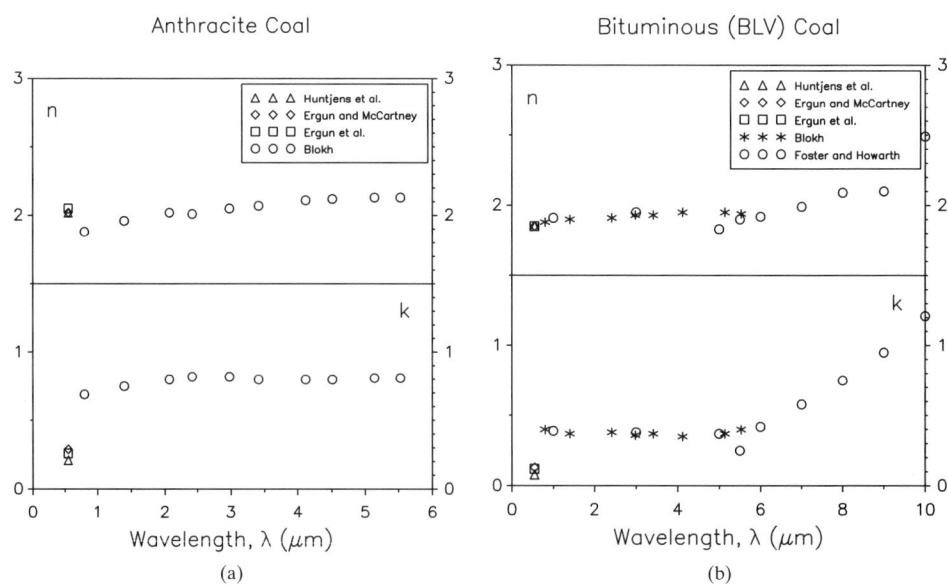

FIGURE 7.26   Complex index of refraction of coal/char particles [237]: (*a*) Anthracite; (*b*) bituminous (BLV).

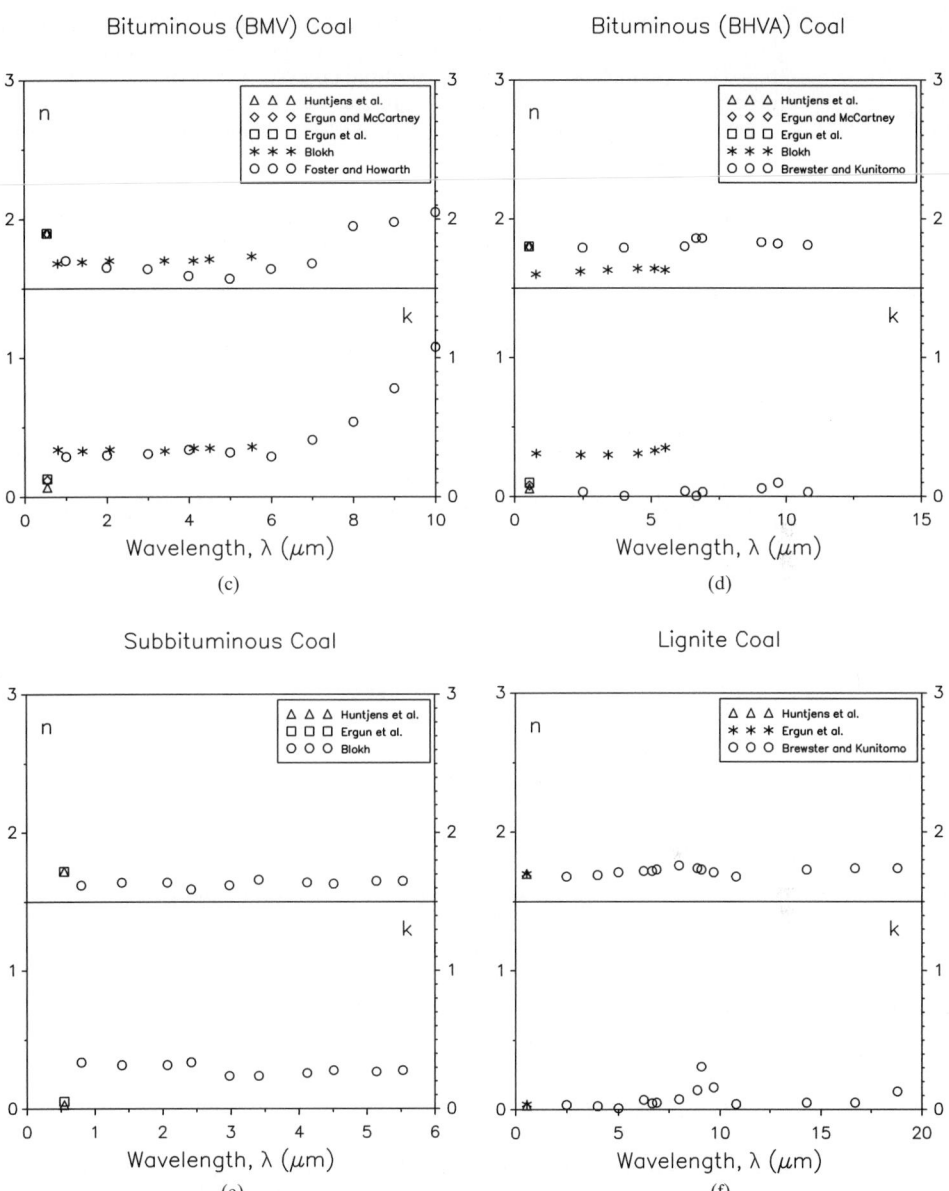

FIGURE 7.26 (*Continued*)   Complex index of refracton of coal/char particles [237]: (*c*) bituminous (BMV); (*d*) bituminous (BHVA); (*e*) subbituminous; (*f*) lignite.

The effective emissivity and transmissivity of coal flames were measured in situ, although the scattering effects were not accounted for in detail [232–236]. These data are best suited for use in zonal method calculations and cannot be employed readily in differential models of the radiative transfer equation. Therefore, they cannot be considered very useful for the present or next-generation modeling attempts. However, it is possible to use this information to validate the trends in different types of experiments performed with similar coals. For example, the data reported by Solomon et al. [236] seem to support the findings of Brewster and Kunitomo [231] in that the imaginary part of the complex index of refraction is much smaller than those previously reported in the literature.

Manickavasagam and Mengüç [229] performed experiments to determine the effective index of refraction of coal/char particles. The particles were suspended in a KBr matrix and measured wide-angle transmission of radiation within the wavelength interval of 2 to 22 μm. The experimental data included transmitted as well as part of the scattered light in a forward cone. These data were used in an inverse analysis to arrive at the best-match spectral index of refraction values, which includes the effects of inhomogeneities and particle irregularities. To validate this approach, additional scattering experiments were performed at a wavelength of 10.6 μm [237, 238, 239], and the corresponding index of refraction was evaluated using a different type of inverse analysis.

Figure 7.27 depicts the results from the experiments of Brewster and Kunitomo [231] and Manickavasagam and Mengüç [229]. It is important to note the similarity in the spectral variation of the extinction index k. Here, the Manickavasagam and Mengüç data need a careful interpretation: *if* the real part of the coal refractive index is assumed 1.8 *and* the imaginary part has the spectral profile depicted in the figure, then the coal radiative properties calculated using the Lorenz-Mie theory yield the experimental results. Even though these refractive index values are not true physical values for coal, the radiative properties calculated with the spherical-particle assumption would be accurate.

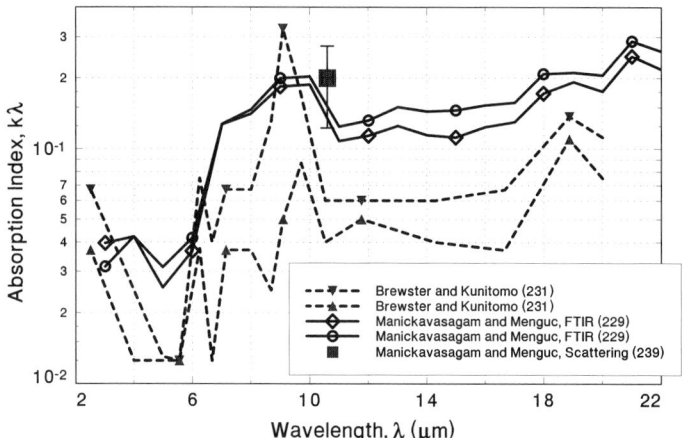

FIGURE 7.27   Absorption index (k_λ) of coal/char particles determined by including the scattering effects.

Fly Ash Particles.   When coal particles burn in a combustion system, the ash present in coal coalesces into small micron- and submicron-sized particles and are carried throughout the system [240]. It has been shown that the shape of the fly ash particles in combustion chambers is primarily spherical. This suggests that, if the complex index of refraction of fly ash particles is known, the Lorenz-Mie theory can be used to determine the required radiative properties.

Fly ash particles are comprised of varying amounts of oxides of silicon (SiO_2), iron (Fe_2O_3), calcium (CaO), and aluminum (Al_2O_3). It may be postulated, then, that the corresponding complex index of refraction is a function of the relative amount of each of these constituents. Depending on the coal rank, type, and flame characteristics, the refractive index of fly ash particles formed may vary, even within the same flame. For example, Wall et al. [241] reported a range of values for fly ash particles, varying from $m = n - ik = 1.43-0.307i$ to $m = 1.50-0.005i$.

Blokh [230] reported similar values for the imaginary part of the refractive index, though the real part reported showed some differences.

Several other researchers reported the refractive indices of fly ash particles determined from different techniques. Lowe et al. [242] and Gupta and Wall [243, 244] reported refractive index values obtained from in situ measurements of fly ash emissivities at the flue gas exit of a pulverized coal burner. Lowe et al. [242] found values of k ranging from 0.01 to 0.02 for the imaginary part; Gupta and Wall's results were smaller by a factor of two. In obtaining these values, the effect of residual carbon within the experimental control volume was not considered, and, in data reduction, the effect of scattering was neglected.

Fly ash optical properties were measured by Wyatt [245] using single, levitated particles, and values of k between 0 and 0.008 were reported. However, a later study indicated that this technique was less reliable for measuring the imaginary part if it was very small, as in the case of fly ash [246].

Extensive studies of the refractive index of fly ash were performed first by Goodwin [247] and later by Ghosal [248] and Ebert [249] at Stanford University. Goodwin used polished wafer layers of several different natural and synthetic coal slags in the experiments and measured transmittance and near-normal reflectance of radiation within the 1–12 μm wavelength range. He deduced the fly ash complex index of refraction using Fresnel reflection theory, as a function of wavelength, temperature, and composition (see Fig. 7.28). Although this is an ex-situ technique, the data obtained can be considered quite reliable because it was possible to prepare uniform samples with polished surfaces, and the structure of fly ash is unlikely to be affected by the wafer preparation.

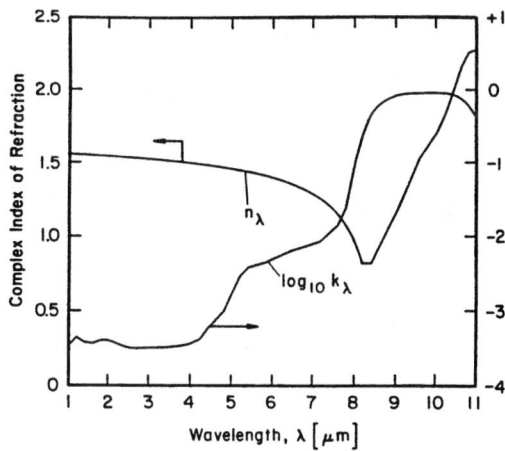

FIGURE 7.28   Typical spectral refractive index data for fly ash [247].

On the other hand, any uncertainty in the composition and complex refractive index of fly ash particles is likely to be less than the uncertainty in the volume fraction distribution of fly ash particles. It was shown that, using detailed spectral data for fly ash particles, one may obtain significantly different results for radiative flux from the flame to the walls of large scale pulverized-coal furnace compared to the results predicted using wavelength-independent properties; however, the difference is lost if the accurate number densities of fly ash and soot present near the walls are not available [52, 250].

The temperature has an insignificant effect on the radiative properties of fly ash particles [247]. Yet, if the medium temperature varies, the spectral variation of fly ash particles becomes more important because of the shift in the peak wavelength of blackbody emission. This shift was shown to affect the divergence of the radiative flux term significantly [250].

Recently, Ebert extended Goodwin's approach and measured the optical properties of molten slags at 1600°C. Figure 7.29 depicts the spectral variation of n_λ and k_λ for three different samples over a spectral wavelength range of 1 to 13 μm. Ebert suggested that the model (which is shown by the lines in the figure) is reliable if three conditions are met: (1) SiO_2 weight percentage is less than 95 percent, (2) the sum of weight percentages of SiO_2, Al_2O_3, CaO, Fe_2O_3, MgO, and TiO_2 is 80 percent or higher, and (3) the weight percentage of Fe_2O_3 is less than 30 percent. It is important to note that the spread shown in the k_λ values in Fig. 7.29 within the spectral range of 3 to 8 μm is roughly the same spread reported by Goodwin for a temperature range of 295 to 1295 K.

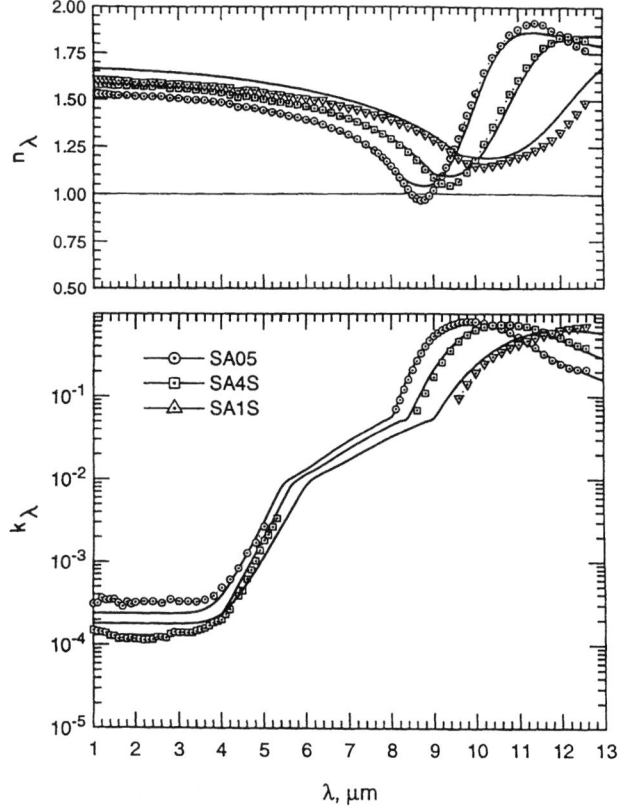

FIGURE 7.29 Spectral refractive index data for three different molten slags at 1600°C [249].

Radiative Properties of Porous Materials

Most porous materials have very small mean free paths for radiation and can be treated as opaque substances. In that case, radiation can be treated as a boundary effect, as for any opaque solid. However, there are at least two important exceptions.

The first is composite materials, which have long fibers imbedded in a matrix of transparent material such as epoxy resin. The properties of these materials are important because of the need for modeling of the radiative transfer in the infrared-assisted curing process for thermosets and the high-temperature molding process for some thermoplastics. The second exception is the class of materials known as reticulated porous ceramics, which have an open pore structure and porosities greater than 80 percent. These materials are used in various high-temperature applications, and radiative transfer through the pore structure is important.

Properties of Composite Fiber/Matrix Systems.   The volume fraction of fibers in most filament-wound structures and in laid-up composite layers can exceed 60 percent. For carbon fibers at this volume fraction, the composite has such a high absorption coefficient that it can be treated as opaque [251]; however, for glass-epoxy composites, the absorption coefficient is small enough, particularly in some spectral windows, that radiation must be treated as a volumetric effect. The situation is complicated by the high fiber volume fraction, which causes the fiber absorption and scattering to be in the dependent regime; however, some data are presented in Ref. 251.

Properties of Reticulated Ceramics.   Reticulated ceramics are used as flame holders for the combustion of gaseous and liquid fuels and in high-temperature solar collectors. Because of their open structure and high porosity (above 80 percent), these materials have a small effective extinction coefficient relative to most porous materials.

Calculation of radiative properties through modeling of radiation absorption and reflection from the ceramic structure has not yet been attempted. Instead, the porous medium is treated as a homogeneous absorbing and scattering medium. For such a treatment, it is necessary to know the effective absorption and scattering coefficients of the medium as well as the scattering phase function.

Hsu et al. [88], during measurement of the thermal conductivity of porous partially stabilized zirconia, used a two-flux radiation model to infer the effective radiative extinction coefficient as a function of pore size. They present a correlation of the data of the form

$$\beta = 1340 - 1540d + 527d^2 \tag{7.195}$$

where β is the mean (spectrally averaged for long wavelength radiation) extinction coefficient in m^{-1} and d is the actual pore diameter in mm. This correlation was for pore sizes in the range $0.3 \text{ mm} \leq d \leq 1.5 \text{ mm}$. The data were collected for temperatures in the range $290 \leq T \leq 890$ K; no temperature dependence was observed. Additionally, Hsu and Howell present a geometric optics prediction of the extinction coefficient versus pore diameter that fits the data very well:

$$\beta = (3/d)(1 - p) \tag{7.196}$$

which applies for $d \geq 0.6$ mm. Here, p is the porosity of the sample, which varied from 0.87 at large pore diameters to 0.84 at the smallest diameters. The method used to obtain these correlations required the assumption of isotropic scattering, and it was not possible to determine independent values of albedo or scattering or absorption coefficients.

Hale and Bohn [252] measured the scattered radiation from a finite sample of reticulated alumina from an incident laser beam at 488 nm. They then matched Monte Carlo predictions of the scattered radiation calculated from various values of extinction coefficient and scattering albedo and chose the values that best matched the experimental data for reticulated alumina samples of 10, 20, 30, and 65 ppi. A scattering albedo of 0.999 and an assumed isotropic scattering phase function reproduced the measured data for all pore sizes. The large reported albedo value indicates that alumina is very highly scattering and that radiative absorption is extremely small for this material.

Hendricks and Howell [253] measured the spectral normal transmittance and normal hemispherical reflectance of three sample thicknesses each of reticulated partially stabilized zirconia and silicon carbide at pore sizes of 10, 20, and 65 ppi. The measurements covered a spectral range of 400–500 nm. They used an inverse discrete ordinates method to find the spectrally dependent absorption and scattering coefficients as well as the constants appropriate for use in the Henyey-Greenstein approximate phase function. Both materials showed best agreement with experimental data when this anisotropic scattering phase function was used. The data for absorption and scattering coefficient for silicon carbide were fairly independent of wavelength, which is the same as the spectral behavior of pure silicon carbide in solid form (see App. A and Refs. 11, 12, 16, and 254). For partially stabilized zirconia, the properties showed a significant change across the range 2500 to 3000 nm but were wavelength-independent on each side of this value. The spectral characteristics of pure and stabilized zirconia do not show this behavior [12]. The phase function in particular shows a radical change in behavior for the reticulated ceramics across this spectral range for PSZ. Scattering albedos varied with wavelength in the range of 0.81–0.999 for zirconia, with some variation with pore diameter, and 0.55–0.81 for silicon carbide.

Integration of the spectral results of Hendricks and Howell [253] provides mean extinction coefficient data that can be compared with that of Hsu et al. [88]. The Hale and Bohn [252] data are only for the wavelength of 488 nm. Hendricks and Howell [253] found that a modified geometrical optics relation also fit the data for the spectrally integrated extinction coefficient of both zirconia and silicon carbide. They recommend the relation

$$\beta = (4.4/d)(1 - p) \tag{7.197}$$

where again β is in m^{-1} and d is the actual pore diameter in mm.

The data presented in Fig. 7.30 obviously have similar characteristics, regardless of the material. The Hale and Bohn data are plotted based on the pore diameter calculated as the inverse of the ppi values reported; however, measured pore sizes are generally smaller than nominal sizes computed in this way [255]. It may be possible to collapse all of the data for extinction coefficient onto a single curve if the Hale and Bohn data are adjusted to actual pore size rather than a pore size calculated from nominal ppi values. This was not attempted, because actual pore size data were not reported in Ref. 252. Even if the data can be collapsed in this way, it will be necessary to have additional data for the scattering albedo of the material so that the individual scattering and absorption coefficients can be recovered from the extinction coefficient data.

Mital et al. [256] have measured the radiative extinction coefficient and scattering albedo for five different porous ceramics in the temperature range from 1200–1400 K, assuming a gray isotropically scattering medium. Doermann and Sacadura [257] have proposed a method for predicting the radiative absorption and scattering coefficients and the phase function of open-celled materials based on the structure of the solid. Recently these authors have presented a comprehensive review of the subject [258].

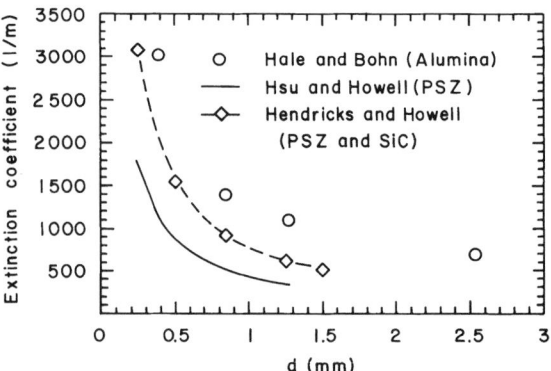

FIGURE 7.30   Extinction coefficient versus pore diameter for alumina, partially stabilized zirconia, and silicon carbide.

Radiative Properties of Semitransparent Materials

Often, layers of materials with differing radiative properties are adjacent to one another, and the transmission of energy through the layers is needed. If the layers have a thickness that is large compared with the wavelength of the radiation, as is the case with sheets of window glass, then a net radiation approach (geometric optics) can be used. If the layer thickness is small compared with the wavelength, then the effects of wave interference (coherence) must be considered, and, for very thin layers, a microscopic approach that considers material structure to the molecular scale may be necessary. For layers intermediate between the geometric optics and coherent optics regime, an analysis using partial coherence effects must be used.

Geometric Optics Results for Cold Plates.   For parallel plates of thickness L with surface reflectivity ρ and transmittance $\tau = \exp(-aL)$, multiple internal reflections occur at both plate-air interfaces within the parallel plate, causing portions of the radiation to have quite long paths within the plates and thus increased absorption and reflection from the primary surface. For a single plate that is cold enough that emission of radiation can be ignored and with $L \gg \lambda$, the fractions of incident energy on one face that are absorbed A, reflected R, and transmitted T, are

$$A = (1 - \rho)(1 - \tau)/(1 - \rho\tau) \tag{7.198}$$

$$R = \rho(1 + \tau T) \tag{7.199}$$

$$T = \tau(1 - \rho)^2/(1 - \rho^2\tau^2) \tag{7.200}$$

Siegel and Howell [1] treat the transmittance, absorptance, and reflectance of multiple parallel plates, each with the same or differing ρ and τ. The normal reflectance for smooth dielectric materials can be estimated from Eq. 7.40.

In Eqs. 198–200, the reflectance can be found for any angle of incidence θ by using the Fresnel relations [Eq. 7.41] for reflectance $\rho(\theta)$ and correcting the transmittance τ with the angle of refraction χ to give $\tau = \exp(-aL/\cos\chi) = \exp(-aL/n\cos\theta)$. This requires knowledge of the refractive index of the plate material.

Geometric Optics Results with Emission.   When the temperature of a semitransparent layer is large, emission of radiation becomes significant, and the problem of radiative transfer becomes more complex. The change in refractive index at each interface causes total internal reflection of radiation in the medium with higher refractive index at the boundary. This effect must be treated in the RTE at the boundary of the medium, and diffuse boundary conditions are no longer correct for the exact solution of this type of problem. Various approaches have been attempted.

Spuckler and Siegel [259, 260] have treated single and multiple semitransparent layers, and they have presented two-flux solutions to steady problems with internal radiation and conduction in the semitransparent layer with convective and radiative loss or gain at the surface [261] and transient problems with similar conditions [262].

Thin Films.   When geometric optics cannot be applied, coherence effects require that electromagnetic wave theory be applied until extremely thin films are treated. For example, the normal reflectance R of a thin nonabsorbing film of refractive index n_1 on a substrate with refractive index n_2 is found to be

$$R = (r_1^2 + r_2^2 + 2r_1 r_2 \cos\gamma_1)/(1 + r_1^2 r_2^2 + 2r_1 r_2 \cos\gamma_1) \tag{7.201}$$

where $\gamma_1 = 4\pi n_1 L/\lambda_o$ and r_1 and r_2 are the normal reflectances of the air-coating and coating-substrate interfaces, respectively, and are given by $r_1 = (n_1 - n_2)/(n_1 + n_2)$ and $r_2 = (n_2 - n_s)/(n_2 + n_s)$. For the special case of a layer thickness that is ¼ wavelength (in the medium) in thickness so that $\gamma_1 = 4\pi n_1 L/\lambda_o = \pi$, Eq. 7.201 reduces to

$$R = [(n_s n_2 - n_1^2)/(n_s n_2 + n_1^2)]^2 \tag{7.202}$$

If the refractive index of the coating is made equal to $(n_s n_2)^{1/2}$, then Eq. 7.202 reduces to $R = 0$. Such a film thus produces an antireflective coating. This result will apply only for normal incidence and at a particular wavelength.

Films that are greater than a few wavelengths in thickness but not yet in the geometric optics region are treated by Chen and Tien [263], and a recent review of thin film and partial coherence effects as well as very thin film effects is in Chen [264].

COMBINED MODES WITH RADIATION

Almost all real engineering problems involving radiative transfer have present other energy transfer modes (conduction, convection, advection) and/or sources or sinks of energy (nuclear reactions and chemical reactions such as combustion). These all affect the temperature of a medium through the energy equation, and thus radiation is nearly always coupled to these other energy forms. Radiative transfer depends on the fourth power of the absolute temperature, while other energy transfer modes depend on local temperature gradients (conduction) or temperature differences (convection). Therefore, we expect a general energy equation to be nonlinear and contain often both integrals of fourth-power temperature for radiative terms and second partial derivatives of temperature for conduction. Such nonlinear integro-partial differential equations do not have the wealth of analytical and numerical solution methods that have been developed for simpler forms. Here, we present a brief review of contemporary methods for solution of combined-mode problems.

The General Energy Equation

A general energy conservation equation is of the form

$$\rho c_p \frac{DT}{D\tau} = \beta T \frac{DP}{D\tau} + \vec{\nabla} \cdot (k\vec{\nabla}T - \vec{q}_{\text{rad}}) + S + \Phi_d \tag{7.203}$$

where $D/D\tau$ is the substantial derivative, β is the temperature coefficient of volume expansion, S is the volumetric source term, and Φ_d is the viscous dissipation. To solve for the temperature distribution in a participating medium, the radiative flux divergence $\vec{\nabla} \cdot \vec{q}_{\text{rad}}$ must be known.* Any of the methods discussed in previous sections (zonal method, multiflux approximations, differential or moment methods, Monte Carlo, discrete ordinate, finite element, finite volume, and so on) can be used along with Eq. 7.105 to evaluate the radiative flux divergence. However, $\vec{\nabla} \cdot \vec{q}_{\text{rad}}$ is itself temperature-dependent, so solution of the energy equation and the radiative flux divergence are coupled.

General CFD solvers are available for the energy equation, but most have at best approximate forms for the radiative flux divergence. The method of choice in commercial codes is often the first-order spherical harmonics (P_1) approximation, because it is in straightforward differential equation form and is easily included in existing numerical codes. However, as noted in the discussion of this method, it is prone to serious errors in many geometries and over wide ranges of parameters so that solutions using this method are suspect. However, it is difficult to couple more general radiative transfer solutions to the energy equation for a number of reasons. One is that the gridding for accurate solution of the RTE may be greatly different from that needed, for example, for resolving a boundary layer flow. If the boundary-

* Here, subscript rad is used to separate radiative flux from conductive, convective, or total heat flux.

layer-generated grid is used for the RTE solution, the time required for solution of the RTE may be prohibitive; yet such gridding is necessary to resolve convective effects.

Recently, Burns and Howell [110] and Farmer and Howell [101] have developed solutions for $\vec{\nabla} \cdot \vec{q}_{rad}$ on an externally generated finite element grid so that existing energy equation solvers could be directly accessed; the former used a direct finite-element solution to the RTE, while the latter used Monte Carlo. Both include spectral properties and anisotropic scattering.

Interaction with Conduction and Convection

For problems where convection and/or conduction within a medium are coupled with radiative transfer, a number of methods and solutions are available in the literature. Most such solutions are for simple geometries (e.g., free convection in the annulus between two concentric horizontal cylinders at different temperatures).

Simplified but useful solutions can be obtained for some cases. For example, if the medium is optically thick at all important wavelengths, then the local radiative flux can be related to the gradient in the local medium emissive power, or

$$q_{rad,x} = -\left(\frac{4}{3\kappa_R}\right)\frac{\partial E}{\partial x} = -\left(\frac{4\sigma}{3\kappa_R}\right)\frac{\partial T^4}{\partial x} = -\left(\frac{16\sigma T^3}{3\kappa_R}\right)\frac{\partial T}{\partial x} = -k_r\frac{\partial T}{\partial x} \qquad (7.204)$$

where $k_r = 16\sigma T^3/3\kappa_R$ is a radiative conductivity, and κ_R is the Rosseland mean absorption coefficient defined by Eq. 7.178.

If portions of the spectrum are transparent, then this method will fail unless the problem is segmented into transparent and optically thick regions, and the results for the entire spectrum are added.

Equation 7.204 has the same form as the Fourier law for conduction heat transfer. However, the equivalent conductivity is highly temperature dependent. For a medium such as a high-temperature solid where convective effects and sources are absent, the general energy equation (Eq. 7.203) reduces to

$$\vec{\nabla} \cdot (k\vec{\nabla}T - \vec{q}_{rad}) = \vec{\nabla} \cdot (k\vec{\nabla}T + k_r\vec{\nabla}T) = \vec{\nabla} \cdot [(k + k_r)\vec{\nabla}T] = 0 \qquad (7.205)$$

Solution of the combined conduction/radiation problem in this case reduces to solution of a conduction problem with temperature-dependent thermal conductivity.

For most problems such as determining the heat transfer characteristics of an industrial furnace or power plant steam generator, simplifying assumptions are simply not justified, as radiation, advection, and conduction are all important, and the effects of spectral properties and anisotropic scattering must also be included. These effects plus compact real geometries make such problems tax even the largest computers.

Efficient treatment of complete combined mode problems is an area of contemporary research, and the texts by Siegel and Howell [1], Modest [3], and Özişik [6] give a good introduction.

Interaction with Combustion and Turbulence

In combustion chambers, radiation transfer from hot combustion products to cooler walls and/or gases/particles depends strongly on the concentration of particles and combustion gases as well as their temperatures. In a typical practical system, these properties fluctuate because of the presence of the turbulent flow field. Radiation transfer affects the fate of particles/droplets at different locations in the flame, which, in turn, influences their concentration distributions. The effect of radiation transfer also depends on the thermophysical properties of the medium, which vary with the degree of interactions. The variations in thermophysical properties, concentrations, and temperatures result in local radiation intensity distributions as well as divergence of radiative flux (i.e., radiative sources and sinks) that oscillate as a function of time. These oscillations, in turn, affect the chemical reactions as well as soot formation/oxidation mechanisms. This cycle is the main impetus for radiation turbulence and radiation combustion interactions.

The first attempt to include the effect of radiation on flow and temperature fields in a turbulent environment was made by Townsend [265]. More recent studies include those of Faeth, Gore, Sivathanu, and their coworkers [266–273], Grosshandler and Joulain [274], Song and Viskanta [275], Yuen et al. [276], and Adams and Smith [277].

Recently, McDonough et al. [278] and Mengüç et al. [279] have approached the problem from a different direction and modeled the fluctuations of the species important for radiation-turbulence interactions using chaotic maps, which appears to be a promising technique.

Although several significant advances have been made in understanding the turbulence-radiation interactions in gaseous flames (especially by Faeth and his coworkers), we still need more fundamental approaches that allow the study of these interactions starting from first principles. There is also a need to employ more sophisticated turbulence simulation techniques along with accurate and reliable radiation and chemical kinetics models to develop a thorough understanding of these complex phenomena and their interactions.

CLOSING REMARKS

In this chapter, we presented a general overview of radiative heat transfer. A number of practical models were included for the solution of the radiative transfer equation and to calculate the required radiative properties of particles, combustion gases, and surfaces. Even though the material presented can allow the reader to tackle a radiative transfer problem, it is not possible to claim that our coverage of the subject was comprehensive. We tried to list most significant references, and the reader is encouraged to consult the literature for more detailed and the most up-to-date analyses and data.

Within the constraints of this chapter, we were forced not to include a number of important topics, including the discussions on dependent-independent scattering, laser-material interactions, and the effect of radiation on manufacturing processes, among others. These are very important subjects; however, it will be difficult to include them in a short chapter like this. Again, the reader is referred to the literature for more detailed discussions of these problems (see Refs. 1, 2, and 3, and the references therein).

APPENDIX A: RADIATIVE PROPERTY TABLES

TABLE A.7.1 Selected Surface Properties

Surface	Temperature, °C (°F)	$\varepsilon_\perp$	Surface	Temperature, °C (°F)	$\varepsilon_\perp$
Metals					
Aluminum:			Heat resistant oxidized	200 (392)	0.639
Bright rolled	170 (338)	0.039	Cast iron, bright	93 (200)	0.21
Bright rolled	500 (932)	0.050	Cast iron, oxidized	93 (200)	0.61
Polished, 98 percent pure	93 (200)	0.05	Black iron oxide	93 (200)	0.56
Aluminum paint	100 (212)	0.20–0.40	Lead:		
Bismuth, bright	80 (176)	0.340	Pure, polished	37.8 (100)	0.05
Chrome, polished	150 (302)	0.058	Gray, oxidized	20 (68)	0.28
Copper polished	20 (68)	0.030	Mercury, pure clean	93 (200)	0.12
Lightly oxidized	20 (68)	0.037	Molybdenium, polished	93 (200)	0.06
Scraped	20 (68)	0.070	Nickel		
Black, oxidized	20 (68)	0.78	Bright matte	100 (212)	0.041
Oxidized	131 (268)	0.76	Polished	100 (212)	0.045
Gold, polished	130 (266)	0.018	Platinum:		
Polished	400 (752)	0.022	Pure, polished	93 (200)	0.05
Iron:			Black	93 (200)	0.93
Pure, polished	93 (200)	0.06	Silumin, cast polished	150 (302)	0.186
Bright etched	150 (302)	0.128	Silver	20 (68)	0.020
Bright abrased	20 (68)	0.24	Tin, bright tinned iron sheet	37.8 (100)	0.08
Red, rusted	20 (68)	0.61	Tungsten, polished	93 (200)	0.04
Hot rolled	20 (68)	0.77	Zinc:		
Hot rolled	130 (266)	0.60	Matte	93 (200)	0.21
Heavily crusted	20 (68)	0.85	Gray, oxidized	20 (68)	0.23–0.28
Heat-resistant oxidized	80 (176)	0.613			
Alloys					
Brass					
Polished	37.8 (100)	0.05	Manganin, bright rolled	118 (245)	0.048
Freshly rubbed with emery	37.8 (100)	0.21			
Dull	37.8 (100)	0.22			
Oxidized	37.8 (100)	0.46			
Paints					
Varnish, dark glossy	37.8 (100)	0.89	Enamel	20 (68)	0.85–0.95
Lacquer, clear on bright			Bakelite	80 (176)	0.935
copper	37.8 (100)	0.07	Gold enamel	93 (200)	0.37
White on clean copper,			Red lead paint	100 (212)	0.93
thin coat	37.8 (100)	0.85	Oil on polished iron:		
White	100 (212)	0.925	0.0008 in thick	37.8 (100)	0.22
Black matte	80 (176)	0.970	0.0080 in thick	37.8 (100)	0.81
Pigments					
Acetylene soot	51.7 (125)	0.99	Red (Fe_2O_3)	51.7 (125)	0.96
Camphor soot	51.7 (125)	0.98	Green (Cr_2O_3)	51.7 (125)	0.95
Lampback	51.7 (125)	0.94	White (Al_2O_3)	51.7 (125)	0.98
Candle soot	51.7 (125)	0.95	White (MgO)	51.7 (125)	0.97
Platinum black	51.7 (125)	0.91			

TABLE A.7.1   Selected Surface Properties (*Continued*)

Surface	Temperature, °C (°F)	$\varepsilon_\perp$	Surface	Temperature, °C (°F)	$\varepsilon_\perp$
		Miscellaneous			
Asbestos paper	260 (500)	0.94	Quartz, fused, rough	37.8 (100)	0.93
Brick, mortar, plaster	20 (68)	0.93	Refractory brick:		
Clay, fired	70 (158)	0.91	Ordinary	1093 (2000)	0.59
Corundum, emery rough	80 (176)	0.855	White	1093 (2000)	0.29
Glass	90 (194)	0.940	Dark chrome	1093 (2000)	0.98
Hoar frost (0.1–0.2 mm thick)	37.8 (100)	0.98	Rubber, gray, soft, rough	37.8 (100)	0.86
Ice:			Sandstone	37.8 (100)	0.83
Smooth, water	0 (32)	0.966	Tar paper	20 (68)	0.93
Rough crystals	0 (32)	0.985	Velvet, black	37.8 (100)	0.97
Limestone	37.8 (100)	0.95	Water (0.1 mm or more thick)	37.8 (100)	0.96
Marble, white	37.8 (100)	0.95	Wood:		
Mica	37.8 (100)	0.75	Beech	70 (158)	0.935
Paper	95 (203)	0.82	Oak, planed	37.8 (100)	0.90
Plaster of paris (0.5 mm)	37.8 (100)	0.91	Spruce, sanded	37.8 (100)	0.82
Porcelain	20 (68)	0.92–0.94			

TABLE A.7.2   Selected Surface Emittance and Absorptances

Material	Emittance (at temperature in K)	Absorptance (at solar temperature)
Aluminum:		
polished	0.102 (573), 0.130 (773), 0.113 (873)	0.09–0.10
anodized	0.842 (296), 0.720 (484), 0.669 (574)	0.12–0.16
with SiO_2 coating	0.366 (263), 0.384 (293), 0.378 (324)	0.11
Carbon black in acrylic binder	0.83 (278)	0.94
Copper, polished	0.041 (338), 0.036 (463), 0.039 (803)	0.35
Gold	0.025 (275), 0.040 (468), 0.048 (668)	0.20–0.23
Iron	0.071 (199), 0.110 (468), 0.175 (668)	0.44
Magnesium oxide	0.73 (380), 0.68 (491), 0.53 (755)	0.14
Nickel	0.10 (310), 0.10 (468), 0.12 (668)	0.36–0.43
Paints:		
Parsons black	0.981 (240), 0.981 (462)	0.98
Acrylic white	0.90 (298)	0.26
White (ZnO)	0.929 (295), 0.926 (478), 0.889 (646)	0.12–0.18

TABLE A.7.3   Real and Imaginary Part of Complex Index of Refraction of Selected Metals at Room Temperature

Aluminum			Beryllium			Copper			Chromium		
λ (μm)	n	k	λ (μm)	n	k	λ (μm)	n	k	λ (μm)	n	k
0.10	0.03	0.79	0.10	0.30	1.07	0.10	1.09	0.73	0.109	1.08	0.69
0.21	0.13	2.39	0.21	0.85	2.64	0.19	0.96	1.37	0.163	0.66	1.23
0.31	0.29	3.74	0.31	2.47	3.08	0.24	1.38	1.80	0.216	0.97	1.74
0.40	0.49	4.86	0.41	2.95	3.14	0.26	1.53	1.71	0.258	0.86	2.13
0.50	0.77	6.08	0.52	3.03	3.18	0.29	1.46	1.64	0.409	1.54	3.71
0.60	1.02	7.26	0.69	3.47	3.23	0.48	1.15	2.50	0.512	2.75	4.46
0.65	1.47	7.79	1.03	3.26	3.96	0.54	1.04	2.59	0.700	3.84	4.37
0.70	1.83	8.31	3.10	2.07	12.6	0.59	0.47	2.81	0.984	4.50	4.28
0.80	2.80	8.45	6.20	3.66	26.7	0.69	0.21	4.05	1.11	4.53	4.30
1.13	1.20	11.2	12.0	11.3	50.1	0.83	0.26	5.26	1.88	3.96	5.95
1.50	1.38	15.4	21.0	19.9	77.1	1.30	0.51	6.92	2.07	4.01	6.48
2.00	2.15	20.7	31.0	37.4	110.0	2.00	0.85	10.6	3.65	2.89	12.0
4.00	6.43	39.8	62.0	86.1	157.0	3.10	1.59	16.5	6.89	8.73	25.4
7.00	14.0	66.2				6.20	5.23	33.0	13.8	11.8	33.9
10.0	25.3	89.8				9.54	10.8	47.5	20.7	21.2	42.0
20.0	60.7	147.0							31.0	14.9	65.2
32.0	103.0	208.0									

Gold			Iron			Molybdenium			Nickel		
λ (μm)	n	k	λ (μm)	n	k	λ (μm)	n	k	λ (μm)	n	k
0.10	1.20	0.84	0.108	0.93	0.84	0.10	1.26	0.92	0.12	0.95	0.87
0.15	1.45	1.11	0.113	0.91	0.83	0.11	1.05	0.77	0.17	1.03	1.27
0.17	1.52	1.07	0.12	0.87	0.91	0.14	0.65	1.41	0.20	1.00	1.54
0.20	1.42	1.12	0.15	0.94	1.18	0.20	0.81	2.50	0.26	1.53	2.11
0.30	1.81	1.92	0.25	1.14	1.87	0.28	2.39	3.88	0.30	1.74	2.00
0.32	1.84	1.90	0.41	1.88	3.12	0.38	3.06	3.18	0.31	1.72	1.98
0.34	1.77	1.85	0.54	2.65	3.34	0.40	3.03	3.22	0.39	1.61	2.33
0.41	1.64	1.96	0.59	2.80	3.34	0.52	3.59	3.78	0.52	1.71	3.06
0.48	1.24	1.80	0.83	3.05	3.77	0.54	3.79	3.61	0.69	2.14	4.00
0.56	0.31	2.88	1.00	3.23	4.35	0.56	3.76	3.41	1.03	2.85	5.10
0.69	0.16	3.80	2.50	4.13	8.59	0.61	3.68	3.49	2.76	4.20	10.2
0.89	0.21	5.88	5.00	4.59	15.4	0.65	3.74	3.58	3.10	3.84	11.4
1.03	0.27	7.07	10.0	5.81	30.4	0.73	3.84	3.51	4.43	4.30	16.0
1.51	0.54	9.58	20.0	9.87	60.1	0.83	3.53	3.30	5.64	4.11	20.2
3.10	1.73	19.2	33.3	22.5	100.0	1.03	2.44	4.22	10.3	7.11	38.3
6.20	5.42	37.5	50.0	45.7	141.0	2.14	1.34	11.3	12.4	9.54	45.8
9.92	12.2	54.7	80.0	75.2	158.0	5.17	3.61	30.0			
						10.3	13.4	58.4			
						12.4	18.5	68.5			

TABLE A.7.3 Real and Imaginary Part of Complex Index of Refraction of Selected Metals at Room Temperature (*Continued*)

Platinum			Silver			Tungsten			Silicon carbide		
λ (μm)	n	k	λ (μm)	n	k	λ (μm)	n	k	λ (μm)	n	k
0.12	1.36	1.18	0.11	1.28	0.56	0.105	1.18	1.48	0.13	1.46	2.21
0.14	1.43	1.14	0.12	1.24	0.57	0.115	1.29	1.39	0.14	1.60	2.15
0.15	1.47	1.15	0.13	1.18	0.55	0.12	1.22	1.33	0.16	2.59	2.87
0.16	1.46	1.19	0.16	0.94	0.83	0.16	0.93	2.06	0.19	4.05	1.42
0.17	1.49	1.22	0.26	1.34	1.35	0.22	2.43	3.70	0.25	3.16	0.26
0.20	1.19	1.40	0.30	1.52	0.99	0.25	3.40	2.85	0.32	2.92	0.01
0.23	1.36	1.61	0.32	0.93	0.50	0.29	3.07	2.31	0.41	2.75	0.00
0.41	1.75	2.92	0.40	0.17	1.95	0.31	2.95	2.43	0.50	2.68	0.00
0.54	2.10	3.67	0.56	0.12	3.45	0.36	3.32	2.70	0.69	2.62	—
0.69	2.51	4.43	0.69	0.14	4.44	0.38	3.45	2.49	0.83	2.60	—
1.03	3.55	5.92	1.03	0.23	6.99	0.40	3.39	2.41	2.00	2.57	0.00
1.55	5.38	7.04	2.00	0.65	12.2	0.44	3.30	2.49	4.00	2.52	0.00
1.77	5.71	6.83	5.17	3.73	31.3	0.67	3.76	2.95	6.67	2.33	0.02
1.91	5.52	6.66	9.92	13.1	53.7	0.71	3.85	2.86	9.80	1.29	0.01
3.10	2.81	11.4				0.77	3.67	2.68	10.40	0.09	0.63
6.20	5.90	24.0				1.03	3.00	3.64	10.81	0.06	1.57
9.54	9.91	36.7				1.29	3.15	4.41	11.9	0.16	4.51
12.4	13.2	44.7				1.35	3.14	4.45	12.6	8.74	18.4
						1.46	2.80	4.33	12.7	17.7	6.03
						2.14	1.18	8.44	13.1	7.35	0.27
						3.10	1.94	13.2	15.4	4.09	0.02
						3.65	1.71	15.7	25.0	3.34	—
						6.89	4.72	31.5			
						10.3	10.1	46.4			
						17.7	26.5	73.8			
						24.8	46.5	93.7			

APPENDIX B: RADIATION CONFIGURATION FACTORS

Element-Element Factors

1. Elemental area of any length z to infinitely long parallel strip of differential width; plane containing element does not intersect strip

Reference: Siegel and Howell [1]

Governing equation: $dF_{d1-d2} = \dfrac{\cos \beta \, d\beta}{2}$

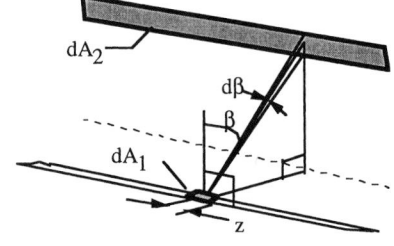

2. Parallel differential strip elements in intersecting planes

Reference: Siegel and Howell [1]

Definition: $Y = y/x$

Governing equation:

$$dF_{d1-d2} = \frac{Y \sin^2 \phi \, dY}{2(1 + Y^2 - 2Y \cos \phi)^{3/2}}$$

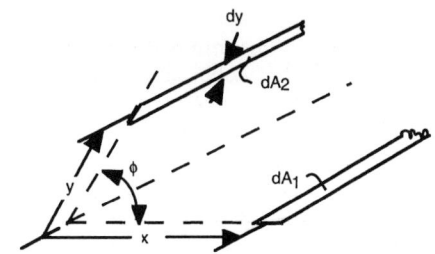

3. Two elemental areas lying on parallel lines

Reference: Sparrow and Eckert [302]

Definition: $X = x/r$

Governing equation:

$$dF_{d1-d2} = \frac{\sin \delta[(1 + X) \cos \delta - 1]}{2[X^2 + 2(X+1)(1 - \cos \delta)]^{3/2}} \, dX$$

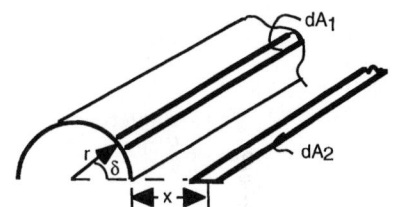

4. Strip of finite length and of differential width, to differential strip of same length on parallel generating line

Reference: Jakob [293]

Definition: $B = b/S$

Governing equation: $dF_{d1-d2} = \frac{\cos \phi \, d\phi}{\pi} \tan^{-1} B$

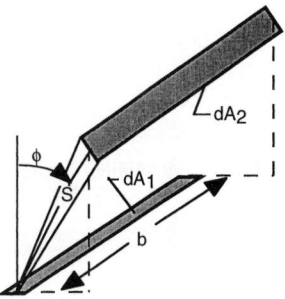

5. Differential ring element to opposed ring element on coaxial disk

Reference: Sparrow and Gregg [303]

Definitions: $R = r_2/r_1$; $H = h/r_1$; $Y = H^2 + R^2 + 1$

Governing equation: $dF_{d1-d2} = \frac{2RH^2Y}{[Y^2 - 4R^2]^{3/2}} \, dR$

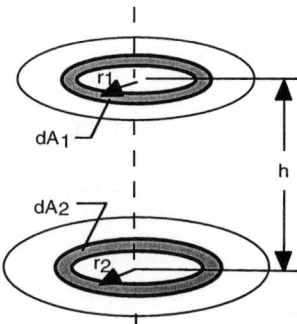

6. Differential ring element on circular disk to opposed coaxial ring element on coaxial disk separated by coaxial cylinder

Reference: Masuda [298]

Definitions: $R_1 = r_1/r$; $R_2 = r_2/r$; $H = h/r$; $X = R_1^2 + R_2^2 + H^2$; $\Phi = \cos^{-1}(1/R_1) + \cos^{-1}(1/R_2)$

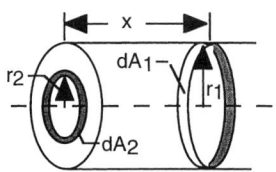

Governing equation:

$$dF_{d1-d2} = \frac{4H^2 R_2 dR_2}{\pi(X^2 - 4R_1^2 R_2^2)}\left\{\frac{R_1 R_2 \sin\Phi}{(X - 2R_1 R_2 \cos\Phi)} + \frac{X}{(X^2 - 4R_1^2 R_2^2)^{1/2}}\tan^{-1}\left[\left(\frac{X + 2R_1 R_2}{X - 2R_1 R_2}\right)^{1/2}\tan\frac{\Phi}{2}\right]\right\}$$

7. Circumferential ring element on interior of right circular cylinder to coaxial ring element on base

Reference: Sparrow et al. [301]

Definitions: $X = x/r_1$; $R = r_2/r_1$

Governing equation:

$$dF_{d1-d2} = \frac{2XR(1 + X^2 - R^2)}{[(1 + X^2 + R^2)^2 - 4R^2]^{3/2}}\,dR$$

8. Ring element on exterior of tube to ring element on coaxial annular element on circular fin

Reference: Masuda [298]

Definitions: $H = h/r_1$; $R = r_2/r_1$; $Z = 1 + R^2 + L^2$

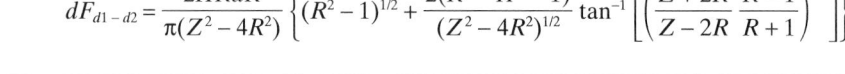

Governing equation:

$$dF_{d1-d2} = \frac{2HRdR}{\pi(Z^2 - 4R^2)}\left\{(R^2 - 1)^{1/2} + \frac{2(R^2 - H^2 - 1)}{(Z^2 - 4R^2)^{1/2}}\tan^{-1}\left[\left(\frac{Z + 2R}{Z - 2R}\frac{R - 1}{R + 1}\right)^{1/2}\right]\right\}$$

9. Two ring elements on the interior of a right circular cylinder

References: Hottel and Keller [291]; Sydnor [305]

Definition: $X = x/2r$

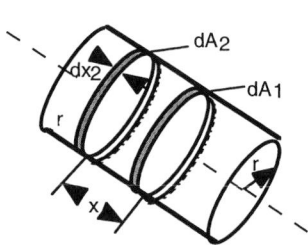

Governing equation: $dF_{d1-d2} = \left[1 - \frac{2X^3 + 3X}{2(X^2 + 1)^{3/2}}\right]dX_2$

Element-Area Factors

1. Differential element of any length to semiinfinite plane. Plane containing element and receiving semiinfinite plane intersect at angle ϕ at edge of semiinfinite plane

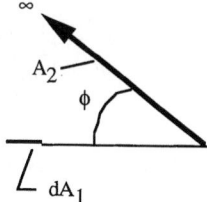

Reference: Hamilton and Morgan [289]

Governing equation: $F_{d1-2} = \dfrac{1}{2}(1 + \cos \phi)$

---

2. Differential tilted planar element dA_1 to disk A_2. Element lies on normal to disk passing through disk center.

References: Naraghi and Chung [300]; Naraghi [299]

Definitions: $H = h/r$; $X = (1 - H^2 \cot^2 \theta)^{1/2}$

Governing equations:

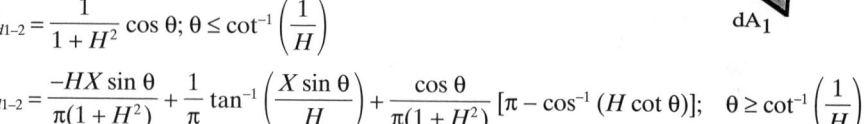

$$F_{d1-2} = \frac{1}{1+H^2} \cos\theta; \ \theta \le \cot^{-1}\left(\frac{1}{H}\right)$$

$$F_{d1-2} = \frac{-HX\sin\theta}{\pi(1+H^2)} + \frac{1}{\pi}\tan^{-1}\left(\frac{X\sin\theta}{H}\right) + \frac{\cos\theta}{\pi(1+H^2)}\left[\pi - \cos^{-1}(H\cot\theta)\right]; \quad \theta \ge \cot^{-1}\left(\frac{1}{H}\right)$$

---

3. Plane element to interior of coaxial right circular cylinder

Reference: Chung and Sumitra [284]

Definitions: $H = h/x$; $R = r/x$

Governing equations:

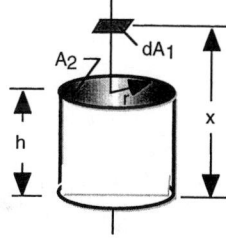

for $H < 1$: $F_{d1-2} = \dfrac{1}{1+R^2} - \dfrac{(1-H)^2}{(1-H)^2 + R^2}$

for $H \ge 1$: $F_{d1-2} = \dfrac{1}{1+R^2}$

---

4. Element on plane to exterior of right circular cylinder of finite length. Plane does not intersect cylinder

Reference: Leuenberger and Person [297]

Definitions: $S = s/r$; $X = x/r$; $Y = y/r$; $H = h/r$; $A = X^2 + Y^2 + S^2$; $B = S^2 + X^2$

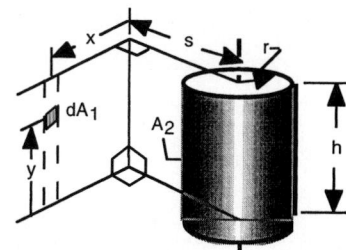

Governing equation:

$$F_{d1-2} = \frac{S}{B}\left(1 - \frac{1}{2\pi}\left\{\cos^{-1}\left(\frac{Y^2 - B + 1}{A - 1}\right) + \cos^{-1}\left[\frac{(H-Y)^2 - B + 1}{(H-Y)^2 + B - 1}\right]\right\}\right.$$

$$- Y\left[\frac{A+1}{((A-1)^2 + 4Y^2)^{1/2}}\right]\cos^{-1}\left[\frac{Y^2 - B + 1}{(A-1)B^{1/2}}\right]$$

$$- (H-Y)\left[\frac{(H-Y)^2 + B + 1}{\{[(H-Y)^2 + B - 1]^2 + 4(H-Y)^2\}^{1/2}}\right]\cos^{-1}\left\{\frac{(H-Y)^2 - B + 1}{[(H-Y)^2 + B - 1]B^{1/2}}\right\} + H\cos^{-1}\frac{1}{B^{1/2}}\right)$$

5. Differential element of any length to surface generated by a line of infinite length parallel to the plane of the element, and moved parallel to itself. Plane of element does not intersect surface

References: Hottel [290]; Hamilton and Morgan [289]

Governing equation: $F_{d1-2} = \frac{1}{2}(\sin \phi_2 - \sin \phi_1)$

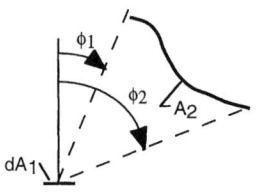

6. Infinitely long strip element to infinitely long parallel cylinder; $r < y$

Reference: Feingold and Gupta [286]

Definitions: $X = x/r$; $Y = y/r$

Governing equation: $F_{d1-2} = \dfrac{Y}{X^2 + Y^2}$

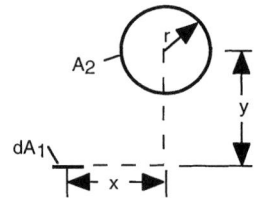

7. Spherical point source to sphere

Reference: Chung and Sumitra [284]

Definition: $R = r/h$

Governing equation: $F_{d1-2} = \frac{1}{2}[1 - (1 - R^2)^{1/2}]$

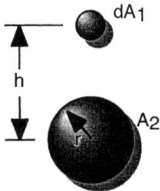

Area-Area Factors

1. Two infinitely long parallel plates of different widths; centerlines of plates are connected by perpendicular between plates

Reference: Wong [306]

Definitions: $B = b/a$; $C = c/a$

Governing equation:

$$F_{1-2} = \frac{1}{2B} \{[(B + C)^2 + 4]^{1/2} - [(C - B)^2 + 4]^{1/2}\}$$

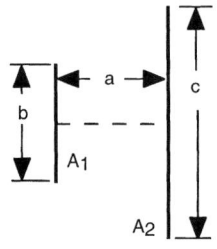

2. Infinitely long enclosure formed by three planar or convex surfaces

Reference: Siegel and Howell [1]

Governing equation: $F_{1-2} = \dfrac{L_1 + L_2 - L_3}{2L_1}$

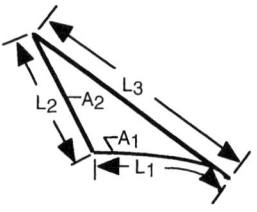

3. Infinite plane to row of parallel cylinders, or n rows of inline cylinders

References: Hottel [290]; Kuroda and Munakata [296]

Definition: $D = d/b$

Governing equations:

For $n = 1$: $F_{1-2} = 1 - (1 - D^2)^{1/2} + D \tan^{-1} \left(\dfrac{1 - D^2}{D^2} \right)^{1/2}$

For $n > 1$: $F_{1-n \text{ rows}} = 1 - (1 - F_{1-2})^n$

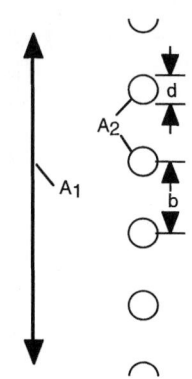

4. Top surface of finite rectangle tilted relative to an infinite plane

Reference: Siegel and Howell [1]

Governing equation: $F_{1-2} = \dfrac{1 - \cos \eta}{2}$

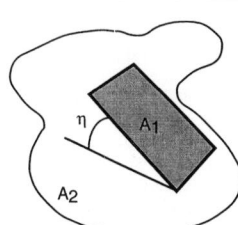

5. Coaxial parallel squares of different edge length

Reference: Crawford [285]

Definitions: $A = a/c$; $B = b/a$;
$X = A(1 + B)$; $Y = A(1 - B)$

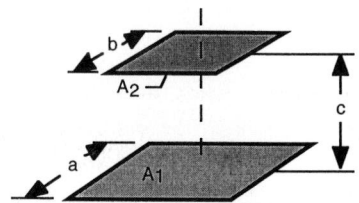

Governing equations:

$$F_{1-2} = \frac{1}{\pi A^2} \left\{ \ln \frac{[A^2(1 + B^2) + 2]^2}{(Y^2 + 2)(X^2 + 2)} + (Y^2 + 4)^{1/2} \left[Y \tan^{-1} \frac{Y}{(Y^2 + 4)^{1/2}} - X \tan^{-1} \frac{X}{(Y^2 + 4)^{1/2}} \right] \right.$$

$$\left. + (X^2 + 4)^{1/2} \left[X \tan^{-1} \frac{X}{(X^2 + 4)^{1/2}} - Y \tan^{-1} \frac{Y}{(X^2 + 4)^{1/2}} \right] \right\}$$

If $A < 0.2$, then $F_{1-2} = \dfrac{(AB)^2}{\pi}$

6. Rectangle to rectangle in a parallel plane; all boundaries are parallel or perpendicular to x and ξ boundaries

References: Gross et al. [288]; also see Boeke and Wall [280]; Chekhovskii et al. [283]; Hsu [292]

Definitions: $X = x/z$; $N = \eta/z$; $Y = y/z$; $S = \xi/z$; $\alpha_{li} = S_l - X_i$; $\beta_{kj} = N_k - Y_j$

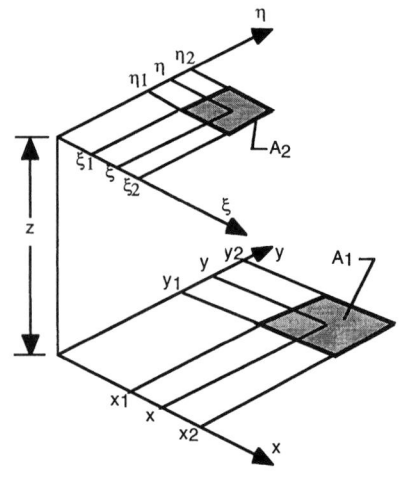

Governing equation:

$$F_{1\text{-}2} = \frac{1}{(X_2 - X_1)(Y_2 - Y_1)} \sum_{l=1}^{2}\sum_{k=1}^{2}\sum_{j=1}^{2}\sum_{i=1}^{2}\left[(-1)^{(i+j+k+l)}G(\alpha_{li}, \beta_{jk})\right]$$

where $G(\alpha_{li}, \beta_{jk}) = \dfrac{1}{2\pi}\left\{\alpha_{li}(1 + \beta_{kj}^2)^{1/2}\tan^{-1}\left[\dfrac{\alpha_{li}}{(1 + \beta_{kj}^2)^{1/2}}\right] - \beta_{kj}\tan^{-1}\beta_{kj}\right.$

$$\left. + \beta_{jk}(1 + \alpha_{li}^2)^{1/2}\tan^{-1}\left[\frac{\beta_{kj}}{(1 + \alpha_{li}^2)^{1/2}}\right] - \frac{1}{2}\,\alpha_{li}^2\ln\alpha_{li}^2 + \frac{1}{2}\ln(1 + \beta_{kj}^2) - \frac{1}{2}\ln(1 + \alpha_{li}^2 + \beta_{kj}^2)\right\}$$

---

7. Rectangle to rectangle in a perpendicular plane; all boundaries are parallel or perpendicular to x and ξ boundaries

References: Gross et al. [288]; also see Boeke and Wall [280]; Chekhovskii et al. [283]

Definitions: $\alpha_{li} = (\xi_l^2 + x_i^2)^{1/2}$; $\beta_{jk} = y_j - \eta_k$

Governing equation:

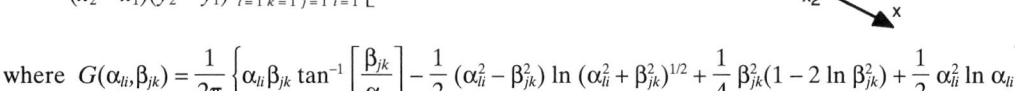

$$F_{1\text{-}2} = \frac{1}{(x_2 - x_1)(y_2 - y_1)} \sum_{l=1}^{2}\sum_{k=1}^{2}\sum_{j=1}^{2}\sum_{i=1}^{2}\left[(-1)^{(i+j+k+l)}G(\alpha_{li},\beta_{jk})\right]$$

where $G(\alpha_{li},\beta_{jk}) = \dfrac{1}{2\pi}\left\{\alpha_{li}\beta_{jk}\tan^{-1}\left[\dfrac{\beta_{jk}}{\alpha_{li}}\right] - \dfrac{1}{2}\,(\alpha_{li}^2 - \beta_{jk}^2)\ln(\alpha_{li}^2 + \beta_{jk}^2)^{1/2} + \dfrac{1}{4}\,\beta_{jk}^2(1 - 2\ln\beta_{jk}) + \dfrac{1}{2}\,\alpha_{li}^2\ln\alpha_{li}\right\}$

---

8. Disk to parallel disk of unequal radius

References: Keene [295]; Hottel [290]; Hamilton and Morgan [289]; Leuenberger and Person [297]

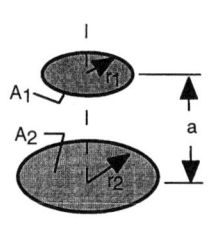

Definitions: $R_1 = r_1/a$; $R_2 = r_2/a$; $X = 1 + \dfrac{1 + R_2^2}{R_1^2}$

Governing equation: $F_{1\text{-}2} = \dfrac{1}{2}\left\{X - \left[X^2 - 4\left(\dfrac{R_2}{R_1}\right)^2\right]^{1/2}\right\}$

9. Exterior of infinitely long cylinder to unsymmetrically placed infinitely long parallel plate; $r < a$

References: Feingold and Gupta [286]; Hamilton and Morgan [289]

Definitions: $B_1 = b_1/a$; $B_2 = b_2/a$

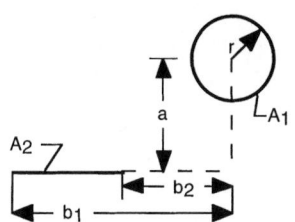

Governing equation: $F_{1\text{-}2} = \dfrac{1}{2\pi}\,(\tan^{-1} B_1 - \tan^{-1} B_2)$

10. Concentric cylinders of infinite length

Reference: Hottel [290]

Definition: $D = D_1/D_2$

Governing equations: $F_{1\text{-}2} = 1$; $F_{2\text{-}2} = 1 - D$; $F_{2\text{-}1} = D$

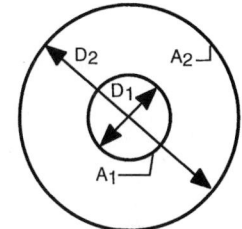

11. Infinite parallel cylinders of different radius

References: Felske [287]; Juul [294]

Definitions: $R = r_2/r_1$; $S = s/r_1$; $C = 1 + R + S$

Governing equations:

Approximate (within 3.9 percent),

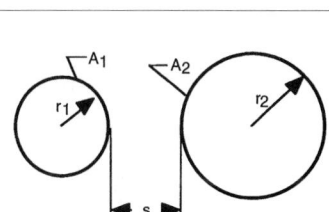

Felske: $F_{1\text{-}2} = \dfrac{C}{\pi}\,\sin^{-1}\left(\dfrac{1}{C}\right) \times \sin^{-1}\left(\dfrac{R}{C}\right)$

Exact solution, Juul:

$$F_{1\text{-}2} = \dfrac{1}{2\pi}\left\{\pi + [C^2 - (R+1)^2]^{1/2} - [C^2 - (R-1)^2]^{1/2} + (R-1)\cos^{-1}\left[\left(\dfrac{R}{C}\right) - \left(\dfrac{1}{C}\right)\right]\right.$$

$$\left. - (R+1)\cos^{-1}\left[\left(\dfrac{R}{C}\right) + \left(\dfrac{1}{C}\right)\right]\right\}$$

12. Outer surface of cylinder to disk at end of cylinder

References: Sparrow et al. [304]; Naraghi and Chung [300]; Masuda [298]

Definitions: $R = r_1/r_2$; $H = h/r_2$; $A = H^2 + R^2 - 1$; $B = H^2 - R^2 + 1$

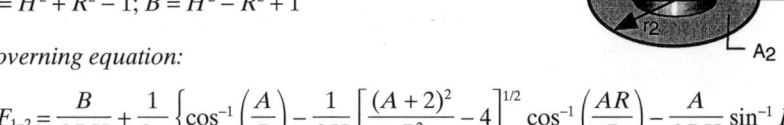

Governing equation:

$$F_{1\text{-}2} = \dfrac{B}{8RH} + \dfrac{1}{2\pi}\left\{\cos^{-1}\left(\dfrac{A}{B}\right) - \dfrac{1}{2H}\left[\dfrac{(A+2)^2}{R^2} - 4\right]^{1/2}\cos^{-1}\left(\dfrac{AR}{B}\right) - \dfrac{A}{2RH}\sin^{-1} R\right\}$$

13. Inner surface of right circular cylinder to itself

   Reference: Buschman and Pittman [282]

   Definition: $H = h/2r$

   Governing equation: $F_{1-1} = 1 + H - (1 + H^2)^{1/2}$

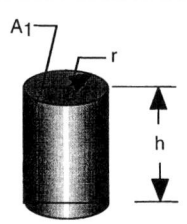

14. Base of right circular cylinder to inside surface of cylinder

   References: Leuenberger and Person [297];
   Buschman and Pittman [282]; Buraczewski [281]

   Definition: $H = h/2r$

   Governing equation: $F_{1-2} = 2H[(1 + H^2)^{1/2} - H]$

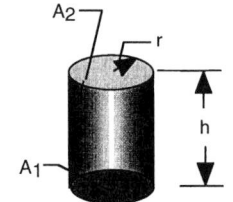

15. Sphere to coaxial disk

   References: Feingold and Gupta [286];
   Naraghi and Chung [300]

   Definitions: $R = r/h$

   Governing equation: $F_{1-2} = \dfrac{1}{2}\left[1 - \dfrac{1}{(1 + R^2)^{1/2}}\right]$

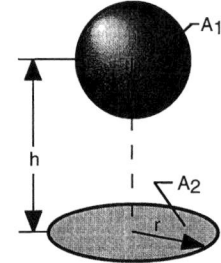

NOMENCLATURE

Symbol, Definition, Unit

A	Area (m^2); fraction absorbed by slab
a_i	Coefficients of the Legendre polynomial expansion of scattering phase function, Eq. 7.100
b_i	Coefficients of the polynomial expansion of the WSGG models, Eq. 7.176
C	Cross sections for particle absorption, extinction, scattering (m^2)
C_1, C_2	Radiation constants, Eq. 7.4
C_3	Constant in Wien's law, Eq. 7.10
c	Speed of light (m/s)
c_p	Specific heat (kJ/kg – K)
D	Particle diameter (μm)
d	Differential; pore diameter (mm)
e	Eccentricity factor for ellipsoidal particles ($= D_{\text{minor}}/D_{\text{major}}$), Eqs. 7.191, 7.192
e_b	Planck's blackbody emissive power (W/m^2)

$F_{0-n\lambda T}$	Fractional blackbody function of first kind
F	Configuration factor
f_v	Volume fraction of particles (m^3/m^3)
f, g	Constants in phase function, Eqs. 7.102, 7.103
G	Incident radiation, Eq. 7.109
h	Planck's constant (J – s)
I	Radiation intensity (W/m^2 – sr)
K	Constant
k	Imaginary part of the complex index of refraction; Boltzmann constant (J/K); conductivity (W/m – K)
L_e	Effective path length, (m)
l	Path
M	Radiation intensity in upper/lower hemispheres, Eqs. 7.112, 7.113
m	Complex index of refraction, $m = n - ik$
N	Particle number density ($\#/m^3$); number of surfaces in enclosure; number of Monte Carlo samples
n	Real part of the complex index of refraction
$\hat{n}$	Surface normal
P	Total pressure (atm)
p	Partial pressure (atm)
Q	Efficiency factors for particle absorption, extinction, scattering (m^2), Eq. 7.180
q	Radiative flux (W/m^2)
R	Random number; fraction reflected from slab
r	Radius (m)
r_e	Electrical resistivity (ohm – cm)
S	Event counter in Monte Carlo analysis; volumetric source term (kW/m^3)
T	Temperature (K); fraction transmitted through slab
V_p	Volume of the computational cell (m^3)
w	Weights in discrete ordinates approximation; energy per sample in Monte Carlo analysis
x	Size parameter ($= \pi D/\lambda$)
x, y, z	Cartesian coordinates (m)

Greek Letters

α	Absorptivity; band strength parameter
β	Extinction coefficient (m^{-1}); coefficient of volume expansion, Eq. 7.197
γ	Constant, Eqs. 7.193 and 7.194
δ_{kj}	Kronecker delta function
ε	Emissivity
η	Direction cosine, Eq. 7.111; wave number ($= 1/\lambda$) [cm^{-1}]
Θ	Scattering angle
θ	Zenith angle
κ	Absorption coefficient (m^{-1})

λ	Wavelength (μm; nm)
μ	Direction cosine, Eq. 7.111
ν	Frequency (sec^{-1})
ξ	Direction cosine, Eq. 7.111; $C_2/n\lambda T$, Eq. 7.14
ρ	Reflectivity; density (kg/m^3)
σ	Stefan-Boltzmann constant, Eq. 7.13; scattering coefficient (m^{-1})
τ	Optical thickness; transmissivity
Φ	Scattering phase function, Eq. 7.99; function given by Eq. 7.157
ϕ	Azimuthal angle
Ψ	Function given by Eq. 7.156
χ	Angle of refraction
Ω	Solid angle (sr)
ω	Single scattering albedo; bandwidth parameter (cm^{-1})

Subscripts

abs	Absorption
b	Blackbody
ext	Extinction
g	Gas
i	Incident; incoming
m	Gaussian quadrature index, Eqs. 7.132, 7.133
o	Outgoing
P	Planck mean value
R	Rosseland mean value
r	Radial component; reflected
rad	Radiative
s	Source; substrate
sca	Scattering
z	Axial component
λ	Spectral (wavelength-dependent)
η	Spectral (wave-number-dependent)
ν	Spectral (frequency-dependent)
0	In vacuum

Superscripts

d	Diffuse
s	Specular
2	Function of two independent parameters, Eq. 7.53
3	Function of three independent parameters, Eqs. 7.49–7.52
$^-$	Normalized or averaged coefficients
$\rightarrow$	Vector
$'$	Directional dependent properties
$+, -$	Upper, lower hemispheres in DP_1 approximation, Eqs. 7.112, 7.113

Others

RTE   Radiative transfer equation

∇   Divergence operator

REFERENCES

1. R. Siegel and J. R. Howell, *Thermal Radiation Heat Transfer,* 3d ed., Hemisphere/Taylor and Francis, Washington, D.C., 1992.

2. M. Q. Brewster, *Thermal Radiative Transfer and Properties,* John Wiley and Sons, New York, 1992.

3. M. M. Modest, *Radiative Heat Transfer,* McGraw-Hill, New York, 1993.

4. R. Siegel, "Two-Flux Method for Transient Radiative Transfer in a Semi-Transparent Layer," *International Journal of Heat Mass Transfer,* 39(5), pp. 1111–1115, 1996.

5. M. Planck, *The Theory of Heat Radiation,* Dover, New York, 1959.

6. M. N. Özişik, *Radiative Transfer and Interactions with Conduction and Convection,* John Wiley and Sons, New York, 1973.

7. N. P. Fox and J. E. Martin, "A Further Intercomparison of Two Cryogenic Radiometers," in *Optical Radiation Measurements II,* SPIE vol. 119, pp. 227–235, 1989.

8. S. L. Chang and K. T. Rhee, "Blackbody Radiation Functions," *International Communications in Heat and Mass Transfer,* vol. 11, p. 451, 1984.

9. Center for Information and Numerical Data Analysis and Synthesis (CINDAS), Purdue University, 2595 Yeager Rd., W. Lafayette, IN 47906.

10. Optical Properties of Solids and Liquids (OPTROP), Sandia National Laboratory, Div. 1824, P.O. Box 5800, Albuquerque, NM 87185.

11. Y. S. Touloukian and D. P. DeWitt, "Thermal Radiative Properties, Metallic Elements and Alloys," vol. 7 in Y. S. Touloukian and C. Y. Ho (eds.), *Thermophysical Properties of Matter,* IFI/Plenum, New York, 1970.

12. Y. S. Touloukian and D. P. DeWitt, "Thermal Radiative Properties, Nonmetallic Solids," vol. 8, in Y. S. Touloukian and C. Y. Ho eds., *Thermophysical Properties of Matter,* IFI/Plenum, New York, 1970.

13. E. D. Palik (ed.), *Handbook of Optical Constants of Solids,* Academic Press, Orlando, 1985.

14. E. D. Palik (ed.), *Handbook of Optical Constants of Solids II,* Academic Press, Orlando, 1991.

15. W. J. Tropf, M. E. Thomas, and T. J. Harris, "Properties of Crystals and Glasses," in M. Bass (ed.), *Handbook of Optics,* vol. 2, McGraw-Hill, New York, 1995.

16. R. A. Paquin, "Properties of Metals," in M. Bass (ed.), *Handbook of Optics,* vol. 2, McGraw-Hill, New York, 1995.

17. P. M. Amirtharaj and D. G. Seiler, "Optical Properties of Semiconductors," in M. Bass (ed.), *Handbook of Optics,* vol. 2, McGraw-Hill, New York, 1995.

18. M. Born and E. Wolf, *Principles of Optics,* 6th ed., Pergamon Press, 1986.

19. S. Chandrasekhar, *Radiative Transfer,* Oxford, 1950 (Dover ed. 1960).

20. J. R. Howell, *A Catalog of Radiation Configuration Factors,* McGraw-Hill, New York, 1982. 2d ed. available on the web at http://www.me.utexas.edu/howe

21. E. F. Sowell and P. F. O'Brien, "Efficient Computation of Radiant-Interchange Factors within an Enclosure," *ASME Journal of Heat Transfer,* 49(3), pp. 326–328, 1972.

22. M. E. Larsen and J. R. Howell, "Least-Squares Smoothing of Direct Exchange Areas in Zonal Analysis," *ASME Journal of Heat Transfer,* 108, pp. 239–242, 1986.

23. J. van Leersum, "A Method for Determining a Consistent Set of Radiation View Factors from a Set Generated by a Nonexact Method," *International Journal of Heat and Fluid Flow,* 10(1), p. 83, 1989.

24. R. P. Taylor, R. Luck, B. K. Hodge, and W. G. Steele, "Uncertainty Analysis of Diffuse-Gray Radiation Enclosure Problems," *AIAA Journal of Thermophysics and Heat Transfer,* 9(1), pp. 63–69, Jan.–March 1995.

25. M. Oguma and J. R. Howell, "Solution of Two-Dimensional Blackbody Inverse Radiation by an Inverse Monte Carlo Method," *Proc. 4th ASME/JSME Joint Symposium,* Maui, March, 1995.

26. V. Harutunian, J. C. Morales, and J. R. Howell, "Radiation Exchange within an Enclosure of Diffuse-Gray Surfaces: The Inverse Problem," *Proc. ASME/AIChE National Heat Trans. Conf.,* Portland, August, 1995.

27. K. Kudo, A. Kuroda, A. Eid, T. Saito, and M. Oguma, "Solution of the Inverse Load Problems by the Singular Value Decomposition," in M. P. Mengüç (ed.), *Radiative Transfer—I: Proceedings of the First International Symposium on Radiative Transfer,* pp. 568–578, Begell House, New York, 1996.

28. J. C. Morales, V. Harutunian, M. Oguma, and J. R. Howell, "Inverse Design of Radiating Enclosures with an Isothermal Participating Medium," in M. P. Mengüç (ed.), *Radiative Transfer—I: Proceedings of the First International Symposium on Radiative Transfer,* pp. 579–593, Begell House, New York, 1996.

29. A. K. Oppenheimer, *ASME Journal of Heat Transfer,* 65, p. 725, 1956.

30. J. Joseph, W. J. Wiscombe, and J. A. Weinman, "The Delta-Eddington Approximation for Radiative Flux Transfer," *Journal of Atmospheric Sciences,* 33, p. 2452, 1976.

31. A. L. Crosbie and G. W. Davidson, "Dirac-Delta Function Approximations to the Scattering Phase Function," *Journal of Quantitative Spectroscopy and Radiative Transfer,* 33, p. 391, 1985.

32. M. P. Mengüç and S. Subramaniam, "A Step Phase Function Approximation for the Experimental Determination of the Effective Scattering Phase Function from the Experiments," *Journal of Quantitative Spectroscopy and Radiative Transfer,* 43, pp. 253–265, 1990.

33. B. G. Carlson and K. D. Lathrop, "Transport Theory—The Method of Discrete Ordinates," in H. Greenspan, C. N. Kelber, and D. Okrent (eds.), *Computing Methods in Reactor Physics,* Gordon and Breach, New York, 1968.

34. R. Viskanta and M. P. Mengüç, "Radiation Heat Transfer in Combustion Systems," *Progress in Energy and Combustion Science,* 13, pp. 97–160, 1987.

35. H. C. Hottel and A. F. Sarofim, *Radiative Transfer,* McGraw-Hill, New York, 1967.

36. J. J. Noble, "The Zone Method: Explicit Matrix Relations For Total Exchange Areas," *International Journal of Heat Mass Transfer,* 18, p. 261, 1975.

37. T. F. Smith, K.-H. Byun, and M. J. Ford, C. L. Tien, V. P. Carey, and J. K. Ferrell (eds.), *Heat Transfer—1986,* vol. 2, pp. 803–808, Hemisphere, Washington, D.C., 1986.

38. K. H. Byun and T. F. Smith, "Development of Zone Method for Linearly Anisotropic Scattering Media," *Journal of Quantitative Spectroscopy and Radiative Transfer,* 40(5), pp. 591–604, 1988.

39. M. E. Larsen and J. R. Howell, "The Exchange Factor Method: An Alternative Zonal Formulation of Radiating Enclosure Analysis," *ASME Journal of Heat Transfer,* 107, pp. 936–942, 1985.

40. M. H. N. Naraghi and B. T. F. Chung, "A Unified Matrix Formulation for the Zone Method: A Stochastic Approach," *International Journal of Heat and Mass Transfer,* 28, pp. 245–251, 1985.

41. M. H. N. Naraghi, B. T. F. Chung, and B. Litkouhi, "A Continuous Exchange Factor Method for Radiative Exchange in Enclosures with Participating Media," *ASME Journal of Heat Transfer,* 110, pp. 456–462, 1988.

42. M. H. N. Naraghi and M. Kassemi, "Radiative Transfer in Rectangular Enclosures: A Discretized Exchange Factor Solution," *ASME Proceedings,* vol. 1, pp. 259–267, 1988.

43. M. H. N. Naraghi and B. Litkouhi, "Discrete Exchange Factor Solution of Radiative Heat Transfer in Three-Dimensional Enclosures," *ASME National Heat Transfer Conference Proceedings,* HTD-vol. 106, pp. 221–229, 1989.

44. K. N. Liou and S. C. Ou, "Infrared Radiative Transfer in Finite Cloud Layers," *Journal of Atmospheric Sciences,* 36, pp. 139–169, 1979.

45. Y. Bayazitoglu and J. Higenyi, "Higher Order Differential Equations of Radiative Transfer: P_3-Approximation," *AIAA Journal,* 14, p. 424, 1979.

46. J. Higenyi and Y. Bayazitoglu, "Differential Approximation of Radiative Heat Transfer in a Gray Medium: Axially Symmetric Radiation Field," *ASME Journal of Heat Transfer,* 102, pp. 719–723, 1980.

47. R. K. Ahluwalia and K. H. Im, "Combined Conduction Convection, Gas, and Particles Radiation in MHD Diffusers," *International Journal of Heat and Mass Transfer,* 24, pp. 1421–1430, 1991.

48. A. C. Ratzel III and J. R. Howell, "Two-Dimensional Radiation in Absorbing-Emitting-Scattering Media Using the P-N Approximation," *ASME Journal of Heat Transfer*, 105, p. 333, 1983.

49. M. P. Mengüç and R. Viskanta, "Radiative Transfer in Three-Dimensional Rectangular Enclosures Containing Inhomogeneous, Anisotropically Scattering Media," *Journal of Quantitative Spectroscopy and Radiative Transfer*, 33, pp. 533–549, 1985.

50. M. P. Mengüç and R. Viskanta, "Radiative Transfer in Axisymmetric, Finite Cylindrical Enclosures," *ASME Journal of Heat Transfer*, 108, pp. 271–276, 1986.

51. M. P. Mengüç and R. Viskanta, "Comparison of Radiative Transfer Approximations for Highly Forward Scattering Planar Medium," *Journal Quantitative Spectroscopy and Radiative Transfer*, 29, pp. 381–394, 1983.

52. M. P. Mengüç and R. Viskanta, "Effect of Fly-Ash Particles on Spectral and Total Radiation Blockage," *Combustion Science and Technology*, 60, pp. 97–115, 1988.

53. M. F. Modest, "Modified Differential Approximation for Radiative Transfer in General Three-Dimensional Media," *AIAA Journal of Thermophysics and Heat Transfer*, 3, pp. 283–288, 1989.

54. D. B. Olfe, "A Modification of a Modified Differential Approximation for Radiative Transfer," *AIAA Journal*, 5, pp. 638–643, 1967.

55. M. P. Mengüç and R. K. Iyer, "Modeling of Radiative Transfer Using Multiple Spherical Harmonics Approximations," *Journal of Quantitative Spectroscopy and Radiative Transfer*, 39, pp. 445–461, 1988.

56. R. K. Iyer and M. P. Mengüç, "Quadruple Spherical Harmonics Approximations for Radiative Transfer in Two-Dimensional Rectangular Enclosures," *AIAA Journal of Thermophysics and Heat Transfer*, 3, p. 266, 1989.

57. B. G. Carlson and C. E. Lee, "Mechanical Quadrature and the Transport Equation," Rep. no. *LA-2573*, Los Alamos Scientific Laboratory, Los Alamos, New Mexico, 1961.

58. K. D. Lathrop, "THREETRAN: A Program to Solve the Multigroup Discrete Ordinates Transport Equation in (x,y,z) Geometry," Los Alamos Scientific Laboratory, Rep. nos. UC-32, UC-79, 1976.

59. W. A. Rhoades, D. B. Simpson, R. L. Childs, and W. W. Engle, "The DOT-IV Two-Dimensional Discrete Ordinates Transport Code with Space Dependent Mesh and Quadrature," ORNL/TM-6529, 1979.

60. W. A. Rhoades and R. L. Childs, "An Updated Version of the DOT-IV One- and Two-Dimensional Neutron/Photon Transport Code with Space Dependent Mesh and Quadrature," ORNL/5851, 1982.

61. E. E. Lewis and W. E. Miller, Jr., *Computational Methods of Neutron Transport*, Wiley, New York, 1984.

62. E. E. Khalil and J. S. Truelove, "Calculation of Radiative Heat Transfer in a Large Gas-Fired Furnace," *Letters in Heat and Mass Transfer*, 4, pp. 353–365, 1977.

63. W. A. Fiveland, "A Discrete-Ordinates Method for Predicting Radiative Heat Transfer in Axisymmetric Enclosures," ASME paper No. 82-HT-20, ASME, New York, 1982.

64. W. A. Fiveland, "Discrete-Ordinate Solutions of the Radiative Transfer Equation for Rectangular Enclosures," *ASME Journal of Heat Transfer*, 106, p. 699, 1984.

65. W. A. Fiveland, "Three-Dimensional Radiative Heat-Transfer Solutions by the Discrete-Ordinates Method," *AIAA Journal of Thermophysics and Heat Transfer*, 2, pp. 309–316, 1988.

66. A. Yücel and M. L. Williams, "Azimuthally Dependent Radiative Transfer in Cylindrical Geometry," in A. M. Smith and T. F. Smith (eds.), *Fundamentals and Applications of Radiation Heat Transfer*, ASME HTD-vol. 72, pp. 29–35, ASME, New York, 1987.

67. J. S. Truelove, "Discrete-Ordinates Solution of the Radiation Transport Equation," *ASME Journal of Heat Transfer*, 109(4), pp. 1048–1051, 1987.

68. J. S. Truelove, "Three-Dimensional Radiation in Absorbing-Emitting-Scattering Media Using the Discrete-Ordinates Approximation," *Journal of Quantitative Spectroscopy and Radiative Transfer*, 39, pp. 27–31, 1988.

69. A. S. Jamaluddin and P. J. Smith, "Predicting Radiative Transfer in Rectangular Enclosures Using the Discrete Ordinates Method," *Combustion Science and Technology*, 62, p. 173, 1988.

70. T.-K. Kim and H. Lee, "Effect of Anisotropic Scattering on Radiative Heat Transfer in Two-Dimensional Rectangular Enclosures," *International Journal of Heat and Mass Transfer*, 31(8), pp. 1711–1721, 1988.

71. T.-K. Kim and H. Lee, "Radiative Heat Transfer in Two-Dimensional Anisotropic Scattering Media," *Journal of Quantitative Spectroscopy and Radiative Transfer,* 42(3), pp. 225–238, 1989.

72. P. D. Jones and Y. Bayazitoglu, "Coordinate Systems for the Radiative Transfer Equation in Curvilinear Media," *Journal of Quantitative Spectroscopy and Radiative Transfer,* 48, pp. 427–440, 1992.

73. N. E. Wakil and J. F. Sacadura, "Some Improvements of the Discrete Ordinates Method for the Solution of the Radiative Transport Equation in Multidimensional Anisotropically Scattering Media," in HTD vol. 203, pp. 119–127, ASME, New York, 1992.

74. W. A. Fiveland and J. P. Jessee, "A Finite Element Formulation of the Discrete-Ordinate Method For Multidimensional Geometries," in *Radiative Heat Transfer: Current Research,* ASME HTD no. 244, New York, 1993.

75. S. Jendoubi, H. Lee, and T.-K. Kim, "Discrete Ordinates Solution for Radiatively Participating Media in a Cylindrical Enclosure," *AIAA Journal of Thermophysics and Heat Transfer,* 7(2), pp. 213–219, 1993.

76. J. C. Chai and S. V. Patankar, "Evaluation of Spatial Differencing Practices for the Discrete Ordinates Method," *AIAA Journal of Thermophysics and Heat Transfer,* vol. 8, pp. 140–144, 1994.

77. W. A. Fiveland and J. P. Jessee, "A Comparison of Discrete Ordinates Formulations for Radiative Heat Transfer in Multidimensional Geometries," *Journal of Thermophysics and Heat Transfer,* 9(1), pp. 47–54, 1995.

78. R. Koch, W. Krebs, S. Wittig, and R. Viskanta. "Discrete Ordinates Quadrature Schemes for Multidimensional Radiative Transfer in Furnaces," *Journal of Quantitative Spectroscopy and Radiative Transfer,* 53, pp. 353–372, 1995.

79. W. Krebs, S. Wittig, and R. Viskanta, "A Parabolic Formulation of the Discrete Ordinates Method for the Treatment of Complex Geometries" in M. P. Mengüç (ed.), *Radiative Transfer—I: Proceedings of the First International Symposium on Radiative Transfer,* pp. 355–371, Begell House, New York, 1996.

80. R. Vaillon, M. Lallemand, and D. Lemonnier, "Radiative Equilibrium in Axisymmetric Semitransparent Gray Shells Using the Discrete Ordinates Method," in M. P. Mengüç (ed.), *Radiative Transfer—I: Proceedings of the First International Symposium on Radiative Transfer,* pp. 62–74, Begell House, New York, 1996.

81. K.-B. Cheong and T.-H. Song, "Application of the Second Order Discrete Ordinate Method to a Radiation Problem in a Square Geometry," in M. P. Mengüç (ed.), *Radiative Transfer—I: Proceedings of the First International Symposium on Radiative Transfer,* pp. 75–91, Begell House, New York, 1996.

82. K. Stamnes, S-C. Tsay, W. Wiscombe, and K. Jayaweera, "Numerically Stable Algorithm for Discrete-Ordinate Method Radiative Transfer in Multiple Scattering and Emitting Layered Media," *Applied Optics,* 27, pp. 2502–2509, 1988.

83. J. P. Jessee, W. A. Fiveland, L. H. Howell, P. Colella, and R. B. Pember, "An Adaptive Mesh Refinement Algorithm for the Discrete Ordinates Method," in R. D. Skocypec, S. T. Thynell, D. A. Kaminski, A. M. Smith, and T. Tong (eds.), *Solution Methods for Radiative Transfer in Participating Media,* ASME HTD vol. 325, ASME, New York, 1996.

84. N. Selçuk and N. Kayakol, "Evaluation of Angular Quadrature and Spatial Differencing Schemes for Discrete Ordinates Method in Rectangular Furnaces," in R. D. Skocypec, S. T. Thynell, D. A. Kaminski, A. M. Smith, and T. Tong (eds.), *Solution Methods for Radiative Transfer in Participating Media,* ASME HTD vol. 325, ASME, New York, 1996.

85. N. Selçuk and N. Kayakol, "Evaluation of Discrete Ordinates Method for Radiative Transfer in Rectangular Furnaces," *International Journal of Heat and Mass Transfer,* 40(2), pp. 213–222, 1997.

86. Z.-Q. Tan and J. R. Howell, "New Numerical Method for Radiative Transfer in Nonhomogeneous Participating Media," *AIAA Journal Thermophysics Heat Transfer,* 4(4), pp. 419–424, 1990.

87. Z.-Q Tan and J. R. Howell, "Combined Radiation and Natural Convection in a Square Enclosure with Participating Medium," *International Journal of Heat Mass Transfer,* 34(3), pp. 785–793, 1991.

88. P. Hsu, Z. Tan, and J. R. Howell, "Application of the YIX Method to the Radiative Heat Transfer Within a Mixture of Highly Anisotropic Scattering Particles and Non-Gray Gas," *Developments in Radiative Heat Transfer,* ASME HTD-vol. 203, pp. 285–300, 1992.

89. P. F. Hsu, Z. Tan, and J. R. Howell, "Radiative Transfer by the YIX Method in Nonhomogeneous, Scattering, and Non-Gray Media," *AIAA Journal of Thermophysics Heat Transfer,* 7(3), pp. 487–495, 1993.

90. P. Hsu and Z. Tan, "The Radiative and Combined Mode Heat Transfer within the L-Shaped Non-homogeneous and Nongray Participating Media" in R. D. Skocypec, S. T. Thynell, D. A. Kaminski, A. M. Smith, and T. Tong (eds.), *Solution Methods for Radiative Transfer in Participating Media,* ASME HTD vol. 325, ASME, New York, 1996.

91. R. P. Gupta, T. F. Wall, and J. S. Truelove, "Radiative Scatter by Fly Ash in Pulverized-Coal-Fired Furnaces: Application of the Monte Carlo Method to Anisotropic Scatter," *International Journal of Heat and Mass Transfer,* 26, pp. 1649–1660, 1983.

92. J. R. Howell, "Application of Monte Carlo to Heat Transfer Problems," T. F. Irvine, Jr. and J. P. Hartnett (eds.), *Advances Heat Transfer,* vol. 5, Academic Press, New York, pp. 1–54, 1968.

93. A. Haji-Sheikh, "Monte Carlo Methods," in W. J. Minkowycz, E. M. Sparrow, R. H. Pletcher, and G. E. Schneider (eds.), *Handbook of Numerical Heat Transfer,* John Wiley, New York, 1988.

94. J. R. Howell, "Thermal Radiation in Participating Media: The Past, the Present, and Some Possible Futures," *ASME Journal of Heat Transfer,* 110, pp. 1220–1229, 1988.

95. H. Taniguchi and M. Funazu, "The Radiative Transfer of Gas in a Three-Dimensional System Calculated by Monte Carlo Method," *Bulletin of JSME,* 13, p. 458, 1970.

96. X. C. Xu, "Mathematical Modeling of Three-Dimensional Heat Transfer from the Flame in Combustion Chambers," *18th Symposium (International) on Combustion,* the Combustion Institute, pp. 1919–1925, 1981.

97. W. Richter, "Scale-Up and Advanced Performance Analysis of Boiler Combustion Chambers," ASME paper No. 85-WA/HT-80, ASME, New York, 1985.

98. J. T. Farmer and J. R. Howell, "Monte Carlo Prediction of Radiative Heat Transfer in Inhomogeneous, Anisotropic, Non-gray Media," *AIAA Journal of Thermophysics and Heat Transfer,* 8(1), pp. 133–139, 1994.

99. J. T. Farmer and J. R. Howell, "Monte Carlo Algorithms for Predicting Radiative Heat Transport in Optically Thick Participating Media," *Proc. 10th International Heat Transfer Conference,* Brighton, England, 1994.

100. J. T. Farmer and J. R. Howell, "Hybrid Monte Carlo/Diffusion Method for Enhanced Solution of Radiative Transfer in Optically Thick Media," AIAA/ASME Heat Transfer Conf., Colorado Springs, June, 1994.

101. J. T. Farmer, *Improved Algorithms for Monte Carlo Analysis of Radiative Heat Transfer in Complex Participating Media,* Ph.D. Dissertation, University of Texas, Austin, August, 1995.

102. S. H. Chan, "Numerical Methods for Multidimensional Radiative Transfer Analysis in Participating Media," *Annual Review of Numerical Fluid Mechanics and Heat Transfer,* vol. 1, Hemisphere, New York, pp. 305–350, 1987.

103. G. D. Raithby and E. H. Chui, "A Finite-Volume Method for Predicting Radiant Heat Transfer in Enclosures with Participating Media," *ASME Journal of Heat Transfer,* vol. 112, p. 415, 1990.

104. M. M. Razzaque, D. E. Klein, and J. R. Howell, "Finite Element Solution of Radiative Heat Transfer in a Two-Dimensional Rectangular Enclosure with Gray Participating Media," *ASME Journal of Heat Transfer,* vol. 105, pp. 933–934, 1983.

105. M. M. Razzaque, J. R. Howell, and D. E. Klein, "Coupled Radiative and Conductive Heat Transfer in a Two-Dimensional Rectangular Enclosure with Gray Participating Media Using Finite Elements," *ASME Journal of Heat Transfer,* vol. 106, pp. 613–619, 1984.

106. T. J. Chung and J. Y. Kim, "Two-Dimensional Combined-Mode Heat Transfer by Conduction, Convection, and Radiation in Emitting, Absorbing, and Scattering Media—Solution by Finite Elements," *ASME Journal of Heat Transfer,* vol. 106, pp. 448–452, 1984.

107. C. N. Sökmen and M. M. Razzaque, "Finite Element Analysis of Conduction-Radiation Heat Transfer in an Absorbing-Emitting and Scattering Medium Contained in an Enclosure with Heat Flux Boundary Conditions," ASME HTD-vol. 81, pp. 17–23, 1987.

108. Z.-Q. Tan, "Radiative Heat Transfer in Multidimensional Emitting, Absorbing, and Anisotropic Scattering Media—Mathematical Formulation and Numerical Method," *ASME Journal of Heat Transfer,* 111, pp. 141–147, 1989.

109. S. P. Burns, J. R. Howell, and D. E. Klein, "Finite Element Solution for Radiative Transfer with Nongray, Nonhomogeneous Radiative Properties," in *Proc. ASME/AIChE National Heat Trans. Conf.,* Portland, August, 1995.

110. S. P. Burns, J. R. Howell, and D. E. Klein, "Solution of Natural Convection-Radiation in a Horizontal, Cylindrical Annulus Using the Finite Element Method," *Proc. 9th Int. Conf. on Numerical Meths. in Thermal Probs.,* Atlanta, July 17–21, 1995.

111. H. A. J. Vercammen and G. F. Fromment, "An Improved Zone Method Using Monte Carlo Techniques for the Simulation of Radiation in Industrial Furnaces," *International Journal of Heat and Mass Transfer,* vol. 23, pp. 329–337, 1984.

112. D. K. Edwards, "Hybrid Monte Carlo Matrix-Inversion Formulation of Radiation Heat Transfer with Volume Scattering" in C. K. Law, Y. Jaluria, W. W. Yuen, and K. Miyasaka (eds.), *Heat Transfer in Fire and Combustion Systems,* ASME, New York, pp. 273–278, 1985.

113. F. C. Lockwood and N. G. Shah, "A New Radiation Solution Method for Incorporation in General Combustion Prediction Procedures," *Eighteenth Symposium (International) on Combustion,* The Combustion Institute, Pittsburgh, pp. 1405–1414, 1981.

114. N. Selçuk and N. Kayakol, "Evaluations of Discrete Transfer Model for Radiative Transfer in Combustors," in M. P. Mengüç (ed.), *Radiative Transfer—I: Proceedings of the First International Symposium on Radiative Transfer,* pp. 127–138, Begell House, New York, 1996.

115. H. Taniguchi, K. Kudo, H. Hayasaka, W.-J Yang, and H. Tashiro, "Thermal Behavior in Furnaces of Complex Geometry," in T. C. Min and J. L. S. Chen (ed.), *Fundamentals in Thermal Radiation Heat Transfer,* ASME, New York, 1984, pp. 29–36, 1984.

116. H. Taniguchi, W.-J Yang, K. Kudo, H. Hayasaka, M. Oguma, A. Kusama, I. Nakamachi, and N. Okigami, "Radiant Transfer in Gas Filled Enclosures by Radiant Energy Absorption Distribution Method," in C. L. Tien, V. P. Carey, and J. K. Ferrell (eds.), *Heat Transfer—1986,* vol. 2, pp. 757–762, 1986.

117. W.-J. Yang, H. Taniguchi, and K. Kudo, "Radiative Heat Transfer by the Monte Carlo Method," T. F. Irvine and J. P. Hartnett (eds.), *Advances in Heat Transfer,* Academic Press, New York, vol. 27, pp. 1–215, 1995.

118. F. R. Steward and K. N. Tennankore, "Towards a Finite Difference Solution Coupled with the Zone Method for Radiative Transfer for a Cylindrical Combustion Chamber," *Journal of the Institute of Energy,* 52, p. 107, 1979.

119. W. T. Tong and R. D. Skocypec, "Summary on Comparison of Radiative Heat Transfer Solutions for a Specified Problem," in *Developments in Radiative Heat Transfer,* ASME HTD-vol. 203, pp. 253–258, 1992.

120. R. L. Hoover, L. Weiming, A. Benmalek, and T. W. Tong, "Sn Solutions for Radiative Heat Transfer in an L-Shaped Participating Medium," in R. D. Skocypec, S. T. Thynell, D. A. Kaminski, A. M. Smith, and T. Tong (eds.), *Solution Methods for Radiative Transfer in Participating Media,* ASME HTD vol. 325, ASME, New York, 1996.

121. P.-F. Hsu and Z. Tan, "Recent Benchmarkings of Radiative Heat Transfer within Nonparticipating Media and the Improved YIX Method," in M. P. Mengüç (ed.), *Radiative Transfer—I: Proceedings of the First International Symposium on Radiative Transfer,* pp. 107–126, Begell House, New York, 1996.

122. J. C. Henson, W. M. G. Malalasekera, and J. C. Dent, "Comparison of the Discrete Transfer and Monte Carlo Methods for Radiative Heat Transfer in Three-Dimensional, Nonhomogeneous Participating Media," in R. D. Skocypec, S. T. Thynell, D. A. Kaminski, A. M. Smith, and T. Tong (eds.), *Solution Methods for Radiative Transfer in Participating Media,* ASME HTD vol. 325, ASME, New York, 1996.

123. P. Cheng, "Exact Solutions and Differential Approximation for Multidimensional Radiative Transfer in Cartesian Coordinate Configuration," *Progress in Astronautics and Aeronautics,* vol. 31, pp. 269–308, 1972.

124. A. L. Crosbie and L. C. Lee, "Relation between Multidimensional Radiative Transfer in Cylindrical and Rectangular Coordinates with Anisotropic Scattering," *Journal of Quantitative Spectroscopy and Radiative Transfer,* 38, pp. 231–241, 1987.

125. N. Selçuk, "Exact Solutions for Radiative Heat Transfer in Box-Shaped Furnaces," *ASME Journal of Heat Transfer,* 107, pp. 648–655, 1985.

126. N. Selçuk and Z. Tahiroglu, "Exact Numerical Solution for Radiative Heat Transfer in Cylindrical Furnaces," *International Journal of Numerical Methods in Engineering,* 26, pp. 1201–1212, 1988.

127. C. L. Tien, "Thermal Radiation Properties of Gases," in T. F. Irvine Jr. and J. P. Hartnett (eds.), *Advances in Heat Transfer,* vol. 5, Academic Press, New York, pp. 253–324, 1968.

128. C. B. Ludwig, W. Malkmus, J. G. Reardon, and J. A. L. Thomson, *Handbook of Infrared Radiation from Combustion Products,* NASA SP-3080, Washington, D.C., 1973.

129. D. K. Edwards, "Molecular Gas Band Radiation," in T. F. Irvine Jr. and J. P. Hartnett (eds.), *Advances in Heat Transfer,* vol. 12, Academic Press, New York, pp. 115–193, 1976.

130. S. N. Tiwari, "Models for Infrared Atmospheric Radiation," in B. Saltzmann (ed.), *Advances in Geophysics,* Academic Press, New York, vol. 20, pp. 1–85, 1978.

131. A. F. Sarofim and H. C. Hottel, *Heat Transfer,* vol. 6, pp. 199–217. Hemisphere Publishing Corp., Washington, D.C., 1978.

132. M. P. Mengüç and R. Viskanta, "An Assessment of Spectral Radiative Heat Transfer Predictions for a Pulverized-Coal Fired Furnace," *Heat Transfer—1986,* 2, pp. 815–820, 1986.

133. D. K. Edwards and W. A. Menard, "Comparison of Models for Correlation of Total Band Absorption," *Applied Optics,* 3, pp. 621–625, 1964.

134. N. Lallemant and R. Weber, "A Computationally Efficient Procedure for Calculating Gas Radiative Properties Using the Exponential Wide Band Model," *International Journal of Heat and Mass Transfer,* 39(15), pp. 3273–3286, 1996.

135. J. D. Felske and C. L. Tien, "A Theoretical Closed Form Expression for the Total Band Absorptance of Infrared-Radiating Gases," *ASME Journal of Heat Transfer,* vol. 96, pp. 155–158, 1974.

136. M. Abramowitz and I. A. Stegun, *Handbook of Mathematical Functions,* Dover, New York, 1972.

137. D. K. Edwards and S. J. Morizumi, "Scaling Vibration-Rotation Band Parameters for Nonhomogeneous Gas Radiation," *Journal of Quantitative Spectroscopy and Radiative Transfer,* 10, pp. 175–188, 1970.

138. J. D. Felske and C. L. Tien, "A Simple Correlation Scheme for the Emissivity of Luminous Flames," *Combustion Science and Technology,* 7, p. 25, 1973.

139. B. Leckner, "Spectral and Total Emissivity of Water Vapor and Carbon Dioxide," *Combustion and Flame,* 19, p. 33, 1972.

140. A. T. Modak, "Radiation from Products of Combustion," *Fire Research,* 1, p. 339, 1979.

141. F. R. Steward and Y. S. Kocaefe, "Total Emissivity and Absorptivity Models for Carbon Dioxide, Water Vapor, and Their Mixtures," in C. L. Tien, V. P. Carey, and J. K. Ferrell (eds.), *Heat Transfer— 1986,* vol. 2, Hemisphere, Washington, D.C., pp. 735–740, 1986.

142. P. B. Taylor and P. J. Foster, "The Total Emissivities of Luminous and Nonluminous Flames," *International Journal of Heat Mass Transfer,* 17, p. 1591, 1974.

143. T. F. Smith, Z. F. Shen, and J. N. Friedman, "Evaluation of Coefficients for the Weighted Sum of Gray Gases Model," *ASME Journal of Heat Transfer,* 104, p. 602, 1982.

144. I. H. Farag, U. Grigull, E. Hahne, K. Stephan, and J. Straub (eds.), *Heat Transfer—1982,* vol. 2, Hemisphere, Washington, D.C., pp. 489–492, 1982.

145. A. Copalle and P. Vervisch, "The Total Emissivities of High-Temperature Flames," *Combustion and Flame,* vol. 49, p. 101, 1983.

146. M. F. Modest, "The Weighted Sum of Gray Gases Model for Arbitrary Solution Methods in Radiative Transfer," *ASME Journal of Heat Transfer,* 113, pp. 650–656, 1991.

147. M. Denison and B. W. Webb, "An Absorption-Line Blackbody Distribution Function for Efficient Calculation of Total Gas Radiative Transfer," *Journal of Quantitative Spectroscopy and Radiative Transfer,* 50, pp. 499–510, 1993.

148. M. Denison and B. W. Webb, "A Spectral Line Based Weighted-Sum-of-Gray-Gases Model for Arbitrary RTE Solvers," *ASME Journal of Heat Transfer,* 115, pp. 1004–1012, 1993.

149. M. K. Denison and B. W. Webb, "*k*-Distributions and Weighted-Sum-of-Gray-Gases—A Hybrid Model," *Heat Transfer—1994,* vol. 2, pp. 19–24, 1994.

150. M. K. Denison and B. W. Webb, "Development and Application of an Absorption-Line Blackbody Distribution Function for CO_2," *International Journal of Heat and Mass Transfer,* 38, pp. 1813–1821, 1995.

151. M. K. Denison and B. W. Webb, "The Spectral Line-Based Weighted-Sum-of-Gray-Gases Model in Non-Isothermal Non-Homogeneous Media," *ASME Journal of Heat Transfer,* 117, pp. 359–365, 1995.

152. M. K. Denison and B. W. Webb, "The Spectral-Line Weighted-Sum-of-Gray-Gases Model for H_2O/CO_2 Mixtures," *ASME Journal of Heat Transfer,* 117, pp. 788–792, 1995.

153. M. K. Denison and B. W. Webb, "The Spectral Line-Based Weighted-Sum-of-Gray-Gases Model—A Review," in M. P. Mengüç (ed.), *Radiative Transfer—I: Proceedings of the First International Symposium on Radiative Transfer,* Begell House, New York, pp. 193–208, 1996.

154. G. Parthasarathy, S. C. Chai, and S. Patankar, "A Simple Approach to Non-Gray Gas Modelling," *Symposium on Thermal Science and Proceedings of the Engineering in Honor of Chancellor Chang-Lin Tien,* November, 1995.

155. P. Y. C. Lee, G. D. Raithby, and K. G. T. Hollands, "Reordering the Absorption Coefficient within the Wide Band for Predicting Gaseous Radiant Exchange," *ASME Journal of Heat Transfer,* 118(2), pp. 394–400, 1996.

156. L. S. Rothman, R. R. Gamache, R. H. Tipping, C. P. Rinsland, M. A. H. Smith, D. Chris Brenner, V. Malathy Devi, J. M. Flaud, C. Camy-Peyret, A. Perrin, A. Goldman, S. T. Massie, and L. R. Brown, "The HITRAN Molecular Database: Editions of 1991 and 1992," *Journal of Quantitative Spectroscopy and Radiative Transfer,* 48(5/6), pp. 469–507, 1992.

157. T. H. Song, "Comparison of Engineering Models of Nongray Behavior of Combustion Products," *International Journal of Heat and Mass Transfer,* 36, pp. 3975–3982, 1993.

158. A. Soufiani and E. Djavdan, "A Comparison between Weighted Sum of Gray Gases and Statistical Narrow Band Radiation Models for Combustion Applications," *Combustion and Flame,* 97, pp. 240–250, 1994.

159. L. Pierrot, A. Soufiani, and J. Taine, "Accuracy of Various Gas IR Radiative Property Models Applied to Radiative Transfer in Planar Media," in M. P. Mengüç (ed.), *Radiative Transfer—I: Proceedings of the First International Symposium on Radiative Transfer,* Begell House, New York, pp. 209–227, 1996.

160. K. C. Tang and M. Q. Brewster, "*K*-Distribution Analysis of Gas Radiation with Non-gray, Emitting, Absorbing, and Anisotropic Scattering Particles," in S. T. Thynell et al. (eds.), *Developments in Radiative Heat Transfer,* ASME-HTD-vol. 203, pp. 311–320, 1992.

161. R. M. Goody, R. West, L. Chen, and D. Chrisp, "The Correlated-*k* Method for Radiation Calculations in Nonhomogeneous Atmospheres," *Journal of Quantitative Spectroscopy and Radiative Transfer,* 42, pp. 539–550, 1989.

162. O. Marin and R. O. Buckius, "Wideband Correlated-*k* Method Applied to Absorbing, Emitting, and Scattering Media," *AIAA Journal of Thermophysics and Heat Transfer,* 10(2), pp. 364–371, 1996.

163. D. K. Edwards and A. Balakrishnan, "Slab Absorptance for Molecular Gas Radiation," *Journal of Quantitative Spectroscopy and Radiative Transfer,* 12, p. 1379, 1972.

164. D. K. Edwards and A. Balakrishnan, "Thermal Radiation by Combustion Gases," *International Journal of Heat Mass Transfer,* 16, p. 25, 1973.

165. W. L. Grosshandler, "Radiation from Nonhomogeneous Combustion Products," *International Journal of Heat Mass Transfer,* 23, p. 1447, 1980.

166. W. L. Grosshandler and A. T. Modak, "Radiation from Nonhomogeneous Combustion Products," *Eighteenth Symposium (International) on Combustion,* the Combustion Institute, Pittsburgh, pp. 601–609, 1981.

167. M. Kerker, *The Scattering of Light,* Academic Press, New York, 1969.

168. C. F. Bohren and E. R. Huffman, *Absorption and Scattering of Light by Small Particles,* Wiley, New York, 1983.

169. W. J. Wiscombe and L. Mugnai, *Single Scattering from Nonspherical Chebyshev Particles: A Compendium of Correlations,* NASA Reference Publication 1157, Washington, D.C., 1986.

170. A. Mugnai and W. J. Wiscombe, "Scattering for Nonspherical Chebyshev Particles. 3: Variability in Angular Scattering Patterns," *Applied Optics,* 28, pp. 3061–3073, 1989.

171. M. L. Mishchenko, L. D. Travis, and D. W. Mackowski, "T-Matrix Computations of Light Scattering by Non-spherical Particles: A Review," *Journal of Quantitative Spectroscopy and Radiative Transfer,* 55(5), pp. 535–575, 1996.

172. E. M. Purcell and C. R. Pennypacker, "Scattering and Absorption by Nonspherical Dielectric Grains," *Astrophysics Journal,* 186, p. 705, 1973.

173. B. T. Draine, "The Discrete-Dipole Approximation and its Application to Interstellar Graphite Grains," *The Astrophysical Journal,* 333, pp. 848–872, 1988.

174. S. Manickavasagam and M. P. Mengüç, "Scattering Matrix Elements of Fractal Like Soot Agglomerates," *Applied Optics,* 36(6), pp. 1337–1351, 1997.

175. R. O. Buckius and D. C. Hwang, "Radiation Properties of Polydispersions: Application to Coal," *ASME Journal of Heat Transfer,* Vol. 102, p. 99, 1980.

176. R. Viskanta, A. Ungan, and M. P. Mengüç, "Predictions of Radiative Properties of Pulverized-Coal and Fly-Ash Polydispersions," ASME paper No. 81-HT-24, 1981.

177. M. P. Mengüç and R. Viskanta, "On the Radiative Properties of Polydispersions: A Simplified Approach," *Combustion Science and Technology,* 44, p. 143, 1985.

178. K. H. Im and R. K. Ahluwalia, "Radiation Properties of Coal Combustion Products," *International Journal of Heat and Mass Transfer,* 36, pp. 293–302, 1993.

179. C. Kim and N. Lior, "Easily Computable Approximations for the Spectral Radiative Properties of Particle Gas Components and Mixtures in Pulverized Coal Combustors," *Fuel,* 74(12), pp. 1891–1902, 1995.

180. R. D. Skocypec and R. O. Buckius, "Total Hemispherical Emittances for CO_2 or H_2O Including Particulate Scattering," *International Journal of Heat Mass Transfer,* 27, p. 1, 1984.

181. R. D. Skocypec, D. V. Walters, and R. O. Buckius, "Total Hemispherical Emittances for Isothermal Mixtures of Combustion Gases and Scattering Particulate," *Combustion Science and Technology,* 47, p. 239, 1986.

182. D. G. Goodwin and J. L. Ebert, "Rigorous Bounds on the Radiative Interactions Between Real Gases and Scattering Particles," *Journal of Quantitative Spectroscopy and Radiative Transfer,* 37, pp. 501–508, 1987.

183. R. D. Skocypec and R. O. Buckius, "Comments on 'Rigorous Bounds on the Radiative Interactions Between Real Gases and Scattering Particles' by D. G. Goodwin and J. L. Ebert," *Journal of Quantitative Spectroscopy and Radiative Transfer,* 37, pp. 509–511, 1987.

184. S. A. Self, "Comment on 'Rigorous Bounds on the Radiative Interactions Between Real Gases and Scattering Particles' by D. G. Goodwin and J. L. Ebert," *Journal of Quantitative Spectroscopy and Radiative Transfer,* 37, pp. 513–514, 1987.

185. T. T. Charalampopoulos, "Morphology and Dynamics of Agglomerated Particulates in Combustion Systems Using Light Scattering Techniques," *Progress in Energy and Combustion Science,* 18, pp. 13–45, 1992.

186. A. Selamet and V. S. Arpaci, "Rayleigh Limit-Penndorf Extension," *International Journal of Heat and Mass Transfer,* 32, pp. 1809–1820, 1989.

187. A. Selamet, "Visible and Infrared Sensitivity of Rayleigh Limit and Penndorf Extension to Complex Index of Refraction of Soot," *International Journal of Heat and Mass Transfer,* 35, pp. 3479–3484, 1992.

188. R. Siegel, "Radiative Behavior of a Gas Layer Seeded with Soot," *NASA TN D-8278,* Washington, D.C., 1976.

189. S. C. Lee and C. L. Tien, "Effect of Soot Shape on Soot Radiation," *Journal of Quantitative Spectroscopy and Radiative Transfer,* 29, pp. 259–265, 1983.

190. D. W. Mackowski, R. A. Altenkirch, and M. P. Mengüç, "Extinction and Absorption Coefficients of Cylindrically Shaped Soot Particles," *Combustion Science and Technology,* 53, p. 399, 1987.

191. D. W. Mackowski, R. A. Altenkirch, M. P. Mengüç, and K. Saito, "Radiative Properties of Chain-Agglomerated Soot Formed in Hydrocarbon Diffusion Flames," *Twenty-Second Symposium (International) on Combustion,* The Combustion Institute, pp. 1263–1269, 1988.

192. T. T. Charalampopoulos and D. W. Hahn, "Extinction Efficiencies of Elongated Soot Particles," *Journal of Quantitative Spectroscopy and Radiative Transfer,* 42, pp. 219–224, 1989.

193. B. T. Draine and P. J. Flatau, "The Discrete-Dipole Approximation for Scattering Calculations," *Journal of Optical Society of America A,* 11, 1491–1499, 1994.

194. B. M. Vaglieco, O. Monda, F. E. Corcione, and M. P. Mengüç, "Optical and Radiative Properties of Particulates at Diesel Engine Exhaust," *Combustion Science and Technology,* 102, pp. 283–299, 1994.

195. Z. Ivezic and M. P. Mengüç, "An Investigation of Dependent/Independent Scattering Regimes for Soot Particles Using Discrete Dipole Approximation," *International Journal of Heat Mass Transfer,* 39, pp. 811–822, 1996.

196. D. W. Mackowski, "Calculation of Total Cross Sections of Multiple-Sphere Clusters," *Journal of Optical Society of America A,* 11, pp. 2851–2861, 1994.

197. K. A. Fuller, "Scattering and Absorption Cross Sections of Compounded Spheres. I: Theory for External Aggregation," *Journal of Optical Society of America A,* 11, pp. 3251–3260, 1994.

198. A. R. Jones, "Electromagnetic Wave Scattering by Assemblies of Particles in the Rayleigh Approximation," *Proceedings of Royal Society of London,* A.366, pp. 111–127. (Corrections in A.375, pp. 453–454, 1979.)

199. A. R. Jones, "Scattering Efficiency Factors for Agglomerates of Small Spheres," *Journal Phys. D: Appl. Phys.* 12, pp. 1661–1672, 1979.

200. J. D. Felske, P. F. Hsu, and J. C. Ku, "The Effect of Soot Particle Optical Inhomogeneity and Agglomeration on the Analysis of Light Scattering Measurements in Flames," *Journal of Quantitative Spectroscopy and Radiative Transfer,* 35, p. 447, 1986.

201. B. L. Drolen and C. L. Tien, "Absorption and Scattering of Agglomerated Soot Particulate," *Journal of Quantitative Spectroscopy and Radiative Transfer,* 37, pp. 433–448, 1987.

202. S. Kumar and C. L. Tien, "Effective Diameter of Agglomerates for Radiative Extinction and Scattering," *Combustion Science and Technology,* 66, pp. 199–216, 1989.

203. J. C. Ku and K-H Shim, "The Effects of Refractive Indices, Size Distribution, and Agglomeration on the Diagnostics and Radiative Properties of Flame Soot Particles," in W. L. Grosshandler and H. G. Semerjian (eds.), *Heat and Mass Transfer in Fires and Combustion Systems,* ASME HTD-vol. 148, ASME, New York, 1990.

204. J. C. Ku and K. H. Shim, "A Comparison of Solutions for Light Scattering and Absorption by Agglomerated or Arbitrarily-Shaped Particles," *Journal of Quantitative Spectroscopy and Radiative Transfer,* 47, pp. 201–220, 1992.

205. T. T. Charalampopoulos and H. Chang, "Agglomerate Parameters and Fractal Dimension of Soot Using Light Scattering: Effects on Surface Growth," *Combustion and Flame,* 87, pp. 89–99, 1991.

206. M. F. Iskander, H. Y. Chen, and J. E. Penner, "Optical Scattering and Absorption by Branched-Chains of Aerosols," *Applied Optics,* 28, pp. 3083–3091, 1989.

207. R. A. Dobbins and C. M. Megaridis, "Morphology of Flame-Generated Soot as Determined by Thermophoretic Sampling," *Langmuir,* 3, pp. 254–259, 1987.

208. R. D. Mountain and D. G. Mulholland, "Light Scattering from Simulated Smoke Agglomerates," *Langmuir,* 4, pp. 1321–1326, 1988.

209. R. A. Dobbins and C. M. Megaridis, "Absorption and Scattering of Light by Polydisperse Aggregates," *Applied Optics,* 30, pp. 4747–4754, 1992.

210. Ü. Ö. Köylü and G. M. Faeth, "Radiative Properties of Flame Generated Soot," *ASME Journal of Heat Transfer,* 115, pp. 409–417, 1993.

211. Ü. Ö. Köylü and G. M. Faeth, "Optical Properties of Soot in Buoyant Laminar Diffusion Flames," *ASME Journal of Heat Transfer,* 116, pp. 971–979, 1994.

212. T. L. Farias, M. G. Carvalho, Ü. Ö. Köylü, and G. M. Faeth, "Computational Evaluation of Approximate Rayleigh-Debye-Gans/Fractal Aggregate Theory for the Absorption and Scattering Properties of Soot," *ASME Journal of Heat Transfer,* 117, pp. 152–159, 1995.

213. T. L. Farias, Ü. Ö. Köylü, and M. G. Carvalho, "Effects of Polydispersity of Aggregates and Primary Particles on Radiative Properties of Simulated Soot," *Journal of Quant. Spectrosc. Radiative Transfer,* 55, p. 357, 1995.

214. S. Manickavasagam, R. Govindan, and M. P. Mengüç, "Estimation of the Morphology of Soot Agglomerates by Measuring their Scattering Matrix Elements," in K. Annamalai et al. (eds.), ASME HTD-Vol. 352, pp. 29–32, 1997.

215. S. Chippet and W. A. Gray, "The Size and Optical Properties of Soot Particles," *Combustion and Flame,* 31, p. 149, 1978.

216. S. C. Lee and C. L. Tien, "Optical Constants of Soot in Hydrocarbon Flames," *Eighteenth Symposium (International) on Combustion,* Combustion Institute, p. 1159, 1981.

217. C. L. Tien and S. C. Lee, "Flame Radiation," *Progress in Energy and Combustion Sciences,* 8, pp. 41–59, 1982.

218. Z. G. Habib and P. Vervisch, "On the Refractive Index of Soot at Flame Temperature," *Combustion Science and Technology,* 59, pp. 261–274, 1988.

219. W. D. Erickson, G. C. Williams, and H. C. Hottel, "A Light Scattering Method for Soot Benzene-Air Flame," *Combustion and Flame,* 8, p. 127, 1970.

220. E. A. Powell and B. T. Zinn, "In Situ Measurements of the Complex Refractive Index of Combustion Generated Particles," *Prog. Astro. Aero.,* 92, p. 238, 1984.

221. T. T. Charalampopoulos and J. D. Felske, "Refractive Indices of Particles as Deduced from In Situ Laser Scattering Measurements," *Combustion and Flame,* 68, pp. 283–294, 1987.

222. T. T. Charalampopoulos and H. Chang, "Refractive Indices of Soot Particles Deduced from In-Situ Laser Light Scattering Measurements," *Combustion and Flame,* 68, p. 283, 1987.

223. T. T. Charalampopoulos and H. Chang, "In-Situ Optical Properties of Soot Particles in the Wavelength Range from 340 to 600 nm," *Combustion Science and Technology,* 59, p. 401, 1988.

224. B. M. Vaglieco, F. Beretta, and A. D'Alessio, "In Situ Evaluation of the Soot Refractive Index in the UV-Visible from the Measurement of the Scattering and Extinction Coefficients in Rich Flames," *Combustion and Flame,* 79, pp. 259–271, 1990.

225. W. H. Dalzell and A. F. Sarofim, "Optical Constants of Soot and Their Application to Heat Flux Calculations," *ASME Journal of Heat Transfer,* 91, p. 100, 1969.

226. D. Janzen, "The Refractive Index of Colloidal Carbon," *Journal of Colloidal and Interface Science,* 69(3), pp. 436–447, 1979.

227. M. P. Mengüç and B. W. Webb, "Radiative Heat Transfer," in L. D. Smoot (ed.), *Fundamentals of Coal Combustion: Clean and Efficient Use,* Elsevier Science, New York, pp. 375–430, 1993.

228. T. Fletcher, J. Ma, and B. W. Webb, "Soot in Coal Combustion Systems," *Progress in Energy and Combustion Systems,* 23(3), 283, 1997.

229. S. Manickavasagam and M. P. Mengüç, "Effective Optical and Radiative Properties of Coal Particles as Determined from FT-IR Spectroscopy Experiments," *Energy and Fuel,* 7(6), pp. 860–869, 1993.

230. A. G. Blokh, *Heat Transfer in Steam Boilers,* Hemisphere, Washington, D.C., 1988.

231. M. Q. Brewster and T. Kunitomo, "The Optical Constants of Coal, Char, and Limestone," *ASME Journal of Heat Transfer,* 106, pp. 678–683, 1984.

232. W. L. Grosshandler and S. L. P. Monterio, "On the Spectral Emissivity of Pulverized Coal and Char," *ASME Journal of Heat Transfer,* 104, pp. 587–593, 1981.

233. P. E. Best, R. M. Carangelo, J. R. Markham, and P. R. Solomon, "Extension of Emission-Transmission Technique to Particulate Samples Using FT-IR," *Combustion and Flame,* 66, pp. 47–66, 1986.

234. P. R. Solomon, R. M. Carangelo, P. E. Best, J. R. Markham, and D. G. Hamblen, "The Spectral Emittance of Pulverized Coal and Char," *Twenty-First Symposium (International) on Combustion,* Combustion Institute, pp. 437–446, 1986.

235. P. R. Solomon, R. M. Carangelo, P. E. Best, J. R. Markham, and D. G. Hamblen, "Analysis of Particle Emittance, Composition, Size, and Temperature by FT-IR Emission/Transmission Spectroscopy," *Fuel,* 66, pp. 897–908, 1987.

236. P. R. Solomon, P. L. Chien, R. M. Carangelo, P. E. Best, and J. R. Markham, "Application of FT-IR Emission/Transmission (E/T) Spectroscopy to Study Coal Combustion Phenomena," *Twenty-Second Symposium (International) on Combustion,* Combustion Institute, pp. 211–221, 1988.

237. S. Manickavasagam, *Effective Radiative and Optical Properties of Coal and Char Particles,* Ph.D. Thesis, University of Kentucky, 1993.

238. M. P. Mengüç, S. Manickavasagam, and D. A. D'sa, "Determination of Radiative Properties of Pulverized Coal Particles from Experiments," *Fuel,* 73(4), pp. 613–625, 1994.

239. S. Manickavasagam and M. P. Mengüç, "Effective Radiation Properties of Coal Particles in Flames at λ = 10.6 μm," in B. Farouk, M. P. Mengüç, R. Viskanta, C. Presser, and S. Chellaiah (eds.), *Heat Transfer in Fire and Combustion Systems–1993,* ASME HTD-vol. 250, pp. 145–157.

240. R. C. Flagan and S. K. Friedlander, "Particle Formation in Pulverized Coal Combustion," in D. T. Shaw (ed.), *Recent Developments in Aerosol Science,* John Wiley and Sons, New York, 1978.

241. T. F. Wall, A. Lowe, L. J. Wibberley, T. Mai-Viet, and R. P. Gupta, "Fly Ash Characteristics and Radiative Heat Transfer in Pulverized-Coal-Fired Furnaces," *Combustion Science and Technology,* 26, pp. 107–121, 1981.

242. A. Lowe, I. M. Stewart, and T. F. Wall, "The Measurement and Interpretation of Radiation from Fly Ash Particles in Large Pulverized Coal Flames," *Seventeenth Symposium (International) on Combustion,* Combustion Institute, Pittsburgh, pp. 105–112, 1979.

243. R. P. Gupta and T. F. Wall, "The Complex Refractive Index of Particles," *Journal of Physics D: Applied Physics,* 14, pp. L95–L98, 1981.

244. R. P. Gupta and T. F. Wall, "The Optical Properties of Fly Ash in Coal Fired Furnaces," *Combustion and Flame,* 61, pp. 145–151, 1985.

245. P. J. Wyatt, "Some Chemical, Physical, and Optical Properties of Fly Ash Particles," *Applied Optics,* 19, pp. 975–983, 1980.

246. E. Marx, "Data Analysis for Size and Refractive Index Determination from Light Scattering by Single Spheres," *Aerosol Science and Technology,* 2, p. 190, 1983.

247. D. G. Goodwin, *Infrared Optical Constants of Coal Slags,* Ph.D. Thesis, Stanford University, 1986.

248. S. Ghosal, *Optical Characterization of Coal Fly Ashes and Infrared Extinction Measurements of Ash Suspensions,* Ph.D. Thesis, Stanford University, 1993.

249. J. Ebert, *Infrared Optical Properties of Coal Slag at High Temperatures,* Ph.D. Thesis, Stanford University, 1994.

250. M. P. Mengüç and S. Subramaniam, "Radiative Transfer Through an Inhomogeneous Fly-Ash Cloud," *Numerical Heat Transfer, Part A: Applications,* 21, pp. 261–273, 1992.

251. B.-C. Chern, T. J. Moon, and J. R. Howell, "Dependent Radiative Transfer Regime for Unidirectional Fiber Composites Exposed to Normal Incident Radiation," 4th ASME/JSME Joint Symposium, Maui, 1995.

252. M. J. Hale and M. S. Bohn, "Measurement of the Radiative Transport Properties of Reticulated Alumina Foams," ASME Paper 92-V-842, August, 1992.

253. T. Hendricks and J. R. Howell, "Absorption/Scattering Coefficients and Scattering Phase Functions in Reticulated Porous Ceramics," *ASME Journal of Heat Transfer,* 116, pp. 79–87, 1996.

254. F. P. Incropera and D. P. DeWitt, *Fundamentals of Heat and Mass Transfer,* 4th ed., Wiley, New York, 1996.

255. J. R. Howell, M. J. Hall, and J. L. Ellzey, "Combustion of Hydrocarbon Fuels within Porous Inert Media," *Progress in Energy and Combustion Sciences,* 22, pp. 121–145, 1996.

256. R. Mital, J. P. Gore, and R. Viskanta, "Measurements of Radiative Properties of Cellular Ceramics at High Temperatures," *AIAA Journal of Thermophysics and Heat Transfer,* 10(1), pp. 33–38, January–March, 1996.

257. D. Doermann and J. F. Sacadura, "Heat Transfer in Open-Cell Foam Insulation," *ASME Journal of Heat Transfer,* 118, pp. 88–93, 1996.

258. D. Doermann and J. F. Sacadura, "Thermal Radiation Properties of Dispersed Media: Theoretical Prediction and Experimental Characterization," *Radiative Transfer—II: Proceedings of the Second International Symposium on Radiative Transfer,* M. P. Mengüç (ed.), Begell House, New York, 1998 (in press).

259. C. M. Spuckler and R. Siegel, "Refractive Index and Scattering Effects on Radiative Behavior of a Semitransparent Layer," *AIAA Journal of Thermophysics and Heat Transfer,* 7(2), pp. 302–310, 1993.

260. C. M. Spuckler and R. Siegel, "Refractive Index and Scattering Effects on Radiation in a Semi-transparent Laminated Layer," *AIAA Journal of Thermophysics and Heat Transfer,* 8(2), pp. 193–201, 1994.

261. R. Siegel and C. M. Spuckler, "Temperature Distributions on Semitransparent Coatings," *AIAA Journal of Thermophysics and Heat Transfer,* 10(1), pp. 39–46, 1996.

262. R. Siegel, "Two-Flux Method with Green's Function Method for Transient Radiative Transfer in a Semitransparent Layer," *Radiative Transfer—I: Proceedings of the First International Symposium on Radiative Transfer,* M. P. Mengüç (ed.), pp. 473–487, Begell House, New York, 1996.

263. G. Chen and C. L. Tien, "Partial Coherence Theory of Thin Film Properties," *ASME Journal of Heat Transfer,* 114, pp. 636–643, 1992.

264. G. Chen, "Heat Transfer in Micro- and Nano-scale Devices," Chap. 1, *Annual Reviews of Heat Transfer,* vol. 7, pp. 1–57, 1995.

265. A. A. Townsend, *Journal of Fluid Mechanics,* 3, p. 361, 1958.

266. G. M. Faeth, S. M. Jeng, and J. Gore in C. K. Law et al. (eds.), *Heat Transfer in Fire and Combustion Systems,* ASME, New York, 1985, HTD-vol. 45, pp. 137–151, 1985.

267. J. P. Gore, S.-M. Jeng, and G. M. Faeth, "Spectral and Total Radiation Properties of Turbulent Hydrogen/Air Diffusion Flames," *ASME Journal of Heat Transfer,* 110, pp. 173–181, 1987.

268. G. M. Faeth, J. P. Gore, S. G. Chuech, and S. M. Jeng, "Radiation From Turbulent Diffusion Flames," in C. L. Tien and T. C. Chawla (eds.), *Annual Reviews of Numerical Fluid Mechanics and Heat Transfer,* Hemisphere, New York, pp. 1–38, 1989.

269. M. E. Kounalakis, Y. R. Sivathanu, and G. M. Faeth, "Infrared Radiation Statistics of Nonluminous Turbulent Diffusion Flames," *ASME Journal of Heat Transfer,* 113, pp. 437–445, 1991.

270. Y. R. Sivathanu, M. E. Kounalakis, and G. M. Faeth, "Soot and Continuum Radiation Statistics of Luminous Turbulent Diffusion Flames," *Twenty-Third Symposium (International) on Combustion,* Combustion Institute, Pittsburgh, pp. 1543–1550, 1991.

271. J. P. Gore, U.-S. Ip, and Y. R. Sivathanu, "Coupled Structure and Radiation Analysis of Acetylene/Air Flames," *ASME Journal of Heat Transfer,* 114, pp. 487–493, 1992.

272. Y. R. Sivathanu and J. P. Gore, "Transient Structure and Radiation Properties of Strongly Radiating Buoyont Flames," *ASME Journal of Heat Transfer,* 114, pp. 659–665, 1992.

273. Y. R. Sivathanu and J. P. Gore, "Coupled Radiation and Soot Kinetics Calculations in Laminar Acetylene/Air Diffusion Flames," *Combustion and Flame,* 97, pp. 161–172, 1994.

274. W. L. Grosshandler and P. Joulain, "The Effect of Large-Scale Fluctuations on Flame Radiation," *Progress in Astronautics and Aeronautics,* 105, Part II, AIAA, Washington, pp. 123–152, 1986.

275. T.-H. Song and R. Viskanta, *Journal of Thermophysics and Heat Transfer,* 1, p. 56, 1987.

276. W. W. Yuen, A. K. Ma, and E. E. Takara, "Turbulence-Radiation Interactions in Non-Gray, Absorbing, Emitting, and Isotropically Scattering Medium," in A. M. Kanury and M. Q. Brewster (eds.), *Heat Transfer in Fire and Combustion Systems,* ASME-HTD vol. 199, pp. 53–62, 1992.

277. B. R. Adams and P. J. Smith, "Modeling Effects of Soot and Turbulence-Radiation Coupling on Radiative Transfer in an Industrial Furnace," in Y. Bayazitoglu et al. (eds.), *Radiative Heat Transfer: Current Research,* ASME-HTD vol. 276, pp. 177–190, 1994.

278. J. M. McDonough, D. Wang, and M. P. Mengüç, "Radiation-Turbulence Interactions in Flames Using Additive Turbulent Decomposition," in M. P. Mengüç (ed.), *Radiative Transfer—I: Proceedings of the First International Symposium on Radiative Transfer,* Begell House, New York, pp. 421–439, 1996.

279. M. P. Mengüç, J. M. McDonough, S. Manickavasagam, S. Mukerji, S. Swabb, and S. Ghosal, "Chaotic Fluctuations of Soot Particles in Turbulent Diffusion Flames: Experimental Data and Logistic Map Models," in M. P. Mengüç, K. Ball, and O. Ezekoye (eds.), *Symposium on Fire and Combustion Systems,* ASME-HTD vol. 335, pp. 271–280, 1996.

280. Willem Boeke and Lars Wall, "Radiative Exchange Factors in Rectangular Spaces for the Determination of Mean Radiant Temperatures," *Build. Serv. Engng.,* 43, pp. 244–253, March, 1976.

281. Czeslaw Buraczewski, "Contribution to Radiation Theory Configuration Factors for Rotary Combustion Chambers," *Pol. Akad. Nauk Pr. Inst. Masz Przeplyw,* 74, pp. 47–73 (in Polish), 1977.

282. Albert Buschman, Jr. and Claud M. Pittman, "Configuration Factors for Exchange of Radiant Energy Between Axisymmetrical Sections of Cylinders, Cones, and Hemispheres and Their Bases," *NASA TN* D-944, 1961.

283. I. R. Chekhovskii, V. V. Sirotkin, Yu. V. Chu-Dun-Chu, and V. A. Chebanov, "Determination of Radiative View Factors for Rectangles of Different Sizes," *High Temp.,* July 1979. (Trans. of Russian original, 17(1) January–February, 1979.)

284. B. T. F. Chung and P. S. Sumitra, "Radiation Shape Factors from Plane Point Sources," *J. Heat Transfer,* 94(3), pp. 328–330, August, 1972.

285. M. Crawford, "Configuration Factor between Two Unequal, Parallel, Coaxial Squares," Paper no. *ASME* 72-WA/HT-16, 1972.

286. A. Feingold and K. G. Gupta, "New Analytical Approach to the Evaluation of Configuration Factors in Radiation from Spheres and Infinitely Long Cylinders," *J. Heat Transfer,* 92(1), pp. 69–76, February, 1970.

287. J. D. Felske, personal communication, August 25, 1981.

288. U. Gross, K. Spindler, and E. Hahne, "Shape Factor Equations for Radiation Heat Transfer between Plane Rectangular Surfaces of Arbitrary Position and Size with Rectangular Boundaries," *Lett. Heat Mass Transfer,* 8, pp. 219–227, 1981.

289. D. C. Hamilton and W. R. Morgan, "Radiant-Interchange Configuration Factors," *NASA TN* 2836, 1952.

290. H. C. Hottel, "Radiant Heat Transmission between Surfaces Separated by Nonabsorbing Media," *Trans. ASME,* 53, FSP-53-196, pp. 265–273, 1931.

291. H. C. Hottel and J. D. Keller, *Trans. ASME,* 55, IS-55-6, pp. 39–49, 1933.

292. C.-J. Hsu, "Shape Factor Equations for Radiant Heat Transfer between Two Arbitrary Sizes of Rectangular Planes," *Can. J. Chem. Eng.,* 45(1), pp. 58–60, 1967.

293. M. Jakob, *Heat Transfer,* 2, John Wiley & Sons, New York, 1957.

294. N. H. Juul, "View Factors in Radiation between Two Parallel Oriented Cylinders," *J. Heat Transfer,* 104, p. 235, 1982.

295. H. B. Keene, "Calculation of the Energy Exchange between Two Fully Radiative Coaxial Circular Apertures at Different Temperatures," *Proc. Roy. Soc.,* vol. LXXXVIIIA, pp. 59–60, 1913.

296. Z. Kuroda and T. Munakata, "Mathematical Evaluation of the Configuration Factors between a Plane and One or Two Rows of Tubes," *Kagaku Sooti* (Chemical Apparatus, Japan), pp. 54–58, November, 1979 (in Japanese).

297. H. Leuenberger and R. A. Person, "Compilation of Radiation Shape Factors for Cylindrical Assemblies," Paper no. 56-A-144, *ASME,* November, 1956.

298. H. Masuda, "Radiant Heat Transfer on Circular-Finned Cylinders," *Rep. Inst. High Speed Mechanics,* Tohoku Univ., 27(225), pp. 67–89, 1973. (See also *Trans. JSME,* 38, pp. 3229–3234, 1972.)

299. M. H. N. Naraghi, *Radiation Configuration Factors between Disks and Axisymmetric Bodies,* M.S. Thesis, Department of Mechanical Engineering, University of Akron, 1981.

300. M. H. N. Naraghi and B. T. F. Chung, "Radiation Configuration between Disks and a Class of Axisymmetric Bodies," *J. Heat Transfer,* 104(3), pp. 426–431, August, 1982.

301. E. M. Sparrow, L. U. Albers, and E. R. G. Eckert, "Thermal Radiation Characteristics of Cylindrical Heat Transfer," *J. Heat Transfer,* 84(1), pp. 73–81, 1962.

302. E. M. Sparrow and E. R. G. Eckert, "Radiant Interaction between Fin and Base Surfaces," *J. Heat Transfer,* 84(1), pp. 12–18, February, 1962.

303. E. M. Sparrow and J. L. Gregg, "Radiant Interchange between Circular Disks Having Arbitrarily Different Temperatures," *J. Heat Transfer,* 83(4), pp. 494–502, November 1961.

304. E. M. Sparrow, G. B. Miller, and V. K. Jonsson, "Radiative Effectiveness of Annular-Finned Space Radiators, including Mutual Irradiation between Radiator Elements," *J. Aerospace Sci.,* 29(11), pp. 1291–1299, 1962.

305. C. L. Sydnor, "A Numerical Study of Cavity Radiometer Emissivities," *NASA Contractor Rept. 32-1462,* Jet Propulsion Laboratory, February 15, 1970.

306. H. Y. Wong, *Handbook of Essential Formulae and Data on Heat Transfer for Engineers,* Longman Group, London, 1977.

CHAPTER 8
MICROSCALE TRANSPORT PHENOMENA

A. Majumdar
University of California at Berkeley

INTRODUCTION

Heat transfer involves the transport of energy from one place to another by energy carriers. In gas phase, gas molecules carry energy either by random molecular motion (diffusion) or by an overall drift of the molecules in a certain direction (advection). In liquids, too, energy can be transported by diffusion and advection of molecules. In solids, however, energy is transported by phonons, electrons, or photons. A *phonon* is a quantum of crystal vibrational energy and dominates heat conduction in insulators and semiconductors. Electrons dominate energy transport in metals. *Photons* are quanta of electromagnetic energy and can transport energy in solids as well as interact with photons and phonons to render radiative properties of solids. The only mode of energy transport in vacuum is by photons.

The study of heat transfer at macroscales by all these energy carriers is well established and is based on continuum theories. For example, energy diffusion in gases, liquids, and solids is usually studied by the Fourier law of heat conduction. Although the physics of all the energy carriers are vastly different, what seems remarkable is the applicability of the Fourier law of heat conduction for any energy carrier—molecules, phonons, electrons, and even photons under the optically thick limit. So, there must be some universality in the transport of all these energy carriers at the macroscopic or continuum scales. The question that one could ask is the following. Does this universality break down at some length scale or is it universal over all length scales? This chapter will attempt to answer this question.

The second fact that is implicit in macroscopic or continuum laws is the idea of local thermodynamic equilibrium. For example, when we write the Fourier law of heat conduction, it is inherently assumed that one can define a temperature at any point in space. This is a rather severe assumption since temperature can be defined only under thermodynamic equilibrium. The question that we might ask is the following. If there is thermodynamic equilibrium in a system, then why should there be any net transport of energy? Thus, we implicitly resort to the concept of local thermodynamic equilibrium, where we assume that thermodynamic equilibrium can be defined over a volume which is much smaller than the overall size of the system. What happens when the size of the object becomes on the order of this volume? Obviously, the macroscopic or continuum theories break down and new laws based on nonequilibrium thermodynamics need to be formulated. This chapter focuses on developing more generalized theories of transport which can be used for nonequilibrium conditions. This involves going to the root of the macroscopic or continuum theories.

Nonequilibrium conditions can be created not only by size restrictions, but also by short time scales. For example, the response time for electrons in metals is much shorter than that of the crystal vibrations or phonons. Hence, if we heat a metal by a sufficiently short pulse of energy, only the electrons will be energized leaving the phonons relatively untouched. This can create nonequilibrium between the electrons and phonons in a metal which can also lead to nonequilibrium phenomena. This chapter addresses such nonequilibrium conditions.

TIME AND LENGTH SCALES

It is clear that continuum and noncontinuum effects must be demarcated by certain characteristic time and length scales of the energy carriers. The smallest length scale for each of the energy carriers in a solid—phonon, electron, or photon—is its wavelength λ. For electrons in a metal that are involved in energy transport, this is typically on the order of 1 to 10 Å. For phonons in semiconductors and insulators, energy is spread over a whole range of wavelengths. However, the wavelength range dominant in energy is typically on the order of 3 to 20 Å at room temperature, which can increase to the 1 μm range as the temperature is reduced. Photons in the visible spectrum fall in the range of 0.4 to 0.7 μm. The near to far infrared span 0.7 to about 50 μm. For gas molecules, the wavelength is too short for wave effects to be important.

Scattering of energy carriers poses resistance to energy transport. In its absence, energy transport will occur at infinite conductivity or conductance. Therefore, it is necessary to study the scattering process in more detail. Associated with scattering are several characteristic time and length scales. These are illustrated in Table 8.1. Consider the timescales first.

TABLE 8.1 Characteristic Time- and Length Scales and Corresponding Transport Phenomena of Energy Carriers

Time scales	Length scales			
	Wavelength λ	Mean free path ℓ	Relaxation length ℓ_r	Diffusion length ℓ_d
Collision time τ_c		Wave transport		
Mean free time τ	Wave transport	Microscopic particle transport theories		
Relaxation time τ_r				
Diffusion time τ_d				Macroscopic or continuum transport

The smallest timescale is the collision time or duration of collision τ_c. Collisions are normally considered instantaneous in classical physics. However, this is not entirely true and there is a finite collision time during wave scattering. This is on the order of the wavelength of the carrier divided by the propagation speed. For electrons in a metal, this is about 10^{-15} s (1 fs) whereas for phonons this is about 10^{-13} s (100 fs). The next timescale is the average time between collision τ or the mean free time. For timescales $t \leq \tau$, carriers travel ballistically and the evolution of the system depends strongly on the details of the initial state. Note that τ is not the relaxation time since it takes several collisions to reach equilibrium. Generally $\tau \gg \tau_c$, although this breaks down in certain cases. For electrons in a metal at room temperature, τ is on the order of 10^{-14} s whereas for phonons in solids it is on the order of 10^{-11} s. For gas molecules, the mean free time is given as ℓ/v where ℓ is the mean free path and v is the molecule velocity. The mean free path depends on the number density ρ and collision cross-sectional area σ and follows the relation $\ell \approx (\rho\sigma)^{-1}$ and is about 0.2 to 0.3 μm for molecules in air at atmospheric pressure. The root-means-square velocity for gas molecules depends on the temperature T and the molecule mass m and is given by the relation $v = \sqrt{3k_BT/m}$, where k_B is the Boltzmann constant. The velocity of oxygen and nitrogen molecules in air at room temperature is about 400 to 500 m/s making the mean free time about 10^{-10} to 10^{-9} s. In liquids, the mean free path is on the order of atomic dimensions or 10^{-10} m whereas the speeds of propagation are about 10^3 m/s. Hence, the mean free time is about 10^{-13} s. Collision-induced equilibrium gives rise to the relaxation time τ_r. Relaxation times are associated with local thermodynamic equilibrium. Since equilibrium is achieved in 5 to 20 collisions, $\tau_r > \tau$. Note that momentum and energy relaxation times of a system can be different. For example, an elastic collision of an electron can change its momentum but not its energy. Hence, energy relaxation times τ_{re} are typically longer than momentum relaxation times τ_{rm}. The last timescale is the diffusion time τ_d, which is on the order of $\tau_d \approx L^2/\alpha$ where L is the size of the object and α is the thermal diffusivity. This depends on particle speed v and mean free time as $\alpha \approx v^2\tau$. Clearly, $\tau_d \approx t_b^2/\tau$ where t_b is the time it takes for the particle to ballistically travel the distance L at speed v. For ballistic transport over a distance L, of course, $\tau \approx \tau_d$.

Note that the diffusion timescale τ_d contains a characteristic size of an object L. This ties in the length scale of the problem. Similarly, the mean free path ℓ can be associated with mean free time between collisions τ by the relation $\ell = v\tau$. Note that this is a statistical quantity since collision distances are not fixed. However, the probability p that a particle emerging from a collision travels a distance x without a collision is related to the mean free path as $p = \exp(-x/\ell)$. Associated with the relaxation time τ_r is a length scale ℓ_r, which is the characteristic size of a volume over which local thermodynamic equilibrium can be defined. Generally, the hierarchy of the length scales is $\lambda < \ell < \ell_r$.

Depending on time t and length L scales of interest, different transport laws can be used. Consider first the smallest length and timescales. For all objects that are comparable in size to the wavelength of the energy carrier, energy transport involves wave phenomena such as reflection, refraction, diffraction, tunneling, and so forth. When the timescale of interest t is of the order collision timescale τ_c, then, again, time-dependent wave mechanics must be used. For electrons and phonons, the Schrodinger equation must be solved to study such phenomena, whereas for photons, Maxwell's equations must be solved. We will not discuss wave transport in this chapter. Instead we will study the intermediate regime of length scales larger than the wavelength of the energy carrier and smaller than the macroscopic scales where continuum theories are valid. This intermediate length scale deals with scattering of energy carriers.

Now consider the next larger length and timescales ℓ or ℓ_r and τ or τ_r. When $L \approx \ell$, ℓ_r and $t \gg \tau$, τ_r, transport is ballistic in nature and local thermodynamic equilibrium cannot be defined. This transport is nonlocal in space. One has to resort to time-averaged statistical particle transport equations. On the other hand, if $L \gg \ell$, ℓ_r and $t \approx \tau$, τ_r, then approximations of local thermodynamic equilibrium can be assumed over space although time-dependent terms cannot be averaged. The nonlocality is in time but not in space. When both $L \approx \ell$, ℓ_r and $t \approx \tau$, τ_r, statistical transport equations in full form should be used and no spatial or temporal averages can be made. Finally, when both $L \gg \ell$, ℓ_r and $t \gg \tau$, τ_r, local thermodynamic equilibrium can be applied over space and time leading to macroscopic transport laws such as the Fourier law of heat conduction.

Let us consider the last case first since that is the easiest one and it also ties the microscopic transport characteristics to the macroscopic world.

KINETIC THEORY

Formulation

The kinetic theory of transport phenomena is the most elementary and perhaps the first step toward understanding more complex transport theories [1]. Consider a plane z, across which particles travel carrying mass and kinetic energy. Consider two fictitious planes at $z + \ell_z$ and

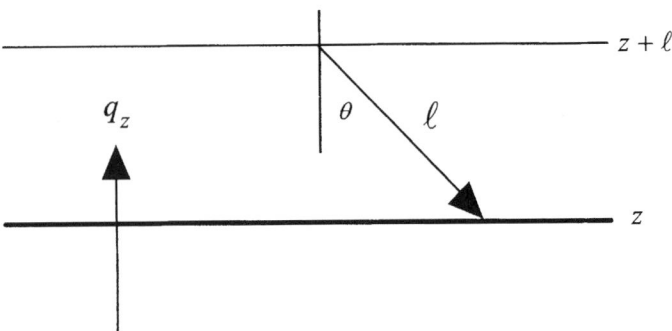

FIGURE 8.1   Schematic diagram showing energy flux across a z plane used in kinetic theory.

$z - \ell_z$ on either side of the z plane as shown in Fig. 8.1. Here, ℓ_z is the z component of the mean free path ℓ which makes an angle θ from the direction perpendicular to z. On average, the particles moving down from $z + \ell_z$ contain an energy density u that is characteristic of the location $u(z + \ell_z)$, whereas those moving up from $z - \ell_z$ have characteristic energy density $u(z - \ell_z)$. If the particles move with a characteristic velocity v then the next flux of energy in the positive z direction is

$$q_z = \tfrac{1}{2} v_z [u(z - \ell_z) - u(z + \ell_z)] \tag{8.1}$$

where v_z is the z component of the velocity and the factor $\tfrac{1}{2}$ is used since only half of the total number of particles at each location move up from $z - \ell_z$ or down from $z + \ell_z$. Using Taylor expansion and keeping only the first order terms, one gets

$$q_z = -v_z \ell_z \frac{du}{dz} = -(\cos^2 \theta) v \ell \frac{du}{dz} \tag{8.2}$$

where it is assumed that $\ell_z = \ell \cos \theta$ and $v_z = v \cos \theta$. Averaging over the whole hemisphere of solid angle 2π, one gets

$$q_z = -v \ell \frac{du}{dz} \left[\frac{1}{2\pi} \int_{\varphi = 0}^{2\pi} \int_{\theta = 0}^{\pi/2} \cos^2 \theta \sin \theta \, d\theta \, d\varphi \right] = -\frac{1}{3} v \ell \frac{du}{dz} \tag{8.3}$$

where φ is the azimuthal angle and θ is the polar angle and $\sin \theta \, d\theta \, d\varphi$ is the elemental solid angle. Assuming local thermodynamic equilibrium such that u is a function of temperature, one can write the flux as

$$q_z = -\frac{1}{3} v \ell \frac{du}{dT} \frac{dT}{dz} = -\frac{1}{3} C v \ell \frac{dT}{dz} \tag{8.4}$$

This is the Fourier law of heat conduction with the thermal conductivity being $k = C v \ell / 3$. Note that we have not made any assumption of the type of energy carrier and, hence, this is a universal law for all energy carriers. The only assumption made is that of local thermodynamic equilibrium such that the energy density u at any location is a function of the local temperature.

The characteristics of the energy carrier are included in the heat capacity C, velocity v, and the mean free path ℓ. Neglecting photons for now, the thermal conductivity can be written as

$$k = \frac{1}{3}[(Cv\ell)_l + (Cv\ell)_e] \tag{8.5}$$

where the first term is the lattice contribution and the second term comes from electrons. In the case of electrons in a metal, the electron contribution is dominant, whereas for semiconductors and insulators, the phonon contribution is dominant.

For gas molecules, the heat capacity is a constant equal to $C = (n/2)\rho k_B$ where n is the number of degrees of freedom for molecule motion, ρ is the number density, and k_B is the Boltzmann constant. The rms speed of molecules is given as $v = \sqrt{3k_B T/m}$, whereas the mean free path depends on collision cross section and number density as $\ell \approx (\rho\sigma)^{-1}$. When they are put together, one finds that the thermal conductivity of a gas is independent of ρ and therefore independent of the gas pressure. This is a classic result of kinetic theory. Note that this is valid only under the assumption that the mean free path is limited by intermolecular collision.

Thermal Conductivity of Crystalline and Amorphous Solids

Since the thermal conductivity depends on C, v, and ℓ, let us investigate the characteristics of these quantities. The electron heat capacity in a metal varies linearly with temperature $C = (\pi^2 \rho k_B^2/2E_F)T$ where E_F is the Fermi energy of a metal and ρ is the electron number density. This is a consequence of the free electron theory of metals in which only electrons within an energy range $k_B T$ around the Fermi energy E_F are responsible for transport phenomena [2]. Here, k_B is the Boltzmann constant (1.38×10^{-23} J/K) and T is the absolute temperature. The Fermi energy of most metals falls in the range of 3 to 10 eV whereas the thermal energy $k_B T$ is 0.026 eV at room temperature. Hence, only a small fraction of all the electrons in a metal contribute to energy transport in metals.

The velocity relevant for transport is the Fermi velocity of electrons. This is typically on the order of 10^6 m/s for most metals and is independent of temperature [2]. The mean free path ℓ can be calculated from $\ell = v_F\tau$ where τ is the mean free time between collisions. At low temperature, the electron mean free path is determined mainly by scattering due to crystal imperfections such as defects, dislocations, grain boundaries, and surfaces. Electron-phonon scattering is frozen out at low temperatures. Since the defect concentration is largely temperature independent, the mean free path is a constant in this range. Therefore, the only temperature dependence in the thermal conductivity at low temperature arises from the heat capacity which varies as $C \propto T$. Under these conditions, the thermal conductivity varies linearly with temperature as shown in Fig. 8.2. The value of k, though, is sample-specific since the mean free path depends on the defect density. Figure 8.2 plots the thermal conductivities of two metals. The data are the best recommended values based on a combination of experimental and theoretical studies [3].

As the temperature is increased, electron-phonon scattering becomes dominant. The mean free path for such scattering varies as $\ell \propto T^{-n}$ with n larger than unity. The mean free path of electrons at room temperature is typically on the order of 100 Å. The mean free path depends on the material but is independent of the sample, since electron-phonon scattering is an intrinsic process. As a result of electron-phonon scattering, thermal conductivity of metals decreases at higher temperatures.

Although the lattice heat capacity in a metal is much larger than its electronic contribution, the Fermi velocity of electrons (typically 10^6 m/s) is much larger than the speed of sound (about 10^3 m/s). Due to the higher energy carrier speed, the electronic contribution to the thermal conductivity turns out to be more dominant than the lattice contribution. For a semiconductor, however, the velocity is not the Fermi velocity but equal to the thermal velocity of the electrons or holes in the conduction or valence bands, respectively. This can be approximated as $v \approx \sqrt{3k_B T/m^*}$, where m^* is the effective electron mass in the conduction band or hole mass in the valence band. This is on the order of 10^5 m/s at room temperature. In addition, the number density of conduction band electrons in a semiconductor is much less than

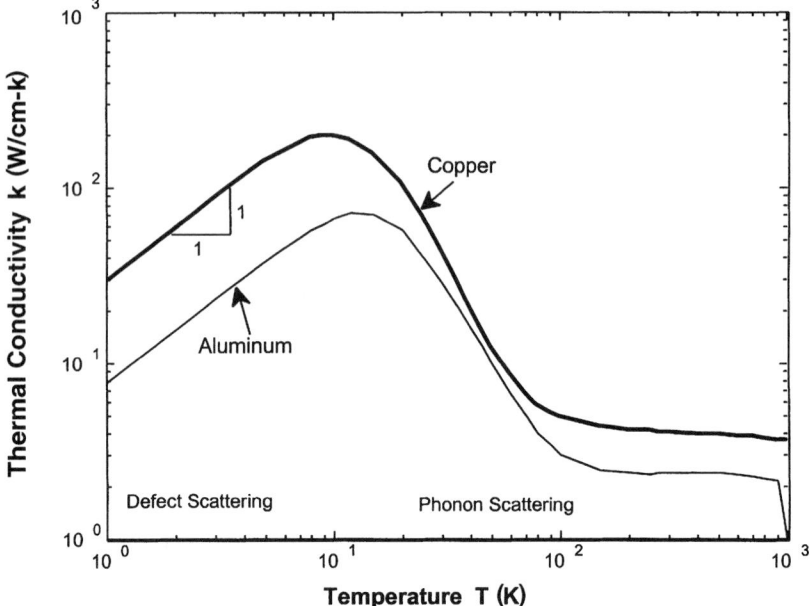

FIGURE 8.2   Thermal conductivity of aluminum and copper as a function of temperature [3]. Note that at low temperature, the thermal conductivity increases linearly with temperature. In this regime, defect scattering dominates and the mean free path is independent of temperature. The thermal conductivity in this regime depends on the purity of the sample. The linear behavior arises from the linear relation between the electronic heat capacity and temperature. As the temperature is increased, phonon scattering starts to dominate and the mean free path reduces with increasing temperature. To a large extent, the thermal conductivity of a metal is independent of the purity of the sample.

the electron density in metals. Hence, the electronic heat capacity is also lower than that of metals. This leads to the fact that electrons play an insignificant role in heat conduction in semiconductors. Therefore, as far as heat conduction is concerned, semiconductors and insulators fall in the same class of materials.

Phonons are quanta of crystal vibration [2,4]. The physics of phonons is quite similar to that of photons in that they follow Bose-Einstein statistics. However, there are some key differences, namely: (1) phonons have a lower cut-off in wavelength and upper cut-off in frequency whereas photon wavelength and frequency are not limited; (2) phonons can have longitudinal polarization whereas photons are transverse waves; (3) phonon-phonon interaction can emit or annihilate phonons and thereby restore thermodynamic equilibrium. Despite these differences, heat conduction by phonons can be studied as a radiative transfer problem.

Figure 8.3 shows the thermal conductivity of crystalline diamond samples with different defect concentrations [5]. At low temperatures, the thermal conductivities of all the samples are nearly equal and follow the T^3 behavior [2,4,6]. This arises from the Planck distribution of phonons at low temperatures. The dominant phonon wavelength at low temperatures can be very large as suggested by the relation $\lambda_{\text{dom}} = h v_s / 3 k_B T$. This is essentially the Wien's displacement law applied to phonons. Hence, the dominant wavelength at low temperatures can be much larger than crystal imperfections such as point defects, dislocations, and grain boundaries. Therefore, mean free path is not limited by the defect scattering but by the size of the crystal. Hence, the mean free path at very low temperatures is temperature-independent.

Phonon velocity is constant and is the speed of sound for acoustic phonons. The only temperature dependence comes from the heat capacity. Since at low temperature, photons and phonons behave very similarly, the energy density of phonons follows the Stefan-Boltzmann relation $\sigma T^4/v_s$, where σ is the Stefan-Boltzmann constant for phonons. Hence, the heat capacity follows as $C \propto T^3$ since it is the temperature derivative of the energy density. However, this T^3 behavior prevails only below the Debye temperature which is defined as $\theta_D = \hbar\omega_D/k_B$. The Debye temperature is a fictitious temperature which is characteristic of the material since it involves the upper cutoff frequency ω_D which is related to the chemical bond strength and the mass of the atoms. The temperature range below the Debye temperature can be thought as the quantum requirement for phonons, whereas above the Debye temperature the heat capacity follows the classical Dulong-Petit law, $C = 3\eta k_B$ [2,4] where η is the number density of atoms. The thermal conductivity well below the Debye temperature shows the T^3 behavior and is often called the *Casimir limit*.

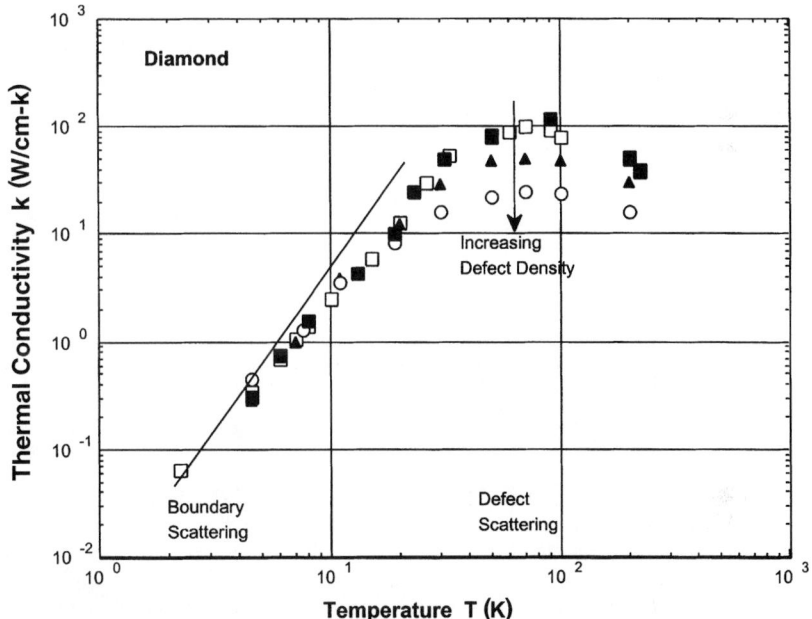

FIGURE 8.3   Thermal conductivity of diamond as a function of temperature [5]. The solid line represents the T^3 behavior.

As the temperature is increased, λ_{dom} becomes comparable to the defect sizes and, hence, the defect concentration in the crystal determines the thermal conductivity. This is the regime near the peak of the thermal conductivity which occurs when the temperature is on the order of $T \approx \theta_D/10$. Although defect scattering is temperature-independent, phonon-phonon interactions are highly temperature-dependent. There are two types of phonon-phonon interactions, namely, normal and Umklapp. *Normal* or *N-scattering* conserves energy and momentum during collision of two or more phonons, whereas *Umklapp* or *U-scattering* conserves energy but does not conserve momentum. Although normal scattering does not directly pose any resistance, it distributes the phonon energy to higher frequencies.

As the temperature is increased beyond $\theta_D/10$, C becomes a constant ($=3\eta k_B$). Also, the phonon density at high frequency and large wave vectors becomes sufficiently high that phonon-phonon Umklapp scattering dominates and determines the phonon mean free path.

Hence, the mean free path decreases drastically with temperature which results in a sharp drop in thermal conductivity with increasing temperature.

Figure 8.4 shows the thermal conductivity for quartz (crystalline SiO_2) and amorphous silica (a-SiO_2) [7]. The quartz data follows the T^3 behavior at low temperature, peaks at about 10 K, and then drops with increasing temperature. As discussed before, this is the expected trend for a crystalline solid. However, amorphous silica behaves very differently. The value of the thermal conductivity is much lower than that of the crystalline sample for all values of temperature. In addition, the temperature dependence of the conductivity is also vastly different. Hence, the model proposed for crystalline solids cannot be applied for such a case. Note that the relation $k = Cv\ell/3$ is still valid although the heat capacity and the mean free path cannot be determined by relations used for crystalline solids.

In 1911, Einstein proposed a model for heat conduction in amorphous solids. In this model, he assumed that all the atoms vibrate as harmonic oscillators at the same frequency ω_E. In addition, he also assumed that a particular oscillator (or atom) is coupled to only first, second, and third nearest neighbors. Hence, the vibrational energy of the oscillator can only be transferred to these atoms. A further assumption was that the phases of these oscillators were uncorrelated and were completely random. Using these assumptions, he derived the thermal conductivity to be

$$k_E = \frac{k_B^2}{\hbar} \frac{\eta^{1/3}}{\pi} \theta_E \frac{x^2 e^x}{(e^x - 1)^2} \tag{8.6}$$

where η is the number density of oscillators, θ_E is the Einstein temperature, and $x = \theta_E/T$. The Einstein temperature is defined as $\theta_E = \hbar w_E/k_B$. Unfortunately, the predictions of this theory

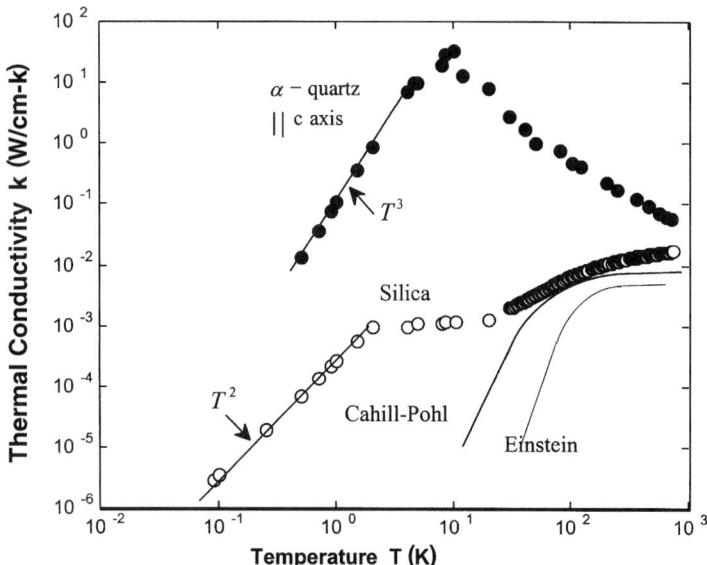

FIGURE 8.4 Thermal conductivity of quartz and glassy silica as a function of temperature [7]. The quartz thermal conductivity exhibits a T^3 behavior at low temperature, a peak at about 10 K, then reduction at higher temperatures. This is typical of a crystalline solid. For amorphous glass the thermal conductivity increases as T^2 plateaus between 1 to 10 K and then increases monotonically with temperature. Also plotted are the predictions of the Cahill-Pohl and Einstein models. The Cahill-Pohl model provides accurate predictions for temperatures higher than 50 K but cannot predict the low temperature behavior. The Einstein model predictions are much lower than the measured values.

were far below those of the measured data as can be observed in Fig. 8.4. The major flaw in this theory was in the assumption that the phases of the neighboring oscillators were uncorrelated. The success of the Debye theory was based on the fact that it considered the coherence of a crystal wave for a distance on the order of a mean free path.

Cahill and Pohl [8,9] recently developed a hybrid model which has the essence of both the localized oscillators of the Einstein model and coherence of the Debye model. In the Cahill-Pohl model, it was assumed that a solid can be divided into localized regions of size $\lambda/2$. These localized regions were assumed to vibrate at frequencies equal to $\omega = 2\pi v_s/\lambda$ where v_s is the speed of sound. Such an assumption is characteristic of the Debye model. The mean free time of each oscillator was assumed to be one-half the period of vibration or $\tau = \pi/\omega$. This implies that the mean free path is equal to the size of the region or $\lambda/2$. Using these assumptions, they derived the thermal conductivity to be

$$k_{CP} = \left(\frac{\pi}{6}\right)^{1/3} k_B \eta^{2/3} \sum_i v_{si} \left(\frac{T}{\theta_i}\right)^2 \int_0^{\theta_i/T} \frac{x^3 e^x}{(e^x - 1)^2} \, dx \tag{8.7}$$

where $x = \theta_i/T$ and $\theta_i = v_{si}(\hbar/k_B)(6\pi\eta)^{1/3}$ is a characteristic temperature equivalent to the Debye and Einstein temperatures. The summation is over the two transverse and one longitudinal polarizations for which the speeds of sound can be different. Note that the agreement between the Cahill-Pohl model and the measured thermal conductivity above 50 K is excellent. The validity of the model is further verified by comparing the measured and predicted thermal conductivities of several amorphous solids at 300 K. Below 50 K, there exists a plateau in the thermal conductivity and a sharp drop at temperatures below 10 K. In the limit $T \ll \theta_i$, the integral in Eq. 8.7 becomes a constant, and the thermal conductivity is expected to vary as $k_{CP} \propto T^2$. Although the trend seems to be correct, the measured values are higher than the predictions by an order of magnitude. This is due to the fact that the Cahill-Pohl model assumes the mean free path to be on the order of the phonon wavelength. At low temperatures ($T < 1$ K), however, the mean free path of the dominant phonons can be larger than the wavelength by a factor of 100.

Although the kinetic theory has been successfully applied to predict the thermal conductivity, it cannot be used under nonequilibrium conditions. For such cases, the Boltzmann transport theory is required.

BOLTZMANN TRANSPORT THEORY

General Formulation

Local thermodynamic equilibrium in space and time is inherently assumed in the kinetic theory formulation. The length scale that is characteristic of this volume is ℓ_r, whereas the timescale is τ_r. When either $L \approx \ell$, ℓ_r or $t \approx \tau$, τ_r or both, the kinetic theory breaks down because local thermodynamic equilibrium cannot be defined within the system. A more fundamental theory is required. The Boltzmann transport equation is a result of such a theory. Its generality is impressive since macroscopic transport behavior such as the Fourier law, Ohm's law, Fick's law, and the hyperbolic heat equation can be derived from this in the macroscale limit. In addition, transport equations such as equation of radiative transfer as well as the set of conservation equations of mass, momentum, and energy can all be derived from the Boltzmann transport equation (BTE). Some of the derivations are shown here.

In a general form, the BTE can be written as the following [4]:

$$\frac{\partial f}{\partial t} + \mathbf{v} \cdot \nabla f + \mathbf{F} \cdot \frac{\partial f}{\partial \mathbf{p}} = \left(\frac{\partial f}{\partial t}\right)_{\text{scat}} \tag{8.8}$$

where $f(\mathbf{r}, \mathbf{p}, t)$ is the statistical distribution function of an ensemble of particles, which varies with time t, particle position vector $\mathbf{r}$, and momentum vector $\mathbf{p}$. $\mathbf{F}$ is the force applied to the particles. The terms on the left side are called *drift terms,* whereas that on the right is the *scattering term.* The BTE applies to all ensembles of particles—electrons, ions, phonons, photons, gas molecules, and so forth—that follow a certain statistical distribution. Of all these, only charged particles such as electrons and ions can usually encounter an appreciable force **F** due to electric and magnetic fields. In the case of electrons under an electric field, for example, the third term on the left side can be written as

$$-\frac{e\mathbf{E}}{\hbar} \cdot \frac{\partial f}{\partial \mathbf{k}} \qquad (8.9)$$

where **E** is the electric field vector, e is the electron charge, and **k** is the electron wave vector.

The right-hand side of Eq. 8.8 is the rate of change of the distribution due to collisions or scattering. This is the term which restores equilibrium. In its most rigorous form, this is very complicated since collisions transfer particles from one set of $(\mathbf{r}',\mathbf{p}')$ coordinates to another set of $(\mathbf{r},\mathbf{p})$ coordinates. This can be written as

$$\left(\frac{\partial f}{\partial t}\right)_{\text{scat}} = \sum_{\mathbf{p}'} [W(\mathbf{p},\mathbf{p}')f(\mathbf{p}') - W(\mathbf{p}',\mathbf{p})f(\mathbf{p})] \qquad (8.10)$$

where $W(\mathbf{p},\mathbf{p}')$ is the scattering rate from state $\mathbf{p}'$ to $\mathbf{p}$. The first term in the summation is from scattering from $\mathbf{p}'$ state to $\mathbf{p}$ and the second term is vice versa. The scattering rates W are often nonlinear functions of $\mathbf{p}$ which make it quite difficult to solve the BTE. However, a simplification is often made through the relaxation-time approximation which assumes that

$$\left(\frac{\partial f}{\partial t}\right)_{\text{scat}} = \frac{f_0 - f}{\tau(\mathbf{r},\mathbf{p})} \qquad (8.11)$$

where f_0 is the equilibrium distribution and $\tau(\mathbf{r},\mathbf{p})$ is the relaxation time as a function of position and momentum. This approximation linearizes the BTE and implies that if a system is thrown out of equilibrium such that $f - f_0$ is nonzero, then collisions restore equilibrium with the dynamics following an exponential decay $f - f_0 \approx \exp(-t/\tau)$. Thus, the BTE under the relaxation-time approximation becomes

$$\frac{\partial f}{\partial t} + \mathbf{v} \cdot \nabla f + \mathbf{F} \cdot \frac{\partial f}{\partial \mathbf{p}} = \frac{f_0 - f}{\tau(\mathbf{r},\mathbf{p})} \qquad (8.12)$$

The equilibrium distribution could be of any type—Maxwell-Boltzmann for gas molecules, Fermi-Dirac for electrons, and Bose-Einstein for photons and phonons.

To study energy transport by particles, it is necessary to solve the BTE to determine the distribution function $f(\mathbf{r},\mathbf{p},t)$. Once found, the rate of energy flow per unit area or the energy flux can then be written as

$$\mathbf{q}(\mathbf{r},t) = \sum_{\mathbf{p}} \mathbf{v}(\mathbf{r},t)f(\mathbf{r},\mathbf{p},t)\varepsilon(\mathbf{p}) \qquad (8.13)$$

where $\mathbf{q}(\mathbf{r},t)$ is the energy flux vector, $\mathbf{v}(\mathbf{r},t)$ is the velocity vector, and $\varepsilon(\mathbf{p})$ is the particle energy as a function of particle momentum. Note that the units of $f(\mathbf{r},\mathbf{p},t)$ is number per unit volume per unit momentum. The summation over momentum space can be changed into an integral over momentum

$$\mathbf{q}(\mathbf{r},t) = \int \mathbf{v}(\mathbf{r},t)f(\mathbf{r},\mathbf{p},t)\varepsilon(\mathbf{p}) \, d^3\mathbf{p} \qquad (8.14)$$

The integral can be changed to that over energy with the introduction of a density of states $D(\varepsilon)$. The energy flux vector can then be written as

$$\mathbf{q}(\mathbf{r},t) = \int \mathbf{v}(\mathbf{r},t)f(\mathbf{r},\varepsilon,t)\varepsilon \, D(\varepsilon) \, d\varepsilon \qquad (8.15)$$

Fourier and Ohm's Laws

Although the solution of the BTE is not trivial, several simplifications can be made. If $t \gg \tau$, τ_r is assumed, then the most common simplification is to drop the time-varying term in Eq. 8.12. In addition, if $L \gg \ell$, ℓ_r is assumed, then the gradient term can be approximated as $\nabla f \approx \nabla f_0$ such that in the one-dimensional case, the BTE can be solved to yield

$$f = f_0 - \tau v_x \frac{\partial f_0}{\partial x} \tag{8.16}$$

where v_x is the x component of velocity. This can be called the *quasi-equilibrium* approximation. The only term that contains lack of equilibrium is the scattering term. Local thermodynamic equilibrium is inherently implied by the approximation $df/dx \approx df_0/dx$. However, since the local equilibrium f_0 can be defined only over a length scale ℓ_r the approximation finally boils down to $df/dx \approx \Delta f_0/\ell_r$. This and the timescale approximations are also made in the kinetic theory, and, hence, one should expect the same results. Since the equilibrium distribution is a function of temperature, one can express

$$\frac{\partial f_0}{\partial x} = \frac{df_0}{dT} \frac{\partial T}{\partial x} \tag{8.17}$$

This leads to the energy flux

$$q_x(x) = -\frac{\partial T}{\partial x} \int v_x^2 \tau \frac{df_0}{dT} \varepsilon \, D(\varepsilon) \, d\varepsilon \tag{8.18}$$

The first term containing f_0 drops out since the integral over all the directions becomes zero. Equation 8.18 is the Fourier law of heat conduction with the integral being the thermal conductivity k. If one assumes that the relaxation time and velocity are independent of particle energy, then the integral becomes

$$k = \int v_x^2 \tau \frac{df_0}{dT} \varepsilon \, D(\varepsilon) \, d\varepsilon = v_x^2 \tau \int \frac{df_0}{dT} \varepsilon \, D(\varepsilon) \, d\varepsilon = \frac{1}{3} C v^2 \tau \tag{8.19}$$

This is exactly the kinetic theory result $k = Cv\ell/3$. Similar derivations and conclusions can be made for Fick's law.

Ohm's law is characterized by the relation $\mathbf{J} = \sigma \mathbf{E}$ where $\mathbf{J}$ is the current density vector at any point in space, $\mathbf{E}$ is the electric field vector, and σ is the electrical conductivity. The Fourier law is the energy analog of Ohm's law due to the following reasons. The electric field vector $\mathbf{E}$ can be written as the negative gradient of the electric potential $\mathbf{E} = -\nabla\Phi$ and hence is analogous to the negative gradient of temperature. The energy flux vector $\mathbf{q}$ in the Fourier law is analogous to the current density vector $\mathbf{J}$ in Ohm's law. Using kinetic theory, it can be shown that the electrical conductivity follows the relation

$$\sigma = \frac{\eta_e e^2 \tau_m}{m} \tag{8.20}$$

where η_e is the density of electrons, e is the electron charge, τ_m is the momentum relaxation time, and m is the electron mass. The ratio of the thermal and electrical conductivity can be expressed by the Weidemann-Franz law [2] which is given as follows, where the right side contains only physical constants and is known as the Lorenz number [2].

$$\frac{k}{\sigma T} = \frac{\pi^2}{3}\left(\frac{k_B}{e}\right)^2 = 2.44 \times 10^{-8} \frac{W}{\Omega - K^2} \tag{8.21}$$

Note, however, that electrical conductivity is related to the momentum relaxation time, whereas the thermal conductivity is related to the energy relaxation time. They are usually close at room temperature or at very low temperatures.

Hyperbolic Heat Equation

If the Boltzmann transport equation is multiplied by the factor $v_x \varepsilon D(\varepsilon) d\varepsilon$ on both sides and integrated over energy, then the equation transforms into

$$\frac{\partial q_x}{\partial t} + \int v_x^2 \frac{\partial f}{\partial x} \varepsilon D(\varepsilon) d\varepsilon = -\int \frac{f v_x \varepsilon D(\varepsilon) d\varepsilon}{\tau(x, \varepsilon)} \tag{8.22}$$

The acceleration term is dropped in this equation. Consider the situation that $L \gg \ell, \ell_r$ and $t \approx \tau, \tau_r$. Now make the following assumptions: (1) the relaxation time is independent of particle energy and is a constant; and (2) the quasi-equilibrium assumption is made for the term $\partial f/\partial x = (df_0/dT)(\partial T/\partial x)$. Then Eq. 8.22 becomes

$$\frac{\partial q_x}{\partial t} + \frac{q_x}{\tau} = -\frac{k}{\tau} \frac{\partial T}{\partial x} \tag{8.23}$$

This is the Cattaneo equation [10], which, in combination with the following energy conservation equation,

$$C \frac{\partial T}{\partial t} + \frac{\partial q_x}{\partial x} = 0 \tag{8.24}$$

leads to the hyperbolic heat equation of the form [10]

$$\tau \frac{\partial^2 T}{\partial t^2} + \frac{\partial T}{\partial t} = \frac{k}{C} \frac{\partial^2 T}{\partial x^2} \tag{8.25}$$

The solution of Eq. 8.25 is wavelike, suggesting that the temperature field propagates as a wave. The speed of propagation of this wave is equal to $\sqrt{k/C\tau}$ which also happens to be the speed of the energy carrier, for example, the speed of sound for phonons. So, this model is nonlocal in time but local in space since the temperature represents a spatially localized thermodynamic equilibrium.

Mass, Momentum, and Energy Conservation—Hydrodynamic Equations

The conservation equations that are encountered in fluid mechanics, heat transfer, and electron transport can be derived as different moments of the BTE [11]. Consider a function $\phi(\mathbf{p})$, which is a power of the particle momentum $\phi(\mathbf{p}) = \mathbf{p}^n$ where n is an integer ($n = 0, 1, 2, \ldots$). Its average can be described as

$$\langle \phi(\mathbf{p}) \rangle = \frac{1}{\rho} \int \phi(\mathbf{p}) f(\mathbf{p}) \, d^3\mathbf{p} \tag{8.26}$$

where ρ is the number density of particles. The BTE is now multiplied by $\phi(\mathbf{p})$ and integrated over momentum. In general form, this gives the *moment equation*

$$\frac{\partial (\rho \langle \phi \rangle)}{\partial t} + \frac{1}{m} \nabla \cdot (\rho \langle \mathbf{p}\phi \rangle) - \rho \mathbf{F} \cdot \left\langle \frac{\partial \phi}{\partial \mathbf{p}} \right\rangle = \rho \sum_{\mathbf{p}'} [\langle W(\mathbf{p},\mathbf{p}')\phi(\mathbf{p}') \rangle - \langle W(\mathbf{p}',\mathbf{p})\phi(\mathbf{p}) \rangle] \tag{8.27}$$

Note that the momentum of each particle can be divided into two components as follows

$$\mathbf{p} = \mathbf{p}_d + \mathbf{p}_r \tag{8.28}$$

where $\mathbf{p}_d$ is the average or drift momentum corresponding to a collective motion of particles in response to an external potential gradient, and $\mathbf{p}_r$ is the random component of the momentum which arises due to thermal motion and is responsible for diffusion. Note that $\langle \mathbf{p} \rangle = \mathbf{p}_d$ since the average of the random component over all the momentum space is zero.

The zeroth moment is when $n = 0$ and $\phi(\mathbf{p})$ is a constant. Using this, one gets the continuity or number conservation equation which is

$$\frac{\partial \rho}{\partial t} + \nabla \cdot (\rho \mathbf{v}_d) = So - Si \tag{8.29}$$

where $\mathbf{v}_d$ is the drift velocity, So is the source or rate of generation rate of particles and Si is the sink or removal rate of particles. In the case of fluids, there are no sources or sinks and hence the right side is zero. However, when electrons and holes are considered, the zeroth moment equation can be written for each valley of the electronic structure of a semiconductor or a metal. Intervalley scattering due to electron-electron, electron-hole, electron-photon, or electron-phonon interactions may be responsible for particle exchange between the different valleys and bands. This creates a source or sink in each valley in which case the right side of Eq. 8.29 may be nonzero. However, if all the valleys and bands are considered together, the right side would be zero since charge or mass must be conserved.

The momentum conservation equation is obtained by taking the first moment, $\phi(\mathbf{p}) = \mathbf{p} = m\mathbf{v}$. This yields the following equation

$$\frac{\partial (\rho \mathbf{p}_d)}{\partial t} + \frac{1}{m} \nabla \cdot (\rho \langle \mathbf{pp} \rangle) - \rho \mathbf{F} = \left(\frac{\partial (\rho \mathbf{p})}{\partial t} \right)_{scat} \tag{8.30}$$

The second term is the average of a tensorial quantity. However, since the average of odd powers of $\mathbf{p}_r$ is zero, we get $\langle \mathbf{pp} \rangle = \mathbf{p}_d \mathbf{p}_d + p_r^2 \delta_{ij}$ where δ_{ij} is the unit tensor. The third term of the left side is what is referred to as a *body force term* in fluid mechanics. It is perhaps more appropriate to refer to it as a *potential gradient term* since a thermodynamic force can be written as a gradient of any potential $\mathbf{F} = -\nabla U$. The potential U is a sum of the gravitational potential G, electrochemical potential Φ, and so forth. For electrons in a metal or semiconductor, the force can be due to electric or magnetic fields which can also be expressed as a gradient of a potential. The right side of Eq. 8.30 is the scattering term. Under the relaxation time approximation, the right side can be assumed to follow

$$\left(\frac{\partial (\rho \mathbf{p})}{\partial t} \right)_{scat} = -\frac{\rho \mathbf{p}}{\tau_m} \tag{8.31}$$

where τ_m is the momentum relaxation time. Therefore, the momentum conservation equation becomes

$$\frac{\partial (\rho \mathbf{p}_d)}{\partial t} + \frac{1}{m} \nabla \cdot (\rho \mathbf{p}_d \mathbf{p}_d) + \frac{1}{m} \nabla (\rho p_r^2) = -\rho \nabla (G + \Phi + \cdots) - \frac{\rho \mathbf{p}_d}{\tau_m} \tag{8.32}$$

The third term on the left side has the form of the kinetic energy of the random particle motion and is representative of the pressure of the particles. Therefore, Eq. 8.32 can be rewritten in the following form:

$$\frac{\partial (\rho m \mathbf{v}_d)}{\partial t} + \nabla \cdot (\rho m \mathbf{v}_d \mathbf{v}_d) = -\rho \nabla \left(\frac{P}{\rho} + G + \Phi + \cdots \right) - \frac{\rho m \mathbf{v}_d}{\tau_m} \tag{8.33}$$

The second term on the left side is often referred to as the *advection term*. When this is negligible, Eq. 8.33 under zero acceleration reduces to the form

$$\mathbf{v}_d = -\frac{\tau_m}{m} \nabla \left(\frac{P}{\rho} + G + \Phi + \cdots \right) \tag{8.34}$$

In the case of electron transport where $\Phi = eV + k_B T \ln(\rho)$ is the electrochemical potential, one can derive the familiar drift-diffusion equation [11]:

$$\mathbf{J} = \frac{\rho e^2 \tau_m}{m} \mathbf{E} + \frac{\rho e \tau_m}{m} \nabla (k_B T \ln \rho) \tag{8.35}$$

Here, the first term is the drift term representing Ohm's law with the electrical conductivity being $\sigma = \rho e^2 \tau_m / m$. The second term is the diffusion term which gives rise to thermoelectric effects and current flow due to electron concentration gradients. In the case of fluid transport, neglecting the left side of Eq. 8.34 gives

$$\mathbf{v} = -\frac{\tau_m}{m} \nabla \left(\frac{P}{\rho} \right) \tag{8.36}$$

which is equivalent to the Darcy equation for flows in porous media. It is evident that Eq. 8.33 has the familiar form of the Navier-Stokes equation except for the last term involving collisions. The Navier-Stokes equation can be derived from the BTE using the Chapman-Enskog approximation where the right side of Eq. 8.33 leads to the diffusion term [1].

The energy conservation equation is obtained from Eq. 8.27 if the second moment is taken $\phi(\mathbf{p}) = p^2$ since energy $\varepsilon = p^2/2m$. This yields the following equation:

$$\frac{\partial \xi}{\partial t} + \nabla \cdot \mathbf{J}_\xi = -\rho \mathbf{F} \cdot \mathbf{v}_d + \rho \sum_{\mathbf{p}'} \left[\left\langle W(\mathbf{p},\mathbf{p}') \frac{p'^2}{2m} \right\rangle - \left\langle W(\mathbf{p}',\mathbf{p}) \frac{p^2}{2m} \right\rangle \right] \tag{8.37}$$

where $\xi = \rho\varepsilon$ is the energy density in J/m^3 and $\mathbf{J}_\xi$ is energy flux vector in W/m^2 which can be expressed in general form as

$$\mathbf{J}_\xi = \mathbf{v}_d \xi + \mathbf{q} \tag{8.38}$$

The term $\mathbf{v}_d \xi$ is the advection of energy which comes from the drift contribution, and $\mathbf{q}$ is the heat flux vector due to diffusion which arises from the random motion of the particles. This reduces the energy equation to

$$\frac{\partial \xi}{\partial t} + \nabla \cdot (\mathbf{v}_d \xi) = \rho \mathbf{v}_d \cdot \nabla U - \nabla \cdot \mathbf{q} + \left(\frac{\partial \xi}{\partial t} \right)_{So} - \left(\frac{\partial \xi}{\partial t} \right)_{Si} \tag{8.39}$$

where U is the sum of all the potentials discussed earlier. Here, the scattering term from Eq. 8.37 is divided into an energy source and an energy sink term which are discussed shortly. The first term on the right side is the work done by a force on the particles and, therefore, must appear in the energy conservation equation. To obtain a relation for $\mathbf{q}$, the next higher moment of the BTE needs to be taken. However, closure is often obtained by assuming the Fourier law $\mathbf{q} = -k\nabla T$. But, recalling the fact that the Fourier law is derived under the assumption of quasi-equilibrium in both space and time, this may not always be a valid assumption. A higher-order relation which takes into account nonlocality in time but quasi-equilibrium in space is the Cattaneo equation for heat flux described in Eq. 8.23.

The energy density ξ of a particle system has contribution from entropic motion as well as from the drift and can be written as

$$\xi = \frac{3}{2} \rho k_B T + \frac{1}{2} \rho m v_d^2 \tag{8.40}$$

Note that the factor $3/2$ is valid for particles such as monoatomic gas molecules and electrons, with only three degrees of freedom of motion, each degree possessing an energy of $k_B T/2$. By multiplying the momentum conservation equation (Eq. 8.33) by $\mathbf{v}_d$ and subtracting it out of the energy conservation equation (Eq. 8.39), the thermal energy conservation equation can be derived as

$$\frac{\partial T}{\partial t} + \mathbf{v}_d \cdot \nabla T + \frac{2}{3} T \nabla \cdot \mathbf{v}_d = \frac{2}{3\rho k_B} \nabla \cdot \mathbf{q} + \left(\frac{\partial T}{\partial t} \right)_{So} - \left(\frac{\partial T}{\partial t} \right)_{Si} \tag{8.41}$$

Note that the work term $\rho \mathbf{v}_d \cdot \nabla U$ drops out since work increases mechanical energy but does not increase entropy or temperature of a system. Only when this work is dissipated by scattering, the entropy of the system is raised and the temperature increases. The scattering term in Eq. 8.41 can be written as follows [12]:

$$\left(\frac{\partial T}{\partial t}\right)_{So} - \left(\frac{\partial T}{\partial t}\right)_{Si} = -\left(\frac{T - T_0}{\tau_\varepsilon(T,v)}\right) + \frac{mv_d^2}{3k_B}\left(\frac{2}{\tau_m} - \frac{1}{\tau_\varepsilon}\right) \tag{8.42}$$

where T_0 is a reservoir temperature and τ_e is the energy relaxation time. The first term on the right side is simply the energy relaxation term with respect to an equilibrium temperature T_0. The second term comes from the difference between the momentum and energy relaxation processes. The energy relaxation time is different from the momentum relaxation time since a collision may change the particle momentum but not its energy. Even if both these times were the same, the term would be nonzero. This is the contribution of the kinetic energy of the particle to the temperature rise. Hence, this is the fraction of the work done which is dissipated resulting in entropy generation and temperature rise. If we consider electrons that are energized by the work done by an external electric field, the electron-phonon interactions eventually dissipate this work and result in energy loss to the phonons. Hence, the reservoir temperature is that of the phonons. Note that although the work term $\rho\mathbf{v}\cdot\mathbf{F}$ is not present, the term $(2mv_d^2/3k_B\tau_m)$ in Eq. 8.42 represents the dissipated work which adds thermal energy to the system. The ratio $(\tau_\varepsilon/\tau_m)$ can be called the Prandtl number Pr of the fluid since the fluid diffusivities are inversely proportional to their respective relaxation times.

Equation of Radiative Transfer for Photons and Phonons

Photons and phonons do not follow number conservation as do electrons and molecules. However, they do follow energy conservation. An intensity of photons or phonons can be defined as follows

$$I_k(\mathbf{r},\mathbf{k},s,t) = \mathbf{v}(\mathbf{k},s)f(\mathbf{r},\mathbf{k},s,t)\hbar\omega(\mathbf{k},s) \tag{8.43}$$

where I_k is the intensity with wave vector $\mathbf{k}$, $\mathbf{v}$ is the velocity at wave vector $\mathbf{k}$, s is the polarization, and $\hbar\omega$ is the energy. The intensity can also be defined in terms of frequency ω and angle (θ,ϕ) in polar coordinates corresponding to the direction of vector $\mathbf{k}$ as follows:

$$I_\omega(\mathbf{r},\omega,\theta,\phi,s,t) = \mathbf{v}(\omega,\theta,\phi,s)f(\mathbf{r},\omega,\theta,\phi,t)\hbar\omega D(\omega,s) \tag{8.44}$$

where $D(\omega,s)$ is the density of $\mathbf{k}$ states in the frequency range ω and $\omega + d\omega$. If the BTE of Eq. 8.8 is multiplied by the factor $\mathbf{v}(\omega,\theta,\phi,s)\hbar\omega D(\omega,s)$, the following equation is obtained:

$$\frac{\partial I_\omega(\mathbf{r},\omega,\theta,\phi,s,t)}{\partial t} + \mathbf{v}\cdot\nabla I_\omega(\mathbf{r},\omega,\theta,\phi,s,t) = \left(\frac{\partial I_\omega}{\partial t}\right)_{scat} \tag{8.45}$$

where
$$\left(\frac{\partial I_\omega}{\partial t}\right)_{scat} = \sum_{(\omega',\theta',\phi',s')}\left[\begin{array}{l}W(\omega',\theta',\phi',s'\to\omega,\theta,\phi,s)I_{\omega'}(\mathbf{r},\omega',\theta',\phi',s',t)\\ -W(\omega,\theta,\phi,s\to\omega',\theta',\phi',s')I_\omega(\mathbf{r},\omega,\theta,\phi,s,t)\end{array}\right]$$

$$+ \sum_{(j,\Omega)}[W(j,\Omega\to\omega,\theta,\phi,s)\varepsilon(\mathbf{r},j,\Omega,t) - W(\omega,\theta,\phi,s\to j,\Omega)I_\omega(\mathbf{r},\omega,\theta,\phi,s,t)] \tag{8.46}$$

Here, each W is a scattering rate. It is evident that the scattering term is quite complicated and needs explanation.

Equation 8.45 is the conservation of energy based on the intensity at frequency ω, polarization s, and direction (θ,ϕ). Consider now the first summation in Eq. 8.46. This increases the intensity $I_\omega(\mathbf{r},\omega,\theta,\phi,s,t)$ due to scattering in frequency $\omega'\to\omega$, polarization $s'\to s$, and direction $\theta',\phi'\to\theta,\phi$. The second term is the loss of intensity $I_\omega(\mathbf{r},\omega,\theta,\phi,s,t)$ due to scattering to other frequencies, polarizations, and directions. Note, however, that if photons are considered, then this term represents photon-photon scattering and not scattering between photons and other particles. So, this term accounts for scattering among the particle type, either photon-photon or phonon-phonon, respectively. This is often known as the *in-scattering term* in photon radiative transfer, although scattering is usually considered only in direction $\theta',\phi'\to\theta,\phi$

and not in frequency and polarization. This is because, inelastic photon scattering is normally ignored in engineering calculations unless processes such as Raman scattering are involved. For phonon radiative transfer, however, inelastic scattering such as normal and Umklapp processes are very common and must be accounted for in this term. In addition, such phonon-phonon scattering is often between different phonon polarizations as allowed by phonon energy and momentum conservation during the collision.

The second summation term in Eq. 8.46 is for increase or decrease in intensity $I_\omega(\mathbf{r},\omega,\theta,\phi,s,t)$ due to interactions with other particles. The particle type is given a tag j, and the phase space defined by momentum and direction is given a tag Ω. For example, an energetic electron in a metal or in the conduction band of a semiconductor can drop in energy by emitting a phonon of a certain polarization (e.g., LO-phonon) due to electron-phonon interactions. Here, the electron is given a tag j and the phonon is given a tag Ω. The frequency, direction, and polarization of this phonon is decided by energy and momentum conservation of the scattering process. In photon radiative transfer, this term is often referred to as the *blackbody source term*. This is true for the particular case of blackbody radiation. However, in a device such as a semiconductor laser or a light-emitting diode, photons are not emitted in a blackbody spectrum but within a certain spectral band that is decided by the semiconductor electronic band structure. Hence, this term is kept as a general emission term in Eq. 8.46. Similarly, there is a loss term when a phonon or a photon is absorbed by another particle and removed from the system.

It is clear that in the most general form as described in Eq. 8.45, the scattering terms pose difficulty for solving. Therefore, the relaxation-time approximation is usually made for convenience, in which case the equation of radiative transfer reduces to

$$\frac{\partial I_\omega}{\partial t} + \mathbf{v} \cdot \nabla I_\omega = -\frac{I_\omega}{\tau_s} - \frac{I_\omega}{\tau_a} + \frac{\varepsilon(j,\Omega)}{\tau_e} + \iint_{\omega',\Theta'} \frac{I_{\omega'}(\mathbf{r},\omega',\Theta',t)}{\tau_s} \, d\omega' \, \frac{d\Theta'}{4\pi} \tag{8.47}$$

The first term on the right side is the out-scattering term with τ_s being the scattering relaxation time, the second term is the photon/phonon absorption (or transfer of energy to other particles such as electrons, or photons to phonons or phonons to photons, etc.) where τ_a is the absorption time, the third term is the emission term with $1/\tau_e$ being the emission rate. Here, energy from other particles is converted and contributed to the intensity I_ω. The last term is the in-scattering term from other frequencies and solid angles Θ'. In an even simpler form, the equation of radiative transfer can be written as

$$\frac{\partial I_\omega}{\partial t} + \mathbf{v} \cdot \nabla I_\omega = \frac{I_\omega^0 - I_\omega}{\tau_s} - \frac{I_\omega}{\tau_a} + \frac{\varepsilon(j,\Omega)}{\tau_e} \tag{8.48}$$

where the in-scattering term is totally ignored but it is assumed that $\omega' \rightarrow \omega$ scattering restores equilibrium that is represented by I_ω^0. This is often the assumption made in phonon radiative transfer where interfrequency scattering restores phonon equilibrium.

The equation of radiative transfer will not be solved here since solutions to some approximations of the equation are well known. In photon radiation, it has served as the framework for photon radiative transfer. It is well known that in the optically thin or ballistic photon limit, one gets the heat flux as $q = \sigma(T_1^4 - T_2^4)$ from this equation for radiation between two black surfaces [13]. For the case of phonons, this is known as the Casimir limit. In the optically thick or diffusive limit, the equation reduces to $q = -k_p \nabla T$ where k_p is the photon thermal conductivity. The same results can be derived for phonon radiative transfer [14,15].

NONEQUILIBRIUM ENERGY TRANSFER

The discussion in the previous sections concentrated on transport by a single carrier, that is, heat conduction by electrons or phonons, charge transport by electrons, and energy transport by photons. Relatively little attention was paid to energy transfer processes between the energy carriers. For example, Joule heating occurs due to electron-phonon interactions whereas radiative heating involves photon-electron and electron-phonon interactions. These are examples of what are commonly called *heat generation mechanisms*.

Traditionally, it is assumed that electrons and phonons within a solid are locally under equilibrium such that a heat generation term can be added to the energy conservation equation. For example, Joule heating during electron transport is usually modeled as I^2R where I is the current and R is the electrical resistance. Such a term is added to the energy conservation equation for the whole solid. Such an equation uses a single temperature T to describe the solid at a point **r** and time t. It inherently assumes that there is equilibrium between the electrons and the phonons. However, this is not quite the picture in many cases. The equilibrium between electrons and phonons can be disrupted by several processes. For example, in the presence of a sufficiently high electric field, electrons can be energized and thrown far out of equilibrium from the phonons. Such nonequilibrium conditions can now be achieved in contemporary technology where electronic devices with submicrometer feature sizes undergo high-field transport. In the case of radiative heating in a metal, for example, the electrons are again thrown out of equilibrium from the lattice due to excitations by ultra-short laser pulses that are on the order of 100 fs. Such lasers are now available and are widely used in physics, chemistry, and materials processing. Therefore, it is clear that when modern engineering systems involving transport phenomena become small and fast, the energy dissipation process required by the second law of thermodynamics can take a highly nonequilibrium path. In this section, a close look is taken at microscopic mechanisms of heat generation and dissipation and models are presented to analyze such problems.

Joule Heating in High-Field Electronic Devices

One of the major goals of the electronics industry is to increase the density of devices on a single chip by reducing the minimum size of features. This has two purposes: (1) to miniaturize and increase the functionality of a single chip; and (2) to increase the speed of logic operations. By the year 2001, the minimum feature size will reduce to 0.18 μm and the speed and power density will increase significantly. A single chip in the future is likely to contain both power and logic devices. This will lead to high temperature and temperature gradients within a chip. New materials choices based on electrical characteristics also influence the thermal problem. For example, the close proximity of transistors on a high-density silicon (Si) chip requires the use of dielectric material such as silicon dioxide (SiO_2) for electrical insulation between devices. Since the thermal conductivity of SiO_2 is about 100 times lower than Si, it leads to high temperatures and temperature gradients [16–18].

Simple Transistors.   Figure 8.5 shows schematic diagrams of a metal-oxide-semiconductor field-effect transistor (MOSFET) and metal-semiconductor field-effect transistor (MESFET) [19]. The MOSFET is usually made of silicon (Si) and is the workhorse of all logic devices and microprocessors. MESFETs are usually made of III-V materials such as GaAs and are usually used in high-speed communication devices such as microwave receivers and transmitters. GaAs is preferable for such devices since the electron mobility is higher than in Si. The current-voltage (I-V) characteristics of these devices are also shown. The voltage bias on the gate opens and closes the gate and in effect controls the resistance between the source and the drain. So, the drain current is a strong function of the gate voltage. Other high-electron mobility transistors also operate in a similar fashion except that the electron channel under the gate has different configurations due to clever control and manipulation of material interfaces and properties.

Most of the potential drop between the drain and the source occurs across the gate. So, the characteristic electric field in a device is on the order of V_{ds}/L_g where V_{ds} is the drain-to-source voltage and L_g is the gate length. When a voltage bias of about 2 V is applied across a

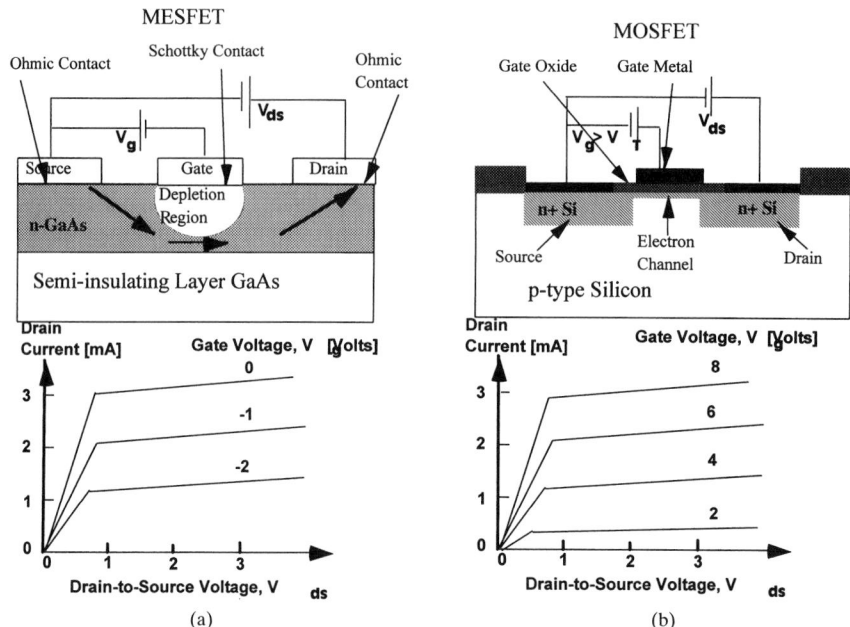

FIGURE 8.5 Schematic diagram and current-voltage (I-V) characteristics of (*a*) metal-semiconductor field-effect transistor (MESFET) and (*b*) metal-oxide semiconductor field-effect transistor (MOSFET).

device with a minimum feature size of 0.2 μm, extremely high electric fields (about 10^7 V/m) are generated. The dynamics of an electron can be expressed as $m^* \dot{\mathbf{v}} = -e\mathbf{E}$ where m^* is the effective electron mass, $\dot{\mathbf{v}}$ is the acceleration vector, and $\mathbf{E}$ is the electric field vector. The electron velocity gained between two collisions is equal to $eE\tau/m^*$ where τ is the average time between collisions. When the electric field is very high, the velocity and the electron energy also becomes very high. Such hot electrons are thrown far out of equilibrium with the lattice vibrations. However, the hot electrons collide with the lattice and at some of these collisions, the electron energy is transferred to the lattice to produce a phonon. The hot electrons do not always follow Ohm's law and, hence, their transport must be studied by the Boltzmann transport equation.

Energy Transfer Processes.   Heat generation occurs by transfer of energy from electrons to phonons. Since Si has two atoms per unit cell, two vibrational modes are present—optical mode and acoustic mode. Similar is the case for GaAs and other III-V materials. Optical phonon energies are higher than that of acoustic phonons. Although electrons interact with both types of phonons, the interactions with optical phonons are restricted to conditions when the electron energy gained from the electric field is higher than the optical phonon energy. So, there exists a critical field beyond which electron-optical phonon interactions can occur. In GaAs for instance, the atomic bond is slightly polar and LO-mode of vibration results in an oscillating dipole which strongly scatters electrons. Hence, electron-LO phonon interaction determines the critical field.

In both Si and GaAs, the critical electric field is on the order of 10^6 V/m. It is clear that in state-of-the-art submicrometer devices with fields on the order of 10^7 V/m, optical phonons will be generated. Although optical phonons interact with hot electrons, their group velocity is very small and hence they do not conduct any heat. So, they eventually decay into acoustic phonons which conduct heat through the device and throughout the package. Therefore, although LO-phonons gain energy from electrons, they must transfer it to acoustic phonons for heat conduction in the solid. Such an energy transfer occurs during scattering of LO-phonons and acoustic phonons which has a characteristic timescale of $\tau_{LO-A} \approx$ 6–10 ps in GaAs [11,20] and about 10 ps in Si [11]. Thus, the electron-LO phonon timescale $\tau_{e-LO} \approx$ 0.1 ps is two orders of magnitude faster than τ_{LO-A}. The timescale of electron-phonon and phonon-phonon interactions can be quite different giving rise to interesting dynamics. Figure 8.6 shows a schematic diagram of the nonequilibrium Joule heating process.

The effect of device temperature on the electrical behavior of the device occurs due to the lattice temperature dependence of the electron scattering rate. When the LO phonon and acoustic phonon temperatures rise, the electron scattering rate increases, thus increasing the electrical resistance or decreasing the carrier mobility. The coupling of electrical and thermal characteristics suggest that these must be analyzed concurrently.

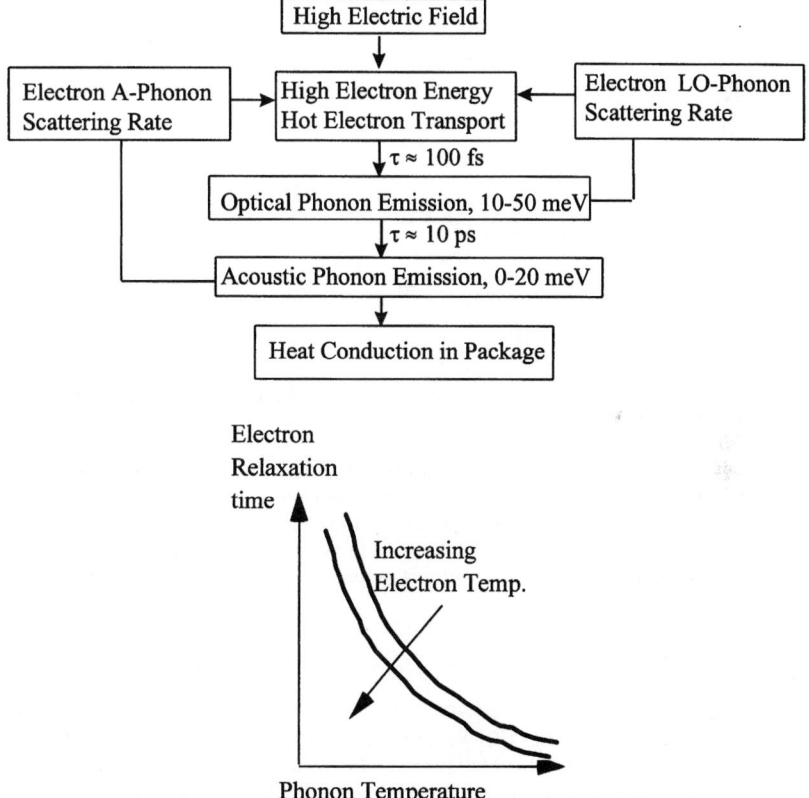

FIGURE 8.6 A flowchart showing the energy transfer mechanisms during Joule heating in high-electric field electronic devices. Note that optical phonons will be emitted only when the electric field is higher than the critical field. Otherwise, hot electrons will directly emit acoustic phonons. The number of phonons (or the phonon temperature) influences the electron scattering rate which in turn changes the device's electrical characteristics. The electron scattering rate depends on both electron and phonon temperatures and follows the qualitative trend shown. The flowchart also shows the typical timescales involved in each process and the energies of phonons.

Governing Equations.   If the problem is to be solved rigorously, the BTE must be solved for electrons in each valley, optical phonons, and acoustic phonons. The distribution function of each of these depends on six variables—three space and three momentum (or energy). The solution to BTE for this complexity becomes very computer intensive, especially due to the fact that the timescales of electron-phonon and phonon-phonon interactions vary by two orders of magnitude. Monte Carlo simulations are sometimes used although this, too, is very time-consuming. Therefore, researchers have resorted mainly to hydrodynamic equations for modeling electron and phonon transport for practical device simulation.

Based on the mechanism of nonequilibrium Joule heating, the governing equations for charge and energy transport are

$$\nabla^2 V = -\frac{e}{\varepsilon_s}(N_D - \rho); \quad \mathbf{E} = -\nabla V \tag{8.49}$$

$$\frac{\partial \rho}{\partial t} + \nabla \cdot (\rho \mathbf{v}) = 0 \tag{8.50}$$

$$\rho m^* \left(\frac{\partial \mathbf{v}}{\partial t} + \mathbf{v} \cdot \nabla \mathbf{v} \right) = -e\rho \mathbf{E} - \nabla(\rho k_B T_e) - \frac{\rho m^* \mathbf{v}}{\tau_m} \tag{8.51}$$

$$\frac{\partial T_e}{\partial t} + \nabla \cdot (\mathbf{v} T_e) = \frac{1}{3} T_e \nabla \cdot \mathbf{v}$$
$$+ \frac{2}{3nk_B} \nabla \cdot (k_e \nabla T_e) - \frac{T_e - T_{LO}}{\tau_{e-LO}} - \frac{T_e - T_A}{\tau_{e-A}} + \frac{m^* v^2}{3k_B}\left(\frac{3}{\tau_m} - \frac{1}{\tau_{e-LO}} - \frac{1}{\tau_{e-A}}\right) \tag{8.52}$$

$$C_{LO}\frac{\partial T_{LO}}{\partial t} = \frac{3\rho k_B}{2}\left(\frac{T_e - T_{LO}}{\tau_{e-LO}}\right) + \frac{\rho m^* v^2}{2\tau_{e-LO}} - C_{LO}\left(\frac{T_{LO} - T_A}{\tau_{LO-A}}\right) \tag{8.53}$$

$$C_A \frac{\partial T_A}{\partial t} = \nabla \cdot (k_A \nabla T_A) + C_{LO}\left(\frac{T_{LO} - T_A}{\tau_{LO-A}}\right) + \frac{3\rho k_B}{2}\left(\frac{T_e - T_A}{\tau_{e-A}}\right) \tag{8.54}$$

The derivation of these equations is described in detail in Refs. 12 and 21. Equation 8.49 is the Poisson equation which satisfies Gauss's law of charge and field. Here V is the potential, ε_s is the dielectric constant of the medium, N_D is the doping concentration, ρ is the electron concentration, and $\mathbf{E}$ is the electric field vector. Equations 8.50 and 8.51 are the electron continuity and energy momentum equations which follow the development in the section entitled "Boltzmann Transport Theory." The momentum equation is quite similar to the Navier-Stokes equation of fluid mechanics. By nondimensionalizing this equation, Lai and Majumdar [22] derived an equivalent electron Reynolds number in terms of device parameters given as $\mathrm{Re} = eV_{ds}\tau_m^2/(m^* L_g^2)$. For most operating conditions of V_{ds} and L_g and values of m^* and τ_m, $\mathrm{Re} \ll 1$ and so the nonlinear convective term can be neglected. In the electron energy conservation equation (Eq. 8.52), the Fourier law of heat conduction has been assumed for the heat flux where k_e is the electron thermal conductivity, T_O is the optical phonon temperature, and τ_{e-o} is the electron energy relaxation time for electron-optical phonon scattering. The right side of Eq. 8.52 contains loss of electron energy to optical phonons, to acoustic phonons, as well as a heat generation term that comes from dissipation of the kinetic energy gained by the electrons. Equation 8.53 represents energy conservation for optical phonons where C_O is the optical phonon heat capacity. The terms on the right side represent energy gain from electrons and loss to acoustic phonons. Note that there is no heat diffusion term due to negligible group velocity of optical phonons. Equation 8.54 represents energy conservation of acoustic phonons where C_A is the acoustic phonon heat capacity, T_A is the acoustic phonon temperature, and the right side contains the heat diffusion term (first term on the right side), the term representing energy gain from the optical phonons, as well as that representing energy gain directly from electrons. Note that all the relaxation times used in the energy equations must be the energy relaxation times which involve inelastic scattering such that energy is transferred during scattering.

The numerical solution of these equations for the case of microelectronic devices have been developed and can be found elsewhere [21,22]. It was found that under steady-state conditions that are realistic for modern devices, the electron temperature is an order of magnitude higher than those of the phonons. This shows the nonequilibrium nature of the Joule heating process. However, the difference between the optical and acoustic phonon temperature is comparatively much smaller. Hence, under steady-state conditions, it is sufficient to group the optical and acoustic phonon temperatures into one equation called the *lattice energy conservation equation*. However, it is important to note that in high-speed devices this is not valid. The phonon-phonon scattering time is about 10 ps in most semiconductors at room temperature. This corresponds to a frequency of 100 GHz which is in the microwave region of the electromagnetic spectrum and is used in wireless communications. Devices that are operated in this regime undergo unsteady heating with a period comparable to the phonon-phonon scattering time. Since electron-optical phonon scattering time is on the order of 100 fs, the electron energy can be efficiently transferred to the optical phonons. However, optical phonons cannot transfer their energy fast enough to the acoustic phonons. Hence, a *phonon bottleneck* is encountered where the optical phonon temperature becomes much higher than the acoustic phonon temperature. This changes the electron scattering rate and thereby changes the electrical performance of the device.

Radiative Heating by Ultrashort Laser Pulses

Pulsed lasers with pulse widths less than 100 fs and as low as 10 fs are now commercially available. It is important to compare the timescale with some characteristic timescales involved in microscale energy transfer processes. As we have noted earlier, the electron-phonon scattering timescale is on the order of 100 fs in most metals and semiconductors at room temperature. This is also the shortest time it takes for the ions in the solid to undergo one vibrational period. In other words, the highest phonon frequency is about 10^{13} Hz which corresponds to 100 fs in time period. Such laser pulse widths are much shorter than the phonon-phonon scattering time at room temperature which is on the order of 10 ps.

If we calculate the period of oscillation of an electron wave function, that is, $t = \lambda/v$, then using λ of 10 Å and v of 10^6 m/s for a Fermi electron, we get the period to be 1 fs. The period of oscillation for a photon in the visible spectrum is also about 1 fs. Therefore, we see that laser pulses of 10 fs contain at least 10 electromagnetic waves and are longer than the time period for wave function of Fermi electrons. Hence, such laser pulses can excite electrons within the duration of a pulse if there are no phonons involved, that is, if the excitation is a direct process. Such pulsed lasers are now widely used to study the electron dynamics in chemical bond formation and dissociation, to investigate carrier relaxation processes in metals and semiconductors, as well as for materials processing with extremely high depth resolution. It is instructive to go through the microscopic energy transfer processes before the governing equations are derived.

Energy Transfer Processes.   The photon energies of such pulsed lasers range from 1 to 5 eV. Using $k_B T$ as the thermal energy of the electrons, we can see that this corresponds to a temperature level of about 10^4 K. Hence, the electrons which absorb these photons and undergo a direct transition in the Brillouin zone are extremely hot for a very short period of time. Note that the electron-electron scattering rate is typically on the order of 10 to 50 fs in metals. In semiconductors, this rate depends on the electron concentration in the conduction band but is also typically in this range. Hence, the distribution of electrons is initially highly nonequilibrium with a large number of electrons at the excited state of 1 to 5 eV due to photon absorption. However, the energy gets quickly redistributed between the electrons such that by about 100 fs the electrons reach the equilibrium Fermi-Dirac distribution. Note, however, that since the electrons do not lose energy to the phonons within this time period, they remain quite energetic. But since equilibrium is established between the electrons, one can define an electron temperature T_e. During this time, the phonons remain at the original ambient temperature T_o. So, as far as the phonons are concerned, they encounter a hot electron reservoir at T_e. Typically, this is about 10^3 to 10^4 K. The temperature difference $T_e - T_o$ drives energy flow from the electron to the phonons.

The most efficient way for the electrons to lose energy is through emission of optical phonons since their wave vectors span the entire Brillouin zone at the same energy. The electron energy, however, must be larger than the optical phonon energy which is typically in the range of 10 to 50 meV. In addition, since optical phonon energies are higher than those of acoustic phonons, optical phonon emission is a faster and more efficient way of energy transfer. When polar interactions are involved, such as in polar bonds in GaAs or other III-V or II-VI materials, the coupling to electrons is even stronger. In such cases, LO-phonons are most efficiently emitted by hot electrons. Optical phonons eventually scatter and emit acoustic phonons which are responsible for lattice heat conduction.

Governing Equations.   To study the dynamics and interactions of photons, electrons, and phonons rigorously from a particle transport viewpoint, the Boltzmann transport equation for all these individual systems must be solved simultaneously. This is a very challenging task and has not been adopted by most investigators. What people have used are the averaged moments of the Boltzmann equation which produce the conservation equations [23–25]. It must be noted, however, that the moments always contain less information than the Boltzmann transport equation and, invariably, an assumption needs to be made at some stage. Otherwise, we have to take higher and higher moments. This is normally called the *closure problem*. However, several assumptions can be made and they seem to work reasonably well for the cases studied in the past.

The energy conservation equations are as follows:

$$\frac{\partial u_c}{\partial t} = \nabla \cdot (k_c \nabla T_c) - \frac{3\rho k_B}{2}\left(\frac{T_c - T_O}{\tau_{c-O}}\right) - \frac{3\rho k_B}{2}\left(\frac{T_c - T_A}{\tau_{c-A}}\right) + S \tag{8.55}$$

$$\frac{\partial U_O}{\partial t} = \frac{3\rho k_B}{2}\left(\frac{T_c - T_O}{\tau_{c-O}}\right) - C_O\left(\frac{T_O - T_A}{\tau_{O-A}}\right) \tag{8.56}$$

$$\frac{\partial u_A}{\partial t} = \nabla \cdot (k_A \nabla T_A) + \frac{3\rho k_B}{2}\left(\frac{T_c - T_A}{\tau_{c-A}}\right) + C_O\left(\frac{T_O - T_A}{\tau_{O-A}}\right) \tag{8.57}$$

Equation 8.55 is the carrier energy conservation equation which involves heat conduction by carriers, energy loss to optical phonons that occur at a timescale of τ_{c-O}, energy loss to acoustic phonons at timescale τ_{c-A}, and a source term containing energy increase due to photon absorption. Note that the assumption of Fourier law is not really valid for short timescale studies but this is an approximation that invokes closure in these hydrodynamic equations. The second equation conserves energy for the optical phonons where the two terms on the right side represent energy gain from electrons and energy loss to acoustic phonons. The third equation is that for acoustic phonons where, again, the Fourier law is used for heat flux. It should be noted that in most of the previous work [23–25], the lattice has been assumed to be a single thermodynamic system. This implicitly assumes equilibrium between optical and acoustic phonons. However, since the timescales for electron–optical phonon (100 fs) and phonon-phonon (10 ps) interactions are different by two orders of magnitude, it is perhaps not a reasonable assumption. Hence, they have been considered as separate systems in this presentation.

The energy densities in the three systems are

$$u_c = \rho E_g + C_c T_c; \quad u_O = C_O T_O; \quad u_A = C_A T_A \qquad (8.58)$$

where ρE_g is the energy density the electrons gain for excitation across the band gap E_g; ρ is the density of the electrons. The source term can be written as

$$S = \frac{\alpha(T_c, \rho) 2J\sqrt{\ln 2}}{t_p \sqrt{\pi}} \left[1 - R(T_{cs}, \rho_s)\right] \exp\left(-\int_0^y \alpha\, dz\right) \exp\left[-4 \ln 2 \left(\frac{t}{t_p}\right)^2\right] \qquad (8.59)$$

where $\alpha(T_c, \rho)$ is the absorption coefficient which is a function of local carrier temperature and density, J is the laser fluence (J/m^2), t_p is the laser pulse width, and R is the surface reflectivity which is a function of the surface carrier temperature and density. The carrier density follows the continuity equation

$$\frac{\partial \rho}{\partial t} + \nabla \cdot (\rho \mathbf{v}) = \frac{S_1}{\hbar\omega} - \gamma\rho^3 \qquad (8.60)$$

where S_1 is the absorption source term that corresponds to direct interband transition that excites an electron to the conduction band and creates an electron-hole pair. Here, $\hbar\omega$ is the photon energy, $\mathbf{v}$ is the carrier velocity, and the last term represents the Auger recombination term. The carrier momentum equation can be written as

$$\rho m^* \left(\frac{\partial \mathbf{v}}{\partial t}\right) = -\nabla(\rho k_B T_e) - \frac{\rho m^* \mathbf{v}}{\tau_m} \qquad (8.61)$$

where the last term contains the momentum relaxation time τ_m.

It is quite evident that the set of conservation equations is nonlinear and highly coupled. They are usually solved numerically which will not be covered here but can be found in the literature.

SUMMARY

This chapter provides an introduction to microscopic transport phenomena with special emphasis on energy transport in solids. The first step in this process is to identify the time- and length scales that are characteristic of the transport process. Based on these scales, the regimes where continuum or macroscopic theories break down are identified. Having done so, the regimes where microscopic transport phenomena occur in both the wave and particle regimes are also indicated. The chapter does not address wave transport issues such as diffraction, refraction, interference, or tunneling, which occur when the size of an object is on the order of the wavelength of an energy carrier or if the phase information of an energy carrier becomes significant during scattering. Hence, this monograph concentrates only on microscopic particle transport theories.

Kinetic theory is introduced and developed as the initial step toward understanding microscopic transport phenomena. It is used to develop relations for the thermal conductivity which are compared to experimental measurements for a variety of solids. Next, it is shown that if the time- or length scale of the phenomena are on the order of those for scattering, kinetic theory cannot be used but instead Boltzmann transport theory should be used. It was shown that the Boltzmann transport equation (BTE) is fundamental since it forms the basis for a vast variety of transport laws such as the Fourier law of heat conduction, Ohm's law of electrical conduction, and hyperbolic heat conduction equation. In addition, for an ensemble of particles for which the particle number is conserved, such as in molecules, electrons, holes, and so forth, the BTE forms the basis for mass, momentum, and energy conservation equations. In cases where particle numbers are not conserved, such as in photons and phonons, the BTE forms the basis for the equation of transfer which span the ballistic to the diffusive transport regimes.

After showing how the BTE forms the fundamental basis for particle transport theories, the monograph describes two case studies: (1) nonequilibrium Joule heating in submicrometer transistors; and (2) nonequilibrium radiative heating by ultrashort laser pulses. Through these case studies, mechanisms and theories of nonequilibrium energy transfer are introduced.

NOMENCLATURE

Symbol, Definition, Units

C	specific heat per unit volume (J/m³-K)
D	density of states
e	electron charge, 1.6×10^{-19} (C)
$\mathbf{E}$	electric field vector (V/m)
E_F	Fermi energy of electrons (eV or J)
f	distribution function of an ensemble of particles
$\mathbf{F}$	force vector (N)
G	gravitational potential (m²/s²)
h	Planck's constant, 6.63×10^{-34} (J-s)
$\hbar$	Planck's constant divided by 2π
I	intensity
j	particle tag
$\mathbf{J}$	flux vector
k	thermal conductivity (W/m-K)
$\mathbf{k}$	wave vector (m⁻¹)
k_B	Boltzmann constant, 1.38×10^{-23} (J/K)
ℓ	characteristic length scale for scattering (m)
L_g	gate width (m)
m	mass (kg)
N_D	doping concentration (m⁻³)
p	momentum (kg-m/s)
P	pressure (N/m²)
q	heat flux (W/m²)
R	reflectivity
Re	electron Reynolds number
s	polarization
S	source function
t	time (s)
T	temperature (K)
u	energy density (J/m³)
U	thermodynamic potential per unit mass (m²/s²)

v velocity (m/s)

V voltage (V)

W scattering rate (1/s)

Greek Letters

α absorption coefficient (m^{-1})

ε energy (J)

λ wavelength of energy carrier (m)

η number density of particles (m^{-3})

ω angular frequency (rad/s)

ρ number density of particles (m^{-3})

σ scattering cross-sectional area (m^2)

       electrical conductivity $[(\Omega m)^{-1}]$

θ polar angle or characteristic temperature (K)

ϕ azimuthal angle

Φ electrochemical potential (m^2/s^2)

τ characteristic time scale for scattering (s)

ξ energy density (J/m^3)

Subscripts

a absorption

A acoustic phonon

c collision

d diffusion, drift

dom   dominant

ds drain-to-source

D Debye, doping

e electron, energy, emission

E Einstein

F Fermi

l lattice

LO longitudinal-optical phonon

m momentum

o equilibrium

r relaxation, random

s sound, scattering

ω frequency

REFERENCES

1. W. G. Vincenti and C. H. Kruger, *Introduction to Physical Gas Dynamics,* Robert Krieger, New York, 1977.
2. C. Kittel, *Introduction to Solid State Physics,* John Wiley & Sons, New York, 1986.

3. W. Powell, C. Y. Ho, and P. E. Liley, *Thermal Conductivity of Selected Materials,* National Bureau of Standards Reference Data Series-8, Washington D.C., 1966.

4. J. M. Ziman, *Electrons and Phonons,* Oxford University Press, London, 1960.

5. R. Berman, E. L. Foster, and J. M. Ziman, *Proc. Roy. Soc. A,* (231):130, 1955.

6. R. Berman, *Thermal Conduction in Solids,* Oxford University Press, Oxford, 1976.

7. D. G. Cahill and R. O. Pohl, "Lattice Vibrations and Heat Transport in Crystals and Glasses," *Ann. Rev. Phys. Chem.* (39): 93–121, 1988.

8. D. Cahill and R. O. Pohl, "Heat Flow and Lattice Vibrations in Glasses," *Solid State Comm.* (70): 927–930, 1989.

9. D. Cahill, S. K. Watson, and R. O. Pohl, "Lower Limit to the Thermal Conductivity of Disordered Crystals," *Phys. Rev. B.* (46): 6131–6140, 1992.

10. D. D. Joseph and L. Preziosi, "Heat Waves," *Rev. Mod. Phys.* (61): 41, 1989.

11. D. K. Ferry, *Semiconductors,* MacMillan, New York, 1991.

12. K. Blotekjaer, "Transport Equations for Electrons in Two-Valley Semiconductors," *IEEE Trans. Electron Dev.* (17): 38–47, 1970.

13. R. Siegel and J. R. Howell, *Thermal Radiation Heat Transfer,* 3d ed., Hemisphere, Washington D.C., 1992.

14. A. Majumdar, "Microscale Heat Conduction in Dielectric Thin Films," *J. Heat Transfer* (115): 7–16, 1993.

15. A. A. Joshi and A. Majumdar, "Transient Ballistic and Diffusive Phonon Heat Transport in Thin Films," *J. Appl. Phys.* (74): 31–39, 1993.

16. K. E. Goodson, M. I. Flik, L. T. Su, and D. A. Antoniadis, "Annealing Temperature Dependence on the Thermal Conductivity of LPCVD Silicon Dioxide Layers," *IEEE Trans. Electron Dev. Lett.* (15): 490–492, 1993.

17. O. W. Kading, H. Skurk, and K. E. Goodson, "Thermal Conduction in Metallized Silicon Dioxide Layers on Silicon," *Appl. Phys. Lett.* (65): 1629–1631, 1994.

18. S. M. Lee and D. G. Cahill, "Heat Conduction in Thin Dielectric Films," *J. Appl. Phys.* (81): 2590–2595, 1997.

19. S. M. Sze, *Physics of Semiconductor Devices,* 2d ed., John Wiley & Sons, New York, 1981.

20. D. von der Linde, J. Kuhl, and H. Klingenburg, "Raman Scattering from Nonequilibrium LO Phonons with Picosecond Resolution," *Phys. Rev. Lett.* (44): 1505, 1980.

21. K. Fushinobu, A. Majumdar, and K. Hijikata, "Heat Generation and Transport in Submicron Semiconductor Devices," *J. Heat Transfer* (117): 25–31, 1995.

22. J. Lai and A. Majumdar, "Concurrent Thermal and Electrical Modeling of Sub-Micrometer Silicon Devices," *J. of Applied Physics* (79): 7353–7361, 1996.

23. H. M. van Driel, "Kinetics of High-Density Plasmas Generated in Si by 1.06 and 0.53-μm Picosecond Laser Pulses," *Phys. Rev. B.* (35): 8166–8176, 1987.

24. T. Q. Qiu and C. L. Tien, "Heat Transfer Mechanisms during Short-Pulse Laser Heating of Metals," *J. Heat Transfer* (115): 835–841, 1993.

25. K. Fushinobu, L. M. Phinney, and N. Tien, "Ultrashort-Pulse Laser Heating of Silicon to Reduce Microstructure Adhesion," *Int. J. Heat Mass Trans.* (39): 3181–3186, 1996.

CHAPTER 9
HEAT TRANSFER IN POROUS MEDIA

Massoud Kaviany
University of Michigan

INTRODUCTION

Examination of heat transfer in natural and engineered porous media relies on the knowledge we have gained in studying heat transfer in plain media. The presence of a permeable solid (which we assume to be rigid and stationary) influences these phenomena significantly. Due to practical limitations, as a general approach we choose to describe these phenomena at a small length scale which is yet larger than the linear dimension of the pore or the linear dimension of the solid particle (for a particle-based porous medium). This requires the use of the local volume-averaging theories. Also, depending on the validity, local mechanical, thermal, and chemical equilibrium or nonequilibrium may be imposed between the fluid (liquid and/or gas) and solid phases.

Figure 9.1 gives a classification of the transport phenomena in porous media based on the single- or two-phase flow through the pores. Figure 9.2 renders these phenomena at the pore level. Description of transport of species, momentum and energy, chemical reaction (endothermic or exothermic), and phase change (solid/liquid, solid/gas, and liquid/gas) at the differential, local phase-volume level and the application of the volume-averaging theories lead to a relatively accurate and yet solvable local description.

The analysis of heat transfer in porous media is required in a large range of applications. The porous media can be *naturally formed* (e.g., rocks, sand beds, sponges, woods) or *fabricated* (e.g., catalytic pellets, wicks, insulations). A review of engineered porous materials is given in Schaefer [1] and the physics and chemistry of porous media is reviewed by Banavar et al. [2]. The applications are in the areas of chemical, environmental, mechanical, and petroleum engineering and in geology. As expected, the range of pore sizes or particle sizes (when considering the solid matrix to be made of consolidated or nonconsolidated particles) is vast and can be of the order of molecular size (ultramicropores with $3 < d < 7$ Å, where d is the average pore size), the order of centimeters (e.g., pebbles, food stuff, debris), or larger. Figure 9.3 gives a classification of the particle size based on measurement technique, application, and statistics. A review of the particle characteristics for particles with diameters smaller than 1 cm is given by Porter et al. [3].

Also shown in Fig. 9.3 is the capillary pressure in a water-air system with the mean radius of curvature equal to the particle radius. It is clear that as the particle size spans over many orders of magnitude, the handling of the radiative heat transfer and the significance of forces such as capillarity and gravity also vary greatly.

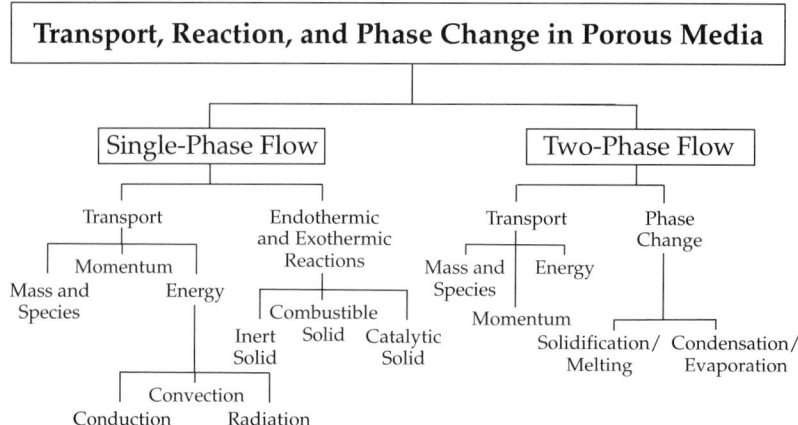

FIGURE 9.1 Aspects of treatment of transport, reaction, and phase change in porous media.

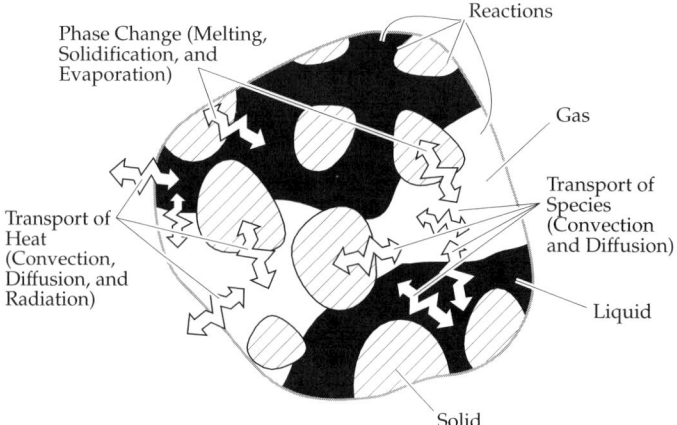

FIGURE 9.2 A rendering of the pore-level transport, reaction, and phase change in porous media.

Other than the particle dimension d, the porous medium has a system dimension L, which is generally much larger than d. There are cases where L is of the order d such as thin porous layers coated on the heat transfer surfaces. These systems with $L/d \simeq O(1)$ are treated by the examination of the fluid flow and heat transfer through a small number of particles, a treatment we call *direct simulation* of the transport. In these treatments, no assumption is made about the existence of the local thermal equilibrium between the finite volumes of the phases. On the other hand, when $L/d \gg 1$ and when the variation of temperature (or concentration) across d is negligible compared to that across L for both the solid and fluid phases, then we can assume that within a distance d both phases are in thermal equilibrium (*local thermal equilibrium*). When the solid matrix structure cannot be fully described by the prescription of solid-phase distribution over a distance d, then a representative elementary volume with a linear dimension larger than d is needed. We also have to extend the requirement of a negligible temperature (or concentration) variation to that over the linear dimension of the representa-

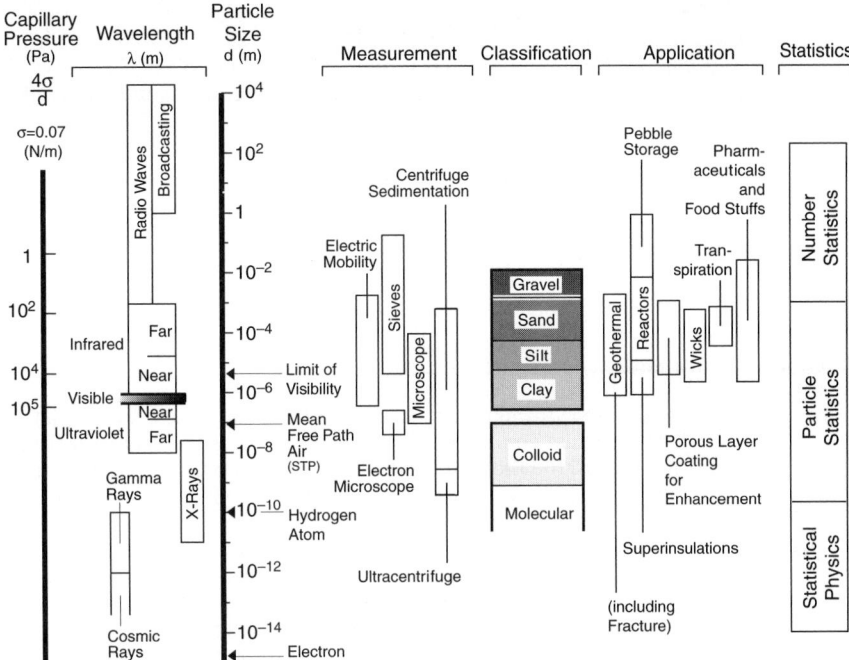

FIGURE 9.3 Particle sizes and their classifications, measurements, and applications.

tive elementary volume ℓ. For some fabricated solid matrices we have $\ell/d \simeq O(1)$, and for natural solid matrices we have $\ell/d \simeq O(10$ or larger$)$. In addition to d, ℓ, and L, a length scale equal to the square root of the permeability is also used. This length scale (called the *Brinkman screening distance*) $K^{1/2}$ is smaller than d and is generally $O(10^{-2}d)$, where K is the *permeability*. The four length scales, with a loose requirement for the presence of local thermal equilibrium (based on the length scale only), are written as

$$K^{1/2} \ll d < \ell \ll L$$

The *solutions* to the conservation equations, marked by the boundary-layer analyses of Cheng and Minkowycz [4] and Vafai and Tien [5] and as reviewed by Nield and Bejan [6] and as part of the general treatment of convective heat transfer by Kaviany [7] are not addressed here. Yet, by undertaking the discussion of principles of the fluid flow, conduction, convection, and radiation for both the single- and two-phase flows in porous media, such treatments must lack the depth that these topics deserve. The *permeability*, the *effective thermal conductivity*, and the *dispersion tensors* are discussed. The governing conservation equations are reached by the method of local volume averaging with these tensors appearing as the effective properties. The materials presented here are excerpts from a monograph on the topic.*

For radiation heat transfer in porous media, the same unit-cell approach is used and the particles in each cell are treated as scatterers. The scattering also becomes dependent when the porosity is not close to unity. The radiation properties are related to the optical properties and the porosity.

---

* *Principles in Heat Transfer in Porous Media,* by M. Kaviany, 2d ed., Springer-Verlag, New York, 1995.

The two-phase flow in porous media (a three-phase system) is approached from the pore-level fluid mechanics. The pertinent forces and their expected contributions are examined, and when available, empirical results are used. After arriving at a set of volume-averaged governing conservation and constitutive equations, some liquid-vapor phase change problems are examined.

Except in the section entitled "Two-Medium Treatment," the local thermal equilibrium is assumed between the phases. A glossary of common words used in transport in porous media is given in the glossary included with this chapter.

SINGLE-PHASE FLOW

Conduction Heat Transfer

Heat conduction through fully-saturated matrices (i.e., a single-phase fluid occupying the pores), as with heat conduction through any heterogeneous media, depends on the structure of the matrix and the thermal conductivity of each phase. One of the most difficult aspects of the analysis of heat conduction through a porous medium is the structural modeling. This is because the representative elementary volumes are three-dimensional and have complicated structures that vary greatly among different porous media. Since the thermal conductivity of the solid phase is generally different than that of the fluid, the manner in which the solid is interconnected influences the conduction significantly. Even when dealing with the nonconsolidated particles, the contact between the particles plays a significant role.

For the analysis of the macroscopic heat flow through heterogeneous media, the local volume-averaged (or *effective*) properties such as the *effective thermal conductivity* $\langle k \rangle = k_e$ are used. These local effective properties such as the *heat capacity* $\langle \rho c_p \rangle$, thermal conductivity $\langle k \rangle$, and radiation absorption and scattering coefficients $\langle \sigma_a \rangle$ and $\langle \sigma_s \rangle$ need to be arrived at from the application of the first principles to the volume over which these local properties are averaged, that is, the representative elementary volume.

As shown later, the local average $\langle \rho c_p \rangle$ is obtained by a simple volume averaging. However, we reiterate that the effective thermal conductivity is expected to depend on the following:

The *thermal conductivity* of each phase; that is, the relative magnitude of k_s/k_f is important.

The *structure* of the solid matrix; that is, the extent of the continuity of the solid phase is very important.

The *contact resistance* between the nonconsolidated particles; that is, the solid surface oxidation and other coatings are all important.

For gases, when the ratio of the *mean free path* and the *average linear pore dimension* (i.e., the *Knudsen number*) becomes large, the bulk gas conductivity cannot be used for the fluid phase.

In the following, after reviewing the requirements for the validity of the assumption of the local thermal equilibrium, the attempt at predicting the effective thermal conductivity is reviewed along with some correlations and comparisons with experimental results.

Local Thermal Equilibrium.   In principle, determination of the thermal conductivity of saturated porous media involves application of the point conduction (energy) equation to a point in the representative elementary volume of the matrix and the integration over this volume. In doing so, we realize that at the pore level there will be a difference ΔT_d between the temperature at a point in the solid and in the fluid. Similarly, across the representative elementary volume, we have a maximum temperature difference ΔT_ℓ. However, we assume that these temperature differences are much smaller than those occurring over the system dimension ΔT_L. Thus, we impose the assumption of local thermal equilibrium by requiring that

$$\Delta T_d < \Delta T_\ell \ll \Delta T_L \qquad (9.1)$$

With this assumed negligible local temperature difference between the phases, we assume that within the local representative elementary volume $V = V_f + V_s$, the solid and fluid phases are in *local thermal equilibrium*. This is stated using the *phase* (or *intrinsic*) and *both-phases* volume averaged temperatures as

$$\langle T \rangle^f \equiv \frac{1}{V_f} \int_{V_f} T_f \, dV = \langle T \rangle^s \equiv \frac{1}{V_s} \int_{V_s} T_s \, dV = \langle T \rangle \equiv \langle T \rangle \frac{1}{V} \int_V T \, dV \qquad (9.2)$$

Note that, although ΔT_d and ΔT_ℓ are small, their gradients in their respective length scales are not small.

For very fast transients and when heat generation exists in the solid or fluid phase, inequality (Eq. 9.1) may not be satisfied and a *two-temperature treatment* should be made. This is further discussed in the section entitled "Two-Medium Treatment" in connection with the local thermal *nonequilibrium* among phases.

Note that in the analysis of the unit cells, we use ΔT_d and ΔT_ℓ, and in dealing with the macroscopic heat transfer of the saturated matrix, we deal with ΔT_L.

Local Volume Averaging. The *porosity* ε is defined as

$$\varepsilon = \frac{V_f}{V_f + V_s} \qquad (9.3)$$

Then the fluid (or solid) temperature within V is decomposed using

$$T_f = \langle T \rangle^f + T_f' \qquad (9.4)$$

where T_f' is the *spatial deviation* component.

Nozad et al. [8] discuss a set of *closure constitutive equations* (or *transformations*) given by

$$T_f' = \mathbf{b}_f \cdot \nabla \langle T \rangle \qquad (9.5)$$
$$T_s' = \mathbf{b}_s \cdot \nabla \langle T \rangle \qquad (9.6)$$

The required choice of a transformation vector $\mathbf{b}$ instead of a scalar also satisfies the tensorial character of the effective thermal conductivity (similar to permeability). Note that $\mathbf{b} = \mathbf{b}(\mathbf{x})$, where $\mathbf{x}$ is within the representative elementary volume, transforms the gradient of local volume-averaged temperature (changing over the length scale L) into deviations changing over length scale ℓ. The *effective thermal conductivity tensor* $\mathbf{K}_e$ is defined, and by using (Eq. 9.5) in the thermal energy equation, we have

$$[\varepsilon(\rho c_p)_f + (1 - \varepsilon)(\rho c_p)_s] \frac{\partial \langle T \rangle}{\partial t} = \nabla \cdot (\mathbf{K}_e \cdot \nabla \langle T \rangle) \qquad (9.7)$$

where $\mathbf{K}_e$ is given by

$$\mathbf{K}_e = [\varepsilon k_f + (1 - \varepsilon) k_s]\mathbf{I} + \frac{k_f - k_s}{V} \int_{A_{fs}} \mathbf{n}_{fs} \mathbf{b}_f \, dA = [\varepsilon k_f + (1 - \varepsilon) k_s]\mathbf{I} + (k_f - k_s)\varepsilon \frac{1}{V_f} \int_{A_{fs}} \mathbf{n}_{fs} \mathbf{b}_f \, dA$$

$$(9.8)$$

A product (of two vectors) such as $\mathbf{n}_{fs} \mathbf{b}_f$ is called a *dyad product* and is a special form of the second-order tensors.

Also, the unit tensor used in (Eq. 9.8) is

$$\mathbf{I} = \begin{bmatrix} 1 & 0 & 0 \\ 0 & 1 & 0 \\ 0 & 0 & 1 \end{bmatrix} \tag{9.9}$$

Determination of **b** for periodic structures is reviewed by Kaviany [9].

Correlations for Effective Conductivity. For packed beds of particles and for the entire range of values of k_s/k_f (larger and smaller than unity) some *empirical* correlations are available for the *isotropic* effective thermal conductivity $\langle k \rangle = k_e$. Three of these are constructed by Krupiczka [10], Kunii and Smith [11], and Zehnder and Schlünder [12]. An extensive review of the literature on the effective thermal conductivity prior to 1960 is given by Krupiczka [10]. The prediction of Krupiczka gives (for $0.2 \leq \varepsilon \leq 0.6$)

$$\frac{k_e}{k_f} = \left(\frac{k_s}{k_f} \right)^{+0.280 - 0.757 \log \varepsilon - 0.057 \log (k_s/k_f)} \tag{9.10}$$

The prediction of Kunii and Smith [11] (for $0.260 \leq \varepsilon \leq 0.476$) gives

$$\frac{k_e}{k_f} = \varepsilon + \frac{(1 - \varepsilon)}{\phi_2 + 4.63(\varepsilon - 0.26)(\phi_1 - \phi_2) + \frac{2}{3}(k_f/k_s)} \tag{9.11}$$

where $\phi_1 = \phi_1(k_f/k_s)$ and $\phi_2 = \phi_2(k_f/k_s)$ and they are monotonically decreasing functions of k_s/k_f. The prediction of Zehnder and Schlünder [12] is

$$\frac{k_e}{k_f} = 1 - (1 - \varepsilon)^{1/2} + \frac{2(1 - \varepsilon)^{1/2}}{1 - (k_f/k_s)B} \left\{ \frac{[1 - (k_f/k_s)]B}{[1 + (k_f/k_s)B]^2} \ln \frac{1}{(k_f/k_s)B} - \frac{B + 1}{2} - \frac{B - 1}{1 - (k_f/k_s)B} \right\} \tag{9.12}$$

where

$$B = 1.25 \left(\frac{1 - \varepsilon}{\varepsilon} \right)^{10/9} \tag{9.13}$$

A modification to the Zehnder-Schlünder correlation is given by Hsu et al. [13].

Hadley [14] combines the Maxwell upper bound with an expression obtained by the introduction of an adjustable function $f_0(\varepsilon)$. In combining these, he uses a weighting function $\alpha_0(\varepsilon)$ which, along with $f_0(\varepsilon)$, is found from the experimental results. While $f_0(\varepsilon)$ changes slightly with ε, α_0 is very sensitive to changes in ε. Hadley's correlation is (for $0 \leq \varepsilon \leq 0.580$)

$$\frac{k_e}{k_f} = (1 - \alpha_0) \frac{\varepsilon f_0 + (k_s/k_f)(1 - \varepsilon f_0)}{1 - \varepsilon(1 - f_0) + (k_s/k_f)\varepsilon(1 - f_0)} + \alpha_0 \frac{2(k_s/k_f)^2(1 - \varepsilon) + (1 + 2\varepsilon)(k_s/k_f)}{(2 + \varepsilon)(k_s/k_f) + 1 - \varepsilon} \tag{9.14}$$

$$f_0 = 0.8, \tag{9.15}$$

where
$\log \alpha_0 = -4.898\varepsilon$ $0 \leq \varepsilon \leq 0.0827$
$\log \alpha_0 = -0.405 - 3.154(\varepsilon - 0.0827)$ $0.0827 \leq \varepsilon \leq 0.298$ (9.16)
$\log \alpha_0 = -1.084 - 6.778(\varepsilon - 0.298)$ $0.298 \leq \varepsilon \leq 0.580$

Prasad et al. [15] performed experiments for k_s/k_f larger and smaller than unity (but for only one order magnitude to either side of unity) and examined the accuracy of the preceding three predictions. In Fig. 9.4, the experimental results of Prasad et al. [15], Nozad [16], and other experimental results reported by her are plotted against these correlations. While for moderate values of k_s/k_f the four predictions are in agreement, for high values, *only* the Hadley correlation can predict the effective thermal conductivity correctly. Another correlation that uses k_s/k_f and ε as parameters is that of Nimick and Leith [17] with accuracies similar to that of Hadley. A correlation for the effective conductivity of rocks based on a geometric model of nonintersecting, oblate spherical pores has been developed by Zimmer-

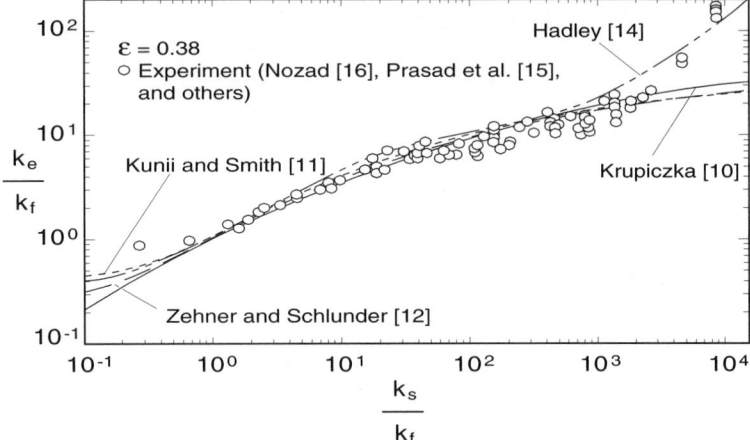

FIGURE 9.4   Comparison of several correlations for the effective thermal conductivity with experimental results from several sources.

man [18]. Correlations for fibrous insulations are given by Stark and Frickle [19]. For two-dimensional structures, some correlations for the bulk effective thermal conductivity are given by Sahraoui and Kaviany [20] and Ochoa-Tapia et al. [21] and by Hsu et al. [22] for two- and three-dimensional periodic structures.

The pore-level structure of the solid and fluid phases significantly influences the effective conductivity, and in general the specifications of ε and k_s/k_f do not suffice. When gases are involved, rarefaction (for small pore size, for low pressure, or near constrictions) should be attended. The thermal conduction of each phase is temperature-dependent (and for gases, also pressure-dependent), and that should also be considered.

Convection Heat Transfer

As we consider simultaneous fluid flow and heat transfer in porous media, the role of the macroscopic (Darcean) and microscopic (pore-level) velocity fields on the temperature field needs to be examined. Experiments have shown that the mere inclusion of $\mathbf{u}_D \cdot \nabla\langle T \rangle$ in the energy equation does not accurately account for all the hydrodynamic effects. The pore-level hydrodynamics also influence the temperature field. Inclusion of the effect of the pore-level velocity nonuniformity on the temperature distribution (called the *dispersion effect* and generally included as a diffusion transport) is the main focus in this section.

This is similar to the Taylor dispersion in tubes, which is the result of area-averaging across the tube. Here a general volume-averaged treatment is presented. The available treatments and closure conditions for periodic structures and the predicted dispersion tensor are examined and comparisons with the experimental results are made. We then look at the correlations available for the dispersion coefficients.

Darcy Equation.   The bulk resistance to flow of an incompressible fluid through a solid matrix, as compared to the resistance at and near the surfaces confining this solid matrix, was first measured by Darcy [23]. Since in his experiment the internal surface area (*interstitial area*) was many orders of magnitude larger than the area of the confining surfaces, the *bulk shear stress resistance* was dominant.

His experiment used nearly uniform size particles that were randomly and loosely packed, that is, a *nonconsolidated, uniform, rigid,* and *isotropic* solid matrix. The macroscopic flow

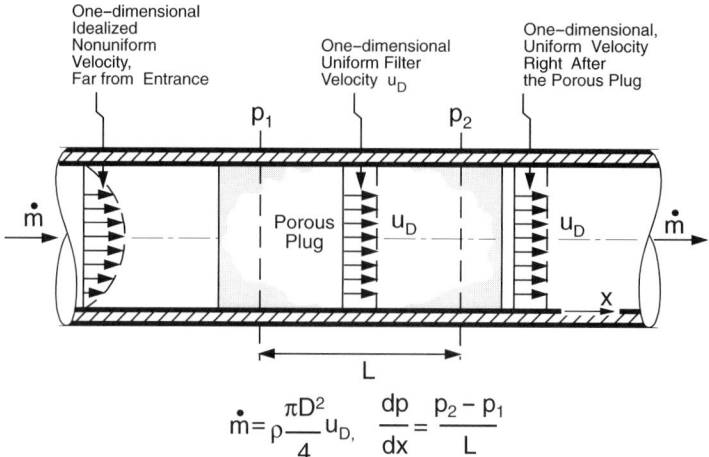

FIGURE 9.5   Determination of filter (or Darcy) velocity.

was steady, one-dimensional, and driven by gravity. A schematic of the flow is given in Fig. 9.5. The mass flow rate of the liquid $\dot{m}$ was measured and the filtration or filter velocity u_D was determined by dividing $\dot{m}$ by the product of the fluid density (assumed incompressible) and the cross-sectional area A of the channel (which was filled with the particles and then the liquid was flown through it). In applying a volumetric force balance to this flow, he discovered that the bulk resistance can be characterized by the viscosity of the Newtonian fluid (fluid parameter) μ and the *permeability* of the solid matrix (solid matrix parameter) K, such that

$$-\frac{dp}{dx} = \frac{\mu}{K} u_D \qquad (9.17)$$

where the dimension of K is in square of length, such as m². Interpretation of Eq. 9.17 has evolved and one unit still used is *darcy*.* One unit of darcy equals 9.87×10^{-13} m². The permeability accounts for the interstitial surface area, the fluid particle path as it flows through the matrix, and other related hydrodynamic characteristics of the matrix. The Darcy model has been examined rather extensively and is not closely followed for liquid flows at high velocities and for gas flows at very low pressures and very high velocities.

At low gas pressures and for small pore size, the mean free path of the gas molecules may be on the order of the pore size and therefore velocity slip occurs (Knudsen effect), resulting in higher permeabilities. However, an increase in the permeability due to an increase in gas pressure has been found in some experiments. Scheidegger [24] discusses the effect of the Knudsen slip, the internal surface roughness, surface absorption, capillary condensation, and molecular diffusion on the measured permeability. By examining these effects at the pore level, it becomes clear that the measured gas and liquid permeabilities can be noticeably different.

For *isotropic media* where the pressure gradient ∇p and the velocity vector $\mathbf{u}_D$ are parallel, Eq. 9.17 is generalized to

$$-\nabla p = \frac{\mu}{K} \mathbf{u}_D \qquad (9.18)$$

---

* One *darcy* is the permeability of a matrix when a cubic sample with each side having a width of 1 cm is used and a fluid with viscosity of 1 centipoise is flown (one-dimensional) through it, resulting in a pressure drop of 1 atmosphere (1.013×10^5 Pa).

For *anisotropic media,* in general, these two vectors are not parallel and a linear transformation can be made using the permeability tensor **K**.

$$-\nabla p = \frac{\mu}{\mathbf{K}}\,\mathbf{u}_D \tag{9.19}$$

A Semiheuristic Momentum Equation.   As much as it is desirable to have one set of governing equations that can describe both the momentum transport through the porous media (K being small) as well as that in the plain media (K being very large), such equations, if they become available, will be too complicated to be of practical use. However, as initiated by Brinkman [25] with his inclusion of a viscous shear stress term (other than the bulk viscous shear stress), which can take into account the shear stresses initiated at the surfaces bounding the porous media (macroscopic shear), attempts are being made to arrive at an equivalent of the Navier-Stokes equation for the description of flow through porous media.

In an effort to extend the Darcy law, the following momentum equation has been suggested by Brinkman and extended to the high-velocity regime for isotropic media and is

$$\frac{\rho_0}{\varepsilon}\left(\frac{\partial \mathbf{u}_D}{\partial t} + \mathbf{u}_D \cdot \nabla \mathbf{u}_D\right) = -\frac{\delta\langle p\rangle^f}{\delta x_i} + \frac{\mu'}{\varepsilon}\nabla^2 \mathbf{u}_D - \frac{\mu}{K}\mathbf{u}_D - \frac{F\varepsilon}{K^{1/2}}\rho\,|\mathbf{u}_D|\mathbf{u}_D \tag{9.20}$$

Note that the viscosity used in both the microscopic and macroscopic viscous terms is the fluid viscosity. Lundgren [26], in giving justification to the Brinkman equation, shows that for spherical particles $\mu' = \mu g(\varepsilon)$, where $0 < g(\varepsilon)$ and $g(1) = 1$ but others have used $\mu' > \mu$.

For the matrices obeying the modified Ergun relation we have

$$\frac{F\varepsilon}{K^{1/2}} = \frac{1.8(1-\varepsilon)}{\varepsilon^3}\frac{1}{d}, \qquad \text{where } K = \frac{1}{180}\frac{\varepsilon^3}{(1-\varepsilon)^2}d^2 \tag{9.21}$$

Some other semiheuristic momentum equations have been recommended, among these is

$$\frac{\rho_0}{\varepsilon}\left(\frac{\partial \mathbf{u}_D}{\partial t} + \mathbf{u}_D \cdot \nabla \mathbf{u}_D\right) = -\nabla p + \rho\mathbf{f}$$

(macroscopic inertial force or macroflow-development term)   (pore pressure gradient)   (body force)

$$+ \frac{\mu}{\varepsilon}\nabla^2\mathbf{u}_D - \frac{\mu}{K}\mathbf{u}_D - \frac{C_E}{K^{1/2}}\rho\,|\mathbf{u}_D|\mathbf{u}_D \tag{9.22}$$

(macroscopic or bulk viscous shear stress diffusion, also called Brinkman viscous term or bounding surface effect)   (microscopic viscous shear stress, Darcy term)   (microscopic intertial force, also called Ergun inertial term or microflow-development term)

where we have used the empirical in place of $F\varepsilon$ Ergun constant C_E.

This is a *semiheuristic volume-averaged* treatment of the flow field. The experimental observations of Dybbs and Edwards [27] show that the macroscopic viscous shear stress diffusion and the flow development (convection) are significant only over a length scale of ℓ from the vorticity generating boundary and the entrance boundary, respectively. However, Eq. 9.22 predicts these effects to be confined to distances of the order of $K^{1/2}$ and Ku_D/ν, respectively. We note that $K^{1/2}$ is smaller than d. Then Eq. 9.22 predicts a macroscopic boundary-layer thickness, which is not only smaller than the representative elementary volume ℓ when $\ell \gg d$, but even smaller than the particle size. However, Eq. 9.22 allows estimation of these macroscopic length scales and shows that for most practical cases, the Darcy law (or the Ergun extension) is sufficient.

Local Volume Averaging.   The principle of volume averaging and the requirement of existence of the local thermal equilibrium between the fluid and solid phases was discussed in the section entitled "Conduction Heat Transfer." In addition to the diffusion time and length scale requirements for the existence of the local thermal equilibrium, the residence timescales (the time it takes for a fluid particle to cover the length scales d, ℓ, and L) must be included in the length and timescale requirements.

Under the assumption of the local thermal equilibrium and by adding the solid- and fluid-phase averaged energy equations, we have

$$[\varepsilon(\rho c_p)_f + (1 - \varepsilon)(\rho c_p)_s]\,\frac{\partial\langle T\rangle}{\partial t} + (\rho c_p)_f \mathbf{u}_D \cdot \nabla\langle T\rangle = (\rho c_p)_f \nabla \cdot (\mathbf{D} \cdot \nabla\langle T\rangle) \qquad (9.23)$$

where $\langle T\rangle^s = \langle T\rangle^f = \langle T\rangle$, and the total diffusivity tensor is

$$\mathbf{D} = \frac{\mathbf{K}_e}{(\rho c_p)_f} + \varepsilon\mathbf{D}^d \qquad (9.24)$$

Here $\mathbf{D}^d$ is the *dispersion tensor.*

As before, a knowledge of $\mathbf{b}$, the transformation vector, leads to the determination of $\mathbf{K}_e$ and $\mathbf{D}^d$. The vector $\mathbf{b}$, which is a function of position only, has a magnitude of the order of the representative elementary volume ℓ and is determined from the differential equation and boundary conditions for T' and its determination is discussed by Kaviany [9].

Correlation for Dispersion Tensor.   Given here are some of the properties of the dispersion tensor.

- $\mathbf{D}$ is a *second-rank* tensor.
- $\mathbf{D}$ is *positive-definite.*
- $\mathbf{D}$ is *symmetrical* for random structures, that is, $D_{ij}^d = D_{ji}^d$.
- The *off-diagonal* elements of $\mathbf{D}$ are zero for isotropic media, that is, $D_{ij}^d = 0$ for $i \neq j$.
- $\mathbf{D}$ is *invariant* to the origin of the coordinate system.
- For *isotropic* media, the *longitudinal* dispersion is given as $D_{\parallel}^d = (\mathbf{u}_D/u_D) \cdot \mathbf{D}^d \cdot (\mathbf{u}_D/u_D)$.
- For *isotropic* media, there is a transverse isotropy in $\mathbf{D}^d$, such that the *transverse* dispersion for $\mathbf{n}$ perpendicular to $\mathbf{u}_D$ is $D_{\perp} = \mathbf{n} \cdot \mathbf{D}^d \cdot \mathbf{n}$.
- The last two can be combined to give $\mathbf{D}^d = \mathbf{nn}D_{\parallel}^d + (\mathbf{I} - \mathbf{nn})D_{\perp}^d$.
- For *periodic* structures, $D_{\perp}^d$ is small (but its magnitude depends on the direction of flow with respect to the symmetry axis of the unit cell).
- It is expected that

$$\mathbf{D}^d = \mathbf{D}^d \left(\text{structure}, \varepsilon, \frac{(\rho c_p)_s}{(\rho c_p)_f}, \frac{k_s}{k_f}, \text{Pr}, \text{Re}_d \right) \qquad (9.25)$$

The high Pe prediction of Koch and Brady [28] is chosen to apply at Pe ≥ 1. The values for $\mathbf{D}^d$ (which is *isotropic*) are those given in Table 9.1 and compared with the experimental result in Fig. 9.6. The hydrodynamic treatment of Koch and Brady involves the leading order analysis of the Stokes (creeping) flow. Therefore, no Reynolds number effects are included. With this in mind, we should interpret their large Pe predictions as being for large Pr (or Sc), not for large Re. This low Re and high Pe restriction is not satisfied in all the experimental data given by Fried and Combarnous [29].

The agreement between the predictions and the experiments is good except for Pe ≤ 20. The prediction of Jeffrey [30] for $\mathbf{D}^d$ does not agree with the experimental results. Note that the improvement over the Maxwell prediction, that is, the Jeffrey results for $\mathbf{K}_e/(\rho c_p)_f = \varepsilon\alpha_f\mathbf{I} + \mathbf{D}^d$, is for *dilute* concentration, that is, is valid only for $\varepsilon \to 1$.

TABLE 9.1   Leading Order Behavior of the Dispersion Coefficient for Various Peclet Number Regimes [28]

$D_\parallel^d/\alpha_f$	$D_\perp^d/\alpha_f$
(a) $\dfrac{3(k_s/k_f-1)}{k_s/k_f+2}(1-\varepsilon)+\dfrac{2^{1/2}}{15}\dfrac{\text{Pe}^2}{(1-\varepsilon)^{1/2}}$	(a) $\dfrac{3(k_s/k_f-1)}{k_s/k_f}(1-\varepsilon)+\dfrac{2^{1/2}}{60}\dfrac{\text{Pe}^2}{(1-\varepsilon)^{1/2}}$
(b) $\left[\dfrac{3}{4}\dfrac{\text{Pe }K^{1/2}}{R^{1/2}}-2-\dfrac{3}{2}\dfrac{R^{1/2}}{\text{Pe }K^{1/2}}+\dfrac{3R}{\text{Pe}^2 K}\right.$ $\left.+\left(\dfrac{3R^{1/2}}{\text{Pe }K^{1/2}}-\dfrac{3R^{3/2}}{\text{Pe}^3 K^{3/2}}\right)\times\ln\left(\dfrac{\text{Pe }K^{1/2}}{R^{1/2}}+1\right)\right]\left(\dfrac{R}{K}\right)^{1/2}$	(b) $\left[\dfrac{1}{4}+\dfrac{3}{4}\dfrac{R^{1/2}}{\text{Pe }K^{1/2}}-\dfrac{3}{4}\dfrac{R}{\text{Pe}^2 K}\right.$ $\left.+\left(\dfrac{3}{2}\dfrac{R^{3/2}}{\text{Pe}^3 K^{3/2}}-\dfrac{3}{4}\dfrac{R^{1/2}}{\text{Pe }K^{1/2}}\right)\times\ln\left(\dfrac{\text{Pe }K^{1/2}}{R^{1/2}}+1\right)\right]\left(\dfrac{R}{K}\right)^{1/2}$
(c) $\tfrac{3}{4}\text{Pe}$	(c) $\dfrac{3(2)^{1/2}}{8}(1-\varepsilon)^{1/2}$
(d) $\tfrac{3}{4}\text{Pe}+\tfrac{1}{6}\pi^2(1-\varepsilon)\text{ Pe ln Pe}+\dfrac{C}{15}(1+\gamma)^2\dfrac{\alpha_f(1-\varepsilon)\text{ Pe}^2}{\alpha_s}$	(d) $\dfrac{63(2)^{1/2}}{320}(1-\varepsilon)^{1/2}\text{ Pe}$

Note: Pe $=u_D R/\alpha_f$, C, and γ are defined in text.
(a) For Pe $\ll (1-\varepsilon)^{1/2}\ll 1$, Pe $K^{1/2}/R^{1/2}\ll 1$; (b) For $(1-\varepsilon)^{3/4}\ll$ Pe $\ll 1$; (c) For $(1-\varepsilon)^{1/2}\ll$ Pe $\ll 1$, Pe $K^{1/2}/R^{1/2}\gg 1$; (d) For Pe $\gg 1$, the last term in $D_\parallel^d/\alpha_f$ is for mass transfer only.

Also note that up to Pe = 1, the effective molecular diffusion dominates. Therefore, the very low Pe results given in Table 9.1 do not make any significant contribution to the total diffusivity. With this in mind, and for Pr characteristics of gases (<1), the significant dispersion contribution will be associated with Re > 1. This, in principle, requires inclusion of the inertial term in the momentum equation. However, as is shown later, the experimental results indicate that even for the inertial regime, Pe remains the only parameter on which $\mathbf{D}^d$ depends.

Tables 9.2 and 9.3 list some of the closed-form solutions and correlations for the longitudinal and transverse dispersion coefficients for ordered and disordered media.

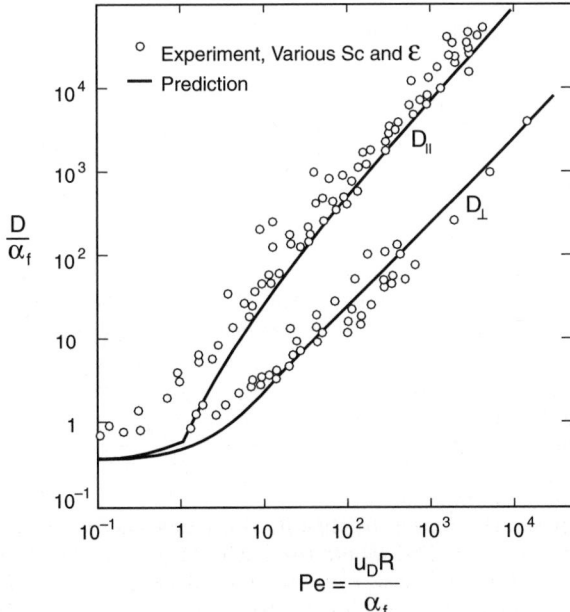

FIGURE 9.6   The prediction of Koch and Brady (for random arrangement of spheres) [28] for longitudinal and transverse total effective thermal diffusivity, compared with the experimental results of many investigators as presented in Fried and Combarnous [29].

TABLE 9.2 Some Closed-Form Solutions for Total Diffusivity D and Dispersion Coefficient D^d

Constraints	$D_{\parallel}/\alpha_f$	$D_{\perp}/\alpha_f$
(a) Re $\gg$ 1, disordered media	$\dfrac{D_{\parallel}}{\alpha_f} = \dfrac{\mathrm{Pe}}{2}$ $\;(\mathrm{Pe} = 2u_D R/\alpha_f)$	$\dfrac{D_{\perp}}{\alpha_f} = \dfrac{\mathrm{Pe}}{11}$
(b) Disordered media	$\dfrac{D^d_{\parallel}}{\alpha_f} = \left(\lambda + \dfrac{3}{4} - 0.577\right)\dfrac{\mathrm{Pe}}{6}$ $(\mathrm{Pe} = u_D\ell/\varepsilon\alpha_f;\ \lambda$ depends on the bed length$)$	$\dfrac{D^d_{\perp}}{\alpha_f} = \dfrac{3\mathrm{Pe}}{16}$
(c) Stokes flow, $\ell/R = 5$, disordered media	$\dfrac{D^d_{\parallel}}{\alpha_f} = \dfrac{\mathrm{Pe}}{6}\left(\ln 1.22\,\mathrm{Pe} - \dfrac{17}{12} - \dfrac{1}{200}\,\mathrm{Pe}\right)$ $\dfrac{D^d_{\parallel}}{\alpha_f} = \dfrac{\mathrm{Pe}^2}{15},\qquad \mathrm{Pe} \ll 1$ $(\mathrm{Pe} = u_D\ell/\varepsilon\alpha_f;\ \ell$ is the average channel length$)$	$\dfrac{D^d_{\perp}}{\alpha_f} = \dfrac{3}{16}\,\mathrm{Pe} + \dfrac{1}{1000}\,\mathrm{Pe}^2$ $\dfrac{D^d_{\perp}}{\alpha_f} = \dfrac{1}{40}\,\mathrm{Pe}^2,\qquad \mathrm{Pe} \ll 1$
(d) $k_f \to \infty$ (uniform T in fluid, but not in solid), disordered media	$\dfrac{D^d_{\parallel}}{\alpha_f} = \dfrac{C^2\varepsilon^2(1-\varepsilon)}{[\varepsilon + C(1-\varepsilon)]^3} \times \left[\dfrac{\mathrm{Pe}^2}{15C}\dfrac{\alpha_f}{\alpha_s} + \dfrac{\mathrm{Pe}}{3}\dfrac{u_D(\rho c_p)_f}{\varepsilon h_{sf}}\right]$ $[\mathrm{Pe} = u_D R/(\varepsilon\alpha_f)],\ C = (\rho c_p)_s/(\rho c_p)_f,\ h_{sf}$ is the interstitial heat transfer coefficient$)$	
(e) Table 9.1, Stokes flow, disordered media	Table 9.1	Table 9.1

Note: (a) Aris and Amundson [31]; (b) De Josselin De Jong [32]; (c) Saffman [33]; (d) Horn [34]; (e) Koch and Brady [28].

The nonuniformities in the phase distributions at and near the bounding surface and their effects on the fluid flow and heat transfer are most significant if the primary heat transfer is through these surfaces. In analyzing the variation of the dispersion coefficient at and near these surfaces, the following should be considered.

- Both $D^d_{\perp}$ and $D^d_{\parallel}$ are expected to vanish at the bounding solid surface.

- Near the bounding surface (fluid or solid), the porosity distribution is nonuniform.

- For packed beds of spheres, the porosity is generally larger near the bounding solid surface, resulting in an increase in the local velocity (component parallel to the surface) and the local Reynolds number.

- Since the Darcean velocity is taken to be uniform, it does not allow for the local variations in $D_{\parallel}(\mathrm{Pe})$ and therefore does not lead to vanishing $D_{\parallel}$ and $D_{\perp}$ on the solid surface. Inclusion of the porosity variation near the surface results in an increase in $D_{\perp}(\mathrm{Pe})$ and $D_{\parallel}(\mathrm{Pe})$. Inclusion of the macroscopic shear stress (Brinkman) term $\mu'\nabla^2\mathbf{u}$ insures that $D_{\parallel} = D_{\perp} = 0$ at the surface. But since the Brinkman screening length is much smaller than the pore size, this inclusion does not account for the actual effect of the pore-level hydrodynamics on the local variation of $D_{\parallel}$ and $D_{\perp}$.

- Inclusion of the pore-level (or particle-based) hydrodynamics along with the appropriate volume averaging allows for the inclusion of the local variation of $D_{\perp}$ and $D_{\parallel}$ into the energy equations. In principle, these variations can only be included if the change from the bulk value to zero at the surface takes place over several representative elementary volumes. Otherwise it will not be in accord with the volume averaging.

- This leaves the rigorous continuum treatment of the near-surface hydrodynamics heat transfer to be rather impossible. The alternative has been the use of various area averages introduced into the volume-averaged equations. These area-averaging-continuum descrip-

TABLE 9.3 Various Correlations for Total Diffusivity Coefficient D

Constraints	$D_\parallel/\alpha_f$	$D_\perp/\alpha_f$
(a) Disordered media		$D_\perp/\alpha_f = \mathrm{Pe}/(5 \sim 15)$ $(\mathrm{Pe} = u_D d/\alpha_f)$
(b) Disordered media	$D_\parallel/\alpha_f = k_s/k_f + (0.7 \sim 0.8)\,\mathrm{Pe}$ $(\mathrm{Pe} = u_D d/\alpha_f)$	$D_\perp/\alpha_f = k_e/k_f + (0.1 \sim 0.3)\,\mathrm{Pe}$ $(\mathrm{Pe} = u_D d/\alpha_f)$
(c) Re < 50 Disordered media	$D_\parallel/\alpha_f = 0.73 + \dfrac{0.5\mathrm{Pe}}{1 + (0.97/\mathrm{Pe})}$ $[\mathrm{Pe} = u_D d/(\varepsilon\alpha_f)]$	
(d) Disordered media	$D_\parallel/\alpha_f = k_e/k_f + 0.8\mathrm{Pe}$ $(\mathrm{Pe} = u_D d/\alpha_f)$	
(e) Disordered media	$D_\parallel/\alpha_f = k_e/k_f + 0.5\mathrm{Pe}^{(1 \sim 1.2)}$ for $\mathrm{Pe} \leq 10^4$; $D_\parallel/\alpha_f = 1.8\mathrm{Pe}$ for $\mathrm{Pe} \geq 10^4$ $(\mathrm{Pe} = u_D d/\alpha_f)$	
(f) Two-dimensional ordered media	$D_\parallel/\alpha_f = 1 + 0.128(\mathrm{Pe}_\ell - 1)$ for $1 < \mathrm{Pe}_\ell < 10$, $\varepsilon = 0.8$ $D_\parallel/\alpha_f = 1 + 0.071(\mathrm{Pe}_\ell - 1)$ for $1 < \mathrm{Pe}_\ell < 10$, $\varepsilon = 0.9$ $D_\parallel/\alpha_f = 0.019\mathrm{Pe}_\ell^{1.82}$ for $10 \leq \mathrm{Pe}_\ell \leq 10^3$, $\varepsilon = 0.8$ $D_\parallel/\alpha_f = 0.009\mathrm{Pe}_\ell^{1.86}$ for $10 \leq \mathrm{Pe}_\ell \leq 10^3$, $\varepsilon = 0.9$	

(a) Baron [35]; (b) Yagi et al. [36], also Schertz and Bischoff [37]; (c) Edwards and Richardson [38]; (d) Vortmeyer [39]; (e) Bear [40]; (f) Sahraoui and Kaviany [41].

tions do not describe the flow and heat transfer accurately, but contain a few adjustable constants, which enable them to match specific experimental data.

- An alternative is the direct simulation of the flow and heat transfer at and near the bounding surfaces. Because of the computational limitations, only simple periodic structures can be analyzed.

Radiation Heat Transfer

We first discuss the equation of radiative transfer and the effective radiative properties. The spectral radiative behavior of a single isolated particle (i.e., independent scattering) is considered. Then the limits, based on large and small particle sizes, are discussed and various predictions are compared. Next the ranges of validity of the independent and dependent scattering are discussed, along with variative properties. Two simplified methods, the geometric layered and the effective radiant conductivity, are also discussed.

- In the presence of a solid matrix, the absorption/emission/scattering from gases is generally masked by the absorption/emission/scattering from the matrix [42]. If the gas contribution is significant, it will be included as if independent of the matrix contributions, that is, the spectral or band contributions will be superimposed,

such as
$$\langle \sigma_{\lambda a} \rangle = \langle (\sigma_{\lambda a})_s \rangle + \langle (\sigma_{\lambda a})_g \rangle \qquad (9.26)$$

This relation is strictly true only if the assumption of independent scattering holds.

- For liquids, the temperature difference existing in the system is generally small and the liquids are highly absorbing in the infrared wavelength range. Therefore, in general, no radiation heat transfer is considered in dealing with fully or partially liquid-saturated matrices. However, when it is necessary to include radiation heat transfer with liquids, the relative index of refraction n used for the properties is n_s/n_f (i.e., for gases $n_f \approx 1$, but *not* for liquids).

- Because the source of thermal radiation energy is generally nonpolarized (gas or solid surface) and because the waves undergo substantial reflections in the interstices of the matrix, the thermal radiation is conceived as being nonpolarized.

Equation of Radiative Transfer and Effective Radiative Properties.   The fundamentals of radiation heat transfer in absorbing/emitting/scattering media have been given by Chandrasekhar [43], Ozisik [44], Siegel and Howell [45, 46], Brewster [47], and Modest [48]. Some of the principles are briefly given herein. Their approach treats the solid-fluid phases as a single continuum. Therefore, the following applies to *heterogeneous* (solid and fluid phases are present simultaneously) differential elements. A schematic showing the coordinate system for a *plane-parallel* geometry (which is the geometry used through most of this chapter) is given in Fig. 9.7. The unit vector in the beam direction is given by *s* and the length of the position vector is given by *S*. The incident beam is shown with subscript *i* and the incident solid angle is shown by $d\Omega_i$.

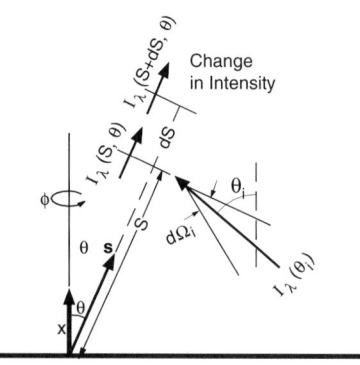

Change in Intensity

FIGURE 9.7   A schematic of the coordinate system.

- It is assumed that the particle size is much smaller than the linear size of the system. Then, the radiative properties are averaged over a representative elementary volume with a linear dimension ℓ, such that $d \ll \ell \ll L$.

- The matrix-fluid system is treated as a continuum by assuming that the local thermal equilibrium (as discussed in the section on continuum treatment) exists in accord with the treatment of conduction and convection.

- Azimuthal symmetry is assumed so that $I_\lambda(\theta, \phi) = I_\lambda(\theta)$.

- The spectral (indicated by subscript λ) radiation intensity I_λ is the radiation energy in the direction θ per unit time, per unit projected area, per unit solid angle, and per interval $d\lambda$ around λ. Then I_λ is given in W/(m²-sr-μm).

- The solid angle differential is $d\Omega = \sin\theta\, d\phi\, d\theta$, when θ is the polar angle and ϕ is the azimuthal angle. Since an azimuthal symmetry is assumed in the following, the differential solid angle is taken as $d\Omega = 2\pi \sin\theta\, d\theta = -2\pi\, d\cos\theta$.

- The absorbed energy is $dI_\lambda = -\sigma_{\lambda a} I_\lambda\, dS$, where again S is the path length and $\sigma_{\lambda a}$ is the absorption coefficient (1/m). Figure 9.7 shows dS and $dI_\lambda = I_\lambda(S + dS, \theta) - I_\lambda(S, \theta)$.

- The equation of radiative transfer for radiation in a direction θ becomes

$$\frac{\partial I_\lambda(S)}{\partial S} = -\langle \sigma_{\lambda a} \rangle I_\lambda(S) + \langle \sigma_{\lambda a} \rangle I_{\lambda b}[T(S)] - \langle \sigma_{\lambda s} \rangle I_\lambda(S) + \frac{\langle \sigma_{\lambda s} \rangle}{2} \int_{-1}^{1} I_\lambda(S, \theta_i) \langle \Phi_\lambda \rangle (\theta_i \rightarrow \theta)\, d\cos\theta_i$$

(9.27)

The arrow indicates from the incident to the scattered direction and θ_o is the angle between the incident (θ_i) and scattered (θ) beam. When rewritten, we have

$$\frac{\partial I_\lambda}{\partial S} = \langle\sigma_{\lambda a}\rangle I_{\lambda b} - (\langle\sigma_{\lambda a}\rangle + \langle\sigma_{\lambda s}\rangle)I_\lambda + \frac{\langle\sigma_{\lambda s}\rangle}{2} \int_{-1}^{1} I_\lambda \langle\Phi_\lambda\rangle(\mu_i \to \mu)\, d\mu_i \tag{9.28}$$

where we have used $\mu = \cos\theta$.

- The spectral radiative heat flux in the direction normal to the parallel slab faces is found from the directional spectral intensity $I_\lambda(\theta)$ by noting that I_λ is per unit projected area ($dA \cos\theta$) and is in the θ direction. Then the contribution from all directions to the normal heat flux is

$$q_{\lambda r} = 2\pi \int_{-1}^{1} I_\lambda \cos\theta\, d\cos\theta \tag{9.29}$$

where $I_\lambda \cos\theta$ is the spectral directional emissive power. In vectorial form, we have

$$\mathbf{q}_{\lambda r} = 2\pi \int_{-1}^{1} \mathbf{s} I_\lambda(S, \mathbf{s})\, d\mu \tag{9.30}$$

- The divergence of the total radiative heat flux, which is used in the energy equation, is found from the radiative transfer equation by its integration over $\int_{4\pi} d\Omega$ and $\int_0^\infty d\lambda$. This gives

$$\int_0^\infty \int_{4\pi} \nabla \cdot \mathbf{s} I_\lambda\, d\Omega\, d\lambda = \nabla \cdot \mathbf{q}_r = 4\pi \int_0^\infty \langle\sigma_{\lambda a}\rangle I_{\lambda b}[T(S)]\, d\lambda - 2\pi \int_0^\infty \langle\sigma_{\lambda a}\rangle \int_{-1}^{1} I_\lambda(S, \theta)\, d\cos\theta\, d\lambda \tag{9.31}$$

or

$$\nabla \cdot \mathbf{q}_r = 4\pi \int_0^\infty \langle\sigma_{\lambda a}\rangle I_{\lambda b}(S)\, d\lambda - 2\pi \int_0^\infty \langle\sigma_{\lambda a}\rangle \int_{-1}^{1} I_\lambda(S, \theta)\, d\mu\, d\lambda \tag{9.32}$$

When no other mode of heat transfer is present and the emitted and absorbed energy are equal, then $\nabla \cdot \mathbf{q}_r = 0$, and the state of radiative equilibrium exists.

Radiative Properties of a Single Particle.   In this section, we treat scattering from a single (i.e., single scatterers) particle. Then, we relate the radiation properties obtained for the individual particles to that for the collection of such elements in the representative elementary volume. Figure 9.8 gives a classification for single-particle scattering, where variation in optical properties, size, shape, and the incident radiation are considered. In terms of the theoretical treatments, the most important distinction, which also leads to significant variation in the rigor of treatment, is that based on size of the spherical particles. For large size parameter $\alpha_R = \pi d/\lambda$ geometric optics can be used. The applicability of various theories, i.e., the Rayleigh, Mie, and geometric optic theories, is discussed. In approaching the theoretical treatment of scattering from particles, we consider cases where the following simplifications can apply: (1) constant optical properties n_s and κ_s within the scatterers, where n_s and κ_s are the solid index of refraction and extinction, respectively, and, (2) smooth scattering surface. The optical properties for solids are strongly wavelength-dependent. We examine this spectral behavior before considering the incidence of electromagnetic waves upon an isolated solid particle.

Wavelength Dependence of Optical Properties.   The relationship between the optical properties n and κ and the other molecular-crystalline properties of the solid are discussed by Siegel and Howell [45]. The theoretical treatments of the prediction of these properties are also discussed by them. Here, we examine some of the limited experimental data on $n(\lambda)$ and $\kappa(\lambda)$ for solids.

Figure 9.9 shows the measured wavelength-dependence of n_s for materials used in visible and infrared optics. We have included these as examples (and because their spectral behavior is studied most extensively) and not because of their common use in heat transfer in porous

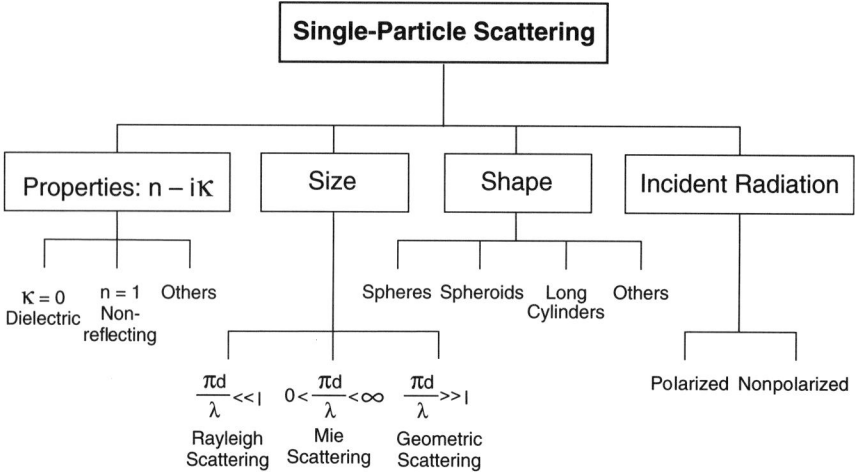

FIGURE 9.8   Parameters influencing scattering from a single particle.

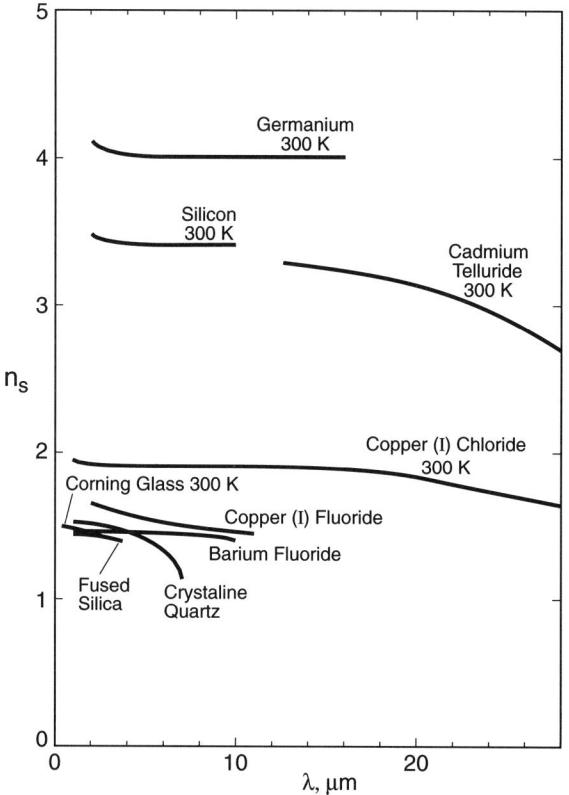

FIGURE 9.9   Variation of index of refraction of some solids used in optics (visible and infrared), with respect to wavelength. This variation is generally negligible for these materials.

media. For these optical materials, the wavelength-dependence is not very strong (in as far as the radiative heat transfer is concerned), except for cadmium, telluride, and crystalline quartz. Because of their near-room-temperature applications, the measured values are for 300 K.

Figure 9.10a and b give $n_s(\lambda)$ and $\kappa_s(\lambda)$ for nonmetallic solids. Again most of the data is for near room temperature. For $n_s(\lambda)$, other than Ge and Si, the other materials shown exhibit strong wavelength-dependence (Fig. 9.10a). The variations in the region $\lambda > 10$ μm, corresponding to below-room-temperature applications, are in general as significant as they are for $\lambda < 10$ μm. The high-temperature applications require data for $\lambda \simeq 1$ μm. Note the variations in $\kappa_s(\lambda)$ shown in Fig. 9.10b. For large particles, the values of κ_s as small as 10^{-5} can result in significant absorption because the attenuation is a function of the product $\kappa_s \alpha_R$. Therefore, reliable data for $\kappa_s(\lambda)$ are very important in determining the transmission through beds. The wavelength-dependence of κ_s is very strong and should be included in the radiative heat transfer analysis.

Figure 9.11a and b give $n_s(\lambda)$ and $\kappa_s(\lambda)$ for metallic solids. Note that n_s does not increase monotonically with λ (through the spectrum shown) for all the materials in Fig. 9.11a, even though Al, Cu, Ag, and Au do show monotonic increase of κ_s with λ. Therefore, for high-temperature applications where $\lambda = O(1$ μm$)$, extreme care should be used in order to properly account for variation of n_s with λ. Figure 9.11b shows that except for titanium below

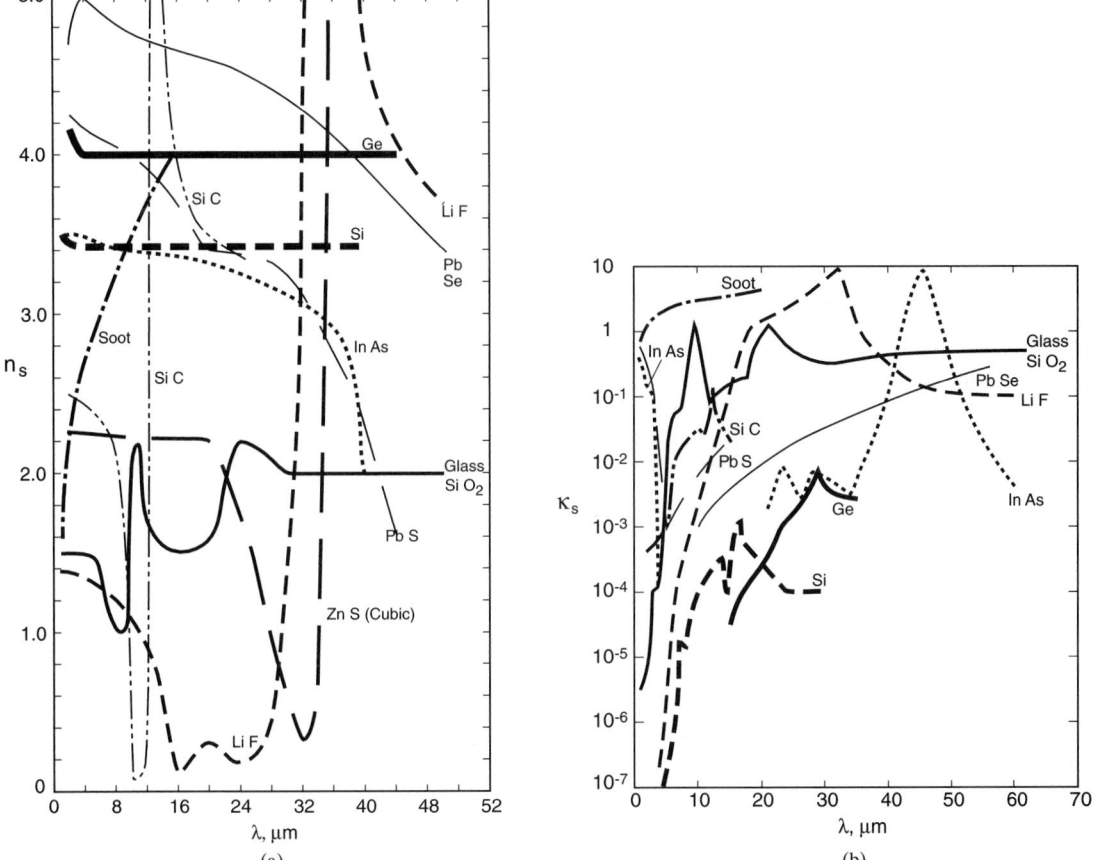

FIGURE 9.10 (a) Variation of index of refraction for some nonmetallic solids, with respect to the wavelength; (b) variation of index of extinction for some nonmetallic solids, with respect to the wavelength.

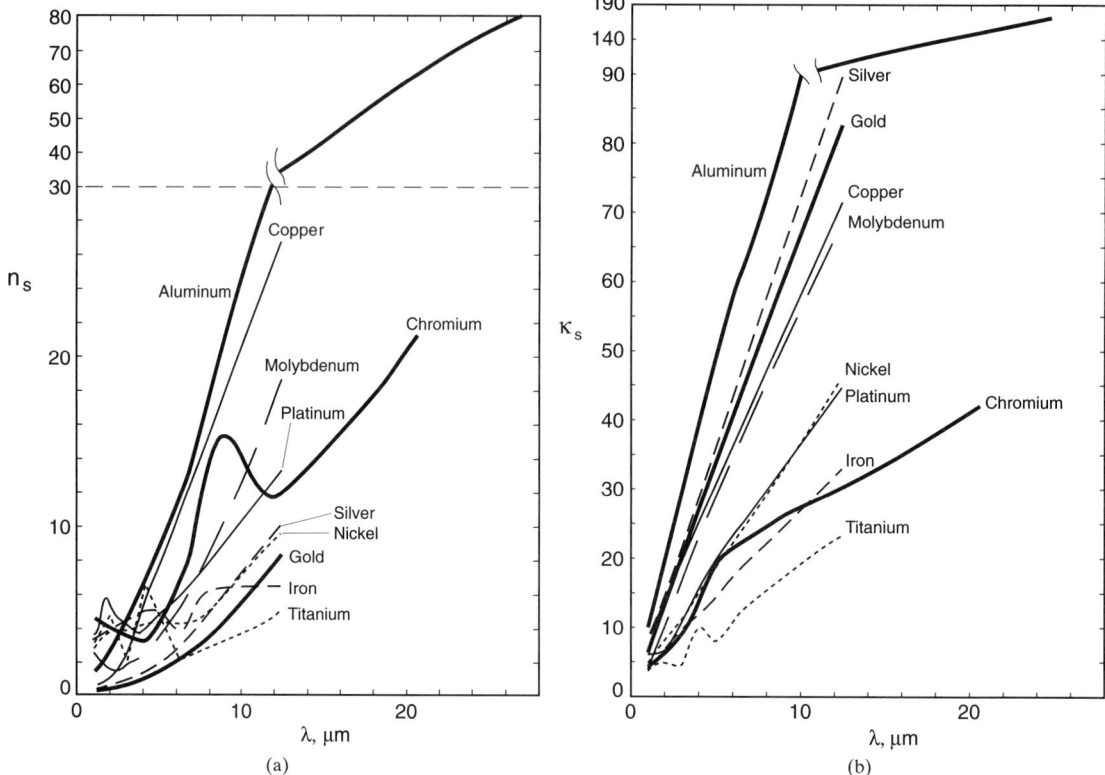

FIGURE 9.11   (*a*) Variation of index of refraction for some metallic solids, with respect to the wavelength; (*b*) variation of index of extinction for some metallic solids, with respect to the wavelength.

$\lambda = 5$ μm, κ_s increases monotonically with λ for metals. Extensive documentation of the optical properties of metals can be found in Weaver [49].

As was mentioned, when the fluid phase is a liquid, we do not expect the radiation heat transfer to be significant, although there are exceptions for some high-temperature applications. However, in Table 9.5, the index of refraction of some liquids is given for the sake of completeness. In treatment of scattering from particles, the *relative index of refraction,* that is, n_s/n_f, is the significant parameter, and as expected when dealing with liquids, this ratio can be substantially different from n_s.

Comparison of Predictions.   We expect the Mie theory to be applicable for all values of n, κ, and size parameter α_R. The Rayleigh theory is applicable for small α_R and small values of $|m\alpha_R|$, $m = n - i\kappa$, and the geometric treatment is expected to be valid for $\alpha_R \gg 1$. Here, we consider spherical particles only. The optical properties are expected to be wavelength-dependent. Van de Hulst [50] gives the classifications for the case of $\kappa_s = 0$. His results are plotted in a diagram named after him and this diagram is shown in Fig. 9.12. He gives the asymptotic relations for the *extinction efficiency* η_{ex} so that the necessity of carrying out the full Mie solution can be avoided. The extinction efficiency η_{ex} is the sum of scattering and absorption efficiencies.

Some of these asymptotes are shown in the van de Hulst diagram. However, because of faster computers and improved subroutines, carrying out a full Mie solution is no longer as prohibitive a task as it once was. The problem lies more in making practical use of it, because no

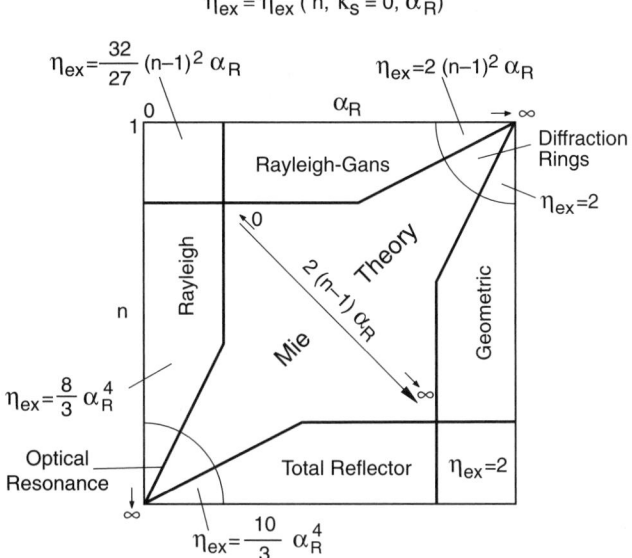

$$\eta_{ex} = \eta_{ex}(n, \kappa_s = 0, \alpha_R)$$

FIGURE 9.12 The α_R–n plane (van de Hulst diagram) showing the various asymptotes for prediction of the extinction efficiency based on the Mie theory. The results are for $\kappa_s = 0$.

method of solution can handle the sharp forward peak produced for large particles. Thus, this peak has to be truncated for geometric size particles and the phase function renormalized to ensure energy conservation. The computation involved increases with increasing α_R. However, for very large values of α_R, the theory of geometric scattering provides a convenient alternative.

For small particles, the Rayleigh theory can be used. Although this does not result in a substantial savings in computation over the Mie theory, it provides a closed-form solution. Here, we compute the *scattering efficiency* $\eta_{\lambda s}(\lambda)$, the *absorption efficiency* $\eta_{\lambda a}(\lambda)$, the *asymmetry parameter* $g_\lambda(\lambda)$, and the *phase function* $\Phi_\lambda(\lambda)$ for a 0.2-mm sphere using the available experimental results for $n_s(\lambda)$ and $\kappa_s(\lambda)$ for glass. The results are shown in Fig. 9.13. The computations are based on the Rayleigh, Mie, and geometric treatments (see Fig. 9.13). Then comparisons among the results of these three theories show the limit of applicability of the Rayleigh and geometric treatments.

Effective Radiative Properties: Dependent and Independent.   The properties of an isolated single particle were discussed in the previous section. However, the equation of radiative transfer (Eq. 9.28) requires knowledge of the radiative properties of the medium, that is, $\langle\sigma_a\rangle$, $\langle\sigma_s\rangle$, and $\langle\Phi_\lambda\rangle$. The scattering and absorption are called *dependent* if the scattering and absorbing characteristics of a particle in a medium are influenced by neighboring particles and are called *independent* if the presence of neighboring particles has no effect on absorption and scattering by a single particle. The assumption of independent scattering greatly simplifies the task of obtaining the radiative properties of the medium. Also, many important applications lie in the independent regime; therefore, the independent theory and its limits will be examined in detail in this section.

In obtaining the properties of a packed bed, the independent theory assumes the following.

- *No interference* occurs between the scattered waves (far-field effects). This leads to a limit on the minimum value of C/λ, where C is the average interparticle clearance. However, most packed beds are made of large particles and can therefore be assumed to be above any such limit.

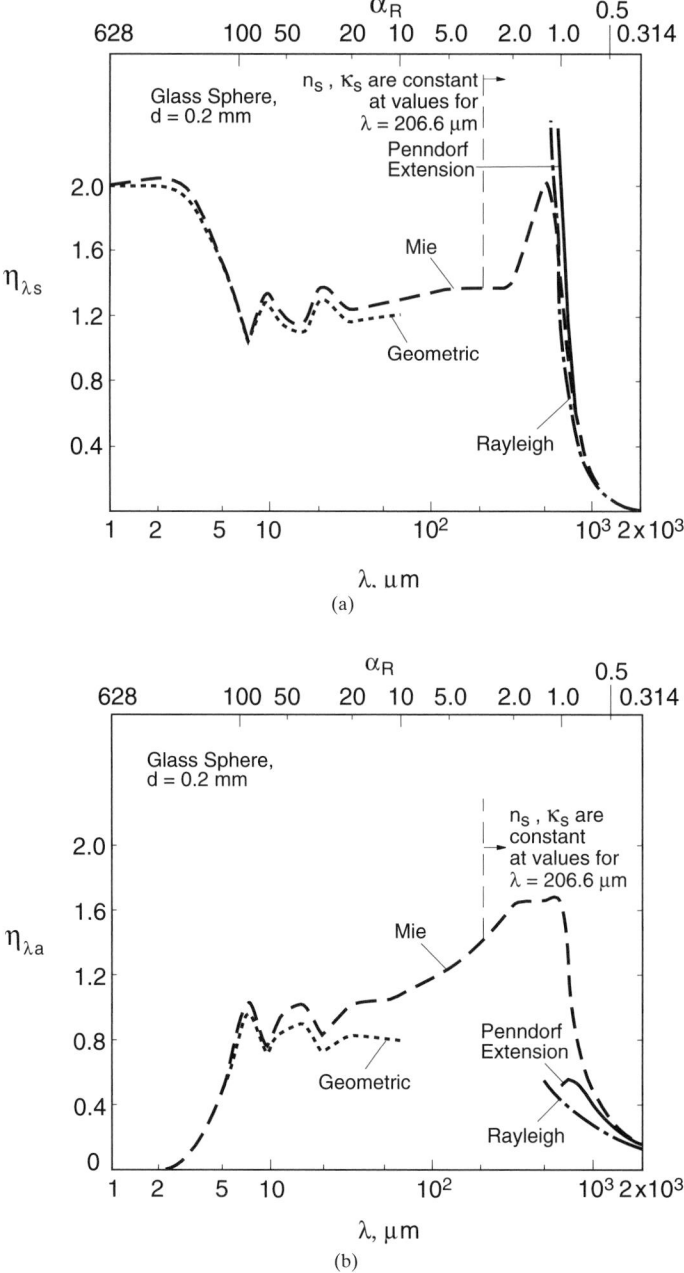

FIGURE 9.13 (*a*) Variation of the spectral scattering efficiency with respect to wavelength for a glass spherical particle of diameter 0.2 mm. When appropriate, the Rayleigh, Mie, and geometrical treatments are shown. Also shown is the Penndorf extension; (*b*) same as (*a*), except for the variation of the spectral absorption efficiency.

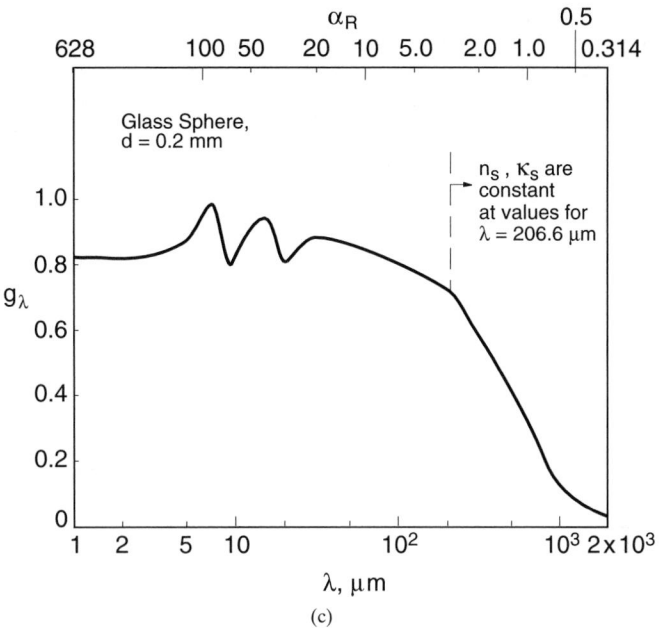

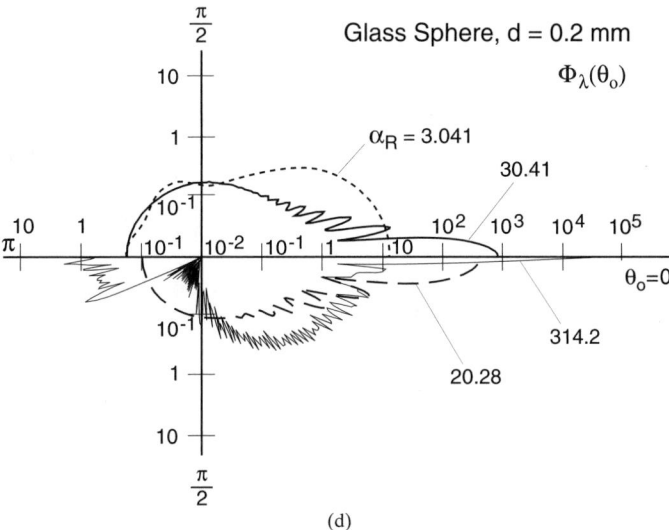

FIGURE 9.13 (*Continued*)   (*c*) same as (*a*), except for the variation of the spectral asymmetry factor; (*d*) same as (*a*), except for the distribution of the phase function.

- *Point scattering* occurs, that is, the distance between the particles is large compared to their size. Thus, a representative elementary volume containing many particles can be found in which there is no multiple scattering and each particle scatters as if it were alone. Then this small volume can be treated as a single scattering volume. This leads to a limit on the porosity.
- The *variation of intensity* across this elemental volume is not large.

Then the radiative properties of the particles can be averaged across this small volume by adding their scattering (absorbing) cross sections. The total scattering (absorbing) cross sections divided by this volume gives the scattering (absorbing) coefficient. The phase function of the single scattering volume is the same as that for a single particle.

Using the number of the scatterers per unit volume N_s (particles/m³) and assuming independent scattering from each scatterer, the spectral scattering coefficient for uniformly distributed monosize scatterers is defined as

$$\langle \sigma_{\lambda s} \rangle = N_s A_{\lambda s} \tag{9.33}$$

Similarly, $\langle \sigma_{\lambda a} \rangle = N_s A_{\lambda a}$ and $\langle \sigma_{\lambda ex} \rangle = \langle \sigma_{\lambda s} \rangle + \langle \sigma_{\lambda a} \rangle$. For spherical particles, the volume of each particle is $4\pi R^3/3$, and in terms of porosity ε, we have

$$\tfrac{4}{3}\pi N_s R^3 = 1 - \varepsilon \qquad \text{or} \qquad N_s = \frac{3}{4\pi} \frac{1-\varepsilon}{R^3} \tag{9.34}$$

Then we have
$$\langle \sigma_{\lambda s} \rangle = \frac{3}{4\pi} \frac{(1-\varepsilon)}{R^3} A_{\lambda s} \tag{9.35}$$

or
$$\langle \sigma_{\lambda s} \rangle = \frac{3}{4} \frac{(1-\varepsilon)}{R} \eta_{\lambda s} \tag{9.36}$$

When the particle diameter is not uniform, we can describe the distribution $N_s(R)\,dR$, that is, the number of particles with a radius between $R + dR$ per unit volume (number density). Note that $N_s(R)\,dR$ has a dimension of particles/m³. Then, assuming independent scattering, we can define the average spectral scattering coefficient as

$$\langle \sigma_{\lambda s} \rangle = \int_0^\infty \eta_{\lambda s}(R)\pi R^2 N_s(R)\,dR \tag{9.37}$$

A similar treatment is given to the absorption and scattering coefficients. The volumetric size distribution function satisfies

$$N_s = \int_0^\infty N_s(R)\,dR \tag{9.38}$$

where N_s is the average number of scatterers per unit volume.

Whenever the particles are placed close to each other, it is expected that they interact. One of these interactions is the *radiation* interaction, in particular, the extent to which the scattering and absorption of radiation by a particle is influenced by the presence of the neighboring particles. This influence is classified by two mechanisms: the coherent addition, which accounts for the phase difference of the superimposed far-field-scattered radiations and the disturbance of the internal field of the individual particle due to the presence of other particles [51]. These interactions among particles can in principle be determined from the Maxwell equations along with the particle arrangement and interfacial conditions. However, the complete solution is very difficult, and, therefore, approximate treatments, that is, modeling of the interactions, have been performed. This analysis leads to the prediction of the extent of interactions, that is, dependency of the scattering and absorption of individual particles on the presence of the other particles. One possible approach is to solve the problem of scattering by a collection of particles and attempt to obtain the radiative properties of the medium from it. However the collection cannot in general be assumed to be a single-scattering volume. For closely packed particles, even a small collection of particles is not a single-scattering volume. Thus, some sort of a regression method might be required to obtain the dependent properties of the medium. For Rayleigh scattering-absorption of dense concentration of small particles, the interaction has been analyzed by Ishimaru and Kuga [52], Cartigny et al. [53], and Drolen and Tien [54].

Hottel et al. [55] were among the first to examine the interparticle radiation interaction by measuring the bidirectional reflectance and transmittance of suspensions and comparing them with the predictions based on Mie theory, that is, by examining $(\eta_{\lambda ex})_{exp}/(\eta_{\lambda ex})_{Mie}$. They used visible radiation and a small concentration of small particles. An arbitrary criterion of 0.95 has been assigned. Therefore, if this ratio is less than 0.95, the scattering is considered dependent (because the interference of the surrounding particles is expected to redirect the scattered energy back to the forward direction).

Hottel et al. [55] identified the limits of independent scattering as $C/\lambda > 0.4$ and $C/d > 0.4$ (i.e., $\varepsilon > 0.73$). Brewster and Tien [56] and Brewster [57] also considered larger particles (maximum value of $\alpha_R = 74$). Their results indicated that no dependent effects occur as long as $C/\lambda > 0.3$, even for a close pack arrangement ($\varepsilon = 0.3$). It was suggested by Brewster [57] that the point-scattering assumption is only an artifice necessary in the derivation of the theory and is not crucial to its application or validity. Thereafter, the C/λ criteria for the applicability of the theory of independent scattering was verified by Yamada et al. [58] ($C/\lambda > 0.5$), and Drolen and Tien [54]. However, Ishimaru and Kuga [52] note dependent effects at much higher values of C/λ. In sum, these experiments seem to have developed confidence in application of the theory of independent scattering in packed beds consisting of large particles, where C/λ almost always has a value much larger than the mentioned limit of the theory of independent scattering. Thus, the approach of obtaining the radiative properties of the packed beds from the independent properties of an individual particle has been applied to packed beds without any regard to their porosity [54, 57]. However, as is shown later, all these experiments were similar in design and most of these experiments used suspensions of small transparent latex particles. Only in the Brewster experiment was a close packing of large semitransparent spheres considered.

Figure 9.14 shows a map of independent/dependent scattering for packed beds and suspensions of spherical particles [59]. The map is developed based on available experimental results. The experiments are from several investigators, and some of the experiments are reviewed later. The results show that for relatively high temperatures in most packed beds, the scattering of thermal radiation can be considered independent.

The rhombohedral lattice arrangement gives the maximum concentration for a given interparticle spacing. This is assumed in arriving at the relation between the average interparticle clearance C and the porosity. This relation is

$$\frac{C}{d} = \frac{0.905}{(1-\varepsilon)^{1/3}} - 1 \quad \text{or} \quad \frac{C}{\lambda} = \frac{\alpha_R}{\pi}\left[\frac{0.905}{(1-\varepsilon)^{1/3}} - 1\right] \tag{9.39}$$

where $C/\lambda > 0.5$ (some suggest 0.3) has been recommended for independent scattering (based on the experimental results). The total interparticle clearance should include the average distance from a point on the surface of one particle to the nearest point on the surface of the adjacent particle in a close pack. This average close-pack separation should be added to the interparticle clearance C obtained when the actual packing is referred to a rhombohedral packing ($\varepsilon = 0.26$). This separation can be represented by $a_1 d$ where a_1 is a constant ($a_1 \approx 0.1$). Therefore, we suggest that the condition for independent scattering be modified to

$$C + 0.1d > 0.5\lambda \tag{9.40}$$

where C is given earlier [60]. This is also plotted in Fig. 9.14. As expected for $\varepsilon \to 1$, this correction is small, while for $\varepsilon \to 0.26$, it becomes significant.

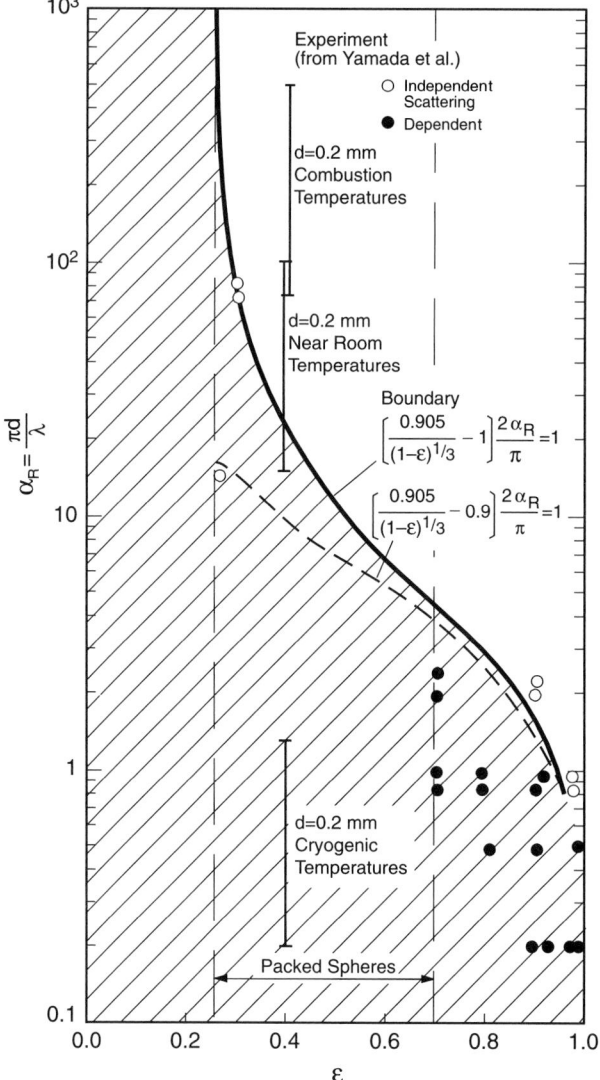

FIGURE 9.14 Experimental results for dependent versus independent scattering shown in the α_R-ε plane. Also shown are two empirical boundaries separating the two regimes.

In Fig. 9.13, the size parameters associated with a randomly packed bed of 0.2-mm-diameter spheres at very high (combustion), intermediate (room temperature), and very low (cryogenic) temperatures are also given. Note that based on Eq. 9.40 only the first temperature range falls into the dependent scattering regime (for $d = 0.2$ mm and $\varepsilon = 0.4$). Table 9.4 gives the range of temperatures, wavelengths, and size parameters for the 0.2-mm sphere considered.

Singh and Kaviany [61] examine dependent scattering in beds consisting of large particles (geometric range) by carrying out Monte Carlo simulations. They argue that the C/λ criterion

TABLE 9.4   Size Parameter $\alpha_R = \pi d/\lambda$ for a 0.2-mm-diameter Particle

λT (μm-K) $(F_{0-\lambda T})^*$	1,888 (0.05)	2,898 (0.25)	12,555 (0.95)
T (K) = 4	$\lambda = 472(\mu m)/(\pi d/\lambda) = 1.33$	724/0.867	3139/0.200
300	$6.29/10^2$	9.66/65	41.9/15
1500	$1.26/2.00 \times 10^2$	$1.93/3.25 \times 10^2$	8.37/75

$$* \, F_{0-\lambda T} = \frac{15}{\pi^4} \sum_{i=1}^{4} \frac{e^{-ix}}{i} \left(x^3 + \frac{3x^2}{i} + \frac{6x}{i^2} + \frac{6}{i^3} \right) \quad \text{for } x = \frac{14{,}388(\mu m\text{-}K)}{\lambda T}$$

only accounts for the far-field effects and that the porosity of the system is of critical importance if near-field effects are to be considered. According to the regime map shown in Fig. 9.14, a packed bed of large particles should lie in the independent range. This is because a very large diameter ensures a large value of C/λ even for small porosities. However, Singh and Kaviany [61] show dependent scattering for very large particles in systems with low porosity. The transmittance through packed beds of different porosities and at different values of τ_{ind} was calculated by the method of discrete ordinates using a 24-point gaussian quadrature.

They show that the independent theory gives good predictions for the bulk behavior of highly porous systems ($\varepsilon \geq 0.992$) for all cases considered. Two distinct dependent scattering effects were identified. The multiple scattering of the reflected rays increases the effective scattering and absorption cross sections of the particles. This results in a decrease in transmission through the bed. The transmission through a particle in a packed bed results in a decrease in the effective cross sections, resulting in an increase in the transmission through a bed. For opaque particles, only the multiple scattering effect is found, while for transparent and semitransparent particles, both of these effects are found and tend to oppose each other.

In conclusion, we note that both the C/λ criterion and the porosity criterion must be satisfied before the independent theory can be used with confidence.

Some Relations for Effective Radiative Properties from Independent Scattering. If the scatterers behave independently, a simple volume integration over the particle concentration distribution results. We now look back at the scattering property of spherical particles as predicted by Rayleigh, Mie, and geometric analyses. We assume independent scattering. Assuming that a continuum treatment of radiation in solid-fluid systems using independent scattering is possible, we use volume averaging over a representative elementary volume to average over the scatterers as in Eq. 9.37.

Tables 9.5 and 9.6 give some approximations for $\langle \sigma_{\lambda s} \rangle \langle \sigma_{\lambda a} \rangle$ and $\langle \Phi_\lambda \rangle (\theta_0)$ along with the applicable constraints. The wavelength λ is that for the wave traveling in the fluid, and if $n_f \neq 1$, then $\lambda = \lambda_0/n_f$, where λ_0 is for travel in vacuum. Table 9.6 shows the various approximations used to represent the phase function $\langle \Phi_\lambda \rangle (\theta_0)$ in terms of Legendre polynomials.

The scattering-absorption of incident beams by a long circular cylinder has also been studied by van de Hulst [50]. He also considers other particle shapes. Wang and Tien [62], Tong and Tien [63], and Tong et al. [64] consider fibers used in insulations. They use the efficiencies derived by van de Hulst [50] and examine the effects of κ_s and d on the overall performance of the insulations. The effect of fiber orientation on the scattering-phase function of the medium is discussed by Lee [65]. The effective radiative properties of a fiber-sphere composite is predicted by Lee et al. [66].

For small particles, a simplified approach to modeling the spectral scattering and absorption coefficient is given by Mengüç and Viskanta [71].

Approximate Geometric, Layered Model. Using geometric optics (radiation size parameter α_R, larger than about five) and the concept of view factor, the emission, transmission, and reflection of periodically arranged, diffuse, opaque particles has been modeled by Mazza et

TABLE 9.5   Volume Averaging of Radiative Properties: Independent Scattering

Constraints	$\langle\sigma_{\lambda s}\rangle(1/m), \langle\sigma_{\lambda a}\rangle(1/m), \langle\Phi_\lambda\rangle(\theta_0)$
(a) Large opaque specularly reflecting spherical particles $2\pi R/\lambda > 5$	$$\langle\sigma_{\lambda sr}\rangle = \pi\rho_\lambda \int_0^\infty R^2 N_s(R)\,dR$$ $N_s(R)\,dR$ is the number density of particles having radius between R and $R + dR$, ρ_λ is the hemispherical reflectivity $$\langle\sigma_{\lambda ar}\rangle = \pi(1 - \rho_\lambda) \int_0^\infty R^2 N_s(R)\,dR$$ $$\langle\Phi_{\lambda r}\rangle(\theta_0) = \frac{\rho_\lambda'}{\rho_\lambda}$$ $\rho_\lambda'[(\pi - \theta_i)/2]$ is the directional specular reflectivity for incident angle θ_i
(b) Large opaque diffusely reflecting spherical particles $2\pi R/\lambda > 5$	$$\langle\sigma_{\lambda sr}\rangle = \pi\rho_\lambda \int_0^\infty R^2 N_s(R)\,dR$$ $$\langle\sigma_{\lambda ar}\rangle = \pi(1 - \rho_\lambda) \int_0^\infty R^2 N_s(R)\,dR$$ $$\langle\Phi_{\lambda r}\rangle(\theta_0) = \frac{8}{3\pi}(\sin\theta_0 - \theta_0\cos\theta_0)$$
(c) Large spherical particles, diffraction contribution, $2\pi R/\lambda > 20$	$$\langle\sigma_{\lambda sd}\rangle = \pi \int_0^\infty R^2 N_s(R)\,dR$$ $$\langle\Phi_{\lambda d}\rangle(\theta_0) = \frac{4J_1^2[(2\pi R/\lambda)\sin\theta_0]}{\sin^2\theta_0}$$ J_1 is the Bessel function of first order and first kind, diffraction contribution
(d) Small spherical particles (extension of Rayleigh's scattering) limits are given in Ku and Felske [67] and Selamet and Arpaci [68]	$$\langle\sigma_{\lambda s}\rangle = \frac{128\pi^5}{3z_1^2\lambda^4}\{[(n^2+\kappa^2)^2 + n^2 - \kappa^2 - 2]^2 + 36n^2\kappa^2\}$$ $$\times\left\{\int_0^\infty R^6 N_s(R)\,dR + \frac{24\pi^2}{5z_1\lambda^2}[(n^2+\kappa^2)^2 - 9]\right.$$ $$\times \int_0^R R^8 N_s(R)\,dR - \frac{64n\kappa\pi^3}{z_1\lambda^3}\int_0^R R^9 N_s(R)\,dR$$ $$\langle\sigma_{\lambda e}\rangle = \frac{48n\kappa\pi^2}{z_1^2\lambda}\int_0^\infty R^3 N_s(R)\,dR + \left\{\frac{4}{15} + \frac{20}{3z_2} + \frac{4.8}{z_1^2}[7(n^2+\kappa^2) + 4(n^2 - \kappa^2 - 5)]\right\}$$ $$\times \frac{8n\kappa\pi^4}{\lambda^3}\int_0^\infty R^5 N_s(R)\,dR + \frac{128\pi^5}{3z_1^2\lambda^4}$$ $$\times \{[(n^2+\kappa^2)^2 + n^2 - \kappa^2 - 2]^2 - 36n^2\kappa^2\} \times \int_0^R R^6 N_s(R)\,dR$$ where $z_1 = (n^2 + \kappa^2)^2 + 4(n^2 - \kappa^2) + 4$ $z_2 = 4(n^2 + \kappa^2)^2 + 12(n^2 - \kappa^2) + 9$

Note: Wavelength λ is for waves traveling in the fluid: $m = n - i\kappa = n_s/n_f - i\kappa_s/n_f$; $\theta_0 = 0$ for forward-scattered beam and π for backward. (a) Siegel and Howell [45]; (b) Siegel and Howell [45]; (c) van de Hulst [50]; (d) Penndorf [69].

TABLE 9.5   Volume Averaging of Radiative Properties: Independent Scattering (*Continued*)

Constraints	$\langle\sigma_{\lambda s}\rangle(1/m), \langle\sigma_{\lambda a}\rangle(1/m), \langle\Phi_\lambda\rangle(\theta_0)$
(e) Small spherical particles, $2\pi R/\lambda < 0.6/n$ (Rayleigh scattering)	$\langle\sigma_{\lambda s}\rangle = \dfrac{128\pi^5}{3\lambda^4}\left\|\dfrac{m^2-1}{m^2+2}\right\|^2 \displaystyle\int_0^\infty R^6 N_s(R)\,dR$ $\langle\Phi_\lambda\rangle(\theta_0) = \frac{3}{4}(1 + \cos^2\theta_0)$
(f) $d <$ interparticle spacing $< \lambda$, $2\pi R/\lambda \ll 1$ (Lorentz-Lorenz scattering)	$\langle\sigma_{\lambda s}\rangle = \dfrac{24\pi^3}{\lambda^4 N_s}\left\|\dfrac{m^2-1}{m^2+2}\right\|^2$ Independent of R, N_s is the number density of scatterers
(g) Spherical particles, interparticle spacing $\gg \lambda$, $2\pi R/\lambda \ll 1$, random arrangement Same with $n \to \infty$	$\langle\sigma_{\lambda s}\rangle = \dfrac{8}{3}\dfrac{\pi^3}{\lambda^4 N_s}\|m^2-1\|^2$ Independent of R $\langle\sigma_{\lambda s}\rangle = \dfrac{160\pi^5}{3\lambda^4}\displaystyle\int_0^\infty R^6 N_s(R)\,dR + \dfrac{256\pi^7}{5\lambda^6}\displaystyle\int_0^\infty R^8 N_s(R)\,dR$ $\langle\Phi\rangle(\theta_0) = \frac{3}{5}[(1 - \frac{1}{2}\cos\theta_0)^2 + (\cos\theta_0 - \frac{1}{2})^2]$ Small spherical particles such that R^8 term is negligible
(h) Nonspherical (Rayleigh-ellipsoid approximation)	$\langle\sigma_{\lambda a}\rangle = \dfrac{\pi}{\lambda}\left(\dfrac{\overline{V}}{\overline{A}}\right)\mathrm{Im}\left[\dfrac{2n}{n-1}(\log n - i\kappa)\right]$ $\overline{V}/\overline{A}$ is the average diameter

(e) van de Hulst [50]; (f) Siegel and Howell [45]; (g) van de Hulst [50]; (h) Bohren and Huffman [70].

al. [76]. For a one-dimensional radiative transfer through a porous medium, with fluxes q_j^+ and q_j^- arriving and leaving from layer j from the left side and fluxes q_{j+1}^+ and q_{j+1}^- leaving and arriving from the right side, the radiative heat flux (across an area A) is given by

$$q_j^- A = \langle T_r\rangle q_{j+1}^- A + \langle\rho_r\rangle q_j^+ A + \langle\varepsilon_r\rangle\sigma T_j^4 A \tag{9.41}$$

$$q_{j+1}^+ A = \langle T_r\rangle q_j^+ A + \langle\rho_r\rangle q_{j+1}^- A + \langle\varepsilon_r\rangle\sigma T_j^4 A \tag{9.42}$$

The effective radiative properties, that is, effective transmissitivity $\langle T_r\rangle$, effective reflectivity $\langle\rho_r\rangle$, and effective emissivity (assumed equal to absorptivity) $\langle\varepsilon_r\rangle$ are determined for various two-dimensional arrangements of spherical particles. Emerging correlations, relating these effective properties to the particle surface emissivity ε_r and medium porosity ε, do not appear to depend significantly on the arrangement. These correlations obtained by Mazza et al. [76] are

$$\langle T_r\rangle = 1 - \langle\rho_r\rangle - \langle\varepsilon_r\rangle \tag{9.43}$$

$$\langle\rho_r\rangle = \frac{\pi(1 - \varepsilon_r)a_1 N_s d^2}{2} \tag{9.44}$$

$$\langle\varepsilon_r\rangle = \frac{\pi\varepsilon_r a_2 N_s d^2}{2} \tag{9.45}$$

$$a_1 = \frac{2}{3[1 + a_3(N_s d^2)^{1.41}]^{1/2}} \tag{9.46}$$

TABLE 9.6 Phase Function in Terms of Legendre Polynomials

Approximations for $\langle\Phi_\lambda\rangle(\theta_0)$

(a) Spherical particles, independent scattering

$$\langle\Phi_\lambda\rangle(\theta_0) = \sum_{i=0}^{N} (2i+1)A_i P_i(\cos\theta_0)$$

where $\cos\theta_0 = \cos\theta_i\cos\theta + \sin\theta_i\sin\theta\cos(\phi-\phi_i)$, $A_i = (1/2)\int_0^\pi \Phi_\lambda(\theta_0)P_i(\cos\theta_0)\,d\cos\theta_0$, P_i is the Legendre polynomial of degree i, and $\Phi_\lambda(\theta_0)$ is the exact phase function obtained through the Mie scattering analysis

- For isotropic scattering $A_0 = 1$, $A_i = 0$
- For Rayleigh scattering $A_0 = 1$, $A_2 = \frac{1}{10}$, $A_i = 0$
- For linear-isotropic scattering $A_0 = 1$, $-\frac{1}{3} \le A_1 \le \frac{1}{3}$, $A_i = 0$

Strong forward scattering

$$\langle\Phi_\lambda\rangle(\theta_0) = 2f_\lambda\delta(1-\cos\theta_0) + (1-f_\lambda)\sum_{i=0}^{M}(2i+1)\hat{A}_i P_i(\cos\theta_0)$$

where $f_\lambda = A_{M+1}$, $\hat{A}_i = (A_i - f_\lambda)/(1-f_\lambda)$, $i = 0, 1, \ldots, M$ and δ are the Dirac delta, $(M+1)/2$ is the order of approximation for a spike in the forward direction $(\delta - M)$ approximation

(b) Strongly forward scattering

$$\langle\Phi_\lambda\rangle(\theta_0) = 2f_\lambda\delta(1-\cos\theta_0) + (1-f_\lambda)(1+3\hat{A}_1\cos\theta_0)$$

is the Delta-Eddington approximation

$$f_\lambda = \begin{cases} A_2, & A_2 \ge (3A_1-1)/2 \\ (A_3-1)/2, & \text{else} \end{cases} \quad \text{with } \hat{A}_1 = \frac{A_1-A_2}{1-A_2}$$

(c) Not very accurate [72]. For the two-flux model an approximation (linear isotropic) is forward scattered,

$$f_\lambda = \frac{1}{2}\int_0^{\pi/2}\langle\Phi_\lambda\rangle(\theta_0)\,d\cos\theta_0 \simeq \frac{1}{2} + \frac{1}{2}\sum_{i=0}^{\infty}\frac{(-1)^i A_{2i+1}(2i)!}{2^{2i+1}i!(i+1)!}$$

backward scattered,

$$b_\lambda = \frac{1}{2}\int_{-\pi/2}^{0}\langle\Phi_\lambda\rangle(\theta_0)\,d\cos\theta_0 = 1 - f_\lambda$$

(d) Strongly forward scattering

$$\langle\Phi_\lambda\rangle(\cos\theta_0) = 2f_\lambda\delta(1-\cos\theta_0) + (1-f_\lambda)\frac{1-g_\lambda^2}{(1+g_\lambda^2-2\cos\theta_0)^{3/2}}$$

is the δ-Henyey-Greenstein approximation,

$$g_\lambda = \frac{A_1-A_2}{1-A_1} \quad \text{and} \quad f_\lambda = \frac{A_2-A_1^2}{1-2A_1+A_2}$$

(a) Chu and Churchill [73]; (b) Wiscombe [74], McKellar and Box [75]; (c) Lee and Buckius [72]; (d) McKellar and Box [75].

$$a_2 = \frac{1}{[1 + a_4(N_s d^2)^{1.74}]^{1/2}} \qquad (9.47)$$

$$a_3 = \frac{1.46\varepsilon_r + 0.484}{1 + 0.16\varepsilon_r} \qquad (9.48)$$

$$a_4 = \frac{1.967\varepsilon_r + 0.00330}{1 + 0.07\varepsilon_r} \qquad (9.49)$$

where N_s is the number of scatterers per unit area.

The correlation applies to $0.630 < N_s d^2 < 1.155$, where the upper limit corresponds to the closest two-dimensional packing of spheres.

The variational upper and lower bounds for the effective emissivity of randomly arranged particles has been obtained by Xia and Strieder [77, 78].

Approximate Radiant Conductivity Model. The radiative heat transfer for a one-dimensional, plane geometry with emitting particles under the steady-state condition is given by [79]:

$$q_r = \frac{F\sigma}{[(1 + \rho_w)/(1 - \rho_w)] + L/d} (T_1^4 - T_2^4) \qquad (9.50)$$

where F is called the *radiative exchange factor* and the properties are assumed to be wavelength-independent. If $\rho_w = 0$ and the bed is several particles deep, then the first term of the denominator can be neglected. Then, for $T_1 - T_2 < 200$ K, a radiant conductivity is defined [59]:

$$\mathbf{q}_r = -k_r \nabla T, \; k_r = 4Fd\sigma T_m^3 \qquad (9.51)$$

The approach has many limitations, but the single most important limitation is that the value of F cannot be easily calculated. Of all the methods, the Monte Carlo method can be used for calculating F for semitransparent particles. The value of F also depends upon the value of the conductivity of the solid phase. In the Kasparek experiment [79] infinite conductivity is assumed, which is justified for metals. Similarly, the case of zero conductivity can be easily treated by considering the rays to be emitted from the same point at which they were absorbed. However, the intermediate case, that is, when the conductivity is comparable to the radiant conductivity, shows a strong dependence of radiant conductivity on the solid conductivity. The extent of this dependence may be seen by comparing the difference in the values of F in Table 9.7 corresponding to low and high emissivities. If the conductivity was small, all the F values would be close to those obtained for the $\varepsilon_r = 0$ case. Thus, a simple tabulation of F as in Table 9.7 is of limited use. On the other hand, this approach is simple.

TABLE 9.7   Radiation Exchange Factor F ($\varepsilon = 0.4$)

Model	Emissivity				
	0.2	0.35	0.60	0.85	1.0
Two-flux (diffuse)	0.88	0.91	1.02	1.06	1.11
Two-flux (specular)	1.11	1.11	1.11	1.11	1.11
Discrete ordinate (diffuse)	1.09	1.15	1.25	1.38	1.48
Discrete ordinate (specular)	1.48	1.48	1.48	1.48	1.48
Argo and Smith	0.11	0.21	0.43	0.74	1.00
Vortmeyer	0.25	0.33	0.54	0.85	1.12
Kasparek (experiment)	—	0.54	—	1.02	—
Monte Carlo (diffuse)	0.32	0.45	0.68	0.94	1.10
Monte Carlo (specular)	0.34	0.47	0.69	0.95	1.10

Determination of F.   Many different models are available for the prediction of F, and these are reviewed by Vortmeyer [79]. Here, the main emphasis is on examining the validity of the radiant conductivity approach by comparing the results of some of these models with the Monte Carlo simulations and with the available experimental results.

A solution to this problem based on the two-flux model is given by Tien and Drolen [59]:

$$F = \frac{2}{d(\overline{\sigma}_{\lambda a} + 2\overline{\sigma}_{\lambda s})} \tag{9.52}$$

which can be written as

$$F = \frac{2}{3(1 - \varepsilon)(\eta_{\lambda a} + 2B\eta_{\lambda s})} \tag{9.53}$$

For isotropic scattering, $B = 0.5$ and Eq. 9.53 becomes independent of the particle emissivity (for large particles).

The low and high conductivity limits of this problem have been explored experimentally [79] and by the Monte Carlo method [61]. In the low-conductivity asymptote, the rays are considered to be emitted from the same point on the sphere at which they were absorbed. In the high-conductivity asymptote, an individual sphere is assumed to be isothermal and a ray absorbed by the sphere is given an equal probability of being emitted from anywhere on the sphere surface. This results in an increase in the radiant conductivity, because the rays absorbed on one side can be emitted from the other side thus bypassing the radiative resistance. In the general problem, the solid and the radiant conductivities can have arbitrary magnitudes. Then, the radiative heat flux q_r for this one-dimensional, plane geometry is given by Eq. 9.50. The radiant conductivity k_r is given by Eq. 9.51, where

$$F = F(k_s^*, \varepsilon_r, \varepsilon) \tag{9.54}$$

and T_m is the mean temperature. The dimensionless solid conductivity k_s^* is defined as

$$k_s^* = \frac{k_s}{4d\sigma T_m^3} \tag{9.55}$$

Within the bed, the radiation is treated by combining the ray tracing with the Monte Carlo method. The conduction through the spheres is allowed by solving for the temperature distribution in a representative sphere for each particle layer in the bed.

The results for $\varepsilon = 0.476$ and various values of ε_r and k_s^* have been obtained for both diffusive and specular surfaces. The results are shown in Fig. 9.15a and b. The results for both surfaces are nearly the same. Both low and high k_s^* asymptotes are present. The low k_s^* asymptotes are reached for $k_s^* < 0.10$ and the high k_s^* asymptote is approached for $k_s^* > 10$. There is a monotonic increase with ε_r, that is, as absorption increases, the radiant conductivity increases for high k_s^*.

The results of Fig. 9.15a and b have been correlated using [9]

$$F = a_1\varepsilon_r \tan^{-1}\left(a_2 \frac{k_s^{*a_3}}{\varepsilon_r}\right) + a_4 \tag{9.56}$$

for given ε. The best-fit values of the constants are given in Table 9.8.

TABLE 9.8   Constants in the Exchange Factor Correlation ($\varepsilon = 0.476$)

	Specular	Diffuse
a_1	0.5711	0.5756
a_2	1.4704	1.5353
a_3	0.8237	0.8011
a_4	0.2079	0.1843

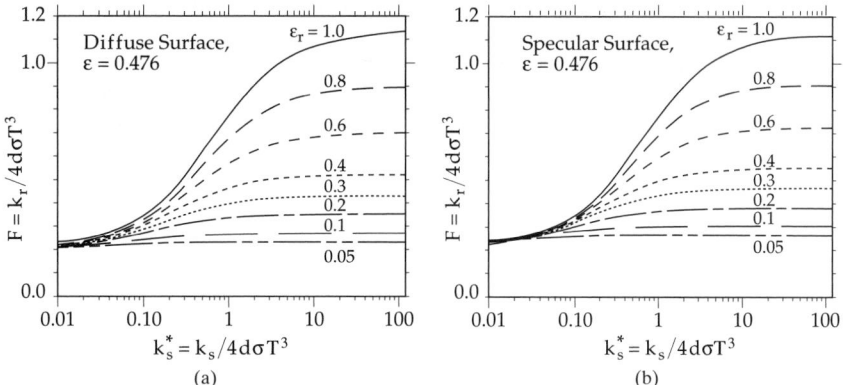

FIGURE 9.15 Effect of dimensionless solid conductivity on the dimensionless radiant conductivity for (*a*) diffuse particle surface and (*b*) specular particle surface [9].

The computer-intensive nature of the problem prevented a thorough sweep of the porosity range as an independent variable. However, the effect of the porosity in the high-conductivity limit has been discussed by Singh and Kaviany [61]. For example, by decreasing the porosity from 0.6 to 0.5, the magnitude of F changes from 0.47 to 0.51 for $\varepsilon_r = 0.35$ (specular surfaces) and from 0.94 to 0.97 for $\varepsilon_r = 0.85$ (diffuse surfaces). In practical packed beds, the porosity ranges between 0.3 to 0.6 with a value of 0.4 for randomly arranged, loosely packed monosized spheres. Therefore, the sensitivity of the radiant conductivity with respect to the porosity (as compared to other parameters) is not expected to be very significant.

The variational upper bound on the radiant conductivity, including the conduction through the particle, has been predicted by Wolf et al. [80].

Summary.   In conclusion, some suggestions are made on how to model the problem of radiative heat transfer in porous media. First, we must choose between a direct simulation and a continuum treatment. Wherever possible, continuum treatment should be used because of the lower cost of computation. However, the volume-averaged radiative properties may not be available in which case continuum treatment cannot be used. Except for the Monte Carlo techniques for large particles, direct simulation techniques have not been developed to solve but the simplest of problems. However, direct simulation techniques should be used in case the number of particles is too small to justify the use of a continuum treatment and as a tool to verify dependent scattering models.

If the continuum treatment is to be employed, we must first identify the elements that make up the system. The choice of elements might be obvious (as in the case of a packed bed of spheres) or some simplifying assumptions might have to be made. Common simplifying assumptions are assuming the system to be made up of cylinders of infinite length (for fibrous media) or assuming arbitrary convex-surfaced particles to be spheres of equivalent cross section or volume. Then the properties of an individual particle can be determined. If the system cannot be broken down into elements, then we have no choice but to determine its radiative properties experimentally.

On the other hand, if we can treat the system as being made up of elements, then we must identify the system as independent or dependent. In theory, all systems are dependent, but if the deviation from the independent theory is not large, the assumption of independent scattering should be made. The range of validity of this assumption can be approximately set at $C/\lambda > 0.5$ and $\varepsilon > 0.95$. If the problem lies in the independent range, then the properties of the bed can be readily calculated, and the equation of transfer can be solved.

However, if the system is in the dependent range, some modeling of the extent of dependence is necessary to get the properties of the packed bed. Models for particles in the Rayleigh range and the geometric range are available. However, no approach is yet available for particles of arbitrary size, and experimental determination of properties is again necessary.

An approximate, geometric, layered model can be used for large particles and the method is described here.

Finally, we note that the thermal conductivity of the solid phase influences the radiation properties. When using the radiant conductivity model, the results show that k_r can increase by fivefold for $k_s \rightarrow \infty$ as compared to that for $k_s \rightarrow 0$ (for $\varepsilon_r = 1$ and typical porosities).

Two-Medium Treatment

In this section, we examine the single-phase flow through solid matrices where the assumption of the local thermal equilibrium between the phases is not valid, i.e., $\langle T \rangle^f / \langle T \rangle^s$. When there is a significant heat generation occurring in any one of the phases (solid or fluid), that is, when the primary heat transfer is by heat generation in a phase and the heat transfer through surfaces bounding the porous medium is less significant, then the local (finite and small) volumes of the solid and fluid phases will be far from the local thermal equilibrium. Also, when the temperature at the bounding surface changes significantly with respect to time, then in the presence of an interstitial flow and when solid and fluid phases have significantly different heat capacities and thermal conductivities, the local rate of change of temperature for the two phases will not be equal.

In the two-medium treatment of the single-phase flow and heat transfer through porous media, no local thermal equilibrium is assumed between the fluid and solid phases, but it is assumed that each phase is continuous and represented with an appropriate effective total thermal conductivity. Then the thermal coupling between the phases is approached either by the examination of the microstructure (for simple geometries) or by empiricism. When empiricism is applied, simple two-equation (or two-medium) models that contain a modeling parameter h_{sf} (called the *interfacial convective heat transfer coefficient*) are used. As is shown in the following sections, only those empirical treatments that contain not only h_{sf} but also the appropriate effective thermal conductivity tensors (for both phases) and the dispersion tensor (in the fluid-phase equation) are expected to give reasonably accurate predictions.

We begin with the phase volume averaging of the energy equations, which shows how the fluid phase dispersion as well as the other convective and conductive effects appear as the coupling coefficients in the energy equations. Then, these coefficients, including the interfacial heat convection coefficient, are evaluated for a simple porous medium, that is, capillary tubes. Then, we examine the existing heuristic two-medium treatments and show that most of them are inconsistent with the results of the local phase volume averaging. Also, in order to examine the cases where the assumptions made in the phase-averaged treatments do not hold, we examine pointwise solutions to a periodic flow. Finally, the chemical reaction in the fluid phase and departure from local thermal equilibrium is examined in an example of premixed combustion in a two-dimensional porous media. For this problem, the results of pointwise (i.e., direct simulation), single- and two-medium treatments is compared for the flame speed and flame structure.

Local Volume Averaging.   The local volume-averaging treatment leading to the coupling between the energy equation for each phase is formulated by Carbonell and Whitaker [81] and is given in Zanotti and Carbonell [82], Levec and Carbonell [83], and Quintard et al. [84]. Their development for the transient heat transfer with a steady flow is reviewed here. Some of the features of their treatment are discussed first.

- For the transient behavior, it is assumed that the penetration depth (in the fluid and solid phases) is larger than the linear dimension of the representative elementary volume. This is required in order to volume-average over the representative elementary volume while satisfying that ΔT_ℓ over this volume is much smaller than that over the system ΔT_L, that is, not all the temperature drop occurs within the representative elementary volume. If ΔT_ℓ is nearly equal to ΔT_L, then the direct simulation of the heat transfer over length ℓ has to be performed. Except for very fast transients, the time for the penetration over ℓ, that is, ℓ^2/α, is much smaller than the timescales associated with the system transients of interest.

- Each phase is treated as a continuum. The phase volume-averaged total thermal diffusivity tensor will be determined for each phase.

- Closure constitutive equations are developed similar to those used when the existence of the local thermal equilibrium was assumed. This requires relating the disturbances in the temperature fields to the gradients of the volume-averaged temperatures and to the difference between the phase volume-averaged temperatures.

After the formal derivations, the energy equation for each phase ($\langle T\rangle^f$ and $\langle T\rangle^s$) can be written in a more compact form by defining the following coefficients. Note that both the hydrodynamic dispersion, that is, the influence of the presence of the matrix on the flow (no-slip condition on the solid surface), as well as the interfacial heat transfer need to be included. The total thermal diffusivity tensors $\mathbf{D}_{ff}$, $\mathbf{D}_{ss}$, $\mathbf{D}_{fs}$, and $\mathbf{D}_{sf}$ and the interfacial convective heat transfer coefficient h_{sf} are introduced. The total thermal diffusivity tensors include both the effective thermal diffusivity tensor (stagnant) as well as the hydrodynamic dispersion tensor. A total convective velocity $\mathbf{v}$ is defined such that the two-medium energy equations become

$$\frac{\partial \langle T\rangle^f}{\partial t} + \mathbf{v}_{ff}\cdot\nabla\langle T\rangle^f + \mathbf{v}_{fs}\cdot\nabla\langle T\rangle^s = \nabla\cdot\mathbf{D}_{ff}\cdot\nabla\langle T\rangle^f + \nabla\cdot\mathbf{D}_{fs}\cdot\nabla\langle T\rangle^s + \frac{A_{fs}}{V_f(\rho c_p)_f}\,h_{sf}(\langle T\rangle^s - \langle T\rangle^f)$$

$$\text{(9.57)}$$

$$\frac{\partial \langle T\rangle^s}{\partial t} + \mathbf{v}_{sf}\cdot\nabla\langle T\rangle^f + \mathbf{v}_{ss}\cdot\nabla\langle T\rangle^s = \nabla\cdot\mathbf{D}_{sf}\cdot\nabla\langle T\rangle^f + \nabla\cdot\mathbf{D}_{ss}\cdot\nabla\langle T\rangle^s + \frac{A_{fs}}{V_s(\rho c_p)_s}\,h_{sf}(\langle T\rangle^f - \langle T\rangle^s)$$

$$\text{(9.58)}$$

As is discussed later, h_{sf} is also used as an overall convection heat transfer coefficient. When h_{sf} is determined experimentally, it is important to note whether the complete form of Eqs. 9.57 and 9.58 are used for its evaluation. The use of oversimplified versions of Eqs. 9.57 and 9.58 results in the inclusion of the neglected terms into h_{sf}. This simplification results in values for h_{sf} that are valid only for those particular experiments.

Interfacial Heat Transfer Coefficient $\mathbf{h_{sf}}$.   In the earlier treatments of transient heat transfer in packed beds, various heuristic models were used instead of the two equations given by Eqs. 9.57 and 9.58. Wakao and Kaguei [85] give the history of the development in this area. In the following, some of these models, which all use an interfacial convection heat transfer coefficient h_{sf}, are discussed. The distinction should be between h_{sf} found from the energy Eqs. 9.57 and 9.58, and that found from the simplified forms of energy equations. Since these different models are used in the determination of h_{sf}, the literature on the reported value of h_{sf} is rather incoherent. Wakao and Kaguei [85] have carefully examined these reported values and classified the modeling efforts.

It should be noted that h_{sf} for a heated single particle in an otherwise uniform temperature field is expected to be significantly different than that for particles in packed beds. Also, since, in general, the thermal conductivity of the solid is not large enough to lead to an isothermal surface temperature, the conductivity of the solid also influences the temperature field around it. Therefore, the interstitial convection heat transfer coefficient obtained from a given fluid-solid combination is not expected to hold valid for some other combinations.

The coefficients in Eqs. 9.57 and 9.58 have been computed for some geometry and range of parameters [84]. Simplified h_{sf}-based models can still be used, and we review some of these heuristic models. However, their inadequacy to explain the process and their limitations cannot be overemphasized.

   Models Based on h_{sf}.   There are many h_{sf}-based models appearing in the literature. Three such models are given here [85]. These are generally for the one-dimensional Darcean flow and heat transfer and for packed beds of spherical particles.

1. *Schumann Model.*   This is the simplest and the least accurate of all models. The two equations are given as

$$\frac{\partial \langle T \rangle^f}{\partial t} + \langle u \rangle^f \frac{\partial \langle T \rangle^f}{\partial x} = \frac{h_{sf} A_0}{\varepsilon (\rho c_p)_f} (\langle T \rangle^s - \langle T \rangle^f) \qquad (9.59)$$

$$\frac{\partial \langle T \rangle^s}{\partial t} = -\frac{h_{sf} A_0}{(1-\varepsilon)(\rho c_p)_s} (\langle T \rangle^s - \langle T \rangle^f) \qquad (9.60)$$

where $A_0 = A_{fs}/V$ is the specific surface area and $u_p = \langle u \rangle^f$ is the average pore velocity. No account is made of the axial conduction and the dispersion in the solid energy equation. This model is for transient problems only.

2. *Continuous-Solid Model.*   In this model, the axial conduction, in both phases, is included through the use of effective thermal conductivities k_{fe} and k_{se}. This gives

$$\frac{\partial \langle T \rangle^f}{\partial t} + \langle u \rangle^f \frac{\partial \langle T \rangle^f}{\partial x} = \frac{\langle k \rangle^f}{\varepsilon (\rho c_p)_f} \frac{\partial^2 \langle T \rangle^f}{\partial x^2} + \frac{h_{sf} A_0}{\varepsilon (\rho c_p)_f} (\langle T \rangle^s - \langle T \rangle^f) \qquad (9.61)$$

$$\frac{\partial \langle T \rangle^s}{\partial t} = \frac{\langle k \rangle^s}{(1-\varepsilon)(\rho c_p)_s} \frac{\partial^2 \langle T \rangle^s}{\partial x^2} - \frac{h_{sf} A_0}{(1-\varepsilon)(\rho c_p)_s} (\langle T \rangle^s - \langle T \rangle^f) \qquad (9.62)$$

No account is made of the dispersion and $\langle k \rangle^f$, $\langle k \rangle^s$, and h_{sf} are to be determined experimentally.

3. *Dispersion-Particle-Based Model.*   This is an improvement over the continuous-solid model and allows for dispersion. The results are

$$\frac{\partial \langle T \rangle^f}{\partial t} + \langle u \rangle^f \frac{\partial \langle T \rangle^f}{\partial x} = \frac{1}{\varepsilon} \left(\frac{\langle k \rangle}{(\rho c_p)_f} + D_{xx}^d \right) \frac{\partial^2 \langle T \rangle^f}{\partial x^2} + \frac{h_{sf} A_0}{\varepsilon (\rho c_p)_f} (T_{sf} - \langle T \rangle^f) \qquad (9.63)$$

$$\frac{\partial T_s}{\partial t} = \frac{\langle k \rangle^s}{(\rho c_p)_s} \frac{1}{r^2} \frac{\partial}{\partial r} \left(r^2 \frac{\partial T_s}{\partial r} \right) \qquad (9.64)$$

$$-k_s \frac{\partial T_s}{\partial r} = h_{sf}(T_{sf} - \langle T \rangle^f) \qquad \text{on } A_{fs} \qquad (9.65)$$

where $T_{sf} = T_s$ on A_{fs}. Wakao and Kaguei [85] suggest $D_{\parallel}^d/\alpha_f = 0.5$ Pe with Pe $= \varepsilon u_p d/\alpha_f$. Note that the bed effective thermal conductivity k is included in the fluid-phase equation [85]. Also note that the suggested coefficient for Pe in the expression for the dispersion is smaller than that given in the section entitled "Convection Heat Transfer," where the presence of the local thermal equilibrium was assumed. This particle-based model is the most accurate among the three and is widely used. This model is for transient problems only.

Experimental Determination of h_{sf}.   Wakao and Kaguei [85] have critically examined the experimental results on h_{sf} and have selected experiments (steady-state and transient) which they found to be reliable. They have used Eqs. 9.59 through 9.65 for the evaluation of h_{sf}. This is a rather indirect method of measuring h_{sf}, and, as was mentioned, the results depend on the model used. They have found the following correlation for h_{sf} for spherical particles (or the dimensionless form of it, the Nusselt number)

$$\mathrm{Nu}_d = \frac{h_{sf}d}{k_f} = 2 + 1.1\mathrm{Re}^{0.6}\,\mathrm{Pr}^{1/3} \tag{9.66}$$

for spherical particles where $\mathrm{Re} = \varepsilon u_p d/\nu = u_D d/\nu$. Equation 9.66 gives a $\mathrm{Re} \to 0$ asymptote of $h_{sf}d/k_f = 2$, which is more reasonable than $h_{sf} \to 0$ found when models other than Eqs. 9.63 to 9.65 are used. It should be mentioned that the measurement of h_{sf} becomes more difficult and the experimental uncertainties become much higher as $\mathrm{Re} \to 0$. Figure 9.16 shows the experimental results compiled by Wakao and Kaguei [85] and their proposed correlation. Note also that at low Re, the interfacial convection heat transfer is insignificant compared to the other terms in the energy equations, and, therefore, the suggested $\mathrm{Re} \to 0$ asymptote cannot be experimentally verified.

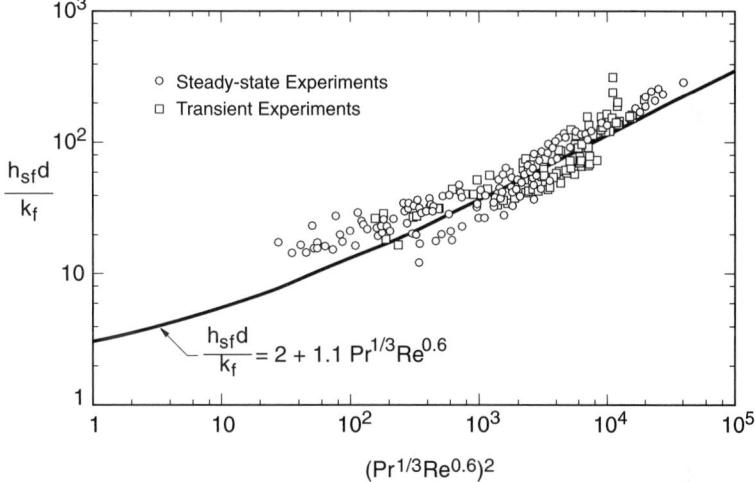

FIGURE 9.16 Experimental results compiled from many sources by Wakao and Kaguei [85] (for steady-state and transient experiments). Also given is their proposed correlation.

The steady-state results are for the heated spheres (the analogous mass transfer is the sublimation of spherical particles).

For ceramic foams, with air as the fluid, Yunis and Viskanta [86] have indirectly measured Nu_d and obtained correlations with Re_d as the variable. They obtain a lower value for the power Re_d. The interfacial heat transfer is also discussed in detail by Kaviany [7].

TWO-PHASE FLOW

In this section, the hydrodynamics and heat transfer of the two-phase (liquid-gas) flow in porous media is addressed. First the volume-averaged momentum equation (for each phase) is considered. The elements of the hydrodynamics of three-phase systems (solid-liquid-gas) are discussed. Then the energy equation and the effective properties are reviewed.

Momentum Equations for Liquid-Gas Flow

The hydrodynamics of two-phase flow in porous media is in part controlled by the dynamics of the liquid-gas-solid contact line. This is in turn determined by the interfacial tensions, the static contact angle, the moving contact angle, and the van der Waals interfacial-layer forces. We need to examine the interfacial tension between a liquid and another fluid. For the case of a static equilibrium at this interface, we can examine the effect of the curvature for the simple problem of ring formation between spheres (and cylinders). For dynamic aspects, we need to examine the combined effect of capillarity and buoyancy by discussing the rise of a bubble in a capillary tube. Then, we should consider more realistic conditions and examine the effects of various factors on the phase distributions and the existing results for the phase distributions in flow through packed beds. The moving contact line and the effects of solid surface tension and the surface roughness and heterogeneities should also be discussed. For the perfectly wetting liquids at equilibrium, a thin extension of the liquid is present on the surface. After the phase-volume averaging of the momentum equation, we discuss the various coefficients that appear in the two momentum equations (one for the wetting phase and one for the nonwetting phase). The coefficients are generally determined empirically, because of the complexity of the phase distributions and their strong dependence on the local saturation. The capillary pressure, phase permeabilities, liquid-gas interfacial drag (due to the difference in the local phase velocities), and the surface tension gradient-induced shear at the liquid-gas interface are discussed in some detail. The special transient problem of immiscible displacement [9] is not examined here. In the following paragraphs, we review some of the definitions used in two-phase flow through porous media and identify the key variables influencing the hydrodynamics.

When compared to the single-phase flows, the two-phase flow in porous media has one significant peculiarity and that is the wetting of the surface of the matrix by one of the fluid phases. Although here the attention is basically on a liquid-phase wetting the surface and a gaseous phase being the nonwetting phase, in some applications the two phases can be two liquids where one preferentially wets the surface. The presence of a curvature at the liquid-gas interface results in a difference between the local gaseous and liquid-phase pressures (capillary pressure). This difference in pressure depends on the fraction of the average pore volume (or porosity of the representative elementary volume) occupied by the wetting phase. This fraction is called the *saturation* and is given as

$$\frac{\varepsilon_\ell}{\varepsilon} \equiv \text{saturation} = s = \frac{\text{fraction of the volume occupied by the wetting phase}}{\text{porosity}} \tag{9.67}$$

As with the single-phase flows, the fractions of the representative elementary volume occupied by the liquid and gas phases are

$$\varepsilon_\ell(\mathbf{x}) = \frac{V_\ell}{V} = \varepsilon s \tag{9.68}$$

$$\varepsilon_g(\mathbf{x}) = \frac{V_g}{V} = \varepsilon(1 - s) \tag{9.69}$$

$$\varepsilon_s + \varepsilon_\ell + \varepsilon_g = 1 \tag{9.70}$$

$$V_s + V_\ell + V_g = V \tag{9.71}$$

$$\varepsilon_s = 1 - \varepsilon \tag{9.72}$$

The subscript ℓ refers to the liquid or wetting phase and g refers to the gaseous or nonwetting phase.

As with the fluid dynamics of two-phase flows in plain media, when the two phases do not have the same interstitial velocity, there will be an interfacial drag whose determination requires a knowledge of the interfacial area $A_{g\ell}$ as well as the local flow field in each phase. This interfacial drag is expected to be important only at high flow rates.

In transient two-phase flows, one phase replaces the other and the dynamics of the wetting-dewetting of the surface, which is influenced by the fluid-fluid interfacial tension, solid-fluid interfacial tensions, and the solid-surface forces, must be closely examined. The research on the dynamics of the contact line (fluid-fluid-solid contact line) has been advanced in the last decade.

Based on this, we expect the following parameters (variables) to influence the dynamics of two-phase flow in porous media.

- *Surface tension.*   Assuming that a membrane stretches over each interface, the magnitudes of the interfacial tension between each pair of phases are the fluid-fluid interfacial tension $\sigma_{g\ell}$, the wetting fluid-solid interfacial tension $\sigma_{\ell s}$, and the nonwetting fluid-solid interfacial tension σ_{gs}. When in static equilibrium, the vectorial force balance at the line of contact (the law of Neumann triangle, Ref. 87) gives

$$\sigma_{g\ell} + \sigma_{\ell s} + \sigma_{gs} = \mathbf{0} \tag{9.73}$$

at contact line. The static mechanical equilibrium of the g-ℓ surface is given by the Young-Laplace equation

$$p_c = p_g - p_\ell = \sigma_{g\ell}\left(\frac{1}{r_1} + \frac{1}{r_2}\right) \equiv \sigma\left(\frac{1}{r_1} + \frac{1}{r_2}\right) \equiv 2H\sigma \qquad \text{on } A_{g\ell} \tag{9.74}$$

where p_c is the capillary pressure and r_1 and r_2 are the two principal radii of curvature of $A_{g\ell}$ and where for simplicity we have used $\sigma_{g\ell} \equiv \sigma$. The mean curvature of the interface H is defined as

$$H \equiv \frac{1}{2}\left(\frac{1}{r_1} + \frac{1}{r_2}\right) \tag{9.75}$$

- *Wettability.*   The extent to which the wetting phase spreads over the solid surface. The angle, measured in the wetting phase, between the solid surface and the g-ℓ interface, is called the *contact angle* θ_c where $\theta_c = 0$ corresponds to complete wetting. Presence of surface roughness, adsorbed surface layers, or surfactants influence θ_c significantly.
- *Matrix structure.*   The size, dimensionality, pore coordinate number, and topology of the matrix influence the phase distributions significantly.
- *Viscosity ratio.* μ_g/μ_ℓ influences the relative flow rates directly and indirectly through the interfacial shear stress. In fast transient flows (e.g., immiscible displacement), depending on whether the viscosity of the displacing fluid is larger than that of the displaced fluid, or vice versa, different displacement frontal behaviors are found.
- *Density ratio.* ρ_g/ρ_ℓ, in addition to the body force, signifies the relative importance of the inertial force for the two phases.
- *Saturation.*   This is the extent to which the wetting phase occupies (averaged over the representative elementary volume) the pore space. At very low saturations the wetting phase becomes disconnected (or immobile). At very high saturations, the nonwetting phase becomes disconnected.

In addition, the presence of temperature and concentration gradients results in interfacial tension gradients and influences the phase distributions and flow rates. In dynamic systems, the history of the flows and the surface conditions also play a role and lead to the observed hysteresis in the phase distributions.

In order to arrive at a local volume-averaged momentum equation for each phase, the effect of the preceding parameters on the microscopic hydrodynamics must be examined. This is done to an extent through the particular forces that appear in the momentum equations.

Now, by including the microscopic inertial and macroscopic inertial terms and by introducing $K_{\ell g1}$, $K_{\ell g2}$, $K_{g\ell1}$, and $K_{g\ell2}$ as the coefficients in the liquid-gas interfacial drag forces, and by assuming that this drag is proportional to the difference in the phase velocities and that for cocurrent flows $|\langle u_j \rangle^g| > |\langle u_j \rangle^\ell|$, we have the following pair of momentum equations for two-phase flow in porous media.

Liquid phase

$$\frac{\rho_\ell}{\varepsilon s}\left(\frac{\partial \langle \mathbf{u}_\ell \rangle}{\partial t} + \langle \mathbf{u}_\ell \rangle \cdot \nabla \langle \mathbf{u}_\ell \rangle \right) = -\nabla \langle p \rangle^\ell + \rho_\ell \mathbf{g} - \frac{\mu_\ell}{K_\ell}\langle \mathbf{u}_\ell \rangle - \frac{\rho_\ell}{K_{\ell i}}|\langle \mathbf{u}_\ell \rangle|\langle \mathbf{u}_\ell \rangle$$

| (ℓ-phase macroscopic inertial force) | (ℓ-phase pore pressure gradient) | (ℓ-phase body force) | (microscopic interfacial ($A_{\ell s}$) shear stress) | (microscopic inertial force) |

$$+ [K_{\ell g1}|\langle \mathbf{u} \rangle^g - \langle \mathbf{u} \rangle^\ell| + K_{\ell g2}(\langle \mathbf{u} \rangle^g - \langle \mathbf{u} \rangle^\ell)^2]\frac{\langle \mathbf{u} \rangle^\ell}{|\langle \mathbf{u} \rangle^\ell|} + \mu_\ell \frac{K_{\ell \Delta \sigma}}{K_\ell}\nabla \sigma \quad (9.76)$$

(microscopic interfacial ($A_{\ell g}$) shear stress)

(microscopic interfacial ($A_{\ell g}$) surface tension gradient force)

Gas phase

$$\frac{\rho_g}{\varepsilon(1-s)}\left(\frac{\partial \langle \mathbf{u}_g \rangle}{\partial t} + \langle \mathbf{u}_g \rangle \cdot \nabla \langle \mathbf{u}_g \rangle \right) = -\nabla \langle p \rangle^g + \rho_g \mathbf{g} - \frac{\mu_g}{K_g}\langle \mathbf{u}_g \rangle - \frac{\rho_g}{K_{gi}}|\langle \mathbf{u}_g \rangle|\langle \mathbf{u}_g \rangle$$

$$+ [K_{g\ell1}|\langle \mathbf{u} \rangle^g - \langle \mathbf{u} \rangle^\ell| + K_{\ell g2}(\langle \mathbf{u} \rangle^g - \langle \mathbf{u} \rangle^\ell)^2]\frac{\langle \mathbf{u} \rangle^g}{|\langle \mathbf{u} \rangle^g|} + \mu_g \frac{K_{g\Delta \sigma}}{K_g}\nabla \sigma \quad (9.77)$$

where we have assumed that all the coefficients are isotropic. This assumption simplifies the preceding equations and is justified because presently only the simple isotropic coefficients are available.

Note that from the definition of the phase averaging, that is, $\langle \mathbf{u} \rangle^\ell = (1/V_\ell)\int_{V_\ell} \mathbf{u}\, dV$, etc., we have

$$\langle \mathbf{u} \rangle^\ell = \frac{\langle \mathbf{u}_\ell \rangle}{\varepsilon s} \quad \text{and} \quad \langle \mathbf{u} \rangle^g = \frac{\langle \mathbf{u}_g \rangle}{\varepsilon(1-s)} \quad (9.78)$$

These momentum equations are solved along with the continuity equations and the appropriate boundary conditions. These are discussed and given in Ref. 9.

Some of the correlations for the coefficients appearing in the momentum equations are given in Tables 9.9 through 9.12.

Local Volume Averaging of Energy Equation

The principles of the local volume averaging as applied to the conduction equation, the single-phase flow convection equation, and the two-phase flow momentum equation is now applied to the two-phase flow energy equation. The concept is that developed by Whitaker [101], where the extensive derivations are given. We expect to arrive at a local volume-averaged energy equation in which the *effective thermal conductivity* is the combined contribution of the three phases (s, ℓ, and g) to the molecular conduction, and the *thermal dispersion* is the combined dispersion in the ℓ and g phases. We consider the general case of transient temperature fields with a local heat generation $\dot{s}$ and the ℓ-g phase change $\dot{n}$. The closure conditions lead to equations for the transformation vectors. For simplicity, we consider the simple case of

TABLE 9.9   Correlations for Capillary Pressure

Constraints	Correlation
(a) Water-air-sand	$\langle p_c \rangle = \dfrac{\sigma}{(K/\varepsilon)^{1/2}} \left[0.364(1 - e^{-40(1-s)}) + 0.221(1-s) + \dfrac{0.005}{s - 0.08} \right]$
(b) Water-air-soil and sandstones	$s = s_{\mathrm{ir}} + \dfrac{1 - s_{\mathrm{ir}} - s_{\mathrm{ir}\,g}}{\left[1 + \left(a_1 \dfrac{\langle p_c \rangle}{\rho_\ell g} \right)^n \right]^{1 - 1/n}}$
	where $n > 1$, a_1 is a constant, n and a_1 depend on the matrix and the drainage or imbibition process
(c) Imbibition, nonconsolidated sand, from Leverett [88] data of water-air	$\langle p_c \rangle = \dfrac{\sigma}{(K/\varepsilon)^{1/2}} \left[1.417(1 - S) - 2.120(1 - S)^2 + 1.263(1 - S)^3 \right]$
	where $S = \dfrac{s - s_{\mathrm{ir}}}{1 - s_{\mathrm{ir}} - s_{\mathrm{ir}\,g}}$
(d) Drainage, oil-water in sandstone	$\langle p_c \rangle = \dfrac{\sigma}{(K/\varepsilon)^{1/2}} \left[a_1 - a_2 \ln (s - s_{\mathrm{ir}}) \right]$
	where $a_1 = 0.30$, $a_2 = 0.0633$, $s_{\mathrm{ir}} = 0.15$

(a) Scheidegger [24] from Leverett [88] experiment; (b) van Genuchten [89]; (c) Udell [90]; (d) Pavone [91].

TABLE 9.10   Correlations for Relative Permeabilities

Constraints	Correlation
(a) Sandstones and limestones, oil-water	$K_{r\ell} = S^4$, $K_{rg} = (1 - S)^2(1 - S^2)$
(b) Nonconsolidated sand, well sorted	$K_{r\ell} = S^3$, $K_{rg} = (1 - S)^3$
(c) Nonconsolidated sand, poorly sorted	$K_{r\ell} = S^{3.5}$, $K_{rg} = (1 - S)^2(1 - S^{1.5})$
(d) Connected sandstone, limestone, rocks	$K_{r\ell} = S^4$, $K_{rg} = (1 - S)^2(1 - S^2)$
(e) Sandstone-oil-water	$K_{r\ell} = S^3$, $K_{rg} = 1 - 1.11S$
(f) Soil-water-gas	$K_{rg} = (1 - s_{\mathrm{ir}} - s_{\mathrm{ir}\,g} - s)^{1/2}\{(1 - s^{1/m})^m - [1 - (1 - s_{\mathrm{ir}} - s_{\mathrm{ir}\,g})^{1/m}]^m\}^2$
	where m is found from experiments
(g) Glass spheres-water (water vapor)	$K_{r\ell} = S^3$
	$K_{rg} = 1.2984 - 1.9832S + 0.7432S^2$
(h) Trickling flow in packed bed	$K_{r\ell} = S^{2.0}$ for increasing liquid flow rate
	$K_{r\ell} = \begin{cases} S^{2.9} & S \geq 0.2 \\ 0.25 S^{2.0} & S < 0.2 \end{cases}$ for decreasing liquid flow rate
	$K_{rg} = (1 - s)^n$
	where
	$n = n(\mathrm{Re}_g$, increase or decrease in $\langle u \rangle^\ell)$
	$n = 4.8$ has been suggested by Saez and Carbonell [92]

(a) Corey given by Wyllie [93]; (b)–(d), Wyllie [93]; (e) Scheidegger [24]; (f) Mualem [94] given in Delshad and Pope [95]; (g) Verma et al. [96]; (h) Levec et al. [97].

TABLE 9.11 Correlations for Microscopic Inertial Coefficients

Constraints	Correlation
(a) Packed beds made of large spheres, air-water flow, no net liquid flow	$K_{rgi} = K_{rg} = \left(\dfrac{1-s}{0.83}\right)^3,\qquad 0.17 \le s \le 1$ $K_{rgi} = K_{rg} = 1,\qquad s \le 0.17$
(b) Cocurrent trickle flow in packed beds	$K_{r\ell i} = K_{r\ell},\qquad K_{rgi} = K_{rg}$ where $K_{r\ell}$ and K_{rg} are as before
(c) Packed beds made of large particles	$K_{r\ell i} = s^5,\qquad 0 \le s \le 0.7$ $K_{rgi} = (1-s)^6,\qquad 0 \le s \le 0.7$ $K_{rgi} = 0.1(1-s)^4,\qquad 0.7 \le s \le 1$
(d) Cocurrent and countercurrent flow	$K_{r\ell i} = K_{r\ell} = s^3$ Bubbly and slug flow, $0.6 \le s \le 1$: $K_{rgi} = \left[\dfrac{1-\varepsilon}{1-\varepsilon(1-s)}\right]^{2/3}(1-s)^3$ Annular flow, $0 \le s \le 0.26$: $K_{rgi} = \left[\dfrac{1-\varepsilon}{1-\varepsilon(1-s)}\right]^{2/3}(1-s)^2$ Transition flow, $0.26 \le s \le 0.6$: K_{rgi} is a smooth function with these two asymptotes

(a) Tutu et al. [98]; (b) Saez and Carbonell [92]; (c) Schulenberg and Müller [99]; (d) Tung and Dhir [100].

TABLE 9.12 Correlations for Liquid-Gas Interfacial Drag Coefficients

Constraints	Correlations
(a) No net liquid flow, packed bed of large spheres, bubbly flow, $s \to 1$	$K_{g\ell1} = 0,\qquad K_{g\ell2} = \dfrac{3\rho_\ell C_D}{4d\varepsilon^2(1-s)}$ where $C_D = C_D(s)$
(b) Packed bed of large spheres, glass (ethanol water solutions) air, large $\langle u \rangle^\ell$, $\langle u \rangle^g$, and $s > 0.5$	$K_{g\ell1} = 0$ $K_{g\ell2} = -\dfrac{\rho_\ell(\rho_\ell - \rho_g)gK\varepsilon^2}{s\sigma}\dfrac{C_E}{K^{1/2}}W(s)$ $K_{\ell g1} = 0,\qquad K_{g\ell2} = -K_{\ell g2}\dfrac{s}{(1-s)}\dfrac{\rho_\ell}{\rho_g}$ for $|\langle \mathbf{u} \rangle^g| > |\langle \mathbf{u} \rangle^\ell|$ where $\dfrac{C_E}{K^{1/2}} = \dfrac{1.75(1-\varepsilon)}{d\varepsilon^3},\qquad W(s) = 350s^7(1-s)$
(c) Packed bed of large spheres, for $s < 0.7$, see Ref. 100	$1 \le s \le 0.7$ $K_{\ell g1} = -\dfrac{a_1 v_\ell \varepsilon(\rho_\ell - \rho_g)}{d_b^2[1 - (\rho_g/\rho_\ell)]}$ $K_{\ell g2} = -\dfrac{a_2[s + (\rho_g/\rho_\ell)(1-s)]s\varepsilon(\rho_\ell - \rho_g)}{d_b[1 - (\rho_g/\rho_\ell)]}$ where $d_b = 1.35\left[\dfrac{\sigma}{g(\rho_\ell - \rho_g)}\right]^{1/2}$ and $a_1 = a_1(s),\qquad a_2 = a_2(s)$ $K_{g\ell1} = -K_{\ell g1}\dfrac{s}{1-s},\qquad K_{g\ell2} = -K_{\ell g2}\dfrac{s}{1-s}$

(a) Tutu et al. [98]; (b) Schulenberg and Müller [99]; (c) Tung and Dhir [100].

a unit cell in a periodic fluid-solid structure. The energy equation given in the last section contains terms that are similar to those that were previously labeled as the effective thermal conductivity and thermal dispersion.

The *total conductivity tensor* **D** is defined as

$$\mathbf{D} = \frac{\mathbf{K}_e}{(\rho c_p)_\ell} + \mathbf{D}^d \qquad (9.79)$$

By using these in the thermal energy equation we have

$$[(1 - \varepsilon)(\rho c_p)_s + \varepsilon s(\rho c_p)_\ell + \varepsilon(1 - s)\langle\rho\rangle^g c_{p_g}] \frac{\partial\langle T\rangle}{\partial t} + [(\rho c_p)_\ell\langle\mathbf{u}_\ell\rangle + (\rho c_p)_g\langle\mathbf{u}_g\rangle] \cdot \nabla\langle T\rangle + \Delta i_{\ell g}\langle\dot{n}\rangle$$

$$= \nabla \cdot [\mathbf{K}_e + (\rho c_p)_\ell\mathbf{D}^d] \cdot \nabla\langle T\rangle + \langle\dot{s}\rangle \quad (9.80)$$

Effective Thermal Conductivity

Then we expect a functional relationship for the effective thermal conductivity tensor, of the form

$$\mathbf{K}_e = \mathbf{K}_e[k_s, k_\ell, k_g, a_\ell(\mathbf{x}), a_g(\mathbf{x})] \qquad (9.81)$$

However, $\mathbf{K}_e$ can be given in terms of the more readily measurable quantities such as

$$\mathbf{K}_e = \mathbf{K}_e(k_s, k_\ell, \; k_g, \mathbf{u}_g, \mathbf{u}_\ell, \sigma, \frac{\mu_g}{\mu_\ell}, \frac{\rho_g}{\rho_\ell}, s, \theta_c, \varepsilon, \text{solid structure, history}) \qquad (9.82)$$

This replacement of the variables is done noting that the phase distributions depend on the velocity field and so forth. We also expect two asymptotic behaviors for $\mathbf{K}_e$, which for isotropic phase distributions are given as the following:

$$\text{for } s \to 1 \qquad \mathbf{K}_e = k_{e(s-\ell)}\mathbf{I} = k_e(s=1)\mathbf{I}$$

$$\text{for } s \to 0 \qquad \mathbf{K}_e = k_{e(s-g)}\mathbf{I} = k_e(s=0)\mathbf{I}$$

$$k_e = k_e(k_s, k_f, \varepsilon, \text{solid structure}) \qquad (9.83)$$

Since in two-phase flow and heat transfer in porous media for any direction, the bulk effective thermal conductivity is generally much smaller than the bulk thermal dispersion, the available studies on $\mathbf{K}_e$ are limited. In the following, we briefly discuss the anisotropy of $\mathbf{K}_e$ and then review the available treatments.

Presently no rigorous solutions for $k_{e\|}$ and $k_{e\perp}$ are available. Although not attempted, one of the readily solvable problems would be that of the periodically constricted tube introduced by Saez et al. [102]. For this simple unit-cell phase distribution (which is an approximation to the simple-cubic arrangement of monosize spheres), we expect

$$k_{e\|} \gg k_{e\perp} \qquad (9.84)$$

because of the smaller thermal conductivity of the gas phase.

Most of the reported experimental results for $\mathbf{K}_e(s)$ are obtained for $\mathbf{u} \neq 0$, and, therefore, are the results of the simultaneous evaluation of $\mathbf{K}_e(s)$ and $\mathbf{D}^d(s)$. In these experiments, generally, $\mathbf{K}_e \ll (\rho c_p)_\ell\mathbf{D}^d$. An exception is the experiment of Somerton et al. [103], where only $\mathbf{K}_e = \mathbf{I}k_e(s)$ was determined but with no examination of the anisotropy. Another one is that of Matsuura et al. [104], where the velocity was reduced sufficiently to allow for the determination of $k_e(s)$ with some accuracy; again, no directional dependence was considered.

Table 9.13 summarizes some of the available empirical correlations for $k_{e\|}(s)$ and $k_{e\perp}(s)$ for packed beds of spherical particles. The experiments of Specchia and Baldi [105], Hashimoto

TABLE 9.13   Correlations for Two-Phase Effective Thermal Conductivity

Constraints	Correlation
(a) Alumina spheres-water-air	$$\dfrac{k_{e\perp}}{k_\ell} = a_1$$ where a_1 is determined experimentally
(b) Glass spheres-water-air ($\varepsilon = 0.375$)	$$\dfrac{k_{e\perp}}{k_\ell} = a_1, \qquad 0.13 < s < 0.6$$ where a_1 is determined experimentally
(c) Glass spheres-water-air ($\varepsilon = 0.4$)	$$\dfrac{k_{e\perp}}{k_\ell} = 1.5, \qquad 0.08 < s < 0.21$$
(d) Glass and ceramic spheres-water-air	$$\dfrac{k_{e\perp}}{k_\ell} = \dfrac{k_e(s=0)}{k_\ell}$$
(e) Nonconsolidated sands-brine-air (moist), no directional dependence investigated	$$\dfrac{k_{e\parallel}}{k_\ell} = \dfrac{k_{e\perp}}{k_\ell} = \dfrac{k_e(s=0)}{k_\ell} + s^{1/2}\left[\dfrac{k_e(s=1) - k_e(s=0)}{k_\ell}\right]$$
(f) Disordered porous media, not experimentally verified	$$\dfrac{k_{e\parallel}}{k_\ell} = s\,\dfrac{k_e(s=1)}{k_\ell} + (1-s)\,\dfrac{k_e(s=0)}{k_\ell}$$
(g) Disordered porous media, not experimentally verified	$$\dfrac{k_{e\perp}}{k_\ell} = \left[\dfrac{k_e(s=1)}{k_\ell}\right]^s\left[\dfrac{k_e(s=0)}{k_\ell}\right]^{1-s}$$

(a) Weekman and Myers [107]; (b) Hashimoto et al. [106]; (c) Matsuura et al. [104]; (d) Specchia and Baldi [105]; (e) Somerton et al. [103] and Udell and Fitch [109]; (f) parallel arrangement; (g) geometric mean.

et al. [106], and Weekman and Myers [107] are not of high enough accuracies at low velocities, and, therefore, their values for a_1 are not expected to be accurate [104]. Also, Matsuura et al. [108] could not resolve the saturation-dependence of k_e in their experiments. Somerton et al. [103] do find a saturation dependence as given in Table 9.13. Also shown in the table are the estimates based on the parallel arrangement and geometric mean.

Examination of Table 9.13 shows that further study of the parameters influencing $\mathbf{K}_e$ is needed. This includes the effects of the wettability and the significance of the expected hysteresis in $\mathbf{K}_e(s)$.

Thermal Dispersion

Under steady-state conditions without a heat source, we can rewrite Eq. 9.80 as

$$\left[\langle \mathbf{u}_\ell \rangle + \frac{(\rho c_p)_g}{(\rho c_p)_\ell}\langle \mathbf{u}_g \rangle\right] \cdot \nabla T = \nabla \cdot \left[\frac{\mathbf{K}_e}{(\rho c_p)_\ell} + \mathbf{D}_\ell^d + \frac{(\rho c_p)_g}{(\rho c_p)_\ell}\mathbf{D}_g^d\right] \cdot \nabla T \tag{9.85}$$

$\mathbf{D}_\ell^d$ and $\mathbf{D}_g^d$ depend on the pore-level velocity and phase distributions (in their respective phases). These phase velocity distributions are coupled at $A_{\ell g}$, and therefore, the motion of one phase influences the other. The general relations for $\mathbf{D}_\ell^d$ and $\mathbf{D}_g^d$ are of the form

$$\mathbf{D}_\ell^d = \mathbf{D}_\ell^d[k_s, k_\ell, k_g, a_\ell(\mathbf{x}), a_g(\mathbf{x}), (\rho c_p)_\ell, (\rho c_p)_g, \mathbf{u}_\ell, \mathbf{u}_g, \mu_\ell, \mu_g, \sigma, \theta_c, \text{history}] \tag{9.86}$$

A similar relationship can be given for $\mathbf{D}_g^d$. Note that the various flow regimes (such as the trickle, pulsing, and bubble regimes in the cocurrent and countercurrent downflow in packed beds of spheres are all represented by a_ℓ and a_g. We expect to recover the two asymptotic behaviors, that is,

$$\text{for } s \to 1: \quad \mathbf{D}_\ell^d = \mathbf{D}_\ell^d(s=1) = \varepsilon \mathbf{D}^d, \qquad \mathbf{D}_g^d = 0$$

$$\text{for } s \to 0: \quad \mathbf{D}_\ell^d = 0, \qquad \mathbf{D}_g^d = \mathbf{D}_g^d(s=0) = \varepsilon \mathbf{D}^d \tag{9.87}$$

where $\mathbf{D}_\ell^d(s=1)$ and $\mathbf{D}_g^d(s=0)$ are the single-phase flow thermal dispersion tensor (multiplied by ε) given in the section on convection heat transfer. From the results of that previous section, we know that

$$\mathbf{D}^d(s=0 \quad \text{or} \quad s=1) = \mathbf{D}^d\left(\text{Pe}, \frac{k_s}{k_f}, \varepsilon, \text{solid structure}\right) \tag{9.88}$$

with a weak dependence on Pr.

Table 9.14 lists existing correlations for the lateral dispersion coefficient for two-phase flow.

TABLE 9.14 Correlations for Two-Phase Flow Lateral Dispersion Coefficient

Constraints	Correlation
(a) Cocurrent, downward flow in packed bed of spheres, air-water-glass, wall heating	$\dfrac{D_\perp^d}{\alpha_\ell} = 0.00174\,\text{Pe}_\ell + 0.172\,\text{Pe}_g$ $\text{Pe}_\ell = \dfrac{\langle u_\ell\rangle d}{\alpha_\ell}, \qquad \text{Pe}_g = \dfrac{\langle u_g\rangle d}{\alpha_g}$
(b) Glass and ceramic spheres and ceramic rings, with water-air, wall heating	$D_\perp^d = D_{g\perp}^d + D_{\ell\perp}^d$ $D_{g\perp}^d = D_{g\perp}^d(s=0), \qquad D_{g\perp}^d \ll D_{\ell\perp}^d$ $\dfrac{\langle u_\ell\rangle d}{D_{\ell\perp}^d} = 338\text{Re}_\ell^{0.67} H^{0.29}\left(\dfrac{dA_0}{\varepsilon}\right)^{-2.7}$ where $H(\text{Re}_\ell, \text{etc.})$ is the liquid holdup, A_0 is the specific surface area, $\text{Re}_\ell = \langle u_\ell\rangle d/\nu_\ell$, and H is correlated by Saez and Carbonell [92] and Crine [110]
(c) Alumina spheres-water-air, nonuniform inlet temperature	For large Pe_ℓ $\dfrac{D_\perp}{\alpha_\ell} = a_1 f\,\text{Pe}_\ell, \qquad \text{Pe}_\ell = \dfrac{\langle u_\ell\rangle d}{\alpha_\ell}$
(d) Glass spheres-air-water, $\varepsilon = 0.4$, nonuniform inlet temperature, trickle flow	$\dfrac{D_\perp^d}{\alpha_\ell} = 0.35\text{Pe}_\ell^{0.8}, \qquad 50 \le \text{Pe}_\ell \le 3000$ $\text{Pe}_\ell = \dfrac{\langle u\rangle^\ell}{\alpha_\ell}\dfrac{d\varepsilon s}{1-\varepsilon}$
(e) Glass and alumina spheres-water (or water glycerin) air, cocurrent, downflow, wall heating	$\dfrac{D_\perp}{\alpha_\ell} = a_1\,\text{Pe}_\ell + 0.095\text{Pe}_g\dfrac{k_g}{k_\ell}$ $\text{Pe}_\ell = \dfrac{\langle u_\ell\rangle d}{\alpha_\ell}, \qquad \text{Pe}_g = \dfrac{\langle u_g\rangle d}{\alpha_g}$ α_g is based on vapor-air properties $a_1 = a_1(\text{Re}_\ell, d_t/d, \text{etc.})$; d_t is tube diameter

(a) Weekman and Myers [107]; (b) Specchia and Baldi [105]; (c) Crine [110]; (d) Saez [111]; (e) Hashimoto et al. [106].

PHASE CHANGE

In this section, we examine evaporation and condensation in porous media in detail and briefly review melting and solidification in porous media. The heat supply or removal causing these to occur is generally through the bounding surfaces and these surfaces can be impermeable or permeable. We begin by considering condensation and evaporation adjacent to vertical impermeable surfaces. These are the counterparts of the film condensation and evaporation in plain media. The presence of the solid matrix results in the occurrence of a two-phase flow region governed by gravity and capillarity. The study of this two-phase flow and its effect on the condensation or evaporation rate (i.e., the heat transfer rate) has begun recently. Evaporation from horizontal impermeable surfaces is considered next. Because the evaporation is mostly from thin-liquid films forming on the solid matrix (in the evaporation zone), the evaporation does not require a significant superheat. The onset of dryout, that is, the failure of the gravity and capillarity to keep the surface wet, occurs at a critical heat flux but only small superheat is required. We examine the predictions of the critical heat flux and the treatment of the vapor-film and the two-phase regions. We also examine the case of thin porous-layer coating of horizontal surfaces and review the limited data on the porous-layer thickness dependence of the heat flow rate versus the superheat curve. Then, we turn to permeable bounding surfaces and examine the moving condensation front occurring when a vapor is injected in a liquid-filled solid matrix. Finally, we examine the heating at a permeable bounding surface where the surface temperature is below the saturation temperature and the resulting surface and internal evaporations result in the gradual drying of the surface. The melting and solidification of single and multicomponent systems are discussed as part of phase change in condensed phase.

Condensation at Vertical Impermeable Bounding Surfaces

When a vertical impermeable surface bounding a semi-infinite vapor-phase domain is cooled below the saturation temperature and when nucleation sites are present, condensation begins, and if the condensed phase (liquid) wets the surface perfectly, the film-condensation flow occurs. For phase-density buoyant flows (i.e., flows caused by the density difference between phases and in presence of the gravity field), if the vapor is in the pores of a solid matrix bounded by this vertical surface, the liquid flows through the matrix both along the gravity vector (due to buoyancy) and perpendicular to it (due to capillarity). A single-phase region (liquid-film region) must be present adjacent to the wall where the liquid is subcooled up to the edge of this film (a distance δ_ℓ from the surface). Beyond δ_ℓ, the capillarity results in a two-phase region that is nearly isothermal. For the phase-density buoyant film condensation flow in plain media, the film thickness is generally small. δ_ℓ increases as $x^{1/4}$ with x taken along the gravity vector and measured from the leading edge of the film. This is characteristic of natural convection in plain media. For condensation in porous media, the local film thickness can be much larger than the pore (or particle) size, which will allow for a local volume-averaged treatment of this fluid-solid heterogeneous system. However, when the local film thickness is smaller or comparable in size to the pore size, the local volume-averaged treatments will no longer be applicable. Figure 9.17 depicts the local flow of liquid for cases where $\delta_\ell/d < 1$, $\delta_\ell/d \simeq 1$, and $\delta_\ell/d > 1$. As was mentioned, for cases with a large δ_ℓ/d, a two-phase region also exists extending from $y = \delta_\ell$ to $y = \delta_{\ell g}$.

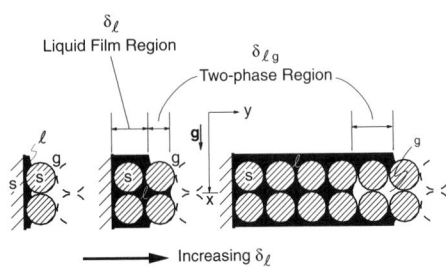

FIGURE 9.17 Condensation at an impermeable vertical surface for cases with $\delta_\ell/d < 1$, $\delta_\ell/d \simeq 1$, and $\delta_\ell/d > 1$.

The case of combined buoyant-forced (i.e., applied external-pressure gradient film condensation flow in porous media) has been examined by Renken et al. [112], and here we examine the case when there is no external pressure gradient.

In the following, we only examine in sufficient detail the fluid flow and heat transfer for thick liquid films, that is, $\delta_\ell/d > 1$.

In order to study two-phase flow and heat transfer for this phase change problem, we assume that the local volume-averaged conservation equations (including the assumption of local thermal equilibrium) are applicable. For this problem, we note the following.

- The liquid-film and two-phase regions each must contain many pores, that is, for gravity-capillarity-dominated (Bo $\simeq$ 1) and capillarity-dominated (Bo < 1) flows, we require

$$\frac{\delta_\ell}{d} \gg 1 \quad \text{and} \quad \frac{\delta_{\ell g}}{d} \gg 1 \quad \text{for Bo} = \frac{g(\rho_\ell - \rho_g)K/\varepsilon}{\sigma} \ll 1 \qquad (9.89)$$

$$\frac{\delta_\ell}{d} \gg 1 \quad \text{for Bo} \leq 1 \qquad (9.90)$$

For the gravity-dominated flows, the two-phase region is absent. This large Bond number asymptote is discussed later.

- The liquid viscosity varies with temperature $\mu_\ell = \mu_\ell(T)$ and can be included in the analysis. However, when μ_ℓ is evaluated at the average film temperature, this variation can be represented with sufficient accuracy.

- The solid structure and the hydrodynamic nonuniformities can cause large variations of ε, K, $k_{\ell\perp}$, and $D_\perp^d$ near the bounding surface. Special attention should be given to these nonuniformities. For the continuum treatment, we require that

$$(\nabla \varepsilon)d \ll 1 \qquad (9.91)$$

that is, a large porosity variation near the bounding surface requires a special treatment.

- The boundary-layer treatment is assumed to be justifiable, that is, $\delta_\ell/L \ll 1$.

- At and near the bounding surface, the lateral effective thermal conductivity $k_{\ell\perp}(y = 0)$ depends on the thermal conductivity of the bounding surface (in addition to the other parameters affecting the bulk values of $k_{\ell\perp}$). Therefore, in using

$$\left(k_{\ell\perp} \frac{\partial T}{\partial y} \right)_{y=0} \simeq \dot{m}_{g\ell} i_{\ell g} \qquad (9.92)$$

a special attention should be given to evaluation of $k_{\ell\perp}(y = 0)$, whenever the experimental results for $T(x, y)$ are used for the evaluation of $(\partial T/\partial y)_{y=0}$. Generally, the local condensation rate $\dot{m}_{g\ell}$ is measured instead of $T(x, y)$. This is because δ_ℓ is generally small, and, therefore, the accuracy of $T(x, 0 < y < \delta_\ell)$ is not high enough to result in acceptable gradients.

- The gaseous phase is assumed to be a simple single-component system. The presence of noncondensable gases, which results in a significant reduction in the condensation rate (because of their accumulation near $y = \delta_\ell$ and their resistance to the flow of vapor toward this interface), can be addressed by the inclusion of the species conservation equation (i.e., the mass diffusion equation). For the $y \geq \delta_\ell + \delta_{\ell g}$ domain, the single-phase mass diffusion can be used. For the two-phase region, a mass diffusion equation for the gaseous phase can be written by noting the anisotropy of the effective mass diffusion and the dispersion tensors. As is discussed later, the determination of the saturation distribution for this condensation problem is not yet satisfactorily resolved, and, therefore, the analysis of the effect of the noncondensables has not yet been carried out with any reasonable accuracy. Since the temperature gradient in the two-phase region is negligibly small, there is a discontinuity in the

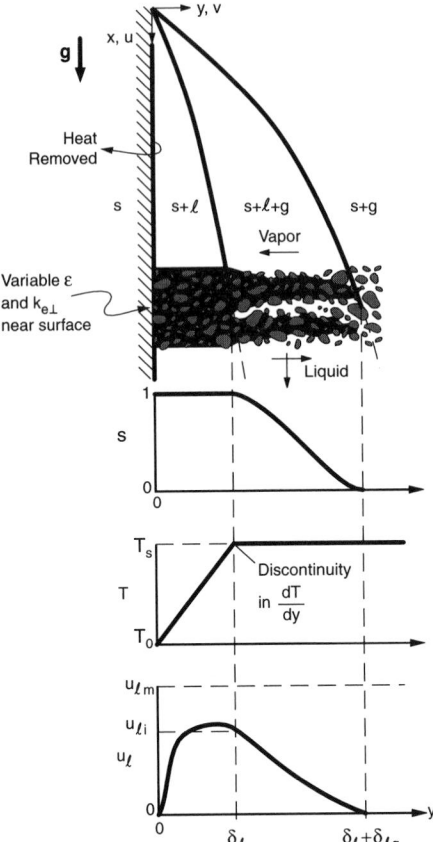

FIGURE 9.18 Film condensation at a vertical impermeable surface with $\delta_\ell/d > 1$ and $\delta_{\ell g}/d > 1$. Distributions of saturation, temperature, and liquid phase velocity are also depicted.

gradient of the temperature at δ_ℓ (Fig. 9.18). Then, following the standard procedures, the analytical treatment is based on the separation of the domains. Here, there are three domains, namely, $0 \leq y \leq \delta_\ell(x)$, $\delta_\ell(x) \leq y \leq \delta_\ell + \delta_{\ell g}(x)$, and $\delta_\ell + \delta_{\ell g}(x) \leq y \leq \infty$. In the following, we treat the first two domains; the third domain is assumed to have uniform fields.

Liquid-Film Region. The single-phase flow and heat transfer in this region can be described by the continuity equation, the momentum equation in Eqs. 9.76 and 9.77, and the energy equation (Eq. 9.80). We deal only with the volume-averaged velocities, such as $\langle u_\ell \rangle = u_\ell$; therefore, we drop the averaging symbol from the superficial (or Darcean) velocities. For the two-dimensional steady-state boundary-layer flow and heat transfer, we have (the coordinates are those shown in Fig. 9.18) the following:

$$\frac{\partial u_\ell}{\partial y} + \frac{\partial v_\ell}{\partial x} = 0 \tag{9.93}$$

$$\frac{\rho_\ell}{\varepsilon}\left(u_\ell \frac{\partial u_\ell}{\partial x} + v_\ell \frac{\partial u_\ell}{\partial y} \right) = \frac{\mu_\ell(x, y)}{\varepsilon} \frac{\partial^2 u_\ell}{\partial y^2} - \frac{\mu_\ell(x, y)}{K} u_\ell + g(\rho_\ell - \rho_g) \tag{9.94}$$

$$u_\ell \frac{\partial T}{\partial x} + v_\ell \frac{\partial T}{\partial y} = \frac{\partial}{\partial y}\left[\frac{k_{\ell\perp}(y)}{(\rho c_p)_\ell} + \varepsilon D_\perp^d(x, y) \right] \frac{\partial T}{\partial y} \tag{9.95}$$

The boundary conditions are

$$u_\ell = v_\ell = 0, \qquad T = T_0 \qquad \text{at } y = 0$$

$$u_\ell = u_{\ell i}, \qquad T = T_s \qquad \text{at } y = \delta_\ell \qquad (9.96)$$

T_s is the saturation temperature.

Since the liquid velocity at δ_ℓ, $u_{\ell i}$ is not known, an extra boundary condition is needed. For single-phase flows using the Brinkman treatment, we have

$$(\mu'_\ell)_{\delta_\ell^-} \left.\frac{\partial u_\ell}{\partial y}\right|_{\delta_\ell^-} = (\mu'_\ell)_{\delta_\ell^+} \left.\frac{\partial u_\ell}{\partial y}\right|_{\delta_\ell^+} \qquad \text{at } y = \delta_\ell \qquad (9.97)$$

This allowed us to make the transition at $y = \delta_\ell$ due to the solid matrix structural change (e.g., discontinuity in permeability), where μ'_ℓ depends on the matrix structure. Presently, we do not have much knowledge about μ'_ℓ for two-phase flows (even though we have assumed that the vapor shear is not significant because at $y = \delta_\ell$ we have $s = 1$). The simplest, but not necessarily an accurate assumption, is that of $(\mu'_\ell)_{\delta_\ell^-} = (\mu'_\ell)_{\delta_\ell^+}$.

The maximum velocity possible in the liquid phase is found by neglecting the macroscopic inertial and viscous terms. The result is

$$u_{\ell m} = \frac{K}{\mu_\ell} g(\rho_\ell - \rho_g) \qquad (9.98)$$

where, since this idealized flow is one-dimensional, as μ_ℓ decreases with increase in y, $u_{\ell m}$ increases with a maximum at $y = \delta_\ell$. However, in practice, u_ℓ does not reach $u_{\ell m}$ because of the lateral flow toward the two-phase region. Also, in most cases of practical interest, $\delta_\ell/d = O(1)$, and, therefore, the velocity no-slip condition at $y = 0$ causes a significant flow retardation throughout the liquid-film region. The velocity reaches its maximum value at $y = \delta_\ell$ if the shear stress $\mu_\ell \partial u_\ell/\partial y$ at δ_ℓ is zero; otherwise it peaks at $y < \delta_\ell$. This is also depicted in Fig. 9.18.

The convection heat transfer in the liquid-film region is generally negligible [113], therefore, Eq. 9.95 can be written as

$$D_\perp \frac{\partial T}{\partial y} \equiv \left[\frac{k_{\ell\perp}}{(\rho c_p)_\ell} + \varepsilon D_\perp^d \right] \frac{\partial T}{\partial y} = \text{constant} \qquad (9.99)$$

For $k_\ell < k_s$, $D_\perp$ is generally smallest near the bounding surface (A_f/A_t is largest) resulting in an expected significant deviation from the linear temperature distribution found in the film condensation in plain media.

Two-Phase Region. The two-phase flow and heat transfer are given by the continuity equations for the ℓ and g phases, the momentum equations (Eqs. 9.76 and 9.77), and the energy equation (Eq. 9.80). The two-phase region is assumed to be isothermal by neglecting the effect of the curvature (i.e., saturation) on the thermodynamic equilibrium state. This is justifiable, except for the very small pores (large p_c). For the steady-state flow considered here, we have (for the assumed isotropic phase permeabilities)

$$\frac{\partial u_\ell}{\partial x} + \frac{\partial v_\ell}{\partial y} = 0 \qquad (9.100)$$

$$\frac{\partial u_g}{\partial x} + \frac{\partial u_g}{\partial y} = 0 \qquad (9.101)$$

$$\frac{\rho\ell}{\varepsilon s} \left(u_\ell \frac{\partial u_\ell}{\partial x} + v_\ell \frac{\partial u_\ell}{\partial y} \right) = -\frac{\partial p_\ell}{\partial x} + \rho_\ell g - \frac{\mu_\ell}{K K_{r\ell}} u_\ell \qquad (9.102)$$

$$\frac{\rho\ell}{\varepsilon s}\left(u_\ell \frac{\partial v_\ell}{\partial x} + v_\ell \frac{\partial v_\ell}{\partial y}\right) = -\frac{\partial p_\ell}{\partial y} - \frac{\mu_\ell}{K K_{r\ell}} v_\ell \tag{9.103}$$

$$\frac{\rho_g}{\varepsilon(1-s)}\left(u_g \frac{\partial u_g}{\partial x} + v_g \frac{\partial u_g}{\partial y}\right) = -\frac{\partial p_g}{\partial x} + \rho_g g - \frac{\mu_g}{K K_{rg}} u_g \tag{9.104}$$

$$\frac{\rho_g}{\varepsilon(1-s)}\left(u_g \frac{\partial v_g}{\partial x} + v_g \frac{\partial v_g}{\partial y}\right) = -\frac{\partial p_g}{\partial y} - \frac{\mu_g}{K K_{rg}} v_g \tag{9.105}$$

along with $$p_g - p_\ell = p_c(s, \text{etc.}) \tag{9.106}$$

When the Leverett idealization (Table 9.9) is used, Eq. 9.106 reduces to $p_c = p_c(s)$. The convective terms in Eqs. 9.102 to 9.105 can be significant when the effect of thickening of δ_ℓ and $\delta_{\ell g}$ and the effect of g and p_c tend to redistribute the phases along the x axis (flow development effects). If the capillarity is more significant than the gravity, that is, $\sigma/[g(\rho_\ell - \rho_g)K/\varepsilon]$ is larger than unity, then we expect larger $\delta_{\ell g}$, and vice versa. The overall energy balance yields

$$\int_0^x \left(k_{\ell\perp}\frac{\partial T}{\partial y}\right)_{y=0} dx = \int_0^{\delta_\ell} \rho_\ell i_{\ell g} u_\ell \, dy + \int_{\delta_\ell}^{\delta_{\ell g}} \rho_g i_{\ell g} u_\ell \, dy + \int_0^{\delta_\ell} (\rho c_p)_\ell u_\ell (T_s - T) \, dy \tag{9.107}$$

where T_s is the saturation temperature. The last term on the right-hand side makes a negligible contribution to the overall heat transfer.

Large Bond Number Asymptote.   Although for the cases where d is small enough to result in $\delta_\ell/d \gg 1$ the capillarity will also become important, we begin by considering the simple case of negligible capillarity. As is shown later, the capillary pressure causes lateral flow of the liquid, thus tending to decrease δ_ℓ. However, the presence of the lateral flow also tends to decrease the longitudinal velocity in the liquid-film region, and this tends to increase δ_ℓ. The sum of these two effects makes for a δ_ℓ, which may deviate significantly from the large Bond number asymptotic behavior. Therefore, the limitation of the large Bond number asymptote, especially its overprediction of u_ℓ, should be kept in mind.

Assuming that Bo $\to \infty$, we replace the boundary condition on u_ℓ at location δ_ℓ with a zero shear stress condition (i.e., $\delta_{\ell g} = 0$ and only two regions are present). In addition, we have the initial conditions

$$u_\ell = v_\ell = 0, \qquad T = T_s \qquad \text{at } x = 0 \tag{9.108}$$

Next, we examine variations in $\mu_\ell(T)$, $\varepsilon(y)$, and $D_\perp(\text{Pe}_\ell, y)$, where $\text{Pe}_\ell = \bar{u}_\ell d/\alpha_\ell$. The variation in μ_ℓ can be nearly accounted for by using $\mu_\ell[(T_s + T_0)/2]$ in Eq. 9.94. For the packed beds of spheres, the variation of ε is significant only for $0 < y < 2d$. White and Tien [114] have included the effect of the variable porosity by using the variation in A_f/A_t. Here, we assume that $\delta_\ell/d \gg 1$, and, therefore, we do not expect the channeling to be significant. For the case of $\delta_\ell/d \simeq O(1)$, this porosity variation must be considered. We note that Pe_ℓ can be larger than unity and that the average liquid velocity $\bar{u}_\ell$ increases with x, and, in general, the variation of $D_\perp$ with respect to y should be included.

A similarity solution is available for Eqs. 9.93 to 9.95 subject to negligible macroscopic inertial and viscous forces, that is, small permeabilities and constant $D_\perp$ [115, 116]. The inertial and viscous forces are included by Kaviany [117] through the regular perturbation of the similarity solution for plain media, that is, the Nusselt solution [113]. The perturbation parameter used is

$$\xi_x = 2\left[\frac{\varepsilon g(\rho_\ell - \rho_g)}{\rho_\ell v_\ell^2}\right]^{-1/2} \frac{\varepsilon x^{1/2}}{K} \tag{9.109}$$

and for large ξ_x, the Darcean flow exists. The other dimensionless parameters are the subcooling parameter $c_{p\ell}(T_s - T_0)/i_{\ell g}$ and Prandtl number $\text{Pr}_\ell = \alpha_\ell/v_\ell$. The results show that, as is the case with the plain media, the film thickness is small. For example, for water with $\mathrm{Pr}_\ell = 10$ and $c_{p\ell}(T_s - T_0)/i_{\ell g} = 0.004$ to 0.2 (corresponding to 2 to 100°C subcooling), the film thickness δ_ℓ is between 1 and 8 mm for $K = 10^{-10}$ m^2 and between 0.1 and 0.8 mm for $K = 10^{-8}$ m^2. By using the Carman-Kozeny relation and $\varepsilon = 0.4$, we find that the latter permeability results in $\delta_\ell/d = 0.03 - 0.25$, which violates the local volume-averaging requirement.

The results of Parmentier [115] and Cheng [116] are (for $\delta_\ell/d \gg 1$) as follows:

$$\frac{2^{1/2}\mathrm{Nu}_x}{\mathrm{Gr}_x^{1/4}} = \frac{2}{\pi^{1/2}} \left[\frac{\mathrm{Pr}_\ell}{\xi_x \, \mathrm{erf} \, (\Delta \, \mathrm{Pr}_\ell^{1/2}/\xi_x^{1/2})} \right]^{1/2} \quad \text{for Bo} > 1 \qquad (9.110)$$

$$\frac{i_{\ell g}}{2c_{p\ell}(T_s - T_0)} + \frac{1}{\pi} = \frac{1}{\pi[\mathrm{erf} \, (\Delta \, \mathrm{Pr}_\ell^{1/2}/\xi_x^{1/2})]^2} \quad \text{for Bo} > 1 \qquad (9.111)$$

where for a given $i_{\ell g}/[c_{p\ell}(T_s - T_0)]$, Pr_ℓ, and ξ_x, Δ, and Nu_x are found, and where

$$\mathrm{Nu} = \frac{qx}{(T_s - T_0)k_{e\perp}}, \qquad \Delta = \frac{\delta_\ell}{x} \left(\frac{\mathrm{Gr}_x}{4} \right)^{1/4}, \qquad \mathrm{Gr}_x = \frac{\varepsilon g(\rho_\ell - \rho_g)x^3}{\rho_\ell v_\ell^2} \qquad (9.112)$$

Note that for small K, the Bond number $[g(\rho_\ell - \rho_g)K/\varepsilon]/\sigma$ is also small; therefore, the capillarity (i.e., the two-phase region) must be included. This is attended to next.

Small Bond Number Approximation. Presently no rigorous solution to the combined liquid-film and two-phase regions is available. However, some approximate solutions are available [118, 119, 120]. The available experimental results [120, 121] are not conclusive as they are either for $\delta_\ell/d \simeq O(1)$ or when they contain a significant scatter.

By considering capillary-affected flows, we expect that $v_g \gg u_g$, since $\partial s/\partial y \gg \partial s/\partial x$. Also, because the inertial force is negligible for the vapor flow, we reduce Eqs. 9.104 and 9.105 to

$$0 = -\frac{\partial p_g}{\partial y} - \frac{\mu_g}{KK_{rg}} v_g \quad \text{or} \quad v_g = -\frac{1}{\mu_g} KK_{rg} \frac{\partial p_g}{\partial y} = -\frac{\dot{m}_{g\ell}}{\rho_g} \qquad (9.113)$$

Note that from Eq. 9.101, we find that v_g is constant along y. For the liquid phase, both u_ℓ and v_ℓ are significant, and v_ℓ changes from a relatively large value at δ_ℓ to zero at $\delta_\ell + \delta_{\ell g}$; therefore, $v_\ell \partial v_\ell/\partial y$ will not be negligibly small. Also, $v_g \partial u_\ell/\partial y$ may not be negligible. Then we have

$$\frac{\rho_\ell}{\varepsilon s} v_\ell \frac{\partial u_\ell}{\partial y} = -\frac{\partial p_\ell}{\partial x} + \rho_\ell g + \frac{\mu_\ell}{KK_{r\ell}} u_\ell \qquad (9.114)$$

$$\frac{\rho_\ell}{\varepsilon s} v_\ell \frac{\partial v_\ell}{\partial y} = -\frac{\partial p_\ell}{\partial y} - \frac{\mu_\ell}{KK_{r\ell}} v_\ell \qquad (9.115)$$

Here, we assume that the gaseous phase hydrostatic pressure is negligible. We note that the approximations made in the evaluations of $K_{r\ell}$ and p_c cause more errors in the determination of u_ℓ and v_ℓ than the exclusion of the inertial terms. Furthermore, we expect $\partial p_\ell/\partial x$ to be small. Then, we can write Eqs. 9.114 and 9.115 as

$$0 = \rho_\ell g - \frac{\mu_\ell}{KK_{r\ell}} u_\ell \quad \text{or} \quad u_\ell = \frac{g}{v_\ell} KK_{r\ell} \qquad (9.116)$$

$$0 = -\frac{\partial p_\ell}{\partial y} - \frac{\mu_\ell}{KK_{r\ell}} v_\ell \quad \text{or} \quad v_\ell = \frac{1}{\mu_\ell} KK_{r\ell} \frac{\partial p_\ell}{\partial y} \qquad (9.117)$$

The velocity distribution given by Eq. 9.116 is that of a monotonic decrease from the value of $u_{\ell i}$ at δ_ℓ to zero at $\delta_\ell + \delta_{\ell g}$. The specific distribution depends on the prescribed $K_{r\ell}(s, \text{etc.})$. The distribution of v_ℓ given by Eq. 9.117 is more complex, because $-\partial p_\ell/\partial y$ increases as s decreases (as $y \to \delta_\ell + \delta_{\ell g}$). Although $\partial p_g/\partial y$ is needed to derive the vapor to δ_ℓ, we note that

$$\frac{\partial p_\ell}{\partial y} = -\frac{\partial p_c}{\partial y} + \frac{\partial p_g}{\partial y} \simeq \frac{\partial p_c}{\partial y} \tag{9.118}$$

From the experimental results on $p_c(s)$, we can conclude that in the two-phase region v_ℓ also decreases monotonically with y. By using Eq. 9.118, we write Eq. 9.117 as

$$v_\ell = \frac{1}{\mu_\ell} K K_{r\ell} \frac{\partial p_c}{\partial y} \tag{9.119}$$

The momentum equations (Eqs. 9.116 and 9.119) can be inserted in Eq. 9.100, and when p_c and $K_{r\ell}$ are given in terms of s, the following saturation equation is obtained:

$$\frac{\partial u_\ell}{\partial x} + \frac{\partial v_\ell}{\partial y} = \frac{gK}{v_\ell} \frac{\partial K_{r\ell}}{\partial x} + \frac{K}{\mu_\ell} \frac{\partial}{\partial y}\left(K_{r\ell} \frac{\partial p_c}{\partial y}\right) \tag{9.120}$$

with
$$
\begin{array}{lll}
s = 1 & \text{at } y = \delta_\ell & \\
s = 0 & \text{at } y = \delta_\ell + \delta_{\ell g} & \\
s = 0 & \text{at } x = 0 &
\end{array}
\tag{9.121}
$$

The evaluation of $u_{\ell i}$, δ_ℓ, and $\delta_{\ell g}$ requires the analysis of the liquid-film region. For a negligible inertial force and with the use of the viscosity evaluated at the average film temperature $\bar{\mu}$ and the definition of $u_{\ell m}$, that is, Eq. 9.98, we can integrate Eqs. 9.94 and 9.96 to arrive at [120]

$$u_\ell = \frac{u_{\ell i} - u_{\ell m} + u_{\ell m} \cosh\left[\delta_\ell/(K/\varepsilon)^{1/2}\right]}{\sinh\left[\delta_\ell/(K/\varepsilon)^{1/2}\right]} \sinh\frac{y}{(K/\varepsilon)^{1/2}} + u_{\ell m}\left[1 - \cosh\frac{y}{(K/\varepsilon)^{1/2}}\right] \tag{9.122}$$

Now, using $(\mu'_\ell)_{\delta_\ell^-} = (\mu'_\ell)_{\delta_\ell^+}$ in Eq. 9.97 and the overall mass balance given by

$$\int_0^x \dot{m}_{g\ell}\, dx = \int_0^{\delta_\ell(x)} \rho_\ell u_\ell\, dy + \int_{\delta_\ell(x)}^{\delta_{\ell g}(x)} \rho_\ell u_\ell\, dy \tag{9.123}$$

Chung et al. [120] solve for the previously mentioned three unknowns. Their experimental and predicted results for $d = 0.35$ mm are shown in Fig. 9.19, where $\mathrm{Ra}_x = \mathrm{Gr}_x\, \mathrm{Pr}_\ell$. We note that in their experiment (a closed system) the condensate collects, that is, the liquid film thickens, at the bottom of the cooled plate. When the plate length is very large, this nonideal lower por-

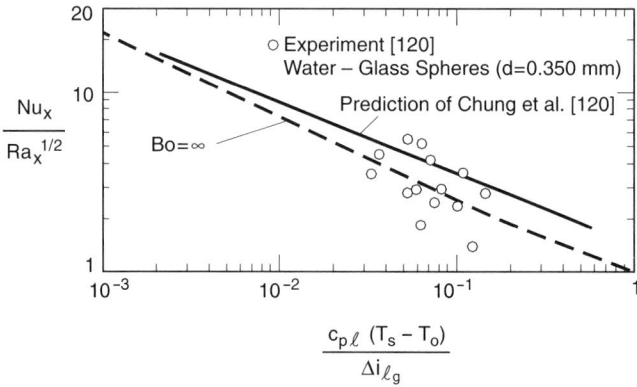

FIGURE 9.19 Prediction and experimental results of Chung et al. [120] for condensation (in a packed bed of spheres) at a bounding vertical impermeable surface. The large Bo results of Cheng [116] are also shown.

tion behavior may be neglected. However, in their experiment with the relatively short plate, this can significantly influence the phase distributions and velocities. The Bond number is 4×10^{-5}. For this size particle and for $c_{p\ell}(T_s - T_0)/i_{\ell g} = 0.1$, they find $u_{\ell i}/u_{\ell m} \simeq 0.06$, that is, the velocity at δ_ℓ is much less than the maximum Darcean velocity given by Eq. 9.98. Note that although the velocity in the liquid phase is so small, the thickness of the liquid-film region is not substantially different than that found for the single-layer (Bo > 1) model of Cheng [116] (as evident from the heat transfer rate). The analysis of Chung et al. [120] shows that $\delta_\ell/d \simeq$ 3.3. They also use $(A_f/A_t)(y)$ in their computation. Note that for this analysis to be meaningful, δ_ℓ/d has to be larger than, say, 10, so that the variation of u_ℓ in $0 \le y \le \delta_\ell$ can be predicted with sufficient accuracy. Therefore, the applicability of the analysis to their experimental condition is questionable.

Evaporation at Vertical Impermeable Bounding Surfaces

For plain media, the film evaporation adjacent to a heated vertical surface is similar to the film condensation. In porous media, we also expect some similarity between these two processes. For the reasons given in the last section, we do not discuss the cases where $\delta_g/d \simeq 1$, where δ_g is the vapor-film region thickness. When $\delta_g/d \gg 1$ and because the liquid flows (due to capillarity) toward the surface located at $y = \delta_g$, we also expect a large two-phase region, that is, $\delta_{g\ell}/d \gg 1$. Then a local volume-averaged treatment can be applied.

The asymptotic solution for Bo > 1, where $\delta_{g\ell} = 0$ (as was $\delta_{\ell g}$) and the similarity solution given there holds. Parmentier [115] has discussed this asymptotic solution. When given in terms of Ra_x, this solution is (given the superheating parameter and Ra_x, then Nu_x and Δ_g are solved simultaneously)

$$\frac{Nu_x}{Ra_x^{1/2}} = \frac{1}{\pi^{1/2} \, erf \, (\Delta_g/2)} \qquad \text{for Bo} \to \infty \qquad (9.124)$$

$$\frac{c_{pg}(T_0 - T_s)}{\Delta i_{\ell g}} = \frac{\pi^{1/2}\Delta_g}{2} \, exp(\Delta^2/2) \, erf \, (\Delta_g/2) \qquad \text{for Bo} \to \infty \qquad (9.125)$$

with
$$Ra_x = \frac{(\rho_\ell - \rho_g)gKx}{\rho_g v_g \alpha_\ell}, \qquad \Delta_g = \frac{\delta_g}{x} \, Ra_x^{1/2} \qquad (9.126)$$

Note that Eqs. 9.110 and 9.111 are given in terms of the perturbation parameter ξ_x, but otherwise are identical to Eqs. 9.124 and 9.125. As we discussed, for small Bond numbers, which is the case whenever $\delta_g/d \gg 1$, a nearly isothermal two-phase region exists. The vapor will be rising due to buoyancy, similar to the falling of the liquid given by Eq. 9.116, except here we include the hydrostatic pressure of the liquid phase. This gives (defining x to be along $-g$)

$$0 = (\rho_\ell - \rho_g)g - \frac{\mu_g}{KK_{rg}} u_g \qquad \text{or} \qquad u_g = \frac{(\rho_\ell - \rho_g)g}{\mu_g} KK_{rg} \qquad (9.127)$$

The lateral motion of the vapor is due to the capillarity, and in a manner similar to Eq. 9.117, we can write

$$0 = -\frac{\partial p_g}{\partial y} - \frac{\mu_g}{KK_{rg}} v_g \qquad \text{or} \qquad v_g = -\frac{1}{\mu_g} KK_{rg} \frac{\partial p_g}{\partial y} \qquad (9.128)$$

The axial motion of the liquid phase is negligible and the lateral flow is similar to that given by Eq. 9.113 and is described as

$$0 = -\frac{\partial p_\ell}{\partial y} - \frac{\mu_\ell}{KK_{r\ell}} v_\ell \qquad \text{or} \qquad v_\ell = \frac{\dot{m}_{\ell g}}{\rho_\ell} = -\frac{1}{\mu_\ell} KK_{r\ell} \frac{\partial p_\ell}{\partial y} \simeq \frac{1}{\mu_\ell} KK_{r\ell} \frac{\partial p_c}{\partial y} \qquad (9.129)$$

The saturation will increase monotonically with y with $s = 0$ at $y = \delta_g$, and $s = 1$ at $y = \delta_{g\ell}$. In principle, we then expect the results of Chung et al. [120] for the condensation to apply to the evaporation. However, since the available experimental results for p_c (drainage versus imbibition) show a hysteresis, and also due to the lack of symmetry in $K_{r\ell}(s)/K_{rg}$, we do not expect a complete analogy.

Evaporation at Horizontal Impermeable Bounding Surfaces

We now consider heat addition to a horizontal surface bounding a liquid-filled porous media from below. When the temperature of the bounding surface is at or above the saturation temperature of the liquid occupying the porous media, evaporation occurs. We limit our discussion to matrices that remain unchanged. When nonconsolidated particles make up a bed, the evaporation can cause void channels through which the vapor escapes [122, 123, 124]. Here, we begin by mentioning a phenomenon observed in some experiments [125] where no significant superheat is required for the evaporation to start, that is, evaporation is through surface evaporation of thin liquid films covering the solid surface (in the evaporation zone). By choosing the surface heating (or external heating), we do not address the volumetrically heated beds (e.g., Ref. 126). For small Bond numbers, for small heat flux, that is, when $(T_0 - T_s) \simeq 0$, the vapor generated at the bottom surface moves upward, the liquid flows downward to replenish the surface, and the surface remains wetted. When a critical heat flux q_{cr} is exceeded, a vapor film will be formed adjacent to the heated surface, and a two-phase region will be present above this vapor-film region. The two-phase region will have an evaporation zone where the temperature is not uniform and a nearly isothermal region where no evaporation occurs. As Bo increases, the role of capillarity diminishes. For Bo > 1, the behavior is nearly the same as when no rigid matrix is present, that is, the conventional pool boiling curve will be nearly observed.

In the following, we examine the experimental results [127], that support these two asymptotic behaviors, that is, Bo > 1, the high-permeability asymptote, and Bo << 1, the low-permeability asymptote. Then, we discuss the low-permeability asymptote using a one-dimensional model [109, 125, 128]. The one-dimensional model is also capable of predicting q_{cr}, that is, the onset of dryout. We note that the hysteresis observed in isothermal two-phase flow in porous media is also found in evaporation-condensation and that the q versus $T_0 - T_s$ curve shows a decreasing q (or $T_0 - T_s$) and an increasing q (or $T_0 - T_s$) branch.

Effect of Bond Number.   In the experiments of Fukusako et al. [127], packed beds of spheres (glass, steel, and aluminum) occupied by fluorocarbon refrigerants were heated from below to temperatures above saturation. Their results for the glass (Freon-11) system with a bed height of 80 mm and for four different particle sizes (Bond numbers) are given in Fig. 9.20. Also shown are their results for pool boiling (no particles). They have not reported the heat flux corresponding to $(T_0 - T_s) < 8°C$ for this solid-fluid system, because all of their results are for $q \geq 10^4$ W/m². For Bo $\to \infty$, that is, with no solid matrix present, the conventional pool boiling curve is obtained, that is, as $(T_0 - T_s)$ increases, after the required superheat for nucleation is exceeded, the nucleate (with a maximum), the transient (with a minimum), and the film boiling regimes are observed. For large particles, this behavior is not significantly altered. However, as the particle size reduces, this maximum and minimum become less pronounced, and for very small particles, they disappear. For the solid-fluid system used in their experiment, this transition appears to occur at Bo ≤ 0.0028. Note that we have only been concerned with the heating from planar horizontal surfaces. For example, the experimental results of Fand et al. [130] for heating of a 2-mm-diameter tube in a packed bed of glass spheres-water ($d = 3$ mm, Bo = 0.003) shows that unlike the results of Fukusako et al. [127], no monotonic increase in q is found for $(T_0 - T_s) > 100°C$. Tsung et al. [131] use a heated sphere in a bed of spheres ($d \geq 2.9$ mm). Their results also show that as d decreases, the left-hand portion of the

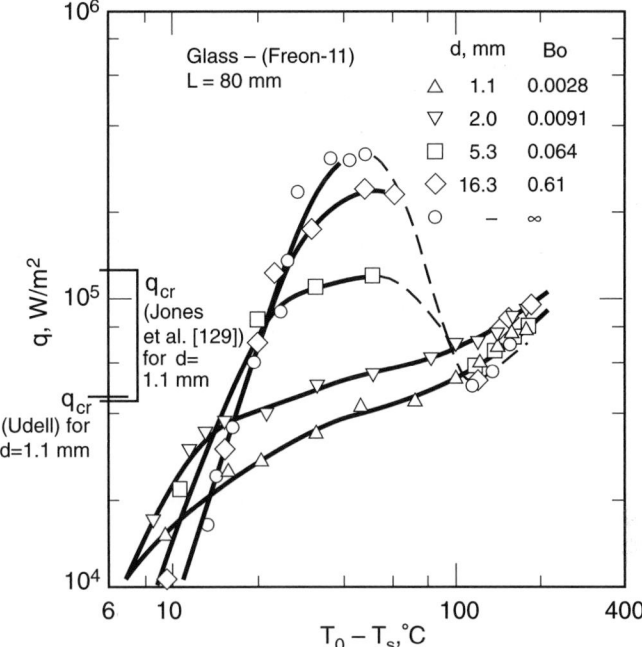

FIGURE 9.20   Experimental results of Fukusako et al. [127] for the bounding surface heating of the liquid-filled beds of spheres to temperatures above saturation. The case of pool boiling (Bo → ∞) is also shown along with the dryout heat flux predicted by Udell [90] and given in the measurement of Jones et al. [129].

q versus ΔT curve moves upward, that is, the required superheat for a given q is smaller in porous media (compared to plain media).

The results given in Fig. 9.20 show that for very small values of Bo (Bo ≤ 0.01), the surface temperature also increases monotonically with the heat flux. This supports the theory of a heat removal mechanism that does not change, unlike that observed in the pool boiling in plain media. For the system shown in Fig. 9.20 and for $d = 1.1$ mm, the transition from the surface-wetted condition to the formation of a vapor film occurs at the critical heat flux

$$q_{cr} = \frac{K\Delta i_{\ell g}(\rho_\ell - \rho_g)g}{v_g}\left[1 + \left(\frac{v_g}{v_\ell}\right)^{1/4}\right]^{-4} = 4.4 \times 10^4 \text{ W/m}^2 \tag{9.130}$$

Jones et al. [129] measure q_{cr} using various fluids. They obtain a range of q_{cr} with the lowest value very close to that predicted by Eq. 9.130. These values of q_{cr} are also shown in Fig. 9.20. Based on the theory of evaporation in porous media given earlier, no vapor film is present until q exceeds q_{cr} (for small Bo). Then, for $q > q_{cr}$, the surface temperature begins to rise, that is, $T_0 - T_s > 0$. Now, by further examination of Fig. 9.20, we note that this low Bond number asymptote—that is, the liquid wetting of the surface for $(T_0 - T_s) = 0$ followed by the simultaneous presence of the vapor film and the two-phase regions in series for $(T_0 - T_s) > 0$—is not distinctly found from the experimental results of Fukusako et al. [127]. Although as Bo → 0, the trend in their results supports this theory; the Bo encountered in their experiment can yet be too large for the realization of this asymptotic behavior.

In Fig. 9.21, q versus ΔT curves are drawn based on the Bo → ∞ and the Bo → 0 asymptotes and the intermediate Bo results of Fukusako et al. [127]. The experimental results of

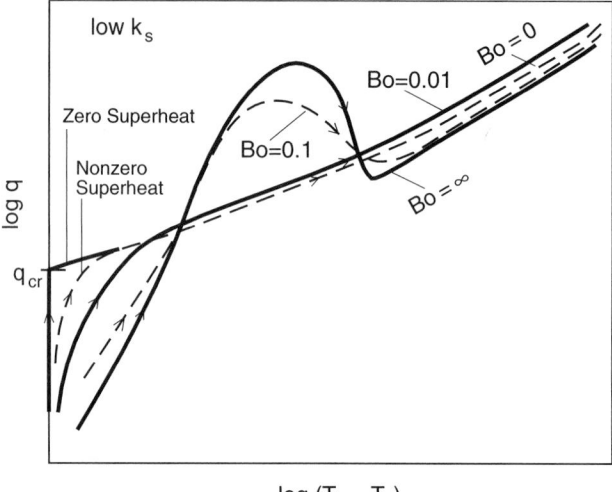

FIGURE 9.21   Effect of the Bond number on the q versus $T_0 - T_s$ curve is depicted based on the Bo $\rightarrow$ 0 and ∞ asymptotes and the experimental results of Fukusako et al. [127]. The solid phase thermal conductivity is low.

Udell [90] for Bo $\rightarrow$ 0 do not allow for the verification of the Bo $\rightarrow$ 0 curve in this figure, that is, the verification of q versus ΔT for Bo $\rightarrow$ 0 is not yet available. Also, the effect of the particle size (Bond number) on the heat flux for $(T_0 - T_s) > 100°C$ is not rigorously tested, and the trends shown are based on the limited results of Fukusako et al. [127] for this temperature range. In examining the q versus ΔT behavior for porous media, we note the following.

- The thermal conductivity of the solid matrix greatly influences the q versus ΔT curve, and the results of Fukusako et al. [127] are for a nonmetallic solid matrix. Later, we examine some of the results for metallic matrices and show that as k_s increases, q increases (for a given $T_0 - T_s$).

- In the experiments discussed earlier, no mention of any hysteresis has been made in the q versus $(T_0 - T_s)$ curve. However, as is shown later, at least for thin porous layer coatings, hysteresis has been found, and in the q decreasing branch, the corresponding $(T_0 - T_s)$ for a given q is much larger than that for an increasing q branch.

- In the behavior depicted in Fig. 9.21, it is assumed that only the particle size is changing, that is, particle shape, porosity, fluid properties, heated surface, and so forth, all remain the same.

- The theoretical zero superheat at the onset of evaporation is not realized, and experiments do show a finite $\partial q / \partial T_0$ as $(T_0 - T_s) \rightarrow 0$.

The analysis for large particle sizes is expected to be difficult. For example, when $(T_0 - T_s) > 0$ and a thin vapor film is found on the heated surface, the thickness of this film will be less than the particle size; therefore, $(T_0 - T_s)$ occurs over a distance less than d. Since $k_s \neq k_s$, this violates the assumption of the local thermal equilibrium. Also, as the particle size increases, boiling occurs with a large range of bubble sizes, that is, the bubbles may be smaller and larger (elongated) than the particle size.

In the following section, we examine the Bo $\rightarrow$ 0 asymptotic behavior by using the volume-averaged governing equations. This one-dimensional analysis allows for an estimation of q_{cr} and the length of the isothermal two-phase region for $q > q_{cr}$.

A One-Dimensional Analysis for Bo << 1.   Figure 9.22 depicts the one-dimensional model for evaporation in porous media with heat addition q from the impermeable lower bounding surface maintained at $T_0 > T_s$, where T_s is the saturation temperature. The vapor-film region has a thickness δ_g, and the two-phase region has a length $\delta_{g\ell}$.

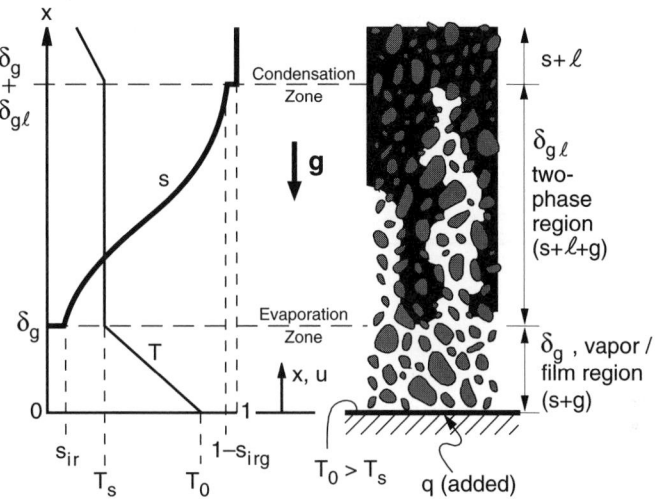

FIGURE 9.22   Evaporation due to the heat addition from below at temperatures above the saturation. The vapor-film region, the two-phase region, and the liquid region, as well as the evaporation and condensation zones are shown. Also shown are the distributions of temperature and saturation within these regions.

For $\delta_g \leq x \leq \delta_g + \delta_{g\ell}$, the saturation is expected to increase monotonically with x. The vapor generated at the evaporation zone (the thickness of this zone is in practice finite but here taken as zero) at $x = \delta_g$, moves upward (buoyancy-driven), condenses (condensation occurs in the condensation zone which is taken to have zero thickness) at $x = \delta_g + \delta_{\ell g}$, and returns as liquid (buoyancy- and capillary-driven). By allowing for irreducible saturations s_{ir} and $s_{ir\,g}$, that is, assuring continuous phase distributions for the two-phase flow, we have to assume an evaporation zone just below $x = \delta_g$ in which s undergoes a step change and evaporation occurs. A similar zone is assumed to exist above $x = \delta_g + \delta_{g\ell}$ over which s undergoes another step change and condensation occurs (condensation zone). Next, we consider cases with $\delta_g > d$ and $\delta_{g\ell} >> d$, where we can apply the volume-averaged governing equations based on bulk properties. For $s < 1$, the liquid will be in a superheated state depending on the local radius of curvature of the meniscus. Therefore, the two-phase region is only approximately isothermal. For steady-state conditions, the heat supplied q is removed from the upper single-phase (liquid) region. Since the heat supplied to the liquid region causes an unstable stratification, natural convection can occur that can influence the two-phase region [125, 132]. In the following one-dimensional analysis, this phenomenon is not considered.

Vapor-Film Region.   The one-dimensional heat conduction for the stagnant vapor-film region is given by

$$q = -k_e(x)\frac{dT}{dx} \tag{9.131}$$

Since $k_g/k_s < 1$, we expect that for the packed beds near the bounding surface the magnitude of k_e will be smaller than the bulk value. Therefore, a nonlinear temperature distribution is expected near this surface. However, if we assume k_e to be constant within δ_g, then we will have

$$q = k_\ell \frac{T_0 - T_s}{\delta_g} \qquad (9.132)$$

where, for a given q, we have $T_0 - T_s$ and δ_g as the unknowns. Generally, $T_0 - T_s$ is also measured, which leads to the determination of δ_g. We note again that between the vapor film and the two-phase region an evaporation zone exists in which the saturation and temperature are expected to change continuously. If a jump in s was allowed across it, it would be inherently unstable and would invade the two adjacent regions intermittently. The condensation zone at $x = \delta_g + \delta_{g\ell}$ is expected to have a similar behavior. The present one-dimensional model does not address the examination of these zones.

Two-Phase Region.   The analysis of the two-phase region is given by Sondergeld and Turcotte [125], Bau and Torrance [128], and more completely by Udell [90] and Jennings and Udell [133]. The vapor that is generated at $x = \delta_g$ and is given by

$$(\rho_g u_g)_{\delta_g} = \frac{q}{\Delta i_{\ell g}} \qquad (9.133)$$

flows upward primarily due to buoyancy. By allowing for the variation in p_g, the momentum equation for the gas phase will be Eq. 9.77, except that the inertial, drag, and surface-tension gradient terms are negligible because of the small Bond number assumption. This gives

$$0 = -\frac{dp_g}{dx} + \rho_g g - \frac{\mu_g}{K K_{rg}} u_g \qquad (9.134)$$

where we have used $u_g = \langle u_g \rangle$, $p_g = \langle p \rangle^g$, and $K_g = K K_{rg}$. Since the net flow at any cross section is zero, we have

$$\rho_g u_g + \rho_\ell u_\ell = 0 \qquad (9.135)$$

as the continuity equation. The momentum equation for the liquid phase (Eq. 9.76) becomes

$$0 = -\frac{dp_\ell}{dx} + \rho_\ell g - \frac{\mu_\ell}{K K_{r\ell}} u_\ell \qquad (9.136)$$

where the local pressure p_g and p_ℓ are related through the capillary pressure (Eq. 9.106).

By using Eqs. 9.134 through 9.136 and Eq. 9.106, we have

$$\frac{dp_c}{dx} = -\frac{q}{K i_{\ell g}} \left(\frac{v_g}{K_{rg}} + \frac{v_\ell}{K_{r\ell}} \right) + (\rho_\ell - \rho_g)g \qquad (9.137)$$

Next, by assuming that the Leverett J function is applicable and that $K_{r\ell}$ and K_{rg} can be given as functions of s only, Eq. 9.137 can be written in terms of the saturation only. Udell [90] uses the p_c correlation given in Table 9.9 and the relative permeabilities suggested by Wyllie [93] as given in Table 9.10. By using these, Eq. 9.137 becomes

$$\frac{\sigma}{(K/\varepsilon)^{1/2}} \frac{dJ}{dx} = -\frac{q}{K i_{\ell g}} \left[\frac{v_g}{(1-S)^3} + \frac{v_\ell}{S^3} \right] + (\rho_\ell - \rho_g)g = \frac{\sigma}{(K/\varepsilon)^{1/2}} \frac{dJ}{ds} \frac{dS}{dx} \qquad (9.138)$$

where, as before,

$$S = \frac{s - s_{ir}}{1 - s_{ir} - s_{ir\,g}} \qquad (9.139)$$

Next, we can translate the origin of x to δ_g, and then by integrating over the two-phase zone, we will have

$$\delta_{g\ell} = \int_0^1 \frac{[\sigma/(K/\varepsilon)^{1/2}](dJ/dS)}{-(q/Ki_{\ell g})\{[v_g/(1-S)^3] + (v_\ell/S^3)\} + (\rho_g - \rho_\ell)g} \, dS \qquad (9.140)$$

Whenever q, the liquid and vapor properties, and K are known, $\delta_{g\ell}$ can be determined from Eq. 9.140 and the saturation distribution can be found from Eqs. 9.138 and 9.139.

Note that when in Eq. 9.137 the viscous and gravity forces exactly balance, the capillary pressure gradient and, therefore, the saturation gradient become zero. For this condition, we have the magnitude $\delta_{g\ell}$ tending to infinity. This is evident in Eq. 9.140. The heat flux corresponding to this condition is called the *critical heat flux* q_{cr}. For $q > q_{cr}$, the thickness of the two-phase region decreases monotonically with q. Figure 9.23 shows the prediction of Udell [90] as given by Eq. 9.140, along with his experimental results for the normalized $\delta_{g\ell}$ as a function of the normalized q. For large q, an asymptotic behavior is observed. The critical heat flux q_{cr} (normalized) is also shown for the specific cases of $v_\ell/v_g = 0.0146$ and Bo $= 5.5 \times 10^{-7}$.

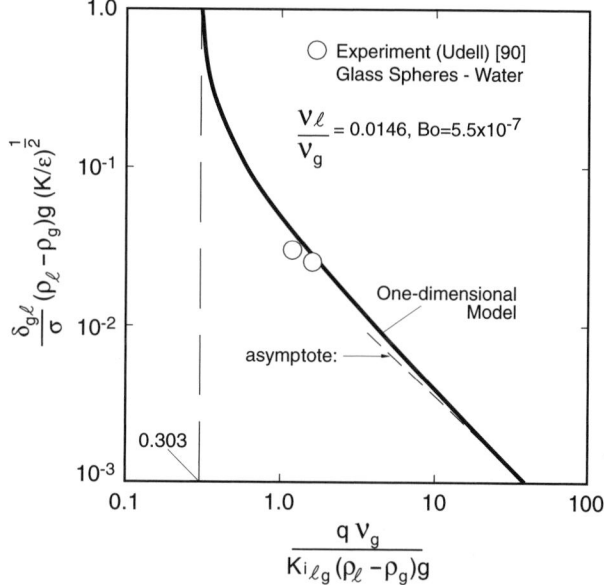

FIGURE 9.23   Variation of the normalized thickness of the two-phase region as a function of the normalized heat flux for evaporation from the heated horizontal surface.

Onset of Film Evaporation.   The saturation at which the saturation gradient is zero (and $\delta_{g\ell} \to \infty$) is found by setting the denominator of Eq. 9.140 to zero, that is,

$$\frac{q_{cr}}{K\Delta i_{\ell g}}\left[\frac{v_g}{(1-S_{cr})^3} + \frac{v_\ell}{S_{cr}^3}\right] = (\rho_\ell - \rho_g)g \qquad (9.141)$$

For this critical reduced saturation S_{cr}, the critical heat flux is given by Eq. 9.130.

Bau and Torrance [128] use a different relative permeability-saturation relation and arrive at a slightly different relation. Jones et al. [129] use a similar treatment and find a relationship for q_{cr} that gives values lower than those predicted by Eq. 9.130 by a factor of approximately 2. It should be noted that these predictions of q_{cr} are estimations and that the effects of wettability, solid matrix structure (all of these studies consider spherical particles only), and surface tension (all of which influence the phase distributions) are included only through the relative permeabilities. These permeabilities, in turn, are given as simple functions of the saturation only. Therefore, the use of realistic and accurate relative permeability relations is critical in the prediction of q_{cr}.

Evaporation at Thin Porous-Layer-Coated Surfaces

Evaporation within and over thin porous layers is of interest in wicked heat pipes and in surface modifications for the purpose of heat transfer enhancement. The case of very thin layers, that is, $\delta/d \simeq 1$ where δ is the porous-layer thickness, has been addressed by Konev et al. [134], Styrikovich et al. [135], and Kovalev et al. [136]. Due to the lack of the local thermal equilibrium in the two-phase region inside the thin porous layer, we do not pursue the analysis for the case of $\delta/d \simeq 1$.

When $\delta/d \gg 1$ but $\delta/(\delta_g + \delta_{g\ell}) < 1$, the two-phase region extends to the plain medium surrounding the porous layer. Presently, no detailed experimental results exist for horizontal surfaces coated with porous layers with $\delta \neq \delta_g + \delta_{g\ell}$. The experimental results of Afgan et al. [137] are for heated horizontal tubes (diameter D) and as is shown in their experiments $\delta < \delta_g + \delta_{g\ell}$. Their porous layers are made by the sintering of metallic particles. The particles are spherical (average diameter $d = 81$ μm) and are fused onto the tube in the process of sintering. From the various porous-layer coatings they use, we have selected the following three cases in order to demonstrate the general trends in their results.

- A layer of thickness $\delta/d \simeq 27$ with $K = 1.4 \times 10^{-10}$ m^2, $\varepsilon = 0.70$, Bo $= 2.6 \times 10^{-5}$, made of stainless steel particles ($k_s = 14$ W/m-K), and coated over a 16-mm-diameter stainless steel tube ($D/\delta = 7.3$).
- A layer of thickness $\delta/d \simeq 6.7$ with $K = 3 \times 10^{-11}$ m^2, $\varepsilon = 0.50$, Bo $= 7.2 \times 10^{-6}$, made of titanium particles ($k_s = 21$ W/m-K), and coated over an 18-mm-diameter stainless steel tube ($D/\delta = 33$).
- A layer of thickness $\delta/d \simeq 5.5$ with $K = 2.0 \times 10^{-12}$ m^2, $\varepsilon = 0.30$, Bo $= 8.9 \times 10^{-7}$, made of stainless steel particles, and coated over a 3-mm-diameter stainless steel tube ($D/\delta = 6.7$).

We have used the mean particle size d of 81 μm and the Carman-Kozeny equation for the calculation of the permeability. The fluid used is water. Their experimental results for these three cases are given in Fig. 9.24. In their experimental results, q is larger in the desaturation

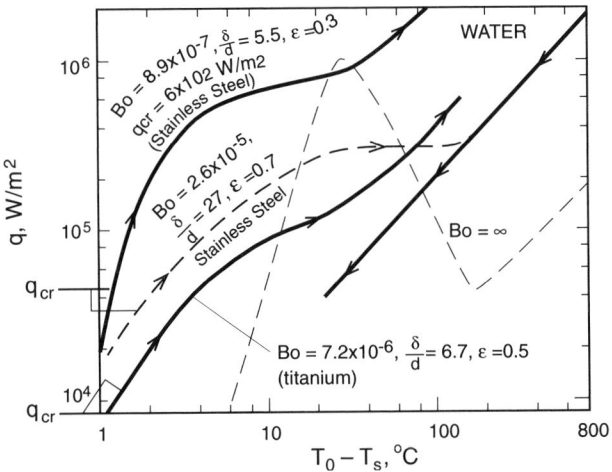

FIGURE 9.24 Experimental results of Afgan et al. [137] for evaporation from tubes coated with porous layers and submerged in a pool of water.

branch, while in the experimental results of Bergles and Chyu [138], q is larger in the saturation branch. In order to examine whether the porous-layer thicknesses used in these experiments are larger than $\delta_g + \delta_{g\ell}$, we apply the prediction of Udell [90] for the thickness of the two-phase region. His results are shown in Fig. 9.23. The asymptote for heat fluxes much larger than the critical heat flux is given by

$$\frac{\delta_{g\ell}(\rho_g - \rho_\ell)g(K/\varepsilon)^{1/2}}{\sigma} \frac{q\nu_g}{K\Delta i_{\ell g}(\rho_\ell - \rho_g)g} = 0.0368 \qquad q \gg q_{cr} \tag{9.142}$$

or

$$\delta_{g\ell} = 0.0368 \frac{\sigma \Delta i_{\ell g}(K\varepsilon)^{1/2}}{q\nu_g} \qquad q \gg q_{cr} \tag{9.143}$$

For $\delta_{g\ell} = \delta$, we have

$$q(\delta_{g\ell} = \delta) = \frac{0.0368\sigma\Delta i_{\ell g}(K\varepsilon)^{1/2}}{\delta\nu_g} \tag{9.144}$$

For those cases presented in Fig. 9.24, we have calculated the required q for $\delta_{g\ell} = \delta$. The values are

$$q\left(\delta_{g\ell} = \delta, \frac{\delta}{d} = 27, \varepsilon = 0.7 \text{ sample}\right) = 1.10 \times 10^6 \text{ W/m}^2$$

$$q\left(\delta_{g\ell} = \delta, \frac{\delta}{d} = 6.7, \varepsilon = 0.5 \text{ sample}\right) = 1.73 \times 10^6 \text{ W/m}^2$$

$$q\left(\delta_{g\ell} = \delta, \frac{\delta}{d} = 5.5, \varepsilon = 0.3 \text{ sample}\right) = 4.23 \times 10^5 \text{ W/m}^2 \tag{9.145}$$

We note that these heat fluxes are lower bounds, because δ is actually occupied by the vapor-film region, evaporation zone, as well as the two-phase region. For the porous layer to contain both of the layers, we need heat fluxes much larger than those given by Eq. 9.145, that is,

$$\delta \geq \delta_g + \delta_{g\ell} \qquad \text{or} \qquad q > q(\delta = \delta_{g\ell}) \tag{9.146}$$

Upon examining the experimental results given in Fig. 9.24, we note that except for the $\delta/d = 5.5$ layer, we have $q < q(\delta = \delta_{g\ell})$, that is, the two-phase region extends beyond the porous layer and into the plain medium.

No rigorous analysis for the case of $\delta < \delta_g + \delta_{g\ell}$ is available. Assuming that the theory of isothermal two-phase is applicable, we postulate that the portion of the two-phase region that is inside the porous layer will be unstable. This instability will be in the form of intermittent drying of this portion, that is, the entire porous layer becoming intermittently invaded by the vapor phase only. When the porous media is dry, there will be a nucleate boiling at the interface of the porous plain medium. Figure 9.25 depicts such an intermittent drying. When the two-phase region extends into the porous layer, the two-phase region will be at the saturation temperature (assuming negligible liquid superheat due to the capillarity). The evaporation takes place in the evaporation zone, just below the two-phase zone. The saturation at $x = \delta$ will be smaller than $1 - s_{ir\,g}$. This saturation is designated by s_m, which is similar to that for thick porous layers discussed in the previous sections. When the porous layer dries out, the evaporation will be at $x = \delta$. The frequency of this transition (i.e., intermittent drying of the porous layer) decreases as the porous-layer thickness increases and should become zero for $\delta > \delta_g + \delta_{g\ell}$. It should be mentioned that, in principle, the theory of evaporation-isothermal two-phase region cannot be extended to thin porous-layer coatings. The previously given arguments are only speculative. The theory of thin porous layers has not yet been constructed.

We now return to Fig. 9.24. For the $\delta/d = 27$ case, we estimate the bulk value of k_e for the vapor-film region ($k_s/k_g = 500$, $\varepsilon = 0.7$) by using Eq. 9.10, and we find k_e to be 0.132 W/m-K.

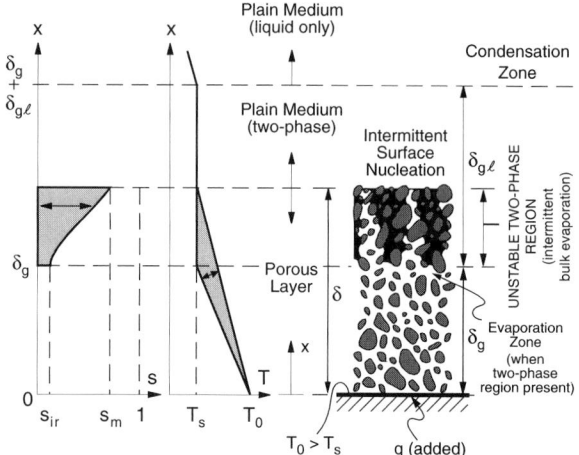

FIGURE 9.25 Evaporation from a horizontal impermeable surface coated with a porous layer with $\delta < \delta_g + \delta_{g\ell}$. The speculated intermittent drying of the layer and the associated temperature and saturation distributions are shown.

We note that the photomicrographs of Afgan et al. [137] show that the particle distribution near the bounding surface is significantly different than that in the bulk. Therefore, this k_e is only an estimate. From Eq. 9.132, we have

$$\delta_g = \frac{k_\ell(T_0 - T_s)}{q} \tag{9.147}$$

For $q = 4 \times 10^4$ W/m^2 and $T_0 - T_s = 100°$C, we have $\delta_g/d = 0.41$, that is, the vapor-film region is less than one particle thick. Then, for $\delta/d = 27$, only part of the two-phase region is in the porous layer.

For the $\delta/d = 5.5$ and 6.7 cases, the vapor-film region thickness is also small and nearly a particle in diameter thick. However, the remaining space occupied by the two-phase region is also very small. Therefore, both the vapor-film and two-phase regions do not lend themselves to the analyses based on the existence of the local thermal equilibrium and the local volume averaging. The two thin porous layers, $\delta/d = 5.5$ and 6.7, result in different heat transfer rates (for a given $T_0 - T_s$), and this difference is also due to the structure of the solid matrix and the value of D/d. For the case of $\delta/d = 5.5$, the one-dimensional analysis predicts that the two-phase region is entirely placed in the porous layer (although the validity of this analysis for such small $\delta_{g\ell}/d$ is seriously questionable). This indicates that the liquid supply to the heated surface is enhanced when the capillary action can transport the liquid through the entire two-phase region. The optimum porous-layer thickness, which results in a small resistance to vapor and liquid flows, a large effective thermal conductivity for the vapor-film region, and possibly some two- and three-dimensional motions, has not yet been rigorously analyzed.

Melting and Solidification

In single-component systems (or pure substances), the chemical composition in all phases is the same. In multicomponent systems, the chemical composition of a given phase changes in response to pressure and temperature changes and these compositions are not the same in all phases. For single-component systems, first-order phase transitions occur with a discontinuity in the first derivative of the Gibbs free energy. In the transitions, T and p remain constant.

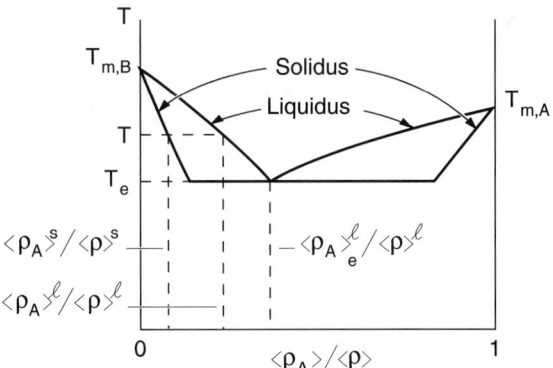

FIGURE 9.26 The thermodynamic equilibrium phase diagram for a binary solid-liquid system. The eutectic temperature and species A mass fraction and a dendritic temperature and liquidus and solidus species A mass fractions are also shown.

In Fig. 9.26, the thermodynamic equilibrium, solid-liquid phase diagram of a binary (species A and B) system is shown for a nonideal solid solution (i.e., miscible liquid but immiscible solid phase). The melting temperatures of pure substances are shown with $T_{m,A}$ and $T_{m,B}$. At the eutectic-point mole fraction, designated by the subscript e, both solid and liquid can coexist at equilibrium. In this diagram the liquidus and solidus lines are approximated as straight lines. A dendritic temperature T and the dendritic mass fractions of species $\langle\rho\rangle^s/\langle\rho\rangle^s$ and $\langle\rho\rangle^\ell/\langle\rho\rangle^\ell$ are also shown. The equilibrium partition ratio k_p is used to relate the solid- and liquid-phase mass fractions of species $\langle\rho\rangle^s/\langle\rho\rangle^s$ and $\langle\rho\rangle^\ell/\langle\rho\rangle^\ell$ on the liquidus and solidus lines at a given temperature and pressure, that is,

$$k_p = \frac{\langle\rho_A\rangle^s/\langle\rho\rangle^s}{\langle\rho_A\rangle^\ell/\langle\rho\rangle^\ell}\bigg|_{T,p} \tag{9.148}$$

A finite, two-phase region (called the *mushy region*) can exist for $k_p < 1$ and corresponds to the case where species A has a limited solubility in the solid phase. For $k_p = 1$, a discrete phase change occurs with no mushy region. This mushy region is a solid-liquid mixture where it is generally assumed that the solid phase is continuous and therefore treated as a permeable solid, that is, a porous medium with the liquid being capable of motion through the porous media.

Figure 9.27 gives a classification of the solid-liquid phase change in porous media. For single-component systems at a given pressure, melting or solidification as a distinct phase change is assumed at a saturation temperature corresponding to the pressure, that is, $T_{m,A}$ for species A. Although at the interface between the solid and liquid this saturation condition holds, the local, bulk phases (i.e., away from the interface) may not be at this temperature, and, therefore, a subcooled liquid (during solidification) or a superheated solid (during melting) may be assumed. For multicomponent systems, in addition to the distinct liquid and solid phases, a mushy (two-phase) region also exists. We note that the solid phase may not only contain the phase-change substance but can contain an inert (not changing phase) solid substance, for example, during solidification of a multicomponent liquid (i.e., molten) when the pore space of a solid matrix is filled with a much higher melting temperature. During this process, the solid fraction increases due to the formation of both a solid phase and a mushy region from the liquid phase (e.g., Ref. 139).

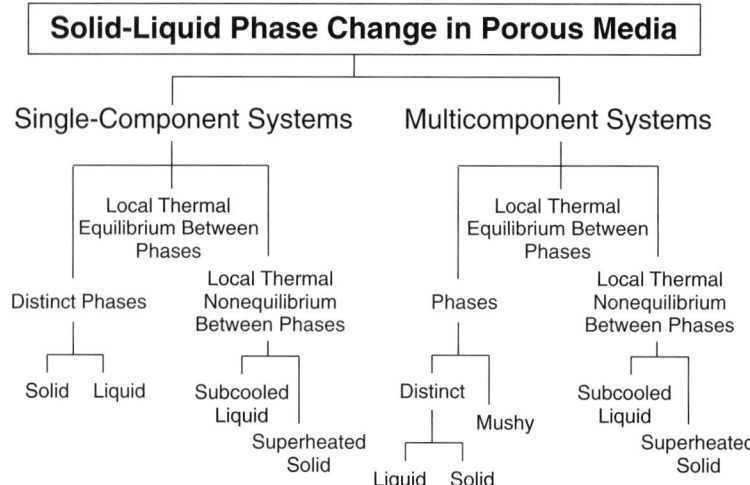

FIGURE 9.27 A classification of solid-liquid phase change in porous media.

A review of melting and solidification of single-component systems follows, as well as a discussion of the multicomponent systems. A more extensive treatment is given by Kaviany [7].

Single-Component Systems. As an example of solid-liquid phase change in porous media, we consider melting of the solid matrix by flow of a superheated liquid through it. The analysis, based on local thermal nonequilibrium between solid and liquid phases, has been performed by Plumb [140] and is reviewed here.

Because of the phase change, the solid-liquid interfacial location changes and this interfacial mass transfer $\langle \dot{n} \rangle$ must be included. The lack of local thermal equilibrium and introduction of local solid and liquid temperature $\langle T \rangle^s$ and $\langle T \rangle^\ell$ can be addressed similar to the two-medium thermal treatment (without phase change). Plumb suggests simplified, semi-heuristic phasic energy equations. These are based on the assumption that the solid is locally at the melting temperature T_m which makes the solid-phase energy equation trivial, and for a one-dimensional transport gives

$$\frac{\partial \varepsilon \langle T \rangle^\ell}{\partial t} - \frac{\langle \dot{n}_\ell \rangle T_m}{\rho_\ell} + \frac{\partial}{\partial x}\, \varepsilon \langle T \rangle^\ell \langle u_\ell \rangle = \frac{\partial}{\partial x}\left[\frac{\langle k \rangle}{(\rho c_p)_\ell} + \varepsilon D^d_{xx}\right]\frac{\partial \langle T \rangle^\ell}{\partial x} - \frac{h_{s\ell}}{(\rho c_p)_\ell}\frac{A_{\ell s}}{V}(\langle T \rangle^\ell - T_m)$$

$$(9.149)$$

$$\langle T \rangle^s = T_m \tag{9.150}$$

$$\langle \dot{n}_\ell \rangle \Delta i_{\ell s} = h_{sf}\frac{A_{\ell s}}{V}(\langle T \rangle^\ell - T_m) \tag{9.151}$$

When Eqs. 9.149 and 9.151 are combined, we have

$$\frac{\partial \langle T \rangle^\ell}{\partial t} + \langle u_\ell \rangle \frac{\partial \langle T \rangle^\ell}{\partial x} = \frac{\partial}{\partial x}\left[\frac{\langle k \rangle}{(\rho c_p)_\ell} + \varepsilon D^d_{xx}\right]\frac{\partial \langle T \rangle^\ell}{\partial x} - \left[\frac{h_{s\ell}}{(\rho c_p)_\ell \varepsilon}\frac{A_{\ell s}}{V} + \langle \dot{n}_\ell \rangle c_{p\ell} \varepsilon\right](\langle T \rangle^\ell - T_m) \quad (9.152)$$

In Eq. 9.152, $\langle k \rangle$ is the effective conductivity D^d_{xx} is the axial dispersion coefficient, and $h_{s\ell}$ is the interfacial heat transfer coefficient.

Using the preceding conservation equations and for an adiabatic system (i.e., no heat losses), subject to a prescribed inlet liquid velocity and liquid superheat $\langle T \rangle_0^\ell - T_m$ flowing into a wettable solid matrix with porosity ε_0, Plumb [140] determines the porosity distribution in the melting front. The approximate melt-front speed is determined from the overall energy balance and by neglecting the axial conduction and is

$$u_F = \frac{\rho_\ell c_{p_\ell}(\langle T \rangle_0^\ell - T_m)\langle u_\ell \rangle_0}{\rho_s \Delta i_{\ell s}\{(1 - \varepsilon)_0 + (\rho_\ell/\rho_s)[c_{p_\ell}(\langle T \rangle_0^\ell - T_m)/\Delta i_{\ell s}]\}} \tag{9.153}$$

The numerical results show that the thickness of the melting front is proportional to the liquid velocity to a power of 0.4. At low velocities, the melt-front thickness can become nearly the same as the pore (or particle) size, and at very low velocities, diffusion dominates the axial heat transfer.

Multicomponent Systems.   Melting and solidification in multicomponent systems is of interest in geological and engineering applications. In solidification, dendritic growth of crystals has been analyzed under the assumption of thermal and chemical equilibrium (as a columnar dendritic growth) and also under the assumption of nonequilibrium, that is, liquid subcooling (as a dispersed equiaxial dendritic growth). The process of solidification of multicomponent liquids is discussed by Kurz and Fisher [141] and a review with the geological applications is given by Hupport [142] and with engineering applications by Beckermann and Viskanta [143]. A more extensive review is given by Kaviany [7] and excerpts of this review are given here. The melting of multicomponent solids is reviewed by Woods [144]. Here, brief reviews of the equilibrium and nonequilibrium treatments are given.

Equilibrium Treatment of Solidification.   As an example of liquid-solid phase change in solid-fluid flow systems with the assumption of local thermal equilibrium imposed, consider the formulation of solid-fluid phase change (solidification/melting or sublimation/frosting) of a binary mixture. For this problem, the equilibrium condition extends to the local thermodynamic equilibrium where the local phasic temperature (thermal equilibrium), pressure (mechanical equilibrium), and chemical potential (chemical equilibrium) are assumed to be equal between the solid and the fluid phases. This is stated as

$$\langle T \rangle^s = \langle T \rangle^f \tag{9.154}$$

$$\langle p \rangle^s = \langle p \rangle^f \tag{9.155}$$

$$\langle \mu_A \rangle^s = \langle \mu_A \rangle^f \tag{9.156}$$

The local volume averaging of the energy equations, with allowance for a phase change in a binary system, has been discussed by Bennon and Incropera [145], Rappaz and Voller [146], Poirier et al. [147], and Hills et al. [148]. In their single-medium treatment, that is, a local volume-averaged description with the assumption of local thermodynamic equilibrium, a distinct interface between the region of solid phase and the region of fluid phase has not been assumed; instead, the fluid-phase volume fraction (i.e., the porosity) is allowed to change continuously. This continuous-medium treatment is in contrast to the multiple-medium treatment which allows for separate solid, fluid, and mushy (i.e., solid-fluid) media (e.g., Refs. 149, 150, 151). Since the explicit tracking of the various distinct interfaces, as defined in the multiple-medium treatment, is not needed in the continuous single-medium treatment, it is easier to implement. Also, for binary systems, the equilibrium temperature depends on the local species concentrations, and due to the variation of the concentration within the medium, the phase transition occurs over a range of temperatures; therefore, the single-medium treatment is even more suitable. In the following, the single-medium treatment of Bennon and Incropera [145] is reviewed. Many simplifications made in the development of this treatment are discussed by Hills et al. [148]. Alternative derivations and assumptions are discussed by Rappaz and Voller [146] and Poirier et al. [147], among others.

Nonequilibrium Treatment of Solidification.   In the following, as examples of nonequilibrium treatments of solidification in a binary system, a kinetic-diffusion controlled dendritic crystal growth and a buoyancy-influenced dendritic crystal growth are examined.

A dispersed-element model for kinetic-diffusion controlled growth.   Assuming that a total number n_s of spherical crystals are nucleated per unit volume at a supercooling of $\Delta T_{sc} = T_m - T_\ell$, then these crystals can grow to final grain radius of R_c

$$R_c = \left(\frac{3}{4\pi n_s}\right)^{1/3} \tag{9.157}$$

The initially spherical crystal will have a dendritic growth and, between the nucleation at time $t = 0$ and the complete growth (end of the solidification) at $t = t_f$, the grain envelope radius R will grow from a radius R_{s_0} (i.e., the initial radius of the nuclei) to R_c. The grain envelope is initially around the crystal, and during the dendritic growth, the envelope contains solid and liquid phases, and finally at the end of the solidification, it will again contain only solid. The content of the grain envelope is treated as a dispersed element and the liquid fraction (or porosity) ε_s of this dispersed element is initially zero and then increases rapidly followed by an eventual decrease, finally becoming zero again. This dispersed element and its surrounding liquid makes a unit-cell model and has been introduced and analyzed by Rappaz and Thévoz [152]. A schematic of this unit-cell model is shown in Fig. 9.28a. The growth of the dispersed element subject to the cooling rate of the unit cell given by the heat transfer rate $q_c 4\pi R_c^2$ with no liquid motion has been analyzed.

The elemental-liquid fraction ε_e and the cell porosity ε_c are defined as

$$\varepsilon_e = 1 - \frac{V_s}{V_e} \qquad \varepsilon_c = 1 - \frac{V_e}{V_c} = 1 - \frac{R^3}{R_c^3} \tag{9.158}$$

The solid volume is

$$V_s = (1 - \varepsilon_e)V_e = (1 - \varepsilon_e)\tfrac{4}{3}\pi R^3 \tag{9.159}$$

The quantities ε_e and ε_c are determined from the heat and mass transfer analyses. An apparent solid radius can be defined as

$$R_s^3 = \frac{V_s}{\tfrac{4}{3}\pi} = (1 - \varepsilon_e)R^3 \qquad \text{or} \qquad R_s = (1 - \varepsilon_e)^{1/3}R \tag{9.160}$$

For convenience, the volume fraction of the solid in the unit cell f_s and the volume fraction of the dispersed element in the unit cell f_c, which are related to ε_e and ε_c, are used in the analysis. These are

$$f_s = \frac{V_s}{V_c} = \frac{R_s^3}{R_c^3} = (1 - \varepsilon_e)(1 - \varepsilon_c) \qquad f_c = \frac{V_e}{V_c} = \frac{R^3}{R_c^3} \tag{9.161}$$

Assuming an equilibrium phase diagram, such as that shown in Fig. 9.28b, the definition of the equilibrium partition ratio k_p was given by Eq. 9.148 and is repeated here.

$$k_p = [(\rho_A/\rho)|_{\ell s}/(\rho_A/\rho)|_{s\ell}]|_{T,p} \tag{9.162}$$

where the solid-liquid interfacial concentrations are $(\rho_A/\rho)_{s\ell}$ and $(\rho_A/\rho)_{\ell s}$ from the solidus and liquidus lines, respectively. The liquidus line mass fraction is related to the temperature using

$$T_{\ell s} - T_{s\ell} = -\gamma\left(\frac{\rho_A}{\rho}\bigg|_{\ell s} - \frac{\rho_A}{\rho}\bigg|_{s\ell}\right) \qquad \text{or} \qquad T_e - T_{m,B} = \gamma\frac{\rho_A}{\rho}\bigg|_e \tag{9.163}$$

where

$$\gamma = \frac{dT_{\ell s}}{d(\rho_A/\rho)_{\ell s}}\bigg|_p = \frac{T_e - T_{m,B}}{(\rho_A/\rho)_e} \tag{9.164}$$

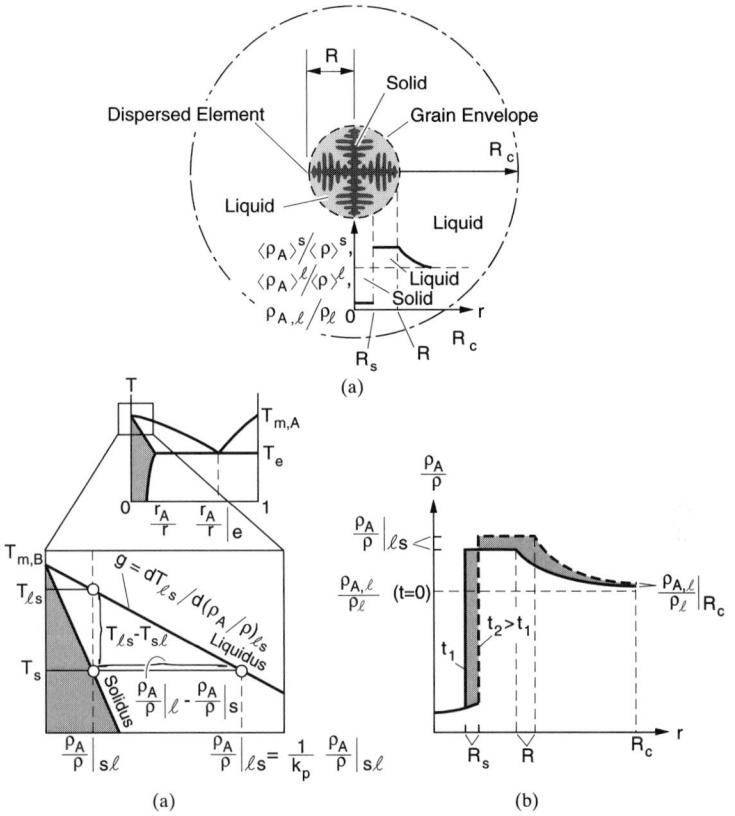

FIGURE 9.28 (*a*) Unit-cell model of the equiaxed dendritic growth of a crystal. The liquid within the grain envelope and within the element are shown as well as the mass fraction distribution of the species *A*; (*b*) the idealized phase diagram; (*c*) the mass fraction distribution of the species *A* for two different elapsed times. (*From Rappaz and Thévos, Ref. 152, reproduced by permission ©1987 Pergamon.*)

The mass fraction difference can also be written as

$$(\rho_A/\rho)_{\ell s} - (\rho_A/\rho)_{s\ell} = (\rho_A/\rho)_{s\ell}\frac{1 - k_p}{k_p} \qquad (9.165)$$

As the crystal grows, the concentration of species *A* in the solid increases slightly. This corresponds to a negligible mass diffusion in the solid ($D_s \to 0$). Other assumptions about the magnitude of D_s (i.e., $D_s \to \infty$ or a finite D_s) do not change the predicted growth rate [152]. The rejected species *A* (i.e., solute) will result in the increase of the concentration of species *A* in the liquid contained in the envelope and in the remainder of liquid in the unit cell. This is depicted in Fig. 28*c* where the concentration distributions in the solid ($0 \le r \le R_s$), in the interelemental liquid ($R_s \le r \le R$), and in the cell liquid ($R < r \le R_c$) are shown for two elapsed times. In the model, the liquid concentration within the element is assumed to be uniform and its magnitude $(\rho_A/\rho)_{\ell s}$ is given by the phase diagram and as a function of T_e. Then, in this model, the dendritic tip concentration and the temperature are $(\rho_A/\rho)_{\ell s}$ and T_e, respectively, and their interrelation is given by Eq. 9.163.

The mass fraction distribution in the liquid region is determined for the species conservation equation which for constant ρ_ℓ and D_ℓ gives

$$\frac{\partial}{\partial t} \frac{\rho_{A,\ell}}{\rho_\ell} = D_\ell \left(\frac{\partial}{\partial r^2} + \frac{2}{r} \frac{\partial}{\partial r} \right) \frac{\rho_{A,\ell}}{\rho_\ell} \qquad R \le r \le R_c \qquad (9.166)$$

the initial and boundary conditions are

$$\frac{\rho_{A,\ell}}{\rho_\ell}(r, 0) = \frac{\rho_{A,\ell}}{\rho_\ell}(t = 0) \qquad R_s \le r \le R_c \qquad (9.167)$$

$$\frac{\partial}{\partial r} \frac{\rho_{A,\ell}}{\rho_\ell} = 0 \qquad r = R_c \qquad (9.168)$$

The concentration at $r = R$ is found from the species balance made over $0 \le r \le R_c$. This is done by the integration of the distributions shown in Fig. 9.28c, that is,

$$\int_0^{R_s} k_p \left. \frac{\rho_A}{\rho} \right|_{\ell s} 4\pi r \, dr + \left. \frac{\rho_A}{\rho} \right|_{\ell s} \tfrac{4}{3}\pi(R^3 - R_s^3) + \int_R^{R_c} \frac{\rho_{A,\ell}}{\rho_\ell} 4\pi r \, dr = \frac{\rho_{A,\ell}}{\rho}(t = 0)\tfrac{4}{3}\pi R_c^3 \qquad (9.169)$$

By differentiating this with respect to time and using Eq. 9.166, we have

$$-4\pi D_\ell R^2 \left. \frac{\partial}{\partial r} \frac{\rho_A}{\rho_\ell} \right|_R = -\frac{4}{3}\pi \left. \frac{d}{dt} \frac{\rho_A}{\rho} \right|_{\ell s} (R^3 - R_s^3) + 4\pi R^2 \frac{dR}{dt}(1 - k_p) \left. \frac{\rho_A}{\rho} \right|_{\ell s} \qquad (9.170)$$

This states that the outward solute flow rate through $r = R$ is determined by the solute ejected due to the solidification and the temperature-caused change in the mass fraction of the interdendritic liquid.

For large solid and liquid conductivities, within the cell the solid, the interdendritic liquid, and the cell liquid can all be in a near-thermal equilibrium. For a temperature change occurring on the boundaries of the cell, the assumption of a uniform temperature within the cell requires that the cell Biot number, that is,

$$\text{Bi} = \frac{\text{Nu}_{dc}^{\text{ext}}}{6} \frac{k_\infty}{\langle k \rangle_{V_c}} \qquad \text{Nu}_{dc}^{\text{ext}} = \frac{q_c 2R_c}{(T_c - T_\infty)k_\infty} \qquad (9.171)$$

be less than 0.1. This condition is assumed to hold (for metals) and a single temperature T_c is used for the unit cell.

The thermal energy balance on the unit cell gives

$$q_c 4\pi R_c^2 = \left[\rho_\ell \Delta i_{\ell s} \frac{df_s}{dt} + \langle \rho c_p \rangle_{V_c} \frac{dT_c}{dt} \right] + \tfrac{4}{3}\pi R_c^3 \qquad (9.172)$$

where $\Delta i_{\ell s}$ is the heat of solidification and

$$\langle \rho c_p \rangle_{V_c} = (1 - \varepsilon_e)(1 - \varepsilon_c)(\rho c_p)_s + [\varepsilon_e(1 - \varepsilon_c) + \varepsilon_c](\rho c_p)_\ell \qquad (9.173)$$

The variation of temperature can be replaced by that of the concentration by assuming that the cell temperature is the equilibrium temperature at the liquidus line. Then using Eq. 9.163, Eq. 9.172 becomes

$$\frac{3q_c}{R_c} = \Delta i_{\ell s} \frac{df_s}{dt} + \langle \rho c_p \rangle_{V_c} \gamma \left. \frac{d}{dt} \frac{\rho_A}{\rho} \right|_{\ell s} \qquad (9.174)$$

The growth rate of the dendritic tip is modeled using the available results for low Péclet number growth in unbounded liquid. The model used by Rappaz and Thévoz [152] is

$$\frac{dR}{dt} = \frac{\gamma D_\ell}{\pi^2(k_p - 1)(\sigma_{\ell s}/\Delta s_{\ell s})(\rho_{A,\ell}/\rho_\ell)(t = 0)} \left(\left. \frac{\rho_A}{\rho} \right|_{\ell s} - \left. \frac{\rho_{A,\ell}}{\rho_\ell} \right|_R \right)^2 \qquad (9.175)$$

This kinetic condition is discussed by Hills and Roberts [153]. The distribution of $\rho_{A,\ell}/\rho_\ell$ in $R \leq r \leq R_c$ as well as the variations of f_s, f_c, $(\rho_A/\rho)_{\ell s}$, and $(\rho_{A,\ell}/\rho_\ell)_R$ with respect to time are determined by solving Eqs. 9.160, 9.169, 9.170, 9.174, and 9.175, simultaneously. This is done numerically and some of their results are reviewed in the following.

The results are for the solidification of Al–Si with an initial silicon concentration $(\rho_{A,\ell}/\rho_\ell)(t = 0) = 0.05$, a eutectic concentration of 0.108, $D_\ell = 3 \times 10^{-9}$ m^2/s, $k_p = 0.117$, $\gamma = -7.0$, $\rho_\ell \Delta i_{\ell s} = 9.5 \times 10^8$ J/m^3, $T_{m,B}$ (aluminum) = 660°C, $T_e = 577$°C, $\sigma_{\ell s}/\Delta s_{s\ell} = 9 \times 10^{-8}$ m-K, and $\langle \rho c_p \rangle_{V_c} = 2.35 \times 10^6$ J/m^3-K. Figure 9.29a shows the results for $\Delta T_{sc} = 0$°C, $R_c = 100$ µm, $R_{s_0} = 1$ µm, and a cooling rate that results in the total solidification of the cell in 10 s.

The variation of temperature T_c and solid and element volume fractions f_s and f_c, respectively, with respect to time, up to the time of the complete solidification, are shown. Since the mass fraction $(\rho_A/\rho)_{\ell s}$ has to increase in order for the growth of the dendrite to begin, as stated by Eq. 9.175, then no change in dT/dt occurs until the point of undershoot is reached, and growth begins with a large and sudden increase in f_c. Since the growth rate of the dendritic tip is larger than the rate of the solidification allowed by the heat removal, the volume fraction of the interdendritic liquid ε_s increases substantially. Therefore, f_s increases much slower than f_c. This increase in f_s decreases $(\rho_A/\rho)_{\ell s}$ and increases T_c, as stated by Eqs. 9.172 and 9.174. The increase in T_c is called *recalescence*. The concentration distributions in the interdendritic liquid and in the cell liquid are shown in Fig. 9.29b for several elapsed times. The elapsed times

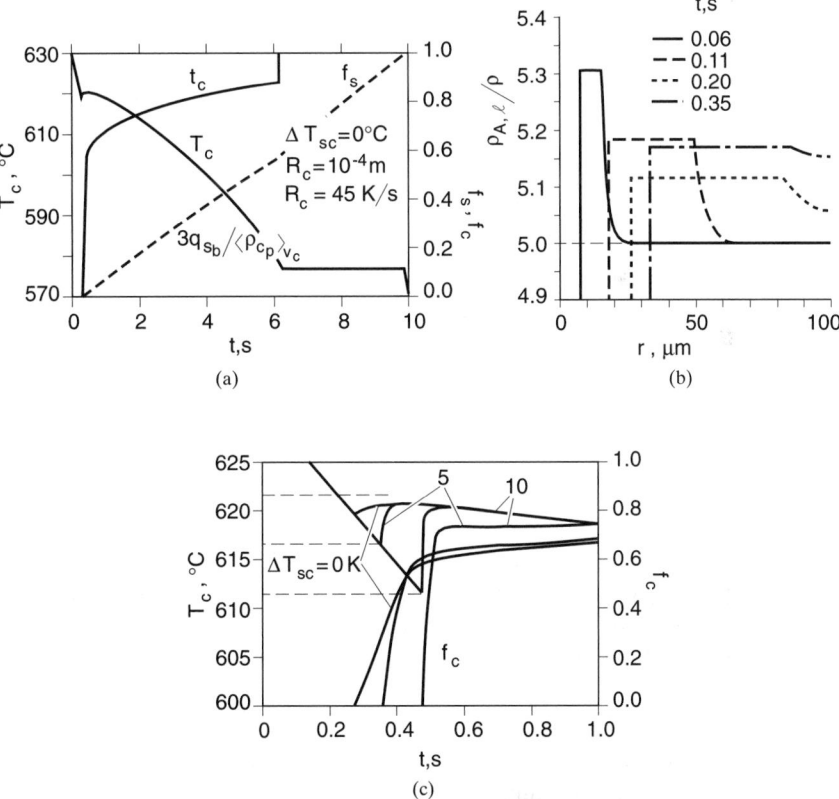

FIGURE 9.29   (a) Computed time variation of temperatures and solid and element volume fractions; (b) radial distribution of the species A in the liquid for four different elapsed times; (c) same as a but for three different supercooling conditions.

correspond to the time of undershoot, shortly after the rise in T_c begins, when T_c reaches a maximum, and shortly after the decrease in T_c begins. For $t > 0.20$ s, the temperature decreases while $(\rho_A/\rho)_{\ell s}$ increases. For an elapsed time slightly larger than 6 s, the element grows to the maximum radius R_c and only the solidification of the interdendritic liquid occurs. This ends when all this liquid is solidified. The undershoot temperature predicted for no supercooling, shown in Fig. 9.29a, will correspond to the supercooling ΔT_{sc} when substantial liquid supercooling exists. Figure 9.29c shows the results for the same conditions as in Fig. 9.29a and b, except 5 and 10°C supercooling are allowed. The results show that the larger the supercooling, the faster the temperature rises after the initial growth (i.e., accelerated recalescence).

Inclusion of buoyant liquid and crystal motions. The unit-cell-based, diffusion-controlled dendritic growth previously discussed has been extended to thermo- and diffusobuoyant convection by using local phase-volume-averaged conservation equations and local thermal and chemical nonequilibrium among the liquid phase, the solid-particles (equiaxed dendritics) phase, and the confining surfaces of the mold. As before, the crystals are assumed to be formed by the bulk nucleation in a supercooled liquid. The liquid temperature $\langle T \rangle^\ell$, the liquid volume fraction ε and its solid particle counterparts $\langle T \rangle^s$ and $1 - \varepsilon$, the velocities $\langle \mathbf{u} \rangle^\ell$ and $\langle \mathbf{u} \rangle^s$, the volumetric solidification rate $\langle \dot{n} \rangle^s$, and the concentration of species A in each phase $\langle \rho_A \rangle^\ell / \rho_\ell$, $\langle \rho_A \rangle^s / \langle \rho \rangle_s$ are all determined from the solution of the local phase-volume-averaged conservation equations. The thermodynamic conditions are applied similar to those in the preceding diffusion treatment, but the growth rate of the dendritic tip is not prescribed. Instead, the interfacial heat and mass transfer is modeled using interfacial Nusselt and Sherwood numbers for the particulate flow and heat transfer. Also, interfacial heat and mass transfer is prescribed as functions of the Reynolds number and is based on the relative velocity and solid particle diameter d.

The local, phase-volume-averaged treatment of flow and heat and mass transfer has been addressed by Voller et al. [154], Prakash [155], Beckermann and Ni [156], Prescott et al. [157], and Wang and Beckermann [158], and a review is given by Beckermann and Viskanta [143]. Here we will not review the conservation equations and thermodynamic reactions. They can be found, along with the models for the growth of bubble-nucleated crystals, in Ref. 158.

Laminar flow is assumed and the modeling of $\langle \mathbf{S} \rangle^\ell$, $\langle \mathbf{S} \rangle^s$, and τ_d are pursued similar to the hydrodynamics of the particulate flow. Since the solid particles are not spherical, the dendritic arms and other geometric parameters should be included in the models. Ahuja et al. [159] develop a drag coefficient for equiaxed dendrites.

The effective media properties $\mathbf{D}_m^\ell$, $\mathbf{D}^\ell$, $\mathbf{D}_m^s$, and $\mathbf{D}^s$, which include both the molecular (i.e., conductive) and the hydrodynamic dispersion components, are also modeled. Due to the lack of any predictive correlations for the nonequilibrium transport, local thermal equilibrium conditions are used. For the interfacial convective transport, the local Nusselt and Sherwood numbers are prescribed. The effect of the solid particles geometry must also be addressed [160].

NOMENCLATURE

Uppercase bold letters indicate that the quantity is a second-order tensor, and lowercase bold letters indicate that the quantity is a vector (or spatial tensor). Some symbols, which are introduced briefly through derivations and, otherwise, are not referred to in the text are locally defined in the appropriate locations and are not listed here. The mks units are used throughout.

a	phase distribution function
a_j	$j = 1, 2, \ldots$, constants
A	area, cross section (m^2)

A_0	volumetric (or specific) surface area (1/m)
A_{sf}	interfacial area between solid and fluid phases (m^2)
Bo	Bond or Eötvös number $\rho_\ell g R^2/\sigma$ where ℓ stands for the wetting phase, also $\rho_\ell g K/\varepsilon/\sigma$
C	average interparticle clearance (m)
C_E	coefficient in Ergun modification of Darcy law
c_p	specific heat capacity (J/kg-°C)
Ca	capillary number where ℓ stands for the wetting phase $\mu_\ell u_{D\ell}/\sigma$
d	pore-level linear length scale (m) or diameter (m)
$\mathbf{d}$	displacement tensor (m)
$\mathbf{D}^d$	dispersion tensor (m^2/s)
$\mathbf{D}^d$	dispersion coefficient (m^2/s)
$\mathbf{D}$	total diffusion tensor (m^2/s)
$\mathbf{D}$	total dispersion coefficient (m^2/s)
D_K	Knudsen mass diffusivity (m^2/s)
$\mathbf{e}$	electric field intensity vector (V/m)
$\mathbf{E}$	strain tensor
f	force (N)
f_δ	van der Waals force (N/m^2)
$\mathbf{f}$	force vector (m/s^2)
F	radiant exchanger factor
g	gravitational constant (m/s^2) or asymmetry parameter
$\mathbf{g}$	gravitational acceleration vector (m/s^2)
h	gap size (m) or height (m)
h_{sf}	solid-fluid interfacial heat transfer coefficient (W/m^2-K)
H	mean curvature of the meniscus $(\frac{1}{2})(1/r_1 + 1/r_2)$ (m) where r_1 and r_2 are the two principal radii of curvature
I	radiation intensity (W/m^2)
$\mathbf{I}$	second-order identity tensor
J	Leverett function
k	conductivity (W/m-K)
k_e	effective conductivity (W/m-K)
k_B	Boltzmann constant 1.381×10^{-23} (J/K)
k_K	Kozeny constant
k_p	equilibrium partition ratio
K	permeability (m^2)
K_{rg}	nonwetting phase relative permeability
$K_{r\ell}$	wetting phase relative permeability
Kn	Knudsen number, λ (mean free path)/C (average interparticle clearance)
$\mathbf{K}$	permeability tensor (m^2)
$\mathbf{K}_e$	effective conductivity tensor (W/m-K)
ℓ	linear length scale for representative elementary volume or unit-cell length (m)

ℓ	length of a period vector (m)
L, L_1, L_2	system dimension, linear length scale (m)
$\dot{m}$	mass flux (kg/m²-s)
$\dot{\mathbf{m}}$	mass flux vector (kg/m²-s)
M	molecular weight (kg/kg·mole)
Ma	Marangoni number $(\partial\sigma/\partial T)(\Delta T)R/(\alpha\mu)$
n	number of molecules per unit volume (molecules/m³)
$\dot{n}_i$	volumetric rate of production of component i (kg/m³-s)
n_r	number of components in the gas mixture
$\mathbf{n}$	normal unit vector
N_A	Avogadro number 6.0225×10^{23} (molecules/mole)
Nu	Nusselt number
p	pressure (Pa)
p_c	capillary pressure (Pa)
P	probability density function
P_i	Legendre polynomial
Pe	Peclet number
Pr	Prandtl number
q	heat flux (W/m²)
r	radial coordinate axis (m), separation distance (m)
$\mathbf{r}$	radial position vector (m)
R	radius (m)
R_g	universal gas constant 8.3144 (kJ/kg·mole-K) $= k_B N_A$
Re_x	Reynolds number ux/ν
s	saturation
s_1	surface saturation
s_{ir}	immobile (or irreducible) wetting phase saturation
$s_{ir\,g}$	immobile (or irreducible) nonwetting saturation
s_g	nonwetting phase saturation $\varepsilon_g/\varepsilon$
$\mathbf{s}$	unit vector
S	reduced (or effective) saturation $(s - s_{ir})/(1 - s_{ir} - s_{ir\,g})$ or path length (m)
$\mathbf{S}$	shear component of stress tensor (Pa)
t	time (s)
$\mathbf{t}$	tangential unit vector
T	temperature (K)
$\mathbf{T}$	stress tensor (Pa)
u, v, w	components of velocity vector in x, y, and z directions (m/s)
$\mathbf{u}$	velocity vector (m/s)
$\mathbf{u}_D$	Darcean (or superficial) velocity vector (m/s)
$\mathbf{u}_F$	Front velocity (m/s)
$\mathbf{u}_p$	pore (interstitial or fluid intrinsic) velocity vector (m/s)
V	volume (m³)

x, y, z	coordinate axes (m)
$\mathbf{x}$	position vector (m)
We	Weber number where ℓ stands for the wetting phase $\rho_\ell u_{D\ell}^2 d/\sigma$

Greek

α	thermal diffusivity (m²/s)
α_R	size parameter
δ	boundary layer thickness (m) or liquid film thickness (m)
δ_j	$j = 1, 2, \ldots$, linear dimension of microstructure (m)
Δ	surface roughness (m)
ε	porosity
ε_r	emissivity
$\eta_{\lambda s}$	scattering efficiency
θ	polar angle (rad)
θ_c	contact angle (rad) measured through the wetting fluid
λ	mean free path (m)
μ	dynamic viscosity (kg/m-s)
ν	kinematic viscosity μ/ρ (m²/s)
ρ	density (kg/m³) or electrical resistivity (ohm/m) or reflectivity
σ	Stefan-Boltzmann constant 5.6696×10^{-8} (W/m²-K⁴)
τ	shear stress (Pa) or tangential or tortuosity
Φ_λ	phase function
φ	scalar

Superscripts

^	Fourier Laplace or other transformation
−	average value
†	transpose
*	dimensionless quantity
'	deviation from volume-averaged value or directional quantity
ℓ	liquid
f	fluid
fs	fluid-solid
g	gas
s	solid
sf	solid-fluid

Subscripts

b	boundary
cr	critical
d	particle
D	Darcy
e	effective
f	fluid-phase

fs	solid-fluid interface
g	gas-phase
gℓ	gas (or nonwetting phase) liquid interface
*gℓ*1, *gℓ*2	gas-liquid drag
*g*Δσ	gas-phase surface tension gradient
gi	gas-phase inertia
h	hydraulic
H	curvature
i	interfacial
ℓ	liquid (or wetting phase), representative elementary volume
ℓf	liquid film
ℓg	liquid-gas
*ℓg*1, *ℓg*2	liquid-gas drag
ℓi	liquid-phase inertia
*ℓ*Δσ	liquid-phase surface tension gradient
L	system
m	mean
n	normal
o	reference
p	pore
pa	porous-ambient
r	relative
s	solid or saturation
s_b	bounding solid surface
sℓ	solid-liquid interface
sf	solid-fluid
sg	solid-gas interface
x, y	*x, y* component

Others

[ ]	matrix
⟨ ⟩	volume average
O( )	order of magnitude

GLOSSARY

absorption coefficient   Inverse of the mean free path that a photon travels before undergoing absorption. The spectral absorption coefficient a_λ is found from $I_\lambda(x) = I_\lambda(0) \times \exp[-\int_0^x \sigma_{\lambda a}(x)\, dx]$, where the beam is traveling along x.

adsorption   Enrichment of one or more components in an interfacial layer.

adsorption isotherm   Variation of the extent of enrichment of one component (amount adsorbed) in the solid-gas interfacial layer with respect to the gas pressure and at a constant temperature. For porous media, the amount adsorbed can be expressed in terms of the gram of gas adsorbed per gram of solid.

Brinkman screening length   A distance $o\,(K^{1/2})$ over which the velocity disturbances, caused by a source, decay; the same as the boundary-layer thickness.

bulk properties   Quantities measured or assigned to the matrix/fluid system without consideration of the existence of boundaries (due to finiteness of the system). Some properties take on different values than their bulk values at or adjacent to these boundaries.

capillary pressure   Local pressure difference between the nonwetting phase and wetting phase (or the pressure difference between the concave and convex sides of the meniscus).

channeling   In packed beds made of nearly spherical particles, the packing near the boundaries is not uniform and the local porosity (if a meaningful representative elementary volume could be defined) is larger than the bulk porosity. When the packed bed is confined by a solid surface and a fluid flows through the bed, this increase in the local porosity (a decrease in the local flow resistance) causes an increase in the local velocity. This increase in the velocity adjacent to the solid boundary is called channeling.

coordination number   The number of contact points between a sphere (or particle of any regular geometry) and adjacent spheres (or particles of the same geometry).

Darcean flow   A flow that obeys $\mathbf{u}_D = -(\mathbf{K}/\mu) \cdot \nabla p$.

dispersion   In the context of heat transfer in porous media and in the presence of a net Darcean motion and a temperature gradient, dispersion is the spreading of heat accounted for separately from the Darcean convection and the effective (collective) molecular conduction. It is a result of the simultaneous existence of temperature and velocity gradients within the pores. Due to the volume averaging over the pore space, this contribution is not included in the Darcean convection, and because of its dependence on $\nabla \langle T \rangle$, it is included in the total effective thermal diffusivity tensor.

drainage   Displacement of a wetting phase by a nonwetting phase. Also called desaturation or dewetting. A more restrictive definition requires that the only force present during draining must be the capillary force.

Dupuit-Forchheimer velocity   Same as pore or interstitial velocity, defined as u_D/ε where u_D is the filter (or superficial or Darcy) velocity and ε is porosity.

effective porosity   The interconnected void volume divided by the total (solid plus total void) volume. The effective porosity is smaller than or equal to porosity.

effective thermal conductivity   Local volume-averaged thermal conductivity used for the fluid-filled matrices along with the assumption of local thermal equilibrium between the solid and fluid phases. The effective thermal conductivity is not only a function of porosity and the thermal conductivity of each phase, but is very sensitive to the microstructure.

extinction coefficient   Sum of the scattering and absorption coefficients $\sigma_s + \sigma_a = \sigma_{\text{ex}}$.

extinction of radiation intensity   Sum of the absorbed and scattered radiation energy, as the incident beam travels through a particle or a collection of particles.

formation factor   Ratio of electrical resistivity of fully saturated matrix (with an electrolyte fluid) to the electrical resistivity of the fluid.

funicular state   Or funicular flow regime. The flow regime in two-phase flow through porous media, where the wetting phase is continuous. The name *funicular* is based on the concept of a continuous wetting phase flowing on the outside of the nonwetting phase and over the solid phase (this two-phase flow arrangement is not realized in practice; instead each phase flows through its individual network of interconnected channels, see Ref. 161).

hydrodynamic dispersion   In the presence of both a net and nonuniform fluid motion and a gradient in temperature, that portion of diffusion or spreading of heat caused by the nonuniformity of velocity within each pore. Also called *Taylor-Aris dispersion*.

hysteresis   Any difference in behavior associated with the past state of the phase distributions. Examples are the hysteresis loop in the capillary pressure-saturation, relative permeability-saturation, or adsorption isotherm-saturation curves. In these curves, depending on whether a given saturation is reached through drainage (reduction in the wetting phase saturation) or by imbibition (increase in the wetting phase saturation), a different value for the dependent variable is found.

imbibition   Displacement of a nonwetting phase with a wetting phase. Also called *saturation, free,* or *spontaneous imbibition.* A more restrictive definition requires that the only force present during imbibition be the capillary force.

immiscible displacement   Displacement of one phase (wetting or nonwetting) by another phase (nonwetting or wetting) without any mass transfer at the interface between the phases (diffusion or phase change). In some cases, a displacement or front develops and right downstream of it the saturation of the phase being displaced is the largest, and behind the front, the saturation of the displacing phase is the largest.

immobile or irreducible saturations s_{ir}   The reduced volume of the wetting phase retained at the highest capillary pressure. For very smooth surfaces, the wetting phase saturation does not reduce any further as the capillary pressure increases. However, for rough and etched surfaces, the irreducible saturation can be zero. The nonwetting phase immobile saturation s_{irg} is found when the capillary pressure is nearly zero and yet some of the nonwetting phase is trapped.

infiltration   Displacement of a wetting (or nonwetting) phase by a nonwetting (or wetting) phase.

intrinsic phase average   For any quantity ψ, the intrinsic phase average over any phase ℓ is defined as

$$\langle \psi_\ell \rangle^\ell = \frac{1}{V_\ell} \int_{V_\ell} \psi_\ell \, dV$$

If ψ_ℓ is a constant, then $\langle \psi_\ell \rangle^\ell = \psi_\ell$. The intrinsic average is useful in analysis of multiphase flow and in dealing with the energy equation.

Knudsen diffusion   When the Knudsen number Kn satisfies

$$\text{Kn} = \frac{\lambda \text{ (mean free path)}}{C \text{ (average interparticle clearance)}} > 10$$

then the gaseous mass transfer in porous media is by the molecular or Knudsen diffusion. In this regime, the intermolecular collisions do not occur as frequently as the molecule-wall (matrix surface) collisions, that is, the motion of molecules is independent of all the other molecules present in the gas.

Laplace or Young-Laplace equation   Equation describing the capillary pressure in terms of the liquid-fluid interfacial curvature $\langle p_c \rangle = \langle p \rangle^g - \langle p \rangle^\ell = 2\sigma H$.

local thermal equilibrium   When the temperature difference between the solid and fluid phases is much smaller than the smallest temperature difference across the system at the level of representative elementary volume, that is, $\Delta T_\ell \ll \Delta T_L$.

macroscopic behavior   System-level (over the entire volume of the porous medium) variations in velocity, temperature, pressure, concentration, and porosity.

matrix or solid matrix   A solid structure with distributed void space in its interior as well as its surface; the solid structure in the porous medium.

mean penetration distance of radiation $(\sigma_a + \sigma_s)^{-1}$ is the inverse of the sum of the absorption and scattering coefficients.

mechanical dispersion   That portion of diffusion or spreading of heat that is due to the presence of the matrix (mechanical dispersion is present only for matrices with random structures) independent of molecular diffusion and in the presence of both a net fluid motion and a gradient in temperature. The tortuous path the fluid particle takes in disordered porous media, as it moves through the matrix, makes it continuously branch out into neighboring conduits, causing spreading of its heat content when a temperature gradient exists.

microscopic behavior Pore-level variations in velocity, temperature, pressure, and concentration. This is different than the micro- or molecular-level variations used in statistical mechanics.

mobility ratio The ratio of flow conductivity of the displacing phase to that of the displaced phase $m = (K_{r\ell}\mu_g)/(K_{rg}\mu_\ell)$. For miscible displacement, the mobility ratio is the viscosity ratio.

molecular diffusion Diffusion or spreading of heat content in the presence of a temperature gradient and absence of a net fluid motion. This molecular diffusion is caused by the Brownian motion of the fluid particles.

optical thickness $\tau = (\sigma_a + \sigma_s)d$ or the number of mean penetration distances a photon encounters as it passes through the particle of diameter d (or as it passes through a finite length d). The optical path length is $\tau(x) = \int_0^x (\sigma_a + \sigma_s)\,dx$.

partial saturation When both liquid and gaseous phases occupy the pores simultaneously, each occupying a portion of the representative elementary volume.

pendular state Or *pendular stage*. The phase distribution at very low saturations (wetting phase), where the wetting phase is distributed in the pores as discrete masses. Each mass is a ring of liquid wrapped around the contact point of adjacent elements of the solid matrix [161].

phase average For any quantity ψ, the phase average over any phase ℓ is defined as

$$\langle\psi_\ell\rangle = \frac{1}{V}\int_{V_\ell} \psi_\ell\,dV$$

If ψ_ℓ is a constant, then $\langle\psi_\ell\rangle = \varepsilon_\ell\psi_\ell$, where $\varepsilon_\ell = V_\ell/V$. In dealing with the single-phase flow, the phase average suffices, otherwise the intrinsic phase average is used.

plain media The domain where no solid matrix is present, that is, the ordinary fluid domain.

porosity Ratio of void volume to total (solid plus void) volume.

reduced or effective saturation The wetting phase saturation normalized using the immobile saturations,

$$\frac{s - s_{\text{ir}}}{1 - s_{\text{ir}} - s_{\text{ir}\,g}}$$

representative elementary volume The smallest differential volume that results in statistically meaningful local average properties such as local porosity, saturation, and capillary pressure. When the representative elementary volume is appropriately chosen, the limited addition of extra pores around this local volume will not change the values of local properties.

saturated matrix A matrix fully filled with one fluid.

saturation The volume fraction of the void volume occupied by a fluid phase $s = \varepsilon_\ell/\varepsilon$, $0 \le s \le 1$. Generally, the wetting phase saturation $\varepsilon_\ell/\varepsilon$ is used.

scattering Interaction between a photon and one or more particles where the photon does not lose its entire energy.

scattering albedo The scattering coefficient divided by the sum of the scattering and absorption coefficients $\sigma_s(\sigma_s + \sigma_a)^{-1}$. For purely scattering media, the albedo is one and for purely absorbing media it is zero.

scattering coefficient Inverse of the mean free path that a photon travels before undergoing scattering. The spectral scattering coefficient $\sigma_{\lambda s}$ is found from $I_\lambda(x) = I_0(x)\exp(-\int_0^x \sigma_{\lambda s}\,dx)$, where the beam travels along x.

scattering by diffraction The change in the direction of motion of a photon as it passes near the edges of a particle.

scattering phase function Scattered intensity in a direction (θ, ϕ), divided by that intensity corresponding to isotropic scattering. This includes the reflected, refracted, and diffracted radiation scattered in any direction.

scattering by reflection The change in the direction of a photon as it collides with a particle and is reflected from the particle surface.

scattering by refraction The change in the direction of a photon as it penetrates through and then escapes from a particle.

specific surface area Or *volumetric surface area*. The surface area of pores (interstitial surface area or surface area between the solid and fluid) per unit volume of the matrix. Direct and inferred methods of measurement are discussed in Scheidegger [24]. In some specific applications, the volume is taken as the volume of the solid phase.

spectral or monochromatic Indicates that the quantity is for a specific wavelength.

superficial velocity Same as Darcy or the filter velocity. It is the volumetric flow rate divided by the surface area (both solid and fluid), so it can be readily used to calculate flow rates.

thermal transpiration When a temperature gradient exists in a porous medium, the gas saturating it flows due to this temperature gradient. The coefficient L in $\rho \mathbf{u}_D = -(L/T)\nabla T$ depends on the gas and local temperature.

tortuosity Traditionally the length of the actual path line between two ports that the fluid particle travels divided by the length of a straight line between these ports. This path is taken by a diffusion (Brownian) motion and is independent of the net velocity. In the modern usage, the tortuosity is found from $\varepsilon(1 + L_t^*) = k_e/k_f$ for $k_s = 0$. L_t^* is also called the tortuosity [81]. The tortuosity tensor is designated by $\mathbf{L}_t^*$.

total thermal diffusivity tensor $\mathbf{D}$, the sum of the effective thermal diffusivity tensor $\mathbf{K}_e/(\rho c_p)_f$ where

$$\mathbf{K}_e/k_f = \mathbf{K}_e/k_f \,(k_s/k_f, \varepsilon, \text{structure})$$

and the dispersion tensor

$$\mathbf{D}^d = \mathbf{D}^d \left(\frac{k_s}{k_f}, \varepsilon, \text{structure, Re, Pr, and } \frac{(\rho c_p)_s}{(\rho c_p)_f} \right)$$

that is, $\mathbf{D} = \mathbf{K}_e/(\rho c_p)_f + \varepsilon \mathbf{D}^d$.

void ratio Ratio of void fraction (porosity) to solid fraction, that is, $\varepsilon(1 - \varepsilon)^{-1}$.

wetting phase The phase that has a smaller contact angle (the contact angle is measured through a perspective phase).

REFERENCES

1. D. W. Schaefer, "Engineered Porous Materials," *MRS Bull.,* (XIX): 14–17, 1994.

2. J. Banavar, J. Koplik, and K. W. Winkler, *Physics and Chemistry of Porous Media,* AIP Conference Proceeding 154, AIP, 1987.

3. H. F. Porter, G. A. Schurr, D. F. Wells, and K. T. Seurau, "Solid Drying and Gas-Solid Systems," in *Perry's Chemical Engineer's Handbook,* McGraw-Hill, New York, 20–79, 1984.

4. P. Cheng and W. J. Minkowycz, "Free Convection about a Vertical Flat Plate Embedded in a Porous Medium with Application to Heat Transfer from a Dike," *J. Geophys. Res.,* (82): 2040–2044, 1977.

5. K. Vafai, and C.-L. Tien, "Boundary and Inertia Effects on Flow and Heat Treansfer in Porous Media," *Int. J. Heat Mass Transfer,* (24): 195–243, 1981.

6. D. A. Nield, and A. Bejan, *Convection in Porous Media,* Springer-Verlag, New York, 1992.

7. M. Kaviany, *Principles of Convective Heat Transfer,* Springer-Verlag, New York, 1994.

8. I. Nozad, R. G. Carbonell, and S. Whitaker, "Heat Conduction in Multi-Phase Systems I: Theory and Experiments for Two-Phase Systems," *Chem. Engng. Sci.,* (40): 843–855, 1985.

9. M. Kaviany, *Principles of Heat Transfer in Porous Media,* 2d ed., Springer-Verlag, New York, 1995.

10. R. Krupiczka, "Analysis of Thermal Conductivity in Granular Materials," *Int. Chem. Engng.,* (7): 122–144, 1967.

11. D. Kunii, and J. M. Smith, "Heat Transfer Characteristics of Porous Rocks," *AIChE J.,* (6): 71–78, 1960.

12. P. Zehnder, and E. U. Schlünder, "Thermal Conductivity of Granular Materials at Moderate Temperatures (in German)," *Chemie. Ingr.-Tech.,* (42): 933–941, 1970.

13. C. T. Hsu, P. Cheng, and K. W. Wong, "Modified Zehnder–Schlünder Models for Stagnant Thermal Conductivity of Porous Media," *Int. J. Heat Mass Transfer,* (37): 2751–2759, 1994.

14. G. R. Hadley, "Thermal Conductivity of Packed Metal Powders," *Int. J. Heat Mass Transfer,* (29): 909–920, 1986.

15. V. Prasad, N. Kladas, A. Bandyopadhaya, and Q. Tian, "Evaluation of Correlations for Stagnant Thermal Conductivity of Liquid-Saturated Porous Beds of Spheres," *Int. J. Heat Mass Transfer,* (32): 1793–1796, 1989.

16. I. Nozad, *An Experimental and Theoretical Study of Heat Conduction in Two- and Three-Phase Systems,* Ph.D. thesis, University of California, Davis, 1983.

17. F. B. Nimick, and J. R. Leith, "A Model for Thermal Conductivity of Granular Porous Media," *ASME J. Heat Transfer,* (114): 505–508, 1992.

18. R. W. Zimmerman, "Thermal Conductivity of Fluid-Saturated Rocks," *J. Pet. Sci. Eng.,* (3): 219–227, 1987.

19. C. Stark, and J. Frickle, "Improved Heat Transfer Models to Fibrous Insulations," *Int. J. Heat Mass Transfer,* (36): 617–625, 1993.

20. M. Sahraoui, and M. Kaviany, "Slip and No-Slip Temperature Boundary Conditions at Interface of Porous, Plain Media: Conduction," *Int. J. Heat Mass Transfer,* (36): 1019–1033, 1993.

21. J. A. Ochoa-Tapia, P. Stroeve, and S. Whitaker, "Diffusion Transport in Two-Phase Media: Spatially Periodic Models and Maxwell's Theory for Isotropic and Anisotropic Systems," *Chem. Engng. Sci.,* (49): 709–726, 1994.

22. C. T. Hsu, P. Cheng, and K. W. Wong, "A Lumped Parameter Model for Stagnant Thermal Conductivity of Spatially Periodic Porous Media," *ASME J. Heat Transfer,* (117): 264–269, 1995.

23. H. Darcy, *Les Fontaines Publiques de la ville de Dijon,* Dalmont, Paris, 1856.

24. A. E. Scheidegger, "Statistical Hydrodynamics in Porous Media," *J. Appl. Phys.,* (25): 994–1001, 1974.

25. H. C. Brinkman, "A Calculation of the Viscous Force Exerted by a Flowing Fluid on a Dense Swarm of Particles," *Appl. Sci. Res.,* (A1): 27–34, 1947.

26. T. S. Lundgren, "Slow Flow Through Stationary Random Beds and Suspensions of Spheres," *J. Fluid Mech.* (51): 273–299, 1972.

27. A. Dybbs and R. V. Edwards, "A New Look at Porous Media Fluid Mechanics—Darcy to Turbulent," in *Fundamentals of Transport Phenomena in Porous Media,* Bear and Corapcioglu, eds., Martinus Nijhoff Publishers, 199–254, 1984.

28. D. L. Koch, and J. F. Brady, "Dispersion in Fixed Beds," *J. Fluid Mech.,* (154): 399–427, 1985.

29. J. J. Fried, and M. A. Combarnous, "Dispersion in Porous Media," *Advances in Hydro. Science,* (7): 169–282, 1971.

30. D. J. Jeffrey, "Conduction Through a Random Suspension of Spheres," *Proc. Roy. Soc.,* London, (A335): 355–367, 1973.

31. R. Aris, and N. R. Amundson, "Some Remarks on Longitudinal Mixing or Diffusion in Fixed Beds," *AIChE J.,* (3): 280–282, 1957.

32. G. De Josselin De Jong, "Longitudinal and Transverse Diffusion in Granular Deposits," *Trans. Amer. Geophys. Union,* (39): 67–74, 1958.

33. P. G. Saffman, "Dispersion Due to Molecular Diffusion and Macroscopic Mixing in Flow Through a Network of Capillaries," *J. Fluid Mech.,* (7): 194–208, 1960.

34. F. J. M. Horn, "Calculation of Dispersion Coefficient by Means of Moments," *AIChE J.,* (17): 613–620, 1971.

35. T. Baron, "Generalized Graphic Method for the Design of Fixed Bed Catalytic Reactors," *Chem. Eng. Prog.,* (48): 118–124, 1952.

36. S. Yagi, D. Kunii, and N. Wakao, "Studies on Axial Effective Thermal Conductivities in Packed Beds," *AIChE J.,* (6): 543–546, 1960.

37. W. W. Schertz, and K. G. Bishoff, "Thermal and Material Transfer in Nonisothermal Packed Beds," *AIChE J.,* (4): 597–604, 1969.

38. M. F. Edwards, and J. E. Richardson, "Gas Dispersion in Packed Beds," *Chem. Engng. Sci.,* (23): 109–123, 1968.

39. D. Vortmeyer, "Axial Heat Dispersion in Packed Beds," *Chem. Engng. Sci.,* (30): 999–1001, 1975.

40. J. Bear, *Dynamics of Fluids in Porous Media,* Dover, New York, 1988.

41. M. Sahraoui, and M. Kaviany, "Slip and No-Slip Temperature Boundary Conditions at Interface of Porous, Plain Media: Convection," *Int. J. Heat Mass Transfer,* (37): 1029–1044, 1994.

42. M. P. Mengüc, and R. Viskanta, "An Assessment of Spectral Radiative Heat Transfer Predictions for a Pulverized Coal-Fired Furnace," in *Proceedings of 8th International Heat and Mass Conference (San Francisco),* (2): 815–820, 1986.

43. S. Chandrasekhar, *Radiation Transfer,* Dover, New York, 1960.

44. M. N. Ozisik, *Radiative Transfer and Interaction with Conduction and Convection,* Werbel and Peck, New York, 1985.

45. R. Siegel, and J. R. Howell, *Thermal Radiation Heat Transfer,* 2d ed., McGraw-Hill, New York, 1981.

46. R. Siegel, and J. R. Howell, *Thermal Radiation Heat Transfer,* 3d ed., Hemisphere, Washington, DC, 1992.

47. M. Q. Brewster, *Thermal Radiative Transfer and Properties,* John Wiley and Sons, New York, 1992.

48. M. F. Modest, *Radiative Heat Transfer,* McGraw-Hill, New York, 1993.

49. *Optical Properties of Metals,* ed. J. H. Weaver, Fachuvnfarmation-szentrum Energie, Physik, Mathematik Gmbh, 1981.

50. H. C. van de Hulst, *Light Scattering by Small Particles,* Dover, New York, 1981.

51. S. Kumar, and C.-L. Tien, "Dependent Scattering and Absorption of Radiation by Small Particles," *ASME J. Heat Transfer,* (112): 178–185, 1990.

52. A. Ishimaru, and Y. Kuga, "Attenuation Constant of a Coherent Field in a Dense Distribution of Particles," *J. Opt. Soc. Amer.,* (72): 1317–1320, 1982.

53. J. D. Cartigny, Y. Yamada, and C.-L. Tien, "Radiative Heat Transfer with Dependent Scattering by Particles: Part 1—Theoretical Investigation," *ASME J. Heat Transfer,* (108): 608–613, 1986.

54. B. L. Drolen, and C.-L. Tien, "Independent and Dependent Scattering in Packed Spheres Systems," *J. Thermophys. Heat Transfer,* (1): 63–68, 1987.

55. H. C. Hottel, A. F. Sarofim, W. H. Dalzell, and I. A. Vasalos, "Optical Properties of Coatings, Effect of Pigment Concentration," *AIAA J.,* (9): 1895–1898, 1971.

56. M. Q. Brewster, and C.-L. Tien, "Radiative Transfer in Packed and Fluidized Beds: Dependent versus Independent Scattering," *ASME J. Heat Transfer,* (104): 573–579, 1982.

57. M. Q. Brewster, "Radiative Heat Transfer in Fluidized Bed Combustors," *ASME* paper no. 83-WA/HT-82, 1983.

58. Y. Yamada, J. D. Cartigny, and C.-L. Tien, "Radiative Transfer with Dependent Scattering by Particles: Part 2—Experimental Investigation," *ASME J. Heat Transfer,* (108): 614–618, 1986.

59. C.-L. Tien, and B. L. Drolen, "Thermal Radiation in Particulate Media with Dependent and Independent Scattering," *Annual Review of Numerical Fluid Mechanics and Heat Transfer,* (1): 1–32, 1987.

60. S. L. Chang, and K. T. Rhee, "Blackbody Radiation Functions," *Int. J. Comm. Heat Mass Transfer,* (11): 451–455, 1984.

61. B. P. Singh, and M. Kaviany, "Independent Theory Versus Direct Simulation of Radiative Heat Transfer in Packed Beds," *Int. J. Heat Mass Transfer,* (34): 2869–2881, 1991.

62. K. Y. Wang, and C.-L. Tien, "Thermal Insulation in Flow Systems: Combined Radiation and Convection Through a Porous Segment," *ASME,* paper no. 83-WA/HT-81, 1983.

63. T. W. Tong, and C.-L. Tien, "Radiative Heat Transfer in Fibrous Insulations—Part 1: Analytical Study," *ASME J. Heat Transfer,* (105): 70–75, 1983.

64. T. W. Tong, Q. S. Yang, and C.-L. Tien, "Radiative Heat Transfer in Fibrous Insulations—Part 2: Experimental Study," *ASME J. Heat Transfer,* (105): 76–81, 1983.

65. S. C. Lee, "Scattering Phase Function for Fibrous Media," *Int. J. Heat Mass Transfer,* (33): 2183–2190, 1990.

66. S. C. Lee, S. White, and J. A. Grzesik, "Effective Radiative Properties of Fibrous Composites Containing Spherical Particles," *J. Thermoph. Heat Transfer,* (8): 400–405, 1994.

67. J. C. Ku, and J. D. Felske, "The Range of Validity of the Rayleigh Limit for Computing Mie Scattering and Extinction Efficiencies," *J. Quant. Spectrosc. Radiat. Transfer,* (31): 569–574, 1984.

68. A. Selamet, and V. S. Arpaci, "Rayleigh Limit Penndorf Extension," *Int. J. Heat Mass Transfer,* (32): 1809–1820, 1989.

69. R. B. Penndorf, "Scattering and Extinction for Small Absorbing and Nonabsorbing Aerosols," *J. Opt. Soc. Amer.,* (8): 896–904, 1962.

70. G. F. Bohren, and D. R. Huffman, *Absorption and Scattering Light by Small Particles,* John Wiley & Sons, New York, 1983.

71. M. P. Mengüc, and R. Viskanta, "On the Radiative Properties of Polydispersions: A Simplified Approach," *Combust. Sci. Technol.,* (44): 143–149, 1985.

72. H. Lee, and R. O. Buckius, "Scaling Anisotropic Scattering in Radiation Heat Transfer for a Planar Medium," *ASME J. Heat Transfer,* (104): 68–75, 1982.

73. C. M. Chu, and S. W. Churchill, "Representation of Angular Distribution of Radiation Scattered by a Spherical Particle," *J. Opt. Soc. Amer.,* (45): 958–962, 1955.

74. W. J. Wiscombe, "The Delta–M Method: Rapid Yet Accurate Flux Calculations for Strongly Asymmetric Phase Functions," *J. Atm. Sci.,* (34): 1408–1422, 1977.

75. B. H. J. McKellar, and M. A. Box, "The Scaling Group of the Radiative Transfer Equation," *J. Atmospheric Sci.,* (38): 1063–1068, 1981.

76. G. D. Mazza, C. A. Berto, and G. F. Barreto, "Evaluation of Radiative Heat Transfer Properties in Dense Particulate Media," *Powder Tech.,* (67): 137–144, 1991.

77. Y. Xia, and W. Strieder, "Variational Calculation of the Effective Emissivity for a Random Bed," *Int. J. Heat Mass Transfer,* (37): 451–460, 1994.

78. Y. Xia, and W. Strieder, "Variational Calculation of the Effective Emissivity for a Random Bed," *Int. J. Heat Mass Transfer,* (37): 451–460, 1994.

79. D. Vortmeyer, "Radiation in Packed Solids," in *Proceedings of 6th International Heat Transfer Conference,* Toronto, (6): 525–539, 1978.

80. J. R. Wolf, J. W. C. Tseng, and W. Strieder, "Radiative Conductivity for a Random Void-Solid Medium with Diffusely Reflecting Surfaces," *Int. J. Heat Mass Transfer,* (33): 725–734, 1990.

81. R. G. Carbonell, and S. Whitaker, "Heat and Mass Transfer in Porous Media," *Fundamentals of Transport Phenomena in Porous Media,* eds. Bear and Corapcioglu, Martinus Nijhoff, 121–198, 1984.

82. F. Zanotti, and R. G. Carbonell, "Development of Transport Equation for Multi-Phase Systems—I–III," *Chem. Engng. Sci.,* (39): 263–278, 279–297, 299–311, 1984.

83. J. Levec, and R. G. Carbonell, "Longitudinal and Lateral Thermal Dispersion in Packed Beds, I–II," *AIChE J.,* (31): 581–590, 591–602, 1985.

84. M. Quintard, M. Kaviany, and S. Whitaker, "Two-Medium Treatment of Heat Transfer in Porous Media: Numerical Results for Effective Properties," *Adv. Water. Resour.* (20): 77–94, 1997.

85. N. Wakao, and S. Kaguei, *Heat and Mass Transfer in Packed Beds,* Gordon and Breach Science, New York, 1982.

86. L. B. Yunis, and R. Viskanta, "Experimental Determination of the Volumetric Heat Transfer Coefficient between Stream of Air and Ceramic Foam," *Int. J. Heat Mass Transfer,* (36): 1425–1434, 1993.

87. R. Defay and I. Prigogine, *Surface Tension and Adsorption* (English edition), Wiley, New York, 1966.

88. M. C. Leverett, "Capillary Behavior in Porous Solids," *Trans. AIME,* (142): 152–169, 1941.

89. M. T. van Genuchten, "A Closed-Form Equation for Predicting the Hydraulic Conductivity of Unsaturated Soils," *Soil Sci. Soc. Amer. J.* (44): 892–898, 1980.

90. K. S. Udell, "Heat Transfer in Porous Media Considering Phase Change and Capillarity—The Heat Pipe Effect," *Int. J. Heat Mass Transfer,* (28): 485–495, 1985.

91. D. Pavone, "Explicit Solution for Free-Fall Gravity Drainage Including Capillary Pressure," in *Multiphase Transport in Porous Media—1989,* ASME FED, vol. 92 (or HTD, vol. 127), 55–62, 1989.

92. A. E. Saez, and R. G. Carbonell, "Hydrodynamic Parameters for Gas-Liquid Co-Current Flow in Packed Beds," *AIChE J.,* (31): 52–62, 1985.

93. M. R. J. Wyllie, "Relative Permeabilities," in *Petroleum Production Handbook,* vol. 2, chap. 25, 25-1–25-14, McGraw-Hill, New York, 1962.

94. D. Mualem, "A New Model for Predicting the Hydraulic Conductivity of Saturated Porous Media," *Water Resour. Res.* (12): 513–522, 1976.

95. M. Delshad, and G. A. Pope, "Comparison of the Three-Phase Oil Relative Permeability Models," *Transp. Porous Media,* (4): 59–83, 1989.

96. A. K. Verma, K. Pruess, C. F. Tsang, and P. A. Withespoon, "A Study of Two-Phase Concurrent Flow of Steam and Water in an Unconsolidated Porous Medium," in *Heat Transfer in Porous Media and Particulate Flows,* ASME HTD, (46): 135–143, 1984.

97. J. Levec, A. E. Saez, and R. G. Carbonell, "The Hydrodynamics of Trickling Flow in Packed Beds, Part II: Experimental Observations," *AIChE J.* (32): 369–380, 1986.

98. N. K. Tutu, T. Ginsberg, and J. C. Chen, "Interfacial Drag for Two-Phase Flow Through High Permeability Porous Beds," in *Interfacial Transport Phenomena,* ASME, 37–44, 1983.

99. T. Schulenberg, and U. Müller, "An Improved Model for Two-Phase Flow through Beds of Coarse Particles," *Int. J. Multiphase Flow* (13): 87–97, 1987.

100. V. X. Tung, and V. K. Dhir, "A Hydrodynamic Model for Two-Phase Flow Through Porous Media," *Int. J. Multiphase Flow,* (14): 47–64, 1988.

101. S. Whitaker, "Simultaneous Heat, Mass, and Momentum Transfer in Porous Media: A Theory of Drying," *Adv. Heat Transfer,* (13): 119–203, 1977.

102. A. E. Saez, R. G. Carbonell, and J. Levec, "The Hydrodynamics of Trickling Flow in Packed Beds, Part I: Conduit Models," *AIChE J.,* (32): 353–368, 1986.

103. W. H. Somerton, J. A. Keese, and S. C. Chu, "Thermal Behavior of Unconsolidated Oil Sands," *SPE J.,* (14): 513–521, 1974.

104. A. Matsuura, Y. Hitcka, T. Akehata, and T. Shirai, "Effective Radial Thermal Conductivity in Packed Beds with Gas-Liquid Down Flow," *Heat Transfer—Japanese Research,* (8): 44–52, 1979.

105. V. Specchia, and G. Baldi, "Heat Transfer in Trickle-Bed Reactors," *Chem. Eng. Commun.,* (3): 483–499, 1979.

106. K. Hashimoto, K. Muroyama, K. Fujiyoshi, and S. Nagata, "Effective Radial Thermal Conductivity in Co-current Flow of a Gas and Liquid Through a Packed Bed," *Int. Chem. Eng.,* (16): 720–727, 1976.

107. V. W. Weekman, and J. E. Meyers, "Heat Transfer Characteristics of Cocurrent Gas-Liquid Flow in Packed Beds," *AIChE J.,* (11): 13–17, 1965.

108. A. Matsuura, Y. Hitake, T. Akehata, and T. Shirai, "Apparent Wall Heat Transfer Coefficient in Packed Beds with Downward Co-Current Gas-Liquid Flow," *Heat Transfer—Japanese Research,* (8): 53–60, 1979.

109. K. S. Udell, and J. S. Fitch, "Heat and Mass Transfer in Capillary Porous Media Considering Evaporation, Condensation and Noncondensible Gas Effects," in *Heat Transfer in Porous Media and Particulate Flows,* ASME HTD, (46): 103–110, 1985.

110. M. Crine, "Heat Transfer Phenomena in Trickle-Bed Reactors," *Chem. Eng. Commun.,* (19): 99–114, 1982.

111. A. E. Saez, *Hydrodynamics and Lateral Thermal Dispersion for Gas-Liquid Cocurrent Flow in Packed Beds,* Ph.D. thesis, University of California-Davis, 1983.

112. K. J. Renken, M. J. Carneiro, and K. Meechan, "Analysis of Laminar Forced Convection Condensation within Thin Porous Coating," *J. Thermophy. Heat Transfer,* (8): 303–308, 1994.

113. J. W. Rose, "Fundamentals of Condensation Heat Transfer: Laminar Film Condensation," *JSME Int. J. Series II,* (31): 357–375, 1988.

114. S. M. White, and C.-L. Tien, "Analysis of Laminar Film Condensation in a Porous Medium," in *Proceeding of the 2d ASME/JSME Thermal Engineering Joint Conference,* 401–406, 1987.

115. E. M. Parmentier, "Two-Phase Natural Convection Adjacent to a Vertical Heated Surface in a Permeable Medium," *Int. J. Heat Mass Transfer,* (22): 849–855, 1979.

116. P. Cheng, "Film Condensation Along an Inclined Surface in a Porous Medium," *Int. J. Heat Mass Transfer,* (24): 983–990, 1981.

117. M. Kaviany, "Boundary-Layer Treatment of Film Condensation in the Presence of a Solid Matrix," *Int. J. Heat Mass Transfer,* (29): 951–954, 1986.

118. A. Shekarriz, and O. A. Plumb, "A Theoretical Study of the Enhancement of Filmwise Condensation Using Porous Fins," *ASME paper no. 86-HT-31,* 1986.

119. A. Majumdar, and C.-L. Tien, "Effects of Surface Tension on Films Condensation in a Porous Medium," *ASME J. Heat Transfer,* (112): 751–757, 1988.

120. J. N. Chung, O. A. Plumb, and W. C. Lee, "Condensation in a Porous Region Bounded by a Cold Vertical Surface," *Heat and Mass Transfer in Frost, Ice, Packed Beds, and Environmental Discharge,* ASME HTD, (139): 43–50, 1990.

121. O. A. Plumb, D. B. Burnett, and A. Shekarriz, "Film Condensation on a Vertical Flat Plate in a Packed Bed," *ASME J. Heat Transfer,* (112): 235–239, 1990.

122. E.R.G. Eckert, R. J. Goldstein, A. I. Behbahani, and R. Hain, "Boiling in an Unconsolidated Granular Medium," *Int. J. Heat Mass Transfer,* (28): 1187–1196, 1985.

123. A. W. Reed, "A Mechanistic Explanation of Channels in Debris Beds," *ASME J. Heat Transfer,* (108): 125–131, 1986.

124. A. K. Stubos, and J.-M. Buchin, "Modeling of Vapor Channeling Behavior in Liquid-Saturated Debris Beds," *ASME J. Heat Transfer,* (110): 968–975, 1988.

125. C. H. Sondergeld, and D. L. Turcotte, "An Experimental Study of Two-Phase Convection in a Porous Media with Applications to Geological Problems," *J. Geophys. Res.,* (82): 2045–2052, 1977.

126. A. S. Naik, and V. K. Dhir, "Forced Flow Evaporative Cooling of a Volumetrically Heated Porous Layer," *Int. J. Heat Mass Transfer,* (25): 541–552, 1982.

127. S. Fukusako, T. Komoriga, and N. Seki, "An Experimental Study of Transition and Film Boiling Heat Transfer in Liquid-Saturated Porous Bed," *ASME J. Heat Transfer,* (108): 117–124, 1986.

128. H. H. Bau, and K. E. Torrance, "Boiling in Low-Permeability Porous Materials," *Int. J. Heat Mass Transfer,* (25): 45–55, 1982.

129. S. W. Jones, M. Epstein, J. D. Gabor, J. D. Cassulo, and S. G. Bankoff, "Investigation of Limiting Boiling Heat Fluxes from Debris Beds," *Trans. Amer. Nucl. Soc.,* (35): 361–363, 1980.

130. R. M. Fand, T. Zheng, and P. Cheng, "The General Characteristics of Boiling Heat Transfer from a Surface Embedded in a Porous Medium," *Int. J. Heat Mass Transfer,* (30): 1231–1235, 1987.

131. V. X. Tsung, V. K. Dhir, and S. Singh, "Experimental Study of Boiling Heat Transfer from a Sphere Embedded in Liquid Saturated Porous Media," in *Heat Transfer in Porous Media and Particulate Flows,* ASME HTD, (46): 127–134, 1985.

132. P. S. Ramesh, and K. E. Torrance, "Stability of Boiling in Porous Media," *Int. J. Heat Mass Transfer,* (33): 1895–1908, 1990.

133. J. D. Jennings, and K. S. Udell, "The Heat Pipe Effect in Heterogeneous Porous Media," in *Heat Transfer in Porous Media and Particulate Flows, ASME HTD,* (46): 93–101, 1985.

134. S. K. Konev, F. Plasek, and L. Horvat, "Investigation of Boiling in Capillary Structures," *Heat Transfer—Soviet Res.,* (19): 14–17, 1987.

135. M. A. Styrikovich, S. P. Malyshenko, A. B. Andrianov, and I. V. Tataev, "Investigation of Boiling on Porous Surfaces," *Heat Transfer—Soviet Res.* (19): 23–29, 1987.

136. S. A. Kovalev, S. L. Solv'yev, and O. A. Ovodkov, "Liquid Boiling on Porous Surfaces," *Heat Transfer—Soviet Res.,* (19): 109–120, 1987.

137. N. M. Afgan, L. A. Jovic, S. A. Kovalev, and V. A. Lenykov, "Boiling Heat Transfer from Surfaces with Porous Layers," *Int. J. Heat Mass Transfer,* (28): 415–422, 1985.

138. A. E. Bergles, and M. C. Chyu, "Characteristics of Nucleate Pool Boiling from Porous Metallic Coatings," *ASME J. Heat Transfer,* (104): 279–285, 1982.

139. M. Singh, and D. R. Behrendt, "Reactive Melt Infiltration of Silicon-Niobium Alloys in Microporous Carbons," *J. Mater. Res.,* (9): 1701–1708, 1994.

140. O. A. Plumb, "Convective Melting of Packed Beds," *Int. J. Heat Mass Transfer,* (37): 829–836, 1994.

141. W. Kurz, and D. J. Fisher, *Fundamentals of Solidifcation,* 3d ed., Trans Tech Publications, Switzerland, 1992.

142. H. E. Hupport, "The Fluid Mechanics of Solidification," *J. Fluid Mech.,* (212): 209–240, 1990.

143. C. Beckermann, and R. Viskanta, "Mathematical Modeling of Transport Phenomena During Alloy Solidification," *Appl. Mech. Rev.,* (46): 1–27, 1993.

144. A. W. Woods, "Fluid Mixing During Melting," *Phys. Fluids,* (A3): 1393–1404, 1991.

145. W. D. Bennon, and F. P. Incropera, "A Continuum Model for Momentum, Heat and Species Transport in Binary Solid-Liquid Phase Change Systems-I., and -II.," *Int. J. Heat Mass Transfer,* (30): 2161–2187, 1987.

146. M. Rappaz, and V. R. Voller, "Modelling of Micro-Macrosegregation in Solidification Processes," *Metall. Trans.,* (21A): 749–753, 1990.

147. D. R. Poirier, P. J. Nandapurkar, and S. Ganesan, "The Energy and Solute Conservation Equations for Dendritic Solidification," *Metall. Trans.,* (22B): 889–900, 1991.

148. R. N. Hills, D. E. Loper, and P. H. Roberts, "On Continuum Models for Momentum, Heat and Species Transport in Solid-Liquid Phase Change Systems," *Int. Comm. Heat Mass Transfer,* (19): 585–594, 1992.

149. M. G. Worster, "Natural Convection in a Mushy Layer," *J. Fluid Mech.,* (224): 335–339, 1991.

150. C.-J. Kim, and M. Kaviany, "A Fully Implicit Method for Diffusion-Controlled Solidification of Binary Alloys," *Int. J. Heat Mass Transfer,* (35): 1143–1154, 1992.

151. F. Vodak, R. Cerny, and P. Prikryl, "A Model of Binary Alloy Solidification with Convection in the Melt," *Int. J. Heat Mass Transfer,* (35): 1787–1791, 1992.

152. M. Rappaz, and Ph. Thévoz, "Solute Model for Equiaxed Dendritic Growth," *Acta Metall.,* (35): 1487–1497, 1987.

153. R. N. Hills, and P. H. Roberts, "A Note on the Kinetic Conditions at a Supercooled Interface," *Int. Comm. Heat Mass Transfer,* (20): 407–416, 1993.

154. V. R. Voller, A. D. Brent, and C. Prakash, "The Modeling of Heat, Mass and Solute Transport in Solidification Systems," *Int. J. Heat Mass Transfer,* (32): 1719–1731, 1989.

155. C. Prakash, "Two-Phase Model for Binary Solid-Liquid Phase Change, Part I: Governing Equations, Part II: Some Illustration Examples," *Num. Heat Transfer,* (18B): 131–167, 1990.

156. C. Beckermann, and J. Ni, "Modeling of Equiaxed Solidification with Convection," *Proceedings, First International Conference on Transport Phenomena in Processing,* ed. S. J. Güceri, Technomic, Lancaster, Pennsylvania, 1992.

157. P. J. Prescott, F. P. Incropera, and D. R. Gaskell, "The Effects of Undercooling, Recalescence and Solid Transport on the Solidification of Binary Metal Alloys," in *Transport Phenomena in Materials Processing and Manufacturing,* ASME HTD, (196), American Society of Mechanical Engineers, New York, 1992.

158. C. Y. Wang, and C. Beckermann, "A Multiphase Micro-Macroscopic Model of Solute Diffusion in Dendritic Alloy Solidification," in *Micro/Macro Scale Phenomena in Solidification,* eds. C. Beckermann et al., ASME HTD, (218): 43–57, American Society of Mechanical Engineers, New York, 1992.

159. S. Ahuja, C. Beckermann, R. Zakhem, P. D. Weidman, and H. C. de Groh III, "Drag Coefficient of an Equiaxed Dendrite Settling in an Infinite Medium," eds. C. Beckermann et al., ASME HTD, (218): 85–91, American Society of Mechanical Engineers, New York, 1992.

160. S. K. Dash, and N. M. Gill, "Forced Convection Heat and Momentum Transfer to Dendritic Structures (Parabolic Cylinder and Paraboloids of Revolution)," *Int. J. Heat Mass Transfer,* (27): 1345–1356, 1984.

161. F. A. L. Dullien, *Porous Media: Fluid Transport and Pore Structure,* Academic, New York, 1979.

CHAPTER 10
NONNEWTONIAN FLUIDS

J. P. Hartnett
University of Illinois at Chicago

Y. I. Cho
Drexel University

INTRODUCTION

Overview

It is well known that the addition of small quantities of a high-molecular-weight polymer to a solvent results in a viscoelastic fluid possessing both viscous and elastic properties. Toms [1] and Mysels [2] discovered that the friction drag of such a viscoelastic fluid under turbulent flow conditions is lower than the value associated with the pure solvent. This initiated a great deal of interest in the use of small amounts of polymers in various transport systems of liquids. The possible areas of application of drag reduction include the transport of liquids or liquid-solid mixtures in pipelines, fire fighting systems, torpedoes and ships, and rotating surfaces of hydraulic machines.

An understanding of the heat transfer behavior of these nonnewtonian fluids is important inasmuch as most of the industrial chemicals and many fluids in the food processing and bio-chemical industries are viscoelastic in nature and undergo heat exchange processes either during their preparation or in their application.

Classification of Nonnewtonian Fluids

Fluids treated in the classical theory of fluid mechanics and heat transfer are the ideal fluid and the newtonian fluid. The former is completely frictionless, so that shear stress is absent, while the latter has a linear relationship between shear stress and shear rate. Unfortunately, the behavior of many real fluids used in the mechanical and chemical industries is not adequately described by these models. Most real fluids exhibit nonnewtonian behavior, which means that the shear stress is no longer linearly proportional to the velocity gradient.

Metzner [3] classified fluids into three broad groups:

1. Purely viscous fluids
2. Viscoelastic fluids
3. Time-dependent fluids

This classification of fluids is essentially the same as that of Skelland [4].

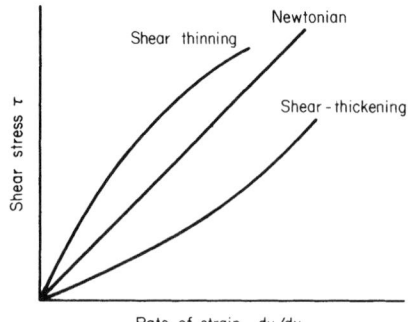

FIGURE 10.1 Flow curves for newtonian fluid and shear-thinning and shear-thickening nonnewtonian fluids [4].

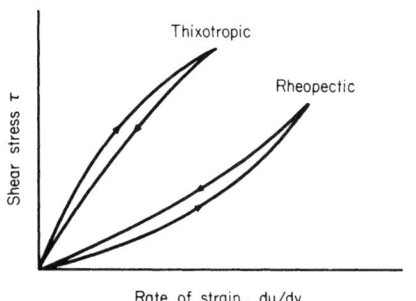

FIGURE 10.2 Flow curves for thixotropic and rheopectic fluids in continuous experiments [4].

Newtonian fluids are a subclass of purely viscous fluids. Purely viscous nonnewtonian fluids can be divided into two categories: (1) shear-thinning fluids, and (2) shear-thickening fluids. Such fluids can be described by a constitutive equation of the general form

$$\tau_{ij} = \eta_{ij}(\text{I, II, III})d_{ij} \qquad (10.1)$$

where η is the viscosity of the fluid. Here η is a decreasing function of the invariants I, II, and III of the strain tensor d_{ij} for shear-thinning fluids and an increasing function of those invariants for shear-thickening fluids. Characteristic flow curves of shear-thinning and shear-thickening fluids are shown in Fig. 10.1. Most nonnewtonian fluids used in the study of drag and heat transfer are shear-thinning. Studies of shear-thickening fluids [5] are relatively rare.

While the stress tensor component τ_{ij} for purely viscous fluids can be determined from the instantaneous values of the rate of deformation tensor d_{ij}, the past history of deformation together with the current value of d_{ij} may become an important factor in determining τ_{ij} for viscoelastic fluids. Constitutive equations to describe stress relaxation and normal stress phenomena are also needed. Unusual effects exhibited by viscoelastic fluids include rod climbing (Weissenberg effect), die swell, recoil, tubeless siphon, drag, and heat transfer reduction in turbulent flow.

Time-dependent fluids are those for which the components of the stress tensor are a function of both the magnitude and the duration of the rate of deformation at constant temperature and pressure [4]. These fluids are usually classified into two groups—thixotropic fluids and rheopectic fluids—depending on whether the shear stress decreases or increases with time at a given shear rate. Thixotropic and rheopectic behavior are common to slurries and suspensions of solids or colloidal aggregates in liquids. Figure 10.2 shows the general behavior of these fluids.

Material Functions of Nonnewtonian Fluids

It is known that incompressible newtonian fluids at constant temperature can be characterized by two material constants: the density ρ and the viscosity η. The characterization of a purely viscous nonnewtonian fluid using the power law model (or any of the so-called generalized newtonian models) is relatively straightforward. However, the experimental description of an incompressible viscoelastic nonnewtonian fluid is more complicated. Although the density can be measured, the appropriate expression for η poses considerable difficulty. Furthermore there is some uncertainty as to what other properties need to be measured. In general, for viscoelastic fluids it is known that the viscosity is not constant but depends on shear rate, that the normal stress differences are finite and depend on shear rate, and that the stress may also depend on the preshear history. To characterize a nonnewtonian fluid, it is necessary to measure the material functions (apparent viscosity, normal stress differences, etc.) in a relatively simple or standard flow. Standard flow patterns used in characterizing nonnewtonian fluids are the simple shear flow and shear-free flow.

Shear Flow Material Functions.   A simple shear flow is given by the velocity field

$$u = \dot{\gamma}_{xy}y; \quad v = 0; \quad w = 0 \qquad (10.2)$$

Here the absolute value of the velocity gradient $\dot{\gamma}_{xy}$ is called the shear rate. For a newtonian fluid it is known that in this simple shear flow only the shear stress τ_{xy} is nonzero. However, it is possible that all six independent components of the stress tensor may be nonzero for a non-newtonian fluid according to its definition. For simple shearing flow of an isotropic fluid it can be proven [6] that the total stress tensor can have the general form

$$
\begin{array}{ccc}
P + \tau_{xx} & \tau_{yx} & 0 \\
\tau_{xy} & P + \tau_{yy} & 0 \\
0 & 0 & P + \tau_{zz}
\end{array}
\tag{10.3}
$$

However, the pressure and normal stress contributions in normal force measurements on surfaces cannot be separated. Hence, the only quantities of experimental interest are the shear stress and two normal stress differences. Assuming that the flow is in the x direction, the stresses usually used in conjunction with shear flow are as follows.

Shear stress:	τ_{xy}
First normal stress difference:	$N_1 = \tau_{xx} - \tau_{yy}$
Second normal stress difference:	$N_2 = \tau_{yy} - \tau_{zz}$

$$\tag{10.4}$$

Under steady shear flow conditions it is presumed that the shear rate has been constant for such a long time that all the stresses in the fluid are time independent. Therefore, the stresses are only functions of the shear rate. Analogously to the viscosity for newtonian fluids, the apparent viscosity η is defined by the following relations:

$$
\tau_{xy} = \eta(\dot{\gamma})\dot{\gamma}_{xy}
\tag{10.5}
$$

Likewise, the normal stress coefficients Ψ_1 and Ψ_2 can be defined as

$$
\tau_{xx} - \tau_{yy} = \Psi_1(\dot{\gamma})\dot{\gamma}_{yx}^2
\tag{10.6}
$$

$$
\tau_{yy} - \tau_{zz} = \Psi_2(\dot{\gamma})\dot{\gamma}_{yx}^2
\tag{10.7}
$$

Here η, Ψ_1, and Ψ_2 are three important material functions of a nonnewtonian fluid in steady shear flow. Experimentally, the apparent viscosity is the best known material function. There are numerous viscometers that can be used to measure the viscosity for almost all nonnewtonian fluids. Manipulating the measuring conditions allows the viscosity to be measured over the entire shear rate range. Instruments to measure the first normal stress coefficients are commercially available and provide accurate results for polymer melts and concentrated polymer solutions. The available experimental results on polymer melts show that Ψ_1 is positive and that it approaches zero as $\dot{\gamma}$ approaches zero. Studies related to the second normal stress coefficient Ψ_2 reveal that it is much smaller than Ψ_1, and, furthermore, Ψ_2 is negative. For 2.5 percent polyacrylamide in a 50/50 mixture of water and glycerin, $-\Psi_2/\Psi_1$ is reported to be in the range of 0.0001 to 0.1 [7].

Rheological Property Measurements

The viscosity of a newtonian fluid can be significantly affected by such variables as temperature and pressure. The viscosity of a newtonian fluid decreases with an increase in temperature approximately according to the Arrhenius relationship:

$$
\eta = Ae^{-B/T}
\tag{10.8}
$$

where T is the absolute temperature and A and B are constants of the liquid. In general, the greater the viscosity, the stronger the temperature dependence. It is generally assumed that Eq. 10.8 holds for nonnewtonian fluids also, at least for polymeric fluids. Since the viscosity of such nonnewtonian fluids is usually very high, the temperature dependence is strong and great care must be taken in such measurements. The same consideration applies when other rheological properties such as the normal stress coefficients are measured.

With few exceptions, the viscosity of a liquid increases exponentially with isotropic pressure. In most practical applications, the departure from standard atmospheric pressure is very small and can generally be ignored.

The viscosity of nonnewtonian fluids generally depends on the shear rate. In practical applications, the shear rate varies from 10^{-6} to 10^{7} s^{-1}, covering 13 orders of magnitude. To cover this range of shear rate, several different types of viscometers are utilized.

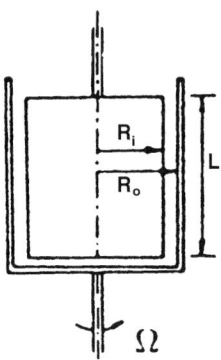

FIGURE 10.3 Schematic of rotating-cylinder viscometer. Torque C measured on inner cylinder (radius R_i). Outer cylinder rotates with angular velocity Ω.

Shear Viscosity. There are three main types of viscometers: the rotational type, the flow-through-constriction type, and the flow-around-obstruction type. The concentric cylinder and the cone-and-plate are the two primary classes of rotational viscometers. The capillary tube is an example of the flow-through-constriction viscometers. Falling-ball or falling-needle viscometers are examples of the flow-around-obstruction type.

The concentric-cylinder viscometer is schematically shown on Fig. 10.3. There are two ways that the rotation can be applied and the torque measured: the first is to drive one member and measure the torque on the same member, while the other is to drive one member and measure the torque on the other. Examples of the first kind are the Haake and Brookfield instruments; examples of the second kind would be the Weissenberg and Rheometrics rheogoniometers.

The analysis of the system shown on Fig. 10.3 for a newtonian fluid yields the working equation (Eq. 10.9)

$$\frac{C}{2\pi R_i^2 L} = \eta \frac{2\Omega}{R_i^2(1/R_i^2 - 1/R_o^2)} \tag{10.9}$$

The left term of Eq. 10.9 is the shear stress on the outer surface of the inner cylinder τ_i. The right term of Eq. 10.9 can be written as $(\eta\dot{\gamma}_i)$, where $\dot{\gamma}_i$ is the shear rate evaluated on the outer surface of the inner cylinder, that is,

$$\dot{\gamma}_i = \frac{2\Omega}{R_i^2(1/R_i^2 - 1/R_o^2)} \tag{10.10}$$

In some cases the shear rate is measured on the inner surface of the outer rotating cylinder (e.g., $\dot{\gamma}_o$). This measurement can be converted to $\dot{\gamma}_i$ by a simple transformation:

$$\frac{\dot{\gamma}_i}{\dot{\gamma}_o} = \left[\frac{R_o}{R_i}\right]^2 \tag{10.11}$$

The viscosity is determined by plotting $\ln[C/2\pi R_i^2 L]$ versus $\ln\{2\Omega/[R_i^2(1/R_i^2 - 1/R_o^2)]\}$. When the value of the abscissa goes to zero, the value of the ordinate will be $\ln\eta$.

If the concentric cylinder viscometer has a narrow gap (i.e., $0.97 \le R_i/R_o < 1$), with the inner cylinder rotating and the outer cylinder stationary, the shearing stress τ_i may be given for any fluid, newtonian or nonnewtonian, by the following expression:

$$\tau = C/2\pi R_o^2 L \tag{10.12}$$

The corresponding shear rate is

$$\dot{\gamma} = R_o\Omega/(R_o - R_i) \tag{10.13}$$

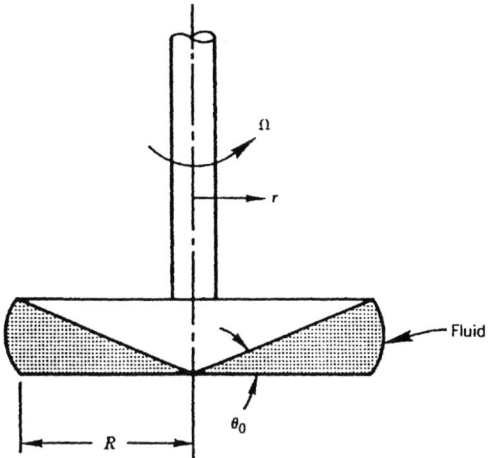

FIGURE 10.4 Schematic of cone-and-plate viscometer. Torque C measured on plate; angular velocity Ω measured on rotating cone.

and the viscosity is given by

$$\eta = C(R_o - R_i)/2\pi R_o^3 \Omega L \qquad (10.14)$$

For wider gaps, the procedure for determining the viscosity of a nonnewtonian fluid becomes more complicated because the velocity depends on the viscosity function [4].

The cone-and-plate viscometer is another type of rotational viscometer and is schematically shown in Fig. 10.4. For any time-independent fluid, the following equations apply if the viscometer cone angle θ_0 is small (1 or 2°):

$$\dot{\gamma} = \Omega/\theta_0 \qquad (10.15)$$

$$\tau = 3C/2\pi R^3 \qquad (10.16)$$

$$\eta = 3C\theta_0/2\pi R^3\Omega \qquad (10.17)$$

It should be noted that special care must be taken in using these rotating viscometers; it is often necessary to correct the results to account for experimental departures from the idealized model (i.e., narrow gap and small cone angle approximation for the concentric-cylinder and cone-and-plate viscometers, respectively).

The capillary tube viscometer is a flow-through-restriction type. When used carefully, it is capable of accuracies of better than 2 percent over its applicable shear-rate range (300 to $4000\,\text{s}^{-1}$). For laminar flow of a nonnewtonian liquid in a capillary tube, it can be shown [4] that the wall shear stress τ_w and the shear rate at the wall $\dot{\gamma}_w$ are given by

$$\tau_w = (R/2)(\Delta P/L) \qquad (10.18)$$

$$\dot{\gamma}_w = \frac{4Q}{\pi R^3}\left(\frac{3}{4} + \frac{1}{4}\frac{d \ln Q}{d \ln [(R/2)(\Delta P/L)]}\right) \qquad (10.19)$$

The viscosity is given by

$$\eta(\dot{\gamma}_w) = \frac{\pi R^4(\Delta P/L)}{8Q\left(\dfrac{3}{4} + \dfrac{1}{4}\dfrac{d \ln Q}{d \ln [(R/2)(\Delta P/L)]}\right)} \qquad (10.20)$$

TABLE 10.1 Generalized Newtonian Models

Model	η	Characteristic time
Power law [18]	$\eta = K\dot\gamma^{n-1}$	None
Bingham [20]	$\eta = \eta_0 + \dfrac{\tau_0}{\dot\gamma} \quad \tau \geq \tau_0^*$ $\dot\gamma = 0 \quad \tau \leq \tau_0$	None
Ellis [21]	$\dfrac{1}{\eta} = \dfrac{1}{\eta_0}\left[1 + \left(\dfrac{\tau}{\tau_{1/2}}\right)^{1/n-1}\right]$	$\dfrac{\eta_0}{\tau_{1/2}}$ †
Powell-Eyring [22]	$\eta = \eta_\infty + (\eta_0 - \eta_\infty)\left(\dfrac{\sinh^{-1} t_p\dot\gamma}{t_p\dot\gamma}\right)$	t_p
Sutterby [23]	$\eta = \eta_0\left(\dfrac{\sinh^{-1} t\dot\gamma}{t\dot\gamma}\right)^{\alpha}$	t
Carreau A [24]	$\eta = \eta_\infty + (\eta_0 - \eta_\infty)[1 + (t\dot\gamma)^2]^{(n-1)/2}$	t

* τ_0 is the yield stress.
† $\tau_{1/2}$ is the value of the shear stress at which $\eta = \eta_0/2$.

Equations 10.19 and 10.20 are derived for a general nonnewtonian fluid. For a simple power-law fluid described by the relation $\eta = K\dot\gamma^{n-1}$ (Table 10.1), these equations can be significantly simplified:

$$\frac{\Delta P}{L}\frac{d}{4} = \left\{K\left(\frac{3n+1}{4n}\right)^n\left(\frac{8U}{d}\right)^n\right\} \tag{10.21}$$

If $\ln[(\Delta P/L)(d/4)]$ is plotted against $\ln[8U/d]$, the slope of the curve yields the power law index n and the ordinate intercept gives $\ln[\{(3n+1)/4n\}^n]$. Some precautions have to be taken during the experiments. Equations 10.18–10.20 were derived assuming a laminar flow. Therefore turbulent flow must be avoided. Viscous heating and end effects can be encountered when using the capillary tube viscometer. The viscous heating effect can be reduced by the use of a constant-temperature bath, while the end effect can be minimized by increasing the length-diameter ratio (>100).

In the case of the *falling-ball viscometer,* details may be found in Ref. 8. Reference 9 provides detailed coverage of the *falling-needle viscometer.*

Normal Stress Coefficients and Oscillatory Viscometric Measurements

Cone-and-Plate Instrument. The fluid to be tested is placed in the gap between the cone and plate. Three measurements are generally made: the torque C on the plate, the total normal force F on the plate, and the pressure distribution $P + \tau_{\theta\theta}$ across the plate. Under the assumptions that (1) inertial effects are negligible, and (2) the angle between the cone and the plate is small (1 to 2°), the first and second normal stress coefficients can be evaluated from the following two equations:

$$\frac{\partial\pi_{\theta\theta}}{\partial\ln r} = -(\Psi_1 + 2\Psi_2)\dot\gamma^2 \tag{10.22}$$

$$\Psi_1 = \frac{2F}{\pi R^2\dot\gamma^2} \tag{10.23}$$

Here, $\pi_{\theta\theta}$ is the pressure, which may be measured by flush-mounted pressure transducers located on the plate, and F is the total force applied on the plate to keep the tip of the cone on the surface of the plate. However, evaluation of both the first and second normal stress coefficients requires the pressure distribution on the plate. Only a few instruments have the capacity to measure both Ψ_1 and Ψ_2.

Oscillatory measurements using the cone-and-plate viscometer are sometimes carried out to demonstrate the elastic behavior of a viscoelastic fluid [10]. The fluid in the viscometer is subjected to an oscillatory strain imposed on the bottom surface while the response of the shearing stress is measured on the top surface. If the phase shift between the input strain and the output stress is 90°, the sample is purely viscous; if it is 0°, the sample is completely elastic. A measured phase shift between 0° and 90° demonstrates that the fluid is viscoelastic.

Thermophysical Properties of Nonnewtonian Fluids

The physical properties of nonnewtonian fluids necessary for the study of forced convection heat transfer are the thermal conductivity, density, specific heat, viscosity, and elasticity. In general these properties must be measured as a function of temperature and, in some instances, of shear rate. In the special case of aqueous polymer solutions it is recommended that all properties except the viscous and elastic properties be taken to be the same as those of water.

This is confirmed by the work of Christiansen and Craig [11], Oliver and Jenson [12], and Yoo [13]. These investigators found that the thermal conductivities of dilute aqueous solutions of Carbopol-934, carboxymethyl cellulose (CMC), polyethylene oxide, and polyacrylamide are no more than 5 percent lower than those of pure water at corresponding temperature. However, Bellet et al. [14] observed substantial decreases in the thermal conductivity measurements for much higher concentrations of aqueous solutions of Carbopol-960 and CMC (i.e., beyond 10 to 15 percent by weight). Lee and Irvine [15] reported that the thermal conductivity of aqueous polyacrylamide solutions was dependent on the shear rate.

Lee et al. [16] measured thermal conductivities of various nonnewtonian fluids at four different temperatures using a conventional thermal conductivity cell. These results, shown in Table 10.2, support the common practice of assuming that the thermal conductivity of aqueous polymer solution is equal to that of pure water of a corresponding temperature if the concentration of the polymer is less than 10,000 wppm (that is, 1 percent by weight).

TABLE 10.2   Data of Thermal Conductivities k_l, W/(m·K)*

Liquid	c, wppm[†]	T, °C			
		20	30	40	50
Water	—	0.593	0.612	0.627	0.645
Polyethylene oxide	100	0.599	0.619	0.630	0.651
(WSR-301)	1,000	0.597	0.619	0.638	0.646
	10,000	0.604	0.624	0.634	0.656
Polyacrylamide	100	0.590	0.602	0.611	0.648
(Separan AP-273)	1,000	0.590	0.609	0.616	0.646
	10,000	0.592	0.610	0.632	0.648
Carboxymethyl	1,000	0.576	0.603	0.632	0.648
cellulose (CMC)	10,000	0.583	0.611	0.637	0.665
Carbopol-960	100	0.585	0.614	0.634	0.648
	1,000	0.595	0.606	0.629	0.651
	10,000	0.616	0.644	0.650	0.679
Attagel-40	1,000	0.594	0.605	0.625	0.650
	10,000	0.604	0.614	0.636	0.645
Polyacrylamide	1,000	0.588	0.604	0.637	0.643
(with 4% NaCl)					

* 1 W/(m·K) = 0.5778 Btu/(h·ft·°F).
[†] wppm = parts per million by weight.

Governing Equations of Nonnewtonian Flow

Conservation Equations.   In the above section, the material functions of nonnewtonian fluids and their measurements were introduced. The material functions are defined under a simple shear flow or a simple shear-free flow condition. The measurements are also performed under or nearly under the same conditions. In most engineering practice the flow is far more complicated, but in general the measured material functions are assumed to hold. Moreover, the conservation principles still apply, that is, the conservation of mass, momentum, and energy principles are still valid. Assuming that the fluid is incompressible and that viscous heating is negligible, the basic conservation equations for newtonian and nonnewtonian fluids under steady flow conditions are given by

Mass: $$(\nabla \cdot V) = 0 \tag{10.24}$$

Momentum: $$\rho \frac{DV}{Dt} = -\nabla P + \nabla \cdot \tau + \rho g \tag{10.25}$$

Energy: $$\rho c_p \frac{DT}{Dt} = k_t \nabla^2 T \tag{10.26}$$

where τ is the shear stress, which will be determined when the constitutive equation of the fluid is specified.

Constitutive Equations.   For a simple shear flow, Eq. 10.5 describes the dependence of shear stress on shear rate. Equation 10.5 can be extended to an arbitrary nonnewtonian flow:

$$\tau_{ij} = \eta d_{ij} \tag{10.27}$$

Here d_{ij} is the rate-of-deformation tensor and η is the nonnewtonian viscosity, a function of the scalar invariant of the rate-of-deformation tensor d_{ij}. In practice, the magnitude of the rate-of-deformation tensor is often used.

$$\dot{\gamma} = \sqrt{\frac{1}{2} \sum \sum d_{ij} d_{ji}} \tag{10.28}$$

In shearing flow, $\dot{\gamma}$ is called the *shear rate*. Many expressions have been proposed to approximate the actual dependence of the viscosity of the magnitude of the rate-of-deformation tensor. Some of the models used to describe the behavior of purely viscous nonnewtonian fluids are listed in Table 10.1.

Among these nonnewtonian fluid models, the power-law model [18] has the simplest viscosity-shear rate relation. For many real fluids this relation generally describes the intermediate shear rate range viscosity very well. It is the most widely used model in nonnewtonian fluid mechanics studies and has proven quite successful in predicting the behavior of a large number of nonnewtonian flows [19]. However, it has several built-in flaws and anomalies. For example, considerable error can occur when the shear rate is very small or very large. In flow over submerged bodies, there usually exist one or more stagnation points. The power-law model predicts an infinite viscosity at the stagnation point. This can cause the drag coefficient to be significantly overpredicted. As the generalized newtonian model becomes more complex such anomalies can be removed; for example, the Carreau model [24] avoids the previously mentioned difficulty and provides sufficient flexibility to fit a wide variety of experimental η-versus-$\dot{\gamma}$ curves. Such fluids are called generalized newtonian fluids. Their viscosity is shear-rate-dependent, but the normal stress differences are negligible.

In flow situations where the elastic properties play a role, viscoelastic fluid models are generally needed. Such models may be linear (e.g., Voigt, Maxwell) or nonlinear (e.g., Oldroyd). In general they are quite complex and will not be treated in this chapter. For further details, interested readers are referred to the textbooks by Bird et al. [6] and Barnes et al. [25].

Use of Reynolds and Prandtl Numbers

Duct flows of nonnewtonian fluids are described by the governing equations (Eq. 10.24–10.26), by the constitutive equation (Eq. 10.27) with the viscosity defined by one of the models in Table 10.1, or by a linear or nonlinear viscoelastic constitutive equation. To compare the available analytical and experimental results, it is necessary to nondimensionalize the governing equations and the constitutive equations. In the case of newtonian flows, a uniquely defined nondimensional parameter, the Reynolds number, is found. However, a comparable nondimensional parameter for nonnewtonian flow is not uniquely defined because of the different choice of the characteristic viscosity.

In the presentation of experimental results describing the fluid mechanics of a power-law nonnewtonian fluid flowing through circular tubes, the five different definitions of the Reynolds number shown in Table 10.3 have been used by various investigators:

1. A generalized Reynolds number Re', introduced by Metzner and Reed
2. A Reynolds number based on the apparent viscosity at the wall, Re_a
3. A generalized Reynolds number Re^+, derived for a power-law fluid from the nondimensional momentum equation
4. A Reynolds number based on the solvent viscosity, Re_s
5. A Reynolds number based on the effective viscosity, Re_{eff}

This use of different Reynolds numbers from one investigator to another makes the comparison of different sets of data quite difficult. The relative merits of the five definitions are discussed below. It was pointed out by Skelland [4] that for fully developed laminar circular-tube flow of nonnewtonian fluids, the wall shear stress τ_w is a unique function of $8U/d$. This may be expressed as

$$\tau_w = K'\left(\frac{8U}{d}\right)^n \qquad (10.29)$$

where K' and n vary with $8U/d$ for most polymeric solutions. It should be noted that K' is not the same as K, the consistency index. To obtain the relationships between these two terms, recast Eq. 10.21 in terms of τ_w and equate the result to Eq. 10.29. This leads to the following relation:

TABLE 10.3 Definitions of Reynolds and Prandtl Numbers: Circular-Tube Flow

	Shear stress–shear rate	Reynolds number	Prandtl number	Pe $(= \rho c_p U d/k_l)$
1	$\tau_w = K'\left(\dfrac{8U}{d}\right)^n$	$Re' = \dfrac{\rho U^{2-n}d^n}{K'8^{n-1}}$	$Pr' = \dfrac{c_p K'(8U/d)^{n-1}}{k_l}$	$Re'\,Pr'$
2	$\tau_w = \eta_a \dot{\gamma}_w$ $\dot{\gamma}_w = \dfrac{3n+1}{4n}\dfrac{8U}{d}$	$Re_a = \dfrac{\rho U d}{\eta_a}$	$Pr_a = \dfrac{\eta_a c_p}{k_l}$	$Re_a\,Pr_a$
3	$\tau_{ij} = K(d_{ij})^n$	$Re^+ = \dfrac{\rho U^{2-n}d^n}{K}$	$Pr^+ = \dfrac{c_p K(U/d)^{n-1}}{k_l}$	$Re^+\,Pr^+$
4	$\eta_s =$ solvent viscosity	$Re_s = \dfrac{\rho U d}{\eta_s}$	$Pr_s = \dfrac{\eta_s c_p}{k_l}$	$Re_s\,Pr_s$
5	$\tau_w = \eta_{eff}\dfrac{8U}{d}$	$Re_{eff} = \dfrac{\rho U d}{\eta_{eff}}$	$Pr_{eff} = \dfrac{\eta_{eff}c_p}{k_l}$	$Re_{eff}\,Pr_{eff}$

$$K' = K\left(\frac{3n+1}{4n}\right)^n \tag{10.30}$$

The dimensionless fully developed pressure drop is given by the Fanning friction factor f, defined by the relation

$$f = \frac{\tau_w}{\frac{1}{2}\rho U^2} \tag{10.31}$$

Metzner and Reed [26] introduced a generalized Reynolds number Re′ such that the Fanning friction factor for fully developed laminar pipe flow is given by

$$f = \frac{16}{\text{Re}'} \tag{10.32}$$

Substituting Eq. 10.21 into Eq. 10.18 and taking note of Eq. 10.31, the generalized Reynolds number Re′ is obtained:

$$\text{Re}' = \frac{\rho U^{2-n}d^n}{K'8^{n-1}} \tag{10.33}$$

This Reynolds number has had wide use because all fully established laminar pipe flow laminar friction factor data for power law fluids lie on the line $f = 16/\text{Re}'$.

A second choice of Reynolds number [27] is based on the apparent viscosity at the wall:

$$\text{Re}_a = \frac{\rho U d}{\eta_a} \tag{10.34}$$

This is a simple modification of the usual definition of Reynolds number for newtonian fluids. The apparent viscosity at the wall is calculated from the following expression:

$$\tau_w = \eta_a \dot{\gamma}_w \tag{10.35}$$

Here $\dot{\gamma}_w$ becomes $[(3n+1)/4n]8U/d$ for established pipe flow. Applying the definition of the Fanning friction factor, it can be shown for the laminar circular-tube flow of a power-law fluid that

$$f = \frac{3n+1}{4n}\frac{16}{\text{Re}_a} \tag{10.36}$$

demonstrating that f is a function not only of Re_a but also of n.

The third approach involving the use of Re+ is sometimes encountered in the study of non-newtonian flow over surfaces such as plates, cylinders, or spheres [28].

$$\text{Re}^+ = \frac{\rho U^{2-n}d^n}{K} \tag{10.37}$$

Earlier investigators studying the drag-reducing phenomenon in viscoelastic fluids often used Re_s and Re_{eff}. The former is generally valid only for dilute polymer solutions, in which case the solution viscosity is quite close to that of the solvent. The use of Re_{eff} seems inappropriate in the study of the drag coefficient because it does not represent any physical property of nonnewtonian fluids, although it produces a unique reference line for experimental friction data in laminar pipe flow:

$$f = \frac{16}{\text{Re}_{\text{eff}}} \tag{10.38}$$

In all five cases the corresponding Prandtl number is defined to be such that the product of the Reynolds and Prandtl numbers yields the Peclet number $\rho c_p U d/k_l$.

In summary, for the experimental or analytical studies of nonnewtonian laminar flow through circular ducts the use of Re' and Re_a is recommended; for the studies of nonnewtonian laminar flow over submerged objects, Re^+ is commonly used.

Use of the Weissenberg Number

In dealing with viscoelastic fluids, especially under turbulent flow conditions, it is necessary to introduce a dimensionless number to take account of the fluid elasticity [29–33]. Either the Deborah or the Weissenberg number, both of which have been used in fluid mechanical studies, satisfies this requirement. These dimensionless groups are defined as follows:

$$\text{De} = \frac{t}{t_F} \tag{10.39}$$

$$\text{Ws} = t\,\frac{U}{d} \tag{10.40}$$

where t is a characteristic time of the fluid and a measure of the elasticity of the fluid t_F is a characteristic time of flow, and U/d is a characteristic shear rate. In this chapter the Weissenberg number will be used to specify the dimensionless elastic effects.

The evaluation of the Weissenberg number requires the determination of the characteristic time of the fluid. This can be accomplished by combining the use of a generalized newtonian model (see Table 10.1) with steady shear viscosity data [21, 34]. The characteristic time of a given sample is obtained by determining the value of t that gives the best fit to the measured viscosity data over the complete shear rate range. Among the various models, the Powell-Eyring model [22] and the Carreau model A [24] were found to be the most suitable for aqueous solutions of polyethylene oxide and polyacrylamide [35–37]. It should be noted that the absolute value of the calculated relaxation time differs from one nonnewtonian model to another. Consequently it is critical that the procedure for determining t be specified when giving numerical values of the Weissenberg number.

LAMINAR NONNEWTONIAN FLOW IN A CIRCULAR TUBE

Velocity Distribution and Friction Factor

For a fully developed nonnewtonian laminar pipe flow, the governing momentum equation can be written as

$$0 = -\frac{dP}{dx} + \frac{1}{r}\frac{d}{dr}(r\tau_{rx}) \tag{10.41}$$

If the power-law model is assumed to describe the viscosity of the fluid, then

$$\tau_{rx} = K\left(\frac{du}{dr}\right)^n \tag{10.42}$$

and the fully developed velocity profile can be shown to be

$$u = u_{\max}\left[1 - \left(\frac{r}{R}\right)^{(n+1)/n}\right] \tag{10.43}$$

where
$$u_{\max} = \left(\frac{\tau_w}{K}\right)^{1/n} \frac{R}{1 + 1/n} \tag{10.44}$$

For values of n less than 1, this gives a velocity that is flatter than the parabolic profile of newtonian fluids. As n approaches zero, the velocity profile predicted by this equation approaches a plug flow profile. Figure 10.5 shows the velocity profile generated by Eq. 10.43 for selected values of the power-law index n. It should be noted that the velocity profiles given in Fig. 5 are valid in the hydrodynamically fully developed region where the entrance effect can be neglected.

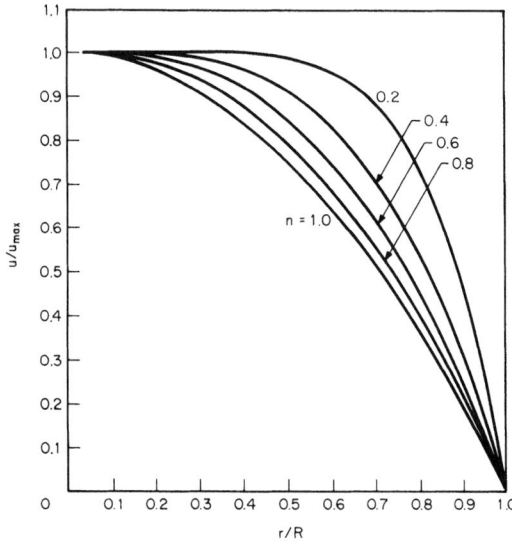

FIGURE 10.5   Velocity profile in fully developed laminar pipe flow for nonnewtonian power-law fluids.

As noted earlier, the Fanning friction factor for fully developed laminar pipe flow of a power-law fluid can be predicted by the following equation:

$$f = \frac{16}{\mathrm{Re}'} \tag{10.32}$$

Experimental measurements of pressure drop for purely viscous nonnewtonian fluids flowing through a circular tube in the fully developed laminar flow region confirm this prediction. In fact, this relationship also applies to fully established flow of viscoelastic fluids through circular tubes as demonstrated by Tung et al. [38]. The reason for this is that there is no mechanism for elasticity to play a role under fully established pipe flow conditions. Equation 10.32 is recommended for the prediction of pressure drop for nonnewtonian fluids, both purely viscous and viscoelastic, in fully established laminar pipe flow.

In the hydrodynamic entrance region where the flow undergoes development of its velocity profile, the governing equations are much more complicated. Bogue [39] calculated the hydrodynamic entrance length using the von Karman integral method for a power-law fluid in laminar pipe flow. Table 10.4 shows the results for four different n values. Experimental studies generally show that nonnewtonian additives, including high-molecular-weight polymers, do not affect the entrance length in the laminar region. Therefore, Table 10.4 is recommended for estimating the hydrodynamic entrance length of purely viscous and viscoelastic fluids in laminar pipe flow.

TABLE 10.4 Hydrodynamic Entrance Length
in Laminar Pipe Flow [39]

n	$L_h/(d\,\mathrm{Re})$
1.00	0.0575
0.75	0.048
0.50	0.034
0.25	0.017

For rectangular channels the hydraulic diameter is
taken as the characteristic length.

Fully Developed Heat Transfer

The fully established laminar heat transfer results for nonnewtonian fluids flowing through a circular tube with a fully developed velocity distribution and constant heat flux boundary condition at the wall can be obtained by solving the following energy equation:

$$\rho c_p u \frac{\partial T}{\partial x} = k_l \frac{1}{r}\frac{\partial}{\partial r}\left(r\frac{\partial T}{\partial r}\right) \tag{10.45}$$

The boundary conditions are

$$\text{At } r = 0 \qquad T = \text{finite}$$

$$\text{At } r = R \qquad -k\left(\frac{\partial T}{\partial r}\right) = q_w'' \tag{10.46}$$

$$\text{At } x = 0 \qquad T = T_{\text{in}}$$

The fully developed velocity profile necessary to solve the preceding equation was calculated for the power-law fluid and presented in Eq. 10.43.

Applying the separation-of-variables technique to solve the preceding partial differential equation, the Nusselt number for the constant heat rate case in the fully developed region can be shown [6] to be given by the following equation:

$$\mathrm{Nu}_\infty = \frac{8(5n+1)(3n+1)}{31n^2 + 12n + 1} \tag{10.47}$$

The Nusselt number for power-law fluids for constant wall heat flux reduces to the newtonian value of 4.36 when $n = 1$ and to 8.0 when $n = 0$. Equation 10.47 is applicable to the laminar flow of nonnewtonian fluids, both purely viscous and viscoelastic, for the constant wall heat flux boundary condition for values of x/d beyond the thermal entrance region. The laminar heat transfer results for the constant wall temperature boundary condition were also obtained by the separation of variables using the fully developed velocity profile. The values of the Nusselt number for $n = 1.0$, ½, and ⅓ calculated by Lyche and Bird [40] are 3.66, 3.95, and 4.18, respectively, while the value for $n = 0$ is 5.80. These values are equally valid for purely viscous and viscoelastic fluids for the constant wall temperature case provided that the thermal conditions are fully established.

Laminar Heat Transfer in the Thermal Entrance Region

The prediction of the local laminar heat transfer coefficient for a power-law fluid in the thermal entrance region of a circular tube was reported by Bird and colleagues [41]. Both the constant wall heat flux and the constant wall temperature boundary condition have been studied. The results can be expressed by the following relationships [42–48].

Local Nusselt number–constant wall heat flux:

$$\mathrm{Nu}_x = 1.41\left(\frac{3n+1}{4n}\right)^{1/3}\mathrm{Gz}^{1/3} \tag{10.48}$$

Local Nusselt number–constant wall temperature:

$$\mathrm{Nu}_x = 1.16\left(\frac{3n+1}{4n}\right)^{1/3}\mathrm{Gz}^{1/3} \tag{10.49}$$

It is interesting to note that the nonnewtonian effect has been taken into account by simply multiplying the corresponding newtonian result by $[(3n + 1)/4n]^{1/3}$. Equations 10.48 and 10.49 may be used to predict the local heat transfer coefficient of purely viscous and viscoelastic fluids in the thermal entrance region of a circular tube. Figure 10.6 shows a typical comparison of the measured local heat transfer coefficient of a viscoelastic fluid with the prediction for a power-law fluid. The good agreement provides evidence to support the applicability of Eq. 10.48 in the case of the constant heat flux boundary condition.

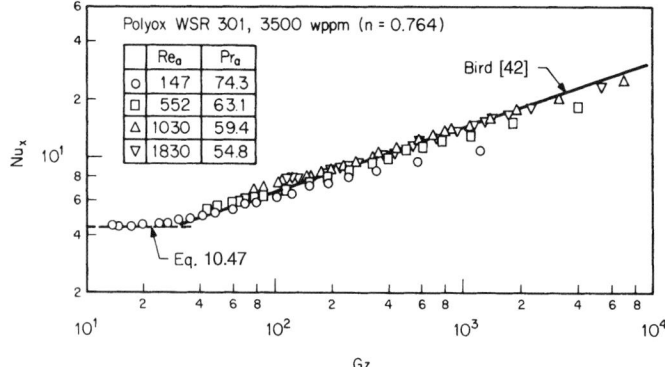

FIGURE 10.6   Experimental results for laminar pipe flow heat transfer for constant wall heat flux boundary conditions [35].

The mean value of the Nusselt number at any position along the tube is equal to 1.5 times the local values given in Eqs. 10.48 and 10.49.

The dimensionless thermal entrance length L_t/d can be estimated using the following expression:

$$L_t/d = 0.04\,\mathrm{Re}\,\mathrm{Pr} \tag{10.50}$$

LAMINAR NONNEWTONIAN FLOW IN A RECTANGULAR DUCT

Velocity Distribution and Friction Factor

A variety of noncircular passage geometries, including the rectangular duct, have been utilized for internal flow applications, for example in compact heat exchangers and solar collectors. The study of the hydrodynamic behavior in a rectangular duct requires a two-dimensional analysis, since the axial velocity even in the fully developed region is a function of two independent variables. The governing equations expressing conservation of mass, momentum, and energy in a rectangular coordinate system under steady-state conditions and in the absence of body forces are

$$\frac{\partial u}{\partial x} + \frac{\partial v}{\partial y} + \frac{\partial w}{\partial z} = 0 \tag{10.51}$$

$$\left(u\frac{\partial u}{\partial x} + v\frac{\partial u}{\partial y} + w\frac{\partial u}{\partial z} \right) = -\frac{1}{\rho}\frac{\partial P}{\partial x} + \frac{1}{\rho}\left[\frac{\partial \tau_{xx}}{\partial x} + \frac{\partial \tau_{xy}}{\partial y} + \frac{\partial \tau_{xz}}{\partial z} \right] \tag{10.52}$$

$$\left(u\frac{\partial v}{\partial x} + v\frac{\partial v}{\partial y} + w\frac{\partial v}{\partial z} \right) = -\frac{1}{\rho}\frac{\partial P}{\partial y} + \frac{1}{\rho}\left[\frac{\partial \tau_{xy}}{\partial x} + \frac{\partial \tau_{yy}}{\partial y} + \frac{\partial \tau_{yz}}{\partial z} \right] \tag{10.53}$$

$$\left(u\frac{\partial w}{\partial x} + v\frac{\partial w}{\partial y} + w\frac{\partial w}{\partial z} \right) = -\frac{1}{\rho}\frac{\partial P}{\partial z} + \frac{1}{\rho}\left[\frac{\partial \tau_{xz}}{\partial x} + \frac{\partial \tau_{yz}}{\partial y} + \frac{\partial \tau_{zz}}{\partial z} \right] \tag{10.54}$$

$$u\frac{\partial T}{\partial x} + v\frac{\partial T}{\partial y} + w\frac{\partial T}{\partial z} = \frac{1}{\rho c_p}\left[\frac{\partial}{\partial x}\left(k\frac{\partial T}{\partial x} \right) + \frac{\partial}{\partial y}\left(k\frac{\partial T}{\partial y} \right) + \frac{\partial}{\partial z}\left(k\frac{\partial T}{\partial z} \right) \right] \tag{10.55}$$

where the stress components of the stress tensor are to be determined using one of the constitutive equations. Equations 10.51–10.55, along with proper boundary conditions and a prescribed constitutive equation, describe the nonnewtonian flow and heat transfer in rectangular ducts.

For hydraulically fully developed flow, the following conditions apply:

$$\frac{\partial u}{\partial x} = 0, \quad v = w = 0, \quad P = P(x), \quad u = u(y, z) \tag{10.56}$$

If it is assumed that the constitutive equation is given by the power law, then

$$\eta = K(\text{II}/2)^{(n-1)/2} \tag{10.57}$$

$$\text{II}/2 = \left\{ 2\left[\left(\frac{\partial u}{\partial x} \right)^2 + \left(\frac{\partial y}{\partial y} \right)^2 + \left(\frac{\partial w}{\partial z} \right)^2 \right] \right\} + \left(\frac{\partial u}{\partial y} + \frac{\partial v}{\partial x} \right)^2 + \left(\frac{\partial u}{\partial z} + \frac{\partial w}{\partial x} \right)^2 + \left(\frac{\partial v}{\partial z} + \frac{\partial w}{\partial y} \right)^2 \tag{10.58}$$

Taking note of Eqs. 10.56 and 10.58, the final equation describing the fully established laminar velocity profile of a power-law fluid flowing through a rectangular duct is given by

$$\frac{1}{K}\frac{dP}{dx} = \frac{\partial}{\partial y}\left\{ \left[\left(\frac{\partial u}{\partial y} \right)^2 + \left(\frac{\partial u}{\partial z} \right)^2 \right]^{(n-1)/2}\frac{\partial u}{\partial y} \right\} + \frac{\partial}{\partial z}\left\{ \left[\left(\frac{\partial u}{\partial y} \right)^2 + \left(\frac{\partial u}{\partial z} \right)^2 \right]^{(n-1)/2}\frac{\partial u}{\partial z} \right\} \tag{10.59}$$

subject to the conditions that the velocity u goes to zero on the boundaries of the flow.

Equation 10.59 was solved by Schechter [49] using a variational principle and by Wheeler and Wissler [50] using a numerical method. Wheeler and Wissler also presented an approximate equation for the square duct geometry. Schechter reported approximate velocity profiles for a power law fluid flowing through rectangular ducts having aspect ratios 0.25, 0.50, 0.75, and 1.0. His results may be expressed as follows:

$$\frac{u(y, z)}{U} = \sum_{i=1}^{6} A_i \sin\left[\alpha_i\pi\frac{(z/a)}{2} + 1 \right] \sin\left[\beta_i\pi\frac{(y/b) + 1}{2} \right] \tag{10.60}$$

where the values of α_i and β_i are shown in Table 10.5, and the values of the coefficients A_i are shown in Table 10.6.

TABLE 10.5   Values of Constants in Eq. 10.60

i	α_i	β_i
1	1	1
2	3	1
3	1	3
4	3	3
5	5	1
6	1	1

TABLE 10.6   Computed Results for Flow in Rectangular Duct[a]

n	α^*	A_1	A_2	A_3	A_4	A_5	A_6
1.00	1.00	2.346	0.156	0.156	0.0289	0.0360	0.0360
0.75	1.00	2.313	0.205	0.205	0.0007	0.0434	0.0434
0.50	1.00	2.263	0.278	0.278	−0.0285	0.0555	0.0555
1.00	0.75	2.341	0.204	0.119	0.0256	0.0498	0.0303
0.75	0.75	2.310	0.235	0.180	0.0001	0.0568	0.0364
0.50	0.75	2.263	0.286	0.267	−0.0277	0.0644	0.0505
1.00	0.50	2.311	0.296	0.104	0.0285	0.0795	0.0303
0.75	0.50	2.288	0.299	0.174	0.0120	0.0811	0.0364
0.50	0.50	2.249	0.312	0.274	−0.0101	0.0780	0.0501
1.00	0.25	2.227	0.503	0.0867	0.0274	0.184	0.0189
0.75	0.25	2.221	0.459	0.160	0.0312	0.160	0.0210
0.50	0.25	2.205	0.407	0.270	0.0257	0.131	0.0364

[a] See Eq. 10.60.

Wheeler and Wissler proposed an approximate equation for the fully developed friction factor for laminar flow of a power-law fluid through a square duct:

$$f \cdot \text{Re}^+ = 1.874 \left(\frac{1.7330}{n} + 5.8606 \right)^n \tag{10.61}$$

Here $\text{Re}^+ = \rho U^{2-n} d_h^n / K$ and $0.4 < n < 1.0$. Chandrupatla and Sastri also reported friction factor results for flow in a square duct [51].

A different approach was taken by Kozicki et al. [52, 53], who generalized the Rabinowitsch-Mooney equation to cover nonnewtonian fluids including the special case of power-law fluids in arbitrary ducts having a constant cross section. These authors introduced a new Reynolds number such that the friction factor for fully developed laminar flow of a power-law fluid through noncircular geometries having constant cross-sectional area is given by a unique equation:

$$f = 16/\text{Re}^* \tag{10.62}$$

where

$$\text{Re}^* = \rho U^{2-n} d_h^n \bigg/ \left[8^{n-1} \left(b^* + \frac{a^*}{n} \right)^n K \right] \tag{10.63}$$

The values of a^* and b^* depend on the geometry of the duct. Table 10.7 presents these values for a rectangular channel as a function of the aspect ratio α^*. It is of interest to note that a^* and b^* are 0.25 and 0.75 for the circular duct, and that the generalized Reynolds number Re^* becomes identical to that proposed by Metzner and Reed [26].

TABLE 10.7   Geometric Constants $a*$ and $b*$ for Rectangular Ducts[a,b]

$\alpha*$	$a*$	$b*$	c	$\alpha*$	$a*$	$b*$	c
1.00	0.2121	0.6771	14.227	0.45	0.2538	0.7414	15.922
0.95	0.2123	0.6774	14.235	0.40	0.2659	0.7571	16.368
0.90	0.2129	0.6785	14.261	0.35	0.2809	0.7750	16.895
0.85	0.2139	0.6803	14.307	0.30	0.2991	0.7954	17.512
0.80	0.2155	0.6831	14.378	0.25	0.3212	0.8183	18.233
0.75	0.2178	0.6870	14.476	0.20	0.3475	0.8444	19.071
0.70	0.2208	0.6921	14.605	0.15	0.3781	0.8745	20.042
0.65	0.2248	0.6985	14.772	0.10	0.4132	0.9098	21.169
0.60	0.2297	0.7065	14.980	0.05	0.4535	0.9513	22.477
0.55	0.2360	0.7163	15.236	0.0	0.5000	1.0000	24.000
0.50	0.2439	0.7278	15.548				

[a] See Eq. 10.63.
[b] Note: $c = 16(a* + b*) = f\,Re$ for newtonian fluid.

The validity of Eq. 10.62 has been confirmed by the experiments of Wheeler and Wissler [50], Hartnett et al. [54], and Hartnett and Kostic [55] for fully developed laminar flow of aqueous polymer solutions in rectangular channels (Fig. 10.7). Given the fact that these solutions are viscoelastic, a number of analytical studies that take elasticity into account predict that the presence of normal forces produces secondary flows [56–60]. However, these analytical studies, along with the previously cited pressure drop measurements, indicate that if such secondary flows exist, they have little effect on the laminar friction factor.

In light of these observations, Eq. 10.62 is recommended for predicting the fully developed friction factor of both purely viscous and viscoelastic fluids in laminar flow through rectangular channels.

Fully Developed Heat Transfer—Purely Viscous Fluids

The solution of Eq. 10.55 describing the conservation of energy requires the solution of the momentum equation for a specified constitutive relationship. The previous section provides this information for a power-law fluid. This section will treat the fully developed heat transfer

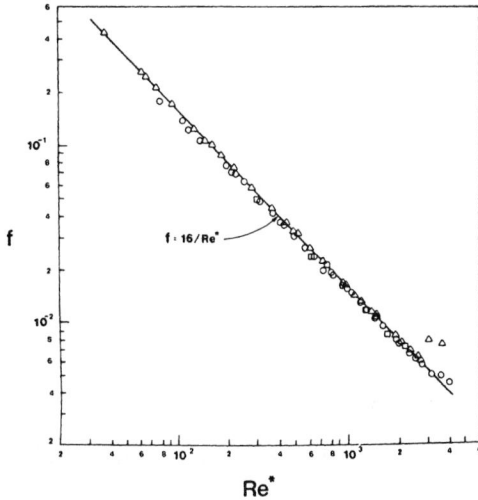

FIGURE 10.7   Experimental friction factor measurements for nonnewtonian fluids in fully established laminar flow through rectangular channels. Results of Wheeler and Wissler [50] (○), Hartnett et al. [54] (△), and Harnett and Kostic [55] (□).

behavior of a purely viscous power-law fluid in laminar flow through a rectangular duct for a variety of thermal boundary conditions.

There are an infinite number of possible thermal boundary conditions describing the temperature and the heat flux that can be imposed on the boundaries of the fluid flowing through a rectangular duct. The heat transfer is strongly dependent on the thermal boundary conditions in the laminar flow regime, but much less dependent in the turbulent flow regime, particularly for fluids with a Prandtl number much larger than unity. This chapter will be restricted mainly to three classes of thermal boundary conditions:

1. Constant temperature imposed on the boundary of the fluid, the so-called T condition
2. Constant axial heat flux with constant local peripheral wall temperature imposed on the boundary of the fluid, the H1 condition
3. Constant heat flux imposed both axially and peripherally on the boundary of the fluid, the H2 condition

If not all of the boundary walls are heated, then the usual nomenclature (i.e., T, H1, and H2) must be modified. Consideration is given here to the thermal boundary conditions: (1) constant temperature imposed on one or more bounding walls with the remaining walls adiabatic; (2) constant heat input per unit length imposed on one or more walls with the associated peripheral wall temperature being constant, while the remaining unheated walls are adiabatic; (3) constant heat input per unit area imposed on one or more walls while the remaining walls are adiabatic. The following examples illustrate the use of the definition:

H1(3L)   thermal boundary condition of the H1 type imposed on three walls (longer version), while one shorter wall is adiabatic

T(2S)   two opposite shorter walls held at constant temperature, while two longer walls are adiabatic

If these terms are used in subscript, such as $Nu_{xH1(3L)}$, it is obvious that this relates to the axially local Nusselt number for the H1(3L) thermal boundary condition. In general, when T, H1, and H2 appear alone, this corresponds to the case where all bounding walls are heated, that is

$$H1 = H1(4)$$

It should be noted that in all cases, the local heat transfer coefficients and the local Nusselt numbers are based on the heated area.

A number of analytical results are available for fully developed Nusselt values for the laminar flow of power law fluids in rectangular channels having aspect ratios ranging from 0 (i.e., plane parallel plates) to 1.0 (i.e., a square duct). Newtonian results ($n = 1$) are available for the T, H1, and H2 boundary conditions for the complete range of aspect ratios. Another limiting case for which many results are available is the slug or plug flow condition, which corresponds to $n = 0$. At other values of n, results are available for plane parallel plates and for the square duct.

Figure 10.8 presents the fully established Nusselt values for the T boundary condition (i.e., constant temperature on all four walls) as a function of the power-law index n with the aspect ratio α^* as a parameter. Many predictions are shown for the plane parallel plates case ($\alpha^* = 0$) covering the range of n values from 0 to 3. The corresponding Nusselt number decreases rather rapidly from a value of 9.87 at $n = 0$ to 7.94 at $n = 0.5$, then decreases more slowly to a value of 7.54 at $n = 1.0$.

In the case of the square duct geometry ($\alpha^* = 1.0$), the Nusselt number also undergoes a large decrease from $n = 0$ to $n = 0.5$ (from 4.918 to 3.184), with the change from $n = 0.5$ to $n = 1.0$ being much more modest (from 3.184 to 2.975). Against this background, with the newtonian and slug flow limits available for all aspect ratios, it is a simple exercise to estimate the

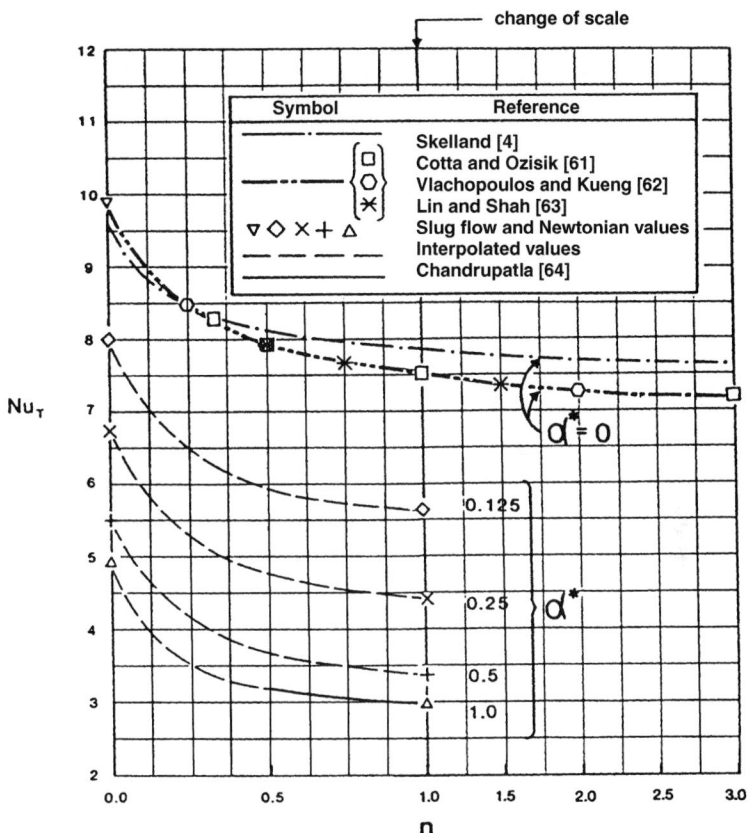

FIGURE 10.8 Fully established laminar Nusselt values for the T boundary condition as a function of the power-law index for various values of the aspect ratio.

fully established Nusselt numbers for the T boundary condition for any aspect ratio for any power-law index n. The dashed lines in Fig. 10.8 represent such estimates.

Turning next to the H1 boundary condition, Fig. 10.9 presents the fully developed Nusselt number predictions for the plane parallel plates case covering the power-law index range from 0 to 3. The available predictions for the square duct, with n varying from 0.5 to 1, are also shown. As in the case of the T boundary condition, the slug flow and newtonian flow limits are also available for the H1 condition for all aspect ratios. As in the constant-temperature case, a large decrease in the Nusselt number occurs for any aspect ratio when n increases from 0 to 0.5, and the subsequent decrease from 0.5 to 1.0 is more gentle. The dashed lines represent estimates of the fully established Nusselt values for intermediate values of the aspect ratio and power-law index.

For the case where the heat flux is constant on all boundary walls—the H2 condition—only results for the square duct and the parallel plates channel are available, as shown in Fig. 10.10. It is recommended that the results shown in Figs. 10.8 and 10.9 be used in conjunction with Fig. 10.10 to estimate values of Nu_{H2} for other aspect ratios.

In many practical applications involving a rectangular duct, other combinations of wall heating may be encountered. In the limiting case of a newtonian fluid, the values reported by

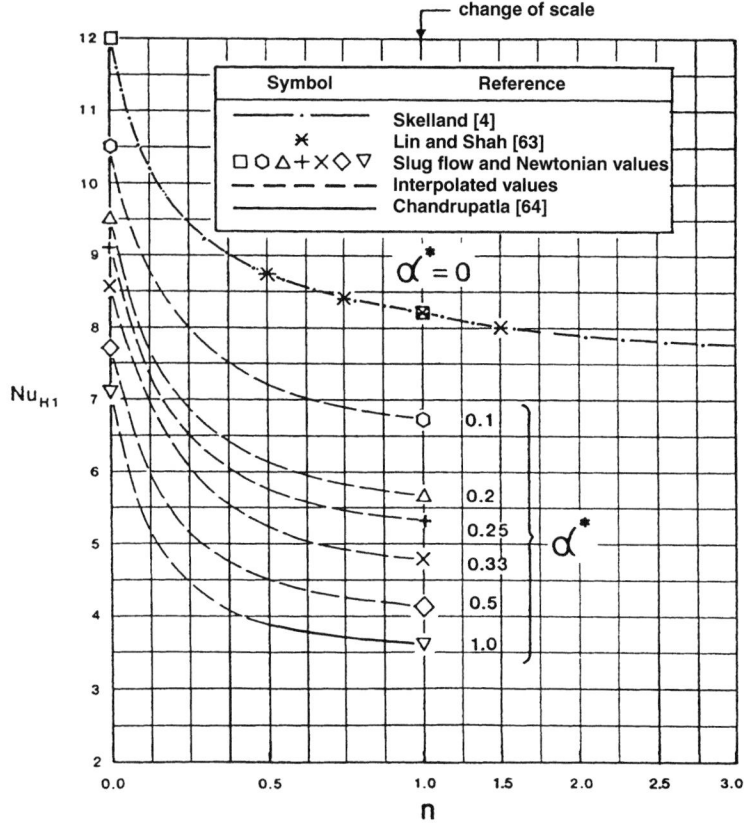

FIGURE 10.9 Fully established laminar Nusselt values for the H1 boundary condition as a function of the power-law index for various values of the aspect ratio.

Shah and London [65] are presented in Fig. 10.11 for the case where one or more walls are held at constant temperature while the remaining walls are adiabatic.

A comparable curve for the case where the H1 condition prevails on one or more walls with the other wall adiabatic is shown as Fig. 10.12. From Figs. 10.11 and 10.12 it is interesting to note for a constant finite value of the aspect ratio that the highest Nusselt number for both the T and H1 boundary conditions occurs for the case where the two long walls are heated (2L). The one-long-wall-heated Nusselt number (1L) falls below the Nusselt value for four heated walls.

The fully established Nusselt numbers for the other limiting case of slug flow, $n = 0$, are available for a wide range of conditions involving the T and H1 boundary conditions on one or more of the bounding walls. Relatively few solutions are available for the H2 boundary condition. Given the fact that the velocity is uniform over the duct cross section for $n = 0$, the analytical solutions are equivalent to the corresponding solutions of the heat conduction equation. Table 10.8 tabulates the resulting fully established Nusselt number for a wide range of thermal conditions for the rectangular duct geometry. Taken together with Figs. 10.11 and 10.12, the upper and lower limits on fully established heat transfer to a pseudoplastic fluid flowing through rectangular channels are established.

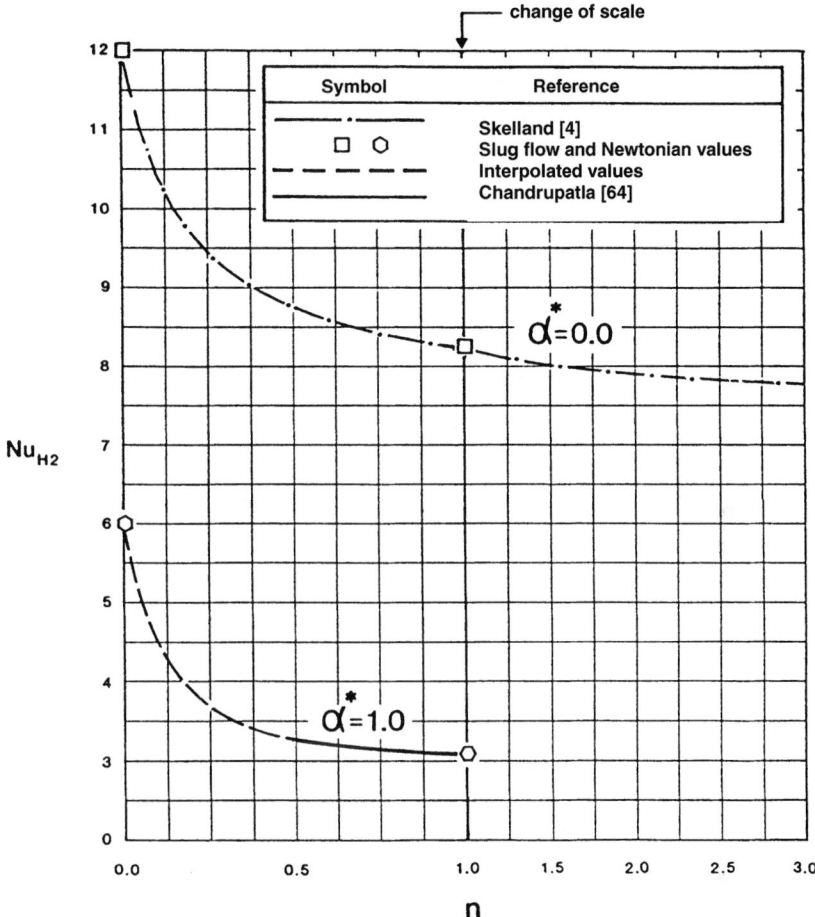

FIGURE 10.10 Fully established laminar Nusselt numbers for the H2 boundary condition as a function of the power-law index for plane parallel plates and for a square channel.

Heat Transfer in the Thermal Entrance Region—Purely Viscous Fluids

Figures 10.13 and 10.14 show the thermally developing Nusselt numbers for the T and H1 boundary conditions, respectively, for purely viscous power-law fluids in laminar flow through rectangular channels. The velocity is assumed to be fully established at the start of heating. Both figures provide results for the limiting cases of plane parallel plates ($\alpha^* = 0$) and the square duct ($\alpha^* = 1.0$) and cover several values of the power-law index n. These figures should provide guidance for estimating the heat transfer performance in the thermal entrance region of rectangular ducts. Such estimates also should be applicable to the case where the velocity and temperature fields are developing simultaneously if the Prandtl number is 50 or greater.

Inspection of Figs. 10.13 and 10.14 reveals that the thermal entrance lengths of purely viscous fluids in laminar flow through rectangular ducts increase as the aspect ratio goes from 0 to 1.0. This is brought out clearly in Fig. 10.15. Taking note of the observation that there is only a modest change in the Nusselt number as the power-law index goes from 0.5 to 1.0, it is

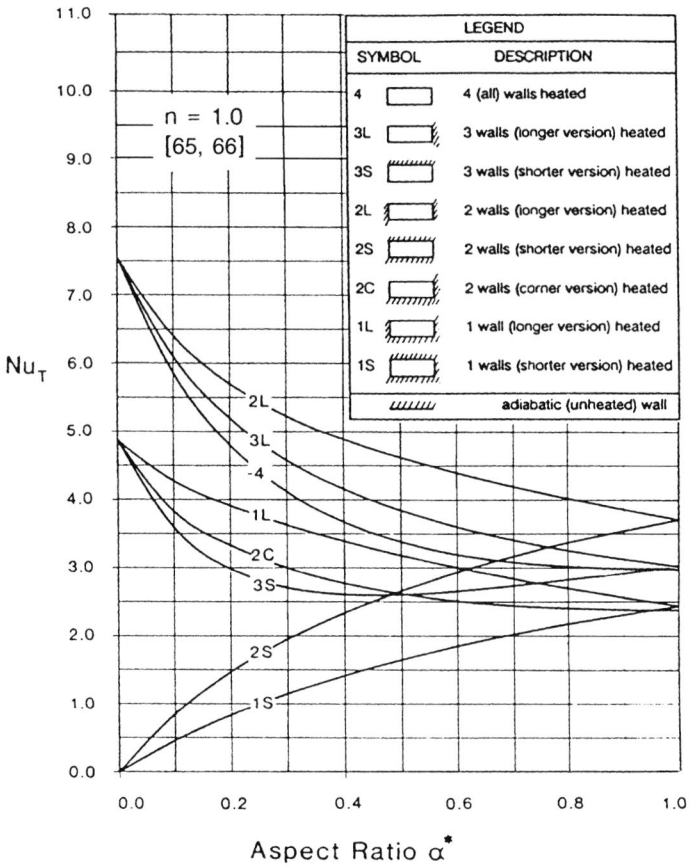

FIGURE 10.11 Fully established laminar Nusselt values for the T boundary condition as a function of the aspect ratio for different combinations of heated and adiabatic walls.

TABLE 10.8 Fully Developed Nusselt Values for Slug Flow, $n = 0$, in Rectangular Channels[a]

α^*	Nu_T				Nu_{HI}				Nu_{H2}			
1.0	4.94	4.11	4.11	2.47	7.11	5.82	5.82	3.56	5.99	—	—	3.00
0.6666	—	—	—	—	7.36	5.07	6.84	3.68	—	—	—	—
0.5	5.48	3.29	5.62	2.74	7.77	4.74	7.60	3.88	—	3.99	—	—
0.4	—	—	—	—	8.18	4.62	8.17	4.09	—	—	—	—
0.3333	—	—	—	—	8.55	4.60	8.63	4.27	—	—	—	—
0.25	6.74	3.29	7.19	3.37	9.12	4.66	9.27	4.56	—	—	—	—
0.20	—	—	—	—	9.54	4.77	9.70	4.77	—	—	—	—
0.1666	—	—	—	—	9.86	4.88	10.03	4.93	—	—	—	—
0.125	7.99	3.74	—	4.00	10.30	5.06	10.46	5.15	—	—	—	—
0.10	—	—	—	—	10.58	5.20	10.728	5.29	—	—	—	—
0.0625	—	4.23	—	—	11.07	5.46	11.18	5.54	—	—	—	—
0.05	—	—	—	—	11.24	5.54	11.34	5.62	—	—	—	—
0.0	9.87	4.94	9.87	4.94	12.00	6.00	12.00	6.00	12.00	6.00	12.00	6.00

[a] ⫽⫽⫽⫽⫽, adiabatic wall.

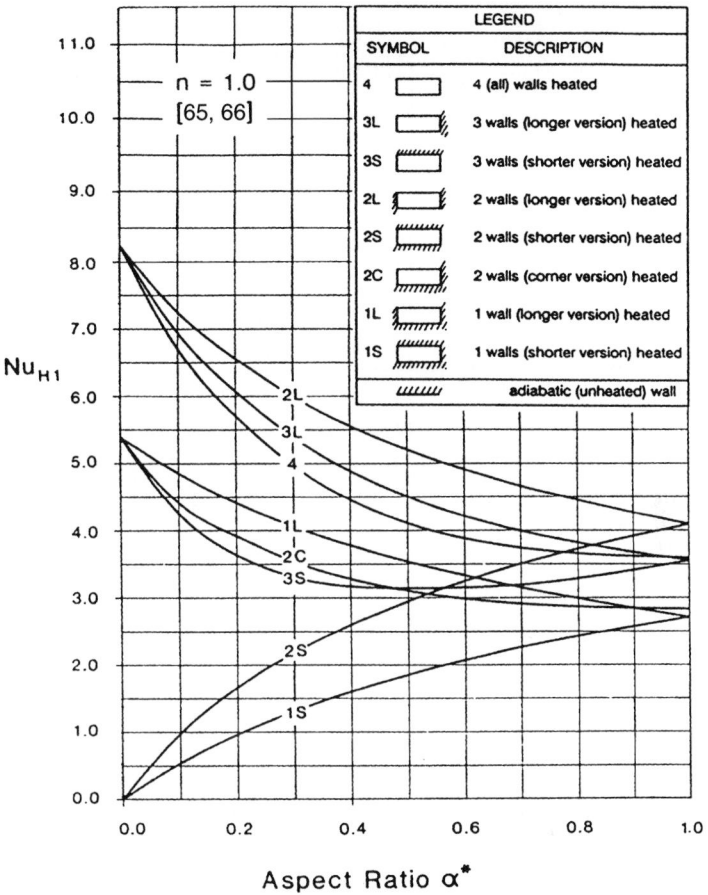

FIGURE 10.12 Fully established laminar Nusselt values for the H1 boundary condition as a function of the aspect ratio for different combinations of heated and adiabatic walls.

recommended that in this range and for aspect ratios equal to or greater than 0.2 the dimensionless thermal entrance length $L_t/[d_h \cdot Re \cdot Pr]$ be taken as 0.055.

Laminar Heat Transfer to Viscoelastic Fluids in Rectangular Ducts

It was pointed out in earlier sections that the friction factor and heat transfer behavior of viscoelastic fluids in laminar flow through circular tubes is the same as the behavior of purely viscous fluids; consequently, these values may be predicted by the power-law model. It was also noted that the power-law model predicts friction factor values in good agreement with measurements for laminar flow of viscoelastic fluids through rectangular channels. Against this background it might be anticipated that the heat transfer behavior of a viscoelastic fluid in established flow through a rectangular channel could be predicted by the power-law formulation. However, this hypothesis turns out to be incorrect, at least for aspect ratios in the range of 0.5 to 1.0. Kostic [69] reported Nusselt numbers for viscoelastic aqueous polyacrylamide

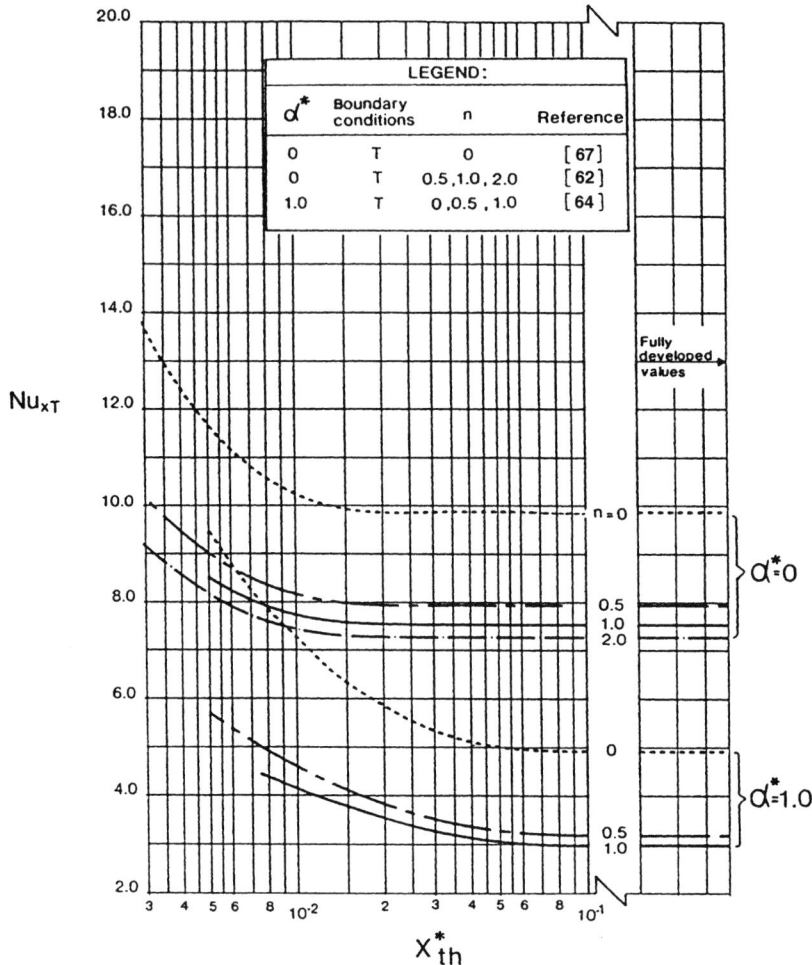

FIGURE 10.13   Thermally developing Nusselt numbers as a function of x_{th}^* and the aspect ratio for the T boundary condition, $n = 0, 0.5, 1.0$, and 2.0.

solutions in laminar flow through a rectangular duct having an aspect ratio of 0.5 with the longer upper and lower walls heated while the shorter side walls were adiabatic. Kostic's experimental results, shown in Fig. 10.16, reveal little difference between the local Nusselt numbers on the upper and lower walls, and both are 2 to 3 times the values predicted for a newtonian or a power-law fluid. These results are consistent with earlier experimental and analytical studies [56–60] that support the contention that the high Nusselt values are due to secondary flows arising from the normal stress differences on the bounding surfaces of the viscoelastic fluid. The influence of these secondary flows on the friction factor is minor, whereas the heat transfer is dramatically increased relative to the values expected for a purely viscous fluid. The available results indicate that the secondary flow effects are greatest in the square duct geometry and that their influence vanishes as the aspect ratio decreases to 0.2. In general, the secondary flow increases as the main flow Reynolds numbers increase.

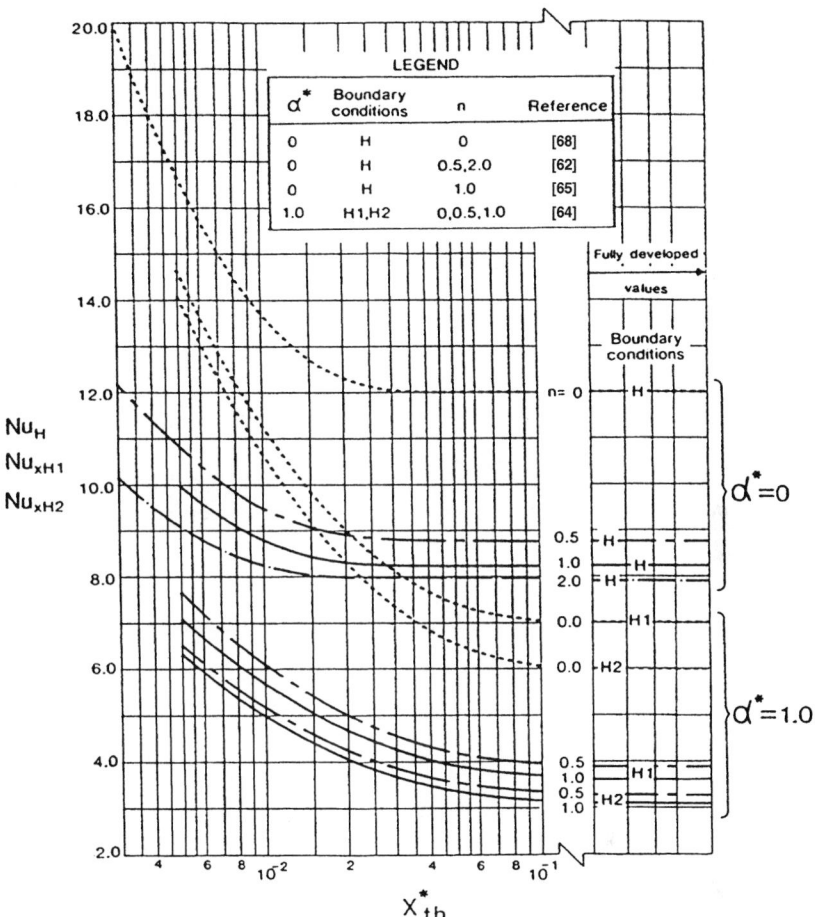

FIGURE 10.14   Thermally developing Nusselt numbers as a function of x_{th}^* and the aspect ratio for the H boundary condition, $n = 0, 0.5, 1.0,$ and 2.0.

Xie [71] added support to the above observations by carrying out studies of aqueous solutions of Carbopol (with deionized water as the solvent) in laminar flow through a rectangular duct having an aspect ratio of 0.5. Three cases were studied (1) top long wall heated, (2) bottom long wall heated, and (3) top and bottom long walls heated symmetrically. The results for the aqueous Carbopol solutions are shown in Figs. 10.17–10.19. In the case where the upper wall is heated and the other walls are adiabatic (Fig. 10.17), the influence of natural convection is minimized, as is evidenced by the baseline water results. The aqueous Carbopol solutions all show higher heat transfer ranging from 25 percent for the 500-wppm solution to 400 percent for the 1200-wppm solution as compared to water. These increases reflect secondary motions that occur in viscoelastic fluids in laminar flow through rectangular ducts. As the polymer concentration increases, the secondary flow increases and the thermal entrance length decreases. When the concentration exceeds 800 wppm, the Nusselt number is almost independent of the Graetz number and appears to be governed primarily by the Reynolds number. The following equation is proposed by Xie for the two largest concentrations:

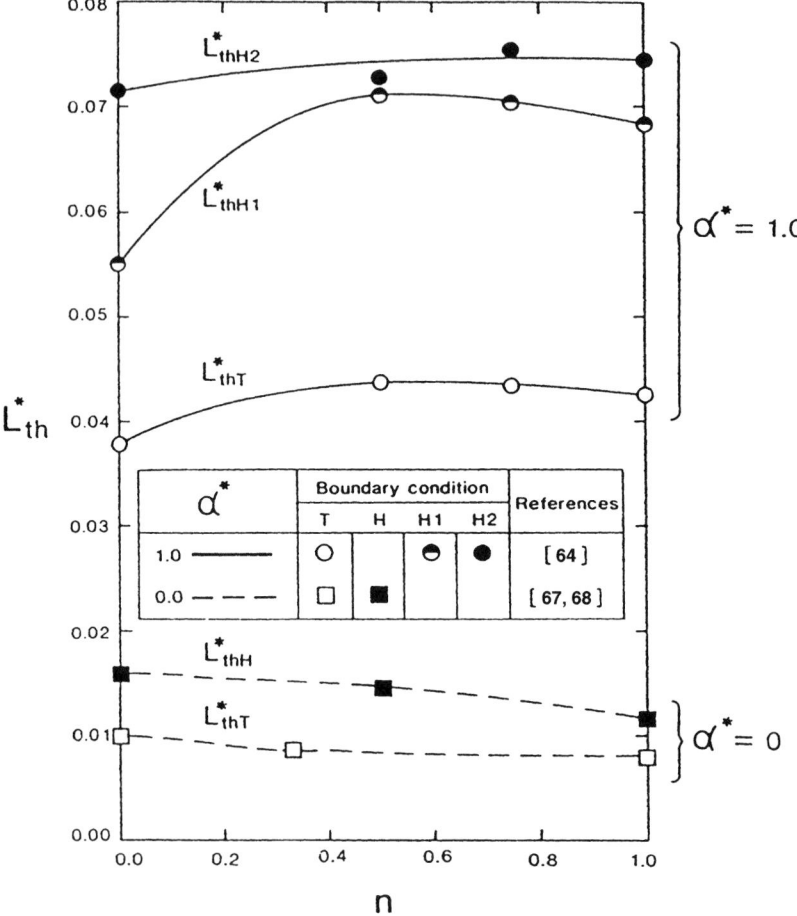

FIGURE 10.15   Thermal entrance lengths for the T and H boundary conditions for power-law fluids.

$$Nu = C_1 \, Re_a^{0.2} \tag{10.64}$$

where $C_1 = 6.0$ for 1000 wppm and $C_1 = 6.7$ for 1200 wppm. Given the closeness of the heat transfer results for these two concentrations, it is suggested that there is an upper asymptotic limit to the Nusselt number. Care must be taken in the use of Eq. 10.64, which may apply only to Carbopol solutions in rectangular ducts of aspect ratio equal to 0.5. Figure 10.18 presents local Nusselt number values for the case where the heated long wall is at the bottom of the channel. The results for water reveal the influence of natural convection on the overall heat transfer performance. Free convection is also evident in the case of the 500-wppm solution, which also reveals enhanced heat transfer that may be ascribed to secondary flows. There is no evidence of natural convection in the 1000- and 1200-wppm solutions indicating that the gravity-induced secondary flow is completely dominated by the secondary flow resulting from the fluid elasticity. These observations are reinforced by Fig. 10.19, which gives experimental heat transfer results for simultaneous heating of the upper and lower long walls with the side walls adiabatic.

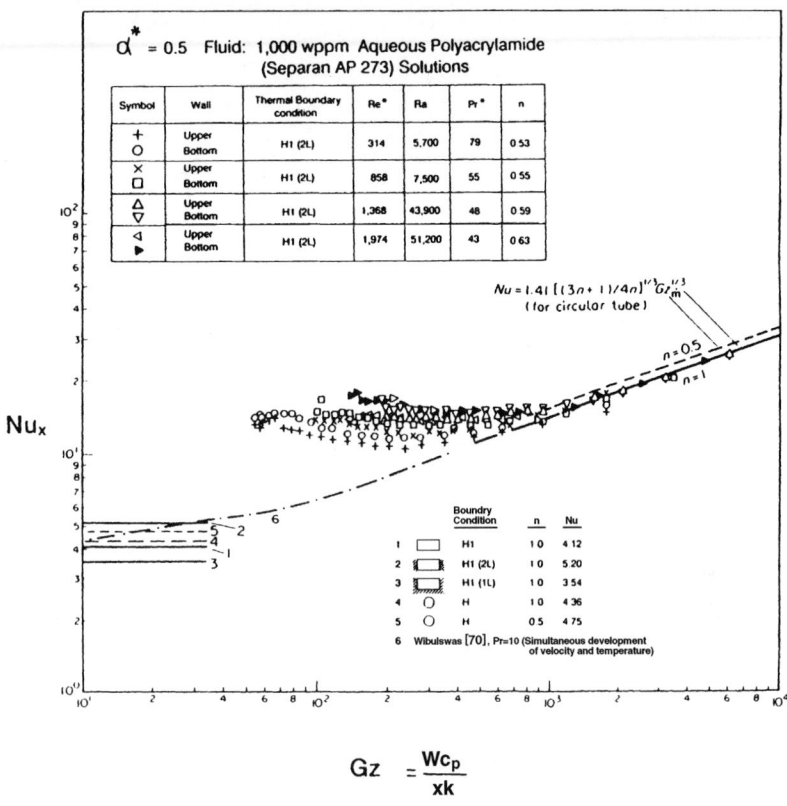

$$Gz = \frac{Wc_p}{xk}$$

FIGURE 10.16 Experimental laminar Nusselt number for aqueous polyacrylamide solutions in rectangular ducts, $\alpha^* = 0.5$, H1 (2L) boundary condition [55].

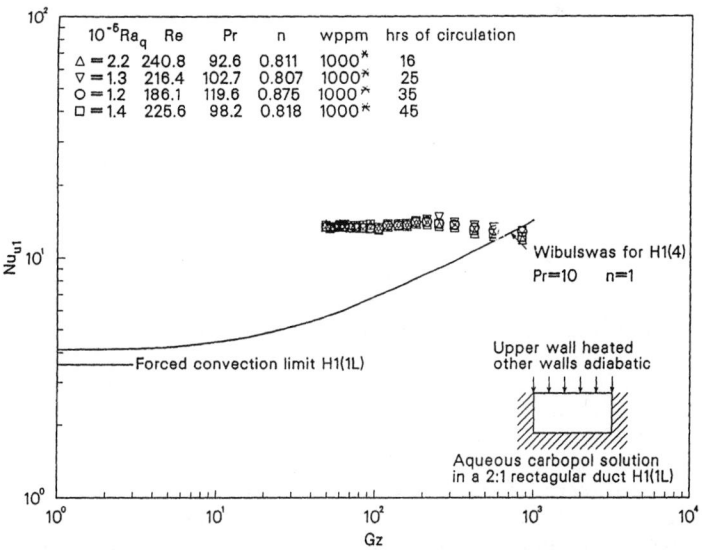

FIGURE 10.17 Laminar heat transfer of Carbopol solutions in a 2:1 rectangular duct with upper wall heated [71].

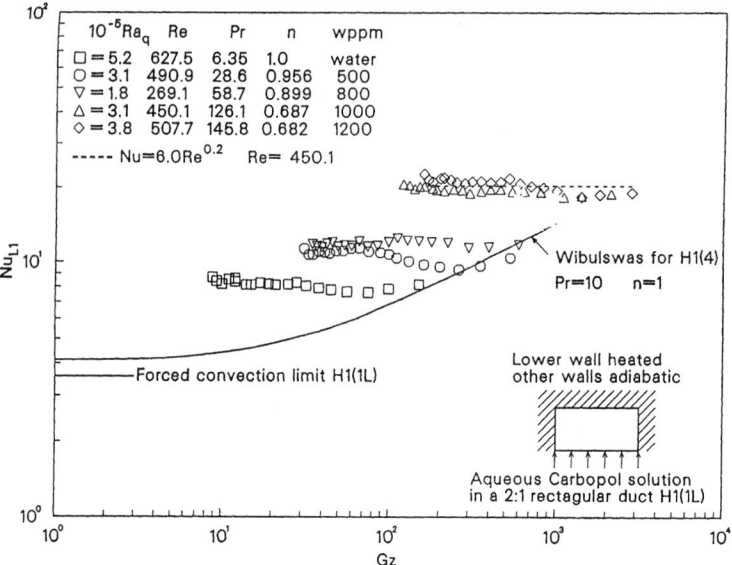

FIGURE 10.18 Laminar heat transfer of Carbopol solutions in a 2:1 rectangular duct with lower wall heated [71].

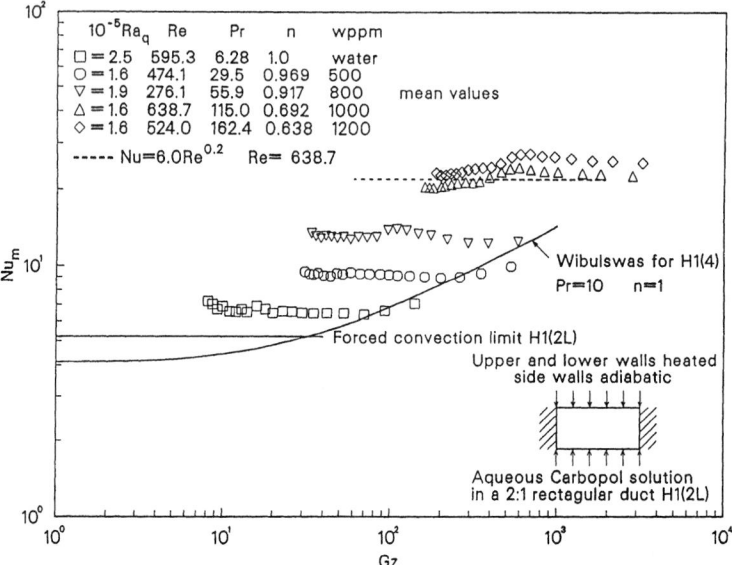

FIGURE 10.19 Mean values of laminar heat transfer of Carbopol solutions in a 2:1 rectangular duct with top and bottom walls heated [71].

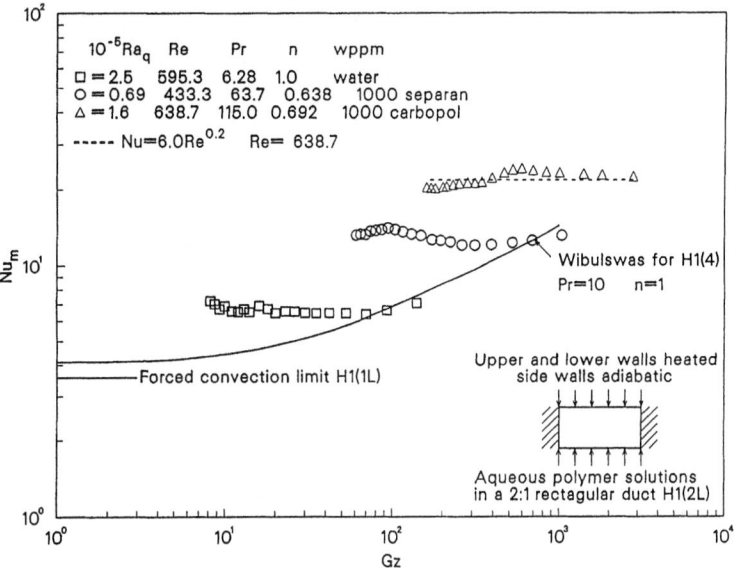

FIGURE 10.20 Laminar heat transfer of water, 1000-wppm Carbopol and 1000-wppm Separan solutions in a 2:1 rectangular duct with upper and lower walls heated [71].

The influence of the polymer-solvent combination on heat transfer may be large. Xie compared the performance of 1000-wppm Carbopol in deionized water with that of a 1000-wppm polyacrylamide in Chicago tap water. Typical results are given in Fig. 10.20 for symmetrical heating of the top and bottom walls with the side walls adiabatic for laminar flow in a rectangular duct having an aspect ratio of 0.5. Surprisingly, the Nusselt numbers for the polyacrylamide solution are only half the values found for the Carbopol solution, which suggests that the Carbopol solution is more elastic. Nevertheless, the polyacrylamide solution yielded Nusselt numbers significantly higher than the corresponding values for a purely viscous fluid. However, it should be noted that the polyacrylamide solution degraded more readily than the Carbopol solution.

TURBULENT FLOW OF PURELY VISCOUS FLUIDS IN CIRCULAR TUBES

Fully Established Friction Factor

A major contribution to the study of purely viscous nonnewtonian fluids in the turbulent flow region was made by Dodge and Metzner [72], who proposed the following turbulent pipe flow correlation to predict the friction factor:

$$1/\sqrt{f} = \frac{4.0}{n^{0.75}} \log \left[\text{Re}' f^{1 - (n/2)} \right] - 0.4/n^{1.2} \tag{10.65}$$

The use of this equation is subject to the restriction that $(\text{Pr } \text{Re}^2) f > 5 \times 10^5$.

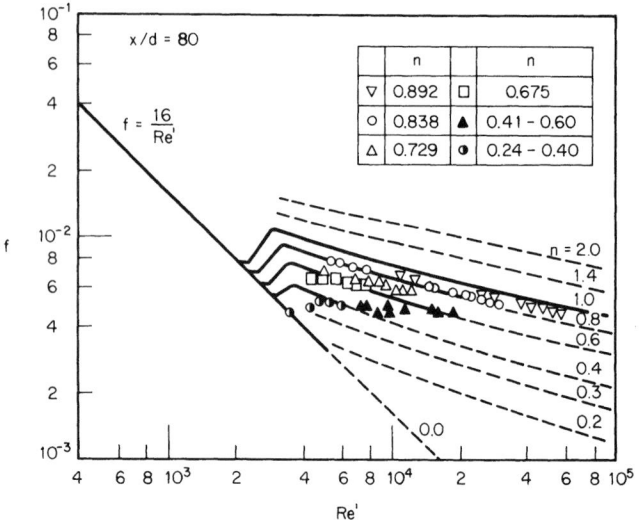

FIGURE 10.21 Experimental pipe flow pressure drop measurements for purely viscous nonnewtonian fluids by Yoo [13]: dashed lines extrapolated from Eq. 10.65.

Experimental measurements [13] for aqueous solutions of Carbopol and slurries of Attagel are in good agreement with the predictions of the Dodge-Metzner equation, as shown in Fig. 10.21.

Subsequently, an explicit equation giving good agreement with Eq. 10.65 was proposed by Yoo [13].

$$f = 0.079 n^{0.675} (\mathrm{Re}')^{-0.25} \qquad (10.66)$$

The hydrodynamic entrance length for purely viscous fluids in turbulent pipe flow is approximately the same as for newtonian fluids, being of the order of 10 to 15 pipe diameters [13].

Heat Transfer

Metzner and Friend [73] measured turbulent heat transfer rates with aqueous solutions of Carbopol, corn syrup, and slurries of Attagel in circular-tube flow. They developed a semi-theoretical correlation to predict the Stanton number for purely viscous fluids as a function of the friction factor and Prandtl number, applying Reichardt's general formulation for the analogy between heat and momentum transfer in turbulent flow:

$$\mathrm{St} = \frac{f/2}{[1.2 + 11.8\sqrt{f/2}\,(\mathrm{Pr} - 1)\,\mathrm{Pr}^{1/3}]} \qquad (10.67)$$

where f is given in Eq. 10.65 or 10.66. The use of Eq. 10.67 is limited to $(\mathrm{Pr}\,\mathrm{Re}^2)f > 5 \times 10^5$ and to a Prandtl number range of 0.5 to 600.

A simple correlation has been given by Yoo [13], who compared his results for Carbopol and Attagel solutions with those of previous investigators. Yoo's empirical equation for predicting turbulent heat transfer for purely viscous fluids is given by

$$\mathrm{St} = 0.0152 \mathrm{Re}_a^{-0.155}\,\mathrm{Pr}_a^{-2/3} \qquad (10.68)$$

This equation describes the available data with a mean deviation of less than 5 percent. It is recommended that Eq. 10.68 be used to predict the heat transfer for purely viscous fluids in turbulent pipe flow for values of the power-law exponent n between 0.2 and 0.9 and over the Reynolds number range from 3,000 to 90,000. The recommended procedure is as follows:

1. Determine the friction factor from Eq. 10.65 or 10.66 as a function of Re'.
2. Convert Re' to Re_a using the following relation: $\text{Re}_a = \text{Re}'\,(3n + 1)/4n$.
3. Use Eq. 10.68 to predict the Nusselt number as a function of Re_a.

The thermal entrance lengths for purely viscous nonnewtonian fluids in turbulent pipe flow are on the order of 10 to 15 pipe diameters, the same order of magnitude as for newtonian fluids [74].

TURBULENT FLOW OF VISCOELASTIC FLUIDS IN CIRCULAR TUBES

Friction Factor and Velocity Distribution

The hydrodynamic behavior of viscoelastic fluids in turbulent pipe flow is quite different from that of the solvent or of a purely viscous nonnewtonian fluid. The friction drag of such a viscoelastic fluid under turbulent flow conditions is substantially lower than the values associated with the pure solvent or with purely viscous nonnewtonian fluids. In general, for turbulent channel flow, this drag reduction increases with higher flow rate, higher polymer molecular weight, and higher polymer concentration. In addition, the diameter of the pipe, the degree of degradation of the polymer, and the chemistry of the solvent are important parameters in the determination of the drag reduction.

It should be noted that the extent of the drag reduction is ultimately limited by a unique asymptote that is independent of the polymer concentration, the solvent chemistry, or the degree of polymer degradation and is solely dependent on the dimensionless axial distance x/d and the Reynolds number [75]. Since polymer concentration, solvent chemistry, and polymer degradation are related to the fluid elasticity, it is postulated that these effects can be incorporated in the dimensionless Weissenberg number and that the friction factor is in general a function of the axial location x/d, the Reynolds number, and the Weissenberg number [31, 37, 76]. However, beyond a certain critical value of the Weissenberg number $(\text{Ws})_f^*$, the friction factor reaches a minimum asymptote value that is dependent solely on the axial distance x/d and the Reynolds number.

In operational terms this can be expressed by the following functional relationships:

$$f = f\left(\frac{x}{d}, \text{Re}_a, \text{Ws}\right) \qquad \text{for Ws} < (\text{Ws})_f^* \qquad (10.69)$$

$$f = f\left(\frac{x}{d}, \text{Re}_a\right) \qquad \text{for Ws} > (\text{Ws})_f^* \qquad (10.70)$$

This behavior can be seen in Fig. 10.22, which shows the fully established turbulent friction factor as a function of Reynolds number Re_a for concentrations ranging from 10 to 1000 wppm of polyacrylamide in Chicago tap water. This series of measurements, which were taken in a tube 1.30 cm in diameter, revealed that the hydrodynamic entrance length varied with concentration, reaching a maximum of 100 pipe diameters at the higher concentrations. Therefore, the friction factors shown in Fig. 22 were measured at values of x/d greater than 100. The asymptotic friction factor is reached at concentrations of approximately 50 wppm of polyacrylamide in tap water for the tube diameter used in the test program [50, 93]. The

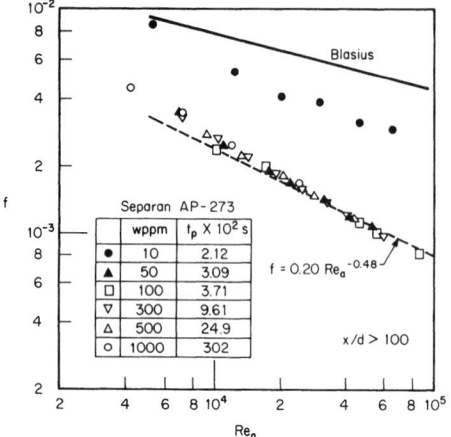

FIGURE 10.22 Fanning friction factor versus Re_a measured in a once-through flow system with polyacrylamide (Separan AP-273) solutions. t_p is the characteristic time calculated from the Powell-Eyring model.

experimental values of the asymptotic friction factors in the turbulent region may be correlated by the following expression [53, 79]:

$$f = 0.20 Re_a^{-0.48} \qquad (10.71)$$

The steady shear viscosity measurements of representative solutions used in the study of the friction factor behavior are given in Fig. 10.23. For concentrations ranging from 50 to 1000 wppm, the viscosity is shear rate dependent. The viscosities for 10-wppm polyacrylamide solutions are relatively independent of shear rate.

Relaxation times can be calculated for each of the polyacrylamide solutions used in the measurements shown in Figs. 10.22 and 10.23. This may be accomplished by combining the experimentally measured viscosity results with an appropriate generalized newtonian model containing relaxation time as a parameter. The Powell-Eyring model [22] has been used to fit the data, and the resulting values of the relaxation time t_p are shown in the tables in the figures. As expected, the relaxation times increase with increasing concentration.

The asymptotic nature of the friction factor is clearly brought out in Fig. 10.24, which shows the measured fully established friction factors taken in three tubes of differing diameters as a function of the Weissenberg number based on the Powell-Eyring relaxation time for fixed values of the Reynolds number for aqueous solutions of polyacrylamide. The critical Weissenberg number for friction $(Ws)_f^*$, is seen to be on the order of 5 to 10. When the Weissenberg number exceeds 10, it is clear that the fully developed friction factor is a function only of the Reynolds number.

Figure 10.25 shows the lower asymptotic values of the fully developed friction factors for highly concentrated aqueous solutions of polyacrylamide and polyethylene oxide as a func-

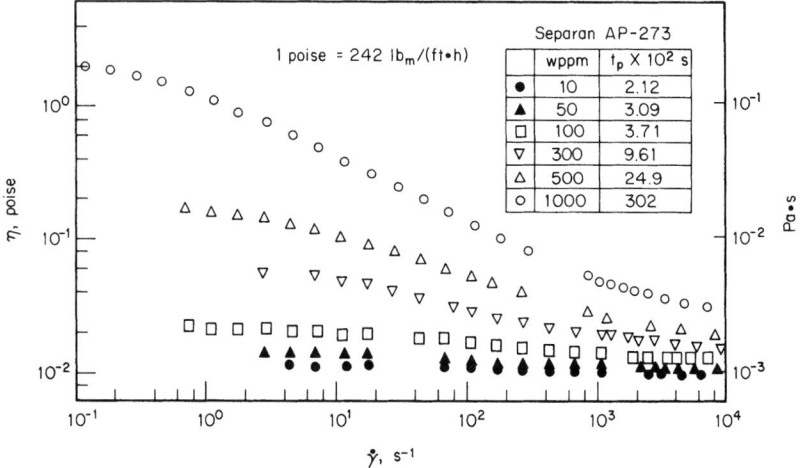

FIGURE 10.23 Steady-shear-viscosity measurements for polyacrylamide (Separan AP-273) solutions from Weissenberg rheogoniometer and capillary-tube viscometer. t_p is the characteristic time calculated from the Powell-Eyring model [37].

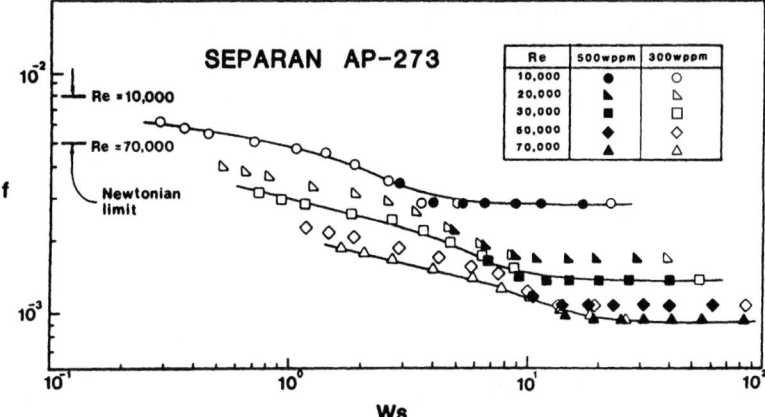

FIGURE 10.24 Fully established friction factors for aqueous polyacrylamide solutions in turbulent pipe flow as a function of the Weissenberg and Reynolds numbers.

tion of the generalized Reynolds number Re′ [79, 108]. These measurements, taken at values of x/d greater than 100, were obtained in tubes of 0.98, 1.30, and 2.25 cm inside diameter. This figure brings out the fact that the laminar flow region extends to values of the Reynolds number on the order of 5000 to 6000. In this laminar flow region the measured friction factors are in excellent agreement with the theoretical prediction, $f = 16/Re′$.

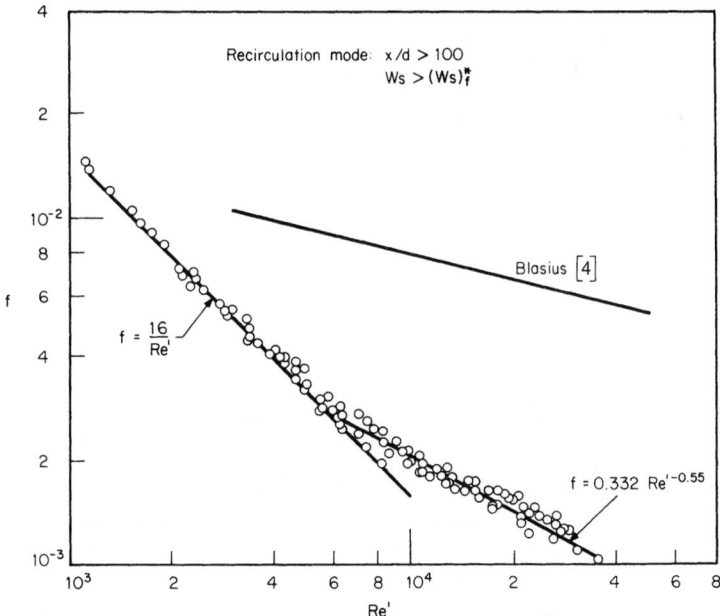

FIGURE 10.25 Fully established friction factor versus Re′ with concentrated polyethylene oxide and polyacrylamide solutions [100].

TABLE 10.9 Various Techniques Used in the Local Velocity Measurements with Dilute Polymer Solutions

References	Method	Polymer
Seyer [80]	Bubble tracer method	Polyacrylamide AP-30, 1000 wppm
Khabakhpasheva and Perepelitsa [81]	Stroboscopic flow visualization	Polyacrylamide, 120 wppm
Rudd [82]	Laser anemometry	Polyacrylamide AP-30, 100 wppm
Arunachalam et al. [83]	Dye injection	Polyethylene oxide coagulant, 5.5 wppm

In the range of Re′ from 6,000 to 40,000, the experimental friction factor measurements may be correlated by the simple expression [37, 77, 79].

$$f = 0.332(\text{Re}')^{-0.55} \tag{10.72}$$

It is recommended that either Eq. 10.71 or Eq. 10.72 be used to predict the fully developed friction factor (that is, for x/d greater than 100) of viscoelastic aqueous polymer solutions in turbulent pipe flow for Reynolds numbers greater than 6000 and for Weissenberg numbers above critical value. The critical Weissenberg number for aqueous polyacrylamide solutions based on the Powell-Eyring relaxation time is on the order of 5 to 10 [50]. In the absence of experimental data for other polymers, this value should be used for other viscoelastic fluids with the appropriate caution.

Direct measurements of the velocity profile for viscoelastic aqueous polymer solutions have been reported by several investigators. Table 10.9 summarizes the techniques and polymers used in obtaining the velocity profiles [80, 83]. The velocity measurements reported by Seyer [80], Khabakhpasheva and Perepelitsa [81], and Rudd [82] shown in Fig. 10.26 are in fairly good agreement. With the predicted velocity profiles obtained from modeling procedures including Prandtl's mixing length model [75], Deissler's continuous eddy diffusivity model [84], and van Driest's damping factor model [85, 86]. These investigations show that the laminar sublayer near the wall is thickened and the velocity distribution in the core region is shifted upward from the newtonian mean velocity profile.

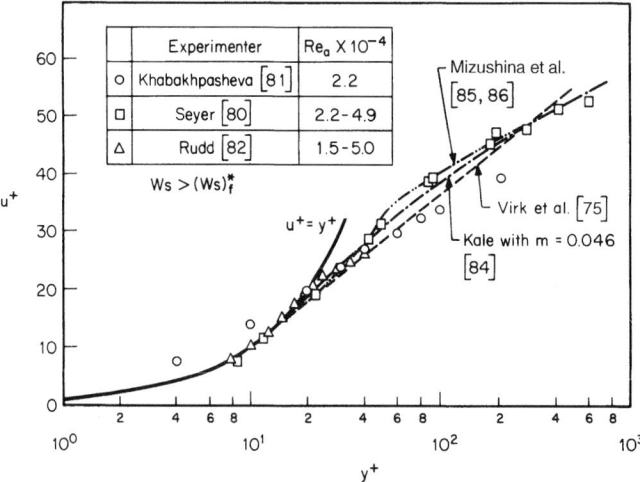

FIGURE 10.26 Experimental measurements of fully established local turbulent velocity profile for the minimum-drag asymptotic case.

It is noteworthy that the use of Pitot tubes and hot-film anemometry, which are applicable to newtonian fluids, is questionable for drag-reducing viscoelastic fluids. The anomalous behavior of Pitot tubes and hot-film probes in these fluids has been observed by many investigators [87–92].

Heat Transfer

Local heat transfer measurements were carried out in the once-through system for the same aqueous polyacrylamide solutions used in the friction factor and viscosity measurements shown in Figs. 10.22 and 10.23 [37, 93]. These heat transfer studies involving a constant heat flux boundary condition required the measurement of the fluid inlet and outlet temperatures and the local wall temperature along the tube. These wall temperatures are presented in terms of a dimensionless wall temperature θ in Fig. 10.27 for four selected concentrations. Here θ is defined as

$$\theta = (T_w - T_b)_{x/d}/(T_w - T_b)_{ex} \tag{10.73}$$

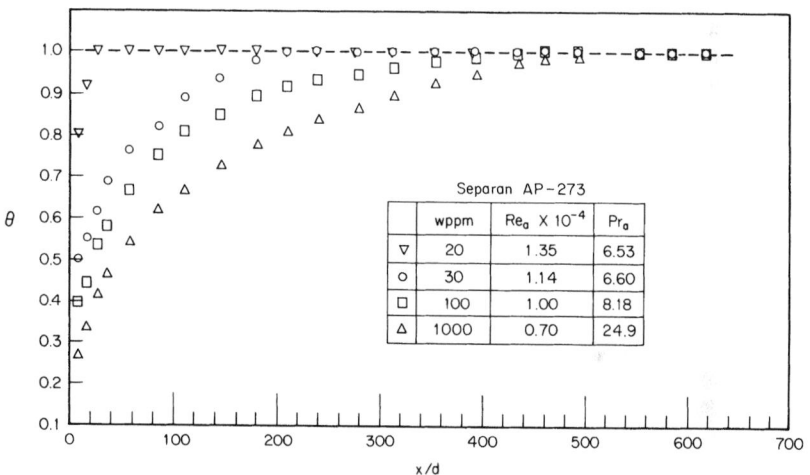

FIGURE 10.27   Thermal entrance length for drag-reducing viscoelastic fluids. Dimensionless wall temperature versus dimensionless axial distance [93].

For a given concentration, the values of x/d associated with values of θ less than unity are referred to as the thermal entrance region; in this region the thermal boundary layer is not fully developed and the heat transfer coefficient is greater than the value in the thermally developed region. Figure 10.27 reveals that the thermal entrance length of the 20-wppm polyacrylamide aqueous solution is almost the same as that of newtonian fluids, which is on the order of 5 to 15 pipe diameters [95–97]. The thermal entrance length increases with increasing concentration (i.e., increasing Weissenberg number), reaching a value of 400 to 500 diameters for the 1000-wppm solutions. It is important to note the long entrance lengths of viscoelastic fluids, which have been overlooked in many studies.

The measured dimensionless heat transfer factors j_H (that is, St $Pr_a^{2/3}$) are shown in Fig. 10.28 as a function of the Reynolds number Re_a for concentrations ranging from 10 to 1000 wppm polyacrylamide [37, 93]. These measurements were made at x/d equal to 430, which corresponds approximately to thermally fully developed conditions as shown in the figure. The asymptotic values of the fully established heat transfer coefficients are reached at a concentration of 500 wppm of polyacrylamide, whereas less than 50 wppm was required to reach

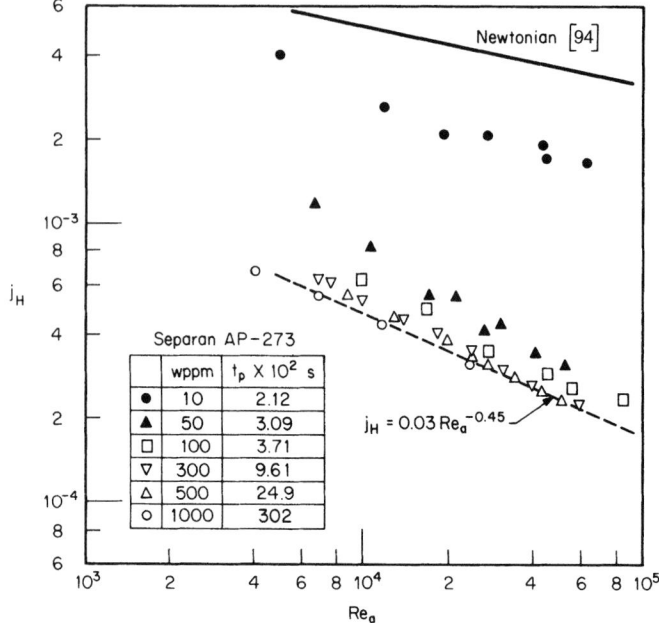

FIGURE 10.28 Turbulent heat transfer results for polyacrylamide solutions measured at $x/d = 430$. t_p is the characteristic time calculated from the Powell-Eyring model.

the asymptotic friction factor values as shown in Fig. 10.22. The fully developed minimum asymptotic heat transfer, which is approximated by the experimental data obtained at x/d equal to 430, is correlated by the following equation [35, 37]:

$$j_H = 0.03 \mathrm{Re}_a^{-0.45} \qquad (10.74)$$

The asymptotic nature of the heat transfer is brought out more vividly in Fig. 10.29, which presents the same data in terms of j_H versus the Weissenberg number based on the Powell-Eyring relaxation time for different values of the Reynolds number [37]. The critical Weissenberg number for heat transfer $(\mathrm{Ws}_p)_h^*$ is approximately 200 to 250, an order of magnitude higher than the critical Weissenberg number for the friction factor $(\mathrm{Ws}_p)_f^*$. Above a Weissenberg number of 250, the dimensionless heat transfer reaches its minimum asymptotic value (Eq. 10.74). Note that this critical Weissenberg value has been established for aqueous polyacrylamide solutions, and appropriate care should be used in applying it to other polymers until additional confirmation is forthcoming.

Values of the asymptotic heat transfer factors j_H in the thermal entrance region are reported for concentrated aqueous solutions of polyacrylamide and polyethylene oxide. The results are shown in Fig. 10.30, as a function of the Reynolds number Re_a. These values were measured in tubes of 0.98, 1.30, and 2.25 cm (0.386, 0.512, and 0.886 in) inside diameter in a recirculating-flow loop. The asymptotic turbulent heat transfer data in the thermal entrance region are seen to be a function of the Reynolds number Re_a and of the axial position x/d. The following empirical correlation is derived from the data [35, 37]:

$$j_H = 0.13(x/d)^{-0.24}(\mathrm{Re}_a)^{-0.45} \qquad (10.75)$$

These same data are shown in Fig. 10.31 as a function of the generalized Reynolds number Re'. Here it may be noted that the laminar data are in excellent agreement with the theoreti-

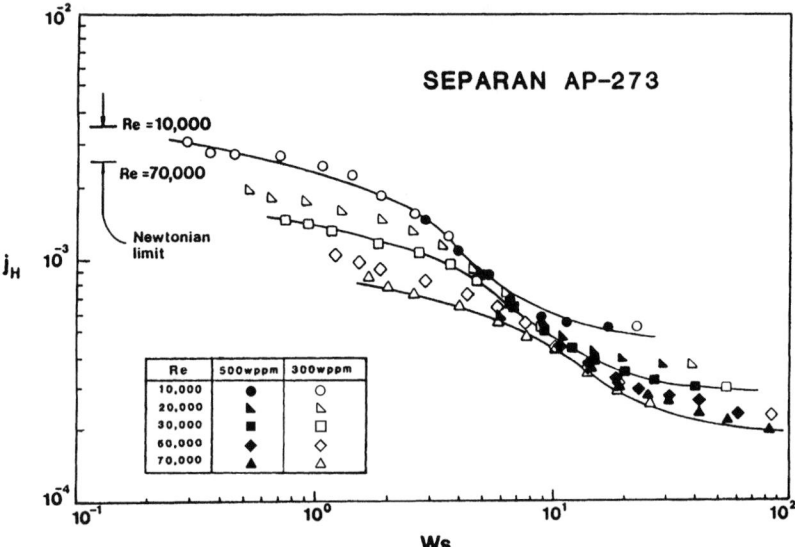

FIGURE 10.29   Fully established dimensionless heat transfer j_H for aqueous polyacrylamide solutions in turbulent pipe flow as a function of the Weissenberg and Reynolds numbers.

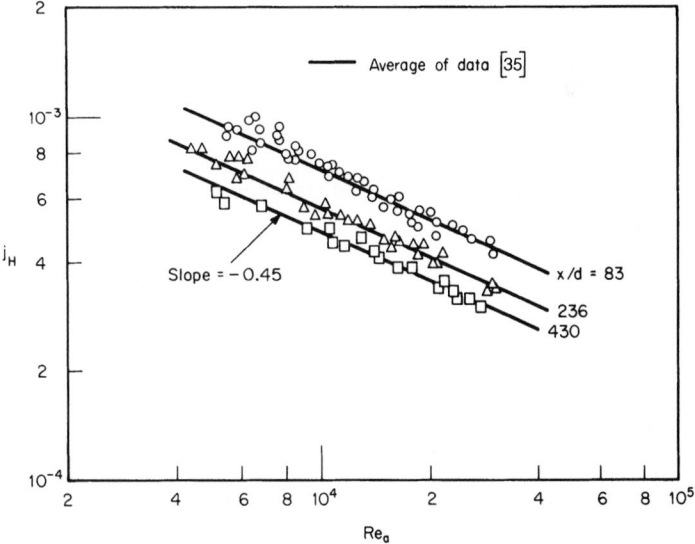

FIGURE 10.30   Experimental results of turbulent heat transfer for concentrated solutions of polyethylene oxide and polyacrylamide in the thermal entrance region.

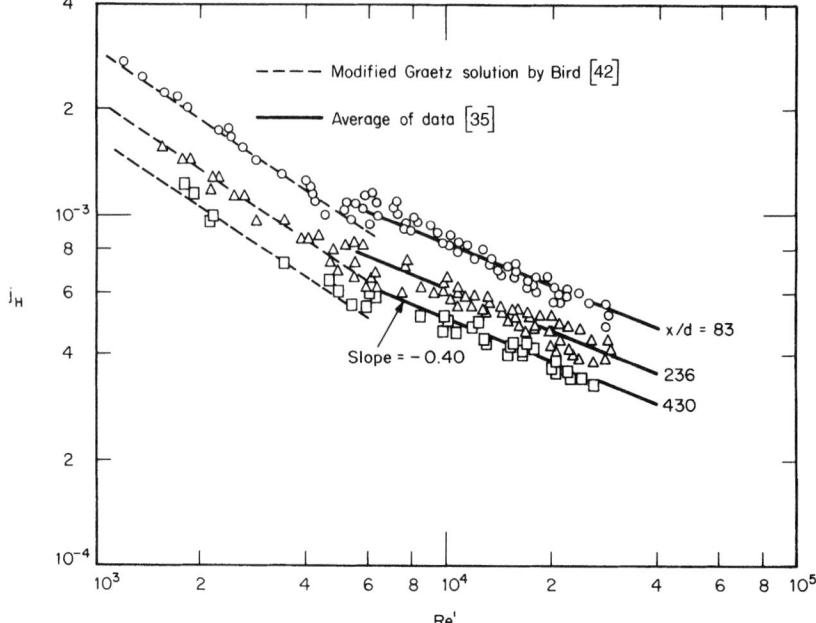

FIGURE 10.31 Experimental results of laminar and turbulent heat transfer for concentrated solutions of polyethylene oxide and polyacrylamide in the thermal entrance region.

cal prediction by Bird [42], lending support to the experimental measurements. Laminar flow extends to a generalized Reynolds number of 5000 to 6000. The empirical correlations resulting from the turbulent flow data given in Fig. 10.31 are [35, 37]

$$j_H = 0.13(x/d)^{-0.3}(\text{Re}')^{-0.4} \qquad \text{for } x/d < 450 \qquad (10.76)$$

$$j_H = 0.02(\text{Re}')^{-0.4} \qquad \text{for } x/d > 450 \qquad (10.77)$$

It is recommended that Eqs. 10.74 and 10.75, or equivalently Eqs. 10.76 and 10.77, be used to predict the heat transfer performance of viscoelastic aqueous polymer solutions for Reynolds numbers greater than 6000 and for values of the Weissenberg number above the critical value for heat transfer. This critical Weissenberg number for heat transfer based on the Powell-Eyring relaxation time is approximately 250 for aqueous polyacrylamide solutions. Appropriate care should be exercised in using this critical value for other viscoelastic fluids.

Degradation

The degradation of the polymer in a viscoelastic polymer solution makes the prediction of the heat transfer and pressure drop extremely difficult, if not impossible, in normal industrial practice. This results from the fact that mechanical degradation, the shearing of the polymer bonds, goes on continuously as the fluid circulates, causing continuous changes in the rheology of the fluid. The elasticity of the fluid is particularly sensitive to this mechanical degradation. These changes in the rheology of the fluid ultimately cause changes in the heat transfer and pressure drop.

Notwithstanding the difficulties of accurately predicting the quantitative effects of degradation on the hydrodynamics and heat transfer, it is nevertheless important to qualitatively understand the process if engineering systems are to be designed to handle such fluids.

Systematic studies have been reported on the heat transfer behavior of degrading polymer solutions with highly concentrated polymer solutions: 1000 wppm of polyacrylamide [36, 37] and 1500 wppm of polyethylene oxide [35]. These studies were conducted in test sections with inside diameters of 2.25 cm ($L/d \approx 280$) and 1.30 cm ($L/d \approx 475$). Heat transfer and pressure drop measurements were carried out at regular time intervals. Although the circulation rate was held approximately constant, periodic flow rate measurements were carried out using the direct weighing and timing method. Fluid samples were removed at regular time intervals from the flow loop for rheological property measurements in the Weissenberg rheogoniometer (WRG) and in the capillary tube viscometer.

Figure 10.32 shows the steady shear viscosity of the polyacrylamide (Separan AP-273) solution as a function of hours of circulation in the flow loop. Chicago tap water was the solvent. This figure brings out very clearly the substantial decrease in the viscosity at low shear rate resulting from the degradation of the polymers, which is accompanied by a decrease in the first normal stress difference and a decrease in the characteristic time [35]. This, in turn, means that a decrease in the Weissenberg number always accompanies degradation. Thus, a circulating aqueous polymer solution experiences a continuing decrease in the Weissenberg number.

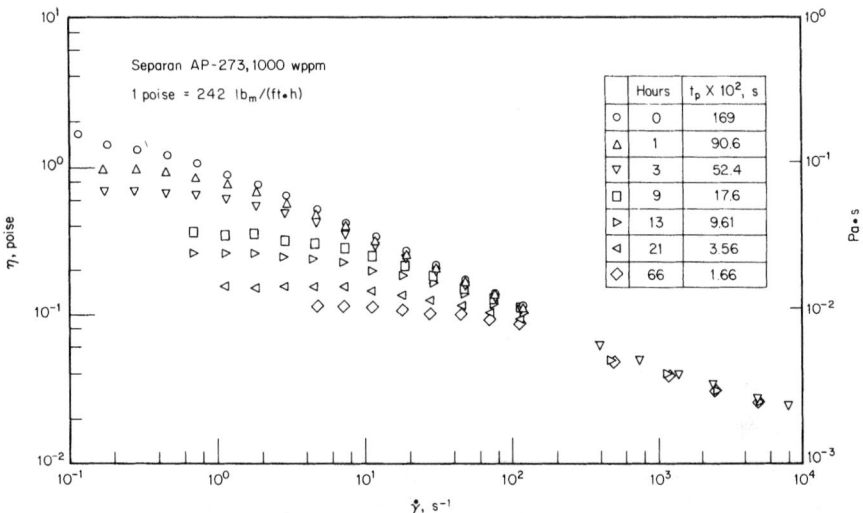

FIGURE 10.32 Degradation effects on steady shear viscosity measurements for polyacrylamide 1000-wppm solution as a function of circulation time.

The Fanning friction factor f and the dimensionless heat transfer coefficient j_H for the polyacrylamide 1000-wppm solution measured at an x/d of 430 and at the Reynolds number equal to 20,000 [37] are presented in Fig. 10.33 as a function of hours of circulation. The dimensionless j_H factor is seen to remain relatively constant at its minimum asymptotic value until some 3 hours have passed. On the other hand, the friction factor does not depart from its asymptotic value until some 30 hours of circulation have occurred. Estimates of the critical Weissenberg number based on the Powell-Eyring model yield values that are in good agreement with those given for the once-through system:

Critical Weissenberg number for friction:

$$(\mathrm{Ws}_p)_f^* = 10 \tag{10.78}$$

Critical Weissenberg number for heat transfer:

$$(\mathrm{Ws}_p)_h^* = 250 \tag{10.79}$$

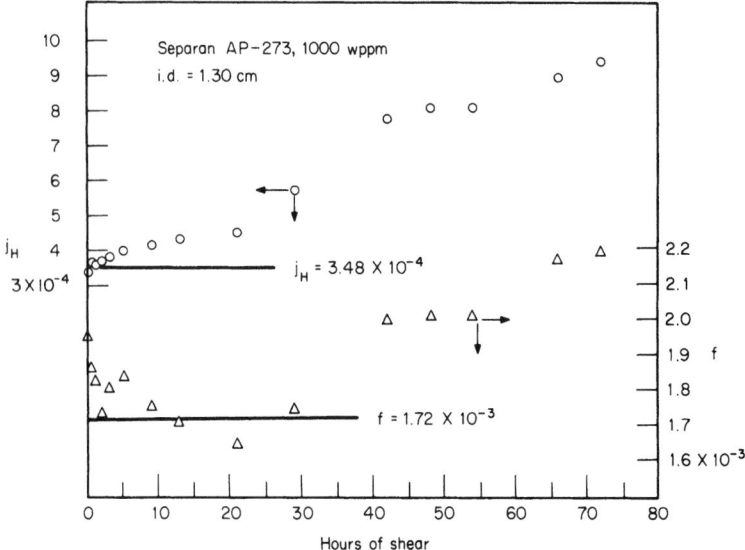

FIGURE 10.33 Fanning friction factor and turbulent heat transfer j versus hours of shear for Reynolds number equal to 20,000 and at $x/d = 430$. Separan AP = 273, 1000 wppm. Solid lines are minimum asymptotic values.

Above the corresponding critical Weissenberg number, the friction factor and the heat transfer remain at their asymptotic values.

A similar degradation test was conducted with a concentrated solution of 1500 wppm polyethylene oxide [35]. Analysis of test results reveals good agreement with the polyacrylamide solutions. In particular, the critical Weissenberg values for the polyethylene oxide solution are of the same order as those for the polyacrylamide solution.

Solvent Effects

When an aqueous solution of a high-molecular-weight polymer is used in a practical engineering system, the solvent is generally predetermined by the system. However, the importance of the solvent on the pressure drop and heat transfer behavior with these viscoelastic fluids has often been overlooked. Since the heat transfer performance in turbulent flow is critically dependent on the viscous and elastic nature of the polymer solution, it is important to understand the solvent effects on the rheological properties of a viscoelastic fluid.

Following the earlier work by Little et al. [98] and Chiou and Gordon [99], Cho et al. [100] measured the rheological properties of the 1000-wppm aqueous solution of polyacrylamide (Separan AP-273) with various solvents: distilled water, tap water, tap water plus acid or base additives, and tap water plus salt. Figure 10.34 presents the steady shear viscosity data over the shear rate ranging from 10^{-2} to $4 \times 10^4 \, \text{s}^{-1}$ using the Weissenberg rheogoniometer and the capillary tube viscometer. The viscosity in the low shear rate of the 1000-wppm polyacrylamide solution with distilled water is greater than that of the polyacrylamide solution with tap water by a factor of 25. However, when the shear rate is increased, the viscosity of the distilled water solution approaches that of the tap water solution. The addition of 100 wppm NaOH to Chicago tap water results in a 100 percent increase of the viscosity in the low-shear-rate range. In contrast, the addition of 4 percent NaCl to the tap water reduces the viscosity of the polyacrylamide solution over the entire range of shear rate by a factor of 4 to 25 depending on the shear rate.

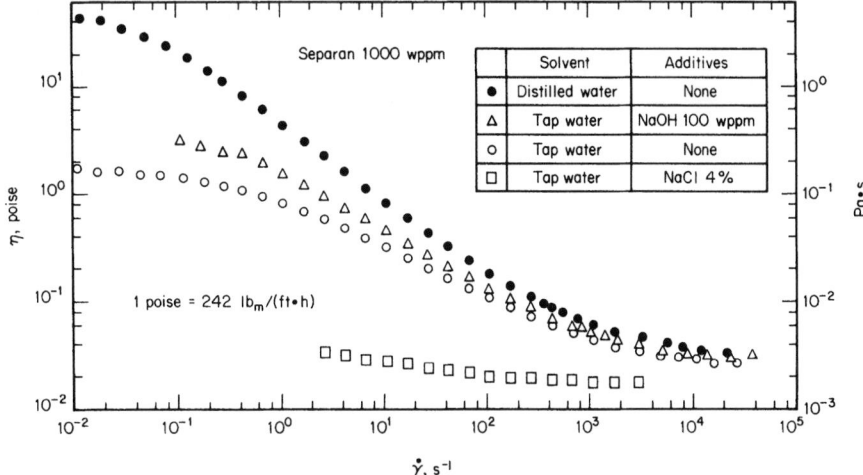

FIGURE 10.34 Steady shear viscosity versus shear rate for polyacrylamide 1000-wppm solutions with four different solvents [100].

The effect on viscosity of the addition on NaOH, NH₄OH, or H₃PO₄ to Chicago tap water has been investigated [100]. The results indicate that for base additives there is an optimum pH number (approximately 10) that maximizes the viscosity of polyacrylamide solutions. For acid additives, an increasing concentration of acid is generally accompanied by a decrease of viscosity. It is noteworthy that similar observations were made with aqueous solutions of polyethylene oxide.

From the above results together with those of other investigators who used distilled water as a solvent [98, 99], it can be concluded that the rheological properties of polymer solutions may be modified by changing the chemistry of the solvent. It follows that the hydrodynamic and heat transfer performance is sensitive to solvent chemistry.

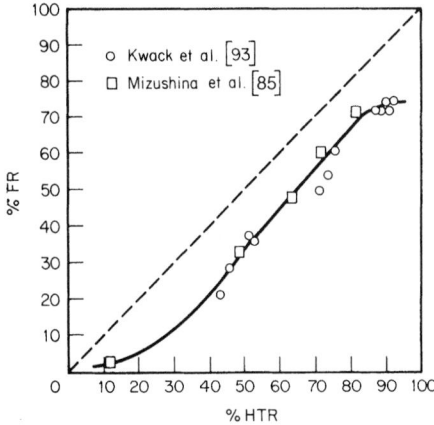

FIGURE 10.35 Comparison of percentage friction reduction and percentage heat transfer reduction.

Failure of the Reynolds-Colburn Analogy

It is well known that for newtonian fluids in turbulent pipe flow, an analogy between momentum and heat transfer can be drawn and expressed in the following form:

$$j_H = f/2 \qquad (10.80)$$

For drag-reducing polymer solutions, there have been many attempts in the literature to formulate and to apply such an analogy [35]. Most of these works attempted to predict turbulent heat transfer rates for drag-reducing fluids from the use of the friction coefficients measurements. To get some insight into the use of the analogy, the measured asymptotic values of the friction factor and the heat transfer are presented in a different form. Figure 10.35 shows the percentage reduction in friction factor resulting from the addition of a long-chained polymer to water plotted against the percent reduction in heat transfer coefficient. Here the reduction is defined as follows:

$$\text{Friction factor reduction} = \text{FR} = (f_S - f_P)/f_S$$

$$\text{Heat transfer reduction} = \text{HTR} = (j_{HS} - j_{HP})/j_{HS}$$

where the subscripts S and P designate the pure solvent and the aqueous polymer solution, respectively.

The solid line in the figure represents the general trend of the experimental observations of Refs. 35 and 85, confirming the fact that the heat transfer reduction always exceeds the friction factor reduction. This contradicts the common assumption of the validity of the Reynolds or Colburn analogy made in a number of heat transfer studies of viscoelastic fluids [101–106].

To further verify the above conclusion on the failure of the analogy between momentum and heat transfer in the case of viscoelastic fluids, the approximate values of the eddy diffusivities of momentum and heat transfer corresponding to the minimum asymptotic cases will be compared. The eddy diffusivity of momentum corresponding to the minimum asymptotic case was calculated by Kale [84] directly from Deissler's continuous eddy diffusivity model:

$$\frac{\epsilon_M}{\nu} = m^2 u^+ y^+ [1 - \exp(-m^2 u^+ y^+)] \qquad y^+ < 150, \quad m = 0.046 \qquad (10.81)$$

where Kale's value of $m = 0.06$ has been changed to 0.046 to conform with the experimental data [35, 37, 79, 107].

Cho and Hartnett [108, 109] calculated the eddy diffusivity of heat for drag-reducing viscoelastic fluids using a successive approximation technique. The result for the minimum asymptotic case can be expressed in the following polynomial equation with respect to y^+:

$$\frac{\epsilon_H}{\nu} = 2.5 \times 10^{-6} y^{+3} \qquad (10.82)$$

A comparison of the calculated eddy diffusivities using Eqs. 10.81 and 10.82 confirms the fact that the eddy diffusivity of heat is much smaller than that of momentum for drag-reducing viscoelastic fluids. This result is consistent with the experimental observation that the thermal entrance length is much longer than the hydrodynamic entrance length for the turbulent pipe flow of drag-reducing viscoelastic fluids. It can be concluded that there is no direct analogy between momentum and heat transfer for drag-reducing viscoelastic fluids in turbulent pipe flows.

TURBULENT FLOW OF PURELY VISCOUS FLUIDS IN RECTANGULAR DUCTS

Friction Factor

The fully established friction factor for turbulent flow of purely viscous nonnewtonian fluids in rectangular channels may be determined by the modified Dodge-Metzner equation [72, 110]:

$$1/\sqrt{f} = \frac{4.0}{n^{0.75}} \log \left[\text{Re}^* f^{1 - (n/2)} \right] - \frac{0.4}{n^{1.2}} \qquad (10.83)$$

Alternatively, the simpler formulation proposed by Yoo [13, 110] may be used:

$$f = 0.079 n^{0.675} (\text{Re}^*)^{-0.25} \qquad \text{for } 0.4 \le n \le 1.0, \quad 5000 < \text{Re}^* < 50,000 \qquad (10.84)$$

Heat Transfer

The fully developed Stanton number for turbulent flow of purely viscous nonnewtonian fluids in rectangular channels may be determined by the modified Metzner-Friend equation [73]:

$$St = \frac{f/2}{[1.2 + 11.8\sqrt{f/2}(Pr - 1)\,Pr^{-1/3}]} \tag{10.85}$$

Equation 10.85 gives values that are within ±10 percent of available data over the range of Reynolds numbers from 5,000 to 60,000.

In the case of turbulent channel flow of purely viscous power-law fluids, the hydrodynamic and thermal entrance lengths can be taken as the same as the corresponding values for a newtonian fluid.

TURBULENT FLOW OF VISCOELASTIC FLUIDS IN RECTANGULAR DUCTS

Friction Factor

The fully established friction factor for turbulent flow of a viscoelastic fluid in a rectangular channel is dependent on the aspect ratio, the Reynolds number, and the Weissenberg number. As in the case of the circular tube, at small values of Ws, the friction factor decreases from the newtonian value. It continues to decrease with increasing values of Ws, ultimately reaching a lower asymptotic limit. This limiting friction factor may be calculated from the following equation:

$$f = 0.2Re_a^{-0.48}$$

where $Re_a = \rho U d_h / \eta_a$.

This is the same equation found for the circular tube and is confirmed by a number of experiments as shown in Fig. 10.36.

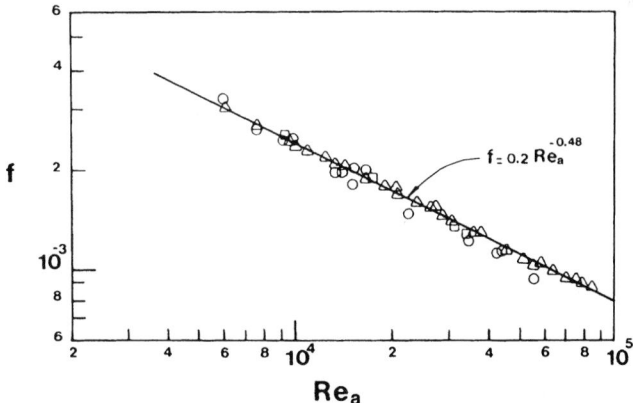

FIGURE 10.36 Measured friction factors of aqueous polyacrylamide solutions in a rectangular duct ($\alpha^* = 0.5$ to 1.0) as a function of the Reynolds number based on the apparent viscosity Re_a. Values are from Kostic and Hartnett [112] ($\bigcirc$), Kwack et al. [111] ($\square$), and Hartnett et al. [113] ($\triangle$).

The behavior of a viscoelastic fluid in turbulent flow in the hydrodynamic entrance region of a rectangular channel can be estimated by assuming that the circular tube results are applicable provided that the hydraulic diameter replaces the tube diameter.

Heat Transfer

Studies of the heat transfer behavior of viscoelastic aqueous polymer solutions have been carried out for turbulent flow in a rectangular channel having an aspect ratio of 0.5. These experimental results obtained with aqueous polyacrylamide solutions are shown in Fig. 10.37, where the minimum asymptotic values of the dimensionless heat transfer coefficient, j_H, are compared with the values reported by Cho and Hartnett for turbulent pipe flow. The turbulent pipe flow results are correlated by

$$j_H^* = \mathrm{Nu}/(\mathrm{Re}^* \, \mathrm{Pr}^{*1/3}) = 0.02\mathrm{Re}^{*-0.4}$$

or alternatively by

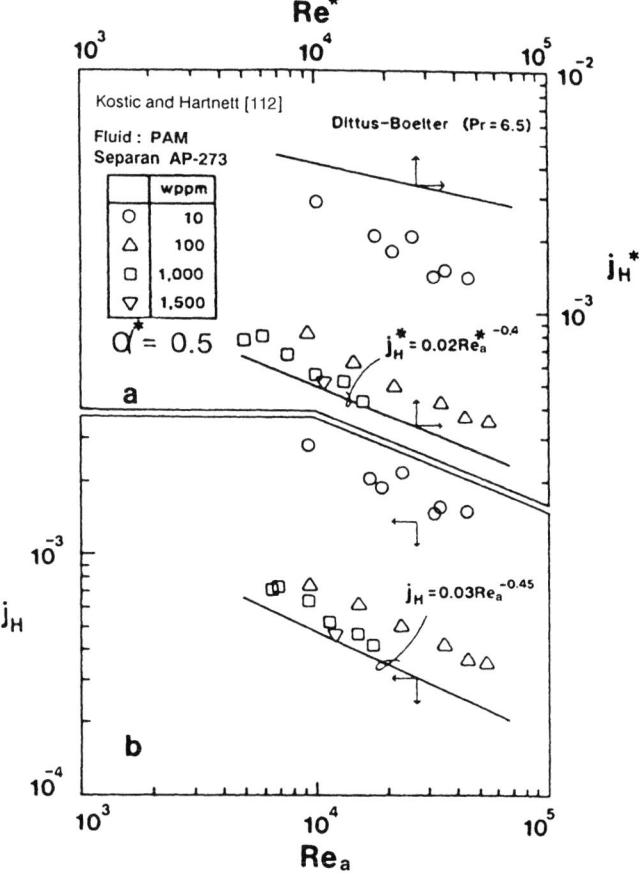

FIGURE 10.37 Heat transfer factor j_H versus Reynolds number Re_a for turbulent flow of aqueous polyacrylamide solution, $\alpha^* = 0.5$ [112].

$$j_H = \text{Nu}/(\text{Re}_a \, \text{Pr}_a^{1/3}) = 0.03\text{Re}_a^{-0.45}$$

Although the rectangular channel data are somewhat higher (5 to 10 percent) than the circular tube correlation equations, it appears that the circular tube predictions may be used for engineering estimates of the asymptotic heat transfer for rectangular ducts having an aspect ratio of approximately 0.5 to 1.0.

In this same spirit, it is proposed that the intermediate values of the heat transfer coefficient lying between the newtonian value and the lower asymptotic limit be estimated from the pipe flow correlation shown in Fig. 10.28 [114]. This approach should give reasonable estimates, at least for aqueous polyacrylamide and polyethylene oxide solutions.

ANOMALOUS BEHAVIOR OF AQUEOUS POLYACRYLIC ACID SOLUTIONS

An exception to the generally observed drag reduction in turbulent channel flow of aqueous polymer solutions occurs in the case of aqueous solutions of polyacrylic acid (Carbopol, from B.F. Goodrich Co.). Rheological measurements taken on an oscillatory viscometer clearly demonstrate that such solutions are viscoelastic. This is also supported by the laminar flow behavior shown in Fig. 10.20. Nevertheless, the pressure drop and heat transfer behavior of neutralized aqueous Carbopol solutions in turbulent pipe flow reveals little reduction in either of these quantities. Rather, these solutions behave like clay slurries and they have been often identified as purely viscous nonnewtonian fluids. The measured dimensionless friction factors for the turbulent channel flow of aqueous Carbopol solutions are in agreement with the values found for clay slurries and may be correlated by Eq. 10.65 or 10.66. The turbulent flow heat transfer behavior of Carbopol solutions is also found to be in good agreement with the results found for clay slurries and may be calculated from Eq. 10.67 or 10.68.

FLOW OVER SURFACES; FREE CONVECTION; BOILING

Page limitations do not permit complete coverage of nonnewtonian fluid mechanics and heat transfer. Readers are referred to the following surveys for more information:

Flow Over Surfaces

R. P. Chhabra, *Bubbles, Drops, and Particles in Non-Newtonian Fluids,* CRC Press, Boca Raton, FL, 1993.

R. P. Chhabra and D. De Kee, *Transport Processes in Bubbles, Drops, and Particles,* Hemisphere, New York, 1992.

D. D. Joseph, *Fluid Dynamics of Viscoelastic Liquids,* Springer-Verlag, New York, 1990.

D. A. Siginer and S. E. Bechtel, "Developments in Non-Newtonian Flows," *AMD,* vol. 191, ASME, 1994.

D. A. Siginer, W. E. VanArsdale, M. C. Altan, and A. N. Alexandrou, "Developments in Non-Newtonian Flows," *AMD,* vol. 175, ASME, 1993.

Z. Zhang, "Numerical and Experimental Studies of Non-Newtonian Fluids in Cross Flow Around a Circular Cylinder," Ph.D. thesis, University of Illinois at Chicago, 1995.

Free Convection

J. L. S. Chen and M. A. Ebadian, "Fundamentals of Heat Transfer in Non-Newtonian Fluids," *HTD,* vol. 174, ASME, 1991.

U. K. Ghosh, S. N. Upadhyay, and R. P. Chhabra, "Heat and Mass Transfer from Immersed Bodies to Non-Newtonian Fluids," *Advances in Heat Transfer* (25): 252–321, 1994.

M. L. Ng, "An Experimental Study on Natural Convection Heat Transfer of Non-Newtonian Fluids from Horizontal Wires," Ph.D. thesis, University of Illinois at Chicago, 1985.

A. V. Shenoy and R. A. Mashelkar, "Thermal Convection in Non-Newtonian Fluids," *Advances in Heat Transfer* (15): 143–225, 1982.

Boiling

Y.-Z. Hu, "Nucleate Pool Boiling from a Horizontal Wire in Viscoelastic Fluids," Ph.D. thesis, University of Illinois at Chicago, 1989.

T.-A. Andrew Wang, "Influence of Surfactants on Nucleate Pool Boiling of Aqueous Polyacrylamide Solutions," Ph.D. thesis, University of Illinois at Chicago, 1993.

Suspensions and Surfactants

K. Gasljevic and E. F. Matthys, "On Saving Pumping Power in Hydronic Thermal Distribution Systems Through the Use of Drag-Reducing Additives," *Energy and Buildings* (20): 45–56, 1993.

E. F. Matthys, "An Experimental Study of Convective Heat Transfer, Friction, and Rheology for Non-Newtonian Fluids: Polymer Solutions, Suspensions of Particulates," Ph.D. thesis, California Institute of Technology, 1985.

Flow of Food Products

S. D. Holdsworth, "Rheological Models Used for the Prediction of the Flow Properties of Food Products: a Literature Review," *Trans. IChemE* (71/C): 139–179, 1993.

Electrorheological Flows

D. A. Siginer, J. H. Kim, S. A. Sherif, and H. W. Coleman, "Developments in Electrorheological Flows and Measurement Uncertainty," *AMD*, vol. 190, ASME, 1994.

NOMENCLATURE

Symbol, Definition, SI Units, English Units

A defined in Eq. 10.8: Pa·s, lb_m/h·ft

A_c cross-sectional area: m^2, ft^2

A_i defined in Eq. 10.60

a half of the longer side of the rectangular duct: m, ft

a^* geometric constant in Kozicki generalized Reynolds number, Eq. 10.63

B defined in Eq. 10.8: °K, °R

b half of the shorter side of the rectangular duct or half of the distance between parallel plates: m, ft

b^* geometric constant in Kozicki generalized Reynolds number, Eq. 10.67

C torque measured on inner cylinder or on plate (see Figs. 10.3 and 10.4: N·m, lb_f·ft

C_1 constant defined by Eq. 10.64

c_p	specific heat at constant pressure: J/(kg·K), Btu/(lb$_m$·°F)
d	tube inside diameter: m, ft
De	Deborah number $= t/t_F$
d_h	hydraulic diameter (equal to d for circular pipe), $4A_c/p$: m, ft
d_{ij}	rate of strain tensor: s^{-1}
F	total force applied on plate (Eq. 10.23): N, lb$_f$
f	Fanning friction factor $= \tau_w/(\rho U^2/2)$
g	acceleration of gravity: m/s^2, ft/s^2
Gr	Grashof number $= \rho^2 g\beta\Delta T d^3/\eta^2$
Gr$_q$	Grashof number $= \rho^2 g\beta q'' d_h''/\eta^2 k_1$
Gz	Graetz number $= Wc_p/k_l x$
h	heat transfer coefficient: W/(m^2·K), Btu/(h·ft^2·°F)
j_H	Colburn heat transfer factor $= \text{St} \cdot \text{Pr}^{2/3}$
K, K'	consistency index in power-law model defined in Table 10.1 and Eq. 10.30: N/(m^2·s^n), lb$_f$/(ft^2·s^n)
k_l	thermal conductivity of liquid: W/(m·K), Btu/(h·ft·°F)
L	tube length: m, ft
L_h	hydrodynamic entrance length in duct flow: m, ft
L_t	thermal entrance length in duct flow: m, ft
N_1	first normal stress difference: N/m^2, lb$_f$/ft^2
N_2	second normal stress difference: N/m^2, lb$_f$/ft^2
Nu$_m$	mean Nusselt number $= h_m d_h/k_l$
Nu$_x$	local Nusselt number $= h_x d_h/k_l$
Nu$_\infty$	fully established Nusselt number, $h_\infty d_h/k_l$
n	power-law index
P	pressure: N/m^2, lb$_f$/ft^2
p	perimeter: m, ft
Pe	Peclet number $= \rho c_p U d_h/k_l$
Pr	Prandtl number $= \eta c_p/k_l$
ΔP	pressure drop along the axial direction: N/m^2, lb$_f$/ft^2
Q	volume flow rate: m^3/s, ft^3/s
q_w''	heat flux at the tube wall: W/m^2, Btu/(h·ft^2)
R	tube radius $= d/2$: m, ft
Ra$_q$	Rayleigh number for constant heat flux boundary condition, Gr$_q$ Pr
Re$^+$	Reynolds number $= \rho U^{2-n} d_h^n/K$
Re$_a$	Reynolds number based on the apparent viscosity at the wall $= \rho U d_h/\eta_a$
Re$'$	Reynolds number defined as $\rho U^{2-n} d_h^n/K' 8^{n-1}$
Re$_s$	Reynolds number based on the solvent viscosity, $\rho U d_h/\eta_s$
Re$_{\text{eff}}$	Reynolds number defined by Eq. 10.38
Re*	Kozicki generalized Reynolds number, Eq. 10.63
r	radial coordinate: m, ft
St	Stanton number $= \text{Nu}/(\text{Re Pr}) = h/\rho U c_p$

T	temperature: K, R
t	characteristic time of the viscoelastic fluid, a measure of elasticity: s
t_F	characteristic time of the flow: s
t_p	characteristic time of a viscoelastic fluid calculated using the Powell-Eyring model (see Table 10.1): s
U	mean velocity in channel flow: m/s, ft/s
u	velocity in the x direction: m/s, ft/s
u^*	friction velocity $= (\tau_w/\rho)^{1/2}$: m/s, ft/s
u^+	normalized velocity $= u/u^*$
V	velocity: m/s, ft/s
v	velocity in the y direction: m/s, ft/s
W	mass flow rate: kg/s, lb_m/s
Ws	Weissenberg number $= tU/d_h$
Ws_p	Weissenberg number based on Powell-Eyring characteristic time, $t_p U/d_h$
w	velocity in the z direction: m/s, ft/s
$(\text{Ws})_f^*$	critical Weissenberg number for friction
$(\text{Ws})_h^*$	critical Weissenberg number for heat transfer
x	axial location along the channel: m, ft
x_t^*	dimensionless distance, $x/(d_h\ \text{Re Pr})$
y	distance normal to the tube wall $= R - r$: m, ft
y	transverse rectilinear coordinate, orthogonal to x and z
y^+	normalized distance from the wall $= yu^*/\nu$
z	transverse rectilinear coordinate, orthogonal to x and y

Greek Symbols

α^*	aspect ratio of rectangular duct: b/a
α_i	defined by Eq. 10.60
β	volumetric coefficient of thermal expansion: $(\text{K})^{-1}$, $(\text{R})^{-1}$
β_i	defined by Eq. 10.60
$\dot{\gamma}$	shear rate: s^{-1}
ε_H	eddy diffusivity of heat: m^2/s, ft^2/s
ε_M	eddy diffusivity of momentum: m^2/s, ft^2/s
η	shear-rate-dependent viscosity: Pa·s, lb_m/(h·ft)
η_{ij}	generalized viscosity: Pa·s, lb_m/(h·ft)
η_a	apparent viscosity: Pa·s, lb_m/(h·ft)
η_0	limiting viscosity at zero shear rate: Pa·s, lb_m/(h·ft)
η_∞	limiting viscosity at infinite shear rate: Pa·s, lb_m/(h·ft)
θ	dimensionless temperature, defined in Eq. 10.38
θ_0	cone angle of cone and plate viscometer, Fig. 10.4
ν	kinematic viscosity: m^2/s, ft^2/s
$\pi_{\theta 0}$	pressure measured by transducer (Eq. 10.22): N/m^2, lb_f/ft^2
ρ	density: kg/m^3, lb_m/ft^3
τ	shear stress: N/m^2, lb_f/ft^2
τ_{ij}	shear stress tensor: N/m^2, lb_f/ft^2

τ_w shear stress at the wall: N/m^2, lb$_f$/ft^2

Ω angular velocity: s^{-1}

Subscripts

a	property based on the apparent viscosity
b	bulk fluid condition
ex	condition at the exit of the tube
H	constant axial heat flux, with peripherally constant wall temperature
in	condition at the inlet of the tube
l	liquid
max	maximum value
w	evaluated at the wall

REFERENCES

1. B. A. Toms, "Some Observations on the Flow of Linear Polymer Solutions through Straight Tubes at Large Reynolds Numbers," *Proc. 1st Int. Cong. Rheol.,* North-Holland, Amsterdam, vol. II, p. 135, 1949.

2. K. J. Mysels, "Flow of Thickened Fluids," U.S. Patent 2,492,173, Dec. 27, 1949.

3. A. B. Metzner, "Heat Transfer in Non-Newtonian Fluids," in *Advances in Heat Transfer,* J. P. Hartnett and T. F. Irvine Jr., eds., vol. 2, p. 357, Academic, New York, 1965.

4. A. H. P. Skelland, *Non-Newtonian Flow and Heat Transfer,* Wiley, New York, 1967.

5. W. H. Suckow, P. Hrycak, and R. G. Griskey, "Heat Transfer to Non-Newtonian Dilatant (Shear-Thickening) Fluids Flowing Between Parallel Plates," *AIChE Symp. Ser.* (199/76): 257, 1980.

6. R. B. Bird, R. C. Armstrong, and O. Hassager, *Dynamics of Polymeric Liquids,* 2d ed., vol. I, p. 212, p. 522, Wiley, New York, 1987.

7. R. Darby, *Viscoelastic Fluids,* Marcel Dekker, New York, 1976.

8. Y. I. Cho, "The Study of non-Newtonian Flows in the Falling Ball Viscometer," Ph.D. thesis, University of Illinois at Chicago, 1979.

9. N. A. Park, "Measurement of Rheological Properties of Non-Newtonian Fluids With the Falling Needle Viscometer," Ph.D. thesis, Mech. Eng. Dept., State Univ. of New York at Stony Brook, 1984.

10. C. Xie, "Laminar Heat Transfer of Newtonian and Non-Newtonian Fluids in a 2:1 Rectangular Duct," Ph.D. thesis, University of Illinois at Chicago, 1991.

11. E. B. Christiansen and S. E. Craig, "Heat Transfer to Pseudoplastic Fluids in Laminar Flow," *AIChE J.* (8): 154, 1962.

12. D. R. Oliver and V. G. Jenson, "Heat Transfer to Pseudoplastic Fluids in Laminar Flow in Horizontal Tubes," *Chem. Eng. Sci.* (19): 115, 1964.

13. S. S. Yoo, "Heat Transfer and Friction Factors for Non-Newtonian Fluids in Turbulent Pipe Flow," Ph.D. thesis, University of Illinois at Chicago, 1974.

14. D. Bellet, M. Sengelin, and C. Thirriot, "Determination of Thermophysical Properties of Non-Newtonian Liquids Using a Coaxial Cylindrical Cell," *Int. J. Heat Mass Transfer* (18): 117, 1975.

15. D.-L. Lee and T. F. Irvine, "Shear Rate Dependent Thermal Conductivity Measurements of Non-Newtonian Fluids," *Experimental Thermal and Fluid Science* (15/1): 16–24, 1997.

16. W. Y. Lee, Y. I. Cho, and J. P. Hartnett, "Thermal Conductivity Measurements of Non-Newtonian Fluids," *Lett. Heat Mass Transfer* (8): 255, 1981.

17. T. T. Tung, K. S. Ng, and J. P. Hartnett, "Pipe Frictions Factors for Concentrated Aqueous Solutions of Polyacrylamide," *Lett. Heat Mass Transfer* (5): 59, 1978.

18. W. Ostwald, "Ueber die Geschwindigkeitsfunktion der viscositat disperser systeme. I.," *Kolloid-Z,* (36): 99–117, 1925.

19. T. F. Irvine Jr. and J. Karni, "Non-Newtonian Fluid Flow and Heat Transfer," in *Handbook of Single-Phase Convective Heat Transfer,* S. Kakac, R. K. Shah, and W. Aung, eds., John Wiley & Sons, New York, 1987.

20. E. C. Bingham, *Fluidity and Plasticity,* McGraw-Hill, New York, 1922.

21. R. B. Bird, "Experimental Tests of Generalized Newtonian Models Containing a Zero-Shear Viscosity and a Characteristic Time," *Can. J. Chem. Eng.* (43): 161, 1965.

22. R. E. Powell and H. Eyring, "Mechanisms for the Relaxation Theory of Viscosity," *Nature* (154): 427, 1944.

23. J. L. Sutterby, "Laminar Converging Flow of Dilute Polymer Solutions in Conical Sections, II," *Trans. Soc. Rheol.* (9): 227, 1965.

24. P. J. Carreau, "Rheological Equations from Molecular Network Theories," Ph.D. thesis, University of Wisconsin, Madison, WI, 1968.

25. H. A. Barnes, J. F. Hutton, and K. Walters, *An Introduction to Rheology,* Elsevier, New York, 1989.

26. A. B. Metzner and J. C. Reed, "Flow of Non-Newtonian Fluids—Correlation of the Laminar Transition, and Turbulent-Flow Regions," *AIChE J.* (1): 434, 1955.

27. M. F. Edwards and R. Smith, "The Turbulent Flow of Non-Newtonian Fluids in the Absence of Anomalous Wall Effects," *J. Non-Newtonian Fluid Mech.* (7): 77, 1980.

28. M. L. Wasserman and J. C. Slattery, "Upper and Lower Bounds on the Drag Coefficient of a Sphere in a Power-Model Fluid," *AIChE J.* (10): 383, 1964.

29. M. Reiner, "The Deborah Number," *Physics Today* (17): 62, 1964.

30. A. B. Metzner, J. L. White, and M. M. Denn, "Constitutive Equation for Viscoelastic Fluids for Short Deformation Periods and for Rapidly Changing Flows: Significance of the Deborah Number," *AIChE J.* (12): 863, 1966.

31. G. Astarita and G. Marrucci, *Principles of non-Newtonian Fluid Mechanics,* McGraw-Hill, New York, 1974.

32. R. R. Huigol, "On the Concept of the Deborah Number," *Trans. Soc. Rheol.* (19): 297, 1975.

33. F. A. Seyer and A. B. Metzner, "Turbulent Flow Properties of Viscoelastic Fluids," *Can. J. Chem. Eng.* (45): 121, 1967.

34. B. Elbirli and M. T. Shaw, "Time Constants from Shear Viscosity Data," *J. Rheol.* (22): 561, 1978.

35. Y. I. Cho and J. P. Hartnett, "Non-Newtonian Fluids in Circular Pipe Flow," in *Advances In Heat Transfer,* T. F. Irvine Jr. and J. P. Hartnett, eds., vol. 15, pp. 59–141, Academic, New York, 1981.

36. K. S. Ng and J. P. Hartnett, "Effects of Mechanical Degradation on Pressure Drop and Heat Transfer Performance of Polyacrylamide Solutions in Turbulent Pipe Flow," in *Studies in Heat Transfer,* T. F. Irvine Jr. et al., eds., p. 297, McGraw-Hill, New York, 1979.

37. E. Y. Kwack, Y. I. Cho, and J. P. Hartnett, "Effect of Weissenberg Number on Turbulent Heat Transfer of Aqueous Polyacrylamide Solutions," *Proc. 7th Int. Heat Transfer Conf.,* Munich, vol. 3, FC11, pp. 63–68, September 1982.

38. T. T. Tung, K. S. Ng, and J. P. Hartnett, "Pipe Friction Factors for Concentrated Aqueous Solutions of Polyacrylamide," *Lett. Heat Mass Transfer* (5): 59, 1978.

39. D. C. Bogue, "Entrance Effects and Prediction of Turbulence in Non-newtonian Flow," *Ind. Eng. Chem.* (51): 874, 1959.

40. B. C. Lyche and R. B. Bird, "The Graetz-Nusselt Problem for a Power Law Non-newtonian Fluid," *Chem. Eng. Sci.* (6): 35, 1956.

41. R. B. Bird, W. E. Stewart, and E. N. Lightfoot, *Transport Phenomena,* Wiley, New York, 1960.

42. R. B. Bird, "Zur Theorie des Wärmeübergangs an nicht-Newtonsche Flüssigkeiten beilaminarer Rohrströmung," *Chem. Ing. Tech.* (315): 69, 1959.

43. M. A. Lévêque, "Les lois de la transmission de la chaleur par convection," *Ann. Mines* (13): 201, 1928.

44. R. L. Pigford, "Nonisothermal Flow and Heat Transfer Inside Vertical Tubes," *Chem. Eng. Prog. Symp. Ser.* (17/51): 79, 1955.

45. A. B. Metzner, R. D. Vaughn, and G. L. Houghton, "Heat Transfer to Non-newtonian Fluids," *AIChE J.* (3): 92, 1957.

46. A. A. McKillop, "Heat Transfer for Laminar Flow of Non-newtonian Fluids in Entrance Region of a Tube," *Int. J. Heat Mass Transfer* (7): 853, 1964.

47. Y. P. Shih and T. D. Tsou, "Extended Leveque Solutions for Heat Transfer to Power Law Fluids in Laminar Flow in a Pipe," *Chem. Eng. Sci.* (15): 55, 1978.

48. S. M. Richardson, "Extended Leveque Solutions for Flows of Power Law Fluids in Pipes and Channels," *Int. J. Heat Mass Transfer* (22): 1417, 1979.

49. R. S. Schechter, "On the Steady Flow of a Non-newtonian Fluid in Cylinder Ducts," *AIChE J.* (7): 445, 1961.

50. J. A. Wheeler and E. H. Wissler, "The Friction Factor–Reynolds Number Relation for the Steady Flow of Pseudoplastic Fluids through Rectangular Ducts," *AIChE J.* (11): 207, 1966.

51. A. R. Chandrupatla and V. M. K. Sastri, "Laminar Forced Convection Heat Transfer of a Non-newtonian Fluid in a Square Duct," *Int. J. Heat Mass Transfer* (20): 1315, 1977.

52. W. Kozicki, C. H. Chou, and C. Tiu, "Non-newtonian Flow in Ducts of Arbitrary Cross-Sectional Shape," *Chem. Eng. Sci.,* vol. 21, pp. 665–679, 1966.

53. W. Kozicki and C. Tiu, "Improved Parametric Characterization of Flow Geometrics," *Can. J. Chem. Eng.,* vol. 49, pp. 562–569, 1971.

54. J. P. Hartnett, E. Y. Kwack, and B. K. Rao, "Hydrodynamic Behavior of Non-Newtonian Fluids in a Square Duct," *J. Rheol.* [30(S)]: S45, 1986.

55. J. P. Hartnett and M. Kostic, "Heat Transfer to a Viscoelastic Fluid in Laminar Flow Through a Rectangular Channel," *Int. J. Heat Mass Transfer* (28): 1147, 1985.

56. A. E. Green and R. S. Rivlin, "Steady Flow of Non-Newtonian Fluids Through Tubes," *Appl. Math.* (XV): 257, 1956.

57. J. A. Wheeler and E. H. Wissler, "Steady Flow of non-Newtonian Fluids in a Square Duct," *Trans. Soc. Rheol.* (10): 353, 1966.

58. A. G. Dodson, P. Townsend, and K. Walters, "Non-Newtonian Flow in Pipes of Non-circular Cross-Section," *Comput. Fluids* (2): 317, 1974.

59. S. Gao, "Flow and Heat Transfer Behavior of non-Newtonian Fluids in Rectangular Ducts," Ph.D. thesis, University of Illinois at Chicago, 1993.

60. P. Payvar, "Heat Transfer Enhancement in Laminar Flow of Viscoelastic Fluids Through Rectangular Ducts," *Int. J. Heat and Mass Transfer* (37): 313–319, 1994.

61. R. M. Cotta and M. N. Ozisik, "Laminar Forced Convection of Power-Law non-Newtonian Fluids Inside Ducts, *Wärme Stoffübertrag* (20): 211, 1986.

62. J. Vlachopoulos and C. K. J. Keung, "Heat Transfer to a Power-Law Fluid Flowing Between Parallel Plates," *AIChE J.* (18): 1272, 1972.

63. T. Lin and V. L. Shah, "Numerical Solution of Heat Transfer to Yield-Power-Law Fluids Flowing in the Entrance Region," *6th Int. Heat Transfer Conf.,* Toronto, vol. 5, p. 317, 1978.

64. A. R. Chandrupatla, "Analytical and Experimental Studies of Flow and Heat Transfer Characteristics of a non-Newtonian Fluid in a Square Duct," Ph.D. thesis, Indian Institute of Technology, Madras, India, 1977.

65. R. K. Shah and A. L. London, "Laminar Flow Forced Convection in Ducts," *Adv. Heat Transfer Suppl. 1,* Academic, New York, 1978.

66. F. W. Schmidt and M. E. Newell, "Heat Transfer in Fully Developed Laminar Flow Through Rectangular and Isosceles Triangular Ducts," *Int. J. Heat Mass Transfer* (10): 1121, 1967.

67. V. Javeri, "Heat Transfer in Laminar Entrance Region of a Flat Channel for the Temperature Boundary Conditions of the Third Kind," *Wärme Stoffübertrag* (10): 127, 1977.

68. R. K. Shah and M. S. Bhatti, "Laminar Convective Heat Transfer in Ducts," in *Handbook of Single Phase Convective Heat Transfer,* S. Kakac, R. K. Shah, and W. Aung eds., p. 3-1, Wiley Interscience, New York, 1987.

69. M. Kostic, "Heat Transfer and Hydrodynamics of Water and Viscoelastic Fluid Flow in a Rectangular Duct," Ph.D. thesis, University of Illinois at Chicago, 1984.

70. P. Wibulswas, "Laminar-Flow Heat-Transfer in Non-circular Ducts," Ph.D. dissertation, Department of Mechanical Engineering, University of London, 1966.

71. C. Xie, "Laminar Heat Transfer of Newtonian and non-Newtonian Fluids in a 2:1 Rectangular Duct," Ph.D. thesis, University of Illinois at Chicago, 1991.

72. D. W. Dodge and A. B. Metzner, "Turbulent Flow of non-Newtonian Systems," *AIChE J.* (5): 189, 1959.

73. A. B. Metzner and P. S. Friend, "Heat Transfer to Turbulent non-Newtonian Fluids," *Ind. Eng. Chem.* (51): 8979, 1959.

74. S. S. Yoo and J. P. Hartnett, "Thermal Entrance Lengths for non-Newtonian Fluid in Turbulent Pipe Flow," *Lett. Heat Mass Transfer* (2): 189, 1975.

75. P. S. Virk, H. S. Mickley, and K. A. Smith, "The Ultimate Asymptote and Mean Flow Structure in Toms' Phenomenon," *Trans. ASME, J. Appl. Mech.* (37): 488, 1970.

76. F. A. Seyer and A. B. Metzner, "Turbulence Phenomena in Drag Reducing Systems," *AIChE J.* (15): 426, 1969.

77. E. Y. Kwack and J. P. Hartnett, "Empirical Correlations of Turbulent Friction Factors and Heat Transfer Coefficients of Aqueous Polyacrylamide Solutions," in *Heat Transfer Science and Technology,* B. X. Wang ed., p. 210, Hemisphere, Washington DC, 1987.

78. A. J. Ghajar and A. J. Azar, "Empirical Correlations for Friction Factor in Drag-reducing Turbulent Pipe Flows," *Int. Comm. Heat Mass Transfer* (15): 705–718, 1988.

79. K. S. Ng, Y. I. Cho, and J. P. Hartnett, *AIChE Symposium Series (19th Natl. Heat Transfer Conference),* no. 199, vol. 76, pp. 250–256, 1980.

80. F. A. Seyer, "Turbulence Phenomena in Drag Reducing Systems," Ph.D. thesis, University of Delaware, Newark, DE, 1968.

81. E. M. Khabakhpasheva and B. V. Perepelitsa, "Turbulent Heat Transfer in Weak Polymeric Solutions," *Heat Transfer Sov. Res.* (5): 117, 1973.

82. M. J. Rudd, "Velocity Measurements Made with a Laser Dopplermeter on the Turbulent Pipe Flow of a Dilute Solution," *J. Fluid Mech.* (51): 673, 1972.

83. V. Arunachalam, R. L. Hummel, and J. W. Smith, "Flow Visualization Studies of a Turbulent Drag Reducing Solution," *Can. J. Chem. Eng.* (50): 337, 1972.

84. D. D. Kale, "An Analysis of Heat Transfer to Turbulent Flow of Drag Reducing Fluids," *Int. J. Heat Mass Transfer* (20): 1077, 1977.

85. T. Mizushina, H. Usui, and T. Yoshida, "Turbulent Pipe Flow of Dilute Polymer Solutions," *J. Chem. Eng. Jpn.* (7): 162, 1974.

86. H. Usui, "Transport Phenomena in Viscoelastic Fluid Flow," Ph.D. thesis, Kyoto University, Kyoto, Japan, 1974.

87. A. B. Metzner and G. Astarita, "External Flows of Viscoelastic Materials: Fluid Property Restrictions on the Use of Velocity-Sensitive Probes," *AIChE J.* (13): 550, 1967.

88. K. A. Smith, E. W. Merrill, H. S. Mickley, and P. S. Virk, "Anomalous Pilot Tube and Hot-Film Measurements in Dilute Polymer Solutions," *Chem. Eng. Sci.* (22): 619, 1967.

89. G. Astarita and L. Nicodemo, "Behavior of Velocity Probes in Viscoelastic Dilute Polymer Solutions," *Ind. Eng. Chem. Fund.* (8): 582, 1969.

90. R. W. Serth and K. M. Kiser, "The Effect of Turbulence on Hot-Film Anemometer Response in Viscoelastic Fluids," *AIChE J.* (16): 163, 1970.

91. N. S. Berman, G. B. Gurney, and W. K. George, "Pilot Tube Errors in Dilute Polymer Solutions," *Phys. Fluids* (16): 1526, 1973.

92. N. A. Halliwell and A. K. Lewkowicz, "Investigation into the Anomalous Behavior of Pilot Tubes in Dilute Polymer Solutions," *Phys. Fluids* (18): 1617, 1975.

93. E. Y. Kwack, Y. I. Cho, and J. P. Hartnett, "Heat Transfer to Polyacrylamide Solutions in Turbulent Pipe Flow: The Once-Through Mode," in *AIChE Symposium Series,* vol. 77, no. 208, pp. 123–130, AIChE, New York, 1981.

94. R. W. Allen and E. R. G. Eckert, "Friction and Heat Transfer Measurements to Turbulent Pipe Flow of Water (Pr = 7 and 8) at Uniform Wall Heat Flux," *Trans. ASME* (86): 301, 1964.

95. R. G. Deissler, "Turbulent Heat Transfer and Friction in the Entrance Regions of Smooth Passages," *Trans. ASME* (77): 1221, 1955.

96. J. P. Hartnett, "Experimental Determination of the Thermal Entrance Length for the Flow of Water and Oil in Circular Pipes," *Trans. ASME* (77): 1211, 1955.

97. V. J. Berry, "Non-uniform Heat Transfer to Fluids Flowing in Conduits," *Appl. Sci. Res.* (A4): 61, 1953.

98. R. C. Little, R. J. Hansen, D. L. Hunston, O. K. Kim, R. L. Patterson, and R. Y. Ting, "The Drag Reduction Phenomenon: Observed Characteristics, Improved Agents and Proposed Mechanisms," *Ind. Eng. Chem. Fund.* (14): 283, 1975.

99. C. S. Chiou and R. J. Gordon, "Low Shear Viscosity of Dilute Polymer Solutions," *AIChE J.* (26): 852, 1980.

100. Y. I. Cho, J. P. Hartnett, and Y. S. Park, "Solvent Effects on the Rheology of Aqueous Polyacrylamide Solutions," *Chem. Eng. Comm.* (21): 369, 1983.

101. K. A. Smith, P. S. Keuroghlian, P. S. Virk, and E. W. Merrill, "Heat Transfer to Drag Reducing Polymer Solutions," *AIChE J.* (15): 294, 1969.

102. G. T. Pruitt, N. F. Whitsitt, and H. R. Crawford, "Turbulent Heat Transfer to Viscoelastic Fluids," Contract No. NA7-369, The Western Company, 1966.

103. C. S. Wells Jr., "Turbulent Heat Transfer in Drag Reducing Fluids," *AIChE J.* (14): 406, 1968.

104. J. C. Corman, "Experimental Study of Heat Transfer to Viscoelastic Fluids," *Ind. Eng. Chem. Process Des. Dev.* (2): 254, 1970.

105. W. A. Meyer, "A Correlation of the Friction Characteristics for Turbulent Flow of Dilute Viscoelastic non-Newtonian Fluids in Pipes," *AIChE J.* (12): 522, 1966.

106. M. Poreh and U. Paz, "Turbulent Heat Transfer to Dilute Polymer Solutions," *Int. J. Heat Mass Transfer* (11): 805, 1968.

107. K. S. Ng, J. P. Hartnett, and T. T. Tung, "Heat Transfer of Concentrated Drag Reducing Viscoelastic Polyacrylamide Solutions," *Proc. 17th Natl. Heat Transfer Conf.,* Salt Lake City, UT, 1977.

108. Y. I. Cho and J. P. Hartnett, "Analogy for Viscoelastic Fluids—Momentum, Heat and Mass Transfer in Turbulent Pipe Flow," *Letters in Heat and Mass Transfer* (7/5): 339–346, 1980.

109. Y. I. Cho and J. P. Hartnett, "Mass Transfer in Turbulent Pipe Flow of Viscoelastic Fluids," *Int. J. Heat and Mass Transfer* (24/5): 945–951, 1981.

110. J. P. Hartnett, "Single Phase Channel Flow Forced Convection Heat Transfer," in *10th International Heat Transfer Conference,* vol. 1, pp. 247–258, Brighton, England, 1994.

111. E. Y. Kwack, Y. I. Cho, and J. P. Hartnett, "Solvent Effects on Drag Reduction of Polyox Solutions in Square and Capillary Tube Flows," *J. Non-Newtonian Fluid Mech.* (9): 79, 1981.

112. M. Kostic and J. P. Hartnett, "Heat Transfer Performance of Aqueous Polyacrylamide Solutions in Turbulent Flow Through a Rectangular Channel," *Int. Commun. Heat Mass Transfer* (12): 483, 1985.

113. J. P. Hartnett, E. Y. Kwack, and B. K. Rao, "Hydrodynamic Behavior of non-Newtonian Fluids in a Square Duct," *J. Rheol.* [30(S)]: S45, 1986.

114. J. P. Hartnett, "Viscoelastic Fluids: A New Challenge in Heat Transfer," *Journal of Heat Transfer* (114): 296–303, 1992.

CHAPTER 11
TECHNIQUES TO ENHANCE HEAT TRANSFER

A. E. Bergles
Rensselaer Polytechnic Institute

INTRODUCTION

General Background

Most of the burgeoning research effort in heat transfer is devoted to analyzing what might be called the "standard situation." However, the development of high-performance thermal systems has also stimulated interest in methods to improve heat transfer. The study of improved heat transfer performance is referred to as heat transfer enhancement, augmentation, or intensification.

The performance of conventional heat exchangers can be substantially improved by a number of enhancement techniques. On the other hand, certain systems, particularly those in space vehicles, may *require* enhancement for successful operation. A great deal of research effort has been devoted to developing apparatus and performing experiments to define the conditions under which an enhancement technique will improve heat (and mass) transfer. Over 5000 technical publications, excluding patents and manufacturers' literature, are listed in a bibliographic report [1]. The recent growth of activity in this area is clearly evident from the yearly distribution of such publications shown in Fig. 11.1. The most effective and feasible techniques have graduated from the laboratory to full-scale industrial use.

The main objective of this chapter is to survey some of the important literature pertinent to each enhancement technique, thus providing guidance for potential users. With the large amount of literature in the field, it is clearly impossible to cite more than representative studies. Wherever possible, correlations for thermal and hydraulic performance will be presented, or key sources of design data will be suggested.

Classification of Heat Transfer Enhancement Techniques

Enhancement techniques can be classified as *passive* methods, which require no direct application of external power, or as *active* schemes, which require external power. The effectiveness of both types depends strongly on the mode of heat transfer, which might range from single-phase free convection to dispersed-flow film boiling. Brief descriptions of passive techniques follow.

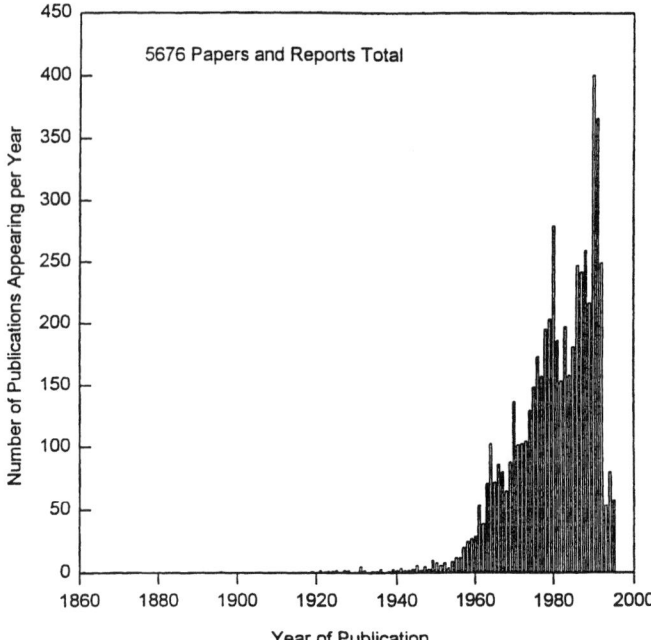

FIGURE 11.1 Citations on heat transfer enhancement versus year of publication (to mid-1995).

Treated surfaces involve fine-scale alternation of the surface finish or coating (continuous or discontinuous). They are used for boiling and condensing; the roughness height is below that which affects single-phase heat transfer.

Rough surfaces are produced in many configurations, ranging from random sand-grain-type roughness to discrete protuberances. The configuration is generally chosen to promote turbulence rather than to increase the heat transfer surface area. The application of rough surfaces is directed primarily toward single-phase flow.

Extended surfaces are routinely employed in many heat exchangers. The development of nonconventional extended surfaces, such as integral inner fin tubing, and the improvement of heat transfer coefficients on extended surfaces by shaping or interrupting the surfaces are of particular interest.

Displaced enhancement devices are inserted into the flow channel so as to indirectly improve energy transport at the heated surface. They are used with forced flow.

Swirl-flow devices include a number of geometric arrangements or tube inserts for forced flow that create rotating and/or secondary flow: inlet vortex generators, twisted-tape inserts, and axial-core inserts with a screw-type winding.

Coiled tubes lead to more compact heat exchangers. The secondary flow leads to higher single-phase coefficients and improvements in most regions of boiling.

Surface-tension devices consist of wicking or grooved surfaces to direct the flow of liquid in boiling and condensing.

Additives for liquids include solid particles and gas bubbles in single-phase flows and liquid trace additives for boiling systems.

Additives for gases are liquid droplets or solid particles, either dilute phase (gas-solid suspensions) or dense phase (fluidized beds).

The active techniques are now described.

Mechanical aids stir the fluid by mechanical means or by rotating the surface. Surface "scraping," widely used for batch processing of viscous liquids in the chemical process industry, is applied to the flow of such diverse fluids as high-viscosity plastics and air. Equipment with rotating heat exchanger ducts is found in commercial practice.

Surface vibration, at either low or high frequency, has been used primarily to improve single-phase heat transfer.

Fluid vibration is the most practical type of vibration enhancement, given the mass of most heat exchangers. The vibrations range from pulsations of about 1 Hz to ultrasound. Single-phase fluids are of primary concern.

Electrostatic fields (dc or ac) are applied in many different ways to dielectric fluids. Generally speaking, electrostatic fields can be directed to cause greater bulk mixing of fluid in the vicinity of the heat transfer surface, which enhances heat transfer. An electrical field and a magnetic field may be combined to provide a forced convection or electromagnetic pumping.

Injection involves supplying gas to a flowing liquid through a porous heat transfer surface or injecting similar fluid upstream of the heat transfer section. Surface degassing of liquids can produce enhancement similar to gas injection. Only single-phase flow is of interest.

Suction involves either vapor removal through a porous heated surface in nucleate or film boiling, or fluid withdrawal through a porous heated surface in single-phase flow.

Two or more of these techniques may be utilized simultaneously to produce an enhancement larger than that produced by only one technique. This simultaneous use is termed *compound enhancement.*

It should be emphasized that one reason for studying enhanced heat transfer is to assess the effect of an inherent condition on heat transfer. Some practical examples include roughness produced by standard manufacturing, degassing of liquids with high gas content, surface vibration resulting from rotating machinery or flow oscillations, fluid vibration resulting from pumping pulsation, and electric fields present in electrical equipment.

Performance Evaluation Criteria

It seems impossible to establish a generally applicable selection criterion for the use of enhancement techniques, since numerous factors influence the ultimate decision. Some of the pertinent considerations are economic: development cost, initial cost, operating cost, maintenance cost, etc.; in addition, other factors such as reliability and safety must be considered. However, to begin the assessment, it is useful to consider the relationship between the thermal and hydraulic performance—particularly in the dominant practical case of single-phase forced convection. The typical heat transfer and flow friction data for "turbulence promoters" shown in Fig. 11.2 illustrate this point. The equivalent plain tube diameter or "envelope diameter" is used in this presentation, as suggested in Ref. 2. The promoters produce a sizable elevation in the heat transfer coefficient at constant velocity; however, there is generally a greater percentage increase in the friction factor.

Common thermal-hydraulic goals include reducing the size of a heat exchanger required for a specified heat duty, increasing the heat duty of an existing heat exchanger, reducing the approach temperature difference for the process streams, or reducing the pumping power. The presence of system and design constraints leads to a number of performance evaluation criteria (PECs). The geometric variables for tube-side flow in a conventional shell-and-tube heat exchanger are tube diameter, tube length, and number of tubes per pass. The primary independent operating variables are the approach temperature difference and the mass flow rate (or velocity). Dependent variables are the heat transfer rate and pumping power (or pressure drop). A PEC is established by selecting for one of the process streams one of the operational variables for the performance objective, in accordance with design constraints on the remaining variables.

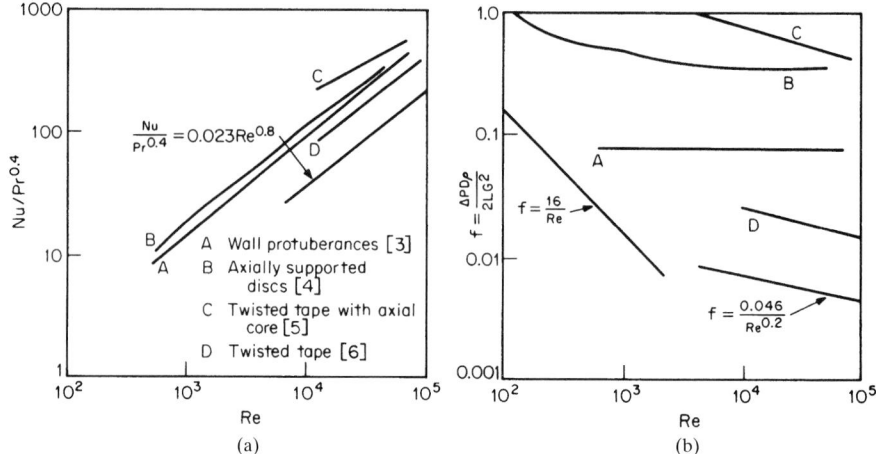

FIGURE 11.2 Typical data for turbulence promoters inserted inside tubes. (*a*) heat transfer data; (*b*) friction data.

Table 11.1 lists PECs for 12 cases of interest concerning enhanced and smooth circular tubes of the same envelope diameter. The table segregates these PECs according to three different geometric constraints:

1. *FG Criteria.* The cross-sectional envelope area and tube length are held constant. The FG criteria may be thought of as a retrofit situation in which there is a one-for-one replacement of plain surfaces with enhanced surfaces of the same basic geometry, for example, tube envelope diameter, tube length, and number of tubes for in-tube flow. The FG-2 criteria have the same objectives as FG-1, but require the enhanced surface design to operate at the same pumping power as the reference smooth tube design. In most cases this

TABLE 11.1 Performance Evaluation Criteria for $D_{ta}/D_{to} = 1$ [7]

		Fixed					Consequences						
Case	Geom.	W	P	q	ΔT_i	Objective	$\dfrac{N_a}{N_o}$	$\dfrac{L_a}{L_o}$	$\dfrac{W_a}{W_o}$	$\dfrac{\text{Re}_{ia}}{\text{Re}_{io}}$	$\dfrac{\mathbf{P}_a}{\mathbf{P}_o}$	$\dfrac{q_a}{q_o}$	$\dfrac{\Delta T_{ia}}{\Delta T_{io}}$
FG-1a	N, L	X	—	—	X	$\uparrow q$	1	1	1	1*	>1	>1	1
FG-1b	N, L	X	—	X	—	$\downarrow \Delta T_i$	1	1	1	1*	1	1	<1
FG-2a	N, L	—	X	—	X	$\uparrow q$	1	1	<1	<1	1	>1	1
FG-2b	N, L	—	X	X	—	$\downarrow \Delta T_i$	1	1	<1	<1	1	1	<1
FG-3	N, L	—	—	X	X	$\downarrow P$	1	1	<1	<1	<1	1	1
FN-1	N	—	X	X	X	$\downarrow L$	1	<1	<1	<1	1	1	1
FN-2	N	X	—	X	X	$\downarrow L$	1	<1	1	1*	<1	1	1
FN-3	N	X	—	X	X	$\downarrow P$	1	<1	1	1*	<1	1	1
VG-1	—	X	X	X	X	$\downarrow NL$	>1[†]	<1	1	<1[†]	1	1	1
VG-2a	$NL^{\ddagger}$	X	X	X	X	$\uparrow q$	>1[†]	<1	1	<1[†]	1	>1	1
VG-2b	$NL^{\ddagger}$	X	X	X	—	$\downarrow \Delta T_i$	>1[†]	<1	1	<1[†]	1	1	<1
VG-3	$NL^{\ddagger}$	X	—	X	X	$\downarrow P$	<1	<1	1	<1[†]	<1	1	1

N, number of tubes; L, tube length.
* For internal roughness. For internal fins, $\text{Re}_{ia}/\text{Re}_{io} = D_{ha}A_{fo}/D_{io}A_{fa}$.
[†] Roughness with high-Pr fluids may not result in $N_a/N_o > 1$ (or $\text{Re}_{ia}/\text{Re}_{io} < 1$).
[‡] The product of N and L is constant in cases VG-2 and 3.

requires the enhanced exchanger to operate at reduced flow rate. The FG-3 criterion seeks reduced pumping power for fixed heat duty.

2. *FN Criteria.* These criteria maintain fixed flow frontal area and allow the length of the heat exchanger to be a variable. These criteria seek reduced surface area (FN-1, FN-2) or reduced pumping power (FN-3) for constant heat duty.

3. *VG Criteria.* In many cases, a heat exchanger is sized for a required thermal duty with specified flow rate. In these situations the FG and FN criteria are not applicable. Because the tube-side velocity must be reduced to accommodate the higher friction characteristics of the enhanced surface, it is necessary to increase the flow area to maintain constant flow rate. This is accomplished by using a greater number of parallel flow circuits. Maintaining a constant exchanger flow rate eliminates the penalty of operating at higher thermal effectiveness encountered in the previous FG and FN cases.

The necessary relations for quantitative formulation of these PECs are summarized in Ref. 7. (A more detailed discussion of PECs is given in Ref. 371.) Evaluation of the objectives is straightforward once the constraints are specified, if the basic heat transfer and flow friction data are available in the range of interest. The calculations can be carried out for any geometry where the data are available; alternatively, using correlations for h and f, the optimum geometry can be determined. Consider the following example.

In proposed ocean thermal energy conversion (OTEC) systems utilizing a closed Rankine cycle, heat exchanger surface is a major consideration, since over half of the plant capital cost is in heat exchangers. The net output is determined primarily by the heat transfer rates in the evaporator and condenser and by the power consumed by the seawater pumps. Since the seawater flow rate is immaterial, FN-1 is an appropriate PEC. The objective can also be interpreted as reduction of NL (or area), and it is convenient to explore the dependence of NL on N. Figure 11.3 presents typical results for the surface area ratio for various types of in-tube enhancement (no outside enhancement) for a baseline OTEC shell-and-tube spray film evaporator [8]. The use of rough tubes, inner fin tubes, or tubes with twisted-tape inserts permits a reduction in surface area by almost 30 percent at $N_a/N_o = 2$. This is very significant, since heat exchanger surface areas in projected OTEC systems are extremely large—on the order of 10,000 $m^2/MW(e)$ net for both evaporator and condenser [372]. Further area reductions can be brought about by enhancing the evaporating side. A study by Webb [13] translates the thermal-hydraulic performance of enhanced tubes into cost advantages for OTEC applications.

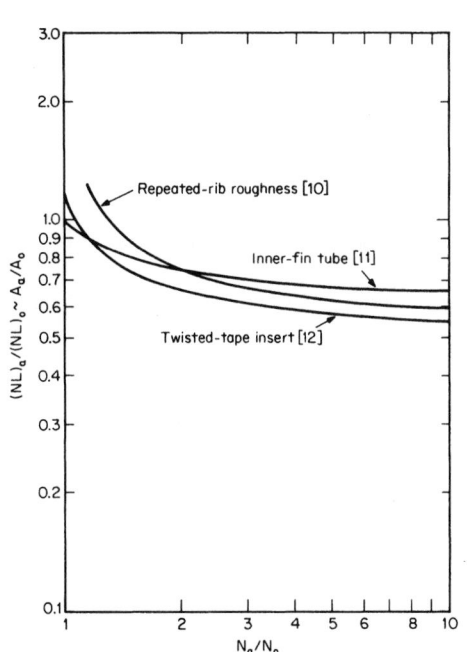

FIGURE 11.3 Composite plots of constant pumping power performance (*VG*-1, $W \neq$ const) of an OTEC evaporator with various in-tube enhancement techniques [9].

The extension of these PECs to two-phase heat transfer is complicated by the dependence of the local heat transfer coefficient on the local temperature difference and/or quality. Heat transfer and pressure drop have been considered in the evaluation of internally finned tubes for refrigerant evaporators [14] and for internally finned tubes, helically ribbed tubes, and spirally fluted tubes for refrigerant condensers [15]. Pumping power has been incorporated into the evaluation of inserts used to elevate subcooled boiling critical heat flux (CHF) [16, 17]. A discussion of the application of enhancement to two-phase systems is given by Webb [373].

These PECs will be used occasionally throughout this chapter to demonstrate the advantages of specific enhanced heat transfer techniques. It should be noted, however, that these thermal-hydraulic comparisons apply only to the basic heat exchanger. A full analysis of the enhancement effect may require consideration of the entire system if flow rates or temperature levels change as a result of the enhancement.

TREATED AND STRUCTURED SURFACES

Boiling

As discussed in Ref. 366, surface material and finish have a strong effect on nucleate and transition pool boiling; thus, optimum surface conditions might be selected for a system operating in these boiling regimes. Certain types of fouling and oxidation, apparently those that improve wettability, produce significant increases in pool-boiling CHF.

A novel technique for promoting nucleate boiling has been proposed by Young and Hummel [18]. Spots of Teflon or other nonwetting material, either on the heated surface or in pits, were found to favor nucleation, as shown in Fig. 11.4. Relatively low wall superheat is required to activate the nonwetting cavities represented by the spots. The spots are placed so that the bubble area of influence includes the whole surface, resulting in a low average superheat. The heat transfer coefficient (at constant heat flux) is increased by factors of 3 to 4. This technique is not effective for refrigerants, as there are no "Freon-phobic" materials [19].

Thin insulating coatings effectively increase rates of heat transfer in pool boiling when heater temperature is the controlling parameter [20]. When surface temperatures are originally in the film-boiling range, a thin coating of Teflon, for example, reduces the fluid-surface interface temperature to the level where the more effective transition or nucleate boiling

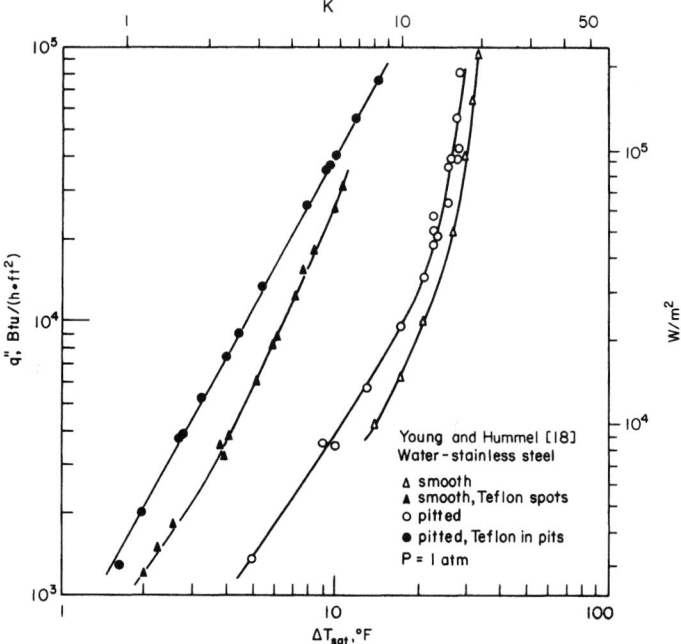

FIGURE 11.4 Influence of surface treatment on saturated pool boiling.

occurs. Bergles and Thompson [21] found large reductions in quench times with scale or oxide coatings, which promote destabilization of film boiling.

With well-wetting fluids (refrigerants, cryogens, organics, alkali liquid metals), doubly reentrant cavities are required to ensure vapor trapping. The probability of having such active nucleation sites present is increased by selective machining, forming, or coating of the surface. Furthermore, large cavities can be created that result in steady-state boiling at low wall super-heat. The surfaces may appear either rough or smooth (as if treated), depending on the man-ufacturing procedure. Hence, the usual "treated" and "rough" classifications are lumped together for purposes of this discussion. Examples of these "structured" boiling surfaces are given in Table 11.2, which is an extension of a table given in Ref. 22.

Wall superheat reductions up to a factor of 10 have been reported with some of these sur-faces. It should be noted that the mechanism of vaporization for these surfaces is different from that for normal cavity boiling. Here the liquid flows via selected paths or channels to the interior, where thin-film evaporation occurs over a large surface area; the vapor is then ejected through other paths by "bubbling" [38, 39]. It should be emphasized that the perfor-mance of these special surfaces is quite sensitive to surface geometry and fluid condition. Additionally, very low temperature differences are involved; hence, it is necessary to be espe-cially careful, or at least consistent, in measuring wall temperatures and saturation tempera-tures (pressures). The first comprehensive comparison of the nucleate boiling performances

TABLE 11.2 Examples of Structured Boiling Surfaces

Category	Report	Procedure	Result
Machined	Kun and Czikk [23]	Cross-grooved and flattened	Regular matrix of reentrant cavities
	Fujikake [24]*	Low fins knurled and compressed	Reentrant cross grooves
	Hwang and Moran [25]	Laser drilling	Regular matrix of slightly reentrant cavities
Formed	Webb [26]*	Standard low-fin tubing with fins bent to reduce gap	Helical circumferential reentrant cavities
	Zatell [27]	As above, with additional variations	Helical circumferential reentrant cavities
	Nakayama et al. [28]*	Rolled, upset, and brushed	Helical circumferential or groove-type reentrant cavities with periodic openings
	Stephan and Mitrovic [29]*	Standard low-fin tubing rolled to form T-shaped fins	Helical circumferential reentrant cavities
Multilayer	Ragi [30]	Stamped sheet with pyramids, open at the top, attached to surface	Regular matrix of reentrant cavities
Coated	Marto and Rohsenow [31]	Poor weld	Irregular matrix of surface and reentrant cavities
	O'Neill et al. [32]*	Sintering or brazing	Irregular matrix of surface and reentrant cavities
	Oktay and Schmeckenbecher [33]	Electrolytic deposition	Irregular matrix that includes reentrant cavities
	Dahl and Erb [34]	Flame spraying	Irregular matrix of surface and reentrant cavities
	Fujii et al. [35]	Particles bonded by plating	Irregular matrix of surface and reentrant cavities
	Janowski et al. [36]*	Metallic coating of a foam substrate	Irregular matrix of surface and reentrant cavities
	Warner et al. [37]	Plasma-deposited polymer	Irregular matrix of surface and reentrant cavities

* Denotes commercial surface.

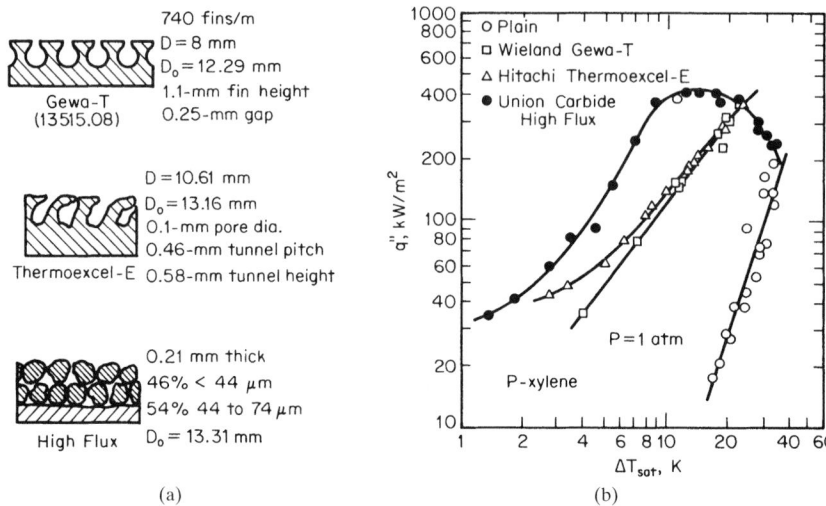

FIGURE 11.5 Pool boiling from smooth and structured surfaces on the same apparatus [40]. (*a*) sketch of cross sections of three enhanced heat transfer surfaces tested; (*b*) boiling curves for three enhanced tubes and smooth tube.

of several structured surfaces was reported recently by Yilmaz et al. [40]. As shown in Fig. 11.5, each of the three surfaces exhibits a boiling curve well to the left of that for a single tube. It is evident that if only low temperature differences are available, high heat fluxes require structured surfaces. Note that the heat flux is based on the area of the equivalent smooth tube for a particular outside diameter.

The previously cited studies do not report boiling curves that exhibit hysteresis, that is, that exhibit different characteristics with increasing heat flux from those with decreasing heat flux. It should be noted, however, that with increasing heat flux, large temperature overshoots and boiling curve hysteresis are common with refrigerants and other highly wetting liquids. Bergles and Chyu [41] provide extensive documentation of such behavior with sintered surfaces; but they note that not all commercial equipment will have start-up problems because of this behavior. Also, in the area of practical application, there is some evidence that tube bundles in a thermosyphon reboiler may exhibit behavior different from that of single tubes with structured surfaces [42]. More recent work suggests that single tubes can be used to anticipate the performance of flow-through bundles [374].

Limited evidence indicates that CHF in pool boiling with structured surfaces is usually as high as or higher than that with plain surfaces [32].

The use of structured surfaces to enhance thin-film evaporation has also been considered recently. Here, in contrast to the flooded-pool experiments noted above, the liquid to be vaporized is sprayed or dripped onto heated horizontal tubes to form a thin film. If the available temperature difference is modest, structured surfaces can be used to promote boiling in the film, thus improving the overall heat transfer coefficient. Chyu et al. [43] found that sintered surfaces yielded nucleate boiling curves similar to those obtained in pool boiling. T-shaped fins did not exhibit low ΔT boiling; however, a threefold convective enhancement was obtained as a result of the increased surface area.

Boiling heat transfer from structured surfaces is a major growth area in enhanced heat transfer. Most of the processes noted in Table 11.2 are covered by patents (see also Refs. 44 and 45), and, as noted, many of the surfaces are offered commercially. In general, the behavior of these surfaces is not yet understood to the point where correlations are available that allow custom production of surfaces for a particular fluid and pressure level. In some cases, however, manufacturers have accumulated sufficient experience to provide optimized surfaces for some of the important applications, for example, flooded-refrigerant evaporators for direct-expansion chillers. A comprehensive discussion of boiling applications of enhancement is given by Thome [375].

Condensing

As noted in Ref. 367, surface treatment for the promotion of dropwise condensation in vapor space environments has been extensively investigated. If dropwise condensation is achieved, the enhancement is 10 to 100 times the filmwise condensation coefficient. Numerous promoters and coatings have been found effective; however, a number of practical problems relate to the method of application, permanence, and compatibility with the rest of the system. Tanasawa's review [47] includes a good discussion of the difficulties that must be overcome if industry is to adopt this condensation process.

It should be noted that the only valid application of dropwise condensation is for steam condensers, since nonwetting substances are not available for most other working fluids. For example, no dropwise condensation promoters have been found for refrigerants (i.e., no dropwise condensation promoters seem to be "Freon-phobic") [46]. The enhancement of dropwise condensation, beyond inducing the process by selection of an effective, durable promoter, is fruitless, since the heat transfer coefficients are already so high.

Glicksman et al. [48] showed that average coefficients for film condensation of steam on horizontal tubes can be improved up to 20 percent by strategically placing horizontal strips of Teflon or other nonwetting material around the tube circumference. The condensate flow is interrupted near the leading edge of a strip and the condensate film is thinner when it re-forms at the downstream edge of the strip.

ROUGH SURFACES

Single-Phase Flow

Surface roughness is usually not considered for free convection since the velocities are commonly too low to cause flow separation or secondary flow. A review [49] of the limited data for free convection from machined or formed roughness with air, water, and oil indicates that increases in heat transfer coefficient of up to 100 percent have been obtained with air. However, the reported increases with liquids are very small. This could be due to inadequate accounting of the radiation in the air experiments.

Surface roughness was one of the first techniques to be considered seriously as a means of enhancing forced-convection heat transfer. Initially, investigators speculated that elevated heat transfer coefficients might accompany the relatively high friction factors characteristic of rough conduits. However, since commercial roughness is not well defined, artificial surface roughness has been employed. Integral roughness may be produced by the traditional manufacturing processes of machining, forming, casting, or welding. Various inserts can also be used to provide surface protuberances. Although the enhanced heat transfer with surface promoters is often due in part to the fin effect, it is difficult to separate the fin contribution from other factors. For the data discussed here, the promoted heat transfer coefficient is referenced to the base or envelope surface area. In view of the infinite number of possible geometric variations, it is not surprising that, even after more than 700 studies [1], no completely unified treatment is available.

The enhancement of deep laminar flow is of particular interest to the food and chemical process industries. Gluck [50] used spiral wire inserts to improve heat transfer to water and various power-law liquids. Nusselt number increases up to 580 percent were obtained. Blumenkrantz and Taborek [51] found that the heating of Alta-Vis-530 in spirally fluted tubes was improved up to 200 percent; however, there was negligible enhancement for cooling. Rozalowski and Gater [52] tested both Alta-Vis-530 and Zerolene SAE-50 in flexible hoses with helical convolutes. Nusselt numbers were increased up to 200 percent for heating and 100 percent for cooling.

Enhancement is widely utilized in plate-type compact heat exchangers. Pescod [53] reported on a study of the improvements obtained through the use of spikes and ripples to enhance nominally laminar flow of air in parallel plate channels of large aspect ratios. Most plate heat exchangers utilize corrugated surfaces for structural reasons as well as for heat transfer enhancement. It is generally agreed that the heat transfer and pressure drop characteristics of commercial corrugated surfaces used in plate exchangers are quite similar for both laminar and turbulent flow.

The diversity of results obtained for turbulent-flow heat transfer to water (composite data for other fluids are similar) in all types of roughened tubes is indicated in Figs. 11.6 and 11.7. Here, the simplest coordinates are chosen for illustrative purposes. All calculations are based on the base area of the tube, with no allowance made for increases in surface area. While the heat transfer coefficients are increased approximately 4 times at the most, friction factors are increased as much as 58 times. Within this matrix of data lie surfaces that are "efficient" as far as both heat transfer and pressure drop are concerned. The PECs noted in the section on per-

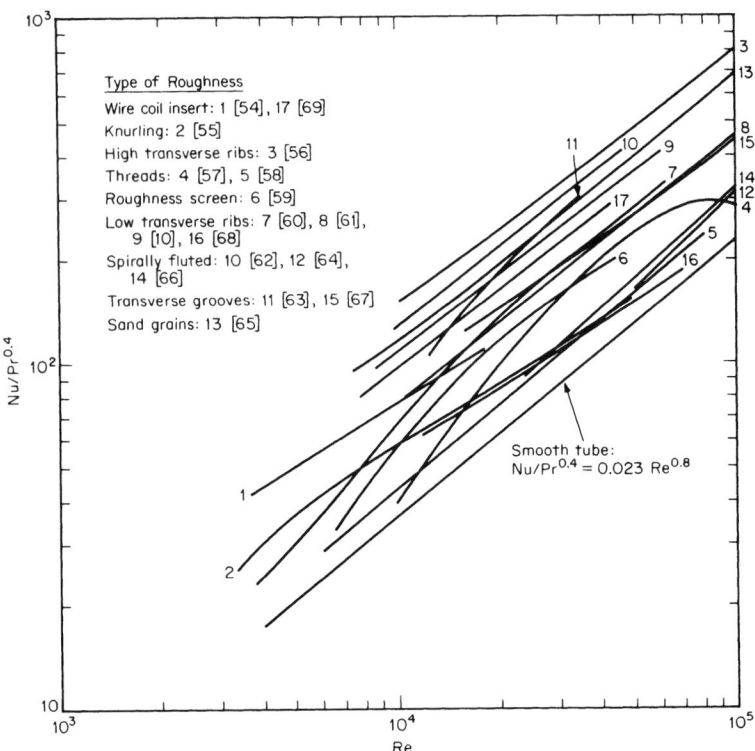

FIGURE 11.6   Summary of heat transfer data for water flowing in internally roughened tubes [9].

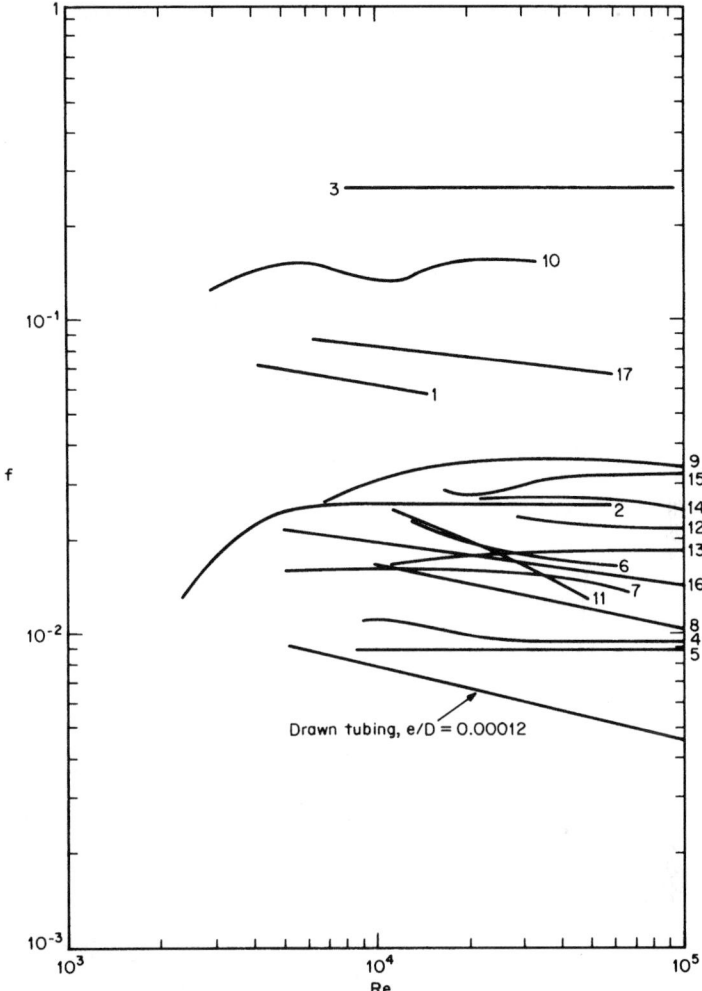

FIGURE 11.7 Summary of friction factor data for water flowing in internally roughened tubes [9].

formance evaluation criteria can be applied to select the best surface for a given application. For example, the transverse-ribbed surface [10] was selected, since its heat transfer and pressure drop characteristics combine to yield one of the best VG-1b ratios.

The ideal situation is to have correlations for h and f that can be introduced into appropriate PECs to obtain the optimum geometry for a particular application. It has been demonstrated that an analogy exists between heat transfer and friction for rough surfaces in turbulent flow; however, the relationship is dependent on the type of roughness. An analogy solution for a sand-grain-type roughness was developed by Dipprey and Sabersky [65]. Recent work has considered surfaces that can be produced commercially. Webb et al. [70] have correlated heat transfer coefficients for various fluids flowing in tubes with transverse repeated-rib roughness. Withers [71, 72] applied this technique to commercial single-helix internally ridged tubes and multiple-helix internally ridged tubes. This similarity correlation method should be valid for any roughness type. It must be recognized, however, that extensive experimental data are required to establish the various functional relations.

A different type of correlation technique has been proposed by Lewis [73]. Basically, the detailed behavior of roughness elements is required: form drag coefficients, heat transfer coefficient distribution, and separation length behind an element. When this information is available, a prediction can be formulated without recourse to data for the actual rough channel. The agreement with experimental data, such as those found in Ref. 10, is surprisingly good, considering the "separate effects" character of the model.

A somewhat simpler, but not fundamental, approach is given in Ref. 376. The mass of data for various types of roughness (Fig. 11.8) is now so large that straightforward power-law correlations for Nusselt number and friction factor can be generated using large computers and statistical analysis software. The power-law correlations developed in this manner are

$$\text{Nu}_a/\text{Nu}_s = \{1 + [2.64\text{Re}^{0.036}(e/D)^{0.212}(p/D)^{-0.21}(\alpha/90)^{0.29}(\text{Pr})^{-0.024}]^7\}^{1/7} \tag{11.1}$$

$$f_a/f_s = \left\{1 + \left[29.1\text{Re}^{(0.67-0.06p/D-0.49\alpha/90)}(e/D)^{(0.37-0.157p/D)}(p/D^{(-1.66\times10^{-6}\,\text{Re}-0.33\alpha/90)}\right.\right.$$

$$\left.\left. \times (\alpha/90^{(4.59+4.11\times10^{-6}\,\text{Re}-0.15p/D)}\left(1+\frac{2.94}{n}\right)\sin\beta\right]^{15/16}\right\}^{16/15} \tag{11.2}$$

where the smooth tube references are

$$\text{Nu}_s = \frac{(f/2)\,\text{Re}\,\text{Pr}}{1.07 + 12.7\sqrt{f/2}(\text{Pr}^{2/3}-1)}$$

and

$$f_s = (1.58\ln\text{Re} - 328)^{-2}$$

The fidelity of this correlation is shown in Fig. 11.9 for 1807 data points, representing a wide range of roughness types and profiles.

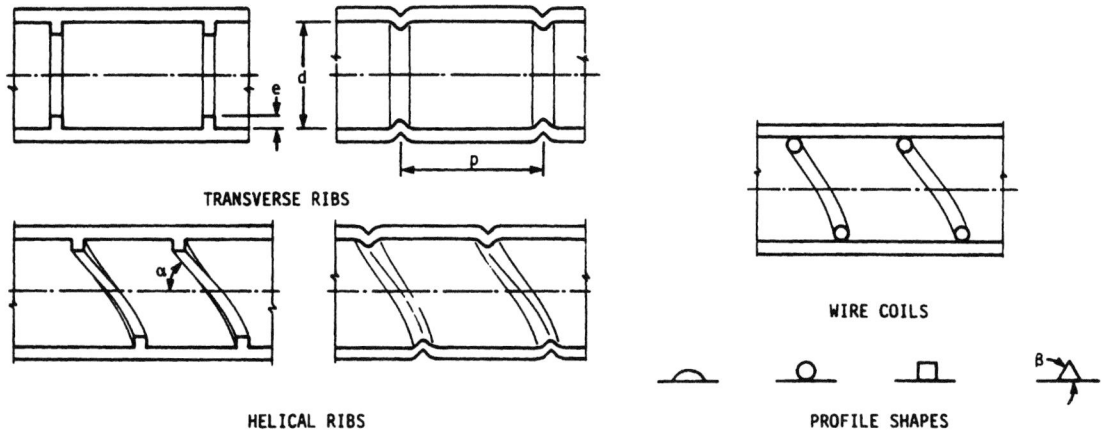

FIGURE 11.8 Various types of tube internal roughness [376].

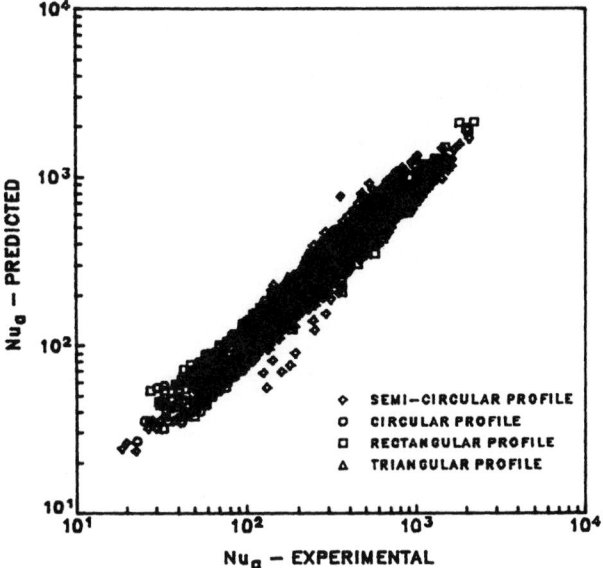

FIGURE 11.9   Performance of heat transfer correlation (Eq. 11.1).

The frequently used annular geometry is generally more suitable for the application of surface roughness. Machined surfaces are relatively easy to produce, and increased friction affects only the inside portion of the wetted surface. The results of Kemeny and Cyphers [74] for a helical groove and a helical protuberance are given in terms of a popular PEC in Fig. 11.10. The grooved surface is not effective in general, although it does tend to improve with increasing Re. The inferior performance of the coiled wire assembly compared to that of the integral protrusion is probably due to poor thermal contact between the wire and the groove.

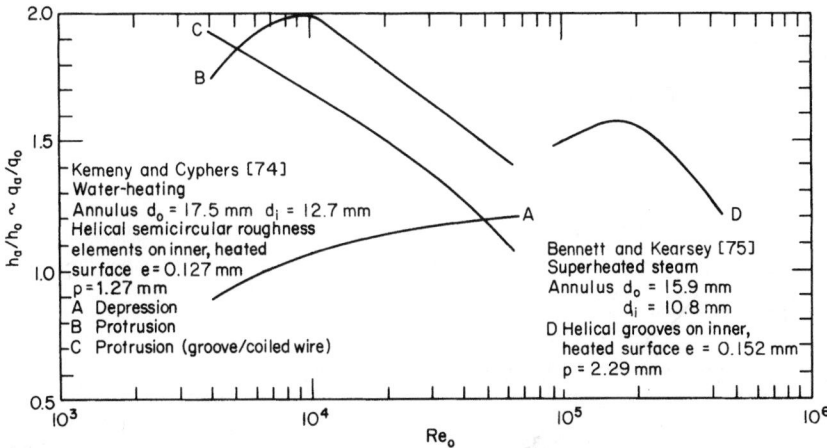

FIGURE 11.10   Performance of annuli with roughness (FG-2a).

The results of Bennett and Kearsey [75] for superheated steam flowing in an annulus are included in Fig. 11.10. The data of Brauer [76] suggest that the optimum P/e for an annular geometry is approximately 3.

Durant et al. [77] summarized an extensive investigation of heat transfer in annuli with the inner heated tube roughened by means of diamond knurls. At equal pumping power, it was found that heat transfer coefficients for the knurled annuli were up to 75 percent higher than those for the smooth annuli.

In commercial gas-cooled nuclear reactors, particularly advanced gas-cooled reactors and gas-cooled fast breeder reactors, increases in core power density have been achieved by artificially roughening the fuel elements. Since bundle experiments are time consuming and expensive, experiments are usually performed with internally roughened tubes or with annuli having electrically heated roughened inner tubes and smooth outer tubes. In these cases it is necessary to transfer the tube or annulus data to fuel bundle conditions. Dalle Donne and Meyer [78] provide a good general description of this area in their discussion of five popular transformation methods. Considering none of these methods to be entirely satisfactory, they propose another method and apply it to two-dimensional rectangular ribs. A discussion of transformation methods and their application to rectangular and trapezoidal ribs is given by Hudina [79]. Subsequent efforts have centered on three-dimensional roughness, since it seems to offer more favorable thermal-hydraulic performance.

Dalle Donne [80] notes that at a Reynolds number of 10^5, for two-dimensional roughness, Stanton numbers are typically increased by a factor of 2 while the corresponding increase in friction factor is 4. He also cites data for two three-dimensional roughnesses that indicate that the Stanton number increased by 3 and 4 with friction factor ratios of 8 and 12, respectively. The first roughness is depicted in Fig. 11.11. Large-scale computer codes such as SAGAP0 [81] are required to accurately predict the thermal-hydraulic behavior of roughened fuel element bundles.

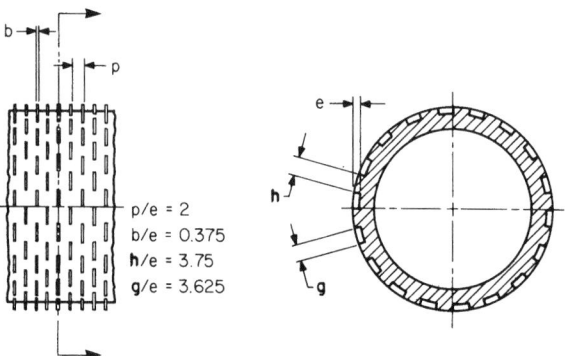

FIGURE 11.11 Three-dimensional roughness with alternate studs [80].

It is common practice in gas-cooled reactor technology to interpret data for rough rods in terms of a merit index $(\mathrm{St}^3/f)^{1/2}$. The average heat flux for a heat exchanger, such as the reactor core, is proportional to this parameter for a given ratio of pumping to thermal power and a given average film temperature drop. For the two-dimensional roughness, then, this ratio is 1.4, while for the best three-dimensional roughness the ratio is 2.3.

Turning to cross flow over bundles of tubes with surface roughness, some work has been done in the context of heat exchanger development for gas-cooled reactors and conventional shell-and-tube heat exchangers. This work is backed up by extensive studies of single cylinders, such as those with pyramid roughness elements tested in air by Achenbach [82]. Nusselt number increases up to about 150 percent were recorded. Zukauskas et al. [83] obtained similar improvements with a single pyramid-roughened cylinder for cross flow of water.

The preceding discussion indicates that certain types of roughness can improve heat transfer performance considerably. Under nonuniform flow or thermal conditions, however, it may be advantageous to roughen only that portion of the heating surface that has a higher heat flux or lower heat transfer coefficient. In many cases the overall pressure drop will not be greatly affected by roughening the hot spot. Any of the foregoing roughness types are then of interest, since they produce large increases in heat transfer coefficient over the smooth-tube value at equal flow rates. The partial roughening technique has been considered for gas-cooled reactors, where the thermal limit is reached only in the downstream portion of the core because of the axial heat flux variation [84]. One scheme for achieving selective roughening in large-diameter pipes involves sandblasting through a smaller tube that is transversed inside the pipe [85].

Boiling

Pool boiling with rough structured surfaces is discussed in the section on treated and structured surfaces. Attention here is focused on the effect of surface roughness on convective flows.

Consider first the gravity-driven flows observed in horizontal-tube spray-film evaporators. Longitudinal ribs or grooves may promote turbulence, but they impede film drainage. Knurled surfaces provide turbulence promotion and may also aid liquid spreading over the surface. When nucleate boiling occurs within the film, it appears that bubble motion is favorably affected by the roughness. Knurling increases coefficients by as much as 100 percent [86]. On the basis of experience with single-phase films (trickle coolers), Newson [87] suggests a longitudinal rib profile for horizontal-tube multiple-effect evaporators.

Annular channels with electrically heated inner tubes have been used to study the effects of roughness on forced-convection boiling. Surface conditions do not appear to significantly alter the boiling curve for reasonably high flow velocities; however, certain surface finishes improve flow-boiling CHF. Durant et al. [77, 88] have demonstrated that there is a substantial increase in subcooled CHF with knurls or threads. It was also suggested that the critical fluxes for the rough tubes were up to 80 percent higher than those for smooth tubes at comparable pumping power. Gomelauri and Magrakvelidze [89] found that for two-dimensional roughness, CHF is dependent on subcooling, with decreases observed at low subcooling and increases up to 100 percent observed at high subcooling. Murphy and Truesdale [90] found that subcooled CHF was decreased 15 to 30 percent with large roughness heights.

For bulk boiling of R-12 in commercial helical-corrugated tubing, Withers and Habdas [91] observed up to a 100 percent increase in heat transfer coefficient and up to a 200 percent increase in CHF. Several investigators have demonstrated that bulk-boiling CHF can be improved by 50 to 100 percent with various other surface modifications: Bernstein et al. [92] (irregular-diameter tubing and slotted helical inserts), Janssen and Kervinen [93] (sandblasting), and Quinn [94] (machined protuberances). Of particular significance for power boilers are the increases in CHF for high-pressure water observed with helical-ribbed tubes. Typical results are given in Fig. 11.12. Pseudo-film boiling is also suppressed with this commercial tubing [96]. Studies of post-CHF, or dispersed-flow film boiling, indicate that roughness elements increase the heat transfer coefficient [94, 97].

Condensing

Medwell and Nicol [98, 99] were among the first to study the effects of surface roughness on condensate films. They condensed steam on the outside of one smooth and three artificially roughened pipes with pyramid-shaped roughness. All were oriented vertically, and the con-

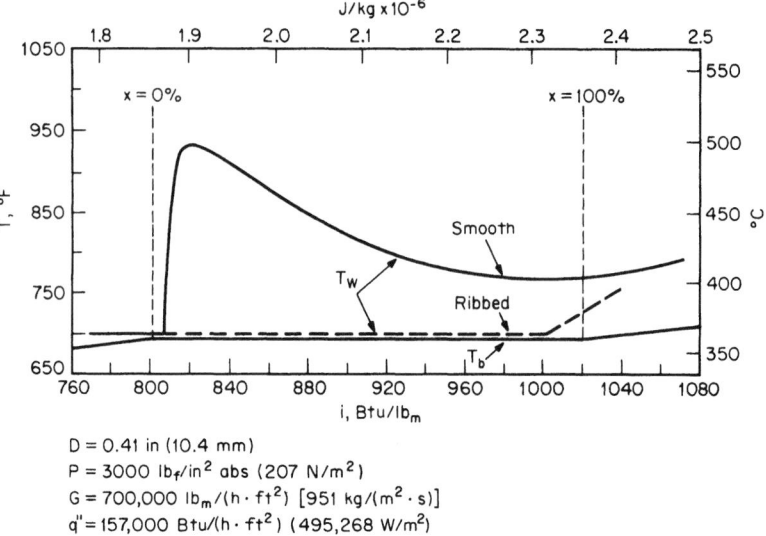

$D = 0.41$ in (10.4 mm)
$P = 3000$ lb_f/in^2 abs (207 N/m^2)
$G = 700,000$ $lb_m/(h \cdot ft^2)$ [951 kg/(m$^2 \cdot$ s)]
$q'' = 157,000$ Btu/(h $\cdot$ ft^2) (495,268 W/m^2)

FIGURE 11.12   Comparison of heat transfer characteristics of smooth and ribbed tubes with once-through boiling of water [95].

densate was drained under gravity alone. The mean heat transfer coefficients were found to increase significantly with roughness height, the values of the roughest tube being almost double those of the smooth tube. Carnavos [100] reported that condensing-side coefficients for knurled tubes were 4 to 5 times the smooth-tube values. Part of this, of course, can be attributed to the area increase.

Cox et al. [101] used several kinds of enhanced tubes to improve the performance of horizontal-tube multiple-effect plants for saline water conversion. Overall heat transfer coefficients (forced convection condensation inside and spray-film evaporation outside) were reported for tubes internally enhanced with circumferential V grooves (35 percent maximum increase in U) and protuberances produced by spiral indenting from the outside (4 percent increase). No increases were obtained with a knurled surface. Prince [102] obtained a 200 percent increase in U with internal circumferential ribs; however, the outside (spray-film evaporation) was also enhanced. Luu and Bergles [15] reported data for enhanced condensation of R-113 in tubes with helical repeated-rib internal roughness. Average coefficients were increased 80 percent above smooth-tube values. Coefficients with deep spirally fluted tubes (envelope diameter basis) were increased by 50 percent.

Random roughness consisting of attached metallic particles (50 percent area density and $e/D = 0.031$) was proposed by Fenner and Ragi [103]. With R-12, the condensing coefficient was increased 300 percent for qualities greater than 0.60, and 140 percent for lower qualities.

EXTENDED SURFACES

Single-Phase Flow

Free Convection.   The topic of finned surfaces in free convection is covered in Sec. C.2 of Ref. 368. From the standpoint of enhancement, much interest has focused on interrupted extended surfaces such as the wire-loop fins used for baseboard hot water heaters or "convectors," finned arrays or "heat sinks" used for cooling electronic components, and serrated fins used in process cooler tube banks. While natural circulation is an important normal or off-design condition in these applications, the flow is basically forced convection due to the chimney effect of the ducting. Hence, it is appropriate to include comments on this subject under the topic of compact heat exchangers with forced convection.

Compact Heat Exchangers. Compact heat exchangers have large surface-area-to-volume ratios, primarily through the use of finned surfaces. An informative collection of articles related to the development of compact heat exchangers is presented by Shah et al. [104].

Compact heat exchangers of the plate-fin, tube and plate-fin, or tube and center variety use several types of enhanced surfaces: offset strip fins, louvered fins, perforated fins, or wavy fins [105]. The flow (usually gases) in these channels is very complex, and few generalized correlations or predictive methods are available. Overall heat exchanger information is often available from manufacturers of surfaces for the automotive, air conditioning, and power industries; however, the air-side coefficients cannot be readily deduced from the published information. In general, many proprietary fin configurations have been developed that have heat transfer coefficients 50 to 100 percent above those of flat fins. The improvements are the result of flow separation, secondary flow, or periodic starting of the boundary layer. The latter is illustrated in Fig. 11.13. It should be emphasized that the tube geometry and arrangement strongly affect the heat transfer and pressure drop. For example, heat transfer coefficients are increased with staggered tubes and pressure drop is reduced with flattened tubes (in the flow direction). Design data for enhanced compact heat exchanger surfaces are given by Kays and London [107]. An overview of mechanisms was presented by Webb [108].

Circular or oval finned-tube banks utilize a variety of enhanced surfaces, as illustrated in Fig. 11.14. With the exception of material pertaining to smooth helical fins, data are rather limited and no generalized correlations presently exist. Webb [108] provides a guide to the literature.

Some complex compact heat exchanger surfaces have been studied using mass transfer methods, for example, naphthalene sublimation [109] and chemical reaction between a surface coating and ammonia added to the air stream [110]. These elegant but tedious methods yield local mass transfer coefficients that can be used to infer heat transfer coefficients by the usual analogy. This detailed information, in turn, should aid in the development of more efficient surfaces. Numerical studies have also yielded useful predictions for laminar flows [111, 112].

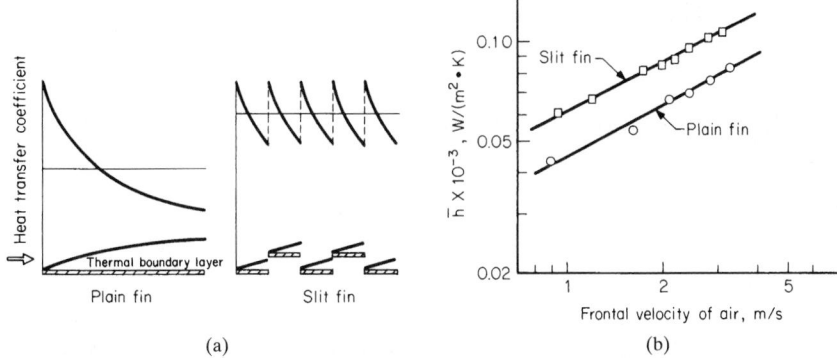

FIGURE 11.13 Principle of interrupted fins and data. (*a*) Local heat transfer coefficients for plain fin and slit fin. (*b*) Average heat transfer coefficients for plain fin and slit fin [106].

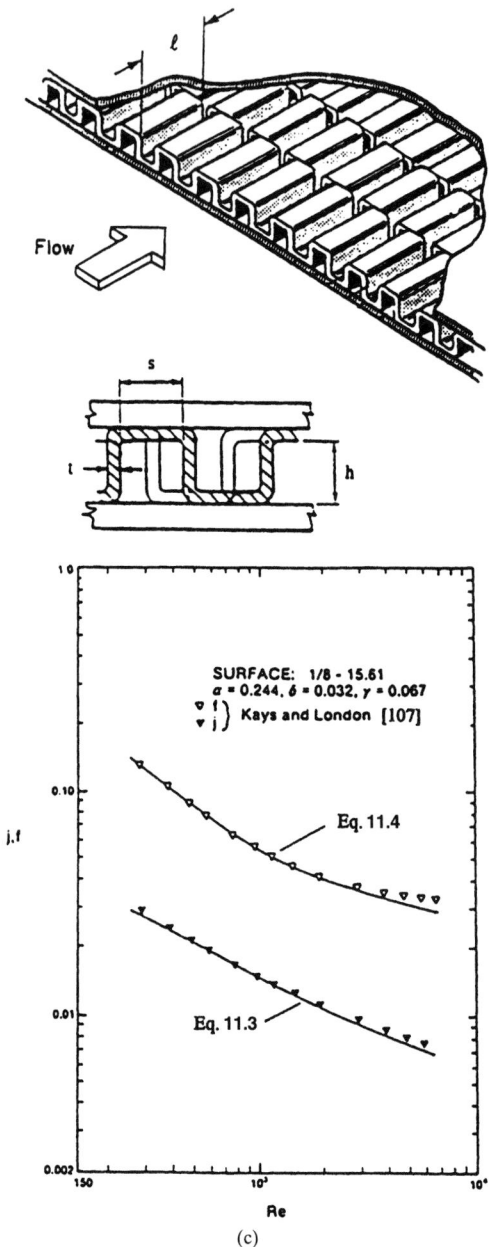

FIGURE 11.13 (*Continued*) (*c*) Geometry of the offset strip and comparison of the predictions with data [107, 377].

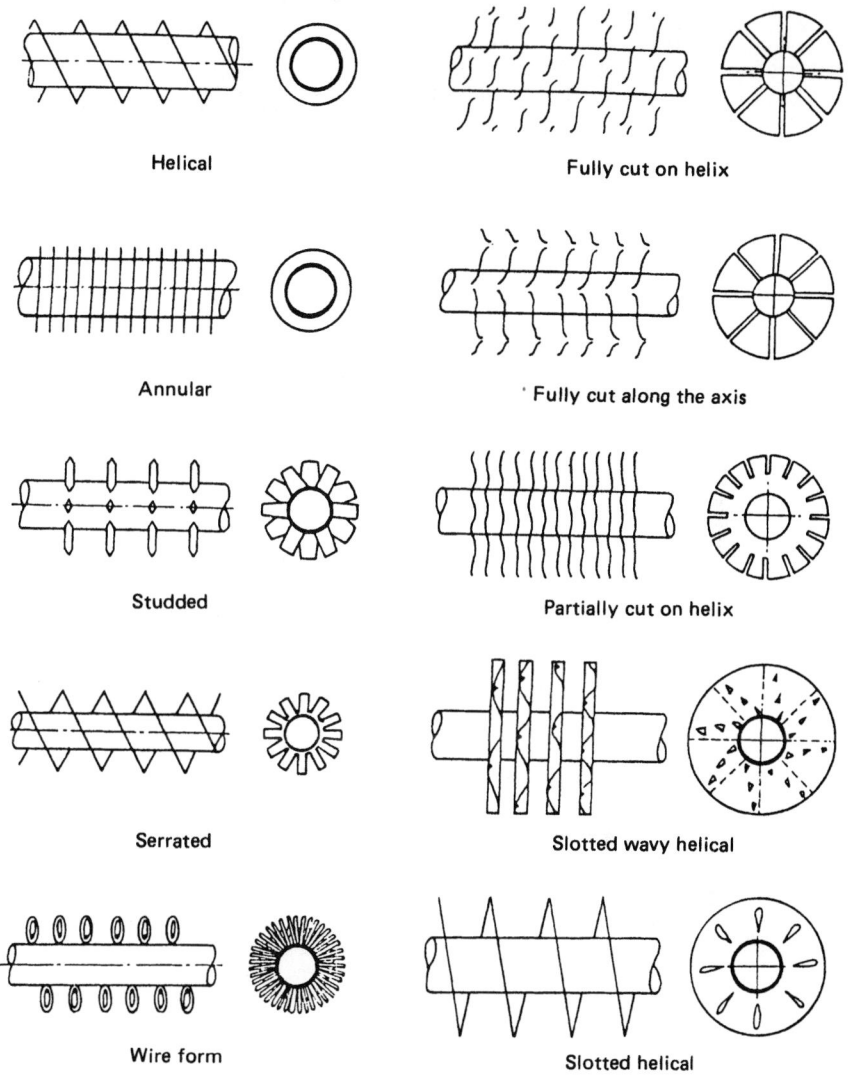

FIGURE 11.14   Enhanced fin configurations for finned-tube banks [105].

The offset strip fin (Fig. 11.13*a*) is widely used in compact heat exchangers. Once again, the data are so numerous that it is possible to develop power-law correlations for heat transfer *j* factor and friction factor [377]:

$$j_h = 0.6522 \mathrm{Re}_h^{-0.5403}\, \alpha^{-0.1541}\delta^{0.1499}\gamma^{-0.0678}[1 + 5.269 \times 10^{-5}\ \mathrm{Re}_h^{1.340}\, \alpha^{0.504}\delta^{0.456}\gamma^{-1.055}]^{0.1} \qquad (11.3)$$

$$f_h = 9.6243 \mathrm{Re}_h^{-0.7422}\, \alpha^{-0.1856}\delta^{0.3053}\gamma^{-0.2659}[1 + 7.669 \times 10^{-8}\ \mathrm{Re}_h^{4.429}\, \alpha^{0.920}\delta^{3.767}\gamma^{0.236}]^{0.1} \qquad (11.4)$$

where j_h, and f_h, and Re_h are based on the hydraulic diameter given by

$$D_h = 4sh\ell/[2(s\ell + h\ell + th) + ts] \qquad (11.5)$$

These equations are based on experimental data for 18 different offset strip fin geometries, and they represent the data continuously in the laminar, transition, and turbulent flow regions, as shown in Fig. 11.13c. The development of accurate power-law correlations for a variety of enhancement configurations is possible when large databases are available.

Internal Flow.   Internally finned circular tubes are commercially available in aluminum and copper (or copper alloys). For laminar flow, the following correlations are available [113]:
  Spiral-Fin Tubes.

$$\frac{Nu_h}{Pr^{1/3}}\left(\frac{D_h}{L}\right)^{1/3}\left(\frac{\mu_b}{\mu_w}\right)^{0.14}\phi = 19.2\left(\frac{b}{p}\right)^{0.5}Re_h^{0.26} \tag{11.6}$$

where
$$\phi = 2.25\frac{1+0.01Gr_h^{1/3}}{\log Re_h} \tag{11.7}$$

  Straight-Fin Tubes.

$$\frac{Nu_h}{Pr^{1/3}}\left(\frac{D_h}{L}\right)^{1/3}\left(\frac{\mu_b}{\mu_w}\right)^{0.14}\phi = 2.43\left(\frac{1}{n}\right)^{0.5}Re_h^{0.46} \tag{11.8}$$

Isothermal Friction Factors for all Tubes:

$$f_h = \frac{16.4(D_h/D)^{1.4}}{Re_h} \tag{11.9}$$

These correlations are based on data for oil in horizontal tubes having approximately uniform temperature (steam heating). Other data for both water and ethylene glycol in both steam-heated and electrically heated tubes are in approximate agreement with the correlations [114]. As noted in Ref. 115, the analytical results for uniformly heated tubes are not in good agreement with data.

The following equations are recommended for turbulent flow in straight- and spiral-fin tubes [116]:

$$Nu_h = 0.023Pr^{0.4}Re_h^{0.8}\left(\frac{A_F}{A_{Fi}}\right)^{0.1}\left(\frac{A_i}{A}\right)^{0.5}(\sec \alpha)^3 \tag{11.10}$$

$$f_h = 0.046Re_h^{-0.2}\left(\frac{A_F}{A_{Fi}}\right)^{0.5}(\sec \alpha)^{0.75} \tag{11.11}$$

These correlations are based on data for air (cooling), water (heating), and ethylene glycol-water (heating). It is noted that fin inefficiency corrections must be incorporated when applying the equations.

Hilding and Coogan [117] provide data for longitudinal interrupted fins, with air as the working fluid. They conclude that regular interruptions of the fins improve the relative heat transfer–flow friction performance in the laminar and transition regions, but that little advantage is apparent in the turbulent region. Several manufacturers now provide this type of surface.

The first analytical study to predict the performance of tubes with straight inner fins for turbulent airflow was conducted by Patankar et al. [118]. The mixing length in the turbulence model was set up so that just one constant was required from experimental data. Expansion of analytical efforts to fluids of higher Prandtl number, tubes with practical contours, and tubes with spiraling fins is still desirable. It would be particularly significant if the analysis could predict with a reasonable expenditure of computer time the optimum fin parameters for a specified fluid, flow rate, etc.

Internally finned tubes can be "stacked" to provide multiple internal passages of small hydraulic diameter. Carnavos [119] demonstrated the large increases in heat transfer coeffi-

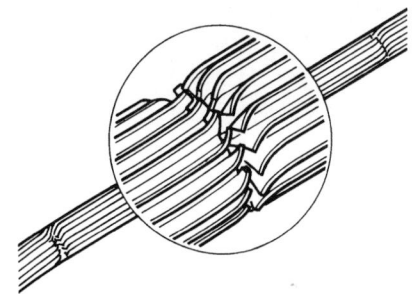

FIGURE 11.15 Close-up of finned tube, showing cut-and-twist construction [121].

cient (based on outer tube nominal area) that can be obtained in these tubes with air flow. Of course, pressure drop is also increased greatly in these tubes. Finned annuli represent one case of considerable practical interest.

Finned concentric annuli, with the fins extending about 85 percent of the way across the gap, have been studied by several investigators in connection with widespread process industry application.

Enhanced fins (interrupted, cut and twisted, perforated) are frequently used [105, 120]. Gunter and Shaw [121] demonstrate that cut-and-twisted finned tubes (Fig. 11.15) have substantially higher coefficients than continuously finned tubes.

Having assembled the available data, Clarke and Winston [122] recommend the correlation curve shown in Fig. 11.16. The Reynolds number is based on the equivalent diameter, L is the length of the finned tube and $\mathscr{P}$ is the wetted perimeter of the channel between two longitudinal fins. In the laminar range (Re < 2000), the curve is valid for both continuous and cut fins if L is interpreted as the distance between cuts, L_c.

Multiple longitudinal-fin tubes are often placed in a single pipe. According to Kern and Kraus [120], the correlation given in Fig. 11.16 is valid, but the entire flow cross section is considered when the hydraulic diameter is evaluated.

Similar longitudinal fins have been used for gas-cooled nuclear reactors. A discussion of this application is given by El-Wakil [123].

Boiling

Low- and medium-fin tubes with external circumferential fins are produced by many manufacturers for pool boiling of refrigerants and organics. Gorenflo [124] reported single-tube boiling of R-11, which indicated that the heat transfer coefficients based on total area were

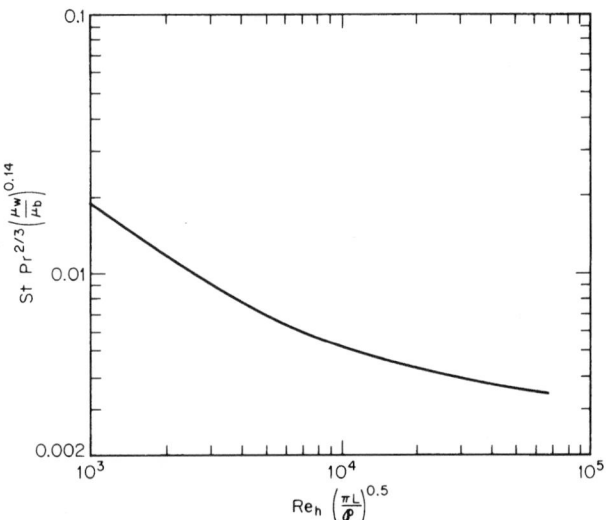

FIGURE 11.16 Correlation of heat transfer data for longitudinally finned tubes in annuli [122].

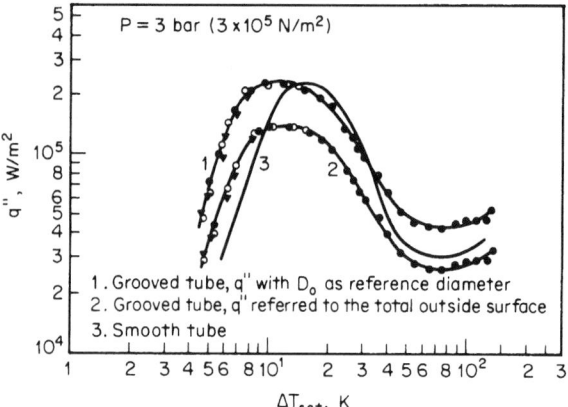

FIGURE 11.17 Boiling curves for R-114, smooth and low-finned tubes [125].

higher for all six tubes tested than for the reference plain tube. As shown in Fig. 11.17, Hesse [125] found that the heat transfer coefficient for pool boiling of R-114 was higher for finned tubes than for a plain tube; however, q''_{cr} was lower. Both comparisons were made on a total area basis. When referenced to the projected area at the maximum outside diameter, the critical heat fluxes were about equal to those of the plain tube. The reduction in local CHF is apparently due to bubble interference between fins. Westwater [126] suggests that the fins may be spaced as close as the departure diameter of a nucleate-boiling bubble (~1.55 mm for R-113).

Katz et al. [127] and Nakajima and Shiozawa [128], for example, found that coefficients on the upper finned tubes in a bundle are higher than coefficients on the lower tubes as a result of bubble-enhanced circulation. Similar results are reported by Arai et al. [129] for a bundle of Thermoexcel-E tubes. It is quite probable that with certain types of enhanced tubes it is sufficient to use the special tubes only in the lower rows since the bubble-enhanced circulation in the upper rows is so high that enhanced tubes are not effective there.

Data for a three-dimensional finned (square, straight-sided fins in a square array) tubular test section were reported by Corman and McLaughlin [130]. Significant reductions in temperature differences were observed in low-flux nucleate boiling; however, temperature differences were larger at high fluxes. The critical heat flux was increased.

A thorough survey of large-scale finned surfaces utilized for pool boiling is given by Westwater [126]. Properly shaped or insulated fins promote very large heat transfer rates if the base temperature is in the film-boiling range.

Circumferential fins with various profiles have been suggested for enhancement of horizontal-tube spray-film evaporation. Area increases are typically 2:1, and in certain cases the fins promote redistribution of the liquid so that thin films are present at the peaks. V-shaped grooves (threads) produce improvements in evaporating heat transfer coefficients of up to 200 percent [131, 132]. An improvement of 200 percent in the overall coefficient was reported by Prince [102] for tubes with straight circumferential flutes outside and shadow ribs inside (condensing). On the basis of analytical results, Sideman and Levin [133] concluded that square-edged grooves should have the best operational characteristics of flow rate and heat transfer. Cox et al. [134] concluded that spirally fluted tubes offered no particular advantage for this service.

With falling film evaporation inside vertical tubes, Thomas and Young [135] found that heat transfer coefficients could be increased by more than a factor of 10 with loosely attached internal fins. While distorted tubes (e.g., doubly fluted and spirally corrugated) have been developed primarily for enhancement of condensation on the outside wall, it is expected that heat transfer coefficients for the evaporating fluid on the inside of the tube should also be increased. This has been confirmed in large-scale tests by Lorenz et al. [136], who obtained a 150 percent increase in vaporization coefficient (external falling film of ammonia) for a doubly fluted tube as compared to the prediction for an equivalent smooth tube. This is greater than the area enhancement of 57 percent.

Tubes with integral or inserted internal fins increase heat transfer rates for horizontal forced convection vaporization of refrigerants by as much as several hundred percent more than smooth-tube values [137–139]. Data from the second study are shown in Fig. 11.18; the heat transfer coefficients are based on the surface area of the smooth tube of the same diameter.

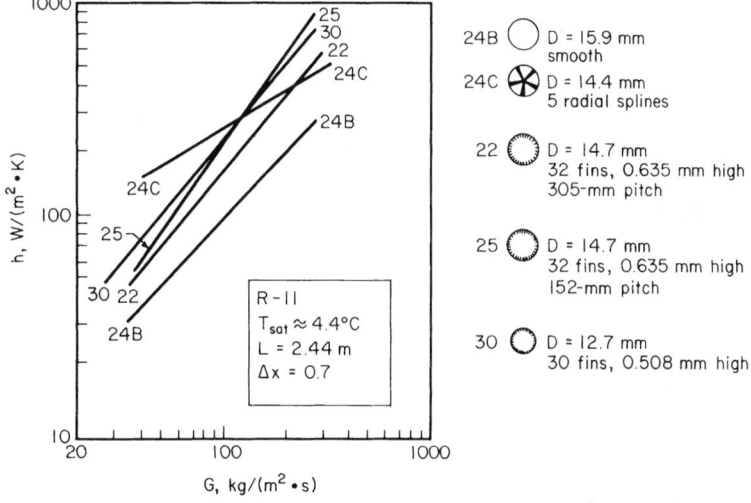

FIGURE 11.18 Heat transfer coefficients for evaporation in internally finned tubes [138].

Microfin tubes—circular tubes with numerous, short, integral fins (see 22, 25, and 30 in Fig. 11.18)—are being widely used in air-conditioning and refrigeration for both small and large units. These enhanced tubes have also been applied to vertical thermosyphon reboilers in the chemical process industry. The use of these tubes for pure conventional refrigerants (including oil effects) has been extensively discussed (e.g., Bergles [380]). The current area of interest is alternate refrigerants.

Microfin tube data for R-123 are reported by Kedzierski [381], and tests of R-134a are described by Eckels et al. [382], these being the generally accepted, chlorine-free substitutes for R-11 and R-12, respectively. The general conclusion is that the enhancement is generally the same for the new refrigerants as for the traditional ones.

Plate-fin heat exchangers are widely used in process heat exchangers. As described in the review by Robertson [140], the various forms of enhanced fins used for single-phase compact heat exchanger cores are also used for evaporators. These include perforated fins, offset strip fins (serrated), and herringbone fins. Both forced convection and falling-film evaporation modes in an offset strip fin compact heat exchanger were tested by Panchal et al. [141] under expected OTEC conditions. A composite heat transfer coefficient of 2,525 Btu/(h·ft^2·°F), or 14,338 W/(m^2·K), was obtained. In the subsequent analysis of Yung et al. [142], the enhancement is attributed to splitting of the film. The thinner film results in a higher heat transfer coefficient than would be obtained with plain fins of the same maximum channel width. This

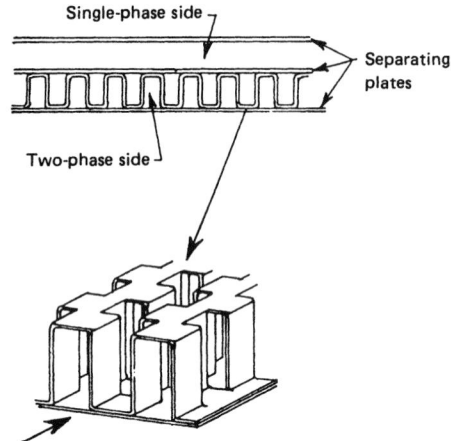

Single-phase side

Separating plates

Two-phase side

FIGURE 11.19 Sections of an offset-strip-fin evaporator [142].

situation is illustrated in Fig. 11.19. This plate-fin heat exchanger also performs well for forced convection (vertical upflow) vaporization [141]. An analysis for this situation was recently presented by Chen et al. [143]. Recognizing the periodic redevelopment of the flow, the local heat transfer coefficient was assumed to be determined by the local ΔT. The computer solution is in good agreement with test data.

Condensing

Surface extensions are widely employed for enhancement of condensation on horizontal or vertical tubes. Consider horizontal arrangements first. Integral low-fin tubing is produced by many manufacturers for shell-and-tube air-conditioning and process condensers. The increased area and thin condensate film near the fin tips result in heat transfer coefficients several times those of a plain tube with the same base diameter. Coefficients based on total area are also higher. However, condensate may bridge the fins and render the enhancement ineffective if the fin spacing is small or if the liquid has a high surface tension. Fins are used with refrigerants and other organic fluids with low surface tension since the condensing side often represents the dominant thermal resistance. Normally, finning would not be used for steam power plant condensers because of the high surface tension of water and the relatively low thermal resistance of the condensing side.

Beatty and Katz [144] proposed the following equation for the average heat transfer coefficient (total area basis) for single low-fin tubes on the basis of data for a variety of fluids:

$$\bar{h} = 0.689\left(\frac{k_f^3 p_f^2 g i_{lg}}{\mu_f \Delta T_{gw}}\right)^{1/4}\left(\frac{A_r}{A_e}\frac{1}{D_r^{0.25}} + 1.3\frac{\eta A_F}{A_e}\frac{1}{L_f^{0.25}}\right) \qquad (11.12)$$

For the copper-finned tubes normally recommended for commercial condensers, the effective surface area A_e is taken to be the total outside surface area. For similar conditions, the mean effective length of a fin is given by $\pi(D_o^2 - D_r^2)/4D_o$. This semiempirical equation is based on the assumption that condensate is readily drained by gravity. Young and Ward [145] suggest that turbulence and vertical-row effects can be included by simply using a different constant for a specific fluid. Rudy and Webb [146] suggested that the success of Eq. 11.16 may be fortuitous in that surface flooding due to condensate retention is compensated for by surface-tension-induced drainage.

There has been much interest in three-dimensional surfaces for horizontal-tube condensers. A finned and machined surface having a three-dimensional character is described by Nakayama et al. [28] and Arai et al. [129]. As shown in Fig. 11.20, the surface resembles a low-fin tube with notched fins. Coefficients (envelope area basis) are as much as 7 times the smooth-tube values. The considerable improvement relative to conventional low fins is apparently due to multidirectional drainage at the fin tips.

Circular pin fins have been tested by Chandran and Watson [147]. Their average coefficients (total area basis) were as much as 200 percent above the smooth-tube values. Square pins have been proposed by Webb and Gee [148]; a 60 percent reduction of fin material as compared to integral-fin tubing is predicted using a gravity drainage model. Notaro [149] described a three-dimensional surface whereby small metal particles are bonded randomly to the surface. The upper portions of the particles promote effective thin-film condensation, and the condensate is drained along the uncoated portion of the tube.

The analytical foundations of contoured surfaces are based on the 1954 paper of Gregorig [150]. The shape suggested by Gregorig is shown in Fig. 11.21. Condensation occurs primarily

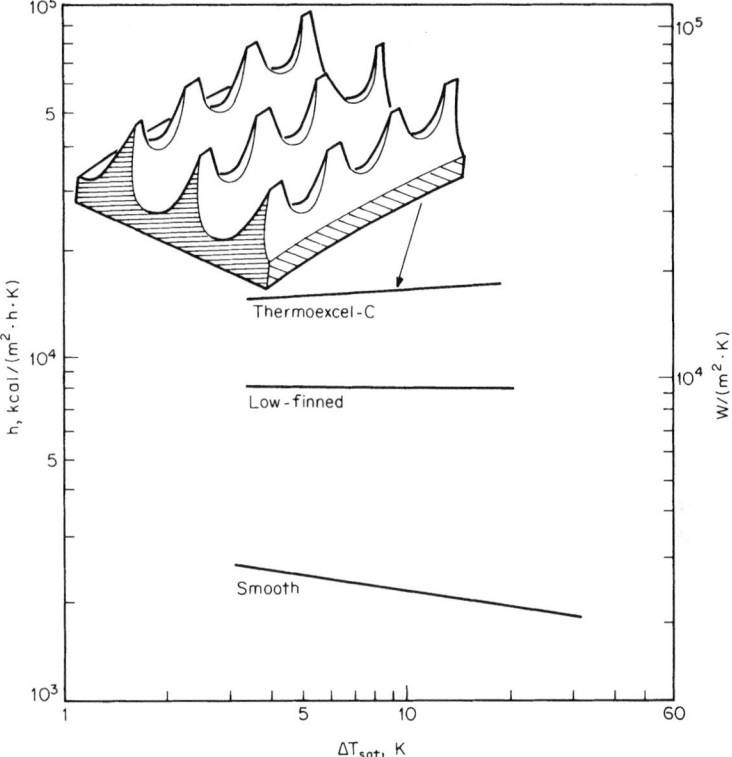

FIGURE 11.20 Performance of enhanced condenser tubes compared to a smooth tube [28].

at the tops of convex ridges. Surface tension forces then pull the condensate into the grooves, which act as drainage channels. The average heat transfer coefficient is substantially greater than that for a uniform film thickness. While manufacturing considerations severely limit the practical realization of such optimum shapes, the surface-tension-driven cross flows are very much in evidence in the finned tubes mentioned above.

Thomas et al. [151] reported a simple resolution of the manufacturing problem. A smooth tube was wrapped with wire so that surface tension pulls the condensate to the base of the wire. The spaces between wires then act as runoff channels. Tests with ammonia indicated a condensing coefficient about 3 times that predicted for the smooth tube.

Several studies considered commercial deep spirally fluted tubes for horizontal single-tube and tube-bundle condensers [152–154]. Commercially available configurations bear some resemblance to the Gregorig profile shown in Fig. 11.21. The derived condensing coefficients (envelope basis) range from essentially no improvement to over 300 percent above the plain-tube values.

Carnavos [155] reported typical *overall* improvements that can be realized with a variety of commercially available enhanced horizontal condenser tubes. The heat flux for single 130-mm-long tubes, in most cases with outside diameters of 19 mm, is plotted in Fig. 11.22 against ΔT_{lm} for 12 tubes qualitatively described in the accompanying table. The overall heat transfer performance gain of the enhanced tubes over the smooth tube is as high as 175 percent. Internal enhancement is a substantial contributor to the overall performance, since the more effective external enhancements produce a large decrease in the shell-side thermal resistance.

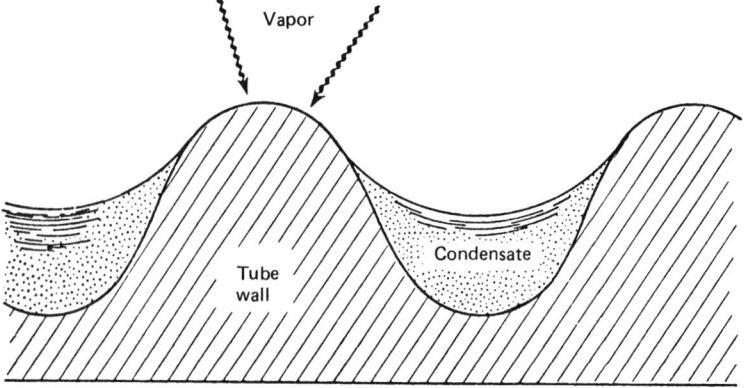

FIGURE 11.21 Profile of the condensing surface developed by Gregorig [150].

These results for a refrigerant, and the data of Marto et al. [154] and Mehta and Rao [156] for steam, provide good practical guidance for the use of enhanced tubes in surface condensers.

Some detailed results are available for condensing in bundles of enhanced tubes. Withers and Young [157] found, for example, that the vertical row effect for corrugated tubes was different from that for bare tubes; in particular, the enhanced tubes were less sensitive to the number of rows.

The enhancement of vertical condensers remains an area of high interest due to potential large-scale power and process industry applications, for example, desalination, reboilers, and OTEC power plants. Tubes with exterior longitudinal fins or flutes, spiral flutes, and flutes on both the interior and exterior (doubly fluted) have been developed and tested. The common objectives are to use the Gregorig effect to create thin-film condensation at the tips of the flutes and to drain effectively.

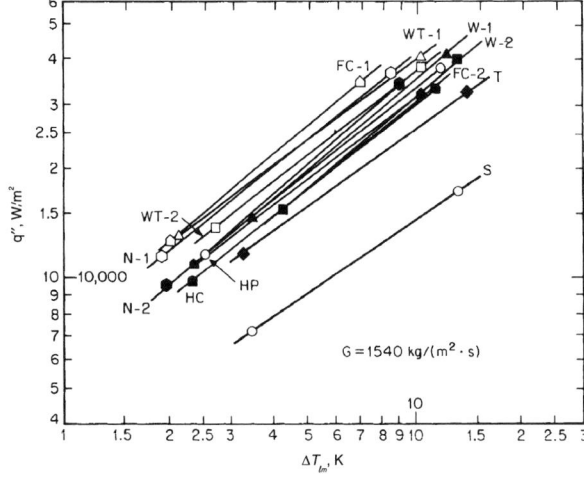

Tube	Outside	Inside
S	smooth	smooth
WT-1	integral helical fins	ribbed inside
WT-2	integral helical fins	ribbed inside
W-1	integral helical fins	plain inside
W-2	integral helical fins	plain inside
HC	interrupted helical fins	plain inside
HP	integral helical fins	plain inside
T	deep spiral flutes	deep spiral flutes
N-1	continuous trapezoidal flutes	helical ribs
N-2	continuous trapezoidal flutes	plain inside
FC-1	trapezoidal pin fins	helical ribs
FC-2	trapezoidal pin fins	plain inside

FIGURE 11.22 Overall performance of condenser tubes tested by Carnavos [155].

Vertical wires, loosely attached and spaced around the tube circumference, provide a simple realization of the desired profile. Thomas reported increases in heat transfer coefficient of up to 800 percent for circular wires [158]. Square wires were found to have a greater condensate-carrying capacity than circular wires of the same dimension [159].

The study of Mori et al. [160] represents a good example of the type of sophisticated analysis that can be performed to obtain the optimum geometry. According to that numerical analysis, the optimum geometry is characterized by four factors: sharp leading edge, gradually changing curvature of the fin surface from tip to root, wide grooves between fins to collect condensate, and horizontal disks attached to the tube to periodically strip off condensate. The recommended geometry is shown in Fig. 11.23. Figure 11.24 presents typical results that illustrate the character of the optimum. The periodic removal of condensate resolves the drainage

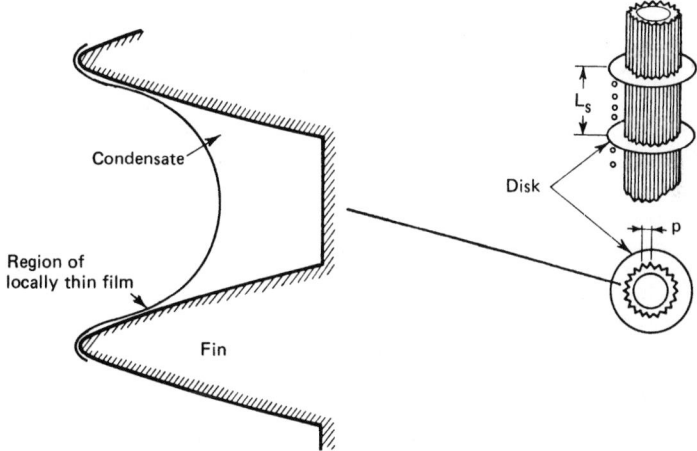

FIGURE 11.23   Recommended flute profile and condensate strippers according to Mori et al. [160].

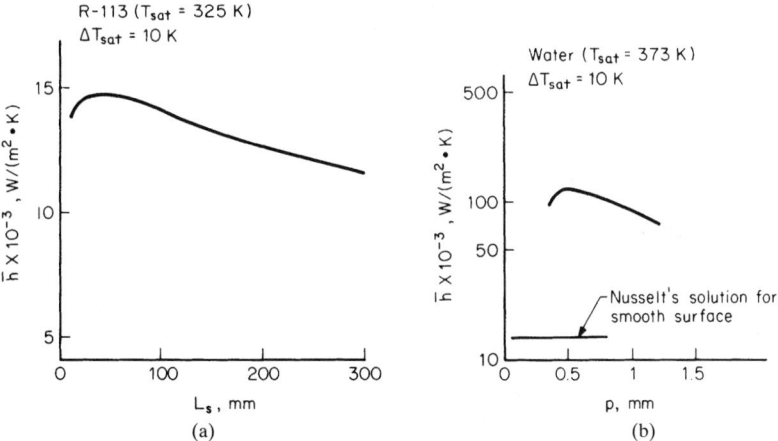

FIGURE 11.24   Optimization of stripper distance and flute pitch for configuration shown in Fig. 11.23 [160]. (*a*) average heat transfer coefficient versus stripper distance; (*b*) average heat transfer coefficient versus flute pitch.

problem with uniform axial geometry. Ideally, the flute size should be changed axially to allow for condensate buildup; however, this compounds the manufacturing difficulty.

Barnes and Rohsenow [161] present a simplified analytical procedure for determining the performance of vertical fluted-tube condensers. For a sine-shaped flute, the average condensing heat transfer coefficient depends on tube geometry approximately as follows:

$$\overline{h} \propto \frac{a^{0.231}}{pL_s^{0.0774}} \tag{11.13}$$

for a given fluid and ΔT. The recommended design procedure includes guidelines for placing the strippers. The design can be easily carried out with a hand calculator.

An important large-scale test of a doubly fluted tube for OTEC condenser service was reported by Lewis and Sather [162]. The tube is the same one noted earlier under enhancement of falling-film evaporation [136]. The ammonia-side heat transfer coefficient was enhanced several times to a value of 3730 Btu/(h·ft²·°F), or 21,181 W/(m²·K).

A major study of condensing on the outside of vertical enhanced tubes has been carried out at Oak Ridge National Laboratory in connection with geothermal Rankine cycle condensers. About 12 tubes were tested with ammonia, isobutane, and various fluorocarbons. The report by Domingo [163] on R-11 concluded that the best surface was the axially fluted tube, followed, in order, by the deep spirally fluted tube, spiral tubes, and roped tubes. The composite (vapor and tube wall) heat transfer coefficient was as much as 5.5 times the smooth-tube value. This high performance was further improved to a factor of 7.2 by using skirts to periodically drain off the condensate.

A guide to the *overall* performance of a wide variety of typical vertical evaporator tubes with condensing outside and vaporization inside is given in Fig. 11.25. This survey by Alexander and Hoffman [164] was specifically directed at vertical-tube evaporators for desalination systems. It is seen that the best surfaces yield increases in overall coefficient up to 200 percent.

The offset-strip-fin compact heat exchanger configuration has also been suggested for OTEC condensers. The flow situation is similar to that of the evaporator noted in Fig. 11.19. A

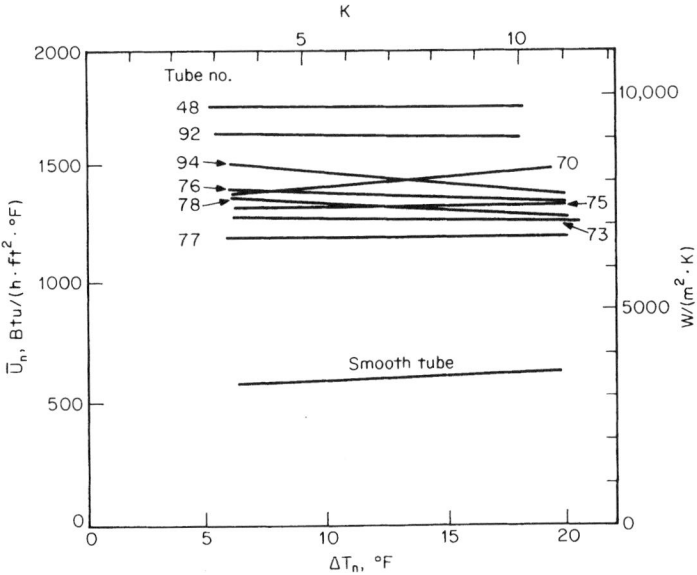

FIGURE 11.25   Heat transfer enhancement in vertical tube evaporator (tubes 48, 70, 94: axial flutes; others: spiral flutes) [164].

composite ammonia-side heat transfer coefficient of 4600 Btu/(h·ft²·°F), or 26,122 W/(m²·K), was reported.

Vrable et al. [165] studied horizontal in-tube forced-flow condensation of R-12 in internally finned tubes. The heat transfer coefficient (envelope basis) was increased by about 200 percent. Reisbig [166] also condensed R-12 (with some oil present) in internally finned tubes having an increased area of up to 175 percent. The nominal heat transfer coefficient was increased by up to 300 percent. Royal and Bergles [167, 168] presented heat transfer and pressure drop data, with correlations, for condensation of steam inside tubes with straight or spiraled fins. Increases of up to 150 percent in average coefficients were observed for complete condensation. Luu and Bergles [169, 170] tested similar tubes with complete condensation of R-113 and found that average coefficients were elevated by up to 120 percent. Grooved tubes were also effective, with increases of 250 percent being reported for longitudinal grooves with steam at high velocities [171] and increases of 100 percent for spiral grooves with R-113 [172]. The relative heat transfer–pressure drop performance of the grooved tubes is considered superior to that of the finned tubes.

The provisionally recommended design equation for steam is [167]

$$\overline{h}_i = 0.0265 \frac{k_l}{D_h} \left(\frac{G_e D_h}{\mu_l} \right)^{0.8} \mathrm{Pr}_l^{0.33} \left[160 \left(\frac{H^2}{lD} \right)^{1.91} + 1 \right] \tag{11.14}$$

where

$$G_e = G \left[(1 - \overline{x}) + \overline{x} \sqrt{\frac{\rho_l}{\rho_g}} \right]$$

For refrigerants the design equation is [169]

$$\overline{h}_i = 0.024 \frac{k_l}{D_h} \left(\frac{G D_h}{\mu_l} \right)^{0.8} \mathrm{Pr}_l^{0.43} \left(\frac{H^2}{lD} \right)^{-0.22} \times \frac{1}{2} \left[\left(\frac{\rho}{\rho_m} \right)_{in}^{0.5} + \left(\frac{\rho}{\rho_m} \right)^{0.5} \right] \tag{11.15}$$

where

$$\frac{\rho}{\rho_m} = 1 + \frac{\rho_l - \rho_g}{\rho_g} \overline{x}$$

Microfin tubes are widely used for in-tube convective condensation, as well as the evaporation applications noted in the previous subsection.

DISPLACED ENHANCEMENT DEVICES

Single-Phase Flow

A rather large variety of tube inserts falls into the category of displaced enhancement devices. The heated surface is left essentially intact, and the fluid flow near the surface is altered by the insert, which might be metallic mesh, static mixer elements, rings, disks, or balls. Laminar heat transfer data for uniform-wall-temperature tubes and uniformly heated tubes are plotted in Figs. 11.26 and 11.27, respectively. The isothermal friction factors are plotted in Fig. 11.28.

Koch [4] employed suspended rings and disks as inserts, as well as tubes packed with Raschig rings and round balls. The disks give maximum enhancement with moderate increases in friction factors, as indicated in Figs. 11.26 and 11.28 (curve d). Enhancement of heat transfer with rings and round balls is quite comparable to that with disks, but rings and balls increase the friction factor by more than 1600 percent (curves c and d). For further comments on packed tubes, see the chapter on heat transfer in fluidized and packed beds.

Sununu [173] and Genetti and Priebe [174] used Kenics static mixers for heating of viscous oils (Figs. 11.26 and 11.28, curves e and f). These mixers consist of 360° segments of twisted tapes; every second element is inverted and the segments are tack-welded together. The

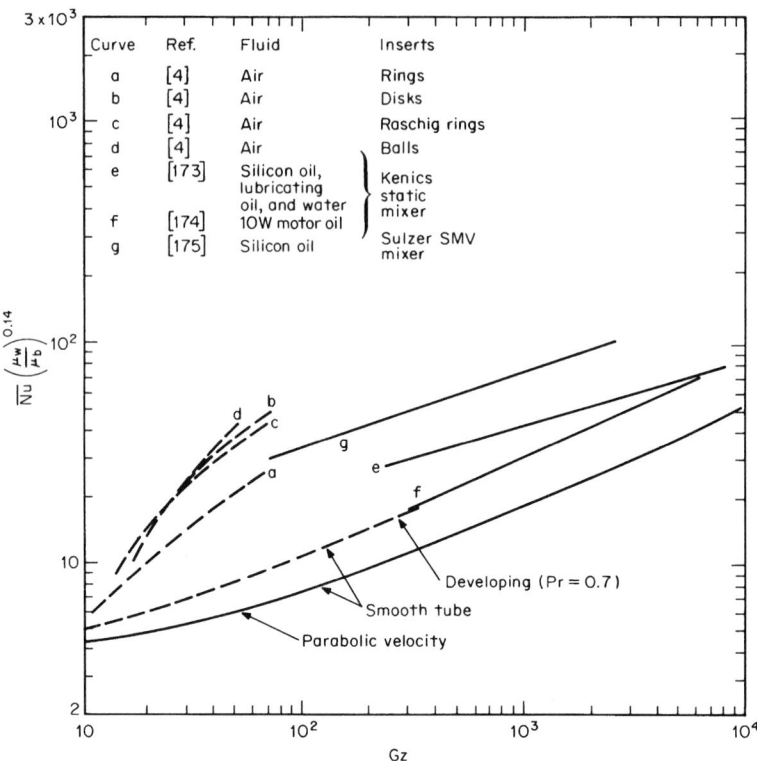

FIGURE 11.26 Representative heat transfer data for displaced promoters in tubes with uniform wall temperature.

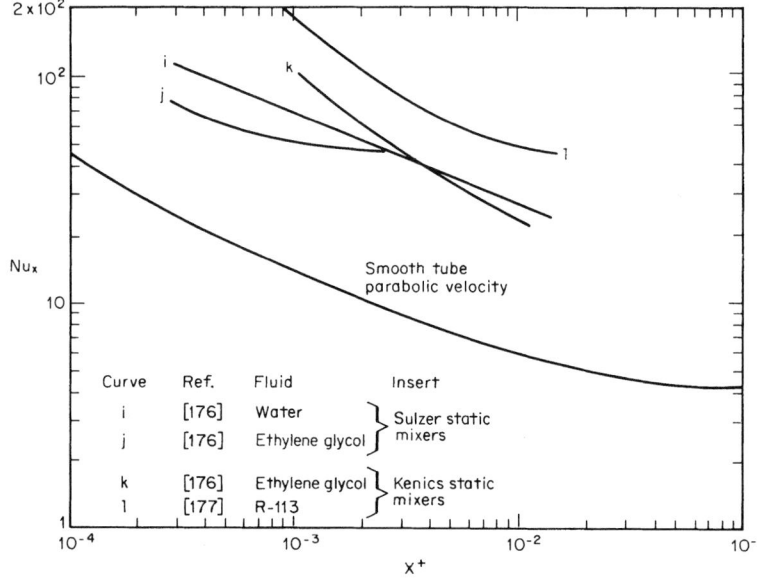

FIGURE 11.27 Representative heat transfer data for displaced promoters with uniform heat flux.

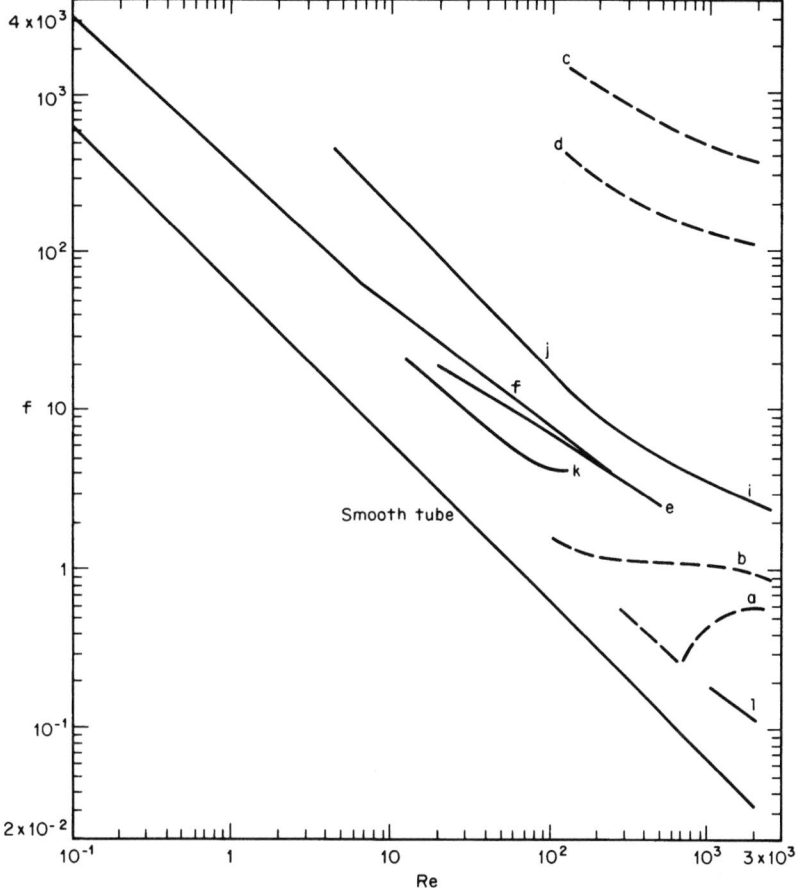

FIGURE 11.28   Isothermal friction factor data corresponding to tests noted in Figs. 11.26 and 11.27.

enhancement of heat transfer is about 150 to 200 percent, but the increase in friction factor is almost 900 percent. Sununu proposed a correlation for Nu, but his data exhibit large scatter about the correlation. Genetti and Priebe have also correlated their heat transfer data, with more success.

Van der Meer and Hoogendoorn [175] used Sulzer mixers for the heating of silicon oil (curves g and h). Each mixer element consists of several layers of corrugated sheet. An increase in heat transfer coefficient of about 400 percent is reported.

In the case of uniformly heated tubes, very high heat transfer coefficients have been obtained with the SMV Sulzer mixer (Fig. 11.27, curves i and j). Comparable heat transfer enhancement is also obtained with Kenics static mixers.

Many companies around the world are involved in the manufacture of static mixers for liquids, to promote either heat transfer or mass transfer. The variety of these mixers, their construction, and other characteristics are described in a comprehensive review article by Pahl and Muschelknautz [178]. There are no broad-based correlations available because of the many geometric arrangements and the strong influence of fluid properties or heating conditions.

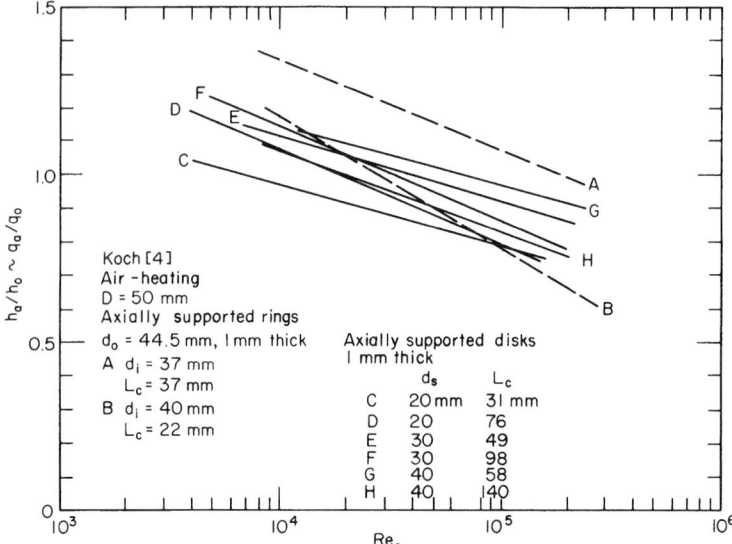

FIGURE 11.29   Performance of tubes with ring or disk inserts (FG-2a).

Similar inserts or packings have been used for turbulent flow; however, this application is usually considered only for short sections since the pressure drop is so high. The problem is illustrated by the results of Koch [4], who placed thin rings or disks in a tube. The typical basic data shown in Fig. 11.2 translate to performance data in Fig. 11.29, where it is seen that rings are effective only in the lower Reynolds number range. These data, as well as the data of Evans and Churchill [179] (disks and streamlined shapes in tubes), Thomas [180] (rings in annuli), and Maezawa and Lock [181] ("Everter" and disk inserts), indicate that these inserts are not particularly effective for turbulent flow.

Mesh or spiral brush inserts were used by Megerlin et al. [182] to enhance turbulent heat transfer in short channels subjected to high heat flux. The largest recorded improvements in turbulent heat transfer coefficients were obtained—up to 8.5 times; however, the pressure drop was up to 2800 times larger. In general, it appears that these displaced enhancement devices are useful in very few practical turbulent situations, for reasons of pressure drop, plugging or fouling, and structural considerations.

Flow Boiling

Janssen and Kervinen [93] reported on bulk boiling CHF with displaced turbulence promoters. Flow-disturbing rings were located on the outer tube of an annular test section. Figure 11.30 shows that critical heat fluxes for quality boiling with the rough liner were as much as 60 percent greater than those for the smooth liner. This is to be expected, since the rings strip the liquid from the inactive surface, thereby increasing the film flow rate on the heated surface. Moeck et al. [183] performed an extensive investigation of CHF for annuli with rough outer tubes. Steam-water mixtures were introduced at the test section inlet so as to obtain high outlet steam qualities. It was found that the critical heat flux increased as the roughness height (1.3 mm maximum) increased and spacing (38 to 114 mm) decreased, with a maximum increase of over 600 percent based on similar inlet conditions. The pressure drop with the

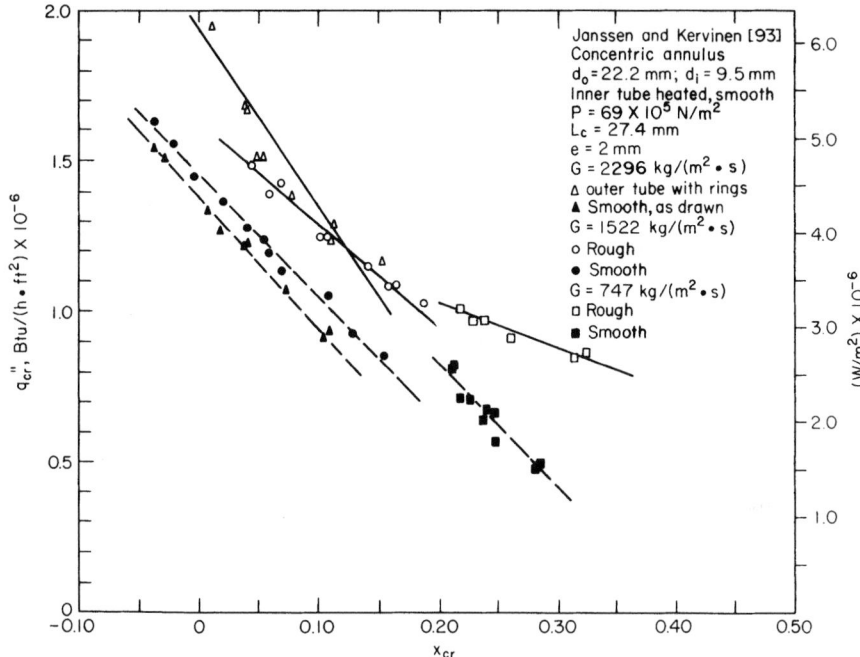

FIGURE 11.30 Effect of displaced promoters on CHF for bulk boiling of water.

most optimum promoter was about 6 times the smooth-annulus value for similar inlet conditions at the critical condition.

Rough liners were also found to produce significant increases in critical power for a simulated boiling-water reactor (BWR) rod bundle [184]. As reported by Quinn [185], rings of stainless steel wire, $e = 1.12$ mm and $p = 25.4$ mm, were spot-welded to the channel wall of a two-rod assembly. Both CHF and film-boiling heat transfer coefficients were improved.

Ryabov et al. [186] summarized a major study of increasing critical power in rod bundles by the use of special spacers, inserts, etc. A comprehensive review of the effects of spacing devices on CHF was presented by Groeneveld and Yousef [187]. Figure 11.31 qualitatively illustrates the expected effect. The majority of the studies cited report beneficial effects of spacing devices on CHF; however, several investigations also report detrimental effects.

Megerlin et al. [182] reported subcooled boiling data for tubes with mesh and brush inserts. Critical heat fluxes were increased by about 100 percent; however, wall temperatures were very high on account of the onset of partial film boiling.

Condensing

Azer et al. [188] reported data for condensation in tubes with Kenics static mixer inserts. Substantial improvements in heat transfer coefficients were reported; however, the increases in pressure drop were very large. A subsequent paper [189] presents a surface renewal model for the condensing heat transfer coefficient. With one experimentally determined constant, the correlation derived from this model is in good agreement with the experimental data.

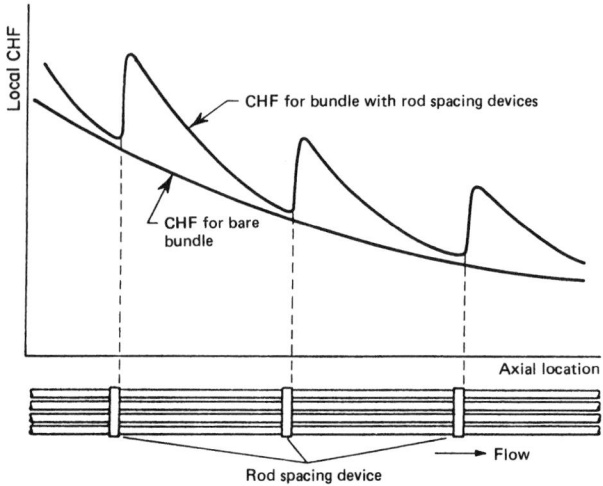

FIGURE 11.31   Effect of rod spacing devices on CHF [187].

SWIRL-FLOW DEVICES

Single-Phase Flow

Swirl-flow devices have been used for more than a century to improve heat transfer in industrial heat exchangers. These devices include inlet vortex generators, twisted-tape inserts, and axial-core inserts with screw-type windings. The enhancement is attributable to several effects: increased path length of flow, secondary flow effects, and, in the case of the tapes, fin effects. Phenomenologically, these devices are part of the general area of confined swirl flows, which also includes curved and rotating systems. The survey by Razgaitis and Holman [190] provides a comprehensive discussion of the entire field. See Sec. H of Ref. 369 for a discussion of curved ducts and coils.

Data for uniform-wall-temperature heating are plotted in Fig. 11.32, and the isothermal friction factors are plotted in Fig. 11.33. Twisted tapes and propellers were used by Koch [4] to heat air (curves a–d). Propellers produce higher heat transfer coefficients than twisted tapes; however, this enhancement is at the expense of a rather large increase in friction factor, as seen in Fig. 11.33. Up to Re = 200, the friction factor for the twisted tape is the same as that for the empty half-tube ($y = \infty$). The twisted-tape data of Marner and Bergles [114] with ethylene glycol exhibit an enhancement of about 300 percent above the smooth-tube values. Swirl at the pipe inlet does not produce any effective enhancement [192].

The following correlation is recommended for fully developed laminar flow in a uniformly heated tube [191]:

$$\text{Nu}_i = 5.172[1 + 5.484 \times 10^{-3} \, \text{Pr}^{0.7} \, (\text{Re}_i/y)^{1.25}]^{0.5} \tag{11.16}$$

Note that the correlation was established for a tape with no heat transfer; considerable increases in heat transfer are predicted with tapes that act as effective fins [193]. The correlation does not seem to be applicable to heating or cooling with a constant wall temperature [114], but more recently a correlation has been proposed for that boundary condition [378]:

$$\text{Nu}_m = 4.612(\{(1 + 0.0951\text{Gz}^{0.894})^{2.5} + 6.413 \times 10^{-9}(\text{Sw} \cdot \text{Pr}^{0.391})^{3.835}\}^{2.0}$$

$$+ 2.132 \times 10^{-14}(\text{Re Ra})^{2.23})^{0.1}\left(\frac{\mu_b}{\mu_s}\right)^{0.14} \tag{11.17}$$

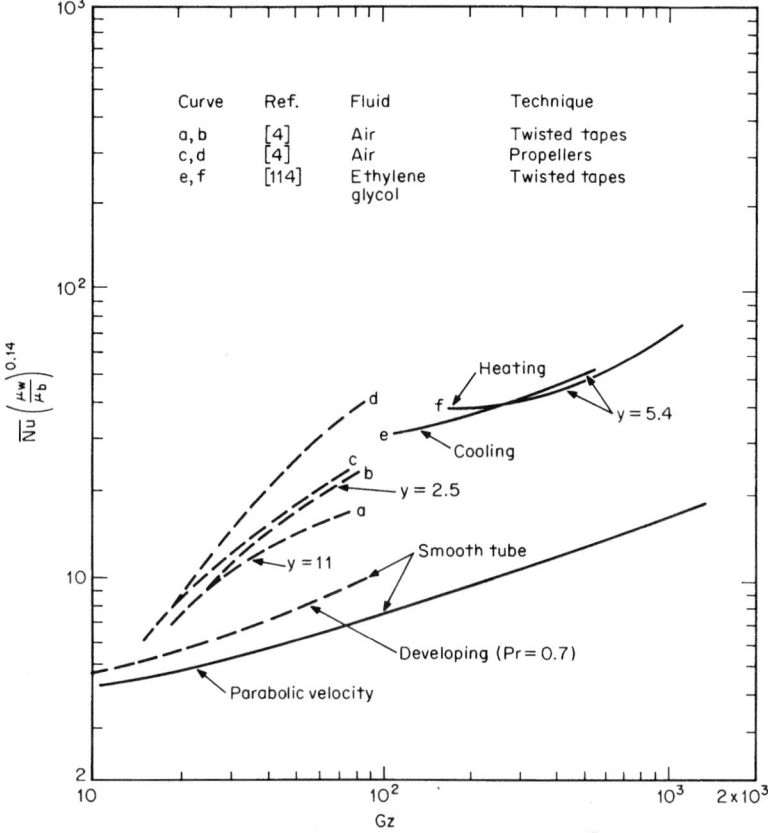

FIGURE 11.32 Representative heat transfer data for swirl-flow devices with uniform wall temperature.

where a new dimensionless parameter $Sw = Re_{sw}/\sqrt{y}$ makes for a good correlation. Strong evidence exists that there is an influence of free convection (horizontal tube) at low Re and low Sw; thus, the term involving Ra is included. A turbulent flow correlation is also available [379].

At Re < 100, the isothermal friction factors can be approximated by the expression for a semicircular tube:

$$f_i = 42.2 Re_i^{-1} \tag{11.18}$$

This line lies slightly below the experimental result for a nontwisted tape (Fig. 11.33, curve r) because the tape thickness is not included in the analysis. There is no substantial increase in friction factor above the empty half-tube results until Re $\cong$ 300. A lengthy set of correlations for the higher Reynolds number region is presented by Shah and London [194].

Turbulent-flow heat transfer in uniformly heated tubes with twisted-tape inserts has been correlated by [195]

$$Nu_h = F\left\{0.023\left[1 + \left(\frac{\pi}{2y}\right)^2\right]^{0.4} Re_h^{0.8} Pr^{0.4} + 0.193\left[\left(\frac{Re_h}{y}\right)^2 \frac{D_h}{D_i} \frac{\Delta\rho}{\rho} Pr\right]^{1.3}\right\} \tag{11.19}$$

This correlation is based on the premise that the average heat transfer coefficient can be represented essentially as a superposition of heat transfer coefficients for spiral convection and

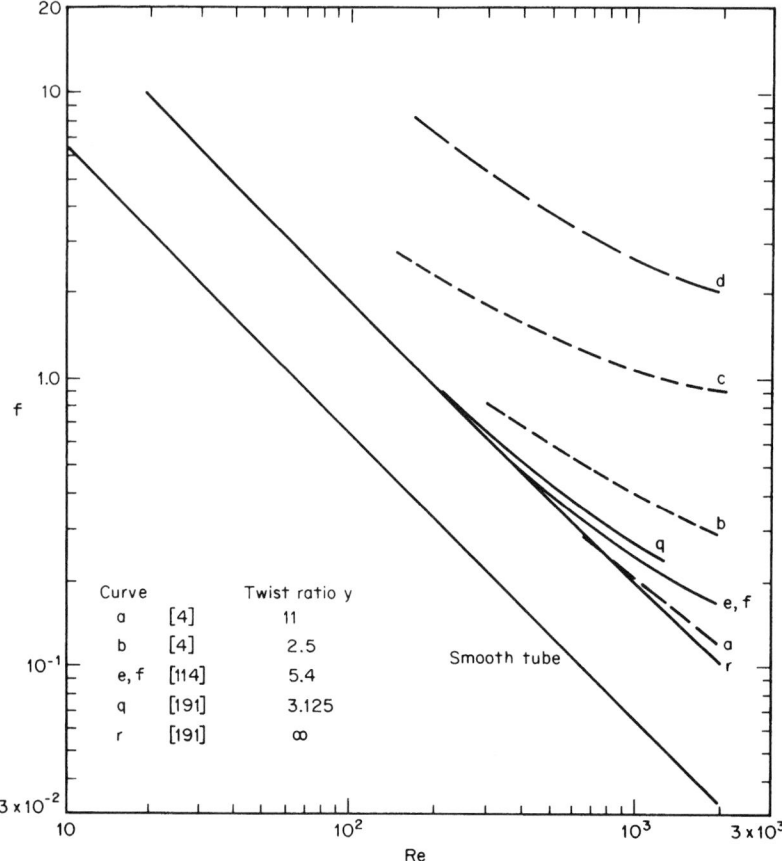

FIGURE 11.33 Isothermal friction factor data for twisted-tape inserts.

centrifugal convection. The fin factor F, which represents the ratio of total heat transfer to the heat transferred by the walls alone, can be estimated from conduction calculations. The value of F is close to unity for a loose tape fit, and may be as high as 1.25 for a tight tape fit. Equation 11.19 is accurate for water heating and cooling (with the second term deleted) and for much gas data as well. An equation specifically for gases, which accounts for large radial temperature gradients, is given by Thorsen and Landis [196]. Isothermal friction factors are given by the following expression [195]:

$$f_{h,\text{iso}} = 0.1276y^{-0.406}\,\text{Re}_h^{-0.2} \tag{11.20}$$

Diabatic friction factors are obtained in the usual manner by applying a viscosity- or temperature-ratio correction. For heating of water at bulk temperatures below 200°F (93.3°C), the following correction to Eq. 11.20 has been suggested [195]:

$$f_h = f_{h,\text{iso}}\left(\frac{\mu_w}{\mu_b}\right)^{0.35(D_h/D)} \tag{11.21}$$

Thorsen and Landis [196] have determined a temperature-ratio correction factor for their correlation of the friction factor data for air.

It is appropriate to conclude this discussion of single-phase data by presenting a constant-pumping-power comparison. Actual friction and heat transfer data from the various investigations are utilized in the computations leading to Fig. 11.34 for air and Fig. 11.35 for water. Because of the diversity in heat transfer and friction data, the performance curves exhibit rather wide scatter; however, the general consensus is that the tapes provide a substantial improvement in performance. Tape twist is, of course, an important parameter; however, even for similar geometries, differences are to be expected due to variations in the fin effect and centrifugal convection effect. The data for axial core assemblies for both air (lines A and B) and water (lines K–M) suggest that the tightest twist ratio is not necessarily the best.

Several studies (e.g., Ref. 197) have considered tapes that do not extend the length of the heated section. For uniformly heated tubes, intermittent tapes do not perform as well as full-

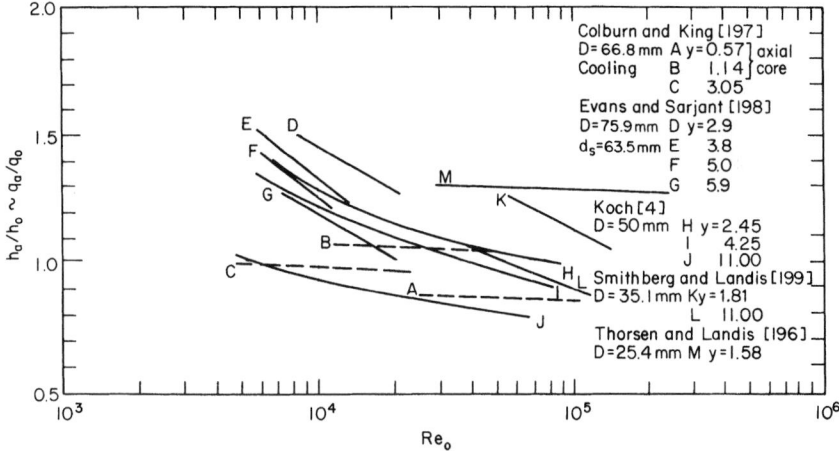

FIGURE 11.34 Performance of twisted-tape inserts with air (FG-2a).

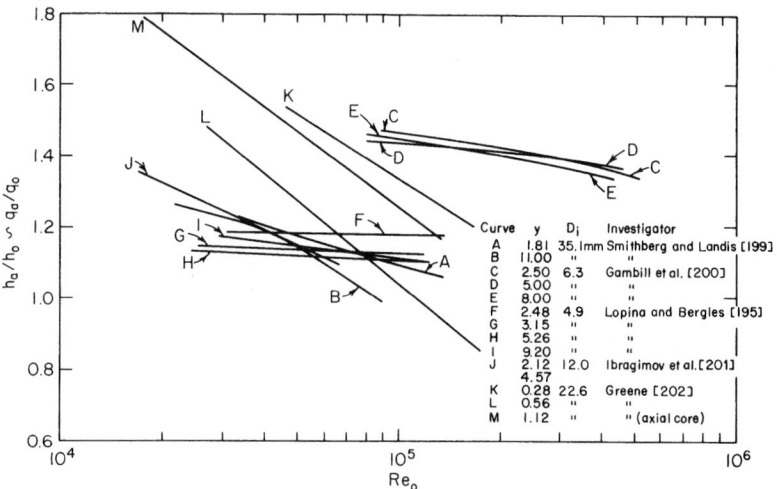

FIGURE 11.35 Performance of twisted-tape inserts with water (FG-2a).

length tapes and are not used. However, intermittent tapes are particularly useful in cases involving nonuniform heat fluxes. The tapes can be placed at the hot-spot location (assuming they can be secured), thus producing the desired improvement in heat transfer with little effect on the overall pressure drop. This technique has been used to eliminate the burnouts caused by degeneration in heat transfer in certain supercritical boiler systems [203].

Boiling

A variety of devices have been proposed to enhance flow boiling by imparting a swirling or secondary motion to the flow. Inlet vortex generators of the spiral ramp or tangential slot variety have been used to accommodate very large heat fluxes for subcooled flow boiling of water. One of the highest fluxes on record, $q''_{cr} = 1.73 \times 10^8$ W/m² or 5.48×10^7 Btu/(h·ft²), has been obtained with this technique by Gambill and Greene [204]. Inlet swirl is effective in increasing CHF for subcooled boiling of water in a tube [205] or in an annulus with a heated inner tube [206].

Twisted tapes are quite popular because of their simplicity and their adaptability to existing heat exchange equipment. They are ideal for hot-spot applications, since a short tape can cure the thermal problem with little effect on the overall pressure drop. Boiling curves for subcooled boiling with twisted tapes are similar to those for empty tubes [207]; however, CHF can be increased by up to 100 percent [200], as shown in Fig. 11.36. Because of a dra-

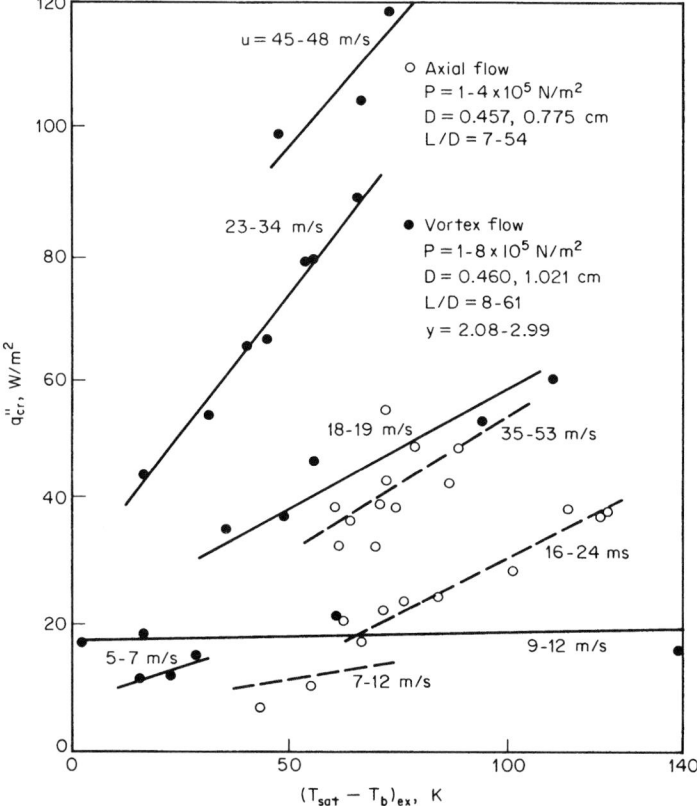

FIGURE 11.36 Influence of twisted-tape inserts on subcooled boiling CHF of water (data from Ref. 200).

matic reduction in the momentum contribution to the pressure drop in swirl flow [207], CHF for swirl flow is higher than that for straight flow at the same test section pumping power. This is demonstrated in Fig. 11.37. Loose-fitting tape inserts have been used by Sephton [208] in tubes that simulated vertical-tube evaporators for seawater desalination. These inserts are also effective for once-through vaporization of cryogenic fluids [209] or steam [210, 211], since all two-phase regimes are beneficially affected. Twisted-tape inserts have also been considered for rod clusters for eventual application to nuclear reactor cores [212].

Coiled-tube vapor generators have advantages in terms of packing and generally higher heat transfer performance. As indicated in the literature survey of Jensen [213], the enhancement of boiling is very sensitive to geometric and flow conditions. Modest improvements in h (circumferential average) for forced convection vaporization are obtained, with an increase in improvement as coil diameter is decreased. In the subcooled region, q_{cr}'' is lower than it is for a comparable straight tube; however, q_{cr}'' or x_{cr} is usually substantially higher than the straight-tube value at outlet qualities of 0.2 and higher (Fig. 11.38). The post-dryout heat transfer coefficient is also increased with helical coils.

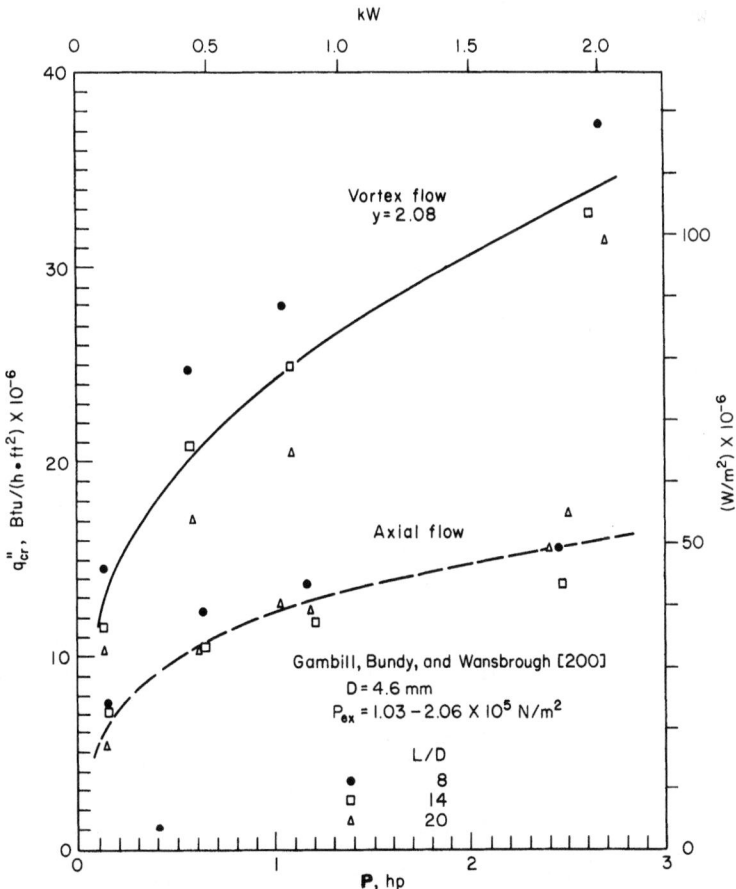

FIGURE 11.37 Dependence of subcooled CHF on pumping power for swirl and axial flow.

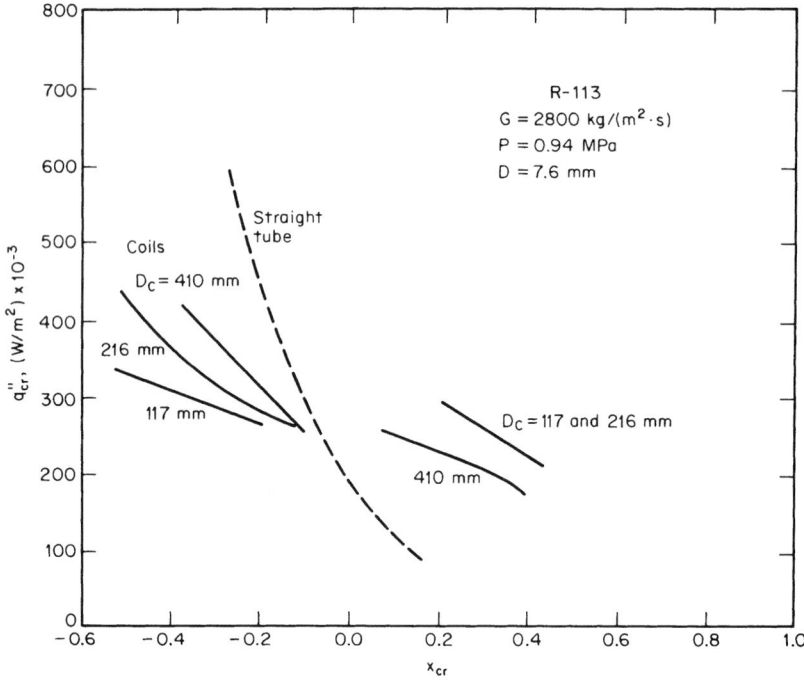

FIGURE 11.38   CHF for helical coils of various diameter compared with straight tube [214].

Condensing

Royal and Bergles [167, 168] found that twisted-tape inserts improved heat transfer coefficients for in-tube condensation of water by 30 percent; however, the pressure drop was quite high. Luu and Bergles [169, 170] report similar results for R-113. The following heat transfer correlations are recommended.

For Steam.

$$\bar{h}_i = 0.0265 \frac{k_l}{D_h}\left(\frac{G_e D_h}{\mu_l}\right)^{0.8} \mathrm{Pr}_l^{0.33}\left[160\left(\frac{H^2}{lD}\right)^{1.91} + 1\right] \tag{11.22}$$

where

$$G_e = G\left(\bar{x}\,\frac{\rho_l}{\rho_g} + 1 - \bar{x}\right)$$

For R-113.

$$\bar{h}_i = 0.024 \frac{k_e}{D_h} F\left(\frac{F_t G D_h}{\mu_l}\right)^{0.8} \mathrm{Pr}_l^{0.43} \times \left[\frac{(\rho/\rho_m)_{\mathrm{in}}^{0.5} + (\rho/\rho_m)_{\mathrm{ex}}^{0.5}}{2}\right] \tag{11.23}$$

where ρ/ρ_m is defined in connection with Eq. 11.15, F is the same factor used with Eq. 11.19, and

$$F_t = \frac{8y^2}{3\pi^2}\left\{\left[\left(\frac{\pi}{2y}\right)^2 + 1\right]^{1.5} - 1\right\}$$

Shklover and Gerasimov [215] used an interesting baffling technique to create a spiraling motion in the vapor condensing outside a tube bundle. With vapor velocities approaching sonic velocities, the condensing coefficients were high; however, no base data were included for comparison.

Condensation in coiled tubes was studied by Miropolskii and Kurbanmukhamedov [216], and condensation in tube bends was studied by Traviss and Rohsenow [217]. In both cases, modest increases in condensation rates were observed relative to straight tubes.

SURFACE-TENSION DEVICES

Within the context of the present classification of enhancement techniques, surface-tension devices are those that involve the application of relatively thick wicking materials to heated surfaces. This technique is distinct from that applied to heat pipes (see the chapter on heat pipes). Wicking is usually considered for situations where coolant is unable to reach the heater surface without the wicking material, for example, the cooling of electronics in aircraft undergoing violent maneuvers or in spacecraft operating in a near-zero-gravity environment. Wicking has also been shown to be effective in enhancing boiling heat transfer from submerged surfaces.

When a heater was completely enclosed with wicking and submerged, Allingham and McEntire [218] found that the heat transfer coefficient for saturated pool boiling was improved at low heat fluxes, but that the reverse was true at moderate fluxes. Costello and Redeker [219] investigated higher heat fluxes and found that the heat flux corresponding to a temperature excursion was only about 10 percent of the normal critical heat flux. It was concluded that proper vapor venting was necessary to avoid blockage of the supplied liquid flow. Subsequent tests [220] indicated that the critical heat flux could be raised by as much as 200 percent when the wicking was not too dense and a narrow channel was maintained at the top for easy escape of vapor. Corman and McLaughlin [221] presented extensive data for wick-augmented surfaces that qualitatively confirm these observations.

Gill [222] spiraled wicking around cylindrical heaters. In all cases, boiling commenced at a superheat of about 1 K, apparently because the wicking provided large nucleation sites. The boiling curve was generally displaced to lower superheat than the normal curve, to a degree dependent on the diameter and pitch of the wicking. No significant change in the critical heat flux was observed with the wicking. The stable film-boiling region was investigated by quenching a copper calorimeter in liquid nitrogen. It appeared that capillary action effectively transported liquid through the vapor film to the heated surface, since the heat transfer coefficient was increased by about 100 percent.

ADDITIVES FOR LIQUIDS

Solid Particles in Single-Phase Flow

Watkins et al. [223] studied suspensions of polystyrene spheres in forced laminar flow of oil. Maximum improvements of 40 percent were observed.

Gas Bubbles in Single-Phase Flow

Tamari and Nishikawa [224] observed increases in average heat transfer coefficient of up to 400 percent when air was injected into either water or ethylene glycol. The injection point was at the base of the vertical heated surface, and up to three injection nozzles were used. Other studies are reviewed by Hart [225], who proposed a correlation to fit his own data as well as the results of other investigators for free convection enhancement.

Kenning and Kao [226] noted heat transfer increases of up to 50 percent when nitrogen bubbles were injected into turbulent water flow. A similar level of enhancement was observed by Baker [227], who created slug flow in small rectangular channels with simulated micro-electronic chips on one of the wide sides.

Surface degassing, which is initiated when wall temperatures are below the saturation temperature, produces an agitation comparable to that of injected bubbles or even of boiling. Behar et al. [228] found that the wall superheat for saturated pool boiling of nitrogen-pressurized *meta*-terphenyl was reduced by as much as 50°F (27.8°C), while in subcooled flow boiling the reduction was as much as 30°F (16.7°C). In general, the surface degassing is effective only at lower heat fluxes; once the nucleate boiling becomes well established, there is negligible reduction in the wall superheat.

Liquid Additives for Boiling

Trace liquid additives have been extensively investigated in pool boiling and, to a lesser extent, in subcooled flow boiling. A great many additives have been investigated, and some have been found to produce substantial heat transfer improvements. With the proper concentration of certain additives (wetting agents, alcohols), increases of about 20 to 40 percent in the heat transfer coefficient for saturated nucleate pool boiling can be realized [229–233]. This occurs in spite of thermodynamic analyses for boiling of binary mixtures that indicate that boiling performance should be decreased [234].

Most additives increase CHF, but the concentration of the additive and the heater geometry have major effects on the enhancement. The typical results of van Stralen et al. [235, 236], as shown in Fig. 11.39, indicate a sharp increase in CHF at some low concentration of 1-pentanol and rather rapid decrease as the concentration is increased. The optimum concen-

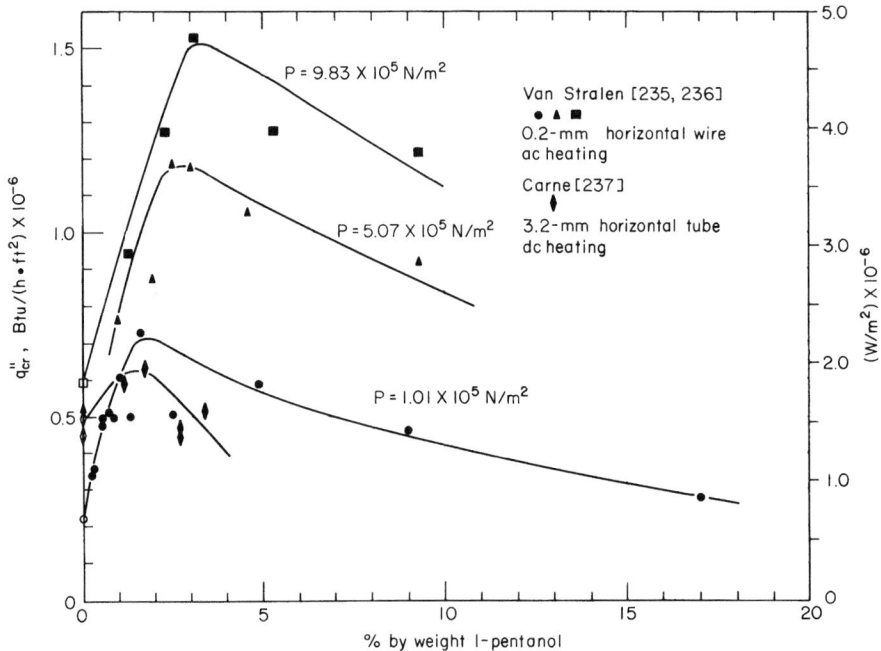

FIGURE 11.39   Dependence of CHF on volatile additive concentration for saturated pool boiling.

tration varies with the mixture, and, to some extent, with the pressure. For a similar water-pentanol system, Carne [237] obtained an increase of only 25 percent in CHF with a 3.2-mm heater; this increase is small compared to the 240 percent increase that van Stralen obtained with a 0.2-mm heater. Van Stralen's extensive program of testing and modeling of mixtures has been extended to film boiling [238]. In this study, it was found that a 4.1 weight-percent mixture of 2-butanone in water improved coefficients by up to 80 percent.

Of related interest are the results of Gannett and Williams [239], who found that small quantities of certain polymers dissolved in water increase nucleate boiling coefficients. As reported by Jensen et al. [240], oily contaminants generally decrease boiling coefficients; however, increases are observed under certain conditions.

In subcooled flow boiling, the improvement (if any) in heat transfer is modest. Leppert et al. [241] found that the main advantage of alcohol-water mixtures was an improvement in smoothness of boiling. The influence of a volatile additive on subcooled flow boiling CHF in tubes has been investigated by Bergles and Scarola [242]. As shown in Fig. 11.40, there is a distinct reduction in CHF at low subcooling with the addition of 1-pentanol. These trends have been confirmed recently [383]. On the other hand, Sephton [243] found that the overall heat transfer coefficient was doubled when a surfactant was added to seawater evaporating in a vertical (upflow) tube.

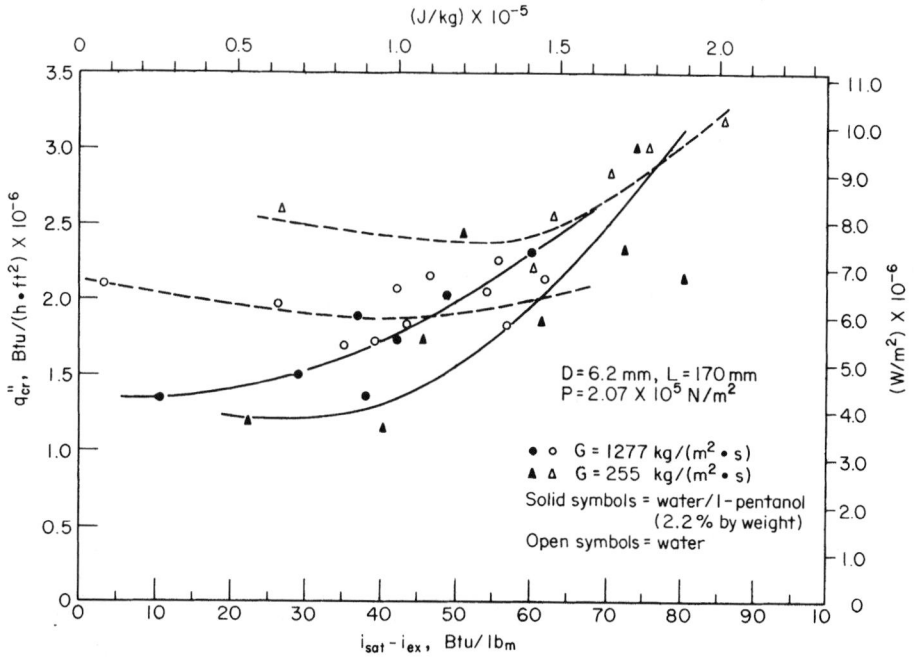

FIGURE 11.40   Influence of volatile additive on subcooled CHF [242].

In general, the improvements in heat transfer and CHF offered by additives are not sufficient to make them useful for practical systems. There are difficulties involved in maintaining the desired concentration, particularly when the additive is volatile. In many cases, the additives, even in small concentrations, are somewhat corrosive and require special piping or seals.

ADDITIVES FOR GASES

Solid Particles in Single-Phase Flow

Dilute gas-solid suspensions have been considered as working fluids for gas turbine and nuclear reactor systems. Solid particles in the micron-to-millimeter size range are dispersed in the gas stream at loading ratios W_s/W_g ranging from 1 to 15. The solid particles, in addition to giving the mixture a higher heat capacity, are highly effective in promoting enthalpy transport near the heat exchange surface. Heat transfer is further enhanced at high temperatures by means of the particle-surface radiation. A summary of typical data for air-solid suspensions is given in Fig. 11.41.

Extensive experimental work was undertaken at Babcock and Wilcox to obtain detailed heat transfer and pressure drop information as well as operating experience. Summary articles by Rhode et al. [245] and Schluderberg et al. [246] elaborate on the conclusions of this work. Heat transfer coefficients for heating were improved by as much as a factor of 10 through the addition of graphite. The suspensions were also shown to be far superior to gas coolants on the basis of pumping power requirements, especially when twisted-tape inserts

Curve	Re	Particle	Size (microns)	Gas	d_p/D
A	18,000	Glass	60	Air	0.0011
B	18,000	Glass	120	Air	0.0027
C	19,000	Sand	230	Air	0.0060
D	19,000	Sand	80	Air	0.0021
E	15,000	Graphite	65	Air	0.0085
F	53,000	Zinc	40	Air	0.0005
G	53,000	Zinc	40	Air	0.0008
H		Graphite			
I	53,000	Zinc	40	Air	0.016
J	15,000	Al₂O₃	65	Air	0.0085
K	13,500	Glass	30	Air	0.0016
L	13,500	Glass	200	Air	0.0111

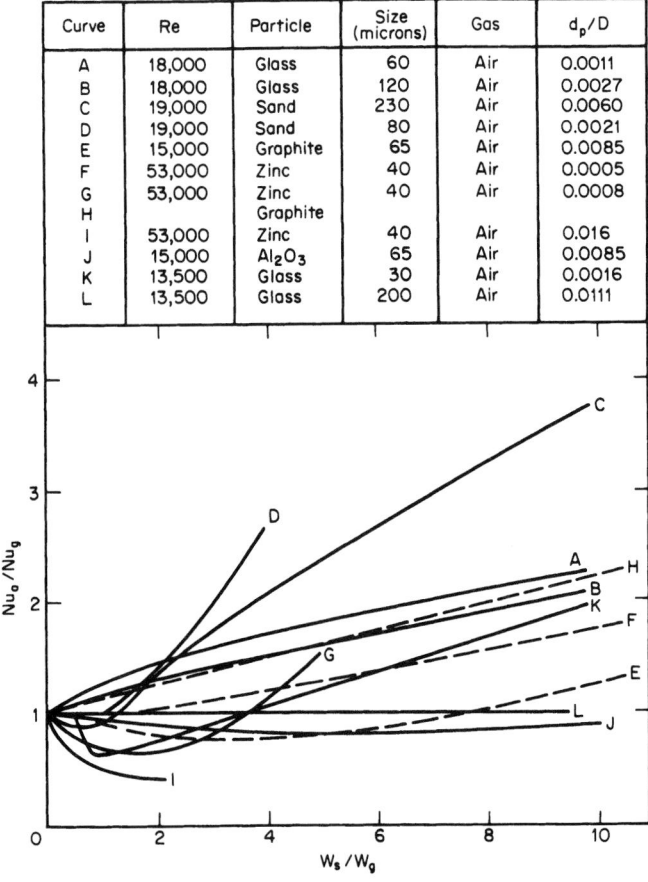

FIGURE 11.41 Composite heat transfer data for air-solid suspensions [244].

were used. There was relatively little settling, plugging, or erosion in the system. With helium suspensions, however, there was serious fouling of the loop coolers, which was attributed to brownian particle motion induced by the temperature gradient.

Abel and coworkers [247] demonstrated that the cold surface deposition is a very serious problem with micronized graphite. This occurred with both helium and nitrogen suspensions and could be alleviated only with very high gas velocities. An economic comparison was presented in terms of a system pumping power–to–heat transfer rate ratio as a function of gas flow rate. This comparison indicated that the pure gas was generally more effective than the suspension at both low and high gas flow rates. In all probability, the loop heater is very effective; however, this gain is offset by the low performance of the cooler.

A comprehensive analysis of much of the data for dilute gas-solid suspension was reported by Pfeffer et al. [248]. Correlations for both heat transfer coefficient and friction factor were developed. These investigators presented a feasibility study of using suspensions as the working fluid in a Brayton space power generation cycle [249]. A subsequent presentation of design information and guide to the literature is given by Depew and Kramer [250].

Fluidized beds represent the other end of the spectrum in terms of solids loading (this subject is covered in the chapter on heat transfer in fluidized and packed beds). The very considerable enhancement of heat transfer coefficients, up to a factor of 20 compared to pure gas flow at the same flow rate, has led to applications in such areas as flue gas heat recovery.

Liquid Drops in Single-Phase Flow

When liquid droplets are added to a flowing gas stream, heat transfer is enhanced by sensible heating of the two-phase mixture, evaporation of the liquid, and disturbance of the boundary layer.

Thomas and Sunderland [251] demonstrated that heat transfer coefficients can be increased by as much as a factor of 30 if a continuous liquid film is formed on the heated surface. A more realistic indication of practical enhancement was provided by Yang and Clark [252], who applied spray cooling to a compact heat exchanger core. The maximum improvement of 40 percent was attributed to formation of a partial liquid film and sensible heating of that film. In general, the large flow rate of liquid required tends to limit practical application of this technique.

MECHANICAL AIDS

Stirring

Attention is now directed toward active techniques that require direct application of external power to create the enhancement. Heating or cooling of a viscous liquid in batch processing is often enhanced by stirrers or agitators built into the tank. Uhl [253] presents a comprehensive survey of this area, including descriptions of hardware and experimental results.

For forced flow in ducts, a "spiralator" has been proposed. This consists of a loose-fitting twisted tape secured at the downstream end in a bearing. Penney [254] found that with heating of corn syrup, the heat transfer coefficient increased by 95 percent at 100 rev/min. A surface "scraper" based on this principle (see next subsection) has been patented [384].

The enhanced convection provided by stirring dramatically improves pool boiling at low superheat. However, once nucleate boiling is fully established, the influence of the improved circulation is small. Pramuk and Westwater [255] found that the boiling curve for methanol was favorably altered for nucleate, transition, and film boiling, with the improvement increasing as agitator speed increased.

Surface Scraping

Close-clearance scrapers for viscous liquids are included in the review by Uhl [253]. An application of scraped-surface heat transfer to air flows is reported by Hagge and Junkhan [256]; a tenfold improvement in heat transfer coefficient was reported for laminar flow over a flat plate. Scrapers were also suggested for creating thin evaporating films. Lustenader et al. [257] outline the technique, and Tleimat [258] presents performance data. The heat transfer coefficients are much higher than those observed for pool evaporation (without nucleate boiling).

Rotating Surfaces

Rotating heat exchanger surfaces occur naturally in rotating electrical machinery, gas turbine rotor blades, and other industrial systems. However, rotation may be deliberately applied to provide active enhancement of heat transfer.

Substantial increases in heat transfer coefficients have been reported for laminar flow in (1) a straight tube rotating about its own axis [259], (2) a straight tube rotating around a parallel axis [260], (3) a rotating circular tube [261], and (4) the rotating curved circular tube [262]. Reference 260 also presents an analysis of turbulent flow and data for both laminar and turbulent flows. Maximum improvements of 350 percent were recorded for laminar flow, but for turbulent flow the maximum increase was only 25 percent.

Tang and McDonald [263] found that when heated cylinders are rotated at high speeds in saturated pools, convective coefficients are so high that boiling can be suppressed. This constitutes an enhancement of pool boiling. Marto and Gray [264] found that critical heat fluxes were elevated in a rotating-drum boiler where the vaporization occurred at the inside of the centrifuged liquid annulus. With proper liquid feed conditions to the heated surface, exit qualities in excess of 99 percent were obtained.

Studies have been reported on condensation in a rotating horizontal disk [265], a rotating vertical cylinder [266], and a rotating horizontal cylinder [267]. Improvements are several hundred percent above the stationary case for water and organic liquids.

Weiler et al. [268] subjected a tube bundle, with nitrogen condensing inside the tube, to high accelerations (essentially normal to the tube axis). At $325g$, the overall heat transfer coefficient was increased by a factor of over 4.

SURFACE VIBRATION

Single-Phase Flow

It has long been recognized that transport processes can be significantly affected by inherent or induced oscillations. In general, sufficiently intense oscillations improve heat transfer; however, decreases in heat transfer have been recorded on both a local and an average basis. A wide range of effects is to be expected as a result of the large number of variables necessary to describe the vibrations and the convective conditions. Most of the research in this area was conducted more than 30 years ago.

In discussing the interactions between vibration and heat transfer, it is convenient to distinguish between vibrations that are applied to the heat transfer surface and those that are imparted to the fluid. The most direct approach is to vibrate the surface mechanically, usually by means of an electrodynamic vibrator or a motor-driven eccentric. In order to achieve an adequate amplitude of vibration, frequencies are generally kept well below 1000 Hz.

The predominant geometry employed in vibrational studies has been the horizontal heated cylinder vibrating either horizontally or vertically. When the ratio of the amplitude to the diameter is large, it is reasonable to assume that the convection process that occurs in the vicinity of the cylinder is quasi-steady. The heat transfer may then be described by conventional correlations for steady convection. In order to achieve the quasi-steady convection, characterized by $a/D_o \gg 1$, it is necessary to use small-diameter cylinders. The data presented in Fig. 11.42 illustrate this situation. These data fall into three rather distinct regions depending on the intensity of vibration: the region of low Re_v, where free convection dominates; a transition region, where free convection and the "forced" convection caused by vibration interact; and finally, the region of dominant vibrational forced convection. The data in this last region are in good agreement with a standard correlation for forced flow normal to a cylinder.

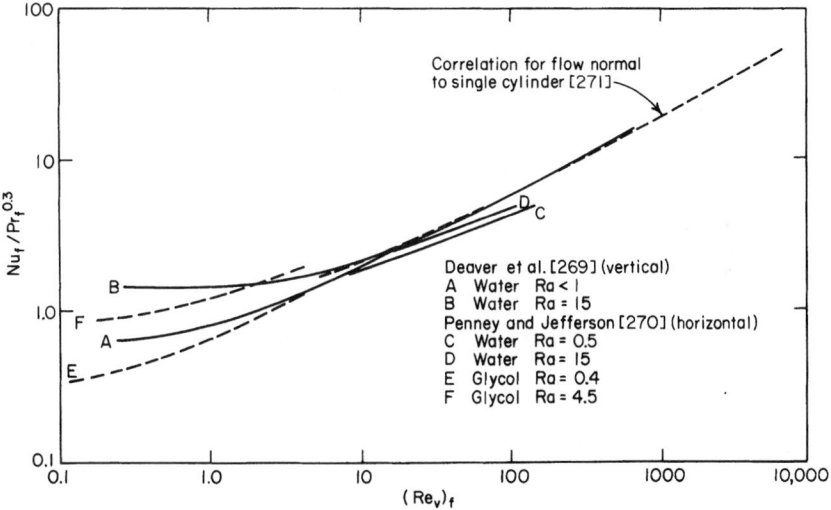

FIGURE 11.42   Influence of mechanical vibration on heat transfer from horizontal cylinders—$a/D_o \gg 1$.

When cylinders of large diameter, typically those found in heat exchange equipment, are used, a different type of behavior is expected. When $a/D_o \le 1$, there is no longer a significant displacement of the cylinder through the fluid to provide enthalpy transport. Natural convection should then dominate. However, where the vibrational intensity reaches a critical value, a secondary flow, commonly called acoustic or thermoacoustic streaming, develops; this flow is able to effect a net enthalpy flux from the boundary layer. Since the coordinates of Fig. 11.42 are inappropriate for description of streaming data, a simple heat transfer coefficient ratio is used in Fig. 11.43 to indicate typical improvements in heat transfer observed under these conditions. The heat transfer coefficient remains at the natural convection value until a critical intensity is reached and then increases with growing intensity. The rate of improvement in heat transfer appears to decrease as Re_v is increased. If these data were plotted on the coordinates of Fig. 11.42, they would lie below the quasi-steady prediction, except at very high Re_v, where they are generally higher.

Several studies have been done concerning the effects of transverse or longitudinal vibrations on heat transfer from vertical plates. Analyses indicate that laminar flow is virtually unaffected; however, experimental observations indicate that turbulent flow is induced by sufficiently intense vibrations. The improvement in heat transfer appears to be rather small, with the largest values of $h_a/h_o < 1.6$ [280].

From an efficiency standpoint, it is important to note that the improvements in heat transfer coefficient with vibration may be quite dramatic, but they are only relative to natural con-

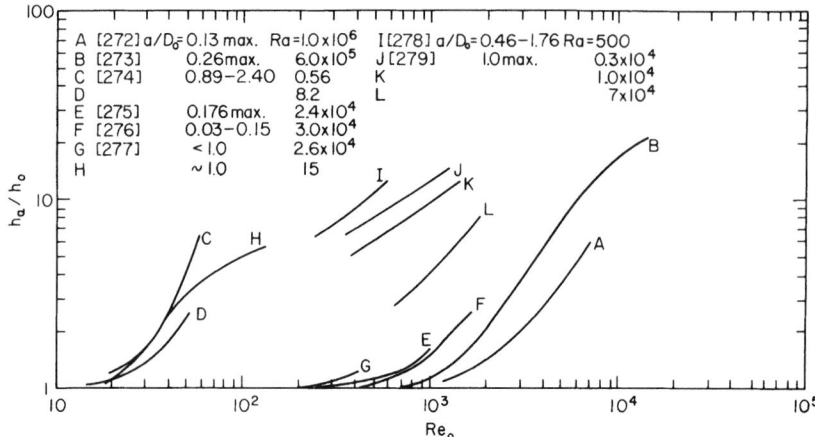

FIGURE 11.43 Influence of mechanical vibration on heat transfer from horizontal cylinders—$a/D_o < 1$.

vection. The average velocities are actually quite low: for example, $4af \approx 1.8$ m/s for the highest-intensity data of Mason and Boelter [273] in Fig. 11.42. For most systems, it would appear to be more convenient and economical to provide steady forced flow to achieve the desired increase in heat transfer coefficient.

Substantial improvements in heat transfer have also been recorded when vibration of the heated surface is used in forced-flow systems. No general correlation has been obtained; however, this is not surprising in view of the diverse geometric arrangements. Figure 11.44 presents representative data for heat transfer to liquids. The effect on heat transfer varies from slight degradation to over 300 percent improvement, depending on the system and the vibrational intensity. One problem is cavitation when the intensity becomes too large. As indicated by curves A and B, the vapor blanketing (*) causes a sharp degradation of heat transfer.

Hsieh and Marsters [288] extended the extensive single-tube experience to a vibrating vertical array of five horizontal cylinders. They found that the average of the heat transfer coef-

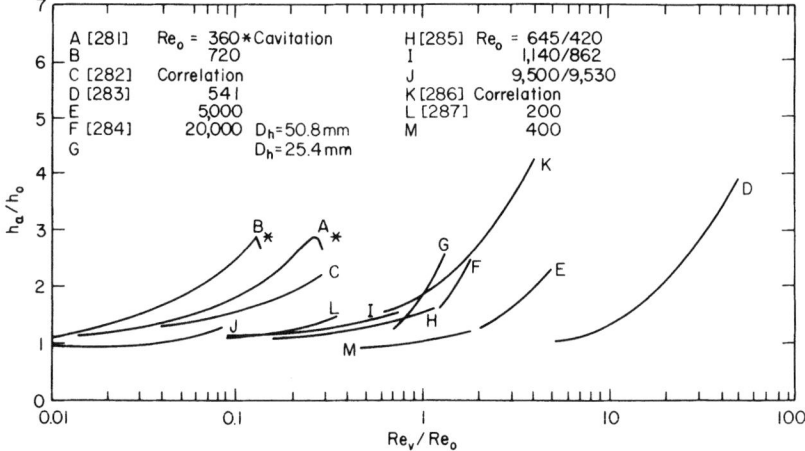

FIGURE 11.44 Effect of surface vibration on heat transfer to liquids with forced flow.

ficients increased by 54 percent at the highest vibrational intensity. The bottom cylinder showed the highest increase; the relatively poor performance of the top cylinders is apparently due to wake interaction.

These experiments indicate that vibrations can be effectively applied to practical heat exchanger geometries; however, economic evaluation is difficult because sufficient data are not available. Apparently, no comparative pressure drop data have been reported for forced flow. In any case, it appears that the overriding consideration is the cost of the vibrational equipment and the power to run it. Ogle and Engel [285] found for one of their runs that about 20 times as much energy was supplied to the vibrator as was gained in improved heat transfer. Even though the vibrator mechanism was not optimized in this particular investigation, the result suggests that heat-surface vibration will not be practical.

Boiling

Experiments by Bergles [279] have established that vibrations have little effect on subcooled or saturated pool boiling. It was found that the coefficients characteristic of single-phase vibrational data govern the entry into boiling conditions. Once boiling is fully established, however, vibration has no discernible effect. The maximum increase in critical heat flux was about 10 percent at an average velocity of 0.25 m/s. Experiments by Parker et al. [289, 290], run over the frequency range from 50 to 2000 Hz, have further confirmed that fully developed nucleate boiling is essentially unaffected by vibration.

Fuls and Geiger [291] studied the effect of enclosure vibration on pool boiling. A slight increase in the nucleate boiling heat transfer coefficient was observed.

Raben et al. [284] reported a study of flow surface boiling with heated-surface vibration. A large improvement was noted at low heat flux, but this improvement decreased with increasing heat flux. This is consistent with those pool-boiling results that indicate no improvement in the region of fully established boiling. Pearce [292] found insignificant changes in bulk-boiling CHF when a boiler tube was vibrated transversely.

Condensing

The few studies in this area include those of Dent [293] and Brodov et al. [294], who both obtained maximum increases of 10 to 15 percent by vibrating a horizontal condenser tube.

FLUID VIBRATION

Single-Phase Flow

In many applications it is difficult to apply surface vibration because of the large mass of the heat transfer apparatus. The alternative technique is then utilized, whereby vibrations are applied to the fluid and focused toward the heated surface. The generators that have been employed range from the flow interrupter to the piezoelectric transducer, thus covering the range of pulsations from 1 Hz to ultrasound of 10^6 Hz. The description of the interaction between fluid vibrations and heat transfer is even more complex than it is in the case of surface vibration. In particular, the vibrational variables are more difficult to define because of the remote placement of the generator. Under certain conditions, the flow fields may be similar for both fluid and surface vibration, and analytical results can be applied to both types of data.

A great deal of research effort has been devoted to studying the effects of sound fields on heat transfer from horizontal cylinders to air. Intense plane sound fields of the progressive or stationary type have been generated by loudspeakers or sirens. The sound fields have been oriented axially and transversely in either the horizontal or vertical plane. With plane transverse fields directed transversely, improvements of 100 to 200 percent over natural convection heat transfer coefficients were obtained by Sprott et al. [295], Fand and Kaye [296], and Lee and Richardson [297].

It is commonly observed that increases in average heat transfer occur at a sound pressure level of about 134 to 140 dB (well above the normal human tolerance of 120 dB), and that these increases are associated with the formation of an acoustically induced flow (acoustic or thermoacoustic streaming) near the heated surface. Large circumferential variations in heat transfer coefficient are present [298], and it has been observed that local improvements in heat transfer occur at intensities well below those that affect the average heat transfer [299]. Correlations have been proposed for individual experiments; however, an accurate correlation covering the limits of free convection and fully developed vortex motion has not been developed.

In general, it appears that acoustic vibrations yield relatively small improvements in heat transfer to gases in free convection. From a practical standpoint, a relatively simple forced-flow arrangement could be substituted to obtain equivalent improvements.

When acoustic vibrations are applied to liquids, heat transfer may be improved by acoustic streaming as in the case of gases. With liquids, though, it is possible to operate with ultrasonic frequencies given favorable coupling between a solid and a liquid. At frequencies of the order of 1 MHz, another type of streaming called *crystal wind* may be developed. These effects are frequently encountered; however, intensities are usually high enough to cause cavitation, which may become the dominant mechanism.

Seely [300], Zhukauskas et al. [301], Larson and London [302], Robinson et al. [303], Fand [304], Gibbons and Houghton [305], and Li and Parker [306] have demonstrated that natural convection heat transfer to liquids can be improved from 30 to 450 percent by the use of sonic and ultrasonic vibrations. In general, cavitation must occur before significant improvements in heat transfer are noted. In spite of these improvements, there appears to be some question regarding the practical aspects of acoustic enhancements. When the difficulty of designing a system to transmit acoustic energy to a large heat transfer surface is considered, it appears that forced flow or simple mechanical agitation will be a more attractive means of improving natural convection heat transfer.

Low-frequency pulsations have been produced in forced convection systems by partially damped reciprocating pumps and interrupter valves. Quasi-steady analyses suggest that heat transfer will be improved in transitional or turbulent flow with sufficiently intense vibrations. However, heat transfer coefficients are usually higher than predicted, apparently due to cavitation. Figure 11.45 indicates the improvements that have been reported for pulsating flow in channels. The improvement is most significant in the transitional range of Reynolds numbers, as might be expected, since the pulsations force the transition to turbulent flow. Interrupter valves are a particularly simple means of generating the pulsations. The valves must be located directly upstream of the heated section to produce cavitation, which appears to be largely responsible for the improvement in heat transfer [310, 311].

A wide variety of geometric arrangements and more complex flow fields are encountered when sound fields are superimposed on forced flow of gases. In general, the improvement is dependent on the relative strengths of the acoustic streaming and the forced flows. The reported improvements in average heat transfer are limited to about 100 percent. Typical data obtained with sirens or loudspeakers placed at the end of a channel are indicated in Fig. 11.46. The improvements in average heat transfer coefficient are generally significant only in the transition range where the vibrating motion acts as a turbulence trigger. The experiments of Moissis and Maroti [318] are important in that they demonstrate practical limitations. Even with high-intensity acoustic vibrations, the gas-side heat transfer coefficient was improved by only 30 percent for a compact heat exchanger core.

The data of Zhukauskas et al. [301] and Larson and London [302] suggest that ultrasonic vibration has no effect on forced convection heat transfer once the flow velocity is raised to

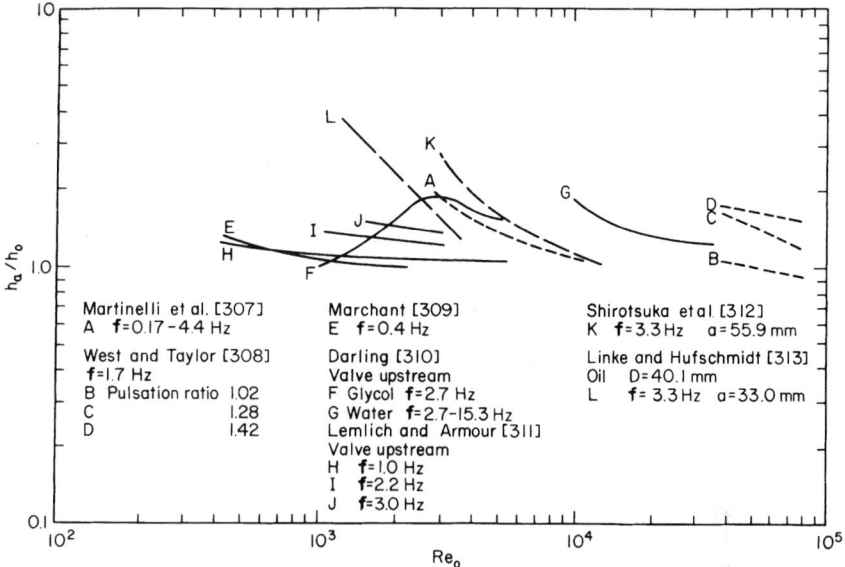

FIGURE 11.45   Effect of upstream pulsations on heat transfer to liquids flowing in pipes.

about 0.3 m/s. However, Bergles [319] demonstrated that lower-frequency vibrations (80 Hz) can produce improvements of up to 50 percent. This experiment was carried out at higher surface temperatures where it was possible to achieve cavitation.

Boiling

The available evidence indicates that fully established nucleate pool boiling is unaffected by ultrasonic vibration, apparently because of the dominance of bubble agitation and attenua-

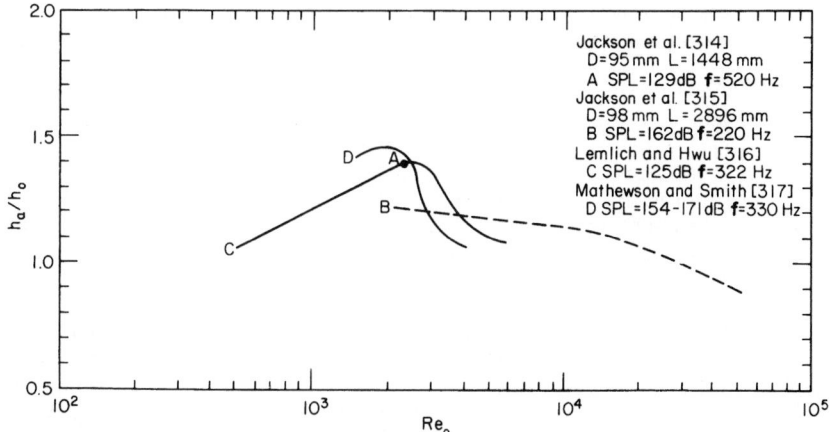

FIGURE 11.46   Influence of acoustic vibrations on heat transfer to air flowing in tubes.

tion of acoustic energy by the vapor [320, 321]. However, enhanced vapor removal can improve CHF by about 50 percent [320, 322]. Transition and film boiling can also be substantially improved, since the vibration has a strong tendency to destabilize film boiling [323].

In channel flow it is usually necessary to locate the transducer upstream or downstream of the test channel, with the result that the sound field is greatly attenuated. Tests with 80-Hz vibrations [319] indicate no improvement of subcooled boiling heat transfer or critical heat flux. Romie and Aronson [324], using ultrasonic vibrations, found that subcooled critical heat flux was unaffected. Even where intense ultrasonic vibrations were applied to the fluid in the immediate vicinity of the heated surface, boiling heat transfer was unaffected [325]. The severe attenuation of the acoustic energy by the two-phase coolant appears to render this technique ineffective for flow boiling systems.

Condensing

Mathewson and Smith [317] investigated the effects of acoustic vibrations on condensation of isopropanol vapor flowing downward in a vertical tube. A siren was used to generate a sound field of up to 176 dB at frequencies ranging from 50 to 330 Hz. The maximum improvement in condensing coefficient was found to be about 60 percent at low vapor flow rates. The condensate film under these conditions was normally laminar; thus, an intense sound field produced sufficient agitation in the vapor to cause turbulent conditions in the film. The effect of the sound field was considerably diminished as the vapor flow rate increased.

ELECTRIC AND MAGNETIC FIELDS

A comprehensive discussion of the fundamental effects electric and magnetic fields have on heat transfer is given in Ref. 370. A magnetic force field retards fluid motion; hence, heat transfer coefficients decrease. On the other hand, if electromagnetic pumping is established with a combined magnetic and electric field, heat transfer coefficients can be increased far above those expected for gravity-driven flows. For example, an analysis by Singer [326] indicates that electromagnetic pumping can increase laminar film condensation rates of a liquid metal by a factor of 10.

Electric fields are particularly effective in increasing heat transfer coefficients in free convection. The configuration may be a heated wire in a concentric tube maintained at a high voltage relative to the wire, or a fine wire electrode may be utilized with a horizontal plate. Reported increases are as much as a factor of 40; however, several hundred percent is normal. Much activity has centered on the application of corona discharge cooling to practical free-convection problems. The cooling of cutting tools by point electrodes was proposed by Blomgren and Blomgren [327], while Reynolds and Holmes [328] have used parallel wire electrodes to improve the heat dissipation of a standard horizontal finned tube. Heat transfer coefficients can be increased by several hundred percent when sufficient electrical power is supplied. It appears, however, that the equivalent effect could be produced at lower capital cost and without the hazards of 10 to 100 kV by simply providing modest forced convection with a blower or fan.

Some very impressive enhancements have been recorded with forced laminar flow. The recent studies of Porter and Poulter [329], Savkar [330], and Newton and Allen [331] demonstrated improvements of at least 100 percent when voltages in the 10-kV range were applied to transformer oil. A typical gas-gas heat exchanger rigged for electrohydrodynamic (EHD) enhancement, on both the tube side and the shell side, is shown in Fig. 11.47. These data show that substantial improvements in the overall heat transfer coefficient are possible. The power expenditure of the electrostatic generator is small, typically only several percent of the pumping power. While it is desirable to take advantage of any naturally occurring electric fields in

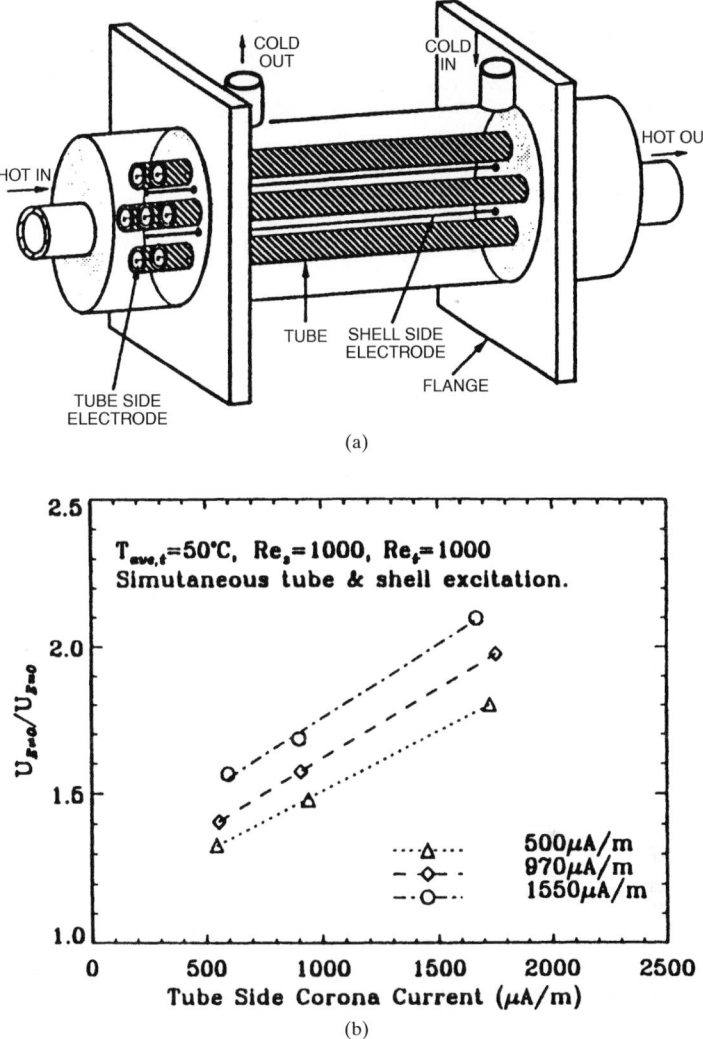

(a)

(b)

FIGURE 11.47   Schematic view of EHD-enhanced heat exchanger, and overall heat transfer coefficient improvement as a function of tube-side corona current. Shell and tube excitation [332].

electrical equipment, enhancement by electrical fields must be considered carefully. Mizushina et al. [333] found that even with intense fields, the enhancement disappeared as turbulent flow was approached in a circular tube with a concentric inner electrode.

The typical effects of electric fields on pool boiling are shown in Fig. 11.48. These data of Choi [334] were taken with a horizontal electrically heated wire located concentrically within a charged cylinder. Because of the large enhancement of free convection, boiling is not observed until relatively high heat fluxes. Once nucleate boiling is initiated, the electric field has little effect. However, CHF is elevated substantially, and large increases in the film-boiling heat transfer coefficient are obtained.

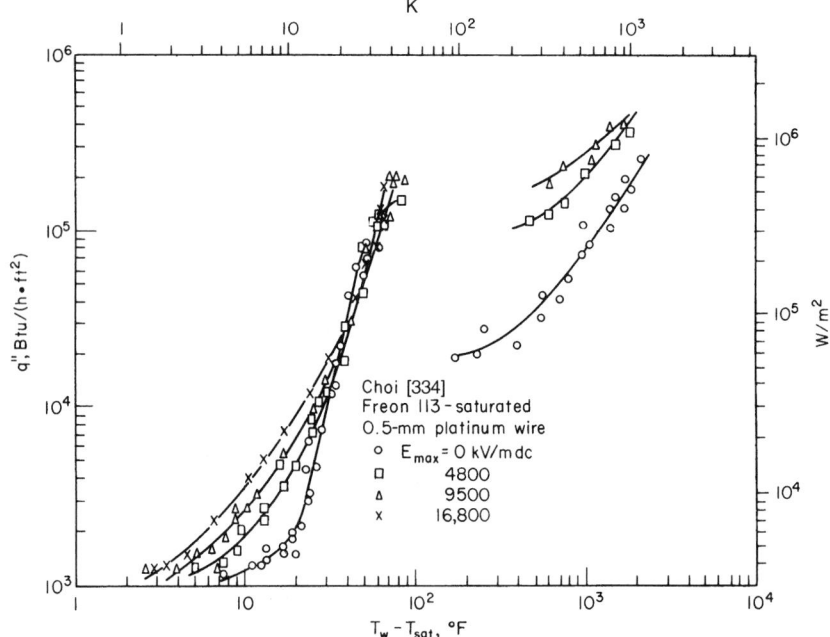

FIGURE 11.48   Influence of electrostatic fields on pool-boiling heat transfer.

Durfee and coworkers [335] conducted an extensive series of tests to evaluate the feasibility of applying EHD to boiling-water nuclear reactors. Tests with water in electrically heated annuli indicated that wall temperatures for flow bulk boiling were slightly reduced through application of the field. Increases in CHF were observed for all pressures, flow rates, and inlet subcoolings, with the improvement falling generally in the 15 to 40 percent range for applied voltages up to 3 kV. On the basis of limited pressure drop data, it was suggested that greater steam energy flow was obtained with the EHD system than with the conventional system at the same pumping power.

Velkoff and Miller [336] investigated the effect of uniform and nonuniform electric fields on laminar film condensation of Freon-113 on a vertical plate. With screen grid electrodes providing a uniform electric field over the entire plate surface, a 150 percent increase in the heat transfer coefficient was obtained with a power expenditure of a fraction of one watt. Choi and Reynolds [337] and Choi [338] recently reported data for condensation of Freon-113 on the outside wall of an annulus in the presence of a radial electric field. With the maximum applied voltage of 30 kV, the average heat transfer coefficients for a 25.4-mm outside diameter by 12.7-mm inside diameter annulus were increased by 100 percent.

EHD has not yet been adopted commercially, largely because of concerns about installing the electrodes and using very high voltages during heat exchanger operation. It has a potential drawback, common to all active techniques, in that an extra system is required (in this case, the electrostatic generator); failure of that system means the enhancement is not obtained.

INJECTION

Injection and suction have been considered primarily in connection with retarding of heat transfer to bodies subject to aerodynamic heating. On the enhancement side, some thought has been given to intensifying heat transfer by injecting gas through a porous heat transfer surface. The bubbling produces an agitation similar to that of nucleate boiling. Gose et al. [339, 340] bubbled gas through sintered or drilled heated surfaces to stimulate nucleate pool and flow boiling. Sims et al. [341] analyzed the pool data and found that Kutateladze's pool boiling relationship correlated the porous-plate data quite well. For their limited forced circulation tests with a sintered pipe, Gose et al. found that heat transfer coefficients were increased by as much as 500 percent in laminar flow and by about 50 percent in turbulent flow. Kudirka [342] found that heat transfer coefficients for flow of ethylene glycol in porous tubes were increased by as much as 130 percent by the injection of air. The practical application of injection appears to be rather limited because of the difficulty of supplying and removing the gas.

Tauscher et al. [348] have demonstrated up to fivefold increases in local heat transfer coefficients by injecting similar fluid into a turbulent tube flow. The effect is comparable to that produced by an orifice plate; in both cases the effect has died out after about 10 L/D.

Bankoff [343] suggested that heat transfer coefficients in film boiling could be substantially improved by continuously removing vapor through a porous heated surface. Subsequent experimental work [344, 345] demonstrated that coefficients could be increased by as much as 150 percent, provided that a porous block was placed on the surface to stabilize the flow of liquid toward the surface. Wayner and Kestin [346] extended this concept to nucleate boiling and found that wall superheats could be maintained at about 3 K (5.4°F) for heat fluxes over 300,000 W/m^2 or 95×10^3 Btu/(h·ft^2). This work was extended by Raiff and Wayner [347]. The need for a porous heated surface and a flow control element appears to limit the application of suction boiling.

SUCTION

Large increases in heat transfer coefficient are predicted for laminar flow [349] and turbulent flow [350] with surface suction. The general characteristics of the latter predictions were confirmed by the experiments of Aggarwal and Hollingsworth [351]. However, suction is difficult to incorporate into practical heat exchange equipment.

The typical studies of laminar film condensation by Antonir and Tamir [352] and Lienhard and Dhir [353] indicate that heat transfer coefficients can be improved by as much as several hundred percent when the film thickness is reduced by suction. This is expected, as the thickness of the condensate layer is the main parameter affecting the heat transfer rate in film condensation.

COMPOUND ENHANCEMENT

Compound techniques are a slowly emerging area of enhancement that holds promise for practical applications since heat transfer coefficients can usually be increased above any of the several techniques acting alone. Some examples for single-phase flows are

Rough tube wall with twisted-tape insert (Bergles et al. [354])

Rough cylinder with acoustic vibrations (Kryukov and Boykov [355])

Internally finned tube with twisted-tape insert (Van Rooyen and Kroeger [356])

Finned tubes in fluidized beds (Bartel and Genetti [357])

Externally finned tubes subjected to vibrations (Zozulya and Khorunzhii [358])

Gas-solid suspension with an electric field (Min and Chao [359])

Fluidized bed with pulsations of air (Bhattacharya and Harrison [360])

It is interesting to note that some compound attempts are unsuccessful. Masliyah and Nandakumar [361], for example, found analytically that average Nusselt numbers for internally finned coiled tubes were lower than they were for plain coiled tubes.

Compound enhancement has also been studied to a limited extent with phase-change heat transfer. For instance, the addition of surface roughness to the evaporator side of a rotating evaporator-condenser increased the overall coefficient by 10 percent [362]. Sephton [208, 243] found that overall coefficients could be doubled by the addition of a surfactant to seawater evaporating in spirally corrugated or doubly fluted tubes (vertical upflow). However, Van der Mast et al. [363] found only slight improvements with a surfactant additive for falling film evaporation in spirally corrugated tubes.

Compound enhancement, as it is used with vapor space condensation, includes rotating finned tubes [147], rotating rough disks [362], and rotating disks with suction [364]. Moderate increases in condensing coefficient are reported. Weiler et al. [268] condensed nitrogen inside rotating tubes treated with a porous coating, which increased coefficients above those for a rotating smooth tube.

PROSPECTS FOR THE FUTURE

This chapter has given an overview of enhanced heat transfer technology, citing representative developments. The literature in enhanced heat transfer appears to be growing faster than the engineering science literature as a whole. At least 10 percent of the heat transfer literature is now directed toward enhancement.

An enormous amount of technology is available; what is needed is technology transfer. Many techniques, and variations thereof, have made the transition from the academic or industrial research laboratory to industrial practice. This development of enhancement technologies must be accelerated. In doing this, however, the "corporate memory" should be retained. The vast literature in the field should be pursued before expensive physical or numerical experiments are started. To facilitate this, bibliographic surveys, such as that in Ref. 1, should be continued. Also, books, such as that of Webb [386], should be consulted.

Enhanced heat transfer will assume greater importance when energy prices rise again. With the current oil and gas "bubbles," there is little financial incentive to save energy. Usually, enhancement is now employed not to save energy costs but to save space. For example, process upgrading, through use of an enhanced heat exchanger that fits a given space, is common. It is expected that the field of enhanced heat transfer will experience another growth phase (refer to Fig. 11.1) when energy concerns are added to volume considerations.

Throughout this whole process, manufacturing methods and materials requirements may be overriding considerations. Can the enhancement be produced in the material that will survive any fouling and corrosion inherent in the environment? Much work needs to be done to define the fouling/corrosion characteristics of enhanced surfaces [385]. Particularly, antifouling surfaces need to be developed.

It should be noted that enhancement technology is still largely experimental, although great strides are being made in analytical/numerical description of the various technologies [386]. Accordingly, it is imperative that the craft of experimentation be kept viable. With the wholesale rush to "technology," laboratories everywhere are being decommissioned. Hands-on experiences in universities are being decreased or replaced by computer skills. Experimentation is still a vital art, needed for direct resolution of transport phenomena in complex enhanced geometries as well as benchmarking of computer codes. As such, experimental skills should continue to be taught, and conventional laboratories should be maintained.

Finally, it is evident that heat transfer enhancement is well established and is used routinely in the power industry, process industry, and heating, ventilation, and air-conditioning. Many techniques are available for improvement of the various modes of heat transfer. Fundamental understanding of the transport mechanism is growing; but, more importantly, design correlations are being established. As noted in Ref. 365, it is appropriate to view enhancement as "second-generation" or "third-generation" heat transfer technology.

This chapter indicates that many enhancement techniques have gone through all of the steps required for commercialization. The prognosis is for the exponential growth curve of Fig. 11.1 to level off, not from a lack of interest, but from a broader acceptance of enchancement techniques in industrial practice.

NOMENCLATURE

Symbol, Definition, SI Units, English Units

Properties are evaluated at the bulk fluid condition unless otherwise noted.

A	heat transfer surface area: m², ft²
A_e	effective heat transfer surface area, Eq. 11.12: m², ft²
A_F	heat transfer surface area of fins: m², ft²
A_f	cross-sectional flow area: m², ft²
A_r	area of unfinned portion of tube: m², ft²
a	vibrational amplitude; amplitude of sinusoidally shaped flute: m, ft
b	stud or fin thickness: m, ft
c_p	specific heat at constant pressure: J/(kg·K), Btu/(lb$_m$·°F)
D, d	tube inside diameter: m, ft
D_c	diameter of coil: m, ft
D_h	hydraulic diameter: m, ft
D_o	outside diameter of circular finned tube or cylinder: m, ft
D_r	root diameter of finned tube: m, ft
d_i	inside diameter of annulus or ring insert: m, ft
d_o	outside diameter of annulus or ring insert: m, ft
d_p	diameter of particles in air-solid suspensions: m, ft
d_s	diameter of spherical packing or disk insert: m, ft
E	electric field strength: V/m, V/ft
e	protrusion height: m, ft
F	fin factor, Eq. 11.19
F_t	convective factor, Eq. 11.23
f	Fanning friction factor $= \Delta PD\rho/2LG^2$
$\mathbf{f}$	vibrational frequency: s⁻¹
G	mass velocity $= W/A_f$ kg/(m²·s), lb$_m$/(h·ft²)
G_e	effective mass velocity, Eq. 11.14: kg/(m²·s), lb$_m$/(h·ft²)
Gr	Grashof number $= g\beta\Delta TD^3/v^2$
Gz	Graetz number $= Wc_p/kL$
g	gravitational acceleration: m/s², ft/s²
$\mathbf{g}$	spacing between protrusions: m, ft
H	fin height: m, ft

h	heat transfer coefficient: W/(m²·K), Btu/(h·ft²·°F); strip fin height: m, ft
$\bar{h}$	mean value of the heat transfer coefficient: W/(m²·K), Btu/h·ft²·°F)
h	protrusion length, Fig. 11.11: m, ft
i	enthalpy: J/kg, Btu/lb$_m$
i_{lg}	enthalpy of vaporization: J/kg, Btu/lb$_m$
j	Colburn j-factor, St Pr$^{2/3}$
k	thermal conductivity: W/(m·K), Btu/(h·ft·°F)
L	channel heated length: m, ft
L_e	finned length between cuts for interrupted fins or between inserts: m, ft
L_f	mean effective length of a fin, Eq. 11.12: m, ft
L_s	distance between condensate strippers: m, ft
l	average space between adjacent fins: m, ft
ℓ	length of one offset module of strip fins: m, ft
N	number of tubes
n	number of fins
Nu	Nusselt number $= hD/k$
$\overline{\text{Nu}}$	mean value of the Nusselt number $= \bar{h}D/k$
P	pressure: N/m², lb$_f$/ft²
P	pumping power: W, Btu/h
$\mathscr{P}$	wetted perimeter of channel between two longitudinal fins, Fig. 11.16: m, ft
ΔP	pressure drop: N/m², lb$_f$/ft²
Pr	Prandtl number $= \mu c_p/k$
p	roughness or flute pitch, Fig. 11.8: m, ft
q	rate of heat transfer: W, Btu/h
q''	heat flux: W/m², Btu/(h·ft²)
q''_{cr}	critical heat flux: W/m², Btu/(h·ft²)
Ra	Rayleigh number = Gr Pr
Re	Reynolds number $= GD/\mu$ (actual G—allowing for any flow blockage—is generally used)
Re$_v$	vibrational Reynolds number $= 2\pi a\mathbf{f}D_o/v$
s	lateral spacing between strip fins: m, ft
St	Stanton number $= h/Gc_p =$ Nu/Re Pr
Sw	Swirl flow parameter Sw $= Re_{sw}\sqrt{y}$
T	temperature: °C, °F
T_{sat}	saturation temperature: °C, °F
ΔT	temperature difference: K, °F
ΔT_{gw}	temperature difference from saturated vapor to wall: K, °F
ΔT_i	heat exchanger inlet temperature difference: K, °F
ΔT_{lm}	log mean temperature difference: K, °F
ΔT_n	shell-side to tube-side exit temperature difference: K, °F
ΔT_{sat}	wall-minus-saturation temperature difference: K, °F
U	overall heat transfer coefficient: W/(m²·K), Btu/(h·ft²·°F)

$\overline{U}_n$ average overall heat transfer coefficient based on nominal tube outside diameter: $W/(m^2 \cdot K)$, $Btu/(h \cdot ft^2 \cdot °F)$

u average axial velocity: m/s, ft/s

W mass flow rate: kg/s, lb_m/s

X^+ dimensionless position $= x/(D \ Re \ Pr)$

x axial position: m, ft; flowing mass quality

Δx quality change along test section

$\overline{x}$ average flowing mass quality

x_{cr} quality at critical heat flux

y twist ratio, tube diameters per 180° tape twist

Greek Symbols

α spiral fin helix angle: rad, deg; aspect ratio for strip fin s/h

β volumetric coefficient of expansion: K^{-1}, R^{-1}; contact angle of rib profile, deg

δ ratio for offset strip fin t/l

γ ratio for offset strip fin t/s

η fin efficiency

μ dynamic viscosity: $N/(m^2 \cdot s)$, $lb_m/(h \cdot ft)$

ν kinematic viscosity: m^2/s, ft^2/s

ρ density: kg/m^3, lb_m/ft^3

$\Delta \rho$ density difference between wall and core fluid: kg/m^2, lb_m/ft^3

ϕ parameter defined in Eq. 11.7

Subscripts

a enhanced heat transfer condition

b evaluated at bulk or mixed-mean fluid condition

cr   at critical heat flow condition

E refers to electrostatic field

ex   condition at outlet of channel

f evaluated at film temperature, $(T_w + T_b)/2$

g based on vapor or gas

h based on hydraulic diameter

i based on maximum inside (envelope) diameter

in   condition at inlet of channel

iso   isothermal

l based on liquid

o nonenhanced data

p particles

s standard condition; refers to solids; refers to shell side

sat   evaluated at saturation condition

sw swirl condition, allows for flow blockage of twisted tape

t refers to tube side

x local value

w evaluated at wall temperature

REFERENCES

1. A. E. Bergles, M. K. Jensen, and B. Shome, Bibliography on "Enhancement of Convective Heat and Mass Transfer," *Heat Transfer Lab Report HTL-23,* Rensselaer Polytechnic Institute, Troy, NY, 1995. Also, "The Literature on Enhancement of Convective Heat and Mass Transfer," *Enhanced Heat Transfer* (4): 1–6, 1996.

2. W. J. Marner, A. E. Bergles, and J. M. Chenoweth, "On the Presentation of Performance Data for Enhanced Tubes Used in Shell-and-Tube Heat Exchangers," *J. Heat Transfer* (105): 358–365, 1983.

3. W. Nunner, "Wärmeübergang und Druckabfall in rauhen Rohren," *Forschungsh. Ver. dt. Ing.* (B22/455): 5–39, 1956. Also, *Atomic Energy Research Establishment (United Kingdom) Lib./Trans.* 786, 1958.

4. R. Koch, "Druckverlust und Wärmeübergang bei verwirbelter Strömung," *Forschungsh. Ver. dt. Ing.* (B24/469): 1–44, 1958.

5. N. D. Greene, *Convair Aircraft,* private communication to W. R. Gambill, May, 1960. Cited in W. R. Gambill and R. D. Bundy, "An Evaluation of the Present Status of Swirl Flow Heat Transfer," *ASME Paper 61-HT-42,* ASME, New York, 1961.

6. R. F. Lopina and A. E. Bergles, "Heat Transfer and Pressure Drop in Tape Generated Swirl Flow," *J. Heat Transfer* (94): 434–442, 1969.

7. R. L. Webb and A. E. Bergles, "Performance Evaluation Criteria for Selection of Heat Transfer Surface Geometries Used in Low Reynolds Number Heat Exchangers," in *Low Reynolds Number Convection in Channels and Bundles,* S. Kakac, R. H. Shah, and A. E. Bergles eds., Hemisphere, Washington, DC, and McGraw-Hill, New York, 1982.

8. L. C. Trimble, B. L. Messinger, H. E. Ulbrich, G. Smith, and T. Y. Lin, "Ocean Thermal Energy Conversion System Study Report," *Proc. 3d Workshop Ocean Thermal Energy Conversion (OTEC), APL/JIIU SR 75-2,* pp. 3–21, August 1975.

9. A. E. Bergles and M. K. Jensen, "Enhanced Single-Phase Heat Transfer for OTEC Systems," *Proc. 4th Conf. Ocean Thermal Energy Conversion (OTEC),* University of New Orleans, New Orleans, LA, pp. VI-41–VI-54, July 1977.

10. R. L. Webb, E. R. G. Eckert, and R. J. Goldstein, "Heat Transfer and Friction in Tubes With Repeated Rib Roughness," *Int. J. Heat Mass Transfer* (14): 601–618, 1971.

11. A. E. Bergles, G. S. Brown Jr., and W. D. Snider, "Heat Transfer Performance of Internally Finned Tubes," *ASME Paper 71-HT-31,* ASME, New York, 1971.

12. E. Smithberg and F. Landis, "Friction and Forced Convection Heat Transfer Characteristics in Tubes With Twisted Tape Swirl Generators," *J. Heat Transfer* (86): 39–49, 1964.

13. R. L. Webb, "Performance, Cost Effectiveness and Water Side Fouling Considerations of Enhanced Tube Heat Exchangers for Boiling Service With Tube-Side Water Flow," *Heat Transfer Engineering* (3/3–4): 84–98, 1982.

14. G. R. Kubanek and D. L. Miletti, "Evaporative Heat Transfer and Pressure Drop Performance of Internally-Finned Tubes with Refrigerant 22," *J. Heat Transfer* (101): 447–452, 1979.

15. M. Luu and A. E. Bergles, "Augmentation of In-Tube Condensation of R-113 by Means of Surface Roughness," *ASHRAE Trans.* (87/2): 33–50, 1981.

16. W. R. Gambill, R. D. Bundy, and R. W. Wansbrough, "Heat Transfer, Burnout, and Pressure Drop for Water in Swirl Flow Tubes With Internal Twisted Tapes," *Chem. Eng. Prog. Symp. Ser.* (57/32): 127–137, 1961.

17. F. E. Megerlin, R. W. Murphy, and A. E. Bergles, "Augmentation of Heat Transfer in Tubes by Means of Mesh and Brush Inserts," *J. Heat Transfer* (96): 145–151, 1974.

18. R. K. Young and R. L. Hummel, "Improved Nucleate Boiling Heat Transfer," *Chem. Eng. Prog.* (60/7): 53–58, 1964.

19. A. E. Bergles, N. Bakhru, and J. W. Shires, "Cooling of High-Power-Density Computer Components," *EPL Rep. 70712-60,* Massachusetts Institute of Technology, Cambridge, MA, 1968.

20. V. M. Zhukov, G. M. Kazakov, S. A. Kovalev, and Y. A. Kuzmakichta, "Heat Transfer in Boiling of Liquids on Surfaces Coated With Low Thermal Conductivity Films," *Heat Transfer Sov. Res.* (7/3): 16–26, 1975.

21. A. E. Bergles and W. G. Thompson Jr., "The Relationship of Quench Data to Steady-State Pool Boiling Data," *Int. J. Heat Mass Transfer* (13): 55–68, 1970.

22. A. E. Bergles, "Principles of Heat Transfer Augmentation. II: Two-Phase Heat Transfer," in *Heat Exchangers, Thermal-Hydraulic Fundamentals and Design,* S. Kakac, A. E. Bergles, and F. Mayinger eds., pp. 857–881, Hemisphere, Washington, DC, and McGraw-Hill, New York, 1981.

23. L. C. Kun and A. M. Czikk, "Surface for Boiling Liquids," *U.S. Pat. 3,454,081,* July 8, 1969.

24. J. Fujikake, "Heat Transfer Tube for Use in Boiling Type Heat Exchangers and Method of Producing the Same," *U.S. Pat. 4,216,826,* Aug. 12, 1980.

25. U. P. Hwang and K. P. Moran, "Boiling Heat Transfer of Silicon Integrated Circuits Chip Mounted on a Substrate," in *Heat Transfer in Electronic Equipment,* M. D. Kelleher and M. M. Yovanovich eds., HTD vol. 20, pp. 53–59, ASME, New York, 1981.

26. R. L. Webb, "Heat Transfer Surface Having a High Boiling Heat Transfer Coefficient," *U.S. Pat. 3,696,861,* Oct. 10, 1972.

27. V. A. Zatell, "Method of Modifying a Finned Tube for Boiling Enhancement," *U.S. Pat. 3,768,290,* Oct. 30, 1973.

28. W. Nakayama, T. Daikoku, H. Kuwahara, and K. Kakizaki, "High-Flux Heat Transfer Surface Thermoexcel," *Hitachi Rev.* (24): 329–333, 1975.

29. K. Stephan and J. Mitrovic, "Heat Transfer in Natural Convective Boiling of Refrigerants and Refrigerant-Oil-Mixtures in Bundles of T-Shaped Finned Tubes," in *Advances in Enhanced Heat Transfer—1981,* R. L. Webb, T. C. Carnavos, E. L. Park Jr., and K. M. Hostetler eds., HTD vol. 18, pp. 131–146, ASME, New York, 1981.

30. E. Ragi, "Composite Structure for Boiling Liquids and Its Formation," *U.S. Pat. 3,684,007,* Aug. 15, 1972.

31. P. J. Marto and W. M. Rohsenow, "Effects of Surface Conditions on Nucleate Pool Boiling of Sodium," *J. Heat Transfer* (88): 196–204, 1966.

32. P. S. O'Neill, C. F. Gottzmann, and C. F. Terbot, "Novel Heat Exchanger Increases Cascade Cycle Efficiency for Natural Gas Liquefaction," *Advances in Cryogenic Engineering* (17): 421–437, 1972.

33. S. Oktay and A. F. Schmeckenbecher, "Preparation and Performance of Dendritic Heat Sinks," *J. Electrochem. Soc.* (21): 912–918, 1974.

34. M. M. Dahl and L. D. Erb, "Liquid Heat Exchanger Interface Method," *U.S. Pat. 3,990,862,* Nov. 9, 1976.

35. M. Fujii, E. Nishiyama, and G. Yamanaka, "Nucleate Pool Boiling Heat Transfer from Micro-Porous Heating Surfaces," in *Advances in Enhanced Heat Transfer,* J. M. Chenoweth, J. Kaellis, J. W. Michel, and S. Shenkman eds., pp. 45–51, ASME, New York, 1979.

36. K. R. Janowski, M. S. Shum, and S. A. Bradley, "Heat Transfer Surface," *U.S. Pat. 4,129,181,* Dec. 12, 1978.

37. D. F. Warner, K. G. Mayhan, and E. L. Park Jr., "Nucleate Boiling Heat Transfer of Liquid Nitrogen From Plasma Coated Surfaces," *Int. J. Heat Mass Transfer* (21): 137–144, 1978.

38. A. M. Czikk and P. S. O'Neill, "Correlation of Nucleate Boiling From Porous Metal Films," in *Advances in Enhanced Heat Transfer,* J. M. Chenoweth, J. Kaellis, J. W. Michel, and S. Shenkman eds., pp. 53–60, ASME, New York, 1979.

39. W. Nakayama, T. Daikoku, H. Kuwahara, and T. Nakajima, "Dynamic Model of Enhanced Boiling Heat Transfer on Porous Surface—Parts I and II," *J. Heat Transfer* (102): 445–456, 1980.

40. S. Yilmaz, J. J. Hwalck, and J. N. Westwater, "Pool Boiling Heat Transfer Performance for Commercial Enhanced Tube Surfaces," *ASME Paper 80-HT-41,* ASME, New York, July 1980.

41. A. E. Bergles and M.-C. Chyu, "Characteristics of Nucleate Pool Boiling From Porous Metallic Coatings," in *Advances in Enhanced Heat Transfer—1981,* R. L. Webb, T. C. Carnavos, E. L. Park Jr., and K. M. Hostetler eds., HTD vol. 18, pp. 61–71, ASME, New York, 1981.

42. S. Yilmaz, J. W. Palen, and J. Taborek, "Enhanced Surfaces as Single Tubes and Tube Bundles," in *Advances in Enhanced Heat Transfer—1981,* R. L. Webb, T. C. Carnavos, E. L. Park Jr., and K. M. Hostetler eds., HTD vol. 18, pp. 123–129, ASME, New York, 1981.

43. M.-C. Chyu, A. E. Bergles, and F. Mayinger, "Enhancement of Horizontal Tube Spray Film Evaporators," *Proceedings 7th Int. Heat Trans. Conf.,* Hemisphere, Washington, DC, vol. 6, pp. 275–280, 1982.

44. R. L. Webb, G. H. Junkhan, and A. E. Bergles, "Bibliography of U.S. Patents on Augmentation of Convective Heat and Mass Transfer—II. Heat Transfer Lab. Rep." *HTL-32, ISU-ERI-Ames-84257, DE S4014865,* Iowa State University, Ames, IA, September 1980.

45. R. L. Webb, "The Evolution of Enhanced Surface Geometries for Nucleate Boiling," *Heat Transfer Eng.* (2/3–4): 46–49, 1981.

46. S. Iltscheff, "Über einige Versuche zur Erzielung von Tropfkondensation mit fluorierten Kältemitteln," *Kältetech. Klim.* (23): 237–241, 1971.

47. I. Tanawasa, "Dropwise Condensation: The Way to Practical Applications," *Heat Transfer 1978, Proc. 6th Int. Heat Transfer Conf.,* Hemisphere, Washington, DC, vol. 6, pp. 393–405, 1978.

48. L. R. Glicksman, B. B. Mikic, and D. F. Snow, "Augmentation of Film Condensation on the Outside of Horizontal Tubes," *AIChE J.* (19): 636–637, 1973.

49. A. E. Bergles, G. H. Junkhan, and R. L. Webb, "Energy Conservation via Heat Transfer Enhancement," *Heat Transfer Lab. Rep. COO-4649-5,* Iowa State University, Ames, IA, 1979.

50. D. F. Gluck, "The Effect of Turbulence Promotion on Newtonian and Non-Newtonian Heat Transfer Rates," MS thesis, University of Delaware, Newark, DE, 1959.

51. A. R. Blumenkrantz and J. Taborek, "Heat Transfer and Pressure Drop Characteristics of Turbotec Spirally Grooved Tubes in the Turbulent Regime," *Heat Transfer Research, Inc., Rep. 2439-300-7,* HTRI, Pasadena, CA, 1970.

52. G. R. Rozalowski and R. A. Gater, "Pressure Loss and Heat Transfer Characteristics for High Viscous Flow in Convoluted Tubing," *ASME Paper 75-HT-40,* ASME, New York, 1975.

53. D. Pescod, "The Effects of Turbulence Promoters on the Performance of Plate Heat Exchangers," in *Heat Exchangers: Design and Theory Sourcebook,* N. H. Afghan and E. U. Schlünder eds., pp. 601–616, Scripta, Washington, DC, 1974.

54. Z. Nagaoka and A. Watanabe, "Maximum Rate of Heat Transfer With Minimum Loss of Energy," *Proc. 7th Int. Cong. Refrigeration* (3): 221–245, 1936.

55. W. F. Cope, "The Friction and Heat Transmission Coefficients of Rough Pipes," *Proc. Inst. Mech. Eng.* (145): 99–105, 1941.

56. D. W. Savage and J. E. Myers, "The Effect of Artificial Surface Roughness on Heat and Momentum Transfer," *AIChE J.* (9): 694–702, 1963.

57. V. Kolar, "Heat Transfer in Turbulent Flow of Fluids Through Smooth and Rough Tubes," *Int. J. Heat Mass Transfer* (8): 639–653, 1965.

58. V. Zajic, "Some Results on Research of Intensified Water Cooling by Roughened Surfaces and Surface Boiling at High Heat Flux Rates," *Acta Technica CSAV* (5): 602–612, 1965.

59. R. A. Gowen, "A Study of Forced Convection Heat Transfer from Smooth and Rough Surfaces," PhD thesis in chemical engineering and applied chemistry, University of Toronto, Toronto, Canada, 1967.

60. E. K. Kalinin, G. A. Dreitser, and S. A. Yarkho, "Experimental Study of Heat Transfer Intensification Under Condition of Forced Flow in Channels," *Jpn. Soc. Mech. Eng. 1967 Semi-Int. Symp.,* Paper 210, JSME, Tokyo, Japan, September 1967.

61. D. Eissenberg, "Tests of an Enhanced Horizontal Tube Condenser Under Conditions of Horizontal Steam Cross Flow," in *Heat Transfer 1970,* vol. 1, paper HE2.1, Elsevier, Amsterdam, 1970.

62. J. M. Kramer and R. A. Gater, "Pressure Loss and Heat Transfer for Non-Boiling Fluid Flow in Convoluted Tubing," *ASME Paper 73-HT-23,* ASME, New York, 1973.

63. G. Grass, "Verbesserung der Wärmeübertragung an Wasser durch künstliche Aufrauhung der Oberflächen in Reaktoren Wärmetauschern," *Atomkernenergie* (3): 328–331, 1958.

64. A. R. Blumenkrantz and J. Taborek, "Heat Transfer and Pressure Drop Characteristics of Turbotec Spirally Grooved Tubes in the Turbulent Regime," *Heat Transfer Research, Inc., Rep. 2439-300-7,* HTRI, Pasadena, CA, 1970.

65. D. G. Dipprey and R. H. Sabersky, "Heat and Momentum Transfer in Smooth and Rough Tubes at Various Prandtl Numbers," *Int. J. Heat Mass Transfer* (6): 329–353, 1963.

66. A. Blumenkrantz, A. Yarden, and J. Taborek, "Performance Prediction and Evaluation of Phelps Dodge Spirally Grooved Tubes, Inside Tube Flow Pressure Drop and Heat Transfer in Turbulent Regime," *Heat Transfer Research, Inc., Rep. 2439-300-4,* HTRI, Pasadena, CA, 1969.

67. E. C. Brouillette, T. R. Mifflin, and J. E. Myers, "Heat Transfer and Pressure Drop Characteristics of Internal Finned Tubes," *ASME Paper 57-A-47,* ASME, New York, 1957.

68. J. W. Smith, R. A. Gowan, and M. E. Charles, "Turbulent Heat Transfer and Temperature Profiles in a Rifled Pipe," *Chem. Eng. Sci.* (23): 751–758, 1968.

69. P. Kumar and R. L. Judd, "Heat Transfer With Coiled Wire Turbulence Promoters," *Can. J. Chem. Eng.* (8): 378–383, 1970.

70. R. L. Webb, E. R. G. Eckert, and R. J. Goldstein, "Generalized Heat Transfer and Friction Correlations for Tubes With Repeated-Rib Roughness," *Int. J. Heat Mass Transfer* (15): 180–184, 1972.

71. J. G. Withers, "Tube-Side Heat Transfer and Pressure Drop for Tubes Having Helical Internal Ridging With Turbulent/Transitional Flow of Single-Phase Fluid. Pt. 1. Single-Helix Ridging," *Heat Transfer Eng.* (2/1): 48–58, 1980.

72. J. G. Withers, "Tube Side Heat Transfer and Pressure Drop for Tubes Having Helical Internal Ridging with Turbulent/Transitional Flow of Single-Phase Fluid. Pt. 2. Multiple-Helix Ridging," *Heat Transfer Eng.* (2/2): 43–50, 1980.

73. M. J. Lewis, "An Elementary Analysis for Predicting the Momentum and Heat-Transfer Characteristics of a Hydraulically Rough Surface," *J. Heat Transfer* (97): 249–254, 1975.

74. G. A. Kemeny and J. A. Cyphers, "Heat Transfer and Pressure Drop in an Annular Gap With Surface Spoilers," *J. Heat Transfer* (83): 189–198, 1961.

75. A. W. Bennett and H. A. Kearsey, "Heat Transfer and Pressure Drop for Superheated Steam Flowing Through an Annulus With One Roughened Surface," *Atomic Energy Research Establishment 4350,* AERE, Harwell, UK, 1964.

76. H. Brauer, "Strömungswiderstand und Wärmeübergang bei Ringspalten mit rauhen Rohren," *Atomkernenergie* (4): 152–159, 1961.

77. W. S. Durant, R. H. Towell, and S. Mirshak, "Improvement of Heat Transfer to Water Flowing in an Annulus by Roughening the Heated Wall," *Chem. Eng. Prog. Symp. Ser.* (60/61): 106–113, 1965.

78. M. Dalle Donne and L. Meyer, "Turbulent Convective Heat Transfer From Rough Surfaces With Two-Dimensional Rectangular Ribs," *Int. J. Heat Mass Transfer* (20): 583–620, 1977.

79. M. Hudina, "Evaluation of Heat Transfer Performances of Rough Surfaces From Experimental Investigation in Annular Channels," *Int. J. Heat Mass Transfer* (22): 1381–1392, 1979.

80. M. Dalle Donne, "Heat Transfer in Gas Cooled Fast Reactor Cores," *Ann. Nucl. Energy* (5): 439–453, 1978.

81. M. Dalle Donne, A. Martelli, and K. Rehme, "Thermo-Fluid-Dynamic Experiments with Gas-Cooled Bundles of Rough Rods and Their Evaluations With the Computer Code SAGAPØ," *Int. J. Heat Mass Transfer* (22): 1355–1374, 1979.

82. E. Achenbach, "The Effect of Surface Roughness on the Heat Transfer From a Circular Cylinder to the Cross Flow of Air," *Int. J. Heat Mass Transfer* (20): 359–369, 1977.

83. A. Zhukauskas, J. Ziugzda, and P. Daujotas, "Effects of Turbulence on the Heat Transfer of a Rough Surface Cylinder in Cross-Flow in the Critical Range of Re," in *Heat Transfer 1978,* vol. 4, pp. 231–236, Hemisphere, Washington, DC, 1978.

84. G. B. Melese, "Comparison of Partial Roughening of the Surface of Fuel Elements With Other Ways of Improving Performance of Gas-Cooled Nuclear Reactors," *General Atomics 4624,* GA, San Diego, CA, 1963.

85. Heat Transfer Capability, *Mech. Eng.,* Vol. 89, p. 55, 1967.

86. R. B. Cox, A. S. Pascale, G. A. Matta, and K. S. Stromberg, "Pilot Plant Tests and Design Study of a 2.5 MGD Horizontal-Tube Multiple-Effect Plant," *Off. Saline Water Res. Dev. Rep. No. 492,* OSW, Washington, DC, October 1969.

87. I. H. Newson, "Heat Transfer Characteristics of Horizontal Tube Multiple Effect (HTME) Evaporators—Possible Enhanced Tube Profiles," *Proc. 6th Int. Symp. Fresh Water from the Sea* (2): 113–124, 1978.

88. W. S. Durant and S. Mirshak, "Roughening of Heat Transfer Surfaces as a Method of Increasing Heat Flux at Burnout," *E. I. Dupont de Nemours and Co. 380,* DP, Savannah, GA, 1959.

89. V. I. Gomelauri and T. S. Magrakvelidze, "Mechanism of Influence of Two Dimensional Artificial Roughness on Critical Heat Flux in Subcooled Water Flow," *Therm. Eng.* (25/2): 1–3, 1978.

90. R. W. Murphy and K. L. Truesdale, "The Mechanism and the Magnitude of Flow Boiling Augmentation in Tubes with Discrete Surface Roughness Elements (III)," *Raytheon Co. Rep. B12-7294,* Raytheon, Bedford, MA, November 1972.

91. J. G. Withers and E. P. Habdas, "Heat Transfer Characteristics of Helical Corrugated Tubes for Intube Boiling of Refrigerant R-12," *AIChE Symp. Ser.* (70/138): 98–106, 1974.

92. E. Bernstein, J. P. Petrek, and J. Meregian, "Evaluation and Performance of Once-Through, Zero-Gravity Boiler Tubes With Two-Phase Water," *Pratt and Whitney Aircraft Co. 428,* DWAC, Middletown, CT, 1964.

93. E. Janssen and J. A. Kervinen, "Burnout Conditions for Single Rod in Annular Geometry, Water at 600 to 1400 psia," *General Electric Atomic Power 3899,* GEAP, San Jose, CA, 1963.

94. E. P. Quinn, "Transition Boiling Heat Transfer Program," *5th Q. Prog. Rep., General Electric Atomic Power 4608,* GEAP, San Jose, CA, 1964.

95. H. S. Swenson, J. R. Carver, and G. Szoeke, "The Effects of Nucleate Boiling Versus Film Boiling on Heat Transfer in Power Boiler Tubes," *J. Eng. Power* (84): 365–371, 1962.

96. J. W. Ackerman, "Pseudoboiling Heat Transfer to Supercritical Pressure Water in Smooth and Ribbed Tubes," *J. Heat Transfer* (92): 490–498, 1970.

97. A. J. Sellers, G. M. Thur, and M. K. Wong, "Recent Developments in Heat Transfer and Development of the Mercury Boiler for the SNAP-8 System," *Proc. Conf. Application of High Temperature Instrumentation to Liquid-Metal Experiments, Argonne National Laboratory 7100,* pp. 573–632, ANL, Argonne, IL, 1965.

98. J. O. Medwell and A. A. Nicol, "Surface Roughness Effects on Condensate Films," *ASME Paper 65-HT-43,* ASME, New York, 1965.

99. A. A. Nicol and J. O. Medwell, "The Effect of Surface Roughness on Condensing Steam," *Can. J. Chem. Eng.* (44/6): 170–173, 1966.

100. T. C. Carnavos, "An Experimental Study: Condensing R-11 on Augmented Tubes," *ASME Paper 80-HT-54,* ASME, New York, 1980.

101. R. B. Cox, G. A. Matta, A. S. Pascale, and K. G. Stromberg, "Second Report on Horizontal Tubes Multiple-Effect Process Pilot Plant Tests and Design," *Off. Saline Water Res. Dev. Rep. No. 592,* DSW, Washington, DC, May 1970.

102. W. J. Prince, "Enhanced Tubes for Horizontal Evaporator Desalination Process," MS thesis in engineering, University of California, Los Angeles, 1971.

103. G. W. Fenner and E. Ragi, "Enhanced Tube Inner Surface Device and Method," *U.S. Pat. 4,154,293,* May 15, 1979.

104. R. K. Shah, C. F. McDonald, and C. P. Howard, eds., *Compact Heat Exchangers—History, Technological Advancement and Mechanical Design Problems,* HTD vol. 10, ASME, New York, 1980.

105. R. K. Shah, "Classification of Heat Exchangers," in *Thermal-Hydraulic Fundamentals and Design,* S. Kakac, A. E. Bergles, and F. Mayinger eds., pp. 9–46, Hemisphere/McGraw-Hill, New York, 1981.

106. M. Ito, H. Kimura, and T. Senshu, "Development of High Efficiency Air-Cooled Heat Exchangers," *Hitachi Rev.* (20): 323–326, 1977.

107. W. M. Kays and A. L. London, *Compact Heat Exchangers,* 3d ed., McGraw-Hill, New York, 1984.

108. R. L. Webb, "Air-Side Heat Transfer in Finned Tube Heat Exchangers," *Heat Transfer Eng.* (1/3): 33–49, 1980.

109. L. Goldstein Jr. and E. M. Sparrow, "Experiments on the Transfer Characteristics of a Corrugated Fin and Tube Heat Exchanger Configuration," *J. Heat Transfer* (98): 26–34, 1976.

110. S. W. Krückels and V. Kottke, "Untersuchung über die Verteilung des Wärmeübergangs an Rippen und Rippen Rohr-Modellen," *Chem. Ing. Tech.* (42): 355–362, 1970.

111. E. M. Sparrow, B. R. Baliga, and S. V. Patankar, "Heat Transfer and Fluid Flow Analysis of Interrupted-Wall Channels, With Applications to Heat Exchangers," *J. Heat Transfer* (99): 4–11, 1977.

112. S. V. Patankar and C. Prakash, "An Analysis of the Effect of Plate Thickness on Laminar Flow and Heat Transfer in Interrupted-Plate Passages," in *Advances in Enhanced Heat Transfer—1981,* R. L.

Webb, T. C. Carnavos, E. L. Park Jr., and K. M. Hostetler eds., HTD vol. 18, pp. 51–59, ASME, New York, 1981.

113. A. P. Watkinson, D. C. Miletti, and G. R. Kubanek, "Heat Transfer and Pressure Drop of Internally Finned Tubes in Laminar Oil Flow," *ASME Paper 75-HT-41,* ASME, New York, 1975.

114. W. J. Marner and A. E. Bergles, "Augmentation of Highly Viscous Laminar Heat Transfer Inside Tubes With Constant Wall Temperature," *Experimental Thermal and Fluid Science* (2): 252–257, 1989.

115. A. E. Bergles, "Enhancement of Heat Transfer," in *Heat Transfer 1978,* Proceedings of the 6th International Heat Transfer Conference, vol. 6, pp. 89–108, Hemisphere, Washington, DC, 1978.

116. T. C. Carnavos, "Heat Transfer Performance of Internally Finned Tubes in Turbulent Flow," in *Advances in Enhanced Heat Transfer,* pp. 61–67, ASME, New York, 1979.

117. W. E. Hilding and C. H. Coogan Jr., "Heat Transfer and Pressure Loss Measurements in Internally Finned Tubes," in *Symp. Air-Cooled Heat Exchangers,* pp. 57–85, ASME, New York, 1964.

118. S. V. Patankar, M. Ivanovic, and E. M. Sparrow, "Analysis of Turbulent Flow and Heat Transfer in Internally Finned Tube and Annuli," *J. Heat Transfer* (101): 29–37, 1979.

119. T. C. Carnavos, "Cooling Air in Turbulent Flow With Internally Finned Tubes," *Heat Transfer Eng.* (1/2): 41–46, 1979.

120. D. Q. Kern and A. D. Kraus, *Extended Surface Heat Transfer,* McGraw-Hill, New York, 1972.

121. A. Y. Gunter and W. A. Shaw, "Heat Transfer, Pressure Drop and Fouling Rates of Liquids for Continuous and Noncontinuous Longitudinal Fins," *Trans. ASME* (64): 795–802, 1942.

122. L. Clarke and R. E. Winston, "Calculation of Finside Coefficients in Longitudinal Finned-Tube Heat Exchangers," *Chem. Eng. Prog.* (51/3): 147–150, 1955.

123. M. M. El-Wakil, *Nuclear Energy Conversion,* American Nuclear Society, La Grange Park, IL, 1978.

124. D. Gorenflo, "Zum Wärmeübergang bei Blasenverdampfung an Rippenrohren," dissertation, Technische Hochschule, Karlsruhe, Germany, 1966.

125. G. Hesse, "Heat Transfer in Nucleate Boiling, Maximum Heat Flux and Transition Boiling," *Int. J. Heat Mass Transfer* (16): 1611–1627, 1973.

126. J. W. Westwater, "Development of Extended Surfaces for Use in Boiling Liquids," *AIChE Symp. Ser.* (69/131): 1–9, 1973.

127. D. L. Katz, J. E. Meyers, E. H. Young, and G. Balekjian, "Boiling Outside Finned Tubes," *Petroleum Refiner* (34): 113–116, 1955.

128. K. Nakajima and A. Shiozawa, "An Experimental Study on the Performance of a Flooded Type Evaporator," *Heat Transfer Jpn. Res.* (4/4): 49–66, 1975.

129. N. Arai, T. Fukushima, A. Arai, T. Nakajima, K. Fujie, and Y. Nakayama, "Heat Transfer Tubes Enhancing Boiling and Condensation in Heat Exchanger of a Refrigerating Machine," *ASHRAE Trans.* (83/2): 58–70, 1977.

130. J. C. Corman and M. H. McLaughlin, "Boiling Heat Transfer With Structured Surfaces," *ASHRAE Trans.* (82/1): 906–918, 1976.

131. V. N. Schultz, D. K. Edwards, and I. Catton, "Experimental Determination of Evaporative Heat Transfer Coefficients on Horizontal, Threaded Tubes," *AIChE Symp. Ser.* (73/164): 223–227, 1977.

132. R. J. Conti, "Experimental Investigations of Horizontal Tube Ammonia Film Evaporators With Small Temperature Differentials," *Proc. 5th Ocean Thermal Energy Conversion Conf.,* Miami Beach, FL, pp. VI-161–VI-180, 1978.

133. S. Sideman and A. Levin, "Effect of the Configuration on Heat Transfer to Gravity Driven Films Evaporating on Grooved Tubes," *Desalination* (31): 7–18, 1979.

134. R. B. Cox, G. A. Matta, A. S. Pascale, and K. G. Stromberg, "Second Report on Horizontal-Tubes Multiple-Effect Process Pilot Plant Tests and Design," *Off. Saline Water Res. Dev. Prog. Rep. No. 529,* OSW, Washington, DC, May 1970.

135. D. G. Thomas and G. Young, "Thin Film Evaporation Enhancement by Finned Surfaces," *Ind. Eng. Chem. Proc. Des. Dev.* (9): 317–323, 1970.

136. J. J. Lorenz, D. T. Yung, D. L. Hillis, and N. F. Sather, "OTEC Performance Tests of the Carnegie-Mellon University Vertical Fluted-Tube Evaporator," *ANL/OTEC-PS-5.* Argonne National Laboratory, Argonne, IL, July 1979.

137. E. U. Schlünder and M. Chwala, "Örtlicher Wärmeübergang und Druckabfall bei der Strömung verdampfender Kältemittel in innenberippten, waggerechten Rohren," *Kältetech. Klim.* (21/5): 136–139, 1969.

138. G. R. Kubanek and D. L. Miletti, "Evaporative Heat Transfer and Pressure Drop Performance of Internally-Finned Tubes With Refrigerant 22," *J. Heat Transfer* (101): 447–452, 1979.

139. M. Ito and H. Kimura, "Boiling Heat Transfer and Pressure Drop in Internal Spiral-Grooved Tubes," *Bull. JSME* (22/171): 1251–1257, 1979.

140. J. M. Robertson, "Review of Boiling, Condensing and Other Aspects of Two-Phase Flow in Plate Fin Heat Exchangers," in *Compact Heat Exchangers—History, Technological Advances and Mechanical Design Problems,* R. K. Shah, C. F. McDonald, and C. P. Howard eds., HTD vol. 10, pp. 17–27, ASME, New York, 1980.

141. C. B. Panchal, D. L. Hillis, J. J. Lorenz, and D. T. Yung, "OTEC Performance Tests of the Trane Plate-Fin Heat Exchanger," *ANL/OTEC-PS-7,* Argonne National Laboratory, Argonne, IL, April 1981.

142. D. Yung, J. J. Lorenz, and C. Panchal, "Convective Vaporization and Condensation in Serrated-Fin Channels," in *Heat Transfer in Ocean Thermal Energy Conversion [OTEC] Systems,* W. L. Owens, ed., HTD vol. 12, pp. 29–37, ASME, New York, 1980.

143. C. C. Chen, J. V. Loh, and J. W. Westwater, "Prediction of Boiling Heat Transfer in a Compact Plate-Fin Heat Exchanger Using the Improved Local Technique," *Int. J. Heat Mass Transfer* (24): 1907–1912, 1981.

144. K. O. Beatty Jr. and D. L. Katz, "Condensation of Vapors on Outside of Finned Tubes," *Chem. Eng. Prog.* (44/1): 55–70, 1948.

145. E. H. Young and D. J. Ward, "How to Design Finned Tube Shell and Tube Heat Exchangers," *The Refining Engineer,* pp. C-32–C-36, November 1957.

146. T. M. Rudy and R. L. Webb, "Condensate Retention of Horizontal Integral-Fin Tubing," in *Advance in Enhanced Heat Transfer—1981,* R. L. Webb, T. C. Carnavos, E. L. Park Jr., and K. M. Hostetler eds., HTD vol. 18, pp. 35–41, ASME, New York, 1981.

147. R. Chandran and F. A. Watson, "Condensation on Static and Rotating Pinned Tubes," *Trans. Inst. Chem. Eng.* (54): 65–72, 1976.

148. R. L. Webb and D. L. Gee, "Analytical Predictions for a New Concept Spine-Fin Surface Geometry," *ASHRAE Trans.* (85/2): 274–283, 1979.

149. F. Notaro, "Enhanced Condensation Heat Transfer Device and Method," *U.S. Pat. 4,154,294,* May 15, 1979.

150. R. Gregorig, "Hautkondensation an Feingewellten Oberflächen bei Berücksichtigung der Oberflächenspannungen," *Z. Angew. Math. Phys.* (5): 36–49, 1954.

151. A. Thomas, J. J. Lorenz, D. A. Hillis, D. T. Young, and N. F. Sather, "Performance Tests of 1 Mwt Shell and Tube Heat Exchangers for OTEC," *Proc. 6th OTEC Conf.,* Washington, DC, vol. 2, p. 11.1, 1979.

152. A. Blumenkrantz and J. Taborek, "Heat Transfer and Pressure Drop Characteristics of Turbotec Spirally Deep Grooved Tubes in the Turbulent Regime," *Heat Transfer Research, Inc., Rep. 2439-300-7,* HTRI, Pasadena, CA, December 1970.

153. J. Palen, B. Cham, and J. Taborek, "Comparison of Condensation of Steam on Plain and Turbotec Spirally Grooved Tubes in a Baffled Shell-and-Tube Condenser," *Heat Transfer Research, Inc., Rep. 2439-300-6,* HTRI, Pasadena, CA, January 1971.

154. P. J. Marto, R. J. Reilly, and J. H. Fenner, "An Experimental Comparison of Enhanced Heat Transfer Condenser Tubing," in *Advances in Enhanced Heat Transfer,* J. M. Chenoweth, J. Kaellis, J. W. Michel, and S. Shenkman eds., pp. 1–9, ASME, New York, 1979.

155. T. C. Carnavos, "An Experimental Study: Condensing R-11 on Augmented Tubes," *ASME Paper 80-HT-54,* ASME, New York, 1980.

156. M. H. Mehta and M. R. Rao, "Heat Transfer and Frictional Characteristics of Spirally Enhanced Tubes for Horizontal Condensers," in *Advances in Enhanced Heat Transfer,* J. M. Chenoweth, J. Kaellis, J. W. Michel, and S. Shenkman eds., pp. 11–21, ASME, New York, 1979.

157. J. G. Withers and E. H. Young, "Steam Condensing on Vertical Rows of Horizontal Corrugated and Plain Tubes," *Ind. Eng. Chem. Process Des. Dev.* (10): 19–30, 1971.

158. D. G. Thomas, "Enhancement of Film Condensation Rate on Vertical Tubes by Longitudinal Fins," *AIChE J.* (14): 644–649, 1968.

159. D. G. Thomas, "Enhancement of Film Condensation Rate on Vertical Tubes by Vertical Wires," *Ind. Eng. Chem. Fund.* (6): 97–103, 1967.

160. Y. Mori, K. Hijikata, S. Hirasawa, and W. Nakayama, "Optimized Performance of Condensers With Outside Condensing Surfaces," *J. Heat Transfer* (103): 96–102, 1981.

161. C. G. Barnes Jr. and W. M. Rohsenow, "Vertical Fluted Tube Condenser Performance Prediction," *Proc. 7th Int. Heat Trans. Conf.,* Hemisphere, Washington, DC, vol. 5, pp. 39–43, 1982.

162. L. G. Lewis and N. F. Sather, "OTEC Performance Tests of the Carnegie-Mellon University Vertical Fluted-Tube Condenser," *ANL/OTEC-PS-4,* Argonne National Laboratory, Argonne, IL, May 1979.

163. N. Domingo, "Condensation of Refrigerant-11 on the Outside of Vertical Enhanced Tubes," *ORNL/TM-7797,* Oak Ridge National Laboratory, Oak Ridge, TN, August 1981.

164. L. G. Alexander and H. W. Hoffman, "Performance Characteristics of Corrugated Tubes for Vertical Tube Evaporators," *ASME Paper 71-HT-30,* ASME, New York, 1971.

165. D. L. Vrable, W. J. Yang, and J. A. Clark, "Condensation of Refrigerant-12 inside Horizontal Tubes With Internal Axial Fins," in *Heat Transfer 1974,* vol. III, pp. 250–254, Japan Society of Mechanical Engineers, Tokyo, Japan, 1974.

166. R. L. Reisbig, "Condensing Heat Transfer Augmentation Inside Splined Tubes," *ASME Paper 74-HT-7,* ASME, New York, July 1974.

167. J. H. Royal and A. E. Bergles, "Augmentation of Horizontal In-Tube Condensation by Means of Twisted-Tape Inserts and Internally-Finned Tubes," *J. Heat Transfer* (100): 17–24, 1978.

168. J. H. Royal and A. E. Bergles, "Pressure Drop and Performance Evaluation of Augmented In-Tube Condensation," in *Heat Transfer 1978, Proc. 6th Int. Conf.,* vol. 2, pp. 459–464, Hemisphere, Washington, DC, 1978.

169. M. Luu and A. E. Bergles, "Experimental Study of the Augmentation of In-Tube Condensation of R-113," *ASHRAE Trans.* (85/2): 132–145, 1979.

170. M. Luu and A. E. Bergles, "Enhancement of Horizontal In-Tube Condensation of R-113," *ASHRAE Trans.* (86/1): 293–312, 1980.

171. V. G. Rifert and V. Y. Zadiraka, "Steam Condensation Inside Plain and Profiled Horizontal Tubes," *Therm. Eng.* (25/8): 54–57, 1978.

172. Y. Mori and W. Nakayama, "High-Performance Mist Cooled Condensers for Geothermal Binary Cycle Plants," *Heat Transfer in Energy Problems, Proc. Jpn-U.S. Joint Sem.,* Tokyo, pp. 189–196, Sept. 30–Oct. 2, 1980.

173. J. H. Sununu, "Heat Transfer with Static Mixer Systems," *Kenics Corp. Tech. Rep. 1002,* Kenics, Danvers, MA, 1970.

174. W. E. Genetti and S. J. Priebe, "Heat Transfer With a Static Mixer," AIChE paper presented at the Fourth Joint Chemical Engineering Conference, Vancouver, Canada, 1973.

175. T. H. Van Der Meer and C. J. Hoogendoorn, "Heat Transfer Coefficients for Viscous Fluids in a Static Mixer," *Chem. Eng. Sci.* (33): 1277–1282, 1978.

176. W. J. Marner and A. E. Bergles, "Augmentation of Tubeside Laminar Flow Heat Transfer by Means of Twisted-Tape Inserts, Static-Mixer Inserts and Internally Finned Tubes," *Heat Transfer 1978, Proc. 6th Int. Heat Transfer Conf.,* Hemisphere, Washington, DC, vol. 2, pp. 583–588, 1978.

177. S. T. Lin, L. T. Fan, and N. Z. Azer, "Augmentation of Single Phase Convective Heat Transfer With In-Line Static Mixers," *Proc. 1978 Heat Transfer Fluid Mech. Inst.,* pp. 117–130, Stanford University Press, Stanford, CA, 1978.

178. M. H. Pahl and E. Muschelknautz, "Einsatz and Auslegung statischer Mischer," *Chem. Ing. Tech.* (51): 347–364, 1979.

179. L. B. Evans and S. W. Churchill, "The Effect of Axial Promoters on Heat Transfer and Pressure Drop Inside a Tube," *Chem. Eng. Prog. Symp. Ser. 59* (41): 36–46, 1963.

180. D. G. Thomas, "Enhancement of Forced Convection Heat Transfer Coefficient Using Detached Turbulence Promoters," *Ind. Eng. Chem. Process Des. Dev.* (6): 385–390, 1967.

181. S. Maezawa and G. S. H. Lock, "Heat Transfer Inside a Tube With a Novel Promoter," *Heat Transfer 1978, Proc. 6th Int. Heat Transfer Conf.,* Hemisphere, Washington, DC, vol. 2, pp. 596–600, 1978.

182. F. E. Megerlin, R. W. Murphy, and A. E. Bergles, "Augmentation of Heat Transfer in Tubes by Means of Mesh and Brush Inserts," *J. Heat Transfer* (96): 145–151, 1974.

183. E. O. Moeck, G. A. Wilkhammer, I. P. L. Macdonald, and J. G. Collier, "Two Methods of Improving the Dryout Heat-Flux for High Pressure Steam/Water Flow," *Atomic Energy of Canada, Ltd. 2109,* AECL, Chalk River, Canada, 1964.

184. L. S. Tong, R. W. Steer, A. H. Wenzel, M. Bogaardt, and C. L. Spigt, "Critical Heat Flux of a Heater Rod in the Center of Smooth and Rough Square Sleeves, and in Line-Contact With an Unheated Wall," *ASME Paper 67-WA/HT-29,* ASME, New York, 1967.

185. E. P. Quinn, "Transition Boiling Heat Transfer Program," *6th Q. Prog. Rep., General Electric Atomic Power 4646,* GEAP, San Jose, CA, 1964.

186. A. N. Ryabov, F. T. Kamen'shchikov, V. N. Filipov, A. F. Chalykh, T. Yugay, Y. V. Stolyarov, T. I. Blagovestova, V. M. Mandrazhitskiy, and A. I. Yemelyanov, "Boiling Crisis and Pressure Drop in Rod Bundles With Heat Transfer Enhancement Devices," *Heat Transfer Sov. Res.* (9/1): 112–122, 1977.

187. D. C. Groeneveld and W. W. Yousef, "Spacing Devices for Nuclear Fuel Bundles: A Survey of Their Effect on CHF, Post CHF Heat Transfer and Pressure Drop," *Proc. ANS/ASME/NRC Information Topical Meeting on Nuclear Reactor Thermal-Hydraulics, Nuclear Regulatory Commission/CP-0014* (2): 1111–1130, 1980.

188. N. Z. Azer, L. T. Fan, and S. T. Lin, "Augmentation of Condensation Heat Transfer With In-Line Static Mixers," *Proc. 1976 Heat Transfer Fluid Mech. Inst.,* Stanford University Press, Stanford, CA, pp. 512–526, 1976.

189. L. T. Fan, S. T. Lin, and N. Z. Azer, "Surface Renewal Model of Condensation Heat Transfer in Tubes With In-Line Static Mixers," *Int. J. Heat Mass Transfer* (21): 849–854, 1978.

190. R. Razgaitis and J. P. Holman, "A Survey of Heat Transfer in Confined Swirl Flows," in *Future Energy Production Systems, Heat and Mass Transfer Processes,* vol. 2, pp. 831–866, Academic, New York, 1976.

191. S. W. Hong and A. E. Bergles, "Augmentation of Laminar Flow Heat Transfer by Means of Twisted-Tape Inserts," *J. Heat Transfer* (98): 251–256, 1976.

192. F. Huang and F. K. Tsou, "Friction and Heat Transfer in Laminar Free Swirling Flow in Pipes," *Gas Turbine Heat Transfer,* ASME, New York, 1979.

193. A. W. Date, "Prediction of Fully-Developed Flow in a Tube Containing a Twisted Tape," *Int. J. Heat Mass Transfer* (17): 845–859, 1974.

194. R. K. Shah and A. L. London, *Laminar Flow Forced Convection in Ducts,* p. 380, Academic, New York, 1978.

195. R. F. Lopina and A. E. Bergles, "Heat Transfer and Pressure Drop in Tape Generated Swirl Flow of Single-Phase Water," *J. Heat Transfer* (91): 434–442, 1969.

196. R. Thorsen and F. Landis, "Friction and Heat Transfer Characteristics in Turbulent Swirl Flow Subject to Large Transverse Temperature Gradients," *J. Heat Transfer* (90): 87–98, 1968.

197. A. P. Colburn and W. J. King, "Heat Transfer and Pressure Drop in Empty, Baffled, and Packed Tubes. III: Relation Between Heat Transfer and Pressure Drop," *Ind. Eng. Chem.* (23): 919–923, 1931.

198. S. I. Evans and R. J. Sarjant, "Heat Transfer and Turbulence in Gases Flowing Inside Tubes," *J. Inst. Fuel* (24): 216–227, 1951.

199. E. Smithberg and F. Landis, "Friction and Forced Convection Heat Transfer Characteristics in Tubes With Twisted Tape Swirl Generators," *J. Heat Transfer* (86): 39–49, 1964.

200. W. R. Gambill, R. D. Bundy, and R. W. Wansbrough, "Heat Transfer, Burnout, and Pressure Drop for Water in Swirl Flow Tubes With Internal Twisted Tapes," *Chem. Eng. Prog. Symp. Ser.* (57/32): 127–137, 1961.

201. M. H. Ibragimov, E. V. Nomofelov, and V. I. Subbotin, "Heat Transfer and Hydraulic Resistance With the Swirl-Type Motion of Liquid in Pipes," *Teploenergetika* (8/7): 57–60, 1962.

202. N. D. Greene, *Convair Aircraft,* private communication to W. R. Gambill, May 1969, cited in W. R. Gambill and R. D. Bundy, "An Evaluation of the Present Status of Swirl Flow Heat Transfer," *ASME Paper 62-HT-42,* ASME, New York, 1962.

203. B. Shiralker and P. Griffith, "The Effect of Swirl, Inlet Conditions, Flow Direction, and Tube Diameter on the Heat Transfer to Fluids at Supercritical Pressure," *J. Heat Transfer* (42): 465–474, 1970.

204. W. R. Gambill and N. D. Greene, "A Preliminary Study of Boiling Burnout Heat Fluxes for Water in Vortex Flow," *Chem. Eng. Prog.* (54/10): 68–76, 1958.

205. F. Mayinger, O. Schad, and E. Weiss, "Investigations Into the Critical Heat Flux in Boiling," *Mannesmann Augsburg Nürnberg Rep. 09.03.01,* MAN, Nürnberg, Germany, 1966.

206. A. P. Ornatskiy, V. A. Chernobay, A. F. Vasilyev, and S. V. Perkov, "A Study of the Heat Transfer Crisis With Swirled Flows Entering an Annular Passage," *Heat Transfer Sov. Res.* (5/4): 7–10, 1973.

207. R. F. Lopina and A. E. Bergles, "Subcooled Boiling of Water in Tape-Generated Swirl Flow," *J. Heat Transfer* (95): 281–283, 1973.

208. H. H. Sephton, "Interface Enhancement for Vertical Tube Evaporator: A Novel Way of Substantially Augmenting Heat and Mass Transfer," *ASME Paper 71-HT-38,* ASME, New York, 1971.

209. A. E. Bergles, W. D. Fuller, and S. J. Hynek, "Dispersed Film Boiling of Nitrogen With Swirl Flow," *Int. J. Heat Mass Transfer* (14): 1343–1354, 1971.

210. M. Cumo, G. E. Farello, G. Ferrari, and G. Palazzi, "The Influence of Twisted Tapes in Subcritical, Once-Through Vapor Generator in Counter Flow," *J. Heat Transfer* (96): 365–370, 1974.

211. A. Hunsbedt and J. M. Roberts, "Thermal-Hydraulic Performance of a 2MWT Sodium Heated, Forced Recirculation Steam Generator Model," *J. Eng. Power* (96): 66–76, 1974.

212. C. Fouré, C. Moussez, and D. Eidelman, "Techniques for Vortex Type Two-Phase Flow in Water Reactors," *Proc. 3d Int. Conf. Peaceful Uses of Atomic Energy,* United Nations, New York, vol. 8, pp. 255–261, 1965.

213. M. K. Jensen, "Boiling Heat Transfer and Critical Heat Flux in Helical Coils," PhD dissertation, Iowa State University, Ames, IA, 1980.

214. M. K. Jensen and A. E. Bergles, "Critical Heat Flux in Helically Coiled Tubes," *J. Heat Transfer* (103): 660–666, 1981.

215. G. G. Shklover and A. V. Gerasimov, "Heat Transfer of Moving Steam in Coil-Type Heat Exchangers," *Teploenergctika* (10/5): 62–65, 1963.

216. Z. L. Miropolskii and A. Kurbanmukhamedov, "Heat Transfer With Condensation of Steam Within Coils," *Therm. Eng.,* No. 5: 111–114, 1975.

217. D. P. Traviss and W. M. Rohsenow, "The Influence of Return Bends on the Downstream Pressure Drop and Condensation Heat Transfer in Tubes," *ASHRAE Trans.* (79/1): 129–137, 1973.

218. W. D. Allingham and J. A. McEntire, "Determination of Boiling Film Coefficient for a Heated Horizontal Tube in Water Saturated with Material," *J. Heat Transfer* (83): 71–76, 1961.

219. C. P. Costello and E. R. Redeker, "Boiling Heat Transfer and Maximum Heat Flux for a Surface With Coolant Supplied by Capillary Wicking," *Chem. Eng. Prog. Symp. Ser.* (59/41): 104–113, 1963.

220. C. P. Costello and W. J. Frea, "The Role of Capillary Wicking and Surface Deposits in the Attainment of High Pool Boiling Burnout Heat Fluxes," *AIChE J.* (10): 393–398, 1964.

221. J. C. Corman and M. H. McLaughlin, "Boiling Augmentation With Structured Surfaces," *ASHRAE Trans.* (82/1): 906–918, 1976.

222. R. S. Gill, "Pool Boiling in the Presence of Capillary Wicking Materials," SM thesis in mechanical engineering, Massachusetts Institute of Technology, Cambridge, MA, 1967.

223. R. W. Watkins, C. R. Robertson, and A. Acrivos, "Entrance Region Heat Transfer in Flowing Suspensions," *Int. J. Heat Mass Transfer* (19): 693–695, 1976.

224. M. Tamari and K. Nishikawa, "The Stirring Effect of Bubbles Upon the Heat Transfer to Liquids," *Heat Transfer Jpn. Res.* (5/2): 31–44, 1976.

225. W. F. Hart, "Heat Transfer in Bubble-Agitated Systems. A General Correlation," *I&EC Process Des. Dev.* (15): 109–111, 1976.

226. D. B. R. Kenning and Y. S. Kao, "Convective Heat Transfer to Water Containing Bubbles: Enhancement Not Dependent on Thermocapillarity," *Int. J. Heat Mass Transfer* (15): 1709–1718, 1972.

227. E. Baker, "Liquid Immersion Cooling of Small Electronic Devices," *Microelectronics and Reliability* (12): 163–173, 1973.

228. M. Behar, M. Courtaud, R. Ricque, and R. Semeria, "Fundamental Aspects of Subcooled Boiling With and Without Dissolved Gases," *Proc. 3d Int. Heat Transfer Conf.,* AIChE, New York, vol. 4, pp. 1–11, 1966.

229. M. Jakob and W. Linke, "Der Wärmeübergang beim Verdampfen von Flüssigkeiten an senkrechten und waagerechten Flächen," *Phys. Z.* (36): 267–280, 1935.

230. T. H. Insinger Jr. and H. Bliss, "Transmission of Heat to Boiling Liquids," *Trans. AIChE* (36): 491–516, 1940.

231. A. I. Morgan, L. A. Bromley, and C. R. Wilke, "Effect of Surface Tension on Heat Transfer in Boiling," *Ind. Eng. Chem.* (41): 2767–2769, 1949.

232. E. K. Averin and G. N. Kruzhilin, "The Influence of Surface Tension and Viscosity on the Conditions of Heat Exchange in the Boiling of Water," *Izv. Akad. Nauk SSSR Otdel. Tekh. Nauk* (10): 131–137, 1955.

233. A. J. Lowery Jr. and J. W. Westwater, "Heat Transfer to Boiling Methanol—Effect of Added Agents," *Ind. Eng. Chem.* (49): 1445–1448, 1957.

234. J. G. Collier, "Multicomponent Boiling and Condensation," in *Two-Phase Flow and Heat Transfer in the Power and Process Industries,* pp. 520–557, Hemisphere, Washington, DC, and McGraw-Hill, New York, 1981.

235. W. R. van Wijk, A. S. Vos, and S. J. D. van Stralen, "Heat Transfer to Boiling Binary Liquid Mixtures," *Chem. Eng. Sci.* (5): 68–80, 1956.

236. S. J. D. van Stralen, "Heat Transfer to Boiling Binary Liquid Mixtures," *Brit. Chem. Eng.* (I/4): 8–17; (II/4): pp. 78–82, 1959.

237. M. Carne, "Some Effects of Test Section Geometry, in Saturated Pool Boiling, on the Critical Heat Flux for Some Organic Liquids and Liquid Mixtures," in *AIChE Preprint 6 for 7th Nat. Heat Transfer Conf.,* AIChE, New York, August 1964.

238. S. J. D. van Stralen, "Nucleate Boiling in Binary Systems," in *Augmentation of Convective Heat and Mass Transfer,* A. E. Bergles and R. L. Webb eds., pp. 133–147, ASME, New York, 1970.

239. H. J. Gannett Jr. and M. C. Williams, "Pool Boiling in Dilute Nonaqueous Polymer Solutions," *Int. J. Heat Mass Transfer* (11): 1001–1005, 1971.

240. M. K. Jensen, A. E. Bergles, and F. A. Jeglic, "Effects of Oily Contaminants on Nucleate Boiling of Water," *AIChE Symp. Ser.* (75/189): 194–203, 1979.

241. G. Leppert, C. P. Costello, and B. M. Hoglund, "Boiling Heat Transfer to Water Containing a Volatile Additive," *Trans. ASME* (80): 1395–1404, 1958.

242. A. E. Bergles and L. S. Scarola, "Effect of a Volatile Additive on the Critical Heat Flux for Surface Boiling of Water in Tubes," *Chem. Eng. Sci.* (21): 721–723, 1966.

243. H. H. Sephton, "Upflow Vertical Tube Evaporation of Sea Water With Interface Enhancement: Process Development by Pilot Plant Testing," *Desalination* (16): 1–13, 1975.

244. A. E. Bergles, G. H. Junkhan, and J. K. Hagge, "Advanced Cooling Systems for Agricultural and Industrial Machines," *SAE Paper 751183,* SAE, Warrendale, PA, 1976.

245. G. K. Rhode, D. M. Roberts, D. C. Schluderberg, and E. E. Walsh, "Gas-Suspension Coolants for Power Reactors," *Proc. Am. Power Conf.* (22): 130–137, 1960.

246. D. C. Schluderberg, R. L. Whitelaw, and R. W. Carlson, "Gaseous Suspensions—A New Reactor Coolant," *Nucleonics* (19): 67–76, 1961.

247. W. T. Abel, D. E. Bluman, and J. P. O'Leary, "Gas-Solids Suspensions as Heat-Carrying Mediums," *ASME Paper 63-WA-210,* ASME, New York, 1963.

248. R. Pfeffer, S. Rossetti, and S. Lieblein, "Analysis and Correlation of Heat Transfer Coefficient and Friction Factor Data for Dilute Gas-Solid Suspensions," *NASA TN D-3603,* NASA, Cleveland, OH, 1966.

249. R. Pfeffer, S. Rossetti, and S. Lieblein, "The Use of a Dilute Gas-Solid Suspension as the Working Fluid in a Single Loop Brayton Space Power Generation Cycle," *AIChE Paper 49c,* AIChE, New York, presented at 1967 national meeting.

250. C. A. Depew and T. J. Kramer, "Heat Transfer to Flowing Gas-Solid Mixtures," in *Advances in Heat Transfer,* vol. 9, pp. 113–180, Academic Press, New York, 1973.

251. W. C. Thomas and J. E. Sunderland, "Heat Transfer Between a Plane Surface and Air Containing Water Droplets," *Ind. Eng. Chem. Fund.* (9): 368–374, 1970.

252. W.-J. Yang and D. W. Clark, "Spray Cooling of Air-Cooled Compact Heat Exchangers." *Int. J. Heat Mass Transfer* (18): 311–317, 1975.

253. V. W. Uhl, "Mechanically Aided Heat Transfer to Viscous Materials," in *Augmentation of Convective Heat and Mass Transfer,* pp. 109–117, ASME, New York, 1970.

254. W. R. Penney, "The Spiralator—Initial Tests and Correlations," in *AIChE Preprint 16 for 8th Nat. Heat Transfer Conf.,* AIChE, New York, 1965.

255. F. S. Pramuk and J. W. Westwater, "Effect of Agitation on the Critical Temperature Difference for Boiling Liquid," *Chem. Eng. Prog. Symp. Ser.* (52/18): 79–83, 1956.

256. J. K. Hagge and G. H. Junkhan, "Experimental Study of a Method of Mechanical Augmentation of Convective Heat Transfer Coefficients in Air," *HTL-3, ISU-ERI-Ames-74158,* Iowa State University, Ames, IA, November 1974.

257. E. L. Lustenader, R. Richter, and F. J. Neugebauer, "The Use of Thin Films for Increasing Evaporation and Condensation Rates in Process Equipment," *J. Heat Transfer* (81): 297–307, 1959.

258. B. W. Tleimat, "Performance of a Rotating Flat-Disk Wiped-Film Evaporator," *ASME Paper 71-HT-37,* ASME, New York, 1971.

259. J. E. McElhiney and G. W. Preckshot, "Heat Transfer in the Entrance Length of a Horizontal Rotating Tube," *Int. J. Heat Mass Transfer* (20): 847–854, 1977.

260. Y. Mori and W. Nakayama, "Forced Convection Heat Transfer in a Straight Pipe Rotating Around a Parallel Axis," *Int. J. Heat Mass Transfer* (10): 1179–1194, 1967.

261. V. Vidyanidhi, V. V. S. Suryanarayana, and V. C. Chenchu Raju, "An Analysis of Steady Freely Developed Heat Transfer in a Rotating Straight Pipe," *J. Heat Transfer* (99): 148–150, 1977.

262. H. Miyazaki, "Combined Free and Forced Convective Heat Transfer and Fluid Flow in a Rotating Curved Circular Tube," *Int. J. Heat Mass Transfer* (14): 1295–1309, 1971.

263. S. I. Tang and T. W. McDonald, "A Study of Heat Transfer From a Rotating Horizontal Cylinder," *Int. J. Heat Mass Transfer* (14): 1643–1658, 1971.

264. P. J. Marto and V. H. Gray, "Effects of High Accelerations and Heat Fluxes on Nucleate Boiling of Water in an Axisymmetric Rotating Boiler," *NASA TN D-6307,* NASA, Cleveland, OH, 1971.

265. V. B. Astafev and A. M. Baklastov, "Condensation of Steam on a Horizontal Rotating Disk," *Therm. Eng.* (17/9): 82–85, 1970.

266. A. A. Nicol and M. Gacesa, "Condensation of Steam on a Rotating Vertical Cylinder," *J. Heat Transfer* (97): 144–152, 1970.

267. R. M. Singer and G. W. Preckshot, "The Condensation of Vapor on a Horizontal Rotating Cylinder," *Proc. 1963 Heat Transfer Fluid Mech. Inst.,* Stanford University Press, Stanford, CA, pp. 205–221, 1963.

268. D. K. Weiler, A. M. Czikk, and R. S. Paul, "Condensation in Smooth and Porous Coated Tubes Under Multi-g Accelerations," *Chem. Eng. Prog. Symp. Ser.* (62/64): 143–149, 1966.

269. F. K. Deaver, W. R. Penney, and T. B. Jefferson, "Heat Transfer from an Oscillating Horizontal Wire to Water," *J. Heat Transfer* (84): 251–256, 1962.

270. W. R. Penney and T. B. Jefferson, "Heat Transfer From an Oscillating Horizontal Wire to Water and Ethylene Glycol," *J. Heat Transfer* (88): 359–366, 1966.

271. W. H. McAdams, *Heat Transmission,* 3d ed., p. 267, McGraw-Hill, New York, 1954.

272. R. C. Martinelli and L. M. K. Boelter, "The Effect of Vibration on Heat Transfer by Free Convection From a Horizontal Cylinder," *Heat. Pip. Air Cond.* (11): 525–527, 1939.

273. W. E. Mason and L. M. K. Boelter, "Vibration—Its Effect on Heat Transfer," *Pwr. Pl. Eng.* (44): 43–46, 1940.

274. R. Lemlich, "Effect of Vibration on Natural Convective Heat Transfer," *Ind. Eng. Chem.* (47): 1173–1180, 1955: errata (53): 314, 1961.

275. R. M. Fand and J. Kaye, "The Influence of Vertical Vibrations on Heat Transfer by Free Convection From a Horizontal Cylinder," in *International Developments in Heat Transfer,* pp. 490–498, ASME, New York, 1961.

276. R. M. Fand and E. M. Peebles, "A Comparison of the Influence of Mechanical and Acoustical Vibrations on Free Convection from a Horizontal Cylinder," *J. Heat Transfer* (84): 268–270, 1962.

277. A. J. Shine, "Comments on a Paper by Deaver et al.," *J. Heat Transfer* (84): 226, 1962.

278. R. Lemlich and M. A. Rao, "The Effect of Transverse Vibration on Free Convection From a Horizontal Cylinder," *Int. J. Heat Mass Transfer* (8): 27–33, 1965.

279. A. E. Bergles, "The Influence of Heated-Surface Vibration on Pool Boiling," *J. Heat Transfer* (91): 152–154, 1969.

280. V. D. Blankenship and J. A. Clark, "Experimental Effects of Transverse Oscillations on Free Convection of a Vertical, Finite Plate," *J. Heat Transfer* (86): 159–165, 1964.

281. J. A. Scanlan, "Effects of Normal Surface Vibration on Laminar Forced Convection Heat Transfer," *Ind. Eng. Chem.* (50): 1565–1568, 1958.

282. R. Anantanarayanan and A. Ramachandran, "Effect of Vibration on Heat Transfer From a Wire to Air in Parallel Flow," *Trans. ASME* (80): 1426–1432, 1958.

283. I. A. Raben, "The Use of Acoustic Vibrations to Improve Heat Transfer," *Proc. 1961 Heat Transfer Fluid Mech. Inst.,* Stanford University Press, Stanford, CA, pp. 90–97, 1961.

284. I. A. Raben, G. E. Cummerford, and G. E. Neville, "An Investigation of the Use of Acoustic Vibrations to Improve Heat Transfer Rates and Reduce Scaling in Distillation Units Used for Saline Water Conversion." *Off. Saline Water Res. Dev. Prog. Rep. No. 65,* OSW, Washington, DC, 1962.

285. J. W. Ogle and A. J. Engel, "The Effect of Vibration on a Double-Pipe Heat Exchanger," *AIChE Preprint 59 for 6th Nat. Heat Transfer Conf.,* AIChE, New York, 1963.

286. I. I. Palyeyev, B. D. Kachnelson, and A. A. Tarakanovskii, "Study of Process of Heat and Mass Exchange in a Pulsating Stream," *Teploenergetika* (10/4): 71, 1963.

287. E. D. Jordan and J. Steffans, "An Investigation of the Effect of Mechanically Induced Vibrations on Heat Transfer Rates in a Pressurized Water System," *New York Operations Office, Atomic Energy Comm.-2655-1,* AEC, New York, 1965.

288. R. Hsieh and G. F. Marsters, "Heat Transfer From a Vibrating Vertical Array of Horizontal Cylinders," *Can. J. Chem. Eng.* (51): 302–306, 1973.

289. F. C. McQuiston and J. D. Parker, "Effect of Vibration on Pool Boiling," in *ASME Paper 67-HT-49,* ASME, New York, 1967.

290. D. C. Price and J. D. Parker, "Nucleate Boiling on a Vibrating Surface," in *ASME Paper 67-HT-58,* ASME, New York, 1967.

291. G. M. Fuls and G. E. Geiger, "Effect of Bubble Stabilization on Pool Boiling Heat Transfer," *J. Heat Transfer* (97): 635–640, 1970.

292. H. R. Pearce, "The Effect of Vibration on Burnout in Vertical, Two-Phase Flow," *Atomic Energy Research Establishment (United Kingdom) 6375,* AERE, Harwell, UK, 1970.

293. J. C. Dent, "Effect of Vibration on Condensation Heat Transfer to a Horizontal Tube," *Proc. Inst. Mech. Eng.* (184/1): 99–105, 1969–1970.

294. Y. M. Brodov, R. Z. Salev'yev, V. A. Permayakov, V. K. Kuptsov, and A. G. Gal'perin, "The Effect of Vibration on Heat Transfer and Flow of Condensing Steam on a Single Tube," *Heat Trans. Sov. Res.* (9/1): 153–155, 1977.

295. A. L. Sprott, J. P. Holman, and F. L. Durand, "An Experimental Study of the Effects of Strong Progressive Sound Fields on Free-Convection Heat Transfer From a Horizontal Cylinder," *ASME Paper 60-HT-19,* ASME, New York, 1960.

296. R. M. Fand and J. Kaye, "The Influence of Sound on Free Convection From a Horizontal Cylinder," *J. Heat Transfer* (83): 133, 1961.

297. B. H. Lee and P. D. Richardson, "Effect of Sound on Heat Transfer From a Horizontal Circular Cylinder at Large Wavelength," *J. Mech. Eng. Sci.* (7): 127–130, 1965.

298. R. M. Fand, J. Roos, P. Cheng, and J. Kaye, "The Local Heat-Transfer Coefficient Around a Heated Horizontal Cylinder in an Intense Sound Field," *J. Heat Transfer* (84): 245–250, 1962.

299. P. D. Richardson, "Local Details of the Influence of a Vertical Sound Field on Heat Transfer From a Circular Cylinder," *Proc. 3d Int. Heat Transfer Conf.,* AIChE, New York, vol. 3, pp. 71–77, 1966.

300. J. H. Seely, "Effect of Ultrasonics on Several Natural Convection Cooling Systems," master's thesis, Syracuse University, Syracuse, NY, 1960.

301. A. A. Zhukauskas, A. A. Shlanchyauskas, and Z. P. Yaronees, "Investigation of the Influence of Ultrasonics on Heat Exchange Between Bodies in Liquids," *J. Eng. Phys.* (4): 58–61, 1961.

302. M. B. Larson and A. L. London, "A Study of the Effects of Ultrasonic Vibrations on Convection Heat Transfer to Liquids," in *ASME Paper 62-HT-44,* ASME, New York, 1962.

303. G. C. Robinson, C. M. McClude III, and R. Hendricks Jr., "The Effects of Ultrasonics on Heat Transfer by Convection," *Am. Ceram. Soc. Bull.* (37): 399–404, 1958.

304. R. M. Fand, "The Influence of Acoustic Vibrations on Heat Transfer by Natural Convection From a Horizontal Cylinder to Water," *J. Heat Transfer* (87): 309–310, 1965.

305. J. H. Gibbons and G. Houghton, "Effects of Sonic Vibrations on Boiling," *Chem. Eng. Sci.* (15): 146, 1961.

306. K. W. Li and J. D. Parker, "Acoustical Effects on Free Convective Heat Transfer From a Horizontal Wire," *J. Heat Transfer* (89): 277–278, 1967.

307. R. C. Martinelli, L. M. Boelter, E. B. Weinberg, and S. Takahi, "Heat Transfer to a Fluid Flowing Periodically at Low Frequencies in a Vertical Tube," *Trans. ASME* (65): 789–798, 1943.

308. F. B. West and A. T. Taylor, "The Effect of Pulsations on Heat Transfer," *Chem. Eng. Prog.* (48): 34–43, 1952.

309. J. M. Marchant, "Discussion of a Paper by R. C. Martinelli et al.," *Trans. ASME* (65): 796–797, 1943.

310. G. B. Darling, "Heat Transfer to Liquids in Intermittent Flow," *Petroleum* (180): 177–178, 1959.

311. R. Lemlich and J. C. Armour, "Forced Convection Heat Transfer to a Pulsed Liquid," in *AIChE Preprint 2 for 6th Nat. Heat Transfer Conf.,* AIChE, New York, 1963.

312. T. Shirotsuka, N. Honda, and Y. Shima, "Analogy of Mass, Heat and Momentum Transfer to Pulsation Flow From Inside Tube Wall," *Kagaku-Kikai* (21): 638–644, 1957.

313. W. Linke and W. Hufschmidt, "Wärmeübergang bei pulsierender Strömung," *Chem. Ing. Tech.* (30): 159–165, 1958.

314. T. W. Jackson, W. B. Harrison, and W. C. Boteler, "Free Convection, Forced Convection, and Acoustic Vibrations in a Constant Temperature Vertical Tube," *J. Heat Transfer* (81): 68–71, 1959.

315. T. W. Jackson, K. R. Purdy, and C. C. Oliver, "The Effects of Resonant Acoustic Vibrations on the Nusselt Number for a Constant Temperature Horizontal Tube," *International Developments in Heat Transfer,* pp. 483–489, ASME, New York, 1961.

316. R. Lemlich and C. K. Hwu, "The Effect of Acoustic Vibration on Forced Convective Heat Transfer," *AIChE J.* (7): 102–106, 1961.

317. W. F. Mathewson and J. C. Smith, "Effect of Sonic Pulsation on Forced Convective Heat Transfer to Air and on Film Condensation of Isopropanol," *Chem. Eng. Prog. Symp. Ser.* (41/59): 173–179, 1963.

318. R. Moissis and L. A. Maroti, "The Effect of Sonic Vibrations on Convective Heat Transfer in an Automotive Type Radiator Section," *Dynatech Corp. Rep. No. 322,* Dynatech, Cambridge, MA, July 1962.

319. A. E. Bergles, "The Influence of Flow Vibrations on Forced-Convection Heat Transfer," *J. Heat Transfer* (86): 559–560, 1964.

320. S. E. Isakoff, "Effect of an Ultrasonic Field on Boiling Heat Transfer—Exploratory Investigation," in *Heat Transfer and Fluid Mechanics Institute Preprints,* pp. 16–28, Stanford University, Stanford, CA, 1956.

321. S. W. Wong and W. Y. Chon, "Effects of Ultrasonic Vibrations on Heat Transfer to Liquids by Natural Convection and by Boiling," *AIChE J.* (15): 281–288, 1969.

322. A. P. Ornatskii and V. K. Shcherbakov, "Intensification of Heat Transfer in the Critical Region With the Aid of Ultrasonics," *Teploenergetika* (6/1): 84–85, 1959.

323. D. A. DiCicco and R. J. Schoenhals, "Heat Transfer in Film Boiling With Pulsating Pressures," *J. Heat Transfer* (86): 457–461, 1964.

324. F. E. Romie and C. A. Aronson, "Experimental Investigation of the Effects of Ultrasonic Vibrations on Burnout Heat Flux to Boiling Water," *Advanced Technology Laboratories A-123,* ATL, Mountainview, CA, July 1961.

325. A. E. Bergles and P. H. Newell Jr., "The Influence of Ultrasonic Vibrations on Heat Transfer to Water Flowing in Annuli," *Int. J. Heat Mass Transfer* (8): 1273–1280, 1965.

326. R. M. Singer, "Laminar Film Condensation in the Presence of an Electromagnetic Field," in *ASME Paper 64-WA/HT-47,* ASME, New York, 1964.

327. O. C. Blomgren Sr. and O. C. Blomgren Jr., "Method and Apparatus for Cooling the Workpiece and/or the Cutting Tools of a Machining Apparatus," *U.S. Pat. 3,670,606,* 1972.

328. B. L. Reynolds and R. E. Holmes, "Heat Transfer in a Corona Discharge," *Mech. Eng.,* pp. 44–49, October 1976.

329. J. E. Porter and R. Poulter, "Electro-Thermal Convection Effects With Laminar Flow Heat Transfer in an Annulus," in *Heat Transfer 1970,* Proceedings of the 4th International Heat Transfer Conference, vol. 2, paper FC3.7, Elsevier, Amsterdam, 1970.

330. S. D. Savkar, "Dielectrophoretic Effects in Laminar Forced Convection Between Two Parallel Plates," *Phys. Fluids* (14): 2670–2679, 1971.

331. D. C. Newton and P. H. G. Allen, "Senftleben Effect in Insulating Oil Under Uniform Electric Stress," *Letters in Heat and Mass Transfer* (4/1): 9–16, 1977.

332. M. M. Ohadi, S. S. Li, and S. Dessiatoun, "Electrostatic Heat Transfer Enhancement in a Tube Bundle Gas-to-Gas Heat Exchanger," *Enhanced Heat Transfer,* vol. 1, pp. 327–335, 1994.

333. T. Mizushina, H. Ueda, and T. Matsumoto, "Effect of Electrically Induced Convection on Heat Transfer of Air Flow in an Annulus," *J. Chem. Eng. Jpn.* (9/2): 97–102, 1976.

334. H. Y. Choi, "Electrohydrodynamic Boiling Heat Transfer," *Mech. Eng. Rep. 63-12-1,* Tufts University, Meford, MA, December 1961.

335. R. L. Durfee, "Boiling Heat Transfer of Electric Field (EHD)," *At. Energy Comm. Rep. NYO-24-04-76,* AEC, New York, 1966.

336. H. R. Velkoff and J. H. Miller, "Condensation of Vapor on a Vertical Plate With a Transverse Electrostatic Field," *J. Heat Transfer* (87): 197–201, 1965.

337. H. Y. Choi and J. M. Reynolds, "Study of Electrostatic Effects on Condensing Heat Transfer," *Air Force Flight Dynamics Laboratory TR-65-51,* 1966.

338. H. Y. Choi, "Electrohydrodynamic Condensation Heat Transfer," *ASME Paper 67-HT-39,* ASME, New York, 1967.

339. E. E. Gose, E. E. Peterson, and A. Acrivos, "On the Rate of Heat Transfer in Liquids With Gas Injection Through the Boundary Layer," *J. Appl. Phys.* (28): 1509, 1957.

340. E. E. Gose, A. Acrivos, and E. E. Peterson, "Heat Transfer to Liquids with Gas Evolution at the Interface," AIChE, New York, paper presented at AIChE annual meeting, 1960.

341. G. E. Sims, U. Aktürk, and K. O. Evans-Lutterodt, "Simulation of Pool Boiling Heat Transfer by Gas Injection at the Interface," *Int. J. Heat Mass Transfer* (6): 531–535, 1963.

342. A. A. Kudirka, "Two-Phase Heat Transfer With Gas Injection Through a Porous Boundary Surface," in *ASME Paper 65-HT-47,* ASME, New York, 1965.

343. S. G. Bankoff, "Taylor Instability of an Evaporating Plane Interface," *AIChE J.* (7): 485–487, 1961.

344. P. C. Wayner Jr. and S. G. Bankoff, "Film Boiling of Nitrogen With Suction on an Electrically Heated Porous Plate," *AIChE J.* (11): 59–64, 1965.

345. V. K. Pai and S. G. Bankoff, "Film Boiling of Nitrogen With Suction on an Electrically Heated Horizontal Porous Plate: Effect of Flow Control Element Porosity and Thickness," *AIChE J.* (11): 65–69, 1965.

346. P. C. Wayner Jr. and A. S. Kestin, "Suction Nucleate Boiling of Water," *AIChE J.* (11): 858–865, 1965.

347. R. J. Raiff and P. C. Wayner Jr. "Evaporation From a Porous Flow Control Element on a Porous Heat Source," *Int. J. Heat Mass Transfer* (16): 1919–1930, 1973.

348. W. A. Tauscher, E. M. Sparrow, and J. R. Lloyd, "Amplification of Heat Transfer by Local Injection of Fluid into a Turbulent Tube Flow," *Int. J. Heat Mass Transfer* (13): 681–688, 1970.

349. R. B. Kinney, "Fully Developed Frictional and Heat Transfer Characteristics of Laminar Flow in Porous Tubes," *Int. J. Heat Mass Transfer* (11): 1393–1401, 1968.

350. R. B. Kinney and E. M. Sparrow, "Turbulent Flow, Heat Transfer, and Mass Transfer in a Tube with Surface Suction," *J. Heat Transfer* (92): 117–125, 1970.

351. J. K. Aggarwal and M. A. Hollingsworth, "Heat Transfer for Turbulent Flow With Suction in a Porous Tube," *Int. J. Heat Mass Transfer* (16): 591–609, 1973.

352. I. Antonir and A. Tamir, "The Effect of Surface Suction on Condensation in the Presence of a Non-condensible Gas," *J. Heat Transfer* (99): 496–499, 1977.

353. J. Lienhard and V. Dhir, "A Simple Analysis of Laminar Film Condensation With Suction," *J. Heat Transfer* (94): 334–336, 1972.

354. A. E. Bergles, R. A. Lee, and B. B. Mikic, "Heat Transfer in Rough Tubes With Tape-Generated Swirl Flow," *J. Heat Transfer* (91): 443–445, 1969.

355. Y. V. Kryukov and G. P. Boykov, "Augmentation of Heat Transfer in an Acoustic Field," *Heat Trans. Sov. Res.* (5/1): 26–28, 1973.

356. R. S. Van Rooyen and D. G. Kroeger, "Laminar Flow Heat Transfer in Internally Finned Tubes With Twisted-Tape Inserts," in *Heat Transfer 1978,* Proceedings of the 6th International Heat Transfer Conference, vol. 2, pp. 577–581, Hemisphere, Washington, DC, 1978.

357. W. J. Bartel and W. E. Genetti, "Heat Transfer From a Horizontal Bundle of Bare and Finned Tubes in an Air Fluidized Bed," *AIChE Symp. Ser. No. 128* (69): 85–93, 1973.

358. N. V. Zozulya and Y. Khorunzhii, "Heat Transfer From Finned Tubes Moving Back and Forth in Liquid, *Chem. Petroleum Eng.* (9–10): 830–832, 1968.

359. K. Min and B. T. Chao, "Particle Transport and Heat Transfer in Gas-Solid Suspension Flow Under the Influence of an Electric Field," *Nucl. Sci. Eng.* (26): 534–546, 1966.

360. S. C. Bhattacharya and D. Harrison, "Heat Transfer in a Pulsed Fluidized Bed," *Trans. Inst. Chem. Eng.* (54): 281–286, 1976.

361. J. H. Masliyah and K. Nandakumar, "Fluid Flow and Heat Transfer in Internally Finned Helical Coils," *Can. J. Chem. Eng.* (55): 27–36, 1977.

362. C. A. Bromley, R. F. Humphreys, and W. Murray, "Condensation on and Evaporation From Radially Grooved Rotating Disks," *J. Heat Transfer* (88): 80–93, 1966.

363. V. C. Van der Mast, S. M. Read, and L. A. Bromley, "Boiling of Natural Sea Water in Falling Film Evaporators," *Desalination* (18): 71–94, 1976.

364. S. P. Chary and P. K. Sarma, "Condensation on a Rotating Disk With Constant Axial Suction," *J. Heat Transfer* (98): 682–684, 1976.

365. A. E. Bergles, "Heat Transfer Enhancement—The Encouragement and Accommodation of High Heat Fluxes," *J. Heat Transfer* (119): 8–19, 1997.

366. W. M. Rohsenow, "Boiling," in *Handbook of Heat Transfer Fundamentals,* W. M. Rohsenow, J. P. Hartnett, and E. N. Ganic eds., chap. 12, McGraw-Hill, New York, 1985.

367. P. Griffith, "Dropwise Condensation," *Handbook of Heat Transfer Fundamentals,* W. M. Rohsenow, J. P. Hartnett, and E. N. Ganic eds., chap. 11, part 2, McGraw-Hill, New York, 1985.

368. G. D. Raithby and K. G. T. Hollands, "Natural Convection," in *Handbook of Heat Transfer Fundamentals,* W. M. Rohsenow, J. P. Hartnett, and E. N. Ganic eds., chap. 6, McGraw-Hill, New York, 1985.

369. W. M. Kays and H. C. Perkins, "Forced Convection, Internal Flow in Ducts," in *Handbook of Heat Transfer Fundamentals,* W. M. Rohsenow, J. P. Hartnett, and E. N. Ganic eds., chap. 7, McGraw-Hill, New York, 1985.

370. R. Viskanta, "Electric and Magnetic Fields," in *Handbook of Heat Transfer Fundamentals,* W. M. Rohsenow, J. P. Hartnett, and E. N. Ganic eds., chap. 10, McGraw-Hill, New York, 1985.

371. R. M. Nelson and A. E. Bergles, "Performance Evaluation for Tubeside Heat Transfer Enhancement of a Flooded Evaporator Water Chiller," *ASHRAE Transactions* (92/1B): 739–755, 1986.

372. W. H. Avery and C. Wu, *Renewable Energy From the Ocean. A Guide to OTEC,* Oxford University Press, New York, 1994.

373. R. L. Webb, "Performance Evaluation Criteria for Enhanced Tube Geometries Used in Two-Phase Heat Exchangers," in *Heat Transfer Equipment Design,* R. K. Shah, E. C. Subbarao, and R. A. Mashelkar eds., Hemisphere, New York, 1988.

374. M. K. Jensen, R. R. Trewin, and A. E. Bergles, "Crossflow Boiling in Enhanced Tube Bundles," *Two-Phase Flow in Energy Systems,* HTD vol. 220, pp. 11–17, ASME, New York, 1992.

375. J. R. Thome, *Enhanced Boiling Heat Transfer,* Hemisphere, New York, 1990.

376. T. S. Ravigururajan and A. E. Bergles, "Development and Verification of General Correlations for Pressure Drop and Heat Transfer in Single-Phase Turbulent Flow in Enhanced Tubes," *Experimental Thermal and Fluid Science* (13): 55–70, 1996.

377. R. M. Manglik and A. E. Bergles, "Heat Transfer and Pressure Drop Correlations for the Rectangular Offset Strip Fin Compact Heat Exchanger," *Experimental Thermal and Fluid Science* (10): 171–180, 1995.

378. R. M. Manglik and A. E. Bergles, "Heat Transfer and Pressure Drop Correlation for Twisted-Tape Inserts in Isothermal Tubes: Part I, Laminar Flows," *Journal of Heat Transfer* (115): 881–889, 1993.

379. R. M. Manglik and A. E. Bergles, "Heat Transfer and Pressure Drop Correlation for Twisted-Tape Inserts in Isothermal Tubes: Part II, Turbulent Flows," *Journal of Heat Transfer* (115): 890–896, 1993.

380. A. E. Bergles, "Some Perspectives on Enhanced Heat Transfer—Second Generation Heat Transfer Technology," *J. Heat Transfer* (110): 1082–1096, 1988.

381. M. A. Kedzierski, "Simultaneous Visual and Calorimetric Measurements of R-11, R-123, and R-123/Alkybenzene Nucleate Flow Boiling," *Heat Transfer with Alternate Refrigerant,* HTD vol. 243, pp. 27–33, ASME, New York, 1993.

382. S. J. Eckels, T. M. Doerr, and M. B. Pate, "Heat Transfer and Pressure Drop of R-134a and Ester Lubricant Mixtures in a Smooth and a Micro-fin Tube: Part I, Evaporation," *ASHRAE Transactions* (100/2): 265–281, 1994.

383. R. A. Pabisz Jr. and A. E. Bergles, "Using Pressure Drop to Predict the Critical Heat Flux in Multiple Tube, Subcooled Boiling Systems," *Experimental Heat Transfer, Fluid Mechanics and Thermodynamics,* vol. 2, pp. 851–858, Edizioni ETS, Pisa, Italy, 1997.

384. H.-C. Yeh, "Device for Producing High Heat Transfer in Heat Exchanger Tubes," *U.S. Patent 4,832,114,* May 23, 1989.

385. E. F. C. Somerscales and A. E. Bergles, "Enhancement of Heat Transfer and Fouling Mitigation," in *Advances in Heat Transfer,* J. P. Hartnett et al. eds., vol. 30, pp. 197–253, Academic, New York, 1997.

386. R. L. Webb, *Principles of Enhanced Heat Transfer,* Wiley, New York, 1994.

CHAPTER 12

HEAT PIPES

G. P. "Bud" Peterson
Texas A&M University

INTRODUCTION

Passive two-phase heat transfer devices capable of transferring large quantities of heat with a minimal temperature drop were first introduced by Gaugler in 1944 [1]. These devices, however, received little attention until Grover et al. [2] published the results of an independent investigation and first applied the term *heat pipe*. Since that time, heat pipes have been employed in numerous applications ranging from temperature control of the permafrost layer under the Alaska pipeline to the thermal control of electronic components such as high-power semiconductor devices [3].

A classical heat pipe consists of a sealed container lined with a wicking structure. The container is evacuated and backfilled with just enough liquid to fully saturate the wick. When a heat pipe operates on a closed two-phase cycle with only pure liquid and vapor present, the working fluid remains at saturation conditions as long as the operating temperature is between the freezing point and the critical state. As shown in Fig. 12.1, heat pipes consist of three distinct regions: the evaporator or heat addition region, the condenser or heat rejection region, and the adiabatic or isothermal region. Heat added to the evaporator region of the container causes the working fluid in the evaporator wicking structure to be vaporized. The high temperature and corresponding high pressure in this region result in flow of the vapor to the other, cooler end of the container, where the vapor condenses, giving up its latent heat of vaporization. The capillary forces in the wicking structure then pump the liquid back to the evaporator. Similar devices, referred to as two-phase thermosyphons, have no wick but utilize gravitational forces for the liquid return.

In order to function properly, heat pipes require three major components: the case, which can be constructed from glass, ceramic, or metal; a wicking structure, which can be fabricated from woven fiberglass, sintered metal powders, screens, wire meshes, or grooves; and a working fluid, which can vary from nitrogen or helium for low-temperature (cryogenic) heat pipes to lithium, potassium, or sodium for high-temperature (liquid metal) heat pipes. Each of these three components is equally important, with careful consideration given to the material type, thermophysical properties, and compatibility.

The heat pipe container or case provides containment and structural stability. As such, it must be fabricated from a material that is (1) compatible with both the working fluid and the wicking structure, (2) strong enough to withstand the pressure associated with the saturation temperatures encountered during storage and normal operation, and (3) of a high enough thermal conductivity to permit the effective transfer of heat either into or out of the vapor space. In addition to these characteristics, which are primarily concerned with the internal

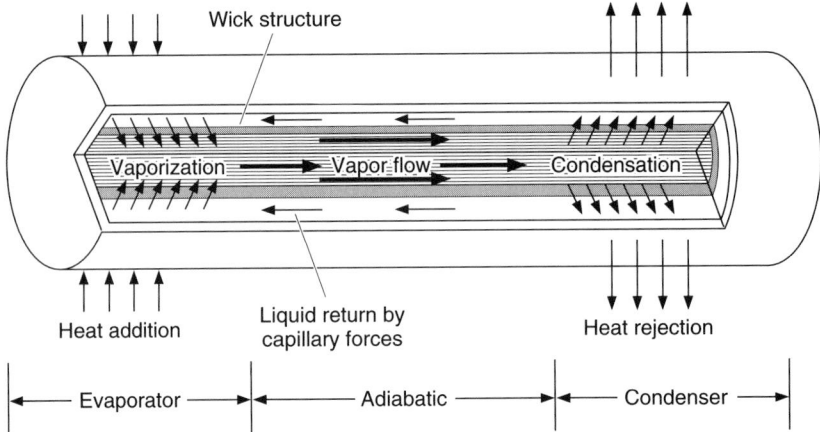

FIGURE 12.1 Heat pipe operation.

effects, the container material must be resistant to corrosion resulting from interaction with the environment and must be malleable enough to be formed into the appropriate size and shape.

The wicking structure has two functions in heat pipe operation: it is the vehicle through which, and provides the mechanism by which, the working fluid is returned from the condenser to the evaporator. It also ensures that the working fluid is evenly distributed over the evaporator surface. In order to provide a flow path with low flow resistance through which the liquid can be returned from the condenser to the evaporator, an open porous structure with a high permeability is desirable. However, to increase the capillary pumping pressure, a small pore size is necessary. Solutions to this apparent dichotomy can be achieved through the use of a nonhomogeneous wick made of several different materials or through a composite wicking structure.

Because the basis for operation of a heat pipe is the vaporization and condensation of the working fluid, selection of a suitable fluid is an important factor in the design and manufacture of heat pipes. Care must be taken to ensure that the operating temperature range is adequate for the application. The most common applications involve the use of heat pipes with a working fluid having a boiling temperature between 250 and 375 K; however, both cryogenic heat pipes (those operating in the 5 to 100 K temperature range) and liquid metal heat pipes (those operating in the 750 to 5000 K temperature range) have been developed and used successfully. Figure 12.2 illustrates the typical operating temperature ranges for some of the various heat pipe fluids. In addition to the thermophysical properties of the working fluid, consideration must be given to the other factors such as the compatibility of the materials and the ability of the working fluid to wet the wick and wall materials [4, 5]. Further criteria for the selection of the working fluids have been presented by Groll et al. [6], Peterson [7], and Faghri [8].

In general, the high heat transfer characteristics, the ability to maintain constant evaporator temperatures under different heat flux levels, and the diversity and variability of evaporator and condenser sizes and shapes make the heat pipe an effective device for use in a wide variety of applications where thermal energy must be transported from one location to another with minimal temperature drop.

FUNDAMENTAL OPERATING PRINCIPLES

Heat pipes and thermosyphons both operate on a closed two-phase cycle and utilize the latent heat of vaporization to transfer heat with very small temperature gradients. Thermosyphons,

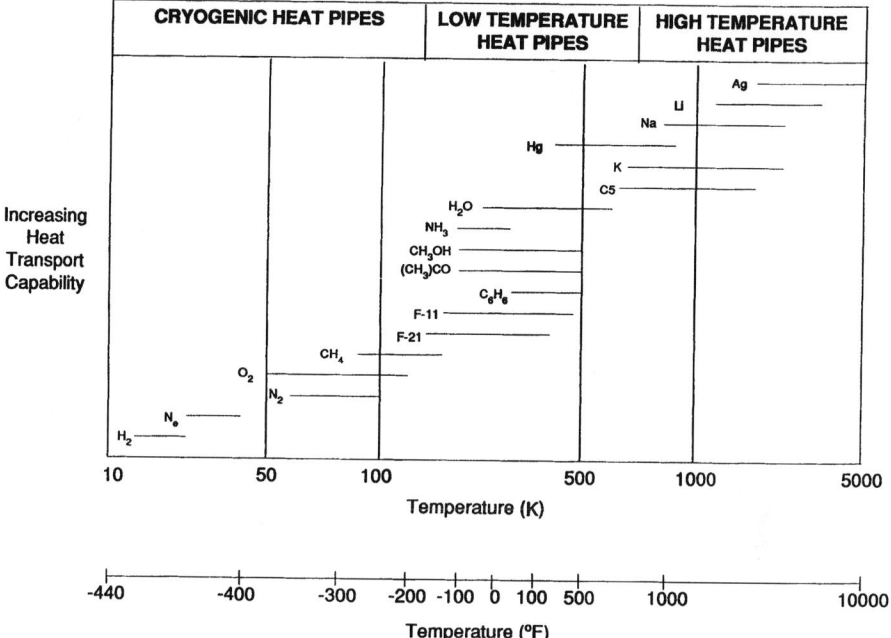

FIGURE 12.2   Heat pipe working fluids [7].

however, rely solely on the gravitational forces to return the liquid phase of the working fluid from the condenser to the evaporator, while heat pipes utilize some sort of capillary wicking structure to promote the flow of liquid from the condenser to the evaporator. As a result of the capillary pumping occurring in this wick, heat pipes can be used in a horizontal orientation, microgravity environments, or even applications where the capillary structure must "pump" the liquid against gravity from the evaporator to the condenser. It is this single characteristic, the dependence of the local gravitational field to promote the flow of the liquid from the condenser to the evaporator, that differentiates thermosyphons from heat pipes [7].

Capillary Limitation

Although heat pipe performance and operation are strongly dependent on shape, working fluid, and wick structure, the fundamental phenomenon that governs the operation of these devices arises from the difference in the capillary pressure across the liquid-vapor interfaces in the evaporator and condenser regions. The vaporization occurring in the evaporator section of the heat pipe causes the meniscus to recede into the wick, and condensation in the condenser section causes flooding. The combined effect of this vaporization and condensation process results in a meniscus radius of curvature that varies along the axial length of the heat pipe as shown in Fig. 12.3a. The point at which the meniscus has a minimum radius of curvature is typically referred to as the "dry" point and usually occurs in the evaporator at the point farthest from the condenser region. The "wet" point occurs at that point where the vapor pressure and liquid pressure are approximately equal or where the radius of curvature is at a maximum. It is important to note that this point can be located anywhere in the condenser or adiabatic sections, but typically is found near the end of the condenser farthest from the evaporator [7].

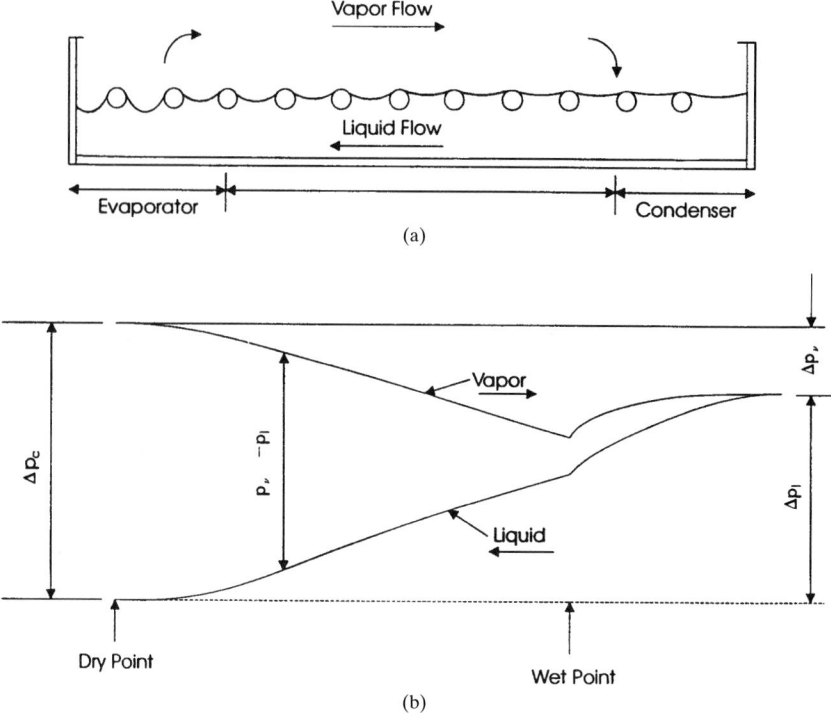

FIGURE 12.3   (*a*) variation of meniscus curvature as a function of axial position [7]; (*b*) typical liquid and vapor pressure distributions in a heat pipe [7].

Figure 12.3*b* illustrates the relationship between the static liquid and static vapor pressures in an operating heat pipe. As shown, the capillary pressure gradient across a liquid-vapor interface is equal to the pressure difference between the liquid and vapor phases at any given axial position. For a heat pipe to function properly, the net capillary pressure difference between the wet and dry points, identified in Fig. 12.3*b*, must be greater than the summation of all the pressure losses occurring throughout the liquid and vapor flow paths. This relationship, referred to as the capillary limitation, can be expressed mathematically as

$$(\Delta P_c)_m \geq \int_{L_{\text{eff}}} \frac{\partial P_v}{\partial x}\, dx + \int_{L_{\text{eff}}} \frac{\partial P_l}{\partial x}\, dx + \Delta P_{PT,e} + \Delta P_{PT,c} + \Delta P_+ + \Delta P_\| \qquad (12.1)$$

where $(\Delta P_c)_m$ = the maximum capillary pressure difference generated within the capillary wicking structure between the wet and dry points

$\dfrac{\partial P_v}{\partial x}$ = the sum of the inertial and viscous pressure drop occurring in the vapor phase

$\dfrac{\partial P_l}{\partial x}$ = the sum of the inertial and viscous pressure drop occurring in the liquid phase

$\Delta P_{PT,e}$ = the pressure gradient across the phase transition in the evaporator
$\Delta P_{PT,c}$ = the pressure gradient across the phase transition in the condenser
ΔP_+ = the normal hydrostatic pressure drop
$\Delta P_\|$ = the axial hydrostatic pressure drop

The first two terms on the right side of this equation, $\partial P_v/\partial x$ and $\partial P_l/\partial x$ represent the summation of viscous and inertial losses in the vapor and liquid flow paths, respectively. The next two, $\Delta P_{PT,e}$ and $\Delta P_{PT,c}$, represent the pressure gradients occurring across the phase transition in the evaporator and condenser and can typically be neglected, and the last two, ΔP_+ and $\Delta P_\parallel$, represent the normal and axial hydrostatic pressure drops. As indicated, when the maximum capillary pressure is equal to or greater than the summation of these pressure drops, the capillary structure is capable of returning an adequate amount of working fluid to prevent dryout of the evaporator wicking structure. When the total capillary pressure across the liquid-vapor interface is not greater than or equal to the summation of all of the pressure drops occurring throughout the liquid vapor flow paths, the working fluid will not be returned to the evaporator, causing the liquid level in the evaporator wicking structure to be depleted, leading to dryout. This condition, referred to as the *capillary limitation,* varies according to the wicking structure, working fluid, evaporator heat flux, and operating temperature. In order to effectively understand the behavior of the vapor and liquid flow in an operating heat pipe, each of the factors contributing to the overall pressure gradient must be clearly understood. The following is a brief explanation of each of these individual terms, summarized from the more detailed explanations presented in Bar-Cohen and Kraus [3] and Peterson [7].

Capillary Pressure. At the surface of a single liquid-vapor interface, a capillary pressure difference, defined as $(P_v - P_l)$ or ΔP_c, exists. This capillary pressure difference can be described mathematically from the Laplace-Young equation,

$$\Delta P_c = \sigma\left(\frac{1}{r_1} + \frac{1}{r_2}\right) \tag{12.2}$$

where r_1 and r_2 are the principal radii of curvature and σ is the surface tension. For many heat pipe wicking structures, the maximum capillary pressure may be written in terms of a single radius of curvature r_c. Using this expression, the maximum capillary pressure between the wet and dry points can be expressed as the difference between the capillary pressure across the meniscus at the wet point and the capillary pressure at the dry point or

$$\Delta P_{c,m} = \left(\frac{2\sigma}{r_{c,e}}\right) - \left(\frac{2\sigma}{r_{c,c}}\right) \tag{12.3}$$

Figure 12.3a illustrates the effect of the vaporization occurring in the evaporator, which causes the liquid meniscus to recede into the wick, and the condensation occurring in the condenser section, which causes flooding of the wick. This combination of meniscus recession and flooding results in a reduction in the local capillary radius $r_{c,e}$, and increases the local capillary radius $r_{c,c}$, respectively, which further results in a pressure difference and, hence, pumping of the liquid from the condenser to the evaporator. During steady-state operation, it is generally assumed that the capillary radius in the condenser or at the wet point r_{cc} approaches infinity, so that the maximum capillary pressure for a heat pipe operating at steady state in many cases can be expressed as a function of only the effective capillary radius of the evaporator wick [7],

$$\Delta P_{c,m} = \left(\frac{2\sigma}{r_{c,e}}\right) \tag{12.4}$$

Values for the effective capillary radius r_c are given in Table 12.1 for some of the more common wicking structures [7]. In the case of other geometries, the effective capillary radius can be found theoretically using the methods proposed by Chi [9] or experimentally using the methods described by Ferrell and Alleavitch [10], Freggens [11], or Tien [12]. In addition, limited information on the transient behavior of capillary structures is also available [13].

Normal Hydrostatic Pressure Drop. There are two hydrostatic pressure drop terms of interest in heat pipes: a normal hydrostatic pressure drop ΔP_+, which occurs only in heat pipes

TABLE 12.1 Effective Capillary Radius for Several Wick Structures [7]

Structure	r_c	Data
Circular cylinder (artery or tunnel wick)	r	r = radius of liquid flow passage
Rectangular groove	ω	ω = groove width
Triangular groove	$\omega/\cos \beta$	ω = groove width β = half-included angle
Parallel wires	ω	ω = wire spacing
Wire screens	$(\omega + d_\omega)/2 = 1/2N$	d = wire diameter N = screen mesh number ω = wire spacing
Packed spheres	$0.41r_s$	r_s = sphere radius

that have circumferential communication of the liquid in the wick, and an axial hydrostatic pressure drop. The first of these is the result of the body force component acting perpendicularly to the longitudinal axis of the heat pipe, and can be expressed as

$$\Delta P_+ = \rho_l g d_v \cos \psi \qquad (12.5)$$

where ρ_l is the density of the liquid, g is the gravitational acceleration, d_v is the diameter of the vapor portion of the pipe, and ψ is the angle the heat pipe makes with respect to the horizontal.

Axial Hydrostatic Pressure Drop. The second hydrostatic pressure drop term is the axial hydrostatic pressure drop, $\Delta P_\parallel$, which results from the component of the body force acting along the longitudinal axis. This term can be expressed as

$$\Delta P_\parallel = \rho_l g L \sin \psi \qquad (12.6)$$

where L is the overall length of the heat pipe.

In a gravitational environment, the normal and axial hydrostatic pressure terms may either assist or hinder the capillary pumping process depending on whether the tilt of the heat pipe promotes or hinders the flow of liquid back to the evaporator (i.e., the evaporator lies either below or above the condenser). In a zero-g environment, both this term and the normal hydrostatic pressure drop term can be neglected because of the absence of body forces.

Liquid Pressure Drop. While the capillary pumping pressure promotes the flow of liquid through the wicking structure, the viscous forces in the liquid result in a pressure drop ΔP_l, which resists the capillary flow through the wick. This liquid pressure gradient may vary along the longitudinal axis of the heat pipe, and hence the total liquid pressure drop can be determined by integrating the pressure gradient over the length of the flow passage [7], or

$$\Delta P_l(x) = -\int_0^x \frac{dP_l}{dx}\, dx \qquad (12.7)$$

where the limits of integration are from the evaporator end to the condenser end ($x = 0$) and dP_l/dx is the gradient of the liquid pressure resulting from frictional drag. Introducing the Reynolds number Re_l and drag coefficient f_l and substituting the local liquid velocity, which is related to the local heat flow, the wick cross-sectional area, the wick porosity ε, and the latent heat of vaporization λ, yields

$$\Delta P_l = \left(\frac{\mu_l}{K A_w \lambda \rho_l} \right) L_{\text{eff}} q \qquad (12.8)$$

TABLE 12.2   Wick Permeability for Several Wick Structures [7]

Structure	K	Data
Circular cylinder (artery or tunnel wick)	$r^2/8$	r = radius of liquid flow passage
Open rectangular grooves	$2\varepsilon(r_{h,l})^2/(f_l\,\mathrm{Re}_l)$	ε = wick porosity w = groove width s = groove pitch δ = groove depth $(r_{h,l}) = 2\omega\delta/(\omega + 2\delta)$
Circular annular wick	$2(r_{h,l})^2/(f_l\,\mathrm{Re}_l)$	$(r_{h,l}) = r_1 - r_2$
Wrapped screen wick	$\dfrac{d^2\varepsilon^3}{122(1-\varepsilon)^2}$	d_ω = wire diameter $\varepsilon = 1 - (1.05\pi N d_w/4)$ N = mesh number
Packed sphere	$\dfrac{r_s^2\varepsilon^3}{37.5(1-\varepsilon)^2}$	r_s = sphere radius ε = porosity (dependent on packing mode)

where L_{eff} is the effective heat pipe length, defined as

$$L_{\mathrm{eff}} = 0.5L_e + L_a + 0.5L_c \tag{12.9}$$

and the wick permeability is given in Table 12.2.

Vapor Pressure Drop.   Mass addition and removal in the evaporator and condenser, respectively, along with the compressibility of the vapor phase, complicate the vapor pressure drop in heat pipes. Applying continuity to the adiabatic region of the heat pipe ensures that for continued operation, the liquid mass flow rate and vapor mass flow rate must be equal. As a result of the difference in the density of these two phases, the vapor velocity is significantly higher than the velocity of the liquid phase. For this reason, in addition to the pressure gradient resulting from frictional drag, the pressure gradient due to variations in the dynamic pressure must also be considered. Chi [9], Dunn and Reay [13], and Peterson [7] have all addressed this problem. The results indicate that on integration of the vapor pressure gradient, the dynamic pressure effects cancel. The result is an expression, which is similar to that developed for the liquid,

$$\Delta P_v = \left(\frac{C(f_v\,\mathrm{Re}_v)\mu_v}{2(r_{h,v})^2 A_v \rho_v \lambda} \right) L_{\mathrm{eff}} q \tag{12.10}$$

where $(r_{h,v})$ is the hydraulic radius of the vapor space and C is a constant that depends on the Mach number.

During steady-state operation, the liquid mass flow rate m_l must equal the vapor mass flow rate m_v at every axial position, and while the liquid flow regime is always laminar, the vapor flow may be either laminar or turbulent. As a result, the vapor flow regime must be written as a function of the heat flux. Typically, this is done by evaluating the local axial Reynolds number in the vapor stream. It is also necessary to determine whether the flow should be treated as compressible or incompressible by evaluating the local Mach number.

Previous investigations summarized by Bar-Cohen and Kraus [3] have demonstrated that the following combinations of these conditions can be used with reasonable accuracy.

$$\mathrm{Re}_v < 2300, \qquad \mathrm{Ma}_v < 0.2$$
$$(f_v\,\mathrm{Re}_v) = 16$$
$$C = 1.00 \tag{12.11}$$

$$\text{Re}_v < 2300, \qquad \text{Ma}_v > 0.2$$

$$(f_v \, \text{Re}_v) = 16$$

$$C = \left[1 + \left(\frac{\gamma_v - 1}{2} \right) \text{Ma}_v^2 \right]^{-1/2} \tag{12.12}$$

$$\text{Re}_v > 2300, \qquad \text{Ma}_v < 0.2$$

$$(f_v \, \text{Re}_v) = 0.038 \left(\frac{2(r_{h,v})q}{A_v \mu_v \lambda} \right)^{3/4}$$

$$C = 1.00 \tag{12.13}$$

$$\text{Re}_v > 2300, \qquad \text{Ma}_v > 0.2$$

$$(f_v \, \text{Re}_v) = 0.038 \left(\frac{2(r_{h,v})q}{A_v \mu_v \lambda} \right)^{3/4}$$

$$C = \left[1 + \left(\frac{\gamma_v - 1}{2} \right) \text{Ma}_v^2 \right]^{-1/2} \tag{12.14}$$

Because the equations used to evaluate both the Reynolds number and the Mach number are functions of the heat transport capacity, it is necessary to first assume the conditions of the vapor flow. Using these assumptions, the maximum heat capacity $q_{c,m}$ can be determined by substituting the values of the individual pressure drops into Eq. 12.1 and solving for $q_{c,m}$. Once the value of $q_{c,m}$ is known, it can then be substituted into the expressions for the vapor Reynolds number and Mach number to determine the accuracy of the original assumption. Using this iterative approach, which is covered in more detail by Chi [9], accurate values for the capillary limitation as a function of the operating temperature can be determined in units of watt·m or watts for $(qL)_{c,m}$ and $q_{c,m}$, respectively.

Other Limitations

While the capillary limitation is the most frequently encountered limitation, there are several other important mechanisms that limit the maximum amount of heat transferred during steady-state operation of a heat pipe. Among these are the viscous limit, sonic limit, entrainment limit, and boiling limit. The capillary wicking limit and viscous limits deal with the pressure drops occurring in the liquid and vapor phases, respectively. The sonic limit results from the occurrence of choked flow in the vapor passage, while the entrainment limit is due to the high liquid vapor shear forces developed when the vapor passes in counterflow over the liquid saturated wick. The boiling limit is reached when the heat flux applied in the evaporator portion is high enough that nucleate boiling occurs in the evaporator wick, creating vapor bubbles that partially block the return of fluid.

In low-temperature applications such as those using cryogenic working fluids, either the viscous limit or capillary limit occurs first, while in high-temperature heat pipes, such as those that use liquid metal working fluids, the sonic and entrainment limits are of increased importance. The theory and fundamental phenomena that cause each of these limitations have been the object of a considerable number of investigations and are well documented by Chi [9], Dunn and Reay [13], Tien [12], Peterson [7], Faghri [8], and the proceedings from the nine International Heat Pipe Conferences held over the past 25 years.

Viscous Limitation. In conditions where the operating temperatures are very low, the vapor pressure difference between the closed end of the evaporator (the high-pressure region) and the closed end of the condenser (the low-pressure region) may be extremely small. Because of this small pressure difference, the viscous forces within the vapor region may prove to be dominant and, hence, limit the heat pipe operation. Dunn and Reay [13] discuss this limit in more detail and suggest the criterion

$$\frac{\Delta P_v}{P_v} < 0.1 \tag{12.15}$$

for determining when this limit might be of concern. Due to the operating temperature range, this limitation will normally be of little consequence in the design of heat pipes for room-temperature applications.

Sonic Limitation.   The sonic limit in heat pipes is analogous to the sonic limit that occurs in converging-diverging nozzles [7], except that in a converging-diverging nozzle, the mass flow rate is constant and the vapor velocity varies because of the changing cross-sectional area, while in heat pipes, the reverse occurs—the area is constant and the vapor velocity varies because of the evaporation and condensation along the heat pipe. Analogous to nozzle flow, decreased outlet pressure, or, in this case, condenser temperature, results in a decrease in the evaporator temperature until the sonic limitation is reached. Any further increase in the heat rejection rate does not reduce the evaporator temperature or the maximum heat transfer capability, but only reduces the condenser temperature, due to the existence of choked flow.
   The sonic limitation in heat pipes can be determined as

$$q_{s,m} = A_v \rho_v \lambda \left(\frac{\gamma_v R_v T_v}{2(\gamma_v + 1)} \right)^{1/2} \tag{12.16}$$

where T_v is the mean vapor temperature within the heat pipe.

Entrainment Limitation.   As a result of the high vapor velocities, liquid droplets may be picked up or entrained in the vapor flow and cause excess liquid accumulation in the condenser and hence dryout of the evaporator wick [14]. This phenomenon requires that, for proper operation, the onset of entrainment in countercurrent two-phase flow be avoided. The most commonly quoted criterion to determine this onset is that the Weber number We, defined as the ratio of the viscous shear force to the force resulting from the liquid surface tension,

$$We = \frac{2(r_{h,w})\rho_v V_v^2}{\sigma} \tag{12.17}$$

be equal to unity. To prevent the entrainment of liquid droplets in the vapor flow, the Weber number must therefore be less than 1.
   By relating the vapor velocity to the heat transport capacity, a value for the maximum transport capacity based on the entrainment limitation may be determined as

$$V_v = \frac{q}{A_v \rho_v \lambda} \tag{12.18}$$

$$q_{e,m} = A_v \lambda \left(\frac{\sigma \rho_v}{2(r_{h,w})} \right)^{1/2} \tag{12.19}$$

where $(r_{h,w})$ is the hydraulic radius of the wick structure, defined as twice the area of the wick pore at the wick-vapor interface divided by the wetted perimeter at the wick-vapor interface. Rice and Fulford [15] developed a somewhat different approach that proposed an expression to define the critical dimensions for wicking structures in order to prevent entrainment.

Boiling Limitation.   At very high radial heat fluxes, nucleate boiling may occur in the wicking structure and bubbles may become trapped in the wick, blocking the liquid return and resulting in evaporator dryout. This phenomenon, referred to as the boiling limit, differs from the other limitations previously discussed in that it depends on the evaporator heat flux as opposed to the axial heat flux [7].

The boiling limit can be found by examining nucleate boiling theory, which is comprised of two separate phenomena—bubble formation and the subsequent growth or collapse of the bubbles. The first of these, bubble formation, is governed by the number and size of nucleation sites on a solid surface; the second, bubble growth or collapse, depends on the liquid temperature and corresponding pressure caused by the vapor pressure and surface tension of the liquid. Utilizing a pressure balance on any given bubble and applying the Clausius-Clapeyron equation to relate the temperature and pressure, an expression for the heat flux beyond which bubble growth will occur may be developed [9]:

$$q_{b,m} = \left(\frac{2\pi L_{\text{eff}} k_{\text{eff}} T_v}{\lambda \rho_v \ln (r_i/r_v)} \right) \left(\frac{2\sigma}{r_n} - \Delta P_{c,m} \right) \tag{12.20}$$

where k_{eff} is the effective thermal conductivity of the liquid-wick combination, given in Table 12.3, r_i is the inner radius of the heat pipe wall, and r_n is the nucleation site radius, which, according to Dunn and Reay [13], can be assumed to be from 2.54×10^{-5} to 2.54×10^{-7} meters for conventional heat pipes [3].

TABLE 12.3   Effective Thermal Conductivity for Liquid-Saturated Wick Structures [7, 9]

Wick structures	k_{eff}
Wick and liquid in series	$\dfrac{k_l k_w}{\varepsilon k_w + k_l(1-\varepsilon)}$
Wick and liquid in parallel	$\varepsilon k_l + k_w(1-\varepsilon)$
Wrapped screen	$\dfrac{k_l[(k_l+k_w) - (1-\varepsilon)(k_l-k_w)]}{[(k_l+k_w) + (1-\varepsilon)(k_l-k_w)]}$
Packed spheres	$\dfrac{k_l[(2k_l+k_w) - 2(1-\varepsilon)(k_l-k_w)]}{(2k_l+k_w) + (1-\varepsilon)(k_l-k_w)}$
Rectangular grooves	$\dfrac{(\omega_f k_l k_w \delta) + \omega k_l(0.185\omega_f k_w + \delta k_l)}{(\omega + \omega_f)(0.185\omega_f k_l + \delta k_f)}$

Once the power level associated with each of the four limitations has been determined as a function of the maximum heat transport capacity, a graphic representation of the operating envelope can be constructed. From this, it is only a matter of selecting the lowest limitation for any given operating temperature to determine the heat transport limitation applicable for a prespecified set of conditions.

DESIGN AND MANUFACTURING CONSIDERATIONS

A number of recent references have focused on the problems associated with the design and manufacture of heat pipes. Most notable are early works by Feldman [16] and Brennan and Kroliczek [17], and more recent ones by Peterson [7] and Faghri [8]. In addition to such factors as cost, size, weight, reliability, fluid inventory, and construction and sealing techniques, the design and manufacture of heat pipes are governed by three operational considerations: the effective operating temperature range, which is determined by the selection of the working fluid; the maximum power the heat pipe is capable of transporting, which is determined by the ultimate pumping capacity of the wick structure (for the capillary wicking limit); and the maximum evaporator heat flux, which is determined by the point at which nucleate boiling occurs.

Working Fluid

Because heat pipes rely on vaporization and condensation to transfer heat, selection of a suitable working fluid is an important factor in the design and manufacture of heat pipes. The operating temperature range must be adequate, and the fluid must be stable over the entire range. For most moderate temperature applications, working fluids with boiling temperatures between 250 and 375 K are required. This includes fluids such as ammonia, Freon 11 or 113, acetone, methanol, and water. For a capillary-wick-limited heat pipe, the characteristics of a good working fluid are a high latent heat of vaporization, a high surface tension, a high liquid density, and a low liquid viscosity. Chi [9] combined these properties into a parameter referred to as the liquid transport factor or figure of merit, which is defined as

$$N_l = \frac{\rho_l \sigma \lambda}{\mu_l} \tag{12.21}$$

This grouping of properties can be used to evaluate various working fluids at specific operating temperatures. The concept of a parameter for evaluating working fluids has been extended by Gosse [18], where it was demonstrated that the thermophysical properties of the liquid-vapor equilibrium state could be reduced to three independent parameters [3].

Along with the importance of the thermophysical properties of the working fluid, consideration must be given to the ability of the working fluid to wet the wick and wall materials as discussed by Peterson [7]. Other important criteria in the selection of the working fluid have been presented by Heine and Groll [19], in whose study a number of other factors including the liquid and vapor pressure, and the compatibility of the materials, are considered.

Wicking Structures

In addition to providing the pumping of the liquid from the condenser to the evaporator, the wicking structure ensures that the working fluid is evenly distributed over the evaporator surface. In order to provide a flow path with low flow resistance through which the liquid can be returned from the condenser to the evaporator, an open porous structure with a high permeability is desirable. However, to increase the capillary pumping pressure, a small pore size is necessary. Solutions to this apparent dichotomy can be achieved through the use of a nonhomogeneous wick made of several different materials or through a composite wicking structure. Udell and Jennings [20] proposed and formulated a model for a heat pipe with a wick consisting of porous media of two different permeabilities oriented parallel to the direction of the heat flux. This wick structure provided a large pore size in the center of the wick for liquid flow and a smaller size pore for capillary pressure.

Composite wicking structures accomplish the same type of effect in that the capillary pumping and axial fluid transport are handled independently. In addition to fulfilling this dual purpose, several wick structures physically separate the liquid and vapor flow. This results from an attempt to eliminate the viscous shear force that occurs during countercurrent liquid-vapor flow.

Materials Compatibility

The formation of noncondensible gases through chemical reactions between the working fluid and the wall or wicking structure, or decomposition of the working fluid, can cause problems with the operation of the heat pipe or with corrosion. For these reasons, careful consideration must be given to the selection of working fluids and wicking and wall materials in order to prevent the occurrence of these problems over the operational life of the heat pipe. The effect of noncondensible gas formation may result in either decreased performance or total failure. Corrosion problems can lead to physical degradation of the wicking structure since solid particles carried to the evaporator wick and deposited there will eventually reduce the wick permeability [21].

Basiulis et al. [22] conducted extensive compatibility tests with several combinations of working fluids and wicking structures, the results of which are summarized in Table 12.4 as well as other investigations by Busse et al. [23]. Other more recent investigations, such as those performed by Zaho et al. [24], in which the compatibility of water and mild steel heat pipes was evaluated; by Roesler et al. [25], in which stainless steel, aluminum, and ammonia combinations were evaluated; and by Murakami and Arai [26], in which a statistical predictive technique for evaluating the long-term reliability of copper-water heat pipes was developed, provide additional insight into the compatibility of various liquid-material combinations that might be used in the thermal control of electronic equipment. Most of the data available are the result of accelerated life tests. Although the majority of the data are based on actual test results, care should be taken to ensure that the tests in which the data were obtained are similar to the application under consideration. Such factors as thermal cycling, mean operating temperature, etc., must be considered.

TABLE 12.4 Working Fluid, Wick and Container Compatibility Data [7, 13]

Material	Water	Acetone	Ammonia	Methanol
Copper	RU	RU	NU	RU
Aluminum	GNC	RL	RU	NR
Stainless steel	GNT	PC	RU	GNT
Nickel	PC	PC	RU	RL
Refrasil	RU	RU	RU	RU

Material	Dow-A	Dow-E	Freon 11	Freon 113
Copper	RU	RU	RU	RU
Aluminum	UK	NR	RU	RU
Stainless steel	RU	RU	RU	RU
Nickel	RU	RL	UK	UK
Refrasil	RU		UK	UK

RU, recommended by past successful usage; RL, recommended by literature; PC, probably compatible; NR, not recommended; NU, not used; UK, unknown; GNC, generation of gas at all temperatures; CNT, generation of gas at elevated temperatures when oxide is present.

These two problems—noncondensible gas generation and corrosion—are only two of the factors to be considered when selecting heat pipe wicks and working fluids. Others include wettability of the fluid-wick combination, strength-to-weight ratio, thermal conductivity and stability, and cost of fabrication.

Heat Pipe Sizes and Shapes

Heat pipes vary in both size and shape, ranging from a 15-m-long monogroove heat pipe developed by Alario et al. [27] for spacecraft heat rejection to a 10-mm-long expandable bellows-type heat pipe developed by Peterson [28] for the thermal control of semiconductor devices. Vapor and liquid flow cross-sectional areas also vary significantly from those encountered in flat-plate heat pipes, which have very large flow areas, to commercially available heat pipes with a cross-sectional area of less than 0.30 mm^2. Heat pipes may be fixed or variable in length and either rigid or flexible for situations where relative motion or vibration poses a problem.

Reliability and Life Tests

Peterson [7] has performed an extensive review of life testing and reliability. This review indicated that many of the early investigations, such as those conducted by Basiulis and Filler [29] and Busse et al. [23] focused on the reliability of various types of material combinations in the intermediate operating temperature range. Long-duration life tests conducted on copper-water heat pipes have been performed by numerous investigators. Most notable among these are the investigations of Kreed et al. [30], which have indicated that this combination, with proper cleaning and charging procedures, can produce heat pipes with expected life spans of tens of years. Other tests [13] have indicated similar results for copper-acetone and copper-methanol combinations. However, in these latter cases, care must be taken to ensure the purity of the working fluid, wick structure, and case materials. Table 12.4 presents a summary of compatibility data obtained from previous tests.

In addition to the use of compatible materials, long-term reliability can be ensured through careful inspection and preparation processes including

- laboratory inspection to ensure that material of high purity (i.e., OFHC copper) is used for the case, end caps, and fill tubes
- appropriate inspection procedures to ensure that the wicking material is made from high quality substances
- inspection and distillation procedures to ensure that the working fluid is of consistently high purity
- fabrication in a clean environment to ensure or eliminate the presence of oils, vapors, etc.
- use of clean solvents during the rinse process [31]

The effects of long-term exposure to elevated temperatures and repeated thermal cycling on heat pipes and thermosyphons can be approximated using a model developed by Baker [32], which utilizes an Arrhenius model to predict the response parameter F,

$$F = Ce^{-A/kT} \tag{12.22}$$

where C = a constant
A = the reaction activation energy
k = the Boltzmann constant
T = the absolute temperature

This model utilized experimental results obtained from the Jet Propulsion Laboratory to predict the rate and amount of hydrogen gas generated over a 20-year lifetime for a stainless steel heat pipe with water as the working fluid. The results of this model, when plotted for several different temperatures, allow the mass of hydrogen generated to be predicted as a function of time.

Although worldwide production of heat pipes designed for applications involving the thermal control of electronic components or devices was in excess of 1,000,000 per year in 1992, it is difficult to calculate a mean time to failure (MTTF) for heat pipes, thermosyphons, and other similar devices due to the relatively small amount of data that exists on actual products in operation. Experience with a wide variety of applications ranging from consumer electronics to industrial equipment has demonstrated that mechanical cleaning of the case and wicking structure with an appropriate solvent, combined with an acidic etch and vacuum bakeout under elevated temperatures, will produce heat pipes free of contaminants that will experience negligible performance degradation (less than 5 percent) over a product lifetime of 10 years [31].

HEAT PIPE THERMAL RESISTANCE

Typically, the temperature drop between the evaporator and condenser of a heat pipe is of particular interest to the designer of heat pipe thermal control systems and is most readily found by utilizing an analogous electrothermal network. Figure 12.4 illustrates the electrothermal analogue for the heat pipe illustrated in Fig. 12.1. As shown, the overall thermal resistance is composed of nine different resistances arranged in a series-parallel combination. These nine resistances can be summarized as follows:

R_{pe} The radial resistance of the pipe wall at the evaporator

R_{we} The resistance of the liquid-wick combination at the evaporator

R_{ie} The resistance of the liquid-vapor interface at the evaporator

R_{ya} The resistance of the adiabatic vapor section

R_{pa} The axial resistance of the pipe wall

R_{wa} The axial resistance of the liquid-wick combination

R_{ic} The resistance of the liquid-vapor interface at the condenser

R_{wc} The resistance of the liquid-wick combination at the condenser

R_{pc} The radial resistance of the pipe wall at the condenser

Estimates of the order of magnitude for each of these resistances indicate that several simplifications can be made [33], that is, the axial resistance of both the pipe wall and the liquid-wick combination may be treated as open circuits and neglected and the liquid-vapor interface resistances and the axial vapor resistance can typically be assumed to be negligible, leaving only the pipe wall radial resistances and the liquid-wick resistances at both the evaporator and condenser.

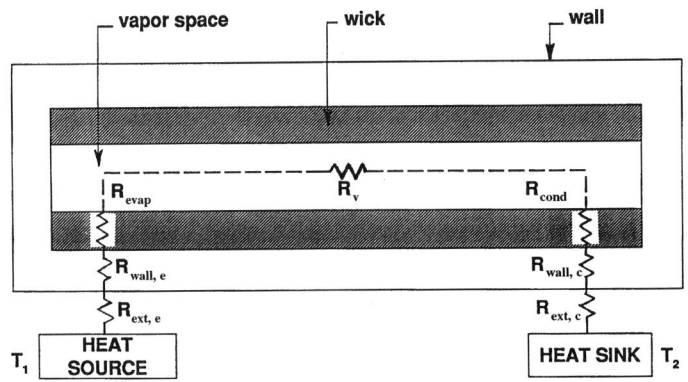

$$\Delta T = T_1 - T_2 = Q \cdot \Sigma R\text{'s}$$

FIGURE 12.4   Electrothermal analogue for a heat pipe [7].

The radial resistances at the pipe wall can be computed from Fourier's law as

$$R_{pe} = \frac{\delta}{k_p A_e} \tag{12.23}$$

for flat plates, where δ is the plate thickness and A_e is the evaporator area, or

$$R_{pe} = \frac{\ln (d_o/d_i)}{2\pi L_e k_p} \tag{12.24}$$

for cylindrical pipes, where L_e is the evaporator length. An expression for the equivalent thermal resistance of the liquid-wick combination in circular pipes is

$$R_{we} = \frac{\ln (d_o/d_i)}{2\pi L_e k_{\text{eff}}} \tag{12.25}$$

where values for the effective conductivity k_{eff} can be found in Table 12.3. For sintered metal materials, an investigation performed by Peterson and Fletcher [34] found that the method presented by Alexander [35], and later by Ferrell et al. [36], for determining the effective thermal conductivity of saturated sintered materials was the most accurate over the widest range of conditions.

The adiabatic vapor resistance, although usually negligible, can be found as

$$R_{va} = \frac{T_v(P_{v,e} - P_{v,c})}{\rho_v \lambda q} \tag{12.26}$$

where $P_{v,e}$ and $P_{v,c}$ are the vapor pressures at the evaporator and condenser. Combining these individual resistances provides a mechanism by which the overall thermal resistance, and, hence, the temperature drop associated with various axial heat fluxes, can be computed.

TYPES OF HEAT PIPES

There are a number of different ways to classify heat pipes, but perhaps the two most important categories are the variable-conductance heat pipes (those in which the magnitude and/or direction of the heat transfer can be controlled) and micro-heat pipes (those that are so small that the mechanisms controlling their operation are significantly different from those in more conventional heat pipes).

Variable-Conductance Heat Pipes

As discussed previously, the presence of noncondensible gases in heat pipes, as in many two-phase cycles, can create a problem due to the partial blockage of the condensing area, since these noncondensible gases are carried to the condenser by the vapor flow and reduce the effective condenser area. This characteristic can, however, be used to control both the direction and amount of heat transferred by the heat pipe. One such method uses a gas-loaded, variable-conductance heat pipe, in which the thermal conductance of the heat pipe varies as a function of the gas front position. As the heat input at the evaporator varies, the vapor temperature changes, causing the gas contained within the reservoir to expand or contract. This expansion and contraction changes the position of the gas front and thereby changes the size of the condenser area, changing the overall conductance, that is, as the heat flux increases, the gas front recedes and the thermal conductance increases due to the larger condenser surface area. In this way, the temperature drop across the evaporator and condenser can be maintained fairly constant even though the evaporator heat flux may fluctuate [7].

While in most applications heat pipes operate in a passive manner, adjusting the heat flow rate to compensate for the temperature difference between the evaporator and condenser [37], several active control schemes have been developed [38]. Most notable among these are: (1) gas-loaded heat pipes with some type of feedback system, (2) excess-liquid heat pipes, (3) vapor-flow-modulated heat pipes, and (4) liquid-flow-modulated heat pipes [9]. In one such pipe, a temperature-sensing device at the evaporator provides a signal to the reservoir heater, which when activated can heat the gas contained in the reservoir, causing it to expand and thereby reducing the condenser area.

Excess-liquid heat pipes operate in much the same manner as gas-loaded heat pipes, but utilize excess working fluid to block portions of the pipe and control the condenser size or prevent reversal of heat transfer. Vapor-flow-modulated heat pipes utilize a throttling valve to control the amount of vapor leaving the evaporator. In this type of control scheme, increased evaporator temperatures result in an expansion of the bellows chamber containing the control fluid. This in turn closes down the throttling valve and reduces the flow of vapor to the condenser. This type of device is typically applied in situations where the evaporator temperature varies and a constant condenser temperature is desired.

Liquid-flow-modulated heat pipes have two separate wicking structures, one to transport liquid from the evaporator to the condenser and one that serves as a liquid trap. As the temperature gradient is reversed, the liquid moves into the trap and starves the evaporator of fluid. In addition to these liquid-vapor control schemes, the quantity and direction of heat transfer can also be controlled through internal or external pumps, or through actual physical contact with the heat sink.

Micro-Heat Pipes

In 1984, Cotter [39] first introduced the concept of very small "micro"-heat pipes incorporated into semiconductor devices to promote more uniform temperature distribution and to improve thermal control. At that time a micro-heat pipe was defined as one "so small that the mean curvature of the liquid-vapor interface is necessarily comparable in magnitude to the reciprocal of the hydraulic radius of the total flow channel." Since this initial introduction, numerous investigations have been conducted on many different types of relatively small heat pipes. Many of these devices were in reality only miniaturized versions of larger, more conventional heat pipes, while others were actually significantly different in their design.

To better understand what the term *micro-heat pipe* implies, Babin and Peterson [40] expressed Cotter's initial definition of a micro-heat pipe mathematically as

$$K \propto \frac{1}{r_h} \qquad (12.27)$$

where K is the mean curvature of the liquid-vapor interface and r_h is the hydraulic radius of the flow channel. Then, by assuming a constant of proportionality of 1 and multiplying both the mean curvature of the liquid-vapor interface and the hydraulic radius by the capillary radius r_c, a dimensionless expression was developed. This expression,

$$\frac{r_c}{r_h} \geq 1 \qquad (12.28)$$

better defines a micro-heat pipe and helps to differentiate between small versions of conventional heat pipes and the more recently developed micro-heat pipes.

The fundamental operating principles of micro-heat pipes are essentially the same as those occurring in larger, more conventional heat pipes. Heat applied to one end of the heat pipe vaporizes the liquid in that region and forces it to move to the cooler end, where it condenses and gives up the latent heat of vaporization. This vaporization and condensation process causes the liquid-vapor interface in the liquid arteries to change continually along the pipe, resulting in a capillary pressure difference between the evaporator and condenser regions. This capillary pressure difference promotes the flow of the working fluid from the condenser back to the evaporator through the triangular corner regions. These corner regions serve as liquid arteries; thus no wicking structure is required. In practical terms, a micro-heat pipe consists of a small noncircular channel that utilizes the sharply angled corner regions as liquid arteries.

Although the initial application proposed by Peterson [41] involved the thermal control of semiconductor devices, a wide variety of other uses have been investigated or proposed. These include the removal of heat from laser diodes and other small, localized heat-generating devices; the thermal control of photovoltaic cells; the removal or dissipation of heat from the leading edge of hypersonic aircraft; and applications involving the nonsurgical treatment of cancerous tissue through either hyper- or hypothermia. Several comprehensive reviews of micro-heat pipes have been conducted in the past several years [7, 41]. These reviews have discussed the wide variety of uses for these devices and have summarized the results of investigations designed to demonstrate the ability to model, design, fabricate, and test heat pipes with an effective cross-sectional radius of less than 100 μm [42–44].

NOMENCLATURE

Symbol, Definition, Units

A	area: m^2
C	constant, defined in text
d	diameter: m
f	drag coefficient, dimensionless
g	gravitational constant, 9.807 m/s^2
k	thermal conductivity: W/m·K
K	wick permeability: m^2
L	length: m
Ma	Mach number, dimensionless
N	screen mesh number: m^{-1}
N_l	liquid figure of merit: W/m^2
P	pressure: N/m^2
q	heat flow rate: W
R	thermal resistance: K/W; or universal gas constant: J/kg·K
Re	Reynolds number, dimensionless
r	radius: m
T	temperature: K
V	velocity: m/s
We	Weber number, dimensionless

Greek Symbols

λ	latent heat of vaporization: J/kg
μ	dynamic viscosity: kg/m·s
ρ	density: kg/m^3
σ	surface tension: N/m

v   vapor

ψ   angle of inclination: degrees or radians

Subscripts

a adiabatic section, air

b boiling

c capillary, capillary limitation, condenser

e entrainment, evaporator section

eff   effective

i inner

l liquid

m maximum

n nucleation

PT phase change

s sonic or sphere

w wire spacing, wick

∥   axial hydrostatic pressure

+   normal hydrostatic pressure

REFERENCES

1. R. S. Gaugler, "Heat Transfer Devices," U.S. Patent 2,350,348, 1944.

2. G. M. Grover, T. P. Cotter, and G. F. Erikson, "Structures of Very High Thermal Conductivity," *J. Appl. Phys.* (35): 1190–1191, 1964.

3. A. Bar-Cohen and A. D. Kraus, *Advances in Thermal Modeling of Electronic Components and Systems*, vol. 1, Hemisphere Publishing Corporation, Washington, DC, pp. 283–336, 1988.

4. G. P. Peterson, X. J. Lu, X. F. Peng, and B. X. Wang, "Analytical and Experimental Investigation of the Rewetting of Circular Channels With Internal V-Grooves," *Int. J. Heat and Mass Transfer* (35/11): 3085–3094, 1992.

5. G. P. Peterson and X. F. Peng, "Experimental Investigation of Capillary Induced Rewetting for a Flat Porous Wicking Structure," *ASME J. Energy Resources Technology* (115/1): 62–70, 1993.

6. M. Groll, W. Supper, and C. J. Savage, "Shutdown Characteristics of an Axial-Grooved Liquid-Trap Heat Pipe Thermal Diode," *J. of Spacecraft* (19/2): 173–178, 1982.

7. G. P. Peterson, *An Introduction to Heat Pipes: Modeling, Testing and Applications*, John Wiley & Sons, Inc., Washington, DC, 1994.

8. A. Faghri, *Heat Pipe Science and Technology*, Taylor & Francis Publishing Company, Washington, DC, 1995.

9. S. W. Chi, *Heat Pipe Theory and Practice*, McGraw-Hill Publishing Company, New York, 1976.

10. J. K. Ferrell and J. Alleavitch, "Vaporization Heat Transfer in Capillary Wick Structures," preprint no. 6, ASME-AIChE Heat Transfer Conf., Minneapolis, MN, 1969.

11. R. A. Freggens, "Experimental Determination of Wick Properties for Heat Pipe Applications," *Proc. 4th Intersoc. Energy Conversion Eng. Conf.,* Washington, DC, pp. 888–897, 1969.

12. C. L. Tien, "Fluid Mechanics of Heat Pipes," *Annual Review of Fluid Mechanics,* 167–186, 1975.

13. P. D. Dunn and D. A. Reay, *Heat Pipes,* 3d ed., Pergamon Press, New York, 1982.

14. G. P. Peterson and B. Bage, "Entrainment Limitations in Thermosyphons and Heat Pipes," *ASME J. Energy Resources Technology* (113/3): 147–154, 1991.

15. G. Rice and D. Fulford, "Influence of a Fine Mesh Screen on Entrainment in Heat Pipes," *Proc. 6th Int. Heat Pipe Conf.,* Grenoble, France, pp. 168–172, 1987.

16. K. T. Feldman, *The Heat Pipe: Theory, Design and Applications,* Technology Application Center, Univ. of New Mexico, Albuquerque, NM, 1976.

17. P. J. Brennan and E. J. Kroliczek, *Heat Pipe Design Handbook,* B&K Engineering, Inc., Towson, MD, 1979.

18. J. Gosse, "The Thermo-Physical Properties of Fluids on Liquid-Vapor Equilibrium: An Aid to the Choice of Working Fluids for Heat Pipes," *Proc. 6th Int. Heat Pipe Conf.,* Grenoble, France, pp. 17–21, 1987.

19. D. Heine and M. Groll, "Compatibility of Organic Fluids With Commercial Structure Materials for Use in Heat Pipes," *Proc. 5th Intl. Heat Pipe Conf.,* Tsukuba, Japan, pp. 170–174, 1984.

20. K. S. Udell and J. D. Jennings, "A Composite Porous Heat Pipe," *Proc. 5th Intl. Heat Pipe Conf.,* Tsukuba, Japan, pp. 41–47, 1984.

21. V. L. Barantsevich, L. V. Barakove, and I. A. Tribunskaja, "Investigation of the Heat Pipe Corrosion Resistance and Service Characteristics," *Proc. 6th Int. Heat Pipe Conf.,* Grenoble, France, pp. 188–193, 1987.

22. A. Basiulis, R. C. Prager, and T. R. Lamp, "Compatibility and Reliability of Heat Pipe Materials," *Proc. 2nd Int. Heat Pipe Conf.,* Bologna, Italy, pp. 357–372, 1976.

23. C. A. Busse, A. Campanile, and J. Loens, "Hydrogen Generation in Water Heat Pipes at 250°C," *First Int. Heat Pipe Conf.,* Stuttgart, Germany, paper no. 4-2, October 1973.

24. R. D. Zaho, Y. H. Zhu, and D. C. Liu, "Experimental Investigation of the Compatibility of Mild Carbon Steel and Water Heat Pipes," *Proc. 6th Int. Heat Pipe Conf.,* Grenoble, France, pp. 200–204, 1987.

25. S. Roesler, D. Heine, and M. Groll, "Life Testing With Stainless Steel/Ammonia and Aluminum/Ammonia Heat Pipe," *Proc. 6th Int. Heat Pipe Conf.,* Grenoble, France, pp. 211–216, 1987.

26. M. Murakami and K. Arai, "Statistical Prediction of Long-Term Reliability of Copper Water Heat Pipes From Acceleration Test Data," *Proc. 6th Int. Heat Pipe Conf.,* Grenoble, France, pp. 2194–2199, 1987.

27. J. Alario, R. Brown, and P. Otterstadt, "Space Constructable Radiator Prototype Test Program," AIAA Paper No. 84-1793, 1984.

28. G. P. Peterson, "Heat Removal Key to Shrinking Avionics," *Aerospace America,* no. 8, October, pp. 20–22, 1987.

29. A. Basiulis and M. Filler, "Operating Characteristics and Long Life Capabilities of Organic Fluid Heat Pipes," AIAA Paper No. 71-408, 1971.

30. H. Kreed, M. Kroll, and P. Zimmermann, "Life Test Investigations with Low Temperature Heat Pipes," *Proc. First Int. Heat Pipe Conf.,* Stuttgart, Germany, paper no. 4-1, October 1973.

31. J. E. Toth and G. A. Meyer, "Heat Pipes: Is Reliability an Issue?," *Proc. IEPS Conf.,* Austin, TX, September 28, 1992.

32. E. Baker, "Prediction of Long Term Heat Pipe Performance From Accelerated Life Tests," *AIAA Journal* (11/9): September 1979.

33. G. A. A. Asselman and D. B. Green, "Heat Pipes," *Phillips Technical Review* (16): 169–186, 1973.

34. G. P. Peterson and L. S. Fletcher, "Effective Thermal Conductivity of Sintered Heat Pipe Wicks," *AIAA J. of Thermophysics and Heat Transfer* (1/3): 36–42, 1987.

35. E. G. Alexander Jr., "Structure-Property Relationships in Heat Pipe Wicking Materials," Ph.D. thesis, North Carolina State University, Dept. of Chemical Engineering, 1972.

36. J. K. Ferrell, E. G. Alexander, and W. T. Piver, "Vaporization Heat Transfer in Heat Pipe Wick Materials," AIAA Paper No. 72-256, 1972.

37. R. I. J. Van Buggenum and D. H. V. Daniels, "Development, Manufacturing and Testing of a Gas Loaded Variable Conductance Heat Pipe," *Proc. 6th Int. Heat Pipe Conf.,* Grenoble, France, pp. 242–249, 1987.

38. Y. Sakuri, H. Masumoto, H. Kimura, M. Furukawa, and D. K. Edwards, "Flight Experiments for Gas-Loaded Variable Conductance Heat Pipe on ETS-III Active Control Package," *Proc. 5th Int. Heat Pipe Conf.,* Tsukuba, Japan, pp. 26–32, 1984.

39. T. P. Cotter, "Principles and Prospects of Micro Heat Pipes," *Proc. 5th Int. Heat Pipe Conf.,* Tsukuba, Japan, pp. 328–335, 1984.

40. B. R. Babin and G. P. Peterson, "Experimental Investigation of a Flexible Bellows Heat Pipe for Cooling Discrete Heat Sources," *ASME Journal of Heat Transfer* (112/3): pp. 602–607, 1990.

41. G. P. Peterson, "An Overview of Micro Heat Pipe Research," *Applied Mechanics Review* (45/5): 175–189, 1992.

42. G. P. Peterson, A. B. Duncan, and M. H. Weichold, "Experimental Investigation of Micro Heat Pipes Fabricated in Silicon Wafers," *ASME J. Heat Transfer* (115/3): 751–756, 1993.

43. G. P. Peterson and A. K. Mallik, "Transient Response Characteristics of Vapor Deposited Micro Heat Pipe Arrays," *ASME J. Electronic Packaging* (117/1): 82–87, 1995.

44. A. K. Mallik and G. P. Peterson, "Steady-State Investigation of Vapor Deposited Micro Heat Pipe Arrays," *ASME J. Electronic Packaging* (117/1): 75–81, 1995.

CHAPTER 13

HEAT TRANSFER IN PACKED AND FLUIDIZED BEDS

Shriniwas S. Chauk and Liang-Shih Fan
The Ohio State University

INTRODUCTION

Fixed and fluidized beds are commonly employed in chemical, biochemical, and petrochemical industries as reactors or vessels for physical operations [1–3]. Successful application of these multiphase systems lies in the accurate characterization of their transport phenomena. Heat transfer study, being an integral part of this, is of paramount importance to engineering practice as exemplified by such operations as drying, nuclear reactor cooling, coal combustion, polymerization reaction, and exothermic or endothermic chemical synthesis. Fluidization systems are typically characterized by thermal uniformity and good heat transfer between the particles, the wall, the immersed surface, and the fluidizing medium, owing to intensive mixing and efficient contact between phases. In addition, the large thermal capacity of the fluidizing medium makes the temperature control of these systems easier. The scale-up of a fluidization system remains a challenging area, particularly when chemical reactions are involved. Heat transfer is affected by system scale-up.

A fluid passing through a bed of particles at a relatively low velocity merely percolates through the interparticle voids. The particles in this case retain their spatial entity and the bed is in the *fixed bed* state. As the flow velocity increases further, the drag exerted on the particles just counterbalances the weight of the bed. At this point the bed is in *incipient fluidization* and the corresponding velocity is called the *minimum fluidization velocity* U_{mf}. As the fluid velocity increases beyond U_{mf}, the bed is in a completely fluidized state.

In a liquid fluidized bed, an increase in liquid velocity beyond U_{mf} causes the bed to expand smoothly and uniformly. The particle concentration is uniform throughout the bed with almost no large-scale voids. This is called *homogeneous fluidization* or *particulate fluidization*.

For gas fluidized beds, based on the operating velocity, fluidization can be generally classified as *dense phase* (lower gas velocity) and *lean phase* (higher gas velocity). Dense-phase fluidization, which is characterized by the presence of an upper bed surface, encompasses particulate fluidization, bubbling fluidization, and turbulent fluidization. The *particulate fluidization* regime is bounded by the minimum fluidization velocity and the minimum bubbling velocity. In particulate fluidization, all the gas passes through the interstitial space between the fluidizing particles without forming bubbles. The bed appears homogeneous, as in a liquid fluidized bed. The *bubbling fluidization* regime is reached with an increase in the gas velocity to U_{mb}. Bubbles form and induce vigorous motion of the particles. In the bubbling fluidization regime, bubble coalescence and breakup take place. The *turbulent fluidization* regime is realized when the gas velocity increases beyond that of the bubbling fluidization regime. In the turbulent fluidization regime, the bubble and emulsion phases may become indistinguishable with increased uniformity of the suspension. The bubbling and turbulent fluidization regimes characterize *heterogeneous fluidization* or *aggregative fluidization*. As the gas velocity increases beyond that of the turbulent fluidization regime, the bubble/void phase eventually disappears and the gas evolves into a continuous phase. In a dense-phase fluidized bed, the particle entrainment rate is low and it increases with increasing gas velocity. As the gas flow rate increases beyond the point corresponding to the disappearance of the bubble/void phase, a drastic increase in the entrainment rate of the particles occurs such that a continuous feeding of particles into the fluidized bed is required to maintain a steady flow of solids. Fluidization at this state, in contrast to dense-phase fluidization, is noted as *lean-phase fluidization*. Lean-phase fluidization encompasses two flow regimes, i.e., fast fluidization and dilute transport. Fast fluidization is characterized by a heterogeneous flow structure, whereas dilute transport is characterized by a homogeneous flow structure. The circulating fluidized bed (CFB) denotes a fluidized bed system in which solid particles circulate between the riser and the downcomer. *Spouting* occurs when the gas is injected vertically at a high velocity through a large orifice at the bottom into the bed, typically one containing Group D particles (with $d_p > 1$ mm). At the center of the spouted bed, the gas jet penetrates the whole bed and carries some particles upward, forming a dilute flow core region. On reaching the top of the core region, particles fall back to the top of the annular dense region located between the core region and the wall. Particles in the annular region move downward in a moving-bed mode and recirculate to the core region, forming a circulatory pattern of solids.

It is well understood that gas fluidization behavior is closely associated with particle properties such as size and density. In order to categorize the wide range of fluidization behavior observed, Geldart [4] proposed a classification of particle behavior. Group A particles have a small mean size of 30 to 100 μm and/or a density less than 1400 kg/m^3. They are relatively easy to fluidize, and uniform bed expansion is observed at the gas velocities immediately beyond U_{mf}. FCC catalyst particles are a typical example of Group A particles. Group B particles have a mean size of 40 to 500 μm and a density ranging from 1400 to 4000 kg/m^3. Their fluidization is characterized by intense bubbling at the gas velocity immediately beyond U_{mf}. Gross solids circulation patterns are developed due to vigorous bubbling. Group C particles are very small in size, typically less than 20 μm. In these particles, the interparticle forces (van der Waals forces) dominate over the hydrodynamic forces. They are extremely difficult to fluidize and gas channeling is their important operating characteristic. Group D particles, which are larger than 1 mm and/or of higher density than Group B particles, show unpredictable fluidization behavior exhibiting severe channeling and the presence of frequently bursting bubbles. These particles mix rather poorly when fluidized. Such a varied particle effect on the hydrodynamics of gas fluidized beds renders it also an important parameter in accounting for heat transfer behavior in fluidized beds.

In gas-solid systems, the governing heat transfer modes include gas-to-particle heat transfer, particle-to-particle heat transfer, and suspension-to-surface heat transfer by conduction, convection, and/or radiation. Numerous heat transfer correlations are available in the literature to predict the heat transfer coefficients for various modes of operation. It is, however, important to note that the heat transfer coefficient defined in a flow system is frequently model specific; therefore, it is necessary to apply the numerical values of these coefficients in the context of the same models that define them. Also, frequently the empirical correlations are defined for very specific or limited flow conditions and geometric arrangements of the flow system. Thus, it is essential that the correlations be used in the specified operating ranges for which they were developed. Selected empirical correlation equations for heat transfer coefficients in fixed and fluidized beds are given in Tables 13.2–13.7. These equations are recommended for use under the range of operating conditions given in the table.

This chapter describes the basic hydrodynamic characteristics and the general modes and mechanisms of heat transfer over a wide range of fluid-solid flow regimes. Since the flow behavior in the bed varies with the geometric configuration, different arrangements of the heat transfer surface result in varied heat transfer performance. Hence, the design considerations for heat transfer are also presented to obtain the optimal heat transfer characteristics. Throughout the text, unless otherwise noted, the correlation equations presented are given in SI units.

HYDRODYNAMICS

The heat transfer characteristics in multiphase systems depend strongly on the hydrodynamics, which vary significantly with particle properties. The particle size, size distribution, and shape affect the particle and fluid flow behavior through particle-fluid and particle-particle interactions. A discussion of the hydrodynamic characteristics of packed and fluidized beds follows.

Packed Beds

In many practical fluidization systems, the presence of unsuspended particles is common and understanding the fluid flow in a packed bed is essential for the study of heat transfer. The important parameter in characterizing the hydrodynamics is the bed voidage in addition to particle properties. Table 13.1 summarizes the voidages of packed beds for different packing arrangements for spherical particles. It is seen that the cubic and rhombohedral arrangements provide the highest and lowest bed voidages, respectively.

TABLE 13.1 Packed Bed Voidage
for Different Packing Arrangements

Packing arrangement	Bed voidage (α)
Spheres—rhombohedral	0.2595
Spheres—tetragonal	0.3019
Spheres—random	0.36–0.43
Spheres—orthorhombic	0.3954
Spheres—cubic	0.4764

Source: From Gabor and Botterill [5].

For packed bed pressure drop estimation, experiments were conducted by Darcy [6] to study viscous flows through homogeneous porous media, which resulted in the well-known Darcy's law relating the fluid velocity to the pressure gradient. However, the Darcy's law does not consider the effects of inertia. For more general description of flows in a packed bed, particularly when the effects of inertia are important, Ergun [7] presented a semiempirical equation covering a wide range of flow conditions, which is known as the Ergun equation [7, 8]. In that approach, the pressure loss was considered to be caused by simultaneous kinetic and viscous energy losses. In Ergun's formulation, four factors contribute to the pressure drop. They are: (1) fluid flow rate, (2) properties of the fluid (such as viscosity and density), (3) closeness (such as porosity) and orientation of packing, and (4) size, shape, and surface of the solid particles. By combining the Blake-Kozeny correlation for laminar flows and Burke-Plummer correlation for turbulent flows, Ergun derived the following correlation for pressure drop through a packed bed of monosized particles:

$$\frac{\Delta p}{H} = 150 \frac{(1-\alpha)^2}{\alpha^3} \frac{\mu U}{\varphi d_p^2} + 1.75 \frac{(1-\alpha)}{\alpha^3} \frac{\rho U^2}{\varphi d_p} \tag{13.1}$$

In Eq. 13.1, φ is the sphericity that is defined as the ratio of the surface area per unit volume of a sphere to the surface area per unit volume of the particle.

Fluidized Beds

When particles are in the fluidized state, the hydrodynamic behavior varies significantly with the flow regimes. Depending on whether liquid or gas is used as the fluidizing fluid, the regime may encompass, as noted in the introduction, particulate, bubbling, turbulent, fast fluidization and dilute transport. The key operating and design variables that affect the hydrodynamics of a fluidized bed include particle properties, fluid velocity, bed diameter, and distributor. Pertinent hydrodynamic characteristics in this regard are briefly described in the following text.

Minimum Fluidization Velocity.   For a gas-solid system the relationship of pressure drop through the bed Δp_b and the superficial gas velocity U for fluidization with uniform particles is illustrated in Fig. 13.1. In the figure, as U increases in the packed bed, Δp_b increases, reaches a peak, and then decreases to a constant. However, as U decreases from a fluidized state, Δp_b follows a different path without passing through the peak. The peak at which the bed is operated is denoted as the minimum fluidization condition and its corresponding superficial gas velocity is defined as the minimum fluidization velocity U_{mf}.

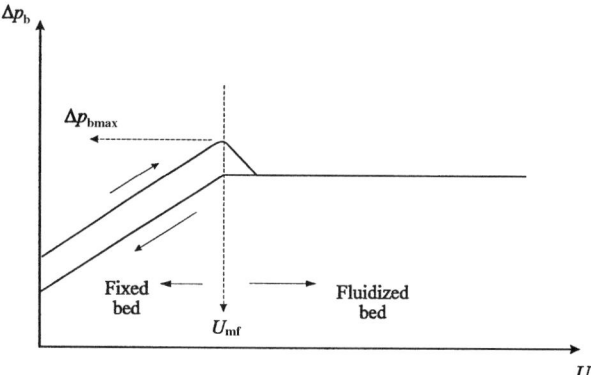

FIGURE 13.1   Determination of U_{mf} from the pressure drop variation with the gas velocity.

For particles with nonuniform properties, the hysteresis effect noted above may also occur; however, the transition of Δp_b from the fixed bed to the fluidized bed is more smooth and U_{mf} can be obtained from the intersection of the Δp_b line for the fixed bed and that for the fluidized bed. Thus, the expression for U_{mf} can be analytically obtained on the basis of the equivalence of the pressure drop for a fixed bed and that for a fluidized bed under the minimum fluidization condition. The pressure drop in the fixed bed can be described by the Ergun equation. Under the minimum fluidization condition, Eq. 13.1 can be written as

$$\frac{\Delta p_b}{H_{\mathrm{mf}}} = 150 \frac{(1-\alpha_{\mathrm{mf}})^2}{\alpha_{\mathrm{mf}}^3} \frac{\mu U_{\mathrm{mf}}}{\varphi^2 d_p^2} + 1.75 \frac{(1-\alpha_{\mathrm{mf}})}{\alpha_{\mathrm{mf}}^3} \frac{\rho U_{\mathrm{mf}}^2}{\varphi d_p} \qquad (13.2)$$

where φ is the sphericity of the particles and H_{mf} is the bed height at minimum fluidization.

Particulate and Aggregative Fluidization. When the fluid phase is liquid, the difference in the densities of fluid and solid is not very large, and the particle size is small, the bed is fluidized homogeneously with an apparent uniform bed structure as the fluid velocity exceeds the minimum fluidization velocity. The fluid passes through the interstitial spaces between the fluidizing particles without forming solids-free "bubbles" or "voids." This state of fluidization characterizes particulate fluidization. In particulate fluidization, the bed voidage can be related to the superficial fluid velocity by the Richardson-Zaki equation. The particulate fluidization occurs when the Froude number at the minimum fluidization is less than 0.13 [9].

When the solids-free "bubbles" or "voids" are present in the bed as in bubbling or turbulent fluidization, the bed of solid particles is fluidized nonhomogeneously. This state of fluidization characterizes aggregative fluidization. The distinct properties of aggregative fluidization are intense mixing of particles, bypassing of fluid through bubbles, and solids entrainment above the bed surface.

Bubbling Fluidization. In a bubbling fluidized bed, the behavior of bubbles significantly affects the flow or transport phenomena in the bed, including solids mixing, entrainment, and heat and mass transfer. Factors that are of relevance to the bubble dynamics include the bubble/jet formation, bubble wake dynamics, and bubble coalescence/breakup. Bubbles are formed due to the inherent instability of gas-solid systems. The instability of a gas-solid fluidized bed is characterized by fast growth in local voidage in response to a system perturbation. Due to the instability in the bed, the local voidage usually grows rapidly into a shape resembling a bubble. Although it is not always the case, the initiation of the instability is usually perceived to be the onset of bubbling, which marks the transition from particulate fluidization to bubbling fluidization.

When gas enters the orifice of the distributor, it can initially form bubbles or jets. The formation of bubbles or jets depends on various parameters including types of particles, fluidization conditions surrounding the orifice, orifice size, and presence of internals or walls in the bed [10]. The initial bubble or jet is then transformed into a chain of bubbles. The jet is defined as an elongated void sufficiently large compared to a bubble, and it extends permanently to some distance from the orifice into the inner bed. In general, bubbles tend to form in the presence of small particles, such as Group A particles [11]; jets tend to form in the presence of large particles, such as Group D particles, when the emulsion phase is not sufficiently fluidized or when internals are present that disrupt the flow of solids to the orifice region [10].

The bubble wake plays an important role in the movement or mixing of solids in the bed and the freeboard. A bubble wake in a single-phase fluid is defined as the streamline-enclosed region beneath the bubble base. Hence, a bubble wake is defined as the region enclosed by streamline of the pseudofluid, which characterizes the emulsion phase, behind the bubble base. In the bed, the wake rises with the bubble and thereby provides an essential means for global solids circulation and induces axial solids mixing. In gas-solid fluidized beds, wake shedding has also been observed. The shed wake fragments are banana-shaped, and the shedding may occur at fairly regular intervals [12].

Bubbles may coalesce to form large bubbles or break up into smaller ones. This bubble interaction leads to a significantly different bubbling behavior in a multibubble system from that in a single-bubble system. The coalescence of bubbles in a gas-solid medium is similar to that of those in a liquid or liquid-solid medium [13]. The coalescence usually takes place with the trailing bubble overtaking the leading bubble through its wake region due to the regional minimum pressure. The breakup of a large bubble may start from an indentation on the upper boundary of the bubble due to the disturbance induced by the relative motion of the particles, which may eventually break the bubble.

In bubbling fluidization, bubble motion becomes increasingly vigorous as the gas velocity increases. This behavior can be reflected in the increase of the amplitude of the pressure fluctuations in the bed. With further increase in the gas velocity, the fluctuation will reach a maximum, decrease, and then gradually level off, as shown in Fig. 13.2, which appears to be typical

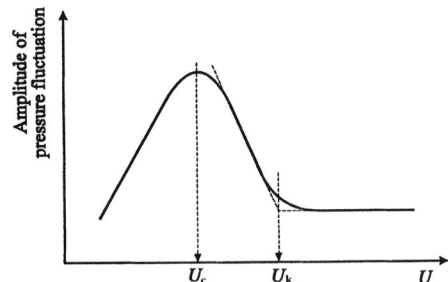

FIGURE 13.2 Variations of pressure fluctuation with the gas velocity for dense-phase fluidized beds with FCC particles (from Yerushalmi and Cankurt [14]).

for Group A particles. This fluctuation variation marks the transition from the bubbling to the turbulent regime.

Turbulent Fluidization. For Group A particles, as shown in Fig. 13.2, the onset velocity of the transition to the turbulent regime is commonly defined as the gas velocity corresponding to the peak U_c, whereas the leveling point U_k may be recognized as the onset of the turbulent regime proper [14]. However, based on direct observation of bed phenomena, appreciable variations in bubble behavior occur at gas velocities around U_c. Specifically, the bubble interaction is dominated by bubble coalescence at gas velocities less than U_c, while it is dominated by bubble breakup at gas velocities greater than U_c.

Due to the existence of bubbles/voids in turbulent fluidized beds, the hydrodynamic behavior of turbulent fluidization under relatively low gas velocity conditions is similar, to certain extent, to that of bubbling fluidization. However, distinct differences exist under relatively high gas velocity conditions, which renders many hydrodynamic and transport correlations developed for the bubbling regime invalid for the turbulent regime. The bubble/void size in the turbulent regime tends to decrease with an increase in the gas velocity due to the predominance of bubble breakup over bubble coalescence. This trend is the opposite of that exhibited in the bubbling regime. However, similarly to the case in the bubbling regime, an increase in the operating pressure at a constant gas velocity or constant excess gas velocity (i.e., $U - U_{mf}$) decreases the bubble size. The bubble/void size in the turbulent bed can eventually be reduced to such a magnitude, with sufficiently high gas velocities and high pressures, that it marks a gradual transition to lean-phase bubble-free fluidization. It is recognized that increasing the operating pressure promotes gas entering the emulsion phase, which eventually results in undistinguishable bubble and emulsion phases in the bed, leading to bubble-free lean-phase fluidization.

The turbulent regime is often regarded as the transition regime from bubbling fluidization to lean-phase fluidization. At relatively low gas velocities, bubbles are present in the turbulent regime, while at relatively high gas velocities in the turbulent regime, the clear boundary of bubbles disappears and the nonuniformity of solids concentration distribution yields distinct gas voids that become less distinguishable as the gas velocity further increases toward lean-phase fluidization.

Entrainment and Elutriation. *Entrainment* refers to the ejection of particles from the dense bed into the freeboard by fluidizing gas. *Elutriation* refers to the separation of fine particles from a mixture of particles, which occurs at all heights of the freeboard, and their ultimate removal from the freeboard. The terms *entrainment* and *elutriation* are sometimes used interchangeably. The carryover rate relates to the quantities of the particles leaving the freeboard. Coarse particles with a particle terminal velocity higher than the gas velocity eventually return to the dense bed, while fine particles eventually exit from the freeboard. The freeboard height required in design consideration is usually higher than the transport disengagement height (TDH), defined as a height beyond which the solids holdup and solids entrainment or carryover rate remain nearly constant.

Particles are ejected into the freeboard via two basic modes: (1) ejection of particles from the bubble roof, and (2) ejection of particles from the bubble wake. The roof ejection occurs when the bubble approaches the surface of the bed and a dome forms on the surface. As the bubble further approaches the bed surface, particles between the bubble roof and the surface of the dome thin out [15]. At a certain dome thickness, eruption of bubbles with pressure higher than the surface pressure takes place, ejecting the particles present on top of the bubble roof to the freeboard. In wake ejection, as the bubble erupts on the surface, the inertia effect of the wake particles traveling at the same velocity as the bubble promptly ejects these particles to the freeboard. The gas leaving the bed surface then entrains the particles ejected to the freeboard.

In bubbling fluidization, especially at a high velocity, bubble coalescence frequently takes place near the bed surface. The coalescence due to an accelerated traveling bubble yields a significant ejection of wake particles from the leading bubble. The wake ejection in this case becomes the dominant source of particles present in the freeboard in bubbling fluidized beds [16]. The wake ejection in the turbulent regime is much more pronounced than that in the bubbling regime due to higher bubble rise velocities.

Circulating Fluidized Beds.   The fast fluidization regime is the principal regime under which the circulating fluidized bed is operated. The fast fluidization regime is characterized by a dense region at the bottom of the riser and a dilute region above it [17]. The interrelationship of the fast fluidization regime with other fluidization regimes in dense-phase fluidization and with the dilute transport regime is reflected in the variations of the pressure drop per unit length of the riser, gas velocity, and solids circulation rate [18].

The axial profile of the cross-sectional averaged voidage in the riser is typically S-shaped [17] as shown in Fig. 13.3 for Group A particles. This profile reflects an axial solids concentration distribution with a dense region at the bottom and a dilute region at the top of the riser. The boundary between the two regions is marked by the inflection point in the profile. An increase in the gas flow rate at a given solids circulation rate reduces the dense region [from (a) to (c) in Fig. 13.3], whereas an increase in the solids circulation rate at a given gas flow rate results in an expansion of the dense region [from (c) to (a) in Fig. 13.3]. When the solids circulation rate is very low and/or the gas velocity is very high, the dilute region covers the entire riser [see curve (d) in Fig. 13.3]. For given gas and solids flow rates, particles of high density or large size yield a low voidage in the bottom of the riser. The axial profile of the voidage or solids concentration given in Fig. 13.3 is influenced not only by the gas velocity, solids circulation rate, and particle properties, but also by the riser entrance and exit geometries. The S-shaped profile shown in Fig. 13.3 is typical for smooth entrance and exit geometries in which the end effects are minimized. The shape of the profile may vary, however, for nonsmooth entrance and exit geometries.

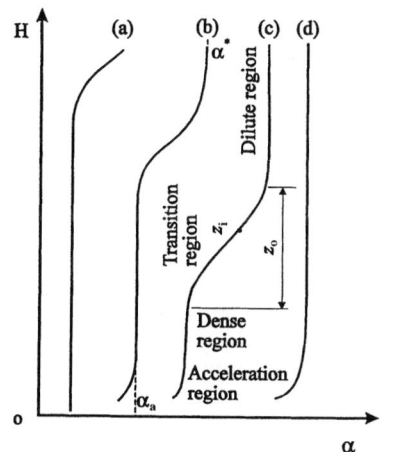

FIGURE 13.3 Typical axial voidage profiles for Group A particles (after Li and Kwauk [17]).

A comprehensive account of radial voidage distribution requires a recognition of the lateral movement of solids in addition to their axial movement. One of the most important and yet least understood aspects of riser hydrodynamics is the lateral solids distribution mechanism [19]. Typical experimental findings for the radial voidage profile are shown in Fig. 13.4. Figure 13.4*a* shows the results for a small CFB unit, while Fig. 13.4*b* shows results for a large unit under similar operating conditions. From Fig. 13.4, it can be seen that the parabolic radial voidage profiles gradually flatten as the height of the axial location increases. Note that both results are time averaged.

The hydrodynamic study at a mesoscopic scale requires the understanding of instantaneous local solids flow structure. The time-variant flow behavior is complex. Analyses of the instantaneous flow structure require recognizing the following factors.

1. Particles migrate to the wall region by means of particle-particle collisions and diffusion, and through particle-wall collision effects that tend to widen the particle velocity distribution in the radial direction.

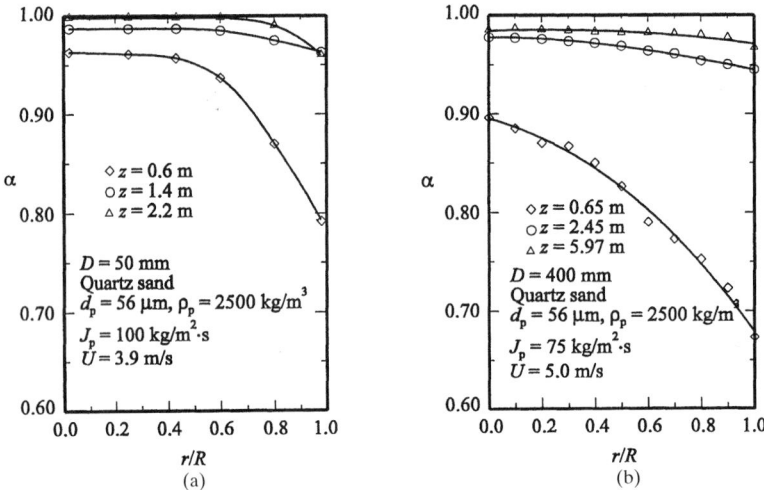

FIGURE 13.4   Typical radial profiles of voidage (after Hartge et al. [20]).

2. No-slip for the gas phase on the wall results in a region of low velocity and low turbulent intensity in the vicinity of the wall.

These factors lead to localized particle accumulation in the wall region. The particle accumulation alters the large-scale motion in the gas-solid flow, which in turn affects the cluster size and motion.

Spouted Beds.   In a spouted bed, gas enters the bed through a jet nozzle of diameter D_i, forming a spout of diameter D_s in the center of the bed. The surrounding annular region forms a downward-moving bed. Particles are entrained into the spout from the bottom and from the sidewall of the spout. Part of the gas seeps into the annular region through the spout wall, whereas the other part leaves the bed from the top of the spout. The particles carried into the spout disengage from the gas in a "solid disengagement fountain" just above the bed and then return to the top of the annular region. Group D particles are commonly used for the spouted bed operation.

In contrast to the minimum fluidization velocity, the minimum spouting velocity U_{msp} depends not only on the particle and gas properties but also on the bed geometry and static bed height. By reducing the gas velocity or by increasing the bed height, spouting can be diminished. U_{msp} can be determined by correlation Eq. 13.5.1, given in Table 13.5 [21]. For $D > 0.4$ m, U_{msp} may be modified by multiplying by a factor of $2D$ [22].

HEAT TRANSFER IN PACKED BEDS

Knowledge of the heat transfer characteristics and spatial temperature distributions in packed beds is of paramount importance to the design and analysis of the packed-bed catalytic or non-catalytic reactors. Hence, an attempt is made in this section to quantify the heat transfer coefficients in terms of correlations based on a wide variety of experimental data and their associated heat transfer models. The principal modes of heat transfer in packed beds consist of conduction, convection, and radiation. The contribution of each of these modes to the overall heat transfer may not be linearly additive, and mutual interaction effects need to be taken into account [23, 24]. Here we limit our discussion to noninteractive modes of heat transfer.

Particle-to-Fluid Heat Transfer

The temperature of the particle surface that is necessary to quantify the heat transfer can be conveniently described in terms of the particle-to-fluid heat transfer coefficient. A considerable amount of study has been carried out to evaluate the particle-to-fluid heat transfer coefficient [e.g., Ref. 25]. The experimental techniques used to measure heat transfer involve either steady-state or unsteady-state conditions. Wakao and Kaguei [1] provide a comprehensive review of the evaluation of the particle-to-fluid heat transfer coefficient. Heat transfer from a single particle in an infinite fluid medium presents the limiting case for heat transfer in packed beds. A simple mathematical treatment of conduction from a sphere (Fourier law) in the absence of convection and/or radiation gives a particle-to-fluid Nusselt number Nu_p of 2. By adding the convective contribution to the overall heat transfer, Ranz and Marshall [26] correlated Nu_p as given by correlation Eq. 13.2.1 in Table 13.2. In a multiparticle system, the heat transfer from particle to fluid and the hydrodynamics are affected by the presence of surrounding particles. Based on the linear velocity, which is higher than the superficial velocity, Ranz and Marshall [26] modified correlation Eq. 13.2.1 for a rhombohedral array (the most dense packing arrangement) of bed particles to give correlation Eq. 13.2.2, where the heat transfer coefficient for a fixed bed is given in terms of particle Reynolds number. Details of this relationship are given in Fig. 13.5, which represents the variation of Nu_p with Re_p. From the figure, it can be seen that correlation Eq. 13.2.2 fits the data well for higher Re_p; however, for lower Re_p the Nusselt number falls below the minimum of 2.

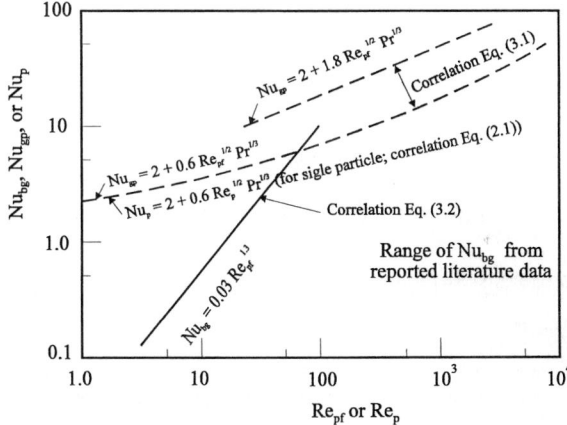

FIGURE 13.5   Particle-to-gas and bed-to-gas heat transfer coefficients under various flow conditions (from Kunii and Levenspiel [2]).

Effective Thermal Conductivity

Although the heat flow and fluid flow in packed beds are quite complex, the heat transfer characteristics can be described by a simple concept of effective thermal conductivity K_e that is based on the assumption that on a macroscopic scale the bed can be described by a continuum. Effective thermal conductivity is a continuum property that depends on temperature, bed material, and structure. It is usually determined by evaluating the steady-state heat flux between two parallel plates separated by a packed bed. The effective thermal conductivity applies very accurately to steady-state heat transfer and to unsteady-state heat transfer if $(t/d_p^2) > 1.94 \times 10^7$ s/m² [27]; in other cases, for unsteady state heat transfer the thermal

TABLE 13.2 Heat Transfer in Packed Beds

	Equation 13.2.1
Investigator	Ranz and Marshall [26]
Type of correlation	Particle-to-fluid heat transfer (single-particle system)
Phases involved	Fluid-solid
Correlation equation	$Nu_p = 2 + 0.6 Re_p^{1/2} Pr^{1/3}$
Range of applicability	$Re_p > 50$

	Equation 13.2.2
Investigator	Ranz and Marshall [26]
Type of correlation	Particle-to-fluid heat transfer (multiparticle system)
Phases involved	Fluid-solid
Correlation equation	$Nu_p = 2 + 1.8 Re_p^{1/2} Pr^{1/3}$
Range of applicability	$Re_p > 50$

	Equation 13.2.3
Investigator	Kunii and Smith [29]
Type of correlation	Effective thermal conductivity of packed bed
Phases involved	Fluid-solid
Model associated	One-dimensional heat transfer model
	Spheres in cubic array
Model equation	$\dfrac{K_e}{K} = (0.7854)(2)\left(\dfrac{K_p}{K_p - K}\right)^2\left(\ln\dfrac{K_p}{K} - \dfrac{K_p - K}{K_p}\right) + 0.2146$

	Equation 13.2.4
Investigator	Kunii and Smith [29]
Type of correlation	Effective thermal conductivity of packed bed
Phases involved	Fluid-solid
Model associated	One-dimensional heat transfer model
	Spheres in orthorhombic array
Model equation	$\dfrac{K_e}{K} = (0.9069)(2)\left(\dfrac{K_p}{K_p - K}\right)^2\left(\ln\dfrac{K_p}{K} - \dfrac{K_p - K}{K_p}\right) + 0.0931$

	Equation 13.2.5
Investigator	Krupiczka [30]
Type of correlation	Effective thermal conductivity of packed bed
Phases involved	Fluid-solid
Model associated	Two-dimensional heat transfer model
	Packed bed consisting of bundle of long cylinders
Model equation	$\log\dfrac{K_e}{K} = \left(0.785 - 0.057\log\dfrac{K_p}{K}\right)\log\dfrac{K_p}{K}$

	Equation 13.2.6
Investigator	Krupiczka [30]
Type of correlation	Effective thermal conductivity of packed bed
Phases involved	Fluid-solid
Model associated	Two-dimensional heat transfer model
	Packed bed consisting of a spherical lattice of spheres
Model equation	$\log\dfrac{K_e}{K} = \left(0.280 - 0.757\log(\alpha) - 0.057\log\dfrac{K_p}{K}\right)\log\dfrac{K_p}{K}$

TABLE 13.2   Heat Transfer in Packed Beds (*Continued*)

Equation 13.2.7	
Investigator	Yagi and Kunii [31]
Type of correlation	Effective thermal conductivity of packed bed
Phases involved	Fluid-solid
Model associated	Accounting the fluid motion (convective contribution)
	Applicable to heat transfer in the direction normal to fluid flow
Correlation equation	$\dfrac{K_{e,conv.}}{K} = \dfrac{K_e}{K} + 0.11 \mathrm{Re}_p\,\mathrm{Pr}$
Range of applicability	$\dfrac{d_p}{D} \leq 0.04$

Equation 13.2.8	
Investigator	Yagi et al. [32]
Type of correlation	Effective thermal conductivity of packed bed
Phases involved	Fluid-solid
Model associated	Accounting the fluid motion (convective contribution)
	Applicable to heat transfer in the radial direction
Correlation equation	$\dfrac{K_{e,conv.}}{K} = \dfrac{K_e}{K} + 0.75 \mathrm{Re}_p\,\mathrm{Pr}$
Range of applicability	$\mathrm{Re}_p < 5400$

Equation 13.2.9	
Investigator	Mohamad et al. [36]
Type of correlation	Packed-bed effective thermal conductivity due to conduction
Phases involved	Fluid-solid
Model equation	$K_{con} = \dfrac{2K}{1-(K/K_p)}\left(\dfrac{\ln(K_p/K)}{1-(K/K_p)} - 1\right)$

Equation 13.2.10	
Investigator	Mohamad et al. [36]
Type of correlation	Packed-bed effective thermal conductivity due to dispersion
Phases involved	Fluid-solid
Model equation	$K_{dis} = 0.0895 \mathrm{Re}_p\,\mathrm{Pr}\,K$

Equation 13.2.11	
Investigator	Mohamad et al. [36]
Type of correlation	Packed-bed effective thermal conductivity due to radiation
Phases involved	Fluid-solid
Correlation equation	$K_{rad} = 0.707 K\left(\dfrac{K_p}{K}\right)^{1.11}\left(\dfrac{4\sigma_b T^3}{2\{[1/(1-\alpha)]-1\}+(1/F_{12})}\dfrac{d_p}{K_p}\right)^{0.96}$
Range of applicability	$20 < K_p/K < 1000$ and $h_r d_p/K < 0.3$

Equation 13.2.12	
Investigator	Li and Finlayson [37]
Type of correlation	Wall-to-bed heat transfer coefficient in packed bed
Phases involved	Fluid-solid
System	Spherical particle-air system
Correlation equation	$\dfrac{h_w d_p}{K} = 0.17 \mathrm{Re}_p^{0.79}$
Range of applicability	$20 < \mathrm{Re}_p < 7600$ and $0.05 < d_p/D < 0.3$

TABLE 13.2 Heat Transfer in Packed Beds (*Continued*)

	Equation 13.2.13
Investigator	Li and Finlayson [37]
Type of correlation	Wall-to-bed heat transfer coefficient in packed bed
Phases involved	Fluid-solid
System	Cylindrical particle-air system
Correlation equation	$\dfrac{h_w d_p}{K} = 0.16 \mathrm{Re}_p^{0.93}$
Range of applicability	$20 < \mathrm{Re}_p < 800$ and $0.03 < d_p/D < 0.2$

	Equation 13.2.14
Investigator	Nasr et al. [38]
Type of correlation	Wall-to-bed heat transfer coefficient in packed bed
Phases involved	Fluid-solid
Comments	Heat transfer augmentation by embedding a heat transfer surface
Correlation equation	$\dfrac{h_w D}{K_e} = 0.53 \left(\dfrac{D}{d_p}\right)^{0.114} \left[\mathrm{Pe}^{0.66(K_p/K)^{-0.0174}} \right]$ where $\mathrm{Pe} = \dfrac{UD}{D_{\mathrm{tem}}}$
Range of applicability	$1 < (D/d_p) < 5$ and $10 < (K_p/K) < 7600$

	Equation 13.2.15
Investigator	Whitaker [39]
Type of correlation	Particle-to-bed radiative heat transfer (radiative flux)
Phases involved	Fluid-solid
Model associated	Treating both particle and bed as gray bodies
Model equation	$J_r = \dfrac{\sigma}{(2/\varepsilon) - 1}\, (T'^4 - T''^4)$

	Equation 13.2.16
Investigator	Wakao and Kato [40]
Type of correlation	Particle-to-bed radiative heat transfer coefficient
Phases involved	Fluid-solid
Model associated	Assumes that any two hemispheres in contact are circumscribed with a diffusely reflecting cylindrical wall. Takes into account the overall view factor
Model equation	$h_r = \dfrac{0.2268}{(2/\varepsilon) - 0.264} \left(\dfrac{T}{100}\right)^3$

	Equation 13.2.17
Investigator	Schotte [41]
Type of correlation	Radiative effective thermal conductivity
Phases involved	Fluid-solid
Model equation	$K_{\mathrm{rad}} = \dfrac{1 - \alpha}{(1/K_p) + (1/K_r)} + \alpha K_r$ where $K_r = 0.229(\alpha)(\varepsilon)(d_p)\, \dfrac{T^3}{10^6}$

response time of each phase needs to be accounted for separately and only after a long time can the effective thermal conductivity be used to approximate the properties of separate phases.

Many theoretical and experimental studies have been carried out on estimation of the effective thermal conductivity (e.g., Ref. 28). These studies broadly consider two approaches: (1) assuming unidirectional heat flow, and (2) assuming two-dimensional heat flow. Using the unidirectional heat flow, Kunii and Smith [29] evaluated the packed-bed effective thermal conductivity by analytical solution. In their interparticle conduction model, they assumed that all the heat transfer occurs in the axial direction and no temperature distribution exists in the radial direction. For spheres in a cubic array ($\alpha = 0.4764$), the effective thermal conductivity with a stagnant fluid is given by model Eq. 13.2.3, while for spheres in orthorhombic array ($\alpha = 0.3954$) with a stagnant fluid, the effective thermal conductivity is given by model Eq. 13.2.4. This model assumes that the heat flow paths are parallel at both microscopic and macroscopic levels. The model is phenomenological and requires the conductivity of bed in a vacuum.

The two-dimensional heat flow model is more realistic than the unidirectional heat flow model. By assuming that the packed bed consists of a bundle of long cylinders, Krupiczka [30] found a numerical solution that gave the effective thermal conductivity of quiescent cylinder bed ($\alpha = 0.215$) as given by model Eq. 13.2.5. By extending the concept to a spherical lattice, Krupiczka [30] expressed the effective thermal conductivity of quiescent bed of spherical particles as given by model Eq. 13.2.6. If the fluid is in motion, then convective contribution augments the heat transfer. Yagi and Kunii [31] accounted for this effect through correlation Eq. 13.2.7, which is applicable to the heat transfer in the direction normal to the fluid flow. For the heat transfer in a radial direction, Yagi et al. [32] recommended correlation Eq. 13.2.8. A detailed survey of various alternate correlations for effective thermal conductivity can be found in Kuzay [33].

Rao and Toor [34] performed a rigorous analysis for a special class of problems where K_e cannot be regarded as a bed continuum property. They showed that the continuum behavior can be assumed only above a certain ratio of radius of the test particle to radius of bed particle. For the cases with radius ratio less than the critical ratio, Rao and Toor proposed a discrete model that satisfactorily explained the lower values of heat transfer rate compared to the continuum model values. Pons et al. [35] investigated the effective thermal conductivity of the packed bed in the presence of a hydriding reaction and found that under high pressure (above the Knudsen transition domain) the wall heat transfer coefficient was extremely high, around 3000 W·m^{-2} K^{-1}, which decreased to 13 W·m^{-2} K^{-1} when the pressure was lowered. Mohamad et al. [36] studied the effective thermal conductivity of a packed bed in a combustor-heater system. Assuming that the porous medium can be treated as a continuum and that K_e is the sum of conductivities due to conduction K_{con}, dispersion K_{dis}, and radiation K_{rad}, Mohamad et al. [36] proposed correlation Eqs. 13.2.9–13.2.11 for predicting the individual effective thermal conductivities.

Wall-to-Bed Heat Transfer

The wall-to-bed heat transfer coefficient has been investigated by many researchers. Examining the published data, Li and Finlayson [37] found that many of the data on h_w had entrance effect. Considering the data that were free from entrance effect, they proposed correlation Eq. 13.2.12 for spherical particle–air systems. They also correlated the data for cylindrical particle-air systems as given by correlation Eq. 13.2.13.

Recently Nasr et al. [38] studied the augmentation of heat transfer by embedding the heat transfer surface in a packed bed. They found that in the presence of particles, the wall-to-bed heat transfer coefficient was up to 7 times greater than that for the case where heat transfer surface was placed in a cross flow. It was shown that the heat transfer coefficient increases with decreasing particle diameter and increasing thermal conductivity of the packing material. They proposed the empirical correlation given by correlation Eq. 13.2.14 for the wall-to-bed heat transfer coefficient in the presence of forced convection.

Radiative Heat Transfer

The contribution of radiative heat transfer becomes significant at temperatures above 600°C. The radiative component of heat transfer is accounted for by linear addition to the conductive and convective heat transfer components. The interactive effects between radiation-conduction and radiation-convection are discussed elsewhere [23, 24].

In a packed bed, the heat transmitted from a particle by radiation is absorbed by surrounding particles and fluid. Hence, measurement of particle-to-bed radiative heat transfer requires an uneven temperature distribution in the bed. The radiant heat transfer flux between two large gray bodies at temperatures T' and T'' is given by model Eq. 13.2.15 [39]. Wakao and Kato [40] proposed a formula for radiative heat transfer coefficient based on unit particle surface area, which includes an overall view factor. They assumed that any two hemispheres in contact were circumscribed with a diffusely reflecting cylindrical wall. After rigorous mathematical analysis, they arrived at model Eq. 13.2.16 for the particle-to-bed radiative heat transfer coefficient. From the contribution of radiant heat transfer to effective thermal conductivity, Schotte [41] derived the radiant effective thermal conductivity as given by model Eq. 13.2.17. Brewster and Tien [42] presented a detailed analysis of dependent versus independent scattering in radiative heat transfer in packed beds. They suggested that interparticle distance is the most important parameter in gauging the importance of dependent scattering. Singh and Kaviany [43] performed modeling of radiative heat transfer in packed beds. They also pointed out that theory of independent scattering fails when interparticle distance and porosity are small.

HEAT TRANSFER IN FLUIDIZED BEDS

Gas-Solid Fluidized Beds

Gas-solid flows involving heat transfer are common in many engineering applications such as petroleum refining, solid fuel combustion, nuclear reactor cooling, and bulk material handling and transport. The understanding of heat transfer characteristics in a gas-solid fluidized bed is of paramount importance because of its widespread application in the industry. This section describes the fundamental mechanisms of heat transfer in gas-solid flows, covering a wide range of operating conditions from dense-phase fluidized beds to spouted beds.

General Modes of Heat Transfer in Gas-Solid Systems. A mechanistic account of suspension-to-surface heat transfer is necessary to accurately quantify heat transfer characteristics. The following section discusses the principal modes and regimes of suspension-to-surface heat transfer.

The heat transfer coefficient between a surface and a gas-solid suspension typically consists of three additive components, viz., particle convection, gas convection, and radiation. Particle convective heat transfer is due to the convective flow of solids from the in-bed region to the region adjacent to the heat transfer surface. Solid particles gain heat by thermal conduction in the hot region, and as they return to the colder region of the bed the heat is dissipated. This is the principal mode of heat transfer in fine particle systems. Gas convective heat transfer is caused by the gas percolating through the bed and also by the gas voids coming in contact with the heat transfer surface. The gas convective component is a significant mode of heat transfer in systems using large particles and high operating pressures. Radiative heat transfer is due to radiant heat transmitted to fluidized particles or solid surfaces from a heat transfer surface at high temperature. The total heat transfer coefficient h can be estimated from the summation of the individual heat transfer coefficients, viz., particle convection h_{pc}, gas convection h_{gc}, and radiation h_r, although the precise relationship may not be linearly additive, i.e.,

$$h \approx h_{pc} + h_{gc} + h_r \qquad (13.3)$$

The heat transfer characteristics are strongly influenced by the operating conditions of the fluidized beds. Different operating conditions such as bubbling and spouting yield a varied bed structure and hence varied heat transfer coefficients. Understanding the governing heat transfer mechanisms is important to the development of simplified heat transfer models and correlations. The relative importance of the heat transfer modes for the suspension-to-surface heat transfer in gas-solid fluidized beds is illustrated in Fig. 13.6 [44], which indicates that the governing mode of heat transfer in the fluidized bed depends on both the particle size and the bed/surface temperature. It can be seen that particle convection is dominant for almost all the conditions, except at low bed temperatures in a bed of large particles, where the gas convection becomes important.

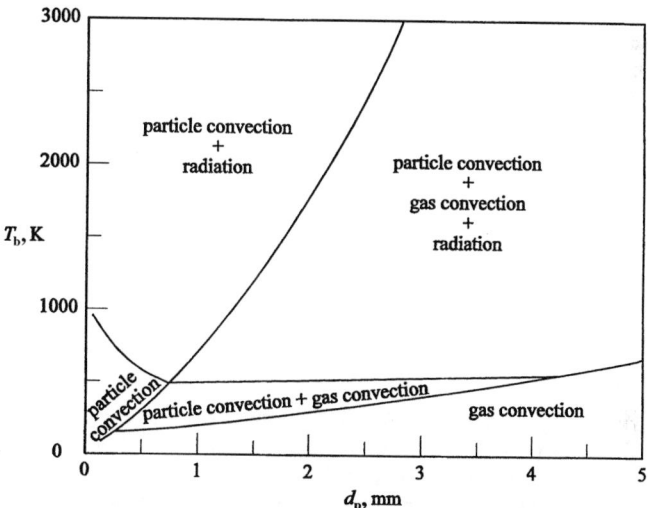

FIGURE 13.6 Heat transfer diagram for various governing modes (from Flamant et al. [44]).

In most dense gas-solid fluidization systems, particle circulation (e.g., induced by the bubbles) is the primary cause of particle convective heat transfer. The heat transfer rate is high when there is an extensive solids exchange between the in-bed region and the region near the heat transfer surface. In light of the relative importance of the particle convective component, more discussion is focused on particle convective heat transfer in the following sections in the context of the heat transfer models and corresponding correlations.

Models for Heat Transfer.   Development of a mechanistic model is essential to the understanding of the heat transfer phenomena in a fluidized system. Models developed for the dense-phase fluidized systems represent a general class and are also applicable to other fluidization systems. Figure 13.7 illustrates the basic heat transfer characteristics in dense-phase fluidization systems that a mechanistic model needs to quantify. The figure shows the variation of heat transfer coefficient with the gas velocity. It can be seen that at a low gas velocity

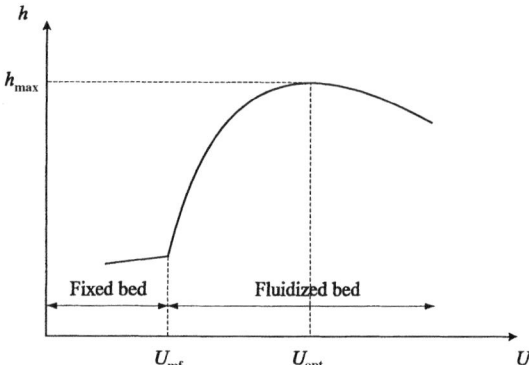

FIGURE 13.7 Typical dependence of the heat transfer coefficient on gas velocity in dense-phase fluidization systems (from Gel'Perin and Einstein [45]).

where the bed is in a fixed state, the heat transfer coefficient is low; with increasing gas velocity, it increases sharply to a maximum value and then decreases. This behavior of the heat transfer coefficient is due to an interplay between the particle convective and gas convective heat transfer that can be explained by mechanistic models given in the following three sections.

The models developed to describe the heat transfer behavior in fluidization systems can be broadly classified into the following three categories: the film model, the single-particle model, and the emulsion phase/packet model. A given model may be more suitable to some fluidization conditions than the others. For example, the film model and the single-particle model are more suitable for a particulate fluidized bed [45] than for a bed containing gas bubbles. The most challenging aspect in applying these models, however, lies in the determination of flow and thermal properties of the region in the vicinity of the heat transfer surface.

Film Model. In a well-fluidized gas-solid system, the in-bed region can be assumed to be isothermal and hence to have negligible thermal resistance. This assumption suggests that the thermal resistance limiting the rate of heat transfer between the bed and the heating surface lies within a narrow gas layer in the vicinity of the heating surface. The film model for fluidized bed heat transfer assumes that the heat is transferred through the thin gas film only by conduction. The heat transfer coefficient in the film model can be expressed as

$$h = \frac{K}{\delta} \tag{13.4}$$

where K is the gas thermal conductivity and δ is the boundary layer thickness, which depends on the velocity and physical properties of the fluid and also on the intensity of motion of the solid particles. The motion of particles erodes the film and reduces its resistive effect. With an increase in the gas velocity, the particles near the surface move more vigorously, but the local concentration of the particles decreases. This interplay results in a maximum in the h-U curve as shown in Fig. 13.7.

The model based on the concept of pure limiting film resistance involves the steady-state concept of the heat transfer process and omits the essential unsteady nature of the heat transfer phenomena observed in many gas-solid suspension systems. The film model discounts the effects of thermophysical properties such as the specific heat of solids and hence would not be able to predict the particle convective component of heat transfer. For estimating the contribution of the particle convective component of heat transfer, the emulsion phase/packet model given in a subsequent section should be used to describe the temperature gradient from the heating surface to the bed.

Single-Particle Model. The single-particle model [46] postulates that the moving solid particles play a significant role in heat transfer by thermal conduction. The model also takes into account the thermal conduction through the gas film at the heating surface. In this model, the contributions from gas convection and bed-to-surface thermal radiation are neglected. The high heat transfer coefficient shown in Fig. 13.7 is due to a high temperature gradient while heating the moving solids. The presence of the maximum in the heat transfer curve is a result of the simultaneous effect of the rise in temperature gradient and the fall in concentration of solid particles.

The simplest model of this kind is the case where an isolated particle surrounded by gas is in contact with or in the vicinity of the heating surface for a certain time, during which the heat transfer between the particle and the heating surface takes place by transient conduction. This model can be extended from a single particle to a single layer of particles at the surface. However, the model requires precise information on the position and the residence time of the particle near the heat transfer surface, and this requirement could limit its wide usage. The model is suitable only when the heat from the heat transfer surface does not penetrate beyond the single-particle layer. The depth of penetration into the bed (δ_{em}) can be estimated from the temperature gradient at the heating surface as

$$\delta_{em} \propto (D_{tem}t_c)^{1/2} \tag{13.5}$$

where D_{tem} is the thermal diffusivity of the emulsion phase and t_c is the average residence time of the particle near or at the heat transfer surface. Equation 13.5 yields

$$\frac{\delta_{em}}{d_p} \propto Fo^{1/2} \tag{13.6}$$

where Fo is the Fourier number ($D_{tem}t_c/d_p^2$). Thus, the single-particle model is suitable only for low Fourier numbers, i.e., large particles with a short contact time. To expand the range of applicability, the heat diffusion equation for multiple particle layers has been solved [27].

Emulsion Phase/Packet Model. In the emulsion phase/packet model, it is perceived that the resistance to heat transfer lies in a relatively thick emulsion layer adjacent to the heating surface. This approach considers the emulsion phase/packets to be a continuous phase. The presence of the maxima in the *h-U* curve can be explained to be due to the simultaneous effect of an increase in the frequency of packet replacement and an increase in the fraction of time for which the heat transfer surface is covered by bubbles/voids. This unsteady-state model reaches its limit when the particle thermal time constant is smaller than the particle contact time determined by the replacement rate for small particles, in which case the heat transfer can be approximated by a steady-state process.

Mickley and Fairbanks [47] treated the packet as a continuum phase and first recognized the significant role of particle heat transfer, since the volumetric heat capacity of the particle is a thousandfold that of the gas at atmospheric conditions. They considered a packet of emulsion phase being swept into contact with the heating surface for a certain period of time. During the contact, the heat is transferred by unsteady-state conduction until the packet is replaced by a fresh one due to circulation, as shown in Fig. 13.8. The heat transfer rate depends on the frequency of the packet replacement at the surface. To simplify the model, the packet of particles and interstitial gas can be regarded as having the

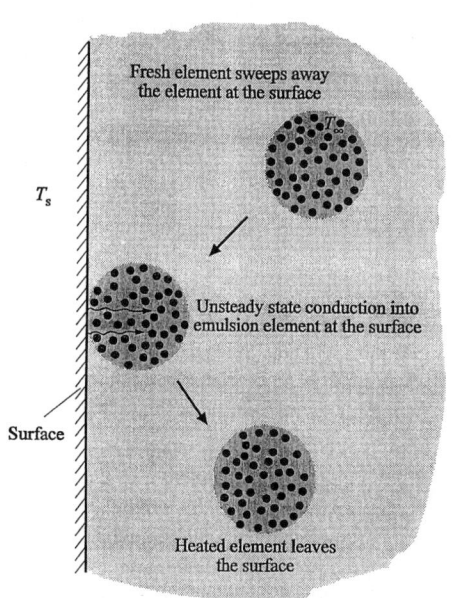

FIGURE 13.8 Conceptual representation of the emulsion contact model of Mickley and Fairbanks [47].

uniform thermal properties of the quiescent suspension. The simplest case is one-dimensional unsteady thermal conduction in a semi-infinite medium. From the model, the instantaneous local heat transfer coefficient is obtained as

$$h_i = \sqrt{\frac{K_{em}c\rho_{em}}{\pi t}} \tag{13.7}$$

Furthermore, the area averaged local heat transfer coefficient can be expressed by

$$h = \frac{1}{A_m} \int_0^{A_m} \left(\int_0^\infty h_i \psi(\tau)\, d\tau \right) dA = \sqrt{K_{em}c\rho_{em}}\, S \tag{13.8}$$

where A_m is the area of the packet in contact with the heating surface, $\psi(\tau)$ represents the frequency of occurrence in time of the packet of age τ, and S is the area mean stirring factor defined as

$$S^{1/2} = \frac{1}{A_m} \int_0^{A_m} \left(\frac{1}{\sqrt{\pi}} \int_0^\infty \frac{\psi(\tau)}{\sqrt{\tau}}\, d\tau \right) dA \tag{13.9}$$

The preceding model successfully explains the role played by the particles in the heat transfer processes occurring in the dense-phase fluidized bed at voidage $\alpha < 0.7$. But it predicts very large values when the contact time of particles with the heating surface decreases. Due to the time-dependent voidage variations near the heating surface, the thermophysical properties of the packet differ from those in the bed, and this difference has not been accounted for in the packet model. Thus, the packet model is accurate only for large Fourier number values.

An important variation of the model of Mickley and Fairbanks is the film penetration model developed by Yoshida et al. [48] by treating packets as a continuum with a finite thickness (δ_{em}). The film penetration theory includes two extremes of emulsion behavior. On one extreme, the packet contacts the heating surface for a short time so that all the heat entering the packet is used to heat up the packet (penetration theory) while none passes through it. On the other extreme, the packet stays at the surface long enough to achieve steady state and simply provides a resistance for heat conduction.

The heat transfer process can be described by a thin layer of emulsion of thickness δ_{em} that is in contact with a heat transfer surface and after a time t_c, is replaced by a fresh element of emulsion from the bulk of the suspension, as shown in Fig. 13.9. Defining the thermal diffusivity of the emulsion phase as

$$D_{tem} = \frac{K_{em}}{\rho_{em}c} \approx \frac{K_{em}}{\rho_p c(1 - \alpha_{mf})} \tag{13.10}$$

the solution for the instantaneous local heat transfer coefficient can be obtained as [48]

$$h_i = \frac{K_{em}}{\sqrt{\pi D_{tem}t}} \left[1 + 2 \sum_{n=1}^\infty \exp\left(-\frac{\delta_{em}^2 n^2}{D_{tem}t} \right) \right]; \qquad \pi \leq \frac{\delta_{em}^2}{D_{tem}t} < \infty \tag{13.11}$$

$$h_i = \frac{K_{em}}{\delta_{em}} \left[1 + 2 \sum_{n=1}^\infty \exp\left(-\frac{\pi^2 D_{tem}t n^2}{\delta_{em}^2} \right) \right]; \qquad 0 < \frac{\delta_{em}^2}{D_{tem}t} \leq \pi \tag{13.12}$$

where the dimensionless group $D_{tem}t/\delta_{em}^2$ characterizes the intensity of renewal. Generally, when $t_c \gg \delta_{em}^2/D_{tem}$, the film theory holds, and when $t_c \ll \delta_{em}^2/D_{tem}$, the penetration theory is valid. The time-averaged heat transfer coefficient h is obtained as

$$h = \int_0^\infty h_i I(t)\, dt \tag{13.13}$$

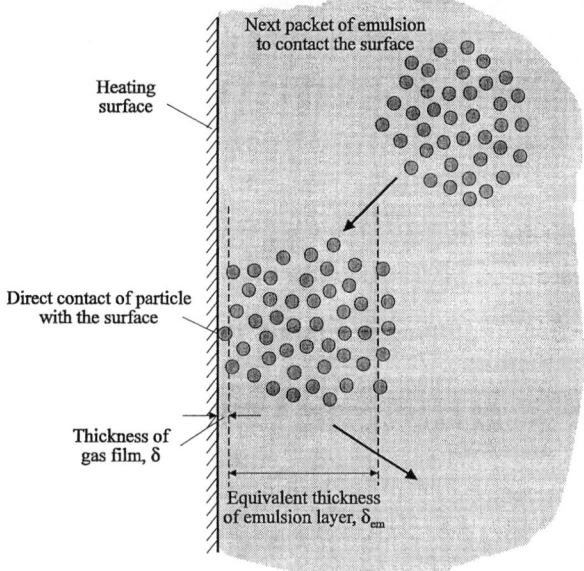

FIGURE 13.9 Conceptual representation of the film penetration model for suspension-to-surface heat transfer (from Yoshida et al. [48]).

where $I(t)$ is the age distribution function, representing the fraction of surface occupied by packets of age between t and $t + dt$. Two commonly used age distribution functions for random surface renewal and for uniform surface renewal are discussed below.

The random surface renewal exists for a surface that is in the in-bed region of a suspension and is continuously contacted by rising bubbles. The age distribution of elements on the surface is represented by that of a continuous-stirred tank reactor (CSTR) as given by

$$I(t) = \frac{1}{t_c} \exp\left(-\frac{t}{t_c}\right) \tag{13.14}$$

Therefore, we obtain the following expressions for the average heat transfer coefficient for two extreme cases as for rapid renewal when $t_c \ll \delta_{em}^2/D_{tem}$, we have

$$h = \frac{K_{em}}{\sqrt{D_{tem}t_c}} \left[1 + 2 \sum_{n=1}^{\infty} \exp\left(-\frac{2\delta_{em}n}{\sqrt{D_{tem}t_c}}\right) \right] \tag{13.15}$$

and for slow renewal when $t_c \gg \delta_{em}^2/D_{tem}$, we have

$$h = \frac{K_{em}}{\delta_{em}} \left(1 + \frac{\delta_{em}^2}{3D_{tem}t_c} \right) \tag{13.16}$$

In the mode of uniform surface renewal, all elements of emulsion contact the surface for the same duration of time; such a situation is encountered in emulsion flowing smoothly past a small heat transfer surface. Here, the age distribution function is represented by that of a plug flow reactor (PFR) as given by

$$I(t) = \begin{cases} 1/t_c & \text{for } 0 < t < t_c \\ 0 & \text{for } t > t_c \end{cases} \tag{13.17}$$

Thus, for rapid renewal, we have

$$h = \frac{2K_{em}}{\sqrt{\pi D_{tem} t_c}} \tag{13.18}$$

and for slow renewal, we have

$$h = \frac{K_{em}}{\delta_{em}} \tag{13.19}$$

It should, however, be noted that the above treatment of the emulsion phase/packet model is suitable for a system with homogeneous emulsion phase (i.e., particulate fluidization). The model needs to be modified when applied to the fluidized bed with a discrete bubble phase.

Heat Transfer in Dense-Phase Fluidized Beds.   The majority of the heat transfer models and correlations derived for gas-solid fluidized beds were originally developed for dense-phase fluidized beds. The following sections discuss the heat transfer coefficients between the suspension (or bed) and the particle, between the suspension (or bed) and the gas, and between the suspension (or bed) and the wall or heat transfer surface.

Particle-to-Gas and Bed-to-Gas Heat Transfer.   The particle-to-gas heat transfer can be quantified by unsteady-state experiments that measure the time required for cold particles of temperature T_{p0}, mass M, and surface area S_p to reach the bed temperature when they are introduced into the bed. The local particle-to-gas heat transfer coefficient h_{gp} can be given by the equation

$$\frac{T_p - T_{p0}}{T_b - T_{p0}} = 1 - \exp\left(-\frac{h_{gp}S_p t}{Mc}\right) \tag{13.20}$$

The particle-to-gas heat transfer coefficient in dense-phase fluidization systems can be determined from correlation Eq. 13.3.1 [2] given in Table 13.3. The correlation indicates that the values of particle-to-gas heat transfer coefficient in a dense-phase fluidized bed lie between those for fixed bed with large isometric particles (with a factor of 1.8 in the second term [49]) and those for the single-particle heat transfer coefficient (with a factor of 0.6 in the second term of the equation).

For suspension (or bed)-to-gas heat transfer in a well-mixed bed of particles, assuming low-Biot number conditions (i.e., negligible internal thermal resistance) and assuming the gas flow to be plug flow, the suspension (or bed)-to-gas heat transfer coefficient h_{bg} can be given by

$$\ln\left(\frac{T - T_b}{T_i - T_b}\right) = -\left(\frac{6h_{bg}(1 - \alpha)}{U\rho c_p d_p}\right)H \tag{13.21}$$

The range of data for the bed-to-gas heat transfer coefficient reported in the literature, which were primarily based on Eq. 13.21, is shown in the shadowed region in Fig. 13.5. From the figure, it can be seen that under high Reynolds numbers ($\mathrm{Re}_{pf} > 100$), values of Nu_{bg} are very close to those of Nu_{gp} determined by correlation Eq. 13.3.1, since the plug flow assumption for the gas phase in the bed is realistic. However, values of Nu_{bg} under low Reynolds numbers ($\mathrm{Re}_{pf} < 100$), as in fine-particle fluidization, are smaller than Nu_{gp} based on correlation Eq. 13.3.1 and are much smaller than the value of 2 for an isolated spherical particle in a stationary condition. Nu_{bg} under this Reynolds number range follows the correlation Eq. 13.3.2. It should, however, be mentioned that this deviation is model dependent rather than being mechanistic because the actual gas-solid contact is much poorer than that portrayed by the plug flow assumption on which Eq. 13.21 is based [2]. The deviation could also be related to the boundary layer reduction due to particle collision, and the generation of turbulence by bubble motion and particle collision [50].

Bed-to-Surface Heat Transfer.   The particle circulation induced by bubble motion plays an important role in the bed-to-surface heat transfer in a dense-phase fluidized bed. This can

TABLE 13.3 Heat Transfer in Dense-Phase Fluidized Beds

Equation 13.3.1	
Investigator	Kunii and Levenspiel [2]
Type of correlation	Particle-to-gas heat transfer coefficient
Phases involved	Gas-solid
Model associated	Gas in plug flow through the bed
Correlation equation	$\mathrm{Nu_{gp}} = \dfrac{h_{gp}d_p}{K} \approx 2 + (0.6 - 1.8)\mathrm{Re}_{pf}^{1/2}\,\mathrm{Pr}^{1/3}$
Range of applicability	$\mathrm{Re}_{pf} > 100$

Equation 13.3.2	
Investigator	Kunii and Levenspiel [2]
Type of correlation	Bed-to-gas heat transfer coefficient
Phases involved	Gas-solid
Correlation equation	$\mathrm{Nu_{bg}} = \dfrac{h_{bg}d_p}{K} = 0.03\mathrm{Re}_{pf}^{1.3}$
Range of applicability	$0.1 < \mathrm{Re}_{pf} < 100$

Equation 13.3.3	
Investigator	Molerus et al. [55, 56]
Type of correlation	Wall-to-bed heat transfer coefficient in bubbling fluidized beds
Phases involved	Gas-solid
Correlation equation	$\dfrac{hL}{K} = \dfrac{0.125(1-\alpha_{mf})(1+A)^{-1}}{1+(K/2c\mu)\{1+BC\}} + 0.165\mathrm{Pr}^{1/3}\,E$

$$\text{where}\quad L = \left[\frac{\mu}{\sqrt{g(\rho_p - \rho_g)}}\right]^{2/3};$$

$$A = 33.3\left\{\sqrt[3]{\left[\frac{U-U_{mf}}{U_{mf}}\right]}\sqrt[3]{\frac{\rho_p c}{Kg}}\,(U-U_{mf})\right\}^{-1};$$

$$B = 0.28(1-\alpha_{mf})^2\left[\frac{\rho_g}{\rho_p - \rho_g}\right]^{0.5};$$

$$C = \left[\sqrt[3]{\frac{\rho_p c}{Kg}}\,(U-U_{mf})\right]^2 \frac{U_{mf}}{(U-U_{mf})};$$

$$E = \left(\frac{\rho_g}{\rho_p - \rho_g}\right)^{1/3}\left[1 + 0.05\left\{\frac{U_{mf}}{(U-U_{mf})}\right\}\right]^{-1}$$

Range of applicability	$\mathrm{Ar} < 10^8$ where $\mathrm{Ar} = \dfrac{d_p^3 g(\rho_p - \rho_g)\rho_g}{\mu^2}$

Equation 13.3.4	
Investigator	Baskakov et al. [57]
Type of correlation	Gas convective heat transfer coefficient
Phases involved	Gas-solid
Correlation equation	$\dfrac{h_{gc}d_p}{K} = 0.009\mathrm{Ar}^{1/2}\,\mathrm{Pr}^{1/3}$
Range of applicability	$0.16\ \text{mm} < d_p < 4\ \text{mm}$

Equation 13.3.5	
Investigator	Denloye and Botterill [58]
Type of correlation	Gas convective heat transfer coefficient
Phases involved	Gas-solid
Correlation equation	$\dfrac{h_{gc}\sqrt{d_p}}{K} = 0.86\mathrm{Ar}^{0.39}$
Range of applicability	$10^3 < \mathrm{Ar} < 2 \times 10^6$ and operating pressure $< 1\ \text{MPa}$

be seen from a study conducted by Tuot and Clift [51] on heat transfer properties around a single bubble rising in a gas-solid suspension. Employing a sensitive probe with low heat capacity and fast response time, these researchers observed that the heat transfer coefficient increased as the bubble rose toward the probe (points A to B on the solid line in Fig. 13.10). The increase results from the particle movement close to the probe surface as the bubble approaches from beneath. As the bubble envelops the probe, the heat transfer coefficient decreases (point C on the solid line) due to the lower thermal conductivity and heat capacity of the gas phase. Further rising of the bubble leads to a peak in the heat transfer coefficient behind the bubble (point D) that is due to high concentration of particles in the wake passing the probe. A relatively slow decay of heat transfer coefficient beyond point D to a new steady value is due to the effect of turbulence in the medium. The dashed line in Fig. 13.10 shows the heat transfer coefficient due to a bubble rising to the side of the probe, and the maximum is again due to the effect of high concentration of particles carried in the wake of the bubble. Thus the bubble wake plays a significant role in particle circulation and hence the heat transfer in gas-solid fluidization.

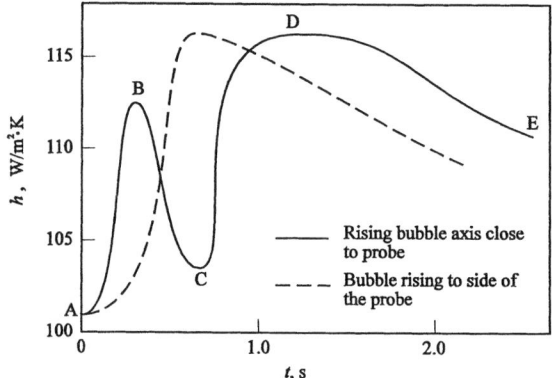

FIGURE 13.10   Probe-to-bed heat transfer coefficient variations in a fluidized bed (from Tuot and Clift [51]).

The bed-to-surface heat transfer consists of three major components, viz., the particle convective component, the gas convective component, and the radiative component. In gas-solid fluidization systems, radiation may be neglected when the bed temperature is less than 400°C. The significance of particle convection and gas convection depends mostly on the types of particles used. As a rule of thumb, particle convection is the dominant mechanism for small particles ($d_p < 400$ μm) and it usually plays a key role for Group A particles. Gas convection becomes important for large particles ($d_p > 1500$ μm) and for high-pressure or high-velocity fluidizations, and it usually plays a key role for Group D particles [52]. For Group B particles, both components are significant.

Particle Convective Component.   Particle convection, caused by the mixing of the particles within the bed, is important for heat transfer from a surface when the surface is in contact with the suspension instead of the void/bubble phase. Thus the heat transfer coefficient due to particle convection can be defined as

$$h_{pc} = \frac{(1 - \alpha_b)}{\text{particle convective heat transfer resistance}} \qquad (13.22)$$

where the particle convective heat transfer resistance can be further divided into the following two series resistances: (1) average packet (particulate phase) resistance $1/h_p$, and (2) film resistance $1/h_f$. Thus, Eq. 13.22 can be expressed by

$$h_{pc} = \frac{(1 - \alpha_b)}{1/h_p + 1/h_f} \tag{13.23}$$

where α_b is the bubble volume fraction, and h_p can be calculated from

$$h_p = \frac{1}{t_c} \int_0^{t_c} h_i \, dt \tag{13.24}$$

where h_i is the instantaneous heat transfer coefficient averaged over the contact area.

Considering the thermal diffusion through an emulsion packet and assuming that the properties of the emulsion phase are the same as those at minimum fluidization, h_i can be expressed by [53]

$$h_i = \left(\frac{K_{em}\rho_p(1 - \alpha_{mf})c}{\pi t} \right)^{1/2} \tag{13.25}$$

Substituting Eq. 13.25 into Eq. 13.24 yields

$$h_p = \frac{2}{\sqrt{\pi}} \left(\frac{K_{em}\rho_p(1 - \alpha_{mf})c}{t_c} \right)^{1/2} \tag{13.26}$$

Assuming that the time fraction needed for the surface to be covered by bubbles equals the bubble volume fraction in the bed, the surface-emulsion phase contact time t_c can be estimated by

$$t_c = \frac{1 - \alpha_b}{f_b} \tag{13.27}$$

where f_b is the bubble frequency at the surface. Equations 13.26 and 13.27 yield

$$h_p = \frac{2}{\sqrt{\pi}} \left(\frac{K_{em}\rho_p(1 - \alpha_{mf})cf_b}{1 - \alpha_b} \right)^{1/2} \tag{13.28}$$

For film resistance, the film heat transfer coefficient can be expressed by

$$h_f = \frac{\xi K}{d_p} \tag{13.29}$$

where ξ is a factor ranging from 4 to 10 [54]. Thus, the particle convective heat transfer component h_{pc} can be calculated from Eqs. 13.23, 13.28, and 13.29.

Bed-to-surface heat transfer in bubbling fluidized beds is influenced by the migration of particles to and from the heat transfer surface. Molerus et al. [55] modeled the bed-to-surface heat transfer coefficient by measuring the particle exchange frequencies using the pulsed-light method. This frequency, along with simultaneously measured heat transfer coefficient, revealed the direct correspondence between particle migration and heat transfer. Molerus et al. [56] proposed the correlation Eq. 13.3.3 for predicting the bed-to-surface heat transfer in bubbling fluidized beds. This correlation accounts for the effect of the thermophysical properties of the gas-solid system and the superficial gas velocity. Figure 13.11 depicts the comparison between the measured heat transfer coefficient for gas-solid systems at ambient conditions and the heat transfer coefficient predicted from the correlation Eq. 13.3.3. This correlation also accounts for the variation in the physical properties of the system, as seen in Fig. 13.12, which shows the effect of operating pressure on heat transfer coefficient. From Fig. 13.12, it can be seen that the heat transfer coefficient increases with the operating pressure.

Gas Convective Component. The gas convective component is caused by the gas percolating through the particulate phase and the gas bubbles coming in contact with the heat transfer surface. For small particles, though the contribution of gas convective component is small in the in-bed region, it could be important in the freeboard region. The gas convective

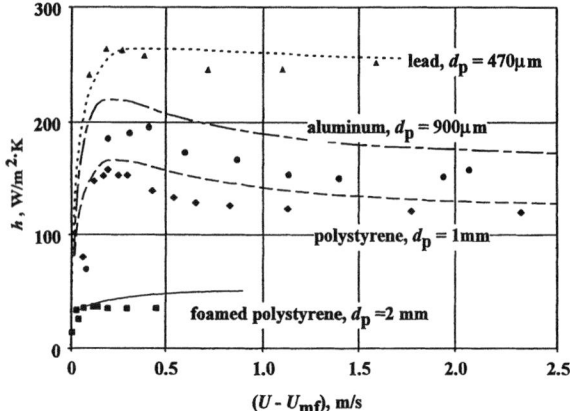

FIGURE 13.11 Comparison between measured and predicted (according to the correlation Eq. 13.3.3) heat transfer coefficients for different solids—air system at ambient conditions. Heat transfer surface: single vertical immersed tube (from Molerus et al. [56]).

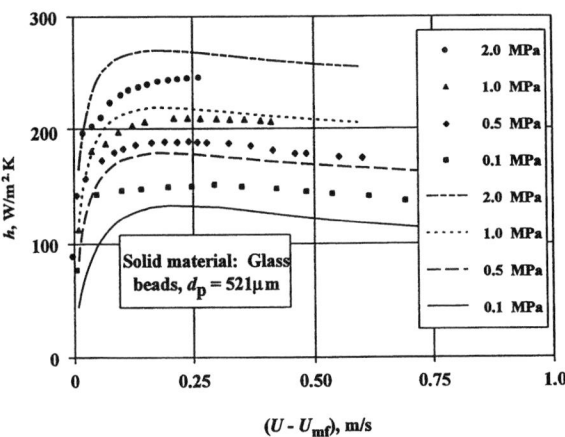

FIGURE 13.12 Comparison between measured and predicted (according to the correlation Eq. 13.3.3) heat transfer coefficients for different glass beads—air system. Heat transfer surface: single vertical immersed tube (from Molerus et al. [56]).

heat transfer coefficient in general varies with the geometry of the heat transfer surface. However, it can be approximated without treating specific surface geometries, as suggested by Baskakov et al. [57] in correlation Eq. 13.3.4 or as proposed by Denloye and Botterill [58] in correlation Eq. 13.3.5. Denloye and Botterill [58] found that the gas convective component becomes a dominant mode of heat transfer as the particle size and the operating pressure increase.

The heat transfer coefficient for the gas convective component can be regarded as comparable to that at incipient fluidization conditions. By assuming $h_{gc} = h_{mf}$, Xavier and Davidson [54] simulated the fluidization system, considering a pseudofluid with the apparent thermal conductivity K_a of the gas-solid medium flowing at the same superficial velocity and the same inlet and outlet temperatures as the gas. They found the temperature distribution in the bed as

$$\frac{T_2 - T_s}{T_1 - T_s} = 4 \sum_{n=1}^{\infty} \frac{1}{\lambda_{n^2}} \exp\left(-4\lambda_{n^2} \frac{K_a L}{\rho c_p U D^2}\right) \tag{13.30}$$

where λ_n are eigenvalues of the eigen equation $J_0(\lambda) = 0$, with the first three being $\lambda_1 = 2.450$, $\lambda_2 = 5.520$, and $\lambda_3 = 8.645$.

Radiative Component.   At high temperatures (above 600°C), radiative heat transfer is significant in many fluidized bed processes such as coal combustion and gasification. If the fluidized bed is treated as a "solid" gray body, the radiative heat transfer coefficient h_r between the fluidized bed at temperature T_b and a heating surface at temperature T_s is defined as

$$h_r = \frac{J_r}{T_b - T_s} = \sigma_b \epsilon_{bs}(T_b^2 + T_s^2)(T_b + T_s) \tag{13.31}$$

where J_r is the radiant heat flux, σ_b is the Stefan-Boltzmann constant, and ϵ_{bs} is the generalized emissivity, which depends on the shape, material property, and emissivity of the radiating and receiving bodies [59]. For two parallel large, perfect gray planes, the generalized emissivity is given as

$$\epsilon_{bs} = (1/\epsilon_b + 1/\epsilon_s - 1)^{-1} \tag{13.32}$$

Due to the multiple surface reflections, the effective bed-to-surface emissivity is larger than the particle emissivity ϵ_p. From Eq. 13.31, it can be seen that the importance of radiative heat transfer significantly increases with the temperature. In general, depending on particle size, h_r increases from being approximately 8 ~ 12 percent of the overall heat transfer coefficient at 600°C to being 20 ~ 30 percent of h at 800°C. Also, increasing the particle size increases the relative radiative heat transfer [60]. Flamant et al. [23] studied the effects and relative importance of parameters such as particle size, particle and wall emissivities, bed and wall temperatures, heat flux direction, and so on, on the heat transfer coefficient in high-temperature gas-solid fluidized beds. Their study indicated that gas convective heat transfer is governed by both wall and bed temperatures, while particle convective heat transfer is mainly affected by the wall temperature.

Effect of Operating Conditions.   When the radiative heat transfer is negligible, the existence of a maximum convective heat transfer coefficient h_{max} is a unique feature of the dense-phase fluidized beds. This phenomenon is distinct for fluidized beds of small particles. For beds of coarse particles, the heat transfer coefficient is relatively insensitive to the gas flow rate once the maximum value is attained.

For a given system, h_{max} depends primarily on the particle and gas properties. For coarse-particle fluidization at $U > U_{mf}$, gas convection is the dominant mode of heat transfer. Thus, h_{max} can be evaluated from the equations for h_{gc}, such as correlation Eq. 13.3.5. On the other hand, h_{max} in a fine-particle bed can be reasonably evaluated from the equations for h_{pc}. In general, h_{max} is a complicated function of $h_{pc,max}$, h_f, and other parameters. An approximation for this functionality was suggested by Xavier and Davidson [54] as

$$\frac{h_{max}}{h_{pc,max}} = \left(\frac{h_f}{h_f + 2h_{pc,max}} + \frac{h_{gc}}{h_{pc,max}}\right)^{0.84} \tag{13.33}$$

The convective heat transfer h_c $(= h_{pc} + h_{gc})$ depends on both the pressure and the temperature. An increase in pressure increases the gas density, yielding a lower U_{mf}. Thus, a pressurized operation enhances the convective heat transfer. h_c is lower under subatmospheric pressure operations than it is under ambient pressure operation due to a lower gas density and a reduction in K_e with decreasing pressure [61]. For fluidized beds with small particles, increasing pressure enhances solids mixing and hence the particle convection [62]. The convective heat transfer increases significantly with pressure for Group D particles; however, in general the pressure effect decreases with decreasing particle size. For Group A and Group B particles, the increase in h_c with pressure is small.

At high temperatures, the decreased gas density causes a decrease in the gas convective component h_{gc}, while the increased gas conductivity at high temperature can increase h_{gc}, K_e, and h_{pc}. For a bed of small particles, the latter is dominant. Thus, a net increase in h_c with increasing temperature can be observed before radiation becomes significant. For Group D particles, h_c decreases with increasing temperature [63]. A higher operating pressure leads to enhanced h_{gc} and h_{pc}. These effects of temperature and pressure on h_{pc}, h_{gc}, and h_{max} are illustrated in Fig. 13.13.

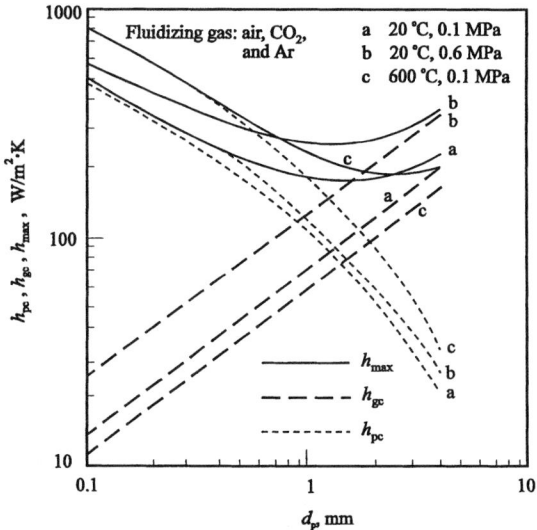

FIGURE 13.13   Effects of temperature and pressure on h_{pc}, h_{gc}, and h_{max} for bed-to-surface heat transfer (from Botterill et al. [64]).

In fluidized bed heat transfer, it is a common practice to use internals such as water-cooled tubes in the bed. Hence, it is important to know the effect of immersed objects on the local fluidization behavior and the local heat transfer characteristics. Here we consider a case in which a horizontal water-cooled tube is placed in a hot fluidized bed [65]. The interference of the immersed tube with the particle circulation pattern leads to an increased average particle residence time at the heat transfer surface. The formation of stagnant zones of particles on the top or nose of the horizontal tube is evidence of such interference. Also, at high gas velocities, gas packets or bubbles are frequently present in the upper part of the horizontal tube. Therefore, there is a significant circumferential variation in the heat transfer coefficient around the tube. The heat transfer coefficient near the nose is considerably lower than that around the sides, indicating a reduction in convective heat transfer due to the stagnant zones near the nose. For large particles for which the gas convective component of heat transfer is significant, the circumferential variations would be somewhat similar, with peaks at the sides, but the variation is less than that with small particles, since the particle convection is no longer dominant. The overall heat transfer coefficient as a result of orientation usually does not differ much from that for a horizontal tube, being slightly lower than that for the vertical tube.

It is noted that most of the models and correlations are developed on the basis of bubbling fluidization. However, most can be extended to the turbulent regime with reasonable error margins. The overall heat transfer coefficient in the turbulent regime is a result of two counteracting effects, one due to the vigorous gas-solid movement that enhances the heat transfer and the other due to the low particle concentration that reduces the heat transfer.

Heat Transfer in Circulating Fluidized Beds.   This section discusses the mechanism of heat transfer in circulating fluidized beds along with the effects of the operating variables on the local and overall heat transfer coefficients.

Mechanism and Modeling.   In a circulating fluidized bed, the suspension-to-wall heat transfer comprises various modes including conduction due to particle clusters contacting the surface or particles sliding along the walls, gas convection to uncovered surface areas, and thermal radiation. Glicksman [66] suggested that the percent of surface area covered by particle clusters is an important parameter in the heat transfer study. The wall-to-bed heat transfer coefficient is a function of the average cluster displacement before breakup. For modeling of the heat transfer mechanism in a circulating fluidized bed, the heat transfer surface is considered to be covered alternately by cluster and dispersed particle phases [67, 68]. Thus, considering that a "packet" represents a "cluster," the packet model developed for dense-phase fluidized beds can be applied.

In circulating fluidized beds, the clusters move randomly and the heat transfer between the surface and clusters occurs via unsteady heat conduction with a variable contact time. The heat transfer due to cluster movement represents the major part of the particle convective component. Heat transfer is also due to gas flow that covers the surface (or a part of surface) and contributes to the gas convective component.

Particle Convective Component.   The particle convection is in general important in the overall bed-to-surface heat transfer. When particles or particle clusters contact the surface, relatively large local temperature gradients are developed. The rate of heat transfer can be enhanced with increased surface renewal rate or decreased cluster residence time in the convective flow of particles in contact with the surface. The particle convective component h_{pc} can be expressed by the following equation, which is an alternative form of Eq. 13.23:

$$h_{pc} = \frac{\delta_c}{1/h_f + 1/h_p} \tag{13.34}$$

Thus, h_{pc} is determined from the wall (film) resistance, $1/h_f$, in series with a transient conduction resistance of homogeneous semi-infinite medium $1/h_p$. By analogy to Eq. 13.29, h_f can be expressed by [69]

$$h_f = \frac{K_f}{\delta^* d_p} \tag{13.35}$$

where δ^* is the dimensionless effective gas layer thickness between wall and cluster (ratio of gas layer thickness to particle size), which is mainly a function of cross-sectionally averaged particle volume fraction [70].

Similarly to Eq. 13.28, h_p can be expressed as [70]

$$h_p = \left(\frac{K_c \rho_p (1 - \alpha_c) c_{pc}}{\pi t_c} \right)^{1/2} \tag{13.36}$$

Equations 13.34–13.36 give h_{pc}.

Gas Convective Component.   The wall-to-bed heat transfer in circulating fluidized beds is greatly influenced by the hydrodynamics near the wall and the thermophysical properties of gas. Wirth [71] studied the effect of particle properties on the heat transfer characteristics in circulating fluidized beds. Their measurements are represented in Fig. 13.14, where the Nusselt number is plotted against the Archimedes number with pressure drop number as the parameter. The Archimedes number and pressure drop number, which accounts for the cross-sectional average solids concentration, characterize the flow dynamics near the wall. From Fig. 13.14 it can be seen that at low Ar, most of the heat transfer occurs by heat conduction in the gas, while at high Ar, gas convection is the dominant mode of heat transfer. Wirth [71] found that particle thermal properties have no influence on heat transfer and proposed correlation Eq. 13.4.1, given in Table 13.4, for predicting the heat transfer in circulating fluidized

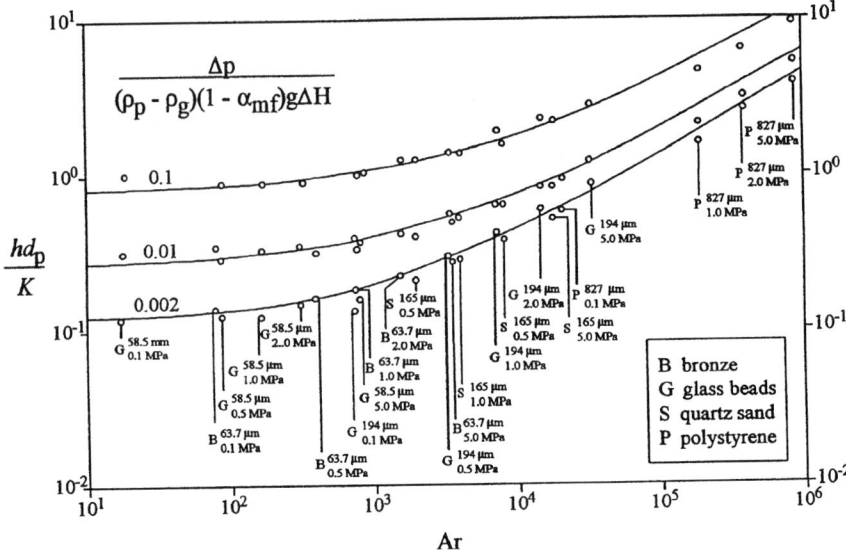

FIGURE 13.14 Heat transfer characteristics in circulating fluidized bed at ambient temperature (from Wirth [71]).

beds. Figure 13.15 shows the comparison between the predicted (from correlation Eq. 13.4.1) and measured heat transfer coefficients at ambient temperature.

In practice, h_{gc} may be evaluated by one of the following approaches:

1. Extended from correlation Eq. 13.3.5 for h_{gc} in dense-phase fluidized beds
2. Approximated as that for dilute-phase pneumatic transport [72, 73]
3. Estimated by the convective coefficient of single-phase gas flow [74]

For high particle concentration on a surface with large dimensions, any one of the approaches listed is reasonable due to the small value of h_{gc}. For low particle concentrations and high temperatures, discrepancies in h_{gc} may exist when these approaches are used.

TABLE 13.4 Heat Transfer in Circulating Fluidized Beds

Equation 13.4.1	
Investigator	Wirth [71]
Type of correlation	Wall-to-suspension heat transfer coefficient
Phases involved	Gas-solid
Model associated	Assumes that heat is transferred simultaneously by gas conduction and gas convection
Correlation equation	$\dfrac{h_{cc}d_p}{K} = 2.85\left(\dfrac{\Delta p}{(\rho_p - \rho_g)(1 - \alpha_{mf})g\Delta H}\right)^{0.5} + 0.00328\text{Re}_w\,\text{Pr}$
Range of applicability	$10^{-3} < \left(\dfrac{\Delta p}{(\rho_p - \rho_g)(1 - \alpha_{mf})g\Delta H}\right) < 0.1$ and $10 < \text{Ar} < 10^6$

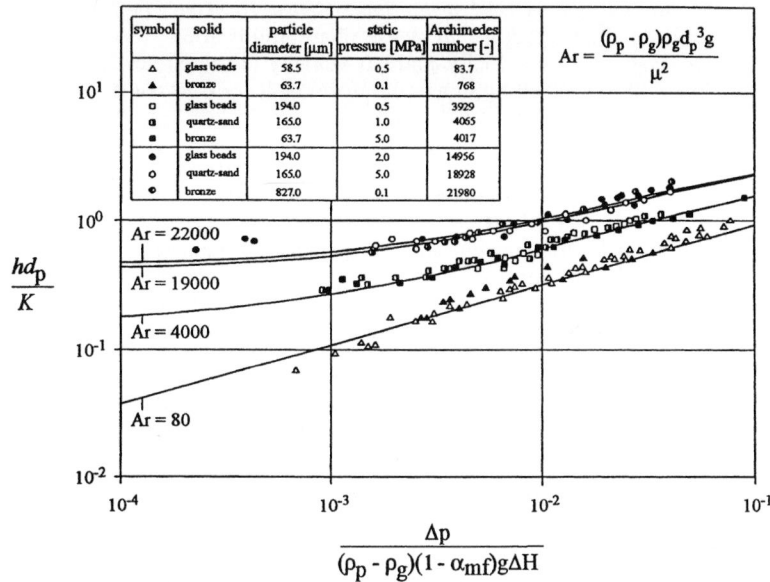

FIGURE 13.15 Comparison between experimentally determined and calculated (correlation Eq. 13.4.1) heat transfer coefficients in circulating fluidized bed at ambient temperature (from Wirth [71]).

Radiative Component. For understanding the radiative heat transfer in a circulating fluidized bed, the bed can be regarded as a pseudogrey body. The radiative heat transfer coefficient is [75]

$$h_r = \frac{\sigma_b(T_b^4 - T_s^4)}{[(1/\epsilon_{sus}) + (1/\epsilon_s) - 1](T_b - T_s)}$$ (13.37)

where ϵ_{sus} is the emissivity of the suspension.

An alternative treatment for radiative heat transfer in a circulating fluidized bed is to consider the radiation from the clusters (h_{cr}) and from the dispersed phase (i.e., the remaining part of gas-solid suspension except clusters h_{dr}) separately [76]

$$h_r = \alpha_c h_{cr} + (1 - \alpha_c)h_{dr}$$ (13.38)

where α_c is the volume fraction of clusters in the bed. These two components can be defined by

$$h_{cr} = \frac{\sigma_b(T_b^4 - T_s^4)}{[(1/\epsilon_c) + (1/\epsilon_s) - 1](T_b - T_s)}$$ (13.39)

$$h_{dr} = \frac{\sigma_b(T_b^4 - T_s^4)}{[(1/\epsilon_d) + (1/\epsilon_s) - 1](T_b - T_s)}$$ (13.40)

The emissivity of the cluster ϵ_c can be determined by Eq. 13.32, and the dispersed phase emissivity ϵ_d is given by [77]

$$\epsilon_d = \sqrt{\frac{\epsilon_p}{(1 - \epsilon_p)B}\left(\frac{\epsilon_p}{(1 - \epsilon_p)B} + 2\right)} - \frac{\epsilon_p}{(1 - \epsilon_p)B}$$ (13.41)

where B is taken as 0.5 for isotropic scattering and 0.667 for diffusely reflecting particles.

Radial and Axial Distributions of Heat Transfer Coefficient. Contrary to the relatively uniform bed structure in dense-phase fluidization, the radial and axial distributions of voidage, particle velocity, and gas velocity in the circulating fluidized bed are considerably nonuniform, resulting in a nonuniform heat transfer coefficient profile in the circulating fluidized bed.

In the axial direction, the particle concentration decreases with height, which leads to a decrease in the cross-sectionally averaged heat transfer coefficient. In addition, the influence of the solids circulation rate is significant at lower bed sections but less significant at upper bed sections, as illustrated in Fig. 13.16. In the radial direction, the situation is more complicated due to the uneven radial distribution of the particle concentration as well as the opposite solids flow directions in the wall and center regions. In general, the coefficient is relatively low and approximately constant in the center region. The coefficient increases sharply toward the wall region. Three representative radial profiles of the heat transfer coefficient with various particle holdups reported by Bi et al. [78] are shown in Fig. 13.16 as described below.

1. When the particle holdup is high, the contribution of h_{pc} plays a dominant role and h_{gc} is less important. The radial distribution of the heat transfer coefficient is nearly parabolic, as shown in Fig. 13.16a.

2. As the gas velocity increases, the solids holdup decreases and thus h_{gc} begins to become as important as h_{pc}. In the center region of the riser, h_{gc} is dominant, and its influence decreases with an increase in the solids holdup along the radial direction toward the wall. In the near wall region, h_{pc} dominates the heat transfer. The contribution of h_{pc} decreases with a decrease in the particle concentration toward the bed center. As a result, a minimum value of h appears at r/R of about $0.5 \sim 0.8$, as indicated in Fig. 13.16b.

3. With further decrease in the particle concentration at $\alpha > 0.93$, h_{gc} becomes dominant except at a region very close to the wall. Thus, the heat transfer coefficient decreases with increasing r/R in most parts of the riser as shown in Fig. 13.16c; this is the same trend as the radial profile of the gas velocity. In the region near the wall, h_{pc} increases sharply, apparently due to the effect of relatively high solids concentration in that region.

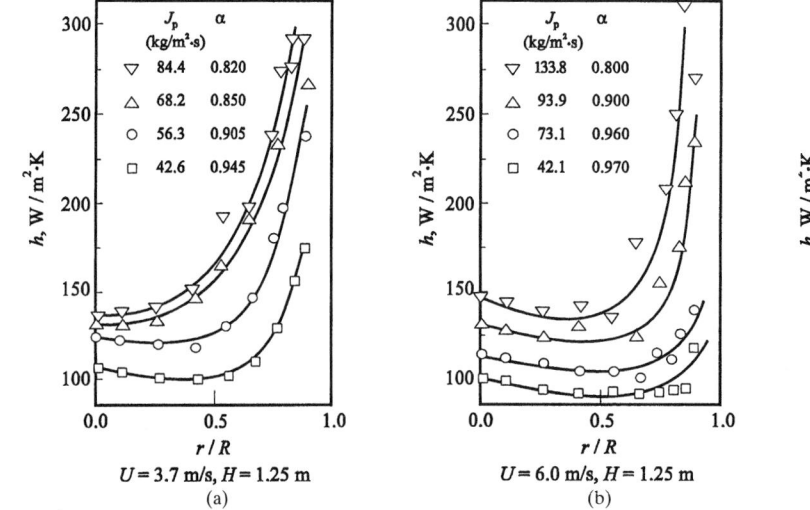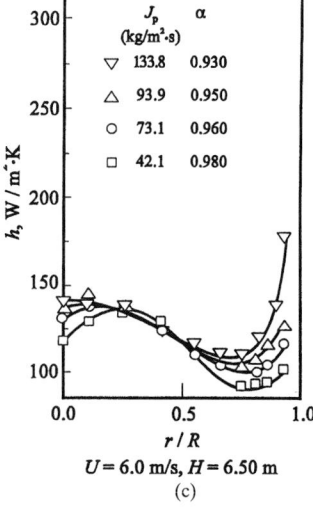

FIGURE 13.16 Radial distributions of overall heat transfer coefficient in a circulating fluidized bed of $d_p = 280$ μm and $\rho_p = 706$ kg/m³ (from Bi et al. [78]).

Effect of Operating Parameters.   The overall heat transfer coefficient can be influenced by the suspension density, solids circulation rate, gas velocity, particle properties, bed temperature, pressure, and dimensions of the heating surface. Basu and Nag [79] presented a critical review on the wall-to-bed heat transfer in circulating fluidized bed boilers. They concluded that the effect of particle size on the heat transfer is insignificant; however, the suspension density shows a dominant effect on the heat transfer coefficient. The overall heat transfer coefficient increases with suspension density [75] and with particle circulation rate. The increase in gas velocity appears to have two counteracting effects, viz., enhancing h_{gc} due to an increased gas convection effect while reducing particle convective heat transfer due to the reduced particle concentration. When h_{pc} dominates (in the near wall region with high particle concentration), h decreases with increasing U. On the other hand, h increases with U if h_{gc} is important (e.g., in the central region where the particle concentration is small). Another reason for the decrease of h in the near wall region is the reduced particle downward velocity caused by increasing U, which results in a prolonged particle-surface contact.

In general, small/light particles can enhance heat transfer. The cluster formation in small/light particle systems contributes to the enhancement of h_{pc}. Also, the gas film resistance can be reduced by fluidizing small particles [80]. When the temperature is lower than 400°C, the effect of bed temperature on the heat transfer coefficient is due to the change of gas properties, while h_r is negligible. At higher temperatures, h would increase with temperature, mainly due to the sharp increase of radiative heat transfer.

Measurements of heat transfer in circulating fluidized beds require use of very small heat transfer probes in order to reduce the interference to the flow field. The dimensions of the heat transfer surface may significantly affect the heat transfer coefficient at any radial position in the riser. All the treatment of circulating fluidized bed heat transfer described above is based on a small dimension for the heat transfer surface. The heat transfer coefficient decreases asymptotically with an increase in the vertical dimension of the heat transfer surface [81]. It can be stated that the large dimensions of the heat transfer surface can prolong the residence time of particles or particle clusters on the surface, resulting in lower renewal frequency and hence a low apparent heat transfer coefficient.

Heat Transfer in Spouted Beds.   The heat transfer behavior in a spouted bed is different from that in the dense-phase and circulating fluidized bed systems due to the inherent differences in their flow structures. The spouted bed is represented by a flow structure that can be characterized by two regions: the annulus and the central spouting region.

Gas-to-Particle Heat Transfer.   The heat transfer phenomena in the annulus and central spouting regions are usually modeled separately. For the central spouting region, the correlation of Rowe and Claxton [82] given by correlation Eq. 13.5.2 in Table 13.5 can be used. In the annulus, the heat transfer can be described using the correlations for fixed beds, for example Littman and Sliva's [83] correlation given by correlation Eq. 13.5.3. Substitution of the corresponding values for the spouted bed into Eqs. 13.5.2, 13.5.3, and 13.20 reveals that the distance required for the gas to travel to achieve a thermal equilibrium with the solids in the annulus region is on the order of magnitude of centimeters, while this distance in the spout region is one or two orders of magnitude larger. An in-depth discussion on the heat transfer between gas and particles in spouted beds, can be found in Mathur and Epstein [84].

The importance of the intraparticle heat transfer resistance is evident for particles with relatively short contact time in the bed or for particles with large Biot numbers. Thus, for a shallow spouted bed, the overall heat transfer rate and thermal efficiency are controlled by the intraparticle temperature gradient. This gradient effect is most likely to be important when particles enter the lowest part of the spout and come in contact with the gas at high temperature, while it is negligible when the particles are slowly flowing through the annulus. Thus, in the annulus, unlike the spout, thermal equilibrium between gas and particles can usually be achieved even in a shallow bed, where the particle contact time is relatively short.

Bed-to-Surface Heat Transfer.   The heat transfer between the bed and the surface in spouted beds is less effective than in fluidized beds. The heat transfer primarily takes place by

TABLE 13.5   Heat Transfer in Spouted Beds

Equation 13.5.1	
Investigator	Mathur and Gishler [21]
Type of correlation	Minimum spouting velocity prediction
Phases involved	Gas-solid
Correlation equation	$U_{msp} = \left(\dfrac{d_p}{D}\right)\left(\dfrac{D_i}{D}\right)^{1/3}\left(\dfrac{2gH_{sp}(\rho_p - \rho_g)}{\rho_g}\right)^{1/2}$
Range of applicability	For $D < 0.4$ m

Equation 13.5.2	
Investigator	Rowe and Claxton [82]
Type of correlation	Gas-to-particle heat transfer coefficient
Phases involved	Gas-solid
Region associated	Central spouting region
Correlation equation	$Nu_{gp} = \dfrac{2}{1 - (1 - \alpha_{cs})^{0.33}} + \dfrac{2}{3\alpha_{cs}}\, Pr^{0.33}\, Re_{pf}^{0.55}$
Range of applicability	$Re_{pf} > 1000$

Equation 13.5.3	
Investigator	Littman and Sliva [83]
Type of correlation	Gas-to-particle heat transfer coefficient
Phases involved	Gas-solid
Region associated	Annulus region
Correlation equation	$Nu_{gp} = 0.42 + 0.35 Re_{pf}^{0.8}$
Range of applicability	$Re_{pf} < 100$

convection. Compared to the fluidized bed, a spouted bed with immersed heat exchangers is less frequently encountered. The bed-to-immersed object heat transfer coefficient reaches a maximum at the spout-annulus interface and increases with the particle diameter due to the convective component of heat transfer [85].

Since the solid particles in the spouted bed are well mixed, their average temperature in different parts of the annulus can be considered to be the same, just as in the case of a fluidized bed. The maximum value of the heat transfer coefficient in the *h-U* plot also exists, similar to the conditions in a dense-phase fluidized bed [84].

Design Considerations for Heat Transfer.   The optimal design considerations for a fluidized bed heat exchanger should consider its heat transfer coefficient and structure properties as given below.

Position and Orientation of Heat Transfer Surface; Intensification of Heat Transfer.   Since the flow behavior in the bed varies spatially, different arrangements of heat transfer surface result in differences in the heat transfer performance. Even for a single surface, different parts of the surface may have quite different heat transfer coefficients. For example, an immersed horizontal tube has a relatively smaller coefficient on the upper surface, due to the possible particle packing and local defluidization in a small area on the top of the tube. The configurations of the heat exchanger tubes, such as horizontal, vertical, slanted, upstream, downstream, sidewall, upward, downward, and so on, are very important for heat transfer, because the local flow field can be varied by changing these factors. The difference in coefficients measured at the different locations inside a dense-phase fluidized bed is not so remarkable compared to the difference obtained from different positions in a circulating fluidized bed. The reason is that the heat transfer coefficient is strongly related to the particle concentration,

TABLE 13.6 Influence of Surface Location and Orientation on Bed-to-Surface Heat Transfer Coefficient in a Circulating Fluidized Combustor (from Grace [86])

Location of heat transfer surface	Orientation	Position in Fig. 13.17	h, W/m²K	Comments
Below secondary air	Horizontal or vertical	A	300–500	Corrosion, erosion, attrition impedes solids lateral mixing
Above secondary air, on wall	Vertical	B	150–250	A preferred location
Above secondary air, suspended	Vertical	C	150–250	Some erosion/attrition, reduces lateral mixing
Extended recycle loop	Horizontal or vertical	D	400–600	Small surface, suitable for big load variation, high cost, needs additional floor space

which is distributed in a more uniform way in a dense bed than in a circulating bed. For similar reasons, the heat transfer coefficients in the dense-phase beds are generally larger than those obtained in circulating fluidized beds.

The influence of surface location and orientation on the bed-to-surface heat transfer coefficient in circulating fluidized bed combustors is summarized in Table 13.6. The geometric construction of the combustor and the heat transfer surface is shown in Fig. 13.17. Besides the location and orientation, differences in local heat transfer can also be found on the heat transfer surface/tube. For example, the upper part of the horizontal tube shows the smallest value for the heat transfer coefficient in dense-phase fluidized beds due to less frequent bubble impacts and the presence of relatively low-velocity particles.

In general, heat transfer can be intensified in the following ways:

1. By considering proper local flow behavior and local heat transfer properties for placement of the heat transfer surface

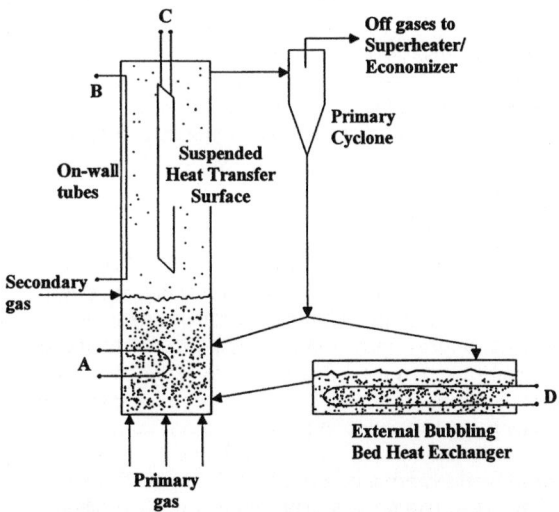

FIGURE 13.17 Immersed surface-to-bed heat transfer in a circulating fluidized bed system (from Grace [86]).

2. By selecting proper configuration, including orientation, for heat transfer

3. By altering the local geometry of the heat transfer surface to intensify the turbulence in the local flow field

4. By using an extended or finned heat transfer surface to increase the area of heat transfer

5. By controlling the fouling and scaling of the heat transfer surface

Structure Properties of Heat Exchanger.   Corrosion, erosion, and mechanical fatigue are the main reasons for the structural failure of heat exchangers. They may occur at the in-bed heat exchanger, waterwall, or in-bed support structure.

The immersed heat exchanger will erode because of the impact of fluidized particles. Compared with other factors, such as corrosion and tube fatigue due to vibrations, wear appears to be the major cause of tube failure in many gas-solid systems. For example, the life of the heat exchanger tube to be used in a multisolids fluidized bed combustor will depend primarily on erosion [87]. The erosion phenomenon of heat exchanger tubes in a fluidized bed is very complex. Sometimes tubes in similar situations may yield entirely different erosion results. It is known, however, that tube erosion is strongly related to the in-bed flow pattern that brings the particles into contact with the surface. Generally, the factors that may influence erosion include particle and surface properties and operating conditions. In dense-phase fluidized beds, the particle impacts are mainly due to the action of bubble wakes because the wake particles possess large kinetic energy. For example, the occurrence of the vertical coalescence of a pair of bubbles just beneath the heat exchanger tube results in the formation of a high-velocity jet of wake particles that strikes the underside of the tube [88]. Thus, any attempt aimed at reducing the bubble size, and hence the kinetic energy, of the wake particles, will be helpful in reducing the erosion of immersed heat exchangers.

The heat exchanger erosion mechanisms for ductile and brittle materials are completely different. A detailed discussion on erosion mechanisms of ductile and brittle materials can be found elsewhere [89]. In the early stages of erosion, the brittle material will form a crack on the surface. Then the formation and propagation of the crack network takes place, yielding material chipping by rodent particles. However, for ductile material, the repeated particle impacts result in the deformation of extruded and forged platelet that reaches a stage of fracture only when it exceeds a local critical strain and is in the final stage of being removed from the surface [90, 91]. The tube materials of interest in most gas-solid suspension systems are all ductile materials.

Some conclusions about surface erosion can be summarized as follows:

1. When the tube is vertical, the erosion rate is less.

2. The erosion rate is smaller for tubes inside a tube bundle than for a single isolated tube [92–93].

3. The erosion rate is strongly influenced by the particle impact velocity, which is caused by the rise and interaction of bubbles in the bed.

4. At high temperature, erosion becomes more complicated due to the involvement of corrosion, deposition, and chemical reactions such as oxidation. The presence of an oxidized layer or deposit may reduce the apparent erosion rate in some cases.

Liquid-Solid Fluidized Beds

In liquid-solid fluidized beds, the presence of solids increases the turbulence in the system and provides additional surface renewal through the thermal boundary layer at the wall. Early studies have indicated that the heat transfer by particle convective mechanism is insignificant and that the convective heat transfer due to turbulent eddies is the principal mode of heat transfer [94]. This distinguished the heat transfer in liquid-solid fluidized beds from that in gas-solid fluidized beds, where particle convective mechanism is dominant. Recently, however, it has been shown that, in conjunction with isotropic fluid microeddies, particles contribute to heat transfer in liquid-solid fluidized beds [95]. In contrast to gas fluidized beds, liquid fluidized beds are generally homogeneous (particulate) and the thermal conductivity of liquid is manyfold more than that of gas. Numerous correlations have been proposed for overall heat transfer in liquid-solid fluidized beds based on a resistance-in-series model considering the near-wall heat transfer resistance and the in-bed heat transfer resistance, which varies with the scale and extent of fluid mixing in the system (e.g., Refs. 96 and 97).

Wall-to-Bed Heat Transfer. The wall-to-bed heat transfer coefficient increases with an increase in liquid flow rate, or equivalently, bed voidage. This behavior is due to the reduction in the limiting boundary layer thickness that controls the heat transport as the liquid velocity increases. Patel and Simpson [94] studied the dependence of heat transfer coefficient on particle size and bed voidage for particulate and aggregative fluidized beds. They found that the heat transfer increased with increasing particle size, confirming that particle convection was relatively unimportant and eddy convection was the principal mechanism of heat transfer. They observed characteristic maxima in heat transfer coefficients at voidages near 0.7 for both the systems.

Recent studies have considered the effects of the in-bed thermal resistance on the overall wall-to-bed heat transfer process. A parabolic radial temperature distribution in the bed indicates a considerable thermal resistance in the in-bed region. Muroyama et al. [96] showed that the contribution of the in-bed thermal resistance relative to the total resistance decreases with increasing bed porosity due to increased bed mobility and radial liquid mixing. For wall-to-bed heat transfer coefficients in liquid-solid fluidized beds of spherical particles, Chiu and Ziegler [98] proposed correlation Eq. 13.7.1, given in Table 13.7. Kang et al. [97] correlated the modified Colburn j factor for heat transfer in liquid fluidized beds, considering the dispersion or mixing of fluidized particles, which appreciably affects the rate of heat transfer. They suggested correlation Eq. 13.7.2 for the wall-to-bed heat transfer coefficient. Both the correlations predict the wall-to-bed heat transfer coefficient satisfactorily in their respective applicability ranges.

Immersed Surface/Particle-to-Bed Heat Transfer. In the design of liquid-solid fluidized beds, the heat transfer between the internals and the bed is also of considerable significance. Macias-Machin et al. [99] studied the heat transfer between a fine immersed wire of the same diameter as the fluidized particles and a liquid fluidized bed. They proposed correlation Eq. 13.7.3 for predicting heat transfer coefficient at low Reynolds numbers ($Re_p < 100$). Kang et al. [97], based on their experiments carried out with a heating source placed at the center of the column, suggested correlation Eq. 13.7.4 for predicting the heat transfer coefficients in the region near the heat transfer surface. Their study reconfirmed the fact stated by Muroyama et al. [96] that when fully fluidized, the heat transfer resistance in the region near the heat transfer surface is more important than the thermal resistance in the in-bed region.

Effective Thermal Conductivity. The effective thermal conductivity signifies the intensity of solids mixing in the interior of the fluidized bed. Muroyama et al. [96] reported that near incipient fluidization the effective thermal conductivity increases sharply with the liquid velocity, passes through a maximum, and then gradually decreases as the liquid velocity is increased. Karpenko et al. [100] reported the effective radial thermal conductivities for liquid fluidized beds of glass and aluminum particles. They obtained correlation Eq. 13.7.5 for predicting the effective thermal conductivity.

TABLE 13.7   Heat Transfer in Liquid-Solid Fluidized Beds

Equation 13.7.1	
Investigator	Chiu and Ziegler [98]
Type of correlation	Wall-to-bed heat transfer coefficient
Phases involved	Liquid-solid
Correlation equation	$\mathrm{Nu}_p = 0.762\mathrm{Re}_m^{0.646}\,\mathrm{Pr}^{0.638}\,U_R^{0.266}(1/\varphi)\left(\dfrac{1-\alpha}{\alpha}\right)$
	where $\quad U_R = U_{\mathrm{mf}}/U_{\mathrm{pt}}$
	$\mathrm{Re}_m = \dfrac{U_L\rho_L}{S_{\mathrm{pc}}(1-\alpha)\mu_L}$
Range of applicability	$0 < \mathrm{Re}_m < 3000$

Equation 13.7.2	
Investigator	Kang et al. [97]
Type of correlation	Wall-to-bed heat transfer coefficient
Phases involved	Liquid-solid
Correlation equation	$j_H = 0.021\mathrm{Pe}_m^{-0.453}$
	where $\quad j_H = \left(\dfrac{\alpha h}{\rho_L c_L U_L}\right)\mathrm{Pr}^{2/3}$
	$\mathrm{Pe}_m = \dfrac{d_p U_L \alpha}{D_p(1-\alpha)}$
Range of applicability	$0 < \mathrm{Re}_{m1} < 3000$
	where $\quad \mathrm{Re}_{m1} = \dfrac{d_p U_L \rho_L}{\mu_L(1-\alpha)}$

Equation 13.7.3	
Investigator	Macias-Machin et al. [99]
Type of correlation	Particle-to-bed heat transfer
Phases involved	Liquid-solid
Correlation equation	$\mathrm{Nu}_p = 1.72 + 2.66(\mathrm{Re}_p/\alpha)^{0.56}\,\mathrm{Pr}^{-0.41}\left(\dfrac{1-\alpha}{\alpha}\right)^{0.29}$
Range of applicability	$0.1 < \mathrm{Re}_p < 100$

Equation 13.7.4	
Investigator	Kang et al. [97]
Type of correlation	Immersed surface-to-bed heat transfer
Phases involved	Liquid-solid
Region associated	Near the heat transfer surface
Correlation equation	$j_{H,\mathrm{surf}} = 0.191\mathrm{Re}_{m1}^{-0.31}$
	where $\quad j_{H,\mathrm{surf}} = \left(\dfrac{\alpha h_{\mathrm{surf}}}{\rho_L c_L U_L}\right)\mathrm{Pr}^{2/3}$
	Re_{m1} is same as defined by Eq. 13.7.2
Range of applicability	$0 < \mathrm{Re}_{m1} < 3000$

Equation 13.7.5	
Investigator	Karpenko et al. [100]
Type of correlation	Effective thermal conductivity
Phases involved	Water/glycerol-solid
Correlation equation	$K_e = 5.05K_{e,\mathrm{max}}\left(\dfrac{\mathrm{Re}_{\mathrm{pf}}}{\mathrm{Re}_{\mathrm{opt}}} - 0.25\right)e^{-1.33\mathrm{Re}_{\mathrm{pf}}/\mathrm{Re}_{\mathrm{opt}}}$
	where $\quad K_{e,\mathrm{max}} = 89.4K\,\mathrm{Ar}^{0.2}\quad$ and $\quad \mathrm{Re}_{\mathrm{opt}} = 0.1\mathrm{Ar}^{0.66}$
Range of applicability	Glycerol concentration (wt%) < 70%

CONCLUDING REMARKS

This chapter presents a brief summary of the hydrodynamic behavior of the packed and fluidized beds and elaborates their heat transfer phenomena. Specifically, the heat transfer mechanisms, models, and characteristics over a wide range of operating conditions for gas-solid and liquid-solid fluidization are described. The particle-to-fluid, wall-to-bed, and immersed surface-to-bed heat transfer properties are discussed in conjunction with the hydrodynamic phenomena including fluidization regimes and their transition.

Packed-bed heat transfer can be conveniently expressed by the concept of effective thermal conductivity, which is based on the assumption that on a macroscale the bed can be described by a continuum. In general, the effective thermal conductivity increases with increasing operating pressure. The wall-to-bed heat transfer coefficient increases with decreasing particle diameter.

In dense-phase gas-solid fluidization systems, particle circulation induced by bubble motion is the primary driving force for bed-to-surface heat transfer. The importance of bubble and bubble wake hydrodynamic characteristics extends to transport phenomena involved in heat transfer and mixing behavior. The significant variations in bubble behavior with gas velocities, column diameter, and particle diameter, and the corresponding significant variations in heat transfer and mixing behavior, generally indicate the shortcomings involved in extrapolating the correlations beyond their range of applicability, specifically the compatible flow regimes. The particle convective heat transfer coefficient typically increases with increasing pressure and decreasing particle size. A pressurized operation also enhances the gas convective heat transfer coefficient. Higher temperatures at which the radiative heat transfer becomes important also favor the overall heat transfer.

Contrary to dense-phase fluidized beds, the radial and axial distributions of voidage, particle velocity, and gas velocity in a circulating fluidized bed are considerably nonuniform, resulting in a nonuniform heat transfer coefficient profile. Since the particle concentration decreases in the axial direction, the heat transfer also decreases. In the radial direction the heat transfer coefficient exhibits a steep profile near the wall, but is almost constant in the center region. The overall heat transfer coefficient increases with suspension density and particle circulation rate.

The heat transfer behavior in a spouted bed is different from that in the dense-phase or circulating fluidized bed system due to the inherent differences in their flow structures. The gas-to-particle heat transfer coefficient in the annulus region is usually an order of magnitude higher than that in the central spout region. The bed-to-surface heat transfer coefficient reaches a maximum at the spout-annulus interface and also increases with the particle diameter.

In liquid-solid fluidized beds, the bed-to-wall heat transfer coefficient increases with an increase in liquid flow rate due to the reduction in thermal boundary layer thickness. The heat transfer coefficient was also found to increase with the particle size. The effective thermal conductivity of liquid fluidized bed increases sharply with liquid velocity beyond minimum fluidization, passes through a maximum near a voidage of 0.7, and then gradually decreases.

Since the flow behavior in a fluidized bed varies in space, different arrangements of heat transfer surface result in differences in the heat transfer performance. Gas velocities, operating pressures, and temperatures have significant effects on enhancement of the heat transfer coefficient. For a given operating condition, the heat transfer coefficient from an immersed surface to a bed is higher than that from column wall to bed. In general, the heat transfer can be intensified by altering the local geometry of the heat transfer surface to increase the turbulence in the local flow field.

NOMENCLATURE

Symbol, Definition

A_m	area of packet in contact with the heating surface
Ar	Archimedes number
B	parameter defined by Eq. 13.41
c	specific heat of particles
c_L	specific heat of liquid
c_p	specific heat at constant pressure
c_{pc}	specific heat of clusters
D	diameter of column
D_i	diameter of jet nozzle
D_p	particle dispersion coefficient
D_s	diameter of spout
D_{tem}	thermal diffusivity of the emulsion phase
d_p	diameter of particle
F_{12}	radiation view factor between two contacting spheres
Fo	Fourier number
f_b	bubble frequency at surface
g	gravitational acceleration
H	height
H_f	expansion bed height
H_{sp}	spouted bed height
h	heat transfer coefficient, bed-to-surface heat transfer coefficient
h_{bg}	bed-to-gas heat transfer coefficient
h_c	convective bed-to-surface heat transfer coefficient
h_{cc}	heat transfer coefficient caused by gas heat conduction and gas heat convection
h_{cr}	radiative heat transfer coefficient of clusters
h_{dr}	radiative heat transfer coefficient of the dispersed phase
h_f	gas-film heat transfer coefficient
h_{gc}	gas convective component of h_c
h_{gp}	particle-to-gas heat transfer coefficient
h_i	instantaneous heat transfer coefficient
h_{mf}	heat transfer coefficient at incipient fluidization condition
h_{max}	maximum value of h
h_p	average heat transfer coefficient between the particulate phase and surface in the absence of gas film resistance
h_{pc}	particle convective component of h_c
$h_{pc,max}$	maximum value of h_{pc}
h_r	radiative heat transfer coefficient
h_{surf}	heat transfer coefficient in the region adjacent to the heater surface

h_w	wall-to-bed heat transfer coefficient
$I(t)$	age distribution function in the film penetration model
J_p	solids recirculation rate or solids flux
J_r	radiant heat flux
j_H	modified Colburn j factor, defined by correlation Eq. 13.7.2
$j_{H,surf}$	modified Colburn j factor in the region adjacent to the heater surface
K	thermal conductivity of gas
K	thermal conductivity of fluid
K_a	apparent thermal conductivity of a gas-solid suspension
K_c	thermal conductivity of clusters
K_{con}	effective thermal conductivity of a fixed bed due to conduction
K_{dis}	effective thermal conductivity of a fixed bed due to dispersion
K_e	effective thermal conductivity of a fixed bed with stagnant fluid
$K_{e,conv.}$	effective thermal conductivity of a fixed bed accounting for the convective contribution due to fluid motion
K_{em}	thermal conductivity of the emulsion phase
$K_{e,max}$	maximum effective radial thermal conductivity, defined by correlation Eq. 13.7.5
K_f	thermal conductivity of gas film
K_p	thermal conductivity of particles
K_{rad}	effective thermal conductivity of a fixed bed due to radiation
L	laminar flow length scale, defined by correlation Eq. 13.3.3
L	length of the column
M	mass of particles
M_p	total mass of particles in the bed
Nu_{bg}	bed-to-gas Nusselt number
Nu_{gp}	particle-to-gas Nusselt number
Nu_p	particle-to-fluid Nusselt number for a single particle
p	total pressure
Δp_b	pressure drop across the bed
p_d	dynamic pressure
Pe	Peclet number, defined by correlation Eq. 13.2.14
Pe_m	modified Peclet number, defined by correlation Eq. 13.7.2
Pr	Prandtl number
r	radial coordinate
R	radius of the bed
R	radius of the particle
Re_m	modified particle Reynolds number, defined by correlation Eq. 13.7.1
Re_{m1}	modified particle Reynolds number, defined by correlation Eq. 13.7.2
Re_{opt}	optimum particle Reynolds number, defined by correlation Eq. 13.7.5
Re_p	particle Reynolds number based on particle diameter and relative velocity
Re_{pf}	particle Reynolds number based on particle diameter and superficial gas velocity

Re_w	particle Reynolds number based on particle diameter and falling velocity of wall strands
S	area mean stirring factor
S_p	surface area of particles
S_{pc}	surface area of particles per unit volume
t	time
t_c	contact time of clusters or the particulate phase and surface
t_e	surface renewal time in the penetration model for bubbles in contact with the emulsion phase
T_1	gas temperature at the inlet of the bed
T_2	gas temperature at the outlet of the bed
T	absolute temperature of gas
T_i	temperature of gas at inlet
T_b	bed temperature
T_p	temperature of particles
T_s	temperature of heating surface
U	superficial gas velocity
U_c	transition velocity between bubbling and turbulent fluidization
U_k	gas velocity corresponding to the pressure fluctuation leveling point in Fig. 13.2
U_L	superficial liquid velocity
U_{mb}	minimum bubbling velocity
U_{mf}	minimum fluidization velocity
U_{msp}	minimum spouting velocity
U_{opt}	superficial gas velocity at $h = h_{max}$
U_{pt}	particle terminal velocity
W	diffusely reflecting wall
z	axial coordinate
z_i	location of inflection point for fast fluidization
z_o	characteristic length of transition region, as shown in Fig. 13.3

Greek Letters

α	bed voidage
α^*	asymptotic voidage in the upper dilute region
α_a	asymptotic voidage in the lower dense region
α_b	volume fraction of bubbles in the bed
α_c	volume fraction of clusters in the bed
α_{cs}	volume fraction in the central spouting region
α_{mf}	bed voidage at minimum fluidization
δ	boundary layer thickness
δ^*	dimensionless gas layer thickness
δ_c	time-averaged fraction of wall area covered by clusters
δ_{em}	layer thickness of emulsion on the surface
ϵ_{bs}	general bed emissivity for bed-to-surface radiation

ϵ_b	emissivity of bed suspension
ϵ_c	emissivity of clusters
ϵ_d	emissivity of dispersed phase
ϵ_p	emissivity of particle surface
ϵ_s	emissivity of heat transfer surface
ϵ_{sus}	emissivity of suspension phase
θ_w	wake angle
ξ	parameter defined by Eq. 13.29
μ	viscosity of gas
μ_L	viscosity of liquid
ρ	density of fluid
ρ_{em}	density of emulsion phase
ρ_g	density of gas
ρ_L	density of liquid
ρ_p	density of particles
σ_b	Stefan-Boltzmann constant
$\psi(t)$	age distribution function in the packet model
φ	sphericity of the particle

REFERENCES

1. N. Wakao and S. Kaguei, *Heat and Mass Transfer in Packed Beds,* Gordon and Breach Science Publishers, New York, 1982.

2. D. Kunii and O. Levenspiel, *Fluidization Engineering,* 2d ed., Butterworth-Heinemann, Boston, 1991.

3. L.-S. Fan and C. Zhu, *Principles of Gas-Solid Flows,* Cambridge University Press, New York, 1998.

4. D. Geldart, "Types of Gas Fluidization," *Powder Tech.* (7): 285, 1973.

5. J. D. Gabor and J. S. M. Botterill, "Heat Transfer in Fluidized and Packed Beds," in *Handbook of Heat Transfer Applications,* Rohsenow, Hartnett, and Ganic eds., McGraw-Hill, New York, 1985.

6. H. Darcy, *Les Fontaines Publiques de la Ville De Dijon,* Victor Dalmon, Paris, 1856.

7. S. Ergun, "Fluid Flow Through Packed Columns," *Chem. Eng. Prog.* (48): 89, 1952.

8. S. Ergun and A. A. Orning, "Fluid Flow Through Randomly Packed Columns and Fluidized Beds," *I&EC* (41): 1179, 1949.

9. R. H. Wilhelm and M. Kwauk, "Fluidization of Solid Particles," *Chem. Eng. Prog.* (44): 201, 1948.

10. L. Massimilla, "Gas Jets in Fluidized Beds," in *Fluidization,* 2d ed., Davidson, Clift, and Harrison eds., Academic Press, London, 1985.

11. P. N. Rowe, H. J. Macgillivray, and D. J. Cheesman, "Gas Discharge From an Orifice Into a Gas Fluidized Bed," *Trans. Instn. Chem. Engrs.* (57): 194, 1979.

12. P. N. Rowe, "Experimental Properties of Bubbles," in *Fluidization,* Davidson and Harrison eds., Academic Press, New York, 1971.

13. L.-S. Fan and K. Tsuchiya, *Bubble Wake Dynamics in Liquid and Liquid-Solid Suspensions,* Butterworths, Boston, 1990.

14. J. Yerushalmi and N. T. Cankurt, "Further Studies of the Regimes of Fluidization," *Powder Tech.* (24): 187, 1979.

15. M. H. Peters, L.-S. Fan, and T. L. Sweeney, "Study of Particle Ejection in the Freeboard Region of a Fluidized Bed With an Image Carrying Probe," *Chem. Eng. Sci.* (38): 481, 1983.

16. S. T. Pemberton, "Entrainment From Fluidized Beds," Ph.D. dissertation, Cambridge University, 1982.

17. Y. Li and M. Kwauk, "The Dynamics of Fast Fluidization," in *Fluidization,* Grace and Matsen eds., Plenum, New York, 1980.

18. D. Bai, Y. Jin, and Z. Yu, "Flow Regimes in Circulating Fluidized Beds," *Chem. Eng. Technol.* (16): 307, 1993.

19. M. Kwauk, *Fluidization: Idealized and Bubbleless, With Applications,* Science Press, Beijing, 1992.

20. E.-U. Hartge, Y. Li, and J. Werther, "Analysis of the Local Structure of the Two-Phase Flow in a Fast Fluidized Bed," in *Circulating Fluidized Bed Technology,* P. Basu ed., Pergamon Press, Toronto, 1986.

21. K. B. Mathur and P. E. Gishler, "A Technique for Contacting Gases With Coarse Solid Particles," *AIChE J.* (1): 157, 1955.

22. A. G. Fane and R. A. Mitchell, "Minimum Spouting Velocity of Scaled-Up Beds," *Can. J. Chem. Eng.* (62): 437, 1984.

23. G. Flamant, J. D. Lu, and B. Variot, "Towards a Generalized Model for Vertical Walls to Gas-Solid Fluidized Beds Heat Transfer—II. Radiative Transfer and Temperature Effects," *Chem. Eng. Sci.* (48/13): 2493, 1993.

24. J. D. Lu, G. Flamant, and B. Variot, "Theoretical Study of Combined Conductive, Convective and Radiative Heat Transfer Between Plates and Packed Beds," *Int. J. Heat Mass Transfer* (37/5): 727, 1994.

25. J. Shen, S. Kaguei, and N. Wakao, "Measurement of Particle-to-Gas Heat Transfer Coefficients From One-Shot Thermal Response in Packed Beds," *Chem. Eng. Sci.* (36): 1283, 1981.

26. W. E. Ranz and W. R. Marshall, "Evaporation from Drops, Part II," *Chem. Eng. Prog.* (48/4): 173, 1952.

27. J. D. Gabor, "Wall-to-Bed Heat Transfer in Fluidized and Packed Beds," *Chem. Eng. Prog. Symp. Ser.* (66/105): 76, 1970.

28. A. B. Duncan, G. P. Peterson, and L. S. Fletcher, "Effective Thermal Conductivity With Packed Beds of Spherical Particles," *J. Heat Tr.* (111): 830, 1989.

29. D. Kunii and J. M. Smith, "Heat Transfer Characteristics of Porous Rocks," *AIChE J.* (6): 71, 1960.

30. R. Krupiczka, "Analysis of Thermal Conductivity in Granular Materials," *Int. Chem. Eng.* (7): 122, 1967.

31. S. Yagi and D. Kunii, "Studies on Effective Thermal Conductivities in Packed Beds," *AIChE J.* (3): 373, 1957.

32. S. Yagi, D. Kunii, and N. Wakao, "Studies on Axial Effective Thermal Conductivities in Packed Beds," *AIChE J.* (6): 543, 1960.

33. T. M. Kuzay, "Effective Thermal Conductivity of Porous Gas-Solid Mixtures," *ASME Winter Ann. Mtg., Paper 80,* Chicago, 1980.

34. S. M. Rao and H. L. Toor, "Heat Transfer From a Particle to a Surrounding Bed of Particles. Effect of Size and Conductivity Ratios," *Ind. Eng. Chem. Res.* (26): 469, 1987.

35. M. Pons, P. Dantzer, and J. J. Guilleminot, "A Measurement Technique and a New Model for the Wall Heat Transfer Coefficient of a Packed Bed of (Reactive) Powder Without Gas Flow," *Int. J. Heat Mass Transfer* (36/10): 2635, 1993.

36. A. A. Mohamad, S. Ramadhyani, and R. Viskanta, "Modeling of Combustion and Heat Transfer in a Packed Bed With Embedded Coolant Tubes," *Int. J. Heat Mass Transfer* (37/8): 1181, 1994.

37. C. H. Li and B. A. Finlayson, "Heat Transfer in Packed Beds—A Reevaluation," *Chem Eng. Sci.* (32): 1055, 1977.

38. K. Nasr, S. Ramadhyani, and R. Viskanta, "An Experimental Investigation on Forced Convection Heat Transfer From a Cylinder Embedded in a Packed Bed," *J. Heat Transfer* (116): 73, 1994.

39. S. Whitaker, "Radiant Energy Transport in Porous Media," *18th Natl. Heat Transfer Conf., ASME, Paper 79-HT-1,* San Diego, CA, 1979.

40. N. Wakao and K. Kato, "Effective Thermal Conductivity of Packed Beds," *J. Chem. Eng. Jpn.* (2): 24, 1969.

41. J. Schotte, "Thermal Conductivity of Packed Beds," *AIChE J.* (6): 63, 1960.

42. M. Q. Brewster and C. L. Tien, "Radiative Heat Transfer in Packed Fluidized Beds: Dependent Versus Independent Scattering," *J. Heat Tr.* (104): 573, 1982.

43. B. P. Singh and M. Kaviany, "Modeling Radiative Heat Transfer in Packed Beds," *Int. J. Heat Mass Transfer* (35/6): 1397, 1992.

44. G. Flamant, N. Fatah, and Y. Flitris, "Wall-to-Bed Heat Transfer in Gas-Solid Fluidized Beds: Prediction of Heat Transfer Regimes," *Powder Tech.* (69): 223, 1992.

45. N. I. Gel'Perin and V. G. Einstein, "Heat Transfer in Fluidized Beds," in *Fluidization,* Davidson and Harrison eds., Academic Press, New York, 1971.

46. S. S. Zabrodsky, "Heat Transfer Between Solid Particles and a Gas in a Non-Uniformly Aggregated Fluidized Bed," *Int. J. Heat & Mass Transfer* (6): 23, 991, 1963.

47. H. S. Mickley and D. F. Fairbanks, "Mechanism of Heat Transfer to Fluidized Beds," *AIChE J.* (1): 374, 1955.

48. K. Yoshida, D. Kunii, and O. Levenspiel, "Heat Transfer Mechanisms Between Wall Surface and Fluidized Bed," *Int. J. Heat & Mass Transfer* (12): 529, 1969.

49. W. E. Ranz, "Friction and Transfer Coefficients for Single Particles and Packed Beds," *Chem. Eng. Prog.* (48): 247, 1952.

50. R. S. Brodkey, D. S. Kim, and W. Sidner, "Fluid to Particle Heat Transfer in a Fluidized Bed and to Single Particles," *Int. J. Heat & Mass Transfer* (34): 2327, 1991.

51. J. Tuot and R. Clift, "Heat Transfer Around Single Bubbles in a Two-Dimensional Fluidized Bed," *Chem. Eng. Prog. Symp. Ser.* (69/128): 78, 1973.

52. V. K. Maskaev and A. P. Baskakov, "Features of External Heat Transfer in a Fluidized Bed of Coarse Particles," *Int. Chem. Eng.* (14): 80, 1974.

53. H. S. Mickley, D. F. Fairbanks, and R. D. Hawthorn, "The Relation Between the Transfer Coefficient and Thermal Fluctuations in Fluidized Bed Heat Transfer," *Chem. Eng. Symp. Ser.* (57/32): 51, 1961.

54. A. M. Xavier and J. F. Davidson, "Heat Transfer in Fluidized Beds: Convective Heat Transfer in Fluidized Beds," in *Fluidization,* 2d ed., Davidson, Clift, and Harrison eds., London: Academic Press, 1985.

55. O. Molerus, A. Burschka, and S. Dietez, "Particle Migration at Solid Surfaces and Heat Transfer in Bubbling Fluidized Beds—I. Particle Migration Measurement Systems," *Chem. Eng. Sci.* (50/5): 871, 1995.

56. O. Molerus, A. Burschka, and S. Dietez, "Particle Migration at Solid Surfaces and Heat Transfer in Bubbling Fluidized Beds—I. Prediction of Heat Transfer in Bubbling Fluidized Beds," *Chem. Eng. Sci.* (50/5): 879, 1995.

57. A. P. Baskakov, O. K. Vitt, V. A. Kirakosyan, V. K. Maskaev, and N. F. Filippovsky, "Investigation of Heat Transfer Coefficient Pulsations and of the Mechanism of Heat Transfer From a Surface Immersed Into a Fluidized Bed," in *Proc. Int. Symposium Fluidization Appl.,* Cepadues-Editions, Toulouse, France, 1974.

58. A. O. O. Denloye and J. M. S. Botterill, "Bed to Surface Heat Transfer in a Fluidized Bed of Large Particles," *Powder Tech.* (19): 197, 1978.

59. A. P. Baskakov, "Heat Transfer in Fluidized Beds: Radiative Heat Transfer in Fluidized Beds," in *Fluidization,* 2d ed, Davidson, Clift, and Harrison eds., Academic Press, London, 1985.

60. A. P. Baskakov, B. V. Berg, O. K. Vitt, N. F. Filippovsky, V. A. Kirakosyan, J. M. Goldobin, and V. K. Maskaev, "Heat Transfer to Objects Immersed in Fluidized Beds," *Powder Tech.* (8): 273, 1973.

61. H.-J. Bock and O. Molerus, "Influence of Hydrodynamics on Heat Transfer in Fluidized Beds," in *Fluidization,* Grace and Matsen eds., Plenum, New York, 1985.

62. V. A. Borodulya, V. L. Ganzha, and V. I. Kovensky, *Nauka I Technika.* Minsk, USSR, 1982.

63. T. M. Knowlton, "Pressure and Temperature Effects in Fluid-Particle System," in *Fluidization VII,* Potter and Nicklin eds., Engineering Foundation, New York, 1992.

64. J. M. S. Botterill, Y. Teoman, and K. R. Yüregir, "Temperature Effects on the Heat Transfer Behaviour of Gas Fluidized Beds," *AIChE Symp. Ser.* (77/208): 330, 1981.

65. J. M. S. Botterill, Y. Teoman, and K. R. Yüregir, "Factors Affecting Heat Transfer Between Gas-Fluidized Beds and Immersed Surfaces," *Powder Tech.* (39): 177, 1984.

66. L. Glicksman, "Circulating Fluidized Bed Heat Transfer," in *Circulating Fluidized Bed Technology II,* P. Basu and J. F. Large eds., Pergamon Press, Oxford, 1988.

67. D. Subbarao and P. Basu, "A Model for Heat Transfer in Circulating Fluidized Beds," *Int. J. Heat & Mass Transfer* (29): 487, 1986.

68. R. L. Wu, J. R. Grace, and C. J. Lim, "A Model for Heat Transfer in Circulating Fluidized Beds," *Chem. Eng. Sci.* (45): 3389, 1990.

69. D. Gloski, L. Glicksman, and N. Decker, "Thermal Resistance at a Surface in Contact With Fluidized Bed Particles," *Int. J. Heat & Mass Transfer* (27): 599, 1984.

70. M. C. Lints and L. R. Glicksman, "Parameters Governing Particle-to-Wall Heat Transfer in a Circulating Fluidized Bed," in *Circulating Fluidized Bed Technology IV,* A. A. Avidan ed., AIChE Publications, New York, 1993.

71. K. E. Wirth, "Heat Transfer in Circulating Fluidized Beds," *Chem. Eng. Sci.* (50/13): 2137, 1995.

72. C. Y. Wen and E. N. Miller, "Heat Transfer in Solid-Gas Transport Lines," *I&EC,* (53): 51, 1961.

73. P. Basu and P. K. Nag, "An Investigation Into Heat Transfer in Circulating Fluidized Beds," *Int. J. Heat & Mass Transfer* (30): 2399, 1987.

74. C. A. Sleicher and M. W. Rouse, "A Convective Correlation for Heat Transfer to Constant and Variable Property Fluids in Turbulent Pipe Flow," *Int. J. Heat & Mass Transfer* (18): 677, 1975.

75. R. L. Wu, J. R. Grace, C. J. Lim, and C. M. H. Brereton, "Suspension-to-Surface Heat Transfer in a Circulating Fluidized Bed Combustor," *AIChE J.* (35): 1685, 1989.

76. P. Basu, "Heat Transfer in High Temperature Fast Fluidized Beds," *Chem. Eng. Sci.* (45): 3123, 1990.

77. M. Q. Brewster, "Effective Absorptivity and Emissivity of Particulate Medium With Applications to a Fluidized Bed," *Trans. ASME, J. Heat Transfer* (108): 710, 1986.

78. H.-T. Bi, Y. Jin, Z. Q. Yu, and D.-R. Bai, "The Radial Distribution of Heat Transfer Coefficients in Fast Fluidized Bed," in *Fluidization VI,* Grace, Shemilt, and Bergougnou eds., Engineering Foundation, New York, 1989.

79. P. Basu and P. K. Nag, "Heat Transfer to Walls of a Circulating Fluidized-Bed Furnace," *Chem. Eng. Sci.* (51/1): 1, 1996.

80. R. L. Wu, C. J. Lim, J. Chaouki, and J. R. Grace, "Heat Transfer From a Circulating Fluidized Bed to Membrane Waterwall Surfaces," *AIChE J.* (33): 1888, 1987.

81. H.-T. Bi, Y. Jin, Z.-Q. Yu, and D.-R. Bai, "An Investigation on Heat Transfer in Circulating Fluidized Bed," in *Circulating Fluidized Bed Technology III,* Basu, Horio, and Hasatani eds., Pergamon Press, Oxford, UK, 1990.

82. P. N. Rowe and K. T. Claxton, "Heat and Mass Transfer From a Single Sphere to a Fluid Flowing Through an Array," *Trans. Instn. Chem. Engrs.* (43): T321, 1965.

83. H. Littman and D. E. Sliva, "Gas-Particle Heat Transfer Coefficient in Packed Beds at Low Reynolds Number," in *Heat Transfer 1970, Paris-Versailles,* CT 1.4, Elsevier, Amsterdam, 1971.

84. K. B. Mathur and N. Epstein, *Spouted Beds,* Academic Press, New York, 1974.

85. N. Epstein and J. R. Grace, "Spouting of Particulate Solids," in *Handbook of Powder Science and Technology,* 2d ed., Fayed and Otten, eds., Chapman and Hall, New York, 1995.

86. J. Grace, "Heat Transfer in Circulating Fluidized Beds," in *Circulating Fluidized Bed Technology,* p. 63, P. Basu ed., Pergamon Press, Toronto, 1986.

87. J. Stringer and A. J. Minchener, "High Temperature Corrosion in Fluidized Bed Combustors," in *Fluidized Combustion: Is It Achieving Its Promise?* vol. 1, p. 255, Institution of Energy, London, 1984.

88. E. K. Levy and F. Bayat, "The Bubble Coalescence Mechanism of Tube Erosion in Fluidized Beds," in *Fluidization VI,* p. 605, Engineering Foundation, Banff, Canada, 1989.

89. I. M. Hutchings, "Surface Impact Damage," in *Tribology in Particulate Technology,* Briscoe and Adams eds., Adam Hilger, Philadelphia, 1987.

90. R. Bellman Jr. and A. Levy, "Erosion Mechanism in Ductile Metals," *Wear* (70): 1, 1981.

91. A. V. Levy, "The Platelet Mechanism of Erosion of Ductile Materials," *Wear* (108): 1, 1986.

92. J. Zhu, J. R. Grace, and C. J. Lim, "Erosion-Causing Particle Impacts on Tubes in Fluidized Beds," in *Fluidization VI,* J. R. Grace, L. W. Shemilt, and M. A. Bergougnou eds., p. 613, Engineering Foundation, New York, 1989.

93. J. Zhu, J. R. Grace, and C. J. Lim, "Tube Wear in Gas Fluidized Beds—I. Experimental Findings," *Chem. Eng. Sci.* (45/4): 1003, 1990.

94. R. D. Patel and J. M. Simpson, "Heat Transfer in Aggregative and Particulate Liquid-Fluidized Beds," *Chem. Eng. Sci.* (32): 67, 1977.

95. M. Magiliotou, Y. M. Chen, and L.-S. Fan, "Bed-Immersed Object Heat Transfer in a Three Phase Fluidized Bed," *AIChE. J.* (34): 1043, 1988.

96. K. Muroyama, M. Fukuma, and A. Yasunishi, "Wall-to-Bed Heat Transfer in Liquid-Solid and Gas-Liquid-Solid Fluidized Beds. Part I: Liquid-Solid Fluidized Beds," *Can. J. Chem. Eng.* (64): 399, 1986.

97. Y. Kang, L. T. Fan, and S. D. Kim, "Immersed Heater-to-Bed Transfer in Liquid-Solid Fluidized Beds," *AIChE J.* (37/7): 1101, 1991.

98. T. M. Chiu and E. N. Ziegler, "Liquid Holdup and Heat Transfer Coefficient in Liquid-Solid and Three-Phase Fluidized Beds," *AIChE J.* (31/9): 1504, 1985.

99. A. Macias-Machin, L. Oufer, and N. Wannenmacher, "Heat Transfer Between an Immersed Wire and a Liquid Fluidized Bed," *Powder Tech.* (66): 281, 1991.

100. A. I. Karpenko, N. I. Syromyatnikov, L. K. Vasanova, and N. N. Galimulin, "Radial Heat Conduction in a Liquid-Fluidized Bed," *Heat Transfer Sov. Res.* (8): 110, 1976.

CHAPTER 14
CONDENSATION

P. J. Marto
Department of Mechanical Engineering,
Naval Postgraduate School,
Monterey, California

INTRODUCTION

Condensation of vapor occurs in a variety of engineering applications. For example, when a vapor is cooled below its saturation temperature, or when a vapor-gas mixture is cooled below its dew point, homogeneous condensation occurs as a fog or cloud of microscopic droplets. Condensation also occurs when vapor comes in direct contact with subcooled liquid such as spraying a fine mist of subcooled liquid droplets into a vapor space or injecting vapor bubbles into a pool of subcooled liquid. The most common type of condensation occurs when a cooled surface, at a temperature less than the local saturation temperature of the vapor, is placed in contact with the vapor. Vapor molecules that strike this cooled surface may stick to it and condense into liquid.

Modes of Condensation

During condensation, the liquid collects in one of two ways, depending on whether it wets the cold surface or not. If the liquid condensate wets the surface, a continuous film will collect, and this is referred to as *filmwise condensation*. If the liquid does not wet the surface, it will form into numerous discrete droplets, referred to as *dropwise condensation*. All surface condensers today are designed to operate in the filmwise mode, since long-term dropwise conditions have not been successfully sustained.

Dropwise condensation is a complex phenomenon that has been studied for over sixty years. It involves a series of randomly occurring subprocesses as droplets grow, coalesce, and depart from a cold surface. The sequence of these subprocesses forms a dynamic "life cycle." The cycle begins with the formation of microscopic droplets that grow very rapidly due to condensation of vapor on them and merge with neighboring droplets. Therefore, they are constantly shifting in position. As a result, rapid surface temperature fluctuations under these droplets occur. This active growth and coalescence continues until larger drops are formed. Although inactive due to condensation, these drops continue to grow due to coalescence with neighboring smaller droplets. Eventually, these large, so-called "dead" drops merge to form a drop that is large enough so that adhesive forces due to surface tension are overcome either by gravity or vapor shear. This very large drop then departs from the surface, sweeping away all condensate in its path, allowing fresh microscopic droplets to begin to grow again and start another cycle.

Condensation Curve

In the last twenty years, it has been demonstrated that, just as with boiling heat transfer, a characteristic condensation curve exists that includes a *dropwise* region, a *filmwise* region, and a *transition* region. Figure 14.1 shows some representative condensation curves for steam at atmospheric pressure [1]. At a fixed vapor velocity, at very low surface subcoolings, drop-wise condensation occurs. Dropwise conditions can persist to relatively large subcoolings (and to very large heat fluxes, near 10 MW/m² for steam). However, at large enough subcool-ings, so much condensate is formed that a relatively thick, continuous liquid film tries to occur (i.e., the rate of formation of condensate exceeds the rate of drop departure). Thus, a maxi-mum heat flux occurs similar to boiling. A transition region follows where the heat flux decreases and approaches the filmwise condensation curve. Further increases in subcooling result in a portion of the condensate actually freezing on the cold surface, and a pseudofilm condensation condition will exist. For steam, this is referred to as *on-ice* or *glacial condensa-tion* [1].

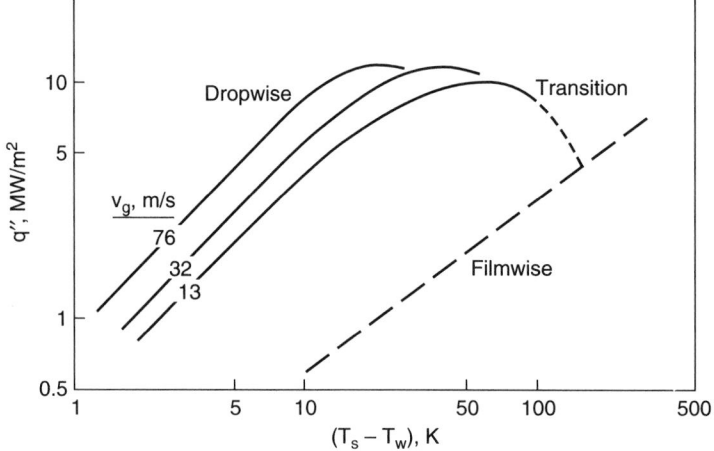

FIGURE 14.1   Condensation curves for steam. (*Adapted from Ref. 1 and printed with permission from Academic Press, Inc., Orlando, FL.*)

Dropwise heat transfer coefficients can be as much as 10–20 times larger than filmwise val-ues during steam condensation at atmospheric pressure on copper surfaces. Under vacuum conditions and for condenser materials with lower thermal conductivities, the dropwise heat transfer coefficient decreases, as shown in Fig. 14.2, making this mode of condensation less attractive. Nevertheless, if a reliable long-term dropwise promoter application technique can be found, a significant economic incentive would exist for design development. In recent years, considerable research has focused on new promoters and on promoter application techniques [2–11], and new breakthroughs may lead to a renewed practical interest in this mode of condensation.

Thermal Resistances

During condensation, thermal resistances exist in the condensate, in the vapor, and across the liquid-vapor interface. These resistances are reflected schematically in Fig. 14.3, which shows the resulting temperature profiles during film condensation on a vertical surface. The dashed

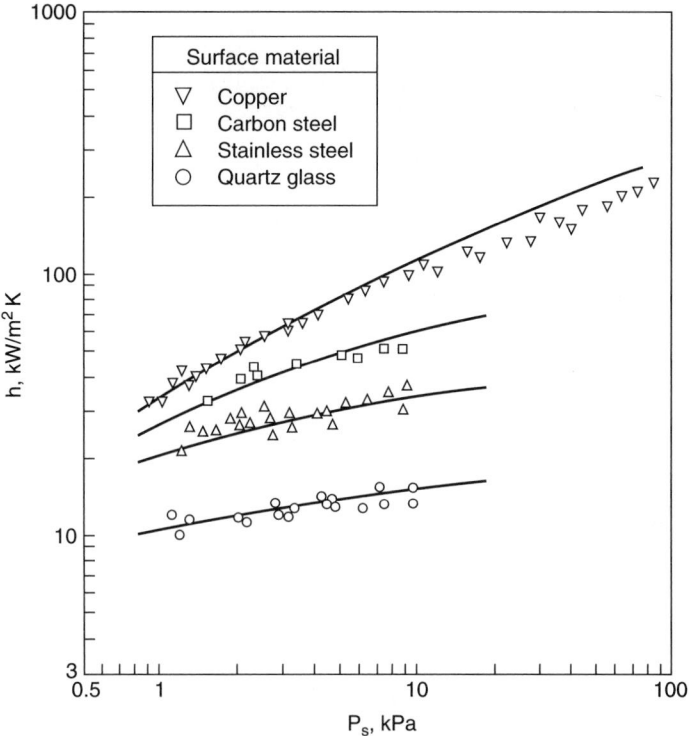

FIGURE 14.2   The influence of surface material and operating pressure on the dropwise condensation heat transfer coefficient of steam. (*Adapted from Ref. 1 and printed with permission from Academic Press, Inc., Orlando, FL.*)

profile denotes the idealized case of a pure vapor with no thermal resistance at the liquid-vapor interface. In reality, the vapor may contain a small amount of noncondensable gas so that the saturation temperature of the vapor far away from the surface is reduced to T_s'. In addition, because of the presence of the gas molecules that concentrate near the interface, the vapor molecules will experience a temperature drop due to a pressure drop caused by their diffusing through the noncondensable gas layer to get to the interface. The resulting vapor interface temperature will be T_{gi}. The influence of noncondensable gases can severely reduce condensation rates; this subject is covered in later sections. An additional temperature drop may exist at the liquid-vapor interface due to the nonequilibrium mass flux of molecules toward and away from the interface during condensation conditions. The theory of interphase mass transfer is reviewed by Tanasawa [1], and an approximate interfacial heat transfer coefficient may be written as

$$h_i = \frac{q''}{(T_{gi} - T_{\ell i})} = \frac{2\sigma}{2 - \sigma} \frac{i_{\ell g}^2 \rho_g}{\sqrt{2\pi R T_g^3}} \tag{14.1}$$

where σ is the condensation coefficient (i.e., the fraction of vapor molecules striking the condensate surface that actually stick and condense on the surface). Recent experimental results indicate that the condensation coefficient is close to 1.0 for condensation of a metal vapor and less than 1.0 (most probably around 0.4) for steam [1]. The interfacial thermal resistance is important only at low pressures and at high condensation rates (where the vapor velocity is high).

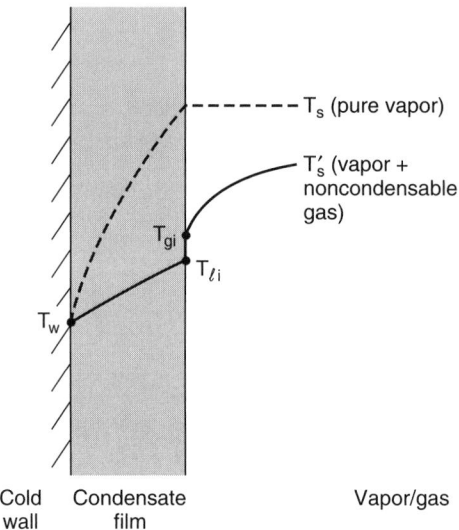

FIGURE 14.3 Temperature distributions during film condensation on a vertical plate.

FILM CONDENSATION ON A VERTICAL PLATE

Approximate Analysis

Laminar Free Convection. When a stagnant vapor condenses on a vertical plate, the motion of the condensate will be governed by body forces, and it will be laminar over the upper part of the plate where the condensate film is very thin. In this region, the heat transfer coefficient can be readily derived following the classical approximate method of Nusselt [12]. Consider the situation depicted in Fig. 14.4 where the vapor is at a saturation temperature T_s and the plate surface temperature is T_w. Neglecting momentum effects in the condensate film, a force balance in the z-direction on a differential element in the film yields

$$\frac{\partial \tau}{\partial y} - \frac{\partial P}{\partial z} + \rho_\ell g = 0 \tag{14.2}$$

A similar force balance in the y-direction gives $\partial P/\partial y = 0$, so that

$$\frac{\partial P}{\partial z} = \frac{dP_g}{dz} = \rho_g g \tag{14.3}$$

Substituting Eq. 14.3 into Eq. 14.2, and integrating from y to δ with the assumption that all fluid properties are constant, yields the shear stress distribution in the film:

$$\tau = \mu_\ell \frac{\partial v_z}{\partial y} = (\rho_\ell - \rho_g)g(\delta - y) \tag{14.4}$$

where the shear stress at $y = \delta$ has been assumed to be zero, since there is no vapor motion. With $v_z = 0$ at $y = 0$, the condensate velocity distribution is therefore

$$v_z = \frac{(\rho_\ell - \rho_g)g}{\mu_\ell} (\delta y - y^2/2) \tag{14.5}$$

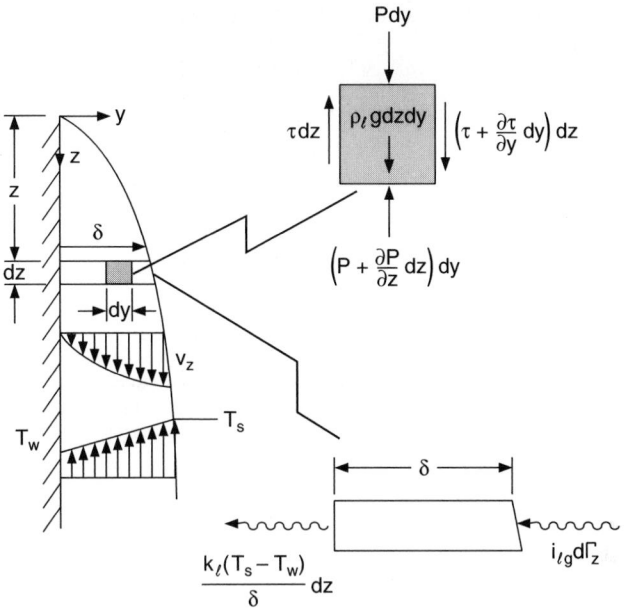

FIGURE 14.4   Model of laminar film condensation on a vertical plate.

The local liquid flow rate (per unit depth) in the film can then be calculated:

$$\Gamma_z = \int_0^\delta \rho_\ell v_z \, dz = \frac{\rho_\ell(\rho_\ell - \rho_g)g\delta^3}{3\mu_\ell} \tag{14.6}$$

Neglecting convection effects in the film (i.e., assuming pure conduction in the film, which yields a linear temperature profile), an energy balance on a differential slice of condensate of width dz (Fig. 14.4) gives

$$\frac{d\Gamma_z}{dz} = \frac{k_\ell(T_s - T_w)}{i_{\ell g}\delta} \tag{14.7}$$

Combining Eq. 14.7 with Eq. 14.6 and assuming the wall temperature T_w to be constant yields the local heat transfer coefficient:

$$h_z = \frac{q''}{(T_s - T_w)} = \frac{k_\ell}{\delta} = \left\{ \frac{k_\ell^3 \rho_\ell(\rho_\ell - \rho_g)g i_{\ell g}}{4\mu_\ell(T_s - T_w)z} \right\}^{1/4} \tag{14.8}$$

which can be converted to an average value between $z = 0$ and L

$$h_m = \frac{1}{L}\int_0^L h_z \, dz = \tfrac{4}{3}h_L = 0.943 \left\{ \frac{k_\ell^3 \rho_\ell(\rho_\ell - \rho_g)g i_{\ell g}}{\mu_\ell(T_s - T_w)L} \right\}^{1/4} \tag{14.9}$$

or in terms of an average Nusselt number

$$\text{Nu}_m = \frac{h_m L}{k_\ell} = 0.943 \left\{ \frac{\rho_\ell(\rho_\ell - \rho_g)g i_{\ell g}L^3}{\mu_\ell(T_s - T_w)k_\ell} \right\}^{1/4} \tag{14.10}$$

The condensation heat transfer coefficient may also be written in terms of the film Reynolds number, Re_z (equal to $4\Gamma_z/\mu_\ell$), where Γ_z is given by Eq. 14.6 or by $\Gamma_z = q''_{avg}L/i_{\ell g}$. With this conversion, Eqs. 14.8 and 14.9 become, respectively,

$$\frac{h_z}{k_\ell}\left(\frac{\mu_\ell^2}{\rho_\ell(\rho_\ell - \rho_g)g}\right)^{1/3} = 1.1\,Re_z^{-1/3} \tag{14.11}$$

and
$$\frac{h_m}{k_\ell}\left(\frac{\mu_\ell^2}{\rho_\ell(\rho_\ell - \rho_g)g}\right)^{1/3} = 1.47\,Re_L^{-1/3} \tag{14.12}$$

In most cases, the vapor density ρ_g is much smaller than the liquid density ρ_ℓ, so the term in brackets in Eqs. 14.11 and 14.12 may be approximated by $(v_\ell^2/g)^{1/3}$.

The Nusselt analysis of laminar film condensation has been shown to be reasonably accurate for a variety of ordinary fluids such as steam and organic vapors, despite the approximations made in the model. Measured heat transfer coefficients are about 15–20 percent higher than predicted values. Numerous studies have been conducted to explain the observed differences. For example, in Eq. 14.9, a correction may be made to the latent heat of evaporation to take into account condensate subcooling:

$$i'_{\ell g} = i_{\ell g} + \tfrac{3}{8}c_{p\ell}(T_s - T_w) \tag{14.13}$$

Rohsenow [13] showed that if the condensate temperature profile was allowed to be nonlinear to account for convection effects in the condensate film, an improved correction term, $i'_{\ell g} = i_{\ell g} + 0.68c_{p\ell}(T_s - T_w)$ results. Another correction pertains to the variation of viscosity with temperature. For the assumed linear temperature profile in the condensate, Drew [14] showed that if $1/\mu_\ell$ is linear in temperature, then the condensate viscosity should be calculated at a reference temperature equal to $T_s - \tfrac{3}{4}(T_s - T_w)$.

Shang and Adamek [15] recently studied laminar film condensation of saturated steam on a vertical flat plate using variable thermophysical properties and found that the Nusselt theory with the Drew [14] reference temperature cited above produces a heat transfer coefficient that is as much as 5.1 percent lower than their more correct model predicts (i.e., the Nusselt theory is conservative).

Condensate Waves and Turbulence.   As the local condensate film thickness (i.e., the film Reynolds number Re_z) increases, the film will become unstable, and waves will begin to grow rapidly. This occurs for $Re_z > 30$. Kapitza [16] has shown that, in this situation, the average film thickness is less than predicted by the Nusselt theory and the heat transfer coefficient increases accordingly. Kutateladze [17] therefore recommends that the following correction be applied to Eq. 14.12:

$$\frac{h_c}{h_m} \approx 0.69\,Re_L^{0.11} \tag{14.14}$$

Butterworth [18] applied Eq. 14.14 to Eq. 14.11 to get

$$\frac{h_z}{k_\ell}\left(\frac{\mu_\ell^2}{\rho_\ell(\rho_\ell - \rho_g)g}\right)^{1/3} = 0.76\,Re_z^{-0.22} \tag{14.15}$$

for $Re_z > 30$. Nozhat [19] recently studied this problem by including the effect of surface tension and free surface curvature in the Nusselt model. With this refinement, he arrived at a correction factor for the Nusselt heat transfer coefficient that may be approximated by

$$\frac{h_c}{h_m} \approx 0.87\,Re_L^{0.07} \tag{14.16}$$

With the above corrections, the presence of waves can easily explain the noted 15–20 percent discrepancy between the Nusselt theory and experimental data.

As the film thickens further, turbulence will develop in the condensate film, and the heat transfer mechanism then undergoes a significant change, since the heat is transferred across the condensate film by turbulent mixing as well as by molecular conduction. For gravity-dominated flow (i.e., natural convection), the transition from laminar-wavy flow to turbulent flow occurs at film Reynolds numbers of about 1600 [18].

Various semiempirical models exist in the literature to predict turbulent film condensation [20–23]. Butterworth [18] recommends the result of Labuntsov [23] for the local coefficient

$$\frac{h_z}{k_\ell}\left(\frac{\mu_\ell^2}{\rho_\ell(\rho_\ell-\rho_g)g}\right)^{1/3} = 0.023\mathrm{Re}_z^{1/4}\,\mathrm{Pr}_\ell^{1/2} \tag{14.17}$$

Using Eqs. 14.11, 14.15, and 14.17, respectively, for the local coefficients in the laminar wave-free ($0 < \mathrm{Re}_z < 30$), laminar-wavy ($30 < \mathrm{Re}_z < 1600$), and turbulent ($\mathrm{Re}_z > 1600$) regions, Butterworth [18] determined an average coefficient from the expression

$$\frac{\mathrm{Re}_L}{h_m} = \int_0^{\mathrm{Re}_L}\frac{d\,\mathrm{Re}_z}{h_z} \tag{14.18}$$

His result is: For $\mathrm{Re}_L < 30$, use Eq. 14.12. For $30 < \mathrm{Re}_L < 1600$,

$$\frac{h_m}{k_\ell}\left(\frac{\mu_\ell^2}{\rho_\ell(\rho_\ell-\rho_g)g}\right)^{1/3} = \frac{\mathrm{Re}_L}{1.08\mathrm{Re}_L^{1.22}-5.2} \tag{14.19}$$

For $\mathrm{Re}_L > 1600$,

$$\frac{h_m}{k_\ell}\left(\frac{\mu_\ell^2}{\rho_\ell(\rho_\ell-\rho_g)g}\right)^{1/3} = \frac{\mathrm{Re}_L}{8750 + 58\mathrm{Pr}_\ell^{-1/2}\,(\mathrm{Re}_L^{3/4}-253)} \tag{14.20}$$

Chun and Kim [24] recommend the following semiempirical average heat transfer coefficient correlation that is valid over a wide range of film Reynolds numbers. For $10 < \mathrm{Re}_L < 3.1 \times 10^4$,

$$\frac{h_m}{k_\ell}\left(\frac{\mu_\ell^2}{\rho_\ell(\rho_\ell-\rho_g)g}\right)^{1/3} = 1.33\mathrm{Re}_L^{-1/3} + 9.56 \times 10^{-6}\,\mathrm{Re}_L^{0.89}\,\mathrm{Pr}_\ell^{0.94} + 8.22 \times 10^{-2} \tag{14.21}$$

This correlation agrees with a variety of data for $1.75 < \mathrm{Pr}_\ell < 5.0$ and is plotted in Fig. 14.5. When viewing this figure, it is clear that Prandtl number is important during turbulent flow conditions. For small Prandtl number fluids, the vertical surface should be as short as possible (i.e., low Re_L) to allow good heat transfer to occur. On the other hand, for large Prandtl number fluids, good heat transfer occurs in the turbulent region (i.e., high Re_L), so the surface should be very long [25].

Laminar Forced Convection.   When the vapor moves in relation to the condensate, a shear stress τ_g will develop at the liquid-vapor interface. At very high vapor velocities, this shear

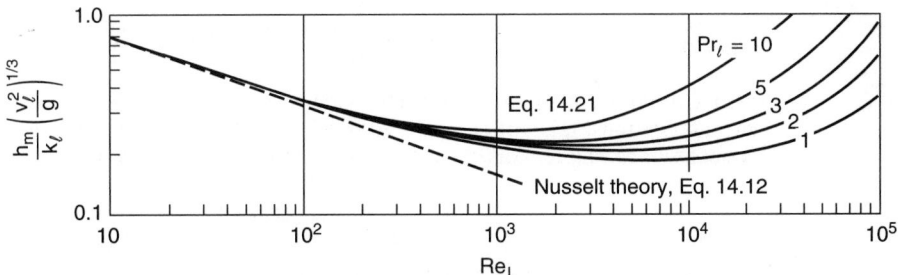

FIGURE 14.5   Average heat transfer coefficients for film condensation on vertical plates.

force can dominate over the gravitational force so that gravitational effects may be completely neglected. The local condensate velocity is then simply

$$v_z = \frac{\tau_g}{\mu_\ell} y \qquad (14.22)$$

If τ_g remains constant, independent of z, a Nusselt-type derivation for heat transfer yields an average Nusselt number

$$\mathrm{Nu}_m = \frac{h_m L}{k_\ell} = 1.04 \left\{ \frac{\rho_\ell \tau_g i_{\ell g} L^2}{\mu_\ell (T_s - T_w) k_\ell} \right\} \qquad (14.23)$$

or, in terms of the film Reynolds number,

$$\frac{h_m}{k_\ell} \left(\frac{\mu_\ell^2}{\rho_\ell (\rho_\ell - \rho_g) g} \right)^{1/3} = 2.2 (\tau_g^*)^{1/2} \, \mathrm{Re}_L^{-1/2} \qquad (14.24)$$

where
$$\tau_g^* = \frac{\rho_\ell \tau_g}{[\rho_\ell (\rho_\ell - \rho_g) \mu_\ell g]^{2/3}} \qquad (14.25)$$

For shear-dominated conditions, the linear temperature distribution correction for subcooling in the condensate film is

$$i'_{\ell g} = i_{\ell g} + \tfrac{1}{3} c_{p\ell} (T_s - T_w) \qquad (14.26)$$

and the reference temperature to evaluate condensate viscosity is $T_s - \tfrac{2}{3}(T_s - T_w)$. See Ref. 26.

When both vapor shear and gravity are important, the average heat transfer coefficient may be approximated by

$$h_m = (h_{\mathrm{sh}}^2 + h_{\mathrm{gr}}^2)^{1/2} \qquad (14.27)$$

where h_{sh} is the average heat transfer coefficient calculated for shear-dominated flow (Eq. 14.24) and h_{gr} is the average value for gravity-dominated flow (Eq. 14.21).

In real situations, the vapor velocity varies with position along the plate, and the interfacial shear stress is not constant since mass is removed due to condensation. The variation in vapor velocity depends upon the condensation rate and any changes in the vapor cross sectional flow area. For moderate condensation rates, the interfacial shear stress may be approximated by:

$$\tau_g = \tfrac{1}{2} f \rho_g v_g^2 \qquad (14.28)$$

where the friction factor f is dependent upon the local vapor Reynolds number, the "waviness" of the film, and any momentum changes due to flow development in the film. As a first approximation, the friction factor f may be estimated by any of the well-known single phase expressions, but this approach ignores the motion of the interface as well as mass transfer effects across the interface due to condensation. Therefore, the results would only be valid at low condensation rates. The friction factor may also be estimated from the adiabatic two-phase flow data of Bergelin et al. [27], shown plotted in Fig. 14.6, where σ_w/σ is the ratio of the surface tension of water to that of the particular fluid being condensed (at the saturation temperature of the condensate). At high condensation rates (i.e., high heat flux), the interfacial shear stress must take into account changes in momentum as the vapor condenses upon the condensate surface. In this case, the shear stress may be represented by

$$\tau_g = (v_g - v_i) \frac{d\Gamma_z}{dz} \approx v_g \frac{d\Gamma_z}{dz} \qquad (14.29)$$

since $v_i \ll v_g$.

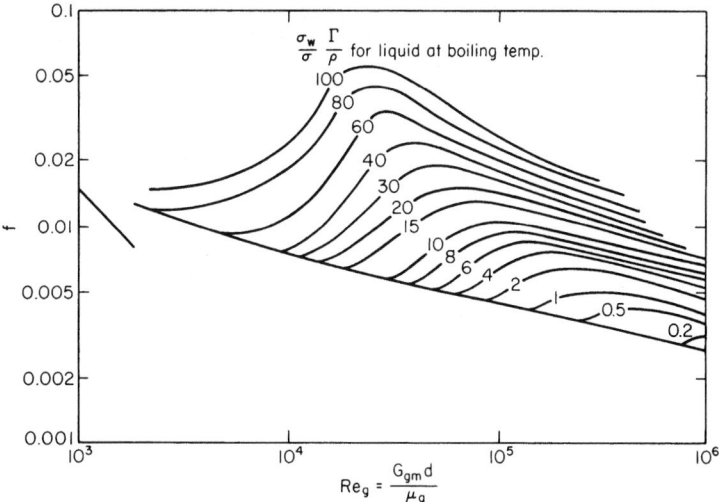

FIGURE 14.6   Friction factor for air flowing in a tube with a liquid layer on the wall [27].

If Eq. 14.29 is now used in Eq. 14.22 (with v_g assumed to be constant), the following simple expression for the average Nusselt number can be derived:

$$\text{Nu}_m = \tilde{\text{Re}}_L^{1/2} \tag{14.30}$$

where $\tilde{\text{Re}}_L = (\rho_\ell v_g L / \mu_\ell)$ is a two-phase Reynolds number involving condensate properties and the vapor velocity. In general, the vapor velocity will not remain constant unless the vapor flow passage geometry is changed to offset the amount of vapor condensed into liquid. The solution is more complicated when the vapor velocity is allowed to change, involving numerical techniques. Rohsenow et al. [21] extended the Nusselt laminar gravity-dominated flow analysis on a vertical plate by including vapor shear stress (assumed to be constant) in the model. These authors also examined the influence of vapor shear stress on the transition from laminar to turbulent flow. For $\tau_g^* < 11$, they arrived at the following expression for the transition film Reynolds number:

$$\text{Re}_{\text{tr}} = 1800 - 246(1 - \rho_g/\rho_\ell)^{1/3}\tau_g^* + 0.667(1 - \rho_g/\rho_\ell)(\tau_g^*)^3 \tag{14.31}$$

In Eq. 14.31, the quantity $(1 - \rho_g/\rho_\ell)$ may be safely taken as unity since ρ_g/ρ_ℓ is very small, except at pressures approaching the critical pressure of the fluid. For $\tau_g^* > 11$, Rohsenow et al. [21] reasoned that a limiting Reynolds number exists, and Butterworth [28] recommends it to be 50. In reality, as shown by the data of Blangetti and Schlünder [29], a distinct laminar to turbulent transition does not exist.

Rohsenow et al. [21] extended the analysis into the turbulent film regime using the heat transfer-momentum analogy. The results for a downward flowing vapor are shown in Fig. 14.7 for $\text{Pr}_\ell = 1.0$ and 10.0. At high vapor velocities, as the dimensionless shear stress τ_g^* increases, the transition to turbulence occurs at smaller values of the film Reynolds number (Eq. 14.31) as represented by the dashed lines. The influence of τ_g^* on both laminar and turbulent film condensation is evident.

Effect of Superheat.   When the vapor is superheated (i.e., $T_g > T_s$) and the cold wall temperature is less than the vapor temperature but greater than the saturation temperature, no condensation occurs. Instead, the vapor is cooled by single-phase free or forced convection

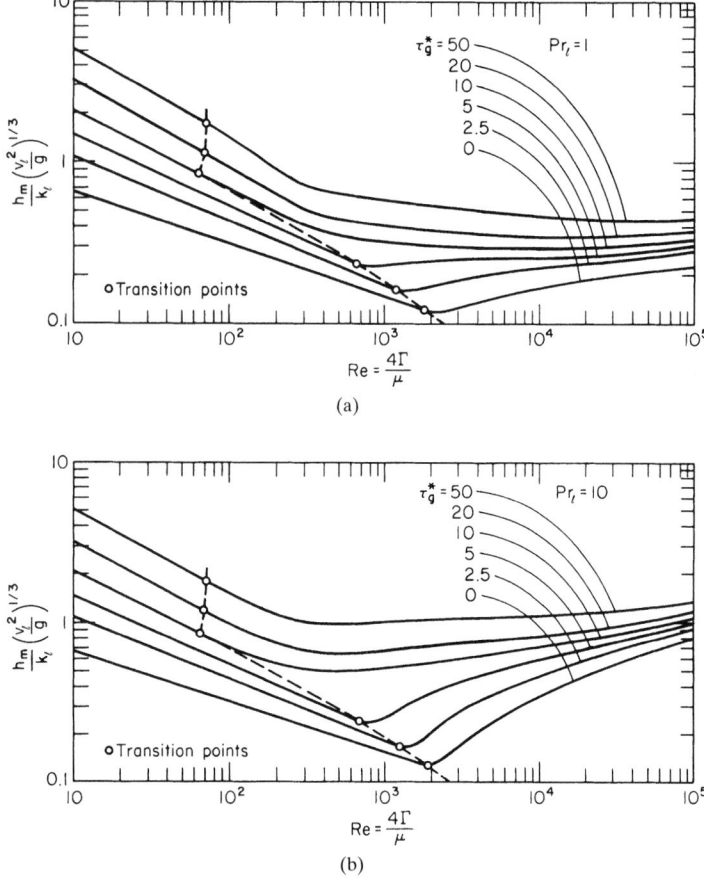

FIGURE 14.7   Effect of turbulence and vapor shear stress during film condensation on a vertical plate [21].

(so-called *dry wall desuperheating*). When the wall temperature is less than the saturation temperature, condensation occurs (so-called *wet wall desuperheating*), and the rate of condensation is slightly increased by the superheat in the vapor. During condensation, this effect of superheat is accounted for in the above analysis by replacing the corrected latent heat of evaporation by

$$i''_{\ell g} = i'_{\ell g} + c_{pg}(T_g - T_s) \tag{14.32}$$

In most practical situations, the increase predicted by Eq. 14.32 is less than a few percent. Miropolskiy et al. [30] found that superheated steam flowing in a tube did not always condense when the wall temperture was less than the saturation temperture. Condensation did not occur unless both the vapor temperature and quality were below certain threshold values.

Boundary Layer Analysis

The boundary layer treatment of laminar film condensation is thoroughly described by Rose [31] and Fujii [32].

Laminar Free Convection.   Sparrow and Gregg [33] were the first to use the boundary layer method to study laminar, gravity-driven film condensation on a vertical plate. They improved upon the approximate analysis of Nusselt by including fluid acceleration and energy convection terms in the momentum and energy equations, respectively. Their numerical results can be expressed as:

$$\frac{\text{Nu}}{\text{Nu}_{\text{Nu}}} = F[H, \text{Pr}_\ell] \tag{14.33}$$

where Nu_{Nu} is the Nusselt number from the Nusselt analysis, and

$$H = c_{p\ell}\Delta T/i_{\ell g} \tag{14.34}$$

For practical values of H and Pr_ℓ, Eq. 14.33 was found to be near unity, indicating that acceleration and convection effects are negligible. Chen [34] included the effect of vapor drag on the condensate motion by using an approximate expression for the interfacial shear stress. He was able to neglect the vapor boundary layer in the process and obtained the results shown in Fig. 14.8. The influence of interfacial shear stress is negligible at Prandtl numbers of ordinary liquids (nonliquid metals, $\text{Pr}_\ell > 1$). Chen [34] was able to represent his numerical results by the approximate (within 1 percent) expression:

$$\frac{\text{Nu}_m}{\text{Nu}_{m,\text{Nu}}} = \left\{ \frac{1 + 0.68\text{Pr}_\ell J_\ell + 0.02\text{Pr}_\ell J_\ell^2}{1 + 0.85J_\ell - 0.15\text{Pr}_\ell J_\ell^2} \right\}^{1/4} \tag{14.35}$$

where

$$J_\ell = \frac{k_\ell \Delta T}{\mu_\ell i_{\ell g}} = \frac{H}{\text{Pr}_\ell} \tag{14.36}$$

Koh et al. [35] solved the boundary layer equations of both the condensate and the vapor using a more accurate representation for the interfacial shear stress. They found a dependence on an additional parameter

$$R = [\rho_\ell \mu_\ell/\rho_g \mu_g]^{1/2} \tag{14.37}$$

but this dependence was negligible. Churchill [36] developed closed-form approximate solutions of the Koh et al. [35] model. He included the effects of acceleration and convection

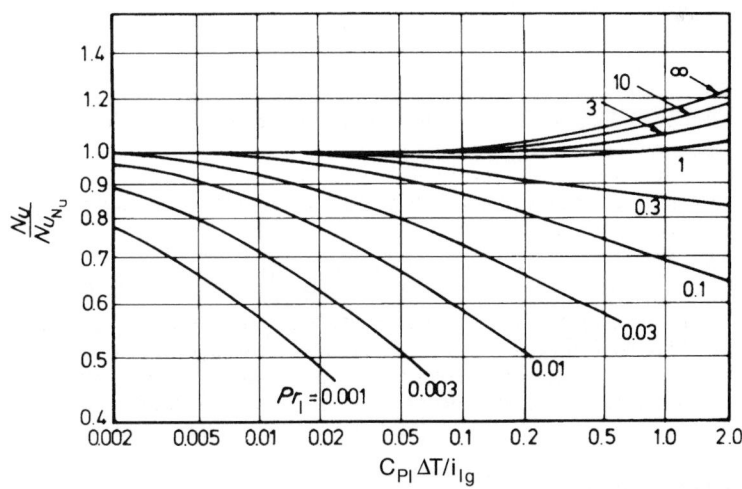

FIGURE 14.8   Influence of vapor drag during laminar free convection condensation on a vertical plate [31]. (*Reprinted with permission from JSME International Journal, Tokyo, Japan.*)

within the condensate, the drag of the vapor, and the curvature of the surface. Thus, his results are applicable also to the outside and inside of vertical tubes.

Laminar Forced Convection. Several investigators have solved the boundary layer equations during forced convection conditions. Cess [37] treated the case where no body force was present (a horizontal plate or a vertical plate where the free stream velocity V_∞ is very large, and the resulting interfacial shear stress dominates the heat transfer). Neglecting the acceleration terms in the momentum equation and the convection terms in the energy equation and using two asymptotic shear stress relationships, he obtained the following approximate asymptotic expressions for the local heat transfer coefficient:

$$\text{Nu}_x\,\tilde{\text{Re}}_x^{-1/2} = \begin{cases} 0.436G^{-1/3} & \text{zero condensation rate limit (i.e., low flux)} \quad (14.38a)\\ 0.5 & \text{infinite condensation rate limit (i.e., high flux)} \quad (14.38b) \end{cases}$$

where

$$G = \left(\frac{k_\ell \Delta T}{\mu_\ell i_{\ell g}}\right)\cdot\left(\frac{\rho_\ell \mu_\ell}{\rho_g \mu_g}\right)^{1/2} = J_\ell \cdot R \tag{14.39}$$

Equation 14.38b, when integrated over the entire length of the plate, yields an average Nusselt number expression identical to Eq. 14.30. Koh [38] solved the same problem more completely by including the acceleration and convection terms and showed that for most practical cases, the effects of acceleration and convection can be safely neglected, just as in the natural convection case. A comparison of his solution to the approximate solution of Cess [37] for high Prandtl number ($\text{Pr}_\ell > 1$) is given in Fig. 14.9. The results show that a definite Prandtl

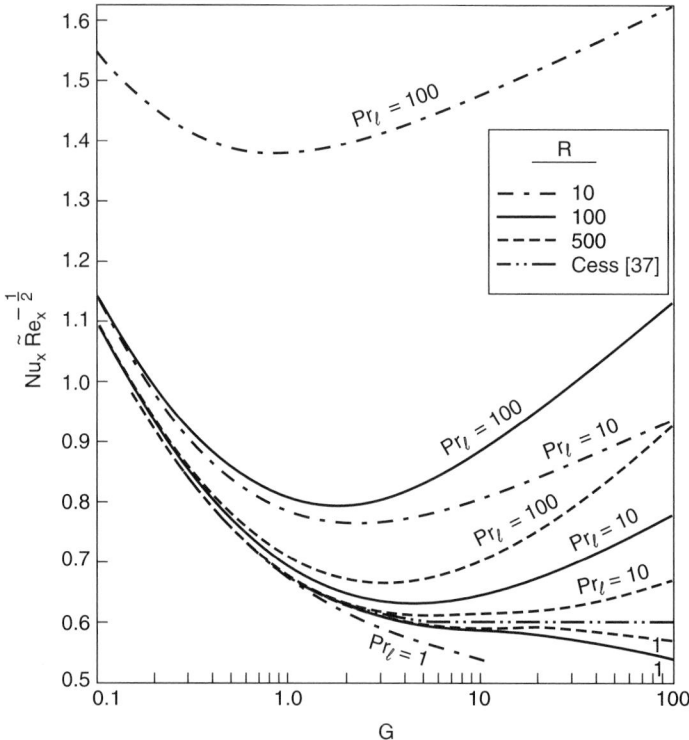

FIGURE 14.9 Influence of acceleration and convection terms during laminar forced convection condensation on a vertical plate [38]. (*Reprinted with permission from Pergamon Press, Tarrytown, New York.*)

number effect occurs. On the other hand, for liquid metals ($Pr_\ell \ll 1$), no such effect is apparent. Solutions for vapor downflow along a vertical flat plate were also obtained by Shekriladze and Gomelauri [39] and Fujii and Uehara [40]. Rose [31] carefully analyzed all the above results and recommends the following expression for the average Nusselt number:

$$Nu_m \, \tilde{Re}_L^{-1/2} = 2K[1 + (\sqrt{2}/3K)^4 F_L]^{1/4} \tag{14.40}$$

where

$$K = 0.436 \left\{ \frac{1.508}{(1 + H/Pr_\ell)^{3/2}} + \frac{1}{G} \right\}^{1/3} \tag{14.41}$$

$$F_L = \frac{\mu_\ell i_{\ell g} g L}{k_\ell \Delta T V_\infty^2} \tag{14.42}$$

Equations 14.40–14.42 reduce to the Nusselt approximate solution (Eq. 14.10) for very low vapor velocities (i.e., $V_\infty \to 0$, $F_L \to \infty$).

Effect of Noncondensable Gas.   As shown in Fig. 14.3, the presence of a noncondensable gas creates an additional thermal resistance to condensation heat transfer due to the required diffusion of the vapor molecules through a gas-rich layer near the surface of the condensate. This additional resistance can reduce the condensation heat transfer rate substantially.

Boundary layer methods have been used to solve this problem. The treatment is more complex than the pure vapor case due to the presence of a gas-vapor boundary layer where the mixture must be treated. In the two-component gas mixture, the local concentrations of the vapor and the noncondensable gas are specified in terms of their mass fractions:

$$W_g = \rho_g/\rho_m \qquad \text{and} \qquad W_G = \rho_G/\rho_m \tag{14.43}$$

where ρ_m is the local density of the mixture, and ρ_g and ρ_G are the local densities of the vapor and the gas, respectively. From these definitions, one can write $W_g + W_G = 1.0$. The solution includes solving the continuity, momentum, and energy equations for the mixture along with the diffusion equation for the noncondensable gas species.

Minkowycz and Sparrow [41] solved this problem under free convection conditions using a similarity transformation. Sparrow et al. [42] conducted a similar analysis under forced convection conditions. Chin et al. [43] modeled both free and forced convection and solved the complete two-phase boundary layer equations using a finite control volume method with an adaptive grid. Figure 14.10 shows their results for a steam-air mixture under both free and forced convection ($V_\infty = 3.05$ m/s) conditions. The serious deterioration in heat transfer under quiescent conditions (using the Nusselt, pure vapor case) is evident in Fig. 14.10(a). A very small concentration of air of 0.1 percent (i.e., $W_{G\infty} = 0.001$) decreases the heat transfer by about 32 percent. Chin et al. [43] show that this effect is more pronounced at smaller values of T_∞ (i.e., at lower operating pressures). The importance of having a vapor sweeping effect is shown in Fig. 14.10(b). In this case the vapor shear created by the steam free stream velocity $V_\infty = 3.05$ m/s causes two effects. It thins the condensate film near the top of the plate, enhancing the heat transfer over the Nusselt case. Secondly, it sweeps the air downstream, reducing the local effect of the noncondensable gas. Excellent agreement with the numerical results of Denny et al. [44] is also shown in Fig. 14.10(b).

For forced convection condensation inside vertical tubes, the bulk concentration of the noncondensable gas outside the boundary layer will increase along the axis of the tube, complicating the above analysis. Wang and Tu [45] provide an approximate solution for this case, assuming that the vapor-gas mixture flows turbulently in the core of the tube. Their results were compared to the experimental data of Borishansky et al. [46] for a steam-nitrogen mixture, and the agreement was good. Forced convection condensation of steam in a vertical tube in the presence of noncondensable gases was recently studied by Hasanein et al. [215].

Rose [47, 48] has suggested an approximate solution for forced convection condensation along a flat plate in the presence of noncondensable gas using an analogy between heat and mass transfer. He points out that, in this situation, the diffusion problem for the vapor-gas

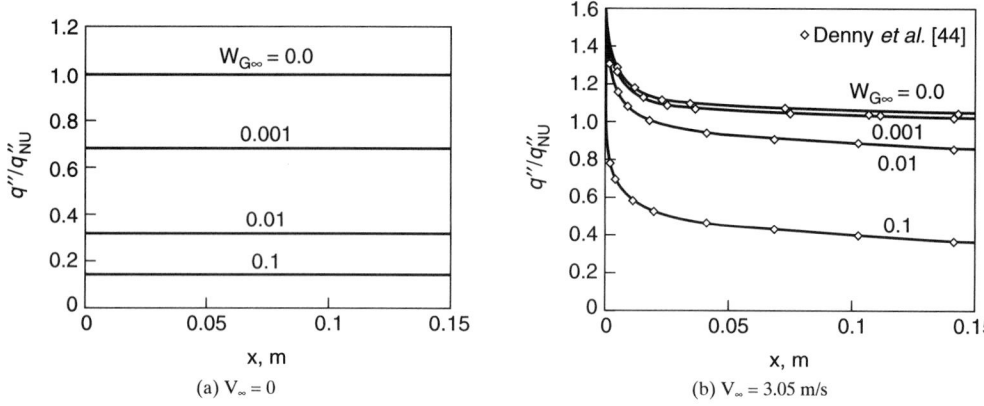

FIGURE 14.10 The influence of noncondensable gas on local heat transfer along a vertical plate [43]. (*Printed with permission from Institution of Chemical Engineers, Rugby, UK.*)

boundary layer is identical to the case of forced convection heat transfer over an isothermal porous plate with surface suction, where the suction velocity v_i has a dependence on local position given by $|v_i| \propto x^{-1/2}$. (Note that $v_i < 0$.) The solution to the heat transfer problem results in the expression:

$$\text{Nu}_x \, \text{Re}_x^{-1/2} = \zeta(\text{Pr})\{1 + 0.941\beta_x^{1.14} \, \text{Pr}^{0.93}\}^{-1} + \beta_x \, \text{Pr} \qquad (14.44)$$

where
$$\beta_x = -(v_i/V_\infty) \, \text{Re}_x^{1/2} \qquad (14.45)$$

$$\text{Re}_x = \frac{\rho V_\infty x}{\mu} \qquad (14.46)$$

and
$$\zeta(\text{Pr}) = \text{Pr}^{1/2} \, (27.8 + 75.9\text{Pr}^{0.306} + 657\text{Pr})^{-1/6} \qquad (14.47)$$

In the above relationships, all the fluid properties are evaluated for the mixture.

Rose [47, 48] applies the analogy by replacing the local Nusselt number in Eq. 14.44 with the local Sherwood number:

$$\text{Sh}_x = \frac{h_D x}{D} \qquad (14.48)$$

where h_D is the local mass transfer coefficient and D is the diffusion coefficient of the mixture (some values of D may be found in Incropera and DeWitt [49]). One must also replace the Prandtl number with the Schmidt number:

$$\text{Sc} = \frac{\nu}{D} \qquad (14.49)$$

The resulting expression relates the local Sherwood number to the local Reynolds number, the Schmidt number, and the parameter β_x. A second equation relating these quantities results from the condition that the interface is impermeable to the noncondensable gas

$$\text{Sh}_x \, \text{Re}_x^{-1/2} = \beta_x \, \text{Sc}/(1 - \omega) \qquad (14.50)$$

where
$$\omega = W_{G\infty}/W_{Gi} \qquad (14.51)$$

Solving the analogous mass-transfer form of Eq. 14.47 (i.e., by replacing Pr by Sc), together with Eq. 14.50, yields an explicit relationship for the interface mass fraction in terms of the interface velocity v_i (through β_x):

$$\omega = \{1 + \beta_x \, \text{Sc} \, (1 + 0.941\beta_x^{1.14} \, \text{Sc}^{0.93})/\zeta\}^{-1} \tag{14.52}$$

When Eq. 14.52 is compared to the numerical results from Sparrow et al. [42] for $\text{Sc} = 0.55$, the agreement is very good. Assuming that the heat transfer at the interface is due entirely to condensation of vapor, an energy balance at the interface yields

$$v_i = -\frac{h_x(T_\infty - T_w)}{\rho_m i_{\ell g}(1 - W_{Gi})} \tag{14.53}$$

Equations 14.52 and 14.53 may be used to solve for the heat transfer coefficient iteratively. The details are provided by Carey [50].

FILM CONDENSATION ON HORIZONTAL SMOOTH TUBES

Single Tube

Laminar Free Convection. Laminar film condensation of a quiescent vapor on an isothermal, smooth horizontal tube, as depicted in Fig. 14.11, may be treated approximately by a Nusselt-type analysis, yielding the following average heat transfer coefficient:

$$\text{Nu}_m = \frac{h_m d_o}{k_\ell} = 0.728\left\{\frac{\rho_\ell(\rho_\ell - \rho_g)g i_{\ell g} d_o^3}{\mu_\ell(T_s - T_{wo})k_\ell}\right\}^{1/4} \tag{14.54}$$

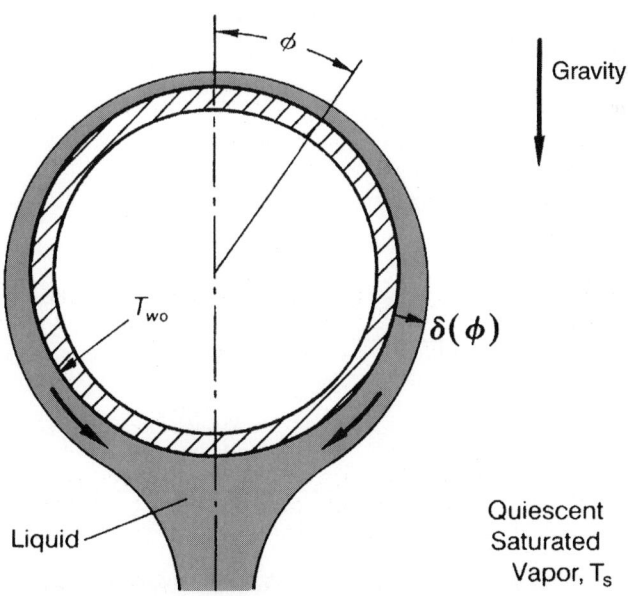

FIGURE 14.11   Laminar film condensation on a horizontal tube.

A convenient alternative in terms of the film Reynolds number Re (equal to $4\Gamma/\mu_\ell$) is

$$\frac{h_m}{k_\ell}\left(\frac{\mu_\ell^2}{\rho_\ell(\rho_\ell - \rho_g)g}\right)^{1/3} = 1.51 \mathrm{Re}^{-1/3} \tag{14.55}$$

where Γ is the mass flow rate of condensate falling off the tube per unit of tube length. Because of a variety of simplifying assumptions, Eq. 14.54 underpredicts measured heat transfer coefficients of ordinary fluids by about 15–20 percent.

Laminar Forced Convection.   When the vapor surrounding a horizontal tube is moving, two important effects influence the heat transfer process: (1) the surface shear stress between the vapor and the condensate film influences the condensate film thickness (therefore, the local vapor flow field must be known), and (2) the effect of vapor separation disturbs the condensate flow downstream of the separation point. Although the vapor flow direction may be oriented in a variety of ways with respect to gravity, most analyses are for vertical downflow, or they neglect the influence of gravity entirely. Rose [31] reviews various analyses of this problem in detail and describes a variety of refinements. Rose [51] points out that, due to complications caused by vapor boundary layer separation, it is questionable whether such refinements are necessary. Rose [51] arrives at the following approximate expression for the average heat transfer coefficient:

$$\mathrm{Nu}_m \, \tilde{\mathrm{Re}}_d^{-1/2} = \frac{0.9 + 0.728 F_d^{1/2}}{(1 + 3.44 F_d^{1/2} + F_d)^{1/4}} \tag{14.56}$$

where

$$\mathrm{Nu}_m = h_m d_o / k_\ell \qquad \tilde{\mathrm{Re}}_d = \rho_\ell V_\infty d_o / \mu_\ell \qquad F_d = \frac{\mu_\ell i_{\ell g} g d_o}{k_\ell \Delta T V_\infty^2} \tag{14.57}$$

Equation 14.56 is compared to data from 12 investigations, using four different fluids, in Fig. 14.12. Equation 14.54 from Nusselt theory is also shown for comparison. It is clear that Eq. 14.56 approaches the Nusselt result at low vapor velocities ($F_d \to \infty$) and gives a reasonable value of the average condensation heat transfer coefficient (greater than the Nusselt theory)

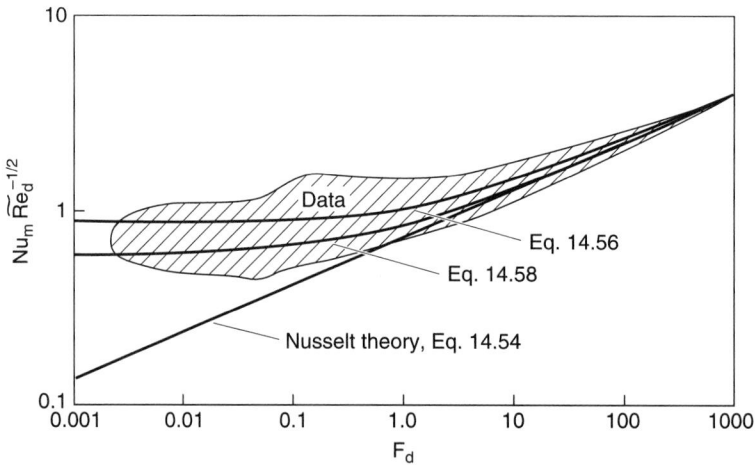

FIGURE 14.12   Comparison of experimental data with several models during condensation on a horizontal tube with vertical vapor downflow. (*Adapted from Ref. 51 and printed with permission from Pergamon Press, Tarrytown, New York.*)

for most practical forced convection situations. Equation 14.58 is a more conservative relationship arrived at by Lee and Rose [52]

$$\mathrm{Nu}_m \, \tilde{\mathrm{Re}}_d^{-1/2} = 0.416[1 + (1 + 9.47F_d)^{1/2}]^{1/2} \tag{14.58}$$

It neglects heat transfer after the separation point and may be used for conservative prediction of forced convection coefficients, as shown in Fig. 14.12.

Tube Bundles

During shell-side condensation in tube bundles, neighboring tubes disturb the vapor flow field and create condensate that flows from one tube to another under the action of gravity and/or vapor shear stress forces. The effects of local vapor velocity and condensate inundation must, therefore, be properly accounted for when calculating the average heat transfer in the bundle. Marto and Nunn [53], Marto [54], and Fujii [55] provide details of these phenomena.

Effect of Vapor Shear. There is a small amount of data for the influence of vapor shear in tube bundles. Nobbs and Mayhew [56] measured steam condensing coefficients during downflow in both staggered and in-line bundles. Kutateladze et al. [57] measured R-21 condensing coefficients during downflow in a staggered bundle. Fujii et al. [58] obtained data for steam flowing downward, upward, and horizontally through both staggered and in-line bundles. Cavallini et al. [59] obtained data for R-11 flowing downward in a staggered bundle. Fujii et al. [58] found little difference between the downward and horizontal data they obtained, but the upward data were as much as 50 percent lower in the range of $0.1 < F_d < 0.5$. They found that the following empirical expression correlated the downward and horizontal data reasonably well:

$$\mathrm{Nu}_m \, \tilde{\mathrm{Re}}_d^{-1/2} = 0.96F_d^{1/5} \tag{14.59}$$

for $0.03 < F_d < 600$.

In a tube bundle, the local vapor velocity must be used to calculate vapor shear effects. However, it is not clear which local value should be used. Butterworth [60] points out that the use of the maximum cross-sectional area would give a conservative prediction, and this velocity is used in Eq. 14.59. Nobbs and Mayhew [56] calculate the velocity based upon a mean flow width given by

$$w = \frac{p_\ell p_t - \pi d_o^2/4}{p_\ell} \tag{14.60}$$

where p_ℓ and p_t are the tube pitches (i.e., centerline to centerline distance) in the longitudinal and transverse directions, respectively, and d_o is the tube diameter.

Effect of Condensate Inundation. In a condenser with quiescent vapor, there is no vapor shear, and condensate flows by gravity onto lower tubes in a bundle. This extra condensate falling on the lower tubes increases the average condensate film thickness around these tubes, and the condensation heat transfer coefficient therefore decreases as one goes further down the bundle.

The Nusselt [12] approximate analysis may be extended to include film condensation on a vertical in-line row of horizontal tubes. If one assumes that all the condensate from a given tube drains as a continuous sheet directly onto the top of the tube below it in a smooth laminar film and that the saturation temperature difference $(T_s - T_{wo})$ remains the same for all the tubes, the average coefficient for a vertical row of N tubes is:

$$\mathrm{Nu}_{mN} = \frac{h_{mN} d_o}{k_\ell} = 0.728 \left\{ \frac{\rho_\ell (\rho_\ell - \rho_g) g i_{\ell g} d_o^3}{\mu_\ell (T_s - T_{wo}) k_\ell N} \right\}^{1/4} \tag{14.61}$$

When Eq. 14.61 is compared to Eq. 14.54, it is clear that the ratio of the average coefficient for N tubes to the average coefficient for a single tube is

$$\frac{h_{mN}}{h_{m1}} = N^{-1/4} \tag{14.62}$$

In terms of the local heat transfer coefficient in the Nth row of the bundle, the Nusselt theory predicts:

$$\frac{h_N}{h_1} = N^{3/4} - (N-1)^{3/4} \tag{14.63}$$

Chen [61] conducted a boundary layer analysis of this problem and included the momentum gain of the condensate in dropping from tube to tube and the condensation that takes place directly on the subcooled condensate film between tubes. His numerical results for the average coefficient of N tubes can be approximated to within 1 percent by:

$$\frac{h_{mN}}{h_{m1}} \cdot N^{1/4} = [1 + 0.2(N-1)\, \mathrm{Pr}_\ell\, J_\ell] \left\{ \frac{1 + 0.68\mathrm{Pr}_\ell\, J_\ell + 0.02\mathrm{Pr}_\ell\, J_\ell^2}{1 + 0.95 J_\ell - 0.15\mathrm{Pr}_\ell\, J_\ell^2} \right\}^{1/4} \tag{14.64}$$

Armbruster and Mitrovic [62] observed that liquid falls from tube to tube in three patterns: discrete droplets, jets or columns, and sheets, depending on the flow rate (i.e., film Reynolds number) and fluid properties. In addition, depending on the tube arrangement and spacing, the condensate may cause ripples, waves, and turbulence to occur in the film; splashing may occur, as well as nonuniform rivulet runoff of condensate because of tube inclination or local vapor velocity effects. As a result, it is impossible to arrive at an analytical expression to describe these complex bundle phenomena. In general, the effect of inundation may be accounted for using

$$\frac{h_{mN}}{h_{m1}} = N^{-s} \tag{14.65}$$

for the average coefficient, and

$$\frac{h_N}{h_1} = N^{1-s} - (1-N)^{1-s} \tag{14.66}$$

for the local value, where s depends on the tube arrangement and pitch-to-diameter ratio, vapor flow direction and velocity, heat flux, and so on. The Nusselt analysis yields $s = \frac{1}{4}$, whereas Kern [63] recommends a less conservative value of $s = \frac{1}{6}$ due to turbulence effects in the film.

An alternative way to account for inundation is to use the expression given by Fuks [64] for the local coefficient:

$$\frac{h_N}{h_1} = \left(\frac{\Gamma_N}{\gamma_N} \right)^{-0.07} \tag{14.67}$$

where Γ_N = total condensate flow rate leaving the Nth tube per unit of tube length
$\qquad \gamma_N$ = condensate flow rate generated on the Nth tube per unit of tube length.

Grant and Osment [65] fit their steam bundle data by Eq. 14.67, but with the exponent 0.07 changed to 0.223.

One way of preventing inundation of condensate on lower tubes is to incline the tube bundle with respect to the horizontal. As the inclination angle increases, a critical value is reached where the condensate no longer drips off the tube but instead clings to the tube and flows to its base. Shklover and Buevich [66] conducted an experimental investigation of steam condensation in an inclined bundle of tubes and recommend an inclination angle of 5°.

Combined Effects of Vapor Shear and Inundation.   In tube bundles, a strong interaction occurs between vapor shear and condensate inundation, and data for these combined effects are limited [54]. The simplest way of handling both phenomena is to separate them and calculate the local heat transfer coefficient for the Nth row as:

$$h_N = h_1 C_N C_{v_g} \tag{14.68}$$

where h_1 is given by Eq. 14.54, C_N is an inundation correction term for the Nth tube (using Eq. 14.66 with $s = \frac{1}{6}$):

$$C_N = N^{5/6} - (1 - N)^{5/6} \tag{14.69}$$

and C_{v_g} is a vapor shear correction term. Berman and Tumanov [67] recommend:

$$C_{v_g} = 1 + 0.0095 \mathrm{Re}_g^{11.8/\sqrt{\mathrm{Nu}_m}} \tag{14.70}$$

provided $\mathrm{Re}_g^{11.8/\sqrt{\mathrm{Nu}_m}} < 50$. The Nusselt number in Eq. 14.70 is calculated from Eq. 14.54. A preferred alternative to the above method is to use Eq. 14.58 for forced convection on a single tube (in place of $h_1 C_{v_g}$) and correct it for condensate inundation using C_N from Eq. 14.69. Then, the local heat transfer coefficient for the Nth row becomes:

$$h_N = 0.416 \frac{k_\ell}{d_o} [1 + (1 + 9.47 F_d)^{1/2}]^{1/2} \, \tilde{\mathrm{Re}}_d^{1/2} [N^{5/6} - (1 - N)^{5/6}] \tag{14.71}$$

McNaught [68] coupled the two phenomena and treated shell-side condensation as two-phase forced convection. He proposed the following relationship for the local coefficient for the Nth row:

$$h_N = (h_{\mathrm{sh}}^2 + h_{\mathrm{gr}}^2)^{1/2} \tag{14.72}$$

where h_{gr} is given by h_1 (from Eq. 14.54) multiplied by C_N (from Eq. 14.69), and, for shear-controlled condensation, h_{sh} is given by:

$$h_{\mathrm{sh}} = a \left[\frac{1}{X_{tt}} \right]^b h_\ell \tag{14.73}$$

In Eq. 14.73, X_{tt} is the Lockhart-Martinelli parameter, defined as

$$X_{tt} = \left(\frac{1 - x}{x} \right)^{0.9} \left(\frac{\rho_g}{\rho_\ell} \right)^{0.5} \left(\frac{\mu_\ell}{\mu_g} \right)^{0.1} \tag{14.74}$$

and h_ℓ is the single-phase forced convection heat transfer coefficient across a bank of tubes, assuming that the liquid phase occupies the total flow area. This is expressed as:

$$h_\ell = C \frac{k_\ell}{d_o} \mathrm{Re}_\ell^m \mathrm{Pr}_\ell^n \tag{14.75}$$

where C, m, and n depend upon the flow conditions through the tube bank. McNaught [68] used the steam data of Nobbs and Mayhew [56] and correlated 90 percent of their data to within ± 25 percent by setting $a = 1.26$ and $b = 0.78$ in Eq. 14.73. Chu and McNaught [69] successfully correlated R-113 data in a tube bundle using Eq. 14.72.

Pressure Loss Considerations.   Vapor pressure losses in flowing through a condenser bundle produce a corresponding decrease in the vapor saturation temperature. If these losses are large due to high vapor velocities, a sizeable loss in temperature-driving potential between the vapor and the coolant can result. Despite the importance of knowing the shell-side pressure loss in condenser design, little information has been published.

For simplicity, shell-side losses may be calculated by using single-phase (i.e., dry-tube) correlations of the form

$$\Delta P = 4 f_m N_t \rho_g \frac{V_m^2}{2} \qquad (14.76)$$

where N_t = number of tube rows
V_m = average maximum vapor velocity in the bundle (i.e., based on minimum flow area)
f_m = friction factor, which depends on V_m

Single phase friction factors in tube bundles may be expressed as:

$$f = \frac{a}{Re^n} \qquad (14.77)$$

where n is near 0.25, and the coefficient a depends on tube bundle geometry. Davidson and Rowe [70] have shown that the above technique successfully predicts pressure losses in a tube bundle. Eissenberg [71] measured friction factor data for air and water in a staggered tube bundle, as shown in Fig. 14.13. Clearly, the dry friction factor is influenced by the amount of water present and the flow direction. More comprehensive methods to calculate pressure losses have been proposed by Grant and Chisholm [72, 73] and Ishihara et al. [74]. These are discussed in more detail by Marto [54].

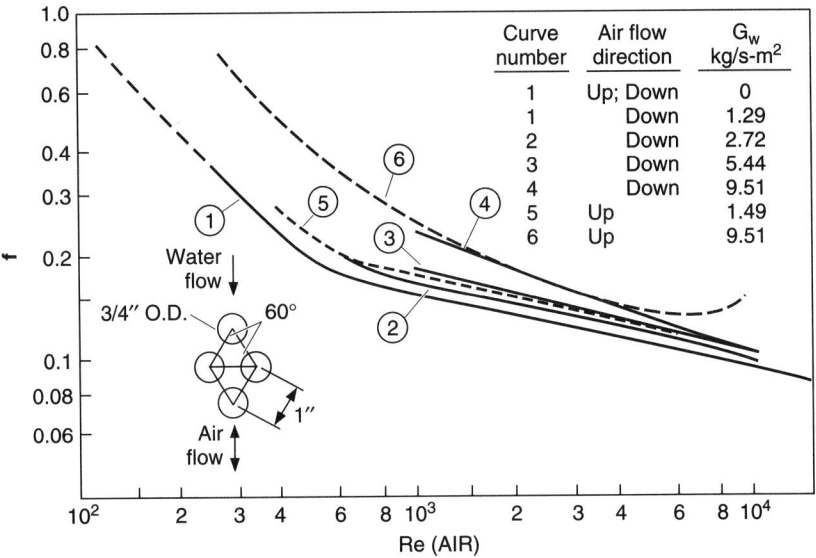

FIGURE 14.13   Effect of two-phase flow on friction factor in a tube bundle [71].

Effect of Noncondensable Gas.   As described earlier, Rose [47, 48] has suggested an approximate method to calculate condensation heat transfer in the presence of a noncondensable gas. For forced convection over a single horizontal tube, he recommends the following relationship (similar to Eq. 14.52 for a flat plate) that relates the mean condensation rate to the free stream conditions:

$$\omega = W_{G\infty}/W_{Gi} = \{1 + 1.75\beta \, Sc^{2/3} \, (1 + \beta \, Sc)\}^{-1} \qquad (14.78)$$

where
$$\beta = -(v_i/V_\infty)\,\mathrm{Re}^{1/2} \tag{14.79}$$

$$\mathrm{Re} = \frac{\rho V_\infty d_o}{\mu} \tag{14.80}$$

and all fluid properties are evaluated for the mixture.

Rose [47, 48] points out that when Eq. 14.78 is used in conjunction with an equation relating the heat transfer rate (i.e., condensation rate) to the temperature difference across the condensate film (an appropriate expression for a single tube might be Eq. 14.56), together with the interface equilibrium condition:

$$W_{Gi} = \frac{P - P_s(T_i)}{P - [1 - M_g/M_G]P_s(T_i)} \tag{14.81}$$

these equations may be solved simultaneously using an iterative technique (while taking $q'' = \dot{m}''i_{\ell g}$), similar to the method described by Carey [50], to arrive at the desired heat transfer rate. Rose [47] compared the predictions from the above method to steam-air measurements of Berman [75] and Mills et al. [76] and found good agreement.

The method of Rose [47, 48] has not been used in practice, however. Instead, for tube bundles, the procedure most widely used is due to Colburn and Hougen [77]. They proposed a point-by-point, trial-and-error method that requires equating the heat transferred locally through the condensate, tube wall, and cooling water film to the sum of the sensible heat transferred by cooling the noncondensable gas and the latent heat deposited on the condensate film due to the amount of vapor transferred by diffusion. An important part of this method is the requirement of knowing the mass transfer coefficient for the diffusion of the vapor through the vapor-gas mixture. This term is evaluated by using the heat and mass transfer analogy together with empirical data for forced-convection gas-side heat transfer.

Berman and Fuks [78] obtained an empirical expression for the mass transfer coefficient in a tube bundle during downward flow of a steam-gas mixture. Their expression can be used to generate an equivalent, noncondensable gas heat transfer coefficient (i.e., giving an added thermal resistance $R_{nc} = 1/h_{nc}$):

$$h_{nc} = \frac{aD}{d_o}\,\mathrm{Re}_g^{1/2}\left(\frac{P}{P - P_g}\right)^b P^{1/3}\left(\frac{\rho_g i_{\ell g}}{T_g}\right)^{2/3}\frac{1}{(T_g - T_s)^{1/3}} \tag{14.82}$$

where, for $\mathrm{Re}_g > 350$, $b = 0.6$ $a = 0.52$ First tube row

$a = 0.67$ Second tube row

$a = 0.82$ Third and later tube rows

and, for $\mathrm{Re}_g < 350$, $b = 0.7$ $a = 0.52$ All tube rows

Fujii [55] recommends the empirical relationship of Fujii and Oda [79] to calculate the ratio of the heat transfer coefficient with air to the coefficient for pure steam in small tube bundles:

$$\frac{h_{\text{air}}}{h_m} = Ae^{-mW_{G\infty}} + Be^{-nW_{G\infty}} \tag{14.83}$$

where
$$A = 0.83(T_s - T_{wo})^{-0.15} + 0.1\ell n(V_\infty) \tag{14.84a}$$

$$B = 0.21(T_s - T_{wo})^{0.25} - 0.09\ell n(V_\infty) \tag{14.84b}$$

$$m = 3.7V_\infty^{-0.12} \tag{14.84c}$$

$$n = 19(T_s - T_{wo})^{0.2}V_\infty^{0.3} \tag{14.84d}$$

and $0 < W_{G\infty} < 0.3,\, 2 < V_\infty < 20$ m/s, $(T_s - T_{wo}) < 15$ K, and $20 < T_s < 40°C$

In general, since the concentration of noncondensable gas increases as a mixture flows through a bundle, the importance of noncondensables increases along the steam flow path.

Computer Modeling. The design of large condensers is a complex problem [53, 54]. The coolant temperature rise from inlet to outlet and vapor pressure losses (causing losses of saturation temperature) in flowing through the bundle affect the local temperature driving force between the vapor and the coolant. The condensation process is influenced by both vapor shear and condensate inundation, which change throughout the bundle. In many applications, too, the vapor distribution into the condenser is nonuniform, and, in some cases, the coolant is not uniformly distributed throughout the tubes due to pressure variations caused by condenser coolant inlet and outlet designs. Noncondensable gases can be a significant problem, especially in large steam power plant condensers that operate at subatmospheric pressure. The noncondensables will tend to collect in the coolest regions of the condenser. Provisions must be made to remove the gases to prevent unusually large concentrations, and gas removal locations must be accurately predicted. Another major problem in condensers is due to fouling deposits on the tubes (predominantly on the coolant side), which can also vary with location.

As a consequence, computer modeling must be used to predict accurately the average shell-side heat transfer coefficient in a tube bundle. The vapor flow must be followed throughout the bundle and within the vapor flow lanes. Knowing the local vapor velocity, pressure and temperature, as well as the distribution of condensate from other tubes (and the local concentration of any noncondensable gases), it is possible to predict a local heat transfer coefficient in the bundle that can then be integrated to arrive at the overall bundle performance. Initial modeling efforts were with simple one-dimensional routines [80, 81, 82, 83]; these were followed by more sophisticated methods [84, 85]. In recent years, Yau and Pouzenc [86] developed a three-dimensional model, and Zhang [87] developed a quasi-three-dimensional model, both of which have been successfully validated against large condenser measurements. The latest modeling efforts are described in Ormiston et al. [210, 211] and Zhang et al. [212, 213].

FILM CONDENSATION ON HORIZONTAL FINNED TUBES

Single Tube

Film condensation on horizontal finned tubes has received considerable attention in recent years [1, 88, 89]. The geometry of a low integral-fin tube with trapezoid-shaped fins is shown in Fig. 14.14(a). During condensation, the fins not only increase the surface area, but the heat transfer coefficient along the fin flanks is increased over the smooth tube value due to a short condensate flow length from fin tip to fin root, as well as to the occurrence of surface tension forces that effectively thin the condensate film. Unfortunately, surface tension also causes condensate to bridge across the space between fins, leading to flooding at the bottom of a tube, Fig. 14.14(b). The influence of surface tension forces on condensation heat transfer is described in detail by Masuda and Rose [90]. Honda et al. [91] arrived at an approximate expression for the flooding angle ϕ_f on a horizontal finned tube with trapezoid-shaped fins for $e > b/2$:

$$\phi_f = \cos^{-1}\left[\frac{4\sigma_\ell \cos\theta}{\rho_\ell g b d_o} - 1\right] \tag{14.85}$$

Equation 14.85 is plotted in Fig. 14.14(b) for a tube with an outside diameter of 21.05 mm and rectangle-shaped fins (i.e., $\theta = 0$). Results are shown for R-113, ethylene glycol, and water. For a given liquid, as fin spacing b decreases, more flooding occurs (i.e., ϕ_f decreases), and, at a critical fin spacing b_c, it is possible for the entire tube to be flooded (i.e., $\phi_f = 0$).

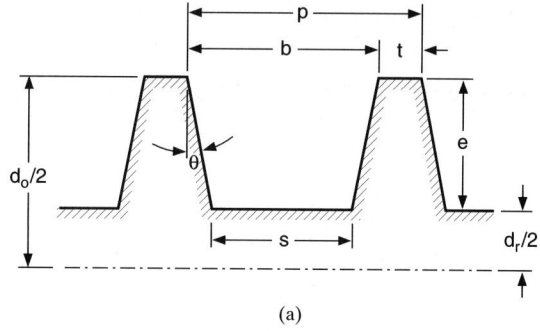

(a)

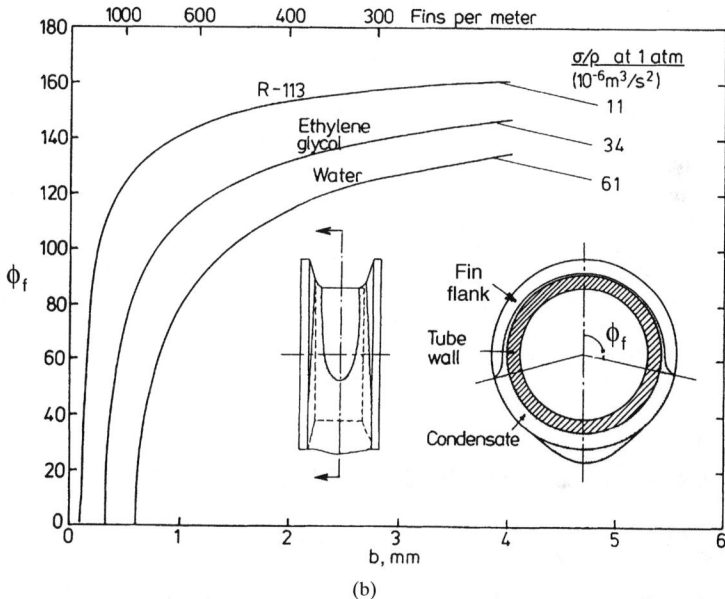

(b)

FIGURE 14.14   Film condensation on a horizontal finned tube. (*a*) Geometric variables. (*b*) Flooding angle variations. (*Adapted from Ref. 88 and printed with permission from Trans. ASME,* Journal of Heat Transfer.)

Beatty and Katz [92] completely neglected surface tension effects and arrived at a very simple expression for a gravity-drainage heat transfer coefficient using Nusselt theory:

$$h_m = 0.689 \left[\frac{\rho_\ell^2 k_\ell^3 g i_{\ell g}}{\mu_\ell \Delta T d_{eq}} \right]^{1/4} \tag{14.86}$$

where

$$\left(\frac{1}{d_{eq}} \right)^{1/4} = 1.30 \eta_f \frac{A_f}{A_{ef}} \frac{1}{\overline{L}^{1/4}} + \frac{A_u}{A_{ef}} \frac{1}{d_r^{1/4}} \tag{14.87a}$$

$$\overline{L} = \pi (d_o^2 - d_r^2)/4d_o \tag{14.87b}$$

$$A_{ef} = \eta_f A_f + A_u \tag{14.87c}$$

and A_f is the surface area of the fin flanks, A_u is the area of the interfin tube surface, and η_f is the fin efficiency. Equation 14.86, which is based on the effective surface area A_{ef}, gives acceptable results for low surface tension fluids and for low fin density tubes. It overpredicts high surface tension fluids and high fin density tubes where condensate flooding from surface tension effects is important.

Various recent fin tube models include surface tension effects [89, 93, 94, 95, 96, 97]. The best model for design purposes, because of its relative simplicity, is due to Rose [95]. His expression for the heat transfer enhancement ratio $\varepsilon_{\Delta T}$ (defined as the average heat transfer coefficient for the finned tube divided by the average value for the smooth tube, both based on the smooth tube surface area at fin root diameter and for the same film temperature difference $(T_s - T_{wo})$) for trapezoidal-shaped fins is [89]:

$$\varepsilon_{\Delta T} = \left(\frac{h_f}{h_s}\right)_{A_r, \Delta T} = \left\{\left\{\left(\frac{d_o}{d_r}\right)^{3/4} t\left[0.281 + \frac{B\sigma d_o}{t^3(\rho_\ell - \rho_g)g}\right]^{1/4}\right.\right.$$

$$+ \frac{\phi_f}{\pi}\left\{\frac{(1-f_f)}{\cos\theta}\frac{(d_o^2 - d_r^2)}{2e_v^{1/4}d_r^{3/4}}\left[0.791 + \frac{B\sigma e_v}{e^3(\rho_\ell - \rho_g)g}\right]^{1/4}\right.$$

$$\left.\left.+ B_1(1-f_s)s\left[(\xi(\phi_f))^3 + \frac{B\sigma d_r}{s^3(\rho_\ell - \rho_g)g}\right]^{1/4}\right\}\right\}\Big/0.728(b+t) \quad (14.88)$$

where

$$\xi(\phi_f) = 0.874 + 0.1991 \times 10^{-2}\phi_f - 0.2642 \times 10^{-1}\phi_f^2 + 0.5530 \times 10^{-2}\phi_f^3 - 0.1363 \times 10^{-2}\phi_f^4 \quad (14.89)$$

$$f_f = \frac{1 - \tan(\theta/2)}{1 + \tan(\theta/2)} \cdot \frac{2\sigma\cos\theta}{\rho_\ell g d_r e} \cdot \frac{\tan(\phi_f/2)}{\phi_f} \quad (14.90a)$$

$$f_s = \frac{1 - \tan(\theta/2)}{1 + \tan(\theta/2)} \cdot \frac{4\sigma}{\rho_\ell g d_r s} \cdot \frac{\tan(\phi_f/2)}{\phi_f} \quad (14.90b)$$

$$e_v = \begin{cases} \dfrac{\phi_f}{\sin\phi_f} \cdot e, & \phi_f \leq \dfrac{\pi}{2} \quad (14.91a) \\[2ex] \dfrac{\phi_f}{2 - \sin\phi_f} \cdot e, & \dfrac{\pi}{2} \leq \phi_f \leq \pi \quad (14.91b) \end{cases}$$

When Eqs. 14.88, 14.89, 14.90, and 14.91 were compared to existing data for finned copper tubes, good agreement was found for $B = 0.143$ and $B_1 = 2.96$. Briggs and Rose [96] extended this model to low conductivity materials by including conduction effects in the fins.

The effect of vapor velocity on finned tube condensation is less than that on a smooth tube [98–100]. Cavallini et al. [101] proposed the following relationship for the average heat transfer coefficient on a finned tube during forced convection conditions:

$$h_m = (h_{st}^2 + h_{fc}^2)^{1/2} \quad (14.92)$$

where h_{st} = heat transfer coefficient under stationary conditions, Eqs. 14.88–14.91 as modified by Briggs and Rose [96]
       h_{fc} = heat transfer coefficient under forced convection conditions

$$= C\frac{k_\ell}{d_o}\text{Re}_{eq}^{0.8}\text{Pr}_\ell^{1/3} \quad (14.93)$$

where $$\text{Re}_{eq} \approx \tilde{\text{Re}}\,(\rho_\ell/\rho_g)^{1/2} \quad (14.94)$$

and $$C = 0.03 + 0.17\left(\frac{t}{b+t}\right) + 0.07\left(\frac{e}{b+t}\right) \quad (14.95)$$

This model correlated the data of Honda [102, 103] for R-113 very well.

Tube Bundles. The average heat transfer coefficient in a bundle of finned tubes is influenced by both vapor shear and condensate inundation, although the effects are not as large as for smooth tubes [88, 102–107]. At low vapor velocities, Webb and Murawski [107] express the local coefficient for the Nth row in terms of the local film Reynolds number:

$$h_N = a \, \text{Re}^{-n} \tag{14.96}$$

Equation 14.96 may be integrated to obtain a bundle-averaged heat transfer coefficient [107]:

$$h_{m,N} = \left[\frac{a}{1-n} (\text{Re}_N - \text{Re}_1) \right] \left[\text{Re}_N^{-n} - \text{Re}_1^{-n} \right] \tag{14.97}$$

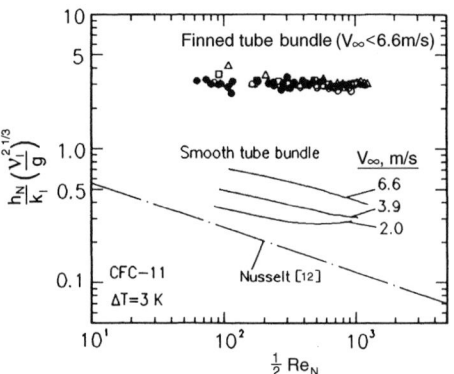

where Re_N and Re_1 are the film Reynolds numbers leaving the Nth and first tube rows, respectively. For smooth tubes, the Nusselt theory predicts that $n = \frac{1}{3}$ (see Eq. 14.55). For finned tubes, $n < 0.1$, presumably since the fins prevent the condensate that drips on a tube from spreading axially [88]. Figure 14.15 compares the finned tube data of Honda et al. [106] and Webb and Murawski [107] to smooth tube predictions, clearly showing the reduced effect of both vapor velocity and inundation for the finned tube bundle. Chu and McNaught [108] measured local heat transfer coefficients in bundles of finned tubes having various fin densities and found that the Rose single tube model [95] has promise in being extended for inundation and vapor velocity effects.

FIGURE 14.15 Variation of Nusselt number with film Reynolds number. (*Adapted from Ref. 106 and printed with permission from United Engineering Trustees, Inc., New York.*)

OTHER BODY SHAPES

The approximate analysis of Nusselt has been applied to a wide variety of geometries. Common assumptions include uniform wall temperature, saturated and quiescent vapor, no interfacial resistance at the liquid-vapor interface, no momentum and convection effects in the condensate, and no variation of properties with temperature. The resulting equations are valid for ordinary liquids (i.e., nonliquid metals) provided $(c_{p\ell} \Delta T / i_{\ell g})$ is less than about 0.2 to 0.5.

Inclined Circular Tubes

Hassan and Jakob [109] analytically studied film condensation on single inclined tubes using Nusselt theory and found that the average heat transfer coefficient for an infinitely long tube was proportional to $(\cos \alpha)^{1/4}$ where α is the inclination angle with respect to the horizontal, Fig. 14.16. Therefore, Eq. 14.54 can be used with g replaced by $(g \cos \alpha)$. Selin [110] confirmed this result with experimental data. Hassan and Jakob [109] also found that, for a finite-length tube of a given length to radius ratio L/r, the mean heat transfer coefficient goes through a maximum at an optimum inclination angle (see Fig. 14.16). For an infinitely long tube, the optimum angle is 0° (i.e., a horizontal tube is best), and, for finite-length tubes, the optimum is greater than zero. For example, for $L/r = 4$, the optimum angle is 60°.

Inclined Upward-Facing Plates

The Nusselt analysis for a vertical plate may be adapted to an inclined upward-facing plate (inclined at an angle α with respect to the horizontal) by replacing g in Eq. 14.10 with $(g \sin \alpha)$. This simple substitution is valid except for plates that are near horizontal (i.e., $\alpha <$

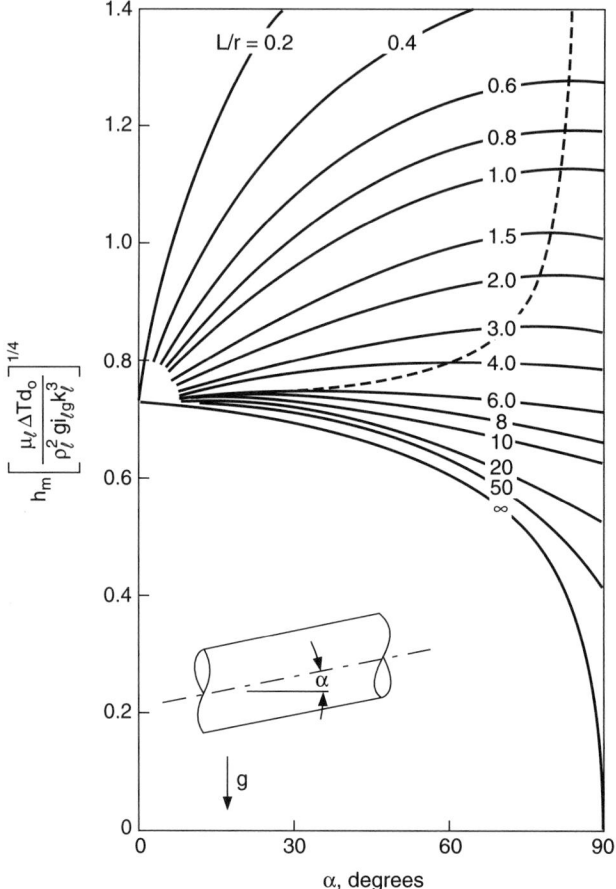

FIGURE 14.16   Effect of inclination angle on the average heat transfer coefficient for circular tubes [109]. (*Reprinted with permission from Trans. ASME, Journal of Heat Transfer.*)

1–2°). Equation 14.10, after modification with the (sin α) term, predicts zero heat transfer when $\alpha = 0$. In reality, there is finite heat transfer even when the plate is horizontal, as discussed below.

Horizontal Upward-Facing Plates and Disks

Nimmo and Leppert [111] used a Nusselt-type analysis to study laminar film condensation on a finite, upward-facing horizontal plate, assuming that the condensate flow was driven by the hydrostatic pressure gradient due to changes in condensate film thickness from the center of the plate to the edges. They arrived at the following approximate expression for the mean Nusselt number:

$$\text{Nu}_m = \frac{h_m L}{k_\ell} \approx 0.82 \left\{ \frac{\rho_\ell^2 g i_{\ell g} L^3}{\mu_\ell \Delta T k_\ell} \right\}^{1/5} \tag{14.98}$$

They found that Eq. (14.98) overpredicted their experimental data for steam by about 20 percent and was influenced by the shape of the edge of the plate (sharp versus rounded), due presumably to surface tension forces that were neglected in their analysis.

Shigechi et al. [112] conducted a boundary layer analysis of this problem and included momentum and convection effects in the condensate film. They obtained different solutions using as a boundary condition various inclination angles of the liquid-vapor interface at the plate edge. Their maximum average Nusselt number was found to agree well with Eq. 14.98. Chiou et al. [214] included surface tension in their model and showed that heat transfer decreases in relation to Eq. 14.98 as the surface tension of the condensate increases.

Chiou and Chang [113] analyzed laminar film condensation on a horizontal upward-facing disk and found that their model overpredicts the theoretical result from Eq. 14.98 by about 25 percent. They attribute this to the fact that on a disk, the surface area increases radially outward, and this makes the condensate film thinner than on a rectangular plate, as studied by Nimmo and Leppert [111].

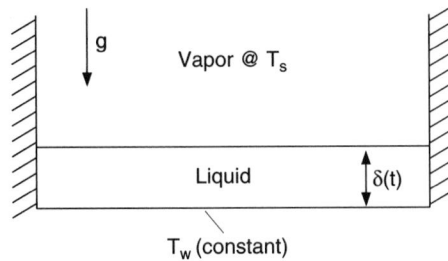

FIGURE 14.17 Condensation on the bottom of a container.

Bottom of a Container

If the condensate is restrained from flowing (i.e., no runoff is allowed), then the condensate film that is collected will vary with time. One such example of this situation is shown in Fig. 14.17, where a container, having insulated walls, has a base temperature that is kept constant at T_w. If the container holds vapor that is supplied at a constant pressure, and therefore temperature $T_s > T_w$, then the transient rate of condensation at the bottom of the container may be approximated by assuming that quasi-steady-state conduction occurs across the film and that an approximately linear temperature distribution exists in the condensate. An energy balance then yields

$$q'' = \frac{k_\ell}{\delta(t)}(T_s - T_w) = \rho_\ell i_{\ell g} \frac{d\delta(t)}{dt} \tag{14.99}$$

Integrating from $\delta = 0$ at $t = 0$ gives the transient film thickness:

$$\delta(t) = \sqrt{\frac{2k_\ell(T_s - T_w)}{\rho_\ell i_{\ell g}} t} \tag{14.100}$$

For this case, the resulting time-dependent heat transfer coefficient is $k_\ell/\delta(t)$. Prasad and Jaluria [114] extended the above simple analysis to the situation where runoff over the plate edges is allowed by conducting a boundary layer analysis of transient film condensation on a horizontal plate.

Horizontal and Inclined Downward-Facing Plates and Disks

Gerstmann and Griffith [115] studied film condensation on the underside of horizontal and inclined surfaces. They predicted the average heat transfer coefficient based upon their observation of the condensate flow as a function of the inclination angle α from the horizontal. On the underside of a perfectly horizontal surface, the condensate is removed as droplets that form due to a Taylor instability created on the condensate surface. At slight inclination angles ($\alpha < 7.5$ degrees), several flow regimes were observed, showing various wave patterns, and, at moderate inclination angles ($\alpha > 20°$), roll waves appeared.

For a horizontal surface, they developed the following expressions:

$$\text{Nu}_m = \begin{cases} 0.69\text{Ra}^{0.20}, & 10^6 < \text{Ra} < 10^8 \\ 0.81\text{Ra}^{0.193}, & 10^8 < \text{Ra} < 10^{10} \end{cases}$$

(14.101a)

(14.101b)

where

$$\text{Nu}_m = \frac{h_m}{k_\ell} \left(\frac{\sigma}{g(\rho_\ell - \rho_g)} \right)^{1/2}$$

(14.102)

$$\text{Ra} = \frac{g\rho_\ell(\rho_\ell - \rho_g)i_{\ell g}}{\mu_\ell \Delta T k_\ell} \left(\frac{\sigma}{g(\rho_\ell - \rho_g)} \right)^{3/2}$$

(14.103)

Equation (101) overpredicted their experimental data for condensation of R-113 by about 10–15 percent.

For slightly inclined surfaces, they arrived at the expression

$$\text{Nu}_m = \frac{0.90\text{Ra}^{1/6}}{(1 + 1.1\text{Ra}^{-1/6})}$$

(14.104)

which agreed with their data to within 10 percent. In Eq. 14.104, Nu_m and Ra are modified from Eqs. 14.102 and 14.103 by replacing g by $(g \cos \alpha)$ in each of the expressions. In both Eqs. 14.101 and 14.104, $k_\ell \Delta T \ll \mu_\ell i_{\ell g}$ and $c_{p\ell} \Delta T \ll i_{\ell g}$. For large inclination angles ($\alpha > 20°$), they modified the Nusselt vertical plate result (Eq. 14.10) by replacing g by $(g \sin \alpha)$ and found that this modified expression overpredicted all their data by about 10 percent.

Yanadori et al. [116] studied film condensation on the underside of small horizontal disks. They found that the heat transfer coefficient reaches an optimum value when the condensing surface diameter is twice as long as the shortest Taylor wavelength $\bar{\lambda}$ of the departing drops, where

$$\bar{\lambda} = 2\pi \sqrt{\frac{\sigma}{g(\rho_\ell - \rho_g)}}$$

(14.105)

General Axisymmetric Bodies

Dhir and Lienhard [117] extended the Nusselt analysis to include a wide variety of body shapes, including axisymmetric bodies (Fig. 14.18). Their result for the local heat transfer coefficient is identical to Eq. 14.8, except g is replaced by g_{ef}, where

$$g_{\text{ef}} = \frac{x(gR)^{4/3}}{\int_0^x g^{1/3} R^{4/3} \, dx}$$

(14.106)

In the above equations, x is the distance along the condensate film measured from the top of the body or from the upper stagnation point, $g(x)$ is the local component of gravity along the flow direction, and $R(x)$ is the local radius of curvature about the vertical axis. Once the local Nusselt number is known, the result may be integrated to get an average expression for the whole body. Some examples follow.

Vertical Cone (of Half-Angle β). In this case, $g(x)$ is constant, equal to $g \cos \beta$ and $R(x) = x \sin \beta$, yielding $g_{\text{ef}} = (7/3)g \cos \beta$, and

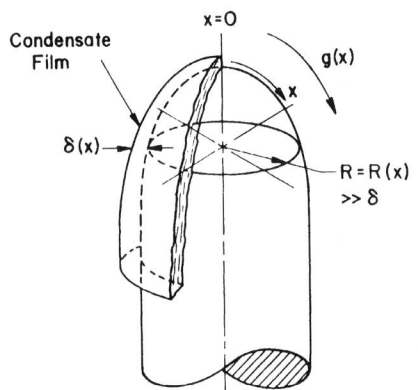

FIGURE 14.18 Condensation on an axisymmetric body [117]. (*Reprinted with permission from Trans. ASME,* Journal of Heat Transfer.)

$$h_x = 0.874 \left\{ \frac{\rho_\ell(\rho_\ell - \rho_g)(g \cos \beta)i_{\ell g}k_\ell^3}{\mu_\ell(T_s - T_w)x} \right\}^{1/4}$$

(14.107)

The average coefficient (for a finite cone of length L) is then:

$$h_m = \frac{1}{\pi \sin \beta L^2} \int_0^L h_x 2\pi x \sin \beta \, dx$$

or
$$h_m = \frac{8}{7} h_L = \left\{ \frac{\rho_\ell(\rho_\ell - \rho_g)(g \cos \beta)i_{\ell g}k_\ell^3}{\mu_\ell(T_s - T_w)L} \right\}^{1/4} \tag{14.108}$$

Sphere (of Diameter d_o).   In this case, $g(x) = g \sin (2x/d_o)$ and $R(x) = (d_o/2) \sin (2x/d_o)$. The result for the average Nusselt number is:

$$\mathrm{Nu}_m = \frac{h_m d_o}{k_\ell} = 0.815 \left\{ \frac{\rho_\ell(\rho_\ell - \rho_g)gi_{\ell g}d_o^3}{\mu_\ell(T_s - T_w)k_\ell} \right\}^{1/4} \tag{14.109}$$

Dhir and Lienhard [118] studied laminar film condensation on two-dimensional isothermal surfaces for which boundary layer similarity solutions exist and found that a similarity solution exists for body shapes that give $g(x) \approx x^n$. Nakayama and Koyama [119] extended the analysis of arbitrarily shaped bodies to include turbulent film condensation.

Horizontal and Inclined Elliptical Cylinders

Karimi [120] and Fieg and Roetzel [121] extended the Nusselt-type analysis of Hassan and Jakob [109] to include condensation on horizontal and inclined elliptical cylinders of various eccentricities. Memory et al. [122] studied both free and forced convection on horizontal elliptical cylinders.

Vertically Oriented Helical Coils

Karimi [120] applied his analysis to solve the reflux condenser geometry shown in Fig. 14.19. In this situation, vapor condenses on the outside of a cooled, helically coiled tube and flows

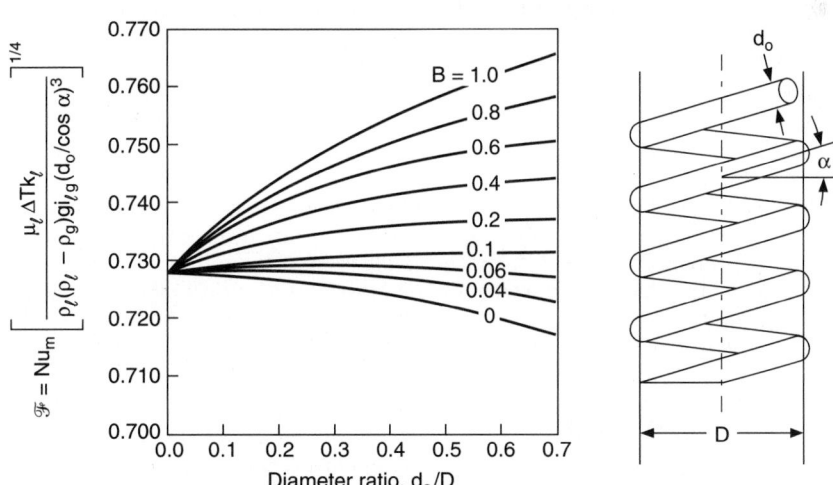

FIGURE 14.19 Average Nusselt number during laminar film condensation on vertical helical coils [120]. (*Reprinted with permission from Pergamon Press, Tarrytown, New York.*)

downward and along the tube axis. In the process, the condensate may be flung off the tube by centrifugal force. Karimi [120] arrived at a numerical solution of this problem

$$\text{Nu}_m = \frac{h_m(d_o/\cos \alpha)}{k_\ell} = \mathcal{F}(d_o/D, \, B) \left[\frac{\rho_\ell(\rho_\ell - \rho_g)gi_{\ell g}(d_o/\cos \alpha)^3}{\mu_\ell(T_s - T_{wo})k_\ell} \right]^{1/4} \tag{14.110}$$

where B, a dimensionless centrifugal parameter, is given by:

$$B = \left(\frac{\rho_\ell - \rho_g}{\rho_\ell} \right) \left(\frac{c_{p\ell}\Delta T}{i_{\ell g}} \right) \left(\frac{\tan^2 \alpha}{\text{Pr}_\ell} \right) \tag{14.111}$$

The function $\mathcal{F}$ is plotted in Fig. 14.19 for various values of (d_o/D) and B.

CONDENSATION WITH ROTATION

When a condensing surface is subjected to a centrifugal force field due to rotation, the resulting body force can significantly influence condensate motion. If the rotational speed is large enough, the local centrifugal acceleration $\omega^2 R$ (where R is the local radius from the axis of rotation) can be much larger than the earth's gravitational acceleration g, making the motion of the condensate independent of the orientation with respect to gravity.

Sparrow and Gregg [123] analyzed laminar film condensation on an isothermal rotating disk situated in a quiescent vapor. Neglecting gravity, they formulated the problem using the complete Navier-Stokes and energy equations and found a similarity solution that yielded the following expression for the local heat transfer coefficient (shown to be constant on the disk), for $\text{Pr}_\ell \geq 1$ and $(c_{p\ell}\Delta T/i_{\ell g}) < 0.1$:

$$\frac{h(\nu_\ell/\omega)^{1/2}}{k_\ell} = 0.904 \left\{ \frac{\text{Pr}_\ell}{c_{p\ell}\Delta T/i_{\ell g}} \right\}^{1/4} \tag{14.112}$$

Equation 14.112 overpredicts the experimental data of Nandapurkar and Beatty [124] for ethanol, methanol, and R-113 by about 25 percent. Sparrow and Gregg [125] subsequently included the effect of induced vapor drag in their analysis and found that this effect was negligible. Sparrow and Hartnett [126] conducted a similar analysis for condensation on the outside of a rotating cone of half-cone angle β and found that

$$\frac{h_{\text{cone}}}{h_{\text{disk}}} = (\sin \beta)^{1/2} \tag{14.113}$$

Condensation inside various shaped tubes rotating about their own axes is described in Refs. 127–129.

Mochizuki and Shiratori [130] investigated condensation of steam on the inside of a vertical tube rotating about an axis parallel to its own. Additional details of this situation are described by Vasiliev and Khrolenok [129].

Condensation of steam on the outside of a vertical pipe rotating around its own axis was experimentally studied by Nicol and Gacesa [131]. At high rotational speeds, where gravity can be neglected, they found that

$$\text{Nu}_m = \frac{h_m d_o}{k_\ell} = 12.26 \text{We}^{0.496} \tag{14.114}$$

where $\text{We} = \rho_\ell \omega^2 d_o^3/4\sigma > 500$. Suryanarayana [132] condensed steam on the outside of a tube rotating about an axis parallel to its own and found that the average heat transfer coefficient could be well correlated by modifying Eq. 14.54 (for laminar free convection on a stationary horizontal tube) in two ways: (1) replacing g by $(\omega^2 R_o)$ where R_o is the radial distance from

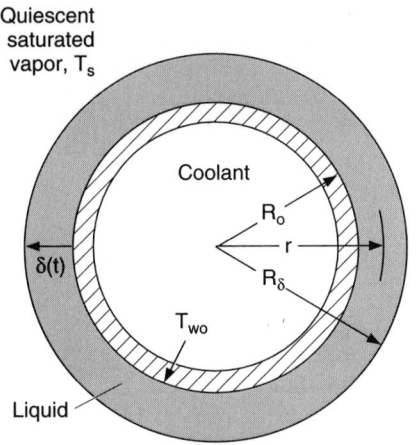

Quiescent
saturated
vapor, T_s

Coolant

R_o

r

R_δ

$\delta(t)$

T_{wo}

Liquid

FIGURE 14.20 Condensation on a tube in zero gravity.

the tube axis to the axis of rotation, and (2) multiplying by a correction term due to the formation of condensate waves, $0.69 Re^{0.11}$ (Eq. 14.14). The result was found to be valid for $1.0 < (\omega^2 R_o/g) < 80$.

ZERO GRAVITY

In the absence of gravity (and other forces acting on the condensate), the condensate formed on a surface will not flow off the surface in a steady manner but will collect in a transient manner, as previously shown.

Consider a cooled circular tube in zero gravity (Fig. 14.20). Making the quasi-steady-state assumption used before, an energy balance yields

$$\frac{q}{L} = \frac{2\pi k_\ell (T_s - T_{wo})}{\ell n(R_\delta/R_o)} = i_{\ell g}\rho_\ell 2\pi R_\delta \frac{dR_\delta}{dt} \qquad (14.115)$$

Integrating from $R_\delta = R_o$ at $t = 0$ yields

$$\frac{1}{2} + \left(\frac{R_\delta}{R_o}\right)^2 \left(\ell n\left(\frac{R_\delta}{R_o}\right) - \frac{1}{2}\right) = \left(\frac{2k_\ell \Delta T}{\rho_\ell i_{\ell g} R_o^2}\right)t \qquad (14.116)$$

Equation 14.116 may be solved for R_δ (and therefore $\delta(t)$) as a function of time, and the resulting heat transfer coefficient is:

$$h = \frac{q}{(2\pi R_o L)\Delta T} = \frac{k_\ell}{R_o \ell n(R_\delta/R_o)} \qquad (14.117)$$

In this situation, it is clear that as the film thickness continues to grow, the heat transfer coefficient becomes smaller and smaller.

In order to stimulate condensate motion under zero-G conditions, other forces must replace the gravitational force. This may be done by centrifugal forces, vapor shear forces, surface tension forces, suction forces, and forces created by an electric field. McEver and Hwangbo [133] and Valenzuela et al. [134] describe how surface tension forces may be used to drain a condenser surface in space. Tanasawa [1] reviews electrohydrodynamics (EHD) enhancement of condensation. Bologa et al. [135] showed experimentally that an electric field deforms the liquid-vapor interface, creating local capillary forces that enhance the heat transfer.

IN-TUBE CONDENSATION

Flow Regimes

During film condensation inside tubes, various flow regimes can occur, depending on the orientation and length of the tube, the heat flux along the tube axis, and relevant fluid properties. For example, when condensation occurs in a long horizontal tube at high condensation rates, the flow passes through various regimes as the fluid proceeds from the inlet (with a quality x near 1.0) to the exit (with $x \leq 0.0$), Fig. 14.21. As a consequence, the heat transfer coefficient and two-phase pressure gradient vary appreciably along the tube axis, depending on the liquid/vapor distribution within the tube.

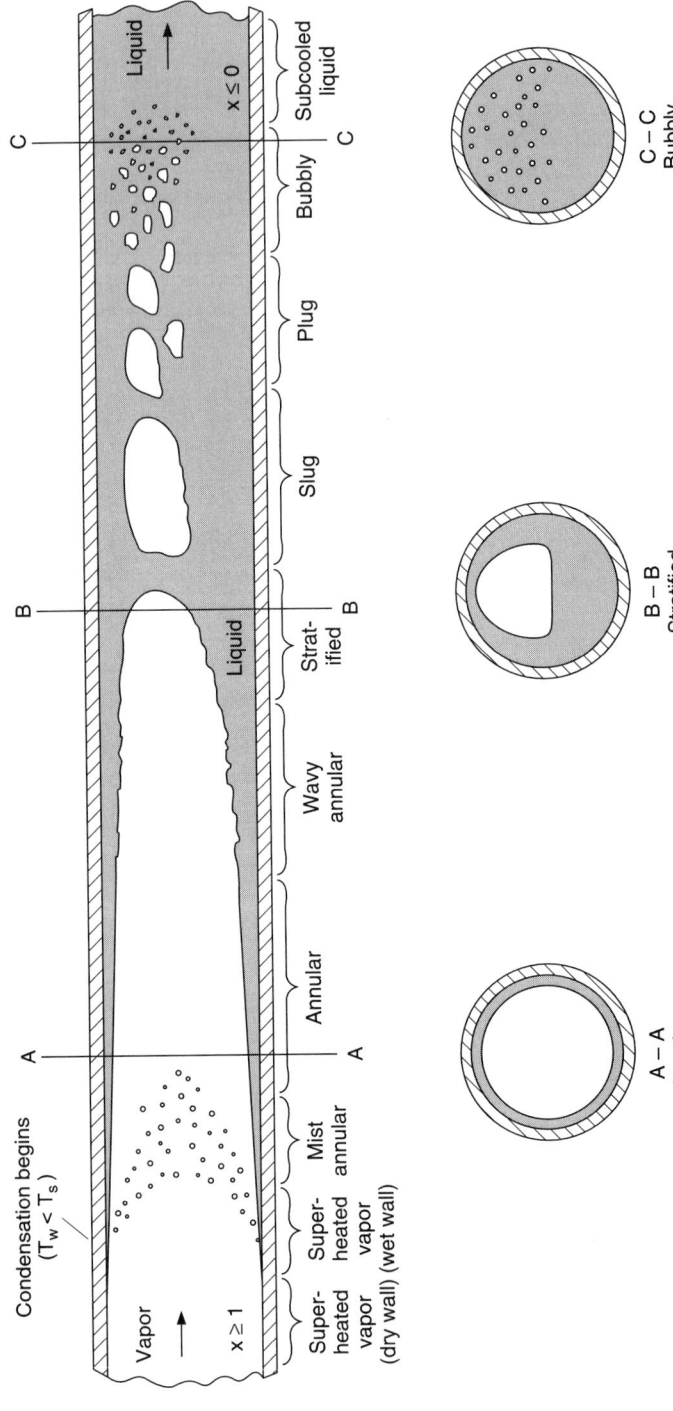

FIGURE 14.21 Flow regime changes during in-tube condensation within a horizontal tube.

If the vapor enters under superheated conditions, the vapor velocity is probably large enough to create turbulent flow. If the tube wall temperature is above the local saturation temperature of the vapor, heat transfer occurs purely by single-phase convection and can be predicted by one of the conventional single-phase turbulent flow correlations (dry wall desuperheating). If the wall temperature is lower than the saturation temperature, condensation starts along the wall while the vapor core is still superheated (wet wall desuperheating). A mist of tiny liquid droplets may occur (mist flow). There is little information available for predicting condensation heat transfer of a flowing superheated vapor [30]. As pointed out previously, a correction may be made to the latent heat of vaporization using Eq. 14.32.

As the vapor velocity slows down due to the mass removed by condensation, the liquid will collect as a thin film along the walls (annular flow). The liquid film will continue to thicken as more condensate collects. Depending upon the orientation of the tube and the magnitude of the vapor shear forces compared to gravitational forces, the liquid may or may not stratify. Waves (or ripples) may form on the liquid film, and eventually, these waves may become large enough to fill the entire cross section of the tube (slug flow). The slugs of liquid will intermittently push plugs of vapor (plug flow) toward the tube exit. Eventually, depending upon the heat flux, it is possible for all the vapor to condense, resulting in single-phase liquid (perhaps subcooled) near the tube exit. In this region, a single-phase correlation may therefore be used, but, due to the large viscosity difference between the liquid and the vapor, the Reynolds number of the liquid will be considerably less than the Reynolds number of the vapor at the same mass flux G, so the flow in this section is likely to be laminar.

Numerous models exist to predict the observed flow patterns [136–138]. One of the simplest, proposed by Breber et al. [136], depends upon the dimensionless vapor mass velocity j_g^*, defined as

$$j_g^* = \frac{xG}{[g\rho_g(\rho_\ell - \rho_g)d_i]^{1/2}} \tag{14.118}$$

and the Lockhart-Martinelli parameter X_{tt} (Eq. 14.74). They reasoned that the different flow patterns depend upon the ratio of shear forces to gravitational forces and the ratio of liquid volume to vapor volume. Their flow pattern criteria are

$$j_g^* > 1.5 \qquad X_{tt} < 1.0 \qquad \text{Mist and annular} \tag{14.119a}$$
$$j_g^* < 0.5 \qquad X_{tt} < 1.0 \qquad \text{Wavy and stratified} \tag{14.119b}$$
$$j_g^* < 0.5 \qquad X_{tt} > 1.5 \qquad \text{Slug} \tag{14.119c}$$
$$j_g^* > 1.5 \qquad X_{tt} > 1.5 \qquad \text{Bubble} \tag{14.119d}$$

Figure 14.22 compares the data of Rahman et al. [137] with the model of Breber et al. [136]. Except for the transition from wavy to stratified-slug flow, the agreement is good.

Vertical Tubes

Condensation in vertical tubes depends on the vapor flow direction and its magnitude. During downflow of vapor, if the vapor velocity is very low, then the condensate flow is controlled by gravity, and the Nusselt results for a vertical flat plate are applicable (unless the tube inside diameter is very small and tube wall curvature effects become important [36]).

The flow may proceed from laminar wave-free to laminar wavy to turbulent conditions, depending on the film Reynolds number (i.e., the heat flux and length of the tube). In this situation, the average heat transfer coefficient may be calculated using Eq. 14.21. If the vapor velocity is very high, then the flow is controlled by vapor shear forces, and annular flow models described in this chapter are applicable.

During upflow of vapor, interfacial shear will retard the drainage of condensate, thicken the condensate film, and decrease heat transfer. Care must be exercised to avoid vapor veloc-

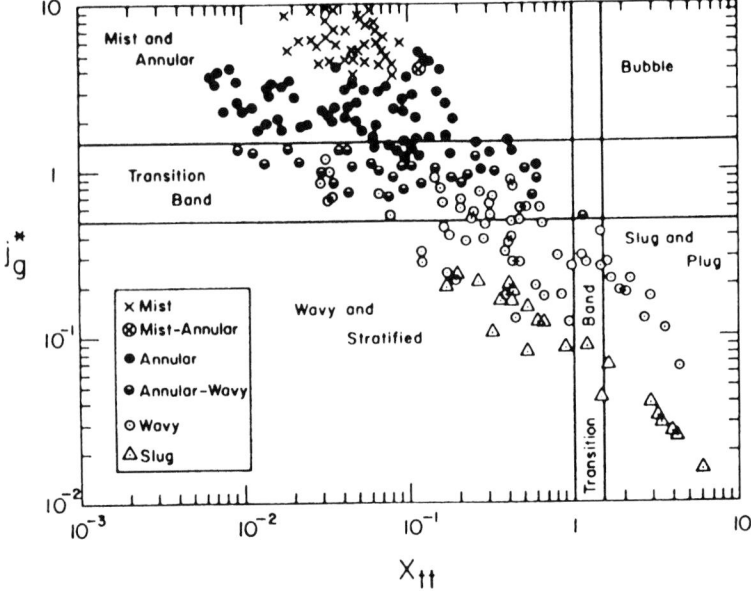

FIGURE 14.22 Comparison of Rahman et al. [137] data with flow regime map of Breber et al. [136]. (*Reprinted with permission from the* Canadian Journal of Chemical Engineering.)

ities that are high enough to cause "flooding," a phenomenon that occurs when vapor shear forces prevent the downflow of condensate. One criterion to predict the onset of flooding is due to Wallis [139], which is based upon air-water measurements

$$(v_g^*)^{1/2} + (v_\ell^*)^{1/2} = C = 0.725 \tag{14.120}$$

where
$$v_g^* = \frac{v_g \rho_g^{1/2}}{[gd_i(\rho_\ell - \rho_g)]^{1/2}} \tag{14.121a}$$

$$v_\ell^* = \frac{v_\ell \rho_\ell^{1/2}}{[gd_i(\rho_\ell - \rho_g)]^{1/2}} \tag{14.121b}$$

The velocities v_g and v_ℓ should be calculated at the bottom of the tube, where they are at their maximum values. Butterworth [140] suggests that C should be corrected for surface tension and tube end effects using the relationship

$$C^2 = 0.53 F_\sigma F_g \tag{14.122}$$

where F_σ is a correction factor for surface tension, equal to $(\sigma/\sigma_w)^{0.5}$, and F_g depends on the geometry of the tube inlet. ($F_g = 1.0$ for a square-ended tube and can be greater than 1.0 for tubes cut at an angle.) The influence of the tube exit geometry on upflow condensation in a vertical tube is discussed by Rabas and Arman [141].

Horizontal Tubes

Stratified. During condensation within horizontal tubes, when the vapor velocity is very low (i.e., j_g^* is less than 0.5), the flow will be dominated by gravitational forces, and stratifica-

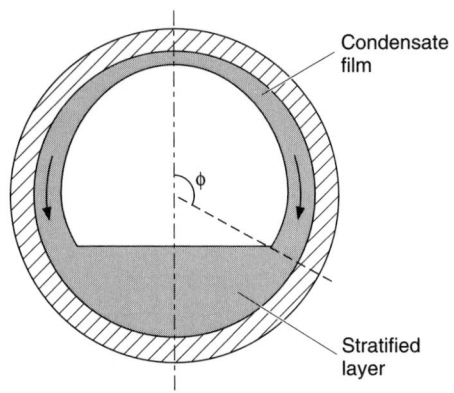

tion of the condensate will occur, as shown idealized in Fig. 14.23. In this case, the condensate forms a thin film on the top portion of the tube walls and drains around the periphery by gravity alone toward the bottom of the tube, where a layer of condensate collects and flows axially due to shear forces. This problem was first studied by Chato [142]. The Nusselt theory is generally valid over the top portion (i.e., the thin film region) of the tube, and heat transfer in the stratified layer is generally negligible. The average heat transfer coefficient over the entire perimeter may therefore be expressed by a modified Nusselt result

FIGURE 14.23 Idealized condensate profile during stratified flow, in-tube condensation within a horizontal tube.

$$h_m = \Omega \left\{ \frac{\rho_\ell(\rho_\ell - \rho_g)gi_{\ell g}k_\ell^3}{\mu_\ell(T_s - T_{wi})d_i} \right\}^{1/4} \qquad (14.123)$$

The coefficient Ω depends on the fraction of the tube circumference that is stratified. Jaster and Kosky [143] have shown that Ω is related to the void fraction of the vapor α_g:

$$\Omega = 0.728\alpha_g^{3/4} \qquad (14.124)$$

where

$$\alpha_g = \frac{1}{1 + [(1-x)/x](\rho_g/\rho_\ell)^{2/3}} \qquad (14.125)$$

Stratified-Wavy. At high vapor velocities, the flow deviates from the idealized situation just described. First of all, heat transfer in the stratified liquid pool at the bottom of the tube may not be negligible. Secondly, axial interfacial vapor shear may influence the motion and heat transfer in the thin film region around the top part of the tube. Dobson [144] studied this more complex situation and reported that stratified-wavy flow exists when $G < 500$ kg/m²s and $\mathrm{Fr}_m < 20$, where Fr_m is a modified Froude number given by:

$$\mathrm{Fr}_m = a \left\{ \frac{\mathrm{Re}_\ell^b}{\mathrm{Ga}^{0.5}} \right\} \left[\frac{1 + 1.09X_{tt}^{0.039}}{X_{tt}} \right]^{1.5} \qquad (14.126)$$

where

$$\mathrm{Re}_\ell = \frac{G(1-x)d_i}{\mu_\ell} \qquad (14.127a)$$

$$\mathrm{Ga} = \frac{\rho_\ell(\rho_\ell - \rho_g)gd_i^3}{\mu_\ell^2} \qquad (14.127b)$$

and

$$a = 0.025 \quad b = 1.59 \quad \text{if } \mathrm{Re}_\ell \le 1250$$
$$a = 1.26 \quad b = 1.04 \quad \text{if } \mathrm{Re}_\ell > 1250$$

Dobson [144] developed an additive model that combined film condensation (i.e., a modified Nusselt analysis) at the top and sidewalls of the tube with forced convection condensation in the stratified pool at the bottom of the tube.

$$\mathrm{Nu} = \mathrm{Nu}_{\text{film}} + \left(1 - \frac{\phi}{\pi}\right)\mathrm{Nu}_{\text{forced}} \qquad (14.128)$$

where ϕ is the angle measured from the top of the tube to the liquid pool (Fig. 14.23) and $(1 - \phi/\pi)$ is the fraction of the tube circumference covered by stratified liquid. This fraction is approximated by:

$$(1 - \phi/\pi) \approx \frac{\arccos\ (2\alpha_g - 1)}{\pi} \tag{14.129}$$

where α_g is the void fraction given in Eq. 14.125.

In Eq. 14.128, the film condensation component is given by

$$\text{Nu}_{\text{film}} = \frac{0.23\text{Re}_{go}^{0.12}}{1 + 1.11X_{tt}^{0.58}} \left\{ \frac{\text{Ga} \cdot \text{Pr}_\ell}{H} \right\}^{1/4} \tag{14.130}$$

where H is given by Eq. 14.34 and

$$\text{Re}_{go} = \frac{Gd_i}{\mu_g} \tag{14.131}$$

The constants in Eq. 14.130 were chosen so that it approaches Eq. 14.54 (with d_o replaced by d_i) at low vapor velocities and at a quality of 1.0. The forced convection component is a modified form of the expression proposed by Traviss et al. [145] for annular flow:

$$\text{Nu}_{\text{forced}} = 0.0195\text{Re}_\ell^{0.8}\ \text{Pr}_\ell^{0.4}\ \sqrt{1.376 + \frac{a}{X_{tt}^b}} \tag{14.132}$$

where, for $\text{Fr}_\ell \leq 0.7$,

$$a = 4.72 + 5.48\text{Fr}_\ell - 1.564\text{Fr}_\ell^2 \tag{14.133a}$$

$$b = 1.773 - 0.169\text{Fr}_\ell \tag{14.133b}$$

and, for $\text{Fr}_\ell > 0.7$,

$$a = 7.242 \qquad b = 1.655 \tag{14.133c}$$

In the above expressions,

$$\text{Fr}_\ell = \frac{(G/\rho_\ell)^2}{gd_i} \tag{14.134}$$

Dobson [144] compared Eq. 14.128 to his experimental data for R-22, R-134a, and two mixtures of R-32/R-125 and found agreement, in general, to within ±15 percent.

Annular.   When the vapor velocity is high enough ($j_g^* > 1.5$), gravitational effects can be neglected, and the condensate collects as a thin annular film around the inside of the tube walls, with no stratification. A significant portion of most condensers operate in this flow regime. There are numerous predictive models described in the literature for annular flow. Laminar flow models predict heat transfer coefficients that are too low, and turbulent models must be used. The most commonly used models are listed in Table 14.1. All models have a form for the local Nusselt number

$$\text{Nu} = \text{Nu}_\ell \cdot F(x) \tag{14.135}$$

where Nu_ℓ is a turbulent flow, single-phase, forced convection Nusselt number for the liquid, and $F(x)$ is a two-phase multiplier that depends on local quality x.

TABLE 14.1 Annular Flow Heat Transfer Models

Akers et al. [146]

$$\text{Nu} = \frac{hd_i}{k_\ell} = C\,\text{Re}_e^n\,\text{Pr}_\ell^{1/3} \tag{14.136}$$

where $\quad C = 0.0265, \qquad n = 0.8 \qquad$ for $\text{Re}_e > 5 \times 10^4$
$\qquad\quad C = 5.03, \qquad\quad n = \frac{1}{3} \qquad$ for $\text{Re}_e < 5 \times 10^4$

$$\text{Re}_e = \frac{G_e d_i}{\mu_\ell} \tag{14.136a}$$

$$G_e = G[(1-x) + x(\rho_\ell/\rho_g)^{1/2}] \tag{14.136b}$$

Boyko and Kruzhilin [147]

$$\text{Nu} = 0.021\text{Re}_{\ell o}^{0.8}\,\text{Pr}_\ell^{0.43}\,[1 + x(\rho_\ell/\rho_g - 1)]^{1/2} \tag{14.137}$$

where $\qquad\qquad\qquad \text{Re}_{\ell o} = \dfrac{Gd_i}{\mu_\ell} \tag{14.137a}$

Cavallini and Zecchin [148]

$$\text{Nu} = 0.05\text{Re}_e^{0.8}\,\text{Pr}_\ell^{0.33} \tag{14.138}$$

where Re_e is calculated from Eq. 14.136a, b

Shah [149]

$$\text{Nu} = \text{Nu}_{\ell o}\left[(1-x)^{0.8} + \frac{3.8x^{0.76}(1-x)^{0.04}}{p_r^{0.38}}\right] \tag{14.139}$$

where $p_r = P/P_c$, and

$$\text{Nu}_{\ell o} = 0.023\text{Re}_{\ell o}^{0.8}\,\text{Pr}_\ell^{0.4} \tag{14.140}$$

Traviss et al. [145]

$$\text{Nu} = \frac{\text{Pr}_\ell\,\text{Re}_\ell^{0.9}}{F_2(\text{Re}_\ell,\,\text{Pr}_\ell)}\,F_1(X_{tt}) \qquad 0.15 < F_1(X_{tt}) < 15 \tag{14.141}$$

where $\qquad\qquad \text{Re}_\ell = \dfrac{G(1-x)d_i}{\mu_\ell} \tag{14.141a}$

$$F_1(X_{tt}) = 0.15\left[\frac{1}{X_{tt}} + \frac{2.85}{X_{tt}^{0.476}}\right] \tag{14.142}$$

and $F_2(\text{Re}_\ell,\,\text{Pr}_\ell)$ is given by

$\quad F_2 = 0.707\text{Pr}_\ell\,\text{Re}_\ell^{0.5} \qquad\qquad\qquad\qquad\qquad \text{Re}_\ell < 50 \tag{14.143a}$

$\quad F_2 = 5\text{Pr}_\ell + 5\ell n[1 + \text{Pr}_\ell\,(0.0964\text{Re}_\ell^{0.585} - 1)] \qquad 50 < \text{Re}_\ell < 1125 \tag{14.143b}$

$\quad F_2 = 5\text{Pr}_\ell + 5\ell n(1 + 5\text{Pr}_\ell) + 2.5\ell n(0.0031\text{Re}_\ell^{0.812}) \qquad \text{Re}_\ell > 1125 \tag{14.143c}$

Fujii [105, 209]

$$\text{Nu} = 0.018[\text{Re}_\ell\,(\rho_\ell/\rho_g)^{1/2}]^{0.9}\left(\frac{x}{1-x}\right)^{0.1x+0.8}\text{Pr}_\ell^{1/3}\,(1 + AH/\text{Pr}_\ell) \tag{14.144}$$

where H is given by Eq. 14.34 and

$$A = 0.071\text{Re}_\ell^{0.1}\left(\frac{\rho_\ell}{\rho_g}\right)^{0.55}\left(\frac{x}{1-x}\right)^{0.2-0.1x}\text{Pr}_\ell^{1/3} \tag{14.144a}$$

All of the in-tube expressions described above are for the local heat transfer coefficient and must be integrated over the length of the tube in order to find an average heat transfer coefficient

$$h_m = \frac{1}{L} \int_0^L h(z) \, dz \tag{14.145}$$

In order to integrate Eq. 14.145, one must know the dependence of the quality x on axial position z. This generally will require subdividing the overall length into a number of subelements of length Δz and following the process from inlet to outlet, using local heat transfer coefficients for each subelement. If the quality is assumed to vary linearly (which, unfortunately, does not occur in many cases), then an average heat transfer coefficient may be approximated by using an average quality $x = 0.5$ in the local expressions listed above. Otherwise, numerical methods must be used.

Palen et al. [150] review additional problems associated with in-tube condensation.

Pressure Losses

During in-tube condensation, the local two-phase pressure gradient, when gravity is neglected, may be written in terms of frictional and acceleration components

$$\frac{dP}{dz} = \left(\frac{dP}{dz} \right)_f + G^2 \frac{d}{dz} \left\{ \frac{(1-x)^2}{\rho_\ell(1-\alpha_g)} + \frac{x^2}{\rho_g \alpha_g} \right\} \tag{14.146}$$

where the two-phase frictional pressure gradient depends on the particular flow regime occurring, and α_g is given by Eq. 14.125. The frictional pressure gradient may be related to the single phase flow of either the liquid or the vapor flowing alone in the tube. Either phase may be assumed to be flowing at their actual respective mass fluxes (e.g., $G_g = xG$) or at the total mass flux. Two-phase frictional multipliers ϕ_ℓ^2, ϕ_g^2, $\phi_{\ell o}^2$, and ϕ_{go}^2 are therefore defined [151]:

$$\left(\frac{dP}{dz} \right)_f = \phi_\ell^2 \left(\frac{dP}{dz} \right)_\ell = \phi_g^2 \left(\frac{dP}{dz} \right)_g = \phi_{\ell o}^2 \left(\frac{dP}{dz} \right)_{\ell o} = \phi_{go}^2 \left(\frac{dP}{dz} \right)_{go} \tag{14.147}$$

where the respective single-phase pressure gradients are

$$\left(\frac{dP}{dz} \right)_\ell = \frac{2f_\ell G^2 (1-x)^2}{\rho_\ell d_i}, \qquad \left(\frac{dP}{dz} \right)_g = \frac{2f_g G^2 x^2}{\rho_g d_i} \tag{14.148a, b}$$

$$\left(\frac{dP}{dz} \right)_{\ell o} = \frac{2f_{\ell o} G^2}{\rho_\ell d_i}, \qquad \left(\frac{dP}{dz} \right)_{go} = \frac{2f_{go} G^2}{\rho_g d_i} \tag{14.148c, d}$$

and where the friction factor f depends on the respective Reynolds number

$$f = 0.079 \text{Re}^{-0.25} \qquad \text{for Re} > 2000 \tag{14.149}$$

Several frictional multiplier correlations have been developed using a separated flow model. These correlations are listed in Table 14.2. Hewitt [151] makes the following tentative recommendations:

1. For $\mu_\ell/\mu_g < 1000$, the Friedel [152] correlation should be used.
2. For $\mu_\ell/\mu_g > 1000$, and $G > 100$ kg/m²s, the Chisholm [153] correlation should be used.
3. For $\mu_\ell/\mu_g > 1000$, and $G < 100$ kg/m²s, the Martinelli [154, 155] correlation (as modified by Chisholm [156]) should be used.

TABLE 14.2   Two-Phase Flow Frictional Multiplier Correlations

Friedel [152]

$$\phi_{\ell o}^2 = E + (3.23FH)/(\mathrm{Fr}^{0.045}\,\mathrm{We}^{0.035}) \tag{14.150}$$

where

$$E = (1-x)^2 + x^2\,\frac{\rho_\ell f_{go}}{\rho_g f_{\ell o}} \tag{14.151}$$

$$F = x^{0.78}(1-x)^{0.224} \tag{14.152}$$

$$H = \left(\frac{\rho_\ell}{\rho_g}\right)^{0.91}\left(\frac{\mu_g}{\mu_\ell}\right)^{0.19}\left(1-\frac{\mu_g}{\mu_\ell}\right)^{0.7} \tag{14.153}$$

$$\mathrm{Fr} = \frac{G^2}{g d_i \rho_h^2} \tag{14.154}$$

$$\mathrm{We} = \frac{G^2 d_i}{\rho_h \sigma} \tag{14.155}$$

$$\rho_h = \frac{\rho_g \rho_\ell}{x\rho_\ell + (1-x)\rho_g} \tag{14.156}$$

Chisholm [153]

$$\phi_{\ell o}^2 = 1 + (Y^2-1)[Bx^{(2-n)/2}(1-x)^{(2-n)/2} + x^{2-n}] \tag{14.157}$$

where

$$Y = \left[\frac{(dP/dz)_{go}}{(dP/dz)_{\ell o}}\right]^{1/2} \tag{14.158}$$

n is the Reynolds number exponent in friction factor relationships (e.g., $n = 0.25$, Eq. 14.149)

For $\quad 0 < Y < 9.5,\quad B = \begin{cases} \dfrac{55}{G^{1/2}} & G \ge 1900\ \mathrm{kg/(m^2 \cdot s)} \\[2mm] \dfrac{2400}{G} & 500 \le G \le 1900\ \mathrm{kg/(m^2 \cdot s)} \\[2mm] 4.8 & G < 500\ \mathrm{kg/(m^2 \cdot s)} \end{cases} \tag{14.159a}$

For $\quad 9.5 < Y < 28,\quad B = \begin{cases} \dfrac{520}{YG^{1/2}} & G \le 600\ \mathrm{kg/(m^2 \cdot s)} \\[2mm] \dfrac{21}{Y} & G > 600\ \mathrm{kg/(m^2 \cdot s)} \end{cases} \tag{14.159b}$

For $\quad Y > 28,\quad B = \dfrac{15{,}000}{Y^2 G^{1/2}} \tag{14.159c}$

Martinelli [154, 155] (as modified by Chisholm [156])

$$\phi_\ell^2 = 1 + \frac{C}{X} + \frac{1}{X^2} \tag{14.160}$$

$$\phi_g^2 = 1 + CX + X^2 \tag{14.161}$$

where

$$X = \left[\frac{(dP/dz)_\ell}{(dP/dz)_g}\right]^{1/2} \tag{14.162}$$

and where the values of the constant C depend on the respective flow regimes associated with the liquid and the vapor

Liquid	Gas	Subscript	C
Turbulent	Turbulent	tt	20
Viscous	Turbulent	vt	12
Turbulent	Viscous	tv	10
Viscous	Viscous	vv	5

Souza et al. [157] measured two-phase pressure drops during turbulent flow of R-12 and R-134a and developed an expression for the two-phase frictional multiplier $\phi_{\ell o}^2$ that successfully predicted their data to within ±10 percent:

$$\phi_{\ell o}^2 = [1.376 + aX_{tt}^{-b}](1 - x)^{1.75} \qquad (14.163)$$

They recommended that a and b should be calculated using Eq. 14.133.

During condensation, because of the mass transfer across the liquid-vapor interface, Sardesai et al. [158] suggest that the following correction be made to the previous correlations:

$$\left(\frac{dP}{dz}\right)_{fc} = \left(\frac{dP}{dz}\right)_f \theta \qquad (14.164)$$

where

$$\theta = \frac{\phi}{1 - \exp(-\phi)} \qquad (14.165)$$

$$\phi = \frac{G(v_g - v_i)}{\tau_g} \qquad (14.166)$$

Groenewald and Kroger [159] studied the effect of mass transfer on turbulent friction during condensation inside tubes. They write the interfacial friction factor as

$$f_i = \beta f_o + f_{tp} \qquad (14.167)$$

where f_o accounts for friction experienced in single-phase gas flow, f_{tp} represents the additional friction due to the formation of waves in the liquid film as well as any vapor separation due to the irregularity of the gas-liquid interface, and β is the single-phase friction enhancement factor due to suction. Their predictions of the frictional pressure drop during condensation of steam in an air-cooled duct agreed to within ±5 percent of experimental data.

During in-tube condensation, since the quality changes along the tube axis, the pressure drop must be calculated in a stepwise manner. As in the case of heat transfer, the tube or channel is divided into a number of short, incremental lengths Δz. The pressure drop over one of these lengths would be

$$\Delta P = \left(\frac{dP}{dz}\right)\Delta z \qquad (14.168)$$

where the gradient (dP/dz) is evaluated using the flow conditions at the midpoint of the length Δz.

Condenser Modeling

In-tube condensers are modeled, generally, in three separate sections: the desuperheater section, the condensing section, and the subcooled section. Different heat transfer and pressure drop correlations are used in each section. The desuperheater and subcooler sections are treated using single-phase flow correlations. In the condensing section, since the local heat transfer coefficient varies along the tube axis and the saturation temperature changes due to pressure losses, a discretization technique must be followed.

For example, to size the condensing section, one can assume magnitudes of d_i, G, and fluid properties. Then, divide the length into equal increments of quality change Δx and calculate the local condensing heat transfer coefficient at the midquality magnitude of each Δx, assuming that the local value is constant over this particular quality range Δx.

The incremental pressure drop ΔP would also be calculated for the assumed flow conditions by using Eq. 14.168 together with Eq. 14.146 and one of the suitable two-phase multiplier models for the frictional pressure gradient. The local saturation temperature along the tube axis can then be found.

An energy balance on one of these incremental lengths yields:

$$\Delta q = h(\pi d_i \Delta z)(T_s - T_{wi}) = G\left(\frac{\pi}{4} d_i^2\right) i_{\ell g} \Delta x$$

or

$$\Delta z = \frac{G d_i \Delta x i_{\ell g}}{4 h \Delta T} \qquad (14.169)$$

The total length of the tube to condense the vapor from $x = 1.0$ to $x = 0$ would be the summation of all the incremental lengths of Δz.

Noncircular Passages

Various investigators have studied in-tube condensation in noncircular passages. Fieg and Roetzel [121] and Chen and Yang [160] analyzed condensation inside elliptical tubes. Kaushik and Azer [161] established an experimental correlation for internally finned tubes. Lee et al. [162] experimentally studied condensation of R-113 within an internally finned tube and a spirally twisted tube and compared performance to that of a smooth tube. Using a modified form of the correlation of Cavallini and Zecchin [148] (Eq. 14.138):

$$\mathrm{Nu} = C \, \mathrm{Re}_e^{0.5} \, \mathrm{Pr}_\ell^{0.33} \qquad (14.170)$$

they obtained the following values of C:

$$C = \begin{cases} 1.05, & \text{smooth tube} \\ 1.65, & \text{spirally twisted tube} \\ 2.59, & \text{internally finned tube} \end{cases}$$

Condensation inside micro-fin tubes has been studied by Schlager et al. [163] and Chamra and Webb [164]. Webb and Yang [165] studied heat transfer and pressure drop of R-12 and R-134a in flat, extruded tubes (having either a smooth wall or micro-fins).

Condensation inside plate condensers has received considerable interest in recent years [166–170]. More information on condensation inside compact heat exchangers may be found in Srinivasan and Shah [171].

DIRECT CONTACT CONDENSATION

Direct contact condensation occurs when vapor condenses directly on a subcooled liquid. It occurs on drops or sprays, on falling films (either supported or unsupported) or jets, and when vapor bubbles are injected into subcooled liquid pools.

Condensation on Drops (Spray Condensers)

When a subcooled liquid at an initial temperature $T_i < T_s$ is sprayed into a vapor space, the atomization process creates many small, near-spherical droplets with initial radii R_i between 50 and 250 μm. As these drops move through the condenser, they grow and heat up due to condensation on their surface. The condensation process is dominated by transient conduction within the drops. Jacobs and Cook [172], assuming a noncirculating, spherical drop, solved the resulting nonlinear, transient conduction problem numerically. For the limiting case of no condensate film resistance (i.e., the drop surface temperature is suddenly changed from T_i to T_s), their numerical result for the time-dependent drop radius agreed with the analytical result:

$$\frac{R(t)}{R_i} = \left\{1 + \text{Ja}\left[1 - \frac{6}{\pi^2}\sum_{n=1}^{\infty}\frac{\exp(-n^2\pi^2\tau)}{n^2}\right]\right\}^{1/3} \tag{14.171}$$

where

$$\text{Ja} = \frac{\rho_\ell c_{p\ell}(T_s - T_i)}{\rho_g i_{\ell g}} \qquad \tau = \text{Fo} = \alpha_\ell\frac{t}{R_i^2} \tag{14.171a, b}$$

Jacobs and Cook [172] showed that the agreement between Eq. 14.171 and the data of Ford and Lekic [173] was acceptable for steam condensing on water droplets for $(T_s - T_i) < 30°C$. For other fluids, the nonlinear conduction problem must be solved.

Sundararajan and Ayyaswamy [174] analyzed the hydrodynamics and the transport phenomena associated with condensation on a single drop moving in a saturated mixture of vapor and noncondensable gas and included internal circulation within the drop that is created by the interfacial shear of the vapor. Excellent agreement was found between their prediction of the transient drop bulk temperature and the data of Kulic and Rhodes [175]. Chang et al. [176] measured the direct contact condensation of R-113 on falling droplets of water, forming an immiscible mixture. They found that the rate of heat transfer was considerably larger than found in miscible liquids. This enhancement was attributed to the observation that the R-113 condensate that collected on the water droplets broke away from the droplet surface (presumably due to interfacial surface forces), thereby thinning the film. Mayinger and Chavez [177] measured direct-contact condensation on an injection spray using pulsed laser holography. Results showed that, at moderate liquid flow rate and relatively high vapor pressure, dramatic changes occur in the spray geometry, considerably reducing the condensation rate. Condensation on a spray of water drops was recently modeled by Sripada et al. [216, 217].

Knowing the time-dependent drop radius $R(t)$, it is possible to estimate the transit time of a droplet and therefore the required distance from the spray nozzle to the condensate pool at the bottom of the condenser in order to size the condenser [178].

Condensation on Jets and Sheets

Hasson et al. [179] studied laminar film condensation on falling jets and sheets using a Graetz-type thermal entrance length solution. For a sheet of thickness 2δ, the local Nusselt number is

$$\text{Nu}_x = \pi^2\left\{\frac{\sum_{n=1}^{\infty}\exp(-4\pi^2(2n-1)^2/\text{Gz})}{\sum_{n=1}^{\infty}(1/(2n-1)^2)\exp(-4\pi^2(2n-1)^2/\text{Gz})}\right\} \tag{14.172}$$

where

$$\text{Nu}_x = \frac{h_x(4\delta)}{k_\ell} \tag{14.173}$$

and

$$\text{Gz} = \text{Re}\,\text{Pr}_\ell\,d_h/x = \frac{(4\delta)^2 u}{\alpha_\ell x} \tag{14.174}$$

For small Graetz numbers that correspond to a very thin sheet, Hasson et al. [179] give

$$\text{Nu}_x = \pi^2 \tag{14.175}$$

and

$$\frac{T_s - T_m(x)}{T_s - T_i} = \frac{8}{\pi^2}\exp(-4\pi^2/\text{Gz}) \tag{14.176}$$

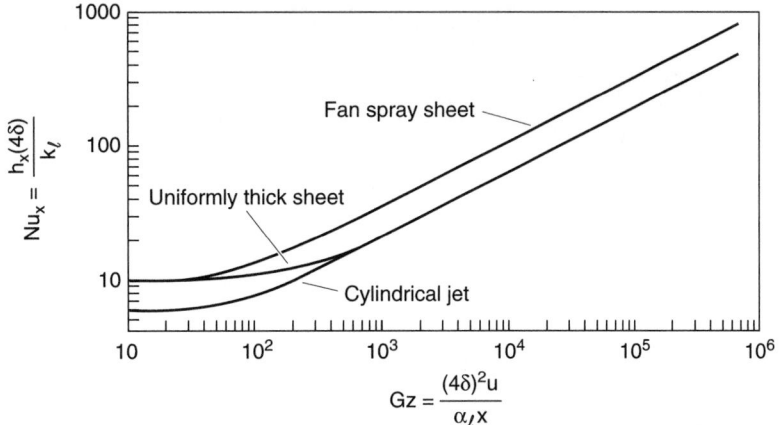

FIGURE 14.24 Local Nusselt number versus Graetz number for jets of various configurations [179]. (*Reprinted with permission from Pergamon Press, Tarrytown, New York.*)

For very large Graetz numbers (i.e., near the sheet entry point),

$$\text{Nu}_x \approx (\text{Gz}/\pi)^{1/2} \tag{14.177}$$

and

$$\frac{T_s - T_m(x)}{T_s - T_i} = 1 - \frac{8}{(\pi \, \text{Gz})^{1/2}} \tag{14.178}$$

Similar results were obtained for cylindrical jets and fan spray sheets, as shown in Fig. 14.24.

Mitrovic and Ricoeur [180] analyzed the hydrodynamics and heat transfer that occur during condensation on freely falling laminar liquid jets of initial diameter d_i and velocity u_i. Their numerical solution for the average Nusselt number is well represented by:

$$\text{Nu}_m = \frac{h_m d_i}{k_\ell} = 4.8 \left\{ \frac{\text{Gz}}{4} \right\}^{0.3} \tag{14.179}$$

where

$$\text{Gz} = \text{Re} \, \text{Pr}_\ell \, d_i/x \tag{14.180a}$$

and

$$\text{Re} = u_i d_i / \nu_\ell \tag{14.180b}$$

In addition, the mean jet temperature can be described by:

$$\frac{T_m(x) - T_i}{T_s - T_i} = 1 - \exp\left[-4.8\left(\frac{4}{\text{Gz}}\right)^{0.7}\right] \tag{14.181}$$

When Eq. 14.181 was compared to the data of Lui et al. [181], the data fell below the correlation, due presumably to the presence of noncondensable gas in the measurements. For small jet lengths (i.e., $x < 70$ mm), the correlation was in good agreement with the measurements of Celata et al. [182].

Condensation on Films

Only a few studies exist for direct contact condensation on a supported film such as gravity-driven flows over inclined trays. Jacobs et al. [183] applied binary boundary layer theory to study the laminar flow of a thin film over a prescribed surface during which condensation occurs on the film. They considered condensation on a coolant film at a temperature T_i falling over a solid, adiabatic sphere of radius R. Over the top part of the sphere, a thermal boundary layer develops in the falling coolant due to condensation on its surface. On the bottom part of the sphere, the entire coolant temperature increases. Jacobs et al. [183] showed that, for a vapor condensing on a coolant of its own liquid, the average Nusselt number on the top part of the sphere up to an angle ϕ can be expressed by

$$\mathrm{Nu}_m(\phi) = 0.58\left\{\frac{\mathrm{Re}^{1/3}\,\mathrm{Pr}^{0.36}}{\phi^{1/2}H^{*0.4}R_m^{1/2}}\right\} \tag{14.182}$$

where

$$\mathrm{Nu}_m(\phi) = \frac{h_m(\phi)(v_\ell^2/g)^{1/3}}{k_\ell} \tag{14.183a}$$

$$\mathrm{Re} = \frac{\dot{m}_f}{2\pi R\mu_\ell} \tag{14.183b}$$

$$R_m = \frac{R}{(v_\ell^2/g)^{1/3}} \tag{14.183c}$$

$$H^* = \frac{c_{p\ell}(T_s - T_i)}{i_{\ell g}} \tag{14.183d}$$

ϕ is the polar angle from the top of the sphere, and $\dot{m}_f$ is the mass flow rate of the film. Equation 14.182 is valid up until the development of the thermal boundary layer is completed. This occurs for a polar angle

$$\phi_D = 15.7\left\{\frac{\mathrm{Re}^{2/3}\,H^{*0.008}R_m^{0.43}}{\mathrm{Pr}^{0.547}}\right\} \tag{14.184}$$

The heat transfer in the lower part of the sphere is highly nonlinear, and no correlation was reported for this region. Karapantsios and Karabelas [184] experimentally examined the influence of flow intermittency on direct contact condensation of a quasistagnant vapor-gas mixture on falling liquid waves. Flow intermittency was found to increase the heat transfer rate by as much as an order of magnitude. Mikielewicz et al. [218] recently included turbulent diffusion effects in studying direct-contact condensation of steam on a horizontal water film.

Condensation of Vapor Bubbles

When vapor bubbles are injected into a subcooled liquid, the individual bubbles will collapse, and this collapse will be controlled by liquid inertia, heat transfer, or both, depending upon the degree of subcooling of the liquid. For heat-transfer-controlled collapse of a stagnant bubble of initial radius R_i, Florshuetz and Chao [185] derived the following expression for the dimensionless time for collapse τ_H:

$$\tau_H = \frac{1}{3}\left[\frac{2}{\beta} + \beta^2 - 3\right] \tag{14.185}$$

where

$$\beta = R(t)/R_i \tag{14.185a}$$

$$\tau_H = \frac{4}{\pi}\,\mathrm{Ja}^2\,\mathrm{Fo} \tag{14.185b}$$

and Ja and Fo are given by Eqs. 14.171a and b, respectively.

For diffusion-controlled collapse of a stagnant bubble, the initial mode and final mode of condensation are given by Moalem-Maron and Zijl [186]:

$$
\beta = \begin{cases} 1 - \dfrac{2}{\sqrt{\pi}}\, \text{Ja Fo}^{1/2} & R(t) \approx R_i \gg (\pi\alpha_\ell t)^{1/2} \\ (1 - 2\, \text{Ja Fo})^{1/2} & R(t) \ll (\pi\alpha_\ell t)^{1/2} \end{cases}
\tag{14.186}
$$

Wittke and Chao [187] considered heat-transfer-controlled condensation on a moving bubble. They assumed that the bubble was a rigid sphere that moved with a constant velocity. They assumed that potential flow theory was valid. Isenberg et al. [188] corrected this model for no slip at the bubble surface and arrived at:

$$
\beta = \left\{ 1 - \frac{3}{2} \left(\frac{K_v}{\pi} \right)^{1/2} \text{Ja Pe}^{1/2}\, \text{Fo} \right\}^{2/3}
\tag{14.187}
$$

where
$$
\text{Pe} = \text{Re Pr}_\ell = \frac{2R_i V_\infty}{\alpha_\ell}
\tag{14.188}
$$

In Eq. 14.187, K_v is a velocity factor for modified potential flow. For a bubble collapsing in its own subcooled liquid, $K_v = 1.0$.

Ullman and Letan [189] studied the effect of noncondensable gas on condensation of bubbles in a subcooled liquid. Bergles and Bar-Cohen [190] provide information on the use of direct contact condensation of bubbles in a so-called submerged condenser. Zangrando and Bharathan [191] established computer models to characterize direct contact condensation of low-density steam (with high noncondensable gas concentrations) on seawater.

Additional information on direct contact condensation may be found in Sideman and Moalem-Maron [192].

CONDENSATION OF MIXTURES

Heat transfer prediction during condensation of mixtures is more difficult than during pure vapor condensation for a variety of reasons. For example, with mixtures, complete or partial condensation can occur depending on whether the coolant temperature is less than the saturation temperature of the more volatile components. Along the condenser, as the less volatile components condense out, the concentration of the more volatile components will increase, and this process creates a vapor temperature decrease that reduces the driving force for condensation through the condenser. Also, the presence of different vapor/gas components introduces mass transfer effects that create an additional thermal resistance that is nonexistent with pure vapors. As a consequence, condensing heat transfer coefficients of mixtures are less than those of single-component pure vapors. Most work with mixtures has dealt with two-component (i.e., binary) mixtures, and little information is known about multicomponent mixtures. Detailed descriptions of mixture processes and calculations may be found in the recent works of Tanasawa [1], Stephan [25], Fujii [32], Hewitt et al. [193], and Webb [194].

Condensation of a vapor in the presence of a noncondensable gas is treated elsewhere in this chapter. Figure 14.3 describes the added thermal resistance that occurs due to mass diffusion of the vapor through a noncondensable, gas-rich layer next to the condensate. The case of two condensing vapors is similar to that depicted in Fig. 14.3. Both vapor components condense, but the more volatile one accumulates at the interface and provides a barrier for the less volatile one, similar to a noncondensable gas. Similar effects are also found with multicomponent mixtures.

The condensate that collects on the cold surface is usually a completely homogeneous, or *miscible,* mixture of components. In general, the relative composition of the liquid components in the condensate is different from the composition in the vapor phase (except for an azeotropic mixture, where the condensate has the same exact molar concentration ratio as the vapor phase) [194]. The film that forms is not necessarily smooth but may show the appearance of streamers (or rivulets), waves, or droplets, depending on the particular mixture and its surface tension (which depends on the local wall temperature) [25, 195, 196]. If the condensate mixture is heterogeneous, or *immiscible* (as can occur when one component, for example, is aqueous and the other is organic), the pattern can be quite complex, looking somewhat like dropwise condensation [25, 193, 197]. These different condensate patterns affect the resulting fluid flow and heat transfer.

Because of the added complexities noted earlier and the important role of mass diffusion during condensation of vapor mixtures, the analysis of these processes is more complex than during condensation of a single-component pure vapor. Boundary layer analysis of these processes for both natural and forced convection condensation is thoroughly described in Fujii [32]. In general, however, boundary layer solutions are limited to the simplest mixtures and geometries [198, 199]. Therefore, for practical design applications, other approximate analytical methods have been used. These are categorized into equilibrium methods and nonequilibrium, or so-called *film,* methods.

Equilibrium Methods

Equilibrium methods, as proposed originally by Silver [200] and extended by Bell and Ghaly [201] and others, all assume that there is local equilibrium between the vapor and the condensate throughout the condenser. Even though condensation is a nonequilibrium process, the gas temperature T_g is assumed to follow a vapor-liquid equilibrium curve at T_g^*, as the vapor mixture is cooled from the mixture dew point T_{dew} to the mixture bubble temperature T_{bub}. These methods therefore require the generation of a cooling or condensation curve (not to be confused with the condensation curve described in Fig. 14.1), as shown in Fig. 14.25,

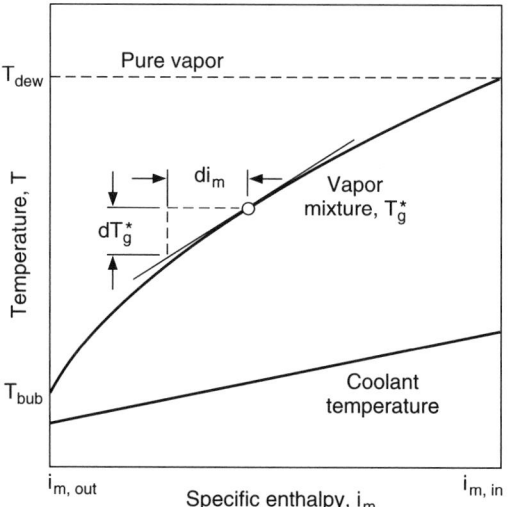

FIGURE 14.25 Equilibrium condensation curve.

which plots condensation-side temperature T_g versus specific enthalpy i_m, or the cumulative heat removal rate $q = \dot{m}(i_{m,\text{in}} - i_m)$, for a mixture. The ideal equilibrium condensation curve approximates the real condensation path. Equilibrium condensation curves may be of the integral type (where the vapor and the liquid are not separated from one another, as might occur, for example, during condensation in a vertical tube) or of the differential type (where the condensate is separated from the vapor, as might occur on the shell side of a shell-and-tube condenser). Webb [194] describes how to calculate these equilibrium curves.

With the Silver [200] method, the local overall heat transfer coefficient from the bulk vapor mixture to the coolant is written as:

$$\frac{1}{U} = \frac{1}{h_c} + R + \frac{1}{h_{\text{ef}}} \tag{14.189}$$

where h_c is the heat transfer coefficient on the coolant side, R is the thermal resistance due to wall conduction and fouling, and h_{ef} is an effective condensing-side heat transfer coefficient, which includes the thermal resistance across the condensate film, as well as the sensible cooling of the gas. The effective coefficient may be expressed as:

$$h_{\text{ef}} = \left[\frac{1}{h_\ell} + \frac{1}{h_g} \left(\frac{q_g''}{q''} \right) \right]^{-1} \tag{14.190}$$

where the ratio (q_g''/q'') is usually written as

$$\frac{q_g''}{q''} = Z = \dot{x}_g c_{pg} \frac{dT_g^*}{di_m} \tag{14.191}$$

In Eq. 14.191, $\dot{x}_g$ is the mass flow fraction of the gas (i.e., $\dot{m}_g/\dot{m}$), c_{pg} is the specific heat of the gas, and dT_g^*/di_m is the local slope of the equilibrium condensation curve (see Fig. 14.25). h_g is calculated for the gas phase flowing along by itself and should be corrected for mass transfer effects using the Ackermann [202] correction factor:

$$h_{gc} = h_g \left(\frac{a}{e^a - 1} \right) \tag{14.192a}$$

where
$$a = \frac{\sum_{i=1}^{n} \dot{n}_i'' \tilde{c}_{pgi}}{h_g} \tag{14.192b}$$

Knowing the equilibrium condensation curve, the local conditions of the mixture, and representative values for h_ℓ and h_g (and thus h_{gc}), h_{ef} can be readily calculated. The total condenser surface area can then be obtained by integration:

$$A_t = \int_{i_{m,\text{out}}}^{i_{m,\text{in}}} \frac{\dot{m} \, di_m}{U(T_g^* - T_c)} \tag{14.193}$$

In practical situations, Eq. 14.193 is replaced by a difference equation, Eq. 14.194:

$$A_t = \sum_{j=1}^{N} \frac{\dot{m} \Delta i_{m,j}}{U_j \Delta T_j} \tag{14.194}$$

This method is approximate not only because of the assumed equilibrium condition between the liquid and the vapor but also because mass transfer effects are ignored. Nevertheless, it is accepted for current industrial design. In cases where the details of the mixture composition at inlet and outlet are important, the more complex nonequilibrium methods must be used.

Nonequilibrium Methods

Nonequilibrium, or film, methods provide physically realistic formulations of the problem that yield more accurate local coefficients at the expense of complexity. Colburn and Hougen [77] developed a trial-and-error solution procedure for condensation of a single vapor mixed with a noncondensable gas. Colburn and Drew [203] extended the method to include condensation of binary vapor mixtures (with no noncondensables). Price and Bell [204] showed how to use the Colburn and Drew [203] method in computer-assisted design.

In recent years, considerable progress has been made to improve further upon this method for use with multicomponent mixtures. Detailed discussions of these methods may be found in Stephan [25], Hewitt et al. [193], and Webb [194]. The procedure of Sardesai et al. [205], which outlines the work of Krishna and Standart [206], is briefly described below.

At any local point along the condenser, the heat flux can be written as

$$q'' = h_{gc}(T_g - T_i) + \sum_{j=1}^{n} \dot{n}_j'' \Delta \tilde{i}_{\ell g, j} \tag{14.195}$$

where

$$\Delta \tilde{i}'_{\ell g, j} = \Delta \tilde{i}_{\ell g, j} + \tilde{c}_{pg, j}(T_g - T_i) \tag{14.196}$$

The heat flux therefore includes three contributions: (1) sensible cooling of the bulk vapor mixture as it moves through the condenser, (2) sensible cooling of the bulk vapor mixture as it flows from the local bulk conditions to the interface (at a temperature T_i), and (3) latent heat of condensation of the various condensing species. The condensation flux of the jth component is given by

$$\dot{n}_j'' = \dot{J}_{jb}'' + \tilde{y}_{jb} \dot{n}_t'' \tag{14.197}$$

where $\dot{J}_{jb}''$ represents a diffusive flux, and $\tilde{y}_{jb} \dot{n}_t''$ represents a convective molar flux.

In order to calculate the diffusive flux, a suitable mass transfer model must be assumed. Two categories of models exist: (1) interactive models (due to Krishna and Standart [206] and Toor [207]), and (2) noninteractive models, known also as *effective diffusivity models*. For the interactive models, the diffusion flux $\dot{J}_{jb}''$ is

$$\dot{J}_{jb}'' = [B][\zeta](\tilde{y}_{jb} - \tilde{y}_{ji}) \tag{14.198}$$

where $[B]$ is a matrix of binary mass transfer coefficients $\beta_{\ell k}$ for all the component pairs and the bulk composition, $[\zeta]$ is a correction matrix that allows for net mass flow on the mass transfer coefficients similar to the Ackermann [202] correction used in Eq. 14.192, and $(\tilde{y}_{jb} - \tilde{y}_{ji})$ is the vapor mole fraction driving force of the jth component. The mass transfer coefficients $\beta_{\ell k}$ are calculated from the Chilton-Colburn analogy:

$$\beta_{\ell k} = \frac{h_g}{\tilde{c}_{pg}} \left[\frac{\mathrm{Pr}}{\mathrm{Sc}_{\ell k}} \right]^{2/3} \tag{14.199}$$

For noninteractive, or effective diffusivity methods, Eq. 14.198 is simplified to

$$\dot{J}_{jb}'' = \lceil B_{\mathrm{ef}} \rfloor \lceil \zeta_{\mathrm{ef}} \rfloor (\tilde{y}_{jb} - \tilde{y}_{ji}) \tag{14.200}$$

where $\lceil B_{\mathrm{ef}} \rfloor$ and $\lceil \zeta_{\mathrm{ef}} \rfloor$ represent diagonal matrices, since each species is assumed to have no interaction with the other species involved. Sardesai et al. [205] compared each of these methods to existing experimental data for ternary systems and found that each method agreed with the experimental data to within about ±10 percent. Since the effective diffusivity method is less complex, it requires less computation time and is consequently the preferred method to use. Webb and McNaught [208] provide a comprehensive, step-by-step design example for a multicomponent mixture where the results of the previously outlined methods are compared.

NOMENCLATURE

A	Coefficient defined by Eq. 14.84a
A	Parameter defined by Eq. 14.144a
A	Surface area, m^2
B	Coefficient defined by Eq. 14.84b
B	Dimensionless parameter defined by Eq. 14.111
B	Function defined by Eq. 14.159
$[B]$	Multicomponent mass transfer coefficient matrix, kmol/(m^2·s)
$\lceil B_{ef} \rfloor$	Diagonal multicomponent mass transfer coefficient matrix, kmol/(m^2·s)
C	Coefficient defined by Eq. 14.95
c_p	Specific heat at constant pressure, J/(kg·K)
$\tilde{c}_p$	Molar specific heat, J/kmol·K
D	Mass transfer coefficient, m^2/s
D	Diameter of helical coil, m
d	Tube diameter, m
E	Dimensionless function defined by Eq. 14.151
e	Radial fin height, m
e_v	Effective mean vertical fin height, m
F	Dimensionless function, defined by Eq. 14.152
F_d	$(\mu_\ell i_{\ell g} g d_o)/(k_\ell \Delta T V_\infty^2)$
F_L	$(\mu_\ell i_{\ell g} g L)/(k_\ell \Delta T V_\infty^2)$
Fo	Fourier number, $\alpha_\ell t/R_i^2$
F_1	Function defined by Eq. 14.142
F_2	Function defined by Eq. 14.143
Fr	Froude number, $G^2/gd_i\rho^2$
Fr$_m$	Modified Froude Number defined by Eq. 14.126
$\mathscr{F}$	Function plotted in Fig. 14.19
f	Friction factor
f_f	Fraction of unflooded part of fin flank blanked by retained condensate at fin root
f_s	Fraction of unflooded part of interfin tube surface blanked by retained consensate at fin root
G	Mass flow rate per unit area, kg/(m^2·s)
G	$J_\ell \cdot R$, Eq. 14.39
g	Gravitational acceleration, m/s^2
Ga	Galileo number, Eq. 14.127b
Gz	Graetz number, Eq. 14.174
H	Dimensionless function defined by Eq. 14.153
H	Phase change number, $c_{p\ell}(T_s - T_w)/i_{\ell g}$
H^*	Modified phase change number, $c_{p\ell}(T_s - T_i)/i_{\ell g}$
h	Heat transfer coefficient, W/(m^2·K)
h_D	Mass transfer coefficient, m/s

h_i	Interfacial heat transfer coefficient, W/(m²·K)
$i_{\ell g}$	Latent heat of evaporation, J/kg
$\tilde{i}_{\ell g}$	Molar latent heat of evaporation, J/kmol
i_m	Specific enthalpy of a mixture, J/kg
J_ℓ	$(k_\ell \Delta T)/(\mu_\ell i_{\ell g})$
Ja	Jakob number, Eq. 14.171a
j''	Diffusive flux, kmol/(m²·s)
j_g^*	Dimensionless vapor mass velocity defined by Eq. 14.118
K	Dimensionless term defined by Eq. 14.41
k	Thermal conductivity, W/(m·K)
L	Length, m
$\overline{L}$	Average condensing length defined by Eq. 14.87b, m
M	Molecular weight, kg/mol
m	Exponent defined by Eq. 14.84c
$\dot{m}$	Mass flow rate, kg/s
$\dot{m}''$	Condensation mass flux, kg/(m²·s)
N	Number of tubes in a vertical row
N_t	Number of tube rows
n	Exponent defined by Eq. 14.84d
$\dot{n}''$	Molar condensing flux, kmol/(m²·s)
Nu	Local Nusselt number on a flat plate at position L, $h_L L/k$
Nu_x	Local Nusselt number on a flat plate at position x, $h_x x/k$
Nu_m	Average Nusselt number $h_m d_o/k$
P	Pressure, N/m²
p	Pitch, m
p_ℓ	Longitudinal pitch, m
p_t	Transverse pitch, m
Pe	Peclet number, Re Pr
Pr	Prandtl number, $\mu c_p/k$
q	Heat transfer rate, W
q''	Heat flux, W/m²
R	Radius, m
R	Gas constant $P/\rho T$, (N·m)/(kg·K)
R	$[(\rho_\ell \mu_\ell)/(\rho_g \mu_g)]^{1/2}$
R	Thermal resistance, (m²·K)/W
R_δ	Radius to the edge of the condensate film, m
r	Radius, m
Ra	Rayleigh number defined by Eq. 14.103
Re	Film Reynolds number, $4\Gamma/\mu_\ell$
Re	Reynolds number, Gd/μ
Re_g	Vapor Reynolds number, Gxd_i/μ_g

Re_{go}	Vapor-only Reynolds number, Gd_i/μ_g
Re_ℓ	Liquid Reynolds number, $G(1-x)d_i/\mu_\ell$
$\mathrm{Re}_{\ell o}$	Liquid-only Reynolds number, $Gd_i/\mu_{\ell o}$
Re_x	Local Reynolds number, $\rho V_\infty x/\mu$
Re_z	Local film Reynolds number, $4\Gamma_z/\mu_\ell$
$\tilde{\mathrm{Re}}_x$	Two-phase Reynolds number, $\rho_\ell v_g x/\mu_\ell$
s	Fin spacing at fin base, m
Sc	Schmidt number, ν/D
Sh_x	Local Sherwood number, $h_D x/D$
T	Temperature, K
T_{bub}	Bubble temperature, K
T_{dew}	Dew point temperature, K
T_g^*	Equilibrium vapor temperature of a mixture, K
ΔT	$(T_s - T_w)$, K
t	Time, s
t	Fin thickness, m
U	Overall heat transfer coefficient, W/(m²·K)
u	Velocity, m/s
V_∞	Free stream velocity, m/s
v	Velocity, m/s
v_i	Velocity component (outward) normal to surface at vapor-liquid interface, m/s
v_z	Velocity in condensate film, m/s
v^*	Dimensionless velocity, defined by Eq. 14.121
W	Mass fraction
w	Mass flow width, defined by Eq. 14.60, m
We	Weber number, $G^2 d_i/\rho_h \sigma$
We	Weber number, $\rho_\ell \omega^2 d_o^3/4\sigma$
X	Lockhart-Martinelli parameter defined by Eq. 14.162
X_{tt}	Lockhart-Martinelli parameter for turbulent-turbulent flow, defined by Eq. 14.74
x	Vapor quality
x	Distance along flow direction, m
Y	Dimensionless pressure gradient ratio defined by Eq. 14.158
y	Distance normal to surface, m
$\tilde{y}$	Mole fraction of component in gas phase
Z	Dimensionless variable defined by Eq. 14.191
z	Distance along the flow direction, m

Greek Symbols

α	Thermal diffusivity, m²/s
α	Inclination angle from the horizontal

α_g	Void fraction of vapor, Eq. 14.125
β	Cone half-angle
β	Suction parameter, $-(v_i/V_\infty)\,\mathrm{Re}^{1/2}$
$\beta_{\ell k}$	Mass transfer coefficient of the $\ell - k$ pair, kmol/(m²·s)
Γ	Liquid film flow rate per perimeter, kg/(s·m)
δ	Liquid film thickness, m
ε	Heat transfer enhancement ratio
ζ	Function defined by Eq. 14.47
$[\zeta]$	High flux correction matrix
$\lceil\zeta_{\mathrm{ef}}\rfloor$	Diagonal high flux correction matrix
η_f	Fin efficiency
θ	Trapezoidal fin angle
θ	Correction factor defined by Eq. 14.165
$\bar{\lambda}$	Shortest Taylor wavelength defined by Eq. 14.105, m
μ	Dynamic viscosity, (N·s)/m²
ν	Kinematic viscosity, m²/s
ξ	Function defined by Eq. 14.89
ρ	Density, kg/m³
ρ_h	Two-phase density defined by Eq. 14.156
σ	Condensation coefficient
σ	Surface tension, N/m
τ	Shear stress, N/m²
τ	Dimensionless time, Eq. 14.171*b*
τ_g	Shear stress at liquid-vapor interface, N/m²
τ_g^*	Dimensionless shear stress, $\rho_\ell \tau_g/[\rho_\ell(\rho_\ell-\rho_g)\mu_\ell g]^{2/3}$
τ_H	Dimensionless time for heat-transfer-controlled collapse of a vapor bubble in a subcooled liquid
ϕ	Polar angle from the top of a sphere
ϕ	Circumferential angle measured from the top of a tube
ϕ	Condensation mass flux factor defined by Eq. 14.166
ϕ^2	Two-phase frictional multiplier, Eq. 14.147
ϕ_f	Flooding angle defined by Fig. 14.14
ω	Angular velocity, rad/s
ω	$W_{G\infty}/W_{gi}$
Ω	Dimensionless coefficient defined by Eq. 14.124

Subscripts

air	Air present
avg	Average
b	Bulk
c	Corrected, critical, coolant

ef	Effective
eq	Equivalent
f	Frictional
f	Fin, fin flank
f	Film
g	Vapor
go	Vapor only
gr	Gravity-dominated
G	Noncondensable gas
h	Hydraulic
i	Interface, inside, initial
in	Inlet
ℓ	Liquid
ℓo	Liquid only
L	Local value at position L
m	Mean/average, maximum, mixture
nc	Noncondensable
N	Vertical row of N tubes, Nth tube
Nu	Nusselt
o	Outside
out	Outlet
r	Fin root
s	Smooth
s	Saturated
sat	Saturated
sh	Shear-dominated
t	Total
tr	Transition
ΔT	Constant $(T_s - T_w)$
u	Unfinned
v	Vapor velocity
w	Wall surface, water
x	Local value at position x
z	Local value at position z
∞	Free stream
δ	Edge of condensate layer

Superscripts

$',''$	Modified, corrected
$''$	Per unit area

REFERENCES

1. I. Tanasawa, "Advances in Condensation Heat Transfer," *Advances in Heat Transfer,* Academic Press, Inc., **21,** pp. 55–139, 1991.

2. D. W. Woodruff and J. W. Westwater, "Steam Condensation on Various Gold Surfaces," *J. Heat Transfer,* **103,** pp. 685–692, 1981.

3. T. G. Sundararaman and T. Venkatram, "Dropwise Condensation Using Newly Developed Promoters on Copper Substrates," *Indian Chem. Engr.,* **23**(4), pp. 35–38, 1981.

4. G. A. O'Neill and J. W. Westwater, "Dropwise Condensation of Steam on Electroplated Silver Surfaces," *Int. J. Heat Mass Transfer,* **27,** pp. 1539–1549, 1984.

5. P. J. Marto, D. J. Looney, J. W. Rose, and A. S. Wanniarachchi, "Evaluation of Organic Coatings for the Promotion of Dropwise Condensation of Steam," *Int. J. Heat Mass Transfer,* **29,** pp. 1109–1117, 1986.

6. D. Zhang, Z. Lin, and J. Lin, "New Surface Materials for Dropwise Condensation," *Proc. 8th Int. Heat Transfer Conf., San Francisco,* **4,** pp. 1677–1682, 1986.

7. C. A. Nash and J. W. Westwater, "A Study of Novel Surfaces for Dropwise Condensation," *Proc. ASME/JSME Thermal Eng. Conf., Honolulu,* **2,** pp. 485–491, 1987.

8. A. S. Gavrish, V. G. Rifert, A. I. Sardak, and V. L. Podbereznyy, "A New Dropwise Condensation Promoter for Desalination and Power Plants," *Heat Transfer Research,* **25**(1), pp. 82–86, 1993.

9. D. Xu, X. Ma, and Z. Long, "Dropwise Condensation on a Variety of New Surfaces," in J. Taborek, J. Rose, and I. Tanasawa (eds.), *Condensation and Condenser Design,* ASME Press, New York, pp. 155–158, 1993.

10. X. Ma, D. Xu, and J. Lin, "A Study of Dropwise Condensation on the Ultra-Thin Polymer Surfaces," *Proc. 10th Int. Heat Transfer Conf., Brighton,* **3,** pp. 359–364, 1994.

11. Q. Zhao and B. M. Burnside, "Dropwise Condensation of Steam on Ion Implanted Condenser Surfaces," *J. Heat Recovery Systems and CHP,* **14**(5), pp. 525–534, 1994.

12. W. Nusselt, "The Condensation of Steam on Cooled Surfaces," *Z. d. Ver. Deut. Ing.,* **60,** pp. 541–546, 569–575, 1916 (Translated into English by D. Fullarton, *Chem. Engr. Funds.,* **1**(2), pp. 6–19, 1982).

13. W. M. Rohsenow, "Heat Transfer and Temperature Distribution in Laminar Film Condensation," *Trans. ASME,* **78,** pp. 1645–1648, 1956.

14. T. B. Drew (personal communication) in W. H. McAdams, *Heat Transmission,* 3d ed., McGraw-Hill, New York, p. 330, 1954.

15. D. Y. Shang and T. Adamek, "Study on Laminar Film Condensation of Saturated Steam on a Vertical Flat Plate for Consideration of Various Physical Factors Including Variable Thermophysical Properties," *Wärme-und Stoffübertragung,* **30,** pp. 89–100, 1994.

16. P. L. Kapitza, "Wave Flow of Thin Layers of Viscous Fluids," *Collected Papers of Kapitza,* Pergamon Press Ltd., New York, **2,** pp. 662–689, 1948.

17. S. S. Kutateladze, *Fundamentals of Heat Transfer,* Adademic Press, Inc., New York, pp. 303–308, 1963.

18. D. Butterworth, "Condensers: Basic Heat Transfer and Fluid Flow," in S. Kakac, A. E. Bergles, and F. Mayinger (eds.), *Heat Exchangers,* Hemisphere Publishing Corp., New York, pp. 289–314, 1981.

19. W. M. Nozhat, "The Effect of Surface Tension and Free Surface Curvature on Heat Transfer Coefficient of Thin Liquid Films," *Proc. ASME/JSME Thermal Eng. Conf., Maui,* **2,** pp. 171–178, 1995.

20. R. A. Seban, "Remarks on Film Condensation with Turbulent Flow," *Trans. ASME,* **76,** pp. 299–303, 1954.

21. W. M. Rohsenow, J. H. Webber, and A. T. Ling, "Effect of Vapor Velocity on Laminar and Turbulent Film Condensation," *Trans. ASME,* **78,** pp. 1637–1643, 1956.

22. H. Uehara, E. Kinosita, and S. Matsuda, "Theoretical Study on Turbulent Film Condensation on a Vertical Plate," *Proc. ASME/JSME Thermal Eng. Conf., Maui,* **2,** pp. 391–397, 1995.

23. D. A. Labuntsov, "Heat Transfer in Film Condensation of Pure Steam on Vertical Surfaces and Horizontal Tubes," *Teploenergetika,* **4**(7), pp. 72–80, 1957.

24. M-H Chun and K-T Kim, "A Natural Convection Heat Transfer Correlation for Laminar and Turbulent Film Condensation on a Vertical Surface," *Proc. ASME/JSME Thermal Eng. Conf., Reno,* **2,** pp. 459–464, 1991.

25. K. Stephan, *Heat Transfer in Condensation and Boiling,* Springer-Verlag, New York, 1992.

26. Y. Mayhew, personal communication, 1986.

27. O. P. Bergelin, P. K. Kegel, F. G. Carpenter, and C. Gazley, "Co-Current Gas Liquid Flow, Part II: Flow in Vertical Tubes, ASME," *Heat Transfer Fluid Mech. Inst.,* pp. 19–28, 1949.

28. D. Butterworth, "Film Condensation of Pure Vapor," in E. U. Schlünder (ed.), *Heat Exchanger Design Handbook,* **2,** Hemisphere Publishing Corp., New York, 1983.

29. F. Blangetti and E. U. Schlünder, "Local Heat Transfer Coefficients on Condensation in a Vertical Tube," *Proc. 6th Int. Heat Transfer Conf., Toronto,* **2,** pp. 437–442, 1978.

30. Z. L. Miropolskiy, R. I. Schneerova, and L. M. Teruakova, "Heat Transfer at Superheated Steam Condensation inside Tubes," *Proc. 5th Int. Heat Transfer Conf., Tokyo,* **3,** pp. 246–249, 1974.

31. J. W. Rose, "Fundamentals of Condensation Heat Transfer: Laminar Film Condensation," *JSME Int'l. J.,* **31,** pp. 357–375, 1988.

32. T. Fujii, *Theory of Laminar Film Condensation,* Springer-Verlag, New York, 1991.

33. E. M. Sparrow and J. L. Gregg, "A Boundary-Layer Treatment of Laminar Film Condensation," *J. Heat Transfer,* **81,** pp. 13–18, 1959.

34. M. M. Chen, "An Analytical Study of Laminar Film Condensation, Part I: Flat Plates," *J. Heat Transfer,* **83,** pp. 48–54, 1961.

35. J. C. Y. Koh, E. M. Sparrow, and J. P. Hartnett, "The Two Phase Boundary Layer in Laminar Film Condensation," *Int. J. Heat Mass Transfer,* **2,** pp. 69–82, 1961.

36. S. W. Churchill, "Laminar Film Condensation," *Int. J. Heat Mass Transfer,* **29,** pp. 1219–1225, 1986.

37. R. D. Cess, "Laminar Film Condensation on a Flat Plate in the Absence of a Body Force," *Z. Angew. Math. Phys.,* **11,** pp. 426–433, 1960.

38. J. C. Y. Koh, "Film Condensation in a Forced-Convection Boundary Layer Flow," *Int. J. Heat Mass Transfer,* **5,** pp. 941–954, 1962.

39. I. G. Shekriladze and V. I. Gomelauri, "Theoretical Study of Laminar Film Condensation of Flowing Vapor," *Int. J. Heat Mass Transfer,* **9,** pp. 581–591, 1966.

40. T. Fujii and H. Uehara, "Laminar Filmwise Condensation on a Vertical Surface," *Int. J. Heat Mass Transfer,* **15,** pp. 217–233, 1972.

41. W. J. Minkowycz and E M. Sparrow, "Condensation Heat Transfer in the Presence of Noncondensables, Interfacial Resistance, Superheating, Variable Properties, and Diffusion," *Int. J. Heat Mass Transfer,* **9,** pp. 1125–1144, 1966.

42. E. M. Sparrow, W. J. Minkowycz, and M. Saddy, "Forced Convection Condensation in the Presence of Noncondensables and Interfacial Resistance," *Int. J. Heat Mass Transfer,* **10,** pp. 1829–1845, 1967.

43. Y. S. Chin, S. J. Ormiston, and H. M. Soliman, "Numerical Solution of the Complete Two-Phase Model for Laminar Film Condensation with a Noncondensable Gas," *Proc. 10th Int. Heat Transfer Conf., Brighton,* **3,** pp. 287–292, 1994.

44. V. E. Denny, A. F. Mills, and V. J. Jusionis, "Laminar Film Condensation from a Steam-Air Mixture Undergoing Forced Flow Down a Vertical Surface," *J. Heat Transfer,* **93,** pp. 297–304, 1971.

45. C-Y Wang and C-J Tu, "Effects of Noncondensable Gas on Laminar Film Condensation in a Vertical Tube," *Int. J. Heat Mass Transfer,* **31,** pp. 2339–2345, 1988.

46. V. M. Borishansky, D. I. Volkov, and N. I. Ivashchenko, "Effects of Noncondensable Gas Content on Heat Transfer in Steam Condensation in a Vertical Tube," *Heat Transfer—Sov. Res.,* **9,** pp. 35–42, 1977.

47. J. W. Rose, "Approximate Equations for Forced-Convection Condensation in the Presence of a Noncondensing Gas on a Flat Plate and Horizontal Tube," *Int. J. Heat Mass Transfer,* **23,** pp. 539–546, 1980.

48. J. W. Rose, "Condensing in the Presence of Noncondensing Gases," in P. J. Marto and R. H. Nunn (eds.), *Power Condenser Heat Transfer Technology,* Hemisphere Publishing Corp., New York, pp. 151–162, 1981.

49. F. P. Incropera and D. P. DeWitt, *Fundamentals of Heat and Mass Transfer,* 2d ed., J. Wiley and Sons, Inc., New York, p. 777, 1985.

50. V. P. Carey, *Liquid-Vapor Phase Change Phenomena,* Hemisphere Publishing Corp., New York, pp. 378–389, 1992.

51. J. W. Rose, "Some Aspects of Condensation Heat Transfer Theory," *Int. Comm. Heat Mass Transfer,* **15,** pp. 449–473, 1988.

52. W. C. Lee and J. W. Rose, "Forced Convection Film Condensation on a Horizontal Tube with and without Non-Condensing Gases," *Int. J. Heat Mass Transfer,* **27,** pp. 519–528, 1984.

53. P. J. Marto and R. H. Nunn (eds.), *Power Condenser Heat Transfer Technology,* Hemisphere Publishing Corp., New York, 1981.

54. P. J. Marto, "Heat Transfer and Two-Phase Flow during Shell-Side Condensation," *Heat Transfer Eng.,* **5,** pp. 31–61, 1984.

55. T. Fujii, "Research Problems for Improving the Performance of Power Plant Condensers," in J. Taborek, J. Rose and I. Tanasawa (eds.), *Condensation and Condenser Design,* ASME Press, pp. 487–498, 1993.

56. D. W. Nobbs and Y. R. Mayhew, "Effect of Downward Vapor Velocity and Inundation on Condensation Rates on Horizontal Tube Banks," Steam Turbine Condensers, National Engineering Laboratory Report No. 619, East Kilbride, Glasgow, pp. 39–52, 1976.

57. S. S. Kutateladze, N. I. Gogonin, A. R. Dorokhov, and V. I. Sosunov, "Film Condensation of Flowing Vapor on a Bundle of Plain Horiontal Tubes," *Thermal Eng.,* **26,** pp. 270–273, 1979.

58. T. Fujii, H. Uehara, K. Hirata, and K. Oda, "Heat Transfer and Flow Resistance in Condensation of Low Pressure Steam Flowing through Tube Banks," *Int. J. Heat Mass Transfer,* **15,** pp. 247–260, 1972.

59. A. Cavallini, S. Frizzerin, and L. Rossetto, "Condensation of R-11 Vapor Flowing Downward outside a Horizontal Tube Bundle," *Proc. 8th Int. Heat Transfer Conf., San Francisco,* **4,** pp. 1707–1712, 1986.

60. D. Butterworth, "Developments in the Design of Shell and Tube Condensers," ASME Winter Annual Meeting, Atlanta, ASME Preprint 77-WA/HT-24, 1977.

61. M. M. Chen, "An Analytical Study of Laminar Film Condensation, Part 2: Single and Multiple Horizontal Tubes," *J. Heat Transfer,* **83,** pp. 55–60, 1961.

62. R. Armbruster and J. Mitrovic, "Patterns of Falling Film Flow over Horizontal Smooth Tubes," *Proc. 10th Int. Heat Transfer Conf., Brighton,* **3,** pp. 275–280, 1994.

63. D. Q. Kern, "Mathematical Development of Loading in Horizontal Condensers," *AIChE J.,* **4,** pp. 157–160, 1958.

64. S. N. Fuks, "Heat Transfer with Condensation of Steam Flowing in a Horizontal Tube Bundle (in Russian)," *Teploenergetika,* **4,** p. 35, 1957 (Translated into English in NEL Report 1041, East Kilbride, Glasgow).

65. I. D. R. Grant and B. D. J. Osment, "The Effect of Condensate Drainage on Condenser Performance," National Engineering Laboratory Report No. 350, East Kilbride, Glasgow, 1968.

66. G. G. Shklover and A. V. Buevich, "Investigation of Steam Condensation in an Inclined Bundle of Tubes," *Thermal Eng.,* **25,** pp. 49–52, 1978.

67. L. D. Berman and Y. A. Tumanov, "Investigation of Heat Transfer in the Condensation of Moving Steam on a Horizontal Tube," *Teploenergetika,* **9,** pp. 77–83, 1962.

68. J. M. McNaught, "Two-Phase Forced Convection Heat Transfer During Condensation on Horizontal Tube Bundles," *Proc. 7th Int. Heat Transfer Conf., Munich,* **5,** pp. 125–131, 1982.

69. C. M. Chu and J. M. McNaught, "Condensation on Bundles of Plain and Low-Finned Tubes—Effects of Vapor Shear and Condensate Inundation," *Proc. 3rd UK Heat Transfer Conf.,* IChemE Symp. Series 129, **1,** pp. 225–232, 1992.

70. B. J. Davidson and M. Rowe, "Simulation of Power Plant Condenser Performance by Computational Methods: An Overview," in P. J. Marto and R. H. Nunn (eds.), *Power Condenser Heat Transfer Technology,* pp. 17–49, Hemisphere Publishing Corp., New York, 1981.

71. D. M. Eissenberg, personal communication, Oak Ridge National Laboratory, Oak Ridge, Tennessee, 1977.

72. I. D. R. Grant and D. Chisholm, "Two-Phase Flow on the Shell-Side of a Segmentally Baffled Shell-and-Tube Heat Exchanger," *J. Heat Transfer,* **101,** pp. 38–42, 1979.

73. I. D. R. Grant and D. Chisholm, "Horizontal Two-Phase Flow Across Tube Banks," *Int. J. Heat Fluid Flow,* **2**(2), pp. 97–100, 1980.

74. K. Ishihara, J. W. Palen, and J. Taborek, "Critical Review of Correlations for Predicting Two-Phase Flow Pressure Drop across Tube Banks," *Heat Transfer Eng.,* **1**(3), pp. 23–32, 1980.

75. L. D. Berman, "Determining the Mass Transfer Coefficient in Calculations on Condensation of Steam Containing Air," *Teploenergetika,* **16,** pp. 68–71, 1969.

76. A. F. Mills, C. Tan, and D. K. Chung, "Experimental Study of Condensation from Steam-Air Mixtures Flowing over a Horizontal Tube: Overall Condensation Rates," *Proc. 5th Int. Heat Transfer Conf., Tokyo,* **5,** pp. 20–23, 1974.

77. A. P. Colburn and O. A. Hougen, "Design of Cooler Condensers for Mixtures of Vapors with Noncondensing Gases," *Ind. Eng. Chem.,* **26,** pp. 1178–1182, 1934.

78. L. D. Berman and S. N. Fuks, "Mass Transfer in Condensers with Horizontal Tubes when Steam Contains Air," *Teploenergetika,* **5,** pp. 66–74, 1958.

79. T. Fujii and K. Oda, "Effect of Air upon the Condensation of Steam Flowing through Tube Bundles (Japanese)," *Trans. JSME,* **50,** pp. 107–113, 1984.

80. E. J. Barsness, "Calculation of the Performance of Surface Condensers by Digital Computer," ASME Paper 63-PWR-2, National Power Conference, Cincinnati, Ohio, 1963.

81. W. H. Emerson, "The Application of a Digital Computer to the Design of Surface Condenser," *The Chemical Engineer,* **228**(5), pp. 178–184, 1969.

82. J. L. Wilson, "The Design of Condensers by Digital Computers," *I. Chem. E. Symp. Ser.,* no. 35, pp. 21–27, 1972.

83. J. A. Hafford, "ORCON1: A Fortran Code for the Calculation of a Steam Condenser of Circular Cross Section," ORNL-TM4248, Oak Ridge National Laboratory, Oak Ridge, Tennessee, 1973.

84. H. Shida, M. Kuragaska, and T. Adachi, "On the Numerical Analysis Method of Flow and Heat Transfer in Condensers," *Proc. 7th Heat Transfer Conf., Munich,* **6,** pp. 347–352, 1982.

85. H. L. Hopkins, J. Loughhead, and C. J. Monks, "A Computerized Analysis of Power Condenser Performance Based upon an Investigation of Condensation," *Condensers: Theory and Practice, I. Chem. E. Symp. Ser.,* no. 75, pp. 152–170, Pergamon Press, London, 1983.

86. K. K. Yau and C. Pouzenc, "Computational Modeling of Power Plant Condenser Performance," *Proc. 3rd UK National Heat Transfer Conf., I Mech. E. Symp. Series 129,* **1,** pp. 217–224, 1992.

87. C. Zhang, "Numerical Modeling Using a Quasi-Three-Dimensional Procedure for Large Power Plant Condensers," *J. Heat Transfer,* **116,** pp. 180–188, 1994.

88. P. J. Marto, "An Evaluation of Film Condensation on Horizontal Intergral-Fin Tubes," *J. Heat Transfer,* **110,** pp. 1287–1305, 1988.

89. J. W. Rose, "Models for Condensation on Horizontal Low-Finned Tubes," *Proc. 4th UK Heat Transfer Conf., I. Mech. E.,* pp. 417–429, 1995.

90. H. Masuda and J. W. Rose, "Static Configuration of Liquid Films on Horizontal Tubes with Low Radial Fins: Implications for Condensation Heat Transfer," *Proc. R. Soc. Lond. A,* **410,** pp. 125–139, 1987.

91. H. Honda, S. Nozu, and K. Mitsumori, "Augmentation of Condensation on Finned Tubes by Attaching a Porous Drainage Plate," *Proc. ASME/JSME Thermal Eng. Joint Conf., Honolulu,* **3,** pp. 289–295, 1983.

92. K. O. Beatty and D. L. Katz, "Condensation of Vapors on Outside of Finned Tubes," *Chem. Eng. Prog.,* **44,** pp. 55–70, 1948.

93. H. Honda and S. Nozu, "A Prediction Method for Heat Transfer During Film Condensation on Horizontal Low Integral-Fin Tubes," *J. Heat Transfer,* **109,** pp. 218–225, 1987.

94. T. Adamek and R. L. Webb, "Prediction of Film Condensation on Horizontal Integral Fin Tubes," *Int. J. Heat Mass Transfer,* **33,** pp. 1721–1735, 1990.

95. J. W. Rose, "An Approximate Equation for the Vapour-Side Heat Transfer Coefficient for Condensation on Low-Finned Tubes," *Int. J. Heat Mass Transfer,* **37,** pp. 865–875, 1994.

96. A. Briggs and J. W. Rose, "Effect of 'Fin Efficiency' on a Model for Condensation Heat Transfer on a Horizontal Integral-Fin Tube," *Int. J. Heat Mass Transfer,* **37,** pp. 457–463, 1994.

97. M. H. Jaber and R. L. Webb, "Steam Condensation on Horizontal Integral-Fin Tubes of Low Thermal Conductivity," *J. Enhanced Heat Transfer,* **3,** pp. 55–71, 1996.

98. I. I. Gogonin and A. R. Dorokhov, "Enhancement of Heat Transfer in Horizontal Shell-and-Tube Condensers," *Heat Transfer-Soviet Research,* **3,** pp. 119–126, 1981.

99. A. G. Michael, P. J. Marto, A. S. Wanniarachchi, and J. W. Rose, "Effect of Vapor Velocity During Condensation on Horizontal Smooth and Finned Tubes," *Heat Transfer for Phase Change,* ASME HTD-Vol. 114, pp. 1–10, 1989.

100. A. Cavallini, B. Bella, A. Longo, and L. Rossetto, "Experimental Heat Transfer Coefficients During Condensation of Halogenated Refrigerants on Enhanced Tubes," *J. Enhanced Heat Transfer,* **2,** pp. 115–125, 1995.

101. A. Cavallini, L. Doretti, N. Klammsteiner, G. Longo, and L. Rossetto, "A New Model for Forced-Convection Condensation on Integral-Fin Tubes," *J. Heat Transfer,* **118,** pp. 689–693, 1996.

102. H. Honda, B. Uchima, S. Nozu, H. Nakata, and E. Torigoe, "Film Condensation of R-113 on In-Line Bundles of Horizontal Finned Tubes," *J. Heat Transfer,* **113,** pp. 479–486, 1991.

103. H. Honda, B. Uchima, S. Nozu, E. Torigoe, and S. Imai, "Film Condensation of R-113 on Staggered Bundle of Horizontal Finned Tubes," *J. Heat Transfer,* **114,** pp. 442–449, 1992.

104. H. Honda, S. Nozu, and Y. Takeda, "A Theoretical Model of Film Condensation in a Bundle of Horizontal Low Finned Tubes," *J. Heat Transfer,* **111,** pp. 525–532, 1989.

105. T. Fujii, "Enhancement to Condensing Heat Transfer—New Developments," *J. Enhanced Heat Transfer,* **2,** pp. 127–138, 1995.

106. H. Honda, M. Takamatsu, and K. H. Kim, "Condensation of CFC-11 and HCFC 123 in In-Line Bundle of Horizontal Finned Tubes," *Proc. Eng. Found. Conf.,* Condensation and Condenser Design, ASME, pp. 543–556, 1993.

107. R. L. Webb and C. G. Murawski, "Row Effect for R-11 Condensation on Enhanced Tubes," *J. Heat Transfer,* **112,** pp. 768–776, 1990.

108. C. M. Chu and J. M. McNaught, "Tube Bundle Effects in Crossflow Condensation on Low-Finned Tubes," *Proc. 10th Int. Heat Transfer Conf., Brighton,* **3,** pp. 293–298, 1994.

109. K. E. Hassan and M. Jakob, "Laminar Film Condensation of Pure Saturated Vapors on Inclined Circular Cylinders," *J. Heat Transfer,* **80,** pp. 887–894, 1958.

110. G. Selin, "Heat Transfer by Condensing Pure Vapors Outside Inclined Tubes," *Int. Devel. Heat Transfer,* pp. 279–289, ASME, New York, 1961.

111. B. G. Nimmo and G. Leppert, "Laminar Film Condensation on a Finite Horizontal Surface," *Proc. 4th Int. Heat Transfer Conf., Paris,* **6,** Cs 2.2, 1970.

112. T. Shigechi, N. Kawae, Y. Tokita, and T. Yamada, "Film Condensation Heat Transfer on a Finite-Size Horizontal Plate Facing Upward," *Heat Transfer-Japanese Research,* **22,** pp. 66–77, 1993.

113. J. S. Chiou and T. B. Chang, "Laminar Film Condensation on a Horizontal Disk," *Wärme-und Stoffübertragung,* **29,** pp. 141–144, 1994.

114. V. Prasad and Y. Jaluria, "Transient Film Condensation on a Finite Horizontal Plate," *Chem. Eng. Commun.,* **13,** pp. 327–342, 1982.

115. J. Gerstmann and P. Griffith, "Laminar Film Condensation on the Underside of Horizontal and Inclined Surfaces," *Int. J. Heat Mass Transfer,* **10,** pp. 567–580, 1967.

116. M. Yanadori, K. Hijikata, Y. Mori, and M. Uchida, "Fundamental Study of Laminar Film Condensation Heat Transfer on a Downward Horizontal Surface," *Int. J. Heat Mass Transfer,* **28,** pp. 1937–1944, 1985.

117. V. Dhir and J. Lienhard, "Laminar Film Condensation on Plane and Axisymmetric Bodies in Nonuniform Gravity," *J. Heat Transfer,* **91,** pp. 97–100, 1971.

118. V. Dhir and J. Lienhard, "Similar Solutions for Film Condensation with Variable Gravity and Body Shape," *J. Heat Transfer,* **93,** pp. 483–486, 1973.

119. A. Nakayama and H. Koyama, "An Integral Treatment of Laminar and Turbulent Film Condensation on Bodies of Arbitrary Geometrical Configuration," *J. Heat Transfer,* **107,** pp. 417–423, 1985.

120. A. Karimi, "Laminar Film Condensation on Helical Reflux Condensers and Related Configurations," *Int. J. Heat Mass Transfer,* **20,** pp. 1137–1144, 1977.

121. G. P. Fieg and W. Roetzel, "Calculation of Laminar Film Condensation in/on Inclined Elliptical Tubes," *Int. J. Heat Mass Transfer,* **37,** pp. 619–624, 1994.

122. S. B. Memory, V. H. Adams, and P. J. Marto, "Free and Forced Convection Laminar Film Condensation on Horizontal Elliptical Tubes," *Int. J. Heat Mass Transfer,* **40,** pp. 3395–3406, 1997.

123. E. M. Sparrow and J. L. Gregg, "A Theory of Rotating Condensation," *J. Heat Transfer,* **81,** pp. 113–120, 1959.

124. S. S. Nandapurkar and K. O. Beatty, "Condensation on a Horizontal Rotating Disk," *Chemical Engr. Prog. Symp. Ser.,* Heat Transfer-Storrs, pp. 129–137, 1959.

125. E. M. Sparrow and J. L. Gregg, "The Effect of Vapor Drag on Rotating Condensation," *J. Heat Transfer,* **82,** pp. 71–72, 1960.

126. E. M. Sparrow and J. P. Hartnett, "Condensation on a Rotating Cone," *J. Heat Transfer,* **83,** pp. 101–102, 1961.

127. P. J. Marto, "Laminar Film Condensation on the Inside of Slender, Rotating Truncated Cones," *J. Heat Transfer,* **95,** pp. 270–272, 1973.

128. P. J. Marto, "Rotating Heat Pipes," in D. E. Metzger and N. H. Afgan (eds.), *Heat and Mass Transfer in Rotating Machinery,* Hemisphere Publishing Corp., New York, pp. 609–632, 1984.

129. L. L. Vasiliev and V. V. Khrolenok, "Heat Transfer Enhancement with Condensation by Surface Rotation," *Heat Recovery Sys. and CHP,* **13,** pp. 547–563, 1993.

130. S. Mochizuki and T. Shiratori, "Condensation Heat Transfer within a Circular Tube under Centrifugal Acceleration Field," *J. Heat Transfer,* **102,** pp. 158–162, 1980.

131. A. A. Nicol and M. Gacesa, "Condensation of Steam on a Rotating, Vertical Cylinder," *J. Heat Transfer,* **92,** pp. 144–152, 1970.

132. N. V. Suryanarayana, "Condensation Heat Transfer under High Gravity Conditions," *Proc. 5th Int. Heat Transfer Conf., Tokyo,* **3,** pp. 279–285, 1974.

133. W. E. McEver and H. Hwangbo, "Surface Tension Effects in a Space Radiator Condenser with Capillary Liquid Drainage," AIAA 18th Thermophysics Conf., Montreal, 1983.

134. J. A. Valenzuela, J. A. McCormick, and J. Thornborrow, "Design and Performance of an Internally Drained Condenser Surface," *Condensation and Condenser Design,* ASME, New York, pp. 557–568, 1993.

135. M. K. Bologa, V. P. Korovkin, and I. K. Savin, "Mechanism of Condensation Heat Transfer Enhancement in an Electric Field and the Role of Capillary Processes," *Int. J. Heat Mass Transfer,* **38,** pp. 175–182, 1995.

136. G. Breber, J. W. Palen, and J. Taborek, "Prediction of Horizontal Tubeside Condensation of Pure Components Using Flow Regime Criteria," *J. Heat Transfer,* **102,** pp. 471–476, 1980.

137. M. M. Rahman, A. M. Fathi, and H. M. Soliman, "Flow Pattern Boundaries During Condensation: New Experimental Data," *Canadian J. Chem. Eng.,* **63,** pp. 547–552, 1985.

138. H. M. Soliman, "Flow Pattern Transitions During Horizontal In-Tube Condensation," *Encycl. of Fluid Mech.,* Gulf Publishing Co., Houston, 1986.

139. G. B. Wallis, "Flooding Velocities for Air and Water in Vertical Tubes," UKAEA Report No. AEEW-R123, 1961.

140. D. Butterworth, "Film Condensation of Pure Vapor," in G. F. Hewitt (ed.), *Handbook of Heat Exchanger Design,* Hemisphere Publishing Corp., New York, 1990.

141. T. J. Rabas and B. Arman, "The Effect of the Exit Condition on the Performance of Intube Condensers," ASME HTD-Vol. 314, pp. 39–47, 1995.

142. J. C. Chato, "Laminar Condensation Inside Horizontal and Inclined Tubes," *ASHRAE J.,* **4,** pp. 52–60, 1962.

143. H. Jaster and P. G. Kosky, "Condensation Heat Transfer in a Mixed Flow Regime," *Int. J. Heat Mass Transfer,* **19,** pp. 95–99, 1976.

144. M. K. Dobson, "Heat Transfer and Flow Regimes During Condensation in Horizontal Tubes," Ph.D. Thesis, University of Illinois, 1994.

145. D. P. Traviss, W. M. Rohsenow, and A. B. Baron, "Forced Convection Condensation inside Tubes: A Heat Transfer Equation for Condenser Design," *ASHRAE Trans., 79*, pp. 157–165, 1972.

146. W. W. Akers, H. A. Deans, and O. K. Crosser, "Condensing Heat Transfer within Horizontal Tubes," *Chem. Eng. Prog. Symp. Series, 55*, pp. 171–176, 1959.

147. L. D. Boyko and G. N. Kruzhilin, "Heat Transfer and Hydraulic Resistance During Condensation of Steam in a Horizontal Tube and in a Bundle of Tubes," *Int. J. Heat Mass Transfer, 10*, pp. 361–373, 1967.

148. A. Cavallini and R. Zecchin, *Proc. 13th Int. Congress Refrigeration,* Washington, D.C., 1971.

149. M. M. Shah, "A General Correlation for Heat Transfer During Film Condensation Inside Pipes," *Int. J. Heat Mass Transfer, 22*, pp. 547–556, 1979.

150. J. W. Palen, R. S. Kistler, and Z. F. Yang, "What We Still Don't Know About Condensation in Tubes," *Condensation and Condenser Design,* ASME, New York, pp. 19–53, 1993.

151. G. F. Hewitt, "Gas-Liquid Flow," in E. U. Schlünder (ed.), *Heat Exchanger Design Handbook,* Hemisphere Publishing Corp., New York, 1983.

152. L. Friedel, "Improved Friction Pressure Drop Correlations for Horizontal and Vertical Two-Phase Pipe Flow," Paper no. E2, European Two-Phase Flow Group Meeting, Ispra, Italy, 1979.

153. D. Chisholm, "Pressure Gradients due to Friction during the Flow of Evaporating Two-Phase Mixtures in Smooth Tubes and Channels," *Int. J. Heat Mass Transfer, 16*, pp. 347–348, 1973.

154. R. W. Lockhart and R. C. Martinelli, "Proposed Correlation of Data for Isothermal Two-Phase Two-Component Flow in Pipes," *Chem. Eng. Prog., 45*(1), pp. 39–48, 1949.

155. R. C. Martinelli and D. B. Nelson, "Prediction of Pressure Drop during Forced-Circulation Boiling of Water," *Trans. ASME, 70*, pp. 695–702, 1948.

156. D. Chisholm, "A Theoretical Basis for the Lockhart-Martinelli Correlation for Two-Phase Flow," *Int. J. Heat Mass Transfer, 10*, pp. 1767–1778, 1967.

157. A. L. Souza, J. C. Chato, J. P. Wattelet, and B. R. Christoffersen, "Pressure Drop During Two-Phase Flow of Pure Refrigerants and Refrigerant-Oil Mixtures in Horizontal Smooth Tubes," ASME HTD-Vol. 243, New York, pp. 35–41, 1993.

158. R. G. Sardesai, R. G. Owen, and D. J. Pulling, "Pressure Drop for Condensation of a Pure Vapor in Downflow in a Vertical Tube," *Proc. 7th Int. Heat Transfer Conf., Munich, 5*, pp. 139–145, 1982.

159. W. Groenewald and D. G. Kroger, "Effect of Mass Transfer on Turbulent Friction During Condensation Inside Ducts," *Int. J. Heat Mass Transfer, 38*, pp. 3385–3392, 1995.

160. C-K Chen and S-A Yang, "Laminar Film Condensation Inside a Horizontal Elliptical Tube with Variable Wall Temperature," *Int. J. Heat and Fluid Flow, 15*, pp. 75–78, 1994.

161. N. Kaushik and N. Z. Azer, "A General Heat Transfer Correlation for Condensation Inside Internally Finned Tubes," *ASHRAE Trans., 94*(2), pp. 261–279, 1988.

162. S. C. Lee, M. Chung, and H. S. Shin, "Condensation Heat Transfer and Pressure Drop Performance of Horizontal Smooth and Internally-Finned Tubes with Refrigerant 113," *Exp. Heat Transfer, Fluid Mech. and Thermodynamics, 2*, pp. 1349–1356, Elsevier Science Publishers, 1993.

163. L. M. Schlager, M. B. Pate, and A. E. Bergles, "Heat Transfer and Pressure Drop During Evaporation and Condensation of R-22 in Horizontal Micro-Fin Tubes," *Int. J. Refrigeration, 12*, pp. 6–14, 1989.

164. L. M. Chamra and R. L. Webb, "Condensation and Evaporation in Micro-Fin Tubes at Equal Saturation Temperatures," *J. Enhanced Heat Transfer, 2*, pp. 219–229, 1995.

165. R. L. Webb and C-Y Yang, "A Comparison of R-12 and R-134a Condensation Inside Small Extruded Aluminum Plain and Micro-Fin Tubes," *Proc. 2nd Vehicle Thermal Management Systems Conf., I. Mech. E., London,* pp. 77–85, 1995.

166. L. L. Tovazhnyanski and P. A. Kapustenko, "Intensification of Heat and Mass Transfer in Channels of Plate Condensers," *Chem. Eng. Commun., 31*, pp. 351–366, 1984.

167. C. B. Panchal, "Condensation Heat Transfer in Plate Heat Exchangers," in J. T. Pearson and J. B. Kitto, Jr. (eds.), *Two-Phase Heat Exchanger Symposium,* HTD-Vol. 44, ASME, New York, pp. 45–52, 1985.

168. F. Chopard, C. Marvillet, and J. Pantaloni, "Assessment of Heat Transfer Performance of Rectangular Channel Geometries: Implications on Refrigerant Evaporator and Condenser Design," *Proc. 3rd UK Nat. Heat Transfer Conference, I. Chem. E. Symp. Ser. 129,* **2,** pp. 725–733, 1992.

169. Z-Z Wang and Z-N Zhao, "Analysis of Performance of Steam Condensation Heat Transfer and Pressure Drop in Plate Condensers," *Heat Transfer Eng.,* **14,** pp. 32–41, 1993.

170. B. Thonon, "Plate Heat Exchangers, Ten Years of Research at GRETh, Part I: Flow Pattern and Heat Transfer in Single Phase and Two Phase Flows," *Rev. Generale de Thermique,* **34,** pp. 77–90A, 1995.

171. V. Srinivasan and R. K. Shah, "Condensation in Compact Heat Exchangers," *J. Enhanced Heat Transfer,* **4,** pp. 237–256, 1997.

172. H. R. Jacobs and D. S. Cook, "Direct Contact Condensation on a Noncirculating Drop," *Proc. 6th Int. Heat Transfer Conf., Toronto,* **2,** pp. 389–393, 1978.

173. J. D. Ford and A. Lekic, "Rate of Growth of Drops during Condensation," *Int. J. Heat Mass Transfer,* **16,** pp. 61–64, 1973.

174. T. Sundararajan and P. S. Ayyaswamy, "Heat and Mass Transfer Associated with Condensation on a Moving Drop: Solutions for Intermediate Reynolds Numbers by a Boundary Layer Formulation," *J. Heat Transfer,* **107,** pp. 409–416, 1985.

175. E. Kulic and E. Rhodes, "Direct Contact Condensation from Air-Steam Mixtures on a Single Droplet," *Can. J. Chem Engrg.,* **55,** pp. 131–137, 1977.

176. C-S Chang, I. Tanasawa, and S. Nishio, "Direct Contact Condensation of an Immiscible Vapor on Falling Liquid Droplets: Experimental Study on Condensation of Freon-R-113 Vapor on Water Droplets," *Proc. ASME/JSME Thermal Eng. Joint Conf., Honolulu,* **3,** pp. 305–310, 1983.

177. F. Mayinger and A. Chavez, "Measurement of Direct Contact Condensation of Pure Saturated Vapor on an Injection Spray by Applying Pulsed Laser Holography," *Int. J. Heat Mass Trans.,* **35,** pp. 691–702, 1992.

178. H. R. Jacobs, "Direct-Contact Condensers," in *Heat Exchanger Design Handbook,* E. U. Schlünder (ed.), pp. 2.6.8, Hemisphere Publishing Corp., New York, 1983.

179. D. Hasson, D. Luss, and R. Peck, "Theoretical Analyses of Vapor Condensation on Laminar Jets," *Int. J. Heat Mass Transfer,* **7,** pp. 969–981, 1964.

180. J. Mitrovic and A. Ricoeur, "Fluid Dynamics and Condensation-Heating of Capillary Liquid Jets," *Int. J. Heat Mass Transfer,* **38,** pp. 1483–1494, 1995.

181. T. L. Lui, H. R. Jacobs, and K. Chen, "An Experimental Study of Direct Condensation on a Fragmenting Circular Jet," *J. Heat Transfer,* **111,** pp. 585–588, 1989.

182. G. P. Celata, M. Cumo, G. E. Farello, and G. Focardi, "A Comprehensive Analysis of Direct Contact Condensation of Saturated Steam on Subcooled Liquid Jets," *Int. J. Heat Mass Transfer,* **32,** pp. 639–654, 1989.

183. H. R. Jacobs, J. A. Bogart, and R. W. Pensel, "Condensation on a Thin Film Flowing over an Adiabatic Sphere," *Proc. 7th Int. Heat Transfer Conf., Munich,* **5,** pp. 89–94, 1982.

184. T. D. Karapantsios and A. J. Karabelas, "Direct-Contact Condensation in the Presence of Noncondensables over Free-Falling Films with Intermittent Liquid Feed," *Int. J. Heat Mass Transfer,* **38,** pp. 795–805, 1995.

185. L. W. Florschuetz and B. T. Chao, "On the Mechanics of Vapor Bubble Collapse," *J. Heat Transfer,* **87,** pp. 209–220, 1965.

186. D. Moalem-Maron and W. Zijl, "Growth Condensation and Departure of Small and Large Bubbles in Pure and Binary Systems," *Chem. Eng. Sci.,* **33,** pp. 1339–1346, 1978.

187. D. D. Wittke and B. T. Chao, "Collapse of Vapor Bubbles with Translatory Motion," *J. Heat Transfer,* **89,** pp. 17–24, 1967.

188. J. Isenberg, D. Moalem-Maron, and S. Sideman, "Direct Contact Heat Transfer with Changes in Phase: Bubble Collapse with Translatory Motion in Single and Two Component Systems," *Proc. 4th Int. Heat Transfer Conf., Paris,* **5,** B.2.5, 1970.

189. A. Ullman and R. Letan, "Effect of Noncondensibles on Condensation and Evaporation of Bubbles," in H. R. Jacobs (ed.), *Proc. 1988 Nat. Heat Transfer Conf.,* **2,** pp. 409–414, ASME Press, 1988.

190. A. E. Bergles and A. Bar-Cohen, "Direct Liquid Cooling of Microelectronic Components," in A. Bar-Cohen and A. D. Kraus (eds.), *Advances in Thermal Modeling of Electronic Components and Systems,* **2,** pp. 278–294, ASME Press, 1990.

191. F. Zangrando and D. Bharathan, "Direct-Contact Condensation of Low-Density Steam on Seawater at High Inlet Noncondensable Concentrations," *J. Heat Transfer.,* **115,** pp. 690–698, 1993.

192. S. Sideman and D. Moalem-Maron, "Direct Contact Condensation," in J. P. Hartnett and T. F. Irvine (eds.), *Adv. in Heat Transfer,* **15,** pp. 228–281, Academic Press, New York, 1982.

193. G. F. Hewitt, G. L. Shires, and T. R. Bott, *Process Heat Transfer,* CRC Press, Inc., Boca Raton, Florida, 1994.

194. D. R. Webb, "Design of Multicomponent Condensers," *Heat Exchanger Design Update,* **2**(1), Begell House Publishers, New York, 1995.

195. T. Fujii, "Overlooked Factors and Unresolved Problems in Experimental Research on Condensation Heat Transfer," *Exp. Thermal and Fluid Science,* **5,** pp. 652–663, 1992.

196. I. Tanasawa, "Recent Advances in Condensation Heat Transfer," *Proc. 10th Int. Heat Transfer Conf., Brighton,* **1,** pp. 297–312, 1994.

197. R. Hashimoto, K. Yanagi, and T. Fujii, "Effects of Condensate Flow Patterns upon Gravity-Controlled Condensation of Ethanol and Water Mixtures on a Vertical Surface," *Heat Transfer-Japanese Research,* **23,** pp. 330–348, 1994.

198. M. Takuma, A. Yamade, T. Matsuo, and Y. Tokita, "Condensation Heat Transfer Characteristics of Ammonia-Water Vapor Mixture on a Vertical Flat Surface," *Proc. 10th Int. Heat Transfer Conf., Brighton,* **3,** pp. 395–400, 1994.

199. W. C. Wang, C. Yu, and B. X. Wang, "Condensation Heat Transfer of a Nonazeotropic Binary Mixture on a Horizontal Tube," *Int. J. Heat Mass Transfer,* **38,** pp. 233–240, 1995.

200. L. Silver, "Gas Cooling with Aqueous Condensation," *Trans. Inst. Chem. Eng.,* **25,** pp. 30–42, 1947.

201. K. J. Bell and M. A. Ghaly, "An Approximate Generalized Design Method for Multicomponent/Partial Condensers," *AIChE Symp. Ser.,* **69**(131), pp. 72–79, 1973.

202. G. Ackermann, "Combined Heat and Mass Transfer in the Same Field at High Temperature and Partial Pressure Differences," *Forsch. Ingenieurwes.,* **8**(382), pp. 1–16, 1937.

203. A. P. Colburn and T. B. Drew, "The Condensation of Mixed Vapors," *Trans. ASChE,* **33,** pp. 197–215, 1937.

204. B. C. Price and K. J. Bell, "Design of Binary Vapour Condensers Using the Colburn-Drew Equations," *AIChEJ Symp. Series,* **7**(138), pp. 267–272, 1974.

205. R. G. Sardesai, R. A. Shock, and D. Butterworth, "Heat and Mass Transfer in Multicomponent Condensation and Boiling," *Heat Transfer Eng.,* **3,** pp. 104–114, 1982.

206. R. Krishna and G. L. Standart, "A Multicomponent Film Model Incorporating a General Matrix Method of Solution to Maxwell-Stephan Equations," *AIChE J.,* **22,** pp. 383–389, 1976.

207. H. L. Toor, "Solution of the Linearized Equations of Multicomponent Mass Transfer," *AIChE J.,* **10,** pp. 448–460, 1964.

208. D. R. Webb and J. M. McNaught, "Condensers," in D. Chisholm (ed.), *Developments in Heat Exchanger Technology,* Applied Science Publishers, London, pp. 71–126, 1980.

209. T. Fujii, H. Honda, and S. Nozu, "Condensation of Fluorocarbon Refrigerants Inside a Horizontal Tube—Proposals of Semi-Empirical Expressions for the Local Heat Transfer Coefficient and the Interfacial Friction Factor," *Refrigeration,* **55**(627), pp. 3–19, 1980 (in Japanese).

210. S. J. Ormiston, G. D. Raithby, and L. N. Carlucci, "Numerical Modeling of Power Station Steam Condensers, Part 1: Convergence Behavior of a Finite-Volume Model," *Num. Heat Transfer, Part B,* **27,** pp. 81–102, 1995.

211. S. J. Ormiston, G. D. Raithby, and L. N. Carlucci, "Numerical Modeling of Power Station Steam Condensers, Part 2: Improvement of Solution Behavior," *Num. Heat Transfer, Part B,* **27,** pp. 103–125, 1995.

212. C. Zhang and Y. Zhang, "Sensitivity Analysis of Heat Transfer Coefficient Correlations on the Predictions of Steam Surface Condensers," *Heat Transfer Eng.,* **15,** pp. 54–63, 1994.

213. C. Zhang, "Local and Overall Condensation Heat Transfer Behavior in Horizontal Tube Bundles," *Heat Transfer Eng.,* **17,** pp. 19–30, 1996.

214. J. S. Chiou, T. B. Chang, and C. K. Chen, "Laminar Film Condensation on a Horizontal Surface with Surface Tension Effect," *J. Heat Transfer,* **118,** pp. 797–799, 1996.

215. H. A. Hasanein, M. S. Kazimi, and M. W. Golay, "Forced Convection In-Tube Steam Condensation in the Presence of Noncondensable Gases," *Int. J. Heat Mass Transfer,* **39,** pp. 2625–2639, 1996.

216. S. Sripada, P. S. Ayyaswamy, and L. J. Huang, "Condensation on a Spray of Water Drops: A Cell Model Study—I. Flow Description," *Int. J. Heat Mass Transfer,* **39,** pp. 3781–3790, 1996.

217. L. J. Huang, P. S. Ayyaswamy, and S. Sripada, "Condensation on a Spray of Water Drops: A Cell Model Study—II. Transport Quantities," *Int. J. Heat Mass Transfer,* **39,** pp. 3791–3797, 1996.

218. J. Mikielewicz, M. Trela, and E. Ihnatowicz, "A Theoretical and Experimental Investigation of Direct-Contact Condensation on Liquid Layer," in G. P. Celata and R. Shah (eds.), *Two-Phase Flow Modelling and Experimentation 1995,* Edizioni ETS, Rome, pp. 221–227, 1995.

CHAPTER 15
BOILING

Geoffrey F. Hewitt
Imperial College of Science, Technology & Medicine

> *"With beaded bubbles winking at the brim"*
>
> —JOHN KEATS, "Ode to a Nightingale"

INTRODUCTION

General Considerations

The processes by which a liquid phase is converted partially or wholly to a vapor phase are of key importance in a wide range of applications including power generation, chemical and petroleum production, air conditioning, refrigeration, and so on. The processes of interconversion between the liquid and vapor phases of water are clearly a vital part of our living environment. In the first two editions of this handbook, Rohsenow [1, 2] defined boiling as "the process of evaporation associated with vapor bubbles in a liquid." However, as we will see, the process of liquid-to-vapor conversion need not be associated with the formation of bubbles. In many situations the evaporation takes place from a liquid surface (for instance, from the surface of the liquid film in annular two-phase flow); for the present purposes, the following definition (a generalized form of that given by Collier and Thome [3]) will be adopted:

> Boiling is defined as being the process of addition of heat to a liquid in such a way that generation of vapor occurs.

The considerable economic importance of boiling processes, coupled with the fascination of the complex phenomena involved, has led to a vast literature on the subject. Within the framework of this short chapter, it would be impossible to even list the work that has been performed. All that can be done is to pick out some of the most salient points (with emphasis on presenting relationships that can be applied by the user of this book). This selection of material is, of course, influenced by the author's interests and background, and it is perhaps important at the outset to apologize for the noninclusion of material that others may have given higher priority!

The reader who wishes to pursue the subject in more depth is referred to the many existing textbooks that have appeared (and will continue to appear) on the subject. Recent examples of these are books by Collier and Thome [3], Carey [4], and Tong and Tang [5]. An interesting general survey on the development of boiling heat transfer and its applications is given by Nishikawa [6], and a number of recent edited volumes arising from significant specialist meetings in the area are also useful sources (Hewitt et al. [7], Dhir and Bergles [8], Chen [9], Gorenflo et al. [10], Celata et al. [11], and Manglik and Krauss [12]); these volumes have been consulted extensively in preparing this chapter and are representative of modern work in the subject area.

Manifestations of Boiling Heat Transfer

In this chapter, the following five manifestations of boiling (i.e., situations in which phase changes occur) are considered:

1. *Pool boiling.* Here, boiling occurs by the generation of vapor from a heated surface in a pool of liquid. Though the liquid may circulate within the pool due to natural convection, mechanically induced circulation of the liquid does not take place.

2. *Cross flow boiling.* In this case, a flow is imposed over the surface. A typical example would be a cylinder with a liquid flow across it in a direction normal to the cylinder axis. At low cross flow velocities, the situation approaches that for pool boiling.

3. *Forced convective boiling in channels.* Here, evaporation of a liquid occurs in flow in a channel (for instance, a round tube). The vapor generated and the remaining liquid form a *two-phase flow* within the tube, and there are strong interactions between this two-phase flow (which can occur in a number of different forms) and the boiling process.

4. *Thin-film evaporation.* Here, a thin film of liquid flows over a heated surface (typically a vertical plate or the inside of a vertical tube) and evaporates. In many situations, this evaporation takes place directly at the surface of the liquid film, without the formation of bubbles at the solid surface. However, at higher heat fluxes, nucleate boiling occurs at the heated surface.

5. *Rewetting of hot surface.* In this case, the liquid phase contacts a hot surface and rewets it, with the accompanying formation of vapor. Examples here are the quenching of hot metal objects in metal forming processes and the rewetting of hot fuel elements in a nuclear reactor following a loss-of-coolant accident.

In the following, relatively simple geometries will be considered. No attempt will be made to deal with complex situations such as multitubular heat exchangers, multirod nuclear fuel elements, and so on. Though such topics are important, the space available in this chapter does not permit them to be covered. Process applications of boiling are dealt with by Hewitt et al. [13] and nuclear applications by Tong and Tang [5].

Structure of This Chapter

The following two sections (starting on p. 15.3 and 15.6) deal with the *generic* issues of phase equilibrium and nucleation and bubble growth, respectively. The remaining sections deal with the various manifestations of boiling as previously described. The first three of these manifestations (pool boiling, cross flow boiling, and forced convective boiling channels) are dealt with in the sections starting on pages 15.30, 15.75, and 15.84, respectively and follow, as nearly as practicable, a consistent pattern. In each case, following a general introduction, the information is presented under the following headings:

Heat Transfer Before the Critical Heat Flux Limit

The Critical Heat Flux Limit

Heat Transfer Beyond the Critical Heat Flux Limit

Under each of these categories are discussions of parametric effects, mechanisms, correlations, predictive models, the effects of multicomponent mixtures and, finally, enhancement or mitigation.

The topics of thin-film evaporation and rewetting of hot surfaces are structured differently, reflecting their special nature.

PHASE EQUILIBRIUM

An essential basis for the study of boiling heat transfer is the thermodynamics of *multiphase* systems. Here, it is normal practice to consider systems at *thermodynamic equilibrium*, in which the temperature of the system is uniform. Of course, as we will see, departures from such thermodynamic equilibrium are important in many instances. In what follows, the thermodynamic equilibrium of a *single-component* material is first considered. In many applications of boiling (particularly in the process and petroleum industries), *multicomponent* mixtures (for example, mixtures of hydrocarbons or refrigerants) are important, and the subject of multicomponent equilibrium is dealt with in the final part of this section.

Single-Component Systems

The relationship between pressure, volume, and temperature for a pure substance is illustrated schematically in Fig. 15.1. The phases indicated are *solid, liquid,* and *vapor*, respectively. The three phases coexist at the triple point (see line in Fig. 15.1 for temperature T_1). For the present purposes, we are concerned with the liquid and vapor regions, the distinction between which disappears at the *critical point* (T_c). The line ABCD represents the pressure-

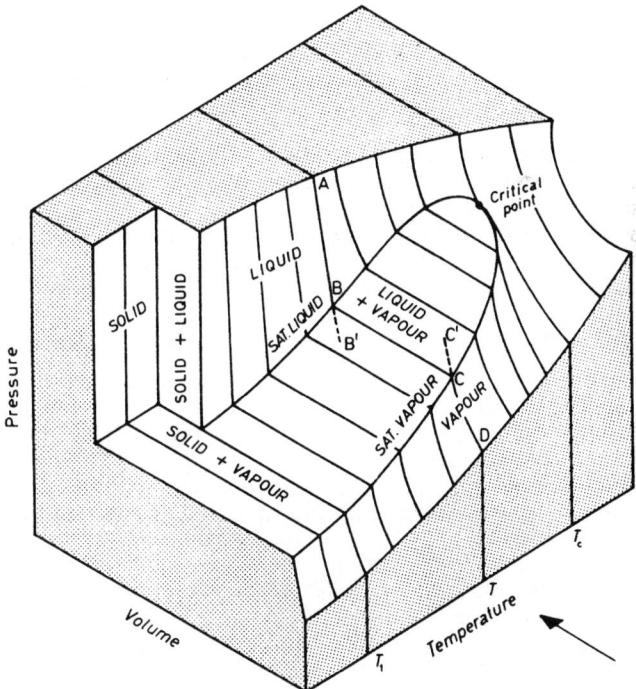

FIGURE 15.1   Pressure-volume-temperature relationship for a pure substance (reprinted from Collier and Thome [3] by permission of Oxford University Press).

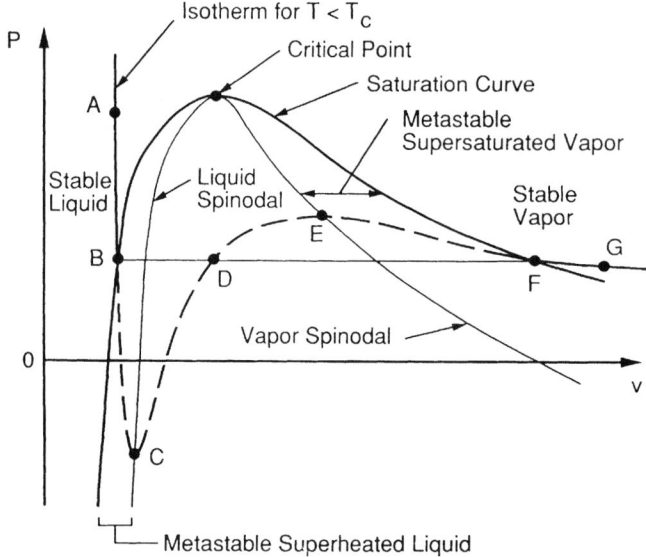

FIGURE 15.2 Pressure-volume relationships in liquid-vapor systems (from Carey [4], with permission from Taylor & Francis, Washington, DC, all rights reserved).

volume relationship for a given temperature (T). In region AB, the substance exists as a liquid; in zone BC, the substance exists as a two-phase mixture of vapor and liquid (in thermodynamic equilibrium); and in zone CD, the substance exists as a vapor. It is possible for the liquid to exist in a *superheated* state at pressures below that corresponding to point B, the limit being indicated by point B'. Similarly, it is possible for the vapor to exist in a *supercooled* or *supersaturated* state to a limit indicated by point C' in Fig. 15.1. In these conditions, the liquid or the vapor is described as being *metastable*. A more detailed description of these metastable states is given in Fig. 15.2.

Here, line ABDFG represents the pressure-volume relationship for a temperature lower than the critical temperature. The liquid phase can exist in a metastable state along line BC and the vapor phase can similarly exist along line FE. The dotted line CDE represents an *unstable* region. Points C and E, representing the limits of the metastable region, are usually referred to as *spinodal* points, and these points have loci (for different isotherms) along the lines labeled Liquid Spinodal and Vapor Spinodal in Fig. 15.2.

An excellent and detailed discussion of the thermodynamics of vapor-liquid equilibrium is given by Carey [4]; it will be sufficient here to state just some of the principal relationships. For the vapor region, the relationship between pressure, volume, and temperature is often represented by the *ideal gas law:*

$$PV = \tilde{N}\tilde{R}T \qquad (15.1)$$

where P is the pressure (Pa), V is the volume (m^3), $\tilde{N}$ is the number of kilogram moles (kmol) of substance, $\tilde{R}$ is the universal gas constant (8314 J/kmol K), and T is the temperature (K). $\tilde{N}$ is given by the ratio of the mass of substance M (kg) to its molecular weight $\tilde{M}$ (kg/kmol). The ideal gas law applies only to low pressures and high temperatures and begins to deviate significantly as the system approaches the critical pressure. There is available a very wide range of equations of state that describe the whole isotherm (line ABCDEFG in Fig. 15.2). One example of these is the van der Waals equation:

$$P = \frac{(\tilde{R}/\tilde{M})T}{v - b} - \frac{a}{v^2} \qquad (15.2)$$

where v is the specific volume (m^3/kg) of the substance and a and b are related to the critical temperature (T_c) and the critical pressure (P_c) by the relationships

$$a = \frac{27(\tilde{R}/\tilde{M})^2 T_c^2}{64 P_c} \tag{15.3}$$

$$b = \frac{(\tilde{R}/\tilde{M}) T_c}{8 P_c} \tag{15.4}$$

Carey [4] suggests the following equation or state in terms of reduced properties ($P_r = P/P_c$, $T_r = T/T_c$, and $v_r = v/v_c$, where v_c is the specific volume at the critical pressure):

$$(P_r + 3T_r^{-\lambda} v_r^{-2})(v_r - \tfrac{1}{3}) = (\tfrac{8}{3}) T_r \tag{15.5}$$

Equation 15.5 reduces to the van der Waals equation for $\lambda = 0$ and to an alternative equation of state (the Bethelot equation) for $\lambda = 1$.

Another useful result from classical thermodynamics is the Claperyon-Clausius equation, which relates the pressure to the temperature along the saturation line as follows:

$$\left(\frac{dP}{dT}\right)_{\text{sat}} = \frac{i_{lg}}{T v_{lg}} \tag{15.6}$$

where i_{lg} is the specific latent heat of vaporization (J/kg) and v_{lg} is the difference in specific volume accompanying phase change (m^3/kg).

Multicomponent Systems

For a single-component system the temperatures at which evaporation and condensation, respectively, occur (on a planar interface) are identical, that is, they correspond to the saturation temperature T_{sat}. In a mixture of fluids, however, the temperature at which the vaporized mixture begins to condense (the so-called *dew point* temperature T_{dew}) is different (higher) than the temperature at which the mixture in liquid form begins to evaporate (the so-called *bubble point* temperature T_{bub}). Thus, the condensate initially formed in condensation is richer in the less volatile component, and the vapor initially formed in evaporation is richer in the more volatile component relative to the original mixture. This behavior can be represented in terms of the *phase equilibrium diagram*. Figure 15.3 shows a typical phase equilibrium diagram for a binary mixture of components A and B. Suppose that we have a liquid in which the mole fraction $\tilde{x}_A = 0.25$ and that this mixture has an initial temperature T_1. If we heat the fluid, vapor begins to form at the bubble point temperature T_{bub}. The vapor formed has a much higher concentration of component A (in this case the mole fraction in the vapor $\tilde{y}_A$ is around 0.7). If, similarly, we start with a vapor of composition $\tilde{y}_A = 0.25$, then it begins to condense at the dew point temperature T_{dew}. In this case, the condensed liquid has a composition x_A of around 0.05. Of course, as will be seen, equilibrium between the bulk phases is often not obtained, although the local compositions of the phases adjacent to the interface are usually very close to equilibrium.

Figure 15.4 shows a phase equilibrium diagram for a somewhat more complex case when an *azeotrope* is formed. At the azeotropic composition, the mole fraction of both components is identical (for equilibrium) in both the vapor and the liquid, and the dew point and bubble point tempera-

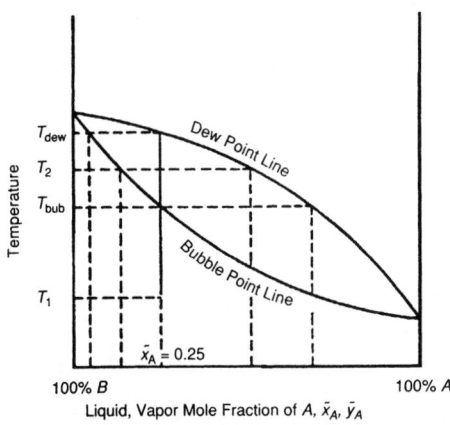

FIGURE 15.3 Phase equilibrium diagrams (for constant pressure) for a binary mixture of components A and B (from Hewitt et al. [13], with permission. Copyright CRC Press, Boca Raton, FL).

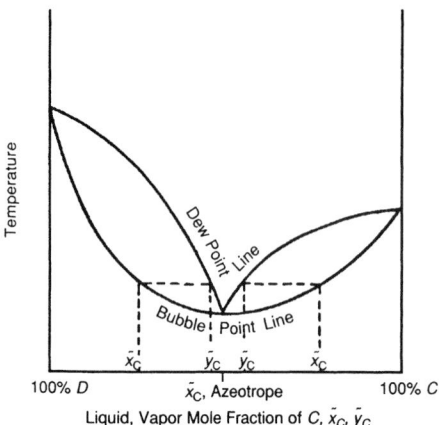

FIGURE 15.4 Phase equilibrium diagram for a binary mixture of components C and D forming a minimum boiling azeotrope (from Hewitt et al. [13], with permission. Copyright CRC Press, Boca Raton, FL).

tures are identical. Thus, the behavior of an azeotrope is similar to that of a single-component fluid. Note that on one side of the azeotropic composition, and for a given temperature, the mole fraction $\tilde{y}_C$ of component C in the vapor is higher than the mole fraction $\tilde{x}_C$ in the liquid and that, on the other side of the azeotropic composition for the same temperature, the liquid mole fraction is higher than that of the vapor. Detailed descriptions of phase equilibrium in multicomponent mixtures are beyond the scope of this present chapter; further information can be obtained in standard textbooks such as that of Prausnitz [14].

NUCLEATION AND BUBBLE GROWTH

Equilibrium of a Bubble

Here we consider a vapor bubble of radius r in equilibrium with a surrounding liquid. The curved surface produces a higher pressure in the bubble than in the surrounding liquid and, to obtain an equilibrium situation, the temperature of the whole system needs to be elevated with respect to T_{sat}. The relationship governing the difference in pressure between the vapor in the bubble and the liquid surroundings is the *Young-Laplace equation*. A detailed derivation of this equation and a description of the underlying physics are given by Carey [4]; here, a rather simpler derivation is presented. For a system of constant volume and temperature, the change in Helmholtz free energy F is given by

$$dF = \sigma dA_i - P_g dV_g - P_l dV_l = 0 \tag{15.7}$$

where σ is the surface tension, P_g and P_l are the vapor and liquid pressures, and dV_g and dV_l are the changes in vapor and liquid volumes, respectively. For a constant-volume system, $dV_g = -dV_l$. For a spherical bubble, it follows that

$$P_g - P_l = \sigma \frac{dA_i}{dV_G} = \sigma \frac{d(4\pi r^2)}{d(\frac{1}{3}\pi r^3)} = \frac{2\sigma}{r} \tag{15.8}$$

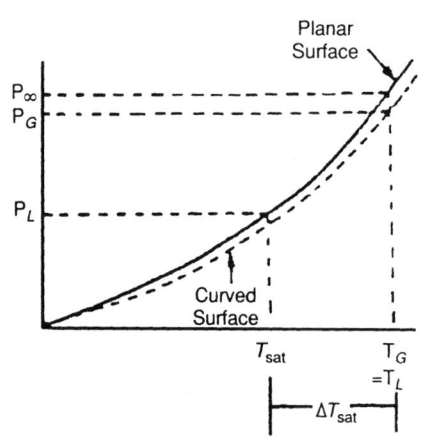

FIGURE 15.5 Relationship between vapor pressure and temperature (from Hewitt et al. [13], with permission. Copyright CRC Press, Boca Raton, FL).

which is the Young-Laplace equation. Note that, in the above derivation, the surface tension is interpreted as an energy per unit surface area (J/m^2).

The next stage is to estimate the difference between the system temperature and the saturation temperature (ΔT_{sat}). The relationship between vapor pressure and temperature is illustrated in Fig. 15.5. The vapor pressure for the curved interface of the bubble (P_g) is slightly different from that for a planar interface at the system temperature (P_∞); the relationship is

$$P_g = P_\infty \exp\left(-\frac{2\sigma v_l \tilde{M}}{r\tilde{R}T}\right) \tag{15.9}$$

where v_l is the specific volume (m^3/kg) of the liquid phase. Introducing the ideal gas law (Eq. 15.1) and expanding the exponential term, we obtain the expression

$$P_\infty - P_g \approx \frac{2\sigma v_l}{r v_g} \tag{15.10}$$

and combining Eqs. 15.8 and 15.10, we have:

$$P_\infty - P_l \approx \frac{2\sigma}{r}\left(1 + \frac{v_L}{v_G}\right) \tag{15.11}$$

This pressure difference can be related to a temperature difference by invoking the Claperyon-Clausius equation (Eq. 15.6). This equation is integrated over the interval of pressure from P_∞ to P_l and the corresponding temperature interval T_{sat} to T_g ($=T_l$) to give

$$\ln\left(\frac{P_\infty}{P_l}\right) = \frac{i_{lg}\tilde{M}}{\tilde{R}}\left(\frac{1}{T_{sat}} - \frac{1}{T_g}\right) \tag{15.12}$$

from which it follows that

$$\Delta T_{sat} = T_l - T_\infty = \frac{\tilde{R}T_g T_{sat}}{i_{lg}\tilde{M}}\ln\left[1 + \frac{P_\infty - P_l}{P_l}\right]$$

$$\approx \frac{\tilde{R}T_g T_{sat}}{i_{lg}\tilde{M}P_l}[P_\infty - P_l] \approx \frac{T_{sat}v_g}{i_{lg}}(P_\infty - P_l) \approx \frac{2\sigma T_{sat}v_g}{i_{lg}r}\left(1 + \frac{v_l}{v_g}\right) \tag{15.13}$$

For $v_l/v_g \ll 1$, we have

$$\Delta T_{sat} = \frac{2\sigma T_{sat}v_g}{i_{lg}r} \tag{15.14}$$

For equilibrium at a temperture that is ΔT_{sat} above the saturation temperature, the equilibrium bubble radius $r*$ is given by

$$r* = 2\sigma T_{sat}v_g/(i_{lg}\Delta T_{sat}) \tag{15.15}$$

Bubbles that are smaller in radius than $r*$ will collapse spontaneously, and bubbles that are bigger will grow. The initiation (*nucleation*) process for bubble growth may under certain circumstances occur within the bulk of the liquid (*homogeneous nucleation*), but more normally it occurs at the interface between the liquid and its containing solid surfaces (*heterogeneous nucleation*). These two cases are discussed in the following two subsections, followed by a discussion of bubble growth and bubble release diameter and frequency.

Homogeneous Nucleation

Statistical molecular fluctuations within a liquid can lead to the formation of microscopic vaporlike regions that have a radius greater than the critical one for bubble growth as given by Eq. 15.15. The formation of such nuclei and their subsequent growth is referred to as *homogeneous nucleation*. There have been two main approaches to the calculation of the temperature required for homogeneous nucleation. These are, respectively, determination of the *thermodynamic limit* and the *kinetic limit*.

The Thermodynamic Limit. In the preceding text (see Fig. 15.2), the limit of the region in which the liquid phase can exist in a metastable state (the *liquid spinodal*) was introduced; this is represented by point C in Fig. 15.2. One view of homogeneous nucleation is that it will occur at the spinodal limit that corresponds (see Fig. 15.2) to the condition (expressed in term of reduced quantities)

$$\left(\frac{\partial P_r}{\partial v_r}\right) = 0 \tag{15.16}$$

From the relationship between reduced properties given by Eq. 15.5, Carey [4] derived the following relationship for the liquid spinodal temperature $(T_r)_s$:

$$(T_r)_s = \left[\frac{(3v_r - 1)^2}{4v_r^3} \right]^{1/(\lambda + 1)}$$ (15.17)

Equations 15.5 and 15.17 can be used to establish the relationship between the reduced temperature and reduced pressure for the spinodal point (and hence for the so-called *thermodynamic limit* for homogeneous nucleation). Carey [4] compared data for homogeneous nucleation with this limit, and his results are illustrated in Fig. 15.6. The results generally lie between those predicted from the van der Waals equation ($\lambda = 0$) and the Berthelot equation ($\lambda = 1$). An empirical expression for the homogeneous nucleation temperature, which is also expressed in terms of reduced properties, is that of Lienhard [15], who relates the reduced homogeneous nucleation temperature $T_{r,n}$ to the reduced saturation temperature $T_{r,\text{sat}}$ as follows:

$$(T_{r,n} - T_{r,\text{sat}}) = 0.905 - T_{r,\text{sat}} + 0.095 T_{r,\text{sat}}^8$$ (15.18)

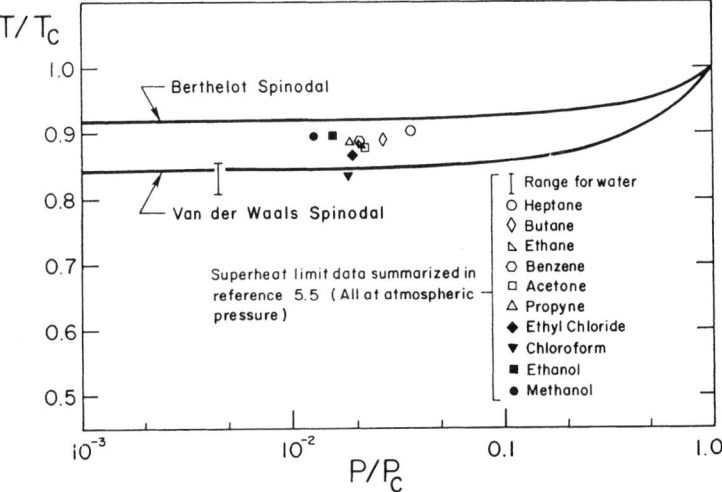

FIGURE 15.6 Comparison of measured homogeneous nucleation temperature with the spinodal temperature calculated from the van der Waals and Berthelot equations of state (from Carey [4], with permission from Taylor & Francis, Washington, DC, all rights reserved).

The Kinetic Limit.   To initiate homogeneous nucleation, a vapor embryo is required that has the critical radius r^*. In principle, the formation of just one such embryo would be sufficient to initiate the nucleation process, but in practice it is found that conditions must be such that J, the number of vapor embryos formed in a unit volume per unit time, has a high value (typically $J > 10^{12}$). Carey [4] derives the following expression for J:

$$J = 1.44 \times 10^{40} \left(\frac{\rho_l^2 \sigma}{\tilde{M}^3} \right)^{1/2} \exp\left\{ \frac{-1.213 \times 10^{24} \sigma^3}{T_l [\eta P_{\text{sat}}(T_l) - P_l]^2} \right\}$$ (15.19)

where η is given by

$$\eta = \exp\left[\frac{P_l - P_{\text{sat}}(T_l)}{\rho_l \tilde{R} T_l} \right]$$ (15.20)

In the above equations, J is the rate at which the embryos are formed $(1/(m^3s))$, $P_{sat}(T_l)$ is the saturation pressure (Pa) corresponding to the liquid temperature T_l (K), σ is the surface tension (N/m), and P_l is the liquid pressure (Pa). Equation 15.19 can be solved to give J as a function of T_b, and, choosing the point at which $J = 10^{12}$ as the limiting value, the homogeneous nucleation temperature T_n can be estimated. For organic fluids, results are in good agreement with the predictions but the values for water (around 300°C) are higher than the highest values (250–280°C) that have been measured.

Analysis of the superheat required for homogeneous nucleation leads to the conclusion that the predicted temperatures are very much higher than those normally required to initiate boiling. The conclusion, therefore, is that it is *heterogeneous* rather than *homogeneous* nucleation that initiates vapor formation in practical boiling processes. This case is discussed in the next section. However, large superheats can exist within liquids before nucleation occurs if the conditions are such as to inhibit heterogeneous nucleation (careful removal of dissolved gases, the use of ultra-smooth surfaces, etc.), and there have been a number of studies (see for instance Merte and Lee [16] and Drach et al. [17]) that indicate that homogeneous nucleation can occur with rapid transient heating.

Heterogeneous Nucleation

As was seen from the preceding discussion, very large superheats are required to nucleate bubbles by the homogeneous nucleation process. For water at atmospheric pressure, a superheat on the order of 200°C (i.e., a liquid temperature of around 300°C) is required for nucleation, and this is clearly much larger than the values commonly observed (typically 10–15°C) for boiling of water from heated surfaces under these conditions. Clearly, then, the surface itself is playing a crucial role in reducing the superheat requirements. If we consider a bubble being formed on a planar solid surface where the contact angle is ϕ (Fig. 15.7a), then the required superheat for homogeneous nucleation is reduced by a factor $f(\phi)$ that is given as follows (Cole [18], Rohsenow [2]):

$$f(\phi) = \left[\frac{2 + 3 \cos \phi - \cos^3 \phi}{4} \right]^{1/2} \tag{15.21}$$

For $\phi = 0$, $f(\phi) = 1$ and the onset of boiling will occur at a superheat identical to that for homogeneous nucleation. On the other hand, if $\phi = 180°$ (its maximum value), then $f(\phi) = 0$ and boiling will be initiated at the surface as soon as the fluid reaches the saturation temperature. However, real contact angles are normally less than 90°; Shakir and Thome [19] report values ranging from 86° for water on a copper surface down to 8° for *n*-propanol on a brass surface.

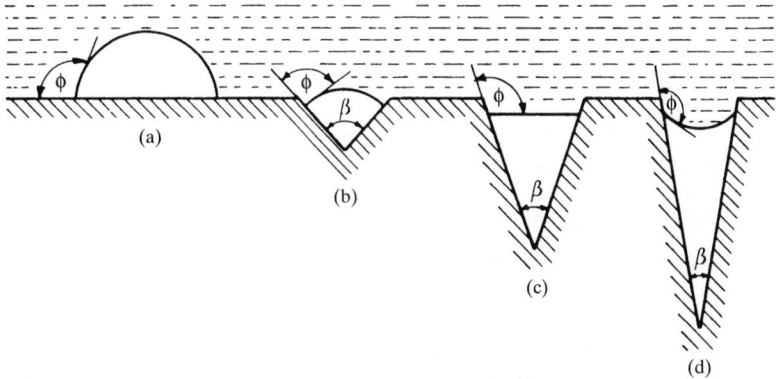

FIGURE 15.7   Formation of bubbles at a solid surface (from Collier and Thome [3], by permission of Oxford University Press).

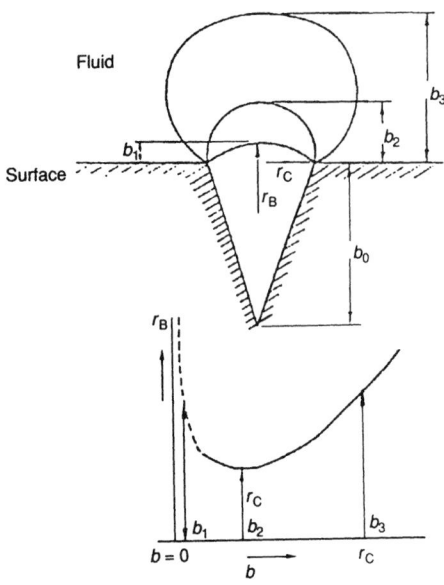

FIGURE 15.8 Bubble growth from an idealized conical cavity (from Hewitt et al. [13], with permission. Copyright CRC Press, Boca Raton FL).

The highest values of ϕ are observed for water on Teflon, where the contact angle is 108°. Even for this extreme value of ϕ, the superheat required for nucleation is still very high (around 290°C for water—Carey [4]). Thus, some other explanation has to be sought for the low values actually observed.

The explanation for the existence of boiling at much lower superheats than predicted by homogeneous nucleation theory is that bubbles are initiated from cavities on the heat transfer surface. Gas or vapor is trapped in these cavities as shown in Fig. 15.7b–d. Once boiling is initiated, these cavities may remain vapor-filled and continue to be active sources for the initiation and growth of bubbles from the surface. The growth process from a conical cavity whose mouth radius is r_c is illustrated in Fig. 15.8.

As the bubble grows from the cavity, it passes through a condition (where the bubble tip is a distance b_2 from the surface) at which the bubble radius is a minimum as shown. This is referred to as the *critical hemispherical condition* and is the condition at which the maximum amount of superheat is required to continue growth. The superheat may be predicted from Eq. 15.14 by substituting $r = r_c$. The larger the heat flux, the larger the superheat on the surface and the smaller the cavities that can be activated. Thus, the number of active cavities increases with heat flux, leading to an increase in heat transfer coefficient with heat flux; Jakob [20] was the first to recognize the connection between heat flux and active site density. Developing an understanding of this relationship has been the main focus of boiling research over the subsequent decades.

Critical to the formation of active nucleation sites is the entrapment of gas into potential sites during the initial wetting of the heat transfer surface during filling of the boiling system with the liquid phase. This process was addressed by Bankoff [21], who considered the motion of an advancing liquid front over a V-groove and suggested that trapping of gas was only possible if

$$\phi_a > \beta \qquad (15.22)$$

where ϕ_a is the *advancing* contact angle and β is the groove angle. If one considers this entrapment process, it does not follow that the residual gas bubble in the cavity has a radius greater than the critical hemispherical radius illustrated in Fig. 15.8. For low contact angles, the radius of the residual bubble trapped in the conical cavity can be smaller than the critical hemispherical radius. This aspect was studied by Lorentz et al. [22], and their results are illustrated in Fig. 15.9.

The ratio of trapped bubble radius to mouth radius increases with cavity angle and contact angle is shown. For water, where the contact angle is high, the bubble growth is likely to be still governed by the critical hemispherical condition, but this need not be necessarily so for organic fluids, where the contact angle is much less.

Detailed studies of the relationship between the actual physical configuration of the boiling surface and the nucleation behavior are reported by Yang and Kim [23] and by Wang and Dhir [24–26]. In these studies, electron microscopy and optical microscopy, respectively, were used to measure the characteristics of the cavities. Yang and Kim [23] assumed conical cavities, but Wang and Dhir [24–26] found that the cavities were more irregular in shape, as illustrated in Fig. 15.10. The area A_c of the cavity mouth could be determined from the microscope pictures, and an effective cavity diameter D_c^* was defined as

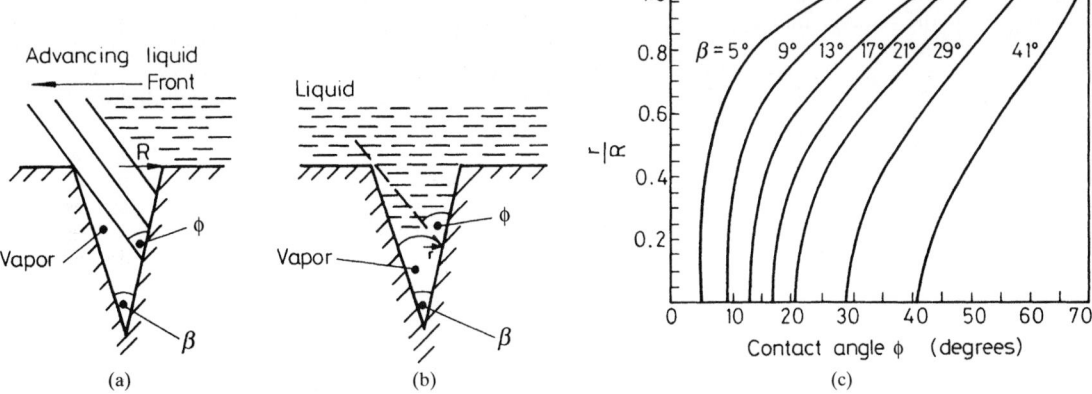

FIGURE 15.9 Vapor/gas trapping at a conical cavity (from Lorentz et al. [22], with permission from Taylor & Francis, Washington, DC. All rights reserved).

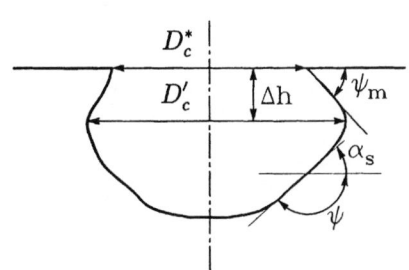

FIGURE 15.10 Definition of cavity characteristics (from Wang and Dhir [24], with permission of ASME).

$$D_c^* = \sqrt{\frac{4A_c}{\pi}} \qquad (15.23)$$

A depth Δh was removed from the surface by polishing and a new equivalent diameter D_c' was determined on the same basis. Thus, the cavity mouth angle ψ_m is then given by

$$\psi_m = \tan^{-1}\frac{D_c^* - D_c'}{2\Delta h} \qquad (15.24)$$

The total number density N_s of all types of cavities present on the copper surface investigated by Wang and Dhir is shown in Fig. 15.11; the cumulative number density N_{as} for sites with $\psi_m < 90°$ is also shown. N_{as} is considerably lower

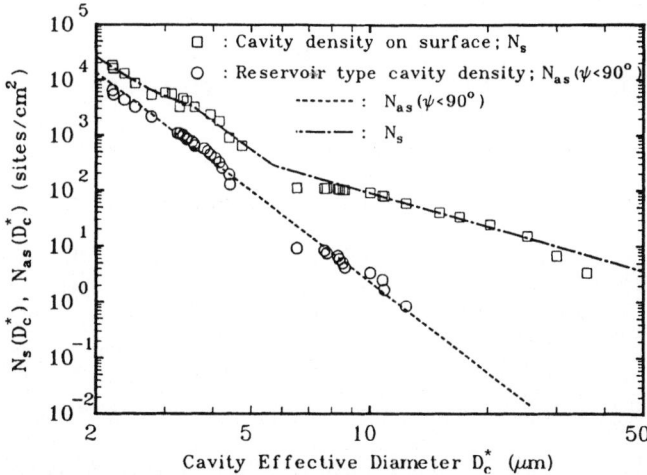

FIGURE 15.11 Cumulative number density distribution of sites on a copper surface (from Wang and Dhir [24], with permission of ASME).

than N_s, indicating the potential for large differences between the total number of sites and the number that are active. In boiling, two conditions are required in order to promote site activity:

1. Adapting the Bankoff [21] criterion, the trapping of vapor or gas within the site depends on the inequality

$$\phi_a > \psi_m \tag{15.25}$$

2. The wall superheat ΔT_{sat} must be greater than the value given by Eq. 15.14. The appropriate radius for use in Eq. 15.14 is given by

$$r = D_c/2 = f_D D_c^*/2 \tag{15.26}$$

where D_c is an effective cavity diameter, which is given by the product of the measured diameter (defined by Eq. 15.23) and a shape factor f_D (accounting for the irregularity of the cavity) that Wang and Dhir found empirically to be 0.89.

Wang and Dhir carried out a series of measurements in which the active site number density N_a was determined by photographing boiling on the copper surface on which the site distribution experiments were carried out. N_a could be determined as function of D_c by measuring the superheat and using Eq. 15.14 to determine the appropriate D_c. These values could be compared with the values estimated from the measurements of the cumulative distributions of N_s and ψ_m just described. The comparisons are shown in Fig. 15.12. Wang and Dhir were able to change the wetting angle from its original value of 90° to lower values of 35° and 18°, respectively, by progressively oxidizing the copper surface. For $\phi = 90°$, the cumulative number density N_a of active sites agrees well with N_{as} (i.e., the number of sites that have $\psi_m < 90°$). However, the number of sites that have $\psi_m < 35°$ and $\psi_m < 18°$ is substantially smaller. Nevertheless, the number distribution of active sites meeting the criteria that $\phi > \psi_m$ for these two cases agrees well with the measured active site number densities from the boiling experiments. Note that, for a given cavity diameter, the number of active sites for $\phi = 18°$ is on the order of 20 times less than for the case where $\phi = 90°$.

The results obtained by Wang and Dhir are consistent with earlier observations by Lorentz et al. [22], who calculated site density from measured superheat values and plotted the site densities as a function of active site radius for organic fluids and for water, respectively. Their results are illustrated in Fig. 15.13.

Here, as we will see, there is a large difference between the number densities for water and for the organic fluids; in this case, the surface has remained the same but the contact angle has been changed by changing the liquid.

The experiments of Wang and Dhir [24–26] and others have demonstrated quantitatively the relationship between the surface characteristics (site number density, internal angles of the cavities, and contact angles) on nucleation processes. These effects have, as we shall see later when discussing nucleate boiling, a profound influence on the overall heat transfer behavior in this region, making the behavior very difficult to predict. In the previous discussion, we have implicitly assumed that the temperature of the liquid phase surrounding the bubble is uniform. However, in real heat transfer situations, there is a temperature gradient away from the surface. This case has been investigated for pool boiling by Hsu [27] and Hsu and Graham [28] and by Bergles and Rohsenow [29], Davis and Anderson [30], Kenning and Cooper [31], and others for forced convective boiling. The situation with a temperature gradient is illustrated in Fig. 15.14. Suppose we have a wide spectrum of cavity sizes on the surface, and the spectrum contains cavities A, B, and C, as shown, that have radii r_A, r_B, and r_C, respectively. If we assume that *the whole of the surface of the bubble must be at a temperature greater than that given by Eq. 15.14* in order to achieve growth, then the requirement for growth is that the temperature at the extremity of the bubble furthest from the wall must be at or above this critical temperature. In Fig. 15.14, line XY represents the critical temperature for a bubble of radius r where

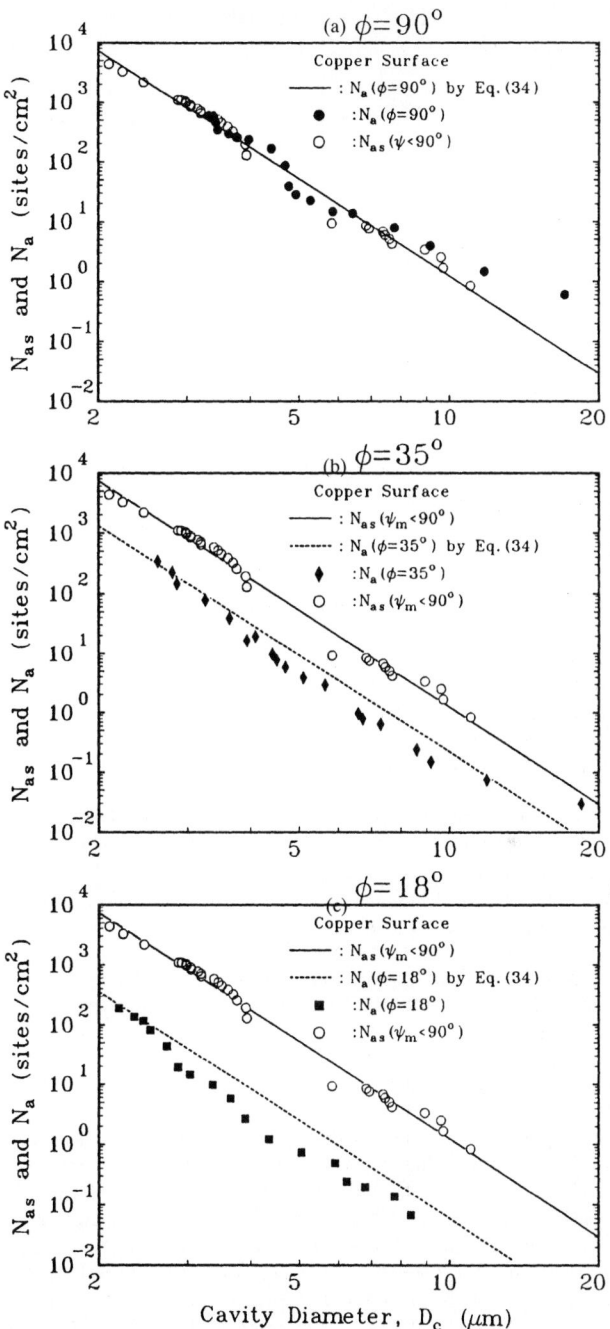

FIGURE 15.12 Comparison of active cavity size distribution estimated from cavity geometry measurements with that measured in boiling for contacting of 90°, 35°, and 18° (from Wang and Dhir [24], with permission of ASME).

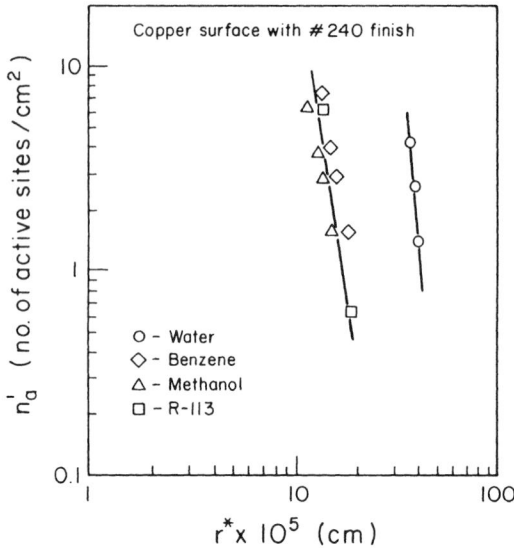

FIGURE 15.13 Number density as a function of calculated cavity radius (adapted by Carey [4] from the results of Lorentz et al. [22], with permission of Taylor & Francis, Washington, DC. All rights reserved).

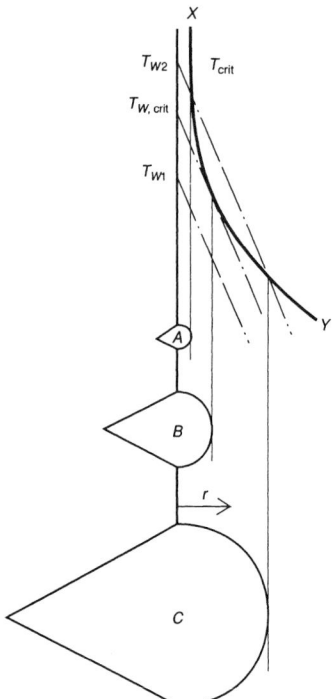

FIGURE 15.14 Bubble nucleation in the presence of a temperature gradient (from Hewitt et al. [13], with permission. Copyright CRC Press, Boca Raton, FL).

(for the critical hemispherical bubble) $r = y$, the distance from the wall at the bubble extremity. The cavities from which nucleation occurs are usually very small, typically on the order of 1 μm in diameter. Thus, the extremity of the bubble is likely to lie in a region where heat transfer is governed by molecular conduction, from which it follows that the temperature gradient from the wall outward is given by

$$\frac{dT}{dy} = \frac{q''}{k_l} \tag{15.27}$$

where q'' is the heat flux (W/m²) and k_l is the liquid thermal conductivity (W/mK). This implies a linear variation of temperature with distance from the wall, and lines showing this variation for a constant heat flux at various values of wall temperature T_W are shown in Fig. 15.14.

Suppose that the wall temperature is T_{W1}; for this case, the temperature never exceeds the value (T_{crit}) for bubble growth (as given by line XY) and no nucleation occurs. If the wall temperature is increased to T_{W2}, then the temperature profile cuts line XY at points corresponding to bubble radii r_A and r_C as shown. Thus, hemispherical bubbles on all cavities in the size range of r_A to r_C will be able to grow. If the wall temperature is $T_{W,crit}$ (Fig. 15.14), then the temperature profile just touches line XY tangentially, corresponding to growth of bubbles of radius r_B. Thus, there is a critical wall temperature $T_{W,crit}$ corresponding to a critical cavity radius ($r_{crit} = r_B$) at which nucleation will start in a surface having a spectrum of nucleation site sizes that includes r_{crit}. Bergles and Rohsenow [29] produced an empirical correlation based on the drawing shown in Fig. 15.14; Davis and Anderson [30] calculated the conditions by equating the slopes of the temperature profile and the T_{crit} line. Thus, from Eq. 15.14:

$$\Delta T_{sat} = T_{crit} - T_{sat} = \frac{2\sigma T_{sat} v_g}{i_{lg} r_{crit}} \tag{15.28}$$

and

$$\frac{dT_{crit}}{dr_{crit}} = \frac{2\sigma T_{sat} v_g}{i_{lg} r_{crit}^2} \tag{15.29}$$

Equating this slope with that given by Eq. 15.27, we have

$$r_{crit} = \left[\frac{2\sigma T_{sat} v_g k_l}{q'' i_{lg}} \right]^{1/2} \tag{15.30}$$

The wall temperature $T_{W,crit}$ is given by

$$T_{W,crit} = T_{crit} + \frac{q'' r_{crit}}{k_l} = T_{sat} + \Delta T_{sat} + \frac{q'' r_{crit}}{k_l} = T_{sat} + \frac{2\sigma T_{sat} v_g}{i_{lg} r_{crit}} + \frac{q'' r_{crit}}{k_l} \tag{15.31}$$

Substituting for r_{crit} from Eq. 15.30, we obtain the following result for the wall superheat required to initiate nucleation:

$$(\Delta T_{sat})_{W,crit} = T_{w,crit} - T_{sat} = \left(\frac{8\sigma T_{sat} v_g q''}{i_{lg} k_l} \right)^{1/2} \tag{15.32}$$

Although Eq. 15.32 is found to give reasonably good results in many cases, deviations may occur due to a number of factors, including:

1. The temperature profile may be distorted by the presence of the bubble. This effect has been studied by Kenning and Cooper [31] and by Frost and Dzakowic [32]. Kenning and Cooper related the bubble surface temperature to the dividing stream line for flow over the bubble and Frost and Dzakowic assumed that the temperature in the undisturbed boundary layer that should be matched with T_{crit} is that at a distance nr_{crit} from the wall. They suggest that

$$n = \left(\frac{c_{pl}\mu_l}{k_l} \right)^2 = Pr_l^2 \tag{15.33}$$

where c_{pl} and μ_l are the specific heat capacity and the dynamic viscosity of the liquid phase and Pr_l is the liquid-phase Prandtl number. This gives

$$(\Delta T_{sat})_{w,crit} = \left(\frac{8\sigma T_{sat} v_g q''}{i_{lg} k_l} \right)^{1/2} Pr_l \tag{15.34}$$

2. The spectrum of cavity sizes may be such that r_{crit} exceeds the maximum cavity radius r_{max}. In this case, the superheat required to initiate nucleation is given by

$$(\Delta T_{sat})_{w,crit} = \frac{2\sigma T_{sat} v_g}{r_{max} i_{lg}} + \frac{q'' r_{max}}{k_l} \tag{15.35}$$

The situation where no cavities of size r_{crit} exist is most likely to occur with well-wetting fluids such as refrigerants. Data of this kind are reviewed by Spindler [39]; typical values for r_{max} are in the range of 0.2 to 0.4 μm, with the values being independent of subcooling and mass flux but reducing with increasing pressure.

Specific studies of nucleation behavior in forced convection are exemplified by the following:

1. Kandlikar [33] studied nucleation in subcooled flow boiling of water. Using a high-powered microscope, he was able to show that the bubbles were nucleated over cavities less than 20 μm in diameter and that the nucleating cavity sizes became smaller as the flow velocity was increased; this is consistent with the preceding derivation if heat fluxes (and hence temperature gradients adjacent to the surface) are increased at the higher velocities.

2. Klausner [34] studied bubble nucleation in stratified flow of refrigerant R113 in a horizontal rectangular channel. The nucleation site density decreased with increasing vapor velocity as illustrated in Fig. 15.15; at a velocity of around 5 m/s, nucleation was totally supressed. Klausner et al. interpreted their data in terms of the relationship between critical radius of nucleation site r_c and number density; the data from this interpretation are shown in Fig. 15.16 and it will be seen, that for these particular conditions, a rather narrow range of site radius applied. The question of suppression of nucleate boiling is discussed further later in this chapter.

3. Hewitt et al. (1963) studied the onset of nucleate boiling in upward annular flow using an annulus test section with a glass outside wall so that nucleation on the inner (steel) heated surface could be observed. Their results are illustrated in Fig. 15.17. Nucleation was observed to occur (for a given mass flux and quality) within a range of heat fluxes indicated by the shaded bands in Fig. 15.17. As will be seen, the heat flux required for nucleation increases with increasing mass flux and quality. Since the heat transfer coefficient increases with these variables (see the following sections), the wall temperature for a given heat flux is reduced and, hence, nucleation is suppressed.

As will be seen, if one considers the evaporation of an initially subcooled fluid in a channel with constant wall heat flux, nucleation is initiated in subcooled boiling once the bulk fluid temperature is high enough to give a wall temperature that is sufficient for nucleation according to the preceding criteria; further along the channel, the high heat transfer coefficients encountered in two-phase flow reduce the wall temperature below that necessary for nucleation and nucleate boiling is suppressed, with the heat transfer mechanism changing to evap-

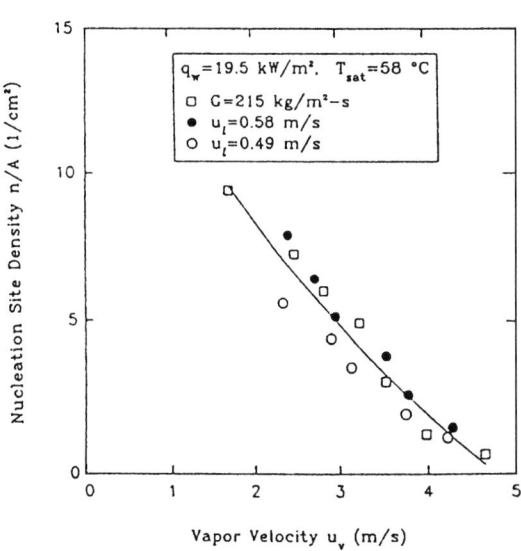

FIGURE 15.15 Variation of nucleation site density with vapor velocity in nucleate boiling in stratified flow of refrigerant R113 (from Klausner et al. [34], with permission from ASME).

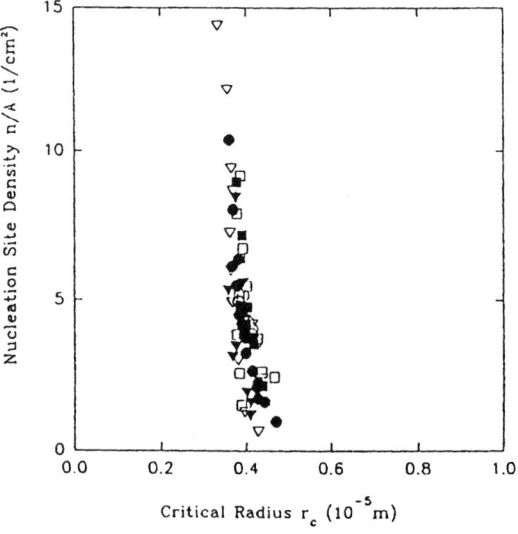

FIGURE 15.16 Number density of active nucleation sites as a function of site radius calculated from the wall superheat using Eq. 15.14 (from Klausner et al. [34], with permission from ASME).

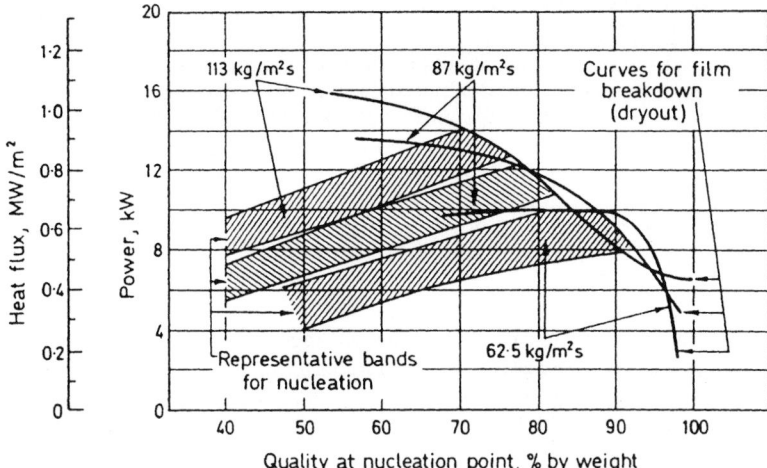

FIGURE 15.17 Nucleation in annular flow (from Hewitt et al. [36], with permission from Elsevier Science Ltd.).

oration at the interfaces in the multiphase flow. Further details of this process are given later in the chapter.

An important aspect of heterogeneous nucleation is that of interaction between adjacent nucleation sites. Nucleation and bubble growth from one site can lead to the initiation of nucleation at adjacent sites. Dramatic demonstrations of site interactions can be obtained using liquid crystal thermography (Kenning and Yan [36], Golobic [37]). Another interesting demonstration of the donor effect from adjacent sites is the observation of "boiling fronts" as reported by Fauser and Mitrovich [38]. In the experiments described by these authors, a horizontal copper tube was heated in a bath of refrigerant 113, the liquid becoming superheated without any nucleation on the tube. Nucleation was initiated at one end of the tube by applying a higher heat flux there and spread along the tube as illustrated in Fig. 15.18 with veloci-

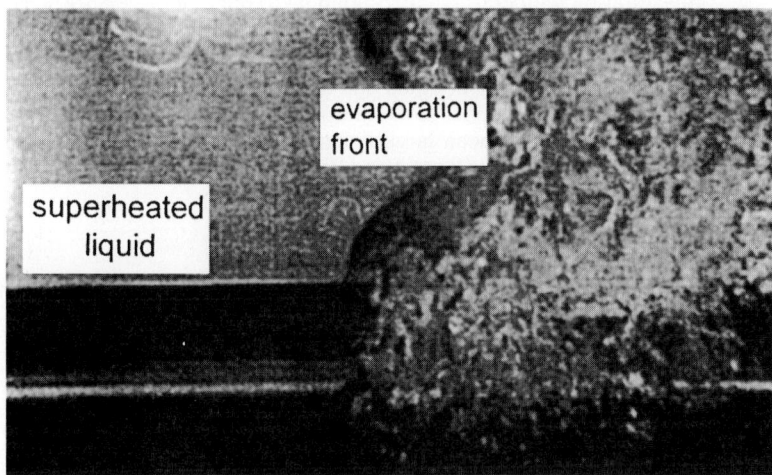

FIGURE 15.18 Propagation of nucleate boiling along a heated copper tube in a pool of refrigerant 113 (from Faucer and Mitrovic [38], with permission from Edizioni ETS).

ties typically in the range of 2–20 m/s, with the velocities increasing with pressure and with wall superheat. The motion of the boiling front led to a transition between natural convection heat transfer to nucleate boiling heat transfer; under some circumstances a transition through nucleate boiling into film boiling was also possible. These propagation phenomena are clearly of great potential importance in considering boiling systems.

Once nucleation has been initiated on a given site, then repeated bubbling from that site may occur even if the heat flux is reduced below the value originally necessary to initiate nucleation; this *hysteresis* effect is discussed further in the section on parametric effects in pool nucleate boiling below.

Bubble Growth

Having nucleated a bubble, the next stage in the boiling process is the growth of that bubble as a result of vaporization of liquid at its interface. Bubble growth and collapse has been of interest for many decades, one of the earliest quantitative studies being that of Lord Rayleigh in 1917 [40]. Comprehensive reviews of bubble growth are given in the books by Carey [4] and Collier and Thome [3] and only a brief treatment is given here. This begins with the consideration of bubble growth in an extensive liquid pool and continues with bubble growth from a surface and bubble growth in binary liquid mixtures.

Bubble Growth in an Extensive Liquid Pool.   Two limiting cases of bubble growth can be recognized:

1. *Inertia-controlled growth.* Here, the growth rate is limited by how rapidly the growing bubble can push back the surrounding liquid. The heat transfer to the interface is very fast and is not a limiting factor. Inertia-controlled growth is typical of the early stages of bubble growth, particularly when the superheat is high. In this region, the growth process was analyzed by Rayleigh [40]; the radius of the bubble $r(t)$ increases linearly with time according to the relationship

$$r(t) = \left\{ \frac{2}{3} \left[\frac{T_l - T_{sat}(P_l)}{T_{sat}(P_l)} \right] \frac{i_{lg}\rho_g}{\rho_l} \right\}^{1/2} t \qquad (15.36)$$

where T_l is the temperature of the liquid pool in which the bubble is growing, $T_{sat}(P_l)$ is the saturation temperature corresponding to the liquid pressure, i_{lg} is the latent heat of vaporization, and ρ_g and ρ_l are the vapor and liquid densities, respectively.

2. *Heat-transfer-controlled growth.* In this case, the growth rate is limited by the transfer of heat between the bulk liquid and the interface where vaporization is occurring. This limiting case usually applies to the later stages of bubble growth when the liquid superheat near the interface has been largely depleted. For this region, the bubble size varies with the square root of time (Plesset and Zwick [41]) and is given by

$$r(t) = \frac{2\Delta T_{sat} k_l}{i_{lg}\rho_g} \left(\frac{3t}{\pi\alpha_l} \right)^{1/2} \qquad (15.37)$$

where k_l is the liquid phase thermal conductivity, ΔT_{sat} is the superheat, and α_l is the liquid phase thermal diffusivity.

Mikic et al. [42] suggest an equation covering both the inertia-controlled and heat-transfer-controlled regions as follows:

$$r^+ = \frac{2}{3} \left[(t^+ + 1)^{3/2} - (t^+)^{3/2} - 1 \right] \qquad (15.38)$$

where: $$r^+ = \frac{r(t)A}{B^2} \qquad (15.39)$$

$$t^+ = \frac{tA^2}{B^2} \tag{15.40}$$

$$A = \left\{ \frac{2[T_l - T_{sat}(P_l)]i_{lg}\rho_g}{\rho_l T_{sat}(P_l)} \right\}^{1/2} \tag{15.41}$$

$$B = \left(\frac{12\alpha_l}{\pi} \right)^{1/2} \left\{ \frac{[T_l - T_{sat}(P_l)]c_{pl}\rho_l}{\rho_g i_{lg}} \right\} \tag{15.42}$$

where c_{pl} is the liquid specific heat capacity. Equation 15.38 reduces to Eq. 15.36 for small values of t^+ and to Eq. 15.37 for large values of t^+.

More recently, numerical calculations have been carried out on bubble growth, and these are not limited by the many assumptions made in deriving Eqs. 15.36 and 15.37. This work is exemplified by the studies of Lee and Merte [43] and Miyatake et al. [44]. In deriving Eq. 15.36, the contribution of surface tension to the bubble internal pressure was ignored; for very tiny bubbles, surface tension becomes important and offsets the pressure difference between the inertia-induced pressure inside the bubble and its surroundings. Thus, there is a short period (of duration t_d) where growth is very slow. Miyatake et al. [44] have produced a new general equation for bubble growth that more accurately covers the whole range than does that of Mikic et al. [42]. The Miyatake et al. equation is as follows:

$$r^+ = \frac{2}{3} \left\{ 1 + \frac{t^+}{3} \exp[-(t^+ + 1)^{1/2}] \right\} [(t^+ + 1)^{3/2} - (t^+)^{3/2} - 1] \tag{15.43}$$

where:

$$r^+ = \frac{A[r(t) - r_c]}{B^2} \tag{15.44}$$

$$A = \left[\frac{2}{3} \frac{\Delta p_o}{\rho_l} \right]^{1/2} \tag{15.45}$$

$$B = \left(\frac{12}{\pi} \right)^{1/2} \left\{ \frac{[T_l - T_{sat}(P_l)]\alpha_l^{1/2}c_{pl}\rho_l}{\rho_g i_{lg}} \right\} \tag{15.46}$$

$$t^+ = \left(\frac{A}{B} \right)^2 \left\{ t - t_d \left[1 - \exp\left[\left(-\frac{t}{t_d} \right)^2 \right] \right] \right\} \tag{15.47}$$

$$r_c = \frac{2\sigma}{\Delta p_o} \tag{15.48}$$

$$t_d = \frac{6r_c}{A} \tag{15.49}$$

$$\Delta p_o = P_{sat}(T_l) - P_l \tag{15.50}$$

There are significant differences between Eqs. 15.38 and 15.43 in the initial stages of growth, though the differences are small in the later stages.

In the above, it is assumed that the bubbles grow only by vaporization at the interface. However, Barthau and Hahne [45] report a case of bubble growth with an *evaporation wave* at the interface of the growing bubble. Such a wave results in a homogeneous flow of vapor and entrained droplets (instead of pure vapor) into the bubble, leading to a dramatic increase in the mass flux into the bubble and a strong increase in the internal pressure. The results obtained by Barthau and Hahne are illustrated in Fig. 15.19. The initial growth proceeds in a manner similar to that predicted from the Plesset and Zwick [41] relationship (Eq. 15.37) but much more rapid growth then occurs as the evaporation wave initiates, leading to the entrainment of small droplets within the bubble.

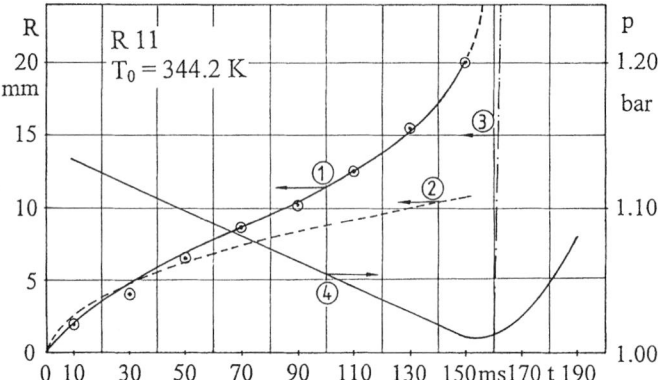

FIGURE 15.19   Bubble growth with the formation of an evaporation wave resulting in entrainment of small droplets into the bubble. 1, experimental data; 2, Plesset and Zwick heat-transfer-controlled solution (Eq. 15.37); 3, Rayleigh solution with $r(t) = 0$ at $t = 0.16$ s; 4, liquid pressure (from Barthau and Hahne [45], with permission).

Bubble Growth From a Surface.   In practice, since heterogeneous rather than homogeneous nucleation is the norm in boiling, bubble growth occurs from a solid surface. An idealized representation of the process is shown in Fig. 15.20. The following stages are envisaged:

1. At $t = 0$, the previous bubble has just departed from the surface, carrying with it the thermal boundary layer. The bulk fluid (at temperature T_∞) is brought into contact with the wall and the thermal boundary layer begins to grow again by transient conduction from the heated surface.

2. During a waiting period t_w, no significant growth of the bubble occurs. During this period, the thermal boundary layer is building up on the surface and it is only after the period t_w that bubble growth can commence.

3. Once the waiting period is over, rapid inertia-controlled bubble growth occurs, the bubble growing in a nearly hemispherical shape as shown in Fig. 15.20c. In this period, a liquid *microlayer* may be left behind that has a thickness near zero at the original nucleation site and a finite thickness at the edge of the hemispherical bubble. The bubble grows as a result of both evaporation at its upper surface (which is in contact with superheated liquid in the displaced boundary layer) and also by evaporation of this microlayer.

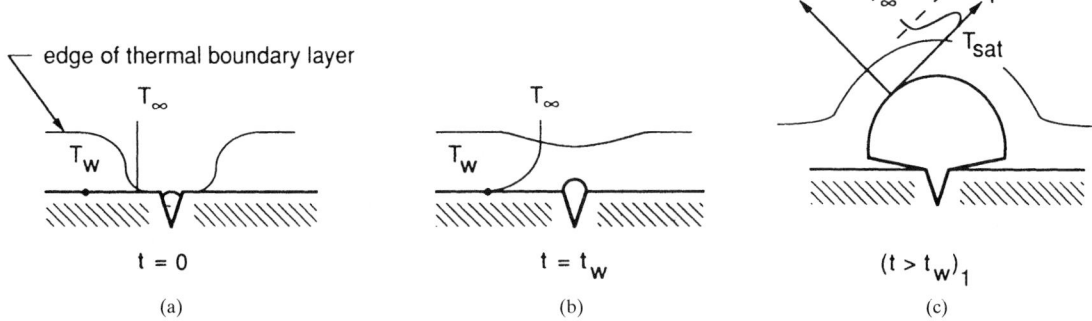

FIGURE 15.20   Stages in bubble growth from a cavity on a heated surface (from Carey [4], with permission from Taylor & Francis, Washington, DC. All rights reserved).

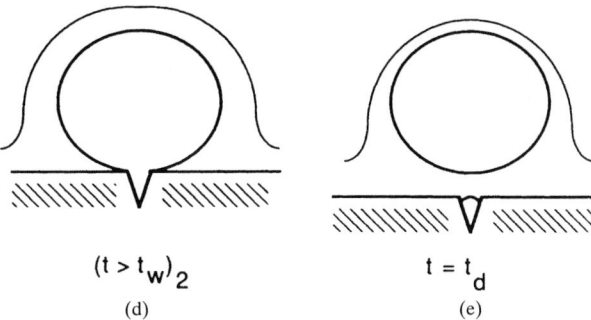

$(t > t_w)_2$

$t = t_d$

(d)

(e)

FIGURE 15.20 (*Continued*) Stages in bubble growth from a cavity on a heated surface (from Carey [4], with permission from Taylor & Francis, Washington, DC. All rights reserved).

4. After the initial rapid growth stage, the growth rate decreases and the bubble growth may become heat-transfer-controlled rather than inertia-controlled; this results in a more spherical bubble as shown in Fig. 15.20d.

5. The bubble is released from the surface at the departure time t_d. The released bubble carries with it a portion of the thermal boundary layer and the cycle is repeated. Figure 15.21 shows this bubble cycle, as well as typical variations of bubble diameter D_b and wall temperature (T_w). Actually, Fig. 15.20d does not fully represent the true picture since the bubbles rapidly outgrow the thermal layer originally surrounding them.

Hsu [27] addressed the problem of prediction of the waiting time t_w by considering transient conduction as illustrated in Fig. 15.22. The treatment is analogous to the steady-state case shown in Fig. 15.14 (and, indeed, was a precursor of it). Following bubble departure, the temperature profile over the thermal boundary layer (of thickness δ) develops with t. At $t = t_1$, the temperature profile becomes tangential to the critical temperature for growth for the bubble of radius r_c (the idealized hemispherical bubble radius on the assumed conical cavity). This condition is reached at $t = t_w$ and the waiting period is over. Hsu [27] and Mikic and Rohsenow

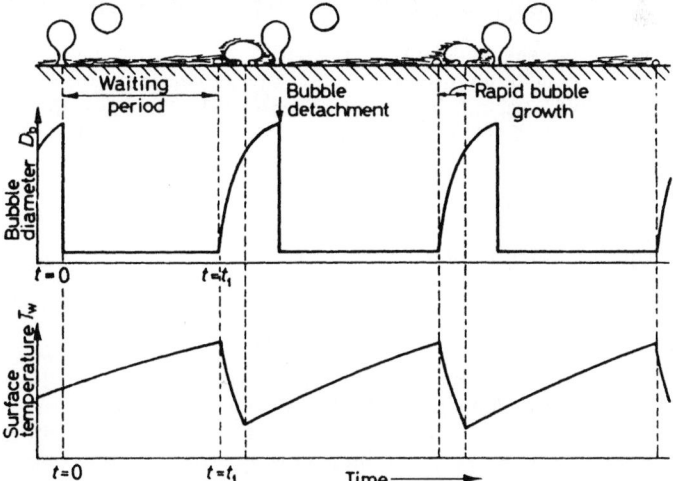

FIGURE 15.21 The bubble growth cycle from a single nucleation site (from Collier and Thome [3], by permission of Oxford University Press).

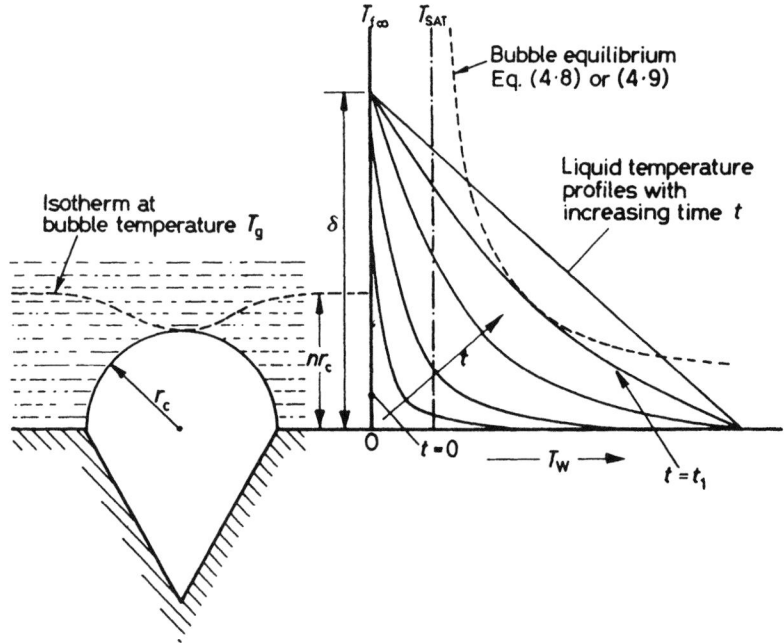

FIGURE 15.22 Transient conduction in thermal boundary layer. Model of Hsu [27] (from Collier and Thome [3], by permission of Oxford University Press).

[47] developed prediction methods for t_w based on the description shown in Fig. 15.22 and on classical transient conduction models. The solution obtained by Mikic and Rohsenow [47] was as follows:

$$t_w = \frac{1}{4\alpha_l} \left\{ \frac{r_c}{\text{erfc}^{-1}\{[T_{\text{sat}}(P_\infty) - T_\infty]/[(T_W - T_\infty)] + [2\sigma T_W(v_g - v_l)]/[(T_W - T_\infty)i_{lg}r_c]\}} \right\}^2 \quad (15.51)$$

Where v_l is the liquid specific volume.

Ignoring the contribution of microlayer evaporation (see Fig. 15.20c), Mikic and Rohsenow [47] considered only heat-transfer-controlled bubble growth (implying spherical bubbles) and obtained the following expression for bubble radius as a function of time for $t > t_w$:

$$r(t) = \frac{2\text{Ja}\sqrt{3\pi\alpha_l(t - t_w)}}{\pi} \left\{ 1 - \frac{T_W - T_\infty}{T_W - T_{\text{sat}}(P_\infty)} \left[\left(1 + \frac{t_w}{t - t_w}\right)^{1/2} - \left(\frac{t_w}{t - t_w}\right)^{1/2} \right] \right\} \quad (15.52)$$

where Ja is the Jakob number given by

$$\text{Ja} = \frac{[T_\infty - T_{\text{sat}}(P_\infty)]c_{pl}\rho_l}{\rho_g i_{lg}} \quad (15.53)$$

For very low pressures (where there is a very large change of specific volume between the liquid and vapor states), bubble growth may be controlled by microlayer evaporation; this situation has been investigated in detail by Cooper and Lloyd [48] and van Stralen et al. [49]. van Stralen et al. [49] proposed the following expression for bubble radius as a function of time:

$$r(t) = \frac{r_1(t)r_2(t)}{r_1(t) + r_2(t)} \quad (15.54)$$

where $r_1(t)$ is a function describing the inertial contribution to bubble growth and $r_2(t)$ is the heat-transfer-controlled contribution. These functions are given as follows:

$$r_1(t) = 0.8165t \left\{ \frac{\rho_g i_{lg}(T_W - T_{sat}(P_\infty)) \exp[-[(t-t_w)/t_d]^{1/2}]}{\rho_l T_{sat}(P_\infty)} \right\}^{1/2} \tag{15.55}$$

$$r_2(t) = 1.9544 \left\{ b^* \exp\left[-\left(\frac{t-t_w}{t_d}\right)^{1/2}\right] + \frac{T_\infty - T_{sat}(P_\infty)}{T_W - T_{sat}(P_\infty)} \right\} Ja \, [\alpha_l(t-t_w)]^{1/2}$$

$$+ 0.3730 \, Pr_l^{-1/6} \left\{ \exp\left[-\left(\frac{t-t_w}{t_d}\right)^{1/2}\right] \right\}^{1/2} Ja \, [\alpha_l(t-t_w)]^{1/2} \tag{15.56}$$

where t_d is the bubble departure time and Pr_l is the liquid-phase Prandtl number, and where

$$b^* = 1.3908 \, \frac{r_2(t_d)}{Ja \, [\alpha_l(t-t_w)]^{1/2}} - 0.1908 Pr_l^{-1/6} \tag{15.57}$$

This analysis assumes that the wall temperature T_W is essentially fixed. However, the wall temperature is known to fluctuate (see Fig. 15.21), and, indeed, there is a complex temperature distribution within the solid that is at least two-dimensional, this temperature distribution developing together with the temperature field in the liquid surrounding the bubble. The processes are conjugated, and this makes the whole bubble growth process extremely complex. Numerical calculations of the evolution of the bubble interface and of the associated liquid and solid temperature fields have been performed by Mei et al. [50, 51] for saturated boiling and by Chen et al. [52] for subcooled boiling. The basis of the model used by these authors is shown in Fig. 15.23. Although the model does contain some empiricism (in particular relating to the extent to which the microlayer evaporation mechanism is contributing), the results are particularly interesting in giving coupled calculations of conduction and velocity fields occurring during the bubble growth cycle. These are exemplified in Fig. 15.24 for a particular time in the cycle ($\tau = 0.5 = t/t_c$, where, in this case, t_c is the time from initiation to collapse of a bubble in subcooled boiling). Figure 15.24a shows the nondimensional isotherms

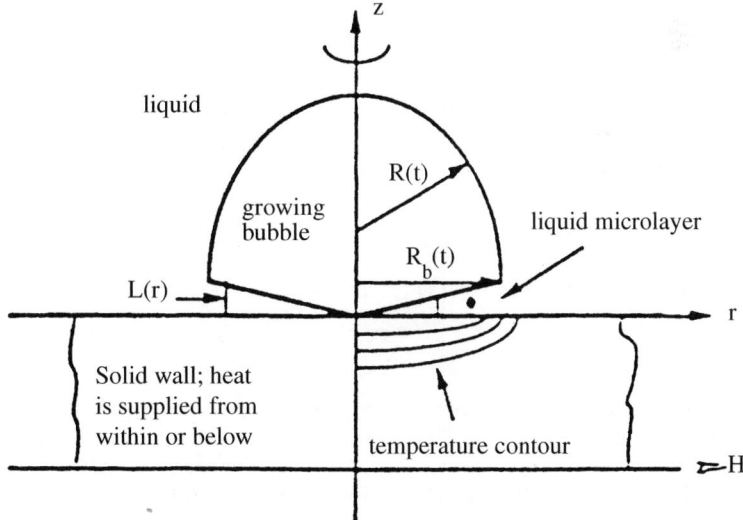

FIGURE 15.23 Basis for numerical model for bubble growth used by Mei et al. [50, 51] and Chen et al. [52] (from Mei et al. [50], with permission from Elsevier Science).

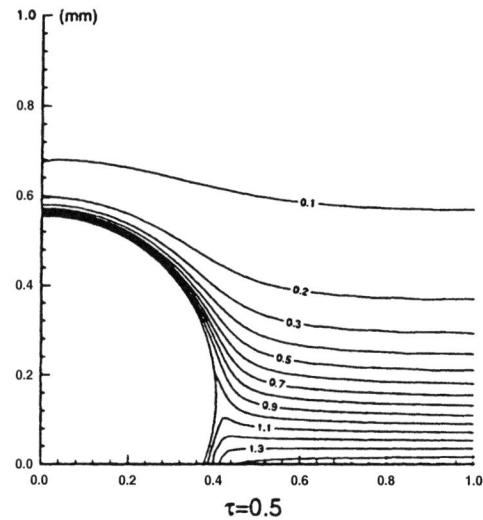

(a) *Isotherms in the liquid phase*

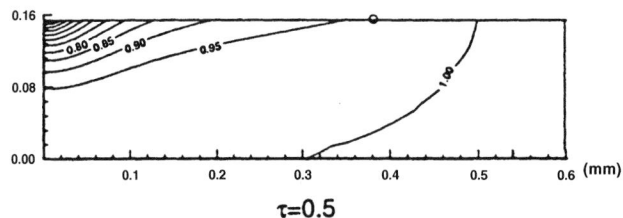

(b) *Isotherms in the solid*

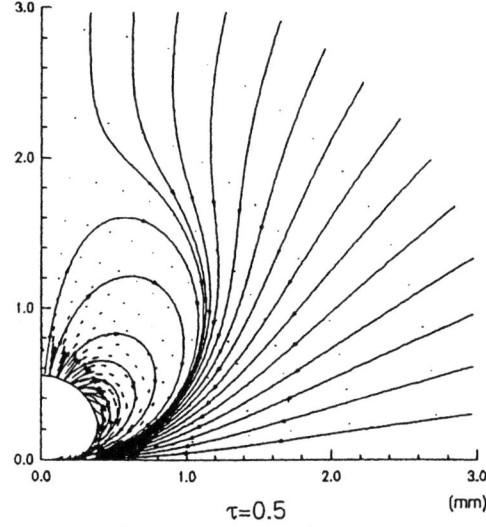

(c) *Velocity vectors and streamlines*

FIGURE 15.24 Typical results from the numerical calculations of bubble growth in subcooled boiling by Chen et al. [52]. Conditions for $\tau = t/t_c = 0.5$ (from Chen et al. [52], with permission from ASME).

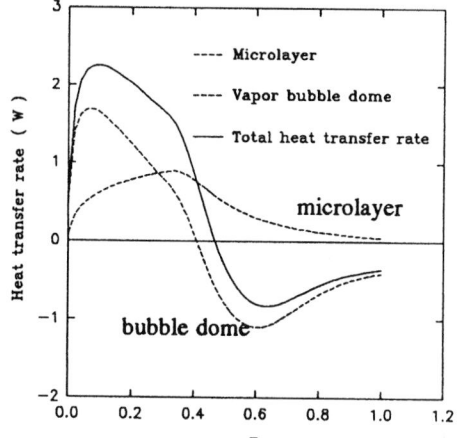

for the liquid, Fig. 15.24*b* the corresponding values for the solid, and Fig. 15.24*c* the associated velocity vectors and streamlines in the liquid phase. The contributions to the overall heat transfer in the growth of a bubble into a subcooled liquid are illustrated for a sample case in Fig. 15.25.

Heat transfer to the bubble occurs both from the microlayer and the bubble dome. The contribution from the microlayer passes through a maximum but always remains positive. In this subcooled boiling case, the contribution to the heat transfer rate to the bubble from the bubble dome passes through a maximum and then becomes negative as the bubble grows through the thermal boundary layer into the subcooled region beyond it and begins to condense and collapse. The model by Chen et al. is in good agreement with growth data obtained by Ellion [53] as shown in Fig. 15.26.

The above analyses assume that the bubble remains in position above the nucleation site from which it arose. However, bubbles may slide along the surface without being released from it and continue to grow during this process. This phenomenon is discussed by Cornwell [54], and observations on sliding bubbles on inclined planes and curved surfaces (carried out using liquid crystal thermography and high-speed video recording) are reported by Yan et al. [55]. Chen [56] analyzed cinefilms of boiling in vertical annular flow obtained by Hewitt et al. [57] and showed that bubbles slid along the surface and grew to a size several times the liquid film thickness before bursting into the continuous vapor phase. In the following text, we return to consideration of such phenomena in discussing the various regimes of boiling.

Bubble Growth in Binary Systems.   In a binary mixture of two components, one of which is more volatile than the other, the bubble growth rate is governed both by heat transfer and by mass transfer in the liquid phase. This is because the more volatile material evaporates first, leaving a preponderance of the less volatile material at the interface, which increases the interfacial saturation temperature and reduces the rate of heat transfer and, hence, of vaporization. Thus, bubble growth in a binary mixture is often much slower than that in a pure component with the same average properties, and this, as we shall see, has a very important effect in boiling.

Bubble growth in binary mixtures is reviewed by Collier and Thome [3]; notable contributions to the subject are those of Scriven [58], van Stralen and Zijl [59], van Stralen [49, 60], Thome [61, 62], and Thome and Davey [63]. The model of van Stralen [60] is illustrated in Fig. 15.27. Bubble growth is considered in a liquid that has a mole fraction $\tilde{x}_o$ of the more volatile material and a superheat (relative to the bubble point temperature—see Fig. 15.3) of ΔT_{sat}. At the actual interface of the growing bubble, the concentration of the more volatile component has fallen to $\tilde{x}$, with the more volatile component being transferred to the interface by diffusion across a nominal diffusion layer of thickness Z_M. The vapor in the bubble has a concentration $\tilde{y}$ in equilibrium with $\tilde{x}$ as shown. However, because of the reduced concentration of volatile component at the interface, the interface temper-

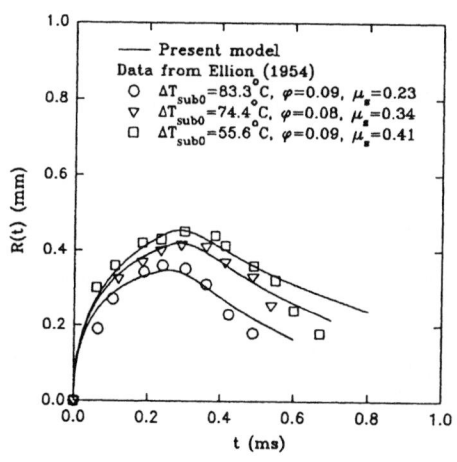

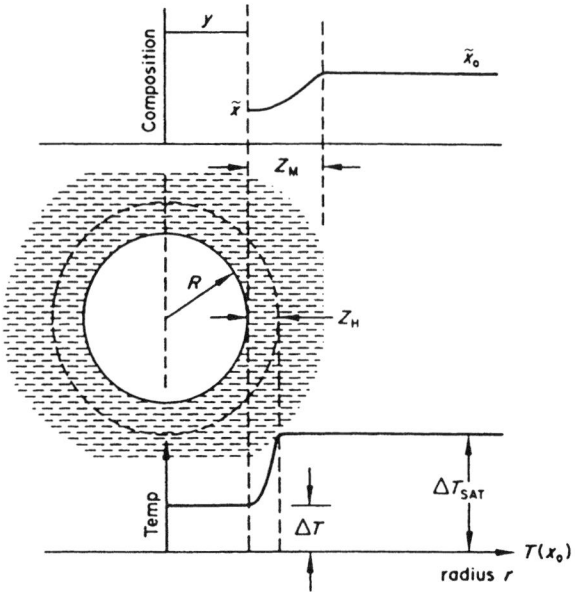

FIGURE 15.27 van Stralen [60] model for bubble growth in binary mixtures (from Collier and Thome [3], by permission of Oxford University Press).

ature is higher by an amount ΔT than the bubble point temperature corresponding to $\tilde{x}_o$. Thus, the driving force for heat transfer across the equivalent thermal layer (thickness Z_H) is reduced from ΔT_o to $(\Delta T_o - \Delta T)$ and the rate of evaporation is consequently reduced. This has a very large effect, as will be seen, on nucleate boiling heat transfer in multicomponent mixtures.

For the heat-transfer-controlled region, the bubble radius at time t can be determined by introducing a correction term Sn (which Thome [61] termed the Scriven number) into Eq. 15.37. Thus:

$$r(t) = \frac{2\Delta T_{\text{sat}} k_l}{i_{lg}\rho_g}\left(\frac{3t}{\pi\alpha_l}\right)^{1/2}\text{Sn} \tag{15.58}$$

where:

$$\text{Sn} = \left[1 - (\tilde{y} - \tilde{x})\left(\frac{\alpha_l}{D}\right)^{1/2}\left(\frac{c_{pl}}{i_{fg}}\right)\left(\frac{\partial T_{\text{bub}}}{\partial \tilde{x}}\right)_p\right]^{-1} \tag{15.59}$$

where D is the diffusion coefficient for the more volatile material in the binary mixture and the gradient $(\partial T_{\text{bub}}/\partial \tilde{x})_p$ can be determined from the equilibrium diagram (see Fig. 15.3). Sn is always less than unity. For the inertia-controlled stage of bubble growth, we can use Eqs. 15.38 or 15.44 with B defined from

$$B = \left(\frac{12\alpha_l}{\pi}\right)^{1/2}\text{Ja Sn} \tag{15.60}$$

Bubble Release Diameter and Frequency

Bubble Departure Diameter. At a certain point in the bubble growth process, the bubble detaches from the surface and the cycle begins again. Clearly the release diameter of the bubble is an important factor in understanding nucleate boiling. For pool boiling, Carey [4] gives a taxonomy of published correlations for bubble departure diameter. A majority of these correlations are expressed in the form of the Bond number Bo, which is defined as follows:

$$\text{Bo} = \frac{g(\rho_l - \rho_g)d_d^2}{\sigma} \qquad (15.61)$$

where d_d is the bubble departure diameter, g is the acceleration due to gravity, and σ is the surface tension. An evaluation of bubble release diameter correlations has been made by Jensen and Memmel [64]; on the basis of the comparisons, these authors recommend a somewhat modified form of the correlation of Kutateladze and Gogonin [65] as follows:

$$\text{Bo}^{1/2} = 0.19(1.8 + 10^5 K_l)^{2/3} \qquad (15.62)$$

where K_l is given by

$$K_l = \left(\frac{\text{Ja}}{\text{Pr}_l}\right)\left\{\left[\frac{g\rho_l(\rho_l - \rho_g)}{\mu_l}\right]\left[\frac{\sigma}{g(\rho_l - \rho_g)}\right]^{3/2}\right\}^{-1} \qquad (15.63)$$

where μ_L is the viscosity of the liquid phase and Ja is the Jakob number, defined, in this case, as

$$\text{Ja} = \frac{(T_W - T_{\text{sat}})c_{pl}\rho_l}{\rho_g i_{lg}} \qquad (15.64)$$

For the boiling of binary mixtures, Thome [62] suggests that the reduction in bubble departure diameter for binary mixtures can be calculated using the relationship

$$\frac{d_{dm}}{d_{di}} = \text{Sn}^{4/5} \qquad (15.65)$$

where d_{dm} is the departure diameter for the mixture and d_{di} is the departure diameter calculated from the above relationships for a fluid having the average physical properties of the mixture.

In forced convective systems, the bubble departure diameter can be critically affected by the presence of a velocity field. Studies of bubble departure diameters in forced convection include those of Al-Hayes and Winterton [66], Winterton [67], Kandlikar et al. [68], and Klausner et al. [69]. The results obtained by Klausner et al. for the probability density function of departure diameter as a function of mass flux and heat flux, respectively, are shown in Figs. 15.28 and 15.29. The most probable departure diameter decreases strongly with increasing flow rate and decreases (less strongly) with decreasing heat flux at a fixed flow rate. It is clear that these velocity effects have to be taken into account in predicting forced convective boiling systems.

Models for bubble departure in flow boiling are given by Klausner et al. [69] and Al-Hayes and Winterton [66]. The Al-Hayes and Winterton model was developed for gas bubble detachment from a surface but has been extended to the case of vapor bubble detachment (in boiling) by Winterton [67]. When inertial forces are negligible, the stability of a bubble on the surface is governed by a balance between three forces on the bubble, resolved parallel to solid surface, namely the buoyancy force F_b, the drag force F_d, and a surface tension force F_s. Al-Hayes and Winterton give the buoyancy force as

$$F_b = \tfrac{1}{3}\rho_l g\pi r^3\{2 + 3\cos\phi - \cos^3\phi\} \qquad (15.66)$$

where r is the bubble radius and ϕ is the equilibrium contact angle. The drag force on the bubble was calculated from

$$F_d = \tfrac{1}{2}C_d\rho u^2 r^2\{\pi - \phi + \cos\phi\sin\phi\} \qquad (15.67)$$

where u is the velocity of the fluid approaching the bubble at a point halfway between the surface and the bubble tip (this can be calculated from standard velocity profiles for either lam-

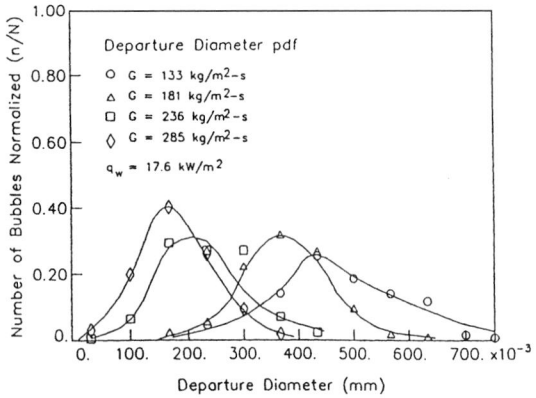

FIGURE 15.28 Probability density function of bubble departure diameter in flow boiling as a function of mass flux for a constant heat flux (from Klausner et al. [69], with permission of Elsevier Science).

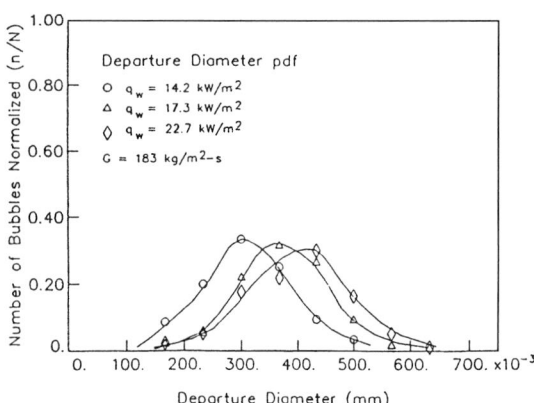

FIGURE 15.29 Probability density function of bubble departure diameter in flow boiling as a function of heat flux at constant mass flux (from Klausner et al. [69], with permission of Elsevier Science).

inar or turbulent flow). C_d is a drag coefficient which Al-Hayes and Winterton calculated from the expressions

$$C_d = 1.22 \qquad \text{for } 20 < \text{Re}_b < 400 \qquad (15.68)$$

$$C_d = 24/\text{Re}_b \qquad \text{for } 4 < \text{Re}_b < 20 \qquad (15.69)$$

where the bubble Reynolds number Re_b is defined as

$$\text{Re}_b = \frac{2\rho_l r u}{\mu_l} \qquad (15.70)$$

The third force involved in the Al-Hayes and Winterton analysis was that due to surface tension (F_s). It is suggested that such a force arises because (just before bubble departure) the contact angle at the furthest upstream point of the bubble is the advancing contact angle ϕ_a and that on the downstream side is the receding contact angle ϕ_r. Al-Hayes and Winterton give the following expression for F_s:

$$F_s = S[\tfrac{1}{2}\pi r\sigma \sin \phi(\cos \phi_a - \cos \phi_r)] \qquad (15.71)$$

The term in brackets is obtained assuming a smooth variation of contact angle around the bubble-to-wall contact. Al-Hayes and Winterton introduced a shape factor S to correct this value for the effect of distortion of the bubbles. S was given by

$$S = \frac{58}{5 + \phi} + 0.14 \qquad (15.72)$$

The balancing of the forces depends on the direction of flow. For downward flow, the drag force is opposed by the buoyancy and surface tension forces, whereas for upward flow, the surface tension force is balanced against the buoyancy and drag forces. For horizontal surfaces, the buoyancy force resolved along the surface is zero and only drag and surface tension forces are included. Thus, the results obtained are sensitive to orientation; for a given bubble size, departure occurs at typically an order of magnitude higher velocity in downward flow than in upward flow. Confirmation of the theory was obtained by measuring the conditions required for detachment of gas bubbles in tube flows at various angles of inclination of the tube. Various contact angles could be obtained depending on the tube material (untreated

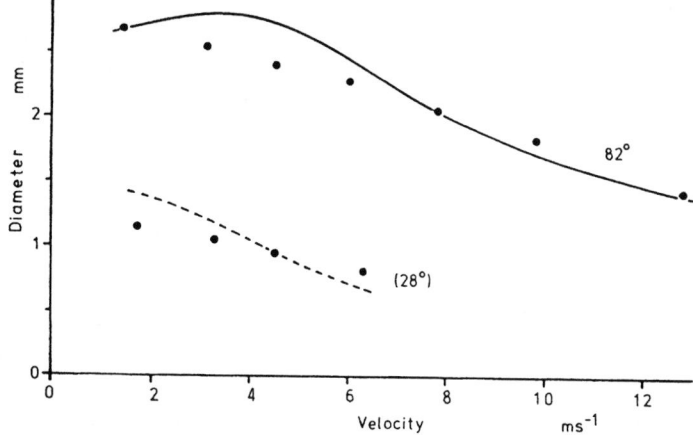

FIGURE 15.30 Conditions necessary for departure of a gas bubble from the surface of a tube in which there is downward flow of ethylene glycol. Tube surfaces have $\phi = 82$, $\phi_a = 99.5$, $\phi_r = 64.5$, and $\phi = 28$, $\phi_a = 46.5$, and $\phi_r = 9.5$, respectively. The advancing contact angles were at the upper (upstream for downflow) side of the bubble (from Al-Hayes and Winterton [66], with permission of Elsevier Science).

glass, treated glass, acrylic resin), and this allowed the testing of the theory for different contact angles. Typical results (for downflow) are shown in Fig. 15.30.

A difficulty with the Al-Hayes and Winterton model is that the values of ϕ, ϕ_a, and ϕ_r need to be known. For a number of boiling experiments, measurements are reported of ϕ, and Winterton [67] shows that reasonable results could be obtained by assuming that $\phi_a = \phi + 10°$ and $\phi_r = \phi - 10°$. Kandlikar and Stumm [374] showed that the Al-Hayes and Winterton model gave excellent results for bubble diameters greater than 800μm but that large discrepancies were observed for smaller bubbles; Kandlikar and Stumm suggest a revised model for this region.

Bubble Frequency. In estimating the heat transfer characteristics in nucleate boiling, the frequency f at which bubbles of diameter d_d depart from a given site is also an important parameter. f is given by

$$f = \frac{1}{t_w + t_g} \tag{15.73}$$

where t_w is the waiting period and t_g is the growth period. Thus, it is not surprising that there is a strong link between the expressions given previously for bubble growth and the departure diameter d_d, and, hence, the frequency. Ivey [70] argued that the product $f^2 d_d$ should be constant if the bubble grows and departs in the inertia-controlled growth regime and that the product $f^{1/2} d_d$ should be constant for heat-transfer-controlled growth. The relationship of Cole [71] for inertia-controlled growth is consistent with this hypothesis:

$$f^2 d_d = \left(\frac{4}{3}\right) \frac{g(\rho_l - \rho_g)}{\rho_l} \tag{15.74}$$

as is that of Mikic and Rohsenow [47] for the heat-transfer-controlled growth case:

$$f^{1/2} d_d = 0.83 \mathrm{Ja} \sqrt{\pi \alpha_l} \tag{15.75}$$

Thus, if d_d is calculated from the relationships given previously, then the bubble frequency can be calculated using Eq. 15.74 or 15.75. A detailed review of alternative expressions for bubble frequency is given by Carey [4].

POOL BOILING

In pool boiling, vapor generation occurs from a surface immersed in a static pool of liquid. The stages in pool boiling from a horizontal cylinder are illustrated in Fig. 15.31. Initially, there is no bubble formation and heat transfer is by natural convection; then, bubbles grow but do not detach from the surface until a much higher bubble population is reached. The process in which vapor generation occurs by nucleation at the surface is known as *nucleate boiling*. Ultimately, the bubbles become so numerous that they coalesce to form a continuous vapor layer on the heated surface; here, the heat transfer process is much less efficient and the surface may become very hot. This region is known as *film boiling*.

The relationship between heat flux and surface temperature in pool boiling is illustrated in Fig. 15.32. A distinction should be made between conditions in which the *wall temperature* is controlled (for instance, by heating with a condensing vapor inside the cylinder) and those in which the *heat flux* is controlled (e.g., by electrical joule heating). With wall temperature control, the relationship between heat flux and wall temperature is illustrated in Fig. 15.32a. Initially, heat transfer is by natural convection; then nucleate boiling starts and a very rapid rise in heat flux occurs with only a small increase in surface temperature. Ultimately, a maximum is reached in heat flux, after which the surface becomes partially dry in the *transition* boiling regime; then the film-boiling regime is entered with a rapid rise in temperature with only a small increase in heat flux. In the case of a heat-flux-controlled situation (e.g., with electrical

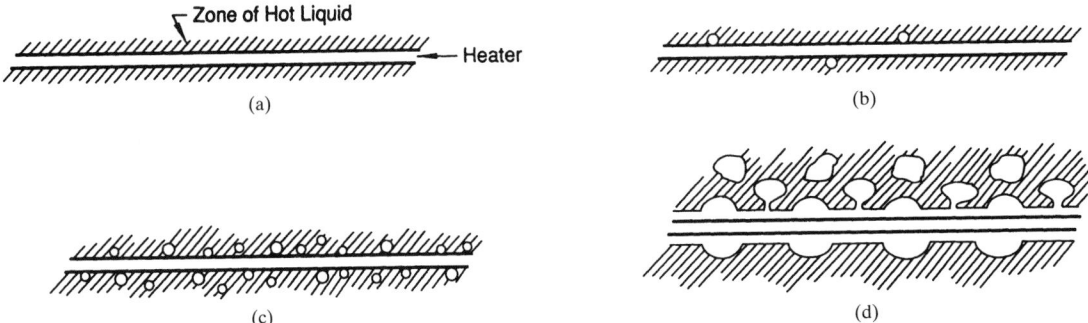

FIGURE 15.31   Stages in subcooled pool boiling as the surface heat flux is increased. (*a*) no bubble formation (natural convection heat transfer); (*b*) bubbles grow but do not detach; (*c*) bubbles increase in number and some detach to condense in the bulk liquid; (*d*) bubbles become so numerous that they coalesce to form a continuous vapor layer on the heated surface (from Hewitt et al. [13], with permission. Copyright CRC Press, Boca Raton, FL).

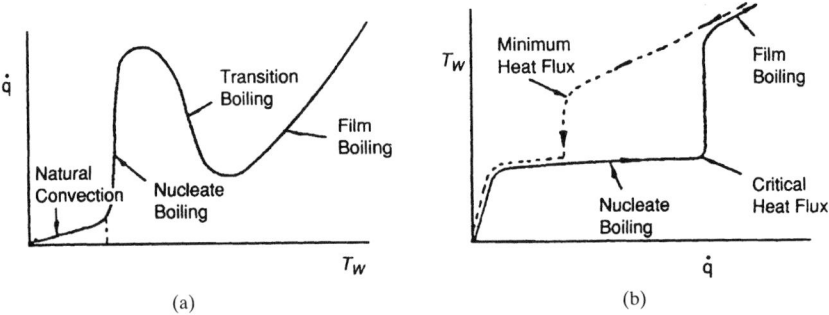

FIGURE 15.32   Pool boiling curves for wall-temperature-controlled and heat-flux-controlled systems (from Hewitt et al. [13], with permission. Copyright CRC press, Boca Raton, FL).

heating), the wall temperature is the dependent variable and is therefore plotted on the y axis. The same form of curve is observed on *increasing* the heat flux: the system passes through the natural convection and nucleate boiling regimes. When the maximum heat flux is reached, there is an excursion into film boiling as shown with a rapid (and often destructive) increase in surface temperature. When the heat flux is *decreased,* the wall temperature decreases until the minimum heat flux is reached, after which there is an excursion back into nucleate boiling. This hysteresis phenomenon is characteristic of electrically heated systems, and it is difficult to obtain transition boiling in such systems. Figure 15.32 presents a somewhat idealized picture. In electrically heated systems, the surface heat flux may be instantaneously different from the steady-state value. This may lead to the conditions passing transiently through transition boiling, even in the imposed (steady-state) heat flux case.

Many applications of pool boiling are transient in nature. Careful experiments on transients in which the wall temperature is either increased or decreased with time are reported by Blum et al. [72]; the results are shown in Fig. 15.33. As will be seen, the boiling curve under transiently changing wall temperature conditions can differ significantly with both the rate of change and the direction of change. These results typify the many unknowns and uncertainties in the pool-boiling process, and the reader should be aware of these from the outset!

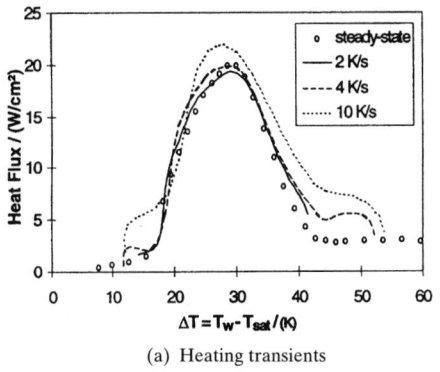

(a) Heating transients

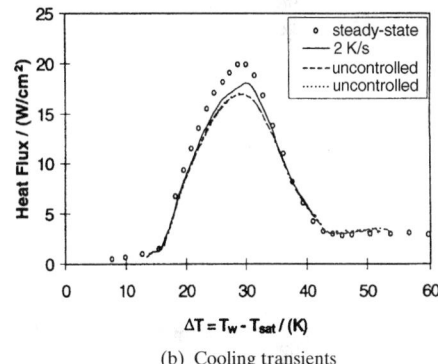

(b) Cooling transients

FIGURE 15.33   Pool-boiling curves obtained under transient heating and cooling compared with pool-boiling curve obtained for steady state (boiling of Fluorinert FC-72) (from Blum et al. [72], with permission).

Pool Boiling Heat Transfer Before the Critical Heat Flux Limit

Nucleate pool boiling has been very widely studied, and it would be impossible in the present context to give a detailed review of the vast amount of work done. Rather, some of the more salient points are addressed. The effect of various system parameters on pool boiling is first discussed, and then the mechanisms of nucleate pool boiling are reviewed with emphasis on recent findings. A number of the most widely used correlations for pool heat transfer are presented, and predictive models are discussed based on a more phenomenological approach. The effect of multicomponent mixtures on nucleate pool boiling and the various methods by which enhanced heat transfer can be obtained are also discussed.

Parametric Effects in Nucleate Pool Boiling

Effect of Pressure.   Results illustrating the effect of pressure on nucleate pool boiling are shown in Fig. 15.34; the superheat required for a given heat flux decreases with increasing pressure as shown. Results for subatmospheric conditions are presented by Schroder et al. [74]; the trends shown in Fig. 15.34 are continued (increasing superheat for reducing pressure), but at the lowest pressure studied (0.1 bar), a new phenomenon is observed. Boiling starts with the creation

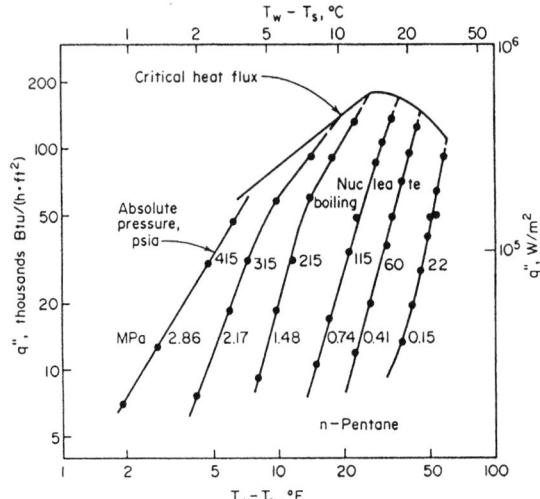

FIGURE 15.34 Effect of pressure on nucleate pool boiling (Cichelli and Bonilla [73]) (from Rohsenow [2], with permission of The McGraw-Hill Companies).

of large bubbles (on the order of 1 cm in size) that collapse and initiate very efficient boiling from a large number of nucleation sites, giving a large reduction in the wall superheat. In fact, once the distributed nucleation is initiated, the superheats are less than those at atmospheric pressure.

Effect of Subcooling. Data on the effect of subcooling (i.e., the difference between the bulk liquid temperature in the pool and the saturation temperature) are reviewed by Judd et al. [75]. Typical results for the variation of wall superheat (difference between wall temperature and saturation temperature) as a function of subcooling, and for a series of fixed heat fluxes, are shown in Fig. 15.35. At low subcoolings, the superheat increases with increasing subcooling; Judd et al. [75] suggest that this is a result of the changes in active site density, average bubble frequency, and the consequential effects on the rate of heat removal from the heated surface. As the subcooling increases further, natural convection begins to be increasingly important and wall superheat decreases with increasing subcooling.

Effect of the Heat Transfer Surface. As we saw from the discussion on pages 15.9–15.18, the number of active nucleation sites on the surface at a given wall superheat and heat flux depends on a variety of factors. First, there is the population of *potentially active* sites, which is a function of the nature and preparation of the surface. Second, there are the *wetting characteristics* associated with the fluid/surface combination. These characteristics are often expressed in terms of the contact angles (ϕ, ϕ_a, and ϕ_r). In general, the heat flux (or heat transfer coefficient) in nucleate boiling heat transfer is strongly dependent on the number of active sites.

As shown in Fig. 15.12, the active site density may decrease very strongly with contact angle for a given distribution of cavity sizes. It is not surprising, therefore, that the prediction of nucleate boiling heat transfer is bedeviled by the problems of predicting surface behavior. The effect of surface finish on nucleate pool boiling is well illustrated by the results of Brown [76] cited by Rohsenow [2] and shown in Fig. 15.36.

Depending on how the surface was pretreated, different results are obtained for heat flux at a given wall superheat. There is even a difference between approximately identical scratches that are circumferential to those that are axial; the experiments were done with the tube in the horizontal position. Investigations of the effect of wetting angle are reported by Jansen et al. [77], who were able to vary the contact angle systematically by using a composite copper/PTFE surface. The results obtained are illustrated in Fig. 15.37. The heat transfer coefficient increases

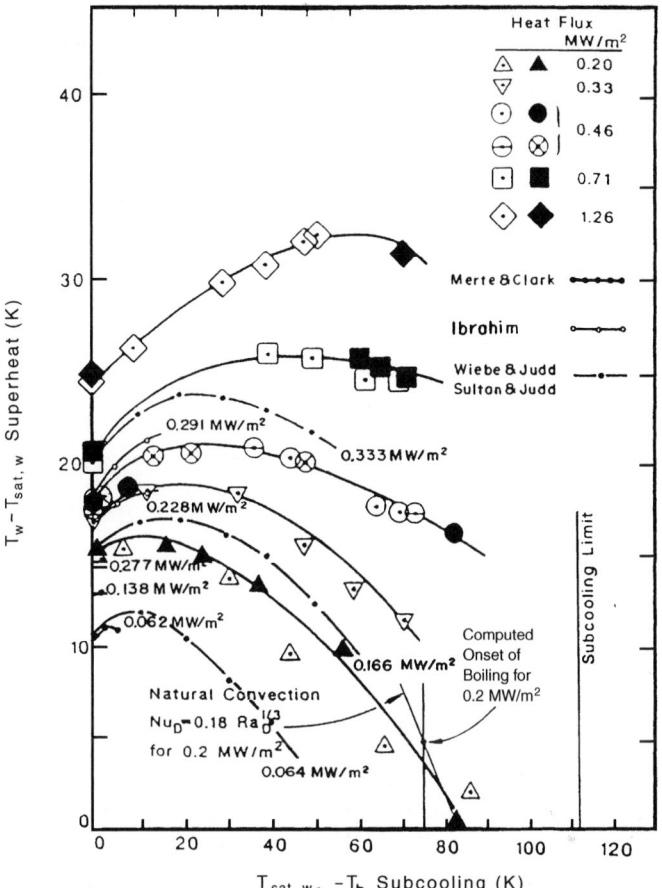

FIGURE 15.35   Effect of subcooling on wall superheat at a given wall heat
flux (from Judd et al. [75], with permission from ASME).

with increasing contact angle at a fixed heat flux and bulk liquid temperature. These observations are consistent with the findings of Wang and Dhir [24–26] as previously discussed. Real-
istically, it seems unlikely that reliable predictions of nucleate pool boiling heat transfer can
ever be made for practical situations in view of these predominant surface effects.

Effect of Wall Thickness.   The results shown in Fig. 15.24 illustrate the strong coupling
that exists between the bubble growth process and the transient developing temperature profile in the solid material under the bubble. The thickness and thermal properties of the solid
material will be expected, therefore, to exert some influence on the boiling process. A major
difficulty in studying this effect is that of decoupling the thermal behavior of the solid from
effects of surface finish. Some elegant experiments that address this problem are described by
Zhou and Bier [78], who studied pool boiling from a horizontal stainless steel tube coated
with various thicknesses of copper. This allows the surface characteristics to be maintained
reasonably constant while the underlying thermal properties are changed. The results
obtained by Zhou and Bier are illustrated in Fig. 15.38.

The thinnest coating used was 2 μm, and the ratio of heat transfer coefficient to the value
obtained with this minimum thickness is plotted as a function of the thickness of the copper

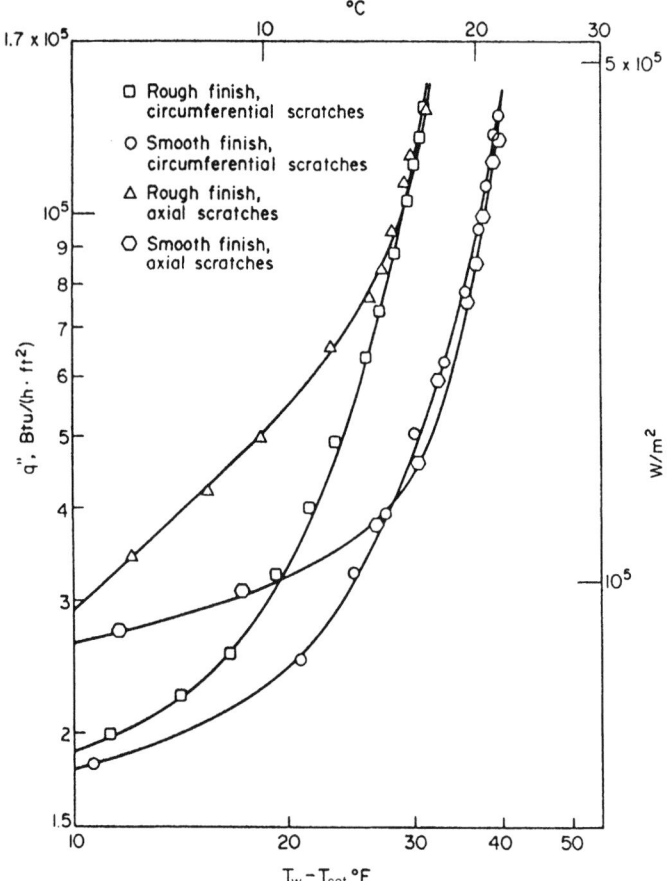

FIGURE 15.36 Effect of surface finish on nucleate pool boiling of water. Results of Brown [76] (from Rohsenow [2], with permission of The McGraw-Hill Companies).

layer for experiments with isopentane and refrigerant R12 (Fig. 15.38). As the thickness of the copper layer increases, the heat transfer coefficient increases toward the limiting value for a thick-walled copper tube. The maximum shift in heat transfer coefficient was on the order of 80% for isopentane and around 40% for refrigerant R12. Zhou and Bier state that these wall thermal property effects are exercised through changes in active site density. Thus, with a copper surface, the underlying solid material can recover its initial temperature profile much more rapidly after bubble departure than can the stainless steel surface, and this will affect the continuing ability to nucleate from a given site. These very significant effects are over and above those associated with the density of potential sites described above.

Effect of Dissolved Gases.   Dissolved gases can have a significant effect on the pool boiling curve; this is illustrated by the results of McAdams et al. [79] illustrated in Fig. 15.39. Boiling incipience occurs at a lower temperature with higher gas contents and the heat flux remains higher at a given temperature difference until temperature differences approaching those for the critical heat flux are reached. A review of the influence of dissolved gases on boiling incipience is given by Bar-Cohen [80].

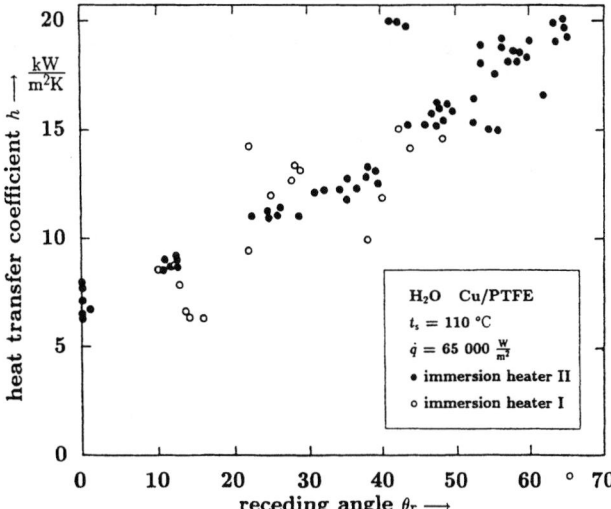

FIGURE 15.37 Effect of contact angle on heat transfer coefficient in nucleate pool boiling with a constant heat flux and subcooling (from Jansen et al. [77], with permission from ASME).

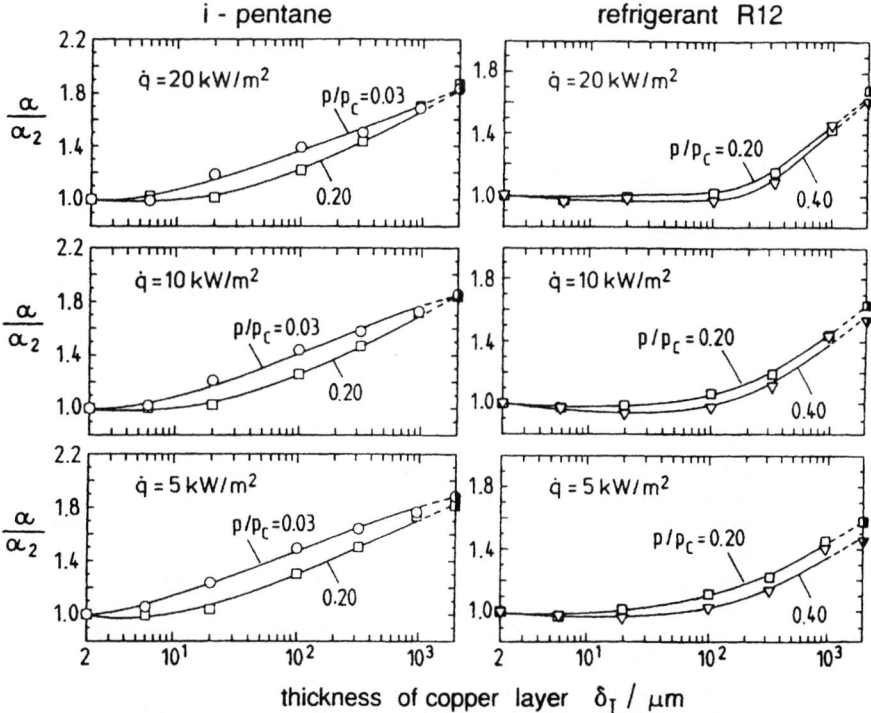

FIGURE 15.38 Pool-boiling heat transfer from a copper-coated stainless steel tube. Ratio of heat transfer coefficient is that for a coating thickness of 2 μm (from Zhou and Bier [78], with permission).

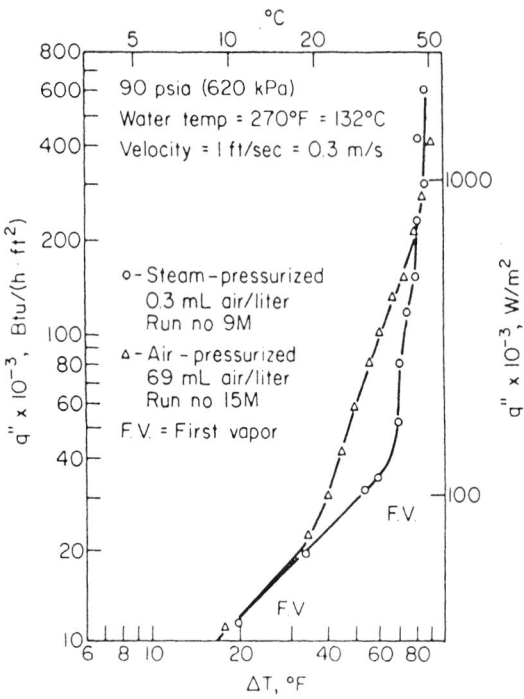

FIGURE 15.39 Influence of dissolved gas content on pool boiling heat transfer. Results of McAdams et al. [70] (from Rohsenow [2], with permission of The McGraw-Hill Companies).

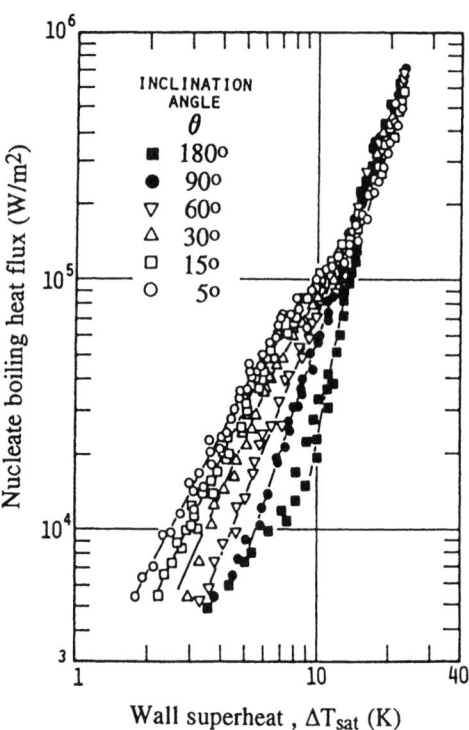

FIGURE 15.40 Effect of angle of inclination on pool boiling from a flat plate (from Nishikawa et al. [82], with permission from ASME).

Effect of Surface Inclination.   The effects of surface inclination on boiling are reviewed in detail by El-Genk [81]. Some very interesting results on boiling from inclined flat surfaces are reported by Nishikawa et al. [82] and are illustrated in Fig. 15.40. The data range from boiling from a downward-facing surface at 5° from the horizontal to boiling from an upward-facing surface (180°). As will be seen, there is a very significant effect of inclination on pool boiling at low superheats, with the upward-facing surface having much greater wall superheats at a given heat flux. However, at higher heat fluxes, the results become independent of the angle of inclination. This is an important finding, implying that the boiling processes are influenced by regions near the wall that are not affected by the angle of inclination.

Effect of Gravity.   The influence of gravitational fields on boiling is important in a number of applications, notably in space power systems where evaporation takes place at near zero gravity. Moreover, the study of the influence of gravity on boiling is of potential interest in evaluating boiling mechanisms. Thus, experiments have been carried out over a range of accelerations with the ratio of the acceleration a to the gravitational acceleration g ranging typically from −1 (negative gravity) to 0 (microgravity) to large positive values (obtained by carrying out boiling experiments in centrifuges). The influence of gravity on pool boiling is reviewed by Merte [83]. The results are exemplified by those of Merte et al. [84], which are illustrated in Fig. 15.41.

With increasing gravity (ratio of acceleration a to gravitational acceleration g greater than unity), the heat transfer is enhanced relative to that for $a/g = 1$. For near-zero and negative gravity, the trend is reversed with an improvement in heat transfer with decreasing gravity. An important point to note is that at high heat fluxes the results become independent of gravitational acceleration. Again, this may indicate that processes are being governed by near-walled phenomena that are unaffected by gravity in this region. Also shown in Fig. 15.41 are lines calculated for natural convection heat transfer without boiling; an important feature here is that, with enhanced gravity, natural convection also increases, and this may provide some explanation of the complexity of the processes. Most experiments on near-zero gravity have been done with apparatus in free fall. A more satisfactory approach is to use an appara-

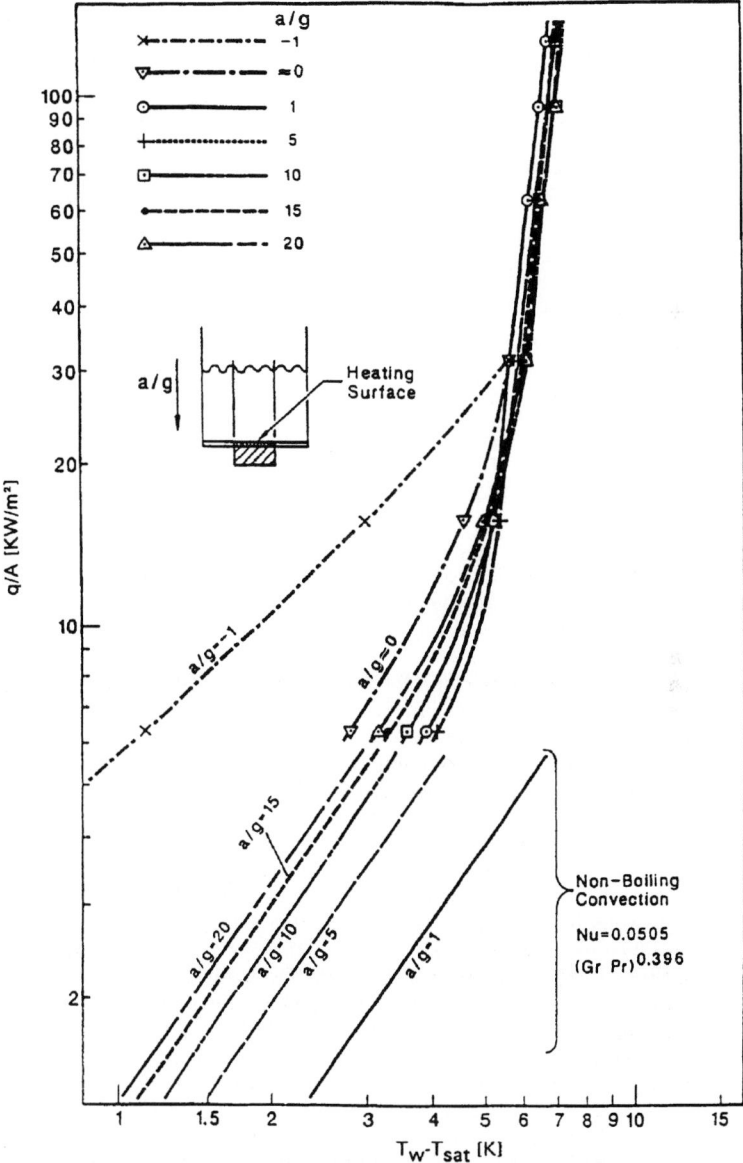

FIGURE 15.41   Influence of gravity on heat transfer in nucleate pool boiling of liquid nitrogen at atmospheric pressure (from Merte et al. [84], with permission).

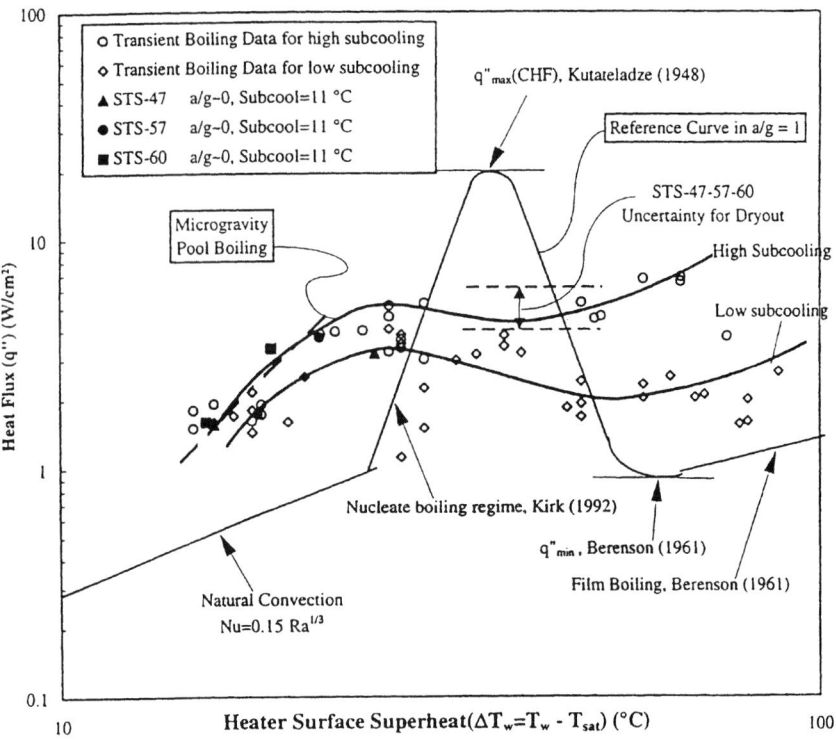

FIGURE 15.42 Pool-boiling curves for the boiling of refrigerant 113 at zero gravity compared with values predicted for normal gravity. Experiments carried out in space shuttle (from Lee and Merte [85] with permission).

tus mounted in an orbiting space craft; such experiments are reported by Lee and Merte [85]. Complete pool-boiling curves were obtained for refrigerant R113 and are illustrated in Fig. 15.42. Again, the heat transfer is enhanced (lower wall superheats at a given heat flux than that calculated for normal gravity) in the nucleate boiling region; for these data, the wall superheat decreases with increasing subcooling, which is analogous to the effects for higher subcooling shown in Fig. 15.35.

Effect of Direction of Change of Heat Flux or Wall Temperature (Hysteresis). Once nucleation is initiated from a given site, then that site tends to remain active even when the heat flux (or wall temperature) is reduced below the value required to initiate the nucleation process. This leads to a hysteresis effect in the curves of heat flux versus wall temperature in pool boiling. A detailed review of hysteresis effects is given by Bar-Cohen [80], and the type of results obtained is exemplified in Fig. 15.43, which shows data obtained by Kim and Bergles [86] for boiling of refrigerant R113 on a copper heater. When the heat flux is increased, the heat transfer occurs by natural convection until boiling is initiated (and probably spreads from the first site by the mechanism shown in Fig. 15.18). This results in a sudden decrease in wall superheat for the given heat flux; a further increase in heat flux leads to an increase in superheat as shown. When the heat flux is decreased, the wall superheat changes smoothly past the point of original initiation, with boiling continuing (with a consequentially lower superheat) until the natural convection region is reached once again. These results are typical of many experiments that demonstrate this effect (see Bar-Cohen [80]).

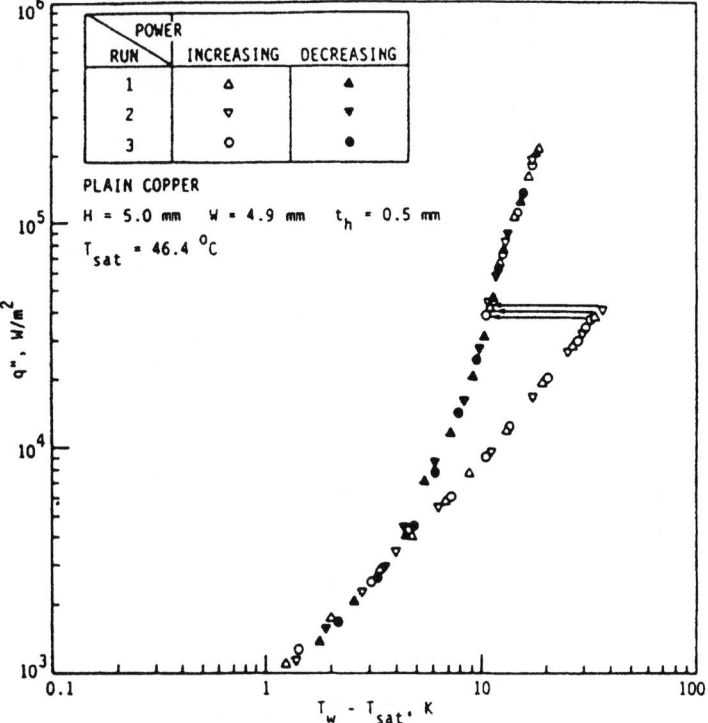

FIGURE 15.43 Hysteresis in the nucleate pool boiling of R113 (from Kim and Bergles [86], with permission from Taylor & Francis, Washington, DC, All rights reserved).

Mechanisms of Nucleate Pool Boiling.   The mechanisms of nucleate pool-boiling heat transfer are a source of endless fascination and are still relatively poorly understood. There have been many thousands of publications on this subject, and only a brief summary can be given within the scope of the present chapter. A thorough survey of the early work in this area is given in the book by Carey [4], and reviews of more recent work are presented by Dhir [87] and Fujita [88].

When boiling data are plotted in terms of system parameters such as gravitational acceleration (see for instance Fig. 15.41), angle of inclination of the surface (see for instance Fig. 15.40), and liquid depth above the boiling surface (Katto et al. [89]), a situation is reached at a certain heat flux where the relationship between heat flux and wall superheat becomes independent of the parameter varied. Clearly, in this situation, the boiling process is governed by near-wall phenomena that are unaffected by the external parameters; this region is termed *fully developed boiling*. At heat fluxes lower than that required for fully developed boiling, the external system parameters can have an influence (as is exemplified in Figs. 15.40 and 15.41) and the evaporation process tends to be associated with the production and release of individual bubbles from individual nucleation sites. This lower-heat-flux region is often referred to as the *isolated bubble region* or as the *partial nucleate boiling region*. It is convenient in the following discussion to treat these regions separately. A number of expressions have been reported in the literature for the heat flux q_{tr}'' at which the transition to fully developed boiling takes place (Fujita [88]). Zuber [90] gives the following expression for the transition:

$$q_{tr}'' = 0.80 \rho_g i_{lg} [g\sigma/(\rho_l - \rho_g)]^{1/4} \qquad (15.76)$$

Moissis and Berenson [91] recognized the influence of contact angle ϕ and suggest the following expression:

$$q''_{tr} = 0.11\rho_g i_{lg}\phi^{1/2}[g\sigma/(\rho_l - \rho_g)]^{1/4} \qquad (15.77)$$

This implies that the transition heat flux increases with increasing contact angle.

Partial Nucleate Boiling (Isolated Bubble) Region. In this region, heat transfer is often considered to be occurring by three separate mechanisms:

1. *Natural convection.* This process is obviously dominant up to the incipience of nucleate boiling. As more and more of the surface becomes covered by bubbles, there is less area over which the natural convection can operate. However, the growth and motion of the bubbles may more than compensate for this by inducing additional convection motions.

2. *Locally enhanced convection coupled with transient conduction.* The bubbles leaving the surface induce the removal of the thermal boundary layer adjacent to the surface and the movement of cool liquid to the surface, which heats up transiently (producing the waiting time t_w—see earlier discussion) before the next bubble is initiated.

3. *Latent heat transport.* Here, the vapor being transported from the surface into the bulk fluid transports heat from the surface in the form of latent heat. This process is particularly important when heat fluxes are high. It may also play a very important role when the situation is such as to promote microlayer evaporation (e.g., at very low pressures).

The total heat flux from the surface is given by

$$q'' = q''_{nc} A_{nc}/A + q''_e A_e/A + q''_t A_t/A \qquad (15.78)$$

where q''_{nc}, q''_e, and q''_t are the heat fluxes for the natural convection, evaporation (latent heat transport), and transient conduction (enhanced convection), respectively, and A_{nc}, A_e, and A_t are the areas of the surface associated with these processes, with A being the total surface area. The usual practice is to take q''_{nc} as equivalent to the normal natural convection heat transfer rate appropriate to the surface and to the geometry, fluid, and wall temperature. It is also usual practice to relate the areas over which the mechanisms occur to a_d, the projected area of the bubble on the surface at the point of bubble departure ($a_d = \pi d_d^2/4$). The microlayer evaporation process would occur over this projected area such that $A_e = N_a a_d$, where N_a is the number of active sites. When the bubble departs from the surface, the bulk liquid is brought into contact with the surface and there is a period of transient conduction as illustrated in Fig. 15.22. It is normally postulated that the area swept by the incoming liquid is greater than the projected area (a_d) of the departing bubble by a factor K. Thus, $A_t = N_a K a_d$. It thus follows that $A_{nc} = (1 - N_a K a_d)$. The difficulty lies in assigning a value to K. In their predictive model for boiling in the isolated bubble region, Mikic and Rohsenow [47] take $K = 4$, whereas Del Valle and Kenning [92] suggest values in the range of 5.8 to 7.5 depending on the heat flux; Judd and Hwang [93] estimate a value of 1.8. Obviously, the value of K can be adjusted arbitrarily to match experimental data.

Although it is convenient to assume the complete removal of the thermal boundary layer over an area $K a_d$ on bubble departure, this is an oversimplification of the process. A very detailed study of the displacement of fluid by a spherical object (for instance a bubble) moving away from a wall is presented by Eames et al. [94], who show that there is a combination of forward displacement and downward displacement as shown in Fig. 15.44. Thus, the actual processes are unlikely to coincide closely with those envisaged by the originators of the transient conduction model. A combination of the type of studies carried out by Eames et al. [94] with the bubble growth predictions of Klausner and coworkers (Mei et al. [50, 51] and Chen et al. [52]) would seem a fruitful area for further research. Another complication is that the influence areas of bubbles being emitted from adjacent sites may overlap as illustrated in Fig. 15.45. This overlapping of the areas of influence was investigated by Kenning and Del Valle [95] and Del Valle and Kenning [92]; it was shown that the overlapping effect had surprisingly

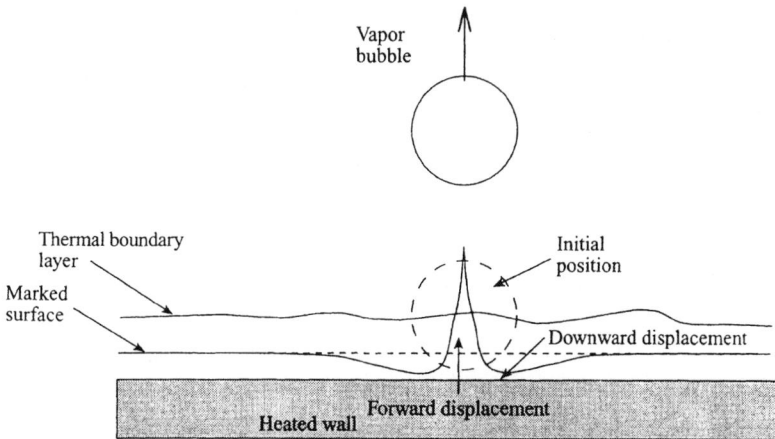

FIGURE 15.44 Fluid displacement on bubble departure from a heated surface (from Eames et al. [94], with permission from Cambridge University Press).

little influence since, though the area available for transient conduction was reduced, the frequency at which the overlapped area was renewed was increased, offsetting the loss of area.

Objective evaluation of the relative roles of the processes included in Eq. 15.78 is very difficult. Paul and Abdel-Khalik [96] and Paul et al. [97] measured bubble departure diameters, bubble departure frequency, and number of active sites for saturated pool boiling of water at atmospheric pressure and were able to estimate the volume of vapor emitted (and hence the contribution of the latent heat transport), as well as the area over which natural convection was occurring and, using a correlation for natural convection, the heat transported by this mechanism. The remaining heat flux was due to enhanced convection due to transient conduction effects as described above. The results are shown in Fig. 15.46, plotted in terms of the fraction of the heat flux that could be ascribed to the respective mechanisms against the total heat flux. At low heat fluxes, natural convection dominated but the contribution from enhanced convection first increased with total heat flux and then decreased. Latent heat transport increased continuously and accounted for all of the transfer at the highest heat fluxes. By this stage, the *fully developed boiling* regime would have been reached (see below).

In another set of experiments aimed at elucidating the contribution of the mechanisms, Judd and Hwang [98] studied the boiling of dichloromethane on a glass surface. In this case, the bubbles condensed at their upper surface in the subcooled liquid and the contribution of latent heat transport (through microlayer evaporation) could not be determined from a measurement of the vapor volume emission. Since the boiling was from a glass surface, Judd and Hwang were able to use optical techniques to measure the thickness of the evaporating microlayer and were able to estimate the contribution of the evaporation of this microlayer to the heat transfer. The contribution of natural convection could be estimated from the observed nonboiling surface area using a normal natural convection correlation and, thus, the value of

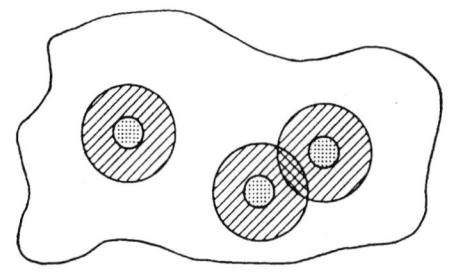

▦ Maximum bubble projected area

▨ Surrounding area of influence

▩ Overlapping areas of influence

☐ Non-boiling area

FIGURE 15.45 Zones of heat transfer in boiling (from Del Valle and Kenning [92], with permission from Elsevier Science).

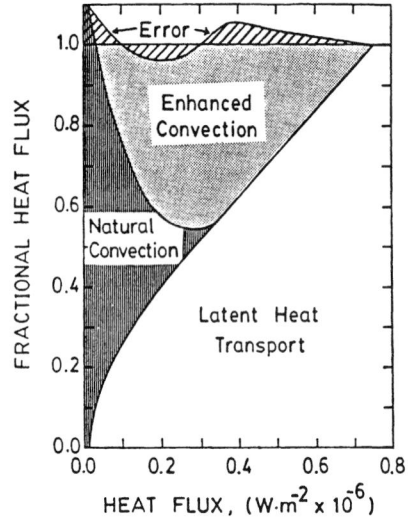

FIGURE 15.46 Relative contributions of latent heat transport, natural convection, and enhanced convection in the saturated pool boiling of water at atmospheric pressure (from Paul et al. [97], with permission from ASME).

q_l'' could be estimated. This led Judd and Hwang to suggest the value $K = 1.8$ as mentioned above. The results obtained in these experiments are illustrated in Fig. 15.47, which shows that at the highest heat flux investigated, microlayer evaporation accounted for about 60 percent of the heat flux. The trends indicated that this fraction would increase with increasing heat flux, which is consistent with the results shown in Fig. 15.46.

It can be seen from the above that, though the principal mechanisms in the partial boiling (isolated bubble) region seemed to have been qualitatively identified, there remain considerable uncertainties about the details of their action under any given set of circumstances and these uncertainties underlie any prediction method based on a mechanistic approach; we will return to this point further in the section on predictive models.

Fully Developed Boiling.   The heat transfer mechanisms in fully developed boiling are even more complex than those in the partial boiling (isolated bubble) regime, and the complexity of the interfacial structure near the heating surface makes observation and measurement of the phenomena extremely difficult. A classical experiment, from which most subsequent work derives, was carried out by Gaertner [99] in 1965. Using photographic techniques, Gaertner was able to observe a number of regimes of boiling, as illustrated in Fig. 15.48.

Initially (Fig. 15.48a), bubbles grow and depart from the individual nucleation centers; this is the partial nucleate boiling or isolated bubble regime previously discussed. At higher heat fluxes, jets of vapor begin to form, and these can combine to form small vapor mushrooms (Fig. 15.48b). At the transition to fully developed boiling, large vapor structures are formed that resemble mushrooms above a liquid layer adjacent to the wall, called the *macrolayer* (Fig. 15.48c). The macrolayer is penetrated by vapor stems that appear to arise from the original nucleation sites. The evaporation process is governed by the macrolayer and its associated vapor stems, which would explain why external features such as acceleration, angle of inclination of the surface, and so forth (see the preceding discussion) have relatively little effect in this region. Drying up of the macrolayer (see Fig. 15.48d) may also represent a mechanism for critical heat flux (see the section on such mechanisms).

An important technique in studying the mechanism of fully developed boiling is to use electrical conductance probes that can be traversed from the bulk fluid to the surface. A tiny needle tip (typically less than 50 μm in diameter) will record the local presence of liquid or vapor, and the fraction of time spent in contact with the vapor gives the void fraction. Typical of the measurements of this type are those of Shoji [100], illustrated in Figs. 15.49 and 15.50.

In fully developed nucleate boiling, the void fraction passes through a peak as shown, and quite different shapes of void fraction variation with distance from the surface are displayed for the critical heat flux, transition boiling, and film boiling regions (we will return to this point in later discussions). Figure 15.50 shows a more detailed interpretation of the nucleate boiling case. Point b represents the macrolayer thickness δ_o at the departure of a vapor mushroom and point a indicates the minimum thickness δ_m of the macrolayer. The thickness $(\delta_o - \delta_m)$ is consumed in the evaporation process between departures of the vapor mushrooms. Shoji [100] was able to show that this liquid consumption accounted for the total heat flux. The thickness δ_o of the macrolayer decreases with increasing heat flux and is typically in the range of 0.05 to 1 mm. A review of relationships for macrolayer thickness is given by Katto [101]. As will be seen from Fig. 15.50, the void fraction corresponding to δ_o is around 0.17, and this represents the ratio of the stem area A_v to the total area A_w.

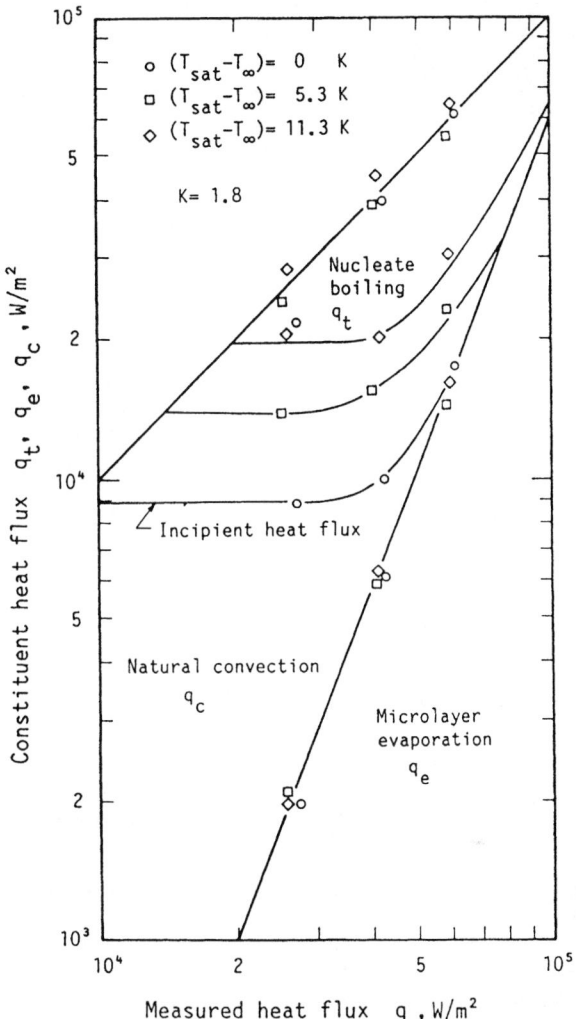

FIGURE 15.47 Contributions to heat transfer in the pool boiling of dichloromethane at 0.5 bar (from Judd and Hwang [98], with permission from ASME).

Katto and Yokoya [102] showed that the volume of the vapor mushrooms increased linearly with time (which is consistent with the hypothesis of the growth being governed in the macrolayer) and obtained the following expression for the "hovering period" τ between vapor mushroom departures:

$$\tau = (3/4\pi)^{1/5}[4(11\rho_l/16 + \rho_g)/g(\rho_l - \rho_g)]^{3/5}v^{1/5} \qquad (15.79)$$

Where v is the volumetric growth rate of the bubble (calculable from the heat flux). The time required for the formation of a new macrolayer with its associated vapor mushroom was very short and the frequency of vapor mushroom departure is therefore $f \approx 1/\tau$.

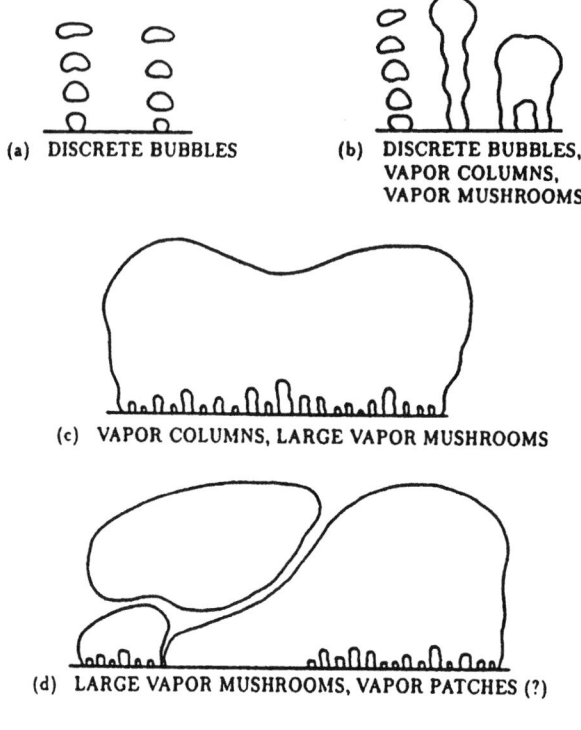

(a) **DISCRETE BUBBLES**

(b) **DISCRETE BUBBLES,
     VAPOR COLUMNS,
     VAPOR MUSHROOMS**

(c) **VAPOR COLUMNS, LARGE VAPOR MUSHROOMS**

(d) **LARGE VAPOR MUSHROOMS, VAPOR PATCHES (?)**

a. Discrete bubble region
b. First transition
c. Vapor mushroom region
d. Second transition region

FIGURE 15.48 Vapor structures in nucleate boiling (from Gaertner [99], with permission from ASME).

If one accepts that the macrolayer formation process is the governing phenomenon in the fully developed boiling region, then there is still the problem of understanding the precise evaporation mechanism associated with the vapor stems. This has been the focus of much recent work that is exemplified by the studies of Dhir and Liaw [103] and Lay and Dhir [104]. The basis of the Dhir and Liaw [103] model is illustrated in Fig. 15.51. The assumption was made that heat transfer occurred by conduction from the solid surface through the liquid phase to the interface, this conduction process occurring mainly in a *thermal layer* adjacent to the interface as shown. The heat transfer process would be influenced by the contact angle between the liquid and the solid surface at the bottom of the vapor stem and, introducing contact angle into the analysis, Dhir and Liaw were able to produce predictions that were consistent with the observed effect of contact angle on fully developed nucleate boiling heat transfer. Lay and Dhir [104] extended this analysis to include the effect of flows induced in the contact region by surface tension and by disjoining pressure (the pressure that tends to maintain an adsorbed layer of liquid on a surface). A similar model is described by Yagov [105]. These models seem capable of providing a reasonably quantitative explanation of the heat transfer phenomena. However, it would be safe to assume that further developments in phenomenological description of fully developed boiling will emerge in the future!

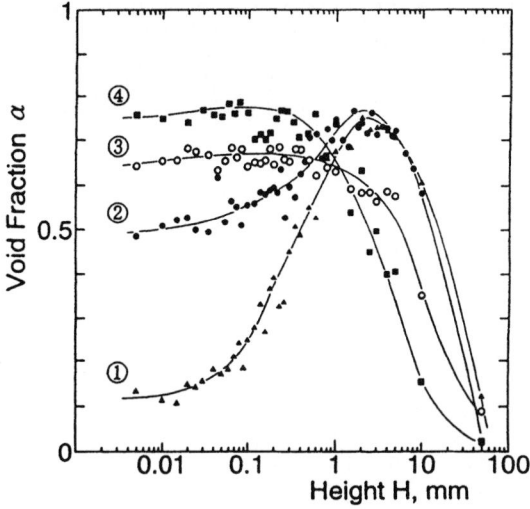

① Nucleate Boilng: ΔT_{sat}= 19.4 K, q=1.33×10⁶ W/m²

② Critical Heat Flux: ΔT_{sat}= 22.2 K, q=1.42×10⁶ W/m²
③ Transition Boiling: ΔT_{sat}= 54.5 K, q=3.98×10⁵ W/m²
④ Film Boiling: ΔT_{sat}=154.1 K, q=1.88×10⁵ W/m²

FIGURE 15.49 Void fraction profiles in boiling of saturated water at atmospheric pressure (from Shoji [100], with permission from ASME).

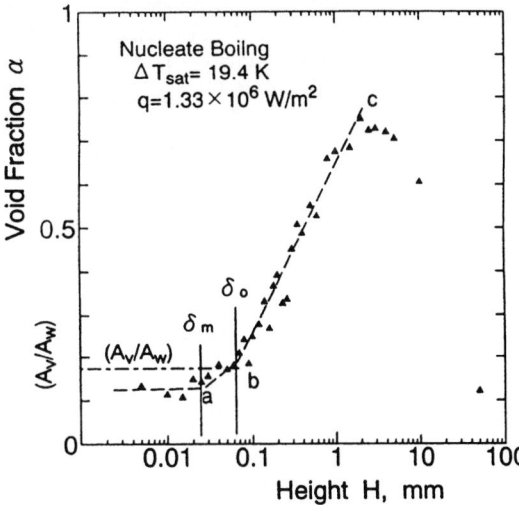

FIGURE 15.50 Macrolayer thicknesses deduced from void fraction measurements (from Shoji [100], with permission from ASME).

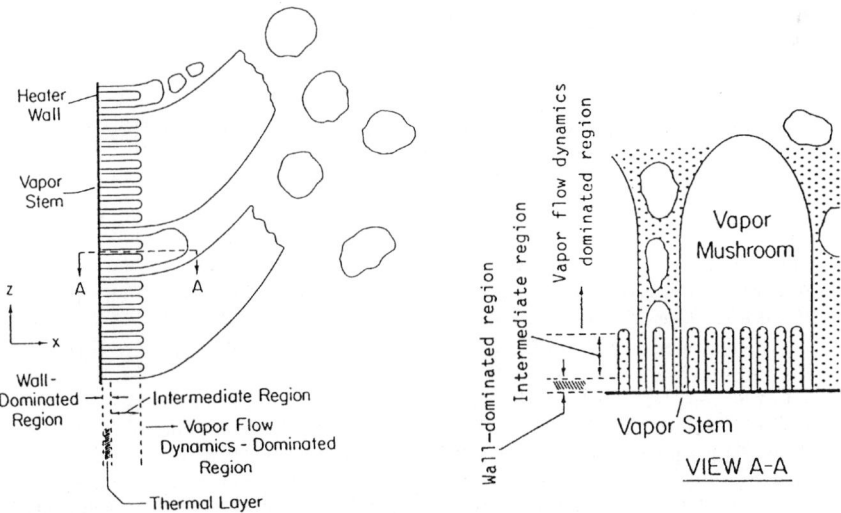

FIGURE 15.51 Basis of model for evaporation of vapor in vapor stems in the macrolayer (from Dhir and Liaw [103], with permission from ASME).

Correlations for Nucleate Pool Boiling.   It is evident from the results for nucleation behavior and from the discussion of parametric effects that the production of correlations for nucleate pool-boiling heat transfer is fraught with difficulty. Even with carefully prepared surfaces, it is difficult to obtain reproducible data and, though the reasons for the differences in nucleation behavior are now better understood, this does not particularly help in the common practical problem of predicting the boiling of an arbitrary fluid on a surface of unknown characteristics. However, the designer is faced with the necessity of calculating heat transfer coefficients for these systems and the use of correlations is inevitable, though the designer should be aware of the uncertainties involved. The following is a brief review of some of the more commonly used correlations. For any given situation, these correlations should be expected to give different predictions for the heat transfer coefficients; a good way (at least approximately) of assessing the uncertainty is to calculate the coefficients using several of the correlations. Unfortunately, there are some systems (for instance, boiling of organic fluids on the outside of tubes with steam condensation on the inside) where the boiling coefficient is the governing one. Here, the uncertainty translates itself directly into capital cost. For such systems, it is wise to take the most conservative values of heat transfer coefficient in calculating the overall surface areas while at the same time arranging control (e.g., of steam pressure) on the heating side in order to prevent overproduction of the vapor.

A wide variety of correlations for pool boiling heat transfer have been published in the literature; only selected ones are given here. The correlation of Forster and Zuber [106] is given for reference since it is used as the basis for nucleate boiling contributions in forced convective heat transfer in the well-known correlation of Chen [107], which is given later in the chapter. Probably the best-known correlation of nucleate pool-boiling heat transfer is that of Rohsenow [108], and this is given for reference since it is still widely used. More recent correlations have tended to be presented in terms of reduced pressure ($P_r = P/P_c$, where P_c is the critical pressure), and correlations of this form probably now represent the best practical route to estimation. In this category, the correlation of Mostinskii [109] and its later developments, the correlation of Cooper [110, 111], and the correlation of Gorenflo [112] as further developed by Leiner and Gorenflo [113, 119] are presented.

The correlation of Forster and Zuber [106] was developed in 1955 in a dimensionless form. It is commonly used in the following dimensional form (where all of the units should be SI):

$$h = \frac{0.00122 \Delta T_{\text{sat}}^{0.24} \Delta P_{\text{sat}}^{0.75} c_{pl}^{0.45} \rho_l^{0.49} k_l^{0.79}}{\sigma^{0.5} i_{lg}^{0.24} \mu_l^{0.29} \rho_g^{0.24}} \tag{15.80}$$

where ΔP_{sat} is the difference in saturation pressure corresponding to the difference between the wall and saturation temperatures (ΔT_{sat}). The problem with the Forster and Zuber correlation is immediately obvious in the light of the preceding discussions; it takes no account of the heater surface/boiling fluid combination. The correlation of Rohsenow [108] does account for surface effects and is given as follows:

$$\frac{c_{pl} \Delta T_{\text{sat}}}{i_{lg}} = C_{SF} \left\{ \frac{q''}{\mu_l i_{lg}} \sqrt{\frac{\sigma}{g(\rho_l - \rho_g)}} \right\}^{0.33} \left[\frac{c_{pl}\mu_l}{k_l} \right]^n \tag{15.81}$$

where C_{SF} is a constant depending on surface finish and fluid. Values for C_{SF} for various fluid/surface combinations are given in Table 15.1.

The exponent n has a value of unity for water and a value of 1.7 for other fluids. Although the Rohsenow correlation explicitly recognizes that, in reality, the surface has a considerable effect, it must be admitted that, in practice, the values of C_{SF} are rarely known and have to be guessed. Where C_{SF} is not known, a value of 0.013 is recommended as a first approximation.

An important class of nucleate boiling correlations are those where the reduced pressure P_r is used as a correlating parameter. This avoids the need for extensive physical property data, and, it can be argued, the fundamental inaccuracy of boiling correlations hardly justifies anything more complex. An early correlation of this type was that of Mostinskii [109], which is in the form

TABLE 15.1   Values of C_{SF} for Various Surface/Fluid Combinations

Liquid-surface combination	C_{SF}
n-pentane on polished copper	0.0154
n-pentane on polished nickel	0.0127
Water on polished copper	0.0128
Carbon tetrachloride on polished copper	0.0070
Water on lapped copper	0.0147
n-pentane on lapped copper	0.0049
n-pentane on emery-rubbed copper	0.0074
Water on scored copper	0.0068
Water on ground and polished stainless steel	0.0080
Water on Teflon-pitted stainless steel	0.0058
Water on chemically etched stainless steel	0.0133
Water on mechanically polished stainless steel	0.0132

Source: Collier and Thome [3], with permission.

$$h = A*(q'')^{0.7}F(P) \qquad (15.82)$$

where $A*$ is given by

$$A* = 3.596 \times 10^{-5}P_c^{0.69} \qquad (15.83)$$

and the pressure function $F(P)$ is given by

$$F(P) = 1.8P_r^{0.17} + 4P_r^{1.2} + 10P_r^{10} \qquad (15.84)$$

Palen et al. [114] suggest that, for design purposes, the last two terms in Eq. 15.84 should be eliminated, giving

$$F(P) = 1.8P_r^{0.17} \qquad (15.85)$$

A further alternative form for $F(P)$ is given by Bier [115]:

$$F(P) = 0.7 + 2P_r\left(4 + \frac{1}{1 - P_r}\right) \qquad (15.86)$$

Cooper [110, 111] analyzed a wide range of boiling data and also gives a penetrating review of existing boiling correlations. He shows that, for a given substance, the physical properties can be represented in the form

$$\text{property} = P_r^k(-\log P_r)^n \times \text{constant} \qquad (15.87)$$

Thus, any correlation for boiling coefficient that involves physical properties can be represented in the form

$$h = \frac{q''}{\Delta T_{sat}} = (q'')^m P_r^q(-\log P_r)^r \times \text{constant} \qquad (15.88)$$

This prompted Cooper [111] to optimize a correlation in the form of Eq. 15.88. His resultant equation is

$$h = 55(q'')^{0.67}P_r^{(0.12 - 0.2 \log R_p)}(-\log P_r)^{-0.55}\tilde{M}^{-1/2} \qquad (15.89)$$

where R_p is the roughness parameter (*glättungstiefe*) in the definition of the German standard DIN 4762 (1960) (μm) and $\tilde{M}$ is the molecular weight (kg/kmol). If R_p is unknown, a value of 1 μm should be assumed.

The generalization in the Cooper correlation in terms of the square root of molecular weight ($\tilde{M}$) is an oversimplification, and considerable errors can sometimes be encountered. It is therefore recommended that an alternative form of reduced pressure correlation developed by Gorenflo and coworkers [116–119] should be used. In its original form [116, 117], the correlation related the heat transfer coefficient h to its value h_o at standard conditions of pressure ($P_r = 0.1$), surface roughness ($R_{po} = 0.4$ μm), and heat flux ($q''_o = 20,000$ W/m²). Values of h_o were tabulated for a wide range of fluids; for instance, for butane $h_o = 3600$ and for water $h_o = 5600$. To obtain coefficients at other conditions, the following equation is used:

$$h = h_o F(P)(q''/q''_o)^n (R_p/R_{po})^{0.133} \tag{15.90}$$

where $F(P)$ is calculated from the expression

$$F(P) = 1.2 P_r^{0.27} + 2.5 P_r + \frac{P_r}{1 - P_r} \tag{15.91}$$

which has a value close to unity at $P_r = 0.1$. The exponent n is calculated as follows:

$$n = 0.9 - 0.3 P_r^{0.3} \tag{15.92}$$

Equations 15.91 and 15.92 apply for all fluids except water and liquid helium. For water, the values of $F(P)$ and n are given by the following equations:

$$F(P) = 1.73 P_r^{0.27} + \left(6.1 + \frac{0.68}{1 - P_r}\right) P_r^2 \tag{15.93}$$

$$n = 0.9 - 0.3 P_r^{0.15} \tag{15.94}$$

The main problem with this correlation is that values of h_o need to be specified. In more recent work, Leiner and Gorenflo [118] and Leiner [119] have developed new nondimensional forms of the correlation that apply for any arbitrary fluid without the need for specification of h_o. The following nondimensional forms of heat transfer coefficient, heat flux, and surface roughness are used:

$$h^* = \frac{h}{P_c(\tilde{R}/\tilde{M}T_c)^{1/2}} \tag{15.95}$$

$$q^* = \frac{q''}{P_c(\tilde{R}T_c/\tilde{M})^{1/2}} \tag{15.96}$$

$$R^* = \frac{R_a}{(\tilde{R}T_c/P_c N_{mol})^{1/3}} \tag{15.97}$$

where $\tilde{R}$ is the universal gas constant (8314 J/kmol K), $\tilde{M}$ is the molecular weight, T_c is the critical temperature, and N_{mol} is the Avogadro number (6.022×10^{26} molecules/kmol). R_a is the arithmetic mean deviation of the surface profile (*Mittenräuhwert*) as defined by ISO4287-1: 1984/DIN4762 (m). The nondimensional correlation has the form

$$h^* = AF'(P)q^{*n}R^{*0.133} \tag{15.98}$$

where n is given by Eq. 15.94 and where the revised pressure function is as follows:

$$F'(P) = 43000^{(n - 0.75)}\left[1.2 P_r^{0.27} + \left(2.5 + \frac{1}{1 - P_r}\right) P_r\right] \tag{15.99}$$

The factor A is given by

$$A = 0.6161 C^{0.1512} K^{0.4894} \tag{15.100}$$

where C and K are given by

$$C = \frac{(\tilde{c}_{pl})_{P_r=0.1}}{\tilde{R}}$$

(15.101)

and

$$K = \frac{T_c \ln P_r}{(1 - T_c)}$$

(15.102)

where $(\tilde{c}_{pl})_{P_r=0.1}$ is the molar specific heat capacity (J/kmol K) at $P_r = 0.1$.

Equation 15.98 was compared with extensive data and had an RMS deviation of around 15%. A three-parameter relationship for A was also investigated but gave little advantage in terms of overall accuracy. If R_a is unknown, then a value of 0.4 μm should be assumed.

Predictive Models for Nucleate Pool Boiling. A natural goal of research on boiling heat transfer has been the prediction, from a consistent physical model, of the boiling heat transfer rate. The preceding sections of this chapter give the reader an impression of the difficulties associated with such predictions! One of the best-known predictive models, dealing specifically with the isolated bubble regime, is that of Mikic and Rohsenow [120]. The bases of this model have been introduced in the context of bubble growth from a surface and in discussing mechanisms for the isolated bubble region (see p. 15.18–15.19). Mikic and Rohsenow neglected the direct contribution due to evaporation (latent heat transport) in Eq. 15.78 and modeled the system in terms of natural convection and the transient heat transfer that occurred after bubble departure. They assumed that this transient heat transfer occurred over a zone of diameter $2d_d$ (equivalent to a factor $K = 4$ on the projected area of the departing bubble—see the section on mechanisms). For bubbles departing at a frequency f from each of the active nucleation sites on the surface, the heat flux (averaged over the whole surface area) is given by Mikic and Rohsenow as

$$q_b'' = 2N_a d_a^2 [T_W - T_{sat}(P_l)](\pi k_l c_{pl} \rho_l f)^{1/2}$$

(15.103)

where N_a is the number of active nucleation sites per unit area. The total heat flux is given by

$$q'' = [1 - N_a K(\pi d_a^2/4)]q_{nc}'' + q_b''$$

(15.104)

where q_{nc}'' is the local heat flux for natural convection (occurring over those areas of the surface not affected by bubbles), which can be calculated from standard natural convection equations. To close the above model, expressions are required for N_a, d_d, and f. Mikic and Rohsenow suggest that the distribution of active nucleation sites on the surface can be expressed in terms of the following relationship:

$$N_a = \left(\frac{r_s}{r}\right)^m$$

(15.105)

where r_s is the radius of the largest cavity on the surface; r_s and the exponent m can be obtained from measured cavity radius distributions or can be estimated by fitting boiling data. r is the minimum radius of site that can be activated at a given wall superheat (Mikic and Rohsenow ignored temperature gradients into the fluid), and this can be calculated using the expressions given in the section on bubble growth. It is thus possible to relate N_a to wall superheat as follows:

$$N_a = r_s^m \left(\frac{i_{lg}\rho_g}{2T\sigma}\right)^m (T_W - T_{sat}(P_l))^m$$

(15.106)

To close the model, Mikic and Rohsenow used the following specific expressions for d_d and f:

$$d_d = a\left[\frac{\sigma}{g(\rho_l - \rho_g)}\right]^{1/2}\left(\frac{\rho_l c_{pl} T_{sat}}{\rho_g i_{lg}}\right)^{5/4}$$

(15.107)

$$fd_a = b\left[\frac{\sigma g(\rho_l - \rho_g)}{\rho_g^2}\right]^{1/4} \tag{15.108}$$

where $a = 1.5 \times 10^{-4}$ for water and 4.65×10^{-4} for other fluids and $b = 0.6$.

A model analogous to that of Mikic and Rohsenow [120] has been recently developed by Benjamin and Balakrishnan [121]. The latter authors took account of microlayer evaporation in the model and also produced a useful relationship between the number of active sites N_a as follows:

$$N_a = 218.8(\theta)^{-0.4}\,\text{Pr}_l^{1.63}\left(\frac{1}{\gamma}\right)[T_W - T_{\text{sat}}(P_l)]^3 \tag{15.109}$$

where $\gamma = (k_{lw}\rho_{lw}c_{plw}/k_l\rho_l c_{pl})^{1/2}$, where the subscript lw indicates fluid properties at the wall and the subscript l refers to bulk properties. θ is given by

$$\theta = 14.4 - 4.5\left(\frac{R_a P_l}{\sigma}\right) + 0.4\left(\frac{R_a P_l}{\sigma}\right)^2 \tag{15.110}$$

A methodology for prediction of fully developed boiling is presented by Lay and Dhir [104] based on modeling of the evaporation processes near the bottom of the vapor stems penetrating the macrolayer (see the section on mechanisms). It was assumed that the number of active nucleation sites was that given by the measurements of Wang and Dhir [24] (see Figs. 15.11 and 15.12 and associated discussion) and that the evaporation took place in a meniscus region at the bottom of the vapor stem, surrounding the nucleation center as illustrated in Fig. 15.52.

Liquid is drawn into this meniscus mainly as a result of the decrease in the radius r_1 in the direction of the axis of the vapor stem. The effect of disjoining pressure were also considered in the analysis but was of less importance. The principal evaporation then occurs in the meniscus region. The analysis was reasonably successful in predicting the fully developed boiling region.

Recent work, exemplified by the papers of Nelson et al. [122] and Sadivasan et al. [123], has focused on very detailed studies of the microprocesses occurring and shows that these are significantly nonlinear. Nelson et al. conclude that "the assumptions used in the older mechanistic models are not valid. Thus, these models should be characterised as very complex correlations." Boiling is a process that involves highly complex interactions between the liquid, vapor, and the solid (heating) surface and, as has been shown by Kenning and Del Valle [95] and others, interactions between adjacent nucleation centers. Thus, the achievement of a true

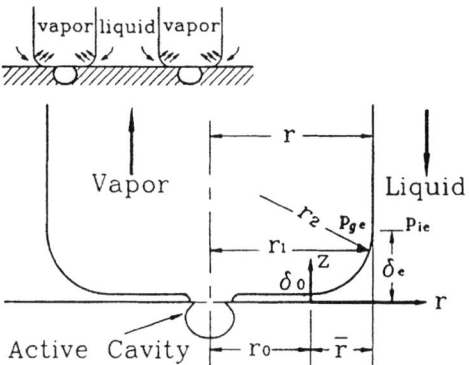

FIGURE 15.52 Evaporation of vapor at base of vapor stem (from Lay and Dhir [104], with permission from ASME).

predictive capability would be expected to remain an elusive goal for many years to come. However, one cannot stress too strongly the importance of working on prediction as a means of developing an understanding of the relative significance of various factors. For the moment, however, practical design will have to depend on the use of correlations of the types described in the section on correlations above.

Effect of Multicomponent Mixtures in Nucleate Pool Boiling. For a binary liquid mixture of composition $\tilde{x}_1$, we may define a heat transfer coefficient h as

$$h = \frac{q''}{(T_W - T_{\text{bub}})} \tag{15.111}$$

where T_{bub} is the bubble point temperature corresponding to $\tilde{x}_1$ (see Fig. 15.3). It might be postulated that, for mixtures of two components, an "ideal" heat transfer coefficient h_{id} could be defined as follows:

$$\frac{1}{h_{id}} = \frac{\tilde{x}_1}{h_1} + \frac{\tilde{x}_2}{h_2} \tag{15.112}$$

where h_1 and h_2 are the heat transfer coefficients for the pure components at the same heat flux. These may be calculated using the correlations given in the section on predictive models. In practice, it is found that h is very much lower than h_{id}, as is exemplified by the results shown in Fig. 15.53 for pool boiling of CF_2Cl_2/SF_6 mixtures. The gross deviation is due to the fact that during bubble growth, the less volatile material is concentrated on the liquid side of the interface, and the effective vapor pressure is much less than for equilibrium. For a mixture with an azeotrope, the coefficient varies with composition as illustrated in Fig. 15.54; when there is no difference between liquid and vapor compositions, at the azeotrope, then the coefficient is that expected from the mean physical properties (see Tolubinskiy and Ostrovskiy [124]).

Commonly, mixtures of three or more components are boiled; Fig. 15.55 shows data reported by Schlunder [125] for the boiling of acetone-methanol-water mixtures. The ratio of measured to ideal heat transfer coefficient (the latter being calculated by extending Eq. 15.112 to include the third component) varies in a complex way with composition; several azeotropic compositions are noted at which the ratio of measured to ideal coefficient approaches unity.

Extensive reviews of multicomponent boiling are given by Shock [126] and Collier and Thome [3].

For multicomponent mixtures, the mechanisms are even more complex, involving mass transfer in both liquid and vapor phases. Detailed measurements of bubble frequency (f), bubble departure diameter (d_d), and number of active nucleation sites (N_a) are reported by Bier and Schmidt [127]; for the propane/n-butane mixtures studied, the bubble departure diameter initially increases (relative to its value for n-butane) with increasing mole fraction of propane and then decreases to a value less than that for pure propane before increasing rapidly with concentration and propane mole fractions greater than about 0.9. The bubble frequency shows the opposite trends. The number of active sites (N_a) passes through a minimum as the mole fraction of propane is increased, the maximum reduction being around a factor of 3. It is clear, therefore, that the effects of having a multicom-

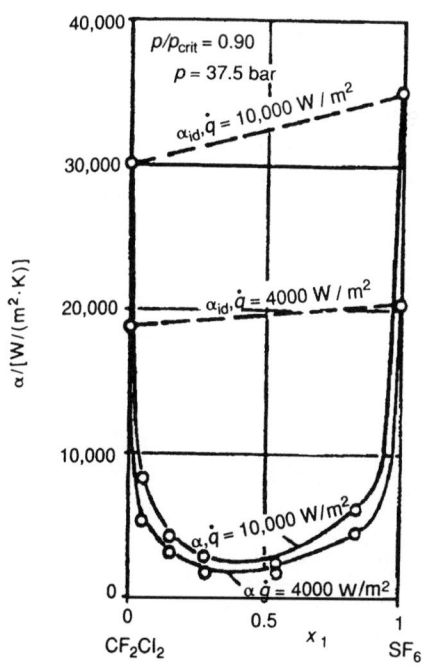

FIGURE 15.53 Comparison of ideal actual heat transfer coefficient for pool boiling of CF_2Cl_2/SF_6 mixtures (from Schlunder [125]).

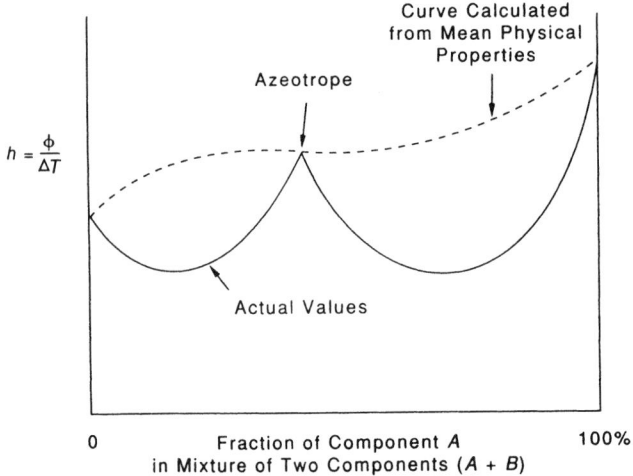

FIGURE 15.54 Schematic diagram of the variation of heat transfer coefficient with composition in a binary mixture forming an azeotrope (from Hewitt et al. [13], with permission. Copyright CRC Press, Boca Raton, FL).

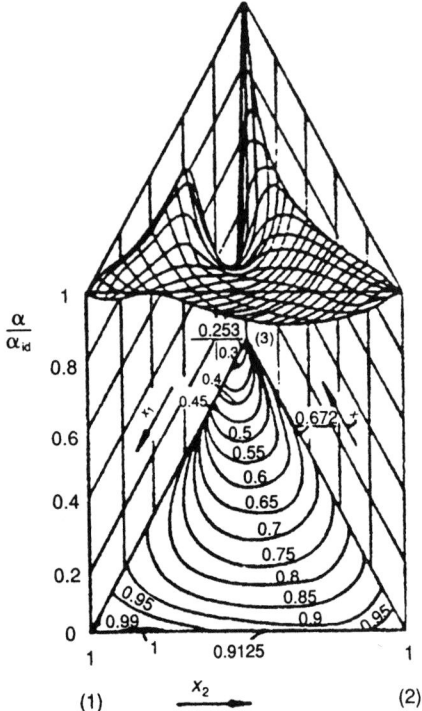

FIGURE 15.55 Boiling data for acetone (1)/methanol (2)/water (3) mixtures (from Schlunder [129], with permission from Taylor & Francis, Washington, DC. All rights reserved).

ponent mixture are highly complex in nature and this makes prediction difficult.

A number of empirical and semiempirical methods have been developed to predict the changes in heat transfer coefficient in pool boiling. Stephan and Korner [128] define an effective temperature driving force for use instead of $(T_W - T_{\text{bub}})$ in Eq. 15.111:

$$\Delta T = \Delta T_I + \Delta T_E \qquad (15.113)$$

where ΔT_I is given by

$$\Delta T_I = \tilde{x}_1 \Delta T_1 + \tilde{x}_2 \Delta T_2 = \tilde{x}_1 \Delta T_1 + (1 - \tilde{x}_1)\Delta T_2 \qquad (15.114)$$

where ΔT_1 and ΔT_2 are the temperature differences for boiling of the pure components at the same heat flux. The additional temperature difference ΔT_E was empirically correlated by the expression

$$\Delta T_E = A(\tilde{y}_{\text{bulk}} - \tilde{x}_{\text{bulk}})\Delta T_I \qquad (15.115)$$

where $\tilde{y}_{\text{bulk}}$ is the vapor composition corresponding to the bulk concentration $\tilde{x}_{\text{bulk}}$ for component 1 at the equilibrium bubble temperature T_{bub}. The coefficient A is given by

$$A = A_o(0.88 + 0.12P) \qquad (15.116)$$

where P is the pressure in bars and A_o varies from binary mixture to binary mixture but has an average value of 1.53.

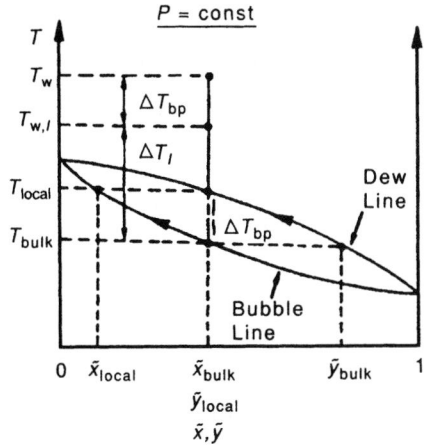

FIGURE 15.56   Thome method for pool boiling of binary mixtures (from Thome [130], with permission from Elsevier Science).

Thome [130] suggests that, at a boiling site, all of the liquid approaching the site may be evaporated such that the local value of $\tilde{y}$ is equal to $\tilde{x}_{\text{bulk}}$. Thus, as illustrated in Fig. 15.56, there is an elevation in temperature ΔT_{bp} corresponding to the difference between the bubble point curve and the dew point curve at the local liquid concentration. Thus

$$\Delta T = \Delta T_l + \Delta T_{bp} \qquad (15.117)$$

A semitheoretical treatment that actually gives results rather close to those of Thome was introduced by Schlunder [125], who obtained the expression

$$\frac{h}{h_{id}} = \frac{1}{1 + (h_{id}/q'')[(T_{S2} - T_{S1})(\tilde{y}_1 - \tilde{x}_1)\{1 - \exp(-B_o q''/\rho_l i_{lg}\beta_l)\}]} \qquad (15.118)$$

where T_{S2} and T_{S1} are the saturation temperatures for components for 2 and 1, respectively, and $\tilde{y}_1$ is the vapor composition in equilibrium with the liquid composition $\tilde{x}_1$ at T_{bub}. β_l is a liquid phase mass transfer coefficient (having a value of approximately 2×10^{-4} m/s), and B_o is a scaling parameter whose value is around unity. The Schlunder analysis represents the data moderately well, but it must still be considered empirical because the constant B_o has to be manipulated in order to fit the data. Gorenflo [131] recommends a value of 10^4 for the ratio B_o/β_l. Fujita et al. [132] carried out a wide range of experiments with the boiling of binary mixtures and compared the data with existing correlations (including those listed above). They suggest the following improved form of the correlation:

$$\frac{h}{h_{id}} = \frac{1}{1 + K_S(\Delta T_{bp}/\Delta T_{id})} \qquad (15.119)$$

where

$$K_S = [1 - \exp(-2.8\Delta T_{id}/\Delta T_S)] \qquad (15.120)$$

where ΔT_{bp} is the boiling range (difference between the dew and bubble point temperatures) as illustrated in Fig. 15.56. ΔT_{id} is defined as follows:

$$\Delta T_{id} = \tilde{x}_1\Delta T_1 + \tilde{x}_2\Delta T_2 \qquad (15.121)$$

where ΔT_1 and ΔT_2 are the wall superheats for the pure components at the same heat flux as the mixture. ΔT_S is the difference in the saturation temperatures of the pure components. This correlation predicted the data within ±20 percent and overall performed better than the other correlations tested.

In contrast to the effect of much larger contents of the second material, the addition of very small amounts of surfactants is known to increase the boiling heat transfer coefficient. Tzan and Yang [133] carried out a wide range of experiments on nucleate pool boiling with surfactant additions and found that there was an optimum concentration of the surfactant beyond which further increase in surfactant concentration reduces the boiling heat transfer coefficient. Work by Wu and Yang [134] suggests that the main effect is to reduce the waiting time between bubble initiations; this increased the bubble frequency (typically by a factor of around 3) and gave a net increase in the heat transfer coefficient in this case.

Enhancement of Pool Nucleate Boiling Heat Transfer.   Study of the enhancement of boiling heat transfer has been one of the fastest-growing areas of research in recent years. The annual publication rate in this area has grown to around 300 papers per year (Bergles [135]).

This growth has been driven by the need to improve boiling heat transfer in high-flux devices (for example, in electronic component cooling) and in reducing the size and cost of equipment in chemical, refrigeration, and other types of plants. Reviews on the subject are presented by Bergles [135] and by Pais and Webb [136]; Bergles presents a classification of the methodologies, dividing these into *passive techniques* (for instance, treated surfaces, rough surfaces, extended surfaces, etc.), *active techniques* (for instance, the use of electrostatic fields or ultrasound), and *compound techniques* (where more than one enhancement methodology is being used at the same time).

By far the most important class of techniques uses enhanced surfaces of various kinds; examples of these are shown in Fig. 15.57. Finned surfaces (Fig. 15.57*a*) increase the rate of boiling heat transfer per unit length of tube compared to a plain tube. However, provided the fin surface is identical in microstructure to the plain tube with which it is being compared, then the increase in heat transfer is accounted for by the increase in surface area (Hubner et al. [138]). The boiling process itself is affected by the provision of *re-entrant* cavities that can be created by mechanical defemation of the surfaces (Fig. 15.57*b*–*e*), and this is the basis of a number of commercial proprietary boiling surfaces. An alternative is to coat the solid surface with a thin, porous layer of particles (Fig. 15.57*f*). This type of surface can produce dramatic improvements in heat transfer as exemplified in Fig. 15.58. Surfaces of this type are now widely used in practical applications. The mechanisms for this large enhancement are discussed by Thome [139] and by Webb and Haider [140]; the principal effect of the enhancement device is to promote bubble formation and emission at much lower superheats, and, in contrast to boiling from plain surfaces as discussed in the section on mechanisms, the boiling process is dominated by latent heat transfer, even at low superheats.

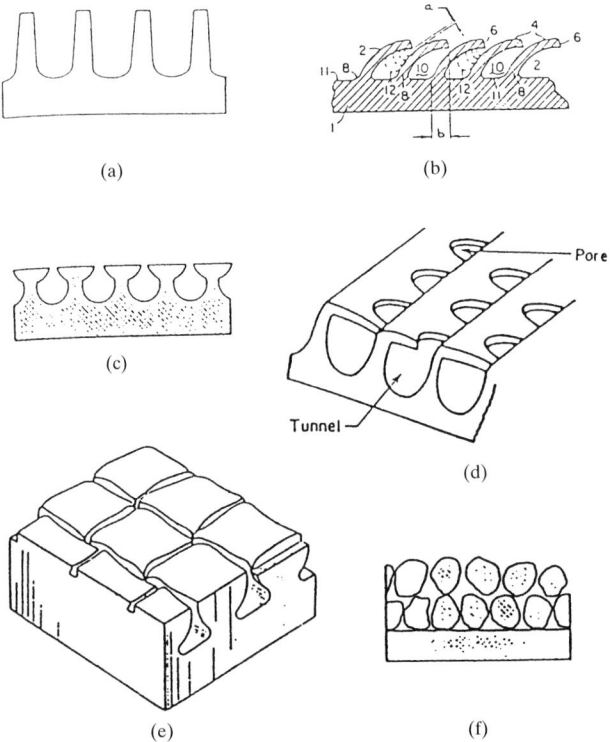

FIGURE 15.57 Enhanced surfaces for boiling (from Webb and Haider [137], with permission from ASME).

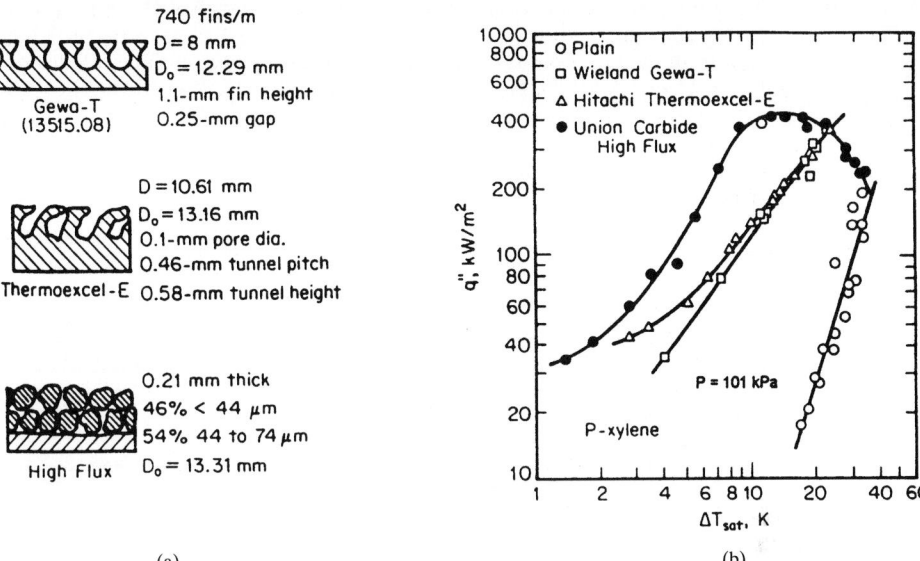

FIGURE 15.58 Improvement in boiling heat transfer using various forms of commercial enhanced surfaces (from Bergles [135], with permission).

There are a number of problems in the application of enhanced surfaces. One of these is that boiling hysteresis (see Fig. 15.43) may have an even more significant role. Hysteresis effects are reviewed by Bar Cohen [80], and the studies of Marto and Leper [141] and Pinto et al. [142] are examples of work on hysteresis with enhanced surfaces. Figure 15.59 shows results obtained by Marto and Leper for boiling on high-flux (porous coated) tubing. The onset of nucleate boiling occurs at around the same superheat as for a plain tube, although once boiling starts, a very large increase in heat transfer coefficient is obtained. This hysteresis effect can have serious implications in the applications of such systems. Tubing of this kind is often applied to refrigerant systems, which frequently have significant amounts of oil dissolved in the liquid. This can significantly reduce the advantage of the enhanced surface (Memory and Marto [143]).

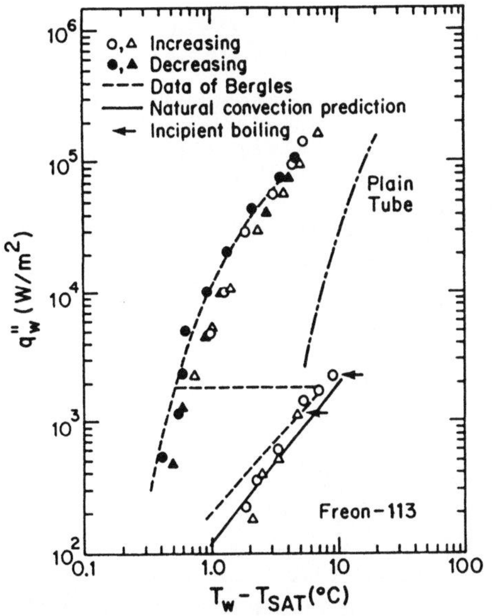

Examples of active techniques are the use of electrohydrodynamic (EHD) enhancement and the use of ultrasound fields. In EHD enhancement, a high voltage (typically 5–25 kV) is applied to the boiling surface and this produces electrically induced secondary motions that can give a very high enhancement in boiling heat transfer. Examples of investigations in this area are those of Ohadi et al. [144] and Zaghdoudi et al. [145]. Ohadi et al. showed that improvement in boiling heat transfer coefficient of a factor of around 3 can be obtained by using an EHD power input corresponding to only 5 percent of the total power transferred from the surface. EHD has obvious safety implications, but by adjusting the surface voltage, it is possible to control the heat transfer rate, and this may be useful in some applications. Boiling in

FIGURE 15.59 Pool boiling of refrigerant R113 from plain and porous coated tubes showing effect of hysteresis (from Marto and Lepere [141], with permission from ASME).

ultrasound fields has been studied by Bonekamp and Bier [146]; they show that significant improvements in boiling heat transfer can be obtained using ultrasound fields, particularly in the case of mixture boiling. Ultrasound fields also diminish the tendency toward the hysteresis effect.

The Critical Heat Flux Limit in Pool Boiling

Parametric Effects in Pool Boiling CHF

Effect of Pressure. The critical heat flux in pool boiling increases and then decreases with increasing pressure (tending to zero at the critical pressure) as illustrated in Fig. 15.60 (see also Fig. 15.34).

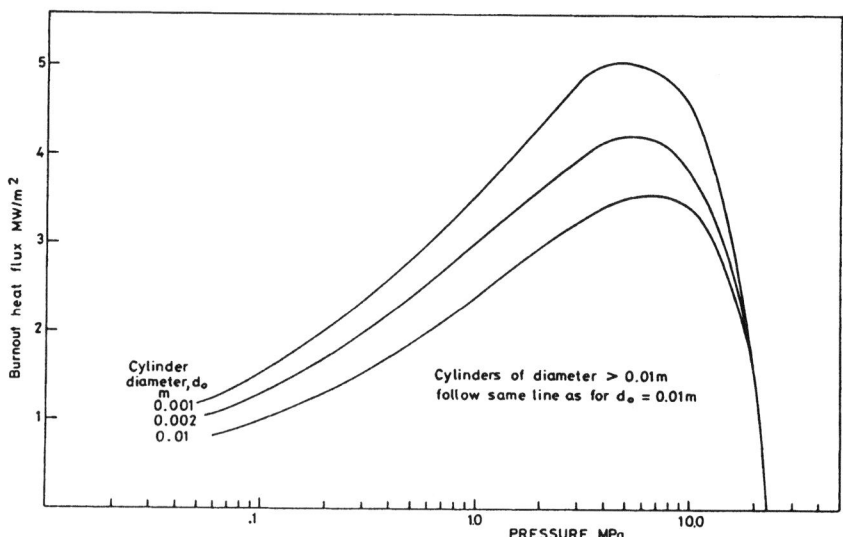

FIGURE 15.60 Effect of pressure and cylinder diameter on burnout heat flux for saturated pool boiling of water from horizontal cylinders (from Hewitt [147], with permission from The McGraw-Hill Companies).

Effect of Subcooling. Critical heat flux increases linearly with subcooling, the effect of subcooling decreasing with increasing pressure. The data are well represented by a correlation from Ivey and Morris [148] that relates the critical heat flux q''_{crit} to its value $(q''_{crit})_{sat}$ for saturated conditions by the expression

$$\frac{q''_{crit}}{(q''_{crit})_{sat}} = 1 + 0.102\left(\frac{\rho_g}{\rho_l}\right)^{0.25} Ja \qquad (15.122)$$

where the Jakob number Ja is given by

$$Ja = \frac{\rho_l c_{pl} \Delta T_{sub}}{\rho_g i_{lg}} \qquad (15.123)$$

At very high subcoolings, the critical heat flux becomes independent of subcooling, this limit being typically reached when the flux is around 2.5 times for the saturation conditions (Elkssabgi and Lienhard [156]). This ratio is probably fluid dependent.

Effect of Surface Finish. As was demonstrated in the section on parametric effects, the microstructure of the surface has a dramatic effect on nucleate boiling heat transfer. Based on results obtained by Berenson [149] in 1960, it has generally been asserted that the critical heat

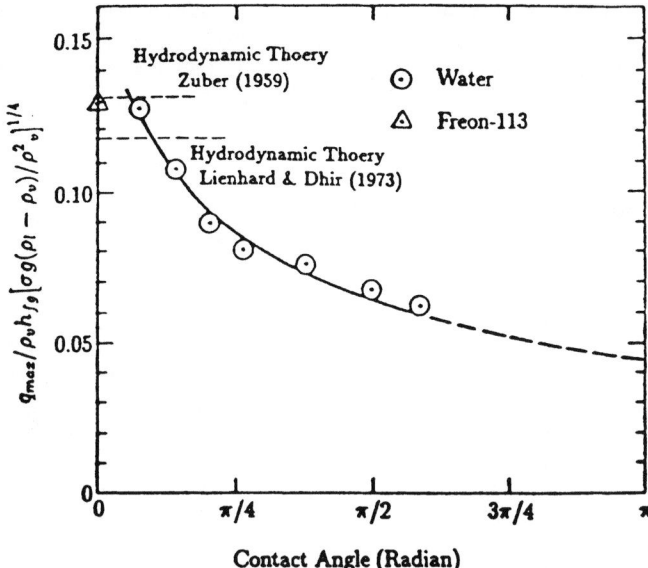

FIGURE 15.61 Variation of critical heat flux with contact angle (from Liaw and Dhir [150], with permission from Taylor & Francis, Washington, DC. All rights reserved).

flux is not significantly affected by surface finish, and this appears to be true for well-wetting fluids (typically with contact angles less than around 20°). However, Liaw and Dhir [150] showed that there was a systematic decrease in critical heat flux with increasing contact angle as shown in Fig. 15.61. This implies the possibility of a change of mechanism, and we will return to this point further in the following text.

Effect of Dissolved Gas. The presence of dissolved gas can lead to a considerable reduction in critical heat flux in pool boiling, as illustrated by the results of Jakob and Fritz [151] shown in Fig. 15.62. The effect of dissolved gases diminishes with decreasing subcooling (increasing fluid temperature), the effect being minimal near saturation conditions.

Effect of Gravity. Critical heat flux decreases with reducing gravitational acceleration (see for instance Fig. 15.42). The hydrodynamic theory of critical heat flux (see the following text section entitled Mechanisms) would suggest that critical heat flux increases with $g^{0.25}$. Experiments over the range of 0.02 to 1.0 times the earth's normal gravity are reported by Siegel and Usiskin [152] and for the range of 1 to 100 times the earth's normal gravity by Adams [153]. For reduced gravity, the results show the predicted variation (q''_{crit} varies with $g^{0.25}$). The enhanced-gravity experiments showed that the exponent fell to around 0.15 in the range from 1 to 10 times normal gravity but rose again to 0.25 for the range from 10 to 100 times normal gravity. Besant and Jones [154] found that the exponent decreased with increasing pressure.

Effect of Heater Size. Heated surfaces of small dimension tend to give higher critical heat fluxes than do surfaces of large size. Lienhard and Dhir [155] suggest a characteristic length scale L' above which the critical heat flux becomes independent of the heater size. L' is given by

$$L' = L\left[\frac{g(\rho_l - \rho_g)}{\sigma}\right]^{0.5} \tag{15.124}$$

where L is the characteristic dimension (the diameter for cylinders, spheres, and circular disks). L' had a value of around 2 for cylinders, around 8 for spheres, and around 15 for circular horizontal disks.

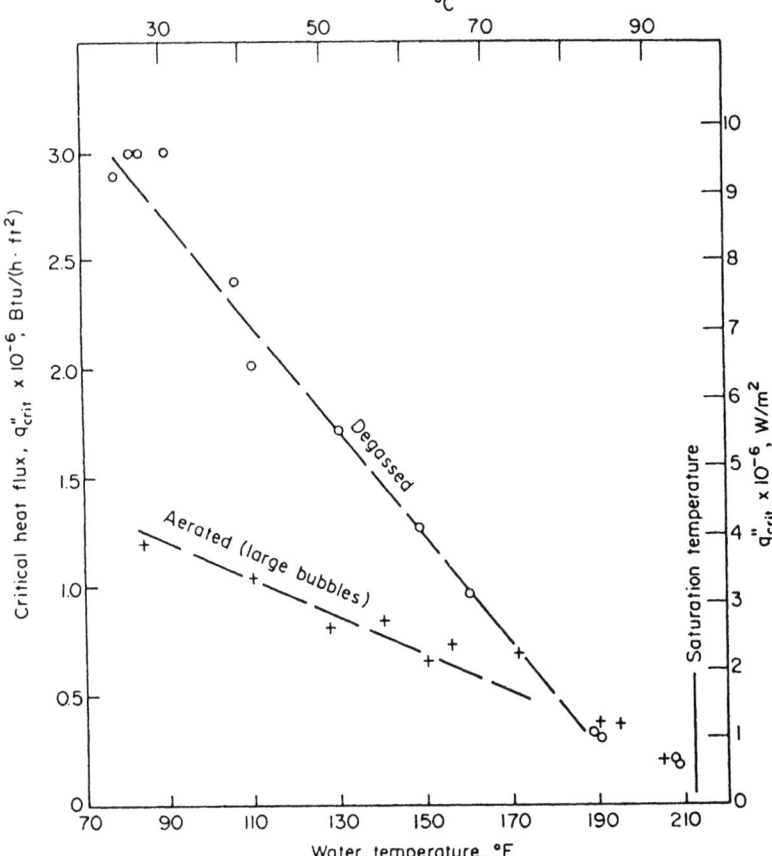

FIGURE 15.62 Results of Jakob and Fritz [151] for the effect of dissolved gases on critical heat flux (from Rohsenow [2], with permission of The McGraw-Hill Companies).

Mechanisms of CHF in Pool Boiling. The mechanism of the critical heat flux phenomenon in pool boiling has been the subject of widespread interest and controversy. Recent reviews relating to mechanisms are presented by Katto [157], Dhir [87], and Bergles [158]. The postulated mechanisms can be approximately classified into four types as follows:

1. *Hydrodynamic instability mechanism.* Here, instabilities occur in the vapor-liquid interfaces leading to the breakdown of the vapor release mechanisms and to vapor accumulation at the surface leading to critical heat flux.

2. *Macrolayer consumption model.* Here it is postulated that the macrolayer formed under the vapor mushrooms in fully developed boiling (see Fig. 15.48) is totally evaporated in the time between the release of the mushroom-shaped bubbles.

3. *Bubble crowding at the heated surface.* In this postulated mechanism, bubbles (or vapor stems in the macrolayer) coalesce, leading to a reduction in the amount of liquid in contact with the wall and, hence, in the overall heat transfer rate, which begins to decrease with increasing wall superheat when this coalescence process begins.

4. *Hot-spot heating.* In this mechanism, a hot spot is formed whose temperature rises to a value at which it cannot be rewetted, thus initiating the CHF transition.

Each of these postulated mechanisms is described in turn and, finally, an attempt is made at an overview of current understanding.

Hydrodynamic Instability Mechanism.   This mechanism was suggested by Zuber (Zuber [159], Zuber et al. [160]); the original Zuber hypothesis was for an infinite flat plate, and the situation is illustrated conceptually in Fig. 15.63.

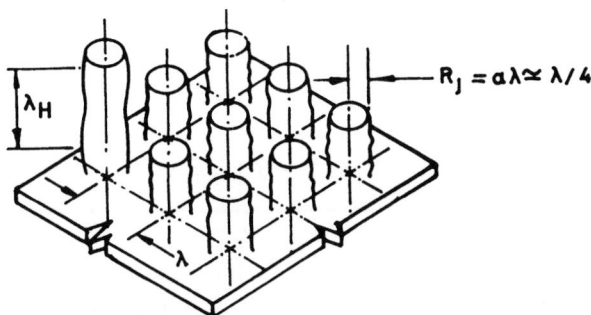

FIGURE 15.63   Representation of Zuber model for vapor escape jets for a horizontal flat plate (from Lienhard and Dhir [155], with permission of ASME).

Zuber postulates that, provided vapor can escape from the layer of bubbles near the surface, thus preventing it from becoming too thick, the liquid phase can penetrate the layer, wetting the surface and preventing overheating leading to the critical heat flux phenomenon. Zuber postulates that the vapor escape mechanism is via the "vapor columns" illustrated in Fig. 15.63. He suggests that these columns occur because the vapor-rich layer adjacent to the surface is fundamentally unstable, i.e., a small disturbance in the interface between the layer and the surrounding liquid is amplified at a rate that depends on the wavelength of the disturbance λ. This phenomenon is known as Taylor instability, and Zuber hypothesized that a rectangular square ray of jets was formed with a pitch λ as shown in Fig. 15.63.

Eventually, the velocity of vapor in the jets becomes so large that the jets themselves become unstable near the interface as a result of Helmholtz instability (of wavelength λ_H, as shown in Fig. 15.63). The breakup of the jets destroys the efficient vapor-removal mechanism, increases vapor accumulation at the interface, and leads to liquid starvation at the surface and to the critical heat flux phenomenon. If jet breakup occurs at a vapor velocity U_H within the jets, the critical heat flux q''_{crit} is given by

$$q''_{\text{crit}} = \rho_g i_{lg} \frac{A_j}{A} U_H \tag{15.125}$$

where A_j is the area occupied by the jets and A is the total surface area. Zuber made the assumption that the jet radius R_j is equal to $\lambda/4$, giving A_j/A as $\pi/16$.

Helmholtz instability theory gives U_H as

$$U_H = \left(\frac{2\pi\sigma}{\rho_g \lambda_H}\right)^{1/2} \tag{15.126}$$

where σ is the surface tension. Various assumptions can be made about λ_H; Zuber assumed that it was equal to the critical Rayleigh wavelength and thus to the circumference of the jet ($\lambda_H = \pi\lambda/2$). Lienhard and Dhir [155, 161] suggested that it is closer to the real physical situation to take $\lambda_H = \lambda$ where λ is the selected value for the Taylor instability wavelength.

Taylor instability theory gives the following value for the wavelength of maximum rate of growth of a disturbance λ_D:

$$\lambda_D = 2\pi \left[\frac{3\sigma}{g(\rho_l - \rho_g)} \right]^{1/2} \tag{15.127}$$

where g is the acceleration due to gravity. The minimum unstable Taylor wavelength λ_c is $\lambda_D/\sqrt{3}$. The form derived for the critical heat flux is

$$q''_{crit} = K \rho_g^{1/2} i_{lg} [\sigma g(\rho_l - \rho_g)]^{1/4} \tag{15.128}$$

where the value given for the constant K depends on the choices made for λ_H and λ. Thus

1. For $\lambda_H = 2\pi R_j = \pi\lambda/2$ and $\lambda = \lambda_D$, $K = 0.119$.
2. For $\lambda_H = 2\pi R_j = \pi\lambda/2$ and $\lambda = \lambda_c$, $K = 0.157$.
3. Zuber hypothesized that K would lie between the values given by choices 1 and 2 and suggested $K = \pi/24 = 0.131$.
4. For $\lambda_H = \lambda_D$ and $\lambda = \lambda_D$, $K = 0.149$.

The final value, due to Lienhard and Dhir, is probably closest to experimental data for flat plates. For a small plate, the number of jets may not be representative of those for an infinite plate, and this effect can lead to either higher or lower critical heat fluxes for small plates, depending on the relationship between λ and the size of the plate (Lienhard and Dhir [161]).

For the case of cylinders, a similar vapor jet formation phenomenon has been postulated to occur as shown in Fig. 15.64. The jets are suggested to have a radius equal to the radius of the cylinder plus the thickness δ of the vapor blanket, as illustrated in Fig. 15.64c. The spacing of the jets depends on the cylinder size; the relationships involved have been investigated by Sun and Lienhard [162]. For small cylinders, the spacing of the jets is approximately λ_D, and the critical Helmholtz wavelength λ_H may be taken as the circumference of the jet (i.e., $2\pi R_j$). For larger cylinders, the spacing increases to approximately 2 jet diameters (Fig. 15.64b) and λ_H is approximately equal to λ_D. The main difficulty in applying Zuber-type analysis to cylinders is the determination of δ, but Sun and Lienhard [162] and Lienhard and Dhir [155, 161] show that δ can be related to the cylinder radius, the relationship being different for small and large cylinders. These relationships are stated on p. 15.63–15.64.

Macrolayer Consumption Model. Although the hydrodynamic instability model agrees well with much of the experimental data, the very extensive photographic studies that have been conducted on boiling (exemplified by those sketched in Fig. 15.48) indicate a quite different pattern of behavior as the critical heat flux is approached. Thus, vapor mushrooms are

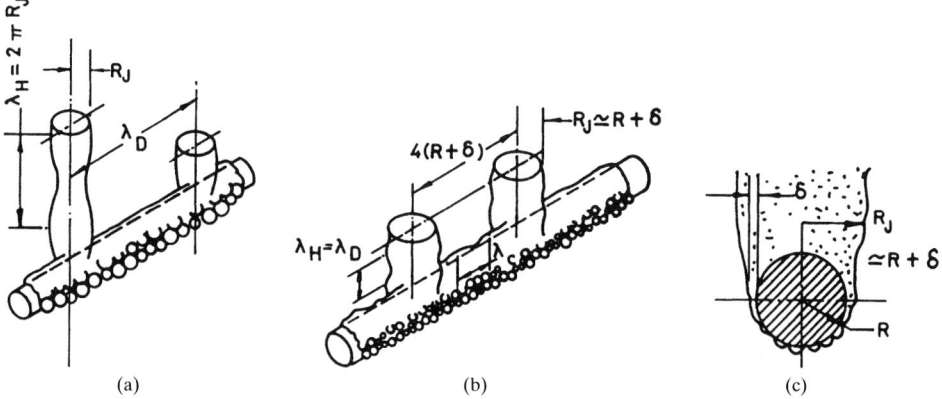

FIGURE 15.64 Vapor escape mechanisms in pool boiling from cylinders. (*a*) small cylinder; (*b*) large cylinder; (*c*) cross section (from Lienhard and Dhir [155], with permission of ASME).

formed on top of the macrolayer as discussed on p. 15.42–15.45. This differs significantly from the picture forming the basis of the hydrodynamic instability model as illustrated in Figures 15.63 and 15.64. Haranura and Katto [163] postulate that the critical heat flux phenomenon occurs when the whole of the macrolayer is consumed during the hovering period τ. The critical heat flux is then given simply by

$$q''_{crit} = (1 - \alpha_M)\delta_o\rho_l i_{lg}/\tau \tag{15.129}$$

where α_M is the void fraction in the macrolayer (equal to A_v/A where A_v is the area of the vapor stems). In this model, relationships are needed for τ, for α_M, and for δ_o; for τ, the Katto and Yokoya [102] relationship (Eq. 15.79) may be used. Haranura and Katto [163] postulated that the length of the vapor stem was limited by Helmholtz instability and that the stem length (macrolayer thickness) δ_o at the point of release of the mushroom-shaped bubble would be given by $\delta_o = \zeta\lambda_H$ where ζ is a factor less than unity. Taking $\zeta = 0.25$, and calculating λ_H using Eq. 15.126, Haranura and Katto obtained the following equation for δ_o:

$$\delta_o = 0.5\pi\sigma[(\rho_l + \rho_g)/\rho_l\rho_g]\alpha_M^2(\rho_g i_{lg}/q'')^2 \tag{15.130}$$

Finally, Haranura and Katto obtained the following empirical relationship for the value of α_M that is required to bring Eq. 15.129 into line with critical heat flux data:

$$\alpha_M = 0.0584(\rho_g/\rho_l)^{0.2} \tag{15.131}$$

Bubble Crowding at Heated Surface. In this mechanism, close packing of bubbles (or vapor stems) leads to a reduction in the heated surface area that is in contact with the liquid phase and, hence, to a fall in heat flux with increasing wall superheat. A model of this type was proposed in 1956 by Rohsenow and Griffith [164], but the method became less popular as the hydrodynamic instability model became more widely accepted. However, Dhir and Liaw [165] have postulated a somewhat analogous model in order to explain the effect of contact angle on critical heat flux as observed by Liaw and Dhir [150] and illustrated in Fig. 15.61. Dhir and Liaw [165] measured void fractions as a function of distance from the wall. The void fraction passed through a maximum value (α_{max}), and this maximum value depended on the heat flux as illustrated in Fig. 15.65.

For well-wetting fluids, the peak void fraction reaches a value close to unity at the critical heat flux condition, and this implies that no liquid may be transferred to the surface under these conditions. For systems with larger contact angles, the peak void fraction was lower than unity, its value at the critical heat flux condition decreasing continuously with increasing contact angle. This implies that, for partially wetting fluids, liquid access to the surface is available at the critical heat flux condition, and this led Dhir and Liaw [165] to postulate the model shown schematically in Fig. 15.66. Assuming an idealized square array of vapor stems, the ratio of the wetted periphery P_s of the stem to the stem spacing L rises with increasing stem diameter D_w and reaches a maximum when the stems begin to merge when $D_w/L = 1$. At this condition, the void fraction would be $\pi/4$. With further increase in D_w, P_s decreases rapidly and, consequently, the heat flux decreases, signifying a critical heat flux condition. The influence of contact angle on critical heat flux (see Fig. 15.61) would presumably arise from the fact that the number of nucleation centers (and hence vapor stems) increases with increasing contact angle.

Hot-Spot Heating. This mechanism has been investigated by Unal et al. [169]. The stages envisaged are that, on departure of the vapor mushroom in the fully developed

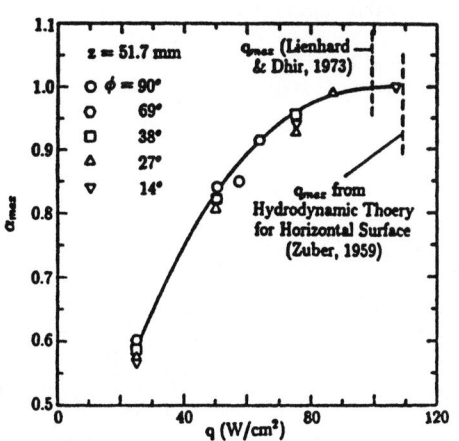

FIGURE 15.65 Dependence of maximum void fraction on heat flux (from Dhir and Liaw [165], with permission of ASME).

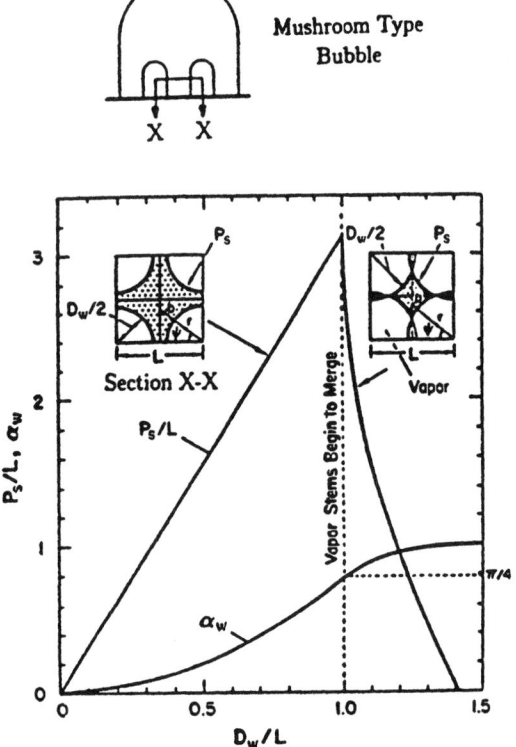

FIGURE 15.66 Variation of stem periphery with steam diameter for an idealized square array of vapor stems (from Dhir and Liaw [165], with permission of ASME).

boiling region, the surface is wetted by fresh liquid and the vapor stems are recreated. At the bottom of the stems, a liquid microlayer is evaporated; the base of the stem is dry and its center temperature rises quickly as a result of the continued heating of the wall. When the next vapor mushroom departs, the dry zone at the base of the vapor stem is rewetted provided that the temperature it has reached during the mushroom hovering period is low enough to allow such rewetting. If the temperature is too high, then rewetting is inhibited and the hot spot becomes permanent and may grow, giving a mechanism for CHF. Using available data for macrolayer thickness, nucleation site density and near-wall void fraction, Unal et al. carried out calculations on the transient heat transfer and demonstrated that, indeed, it seemed possible that the wall temperature at the center of the base of the vapor stem could reach the minimum film boiling temperature (see below), which gave at least a *prima facie* case for this mechanism.

Overview. There still seems to be considerable controversy about the mechanism of critical heat flux in pool boiling. The classical hydrodynamic instability model seems inconsistent with visual observations of the phenomena, though it is extremely difficult to view what is happening in the region close to the surface. Dhir [87] points out that the near-wall void fraction is not well represented by Eq. 15.131 and that instabilities of the vapor stems have not been discerned from visual observations. Furthermore, the macrolayer consumption model is unable to explain the effect of wetting angle on the critical heat flux. The vapor stem merging model of Dhir and Liaw [165] seems appealing for partially wetting fluids, and the observation of near-wall void fractions close to unity for well-wetting fluids seems to support the idea that there is a change in mechanism (perhaps to something like the hydrodynamic instability mechanism) for this case. The hot-spot mechanism of Unal et al. [169] is an interesting one and needs further investigation; however, Dhir [87] argues that the processes are steady-state rather than transient, with the vapor stems remaining in place between successive vapor mushroom departures. This area seems ripe for more detailed investigation.

Correlations for CHF in Pool Boiling.   Most correlations for critical heat flux in pool boiling have been of the form indicated in Eq. 15.128. Although this equation was introduced in the context of the hydrodynamic instability model of Zuber [159, 160], the form of the equation was derived some years earlier by Kutateladze [166]. Thus, the use of the equation is not necessarily associated with any physical model. In the following text, equations will be given for the most usual practical cases of horizontal flat plates and horizontal cylinders; relationships for other shapes are discussed by Lienhard and Dhir [155, 161].

Large Horizontal Flat Plates.   Here, the form of Eq. 15.128 suggested by Lienhard and Dhir [155, 161] is recommended as follows:

$$q''_{crit} = 0.149 \rho_g^{1/2} i_{lg} [\sigma g (\rho_l - \rho_g)]^{1/4} \tag{15.132}$$

The correlation is accurate to about ±20 percent and has the following main limitations:

1. It is for saturated pool boiling only; if the liquid in the pool is subcooled, the critical heat flux is higher.
2. It is applicable only to large plates. The characteristic dimension of the plate L (m) should obey

$$L > \frac{32.6}{[g(\rho_l - \rho_g)/\sigma]^{1/2}} \tag{15.133}$$

   where L is given by the shortest side for a rectangular plate or by the diameter for a circular plate. For smaller plates, Lienhard and Dhir [155, 161] suggest that the critical heat flux could be either higher or lower; they ascribe this (using the hydrodynamic instability theory) to the number of jets that could be accommodated on the plate.
3. Effects of liquid viscosity are not included in Eq. 15.132, although critical heat flux for viscous liquids is higher than that for those with low viscosity. A more detailed correlation taking viscosity effects into account is given by Dhir and Lienhard [167]. To use Eq. 15.132, the viscosity number Vi as defined by

$$Vi = \frac{\rho_l \sigma^{3/4}}{\mu_l g^{1/4} (\rho_l - \rho_g)^{3/4}} \tag{15.134}$$

   should be greater than 400.
4. The correlation does not apply to liquid metals; a discussion of this case is given by Rohsenow [168].

Horizontal Cylinders.   Here, the critical heat flux is given by Leinhard and Dhir [155.161]

$$q''_{crit} = K \rho_g^{1/2} i_{lg} [\sigma g (\rho_l - \rho_g)]^{1/4} \tag{15.135}$$

where the constant K is given by

$$K = 0.118 \qquad \text{for } R' > 1.17 \tag{15.136}$$

$$K = \frac{0.123}{(R')^{1/4}} \qquad \text{for } 1.17 > R' > 0.12 \tag{15.137}$$

where R' is a nondimensional radius defined by

$$R' = R\left[\frac{g(\rho_l - \rho_g)}{\sigma}\right]^{1/2} \tag{15.138}$$

where R is the cylinder radius. The correlation given by Eqs. 15.135–15.138 is accurate to around ±20 percent and has the following main limitations:

1. It does not apply to very small cylinders (i.e., $R' < 0.12$).
2. It applies only for low-viscosity systems (i.e., Vi > 400). A correlation for viscous fluids is described by Dhir and Lienhard [167].
3. The expression will not apply accurately to short cylinders (typically the cylinder should be at least 20 diameters long for the equation to be applied).
4. The correlation is for saturated fluids; the effect of liquid subcooling can be taken into account using Eq. 15.122.
5. The correlation does not apply accurately to liquid metals; again, this case is discussed in some detail by Rohsenow [168].

The correlations given here are also limited to well-wetting fluids; increase in contact angle gives a decrease in critical heat flux as discussed previously.

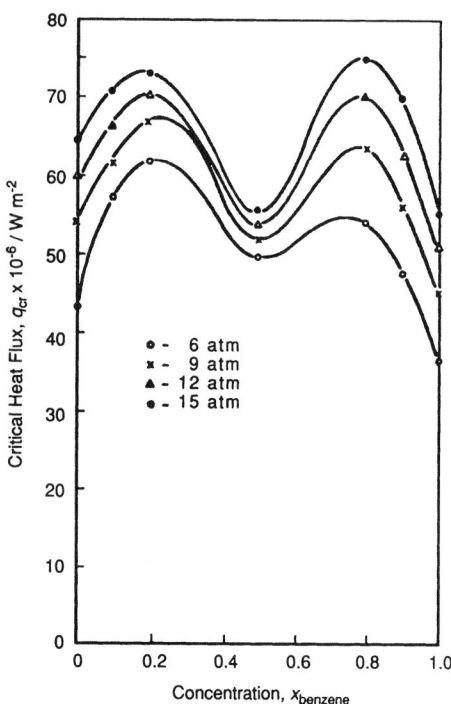

FIGURE 15.67 Variation of pool-boiling critical heat flux with composition and pressure for ethanol/benzene mixtures (from Afgan [170], with permission of Taylor & Francis, Washington, DC. All rights reserved).

Prediction of Pool Boiling CHF. The pool-boiling case is unusual in that correlations and prediction methods are commonly based on mechanistic models. These models were introduced previously; in general, prediction methods based on the hydrodynamic instability model are applicable only for well-wetting fluids. For partially wetting fluids, the model of Dhir and Liaw [165] appears promising, though the alternative interpretation of Unal et al. [169] indicates the remaining uncertainties. One may conclude, therefore, that *prediction* of critical heat flux in pool boiling is still surrounded by mechanistic uncertainties and that recourse must be had to the reasonably well-established correlations (notwithstanding their deficiencies as listed previously).

CHF in Pool Boiling of Multicomponent Mixtures. Critical heat flux in binary and multicomponent mixtures can be very different from that calculated based on the average physical properties of the mixture. Early data in this area are typified by the results of Afgan [170] for the boiling of ethanol/benzene mixtures (which form an azeotrope), which are shown in Fig. 15.67. In contrast to the results obtained for nucleate boiling heat transfer coefficient (see Fig. 15.54), the critical heat flux is *increased* on either side of the azeotrope as shown. Van Stralen [171] noted that the critical heat flux in these early experiments reached a maximum when the value of $(\tilde{y} - \tilde{x})$ (see Fig. 15.56) reached a maximum; this condition also corresponded to the minimum bubble growth rate. Because of mass transfer limitations, the interface temperature is higher than that for equilibrium. This produces what Reddy and Lienhard [172] call *induced subcooling* in the bulk liquid, and these authors suggest that the increased critical heat flux is analogous to the increase in heat flux with subcooling found with single components. Reddy and Lienhard developed a correlation for critical heat flux based on this concept, which (in the form stated by Fujita and Bai [173]) is as follows:

$$q''_{crit} = (q''_{crit})_{id}(1 - 0.170 Ja_e^{0.308})^{-1} \tag{15.139}$$

where $(q''_{crit})_{id}$ is the ideal mixture critical heat flux calculated from Eq. 15.135 with a value of K given by

$$K = K_1 \tilde{x}_1 + K_2 \tilde{x}_2 \tag{15.140}$$

where K_1 and K_2 are estimated (in the case of cylindrical heaters) from Eqs. 15.136 and 15.137 for the respective pure components. Ja_e is given by

$$Ja_e = (\rho_l c_{pl} \Delta T_{bp})/(\rho_g i_{fg}) \tag{15.141}$$

where ΔT_{bp} is the bubble point to dew point temperature difference as illustrated in Fig. 15.56.

More recent work has shown that the enhancement of critical heat flux does not always occur for binary mixtures; indeed, the critical heat flux can sometimes be reduced relative to the ideal value. A possible explanation for this variability arises from the influence of surface tension differences (Marangoni effects). This possibility was first suggested by Hovestreijdt [174] in 1963, and this suggestion has been developed into the form of a correlation by Fujita and Bai [173]. Flows induced by surface tension differences (Marangoni flows) will be expected to increase CHF in so-called *positive mixtures* whose surface tension is decreased with increasing concentration of the more volatile component; *negative mixtures* have the opposite effect and, it is suggested, would decrease the CHF. Fujita and Bai [173] give the following correlation accounting for these effects:

$$\dot{q}''_{crit} = (q''_{crit})_{id}\left[1 - 1.83 \times 10^{-3} \frac{|Ma|^{1.43}}{Ma}\right]^{-1} \tag{15.142}$$

where the Marangoni number Ma is defined as

$$Ma = (\Delta\sigma/\rho_l v_l^2)[\sigma/g(\rho_l - \rho_g)]^{1/2} Pr_l \tag{15.143}$$

where v_l is the liquid kinematic viscosity and $\Delta\sigma$ is defined as

$$\Delta\sigma = \sigma_D - \sigma_B \tag{15.144}$$

where σ_D and σ_B are the surface tensions at the dew point and bubble point corresponding to concentration $\tilde{x}_1$, respectively. Good qualitative and quantitative agreement was obtained using this relationship, including the prediction for azeotropes. It may be possible to include such effects in more analytical models, for example the model of Lay and Dhir [104].

Mitigation of Pool Boiling CHF.   Mitigation of the critical heat flux phenomenon is reviewed in detail by Collier and Thome [3]. Some brief examples of mitigation methods follow:

1. *Finned surfaces.* Here, thick fins or studs are attached to the surface and heat is conducted along them. Near the original surface, film boiling occurs but excessive temperature rises are avoided by thermal conductance along the fins to regions where nucleate boiling prevails. Typical work on this area is that of LeFranc et al. [175] and Haley and Westwater [76].

2. *Electrical fields.* Here, electrohydrodynamic (EHD) methods are used to enhance CHF. Increases of up to around a factor of 5 are possible by this technique. An example of this work is that of Markels and Durfee [177], who obtained an increase in CHF by a factor of 4.5 by the application of 7,000 volts DC to a 9.5-mm tube in pool boiling of isopropanol at atmospheric pressure.

3. *Ultrasonic vibration.* Several authors have investigated pool boiling in ultrasonic fields. An example here is the work of Ornatskii and Shehebakov [178], who observed increases in CHF of between 30 and 80 percent, with the improvement increasing with increasing sub-cooling. A 1-MHz ultrasonic field was employed.

As was discussed in the section on multicomponent mixtures, enhancement of CHF may also be obtained in pool boiling of multicomponent mixtures.

Heat Transfer Beyond the Critical Heat Flux Limit in Pool Boiling

Referring to the schematic pool-boiling diagram in Fig. 15.32, we see that there are two distinct regions of heat transfer behavior in the region beyond the critical (maximum) heat flux. These are, respectively, *transition boiling,* in which the heat flux decreases with increasing wall temperature and, *film boiling,* in which the heat flux begins to increase again with wall temperature. The two regions join at the point of *minimum heat flux* corresponding to a temperature defined as the *minimum film-boiling temperature* (T_{min}). Again, there has been a vast amount of work on post-CHF heat transfer in pool boiling, and it is impossible to even list it in the space available. Recent reviews on transition boiling heat transfer are those of Auracher [179] and Sakiurai and Shiotsu [180], the latter review also dealing in detail with the minimum film-boiling tempera-ture. Film boiling is perhaps the only region of boiling where well-founded theoretical treatment can be made in terms of the governing equations for fluid flow and heat transfer. An excellent presentation of these fundamental relationships is given in the book by Carey [4].

Parametric Effects in Post-CHF Pool Boiling

Effect of Pressure. The effect of pressure on pool boiling has been investigated by Pan and Lin [181], whose results are shown in Fig. 15.68. The effect of pressure is quite complex, changing from the nucleate boiling region into the transition boiling region and finally into the film-boiling region. At a given wall superheat in the transition boiling region, the heat flux decreases with increasing pressure, whereas in the film-boiling region, the heat flux increases

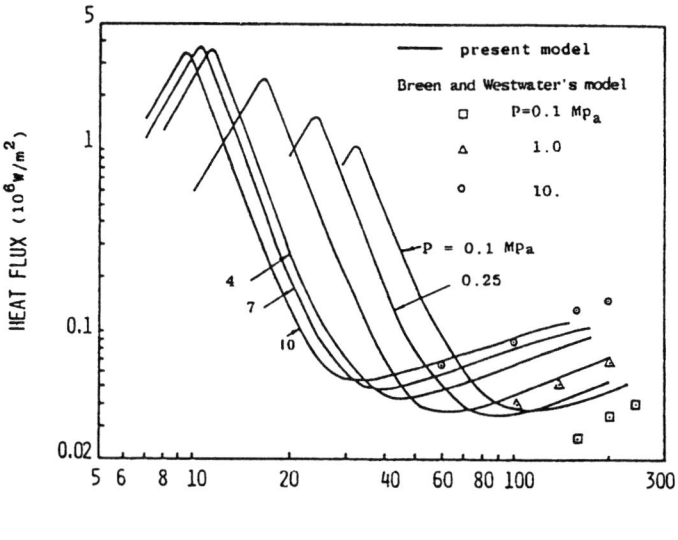

FIGURE 15.68 Effect of pressure on the pool-boiling curve for water on a copper surface (from Pan and Lin [181], with permission from Elsevier Science).

with increasing pressure at a given superheat. The calculated results in Fig. 15.68 show that the wall superheat at the minimum heat flux decreases with pressure. However, since the saturation temperature increases with pressure, the value of T_{min} increases with pressure despite the fall in wall superheat. This is consistent with the results of Sakurai and Shiotsu [180] shown in Fig. 15.69; Sakurai and Shiotsu show that the minimum film boiling temperature for boiling of water from cylinders increases with increasing pressure, approaching the homogeneous nucleation temperature at high pressures.

Effect of Subcooling. Increasing the subcooling increases the heat flux in the transition boiling and film-boiling regions. Results obtained by Tsuchiya [182] illustrating the effect in the transition boiling region are shown in Fig. 15.70. Figure 15.71 shows data for heat transfer coefficient in film boiling; the heat transfer coefficient increases with increasing subcooling and decreases, at a given subcooling, with increasing values of ΔT_{sat}, the wall superheat.

Effect of Surface. The roughness and contact angle of the surface have a significant effect in the transition boiling region but little or no effect in the film-boiling region. A review of surface effects in transition boiling is presented by Auracher [179]. Increasing surface roughness increases the heat flux at a given surface temperature in the transition boiling region. The heat flux tends to decrease with increasing contact angle, and Shoji et al. [183] observed the remarkable transition at high contact angles illustrated in Fig. 15.72; a new form of film boiling was observed with a much lower minimum film-boiling temperature.

Effect of Angle of Surface. El-Genk and Guo [184] investigated the effect of angle of inclination on transition pool boiling of saturated water. They used a transient (quenching) technique and their results are shown in Fig. 15.73. The angle of inclination of the surface was changed systematically from zero (downward-facing) to 90° (vertical). The heat flux at a given wall superheat increased with increasing angle; the most rapid change was between a downward-facing surface and a surface inclined at only 5°. The effect of angle in the film boiling region has been investigated by Nishio et al. [185] for the case of saturated film boiling of liquid helium. These results are illustrated in Fig. 15.74; here the angle is defined relative to an

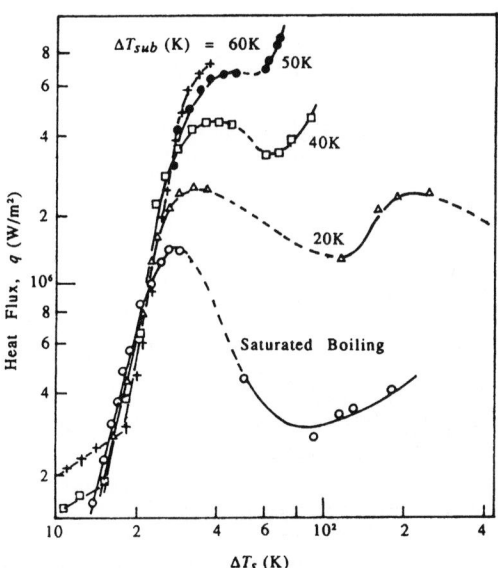

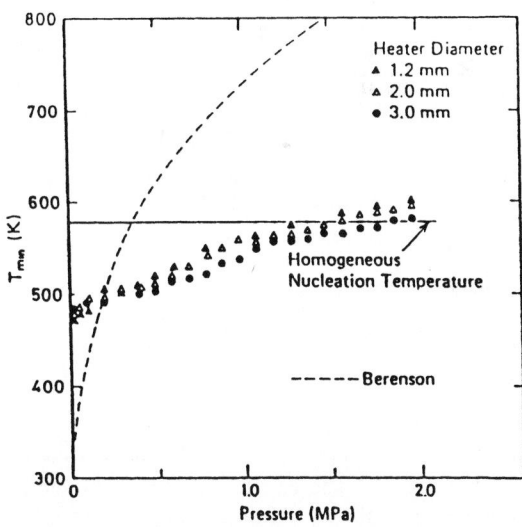

FIGURE 15.69 Effect of subcooling on transition boiling of water at atmospheric pressure on a 20-mm disk heater (results of Tsuchiya [182] quoted by Katto [157]; reproduced with permission from ASME).

FIGURE 15.70 Minimum film boiling temperature for boiling of water from horizontal cylinders as a function of system pressure (from Sakurai and Shiotsu [180], with permission from ASME).

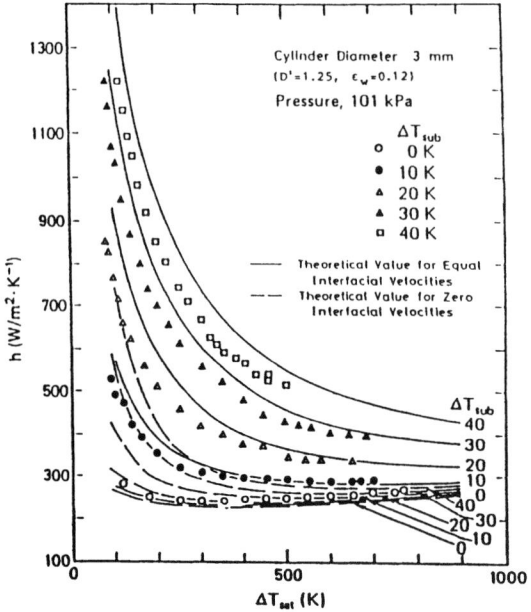

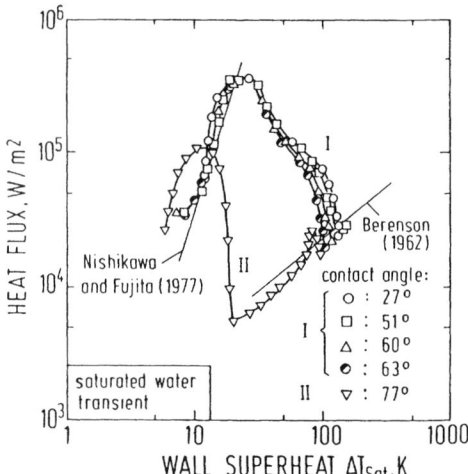

FIGURE 15.72 Effect of contact angle on transition boiling heat transfer. Transient tests with saturated water on 100-mm copper surface (from Shoji et al. [183], with permission from ASME).

FIGURE 15.71 Heat transfer coefficient in film boiling of water from a 3-mm-diameter cylinder (from Sakurai and Shiotsu [180], with permission from Taylor & Francis, Washington, DC. All rights reserved).

upward-facing surface and the results are related to the theoretical value for a vertical surface. The heat transfer coefficient has a peak for the vertical surface, decreasing somewhat with angle for upward-facing surfaces and more sharply with angle for downward-facing surfaces.

Mechanisms of Post-CHF Heat Transfer in Pool Boiling. It is convenient to divide the discussion of mechanisms into three areas, namely transition boiling, film boiling, and minimum film boiling temperature.

Transition Boiling Mechanisms. A detailed review of mechanisms of transition boiling is given by Auracher [179]. A key feature of the transition boiling region is that a fraction of the surface is in contact with the liquid phase; the extent of liquid contact can be determined using electrical methods and typical results are shown in Fig. 15.75. Though the results in Fig. 15.75 show that the fraction of liquid contact is close to unity at the critical heat flux point, it should be pointed out that there is a large discrepancy between data reported by various investigators. The contact fraction falls to near zero at the minimum heat flux (MHF) condition. Thus, the heat flux in transition boiling can be expressed formally in terms of the expression

$$q'' = Fq''_l + (1 - F)q''_g \tag{15.145}$$

where q''_l and q''_g are the average heat fluxes during the periods of liquid contact and vapor contact, respectively, and F is the fraction of liquid contact (as shown in Fig. 15.75).

The concept of partial liquid contact in transition boiling is consistent with the macrolayer theory of fully developed boiling as discussed earlier. In fact, Dhir and Liaw [103] extend their model (as illustrated in Fig. 15.51) into the transition boiling region. Thus there is a continuous development of macrolayer behavior from the fully developed boiling region into the CHF region (where vapor stems begin to coalesce) and through the transition boiling region until the surface is completely covered by vapor at the minimum heat flux (T_{min}) point.

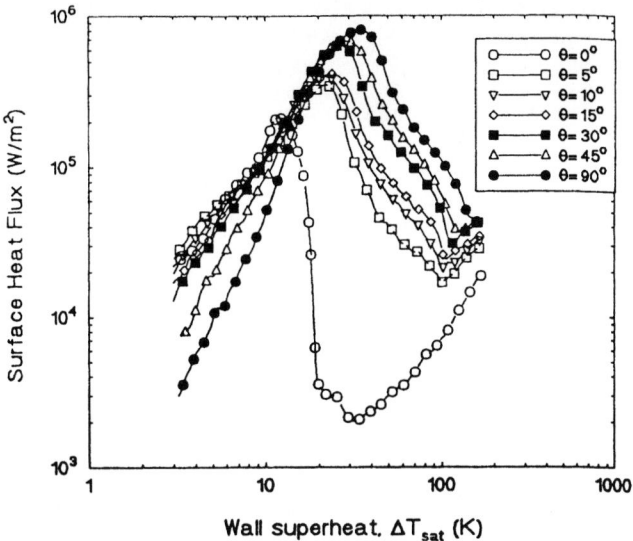

FIGURE 15.73   Effect of surface inclination angle on transition boiling of saturated water. Experiment of transient (quenching) type (from El-Genk and Guo [184], with permission from ASME).

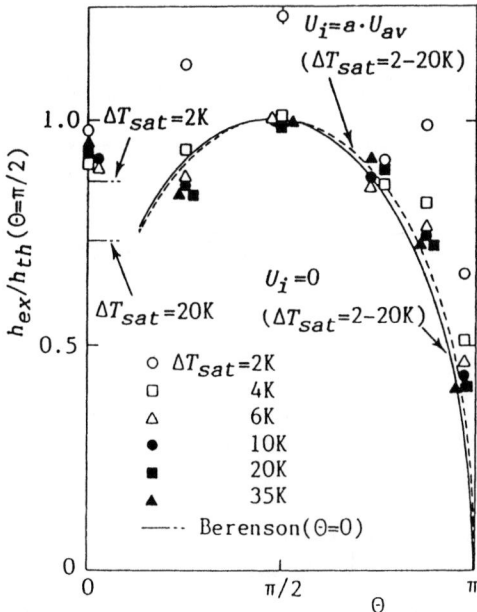

FIGURE 15.74   Effect of angle of inclination in film boiling of liquid helium (angle relative to horizontal) (from Nishio et al. [185]).

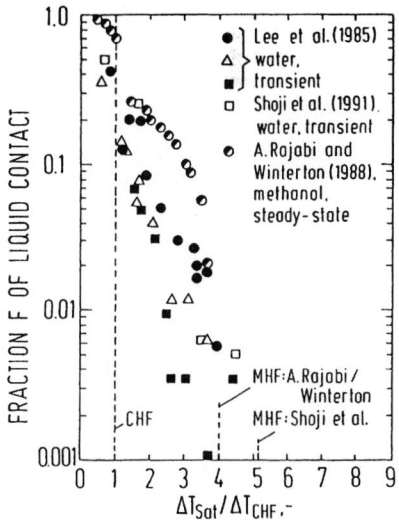

FIGURE 15.75   Fractional liquid contact as a function of surface superheat (from Auracher [179], with permission from ASME).

Film Boiling.   In film boiling, a vapor layer is formed that separates the solid surface from the liquid phase. Thus, film boiling is not dependent on the detailed microstructure of the surface; essentially, the heat transfer process is governed by conduction, convection, and radiation across the vapor layer. The contribution of radiation is, of course, governed by the emissivity of the solid surface, but the radiation component is usually quite small relative to the other components. Thus film boiling is in general a much more predictable mode of heat transfer than is nucleate or transition boiling, and this has led to an enormous amount of work in the area. Figure 15.76 shows various modes of film boiling. For vertical flat plates, spheres, and cylinders, a vapor layer is formed that flows upward over the hot surface as shown in Fig. 15.76a and b.

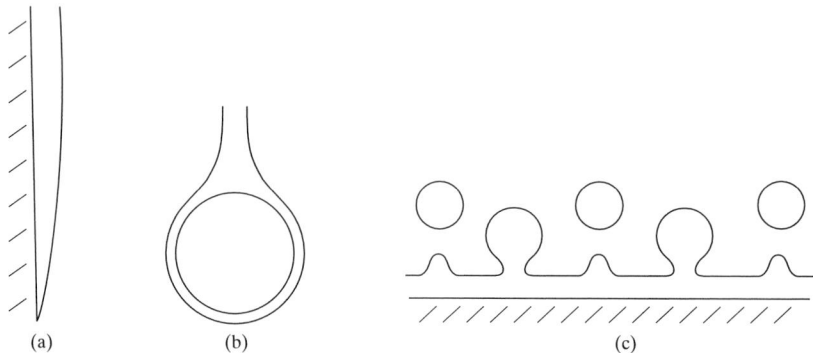

(a)   (b)   (c)

FIGURE 15.76   Modes of film boiling. (*a*) vertical flat plate; (*b*) sphere or cylinder; (*c*) horizontal plate.

These modes of heat transfer shown in Fig. 15.76a and b have some analogies with film condensation (though the flows are in the opposite direction, of course!) and the analytical expressions have some similarity to those derived in condensation. When the boiling occurs from a horizontal plate, the vapor release mechanism is more complex as shown in Fig. 15.76c. A beautiful regular pattern of bubbles is formed on the vapor-liquid interface due to classical Taylor instability. The theoretical background for such instability is given, for instance, by Carey [4]; briefly, the surface separating the vapor and the liquid is unstable to small perturbations. Perturbations of wavelength λ_D ("most dangerous wavelength") are the ones that grow most rapidly and have been usually associated with the bubble behavior shown in Fig. 15.76c. Many of the correlations and prediction methods for film boiling on horizontal surfaces had their origin in the analysis of such instabilities.

Minimum Conditions for Film Boiling.   Zuber [159] suggested that the minimum heat flux $q''_{\min}$ for film boiling (corresponding to the minimum film boiling temperature $T_{\min}$) would occur when the vapor production rate required by the Taylor instability mechanisms sketched in Fig. 15.76 became greater than the generation rate of vapor by the process of conduction, convection, and radiation heat transfer from the surface to the vapor-liquid interface. The following expression was obtained for the minimum heat flux $q''_{\min}$:

$$q''_{\min} = C_2 \rho_g i_{lg} \left[\frac{g\sigma(\rho_l - \rho_g)}{(\rho_l + \rho_g)^2} \right]^{1/4} \tag{15.146}$$

where C_2 is a constant introduced to take account of differences from the linear instability theory; Berenson [186] fitted data for pool-boiling minimum heat flux on a flat plate with C_2 = 0.09. A modified form of Eq. 15.146 has been derived by Lienhard and Wong [187]. However, the Zuber/Berenson form of the equation seems incapable of predicting the effect of pressure on T_{min}, as shown in Fig. 15.70. Sakurai and Shiotsu [180] present a convincing case that T_{min} is related to the temperature T_l that is reached at the liquid-solid interface when the liquid and solid are brought into contact. Sakurai and Shiotsu [180] suggest the following equation for T_l:

$$T_l = 0.92 T_c \{1 - 0.26 \exp(-20 P_r (1 + 1700/P_c)^{-1})\} \tag{15.147}$$

where T_c is the critical temperature, P_c is the critical pressure, and P_r is the reduced pressure ($= P/P_c$). T_l approaches the homogeneous nucleation temperature at high pressures.

Correlations for Post-CHF Heat Transfer in Pool Boiling

Transition Boiling.   An approximate method of predicting transition boiling is simply to linearly interpolate between the critical heat flux (q''_{crit}) and minimum heat flux (q''_{min}) conditions on the boiling curve. Based on a model for transition boiling, Ramilison and Lienhard [189] proposed the following correlation for predicting heat flux in the transition boiling region:

$$Bi^* = 3.74 \times 10^{-6} (Ja^*)^2 K \tag{15.148}$$

where:

$$Bi^* = \frac{(q'' - q''_{fb}) \sqrt{\alpha_s \tau}}{k_s [T_w - T_{sat}(P_l)] K} \tag{15.149}$$

$$Ja^* = \frac{(\rho_s c_{ps})(T_{dfb} - T_w)}{\rho_g i_{lg}} \tag{15.150}$$

$$\tau = \left[\frac{\sigma}{g^3 (\rho_l - \rho_g)} \right]^{1/4} \tag{15.151}$$

$$K = \frac{k_l / \alpha_l^{1/2}}{k_l / \alpha_l^{1/2} + k_s / \alpha_s^{1/2}} \tag{15.152}$$

In the above expressions, q''_{fb} is the film-boiling heat flux predicted at the given wall superheat; ρ_s, c_{ps}, k_s, and α_s are the density, specific heat capacity, thermal conductivity, and thermal diffusivity of the heating wall material. α_l is the thermal diffusivity of the liquid phase, and T_{dfb} is the wall temperature at which liquid contact starts (roughly equivalent to T_{min}). Ramilison and Lienhard [189] gave the following expression for the calculation of T_{dfb}:

$$\frac{T_{dfb} - T_{sat}}{T_{hn} - T_{sat}} = 0.97 \exp(-0.00060 \phi_a^{1.8}) \tag{15.153}$$

where ϕ_a is the advancing contact angle, and T_{hn} is the homogeneous nucleation temperature, which was calculated from the expression

$$T_{hn} = \left[0.932 + 0.077 \left(\frac{T_{sat}}{T_c} \right)^9 \right] T_c \tag{15.154}$$

It should be noted that Eqs. 15.153 and 15.154 differ from expressions given earlier in this chapter for T_{min} and for T_{hn}. However, these equations are given here for consistency with the above correlation.

Film Boiling.   A bewildering range of relationships exists in the literature for film-boiling heat transfer. Usually, these relationships have some basis in theory, though it is also usual for empirical constants to be included to bring the theoretical framework into line with experimental data. Extensive surveys of film-boiling relationships are given by Carey [4] and Tong and Tang [5]. Because of the low heat transfer coefficients encountered in film boiling, the surface temperature can be very high, and this can lead to a significant radiation component in the heat flux. It is usual practice to define a total heat transfer coefficient h as the sum of a convective coefficient h_c and a factor J multiplied by a radiation heat transfer coefficient h_r as follows:

$$h = h_c + Jh_r \tag{15.155}$$

h_r is given by

$$h_r = \left[\frac{\sigma_S}{1/e_s + 1/e_l - 1} \right]\left[\frac{T_w^4 - T_{sat}^4}{T_w - T_{sat}} \right] \tag{15.156}$$

where σ_S is the Stefan-Boltzmann constant and e_s and e_l are the emmisivities of the solid and liquid surfaces (e_l is often close to unity and is not included in many of the expressions for h_r). J is often assigned a value of 0.75, though a more accurate correlation for J for cylinders has been developed by Sakurai and Shiotsu [180] and is presented below. First we consider correlations for h_c. Here, just a few correlations are selected from the literature to cover the most common geometries.

Vertical Flat Plates. Following the earlier work by Bromley [190] on cylinders (see below), Hsu and Westwater [191] derived an expression for laminar film boiling on a vertical plate that is analogous to that for laminar film condensation. The average value of h_c over a plate of height L is given by

$$h_c = 0.943 \left[\frac{g(\rho_l - \rho_g)\rho_g k_g^3 i_{lg}'}{L\mu_g(T_w - T_{sat})} \right]^{1/4} \tag{15.157}$$

where the physical properties are evaluated at a temperature corresponding to $(T_w + T_{sat})/2$ and where i_{lg}' is a corrected latent heat of vaporization (introduced to allow for sensible heating of the vapor film) defined as

$$i_{lg}' = i_{lg}\left[1 + \frac{Cc_{pg}(T_w - T_{sat})}{i_{lg}} \right] \tag{15.158}$$

This corrected form of latent heat is often used in expressions for film boiling; the constant C is assigned various values, the particular value used by Hsu and Westwater being $C = 0.34$. It should be noted that the *local* heat transfer coefficient varies along the vapor film, generally decreasing with increasing length of film.

For a long enough plate, waves begin to appear on the film surface; a detailed analysis taking account of such waves is presented by Bui and Dhir [192]. Another effect is that the film eventually becomes turbulent. We may define a Reynolds number for the vapor film (Re_g) as

$$Re_g = \frac{4\Gamma_{gL}}{\mu_g} \tag{15.159}$$

where Γ_{gL} is the mass flow per unit periphery of the surface at distance L along the surface. Γ_{gL} is given as

$$\Gamma_{gL} = q''L/i_{lg} \tag{15.160}$$

Hsu and Westwater [191] recommend the following expression for h_c for Re_g in the range 800–5000:

$$h_c\left[\frac{\mu_g^2}{k_g^3\rho_g(\rho_l - \rho_g)g} \right]^{1/3} = 0.002Re_g^{0.6} \tag{15.161}$$

Horizontal Flat Plates. For this case, the classical correlation is that of Berenson [186]. This correlation was derived on the basis of a model describing the bubble release mechanism shown in Fig. 15.76c. Berenson obtained the following expression for h_c for these conditions:

$$h_c = 0.425 \left\{ \left[\frac{k_g^3 g \rho_g (\rho_l - \rho_g) i'_{lg}}{\mu_g (T_w - T_{sat})} \right] \left[\frac{g(\rho_l - \rho_g)}{\sigma} \right]^{1/2} \right\}^{1/4} \tag{15.162}$$

where σ is the surface tension. Here, i'_{lg} is obtained from Eq. 15.158 with $C = 0.50$.

An extension of this model to take account of turbulence is described by Klimenko [194].

Horizontal Cylinders. The horizontal cylinder has been the most widely studied case of film boiling. Again, there is a strong analogy with condensation (see Fig. 15.76b) and the classical expression for this case is that due to Bromley [190] as follows:

$$h_c = 0.62 \left[\frac{g(\rho_l - \rho_g)\rho_g k_g^3 i'_{lg}}{D \mu_g (T_w - T_{sat})} \right]^{1/4} \tag{15.163}$$

where D is the cylinder diameter and where i'_{lg} is obtained from Eq. 15.158 with $C = 0.68$.

The Bromley equation (though still widely used) can give significant errors under a variety of conditions. An extensive exercise on the correlation of pool film-boiling heat transfer from cylinders is reported by Sakurai and Shiotsu [180], whose expression for the mean convective heat transfer coefficient is expressed in dimensionless form as follows:

$$\frac{Nu_g}{(1 + 2/Nu_g)} = KM*^{1/4} \tag{15.164}$$

where

$$Nu_g = \frac{h_c D}{k_g} \tag{15.165}$$

and where the parameter K is a function of nondimensional diameter D' defined as follows:

$$D' = D[g(\rho_l - \rho_g)/\sigma]^{1/2} \tag{15.166}$$

The relationship between K and D' is as follows:

$$K = 0.415 D'^{1/4} \qquad \text{for } D' > 6.6 \tag{15.167}$$

$$K = 2.1 D'/(1 + 3D') \qquad \text{for } 1.25 \le D' \le 6.6 \tag{15.168}$$

$$K = 0.75/(1 + 0.28D') \qquad \text{for } 0.14 \le D' < 1.25 \tag{15.169}$$

The parameter $M*$ is given by

$$M* = \{Gr_g \, Pr_g \, i'_{lg}/[c_{pg}(T_w - T_{sat})]\} \psi \tag{15.170}$$

where i'_{lg} is calculated from Eq. 15.158 with $C = 0.5$ and where the vapor Grashof number (Gr_g) and Prandtl number (Pr_g) are calculated from the expressions

$$Gr_g = gD^3(\rho_l - \rho_g)/(\rho_g v_g^2) \tag{15.171}$$

$$Pr_g = c_{pg} \mu_g / k_g \tag{15.172}$$

The parameter ψ in Eq. 15.170 is introduced to take account of subcooling, sensible heating, and the relative motion of the vapor interface. It is given by

$$\psi = \{E^3/[1 + E/(S_p \, Pr_l)]\}/(R \, Pr_l \, S_p)^2 \tag{15.173}$$

where

$$E = (A + C\sqrt{B})^{1/3} + (A - C\sqrt{B})^{1/3} + S_c/3 \tag{15.174}$$

$$A = (1/27)S_c^3 + (1/3)R^2 S_p \, Pr_l \, S_c + (1/4)R^2 S_p \, Pr_l^2 \tag{15.175}$$

$$B = (-4/27)S_c^3 + (2/3)S_p \, Pr_l \, S_c - (32/37)S_p \, Pr_l^2 \, R^2 \tag{15.176}$$

$$C = (1/2)R^2 S_p \, \mathrm{Pr}_l \tag{15.177}$$

$$R = [\rho_g \mu_g/(\rho_l \mu_l)]^{1/2} \tag{15.178}$$

$$\mathrm{Pr}_l = c_{pl}\mu_l/k_l \tag{15.179}$$

$$S_p = c_{pg}(T_w - T_{\mathrm{sat}})/(i'_{lg} \, \mathrm{Pr}_g) \tag{15.180}$$

$$S_c = K_a c_{pl}(T_l - T_{\mathrm{sat}})i'_{lg} \tag{15.181}$$

$$K_a = [0.93\mathrm{Pr}_l^{0.22} + 3.0 \exp(-100S_p \mathrm{Pr}_l \, S_c^{-0.8})][0.45 \times 10^5 \, \mathrm{Pr}_l \, S_c/(1 + 0.45 \times 10^5 \, \mathrm{Pr}_l \, S_c)] \tag{15.182}$$

The justification for such a complex correlation seems to be that it covers a very wide range of fluids (including liquid metals), cylinder diameters, and subcoolings. It was shown to fit a very wide range of data and to perform much better than the Bromley equation (Eq. 15.163) and a variety of other earlier correlations.

Radiation Correction Factor.   Bromley [190] suggested a value of $J = 0.75$ as a multiplier for the radiation heat transfer coefficient in Eq. 15.155. Detailed analytical studies by Sakurai and Shiotsu [180] show that J can vary over a wide range; they fitted the following expression to their analytical results for cylinders:

$$J = F + (1 - F)/(1 + 1.4h_c/h_r) \tag{15.183}$$

where

$$F = [1 - 0.25 \exp(-0.13S_{pr})] \exp(-0.64R^{0.60} \, \mathrm{Pr}_l^{-0.45} \, S_{pr}^{-0.73} S_{cr}^{1.1}) \tag{15.184}$$

where

$$S_{pr} = c_{pg}(T_w - T_{\mathrm{sat}})/(i_{lg} \, \mathrm{Pr}_g) \tag{15.185}$$

$$S_{cr} = c_{pl}(T_l - T_{\mathrm{sat}})/i_{lg} \tag{15.186}$$

If the value of F calculated from Eq. 15.184 is less than 0.19, F should be taken as 0.19.

Prediction of Post-CHF Heat Transfer in Pool Boiling.   Since film boiling often has a reasonably well-established geometry (see Fig. 15.74), it has been the subject of a great deal of effort in prediction. Such prediction methods are reviewed by Carey [4] and more recent prediction activities are reviewed by Sakurai and Shiotsu [180]. These prediction efforts have essentially been built into the previously given empirical correlations for film boiling. For transition boiling, there are greater uncertainties about the precise mechanisms, though models have been developed based on assumptions about the behavior of vapor jets under the macrolayer (Dhir and Liao [165], Shoji [195]).

Post-CHF Pool Boiling Heat Transfer with Multicomponent Mixtures.   In the transition boiling region, Happel and Stephan [196] have shown that the effect of having binary mixtures rather than pure components was qualitatively similar to that observed in nucleate pool boiling as discussed above (p. 15.51–15.53), i.e., causing a net reduction in heat transfer coefficient. The minimum heat flux may be expected to be greater for binary mixtures compared with an equivalent pure fluid. Collier and Thome [3] and Yue and Weber [197]) have shown that, in the film-boiling region, the net effect of the concentration of the less volatile phase at interface is to increase the heat transfer coefficient relative to that expected for the mean properties of the mixture. Thus, equations of the type given earlier underpredict the heat fluxes by typically 30 percent.

Enhancement of Post-CHF Heat Transfer in Pool Boiling.   Thome [139] shows that both transition and film-boiling heat transfer coefficients are increased when porous surfaces of the type discussed previously are used. The influence of high-voltage electrical fields on film-boiling heat transfer was investigated by Verplaetsen and Berghmans [198], who suggest that, typically, a fivefold increase in heat transfer coefficient in the film-boiling region can be achieved with electrically conducting liquids.

CROSS FLOW BOILING

In the pool-boiling situation described previously, boiling occurs from a heated surface mounted in a static pool of liquid. Of course, even in the pool-boiling case, circulation of liquid occurs within the pool, contributing to natural convection heat transfer from the surface. However, there are many situations in which flow across the heated surface occurs, and boiling in these circumstances is usually referred to as *cross flow boiling*. Examples of cross flow boiling would be boiling from a horizontal cylinder or a sphere with a flow passing upward around the object. Though there are practical applications of such cases, the most common example of cross flow boiling is that of boiling in a horizontal bundle where boiling occurs from the surfaces of an array of (usually horizontal) tubes over which the fluid flows in a direction normal to the tube axis. Typical process industry applications are in horizontal thermosiphon and kettle reboilers. In the horizontal thermosiphon reboiler (Fig. 15.77), flow over a heated tube bundle is induced by natural circulation through the loop from the distillation tower as shown. In the kettle reboiler (Fig. 15.78) there is no recirculation through the boiler, the generated vapor leaving the surface of the liquid and passing back to the distillation column as shown. The kettle reboiler has often been considered in terms of a direct analogy with pool boiling, but it is probably more correct to think of it as a case of cross flow boiling since circulation is induced within the liquid pool as a result of boiling in the bundle as illustrated in Fig. 15.79.

Although cross flow boiling in tube bundles is in some ways related to cross flow boiling across single tubes, it also has a relationship to the case of forced convective boiling in channels (which will be dealt with later). As seen from Fig. 15.79, there are vertical "channels" between the rows of tubes, and the flow up these channels may behave in a way that is analogous to flow boiling in tubes. The reviews of Whalley [199] and Jenson [200] relate specifically to external flow boiling and shellside boiling.

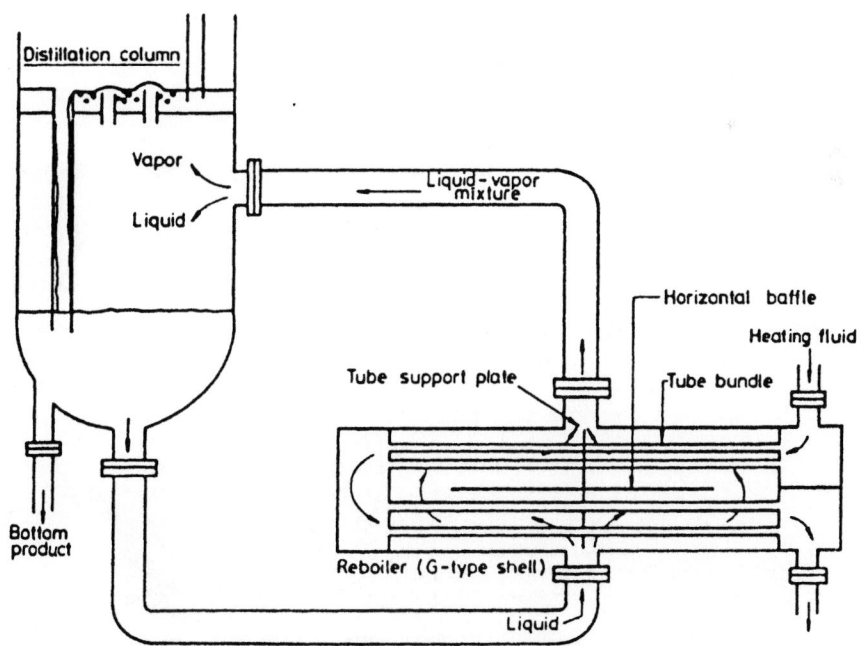

FIGURE 15.77   Horizontal thermosiphon reboiler (from Hewitt et al. [13], with permission. Copyright CRC Press, Boca Raton, FL).

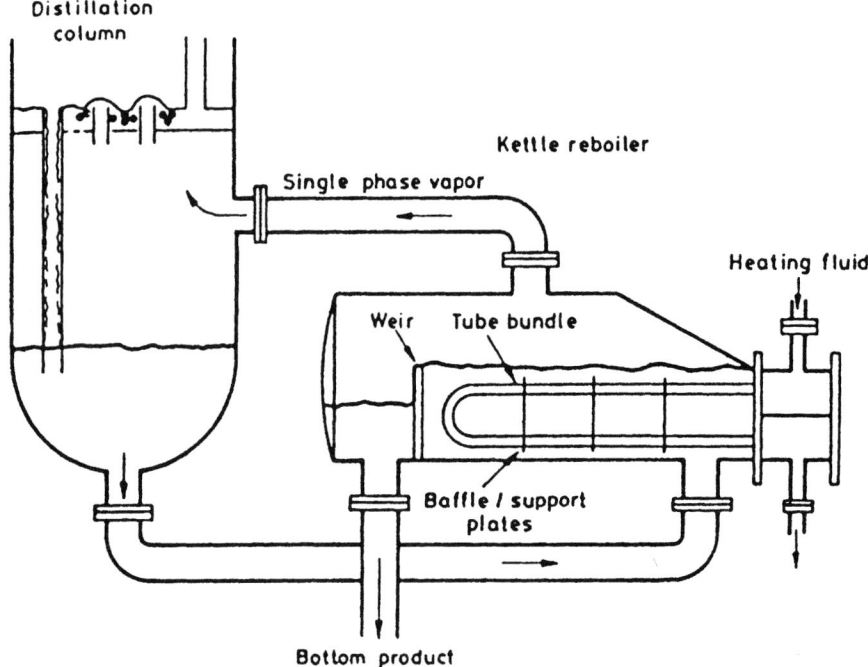

FIGURE 15.78 Kettle reboiler (from Hewitt et al. [13], with permission. Copyright CRC Press, Boca Raton, FL).

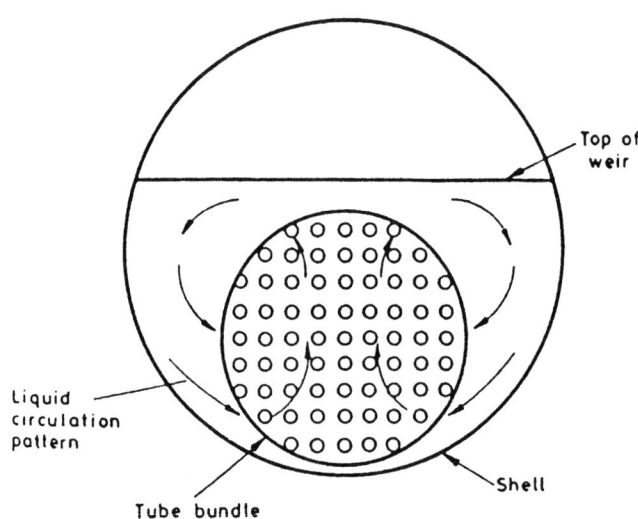

FIGURE 15.79 Induced circulation in a kettle reboiler (from Hewitt et al. [13], with permission. Copyright CRC Press, Boca Raton, FL).

Heat Transfer Below the Critical Heat Flux Limit in Cross Flow Boiling

Pre-CHF Cross Flow Boiling from Single Tubes.   Data for boiling in cross flow over a single tube have been obtained by a number of authors including Yilmaz and Westwater [201], Singh et al. [202, 203], and Fink et al. [204]. It is found that the existence of a cross flow velocity has a considerable effect on the behavior, even at very low velocity. The results obtained are exemplified by those of Fink et al. [204], illustrated in Fig. 15.80. The heat flux at a given wall superheat increases with fluid velocity at low wall superheats but there is much less effect of fluid velocity as the fully developed boiling region is entered at higher wall superheats. Data like those shown in Fig. 15.80 can be predicted by the super-position model of Bergles and Rohsenow [205] in which the heat flux for nucleate boiling at the given wall superheat (calculated using the correlations given previously) is added to the heat flux for single-phase forced convective flow over the tube.

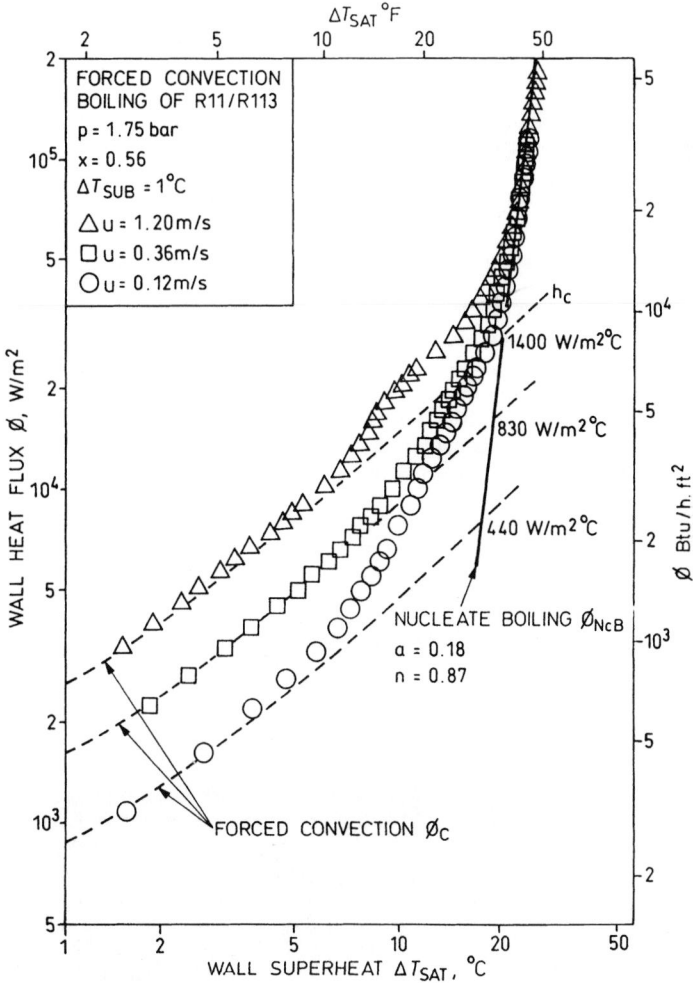

FIGURE 15.80   Effect at cross flow velocity on boiling from a single tube (from Fink et al. [204], with permission from Taylor & Francis, Washington, DC. All rights reserved).

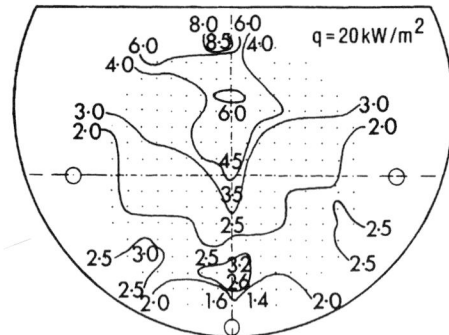

FIGURE 15.81 Contours of heat transfer coefficient (KW/m²K) in a simulated kettle reboiler experiment (from Leong and Cornwell [206], with permission).

Pre-CHF Cross Flow Boiling from Tube Bundles. Cross flow heat transfer in tube bundles is highly complex, particularly when the flow and heat transfer are strongly coupled as in natural circulation in the case of kettle reboilers (see Fig. 15.79). Leong and Cornwell [206] measured the average coefficient on each tube in a model simulating a kettle reboiler. They mounted 241 tubes 19.05 mm in diameter and 25.4 mm long in a square array between vertical pipes in an arrangement simulating a kettle reboiler. Each tube was electrically heated using cartridge heaters and the wall temperatures were determined, giving a value for the heat transfer coefficient. The fluid boiled was refrigerant 113 at atmospheric pressure. Contours of heat transfer coefficient were plotted and are illustrated in Fig. 15.81. The dots represent the positions of the tubes. As will be seen, very large variations in heat transfer coefficient occurred in the bundle, arising from the increase in flow quality. Such large variations do not occur at high mass fluxes and/or high heat fluxes [200]. Nevertheless, the average coefficients for tube bundles in the pre-CHF region are higher than those for single tubes, as is illustrated by the data of Palen et al. shown in Fig. 15.82.

Gorenflo et al. [208] carried out experiments in which the influence on heat transfer of bubbles generated from a lower tube was investigated. Typical results are shown in Fig. 15.83 (the lower tube was actually simulated using a U-shaped heater represented by the black dots in the sketch in the figure). Depending on the equivalent heat flux for the lower tube, the heat transfer coefficient for the upper tube varied considerably as shown. The influence of the lower tube became less at high heat fluxes. The influence of bubbles generated upstream has been investigated experimentally and analytically by Cornwell [209, 210]; bubbles arising from lower tubes impinge on the surfaces of upper tubes and slide around them. Between these sliding bubbles and the surface of the tube is a thin liquid layer that evaporates, contributing considerably to the heat transfer. The calculation of local heat transfer coefficient in tube bundles was considered by Polley et al. [211], who suggested that the coefficient could be calculated from the expression

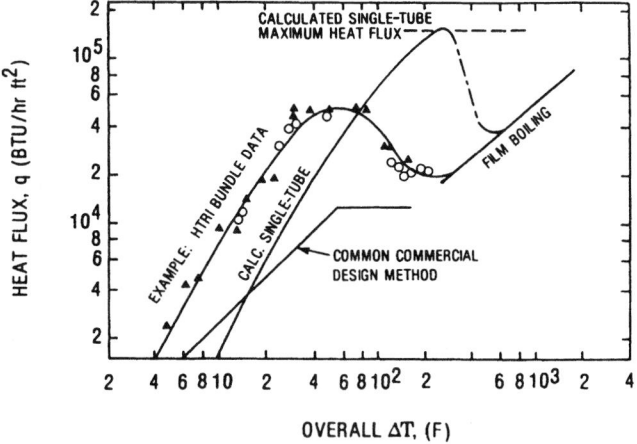

FIGURE 15.82 Comparison of average boiling curve for a tube bundle with boiling curve calculated for a single tube (from Palen et al. [207], with permission).

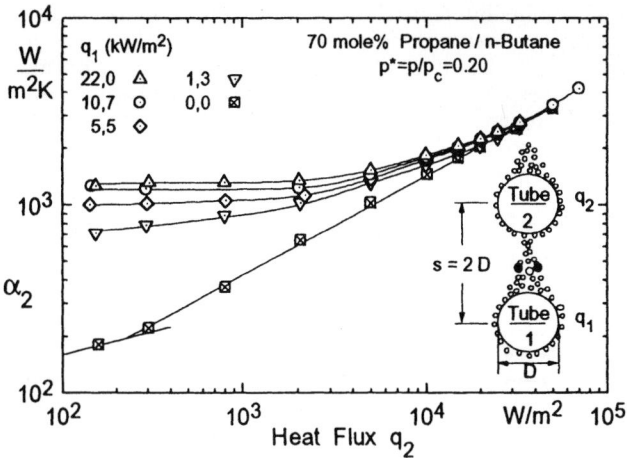

$$h = h_{FC} + h_{NB} \tag{15.187}$$

Here h_{NB} is the nucleate boiling coefficient (calculated from equations analogous to those given previously) and h_{FC} is the forced convective component, which is related to the heat transfer coefficient h_l for the liquid phase flowing alone across the tube bundle by the expression

$$\frac{h_{FC}}{h_l} = \left(\frac{1}{1-\alpha}\right)^{0.744} \tag{15.188}$$

where α is the void fraction (fraction of the free volume in the bundle occupied by the gas phase), which Polley et al. calculated from an equation due to Armand [212] as follows:

$$\alpha = \frac{0.833x}{x + (1-x)\rho_g/\rho_l} \tag{15.189}$$

where x is the flow quality (fraction of the mass flow through the bundle, which is in the vapor phase).

Using Eqs. 15.187–15.189 it is possible to trace the variation of heat transfer coefficient through the bundle. If the heat flux and mass flux are both known, then the local quality x can be calculated from a simple heat balance. If the heat flux has to be calculated (taking into account the heat transfer from the heating fluid inside the tubes and also the local heat transfer coefficient on the boiling side), then the calculation becomes more complex, even if the mass flux is known. However, the mass flux is often governed by natural circulation and is related to the heat flux and vapor generation rate. A fairly complex iterative calculation is then required to establish the heat transfer conditions within the bundle. This approach has been followed by Brisbane et al. [213] and (in a simplified form) by Whalley and Butterworth [214]. Further information about the methods is given by Hewitt et al. [13]. Simplified methods that involve calculating heat transfer to tube bundles in kettle reboilers are presented by Palen [215] and Swanson and Palen [216]. Palen [215] recommends the following expression to obtain the heat transfer coefficient for boiling in a bundle:

$$h = F_b F_c h_{NB} + h_{NC} \tag{15.190}$$

where h_{NC} is the coefficient for natural convection (approximately 250 W/m^{2}K for hydrocarbons and around 1000 W/m^{2}K for water); h_{NC} does not become significant except at very low temperature differences. The factors F_b and F_c in Eq. 15.190 are correction factors to the pool boiling heat transfer coefficient h_{NB} (calculated from the methods given previously) to account for the effect of circulation in the bundle and the influence of multicomponents, respectively. F_c may be calculated by the methods given in the section on the effect of multi-component mixtures (for example, from Eq. 15.118), but Palen suggests that, for design purposes, a simple expression may be used as follows:

$$F_c = \frac{1}{1 + 0.023q''^{0.15}BR^{0.75}} \qquad (15.191)$$

Here q'' is the heat flux and BR is the boiling range (difference between dew point and bubble point temperatures, K). The factor F_b has values typically in the range of 1.0–3.0. At heat fluxes typically above 50 kW/m², F_b is close to unity since the heat transfer is often in the fully developed boiling mode where convection has little effect. However, commercial kettle reboilers and flooded evaporators work typically in the range of 5–30 kW/m² and a typical F_b value for this range would be 1.5. Alternatively, F_b can be calculated from the following approximate formula from Taborek [217]:

$$F_b = 1.0 + 0.1 \left(\frac{0.785 D_b}{C_1 (p_t/D_o)^2 D_o} - 1.0 \right) \qquad (15.192)$$

where D_b and D_o are the bundle and tube diameters and p_t is the tube pitch. The constant C_1 has a value of 1.0 for square and rotated square tube layouts and 0.866 for triangle and rotated triangle layouts.

There has been an increasing interest in boiling from bundles of tubes with enhanced boiling surfaces. Such surfaces were described previously (see for instance Figs. 15.57 and 15.58). The papers of Bergles [135] and Jensen et al. [218] typify recent studies of boiling from bundles of enhanced surface tubing. Thonon et al. [219] report studies of boiling of *n*-pentane on low-fin tubing. Figure 15.84 shows typical results obtained by Jensen et al. [218] for boiling of

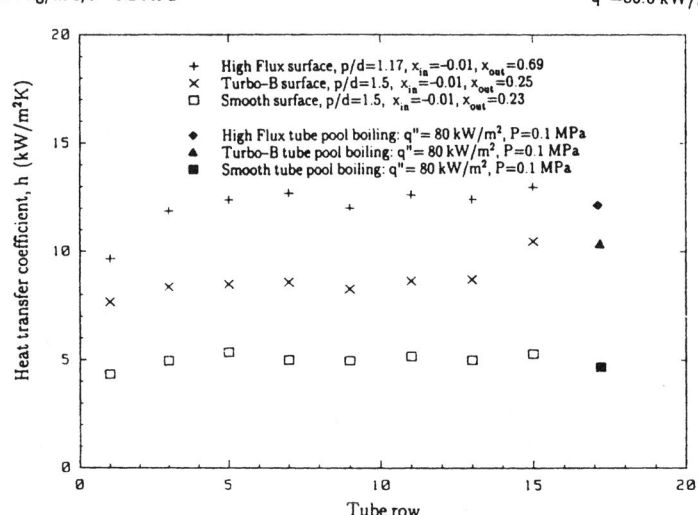

SHELLSIDE BOILING HEAT TRANSFER COEFFICIENTS

G=217 kg/m²s, P=0.2 MPa   q"=80.6 kW/m²

Legend:
+ High Flux surface, p/d=1.17, x_{in}=−0.01, x_{out}=0.69
× Turbo-B surface, p/d=1.5, x_{in}=−0.01, x_{out}=0.25
□ Smooth surface, p/d=1.5, x_{in}=−0.01, x_{out}=0.23

♦ High Flux tube pool boiling: q"= 80 kW/m², P=0.1 MPa
▲ Turbo-B tube pool boiling: q"= 80 kW/m², P=0.1 MPa
■ Smooth tube pool boiling: q"= 80 kW/m², P=0.1 MPa

Heat transfer coefficient, h (kW/m²K) vs Tube row

FIGURE 15.84 Comparison of heat transfer coefficient for boiling refrigerant 113 on smooth and enhanced surfaces in vertical upward flow over a tube bundle (from Jensen et al. [218], with permission from Taylor & Francis, Washington DC. All rights reserved).

refrigerant 113 in a vertical rectangular cross section bundle with electrically heated tubes; this arrangement allows both the heat flux and mass flux to be fixed. Wolverine Turbo-B and Linde High Flux tubes were used and their performance was compared to that of smooth tubes. As will be seen, increases in heat transfer coefficient of the same order as those obtained with single tubes in pool boiling were observed. The data shown in Fig. 15.84 are in the fully developed boiling region and there was little variation from tube to tube.

Critical Heat Flux in Cross Flow Boiling

CHF in Cross Flow Boiling from Single Tubes. Critical heat flux for pool boiling from single tubes was discussed previously. The influence of an upward cross flow over the tubes was investigated by Lienhard and Eichorn [220]; their results are illustrated in Fig. 15.85. In the absence of cross flow, and for small cross flow velocities, the vapor is released from the surface in the form of three-dimensional jets as illustrated (see also Fig. 15.64 for the pool-boiling case). The classical interpretation of the critical heat flux phenomenon is that it occurs when these jets break due to Helmholtz instability and the vapor release mechanism begins to fail, thus allowing the buildup of vapor near the surface (see discussion on p. 15.59–15.60). As the cross flow velocity is increased, there is a transition (as illustrated in Fig. 15.85) to a two-dimensional jet and, subsequently, an increase in critical heat flux with increasing velocity.

A correlation for critical heat flux in cross flow is given by Katto [101] and is as follows:

$$\frac{q''_{crit}}{\rho_l U_\infty i_{lg}} = K(\sigma/\rho_l U_\infty^2 D_o)^{1/3} \tag{15.193}$$

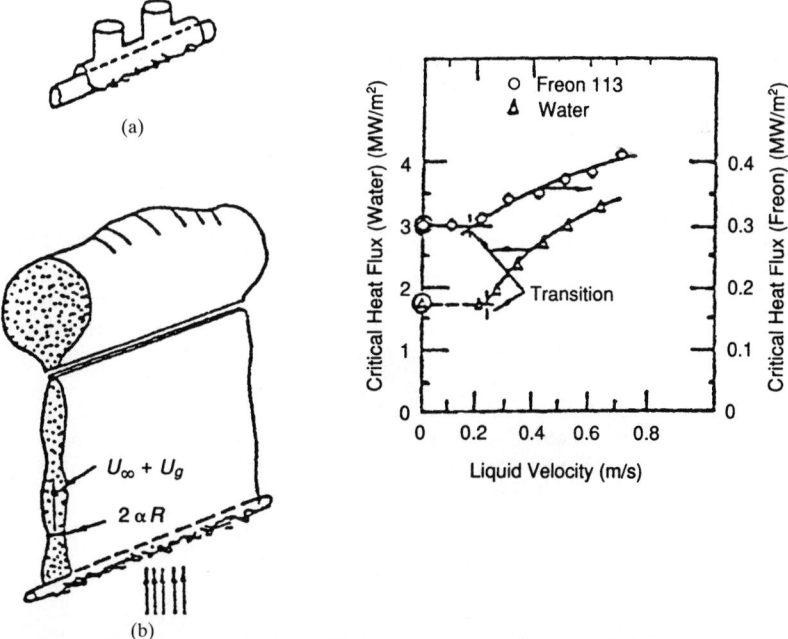

FIGURE 15.85 Critical heat flux in cross flow over cylinders. (*a*) three-dimensional jets; (*b*) two-dimensional jets (from Lienhard and Eichorn [220], with permission from Elsevier Science).

where U_∞ is fluid velocity approaching the tube, D_o is tube diameter, σ is the surface tension, and K is given by

$$K = 0.151(\rho_g/\rho_L)^{0.467}[1 + (\rho_g/\rho_l)]^{1/3} \qquad (15.194)$$

If the value for q''_{crit} is less than that for pool boiling, then the pool-boiling value should be taken (consistent with Fig. 15.85).

A recent study on the influences of mixtures and tube enhancement on cross flow boiling over single tubes is that of Kramer et al. [221].

CHF in Cross Flow Boiling from Tube Bundles. As was shown in Fig. 15.82, the critical heat flux for tube bundles tends to be less than that for single tubes (though the pre-CHF heat transfer coefficients are higher). Critical heat flux in bundles may occur by a number of different mechanisms; obviously, one limiting case would be the critical heat flux mechanisms applicable to pool boiling (see p. 15.58–15.63). Another mechanism might occur if flow into the bottom of the bundle was restricted and the critical heat flux phenomena would be limited by the ingress of liquid at the top of the bundle, the rate of which would be governed by the vapor generation rate (the "flooding" phenomenon). In this latter case, the onset of a critical heat flux phenomenon would occur at the bottom of the bundle, furthest from the point of liquid ingress. However, in most practical situations, liquid ingress at the bottom of the bundle is possible and, as the two-phase flow develops up the bundle, the annular mist flow regime occurs and the limitation is dryout of the liquid film on the surface of the tube (a situation rather similar to that occurring for forced convective boiling in tubes and described in detail on p. 15.123–15.128).

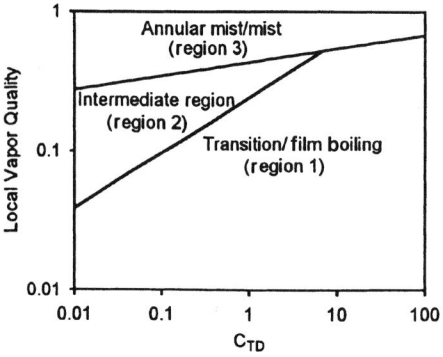

Jensen and Tang [222] have developed a methodology for predicting critical heat flux in cross flow in tube bundles based on two-phase flow regime identification. They present a critical heat flux flow regime map as shown in Fig. 15.86. The map is in terms of local quality (x) and a parameter C_{TD}, which was defined by Taitel and Dukler [223] in evaluating flow patterns for horizontal pipes. C_{TD} is defined as follows:

$$C_{TD} = g\rho_H \frac{(\rho_l - \rho_g)D_o}{G^2} \qquad (15.195)$$

where ρ_H is the homogeneous density, which is related to the densities of the phases and the local quality as follows:

$$\rho_H = \frac{\rho_g \rho_l}{x\rho_l + (1 - x)\rho_g} \qquad (15.196)$$

FIGURE 15.86 Map of critical heat flux regimes for boiling in tube bundles in cross flow (from Jensen and Tang [222], with permission from ASME).

and D_o is the outside diameter of the tubes and G is a mass flux based on the minimum flow area between the tubes.

Jensen and Tang suggest that the annular mist region (where the critical heat flux is governed by film dryout) occurs for qualities greater than x_a given by

$$x_a = 0.432C_{TD}^{0.098} \qquad (15.197)$$

The region in which the mechanisms of critical heat flux are similar to those for pool boiling (region 1 in Fig. 15.86) is bounded by x_a and by a transition quality x_i to an intermediate region (region 2 in Fig. 15.86), with x_i given by

$$x_i = 0.242C_{TD}^{0.396} \qquad (15.198)$$

For region 1 (pool-boiling-type critical heat flux) the critical heat flux for staggered bundles is given by

$$q''_{crit,1} = q''_{crit,t} \exp\left(-0.0322 - \frac{10.1}{\psi^{0.585}}\right) \qquad (15.199)$$

where $q''_{crit,t}$ is the critical heat flux for a single tube in pool boiling for the corresponding conditions and Ψ is given by the expression

$$\Psi = D_o \left[\frac{\rho_H}{\mu_l} \right] \left[\frac{\sigma g(\rho_l - \rho_g)}{\rho_l^2} \right]^{1/4} \tag{15.200}$$

For the annular mist region (region 3), Jensen and Tang suggest the following expression:

$$q''_{crit,3} = 1.97 \times 10^{-5} G i_{lg} C_{TD}^{0.165} \, \text{Re}^{-0.0858} \tag{15.201}$$

where Re is the Reynolds number for the shellside flow ($=GD_o/\mu_l$). If conditions are such that the critical heat flux is occurring in region 2, then the transition qualities x_a and x_l are calculated for the value of C_{TD} estimated from Eq. 15.195, using Eqs. 15.197 and 15.198, respectively. The critical heat fluxes $q''_{crit,a}$ and $q''_{crit,r}$ are calculated from Eqs. 15.201 and 15.199, respectively, corresponding to qualities x_a and x_i, and the value of the critical heat flux in the intermediate region (region 2) is estimated by interpretation as follows:

$$q''_{crit,2} = q''_{crit,i} + (q''_{crit,a} - q''_{crit,i}) \left[\frac{x - x_i}{x_a - x_i} \right] \tag{15.202}$$

Jensen and Tang also give relationships for in-line bundles. Application of Eqs. 15.195–15.202 requires a knowledge of local quality within the bundle. If the mass flux is known, then this can be obtained very simply from a heat balance, but if the mass flux is unknown and has to be calculated (as in kettle reboilers), then recourse must be had to methodologies of the type described by Brisbane et al. [213] and Whalley and Butterworth [214]. As in the case of heat transfer coefficient, simple methods have been developed for prediction of critical heat flux in tube bundles by Palen and coworkers (Palen [215], Palen and Small [224]), who relate the single tube critical heat flux (calculated by the correlations given previously on p. 15.63–15.65) by a simple bundle correction factor Φ_b as follows:

$$q''_{crit} = \Phi_b q''_{crit,t} \tag{15.203}$$

where Φ_b is given by Palen [215] as

$$\Phi_b = \frac{9.74 D_b L}{A} \tag{15.204}$$

where D_b is the bundle diameter, L is the bundle length, and A is the total heat transfer surface area in the bundle. If Eq. 15.204 gives a value higher than unity for Φ_b, then it should be assumed that $\Phi_b = 1.0$.

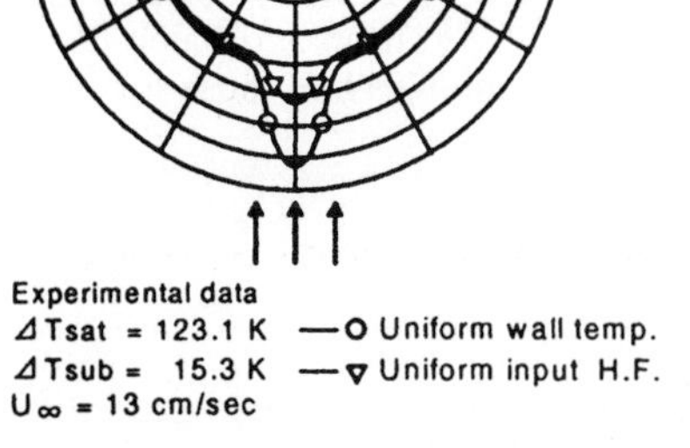

Heat Transfer Beyond the Critical Heat Flux Limit in Cross Flow Boiling

Post-CHF Heat Transfer in Cross Flow Boiling from Single Tubes. There have been relatively few studies of heat transfer in the post-CHF region in the presence of a cross flow. Montasser and Shoji [225] investigated film boiling of refrigerant 113 on a 3.3-cm horizontal heated cylinder. A typical radial distribution of Nusselt number ($=hD_o/k_g$) is shown in Fig. 15.87. The large peak at the stagnation point will be observed (this is where the vapor film is thinnest), and Montasser and Shoji demonstrated the importance of the cross flow velocity on the heat transfer coefficient (which increases with increasing velocity). To obtain a conservative estimate of the heat transfer coefficient in film boiling, therefore, the relationships described earlier (p. 15.71–15.74) for pool boiling can be used.

Nu=3000

2000

1000

Experimental data
Δ Tsat = 123.1 K —O Uniform wall temp.
Δ Tsub = 15.3 K —▽ Uniform input H.F.
U ∞ = 13 cm/sec

FIGURE 15.87 Variation of Nusselt number around tube periphery for film boiling of refrigerant 113 on a horizontal tube (from Montasser and Shoji [225], with permission from Taylor & Francis, Washington, DC. All rights reserved).

An early correlation for the effect of cross flow in film boiling from cylinders was that of Bromley et al. [226], who suggested that, for

$$U_\infty \geq 2\sqrt{gD_o} \tag{15.205}$$

the convective heat transfer coefficient in film boiling (h_c) can be estimated from the expression cone

$$h_c = 2.7 \left[\frac{U_\infty k_g \rho_g (i_{lg} + 0.4 c_{pg} \Delta T_{\text{sat}})}{D_o \Delta T_{\text{sat}}} \right]^{1/2} \tag{15.206}$$

This relationship has not been extensively tested.

Post-CHF Heat Transfer in Cross Flow Boiling from Spheres. There has been a considerable interest in post-CHF heat transfer from spheres (particularly in the transient cooling case) because of the importance of heat transfer to dispersed nuclear fuel materials in postulated nuclear accident situations. Film-boiling heat transfer from spheres and bodies of other shapes is considered theoretically by Witte and Orozco [227] and transient measurements on spheres plunged at a controlled speed into a bath of saturated or subcooled water are reported by Aziz et al. [228]. Transient measurements for spheres in free fall in water are reported by Zvirin et al. [229]. Heat transfer rates in stable film boiling were found to be somewhat higher than the correlation of Witte and Orozco. The most interesting finding was that subcoolings of more than 10 K could cause a transition from stable film boiling to microbubble boiling at temperatures as high as 680°C. The transition to microbubble boiling was accompanied by an order of magnitude increase in heat transfer rate and, sometimes, by the creation of lateral hydrodynamic forces on the spheres.

Post-CHF Heat Transfer in Cross Flow Boiling from Tube Bundles. It is not normal practice to operate tube bundles in the transition or film-boiling regions. However, circumstances can arise where this happens inadvertently. Suppose that the tubeside fluid is at a temperature that is greater than T_{min}, the minimum film-boiling temperature for the shellside fluid. In pre-CHF conditions, the wall temperature (T_w) on the shell side would be much lower than the tubeside fluid temperature due to the fact that heat transfer is producing a temperature drop on the tube side and also in the tube wall material. If, however, the tubeside fluid is introduced before the shellside fluid, then the wall temperature on the shell side may initially be nearly equal to the tubeside fluid temperature (and therefore greater than T_{min}). Thus, when the shellside fluid is introduced, film boiling may be initiated with a lower heat flux, maintaining the tube outer wall temperature at a value greater than T_{min}. Thus, the boiler would underperform considerably; this situation should be borne in mind in operation and design.

A discussion of post-CHF heat transfer in bundles is given by Swanson and Palen [216]. They suggest that for conditions leading to critical heat flux in regions 1 and 2 in Fig. 15.86, the pool-boiling correlations for post-CHF heat transfer described previously could be used to give a conservative estimate of the heat transfer coefficient. Swanson and Palen [216] also observe that for region 3 in Fig. 15.86 (annular mist flows), models like those for the same region in channel flow (see p. 15.134–15.136) might be used.

FORCED CONVECTIVE BOILING IN CHANNELS

Although most of the research on boiling has been conducted (for convenience) with pool-boiling systems, the most important applications are those where boiling occurs in a channel such as a tube in a vertical thermosyphon reboiler or a round tube, or a narrow rectangular passage in a compact heat exchanger, or in longitudinal flow through a bundle of rods as in the fuel elements of a nuclear reactor. The stages of forced convective boiling in a tube are

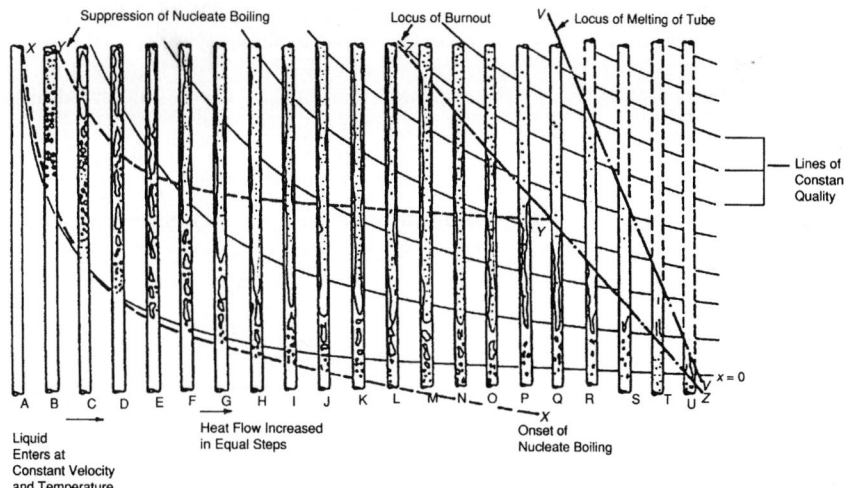

FIGURE 15.88 Stages in the evaporation of a liquid in a tube (from Hewitt et al. [13], with permission. Copyright CRC Press, Boca Raton, FL).

illustrated in Fig. 15.88, which shows the successive patterns of flow and heat transfer as the heat input is increased in equal steps. The heat input at step A is just sufficient to bring the liquid entering the tube at the bottom to the saturation temperature. Case B has twice this heat flux, case C has 3 times this heat flux, etc. The following main features are observed in the diagram:

1. Lines of constant quality x are shown. Quality is normally defined as the ratio of the vapor mass flux G_g to the total mass flux G. However, in an evaporating system, the mass flow rate of vapor has to be calculated from a heat balance. For a channel with cross-sectional area A and periphery S, the enthalpy i at a distance z from the channel inlet is given by

$$i = i_i + \frac{S}{GA} \int_0^z q'' \, dz = x i_g + (1 - x) i_l \tag{15.207}$$

where G is the total mass flux, i_i is the enthalpy of the inlet fluid and i_g and i_l are the saturated enthalpies for the gas and liquid phases, respectively. For flow in a round tube, $S/A = 4/D$, where D is the tube diameter; for uniform heat flux q'', the integral is replaced by $q''z$, where z is the distance from the channel inlet. Local quality can be calculated from the local enthalpy via the relationship

$$x = \frac{i - i_l}{i_g - i_l} = \frac{i - i_l}{i_{lg}} \tag{15.208}$$

where i_{lg} is the latent heat of vaporization. Defining a thermodynamic quality from Eq. 15.208, we see that (in a formal sense) this quality can be negative when the local enthalpy is less than that of the saturated liquid (i.e., conditions are *subcooled*) and greater than unity when the enthalpy is greater than the saturated vapor enthalpy (i.e., the conditions are *superheated*). The actual quality (defined as a ratio of vapor flow to total flow) can be positive in the subcooled region or less than unity in the superheated region, due to the presence of bubbles in subcooled boiling and to the presence of droplets coexisting with superheated vapor, respectively. These thermodynamic nonequilibrium effects are often of great importance, as will be shown below.

2. The onset of nucleate boiling (line XX in Fig. 15.88) occurs above the $x = 0$ line at low heat flux (i.e., there is a net bulk superheat of the liquid) but at qualities less than 0 for high heat fluxes, corresponding to the region of subcooled nucleate boiling. For heat transfer to a single-phase liquid, the wall superheat $(\Delta T_{sat})_W$ can be calculated from

$$(\Delta T_{sat})_W = T_W - T_{sat} = \frac{q''}{h_l} + T_B - T_{sat} \qquad (15.209)$$

where T_B is the bulk temperature and h_l is the heat transfer coefficient for the single-phase liquid. For a constant heat flux in a round tube, T_B can be calculated from the expression

$$T_B = T_{in} + \frac{4q''z}{Gc_{pl}D} \qquad (15.210)$$

where c_{pl} is the specific heat capacity of the liquid phase. To calculate the distance z required for the onset of nucleate boiling, $(\Delta T_{sat})_W$ is equated to the value of $(\Delta T_{sat})_{W,crit}$, which may be calculated from the equations given earlier (for instance, the expression from Davis and Anderson [30], Eq. 15.32).

3. The addition of vapor to the flow leads to the existence of a two-phase vapor-liquid flow in the channel; it is the existence of this flow and its interaction with the heat transfer processes that makes forced convective boiling so different from pool boiling. As more and more vapor enters the flow, the two-phase flow patterns or regimes develop in the following succession:

Bubble flow. Here there is a suspension of bubbles within the liquid continuum.

Slug flow. Here the bubbles have coalesced to form large bubbles (sometimes called Taylor bubbles) that are separated by slugs of liquid, the latter often containing a dispersion of smaller bubbles.

Churn flow. Here the slug flow regime breaks down to form a region in which liquid is carried upward in large waves, between which a liquid film at the wall may fall downward.

Annular flow. Here there is a liquid film carried upward on the walls of the channel, with the vapor flowing in the center of the channel. Usually the vapor contains a dispersion of fine droplets of liquid entrained from the liquid film.

Dispersed droplet flow. In this regime, which can only normally exist in heated systems, the liquid phase is completely dispersed as droplets in the vapor.

A detailed discussion of flow regimes in two-phase flow is beyond the scope of this chapter. A more detailed discussion is given by Hewitt [230].

4. At higher qualities, the heat transfer rate by direct forced convection through the liquid film in annular flow may be so great that it may become impossible to maintain a wall temperature that is sufficiently high to sustain nucleate boiling. Line YY in Fig. 15.88 represents the onset of suppression of nucleate boiling, which, therefore, only exists between XX and YY. The conditions for suppression of nucleate boiling can be calculated in a manner similar of that for calculating its onset. Thus, the wall superheat is given by

$$(\Delta T_{sat})_W = \frac{q''}{h_{FC}} - T_{sat} \qquad (15.211)$$

where h_{FC} is the two-phase forced convective heat transfer coefficient (see the following section). h_{FC} increases with quality so that $(\Delta T_{sat})_W$ falls progressively along the channel. At a given point, corresponding to complete suppression of nucleate boiling, $(\Delta T_{sat})_W$ falls below the value of $(\Delta T_{sat})_{W,crit}$ calculated from, say, Eq. 15.32.

Figure 15.89 shows an expanded view of the regimes of flow and heat transfer in a single tube (which might correspond roughly to tube N in Fig. 15.88). Also shown in Fig. 15.89 is the typical behavior of wall and fluid temperature along the channel. An alternative representation of the various regimes is given in Fig. 15.90, which is plotted in terms of a map relating the regions of heat transfer to heat flux and quality. Finally, Fig. 15.91 shows the variation of heat transfer coefficient with heat flux and quality. Here the fluid temperature is defined as the saturation temperature for saturated conditions and as the mixed mean liquid temperature for subcooled conditions. In this particular diagram, the critical heat flux transition is termed departure from nucleate boiling (DNB) in the region where nucleate boiling is not suppressed by the stage at which the critical phenomenon occurs (see Fig. 15.88). In Fig. 15.91, the heat transfer coefficient is shown as being constant in the saturated nucleate boiling region. However, the coefficient may decline with increasing quality; a more recent flow boiling map showing detailed trends of this type has been published by Kandlikar [375].

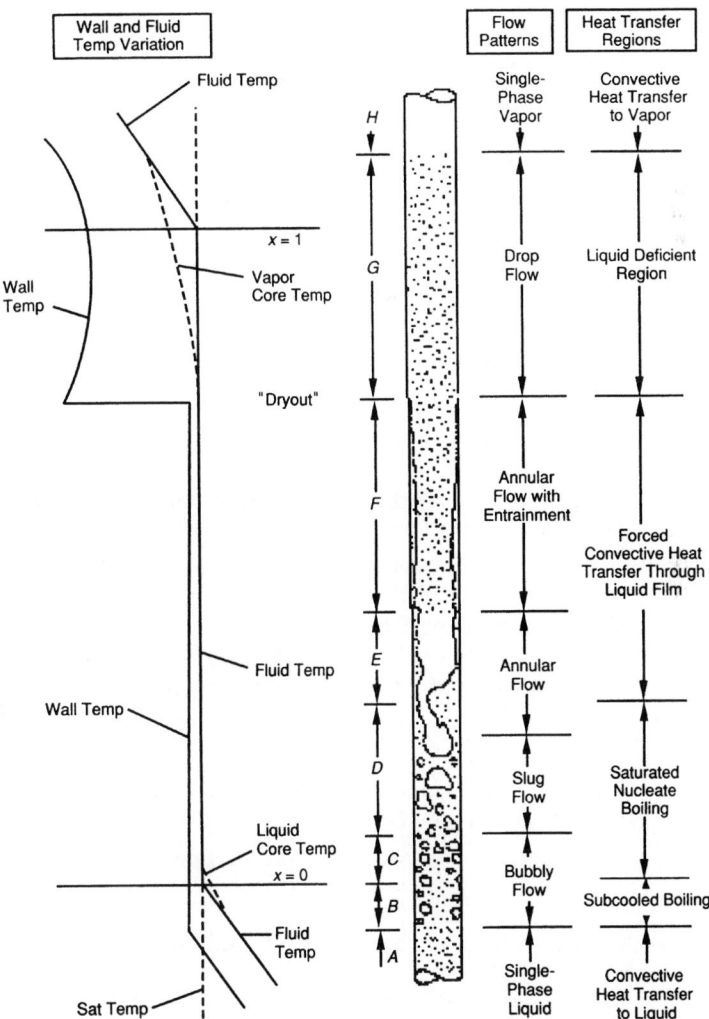

FIGURE 15.89 Flow regimes, heat transfer regimes, and wall and fluid temperatures in forced convective boiling (from Collier and Thome [3], by permission of Oxford University Press).

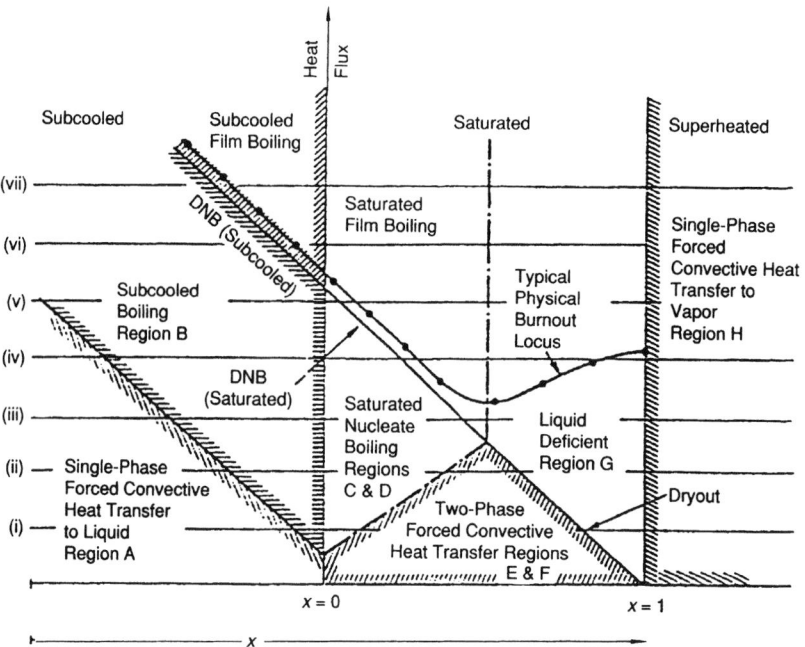

FIGURE 15.90 Regions of operation of the various regimes of heat transfer in terms of heat flux and quality (from Collier and Thome [3], by permission of Oxford University Press).

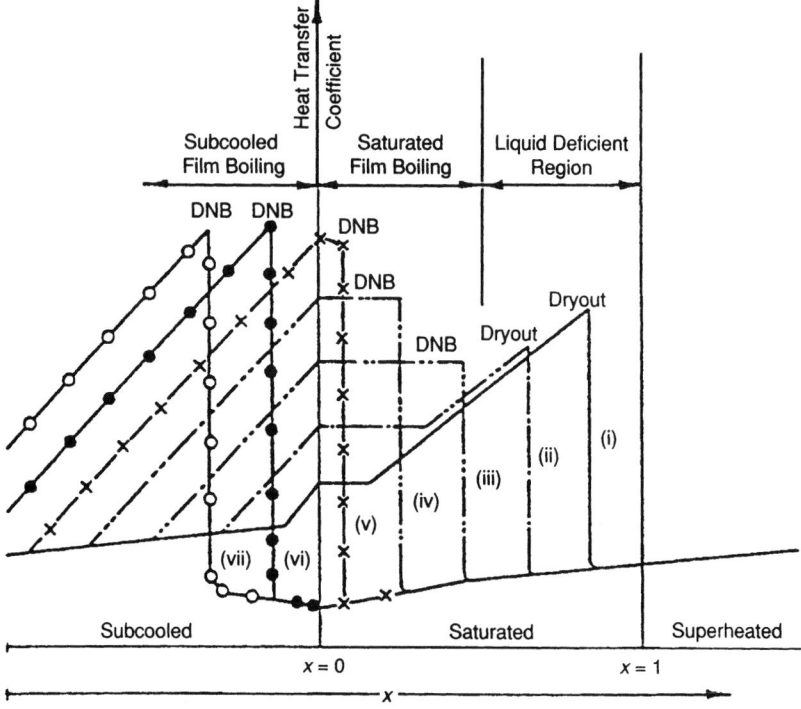

FIGURE 15.91 Variation of heat transfer coefficient with quality at various heat flux levels. Note that the numbers on the curves are cross references to the various levels indicated in Fig. 15.90 (from Collier and Thome [3], by permission of Oxford University Press).

Heat Transfer Below the Critical Heat Flux Limit in Forced Convective Boiling in Channels

Parametric Effects in Pre-CHF Forced Convective Boiling in Channels

Effect of Mass Flux, Heat Flux, and Quality (Subcooling). In general, one must consider the effects of mass flux (G), heat flux (q''), and quality (x) in forced convective boiling to be strongly interrelated so that they have to be considered simultaneously. Here we include subcooling within the overall definition of quality; conditions where the average enthalpy of the bulk fluid is less than that for saturated liquid can be thought of as being at negative quality. There is no sudden change between negative and positive qualities due to the fact that bubbles form at negative qualities due to subcooled boiling and the phenomena are continuous across the zero-quality boundary. The best way of introducing the phenomena is to take some specific examples as follows:

1. Figure 15.92 shows data (plotted in the form of the boiling curve, i.e., as heat flux versus wall superheat) for fixed subcooling at various velocities. For this particular set of data, the results at the lowest velocity agreed well with a pool-boiling curve obtained for the same surface, and the results at high velocity asymptoted to an extrapolation of this pool-boiling curve at high heat fluxes. At lower wall superheats, the results for the higher liquid velocities become dominated by forced convective heat transfer. Although this particular set of data lends credence to the idea of using correlations for fully developed pool-boiling for forced convective situations, it should be stated that this is not always the case.

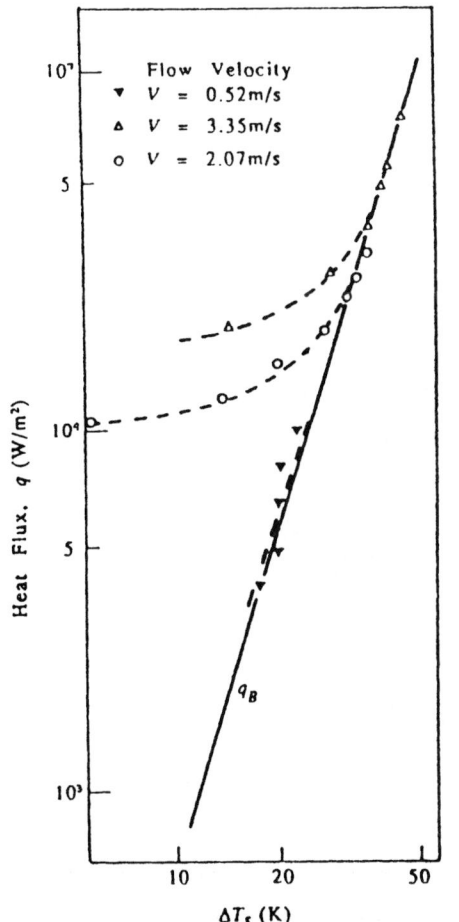

FIGURE 15.92 Subcooled flow boiling curve for fixed subcooling and for a range of velocities (from Bergles and Rohsenow [29], with permission of ASME).

2. An alternative way of presenting the data is to plot the heat transfer coefficient at a given quality (subcooling) as a function of the heat flux with velocity as a parameter. An example of this is shown in Fig. 15.93, which is taken from the paper by Müller-Steinhagen and Jamialahmadi [231]. At low heat fluxes, the data show an approach to a constant heat transfer coefficient (characteristic of the single-phase forced convective region), this coefficient being independent of heat flux. At high heat fluxes, on the other hand, the heat transfer coefficient depends only on heat flux and is independent of velocity. This is consistent with the idea of a fully developed boiling region where heat transfer is dominated by near-wall effects, as in fully developed pool boiling.

3. Figure 15.94 shows data obtained by Robertson and Wadekar [232] for ethanol boiling in a vertical 10-mm internal diameter tube. Here, the heat transfer coefficient is plotted as a function of quality for two mass fluxes with heat flux as a parameter. Again, the data asymptote to a line representing pure forced convection as the quality increases.

4. Kenning and Cooper [233] report data for forced convective boiling of water in a vertical tube. At high qualities, the heat transfer coefficient becomes independent of heat flux and varies approximately linearly with quality as illustrated in Fig. 15.95.

These examples illustrate the interplay of heat flux, quality, and mass flux, and it is this, as we shall see, that presents a particular challenge in correlating or predicting forced convective boiling data.

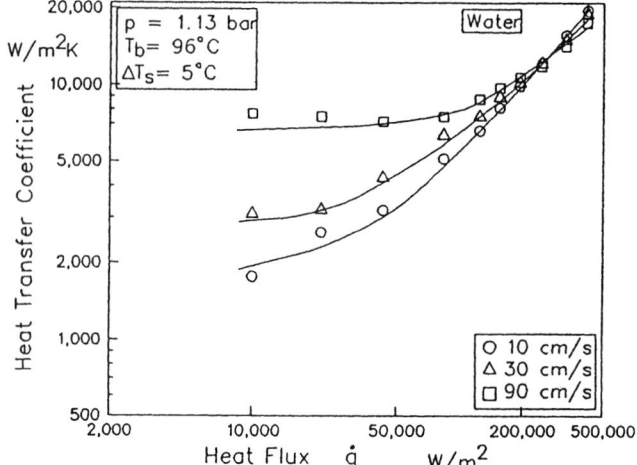

FIGURE 15.93   Influence of heat flux and flow velocity on the heat transfer coefficient for subcooled flow boiling of water (from Müller-Steinhagen and Jamialahmadi [231], with permission of Taylor & Francis, Washington, DC. All rights reserved).

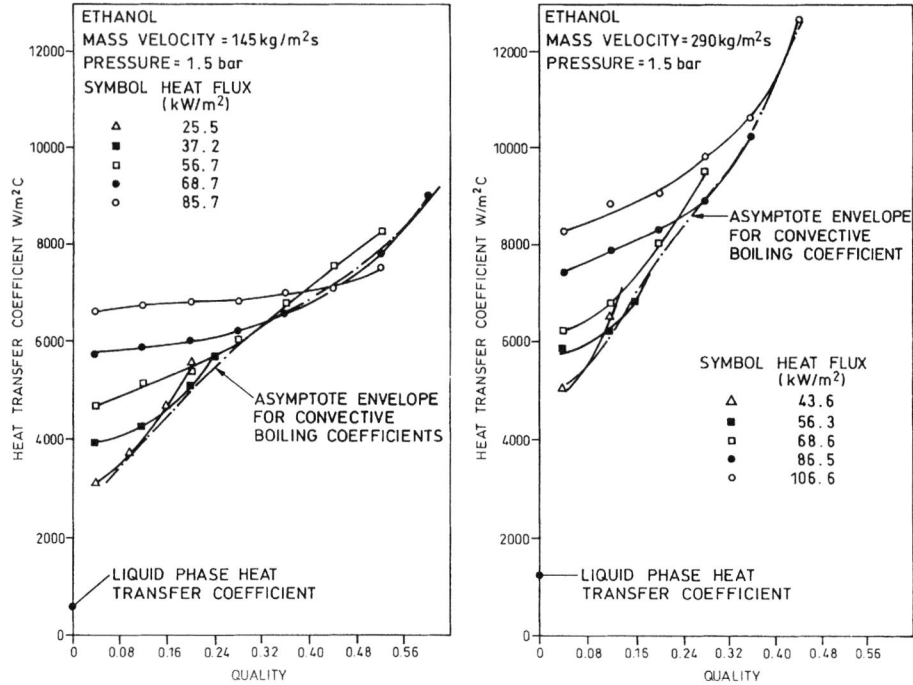

FIGURE 15.94   Heat transfer coefficients for the boiling of ethanol in a vertical 10-mm diameter tube (from Robertson and Wadekar [232]).

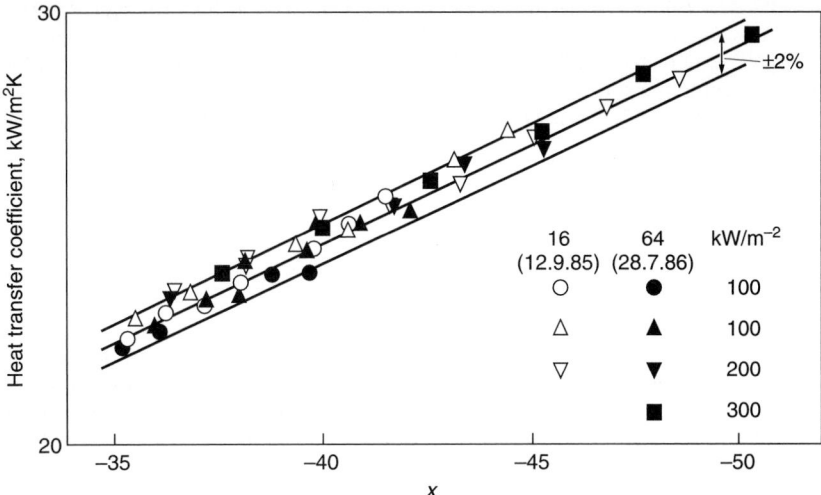

FIGURE 15.95 Variation of heat transfer coefficient with quality in the forced convective region (from Kenning and Cooper [233], with permission of Elsevier Science).

Effect of Surface. As discussed previously, there is a strong effect of heater surface on nucleate boiling heat transfer in pool boiling. However, it seems probable that the effect is less in forced convection; Brown [76] carried out experiments with the same cylindrical heating surface both in pool boiling and in forced convection. For the forced convection tests, the heater was mounted to form the internal surface of an annulus through which water was passed. The pool boiling results are shown in Fig. 15.36. The equivalent forced convection data are shown in Fig. 15.96. Though there are still significant effects of surface roughness, these are much less than in the case of the pool boiling experiments.

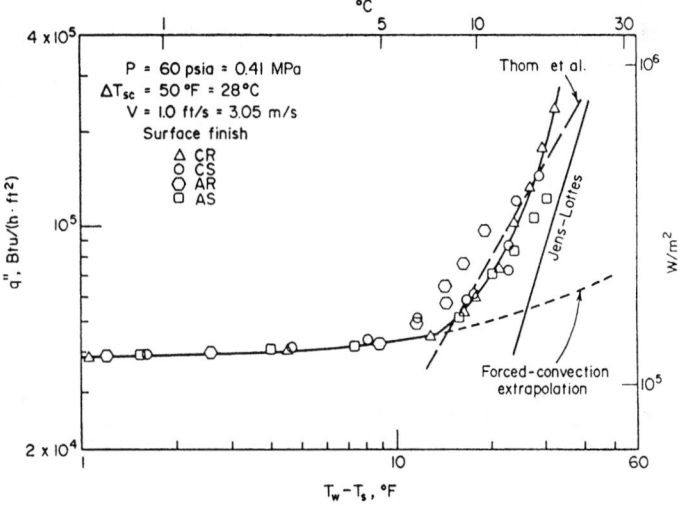

FIGURE 15.96 Effect of surface finish on forced convective boiling (results of Brown [76], from Rohsenow [2], with permission of The McGraw-Hill Companies).

Effect of Gravity. At high fluid velocities, it may be expected that the effect of gravity would be much less than that in the case of pool boiling. In order to evaluate the effects of gravity, Kirk and Merte [234] and Kirk et al. [235] investigated the effect of the orientation of a boiling surface placed in a channel through which refrigerant 113 was flowed. As the orientation was changed from horizontal through to downward-facing, the heat transfer coefficient in forced convective nucleate boiling increased at low velocity (4.1 cm/s) but was unaffected by the heater orientation at higher velocities (32.4 cm/s). These authors argue that the same result would be obtained in microgravity.

Hysteresis. Just as in the case of pool boiling, hysteresis in nucleate boiling heat transfer can occur with forced convection. An early set of data demonstrating such effects was that of Abdelmessih et al. [236], and more recent measurements are typified by those of Bilicki [237]. At a given point in the channel, the onset of nucleate boiling may be delayed until a certain heat flux is achieved. When the heat flux is reduced, nucleate boiling may be retained down to heat fluxes much lower than those required for initiation of boiling. Bilicki carried out experiments in which a bubbly mixture of refrigerant 21 liquid and vapor was fed to the channel. Hysteresis in nucleate boiling still occurred under these circumstances, indicating that the hysteresis effect is governed by near-wall conditions and can still occur despite the existence of vapor in the core of the flow.

Mechanisms for Pre-CHF Heat Transfer in Forced Convective Boiling in Channels. It was convenient to discuss the onset of nucleate boiling and subsequent bubble growth for *both* pool boiling and forced convective boiling. The sections covering these subjects (p. 15.9–15.18) should be consulted for information in this area. Furthermore, in fully developed nucleate boiling in forced convection, the detailed near-wall phenomena are likely to be similar to those encountered in pool boiling, and the reader is referred to p. 15.42–15.45 for a detailed description of these phenomena. In the present section we will deal with the interrelationships between pre-CHF heat transfer and the respective flow regimes, starting with bubbly flow and continuing with slug flow and annular flow.

Bubbly Flow and Subcooled Boiling. The generally accepted picture of void generation in subcooled boiling leading to bubbly flow under bulk boiling (positive thermodynamic quality conditions) is sketched in Fig. 15.97. After a zone of single-phase heat transfer, nucleate boiling begins at a distance z_n from the start of the heated section. Initially (region 1) bubbles are formed locally on the wall and are condensed in the near-wall region before they can penetrate into the core of the flow. Essentially, these bubbles remain located close to their point of original nucleation. However, at a distance z_d, bubbles begin to detach from the wall and mix with the flow, condensing relatively slowly, and form a bubbly two-phase flow. Eventually, at a distance z_{bulk} from the start of the heated section, the thermodynamic equilibrium quality (calculated from Eq. 15.208) reaches zero as the mean bulk fluid enthalpy reaches that of the saturated liquid. However, at this point, there is already significant void fraction (fraction of the pipe occupied by the vapor phase), and this implies that the liquid phase is still subcooled and is in contact with vapor bubbles that are still condensing. It is only after a distance of z_{eq} that the liquid and vapor phases are in equilibrium (i.e., at a condition in which $x > 0$). The distance z_n can be calculated by combining Eqs. 15.209 and 15.210 and, say, the Davis and Anderson [30] expression for the onset of nucleate boiling (Eq. 15.32), which gives the following result:

$$z_n = \frac{GD_{cpl}}{4}\left[\left(\frac{8\sigma T_{\text{sat}}v_g}{k_l i_{lg}q''}\right)^{1/2} - \frac{1}{h_l} + \frac{T_{\text{sat}} - T_{in}}{q''}\right] \qquad (15.212)$$

The next stage is to calculate z_d, the point at which bubbles begin to depart from the heated surface, and a typical model for predicting z_d is that of Saha and Zuber [239]. They suggest the following relationship for the quality at the point of bubble departure $x(z_d)$:

$$x(z_d) = -0.0022 \frac{q''Dc_{pl}}{i_{lg}k_l} \qquad \text{for } \frac{GDc_{pl}}{k_l} < 70000 \qquad (15.213)$$

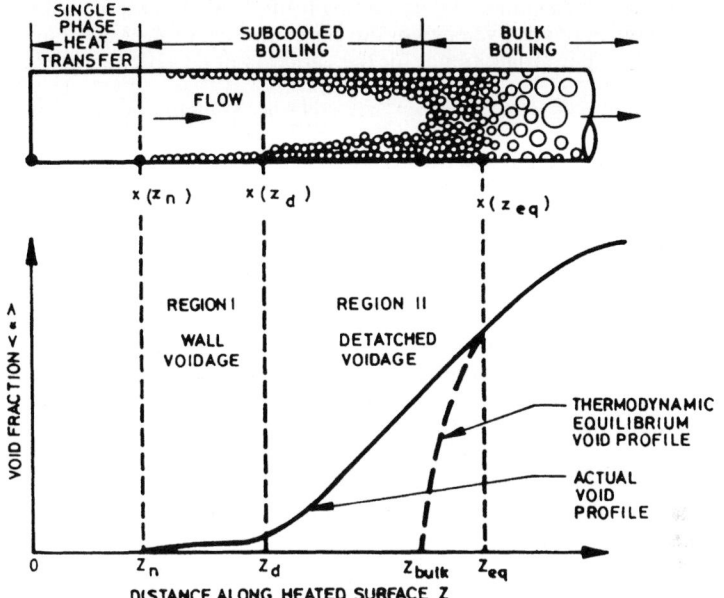

FIGURE 15.97 Void formation in subcooled boiling (from Hewitt [238], with permission of The McGraw-Hill Companies).

$$x(z_d) = -154 \frac{q''}{Gi_{lg}} \quad \text{for} \quad \frac{GDc_{pl}}{k_l} > 70000 \tag{15.214}$$

It should be noted that $x(z_d)$ is negative, indicating that the point of bubble departure occurs while the bulk fluid is still subcooled. z_d is then given by

$$z_d = \frac{GD[i_{lg}x(z_d) + i_l - i_{in}]}{4q''} \tag{15.215}$$

where i_{in} is the inlet enthalpy.

The next step in the calculation is to determine the *actual* quality in the region beyond the bubble detachment. There have been a number of attempts to predict this from mechanistic models in which the rates of evaporation near the wall and condensation in the core of the flow are estimated and the quality is evaluated; surveys of early versions of such models are given by Mayinger [240] and Lahey and Moody [241]. A more recent example of such an approach is that of Zeitoun and Shoukri [242]. A simpler class of methods uses a profile fit; these methods are exemplified by that of Levy [243], who relates the actual quality x_a to the local equilibrium quality x (calculated from Eq. 15.208) and $x(z_d)$ (calculated from Eq. 15.213 or 15.214) as follows:

$$x_a = x - x(z_d) \exp\left[\frac{x}{x(z_d)} - 1\right] \tag{15.216}$$

If the local actual quality x_a is known, then the local void fraction α can be calculated from standard relationships for two-phase void fraction. For example, the relationship of Zuber and Findlay [244] may be employed as follows:

$$\alpha = \frac{x_a\rho_l}{C_o\{x_a\rho_l + [1 - x_a]\rho_g\} + \rho_l\rho_g u_{GU}/G} \tag{15.217}$$

where C_o is a parameter accounting for the distribution of void fraction in the flow and u_{GU} is the mean relative velocity of the gas compared to the bulk fluid velocity. An expression that fits the appropriate trends for the variation of C_o with quality is that of Dix [245]:

$$C_o = \beta\left[1 + \left(\frac{1}{\beta} - 1\right)^b\right] \qquad (15.218)$$

where β is the volumetric flow ratio, which is related to local flow quality by the expression

$$\beta = \frac{x_a}{x_a + [1 - x_a]\rho_g/\rho_l} \qquad (15.219)$$

and b is related (in the correlation by Dix) to density ratio as follows:

$$b = \left(\frac{\rho_g}{\rho_l}\right)^{0.1} \qquad (15.220)$$

The mean relative velocity u_{GU} can be calculated from the following expression by Lahey and Moody [241]:

$$u_{GU} = 2.9\left[\frac{(\rho_l - \rho_g)\sigma g}{\rho_l^2}\right]^{0.25} \qquad (15.221)$$

where σ is the surface tension and g is the acceleration due to gravity.

Slug Flow. For void fractions (fractions of the channel volume occupied by the vapor phase) greater than around 0.3, bubble coalescence leads to the formation of slug flow (see Fig. 15.98), in which large bubbles (often referred to as Taylor bubbles) are formed separated by slugs of liquid. Forced convective evaporative heat transfer in slug flow was studied by Wadekar and Kenning [246], who proposed a model for this region, taking its upper boundary as being given by McQuillan and Whalley [247] in terms of the dimensionless parameter j_g^*:

$$j_g^* = \frac{Gx}{[gD(\rho_l - \rho_g)\rho_g]^{1/2}} > 1.0 \qquad (15.222)$$

Wadekar and Kenning modeled the heat transfer in two regions, namely single-phase convective heat transfer in the liquid slug region and falling film heat transfer in the bubble region. The predicted results are compared with measurements by Kenning and Cooper [233] in Fig. 15.99; the results were chosen to be such that nucleate boiling heat transfer was not occurring and the heat transfer was with the forced convection only. The solid lines in Fig. 15.99 show the predictions from the theory and the dashed lines show the heat transfer coefficient expected for the liquid phase flowing alone in the pipe; all of the lines terminate with the condition of $j_g^* = 1$ (beyond which annular flow will take place). The predictions are in good agreement with the model at low mass fluxes but underpredict the data for high mass fluxes. This could be the result of the breakdown of the slug flow regime into churn flow (where there is a continuous vapor core as in annular flow but where the liquid layer at the wall is traversed by large, upwardly moving waves between which flow reversal occurs in the liquid layer).

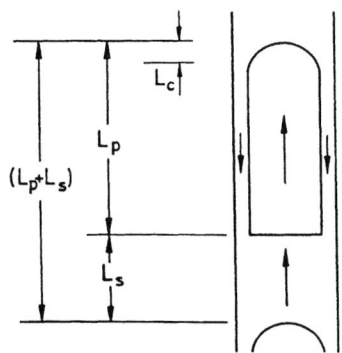

FIGURE 15.98 Bases for description of evaporative heat transfer in slug flow (from Wadekar and Kenning [246], with permission from Taylor & Francis, Washington, DC. All rights reserved).

The case of slug flow in horizontal pipes has been considered by Sun et al. [248]. Here the situation is complicated by circumferential drainage of the liquid film layer between the slugs. The effects of nucleate boiling were also taken into account in this model.

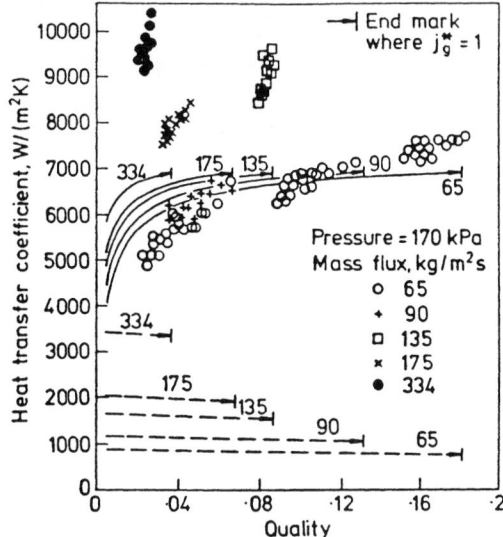

FIGURE 15.99 Comparison of experimental and predicted heat transfer coefficients for slug flow (from Wadekar and Kenning [246], with permission from Taylor & Francis, Washington, DC. All rights reserved).

Annular Flow. In annular flow, nucleate boiling at the wall tends to be strongly suppressed (though not always totally so—see p. 15.16–15.17 for a discussion of nucleation behavior in this regime). Thus the situation is usually regarded as being dominated by convective heat transfer from the heated surface to the interface between the liquid film and the vapor core. This type of heat transfer would be expected to be independent of heat flux, and results such as those shown in Fig. 15.95 tend to confirm this view. However, a completely different point of view was put forward by Messler [249], who suggested that the improvement in heat transfer in forced convection was a result of enhancement rather than suppression of nucleation and bubble growth. Though nucleation at the wall could be suppressed, Messler suggested that secondary nucleation at the interface could occur as a cyclic process as illustrated in Fig. 15.100 (Messler [250]). Bubbles departing from the liquid film would leave behind vapor nuclei that themselves would grow within the superheated film liquid, giving rise to further bubble releases and further nuclei creating a chain reaction as shown in Fig. 15.100. The process could be initiated by drop impact on the surface or by gas bubbles being entrained in wave action on the liquid film interface. It could be argued that with secondary nucleation the heat transfer coefficient would not increase with increasing heat flux as in the case of nucleate boiling with wall nucleation, where the number of nucleation sites increases with the flux. Thus, constancy of the transfer coefficient cannot be taken as a disproof of the secondary nucleation mechanism.

Work at the Harwell Laboratory and at Imperial College in London has focused on investigating the secondary nucleation model by comparing condensation and evaporation heat transfer in fully developed annular steam-water mixture flows. The flow is brought into an equilibrium condition by the use of an adiabatic section upstream of the heat transfer test section. Fluid heating and cooling was used to investigate evaporation and condensation under precisely the same conditions. Although early experiments of this type (Chan [56]) appeared to indicate some differences between evaporation and condensation, later, more accurate experiments showed that there was little difference, as exemplified in Fig. 15.101 (Sun and Hewitt [251]). Although more investigation of these phenomena is required, it seems unlikely that secondary nucleation is the explanation for the heat transfer enhancement in the forced

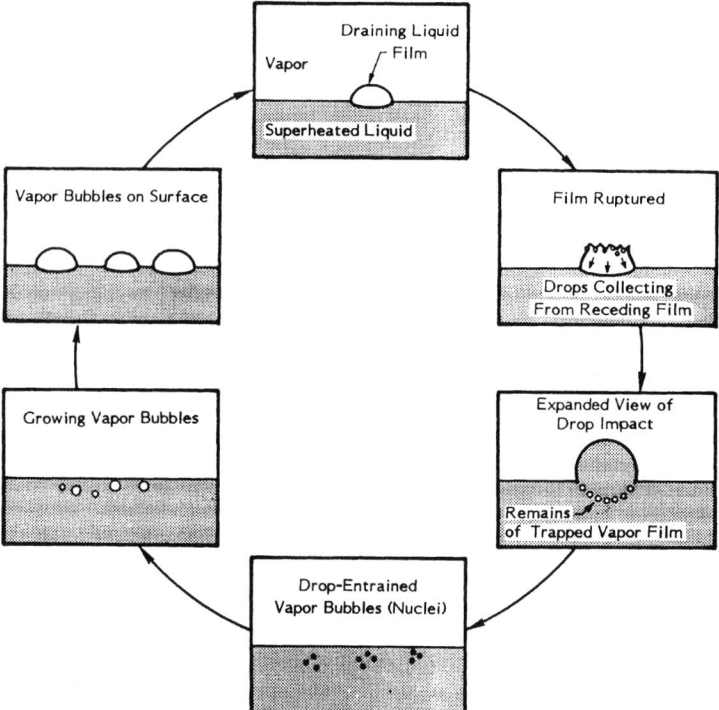

FIGURE 15.100   Chain reaction of secondary nucleations (from Messler [250], with permission from ASME).

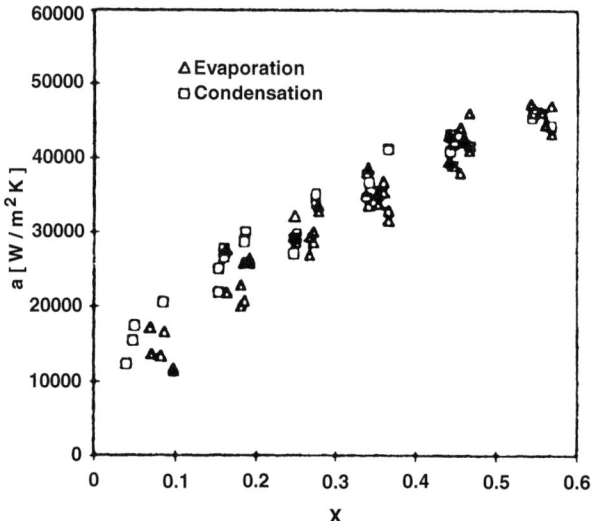

FIGURE 15.101   Comparison of evaporative and condensing heat transfer coefficients under identical conditions (from Sun and Hewitt [251], with permission).

convective region. We shall return to the topic of mechanisms in annular flow heat transfer in discussing predictions later.

Correlations for Pre-CHF Forced Convective Boiling in Channels.   A very wide variety of correlations have been developed over the past five decades for the heat transfer coefficient in forced convective evaporation in channels. Again, it would be impossible to present all of the alternative correlations in detail; rather, examples are given of the various generic types of correlation, an attempt being made to select the correlations corresponding with those most widely used. For much more detailed descriptions and discussions of correlations, the reader is referred to the books of Collier and Thome [3] and Carey [4]. The correlations can be grouped into a number of categories, as follows:

1. *Fluid-specific correlations.* Here the data for a particular fluid are correlated in a simple form in terms of the system parameters. Although such correlations do not attempt to represent the competing mechanisms, they are often useful for quick design purposes.

2. *Correlations for pure forced convection.* At low heat fluxes and/or high mass fluxes, nucleate boiling heat transfer becomes negligible and the heat transfer is by conduction/convection from the wall to the interface where the evaporation is occurring. Such forced convective heat transfer represents an important limiting case.

3. *Power-law interpolation correlations.* Here the total heat flux q'' is calculated from a formula of the type

$$q'' = (q''^{n}_{FC} + q''^{n}_{NB})^{1/n} \tag{15.223}$$

where q'' is the heat flux due to forced convection, given by:

$$q''_{FC} = h_{FC}(T_W - T_B) \tag{15.224}$$

where h_{FC} is the forced convective heat transfer coefficient, T_W is the wall temperature, and T_B is the bulk temperature of the fluid. q''_{NB} is the heat flux due to nucleate boiling, which can be expressed by the equation

$$q''_{NB} = h_{NB}(T_W - T_{\text{sat}}) \tag{15.225}$$

where T_{sat} is the saturation temperature. In the saturated (positive quality) region, T_B is defined as being equal to T_{sat} and, for this region, therefore, one may combine Eqs. 15.223–15.225 to give an alternative form for the parallel interpolation type of correlation:

$$h = (h^{n}_{FC} + h^{n}_{NB})^{1/n} \tag{15.226}$$

where $h = q''/\Delta T_{\text{sat}}$ is the heat transfer coefficient including both forced convective and nucleate boiling contributions.

4. *Suppression correlations.* Here, an attempt is made to correlate the suppression of either the nucleate boiling or the forced convective component and to correct for this suppression in calculating the total heat transfer coefficient.

5. *General empirical correlations.* In this approach, no attempt is made to base the correlations on nucleate pool-boiling correlations combined in some way with forced convective correlations. Rather, the data are correlated independently using a number of dimensionless groups.

In what follows, examples are given of each of the above approaches.
  Fluid-Specific Correlations.   Since water is the most widely boiled fluid in forced convection (i.e., in power generation systems), it was natural that correlations specific to water were developed at an early stage. Perhaps the best known of such correlations is that of Jens and Lottes [252], which relates the wall superheat to the heat flux and pressure as follows

$$\Delta T_{\text{sat}} = 25 q''^{0.25} \exp(-P/62) \tag{15.227}$$

It should be noted that the dimensions of the quantities in Eq. 15.227 are not in SI units; ΔT_{sat} is in K, q'' is in MW/m^2, and P is in bar. An improved equation, giving a closer fit to experimental data, was suggested by Thom et al. [253] as follows:

$$\Delta T_{sat} = 22.5 q''^{0.5} \exp(-P/87) \tag{15.228}$$

where the units are identical to those used for Eq. 15.227. These equations are valid up to pressures around 200 bar and may be used for subcooled boiling and for forced convective boiling when the nucleate boiling contribution is dominant. They are extremely useful in obtaining a rough estimate of temperature differences.

Correlations for Pure Convection. It is usually assumed that if the heat transfer coefficient is independent of heat flux, the mode of heat transfer is pure forced convection (see the preceding section on mechanisms). Earlier correlations for forced convective evaporative heat transfer had the form

$$F = \frac{h_{FC}}{h_l} = a \left(\frac{1}{X_{tt}} \right)^b \tag{15.229}$$

where h_l is the heat transfer coefficient for the liquid phase flowing alone in the pipe and X_{tt} is the Martinelli parameter. h_l is often calculated using the equation of Dittus and Boelter [254] as follows:

$$\mathrm{Nu}_l = \frac{h_l D}{k_l} = 0.023 \mathrm{Re}_l^{0.8} \, \mathrm{Pr}_l^{0.4} \tag{15.230}$$

$$\mathrm{Nu}_l = 0.023 \left[\frac{G(1-x)D}{\mu_l} \right]^{0.8} \left[\frac{\mu_l c_{pl}}{k_l} \right]^{0.4} \tag{15.231}$$

The Martinelli parameter is given by

$$X_{tt} = \left[\frac{(dp/dz)_l}{(dp/dz)_g} \right]^{1/2} \approx \left(\frac{1-x}{x} \right)^{0.9} \left(\frac{\rho_g}{\rho_l} \right)^{0.5} \left(\frac{\mu_l}{\mu_g} \right)^{0.1} \tag{15.232}$$

where $(dp/dz)_l$ and $(dp/dz)_g$ are the pressure gradients for the liquid and vapor phases flowing alone in the pipe. Various values for the constants a and b in Eq. 15.229 have been reported in the literature as follows:

	a	b
Dengler and Addoms [255]	3.5	0.5
Guerrieri and Talty [256]	3.4	0.45
Collier and Pulling [257]	2.5	0.7

The values found by Collier and Pulling [257] are probably the most reliable.

In his correlation for combined forced convective and nucleate boiling heat transfer, Chen [107] derived a graphical relationship between F ($=h_{FC}/h_l$) and $1/X_{tt}$ as illustrated in Fig. 15.102.

Butterworth [258] fitted the graphic relationship with the expression

$$F = \frac{h_{FC}}{h_l} = 2.35 \left[\frac{1}{X_{tt}} + 0.213 \right]^{0.736} \tag{15.233}$$

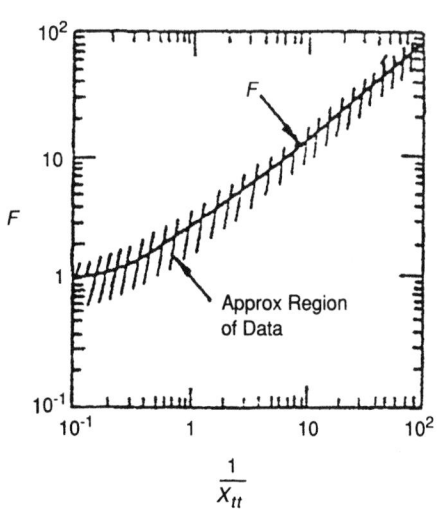

FIGURE 15.102 Graphical correlation of F (Eq. 15.229) as a function of $1/X_{tt}$ (Eq. 15.232) (from Chen [107]).

An alternative correlation in terms of flow quality x and liquid-to-vapor density ratio (ρ_l/ρ_g) is given by Steiner and Taborek [259] as follows:

$$F = \frac{h_{FC}}{h_l} = \left[(1 - x)^{1.5} + 1.9x^{0.6} \left(\frac{\rho_l}{\rho_g} \right)^{0.35} \right]^{1.1} \tag{15.234}$$

Power-Law Interpolation Correlations. Correlations of the form of Eq. 15.223 or 15.226 have been widely used to account for the simultaneous action of forced convection and nucleate boiling heat transfer. The first region considered here is that of forced convective subcooled boiling. Detailed discussions of the various models for this region are given in the books of Collier and Thome [3] and Carey [4]. The relationship between heat flux q'' and wall temperature T_W for a point along the channel at which the bulk fluid temperature is $T_l(z)$ is illustrated in Fig. 15.103. Line ABCI represents forced convective single-phase heat transfer, with q'' being given by

$$q''_{FC} = h_l[T_W - T_l(z)] \tag{15.235}$$

where h_l is calculated from Eq. 15.231 (since the conditions are subcooled, x is taken as 0).

Curve FDIE in Fig. 15.103 is calculated for fully developed nucleate boiling (using, for instance, the correlations for pool nucleate boiling given previously). The two curves intersect at point I, where the heat flux is q''_I. At point E, the heat transfer is dominated by nucleate boiling and the heat flux at this condition (q''_{fdb}) is approximately 1.4 q''_I (Engelberg-Forester and Grief [260]). The onset of boiling occurs when the wall temperature reaches T_{ONB}; this condition can be estimated using the methods given earlier (for instance, from Eq. 15.32). The region between the onset of boiling and the attainment of fully developed nucleate boiling is often termed the *partial boiling region*. Bergles and Rohsenow [29] suggest that in this region, the heat flux due to nucleate boiling should be calculated from the expression

$$q''_{NB} = q''_R - q''_D \tag{15.236}$$

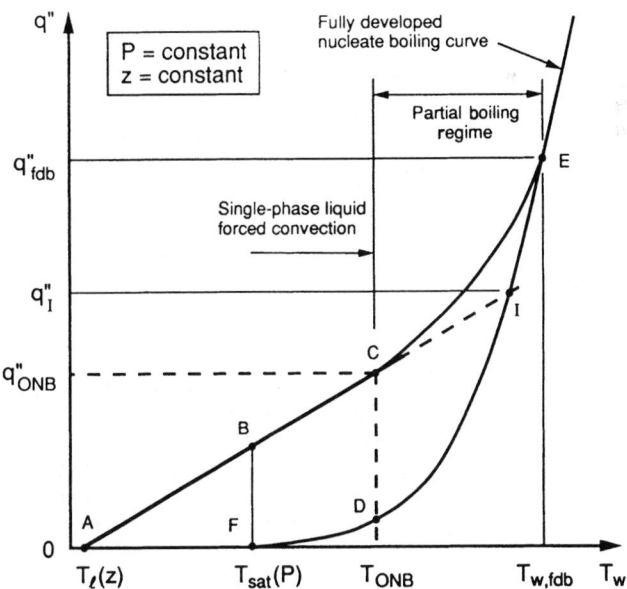

FIGURE 15.103 Relationship between heat flux and wall temperature at fixed bulk liquid temperature in subcooled boiling (from Carey [4], with permission from Taylor & Francis, Washington, DC. All rights reserved).

where q_R'' is the heat flux calculated (for this particular wall temperature T_W) from the Rohsenow [108] correlation (Eq. 15.81) for pool nucleate boiling. q_D'' is the heat flux calculated from the same correlation at the point of onset of nucleate boiling (T_{ONB} in Fig. 15.103). Bergles and Rohsenow further suggest a value of $n = 2$ in Eq. 15.223, which leads to the following expression for the heat flux in the partial boiling region:

$$q'' = \left\{ q_{FC}''^2 + \left[q_R'' \left(1 - \frac{q_D''}{q_R''} \right) \right]^2 \right\}^{1/2}$$
(15.237)

The correlation approach used by Bergles and Rohsenow [29] was extended into the quality region by Bjorge et al. [261].

For the saturated boiling region, the power-law interpolation method has been used by, for instance, Steiner and Taborek [259] and Wattelet [262]. Steiner and Taborek deal with forced convective boiling in the quality region and hence use Eq. 15.226 as a basis. Their expression is as follows:

$$h = (h_{FC}^3 + h_{NB}^3)^{1/3}$$
(15.238)

where h_{FC} is calculated from Eq. 15.234 and h_{NB} is calculated using the correlations of Gorenflo and coworkers [116–119] given previously (see Eqs. 15.90–15.94).

Suppression Correlations.   Probably the most widely used correlation for forced convection boiling heat transfer is that of Chen [107], which is in the simple form:

$$h = h_{FC} + h_{NB}$$
(15.239)

where h_{FC} is calculated as Fh_l, with F being obtained from Fig. 15.102 or from Eq. 15.233. The nucleate boiling coefficient h_{NB} is given as

$$h_{NB} = Sh_{FZ}$$
(15.240)

where h_{FZ} is the nucleate boiling coefficient calculated from the pool boiling correlation of Forster and Zuber [106] (Eq. 15.80) and S is a suppression factor that Chen correlated graphically as a function of the product $\mathrm{Re}_l \, F^{1.25}$, where Re_l is the Reynolds number for the liquid phase flowing alone in the pipe and F is obtained from Fig. 15.102 or Eq. 15.233. The graphic correlation for S is shown in Fig. 15.104 and may be calculated from the expression (Butterworth [258]):

$$S = \frac{1}{1 + 2.53 \times 10^{-6} (\mathrm{Re}_l \, F^{1.25})^{1.17}}$$
(15.241)

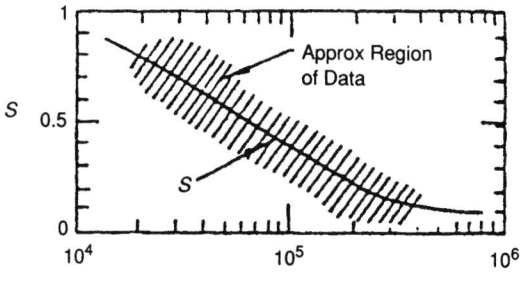

FIGURE 15.104   Graphic correlation for suppression factor S (from Chen [107]).

Although primarily designed for the quality region, the Chen correlation may also be applied to subcooled boiling. In this case, S is calculated from Eq. 15.241 by taking $F = 1$. The heat flux is then calculated from the expression

$$q'' = h_l(T_W - T_B) + h_{NB}(T_W - T_{sat}) \qquad (15.242)$$

Though the assumption by Chen that nucleate boiling is suppressed at increasing mass fluxes and qualities seems consistent with the experimental findings reviewed earlier (see, for instance, Figs. 15.15–15.17), it could also be postulated that as the wall temperature increases, the bubble population on the surface increases and the area available for forced convection is therefore reduced. Wadekar [263] presents a correlation based on the idea that the full nucleate boiling coefficient is retained but the forced convective coefficient is gradually reduced as the area available for forced convection decreases. A similar idea had been put forward previously for subcooled boiling by Bowring [264]. Probably the real situation lies between the extremes of suppression of nucleate boiling and reduction of forced convection. The contribution from forced convection may be gradually reduced as the situation of fully developed nucleate boiling (with the heat transfer controlled by near-wall conditions) is approached. Similarly, the nucleate boiling contribution is suppressed as the point of onset of nucleate boiling is neared. Possibly, there is scope for the development of correlation that includes both effects.

General Empirical Correlations. The correlations described above were either fluid specific or related to correlations for forced convection evaporation and/or pool nucleate boiling, respectively. An alternative approach is to develop correlations based only on the data for forced convective boiling. The most widely used correlation of this form is that of Shah [265], who correlated data for convective flow boiling in both vertical and horizontal pipes in the form

$$\frac{h}{h_l} = fn(\text{Co, Bo, Fr}_{le}) \qquad (15.243)$$

where Co is a convection number, defined as

$$\text{Co} = \left(\frac{1-x}{x}\right)^{0.8}\left(\frac{\rho_g}{\rho_l}\right)^{0.5} \qquad (15.244)$$

The boiling number Bo is defined as

$$\text{Bo} = \frac{q''}{Gi_{lg}} \qquad (15.245)$$

The Froude number for the total flow flowing as liquid (Fr_{le}) is given as

$$\text{Fr}_{le} = \frac{G^2}{\rho_l g D} \qquad (15.246)$$

The original correlation was in graphic form (Fig. 15.105), though Shah [266] provided a set of equations describing the curves. Shah's chart contains three regions as follows:

1. A *pure nucleate boiling regime,* which occurs at high values of Co; in this regime, the value of h/h_l is constant independent of Co.

2. A *bubble suppression regime* in which h/h_l increases with decreasing Co. Line AB represents the limit of pure convective boiling. Note that the curves for given values of Bo asymptote to line AB as Co reduces.

3. Conditions to the right of line AB apply to vertical tubes and also to horizontal tubes for the case where Fr_{le} is greater than around 0.04. In horizontal tubes, however, there is gravitational separation of the phases for $\text{Fr}_{le} > 0.04$, and this results in the upper part of the tube being dry. The third region in the graphical correlation, therefore, is for pure convec-

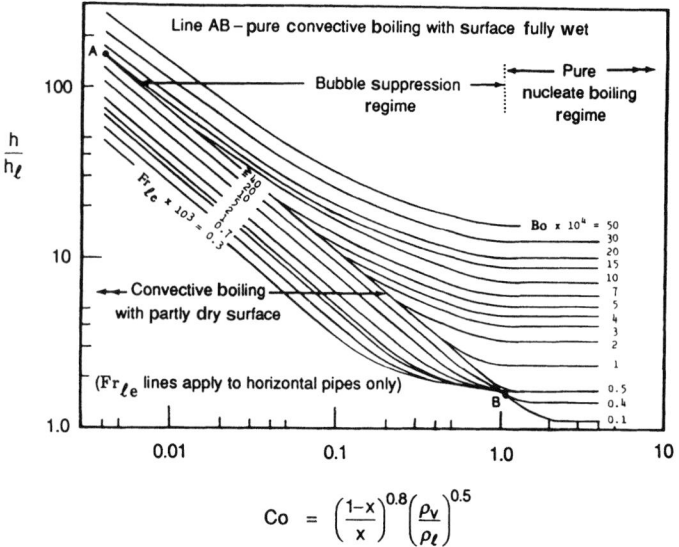

$$Co = \left(\frac{1-x}{x}\right)^{0.8}\left(\frac{\rho_v}{\rho_\ell}\right)^{0.5}$$

FIGURE 15.105   Shah correlation for forced convective boiling (from Shah [265], with permission).

tive boiling with a partially dry surface, the consequential reduction in h/h_l being related to Fr_{le} as shown. Of course, it would be possible for nucleate boiling to occur in a partially wetted horizontal tube, but this is not taken into account in this correlation.

Kandlikar [267] used the same dimensionless groups as Shah [265] (Eqs. 15.244–15.246) and produced a correlation in the following form:

$$\frac{h}{h_l} = C_1\, Co^{C_2}\, (25Fr_{le})^{C_5} + C_3\, Bo^{C_4}\, F_{fl} \tag{15.247}$$

For the case of vertical flows, $C_5 = 0$ and there is no influence of Froude number. The various constants in Eq. 15.247 are given as follows:

	Convective Region	Nucleate Boiling Region
C_1	1.136	0.6683
C_2	−0.9	−0.2
C_3	667.2	1058.0
C_4	0.7	0.7
C_5 (horizontal flows)	0.3	0.3
C_5 (vertical flows)	0	0

The coefficient is calculated for both the convective and nucleate boiling regions using the respective values of C_1–C_5, and the highest calculated coefficient is taken. The value of parameter F_{fl} depends on the fluid used. Kandlikar gives the following list of values:

Fluid	F_{fl}	Fluid	F_{fl}
Water	1.0	R-113	1.30
R-11	1.30	R-114	1.24
R-12	1.50	R-134a	1.63
R-13B1	1.31	R-152a	1.10
R-22	2.20	Nitrogen	4.70
		Neon	3.50

A correlation that also falls in this category is that of Gungor and Winterton [268] (which supercedes an earlier correlation similar in form [269]). For vertical tubes, the correlation has the form

$$\frac{h}{h_l} = \left[1 + 3000 \text{Bo}^{0.86} + 1.12 \left(\frac{x}{1-x} \right)^{0.75} \left(\frac{\rho_l}{\rho_g} \right)^{0.41} \right] \qquad (15.248)$$

and for horizontal tubes, the correlation is modified to include Fr_{le} as follows:

$$\frac{h}{h_l} = \left[1 + 3000 \text{Bo}^{0.86} + 1.12 \left(\frac{x}{1-x} \right)^{0.75} \left(\frac{\rho_l}{\rho_g} \right)^{0.41} \right] \text{Fr}_{le}^{(0.1 - 2\text{Fr}_{le})} \qquad (15.249)$$

The correlation was compared with a wide range of data and was shown to perform better than many of the earlier (more complex) correlations.

Overview. The selection of a forced convective boiling correlation is a matter of personal choice (and often also of the tradition within a particular organization!). The Gungor and Winterton [268] correlation has the advantage of simplicity and performs as well as (and in many cases better than) some of the more complex correlations. The Chen [107] correlation seems nearest to describing the physics of the situation, and the Steiner and Taborek [259] correlation is straightforward and extrapolates to the right limits for forced convection and fully developed nucleate boiling. The Kandlikar [267] correlation has been carefully fitted to a wide range of data and has the advantage of reflecting better some of the observed trends (e.g., reducing heat transfer coefficient with increasing quality). There still seems to be scope to develop a correlation in which the influences of nucleate boiling on forced convection are considered *in addition to* the influences of forced convection on nucleate boiling.

Prediction of Pre-CHF Forced Convective Boiling in Channels. The problems of predicting heat transfer in flow boiling are enormous, and this is why recourse must usually be made to empirical correlations of the type discussed in the preceding text. However, it is useful to pursue prediction methods as a means of gaining better understanding of the phenomena, and a great deal of work has been done in this area. In the following, we first consider the work that has been done on modeling of the forced convective component of the heat transfer, specifically in annular flow. We then consider the prediction of the nucleate boiling component.

Forced Convective Heat Transfer in Annular Flow. Over the past four decades, a great deal of work has been done on the prediction of forced convective heat transfer in annular flow. This work has been done not only in the context of evaporation, but also for the case of condensation. In the simplified presentation here, the following assumptions are made:

1. The liquid film on the channel wall is thin compared to the radius of the channel and the flow in the film can therefore be considered two-dimensional (i.e., equivalent to a film on a flat plate).
2. It is assumed that the interfacial shear stress τ_i is the dominant force on the liquid film and that gravitational effects can be ignored. In reality (and particularly close to the onset of annular flow), the shear stress may vary considerably across the liquid film.
3. The liquid film is assumed to be uniformly distributed around the channel periphery. This is likely to be a reasonable assumption in the case of vertical round tubes; for horizontal tubes, the film tends to be redistributed, with a thicker film at the bottom of the channel and a thinner film at the top. Also, for rectangular and square channels, there is a tendency for the liquid to move toward the corners of the channels under the influence of surface tension.

Analytical studies in which these assumptions are not made are described by Hewitt [270], Hewitt and Hall-Taylor [271], and Sun et al. [272]. However, for most cases of annular flow, the assumptions are reasonably closely obeyed. The system parameters used in the analysis are sketched in Fig. 15.106. The analysis proceeds in the following steps:

1. The flow rate per unit periphery of the liquid film (Γ) is calculated from the expression

$$\Gamma = \frac{G(1-x)(1-E)(\pi D^2/4)}{\pi D} = \frac{GD(1-x)(1-E)}{4} \qquad (15.250)$$

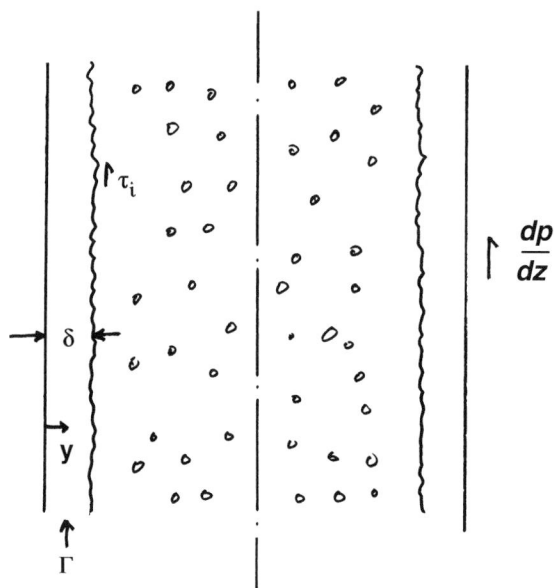

FIGURE 15.106   Parameters of annular flow in a channel.

where E is the fraction of the liquid phase that is entrained as droplets in the gas core. We will return to a discussion of the calculation of entrainment fraction in the context of critical heat flux prediction later.

2. The interfacial shear stress τ_i is calculated. If the thickness of the liquid film is neglected compared to the tube diameter, then τ_i is given by

$$\tau_i = -\frac{D}{4}\left(\rho_c g + \frac{dp}{dz}\right)$$

(15.251)

where dp/dz is the pressure gradient and ρ_c is the density of the vapor/droplet core. The pressure gradient may be estimated from standard correlations (see Hewitt [273]) or from more complex models of the core flow (see, for instance, Owen and Hewitt [274]). The core density is usually estimated by assuming that the droplets and vapor form a homogeneous mixture, which leads to the expression:

$$\rho_c = \frac{\rho_g \rho_l[(1-x)E + x]}{\rho_l x + \rho_g(1-x)E}$$

(15.252)

Often, $\rho_c g$ is small compared to dp/dz.

3. The thickness of the liquid film δ is calculated corresponding to the flow rate per unit periphery Γ. In a general case, this is done iteratively with the use of the following steps:

a. A value of δ is guessed.

b. The velocity profile from $y = 0$ to $y = \delta$ is then calculated by integrating the expression

$$\tau_i = \mu_{\text{eff}} \frac{du}{dy}$$

(15.253)

where μ_{eff} is the effective viscosity in the film.

c. From the calculated velocity profile, Γ is estimated by integrating the velocity profile as follows:

$$\Gamma = \rho_l \int_0^\delta u \, dy \tag{15.254}$$

d. The value of Γ calculated from Eq. 15.254 is compared with the value calculated in step 1. If they do not agree, the value of δ is readjusted and the calculation is repeated until agreement is obtained.

4. The convective heat transfer coefficient $h = q''/(T_W - T_{sat})$ is determined by integrating the expression

$$q'' = -k_{eff} \frac{dT}{dy} \tag{15.255}$$

with boundary conditions $T = T_W$ and $y = 0$ and $T = T_{sat}$ at $y = \delta$. k_{eff} is the effective thermal conductivity in the liquid film.

In laminar flow, $\mu_{eff} = \mu_l$ and $k_{eff} = k_l$ and an explicit solution is possible. Thus the local velocity in the liquid film is given by

$$u = \int_0^y \frac{\tau_i}{\mu_l} \, dy = \frac{\tau_i y}{\mu_l} \tag{15.256}$$

and the flow rate per unit periphery is given by

$$\Gamma = \rho_l \int_0^\delta u \, dy = \rho_l \delta^2 \tau_i / 2\mu_l \tag{15.257}$$

thus giving the value of δ as

$$\delta = (2\Gamma \mu_l / \tau_i \rho_l)^{1/2} \tag{15.258}$$

from which the heat transfer coefficient h can be estimated as

$$h = \frac{k_l}{\delta} = (k_l^2 \tau_i \rho_l / 2\Gamma \mu_l)^{1/2} \tag{15.259}$$

Over most of the range of practical interest, the flow in the liquid film is turbulent and the calculation becomes more complicated since both μ_{eff} and k_{eff} vary with distance from the wall. The normal practice is to express these parameters as follows:

$$\mu_{eff} = \mu_l + \varepsilon_m \rho_l \tag{15.260}$$

$$k_{eff} = k_l + \varepsilon_h \rho_l c_{pl} \tag{15.261}$$

where ε_m and ε_h are eddy diffusivities for momentum and heat, respectively. It is normally assumed that $\varepsilon_m = \varepsilon_h = \varepsilon$. The intergration of Eqs. 15.253 and 15.255 thus requires values for ε. A very large range of relationships is available for turbulence in single-phase flows, and these relationships are often used for the case of annular flow. One of the simplest of these relationships is based on the universal velocity profile, which relates a nondimensional velocity u^+ to a nondimensional distance from the wall y^+. u^* and y^* are defined as follows:

$$u^+ = u/u^* \tag{15.262}$$

$$y^+ = \frac{yu^* \rho_l}{\mu_l} \tag{15.263}$$

where u^* is the friction velocity given by

$$u^* = \sqrt{\tau_i / \rho_l} \tag{15.264}$$

Relationships for ε that are consistent with the universal velocity profile are as follows:

$$\varepsilon = 0 \qquad \text{for } y^+ < 5 \tag{15.265}$$

$$\varepsilon = \frac{\mu_l}{\rho_l}\left(\frac{y^+}{5} - 1\right) \qquad 5 < y^+ < 30 \tag{15.266}$$

$$\varepsilon = \frac{\mu_l}{\rho_l}\left(\frac{y^+}{2.5} - 1\right) \qquad y^+ > 30 \tag{15.267}$$

A somewhat more complex set of relationships for ε was used by Hewitt [270], but the advantage was marginal and the solution required numerical integration. However, even with the simpler expressions for ε (Eqs. 15.265–15.267) the solutions for δ are implicit and require iteration as indicated. However, the solutions for δ can be represented in dimensionless form, relating δ^+ defined by

$$\delta^+ = \frac{u^* \delta \rho_l}{\mu_l} \tag{15.268}$$

to a film Reynolds number Re_f defined as follows:

$$\text{Re}_f = \frac{4\Gamma}{\mu_l} \tag{15.269}$$

Kosky and Staub [275] derived the following explicit relationship between δ^+ and Re_f by fitting the results obtained by the iterative calculation

$$\delta^+ = 0.7071\text{Re}_f^{0.5} \qquad \text{for } \text{Re}_f \leq 50 \tag{15.270}$$

$$\delta^+ = 0.6323\text{Re}_f^{0.5286} \qquad \text{for } 50 \leq \text{Re}_f \leq 1483 \tag{15.271}$$

$$\delta^+ = 0.0504\text{Re}_f^{0.875} \qquad \text{for } \text{Re}_f > 1483 \tag{15.272}$$

The equation for heat transfer can also be nondimensionalized. Combining Eqs. 15.255 and 15.261, we have:

$$q'' = -(k_l + \varepsilon \rho_l c_{pl})\frac{dT}{dy} \tag{15.273}$$

which can be transformed into the nondimensional form

$$1 = \left(\frac{1}{\text{Pr}_l} + \frac{\varepsilon \rho_l}{k_l}\right)\frac{dT^+}{dy^+} \tag{15.274}$$

where Pr_l is the liquid Prandtl number and the nondimensional temperature T^+ is given by

$$T^+ = \frac{c_{pl}\rho_l u^*}{q''}(T_w - T) \tag{15.275}$$

Introducing the relationships for ε given by Eqs. 15.265–15.267, Eq. 15.274 may be integrated to produce explicit relationships for T_δ^+, the value of T^+ at $y = \delta$ ($y^+ = \delta^+$), as follows:

$$T_\delta^+ = \delta^+ \text{Pr}_l \qquad\qquad \text{for } \delta^+ \leq 5 \tag{15.276}$$

$$T_\delta^+ = 5[\text{Pr}_l + \ln\{1 + \text{Pr}_l\,(\delta^+/5 - 1)\}] \qquad \text{for } 5 \leq \delta^+ \leq 30 \tag{15.277}$$

$$T_\delta^+ = 5\left[\text{Pr}_l + \ln\{1 + 5\text{Pr}_l\} + \frac{1}{2}\ln\left(\frac{\delta^+}{30}\right)\right] \qquad \text{for } \delta^+ > 30 \tag{15.278}$$

The heat transfer coefficient is then related to T_δ^+ as follows:

$$h = \frac{c_{pl}(\rho_l \tau_i)^{1/2}}{T_\delta^+}$$

(15.279)

Though the above equations are claimed to give reasonably accurate predictions for condensation [275], they overpredict the heat transfer coefficient in evaporation by typically 50 percent or so. It has often been suggested that the turbulence would be damped near the interface and that the use of equations like Eqs. 15.265–15.267 for eddy diffusivity, which are derived from single-phase flow data, is inappropriate. An approach suggested by Levich [276] is to introduce a damping factor that reduces the eddy diffusivity to zero at the interface. Thus, a modified eddy diffusivity ε' is calculated from the expression

$$\varepsilon' = \varepsilon\left(1 - \frac{y}{\delta}\right)^n$$

(15.280)

where ε is calculated from relationships like Eqs. 15.265–15.267. Various values have been used in the literature for the exponent n, and the results obtained suggest that it is not a universal constant. Typical values for n might be on the order of 1–2; a correlation for n is given by Sun et al. [272].

It is a gross oversimplification to treat heat transfer in annular flow in terms of average properties of the liquid film as described above. In reality, the film interface is highly complex; characteristically, large disturbance waves traverse the interface, whose length is typically on the order of 20 mm and whose height is typically on the order of 5 times the thickness of the thin (substrate) layer between the waves. This wave/substrate system can be modeled using the techniques of computational fluid dynamics (CFD), and recent predictions using these techniques are reported by Jayanti and Hewitt [277]. The calculations suggest that there is recirculation within the wave and that the wave behaves as a package of turbulence traveling over a laminar substrate film. This is illustrated by the results for turbulent viscosity distribution shown in Fig. 15.107.

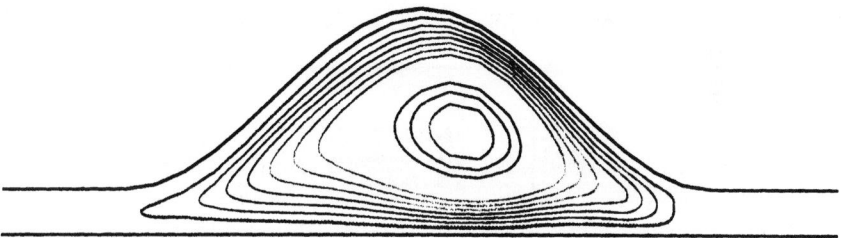

FIGURE 15.107   Turbulent viscosity distribution in a disturbance wave (from Jayanti and Hewitt [277], with permission from Elsevier Science). (Horizontal scale foreshortened.)

The recirculation within the wave distorts the temperature distribution as illustrated in Fig. 15.108. Quantitative comparisons between such calculations and measured heat transfer coefficients indicate reasonable agreement; the method has the advantage of not needing an arbitrary correction in the form, say, of Eq. 15.280. This seems an area of potential fruitful study in the future.

Combined Nucleate Boiling and Forced Convection in Annular Flow.   Even without nucleate boiling, annular flow heat transfer is highly complex (as discussed above). The coexistence of nucleate boiling makes the situation even more difficult, but it is still worth trying to produce a comprehensive model, if only to understand the relative importance of the various variables. Sun et al. [272] describe a model for the nucleate boiling component that, briefly, is on the following lines:

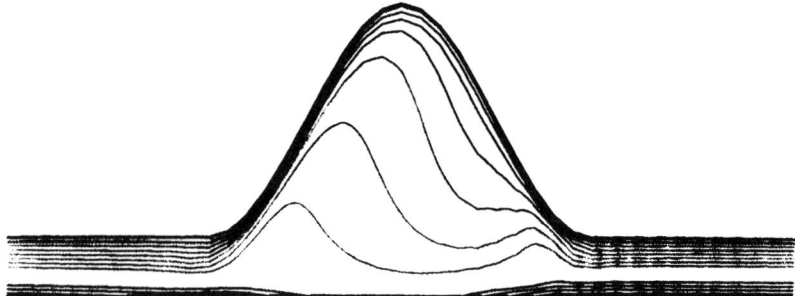

FIGURE 15.108 Temperature distribution in a disturbance wave (from Jayanti and Hewitt [277], with permission from Elsevier Science). (Horizontal scale foreshortened.)

1. The number of active nucleation sites is estimated as a function of heat flux using relationships for heterogeneous nucleation similar to those given earlier.

2. The growth of the bubble on the surface is estimated and the size at which it is swept from the nucleation site is calculated on the basis of a force balance.

3. The released bubble slides up the liquid film, continues to grow (and indeed is calculated to reach a size several times greater than the thickness of the liquid film), and ultimately bursts. The contribution of nucleate boiling is assumed to be that due to the latent heat of vaporization released by the bursting bubbles. A correlation was produced for bubble burst size. Bubble departure and burst sizes for a given mass flux were plotted in terms of vapor velocity as shown in Fig. 15.109, which also shows the liquid film thickness. The convective heat transfer was modeled, taking account of interface damping as mentioned above. Figure 15.110 shows the contributions to the total coefficient from convection and boiling, respectively. As expected, the nucleate boiling is damped out with increasing quality but, simultaneously, the convective contribution increases, giving an approximately constant heat transfer coefficient until the convection is totally suppressed, after which the

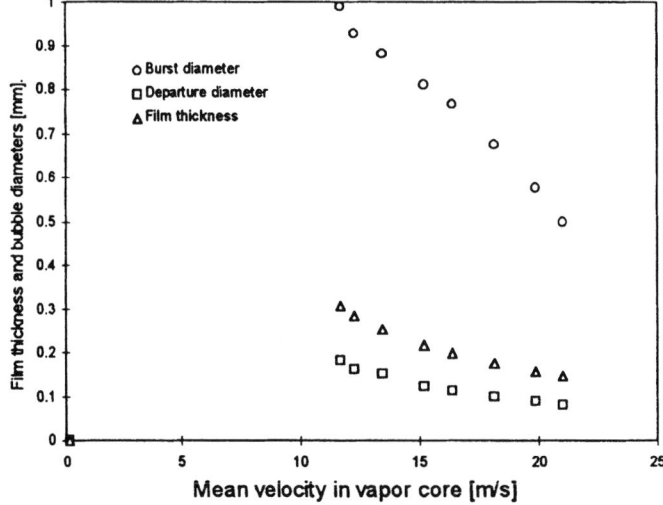

FIGURE 15.109 Typical departure diameters and burst diameters for bubbles formed in nucleate boiling in annular flow (from Sun et al. [272], with permission from Taylor & Francis, Washington, DC. All rights reserved).

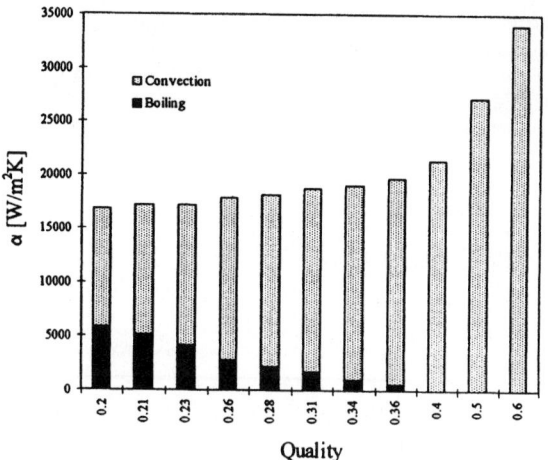

FIGURE 15.110 Contribution of nucleate boiling and forced convective in annular flow evaporation (from Sun et al. [272], with permission from Taylor & Francis, Washington, DC. All rights reserved).

heat coefficient rises with increasing quality. This does, of course, agree with the shape of the curve normally observed, and the predictions were in reasonable agreement with a range of data. Thus, the model (though preliminary) does appear to reflect the physical phenomenon. Again, this is obviously an area for further investigation.

Forced Convective Boiling of Multicomponent Mixtures in Channels.   There is a growing literature on forced convective boiling of multicomponent mixtures; reviews on the area are given by Collier and Thome [3], Carey [4], and Fujita [278]. Where the heat transfer is dominated by nucleate boiling, reductions in the heat transfer coefficient may occur, as in the case of pool boiling, and can be estimated using the methodologies described previously. Results in this category include those of Müller-Steinhagen and Jamialahmadi [231], Fujita and Tsutsui [279], Celata et al. [280], and Steiner [281]. Typical results of this kind are shown in Fig. 15.111. As will be seen, the data lie between the Stephan and Korner [128] and Schulunder [129] methods.

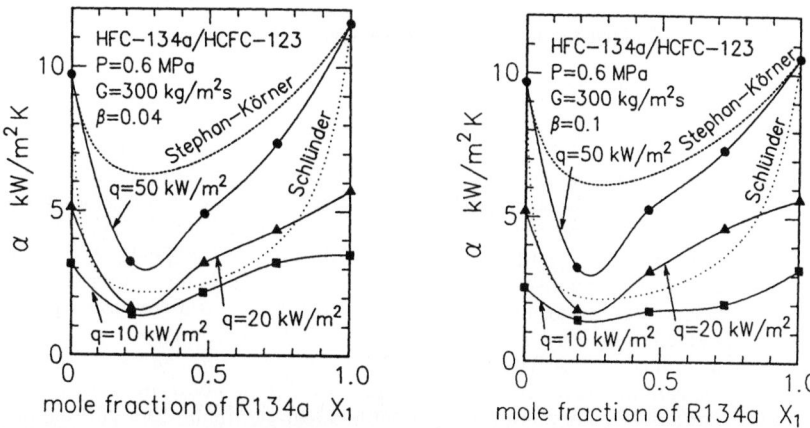

FIGURE 15.111 Variation of heat transfer coefficient with composition in the forced convective boiling of R134a/R123 mixtures (from Fujita and Tsutsui [279], with permission from Taylor & Francis, Washington, DC. All rights reserved).

At high enough qualities and mass fluxes, however, it would be expected that the nucleate boiling would be suppressed and the heat transfer would be by forced convection, analogous to that for the evaporation for pure fluids. Shock [282] considered heat and mass transfer in annular flow evaporation of ethanol water mixtures in a vertical tube. He obtained numerical solutions of the turbulent transport equations and carried out calculations with mass transfer resistance calculated in both phases and with mass transfer resistance omitted in one or both phases. The results for interfacial concentration as a function of distance are illustrated in Fig. 15.112. These results show that the liquid phase mass transfer resistance is likely to be small and that the main resistance is in the vapor phase. A similar conclusion was reached in recent work by Zhang et al. [283]; these latter authors show that mass transfer effects would not have a large effect on forced convective evaporation, particularly if account is taken of the enhancement of the gas mass transfer coefficient as a result of interfacial waves.

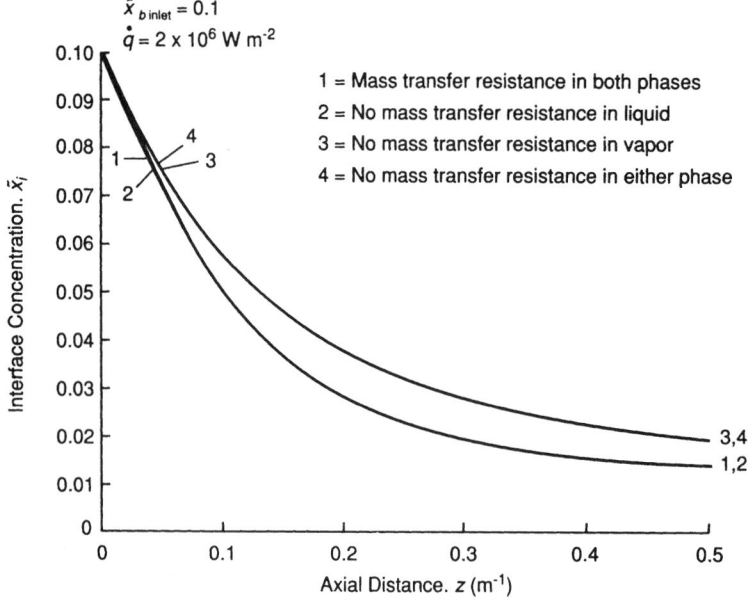

FIGURE 15.112   Axial variation of interface concentration of ethanol in annular flow evaporation of ethanol/water mixtures in a vertical tube (from Shock [282], with permission from Elsevier Science).

The results shown in Fig. 15.112 are for an axisymmetric annular flow. However, many multicomponent evaporation processes occur in horizontal tubes where the liquid film flow is certainly not axisymmetric. Here the film tends to be thicker at the bottom of the tube and circumferential transport of the liquid phase may be limited. Some interesting results relating to this form of evaporation were obtained by Jung et al. [284]. In what had been established as clearly forced convective evaporation without nucleate boiling, the reduction of heat transfer below that for an ideal mixture was still observed as is shown in Fig. 15.113. The relative reduction increased with decreasing quality. Obviously, just as in pool boiling, local concentration of the less volatile component was occurring. Support for this hypothesis is given by measurements of local temperature at the top and bottom of the tube as shown in Fig. 15.114. For a pure component, the wall temperature at the bottom of the tube was higher than at the top; this is consistent with the fact that the liquid layer thickness at the bottom is greater than that at the top, giving a higher heat transfer coefficient at the top. For an R22/R114 mixture,

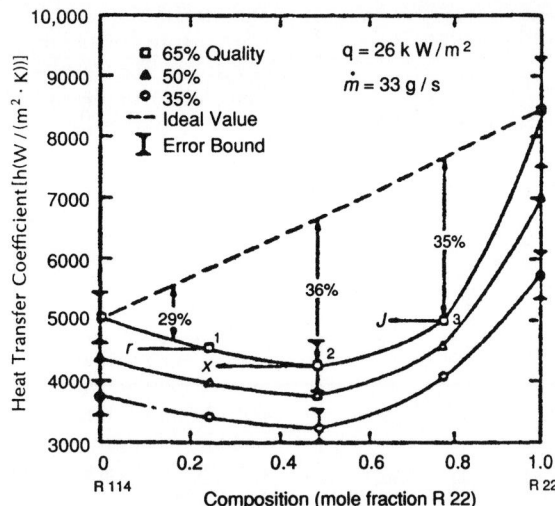

FIGURE 15.113   Average heat transfer coefficient for boiling of R22/R114 mixtures in a 9-mm bore tube (from Jung et al. [284], with permission from Elsevier Science).

on the other hand, the bottom temperature was considerably lower than the top temperature, particularly in the intermediate range of qualities where flow separation is likely to be greatest. What is happening, therefore, is that the liquid flow at the top of the tube is being denuded in the more volatile component and concentrated in the less volatile component, giving an increase in the interface temperature. This was confirmed in the experiments of

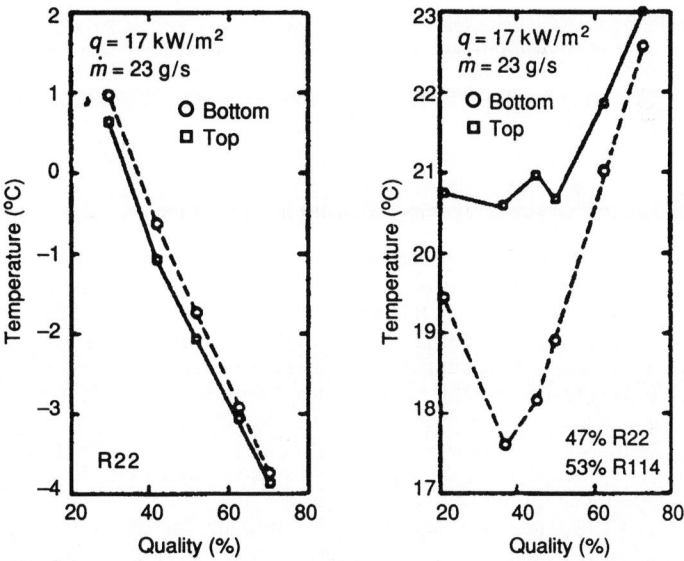

FIGURE 15.114   Circumferential variation of wall temperature for evaporation of pure R22 and for evaporation of an R22/R114 mixture in a horizontal tube (from Jung et al. [284], with permission from Elsevier Science).

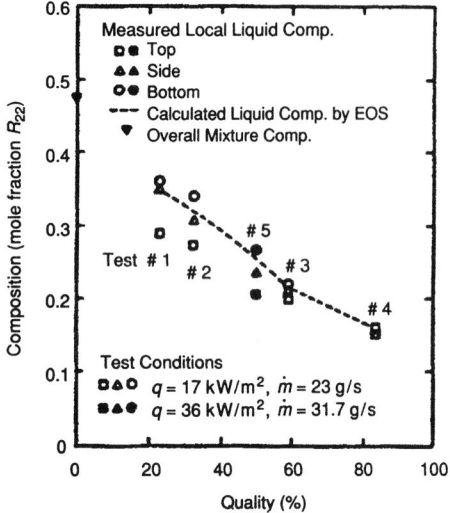

FIGURE 15.115 Circumferential concentration variations in the evaporation of an R22/R114 mixture in a horizontal tube (from Jung et al. [284], with permission from Elsevier Science).

Jung et al. [284] by direct measurements of the concentration of the more volatile material at the top, side, and bottom of the tube. The results shown in Fig. 15.115 confirm the denudation of the more volatile material at the top; the effect reduces with increasing quality, as might be expected because the variation of the flow rate around the periphery decreases with increasing quality.

For cases where both forced convective heat transfer and nucleate boiling are significant, then the power-law interpolation and suppression-type correlations can be employed. The nucleate boiling component is adjusted for the effect of multiple components as described earlier. Yoshida et al. [285] have developed a suppression-type correlation, and Winterton [286] describes the application of the Liu and Winterton [287] correlation to multicomponent mixtures; the Liu and Winterton correlation is of a hybrid type that includes a suppression factor on the nucleate boiling component but that uses the power-law interpolation with $n = 2$ in Eq. 15.226.

Enhancement of Forced Convective Boiling Heat Transfer in Channels. The topic of enhancement of boiling heat transfer in channel flows has met with much less attention than is the case for pool boiling or cross-flow boiling. This is because it is more difficult to produce enhanced surfaces on the inside of tubes. Koyama et al. [288] studied boiling heat transfer inside an 8.37-mm-diameter tube with sixty 0.168-mm-high trapezoidal fins machined on the tube walls with the fins being at a helix angle of 18°. Heat transfer was enhanced (compared to a smooth tube of the same internal diameter) by approximately 100 percent. MacBain and Bergles [289] studied boiling heat transfer in a deep spirally fluted tube and reported enhancement factors in the range of 1.8 to 2.7 in the nucleate boiling regime and 3.3 to 7.8 in the forced convective regime.

Another important area relating to enhancement of boiling heat transfer is the influence of fin design in plate-fin heat exchangers. This topic is reviewed by Carey [4], and this type of work is exemplified by the study of Hawkes et al. [290] on the hydrodynamics of flow with off-set strip fins in plate-fin heat exchangers.

Critical Heat Flux in Forced Convective Boiling in Channels

Just as in pool boiling and cross flow boiling, a critical phenomenon occurs in which the heat transfer process deteriorates. This is signaled by an increase in surface temperature for a small incremental change in the surface heat flux (hence the name *critical heat flux*) or by a reduction in heat flux arising from a small incremental increase in surface temperature in cases where the surface temperature is controlled. As before, we will use the term critical heat flux (CHF) to describe this phenomenon. Other terms used in the literature include *burnout, boiling crisis, departure from nucleate boiling (DNB), dryout,* and *boiling transition.* None of these terms is totally satisfactory (Hewitt [291]), but the term *critical heat flux* is retained for the present section since it has the most currency within the literature, particularly the North American literature.

Critical heat flux has attracted a large amount of attention as a result of its importance as a limiting condition in water-cooled water reactors. The literature is vast and could certainly not be dealt with in detail in the present context. Reviews of earlier publications are given by Hewitt [291, 292] and more recent review material is presented in the books of Collier and Thome [3], Carey [4], and Tong and Tang [5]. Here the objective is to pick out the most salient points; the reader is referred to these earlier reviews for further information.

Parametric Effects in CHF in Forced Convective Boiling in Channels.   Detailed discussions of parametric effects on critical heat flux in flow boiling are given by Hewitt [291] and by Tong and Tang [5]. To illustrate the effect of various parameters on critical heat flux, it is convenient to use the example of upward water flow in a 0.01-m-diameter tube and to employ the correlation of Bowring [293] (see the following text). The calculations were for uniformly heated tubed where the critical condition occurs at the end of the tube. Based on this evaluation, the following results are obtained:

Effect of Subcooling.   The critical heat flux increases approximately linearly with increasing inlet subcooling over a wide range of subcooling, and for constant mass flux, pressure, and tube length as illustrated in Fig. 15.116.

Effect of Tube Length.   The critical heat flux decreases with tube length as shown in Fig. 15.117. The total power input increases with length (also shown in Fig. 15.117). Some data show an asymptotic value of power input over a wide range of tube length, corresponding to a constant outlet quality for the critical condition.

Effect of Pressure.   For fixed mass flux, tube diameter, tube length, and inlet subcooling, the critical heat flux initially increases with increasing pressure and then decreases with pressure approaching zero as the critical pressure is approached.

Effect of Tube Diameter.   For fixed mass flux, length, pressure, and inlet subcooling, the critical heat flux increases with tube diameter, eventually reaching an approximately constant value independent of tube diameter.

Effect of Mass Flux.   Typical results calculated from the Bowring [293] correlation for the effect of mass flux are shown in Fig. 15.118. At low mass fluxes, the critical heat flux increases rapidly with increasing mass flux and tends to approach a constant value.

Effect of Channel Orientation.   For the subcooled boiling region, a study of the effect of channel inclination on critical heat flux is reported by Brusstar and Merte [294]. The channel used by these authors was rectangular in cross section with one side heated. The orientation of this heated surface with respect to the horizontal could be varied between 0 and 360°. At low velocities, a sharp decrease in critical heat flux was observed when the heater surface was downward-facing, as exemplified by the results shown in Fig. 15.119.

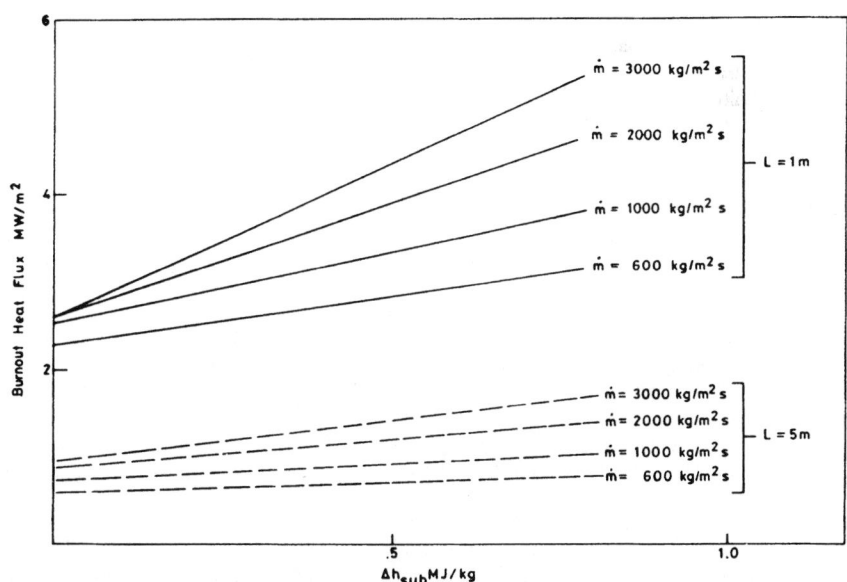

FIGURE 15.116   Relation between critical heat flux and inlet subcooling (calculated from the Bowring [293] correlation for water for a system pressure of 6 MPa and a tube diameter of 0.01 m) (from Hewitt [298], with permission from The McGraw-Hill Companies).

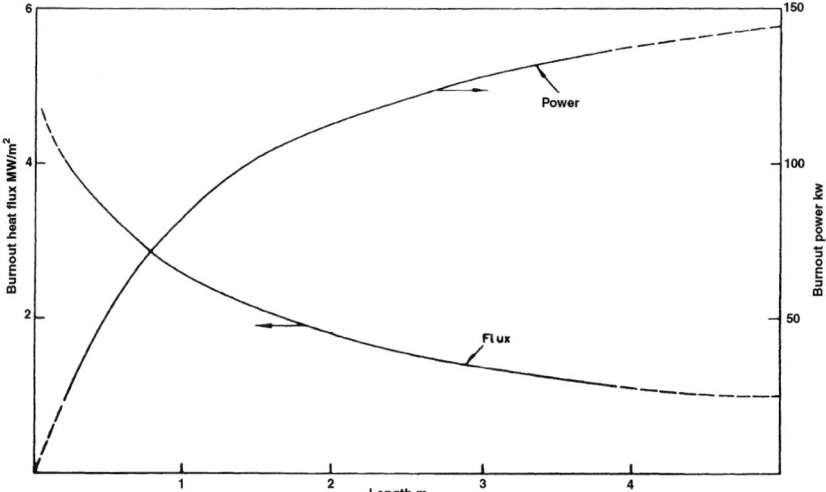

FIGURE 15.117   Effect of tube length in critical heat flux and power input at the CHF condition (calculated from the correlation of Bowring [293] for water for a mass flow of 3000 kg/m²s, a tube diameter of 0.01 m, a pressure of 6 MPa, and zero inlet subcooling) (from Hewitt [291], with permission from The McGraw-Hill Companies).

For higher flow velocities, the minimum in the critical heat flux was much less pronounced. When the quality region is entered, there is a complex interaction between the flow patterns existing within the channel and the heat transfer behavior. A particularly important case is that of evaporation in horizontal evaporator tubes as used, for instance, in many refrigeration and air-conditioning plants. Here, due to the action of gravity, the liquid phase tends to be

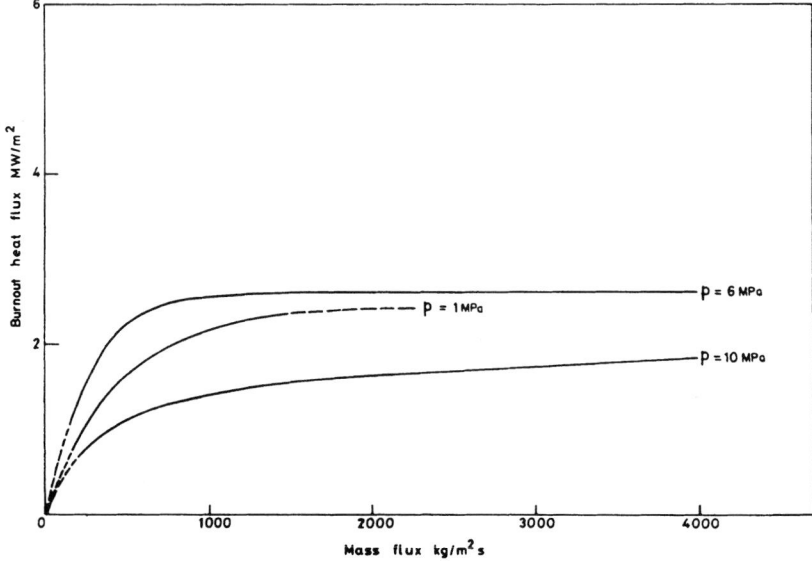

FIGURE 15.118   Effect of mass flux on critical heat flux in upward water flow in a vertical tube (calculated from the Bowring [293] correlation for a tube length of 1 m, a tube diameter of 0.01 m, and zero inlet subcooling) (from Hewitt [291], with permission from The McGraw-Hill Companies).

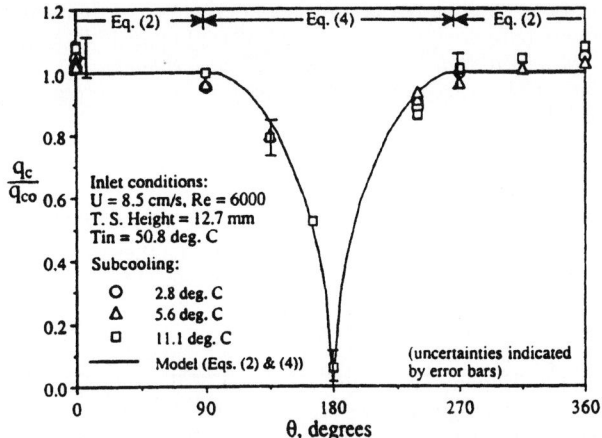

FIGURE 15.119   Effect of heated surface orientation in subcooled flow boiling in a rectangular channel (from Brusstar and Murte [294] with permission).

moved away from the upper surface of the tube, which may become dry (with a consequent reduction in heat transfer performance). Detailed discussions of this phenomenon in terms of flow pattern boundaries for horizontal two-phase flow are given by Ruder et al. [295] and Bar-Cohen et al. [296, 297].

Mechanisms of CHF in Forced Convective Boiling in Channels.   Detailed reviews of critical heat flux mechanisms in forced convective boiling are given by Hewitt [291], Tong and Tang [5], Collier and Thome [3], and Katto [101]. The more commonly accepted mechanisms for the occurrence of critical heat flux in forced convection are as follows:

Dryout Under a Vapor Cloud or Slug Flow Bubble.   Here, a large vapor bubble may be formed on the wall and dryout occurs under it. As discussed by Katto [101], this mechanism may be closely related to mechanisms in pool boiling, as is illustrated in Fig. 15.120. Assuming the existence of a macrolayer underneath the clot, the critical heat flux condition may be initiated if there is sufficient time for the macrolayer to evaporate. In pool boiling, this time is governed by the frequency of release of the large vapor mushrooms (clots), whereas in forced

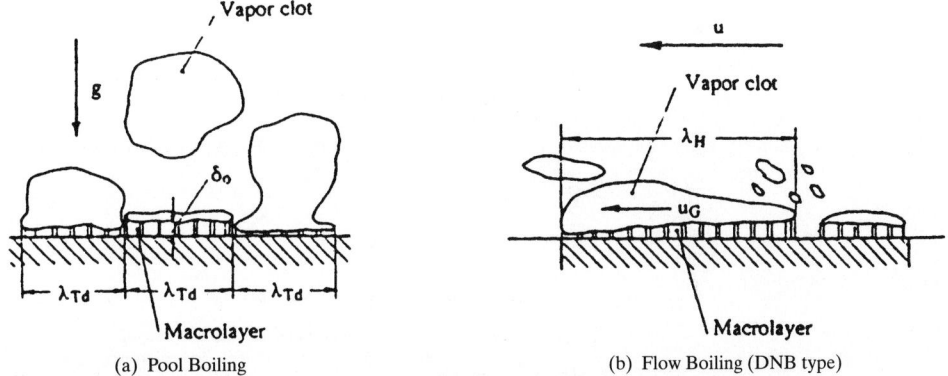

FIGURE 15.120   Behavior of vapor clot and macrolayer in pool and forced convective boiling (from Katto [101], with permission from Taylor & Francis, Washington, DC. All rights reserved).

convective boiling the occurrence of critical heat flux will depend on the time taken for the vapor clot to sweep over the surface (see Fig. 15.120).

Near-Wall Bubble Crowding and Vapor Blanketing.   Here, a layer of vapor bubbles builds up near the wall and this prevents the ingress of liquid to the tube surface, leading to a decrease in efficiency of cooling and to the critical phenomenon.

Hot Spot Growth Under a Bubble.   When bubbles grow and detach from a nucleation center on a solid surface, evaporation of the liquid layer commonly occurs, separating the bubble from the solid surface. This *microlayer evaporation* process is particularly important at low pressures. When a small zone under the bubble becomes dry as a result of this process, its temperature increases, and this increase can, under certain conditions, be sufficient to prevent rewetting of the surface on bubble departure, leading to a permanent hot spot and onset of the critical phenomenon.

Film Dryout.   In the annular flow regime, the critical heat flux condition is reached as a result of film dryout. The film dries out because of the entrainment of droplets from its surface and as a result of evaporation, and despite the redeposition of droplets counteracting the effect of droplet entrainment. Reviews of the extensive experimental studies relating to this mechanism are given by Hewitt [291] and Hewitt and Govan [298]. An example of the experimental evidence demonstrating this mechanism is shown in Fig. 15.121. As the power input to the channel is increased, annular flow begins at the end of the channel, and the flow rate of the annular liquid film decreases with further increases in power input. Eventually, the film flow rate reaches zero at a point corresponding to the critical heat flux (burnout) condition as shown. We will return to a discussion of the prediction of entrainment and deposition phenomena along the channel in reviewing prediction methods later.

It is probable that all of the above mechanisms play a role, their influence depending on the flow and thermodynamic conditions within the channel. Semeria and Hewitt [300] represented the regions of operation of the various mechanisms in terms of the conceptual diagram reproduced in Fig. 15.122. The regions are plotted in terms of mass flux and local quality. As will be seen, the most important mechanism for tubes of reasonable length (where higher qualities will be generated) is that of annular flow dryout.

Correlations for CHF in Forced Convective Boiling in Channels.   The importance of the critical heat flux phenomenon in nuclear reactor design has led to extensive work on correlation of critical heat flux data. The correlations in the literature have taken two main forms as sketched in Fig. 15.123:

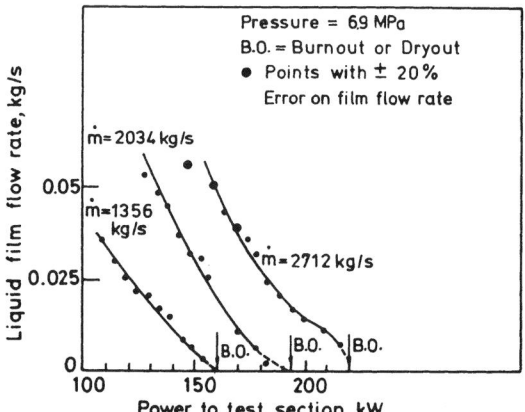

FIGURE 15.121   Variation of film flow rate with input power at the end of a uniformly heated round tube in which water is being evaporated at 6.9 MPa (from Hewitt [299], with permission from Taylor & Francis, Washington, DC. All rights reserved).

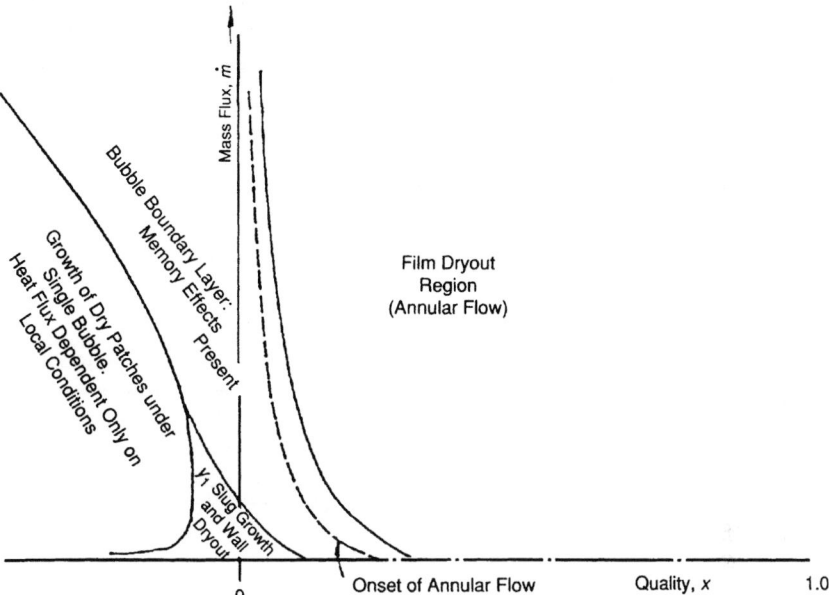

FIGURE 15.122 Tentative map of regimes of operation of various forced convective critical heat flux mechanisms (from Hewitt and Semaria [300], with permission from Taylor & Francis, Washington, DC. All rights reserved).

1. The data for a given fluid, pressure, mass flux, channel cross section, and orientation are found to fall approximately on a single curve of critical heat flux versus quality, the critical phenomena occurring in this (uniformly heated) case at the end of the channel. Thus, data for all lengths and inlet subcoolings are represented by a single line. For the range of data covered, the relationship is often approximately linear and many of the available correlations are in this linear form.

2. The same data that were plotted in the heat flux/quality form can also be plotted in terms of x_{crit} against boiling length $(L_B)_{crit}$, where boiling length is the length between the point where $x = 0$ and the point where a critical phenomenon occurs (usually at the end of the channel for uniform heat flux).

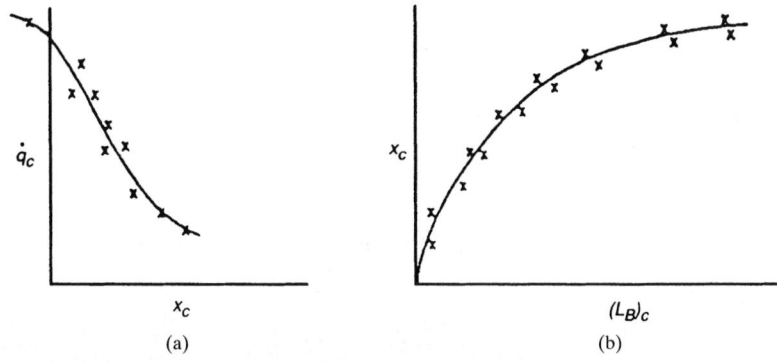

FIGURE 15.123 Bases for correlation of critical heat flux data (from Hewitt et al. [13], with permission. Copyright CRC Press, Boca Raton, FL).

For uniform heat flux, the two forms are essentially equivalent, because if

$$q''_{\text{crit}} = fn(x_{\text{crit}}) \tag{15.281}$$

then it follows that

$$(L_B)_{\text{crit}} = \frac{DGi_{lg}x_{\text{crit}}}{4q''_{\text{crit}}} = fn(x_{\text{crit}}) \tag{15.282}$$

and thus

$$x_{\text{crit}} = fn(L_B)_{\text{crit}} \tag{15.283}$$

For channels with axial variations of heat flux (which is typical of nuclear reactors and industrial boiling systems), Eqs. 15.281 and 15.283 no longer give the same result; in these cases, the critical heat flux condition can occur upstream of the end of the channel in some circumstances. In general, the quality–versus–boiling length representation gives better fit to the data for nonuniformly heated channels, as is illustrated by some data taken by Keeys et al. [301, 302] that are plotted in the flux/quality form in Fig. 15.124 and in the quality/boiling length form in Fig. 15.125. Generally, therefore, it is recommended that if a correlation of the critical heat flux/quality type is used for application to nonuniform heat flux, then it should be transformed into a quality/boiling length relationship using Eq. 15.282 and used in that form. However, this suggestion is less valid at low qualities and in the subcooled region. Extensive discussion of the influence of nonuniform heat flux is given by Tong and Tang [5] and by Hewitt [291].

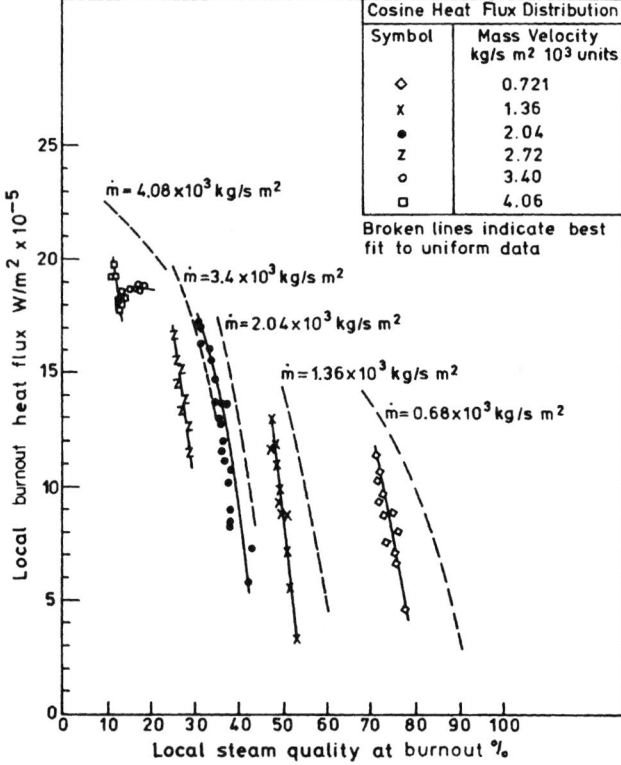

FIGURE 15.124 Comparison of critical heat flux data for tubes with cosine variation of axial heat flux and uniform heat flux: evaporation of water at 6.89 MPa in a 12.7-mm bore tube (from Keeys et al. [301], reproduced by permission of AEA Technology plc).

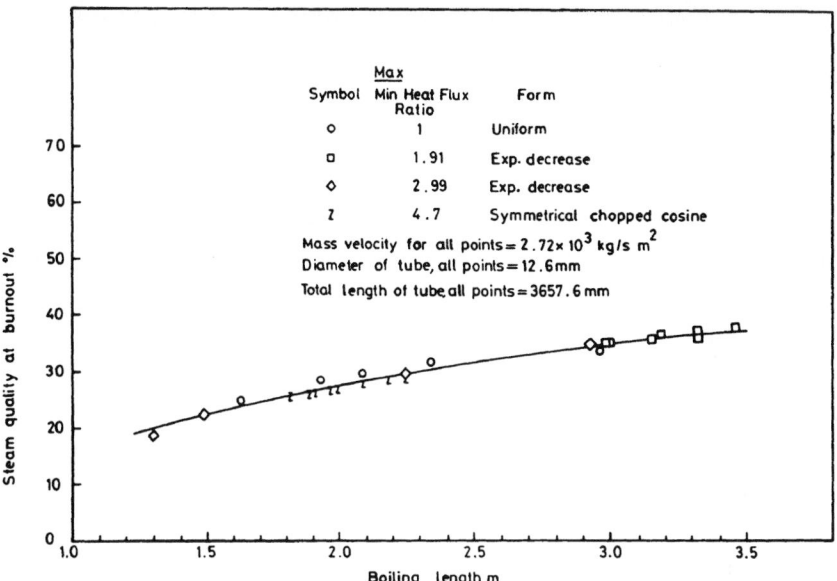

FIGURE 15.125 Data for critical heat flux for various heat flux distributions plotted in the quality–versus–boiling length form: system pressure 6.89 MPa (from Keys et al. [302], reproduced by permission of AEA Technology plc).

Widely used correlations of the flux/quality category include the "standard" tabular presentations of critical heat flux data for water by Groeneveld et al. [303] and Kirillov et al. [304] and the correlations of Thompson and MacBeth [305], Becker [306], and Bowring [293]. Correlations of the boiling length type include those of Bertoletti et al. [307] and Biasi et al. [308]. Generalized empirical correlations that attempt to combine both upstream and local effects include those of Shah [309] and Katto and Ohne [310]. It is clearly impossible to present a comprehensive treatment of critical heat flux correlation in the space available here. The reader is referred to the cited references and to the books of Collier and Thome [3] and Tong and Tang [5] for a more comprehensive presentation. In the following, we will deal only with the case of round tubes; critical heat flux in other geometries such as annuli, rectangular channels, and bundles of nuclear fuel elements is discussed by Hewitt [291] and Tong and Tang [5]. Here, the correlation of Bowring [293] for water upflow in vertical tubes and the more general correlation of Katto and Ohne [310] for vertical upflow will be presented. Though the Bowring correlation is for water only, it can be extended to cover other fluids by scaling methods, and this will be discussed. For horizontal tubes, the critical heat flux can be much lower than for vertical tubes; the correlation of Merrilo [46] is presented for this case.

Bowring [293] Correlation for Upward Flow of Water in Vertical Tubes. This correlation is of a linear flux/quality form and is written as

$$q''_{\text{crit}} = \frac{A' + 0.25DG\Delta i_{\text{sub}}}{C' + L} \tag{15.284}$$

where D is the tube diameter, G is the mass flux, Δi_{sub} is the inlet subcooling (difference between saturated liquid enthalpy and inlet liquid enthalpy), and L is the tube length. The parameters A' and C' are given by the expressions

$$A' = 2.317(0.25i_{lg}DG)F_1/(1 + 0.0143F_2D^{1/2}G) \tag{15.285}$$

$$C' = 0.077F_3DG/[1.0 + 0.347F_4(G/1356)^n] \tag{15.286}$$

where
$$n = 2.0 - 0.5P_R \qquad (15.287)$$

where P_R is given by
$$P_R = P/6.895 \times 10^6 \qquad (15.288)$$

The parameters F_1 to F_4 in Eqs. 15.285 and 15.286 are given for $P_R < 1$ by

$$F_1 = \{P_R^{18.492} \exp[20.8(1 - P_R)] + 0.917\}/1.917 \qquad (15.289)$$

$$F_1/F_2 = \{P_R^{1.316} \exp[2.444(1 - P_R)] + 0.309\}/1.309 \qquad (15.290)$$

$$F_3 = \{P_R^{17.023} \exp[16.658(1 - P_R)] + 0.667\}/1.667 \qquad (15.291)$$

$$F_4/F_3 = P_R^{1.649} \qquad (15.292)$$

For $P_R > 1$, F_1 to F_4 are given by

$$F_1 = P_R^{-0.368} \exp[0.648(1 - P_R)] \qquad (15.293)$$

$$F_1/F_2 = P_R^{-0.448} \exp[0.245(1 - P_R)] \qquad (15.294)$$

$$F_3 = P_R^{0.219} \qquad (15.295)$$

$$F_4/F_3 = P_R^{1.649} \qquad (15.296)$$

The Bowring correlation was based on data for 0.2 MPa < P < 19 MPa, 2 mm < D < 45 mm, 0.15 m < L < 3.7 m, and 136 kg/m²s < G < 18,600 kg/m²s, but it should not be assumed that all combinations of these parameter ranges are covered. Within its range of applicability, the Bowring correlation gives, for round tubes, a standard deviation of about 7 percent when compared with around 3000 data points. The Bowring correlation is for water only, but it may be used in the prediction of critical heat flux for other fluids by using scaling methods, perhaps the most successful of which is that of Ahmad [311]. The basis of this method is that, for given values of $\Delta i_{sub}/i_{lg}$, ρ_l/ρ_g, and L/D, the boiling number Bo_{crit} for the critical condition is a function of a scaling parameter ϕ as follows:

$$Bo_{crit} = q''_{crit}/Gi_{lg} = fn(\psi) \qquad (15.297)$$

where ψ is given by

$$\psi = \left[\frac{GD}{\mu_l}\right]\left[\frac{\mu_l}{\sigma D \rho_l}\right]^{2/3}\left[\frac{\mu_g}{\mu_l}\right]^{1/5} \qquad (15.298)$$

where σ is the surface tension and μ_g and μ_l are the viscosities of the vapor and liquid, respectively. The procedure for establishing the critical heat flux for a nonaqueous fluid is thus:

1. Using steam tables, determine the pressure at which the water/steam density ratio ρ_l/ρ_g is the same as that specified for the fluid being used.

2. Using Eq. 15.298, calculate the ratio for the fluid $(\psi/G)_F$ and the ratio for water $(\psi/G)_W$, respectively. In this calculation, the physical properties for water at the pressure estimated in step 1 are used. The equivalent mass flux for water G_W may then be calculated from

$$G_W = G_F\left[\frac{(\psi/G)_F}{(\psi/G)_W}\right] \qquad (15.299)$$

3. To maintain the condition of equal ratios of inlet subcooling to latent heat of vaporization, the equivalent subcooling for water is calculated as follows:

$$(\Delta i_{sub})_W = (\Delta i_{sub})_F\left[\frac{(i_{lg})_W}{(i_{lg})_F}\right] \qquad (15.300)$$

4. Using the Bowring [293] correlation described above (or any alternative correlation for water data), the critical heat flux is then determined for the known values of L and D and for the values of mass flux (G_W) and inlet subcooling $[(\Delta i_{sub})_W]$ calculated as above.

5. The critical heat flux for the fluid $(q''_{crit})_F$ is then estimated from the value $(q''_{crit})_W$ calculated for water, taking account of the fact that Bo_{crit} is the same for equivalent values of ψ for both water and the fluid. Thus:

$$(q''_{crit})_F = \frac{G_F(i_{lg})_F}{G_W(i_{lg})_W}(q''_{crit})_W \qquad (15.301)$$

A worked example using this method is given by Hewitt et al. [13].

The Correlation of Katto and Ohne [310] for Critical Heat Flux in Upward Flow in Vertical Tubes. The Katto and Ohne correlation is based on data for a wide variety of fluids and is not (like the Bowring [293] correlation) restricted to water, and there is no need to use the scaling procedure in this case to calculate the critical heat flux. The Katto and Ohne correlation is expressed generally in the form:

$$q''_{crit} = XG(i_{lg} + K\Delta i_{sub}) \qquad (15.302)$$

where Δi_{sub} is the inlet subcooling and where X and K are functions of three dimensionless groups as follows:

$$Z' = z/D \qquad (15.303)$$

$$R' = \rho_g/\rho_l \qquad (15.304)$$

$$W' = \left[\frac{\sigma \rho_l}{G^2 z}\right] \qquad (15.305)$$

where z is the distance along the channel ($z = L$ for uniform heat flux where the critical heat flux occurrence is at the end of the channel). Although the basic equation (Eq. 15.302) is quite simple, there is a very complex set of alternative relationships for X and K. The five alternative expressions for X are as follows:

$$X_1 = \frac{CW'^{0.043}}{Z'} \qquad (15.306)$$

where:
$$C = 0.25 \qquad \text{for } Z' < 50 \qquad (15.307)$$

$$C = 0.25 + 0.0009(Z' - 50) \qquad \text{for } 50 < Z' < 150 \qquad (15.308)$$

$$C = 0.34 \qquad \text{for } Z' > 150 \qquad (15.309)$$

and the remaining values of X are given as follows:

$$X_2 = \frac{0.1R'^{0.133}W'^{0.333}}{1 + 0.0031Z'} \qquad (15.310)$$

$$X_3 = \frac{0.098R'^{0.133}W'^{0.433}Z'^{0.27}}{1 + 0.0031Z'} \qquad (15.311)$$

$$X_4 = \frac{0.0384R'^{0.6}W'^{0.173}}{1 + 0.28W'^{0.233}Z'} \qquad (15.312)$$

$$X_5 = \frac{0.234R'^{0.513}W'^{0.433}Z'^{0.27}}{1 + 0.0031Z'} \qquad (15.313)$$

Similarly, there are alternative expressions for K for use in Eq. 15.302. These are as follows:

$$K_1 = \frac{0.261}{CW'^{0.043}} \tag{15.314}$$

$$K_2 = \frac{0.833[0.0124 + (1/Z')]}{R'^{0.133}W'^{0.333}} \tag{15.315}$$

$$K_3 = \frac{1.12[1.52W'^{0.233} + (1/Z')]}{R'^{0.6}W'^{0.173}} \tag{15.316}$$

The methodology for choosing the expression for X and K is given in the following table:

	For $R' < 0.15$		For $R' > 0.15$	
If	$X_1 < X_2$	$X = X_1$	$X_1 < X_5$	$X = X_5$
	$\left.\begin{array}{c} X_1 > X_2 \\ \text{and} \\ X_2 < X_3 \end{array}\right\}$	$X = X_2$	$\left.\begin{array}{c} X_1 > X_5 \\ \text{and} \\ X_5 > X_4 \end{array}\right\}$	$X = X_5$
	$\left.\begin{array}{c} X_1 > X_2 \\ \text{and} \\ X_2 < X_3 \end{array}\right\}$	$X = X_3$	$\left.\begin{array}{c} X_1 > X_5 \\ \text{and} \\ X_5 < X_4 \end{array}\right\}$	$X = X_4$
	$K_1 > K_2$	$K = K_1$	$K_1 > K_2$	$K = K_1$
	$K_1 < K_2$	$K = K_2$	$\left.\begin{array}{c} K_1 < K_2 \\ \text{and} \\ K_2 < K_3 \end{array}\right\}$	$K = K_2$
			$\left.\begin{array}{c} K_1 < K_2 \\ \text{and} \\ K_2 > K_3 \end{array}\right\}$	$K = K_3$

The Katto and Ohne correlation covers tube diameters in the range 0.001 to 0.038 m, values of Z' from 5 to 880, values of R' from 0.0003 to 0.41, and values of W' from 3×10^{-9} to 2×10^{-2}.

The Correlation of Merilo [46] for Horizontal Tubes.   In horizontal tubes, as mentioned earlier, liquid tends to drain to the bottom of the channel under gravity, leading to the critical condition occurring at the top of the channel where the liquid film dries out. This means that the critical heat flux is very much less for horizontal channels than for vertical channels, as is illustrated in Fig. 15.126. This figure contains both data for water and also data for refrigerant 12 that had been scaled through data for water using the Ahmad [311] scaling factor (Eq. 15.298). Merilo observed that the Ahmad scaling factor did not work well for horizontal tubes and proposed an alternative scaling factor that includes the effects of gravity and is defined as follows:

$$\psi_H = \frac{GD}{\mu_l}\left(\frac{\mu_l^2}{\sigma D \rho_l}\right)^{-1.57}\left[\frac{(\rho_l - \rho_g)gD^2}{\sigma}\right]^{-1.05}\left(\frac{\mu_l}{\mu_g}\right)^{6.41} \tag{15.317}$$

ψ_H can be used in precisely the same way as described above for scaling water and refrigerant data for horizontal tubes. However, a correlation for critical heat flux for water flow in a horizontal tube would then be needed, and Merilo [46] suggested a formulation that includes the scaling group within a general correlation for horizontal tubes as follows:

$$\frac{\dot{q}_{crit}}{Gi_{lg}} = 6.18\psi_H^{-0.340}\left(\frac{L}{D}\right)^{-0.511}\left(\frac{\rho_l - \rho_g}{\rho_g}\right)^{1.27}\left(1 + \frac{\Delta i_{sub}}{i_{lg}}\right)^{1.64} \tag{15.318}$$

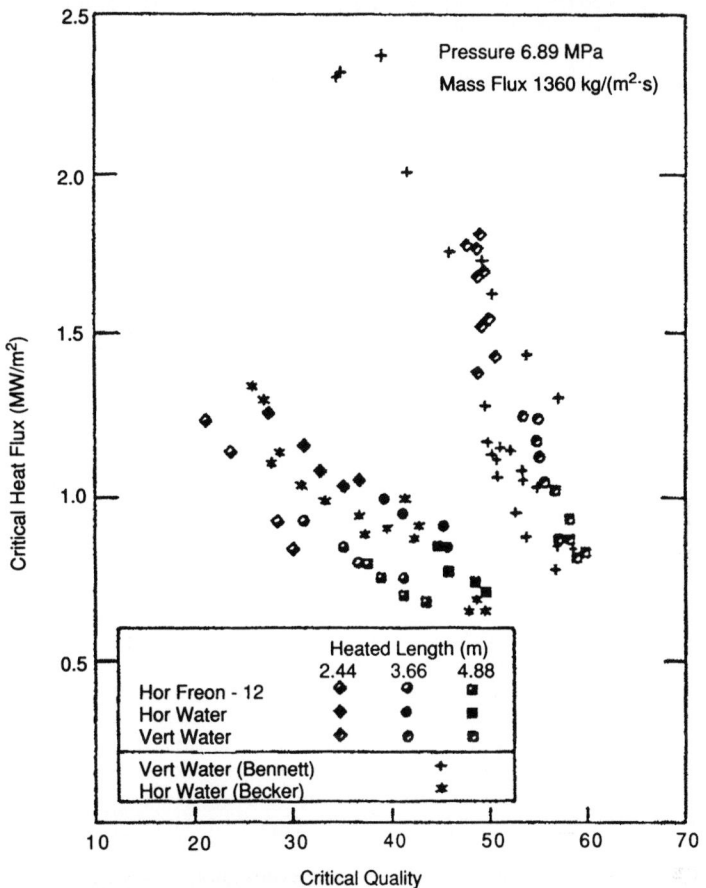

FIGURE 15.126   Comparison of critical heat flux in vertical and horizontal channels (data for refrigerant 12 scaled to that for water using Eq. 15.298) (from Merilo [46], with permission from Elsevier Science.)

Prediction of CHF in Forced Convective Boiling in Channels.   The correlations described in the preceding section can be seen to be essentially multiparameter fits to collections of data. The more data that have to be fitted, the greater the number of fitting parameters required. The difficulty is that there must always be uncertainties in extrapolating such correlations outside the range of data for which they were derived. This has led to attention being focused, over many decades, on predicting the critical heat flux phenomenon. Extensive discussions of such prediction methods are presented by Hewitt [291, 292], Collier and Thome [3], Tong and Tang [5], and Hewitt and Govan [298]. Again, it is beyond the scope of this chapter to deal with this subject in detail. In what follows, therefore, brief summaries are given of the development of predictive methods for the subcooled and low-quality regions and for the annular flow region, respectively.

   Prediction of Critical Heat Flux Under Subcooled and Low-Quality Conditions.   A detailed review of subcooled and low-quality critical heat flux prediction methods is given by Tong and Tang [5]. The same difficulties in interpreting the critical heat flux phenomenon as presented earlier exist in this region, namely the difficulty of understanding the detailed physical behavior in a location close to the surface. Forms of prediction methodology that have been used for forced convective critical heat flux prediction in this region include:

1. *Boundary layer separation models.* In this class of model, the critical heat flux phenomenon is considered to be analogous to the phenomenon of boundary layer separation from a permeable plate through which gas is flowed in a direction normal to the flow over the plate. This mechanism was initially suggested by Kutateladze and Leontiev [312] and was further developed by Tong [313] and others and more recently by Celata et al. [314]. This method of prediction leads to an equation of the form

$$q''_{\text{crit}} = \frac{C_1 i_{lg} \rho_l U}{(\text{Re})^n} \qquad (15.319)$$

where U is the main stream fluid velocity and Re is the main stream fluids Reynolds number, given by $\text{Re} = \rho_l U D/\mu_{ls}$ where μ_{ls} is the saturated liquid viscosity. C_1 and n are fitted parameters; Celata et al. suggest that $n = 0.5$ and that C_1 is given by:

$$C_1 = (0.216 + 4.74 \times 10^{-8} P)\psi \qquad (15.320)$$

where the parameter ψ is related to the thermodynamic equilibrium quality (calculated from Eq. 15.208) as follows:

$$\psi = 0.825 + 0.986x \qquad \text{for } 0 > x > -0.1 \qquad (15.321)$$

$$\psi = 1 \qquad \text{for } x < -0.1 \qquad (15.322)$$

$$\psi = \frac{1}{2 + 30x} \qquad \text{for } x > 0 \qquad (15.323)$$

2. *Bubble crowding models.* In this form of model, the processes of bubble formation at the wall, bubble condensation into the subcooled core at the edge of the wall bubble layer ("bubble boundary layer"), and liquid percolation through the bubble boundary layer are modeled. Many models of this kind have been formulated, but they are typified by that of Weisman and Pei [315], who assumed that there is a limiting void fraction in the bubble boundary layer of 0.82 in which an array of ellipsoidal bubbles can be maintained without significant contact between the bubbles. Weisman and Pei suggest models for evaporation at the channel wall and for condensation at the edge of the boundary layer that allow the calculation of this critical condition. Originally the Weisman and Pei model was only for subcooled boiling, but it has been extended to cover both subcooled and saturated boiling by Hewitt and Govan [298]. This extended model is compared with data from the standard tables of Groeneveld et al. [303] and Kirillov et al. [304] in Fig. 15.127.

3. *Macrolayer evaporation models.* In this class of models, the critical heat flux is considered to be governed by a macrolayer underneath a vapor clot as illustrated in Fig. 15.120. This type of model is analogous to those suggested for pool boiling, and it has been pursued by a number of authors including Katto [101], Mudawwar et al. [316], and Lee and Mudawwar [317].

In general, the processes modeled in the prediction methods for subcooled and low-quality critical heat flux mentioned above are clearly important ones; there seems scope for fundamental work on the precise mechanisms involved in the near-wall region.

Prediction of Critical Heat Flux in Annular Flow. In annular flow, the situation to be modeled is illustrated in Fig. 15.106. There is a thin liquid film on the channel wall that has a flow rate Γ per unit periphery. Droplets are being entrained from this film into the vapor core at a rate $\dot{m}_E$ (mass rate of entrainment per unit peripheral area, kg/m²s) and are being redeposited from the core at a rate $\dot{m}_D$ (kg/m²s). In addition, the liquid film is being evaporated at a rate q''/i_{lg} per unit peripheral area. Thus, the rate of change of Γ with distance is given by

$$\frac{d\Gamma}{dz} = \dot{m}_D - \dot{m}_E - q''/i_{lg} \qquad (15.324)$$

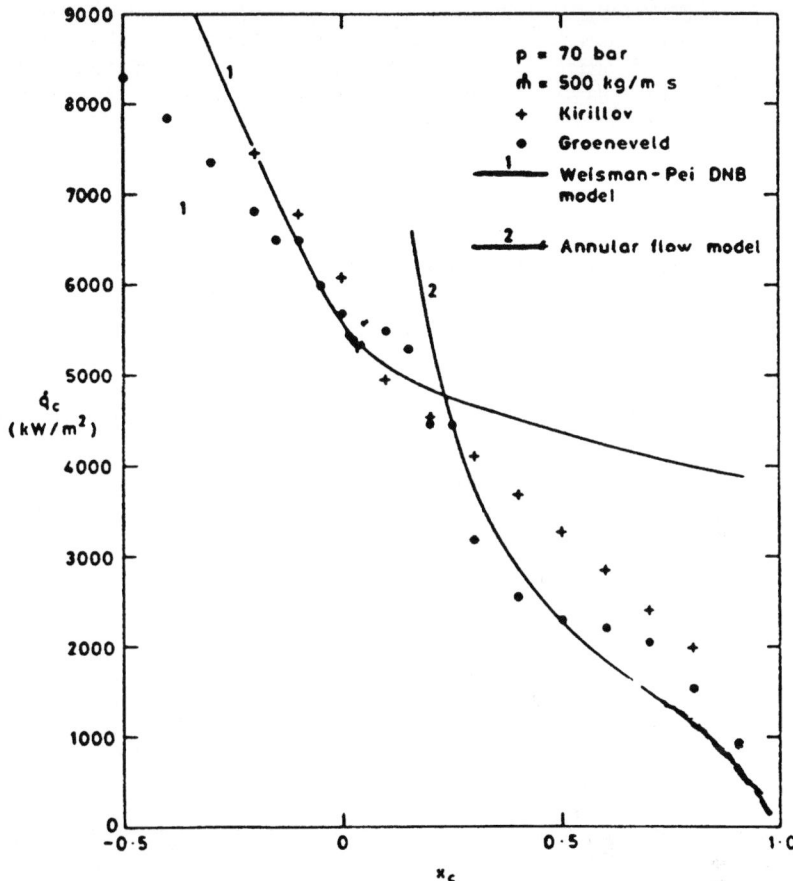

FIGURE 15.127   Comparison of predictions from bubble layer (Weisman and Pei [315]) and annular flow models (from Hewitt and Govan [298], with permission from ASME).

As shown from the results presented previously, the critical heat flux in annular flow occurs when the liquid film dries out on the channel wall. In principle, therefore, the prediction of critical heat flux involves simply integrating Eq. 15.324 along the channel until the point is reached at which $\Gamma = 0$. In order to do this, one needs a boundary value for Γ at the onset of annular flow and also relationships for $\dot{m}_D$ and $\dot{m}_E$, the deposition and entrainment rates. A reasonable choice for the boundary value Γ_a at the onset of annular flow is to assume that the entrainment and deposition processes are at equilibrium at that point (Hewitt and Govan [298]), but the results for the calculation of critical heat flux are not very sensitive to the precise value chosen for Γ_a. Thus, most of the effort in predicting critical heat flux in annular flow has been focused on the development of relationships for $\dot{m}_D$ and $\dot{m}_E$. Extensive reviews are given by Hewitt [291], Collier and Thome [3], and Hewitt and Govan [298]. As the deposition and entrainment rate relationships have gradually evolved over the past two and a half decades, the predictions from the annular flow critical heat flux model have gradually improved and become more general. However, the processes involved are extremely complex and one can foresee that this evolution will continue. $\dot{m}_D$ is often calculated from the relationship

$$\dot{m}_D = kC \tag{15.325}$$

where C is the concentration (mass per unit volume) of the droplets in the gas core and k is a droplet and mass transfer coefficient. C is usually calculated on the basis of there being a homogeneous mixture of droplets and vapor in the core and is thus given by

$$C = \frac{E\rho_l\rho_g(1-x)}{E(1-x)\rho_g + \rho_l^x}$$

(15.326)

where E is the fraction of the liquid phase that is entrained. The earlier correlations for k (see Hewitt [291]) did not take into account the effect of concentration on k, which later work showed to be significant. A more recent correlation is that of Govan [318] (see Hewitt and Govan [298]), which is compared with available data for deposition coefficient in Fig. 15.128. The following equations were given for the calculation of k:

$$k\sqrt{\frac{\rho_g D}{\sigma}} = 0.18 \qquad \text{for } C/\rho_g < 0.3$$

(15.327)

$$k\sqrt{\frac{\rho_g D}{\sigma}} = 0.83(C/\rho_g)^{-0.65} \qquad \text{for } C/\rho_g > 0.3$$

(15.328)

Having established a correlation for k (and hence a means of calculating $\dot{m}_D$), it is possible to deduce values of $\dot{m}_E$ from equilibrium annular flow data where $\dot{m}_D = \dot{m}_E$. This leads to a correlation for $\dot{m}_E$ (Fig. 15.129) in the following form:

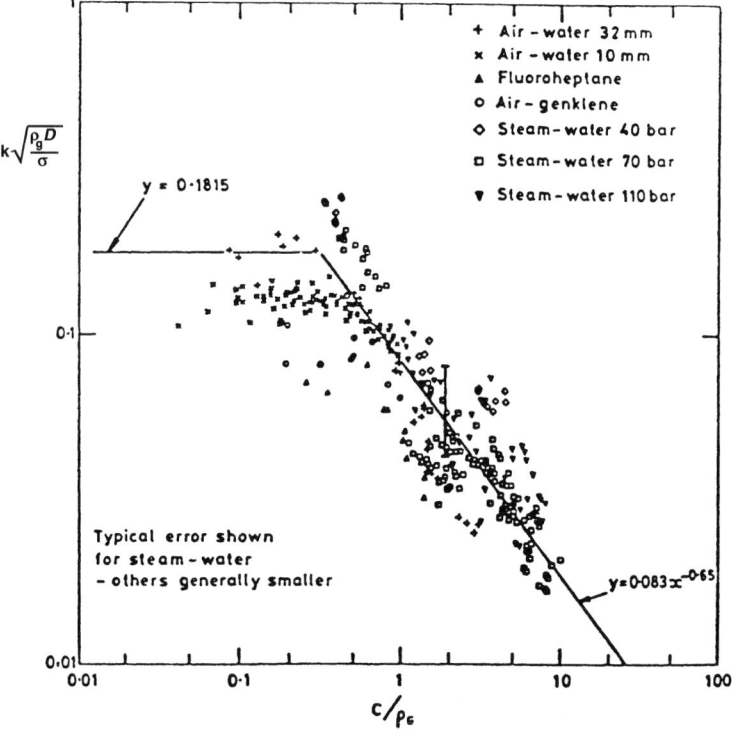

FIGURE 15.128 Correlation for deposition coefficient k (from Hewitt and Govan [298], with permission from ASME).

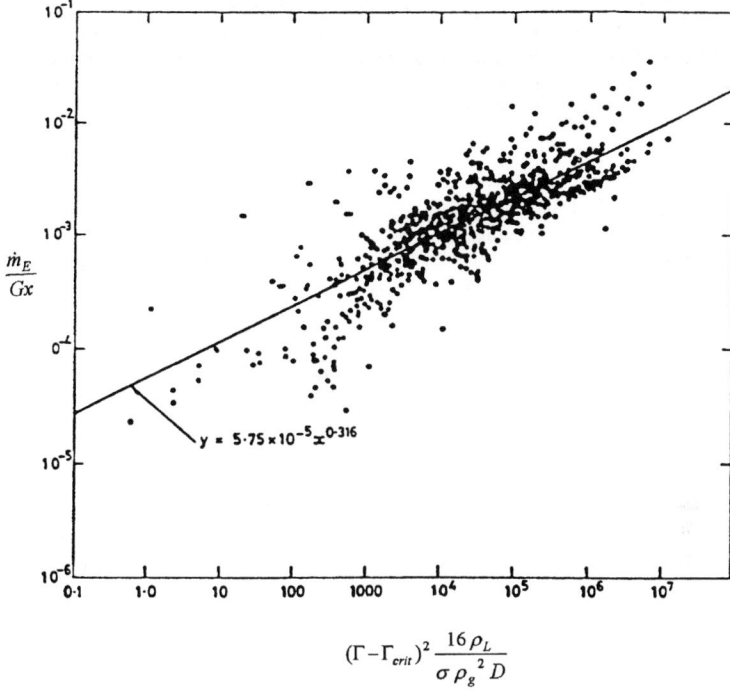

$$(\Gamma - \Gamma_{crit})^2 \frac{16\,\rho_L}{\sigma\,\rho_g^{\,2}\,D}$$

FIGURE 15.129 Correlations for entrainment rate $\dot{m}_E$ (from Hewitt and Govan [298], with permission from ASME).

$$\dot{m}_E/Gx = 5.75 \times 10^{-5}\left[(\Gamma - \Gamma_{crit})^2\,\frac{16\rho_l}{\sigma\rho_g^2 D}\right]^{0.316} \qquad (15.329)$$

where Γ_{crit} is a critical film flow rate for the onset of disturbance waves, which are a necessary condition for entrainment (i.e., $\dot{m}_E$ is 0 below a film flow rate per unit periphery less than Γ_{crit}). Γ_{crit} is given by

$$\frac{4\Gamma_{crit}}{\mu_l} = \exp[5.8504 + 0.4249(\mu_g/\mu_l)(\rho_l/\rho_g)^{1/2}] \qquad (15.330)$$

At high heat fluxes, and particularly at high pressures, the presence of the heat flux may itself influence the entrainment rate. Experiments aimed at evaluating the magnitude of this thermal contribution to entrainment have been carried out by Milashenko et al. [319]; they correlate the extra entrainment rate $\dot{m}_{Et}$ arising from this source by the following expression (which applies only to water):

$$\dot{m}_{Et} = 1.75[10^{-6}q''\rho_g/\rho_l]^{1.3}\,[\Gamma/(\pi D)] \qquad (15.331)$$

The application of annular flow modeling to the prediction of the critical heat flux phenomenon is illustrated here by taking two examples as follows:

1. A severe test of annular flow prediction models is provided by some data obtained by Bennett et al. [320], the results of which are illustrated in Fig. 15.130. In these experiments, film flow rate was measured as a function of distance along the channel and, knowing the local quality, the entrained flow rate could be calculated and is plotted. The results show two

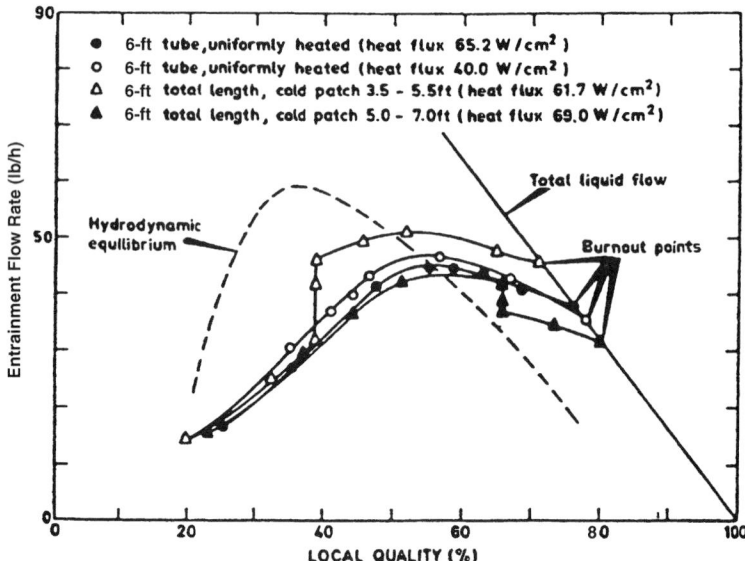

FIGURE 15.130   Variation of entrained droplet flow with quality (from Bennett et al. [320], with permission).

sets of data for uniformly heated tubes of two different lengths and also data for which there was a cold patch (i.e., an unheated length of tube) in which a shift of entrained flow rate was obtained at a fixed quality. The entrained flow rate tended toward the equilibrium value (where $\dot{m}_D = \dot{m}_E$) as shown, and this gave rise to a peak in the entrained flow rate as the data for the heated tube passed through the equilibrium as shown. When the entrained flow rate becomes equal to the total liquid flow rate, there is no liquid in the film and the critical heat flux is reached. Predictions of these data using annular flow modeling are shown in Fig. 15.131, and as stated by Hewitt and Govan [298], the model predicts all of the cases shown, including that for hydrodynamic equilibrium.

2. The predictions of the annular flow model are compared with data from the tables of Groeneveld et al. [303] and Kirillov et al. [304] in Fig. 15.127 (which also presents predictions from a modified version of the Weisman and Pei [315] low-quality CHF model). At high qualities, there is reasonable agreement between the annular flow model and the critical heat flux data, but a more interesting finding shown in this figure is the relationship between the subcooled/low-quality model and the annular flow model. Basically, the critical heat flux prediction selected should be the one that gives the lowest critical heat flux value at a given quality. As will be seen, the two lines cross over each other at a small but positive quality. Taking account of the scatter of the data, no dramatic difference is observed on this change of mechanism.

Annular flow modeling has been used extensively in predicting critical heat flux phenomena in annular flow. It has been used for prediction of critical heat flux in annuli and rod bundles (see Hewitt [291] for a review) and has also been successfully applied to the prediction of transient critical heat flux and to limiting cases of rewetting of a hot surface (see Hewitt and Govan [298]).

CHF in Forced Convective Boiling of Multicomponent Mixtures in Channels.   Reviews of critical heat flux data for the forced convective boiling of mixtures are presented by Collier and Thome [3] and by Celata [321]. In subcooled boiling and low quality, nucleate boiling predominates and similar effects are observed to those seen with pool boiling. This is exemplified

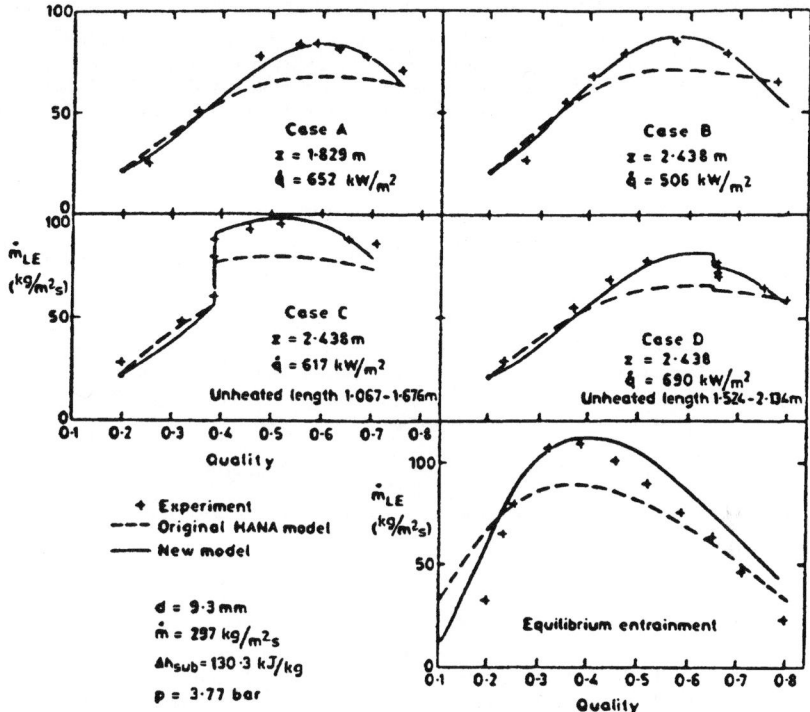

FIGURE 15.131 Predictions of annular flow model compared with entrained flow measurements of Bennett et al. [320] (from Hewitt and Govan [298], with permission from ASME).

by the results obtained by Tolubinski and Matorin [322], which are illustrated in Fig. 15.132. These subcooled boiling data were obtained for benzene-ethanol mixtures that have an azeotrope, and the pattern of results obtained (with a double peak in the critical heat flux) is very similar to the type of result obtained in pool boiling as illustrated in Fig. 15.67. As the quality increases, the critical heat flux behavior approaches a more idealized one with a linear interpolation between the values for the respective components of the mixture. This trend is described by Celata [321] and is well illustrated also by the data of Miyara [323], which are exemplified in Fig. 15.133. The critical heat flux at a given quality decreased with increasing length, tending to an asymptotic value. The data shown in Fig. 15.133 are for both the shortest length (0.058 m) and for the asymptotic figure. As will be seen, the critical heat flux shows a typical peak for the subcooled region but the asymptotic values in the quality region show a more ideal (linear) trend between the two components.

The implication is that correlations similar to those used for pool boiling can be used for multicomponent boiling at low-quality or subcooled conditions. For the annular flow region, predictions based on the models described in the preceding section should give reasonable results based on mean physical properties of the mixture.

Mitigation of CHF in Forced Convective Boiling in Channels.   In systems where the critical heat flux limit presents a serious problem, there is an obvious interest in enhancing its value. Methods of augmentation of critical heat flux in forced convective systems are reviewed by Collier and Thome [3] and by Bergles [324]. Although enhancement in forced convective subcooled boiling has been achieved using electrical fields (Nichols et al. [325]), the main practical means of enhancement rely on the use of swirl or mixing vanes, and tubes with internal ribbing.

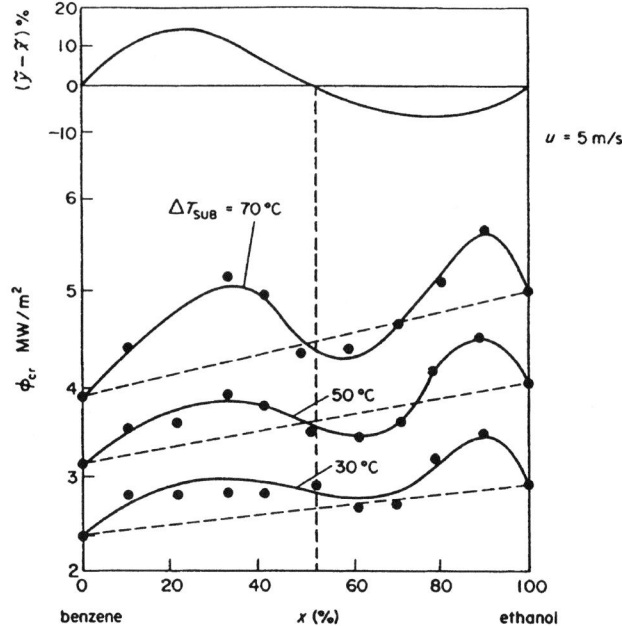

FIGURE 15.132 Critical heat flux in the forced convective subcooled boiling of benzene-ethanol mixtures (from Tolenbinsky and Maturin [322]. Reprinted by permission of John Wiley & Sons, Inc.).

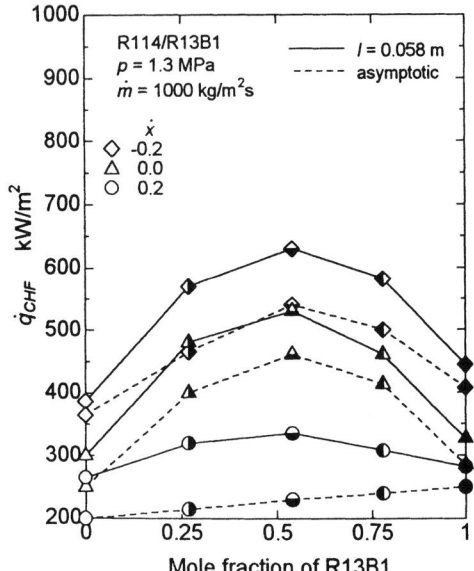

FIGURE 15.133 Variation of critical heat flux with composition at various qualities for boiling of R114/R13B1 mixtures (from Miyara et al. [323], with permission).

Swirling the flow with twisted tapes or mixing vanes can induce the liquid phase to wet the heated surface (thus preventing dryout). The tapes can be continuous along the channel or can consist of a number of shorter sections with plain tube between them. Typical studies using this technique are those of Moeck et al. [326], Nariai and Inasaka [327], and Chung et al. [328]. Although the tape can increase the critical heat flux at a given mass flux, it is also possible to observe a decrease in flux due to capture of liquid by the tape itself; this effect is illustrated in Fig. 15.134.

The use of ordinary surface roughness elements tends to reduce the critical heat flux. However, by introducing helical and fin-form deformations on the tube surface, it is possible to induce swirl and turbulence in the flow, which promotes transfer of the liquid phase to the surface and thus gives enhancement. An example of work of this kind is illustrated in Fig. 15.135, which shows results obtained by Nishikawa et al. [329] using rifled and ribbed tubes.

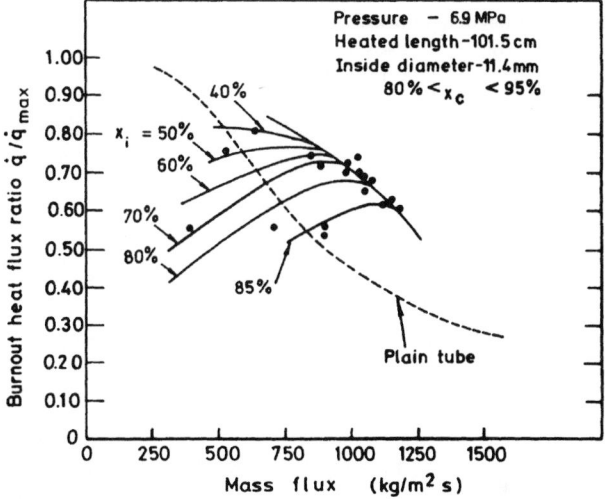

FIGURE 15.134 Fraction of liquid evaporated before onset of critical heat flux as a function of mass velocity for a plain tube and for a tube with helical tape inserts; q''_{max} is the flux corresponding to the evaporation of all the injected water (from Moeck et al. [326], with permission).

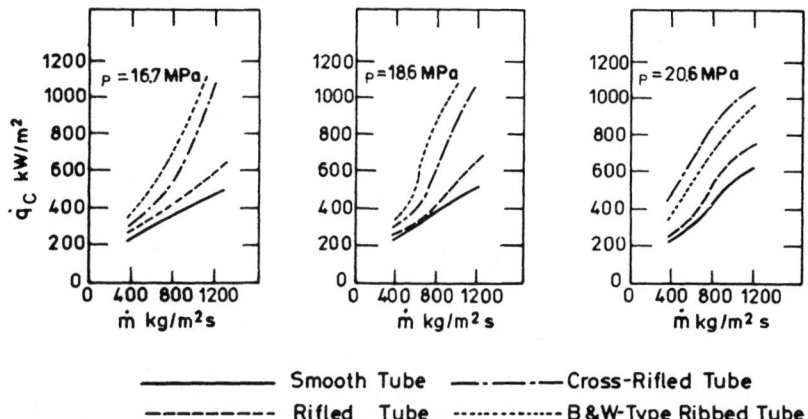

FIGURE 15.135 Critical heat flux data for plain and modified surface tubes (from Nishikawa et al. [329]. Reprinted by permission of John Wiley & Sons, Inc.).

Heat Transfer Beyond the Critical Heat Flux Limit in Forced Convective Boiling in Channels

Because of the importance of estimating temperatures in the postcritical region (mainly in the context of nuclear safety), there has been a vast literature on postcritical heat transfer; again, it is impossible to present it comprehensively within the space available here. The reader is referred to the review articles by Collier [330], Chen and Costigan [331], Ishii [332], Groeneveld [333], and Andreani and Yadigaroglu [334] for further information about the wealth of studies in this area.

Parametric Effects in Post-CHF Heat Transfer in Forced Convective Boiling in Channels. Forced convective post-CHF heat transfer is highly complex and the earlier practice of using pool boiling heat transfer relationships such as those described previously is entirely inappropriate. The complexity is illustrated by some results obtained by Hammouda et al. [335], which are shown in Fig. 15.136. The heat transfer coefficient (defined as the ratio of the heat flux to the wall superheat) initially decreases with increasing quality (decreasing subcooling) (region I). With further increases in equilibrium quality, the heat transfer coefficient rises rapidly with quality (region II) and then becomes relatively constant (region III). Finally, the heat transfer coefficient increases with increasing quality in region IV. As will be seen, the heat transfer coefficient increases with increasing mass flux and is higher with R-134a than with R-12.

Mechanisms of Post-CHF Heat Transfer in Forced Convective Boiling in Channels. The mechanisms of forced convective post-CHF heat transfer are reviewed by Collier and Thome [3], Carey [4], Tong and Tang [5], Ishii [332], Katto [101], and Andreani and Yadigaroglu [334], as well as others. The successive regimes that occur when the critical heat flux transition occurs in the subcooled boiling region are illustrated in Fig. 15.137 which is taken from the paper of Hammouda et al. [335]. Hammouda et al. associated the various regions of heat transfer illustrated in Fig. 15.136 with the regimes shown in Fig. 15.137. Thus, region I is associated with the subcooled inverted annular flow regime and region IV with the dispersed droplet flow regime; the relationships between regions II and III and the flow regime diagram are less clear-cut. When dryout occurs in annular flow, then the sequence of events is different, the system passing immediately to the dispersed droplet flow regime (where the droplet size distribution may be different since the droplets are created differently).

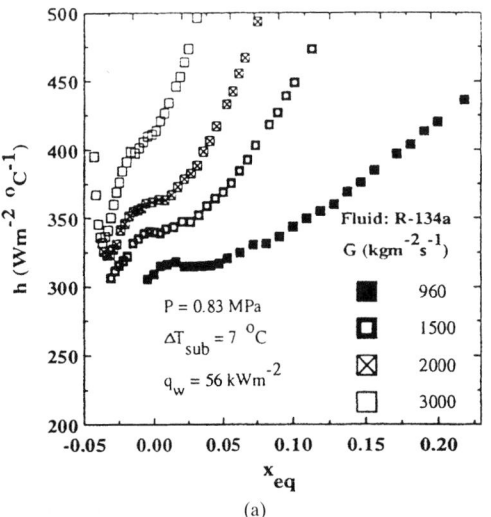
(a)

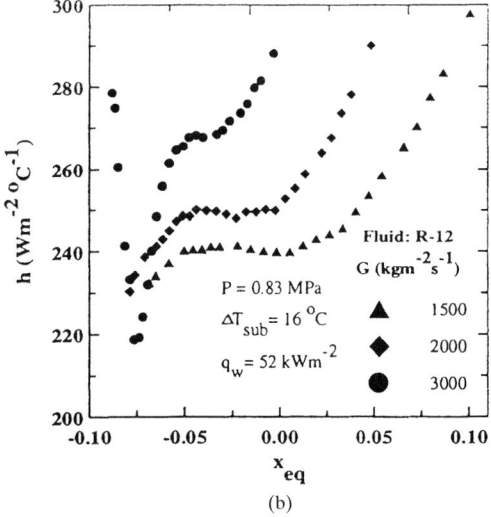
(b)

FIGURE 15.136 Post-CHF heat transfer coefficient as a function of equilibrium quality (from Hammouda et al. [335], with permission from Elsevier Science). (Continued on p. 15.133.)

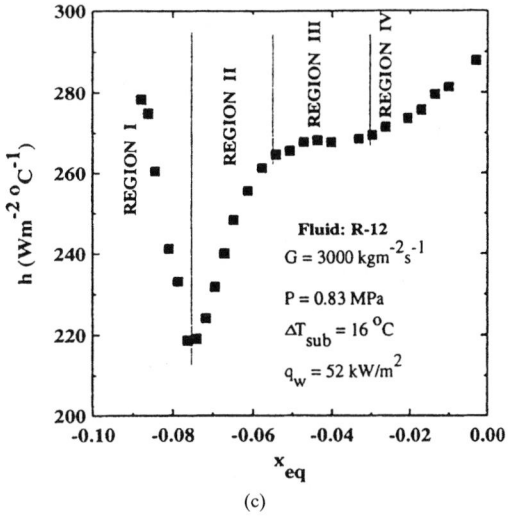

(c)

FIGURE 15.136 (*Continued*) Post-CHF heat transfer coefficient as a function of equilibrium quality (from Hammouda et al. [335], with permission from Elsevier Science).

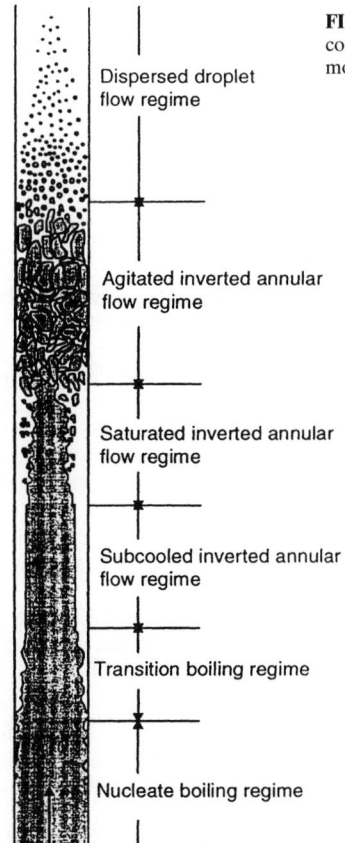

FIGURE 15.137 Flow regimes in low-quality film boiling (from Hammouda et al. [335], with permission from Elsevier Science).

Correlations for Post-CHF Heat Transfer in Forced Convective Boiling in Channels. Despite the complexity of the phenomena involved (as discussed earlier), there have been many attempts at providing general correlations. Here, we give two such correlations that serve as examples of this approach. The previously cited reviews should be consulted for further information.

The first example is that of Leonard et al. [336]. These authors suggest that it would be inappropriate to use the vertical height from the critical heat flux position as a length scale, as was done for the case of the vertical flat plate by Hsu and Westwater [191] (see Eq. 15.157). Rather, they suggested that the equation of Bromley [190] for horizontal cylinders (Eq. 15.163) should be used with the Helmholtz instability wavelength λ_H substituted for the tube diameter as the length scale for the vertical forced convective film boiling case. λ_H is given by

$$\lambda_H = 16.24\left[\frac{\sigma^4 i_{lg}^3 \mu_g^5}{\rho_g(\rho_l - \rho_g)^5 g^5 k_g^5 (T_w - T_{sat})^2}\right]^{1/2} \quad (15.332)$$

and the heat transfer coefficient is thus given by

$$h = 0.62\left[\frac{k_g^3 \rho_g(\rho_l - \rho_g)g i_{lg}}{\mu_g(T_w - T_{sat})\lambda_H}\right]^{1/4} \quad (15.333)$$

The Leonard et al. [336] correlation would apply to the inverted annular flow regimes illustrated in Fig. 15.137.

A correlation for the dispersed droplet flow regime (see Fig. 15.137) was developed by Groeneveld [337] and, for tubes, the correlation is as follows:

$$\frac{hD}{k_g} = 0.00109\left\{\left(\frac{GD}{\mu_g}\right)\left[x + \frac{\rho_g}{\rho_l}(1-x)\right]\right\}^{0.989} \mathrm{Pr}_g^{1.41}\, Y^{-1.15} \qquad (15.334)$$

where k_g, μ_g, and ρ_g are evaluated at the mean of the wall and saturation temperatures but Pr_g is evaluated at the wall temperature. The parameter Y is given by

$$Y = 1 - 0.1\left(\frac{\rho_l}{\rho_g} - 1\right)^{0.4}(1-x)^{0.4} \qquad (15.335)$$

Prediction of Post-CHF Heat Transfer in Forced Convective Boiling in Channels

Inverted Annular Region. The application of two-fluid models to the prediction of heat transfer in the inverted annular film-boiling regime (see Fig. 15.137) has been developed by a number of authors, including Analytis and Yadigaroglu [338] and Hammouda et al. [339]. These models are quite complex and sophisticated and involve:

1. Describing the process of heat transfer from the wall to the vapor-liquid interface and from the interface to the bulk fluid, taking account of convection and radiation.

2. Developing a description of the vapor film, modeling both the interfacial shear stress (affected by the bulk flow) and the wall shear stress.

3. Integrating the model from the critical heat flux point, calculating the development of the temperature profile downstream from this point.

Models of this type are beginning to show rather good agreement with experimental data, but they still involve a wide variety of assumptions with regard to closure relationships.

Model for Dispersed Film Flow Boiling. The dispersed flow film boiling region is an attractive one from the point of view of modeling since it does not involve the highly complex interfaces that are encountered in the inverted annular regime (see Fig. 15.137). Rather, in the dispersed region, there is a suspension of droplets in a vapor continuum and, if the droplets are considered to be spherical, this allows a much more mechanistic approach to be followed. In the earlier prediction methods it was assumed that the heat transfer took place in two steps, that is, from the wall to the vapor phase and then from the (superheated) vapor phase to the liquid droplets. This two-step mechanism was first suggested by Laverty and Rohsenow [340] and was further developed by Forsland and Rohsenow [341]. Work then proceeded simultaneously at AERE (UK) (Bennett et al. [342]) and at MIT (USA) (Forsland and Rohsenow [343]) on the development of the so-called four gradients method. In this method, it is assumed that the vapor is at the saturation temperature at the point of film dry-out and that a droplet size can be specified (in the event, over a reasonable range, the results are not too sensitive to the size selected). Downstream of this point, the droplet diameter, the local (actual) quality, the axial velocity of the drops, and the temperature of the vapor phase are determined by solving four simultaneous first-order differential equations as follows:

1. *Liquid momentum equation.* This considers the acceleration of the droplets as the vapor velocity increases along the channel. The difference in velocity between the vapor and the drops is an important parameter in heat transfer, and it is necessary to include this calculation.

2. *Liquid-mass continuity equation.* This equation describes the evolution of droplet diameter by evaporation. Early versions of the model took into account only of evaporation by heat transfer from the superheated vapor, but later versions took some account of direct evaporation of droplets at the wall.

3. *Vapor continuity equation.* Since the droplets are evaporating by heat transfer, the actual quality of the flow is also changing and the change with length can be estimated from the change in droplet diameter.

4. *Vapor energy equation.* Any heat that is transferred from the wall to the fluid, and that is not used to evaporate the droplets, is absorbed by superheating the carrier vapor. Of course, the higher the superheat, the more rapid the rate of evaporation of the droplets.

If one ignores the contribution of direct droplet-wall heat transfer and assumes that the heat transfer from the wall to the vapor continuous phase is identical to that for the vapor flowing alone, then two limiting cases of the model can be immediately recognized:

1. The heat fed through the channel wall is used entirely in superheating the vapor with no evaporation of the drops. Thus, the vapor and wall temperatures increase along the channel.

2. As long as the drops are present, they evaporate rapidly, absorbing heat from the vapor and maintaining it at saturation temperature. Here, the vapor velocity increases along the channel leading to increasing heat transfer coefficient and decreasing wall temperature until (at $x = 1$) the droplets are completely evaporated.

The wall and vapor temperature profiles associated with these two extreme cases are depicted in Fig. 15.138.

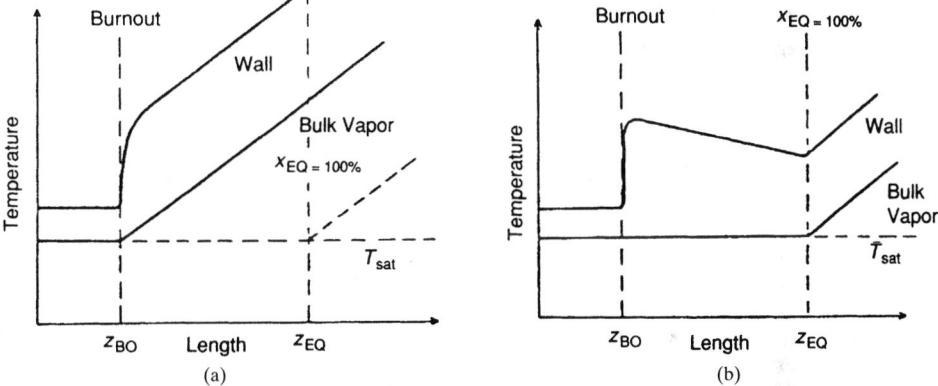

FIGURE 15.138 Limiting cases for postdryout heat transfer in the dispersed flow film-boiling region (from Hewitt et al. [13], with permission. Copyright CRC Press, Boca Raton, FL). (z_{BO} is the point at which dryout occurs and z_{EQ} is the point at which the equilibrium quality reaches unity.)

The actual wall temperature profiles tend to the first of these extremes at low mass fluxes and to the second at high mass fluxes (where the concentration of droplets is sufficient to maintain near-saturation conditions in the vapor). The four-gradient models are remarkably successful in predicting the systematic change from one extreme to the other, the calculated wall temperature profiles agreeing well with those measured.

Since the early development of the four-gradient models in the late 1960s, there has been continuous further development of these models, particularly under the direction of W. M. Rohsenow at MIT. A detailed review of the developments is presented by Andreani and Yadigaroglu [334]. Here, just a few salient features will be mentioned as follows:

1. *Drop-wall interaction.* The earlier models ignored the contribution of drop-wall interaction. The direct evaporation of droplets at the hot channel wall can play an important role when the wall temperatures are relatively low just upstream of the dryout point. Evans et al. [344] made measurements of vapor superheat just downstream of the dryout point and found that for approximately one-third of a meter downstream, the vapor remained at its saturation temperature, indicating that, in this region, the heat flux was being absorbed by direct droplet-wall contacts. Adaptation of the four-gradient models to take account of direct droplet-wall heat transfer is discussed by Ganic and Rohsenow [345], Iloeje et al. [346], and Varone and Rohsenow [347], among others.

2. *Radiation.* At very high wall temperatures, radiation heat transfer from the wall to the droplets can become significant. A typical model describing radiative heat transfer to a droplet-vapor mixture is that of Sun et al. [348].

3. *Turbulence modification.* The presence of a dispersed phase can influence the turbulence in the continuous fluid. Varone and Rohsenow [349] investigated this effect using a suspension of solid particles in air to simulate the vapor-droplet dispersion. Significant reductions of heat transfer coefficient were observed over a range of conditions. However, introduction of a correction to take account of this deterioration gave overpredictions of wall temperature and led Varone and Rohsenow to suggest that two-dimensional calculations were necessary to take account of the distributed heat sink. The concept of two-dimensional interactions between the droplet profile and the vapor temperature profile is being pursued by Andreani and Yadigaroglu [334].

In view of all the above factors, it may be the case that the four-gradient model in its simplest form is performing better than could reasonably be expected. Dispersed flow film boiling is a fascinating area of research and one might expect that this research will continue with the development of models of ever increasing complexity. For design purposes, however, even the simplest forms of the four-gradient model appear to do a better job than empirical correlations. This is well illustrated by considering cases where the heat flux in the postdryout region is nonuniform (Keeys et al. [350]).

Post-CHF Heat Transfer to Multicomponent Mixtures in Forced Convective Boiling in Channels. Studies of post-CHF for multicomponent mixtures are rather rare. Measurements with short tubes (and therefore at low quality) are reported by Auracher and Marroquin [351] and by Miyara et al. [323]. The studies were restricted to the subcooled and low-quality region. Figure 15.139 illustrates boiling curves obtained for pure refrigerant 114 and for a mixture of refrigerant 114 and refrigerant 13B1 ($CBrF_3$) with a mole fraction of 0.54 R13B1.

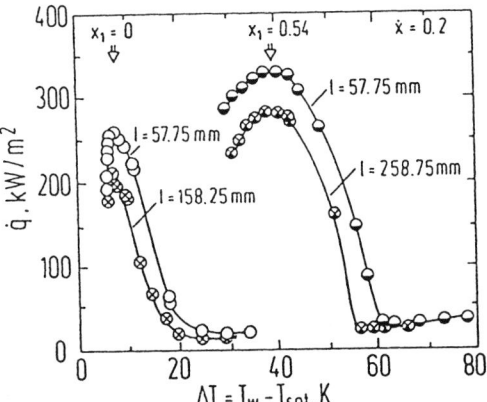

FIGURE 15.139 Comparison of forced convective boiling curves for boiling of R114 and a mixture of R114 and R13B1 in a 17-mm-diameter vertical tube with a mass flux of 1000 kg/m²s. Mixture composition: 0.54 mole fraction R13B1 (from Auracher and Marroquin [351], with permission from Taylor & Francis, Washington DC. All rights reserved).

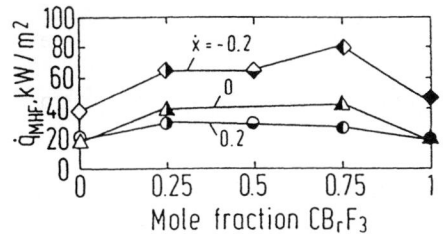

FIGURE 15.140 Effect of composition on minimum heat flux for forced convective boiling of R114/R13B1 (CBrF₃) mixtures in a vertical 17-mm-diameter pipe (from Auracher and Marroquin [351], with permission from Taylor & Francis, Washington, DC. All rights reserved).

Results were obtained at a flow quality of 0.2 and illustrate the enhancement of critical heat flux (as mentioned previously) and demonstrate a significant length effect in the transition boiling region, and also, of course, an effect of the multicomponent mixture. The minimum heat flux becomes insensitive to the mole fraction of R13B1 as the quality increases, as is illustrated in Fig. 15.140.

In the film-boiling region, there does not seem to be a dramatic effect of the multicomponent nature of the fluid (though the curves for the pure and mixed fluids do not overlap) and the effect of upstream length disappears (see Fig. 15.139).

In the dispersed flow film-boiling region, some effect of using fluid mixtures may become apparent since mass transfer effects would influence the rate of evaporation of the droplets. This evaporation rate would be decreased, giving a somewhat increased wall superheat relative to the pure fluids.

Enhancement of Post-CHF Heat Transfer in Forced Convective Boiling in Channels. The use of relatively large longitudinal fins attached to the heat transfer surface may potentially be a means of enhancing heat transfer in the post-CHF region. Clearly, such fins are most conveniently attached to the *outside* of a tube and can be used in parallel flow through rod bundles. Here, the enhancement is by the so-called vapotron principle, with film boiling occurring at the fin root area but with nucleate boiling occurring near the fin tip. Conduction along the fin minimizes the temperatures near the tube surface by this mechanism. A review of information on this form of enhancement is given by Collier and Thome [3].

In nuclear reactors, the fuel elements are often in the form of bundles of cylindrical rods with parallel flow between them. The individual fuel rods are separated from one another by means of supports (grids) placed periodically along the fuel element. In post-CHF heat transfer in such fuel bundles, it has been shown that the spacer grids can lead to a considerable enhancement of heat transfer through their interaction with the liquid phase, causing it to partially redeposit on the fuel elements downstream of the grid. The grids also enhance heat transfer by breaking up droplets and by enhancing the single-phase heat transfer to the vapor. A review of this topic is given by Hochreiter et al. [352].

THIN FILM HEAT TRANSFER

In the preceding section, the case of evaporation of thin films in annular flow was discussed. In this case, the vapor had a dominant role in determining the flow of the film through its influence on the interfacial shear stress. However, there are many situations of industrial importance in which a liquid film is present that falls down the heat transfer surface under the influence of gravity. Here, we can distinguish two cases:

1. The film is surrounded by the vapor phase and the interface is at a temperature close to its saturation temperature.

2. The liquid in the film is heated (or cooled) without any phase change occurring.

The relationships developed for these two cases are different and will be discussed below. Equipment using falling film heat transfer can be classified into *vertical* and *horizontal* systems. The vertical systems can include falling films on the inside or outside of tubes, or alternatively (in plate-type evaporators) on vertical flat plates. Generally, the liquid films are sufficiently thin to be treated as equivalent to the flat plate case for all of these configurations. Another important case is that of falling films on tube banks, as illustrated in Fig. 15.141; the

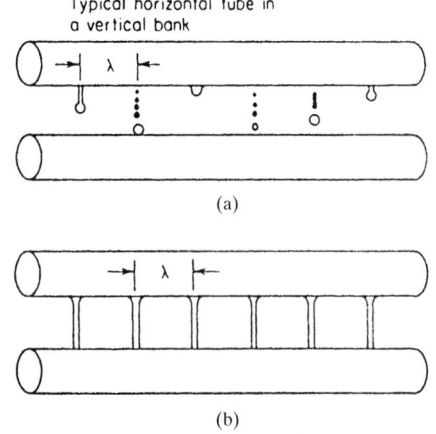

Typical horizontal tube in
a vertical bank

(a)

(b)

FIGURE 15.141   Falling films in tube banks (from Rohsenow [2], with permission of The McGraw-Hill Companies).

liquid can be transferred between the tubes in droplet form (Fig. 15.141a) or in liquid columns (Fig. 15.141b).

In the following, only the vertical flow case will be discussed for reasons of brevity. The reader interested in the tube bundle case can consult, for instance, the work of Gimbutis [353] and Ganic and Mastinaiah [354].

It is convenient in what follows to define a *film Reynolds number* Re_f as follows:

$$Re_f = \frac{4\Gamma}{\mu_l} \tag{15.336}$$

where Γ is the mass rate of liquid flow per unit periphery. For tubes, it is noteworthy that Re_f is the same as the Reynolds number for liquid flow alone in the tube.

The type of flow occurring in a falling film depends on Re_f as follows:

1. For $Re_f < 20$–30, the film is smooth and laminar.

2. For $Re_f > 30$–50, interfacial waves begin to appear on the film that can have a significant influence on hydrodynamic and heat transfer behavior.

3. At Re_f greater than approximately 1600, the film becomes turbulent, with the waves still present.

In the following, the case of evaporating liquid films for both laminar and turbulent flows is first discussed, followed by a discussion on the evaporation of liquid films consisting of multicomponent mixtures. Nucleate boiling can occur in liquid films if heat fluxes are high enough, and this topic is discussed next. There follows a brief discussion of heat transfer to nonevaporating (subcooled) falling films; finally, the important topic of film breakdown is briefly reviewed. As stated previously, all the discussion is focused on vertical falling film systems. The restriction of space in the present context necessitates a rather brief discussion of all these topics; the reader is referred to earlier reviews of this topic given by Hewitt and Hall-Taylor [271] and Fulford [355].

Evaporating Liquid Films: Laminar Flow

The classical treatment of falling liquid films is that of Nusselt [356]. By balancing viscous and gravitational forces, Nusselt obtained the following expression for the film thickness δ:

$$\delta = \left(\frac{3\mu_l\Gamma}{\rho_l^2 g}\right)^{1/3} = \left(\frac{3\mu_l^2}{4\rho_l^2 g}\right)^{1/3} Re_f^{1/3} \tag{15.337}$$

The heat transfer coefficient for laminar flow is given simply by

$$h = \frac{k_l}{\delta} = \left(\frac{4\rho_l^2 g k_l^3}{3\mu_l^2}\right)^{1/3} Re_f^{-1/3} \tag{15.338}$$

It is convenient to express the heat transfer coefficient in a nondimensional format (similar to a Nusselt number) and, in this format, Eq. 15.338 becomes

$$\frac{h}{k_l}\left(\frac{\mu_l^2}{\rho_l^2 g}\right)^{1/3} = \left(\frac{4}{3}\right)^{1/3} Re_f^{-1/3} = 1.1006 Re_f^{-1/3} \tag{15.339}$$

Although Eq. 15.337 fits data for film mean thickness for nonturbulent films, the heat transfer coefficient is enhanced in the wavy region; Chun and Seban [357] give the following expression taking account of this enhancement:

$$\frac{h}{k_l}\left(\frac{\mu_l^2}{\rho_l^2 g}\right)^{1/3} = 0.821 \text{Re}_f^{-0.22} \tag{15.340}$$

which applies for

$$\text{Re}_f < 5800 \text{Pr}_l^{-1.06} \tag{15.341}$$

Extensive work has been done on the study of waves on falling films (see for instance the review by Wasden and Dukler [358]), and information is available on the shape of the waves in fully developed falling film flows. Using this information, Jayanti and Hewitt [359] were able to use a computational fluid dynamics code (CFDS-CFX) to calculate the flow and heat transfer behavior in the waves. The waves can be considered as solitary ones traveling above a substrate. For high ratios of wave height to substrate height, recirculation within the wave begins as illustrated in Fig. 15.142. Temperature profiles in the waves were also calculated; the enhancement of heat transfer could be estimated from these. It was shown that the enhancement due to the waves is not a direct one; rather, the waves transport an increasing fraction of the fluid as Re_f increases and, for a given total liquid flow rate, the substrate region between the waves is thinned, giving a higher heat transfer coefficient in this region.

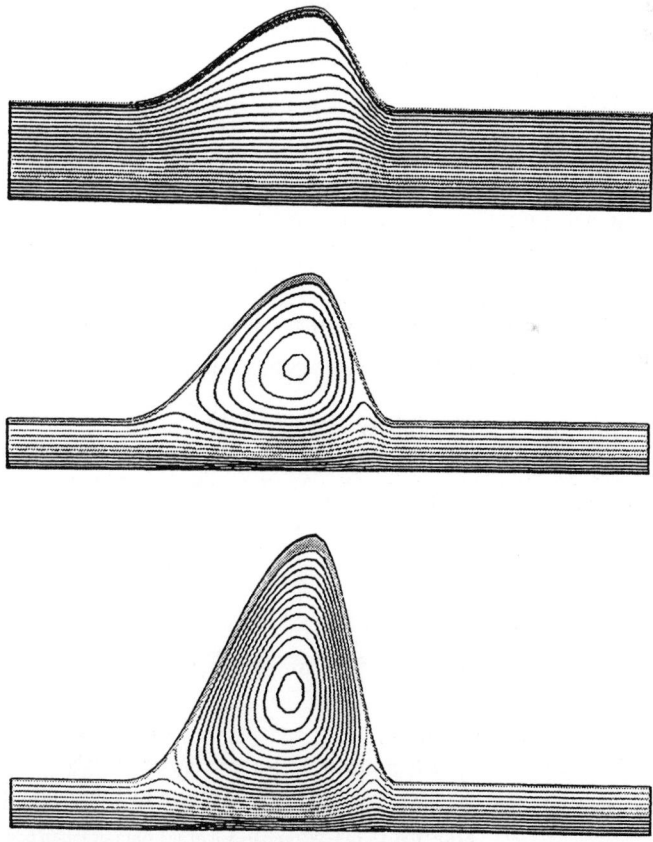

FIGURE 15.142   Flow patterns in falling film waves for wave height to substrate height ratios of 2, 4, and 6 (from Jayanti and Hewitt [359], with permission of Elsevier Science). Note: axial distance foreshortened by a factor of about 400 for presentational purposes.

Evaporating Liquid Films: Turbulent Flow

In turbulent falling films, there is a complex combined effect on heat transfer of the turbulence and interfacial waves. An empirical correlation for this region is that of Chun and Seban [357], which is as follows:

$$\frac{h}{k_l}\left(\frac{\mu_l^2}{\rho_l^2 g}\right)^{1/3} = 3.8 \times 10^{-3}\ \mathrm{Re}_f^{0.4}\ \mathrm{Pr}_l^{0.65} \qquad (15.342)$$

This equation applies for

$$\mathrm{Re}_f > 5800\mathrm{Pr}_l^{-1.06} \qquad (15.343)$$

Evaporating Liquid Films: Multicomponent Mixtures

Since falling films are used extensively in the concentration of solutes in aqueous solution, multicomponent mixture evaporation is very important from an industrial standpoint. Studies of multicomponent falling film evaporation are reported, for instance, by Palen et al. [360]. For such mixtures, the heat transfer coefficient can be below that predicted by Eq. 15.342; a review of data and correlations for this case is presented by Numrich [361], who suggested the following modified form of Eq. 15.342 to fit this data:

$$\frac{h}{k_l}\left(\frac{\mu_l^2}{\rho_l^2 g}\right)^{1/3} = 0.003\mathrm{Re}_f^{0.44}\ \mathrm{Pr}_l^{0.4} \qquad (15.344)$$

Comparisons between Eq. 15.344, Eq. 15.342, and the data of Palen et al. are shown in Fig. 15.143.

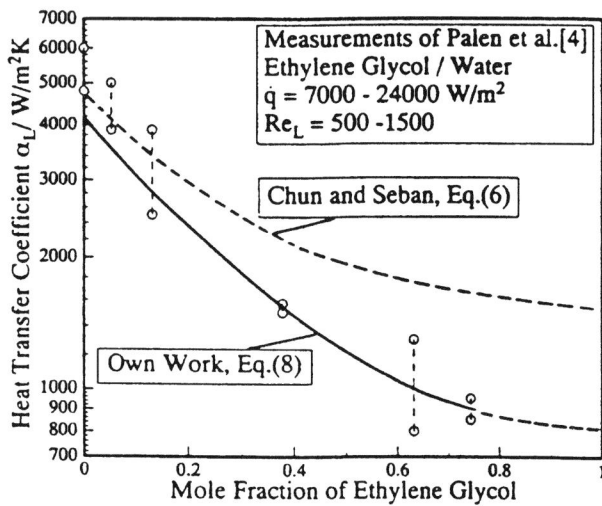

FIGURE 15.143 Comparison of the data of Palen et al. [360] with Eq. 15.342 (Chun and Seban) and Eq. 15.344 (Numrich) (from Numrich [361], with permission of Taylor & Francis, Washington, DC. All rights reserved).

Evaporating Liquid Films With Nucleate Boiling

At high heat fluxes, nucleate boiling can be initiated in falling film flow. It is often inappropriate to allow nucleate boiling to be initiated, particularly when films containing insoluble material are being evaporated. In this case, deposition of the solute may occur around the nucleation centers; to avoid nucleation, the heat fluxes should be kept below those calculated using the methods given earlier.

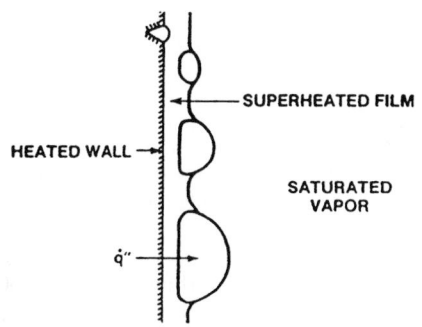

FIGURE 15.144   Bubble nucleation and growth in a falling liquid film (from Cerza [363], with permission from ASME).

When nucleate boiling does occur, correlations similar to those for pool boiling and forced convective boiling may be developed. An example of such a correlation, obtained for fully developed nucleate boiling in falling water films at atmospheric pressure, is that of Fujita and Ueda [362], which is as follows:

$$h = 1.24q''^{0.71} \qquad (15.345)$$

A mechanistic study of nucleate boiling in falling liquid films is reported by Cerza [363]. In a falling film, the bubbles detach from the nucleation site and continue to grow as they fall down the film, as illustrated in Fig. 15.144. Note that the bubble grows to a size much greater than the thickness of the film before bursting. Cerza developed a model based on the evaporation of the microlayer underneath the traveling bubble.

Heat Transfer to a Nonevaporating (Subcooled) Falling Liquid Film

For an evaporating film, the heat transfer coefficient is defined as $q''/(T_w - T_{sat})$ where T_w is the wall temperature and T_{sat} is the saturation temperature. However, if no evaporation is taking place and the film is merely warming up as a result of heat transfer from the wall, then the heat transfer coefficient is defined as $q''/(T_w - T_b)$ where T_b is the bulk (mixed mean) temperature, as conventionally used in single-phase heat transfer. For this case, the heat transfer coefficient for pure laminar flow (not taking account of the effect of waves and turbulence) is given by (Hewitt and Hall-Taylor [270]):

$$h = \frac{q''}{(T_w - T_b)} = \frac{280}{141} \left(\frac{\rho_l^2 g k_l^3}{3 \mu_l \Gamma} \right)^{1/3} = \frac{280}{141} \left(\frac{4 \rho_l^2 g k_l^3}{3 \mu_l^2} \right)^{1/3} \mathrm{Re}_f^{-1/3} \qquad (15.346)$$

An empirical correlation (subsuming the effects of waves and turbulence) for this subcooled film heat transfer case is given by Wilkie [364] as follows:

$$\frac{h\delta}{k} = C_o \, \mathrm{Re}_f^m \, \mathrm{Pr}_l^{0.344} \qquad (15.347)$$

where $C_o = 0.029$ and $m = 0.533$ for $\mathrm{Re}_f > 1600$, $C_o = 0.212 \times 10^{-3}$ and $m = 1.2$ for $1600 > \mathrm{Re}_f > 3200$, and $C_o = 0.181 \times 10^{-2}$ and $m = 0.933$ for $\mathrm{Re}_f > 3200$. The film thickness δ used in Eq. 15.347 is calculated from Eq. 15.337 for $\mathrm{Re}_f < 1600$ and from the following expression (from Feind [365]) for $\mathrm{Re}_f > 1600$:

$$\left(\frac{\delta^3 \rho_l^2 g}{\mu_l^2} \right)^{0.5} = 0.137 \mathrm{Re}_f^{0.75} \qquad (15.348)$$

Equation 15.347 is for *average* heat transfer coefficient. The local value of the coefficient may vary considerably along the film, and measurements demonstrating this are reported by Ulucakli [366].

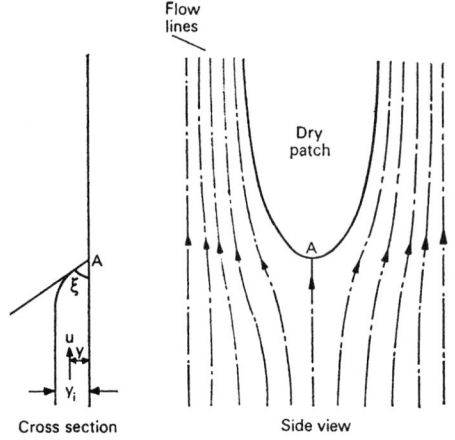

Flow lines

Dry patch

Cross section   Side view

FIGURE 15.145 Forces applying at an idealized dry patch on a liquid film (from Hartley and Murgatroyd [367], with permission from Elsevier Science).

Film Breakdown

A significant problem in the use of falling film evaporation systems is that of maintaining the integrity of the liquid film. The *minimum wetting rate* is required to rewet any dry patches formed on the film and evaporators should be operated above this minimum condition. A review on this topic is given by Hewitt and Hall-Taylor [271]. Typical of the earlier work in this area is that of Hartley and Murgartroyd [367], who considered the situation illustrated in Fig. 15.145.

A balance is made between the dynamic force arising from the change of direction of the liquid film and the surface tension force causing the dry patch to grow. When the dynamic force is greater than the surface tension force, then the dry patch will disappear. This allows the calculation of minimum wetting rate if the contact angle (see Fig. 15.145) is known.

A full review of rewetting behavior is beyond the scope of this present chapter, but several important factors can be mentioned as follows:

1. When surface tension variations can occur along the interface, then breakdown may be caused through the Marangoni effect. This phenomenon is reviewed by Hewitt and Hall-Taylor [271]. The surface tension variation can occur due to variation of heat and mass transfer from point to point along the surface due to the presence of the interfacial waves. This can occur in heat transfer to subcooled single components (though with evaporating systems the interface is maintained at close to the saturation temperature and the effect is largely eliminated). It can also occur when mass transfer is taking place and, in this case, the effect on minimum wetting rate is dependent on the effect of mass transfer on surface tension. If the concentration of one component in the thin film region between the waves leads to a decrease in surface tension, then this is destabilizing and vice versa. A particularly interesting case is that of an azeotrope where the concentration effect is different on one side of the azeotrope than on the other. The effect on minimum wetting rate is illustrated for this case by the work of Norman and Binns [368], whose results are shown in Fig. 15.146. On one side of the azeotropic composition, the minimum wetting rate is reduced as a result of mass transfer, whereas it is increased on the other side.

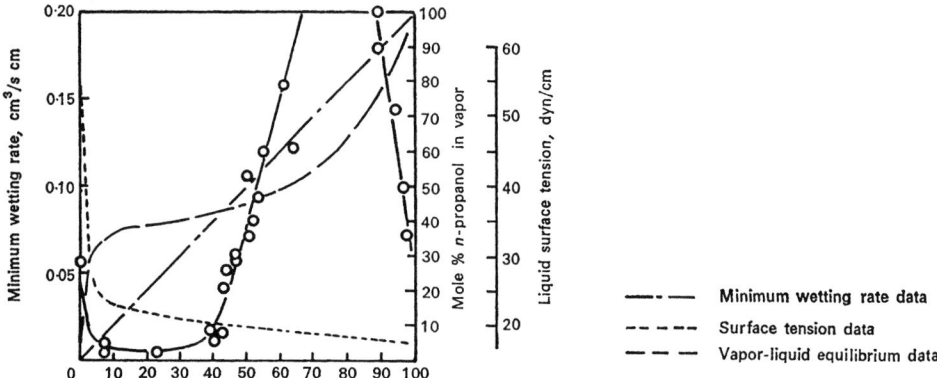

————— Minimum wetting rate data

– – – – – Surface tension data

— — — Vapor-liquid equilibrium data

FIGURE 15.146 Minimum wetting rate for a falling film in the presence of mass transfer systems: water/propanol, forming an azeotrope (from Norman and Binus [368], with permission).

2. As will be seen from Fig. 15.145, wetting behavior is intimately connected with surface forces. These, in turn, are strongly influenced by the presence of adsorbed layers on the surface; rewetting behavior in the presence of such adsorbed layers is discussed, for instance, by Wayner [369].

3. Film breakdown in falling film systems may be triggered by nucleate boiling; evidence for this is given by Alhusseini et al. [370].

Wettability is also affected by the presence of surface contaminants (such as grease), and the problem of wetting must always be addressed in falling film systems.

REWETTING OF HOT SURFACES

When a surface becomes very hot, the process of rewetting (quenching) of that surface by the application of a liquid flow onto it is of considerable industrial importance. The problem of quenching has, of course, been significant in the metallurgical industries for many centuries, in the context of quenching of hot metal objects after casting or forging. The quenching process causes a modification of the crystal structure of the metal and is an important part of obtaining the required properties. Moreover, in the past three decades, quenching processes have been studied extensively in the context of nuclear reactor safety. If the coolant is ejected from a water-cooled nuclear reactor as a result of a loss of coolant accident (LOCA), then emergency cooling water has to be fed to the core, and the rate at which this water can quench the core is of vital importance in the safety assessment of such reactors. There has been a wide range of studies in the nuclear reactor context; the space available here does not allow more than a cursory overview to be made of these. More thorough reviews are given by Collier [371] and by Nelson and Pasamehmetoglu [372].

A useful classification of regimes in quenching is given by Nelson and Pasamehmetoglu [372] and is shown in Fig. 15.147. Basically, four regions were identified:

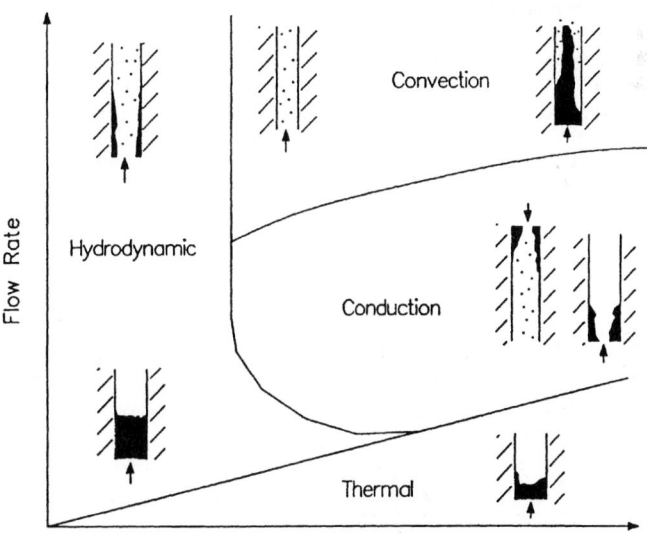

FIGURE 15.147   Classification of quenching regimes (from Nelson and Pasa-mehmetoglu [372], with permission. Copyright CRC Press, Boca Raton, FL).

1. *Hydrodynamic limited rewetting.* This regime occurs at low flow rates and wall temperatures and represents the limiting rate at which fluid can access the hot surface. For instance, in annular flow, the film dries out at zero film flow rate and the propagation of this zero film flow rate condition represents a limiting case for rewetting. The zero film flow rate limit is reached when, upstream of the dryout point, the processes of entrainment, evaporation, and deposition are such that the film flow rate is zero at the dryout point; modeling of annular flow dryout was described earlier. In the case of rewetting, a transient in the inlet flow causes the surface to begin to rewet.

2. *Convection limited.* Here, heat transfer downstream of the rewetting front is sufficient to bring the surface temperature down to a low enough value to allow easy rewetting. Thus, in this case, control of the phenomenon is in convective flow downstream of the rewetting point.

3. *Conduction limited.* Here, the rate of removal of stored energy from the hot wall in front of the quench front is governed by conduction from the hot zone to the colder zone upstream of the front.

4. *Thermal limited.* In the hydrodynamic convection and conduction limited cases, there is usually an amount of fluid entering the system that is in excess of that required to cool the hot surface as a result of evaporation. However, there will exist a thermal limit in which the heat generation in the system exceeds that which can be removed by the liquid being introduced.

The modeling of the quenching phenomenon has usually been focused on the conduction limited case, which is arguably the most important. Models are usually constructed in terms of a coordinate framework that moves at the quench front velocity (u_r). The models take two main forms:

1. *One-dimensional models.* Here, it is assumed that the temperature in the hot wall does not vary with position across the wall; clearly this is a considerable oversimplification, but it may be a reasonable description if the wall is relatively thin.

2. *Two-dimensional models.* Here, the two-dimensional nature of the temperature profiles within the solid wall is taken into account.

Detailed tabular information on the many models of the above two types that have been used is given in the reviews of Collier [371] and Nelson and Pasamehmetoglu [372]. The models differ in their assumption of the distribution of heat transfer coefficients around the quench front and in their specification of the temperature of the surface adjacent to the quench front, the so-called rewetting temperature or sputtering temperature (T_{sp}). Since assumptions about heat transfer coefficients and about the value of T_{sp} can be made independently, these values can be adjusted to give good fit to ranges of experimental data. The simplest model is that of Yamanochi [373], who gives the following expression for u_r:

$$\frac{1}{u_r} = \frac{\rho_w c_{pw}}{2}\left(\frac{\delta_w}{h(z)k_w}\right)^{1/2}\left\{\left[\frac{2(T_w - T_{sp})}{T_{sp} - T_{sat}}\right]^2 - 1\right\}^{1/2} \tag{15.349}$$

where ρ_w, c_{pw}, δ_w, and k_w are the density, specific heat capacity, thickness, and thermal conductivity of the wall material. T_w is the temperature of the hot wall downstream of the quench front. The heat transfer coefficient $h(z)$ was assumed in the work of Yamanochi to be essentially zero downstream of the quench front and to have a fixed value (between 2×10^5 and 10^6 W/m^{2}K) in the region upstream of the quench front. A correlation of $h(z)$ suitable for use with this model is presented by Yu et al. [188].

The mechanism controlling rewetting may change during a particular flow transient. Hewitt and Govan [298] describe calculations on the rewetting of the tube (which had been previously dried out) as a consequence of a flow transient. Their results are typified by those shown in Fig. 15.148. If a value of $T_{sp} = T_{sat} + 80(K)$ is assumed, then the rewetting behavior is

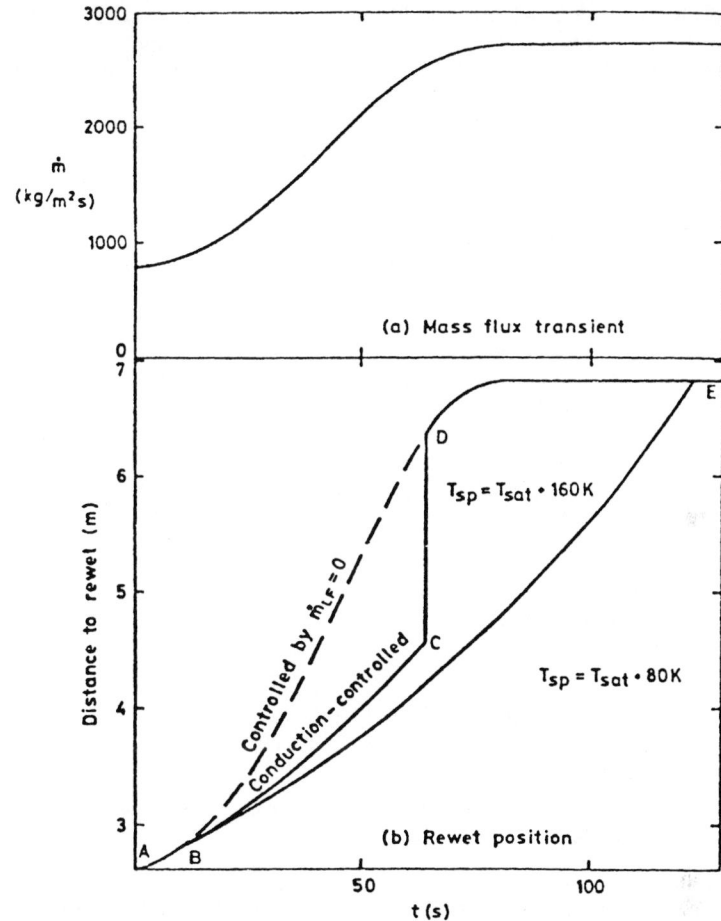

FIGURE 15.148   Movement of quench front calculated for an increasing flow transient (from Hewitt and Govan [298], with permission from ASME).

conduction-controlled throughout. If, on the other hand, a value of $T_{sp} = T_{sat} + 160(K)$ is taken, the rewetting is initially conduction-controlled (along line ABC in Fig. 15.148) but then switches rapidly to hydrodynamic control (film flow rate equal to zero at the quench front) for the remainder of the quench (line CDE).

NOMENCLATURE

Symbol, Definition, Units

a	parameter in Eq. 15.2, m^4/s^2
	acceleration, m^2/s
a_d	projected area of bubble on surface at time of bubble departure, m^2

A	parameter defined by Eq. 15.41 or 15.45
	factor in Eq. 15.98 (given by Eq. 15.100)
	factor in Eq. 15.115
	total surface area, m^2
$A*$	parameter defined by Eq. 15.83, $Pa^{0.69}$
A_e	area of surface associated with evaporation, m^2
A_i	interfacial area, m^2
A_j	area occupied by jets, m^2
A_{nc}	area of surface associated with natural convection, m^2
A_o	factor in Eq. 15.116
A_t	area of surface associated with transient conduction, m^2
A_v	vapor stem area, m^2
A_w	total wall area, m^2
b	parameter in Eq. 15.2, m^3/kg
	distance from surface to edge of bubble (see Fig. 15.8), m
	exponent in Eq. 15.218
B	parameter defined by Eq. 15.42 or 15.46
Bi*	Biot number defined by Eq. 15.149
Bo	Bond number (Eq. 15.61)
	scaling parameter in Eq. 15.118
	boiling number (Eq. 15.245)
Bo_{crit}	boiling number corresponding to critical heat flux (Eq. 15.297)
c_p	specific heat capacity, J/kg K
$\tilde{c}_p$	molar specific heat capacity, J/kmol K
C	factor defined by Eq. 15.101
C_o	distribution parameter (Eqs. 15.217 and 15.218)
	convection number (Eq. 15.244)
C_{SF}	constant in Eq. 15.81
C_{TD}	flow pattern parameter (Eq. 15.195)
d_b	bubble departure diameter, m
D	cylinder diameter, m
	tube inside diameter, m
D'	nondimensional diameter defined by Eq. 15.166
D_b	bubble diameter, m
	tube bundle diameter, m
D_c	effective cavity diameter (Eq. 15.26), m
D_c^*	effective cavity diameter (Eq. 15.23)
D_c^1	cavity equivalent diameter after removal of material (see Fig. 15.10), m
D_o	tube outside diameter, m
D_W	stem diameter, m
e	emissivity
E	fraction of liquid entrained as droplets

f	bubble frequency, 1/s
	frequency of vapor mushroom departure, 1/s
f_D	cavity shape factor (Eq. 15.26)
$f(\phi)$	factor by which nucleation temperature is reduced on a surface with contact angle ϕ
F	Helmholtz free energy, J
	fraction of liquid contact (Eq. 15.145)
	enhancement factor (Eq. 15.229)
F_b	buoyancy force (Eq. 15.66), N
	correction factor for circulation (Eq. 15.192)
F_c	correction factor for boiling range (Eq. 15.191)
F_d	drag force (Eq. 15.67), N
$F(P)$	pressure function (Eq. 15.82)
Fr_{le}	Froude number defined by Eq. 15.246
F_s	surface tension force (Eq. 15.71), N
$F_1–F_4$	parameters in Eqs. 15.285–15.296
g	acceleration due to gravity, m²/s
G	mass flux, kg/m²s
G_F	mass flux for scaling fluid (Eq. 15.299), kg/m²s
Gr	Grashof number
G_W	equivalent mass flux for water (Eq. 15.299), kg/m²s
h	heat transfer coefficient, W/m²K
h^*	dimensionless heat transfer coefficient (Eq. 15.95)
h_c	convective film boiling coefficient, W/m²K
h_{FC}	forced convective heat transfer coefficient, W/m²K
h_{FZ}	nucleate boiling coefficient calculated from Forster and Zuber correlation (Eq. 15.80), W/m²K
h_{id}	ideal heat transfer coefficient (Eq. 15.112), W/m²K
h_{NB}	nucleate boiling heat transfer coefficient, W/m²K
h_o	reference value of heat transfer coefficient (Eq. 15.90), W/m²K
h_r	radiative film boiling coefficient, W/m²K
h_1	heat transfer coefficient for component 1 at given heat flux, W/m²K
h_2	heat transfer coefficient for component 2 at given heat flux, W/m²K
i	enthalpy, J/kg
i_i	inlet enthalpy, J/kg
i_{lg}	latent heat of vaporization, J/kg
i'_{lg}	corrected latent heat of vaporization (Eq. 15.158), J/kg
$i_{l,sat}$	saturated liquid enthalpy, J/kg
j_g^*	Wallis dimensionless parameter (Eq. 15.222)
J	number of vapor embryos formed per unit volume per unit time, 1/m³s
	parameter in Eq. 15.155
Ja	Jakob number (Eq. 15.53)

Ja_e	Jakob number defined by Eq. 15.141
k	thermal conductivity, W/m K
	droplet deposition coefficient, m/s
k_{eff}	effective (turbulent) thermal conductivity, W/m K
K	factor relating area swept by incoming liquid to projected area (a_d) of departing bubble
	factor defined by Eq. 15.102
	constant in Eq. 15.128
	constant in Eq. 15.135
	parameter in Eq. 15.148 (see Eq. 15.152)
	parameter used in Katto and Ohne correlation (Eq. 15.302)
K_l	parameter defined by Eq. 15.63
K_s	parameter in Eq. 15.119 (given by Eq. 15.120)
L	characteristic dimension, m
	stem spacing, m
	plate height, m
	channel length, m
L'	characteristic length scale (Eq. 15.124)
L_B	boiling length, m
$(L_B)_{crit}$	critical boiling length for dryout, m
$\dot{m}_D$	droplet deposition rate, kg/m²s
$\dot{m}_E$	droplet entrainment rate, kg/m²s
$\dot{m}_{Et}$	droplet entrainment rate due to heat flux, kg/m²s
M	mass of substance, kg
$M*$	parameter defined by Eq. 15.170
$\tilde{M}$	molecular weight, kg/kmol
Ma	Marangoni number (Eq. 15.143)
n	factor in Eq. 15.33
	exponent given by Eq. 15.92
	exponent in Eq. 15.233
	exponent in Eq. 15.286
$\tilde{N}$	number of kmols of material, kmol
N_a	number of active sites per unit surface area, 1/m²
N_{as}	number of sites per unit surface area having $\psi_m < 90°$, 1/m²
N_{mol}	Avogadro number (6.022×10^{26}), molecules/kmol
N_s	number of all cavities per unit surface area, 1/m²
Nu	Nusselt number
p_t	Tube pitch, m
P	pressure, Pa
P_c	critical pressure, Pa
P_r	reduced pressure (P/P_c)
Pr	Prandtl number ($c_p\mu/k$)

P_R	pressure relative to 6.895 MPa (Eq. 15.288)
P_s	wetted periphery (see Fig. 15.66), m
P_{sat}	saturation pressure, Pa
P_∞	vapor pressure on planar interface, Pa
q^*	dimensionless heat flux (Eq. 15.96)
q''	heat flux, W/m^2
q''_b	heat flux due to boiling, W/m^2
q''_{crit}	critical heat flux, W/m^2
q''_e	heat flux due to evaporation, W/m^2
q''_{min}	minimum heat flux, W/m^2
q''_{nc}	heat flux due to natural convection (local or average), W/m^2
q''_R	heat flux calculated from Rohsenow equation (Eq. 15.81), W/m^2
q''_t	heat flux due to transient conduction, W/m^2
q''_{tr}	heat flux for transition to fully developed boiling, W/m^2
r	radius of curvature, m
r^*	equilibrium bubble radius (Eq. 15.15), m
r^+	nondimensional radius defined by Eq. 15.39
r_c	conical cavity mouth radius, m
	critical radius given by Eq. 15.48, m
r_{crit}	critical cavity radius, m
r_{max}	maximum cavity radius, m
r_s	radius of largest cavity, m
$r_1(t)$	inertial contribution to bubble growth (Eq. 15.55), m
$r_2(t)$	thermal contribution to bubble growth (Eq. 15.56), m
R	cylinder radius, m
R^*	dimensionless roughness (Eq. 15.97)
$\tilde{R}$	universal gas constant (8314 J/kmol K), J/kmol K
R'	nondimensional radius (Eq. 15.138)
R_a	arithmetic mean deviation of surface profile (*Mittenrauhwert*), m
Re	Reynolds number
Re$_b$	bubble Reynolds number (Eq. 15.70)
Re$_f$	film Reynolds number (Eq. 15.269, Eq. 15.336)
R_j	jet radius (see Fig. 15.64), m
S	shape factor (Eq. 15.72)
	channel periphery, m
	suppression factor (Eq. 15.240)
Sn	Scriven number (Eq. 15.59)
t	time, s
t^+	nondimensional time defined by Eq. 15.40
t_c	time between initiation and collapse of a bubble, s
t_d	time at which bubble departs, s

t_g	growth time, s
t_w	waiting period (for start of bubble growth), s
T	temperature, K
T^+	nondimensional temperature (Eq. 15.275)
T_B	bulk (mixed) temperature in forced convection, K
T_{bub}	bubble point temperature, K
T_c	critical temperature, K
T_{crit}	critical temperature for bubble growth, K
T_{dew}	dew point temperature, K
T_{dfb}	temperature at which liquid contact starts, K
T_{hn}	homogeneous nucleation temperature, K
T_I	temperature at liquid/solid interface on contact, K
$T_{\min}$	minimum temperature, K
T_n	homogeneous nucleation temperature, K
T_r	reduced temperature (T/T_c)
$T_{r,n}$	reduced homogeneous nucleation temperature
$(T_r)_s$	reduced spinoidal temperature
$T_{r,\text{sat}}$	reduced saturation temperature, K
T_{sat}	saturation temperature, K
$T_{\text{sat}}(P_l)$	saturation temperature at pressure P_l, K
$T_{\text{sat}}(P_\infty)$	saturation temperature far from solid surface, K
T_W	wall temperature, K
T_∞	bulk fluid temperature, K
u	velocity
u_{GU}	mean relative velocity (see Eq. 15.221), m/s
U_H	velocity for Helmholtz instability (Eq. 15.126), m/s
U_∞	fluid velocity approaching cylinder, m/s
v	specific volume, m^3/kg
	volumetric growth rate of bubble, m^3/s
v_c	critical specific volume, m^3/kg
v_{lg}	change of specific volume on vaporization, m^3/kg
v_r	reduced specific volume (v/v_c)
V	volume, m^3
Vi	viscosity number (Eq. 15.134)
W'	parameter used in Katto and Ohne correlation (Eq. 15.305)
x	quality
$\tilde{x}$	mole fraction in liquid phase
x_{crit}	critical quality for dryout (CHF)
X	parameter used in Katto and Ohne correlation (Eq. 15.302)
X_{tt}	Martinelli parameter (Eq. 15.232)
y	distance from wall, m
$\tilde{y}$	mole fraction in vapor

y^+	friction distance parameter (Eq. 15.263)
Y	parameter in Groeneveld correlation (Eqs. 15.334 and 15.335)
z	axial distance, m
z_d	distance to onset of bubble detachment, m
z_n	distance to onset of nucleation, m
Z'	parameter used in Katto and Ohne correlation (Eq. 15.308)

Greek Symbols

α	thermal diffusivity
	void fraction
α_m	void fraction in macrolayer
β	angle of cavity (see Fig. 15.9), degrees (radians)
	volume flow ratio (Eq. 15.219)
β_l	liquid phase mass transfer coefficient, m/s
Γ	liquid mass flow per unit periphery, kg/ms
Γ_{crit}	liquid flow per unit periphery for onset of entrainment, kg/ms
Γ_{gL}	mass flow of vapor per unit periphery at distance L along surface, kg/ms
δ	liquid film thickness, m
δ^+	nondimensional film thickness (Eq. 15.268)
δ_m	minimum thickness of macrolayer, m
δ_o	macrolayer thickness at point of departure of vapor mushroom, m
Δi_{sub}	inlet subcooling ($i_{l,sat} - i_i$), J/kg
Δp_o	initial pressure difference between interior and exterior of bubble, Pa
ΔT	temperature difference, K
ΔT_{bp}	elevation in temperature corresponding to difference between bubble point and dew point at local liquid concentration, K
ΔT_E	additional temperature difference (Eq. 15.115), K
ΔT_I	ideal temperature difference (Eq. 15.14), K
ΔT_S	difference in saturation temperatures of pure components, K
ΔT_{sat}	saturation temperature difference, K
ΔT_{sub}	difference between saturation temperature and temperature (subcooling), K
ΔT_1 and ΔT_2	temperature difference for boiling of components 1 and 2 at given heat flux, K
ε	eddy diffusivity, m²/s
ε'	modified eddy diffusivity (Eq. 15.280), m²/s
ε_h	eddy diffusivity for heat, m²/s
ε_m	eddy diffusivity for momentum, m²/s
η	parameter given by Eq. 15.20
θ	parameter defined by Eq. 15.110
λ	parameter in Eq. 15.5
λ_c	minimum unstable Taylor wavelength, m
λ_D	wavelength for maximum growth rate (Eq. 15.127), m
λ_H	wavelength for Helmholtz instability, m
μ	viscosity, Ns/m²

μ_{eff}	effective viscosity, Ns/m^2
ρ	density, kg/m^3
ρ_c	core density (Eq. 15.252), kg/m^3
ρ_H	homogeneous density (Eq. 15.196), kg/m^3
σ	surface tension, N/m (or J/m^2)
σ_S	Stefan-Boltzmann constant (5.669×10^{-8}), W/m^{2}K^4
τ	t/t_c (Fig. 15.24)
	hovering period of vapor mushrooms, s
	dimensionless time (Eq. 15.151)
τ_i	interfacial shear stress, N/m^2
υ	kinematic viscosity, m^2/s
ϕ	contact angle, degrees (radians)
ϕ_a	advancing contact angle, degrees (radians)
Φ_b	bundle correction factor (Eq. 15.204)
ψ	Ahmad scaling parameter (Eq. 15.298)
ψ_H	scaling parameter for horizontal flow (Eq. 15.317)
ψ_m	cavity mouth angle (Eq. 15.24), degrees (radians)
Ψ	parameter defined by Eq. 15.200

Subscripts

A,B,C,D	component A, B, C, D in mixtures
bub	bubble point
bulk	bulk
c	critical
dew	dew point
F	scaling fluid
g	gas (vapor)
i	interfacial
in	inlet
l	liquid
r	reduced (relative to critical)
s	solid
sat	saturation
W	wall
W	water

Superscripts

~	molar value

REFERENCES

1. W. M. Rohsenow, "Boiling," in *Handbook of Heat Transfer,* W. M. Rohsenow and J. P. Hartnett eds., Sec. 13, McGraw-Hill Book Company, New York, 1973.

2. W. M. Rohsenow, "Boiling," in *Handbook of Heat Transfer Fundamentals,* W. M. Rohsenow, J. P. Hartnett, and E. N. Ganic eds., Chap. 12, McGraw-Hill Book Company, New York, 1985.

3. J. G. Collier and J. R. Thome, *Convective Boiling and Condensation,* 3d ed., Clarendon Press, Oxford, UK, 1994.

4. V. P. Carey, *Liquid Vapor Phase Change Phenomena,* Hemisphere Publishing Corporation, Washington, DC, 1992.

5. L. S. Tong and Y. S. Tang, *Boiling Heat Transfer and Two-Phase Flow,* 2d ed., Taylor and Francis, Washington, DC, 1997.

6. K. Nishikawa, "Developments of Boiling Heat Transfer and its Applications," *Proc. 2nd International Conference on Multiphase Flow,* Kyoto, Japan, PL3-1/PL12, April 1995.

7. G. F. Hewitt, J. M. Delhaye, and N. Zuber, eds., *Post-Dryout Heat Transfer,* CRC Press, Boca Raton, FL, 1992.

8. V. K. Dhir and A. E. Bergles, eds., *Pool and External Flow Boiling,* ASME, New York, 1992.

9. J. C. Chen, ed., *Convective Flow Boiling,* Taylor & Francis, Washington, DC, 1996.

10. D. Gorenflo, D. B. R. Kenning, and C. Marvillet, eds., *Proceedings of EUROTHERM Seminar No. 48 (Pool Boiling 2),* Edizioni ETS, Pisa, Italy, 1996.

11. G. P. Celata, P. D. Marco, and A. Mariani, eds., *Proc. 2nd European Thermal Sciences and 14th UIT National Heat Transfer Conference, 1996,* Edizioni ETS, Pisa, Italy, 1996.

12. R. M. Manglik and A. G. Krauss, eds., *Process Enhanced and Multiphase Heat Transfer. A Festschrift for A. E. Bergles,* Begell House, New York, 1996.

13. G. F. Hewitt, G. L. Shires, and T. R. Bott, *Process Heat Transfer,* CRC Press, Boca Raton, FL, 1994.

14. J. M. Prausnitz, *Molecular Thermal Dynamics of Fluid Phase Equilibria,* Prentice-Hall, Englewood Cliffs, NJ, 1969.

15. J. H. Lienhard, "Correlation of the Limiting Liquid Superheat," *Chem. Eng. Sci.* (31): 847–849, 1976.

16. H. Merte and H. S. Lee, *Homogeneous Nucleation in Microgravity at Low Heat Flux: Experiments and Theory,* ASME Paper 95, WA/HT-41, 1995.

17. V. Drach, N. Sack, and J. Fricke, "Surface Effects in Transient Heat Transfer Into Liquid Nitrogen," in *Proc. 2nd European Thermal Sciences and 14th UIT National Heat Transfer Conference, 1996,* G. P. Celata, P. D. Marco, and A. Mariani eds., Edizioni ETS, Pisa, Italy, 1996.

18. R. Cole, "Boiling Nucleation," *Advances in Heat Transfer* (10): 86–164, 1974.

19. S. Shakir and J. R. Thome, "Boiling Nucleation of Mixtures on Smooth and Enhanced Surfaces," in *Proc. 8th Int. Heat Transfer Conference,* San Francisco, CA, vol. 4, pp. 2081–2086, 1986.

20. M. Jakob, *Heat Transfer,* John Wiley and Sons, New York, 1949.

21. S. G. Bankoff, "Entrapment of Gas in the Spreading of a Liquid Over a Rough Surface," *AIChE J.* (4): 24–26, 1958.

22. J. J. Lorenz, B. B. Mikic, and W. M. Rohsenow, *Proc. 5th Int. Heat Transfer Conf.,* Tokyo, Japan, vol. IV, p. 35, 1974.

23. S. R. Yang and R. H. Kim, "A Mathematical Model of the Pool Boiling Nucleation Site Density in Terms of the Surface Characteristics," *Int. J. Heat Mass Transfer* (31): 1127–1135, 1988.

24. C. H. Wang and V. K. Dhir, "On the Prediction of Active Nucleation Sites Including the Effect of Surface Wettability," in *Pool and External Flow Boiling,* V. K. Dhir and A. E. Bergles eds., ASME, New York, 1992.

25. C. H. Wang and V. K. Dhir, "On the Gas Entrapment and Nucleation Site Density During Pool Boiling of Saturated Water," *J. Heat Transfer* (115): 670–679, 1993.

26. C. H. Wang and V. K. Dhir, "Effect of Surface Wettability on Active Nucleation Site Density During Pool Boiling of Water on a Vertical Surface," *J. Heat Transfer* (115): 659–669, 1993.

27. Y. Y. Hsu, "On the Size Range of Active Nucleation Cavities on a Heating Surface," *J. Heat Transfer* (84): 207–213, 1962.

28. Y. Y. Hsu and R. W. Graham, *An Analytical and Experimental Study of the Thermal Boundary Layer and the Ebullition Cycle in Nucleate Boiling,* NASA Report TN-D-594, 1961.

29. A. E. Bergles and W. M. Rohsenow, "The Determination of Forced Convection Surface Boiling Heat Transfer," *J. Heat Transfer* (86): 365–372, 1964.

30. E. J. Davis and G. H. Anderson, "The Incipience of Nucleate Boiling in Forced Convection Flow," *AIChE J.* (12): 774, 1966.

31. D. B. R. Kenning and M. G. Cooper, "Flow Patterns Near Nuclei and the Initiation of Boiling During Forced Convection Heat Transfer," *Proc. Inst. Mech. Eng.* (180/3C): 112, 1965–1966.

32. W. Frost and G. S. Dzakowic, "An Extension of the Method of Predicting Incipient Boiling on Commercially Finished Surfaces," in *ASME/AIChE Heat Transfer Conference,* Paper 67-HT-61, 1967.

33. S. G. Kandlikar, "Bubble Behavior and Departure Bubble Diameter of Bubbles Generated over Nucleating Cavities in Flow Boiling," in *Pool and External Flow Boiling,* V. K. Dhir and A. E. Bergles eds., pp. 447–457, ASME, New York, 1992.

34. J. P. Klausner, L. Z. Zeng, and R. Mei, "Vapor Bubble Nucleation, Growth and Departure in Forced Convection Boiling," in *Pool External Flow Boiling,* V. K. Dhir and A. E. Bergles eds., pp. 453–457, ASME, New York, 1992.

35. G. F. Hewitt, H. A. Kearsey, P. M. C. Lacey, and D. G. Pulling, *Burnout and Nucleation in Climbing Film Flow,* AERE Report R-4374, 1963. [See also *Int. J. Heat Mass Transfer* (8): 793, 1965].

36. D. B. R. Kenning and Y. Yan, "Pool Boiling on a Thin Plate: Features Revealed by Liquid Crystal Thermography," *Int. J. Heat Mass Transfer* (39): 3117–3137, 1996.

37. I. Golobic, E. Pavlovic, S. Strgar, D. B. R. Kenning, and Y. Yan, "Wall Temperature Variations During Bubble Growth on a Thin Plate: Computation and Experimentation," in *Proc. EUROTHERM Seminar No. 48: Pool Boiling 2,* D. Gorenflo, D. B. R. Kenning, and C. Marvillet eds., Edizioni ETS, Pisa, Italy, pp. 25–32, 1996.

38. J. Fauser and J. Mitrovich, "Heat Transfer During Propagation of Boiling Fronts in Superheated Liquids," in *Proc. EUROTHERM Seminar No. 48: Pool Boiling 2,* D. Gorenflo, D. B. R. Kenning, and C. Marvillet eds., Edizioni ETS, Pisa, Italy, pp. 283–290, 1996.

39. K. Spindler, "Flow Boiling," in *Proc. 10th International Heat Transfer Conference,* G. F. Hewitt ed., Brighton, UK, vol. 1, pp. 349–368, 1994.

40. Lord Rayleigh, "On the Pressure Developed in a Liquid During the Collapse of a Spherical Cavity," *Phil. Mag.* (34): 94–98, 1917.

41. M. S. Plesset and S. A. Zwick, "The Growth of Vapour Bubbles in Superheated Liquid," *J. Appl. Phys.* (25): 493–500, 1954.

42. B. B. Mikic, W. M. Rohsenow, and D. Griffith, "On Bubble Growth Rates," *Int. J. Heat Mass Transfer* (13): 657–666, 1970.

43. H. S. Lee and H. Merte, "Spherical Vapour Bubble Growth in Uniformly Superheated Liquids," *Int. J. Heat Mass Transfer* (39): 2427–2447, 1996.

44. O. Miyatake, I. Tanaka, and N. Lior, "A Simple Universal Equation for Bubble Growth in Pure Liquids and Binary Solutions With a Non-Volatile Solute," *Int. J. Heat Mass Transfer* (40): 1477–1584, 1997.

45. G. Barthau and E. Hahne, "Fragmentation of a Vapour Bubble Growing in a Superheated Liquid," in *Proc. EUROTHERM Seminar No. 48: Pool Boiling 2,* G. Gorenflo, D. B. R. Kenning, and C. Marvillet eds., Edizioni ETS, Pisa, Italy, pp. 105–110, 1996.

46. M. Merilo, "Fluid-to-Fluid Modelling and Correlation of Flow Boiling Crisis in Horizontal Tubes," *Int. J. Multiphase Flow* (5): 313, 1979.

47. B. B. Mikic and W. M. Rohsenow, "Bubble Growth Rates in Non-Uniform Temperature Field," *Prog. Heat Mass Transfer* (2): 283–292, 1969.

48. N. G. Cooper and A. J. P. Lloyd, "The Microlayer in Nucleate Pool Boiling," *Int. J. Heat Mass Transfer* (12): 895–913, 1969.

49. S. J. D. van Stalen, M. S. Sohal, R. Cole, and W. M. Sluyter, "Bubble Growth Rates in Pure and Binary Systems: Combined Effect of Relaxation and Evaporation Microlayers," *Int. Heat Mass Transfer* (18): 453–467, 1975.

50. R. Mei, W. Chen, and J. F. Klausner, "Vapour Bubble Growth in Heterogeneous Boiling. I. Formulation," *Int. J. Heat Mass Transfer* (38): 909–919, 1995.

51. R. Mei, W. Chen, and J. F. Klausner, "Vapour Bubble Growth in Heterogeneous Boiling. II. Growth Rate and Thermal Fields," *Int. J. Heat Mass Transfer* (38): 921–934, 1995.

52. W. Chen, R. Mei, and J. F. Klausner, "Vapour Bubble Growth in Highly Subcooled Heterogeneous Boiling," in *Convective Flow Boiling*, J. C. Chen ed., pp. 91–98, Taylor & Francis, Washington, DC, 1996.

53. M. E. Ellion, "A Study of Mechanism of Boiling Heat Transfer," Ph.D. dissertation, California Institute of Technology, Pasadena, CA, Jet Propulsion Laboratory Report, Memo 20-88, 1954.

54. K. Cornwell, "The Influence of Bubbly Flow on Boiling from a Tube in a Bundle," *Int. J. Heat Mass Transfer* (33): 2579–2584, 1990.

55. Y. Yan, D. B. R. Kenning, and K. Cornwell, "Sliding and Sticking Vapour Bubbles under Inclined Plain and Curved Surfaces," *Proc. EUROTHERM Seminar No. 48: Pool Boiling II*, D. Gorenflo, D. B. R. Kenning, and C. Marvillet eds., Edizioni ETS, Pisa, Italy, pp. 189–200, 1996.

56. W. H. G. T. Chan, "Evaporation and Condensation in Annular Vertical Upward Flow of Water-Steam," Ph.D. thesis, Imperial College, London, UK, 1990.

57. G. F. Hewitt, H. A. Kearsey, P. M. C. Lacey, and D. J. Pulling, "Burnout and Film Flow in the Evaporation of Water in Tubes," *Proc. Inst. Mech. Eng.* (180/3C): paper 2, 1965.

58. L. E. Scriven, "On Dynamics of Phase Growth," *Chem. Eng. Sci.* (10): 1–13, 1959.

59. S. J. D. van Stralen and W. Zijl, "Fundamental Developments in Bubble Dynamics," in *Proc. 6th International Heat Transfer Conference*, Toronto, Canada, August 7–11, 1978.

60. S. T. D. van Stralen, "Bubble Growth Rates in Boiling Binary Mixtures," *Brit. Chem. Eng.* (12/3): 390–395, 1967.

61. J. R. Thome, "Nucleate Pool Boiling in Binary Liquids: An Analytical Equation," *AIChE Symp. Ser.* (77/208): 238–250, 1981.

62. J. R. Thome, "Boiling Heat Transfer in Binary Liquid Mixtures," in *NATO Advanced Research Workshop on Advances in Two-Phase Flow and Heat Transfer*, Spitzingsee, Germany, August 31–September 3, 1982.

63. J. R. Thome and G. Davey, "Bubble Growth Rate in Liquid Nitrogen, Argon and Their Mixtures," *Int. J. Heat Mass Transfer* (24): 89–97, 1981.

64. M. K. Jenson and G. J. Memmel, "Evaluation of Bubble Departure Diameter Correlations," in *Proc. 8th Int. Heat Trans. Conf.*, San Francisco, CA, vol. 4, pp. 1907–1912, 1986.

65. S. S. Kutateladze and I. I. Gogonin, "Growth Rate and Detachment Diameter of a Vapour Bubble in Free Convection Boiling of a Saturated Liquid," *High Temperature* (17): 667–671, 1979.

66. R. A. N. Al-Hayes and R. H. S. Winterton, "Bubble Diameter on Detachment in Flowing Liquids," *Int. J. Heat Mass Transfer* (24): 223–230, 1981.

67. R. H. S. Winterton, "Flow Boiling: Prediction of Bubble Departure," *Int. J. Heat Mass Transfer* (27): 1422–1424, 1984.

68. S. G. Kandlikar, V. R. Mizo, and M. D. Cartright, "Investigation of Bubble Departure Mechanism in Subcooled Flow Boiling of Water Using High-Speed Photography," in *Convective Flow Boiling*, J. C. Chen ed., pp. 161–166, Taylor & Francis, Washington, DC, 1996.

69. J. F. Klausner, R. Mei, D. M. Bernhard, and L. Z. Zeng, "Vapour Bubble Departure in Forced Convection Boiling," *Int. J. Heat Mass Transfer* (36): 2271–2279, 1993.

70. H. J. Ivey, "Relationships Between Bubble Frequency, Departure Diameter and Ride Velocity in Nucleate Boiling," *Int. J. Heat Mass Transfer* (10): 1023–1040, 1967.

71. R. Cole, "Photographic Study of Boiling in the Region of Critical Heat Flux," *AIChE J.* (6): 533–542, 1960.

72. J. Blum, R. Hohl, W. Marquardt, and H. Auracher, "Controlled Transient Pool Boiling Experiments—Methodology and Results," in *Proc. EUROTHERM Seminar No. 48: Pool Boiling II*, D. Gorenflo, D. B. R. Kenning, and C. Marvillet eds., Edizioni ETS, Pisa, Italy, pp. 301–310, 1996.

73. M. T. Cichelli and C. F. Bonilla, "Heat Transfer to Liquids Boiling Under Pressure," *Trans. AIChE* (41): 755, 1945.

74. J. J. Schroder, S. McGill, M. Dirbach, and F. Podzelny, "Secondary Nucleation Ambivalent Heat Transfer and Sound Emission in Low Pressure Subcooled Boiling," in *Proc. EUROTHERM Semi-*

nar No 48: Pool Boiling II, D. Gorenflo, D. B. R. Kenning, and C. Marvillet eds., Edizioni ETS, Pisa, Italy, pp. 291–299, 1996.

75. R. L. Judd, H. Merte, and M. E. Ulucaki, "Variation of Superheat With Subcooling in Nucleate Pool Boiling," *J. Heat Transfer* (113): 201–208, 1991.

76. W. Brown, "Study of Flow Surface Boiling," Sc.D. thesis, Massachusetts Institute of Technology, Cambridge, MA, 1967.

77. F. Jansen, M. Mineur, and J. J. Schroder, "The Influence of Wetting Behaviour on Pool Boiling Heat Transfer," in *Pool and External Flow Boiling,* V. K. Dhir and A. E. Bergles eds., pp. 55–61, ASME, New York, 1992.

78. X. Zhou and K. Bier, "Influence of the Heat Conduction Properties of the Wall Material and of the Wall Thickness on Pool Boiling Heat Transfer," in *Proc. EUROTHERM Seminar No. 48: Pool Boiling II,* D. Gorenflo, D. B. R. Kenning, and C. Marvillet eds., Edizioni ETS, Pisa, Italy, pp. 43–54, 1996.

79. W. H. McAdams, W. E. Kennel, C. S. Minden, C. Rudolf, and J. E. Dow, "Heat Transfer at High Rates to Water With Surface Boiling," *Ind. Eng. Chem.* (41): 1945, 1959.

80. A. Bar-Cohen, "Hysteresis Phenomena at the Onset of Nucleate Boiling," in *Pool and External Flow Boiling,* V. K. Dhir and A. E. Bergles eds., pp. 1–14, ASME, New York, 1992.

81. M. S. El-Genk, "A Review of Pool Boiling From Inclined and Downward Facing Flat Surfaces," in *Proc. 2nd European Thermal-Sciences and 14th UIT National Heat Transfer, Rome, 1996,* C. P. Celata, P. D. Marco, and A. Mariani eds., Edizioni ETS, Pisa, Italy, pp. 1591–1600, 1996.

82. K. Nishikawa, Y. Fujita, S. Uchida, and H. Ohta, "Effect of Heating Surface Orientation on Nucleate Boiling Heat Transfer," *ASME-JSME Thermal Engineering Joint Conf.* (1): 129–136, 1983.

83. H. Merte, "Nucleate Pool Boiling in Variable Gravity," in *Progress in Astronautics and Aeronautics, Low Gravity Fluid Dynamics and Transport Phenomena,* J. N. Coster and R. L. Sani eds., vol. 130, pp. 15–72, A.I.A.A., Washington, DC, 1990.

84. H. Merte, H. S. Lee, and J. S. Irvine, "Transient Nucleate Pool Boiling in Microgravity—Some Initial Results," *Microgravity Sci. Tech.* (VII/2): 173–179, 1994.

85. H. S. Lee and H. Merte, "The Pool Boiling Curve in Microgravity," in *34th Aerospace Sciences Meeting & Exhibit,* Reno, NV, Paper A1AA 96-0499, January 1996.

86. C. J. Kim and A. E. Bergles, "Incipient Boiling Behaviour of Porous Boiling Surfaces Used for Cooling of Microelectronic Chips," in *Particulate Phenomena and Multiphase Transport,* T. N. Veziroglu ed., vol. 2, pp. 3–18, Hemisphere Publishing Corporation, Washington, DC, 1988.

87. V. K. Dhir, "Pool Boiling Heat Transfer: Recent Advances and Expectations for the Future," in *Process, Enhanced and Multiphase Heat Transfer: A Festschrift for A. E. Bergles,* R. M. Manglik and A. D. Kraus eds., pp. 99–126, Begell House, New York, 1996.

88. Y. Fujita, "The State of the Art—Nucleate Boiling Mechanism," in *Pool and External Flow Boiling,* V. K. Dhir and A. E. Bergles eds., pp. 83–97, ASME, New York, 1992.

89. Y. Katto, S. Yokoya, and M. Yasunaka, "Mechanism of Boiling Crisis and Transition Boiling in Pool Boiling," in *Proc. 4th Int. Heat Transfer Conf.,* Paris-Versailles, France, vol. V, p. B3.2, 1970.

90. N. Zuber, "Nucleate Boiling. The Region of Isolated Bubbles and the Similarity With Natural Convection," *Int. J. Heat Mass Transfer* (6): 53–78, 1963.

91. R. Moissis and B. J. Berenson, "On the Hydrodynamic Transitions in Nucleate Boiling," *J. Heat Transfer* (85): 221–229, 1963.

92. M. V. H. Del Valle and D. B. R. Kenning, "Subcooled Flow Boiling at High Heat Flux," *Int. J. Heat Mass Transfer* (28): 1907–1920, 1985.

93. R. L. Judd and K. S. Hwang, "A Comprehensive Model for Nucleate Pool Boiling Heat Transfer Including Microlayer Evaporation," *J. Heat Transfer* (98): 623–629, 1976.

94. I. Eames, J. C. R. Hunt, and S. E. Belcher, "Displacement of Invisid Fluids by a Sphere Moving Away From a Wall," *J. Fluid Mech.* (324): 333–353, 1996.

95. D. B. R. Kenning and M. V. H. Del Valle, "Fully Developed Nucleate Boiling: Overlap of Areas of Influence and Interference Between Bubble Sites," *Int. J. Heat Mass Transfer* (24): 1025–1032, 1981.

96. D. D. Paul and S. I. Abdel-Khalik, "A Statistical Analysis of Saturated Nucleate Boiling Along a Heated Wire," *Int. J. Heat Mass Transfer* (26): 509–519, 1983.

97. D. D. Paul, S. M. Ghiaasiaan, and S. I. Abdel-Khalik, "On the Contribution of Various Mechanisms to Nucleate Pool Boiling Heat Transfer," in *Pool and External Flow Boiling,* V. K. Dhir and A. E. Bergles eds., pp. 125–133, ASME, New York, 1992.

98. R. O. Judd and K. S. Hwang, "A Comprehensive Model for Nucleate Pool Boiling Heat Transfer Including Microlayer Evaporation," *J. Heat Transfer* (98): 623–629, 1976.

99. R. F. Gaertner, "Photographic Study of Nucleate Pool Boiling on a Horizontal Surface," *J. Heat Transfer* (87): 17–29, 1965.

100. M. Shoji, "A Study of Steady Transition Boiling of Water: Experimental Verification of Macrolayer Evaporation and Model," in *Pool and External Flow Boiling,* V. K. Dhir and A. E. Bergles eds., pp. 237–242, ASME, New York, 1992.

101. Y. Katto, "Critical Heat Flux Mechanisms," in *Convective Flow Boiling,* J. C. Chen ed., pp. 29–44, Taylor & Francis, Washington, DC, 1996.

102. Y. Katto and S. Yokoya, "Principal Mechanism of Boiling Crisis in Pool Boiling," *Int. J. Heat Mass Transfer* (11): 993–1002, 1968.

103. V. K. Dhir and S. P. Liaw, "Framework for a Unified Model for Nucleate and Transition Pool Boiling," *J. Heat Transfer* (111): 739–746, 1989.

104. J. H. Lay and V. K. Dhir, "Shape of a Vapour Stem During Nucleate Boiling of Saturation Liquids," *J. Heat Transfer* (117): 394–401, 1995.

105. V. V. Yagov, "The Principle Mechanisms for Boiling Contribution in Flow Boiling Heat Transfer," in *Convective Flow Boiling,* J. C. Chen ed., pp. 175–180, Taylor & Francis, Washington, DC, 1996.

106. H. K. Forster and N. Zuber, "Dynamics of Vapour Bubbles and Boiling Heat Transfer," *AIChE J.* (1): 531, 1955.

107. J. C. Chen, "Correlation for Boiling Heat Transfer to Saturated Fluids in Convective Flow," *Ind. Eng. Chem. Proc. Des. Dev.* (5): 322, 1966.

108. W. M. Rohsenow, "A Method of Correlating Heat Transfer Data for Surface Boiling of Liquids," *Trans. ASME* (74): 969, 1952.

109. I. L. Mostinskii, "Calculation of Heat Transfer and Critical Heat Fluxes in Liquids," *Teploenergetika* (10/4): 66, 1963.

110. M. G. Cooper, "Correlation for Nucleate Boiling—Formulation Using Reduced Properties," *Physicochem. Hydrodynam.* (3/2): 89, 1982.

111. M. G. Cooper, "Saturation and Nucleate Pool Boiling—A Simple Correlation," *Inst. Chem. Eng. Symp. Ser.* (86/2): 785, 1984.

112. D. Gorenflo, "Pool Boiling," in *VDI-Warmeatlas,* 6th ed., VDI Verlag, Dusseldorf, Germany, 1990.

113. W. Leiner and D. Gorenflo, "Methods of Predicting the Boiling Curve and a New Equation Based on Thermodynamic Similarity," in *Pool and External Flow Boiling,* V. K. Dhir and A. E. Bergles eds., pp. 99–103, ASME, New York, 1992.

114. J. W. Palen, A. Yarden, and J. Taborek, "Characteristics of Boiling Outside Large Scale Horizontal Multi-Tube Bundles," *AIChE Symp. Ser.* (68/118): 50, 1972.

115. K. Bier, J. Schmedel, and D. Gorenflo, "Influence of Heat Flux and Saturation Pressure on Pool Boiling Heat Transfer to Binary Mixtures," *Chem. Eng. Fundam.* (1/2): 79, 1983.

116. D. Gorenflo, P. Sokol, and S. Caplanis, "Pool Boiling Heat Transfer From Single Plain Tubes to Various Hydrocarbons," *Int. J. Refrig.* (13): 286–292, 1990.

117. D. Gorenflo, "Pool Boiling," in *VDI-Warmeatlas,* 6th ed., VDI Verlag, Dusseldorf, Germany, 1990.

118. W. Leiner and D. Gorenflo, "Methods of Predicting the Boiling Curve and a New Equation Based on Thermodynamic Similarity," in *Pool and External Flow Boiling,* V. K. Dhir and A. E. Bergles eds., pp. 99–103, ASME, New York, 1992.

119. W. Leiner, "Heat Transfer by Nucleate Pool Boiling—General Correlation Based on Thermodynamic Similarity," *Int. J. Heat Mass Transfer* (37): 763–769, 1994.

120. B. B. Mikic and W. M. Rohsenow, "A New Correlation of Pool Boiling Data Including the Effect of Heating Surface Characteristics," *J. Heat Transfer* (91): 245, 1969.

121. R. J. Benjamin and A. R. Balakrishnan, "Nucleate Pool Boiling Heat Transfer of Pure Liquids at Low to Moderate Heat Fluxes," *Int. J. Heat Mass Transfer* (39): 2495–2500, 1996.

122. R. Nelson, D. B. R. Kenning, and M. Shoji, "Nonlinear Dynamics in Boiling Phenomena," *Journal of the Heat Transfer Society of Japan* (35/136): 22–34, 1996.

123. P. Sadasivan, C. Unal, and R. Nelson, "Nonlinear Aspects of High Heat Flux Nucleate Boiling Heat Transfer," *J. Heat Transfer* (117): 981–989, 1995.

124. V. I. Tolubinskiy and Y. N. Ostrovskiy, "Mechanism of Heat Transfer in Boiling of Binary Mixtures," *Heat Transfer Sov. Res.* (1/6): 1969.

125. E. U. Schlunder, "Heat Transfer in Bubble Vapourisation From Liquid Mixtures," *Verfahrenstechnik* (16/9): 692, 1982.

126. R. A. W. Shock, "Boiling in Multicomponent Fluids," *Multiphase Sci. Technol.* (1): 281, 1982.

127. K. Bier and J. Schmidt, "Bubble Formation with Pool Boiling of Pure Fluids and Binary Fluid Mixtures," in *Proc. 2nd European Thermal Sciences and 14th UIT National Heat Transfer Conf., Rome, 1996,* C. P. Celata, P. D. Marco, and A. Mariani eds., Edizioni ETS, Pisa, Italy, pp. 1661–1675, 1996.

128. K. Stephan and P. Preusser, "Heat Transfer in Natural Convection Boiling of Polynary Mixtures," in *Proc. 6th Int. Heat Transfer Conference,* Toronto, Canada, vol. 1, p. 187, 1978.

129. E. U. Schlunder, "Heat Transfer in Nucleate Boiling of Mixtures," in *Proc. 8th Int. Heat Transfer Conference,* San Francisco, CA, vol. 4, p. 2073, 1986.

130. J. R. Thome, "Prediction of Binary Mixture Boiling Heat Transfer Coefficients Using Only Phase Equilibrium Data," *Int. J. Heat Mass Transfer* (26): 965, 1983.

131. D. Gorenflo, "Abschnitt Hab Behaltersieden," in *VDI-Warmeatlas, 7. Aufl.* VDI-Verlag, Dusseldorf, Germany, 1994.

132. Y. Fujita, Q. Bai, and N. Tsutsui, "Heat Transfer of Binary Mixtures in Nucleate Pool Boiling," in *Proc. 2nd European Thermal-Sciences and 14th UIT National Heat Transfer Conference, Rome, 1996,* C. P. Celata, P. D. Marco, and A. Mariani eds., Edizioni ETS, Pisa, Italy pp. 1369–1646, 1996.

133. Y. L. Tzan and Y. M. Yang, "Experimental Study of Surfactant Effects on Pool Boiling Heat Transfer," *J. Heat Transfer* (112): 207–212, 1990.

134. W. T. Wu and Y. M. Yang, "Enhanced Boiling Heat Transfer by Surfactant Additives," in *Pool and External Flow Boiling,* V. K. Dhir and A. E. Bergles eds., ASME, New York, pp. 361–366, 1992.

135. A. E. Bergles, "Enhancement of Pool Boiling," in *Proc. EUROTHERM Seminar No. 48: Pool Boiling 2,* D. Gorenflo, D. B. R. Kenning, and C. Marvillet eds., Edizioni ETS, Pisa, Italy, pp. 111–121, 1996.

136. C. Pais and R. L. Webb, "Literature Survey of Pool Boiling on Enhanced Surfaces," *ASHRAE Transactions* (97/1): 1991.

137. R. L. Webb and I. Haider, "An Analytical Model for Nucleate Boiling on Enhanced Surfaces," in *Pool and External Flow Boiling,* V. K. Dhir and A. E. Bergles eds., pp. 345–360, ASME, New York, 1992.

138. P. Hubner, W. Kunstler, and D. Gorenflo, "Pool Boiling Heat Transfer at Fin Tubes: Influence of Surface Roughness and Shape of the Fins," in *Proc. EUROTHERM Seminar No. 48: Pool Boiling 2,* D. Gorenflo, D. B. R. Kenning, and C. Marvillet eds., Edizioni ETS, Pisa, Italy, pp. 131–139, 1996.

139. J. R. Thome, "Mechanisms of Enhanced Nucleate Pool Boiling," in *Pool and External Flow Boiling,* V. K. Dhir and A. E. Bergles eds., pp. 337–343, ASME, New York, 1992.

140. R. L. Webb and I. Haider, "An Analytical Model for Nucleate Boiling on Enhanced Surfaces," in *Pool and External Flow Boiling,* V. K. Dhir and A. E. Bergles eds., pp. 345–360, ASME, New York, 1992.

141. P. J. Marto and V. J. Lepere, "Pool Boiling Heat Transfer From Enhanced Surfaces to Dielectric Liquids," *J. Heat Transfer* (104): 292–299, 1982.

142. A. D. Pinto, S. Caplanis, P. Sokol, and D. Gorenflo, "Variation of Hysteresis Phenomena With Saturation Pressure and Surface Roughness," in *Pool and External Flow Boiling,* V. K. Dhir and A. E. Bergles eds., pp. 37–47, ASME, New York, 1992.

143. S. P. Memory and P. J. Marto, "The Influence of Oil on Boiling Hysteresis of R114 from Enhanced Surfaces," in *Pool and External Flow Boiling,* V. K. Dhir and A. E. Bergles eds., pp. 63–71, ASME, New York, 1992.

144. M. M. Ohardi, R. A. Papar, A. Kumar, and A. I. Ansari, "Some Observations on EHD-Enhanced Boiling of R-123 in the Presence of Oil Contamination," in *Pool and External Flow Boiling,* V. K. Dhir and A. E. Bergles eds., pp. 387–396, ASME, New York, 1992.

145. M. C. Zaghdoudi, S. Cioulachtjian, J. Bonjour, and M. Lallemand, "Analysis of Hysteresis and Polarity Influence in Nucleate Pool Boiling Under DC Electric Field," in *Proc. EUROTHERM Seminar No. 48: Pool Boiling 2,* D. Gorenflo, D. B. R. Kenning, and C. Marvillet eds., Edizioni ETS, Pisa, Italy, pp. 247–264, 1996.

146. S. Bonekamp and K. Bier, "Influence of Ultrasound on Pool Boiling Heat Transfer to Mixtures of the Refrigerants R-23 and R-134A," in *Proc. EUROTHERM Seminar No. 48: Pool Boiling 2,* D. Gorenflo, D. B. R. Kenning, and C. Marvillet eds., Edizioni ETS, Pisa, Italy, pp. 227–238, 1996.

147. G. F. Hewitt, "Burnout," in *Handbook of Multiphase Systems,* G. Hetsroni ed., chap. 6.4, McGraw-Hill Book Company, New York, 1982.

148. H. J. Ivey and D. J. Morris, *On the Relevance of the Vapour-Liquid Exchange Mechanism for Sub-Cooled Boiling Heat Transfer at High Pressure,* UKAEA Report AEEW-R137, 1962.

149. P. J. Berenson, *Transition Boiling Heat Transfer From a Horizontal Surface,* MIT (Division of Sponsored Research) Report No. 17, 1960.

150. S. P. Liaw and V. K. Dhir, "Effect of Surface Wettability on Transition Boiling Heat Transfer From a Vertical Surface," in *Proc. 8th Int. Heat Transfer Conference,* San Francisco, CA, vol. 4, pp. 2031–2036, 1986.

151. M. Jakob and W. Fritz, "Versuche über der Verdampfungsvorgang," *Forsch. Gebite Ing.* (2): 434, 1931.

152. R. Siegel and C. Usiskin, "Photographic Study of Boiling in Absence of Gravity," *J. Heat Transfer* (81): 3, 1959.

153. J. M. Adams, *A Study of the Critical Flux in an Accelerating Pool Boiling System,* NSF Report G-19697, University of Washington, Seattle, WA, 1962.

154. W. R. Beasant and H. W. Jones, *The Critical Heat Flux in Pool Boiling Under Combined Effect of High Acceleration and Pressure,* UKAEA Report AEEW-R275, 1963.

155. J. H. Lienhard and V. K. Dhir, "Hydrodynamic Prediction of Peak Pool Boiling Heat Fluxes from Finite Bodies," *J. Heat Transfer* (95): 152–158, 1973.

156. Y. Elkassabgi and J. H. Lienhard, "The Peak Pool Boiling Heat Fluxes From Horizontal Cylinders in Sub-Cooled Liquid," *J. Heat Transfer* (110): 479–496, 1988.

157. Y. Katto, "Critical Heat Flux in Pool Boiling," in *Pool and External Flow Boiling,* V. K. Dhir and A. E. Bergles eds., pp. 151–164, ASME, New York, 1992.

158. A. E. Bergles, "What is the Real Mechanism of CHF in Pool Boiling?" in *Pool and External Flow Boiling,* V. K. Dhir and A. E. Bergles eds., pp. 165–170, ASME, New York, 1992.

159. N. Zuber, *Hydrodynamic Aspects of Boiling Heat Transfer,* Report AECU-4439, Atomic Energy Commission (US), 1958.

160. N. Zuber, M. Tribus, and J. W. Westwater, "The Hydrodynamic Crisis in Pool Boiling of Saturated Liquids," in *International Developments in Heat Transfer,* pp. 230–236, ASME, New York, 1963.

161. J. H. Lienhard and V. K. Dhir, *Extended Hydrodynamic Theory of the Peak and Minimum Pool Boiling Heat Fluxes,* NASA Report CR-2270, 1973.

162. K. H. Sun and J. H. Lienhard, "The Peak Pool Boiling Heat Flux on Horizontal Cylinders," *Int. J. Heat Mass Transfer* (13): 1425–1439, 1970.

163. Y. Haramura and Y. Katto, "A New Hydrodynamic Model of Critical Heat Flux, Applicable Widely to Both Pool and Forced Convection Boiling on Submerged Bodies in Saturated Liquids," *Int. J. Heat Mass Transfer* (26): 389–399, 1983.

164. W. M. Rohsenow and P. Griffith, "Correlation of Maximum Heat Flux Data for Boiling of Saturated Liquids," *Chem. Eng. Prog. Symp. Series* (52): 47–49, 1956.

165. V. K. Dhir and S. P. Liaw, "Framework for a Unified Model for Nucleate and Transition Pool Boiling," *J. Heat Transfer* (111): 739–745, 1989.

166. S. S. Kutateladze, "On the Transition to Film Boiling Under Natural Convection," *Kotloturbostroenie* (3): 10, 1948.

167. V. K. Dhir and J. H. Lienhard, "Peak Pool Boiling Heat Flux in Viscous Liquids," *J. Heat Transfer* (96): 71–78, 1974.

168. W. M. Rohsenow, "Pool Boiling," in *Handbook of Multiphase Systems,* G. Hetsroni ed., chap. 6.2, McGraw-Hill Book Company, New York, 1982.

169. C. Unal, P. Sadasivan, and R. M. Nelson, "On the Hot-Spot Controlled Critical Heat Flux Mechanism in Pool Boiling of Saturated Fluids," in *Pool and External Flow Boiling,* V. K. Dhir and A. E. Bergles eds., pp. 193–201, ASME, New York, 1992.

170. N. H. Afgan, "Boiling Heat Transfer and Burnout Heat Flux of Ethyl-Alcohol/Benzene Mixtures," in *Proc. 3rd Int. Heat Transfer Conf.,* vol. 3, p. 175, 1966.

171. S. J. D. van Stralen, "The Mechanism of Nucleate Boiling in Pure Liquids and in Binary Mixtures, Parts I, II, III," *Int. J. Heat Mass Transfer* (9): 995–1046, 1966; (10): 1469, 1967; (11): 1467, 1968.

172. R. P. Reddy and J. H. Lienhard, "The Peak Boiling Heat Flux in Saturated Ethanol-Water Mixtures," *J. Heat Transfer* (111): 480–486, 1989.

173. Y. Fujita and Q. Bai, "Critical Heat Flux of Binary Mixtures in Pool Boiling and its Correlation in Terms of Marangoni Number," *Proc. EUROTHERM Seminar No. 48: Pool Boiling 2,* D. Gorenflo, D. B. R. Kenning, and C. Marvillet eds., Edizioni ETS, Pisa, Italy, pp. 319–326, 1996.

174. J. Hovestreijdt, "The Influence of the Surface Tension Difference on the Boiling of Mixtures," *Chem. Eng. Sci.* (18): 631–639, 1963.

175. J. D. Le Franc, H. Bruchener, P. Domenjoud, and R. Morin, "Improvements Made to the Thermal Transfer of Fuel Elements by Using the Vapotron Process," Paper A/CONF28/P/96, presented at the 3rd International Conference on Peaceful Uses of Atomic Energy, 1964.

176. K. W. Haley and J. W. Westwater, "Boiling Heat Transfer from Single Fins," in *Proc. 3rd Int. Heat Transfer Conf.,* New York, NY, vol. 2, pp. 245–253, 1966.

177. M. Markels and R. L. Durfee, "The Effect of Applied Voltage on Boiling Heat Transfer," AIChE Paper 157, presented at the 55th Annual Meeting, Chicago, IL, December 1962.

178. A. P. Ornatskii and V. K. Sheherbakov, "Intensification of Heat Transfer in the Critical Area With the Aid of Ultrasonics," quoted in Report NYO-9500, Atomic Energy Comission (US), 1960.

179. H. Auracher, "Transition Boiling in Natural Convection Systems," in *Pool and External Flow Boiling,* V. K. Dhir and A. E. Bergles eds., pp. 219–236, ASME, New York, 1992.

180. A. Sakurai and M. Shiotsu, "Pool Film Boiling Heat Transfer and Minimum Film Boiling Temperature," in *Pool and External Flow Boiling,* V. K. Dhir and A. E. Bergles eds., pp. 277–301, ASME, New York, 1992.

181. C. Pan and T. N. Lin, "Marangoni Flow Effect in Nucleate Boiling Near Critical Heat Flux," in *Int. Comm. Heat Mass Transfer* (16): 475–486, 1989.

182. T. Tsuchiya, "A Few Experiments Into Transition Boiling," in *Proc. 7th National Heat Transfer Symposium,* Japan, pp. 93–96, 1970.

183. M. Shoji, L. C. Witte, S. Yokoya, and M. Ohshima, "Liquid Solid Contact and Effects of Surface Roughness and Wettability in Film and Transition Boiling on an Horizontal Large Surface," in *Proc. 9th Int. Heat Transfer Conference,* Jerusalem, Israel, vol. 2, pp. 135–140, 1990.

184. M. S. El-Genk and Z. Guo, "Saturated Pool Boiling From Downward Facing and Inclined Surfaces," in *Pool and External Flow Boiling,* V. K. Dhir and A. E. Bergles eds., pp. 243–249, ASME, New York, 1992.

185. S. Nishio, G. R. Chandratilleke, and T. Ozu, "Natural Convection Film Boiling Heat Transfer (1st Report, Saturated Film Boiling with Long Vapour Film)," *J.S.M.E. Int. J., Series II* (34): 202–211, 1991.

186. P. B. Berenson, "Film Boiling Heat Transfer From a Horizontal Surface," *J. Heat Transfer* (83): 351, 1961.

187. J. H. Lienhard and P. T. Y. Wong, "The Dominant Unstable Wavelength and Minimum Heat Flux During Film Boiling on a Horizontal Cylinder," *J. Heat Transfer* (86): 220–226, 1964.

188. S. K. W. Yu, P. R. Farmer, and M. W. E. Conly, "Methods and Correlations for the Prediction of Quenching Rates on Hot Surfaces," *Int. J. Multiphase Flow* (3): 415–444, 1977.

189. J. M. Ramilison and J. H. Lienhard, "Transition Boiling Heat Transfer and the Film Transition Regime," *J. Heat Transfer* (109): 746–752, 1987.

190. L. A. Bromley, "Heat Transfer in Stable Film Boiling," *Chem. Eng. Prog.* (46): 221–227, 1950.

191. Y. Y. Hsu and J. W. Westwater, "Approximate Theory for Film Boiling on Vertical Surfaces," *AIChE Chem. Eng. Prog. Symp. Ser.* (56/30): 15, 1960.

192. T. D. Bui and V. K. Dhir, "Film Boiling Heat Transfer on an Isothermal Vertical Surface," *J. Heat Transfer* (107): 764–771, 1985.

193. Y. Y. Hsu and J. W. Westwater, "Approximate Theory for Film Boiling on Vertical Surfaces," *AIChE J.* (4): 58, 1958.

194. V. V. Klimenko, "Film Boiling on a Horizontal Plate—New Correlation," *Int. J. Heat Mass Transfer* (24): 69–79, 1981.

195. M. Shoji, "A Study of Steady Transition Boiling of Water: Experimental Verification of Macrolayer Evaporation Model," in *Pool and External Flow Boiling,* V. K. Dhir and A. E. Bergles eds., pp. 237–242, ASME, New York, 1992.

196. O. Happel and K. Stephan, "Heat Transfer From Nucleate Boiling to the Beginning of Film Boiling in Binary Mixtures," *Proc. 5th Int. Heat Transfer Conference,* Tokyo, Japan, Paper B7.8, 1974.

197. P. L. Yue and M. E. Weber, "Minimum Film Boiling Flux of Binary Mixtures," *Trans. Inst. Chem. Eng.* (52): 217–221, 1974.

198. F. Verplaetsen and J. A. Berghmans, "The Influence of an Electric Field on the Heat Transfer Rate During Film Boiling of Stagnant Fluids," in *Proc. EUROTHERM Seminar No. 48: Pool Boiling 2,* D. Gorenflo, D. B. R. Kenning, and C. Marvillet eds., Edizioni ETS, Pisa, Italy, pp. 327–334, 1996.

199. P. B. Whalley, "The Importance of External Flow Boiling," in *Pool and External Flow Boiling,* V. K. Dhir and A. E. Bergles eds., pp. 411–426, ASME, New York, 1992.

200. M. K. Jensen, "Fundamental Issues in Shellside Boiling," in *Pool and External Flow Boiling,* V. K. Dhir and A. E. Bergles eds., pp. 427–437, ASME, New York, 1992

201. S. Yilmaz and J. W. Palen, *Performance of Finned Tube Reboilers in Hydrocarbon Service,* ASME Paper 84-HT-91, 1984.

202. R. L. Singh, J. S. Saini, and H. K. Varma, "Effect of Cross-Flow on Boiling Heat Transfer in Water," *Int. J. Heat Mass Transfer* (26): 1882–1885, 1983.

203. R. L. Singh, J. S. Saini, and H. K. Varma, "Effect of Cross-Flow on Boiling Heat Transfer of Refrigerant-12," *Int. J. Heat Mass Transfer* (28): 512–514, 1985.

204. J. Fink, E. S. Gaddis, and A. Voglpohl, "Forced Convection Boiling of a Mixture of Freon-11 and Freon-113 Flowing Normal to a Cylinder," in *Proc. 7th Int. Heat Transfer Conf.,* Munich, Germany, vol. 4, pp. 207–212, 1982.

205. A. E. Bergles and W. M. Rohsenow, "The Determination of Forced Convection Surface Boiling Heat Transfer," *J. Heat Transfer* (86): 365, 1964.

206. L. S. Leong and K. Cornwell, "Heat Transfer Coefficients in a Reboiler Tube Bundle," *Chemical Engineer* (343): 219–221, 1979.

207. J. W. Palen and W. M. Small, "Kettle and Internal Reboilers," *Hydrocarbon Processing* (43/11): 199–208, 1964.

208. D. Gorenflo, M. Buschmeier, and P. Kaupmann, "Heat Transfer From a Horizontal Tube to Boiling Binary Mixtures With Superimposed Convective Flow," in *Convective Flow Boiling,* J. C. Chen ed., pp. 265–270, Taylor & Francis, Washington, DC, 1996.

209. K. Cornwell, "The Influence of Bubble Flow on Boiling From a Tube in a Bundle," *Int. J. Heat Mass Transfer* (33): 2579–2584, 1990.

210. K. Cornwell, "The Role of Sliding Bubbles in Boiling on Tube Bundles," *Proc. 9th Int. Heat Transfer Conference,* Jerusalem, Israel, vol. 3, pp. 455–460, 1990.

211. G. T. Polley, T. Ralston, and I. D. R. Grant, "Forced Flow Boiling in an Ideal In-Line Tube Bundle," *ASME/AIChE Heat Transfer Conference,* Orlando, FL, Paper 80-HT-46, 1980.

212. A. A. Armand, "The Resistance During the Movement of a Two-Phase System in Horizontal Pipes," *Izv. Vses. Tpel. Inst.* (1): 16, 1946.

213. T. W. C. Brisbane, I. D. R. Grant, and P. B. Whalley, "A Prediction Method for Kettle Reboiler Performance," *ASME/AIChE Heat Transfer Conf.,* Orlando, FL, ASME Paper 80-HT-42, 1980.

214. P. B. Whalley and D. Butterworth, "A Simple Method of Calculating the Recirculating Flow in Vertical Thermasyphon and Kettle Reboilers," *21st ASME/AIChE National Heat Transfer Conference,* Seattle, WA, 1983.

215. J. W. Palen, "Thermal Design of Shell-and-Tube Reboilers," in *Handbook of Heat Exchanger Design,* sec. 3.6.2, Begell House, New York, 1992.

216. L. W. Swanson and J. W. Palen, "Convective Boiling Applications in Shell-and-Tube Heat Exchangers," in *Convective Flow Boiling,* J. C. Chen ed., pp. 45–56, Taylor & Francis, Washington, DC, 1996.

217. J. Taborek, *Basic Design Principles for Process Reboilers,* lecture given at the University of California, Santa Barbara, CA, 1985.

218. M. K. Jensen, R. T. Trewin, and A. E. Bergles, "Cross-Flow Boiling in Enhanced Tube Bundles," in *Pool and External Flow Boiling,* V. K. Dhir and A. E. Bergles eds., pp. 373–379, ASME, New York, 1992.

219. B. Thonon, T. Lang, and G. Schuz, "Pool Boiling of *n*-Pentane on a Bundle of Low Finned Tubes," in *Proc. EUROTHERM Seminar No. 48: Pool Boiling 2,* D. Gorenflo, D. B. R. Kenning, and C. Marvillet eds., Edizioni ETS, Pisa, Italy, pp. 157–164, 1996.

220. J. H. Lienhard and R. Eichhorn, "Peak Boiling Heat Flux on Cylinders in a Cross-Flow," *Int. J. Heat Mass Transfer* (19): 1135, 1976.

221. W. Kramer, H. Auracher, and C. Marvillet, "Pool Boiling Heat Transfer of Ethanol-Water Mixtures on Horizontal Smooth and GEWA-TX Tubes: The Influence of Two-Phase Cross-Flow," in *Proc. EUROTHERM Seminar No. 48: Pool Boiling 2,* D. Gorenflo, D. B. R. Kenning, and C. Marvillet eds., Edizioni ETS, Pisa, Italy, pp. 165–172, 1996.

222. M. K. Jensen and H. Tang, "Correlations for the CHF Condition in Two-Phase Cross Flow Through Multitube Bundles," *J. Heat Transfer* (116): 780–784, 1994.

223. Y. Taitel and A. E. Dukler, "A Model for Predicting Flow Regime Transition in Horizontal and Near-Horizontal Gas-Liquid Flow," *AIChE J.* (22): 47–55, 1976.

224. J. W. Palen and D. W. Small, "A New Way to Design Kettle and Internal Reboilers," *Hydrocarbon Process* (43/11): 199, 1964.

225. O. A. Montasser and M. Shoji, "Effect of Wall Thermal Conditions on Cross-Flow Film Boiling Heat Transfer," in *Convective Flow Boiling,* J. C. Chen ed., pp. 231–236, Taylor & Francis, Washington, DC, 1996.

226. L. A. Bromley, N. R. Leroy, and J. A. Robbers, "Heat Transfer in Forced Convection Film Boiling," *Ind. Eng. Chem.* (45): 2639–2646, 1953.

227. L. C. Witte and J. Orozco, "The Effect of Vapour Velocity Profile Shape on Flow Film Boiling From Submerged Bodies," *J. Heat Transfer* (105): 191–197, 1984.

228. S. Aziz, G. F. Hewitt, and D. B. R. Kenning, "Heat Transfer Regimes in Forced Convection Film Boiling on Spheres," in *Proc. 8th Int. Heat Transfer Conf.,* San Francisco, CA, vol. 5, pp. 2149–2154, 1986.

229. Y. Zvirin, G. F. Hewitt, and D. B. R. Kenning, "Boiling on Free Falling Spheres, Drag and Heat Transfer Coefficients," *Experimental Heat Transfer* (3): 185–214, 1990.

230. G. F. Hewitt, "Flow Regimes," in *Handbook of Multiphase Systems,* G. Hetsroni ed., chap. 2.1, McGraw-Hill Book Company, New York, 1982.

231. H. Muller-Steinhagen and M. Jamialahmandi, "Subcooled Flow Boiling Heat Transfer to Mixtures and Solutions," in *Convective Flow Boiling,* J. C. Chen ed., pp. 277–283, Taylor & Francis, Washington, DC, 1996.

232. J. M. Robertson and V. V. Wadekar, "Vertical Upflow Boiling of Ethanol in a 10 mm Tube," in *Trans. 2nd UK Nat. Heat Transfer Conf.,* Glasgow, UK, vol. 1, pp. 67–77, 1988.

233. D. B. R. Kenning and M. G. Cooper, "Saturated Flow Boiling of Water in Vertical Tubes," *Int. J. Heat Mass Transfer* (32): 445–458, 1989.

234. K. M. Kirk and H. Merte, "A Mixed Natural/Forced Convection Nucleate Boiling Heat Transfer Criteria," in *Proc. 10th Int. Heat Transfer Conf.,* Brighton, UK, pp. 479–484, 1994.

235. K. M. Kirk, H. Merte, and R. Keller, "Low-Velocity Subcooled Nucleate Flow Boiling at Various Orientations," *J. Heat Transfer* (117): 380–386, 1995.

236. A. H. Abdelmessih and A. Fakhri, "Hysteresis Effects in Incipient Boiling Superheat of Freon-11," in *Proc. 5th Int. Heat Transfer Conf.,* Tokyo, Japan, vol. 4, p. 165, 1974.

237. Z. Bilicki, "The Relation Between the Experiment and Theory for Nucleate Forced Boiling," in *Proc. 4th World Conf. on Experimental Heat Transfer, Fluid Mechanics and Thermodynamics,* Brussels, Belgium, vol. 2, pp. 571–578, June 1997.

238. G. F. Hewitt, "Void Fraction," in *Handbook of Multiphase Systems,* G. Hetsroni ed., chap. 2.3, McGraw-Hill Book Company, New York, 1982.

239. P. Saha and N. Zuber, "Point of Net Vapour Generation and Vapour Void Fraction in Subcooled Boiling," in *Proc. 5th Int. Heat Transfer Conf.,* Tokyo, Japan, vol. 4, pp. 175–179, 1974.

240. F. Mayinger, "Subcooled Boiling," in *Two-Phase Flows and Heat Transfer With Application to Nuclear Reactor Design Problems,* J. J. Ginoux ed., pp. 339–410, Hemisphere, Washington, DC, 1978.

241. R. T. Lahey and F. J. Moody, *The Thermal Hydraulics of a Boiling Water Nuclear Reactor,* American Nuclear Society, 1997.

242. O. Zeitoun and M. Shoukri, "On the Net Vapour Generation Phenomenon in Low Pressure and Low Mass Flux Subcooled Flow Boiling," in *Convective Flow Boiling,* J. C. Chen ed., pp. 85–90, Taylor & Francis, Washington, DC, 1996.

243. S. Levy, *Forced Convection Subcooled Boiling—Prediction of Vapour Volumetric Function,* General Electric Co. Report GEAP-5157, 1966.

244. N. Zuber and J. A. Findlay, "Average Volumetric Concentration in Two-Phase Flow Systems," *J. Heat Transfer* (87): 453–468, 1965.

245. G. E. Dix, *Vapour Void Fractions for Forced Convection With Subcooled Boiling at Low Flow Rates,* General Electric Co. Report NEDO-10491, 1971.

246. V. V. Wadekar and D. B. R. Kenning, "Flow Boiling Heat Transfer in Vertical Slug and Churn Flow Region," in *Proc. 9th Int. Heat Transfer Conf.,* Jerusalem, Israel, vol. 3, pp. 449–454, 1990.

247. K. W. McQuillan and P. B. Whalley, "Flow Patterns in Vertical Two-Phase Flow," *Int. J. Multiphase Flow* (11): 161–175, 1985.

248. G. Sun, V. V. Wadekar, and G. F. Hewitt, "Heat Transfer in Horizontal Slug Flow," in *Proc. 10th Int. Heat Transfer Conf.,* Brighton, UK, vol. 6, pp. 271–276, 1994.

249. R. B. Mesler, "An Alternative to the Dengler and Addoms Convection Concept of Forced Convection Boiling Heat Transfer," *AIChE J.* (23): 448–453, 1977.

250. R. B. Mesler, "Improving Nucleate Boiling Using Secondary Nucleation," in *Pool and External Flow Boiling,* V. K. Dhir and A. E. Bergles eds., pp. 43–47, ASME, New York, 1992.

251. G. Sun and G. F. Hewitt, "Experimental Studies on Heat Transfer in Annular Flow," *Heat and Technology* (14/2): 87–93, 1996.

252. W. H. Jens and P. A. Lottes, *Analysis of Heat Transfer, Burnout, Pressure Drop and Density Data for High Pressure Water,* Argonne National Laboratory Report ANL-4627, May 1951.

253. J. R. S. Thom, W. M. Walker, T. A. Fallon, and G. F. S. Reising, "Boiling in Subcooled Water During Flow Up Heated Tubes or Annuli," in *Symposium on Boiling Heat Transfer in Steam Generation Units and Heat Exchangers, Manchester,* IMechE, London, Paper 6, September 1965.

254. F. W. Dittus and L. M. K. Boelter, *Publications on Engineering,* vol. 2, p. 443, University of California, Berkeley, CA, 1930.

255. C. E. Dengler and J. Addoms, "Heat Transfer Mechanism for Vaporisation of Water in a Vertical Tube," *Chem. Eng. Prog. Symp. Ser.* (52/18): 95, 1956.

256. S. A. Guerrieri and R. D. Talty, "A Study of Heat Transfer to Organic Liquids in Single Tube Natural Circulation Vertical Tube Boilers," *Chem. Eng. Prog. Symp. Ser.* (52/18): 69, 1956.

257. J. G. Collier and D. H. Pulling, "Heat Transfer to Two-Phase Gas-Liquid Systems: Part II. Further Data on Steam-Water Mixtures in the Liquid Dispersed Region in an Annulus," *Trans. Inst. Chem. Eng.* (42): 127, 1962.

258. D. Butterworth, *The Correlation of Cross Flow Pressure Drop Data by Means of a Permeability Concept,* UKAEA Report AERE-R9435, 1979.

259. D. Steiner and J. Taborek, "Flow Boiling Heat Transfer in Vertical Tubes Correlated by an Asymptotic Method," *Heat Transfer Eng.* (13/2): 43, 1992.

260. K. Engelberg-Forester and R. Greif, "Heat Transfer to a Boiling Liquid—Mechanism and Correlations," *J. Heat Transfer* (81): 43–53, 1959.

261. R. W. Bjorge, G. R. Hall, and W. M. Rohsenow, "Correlation of Forced Convection Boiling Heat Transfer Data," *Int. J. Heat Mass Transfer* (25): 753–757, 1982.

262. J. P. Wattelet, "Predicting Boiling Heat Transfer in a Small Diameter Round Tube Using an Asymptotic Method," in *Convective Flow Boiling*, J. C. Chen ed., pp. 377–382, Taylor & Francis, Washington, DC, 1996.

263. V. V. Wadekar, "An Alternative Model for Flow Boiling Heat Transfer," in *Convective Flow Boiling*, J. C. Chen ed., pp. 187–192, Taylor & Francis, Washington, DC, 1996.

264. R. W. Bowring, *Physical Model Based on Bubble Detachment and Calculation of Steam Voidage in the Subcooled Region of a Heated Channel,* OECD Holden Reactor Project Report HPR-10, 1962.

265. M. M. Shah, "A New Correlation for Heat Transfer During Boiling Flow Through Pipes," *ASHRAE Trans.* (82/2): 66–86, 1976.

266. M. M. Shah, "Chart Correlation for Saturated Boiling Heat Transfer: Equations and Further Study," *ASHRAE Trans.* (88): 185–196, 1982.

267. S. G. Kandlikar, "A General Correlation for Saturated Two-Phase Flow Boiling Heat Transfer Inside Horizontal and Vertical Tubes," *J. Heat Transfer* (112): 219–228, 1990.

268. A. E. Gungor and R. S. H. Winterton, "Simplified General Correlation for Saturated Flow Boiling and Comparisons of Correlations With Data," *Chem. Eng. Res. Des.* (65): 148–156, 1987.

269. A. E. Gungor and R. H. S. Winterton, "A General Correlation for Flow Boiling in Tubes and Annuli," *Int. J. Heat Mass Transfer* (29/3): 351–358, 1986.

270. G. F. Hewitt, "Reboilers," in *Process Heat Transfer,* G. F. Hewitt, G. L. Shires, and T. R. Bott eds., chap. 14, CRC Press, Boca Raton, FL, 1994.

271. G. F. Hewitt and N. S. Hall-Taylor, *Annular Two-Phase Flow,* Pergamon Press, Oxford, UK, 1970.

272. G. Sun, W. H. G. T. Chan, and G. F. Hewitt, "A General Heat Transfer Model for Two-Phase Annular Flow," in *Convective Flow Boiling,* J. C. Chen ed., pp. 193–198, Taylor & Francis, Washington, DC, 1996.

273. G. F. Hewitt, "Pressure Drop," in *Handbook of Multiphase Systems,* G. Hetsroni ed., chap. 2.2, McGraw-Hill Book Company, New York, 1982.

274. D. G. Owen and G. F. Hewitt, "An Improved Annular Two-Phase Flow Model," in *Proc. 3rd Int. Multiphase Flow Conference,* The Hague, Netherlands, Paper No. C1, 1987.

275. P. G. Kosky and F. W. Staub, "Local Condensing Heat Transfer Coefficients in the Annular Flow Regime," *AIChE J.* (17): 1037–1043, 1971.

276. V. G. Levich, in *Physicochemical Hydrodynamics,* chap. XII, Prentice-Hall, Englewood Cliffs, NJ, 1962.

277. S. Jayanti and G. F. Hewitt, "Hydrodynamics and Heat Transfer in Wavy Annular Gas-Liquid Flow: A Computational Fluid Dynamics Study," *Int. J. Heat Mass Transfer* (40): 2445–2460, 1997.

278. Y. Fujita, "Predictive Methods of Heat Transfer Coefficient and Critical Heat Flux in Mixture Boiling," in *Proc. 4th World Conference on Experimental Heat Transfer, Fluid Mechanics and Thermodynamics,* Brussels, Belgium, vol. 2, pp. 831–842, 1997.

279. Y. Fujita and M. Tsutsui, "Convective Flow Boiling of Binary Mixtures in a Vertical Tube," in *Convective Flow Boiling,* J. C. Chen ed., pp. 259–264, Taylor & Francis, Washington, DC, 1996.

280. G. P. Celata, M. Cumo, and D. Sataro, "Heat Transfer in Vertical Forced Convective Boiling of Binary Mixtures," in *Convective Flow Boiling,* J. C. Chen ed., pp. 251–257, Taylor & Francis, Washington, DC, 1996.

281. D. Steiner, "Forced Convective Vaporization of Saturated Liquid Mixtures," in *Convective Flow Boiling,* J. C. Chen ed., pp. 271–276, Taylor & Francis, Washington, DC, 1996.

282. R. A. W. Shock, "Evaporation of Binary Mixtures in Annular Flow," *Int. J. Multiphase Flow* (2): 411, 1976.

283. L. Zhang, E. Hihara, and T. Satiao, "Boiling Heat Transfer of a Ternary Refrigerant Mixture Inside a Horizontal Smooth Tube," *Int. J. Heat Mass Transfer* (40): 2009–2017, 1997.

284. D. E. Jung, M. McLinden, R. Rademacher, and D. Didion, "Horizontal Flow Boiling Heat Transfer Experiments With a Mixture of R22/R114," *Int. J. Heat Mass Transfer* (32): 131, 1989.

285. S. Yoshida, H. Mori, H. Hong, and T. Matsunaga, "Prediction of Heat Transfer Coefficient for Refrigerants Flowing in Horizontal Evaporator Tubes," *Trans. JAR* (11): 67–78, 1994.

286. R. Winterton, "Extension of a Pool Boiling Based Correlation to Flow Boiling of Mixtures," in *Proc. EUROTHERM Seminar No. 48: Pool Boiling 2,* D. Gorenflo, D. B. R. Kenning, and C. Marvillet eds., Edizioni ETS, Pisa, Italy, pp. 173–180, 1996.

287. Z. Liu and R. H. S. Winterton, "A General Correlation for Saturated and Subcooled Flow Boiling in Tubes and Annuli Based on a Nucleate Pool Boiling Equation," *Int. J. Heat Mass Transfer* (34): 2759–2766, 1991.

288. S. Koyama, J. Yu, S. Momoki, T. Fujii, and H. Honda, "Forced Convective Flow Boiling Heat Transfer of Pure Refrigerants Inside a Horizontal Microfin Tube," in *Convective Flow Boiling,* J. C. Chen ed., pp. 137–142, Taylor & Francis, Washington, DC, 1996.

289. S. M. MacBain and A. E. Bergles, "Heat Transfer and Pressure Drop Characteristics of Forced Convection Evaporation in Deep Spirally Fluted Tubing," in *Convective Flow Boiling,* J. C. Chen ed., pp. 143–148, Taylor & Francis, Washington, DC, 1996.

290. N. E. Hawkes, K. A. Shollenberger, and V. P. Carey, "The Effects of Internal Ribs and Fins on Annular Two Phase Transport in Compact Evaporator Passages," in *Convective Flow Boiling,* J. C. Chen ed., pp. 317–322, Taylor & Francis, Washington, DC, 1996.

291. G. F. Hewitt, "Burnout," in *Handbook of Multiphase Systems,* G. Hetsroni ed., chap. 6.4, McGraw-Hill Book Company, New York, 1982.

292. G. F. Hewitt, "Critical Heat Flux in Flow Boiling," in *Proc. 6th Int. Heat Transfer Conf.,* Toronto, Canada, vol. 6, pp. 143–171, 1978.

293. R. W. Bowring, *A Simple but Accurate Round Tube, Uniform Heat Flux, Dryout Correlation Over the Pressure Range 0.7–17 mn/m²,* UKAEA Report AEEWR789, 1972.

294. M. J. Brusstar and H. Merte, "Effects of Buoyancy on the Critical Heat Flux in Forced Convection," *J. Thermophysics and Heat Transfer* (8): 322–328, 1994.

295. Z. Ruder, A. Bar-Cohen, and P. Griffiths, "Major Parametric Effect on Isothermality in Horizontal Steam Generating Tubes at Low and Moderate Steam Qualities," *Heat and Fluid Flow* (8): 218–227, 1987.

296. A. Bar-Cohen, Z. Ruder, and P. Griffiths, "Thermal and Hydrodynamic Phenomena in Horizontal Uniformly Heated Steam Generating Pipe," *J. Heat Transfer* (109): 739–745, 1987.

297. A. Bar-Cohen, Z. Ruder, and P. Griffiths, "Development and Validation of Boundaries for Circumferential Isothermality in Horizontal Boiler Tubes," *Int. J. Multiphase Flow* (12): 63–77, 1986.

298. G. F. Hewitt and A. H. Govan, "Phenomena and Prediction in Annular Two-Phase Flow," in *Advances in Gas-Liquid Flows,* J. A. Kim, U. S. Rohatgi, and A. Hashemi eds., ASME-FED, vol. 99, HTD-vol. 155, ASME, New York, 1990.

299. G. F. Hewitt, "Experimental Studies of the Mechanisms of Burnout in Heat Transfer to Steam Water Mixtures," in *Proc. 4th Int. Heat Transfer Conf.,* Versailles, France, Paper B6.6, 1970.

300. R. Semeria and G. F. Hewitt, "Aspects of Two-Phase Gas-Liquid Flow," in *International Centre for Heat and Mass Transfer, Seminar on Recent Developments in Heat Exchangers,* Trogir, Yugoslavia, Lecture Session H, September 1972.

301. R. K. F. Keeys, J. C. Ralph, and D. N. Roberts, *Post Burnout Heat Transfer in High Pressure Steam-Water Mixtures in a Tube with Cosine Heat Flux Distribution,* UKAEA Report AERE-R6411, 1970.

302. R. K. F. Keeys, J. C. Ralph, and D. N. Roberts, *The Effect of Heat Flux on Liquid Entrainment in Steam Water Flow in a Vertical Tube at 1000 psia,* UKAEA Report AERE-R6294, 1971.

303. D. C. Groeneveld, S. C. Cheng, and T. Doan, "The CHF Look-Up Table, A Simple and Accurate Method for Predicting Critical Heat Flux," *Heat Transfer Engineering* (7): 46–62, 1986.

304. P. L. Kirillov, V. P. Bobkov, V. N. Vinogradov, V. S. Denisov, A. A. Ivashkevitch, I. B. Katan, E. I. Panuitchev, I. P. Smogalev, and O. B. Salnikova, "On Standard Critical Heat Flux Data for Round Tubes," in *Proc. 4th Int. Topical Meeting on Nuclear Reactor Thermal-Hydraulics (NURETH-4),* Karlsruhe, Germany, pp. 103–108, October 1989.

305. B. Thompson and R. V. Macbeth, *Boiling Water Heat Transfer—Burnout in Uniformly Heated Round Tubes: A Compilation of World Data with Accurate Correlations,* UKAEA Report AEEW-R356, 1964.

306. K. M. Becker, *An Analytical and Experimental Study of Burnout Conditions in Vertical Round Ducts,* AB Atomenergi Report AE178, 1965.

307. S. Bertoletti, G. P. Gaspari, C. Lombardi, G. Peterlongo, M. Silvestri, and F. A. Tacconi, "Heat Transfer Crisis With Steam-Water Mixtures," *Energia Nucleare* (12/3): 121–172, 1965.

308. L. Biasi, G. C. Clerici, S. Garribba, and R. Sara, "Studies on Burnout. Part 3," *Energia Nucleare* (14/9): 530–536, 1967.

309. M. M. Shah, "Improved General Correlation for Critical Heat Flux During Upflow in Uniformly Heated Vertical Tubes," *Heat and Fluid Flow* (8/4): 326–335, 1987.

310. Y. Katto and H. Ohne, "An Improved Version of the Generalised Correlation of Critical Heat Flux for Forced Convection Boiling in Uniformly Heated Vertical Tubes," *Int. J. Heat Mass Transfer* (27/9): 1641–1648, 1984.

311. S. Y. Ahmad, "Fluid-to-Fluid Modelling of Critical Heat Flux: A Compensated Distortion Model," *Int. J. Heat Mass Transfer* (16): 641, 1973.

312. S. S. Kutateladze and A. I. Leontiev, "Some Applications of the Asymptotic Theory of the Turbulent Boundary Layer," in *Proc. 3rd Int. Heat Transfer Conf.,* New York, NY, vol. 3, pp. 1–6, 1966.

313. L. S. Tong, "An Evaluation of the Departure From Nucleate Boiling in Bundles of Reactor Fuel Rods," *Nuclear Sci. Eng.* (33): 7–15, 1968.

314. G. P. Celata, M. Cumo, and A. Mariani, "Assessment of Correlations and Models for the Prediction of CHF in Sub-Cooled Flow Boiling," *Int. J. Heat Mass Transfer* (37/2): 237–255, 1994.

315. J. Weisman and B. S. Pei, "Prediction of CHF in Flow Boiling at Low Qualities," *Int. J. Heat Mass Transfer* (26): 1463, 1983.

316. I. A. Mudawwar, T. A. Incropera, and F. P. Incropera, "Boiling Heat Transfer and Critical Heat Flux in Liquid Films Falling on Vertically-Mounted Heat Sources," *Int. J. Heat Mass Transfer* (30): 2083–2095, 1987.

317. C. H. Lee and I. A. Mudawwar, "A Mechanistic CHF Model for Sub-Cooled Flow Boiling Based on Local Bulk Flow Conditions," *Int. J. Multiphase Flow* (14): 711, 1988.

318. A. H. Govan, "Modelling of Vertical Annular and Dispersed Two-Phase Flow," Ph.D. thesis, University of London, London, UK, 1990.

319. V. I. Milashenko, B. I. Nigmatulin, V. V. Petukhov, and N. I. Trubkin, "Burnout and Distribution of Liquid in Evaporative Channels of Various Lengths," *Int. J. Multiphase Flow* (15): 391–402, 1989.

320. A. W. Bennet, G. F. Hewitt, H. A. Kearsey, R. K. F. Keeys, and D. J. Pulling, "Studies of Burnout in Boiling Heat Transfer," *Trans. Inst. Chem. Eng.* (45): T139–T333, 1967.

321. G. P. Celata, "Pool and Forced Convective Boiling of Binary Mixtures," in *Pool and External Flow Boiling,* V. K. Dhir and A. E. Bergles eds., pp. 397–409, ASME, New York, 1992.

322. V. I. Tolubinsky and P. S. Matorin, "Forced Convection Boiling Heat Transfer Crisis With Binary Mixtures," *Heat Transfer—Soviet Research* (5/2): 98–101, 1973.

323. A. Miyra, A. Marroquin, and H. Auracher, "Critical Heat Flux and Minimum Heat Flux of Film Boiling of Mixtures in Forced Convection Boiling," in *Proc. 4th World Conference on Experimental Heat Transfer, Fluid Mechanics and Thermodynamics,* Brussels, Belgium, vol. 2, pp. 873–880, 1997.

324. A. E. Bergles, "Burnout in Boiling Heat Transfer. Part 1: Pool Boiling Systems," *Nuclear Safety* (16/1): 29–42, 1975.

325. C. R. Nichols, J. M. Spurlock, and M. Markels, *Effect of Electrical Fields on Boiling Heat Transfer,* NYO-2404-5, 1964, and NYO-2404-70, 1965.

326. E. O. Moeck, G. A. Wikhammer, I. P. I. MacDonald, and J. G. Collier, *Two Methods of Improving the Dryout Heat Flux for High Pressure Steam/Water Flow,* AECL2109, 1964.

327. H. Nariai and F. Inasaka, "Critical Heat Flux of Sub-Cooled Flow Boiling in Tube With and Without Internal Twisted Tape Under Uniform and Non-Uniform Heat Flux Conditions," in *Convective Flow Boiling,* J. C. Chen ed., pp. 207–212, Taylor & Francis, Washington, DC, 1996.

328. J. B. Chung, W. P. Baek, and S. H. Chang, "Effects of the Space and Mixing Vanes on Critical Heat Flux for Low Pressure Water and Low Velocities," *Int. Comm. Mass Transfer* (23): 757–765, 1966.

329. K. Nishikawa, T. Fujii, S. Yoshida, and M. Onno, "Flow Boiling Crisis in Grooved Boiler Tubes," *Heat Transfer Japanese Research* (4): 270–274, 1974.

330. J. G. Collier, "Heat Transfer in the Post Burnout Region and During Quenching and Reflooding," in *Handbook of Multiphase Systems,* G. Hetsroni ed., chap. 6.5, McGraw-Hill Book Company, New York, 1982.

331. J. C. Chen and J. Costigan, "Review of Post-Dryout Heat Transfer in Dispersed Two-Phase Flow," in *Post-Dryout Heat Transfer,* G. F. Hewitt, J.-M. Delhaye, and N. Zuber eds., chap. 1, CRC Press, Boca Raton, FL, 1992.

332. M. Ishii, "Flow Phenomena in Post-Dryout Heat Transfer," in *Post-Dryout Heat Transfer,* G. F. Hewitt, J.-M. Delhaye, and N. Zuber eds., chap. 4, CRC Press, Boca Raton, FL, 1992.

333. D. C. Groeneveld, "A Review of Inverted Annular and Low Quality Film Boiling," in *Post-Dryout Heat Transfer,* G. F. Hewitt, J.-M. Delhaye, and N. Zuber eds., chap. 5, CRC Press, Boca Raton, FL, 1992.

334. M. Andreani and G. Yadigaroglu, "Prediction Methods for Dispersed Flow Film Boiling," *Int. J. Multiphase Flow* (20, suppl.): 1–51, 1994.

335. N. Hammouda, D. C. Groeneveld, and S. C. Cheng, "An Experimental Study of Sub-Cooled Film Boiling of Refrigerants in Vertical Up-Flow," *Int. J. Heat Mass Transfer* (39): 3799–3812, 1996.

336. J. E. Leonard, K. H. Sun, and G. E. Dix, "Solar and Nuclear Heat Transfer," *AIChE Symp. Ser.* (73/164): 7, 1977.

337. D. C. Groeneveld, *Post-Dryout Heat Transfer at Reactor Operating Conditions,* AECL-4513, 1973.

338. G. T. Analytis and G. Yadigaroglu, "Analytical Modelling of Inverted Annular Film Boiling," *Nuclear Engineering Design* (99): 201–212, 1987.

339. N. Hammouda, D. C. Groeneveld, and S. C. Cheng, "Two-Fluid Modelling of Inverted Annular Film Boiling," *Int. J. Heat Mass Transfer* (40): 2655–2670, 1997.

340. W. F. Laverty and W. M. Rohsenow, *Film Boiling of Saturated Liquid Flowing Upward Through a Heated Tube: High Vapour Quality Range,* MIT Heat Transfer Lab. Rep. 9857-32, 1964.

341. R. P. Forslund and W. M. Rohsenow, *Thermal Non-Equilibrium in Dispersed Flow Film Boiling in a Vertical Tube,* MIT Heat Transfer Lab. Rep. 75312-44, 1966.

342. A. W. Bennet, G. F. Hewitt, H. A. Kearsey, and R. K. F. Keeys, *Heat Transfer to Steam Water Mixtures Flowing in Uniformly Heated Tubes in Which the Critical Heat Flux has Been Exceeded,* UKAEA Report AERE-R5373, 1967.

343. R. P. Forslund and W. M. Rohsenow, "Dispersed Flow Film Boiling," *J. Heat Transfer* (90): 399–407, 1968.

344. D. Evans, S. W. Webb, and J. C. Chen, "Axially Varying Vapour Superheats in Convective Film Boiling," *J. Heat Transfer* (107): 663–669, 1985.

345. E. M. Ganic and W. M. Rohsenow, "Dispersed Flow Heat Transfer," *Int. J. Heat Mass Transfer* (90): 399–407, 1977.

346. O. C. Iloeje, W. M. Rohsenow, and P. Griffith, *Three-Step Model of Dispersed Flow Heat Transfer (Post CHF Vertical Flow),* ASME Paper 75-WA/HT1, 1975.

347. A. F. Varone and W. M. Rohsenow, "Post-Dryout Heat Transfer Prediction," in *Proc. First Int. Workshop on Fundamental Aspects of Post-Dryout Heat Transfer,* Salt Lake City, UT, NUREG/CP-0060, April 2–4, 1984.

348. K. H. Sun, J. M. Gonzales, and C. L. Tien, *Calculation of Combined Radiation and Convection Heat Transfer in Rod Bundles Under Emergency Cooling Conditions,* ASME Paper 75-HT-64, 1975.

349. A. F. Varone and W. M. Rohsenow, *The Influence of the Dispersed Phase on the Convective Heat Transfer in Dispersed Flow Film Boiling,* MIT Report No. 71999-106, 1990.

350. R. K. F. Keeys, J. C. Ralph, and D. N. Roberts, "Post Burnout Heat Transfer in High Pressure Steam-Water Mixtures in a Tube With Co-Sign Heat Flux Distribution," *Prog. Heat Mass Transfer* (6): 99–118, 1972.

351. H. Auracher and A. Marroquin, "Critical Heat Flux and Minimum Heat Flux of Film Boiling of Binary Mixtures Flowing Upwards in a Vertical Tube," in *Convective Flow Boiling,* J. C. Chen ed., pp. 213–218, Taylor & Francis, Washington, DC, 1996.

352. L. E. Hochreiter, M. J. Loftus, F. J. Erbacher, P. Ihle, and K. Rust, "Post CHF Effects of Spacer Grids and Blockages in Rod Bundles," in *Post-Dryout Heat Transfer,* G. F. Hewitt, J. M. Delhaye, and N. Zuber eds., chap. 3, CRC Press, Boca Raton, FL, 1992.

353. G. Gimbutis, "Heat Transfer in Film Heat Exchangers," in *Proc. 14th Int. Congress of Refrigeration,* Moscow, vol. 2, pp. 1–7, 1975.

354. E. N. Ganic and K. Mastinaiah, "Hydrodynamics and Heat Transfer in Falling Film Flow," in *Low Reynolds Number Heat Exchangers,* S. Kakac, R. K. Shah, and A. E. Bergles eds., Hemisphere Publishing Corporation, Washington, DC, 1982.

355. G. D. Fulford, "The Flow of Liquids in Thin Films," *Adv. Chem. Eng.* (5): 151–236, 1964.

356. W. Nusselt, "Die Oberflaechenkondensateion des Wasserdampfes," *VDI Zeitschrift* (60): 541–546, 569–575, 1916.

357. K. R. Chun and R. A. Seban, "Heat Transfer to Evaporating Liquid Films," *J. Heat Transfer* (91): 391–396, 1971.

358. F. K. Wasden and A. E. Dukler, "Numerical Investigation of Large Wave Interaction on Free Falling Film," *Int. J. Multiphase Flow* (15): 357–370, 1989.

359. S. Jayanti and G. F. Hewitt, "Hydrodynamics and Heat Transfer of Wavy Thin Film Flow," *Int. J. Heat Mass Transfer* (40): 179–190, 1997.

360. J. W. Palen, Q. Wang, and J. C. Chen, "Falling Film Evaporation of Binary Mixtures," *AIChE J.* (40): 207–214, 1994.

361. R. Numrich, "Falling Film Evaporation of Soluble Mixtures," in *Convective Flow Boiling,* J. C. Chen ed., pp. 335–338, Taylor & Francis, Washington, DC, 1996.

362. T. Fujita and T. Ueda, "Heat Transfer to Falling Liquid Films and Film Breakdown—II," *Int. J. Heat Mass Transfer* (21): 109–118, 1978.

363. M. Cerza, "Nucleate Boiling in Thin Falling Liquid Films," in *Pool and External Flow Boiling,* V. K. Dhir and A. E. Bergles eds., pp. 459–466, ASME, New York, 1992.

364. W. Wilke, "Warmeubergang an Rieselfilme," in *VDI Forschungsh.,* no. 490, Dusseldorf, Germany, 1962.

365. K. Feind, "Stromungsuntersuchengen bei Gegenstrom von Rieselfilmen und Gas in Lotrechten Rohren," in *VDI Forschungsh.,* no. 481, Dusseldorf, Germany, 1960.

366. E. Ulucakli, "Heat Transfer in a Sub-Cooled Falling Liquid Film," in *Convective Flow Boiling,* J. C. Chen ed., pp. 329–334, Taylor & Francis, Washington, DC, 1996.

367. D. E. Hartley and W. Murgatroyd, "Criteria for the Breakup of Thin Liquid Layers Flowing Isothermally Over Solid Surfaces," *Int. J. Heat Mass Transfer* (7): 1003, 1964.

368. W. S. Norman and D. T. Binns, "The Effect of Surface Changes on the Minimum Wetting Rates in a Wetted Rod Distillation Column," *Trans. Inst. Chem. Eng.* (38): 294, 1960.

369. P. C. Wayner, "Evaporation and Stress in the Contact Line Region," in *Pool and External Flow Boiling,* V. K. Dhir and A. E. Bergles eds., pp. 251–256, ASME, New York, 1992.

370. A. A. Alhusseini, B. C. Hoke, and J. C. Chen, "Critical Heat Flux in Falling Films Undergoing Nucleate Boiling" in *Convective Flow Boiling,* J. C. Chen ed., p. 339–350, Taylor & Francis, Washington, DC, 1996.

371. J. G. Collier, "Heat Transfer During Quenching and Reflooding," in *Handbook of Multiphase Systems,* G. Hetsroni ed., sec. 6.5.2, McGraw-Hill Book Company, New York, 1982.

372. R. A. Nelson and K. O. Pasamehmetoglu, "Quenching Phenomena," in *Post-Dryout Heat Transfer,* G. F. Hewitt, J.-M. Delhaye, and N. Zuber eds., chap. 2, CRC Press, Boca Raton, FL, 1992.

373. A. Yamanouchi, "Effect of Core Spray Cooling in Transient State after Loss of Coolant Accident," *J. Nucl. Sci. Tech.* (5): 547–558, 1968.

374. S. G. Kandlikar and B. J. Stumm, "A Control Volume Approach for Investigating Forces on a Departing Bubble under Subcooled Flow Boiling," *J. Heat Transfer* (117): 990–997, 1995.

375. S. G. Kandlikar, "Development of a Flow Boiling Map for Subcooled and Saturated Flow Boiling of Different Fluids inside Circular Tubes," *J. Heat Transfer* (113): 190–200, 1991.

CHAPTER 16
MEASUREMENT OF TEMPERATURE AND HEAT TRANSFER

R. J. Goldstein
University of Minnesota

P. H. Chen
National Taiwan University

H. D. Chiang
Industrial Technology Research Institute
Energy & Resources Laboratories
Chutung, Hsinchu, Taiwan

INTRODUCTION

In heat transfer studies, many different properties or quantities may be measured. These include thermodynamic properties, transport properties, velocities, mass flow, and concentration. There are, however, two key quantities that must be measured in almost every heat transfer experiment or project, be it for research purposes or for a specific application. These are temperature and the quantity of heat transferred. The present chapter will discuss the means of measuring these quantities. First, we review definitions of temperature and temperature scales and then discuss specific instruments that are used to measure temperature. The heart of a temperature measuring instrument (or thermometer) consists of a sensor (or transducer) that has a unique response to changes in its temperature, be it a change in length, an EMF change, or a change in electrical resistance. The output of the sensor is used as a direct indicator or converted through signal processing to produce a reading of the temperature. Next, calibration techniques are introduced as a means of assuring the accuracy of various temperature measuring instruments. In practice, however, an accurate instrument does not guarantee an accurate result during actual temperature measurement, the reason being that there are at least two additional sources of error:

1. The environmental influence resulting from installation of the sensor
2. The quality system that is associated with the measurement process, which includes the technical capability of the operator and adherence to the standard operating procedure, among other factors.

These topics will be addressed in more detail later. A later section covers heat flux measurement techniques. We close the chapter with a discussion of mass transfer analogies that are used in experiments to simulate heat transfer.

TEMPERATURE MEASUREMENT

Basic Concepts and Definitions

Temperature measurement is important in virtually every heat transfer study and for many other systems as well. The temperature difference is, of course, the driving force that causes heat to be transferred from one system to another. In addition, knowledge of temperature is often required for determining the properties of systems in equilibrium.

Temperature is a thermodynamic property and is defined for systems in equilibrium. Even with nonequilibrium systems—e.g., in a medium in which there is a temperature gradient—local measurements are generally assumed to give a temperature that can help determine local thermodynamic properties by assuming local quasiequilibrium.

From the zeroth law of thermodynamics, we know that two systems that are in thermal equilibrium with a third system are in thermal equilibrium with one another and, by definition, have the same temperature. The zeroth law is not only important in defining systems that have the same temperature, but it also provides the basic principle behind thermometry; one measures temperatures of different systems by thermometers that are, in turn, compared to some standard temperature systems or standard thermometers.

Initial concepts of temperature came from the physical sensation of the relative hotness or coldness of bodies. This sensation of warmth or cold is so subjective relative to our immediate prior exposure that it is difficult to use for anything but simple qualitative comparison. The need to assign a quantitative value to temperature leads to the definition of a temperature scale. The concept of fixed points of temperature arises from the observation that there exist some systems in nature that always exhibit the same temperatures. The scientific or thermodynamic definition of temperature comes from Kelvin, who defined the ratio of the thermodynamic or absolute temperatures of two systems as being equal to the ratio of the heat added to the heat rejected for a reversible heat engine operated between the systems. This unique temperature scale requires only one fixed point, the triple point of water, for its definition.

The use of an idealized (Carnot) engine for measuring temperature or the ratio of two temperatures is not practical for most systems, though such an engine is approximated in some devices at very low temperature. The definition of temperature from a Carnot cycle leads to the concept of ideal gas systems for thermodynamic temperature measurement. These are approached by special low-pressure, high-precision gas thermometers. The thermodynamic (or ideal gas) temperature scale is elegant in its definition, but, to measure temperatures, one must, in principle, set up a thermodynamic (or statistical mechanical) system that can relate the unknown equilibrium states back to the reference state and then calculate their thermodynamic temperatures. Such procedures are complex, costly, and difficult to carry out experimentally, and practical measurements can be made with greater precision and reproducibility using nonthermodynamic thermometers calibrated against, or traceable to, an internationally accepted temperature scale.

The definition of a practical temperature scale requires fixed points (i.e., quantitatively defined and easily reproduced temperatures), and one or more instruments (thermometers) that are sensitive primarily to temperature and not other variables. In addition, an interpolation equation is required to measure temperatures between the fixed points.

Fixed points are temperatures that are relatively easy to reproduce given a reasonable apparatus and sufficient care. They usually involve two- or three-phase thermodynamic equilibrium points of pure substances such as the freezing point of a liquid, the boiling point of a liquid, or a triple point where solid, liquid, and vapor are in equilibrium. For a pure substance, the triple point is a uniquely fixed temperature, while the freezing point usually has only a slight dependence on pressure. Liquid-vapor equilibrium temperatures are dependent on pressure and thus are not as convenient except at temperatures where other equilibrium points are not readily available.

The Kelvin or thermodynamic temperature scale uses a single fixed point to define the size of a degree—the triple point of water at 273.16 Kelvin (K). Special low-pressure, constant-volume gas thermometers are used in a handful of laboratories to approximate the thermodynamic temperatures of other fixed points.

Many different thermometers have been used in temperature measurements. The most common types are those in which the temperature-dependent measured variable is: (1) volume or length of a system, as with liquid-in-glass thermometers, (2) electrical resistance (platinum and other resistance thermometers, including thermistors), (3) electromotive force (EMF)—particularly as used in thermocouples, and (4) radiation emitted by a surface, as in various types of pyrometers that are used primarily with high-temperature systems. These thermometers as well as some others will be described below.

In principle, any device that has one or more physical properties uniquely related to temperature in a reproducible way can be used as a thermometer. Such a device is usually classified as either a primary or secondary thermometer. If the relation between the temperature and the measured physical quantity is described by an exact physical law, the thermometer is referred to as a primary thermometer; otherwise, it is called a secondary thermometer. Examples of primary thermometers include special low-pressure gas thermometers that behave according to the ideal gas law and some radiation-sensitive thermometers that are based upon the Planck radiation law. Resistance thermometers, thermocouples, and liquid-in-glass thermometers all belong to the category of secondary thermometers. Ideally, a primary thermometer is capable of measuring the thermodynamic temperature directly, whereas a secondary thermometer requires a calibration prior to use. Furthermore, even with an exact calibration at fixed points, temperatures measured by a secondary thermometer still do not quite match the thermodynamic temperature; these readings are calculated from interpolation formulae, so there are differences between these readings and the true thermodynamic temperatures. Of course, the better the thermometer and its calibration, the smaller the deviation would be.

Well-designed low-pressure gas thermometers can be used to determine (really approximate) the thermodynamic temperature. However, from a practical standpoint, where precision and simplicity in the implementation and transfer are the major considerations, secondary thermometers were chosen as the defining standard thermometers* for a practical temperature scale. This scale was defined by the use of fixed reference points whose thermodynamic temperatures were determined from gas thermometer measurements. The International Committee of Weights and Measures (Comite International des Poids et Mesures, CIPM) is responsible for developing and maintaining the scale.

National standard laboratories, such as the National Institute of Standards and Technology (NIST) in the United States, implement and maintain the practical temperature scale for their respective countries. They also help in the transfer of the scale by calibrating the defining standard thermometers. These defining standard thermometers are costly to maintain and are primarily used in temperature calibration laboratories in industry or universities. They are directly or indirectly used for calibration of thermometers used in actual applications.

Standards and Temperature Scales

The standardized scale now used in temperature measurement is the International Temperature Scale of 1990 (ITS-90) [1–3]. ITS-90 has been designed to give values as close to the corresponding thermodynamic temperatures as practically possible. It covers the range of

---

* See Table 16.1 for definition of different thermometers.

TABLE 16.1 Definitions of Different Thermometers

Standard thermometer types	
Primary thermometer	One that uses a physical law to define an exact relation between the temperature and the measured physical quantity.
Secondary thermometer	One that does not have an exact relationship between the temperature and measured quantity; calibration is required.
Defining standard thermometer	A thermometer used in the definition of ITS-90; calibration performed at defining fixed points or against another defining standard.
Transfer standard thermometer	An intermediate standard used to minimize the use and drift of a defining standard thermometer.
Working standard thermometer	A standard used in calibration that is the same type of thermometer as the thermometers to be calibrated. Calibration against a defining standard thermometer or a transfer standard thermometer is required at periodic intervals to ensure accuracy.
Thermometer types with abbreviations	
DWRT	Double wavelength radiation thermometer
PRT	Platinum resistance thermometer
RTD	Resistance temperature detector
SPRT	Standard platinum resistance thermometer
Type S thermocouple	Platinum-10% rhodium vs. platinum thermocouple
Type R thermocouple	Platinum-13% rhodium vs. platinum thermocouple
Type B thermocouple	Platinum-30% rhodium vs. platinum-6% rhodium thermocouple
Type T thermocouple	Copper-constantan thermocouple
Type J thermocouple	Iron-constantan thermocouple
Type E thermocouple	Chromel-constantan thermocouple
Type K thermocouple	Chromel-Alumel thermocouple
Type N thermocouple	Nicrosil-Nisil thermocouple

temperatures from 0.65 K to the highest measurable temperature in practice. Temperature in the ITS-90 scale is presented in International Kelvin Temperature T_{90} and International Celsius Temperature t_{90} with units Kelvin (K) and degrees Celsius (°C), respectively. The temperatures are related by:

$$t_{90} = T_{90} - 273.15 \text{ K} \tag{16.1}$$

Note that a temperature difference has the same numerical value when expressed in either of these units.

$$\Delta t_{90}(°C) = \Delta T_{90}(K) \tag{16.2}$$

The ITS-90 scale is the latest development that evolved from a number of earlier international temperature scales. The first of these scales was the normal hydrogen scale adopted by the CIPM in 1887. This temperature scale was based on hydrogen gas thermometer measurements using the freezing and boiling points of water of 0°C and 100°C as the two defining fixed points. Mercury-in-glass thermometers were used as transfer standards for distribution to other laboratories. During the last century, there have been three earlier major changes (ITS-27, ITS-48, and IPTS-68), one amendment [IPTS-68(75)] and one extension (EPT-76) in the definition of the international temperature scale. For a more detailed historical background, refer to Refs. 2 and 5.

The ITS-90 scale includes 14 defining fixed points ranging from the triple point of hydrogen (13.8033 K) to the normal freezing point of copper (1357.77 K) (see Table 16.2). Four temperature ranges are defined; for two of these, a defining standard thermometer and the specific forms of interpolation or calibration equations are specified [1]. In brief, T_{90} between 0.65 K and 5.0 K is defined in terms of (1) the vapor-pressure versus temperature relationship of $_3$He and $_4$He. (2) Between 3.0 K and the triple point of neon (24.5561 K), T_{90} is defined by a helium gas thermometer. (3) A standard platinum resistance thermometer (SPRT) is used for the definition of T_{90} from the triple point of equilibrium hydrogen (13.8033 K) up to the freezing point of silver (1234.93 K). Above the freezing point of silver (1234.93 K), (4) T_{90} is defined in terms of the Planck radiation law and a defining fixed point. Table 16.3 gives estimates of the one standard deviation uncertainty in the values of thermodynamic temperature T_1 and in the current best practical realizations T_2 of the defining fixed points of the ITS-90 [2]. The definition of the ITS-90 is now sufficiently precise that it should be reproducible to better than 1.0 mK between 0.65 K and the freezing point of aluminum (933.473 K), with a slight degradation up to the freezing point of silver (1234.93 K) [4]. It is also believed that the ITS-90 is now a close approximation (to within 1 mK) of the thermodynamic temperature scale at all temperatures [4].

One of the guiding principles underlying the establishment of the ITS-90 is that it should provide the user with as much choice in its realization as is compatible with an accurate and reproducible scale. The design of the ITS-90 thus includes four generally overlapping main ranges (as described above with their respective interpolation instruments) and many subranges, as shown schematically in Fig. 16.1. Thus, if an SPRT is to be calibrated over the entire high temperature range from 273.16 to 1234.93 K, all seven fixed points in that range must be used. However, if our application range is limited only to between 273.16 and 500 K, then the SPRT need only be calibrated at four fixed points. Under the design of ITS-90, an SPRT may be calibrated from 0 to 30°C using just two fixed points—the triple point of water (0.01°C) and the melting point of gallium (29.7646°C). This last calibration provides the simplest

TABLE 16.2   The Defining Fixed Points of the ITS-90 [1]

Number	Temperature		Substance*	State†
	T_{90}/K	t_{90}/°C		
1	13.8033	−259.3467	e-H_2	tp
2	24.5561	−248.5939	Ne	tp
3	54.3584	−218.7916	O_2	tp
4	83.8058	−189.3442	Ar	tp
5	234.3156	−38.8344	Hg	tp
6	273.16	0.01	H_2O	tp
7	302.9146	29.7646	Ga	mp
8	429.7485	156.5985	In	fp
9	505.078	231.928	Sn	fp
10	692.677	419.527	Zn	fp
11	933.473	660.323	Al	fp
12	1234.93	961.78	Ag	fp
13	1337.33	1064.18	Au	fp
14	1357.77	1084.62	Cu	fp

* All substances are of natural isotopic composition; e-H_2 is hydrogen at the equilibrium concentration of the ortho- and para-molecular forms.

† For complete definitions and advice on the realization of these various states, see Ref. 2. The symbols have the following meaning: tp: triple point (temperature at which the solid, liquid, and vapour phases are in equilibrium); mp, fp: melting point, freezing point (temperature at a pressure of 101 325 Pa at which the solid and liquid phases are in equilibrium).

TABLE 16.3 Estimates of the (1σ) Uncertainty in the Values of Thermodynamic Temperature ΔT_1 and in the Current (1990) Best Practical Realizations ΔT_2 of the Defining Fixed Points of the ITS-90 [2]

Fixed points	T_{90}/K	ΔT_1/mK	ΔT_2/mK
H_2 tp	13.8033	0.5	0.1
Ne tp	24.5561	0.5	0.2
O_2 tp	54.3584	1	0.1
Ar tp	83.8058	1.5	0.1
Hg tp	234.3156	1.5	0.05
H_2O tp	273.16	0	0.02
Fixed points	t_{90}/°C	ΔT_1/mK	ΔT_2/mK
Ga mp	29.7646	1	0.05
In fp	156.5985	3	0.1
Sn fp	231.928	5	0.1
Zn fp	419.527	13	0.1
Al fp	660.323	25	0.3
Ag fp	961.78	40	1*, 10†
Au fp	1 064.18	50	10†
Cu fp	1 084.62	60	15†

* For platinum resistance thermometry.
† For radiation thermometry.

means of achieving the highest accuracy thermometry over the room-temperature range. There are two advantages to this design: (1) The user needs only set up calibration fixed points for the temperature range of interest, and (2) by not requiring the defining standard thermometer to be significantly and unnecessarily heated above (or cooled below) the temperature of normal use, the thermometer would be maintained under the best possible operating conditions. This useful flexibility in the ITS-90 will, of course, introduce some subrange inconsistency (see Fig. 16.2) and increased nonuniqueness (see Fig. 16.3) when compared with a scale with no overlapping ranges and subranges. However, such uncertainties are too small to be of concern to most engineers and scientists.

Two common temperature scales having different-sized degrees are widely used: the Celsius and Fahrenheit scales. The Celsius scale is essentially described by the ITS-90 and Eq. 16.1, as noted above. The Fahrenheit scale (°F) has an analogous absolute temperature scale, the Rankine scale (R), with a value of zero at absolute zero. Temperatures in the Rankine scale can be determined from the Kelvin temperature by

$$T(R) = 1.8T(K) \tag{16.3}$$

The Fahrenheit scale has the same-sized degree as the Rankine scale and can be determined from

$$t(°F) = T(R) - 459.67 \tag{16.4}$$

In accordance with the historical definitions, this leads to a value of 32°F at the freezing point of water and 212°F at the boiling point of water at standard atmospheric pressure. Temperatures in °C and °F are related by

$$t(°F) = (9/5)t(°C) + 32 \tag{16.5}$$

It should be noted that the Fahrenheit and Rankine scales are, with few exceptions, now used only in the United States.

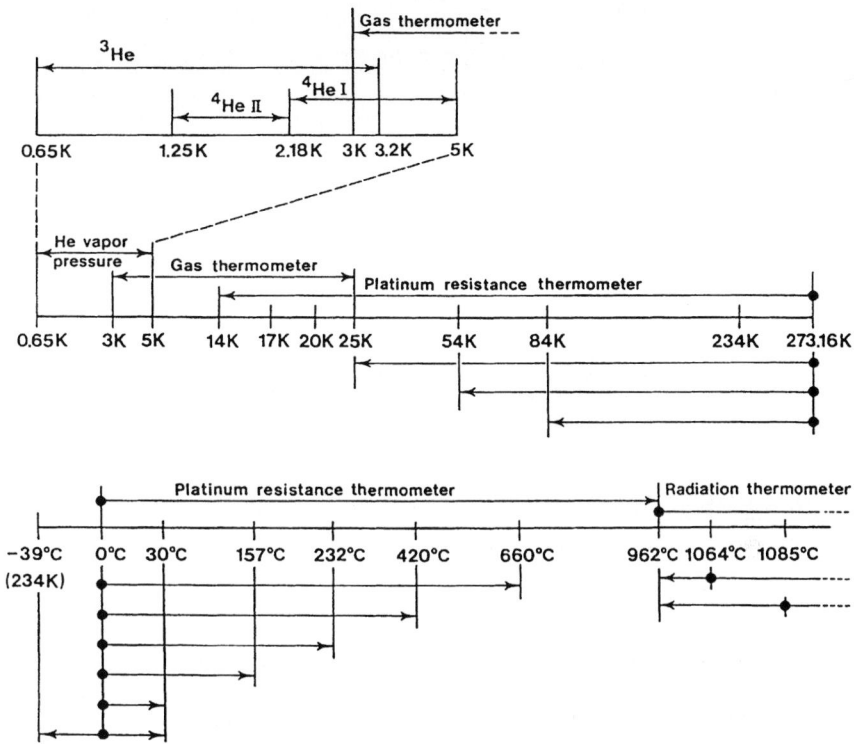

FIGURE 16.1   Schematic representation [2] of the ranges, subranges, and interpolation instruments of ITS-90. (The temperatures shown are approximate only.)

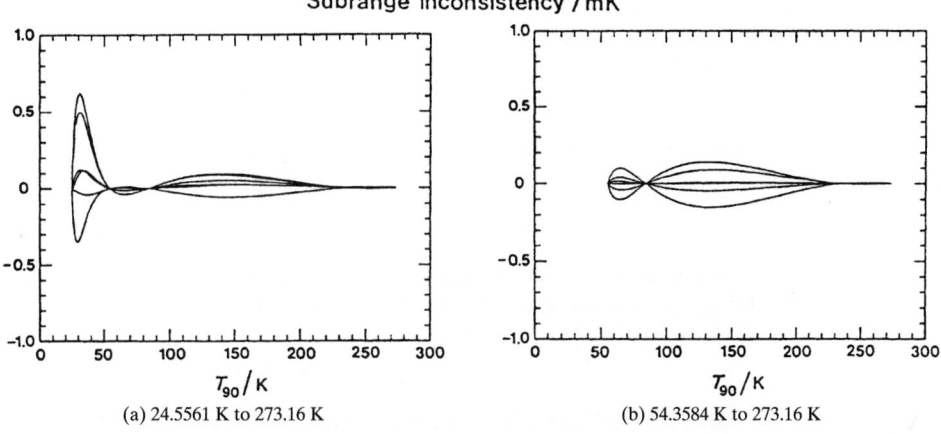

FIGURE 16.2   The subrange inconsistency of the ITS-90 for a number of SPRTs [2].

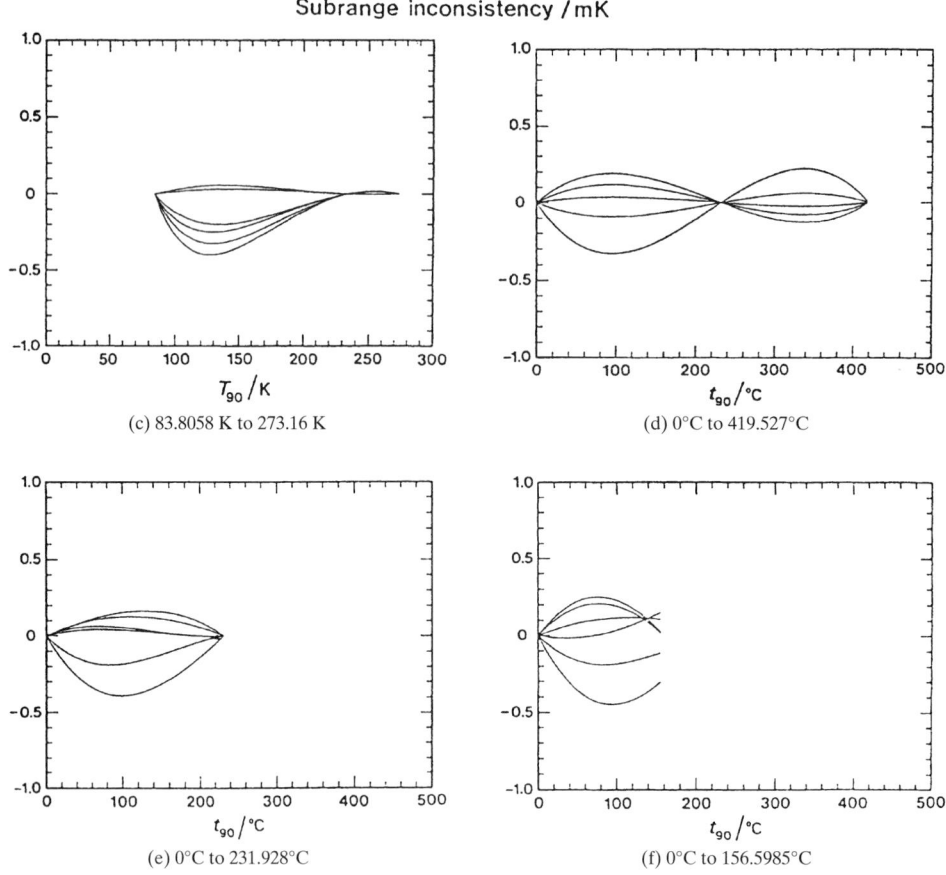

Subrange inconsistency / mK

(c) 83.8058 K to 273.16 K

(d) 0°C to 419.527°C

(e) 0°C to 231.928°C

(f) 0°C to 156.5985°C

FIGURE 16.2 (*Continued*)   The subrange inconsistency of the ITS-90 for a number of SPRTs [2].

Sensors

Introduction.   Historically, low-pressure, constant-volume gas thermometers were the only primary thermometers that had the accuracy required for determining the temperatures of defining fixed points. Recent advances in thermometry have resulted in the development of several types of primary thermometers capable of accurate thermodynamic temperature measurements [5–6]. A list of present-day primary thermometers capable of thermodynamic temperature measurement includes:

1. Gamma-ray anisotropy thermometers—applicable from 10 μK to 1 K

2. Magnetic thermometers—1 mK to 100 K

3. Noise thermometers:

   • Josephson-junction type—1 mK to 1 K
   • "Normal" type—2 to 2000 K

4. Dielectric-constant gas thermometers—3 to 50 K

5. Acoustic gas thermometers—1 to 1000 K

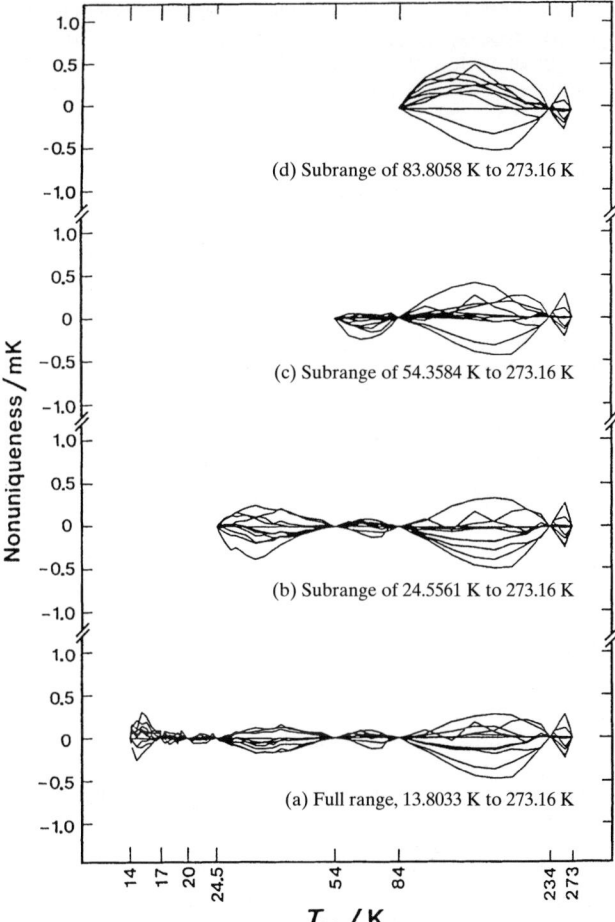

FIGURE 16.3   The nonuniqueness ΔT of the ITS-90 in the range 13.8–273.16 K for a set of eleven SPRTs [2].

6. Constant-volume gas thermometers—1 to 1400 K
7. Total radiation thermometers—100 to 700 K
8. Spectral radiation thermometers—above 700 K

The total and spectral radiation thermometers are not really primary thermometers, since a one-point (or more) calibration is needed for thermodynamic temperature measurements. For that matter, magnetic thermometers also belong to this class—pseudoprimary thermometers.

Almost all thermometers used in practical applications belong to the class of secondary thermometers. They can be categorized according to their temperature-dependent function, represented by the type of transducer or output they employ. They include systems whose temperature is indicated by:

1. *Changes in dimension.* With such devices, a change in physical dimension occurs with a change in temperature. In this category are liquid-in-glass or other fluid-expansion thermometers, bimetallic strips, and others.

2. *Electrical effects.*   Electrical methods are convenient because an electrical signal can be easily processed. Resistance thermometers (including thermistors) and thermocouples are the most widely used. Other electrical methods include: noise thermometers using the Johnson noise as a temperature indicator; resonant-frequency thermometers, which rely on the temperature dependence of the resonant frequency of a medium, including nuclear quadrupole resonance thermometers, ultrasonic thermometers, and quartz thermometers; and semiconductor-diode thermometers, where the relation between temperature and junction voltage at constant current is used.

3. *Radiation.*   The thermal radiation emitted by a body is a function of the temperature of the body; hence, measurement of the radiant energy can be used to indicate the temperature. Commonly employed sensors in this category are optical thermometers, infrared scanners, spectroscopic techniques, and total-radiation calorimeters.

4. *Other methods.*   In addition to the systems mentioned above, there are many others. Commonly encountered ones include: optical methods in which the index-of-refraction variation is determined and, from this, the temperature; and liquid-crystal or other contact thermographic methods in which the color of cholesteric liquid crystals or thermally sensitive materials are determined by the temperature.

Different devices have their own temperature range of operation—some suitable for low-temperature measurements, some only applicable at high temperature, and others in between. No single device is applicable over the entire range of temperatures in the physical environment. Even ITS-90 requires three different defining standard thermometers: a gas thermometer, a standard platinum resistance thermometer (SPRT), and a radiant energy sensor. Figure 16.4 gives an approximate range of operation for some commonly used practical thermometers, and Table 16.4 summarizes the relative merits of four common insertion-type thermometers.

Accuracy of a defining standard thermometer in reproducing the ITS-90 temperature is achieved through calibration at the appropriate fixed points. Defining standard thermometers calibrated according to the specifications of ITS-90 are normally done at a government standards laboratory. Other thermometers are calibrated by comparison against defining

TABLE 16.4   Advantages and Disadvantages of Various Insertion- or Probe-Type Sensors

Sensor	Advantages	Disadvantages
Mercury-in-glass thermometers	1. Stable 2. Cheap 3. Good accuracy	1. Slow time response 2. Automation and remote sensing impractical
Thermocouples	1. Wide temperature range 2. Fast response 3. Ease of remote sensing 4. Reasonable cost 5. Rugged 6. Small	1. Need reference junction 2. Do not have extreme sensitivity and stability 3. Fair accuracy
Thermistors	1. Fast response 2. Ease of remote sensing 3. Low cost 4. Rugged	1. Highly nonlinear 2. Small temperature range 3. Limited interchangeability
Platinum resistance thermometers	1. Good linearity 2. Wide temperature range 3. Very good stability even at high temperature 4. Very good precision and accuracy	1. Slow response 2. Vibration and shock fragility 3. High cost

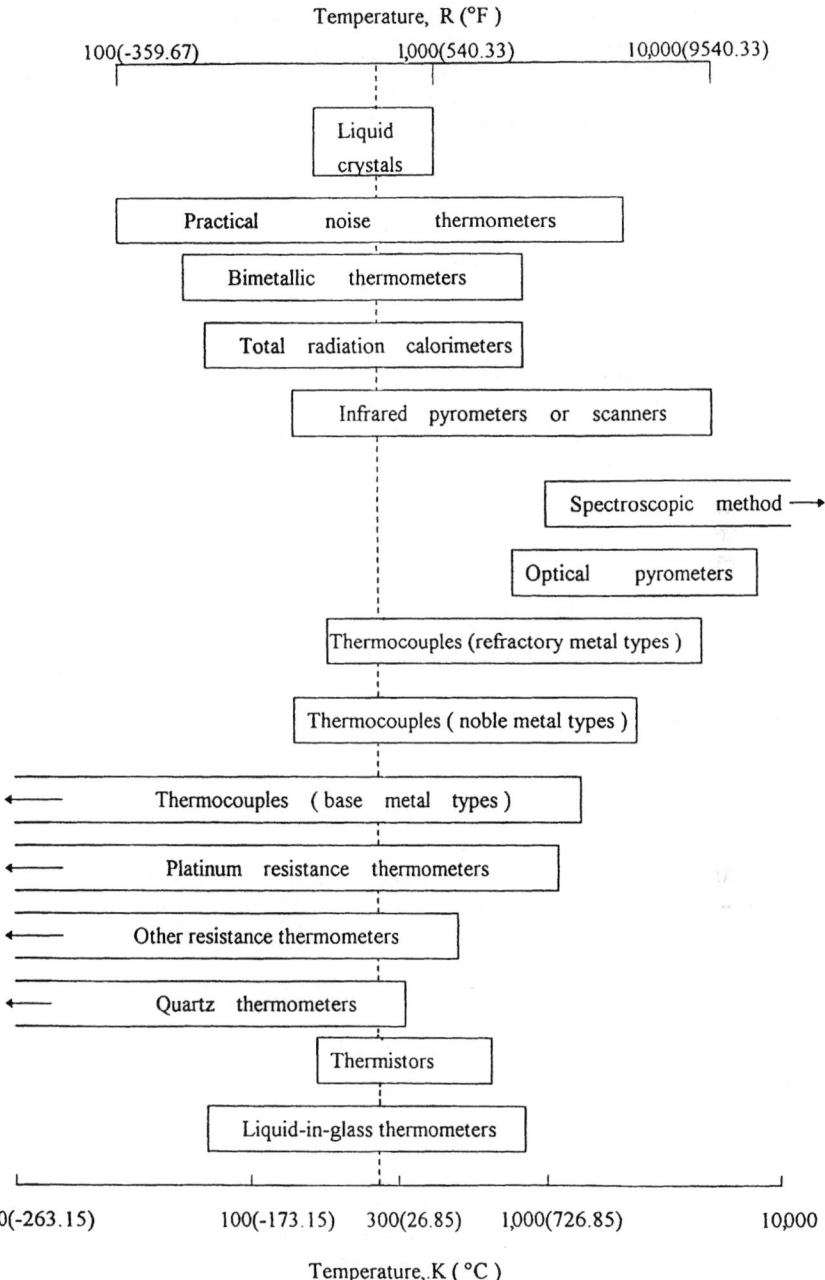

FIGURE 16.4 Temperature ranges for commonly used secondary thermometers.

standard thermometers in constant-temperature environments—a well-stirred constant-temperature liquid bath or special furnace. Comparisons are generally made at a modest number of temperatures with intermediate values described by a correlating equation for the specific thermometer being used. Easily established fixed points, such as the triple point of water or the water-ice equilibrium point (the latter is no longer a defining fixed point for ITS-90), are used for precision thermometers to see if there has been any drift in the correlation. Because of the fragility of SPRTs and other defining standard thermometers, thermometers for many applications are often calibrated against working standard thermometers that have been previously calibrated against a transfer standard thermometer (see definitions in Table 16.1) or a defining standard thermometer which is traceable to NIST. A working standard thermometer is usually of the same type as the thermometers to be calibrated to reduce the instrumentation requirements; a transfer standard thermometer serves as an intermediate standard to minimize the use and drift of the defining standard thermometer. All standard thermometers should be calibrated periodically to ensure their accuracy. In the United States, NIST provides a wide range of calibration services for standard thermometers [7].

In the following sections, some of the commonly used thermometers are reviewed. Also included are some statements about the accuracy (precision, stability, etc.) of individual sensing methods. The number quoted as the accuracy is the "ideal," which is the accuracy a thermometer is capable of producing, but is not the same as the accuracy of the actual measurement, which involves more than just the thermometer.

Liquid-in-Glass Thermometers

Definitions and Principles.   A typical liquid-in-glass thermometer consists of a liquid in a glass bulb attached to a glass stem, with the bulb and stem system sealed against the environment. The volume of the liquid in the bulb depends on the liquid temperature, and, with almost all real thermometers, both the volume of the liquid and the cavity itself increase with increasing temperature. The increase in the volume of the cavity is a secondary effect in the measurement. The volume of the liquid is indicated by the height of liquid in the stem. A scale is provided on the stem to facilitate the reading of the temperature of the bulb. The volume above the liquid into which it can expand is either evacuated or filled with a dry inert gas. An expansion chamber is provided at the top end of the stem to protect the thermometer in case of overheating. Liquid-in-glass thermometers for use at high temperatures have an auxiliary scale for calibration at the ice-point temperature and a contraction chamber to reduce the stem length. A schematic of a high-temperature thermometer is shown in Fig. 16.5.

Liquid-in-glass thermometers are almost exclusively used to determine the temperature of fluids that are relatively uniform—that is, they contain no large temperature gradients. Conduction along the glass stem can affect the temperature of the glass as well as that of the liquid. Therefore, thermometers are usually calibrated for a specified depth of immersion. They should then show the correct temperature when inserted to that level in the fluid whose temperature is to be measured. When a thermometer is used in situations where the immersion is other than that for which it was designed and calibrated, a stem correction [8] should be applied.

Three immersion types are available: (1) partial immersion—the bulb and a specified portion of the stem are exposed to the fluid whose temperature is being measured,

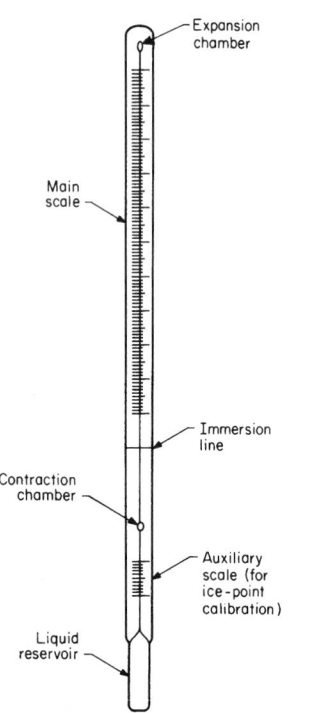

FIGURE   16.5   High-temperature   liquid-in-glass thermometer.

(2) total immersion—the bulb and all of the liquid column are in the fluid, and (3) complete immersion—the entire thermometer is immersed in the fluid whose temperature is being measured.

The most widely used liquid-in-glass thermometer is the mercury-in-glass thermometer, which can be used between −38.9 and about 300°C. The upper limit can be extended to about 540°C by filling the space above the mercury with nitrogen. The lower limit can be extended to −56°C by alloying the mercury with thallium.

For temperatures down to −200°C, organic liquids can be used as the working fluid. Thermometers filled with organic liquids, however, are not considered as reliable as mercury-filled thermometers.

Accuracy. Mercury-in-glass thermometers are relatively inexpensive and can be obtained in a wide variety of accuracy and temperature ranges. For example, between 0 and 100°C, thermometers with a 0.1°C graduation interval are readily available. Factors that affect the accuracy of the thermometer reading include changes in volume of the glass bulb under thermal stress, pressure effects, and response lag. With proper calibration by NIST [9, 10] or traceable to NIST, an accuracy of from 0.01 to 0.03°C can be achieved. Table 16.5 summarizes

TABLE 16.5 Expected Accuracies of Various Mercury-in-Glass Thermometers

Total-Immersion Thermometers (°C) [8]

Temperature range, °C	Graduation interval	Tolerance	Accuracy
Thermometer graduated under 150°C			
0 up to 150	1.0 or 0.5	0.5	0.1 to 0.2
0 up to 150	0.2	0.4	0.02 to 0.05
0 up to 100	0.1	0.3	0.01 to 0.03
Thermometers graduated under 300°C			
0 up to 100	1.0 or 0.5	0.5	0.1 to 0.2
Above 100 up to 300		1.0	0.2 to 0.3
0 up to 100	0.2	0.4	0.02 to 0.05
Above 100 up to 200		0.5	0.05 to 0.1
Thermometers graduated above 300°C			
0 up to 300	0.2	2.0	0.2 to 0.5
Above 300 up to 500		4.0	0.5 to 1.0
0 up to 300	1.0 or 0.5	2.0	0.1 to 0.5
Above 300 up to 500		4.0	0.2 to 0.5

Partial-Immersion Thermometers (°C) [8]

Temperature range, °C	Graduation interval*	Tolerance	Accuracy[†]
Thermometers not graduated above 150°C			
0 up to 100	1.0 or 0.5	1.0	0.1 to 0.3
0 up to 150	1.0 or 0.5	1.0	0.1 to 0.5
Thermometers not graduated above 300°C			
0 up to 100	1.0	1.0	0.1 to 0.3
Above 100 up to 300	1.0	1.5	0.5 to 1.0
Thermometers graduated above 300°C			
0 up to 300	2.0 or 1.0	2.5	0.5 to 1.0
Above 300 up to 500		5.0	1.0 to 2.0

  * Partial-immersion thermometers are sometimes graduated in smaller intervals than shown in these tables, but this in no way improves the performance of the thermometers, and the listed tolerances and accuracies still apply.

  [†] The accuracies shown are attainable only if emergent stem temperatures are closely known and accounted for.

the tolerances and the expected accuracy for various mercury-in-glass thermometers calibrated by NIST. Guidelines for acceptance for calibration by NIST can be found in Ref. 8. Total-immersion types are generally more accurate than partial-immersion types. For accurate measurement, thermometers should be checked periodically at a reference point (normally, the ice point) for changes in calibration. In general, stem correction is not significant for applications below 100°C. At higher temperatures, accuracy can only be assured with a proper stem correction [8].

Resistance Thermometers

Principles and Sensors.   The electrical resistance of a material is a function of temperature. This functional dependence is the basis for the operation of resistance thermometers. A resistance thermometer includes a properly mounted and protected resistor and a resistance-measuring instrument. The resistance values are then converted to temperature readings.

Resistive materials used in thermometry include platinum, copper, nickel, rhodium-iron, and certain semiconductors known as *thermistors*. Sensors made from platinum wires are called *platinum resistance thermometers* (PRTs) and, though expensive, are widely used. They have excellent stability and the potential for high-precision measurement. The temperature range of operation is from −260 to 1000°C. Other resistance thermometers are less expensive than PRTs and are useful in certain situations. Copper has a fairly linear resistance-temperature relationship, but its upper temperature limit is only about 150°C, and because of its low resistance, special measurements may be required. Nickel has an upper temperature limit of about 300°C, but it oxidizes easily at high temperature and is quite nonlinear. Rhodium-iron resistors are used in cryogenic temperature measurements below the range of platinum resistors [11]. Generally, these materials (except thermistors) have a positive temperature coefficient of resistance—the resistance increases with temperature.

Thermistors are usually made from ceramic metal oxide semiconductors, which have a large negative temperature coefficient of electrical resistance. *Thermistor* is a contraction of thermal-sensitive-resistor. The recommended temperature range of operation is from −55 to 300°C. The popularity of this device has grown rapidly in recent years. Special thermistors for cryogenic applications are also available [12].

Among the different resistance-type thermometers, we shall consider only PRTs and thermistors in greater detail.

PRTs.   PRTs are the most widely used temperature-measuring devices for high-precision applications. The precision of PRTs is due to: (1) the very high degree of purity of platinum that can be obtained for the manufacture of platinum wires, (2) the reproducibility of that purity from batch to batch, and (3) the inertness of pure platinum. These characteristics allow a high degree of stability and repeatability in PRTs.

As mentioned above, a special PRT that satisfies the specifications of ITS-90 is the defining standard thermometer (called the standard platinum resistance thermometer, SPRT) for the range of temperatures from 13.80 to 1234.93 K (961.78°C). According to ITS-90 [2], an acceptable SPRT must be made from sufficiently pure and strain-free platinum wire, and it must satisfy at least one of the following two relations:

$$\Pi(29.7646°C) > 1.11807 \qquad (16.6a)$$

$$\Pi(-38.8344°C) < 0.844235 \qquad (16.6b)$$

This implies a thermal coefficient of resistance greater than about 0.004/°C. An acceptable high-temperature SPRT must also satisfy the relation:

$$\Pi(961.78°C) > 4.2884 \qquad (16.6c)$$

The factor Π is defined as the ratio of the resistance $R(T_{90})$ at a temperature T_{90} to the resistance $R(273.16 \text{ K})$ at the triple point of water,

$$\Pi(T_{90}) = R(T_{90})/R(273.16 \text{ K}) \qquad (16.7)$$

The choice of an SPRT as the standard thermometer stems from the extremely high stability, repeatability, and accuracy that can be achieved by an SPRT in a strain-free, annealed state. The high cost of calibrating and maintaining an SPRT, the care needed in proper handling, the slow time-response, and the small change in resistance per degree are factors that make it a standard device rather than one to be used in most measurements. Figures 16.6 and 16.7 show the construction of typical designs for SPRTs and high-temperature SPRTs, respectively. Figure 16.8 shows the sensing element for a PRT standard with a more rugged design. However, its stability and precision are not as good as the units shown in Figs. 16.6 and 16.7; thus, it is only recommended as a transfer or working standard. Further information on SPRTs is available in the many papers of Ref. 13.

Besides SPRTs, PRTs are available in many different forms and sizes for a variety of industrial applications [14]. The basic design of an industrial PRT involves a length of platinum wire wound on an inert supporting material with proper insulation to prevent shorting. Figure 16.9 shows the construction of some industrial PRTs.

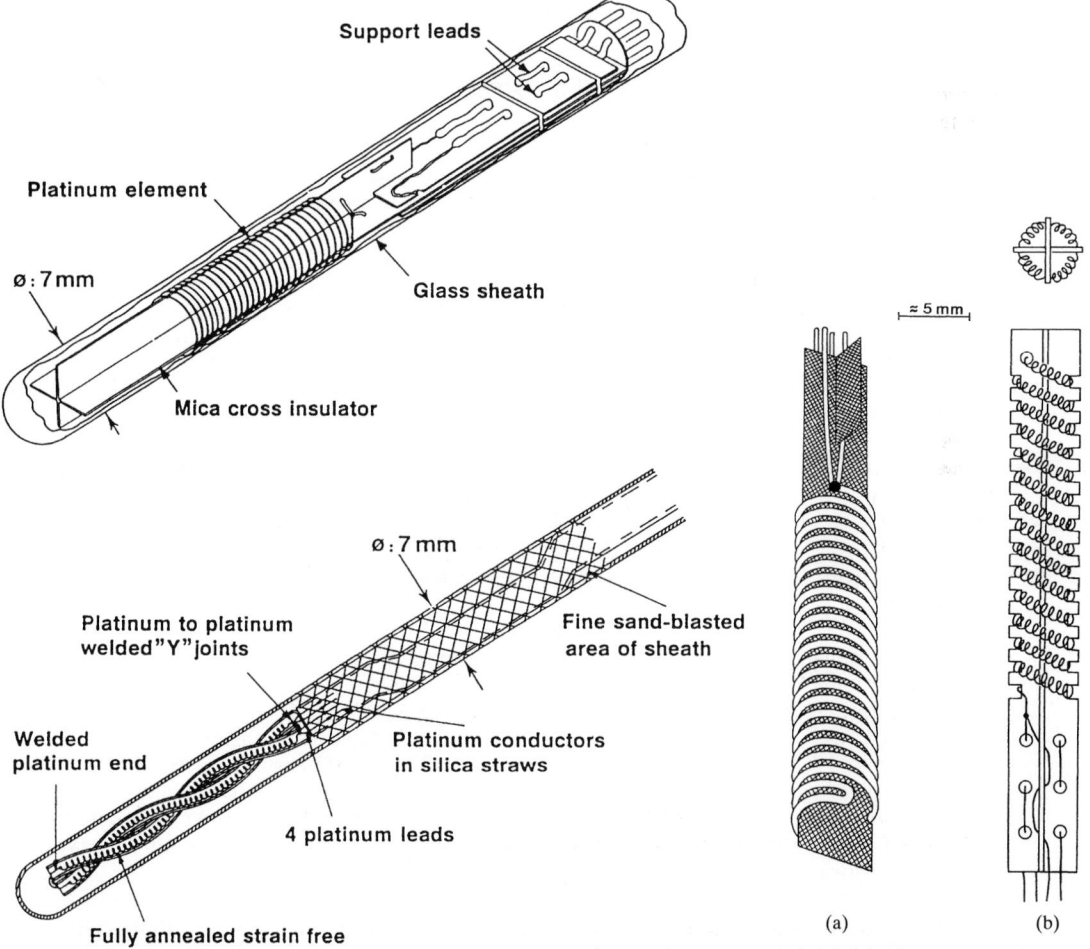

FIGURE 16.6   Typical designs of 25 Ω long-stem SPRTs [2].

FIGURE 16.7   Typical designs of high temperature SPRTs, (a) R_{tp} = 0.25 Ω, (b) R_{tp} = 2.5 Ω [2].

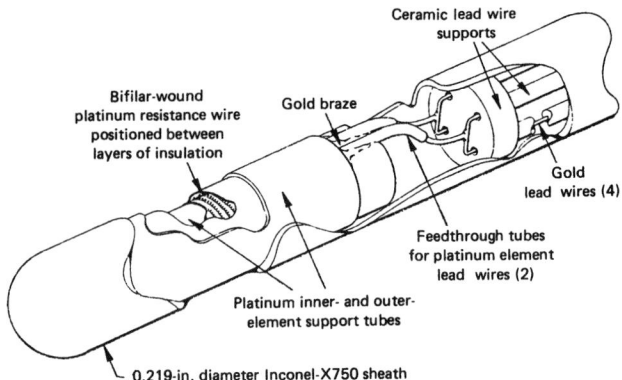

FIGURE 16.8 PRT design typical used as transfer or working standard thermometer. (*Courtesy of Rosemount, Inc.*)

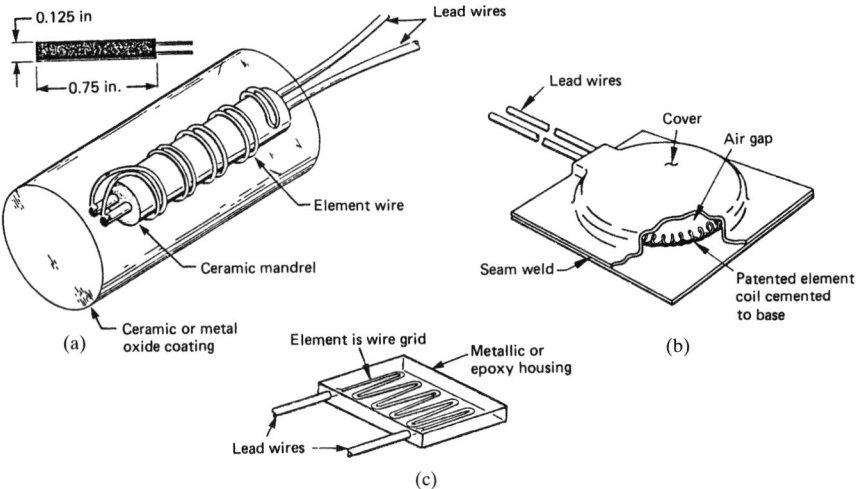

FIGURE 16.9 Industrial PRTs: (*a*) rod type, (*b*) disk type, (*c*) strain-gauge type. (*Courtesy of Rosemount, Inc.*)

The sensitivity of a PRT depends on its nominal resistance—the resistance measured at the ice point R_0. The sensitivity is measured by the change in resistance per degree change in temperature. A typical industrial PRT having an R_0 of 100 Ω has a sensitivity of about 0.4 Ω/°C; as comparison, SPRTs usually have an R_0 of 25 Ω with a sensitivity of only 0.1 Ω/°C. Industrial PRTs have a range of R_0 from 10 to 2000 Ω. For a custom-made sensor, an R_0 of 10,000 Ω or greater can be achieved. The choice of R_0 normally depends on the operating temperature of the sensor; a PRT with a low R_0 would be used in high-temperature applications to avoid shorting through the insulation, while a large R_0 is normally chosen for cryogenic applications to increase the sensitivity.

The platinum wire used in PRTs generally has a diameter between 7.6 and 76 μm. The wire resistance is about 4 Ω/cm for 17-μm wire. The choice of wire diameter would depend on the sensitivity and physical size of the sensor.

The resistance of a wire is a function of temperature, deformation, and impurities in the wire. The temperature effect is the one sought; the other two introduce uncertainties (errors).

The effect of deformation is minimized in an SPRT by a strain-free mounting (thus requiring care in handling to avoid straining the wire) and annealing (to remove the strains). PRTs with compact insulation and support have some inherent uncertainty in repeatability due to mechanical strain. Additional deformation can be caused by vibration and mechanical or thermal shock. These should be avoided in applications if the stability of the PRT is important.

PRTs are often used to measure temperatures of gases, liquids, and granular solids. They are usually limited to steady-state or relatively slow transient measurements (compared to thermistors or thermocouples) as a result of constraints imposed by the construction, mainly of physical size and mass. For the same reason, PRTs are not generally used to measure local values in systems with significant temperature gradients. Commercially available PRTs with a small measuring volume have typical sizes of 1.5 mm in diameter by 15.0 mm in length for a rod-type sensor and 0.76 mm in thickness with a square surface of 6.4-mm-long sides for a disc-type sensor.

The common types of insulation used in PRTs include mica, ceramic, glass, epoxy resin, and aluminum oxide (Al_2O_3). Mica and ceramic are used in SPRTs for both insulation and support. In other PRTs, ceramic glazes or glass or epoxy coatings as well as Al_2O_3 packing are frequently used with metal supports. In situations where better response time is required, an open-end-type construction is used with the platinum wire wound on top of an insulation coating and with no insulation on the outside other than a thin coating on the wire.

The two ends of the platinum wire in a PRT are connected to lead wires. Two-, three-, or even four-lead configurations are available. The choice would depend on the accuracy required, which is discussed in the subsection on resistance measurement, after the section on thermistors.

Thermistors. Thermistors [15–16] are made from semiconductors that have large temperature coefficients (change in resistance with temperature) as compared to metal-wire resistance sensors. Although positive and negative temperature coefficient thermistors are available, only the latter are normally used as temperature sensors. The most widely used thermistor sensors are made of metal oxides, including oxides of manganese, nickel, cobalt, copper, iron, and titanium. Their high resistance allows for remote sensing capability, as the lead resistance can be safely ignored.

Commercial thermistors are available in two major classes: (1) embedded-lead types, including bead (glass-coated or uncoated), glass-probe, and glass-rod thermistors; and (2) metal-contact types, including thermistor discs, chips, rods, washers, and so on. This division is based on the means of attaching the lead wires. Thermistors of the first type use direct sintering of metal oxide onto the lead wires. They exhibit high stability, are available in units of very small size and fast response, and can be used at temperatures up to 300°C. Those of the second type have the lead wires soldered on after the thermistor is fabricated. They are generally not usable at temperatures above 125°C and have slower response times than the first type. Table 16.6 gives a summary of the characteristics of thermistors [17].

Figure 16.10 shows the schematic of some thermistor sensors. The characteristic rating for a thermistor is given by its resistance at 25°C. Unlike PRTs, which have a rather limited range of resistance values, commercially available thermistor probes have a large range of characteristic resistance, from as low as 30 Ω to as high as 20 MΩ. The physical size of a thermistor could be as small as 0.075 mm in diameter on lead wires of 0.018 mm diameter for a bead thermistor. Metal-contact thermistors are generally bigger, starting from 0.25 mm in thickness and 1.0 mm in diameter for disc types and from 0.5 mm in diameter and 5 mm in length for rod-type thermistors.

Resistance Measurement. The common methods of resistance measurement in resistance thermometry are the bridge method and the potentiometric method. Basically, the bridge method uses the resistance sensor together with a variable resistor and two fixed resistors to form the four legs of a conventional Wheatstone bridge circuit. On the other hand, the potentiometric method, also called a half bridge, connects the resistance sensor in series with a known resistor.

TABLE 16.6 Characteristics of Thermistors [17]

Embedded-Lead Thermistors (Beads, Probes)	
Advantages	Disadvantages
1. High stability, reliability 2. Available in very small (0.1 mm) diameter 3. Fast response 4. Can be used at high temperatures (300°C)	1. Relatively high cost for close tolerances and interchangeability 2. Resistor network padding or use of matched pairs required 3. Comparatively low dissipation constants
Metal-Contact Thermistors (Disks, Wafers, Chips)	
Advantages	Disadvantages
1. Close tolerances and interchangeability at relatively low cost 2. Comparatively high dissipation constants 3. Reasonably fast response times for uncoated units—particularly for 1 mm × 1 mm × ½ mm thick units	1. Difficult to obtain reliable electrodes 2. Reliability and stability problems 3. Size limited to about 1 mm diameter (or 1 mm × 1 mm) uncoated; about 1.5 mm diameter coated 4. Cannot be used at high temperatures—requires cold sterilization 5. Comparatively long response times for coated units

The basic difference between the bridge method and the potentiometric method is that, in the bridge method the resistance is measured directly by comparing the sensor's resistance with that of a known resistor, while, in the potentiometric method, the measurement of the voltage drop is used to determine the sensor's resistance. The bridge method is inherently tedious and is not easily automated. In the potentiometric circuit, a known fixed resistor and constant voltage source are used. From a measurement of the voltage drop across the fixed resistor, the sensor's resistance can be determined. This system is simple and versatile and is widely used. Figure 16.11 shows a simple circuit used to determine resistance of a thermistor using a constant voltage source. Variations to this basic circuit can be found in Ref. 18.

In precision measurements, for SPRTs in particular, the simple potentiometric circuit shown in Fig. 16.11 is not sufficient. With this potentiometric method, a problem arises from the difficulty in maintaining a constant voltage source and measuring it to an acceptable accu-

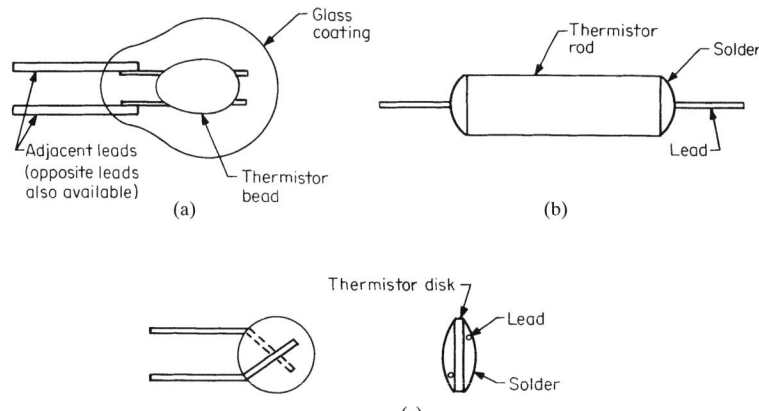

FIGURE 16.10 Thermistors: (*a*) glass-coated bead type, (*b*) rod type, (*c*) disk type.

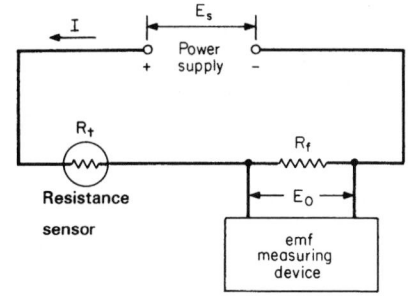

FIGURE 16.11 Resistance thermometer circuit with constant voltage source.

racy (and the inability to account for lead-wire resistance with a two-lead sensor). For high-precision measurements of a sensor with four lead wires, a Mueller bridge [19, 20] can be used. A schematic of the bridge is shown in Fig. 16.12. Its advantage over the conventional Wheatstone bridge is the provision for interchanging the leads (c and C with t and T) so that the average of the two readings is independent of the lead resistance. Referring to Fig. 16.12, at null condition, the equation of balance is

$$R_{D1} + R_C = R_X + R_T \qquad (16.8)$$

where R_C and R_T are the resistance of leads C and T, respectively, and R_{D1} is the value of R_D required for zero current in the galvanometer with these connections to the SPRT. After interchanging the leads c and C with t and T (reverse connection, see Fig. 16.12) and rebalancing (changing only R_D)

$$R_{D2} + R_T = R_X + R_C \qquad (16.9)$$

From Eqs. 16.8 and 16.9, the unknown resistance of the SPRT R_X can be found:

$$R_X = \frac{R_{D1} + R_{D2}}{2} \qquad (16.10)$$

To eliminate the error caused by spurious EMFs, the bridge current is reversed (not shown on Fig. 16.12)—that is, four bridge balances are made in the order $NRRN$ (where $N =$ normal and $R =$ reverse connections of the bridge) for each determination of R_X.

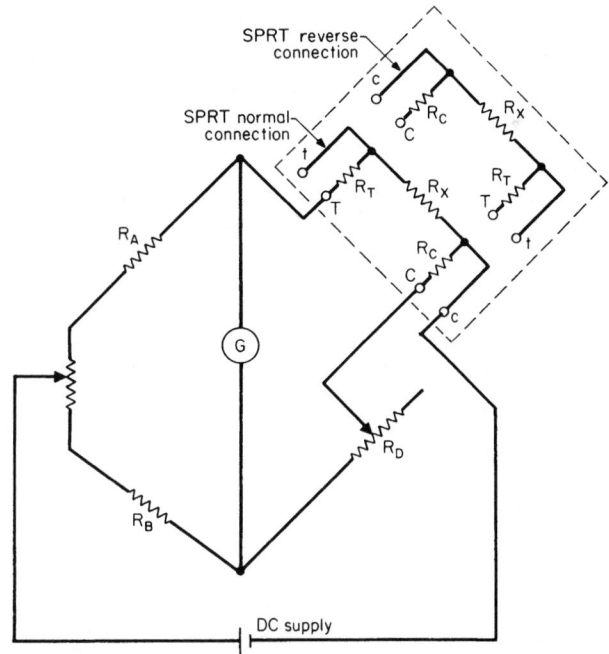

FIGURE 16.12 Mueller bridge circuit for precision resistance measurement.

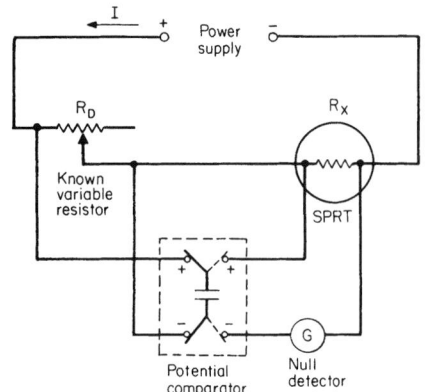

FIGURE 16.13 Potentiometric circuit for resistance measurement with comparator.

Conventional potentiometric methods for precision thermometry utilize a potentiometer [21] in measuring the resistance. The procedure is tedious, requiring four balancings for each reading. A modification using an isolating potential comparator requires only two balancings [21]. A schematic of a potentiometric circuit with an isolating potential comparator (for an SPRT or a thermistor standard, both with four lead wires) is shown in Fig. 16.13. A variable resistor (precalibrated so that its resistance is known "exactly") is used. The voltage drop across this resistor is compared to that across the SPRT by a potential comparator with provision for current reversal. The potential comparator basically consists of a high-quality capacitor that is successively connected to the leads of the SPRT and those of the known resistor by a high-speed double-pole double-throw switch (commonly called a *chopper*). If the resistance of the known resistor is not equal to the resistance of the SPRT, the capacitor will experience a process of successive charging and discharging, thus causing a current to flow through the galvanometer (or null detector). The resistance of the SPRT is given by the resistance of the known resistance under null condition (no current through the galvanometer). A provision for current reversal (not shown on Fig. 16.13) is used to eliminate the error caused by spurious EMFs.

For the methods described above, which use a DC power source, two to four balancings are required for each reading, and automation is difficult. The development of AC bridges eliminates the need of current reversal and permits automation in precision thermometry [22, 23].

PRTs can have either two, three, or four lead wires. Besides the Mueller bridge setup, two-wire and three-wire bridges are available [24]. A two-wire (Wheatstone) bridge is the least accurate, because the lead-wire resistances are not accounted for. A three-wire bridge (shown in Fig. 16.14) allows for the compensation of lead-wire resistances by adding the resistance of the third lead wire to the leg of the bridge, where the known variable resistor resides. Hence, if the resistances of the two lead wires (one connected to one leg and the second to the other leg) are the same, they will effectively cancel each other out. However, the resistances of the two lead wires would normally not be equal if for no other reason than the presence of the temperature gradient along the lead wires. With potentiometric circuits, only two- or four-lead-wire sensors are used. A two-lead-wire sensor is less accurate because of the unknown lead-wire resistances.

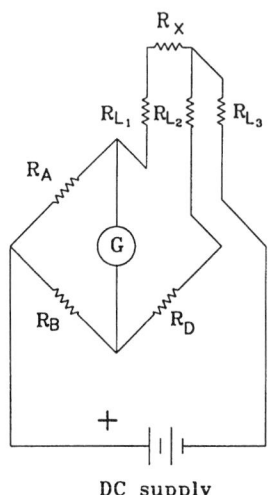

FIGURE 16.14 A circuit for resistance measurement of 3-wire PRTs.

Resistance-Temperature Conversion. A measuring circuit (bridge or potentiometer) can provide the resistance of the sensor in a resistance thermometer. The next step is to convert the resistance reading into a value of sensor temperature.

In the definition of ITS-90, interpolation formulas are provided for the calibration of SPRTs. These formulas are rather involved, including reference functions and deviation functions. For temperature above 0°C, the reference function is a 9th-order polynomial with fixed coefficients and the deviation function is a cubic polynomial with four constants, determined by calibration at the triple point of water (0.01°C) and the freezing points of tin (231.928°C), zinc (419.527°C), aluminum (660.323°C), and silver (961.78°C). These equations are complex and usually of interest only to specialists working at national laboratories. Hence, they will not be reproduced here; Ref. 3 treats this subject in detail.

For industrial PRTs, standard R_0 of 100 Ω is usually adopted, which enables the specification of a standard resistance-versus-temperature curve as well as interchangeability tolerance. The specification of the American Society for Testing and Materials (ASTM) [14] for values of R at different t are given in Table 16.7(a). Table 16.7(b) gives the expected maximum deviations from the standard R-versus-t relationship for two different classes of commercially available PRTs.

The resistance of a thermistor varies approximately exponentially with temperature. Sample resistance-temperature characteristics of thermistors are shown in Fig. 16.15. The relationship for an idealized thermistor can be written as

$$R = a \exp \frac{b}{t}$$

(16.11)

The existence of impurities causes a deviation from the simple exponential relation given in Eq. 16.11. Accurate representation is given by the addition of higher powers of $1/t$ to the exponential term in Eq. 16.11.

TABLE 16.7(a)   Resistance versus Temperature for Standard Industrial Platinum Resistance Thermometers [14]

Temperature, °C	Resistance, Ω	Temperature, °C	Resistance, Ω
−200	18.52	80	130.90
−180	27.10	100	138.51
−160	35.54	150	157.33
−140	43.88	200	175.86
−120	52.11	250	194.10
−100	60.26	300	212.05
−80	68.33	350	229.72
−60	76.33	400	247.09
−40	84.27	450	264.18
−20	92.16	500	280.98
0	100.00	550	297.49
20	107.79	600	313.71
40	115.54	650	329.64
60	123.24		

TABLE 16.7(b)   Classification Tolerances [14]

Temperature t, °C	Grade A		Grade B	
	°C	Ω	°C	Ω
−200	0.47	0.20	1.1	0.47
−100	0.30	0.12	0.67	0.27
0	0.13	0.05	0.25	0.10
100	0.30	0.11	0.67	0.25
200	0.47	0.17	1.1	0.40
300	0.64	0.23	1.5	0.53
400	0.81	0.28	1.9	0.66
500	0.98	0.33	2.4	0.78
600	1.15	0.37	2.8	0.88
650	1.24	0.40	3.0	0.94

The table represents values for 3-wire and 4-wire PRTs. Caution must be exercised with 2-wire PRTs because of possible errors caused by connecting wires. Tabulated values are based on elements of 100.0 Ω (nominal) at 0°C.

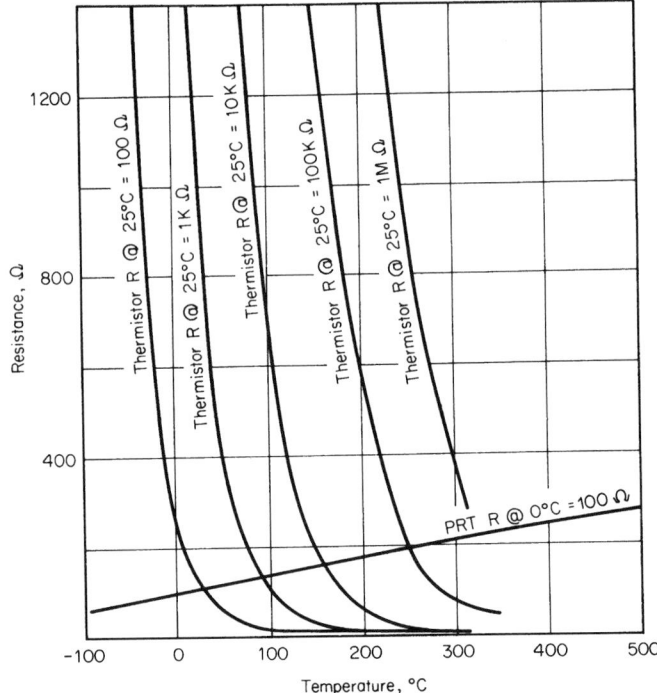

FIGURE 16.15 Resistance-temperature characteristics of thermistors [15].

For most applications, an alternative is employed. Recall that, in measuring the resistance of a thermistor, a fixed resistor is normally connected in series with the sensor. If a constant-voltage source (E_s) is used, the circuit current is inversely proportional to the total resistance. Then the relationship between the measured voltage drop across the fixed resistor and the thermistor temperature can be almost linear over a range of temperature. The linear part of this curve can be shifted along the temperature scale by changing the value of the fixed resistor.

$$E_0 = IR_f = \frac{E_s}{R_t + R_f} R_f \tag{16.12}$$

where R_t and R_f, as shown in Fig. 16.11, denote the resistances of a fixed resistor and of a thermistor, respectively.

The resistance of the thermistor is normally expressed in terms of its resistance at a standard reference temperature of 25°C, R_{25}, by introducing a ratio γ called the resistance-ratio characteristic:

$$\gamma = \frac{R_t}{R_{25}} \tag{16.13}$$

The dependence of γ on t would come from Eq. 16.11, or some variation of it, for the particular thermistor used. Since R_{25} is a known fixed value, the resistance ratio $s = R_{25}/R_f$ is a circuit constant. With the introduction of γ and s, Eq. 16.12 becomes

$$\frac{E_0}{E_s} = \frac{R_f}{R_t + R_f} = \frac{1}{1 + R_t/R_f} = \frac{1}{1 + s\gamma} = F(t) \tag{16.14}$$

The function $F(t)$ defines the curve mentioned above. By varying the parameter s, a family of such curves can be generated. They can be used to define the standard curves for different thermistor materials.

These curves can be almost linear over a range of 40 to 60°C at temperatures above 0°C and over a range of about 30°C for temperatures below 0°C. The choice of the value for the fixed resistor R_f is such that the operating temperature range of the thermistor would coincide with the linear portion of $F(t)$, thus simplifying the data reduction process. Actual $F(t)$ curves for different thermistor materials are given in Ref. 18.

Reliability.   A discussion of the reliability of a sensor requires the use of terms like precision, accuracy, and stability. These are popular but not always well defined. Here we shall give brief descriptions of these terms. A more detailed discussion on reliability and procedures will be covered in the section on calibration.

Within its specified temperature range of operation, an SPRT has the best precision (used here interchangeably with repeatability—the ability to reproduce the same reading for the same conditions) as a temperature sensor. Thus, it is used as the defining standard thermometer for ITS-90. The precision of the best SPRT around room temperature is of the order of 0.01 mK. The ability of an SPRT to realize ITS-90 temperature (its accuracy) at the calibration points (defining fixed points) can be better than 1 mK. At temperatures other than the calibration points, there is an additional error due to the interpolation process; this is also of the order of 1 mK under the best conditions.

When using an SPRT, high precision and accuracy are difficult to maintain. The delicate construction of an SPRT is vulnerable to mechanical strains. An audible tapping of the sheath of an SPRT could produce an increase as large as 100 $\mu\Omega$ in R_0 (corresponding to a 1-mK change in temperature if R_0 is 25 Ω). The cost and fragility of a calibrated SPRT plus the need to have high-precision auxiliary equipment for resistance measurement makes the SPRT a defining standard thermometer rather than a normal laboratory sensor. NIST provides a calibration service for SPRTs [25].

For industrial applications, PRTs are also called resistance temperature detectors (RTDs). For applications in the range from 0 to 300°C, these commercially available RTDs are capable of an accuracy of about 0.05°C, and they can usually maintain their accuracy to within 0.2°C for two or more years. If these RTDs are first annealed before calibration they are capable of maintaining an accuracy of 0.01°C [26].

Thermistors, on the other hand, are capable of a higher sensitivity or resolution (the ability to detect temperature change), due to their high resistance values and high temperature coefficients of resistance. Thermistors are now increasingly popular in the biomedical community; however, they were not popular in the early years because of limitations of stability (ability to maintain accuracy over a period of time) or drift (ability to resist change in reading over a period of time). Low-drift, fast-response thermistors were only possible with the development of bead-type (embedded-lead) thermistors [27]. Thermistor standards have been developed for use at temperatures between 0 and 100°C [28, 29] and 100 and 200°C [29]. These are bead-in-glass, probe-type thermistors assembled into thin-wall stainless steel housings with $R_{25} = 4000 \, \Omega$ for the former and $R_{125} = 24{,}000 \, \Omega$ for the latter. When properly calibrated, they are capable of temperature errors of less than 1 mK at 25°C and 5 mK at 125°C, respectively. The stability of the low-temperature thermistor standard is guaranteed to be within 5 mK/yr, and actual field experience shows most units drift less than 1 mK after several years of use. The combination of this stability with rugged design (good vibration and mechanical shock resistance) yields a good transfer standard thermometer near room temperatures.

Recent development of so-called super-stable disc thermistors have resolved the stability problem of early disc thermistors [30, 31]. Though still slightly less stable than bead-type thermistors, they are already capable of maintaining a stability of 2–3 mK/yr [31].

Thermoelectric Thermometers

Basic Principles.   Thermocouples are widely used for temperature measurement. Advantages of thermocouples include simplicity in construction (they are formed by only two wires

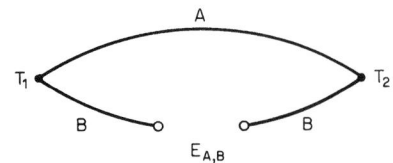

FIGURE 16.16   Simple thermoelectric circuit.

joined at their ends), ease of remote measurement, flexibility in construction (the size of the measuring junction can be made large or small according to the needs in life expectancy, drift, local measurement, and response time), simplicity in operation and signal processing (there is negligible self-heating, and the electrical output can be recorded directly), and availability of thermocouple wires at a nominal cost.

The operation principle of a thermocouple is described by the Seebeck effect: When two dissimilar materials are joined together at two junctions and these junctions are maintained at different temperatures, an electromotive force (EMF) exists across the two junctions.

An elementary thermocouple circuit is shown in Fig. 16.16. The EMF generated in this circuit is a function of the materials used and the temperatures of the junctions. It is useful to describe briefly the basic thermoelectric phenomena or effects that are related to the Seebeck effect and are present in thermocouple measurements. They include two well-known irreversible phenomena—Joule heating and thermal conduction—and two reversible phenomena—the Peltier effect and the Thompson effect.

Joule heating is the energy dissipation that occurs with an electric current flowing through a resistor and has magnitude I^2R. Thermal conduction is often quantitatively determined from Fourier's law: the heat conduction in a material is proportional to the temperature gradient.

The Peltier effect relates to reversible heat absorption or heat rejection at the junction between two dissimilar materials through which an electric current flows. The quantity of heat rejected or received is proportional to the electric current:

$$\dot{Q}_p = \pi_{AB}I \tag{16.15}$$

where π_{AB}, the Peltier coefficient for the junction, is a function of the temperature and the materials (A and B) used. Note that, if the current were reversed, the sign of the heat transfer would also change.

The Thompson effect refers to a heat addition or rejection per unit length of a conductor. The heat addition is proportional to the product of the electric current and the temperature gradient along the conductor. The rate of transfer of Thompson heat per unit length of a wire (conductor) is given by $\sigma_T dT/dx$, where σ_T is the Thompson coefficient. Integration over the entire length of a wire gives the Thompson heat:

$$\dot{Q}_T = \int_{\text{along wire}} \left(\sigma_T I \frac{dT}{dx} \right) dx = I \cdot \int_{\text{along wire}} \sigma_T \, dT \tag{16.16}$$

The direction or sign of this heat flow depends on the direction of the current relative to the temperature gradient. The magnitude of the Thompson coefficient is a function of the material (assumed homogeneous) in the conductor and the temperature level.

A simple and convenient means of determining the EMF generated in complex thermocouple circuit can be derived from the relations of irreversible thermodynamics. The result of this analysis [32] is that the zero-current EMF for a single homogeneous wire of length dx is

$$\left(\frac{dE}{dx} \right)_{I=0} = -S^* \frac{dT}{dx} \tag{16.17}$$

and, over a finite-length wire,

$$\Delta E = -\int_0^L S^* \frac{dT}{dx} \, dx = -\int_{\text{along wire length}} S^* \, dT \tag{16.18}$$

S^*, the entropy transfer parameter, is the ratio of the electric current-driven entropy transport to the electric current transport itself in the conductor. It should be noted that this ratio

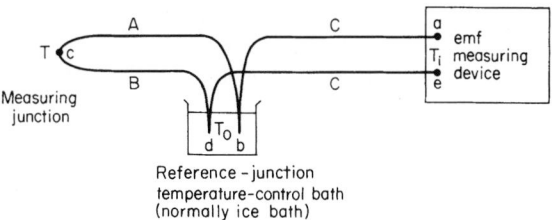

FIGURE 16.17   Basic circuit for a single thermocouple.

exists even when the current is zero. S^* is a function solely of the material composition and the temperature. Equation 16.17 can be integrated over any circuit to give the net EMF. In doing this, individual wires are generally assumed to be homogeneous.

Consider a basic thermocouple circuit with one measuring junction, as shown in Fig. 16.17. The two thermoelements A and B are joined at point c to form the measuring junction at temperature T. The thermoelements are connected to wires C at points b and d, both immersed in an ice bath (liquid water and ice in equilibrium) at T_0. The two wires C are connected to the input of an EMF measuring device. The input ports, a and e, are maintained at temperature T_i. Applying Eq. 16.18 over the various legs of the circuit gives

$$E_a - E_b = \int_{T_i}^{T_0} S_C^* \, dT \qquad E_c - E_d = \int_{T}^{T_0} S_B^* \, dT$$

$$E_b - E_c = \int_{T_0}^{T} S_A^* \, dT \qquad E_d - E_e = \int_{T_0}^{T_i} S_C^* \, dT \qquad (16.19)$$

$$E_a - E_e = \int_{T_0}^{T} (S_A^* - S_B^*) \, dT = E_{AB} \qquad (16.20)$$

The quantity $(S_A^* - S_B^*)$ is called the Seebeck coefficient α_{AB} and is a function of material A, material B, and temperature. Equation 16.20 is often presented in the form

$$E_{AB} = \int_{T_0}^{T} \alpha_{AB} \, dT \qquad (16.21)$$

Equation 16.18 is particularly useful in deriving the net EMF of a complex thermoelectric circuit. In many reference works, three laws—the law of homogeneous materials, the law of intermediate materials, and the law of intermediate temperature—are used to show how the EMF at a measuring device is affected by various lead wires from a thermocouple junction [33]. These laws can be derived from Eq. 16.17 but are generally more difficult to apply in a complex circuit.

Thermoelectric Circuits.   A typical circuit for a single thermocouple of materials A and B is shown in Fig. 16.17. The reference temperature (at which junctions b and d are maintained) is usually the ice point, 0°C. The connecting wires C are usually copper wires. Note that, according to Eq. 16.20, the connecting (copper) wires C should not affect the EMF E_{AB}, which, for given materials A and B, is just a function of the temperature T.

With more than one thermocouple, several different circuits with varying degrees of precision are available. The circuit given in Fig. 16.18(a) can be used for precision work—with each reference junction at the ice point. To avoid using multiple reference junctions, a circuit with only one reference junction and a uniform temperature zone box can be used. See Fig. 16.18(b). Alternatively, if better precision is required and the calibrations of the individual thermocouples are not identical, a variant of the apparatus in Fig. 16.18(b) can be used. With this, the junction in the ice bath is treated as a regular junction and in this way is used to determine the temperature of the selector that serves as the uniform temperature zone. Then the

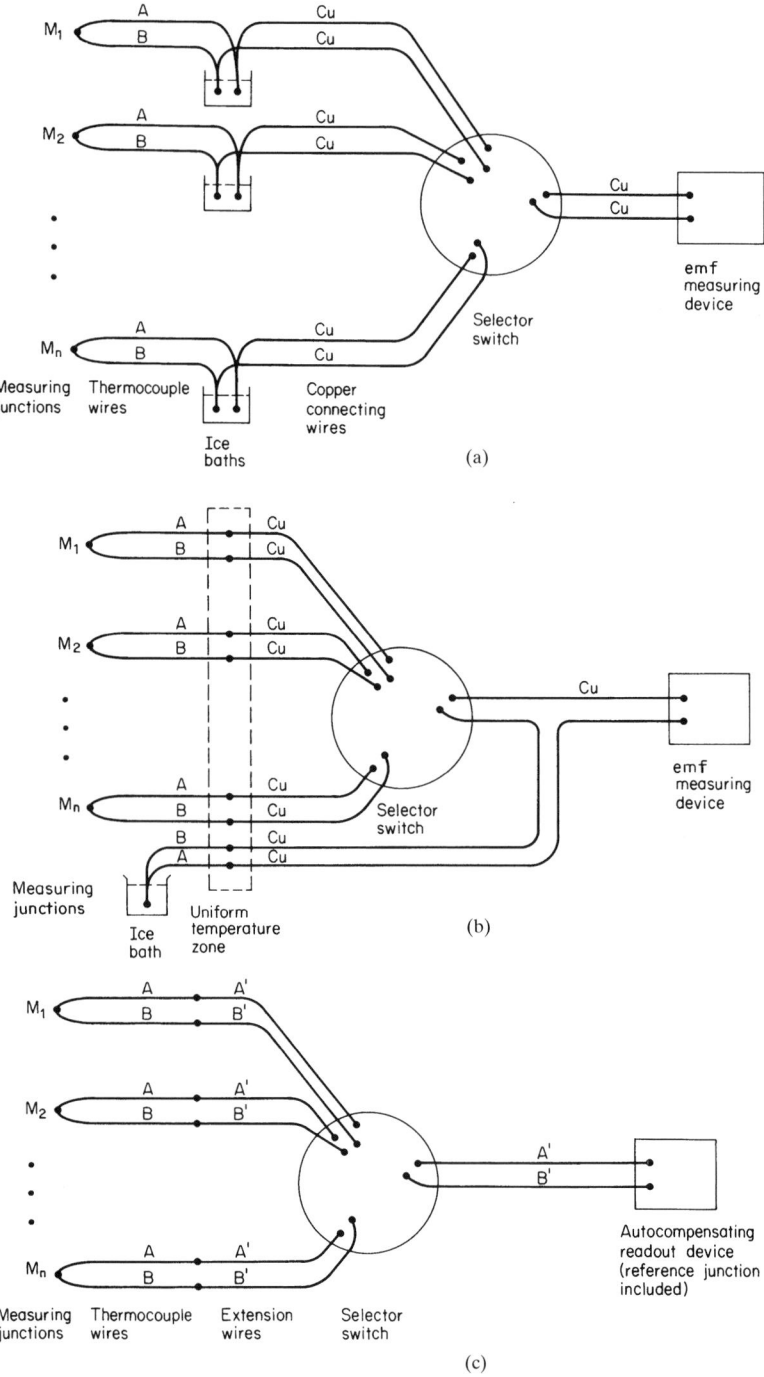

FIGURE 16.18   Measuring circuits for thermocouples: (*a*) ideal circuit for multiple thermocouples, (*b*) zone-box circuit with only one reference junction, (*c*) typical circuit for industrial use—no ice junction.

temperature (not the equivalent EMF) of the selector switch can be subtracted from (added to) the temperature readings of the individual junctions. Finally, a typical industrial thermocouple circuit using extension wires that approximate the composition of the thermocouple wires is shown in Fig. 16.18(*c*).

Some different thermocouple circuits are shown in Figs. 16.19(*a*) and (*b*). Figure 16.19(*a*) shows a thermopile, which has a number of thermocouples in series. This circuit amplifies the output by the number of thermocouples in series. The amplification permits detection of small temperature differences, and such a circuit is often used in heat flux gauges. Figure 16.19(*b*) shows a parallel arrangement for sensing average temperature level; one reading gives the arithmetic mean of the temperatures sensed by a number of individual junctions.

Thermocouple Materials.   Thermoelectric properties of different materials can be represented by their respective Seebeck coefficients. Usually this is done with reference to Eq. 16.20, assuming a standard reference material (material *B*) and a standard reference temperature T_0 (usually 0°C; see Fig. 16.17).

In principle, any two different materials can be used in a thermocouple combination. Of the enormous number of possible combinations, only a few are commonly used. Based on general usage and recognition by the American National Standard Institute (ANSI), a standard was established [34] that designates eight thermocouples as standard types. Table 16.8 gives the designations, popular names, materials (with color codes), temperature ranges, and Seebeck coefficients of these eight standardized thermocouples. Three of these (types B, R, and S) are referred to as noble metal thermocouples, since they contain platinum and platinum-rhodium combinations; the other five (types E, J, K, N, and T) are called base metal thermocouples.

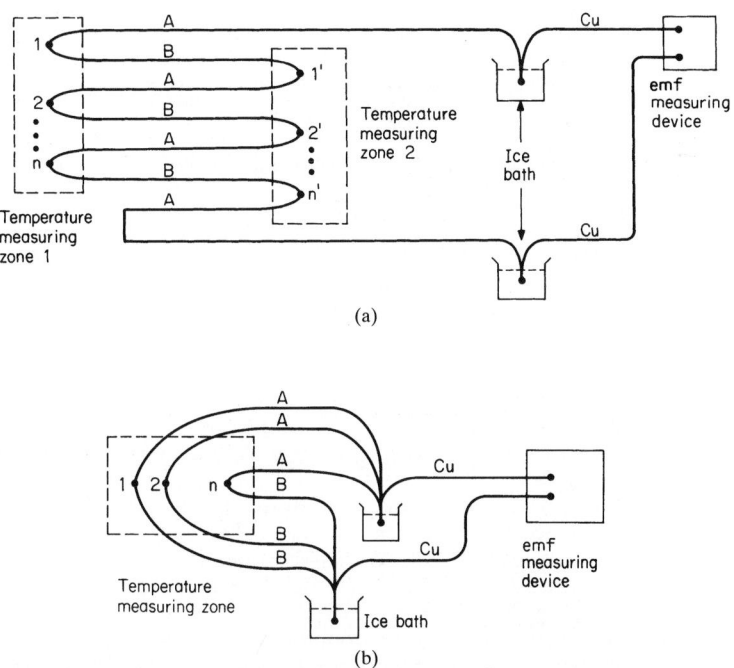

FIGURE 16.19   Thermocouples in series and parallel circuits: (*a*) thermopile for magnifying EMF when measuring small temperature differences, and (*b*) parallel arrangement for sensing average temperature level.

TABLE 16.8 Standard Thermocouple Types

SLD[1]	Popular name	Materials (color code)[2] (positive material appears first)	Typical temperature range[3]	Seebeck coefficient at 100°C (212°F)[4], μV/°C
S	—	Platinum-10% rhodium vs. platinum	−50 to 1767°C	7.3
R	—	Platinum-13% rhodium vs. platinum	−50 to 1767°C	7.5
B	—	Platinum-30% rhodium vs. platinum-6% rhodium	0 to 1820°C	0.9
T	Copper-constantan	Copper (blue) vs. a copper-nickel alloy[5] (red)	−270 to 400°C	46.8
J	Iron-constantan	Iron (white) vs. a slightly different copper-nickel alloy[6] (red)	−210 to 760°C[9]	54.4
E	Chromel-constantan	Nickel-chromium alloy[7] (purple) vs. a copper-nickel alloy[5] (red)	−270 to 1000°C	67.5
K	Chromel-Alumel	Nickel-chromium alloy[7] (yellow) vs. nickel-aluminum alloy[8] (red)	−270 to 1372°C	41.4
N	Nicrosil-Nisil	Nickel-chromium-silicon alloy[10] (orange) vs. Nickel-chromium-magnesium alloy[10] (red)	−270 to 1300°C	29.6

[1] SLD stands for Standardized Letter Designation. The letter designation is for the combined thermocouple with each individual thermoelement designated by P or N for positive or negative legs, respectively—for example, SN stands for platinum, TP stands for copper, and so on.

[2] The color codes given in parentheses are the color of the duplex-insulated wires [34]. Color codes are not available for the noble metal types (S, R, and B). For the base metal types (T, J, E, and K), the overall insulation color is brown.

[3] These temperature ranges are taken from Ref. 42.

[4] Conversion to μV/°F is by dividing the figures by 1.8.

[5] This copper-nickel alloy is the same for both EN and TN, often referred to as Adams constantan or, sometimes, constantan.

[6] This copper-nickel alloy is used in JN. It is similar to, but not always interchangeable with, EN and N. By SAMA specifications, this substance is often referred to as *SAMA constantan,* but it is also loosely called *constantan.*

[7] EP and KP is a nickel-chromium alloy that is usually referred to by its trade name, Chromel (Hoskins Manufacturing Co.).

[8] KN is a nickel-aluminum alloy usually referred to by its trade name, Alumel (Hoskins Manufacturing Co.).

[9] Even though EMF-temperature values are available up to 1200°C [42], thermophysical properties of type J thermocouples are not stable above 760°C.

[10] See Ref. 37 for details.

The letter designations for thermocouples were originally assigned by the Instrument Society of America (ISA) and accepted as an American Standard in ANSI-C96.1-1964. They are not used to specify the precise chemical composition of the materials but only their thermoelectric properties. Hence, the composition of a given type of thermocouple may be different and still be acceptable by the standard as long as the combined thermoelectric properties remain within specified tolerances—see Table 16.9 [35].

Some wire sizes and their corresponding upper temperature limits for each type of standard thermocouple as recommended by the ASTM [36] are summarized in Table 16.10. These limits are for protected thermocouples in conventional closed-end protecting tubes. When a standard type of thermocouple is used beyond the recommended temperature ranges given in Table 16.10, accuracy and reliability may be compromised. Commercially available thermocouple wires are available in many more gauge sizes than those indicated in Table 16.10.

TABLE 16.9   Tolerances on Initial Values of EMF versus Temperature [35]

Thermocouple type	Temperature range		Tolerances-reference junction 0°C (32°F)			
			Standard tolerances		Special tolerances	
	°C	°F	°C (whichever is greater)	°F	°C (whichever is greater)	°F
T	0 to 370	32 to 700	±1 or ±0.75%	Note 2	±0.5 or 0.4%	Note 2
J	0 to 760	32 to 1400	±2.2 or ±0.75%		±1.1 or 0.4%	
E	0 to 870	32 to 1600	±1.7 or ±0.5%		±1 or ±0.4%	
K or N	0 to 1260	32 to 2300	±2.2 or ±0.75%		±1.1 or ±0.4%	
R or S	0 to 1480	32 to 2700	±1.5 or ±0.25%		±0.6 or ±0.1%	
B	870 to 1700	1600 to 3100	±0.5%		±0.25%	
T*	−200 to 0	−328 to 32	±1 or ±1.5%		†	
E*	−200 to 0	−328 to 32	±1.7 or ±1%		†	
K*	−200 to 0	−328 to 32	±2.2 or ±2%		†	

* Thermocouples and thermocouple materials are normally supplied to meet the tolerances specified in the table for temperatures above 0°C. The same materials, however, may not fall within the tolerances given for temperatures below 0°C in the second section of the table. If materials are required to meet the tolerances stated for temperatures below 0°C the purchase order must so state. Selection of materials usually will be required.

† Special tolerances for temperatures below 0°C are difficult to justify due to limited available information. However, the following values for Types E and T thermocouples are suggested as a guide for discussion between purchaser and supplier:

Type E −200 to 0°C ±1°C or ±0.5% (whichever is greater)
Type T −200 to 0°C ±0.5°C or ±0.8% (whichever is greater)

Initial values of tolerance for Type J thermocouples at temperatures below 0°C and special tolerances for Type K thermocouples below 0°C are not given due to the characteristics of the materials.

Note: 1—Tolerances in this table apply to new essentially homogeneous thermocouple wire, normally in the size range 0.25 to 3 mm in diameter (No. 30 to No. 8 Awg) and used at temperatures not exceeding the recommended limits of Table 16.10. If used at higher temperatures these tolerances may not apply.

Note: 2—The Fahrenheit tolerance is 1.8 times larger than the °C tolerance at the equivalent °C temperature. Note particularly that percentage tolerances apply only to temperatures that are expressed in °C.

Note: 3—*Caution:* Users should be aware that certain characteristics of thermocouple materials, including the emf versus temperature relationship may change with time in use; consequently, test results and performance obtained at time of manufacture may not necessarily apply throughout an extended period of use. Tolerances given in this table apply only to new wire as delivered to the user and *do not allow for changes in characteristics with use.* The magnitude of such changes will depend on such factors as wire size, temperature, time of exposure, and environment. It should be further noted that due to possible changes in homogeneity, attempting to recalibrate *used* thermocouples is likely to yield irrelevant results, and is not recommended. However, it may be appropriate to compare used thermocouples in-situ with new or known good ones to ascertain their suitability for further service under the conditions of the comparison.

TABLE 16.10   Recommended Upper Temperature Limits for Protected Thermocouples [36]

Thermocouple type	Upper temperature limit for various wire sizes (awg), °C (°F)					
	No. 8 gage, 3.25 mm (0.128 in.)	No. 14 gage, 1.63 mm (0.064 in.)	No. 20 gage, 0.81 mm (0.032 in.)	No. 24 gage, 0.51 mm (0.020 in.)	No. 28 gage, 0.033 mm (0.013 in.)	No. 30 gage, 0.25 mm (0.010 in.)
T	—	370 (700)	260 (500)	200 (400)	200 (400)	150 (300)
J	760 (1400)	590 (1100)	480 (900)	370 (700)	370 (700)	320 (600)
E	870 (1600)	650 (1200)	540 (1000)	430 (800)	430 (800)	370 (700)
K and N	1260 (2300)	1090 (2000)	980 (1800)	870 (1600)	870 (1600)	760 (1400)
R and S	—	—	—	1480 (2700)	—	—
B	—	—	—	1700 (3100)	—	—

Note: This table gives the recommended upper temperature limits for the various thermocouples and wire sizes. These limits apply to protected thermocouples; that is, thermocouples in conventional closed-end protecting tubes. They do not apply to sheathed thermocouples having compacted mineral oxide insulation. Properly designed and applied sheathed thermocouples may be used at temperatures above those shown in the tables. Other literature sources should be consulted.

Some environmental limitations of the standard thermocouple materials compiled by ASTM [36] are reproduced in Table 16.11. The thermal EMF of standard thermoelements relative to platinum is shown in Fig. 16.20 [36]. Seebeck coefficients (first derivative of thermal EMF with respect to temperature) for each of the standard thermocouples as a function of temperature are tabulated in Table 16.12.

Type-S thermocouples (platinum-10 percent rhodium versus platinum) were introduced in 1886 by Le Chatelier. Because of their stability and reproducibility, they became an early standard and were used as the defining instrument for IPTS-68 between 630.74 and

TABLE 16.11   Environmental Limitations of Thermoelements [36]

Thermoelement	Environmental recommendations and limitations (See Notes)
JP	For use in oxidizing, reducing, or inert atmospheres or in vacuum. Oxidizes rapidly above 540°C (1000°F). Will rust in moist atmospheres as in subzero applications.
	Stable to neutron radiation transmutation. Change in composition is only 0.5% (increase in manganese) in 20-year period.
JN, TN, EN	Suitable for use in oxidizing, reducing, and inert atmospheres or in vacuum. Should not be used unprotected in sulfurous atmospheres above 540°C (1000°F).
	Composition changes under neutron radiation since copper content is converted to nickel and zinc. Nickel content increases 5% in 20-year period.
TP	Can be used in vacuum or in oxidizing, reducing, or inert atmospheres. Oxidizes rapidly above 370°C (700°F). Preferred to Type JP element for subzero use because of its superior corrosion resistance in moist atmospheres.
	Radiation transmutation causes significant changes in composition.
	Nickel and zinc grow into the material in amounts of 10% each in a 20-year period.
KP, EP	For use in oxidizing or inert atmospheres. Can be used in hydrogen or cracked ammonia atmospheres if dew point is below −40°C (−40°F). Do not use unprotected in sulfurous atmospheres above 540°C (1000°F).
	Not recommended for service in vacuum at high temperatures except for short time periods because preferential vaporization of chromium will alter calibration. Large negative calibration shifts will occur if exposed to marginally oxidizing atmospheres in temperature range 815 to 1040°C (1500 to 1900°F).
	Quite stable to radiation transmutation. Composition change is less than 1% in 20-year period.
NP	Same general use as type KP, except less affected by sulfurous atmospheres because of the silicon addition. Best used in oxidizing or neutral atmospheres.
KN	Can be used in oxidizing or inert atmospheres. Do not use unprotected in sulfurous atmospheres as intergranular corrosion will cause severe embrittlement.
	Relatively stable to radiation transmutation. In 20-year period, iron content will increase approximately 2%. The manganese and cobalt contents will decrease slightly.
NN	Can be used in oxidizing or inert atmospheres. Do not use unprotected in sulfurous atmospheres as intergranular corrosion will cause severe embrittlement.
	Relatively stable to radiation transmutation. In 20-year period, iron content will increase approximately 2%. The manganese and cobalt contents will decrease slightly.
RP, SP, SN, RN, BP, BN	For use in oxidizing or inert atmospheres. Do not use unprotected in reducing atmospheres in the presence of easily reduced oxides, atmospheres containing metallic vapors such as lead or zinc, or those containing nonmetallic vapors such as arsenic, phosphorus, or sulfur. Do not insert directly into metallic protecting tubes. Not recommended for service in vacuum at high temperatures except for short time periods.
	Types RN and SN elements are relatively stable to radiation transmutation. Types BP, BN, RP, and SP elements are unstable because of the rapid depletion of rhodium. Essentially, all the rhodium will be converted to palladium in a 10-year period.

Note: 1—Refer to Table 16.10 for recommended upper temperature limits.
Note: 2—Stability under neutron radiation refers to chemical composition of thermoelement, not to stability of thermal EMF.
Note: 3—Radiation transmutation rates are based on exposure to a thermal neutron flux of 1×10^{14} neutrons/cm²s. See W. E. Browning, Jr. and C. E. Miller, Jr., "Calculated Radiation Induced Changes in Thermocouple Composition," *Temperature: Its Measurement and Control in Science and Industry,* vol. C, pt. 2, Rheinhold, New York, 1962, p. 271.

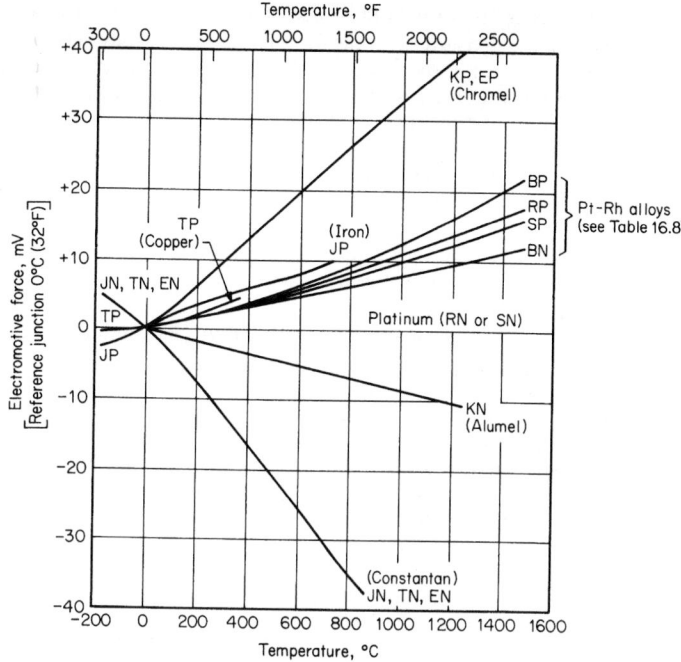

FIGURE 16.20 Electromotive force of thermoelements relative to platinum [36].

1064.43°C. However, after the availability of high-temperature SPRTs, type-S thermocouples are no longer used as a defining standard in the latest temperature scale, ITS-90.

Some early SP (the second capital letter designates the polarity of the thermoelement: P = positive, N = negative—see Table 16.8) thermoelements were found to contain iron impurities from the use of impure rhodium. Later, when purer rhodium became available, 13 percent

TABLE 16.12 Nominal Seebeck Coefficients (Thermoelectric Power), α_{AB} ($\mu V/°C$) [36]

Temperature, °C	Thermocouple						
	E Chromel- constantan	J Iron- constantan	K Chromel- Alumel	R Pt-13% Rh vs. Pt	S Pt-10% Rh vs. Pt	T Copper- constantan	B Pt-30% Rh vs. Pt-6% Rh
−190	27.3	24.2	17.1			17.1	
−100	44.8	41.4	30.6			28.4	
0	58.5	50.2	39.4			38.0	
200	74.5	55.8	40.0	8.8	8.5	53.0	2.0
400	80.0	55.3	42.3	10.5	9.5		4.0
600	81.0	58.5	42.6	11.5	10.3		6.0
800	78.5	64.3	41.0	12.3	11.0		7.7
1000			39.0	13.0	11.5		9.2
1200			36.5	13.8	12.0		10.3
1400				13.8	12.0		11.3
1600					11.8		11.6

rhodium instead of the usual 10 percent had to be used in the platinum-rhodium alloy to match the previous standard thermal EMF—hence the development of the type-R thermocouple.

Type-B thermocouples have better stability and greater mechanical strength and can be used at higher temperatures than types R and S. The small Seebeck coefficients of these thermocouples make them unattractive for low-temperature measurements. The particularly small Seebeck coefficients of type-B thermocouples at low temperatures mean the actual temperature of the reference junction does not greatly affect the EMF as long as it is between 0 and 50°C.

In order to reduce cost and obtain higher EMFs, base metal thermocouples were introduced. Iron and nickel were good candidates. However, pure nickel was found to be very brittle after oxidation; to avoid this, a copper-nickel alloy commonly referred to as *constantan* was introduced. The iron-constantan combination is designated type J. Later, type K was developed for use at higher temperatures than type J, and type T was introduced particularly for measurements below 0°C. Unlike the other three, the type-E thermocouple does not have a well-documented history. It was formed by a combination of KP and TN. Type-E thermocouples have the largest Seebeck coefficient among the letter-designated thermocouples, which can be important especially in differential temperature measurements. Type-N thermocouple is the newest member of the standardized thermocouples. Its letter designation is only available to the general public since 1993 [36], although its thermoelectric properties have been well documented since the late 1970s [37]. Type-N thermocouples have excellent resistance to preferential oxidation in oxidizing and reducing atmosphere at temperatures above 1000°C and are preferable over type K at high temperature. Their respective environmental limitations are covered in detail in Table 16.11.

In situations when mechanical strength between measuring junction and reference junction is required (e.g., when very fine wires are used as the measuring junction for fast response time) or when cost reduction is a significant factor (e.g., when using noble metal thermocouples), extension wires can be used between the measuring and reference junctions. Extension wires are wires having similar thermoelectric properties to those of the thermocouple wires to which they are attached. However, they can increase the deviation of the thermocouple from the standard tables due to improper combinations of thermoelements and extension wires. A deviation of as much as 5°C—more than twice the commercial tolerance limits shown in Table 16.9—can be introduced. Additional deviations can also arise if the temperatures of the two thermoelement/extension-wire junctions are different. One should avoid extension wires unless absolutely necessary; if used, care should be exercised in the selection of extension wires to ensure proper matching and the equalization of thermoelement/extension-wire junction temperatures. Detailed information on extension wires can be found in Ref. 36. Once the wires have been brought out to the reference junction, copper wires can be used for connection to the EMF-measuring device.

Thermocouple materials can be divided into four categories: (1) noble metals, (2) base metals, (3) refractory metals, and (4) nonmetal types. Of the eight letter-designated thermocouples, three are from the first category, and five are from the second. In addition to the eight designated thermocouple combinations, other nonstandard combinations are available commercially. Over the years, hundreds of combinations have been investigated for various applications, many of which never went beyond the research stage. Refractory metal thermocouples (e.g., tungsten-rhenium alloys) are normally used for high-temperature applications. Carbides, graphites, and ceramics are the main nonmetal thermoelements; they are not very popular, because they are bulky and have poor reproducibility. Nonstandard thermocouples are described in Refs. 38–41.

Thermocouple Components and Fabrication. A thermocouple measurement assembly includes a sensing element assembly, extension wires (when used), reference junction, connecting wires, an EMF-measuring device (possibly with signal-processing equipment), and other hardware needed for applications in adverse environments such as protection tubes, connectors, adapters, and so on. Each of the above components will be discussed in the following paragraphs.

The sensing element assembly includes two thermoelements properly joined together to form the measuring junction with appropriate electrical insulation to avoid shorting the two thermoelements except at the measuring junction.

Commercial off-the-shelf thermocouple wires are available in different forms: bare wires, insulated wires, and sheathed wires. Bare wires come on spools in a wide range of gauge sizes. They have to be individually matched, fabricated, and insulated before use.

Insulated wires come as single-insulated thermoelements or double-insulated duplex wires. Duplex wires can be obtained with a stainless steel overbraid for wear and abrasion protection. Table 16.13 lists characteristics of insulations used with thermocouple wires.

Sheathed wires have two bare wires embedded within a sheath packed with ceramic insulation. The crushed ceramic insulation rapidly absorbs moisture. To avoid deterioration due to moisture and contamination, the wires should be properly sealed at both ends at all times, and clean tools should be used during fabrication. If the wires have to be left unsealed temporarily, they should be heated in an oven at about 100°C or higher to remove moisture.

Fabrication of a thermocouple [34] requires some skill and familiarity, especially when using small-diameter wires. The measuring junction should be a joint of good thermal and electrical contacts produced without destroying the thermoelectric properties of the wires at the junction. For applications below 500°C, silver solder with borax flux is sufficient for most base metal types, whereas junctions formed by welding are recommended for use above

TABLE 16.13 Insulation Characteristics [36]

Insulation	Lower temperature limit, °C (°F)	Continuous use temperature limit, °C (°F)	Single exposure temperature limit, °C (°F)	Moisture resistance	Abrasion resistance
Cotton	—	95 (200)	95 (200)	Poor	Fair
Enamel and cotton	—	95 (200)	95 (200)	Fair	Fair
Polyvinyl chloride	−40 (−40)	105 (220)	105 (220)	Excellent	Good
Type R[a]	−55 (−67)	125 (257)	125 (257)	Excellent	Good
Nylon[b]	−55 (−67)	125 (260)	125 (260)	Poor	Good
Teflon[b]	−55 (−67)	205 (400)	315 (600)	Excellent	Good
Kapton[b]	−260 (−436)	260 (500)	260 (500)	Excellent	Excellent
B fiber[a]	—	260 (500)	260 (500)	Fair	Excellent
Teflon and fiberglass[c]	—	315 (600)	370 to 540 (700 to 1000)	Excellent to 600°F	Good
Fiberglass-varnish or silicone impregnation E[d]	—	480 (900)	540 (1000)	Fair to 400°F, poor above 400°F	Fair to 400°F, poor above 400°F
Fiberglass nonimpregnated S[e]	—	540 (1000)	650 (1200)	Poor	Fair
Asbestos[f] and fiberglass with silicone[g]	—	480 (900)	650 (1200)	Fair to 400°F	Fair to 400°F, poor above 400°F
Felted asbestos	—	540 (1000)	650 (1200)	Poor	Poor
Asbestos over asbestos	—	540 (1000)	650 (1200)	Poor	Poor
Refrasil[h]	—	870 (1600)	1100 (2000)	Very poor	Very poor
Ceramic fibers (for example, Nextel[i] and Cefir[j])	—	1000 (1830)	1370 (2500)	Very poor	Poor

[a] Type R and B fiber trademark of Thermo Electric's Thermoplastic Elastomer and Polyimide fiber, respectively.
[b] Nylon, Teflon, and Kapton are trademarks of the E. I. duPont Co.
[c] The Teflon vaporizes at 315°C (600°F) with possible toxic effects.
[d] E = Electrical grade fiberglass.
[e] S = Structural grade fiberglass.
[f] Asbestos is hazardous to our health and environment. The users are encouraged to use alternate material.
[g] Individual wires are asbestos and overbraid is fiberglass.
[h] Trademark of the H. I. Thompson Company.
[i] Trademark of the 3M Corporation. Nextel 312 and Nextel 440 can be used up to 1200°C (2200°F) and 1370°C (2500°F), respectively.
[j] Trademark of Thermo Electric Company.

TABLE 16.14 Summary of Methods for Joining of Bare-Wire Thermocouples [34]

Type	Materials	Lower melting	Silver brazing*	Welding Flux†	Welding Gas	Welding Arc	Resistance welding	Plasma arc or tungsten inert gas	Butt welding‡
T	Copper-constantan	TP	✓	Borax	✓	✓	N.R.	✓	N.R.
J	Iron-constantan	JN	✓	Borax	✓	✓	✓	✓	✓
E	Chromel-constantan	EN	✓	Fluorspar	✓	✓	N.R.	✓	✓
K	Chromel-Alumel	KN	✓	Fluorspar	✓	✓	✓	✓	✓
R	Pt-13% Rh vs. Pt	RN	N.R.	None	✓	✓	✓	✓	—
S	Pt-10% Rh vs. Pt	SN	N.R.	None	✓	✓	✓	✓	—
B	Pt-30% Rh vs. Pt-6% Rh	BN	N.R.	None	✓	✓	✓	✓	—

* Only recommended for use below 500°C for base metal thermocouple.
† Boric acid also recommended for types J, E, and K.
‡ Recommended for 8 through 20 AWG wires.
N.R.—Not recommended.
Reprinted from ANSI MC96.1-1982 with the permission of the copyright holder. Copyright © Instrument Society of America. Users should refer to a complete copy of the standard available from the publisher. ISA, 67 Alexander Drive, P.O. Box 12277, Research Triangle Park, North Carolina 27709.

500°C. Common welding methods include gas, electric arc, resistance, butt, tungsten-inert gas, and plasma-arc welding. Table 16.14 summarizes fabrication methods for bare wires. If present, the insulation and/or sheath should first be removed and the wires properly cleaned before the joining process. For the removal of sheath materials, commercially available stripping tools can be used. Figure 16.21 shows pairs of wires prepared for different types of welding to form junctions.

Bare wires or insulated wires are normally fabricated into butt-welded or beaded (welded or soldered) junctions. Sheathed wires can be fabricated into these exposed types as well as protected (grounded or ungrounded) types. Figure 16.22 shows different thermocouple junctions.

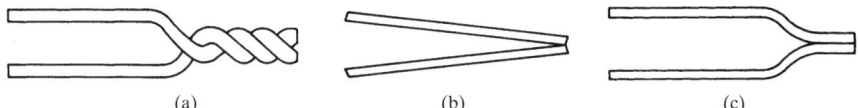

(a)   (b)   (c)

FIGURE 16.21 Preparation of thermocouple wires: (*a*) wire placement for gas and arc welding of base metal thermocouples, (*b*) wire placement for arc welding of all standardized thermocouples, and (*c*) wire placement for resistance welding.

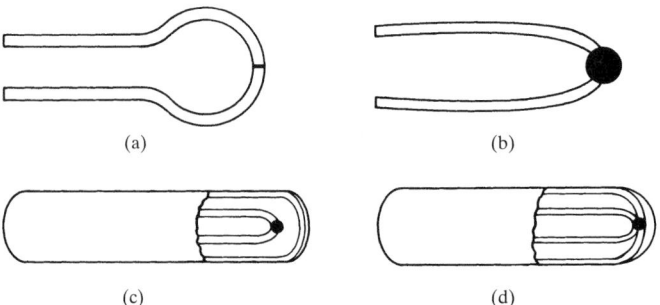

(a)   (b)

(c)   (d)

FIGURE 16.22 Thermocouple junctions: (*a*) butt-welded junction, (*b*) beaded junction, (*c*) protected junction, ungrounded, (*d*) protected junction, grounded.

At the reference junction, each thermoelement (or extension wire) is soldered to a separate copper connecting wire. These copper wires are then connected to the input terminals of an EMF-measuring device. Depending on the application and accuracy required, different EMF-measuring devices are available: (1) direct EMF readout, normally a precision digital voltmeter, and (2) automatic compensating readout, giving a reading directly in temperature, available in both analog and digital off-the-shelf devices with built-in compensation to account for variations in the reference-junction temperature to be used with a specified type of thermocouple. The choice between the two depends primarily on the required accuracy in the measurements. The automatic compensator is convenient to use, but accuracy beyond the commercial limits should not be expected. On the other hand, the better accuracy achievable using the direct EMF readout requires calibration and calculation of temperature from the EMF.

The EMF measured by a readout device reflects the difference between the temperatures of the measuring junction and the reference junction; thus, the temperature at the reference junction must be properly controlled. It is usually maintained at the freezing point of water (0°C) by using an ice bath (Fig. 16.23). The ice bath should be a mixture of crushed ice and pure water in a Dewar flask with considerably more ice than liquid. It can be very accurate but not too convenient in some applications, since frequent replacement of melted ice is required. With a thermoelectric refrigeration system [36], ice can be maintained by cooling the bath of water with thermoelectric cooling elements.

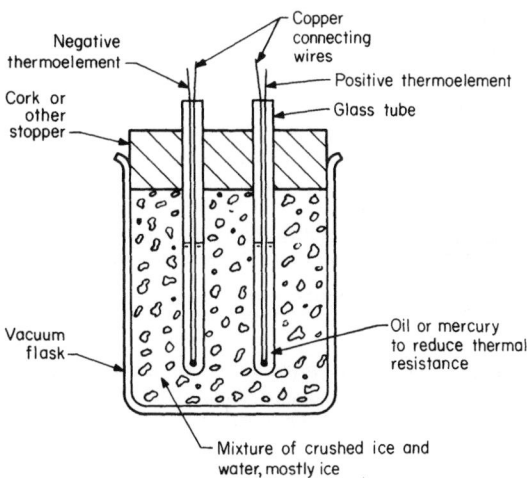

FIGURE 16.23   Ice bath for thermocouple reference junction.

In contrast to direct EMF-readout systems, electrical compensation can be used [36]. The ice bath is replaced by an electrical circuit that has a temperature-sensitive resistor. The circuit is preset to compensate for variation in the reference junction temperature.

Another method used to control the reference junction temperature is to use a constant-temperature oven. The oven will maintain the reference junction at a fixed temperature that is above the highest anticipated ambient temperature. Corrections to the standard reference EMF values must be applied when using this method.

For applications in adverse environments such as measurement in high-temperature furnaces, combustion chambers, and nuclear reactor cores, a protection tube is normally used with the thermocouple. Detailed descriptions on the selection of protection tubes and typical industrial high-temperature thermocouple assemblies are given in Ref. 36.

Reference Tables and Reliability.   For the eight letter-designated thermocouples (S, R, B, T, J, E, N, and K), standard temperature versus EMF reference functions and tables are available, both from ASTM [35] and NIST [42]. These reference functions, together with the appropriate tolerances (Table 16.9), serve as the standard for industry in the manufacture of thermocouple wires.

Where accuracy of measurement is not vital, commercially available electrical compensators (which assume the outputs of the thermocouples used will match the standard function within the tolerances stated in Table 16.9) are sufficient for direct readout of temperatures. With proper selection, accuracy of at least 2.2°C (or 0.75 percent of readings) for standard wires and 1.1°C (or 0.4 percent of readings) for wires of special tolerance can be achieved without calibration (Table 16.9). However, the precision or reproducibility of a single thermocouple is usually much better than these limits.

When accuracy greater than the tolerances in Table 16.9 is required, the wires must be calibrated. Normally, this requires comparison calibration of sample thermocouples taken from each spool to account for spool variability. Typically, two thermocouples—one fabricated from the beginning and the other from the end of the spool—are calibrated to determine an average calibration for the entire spool. If the deviation between the two calibrations is not within the required uncertainty, a third thermocouple fabricated from the center of the spool should be used. If the results are still unsatisfactory, then each thermocouple should be calibrated individually, or a different spool should be used.

Transfer or working-standard thermocouples (including connecting wires—see Fig. 16.17) are individually calibrated by comparison calibration against a defining standard thermometer (such as an SPRT) or another transfer standard thermometer (usually a thermocouple).

The type-S thermocouple, though no longer used as a defining standard for ITS-90, is still a reasonably accurate transfer standard thermometer. The precision of a type-S thermocouple at temperatures between 600 to 1000°C is about 0.02°C, and its accuracy is about 0.2 to 0.3°C. At lower temperatures (between about 0 and 200°C), a base-metal-type thermocouple (e.g., type T) is capable of a precision of about 0.01°C and an accuracy of 0.1°C.

The calibration and use of base metal thermocouples at temperatures above about 300°C will produce inhomogeneities in the wires, which can change the calibration itself [43]. The usual practice to overcome this dilemma for application at high temperature is to calibrate sample thermocouples to obtain the calibration for the remainder of the spool of wire and discard the calibrated thermocouples.

Factors that can affect thermocouple reliability include wire inhomogeneity and mechanical strain. Wire inhomogeneity (short range) can be detected by observing whether a spurious EMF is generated when a local heat source is applied to thermocouple wire [44]. Short-range inhomogeneities are small and usually not significant in most wires manufactured today. Medium- and longer-range inhomogeneities—changes in calibration from thermocouple to thermocouple or from one end of a spool of wire to the other—are normally small and gradual. They are compensated by the calibration method outlined above (for spool variability).

Mechanical strains in the wires can be removed by annealing. Commercial base-metal thermocouple wire is usually annealed by the manufacturer. Most noble metal thermocouple wires can also be obtained in an annealed form. Annealing by the user is needed only when high accuracy is sought [45].

The inability to control precisely the reference junction temperature or to obtain accurate compensation would also affect thermocouple reliability. In the case of an ice bath, factors that can cause the junction temperature to depart from 0°C include nonuniformity in the temperature of the ice-water mixture, small depth of immersion, insufficient ice, and large wire sizes (conduction effects). Reference 46 describes various sources of errors in an ice bath.

Radiation Thermometer

Introduction.   The relationship between the radiant energy emitted by an ideal (black) body and its temperature is described by Plancks radiation law. Radiation thermometers measure the radiation emitted by a body to determine its temperature.

Radiation thermometers can be sensitive to radiation in all wavelengths (total-radiation thermometers) or only to radiation in a band of wavelengths (spectral-radiation thermometers). Thermocouple and thermopile junctions or a calorimeter are the usual detectors in a total-radiation thermometer. For spectral systems, the classification is normally based on the effective wavelength or wavelength band used—as determined, for example, by a filter, which allows only near-monochromatic radiation to reach the detector, or by the use of a detector sensitive only to radiation in a specific wavelength band. Radiation thermometers utilize the visible portion of the radiation spectrum, infrared thermometers or scanners measure infrared radiation, and spectroscopic thermometers operate with radiation that is normally of shorter wavelength than the other two methods.

Thermal Radiation.   The wavelength distribution of an ideal thermal radiation emitter (a blackbody) in a vacuum is given by the Planck distribution function

$$e_{b\lambda} = \frac{c_1 \lambda^{-5}}{\exp(c_2/\lambda T) - 1} \tag{16.22}$$

where $e_{b\lambda}$ = monochromatic blackbody emissive power, energy/(time·area·wavelength interval)
λ = wavelength, μm
$c_1 = (3.741832 \pm 0.000020) \times 10^8$ W·μm⁴/m²
$c_2 = (1.438786 \pm 0.000045) \times 10^4$ μm·K

Integration of Eq. 16.22 over all wavelengths gives the total thermal radiation emitted by a blackbody

$$e_b = \int_0^\infty e_{b\lambda} = \sigma T^4 \tag{16.23}$$

where e_b = emissive power of a blackbody at temperature T, W/m²
σ = Stefan-Boltzmann constant $(5.67032 \pm 0.00071) \times 10^{-8}$ W/(m²·K⁴)

Thermal radiation emitted by a real body has an irregular wavelength dependence (see Ref. 47). The emissivity is defined to relate the emissive power of a real body, e_λ or e, to that of a blackbody:

$$e_\lambda = \varepsilon_\lambda e_{b\lambda} \tag{16.24a}$$

$$e = \varepsilon e_b \tag{16.24b}$$

where ε_λ and ε are the monochromatic emissivity and the total emissivity of the body, respectively. A gray body is defined as one for which ε_λ is independent of wavelength; then $\varepsilon_\lambda = \varepsilon$.

Principles of Radiation Thermometers.   A detailed discussion on opto-electronic temperature measuring systems for radiance thermometers can be found in Ref. 48. In the USA, ITS-90 above the gold point is maintained by NIST [49]. A classical radiation thermometer is shown in Fig. 16.24. Radiation from the object whose temperature is to be measured is

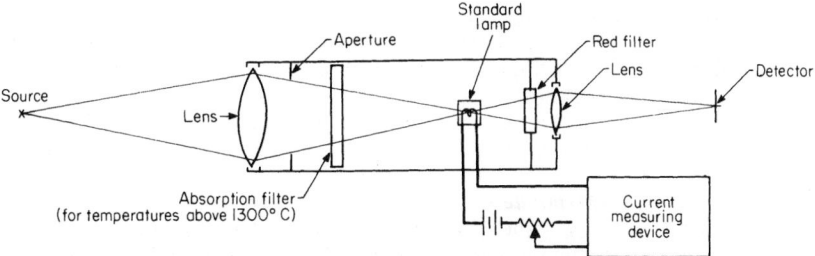

FIGURE 16.24   Schematic of an optical pyrometer.

focused on a standard adjustable-brightness lamp. The power supply to the lamp filament can be adjusted to change the brightness of the filament. This brightness is then compared with the brightness of the incoming radiation.

The wavelength at which $e_{b\lambda}$ is a maximum, λ_{max}, can be determined from Wien's displacement law, which can be derived from Eq. 16.22:

$$\lambda_{max} T = 2897.8 \ \mu mK \tag{16.25}$$

At the gold point (the freezing point of gold), 1337.58 K, $\lambda_{max} = 2.17 \ \mu m$. Although this is well into the infrared region, a substantial fraction of the emitted radiation at the gold point is in the visible region. The effective wavelength of a radiation thermometer, determined by a filter (usually red), is often taken to be 0.655 μm.

The emissive power is determined by comparing the incoming radiation from the object with the radiation from the standard lamp. The filament current is adjusted until the filament image just disappears in the image of the test body; such a condition is similar to a null condition in resistance measurement with a bridge. Prior calibration of the radiation thermometer to determine the filament current setting and the corresponding filament temperature is used to calculate the temperature of the test body from the current setting. As a real body emits less thermal radiation than a blackbody at the same temperature, the temperature reading of a radiation thermometer must be corrected. The measured emissive power can be represented by a blackbody at temperature T_m corresponding to the calibrated-filament lamp reading.

For a monochromatic measurement, the spectral radiance temperature T_λ can be derived from Eqs. 16.22 and 16.24(a),

$$\exp \frac{c_2}{\lambda T_\lambda} - 1 = \varepsilon_\lambda \left(\exp \frac{c_2}{\lambda T_m} - 1 \right) \tag{16.26}$$

Rearranging,

$$\exp \frac{c_2}{\lambda T_\lambda} = \varepsilon_\lambda \exp \frac{c_2}{\lambda T_m} + (1 - \varepsilon_\lambda) \tag{16.27}$$

Neglecting $1 - \varepsilon_\lambda$ and taking the natural logarithm of both sides,

$$\frac{1}{T_\lambda} - \frac{1}{T_m} = \frac{\lambda}{c_2} \ln \varepsilon_\lambda \approx 4.55 \times 10^{-5} \ln \varepsilon_\lambda \qquad \text{for } \lambda = 0.655 \ \mu m \tag{16.28}$$

With ε_λ known and T_m having been measured, Eq. 16.28 can be solved for the unknown temperature, T [50]. Note that $T = T_m$, and, the smaller the value of ε_λ, the greater the temperature correction. This correction is small unless the emissivity is considerably less than 1.0.

If the total emissive power were used, then

$$e_b(T_m) = e(T_\lambda) = \varepsilon e_b(T)$$

$$\sigma T_m^4 = \varepsilon \sigma T_\lambda^4$$

$$T_\lambda = T_m \left(\frac{1}{\varepsilon} \right)^{1/4} \tag{16.29}$$

Similarly, as an approximation to Eq. 16.28, the variation of $e_{b\lambda}$ with T can be assumed to follow a power law at least over a limited temperature range:

$$e_{b\lambda} = c_3 T_\lambda^n \tag{16.30}$$

Then,

$$e_{b\lambda}(T_m) = e_\lambda(T_\lambda) = \varepsilon_\lambda e_{b\lambda}(T_\lambda)$$

$$c_3 T_m^n = \varepsilon_\lambda c_3 T_\lambda^n$$

$$T_\lambda = T_m \left(\frac{1}{\varepsilon_\lambda} \right)^{1/n} \tag{16.31}$$

Note that for $\lambda = 0.655 \ \mu m$ and $T = 1340$ K, n is about 16.

Table 16.15 [51] shows the variation in ε_λ for different materials. For oxides, ε_λ has a strong dependence on surface roughness, which leads to large uncertainty [51]. Care should be exercised to ensure that the surface condition of the object whose temperature is being measured is such that the emissivity is known with a reasonable certainty. Sometimes a small hole with a depth of about five diameters is made in the object being studied, and the pyrometer is focused on this hole. Internal reflections in the hole cause it to approximate a black surface.

Reliability of Radiation Thermometers.   Calibration of radiation thermometers at NIST is accomplished by focusing the radiance sensor at a blackbody furnace with known temperature. This blackbody furnace is previously calibrated by comparison calibration against a standard lamp, which, in turn, is calibrated at the gold point [52]. With calibration performed at NIST, the accuracy of a radiation thermometer is within 0.4°C at the gold point, within about 2°C at 2200°C and about 10°C at 4000°C.

Other Radiation Thermometers.   Traditionally, radiation thermometers used a red filter to achieve a monochromatic comparison between the incoming radiation and that of a standard lamp. With the development of photomultipliers, better precision and automation are possible, including direct detection—eliminating the use of a standard lamp in actual applications.

The major problem in using a single wavelength radiation thermometer to measure the surface temperature is the unknown emissivity of the measured surface. The emissivity is the major parameter in the spectral radiance temperature equation (Eq. 16.28) for the temperature evaluation. Objects encountered for temperature measurements are often oxidized metal surfaces, molten metal, or even semitransparent materials. On these surfaces, the emissivity is usually affected by the surface temperature and the manufacturing process for these materials.

To reduce the error in the temperature evaluation caused by the uncertainty of the emissivity, radiation measurements for two or multiple distinct wavelengths may resolve the problem. For each wavelength, both spectral radiance temperature equations can be respectively written as

$$\frac{1}{T_m} = \frac{1}{T_{\lambda_1}} + \frac{\lambda_1}{c_2} \ln (\varepsilon_{\lambda_1}) \tag{16.32}$$

and

$$\frac{1}{T_m} = \frac{1}{T_{\lambda_2}} + \frac{\lambda_2}{c_2} \ln (\varepsilon_{\lambda_2}) \tag{16.33}$$

TABLE 16.15(a)   Spectral Emissivity of Materials ($\lambda = 0.65$ μm): Unoxidized Surfaces [51]

Element	Solid	Liquid	Element	Solid	Liquid
Beryllium	0.61	0.61	Thorium	0.36	0.40
Carbon	0.80–0.93		Titanium	0.63	0.65
Chromium	0.34	0.39	Tungsten	0.43	
Cobalt	0.36	0.37	Uranium	0.54	0.34
Columbium	0.37	0.40	Vanadium	0.35	0.32
Copper	0.10	0.15	Yttrium	0.35	0.35
Erbium	0.55	0.38	Zirconium	0.32	0.30
Gold	0.14	0.22	Steel	0.35	0.37
Iridium	0.30		Cast iron	0.37	0.40
Iron	0.35	0.37	Constantan	0.35	
Manganese	0.59	0.59	Monel	0.37	
Molybdenum	0.37	0.40	Chromel P		
Nickel	0.36	0.37	(90Ni-10Cr)	0.35	
Palladium	0.33	0.37	80Ni-20Cr	0.35	
Platinum	0.30	0.38	60Ni-24Fe-16Cr	0.36	
Rhodium	0.24	0.30	Alumel		
Silver	0.07	0.07	(95Ni; bal. Al, Mn, Si)	0.37	
Tantalum	0.49		90Pt-10Rh	0.27	

TABLE 16.15(b) Spectral Emissivity of Materials ($\lambda = 0.65$ μm): Oxides [51]

Material*	Range of observed values	Probable value for the oxide formed on smooth metal
Aluminum oxide	0.22 to 0.40	0.30
Beryllium oxide	0.07 to 0.37	0.35
Cerium oxide	0.58 to 0.80	
Chromium oxide	0.60 to 0.80	0.70
Cobalt oxide		0.75
Columbium oxide	0.55 to 0.71	0.70
Copper oxide	0.60 to 0.80	0.70
Iron oxide	0.63 to 0.98	0.70
Magnesium oxide	0.10 to 0.43	0.20
Nickel oxide	0.85 to 0.96	0.90
Thorium oxide	0.20 to 0.57	0.50
Tin oxide	0.32 to 0.60	
Titanium oxide		0.50
Uranium oxide		0.30
Vanadium oxide		0.70
Yttrium oxide		0.60
Zirconium oxide	0.18 to 0.43	0.40
Alumel (oxidized)		0.87
Cast iron (oxidized)		0.70
Chromel P (90Ni-10Cr) (oxidized)		0.87
80Ni-20Cr (oxidized)		0.90
60Ni-24Fe-16Cr (oxidized)		0.83
55Fe-37.5Cr-7.5Al (oxidized)		0.78
70Fe-23Cr-5Al-2Co (oxidized)		0.75
Constantan (55Cu-45Ni) (oxidized)		0.84
Carbon steel (oxidized)		0.80
Stainless steel (18-8) (oxidized)		0.85
Porcelain	0.25 to 0.50	

* The emissivity of oxides and oxidized metals depends to a large extent upon the roughness of the surface. In general, higher values of emissivity are obtained on the rougher surfaces.

In ratio pyrometers [5], a device is designed to measure the spectral radiance temperatures T_{λ_1} and T_{λ_2}, at two wavelengths λ_1 and λ_2. Then, the true temperature T_m is determined from the ratio temperature T_{ra} and is given by

$$\frac{1}{T_m} = \frac{1}{T_{ra}} + \frac{\Lambda}{c_2} \ln{(\varepsilon_r)} \tag{16.34}$$

where the ratio temperature is defined as

$$\frac{1}{T_{ra}} = \Lambda \left(\frac{1}{\lambda_1 T_{\lambda_1}} - \frac{1}{\lambda_2 T_{\lambda_2}} \right) \tag{16.35}$$

Λ is the effective wavelength, given by

$$\Lambda = \left(\frac{\lambda_1 \lambda_2}{\lambda_2 - \lambda_1} \right) \tag{16.36}$$

and the emissivity ratio ε_r is expressed as

$$\varepsilon_r = \left(\frac{\varepsilon_{\lambda 1}(\lambda_1, T_m)}{\varepsilon_{\lambda 2}(\lambda_2, T_m)} \right) \tag{16.37}$$

However, the emissivity ratio ε_r is still an unknown quantity for some materials. Therefore, the ratio pyrometer is commonly used for temperature measurements on materials that can be assumed to be gray with ε_r assumed equal to unity.

To overcome the problems faced by the single-wavelength radiation thermometer and the ratio pyrometer, a double-wavelength radiation thermometer (DWRT) measures the spectral radiance itself at two distinct wavelengths for surface temperature evaluation. For this method to be used, the emissivity compensation function $\varepsilon_{\lambda 1} = f(\varepsilon_{\lambda 2})$ must be defined. A detailed description of the principle for DWRT can be found in Ref. 53. When the emissivity relation $\varepsilon_{\lambda 1} = f(\varepsilon_{\lambda 2})$ at two distinct wavelengths $\varepsilon_{\lambda 1} = f(\varepsilon_{\lambda 2})$ is established, the true temperature on the measured surface can be determined from the inferred temperature, which is defined as

$$\frac{1}{T_{\text{inf}}} = \frac{c_4}{T_{\lambda_1}} + \frac{c_5}{T_{\lambda_2}} + c_6 \tag{16.38}$$

where the constants c_4, c_5, and c_6 are empirical constants determined from the specific emissivity compensation function, which varies with the particular algorithm being used. In practical applications, an emissivity compensation function should be established for a particular material in a specific range of operation. The values of c_4, c_5, and c_6 for aluminum alloys are found in Ref. 54.

Infrared thermometers are similar to radiation thermometers, the main difference being that an infrared thermometer uses a detector sensitive to a band of radiation in the infrared region. A band filter is often used. An infrared scanner is an infrared detector coupled with a scanning mechanism to produce a two-dimensional output display. In order to eliminate the errors due to thermal noise (infrared radiation from the casing, etc.), the detector has to be cooled and properly sealed. Many infrared scanners used in airborne measuring systems for thermal mapping are sensitive to a portion of the spectrum is which the atmosphere has low absorption, 8 to 14 µm. Other commonly found units for ground-level application (often smaller in physical size) are sensitive in the 3- to 5-µm band of radiation. More information on infrared scanners and remote sensing is available in Refs. 55 and 56.

Spectroscopic methods depend on the spectral line intensity emitted by the media of interest. These techniques have been used for temperature measurement in high-temperature gases [48]. The wavelength involved is generally shorter than those in the infrared band.

Optical Thermometers

Introduction. An optical thermometer usually consists of a light source, often an excitation laser; an optic sensor for capturing the spectral characteristics of the reflected light; and an optic fiber for transmitting the optical signal. In some cases, the optic fiber is itself used as the optic sensor. An overview of fiberoptic thermometry is given in Ref. 57. In order to provide temperature-dependent optic characteristics on the test surface, light reflection materials or light tracers are usually attached or doped on the test surface during practical applications. Optical thermometers can be designed to measure the following temperature-dependent optical characteristics:

1. Index of refraction: schlieren system, shadowgraph system, interferometer
2. Fluorescence: photoluminescence sensors, fluorescence thermometry
3. Optical absorption
4. Reflectivity

Details of optical thermometers for measuring each temperature-dependent characteristic will be described as follows.

Index of Refraction. The refraction of index of a fluid is usually a function of the thermodynamic state, often only the density. According to the Lorenz-Lorentz equation, the relation between the index of refraction and temperature is given by

$$\frac{n_i^2 - 1}{\rho(n_i^2 + 2)} = \frac{N}{M} \tag{16.39}$$

where N is the molar refractivity, M denotes the molecular weight, ρ is the density, and n_i is the refractive index at a particular position.

For gases with $n_i \sim 1$, Eq. 16.39 can be simply reduced to the Gladstone-Dale equation:

$$\frac{(n_i - 1)}{\rho} = \frac{3N}{2M} = C \tag{16.40}$$

where C is called the Gladstone-Dale constant. In practical applications, usually the value of C is replaced by the index of refraction at the standard condition n_0. Thus, Eq. 16.40 becomes

$$n_i - 1 = \frac{\rho}{\rho_0}(n_{i,0} - 1) \tag{16.41}$$

where the subscript 0 denotes a state at the standard condition.

In a single component system, the density of a gas can be determined from the equation of state. If the pressure in the test section is kept at a constant value, the temperature of the gas is given by

$$T = \frac{n_{i,0} - 1}{n_i - 1} \frac{P}{P_0} T_0 \tag{16.42}$$

In this section, three optical techniques are introduced: schlieren, shadowgraph, and interferometric. These three techniques are described in detail in Refs. 58 and 59. Although these three optic techniques depend on the variation of the index of refraction with the position in a transparent medium in the test section through which a light beam passes, quite different quantities are measured with each one. Interferometers measure the differences in the optical path lengths between two light beams. The schlieren and shadowgraph systems can provide the first and second derivatives of the index of refraction, respectively.

Schlieren System.   With a schlieren system, the first derivative of index refraction is determined and given by

$$\frac{\partial \rho}{\partial y} = \frac{\rho_0}{n_{i,0} - 1} \frac{\partial n_i}{\partial y} \tag{16.43}$$

If the pressure is a constant and the ideal gas equation of state holds for the gas medium in the test section, the relationship between the first derivatives in the index of refraction and the index of temperature can be expressed as

$$\frac{\partial n_i}{\partial y} = -\frac{n_{i,0} - 1}{T} \frac{\rho}{\rho_{i,0}} \frac{\partial T}{\partial y} \tag{16.44}$$

A schematic view of a typical schlieren system using lenses is shown in Fig. 16.25. For simplicity in this discussion, the index of reflection varies only in the y direction in the test section that is surrounded by the ambient air. Since the velocity of a light beam varies with the index of refraction in the medium, the angular deflection of the light beam in the yz plane occurs when the light beam passes through the disturbed region with a variation in the index of refraction in the test section. If the angular deflection is small, the angular deflection of a light beam is given by

$$\alpha = \frac{1}{n_{i,a}} \int \frac{\partial n_i}{\partial y} dz \tag{16.45}$$

where $n_{i,a}$ is the index of refraction of the ambient air.

The schlieren system is a device that is used to measure or indicate the angular deflection α, as illustrated in Fig. 16.25. In the schlieren system, a light source located at the focus of the lens L_1 provides a parallel light beam passing through the test section. The deflected light beam, when a disturbance in the test section is present, is marked by the dashed line. The light

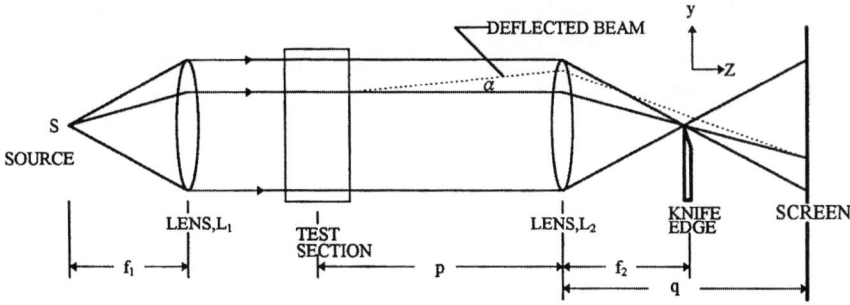

FIGURE 16.25   Schematic view of a schlieren system.

is collected by the second lens L_2, at whose focus a knife edge is installed, and then enters onto a screen. The screen is placed at the conjugate focus of the test section, or, as is often the case, a camera is placed after the knife edge and is focused on the center of the test section. Focusing mirrors are commonly used in place of the lenses shown in Fig. 16.25.

Shadowgraph System.   In a shadowgraph system, the second derivative of index refraction is measured and given by

$$\frac{\partial^2 \rho}{\partial y^2} = \frac{\rho_0}{n_{i,0} - 1} \frac{\partial^2 n_i}{\partial y^2} \tag{16.46}$$

If the pressure is assumed to be constant and the ideal gas equation of state holds, then

$$\frac{\partial^2 n_i}{\partial y^2} = C\left[-\frac{\rho}{T} \frac{\partial^2 T}{\partial y^2} + \frac{2\rho}{T^2} \left(\frac{\partial T}{\partial y}\right)^2 \right] \tag{16.47}$$

A schematic view of a typical shadowgraph system with parallel light beams is shown in Fig. 16.26. The shadowgraph system measures the displacement of the disturbed light beam rather than the angular deflection, as in a schlieren system. However, the linear displacement of the light beam is usually small and difficult to measure. Instead, the contrast is shown in shadowgraphs of the flow or heat transfer fields being investigated.

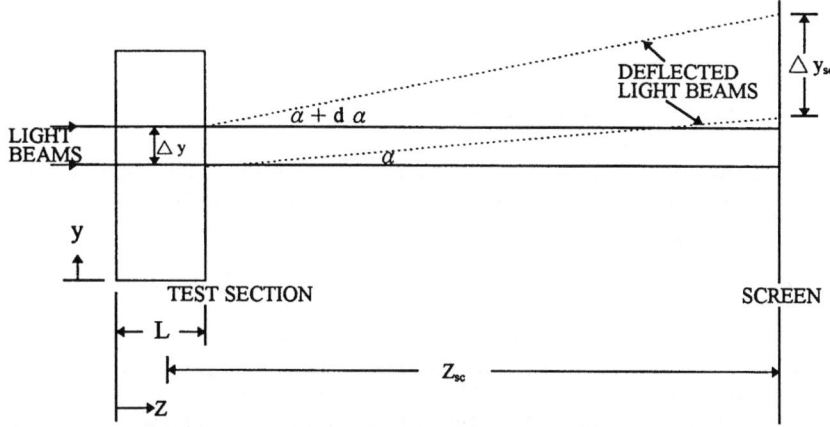

FIGURE 16.26   Optical path of light beam for a shadowgraph system.

As parallel light beams enter a test section in which the index of refraction of medium varies with the position, the light beams leaving the test section are deflected by an angle α, which is a function of vertical location y. The relationship between the illumination on the screen $\mathbf{I}_{sc}$ and the initial illumination intensity $\mathbf{I}_i$ is given by

$$\mathbf{I}_{sc} = \frac{\Delta y}{\Delta y_{sc}} \mathbf{I}_i \tag{16.48}$$

where Δy and Δy_{sc} denotes intervals between two light beams through the test section and on the screens, respectively. If the distance between the test section and the screen is z_{sc}, then

$$\Delta y_{sc} = \Delta y + z_{sc} d\alpha \tag{16.49}$$

Combining Eq. 16.49 with Eq. 16.48, the contrast becomes

$$\frac{\Delta \mathbf{I}}{\mathbf{I}_i} = \frac{\mathbf{I}_{sc} - \mathbf{I}_i}{\mathbf{I}_i} = \frac{\Delta y}{\Delta y_{sc}} - 1 \cong -z_{sc} \frac{\partial \alpha}{\partial y} \tag{16.50}$$

Substituting Eq. 16.45 into Eq. 16.50 yields

$$\frac{\Delta \mathbf{I}}{\mathbf{I}_i} = -\frac{z_{sc}}{n_{i,a}} \int \frac{\partial^2 n_i}{\partial y^2}\, dz \tag{16.51}$$

If the pressure in the test section is assumed to be constant and the ideal gas law holds, Eq. 16.51 becomes

$$\frac{\Delta \mathbf{I}}{\mathbf{I}_i} = -\frac{z_{sc}}{n_{i,a}} \int C \left[-\frac{\rho}{T} \frac{\partial^2 T}{\partial y^2} + \frac{2\rho}{T^2} \left(\frac{\partial T}{\partial y} \right)^2 \right] dz \tag{16.52}$$

Often a system similar to that shown in Fig. 16.25—or with focusing mirrors—is used for shadowgraph studies. Note that when shadowgraphs are taken, the knife edge (in Fig. 16.25) would be removed and the screen would not be placed at the conjugate focus of the test section.

Interferometer.   A schematic view of a Mach-Zehnder interferometer is shown in Fig. 16.27. Two parallel beams of light are provided by a monochromatic light source with a lens and a splitter plate. If both light beams 1 and 2 pass through homogeneous media, the recombined beam shown on the screen is uniformly bright. Once a disturbance caused by the temperature variation is present in the test section, the difference in the optical path lengths of both light beams is no longer zero. Thus, the initially bright field will have fringes occurring on the screen.

If a heat transfer field is two-dimensional and the variation in the index of refraction only occurs perpendicular to the light beam direction, the fringe shift or difference in the optical path length (measured in vacuum wavelength) ε can be expressed as

$$\varepsilon = L\{n_i - n_{i,0}\}/\lambda_o \tag{16.53}$$

where L is the length of the test section in which the refractive index varies due to the change in density or temperature, n_i is the refractive index field to be measured, $n_{i,0}$ is the reference refractive index in reference beam 2, and λ is the wavelength of the monochromatic light source.

In a single-component system, the density of a gas can be determined from the ideal gas equation of state. If the pressure in the test section is kept at a constant value and the index of refraction is only a function of density, the temperature distribution in the test section can be evaluated from the equation, given by

$$T(x, y) = \left[\frac{\lambda_o R_u}{PCLM} \varepsilon + \frac{1}{T_0} \right]^{-1} \tag{16.54}$$

where R_u denotes the universal gas constant, T_0 is the reference temperature, and P is the absolute pressure in the test section.

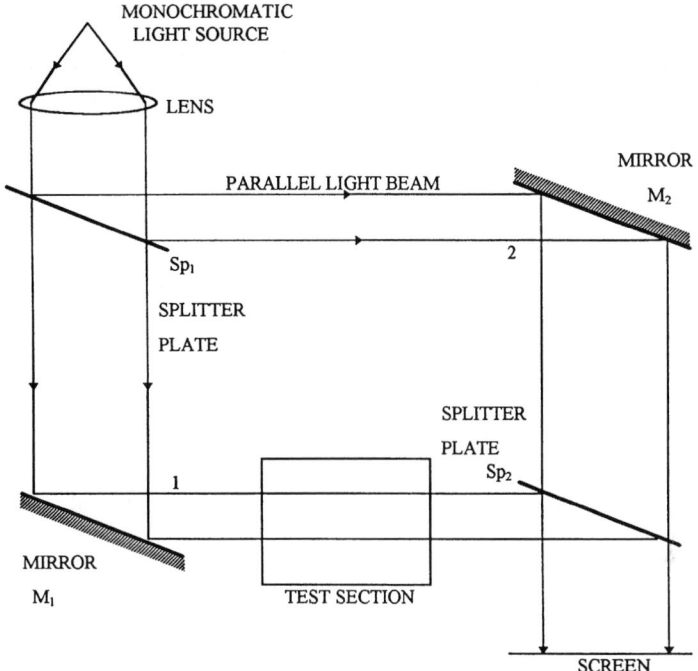

FIGURE 16.27 Schematic view of Mach-Zehnder interferometer.

Holographic interferometers are also commonly used in both flow and heat transfer measurements. The basic principle of a holographic interferometer is the same as that of an interferometer except that the fringe pattern between the object beam and reference beam is shown on a hologram plate. The beam emitted from a laser is split and expanded into two parallel beams: the reference beam and the object beam. After passing through the test section, initially uniform in temperature, the object beam interferes with the reference beam, which bypasses the test section. Both wavefronts then combine to form a single wavefront, known as the *comparison wavefront,* which is then recorded on a hologram. The hologram is then developed and repositioned back to the same location. As the developed hologram is illuminated by the same reference beam, the reconstructed beam as observed from the hologram is the same as the object beam before the hologram is developed. To record the temperature field in the test section where the heat transfer starts to take place, the object beam is emitted from the laser and travels through the test section and interferes with the reconstructed wavefront beam to form the fringe pattern. For simultaneously measuring the heat and mass transfer, a double-wavelength holographic interferometer is described in Refs. 60 and 61.

In the aforementioned optical techniques, the light beam essentially integrates the variation in index of refraction or its gradients along the path of the beam and is usually not useful for local measurements in a three-dimensional temperature field. This difficulty can be overcome with multiple views or with holography.

Laser-Induced Fluorescence.   The fluorescence lifetime of a rare-earth-doped ceramic phosphor decreases with an increase in temperature. This relationship permits the fluorescence to be used in noncontact and remote temperature measurements. After a rare-earth-doped ceramic phosphor is excited by a delta-function excitation, electrons will be elevated from the valence band to the conduction band. The material fluorescence results from the energy released by electrons moving from the conduction band back to the original valence band.

The fluorescence amplitude of the ceramic material is expressed as

$$G(t) = G_0 \exp(-\theta/\tau) \exp[-\chi(\theta/\tau)^\beta] \qquad (16.55)$$

where $G(t)$ denotes the fluorescence amplitude, G_0 is the initial amplitude, θ is the time, τ denotes the lifetime of fluorescence, and χ and β are the two dimensionless scaling parameters [62]. The parameter β depends on the energy mechanism of the fluorescence material being used: $\beta = 0.5$ for dipole-dipole interactions and 0.3 for quadrupole-quadrupole interactions. Except for β, the other parameters (G_0, τ, and χ) are all dependent on the temperature. The typical lifetime of fluorescence ranges from 1 μs to 1 ms, and the amplitude of fluorescence decreases with an increase in temperature. To employ the relationship between the temperature and lifetime of excited fluorescence, a fiber-optic thermometer [63] can be used to sense the decay time of the excited fluorescence doped in a phosphate glass. The excited fluorescence light is first transmitted in the fiber, and then the light intensity is detected by a silicon diode within a nanosecond response time. A linear relation between the decay time of fluorescence and the temperature is illustrated in Ref. 63. In fluorescent decay-time thermometers, both rare-earth-doped and transition-metal-doped phosphors are the popular materials being used. Table 16.16 [64] lists the phosphor properties of different decay-time thermometers.

In addition to measuring the light intensity at a particular wavelength of fluorescence, an improved approach for determining surface temperature is based on the ratio of the intensities of fluorescence emitted from the doped material at two different wavelengths when the material is subject to the same excited laser light. The intensity ratio is strongly dependent on the temperature. Chyu and Bizzak [65] applied the Nd:YAG laser to excite a test surface coated with phosphor ceramic material $Eu^{+3}:La_2O_2S$ and then recorded the reflected fluorescence image. They used two optic filters to measure the spectrum intensity at two particular wavelengths: one at 512 nm that is sensitive to temperature change in the room temperature range and the other at 620 nm that is insensitive to temperature change in the room temperature. A correlation can be established between the temperature and the intensity ratio from these two emission spectrums at different wavelengths. Once the correlation is established, laser-induced fluorescence can be used to indicate the surface temperature on a test piece coated with the particular phosphor transducer. Due to the short lifetime of fluorescence and the weak intensity in the 512-nm fluorescence emission signal, an improvement of the camera

TABLE 16.16   Phosphor Properties of the Decay-Time Thermometers [64]

Phosphor	Emission line, nm	Absorption bands, nm	Excitation device	Decay time @T_{amb}, μs	Sensitivity, μs/°C	Temperature range, °C
$Eu:La_2O_2S$	514	337, 355	NV N_2 Laser Nd:YAG laser	7	0.2	0–100
$Eu:Y_2O_3$	611	396, 466	Pulsed N_2 + Dye laser	1000	2	500–900
Nd:Glass	1054	750, 805, 880	IR-LED Laser diode	240	0.2	−50–350
Nd:YAG	1064	750, 810, 880	IR-LED Laser diode	250	0.05	0–800
$Cr:Al_2O_3$	694	400, 550	He-Ne laser, LED	2900	5	100–500
$Cr:BeAl_2O_4$	680	420, 580	He-Ne laser, LED	300	0.5	0–200
Cr:YAG	689	430, 600	He-Ne laser, Laser diode	1450	10	−100–300
$Mn:MgGeF_6$	660	290, 420	Xenon flash lamp	3500	6	−60–400

and the data acquisition was proposed in Chyu and Bizzak [65] to increase the signal-to-noise ratio of the measuring system.

For a typical measurement, the phosphor transducer is mounted on the test surface, and a pulse laser is used to excite the transducer. Then, the time evolution of the phosphors fluorescence spectrum is captured by an optic filter and an optoelectronic device. The fluorescence thermometer should be carefully calibrated before application in temperature measurements.

Optical Absorption.   For semiconductor materials (e.g., GaAs), the optical absorption can be significantly affected by a change in the temperature. The optical absorption H can be empirically expressed as

$$H = H_0 \exp[c_7(h_p\upsilon - E_g)/k_B T] \tag{16.56}$$

where c_7 is a material constant insensitive to temperature, h_p is Plancks constant, k_B is Boltzmann constant, υ is the light frequency, E_g is the band gap energy, H_0 is the initial amplitude of emitted light from the light source. When the light-emitting energy $h_p\upsilon$ at a fixed frequency is close to the band gap energy E_g, the optical absorption ratio H/H_0 is very sensitive to changes in the thermal energy $k_B T$. Figure 16.28 shows a schematic view of optical absorption thermometry. A review of such an optical temperature sensor is given in Ref. 66.

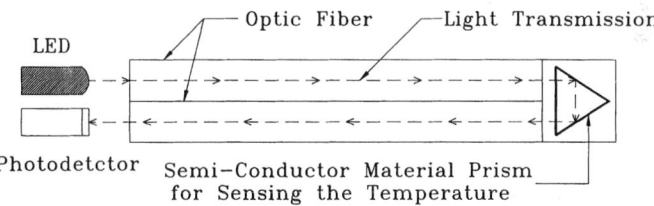

FIGURE 16.28   Optical absorption thermometer.

Reflectivity.   Since the reflectivity of pure materials is temperature-dependent, it can be used for the determination of surface temperature. However, the reflectivity is a weak function of temperature. Thus, a small error in the measurement of the intensity of light reflection can cause a large uncertainty in the temperature measurement [67]. A different measurement technique [68] can minimize the uncertainty of the noncontact temperature measurement within a range of 0.5°C. In that study, the surface temperature of the measured material is determined by measuring the difference in reflection intensity between a probe beam and a reference beam. A further application of this kind of reflection measurement is employed to investigate the transient optic reflectivity of an He-Ne laser that is reflected from heated thin polysilicon films [69]. A relation between the transient temperature history and the reflection index of thin polysilicon films is established this way. With this information, the remote sensing of temperature variation during the fabrication process of integrated circuits becomes possible. The transient temperature variations in microsecond or picosecond time domains can be measured using this reflection measurement because the reflection index of materials changes as rapidly as the change in temperature.

Being optical techniques, these systems essentially have zero inertia. As no probe is used, they would not influence a flowing fluid whose temperature is being studied. However, they do require a transparent medium with windows to permit a light beam or beams to enter and pass through the medium.

Surface Temperature Measurement

Liquid Crystal.   Many liquid crystals have a wavelength-dependent reflectivity that varies with temperature. The helical structure of a liquid crystal reflects a wider light spectrum wavelength range at higher temperature. Since the reflected light spectrum of the liquid crystal varies with temperature, the characteristic of the detectable color variation also varies with surface temperature when liquid crystals are attached onto a test piece. These characteristics allow liquid crystals to be used to measure the transient temperature variation of a test surface. A brief review of thermochromic liquid crystals is presented in Ref. [70].

In general, the term liquid crystal is used to describe an intermediate phase between liquid and solid occurring in some organic compounds. The phase of liquid crystal can be divided into two mesophases: smectic and nematic. Nematic liquid crystals can be further divided as chiral nematic or archiral nematic. In chiral nematic liquid crystals, sterol-related compounds are called cholesteric, and non-sterol-based compounds are termed chiral nematic. For heat transfer applications, encapsulated forms of chiral nematic [71] or the composite liquid crystal sheets of the cholesteric type [72] are commonly used. Recently, the application of micro-encapsulated liquid crystals has become more popular in heat transfer measurements because of the fast response and easy paintbrush or spray application to the test surface.

Since the temperature on the test surface is determined from the reflection color on the liquid crystal surface, the color sensation for the reflected light from a liquid crystal surface becomes the essential element that affects the uncertainty in temperature or heat transfer measurements. Several parameters affect the color sensation: the spectral characteristics of the incident light that illuminates the liquid crystal surface, the helical structure of the liquid crystal, the incidence and reflection angle of light, and the color capturing device [73].

Figure 16.29 shows a schematic view of a test rig using the thermochromic liquid crystal technique to determine the heat transfer coefficient over the test surface coated with the liquid crystal. A heated air stream is suddenly supplied to the test section, and transient heat transfer takes place between the heated air stream and the test surface coated with the liquid crystal. As the temperature of the liquid crystal surface gradually rises to a value at which the reflection light from the liquid crystal surface changes color, the transient temperature-dependent color pattern reflected from the liquid crystal surface can be recorded by an image capturing device. Through an appropriate analysis, the temperature and heat transfer information over the test surface can be deduced from the recorded luminance signal. Several analytical procedures for obtaining the temperature information from the recorded signal have been proposed. A method to deduce temperature information from the light intensity history

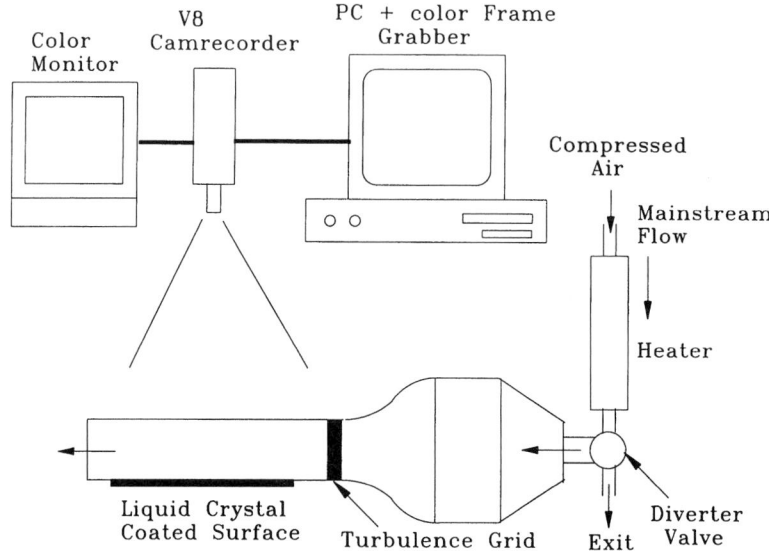

FIGURE 16.29 Schematic view of the liquid-crystal temperature measurements.

of the liquid crystal surface was employed in Ref. 74. A method to determine temperature from the luminance signal of the surface color was used in Ref. 75. A hue-capturing technique for transient heat transfer measurements is described in Ref. 73. This hue-capturing technique was employed in Ref. 76 to determine the heat transfer over curved surfaces on which a mixture of three narrow-band liquid crystals was sprayed [76].

The isothermal lines on the test surface can be evaluated from the measured dominant wavelength distribution reflected from the liquid crystal surface. In a transient test, the surface temperature can be evaluated from the one-dimensional conduction equation in a semi-infinite domain:

$$\frac{\partial^2 T}{\partial x^2} = \alpha_s \frac{\partial T}{\partial \theta} \qquad (16.57)$$

which is subject to the initial condition

$$\text{at } \theta = 0, \, T = T_i \qquad (16.58)$$

and the boundary conditions

$$\text{at } x = 0, \, -k \frac{\partial T}{\partial x} = h(T_{re} - T_w); \text{ at } x, \, T = T_i \qquad (16.59)$$

When the given heat conduction equation is solved, the temperature on the surface coated with liquid crystal is obtained as

$$\frac{T_w - T_i}{T_{re} - T_i} = 1 - \exp\left[\frac{h^2\alpha_s\theta}{k^2}\right]\text{erfc}\left[\frac{h\sqrt{\alpha_s\theta}}{k}\right] \qquad (16.60)$$

where α_s and k, respectively, denote the thermal diffusivity and conductivity of the liquid crystal; and T_w, T_i, and T_{re} denote the wall temperature, the initial temperature, and the reference fluid temperature for driving the convection heat transfer, respectively. However, it is physically impossible to provide a step temperature rise in the fluid flow. Therefore, Duhamel's principle can be applied to obtain the wall temperature with a recorded temperature rise, shown as

$$T_w - T_i = \sum_{j=1}^{N}\left\{1 - \exp\left[\frac{h^2\alpha_s(\theta - \theta_j)}{k^2}\right]\text{erfc}\left[\frac{h\sqrt{\alpha_s(\theta - \theta_j)}}{k}\right]\right\}(T_{re,j} - T_i) \qquad (16.61)$$

where $T_{re,j}$ represents the time-dependent reference fluid temperature at θ_j.

In order to use the surface color to determine the temperature, one needs to know only the response time θ for the surface temperature T_w to reach the threshold value at which the liquid crystal surface changes color. Once the response time θ for the surface color change is known, the only remaining unknown in Eq. 16.61 is the heat transfer coefficient h. However, the local intensity of illumination may vary with location on the test surface due to the incidence angle of the incoming light, or it may vary with the difference in the lighting intensity between calibration and actual experiment. This may cause some error in the determination of the response time θ from the reflection intensity from one single color. Consequently, an error in determining the local heat transfer coefficient results from the uncertainty in the evaluation of the surface temperature from the light intensity of a single color, as shown in Eq. 16.61. To overcome this problem, two schemes have been proposed to calculate the local heat transfer coefficient. One is to correlate the hue of luminance signal from the liquid crystal surface with the surface temperature [73], and the other is to determine the local heat transfer coefficient ratio directly by matching the full local intensity histories at two particular locations on the test surface [77].

To represent the color, three different colorimetric coordinate systems can be used for the signals: three primary colors (red, green, and blue); hue, saturation, and intensity (H, S, I)

[73]; or intensity and two different color signals (I, B-I, R-I) [71]. The values of H, S, and I can be calculated from the R, G, and B signals. For the NTSC (National Television Standards Committee) broadcasting standards, the luminance (or intensity) signal **I** can be expressed as

$$\mathbf{I} = \mathbf{I}_l * \int_{-\infty}^{\infty} e(\lambda)R_s(\lambda,\, T)(0.299r(\lambda) + 0.587g(\lambda) + 0.144b(\lambda))\, d\lambda \qquad (16.62)$$

where $\mathbf{I}_l$ is the local lighting intensity, $e(\lambda)$ denotes the normalized illumination spectrum, $R_s(\lambda,\, T)$ denotes the surface reflectance, and $r(\lambda)$, $g(\lambda)$, and $b(\lambda)$ represent the filter function for the red, green, and blue signals, respectively. A detailed discussion of the use of luminance time history for each pixel of the recorded image is described in Ref. 77. In addition, another approach for using the hue information for the evaluation of the surface temperature is presented in Ref. 73.

For liquid crystal heat transfer measurements, the calibration process requires matching the values for the reflection intensity or hue from the liquid crystal surface with the temperature using the same illuminating system and color-capturing device. Calibrations between three different hue definitions with the temperature have been performed and reported in Ref. 70. The definition based on the RGB triangle is the best among the three hue definitions tested in this study. It is worthwhile to note that the reflection wavelength is dependent on the incident and reflective angle of the illuminating light. Therefore, the view angle of the color-capturing device and the incident angle of the illuminating light should be the same for both the calibration and measuring processes.

Melting of Surface Coating.   A thin layer of a coating material with a low-temperature melting point is used to coat the surface to be investigated. Some materials with a low melting point, near 40°C, are commercially available [78]. When a heated air flow is suddenly supplied into the test section, the wall temperature is obtained from Eq. 16.60. As the heat flux from the heated fluid flow raises the test surface temperature above the melting point of the coating material, the color of the coating material changes and is recorded by an image-capturing device such as a CCD camera. Then, at any surface point, the heat transfer coefficient can be determined by measuring the required time for the wall to reach the phase-change temperature. For a fast data acquisition system, a motor-driven camera is employed to record the melting pattern [79]. Note that the coating material should be thin enough to avoid long thermal response time in the material itself.

Other Thermometers.   Among the many other types of thermometers, we will briefly discuss the following: bimetallic thermometers, noise thermometers, resonant-frequency thermometers, and semiconductor diode thermometers.

Bimetallic thermometers measure temperature by the change in physical dimension of the sensor. They have often been used [80]. The sensor consists of a composite strip of material, normally in a helical shape, formed by two different metals. Differences in the thermal expansion of the two metals cause the curvature of the strip to be a function of temperature. The strip is used as a temperature indicator with a self-contained scale. Bimetallic thermometers have been used at temperatures from −185 to 425°C. The precision and accuracy of such thermometers are described in Ref. 81.

With noise thermometers, the Johnson noise fluctuation generated in a resistor is the temperature indicator [82]. The random noise fluctuation in a frequency band is related to the temperature and resistance of the resistor, thus ideally enabling the noise thermometer to measure absolute temperature. Noise thermometers have a wide temperature range, from cryogenic temperatures up to about 1500°C. They have a long integration time—between 1 and 10 h for thermodynamic temperature measurements and on the order of seconds for an industrial sensor. Problems encountered in their use include noise of nonthermal origin and the stability of the measuring instrumentation during integration.

In a resonant-frequency thermometer, the resonance frequency of the medium serves as the temperature indicator. Included in this category are nuclear quadrupole resonance thermometers, quartz thermometers, and ultrasonic thermometers. These thermometers usually come complete with their frequency measuring (counting) device. The improvement in modern frequency counting allows high-precision measurements.

Nuclear quadrupole resonance thermometers [83] can be used between 20 and 400 K. In the ultralow temperature range below 1 K, a direct measurement of temperature is feasible by using the spectrometer to measure the intensity ratio of magnetic resonance lines [84]. A precision and accuracy of 1 mK can be achieved. However, these thermometers are quite expensive and are therefore used most often as transfer standards.

For a quartz thermometer, the resonant frequency of a quartz crystal resonator is strongly related to the temperature variation. With high resolution, the temperature change can be directly determined from the frequency change of a quartz crystal thermometer. A quartz thermometer developed for use between −80 and 250°C [85] has a resolution of 0.1 mK. If used at the same temperature, a comparable precision can be achieved. However, with temperature cycling, hysteresis can reduce its repeatability. An accuracy of 0.05°C can be achieved with calibration. Nevertheless, the temperature resolution for the quartz resonator is found to be less accurate at lower temperatures. Over the temperature range from 4.2 to 400 K, the temperature resolution with the resonant frequency change for a YS cut quartz crystal thermometer drops from 1 kHz/K at 300 K to 80 Hz/K at 4.2 K [86].

Ultrasonic thermometers [87] measure the temperature based on sound velocity. They are mainly used for measurements at high temperatures (300 up to 2000 K).

When a PN-junction diode is subjected to a constant forward bias current, its junction voltage is inversely proportional to the absolute temperature [88]. This is the physical basis for semiconductor diode thermometers. They can be used from liquid helium temperature to 200°C. Germanium, silicon, and gallium arsenide diodes have been used as thermometers. Due to the difficulty in maintaining a uniform manufacturing process, the interchangeability of diode thermometers is a significant problem. This variation from unit to unit can be accommodated by adjustment in the signal conditioning circuit, which normally comes as part of the thermometer.

Local Temperature Measurement

Introduction.   In choosing a thermometer, consideration must be given to the specific environment in which the temperature is being measured (see Ref. 89). Many measurements are taken in systems in which the temperature varies with position and perhaps also with time. The presence of a sensor and/or its connecting leads can change the temperature field being studied. Factors that affect the suitability and accuracy of a particular thermometer for making a temperature measurement include the size and the physical characteristics (thermal properties) of the sensing element.

Small size is important if there are steep gradients of temperature in the medium being studied. Even with a small sensor such as a thermocouple junction, conduction along the lead wires can affect the temperature of the junction. The leads near a sensor should pass, when possible, through an isothermal region. With a thermometer well, conduction along the well should also be considered.

Steady Temperature Measurement in Solids.   In measuring internal temperatures of a solid, the leads to the temperature sensor should follow an isothermal path. In addition, there should be good thermal contact between the sensor and the surrounding solid. This can be provided by a thermally conducting paste or, in a thermometer well, a liquid.

Often, the knowledge of a solid's surface temperature is required. Figure 16.30 [33] shows common methods of thermocouple attachment to surfaces. The junction can be held close to the surface by soldering, welding, using contacting cement, or simply by applying pressure. The lead wires should be held in good contact with an isothermal portion of the surface for the order of 20 wire diameters to avoid lead conduction errors. Temperatures can also be measured at different locations below the surface, and extrapolation used to determine the temperature at the surface.

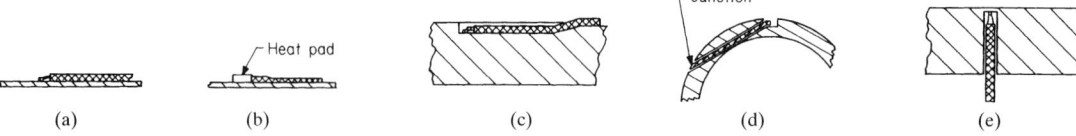

FIGURE 16.30 Common method of thermocouple attachment to the surface [36]: (*a*) junction mounted directly on surface, (*b*) junction in heating collecting pad, (*c*) junction mounted in groove, (*d*) junction mounted in chordal hole in tube wall, and (*e*) junction mounted from rear of surface.

Temperature Measurement in Fluids. The definitions of temperature and temperature scale come from thermodynamic considerations; the meaning of temperature is strictly valid only for a state of thermodynamic equilibrium. In most real situations, nonuniformities are present. We generally assume the fluid is in local thermodynamic equilibrium and the properties of the fluid are interrelated through the material's equilibrium equation of state. The local temperature can be determined from a probe that is small compared to the region over which the temperature varies by any significant amount. Problems arise in special cases such as rarefied gases as well as regions with very large gradients, as in shock waves.

A temperature-sensing element measures its own temperature. Ideally, it is in thermal equilibrium with the surrounding region and does not affect the local temperature because of its presence. Often this ideal situation is not met, and factors that must be considered include heat transfer to or from the sensor by radiation and conduction, conversion of kinetic energy into internal energy in a flow surrounding the probe, and convective heat transfer from the surrounding medium, including that due to a temperature variation in temperature.

An energy balance can be made on the temperature sensor to take into account the energy flows to and from it. This could indicate that its temperature is the same as that of the surrounding medium or that corrections should be made. For example, the principal heat transfer mechanisms might be convection from the adjacent fluid, conduction along the leads, and thermal radiation transport. If the conduction and radiation heat flows are small and/or the sensor is in good thermal contact (high heat transfer coefficient) with the immediately surrounding fluid, the sensor's temperature can closely approximate that of the adjacent fluid.

Radiation shields [90] can be used to isolate a probe from a distant medium so that there will be relatively little radiation heat transfer to it; at the same time, they do not interfere with good thermal contact between the probe and the surrounding fluid. Designs of thermometer probes for gas temperature measurement are described in Refs. 91 and 92. Analyses to account for some uncertainties in probe measurement can be found in Ref. 93.

When a probe is immersed in a flowing fluid, the flow comes to rest in the immediate vicinity of the probe. In this deceleration, kinetic energy is converted into internal energy, which can significantly increase the fluid temperature. Although this change in temperature is generally small in liquid flows, it can be significant in gas flows. The total and static temperatures of a gas with velocity V and (constant) specific heat c_p are related by the equation

$$T_t = T_{st} + \frac{V^2}{2c_p} \tag{16.63}$$

The static temperature T_{st} would be observed by a thermometer moving along with the flow, while the total temperature T_t would be attained by the fluid following adiabatic conversion of the kinetic energy into internal energy. As a flow is brought to rest at a real probe, the temperature generally is not equal to the total temperature other than in an idealized case or in specially designed probes [90, 91]. Often, as a result of dissipative processes (conduction, viscosity), the temperature is some value less than T_t. The temperature at an adiabatic surface is called the recovery temperature T_r, which is

$$T_r = T_{st} + rT_d \tag{16.64}$$

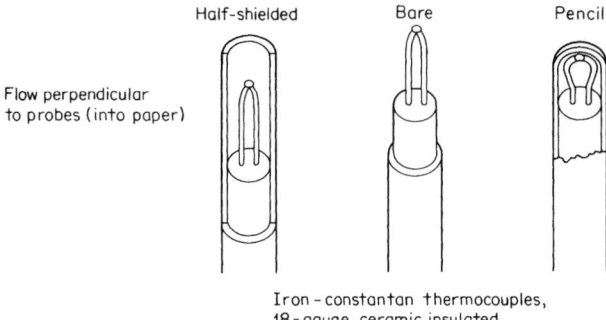

FIGURE 16.31 Thermocouple probes for measuring temperatures of flowing fluids.

where r is the recovery factor, and T_d is the dynamic temperature $T_t - T_{st}$. With a laminar boundary layer flowing along a flat surface parallel to the flow, the recovery factor is equal to the square root of the Prandtl number [90]. In general, the recovery factor around a temperature probe is not uniform and often must be measured if accurate results are required.

Specific values of the recovery (dynamic correction) factors for the three probe geometries indicated in Fig. 16.31 have been measured [94]. The results are reproduced in Fig. 16.32. Note that the "half-shielded" probe has a relatively constant recovery factor of about 0.96 over the range of conditions studied.

Transient Temperature Measurements. The response of a thermometer to a change in the temperature of its environment depends on the physical properties of the sensor and the dynamic properties of the surrounding environment. A standard approach is to determine a time constant for the probe, assuming the probe is small enough or its conductivity is high enough that, as a first approximation, the temperature within the probe is uniform. Neglecting radiation and lead-wire conduction, the increase in energy stored within the probe would be equal to the net heat transfer convected into it:

$$mc_p \frac{dT}{d\theta} = hA(T_f - T) \qquad (16.65)$$

where m, c_p, and A are mass, specific heat, and surface area of the sensor, respectively; T and T_f are the temperatures of the sensor and the surrounding material (often a fluid), respectively; and h is the heat transfer coefficient between the sensor and its surroundings. A time constant θ_c can be defined in the form

$$\theta_c = \frac{mc_p}{hA} = \frac{\text{thermal capacitance of the sensor}}{\text{thermal conductance of the fluid}} \qquad (16.66)$$

The response of such a sensor has been analyzed for a number of fluid temperature transients—in particular, a step change, a ramp change (linear increase with time), and a periodic change. For a step change in fluid temperature, θ_c is the time for the sensor to have changed its temperature reading T so that $(T_f - T)$ is equal to e^{-1} times the original temperature difference. For ramp and periodic changes, the time constant is (after the initial transient has faded) the time the sensor lags its environment. Examples of error estimates for transient temperature measurement in solids can be found in Refs. 95 to 98. Reference 93 provides an analysis of sensor response to temperature transients in unsteady fluid flows.

The above analyses provide error estimates in measurement of time-varying temperatures. However, the best practice is to reduce the potential lag by minimizing the time constant, usu-

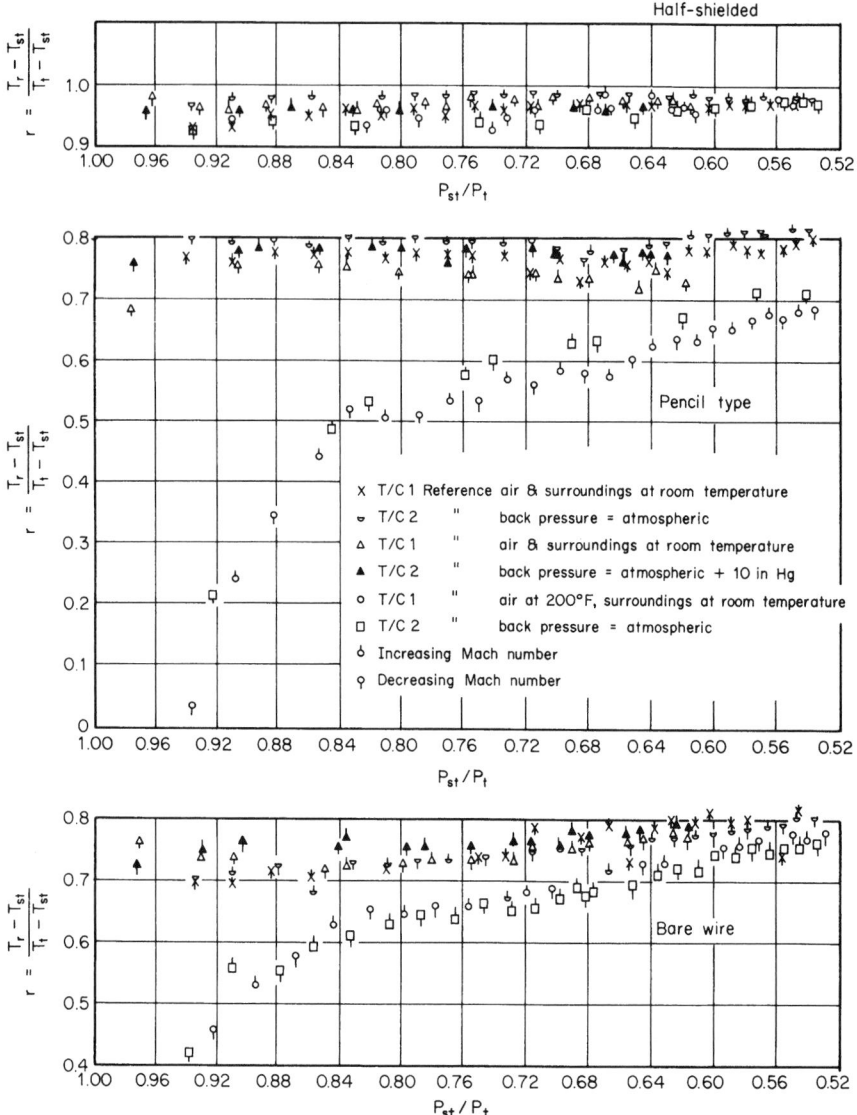

FIGURE 16.32 Dynamic correction fractors for temperature measurement in air flow [94].

ally by having a small sensor. One can also sometimes reduce the thermal resistance (i.e., increase h) to the surroundings.

Calibration of Thermometers and Assurance of Measurements

Basic Concepts and Definitions. The task of relating thermometer output (i.e., magnitude of the variable dependent on temperature) to its temperature is achieved through calibration. Two general means of calibration are available: (1) fixed-point calibration, and (2) comparison calibration. ITS-90 provides 14 defining fixed points to which unique values of temperature are assigned (Table 16.2). These points are used for fixed-point calibration of defining standard thermometers and could also be used, though it is seldom done, in calibration of other standard thermometers.

The second means of calibration compares the output of the sensor to be calibrated to that of a standard thermometer, and it is called *comparison calibration.* The temperature of the sensor and of the standard thermometer at the time of calibration should be identical. This temperature, called the *calibration temperature,* does not have to be a fixed point; it can be any temperature as long as it can be determined by the standard thermometer.

Any experimental procedure contains uncertainties, and an error analysis is essential to attach significance to the results. The calibration process is no exception. The reliability of a particular temperature sensor is only assured after the calibration and accompanying error estimate are completed.

The terms *error* and *uncertainty* quantify the accuracy and precision of a sensor. *Error* usually indicates the difference between the measured and the actual values. *Uncertainty* normally refers to the inability to pinpoint the value; it is the range of readings obtained when repeatedly measuring the same temperature.

The terms *accuracy, precision* (or *repeatability*), *stability* (or *drift*), *hysteresis, dead-band,* and *interchangeability* all relate to the reliability of temperature measurements. *Accuracy* indicates the closeness of a thermometer's reading to the actual ITS-90 value. The terms *precision* or *repeatability* indicate the ability of a thermometer to reproduce the same readings at the same temperature or the ability to reproduce the same calibration [99]. *Stability* is a measure of how well a thermometer could maintain the same reading for a given condition over a period of time, while *drift* is a measure of the degree of departure from the original reading over time. *Hysteresis* is a description of the different readings from a thermometer when its temperature is raised to the desired condition versus that when its temperature is lowered to the desired condition. The term *dead-band* is a measure of the extent of temperature change needed to induce a change in the thermometer's output. Finally, *interchangeability* is a measure of how closely a replacement thermometer's readings would match a similar thermometer's without recalibration.

For a themometer used in the field, its reliability can be achieved by the traceability requirement. Traceability is a term describing the link between the accuracy of the thermometer's readings to the ITS-90 temperature scale maintained in a national laboratory. In a calibration or testing laboratory, traceability of results (be it thermometer calibrations or temperature measurements) can be achieved through a laboratory accreditation program. This will be described in more detail later.

Choosing a Temperature Measuring System.   As some previous examples have indicated, actual application first involves the choice of a temperature measuring system, which includes: (1) a sensor that converts the temperature at the location of the sensor to some other physical quantities (e.g., EMFs), (2) an electrical system that transmits the EMFs to a signal processor, (3) a signal processor that converts the EMFs back to corresponding temperatures, and (4) a recorder that stores temperature readings for future analysis.

One should consider the following when choosing a temperature sensor:

1. The spatial variation of the temperature field to be studied, which determines the size and position of the sensors that are best suited for the situation.

2. The transient characteristics of the temperature variation at the measurement location, which determine the time response needed. Too large a time constant would lead to a failure to pick up the temperature fluctuations, while too small a time constant could lead to excessive noise, which is also undesirable.

3. If surface temperature measurements are needed, then the means of attaching the sensor to the surface is also a critical factor in the accuracy of measurements.

In addition to the sensor, the remainder of the measurement system must be chosen. Considerations include:

1. What funds are available?
2. Are the temperature readings for reference only or used as control variables?
3. Is automatic data gathering needed, or will manual recording suffice?
4. When an automatic system is planned, is a centrally located control room a requirement, or will local data-loggers suffice?
5. How is the system to be maintained and serviced?
6. Who is going to perform the actual tasks? Is he/she qualified or properly trained?

In summary, the following three factors are needed before the accuracy of actual temperature measurements can be assured:

1. The accuracy of the temperature measuring system
2. The environmental influence resulting from installation of the sensor
3. The quality system that is associated with the measurement process, and the reporting of the results

The accuracy of the temperature measuring system is achieved through calibration, which will be discussed next.

Calibration Systems and Procedures. In calibrating a temperature sensor, the sensor should be maintained at a known temperature while its output signal (e.g., EMF or resistance) is measured. The following description is only meant to be a brief introduction. Interested readers are referred to Refs. 7, 9, 25, 40, 52, and 100.

Fixed-point calibration has the highest accuracy, but to achieve such accuracy requires great precautions and is very time-consuming. Most freezing- and triple-point cells are difficult to maintain and usually cannot accommodate more than one sensor at a time. Thus, fixed-point calibrations are usually carried out only for high-precision thermometers and usually only at national laboratories. However, easily maintained water triple-point (or ice-point) systems are used in many laboratories to correct thermometer drift, particularly for SPRTs.

Constant-temperature baths (or furnaces) and defining standard thermometers (such as SPRTs that were previously calibrated by a national laboratory) or a transfer standard thermometer are generally used when calibrating temperature measuring devices for laboratory experiments. In an industrial setting, when calibration of the thermometers used for actual production measurement is required, comparison calibration against a working standard thermometer is used. The working standard is usually the same type of thermometer as the thermometers to be calibrated to reduce the instrumentation requirements; this standard should have been previously calibrated against a defining or transfer standard thermometer.

The defining standard thermometers are thermometers specified for ITS-90. They should be calibrated by NIST at regular time intervals, and are then used to calibrate transfer or working standard thermometers. Transfer standard thermometers serve as intermediate standards to reduce the use and drift of the defining standard thermometers.

Constant-temperature baths or furnaces are needed to maintain the uniform temperature environment for comparison calibration. Stirred-liquid baths and temperature controllers have good characteristics as constant-temperature media due to their ability to maintain temperature uniformity [100]. The liquids used include refrigerants, water, oils, molten tin, and molten salt. Water can be used for temperatures between 0 and 100°C and oils above 300°C. Refrigeration units are available for commercial constant-temperature baths with alcohol as the working fluid and temperatures as low as −80°C. At NIST, special cryostats [8, 25] with liquid nitrogen or liquid helium as the coolant are used for comparison calibration at low temperatures. A molten tin bath is used at NIST for calibration between 315 and 540°C. A molten salt bath [101] can be used for calibration in a temperature range similar to that for the tin baths. Air-fluidized solid baths for calibration up to 1100°C have been developed [102]. Gas furnaces are used for high-temperature calibrations. A furnace has been used for calibration of high-temperature thermocouples up to 2200°C [103]. However, the temperature in a furnace is not as uniform as that in a liquid bath because of the relatively poor heat transfer characteristics of gases. To ensure uniformity, a copper block is sometimes used as a holder for the thermometer to be calibrated and the standard thermometer. The copper block can be partially insulated from the surroundings to reduce the effect of small fluctuations in the surrounding temperature, and the positions of the two thermometers are interchanged to eliminate the influence of spatial variations in temperature.

A typical procedure for comparison calibration of a thermometer is as follows. First, the temperature range of operation for the thermometer is decided. Then, an interpolation formula calibration equation is selected, usually a polynomial. Third, a number of calibration temperatures are chosen; these are distributed over the desired temperature range of calibration. For comparison calibration, the number of calibration points, usually at regular temperature intervals, is greater than the unknown coefficients in the interpolation formulae to enable least-squares determination. The fourth step is to measure the thermometer output while it is maintained at each of the calibration temperatures and while the actual temperature is determined with a standard thermometer maintained in the same temperature environment. From the thermometer outputs and the corresponding temperatures, the unknown coefficients in the calibration equation are determined.

Although the thermometer can be calibrated separately from the measuring instrument (e.g., a digital voltmeter for the measurement of EMF from a thermocouple) that is used to read the thermometer output, generally the measuring instrument and thermometer are a unit and should be calibrated together. With the advent of personal computers and multichannel data-loggers plus A/D converters, the whole data collection path should be calibrated together as a system.

Error Analysis and Measurement Assurance.   Sources of error in a calibration include: (1) difficulty in maintaining the fixed points, (2) accuracy of the standard thermometer, (3) uniformity of the constant temperature medium, (4) accuracy in the signal-reading instrument used, (5) stability of each of the components, (6) hysteresis effects, (7) interpolation uncertainty, and (8) operator error. Techniques for error analysis are described in a number of papers on experimental measurement [104, 105].

Reducing errors to attain high accuracy is difficult. Today, the highest accuracy for the realization of ITS-90 is maintained in a handful of laboratories, mainly National Standard Laboratories, by the Measurement Assurance Program (MAP) [106]. Through the MAP program, SPRTs circulate among a number of standards laboratories, each of which recalibrates the thermometers and then returns them to NIST for another recalibration. This procedure permits the detection of any drift in the whole calibration system at the participating laboratories.

The accuracy of a thermometer used in the field can be assured through the traceability requirement. *Traceability* is a term describing the link between the accuracy of thermometer readings to a national laboratory that is in the MAP program [106], which would have proven accuracy in the realization of ITS-90. Since it is usually impractical to send the actual thermometer to, say, NIST, the usual procedure is to perform a comparison calibration against a working standard thermometer, which is then sent to a calibration laboratory that has traceability to NIST. At NIST, the working standard thermometer is often still calibrated by comparison against a transfer standard thermometer (which may or may not be an SPRT).

Most temperature calibration laboratories do not participate in the MAP program. The traceability of measurements can best be assured through laboratory accreditation, which is the subject of the next section.

Laboratory Accreditation—ISO Guide 25.   Accreditation is a process that is administered by an organization (the Accreditation Body) to verify through the use of preapproved auditors that the applicant satisfies the minimum requirements set forth. The accreditation of laboratories is governed by an ISO standard, namely, the ISO Guide 25:1990, "General Requirements for the Competence of Calibration and Testing Laboratories" [108]. This standard requires the laboratory to have a quality system plus a proof of competency for the measurements. To put it simply, the laboratory has to preannounce its capability and then prove it to an external auditor (i.e., it may provide more than what a particular customer may need), and, of course, there is a price attached. The higher the accuracy, usually the higher the cost.

Each country has its own accreditation body for laboratories. In the United States, NIST is administering the National Voluntary Laboratory Accreditation Program (NVLAP) [109]. The criteria for accreditation used by NVLAP is consistent with ISO Guide 25, which involves the following items: (1) quality system, (2) staff competence and training, (3) facilities and equipment, (4) calibration and traceability, (5) test methods and procedures, (6) recordkeeping, and (7) test reports.

To satisfy the requirements of ISO Guide 25, the laboratory must first have answers to the following questions:

1. What are the standard operating procedures (SOPs) of the laboratory?
2. How is the accuracy of results assured?
3. Who is in charge of the laboratory? Does he or she have the full support of management to ensure that the laboratory will function independently, without financial concerns or administrative interference?
4. What is the organizational structure? Are the different job assignments adequately staffed? And are the interfaces of the various tasks properly defined and explained to all concerned?
5. Has the staff been properly trained for the job?
6. What happens when there are problems or customers complain?
7. When happens when the equipment is found to deviate?
8. Is there a plan for continuous improvement of service?

After a laboratory in the United States completes its preparation, it can then apply for accreditation to NVLAP. Auditors will be sent by NVLAP to verify that the laboratory satisfies the minimum requirements set forth in the ISO Guide 25.

To put all of the above in terms that may be easier to remember, one needs to consider these three components that make a laboratory work with consistency:

1. A program to ensure the accuracy of the measurements is maintained.
2. A documentation control system that keeps track of all procedures, records, and reports.
3. A personnel competency training program to match the staff's skill to the tasks.

HEAT FLUX MEASUREMENT

Basic Principles

Unlike temperature, heat is not a thermodynamic property. The definition of heat comes from the first law of thermodynamics, which, for a closed, fixed-mass system, can be written

$$Q = W + \Delta U \tag{16.67}$$

The term Q was introduced into this energy balance when considering the interaction of a closed system with a surrounding that is not in thermal equilibrium (not at the same temperature) with the system itself. Heat can thus be defined as energy transported between two systems due to the temperature difference between the two systems. Note that, in Eq. 16.67, Q is the heat added to the system and W is the work done by the system. From the second law of thermodynamics, we know that the heat transferred between two systems goes from the one with the higher temperature to the one with the lower temperature.

Measurements of heat transfer are usually based on one or more of a small number of physical laws. For many measurements, the first law, Eq. 16.67, is used directly. This is true for some heat flux gauges, such as transient ones that normally have $W = 0$, where the measurement of ΔU would then give Q. Some experimental systems achieve a constant heat flux boundary condition by electric heating. Then the energy input is a work term in the form of electrical energy. If a steady state is achieved, there is no change in internal energy, and the work (electrical energy) added is equal to the heat transferred from the electric element.

With an open system, the first law for steady flow through a control volume with one exit and one entrance can be written in simplified form to obtain the net heat flow through the boundaries of the control volume:

$$\dot{Q} = \dot{W} + \dot{m}(i_{\text{exit}} - i_{\text{ent}}) \tag{16.68}$$

where i is enthalpy per unit mass. This equation is often used to determine the net heat (or energy) flow to a fluid passing through some system, such as heat exchanger, nuclear reactor, or test apparatus.

Some heat flow measuring instruments are based on Fourier's law. For a one-dimensional system, Fourier's law in rectangular coordinates is

$$q = \frac{\dot{Q}}{A} = -k \frac{\partial T}{\partial x} \tag{16.69}$$

Thus, the first law of thermodynamics provides the definition of heat flow, and it, along with Fourier's law, provides the basis for most instrumentation used to measure heat transfer or heat flux. Specific systems using these laws are described below.

Methods

Introduction. Most heat-flux-measuring devices can be placed into one of four categories [110]: (1) the thermal resistance type, where measurement of temperature drop gives an indication of the heat flux from Fourier's law (e.g., the sandwich type or circular foil type); (2) the thermal capacitance or calorimetric type, where a heat balance on the device yields the desired heat flow (e.g., the wall-heating (cooling) type or temperature-transient type); (3) the energy input or output type, where a direct measurement of the energy input or output on the heat flux instrument is required; and (4) a temperature gradient is measured in the fluid adjacent to the test surface. For reliable results, heat-flux-measuring systems should be calibrated before use.

With a thermal resistance heat flux sensor, the presence of the instrument in the environment will disturb the temperature field somewhat and introduce an error in the measurement. Wall-heating systems require a heat source (or sink) and an appropriate heat balance equation to determine the heat flux. The temperature-transient types require a measurement of the temperature variation with time. The energy input or output types require good control or measurement of the temperature of the heat flux instrument. For the fourth type, the properties of the fluid are required. A brief discussion of different types of heat flux sensors is given below.

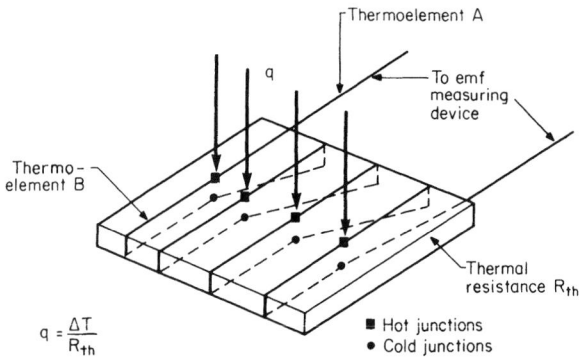

FIGURE 16.33   Sandwich-type heat flux gauge.

Thermal Resistance Gauges

Sandwich Type.   When heat is conducted through a thin slab of material (a thermal resistance), there is a temperature drop across the material. A measurement of the temperature difference across the slab can be used as a direct indication of the heat flux [111]. A schematic of a sandwich-type gauge is shown in Fig. 16.33. The steady one-dimensional heat flux through the material is related to the temperature difference ΔT by

$$q = \frac{k}{\delta} \Delta T \tag{16.70}$$

or

$$q = \frac{\Delta T}{R_{th}} \tag{16.71}$$

where k and δ are the thermal conductivity and thickness of the material, respectively, and R_{th} is the thermal resistance δ/k. Often, k and δ are difficult to measure directly, and, thus, thermal resistance R_{th} is determined from a calibration of the heat flux gauge. Even for a finite-size system where the temperature and/or R_{th} are not perfectly uniform along the gauge, Eq. 16.71 can still be used with R_{th} determined by calibration.

The temperature drop across the thermal resistance is usually measured with a multijunction thermocouple (thermopile) to increase the sensitivity of the device. The sensitivity of a heat flux sensor depends on the slab material and slab thickness, which essentially determine R_{th}, and the number of junctions in the thermopile, which determines the output EMF as a function of ΔT.

Heat flux sensors having a large sensitivity (output per unit heat flux) generally have a low maximum allowable heat flux and relatively large physical size (thus slower response time). For example, one commercial unit with a sensitivity of 60 µV per W/m² has a response time of a few seconds [112]; whereas a unit that can accommodate heat fluxes up to 6×10^5 W/m² has a sensitivity of 0.006 µV per W/m² and a response time of less than one second [113].

Circular Foil Heat Flux Gauge.   The instruments (often called Gardon heat flux gauges [114]) shown in Fig. 16.34 are also based on Fourier's law. A copper heat sink is installed in the wall of the measuring site with a thin constantan disc mounted over it. A small copper wire is attached to the center of the constantan disc. Another copper wire attached to the copper heat sink completes a thermocouple circuit. Heat flow to the constantan disc is conducted radially outward to the copper heat sink, creating a temperature difference between the center and edge of the disc. The copper and constantan (other materials could be used) act as a thermocouple pair to measure this temperature difference.

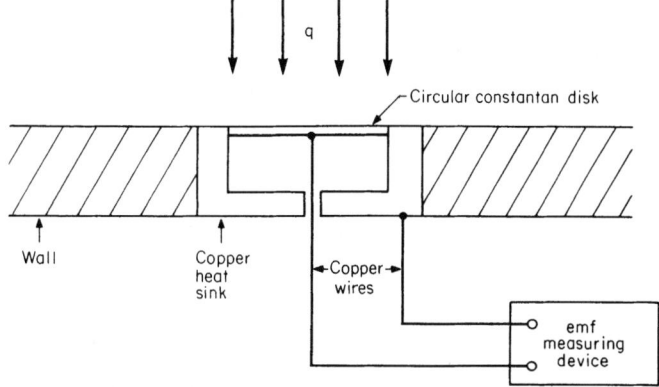

FIGURE 16.34   Gardon-type heat flux gauge.

Such gauges are used for measurement of convection and/or radiation heat fluxes. Although ideally the relationship between the heat flux and sensor temperature difference can be analyzed, calibration of the gauge is almost always necessary. More information is available in Refs. 115 and 116.

Calorimetric Gauges

Wall-Heating (Cooling) Type.   A wall-heating system employing a heat source (or heat sink) can be used not only to measure the heat flow from a surface but also to control the thermal boundary conditions. Two distinct types of systems can be considered: one, where the heat source or sink is behind the wall to which heat is being transferred; and the other, where the system is placed directly on the surface itself. The first type has often been used to obtain the heat flux. The rear of a wall can be sectioned off into individual regions where a forced liquid flow [117] or condensing vapor is used to measure the local heat flow. The mass flow and change of enthalpy of the fluid flowing to the rear face of the wall are used (Eq. 16.68) to determine the heat input to each individual region of the front surface. Other systems measure the total heat loss from the surface by having a single fluid region behind the wall and measuring the mass flow and enthalpy change to that fluid. Sometimes the temperature drop across the wall, and a calculated thermal resistance can be used as a check on the total heat flow.

For some systems, heat input to different regions of the surface is adjusted to obtain a specific boundary condition—often to approximate an isothermal wall. This is most easily done with a number of small heating elements that can be individually adjusted to maintain a constant wall temperature. The local heat flux, or at least the heat flux averaged over the size of each individual heater, can be determined from the power input to the heaters. These heaters can be placed quite close to the surface.

If a constant heat flux boundary condition is required, an electrical heating element, often a thin, metallic foil, can be stretched over an insulated wall. The uniform heat flux is obtained by Joule heating. If the wall is well insulated, then, under steady-state conditions, all of the energy input to the foil goes to the fluid flowing over the wall. Thermocouples attached to the wall beneath the heater can be used to measure local surface temperature. From the energy dissipation per unit time and area, the local surface temperature, and the fluid temperature, the convective heat transfer coefficient can be determined. Corrections to the total heat flow (e.g., due to radiation heat transfer or wall conduction) may have to be made.

Temperature-Transient Type.   With a temperature-transient gauge, the time history of temperature (an indicator of the change in internal energy) is used to determine the heat flux. Assuming two-dimensional (rectangular coordinates) heat flow, the governing equation for the temperature within a homogeneous gauge is

$$\alpha_s\left(\frac{\partial^2 T}{\partial x^2} + \frac{\partial^2 T}{\partial y^2}\right) = \frac{\partial T}{\partial \theta} \tag{16.72}$$

where α_s is the thermal diffusivity. Equation 16.72 can readily be solved to give a relationship between the temperature and the heat flux to the surface.

Various temperature-transient-type gauges are available [118]. They can be categorized by the boundary conditions used: (1) the semi-infinite type, where the heat flux is derived from the solution of Eq. 16.72 for heat flow into a semi-infinite conductor; and (2) the finite-thickness type, where the wall beneath the gauge is assumed to be insulated and the heat flux is determined by the temporal variation of the (assumed uniform) sensor temperature.

The solution to Eq. 16.72 for transient conduction in a semi-infinite solid is utilized for the design of the thin-film gauge. Such a thin-film heat flux gauge consists of a thin—usually metallic, often platinum—film attached on the wall to (from) which the heat flows. The thin-film platinum resistance thermometer is fabricated using a semiconductor manufacturing process such as vapor deposition or painting. To construct the thin-film heat flux gauge, the platinum is painted on silicon substrate, and a protective coating is applied to the gauge using a vapor deposition method. For a typical thin-film heat flux gauge, the thickness of platinum is around 1000 Å, and the response time is at an order of magnitude of 10^{-8} sec [119].

However, Eq. 16.72 may involve some errors when a long-period calibration is performed on a rectangular thin-film heat flux gauge, shown in Fig. 16.35. Thus, a three-dimensional heat diffusion equation should be used. The boundary and initial conditions for the rectangular heat flux gauge are expressed as:

at $\theta = 0$, $T = T_i$

at $x = \pm b$ for $-a < y < a$ and $y = \pm a$ for $-b < x < b$; adiabatic

at $\theta > 0$ for $-a < y < a$ and $-b < x < b$, $q \neq 0$ \hfill (16.73)

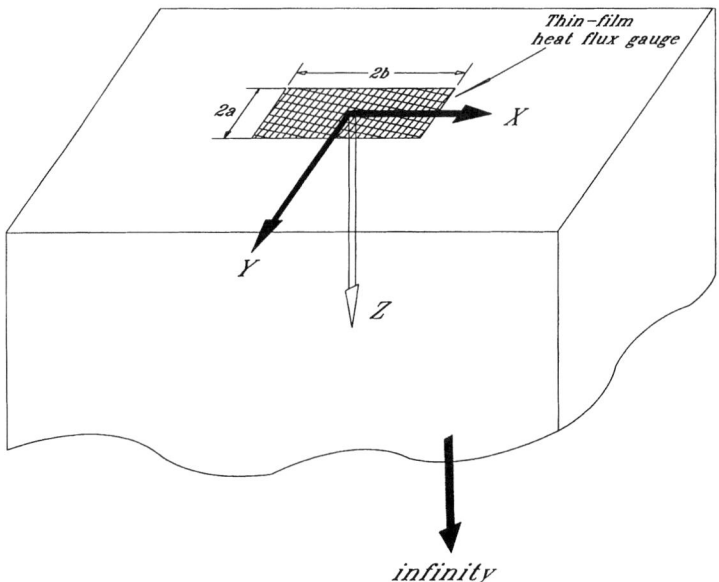

FIGURE 16.35 Geometry for determining the temperature response of a thin-film heat flux gauge attached on an infinitely long surface [120].

By solving the three-dimensional heat diffusion equation on the surface at $x = 0$, the average temperature for the rectangular heat flux gauge with an area of $4ab$ [120] is

$$\overline{T} = T_i + (q\alpha_s/abk) \int_0^\theta \{a - (\alpha_s(\theta - \xi))^{1/2}[(1/\pi)^{1/2} - i \operatorname{erfc} (a/(\alpha_s(\theta - \xi))^{1/2})]\}$$

$$\times \{b - (\alpha_s(\theta - \xi))^{1/2}[(1/\pi)^{1/2} - i \operatorname{erfc} (b/(\alpha_s(\theta - \xi))^{1/2})]\} \times (1/(\pi\alpha_s(\theta - \xi)))^{1/2} d\xi \quad (16.74)$$

For a thin-film heat flux gauge, the calibration of the physical properties k, c_p, and ρ of each layer of the heat flux gauge are crucial for the accurate measurement of heat flux on the test surface. A calibration procedure was described in Keltner et al. [120]. Using the heat flux gauge itself as the heat source, the gauge was heated by supplying a step, a ramp, or a sinusoidal current into the gauge. Then, the change in resistance was measured. The temperature was evaluated from the measured resistance variation. Through a comparison between the measured temperature response and numerical predictions based on a one-dimensional transient conduction model for a sudden step change in wall temperature, the thermal impedance and conductivity of the gauge were obtained. To prevent heat loss from the gauge to the test surface, a multilayer thin film gauge with an insulated enamel layer was developed [121].

Finite-thickness-type gauges include slug (or plug) calorimeters and thin-wall (or thin-skin) calorimeters. They assume the gauge is exposed to a heat flux on the front surface and is insulated on the back. The slug calorimeter consists of a small mass of high-conductivity material inserted into the insulated wall. A thin-wall calorimeter covers a large (or the entire) surface of a well-insulated wall. Both calorimeters assume that the temperature within the gauge is uniform; thus, the energy balance equations for a slug (plug) calorimeter [Fig. 16.36(a)] and thin-wall calorimeter [Fig. 16.36(b)] are, respectively

$$q = \frac{mc_p}{A} \frac{dT_s}{d\theta} \quad (16.75a)$$

or

$$q = \rho\delta c_p \frac{dT_s}{d\theta} \quad (16.75b)$$

where m and A are the mass and surface area of the slug (plug); ρ and δ are the density and thickness of the wall; and c_p and T_s are the specific heat and mean temperature of the slug or wall.

The time derivative of the mean temperature is needed to determine the heat flux. This is obtained by using either a thermocouple to measure the back surface temperature of the sensor or the sensor as a resistance thermometer to measure its average temperature [118, 122–124]. Since the mean temperature increases as long as the heat flow is positive, these sensors generally are limited to short-duration measurement of transient heat flux, as is also true for the thin-film calorimeter.

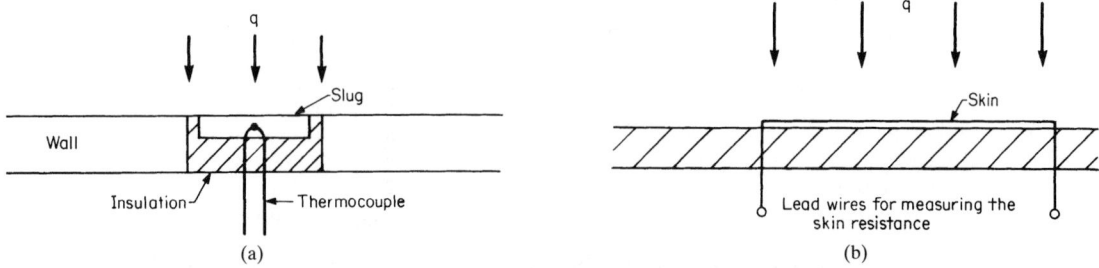

FIGURE 16.36   Calorimeter: (a) slug (plug) calorimeter, and (b) thin-wall (thin-skin) calorimeter.

Uncertainties in the use of these sensors include the temperature nonuniformity across the sensor's thickness at high heat flux, the edge correction for localized gauges, and the disturbance of the temperature field caused by the presence of the sensor [118, 122–124].

MEASUREMENT BY ANALOGY

Introduction

Analogies have been widely used to study heat transfer. An analog system is often simpler to construct than a heat transfer test apparatus. In addition, analog systems can often be set up to avoid secondary effects (e.g., conduction) that tend to introduce errors in temperature and heat transfer measurements.

Electric networks have been used to describe radiation heat transfer. Because electric networks have commonly available solutions, this analogy is useful. It also permits the use of an analog computer for solving complex problems. Similarly, conduction systems have been studied using small analog models made of various materials, including conducting paper.

Numerical analysis using high-speed digital computers has taken the place of the above analogies in many situations that require accurate analysis. Many conduction and radiation problems with known physical properties are amenable to computer modeling and solution. For this reason, the analogies for conduction and radiation heat transfer, though still used as teaching tools, will not be discussed here.

Computer modeling of convection has had mixed success. Many convection problems, particularly those involving laminar flow, can readily be solved by special computer programs. However, in situations where turbulence and complex geometries are involved, computer analysis and modeling are still under development. Mass transfer analogies can play a key role in the study of convective heat transfer processes. Two mass transfer systems, the sublimation technique and the electrochemical technique, are of particular interest because of their convenience and advantages relative to direct heat transfer measurements.

The principal governing equations in convective heat transfer are the continuity equation, the momentum equations, and the energy equation. In a mass transfer system involving a two-component single-phase medium, the energy equation is replaced by the species diffusion equation. For the analogy between heat and mass transfer to be valid, the energy and species diffusion equations have to be similar in both form and boundary conditions. The conditions for similarity can be readily derived [125]. For laminar flow, the Prandtl number must be equal to the Schmidt number, and there must be similarity in the boundary conditions. With turbulence, the energy and species diffusion equations both have an additional term involving a turbulent Prandtl number and a turbulent Schmidt number, respectively. Fortunately, experimental evidence suggests an equality between these two turbulent quantities in many flows.

In most heat transfer systems, the component of velocity normal to the active boundaries is zero, while, for the corresponding mass transfer system, this may not be the case. However, the magnitude of this normal velocity is usually sufficiently small that the analogy is not affected [125].

The advantages of using a mass transfer system to simulate a heat transfer system include the potential for improved accuracy of measurement and control of boundary conditions. For example, electric current and mass changes can generally be measured with greater accuracy than heat flux. Also, while adiabatic walls are an ideal that, at best, we can only approach, impermeable walls are an everyday reality. Thus, mass transfer systems are gaining popularity in precision experimental studies.

In convective heat transfer, knowledge of the heat transfer coefficient h is often required:

$$h = \frac{q}{\Delta T} \tag{16.76}$$

where ΔT is the driving force for heat transfer. For mass transfer (of component i), a mass transfer coefficient $h_{D,i}$ can be defined:

$$h_{D,i} = \frac{j_i}{\Delta C_i} \qquad (16.77)$$

where j_i is the mass flux of component i and the concentration difference ΔC_i is the driving force for mass transfer.

The dimensionless forms of the transfer coefficients are the Nusselt number Nu and the Sherwood number Sh for the heat and mass transfer processes, respectively:

$$\mathrm{Nu} = \frac{hL}{k} \qquad \mathrm{Sh} = \frac{h_{D,i}L}{D} \qquad (16.78)$$

Each of these is a ratio of a convective transfer rate to the corresponding diffusion rate of transfer. Dimensionless analysis indicates that, for fixed geometry and constant properties, the Nusselt number and the Sherwood number depend on the Reynolds number (forced convection), Rayleigh number (natural convection), flow characteristics, Prandtl number (heat transfer), and Schmidt number (mass transfer).

Sublimation Technique

A comprehensive review of the naphthalene sublimation technique is given in Ref. 126. The naphthalene sublimation technique, commonly employed to measure convective transport phenomena, has several advantages over direct heat transfer measurement techniques. These advantages are: more detailed mass transfer distribution over the test piece (typically thousands of data measured points), avoidance of heat conduction and radiation loss, and better control on boundary conditions.

In typical applications, pure solid naphthalene is melted and poured into a mold so it will have the desired shape such as a flat plate [127], a circular cylinder [128], or a turbine blade [129]. For average mass transfer measurements on a test surface, the section coated with naphthalene can be weighed before and after exposure to air flow to determine the mass transfer rate. Local mass transfer coefficients can be determined from the sublimation depth, which is the difference in surface profiles, measured using a profilometer, before and after each test run. Once the vapor density of naphthalene is known, the local mass transfer coefficient h_D can be evaluated from the following expression:

$$h_D = \frac{\rho_s L_{sb}}{\rho_{v,w}\Delta\theta} \qquad (16.79)$$

where ρ_s denotes the solid density of naphthalene, L_{sb} is the naphthalene sublimation depth, $\rho_{v,w}$ is the vapor density of naphthalene over the test piece surface, and $\Delta\theta$ is the time the test piece is exposed to the air stream.

As seen from Eq. 16.79, the measurement requires knowledge of the physical properties of naphthalene including its vapor pressure. The same properties of naphthalene are listed in Table 16.17.

TABLE 16.17   Physical Properties of Naphthalene

Properties	Value
Molecular weight	128.7
Melting point, °C [130]	80.3
Normal boiling point, °C (in air at 1 atm.) [130]	218
Solid density, kgm⁻³ (at 20°C) [131]	1175

An empirical correlation given in Ref. 132 has been commonly used to determine the vapor pressure of naphthalene:

$$T_{v,w} \log P_{v,w} = 0.5 c_8 + \sum_{s=9}^{11} c_s E_s(c_{12}) \tag{16.80}$$

where $T_{v,w}$ and $P_{v,w}$, respectively, denote the absolute temperature and pressure of vapor naphthalene. Constants and other parameters are described as follows:

$$c_8 = 301.6247$$

$$c_9 = 791.4937$$

$$c_{10} = -8.25336$$

$$c_{11} = 0.4043$$

$$c_{12} \doteq (T_{v,w} - 287)/57$$

$$E_9(c_{12}) = c_5$$

$$E_{10}(c_{12}) = 2c_{12}^2 - 1$$

$$E_{11}(c_{12}) = 4c_{12}^3 - 3c_{12} \tag{16.81}$$

In addition, the dimensionless form of the mass transfer coefficient, the Sherwood number, requires the diffusion coefficient of naphthalene in air. Empirical correlations [133] fit from the measured data of the naphthalene diffusion coefficient [134, 135] for D_{naph} and Sc of naphthalene are respectively given by

$$D_{naph} = 0.0681 \left(\frac{T_{v,w}}{298.16}\right)^{1.93} \left(\frac{760}{P_{atm}}\right) \quad [\text{cm}^2/\text{s}] \tag{16.82}$$

$$Sc = 2.28 \left(\frac{T_{v,w}}{298.16}\right)^{-0.1526} \tag{16.83}$$

Recent results for the dependence of D_{naph} on temperature are reported in Ref. 136.

A computer-controlled data acquisition system allows many data points to be taken at designated positions. For a typical profile measurement, it may take an hour to measure the naphthalene sublimation depth at several thousand measured locations. The extra sublimation loss during the profile measurements should be taken into account in order to reduce the measurement errors.

Electrochemical Technique

The sublimation method is used for mass transfer measurements in air flows. For measurement in some liquids, the electrochemical technique can be used.

Systematic studies with the mass transfer process in an electrochemical system date back to the 1940s [137, 138]. Later investigators extended the use of the method to both natural and forced convection flows. Extensive bibliographies of natural and forced convection studies using the electrochemical technique are available [139, 140]. Convenient sources of information on the general treatment of electrochemical transport phenomena can be found in Refs. 141 and 142.

The working fluid in an electrochemical system is the electrolyte. When an electric potential is applied across two electrodes in a system, the positive ions of the electrolyte will move toward the cathode, while the negative ions move toward the anode. The movement of the ions is controlled by (1) migration due to the electric field, (2) diffusion because of the ion density gradient, and (3) convection if the fluid is in motion. Fluid motion can be driven by the pressure drop in forced flow or by the density gradient in natural convection.

With heat transfer, convection and diffusion processes are present, but there is no equivalent to migration. In order to use ion transport as an analog to the heat transfer process, the ion migration has to be made negligible. This is done by introducing a second electrolyte, the so-called supporting electrolyte. It is normally in the form of an acid or base with a concentration of the order of 30 times that of the active electrolyte and selected so that its ions do not react at the electrodes at the potential used in the experiment. This supporting electrolyte tends to neutralize the charge in the bulk of the fluid. The addition of a supporting electrolyte does not significantly affect the transport phenomena in forced flow. However, it introduces an additional density gradient into the buoyancy force term for natural convection. Therefore, the analogy between heat and mass transfer in natural convection flow does not rigorously apply, and the effect of this additional gradient must be considered in applying the results of a mass transfer study.

Among the more commonly used electrolytic solutions are:

1. Copper sulfate-sulfuric acid solution: $CuSO_4$-H_2SO_4-H_2O

2. Potassium ferrocyanide-potassium ferricyanide-sodium hydroxide solution: $K_3[Fe(CN)_6]$-$K_4[Fe(CN)_6]$-$NaOH$-H_2O

With a copper sulfate solution, copper is dissolved from the anode and deposited on the cathode. For the other solution (also known as redox couples solution), only current transfer occurs at the electrodes. The respective reactions at the electrodes are:

$$Cu^{++} + 2e \underset{anode}{\overset{cathode}{\rightleftharpoons}} Cu$$

$$[Fe(CN)_6]^3 + e \underset{anode}{\overset{cathode}{\rightleftharpoons}} [Fe(CN)_6]^4$$

Typical transport properties for the above systems are listed in Table 16.18. Note that they are high-Schmidt number (analogous to Prandtl number) fluids.

The mass transfer coefficient for species i ($h_{D,i}$) is defined in the usual manner:

$$h_{D,i} = \frac{(\dot{N}_i'')_{DC}}{\Delta C_i} \tag{16.84}$$

where $(\dot{N}_i'')_{DC}$ is the transfer flux of species i in kgmol/(s·m²) due to diffusion and convection, and ΔC_i is the concentration difference of species i in kgmol/m³ across the region of interest.

The total mass flux can be determined from the electric current using the basic electrochemical relations. With the introduction of the supporting electrolyte, diffusion and convec-

TABLE 16.18 Transport Properties of Typical Electrolyte Solutions

Solution	Density ρ, g/cm³	Viscosity μ, N·s/m²	Diffusion coefficient of active species, $D \times 10^5$, cm²/s		Schmidt number Sc = $\mu/\rho D$
A	1.095	0.0124	$CuSO_4$	0.648	1750
B	1.095	0.0139	$K_3[Fe(CN)_6]$	0.537	~2500
			$K_4[Fe(CN)_6]$	0.460	

Solution A: $CuSO_4$: 0.05 gmol H_2SO_4 1.5 gmol
Solution B: $K_3[Fe(CN)_6]$: 0.05 gmol $NaOH$ 1.9 gmol
$K_4[Fe(CN)_6]$: 0.05 gmol

tion are the prime contributors to the total mass flux, while the migration effect is accounted for as a correction. The concentration difference ΔC_i is determined from the bulk and surface concentrations. The bulk concentration is determined through chemical analysis of the solution; however, the surface concentration is an unknown. This is resolved by the use of the limiting current condition [141]. As the voltage across the system is increased, the current increases monotonically until a plateau in the graph of current versus voltage occurs. At this limiting current, the surface concentration of the active species at one electrode is zero.

ACKNOWLEDGMENTS

We would like to thank Dr. Haiping Wang for his careful proofreading of the final draft of this chapter. His feedback improved the quality of the book.

NOMENCLATURE

Symbols, Definitions, SI Units

A	Area, m^2
a, b	Constants used in Eq. 16.11
C	Gladstone-Dale constant
C_i	Mass concentration of species i, kg/m^3
$\mathbf{C}_i$	Molar concentration of species i in an electrochemical system, $kgmol/m^3$
c_p	Specific heat at constant pressure, $J/(kgK)$
c_1, c_2	Constants defined in Eq. 16.22
c_3	Constant defined in Eq. 16.30
$c_4 - c_6$	Constants defined in Eq. 16.38
c_7	Material constant defined in Eq. 16.56
$c_8 - c_{12}$	Constants defined in Eq. 16.80
D	Diffusion coefficient, m^2/s
D_{naph}	Diffusion coefficient, m^2/s
E	Electric potential, or thermoelectric EMF, V
E_{AB}	Electric potential of thermocouple circuit with materials A and B, V
$E_a - E_e$	Electric potential at junctions a to e in Fig. 16.17, V
E_g	Band gap energy, erg
E_s, E_0	Electric potential used in Eq. 16.12, V
e	Emissive power, W/m^2
e_b	Blackbody emissive power, W/m^2
F	Function defined in Eq. 16.14
G	Amplitude of fluorescence
G_0	Initial amplitude of fluorescence
H	Optical absorption
H_0	Initial amplitude of emitted light
h	Heat transfer coefficient, $W/(m^2K)$
h_p	Planck's constant $= 6.6262 \times 10^{-27}$, ergs

h_D	Mass transfer coefficient, m/s
$h_{D,i}$	Mass transfer coefficient for species i, m/s
I	Electric current, A
$\mathbf{I}$	Luminance (or intensity) signal
$\mathbf{I}_l$	Local lighting intensity
$\mathbf{I}_i$	Initial illumination intensity in the shadowgraph system
$\mathbf{I}_{sc}$	Illumination intensity on the screen in the shadowgraph system
i_{exit}, i_{ent}	Enthalpy per unit mass at exit and entrance, J/kg
j_i	Mass flux of species i, kg/(sm^2)
k	Thermal conductivity, W/(mK)
k_B	Boltzmann constant $= 1.3806 \times 10^{-16}$ erg/K
L	Length, m
L_{sb}	Naphthalene sublimation depth, m
M	Molecular weight, kg/kgmol
m	Mass, kg
$\dot{m}$	Mass flux, kg/s
$(\dot{N}_i^m)_{DC}$	Diffusion and convection flux for species i in an electrochemical system, kgmol/(sm^2)
N	Molar refractivity
Nu	Nusselt number
n	Exponential defined in Eq. 16.30
n_i	Index of refraction
P	Pressure, Pa (N/m^2)
$P_{v,w}$	The vapor pressure of naphthalene, Pa
Q	Heat, J
$\dot{Q}$	Heat flow rate, W
$\dot{Q}_p$	Peltier heat, W
$\dot{Q}_T$	Thompson heat, W
q	Heat flux, W/m^2
q_e, q_r	External energy source input and radiative flux input, W/m^2
R	Electrical resistance, Ω
R_C, R_D, R_T, R_X	Resistance used in Eqs. 16.8–16.10, Ω
R_f, R_t	Resistance of a fixed resistor and thermistor, Ω
R_s	Surface reflectance
R_{th}	Thermal resistance, ($^\circ$C m^2)/W
R_u	Universal gas constant
$R_0, R_{25}, R_{100}, R_{125}$	Electrical resistance at 0°C, 25°C, 100°C, and 125°C, Ω
r	Recovery factor
S	Optical path length
S^*	Entropy transfer parameter, V/$^\circ$C
S_A^*–S_D^*	Entropy transfer parameters for materials A to D, V/$^\circ$C
Sh	Sherwood number
s	Resistance ratio
T	Absolute temperature, K

T_d, T_r, T_{st}, T_t	Dynamic temperature, recovery temperature, static temperature, and total temperature in Eqs. 16.63 and 16.64, K or °C
T_f	Fluid temperature, K or °C
T_i, T_{re}, T_w	Initial temperature, reference temperature, and wall fluid temperature in Eq. 16.60, K
T_{inf}	Inferred temperature, K
T_m	Blackbody temperature corresponding to the pyrometer-measured radiant energy, K
T_{ra}	Ratio temperature defined in Eq. 16.35
T_s	Temperature of a heat flux sensor, K or °C
$T_{v,w}$	Temperature of vapor naphthalene, K
T_λ	Spectral radiation temperature, K
T_0, T_e, T_i	Junction temperatures in Fig. 16.17, K or °C
T_{90}	International Kelvin temperature, K
t	Temperature, °C °F
t_{90}	International Celsius temperature, °C
U	Internal energy, J
V	Velocity, m/s
W	Work, J
$\dot{W}$	Rate of work, W
x, y, z	Rectangular coordinates, m

Greek Symbols

α	Deflected angle of light beam
α_{AB}	Seebeck coefficient, V/°C
α_s	Thermal diffusivity, m²/s
β; χ	Scaling parameters defined in Eq. 16.55
Δ	Finite increment
δ	Thickness, m
ε	Emissivity
ε_r	Emissivity ratio defined in Eq. 16.37
γ	Resistance-ratio defined in Eq. 16.13
Λ	Effective wavelength defined in Eq. 16.36
λ	Wavelength, m
λ_{max}	Wavelength at which $e_{b\lambda}$ is maximum, m
Π	Ratio of resistance defined in Eq. 16.7
π_{AB}	Peltier coefficient, W/A
θ	Time, s
θ_c	Time constant defined in Eq. 16.66, s
ρ	Density, kg/m³
ρ_s	Density of solid naphthalene, kg/m³
$\rho_{v,w}$	Vapor density of naphthalene, kg/m³
σ	Stefan-Boltzmann constant, W/(m²K⁴)

σ_T	Thompson coefficient, W/(A°C)
τ	Lifetime of fluorescence in Eq. 16.55, s
υ	Light frequency, Hz

Subscripts

a	Ambient
sc	Screen in the shadowgraph system
st	Static
t	Total
λ	Monochromatic value, evaluated at wavelength λ
0	Standard (or reference) condition

LIST OF ABBREVIATIONS

ANSI	American National Standard Institute
ASTM	American Society for Testing and Materials
CIPM	International Committee of Weights and Measures
DWRT	Double-Wavelength Radiation Thermometer
IPTS	International Practical Temperature Scale
IPTS-68	International Practical Temperature Scale of 1968
ISA	Instrument Society of America
ISO	International Organization of Standardization
ITS	International Temperature Scale
ITS-90	International Temperature Scale of 1990
NBS	National Bureau of Standards
NIST	National Institute of Standards and Technology (formally NBS)
NTSC	National Television Standards Committee
NVLAP	National Voluntary Laboratory Accreditation Program

REFERENCES

1. The International Temperature Scale of 1990 (ITS-90), *Metrologia,* vol. 27, pp. 3–10, 1990.

2. Supplementary Information for the International Temperature Scale of 1990, Bureau International des Poids et Mesures, Sevres, France, 1990.

3. B. W. Mangum and G. T. Furukawa, "Guidelines for Realizing the International Temperature Scale of 1990 (ITS-90)," NIST Technical Note 1265, 1990.

4. C. A. Swenson, "From the IPTS-68 to the ITS-90," in *Temperature: Its Measurements and Control in Science and Industry,* vol. 6, pt. 1, pp. 1–7, American Institute of Physics, New York, 1992.

5. T. J. Quinn, *Temperature,* 2d ed., Academic Press, London, 1990.

6. R. L. Rusby, R. P. Hudson, M. Durieux, J. F. Schooley, P. P. M. Steur, and C. A. Swenson, "A Review of Progress in the Measurement of Thermodynamic Temperature," in *Temperature: Its Measurement and Control in Science and Industry,* vol. 6, pt. 1, pp. 9–14, American Institute of Physics, New York, 1992.

7. NIST Calibration Services Users Guide/Office of Physical Measurement Services, National Institute of Standards and Technology 1991, NIST Special Publication 250, Gaithersburg, Maryland.

8. J. A. Wise, "Liquid-in-Glass Thermometry," NBS Monograph 150, National Bureau of Standards, 1976.

9. J. A. Wise and R. J. Soulen, Jr., "Thermometer Calibration: A Model for State Calibration Laboratories," NBS Monograph 174, 1986.

10. J. A. Wise, "A Procedure for the Effective Recalibration of Liquid-in-Glass Thermometers," NIST Special Publication 819, 1991.

11. G. Schuster, "Temperature Measurement with Rhodium-Iron Resistors below 0.5 K," in *Temperature: Its Measurement and Control in Science and Industry,* vol. 6, pt. 1, pp. 449–451, American Institute of Physics, New York, 1992.

12. W. F. Schlooser and R. H. Munnings, "Thermistors as Cryogenic Thermometers," in *Temperature: Its Measurement and Control in Science and Industry,* vol. 4, pt. 2, pp. 795–801, Instrument Society of America, Pittsburgh, 1972.

13. *Temperature: Its Measurement and Control in Science and Industry,* vol. 6, pt. 1, sec. 4, American Institute of Physics, New York, 1992.

14. Standard Specification for Industrial Platinum Resistance Thermometers, ASTM Standard E1137-95, 1995.

15. M. Sapoff, "Thermistors for Resistance Thermometry," *Measurements and Control,* 14, 2, pp. 110–121, 1980.

16. Specification for Thermistor Sensors for Clinical Laboratory Temperature Measurements, ASTM Standard E879-93, 1993.

17. M. Sapoff, "Thermistors: Part 2—Manufacturing Techniques," *Measurements and Control,* 14, 3, pp. 112–117, 1980.

18. M. Sapoff, "Thermistors: Part 4—Optimum Linearity Techniques," *Measurements and Control,* 14, 5, pp. 112–119, 1980.

19. E. F. Mueller, "Precision Resistance Thermometry," in *Temperature: Its Measurement and Control in Science and Industry,* vol. 1, pp. 162–179, Reinhold, New York, 1941.

20. J. P. Evans, "An Improved Resistance Thermometer Bridge," in *Temperature: Its Measurement and Control in Science and Industry,* vol. 3, pt. 1, pp. 285–289, Reinhold, New York, 1962.

21. T. M. Dauphinee, "Potentiometric Methods of Resistance Measurement," in *Temperature: Its Measurement and Control in Science and Industry,* vol. 3, pt. 1, pp. 269–283, Reinhold, New York, 1962.

22. R. D. Cutkosky, "Automatic Resistance Thermometer Bridges for New and Special Applications," in *Temperature: Its Measurement and Control in Science and Industry,* vol. 5, pt. 2, pp. 711–712, American Institute of Physics, New York, 1982.

23. N. L. Brown, A. J. Fougere, J. W. McLeod, and R. J. Robbins, "An Automatic Resistance Thermometer Bridge," in *Temperature: Its Measurement and Control in Science and Industry,* vol. 5, pt. 2, pp. 719–727, American Institute of Physics, New York, 1982.

24. S. Anderson and D. Myhre, "Resistance Temperature Detectors—A Practical Approach to Application Analysis," *Rosemount Rep. 108123,* 1981.

25. B. W. Mangum, "Platinum Resistance Thermometer Calibrations," NBS Special Publication 250-22, 1987.

26. H. M. Hashemain and K. M. Petersen, "Achievable Accuracy and Stability of Industrial RTDs," in *Temperature: Its Measurement and Control in Science and Industry,* vol. 6, pt. 1, pp. 427–432, American Institute of Physics, New York, 1992.

27. S. D. Wood, B. W. Mangum, J. J. Filliben, and S. B. Tillett, "An Investigation of the Stability of Thermistors," *J. Res. (NBS),* 83, pp. 247–263, 1978.

28. M. Sapoff and H. Broitman, "Thermistors-Temperature Standards for Laboratory Use," *Measurements & Data,* 10, pp. 100–103, 1976.

29. W. R. Siwek, M. Sapoff, A. Goldberg, H. C. Johnson, M. Botting, R. Lonsdorf, and S. Weber, "A Precision Temperature Standard Based on the Exactness of Fit of Thermistor Resistance-Temperature Data Using Third-Degree Polynomials," in *Temperature: Its Measurement and Control in Science and Industry,* vol. 6, pt. 1, pp. 491–496, American Institute of Physics, New York, 1992.

30. W. R. Siwek, M. Sapoff, A. Goldberg, H. C. Johnson, M. Botting, R. Lonsdorf, and S. Weber, "Stability of NTC Thermistors," in *Temperature: Its Measurement and Control in Science and Industry,* vol. 6, pt. 1, pp. 497–502, American Institute of Physics, New York, 1992.

31. J. A. Wise, "Stability of Glass-Encapsulated Disc-Type Thermistors," in *Temperature: Its Measurement and Control in Science and Industry,* vol. 6, pt. 1, pp. 481–484, American Institute of Physics, New York, 1992.

32. M. W. Zemansky, *Heat and Thermodynamics,* 4th ed., pp. 298–309, McGraw-Hill, New York, 1957.

33. W. F. Roeser, "Thermoelectric Thermometry," in *Temperature: Its Measurement and Control in Science and Industry,* vol. 1, pp. 180–205, Reinhold, New York, 1941.

34. American National Standard for Temperature Measurement Thermocouples, ANSI-MC96-1 1982, Instrument Society of America (sponsor), 1982.

35. Temperature Electromotive Force (EMF) Tables for Standardized Thermocouples, ASTM Standard E230-93, 1993.

36. *ASTM Manual MNL 12: Manual on the Use of Thermocouples in Temperature Measurement,* 4th ed., American Society for Testing and Materials, 1993.

37. N. A. Burley, R. L. Powell, G. W. Burns, and M. G. Scroger, "The Nicrosil vs. Nisil Thermocouple: Properties and Thermoelectric Reference Data," NS Monograph 161, 1978.

38. P. A. Kinzie, *Thermocouple Temperature Measurement,* Wiley, New York, 1973.

39. E. D. Zysk and A. R. Robertson, "Newer Thermocouple Materials," in *Temperature: Its Measurement and Control in Science and Industry,* vol. 4, pt. 3, pp. 1697–1734, Instrument Society of America, Pittsburgh, 1972.

40. G. W. Burns and M. G. Scroger, "The Calibration of Thermocouples and Thermocouple Materials," NIST Special Publication 250-35, National Institute of Standards and Technology, 1989.

41. G. W. Burns, G. F. Strouse, B. M. Liu, and B. W. Mangum, "Gold versus Platinum Thermocouples: Performance Data and an ITS-90-Based Reference Function," in *Temperature: Its Measurement and Control in Science and Industry,* vol. 6, pt. 1, pp. 531–536, American Institute of Physics, New York, 1992.

42. G. W. Burns, M. G. Scroger, G. F. Strouse, M. C. Croarkin, and W. F. Guthrie, Temperature-Electromotive Force Reference Functions and Tables for the Letter-Designated Thermocouple Types Based on the ITS-90, NIST Monograph 175, National Institute of Standards and Technology, 1993.

43. A. Mossman, J. L. Horton, and R. L. Anderson, "Testing of Thermocouples for Inhomogeneities: A Review of Theory, with Examples," in *Temperature: Its Measurement and Control in Science and Industry,* vol. 5, pt. 2, pp. 923–929, American Institute of Physics, New York, 1982.

44. R. L. Powell, "Thermocouple Thermometry," in E. R. G. Eckert and R. J. Goldstein (eds.), *Measurements in Heat Transfer,* 2d ed., pp. 112–115, McGraw-Hill, 1971.

45. G. W. Burns and M. G. Scroger, "NIST Measurement Services: The Calibration of Thermocouples and Thermocouple Materials," NIST Special Publication 250-35, National Institute of Standards and Technology, 1989.

46. F. R. Caldwell, "Temperature of Thermocouple Reference Junctions in an Ice Bath," *J. Res. (NBS),* 69c, pp. 256–262, 1965.

47. E. R. G. Eckert, C. L. Tien, and D. K. Edwards, "Radiation," in W. M. Rohsenow, J. P. Hartnett, and E. N. Ganic (eds.), *Handbook of Heat Transfer Fundamentals,* Chap. 14, McGraw-Hill, New York, 1985.

48. T. D. McGee, *Principles and Methods of Temperature Measurement,* John Wiley & Sons, New York, 1988.

49. K. D. Mielenz, R. D. Saunders, A. C. Parr, and J. J. Hsia, "The 1990 NIST Scales of Thermal Radiometry," *J. Res. (NIST),* 95, 6, pp. 621–629, 1990.

50. R. P. Benedict, *Fundamentals of Temperature: Pressure and Flow Measurements,* 2d ed., pp. 144–146, 265–273, Wiley, New York, 1977.

51. W. F. Roeser and H. T. Wensel in *Temperature: Its Measurement and Control in Science and Industry,* vol. 1, p. 1313, Reinhold, New York, 1941.

52. Radiance Temperature Calibrations, NIST Special Publication 250-7, National Institute of Standards and Technology, 1996.

53. D. P. DeWitt, "Advances and Challenges in Radiation Thermometry," in G. F. Hewitt (ed.) *Heat Transfer 1994: Proceedings of the Tenth International Heat Transfer Conference,* vol. 1, pp. 205–222, 1994.

54. B. K. Tsai, D. P. Dewitt, and G. J. Dail, "Dual Wavelength Radiation Thermometry for Aluminum Alloys," *Measurement,* 11, pp. 211–221, 1993.

55. R. D. Hudson, Jr., *Infrared System Engineering,* Wiley, New York, 1969.

56. R. E. Engelhardt and W. A. Hewgley, "Thermal and Infrared Testing," in Non-Destructive Testing, NASA Rep. SP-5113, pp. 119–140, 1973.

57. K. A. Wickersheim, "Fiberoptic Thermometry: An Overview," in *Temperature: Its Measurement and Control in Science and Industry,* vol. 6, pt. 2, pp. 711–714, American Institute of Physics, New York, 1992.

58. R. J. Goldstein, "Optical Techniques for Temperature Measurement," in E. R. G. Eckert and R. J. Goldstein (eds.), *Measurement in Heat Transfer,* 2d ed., pp. 241–293, McGraw-Hill, New York, 1976.

59. R. J. Goldstein and T. H. Kuehn, "Optical System for Flow Measurement: Shadowgraph, Schlieren, and Interferometric Techniques," in R. J. Goldstein (ed.), *Fluid Dynamics Measurements,* 2d ed., pp. 451–508, Taylor & Francis, Washington, 1996.

60. F. Mayinger, "Image-Forming Optical Techniques in Heat Transfer: Revival by Computer-Aided Data Processing," *ASME J. of Heat Transfer,* 115, pp. 824–834, 1993.

61. F. Mayinger, "Modern Electronics in Image-Processing and in Physical Modeling—A New Challenge for Optical Techniques," *Heat Transfer 1994: Proceedings of the Tenth International Heat Transfer Conference,* in G. F. Hewitt (ed.), pp. 61–79, 1994.

62. L. J. Dowell, "Fluorescence Thermometry," *Appl. Mech. Rev.,* 45, 7, pp. 253–260, 1992.

63. K. T. V. Grattan, J. D. Manwell, S. M. L. Sim, and C. A. Wilson, "Fibre-Optic Temperature Sensor with Wide Temperature Range Characteristics," *IEE PROCEEDINGS,* 134, 5, pp. 291–294, 1987.

64. V. Pernicola and L. Crovine, "Two Fluorescent Decay-Time Thermometers," in *Temperature: Its Measurement and Control in Science and Industry,* vol. 6, pt. 2, pp. 725–730, American Institute of Physics, New York, 1992.

65. M. K. Chyu and D. J. Bizzak, "Surface Temperature Measurement Using a Laser-Induced Fluorescence Thermal Imaging System," *ASME J. of Heat Transfer,* 116, pp. 263–266, 1994.

66. T. V. Samulski, "Fiberoptic Thermometry: Medical and Biomedical Applications," in *Temperature: Its Measurement and Control in Science and Industry,* vol. 6, pt. 2, pp. 1185–1189, American Institute of Physics, New York, 1992.

67. J. M. C. England, N. Zissis, P. J. Timans, and H. Ahmed, "Time-Resolved Reflectivity Measurements of Temperature Distributions During Swept-Line Electron-Beam Heating of Silicon," *Journal of Applied Physics,* 70, 1, pp. 389–397, 1991.

68. T. Q. Qiu, C. P. Grigoropoulos, and C. L. Tien, "Novel Technique for Noncontact and Microscale Temperature Measurements," *Experimental Heat Transfer,* 6, pp. 231–241, 1993.

69. X. Xu, C. P. Grigoropoulos, and R. E. Russo, "Transient Temperature During Pulsed Excimer Laser Heating of Thin Polysilicon Films Obtained by Optical Reflectivity Measurement," *ASME J. of Heat Transfer,* 117, pp. 17–24, 1995.

70. J. L. Hay and D. K. Hollingsworth, "A Comparison of Trichromic Systems for Use in the Calibration of Polymer-Dispersed Thermochromic Liquid Crystals," *Experimental Thermal and Fluid Science,* 12, 1, pp. 1–12, 1996.

71. K.-H. Platzer, C. Hirsch, D. E. Metzger, and S. Wittig, "Computer-Based Areal Surface Temperature and Local Heat Transfer Measurements with Thermochromic Liquid Crystals (TLC)," *Experiments in Fluids,* 13, pp. 26–32, 1992.

72. S. A. Hippensteele, L. M. Russell, and F. S. Stepke, "Evaluation of a Method for Heat Transfer Measurements and Thermal Visualization Using a Composite of a Heater Element and Liquid Crystals," *ASME J. of Heat Transfer,* 105, pp. 184–189, 1983.

73. C. Camci, K. Kim, and S. A. Hippensteele, "A New Hue Capturing Technique for the Quantitative Interpretation of Liquid Crystal Images Used in Convective Heat Transfer Studies," *ASME J. of Turbomachinery,* 114, pp. 512–518, 1992.

74. Z. Wang, P. T. Ireland, and T. V. Jones, "A Technique for Measuring Convective Heat-Transfer at Rough Surfaces," ASME Paper 90-GT-300, 1990.

75. R. S. Bunker, D. E. Metzger, and S. Wittig, "Local Heat Transfer in Turbine Disk Cavities. Part I: Rotor and Stator Cooling with Hub Injection of Coolant," *ASME J. of Turbomachinery,* 114, pp. 211–220, 1992.

76. C. Camci, K. Kim, S. A. Hippensteele, and P. E. Poinsatte, "Evaluation of a Hue-Capturing-Based Transient Liquid Crystal Method for High-Resolution Mapping of Convective Heat Transfer on Curved Surfaces," *ASME J. of Heat Transfer,* 115, pp. 311–318, 1993.

77. Z. Wang, P. T. Ireland, and T. V. Jones, "An Advanced Method of Processing Liquid Crystal Video Signals from Transient Heat Transfer Experiments," *ASME J. of Turbomachinery,* 117, pp. 184–188, 1995.

78. D. E. Metzger and R. S. Bunker, "Local Heat Transfer in Internally Cooled Turbine Airfoil Leading Edge Regions: Part II—Impingement Cooling with Film Coolant Extraction," *ASME J. of Turbomachinery,* 112, pp. 459–466, 1990.

79. D. E. Metzger and D. E. Larson, "Use of Melting Point Surface Coating for Local Convection Heat Transfer Measurements in Rectangular Channel Flows with 90° Turns," *ASME J. of Heat Transfer,* 108, pp. 48–54, 1986.

80. Bimetallic Thermometers, SAMA Std. PMC-4-1-1962, 1962.

81. W. D. Huston, "The Accuracy and Reliability of Bimetallic Temperature Measuring Elements," in *Temperature: Its Measurement and Control in Science and Industry,* vol. 3, pt. 2, pp. 949–957, Reinhold, New York, 1962.

82. T. V. Blalock and R. L. Shepard, "A Decade of Progress in High-Temperature Johnson Noise Thermometry," in *Temperature: Its Measurement and Control in Science and Industry,* vol. 5, pt. 2, American Institute of Physics, New York, 1982.

83. A. Ohte and H. Iwaoka, "A New Nuclear Quadrupole Resonance Standard Thermometer," in *Temperature: Its Measurement and Control in Science and Industry,* vol. 5, pt. 2, pp. 1173–1180, American Institute of Physics, New York, 1982.

84. P. M. Anderson, N. S. Sullivan, and B. Andraka, "Nuclear Quadrupole Resonance Spectroscopy for Ultra-Low-Temperature Thermometry," in *Temperature: Its Measurement and Control in Science and Industry,* vol. 6, pt. 2, pp. 1013–1016, American Institute of Physics, New York, 1992.

85. A. Benjaminson and F. Rowland, "The Development of the Quartz Resonator as a Digital Temperature Sensor with a Precision of 1×10^{-4}," in *Temperature: Its Measurement and Control in Science and Industry,* vol. 3, pt. 1, pp. 701–708, Reinhold, New York, 1962.

86. K. Agatsuma, F. Uchiyama, T. Ohara, K. Tukamoto, H. Tateishi, S. Fuchino, Y. Nobue, S. Ishigami, M. Sato, and H. Sugimoto, *Advanced in Cryogenic Engineering,* 35, pp. 1563–1571, Plenum Press, New York, 1990.

87. L. C. Lynnworth, "Temperature Profiling Using Multizone Ultrasonic Waveguides," in *Temperature: Its Measurement and Control in Science and Industry,* vol. 5, pt. 2, pp. 1181–1190, American Institute of Physics, New York, 1982.

88. R. W. Treharne and J. A. Riley, "A Linear-Response Diode Temperature Sensor," *Instrum. Technol.,* 25, 6, pp. 59–61, 1978.

89. H. D. Baker, E. A. Ryder, and N. H. Baker, *Temperature Measurement in Engineering,* vol. H, Omega Press, Stanford, Connecticut, 1975.

90. E. R. G. Eckert and R. M. Drake, Jr., *Analysis of Heat and Mass Transfer,* pp. 417–422, 694–695, McGraw-Hill, New York, 1972.

91. R. J. Moffat, "Gas Temperature Measurements," in *Temperature: Its Measurement and Control in Science and Industry,* vol. 3, pt. 2, pp. 553–571, Reinhold, New York, 1962.

92. S. J. Green and T. W. Hunt, "Accuracy and Response of Thermocouples for Surface and Fluid Temperature Measurement," in *Temperature: Its Measurement and Control in Science and Industry,* vol. 3, pt. 2, pp. 695–722, Reinhold, New York, 1962.

93. E. M. Sparrow, "Error Estimates in Temperature Measurements," in E. R. G. Eckert and R. J. Goldstein (eds.), *Measurement in Heat Transfer,* 2d ed., pp. 1–24, McGraw-Hill, New York, 1976.

94. R. P. Benedict, "Temperature Measurement in Moving Fluids," ASME Paper 59A-257, 1959.

95. C. D. Henning and R. Parker, "Transient Response of an Intrinsic Thermocouple," *ASME J. of Heat Transfer,* 89, pp. 146–154, 1967.

96. J. V. Beck and H. Hurwicz, "Effect of Thermocouple Cavity on Heat Sink Temperature," *ASME J. of Heat Transfer,* 82, pp. 27–36, 1960.

97. J. V. Beck, "Thermocouple Temperature Disturbances in Low Conductivity Materials," *ASME J. of Heat Transfer,* 84, pp. 124–132, 1962.

98. R. C. Pfahl, Jr. and D. Dropkin, "Thermocouple Temperature Perturbations in Low Conductivity Materials," ASME Paper 66-WA/HT-8, 1966.

99. "Use of the Terms *Precision* and *Accuracy* as Applied to Measurement of a Property of a Material," ASTM Std. E-177, Annual Standard of ASTM, p. 41, 1981.

100. G. N. Gray and H. C. Chandon, "Development of a Comparison Temperature Calibration Capability," in *Temperature: Its Measurement and Control in Science and Industry,* vol. 4, pt. 2, pp. 1369–1378, Instrument Society of America, Pittsburgh, 1972.

101. Temperature Measurement Instruments and Apparatus, ASME-PTC 19.3-1974, supplement to ASME performance test codes, 1974.

102. H. K. Staffin and C. Rim, "Calibration of Temperature Sensors between 538°C (1000°F) and 1092°C (2000°F) in Air Fluidized Solids," in *Temperature: Its Measurement and Control in Science and Industry,* vol. 4, pt. 2, pp. 1359–1368, Instrument Society of America, Pittsburgh, 1972.

103. D. B. Thomas, "A Furnace for Thermocouple Calibrations to 2200°C," *J. Res. (NBS),* 66c, pp. 255–260, 1962.

104. R. B. Abernethy, R. P. Benedict, and R. B. Dowdell, "ASME Measurement Uncertainty," *ASME J. of Fluids Engineering,* 107, pp. 161–164, 1985.

105. R. J. Moffat, "Describing the Uncertainties in Experimental Results," *Experimental Thermal and Fluid Science,* 1, pp. 3–17, 1988.

106. G. F. Strouse and B. W. Mangum, "NIST Measurement Assurance of SPRT Calibrations on the ITS-90: A Quantitative Approach," in *Proceedings of the 1993 Measurement Science Conference,* Session 1-D, January 20–24, 1993.

107. J. H. Garside, "Development of Laboratory Accreditation," in *Proceedings of the Chinese National Laboratory Accreditation Annual Conference and Laboratory Accreditation Symposium,* Taipei, Taiwan, 1995.

108. ISO Guide 25: General Requirements for the Competence of Calibration and Testing Laboratories, International Organization for Standardization, 1990.

109. National Voluntary Laboratory Accreditation Program (NVLAP), administered by NIST.

110. T. E. Diller, "Advances in Heat Flux Measurement," *Advances in Heat Transfer,* 23, pp. 279–367, 1993.

111. N. E. Hager, Jr., "Thin Foil Heat Meter," *Rev. Sci. Instrum.,* 36, pp. 1564–1570; and 1965.62. Temperature Measurement Instruments and Apparatus, ASME-PTC 19.3-1974, supplement to ASME performance test codes, 1974.

112. Hy-Cal Engineering, sale brochure, Santa Fe, California.

113. RdF Corporation sale brochure, Hudson, New Hampshire.

114. R. Gardon, "An Instrument for the Direct Measurement of Thermal Radiation," *Rev. Sci. Instrum.,* 24, pp. 366–370, 1953.

115. "Standard Method for Measurement of Heat Flux Using a Copper-Constantan Circular Foil Heat Flux Gauge," ASTM Std. E-511, Annual Standard of ASTM, Pt. 41, 1981.

116. N. R. Keltner and M. W. Wildin, "Transient Response of Circular Foil Heat-Flux Gauges to Radiative Fluxes," *Rev. Sci. Instrum.,* 46, pp. 1161–1166, 1975.

117. Standard Method for Measuring Heat Flux Using A Water-Cooled Calorimeter, *ASTM Std. E-422,* Annual Standard of ASTM, Pt. 41, 1981.

118. D. L. Schultz and T. V. Jones, "Heat Transfer Measurements in Short-Duration Hypersonic Facilities," AGAR Dograph No. 165, 1973.

119. M. G. Dunn and A. Hause, "Measurement of Heat Flux and Pressure in a Turbine Blade," *ASME J. of Engineering for Power,* 104, pp. 215–223, 1982.

120. N. R. Keltner, B. L. Bainbridge, and J. V. Beck, "Rectangular Heat Source on a Semi-infinite Solid—An Analysis for a Thin-Film Heat Flux Gage Calibration," *ASME J. of Heat Transfer,* 110, pp. 42–48, 1988.

121. J. E. Doorley, "Procedures for Determining Surface Heat Flux Using Thin Film Gages on a Coated Metal Model in a Transient Test Facility," *ASME J. of Turbomachinery,* 110, pp. 242–250, 1988.

122. Standard Method for Measuring Heat Transfer Rate Using a Thermal Capacitance (Slug) Calorimeter, ASTM Std. E0457, Annual Standard of ASTM, pt. 41, 1981.

123. Standard Method for Design and Use of a Thin-Skin Calorimeter for Measuring Heat Transfer Rate, ASTM Std. E-459, Annual Standard of ASTM, pt. 41, 1981.

124. C. S. Landram, Transient Flow Heat Transfer Measurements Using the Thin-Skin Method, *ASME J. of Heat Transfer,* 96, pp. 425–426, 1974.

125. E. R. G. Eckert, "Analogies to Heat Transfer Processes," in E. R. G. Eckert and R. J. Goldstein (eds.), *Measurement in Heat Transfer,* 2d ed., pp. 397–423, McGraw-Hill, New York, 1976.

126. R. J. Goldstein and H. H. Cho, "A Review of Mass (Heat) Transfer Measurement Using Naphthalene Sublimation," *Experimental Thermal and Fluid Science,* 10, pp. 416–434, 1995.

127. R. J. Goldstein, M. K. Chyu, and R. C. Hain, "Measurement of Local Mass Transfer on a Surface in the Region of the Base of a Protruding Cylinder with a Computer-Controlled Data Acquisition System," *Int. J. Heat Mass Transfer,* 28, pp. 977–985, 1985.

128. J. Karni and R. J. Goldstein, "Surface Injection Effect on Mass Transfer From a Cylinder in Crossflow: A Simulation of Film Cooling in the Leading Edge Region of a Turbine Blade," *ASME J. of Turbomachinery,* 112, pp. 418–427, 1990.

129. P. H. Chen and R. J. Goldstein, "Convective Transport Phenomena on the Suction Surface of a Turbine Blade," *ASME J. of Turbomachinery,* 114, pp. 418–427, 1992.

130. A. P. Kudchadker, S. A. Kudchadker, and R. C. Wilhoit, *Naphthalene,* American Petroleum Institute, Washington, 1978.

131. J. A. Dean, *Handbook of Organic Chemistry,* McGraw-Hill, New York, pp. 1-308–1-309, 1987.

132. D. Ambrose, I. J. Lawrenson, and C. H. S. Sprake, "The Vapour Pressure of Naphthalene," *J. Chem. Thermodynamics,* 7, pp. 1173–1176, 1975.

133. H. H. Cho, M. Y. Jabbari, and R. J. Goldstein, "Mass Transfer with Flow Through an Array of Rectangular Cylinders," *ASME J. of Heat Transfer,* 116, pp. 904–911, 1994.

134. K. Cho, T. T. Irvine, Jr., and J. Karni, "Measurement of the Diffusion Coefficient of Naphthalene into Air," *Int. J. Heat Mass Transfer,* 35, 4, pp. 957–966, 1992.

135. L. Caldwell, "Diffusion Coefficient of Naphthalene in Air and Hydrogen," *J. Chem. Eng. Data,* 29, pp. 60–62, 1984.

136. P.-H. Chen, J.-M. Miao, and C.-S. Jian, "Novel Technique for Investigating the Temperature Effect on the Diffusion Coefficient of Naphthalene into Air," *Rev. Sci Instrum.,* 67, pp. 2831–2836, 1996.

137. J. N. Agar, "Diffusion and Convection at Electrodes," *Discussion Faraday Soc.,* 1, pp. 26–37, 1947.

138. C. Wagner, "The Role of Natural Convection in Electrolytic Processes," *Trans. Electrochem. Soc.,* 95, pp. 161–173, 1949.

139. H. D. Chiang and R. J. Goldstein, "Application of the Electrochemical Mass Transfer Technique to the Study of Buoyancy-Driven Flows," *Proc. 4th Int. Symposium on Transport Phenomena (ISTP-4)—Heat and Mass Transfer,* Australia, 1991.

140. A. A. Wragg, "Applications of the Limiting Diffusion Current Technique in Chemical Engineering," *Chem. Eng. (London),* January 1977, pp. 39–44, 1977.

141. J. R. Selman and C. W. Tobias, "Mass-Transfer Measurements by the Limiting-Current Technique," *Advances in Chemical Engineering,* 10, pp. 211–318, 1978.

142. J. Newman, *Electrochemical Systems,* 1st ed., Prentice-Hall, Englewood Cliffs, New Jersey, 1973.

CHAPTER 17
HEAT EXCHANGERS

R. K. Shah* and D. P. Sekulić
University of Kentucky

INTRODUCTION

A *heat exchanger* is a device that is used for transfer of thermal energy (enthalpy) between two or more fluids, between a solid surface and a fluid, or between solid particulates and a fluid, at differing temperatures and in thermal contact, usually without external heat and work interactions. The fluids may be single compounds or mixtures. Typical applications involve heating or cooling of a fluid stream of concern, evaporation or condensation of a single or multicomponent fluid stream, and heat recovery or heat rejection from a system. In other applications, the objective may be to sterilize, pasteurize, fractionate, distill, concentrate, crystallize, or control process fluid. In some heat exchangers, the fluids exchanging heat are in direct contact. In other heat exchangers, heat transfer between fluids takes place through a separating wall or into and out of a wall in a transient manner. In most heat exchangers, the fluids are separated by a heat transfer surface, and ideally they do not mix. Such exchangers are referred to as the *direct transfer type,* or simply *recuperators.* In contrast, exchangers in which there is an intermittent heat exchange between the hot and cold fluids— via thermal energy storage and rejection through the exchanger surface or matrix—are referred to as the *indirect transfer type* or *storage type,* or simply *regenerators.* Such exchangers usually have leakage and fluid carryover from one stream to the other.

A heat exchanger consists of heat exchanging elements such as a core or a matrix containing the heat transfer surface, and fluid distribution elements such as headers, manifolds, tanks, inlet and outlet nozzles or pipes, or seals. Usually there are no moving parts in a heat exchanger; however, there are exceptions such as a rotary regenerator (in which the matrix is mechanically driven to rotate at some design speed), a scraped surface heat exchanger, agitated vessels, and stirred tank reactors.

The heat transfer surface is a surface of the exchanger core that is in direct contact with fluids and through which heat is transferred by conduction. The portion of the surface that also separates the fluids is referred to as the *primary* or *direct surface.* To increase heat transfer area, appendages known as fins may be intimately connected to the primary surface to provide *extended, secondary,* or *indirect surface.* Thus, the addition of fins reduces the thermal resistance on that side and thereby increases the net heat transfer from/to the surface for the same temperature difference. The heat transfer coefficient can also be higher for fins.

A gas-to-fluid heat exchanger is referred to as a *compact heat exchanger* if it incorporates a heat transfer surface having a surface area density above about 700 m²/m³ (213 ft²/ft³) on at

---

* Current address: Delphi Harrison Thermal Systems, Lockport, New York.

least one of the fluid sides, which usually has gas flow. It is referred to as a *laminar flow heat exchanger* if the surface area density is above about 3000 m²/m³ (914 ft²/ft³), and as a *micro-heat exchanger* if the surface area density is above about 10,000 m²/m³ (3050 ft²/ft³). A liquid/two-phase fluid heat exchanger is referred to as a *compact heat exchanger* if the surface area density on any one fluid side is above about 400 m²/m³ (122 ft²/ft³). A typical process industry shell-and-tube exchanger has a surface area density of less than 100 m²/m³ on one fluid side with plain tubes and 2–3 times that with the high-fin-density, low-finned tubing. Plate-fin, tube-fin, and rotary regenerators are examples of compact heat exchangers for gas flows on one or both fluid sides, and gasketed and welded plate heat exchangers are examples of compact heat exchangers for liquid flows.

CLASSIFICATION OF HEAT EXCHANGERS

Heat exchangers may be classified according to transfer process, construction, flow arrangement, surface compactness, number of fluids and heat transfer mechanisms as shown in Fig. 17.1 modified from Shah [1] or according to process functions as shown in Fig. 17.2 [2]. A brief description of some of these exchangers classified according to construction is provided next along with their selection criteria. For further general description, see Refs. 1–4.

Shell-and-Tube Exchangers

The tubular exchangers are widely used in industry for the following reasons. They are custom designed for virtually any capacity and operating conditions, such as from high vacuums to ultra-high pressures (over 100 MPa or 15,000 psig), from cryogenics to high temperatures (about 1100°C, 2000°F), and any temperature and pressure differences between the fluids, limited only by the materials of construction. They can be designed for special operating conditions: vibration, heavy fouling, highly viscous fluids, erosion, corrosion, toxicity, radioactivity, multicomponent mixtures, and so on. They are the most versatile exchangers made from a variety of metal and nonmetal materials (graphite, glass, and Teflon) and in sizes from small (0.1 m², 1 ft²) to super-giant (over 100,000 m², 10^6 ft²). They are extensively used as process heat exchangers in the petroleum-refining and chemical industries; as steam generators, condensers, boiler feed water heaters, and oil coolers in power plants; as condensers and evaporators in some air-conditioning and refrigeration applications; in waste heat recovery applications with heat recovery from liquids and condensing fluids; and in environmental control.

Shell-and-tube exchangers are basically noncompact exchangers. Heat transfer surface area per unit volume ranges from about 50 to 100 m²/m³ (15 to 30 ft²/ft³). Thus, they require a considerable amount of space, support structure, and capital and installation costs. As a result, overall they may be quite expensive compared to compact heat exchangers. The latter exchangers have replaced shell-and-tube exchangers in those applications today where the operating conditions permit such use. For the equivalent cost of the exchanger, compact heat exchangers will result in high effectiveness and be more efficient in energy (heat) transfer.

Shell-and-tube heat exchangers are classified and constructed in accordance with the widely used Tubular Exchanger Manufacturers Association (TEMA) standards [5], DIN and other standards in Europe and elsewhere, and ASME Boiler and Pressure Vessel Codes. TEMA has developed a notation system to designate the main types of shell-and-tube exchangers. In this system, each exchanger is designated by a three-letter combination, the first letter indicating the front-end head type, the second the shell type, and the third the rear-end head type. These are identified in Fig. 17.3. Some of the common shell-and-tube exchangers are BEM, BEU, BES, AES, AEP, CFU, AKT, and AJW. Other special types of commercially available shell-and-tube exchangers have front-end and rear-end heads different from those in Fig. 17.3; these exchangers may not be identifiable by the TEMA letter designation.

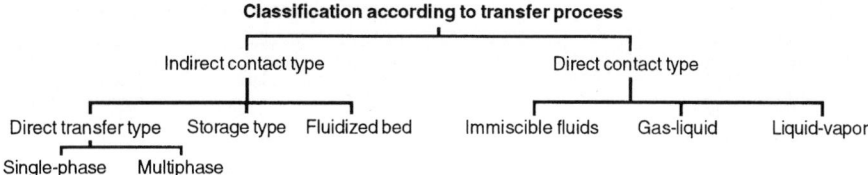

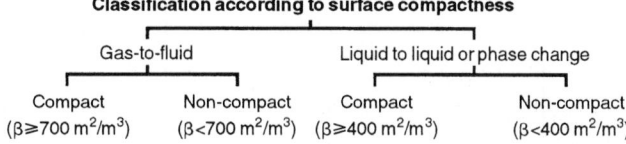

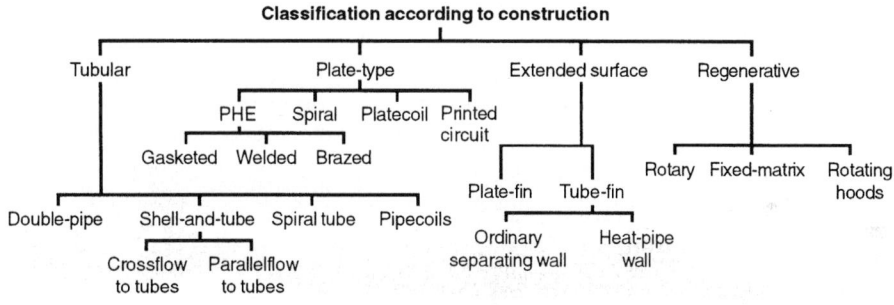

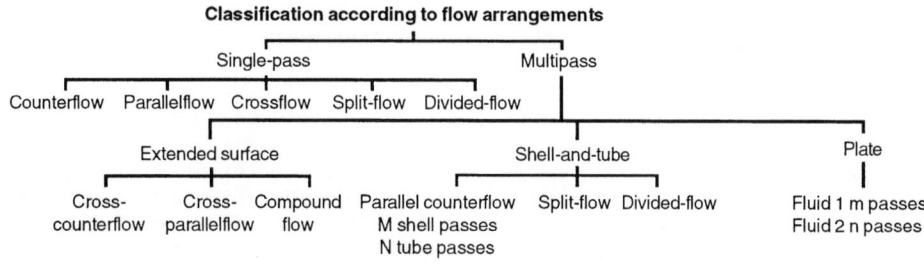

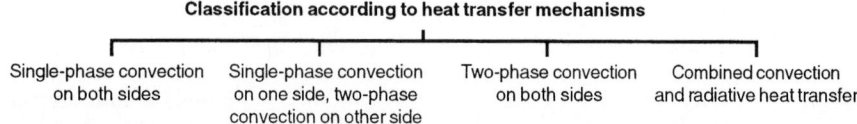

FIGURE 17.1 Classification of heat exchangers.

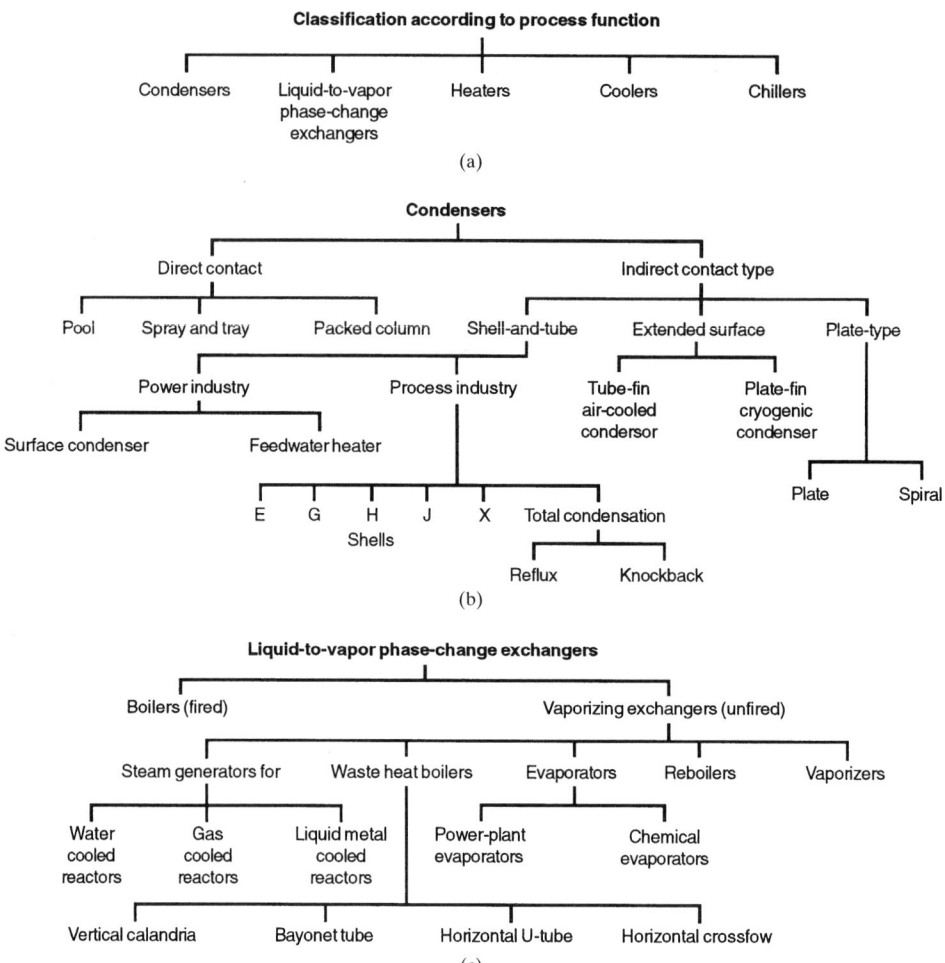

FIGURE 17.2 (*a*) Classification according to process function. (*b*) Classification of condensers. (*c*) Classification of liquid-to-vapor phase-change exchangers.

The three most common types of shell-and-tube exchangers are fixed tubesheet design, U-tube design, and the floating head type. In all types, the front-end head is stationary, while the rear-end head could be either stationary or floating depending upon the thermal stresses in the shell, tube, or tubesheet due to temperature differences as a result of heat transfer.

The exchangers are built in accordance with three mechanical standards that specify design, fabrication, and materials of unfired shell-and-tube heat exchangers. Class R is for generally severe requirements of petroleum and related processing applications. Class C is for generally moderate requirements for commercial and general process applications. Class B is for chemical process service. The exchangers are built to comply with the applicable ASME (American Society of Mechanical Engineers) Boiler and Pressure Vessel Code Section VIII, or other pertinent codes and/or standards. The TEMA standards supplement and define the ASME code for heat exchanger applications. In addition, the state and local codes applicable to the plant location must also be met. In this chapter, we use the TEMA standards, but there are other standards such as DIN 28 008.

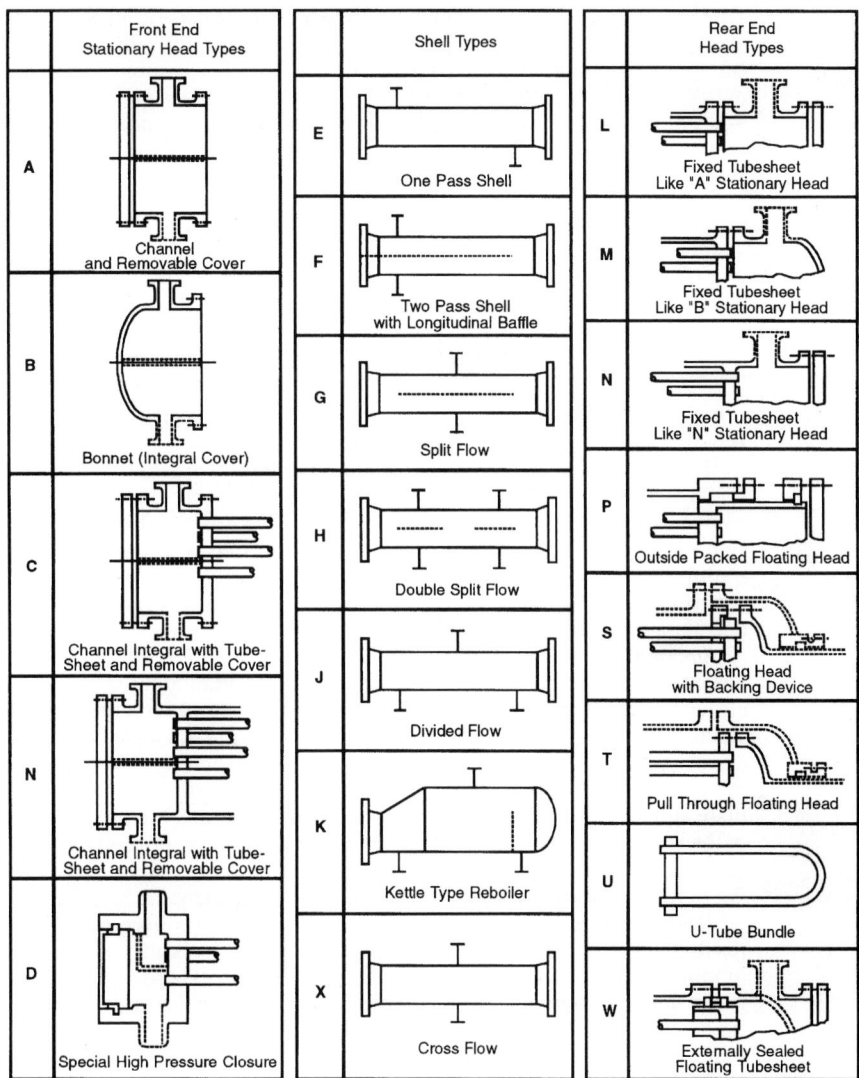

FIGURE 17.3   Standard front-end head, shell-type, and rear-end head types, from TEMA [5].

The TEMA standards specify the manufacturing tolerances for various mechanical classes, the range of tube sizes and pitches, baffling and support plates, pressure classification, tubesheet thickness formulas, and so on, and must be consulted for all these details.

Criteria for Mechanical Selection

Shells.   Shells are generally made from standard pipes for shell diameters less than 610 mm (2 ft) and by rolling and welding plates to the desired diameters for larger sizes. The shell diameters range from less than 50 mm (2 in) to 6.10 m (20 ft) for special applications. The E shell (see Fig. 17.3) is a single-pass shell that is economical and usually has the most efficient thermal arrangement (i.e., it has the highest mean temperature difference correction factor

F). However, if the F factor is low enough to require two E shells in series for multipass tube-side exchangers, the F shell (a two-pass shell; a counterflow unit) can be used as an equivalent but more economical unit. However, the F shell baffle is subject to fluid and thermal leakage across the longitudinal baffle, so it must be carefully designed and constructed. It also provides more problems in removing or replacing the tube bundle. The F shell is used for single-phase applications. If the pressure drop in an F shell is limiting, a split-flow G or H shell can be used with some sacrifice in the F factor. The G shell is used in many applications, with the shellside thermosiphon and forced convective boiling as one of the common applications. If shellside pressure drop becomes limiting, the divided-flow J shell is used; however, there is some loss in the thermal efficiency (a lower F factor). The J shell is commonly used in vacuum condensing applications. The X shell is used for large shell flows or for the lowest shellside Δp for a given flow rate. In the X shells, full-size support plates are used to prevent tube vibration. For high flow rates (inlet velocity), alternately H or J shells with two inlet nozzles are used. The G and H shells are seldom used for shellside single-phase applications, since there is no advantage over E or X shells. They are used for thermosiphon reboilers, condensers, and other phase-change applications. The K shell is exclusively used for vaporization of liquid on the shell side.

The type of shell shown in Fig. 17.3 has either one or two shell passes per shell. Because of the high cost of the shell compared to tubes, three or four shell passes in a shell could be made by the use of longitudinal baffles with positive sealing. However, such multipassing will reduce the flow area compared to a single-pass unit with possibly higher Δp. Alternatively, multiple shells in series are used in some applications (such as up to six shells in series in heat recovery trains) for increased effectiveness, part-load operation, spare bundle requirement, and shipping and handling requirements.

Stationary Heads.   These are used to get the tubeside fluid into the tubes. There are two basic types of stationary (front-end) heads: the bonnet and the channel. The bonnet (B) has either a side-entering or end-entering nozzle and is used for generally clean tubeside fluids; it has fewer joints (and hence is less expensive than the A head) but does require breaking the piping joints in order to clean or inspect the tubes. The channel head can be removable (A) or integral with the tubesheet (C and N). It has side-entering nozzles and a removable cover plate allowing easy access to the tubes without disturbing the piping. While the shell is flanged in the C head, it is welded in the N head to eliminate any potential leak between the shell and tubes. The D head has a special high pressure enclosure and is used in feedwater heaters having tubeside pressures 10–40 MPa (1500–6000 psig) range.

Rear-End Heads.   Shells and tubes are exposed to different temperatures in operation, resulting in thermal stresses that can cause bending, buckling, or fracture of the tubes or shell or failure of tube-to-tubesheet joints. This thermal stress problem can be further compounded if the shell and tube materials are different, or residual stresses remain after the exchanger fabrication. Proper rear head design can minimize/eliminate these thermal stresses, and the specific design depends upon the thermal stresses in the operation.

The fixed tubesheet (L, M, or N) is a rigid design and permits differential thermal expansion to moderate inlet temperature differences (< 56°C or 100°F) between the tubes and shell. Use of a shell expansion joint can raise this temperature difference limit to 83°C (150°F). Any number of tube passes can be used. However, the shell side can only be chemically cleaned. Individual tubes can be replaced. These heads allow the least clearance between the shell and the tube bundle (10–12 mm, 0.4–0.5 in), thus minimizing the bundle-to-shell bypass flow. Fixed tubesheet exchangers are used for low temperatures (315°C or 600°F max) coupled with low pressures (2100 kPa gauge or 300 psig max). This is a low-cost exchanger but slightly higher in cost than the U-tube exchanger.

The U-tube head (U) is a very simple design requiring a bundle of U tubes, only one tubesheet, no expansion joints, and no rear-end head at all, allowing easy removal of the bundle. The thermal stress problem is eliminated because each tube is free to expand/contract independently. In this design, individual tube replacement is not possible except in the outer rows, and an even number of tube passes is required. Some tubes are lost in the center due to tube bend limit, and tubeside mechanical cleaning of the bends is difficult. Flow-induced vibration could be a problem for tubes in the outermost row. It is the lowest-cost design because there is no need for the second tubesheet.

The outside packed floating head (P) provides for expansion and can be designed for any number of passes. Shell and tube fluids cannot mix if gaskets or packing develop leaks, since the leak is to the atmosphere; however, very toxic fluids are not used. This P head requires a larger bundle-to-shell clearance, and sealing strips are used in some designs to block the bundle-to-shell bypass flow partially. This design allows only an even number of tube passes. For a given amount of surface area, it requires a larger shell diameter compared to the L, M, or N head design. This is a high-cost design.

The split-ring floating head (S) has the tubesheet sandwiched between a removable split ring and the cover, which has a larger diameter than the shell. This permits a smaller clearance between the shell and bundle, and the sealing strip is required for only selected applications. On account of the floating head location, the minimum outlet baffle spacing is the largest of any design. Gasket failure is not visible and allows mixing of tube and shell fluids. To remove the bundle or clean the tubes, both ends of the exchanger must be disassembled. Cleaning costs somewhat more than for the pull-through type (T), and the exchanger cost is relatively high.

The pull-through floating head (T) can be removed from the shell by disassembling the stationary head. Because of the floating-head flange bolting, this design has the largest bundle-to-shell clearance, and thus sealing strips are necessary. Even-numbered multipassing is imposed. Again, gasket leakage allows mixing of shell and tube fluids and is not externally visible. Cost is relatively high.

The packed floating head with lantern ring (W) has the lantern ring packing compressed by the rear head bolts. Bundle-to-shell clearance is relatively small. A single- or two-pass arrangement is possible. Potential leakage of either shell or tube fluid is to the atmosphere; however, mixing of these two fluids is possible in the leakage area. Hence, this design is used for benign fluids and low to very moderate pressures and temperatures. The bundle is easily removed, but this design is not recommended on account of severe thermal fluctuations, which can loosen the packing. This floating head design is the lowest in cost. The design features of shell-and-tube exchangers with various rear heads are summarized in Table 17.1.

Baffles. Longitudinal baffles are used in the shell to control the overall flow direction of the shell fluid as in F, G, and H shells. The transverse baffles may be classified as plate baffles and axial-flow baffles (rod, NEST, etc.). The plate baffles (see Fig. 17.4) are used to support the tubes, to direct the fluid in the tube bundle at about 90° to the tubes, and to increase the turbulence of the shell fluid. The rod (and other axial-flow) baffles (see Fig. 17.5) are used to support the tubes, to have the fluid flowing axially over the tubes, and to increase the turbulence of the shell fluid. Flow-induced vibration is virtually eliminated in rod and other axial-flow baffles having axial shellside flows.

Plate baffles can be segmental with or without tubes in the window, multisegmental, or disk-and-doughnut (see Fig. 17.4). The single-segmental baffle is most common and is formed by cutting a segment from a disk. As shown in Fig. 17.4, the cuts are alternately 180° apart and cause the shell fluid to flow back and forth across the tubes more or less perpendicularly. Baffle cut and baffle spacing are selected from the shellside fouling and Δp considerations. In fouling situations, the baffle cut should be below 25 percent. Baffle spacing is chosen to avoid tube vibrations and optimize heat transfer/pressure drop by keeping about the same flow area in the crossflow and window zone. The direction of the baffle cut is preferred horizontal for very viscous liquids for better mixing (as shown for segmental baffles in Fig. 17.4). The direction of baffle cut is selected vertical for shellside consideration (for better drainage), evaporation/boiling (to promote more uniform flow), solids entrained in liquid, and multipassing on the shellside. One disadvantage of the segmental types is the flow bypassing that occurs in the annular spaces (clearances) between the tube bundle and shell. If the pressure drop is too high or more tube supports are needed to prevent vibration, the segmental baffle is further subdivided into a double-segmental or triple-segmental arrangement. An alternate method of

TABLE 17.1 Design Features of Shell-and-Tube Heat Exchangers [1]

Design features	Fixed tubesheet	Return bend (U-tube)	Outside-packed stuffing box	Outside-packed latern ring	Pull-through bundle	Inside split backing ring
TEMA rear-head type	L, M, N	U	P	W	T	S
Tube bundle removable	No	Yes	Yes	Yes	Yes	Yes
Spare bundles used	No	Yes	Yes	Yes	Yes	Yes
Provides for differential movement between shell and tubes	Yes, with bellows in shell	Yes	Yes	Yes	Yes	Yes
Individual tubes can be replaced	Yes	Yes[a]	Yes	Yes	Yes	Yes
Tubes can be chemically cleaned, both inside and outside	Yes	Yes	Yes	Yes	Yes	Yes
Tubes can be mechanically cleaned on inside	Yes	With special tools	Yes	Yes	Yes	Yes
Tubes can be mechanically cleaned on outside	Yes	Yes[b]	Yes[b]	Yes[b]	Yes[b]	Yes[b]
Internal gaskets and bolting are required	No	No	No	No	Yes	Yes
Double tubesheets are practical	Yes	Yes	Yes	No	No	No
Number of tubesheet passes available	Any	Any even number	Any[c]	One or two[d]	Any[e]	Any[e]
Approximate diametral clearance (mm) (Shell ID–D_{otl})	11–18	11–18	25–50	15–35	95–160	35–50
Relative costs in ascending order, least expensive = 1	2	1	4	3	5	6

[a] Only those in outside rows can be replaced without special designs.
[b] Outside mechanical cleaning possible with square or rotated square pitch, or wide triangular pitch.
[c] Axial nozzle required at rear end for odd number of passes.
[d] Tube-side nozzles must be at stationary end for two passes.
[e] Odd number of passes requires packed gland or bellows at floating head.

improving tube support for vibration prevention is to eliminate tubes in the window zone, in which case intermediate support baffles can be used. This design requires a larger shell to contain the same number of tubes in a segmental baffle exchanger, but a lower pressure drop and improved heat transfer (due to improved flow distribution and less fouling) can help to reduce this diameter increase.

Tube Pitch and Layout. The selection of tube pitch (see Fig. 17.6 for the definition) is a compromise between a close pitch for increased shellside heat transfer and surface compactness, and an open pitch for decreased shellside plugging and ease in shellside cleaning. In most shell-and-tube exchangers, the ratio between tube pitch and outside diameter ratio varies from 1.25 to 2.00. The recommended ligament width depends upon the tube diameter and pitch; the values are provided by TEMA [5]; the minimum value is 3.18 mm (1/8 in) for clean services and 6.35 mm (1/4 in) where mechanical cleaning is required.

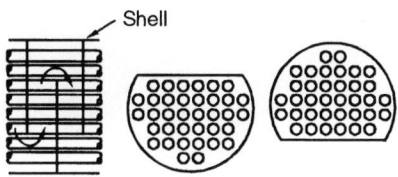

(a) Single-segmental baffle

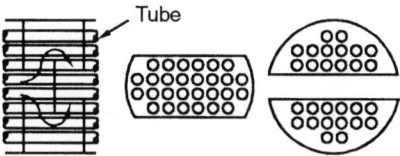

(b) Double-segmental baffle

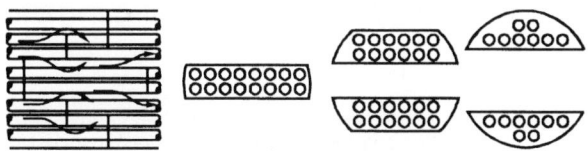

(c) Triple-segmental baffle

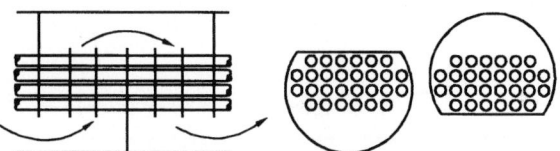

(d) No-tubes-in-window segmental baffle

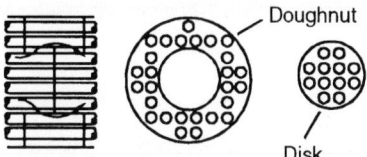

(e) Disk-and-doughnut baffle

FIGURE 17.4 Plate baffle types: (*a*) single-segmental baffle, (*b*) double-segmental baffle, (*c*) triple-segmental baffle, (*d*) no-tubes-in-window segmental baffle, and (*e*) disc-and-doughnut baffle.

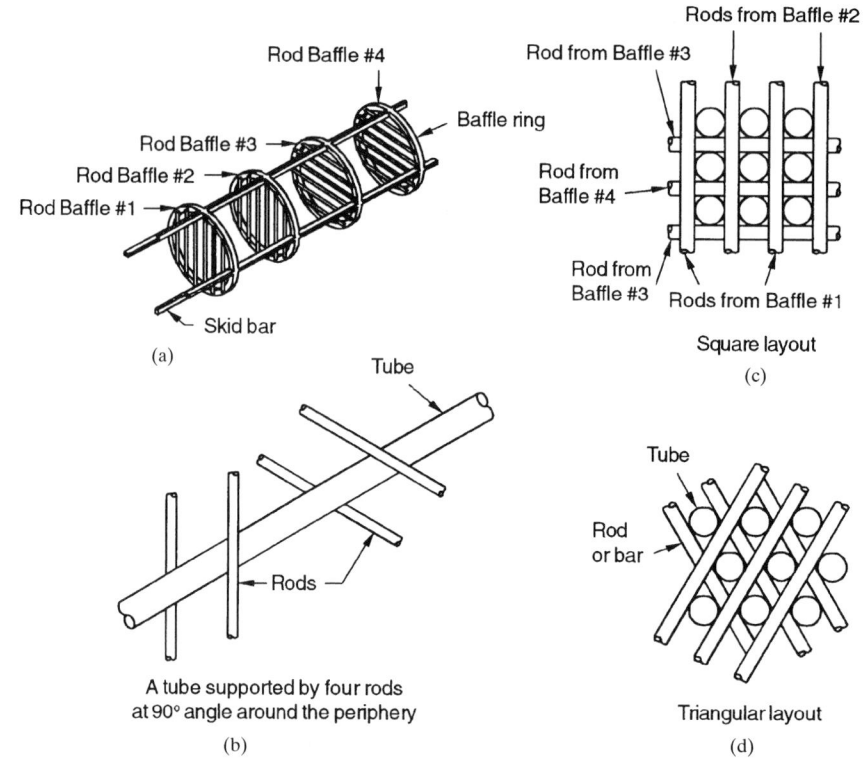

FIGURE 17.5 Rod baffle supports: (*a*) four rod baffles supported by skid bars (no tubes shown), (*b*) a tube supported by four rods at 90° angle around the periphery, (*c*) a square layout of tubes with rods, (*d*) a triangular layout of tubes with rods.

	30° Triangular Staggered Array	60° Rotated Triangular Staggered Array	90° Square Inline Array	45° Rotated Square Staggered Array
Transverse tube pitch X_t	p_t	$\sqrt{3}\, p_t$	p_t	$\sqrt{2}\, p_t$
Longitudinal tube pitch X_ℓ	$\left(\sqrt{3}/2\right)p_t$	$p_t/2$	p_t	$p_t/\sqrt{2}$
Ratio of minimum free flow area to frontal area, $A_o/A_{fr} = \sigma$	$\dfrac{p_t - d_o}{p_t}$	$\dfrac{\sqrt{3}\, p_t - d_o}{\sqrt{3}\, p_t}$ for $\dfrac{p_t}{d_o} \geq 3.732$ $\dfrac{\sqrt{2}\, p_t - d_o}{\sqrt{3}\, p_t}$ for $\dfrac{p_t}{d_o} \leq 3.732$	$\dfrac{p_t - d_o}{p_t}$	$\dfrac{\sqrt{2}\, p_t - d_o}{\sqrt{2}\, p_t}$ for $\dfrac{p_t}{d_o} \geq 1.707$ $\dfrac{\sqrt{2}\, p_t - d_o}{\sqrt{2}\, p_t}$ for $\dfrac{p_t}{d_o} \leq 1.707$

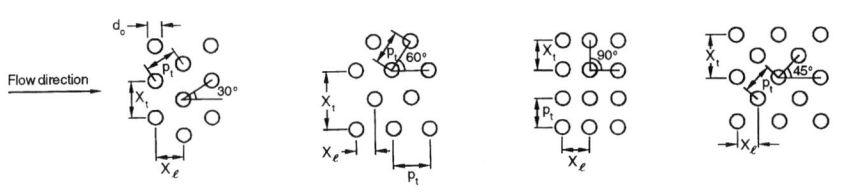

FIGURE 17.6 Nomenclature and geometrical properties of tube banks common in shell-and-tube exchangers.

Two standard types of tube layouts are the square and the equilateral triangle, as shown in Fig. 17.6. The equilateral tube layout can be oriented at a 30° or 60° angle to the flow direction. For the square tube layout, it is 45° and 90°. Note that the 30° and 45° arrangements are staggered, and the 60° and 90° are inline.

For identical tube pitch and flow rates, the tube layouts in decreasing order of shellside heat transfer coefficient and pressure drop are: 30°, 45°, 60°, and 90°, with the 90° layout having the lowest heat transfer coefficient and pressure drop.

The square tube layout (90° or 45°) is used when jet or mechanical cleaning is necessary on the shell side. The triangular tube layout is generally used in the fixed tubesheet design because there is no need for cleaning. It provides a more compact arrangement, usually resulting in a smaller shell, and the strongest header sheet for a specified shellside flow area. Hence, it is preferred when the operating pressure difference between two fluids is large. When mechanical cleaning is required, the 45° layout is preferred for laminar or turbulent flow of a single-phase fluid and for condensing fluid on the shell side. If the pressure drop is constrained on the shell side, the 90° layout is used for turbulent flow. For boiling applications, the 90° layout, which provides vapor escape lanes, is preferred. However, if mechanical cleaning is not required, the 30° layout is preferred for single-phase laminar or turbulent flow and condensing applications involving high ΔT range (a mixture of condensibles). The 60° layout is preferred for condensing applications involving high ΔT range (generally pure vapor condensation) and for boiling applications. Horizontal tube bundles are used for shellside condensation or vaporization.

Tubes. The tubes are either plain or finned with low fins (0.79–1.59 mm or 0.031–0.063 in) or high fins (generally 15.88 to 19.05 mm or 0.63 to 0.75 in) with 630–1260 fins/m (16–32 fins/in). Consult manufacturers' catalogs for dimensions. The plain tubes range 6.35–50.8 mm (0.25–2 in) in outside diameter. For small exchangers of less than 203 mm (8 in) shell diameter, smaller tubes and pitches are used, but these exchangers fall outside the range of TEMA standards. For mechanical cleaning, the smallest practical tube diameter is 19.05 mm (3/4 in). The tube diameter and length are based on the type of cleaning to be used. If a drilling operation is required, the minimum tube diameter considered is 19.05 or 25.4 mm (3/4 or 1 in), and the maximum tube length is 4.9 m (16 ft). Longer exchangers made with plain tubes are up to 30 m (100 ft), the length limited by the ability to handle such long exchangers in the shop and field.

Tubes are fastened to tubesheets by welding, mechanical rolling, or both. However, these joints are susceptible to thermal and pressure stresses and may develop leaks. In those instances where mixing of the shell and tube fluids would result in corrosive or other hazardous conditions, special designs such as double tubesheets are used, with the space between the tubesheets vented. A double tubesheet can be used only in the following rear head designs: fixed tubesheets (L, M, or N) and the outside packed head (P). Bimetal tubes are used when corrosive conditions of shell and tube fluids require the use of different metals.

Pass Arrangements. The number of tube passes per exchanger can range from 1 to 16. If more than one pass is used, some loss in thermal efficiency results because of the effect of flow pattern on the mean temperature difference. A design for large numbers of passes results from the need to compensate for low flow rates or the need to maintain high velocities to reduce fouling and get good heat transfer. However, large temperature changes in the tube fluid can, by thermal expansion, cause the floating tubesheets to cock and bind. The passes should be so arranged as to minimize the number of lanes between the passes that are in the same direction as the shell fluid flow. Also, the passes should be arranged so that the tube side can be drained and vented.

Shell Nozzles and Impingement Methods. Whenever a high-velocity two-phase flow enters the shell, some type of impingement protection is required to avoid tube erosion and vibration; some examples are shown in Fig. 17.7: annular distributors (*d*), impingement plates (*a, c*), and impingement rods (*b*). The nozzles must also be sized with the understanding that the tube bundle will partially block the opening. In order to provide escape area with impingement plates, some tubes may have to be removed. The annular distributor is an excellent design that allows any orientation of the nozzles, provides impingement protection, and

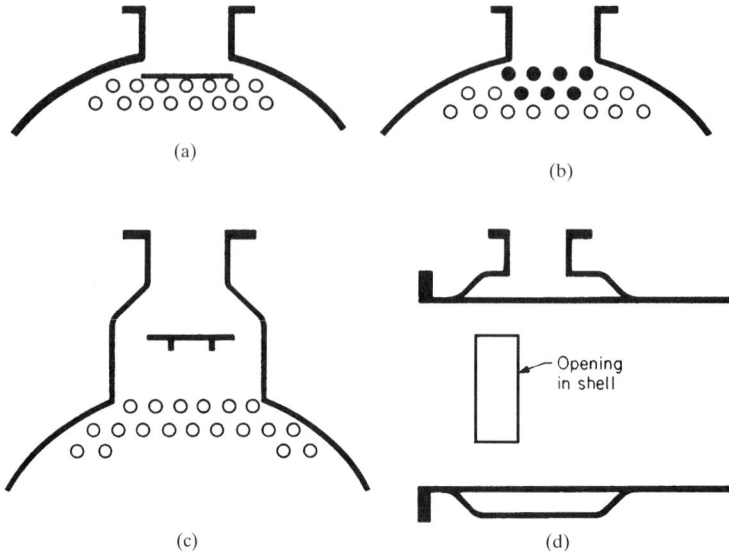

FIGURE 17.7 Impingement protection designs: (*a*) impingement plate, (*b*) impingement rods, (*c*) nozzle impingement baffle, (*d*) vapor belt.

allows baffling closer to the tubesheets and thus higher velocities; however, it is a very expensive design. Dummy rods or extra-heavy walled tubes near the nozzles are also good impingement devices.

Drains and Vents. All exchangers need to be drained and vented; therefore, care should be taken to locate and size drains and vents properly. The proper location depends upon the exchanger design and orientation. Additional openings may be required for instruments such as pressure and temperature sensors.

Selection Procedure. The selection procedure for a specific design of exchanger involves the consideration of many and often conflicting requirements of process conditions, operation, and maintenance. Depending upon the relative importance of these factors as determined by the designer, one or several designs may be selected for evaluation.

Selecting Tubeside Fluid. The choice of the fluid to be on the tube side will influence the selection of the type of exchanger and require evaluation of the following factors to arrive at a satisfactory compromise.

Cleanability. The shell is difficult to clean and requires the cleaner fluid.

Corrosion. Corrosion or process cleanliness may dictate the use of expensive alloys; therefore, the more corrosive fluids are placed inside the tubes in order to save the cost of an alloy shell.

Pressure. High-pressure shells, because of their diameters, are thick-walled and expensive; therefore, high-pressure fluids are placed in the tubes.

Temperatures. The high-temperature fluid should be inside the tubes. High temperatures reduce the allowable stresses in materials, and the effect is similar to high pressure in determining shell thickness. Furthermore, safety of personnel may require the additional cost of insulation if the high-temperature fluid is in the shell.

Hazardous or expensive fluids. The more hazardous or expensive fluid should be placed on the tighter side of the exchanger, which is the tube side of some types of exchangers.

Quantity.   A better overall design may be obtained when the smaller quantity of fluid (i.e., the fluid with lower mass flow rate) is placed in the shell. This choice may be to avoid multipass construction with consequent loss of exchanger effectiveness (or the F factor) or to obtain turbulent flow in the shell at low Reynolds numbers.

Viscosity.   The critical Reynolds number for turbulent flow on the shell side is about 200; hence, when the fluid flow in the tubes is laminar, it may be turbulent if that same fluid is placed in the shell. However, if the flow is still laminar when in the shell, then it is best to place the fluid back inside the tubes, as it will be somewhat easier to predict both heat transfer and flow distribution.

Pressure drop.   If the pressure drop of one fluid is critical and must be accurately predicted, then place that fluid inside the tubes. Pressure drop inside tubes can be calculated with less error, as the pressure drop in the shell will deviate widely from theoretical values depending upon the shell leakage clearances in the particular exchanger.

Selecting Shell and Head.   The E shell is the best arrangement; however, if shellside pressure drop is too high, a divided-flow J or G shell may be used. The F shell is a possible alternative when a temperature cross occurs and more than one shell pass is required. Accessibility to the tubes governs the selection of the stationary head, while thermal stress, need for cleaning, possible gasket problems, leakage, plant maintenance experience, and cost are factors influencing the rear head selection. See the earlier section on criteria for mechanical selection for comments on specific heads.

Selecting Tube Size and Layout.   The best ratio of heat transfer to pressure drop is obtained with the smallest-diameter tubes; however, the minimum size is determined by the ability to clean the tubes. Pressure drop, tube vibration, tubesheet joints, and cost are several factors limiting the minimum size. Also, a reasonable balance between the tubeside and shellside heat transfer coefficients is desired. The ligament between tubes is governed by the pitch ratio and tube size selected; however, for tubesheet strength, drilling tolerances, and the ability to roll a tight tube joint, a minimum ligament of 3.2 mm (1/8 in) to 6.4 mm (1/4 in) is recommended; the more conservative design uses larger ligaments. The pitch ratio and ligament thickness also affect the shellside fluid velocity and hence the heat transfer and pressure drop.

Tube layouts are either triangular or square; the choice usually depends on the need for shellside cleaning. The square pitch is particularly suitable for cleaning; however, a larger triangular pitch can also be used. For example, a 25.4-mm (1-in) tube on a 34.9-mm (1-3/8-in) triangular pitch will have essentially the same tube count, shell velocities, and heat transfer coefficients; it will also have almost the same clearances for cleaning as a 25.4-mm (1-in) tube on a 31.8-mm (1-1/4-in) square pitch, but the 9.5-mm (3/8-in) ligament will be 50 percent stronger. Other factors including number of tubes and heat transfer for different flow angles (30°, 45°, etc.) are discussed above.

Selecting Baffles.   The segmental baffle is commonly used unless problems of pressure drop, tube vibration, or tube support dictate the use of double, triple, or rod baffling or a no-tube-in-window configuration. Note that these alternate choices also seriously affect the reliability of the correlations for heat transfer and pressure drop. The segmental baffles are spaced at a minimum distance of 50.8 mm (2 in) or $0.2D_s$, whichever is larger, and a maximum spacing of D_s. The baffle cut also depends linearly on the baffle spacing and should be 20 percent of D_s at $0.2D_s$ spacing and 33 percent at D_s. The maximum spacing is also determined by the need for tube support. The TEMA maximum unsupported length depends on tube size and material but ranges from 50 to 80 tube diameters (see TEMA standards for specific values). This maximum length usually occurs at the ends of the exchanger in the window area of the first or last baffle, since the end baffle spacing generally is greater (due to nozzle location) than the central baffle. Baffle spacing is selected to obtain a high velocity within a pressure drop limit.

Selecting Nozzles.   Nozzle sizes and impingement devices are related by the TEMA rule-of-thumb value of ρV^2 where ρ is the shell fluid density and V is the shell fluid velocity at the nozzle exit and entrance to the shell side. If $\rho V^2 > 2250$ kg/(m·s²) or 1500 lbm/(ft·s²) for non-corrosive or nonabrasive single-phase fluids, or 750 kg/(m·s²) [500 lbm/(ft·s²)] for other liquids, any vapor-liquid mixture or saturated vapor, then an impingement device is needed. There are several possible configurations indicated in Fig. 17.7. Also the entrance into the tube bundle should have a ρV^2 less than 6000 kg/(m·s²) or 4000 lbm/(ft·s²). The entrance area is the total free area between a nozzle and the projected area on the tube bundle. Meeting these requirements may require removal of some tubes. Usually such dimensions and area are not available until the mechanical drawings have been made. In the design stage, an estimate of these effects is made or a final check calculation is made based on final drawings if the shell pressure drops are marginal. Nozzle locations with respect to the shell flange are governed by pressure vessel codes.

Selecting Tube Passes. The number of tube passes is kept as low as possible in order to get simple head and tubesheet designs. For even numbers of multiple pass designs, no off-center nozzle on the floating head is required. The flow quantity and the desired minimum tubeside velocity determine the number of tubes per pass; and the total area and tube length then fix the number of passes for the desired performance. However, the number of tube passes must be an even integer; hence, the tube length is variable.

Newer Designs of Shell-and-Tube Exchangers

In a conventional shell-and-tube exchanger, transverse plate baffles are used to support the tubes and direct the shellside stream to flow across the tubes. However, it results in the shell-side flow that wastes pressure drop in turning back and forth without yielding the corresponding heat transfer. The high turnaround pressure drop also results in more leakage flow (shell-to-baffle and baffle-to-tube), lower crossflow, and subsequent lower heat transfer coefficients. The transverse baffles create dead spots or recirculation zones that could promote fouling. Various leakage and bypass flows on the shell side reduce the mean temperature difference in the exchanger and the performance of the exchanger; a very high exchanger effectiveness may not be achievable in this type of exchanger regardless of a large increase in the surface area.

Some of these problems can be eliminated by modifying the shellside design to achieve axial or longitudinal flows; one construction with rod baffles is shown in Fig. 17.5. Such designs require different ways to support the tubes and may virtually eliminate the flow-induced tube vibration problem. Usually heat transfer rate per unit pressure drop is high in such designs; but on the absolute scale, both heat transfer rate and pressure drops are low. As a result, the exchanger usually ends up with a relatively large shell length-to-diameter ratio. In addition to rod and NEST baffle types, several new designs have been developed to induce axial flows, as shown in Fig. 17.8. Figure 17.8*a* shows a design with a full circle baffle with

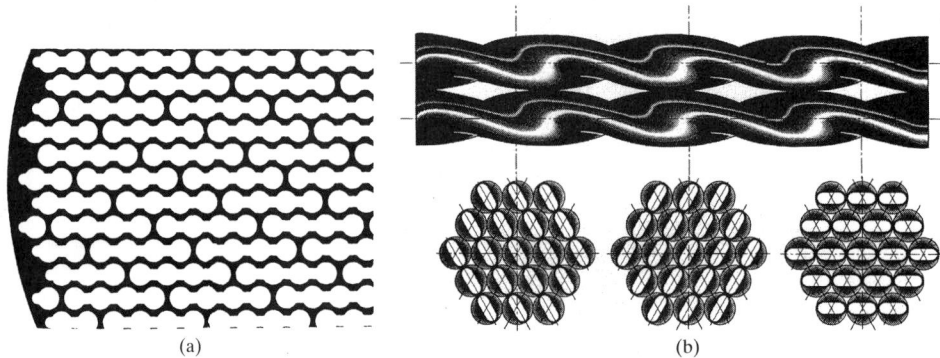

(a)   (b)

FIGURE 17.8 (*a*) Axial flow baffle, courtesy of Brown Fintube Company, Houston, Texas, (*b*) a twisted tube exchanger, courtesy of ABB Lummus Heat Transfer, Bloomfield, New Jersey.

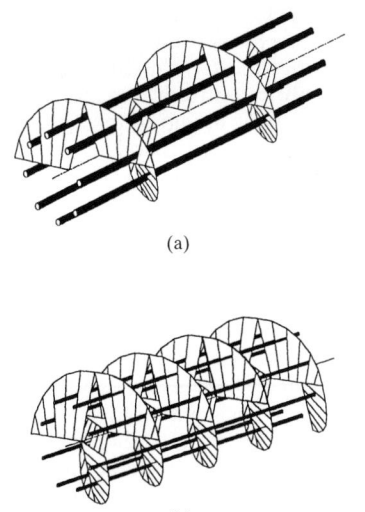

(a)

(b)

FIGURE 17.9 A helical baffle shell-and-tube exchanger: (a) single helix, (b) double helix, courtesy of ABB Lummus Heat Transfer, Bloomfield, New Jersey.

extra space for shellside fluid flow. Figure 17.8b delineates a design with twisted flattened tubes that would yield about 40 percent higher heat transfer coefficient than the conventional shell-and-tube exchanger for the same pressure drop. Plain tubes may be interspersed between twisted tubes for greater design flexibility.

An alternative to conventional and axial flow shell-and-tube exchangers is an exchanger with helical shellside flow. It can be either a single-helix baffle, as shown in Fig. 17.9a, or a double-helix baffle as shown in Fig. 17.9b. There are several variations of angled baffle exchangers available commercially. The helical flow reduces the shellside flow turning losses and fouling tendency compared to a conventional shell-and-tube exchanger, but introduces radial variations in shellside mass flow rate and temperature variations that can be overcome by a radial variation in the tube pitch design.

Compact Heat Exchangers

As defined earlier, compact heat exchangers are characterized by a large heat transfer surface area per unit volume of the exchanger, resulting in reduced space, weight, support structure, and footprint; reduced energy requirement and cost; improved process design, plant layout, and processing conditions; and low fluid inventory compared to conventional designs such as shell-and-tube exchangers. Extremely high heat transfer coefficients h are achievable with small-hydraulic-diameter flow passages with gases, liquids, and two-phase flows. A typical plate heat exchanger has about two times the heat transfer coefficient (h) or overall heat transfer coefficient (U) of a shell-and-tube exchanger for water/water applications. Basic constructions of gas-to-gas compact heat exchangers are plate-fin, tube-fin, and all prime surface recuperators (includes polymer film and laminar flow exchangers) and compact regenerators; basic flow arrangements of two fluids are single-pass crossflow, counterflow, and multipass cross-counterflow. The last two flow arrangements can yield very high exchanger effectiveness or very small temperature differences between fluid streams and very small pressure drops compared to shell-and-tube exchangers. Basic constructions for liquid-to-liquid and liquid-to-phase-change-fluid compact exchangers are: gasketed and welded plate-and-frame, welded stacked plate (without frames), spiral plate, printed circuit, and dimple plate heat exchangers.

Gas-to-Fluid Exchangers.   The unique characteristics of compact extended (plate-fin and tube-fin) surface exchangers, as compared with the conventional shell-and-tube exchangers, are: (1) there are many surfaces available with different orders of magnitude of surface area density; (2) there is flexibility in distributing surface area on the hot and cold sides as warranted by design considerations; and (3) there is generally substantial cost, weight, or volume savings.

The important design and operating considerations for compact *extended* surface exchangers are: (1) usually at least one of the fluids is a gas or specific liquid that has low h; (2) fluids must be clean and relatively noncorrosive because of small-hydraulic-diameter (D_h) flow passages and no easy techniques for mechanically cleaning them; (3) the fluid pumping power (i.e., pressure drop) design constraint is often as equally important as the heat transfer rate; (4) operating pressures and temperatures are somewhat limited compared to shell-and-tube exchangers due to joining of the fins to plates or tubes such as brazing and mechanical expansion; (5) with the use of highly compact surfaces, the resultant shape of a gas-to-fluid exchanger is one having a large frontal area and a short flow length; the header design of a compact heat exchanger is thus important for a uniform flow distribution among the very large number of small flow passages; and (6) the market potential must be large enough to warrant the sizable manufacturing research and tooling costs for new forms to be developed.

Some of the advantages of plate-fin exchangers over conventional shell-and-tube exchangers are as follows. Compact heat exchangers, generally fabricated from thin metallic plates, yield large heat transfer surface area per unit volume (β), typically up to ten times greater than the 50 to 100 m^2/m^3 provided by a shell-and-tube exchanger for general process application and from 1000 to 6000 m^2/m^3 for highly compact gas-side surfaces. Compact liquid or two-phase fluid side surfaces have a ratio ranging from 400 to 600 m^2/m^3. A compact exchanger provides a tighter temperature control and thus is useful for heat sensitive materials. It improves the product (e.g., refining fats from edible oil) and quality (such as a catalyst bed). Also, a compact exchanger provides rapid heating or cooling of a process stream, thus further improving the product quality. The plate-fin exchangers can accommodate multiple (up to 12 or more) fluid streams in one exchanger unit with proper manifolding, thus allowing process integration and cost-effective compact solutions.

The major limitations of plate-fin and other compact heat exchangers are as follows. Plate-fin and other compact heat exchangers have been and can be designed for high-temperature applications (up to about 850°C or 1550°F), high-pressure applications (over 20 MPa or 3000 psig), and moderate fouling applications. However, applications usually do not involve both high temperature and high pressure simultaneously. Highly viscous liquids can be accommodated in the plate-fin exchangers with a proper fin height; fibrous or heavy fouling fluids are not used in the plate-fin exchangers because mechanical cleaning in general is not possible. However, these liquids can be readily accommodated in plate heat exchangers. Most of the plate-fin heat exchangers are brazed. At the current state-of-the-art, the largest size exchanger that can be brazed is about $1.2 \times 1.2 \times 6$ m ($4 \times 4 \times 20$ ft). While plate-fin exchangers are brazed in a variety of metals including aluminum, copper, stainless steels, nickel, and cobalt-based superalloys, the brazing process is generally of proprietary nature and is quite expensive to set up and develop. The plate-fin exchanger is readily repairable if leaks occur at the external border seams.

Fouling is one of the major potential problems in compact heat exchangers (except for plate-and-frame heat exchangers), particularly having a variety of fin geometries or very fine circular or noncircular flow passages that cannot be cleaned mechanically. Chemical cleaning may be possible; thermal baking and subsequent rinsing is possible for small-size units. Hence, extended surface compact heat exchangers may not be used in heavy fouling applications. Nonfouling fluids are used where permissible, such as for clean air or gases, light hydrocarbons, and refrigerants.

Other important limitations of compact heat exchangers are as follows. With a high-effectiveness heat exchanger and/or large frontal area, flow maldistribution could be another serious problem. More accurate thermal design is required, and a heat exchanger must be considered a part of a system. Due to short transient times, a careful design of controls is required for startup of some compact heat exchangers compared to shell-and-tube exchangers. Flow oscillation could be a problem for some compact heat exchangers. No industry standards or recognized practice for compact heat exchangers are yet available, particularly for the power and process industry (note that this is not a problem for aircraft, vehicular, and marine transportation industries). Structural integrity is required to be examined on a case-by-case basis utilizing standard pressure vessel codes.

Liquid-to-Liquid Exchangers.   Liquid-to-liquid and phase-change exchangers are plate-and-frame and welded PHE, spiral plate, and printed circuit exchangers. Some of these are described in some detail later in this section. Some compact heat exchangers and their applications are now summarized.

Plate-Fin Exchangers.   This type of exchanger has corrugated fins (having triangular and rectangular cross sectional shapes most common) sandwiched between parallel plates (referred to as *plates* or *parting sheets*), as shown in Fig. 17.10. Sometimes fins are incorpo-

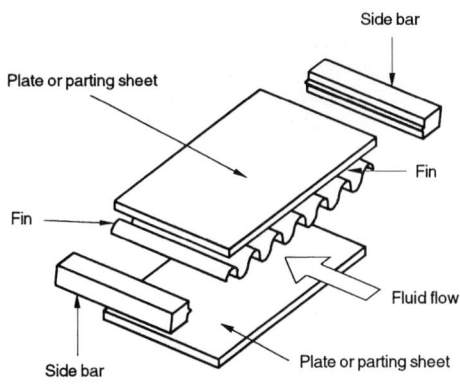

FIGURE 17.10 A plate-fin assembly.

rated in a flat tube with rounded corners (referred to as a *formed tube*), thus eliminating a need for the side bars. If liquid or phase-change fluid flows on the other side, the parting sheet is usually replaced by a flat tube with or without inserts/webs (Fig. 17.11). Other plate-fin constructions include drawn-cup (see Fig. 17.12) or tube-and-center configurations. Fins are die- or roll-formed and are attached to the plates by brazing, soldering, adhesive bonding, welding, or extrusion. Fins may be used on both sides in gas-to-gas heat exchangers. In gas-to-liquid applications, fins are usually used only on the gas side; if employed on the liquid side, they are used primarily for structural strength and flow mixing purposes. Fins are also sometimes used for pressure containment and rigidity.

Plate fins are categorized as (1) plain (i.e., uncut) and straight fins such as plain triangular and rectangular fins, (2) plain but wavy fins (wavy in the main fluid flow direction), and (3) interrupted fins such as offset strip, louver, and perforated. Examples of commonly used fins are shown in Fig. 17.13.

Plate-fin exchangers have been built with a surface area density of up to 5900 m²/m³ (1800 ft²/ft³). There is a total freedom of selecting fin surface area on each fluid side, as required by the design, by varying fin height and fin density. Although typical fin densities are 120 to 700 fins/m (3 to 18 fins/in), applications exist for as many as 2100 fins/m (53 fins/in). Common fin thicknesses range from 0.05 to 0.25 mm (0.002–0.01 in). Fin heights range from 2 to 25 mm (0.08–1.0 in). A plate-fin exchanger with 600 fins/m (15.2 fins/in) provides about

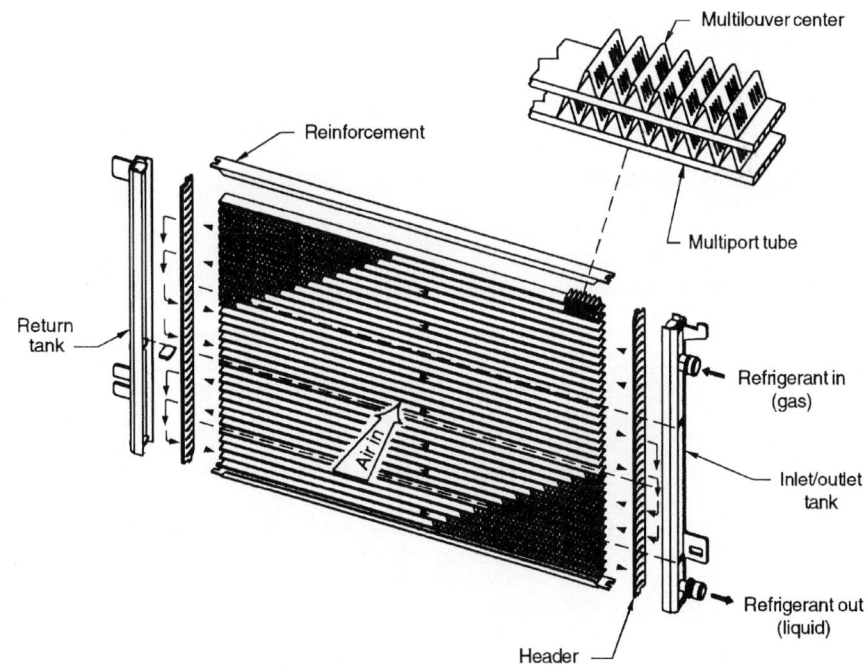

FIGURE 17.11 Flat webbed tube and multilouver fin automotive condenser, courtesy of Delphi Harrison Thermal Systems, Lockport, New York.

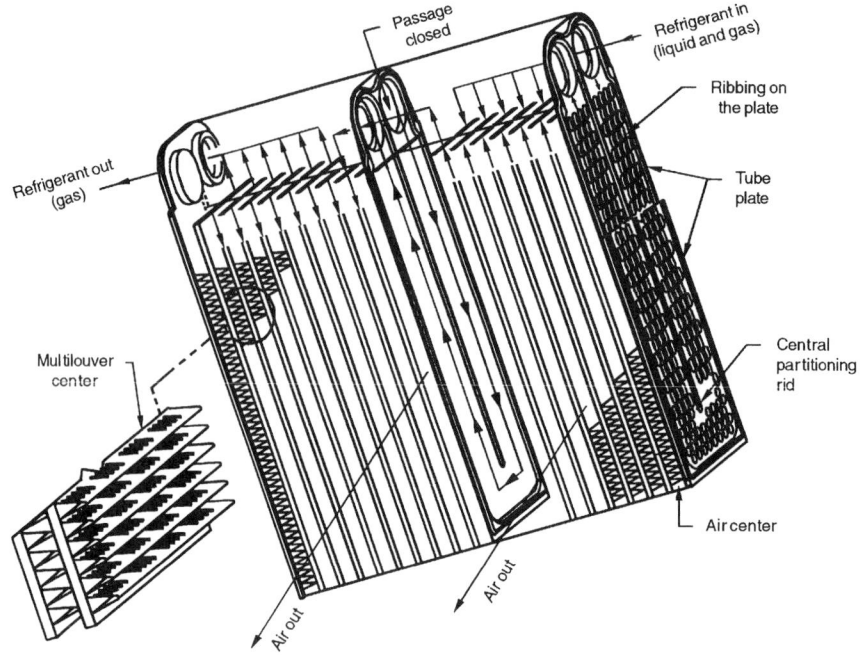

FIGURE 17.12 U-channel ribbed plates and multilouver fin automotive evaporator, courtesy of Delphi Harrison Thermal Systems, Lockport, New York.

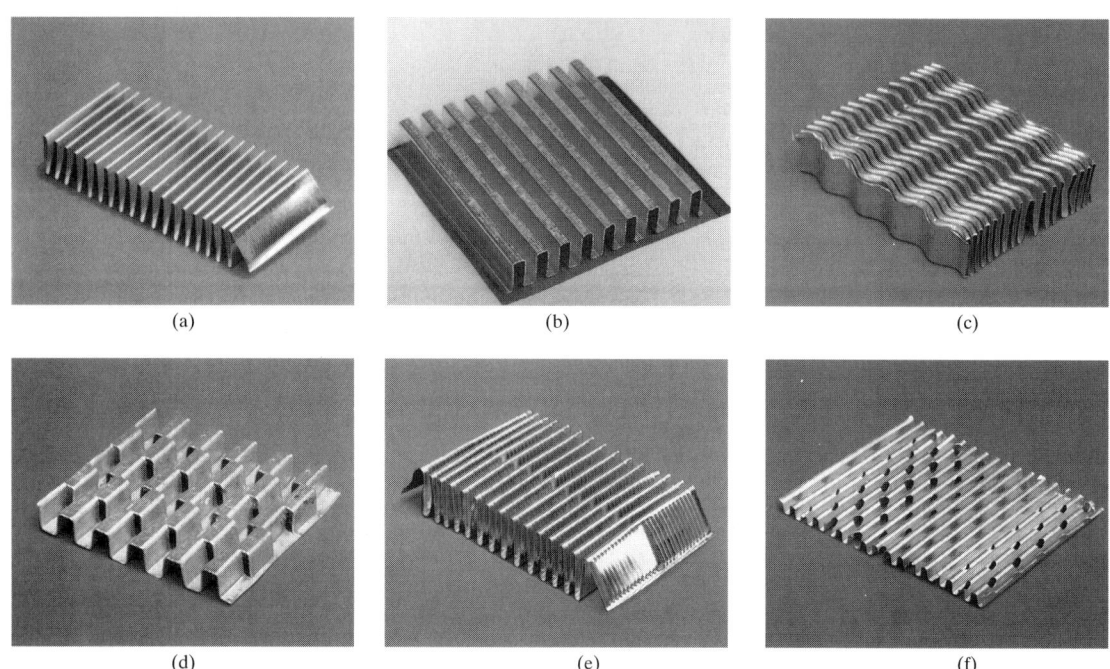

FIGURE 17.13 Fin geometries for plate-fin heat exchangers: (*a*) plain triangular fin, (*b*) plain rectangular fin, (*c*) wavy fin, (*d*) offset strip fin, (*e*) multilouver fin, and (*f*) perforated fin.

1300 m^2 ($400 \text{ ft}^2/\text{ft}^3$) of heat transfer surface area per cubic meter volume occupied by the fins. Plate-fin exchangers are manufactured in virtually all shapes and sizes and are made from a variety of materials.

Plate-fin exchangers are widely used in electric power plants (gas turbine, steam, nuclear, fuel cell), propulsive power plants (automobile, truck, airplane), thermodynamic cycles (heat pump, refrigeration), and in electronics, cryogenics, gas-liquefaction, air-conditioning, and waste heat recovery systems.

Tube-Fin Exchangers.   In this type of exchanger, round and rectangular tubes are the most common, although elliptical tubes are also used. Fins are generally used on the outside, but they may be used on the inside of the tubes in some applications. They are attached to the tubes by a tight mechanical (press) fit, tension winding, adhesive bonding, soldering, brazing, welding, or extrusion. Depending upon the fin type, the tube-fin exchangers are categorized as follows: (1) an individually finned tube exchanger or simply a *finned tube exchanger,* as shown in Figs. 17.14*a* and 17.15, having normal fins on individual tubes; (2) a tube-fin exchanger having flat (continuous) fins, as shown in Figs. 17.14*b* and 17.16; the fins can be plain, wavy, or interrupted, and the array of tubes can have tubes of circular, oval, rectangular, or other shapes; and (3) longitudinal fins on individual tubes. The exchanger having flat (continuous) fins on tubes has been variously referred to as a *plate-fin and tube, plate-finned tube,* and *tube in plate-fin exchanger* in the literature. In order to avoid confusion with a plate-fin exchanger defined above, we will refer to this type as a tube-fin exchanger having flat (plain, wavy, or interrupted) fins. Individually finned tubes are probably more rugged and practical in large tube-fin exchangers. Shell-and-tube exchangers sometimes employ low finned tubes to increase the surface area on the shellside when the shellside heat transfer coefficient is low compared to the tubeside coefficient. The exchanger with flat fins is usually less expensive on a unit heat transfer surface area basis because of its simple and mass-production-type construction features. Longitudinal fins are generally used in condensing applications and for viscous fluids in double pipe heat exchangers.

Tube-fin exchangers can withstand high pressures on the tube side. The highest temperature is again limited by the type of bonding, materials employed, and material thickness. Tube-fin exchangers with an area density of about $3300 \text{ m}^2/\text{m}^3$ ($1000 \text{ ft}^2/\text{ft}^3$) are commercially available. On the fin side, the desired surface area can be employed by using the proper fin

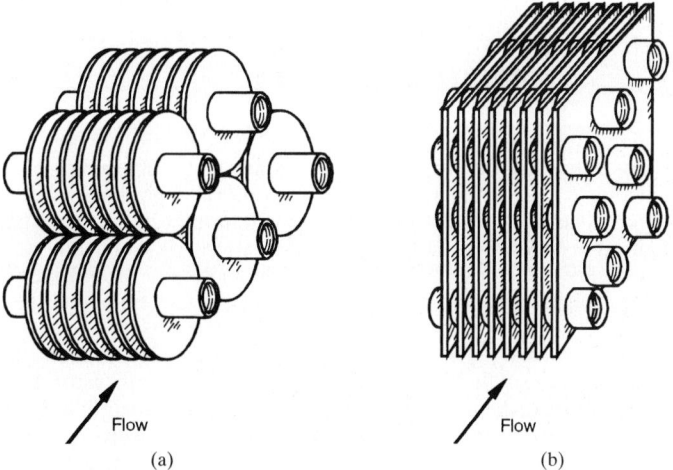

Flow   Flow

(a)   (b)

FIGURE 17.14   (*a*) Individually finned tubes; (*b*) flat or continuous fins on an array of tubes.

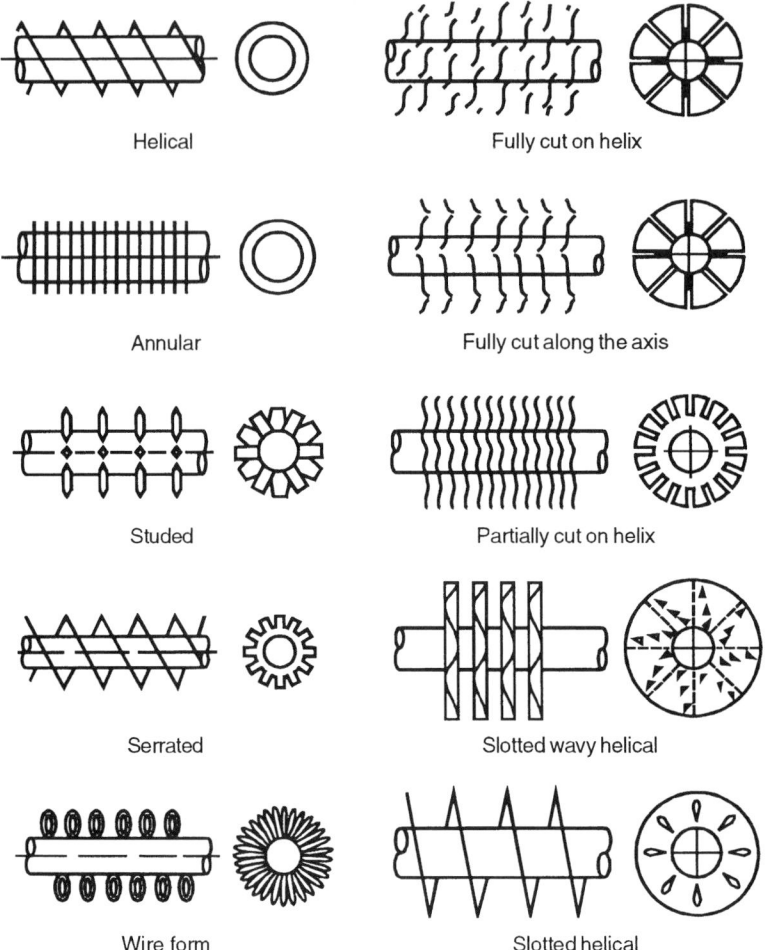

Helical

Fully cut on helix

Annular

Fully cut along the axis

Studed

Partially cut on helix

Serrated

Slotted wavy helical

Wire form

Slotted helical

FIGURE 17.15 Individually finned tubes.

density and fin geometry. The typical fin densities for flat fins vary from 250 to 800 fins/m (6–20 fins/in), fin thicknesses vary from 0.08 to 0.25 mm (0.003–0.010 in), and fin flow lengths from 25 to 250 mm (1–10 in). A tube-fin exchanger having flat fins with 400 fins/m (10 fins/in) has a surface area density of about 720 m^2/m^3 (220 ft^2/ft^3). These exchangers are extensively used as condensers and evaporators in air-conditioning and refrigeration applications, as condensers in electric power plants, as oil coolers in propulsive power plants, and as air-cooled exchangers (also referred to as a fin-fan exchanger) in process and power industries.

Regenerators. The regenerator is a storage type exchanger. The heat transfer surface or elements are usually referred to as a matrix in the regenerator. In order to have continuous operation, either the matrix must be moved periodically into and out of the fixed streams of gases, as in a *rotary* regenerator (Fig. 17.17*a*), or the gas flows must be diverted through valves to and from the fixed matrices as in a *fixed-matrix* regenerator (Fig. 17.17*b*). The latter is also sometimes referred to as a *periodic-flow regenerator* or a *reversible heat accumulator.* A third type of regenerator has a fixed matrix (in the disk form) and the fixed stream of gases, but the

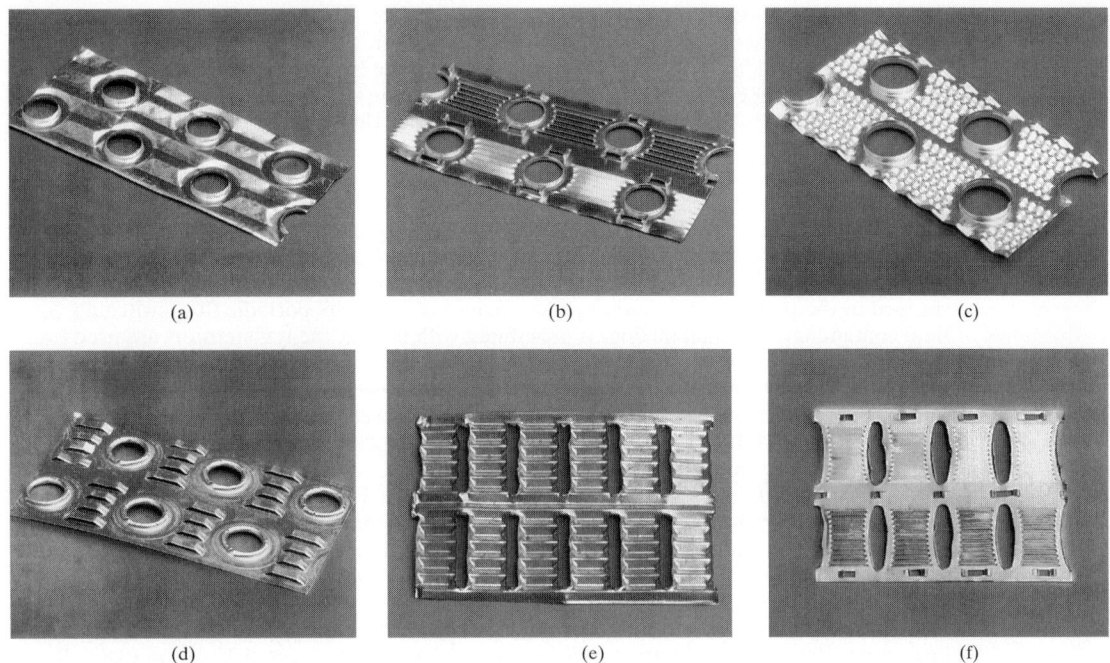

FIGURE 17.16 Flat or continuous fins on an array of tubes: (*a*) wavy fin, (*b*) multilouver fin, (*c*) fin with structured surface roughness (circular dimples), (*d*) parallel louver fin; all four fins with staggered round tubes, (*e*) wavy fin on inline flat tubes, and (*f*) multilouver fin with inline elliptical tubes.

gases are ducted through rotating hoods (headers) to the matrix as shown in Fig. 17.17*c*. This Rothemuhle regenerator is used as an air preheater in some power generating plants. The thermodynamically superior counterflow arrangement is usually employed for regenerators.

The *rotary regenerator* is usually a disk type in which the matrix (heat transfer surface) is in a disk form and fluids flow axially. It is rotated by a hub shaft or a peripheral ring gear drive. For a rotary regenerator, the design of seals to prevent leakages of hot to cold fluids and vice versa becomes a difficult task, especially if the two fluids are at significantly differing pressures. Rotating drives also pose a challenging mechanical design problem.

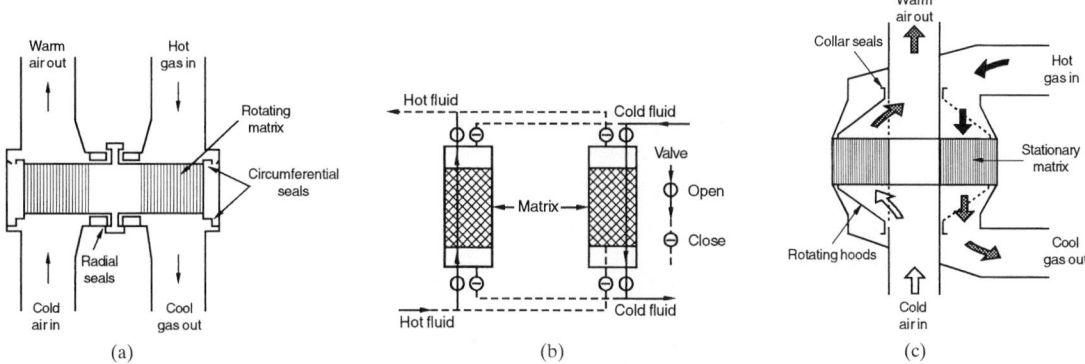

FIGURE 17.17 Regenerators: (*a*) rotary, (*b*) fixed-matrix, and (*c*) rotating hoods.

Major advantages of rotary regenerators follow: For a highly compact regenerator, the cost of the regenerator surface per unit of heat transfer area is usually substantially lower than that for the equivalent recuperator. Another important advantage for a counterflow regenerator compared to a counterflow recuperator is that the design of inlet and outlet headers to distribute the hot and cold gases in the matrix is simple. This is because both fluids flow in different sections (separated by radial seals) of a rotary regenerator. The matrix surface has self-cleaning characteristics for low gas-side fouling because the hot and cold gases flow alternately in the opposite directions in the same fluid passage. Compact surface area density and the counterflow arrangement make the regenerator ideally suited for gas-to-gas heat exchanger applications requiring high exchanger effectiveness, generally exceeding 85 percent. A major disadvantage of a regenerator is an unavoidable carryover of a small fraction of the fluid trapped in the passage to the other fluid stream just after the periodic flow switching. Since fluid contamination (small mixing) is prohibited with liquids, the regenerators are used exclusively for gas-to-gas heat or energy recovery applications. Cross contamination can be minimized significantly by providing a purge section in the disk and using double-labyrinth seals.

Rotary regenerators have been designed for surface area density of up to about 6600 m^2/m^3 (2000 ft^2/ft^3) and exchanger effectiveness exceeding 85 percent for a number of applications. They can employ thinner-stock material, resulting in the lowest amount of material for a given exchanger effectiveness and pressure drop of any heat exchanger known today. The metal rotary regenerators have been designed for continuous operating temperatures up to about 790°C (1450°F) and ceramic matrices for higher-temperature applications; these regenerators are designed up to 400 kPa or 60 psi pressure difference between hot and cold gases. Plastic, paper, and wool are used for regenerators operating below 65°C (150°F) temperatures and one atmospheric pressure. Typical regenerator rotor diameters and rotational speeds are as follows: up to 10 m (33 ft) and 0.5–3 rpm for power plant regenerators, 0.25 to 3 m (0.8 to 9.8 ft) and up to 10 rpm for air-ventilating regenerators, and up to 0.6 m (24 in) and up to 18 rpm for vehicular regenerators. Refer to Shah [1] for the description of the *fixed-matrix regenerator,* also referred to as a *periodic-flow, fixed-bed, valved,* or *stationary* regenerator.

Plate-Type Exchangers.   These exchangers are usually built of thin plates (all prime surface). The plates are either smooth or have some form of corrugations, and they are either flat or wound in an exchanger. Generally, these exchangers cannot accommodate very high pressures, temperatures, and pressure and temperature differentials. These exchangers may be further classified as plate, spiral plate, lamella, and platecoil exchangers, as shown in Fig. 17.1. The plate heat exchanger, being the most important of these, is described next.

The *plate-and-frame* or *gasketed plate heat exchanger* (PHE) consists of a number of thin rectangular corrugated or embossed metal plates sealed around the edges by gaskets and held

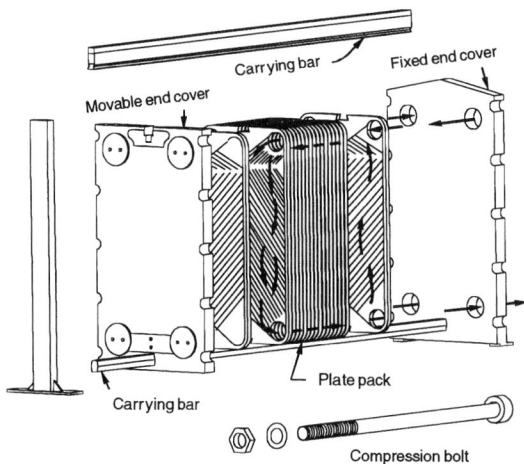

FIGURE 17.18   A plate-and-frame or gasketed plate heat exchanger.

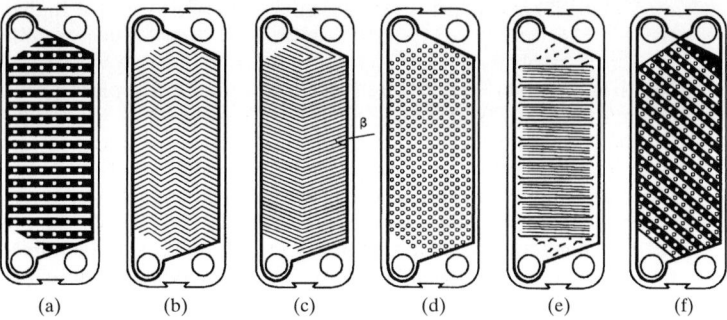

FIGURE 17.19 Plate patterns: (*a*) washboard, (*b*) zig-zag, (*c*) chevron or herring-bone, (*d*) protrusions and depressions, (*e*) washboard with secondary corrugations, and (*f*) oblique washboard.

together in a frame as shown in Fig. 17.18. The plate pack with fixed and movable end covers is clamped together by long bolts, thus compressing the gaskets and forming a seal. Typical plate geometries (corrugated patterns) are shown in Fig. 17.19. Sealing between the two fluids is accomplished by elastomeric molded gaskets (typically 5 mm or 0.2 in thick) that are fitted in peripheral grooves mentioned earlier. The most conventional flow arrangement is 1 pass–1 pass counterflow with all inlet and outlet connections on the fixed end cover. By blocking flow through some ports with proper gasketing, either one or both fluids could have more than one pass. Also, more than one exchanger can be accommodated in a single frame with the use of intermediate connector plates such as up to five "exchangers" or sections to heat, cool, and regenerate heat between raw milk and pasteurized milk in a milk pasteurization application.

Typical plate heat exchanger dimensions and performance parameters are given in Table 17.2 [1]. Any metal that can be cold-worked is suitable for PHE applications. The most com-

TABLE 17.2 Plate-and-Frame Exchanger Geometrical, Operational, and Performance Parameters [1]

Unit	
Maximum surface area, m^2	2500
Number of plates	3 to 700
Port size, mm	up to 400
Plates	
Thickness, mm	0.5 to 1.2
Size, m^2	0.03 to 3.6
Spacing, mm	1.5 to 5
Width, mm	70 to 1200
Length, m	0.6 to 5
Operation	
Pressure, MPa	0.1 to 2.5
Temperature, °C	−40 to 260
Maximum port velocity, m/s	6
Channel flow rates, m^3/h	0.05 to 12.5
Maximum unit flow rate, m^3/h	2500
Performance	
Temperature approach, °C	as low as 1
Heat exchanger effectiveness, %	up to 93%
Heat transfer coefficients for water-water duties, W/m^{2}K	3000–7000

mon plate materials are stainless steel (AISI 304 or 316) and titanium. Plates made from Incoloy 825, Inconel 625, and Hastelloy C-276 are also available. Nickel, cupronickel, and monel are rarely used. Carbon steel is not used due to low corrosion resistance for thin plates. The heat transfer surface area per unit volume for plate exchangers ranges from 120 to 660 m^2/m^3 (37 to 200 ft^2/ft^3).

Since plate heat exchangers are mainly used for liquid-to-liquid heat exchange applications, the characteristics of these exchangers will be briefly summarized here. The most significant characteristic of a gasketed PHE is that it can easily be taken apart into its individual components for cleaning, inspection, and maintenance. The heat transfer surface area can be readily changed or rearranged through the flexibility of the number of plates, plate type, and pass arrangements. The high turbulence due to plates reduces fouling to about 10 to 25 percent that of a shell-and-tube exchanger. Because of the high heat transfer coefficients, reduced fouling, absence of bypass and leakage streams, and pure counterflow arrangements, the surface area required for a plate exchanger is 1/2 to 1/3 that of a shell-and-tube exchanger for a given heat duty. This would reduce the cost, overall volume, and maintenance space for the exchanger. Also, the gross weight of a plate exchanger is about 1/6 that of an equivalent shell-and-tube exchanger. Leakage from one fluid to the other cannot take place unless a plate develops a hole. Since the gasket is between the plates, any leakage from the gaskets is to the outside of the exchanger. The residence time for fluid particles on a given side is approximately the same for uniformity of heat treatment in applications such as sterilizing, pasteurizing, and cooking. There are no significant hot or cold spots in the exchanger that could lead to the deterioration of heat-sensitive fluids. The volumes of fluids held up in the exchanger are small. This is important with expensive fluids for faster transient response and a better process control. Finally and most importantly, high thermal performance can be achieved in plate exchangers. The high degree of counterflow in PHEs makes temperature approaches of up to 1°C (2°F) possible. The high thermal effectiveness (up to about 93 percent) facilitates economical low-grade heat recovery. Flow-induced vibration, noise, high thermal stresses, and entry impingement problems of shell-and-tube heat exchangers do not exist for plate heat exchangers.

Plate heat exchangers are most suitable for liquid-liquid heat transfer duties that require uniform and rapid heating or cooling, as is often the case when treating thermally sensitive fluids. Special plates capable of handling two-phase fluids (e.g., steam condensation) are available. PHEs are not suitable for erosive duties or for fluids containing fibrous materials. In certain cases, suspensions can be handled; but, to avoid clogging, the largest suspended particle should be at most one-third the size of the average channel gap. Viscous fluids can be handled, but extremely viscous fluids lead to flow maldistribution problems, especially on cooling.

Some other inherent limitations of the plate heat exchangers are due to the plates and gaskets as follows. The plate exchanger is used for a maximum pressure of about 2.5 MPa gauge (360 psig) but usually below 1.0 MPa gauge (150 psig). The gasket materials (except for the recent Teflon-coated type) restrict the use of PHEs in highly corrosive applications; they also limit the maximum operating temperature to 260°C (500°F) but usually below 150°C (300°F) to avoid the use of expensive gasket materials. Gasket life is sometimes limited. Frequent gasket replacement may be needed in some applications. Pinhole leaks are hard to detect. For equivalent flow velocities, pressure drop in a plate exchanger is very high compared to a shell-and-tube exchanger. However, the flow velocities are usually low, and plate lengths are short, so the resulting pressure drops are generally acceptable. Some of the largest units have a total surface area of about 2,500 m^2 (27,000 ft^2) per frame. Large differences in fluid flow rates of two streams cannot be handled in a PHE.

Gasketed plate-and-frame heat exchangers are most common in the dairy, beverage, general food processing, and pharmaceutical industries where their ease of cleaning and the thermal control required for sterilization/pasteurization makes them ideal. They are used in the synthetic rubber industry, paper mills, and petrochemical plants. In addition, they are also used in the process industry for water-water duties (heating, cooling, and temperature control) with stainless steel construction when rather high pressure drops are available.

One of the limitations of gasketed plate heat exchanger is the presence of the gaskets, which restricts their use to compatible fluids (with respect to the gasket material) and limits operating temperatures and pressures. In order to overcome this limitation, a number of *welded plate heat exchanger* designs have surfaced, with a welded pair of plates for one or both fluid sides. However, the disadvantage of such a design is the loss of disassembling flexibility on the fluid sides where the welding is done. Essentially, welding is done around the complete circumference where the gasket is normally placed. A *stacked plate heat exchanger* is another welded plate heat exchanger design from Pacinox in which rectangular plates are stacked and welded at the edges. The physical size limitations of PHEs (1.2 m wide × 4 m long max., 4 × 13 ft) are considerably extended to 1.5 m wide × 20 m long (5 × 66 ft) in this exchanger. A maximum surface area of over 10,000 m^2 or 100,000 ft^2 can be accommodated in one unit. The potential maximum operating temperature is 815°C (1500°F), with an operating pressure of up to 20 MPa (3000 psig) when the stacked plate assembly is placed in a cylindrical pressure vessel. For operating pressures below 2 MPa (300 psig) and operating temperatures below 200°C (400°F), the plate bundle is not contained in a pressure vessel but is bolted between two heavy plates. Some of the applications of this exchanger are catalytic reforming, hydrosulfurization, crude distillation, synthesis converter feed effluent exchanger for methanol, propane condenser, and so on.

A vacuum *brazed plate heat exchanger* is a compact PHE for high-temperature and high-pressure duties, and it does not have gaskets, tightening bolts, frame bars, or carrying and guide bars. It simply consists of stainless steel plates and two end plates brazed together. The brazed unit can be mounted directly on piping without brackets and foundations.

A number of other plate heat exchanger constructions have been developed to address some of the limitations of the conventional PHEs. A double-wall PHE is used to avoid mixing of the two fluids. A wide-gap PHE is used for fluids having high fiber content or coarse particles. A graphite PHE is used for highly corrosive fluids. A flow-flex exchanger has plain fins on one side between plates, and the other side has conventional plate channels and is used to handle asymmetric duties (flow rate ratio of 2 to 1 and higher). A design guide for the selection of these exchangers is presented in Table 17.3, which takes into consideration fluids, operating cost, and maintenance cost [6].

The *printed circuit heat exchanger,* as shown in Fig. 17.20, has only primary heat transfer surface as PHEs. Fine grooves are made in the plate by using the same techniques as those employed for making printed electrical circuits. Very high surface area densities (1000–5000 m^2/m^3 or 300–1520 ft^2/ft^3) are achievable. A variety of materials such as 316 SS, 316L SS, 304 SS, 904L SS, cupronickel, monel, nickel, and superalloys can be used. This exchanger has been successfully used with relatively clean gases, liquids, and phase-change fluids in chemical processing, fuel processing, waste heat recovery, and refrigeration industries. Again, this exchanger is a new construction with limited current special applications.

FIGURE 17.20 A segment of a printed circuit heat exchanger, courtesy of Heatric Ltd, Dorset, UK.

EXCHANGER SINGLE-PHASE HEAT TRANSFER AND PRESSURE DROP ANALYSIS

Our objective is to develop relationships between the heat transfer rate q exchanged between two fluids, heat transfer surface area A, heat capacity rates C of individual fluid streams, overall heat transfer coefficient U, and fluid terminal temperatures. In this section, starting with idealizations for heat exchanger analysis and the thermal circuit associated with a two-fluid exchanger, ε-NTU, P-NTU, and MTD methods used for an exchanger analysis are presented,

TABLE 17.3   A Guide for Selection of Plate Heat Exchangers [6]

	Std. PHE	Flow-Flex PHE	Wide-Gap PHE	Double Wall PHE	Semi-welded PHE	Diabon F graphite PHE	Brazed PHE	Fully welded PHE
Performance data								
Pressure range, full vacuum to MPa (psi)	2.5 (355)	2.0 (285)	0.9 (130)	2.5 (355)	2.5 (355)	0.6 (85)	3.0 (427)	4.0 (570)
Temperature range								
°C (Min)	−30	−30	−30	−30	−30	0	−195	−50
(Max)	+180	+180	+180	+180	+180	+140	+225	+350
Service								
Liquid/liquid	1	1	1	1	1	1	1	1
Gas/liquid	1–3*	1–3*	1–3*	1–3*	1–3*	1–3*	1	1–3*
Gas/gas	1–3*	1–3*	1–3*	1–3*	1–3*	1–3*	1–3*	1–3*
Condensation	1–3*	1–3*	1–3*	1–3*	1–3*	1–3*	1	1–3*
Vaporization	1–3*	1–3*	1–3*	1–3*	1–3*	1–3*	1	1–3*
Nature of media								
Corrosive	1	1	1	1	1	1	3	1
Aggressive	3	3	3	3	1	1	4	1
Viscous	1	1	1	1	1	1	3	1
Heat-sensitive	1	1	1	1	1	1	1	1
Hostile reaction	3	3	3	1	2	3	4	2
Fibrous	4	3	1	4	4	4	4	4
Slurries and suspension	3	2	2	3	3	3	4	4
Fouling	3	2	2	3	3	3	3	3
Inspection								
Corrosion	A	A	A	A	B	A	C	C
Leakage	A	A	A	A	A	A	C	C
Fouling	A	A	A	A	B	A	C	C
Maintenance								
Mechanical cleaning	A	A	A	A	B	A	C	C
Modification	A	A	A	A	A	A	C	C
Repair	A	A	A	A	A	A	C	C

1 = Usually best choice
2 = Often best choice
3 = Sometimes best choice
4 = Seldom best choice
A = Both sides
B = One side
C = No side
* Depending on operating pressures, gas/vapor density, etc.

followed by the fin efficiency concept and various expressions. Finally, pressure drop expressions are outlined for various single-phase exchangers.

Heat Transfer Analysis

Idealizations for Heat Exchanger Analysis.   The energy balances, the rate equations, and the subsequent analyses are subject to the following idealizations.

1. The heat exchanger operates under steady-state conditions (i.e., constant flow rate, and fluid temperatures at the inlet and within the exchanger independent of time).
2. Heat losses to the surroundings are negligible (i.e., the heat exchanger is adiabatic).
3. There are no thermal energy sources and sinks in the exchanger walls or fluids.
4. In counterflow and parallelflow exchangers, the temperature of each fluid is uniform over every flow cross section. From the temperature distribution point of view, in crossflow exchangers each fluid is considered mixed or unmixed at every cross section depending upon the specifications. For a multipass exchanger, the foregoing statements apply to each pass depending on the basic flow arrangement of the passes; the fluid is considered mixed or unmixed between passes.
5. Either there are no phase changes in the fluid streams flowing through the exchanger or the phase changes (condensation or boiling) occur under one of the following conditions: (a) phase change occurs at a constant temperature as for a single component fluid at constant pressure; the effective specific heat for the phase-changing fluid is infinity in this case, and hence $C_{max} \to \infty$; (b) the temperature of the phase-changing fluid varies linearly with heat transfer during the condensation or boiling. In this case, the effective specific heat is constant and finite for the phase-changing fluid.
6. The specific heats (as well as other fluid properties implicitly used in NTU) of each fluid are constant throughout the exchanger.
7. The velocity and temperature at the entrance of the heat exchanger on each fluid side are uniform.
8. For an extended surface exchanger, the overall extended surface efficiency η_o is considered uniform and constant.
9. The individual and overall heat transfer coefficients are constant (independent of temperature, time, and position) throughout the exchanger, including the case of phase-changing fluid in idealization 5.
10. The heat transfer surface area is distributed uniformly on each fluid side. In a multipass unit, heat transfer surface area is equal in each pass.
11. For a plate-baffled shell-and-tube exchanger, the temperature rise (or drop) per baffle pass is small compared to the overall temperature rise (or drop) of the shell fluid in the exchanger so that the shell fluid can be treated as mixed at any cross section. This implies that the number of baffles is large.
12. The fluid flow rate is uniformly distributed through the exchanger on each fluid side in each pass. No flow maldistribution, flow stratification, flow bypassing, or flow leakages occur in any stream. The flow condition is characterized by the bulk (or mean) velocity at any cross section.
13. Longitudinal heat conduction in the fluid and in the wall is negligible.

Idealizations 1 through 4 are necessary in a theoretical analysis of steady-state heat exchangers. Idealization 5 essentially restricts the analysis to single-phase flow on both sides or on one side with a dominating thermal resistance. For two-phase flows on both sides, many of the foregoing idealizations are not valid, since mass transfer in phase change results in variable properties and variable flow rates of each phase, and the heat transfer coefficients vary significantly. As a result, the heat exchanger cannot be analyzed using the theory presented here.

If idealization 6 is not valid, divide the exchanger into small segments until the specific heats and/or other fluid properties can be treated as constant. Refer to the section on maldistribution later if idealization 7 is violated. Some investigation is reported in the literature when idealizations 8 through 13 are not valid; this is summarized in the following section. If any of these idealizations are not valid for a particular exchanger application, the best solution is to work directly with either local energy balances and the overall rate equations (see Eqs. 17.1 and 17.2) or their modified form by including a particular effect, and to integrate or numerically analyze them over a small exchanger segment in which all of the idealizations are valid.

Thermal Circuit. In order to develop relationships among variables for various exchangers, consider a two-fluid exchanger (in Fig. 17.21 a counterflow exchanger is given as an example). Two energy conservation differential equations for an overall adiabatic two-fluid exchanger with any flow arrangement are

$$dq = q''dA = -C_h dT_h = \pm C_c dT_c \tag{17.1}$$

Here dq is heat transfer rate from the hot to cold fluid across the surface area dA; C_h and C_c are the heat capacity rates for the hot and cold fluids, and the $\pm$ sign depends on whether dT_c is increasing or decreasing with increasing dA. The overall rate equation on a local basis is

$$dq = q''dA = U(T_h - T_c)_{loc} dA = U\Delta T dA \tag{17.2}$$

where U is the overall heat transfer coefficient.

Integration of Eqs. 17.1 and 17.2 across the exchanger surface area results in overall energy conservation and rate equations as follows.

$$q = C_h(T_{h,i} - T_{h,o}) = C_c(T_{c,o} - T_{c,i}) \tag{17.3}$$

and

$$q = UA\Delta T_m = \Delta T_m / \mathbf{R}_o \tag{17.4}$$

Here ΔT_m is the true mean temperature difference dependent on the exchanger flow arrangement and degree of fluid mixing within each fluid stream. The inverse of the overall thermal conductance UA is referred to as the overall thermal resistance $\mathbf{R}_o$, which consists of component resistances in series as shown in Fig. 17.22 as follows.

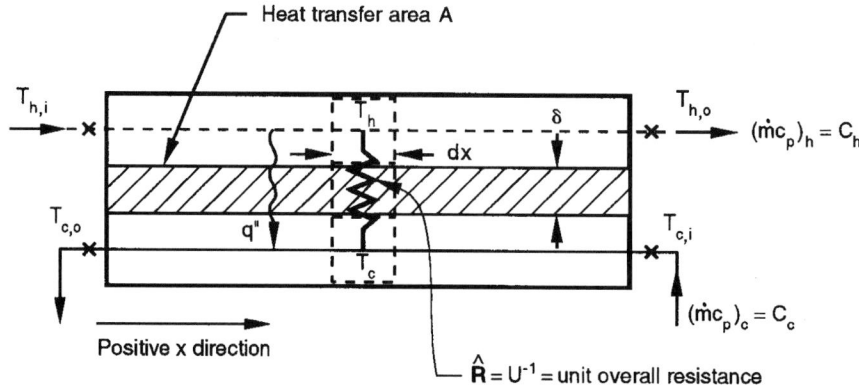

FIGURE 17.21 Nomenclature for heat exchanger variables.

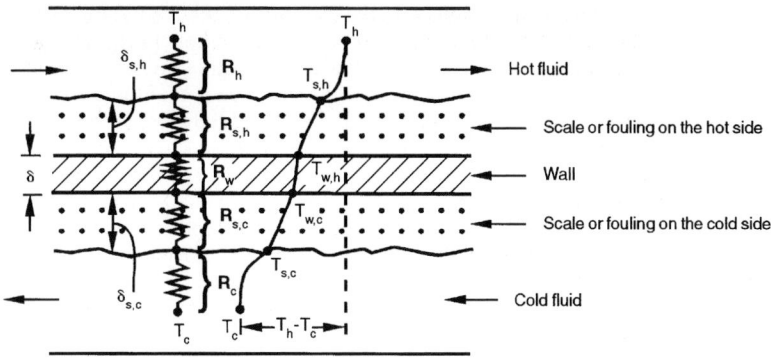

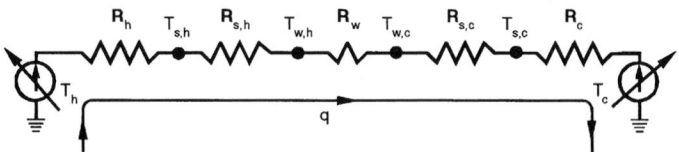

FIGURE 17.22   Thermal circuit for heat transfer in an exchanger.

$$\mathbf{R}_o = \mathbf{R}_h + \mathbf{R}_{s,h} + \mathbf{R}_w + \mathbf{R}_{s,c} + \mathbf{R}_c \qquad (17.5)$$

where the subscripts h, c, s, and w denote hot, cold, fouling (or scale), and wall, respectively. In terms of the overall and individual mean heat transfer coefficients, Eq. 17.5 is represented as

$$\frac{1}{U_m A} = \frac{1}{(\eta_o h_m A)_h} + \frac{1}{(\eta_o h_s A)_h} + \mathbf{R}_w + \frac{1}{(\eta_o h_s A)_c} + \frac{1}{(\eta_o h_m A)_c} \qquad (17.6)$$

where η_o is the total surface effectiveness of an extended (fin) surface and is related to the fin efficiency η_f, surface area A_f, and the total (primary plus secondary or finned) surface area A as defined in Eq. 17.24. Note that no fins are shown in the upper sketch of Fig. 17.22; however, η_o is included in the aforementioned various resistance terms in order to make them most general. For all prime surface exchangers (i.e., having no fins or extended surface), $\eta_{o,h}$ and $\eta_{o,c}$ are unity.

A comparison of each respective term of Eqs. 17.5 and 17.6 defines the value of individual thermal resistances. Note $U = U_m$ and $h = h_m$ in Eq. 17.6, since we have idealized constant and uniform individual and overall heat transfer coefficients (the same equation still holds for local U and h values).

The wall thermal resistance $\mathbf{R}_w$ of Eq. 17.5 or 17.6 is given by

$$\mathbf{R}_w = \begin{cases} \delta/A_w k_w & \text{for flat walls with a single layer} \\[2mm] \dfrac{\ln (d_o/d_i)}{2\pi k_w L N_t} & \text{for circular tubes with a single-layer wall} \\[2mm] \dfrac{1}{2\pi L N_t}\left[\sum_j \dfrac{\ln (d_{j+1}/d_j)}{k_{w,j}}\right] & \text{for circular tubes with a multiple-layer wall} \end{cases} \qquad (17.7)$$

where δ is the wall (plate) thickness, A_w is the total wall area for conduction, k_w is the thermal conductivity of the wall material, d_o and d_i are the tube outside and inside diameters, L is the exchanger length, and N_t is the number of tubes. A flat or plain wall is generally associated with a plate-fin or an all-prime surface plate heat exchanger. In this case, $A_w = L_1 L_2 N_p$ where L_1, L_2, and N_p are the length, width, and total number of separating plates.

If there is any contact or bond resistance present between the fin and tube or plate on the hot or cold fluid side, it is included as an added thermal resistance on the right side of Eq. 17.5 or 17.6. For a heat pipe heat exchanger, additional thermal resistances associated with the heat pipe should be included on the right side of Eq. 17.5 or 17.6; these resistances are evaporator resistance at the evaporator section of the heat pipe, viscous vapor flow resistance inside heat pipe (very small), internal wick resistance at the condenser section of the heat pipe, and condensation resistance at the condenser section.

If one of the resistances on the right-hand side of Eq. 17.5 or 17.6 is significantly higher than the other resistances, it is referred to as the *controlling resistance;* it is considered significantly dominant when it represents more than 80 percent of the total resistance. A reduction in the controlling thermal resistance will have more impact in reducing the exchanger surface area A requirement compared to the reduction in A due to the reduction in other thermal resistances.

UA in Eq. 17.6 may be defined in terms of hot or cold fluid side surface area or wall conduction area as

$$UA = U_h A_h = U_c A_c = U_w A_w \tag{17.8}$$

The knowledge of wall temperature in a heat exchanger is essential to determine localized hot spots, freeze points, thermal stresses, local fouling characteristics, or boiling/condensing coefficients. Based on the thermal circuit of Fig. 17.22, when $\mathbf{R}_w$ is negligible, $T_{w,h} = T_{w,c} = T_w$ is computed from

$$T_w = \frac{T_h + [(\mathbf{R}_h + \mathbf{R}_{s,h})/(\mathbf{R}_c + \mathbf{R}_{s,c})]T_c}{1 + [(\mathbf{R}_h + \mathbf{R}_{s,h})/(\mathbf{R}_c + \mathbf{R}_{s,c})]} \tag{17.9}$$

When there is no fouling on either side ($\mathbf{R}_{s,h} = \mathbf{R}_{s,c} = 0$), Eq. 17.10 reduces to

$$T_w = \frac{T_h/\mathbf{R}_h + T_c/\mathbf{R}_c}{1/\mathbf{R}_h + 1/\mathbf{R}_c} = \frac{(\eta_o hA)_h T_h + (\eta_o hA)_c T_c}{(\eta_o hA)_h + (\eta_o hA)_c} \tag{17.10}$$

Here T_h, T_c, and T_w are local temperatures in this equation.

The ε-NTU, P-NTU, and MTD Methods

If we consider the fluid outlet temperatures or heat transfer rate as dependent variables, they are related to independent variable/parameters of Fig. 17.21 as follows.

$$T_{h,o}, T_{c,o}, \text{ or } q = \phi\{T_{h,i}, T_{c,i}, C_h, C_c, U, A, \text{ flow arrangement}\} \tag{17.11}$$

Six independent and three dependent variables of Eq. 17.11 for a given flow arrangement can be transferred into two independent and one dependent dimensionless groups; three different methods for design and analysis of heat exchangers are presented in Table 17.4 based on the choice of three dimensionless groups. The relationship among three dimensionless groups is derived by integrating Eqs. 17.1 and 17.2 across the surface area for a specified exchanger flow arrangement. Such expressions are presented later in Table 17.6 for industrially most important flow arrangements. Note that there are other methods such as ψ-P [7] and P_1-P_2 charts [8] in which the important dimensionless groups of three methods of Table 17.4 are delineated; using these charts, the solutions to the rating and sizing problems of heat exchanger design can be obtained graphically straightforward without any iterations. However, the description of these methods will not add any more information from a designer's

TABLE 17.4 General Functional Relationships and Dimensionless Groups for ε-NTU, P-NTU, and LMTD Methods

ε-NTU method	P-NTU method	LMTD method
$q = \varepsilon C_{min}(T_{h,i} - T_{c,i})$	$q = P_1 C_1 \lvert T_{2,i} - T_{1,i}\rvert$	$q = UAF\,\Delta T_{lm}$
$\varepsilon = \phi(\text{NTU}, C^*, \text{flow arrangement})$	$P_1 = \phi(\text{NTU}_1, R_1, \text{flow arrangement})$	$F = \phi(P_1, R_1, \text{flow arrangement})^{\dagger}$
$\varepsilon = \dfrac{C_h(T_{h,i} - T_{h,o})}{C_{min}(T_{h,i} - T_{c,i})} = \dfrac{C_c(T_{c,o} - T_{c,i})}{C_{min}(T_{h,i} - T_{c,i})}$	$P_1 = \dfrac{T_{1,o} - T_{1,i}}{T_{2,i} - T_{1,i}}$	$F = \dfrac{\Delta T_m}{\Delta T_{lm}}$
$\text{NTU} = \dfrac{UA}{C_{min}} = \dfrac{1}{C_{min}}\displaystyle\int_A U\,dA$	$\text{NTU}_1 = \dfrac{UA}{C_1} = \dfrac{\lvert T_{1,o} - T_{1,i}\rvert}{\Delta T_m}$	$\text{LMTD} = \Delta T_{lm} = \dfrac{\Delta T_1 - \Delta T_2}{\ln\,(\Delta T_1 / T_2)}$
$C^* = \dfrac{C_{min}}{C_{max}} = \dfrac{(\dot{m}c_p)_{min}}{(\dot{m}c_p)_{max}}$	$R_1 = \dfrac{C_1}{C_2} = \dfrac{T_{2,i} - T_{2,o}}{T_{1,o} - T_{1,i}}$	$\Delta T_1 = T_{h,i} - T_{c,o} \qquad \Delta T_2 = T_{h,o} - T_{c,i}$

† P_1 and R_1 are defined in the P-NTU method.

viewpoint since the closed-form solutions are presented in Table 17.6 later, and hence these methods will not be discussed. Now we will briefly describe the three methods of Table 17.4.

The ε-NTU Method.   In this method, the heat transfer rate from the hot fluid to the cold fluid in the exchanger is expressed as

$$q = \varepsilon C_{min}(T_{h,i} - T_{c,i}) \tag{17.12}$$

Here the exchanger effectiveness ε is an efficiency factor. It is a ratio of the actual heat transfer rate from the hot fluid to the cold fluid in a given heat exchanger of any flow arrangement to the maximum possible heat transfer rate q_{max} thermodynamically permitted. The q_{max} is obtained in a counterflow heat exchanger (recuperator) of infinite surface area operating with the fluid flow rates (heat capacity rates) and fluid inlet temperatures equal to those of an actual exchanger (constant fluid properties are idealized). As noted in Table 17.4, the exchanger effectiveness ε is a function of NTU and C^* in this method. The number of transfer units NTU is a ratio of the overall conductance UA to the smaller heat capacity rate C_{min}. NTU designates the dimensionless heat transfer size or thermal size of the exchanger. It may be interpreted as the C_{min} fluid dimensionless residence time, a temperature ratio, or a modified Stanton number [9]. The heat capacity rate ratio C^* is simply a ratio of the smaller to the larger heat capacity rate for the two fluid streams. Note that $0 \le \varepsilon \le 1$, $0 \le \text{NTU} \le \infty$, and $0 \le C^* \le 1$. Graphical results are provided in Figs. 17.23 and 17.24 for a counterflow and an unmixed-unmixed crossflow exchanger as an illustration. The results for many others can be obtained from the closed-form expressions of Table 17.6.

The P-NTU Method.   This method represents a variant of the ε-NTU method. The ε-NTU relationship is different depending on whether the shell fluid is the C_{min} or C_{max} fluid in the (stream asymmetric) flow arrangements commonly used for shell-and-tube exchangers. In order to avoid possible errors and confusion, an alternative is to present the temperature effectiveness P as a function of NTU and R, where P, NTU, and R are defined consistently either for the fluid 1 side or fluid 2 side; in Table 17.4, they are defined for the fluid 1 side (regardless of whether that side is the hot or cold fluid side or the shell or tube side). Closed-form $P_1 - \text{NTU}_1$ expressions for many industrially useful heat exchanger flow arrangements are provided in Table 17.6, where the fluid 1 side is clearly identified in the sketch for each flow arrangement; it is the shell side in a shell-and-tube exchanger. Note that

$$q = P_1 C_1 \lvert T_{2,i} - T_{1,i}\rvert = P_2 C_2 \lvert T_{1,i} - T_{2,i}\rvert \tag{17.13}$$

$$P_1 = P_2 R_2 \qquad P_2 = P_1 R_1 \tag{17.14}$$

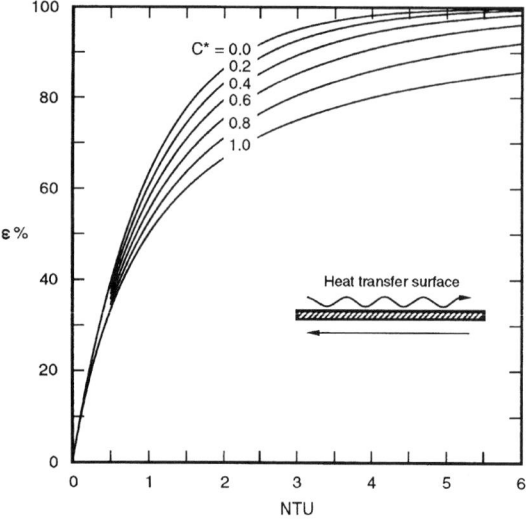

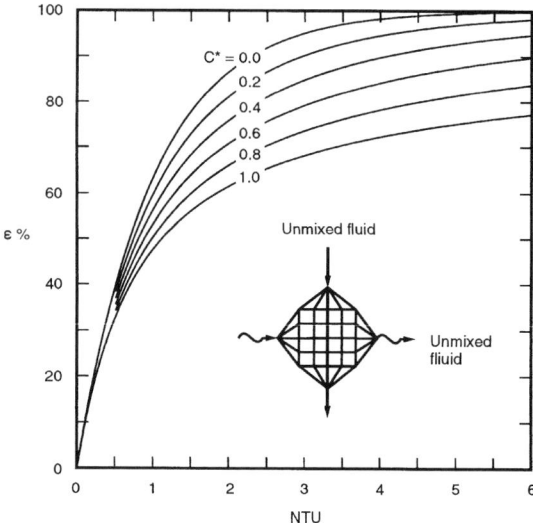

FIGURE 17.23   Heat exchanger effectiveness ε as a function of NTU and C^* for a counterflow exchanger.

FIGURE 17.24   Heat exchanger effectiveness ε as a function of NTU and C^* for a crossflow exchanger with both fluids unmixed.

$$\text{NTU}_1 = \text{NTU}_2 R_2 \qquad \text{NTU}_2 = \text{NTU}_1 R_1 \qquad (17.15)$$

$$\text{and} \qquad R_1 = 1/R_2 \qquad (17.16)$$

P-NTU results for one of the most common 1-2 TEMA E shell-and-tube exchanger are shown in Fig. 17.25.

In Table 17.6, P-NTU-R results are provided for (1) single-pass counterflow and parallel-flow exchangers, (2) single-pass crossflow exchangers, (3) shell-and-tube exchangers, (4) heat exchanger arrays, and (5) plate heat exchangers. For additional plate exchanger flow arrangements, refer to Ref. 10. For the results of two-pass cross-counterflow or cross-parallelflow exchangers, refer to Ref. 7 for some flow arrangements, and Bačlić [11] for 72 different flow arrangements. Results for some three-pass cross-counterflow and some compound multipass crossflow arrangements are presented in Ref. 7.

The MTD Method.   In this method, the heat transfer rate from the hot fluid to the cold fluid in the exchanger is given by

$$q = UAF\Delta T_{lm} \qquad (17.17)$$

where the log-mean temperature difference correction factor F is a ratio of true (actual) mean temperature difference (MTD) to the log-mean temperature difference (LMTD)

$$\text{LMTD} = \Delta T_{lm} = \frac{\Delta T_1 - \Delta T_2}{\ln(\Delta T_1/\Delta T_2)} \qquad (17.18)$$

Here ΔT_1 and ΔT_2 are defined as

$$\Delta T_1 = T_{h,i} - T_{c,o} \qquad \Delta T_2 = T_{h,o} - T_{c,i} \qquad \text{for all flow arrangements except parallelflow} \qquad (17.19)$$

$$\Delta T_1 = T_{h,i} - T_{c,i} \qquad \Delta T_2 = T_{h,o} - T_{c,o} \qquad \text{for parallelflow} \qquad (17.20)$$

The LMTD represents a true mean temperature difference for counterflow and parallelflow arrangements under the idealizations listed below. Thus the LMTD correction factor F rep-

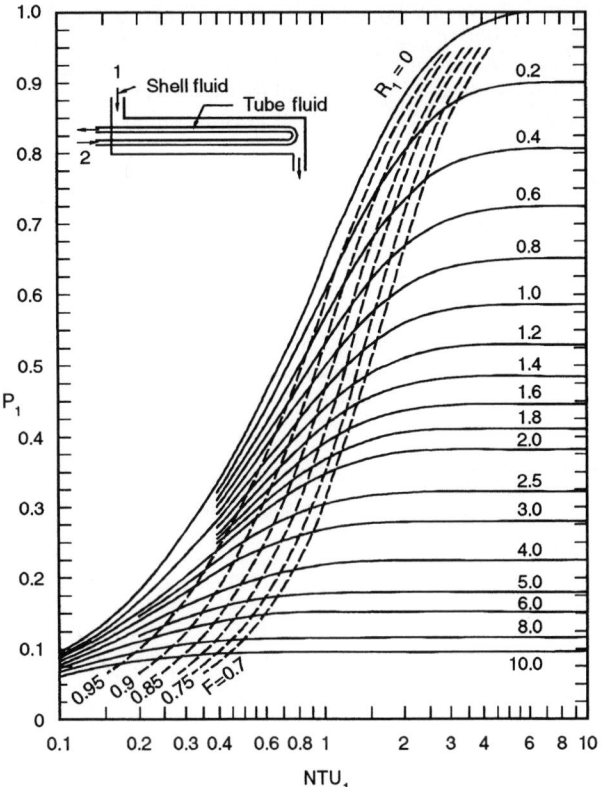

FIGURE 17.25 Temperature effectiveness P_1 as a function of NTU_1 and R_1 for the 1-2 TEMA E shell-and-tube exchanger with shell fluid mixed.

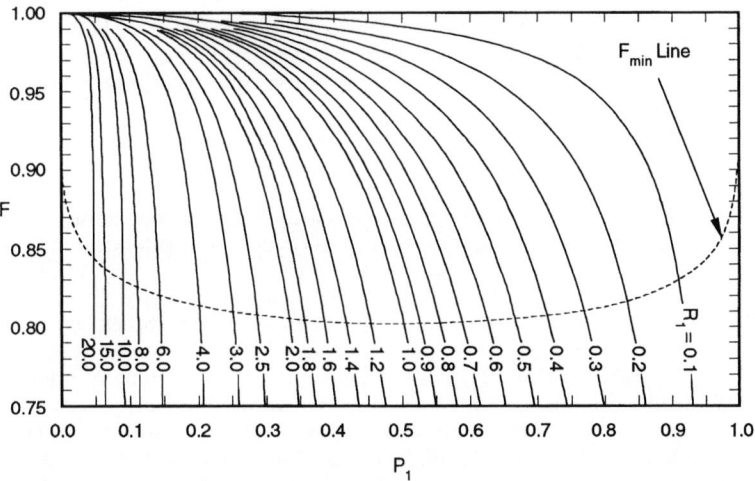

FIGURE 17.26 The LMTD correction factor F as a function of P_1 and R_1 for the 1–2 shell-and-tube exchanger (TEMA E) with shell fluid mixed.

resents a degree of departure for the MTD from the counterflow LMTD; it does not represent the effectiveness of a heat exchanger. For a given flow arrangement, it depends on two dimensionless groups: P_1 and R_1 or P_2 and R_2. (See Fig. 17.26.)

The relationships among the dimensionless groups of the ε-NTU, P-NTU, and MTD methods are presented in Table 17.5. It must be emphasized that closed-form formulas for F factors are available only for (1) a crossflow exchanger with the tube fluid unmixed, shell fluid mixed, (2) a crossflow exchanger with the tube fluid mixed, shell fluid unmixed, and (3) a 1–2 parallel counterflow exchanger (TEMA E). For all other exchanger flow arrangements, one must calculate P_1 first for given NTU_1 and R_1 and use the relationship in Table 17.5 to get the F factor. This is the reason Table 17.6 represents P_1-NTU_1-R_1 formulas, and not the formulas for F factors.

TABLE 17.5 Relationships between Dimensionless Groups of the P-NTU_1, LMTD, and ψ-P Methods, and Those of the ε-NTU Method

$$P_1 = \frac{C_{min}}{C_1} \, \varepsilon = \begin{cases} \varepsilon & \text{for } C_1 = C_{min} \\ \varepsilon C^* & \text{for } C_1 = C_{max} \end{cases}$$

$$R_1 = \frac{C_1}{C_2} = \begin{cases} C^* & \text{for } C_1 = C_{min} \\ 1/C^* & \text{for } C_1 = C_{max} \end{cases}$$

$$NTU_1 = NTU \frac{C_{min}}{C_1} = \begin{cases} NTU & \text{for } C_1 = C_{min} \\ NTU \, C^* & \text{for } C_1 = C_{max} \end{cases}$$

$$F = \frac{NTU_{cf}}{NTU} = \frac{1}{NTU(1-C^*)} \ln\left[\frac{1-C^*\varepsilon}{1-\varepsilon}\right] \xrightarrow{C^*=1} \frac{\varepsilon}{NTU(1-\varepsilon)}$$

$$F = \frac{1}{NTU_1(1-R_1)} \ln\left[\frac{1-R_1P_1}{1-P_1}\right] \xrightarrow{R_1=1} \frac{P_1}{NTU_1(1-P_1)}$$

$$\psi = \frac{\varepsilon}{NTU} = \frac{P_1}{NTU_1} = \frac{FP_1(1-R_1)}{\ln[(1-R_1P_1)/(1-P_1)]} \xrightarrow{R_1=1} F(1-P_1)$$

Fin Efficiency and Extended Surface Efficiency

Extended surfaces have fins attached to the primary surface on one side of a two-fluid or a multifluid heat exchanger. Fins can be of a variety of geometries—plain, wavy, or interrupted—and can be attached to the inside, outside, or both sides of circular, flat, or oval tubes or parting sheets. Fins are primarily used to increase the surface area (when the heat transfer coefficient on that fluid side is relatively low) and consequently to increase the total rate of heat transfer. In addition, enhanced fin geometries also increase the heat transfer coefficient compared to that for a plain fin. Fins may also be used on the high heat transfer coefficient fluid side in a heat exchanger primarily for structural strength purposes (for example, for high-pressure water flow through a flat tube) or to provide a thorough mixing of a highly viscous liquid (such as for laminar oil flow in a flat or a round tube). Fins are attached to the primary surface by brazing, soldering, welding, adhesive bonding, or mechanical expansion (press fit) or extruded or integrally connected to the tubes. Major categories of extended surface heat exchangers are plate-fin (Fig. 17.10) and tube-fin (Fig. 17.14) exchangers. Note that shell-and-tube exchangers sometimes employ individually finned tubes—low finned tubes (similar to Fig. 17.14a but with low-height fins).

The concept of fin efficiency accounts for the reduction in temperature potential between the fin and the ambient fluid due to conduction along the fin and convection from or to the fin surface depending on the fin cooling or heating situation. The fin temperature effectiveness or *fin efficiency* is defined as the ratio of the actual heat transfer rate through the fin base

TABLE 17.6 P_1–NTU_1 Formulas and Limiting Values P_1 and $R_1 = 1$ and $NTU_1 \to \infty$ for Various Exchanger Flow Arrangements*

Flow arrangement	Eq. no.	General formula	Value for $R_1 = 1$	Value for $NTU_1 \to \infty$
Counterflow exchanger, stream symmetric	I.1.1	$P_1 = \dfrac{1 - \exp[-NTU_1(1 - R_1)]}{1 - R_1\exp[-NTU_1(1 - R_1)]}$	$P_1 = \dfrac{NTU_1}{1 + NTU_1}$	$P_1 \to 1$ for $R_1 \le 1$ \ $P_1 \to 1/R_1$, for $R_1 \ge 1$
	I.1.2	$NTU_1 = \dfrac{1}{(1 - R_1)}\ln\left[\dfrac{1 - R_1P_1}{1 - P_1}\right]$	$NTU_1 = \dfrac{P_1}{1 - P_1}$	$NTU_1 \to \infty$
	I.1.3	$F = 1$	$F = 1$	$F = 1$
Parallelflow exchanger, stream symmetric	I.2.1	$P_1 = \dfrac{1 - \exp[-NTU_1(1 + R_1)]}{1 + R_1}$	$P_1 = \dfrac{1}{2}[1 - \exp(-2NTU_1)]$	$P_1 \to \dfrac{1}{1 + R_1}$
	I.2.2	$NTU_1 = \dfrac{1}{1 + R_1}\ln\left[\dfrac{1}{1 - P_1(1 + R_1)}\right]$	$NTU_1 = \dfrac{1}{2}\ln\left[\dfrac{1}{1 - 2P_1}\right]$	$NTU_1 \to \infty$
	I.2.3	$F = \dfrac{(R_1 + 1)\ln[(1 - R_1P_1)/(1 - P_1)]}{(R_1 - 1)\ln[1 - P_1(1 + R_1)]}$	$F = \dfrac{2P_1}{(P_1 - 1)\ln(1 - 2P_1)}$	$F \to 0$
Single-pass crossflow exchanger, both fluids unmixed, stream symmetric	II.1	$P_1 = 1 - \exp(NTU_1)$ $-\exp[-(1 + R_1)NTU_1]$ $\times \displaystyle\sum_{n=1}^{\infty} R_1^n P_n(NTU_1)$ $P_n(y) = \dfrac{1}{(n + 1)!}\displaystyle\sum_{j=1}^{n}\dfrac{(n + 1 - j)}{j!}\,y^{n+j}$	Same as Eq. (II.1) with $R_1 = 1$	
Single-pass crossflow exchanger, fluid 1 unmixed, fluid 2 mixed	II.2.1	$P_1 = [1 - \exp(-KR_1)]/R_1$ $K = 1 - \exp(-NTU_1)$	$P_1 = 1 - \exp(-K)$	$P_1 \to \dfrac{1 - \exp(-R_1)}{R_1}$
	II.2.2	$NTU = \ln\left[\dfrac{1}{1 + (1/R_1)\ln(1 - R_1P_1)}\right]$	$NTU_1 = \ln\left[\dfrac{1}{1 + \ln(1 - P_1)}\right]$	$NTU_1 \to \infty$
	II.2.3	$F = \dfrac{\ln[(1 - R_1P_1)/(1 - P_1)]}{(R_1 - 1)\ln[1 + (1/R_1)\ln(1 - R_1P_1)]}$	$F = \dfrac{P_1}{(P_1 - 1)\ln[1 + \ln(1 - P_1)]}$	$F \to 0$

*Table condensed from R. K. Shah and A. Pignotti, *Basic Thermal Design of Heat Exchangers*, National Science Foundation Report, Int-8601771, 1988. In this table, all variables except P_1, R_1, NTU_1, and F are local or dummy variables not necessarily related to those defined in the nomenclature.

TABLE 17.6 P_1–NTU_1 Formulas and Limiting Values P_1 and $R_1 = 1$ and $NTU_1 \to \infty$ for Various Exchanger Flow Arrangements (*Continued*)

Flow arrangement	Eq. no.	General formula	Value for $R_1 = 1$	Value for $NTU_1 \to \infty$
Single-pass crossflow exchanger, fluid 1 mixed, fluid 2 unmixed	II.3.1	$P_1 = 1 - \exp(-K/R_1)$ $K = 1 - \exp(-R_1 NTU_1)$	$P_1 = 1 - \exp(-K)$ $K = 1 - \exp(-NTU_1)$	$P_1 \to 1 - \exp(-1/R_1)$
	II.3.2	$NTU_1 = \dfrac{1}{R_1} \ln\left[\dfrac{1}{1+R_1 \ln(1-P_1)}\right]$	$NTU_1 = \ln\left[\dfrac{1}{1+\ln(1-P_1)}\right]$	$NTU_1 \to \infty$
	II.3.3	$F = \dfrac{\ln\,(1-R_1 P_1)/(1-P_1)}{(1-1/R_1)\ln[1+R_1\ln(1-P_1)]}$	$F = \dfrac{P_1}{(P_1-1)\ln[1+\ln(1-P_1)]}$	$F \to 0$
Single-pass crossflow exchanger, both fluids mixed, stream symmetric	II.4	$P_1 = \left[\dfrac{1}{K_1} + \dfrac{R_1}{K_2} - \dfrac{1}{NTU_1}\right]^{-1}$ $K_1 = 1 - \exp(-NTU_1)$ $K_2 = 1 - \exp(-R_1 NTU_1)$	$P_1 = \left[\dfrac{2}{K_1} - \dfrac{1}{NTU_1}\right]^{-1}$	$P_1 \to \dfrac{1}{1+R_1}$
1-2* TEMA E shell-and-tube exchanger, shell fluid mixed, stream symmetric	III.1.1	$P_1 = \dfrac{2}{1+R_1+E\coth(ENTU_1/2)}$ $E = [1+R_1^2]^{1/2}$	$P_1 = \dfrac{1}{1+\coth(NTU_1/\sqrt{2})/\sqrt{2}}$	$P_1 \to \dfrac{2}{1+R_1+E}$
	III.1.2	$NTU_1 = \dfrac{1}{E}\ln\left[\dfrac{2-P_1(1+R_1-E)}{2-P_1(1+R_1+E)}\right]$	$NTU_1 = \ln\left[\dfrac{2-P_1}{2-3P_1}\right]$	$NTU_1 \to \infty$
	III.1.3	$F = \dfrac{E\ln[(1-R_1P_1)/(1-P_1)]}{(1-R_1)\ln\left[\dfrac{2-P_1(1+R_1-E)}{2-P_1(1+R_1+E)}\right]}$	$F = \dfrac{P_1/(1-P_1)}{\ln[(2-P_1)/(2-3P_1)]}$	$F \to 0$

	III.2	$$P_1 = \frac{1}{R_1}\left[1 - \frac{(2-R_1)(2E+R_1 B)}{(2+R_1)(2E-R_1/B)}\right]$$ $E = \exp(NTU_1)$ $B = \exp(-NTU_1 R_1/2)$ same as 1-1 J shell, Eq. (III.10)	$$P_1 = \frac{1}{2}\left[1 - \frac{1+E^{-2}}{2(1+NTU_1)}\right]$$ for $R_1 = 2$ $P_1 \to \dfrac{2}{2+R_1}$ for $R_1 \le 2$ $P_1 \to \dfrac{1}{R_1}$ for $R_1 \ge 2$

1-2 TEMA E shell-and-tube exchanger, shell fluid divided into two streams individually mixed

	III.3	$$P_1 = \frac{1}{R_1}\left[1 - \frac{C}{AC+B^2}\right]$$ $A = X_1(R_1+\lambda_1)(R_1-\lambda_2)/2\lambda_1 - X_3\delta$ $\quad - X_2(R_1+\lambda_2)(R_1-\lambda_1)/2\lambda_2 + 1/(1-R_1)$ $B = X_1(R_1-\lambda_2) - X_2(R_1-\lambda_1) + X_3\delta$ $C = X_2(3R_1+\lambda_2) - X_1(3R_1+\lambda_1) + X_3\delta$ $X_i = \exp(\lambda_i NTU_1/3)/2\delta, \quad i = 1, 2, 3$ $\delta = \lambda_1 - \lambda_2$ $\lambda_1 = -\dfrac{3}{2} + \left[\dfrac{9}{4} + R_1(R_1-1)\right]^{1/2}$ $\lambda_2 = -\dfrac{3}{2} - \left[\dfrac{9}{4} + R_1(R_1-1)\right]^{1/2}$ $\lambda_3 = -R_1$	Same as Eq. (III.3) with $R_1 = 1$ $A = -\exp(-NTU_1)/18$ $\quad - \exp(NTU_1/3)/2$ $\quad + (NTU_1+5)/9$ $P_1 \to 1$ for $R_1 \le 1$ $P_1 \to \dfrac{1}{R_1}$ for $R_1 \ge 1$

1-3 TEMA E shell-and-tube exchanger, shell and tube fluids mixed, one parallelflow and two counterflow passes

	III.4	$$P_1 = 4[2(1+R_1) + DA + R_1 B]^{-1}$$ $A = \coth(DNTU_1/4)$ $B = \tanh(R_1 NTU_1/4)$ $D = [4+R_1^2]^{1/2}$	$$P_1 = 4[4+\sqrt{5}A+B]^{-1}$$ $A = \coth(\sqrt{5}NTU_1/4)$ $B = \tanh(NTU_1/4)$ $$P_1 \to \frac{4}{2(1+2R_1)+D-R_1}$$

1-4 TEMA E shell-and-tube exchanger, shell and tube fluids mixed

* 1-2 means one pass on shellside, two passes on tubeside.

TABLE 17.6 P_1–NTU$_1$ Formulas and Limiting Values P_1 and $R_1 = 1$ and NTU$_1 \to \infty$ for Various Exchanger Flow Arrangements (*Continued*)

Flow arrangement	Eq. no.	General formula	Value for $R_1 = 1$	Value for NTU$_1 \to \infty$
Same as III.4 with $n \to \infty$	III.5	Eq. (II.4) applies in this limit with $n \to \infty$	same as for Eq. (II.4)	same as for Eq. (II.4)
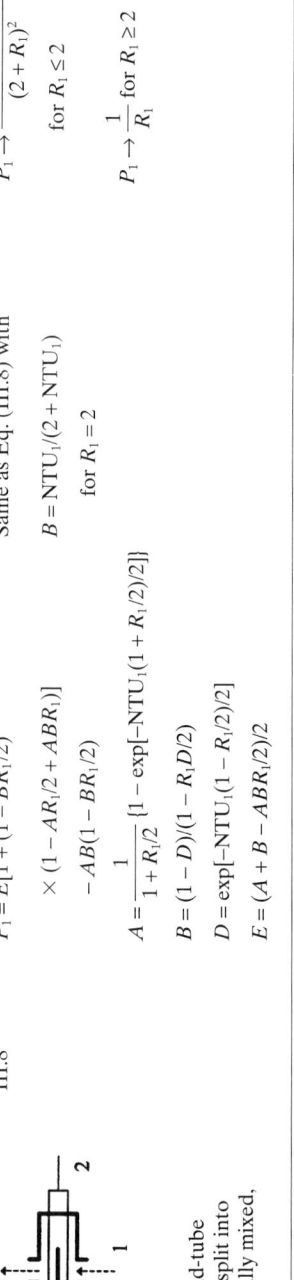 1-1 TEMA G shell-and-tube exchanger; tube fluid split into two streams individually mixed, shell fluid mixed; stream symmetric	III.6	$P_1 = A + B - AB(1 + R_1) + R_1 AB^2$ $A = \dfrac{1}{1 + R_1} \{1 - \exp[-\text{NTU}_1(1 + R_1)/2]\}$ $B = (1 - D)/(1 - R_1 D)$ $D = \exp[-\text{NTU}_1(1 - R_1)/2]$	Same as for Eq. (III.6) with $B = \text{NTU}_1/(2 + \text{NTU}_1)$ for $R_1 = 1$	$P_1 \to 1$ for $R_1 \le 1$ $P_1 \to \dfrac{1}{R_1}$ for $R_1 \ge 1$
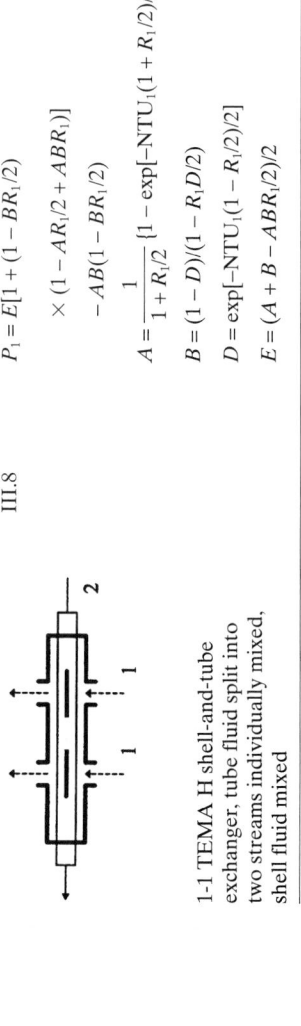 Overall counterflow 1-2 TEMA G shell-and-tube exchanger; shell and tube fluids mixed in each pass at a cross section	III.7	$P_1 = (B - \alpha^2)/(A + 2 + R_1 B)$ $A = -2R_1(1 - \alpha)^2/(2 + R_1)$ $B = [4 - \beta(2 + R_1)]/(2 - R_1)$ $\alpha = \exp[-\text{NTU}_1(2 + R_1)/4]$ $\beta = \exp[-\text{NTU}_1(2 - R_1)/2]$	$P_1 = \dfrac{1 + 2\text{NTU}_1 - \alpha^2}{4 + 4\text{NTU}_1 - (1 - \alpha)^2}$ for $R_1 = 2$ $\alpha = \exp(-\text{NTU}_1)$	$P_1 \to \dfrac{2 + R_1}{R_1^2 + R_1 + 2}$ for $R_1 \le 2$ $P_1 \to \dfrac{1}{R_1}$ for $R_1 \ge 2$
1-1 TEMA H shell-and-tube exchanger, tube fluid split into two streams individually mixed, shell fluid mixed	III.8	$P_1 = E[1 + (1 - BR_1/2)]$ $\quad \times (1 - AR_1/2 + ABR_1)]$ $\quad - AB(1 - BR_1/2)$ $A = \dfrac{1}{1 + R_1/2} \{1 - \exp[-\text{NTU}_1(1 + R_1/2)/2]\}$ $B = (1 - D)/(1 - R_1 D/2)$ $D = \exp[-\text{NTU}_1(1 - R_1/2)/2]$ $E = (A + B - ABR_1/2)/2$	Same as Eq. (III.8) with $B = \text{NTU}_1/(2 + \text{NTU}_1)$ for $R_1 = 2$	$P_1 \to \dfrac{4(1 + R_1) - R_1^2}{(2 + R_1)^2}$ for $R_1 \le 2$ $P_1 \to \dfrac{1}{R_1}$ for $R_1 \ge 2$

17.38

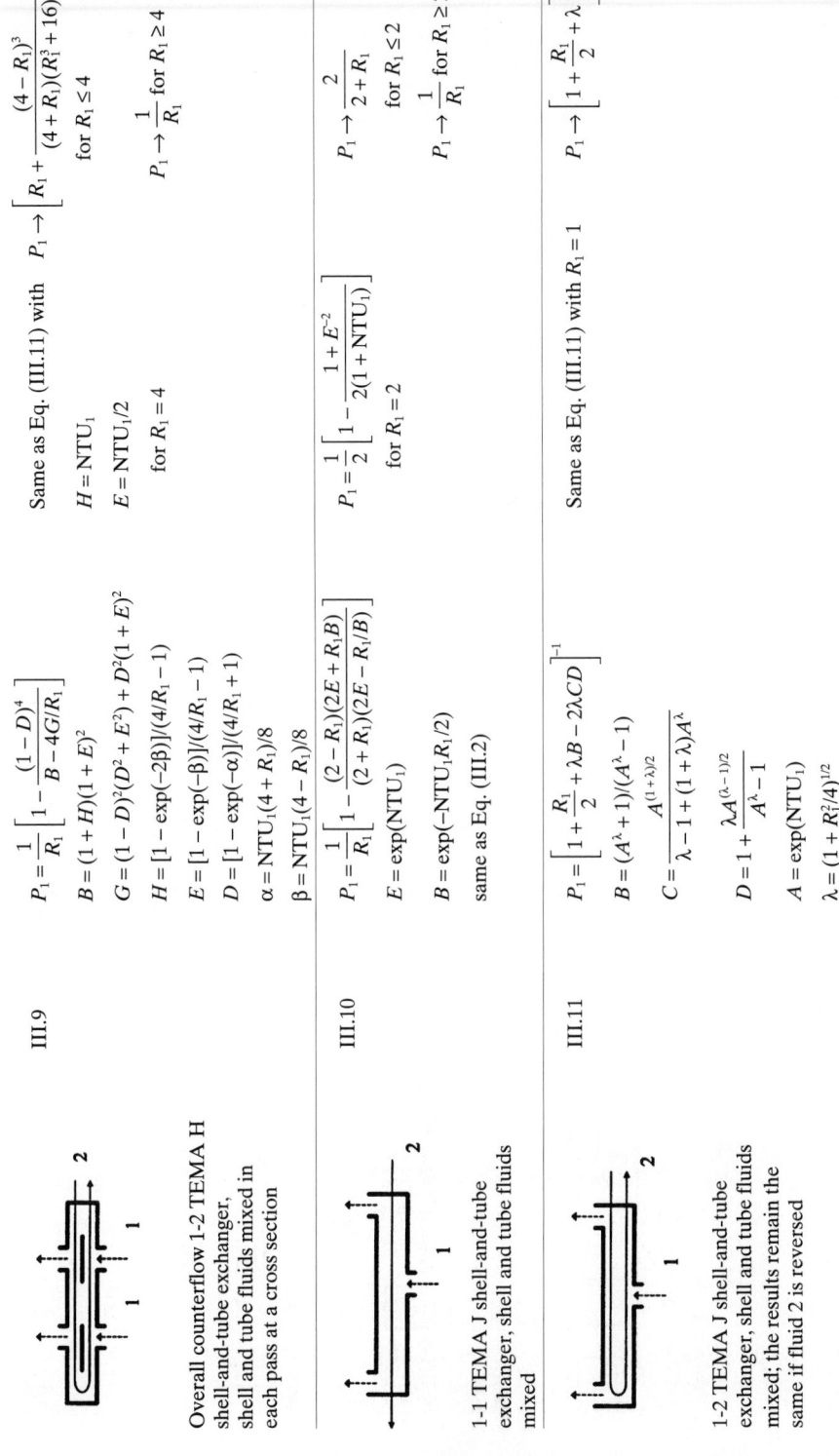

III.9

$$P_1 = \frac{1}{R_1}\left[1 - \frac{(1-D)^4}{B - 4G/R_1}\right]$$

$B = (1+H)(1+E)^2$

$G = (1-D)^2(D^2+E^2) + D^2(1+E)^2$

$H = [1 - \exp(-2\beta)]/(4/R_1 - 1)$

$E = [1 - \exp(-\beta)]/(4/R_1 - 1)$

$D = [1 - \exp(-\alpha)]/(4/R_1 + 1)$

$\alpha = NTU_1(4 + R_1)/8$

$\beta = NTU_1(4 - R_1)/8$

Same as Eq. (III.11) with $P_1 \to \left[R_1 + \dfrac{(4 - R_1)^3}{(4 + R_1)(R_1^3 + 16)}\right]^{-1}$

for $R_1 \le 4$

$P_1 \to \dfrac{1}{R_1}$ for $R_1 \ge 4$

Overall counterflow 1-2 TEMA H shell-and-tube exchanger, shell and tube fluids mixed in each pass at a cross section

---

III.10

$$P_1 = \frac{1}{R_1}\left[1 - \frac{(2-R_1)(2E+R_1B)}{(2+R_1)(2E-R_1/B)}\right]$$

$E = \exp(NTU_1)$

$B = \exp(-NTU_1R_1/2)$

same as Eq. (III.2)

$H = NTU_1$

$E = NTU_1/2$

for $R_1 = 4$

$$P_1 = \frac{1}{2}\left[1 - \frac{1+E^{-2}}{2(1+NTU_1)}\right]$$

for $R_1 = 2$

$P_1 \to \dfrac{2}{2 + R_1}$

for $R_1 \le 2$

$P_1 \to \dfrac{1}{R_1}$ for $R_1 \ge 2$

1-1 TEMA J shell-and-tube exchanger, shell and tube fluids mixed

---

III.11

$$P_1 = \left[1 + \frac{R_1}{2} + \lambda B - 2\lambda CD\right]^{-1}$$

$B = (A^\lambda + 1)/(A^\lambda - 1)$

$C = \dfrac{A^{(1+\lambda)/2}}{\lambda - 1 + (1+\lambda)A^\lambda}$

$D = 1 + \dfrac{\lambda A^{(\lambda-1)/2}}{A^\lambda - 1}$

$A = \exp(NTU_1)$

$\lambda = (1 + R_1^2/4)^{1/2}$

Same as Eq. (III.11) with $R_1 = 1$

$P_1 \to \left[1 + \dfrac{R_1}{2} + \lambda\right]^{-1}$

1-2 TEMA J shell-and-tube exchanger, shell and tube fluids mixed; the results remain the same if fluid 2 is reversed

TABLE 17.6 P_1–NTU_1 Formulas and Limiting Values P_1 and $R_1 = 1$ and $NTU_1 \to \infty$ for Various Exchanger Flow Arrangements (*Continued*)*

Flow arrangement	Eq. no.	General formula	Value for $R_1 = 1$	Value for $NTU_1 \to \infty$
	III.12	$P_1 = \left[1 + \dfrac{R_1}{4}\left(\dfrac{1+3E}{1+E}\right) + \lambda B - 2\lambda CD\right]^{-1}$ $B = \dfrac{A^\lambda + 1}{A^\lambda - 1}$ $C = \dfrac{A^{(1+\lambda)/2}}{\lambda - 1 + (1+\lambda)A^\lambda}$ $D = 1 + \dfrac{\lambda A^{(\lambda-1)/2}}{A^\lambda - 1}$ $A = \exp(NTU_1)$ $E = \exp(R_1 NTU_1/2)$ $\lambda = (1 + R_1^2/16)^{1/2}$	Same as Eq. (III.12) with $R_1 = 1$	$P_1 \to \left[1 + \dfrac{3R_1}{4} + \lambda\right]^{-1}$
1-4 TEMA J shell-and-tube exchanger, shell and tube fluids mixed				
Limit of 1-n TEMA J shell-and-tube exchangers for $n \to \infty$; shell and tube fluids mixed; stream symmetric	III.13	Eq. (II.4) applies in this limit.	Same as for Eq. (II.4)	Same as for Eq. (II.4)
	IV.1.1	$P_1 = 1 - \displaystyle\prod_{i=1}^{n}(1 - P_{1,A_i})$	Same as Eq. (IV.1.1)	Same as Eq. (IV.1.1)
	IV.1.2	$\dfrac{1}{R_1} = \displaystyle\sum_{i=1}^{n}\dfrac{1}{R_{1,A_i}}$	$1 = \displaystyle\sum_{i=1}^{n}\dfrac{1}{R_{1,A_i}}$	Same as Eq. (IV.1.2)
Parallel coupling of n exchangers; fluid 2 split arbitrarily into n streams	IV.1.3	$NTU_1 = \displaystyle\sum_{i=1}^{n} NTU_{1,A_i}$	Same as Eq. (IV.1.3)	$NTU_1 \to \infty$
	IV.2.1	$P_1 = \dfrac{\prod_{i=1}^{n}(1 - R_1 P_{1,A_i}) - \prod_{i=1}^{n}(1 - P_{1,A_i})}{\prod_{i=1}^{n}(1 - R_1 P_{1,A_i}) - R_1 \prod_{i=1}^{n}(1 - P_{1,A_i})}$	$P_1 = \dfrac{\sum_{i=1}^{n}(P_{1,A_i})/(1 - P_{1,A_i})}{1 + \sum_{i=1}^{n}(P_{1,A_i})/(1 - P_{1,A_i})}$	Same as Eq. (I.1.1) counterflow
	IV.2.2	$R_1 = R_{1,A_i},\; i = 1, \ldots, n$	$1 = R_{1,A_i},\; i = 1, \ldots, n$	Same as Eq. (IV.2.2)
Series coupling of n exchangers, overall counterflow arrangement; stream symmetric if all A_i are stream-symmetric	IV.2.3	$NTU_1 = \displaystyle\sum_{i=1}^{n} NTU_{1,A_i}$	Same as for Eq. (IV.2.3)	$NTU_1 \to \infty$
	IV.2.4	$F = \dfrac{1}{NTU_1}\displaystyle\sum_{i=1}^{n} NTU_{1,A_i} F_{A_i}$	Same as for Eq. (IV.2.4)	Same as Eq. (IV.2.4)
	IV.3.1	$P_1 = \dfrac{1}{1 + R_1}\left\{1 - \displaystyle\prod_{i=1}^{n}\left[1 - (1 + R_1)P_{1,A_i}\right]\right\}$	$P_1 = \dfrac{1}{2}\left\{1 - \displaystyle\prod_{i=1}^{n}\left[1 - 2P_{1,A_i}\right]\right\}$	Same as Eq. (IV.3.1)
	IV.3.2	$R_1 = R_{1,A_i},\; i = 1, \ldots, n$	$1 = R_{1,A_i},\; i = 1, \ldots, n$	Same as Eq. (IV.3.2)
Series coupling of n exchangers, overall parallelflow arrangement; stream symmetric if all A_i are stream-symmetric	IV.3.3	$NTU_1 = \displaystyle\sum_{i=1}^{n} NTU_{1,A_i}$	Same as Eq. (IV.3.3)	$NTU_1 \to \infty$

Flow arrangement	Eq. no.	General formula	Value for $R_1 = 1$ unless specified differently	Value for $NTU_1 \to \infty$
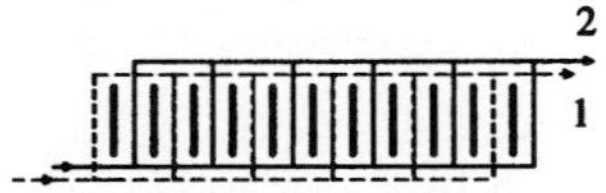 1 pass-1 pass parallelflow plate exchanger, stream symmetric	V.1	$P_1 = A$ $A = P_p(NTU_1, R_1)$ P_1 same as in Eq. (I.2)	$P_1 = \frac{1 - \exp(-2NTU_1)}{2}$	$P_1 = \frac{1}{1 + R_1}$
 1 pass-1 pass counterflow plate exchanger, stream symmetric	V.2	$P_1 = B$ $B = P_c(NTU_1, R_1)$ P_1 same as in Eq. (I.1)	$P_1 = \frac{NTU_1}{1 + NTU_1}$	$P_1 = \begin{cases} 1 & \text{for } R_1 \le 1 \\ \frac{1}{R_1} & \text{for } R_1 > 1 \end{cases}$
 1 pass-2 pass plate exchanger	V.3	$P_1 = \frac{1}{2}(A + B - \frac{1}{2}ABR_1)$ $A = P_p(NTU_1, R_1/2)$ $B = P_c(NTU_1, R_1/2)$	Same as Eq. (V.3) with $B = \frac{NTU_1}{1 + NTU_1}$ for $R_1 = 2$	$P_1 = \begin{cases} \frac{2}{2 + R_1} & \text{for } R \le 2 \\ \frac{1}{R_1} & \text{for } R_1 > 2 \end{cases}$
 1 pass-3 pass plate exchanger with two end passes in parallelflow	V.4	$P_1 = \frac{1}{3}[B + A(1 - R_1B/3)(2 - R_1A/3)]$ $A = P_p(NTU_1, R_1/3)$ $B = P_c(NTU_1, R_1/3)$	Same as Eq. (V.4) with $B = \frac{NTU_1}{1 + NTU_1}$ for $R_1 = 3$	$P_1 = \begin{cases} \frac{9 + R_1}{(3 + R_1)^2} & R_1 \le 3 \\ \frac{1}{R_1} & R_1 > 3 \end{cases}$
 1 pass-3 pass plate exchanger with two end passes in counterflow	V.5	$P_1 = \frac{1}{3}[A + B(1 - R_1A/3)(2 - R_1B/3)]$ $A = P_p(NTU_1, R_1/3)$ $B = P_c(NTU_1, R_1/3)$	Same as Eq. (V.5) with $B = \frac{NTU_1}{1 + NTU_1}$ for $R_1 = 3$	$P_1 = \begin{cases} \frac{9 - R_1}{9 + 3R_1} & R_1 \le 3 \\ \frac{1}{R_1} & R_1 > 3 \end{cases}$

* In all formulas of plate heat exchangers with the number of thermal plates $N \to \infty$ (equation numbers starting with V), the single-pass parallelflow and counterflow temperature effectivenesses are presented in implicit forms. Their explicit forms are as follows, with x and y representing the appropriate values of the number of transfer units and heat capacity rate ratios, respectively.

Single-pass parallelflow

$$P_p(x, y) = \frac{1 - \exp[-x(1 + y)]}{1 + y}$$

$$P_p(x, 1) = [1 - \exp(-2x)]/2$$

$$P_p(\infty, y) = 1/(1 + y)$$

Single-pass counterflow

$$P_c(x, y) = \frac{1 - \exp[-x(1 - y)]}{1 - y\exp[-x(1 - y)]}$$

$$P_c(x, 1) = x/(1 + x)$$

$$P_c(\infty, y) = \begin{cases} 1 & \text{for } y < 1 \\ 1/y & \text{for } y > 1 \end{cases}$$

TABLE 17.6 P_1–NTU_1 Formulas and Limiting Values P_1 and $R_1 = 1$ and $NTU_1 \rightarrow \infty$ for Various Exchanger Flow Arrangements (*Continued*)

Flow arrangement	Eq. no.	General formula	Value for $R_1 = 1$ unless specified differently	Value for $NTU_1 \rightarrow \infty$
1 pass-4 pass plate exchanger	V.6	$P_1 = (1 - Q)/R_1$ $Q = (1 - AR_1/4)^2(1 - BR_1/4)^2$ $A = P_p(NTU_1, R_1/4)$ $B = P_c(NTU_1, R_1/4)$	Same as Eq. (V.6) with $B = \dfrac{NTU_1}{1 + NTU_1}$ for $R_1 = 4$	$P_1 = \begin{cases} \dfrac{16}{(4 + R_1)^2} & R_1 \leq 4 \\ \dfrac{1}{R_1} & R_1 > 4 \end{cases}$
2 pass-2 pass plate exchanger with overall parallelflow and individual passes in parallelflow, stream symmetric	V.7	P_1 same as in Eq. (V.1)	Same as Eq. (V.1)	Same as Eq. (V.1)
2 pass-2 pass plate exchanger with overall parallelflow and individual passes in counterflow, stream symmetric	V.8	$P_1 = B[2 - B(1 + R_1)]$ $B = P_c(NTU_1/2, R_1)$	Same as Eq. (V.8) with $B = \dfrac{NTU_1}{2 + NTU_1}$ for $R_1 = 1$	$P_1 = \begin{cases} 1 - R_1 & R_1 \leq 1 \\ \dfrac{R_1 - 1}{R_1^2} & R_1 > 1 \end{cases}$
2 pass-2 pass plate exchanger with overall counterflow and individual passes in parallelflow, stream symmetric	V.9	$P_1 = \dfrac{2A - A^2(1 + R_1)}{1 - R_1 A^2}$ $A = P_p(NTU_1/2, R_1)$	Same as Eq. (V.9)	$P_1 = \dfrac{1 + R_1}{1 + R_1 + R_1^2}$
2 pass-2 pass plate exchanger with overall counterflow and individual passes in counterflow, stream symmetric	V.10	P_1 same as Eq. (V.2)	Same as Eq. (V.2)	Same as Eq. (V.2)

Configuration	Eq.	Equations	Limiting values
 2 pass-3 pass plate exchanger with overall parallelflow	V.11	$P_1 = A + B - (2/9 + D/3)(A^2 + B^2)$ $\quad - (5/9 + 4D/3)AB$ $\quad + D(1 + D)AB(A + B)/3$ $\quad - D^2A^2B^2/9$ $A = P_p(\text{NTU}_1/2, D)$ $B = P_c(\text{NTU}_1/2, D)$ $D = 2R_1/3$	Same as Eq. (V.11) with $B = \dfrac{\text{NTU}_1}{2 + \text{NTU}_1}$ for $R_1 = \dfrac{3}{2}$ $P_1 = \begin{cases} \dfrac{9 - 2R_1}{9 + 6R_1} & R_1 \le \dfrac{3}{2} \\[3mm] \dfrac{4R^2 + 2R_1 - 3}{2R_1^2(3 + 2R_1)} & R_1 > \dfrac{3}{2} \end{cases}$
 2 pass-3 pass plate exchanger with overall counterflow	V.12	$P_1 = (A + 0.5B + 0.5C + D)/R_1$ $A = \dfrac{2R_1EF^2 - 2EF + F - F^2}{2R_1E^2F^2 - E^2 - F^2 - 2EF + E + F}$ $B = A(E - 1)/F;\ C = (1 - A)/E$ $D = R_1E^2C - R_1E + R_1 - C/2$ $E = 1/(2R_1G/3);\ F = 1/(2R_1H/3)$ $G = P_c(\text{NTU}_1/2, 2R_1/3)$ $H = P_p(\text{NTU}_1/2, 2R_1/3)$	Same as Eq. (V.12) with $G = \dfrac{\text{NTU}_1}{2 + \text{NTU}_1}$ for $R_1 = \dfrac{3}{2}$ $P_1 = \begin{cases} \dfrac{27 + 12R_1 - 4R_1^2}{27 + 12R_1 + 4R_1^2} & R_1 \le \dfrac{3}{2} \\[3mm] \dfrac{1}{R_1} & R_1 > \dfrac{3}{2} \end{cases}$
 2 pass-4 pass plate exchanger with overall parallelflow	V.13	$P_1 = 2D - (1 + R_1)D^2$ $D = (A + B - ABR_1/2)/2$ $A = P_p(\text{NTU}_1/2, R_1/2)$ $B = P_c(\text{NTU}_1/2, R_1/2)$	Same as Eq. (V.13) with $B = \dfrac{\text{NTU}_1}{2 + \text{NTU}_1}$ for $R_1 = 2$ $P_1 = \begin{cases} \dfrac{4}{(2 + R_1)^2} & R_1 \le 2 \\[3mm] \dfrac{R_1 - 1}{R_1^2} & R_1 > 2 \end{cases}$
 2 pass-4 pass plate exchanger with overall counterflow	V.14	$P_1 = \dfrac{2D - (1 + R_1)D^2}{1 - D^2R_1}$ $D = (A + B - ABR_1/2)/2$ $A = P_p(\text{NTU}_1/2, R_1/2)$ $B = P_c(\text{NTU}_1/2, R_1/2)$	Same as Eq. (V.14) with $B = \dfrac{\text{NTU}_1}{2 + \text{NTU}_1}$ for $R_1 = 2$ $P_1 = \begin{cases} \dfrac{4}{4 + R_1^2} & R_1 \le 2 \\[3mm] \dfrac{1}{R_1} & R_1 > 2 \end{cases}$

divided by the maximum possible heat transfer rate through the fin base, which would be obtained if the entire fin was at the base temperature (i.e., its material thermal conductivity was infinite). Since most of the real fins are thin, they are treated as one-dimensional (1-D) with standard idealizations used for the analysis [12]. This 1-D fin efficiency is a function of the fin geometry, fin material thermal conductivity, heat transfer coefficient at the fin surface, and the fin tip boundary condition; it is not a function of the fin base or fin tip temperature, ambient temperature, and heat flux at the fin base or fin tip.

The expressions for 1-D fin efficiency formulas for some common fins are presented in Table 17.7. For other fin geometries, refer to Refs. 13 and 14. The fin efficiencies for straight (first and third from the top in Table 17.7) and circular (seventh from the top in Table 17.7) fins of uniform thickness δ are presented in Fig. 17.27 ($r_e/r_o = 1$ for the straight fin).

The fin efficiency for flat fins (Fig. 17.14b) is obtained by a sector method [15]. In this method, the rectangular or hexagonal fin around the tube (Fig. 17.28a and b) or its smallest symmetrical section is divided into N sectors. Each sector is then considered as a circular fin with the radius $r_{e,i}$ equal to the length of the centerline of the sector. The fin efficiency of each sector is subsequently computed using the circular fin formula of Table 17.7. The fin efficiency η_f for the whole fin is then the surface area weighted average of $\eta_{f,i}$ of each sector.

$$\eta_f = \frac{\sum_{i=1}^{N} \eta_{f,i} A_{f,i}}{\sum_{i=1}^{N} A_{f,i}} \tag{17.21}$$

Since the heat flow seeks the path of least thermal resistance, actual η_f will be equal or higher than that calculated by Eq. 17.21; hence Eq. 17.21 yields a somewhat conservative value of η_f.

The η_f values of Table 17.7 or Eq. 17.21 are not valid in general when the fin is thick, is subject to variable heat transfer coefficients or variable ambient fluid temperature, or has temperature depression at the base. For a thin rectangular fin of constant cross section, the fin efficiency as presented in Table 17.7 is given by

$$\eta_f = \frac{\tanh(m\ell)}{m\ell} \tag{17.22}$$

where $m = [2h(1 + \delta_f/\ell_f)/k_f\delta_f]^{1/2}$.

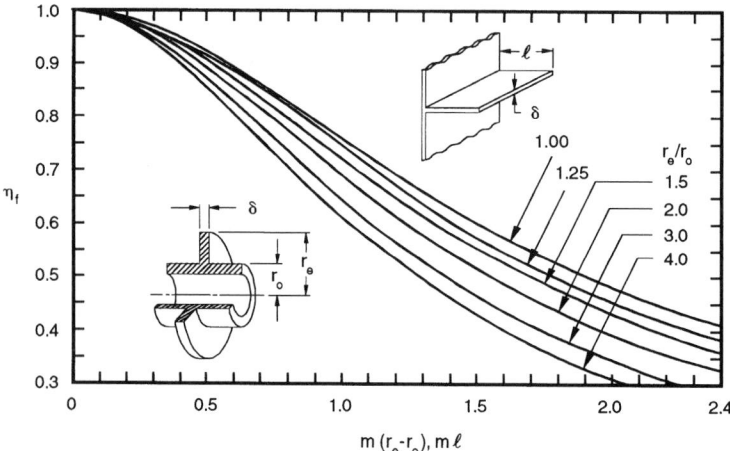

FIGURE 17.27 Fin efficiency of straight and circular fins of uniform thickness.

TABLE 17.7 Fin Efficiency for Plate-Fin and Tube-Fin Geometries of Uniform Fin Thickness

Geometry	Fin efficiency formula $m_i = \left[\dfrac{2h}{k_f\delta_i}\left(1+\dfrac{\delta_i}{\xi}\right)\right]^{1/2}$ $E_i = \dfrac{\tanh(m_i\ell_i)}{m_i\ell_i}$ $i = 1, 2, 3, 4$
 Plain, wavy, or offset strip fin of rectangular cross section	$\eta_f = E_1$ $\ell_1 = \dfrac{b}{2} - \delta_1 \qquad \delta_1 = \delta$
 Triangular fin heated from one side	$\eta_f = \dfrac{hA_1(T_0 - T_\infty)\dfrac{\sinh(m_1\ell_1)}{m_1\ell_1} + q_e}{\cosh(m_1\ell_1)\left[hA_1(T_0 - T_\infty) + q_e\dfrac{T_0 - T_\infty}{T_1 - T_\infty}\right]} \qquad \delta_1 = \delta$
 Plain, wavy, or louver fin of triangular cross section	$\eta_f = E_1$ $\ell_1 = \dfrac{\ell}{2} \qquad \delta_1 = \delta$
 Double sandwich fin	$\eta_f = \dfrac{E_1\ell_1 + E_2\ell_2}{\ell_1 + \ell_2}\,\dfrac{1}{1 + m_1^2 E_1 E_2 \ell_1 \ell_2}$ $\delta_1 = \delta \qquad \delta_2 = \delta_3 = \delta + \delta_s$ $\ell_1 = b - \delta + \dfrac{\delta_s}{2} \qquad \ell_2 = \ell_3 = \dfrac{p_f}{2}$
 Triple sandwich fin	$\eta_f = \dfrac{(E_1\ell_1 + 2\eta_{f24}\ell_{24})/(\ell_1 + 2\ell_2 + \ell_4)}{1 + 2m_1^2 E_1 \ell_1 \eta_{f24}\ell_{24}}$ $\eta_{f24} = \dfrac{(2E_2\ell_2 + E_4\ell_4)/(2\ell_2 + \ell_4)}{1 + m_2^2 E_2 E_4 \ell_2 \ell_4/2} \qquad \ell_{24} = 2\ell_2 + \ell_4$ $\delta_1 = \delta_4 = \delta \qquad \delta_2 = \delta_3 = \delta + \delta_s$ $\ell_1 = b - \delta + \dfrac{\delta_s}{2} \qquad \ell_2 = \ell_3 = \dfrac{p_f}{2} \qquad \ell_4 = \dfrac{b}{2} - \delta + \dfrac{\delta_s}{2}$
 Pin fin	$\eta_f = \dfrac{\tanh(m\ell)}{m\ell}$ $\ell = \dfrac{b}{2} - d_o \qquad m = \left(\dfrac{4h}{k_f d_o}\right)^{1/2} \qquad \delta = \dfrac{d_o}{2}$
 Circular fin	$\eta_f = \begin{cases} a(m\ell_e)^{-b} & \text{for } \Phi > 0.6 + 2.257(r^*)^{-0.445} \\[4pt] \dfrac{\tanh \Phi}{\Phi} & \text{for } \Phi \le 0.6 + 2.257(r^*)^{-0.445} \end{cases}$ $a = (r^*)^{-0.246} \qquad \Phi = m\ell_e(r^*)^n \qquad n = \exp(0.13m\ell_e - 1.3863)$ $b = \begin{cases} 0.9107 + 0.0893r^* & \text{for } r^* \le 2 \\ 0.9706 + 0.17125 \ln r^* & \text{for } r^* > 2 \end{cases}$ $m = \left(\dfrac{2h}{k_f\delta}\right)^{1/2} \qquad \ell_e = \ell_f + \dfrac{\delta}{2} \qquad r^* = \dfrac{d_e}{d_o}$
 Studded fin	$\eta_f = \dfrac{\tanh(m\ell_e)}{m\ell_e}$ $m = \left[\dfrac{2h}{k_f\delta}\left(1+\dfrac{\delta}{w}\right)\right]^{1/2} \qquad \ell_e = \ell_f + \dfrac{\delta}{2} \qquad \ell_f = \dfrac{(d_e - d_o)}{2}$
Rectangular fin over circular tubes	See the text.

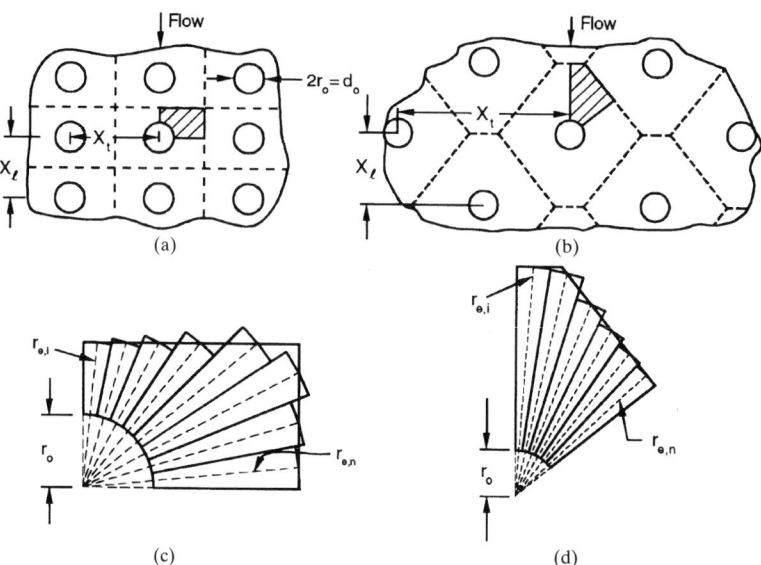

FIGURE 17.28 Flat fin over (*a*) an inline and (*b*) staggered tube arrangement; the smallest representative shaded segment of the fin for (*c*) an inline and (*d*) a staggered tube arrangement.

For a thick rectangular fin of constant cross section, the fin efficiency (a counterpart of Eq. 17.22) is given by Huang and Shah [12] as

$$\eta_f = \frac{(\text{Bi}^+)^{1/2}}{\alpha_f^* \, \text{Bi}} \, \tanh \left[\alpha_f^* (\text{Bi}^+)^{1/2} \right] \tag{17.23}$$

where $\text{Bi}^+ = \text{Bi}/(1 + \text{Bi}/4)$, $\text{Bi} = h\delta_f/2k_f$, $\alpha_f^* = 2\ell/\delta_f$. Equation 17.22 is accurate (within 0.3 percent) for a thick rectangular fin of $\eta_f > 80$ percent; otherwise use Eq. 17.23 for a thick fin.

The nonuniform heat transfer coefficient over the fin surface can lead to significant error in η_f [12] compared to that for a uniform h over the fin surface. However, generally h is obtained experimentally by considering a constant (uniform) value of h over the fin surface. Hence, such experimental h will not introduce significant errors in η_f while designing a heat exchanger, particularly for $\eta_f > 80$ percent. However, one needs to be aware of the impact of nonuniform h on η_f if the heat exchanger test conditions and design conditions are significantly different. Nonuniform ambient temperature has less than a 1 percent effect on the fin efficiency for $\eta_f > 60$ percent and hence can be neglected. The longitudinal heat conduction effect on the fin efficiency is less than 1 percent for $\eta_f > 10$ percent and hence can be neglected. The fin base temperature depression increases the total heat flow rate through the extended surface compared to that with no fin base temperature depression. Hence, neglecting this effect provides a conservative approach for the extended surface heat transfer. Refer to Huang and Shah [12] for further details on the foregoing effects and modifications to η_f for rectangular fins of constant cross sections.

In an extended surface heat exchanger, heat transfer takes place from both the fins ($\eta_f < 100$ percent) and the primary surface ($\eta_f = 100$ percent). In that case, the total heat transfer rate is evaluated through a concept of total surface effectiveness or *extended surface efficiency* η_o defined as

$$\eta_o = \frac{A_p}{A} + \eta_f \frac{A_f}{A} = 1 - \frac{A_f}{A} (1 - \eta_f) \tag{17.24}$$

where A_f is the fin surface area, A_p is the primary surface area, and $A = A_f + A_p$. In Eq. 17.24, the heat transfer coefficients over the finned and unfinned surfaces are idealized to be equal. Note that $\eta_o \geq \eta_f$ and η_o is always required for the determination of thermal resistances of Eq. 17.6 in heat exchanger analysis.

Extensions of the Basic Recuperator Thermal Design Theory

Nonuniform Overall U.   One of the idealizations involved in all of the methods listed in Table 17.4 is that the overall heat transfer coefficient between two fluids is uniform throughout the exchanger and invariant with time. However, the local heat transfer coefficients on each fluid side can vary slightly or significantly due to two effects: (1) changes in the fluid properties or radiation as a result of a rise in or drop of fluid temperatures, and (2) developing thermal boundary layers (referred to as the *length effect*). The first effect due to fluid property variations (or radiation) consists of two components: (1) distortion of velocity and temperature profiles at a given flow cross section due to fluid property variations—this effect is usually taken into account by the so-called property ratio method, with the correction scheme of Eqs. 17.109 and 17.110, and (2) variations in the fluid temperature along the axial and transverse directions in the exchanger depending on the exchanger flow arrangement; this effect is referred to as the *temperature effect*. The resultant axial changes in the overall mean heat transfer coefficient can be significant; the variations in U_{local} could be nonlinear depending on the type of fluid. While both the temperature effect and the thermal entry length effect could be significant in laminar flows, the latter effect is generally not significant in turbulent flow except for low Prandtl number fluids.

It should be mentioned that, in general, the local heat transfer coefficient in a heat exchanger is also dependent upon variables other than the temperature and length effects such as flow maldistribution, fouling, and manufacturing imperfections. Similarly, the overall heat transfer coefficient is dependent upon heat transfer surface geometry, individual Nu (as a function of relevant parameters), thermal properties, fouling effects, temperature variations, temperature difference variations, and so on. However, we will concentrate only on nonuniformities due to temperature and length effects in this section.

In order to outline how to take into account the temperature and length effects, specific definitions of local and mean overall heat transfer coefficients are summarized in Table 17.8 [18]. The three mean overall heat transfer coefficients are important: (1) the traditional U_m defined by Eq. 17.6 or 17.25, (2) $\tilde{U}$ that takes into account only the temperature effect; and (3) $\hat{U}$ that takes into account both effects, with κ providing a correction for the length effect. Note that $U_m(T)$ is traditionally (in the rest of this chapter) defined as

$$\frac{1}{U_m A} = \frac{1}{(\eta_o h_m A)_h} + \mathbf{R}_w + \frac{1}{(\eta_o h_m A)_c} \tag{17.25}$$

where h_m is the mean heat transfer coefficient averaged over the heat transfer surface; $h_{m,h}$ and $h_{m,c}$ are evaluated at the reference temperature T_m for fluid properties; here T_m is usually the arithmetic mean of inlet and outlet fluid temperatures on each fluid side.

Temperature Effect.   In order to find whether the variation in UA is significant with the temperature changes, first evaluate UA at the two ends of a counterflow exchanger or a hypothetical counterflow for all other exchanger flow arrangements. If it is determined that the variations in UA are significant for these two points, evaluate the mean value $\tilde{U}$ by integrating the variations in UA by a three-point Simpson method [17, 18] as follows [16]; note that this method also takes into account the variations in c_p with temperature.

1. Hypothesize the given exchanger as a counterflow exchanger and determine individual heat transfer coefficients and enthalpies at three points in the exchanger: inlet, outlet, and a third point designated with a subscript 1/2 within the exchanger. This third point—a central point on the $\ln \Delta T$ axis—is determined by

TABLE 17.8 Definitions of Local and Mean Overall Heat Transfer Coefficients

Symbol	Definition	Meaning	Comments
U	$U = \dfrac{dq}{dA\,\Delta T}$	Local heat flux per unit of local temperature difference	This is the basic definition of the *local* overall heat transfer coefficient.
U_m	$\dfrac{1}{U_m A} = \dfrac{1}{(\eta_o h_m A)_h} + \mathbf{R}_w + \dfrac{1}{(\eta_o h_m A)_c}$	Overall heat transfer coefficient defined using area average heat transfer coefficients on both sides	Individual heat transfer coefficients should be evaluated at respective reference temperatures (usually arithmetic mean of inlet and outlet fluid temperatures on each fluid side).
$\breve{U}$	$\breve{U} = \dfrac{1}{A} \displaystyle\int_A U(A)\, dA$	Overall heat transfer coefficient averaged over: Heat transfer surface area	Overall heat transfer coefficient is either a function of: (1) local position only (laminar gas flow) $\breve{U}$, (2) temperature only (turbulent liquid flow) $\tilde{U}$, or (3) both local position and temperature (a general case) $\overline{\overline{U}}$. $U(T)$ in $\tilde{U}$ represents a position average overall heat transfer coefficient evaluated at a local temperature. Integration should be performed numerically and/ or can be approximated with an evaluation at three points. The values of the correction factor κ are presented in Fig. 17.29.
$\tilde{U}$	$\tilde{U} = (\ln \Delta T_b - \ln \Delta T_a)\left[\displaystyle\int_{\ln \Delta T_a}^{\ln \Delta T_b} \dfrac{d(\ln \Delta T)}{U(T)}\right]^{-1}$	Temperature range	
$\overline{\overline{U}}$	$\overline{\overline{U}} = \kappa \tilde{U}$	Local position and temperature range	

$$\Delta T_{1/2}^* = (\Delta T_1 \Delta T_2)^{1/2} \tag{17.26}$$

where $\Delta T_1 = (T_h - T_c)_1$ and $\Delta T_2 = (T_h - T_c)_2$ (subscripts 1 and 2 denote terminal points).

2. In order to consider the temperature-dependent specific heats, compute the specific enthalpies i of the C_{max} fluid (with a subscript j) at the third point (referred with 1/2 as a subscript) within the exchanger from the following equation using the known values at each end of a real or hypothetical counterflow exchanger

$$i_{j,1/2} = i_{j,2} + (i_{j,1} - i_{j,2}) \frac{\Delta T_{1/2}^* - \Delta T_2}{\Delta T_1 - \Delta T_2} \tag{17.27}$$

where $\Delta T_{1/2}^*$ is given by Eq. 17.26. If $\Delta T_1 = \Delta T_2$ (i.e., $C^* = 1$), the quotient in Eq. 17.27 becomes 1/2. If the specific heat does not vary significantly, Eq. 17.27 could also be used for the C_{min} fluid. However, when it varies significantly, as in a cryogenic heat exchanger, the third point calculated for the C_{max} and C_{min} fluid separately by Eq. 17.27 will not be physically located close enough to the others. In that case, compute the third point for the C_{min} fluid by the energy balance as follows:

$$[\dot{m}(i_i - i_{1/2})]_{C_{max}} = [\dot{m}(i_{1/2} - i_o)]_{C_{min}} \tag{17.28}$$

Subsequently, using the equation of state or tabular/graphic results, determine the temperature $T_{h,1/2}$ and $T_{c,1/2}$ corresponding to $i_{h,1/2}$ and $i_{c,1/2}$. Then

$$\Delta T_{1/2} = T_{h,1/2} - T_{c,1/2} \tag{17.29}$$

3. The heat transfer coefficient $h_{j,1/2}$ on each fluid side at the third point is evaluated at the following corrected reference temperature for a noncounterflow exchanger.

$$T_{j,1/2,\text{corr}} = T_{j,1/2} + \frac{3}{2}(-1)^j(T_{h,1/2} - T_{c,1/2})\frac{1-F}{1+R_j^{2/3}} \qquad (17.30)$$

In Eq. 17.30, the subscript $j = h$ or c (hot or cold fluid), the exponent $j = 1$ or 2, respectively, for the subscript $j = h$ or c, F is the log-mean temperature difference correction factor, and $R_h = C_h/C_c$ or $R_c = C_c/C_h$. The temperatures $T_{h,1/2,\text{corr}}$ and $T_{c,1/2,\text{corr}}$ are used only for the evaluation of fluid properties to compute $h_{h,1/2}$ and $h_{c,1/2}$. The foregoing correction to the reference temperature $T_{j,1/2}$ results in the cold fluid temperature being increased and the hot fluid temperature being decreased.

Calculate the overall conductance at the third point by

$$\frac{1}{U_{1/2}A} = \frac{1}{\eta_{o,h}h_{h,1/2}A_h} + \mathbf{R}_w + \frac{1}{\eta_{o,c}h_{c,1/2}A_c} \qquad (17.31)$$

Note that η_f and η_o can be determined accurately at local temperatures.

4. Calculate the apparent overall heat transfer coefficient at this point.

$$U_{1/2}^*A = U_{1/2}A\frac{\Delta T_{1/2}}{\Delta T_{1/2}^*} \qquad (17.32)$$

5. Knowing the heat transfer coefficient at each end of the exchanger evaluated at the respective actual temperatures, compute overall conductances according to Eq. 17.31 and find the mean overall conductance for the exchanger (taking into account the temperature dependency of the heat transfer coefficient and heat capacities) from the following equation (Simpson's rule):

$$\frac{1}{\tilde{U}A} = \frac{1}{6}\frac{1}{U_1A} + \frac{2}{3}\frac{1}{U_{1/2}^*A} + \frac{1}{6}\frac{1}{U_2A} \qquad (17.33)$$

6. Finally, the true mean heat transfer coefficient that also takes into account the laminar flow entry length effect is given by:

$$\overline{\overline{U}}A = \tilde{U}A \cdot \kappa \qquad (17.34)$$

where the entry length effect factor $\kappa \leq 1$ is given in Fig. 17.29.

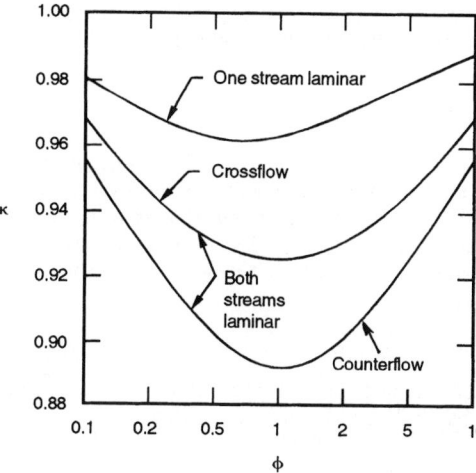

FIGURE 17.29   The length effect correction factor κ for one or both laminar streams as a function of ϕ [17].

Shah and Sekulić [16] recently conducted an analysis of the errors involved with various U averaging methods. They demonstrated that none of the existing methods, including the Roetzel method presented here, can accurately handle a nonlinear temperature variation of U for the surface area determination. The only plausible method in such a case is the numerical approach [16].

If the fluid properties or heat transfer coefficients vary significantly and/or other idealizations built into the ε-NTU or MTD methods are not valid, divide the exchanger into many small segments, and analyze individual small segments with energy balance and rate equations. In such individual small segments, h and other quantities are determined using local fluid properties.

Length Effect. The heat transfer coefficient can vary significantly in the entrance region of the laminar flow. For hydrodynamically developed and thermally developing flow, the local and mean heat transfer coefficients h_x and h_m for a circular tube or parallel plates are related as [19]

$$h_x = \frac{2}{3} h_m (x^*)^{-1/3} \tag{17.35}$$

where $x^* = x/(D_h \, \mathrm{Re} \, \mathrm{Pr})$. Using this variation in h on one or both fluid sides, counterflow and crossflow exchangers have been analyzed and the correction factor κ is presented in Fig. 17.29 [17, 18] as a function of ϕ_1 where

$$\phi_1 = \eta_{o,2} h_{m,2} A_2 \left[\frac{1}{\eta_{o,1} h_{m,1} A_1} + \mathbf{R}_w \right] \tag{17.36}$$

The value of κ is 0.89 when the exchanger has the thermal resistances approximately balanced and $\mathbf{R}_w = 0$, $\phi_1 = (\eta_o hA)_2/(\eta_o hA)_1 = 1$. Thus when variation in the heat transfer coefficient due to thermal entry length effect is considered, $\overline{U} \leq \tilde{U}$ or U_m. The reason for this can be explained easily if one considers thermal resistances connected in series for the problem. For example, consider a very simplified problem with the heat transfer coefficient on each side varying from 80 to 40 W/m²K from entrance to exit and $A_1 = A_2$, $\mathbf{R}_w = 0$, and $\eta_{o,1} = \eta_{o,2} = 1$. In this case, $h_{m,1} = h_{m,2} = 60$ W/m²K and $U_m = 30$ W/m²K. However, at each end of this counterflow exchanger, $U_1 = U_2 = 26.67$ W/m²K (since $1/U = 1/80 + 1/40$). Hence $\overline{U} = (U_1 + U_2)/2 = 26.67$ W/m²K. Thus $\overline{U}/U_m = 26.67/30 = 0.89$.

Combined Effect. Specific step-by-step procedures are presented in Ref. 16 to take into account the combined temperature and length effects for counterflow, crossflow, and 1-2N TEMA E shell-and-tube exchangers. In a broader sense, the effect due to fluid property variations (or radiation) consists of two components: (1) distortion of velocity and temperature profiles at a given flow cross section due to fluid property variations, and (2) variations in fluid temperature along the axial and transverse directions in the exchanger. In general, most of the correlations for heat transfer coefficient are experimentally derived at almost constant fluid properties (because generally small temperature differences are maintained during the experiments) or are theoretically/numerically obtained for constant fluid properties. When the temperature differences between the fluid and wall (heat transfer surface) are large, the fluid properties will vary considerably across a given cross section at a local x and will distort both velocity and temperature profiles [20]. In that case, the dilemma is whether to use fluid bulk mean temperature, wall temperature or something in between for fluid properties to determine hs for the constant property correlations. Unless a specific heat transfer correlation includes this effect, it is commonly taken into account by a property ratio method [20] using both fluid bulk mean temperatures and wall temperature. Hence, it must be emphasized that the local heat transfer coefficients at specific points needed in the Simpson method of integration must first be corrected for the local velocity and temperature profile distortions by the property ratio method and then used as local hs for the integration. The net effect on $\tilde{U}$ due to these two temperature effects can be significant, and $\tilde{U}$ can be considerably higher or lower compared to U_m at constant properties.

The individual heat transfer coefficients in the thermal entrance region could be generally high. However, in general they will have less impact on the overall heat transfer coefficient. This is because, when computing U_{loc} by Eq. 17.25, with U_m and h_ms replaced by corresponding local values, its impact will be diminished due to the presence of the other thermal resistances in the series that are controlling (i.e., having low hA). It can also be seen from Fig. 17.29 that the reduction in U_m due to the entry length effect is at most 11 percent, i.e., $\kappa = 0.89$. Usually the thermal entry length effect is significant for laminar gas flow in a heat exchanger.

Unequal Heat Transfer Area in Individual Exchanger Passes. In a multipass exchanger, it may be preferable to have different heat transfer surface areas in different passes to optimize the exchanger performance. For example, if one pass has two fluids in counterflow and the second pass has two fluids in parallelflow, the overall exchanger performance for a specified total surface area will be higher if the parallelflow pass has a minimum amount of surface area.

Roetzel and Spang [21] analyzed 1-2, 1-3, and 1-2N TEMA E exchangers for unequal heat transfer area in counterflow and parallelflow passes, with the shell inlet either at the stationary head or at the floating head. For a 1-2 TEMA E exchanger, they obtained the following expression for tubeside P_t, NTU_t, and R_t.

$$\frac{1}{P_t} = \nu + R_t + \frac{1}{NTU_t}\frac{m_1 e^{m_1} - m_2 e^{m_2}}{e^{m_1} - e^{m_2}} \tag{17.37}$$

where
$$m_1, m_2 = \frac{NTU_t}{2}\{\pm[(R_t + 2\nu - 1)^2 + 4\nu(1 - \nu)]^{1/2} - (R_t + 2\nu - 1)\} \tag{17.38}$$

$$\nu = \frac{NTU_{pf}}{NTU_t}, \qquad R_t = \frac{C_t}{C_s} \tag{17.39}$$

Here the NTU_{pf} represents the NTU of the parallelflow pass, and NTU_t is the total NTU of the exchanger on the tube side.

Roetzel and Spang [21] showed that Eq. 17.37 represents an excellent approximation for a 1-2N exchanger for $NTU_t \leq 2$ with ν not close to zero. If ν is close to zero, the appropriate formulas are given in Ref. 21. Refer to Ref. 21 for formulas for unequal passes for 1-3 and 1-2N exchangers. The following are the general observations that may be made from the above results.

- As expected, F factors are higher for $\mathbf{K} > 1.0$ compared to the $\mathbf{K} = 1$ (balanced pass) case for given P and R, where $\mathbf{K} = (UA)_{cf}/(UA)_{pf} = (1 - \nu)/\nu$ and the subscripts cf and pf denote counterflow and parallelflow passes, respectively.
- As $\mathbf{K}$ increases, P increases for specified F (or NTU) and R.
- The F factors for the 1-2 exchanger are higher than those for the 1-4 exchanger for specified values of P, R, and $\mathbf{K}$.
- As the number of passes is increased, the F factors (or P) continue to approach to a crossflow exchanger with both fluids mixed, and the advantage of unbalanced passes over balanced passes becomes negligible.
- Although not specifically evaluated, the unbalanced UA (i.e., $\mathbf{K} > 1$) exchanger will have higher total tubeside pressure drop and lower tubeside h compared to those for the balanced UA (i.e., $\mathbf{K} = 1$) exchanger.

Since the analysis was based on the value of $\mathbf{K} = U_{cf}A_{cf}/U_{pf}A_{pf}$, it means that not only the influence of unequal tube pass area can be taken into account, but also the unequal tube side heat transfer coefficient can be taken into account. Similarly, it should be emphasized that the results for nonuniform UA presented in the preceding subsection, if properly interpreted, can also apply for unequal surface areas in different passes. As noted above, higher exchanger performance can be achieved with higher values of **K** and **K** = U_{cf}/U_{pf} for equal pass areas. Hence, the shell inlet nozzle should be located at the stationary head when heating the tube fluid and at the floating head when cooling the tube fluid. This is because higher temperatures mean higher heat transfer coefficients. It should be emphasized that U_{cf} and U_{pf} represent mean values of U across the counterflow and parallelflow tube passes and not at the inlet and outlet ends.

Spang et al. [22] and Xuan et al. [23] have analyzed 1-N TEMA G (split flow) and 1-N TEMA J (divided flow) shell-and-tube exchangers, respectively, with an arbitrary number of passes N, arbitrary surface area (NTU$_i$) in each pass, and arbitrary locations of inlet and outlet shellside nozzles in the exchangers. Bačlić et al. [24] have analyzed two-pass cross-counterflow heat exchanger effectiveness deterioration caused by unequal distribution of NTU between passes.

Finite Number of Baffles.   Idealization 11 (see p. 17.27) indicates that the number of baffles used is very large and can be assumed to approach infinity. Under this idealization, the temperature change within each baffle compartment is very small in comparison with the total temperature change of the shell fluid through the heat exchanger. Thus the shell fluid can be considered as uniform (perfectly mixed) at every cross section (in a direction normal to the shell axis). It is with this model that the mean temperature difference correction factor for exchanger effectiveness is normally derived for single-phase exchangers. In reality, a finite number of baffles are used, and the condition stated above can be achieved only partially. Shah and Pignotti [25] have made a comprehensive review and obtained new results as appropriate; they arrived at the following specific number of baffles beyond which the influence of the finite number of baffles on the exchanger effectiveness is not significantly larger than 2 percent.

- $N_b \geq 10$ for 1-1 TEMA E counterflow exchanger
- $N_b \geq 6$ for 1-2 TEMA E exchanger for NTU$_s \leq 2$, $R_s \leq 5$
- $N_b \geq 9$ for 1-2 TEMA J exchanger for NTU$_s \leq 2$, $R_s \leq 5$
- $N_b \geq 5$ for 1-2 TEMA G exchanger for NTU$_s \leq 3$
- $N_b \geq 11$ for 1-2 TEMA H exchanger for NTU$_s \leq 3$

For 1-N TEMA E exchangers, the exchanger effectiveness will depend on the combination of the number of baffles and tube passes [25].

Shell Fluid Bypassing.   Various clearances are required for the construction of a plate-baffled shell-and-tube exchanger. The shell fluid leaks or bypasses through these clearances with or without flowing past the tubes. Three clearances associated with a plate baffle are tube-to-baffle hole clearance, bundle-to-shell clearance, and baffle-to-shell clearance. Various leakage streams associated with these clearances are identified elsewhere.

Gardner and Taborek [26] have summarized the effect of various bypass and leakage streams on the mean temperature difference. As shown in Fig. 17.30, the baffle-to-shell leakage stream E experiences practically no heat transfer; the bundle-to-shell bypass stream C indicates some heat transfer, and the crossflow stream B shows a large temperature change and a possible pinch or temperature cross ($T_{B,0} > T_{t,0}$). The mixed mean outlet temperature T_{s0} is much lower than the B stream outlet temperature $T_{B,0}$, thus resulting in an indicated temperature difference larger than is actually present; the overall exchanger performance will be lower than the design value. Since the bypass and leakage streams can exceed 30 percent of the total flow, the effect on the mean temperature difference can be very large, especially for close temperature approaches. The Bell-Delaware method of designing shell-and-tube exchangers that includes the effect of leakage and bypass streams is described on p. 17.113.

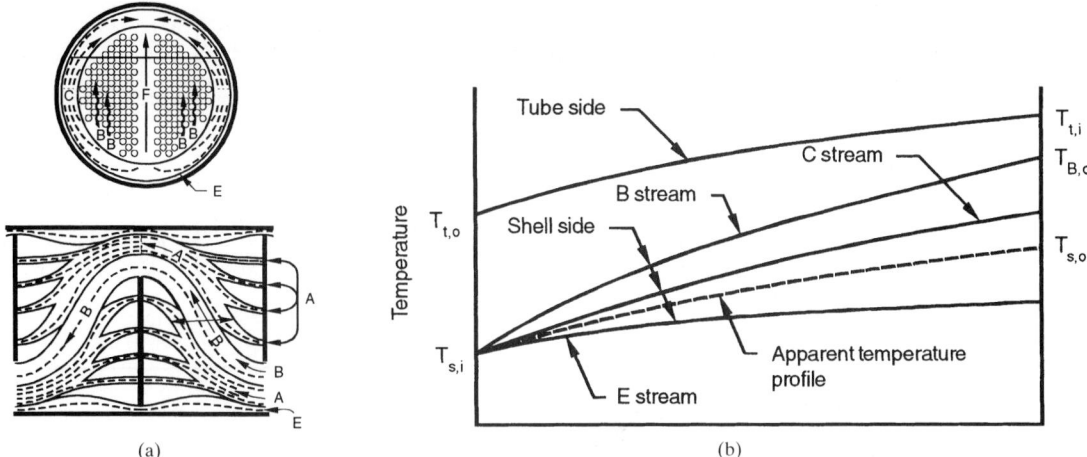

FIGURE 17.30   Effect of bypass and leakage streams on the temperature profile of a shell-and-tube exchanger: (a) streams, (b) temperature profiles.

Longitudinal Wall Heat Conduction Effects.   All three methods discussed in the preceding sections are based on the idealizations of zero longitudinal heat conduction both in the wall and in the fluid in the flow direction. Longitudinal heat conduction in the fluid is negligible for $Pe > 10$ and $x^* \geq 0.005$ [19], where $Pe = Re\, Pr$ and $x^* = x/(D_h\, Re\, Pr)$. For most heat exchangers, except for liquid metal exchangers, Pe and x^* are higher than the above indicated values, and hence longitudinal heat conduction in the fluid is negligible.

Longitudinal heat conduction in the wall reduces the exchanger effectiveness and thus reduces the overall heat transfer performance. The reduction in the exchanger performance could be important and thus significant for exchangers designed for effectivenesses greater than about 75 percent. This would be the case for counterflow and single-pass crossflow exchangers. For high-effectiveness multipass exchangers, the exchanger effectiveness per pass is generally low, and thus longitudinal conduction effects for each pass are generally negligible. The influence of longitudinal wall heat conduction on the exchanger effectiveness is dependent mainly upon the longitudinal conduction parameter $\lambda = k_w A_k / L C_{min}$ (where k_w is the wall material thermal conductivity, A_k is the conduction cross-sectional area, and L is the exchanger length for longitudinal conduction). It would also depend on the convection-conductance ratio $(\eta_o hA)^*$, a ratio of $\eta_o hA$ on the C_{min} to that on the C_{max} side, if it varies significantly from unity. The influence of longitudinal conduction on ε is summarized next for counterflow and single-pass crossflow exchangers.

Kroeger [27] analyzed extensively the influence of longitudinal conduction on counterflow exchanger effectiveness. He found that the influence of longitudinal conduction is the largest for $C^* = 1$. For a given C^*, increasing λ decreases ε. Longitudinal heat conduction has a significant influence on the counterflow exchanger size (i.e., NTU) for a given ε when $NTU > 10$ and $\lambda > 0.005$. Kroeger's solution for $C^* = 1$, $0.1 \leq (\eta_o hA)^* \leq 10$, and $NTU \geq 3$ is as follows:

$$\varepsilon = 1 - \frac{1}{1 + NTU \dfrac{1 + \lambda[\lambda NTU/(1 + \lambda NTU)]^{1/2}}{1 + \lambda NTU}} \tag{17.40}$$

The results for $1 - \varepsilon$ from this equation are presented in Fig. 17.31a.

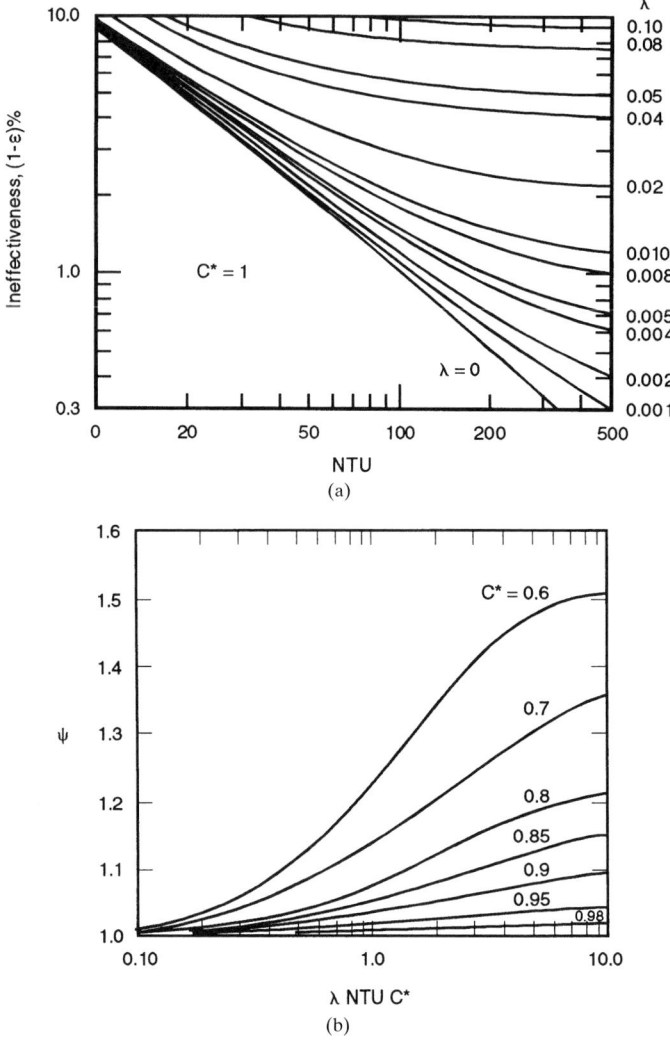

FIGURE 17.31 (a) Counterflow exchanger ineffectiveness as a function of NTU and λ for $C^* = 1.0$, (b) the parameter ψ for Eq. 17.41.

Kroeger [27] also obtained the detailed results for $1 - \varepsilon$ for $0.8 \leq C^* \leq 0.98$ for the counterflow exchanger. He correlated all his results for $1 - \varepsilon$ for $0.8 \leq C^* \leq 1$ as follows:

$$1 - \varepsilon = \frac{1 - C^*}{\psi \exp(r_1) - C^*} \tag{17.41}$$

where

$$r_1 = \frac{(1 - C^*)\text{NTU}}{1 + \lambda \text{NTU} C^*} \tag{17.42}$$

In Eq. 17.41 the parameter ψ is a function of λ, C^*, and NTU

$$\psi = f(\alpha, C^*) \tag{17.43}$$

where

$$\alpha = \lambda \text{NTU} C^* \tag{17.44}$$

The parameter ψ is given in Fig. 17.31b and Ref. 27. For $0.5 < (\eta_o hA)^*/C^* \le 2$, the error introduced in the ineffectiveness is within 0.8 percent and 4.7 percent for $C^* = 0.95$ and 0.8, respectively.

For a crossflow exchanger, temperature gradients in the wall exist in the x and y directions (two fluid flow directions). As a result, two longitudinal conduction parameters λ_h and λ_c are used to take into account the longitudinal conduction effects in the wall. Detailed tabular results are presented in Ref. 15, as reported by Chiou, on the effect of λ_h and λ_c on the exchanger ε for an unmixed-unmixed crossflow exchanger.

ε-NTU$_o$ and Λ-Π Methods for Regenerators

Heat transfer analysis for recuperators needs to be modified for regenerators in order to take into account the additional effects of the periodic thermal energy storage characteristics of the matrix wall and the establishment of wall temperature distribution dependent on $(hA)_h$ and $(hA)_c$. These two effects add two additional dimensionless groups to the analysis to be discussed in the following subsection. All idealizations, except for numbers 8 and 11, listed on p. 17.27, are also invoked for the regenerator heat transfer analysis. In addition, it is idealized that regular periodic (steady-state periodic) conditions are established; wall thermal resistance in the wall thickness (transverse) direction is zero, and it is infinity in the flow direction; no mixing of the fluids occurs during the switch from hot to cold flows or vice versa; and the fluid carryover and bypass rates are negligible relative to the flow rates of the hot and cold fluids. Note that negligible carryover means the dwell (residence) times of the fluids are negligible compared to the hot and cold gas flow periods.

ε-NTU$_o$ and Λ-Π Methods.   Two methods for the regenerator heat transfer analysis are the ε-NTU$_o$ and Λ-Π methods [28]. The dimensionless groups associated with these methods are defined in Table 17.9, the relationship between the two sets of dimensionless groups is presented in Table 17.10a, and these dimensionless groups are defined in Table 17.10b for rotary and fixed-matrix regenerators. Notice that the regenerator effectiveness is dependent on four dimensionless groups, in contrast to the two parameters NTU and C^* for recuperators (see Table 17.4). The additional parameters C_r^* and $(hA)^*$ for regenerators denote the dimensionless heat storage capacity rate of the matrix and the convection-conductance ratio of the cold and hot fluid sides, respectively.

Extensive theory and results in terms of the Λ-Π method have been provided by Hausen [29] and Schmidt and Willmott [30]. The ε-NTU$_o$ method has been used for rotary regenerators and the Λ-Π method for fixed-matrix regenerators. In a rotary regenerator, the outlet fluid temperatures vary across the flow area and are independent of time. In a fixed-matrix regenerator, the outlet fluid temperatures vary with time but are uniform across the flow area at any instant of time.* In spite of these subtle differences, if the elements of a regenerator (either rotary or fixed-matrix) are fixed relative to the observer by the selection of the appropriate coordinate systems, the heat transfer analysis is identical for both types of regenerators for arriving at the regenerator effectiveness.

In the Λ-Π method, several different designations are used to classify regenerators depending upon the values of Λ and Π. Such designations and their equivalent dimensionless groups of the ε-NTU$_o$ method are summarized in Table 17.11.

---

* The difference between the outlet temperatures of the heated air (cold fluid) at the beginning and end of a given period is referred to as the temperature swing δT.

TABLE 17.9 General Functional Relationships and Basic Definitions of Dimensionless Groups for ε-NTU$_o$ and Λ-Π Methods for Counterflow Regenerators

ε-NTU$_0$ method	Λ-Π method*
$q = \varepsilon C_{min}(T_{h,i} - T_{c,i})$	$Q = \varepsilon_h C_h \mathscr{P}_h(T_{hi} - T_{ci}) = \varepsilon_c C_c \mathscr{P}_c(T_{hi} - T_{ci})$
$\varepsilon = \phi\{NTU_0, C^*, C_r^*, (hA)^*\}$	$\varepsilon_r, \varepsilon_h, \varepsilon_c = \phi(\Lambda_m, \Pi_m, \gamma, R^*)$
$\varepsilon = \dfrac{C_h(T_{h,i} - T_{h,o})}{C_{min}(T_{h,i} - T_{c,i})} = \dfrac{C_c(T_{c,o} - T_{c,i})}{C_{min}(T_{h,i} - T_{c,i})}$	$\varepsilon_h = \dfrac{Q_h}{Q_{max,h}} = \dfrac{C_h \mathscr{P}_h(T_{h,i} - \overline{T}_{h,o})}{C_h \mathscr{P}_h(T_{h,i} - T_{ci})} = \dfrac{T_{h,i} - \overline{T}_{h,o}}{T_{h,i} - T_{ci}}$
$NTU_0 = \dfrac{1}{C_{min}}\left[\dfrac{1}{1/(hA)_h + 1/(hA)_c}\right]$	$\varepsilon_c = \dfrac{Q_c}{Q_{max,c}} = \dfrac{C_c \mathscr{P}_c(\overline{T}_{c,o} - T_{ci})}{C_c \mathscr{P}_c(T_{h,i} - T_{ci})} = \dfrac{\overline{T}_{c,o} - T_{c,i}}{T_{h,i} - T_{c,i}}$
$C^* = \dfrac{C_{min}}{C_{max}}$	$\varepsilon_r = \dfrac{Q_h + Q_c}{Q_{max,h} + Q_{max,c}} = \dfrac{2Q}{Q_{max,h} + Q_{max,c}}$
$C_r^* = \dfrac{C_r}{C_{min}}$	$\dfrac{1}{\varepsilon_r} = \dfrac{1}{2}\left(\dfrac{1}{\varepsilon_h} + \dfrac{1}{\varepsilon_c}\right)$
$(hA)^* = \dfrac{hA \text{ on the } C_{min} \text{ side}}{hA \text{ on the } C_{max} \text{ side}}$	$\dfrac{1}{\Pi_m} = \dfrac{1}{2}\left(\dfrac{1}{\Pi_h} + \dfrac{1}{\Pi_c}\right)$ $\quad$ $\dfrac{1}{\Lambda_m} = \dfrac{1}{2\Pi_m}\left(\dfrac{\Pi_h}{\Lambda_h} + \dfrac{\Pi_c}{\Lambda_c}\right)$
	$\gamma = \dfrac{\Pi_c/\Lambda_c}{\Pi_h/\Lambda_h}$ $\quad$ $R^* = \dfrac{\Pi_h}{\Pi_c}$ $\quad$ $\Lambda_h = \dfrac{(hA)_h}{C_h}$
	$\Lambda_c = \dfrac{(hA)_c}{C_c}$ $\quad$ $\Pi_h \approx \left(\dfrac{hA}{C_r}\right)_h$ $\quad$ $\Pi_c \approx \left(\dfrac{hA}{C_r}\right)_c$

* P_h and P_c represent hot-gas and cold-gas flow periods, respectively, in seconds.

TABLE 17.10(a) Relationship between Dimensionless Groups of ε-NTU$_0$ and Λ-Π Methods for $C_c = C_{min}$[†]

$$\varepsilon = \varepsilon_c = \frac{\varepsilon_h}{\gamma} = (\gamma + 1)\frac{\varepsilon_r}{2\gamma} \text{ for } C_c = C_{min}$$

ε-NTU$_0$	Λ-Π
$NTU_0 = \dfrac{\Lambda_m(1 + \gamma)}{4\gamma} = \dfrac{\Lambda_c/\Pi_c}{1/\Pi_h + 1/\Pi_c}$	$\Lambda_h = C^*\left[1 + \dfrac{1}{(hA)^*}\right]NTU_0$
$C^* = \gamma = \dfrac{\Pi_c/\Lambda_c}{\Pi_h/\Lambda_h}$	$\Lambda_c = [1 + (hA)^*]NTU_0$
$C_r^* = \dfrac{\Lambda_m(1 + \gamma)}{2\gamma\Pi_m} = \dfrac{\Lambda_c}{\Pi_c}$	$\Pi_h = \dfrac{1}{C_r^*}\left[1 + \dfrac{1}{(hA)^*}\right]NTU_0$
$(hA)^* = \dfrac{1}{R^*} = \dfrac{\Pi_c}{\Pi_h}$	$\Pi_c = \dfrac{1}{C_r^*}[1 + (hA)^*]NTU_0$

[†] If $C_h = C_{min}$, the subscripts c and h in this table should be changed to h and c, respectively.

TABLE 17.10(b)   Working Definitions of Dimensionless Groups for Regenerators in Terms of Dimensional Variables of Rotary and Fixed-Matrix Regenerators for $C_c = C_{min}$[†]

Dimensionless group	Rotary regenerator	Fixed-matrix regenerator
NTU_0	$\dfrac{h_c A_c}{C_c} \dfrac{h_h A_h}{h_h A_h + h_c A_c}$	$\dfrac{h_c A}{C_c} \dfrac{h_h \mathcal{P}_h}{h_h \mathcal{P}_h + h_c \mathcal{P}_c}$
C^*	$\dfrac{C_c}{C_h}$	$\dfrac{C_c \mathcal{P}_c}{C_h \mathcal{P}_h}$
C_r^*	$\dfrac{M_w c_w \omega}{C_c}$	$\dfrac{M_w c_w}{C_c \mathcal{P}_c}$
$(hA)^*$	$\dfrac{h_c A_c}{h_h A_h}$	$\dfrac{h_c \mathcal{P}_c}{h_h \mathcal{P}_h}$
$\dfrac{1}{\Lambda_m}$	$\dfrac{C_c + C_h}{4}\left(\dfrac{1}{h_h A_h} + \dfrac{1}{h_c A_c}\right)$	$\dfrac{C_c \mathcal{P}_c + C_h \mathcal{P}_h}{4A}\left(\dfrac{1}{h_h \mathcal{P}_h} + \dfrac{1}{h_c \mathcal{P}_c}\right)$
$\dfrac{1}{\Pi_m}$	$\dfrac{M_w c_w \omega}{2}\left(\dfrac{1}{h_h A_h} + \dfrac{1}{h_c A_c}\right)$	$\dfrac{M_w c_w}{2A}\left(\dfrac{1}{h_h \mathcal{P}_h} + \dfrac{1}{h_c \mathcal{P}_c}\right)$
γ	$\dfrac{C_c}{C_h}$	$\dfrac{C_c \mathcal{P}_c}{C_h \mathcal{P}_h}$
R^*	$\dfrac{h_h A_h}{h_c A_c}$	$\dfrac{h_h \mathcal{P}_h}{h_c \mathcal{P}_c}$

[†] If $C_h = C_{min}$, the subscripts c and h in this table should be changed to h and c, respectively. The definitions are given for one rotor (disk) of a rotary regenerator or for one matrix of a fixed-matrix regenerator. $\mathcal{P}_h$ and $\mathcal{P}_c$ represent hot-gas and cold-gas periods, respectively, s. ω is rotational speed, rev/s.

TABLE 17.11   Designation of Various Types of Regenerators Depending upon the Values of Dimensionless Groups

Terminology	Λ-Π method	ε-NTU_0 method
Balanced regenerators	$\Lambda_h/\Pi_h = \Lambda_c/\Pi_c$ or $\gamma = 1$	$C^* = 1$
Unbalanced regenerators	$\Lambda_h/\Pi_h \neq \Lambda_c/\Pi_c$	$C^* \neq 1$
Symmetric regenerators	$\Pi_h = \Pi_c$ or $R^* = 1$	$(hA)^* = 1$
Unsymmetric regenerators	$\Pi_h \neq \Pi_c$	$(hA)^* \neq 1$
Symmetric and balanced regenerators	$\Lambda_h = \Lambda_c, \Pi_h = \Pi_c$	$(hA)^* = 1, C^* = 1$
Unsymmetric but balanced regenerators	$\Lambda_h/\Pi_h = \Lambda_c/\Pi_c$	$(hA)^* \neq 1, C^* = 1$
Long regenerators	$\Lambda/\Pi > 5$	$C_r^* > 5$

A closed-form solution for a *balanced and symmetric* counterflow regenerator [$C^* = 1$, $(hA)^* = 1$] has been obtained by Bačlić [31], valid for all values of C_r^*, as follows.

$$\varepsilon = C_r^* \frac{1 + 7\beta_2 - 24\{B - 2[R_1 - A_1 - 90(N_1 + 2E)]\}}{1 + 9\beta_2 - 24\{B - 6[R - A - 20(N - 3E)]\}} \tag{17.45}$$

where $B = 3\beta_3 - 13\beta_4 + 30(\beta_5 - \beta_6)$
$R = \beta_2[3\beta_4 - 5(3\beta_5 - 4\beta_6)]$
$A = \beta_3[3\beta_3 - 5(3\beta_4 + 4\beta_5 - 12\beta_6)]$
$N = \beta_4[2\beta_4 - 3(\beta_5 + \beta_6)] + 3\beta_5^2$
$E = \beta_2\beta_4\beta_6 - \beta_2\beta_5^2 - \beta_3^2\beta_6 + 2\beta_3\beta_4\beta_5 - \beta_4^3 \tag{17.46}$
$N_1 = \beta_4[\beta_4 - 2(\beta_5 + \beta_6)] + 2\beta_5^2$
$A_1 = \beta_3[\beta_3 - 15(\beta_4 + 4\beta_5 - 12\beta_6)]$
$R_1 = \beta_2[\beta_4 - 15(\beta_5 - 2\beta_6)]$
$\beta_i = V_i(2NTU_o, 2NTU_o/C_r^*)/(2NTU_o)^{i-1}, \quad i = 2, 3, \ldots, 6$

and

$$V_i(x, y) = \exp[-(x + y)] \sum_{n=i-1}^{\infty} \binom{n}{i-1}(y/x)^{n/2} I_n(2\sqrt{xy}) \tag{17.47}$$

In these equations, all variables and parameters are local except for NTU_o, C_r^*, and ε. Here I_n represents the modified Bessel function of the first kind and nth order. Shah [32] has tabulated the effectiveness of Eq. 17.46 for $0.5 \leq NTU_o \leq 500$ and $1 \leq C_r^* \leq \infty$.

Extensive numerical results have been obtained by Lambertson as reported in Ref. 20 for a counterflow regenerator and Theoclitus and Eckrich [33] for a parallelflow regenerator for a wide range of NTU_o, C^*, and C_r^*. Their results for $C^* = 1$ are presented in Figs. 17.32 and 17.33. Note that longitudinal heat conduction in the wall is neglected in these results since infinite thermal resistance is specified for the matrix in the flow direction.

Razelos, as reported in Refs. 15 and 29, proposed the following approximate procedure to calculate the counterflow regenerator effectiveness ε for unbalanced and unsymmetric regenerators for $C_r^* \geq 1$, $0.25 \leq (hA)^* \leq 4$, and the complete range of C^* and NTU_o. For the known values of NTU_o, C^*, and C_r^*, calculate appropriate values of NTU_o and C_r^* for an equivalent balanced regenerator ($C^* = 1$), designated with a subscript m, as follows:

$$NTU_{o,m} = \frac{2NTU_o C^*}{1 + C^*} \tag{17.48}$$

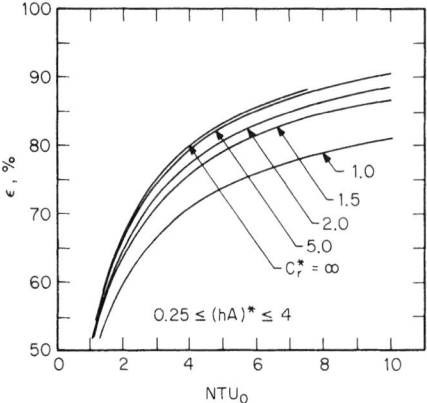

FIGURE 17.32 Counterflow regenerator effectiveness as a function of NTU_o and C_r^* for $C^* = 1$ [20].

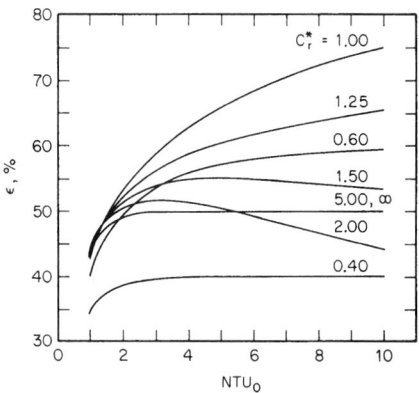

FIGURE 17.33 Parallelflow regenerator effectiveness as a function of NTU_o and C_r^* for $C^* = 1$ and $(hA)^* = 1$ [33].

$$C^*_{r,m} = \frac{2C^*_r C^*}{1 + C^*} \tag{17.49}$$

Then the equivalent balanced regenerator effectiveness ε_r is given by $\varepsilon_r = \varepsilon$ in Eq. 17.45 using the above $\text{NTU}_{o,m}$ and $C^*_{r,m}$ for NTU_o and C^*_r in Eq. 17.45. For $C^*_{r,m} < 1$, the regenerator effectiveness can be obtained from Hausen's effectiveness chart in Figs. 13–16 of Ref. 29 or Fig. 5.4 of Ref. 30 using $\Lambda = 2\text{NTU}_{o,m}$ and $\Pi = 2\text{NTU}_{o,m}/C^*_{r,m}$.

Finally, calculate the desired regenerator effectiveness ε from

$$\varepsilon = \frac{1 - \exp\{\varepsilon_r(C^{*2} - 1)/[2C^*(1 - \varepsilon_r)]\}}{1 - C^* \exp\{\varepsilon_r(C^{*2} - 1)/[2C^*(1 - \varepsilon_r)]\}} \tag{17.50}$$

where $\varepsilon_r = \varepsilon$ of Eq. 17.45 using $\text{NTU}_{o,m}$ and $C^*_{r,m}$ of Eqs. 17.48 and 17.49 for NTU_o and C^*_r in Eq. 17.45.

Longitudinal Heat Conduction in Wall. Longitudinal heat conduction in the wall was neglected in deriving the results of the preceding section. However, it may not be negligible, particularly for a high-effectiveness regenerator having a short flow length L and resultant large temperature gradient in the axial direction. It reduces the regenerator effectiveness and the overall heat transfer rate. For example, for regenerators designed for $\varepsilon > 85$ percent, a 1 percent reduction in ε would reduce gas turbine power plant efficiency by about 1 to 5 percent depending upon the load conditions, which could translate into a significant economic penalty. The reduction in ε due to longitudinal conduction in the wall can be 1 percent or higher and hence must be properly considered in the design. Based on extensive numerical results by Bahnke and Howard [20, 34], this effect can be taken into account by an additional parameter λ, referred to as the longitudinal conduction parameter:

$$\lambda = \frac{k_w A_{k,t}}{L C_{\min}} \tag{17.51}$$

where k_w is the thermal conductivity of the matrix wall, and $A_{k,t}$ is the total solid area for longitudinal conduction

$$A_{k,t} = A_{k,c} + A_{k,h} = A_{fr} - A_o = A_{fr}(1 - \sigma) \tag{17.52}$$

Bahnke and Howard's results for $C^* = 1$ can be accurately expressed by

$$\varepsilon = C_\lambda \varepsilon_{\lambda = 0} \tag{17.53}$$

where $\varepsilon_{\lambda = 0}$ is given by Eq. 17.45 and C_λ is given by

$$C_\lambda = \frac{1 + \text{NTU}_o}{\text{NTU}_o} \left[1 - \frac{1}{1 + \text{NTU}_o(1 + \lambda\phi)/(1 + \lambda\text{NTU}_o)} \right] \tag{17.54}$$

where
$$\phi = \left[\frac{\lambda\text{NTU}_o}{1 + \lambda\text{NTU}_o} \right]^{1/2} \tanh \left[\frac{\text{NTU}_o}{\{\lambda\text{NTU}_o/(1 + \lambda\text{NTU}_o)\}^{1/2}} \right] \tag{17.55}$$

$$\approx \left[\frac{\lambda\text{NTU}_o}{1 + \lambda\text{NTU}_o} \right]^{1/2} \qquad \text{for } \text{NTU}_o \geq 3 \tag{17.56}$$

Bahnke and Howard's results for $C^* = 1$ and $C^*_r > 5$ are the same as those shown in Fig. 17.31a provided that the abscissa NTU is replaced by NTU_o. The regenerator effectiveness due to longitudinal conduction decreases with increasing values of λ and C^*, with the maximum effect at $C^* = 1$.

For $C^* < 1$, use the following Razelos method to account for the effect of longitudinal conduction in the wall.

1. Compute $NTU_{o,m}$, $C_{r,m}^*$ and $\varepsilon_{r,\lambda=0}$ for an equivalent balanced regenerator using Eqs. 17.48, 17.49, and 17.45, respectively.
2. Compute C_λ from Eq. 17.54 using $NTU_{o,m}$ and λ.
3. Calculate $\varepsilon_{r,\lambda\neq0} = C_\lambda \varepsilon_{r,\lambda=0}$.
4. Finally, ε is determined from Eq. 17.50, with ε_r replaced by $\varepsilon_{r,\lambda\neq0}$.

This procedure yields ε that is accurate within 1 percent for $1 \leq NTU_o \leq 20$ for $C_r^* \geq 1$ when compared to Bahnke and Howard's results.

Influence of Transverse Heat Conduction in Wall. The thermal resistance for heat conduction in the wall thickness direction is considered zero in all of the preceding ε-NTU_o results. This is a good idealization for metal matrices with thin walls. For most rotary regenerators, the thermal resistance in the transverse direction is negligible except possibly for ceramic regenerators.

The wall thermal resistance is evaluated separately during the hot-gas and cold-gas flow periods, since there is no continuous heat flow from the hot gas to the cold gas in the regenerator. Based on the unit area, it is given by [29]

$$\hat{\mathbf{R}}_w = \mathbf{R}_w A = \frac{\delta}{6k_w} \Phi^* \tag{17.57}$$

so that the effective heat transfer coefficients during the hot- and cold-gas flow periods (designated by a superscript bar) are

$$\frac{1}{\overline{h_h}} = \frac{1}{h_h} + \frac{\delta}{6k_w}\Phi^* \qquad \frac{1}{\overline{h_c}} = \frac{1}{h_c} + \frac{\delta}{6k_w}\Phi^* \tag{17.58}$$

where δ is the wall thickness and Φ^* for a plain wall is given by

$$\Phi^* = \begin{cases} 1 - \tfrac{1}{15}Z & \text{for } Z \leq 5 \tag{17.59} \\ 2.142[0.3 + 2Z]^{-1/2} & \text{for } Z > 5 \tag{17.60} \end{cases}$$

where $Z = (Bi_h/\Pi_h) + (Bi_c/\Pi_c)$ and $Bi_h = h_h(\delta/2)/k_w$, $\Pi_h = h_h A_h/C_{r,h}$, Bi_c and Π_c are defined in a similar manner. The range of Φ^* for Eq. 17.59 is 2/3 to 1, and for Eq. 17.60 from 0 to 2/3. When $Bi \to 0$, the transverse thermal resistance $\mathbf{R}_w$ of Eq. 17.57 approaches zero. Equations 17.59 and 17.60 are valid for Bi_h and Bi_c lower than 2. For $Bi > 2$, use the numerical results of Heggs et al. [35]. The accuracy of Eqs. 17.59 and 17.60 decreases with increasing Bi/Π and decreasing C_r^*.

The prediction of the temperature swing δT in a fixed-matrix regenerator will not be accurate by the foregoing approximate method. The numerical analysis of the type made by Heggs et al. [35] is essential for accurate δT determination. It may be noted that the δT values of Table 1 of Ref. 35 had a typing error and all should be multiplied by a factor of 10; also all the charts in Fig. 1 of Ref. 35 are poorly drawn, as a result of which the δT values shown are approximate.

Fluid Pressure Leakage and Carryover. In rotary regenerators, fluid mixing from the cold to hot gas stream and vice versa occurs due to fluid pressure and carryover leakages. A comprehensive gas flow network model of pressure leakage and carryover is presented by Shah and Skiepko [36] as shown in Fig. 17.34. A rotary regenerator disk with its housing and radial, peripheral, and axial seals (to prevent leakages) is designated by a boundary indicating an *actual regenerator* in Fig. 17.34; and the regenerator disk or matrix with no leakage streams within its boundary is designated the *internal regenerator;* the hot and cold gas inlet faces are designated the *hot and cold ends.* The high-pressure cold gas (air) can leak through the low-pressure hot gas in a number of ways due to the pressure difference. Also, due to the pressure drop on each gas side in the regenerator, the inlet pressures are going to be higher than the outlet pressures on the respective sides, and hence the cold and hot gases can bypass the

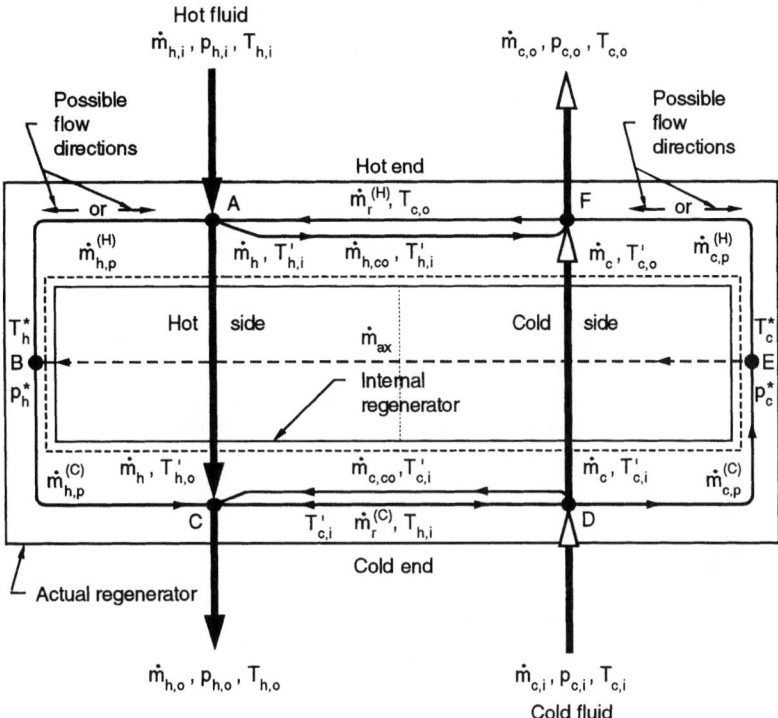

FIGURE 17.34 Regenerator gas flow model with leakages [36].

regenerator matrix on individual sides through the gap between the disk and housing. Various flow leakage streams due to the pressure differences are minimized through the use of radial, peripheral, and axial seals. Because of the mechanical design considerations, there will be finite clearances between the housing and the seals; these clearances will set the leakage flow rates depending on the operating pressures, flow rates and fluid properties. Various leakage flow rate terms designated in Fig. 17.34 are defined in Table 17.12 and are as follows.

- Pressure leakages
  - due to a part of higher pressure gas stream passing through the sealing system, and entering into the lower pressure gas stream: $\dot{m}_r^{(H)}$, $\dot{m}_r^{(C)}$, $\dot{m}_{ax}$.
  - due to a flow bypass from inlet to outlet (through the gap between the housing and disk) on each gas side associated with the pressure drop in the matrix: $\dot{m}_{h,p}^{(H)}$, $\dot{m}_{c,p}^{(H)}$, $\dot{m}_{h,p}^{(C)}$, $\dot{m}_{c,p}^{(C)}$.
- Carryover leakages, when a part of one gas stream trapped into void volumes of the matrix at the end of the period is carried into the other gas stream at the beginning of the following period: $\dot{m}_{h,co}$, $\dot{m}_{c,co}$.

Clearances associated with seals and pressure leakages are considered orifices, and the leakage flow rates are computed using the following ASME orifice formulas with the known seal gap flow areas $A_{o,s}$ as follows.

$$\dot{m}_{\text{seal}} = C_d A_{o,s} Y \sqrt{2\rho\Delta p} \qquad (17.61)$$

Here, the coefficient of discharge $C_d = 0.80$ [36], expansion factor $Y = 1$, and the specific values of Δp and ρ at inlet for each leakage are given in Table 17.12.

TABLE 17.12   Rotary Regenerator Pressure Leakage and Carryover Flow Rates, Δp and Inlet Density for the Orifice Analysis[†]

Leakage terms	Symbols	Pressure drops	Density
Flows through radial seal clearances:			
Flow of the higher pressure cold gas at the hot end	$\dot{m}_r^{(H)}$	$p_{c,o} - p_{h,i}$	$\rho_{c,o}$
Flow of the higher pressure cold gas at the cold end	$\dot{m}_r^{(C)}$	$p_{c,i} - p_{h,o}$	$\rho_{c,i}$
Flows through peripheral seal clearances:			
Flows at the hot end of the disc face			
Flow around inlet to the lower pressure hot gas zone	$\dot{m}_{h,p}^{(H)}$	$p_{h,i} - p_h^*$ $p_h^* - p_{h,i}$	$\rho_{h,i}$ if $p_{h,i} > p_h^*$ ρ_h^* if $p_{h,i} < p_h^*$
Flow around outlet from the higher pressure cold gas zone	$\dot{m}_{c,p}^{(H)}$	$p_{c,o} - p_c^*$ $p_c^* - p_{c,o}$	$\rho_{c,o}$ if $p_{c,o} > p_c^*$ ρ_c^* if $p_{c,o} < p_c^*$
Flows at the cold end of the disc face			
Flow around outlet from the lower pressure hot gas zone	$\dot{m}_{h,p}^{(C)}$	$p_h^* - p_{h,o}$	ρ_h^*
Flow around inlet to the higher pressure cold gas zone	$\dot{m}_{c,p}^{(C)}$	$p_{c,i} - p_c^*$	$\rho_{c,i}$
Flow through axial seal clearances:	$\dot{m}_{ax}$	$p_c^* - p_h^*$	ρ_c^*
Gas carryover:			
Carryover of the lower pressure hot gas into cold gas	$\dot{m}_{h,co}$		$\bar{\rho}_h$
Carryover of the higher pressure cold into hot gas	$\dot{m}_{c,co}$		$\bar{\rho}_c$

[†] p_h^* and ρ_h^* are pressure and density at Point B in Fig. 17.34; $\bar{\rho}$ is an average density from inlet to outlet.

Several models have been presented to compute the carryover leakage [15, 36], with the following model as probably the most representative of industrial regenerators.

$$\dot{m}_{co} = A_{fr} \mathbf{N} \bar{\rho} \left[\sum_{i=1}^{n} (L_i \sigma_i) + \Delta L \right] \qquad (17.62)$$

where $\mathbf{N}$ is the rotational speed (rev/s) of the regenerator disk, $\bar{\rho}$ is the gas density evaluated at the arithmetic mean of inlet and outlet temperatures, and σ_i and L_i represent the porosity and height of several layers of the regenerator (use σ and L for uniform porosity and a single layer of the matrix) and ΔL represents the height of the header.

Equations 17.61 and 17.62 represent a total of nine equations (see Table 17.12 for nine unknown mass flow rates) that can be solved once the pressures and temperatures at the terminal points of the regenerator of Fig. 17.34 are known. These terminal points are known once the rating of the internal regenerator is done and mass and energy balances are made at the terminal points based on the previous values of the leakage and carryover flow rates. Refer to Shah and Skiepko [36] for further details.

Single-Phase Pressure Drop Analysis

Fluid pumping power is a design constraint in many applications. This pumping power is proportional to the pressure drop in the exchanger in addition to the pressure drops associated with inlet and outlet headers, manifolds, tanks, nozzles, or ducting. The fluid pumping power P associated with the core frictional pressure drop in the exchanger is given by

$$P = \frac{\dot{m}\Delta p}{\rho} \approx \begin{cases} \dfrac{1}{2g_c} \dfrac{\mu}{\rho^2} \dfrac{4L}{D_h} \dfrac{\dot{m}^2}{D_h A_o} f\,\text{Re} & \text{for laminar flow} & (17.63a) \\[3mm] \dfrac{0.046}{2g_c} \dfrac{\mu^{0.2}}{\rho^2} \dfrac{4L}{D_h} \dfrac{\dot{m}^{2.8}}{A_o^{1.8} D_h^{0.2}} & \text{for turbulent flow} & (17.63b) \end{cases}$$

Only the core friction term is considered in the right-hand side approximation for discussion purposes. Now consider the case of specified flow rate and geometry (i.e., specified $\dot{m}$, L, D_h, and A_o). As a first approximation, f Re in Eq. 17.63a is constant for fully developed laminar flow, while $f = 0.046\text{Re}^{-0.2}$ is used in deriving Eq. 17.63b for fully developed turbulent flow. It is evident that P is strongly dependent on ρ (P $\propto 1/\rho^2$) in laminar and turbulent flows and on μ in laminar flow, and weakly dependent on μ in turbulent flow. For high-density, moderate-viscosity liquids, the pumping power is generally so small that it has only a minor influence on the design. For a laminar flow of highly viscous liquids in large L/D_h exchangers, pumping power is an important constraint; this is also the case for gases, both in turbulent and laminar flow, because of the great impact of $1/\rho^2$.

In addition, when blowers and pumps are used for the fluid flow, they are generally head-limited, and the pressure drop itself can be a major consideration. Also, for condensing and evaporating fluids, the pressure drop affects the heat transfer rate. Hence, the Δp determination in the exchanger is important. As shown in Eq. 17.177, the pressure drop is proportional to D_h^{-3} and hence it is strongly influenced by the passage hydraulic diameter.

The pressure drop associated with a heat exchanger consists of (1) core pressure drop and (2) the pressure drop associated with the fluid distribution devices such as inlet and outlet manifolds, headers, tanks, nozzles, ducting, and so on, which may include bends, valves, and fittings. This second Δp component is determined from Idelchik [37] and Miller [38]. The core pressure drop may consist of one or more of the following components depending upon the exchanger construction: (1) friction losses associated with fluid flow over heat transfer surface; this usually consists of skin friction, form (profile) drag, and internal contractions and expansions, if any; (2) the momentum effect (pressure drop or rise due to fluid density changes) in the core; (3) pressure drop associated with sudden contraction and expansion at the core inlet and outlet; and (4) the gravity effect due to the change in elevation between the inlet and outlet of the exchanger. The gravity effect is generally negligible for gases. For vertical flow through the exchanger, the pressure drop or rise ("static head") due to the elevation change is given by

$$\Delta p = \pm \frac{\rho_m g L}{g_c} \tag{17.64}$$

Here the "+" sign denotes vertical upflow (i.e., pressure drop), the "−" sign denotes vertical downflow (i.e., pressure rise or recovery). The first three components of the core pressure drop are now presented for plate-fin, tube-fin, regenerative, and plate heat exchangers. Pressure drop on the shellside of a shell-and-tube heat exchanger is presented in Table 17.31.

Plate-Fin Heat Exchangers.   For the plate-fin exchanger (Fig. 17.10), all three components are considered in the core pressure drop evaluation as follows.

$$\frac{\Delta p}{p_i} = \frac{G^2}{2g_c}\frac{1}{p_i \rho_i}\left[(1 - \sigma^2 + K_c) + f\frac{L}{r_h}\rho_i\left(\frac{1}{\rho}\right)_m + 2\left(\frac{\rho_i}{\rho_o} - 1\right) - (1 - \sigma^2 - K_e)\frac{\rho_i}{\rho_o}\right] \tag{17.65}$$

where f is the Fanning friction factor, K_c and K_e are flow contraction (entrance) and expansion (exit) pressure loss coefficients, and σ is a ratio of minimum free flow area to frontal area. K_c and K_e for four different long ducts are presented by Kays and London [20] as shown in Fig. 17.35 for which flow is fully developed at the exit. For partially developed flows, K_c is lower and K_e is higher than that for fully developed flows. For interrupted surfaces, flow is never of the fully developed boundary-layer type. For highly interrupted fin geometries, the entrance and exit losses are generally small compared to the core pressure drop, and the flow is well mixed; hence, K_c and K_e for Re $\rightarrow \infty$ should represent a good approximation. The entrance and exit losses are important at low values of σ and L (short cores), at high values of Re, and for gases; they are negligible for liquids. The mean specific volume v_m or $(1/\rho)_m$ in Eq. 17.65 is given as follows: for liquids with any flow arrangement, or for a perfect gas with $C^* = 1$ and any flow arrangement (except for parallelflow),

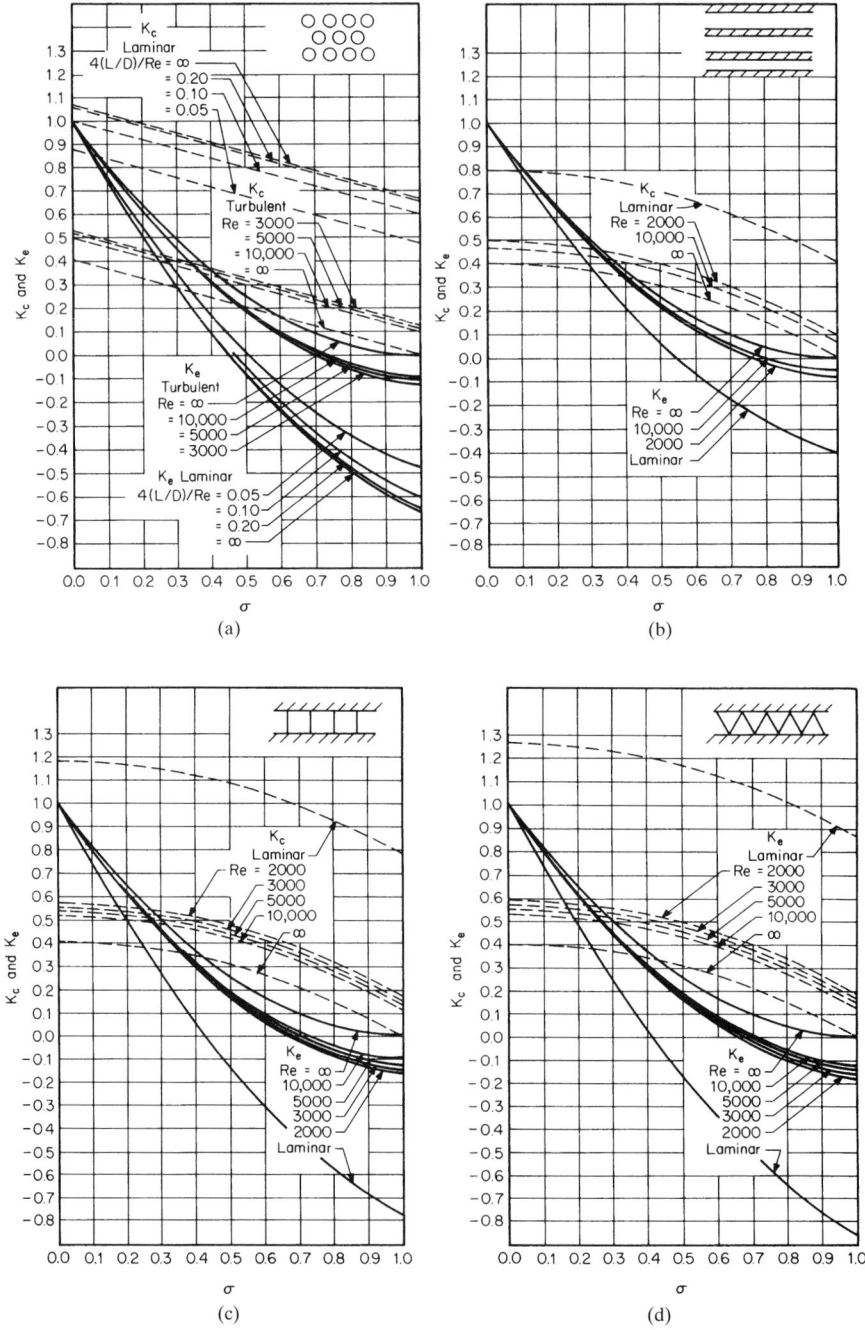

FIGURE 17.35 Entrance and exit pressure loss coefficients: (*a*) circular tubes, (*b*) parallel plates, (*c*) square passages, and (*d*) triangular passages [20]. For each of these flow passages, the fluid flows perpendicular to the plane of the paper into the flow passages.

$$\left(\frac{1}{\rho}\right)_m = v_m = \frac{v_i + v_o}{2} = \frac{1}{2}\left(\frac{1}{\rho_i} + \frac{1}{\rho_o}\right) \tag{17.66}$$

where v is the specific volume in m^3/kg.

For a perfect gas with $C^* = 0$ and any flow arrangement,

$$\left(\frac{1}{\rho}\right)_m = \frac{\tilde{R}}{p_{ave}} T_{lm} \tag{17.67}$$

Here $\tilde{R}$ is the gas constant in $J/(kg\ K)$, $p_{ave} = (p_i + p_o)/2$, and $T_{lm} = T_{const} + \Delta T_{lm}$, where T_{const} is the mean average temperature of the fluid on the other side of the exchanger; the log-mean temperature difference ΔT_{lm} is defined in Table 17.4. The core frictional pressure drop in Eq. 17.65 may be approximated as

$$\Delta p \approx \frac{4fLG^2}{2g_c D_h}\left(\frac{1}{\rho}\right)_m \approx \frac{4fLG^2}{2g_c \rho_m D_h} \tag{17.68}$$

Tube-Fin Heat Exchangers.   The pressure drop inside a circular tube is computed using Eq. 17.65 with proper values of f factors (see equations in Tables 17.14 and 17.16) and K_c and K_e from Fig. 17.35 for circular tubes.

For flat fins on an array of tubes (see Fig. 17.14b), the components of the core pressure drop (such as those in Eq. 17.65) are the same with the following exception: the core friction and momentum effect take place within the core with $G = \dot{m}/A_o$, where A_o is the minimum free flow area within the core, and the entrance and exit losses occur at the leading and trailing edges of the core with the associated flow area A_o' so that

$$\dot{m} = GA_o = G'A_o' \qquad \text{or} \qquad G'\sigma' = G\sigma \tag{17.69}$$

where σ' is the ratio of free flow area to frontal area at the fin leading edges and is used in the evaluation of K_c and K_e from Fig. 17.35. The pressure drop for flow normal to a tube bank with flat fins is then given by

$$\frac{\Delta p}{p_i} = \frac{G^2}{2g_c}\frac{1}{p_i\rho_i}\left[f\frac{L}{r_h}\rho_i\left(\frac{1}{\rho}\right)_m + 2\left(\frac{\rho_i}{\rho_o} - 1\right)\right] + \frac{G'^2}{2g_c}\frac{1}{p_i\rho_i}\left[(1 - \sigma'^2 + K_c) - (1 - \sigma'^2 - K_e)\frac{\rho_i}{\rho_o}\right] \tag{17.70}$$

For individually finned tubes as shown in Fig. 17.14a, flow expansion and contraction take place along each tube row, and the magnitude is of the same order as that at the entrance and exit. Hence, the entrance and exit losses are generally lumped into the core friction factor. Equation 17.65 for individually finned tubes then reduces to

$$\frac{\Delta p}{p_i} = \frac{G^2}{2g_c}\frac{1}{p_i\rho_i}\left[f\frac{L}{r_h}\rho_i\left(\frac{1}{\rho}\right)_m + 2\left(\frac{\rho_i}{\rho_o} - 1\right)\right] \tag{17.71}$$

Regenerators.   For regenerator matrices having cylindrical passages, the pressure drop is computed using Eq. 17.65 with appropriate values of f, K_c, and K_e. For regenerator matrices made up of any porous material (such as checkerwork, wire, mesh, spheres, or copper wool), the pressure drop is calculated using Eq. 17.71, in which the entrance and exit losses are included in the friction factor f.

Plate Heat Exchangers.   Pressure drop in a plate heat exchanger consists of three components: (1) pressure drop associated with the inlet and outlet manifolds and ports, (2) pressure drop within the core (plate passages), and (3) pressure drop due to the elevation change. The pressure drop in the manifolds and ports should be kept as low as possible (generally <10 percent, but it is found as high as 25–30 percent or higher in some designs). Empirically, it is calculated as approximately 1.5 times the inlet velocity head per pass. Since the entrance and exit losses in the core (plate passages) cannot be determined experimentally, they are included in the friction factor for the given plate geometry. The pressure drop (rise) caused by the elevation change for liquids is given by Eq. 17.64. Hence, the pressure drop on one fluid side in a plate heat exchanger is given by

$$\Delta p = \frac{1.5G^2 N}{2g_c \rho_i} + \frac{4fLG^2}{2g_c D_e}\left(\frac{1}{\rho}\right)_m + \left(\frac{1}{\rho_o} - \frac{1}{\rho_i}\right)\frac{G^2}{g_c} \pm \frac{\rho_m g L}{g_c} \tag{17.72}$$

where N is the number of passes on the given fluid side and D_e is the equivalent diameter of flow passages (usually twice the plate spacing). Note that the third term on the right side of the equality sign of Eq. 17.72 is for the momentum effect, which generally is negligible for liquids.

SINGLE-PHASE SURFACE BASIC HEAT TRANSFER AND FLOW FRICTION CHARACTERISTICS

Accurate and reliable surface heat transfer and flow friction characteristics are a key input to the exchanger heat transfer and pressure drop analyses or to the rating and sizing problems (see Fig. 17.36). After presenting the associated nondimensional groups, we will present important experimental methods, analytical solutions, and empirical correlations for some important exchanger geometries.

The dimensionless heat transfer and fluid flow friction (pressure drop) characteristics of a heat transfer surface are simply referred to as the surface basic characteristics or surface basic

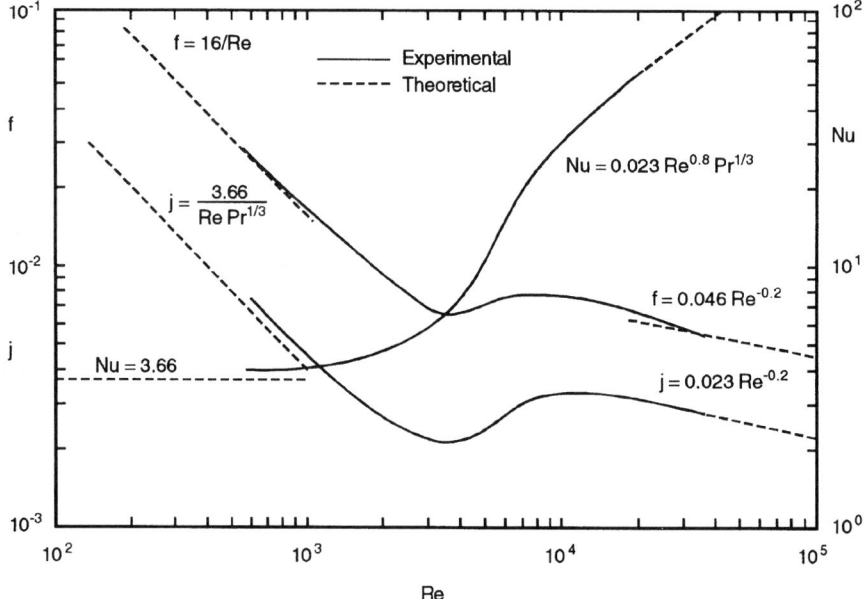

FIGURE 17.36 Basic heat transfer and flow friction characteristics for air flow through a long circular tube.

data.* Generally, the dimensionless experimental heat transfer characteristics are presented in terms of the Colburn factor $j = \mathrm{St}\,\mathrm{Pr}^{2/3}$ versus Reynolds number Re, and the theoretical characteristics in terms of Nusselt number Nu versus Re or $x^* = x/(D_h\,\mathrm{Re}\,\mathrm{Pr})$. The dimensionless pressure drop characteristics are presented in terms of the Fanning friction factor f versus Re or modified friction factor per tube row f_{tb} versus Re_d. These and other important dimensionless groups used in presenting and correlating internal flow forced convection heat transfer are summarized in Table 17.13 with their definition and physical meanings. Where applicable, the hydraulic diameter D_h is used as a characteristic dimension in all dimensionless groups for consistency. However, it must be emphasized that the hydraulic diameter or any other characteristic length does not represent a universal characteristics dimension. This is because the three-dimensional boundary layer and wake effects in noncircular continuous/interrupted flow passages cannot be correlated with a single-length dimension. For some of the dimensionless groups of Table 17.13, a number of different definitions are used in the literature; the user should pay particular attention to the specific definitions used in any research paper before using specific results. This is particularly true for the Nusselt number (where many different temperature differences are used in the definition of h), and for f, Re, and other dimensionless groups having characteristic dimensions different from D_h.

Since the majority of basic data for compact surfaces are obtained experimentally, the dimensionless heat transfer and pressure drop characteristics of these surfaces are presented in terms of j and f versus Re. As an example of these correlating groups, basic heat transfer and flow friction characteristics for air flow in a long circular tube are presented in Fig. 17.36. This figure shows three flow regimes: laminar, transition, and turbulent. This is characteristic of fully developed flow in continuous flow-passage geometries such as a long circular tube and triangular tube. Generally, the compact interrupted surfaces do not have a sharp dip in the transition region (Re ~ 1500–10,000) as shown for the circular tube. Notice that there is a parallel behavior of j versus Re and f versus Re curves, but no such parallelism exists between Nu and f versus Re curves. The parallel behavior of j and f versus Re is useful for: (1) identifying erroneous test data for some specific surfaces for which parallel behavior is expected but indicated otherwise by test results (see Fig. 17.38); (2) identifying specific flow phenomena in which the friction behavior is different from the heat transfer behavior (such as rough surface flow for friction and boundary-layer-type flow for heat transfer in a turbulent flow regime for highly interrupted fin geometries); and (3) predicting the f factors for an unknown surface when the j factors are known by some predictive method. It should be remembered that j versus Re can readily be converted to Nu versus Re curves or vice versa, because $j = \mathrm{Nu}\,\mathrm{Pr}^{-1/3}/\mathrm{Re}$ by definition. Because the values of j, f, and Re are dimensionless, the test data are applicable to surfaces of any hydraulic diameter, provided that complete geometric similarity is maintained.

The limitations of the j versus Re plot, commonly used in presenting compact heat exchanger surface basic data, should be understood. In fully developed laminar flow, as will be discussed, the Nusselt number is theoretically constant, independent of Pr (and also Re). Since $j = \mathrm{St}\,\mathrm{Pr}^{2/3} = \mathrm{Nu}\,\mathrm{Pr}^{-1/3}/\mathrm{Re}$, then j will be dependent upon Pr in the fully developed laminar region, and hence the j factors presented in Chap. 7 of Kays and London [20] for gas flows in the fully developed laminar region should be first converted to a Nusselt number (using Pr = 0.70), which can then be used directly for liquid flows as constant property results. Based on theoretical solutions for thermally developing laminar flow (to be discussed), Nu $\propto$ $(x^*)^{-1/3}$. This means Nu $\mathrm{Pr}^{-1/3}$ is independent of Pr, and, hence, j is independent of Pr for thermally developing laminar flows.† For fully developed turbulent flow, Nu $\propto \mathrm{Pr}^{0.4}$, and, hence,

---

* We will not use the terminology *surface performance data,* since *performance* in industry means a dimensional plot of heat transfer rate and pressure drop as a function of the fluid flow rate for an exchanger. Note that we need to distinguish between the performance of a surface geometry and the performance of a heat exchanger.

† If a slope of −1 for the log j-log Re characteristic is used as a criterion for fully developed laminar flow, none of the surfaces (except for long smooth ducts) reported in Chap. 10 of Kays and London [20] would qualify as being in a fully developed laminar condition. Data for most of these surfaces indicate thermally developing flow conditions for which j is almost independent of Pr as indicated (as long as the exponent on Pr is about −⅓), and hence the j-Re characteristic should not be converted to the Nu-Re characteristic for the data of Chap. 10, Ref. 20.

TABLE 17.13   Important Dimensionless Groups for Internal Flow Forced Convection Heat Transfer and Flow Friction Useful in Heat Exchanger Design

Dimensionless groups	Definitions and working relationships	Physical meaning and comments
Reynolds number	$\mathrm{Re} = \dfrac{\rho V D_h}{\mu} = \dfrac{G D_h}{\mu}$	A flow modulus proportional to the ratio of inertia force to viscous force
Fanning friction factor	$f = \dfrac{\tau_w}{(\rho V^2/2g_c)}$ $f = \Delta p^* \dfrac{r_h}{L} = \dfrac{\Delta p}{(\rho V^2/2g_c)} \dfrac{r_h}{L}$	The ratio of wall shear (skin frictional) stress to the flow kinetic energy per unit volume; commonly used in heat transfer literature
Apparent Fanning friction factor	$f_{\mathrm{app}} = \Delta p^* \dfrac{r_h}{L}$	Includes the effects of skin friction and the change in the momentum rates in the entrance region (developing flows)
Incremental pressure drop number	$K(x) = (f_{\mathrm{app}} - f_{fd}) \dfrac{L}{r_h}$ $K(\infty) = \text{constant for } x \to \infty$	Represents the excess dimensionless pressure drop in the entrance region over that for fully developed flow
Darcy friction factor	$f_D = 4f = \Delta p^* \dfrac{D_h}{L}$	Four times the Fanning friction factor; commonly used in fluid mechanics literature
Euler number	$\mathrm{Eu} = \Delta p^* = \dfrac{\Delta p}{(\rho V^2/2g_c)}$	The pressure drop normalized with respect to the dynamic velocity head
Dimensionless axial distance for the fluid flow problem	$x^+ = \dfrac{x}{D_h \, \mathrm{Re}}$	The ratio of the dimensionless axial distance (x/D_h) to the Reynolds number; useful in the hydrodynamic entrance region
Nusselt number	$\mathrm{Nu} = \dfrac{h}{k/D_h} = \dfrac{q'' D_h}{k(T_w - T_m)}$	The ratio of the convective conductance h to the pure molecular thermal conductance k/D_h
Stanton number	$\mathrm{St} = \dfrac{h}{G c_p}$ $\mathrm{St} = \dfrac{\mathrm{Nu}}{\mathrm{Pe}} = \dfrac{\mathrm{Nu}}{\mathrm{Re} \, \mathrm{Pr}}$	The ratio of convection heat transfer (per unit duct surface area) to amount virtually transferable (per unit of flow cross-sectional area); no dependence upon any geometric characteristic dimension
Colburn factor	$j = \mathrm{St} \, \mathrm{Pr}^{2/3} = (\mathrm{Nu} \, \mathrm{Pr}^{-1/3})/\mathrm{Re}$	A modified Stanton number to take into account the moderate variations in the Prandtl number for $0.5 \lesssim \mathrm{Pr} \lesssim 10.0$ in turbulent flow
Prandtl number	$\mathrm{Pr} = \dfrac{\nu}{\alpha} = \dfrac{\mu c_p}{k}$	A fluid property modulus representing the ratio of momentum diffusivity to thermal diffusivity of the fluid
Péclet number	$\mathrm{Pe} = \dfrac{\rho c_p V D_h}{k} = \dfrac{V D_h}{\alpha} = \mathrm{Re} \, \mathrm{Pr}$	Proportional to the ratio of thermal energy convected to the fluid to thermal energy conducted axially within the fluid; the inverse of Pe indicates relative importance of fluid axial heat conduction
Dimensionless axial distance for the heat transfer problem	$x^* = \dfrac{x}{D_h \, \mathrm{Pe}} = \dfrac{x}{D_h \, \mathrm{Re} \, \mathrm{Pr}}$	Useful in describing the thermal entrance region heat transfer results
Graetz number	$\mathrm{Gz} = \dfrac{\dot{m} c_p}{k L} = \dfrac{\mathrm{Pe} \, P}{4L} = \dfrac{P}{4 D_h} \dfrac{1}{x^*}$ $\mathrm{Gz} = \pi/(4x^*) \text{ for a circular tube}$	Conventionally used in the chemical engineering literature related to x^* as shown when the flow length in Gz is treated as a length variable

$j \propto Pr^{0.07}$. Thus, j is again dependent on Pr in the fully developed turbulent region.* All of the foregoing comments apply to either constant-property theoretical solutions, or almost-constant-property (low-temperature difference) experimental data. The influence of property variations (see p. 17.88) must be taken into account by correcting the aforementioned constant property j or Nu when designing a heat exchanger.

Experimental Methods

Primarily, three different test techniques are used to determine the surface heat transfer characteristics. These techniques are based on the steady-state, transient, and periodic nature of heat transfer modes through the test sections. We will cover here the most common steady-state techniques used to establish the j versus Re characteristics of a recuperator surface. Different data acquisition and reduction methods are used depending upon whether the test fluid is a gas (air) or a liquid. The method used for liquids is generally referred to as the Wilson plot technique. Refer to Ref. 15 for the transient and periodic techniques. Generally, the isothermal steady-state technique is used for the determination of f factors. These test techniques are now described.

Steady-State Test Technique for Gases.   Generally, a crossflow heat exchanger is employed as a test section. On one side, a surface for which the j versus Re characteristic is known is employed; a fluid with high heat capacity rate flows on this side. On the other side of the exchanger, a surface for which the j versus Re characteristic is to be determined is employed; the fluid which flows over this unknown-side surface is preferably one that is used in a particular application of the unknown-side surface. Generally, air is used on the unknown side; while steam, hot water, chilled water, or oils having high hAs are used on the known side. A typical test setup used by Kays and London [20] is shown in Fig. 17.37 to provide some ideas on the air-side (unknown-side) components of the test rig. For further details, refer to Ref. 40.

In the experiments, the fluid flow rates on both sides of the exchanger are set constant at predetermined values. Once the steady-state conditions are achieved, fluid temperatures upstream and downstream of the test section on both sides are measured, as well as all pertinent measurements for the determination of the fluid flow rates. The upstream pressure and pressure drop across the core on the unknown side are also recorded to determine the "hot" friction factors.[†] The tests are repeated with different flow rates on the unknown side to cover the desired range of the Reynolds number.

---

* In 1933 Colburn [39] proposed $j = St\ Pr^{2/3}$ as a correlating parameter to include the effect of Prandtl number based on the then available data for turbulent flow. Based on presently available experimental data, however, the j factor is clearly dependent on Pr for fully developed turbulent flow and for fully developed laminar flow, but not for ideal developing laminar flows.
† The friction factor determined from the Δp measurement taken during the heat transfer testing is referred to as the hot friction factor.

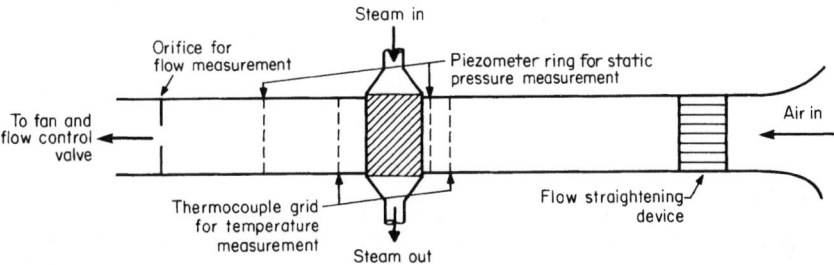

FIGURE 17.37   Schematic of a steam-to-air steady-state heat transfer test rig.

In order to determine the j factor on the unknown side, the exchanger effectiveness ε is determined from the temperature measurements, and the heat capacity rate ratio C^* is determined from individual flow measurements and specific heats. NTU is subsequently computed from the appropriate ε-NTU relationship for the test core flow arrangement (such as Eq. II.1 in Table 17.6). Generally, the test section is a new exchanger core, and fouling resistances are negligible; $\eta_o hA$ on the unknown side is determined from the following thermal resistance equation where UA is found from NTU:

$$\frac{1}{UA} = \frac{1}{(\eta_o hA)_{\text{unknown side}}} + \mathbf{R}_w + \frac{1}{(\eta_o hA)_{\text{known side}}} \tag{17.73}$$

Once the surface area and the geometry are known for the extended surface (if any), h and η_o are computed iteratively using Eqs. 17.73, 17.24, and 17.22 (or an appropriate expression for η_f). Then the j factor is calculated from its definition. The Reynolds number on the unknown side for the test point is determined from its definition for the known mass flow rate and temperature measurements.

The test core is designed with two basic considerations in mind to reduce the experimental uncertainty in the j factors: (1) the magnitudes of thermal resistances on each side as well as of the wall, and (2) the range of NTU.

The thermal resistances in a heat exchanger are related by Eq. 17.73. To reduce the uncertainty in the determination of the thermal resistance of the unknown side (with known overall thermal resistance $1/UA$), the thermal resistances of the exchanger wall and the known side should be kept as small as possible by design. The wall thermal resistance is usually negligible when one of the fluids in the exchanger is air. This may be further minimized through the use of a thin material with high thermal conductivity. On the known side, the thermal resistance is minimized by the use of a liquid (hot or cold water) at high flow rates or a condensing steam (to achieve a high h) or extending the surface area. Therefore, the thermal boundary condition achieved during steady-state testing is generally a close approach to a uniform wall temperature condition.

The NTU range for testing is generally restricted between 0.5 and 3 or between 40 and 90 percent in terms of the exchanger effectiveness. In order to understand this restriction and point out precisely the problem areas, consider the test fluid on the unknown side to be cold air being heated in the test section and the fluid on the known side to be hot water (steam replaced by hot water and its flow direction reversed in Fig. 17.37 to avoid air bubbles). The high NTU occurs at low air flows for a given test core. Both temperature and mass flow rate measurements become more inaccurate at low air flows, and the resultant heat unbalances $(q_w - q_a)/q_a$ increase sharply at low air flows with decreasing air mass flow rate. In this subsection, the subscripts w and a denote water and air sides, respectively. Now, the exchanger effectiveness can be computed in two different ways:

$$\varepsilon = \frac{q_a}{C_a(T_{w,i} - T_{a,i})} \quad \text{or} \quad \varepsilon = \frac{q_w}{C_a(T_{w,i} - T_{a,i})} \tag{17.74}$$

Thus, a large variation in ε will result at low air flows depending upon whether it is based on q_a or q_w. Since ε-NTU curves are very flat at high ε (high NTU), there is a very large error in the resultant NTU, h, and j. The j versus Re curve drops off consistently with decreasing Re, as shown by a dashed line in Fig. 17.38. This phenomenon is referred to as *rollover* or *drop-off* in j. Some of the problems causing the rollover in j are the errors in temperature and air mass flow rate measurements as follows:

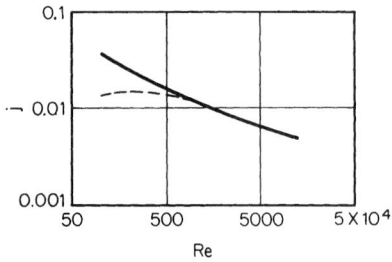

FIGURE 17.38   The rollover phenomenon for j versus Re characteristic of a heat exchanger surface at low airflows. The dashed curve indicates the rollover phenomenon; the solid curve represents the accurate characteristics.

1. Basically, the heat transfer coefficients associated with the thermocouple junction or resistance thermometer are quite low at low air flows. Hence, what we measure is the junction temperature and not the ambient temperature. Thus, the measured air temperature downstream of the test core $T_{a,o}$ may be too low due to heat conduction along the thermocouple wire. This heat conduction error is not so pronounced for the upstream temperature measurement, since air is at a lower temperature. The measured air temperature upstream of the test core $T_{a,i}$ may be too high due to the radiation effect from the hot core and the hot walls of the wind tunnel because of heat conduction in the duct wall from the hot test core. This error is negligible for the core downstream, since the duct walls are at about the same temperature as the outlet air. Both the aforementioned errors in $T_{a,i}$ and $T_{a,o}$ will decrease the calculated q_a.

2. At low airflows, temperature stratification in the vertical direction would be a problem both upstream and downstream of the test core. Thus it becomes difficult to obtain true bulk mean temperatures $T_{a,i}$ and $T_{a,o}$.

3. On the water side, the temperature drop is generally very small, and hence it will require very accurate instrumentation for ΔT_w measurements. Also, care must be exercised to ensure good mixing of water at the core outlet before ΔT_w is measured.

4. There are generally some small leaks in the wind tunnel between the test core and the point of air mass flow rate measurement. These leaks, although small, are approximately independent of the air mass flow rate, and they represent an increasing fraction of the measured flow rate $\dot{m}_a$ at low air flows. A primary leak test is essential at the lowest encountered test airflow before any testing is conducted.

5. Heat losses to the ambient are generally small for a well-insulated test section. However, they could represent a good fraction of the heat transfer rate in the test section at low airflows. A proper calibration is essential to determine these heat losses.

6. For some test core surfaces, longitudinal heat conduction in the test core surface wall may be important and should be taken into account in the data reduction.

The first five factors cause heat imbalances $(q_w - q_a)/q_a$ to increase sharply at decreasing low air flow rates. In order to minimize or eliminate the rollover in j factors, the data should be reduced based on $q_{ave} = (q_w + q_a)/2$, and whenever possible, by reducing the core flow length (i.e., reducing NTU) by half and then retesting the core.

The uncertainty in the j factors obtained from the steady-state tests ($C^* \approx 0$ case) for a given uncertainty in Δ_2 ($=T_s - T_{a,o}$ or $T_{w,o} - T_{a,o}$) with T_s as condensing steam temperature is given by [15, 40]

$$\frac{d(j)}{j} = \frac{d(\Delta_2)}{\Delta_0} \frac{\text{ntu}_c}{\text{NTU}} \frac{e^{\text{NTU}}}{\text{NTU}} \qquad (17.75)$$

Here $\Delta_0 = T_{w,i} - T_{a,i}$ and $\text{ntu}_c/\text{NTU} \approx 1.1$. Thus, a measurement error in the outlet temperature difference [i.e., $d(\Delta_2)$] magnifies the error in j by the foregoing relationship both at high NTU (NTU > 3) and low NTU (NTU < 0.5). The error at high NTU due to the errors in Δ_2 and other factors was discussed above. The error at low NTU due to the error in Δ_2 can also be significant. Hence a careful design of the test core is essential for obtaining accurate j factors.

In addition to the foregoing measurement errors, incorrect j data are obtained for a given surface if the test core is not constructed properly. The problem areas are poor thermal bonds between the fins and the primary surface, gross blockage (gross flow maldistribution) on the air side or water (steam) side, and passage-to-passage nonuniformity (or maldistribution) on the air side. These factors influence the measured j and f factors differently in different Reynolds number ranges. Qualitative effects of these factors are presented in Fig. 17.39 to show the trends. The solid lines in these figures represent the j data of an ideal core having a perfect thermal bond, no gross blockage, and perfect uniformity. The dashed lines represent what happens to j factors when the specified imperfections exist. It is imperative that a detailed air temperature distribution be measured at the core outlet to ensure none of the foregoing problems are associated with the core.

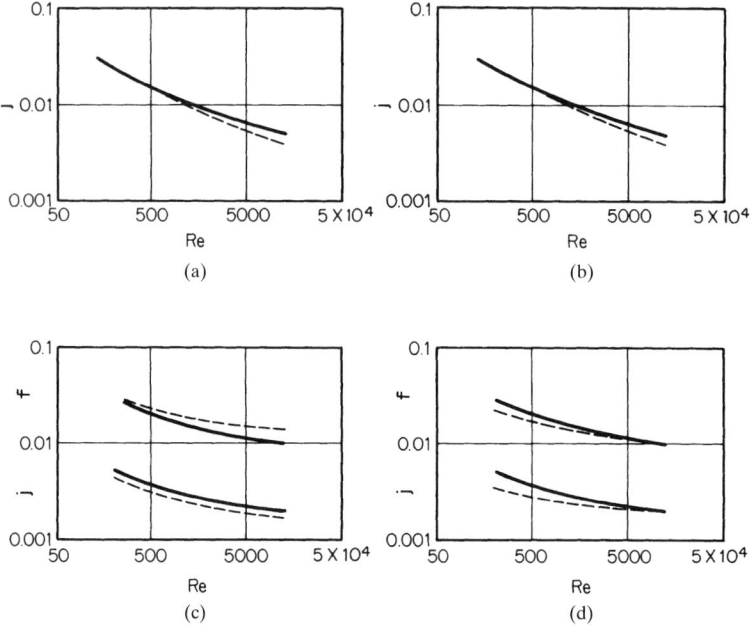

FIGURE 17.39 The influence on measured j data due to (a) poor thermal bond between fins and primary surface, (b) water- (steam-) side gross blockage, (c) air-side blockage, and (d) air-side passage-to-passage nonuniformity. The solid lines are for the perfect core, the dashed lines for the specified imperfect core.

The experimental uncertainty in the j factor for the foregoing steady-state method is usually within ±5 percent when the temperatures are measured accurately to within ±0.1°C (0.2°F) and none of the aforementioned problems exist in the test core. The uncertainty in the Reynolds number is usually within ±2 percent when the mass flow rate is measured accurately within ±0.7 percent.

Wilson Plot Technique for Liquids. In order to obtain highly accurate j factors, one of the considerations for the design of a test core in the preceding method was to have the thermal resistance on the test fluid (gas) side dominant (i.e., the test fluid side thermal conductance $\eta_o hA$ significantly lower compared to that on the other known side). This is achieved by either steam or hot or cold water at high mass flow rates on the known side. However, if the test fluid is water or another liquid and it has a high heat transfer coefficient, it may not represent a dominant thermal resistance, even if condensing steam is used on the other side. This is because the test fluid thermal resistance may be of the same order of magnitude as the wall thermal resistance. Hence, for liquids, Wilson [41] proposed a technique to obtain heat transfer coefficients h or j factors for turbulent flow in a circular tube.

In this method, liquid (test fluid, unknown side, fluid 1) flows on one side for which j versus Re characteristics are being determined, condensing steam, liquid, or air flows on the other side (fluid 2), for which we may or may not know the j versus Re characteristics. The fluid flow rate on the fluid 2 side and the log-mean average temperature *must* be kept constant (through iterative experimentation) so that its thermal resistance and C_2 in Eq. 17.79 are truly constant. The flow rate on the unknown (fluid 1) side is varied systematically. The fluid flow rates and temperatures upstream and downstream of the test core on each fluid side are measured for each test point. Thus when ε and C^* are known, NTU and UA are computed.

For discussion purposes, consider the test fluid side to be cold and the other fluid side to be hot. UA is given by

$$\frac{1}{UA} = \frac{1}{(\eta_o hA)_c} + \mathbf{R}_{s,c} + \mathbf{R}_w + \mathbf{R}_{s,h} + \frac{1}{(\eta_o hA)_h} \qquad (17.76)$$

Note that $\eta_o = 1$ on the fluid side, which does not have fins. For fully developed turbulent flow through constant cross-sectional ducts, the Nusselt number correlation is of the form

$$\text{Nu} = C_o\, \text{Re}^a\, \text{Pr}^{0.4}\, (\mu_w/\mu_m)^{-0.14} \qquad (17.77)$$

where C_o is a constant and $a = 0.8$ for the Dittus-Boelter correlation. However, note that a is a function of Pr, Re, and the geometry. For example, a varies from 0.78 at Pr = 0.7 to 0.90 at Pr = 100 for Re = 5×10^4 for a circular tube [15]; it also varies with Re for a given Pr. Theoretically, a will vary depending on the tube cross-sectional geometry, particularly for augmented tubes, and is not known a priori. Wilson [41] used $a = 0.82$. The term $(\mu_w/\mu_m)^{-0.14}$ takes into account the variable property effects for liquids; for gases, it should be replaced by an absolute temperature ratio function (see Eq. 17.109). By substituting the definitions of Re, Pr, and Nu in Eq. 17.77 and considering the fluid properties as constant,

$$h_c A_c = A_c (C_o k^{0.6} \rho^{0.82} c_p^{0.4} \mu^{-0.42} D_h^{-0.18})_c V^{0.82} = C_1' V^{0.82} = C_1 V^{0.82}/\eta_{o,c} \qquad (17.78)$$

The test conditions are maintained such that the fouling (scale) resistances $\mathbf{R}_{s,c}$ and $\mathbf{R}_{s,h}$ remain approximately constant though not necessarily zero, although Wilson [41] had neglected them. Since h is maintained constant on the fluid 2 side, the last four terms on the right side of the equality sign of Eq. 17.76 are constant—let us say equal to C_2. Now, substituting Eq. 17.78 in Eq. 17.76, we get

$$\frac{1}{UA} = \frac{1}{C_1 V^{0.82}} + C_2 \qquad (17.79)$$

Equation 17.79 has the form $y = mx + b$ with $y = 1/UA$, $m = 1/C_1$, $x = V^{-0.82}$, and $b = C_2$. Wilson plotted $1/UA$ versus $V^{-0.82}$ on a linear scale as shown in Fig. 17.40. The slope $1/C_1$ and the intercept C_2 are then determined from this plot. Once C_1 is known, h_c from Eq. 17.78 and hence the correlation given by Eq. 17.77 is known.

For this method, the Re exponent of Eq. 17.77 should be known and both resistances on the right side of Eq. 17.79 should be of the same order of magnitude. If C_2 is too small, it could end up negative in Fig. 17.40, depending on the slope due to the scatter in the test data; in this case, ignore the Wilson plot technique and use Eq. 17.76 for the data reduction using the best estimate of C_2. If C_2 is too large, the slope $1/C_1$ will be close to zero and will contain a large experimental uncertainty. If $\mathbf{R}_w$ or $\mathbf{R}_{s,h}$ is too high, $\mathbf{R}_h = 1/(\eta_o hA)_h$ must be kept too low so that C_2 is not very large. However, if $\mathbf{R}_h$ is too low and the hot fluid is a liquid or gas, its temperature drop may be difficult to measure accurately. C_2 can be reduced by increasing h on that side.

The limitations of the Wilson plot technique may be summarized as follows. (1) The fluid flow rate and its log-mean average temperature on the fluid 2 side *must* be kept constant so that C_2 is a constant. (2) The Re exponent in Eq. 17.77 is presumed to be known (such as 0.82 or 0.8). However, in reality it is a function of Re, Pr, and the geometry itself. Since the Re exponent is not known a priori, the Wilson plot technique *cannot* be utilized to determine the constant C_o of Eq. 17.77 for most heat transfer surfaces. (3) All the test data must be in one flow region (e.g., turbulent flow) on fluid 1 side, or the Nu correlation must be expressed by an

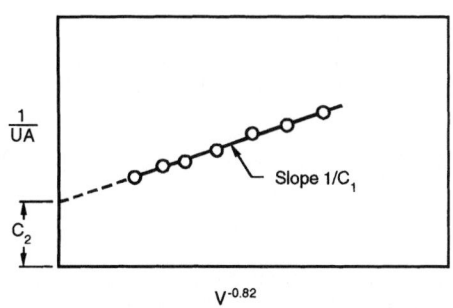

FIGURE 17.40   Original Wilson plot of Eq. 17.79.

explicit equation with only one unknown constant, such as Eq. 17.77 for known exponent a. (4) Fluid property variations and the fin thermal resistance are not taken into consideration on the unknown fluid 1 side. (5) Fouling on either fluid side of the exchanger *must* be kept constant so that C_2 remains constant in Eq. 17.79. Shah [42] discusses how to relax all of the above limitations of the Wilson plot technique except for the third limitation (one flow region for the complete testing); this will be discussed later.

In the preceding case of Eq. 17.79, unknowns are C_1 (means unknown C_o) and C_2. Alternatively, it should be emphasized that if $\mathbf{R}_{s,c}$, $\mathbf{R}_w$, and $\mathbf{R}_{s,h}$ are known a priori, then an unknown C_2 means that only its C_o and a for fluid 2 are unknown. Thus the heat transfer correlation on fluid 2 side can also be evaluated using the Wilson plot technique if the exponents on Re in Eq. 17.77 are known on both fluid sides. The Wilson plot technique thus represents a problem with two unknowns.

For a more general problem (e.g., a shell-and-tube exchanger), consider the Nu correlation on the tube side as Eq. 17.77 with $C_o = C_t'$ and on the shell side as Eq. 17.77 with $C_o = C_s'$ and the Re exponent as d, we can rewrite Eq. 17.76 as follows after neglecting $\mathbf{R}_{s,t} = \mathbf{R}_{s,s} = 0$ for a new/clean exchanger.

$$\frac{1}{UA} = \frac{1}{C_t[\mathrm{Re}^a \, \mathrm{Pr}^{0.4} \, Ak/D_h]_t(\mu_w/\mu_m)_t^{-0.14}} + \mathbf{R}_w + \frac{1}{C_s[\mathrm{Re}^d \, \mathrm{Pr}^{0.4} \, Ak/D_h]_s(\mu_w/\mu_m)_s^{-0.14}} \quad (17.80)$$

where $C_t = \eta_{o,t}C_t'$ and $C_s = \eta_{o,s}C_s'$. Thus, the more general Wilson plot technique has five unknowns (C_t, C_s, a, d, and $\mathbf{R}_w$); Shah [42] discusses the solution procedure.

As mentioned earlier, if one is interested in determining a complete correlation on one fluid side (such as the tube side, Eq. 17.77 without either knowing or not being concerned about the correlation on the other (such as the shell side), it represents a three unknown (C_t, a, and C' of Eq. 17.81) problem. The following procedure is suggested.

1. If the j or Nu versus Re characteristics on the shell side are accurately known, back-calculate the tubeside h from Eq. 17.76 with all other terms known (here, subscripts $c = t$ and $h = s$).

2. If the j or Nu versus Re characteristics on the shell side are *not* known, then the shellside mass flow rate (Reynolds number) and log-mean average temperature *must* be kept constant during the testing. In this case, Eq. 17.80 is manipulated as follows.

$$\left[\frac{1}{UA} - \mathbf{R}_w\right]\left[\frac{\mu_w}{\mu_m}\right]_s^{-0.14} = \frac{1}{C_t}\left\{\frac{(\mu_w/\mu_m)_s^{-0.14}/(\mu_w/\mu_m)_t^{-0.14}}{[\mathrm{Re}^a \, \mathrm{Pr}^{0.4} \, Ak/D_h]_t}\right\} + C' \quad (17.81)$$

where

$$C' = \frac{1}{C_s[\mathrm{Re}^d \, \mathrm{Pr}^{0.4} \, Ak/D_h]_s} = \frac{1}{(\eta_o hA)_s} \quad (17.82)$$

Equation 17.81 has three unknowns, C_t, a, and C', and it represents a variant of the Briggs and Young method [43] for the three-unknown problem. These constants are determined by two successive linear regressions iteratively. The modified Wilson plot of Eq. 17.81 is shown in Fig. 17.41 considering a as known (guessed). In reality, a single plot as shown in Fig. 17.41 is not sufficient. It will require an internal iterative scheme by assuming C' or using it from the previous iteration, computing Nu_s and hence h_s, determining T_w with the measured q, and finally calculating the viscosity ratio functions of Eq. 17.81. Iterations of regression analyses are continued until the successive values of C_t converge within the desired accuracy. Now, with known C', Eq. 17.80 is rearranged as follows.

$$\left[\frac{1}{UA} - \mathbf{R}_w - \frac{C'}{(\mu_w/\mu_m)_s^{-0.14}}\right] \times [\mathrm{Pr}^{0.4} \, Ak/D_h](\mu_w/\mu_m)_t^{-0.14} = \frac{1}{C_t \, \mathrm{Re}_t^a} \quad (17.83)$$

Substituting y_t for the left side of Eq. 17.83 and taking logarithms:

$$\ln(1/y_t) = a \ln(\mathrm{Re}_t) + \ln(C_t) \quad (17.84)$$

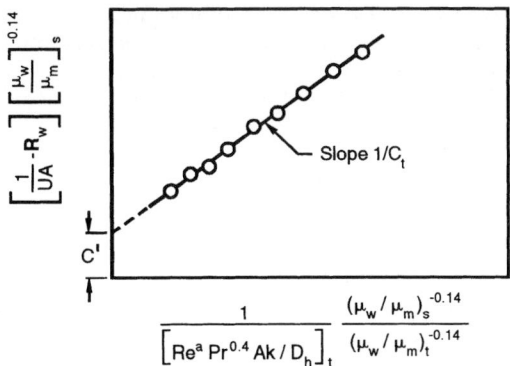

FIGURE 17.41   A tubeside Wilson plot of Eq. 17.81.

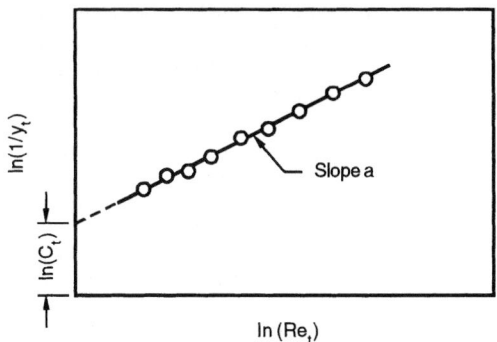

FIGURE 17.42   A tubeside Wilson plot of Eq. 17.84 where y_t is defined by the left side of Eq. 17.83.

Since Eq. 17.84 has a form $Y = mX + b$, C_t and a can be determined from the modified Wilson plot as shown in Fig. 17.42. Note that, here, an internal iterative scheme is not required for the viscosity ratio functions because the shellside C' (correlation) needed to compute the wall temperature is already known from the previous step. Iterations of the modified Wilson plots of Figs. 17.41 and 17.42 are continued until C_t, a, and C' converge within the desired accuracy.

For an accurate determination of C_t and a through the solution of Eq. 17.84, the thermal resistance for the tube side should be dominant for all test points for y_t (of Eq. 17.84) to remain positive. In practice, the purpose of using this modified technique is to determine the tube-side h when its thermal resistance is *not* dominant. If it would have been dominant, use Eq. 17.76 to back-calculate h. If the tube-side resistance cannot be made dominant due to the limitations of test equipment, this method will not yield an accurate tube-side correlation. Hence, a careful design of testing is essential before starting any testing.

If all test points are not in the same flow regime (such as in turbulent flow) for the unknown side of the exchanger using the Wilson plot technique or its variant, use the method recommended in Refs. 15 and 42 to determine h or Nu on the unknown side.

Test Technique for Friction Factors.   The experimental determination of flow friction characteristics of compact heat exchanger surfaces is relatively straightforward. Regardless of the core construction and the method of heat transfer testing, the determination of f is made under steady fluid flow rates with or without heat transfer. For a given fluid flow rate on the unknown side, the following measurements are made: core pressure drop, core inlet pressure and temperature, core outlet temperature for hot friction data, fluid mass flow rate, and the core geometric properties. The Fanning friction factor f is then determined from the following equation:

$$f = \frac{r_h}{L} \frac{1}{(1/\rho)_m} \left[\frac{2g_c \Delta p}{G^2} - \frac{1}{\rho_i}(1 - \sigma^2 + K_c) - 2\left(\frac{1}{\rho_o} - \frac{1}{\rho_i}\right) + \frac{1}{\rho_o}(1 - \sigma^2 - K_e)\right] \qquad (17.85)$$

This equation is an inverted form of the core pressure drop in Eq. 17.65. For the isothermal pressure drop data, $\rho_i = \rho_o = 1/(1/\rho)_m$. The friction factor thus determined includes the effects of skin friction, form drag, and local flow contraction and expansion losses, if any, within the core. Tests are repeated with different flow rates on the unknown side to cover the desired range of the Reynolds number. The experimental uncertainty in the f factor is usually within ±5 percent when Δp is measured accurately within ±1 percent.

Generally, the Fanning friction factor f is determined from isothermal pressure drop data (no heat transfer across the core). The hot friction factor f versus Re curve should be close to the isothermal f versus Re curve, particularly when the variations in the fluid properties are small, that is, the average fluid temperature for the hot f data is not significantly different from the wall temperature. Otherwise, the hot f data must be corrected to take into account the temperature-dependent fluid properties.

Analytical Solutions

Flow passages in most compact heat exchangers are complex with frequent boundary layer interruptions; some heat exchangers (particularly the tube side of shell-and-tube exchangers and highly compact regenerators) have continuous flow passages. The velocity and temperature profiles across the flow cross section are generally fully developed in the continuous flow passages, whereas they develop at each boundary layer interruption in an interrupted surface and may reach a periodic fully developed flow. The heat transfer and flow friction characteristics are generally different for fully developed flows and developing flows. Analytical results are discussed separately next for developed and developing flows for simple flow passage geometries. For complex surface geometries, the basic surface characteristics are primarily obtained experimentally, as discussed in the previous section; the pertinent correlations are presented in the next subsection.

Analytical solutions for developed and developing velocity/temperature profiles in constant cross section circular and noncircular flow passages are important when no empirical correlations are available, when extrapolations are needed for empirical correlations, or in the development of empirical correlations. Fully developed laminar flow solutions are applicable to highly compact regenerator surfaces or highly compact plate-fin exchangers with plain uninterrupted fins. Developing laminar flow solutions are applicable to interrupted fin geometries and plain uninterrupted fins of short lengths, and turbulent flow solutions to not-so-compact heat exchanger surfaces.

Three important thermal boundary conditions for heat exchangers are ⓣ, ⓗ₁, and ⓗ₂. The ⓣ boundary condition refers to constant wall temperature, both axially and peripherally throughout the passage length. The wall heat transfer rate is constant in the axial direction, while the wall temperature at any cross section is constant in the peripheral direction for the ⓗ₁ boundary condition. The wall heat transfer rate is constant in the axial direction as well as in the peripheral direction for the ⓗ₂ boundary condition. The ⓗ₁ boundary condition is realized for highly conductive materials where the temperature gradients in the peripheral direction are at a minimum; the ⓗ₂ boundary condition is realized for very poorly conducting materials for which temperature gradients exist in the peripheral direction. For intermediate thermal conductivity values, the boundary condition will be in between that of ⓗ₁ and ⓗ₂. In general, $Nu_{H1} > Nu_T$, $Nu_{H1} \geq Nu_{H2}$, and $Nu_{H2} \gtrless Nu_T$.

The heat transfer rate in the laminar duct flow is very sensitive to the thermal boundary condition. Hence, it is essential to carefully identify the thermal boundary condition in laminar flow. The heat transfer rate in turbulent duct flow is insensitive to the thermal boundary condition for most common fluids ($Pr > 0.7$); the exception is liquid metals ($Pr < 0.03$). Hence, there is generally no need to identify the thermal boundary condition in turbulent flow for all fluids except liquid metals.

Fully Developed Flows

Laminar Flow. Nusselt numbers for fully developed laminar flow are constant but depend on the flow passage geometry and thermal boundary conditions. The product of the Fanning friction factor and the Reynolds number is also constant but dependent on the flow passage geometry. Fully developed laminar flow problems are analyzed extensively in Refs. 19 and 44; most of the analytical solutions are also presented in closed-form equations in Ref. 44. Solutions for some technically important flow passages are presented in Table 17.14. The following observations may be made from this table: (1) There is a strong influence of flow passage geometry on Nu and f Re. Rectangular passages approaching a small aspect ratio exhibit the highest Nu and f Re. (2) Three thermal boundary conditions have a strong influence on the Nusselt numbers. (3) As $Nu = hD_h/k$, a constant Nu implies the convective heat transfer coefficient h independent of the flow velocity and fluid Prandtl number. (4) An increase in h can be best achieved either by reducing D_h or by selecting a geometry with a low aspect ratio rectangular flow passage. Reducing the hydraulic diameter is an obvious way to increase exchanger compactness and heat transfer, or D_h can be optimized using well-known heat transfer correlations based on design problem specifications. (5) Since $f\,\mathrm{Re} = \text{constant}$, $f \propto 1/\mathrm{Re} \propto 1/V$. In this case, it can be shown that $\Delta p \propto V$. Many additional analytical results for fully developed laminar flow ($\mathrm{Re} \le 2000$) are presented in Refs. 19 and 44. For most channel shapes, the mean Nu and f will be within 10 percent of the fully developed value if $L/D_h > 0.2\mathrm{Re}\,\mathrm{Pr}$. The entrance effects, flow maldistribution, free convection, property variation, fouling, and surface roughness all affect fully developed analytical solutions as shown in Table 17.15. Hence, in order to consider these effects in real plate-fin plain fin geometries having fully developed flows, it is best to reduce the magnitude of the analytical Nu by a minimum of 10 percent and increase the value of the analytical $f\,\mathrm{Re}$ by a minimum of 10 percent for design purposes.

Analytical values of L_{hy}^{+} and $K(\infty)$ are also listed in Table 17.14. The hydrodynamic entrance length L_{hy} [dimensionless form is $L_{hy}^{+} = L_{hy}/(D_h\,\mathrm{Re})$] is the duct length required to

TABLE 17.14   Solutions for Heat Transfer and Friction for Fully Developed Laminar Flow through Specified Ducts [19]

Geometry ($L/D_h > 100$)		Nu_{H1}	Nu_{H2}	Nu_T	$f\,\mathrm{Re}$	$\dfrac{j_{H1}}{f}$ *	$K(\infty)$[†]	L_{hy}^{+}[‡]
(triangular, 2b, 2a)	$\dfrac{2b}{2a} = \dfrac{\sqrt{3}}{2}$	3.014	1.474	2.39	12.630	0.269	1.739	0.04
(triangular 60°, 2b, 2a)	$\dfrac{2b}{2a} = \dfrac{\sqrt{3}}{2}$	3.111	1.892	2.47	13.333	0.263	1.818	0.04
(square, 2b, 2a)	$\dfrac{2b}{2a} = 1$	3.608	3.091	2.976	14.227	0.286	1.433	0.090
(hexagon)		4.002	3.862	3.34	15.054	0.299	1.335	0.086
(rectangle, 2b, 2a)	$\dfrac{2b}{2a} = \dfrac{1}{2}$	4.123	3.017	3.391	15.548	0.299	1.281	0.085
(circle)		4.364	4.364	3.657	16.000	0.307	1.25	0.056
(rectangle, 2b, 2a)	$\dfrac{2b}{2a} = \dfrac{1}{4}$	5.331	2.94	4.439	18.233	0.329	1.001	0.078
(rectangle, 2b, 2a)	$\dfrac{2b}{2a} = \dfrac{1}{6}$	6.049	2.93	5.137	19.702	0.346	0.885	0.070
(rectangle, 2b, 2a)	$\dfrac{2b}{2a} = \dfrac{1}{8}$	6.490	2.94	5.597	20.585	0.355	0.825	0.063
(parallel plates)	$\dfrac{2b}{2a} = 0$	8.235	8.235	7.541	24.000	0.386	0.674	0.011

* $j_{H1}/f = \mathrm{Nu}_{H1}\,\mathrm{Pr}^{-1/3}/(f\,\mathrm{Re})$ with $\mathrm{Pr} = 0.7$. Similarly, values of j_{H2}/f and j_T/f may be computed.
[†] $K(\infty)$ for sine and equilateral triangular channels may be too high [19]; $K(\infty)$ for some rectangular and hexagonal channels is interpolated based on the recommended values in Ref. 19.
[‡] L_{hy}^{+} for sine and equilateral triangular channels is too low [19], so use with caution. L_{hy}^{+} for rectangular channels is based on the faired curve drawn through the recommended value in Ref. 19. L_{hy}^{+} for a hexagonal channel is an interpolated value.

TABLE 17.15 Influence of Increase of Specific Variables on Laminar Theoretical Friction Factors and Nusselt Numbers.

Variable	f	Nu
Entrance effect	Increases	Increases
Passage-to-passage nonuniformity	Decreases slightly	Decreases significantly
Gross flow maldistribution	Increases sharply	Decreases
Free convection in a horizontal passage	Increases	Increases
Free convection with vertical aiding flow	Increases	Increases
Free convection with vertical opposing flow	Decreases	Decreases
Property variation due to fluid heating	Decreases for liquids and increases for gases	Increases for liquids and decreases for gases
Property variation due to fluid cooling	Increases for liquids and decreases for gases	Decreases for liquids and increases for gases
Fouling	Increases sharply	Increases slightly
Surface roughness	Affects only if the surface roughness height profile is nonnegligible compared to D_h	Affects only if the surface roughness height profile is nonnegligible compared to D_h

achieve a maximum channel section velocity of 99 percent of that for fully developed flow when the entering fluid velocity profile is uniform. Since the flow development region precedes the fully developed region, the entrance region effects could be substantial, even for channels having fully developed flow along a major portion of the channel. This increased friction in the entrance region and the change of momentum rate is taken into account by the incremental pressure drop number $K(\infty)$ defined by

$$\Delta p = \left[\frac{4 f_{fd} L}{D_h} + K(\infty) \right] \frac{G^2}{2 g_c \rho} \tag{17.86}$$

where the subscript fd denotes the fully developed value.

Transition Flow.   The initiation of transition to turbulent flow, the lower limit of the critical Reynolds number (Re_{cr}), depends on the type of entrance (e.g., smooth versus abrupt configuration at the exchanger flow passage entrance) in smooth ducts. For a sharp square inlet configuration, Re_{cr} is about 10–15 percent lower than that for a rounded inlet configuration. For most exchangers, the entrance configuration would be sharp. Some information on Re_{cr} is provided by Ghajar and Tam [45]. The lower limits of Re_{cr} for various passages with a sharp square inlet configuration vary from about 2000 to 3100 [46]. The upper limit of Re_{cr} may be taken as 10^4 for most practical purposes.

Transition flow and fully developed turbulent flow Fanning friction factors for a circular duct are given by Bhatti and Shah [46] as

$$f = A + B \, \text{Re}^{-1/m} \tag{17.87}$$

where $A = 0.0054, B = 2.3 \times 10^{-8}, m = -2/3$ for $2100 \leq \text{Re} \leq 4000$
$A = 0.00128, B = 0.1143, m = 3.2154$ for $4000 \leq \text{Re} \leq 10^7$

Equation 17.87 is accurate within ±2 percent [46]. The transition flow f data for noncircular passages are rather sparse; Eq. 17.87 may be used to obtain fair estimates of f for noncircular flow passages (having no sharp corners) using the hydraulic diameter as the characteristic dimension.

The transition flow and fully developed turbulent flow Nusselt number correlation for a circular tube is given by Gnielinski as reported in Bhatti and Shah [46] as

$$\text{Nu} = \frac{(f/2)(\text{Re} - 1000)\,\text{Pr}}{1 + 12.7(f/2)^{1/2}(\text{Pr}^{2/3} - 1)} \qquad (17.88)$$

which is accurate within about ±10 percent with experimental data for $2300 \le \text{Re} \le 5 \times 10^6$ and $0.5 \le \text{Pr} \le 2000$. For higher accuracies in turbulent flow, refer to the correlations by Petukhov et al. reported by Bhatti and Shah [46]. Churchill as reported in Bhatti and Shah [46] provides a correlation for laminar, transition, and turbulent flow regimes in a circular tube for $2100 < \text{Re} < 10^6$ and $0 < \text{Pr} < \infty$. Since no Nu and j factors are available for transition flow for noncircular passages, Eq. 17.88 may be used to obtain a fair estimate of Nu for noncircular passages (having no sharp corners) using D_h as the characteristic dimension.

Turbulent Flow.   A compendium of available f and Nu correlations for circular and noncircular flow passages are presented in Ref. 46. Table 17.16 is condensed from Ref. 46, summarizing the most accurate f and Nu correlations for smooth circular and noncircular passages.

It is generally accepted that the hydraulic diameter correlates Nu and f for fully developed turbulent flow in circular and noncircular ducts. This is true for the results accurate to within ±15 percent for most noncircular ducts. Exceptions are for those having sharp-angled corners in the flow passage or concentric annuli with inner wall heating. In these cases, Nu and f could be lower than 15 percent compared to the circular tube values. Table 17.16 can be used for more accurate correlations of Nu and f for noncircular ducts.

Roughness on the surface causes local flow separation and reattachment. This generally results in an increase in the friction factor as well as the heat transfer coefficient. A roughness element has no effect on laminar flow, unless the height of the roughness element is not negligible compared to the flow cross section size. However, it exerts a strong influence on turbulent flow. Specific correlations to account for the influence of surface roughness are presented in Refs. 46 and 47.

A careful observation of accurate experimental friction factors for all noncircular smooth ducts reveals that ducts with laminar f Re < 16 have turbulent f factors lower than those for the circular tube, whereas ducts with laminar f Re > 16 have turbulent f factors higher than those for the circular tube [48]. Similar trends are observed for the Nusselt numbers. If one is satisfied within ±15 percent accuracy, Eqs. 17.87 and 17.88 for f and Nu can be used for noncircular passages with the hydraulic diameter as the characteristic length in f, Nu, and Re; otherwise, refer to Table 17.16 for more accurate results for turbulent flow.

Hydrodynamically Developing Flows

Laminar Flow.   Based on the solutions for laminar boundary layer development over a flat plate and fully developed flow in circular and some noncircular ducts, f_{app} Re can be correlated by the following equation:

$$f_{app}\,\text{Re} = 3.44(x^+)^{-0.5} + \frac{K(\infty)/(4x^+) + f\,\text{Re} - 3.44(x^+)^{-0.5}}{1 + C'(x^+)^{-2}} \qquad (17.89)$$

where the values of $K(\infty)$, f Re, and C' are given in Table 17.17 for three geometries. Here f_{app} is defined the same way as f (see the nomeclature), but Δp includes additional pressure drop due to momentum change and excess wall shear between developing and developed flows.

Turbulent Flow. f_{app} Re for turbulent flow depends on Re in addition to x^+. A closed-form formula for f_{app} Re is given in Refs. 46 and 48 for developing turbulent flow. The hydrodynamic entrance lengths for developing laminar and turbulent flows are given by Refs. 44 and 46 as

$$\frac{L_{hy}}{D_h} = \begin{cases} 0.0565\text{Re} & \text{for laminar flow (Re} \le 2100) \\ 1.359\text{Re}^{1/4} & \text{for tubulent flow (Re} \ge 10^4) \end{cases} \qquad (17.90)$$

TABLE 17.16   Fully Developed Turbulent Flow Friction Factors and Nusselt Numbers (Pr > 0.5) for Technically Important Smooth-Walled Ducts [44]

Duct geometry and characteristic dimension	Recommended correlations[†]
Circular $D_h = 2a$	Friction factor correlation for $2300 \leq \mathrm{Re} \leq 10^7$ $$f = A + \frac{B}{\mathrm{Re}^{1/m}}$$ where $A = 0.0054$, $B = 2.3 \times 10^{-8}$, $m = -\frac{2}{3}$ for $2100 \leq \mathrm{Re} \leq 4000$ and $A = 1.28 \times 10^{-3}$, $B = 0.1143$, $m = 3.2154$ for $4000 \leq \mathrm{Re} \leq 10^7$ Nusselt number correlation by Gnielinski for $2300 \leq \mathrm{Re} \leq 5 \times 10^6$: $$\mathrm{Nu} = \frac{(f/2)(\mathrm{Re} - 1000)\,\mathrm{Pr}}{1 + 12.7(f/2)^{1/2}(\mathrm{Pr}^{2/3} - 1)}$$
Flat $D_h = 4b$	Use circular duct f and Nu correlations. Predicted f are up to 12.5% lower and predicted Nu are within ±9% of the most reliable experimental results.
Rectangular $D_h = \dfrac{4ab}{a+b}$, $\alpha^* = \dfrac{2b}{2a}$ $\dfrac{D_1}{D_h} = \frac{2}{3} + \frac{11}{24}\,\alpha^*(2 - \alpha^*)$	f factors: (1) substitute D_1 for D_h in the circular duct correlation, and calculate f from the resulting equation. (2) Alternatively, calculate f from $f = (1.0875 - 0.1125\alpha^*)f_c$ where f_c is the friction factor for the circular duct using D_h. In both cases, predicted f factors are within ±5% of the experimental results. Nusselt numbers: (1) With uniform heating at four walls, use circular duct Nu correlation for an accuracy of ±9% for $0.5 \leq \mathrm{Pr} \leq 100$ and $10^4 \leq \mathrm{Re} \leq 10^6$. (2) With equal heating at two long walls, use circular duct correlation for an accuracy of ±10% for $0.5 < \mathrm{Pr} \leq 10$ and $10^4 \leq \mathrm{Re} \leq 10^5$. (3) With heating at one long wall only, use circular duct correlation to get approximate Nu values for $0.5 < \mathrm{Pr} < 10$ and $10^4 \leq \mathrm{Re} \leq 10^6$. These calculated values may be up to 20% higher than the actual experimental values.
Equilateral triangular $D_h = 2\sqrt{3}a = 4b/3$ $D_1 = \sqrt{3}a = 2b/3\sqrt{3}$	Use circular duct f and Nu correlations with D_h replaced by D_1. Predicted f are within +3% and −11% and predicted Nu within +9% of the experimental values.

[†] The friction factor and Nusselt number correlations for the circular duct are the most reliable and agree with a large amount of the experimental data within ±2% and ±10% respectively. The correlations for all other duct geometries are not as good as those for the circular duct on an absolute basis.

Thermally Developing Flows

Laminar Flow.   Thermal entry length solutions with developed velocity profiles are summarized in Refs. 19 and 44 for a large number of practically important flow passage geometries with extensive comparisons.

Shah and London [19] proposed the following correlations for thermal entrance solutions for circular and noncircular ducts having laminar developed velocity profiles and developing temperature profiles.

$$\mathrm{Nu}_{x,\mathrm{T}} = 0.427(f\,\mathrm{Re})^{1/3}(x^*)^{-1/3} \tag{17.91}$$

$$\mathrm{Nu}_{m,\mathrm{T}} = 0.641(f\,\mathrm{Re})^{1/3}(x^*)^{-1/3} \tag{17.92}$$

TABLE 17.16 Fully Developed Turbulent Flow Friction Factors and Nusselt Numbers ($Pr > 0.5$) for Technically Important Smooth-Walled Ducts [44] (*Continued*)

Duct geometry and characteristic dimension	Recommended correlations[†]
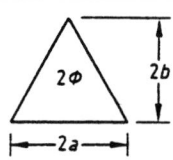 Isosceles triangular $D_h = \dfrac{4ab}{a + \sqrt{a^2 + b^2}}$ $\dfrac{D_g}{D_h} = \dfrac{1}{2\pi}\left[3 \ln \cot \dfrac{\theta}{2}\right.$ $\left. + 2 \ln \tan \dfrac{\phi}{2} - \ln \tan \dfrac{\theta}{2}\right]$ where $\theta = (90° - \phi)/2$	For $0 < 2\phi < 60°$, use circular duct f and Nu correlations with D_h replaced by D_g; for $2\phi = 60°$, replace D_h by D_1 (see previous geometry); and for $60° < 2\phi \leq 90°$ use circular duct correlations directly with D_h. Predicted f and Nu are within +9% and −11% of the experimental values. No recommendations can be made for $2\phi > 90°$ due to lack of the experimental data.
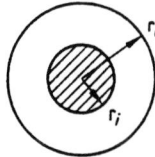 Concentric annular $D_h = 2(r_o - r_i),$ $r^* = \dfrac{r_i}{r_0}$ $\dfrac{D_1}{D_h} = \dfrac{1 + r^{*2} + (1 - r^{*2})/\ln r^*}{(1 - r^*)^2}$	f factors: (1) Substitute D_1 for D_h in the circular duct correlation, and calculate f from the resulting equation. (2) Alternatively, calculate f from $f = (1 + 0.0925r^*)f_c$ where f_c is the friction factor for the circular duct using D_h. In both cases, predicted f factors are within ±5% of the experimental results. Nusselt Numbers: In all the following recommendations, use D_h with a wetted perimeter in Nu and Re: (1) Nu at the outer wall can be determined from the circular duct correlation within the accuracy of about ±10% regardless of the condition at the inner wall. (2) Nu at the inner wall cannot be determined accurately regardless of the heating/cooling condition at the outer wall.

[†] The friction factor and Nusselt number correlations for the circular duct are the most reliable and agree with a large amount of the experimental data within ±2% and ±10% respectively. The correlations for all other duct geometries are not as good as those for the circular duct on an absolute basis.

$$\mathrm{Nu}_{x,\mathrm{H1}} = 0.517(f\,\mathrm{Re})^{1/3}(x^*)^{-1/3} \tag{17.93}$$

$$\mathrm{Nu}_{m,\mathrm{H1}} = 0.775(f\,\mathrm{Re})^{1/3}(x^*)^{-1/3} \tag{17.94}$$

where f is the Fanning friction factor for fully developed flow, Re is the Reynolds number, and $x^* = x/(D_h\,\mathrm{Re}\,\mathrm{Pr})$. For interrupted surfaces, $x = \ell_{ef}$. Equations 17.91–17.94 are recommended for $x^* < 0.001$.

The following observations may be made from Eqs. 17.91–17.94 and solutions for laminar flow surfaces having developing temperature profiles given in Refs. 19 and 44: (1) the influence of thermal boundary conditions on the convective behavior appears to be of the same order as that for fully developed flow, (2) since Nu $\propto (x^*)^{-1/3} = [x/(D_h\,\mathrm{Re}\,\mathrm{Pr})^{-1/3}]$, then Nu $\propto$ $\mathrm{Re}^{1/3} \propto V^{1/3}$—therefore h varies as $V^{1/3}$, (3) since the velocity profile is considered fully devel-

TABLE 17.17 $K(\infty)$, $f\,Re$, and C' for Use in Eq. 17.89 [19]

	$K(\infty)$	$f\,Re$	C'
α^*		Rectangular ducts	
1.00	1.43	14.227	0.00029
0.50	1.28	15.548	0.00021
0.20	0.931	19.071	0.000076
0.00	0.674	24.000	0.000029
2Φ		Equilateral triangular duct	
60°	1.69	13.333	0.00053
r^*		Concentric annular ducts	
0	1.25	16.000	0.000212
0.05	0.830	21.567	0.000050
0.10	0.784	22.343	0.000043
0.50	0.688	23.813	0.000032
0.75	0.678	23.967	0.000030
1.00	0.674	24.000	0.000029

oped, $\Delta p \propto V$ as noted earlier; (4) the influence of the duct shape on thermally developed Nu is not as great as that for the fully developed Nu.

The theoretical ratio Nu_m/Nu_{fd} is shown in Fig. 17.43 for several passage geometries having constant wall temperature boundary conditions. Several observations may be made from this figure. (1) The Nusselt numbers in the entrance region and hence the heat transfer coefficients could be 2–3 times higher than the fully developed values depending on the interruption length. (2) At $x^* \approx 0.1$, the local Nusselt number approaches the fully developed value, but the value of the mean Nusselt number can be significantly higher for a channel of length $\ell_{ef}^* = x^* = 0.1$. (3) The order of increasing Nu_m/Nu_{fd} as a function of channel shape at a given x^*

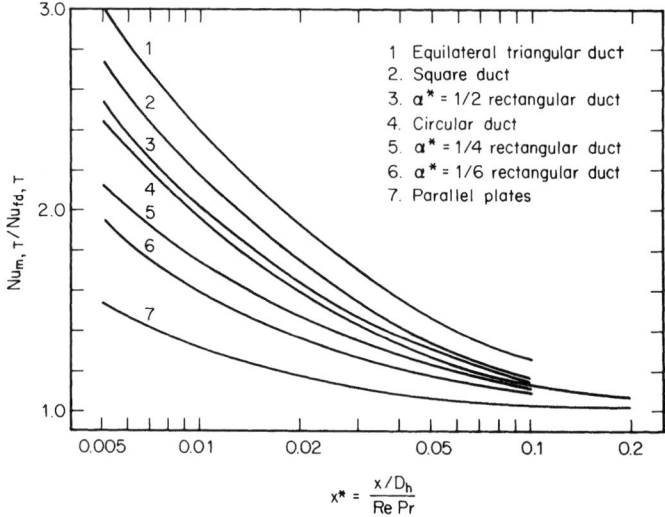

FIGURE 17.43 The ratio of laminar developing to developed Nu for different ducts; the velocity profile developed for both Nu's.

is the opposite of Nu_{fd} in Table 17.14. For a highly interrupted surface, a basic inferior passage geometry for fully developed flow (such as triangular) will not be penalized in terms of low Nu or low h in developing flow. (4) A higher value of Nu_m/Nu_{fd} at $x^* = 0.1$ means that the flow channel has a longer entrance region.

Turbulent Flow.   The thermal entry length solutions for smooth ducts for several cross-sectional geometries have been summarized [46]. As for laminar flow, the Nusselt numbers in the thermal region are higher than those in the fully developed region. However, unlike laminar flow, $Nu_{x,T}$ and $Nu_{x,H1}$ are very nearly the same for turbulent flow. The local and mean Nusselt numbers for a circular tube with Ⓣ and Ⓗ boundary conditions are [46]:

$$\frac{Nu_x}{Nu_\infty} = 1 + \frac{c}{10(x/D_h)} \qquad \frac{Nu_m}{Nu_\infty} = 1 + \frac{c}{x/D_h} \qquad (17.95)$$

where Nu_∞ stands for the fully developed Nu_T or Nu_H derived from the formulas in Table 17.16, and

$$c = \frac{(x/D_h)^{0.1}}{Pr^{1/6}} \left(0.68 + \frac{3000}{Re^{0.81}} \right) \qquad (17.96)$$

This correlation is valid for $x/D_h > 3$, $3500 < Re < 10^5$, and $0.7 < Pr < 75$. It agrees within ±12 percent with the experimental measurements for $Pr = 0.7$.

Simultaneously Developing Flows

Laminar Flow.   In simultaneously developing flow, both the velocity and temperature profiles develop in the entrance region. The available analytical solutions are summarized in Refs. 19 and 44. The theoretical entrance region Nusselt numbers for simultaneously developing flow are higher than those for thermally developing and hydrodynamically developed flow. These theoretical solutions do not take into account the wake effect or secondary flow effect that are present in flow over interrupted heat transfer surfaces. Experimental data indicate that the interrupted heat transfer surfaces do not achieve higher heat transfer coefficients predicted for the simultaneously developing flows. The results for thermally developing flows (and developed velocity profiles) are in better agreement with the experimental data for interrupted surfaces and hence are recommended for design purposes.

Turbulent Flow.   The Nusselt numbers for simultaneously developing turbulent flow are practically the same as the Nusselt numbers for the thermally developing turbulent flow [46]. However, the Nusselt numbers for simultaneously developing flow are sensitive to the passage inlet configuration.

Table 17.18 summarizes the dependence of Δp and h on V for developed and developing laminar and turbulent flows. Although these results are for the circular tube, the general functional relationship should be valid for noncircular ducts as a first approximation.

TABLE 17.18 Dependence of Pressure Drop and Heat Transfer Coefficient on the Flow Mean Velocity for Internal Flow in a Constant Cross-Sectional Duct

Flow type	$\Delta p \propto V^p$ Laminar	$\Delta p \propto V^p$ Turbulent	$h \propto V^q$ Laminar	$h \propto V^q$ Turbulent
Fully developed	V	$V^{1.8}$	V^0	$V^{0.8}$
Hydrodynamically developing	$V^{1.5}$	$V^{1.8}$	—	—
Thermally developing	V	$V^{1.8}$	$V^{1/3}$	$V^{0.8}$
Simultaneously developing	$V^{1.5}$	$V^{1.8}$	$V^{1/2}$	$V^{0.8}$

Experimental Correlations

Analytical results presented in the preceding section are useful for well-defined constant cross-sectional surfaces with essentially unidirectional flows. The flows encountered in heat exchangers are generally very complex, having flow separation, reattachment, recirculation, and vortices. Such flows significantly affect Nu and f for the specific exchanger surfaces. Since no analytical or accurate numerical solutions are available, the information is derived experimentally. Kays and London [20] and Webb [47] presented many experimental results reported in the open literature. In the following, empirical correlations for only some important surfaces are summarized due to space limitations.

A careful examination of all good data that are published has revealed the ratio $j/f \leq 0.25$ for strip fin, louver fin, and other similar interrupted surfaces. This can be approximately justified as follows. The flow develops along each interruption in such a surface. Based on the Reynolds analogy for fully developed turbulent flow over a flat plate, in the absence of form drag, j/f should be 0.5 for $Pr \approx 1$. Since the contribution of form drag being of the same order of magnitude as the skin friction in developing laminar flows for such an interrupted surface, j/f will be about 0.25. Published data for strip and louver fins are questionable if $j/f > 0.3$. All pressure and temperature measurements and possible sources of flow leaks and heat losses must be checked thoroughly for all those basic data having $j/f > 0.3$ for strip and louver fins.

Bare Tubebanks.   One of the most comprehensive correlations for crossflow over a plain tubebank is presented by Zukauskas [49] as shown in Figs. 17.44 and 17.45 for inline (90° tube layout) and staggered arrangement (30° tube layout) respectively, for the Euler number. These results are valid for the number of tube rows above about 16. For other inline and staggered tube arrangements, a correction factor χ is obtained from the inset of these figures to compute Eu. Zukauskas [49] also presented the mean Nusselt number $Nu_m = h_m d_o / k$ as

$$Nu_m = F_c (Nu_m)_{16\,rows} \tag{17.97}$$

Values of Nu_m for 16 or more tube rows are presented in Table 17.19 for inline (90° tube layout, Table 17.19a) and staggered (30° tube layout, Table 17.19b) arrangements. For all expressions in Table 17.19, fluid properties in Nu, Re_d, and Pr are evaluated at the bulk mean temperature and for Pr_w at the wall temperature. The tube row correction factor F_c is presented in Fig. 17.46 as a function of the number of tube rows N_r for inline and staggered tube arrangements.

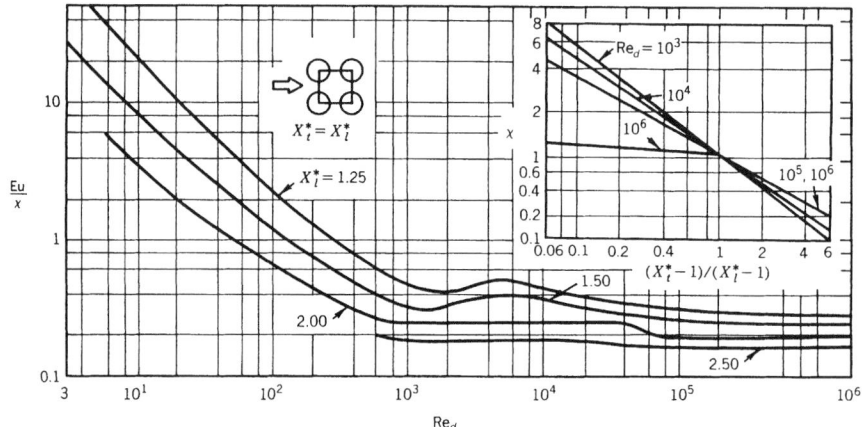

FIGURE 17.44   Friction factors for the inline tube arrangements for $X_t^* = 1.25, 1.5, 2.0,$ and 2.5 where $X_t^* = X_t/d_o$ and $X_l^* = X_l/d_o$ [49].

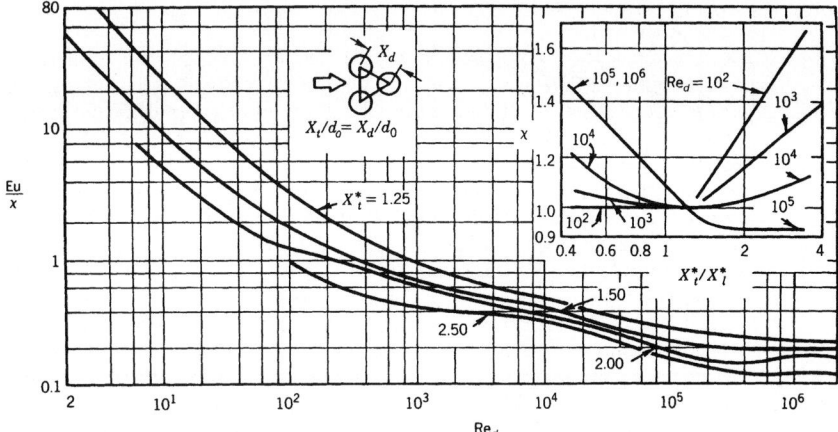

FIGURE 17.45   Friction factors for the staggered tube arrangements for $X_t^* = 1.25$, 1.5, 2.0, and 2.5 where $X_\ell^* = X_\ell/d_o$ and $X_t^* = X_t/d_o$ [49].

Plate-Fin Extended Surfaces

Offset Strip Fins.   This is one of the most widely used enhanced fin geometries (Fig. 17.47) in aircraft, cryogenics, and many other industries that do not require mass production. This surface has one of the highest heat transfer performances relative to the friction factor. Extensive analytical, numerical, and experimental investigations have been conducted over the last 50 years. The most comprehensive correlations for j and f factors for the offset strip fin geometry are provided by Manglik and Bergles [50] as follows.

$$j = 0.6522 \mathrm{Re}^{-0.5403} \left(\frac{s}{h'}\right)^{-0.1541} \left(\frac{\delta_f}{\ell_f}\right)^{0.1499} \left(\frac{\delta_f}{s}\right)^{-0.0678} \left[1 + 5.269 \times 10^{-5}\, \mathrm{Re}^{1.340} \left(\frac{s}{h'}\right)^{0.504} \left(\frac{\delta_f}{\ell_f}\right)^{0.456} \left(\frac{\delta_f}{s}\right)^{-1.055}\right]^{0.1}$$

$$(17.98)$$

TABLE 17.19(a)   Heat Transfer Correlations for Inline Tube Bundles for $n > 16$ [49]

Recommended correlations	Range of Re_d
$\mathrm{Nu} = 0.9 \mathrm{Re}_d^{0.4}\, \mathrm{Pr}^{0.36}\, (\mathrm{Pr}/\mathrm{Pr}_w)^{0.25}$	$10^0 - 10^2$
$\mathrm{Nu} = 0.52 \mathrm{Re}_d^{0.5}\, \mathrm{Pr}^{0.36}\, (\mathrm{Pr}/\mathrm{Pr}_w)^{0.25}$	$10^2 - 10^3$
$\mathrm{Nu} = 0.27 \mathrm{Re}_d^{0.63}\, \mathrm{Pr}^{0.36}\, (\mathrm{Pr}/\mathrm{Pr}_w)^{0.25}$	$10^3 - 2 \times 10^5$
$\mathrm{Nu} = 0.033 \mathrm{Re}_d^{0.8}\, \mathrm{Pr}^{0.4}\, (\mathrm{Pr}/\mathrm{Pr}_w)^{0.25}$	$2 \times 10^5 - 2 \times 10^6$

TABLE 17.19(b)   Heat Transfer Correlations for Staggered Tube Bundles for $n > 16$ [49]

Recommended correlations	Range of Re_d
$\mathrm{Nu} = 1.04 \mathrm{Re}_d^{0.4}\, \mathrm{Pr}^{0.36}\, (\mathrm{Pr}/\mathrm{Pr}_w)^{0.25}$	$10^0 - 5 \times 10^2$
$\mathrm{Nu} = 0.71 \mathrm{Re}_d^{0.5}\, \mathrm{Pr}^{0.36}\, (\mathrm{Pr}/\mathrm{Pr}_w)^{0.25}$	$5 \times 10^2 - 10^3$
$\mathrm{Nu} = 0.35 (X_t^*/X_l^*)^{0.2}\, \mathrm{Re}_d^{0.6}\, \mathrm{Pr}^{0.36}\, (\mathrm{Pr}/\mathrm{Pr}_w)^{0.25}$	$10^3 - 2 \times 10^5$
$\mathrm{Nu} = 0.031 (X_t^*/X_l^*)^{0.2}\, \mathrm{Re}_d^{0.8}\, \mathrm{Pr}^{0.36}\, (\mathrm{Pr}/\mathrm{Pr}_w)^{0.25}$	$2 \times 10^5 - 2 \times 10^6$

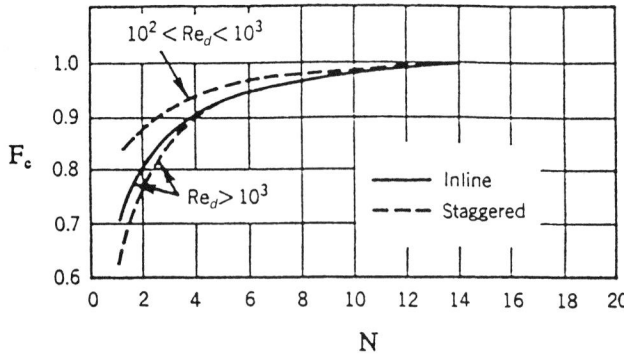

FIGURE 17.46 A correction factor F_c to take into account the tube-row effect for heat transfer for flow normal to bare tubebanks.

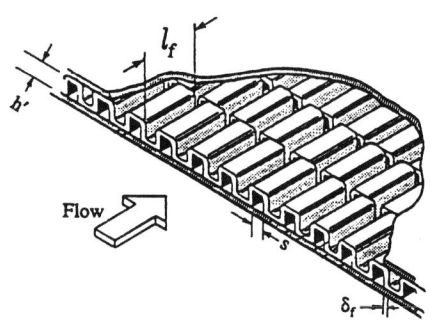

FIGURE 17.47 An offset strip fin geometry.

$$f = 9.6243 \text{Re}^{-0.7422} \left(\frac{s}{h'}\right)^{-0.1856} \left(\frac{\delta_f}{\ell_f}\right)^{0.3053} \left(\frac{\delta_f}{s}\right)^{-0.2659} \left[1 + 7.669 \times 10^{-8} \, \text{Re}^{4.429} \left(\frac{s}{h'}\right)^{0.920} \left(\frac{\delta_f}{\ell_f}\right)^{3.767} \left(\frac{\delta_f}{s}\right)^{0.236}\right]^{0.1}$$

(17.99)

where
$$D_h = 4A_o/(A/\ell_f) = 4sh'\ell_f/[2(s\ell_f + h'\ell_f + \delta_f h') + \delta_f s]$$
(17.100)

Geometrical symbols in Eq. 17.100 are shown in Fig. 17.47.

These correlations predict the experimental data of 18 test cores within ±20 percent for $120 \leq \text{Re} \leq 10^4$. Although all experimental data for these correlations are obtained for air, the j factor takes into consideration minor variations in the Prandtl number, and the above correlations should be valid for $0.5 < \text{Pr} < 15$.

The heat transfer coefficients for the offset strip fins are 1.5 to 4 times higher than those of plain fin geometries. The corresponding friction factors are also high. The ratio of j/f for an offset strip fin to j/f for a plain fin is about 80 percent. If properly designed, the offset strip fin would require substantially lower heat transfer surface area than that of plain fins at the same Δp, but about a 10 percent larger flow area.

Louver Fins. Louver or multilouver fins are extensively used in auto industry due to their mass production manufacturability and lower cost. It has generally higher j and f factors than those for the offset strip fin geometry, and also the increase in the friction factors is in general higher than the increase in the j factors. However, the exchanger can be designed for higher heat transfer and the same pressure drop compared to that with the offset strip fins by a proper selection of exchanger frontal area, core depth, and fin density. Published literature and correlations on the louver fins are summarized by Webb [47] and Cowell et al. [51], and the understanding of flow and heat transfer phenomena is summarized by Cowell et al. [51]. Because of the lack of systematic studies reported in the open literature on modern louver fin geometries, no correlation can be recommended for the design purpose.

Other Plate-Fin Surfaces. Perforated and pin fin geometries have been investigated, and it is found that they do not have superior performance compared to offset strip and louver fin geometries [15]. Perforated fins are now used only in a limited number of applications. They are used as "turbulators" in oil coolers and in cryogenic air separation exchangers as a replacement to the existing perforated fin exchangers; modern cryogenic air separation exchangers use offset strip fin geometries. Considerable research has been reported on vortex generators using winglets [52, 53], but at present neither definitive conclusions are available on the superiority of these surfaces nor manufactured for heat exchanger applications.

Tube-Fin Extended Surfaces.   Two major types of tube-fin extended surfaces are: (1) individually finned tubes, and (2) flat fins (also sometimes referred to as plate fins), with or without enhancements/interruptions on an array of tubes as shown in Fig. 17.14. An extensive coverage of the published literature and correlations for these extended surfaces is provided by Webb [47] and Kays and London [20]. Empirical correlations for some important geometries are summarized below.

Individually Finned Tubes.   In this fin geometry, helically wrapped (or extruded) circular fins on a circular tube as shown in Fig. 17.14a, is commonly used in process and waste heat recovery industries. The following correlation for *j* factors is recommended by Briggs and Young (see Webb [47]) for individually finned tubes on staggered tubebanks.

$$j = 0.134\text{Re}_d^{-0.319} \, (s/\ell_f)^{0.2}(s/\delta_f)^{0.11} \tag{17.101}$$

where ℓ_f is the radial height of the fin, δ_f is the fin thickness, $s = p_f - \delta_f$ is the distance between adjacent fins, and p_f is the fin pitch. Equation 17.101 is valid for the following ranges: $1100 \leq \text{Re}_d \leq 18{,}000$, $0.13 \leq s/\ell_f \leq 0.63$, $1.01 \leq s/\delta_f \leq 6.62$, $0.09 \leq \ell_f/d_o \leq 0.69$, $0.011 \leq \delta_f/d_o \leq 0.15$, $1.54 \leq X_t/d_o \leq 8.23$, fin root diameter d_o between 11.1 and 40.9 mm, and fin density N_f ($=1/p_f$) between 246 and 768 fins per meter. The standard deviation of Eq. 17.101 with experimental results was 5.1 percent.

For friction factors, Robinson and Briggs (see Webb [47]) recommended the following correlation.

$$f_{tb} = 9.465\text{Re}_d^{-0.316} \, (X_t/d_o)^{-0.927}(X_t/X_d)^{0.515} \tag{17.102}$$

Here $X_d = (X_t^2 + X_l^2)^{1/2}$ is the diagonal pitch and X_t and X_l are the transverse and longitudinal tube pitches, respectively. The correlation is valid for the following ranges: $2000 \leq \text{Re}_d \leq 50{,}000$, $0.15 \leq s/\ell_f \leq 0.19$, $3.75 \leq s/\delta_f \leq 6.03$, $0.35 \leq \ell_f/d_o \leq 0.56$, $0.011 \leq \delta_f/d_o \leq 0.025$, $1.86 \leq X_t/d_o \leq 4.60$, $18.6 \leq d_o \leq 40.9$ mm, and $311 \leq N_f \leq 431$ fins per meter. The standard deviation of Eq. 17.102 with correlated data was 7.8 percent.

For crossflow over low-height finned tubes, Rabas and Taborek [54], Ganguli and Yilmaz [55], and Chai [56] have assessed the pertinent literature. A simple but accurate correlation for heat transfer is given by Ganguli and Yilmaz [55] as

$$j = 0.255\text{Re}_d^{-0.3} \, (d_e/s)^{-0.3} \tag{17.103}$$

A more accurate correlation for heat transfer is given by Rabas and Taborek [54]. Chai [56] provides the best correlation for friction factors:

$$f_{tb} = 1.748\text{Re}_d^{-0.233} \left(\frac{\ell_f}{s}\right)^{0.552}\left(\frac{d_o}{X_t}\right)^{0.599}\left(\frac{d_o}{X_l}\right)^{0.1738} \tag{17.104}$$

This correlation is valid for $895 < \text{Re}_d < 713{,}000$, $20 < \theta < 40°$, $X_t/d_o < 4$, $N_r \geq 4$, and θ is the tube layout angle. It predicts 89 literature data points within a mean absolute error of 6 percent; the range of actual error is from −16.7 to 19.9 percent.

Plain Flat Fins on a Staggered Tubebank.   This geometry, as shown in Fig. 17.14b, is used in the air-conditioning/refrigeration industry for cost considerations as well as where the pressure drop on the fin side prohibits the use of enhanced/interrupted flat fins. An inline tubebank is generally not used unless very low finside pressure drop is the essential requirement. The heat transfer correlation for Fig. 17.14b for flat plain fins on staggered tubebanks is provided by Gray and Webb (see Webb [47]) as follows for four or more tube rows with the subscript 4.

$$j_4 = 0.14\text{Re}_d^{-0.328} \, (X_t/X_l)^{-0.502}(s/d_o)^{0.031} \tag{17.105}$$

For the number of tube rows N_r from 1 to 3, the *j* factor is lower and is given by

$$\frac{j_N}{j_4} = 0.991[2.24\text{Re}_d^{-0.092} \, (N_r/4)^{-0.031}]^{0.607(4-N_r)} \tag{17.106}$$

Gray and Webb hypothesized the friction factor consisting of two components—one associated with the fins and the other associated with the tubes as follows.

$$f = f_f \frac{A_f}{A} + f_t \left(1 - \frac{A_f}{A}\right)\left(1 - \frac{\delta_f}{p_f}\right)$$

(17.107)

where

$$f_f = 0.508 \mathrm{Re}_d^{-0.521} (X_t/d_o)^{1.318}$$

(17.108)

and f_t (defined the same way as f) is the Fanning friction factor associated with the tube and can be determined from Eu of Fig. 17.45 as $f_t = \mathrm{Eu}\, N_r (X_t - d_o)/\pi d_o$. Equation 17.107 correlated 90 percent of the data for 19 heat exchangers within ±20 percent. The ranges of dimensionless variables of Eqs. 17.107 and 17.108 are $500 \le \mathrm{Re} \le 24{,}700$, $1.97 \le X_t/d_o \le 2.55$, $1.7 \le X_l/d_o \le 2.58$, and $0.08 \le s/d_o \le 0.64$.

Influence of Temperature-Dependent Fluid Properties

One of the basic idealizations made in the theoretical solutions for Nu and f is that the fluid properties remain constant throughout the flow field. Most of the experimental j and f data obtained in the preceding section involve small temperature differences so that the fluid properties generally do not vary significantly. In certain heat exchanger applications, fluid temperatures vary significantly. At least two questions arise: (1) Can we use the j and f data obtained for air at 50 to 100°C (100 to 200°F) for air at 500 to 600°C (900 to 1100°F)? (2) Can we use the j and f data obtained with air (such as all data in Ref. 20) for water, oil, and viscous liquids? The answer is yes, by modifying the constant-property j and f data to account for variations in the fluid properties within a heat exchanger. The property ratio method is the most commonly used technique to take into account the fluid property variations in the heat exchanger. In this method, the Nusselt number and friction factors for the variable fluid property case are related to the constant-property values for gases and liquids as follows:

For gases:

$$\frac{\mathrm{Nu}}{\mathrm{Nu}_{cp}} = \left(\frac{T_w}{T_m}\right)^n \qquad \frac{f}{f_{cp}} = \left(\frac{T_w}{T_m}\right)^m$$

(17.109)

For liquids:

$$\frac{\mathrm{Nu}}{\mathrm{Nu}_{cp}} = \left(\frac{\mu_w}{\mu_m}\right)^n \qquad \frac{f}{f_{cp}} = \left(\frac{\mu_w}{\mu_m}\right)^m$$

(17.110)

Here the subscript cp refers to the constant property variable, and all temperatures in Eq. 17.109 are *absolute*. All of the properties in the dimensionless groups of Eqs. 17.109 and 17.110 are evaluated at the *bulk mean* temperature. The values of the exponents n and m for fully developed laminar and turbulent flows in a circular tube are summarized in Table 17.20 for heating and cooling situations. These correlations, Eqs. 17.109 and 17.110, with exponents from Table 17.20a and b, are derived for the constant heat flux boundary condition. The variable-property effects are generally not important for fully developed flow having constant wall temperature boundary condition, since T_m approaches T_w for fully developed flow. Therefore, in order to take into account minor influence of property variations for the constant wall temperature boundary condition, the correlations of Eqs. 17.109 and 17.110 are adequate.

The Nu and f factors are also dependent upon the duct cross-sectional shape in laminar flow, and are practically independent of the duct shape in turbulent flow. The influence of variable fluid properties on Nu and f for fully developed laminar flow through rectangular ducts has been investigated by Nakamura et al. [57]. They concluded that the velocity profile is strongly affected by the μ_w/μ_m ratio, and the temperature profile is weakly affected by the μ_w/μ_m ratio. They found that the influence of the aspect ratio on the correction factor $(\mu_w/\mu_m)^m$ for the friction factor is negligible for $\mu_w/\mu_m < 10$. For the heat transfer problem, the Sieder-Tate correlation ($n = -0.14$) is valid only in the narrow range of $0.4 < \mu_w/\mu_m < 4$.

TABLE 17.20(a) Property Ratio Method Exponents of Eqs. 17.109 and 17.110 for Laminar Flow

Fluid	Heating	Cooling
Gases	$n = 0.0$, $m = 1.00$ for $1 < T_w/T_m < 3$	$n = 0.0$, $m = 0.81$ for $0.5 < T_w/T_m < 1$
Liquids	$n = -0.14$, $m = 0.58$ for $\mu_w/\mu_m < 1$	$n = -0.14$, $m = 0.54$ for $\mu_w/\mu_m < 1$

TABLE 17.20(b) Property Ratio Method Correlations or Exponents of Eqs. 17.109 and 17.110 for Turbulent Flow

Fluid	Heating	Cooling
Gases	$\mathrm{Nu} = 5 + 0.012\mathrm{Re}^{0.83}\,(\mathrm{Pr} + 0.29)(T_w/T_m)^n$ $n = -[\log_{10}(T_w/T_m)]^{1/4} + 0.3$ for $1 < T_w/T_m < 5$, $0.6 < \mathrm{Pr} < 0.9$, $10^4 < \mathrm{Re} < 10^6$ and $L/D_h > 40$ $m = -0.1$ for $1 < T_w/T_m < 2.4$	$n = 0$ $m = -0.1$ (tentative)
Liquids	$n = -0.11*$ for $0.08 < \mu_w/\mu_m < 1$ $f/f_{cp} = (7 - \mu_w/\mu_m)/6^\dagger$ or $m = 0.25$ for $0.35 < \mu_w/\mu_m < 1$	$n = -0.25*$ for $1 < \mu_w/\mu_m < 40$ $m = 0.24^\dagger$ for $1 < \mu_w/\mu_m < 2$

* Valid for $2 \le \mathrm{Pr} \le 140$, $10^4 \le \mathrm{Re} \le 1.25 \times 10^5$.
† Valid for $1.3 \le \mathrm{Pr} \le 10$, $10^4 \le \mathrm{Re} \le 2.3 \times 10^5$.

Influence of Superimposed Free Convection

The influence of superimposed free convection over pure forced convection flow is important when the flow velocity is low, a high temperature difference $(T_w - T_m)$ is employed, or the passage geometry has a large hydraulic diameter D_h. The effect of the superimposed free convection is generally important in the laminar flow of a noncompact heat exchanger; it is quite negligible for compact heat exchangers [19], and hence it will not be covered here. The reader may refer to Ref. 58 for further details.

It should be emphasized that, for laminar flow of liquids in tubes, the influence of viscosity and density variations (buoyancy or free convection effects) must be considered simultaneously for heat exchanger applications. Some correlations and work in this area have been summarized by Bergles [59].

TWO-PHASE HEAT TRANSFER AND PRESSURE DROP CORRELATIONS

Flow Patterns

The *flow pattern* depicts a distinct topology (regarding the spatial and temporal distributions of vapor and liquid phases) of two-phase flow and greatly influences the resulting phenomena of heat transfer and friction. An important feature of a particular flow pattern is the direct relationships of the heat transfer and pressure drop characteristics to the pattern type, leading to an easy identification of important macroscopic heat transfer modes. Consequently, an approach to the selection of appropriate heat transfer and/or pressure drop correlations has to be preceded by an identification of the involved flow patterns.

A particular flow pattern depends upon the flow passage geometry, its orientation, relative magnitudes of flow rates of fluid phases, fluid properties, boundary conditions, and so on. For a heat exchanger designer, several characteristic geometric settings are of special interest: (1) two-phase flow patterns in vertical, horizontal, and inclined flow passages (upward or downward internal flow); and (2) flow patterns in vertical and horizontal flow passages of the shell side of a shell-and-tube heat exchanger (i.e., external crossflow—shellside flow).

Internal Flow

Vertical Ducts. Typical flow patterns in upward vertical two-phase flow in a tube are presented in Fig. 17.48a. At low vapor qualities and low mass flow rates, the flow usually obeys the bubbly flow pattern. At higher vapor qualities and mass flow rates, slug or plug flow replaces the bubbly flow pattern. Further increase in vapor quality and/or mass flow rates leads to the appearance of the churn, annular, and wispy annular flow patterns.

The prediction of a flow pattern is a very important step in the analysis of a two-phase flow. In a graphical form, flow-pattern maps provide an empirical quantitative set of criteria for predicting flow conditions for a given flow pattern to occur; also, flow pattern maps delineate the boundaries of transitions from one flow regime to another. In Fig. 17.49, a flow pattern map is presented for a vertical upward flow [60, 113]. The map offers only a rough estimation of all the pattern transitions in terms of the so-called superficial momentum fluxes of the vapor and liquid (i.e., $\rho_v j_v^2 = G^2 x^2 / \rho_v$ versus $\rho_l j_l^2 = G^2 (1-x)^2 / \rho_l$). Note that loci of points correspond to the precisely defined pairs of superficial vapor and liquid momentum fluxes, but the corresponding transition boundaries should actually be interpreted as transition regions. The presented map, although not entirely reliable and less sophisticated than some subsequently developed for various situations, still provides a guide for engineering use. Taitel et al. [61] developed semiempirical models for transition between the flow patterns of steady upward gas-liquid flow in vertical tubes. Weisman and Kang [62] developed empirical flow pattern maps for both vertical and upward inclined tubes.

Horizontal Ducts. The topology of flow patterns in horizontal tubes with a circular cross section has different although analogous structure to that for the upward vertical two-phase flow. In Fig. 17.48b, the major types of the horizontal tube flow patterns are presented.

A flow pattern map for two-phase horizontal tube flow is given in Fig. 17.50. This map, developed by Taitel and Dukler [63], presents the transitions between various flow patterns in terms of three parameters K, F, and T as a function of the Martinelli parameter X, the ratio of the frictional pressure gradients for the gas and liquid phases flowing alone in the duct. All these parameters are defined in the figure caption. The transition between dispersed bubbly flow and intermittent flow (i.e., plug and slug flows, see Fig. 17.48b) is defined with a T versus X curve in Fig. 17.50. The transition between wavy (stratified) and both intermittent and annular (dispersed) flows is defined as an F versus X curve. The transition between stratified (smooth) and wavy flow is defined as a K versus X curve. Finally, the transition between either dispersed bubbly or intermittent flow patterns and annular flow is defined with a constant X line ($X = 1.6$). In addition to the Taitel and Dukler semiempirical map, a number of other flow pattern maps have been proposed, and the corresponding studies of the flow pattern transitions for horizontal two-phase flows have been conducted. An analysis of flow patterns for the entire range of duct inclinations is given by Barnea [64, 65]. An overview of the interfacial and structural stability of separated flow (both stratified and annular) is presented in Ref. 66. The most recent review and flow pattern map for boiling is provided by Kattan et al. [67].

External Flow (Shell Side).

Two-phase flow patterns for flow normal to tube bundles (crossflow), such as on the shell side of a shell-and-tube heat exchanger, are much more complex than those inside a plain circular tube. Consequently, prediction of flow patterns in such situations is very difficult. It is important to note that two-phase shellside flow patterns are substantially less analyzed than those for internal flows. A review of the shellside flow pattern is presented by Jensen [68]. The dominant flow patterns (see Fig. 17.51 [69]) may be assessed

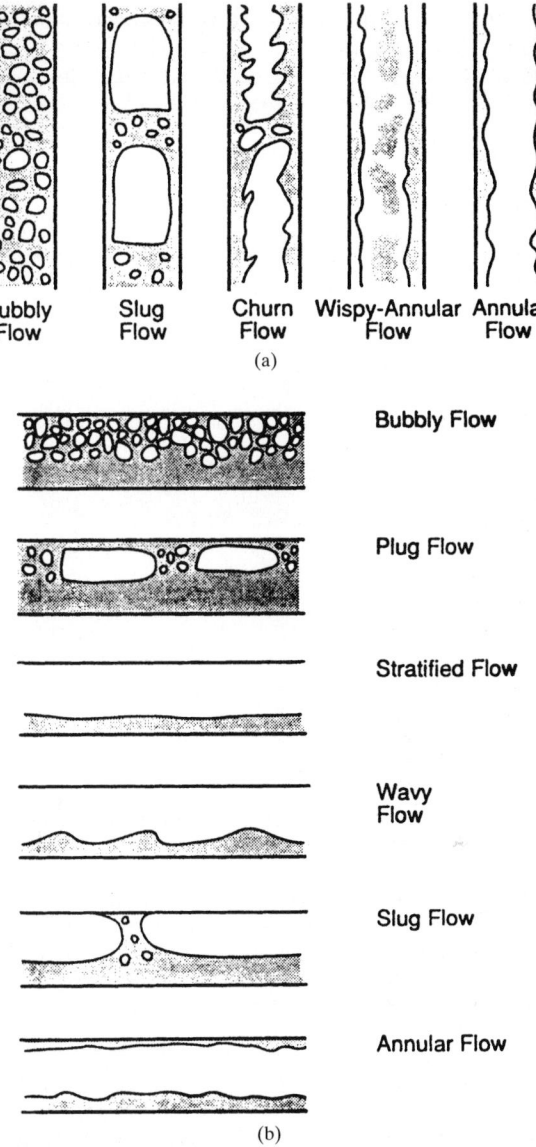

FIGURE 17.48   Flow patterns: (*a*) vertical cocurrent two-phase flow, (*b*) horizontal cocurrent two-phase flow [76].

using the flow pattern map presented in Fig. 17.52 by Grant and Chrisholm [70]. Again, as in the case of flow pattern maps for internal flows, the actual transitions are not abrupt as presented in Fig. 17.52. The flow pattern transitions are given as functions of liquid and gas velocities based on minimum cross-sectional areas and modified by a fluid property parameter to take into account property variations.

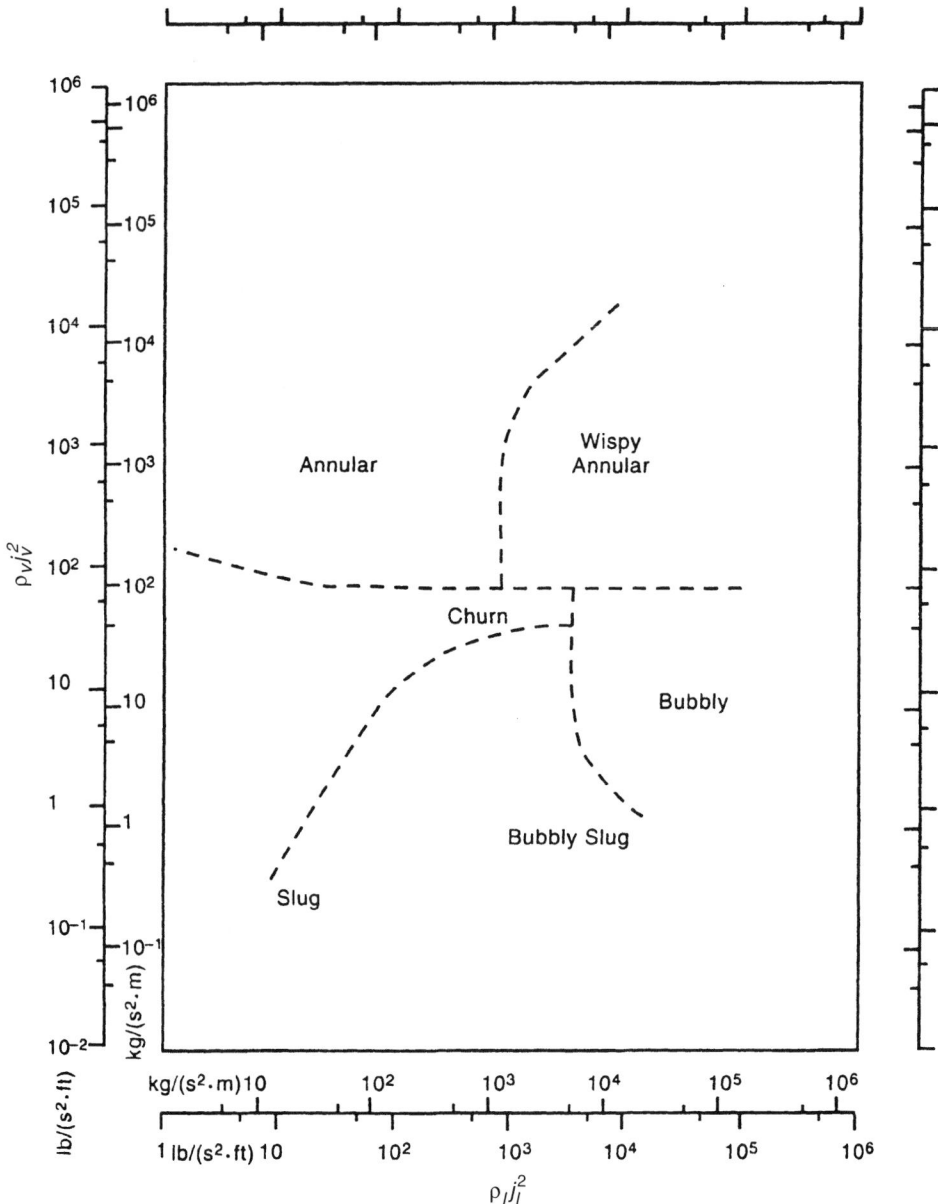

FIGURE 17.49   Flow pattern map for vertical, upward two-phase flow [60].

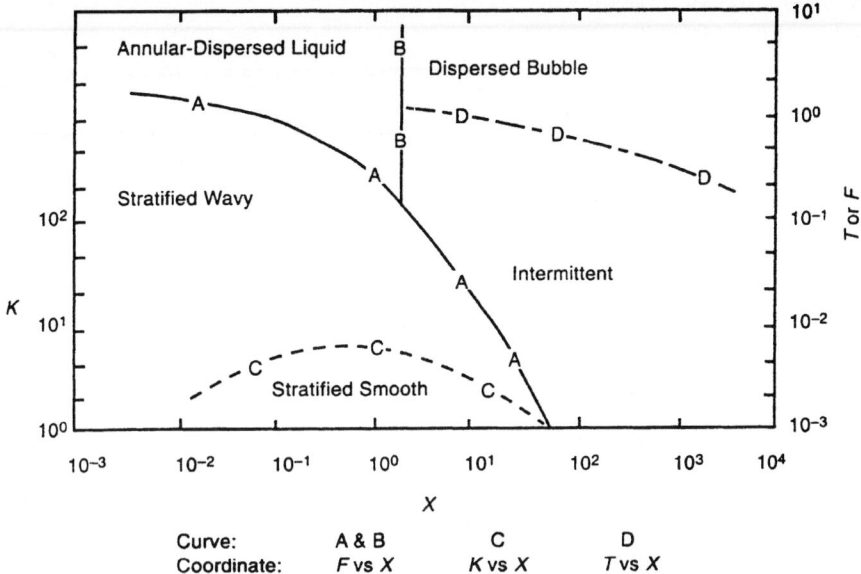

FIGURE 17.50 Flow pattern map for horizontal two-phase flow [63]. $X = [(dp/dz)_l/(dp/dz)_v]^{1/2}$, $K = F(d_i j_l/\nu_l)^{1/2}$, $F = [\rho_v/(\rho_l - \rho_v)]^{1/2} j_v/(d_i g \cos\theta)^{1/2}$, and $T = \{(dp/dz)_l/[(\rho_l - \rho_v)g \cos\theta]\}^{1/2}$ [63].

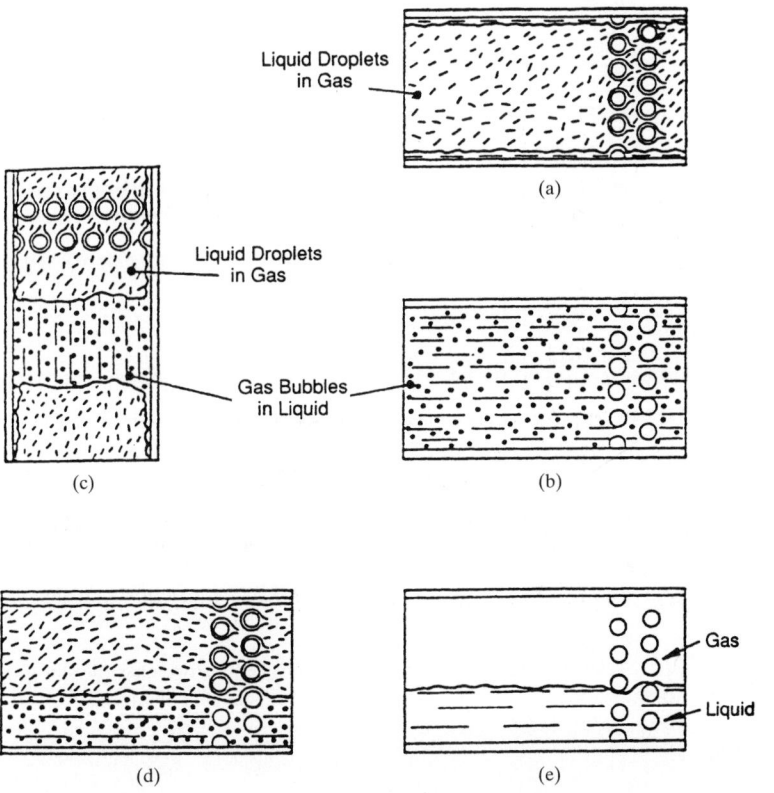

FIGURE 17.51 Flow patterns in tube bundles: (*a*) spray flow, (*b*) bubbly flow (vertical and horizontal), (*c*) chugging flow (vertical), (*d*) stratified spray flow, (*e*) horizontal stratified flow as defined by Grant and reported in Ref. 69.

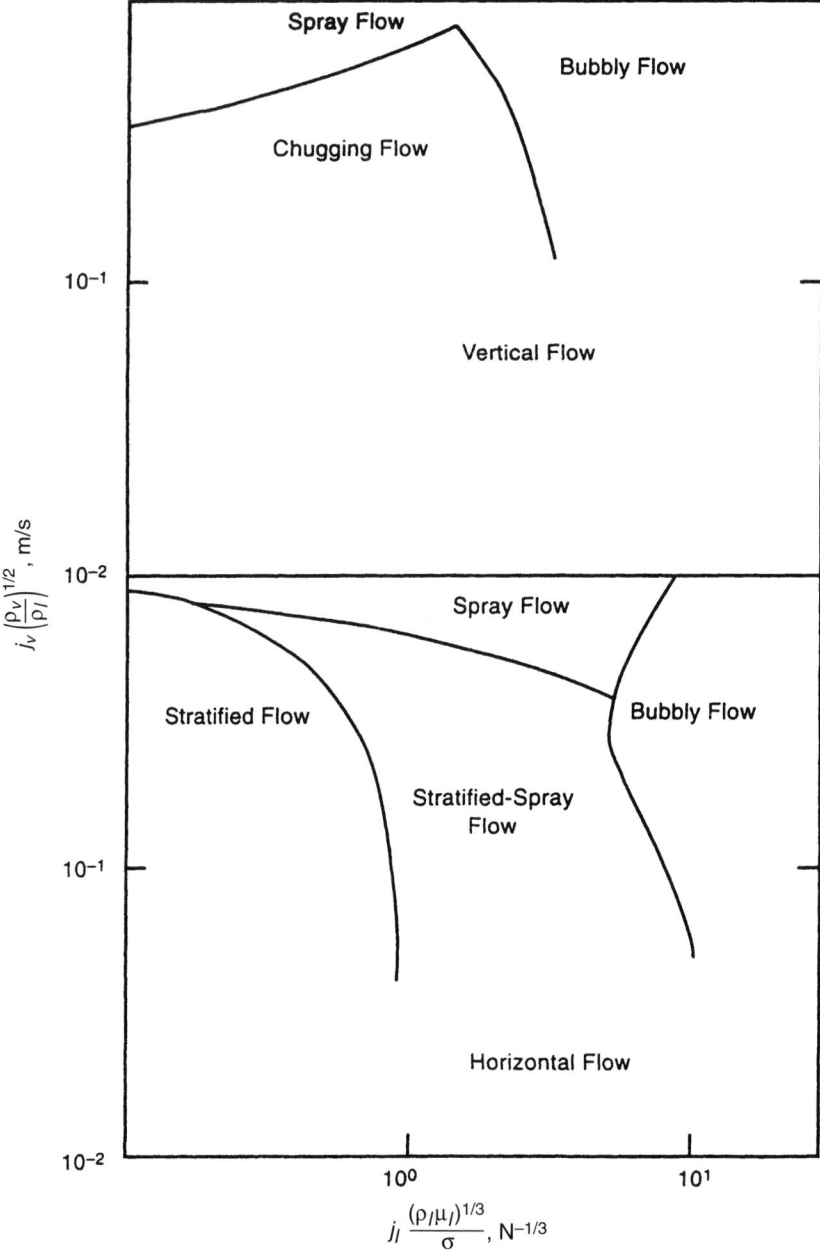

FIGURE 17.52 A flow pattern map for shellside two-phase flow by Grant, as reported in [71].

Two-Phase Pressure Drop Correlations

As a rule, the pressure drop for a two-phase flow is difficult to predict with good accuracy because of the presence of two phases, which results into various pressure drop components of individual phases and their interactions to well-understood, single-phase flows. The total pressure drop in a two-phase flow can be calculated as follows:

$$\Delta p = \Delta p_s + \Delta p_f + \Delta p_m + \Delta p_g \tag{17.111}$$

Major contributions to the total pressure drop (i.e., the various terms on the right-hand side of the equality sign of Eq. 17.111) depend on losses caused by friction and momentum changes along the two-phase fluid flow path. The list may include the following contributions: (1) pressure drop due to various singularities along the flow path such as an abrupt change in the free flow area, bends, and valves Δp_s (details regarding calculation of this pressure drop are provided in [71]), (2) two-phase friction loss Δp_f, (3) hydrostatic loss (i.e., the gravity loss or static head) Δp_g, and (4) momentum change loss (caused by acceleration or deceleration of the flow) Δp_m. The three dominant contributions are friction (the most difficult to determine accurately), momentum change (can be sizable in both vaporizers and condensers, primarily in vacuum operation), and hydrostatic effects (important only in a nonhorizontal flow). In most geometries, entrance and exit losses of Δp_s are difficult to measure. Hence, in two-phase or multiphase flow, they are often lumped into friction losses.

Analytical expressions for the three dominant pressure drop contributions will be discussed next. The correlations will be presented separately for internal (in-tube) and external (shellside—tube bundle) two-phase flows.

Intube Pressure Drop. The two-phase friction pressure drop can be estimated from the corresponding pressure drop for single-phase flow (with total two-phase fluid flowing as vapor or liquid) and multiplying that pressure drop magnitude with the so-called two-phase friction multiplier denoted as φ^2. The two-phase friction multiplier should be defined for corresponding hypothetical single-phase flows assuming that mass velocities are either equal to the actual respective mass velocities $[G_l = (1 - x)G, G_v = xG]$, or to the total mass velocity G. For example, the frictional multiplier φ_{lo}^2 represents the ratio of the two-phase frictional pressure gradient and the single-phase (liquid) pressure gradient for a flow with the same total mass velocity as liquid [71]. Therefore, the friction pressure drop can be presented as:

$$\Delta p_f = \Delta p_{f,lo}\varphi_{lo}^2 = f_{lo}\frac{2L}{D_h}\frac{G^2}{g_c\rho_l}\varphi_{lo}^2 \quad \text{or} \quad \Delta p_f = \Delta p_{f,vo}\varphi_{vo}^2 = f_{vo}\frac{2L}{D_h}\frac{G^2}{g_c\rho_v}\varphi_{vo}^2 \tag{17.112}$$

where f_{lo} and f_{vo} represent the single-phase Fanning friction factor (the total mass flow rate as liquid or vapor, respectively, f_{lo} equal to $16/\text{Re}_{lo}$ for $\text{Re}_{lo} = GD_h/\mu_l < 2000$, and $f_{lo} = 0.079(\text{Re}_{lo})^{-0.25}$ for $\text{Re}_{lo} > 2000$). In Table 17.21, the two most reliable correlations for the friction multiplier are presented, a Friedel correlation [71, 72] for vertical upward and horizontal flow and $\mu_l/\mu_v < 1000$, and the Chisholm correlation [71, 73] for $\mu_l/\mu_v > 1000$ and $G > 100$ kg/m²s. An empirically determined standard deviation for Friedel correlation [72] can be fairly large, up to 50 percent for two-component flows compared with a data bank of 25,000 data points [71]. The standard deviation is smaller for single-component flows, up to 30 percent. However, for small mass fluxes ($G < 100$ kg/m²s), the two correlations given in Table 17.21 are not accurate. The best correlation available for this range of mass fluxes is the well-known Lockhart-Martinelli correlation [74]. This correlation uses φ_l^2 and φ_v^2, fractional multipliers for vapor and liquid phases, based on the single-phase pressure gradients defined by

$$\Delta p_f = \Delta p_{f,l}\varphi_l^2 = \Delta p_{f,v}\varphi_v^2 \tag{17.113}$$

where φ_l^2 and φ_v^2 are given by Ref. 74 as follows for $\mu_l/\mu_v > 1000$ and $G < 100$ kg/m²s:

$$\varphi_l^2 = \frac{(dp/dz)}{(dp/dz)_l} = 1 + \frac{c}{X} + \frac{1}{X^2} \tag{17.113a}$$

TABLE 17.21   Frictional Multiplier Correlations*

Correlation	Parameters	Ref.
Friedel correlation $\varphi_{lo}^2 = E + 3.23 \dfrac{FH}{\mathrm{Fr}^{0.045}\,\mathrm{We}^{0.035}}$	$E = (1-x)^2 + x^2\,\dfrac{\rho_l}{\rho_v}\,\dfrac{f_{vo}}{f_{lo}} \qquad F = x^{0.78}(1-x)^{0.24}$ $H = \left(\dfrac{\rho_l}{\rho_v}\right)^{0.91}\left(\dfrac{\mu_v}{\mu_l}\right)^{0.19}\left(1 - \dfrac{\mu_v}{\mu_l}\right)^{0.7}$ $\mathrm{Fr} = \dfrac{G^2}{gD_h\rho_{\mathrm{hom}}^2} \qquad \mathrm{We} = \dfrac{G^2 D_h}{\sigma\rho_{\mathrm{hom}}}$ $\dfrac{1}{\rho_{\mathrm{hom}}} = \dfrac{x}{\rho_v} + \dfrac{1-x}{\rho_l}; \qquad \dfrac{\mu_l}{\mu_v} < 1000$	71
Chisholm correlation $\varphi_{lo}^2 = 1 + (Y^2 - 1)[Bx^{n^*}(1-x)^{n^*} + x^{2-n}]$ $n^* = \dfrac{2-n}{2}$	$Y = (\Delta p_{f,vo}/\Delta p_{l,lo})^{1/2}, \quad n = \tfrac{1}{4}\ [\text{exponent in } f = C\,\mathrm{Re}^n]$ $B = \begin{cases} 4.8 & G < 500 \\ 2400/G & 500 \le G < 1900 \\ 55/G^{1/2} & G \ge 1900 \end{cases}$ for $Y \le 9.5$ $B = \begin{cases} 520/(YG^{1/2}) & G \le 600 \\ 21/Y & G > 600 \end{cases}$ for $9.5 < Y < 28$ $B = 15{,}000/(Y^2 G^{1/2})$ for $Y > 28$ $\dfrac{\mu_l}{\mu_v} > 1000;\ G > 100$	71

* The parameters E, F, H, Y, and B are local for this table as defined; other variables in SI units.

$$\varphi_v^2 = \frac{(dp/dz)}{(dp/dz)_v} = 1 + cX + X^2 \tag{17.113b}$$

In Eqs. 17.113a and 17.113b, the value of c depends on the single-phase regime for liquid and vapor streams as follows: (1) if both liquid and vapor phases are turbulent, $c = 20$; (2) if the liquid phase is turbulent and the vapor phase is viscous (laminar), $c = 10$; (3) if the liquid phase is viscous and the vapor phase is turbulent, $c = 12$; and (4) if both liquid and vapor phases are viscous, $c = 5$.

The momentum pressure drop can be calculated integrating the momentum balance equation [75], thus obtaining:

$$\Delta p_m = \frac{G^2}{g_c}\left[\left(\frac{x^2}{\alpha\rho_v} + \frac{(1-x)^2}{(1-\alpha)\rho_l}\right)_{z=z_2} - \left(\frac{x^2}{\alpha\rho_v} + \frac{(1-x)^2}{(1-\alpha)\rho_l}\right)_{z=z_1}\right] \tag{17.114}$$

where α represents the void fraction of the vapor (gas) phase. For the homogeneous model, the two-phase flow behaves like a single phase, and the vapor and liquid velocities are equal. A number of correlations for the void fraction exist [76]. An empirical correlation whose general form is valid for several frequently used models is given by Butterworth as reported in [76]:

$$\alpha = \left[1 + A\left(\frac{1-x}{x}\right)^p\left(\frac{\rho_v}{\rho_l}\right)^q\left(\frac{\mu_l}{\mu_v}\right)^r\right]^{-1} \tag{17.115}$$

where the constants A, p, q, and r depend on the two-phase model and/or empirical data chosen. These constants for a nonhomogeneous model, based on steam-water data, are [76]: $A = 1$, $p = 1$, $q = 0.89$, and $r = 0.18$. For the homogeneous model, $A = p = q = 1$ and $r = 0$. The Lockhart and Martinelli model assumes $A = 0.28$, $p = 0.64$, $q = 0.36$, and $r = 0.07$. For engineering design calculations, the homogeneous model yields the best results where the slip velocity between the phases is small (for bubbly or mist flows).

Finally, the pressure drop caused by the gravity (hydrostatic) effect is:

$$\Delta p_g = \pm \frac{g}{g_c} \sin \theta \int_0^L [\alpha \rho_v + (1 - \alpha)\rho_l] \, dz \tag{17.116}$$

Note that the negative sign (i.e., the pressure recovery) stands for a downward flow in an inclined or vertical fluid flow.

Shellside Pressure Drop.   Surprisingly little attention has been devoted in engineering literature to estimate two-phase pressure drop on the shell side of shell-and-tube heat exchangers [77, 78]. In engineering practice, the estimation of the two-phase flow pressure drop can be performed in some situations using modified single-phase flow correlations. This approach is, however, highly unreliable.

In Table 17.22, two correlations are presented for shellside two-phase flow pressure drop estimation, based on modifications of the internal flow correlations. The first correlation uses the modified Chrisholm correlation [69, 79], and the second one [80] employs the modified Lockhart-Martinelli correlation. The first correlation is for horizontal crossflow (crossflow in a baffled horizontal heat exchanger with horizontal or vertical baffle cuts). The second one is for vertical crossflow (upflow in a horizontal tube bundle).

Heat Transfer Correlations for Condensation

The objective of this subsection is to present the most important condensation correlations for design of heat transfer equipment.

Condensation represents a class of vapor-liquid phase change phenomena that usually take place when vapor is cooled below its saturation temperature at a given pressure. In the case of condensation on a heat exchanger surface, the heat transfer interaction between the bulk of the vapor and the surface involves heterogeneous nucleation that leads to a formation

TABLE 17.22   Shellside Two-Phase Pressure Drop Correlations

Correlation	Parameters						Ref.
Chisholm correlation: $\Delta p_f = \Delta p_{f,lo} \varphi_{lo}^2$	Orientation	Flow pattern		n	B		
$\varphi_{lo}^2 = 1 + (Y^2 - 1)[Bx^{(2-n)/2}(1-x)^{(2-n)/2} + x^{2-n}]$	Vertical ↑↓	Spray and bubble		0.37	1		69
	Horizontal	Spray and bubble		0.46	0.75		79
$Y = (\Delta p_{f,vo} / \Delta p_{l,lo})^{1/2}$	Horizontal	Stratified and spray		0.46	0.25		
	Window zone flow						
See Eq. 17.112 for Δp_{fvo} and Δp_{flo}	Vertical ↑↓			0	0.25		
	Horizontal			0	$2/(Y+1)$		

Correlation	Parameters						Ref.
Modified Lockhart-Martinelli correlation: $\Delta p_f = \Delta p_{fl} \varphi_l^2$ For $\mathrm{Fr}_l > 0.15$:	$C = c_1 \, \mathrm{Fr}_l^{c_2} \ln X_{tt} + c_3 \, \mathrm{Fr}_l^{c_4}$						
$\varphi_l^2 = 1 + \dfrac{C}{X_{tt}} + \dfrac{c_5}{X_{tt}^2}$ For $\mathrm{Fr}_l \le 0.15$:	Flow pattern	c_1	c_2	c_3	c_4	c_5	80
$\varphi_l^2 = 1 + \dfrac{8}{X_{tt}} + \dfrac{1}{X_{tt}^2}$	Bubble	0.036	1.51	7.79	−0.057	0.774	
	Slug ($\mathrm{Fr}_l > 1.15$)	2.18	−0.643	11.6	0.233	1.09	
$X_{tt} = \left(\dfrac{1-x}{x}\right)^{0.9} \left(\dfrac{\rho_v}{\rho_l}\right)^{0.5} \left(\dfrac{\mu_l}{\mu_v}\right)^{0.1}$	Spray ($\mathrm{Fr}_l > 1.15$)	0.253	−1.50	12.4	0.207	0.205	
$X_{tt} < 0.2$							

of liquid droplets (dropwise condensation) and/or a liquid layer (filmwise condensation) between the surface and the condensing vapor. The dropwise condensation is desirable because the heat transfer coefficients are an order of magnitude higher than those for filmwise condensation. Surface conditions, though, are difficult for sustaining dropwise condensation. Hence, this mode is not common in practical applications. The heat transfer correlations presented in this section will deal primarily with filmwise condensation (also classified as surface condensation). Refer to Chap. 12 and Refs. 75, 76, 81, and 82 for additional information.

Heat transfer coefficients for condensation processes depend on the condensation models involved, condensation rate, flow pattern, heat transfer surface geometry, and surface orientation. The behavior of condensate is controlled by inertia, gravity, vapor-liquid film interfacial shear, and surface tension forces. Two major condensation mechanisms in film condensation are gravity-controlled and shear-controlled (forced convective) condensation in passages where the surface tension effect is negligible. At high vapor shear, the condensate film may became turbulent.

Now we will present separately heat transfer correlations for external and internal filmwise condensation.

Heat Transfer Correlations for External Condensation. Although the complexity of condensation heat transfer phenomena prevents a rigorous theoretical analysis, an external condensation for some simple situations and geometric configurations has been the subject of a mathematical modeling. The famous pioneering Nusselt theory of film condensation had led to a simple correlation for the determination of a heat transfer coefficient under conditions of gravity-controlled, laminar, wave-free condensation of a pure vapor on a vertical surface (either flat or tube). Modified versions of Nusselt's theory and further empirical studies have produced a list of many correlations, some of which are compiled in Table 17.23.

Vertical Surfaces. Condensation heat transfer coefficients for external condensation on vertical surfaces depend on whether the vapor is saturated or supersaturated; the condensate film is laminar or turbulent; and the condensate film surface is wave-free or wavy. Most correlations assume a constant condensation surface temperature, but variable surface temperature conditions are correlated as well as summarized in Table 17.23. All coefficients represent mean values (over a total surface length), that is, $\bar{h} = (1/L) \int_0^L h_{\text{loc}} \, dx$.

The first two correlations in Table 17.23 for laminar condensation of saturated vapor with negligible interfacial shear and wave-free condensate surface are equivalent, the difference being only with respect to the utilization of a condensate Reynolds number based on the condensation rate evaluated at distance L. If the assumption regarding the uniformity of the heat transfer surface temperature does not hold, but condensation of a saturated vapor is controlled by gravity only, the heat transfer surface temperature can be approximated by a locally changing function as presented in Table 17.23 (third correlation from the top). This results into a modified Nusselt correlation, as shown by Walt and Kröger [83]. It is important to note that all heat transfer correlations mentioned can be used for most fluids regardless of the actual variation in thermophysical properties as long as the thermophysical properties involved are determined following the rules noted in Table 17.23.

A presence of interfacial waves increases the heat transfer coefficient predicted by Nusselt theory by a factor up to 1.1. An underprediction of a heat transfer coefficient by the Nusselt theory is more pronounced for larger condensate flow rates. For laminar condensation having both a wave-free and wavy portion of the condensate film, the correlation based on the work of Kutateladze as reported in [81] (the fourth correlation from the top of Table 17.23) can be used as long as the flow is laminar.

Film turbulence (the onset of turbulence characterized by a local film Reynolds number range between 1600 and 1800) changes heat transfer conditions depending on the magnitude of the Pr number. For situations when the Prandtl number does not exceed 10, a mean heat transfer coefficient may be calculated using the correlation provided by Butterworth [81] (the fifth correlation from the top of Table 17.23). An increase in the Pr and Re numbers causes an

TABLE 17.23 Heat Transfer Correlations for External Condensation on Vertical Surfaces

Vapor condition*	Liquid-vapor interface	Condensation surface	Correlation	Ref.	Comment‡
Saturated vapor	Laminar wave-free	$T_w = $ const.	$\bar{h} = 0.943\left[\dfrac{k_l^3 \rho_l(\rho_l - \rho_v)g i_{lv}}{\mu_l(T_{sat} - T_w)L}\right]^{1/4}$	81	i_{lv}, ρ_v @ T_{sat}; $k_l = [(k_l)_{T_w} + (k_l)_{T_{sat}}]/2$; $\rho_l = [(\rho_l)_{T_w} + (\rho_l)_{T_{sat}}]/2$; $\mu_l = \dfrac{3\mu_{l,T_w} + \mu_{l,T_{sat}}}{4}$; $Re_L = \dfrac{4\Gamma_L}{\mu_l} < 1600$; $\Gamma_L = \dfrac{\dot{m}}{s}$†
			$\bar{h} = 1.47\,\dfrac{k_l}{Re_L^{1/3}}\left[\dfrac{\mu_l^2}{\rho(\rho_l - \rho_v)g}\right]^{-1/3}$	81	
		$T_w = T_{sat} - az^n$ $0 < n < 3$	$\bar{h} = \dfrac{4}{3-n}\left(\dfrac{n+1}{4}\right)^{1/4}\left[\dfrac{k_l^3 \rho(\rho_l - \rho_g)g i_{lv}}{\mu(T_{sat} - T_w)_{z=L}L}\right]^{1/4}$	83	
	Laminar and both wave-free and wavy	$T_w = $ const.	$\bar{h} = \dfrac{Re_L\, k_l[\mu_l^2/[\rho(\rho_l - \rho_v)g]]^{-1/3}}{1.08 Re_L^{1.22} - 5.2}$	81	
	Laminar wave-free, wavy and turbulent		$\bar{h} = \dfrac{Re_L\, k_l[\mu_l^2/[\rho(\rho_l - \rho_v)g]]^{-1/3}}{8750 + 58Pr^{-0.5}\,(Re_L^{0.75} - 253)}$	81	$Pr \leq 10$
Superheated vapor	Laminar, wavy-free		$\bar{h} = \bar{h}_{sat}\left[1 + \dfrac{c_{p,v}(T_v - T_{sat})}{i_{lv}}\right]^{1/4}$	84	$h_{sat} = \bar{h}$ as given above

* Negligible vapor velocity.
† s is the plate width or tube perimeter for a tube.
‡ Thermophysical properties are taken as indicated for all correlations unless indicated otherwise.

increase in the heat transfer coefficient in the turbulent region. The correlation given in Table 17.23 tends to overpredict the heat transfer coefficient for Pr > 10. For further details about treating the existence of turbulence, see Refs. 76 and 78.

It must be noted that predictions of heat transfer coefficients in all mentioned situations may be treated, as a rule, as conservative as long as the correlation is based on the Nusselt theory. Two important additional phenomena, though, are *not* included: vapor superheat and vapor shear effects. The influence of superheating can be included (although the effect is usually small) by the sixth correlation from the top in Table 17.23.

An interfacial shear may be very important in so-called shear-controlled condensation because downward interfacial shear reduces the critical Re number for onset of turbulence. In such situations, the correlations must include interfacial shear stress, and the determination of the heat transfer coefficient follows the Nusselt-type analysis for zero interfacial shear [76]. According to Butterworth [81], data and analyses involving interfacial shear stress are scarce and not comprehensive enough to cover all important circumstances. The calculations should be performed for the local heat transfer coefficient, thus involving step-by-step procedures in any condenser design. The correlations for local heat transfer coefficients are presented in [81] for cases where interfacial shear swamps any gravitational forces in the film or where both vapor shear and gravity are important.

Horizontal and Inclined Surfaces.   The Nusselt theory of gravity-controlled film condensation can easily be applied to horizontal or inclined surfaces. Correlations for horizontal single tubes and tube bundles are given in Table 17.24. The first two correlations in the table are valid for negligible vapor shear effect. The correlations predict the mean heat transfer coefficient around the tube circumference at a given location along the tube. The last correlation is for condensing superheated vapor. The correlations for a single tube are conservative. They generally underpredict the heat transfer coefficients by up to 20 percent.

When vapor is moving at a large approaching velocity, the shear stress between the vapor and the condensate surface must be taken into account (i.e., shear forces are large compared to gravity force). A good review of the work devoted to this problem is found in Rose [85], who provided a detailed discussion of film condensation under forced convection. In Table 17.24, a correlation derived by Fuji et al. [86] and suggested by Butterworth [81] is included for the vapor shear effect. The same equation can be applied for a tube bundle. In such a situation, the approach velocity u_v should be calculated at the maximum free-flow area cross section within the bundle.

Film condensation in tube bundles (more commonly used in shell-and-tube heat exchangers) characterize more complex physical conditions compared to condensation on a single tube. The gravity-controlled and surface-shear-stress-influenced condensate films must be modeled in different ways to accommodate combined influences of condensate drain to lower tubes (i.e., condensate inundation) and shear effects. Such a correlation, the fourth correlation from the top of Table 17.24, was proposed by Kern and modified by Butterworth [81].

In the absence of vapor shear effects, the heat transfer coefficient around the lower tubes in a bundle should decrease. However, in general, it is difficult to predict the actual value in a tube bundle depending on the influence of vapor and condensate velocities, turbulence effects, vapor flow direction, tube bundle layout, pressure, heat transfer surface conditions, and so on.

Heat Transfer Correlations for Internal Condensation.

Internal condensation processes are complex because a simultaneous motion of both vapor and condensate takes place (in addition to phase change phenomena) in a far more complex manner than for unconfined external condensation. The flow regime can vary substantially. Characteristics of a particular flow pattern involved are extremely important in describing particular heat transfer conditions. Correspondingly, to predict with confidence the heat transfer coefficient for internal film condensation appears to be even more difficult than for external condensation.

Convective condensation in horizontal and vertical tubes is most important with two flow patterns: annular film flow and stratified flow.

TABLE 17.24 Heat Transfer Correlations for External Condensation on Horizontal and Inclined Tubes and Tube Bundles

Vapor condition	Liquid-vapor interface	Condensation surface	Correlation	Ref.	Comment
Saturated vapor	Laminar, no vapor shear	T_w = const. Horizontal, single tube ($\theta = 0$); inclined tube ($\theta \neq 0$)	$\bar{h} = 0.728\left[\dfrac{k_l^3 \rho_l(\rho_l - \rho_v)g\cos\theta\, i_{lv}}{\mu_l(T_{sat} - T_w)d_o}\right]^{1/4}$	81	i_{lv}, ρ_v @ T_{sat}; $k_l = [(k_l)_{T_w} + (k_l)_{T_{sat}}]/2$; $\rho_l = [(\rho_l)_{T_w} + (\rho_l)_{T_{sat}}]/2$; $\mu_l = \dfrac{3\mu_{l,T_w} + \mu_{l,T_{sat}}}{4}$
		q_w = const. Horizontal single tube	$\bar{h} = 0.70\left[\dfrac{k_l^3 \rho_l(\rho_l - \rho_v)g i_{lv}}{\mu_l(T_{sat} - T_w)d_o}\right]^{1/4}$	86	
	Laminar, vapor shear exists	T_w = const. Single tube	$\bar{h} = 0.728\xi\left[\dfrac{k_l^3 \rho_l(\rho_l - \rho_v)g i_{lv}}{\mu_l(T_{sat} - T_w)d_o}\right]^{1/4}$	86	$1 < \xi \le 1.4\left[\dfrac{u_v^2(T_{sat} - T_w)k_l}{\mu_l i_{lv}g d_o}\right]^{1/20} < 1.7$
	Laminar	T_w = const. Horizontal tube bundle	$\bar{h} = \bar{h}_{single,top}aN^{-b}$	81	$a = 1, b = \tfrac{1}{6}$, for $N < 10$; $a = 1.24, b = \tfrac{1}{4}$, for $N \ge 10$
Superheated vapor	Laminar		$\bar{h} = \bar{h}_{sat}\left[1 + \dfrac{c_{p,v}(T_v - T_{sat})}{i_{lv}}\right]^{1/4}$	84	$h_{sat} = \bar{h}$ as given above

Note: $\bar{h}$ is the mean coefficient around the tube perimeter; u_v is the vapor approach velocity.

TABLE 17.25   Heat Transfer Correlations for Internal Condensation in Horizontal Tubes

Stratification conditions	Correlation	Ref.
Annular flow* (Film condensation)	$h_{\mathrm{loc}} = h_l\left[(1-x)^{0.8} + \dfrac{3.8x^{0.76}(1-x)^{0.04}}{(p/p_c)^{0.38}}\right]$	
	where $h_l = 0.023\,\dfrac{k_l}{d_i}\,\mathrm{Re}_l^{0.8}\,\mathrm{Pr}_l^{0.4}$	88
	$\mathrm{Re}_l = \dfrac{Gd_i}{\mu_l},\quad G = \text{total mass velocity (all liquid)}$	
	$100 \le \mathrm{Re}_l \le 63{,}000$	
	$0 \le x \le 1 \qquad 1 \le \mathrm{Pr}_l \le 13$	
Stratified flow	$\bar{h} = 0.728\left[1 + \dfrac{1-x}{x}\left(\dfrac{\rho_v}{\rho_l}\right)^{2/3}\right]^{-(3/4)}\left[\dfrac{k_l^3\rho_l(\rho_l - \rho_g)gi'_{lv}}{\mu_l(T_{\mathrm{sat}} - T_w)d_i}\right]^{1/4}$	76
	where: $i'_{lv} = i_{lv} + 0.68c_{p,l}(T_{\mathrm{sat}} - T_w)$	87

* Valid for horizontal, vertical or inclined tubes of diameters ranging from 7 to 40 mm.

Horizontal Surfaces.   For annular film flow, the Nusselt-theory-based correlations usually fail to provide acceptable predictions. This type of flow is shear-dominated flow. This problem has been a subject of extensive research, and numerous correlations can be found in literature [78]. The correlation given by Shah [88] in Table 17.25 is the best, as it is valid for a wide range of fluids and flow conditions. The mean deviation for 474 data points analyzed was found to be 15 percent.

In stratified flow, the stratified layer at the lower part of the tube free-flow area is influenced primarily by shear effects, while a thin film covers the upper portions of the inner tube wall and stratifies under the influence of gravity. The heat transfer conditions in two regions are quite different, but it is a standard practice to correlate heat transfer based on the entire perimeter. In Table 17.25, a correlation based on the modified Nusselt theory is given for stratified flow, developed by Chato [87] and modified by Jaster and Kosky, as reported by Carey [76]. Consult Carey [76] and Butterworth [81] for a detailed analysis of related phenomena. The most recent condensation correlations are given by Dobson and Chato [89].

Vertical Surfaces.   If the laminar flow direction is downward and gravity-controlled, heat transfer coefficient for internal condensation inside vertical tubes can be predicted using the correlations for external film condensation—see Table 17.23. The condensation conditions usually occur under annular flow conditions. Discussion of modeling of the downward internal convective condensation is provided in Ref. 76.

For the interfacial shear-controlled flows, annular film flow pattern is established, and the tube orientation is irrelevant. Consequently, the correlations for annular condensation in horizontal tubes can be applied for vertical internal downward flows as well—see Table 17.25.

For an upward flow direction, the shear forces may influence the downward-flow of the condensate, causing an increase of the condensate film thickness. Therefore, the heat transfer coefficient under such conditions shall decrease up to 30 percent compared to the result obtained using the same correlation as the upward-flowing vapor. If the vapor velocity increases substantially, the so-called flooding phenomenon may occur. Under such condition, the shear forces completely prevent the downward condensate flow and flood (block) the tube with the condensate. Prediction of the flooding conditions is discussed by Wallis, as reported by Butterworth [81].

Heat Transfer Correlations for Condensation Under Special Conditions.   In a number of practical engineering situations, condensation phenomena may occur under quite different conditions compared to the ones discussed in the preceding subsections. This includes noncircular compact heat exchanger passages, augmented tube geometries, condensation of multi-component vapor mixtures, presence of noncondensable gases, surface tension driven flows, and so on. Also, the appearance of dropwise condensation radically changes heat transfer performance. In all these situations, the conservative approach cannot be applied. For detailed discussion of these factors, one may consult Refs. 76, 78, 81, and 90–93 for details.

Heat Transfer Correlations for Boiling

Vaporization (boiling and evaporation) phenomena have been extensively investigated and reported in the literature [75, 76, 94, 95]. This section is devoted primarily to forced convective boiling and critical heat flux correlations important for heat exchanger design. These correlations are considered for the intube and shellside of a shell-and-tube heat exchanger. Vaporization in both geometries is a very complex process, and the empirical data are the primary source for engineering heat transfer correlations. Over the years, a large number of correlations have been developed; for example, well over 30 correlations are available for saturated flow boiling [96]. We will present now the most accurate correlations (based on experimental data for many fluids) for both intube and shellside forced convective boiling.

Heat Transfer Coefficient Correlations

Intube Forced (Flow) Saturated Boiling. The correlation proposed by Kandlikar [96] is based on empirical data for water, refrigerants, and cryogens. The correlation consists of two parts, the convective and nucleate boiling terms, and utilizes a fluid-dependent parameter.

TABLE 17.26   Heat Transfer Correlations for Boiling

Geometry	Correlation

$$\frac{h}{h_l} = c_1 \, \text{Co}^{c_2} \, (25\text{Fr}_{lo})^{c_5} + c_3 \, \text{Bo}^{c_4} \, F_{fl}$$

$$h_l = 0.023 \text{Re}_l^{0.8} \, \text{Pr}_l^{0.4} \left(\frac{k_l}{d_i}\right) \qquad \text{Co} = \left(\frac{1-x}{x}\right)^{0.8}\left(\frac{\rho_v}{\rho_l}\right)^{0.5}$$

Intube [96]
$$\text{Re}_l = \frac{G d_i (1-x)}{\mu_l} \qquad \text{Fr}_{lo} = \frac{G^2}{g d_i \rho_l^2} \qquad \text{Bo} = \frac{q''}{G i_{lv}}$$

c_i	Convective boiling	Nucleate boiling	Fluid	F_{fl}
c_1	1.1360	0.6683	H_2O	1.00
c_2	−0.9	−0.2	R-11	1.30
c_3	667.2	1058.0	R-12	1.50
c_4	0.7	0.7	R-22	2.20
c_5	0.3	0.3	R-113	1.30
Note: $c_5 = 0$ for vertical tubes			R-114	1.24
and $c_5 = 0$ for horizontal tubes			N_2	4.70
for $\text{Fr}_{lo} > 0.04$.			Ne	3.50

Shellside [68]

$h = F h_{\text{conv}} + S h_{NB}$

h_{conv}, single-phase correlation of a type $\text{Nu} = C \, \text{Re}^n \, \text{Pr}^{0.34}$

h_{NB}, a nucleate pool boiling correlation [95]

$F = (\varphi_l^2)^{n/(2-m)}$, m from a single-phase correlation $f = C' \, \text{Re}^{-m}$

$$S = \frac{k}{F} \, h_{\text{conv}} X_o \left[1 - \exp\left(-\frac{F h_{\text{conv}} X_o}{k}\right)\right] \qquad X_o = 0.041 \left[\frac{\sigma}{g(\rho_l - \rho_v)}\right]^{0.5}$$

φ_l^2, friction multiplier, see Table 17.22

For single-phase correlations, refer to the major section starting on p. 17.66.

The correlation is given in Table 17.26 and can be applied for either vertical (upward and downward) or horizontal intube flow. A mean deviation of slightly less than 16 percent with water and 19 percent with refrigerants has been reported.

Shellside Forced Boiling. Additional turbulence effects, complex flow fields, and drag effects can significantly increase heat transfer coefficients on the shell side of a tube bundle. A detailed review of the topic is given in Ref. 68.

As in the case of the intube boiling, the shellside boiling is controlled by a trade-off between the convective and nucleate boiling. This has been the reason why a correlation type originally developed for intube boiling has extensively been used for shellside boiling conditions as well. It should be pointed out that the prediction of the heat transfer coefficient on the shell side can easily reach an error margin of 40 percent [68]. One such correlation is presented in Table 17.26 [68].

Critical Heat Flux. The importance of the critical heat flux (CHF) for engineering practice cannot be overemphasized. A sharp increase of the wall temperature caused by the onset of critical conditions can lead to failure of heat transfer equipment. This is the reason for a large number of correlations in the literature. An instructive overview of the topic is provided in Ref. 76. The critical heat flux conditions in a single tube or on the shell side of the tube bundle are not the same, and the suggestions regarding the use of the most precise correlations are given below.

Intube Forced (Flow) Boiling. The prediction of the onset of the internal boiling flow is more accurate (the root-mean-square error of correlations is reported to be usually between 7 and 15 percent). The correlation presented in Table 17.27, along with the direction concerning the application for vertical uniformly heated tubes, is proposed by Katto and Ohno [97], as reported by Carey [76]. An explicit information regarding the accuracy of this correlation is not available.

TABLE 17.27 CHF Correlation for a Vertical, Uniformly Heated Tube [97]

CHF correlation
$$q''_{crit} = q''_{co}\left(1 + K_K \frac{i_{l,sat} - i_{in}}{i_{lv}}\right)$$

$\rho^* = \rho_v/\rho_l \geq 0.15$	$\rho^* = \rho_v/\rho_l < 0.15$
$q''_{co} = q''_{co1}$ for $q''_{co1} < q''_{co5}$	$q''_{co} = q''_{co1}$ for $q''_{co1} < q''_{co2}$
$q''_{co} = q''_{co5}$ for $q''_{co1} > q''_{co5} > q''_{co4}$	$q''_{co} = q''_{co2}$ for $q''_{co1} > q''_{co2} < q''_{co3}$
$q''_{co} = q''_{co4}$ for $q''_{co1} > q''_{co5} \leq q''_{co4}$	$q''_{co} = q''_{co3}$ for $q''_{co2} \geq q''_{co3}$
$K_K = K_{K1}$ for $K_{K1} > K_{K2}$	$K_K = K_{K1}$ for $K_{K1} > K_{K2}$
$K_K = K_{K2}$ for $K_{K1} \leq K_{K2} < K_{K3}$	$K_K = K_{K2}$ for $K_{K1} \leq K_{K2}$
$K_K = K_{K3}$ for $K_{K1} \leq K_{K2} \geq K_{K3}$	

$$K_{K1} = \frac{1.043}{4C_K \, \mathrm{We}_K^{-0.043}} \quad K_{K2} = \frac{5}{6}\frac{0.0124 + d_i/L}{(\rho^*)^{0.133}\,\mathrm{We}_K^{-1/3}} \quad K_{K3} = 1.12 \frac{1.52\mathrm{We}_K^{-0.233} + d_i/L}{(\rho^*)^{0.6}\,\mathrm{We}_K^{-0.173}}$$

$$q''_{co1} = C_K Gi_{lv}\,\mathrm{We}_K^{-0.043}\,(L/d_i)^{-1} \quad q''_{co2} = 0.1Gi_{lv}(\rho^*)^{0.133}\,\mathrm{We}_K^{-1/3}\left[\frac{1}{1 + 0.0031(L/d_i)}\right]$$

$$q''_{co3} = 0.098Gi_{lv}(\rho^*)^{0.133}\,\mathrm{We}_K^{-0.433}\left[\frac{(L/d_i)^{0.27}}{1 + 0.0031(L/d_i)}\right] \quad q''_{co4} = 0.0384Gi_{lv}(\rho^*)^{0.6}\,\mathrm{We}_K^{-0.173}\left[\frac{1}{1 + 0.28\mathrm{We}_K^{-0.233}(L/d_i)}\right]$$

$$q''_{co5} = 0.234Gi_{lv}(\rho^*)^{0.513}\,\mathrm{We}_K^{-0.433}\left[\frac{(L/d_i)^{0.27}}{1 + 0.0031(L/d_i)}\right] \quad \mathrm{We}_K = \frac{G^2 L}{\sigma\rho_l}$$

$$C_K = 0.25 \quad \text{for } \frac{L}{d_i} < 50; \quad C_K = 0.25 + 0.0009\left[\left(\frac{L}{d_i}\right) - 50\right] \quad \text{for } 50 \leq \frac{L}{d_i} \leq 150; \quad C_K = 0.34 \quad \text{for } \frac{L}{d_i} > 150$$

Shellside Forced Boiling.   As is the case with the heat transfer coefficient, an accurate prediction of the CHF for shellside boiling is much more difficult than for internal flow. Available experimental data suggest that a correlation presented by Palen and Small in graphical form [98] is very conservative, but it is still the only one available for general use [68].

THERMAL DESIGN FOR SINGLE-PHASE HEAT EXCHANGERS

Exchanger Design Methodology

The problem of heat exchanger design is complex and multidisciplinary [99]. The major design considerations for a new heat exchanger include: process/design specifications, thermal and hydraulic design, mechanical design, manufacturing and cost considerations, and trade-offs and system-based optimization as shown in Fig. 17.53, with possible strong interactions among these considerations as indicated by double-sided arrows. The thermal and hydraulic designs are mainly analytical and the structural design is to some extent. Most of the other major design considerations involve qualitative and experience-based judgments, trade-offs, and compromises. Therefore, there is no unique solution to designing a heat exchanger for given process specifications. Further details on this design methodology are given by Ref. 99.

Two most important heat exchanger design problems are the rating and sizing problems. Determination of heat transfer and pressure drop performance of either an existing exchanger or an already sized exchanger is referred to as the *rating problem*. The objective here is to verify vendor specifications or determine the performance at off-design conditions. The rating problem is also sometimes referred to as the performance problem. In contrast, the design of a new or existing-type exchanger is referred to as the *sizing problem*. In a broad sense, it means the determination of the exchanger construction type, flow arrangement, heat transfer surface geometries and materials, and the physical size of an exchanger to meet the specified heat transfer and pressure drops. However, from the viewpoint of quantitative thermal-hydraulic analysis, we will consider that the selection of the exchanger construction type, flow arrangement, and materials has already been made. Thus, in the sizing problem, we will determine here the physical size (length, width, height) and surface areas on each side of the exchanger. The sizing problem is also sometimes referred to as the design problem.

The step-by-step solution procedures for the rating and sizing problems for counterflow and crossflow single-pass plate-fin heat exchangers have been presented with a detailed illustrative example by Ref. 100. Shah [101] presented further refinements in these procedures as well as step-by-step solution procedures for two-pass cross-counterflow plate-fin exchangers and single-pass crossflow and two-pass cross-counterflow tube-fin exchangers. Also, step-by-step solution procedures for the rating and sizing problems for rotary regenerators [32], heat pipe heat exchangers [102], and plate heat exchangers [103] are available. The well-established step-by-step solution procedures for two-fluid heat exchangers cannot be extended easily when more than two fluids are involved, such as in three-fluid or multifluid heat exchangers [104]. As an illustration, the step-by-step solution procedures will be covered here for a two-fluid single-pass crossflow exchanger.

Extended Surface Heat Exchangers

Rating Problem for a Crossflow Plate-Fin Exchanger.   We will present here a step-by-step solution procedure for the rating problem for a crossflow plate-fin exchanger. Inputs to the rating problem for a two-fluid exchanger are: the exchanger construction, flow arrangement and overall dimensions, complete details on the materials and surface geometries on both sides including their nondimensional heat transfer and pressure drop characteristics (j and f

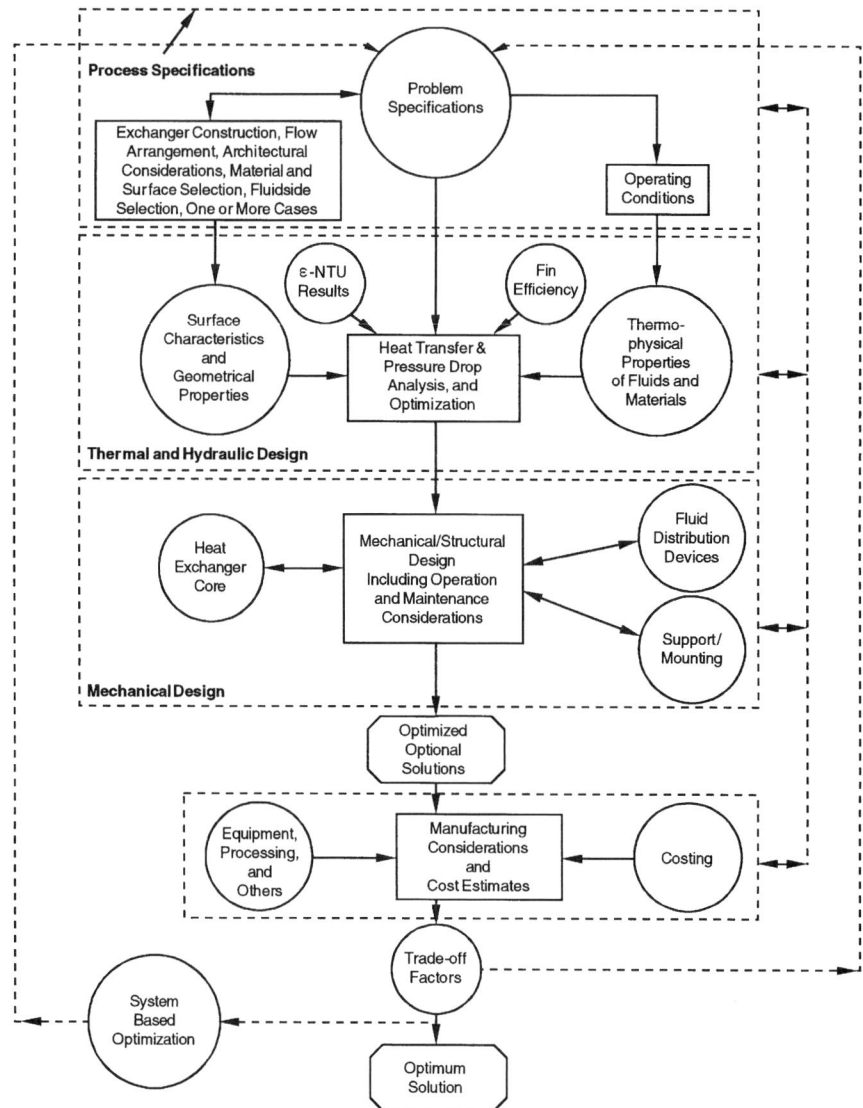

FIGURE 17.53   Heat exchanger overall design methodology.

versus Re), fluid flow rates, inlet temperatures, and fouling factors. The fluid outlet temperatures, total heat transfer rate, and pressure drops on each side of the exchanger are then determined as the rating problem solution.

1. Determine the surface geometrical properties on each fluid side. This includes the minimum free flow area A_o, heat transfer surface area A (both primary and secondary), flow
   lengths L, hydraulic diameter D_h, heat transfer surface area density β, the ratio of minimum free flow area to frontal area σ, fin length and fin thickness δ_f for fin efficiency determination, and any specialized dimensions used for heat transfer and pressure drop
   correlations.

2. Compute the fluid bulk mean temperature and fluid thermophysical properties on each fluid side. Since the outlet temperatures are not known for the rating problem, they are estimated initially. Unless it is known from past experience, assume an exchanger effectiveness as 60–75 percent for most single-pass crossflow exchangers or 80–85 percent for single-pass counterflow exchangers. For the assumed effectiveness, calculate the fluid outlet temperatures.

$$T_{h,o} = T_{h,i} - \varepsilon (C_{\min}/C_h)(T_{h,i} - T_{c,i}) \tag{17.117}$$

$$T_{c,o} = T_{c,i} + \varepsilon (C_{\min}/C_c)(T_{h,i} - T_{c,i}) \tag{17.118}$$

Initially, assume $C_c/C_h = \dot{m}_c/\dot{m}_h$ for a gas-to gas exchanger, or $C_c/C_h = \dot{m}_c c_{p,c}/\dot{m}_h c_{p,h}$ for a gas-to-liquid exchanger with very approximate values of c_p for the fluids in question.

For exchangers with $C^* \geq 0.5$ (usually gas-to-gas exchangers), the bulk mean temperatures on each fluid side will be the arithmetic mean of the inlet and outlet temperatures on each fluid side [100]. For exchangers with $C^* < 0.5$ (usually gas-to-gas exchangers), the bulk mean temperature on the $C_{\max}$ side will be the arithmetic mean of inlet and outlet temperatures; the bulk mean temperature on the $C_{\min}$ side will be the log-mean average temperature obtained as follows:

$$T_{m,C_{\min}} = T_{m,C_{\max}} \pm \Delta T_{lm} \tag{17.119}$$

where ΔT_{lm} is the log-mean temperature difference based on the terminal temperatures (see Eq. 17.18); use the plus sign only if the $C_{\min}$ side is hot.

Once the bulk mean temperatures are obtained on each fluid side, obtain the fluid properties from thermophysical property software or handbooks. The properties needed for the rating problem are μ, c_p, k, Pr, and ρ. With this c_p, one more iteration may be carried out to determine $T_{h,o}$ or $T_{c,o}$ from Eq. 17.117 or 17.118 on the $C_{\max}$ side and, subsequently, T_m on the $C_{\max}$ side. Refine fluid properties accordingly.

3. Calculate the Reynolds number $Re = GD_h/\mu$ and/or any other pertinent dimensionless groups (from the basic definitions) needed to determine the nondimensional heat transfer and flow friction characteristics (e.g., j or Nu and f) of heat transfer surfaces on each side of the exchanger. Subsequently, compute j or Nu and f factors. Correct Nu (or j) for variable fluid property effects [100] in the second and subsequent iterations from the following equations.

For gases:
$$\frac{\text{Nu}}{\text{Nu}_{cp}} = \left[\frac{T_w}{T_m} \right]^n \qquad \frac{f}{f_{cp}} = \left[\frac{T_w}{T_m} \right]^m \tag{17.120}$$

For liquids:
$$\frac{\text{Nu}}{\text{Nu}_{cp}} = \left[\frac{\mu_w}{\mu_m} \right]^n \qquad \frac{f}{f_{cp}} = \left[\frac{\mu_w}{\mu_m} \right]^m \tag{17.121}$$

where the subscript cp denotes constant properties, and m and n are empirical constants provided in Table 17.20a and 17.20b. Note that T_w and T_m in Eqs. 17.120 and 17.121 and in Table 17.20a and 17.20b are absolute temperatures, and T_w is computed from Eq. 17.9.

4. From Nu or j, compute the heat transfer coefficients for both fluid streams.

$$h = \text{Nu } k/D_h = jGc_p \text{ Pr}^{-2/3} \tag{17.122}$$

Subsequently, determine the fin efficiency η_f and the extended surface efficiency η_o:

$$\eta_f = \frac{\tanh ml}{ml}$$

where
$$m^2 = \frac{h\mathbf{P}}{k_f A_k} \tag{17.123}$$

where **P** is the wetted perimeter of the fin surface.

$$\eta_o = 1 - (1 - \eta_f)A_f/A \tag{17.124}$$

Also calculate the wall thermal resistance $\mathbf{R}_w = \delta/A_w k_w$. Finally, compute overall thermal conductance UA from Eq. 17.6, knowing the individual convective film resistances, wall thermal resistances, and fouling resistances, if any.

5. From the known heat capacity rates on each fluid side, compute $C^* = C_{min}/C_{max}$. From the known UA, determine NTU $= UA/C_{min}$. Also calculate the longitudinal conduction parameter λ. With the known NTU, C^*, λ, and the flow arrangement, determine the crossflow exchanger effectiveness (from either closed-form equations of Table 17.6 or tabular/graphical results from Kays and London [20].

6. With this ε, finally compute the outlet temperatures from Eqs. 17.117 and 17.118. If these outlet temperatures are significantly different from those assumed in step 2, use these outlet temperatures in step 2 and continue iterating steps 2–6 until the assumed and computed outlet temperatures converge within the desired degree of accuracy. For a gas-to-gas exchanger, one or two iterations may be sufficient.

7. Finally compute the heat duty from

$$q = \varepsilon C_{min}(T_{h,i} - T_{c,i}) \tag{17.125}$$

8. For the pressure drop calculations, first we need to determine the fluid densities at the exchanger inlet and outlet (ρ_i and ρ_o) for each fluid. The mean specific volume on each fluid side is then computed from Eq. 17.66.

    Next, the entrance and exit loss coefficients K_c and K_e are obtained from Fig. 17.35 for known σ, Re, and the flow passage entrance geometry.

    The friction factor on each fluid side is corrected for variable fluid properties using Eq. 17.120 or 17.121. Here, the wall temperature T_w is computed from

$$T_{w,h} = T_{m,h} - (\mathbf{R}_h + \mathbf{R}_{s,h})q \tag{17.126}$$

$$T_{w,c} = T_{m,c} + (\mathbf{R}_c + \mathbf{R}_{s,c})q \tag{17.127}$$

where the various resistance terms are defined by Eq. 17.6.

    The core pressure drops on each fluid side are then calculated from Eq. 17.65. This then completes the procedure for solving the rating problem.

Sizing Problem for Plate-Fin Exchangers.   As defined earlier, we will concentrate here to determine the physical size (length, width, and height) of a single-pass crossflow exchanger for specified heat duty and pressure drops. More specifically inputs to the sizing problem are surface geometries (including their nondimensional heat transfer and pressure drop characteristics), fluid flow rates, inlet and outlet fluid temperatures, fouling factors, and pressure drops on each side.

    For the solution to this problem, there are four unknowns—two flow rates or Reynolds numbers (to determine correct heat transfer coefficients and friction factors) and two surface areas—for the two-fluid crossflow exchanger. Equations 17.128, 17.129, 17.130 for $q = 1, 2$, and 17.132 are used to solve iteratively the surface areas on each fluid side: UA in Eq. 17.128 is determined from NTU computed from the known heat duty or ε and C^*; G in Eq. 17.130 represents two equations for fluids 1 and 2 [101]; and the volume of the exchanger in Eq. 17.132 is the same based on the surface area density of fluid 1 (hot) or fluid 2 (cold).

$$\frac{1}{UA} \approx \frac{1}{(\eta_o hA)_h} + \frac{1}{(\eta_o hA)_c} \tag{17.128}$$

Here, we have neglected the wall and fouling thermal resistances. Defining $ntu_h = (\eta_o hA)_h/C_h$ and $ntu_c = (\eta_o hA)_c/C_c$, Eq. 17.128 in nondimensional form is given by

$$\frac{1}{NTU} = \frac{1}{ntu_h(C_h/C_{min})} + \frac{1}{ntu_c(C_c/C_{min})} \tag{17.129}$$

$$G_q = \left[\frac{2g_c\Delta p}{Deno}\right]_q^{1/2} \qquad q = 1, 2 \tag{17.130}$$

where

$$(Deno)_q = \left[\frac{f}{j}\frac{ntu}{\eta_o}Pr^{2/3}\left(\frac{1}{\rho}\right)_m + 2\left(\frac{1}{\rho_o} - \frac{1}{\rho_i}\right) + (1 - \sigma^2 + K_c)\frac{1}{\rho_i} - (1 - \sigma^2 - K_e)\frac{1}{\rho_o}\right]_q \tag{17.131}$$

$$V = \frac{A_1}{\alpha_1} = \frac{A_2}{\alpha_2} \tag{17.132}$$

In the iterative solutions, one needs ntu_h and ntu_c to start the iterations. These can be determined either from the past experience or by estimations. If both fluids are gases or liquids, one could consider that the design is balanced (i.e., the thermal resistances are distributed approximately equally on the hot and cold sides). In that case, $C_h = C_c$, and

$$ntu_h \approx ntu_c \approx 2NTU \tag{17.133}$$

Alternatively, if we have liquid on one side and gas on the other side, consider 10 percent thermal resistance on the liquid side, i.e.

$$0.10\left(\frac{1}{UA}\right) = \frac{1}{(\eta_o hA)_{liq}} \tag{17.134}$$

Then, from Eqs. 17.128 and 17.129 with $C_{gas} = C_{min}$, we can determine the ntu on each side as follows.

$$ntu_{gas} = 1.11NTU, \qquad ntu_{liq} = 10C^*NTU \tag{17.135}$$

Also note that initial guesses of η_o and j/f are needed for the first iteration to solve Eq. 17.131. For a good design, consider $\eta_o = 0.80$ and determine approximate value of j/f from the plot of j/f versus Re curve for the known j and f versus Re characteristics of each fluid side surface. The specific step-by-step design procedure is as follows.

1. In order to compute the fluid bulk mean temperature and the fluid thermophysical properties on each fluid side, determine the fluid outlet temperatures from the specified heat duty.

$$q = (\dot{m}c_p)_h(T_{h,i} - T_{h,o}) = (\dot{m}c_p)_c(T_{c,o} - T_{c,i}) \tag{17.136}$$

or from the specified exchanger effectiveness using Eqs. 17.117 and 17.118. For the first time, estimate the values of c_p.

For exchangers with $C^* \geq 0.5$, the bulk mean temperature on each fluid side will be the arithmetic mean of inlet and outlet temperatures on each side. For exchangers with $C^* < 0.5$, the bulk mean temperature on the C_{max} side will be the arithmetic mean of the inlet and outlet temperatures on that side, the bulk mean temperature on the C_{min} side will be the log-mean average as given by Eq. 17.119. With these bulk mean temperatures, determine c_p and iterate one more time for the outlet temperatures if warranted. Subsequently, determine μ, c_p, k, Pr, and ρ on each fluid side.

2. Calculate C^* and ε (if q is given) and determine NTU from the ε-NTU expression, tables, or graphical results for the selected crossflow arrangement (in this case, it is unmixed-unmixed crossflow, Table 17.6). The influence of longitudinal heat conduction, if any, is ignored in the first iteration, since we don't know the exchanger size yet.

3. Determine ntu on each side by the approximations discussed with Eqs. 17.133 and 17.135 unless it can be estimated from the past experience.

4. For the selected surfaces on each fluid side, plot j/f versus Re curve from the given surface characteristics, and obtain an approximate value of j/f. If fins are employed, assume $\eta_o = 0.80$ unless a better value can be estimated.

5. Evaluate G from Eq. 17.130 on each fluid side using the information from steps 1–4 and the input value of Δp.

6. Calculate Reynolds number Re, and determine j and f for this Re on each fluid side from the given design data for each surface.

7. Compute h, η_f, and η_o using Eqs. 17.122–17.124. For the first iteration, determine U_1 on the fluid 1 side from the following equation derived from Eqs. 17.6 and 17.132.

$$\frac{1}{U_1} = \frac{1}{(\eta_o h)_1} + \frac{1}{(\eta_o h_s)_1} + \frac{\alpha_1/\alpha_2}{(\eta_o h_s)_2} + \frac{\alpha_1/\alpha_2}{(\eta_o h)_2} \tag{17.137}$$

where $\alpha_1/\alpha_2 = A_1/A_2$, $\alpha = A/V$ and V is the exchanger total volume, and subscripts 1 and 2 denote the fluid 1 and 2 sides. For a plate-fin exchanger, α's are given by [20, 100]:

$$\alpha_1 = \frac{b_1 \beta_1}{b_1 + b_2 + 2\delta} \qquad \alpha_2 = \frac{b_2 \beta_2}{b_1 + b_2 + 2\delta} \tag{17.138}$$

Note that the wall thermal resistance in Eq. 17.137 is ignored in the first iteration. In the second and subsequent iterations, compute U_1 from

$$\frac{1}{U_1} = \frac{1}{(\eta_o h)_1} + \frac{1}{(\eta_o h_s)_1} + \frac{\delta A_1}{k_w A_w} + \frac{A_1/A_2}{(\eta_o h_s)_2} + \frac{A_1/A_2}{(\eta_o h)_2} \tag{17.139}$$

where the necessary geometry information A_1/A_2 and A_1/A_w is determined from the geometry calculated in the previous iteration.

8. Now calculate the core dimensions. In the first iteration, use NTU computed in step 2. For subsequent iterations, calculate longitudinal conduction parameter λ and other dimensionless groups for a crossflow exchanger. With known ε, C^*, and λ, determine the correct value of NTU using either a closed-form equation or tabulated/graphical results [10]. Determine A_1 from NTU using U_1 from previous step and known C_{min}.

$$A_1 = \text{NTU} C_{min}/U_1 \tag{17.140}$$

and hence
$$A_2 = (A_2/A_1)A_1 = (\alpha_2/\alpha_1)A_1 \tag{17.141}$$

A_o is derived from known $\dot{m}$ and G as

$$A_{o,1} = (\dot{m}/G)_1 \qquad A_{o,2} = (\dot{m}/G)_2 \tag{17.142}$$

so that
$$A_{fr,1} = A_{o,1}/\sigma_1 \qquad A_{fr,2} = A_{o,2}/\sigma_2 \tag{17.143}$$

where σ_1 and σ_2 are generally specified for the surface or can be computed for plate-fin surfaces from [20, 100]:

$$\sigma_1 = \frac{b_1 \beta_1 D_{h,1}/4}{b_1 + b_2 + 2\delta} \qquad \sigma_2 = \frac{b_2 \beta_2 D_{h,2}/4}{b_1 + b_2 + 2\delta} \tag{17.144}$$

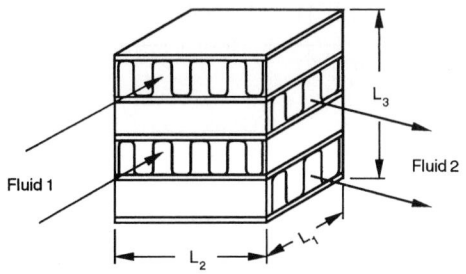

FIGURE 17.54 A single-pass crossflow exchanger.

Now compute the fluid flow lengths on each side (see Fig. 17.54) from the definition of the hydraulic diameter of the surface employed on each side.

$$L_1 = \left(\frac{D_h A}{4A_o}\right)_1 \qquad L_2 = \left(\frac{D_h A}{4A_o}\right)_2 \qquad (17.145)$$

Since $A_{fr,1} = L_2 L_3$ and $A_{fr,2} = L_1 L_3$, we can obtain

$$L_3 = \frac{A_{fr,1}}{L_2} \quad \text{or} \quad L_3 = \frac{A_{fr,2}}{L_1} \qquad (17.146)$$

Theoretically, L_3 calculated from both expressions of Eq. 17.146 should be identical. In reality, they may differ slightly due to the round-off error. In that case, consider an average value for L_3.

9. Now compute the pressure drop on each fluid side, after correcting f factors for variable property effects, in a manner similar to step 8 of the rating problem for the crossflow exchanger.

10. If the calculated values of Δp are close to input specifications, the solution to the sizing problem is completed. Finer refinements in the core dimensions such as integer numbers of flow passages may be carried out at this time. Otherwise, compute the new value of G on each fluid side using Eq. 17.65 in which Δp is the input specified value and f, K_c, K_e, and geometrical dimensions are from the previous iteration.

11. Repeat (iterate) steps 6–10 until both heat transfer and pressure drops are met as specified. It should be emphasized that, since we have imposed no constraints on the exchanger dimensions, the above procedure will yield L_1, L_2, and L_3 for the selected surfaces such that the design will meet the heat duty and pressure drops on both fluid sides exactly.

Shell-and-Tube Heat Exchangers

The design of a shell-and-tube heat exchanger is more complex than the plate-fin and tube-fin exchangers. There are many variables associated with the geometry (i.e., shell, baffles, tubes, front and rear end, and heads) and operating conditions including flow bypass and leakages in a shell-and-tube heat exchanger [5]. There are no systematic quantitative correlations available to take into account the effect of these variables on the exchanger heat transfer and pressure drop. As a result, the common practice is to presume the geometry of the exchanger and determine the tube (shell) length for the sizing problem or do the rating calculations for the given geometry to determine the heat duty, outlet temperatures, and pressure drops. Hence, effectively, the rating calculations are done for the determination of the heat duty or the exchanger length; in both cases, the basic exchanger geometry is specified. The design calculations are essentially a series of iterative rating calculations made on an assumed design and modified as a result of these calculations until a satisfactory design is achieved.

The following is a step-by-step procedure for the "sizing" problem in which we will determine the exchanger (shell-and-tube) length.

The key steps of the thermal design procedure for a shell-and-tube heat exchanger are as follows:

1. For a given (or calculated) heat transfer rate (required duty), compute (or select) the fluid streams inlet and/or outlet temperatures using overall energy balances and specified (or selected) fluid mass flow rates.

2. Select a preliminary flow arrangement (i.e., a type of the shell-and-tube heat exchanger based on the common industry practice).

TABLE 17.28   Shell-and-Tube Overall Heat Transfer Coefficient, Modified from Ref. 115

Cold-side fluid	Hot-side fluid U, W/(m²K)*							
	Gas @ 10^5 Pa	Gas @ 2×10^6 Pa	Process H_2O	Organic liquid[†]	Viscous liquid[‡]	Condensing steam	Condensing hydrocarbon	Condensing hydrocarbon and inert gas
Gas @ 10^5 Pa	55	93	102	99	63	107	100	86
Gas @ 2×10^6 Pa	93	300	429	375	120	530	388	240
H_2O, treated	105	484	938	714	142	1607	764	345
Organic liquid[†]	99	375	600	500	130	818	524	286
High-viscosity liquid[‡]	68	138	161	153	82	173	155	124
H_2O, boiling	105	467	875	677	140	1432	722	336
Organic liquid, boiling[§]	99	375	600	500	130	818	524	286

* Based on data given in [G.F. Hewit, A.R. Guy, and R. Marsland, Heat Transfer Equipment, Ch. 3 in *A User Guide on Process Integration for the Efficient Use of Energy*, eds. B. Linnhoff et al., The IChemE, Rugby, 1982]. Any such data, including the data given in this table, should be used with caution. The numbers are based on empirical data and should be considered as mean values for corresponding data ranges. Approximate values for boiling and condensation are given for convenience.
  [†] Viscosity range 1 to 5 mPa s.
  [‡] Viscosity range > 100 mPa s.
  [§] Viscosity typically < 1 mPa s.

3. Estimate an overall heat transfer coefficient using appropriate empirical data (see, for example, Table 17.28).

4. Determine a first estimate of the required heat transfer area using Eq. 17.17 (i.e., using a first estimate of the log-mean temperature difference ΔT_{lm} and the correction factor F, the estimated overall heat transfer coefficient U, and given heat duty q). Good design practice is to assume $F = 0.8$ or a higher value based on past practice. Based on the heat transfer area, the mass flow rates, and the process conditions, select suitable types of exchangers for analysis (see Refs. 106 and 109). Determine whether a multipass exchanger is required.

5. Select tube diameter, length, pitch, and layout. Calculate the number of tubes, the number of passes, shell size, and baffle spacing. Select the tentative shell diameter for the chosen heat exchanger type using manufacturer's data. The preliminary design procedure presented on p. 17.116 can be used to select these geometrical parameters.

6. Calculate heat transfer coefficients and pressure drops using the Bell-Delaware Method [105] or the stream analysis method [106].

7. Calculate a new value of the overall heat transfer coefficient.

8. Compare the calculated values for the overall heat transfer coefficient (obtained in step 7) with the estimated value of the overall heat transfer coefficient (step 3), and similarly calculated pressure drops (obtained in step 6) with allowable values for pressure drops.

9. Inspect the results and judge whether the performance requirements have been met.

10. Repeat, if necessary, steps 5 to 9 with an estimated change in design until a final design is reached that meets, for instance, specified q and Δp, requirements. If it cannot, then one may need to go back to step 2 for iteration. At this stage, an engineer should check for meeting TEMA standards, ASME Pressure Vessel Codes (and/or other pertinent standards and/or codes as appropriate), potential operating problems, cost, and so on; if the design change is warranted, iterate steps 5 to 9 until the design meets thermal/hydraulic and other requirements.

This step-by-step procedure is consistent with overall design methodology and can be executed as a straightforward manual method or as part of a computer routine. Although the actual design has been frequently carried out using available sophisticated commercial software, a successful designer ought to know all the details of the procedure in order to interpret and assess the results from the commercial software.

The central part of thermal design procedure involves determination of heat transfer and pressure drops. A widely utilized, most accurate method in the open literature is the well-known Bell-Delaware method [105] that takes into account various flow characteristics of the complex shellside flow. The method was developed originally for design of fully tubed E-shell heat exchangers with nonenhanced tubes based on the experimental data obtained for an exchanger with geometrical parameters closely controlled. It should be noted that this method can be applied to the broader range of applications than originally intended. For example, it can be used to design J-shell or F-shell heat exchangers. Also, an external low-finned tubes design can easily be considered [105, 106].

Bell-Delaware Method. Pressure drop and heat transfer calculations (the step 6 of the above thermal design procedure) constitute the key part of design. Tubeside calculations are straightforward and should be executed using available correlations for internal forced convection. The shellside calculations, however, must take into consideration the effect of various leakage streams (A and E streams in Fig. 17.30) and bypass streams (C and F streams in Fig. 17.30) in addition to the main crossflow stream B through the tube bundle. Several methods have been in use over the years, but the most accurate method in the open literature is the above mentioned Bell-Delaware method. This approach is based primarily on limited experimental data. The set of correlations discussed next constitutes the core of the Bell-Delaware method.

Heat Transfer Coefficients. In this method, an actual heat transfer coefficient on the shellside h_s is determined, correcting the ideal heat transfer coefficient h_{ideal} for various leakage and bypass flow streams. The h_{ideal} is determined for pure crossflow in an ideal tubebank, assuming the entire shellside stream flows across the tubebank at or near the centerline of the shell. The correction factor is defined as a product of five correction factors $J_1, J_2, \ldots J_5$ that take into account, respectively, the effects of:

- Baffle cut and baffle spacing ($J_1 = 1$ for an exchanger with no tubes in the window and increases to 1.15 for small baffle cuts and decreases to 0.65 for large baffle cuts)
- Tube-to-baffle and baffle-to-shell leakages (A and E streams, Fig. 17.30); a typical value of J_2 is in the range of 0.7–0.8
- Tube bundle bypass and pass partition bypass (C and F streams, Fig. 17.30); a typical value of J_3 is in the range 0.7–0.9
- Laminar flow temperature gradient buildup (J_4 is equal to 1.0 except for shellside Reynolds numbers smaller than 100)
- Different central versus end baffle spacings (J_5 usually ranges from 0.85 to 1.0)

A complete set of equations and parameters for the calculation of the shellside heat transfer coefficient is given in Tables 17.29 and 17.30.

A combined effect of all five corrections can reduce the ideal heat transfer coefficient by up to 60 percent. A comparison with a large number of proprietary experimental data indicates the shellside h predicted using all correction factors is from 50 percent too low to 200 percent too high with a mean error of 15 percent low (conservative) at all Reynolds numbers.

Pressure Drops. Shellside pressure drop has three components: (1) pressure drop in the central (crossflow) section Δp_c, (2) pressure drop in the window area Δp_w, and (3) pressure drop in the shell side inlet and outlet sections, $\Delta p_{i\text{-}o}$. It is assumed that each of the three components is based on the total flow and that each component can be calculated by correcting the corresponding ideal pressure drops.

The ideal pressure drop in the central section Δp_{bi} assumes pure crossflow of the fluid across the ideal tube bundle. This pressure drop should be corrected for: (a) leakage streams (A and E, Fig. 17.30; correction factor R_ℓ), and (b) bypass flow (streams C and F, Fig. 17.30;

TABLE 17.29   The Heat Transfer Coefficient on the Shell Side, Bell-Delaware Method

Shell-side heat transfer coefficient h_s

$$h_s = h_{ideal}J_1J_2J_3J_4J_5$$

$$h_{ideal} = j_i c_p G_s\, \text{Pr}_s^{-2/3}\, \phi_s$$

$$\phi_s = \begin{cases} (\mu_s/\mu_w)^{0.14} & \text{for liquid} \\ 1 & \text{for gas (cooled)} \\ (T_s/T_w)^{0.25} & \text{for gas (heated)} \end{cases}$$

$$j_i = j_i(\text{Re}_s, \text{tube layout, pitch})$$

$$\mu_w = \mu_{T_w},\ T_s = \overline{T}_s,\ T_s \text{ and } T_w \text{ in } [K]$$

$$G_s = \frac{\dot{m}_s}{A_{mb}} \quad \text{Re}_s = \frac{d_o G_s}{\mu_s}$$

$j_i = j$ from Figs. 17.55–17.57 or alternately from correlations as those given in Table 17.19*

$$J_1 = 0.55 + 0.72F_c$$

F_c from Table 17.30

$$J_2 = 0.44(1 - r_s) + [1 - 0.44(1 - r_s)]\exp(-2.2r_{lm})$$

$$r_s = \frac{A_{sb}}{A_{sb} + A_{tb}} \quad r_{lm} = \frac{A_{sb} + A_{tb}}{A_{mb}}$$

A_{sb}, A_{tb}, A_{mb} from Table 17.30

$$r_b = \frac{A_{ba}}{A_{mb}} \quad N_{ss}^+ = \frac{N_{ss}}{N_{tcc}}$$

$$J_3 = 1 \qquad\qquad\qquad \text{for } N_{ss}^+ \geq \tfrac{1}{2}$$
$$J_3 = \exp\{-Cr_b[1 - (2N_{ss}^+)^{1/3}]\} \quad \text{for } N_{ss}^+ < \tfrac{1}{2}$$

A_{ba}, N_{ss}, N_{tcc} from Table 17.30
$C = 1.35$ for $\text{Re}_s \leq 100$
$C = 1.25$ for $\text{Re}_s > 100$

$$J_4 = \begin{cases} 1 & \text{Re}_s > 100 \\ (10/N_c)^{0.18} & \text{Re}_s \leq 20 \end{cases}$$

$N_c = N_{tcc} + N_{tcw}$
N_{tcw} from Table 17.30
Linear interpolation for $20 < \text{Re}_s < 100$

$$J_5 = \frac{N_b - 1 + (L_i^+)^{(1-n)} + (L_o^+)^{(1-n)}}{N_b - 1 + L_i^+ + L_o^+}$$

$$L_i^+ = \frac{L_{bi}}{L_{bc}} \quad L_o^+ = \frac{L_{bo}}{L_{bc}} \quad N_b = \frac{L_{ti}}{L_{bc}} - 1$$

L_{bi}, L_{bo}, L_{bc}, and L_{ti} from Table 17.30
$n = 0.6$ (turbulent flow)

* A number of accurate correlations such as those given in Table 17.19 are available. Traditionally, the diagrams such as those given in Figs. 17.55–17.57 have been used in engineering practice.

correction factor R_b). The ideal window pressure drop Δp_w has also to be corrected for both baffle leakage effects. Finally, the ideal inlet and outlet pressure drops $\Delta p_{i\text{-}o}$ are based on an ideal crossflow pressure drop in the central section. These pressure drops should be corrected for bypass flow (correction factor R_b) and for effects of uneven baffle spacing in inlet and outlet sections (correction factor R_s). Typical correction factor ranges are as follows:

- Baffle leakage effects (i.e., tube-to-baffle and baffle-to-shell leakages, A and E streams, Fig. 17.30); a typical value of R_ℓ is in the range of 0.4–0.5
- Tube bundle and pass partition bypass flow effects (i.e., streams C and F, Fig. 17.30); a typical value of R_b is in the range of 0.5–0.8
- The inlet and outlet baffle spacing effects correction factor R_s, in the range of 0.7–1

The complete set of equations, including the correcting factors, is given in Table 17.31.

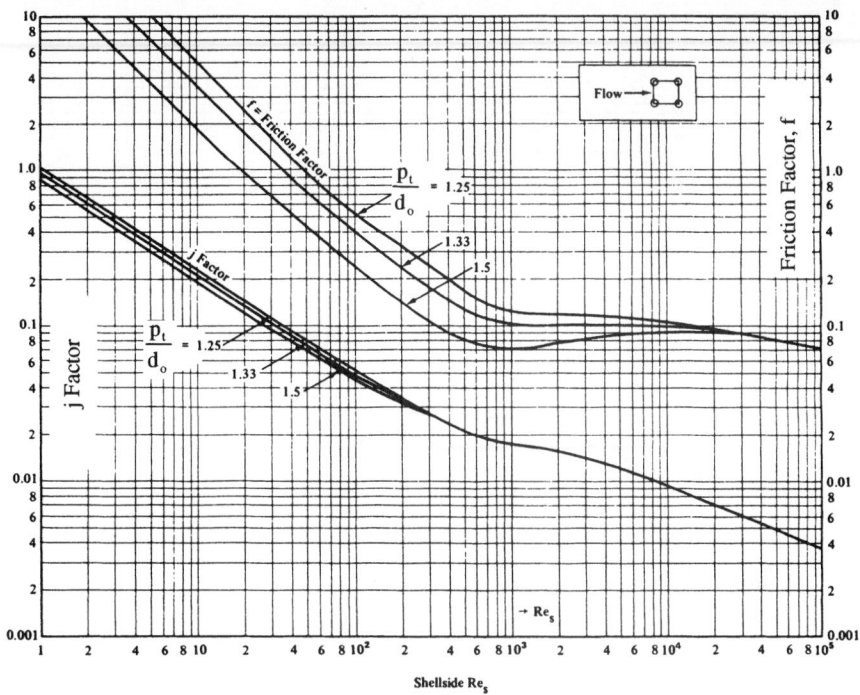

FIGURE 17.55 Colburn factors and friction factors for ideal crossflow in tube bundles, 90° inline layout [106].

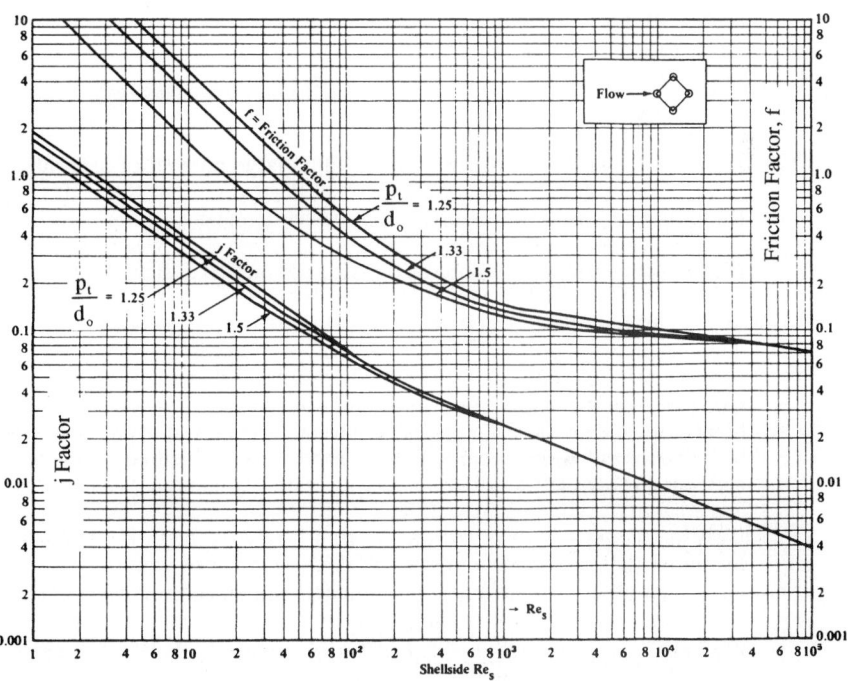

FIGURE 17.56 Colburn factors and friction factors for ideal crossflow in tube bundles, 45° staggered layout [106].

17.115

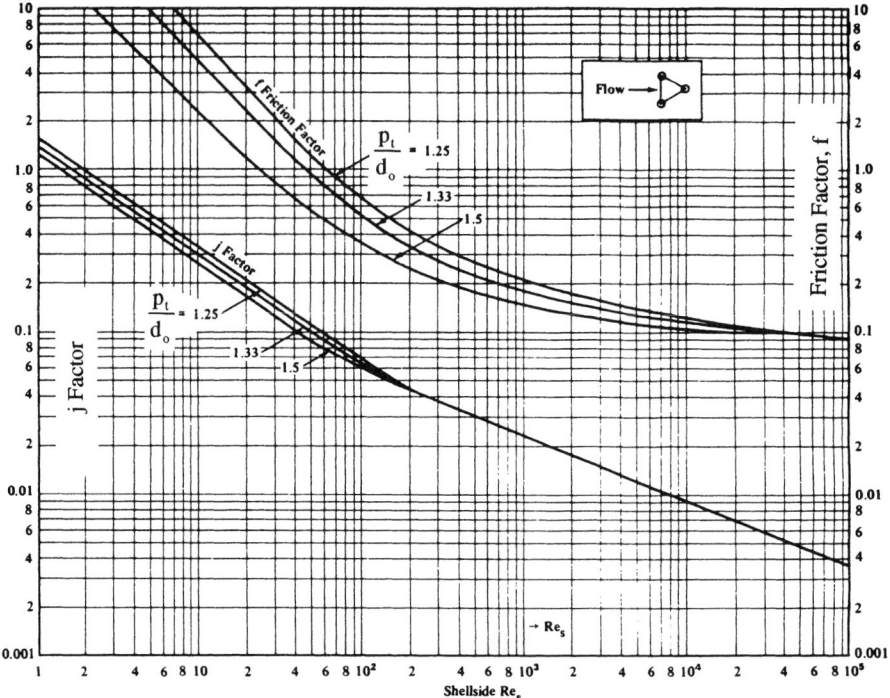

FIGURE 17.57   Colburn factors and friction factors for ideal crossflow in tube bundles, 30° staggered layout [106].

The combined effect of pressure drop corrections reduces the ideal total shellside pressure drop by 70–80 percent. A comparison with a large number of proprietary experimental data indicate shellside Δp from about 5 percent low (unsafe) at $Re_s > 1000$ to 100 percent high at $Re_s < 10$. The tubeside pressure drop is calculated using Eq. 17.65 for single-phase flow.

Preliminary Design.   A state-of-the-art approach to design of heat exchangers assumes utilization of computer software, making any manual method undoubtedly inferior. For a review of available computer software, consult Ref. 107. The level of sophistication of the software depends on whether the code is one-, two-, or three-dimensional. The most complex calculations involve full-scale CFD (computational fluid dynamics) routines. The efficiency of the software though is not necessarily related to the complexity of the software because of a need for empirical data to be incorporated into design and sound engineering judgment due to the lack of comprehensive empirical data. The design of shell-and-tube heat exchangers is more accurate for a variety of fluids and applications by commercial software than any other heat exchanger type [108] because of its verification by extensive experimental data.

A successful design based on the Bell-Delaware method obviously depends to a great extent on the experience and skills of the designer. An important component of the experience is an ability to perform a preliminary estimate of the exchanger configuration and its size. A useful tool in accomplishing this task is an approximate sizing of a shell-and-tube heat exchanger. Brief details of this procedure according to Ref. 109 follow. The procedure is based on the MTD method.

TABLE 17.30 Shell-and-Tube Geometric Characteristics to Accompany Tables 17.29 and 17.31

Shellside geometry*

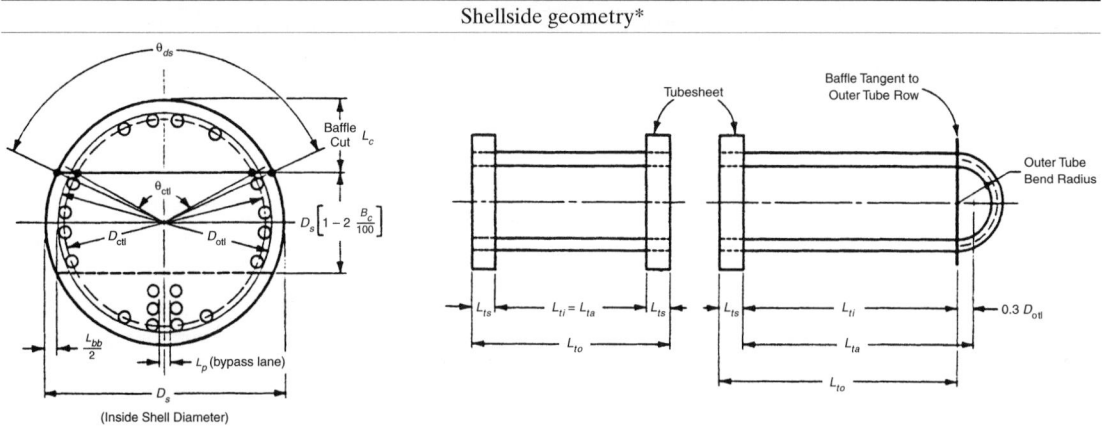

$$A_{mb} = L_{bc}\left[L_{bb} + \frac{D_{ctl}}{p_{t,\text{eff}}} (p_t - d_o) \right] \qquad D_{ctl} = D_{otl} - d_o \qquad p_{t,\text{eff}} = \xi p_t \qquad \begin{array}{l} \xi = 1 \text{ for } 30^\circ \text{ and } 90^\circ \\ \xi = 0.707 \text{ for } 45^\circ \text{ layout} \end{array}$$

$$F_c = 1 - 2F_w \qquad F_w = \frac{\theta_{ctl}}{360^\circ} - \frac{\sin \theta_{ctl}}{2\pi} \qquad \theta_{ctl} = 2 \cos^{-1}\left[\frac{D_s}{D_{ctl}}\left(1 - 2\frac{B_c}{100}\right) \right] \qquad B_c = \frac{\text{Baffle cut}}{D_s} \times 100$$

$$A_{sb} = \pi D_s \frac{L_{sb}}{2} \frac{360^\circ - \theta_{ds}}{360^\circ} \qquad \text{see Ref. 5 for allowable } L_{sb} \text{ and } L_{tb} \qquad \theta_{ds} = 2 \cos^{-1}\left[1 - 2\frac{B_c}{100} \right]$$

$$A_{tb} = \frac{\pi}{4}\left[(d_o + L_{tb})^2 - d_o^2 \right] N_t (1 - F_w) \qquad A_{ba} = L_{bc}[(D_s - D_{otl}) + L_{pl}] \qquad L_{pl} = \begin{cases} 0 & \text{standard} \\ \tfrac{1}{2}d_o & \text{estimation} \end{cases}$$

$$N_{tcc} = \frac{D_s}{X_\ell}\left(1 - 2\frac{B_c}{100}\right) \qquad N_{tcw} = 0.8\frac{L_{cp}}{X_\ell} \qquad N_{ss} = 1 \text{ per 4 or 6 tube rows crossed}$$

$$A_w = A_{wg} - A_{wt} \qquad A_{wg} = \frac{\pi}{4} D_s^2\left(\frac{\theta_{ds}}{360^\circ} - \frac{\sin \theta_{ds}}{2\pi} \right) \qquad A_{wt} = N_t F_w \frac{\pi d_o^2}{4}$$

Note: Specification of the shell-side geometry provided in this table follows (with a few exceptions) the notation adopted in Ref. 7. Somewhat different approach is provided in Ref. 105. Refer to Ref. 106 for further details.

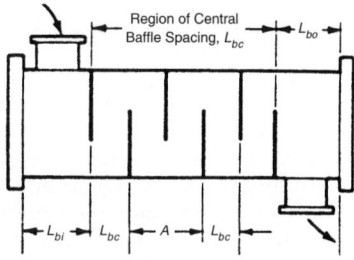

---

* A proper set of units should be used for calculating data in Tables 17.29, 17.30, and 17.31. If using SI units, refer for further details to Ref. 106; if using U.S. Engineering units, refer to Ref. 5.

TABLE 17.31 Shellside Pressure Drop, Bell-Delaware Method

Shellside pressure drop Δp_s^*	
$\Delta p_s = \Delta p_c + \Delta p_w + \Delta p_{i-o}$	

$\Delta p_c = \Delta p_{bi}(N_b - 1)R_b R_l$

$N_b = \dfrac{L_{ti}}{L_{bc}} - 1$

$\Delta p_{bi} = 2fN_{tcc}\dfrac{G_s^2}{g_c\rho_s}\phi_s$

L_{ti}, L_{bc} from Table 17.30;

$f = f(\text{Re}_s, \text{tube layout, pitch})$

f from Figs. 17.55–17.57[†]
$\text{Re}_s, G_s, \phi_s, N_{ss}^+$ defined in
Table 17.29

$R_b = \exp\{-\mathbf{D}r_b[1 - (2N_{ss}^+)^{1/3}]\}$ for $N_{ss}^+ < 1/2$

N_{tcc} from Table 17.30
r_b, r_{lm}, r_s from Table 17.29

$R_l = \exp[-1.33(11 + r_s)(r_{lm})^p]$

$p = [-0.15(1 + r_s) + 0.8]$
$R_b = 1$ at $N_{ss}^+ \geq 1/2$

$\mathbf{D} = 4.5$ for $\text{Re}_s \leq 100$; $\mathbf{D} = 3.7$ for $\text{Re}_s > 100$

$\Delta p_w = \begin{cases} N_b(2 + 0.6N_{tcw})\dfrac{G_w^2}{2g_c\rho_s}R_l & \text{for } \text{Re}_s \geq 100 \\[3mm] N_b\left[26\dfrac{G_w\mu_s}{\rho_s}\left(\dfrac{N_{tcw}}{p_t - d_o} + \dfrac{L_{bc}}{D_w^2}\right) + 2(10^{-3})\dfrac{G_w^2}{2g_c\rho_s}\right]R_l & \text{for } \text{Re} < 100 \end{cases}$

N_{tcw}, L_{bc}
from Table 17.30

$G_w = \dfrac{\dot{m}_s}{(A_{mb}A_w)^{1/2}}$

A_{mb}, A_w from Table 17.30

$D_w = \dfrac{4A_w}{\pi d_o F_w N_t + \pi D_s \dfrac{\theta_{ds}}{360}}$

$\Delta p_{i-o} = \Delta p_{bi}\left(1 + \dfrac{N_{tcw}}{N_{tcc}}\right)R_b R_s$

$N_{tcc}, L_{bo}, L_{bi},$ and L_{bc}
from Table 17.30

$R_s = \left(\dfrac{L_{bc}}{L_{bo}}\right)^{2-n} + \left(\dfrac{L_{bc}}{L_{bi}}\right)^{2-n}$ $n = \begin{cases} 1.0 & \text{laminar flow} \\ 0.2 & \text{turbulent flow} \end{cases}$

* Note regarding the units: Δp in Pa or psi; A_{mb} and A_w in mm² or in²; $p_t, d_o,$ and D_w in mm or in. See
notes in Table 17.30.
† A number of accurate correlations such as those given in Table 17.19 are available. Traditionally,
the diagrams such as those given in Figs. 17.55–17.57 have been used in engineering practice.

1. Determine the heat load. If both streams are single phase, calculate the heat load q using
   Eq. 17.3. If one of the streams undergoes a phase change, calculate $q = \dot{m}i$ where $\dot{m}$ = mass
   flow rate of that stream and i = specific enthalpy of phase change.

2. Determine the logarithmic mean temperature difference using Eq. 17.18.

3. Estimate the log-mean temperature difference correction factor F. For a single TEMA E
   shell with an arbitrary even number of tubeside passes, the correction factor should be
   $F > 0.8$. The correction factor F should be close to 1 if one stream changes its temperature
   only slightly in the exchanger. F should be close to 0.8 if the outlet temperatures of the two
   streams are equal. Otherwise, assume $F = 0.9$.

4. Estimate the overall heat transfer coefficient (use Table 17.28 with judgment or estimate
   the individual heat transfer coefficients and wall resistance [109], and afterwards calculate
   the overall heat transfer coefficient using Eq. 17.6).

5. Calculate the total outside tube heat transfer area (including fin area) using $A = A_p + A_f$.

6. Determine the set of heat exchanger dimensions that will accommodate the calculated
   total heat transfer area for a selected shell diameter and length using the diagram given
   in Fig. 17.58. The diagram in Fig. 17.58 corresponds to plain tubes with a 19-mm outside

diameter on a 23.8-mm equilateral triangular tube layout. The extension of this diagram to other shell/bundle/tube geometries requires determination of a corrected effective total heat transfer area using the procedure outlined in Ref. 109. The abscissa in Fig. 17.58 is the effective tube length of a single straight section. The effective length is from tubesheet to tubesheet for a straight tube exchanger and from tubesheet to tangent line for a U-tube bundle. The dashed lines show the approximate locus of shells with a given effective tube length-to-shell diameter ratio. The solid lines are the inside diameters of the shell. The proper selection of the combination of parameters and the effective tube length depends on the particular requirements and given conditions and is greatly influenced by the designer's experience. For a good design, the L/D ratio for the shell is kept between 6 and 15 to optimize the cost of the shell (diameter) and the tubeside pressure drop (tube length).

The thermal design and some aspects of the mechanical design of a shell-and-tube heat exchanger are empirically based, as discussed above. However, there are many criteria for mechanical selection [5], many experience-based criteria that can avoid or minimize operating problems [155], and other design considerations such as identification of thermodynamic irreversibilities [15, 110], thermoeconomic considerations [111], system optimization, and process integration [112]. In industrial applications, thermoeconomic optimization should be

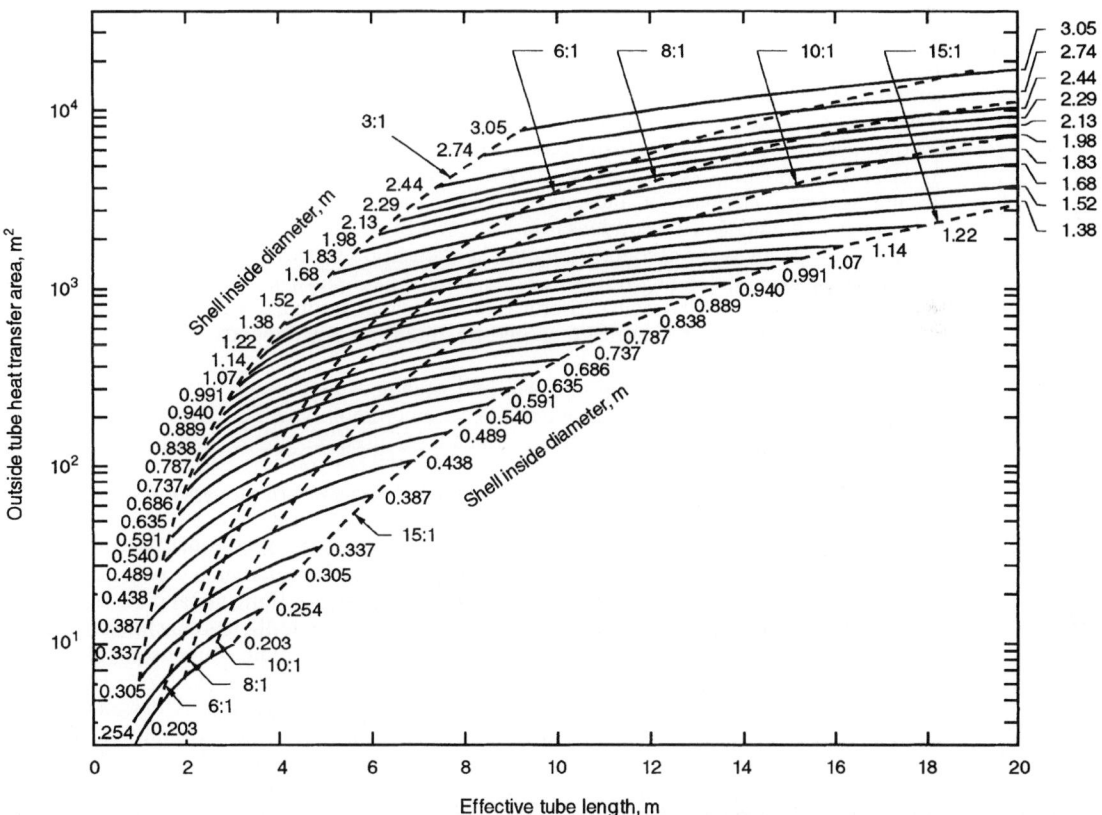

FIGURE 17.58 Heat transfer area as a function of the tube length and shell inside diameter for 19.0-mm outside diameter plain tubes on a 23.8-mm equilateral tube layout, fixed tubesheet, one tubeside pass, and fully tubed shell [109].

carried out at the system level, but individual irreversibilities of the heat exchanger expressed in terms of their monetary values must be identified [15]. All these clearly demonstrate the complexity of heat exchanger thermal design.

THERMAL DESIGN FOR TWO-PHASE HEAT EXCHANGERS

Most common heat exchangers operating under two-phase and multiphase flow conditions are condensers and vaporizers. See Fig. 17.2 for further classification.

The variety of phase-change conditions, the diversity of heat exchanger constructions, and the broad ranges of operating conditions prevent a thorough and complete presentation of design theory and design considerations in a limited space. The objective of this section, though, is to summarize the key points regarding thermal design and to present design guidelines for the most frequently utilized two-phase flow heat exchangers.

Condensers

In a condenser, the process stream (single component or multicomponent with or without noncondensable gases) is condensed to a liquid with or without desuperheating and/or subcooling. The diversity of major design features of various condensers is very broad, as can be concluded from many different applications presented in Fig. 17.2b. Consequently, various aspects of condenser operation as well as their various design characteristics cannot be presented in a unified fashion. Important aspects of condenser operation involve, but are not restricted to: (1) the character of the heat transfer interaction (direct or indirect contact type); (2) the geometry of the heat transfer equipment (shell-and-tube, extended surface, plate, and so on); (3) the number of components in the condensing fluid (single or multicomponent); (4) desuperheating, condensation, and subcooling; and (5) the presence of noncondensable gas in the condensing fluid (partial condensation).

Primary objectives for accomplishing the condensation process vary depending on a particular application, but common features of a vapor-liquid phase-change lead to certain general similarities in thermal design procedure. Nonetheless, thermal design of a condenser does not necessarily follow a standardized procedure, and it greatly depends on a condenser type and the factors mentioned above.

In indirect contact type condensers, two fluid streams are separated by a heat transfer surface. A shell-and-tube condenser is one of the most common type. For example, surface condensers are the turbine exhaust steam condensers used in power industry. In another condenser, a boiler feedwater is heated with a superheated steam on the shell side, causing desuperheating, condensing, and subcooling of the steam/water. In process industry, condensation of either single or multicomponent fluids (with or without noncondensable gases) may occur inside or outside the tubes, the tubes being either horizontal or vertical. Extended surface condensers are used both in power and process industries (including cryogenic applications) and are designed either as tube-fin or plate-fin exchangers. If the metal plate substitutes a tube wall to separate the two fluids (the condensing vapor and the coolant) in all primary surface condensers, the resulting design belongs to the family of plate condensers (plate-and-frame, spiral plate, and printed circuit heat exchangers).

In direct contact condensers, a physical contact of the working fluids (a saturated or superheated vapor and a liquid) occurs, allowing for the condensation to be accomplished simultaneously with the mixing process. The fluids can be subsequently separated only if they are immiscible. Direct contact is generally characterized with a very high heat transfer rate per unit volume. The classification of indirect and direct contact heat exchangers is discussed in more detail in Ref. 2.

Thorough discussion of various topics related to condensers and their characteristics is provided in Refs. 113–115.

Indirect Contact Type Condensers

Thermal Design.   Sizing or rating of an indirect contact condenser involves the very same heat transfer rate equation, Eq. 17.4, that serves as a basis for the thermal design of a single-phase recuperator. In the case of a condenser, however, both the overall heat transfer coefficient and the fluid temperature difference vary considerably along and across the exchanger. Consequently, in the design of a condenser, the local heat transfer rate equation, Eq. 17.2:

$$dq = U\Delta T dA \tag{17.147}$$

may be supplemented with an approximate equation:

$$q = \check{U}\Delta T_m A \tag{17.148}$$

where
$$\check{U} = \frac{1}{A}\int_A U\, dA \tag{17.149}$$

and/or
$$\Delta T_m = \frac{q}{\int_A U\, dA} \tag{17.150}$$

or alternately, the integration of Eq. 17.147 must be rigorously executed.

Now, the problem is how to determine the mean overall heat transfer coefficient and the corresponding mean temperature difference, Eqs. 17.149 and 17.150.

In practice, calculation has to be performed by dividing the condenser's total heat transfer load in an appropriate number of heat duty zones and subsequently writing auxiliary energy balances based on enthalpy differences for each zone. One must simultaneously establish the corresponding temperature variation trends, corresponding zonal mean overall heat transfer coefficients, and mean temperature differences. As a result, one can calculate the heat transfer surface for each zone using Eq. 17.148. Total heat transfer area needed for design is clearly equal to the sum of the heat transfer areas of all zones. In a limit, for a very large number of zones, the total heat transfer area is equal to:

$$A = \int_q \frac{dq}{U\Delta T} \tag{17.151}$$

Modern computer codes for designing heat exchangers evaluate Eq. 17.151 numerically, utilizing local overall heat transfer coefficients and local fluid temperature differences.

A method based on this simple set of propositions leads to the formulation of the thermal evaluation method as suggested by Butterworth [113]. This method is convenient for a preliminary design of E- and J-type shell-and-tube condensers. The complete design effort must include a posteriori the determination of pressure drop and corresponding corrections of saturation temperature and should ultimately end with an economic assessment based on, say, capital cost. The thermal evaluation method can be summarized for the shell side of a shell-and-tube condenser having a single tube pass as follows:

1. Construct an exchanger operating diagram. The plot provides the local shellside fluid equilibrium temperature T_s as a function of the corresponding fluid specific enthalpy (see Fig. 17.59). A correlation between the shellside and tubeside fluid enthalpies is provided by the enthalpy balance, therefore the tubeside temperature dependence T_t can be presented as well. The local equilibrium temperature is assumed to be the temperature of the stream well mixed at the point in question. Note that this step does not involve an estimation of the overall heat transfer coefficient.

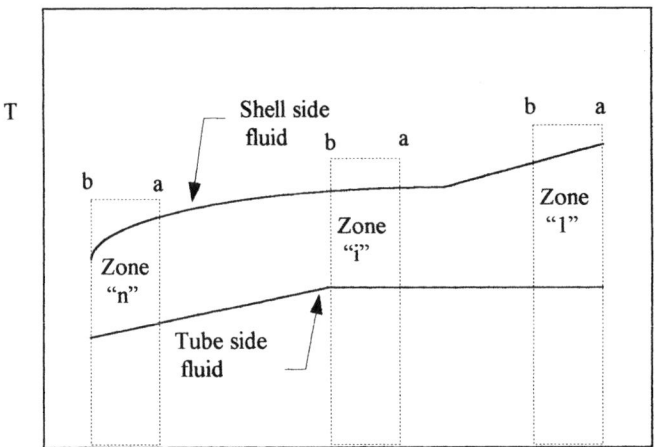

FIGURE 17.59   Operating diagram of a condenser.

2. Divide the exchanger operating diagram into N zones, $\{a, b\}_i$, for which both corresponding temperatures vary linearly with the shellside enthalpy. Here, a_i and b_i denote terminal points of the zone i.

3. Determine logarithmic mean temperature differences for each zone:

$$\Delta T_m = \Delta T_{lm,i} = \frac{\Delta T_{a,i} - \Delta T_{b,i}}{\ln (\Delta T_{a,i}/\Delta T_{b,i})} \tag{17.152}$$

4. Calculate the overall heat transfer coefficient for each zone using an appropriate set of heat transfer correlations and an appropriate correlation from Table 17.32. More specifically, if a linear dependence between U and A can be assumed, an arithmetic mean between the terminal U values should be used as a mean value. If both U and T vary linearly with q, the mean U value should be calculated from a logarithmic mean value of the $U\Delta T$ product as indicated in Table 17.32. Next, if both $1/U$ and T vary linearly with q, the third equation for the mean U value from Table 17.32 should be used. Finally, if U is not a linear function of either A or q, the mean value should be assessed following the procedure described in the section starting on p. 17.47.

TABLE 17.32   Mean (Zonal) Overall Heat Transfer Coefficient

Conditions	Mean overall heat transfer coefficient
$\check{U}$ vs. A linear within the zone $a - b$	$\check{U} = \dfrac{U_a + U_b}{2}$
$\check{U}$ and ΔT vs. A linear within the zone $a - b$	$\check{U} = \dfrac{U_a \Delta T_b - U_b \Delta T_a}{\Delta T_{lm} \ln \left(\dfrac{U_a \Delta T_b}{U_b \Delta T_a} \right)}$
$\dfrac{1}{\check{U}}$ and ΔT vs. A linear within the zone $a - b$	$\dfrac{1}{\check{U}} = \dfrac{1}{U_a} \left(\dfrac{\Delta T_{lm} - \Delta T_b}{\Delta T_a - \Delta T_b} \right) + \dfrac{1}{U_b} \left(\dfrac{\Delta T_a - \Delta T_{lm}}{\Delta T_a - \Delta T_b} \right)$
$\check{U}$ vs. A nonlinear within the zone $a - b$	See text on p. 17.47

5. Calculate heat transfer area for each zone:

$$A_i = \frac{\dot{m}_s \Delta i_i}{\breve{U}_i \Delta T_{lm,i}}$$
(17.153)

6. The total heat transfer area is then:

$$A = \sum_{i=1}^{i=N} A_i$$
(17.154)

This procedure is applicable to either countercurrent or cocurrent condensers (the difference being only the enthalpy balances in formal writing). The use of the exchanger operating diagram can also be utilized for shellside E-type condensers with more than one tube pass (i.e., 2, 4, and more passes); see Ref. 113 for details. As it was already pointed out, this method does not cover the complete set of design requirements (i.e., the pressure drop considerations must be included into the analysis). The preliminary design obtained by using the described method should be corrected as necessary, repeating the procedure for different assumed geometries, calculating the pressure drops, and evaluating mechanical and economic aspects of the design.

A modern approach to the design of condensers inevitably involves the use of complex numerical routines. An overview of numerical methods is provided in Ref. 117.

Overall Design Considerations and Selection of Condenser Types.   Regardless of the particular thermal design method involved, a designer should follow an overall design procedure as outlined by Mueller for preliminary sizing of shell-and-tube condensers [114]: (1) determine a suitable condenser type following specific selection guidelines (see Table 17.33), (2) determine the heat load, (3) select coolant temperatures and calculate mean temperatures, (4) estimate the overall heat transfer coefficient, (5) calculate the heat transfer area, (6) select geometric characteristics of heat transfer surfaces (e.g., for a shell-and-tube heat exchanger, select the tube size, pitch, length, the number of tubes, shell size, and baffling), (7) compute pressure drops on both sides, and (8) refine the sizing process in an iterative procedure (as a rule using a computer). The final design has to be accompanied by mechanical design and thermoeconomic optimization.

Pressure drops on both sides of a condenser are usually externally imposed constraints and are calculated using the procedures previously described (see text starting on p. 17.62 for single-phase and p. 17.95 for two-phase). However, such calculated pressure drops for two-phase flow have a much larger uncertainty than those for single-phase conditions.

Comprehensive guidelines regarding the condenser selection process are given in Ref. 114 and are briefly summarized in Table 17.33. Most tubeside (condensation on tubeside) condensers with horizontal tubes are single-pass or two-pass shell-and-tube exchangers. They are acceptable in partial condensation with noncondensables. The tube layout is governed by the coolant side conditions. Tubeside condensers with vertical downflow have baffled shell sides, and the coolant flows in a single-pass countercurrent to the vapor. The vapor in such settings condenses, usually with an annular flow pattern. If the vapor condenses in upflow, the important disadvantage may be the capacity limit influenced by flooding. Shellside condensers with horizontal tubes can be baffled or the crossflow type. In the presence of noncondensables, the baffle spacing should be made variable. If the shellside pressure drop is a severe constraint, J-shell and X-shell designs are preferable. Tubes on the vapor side are often enhanced with low-height fins. The tube side can have multipasses. Vertical shellside condensers usually do not have baffled shell sides, and as a rule, vapor is in downflow.

Design procedures for condensers with noncondensables and multicomponent mixtures are summarized in Ref. 2.

Direct Contact Condensers

Thermal Design.   A unified approach to the design of direct contact condensers does not exist. A good overview of direct contact condensation phenomena is provided in Ref. 115.

TABLE 17.33 Selection Guidelines for Condensers (Modified from Ref. 112)

	Tubeside condensation						Shellside condensation								Direct contact condensation	
	Horizontal		Vertical				Horizontal				Vertical					
			Downflow		Upflow		Cross "X"		Baffled		Downflow		Upflow			
	Possible slugging of condensate for total condensation.		*Handles dirty or polymerizing vapors.*		*Flooding. Hot condensate return and removal of small amounts of low boilers for total condensation.*		*Possible venting problems for total condensation.*		*Multipassing and variable baffle spacing can be used for a partial condensation case.*		*Multipassing and variable baffle spacing and falling-water film can be used for a partial condensation case.*		*For partial condensation, no known practical application.*		*Design information is limited. If coolant is condensate, then an external cooler is required. With water coolant, there may be pollution problems. Applicable for very low or intermediate pressure, very corrosive vapor, and severe or moderate coolant fouling.*	
	Total	Partial	Total	Partial	Total	Partial	Total	Partial	Total	Partial	Total	Partial	Total	Partial	Total	Partial
SCV¹	G²,³	G	G	G	F	X	G	P	G	P	G	F	G	P	G	X
MCV	F	P	G	G	F	X	G	F	G	P	G	P	F	X	P	F
SC	P	G	G	G	X	X	F	G	P	G	F	F	X	X	X	F
Δp high	G	G	G	G	X	P	G	G	G	G	G	G	X	X	X	X
Δp low	P	P	P	G	G	F	G	G	F	P	G	G	F	F	G	G
C liquid	G	G	G	G	G	G	G	G	G	G	G	G	G	G	G	G
C gas	G	G	G	G	G	G	G	G	G	G	G	G	G	G	X	X
C boiling	G	G	G	G	G	G	X	X	X	X	G	G	G	G	X	X

Descriptive notes at top of major headings:

Tubeside condensation: High and intermediate pressures. Noncorrosive or corrosive vapor. High temperature (> 400°C). Severe vapor fouling. Good control of venting. Poor for pressures below 3 kPa.

— Horizontal: *Possible slugging of condensate for total condensation.*
— Vertical: *Tubesheet vents required.*

Shellside condensation: Intermediate pressure. Noncorrosive vapor. Severe coolant or moderate vapor fouling. Can handle freezing condensate. Tube vibration problems. Multipassing and variable baffle spacing can be used for total condensation. Good venting control.

— Horizontal: *Can use finned or enhanced tubes.*
— Vertical: *Can use a falling-water film but possible venting problems for total condensation.*

Direct contact condensation: Design information is limited. If coolant is condensate, then an external cooler is required. With water coolant, there may be pollution problems. Applicable for very low or intermediate pressure, very corrosive vapor, and severe or moderate coolant fouling.

¹SCV - Single-component vapor; MCV - Multicomponent vapor; SC - Subcooled condensate; Δp - Pressure drop; C - Coolant.

²Acceptability: G - good; F - fair; P - poor; X - not acceptable or not recommended.

³ Predictability: ☺ - average (~25%); ☺ - fair (< 50%); ☺ - poor (50+%); ☒ - no method or not recommended.

Physical conditions greatly depend on the aggregate state of the continuous phase (vapor in spray and tray condensers, liquid in pool-type condensers, and liquid film on the solid surface in packed bed condensers). Design of the most frequently used spray condensers, featuring vapor condensation on the water droplets, depends on the heat and mass transfer phenomena involved with saturated vapor condensation in the presence of the subcooled liquid droplets of changing mass. The process is very complex. For further details on the problems involved, consult Refs. 116 and 118. Such designs involve a substantial input of empirical data.

The key process variables are the time required for a spray drop of a particular size to reach prescribed distance and the quantity of heat received by droplets from the vapor. The initial size of a droplet obviously influences the size of a heat exchanger. Subsequent transient heat and mass transfer processes of vapor condensation on a droplet of changing size has a key role in the exchanger operation. Initial droplet sizes and their distribution is controlled by design of spray nozzles. Thermofluid phenomena models involve a number of idealizations; the following are important: (1) heat transfer is controlled by transient conduction within the droplet as a solid sphere, (2) droplet size is uniform and surface temperature equal to the saturation temperature, and (3) droplets are moving relative to the still vapor. Although these idealizations seem to be too radical, the models developed provide at least a fair estimate for the initial design. In Table 17.34, compiled are the basic relations important for contact condensation of saturated vapor on the coolant liquid. Generally, guidelines for design or rating a direct contact condenser do not exist and each design should be considered separately. A good overview of the calculations involved is provided in Ref. 118.

TABLE 17.34   Direct Contact Condensation Thermofluid Variables

	Correlation	Parameters
Liquid drop residence time	$\mathrm{Fo} = -\dfrac{1}{\pi^2} \ln\left[1 - \left(\dfrac{T - T_i}{T_{\mathrm{sat}} - T_i}\right)^2\right]$	$\mathrm{Fo} = \dfrac{4\alpha\tau}{D^2} \qquad \alpha = \dfrac{k_l}{\rho_l c_{p,l}}$
Drop travel distance, m	$L = 0.06 \dfrac{D^{1.84}}{\Gamma} (V_i^{0.84} - V^{0.84})$	$\Gamma = v_v^{0.84} \dfrac{\rho_v}{\rho_l}$
Drop velocity, m/s	$V = \left(V_i^{-0.16} + 3.23 \dfrac{\Gamma\tau}{D^{1.84}}\right)^{-1/0.16}$	
Heat transfer rate	$q = (\dot{m}c_p)_l (T - T_i)$	
Condensate mass flow rate	$\dot{m}_v = \dfrac{q}{i_{lv}}$	

Vaporizers

Heat exchangers with liquid-to-vapor phase change constitute probably the most diverse family of two-phase heat exchangers with respect to their functions and applications (see Fig. 17.2). We will refer to them with the generic term *vaporizer* to denote any member of this family. Therefore, we will use a single term to denote boilers, steam and vapor generators, reboilers, evaporators, and chillers. Design methodologies of these vaporizers differ due to construction features, operating conditions, and other design considerations. Hence, we will not be able to cover them here but will emphasize only a few most important thermal design topics for evaporators.

Thermal Design.   The key steps of an evaporator thermal design procedure follow the heat exchanger overall design methodology. For a two-phase liquid-vapor heat exchanger, the procedure must accommodate the presence of phase change and corresponding variations of local heat transfer characteristics, the same two major features discussed for condensers. The procedure should, at least in principle, include the following steps:

1. Select an appropriate exchanger type following the analysis of the vaporizer function, and past experience if any. The selection influences both heat transfer and nonheat transfer factors such as: heat duty, type of fluids, surface characteristics, fouling characteristics, operating conditions (operating pressure and design temperature difference), and construction materials. For example, a falling-film evaporator should be used at pressures less than 1 kP (0.15 psi). At moderate pressures (less than 80 percent of the corresponding reduced pressure), the selection of a vaporizer type does not depend strongly on the pressure, and other criteria should be followed. For example, if heavy fouling is expected, a vertical tubeside thermosiphon may be appropriate.

2. Estimate thermofluid characteristics of liquid-vapor phase change and related heat transfer processes such as circulation rate in natural or forced internal or external fluid circulation, pressure drops, and single- and two-phase vapor-liquid flow conditions. The initial analysis should be based on a rough estimation of the surface area from the energy balance.

3. Determine local overall heat transfer coefficient and estimate corresponding local temperature difference (the use of an overall logarithmic mean temperature difference based on inlet and outlet temperatures is, in general, not applicable).

4. Evaluate (by integration) the total heat transfer area, and subsequently match the calculated area with the area obtained for a geometry of the selected equipment.

5. Evaluate pressure drops. The procedure is inevitably iterative and, in practice, ought to be computer-based.

6. Determine design details such as the separation of a liquid film from the vapor (i.e., utilization of baffles and separators).

Important aspects in thermal design of evaporators used in relation to concentration and crystallization in the process/chemical industry can be summarized as follows [119]:

1. The energy efficiency of the evaporation process (i.e., the reduction of steam consumption by adequate preheating of feed by efficient separation, managing the presence of noncondensable gases, avoiding high concentrations of impurities, and proper selection of take-off and return of the liquid)

2. The heat transfer processes

3. The means by which the vapor and liquid are separated

Preliminary thermal design is based on the given heat load, estimated overall heat transfer coefficient, and temperature difference between the saturation temperatures of the evaporating liquid and condensing vapor. The guidelines regarding the preliminary estimation of the magnitude of the overall heat transfer coefficient are provided by Smith [119]; also refer to Table 17.28 for shell-and-tube heat exchangers.

Problems that may be manifested in the operation of evaporators and reboilers are numerous: (1) corrosion and erosion, (2) flow maldistribution, (3) fouling, (4) flow instability, (5) tube vibration, and (6) flooding, among others. The final design must take into account some or all of these problems in addition to the thermal and mechanical design.

A review of thermal design of reboilers (kettle, internal, and thermosiphon), and an overview of important related references is provided by Hewitt at al. [115]. It should be pointed out that a computer-based design is essential. Still, one must keep in mind that the results greatly depend on the quality of empirical data and correlations. Thermal design of kettle and internal reboilers, horizontal shellside and vertical thermosiphon reboilers, and the useful guidelines regarding the special design considerations (fouling, flow regime consideration, dryout, overdesign, vapor separation, etc.) are provided in Ref. 2.

Finally, it should be noted that nuclear steam generators and waste heat boilers, although working in different environments, both represent modern unfired steam raisers (i.e., steam generators) that deserve special attention. High temperatures and operating pressures, among the other complex issues, impose tough requirements that must be addressed in design. The basic thermal design procedure, though, is the same as for other vapor-liquid heat exchangers [120, 121].

FLOW-INDUCED VIBRATION

In a tubular heat exchanger, interactions between fluid and tubes or shell include the coupling of fluid flow-induced forces and an elastic structure of the heat exchanger, thus causing oscillatory phenomena known under the generic name *flow-induced vibration* [122]. Two major types of flow-induced vibration are of a particular interest to a heat exchanger designer: tube vibration and acoustic vibration.

Tube vibrations in a tube bundle are caused by oscillatory phenomena induced by fluid (gas or liquid) flow. The dominant mechanism involved in tube vibrations is the *fluidelastic instability* or *fluidelastic whirling* when the structure elements (i.e., tubes) are shifted elastically from their equilibrium positions due to the interaction with the fluid flow. The less dominant mechanisms are vortex shedding and turbulent buffeting.

Acoustic vibrations occur in fluid (gas) flow and represent standing acoustic waves perpendicular to the dominant shellside fluid flow direction. This phenomenon may result in a loud noise.

A key factor in predicting eventual flow-induced vibration damage, in addition to the above mentioned excitation mechanisms, is the natural frequency of the tubes exposed to vibration and damping provided by the system. Tube vibration may also cause serious damage by fretting wear due to the collision between the tube-to-tube and tube-to-baffle hole, even if resonance effects do not take place.

Flow-induced vibration problems are mostly found in tube bundles used in shell-and-tube, duct-mounted tubular and other tubular exchangers in nuclear, process, and power industries. Less than 1 percent of such exchangers may have potential flow-induced vibration problems. However, if it results in a failure of the exchanger, it may have a significant impact on the operating cost and safety of the plant.

This subsection is organized as follows. The tube vibration excitation mechanism (the fluidelastic whirling) will be considered first, followed by acoustic phenomena. Finally, some design-related guidelines for vibration prevention will be outlined.

Tube Vibration

Fluidelastic Whirling.   A displacement of a tube in a tube bundle causes a shift of the flow field, and a subsequent change of fluid forces on the tubes. This change can induce instabilities, and the tubes will start vibrating in oval orbits. These vibrations are called the fluidelastic whirling (or the fluidelastic instability). Beyond the critical intertube flow velocity V_{crit}, the amplitude of tube vibrations continues to increase exponentially with increasing flow velocity. This phenomenon is recognized as the major cause of tube vibrations in the tubular heat exchangers. The critical velocity of the complex phenomenon is correlated semiempirically as follows [122].

$$\frac{V_{crit}}{f_n d_o} = \mathbf{C}\left(\frac{\delta_o M_{eff}}{\rho_s d_o^2}\right)^a \tag{17.155}$$

where δ_o represents the logarithmic decrement, ρ_s is shellside fluid density, and M_{eff} is the virtual mass or the effective mass per unit tube length given by

$$M_{\text{eff}} = \frac{\pi}{4}(d_o^2 - d_i^2)\rho_t + \frac{\pi}{4}d_i^2\rho_{f,t} + \frac{\pi}{4}d_o^2 C_m \rho_s \tag{17.156}$$

The effective mass per unit tube length, M_{eff}, includes the mass of the tube material per unit length, the mass of the tubeside fluid per unit tube length, and the hydrodynamic mass per unit tube length (i.e., the mass of the shellside fluid displaced by a vibrating tube per unit tube length). In the hydrodynamic mass per unit tube length, ρ_s is the shellside fluid density and C_m is the added mass (also virtual mass or hydrodynamic mass) coefficient provided in Fig. 17.60. In addition to the known variables, two additional coefficients ought to be introduced: the coefficient **C** (also referred to as the *threshold instability constant* [130, 154] or *fluidelastic instability parameter* [154]) and the exponent a. Both coefficient **C** and exponent a can be obtained by fitting experimental data for the critical flow velocity as a function of the so-called *damping parameter* (also referred to as *mass damping*), the bracketed quantity on the right side of Eq. 17.155. The coefficient **C** is given in Table 17.35 as a function of tube bundle layout under the condition that exponent a take the value of 0.5 as predicted by the theory of fluidelastic instability developed by Connors [124] and for the damping parameter greater than 0.7. If the damping parameter is smaller than 0.7, a least-square curve fit of available data gives $\mathbf{C}_{\text{mean}} = 3.9$, and $a = 0.21$ (the statistical lower bound $\mathbf{C}_{90\%}$ being 2.7) [122]. The smallest coefficient $\mathbf{C}_{\text{mean}}$ in Table 17.35 corresponds to the 90° tube layout, thus implying this layout has the smallest critical velocity (the worst from the fluidelastic whirling point of view) when other variables remain the same. The same correlation with different coefficients and modified fluid density can be applied for two-phase flow [125]. It should be noted that the existing models cannot predict the fluidelastic whirling with the accuracy better than the one implied by a standard deviation of more than 30 percent of existing experimental data [126].

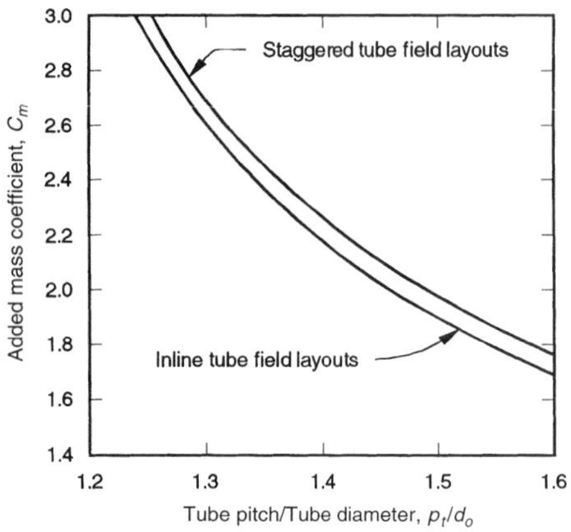

FIGURE 17.60 Upper bound of added mass coefficient C_m [128].

TABLE 17.35 Coefficient **C** in Eq. 17.155 [122]

	Tube pattern					
C	30°	60°	45°	90°	All	Single tube row
$\mathbf{C}_{\text{mean}}$	4.5	4.0	5.8	3.4	4.0	9.5
$\mathbf{C}_{90\%}$	2.8	2.3	3.5	2.4	2.4	6.4

Acoustic Vibrations

This mechanism produces noise and generally does not produce tube vibration. It is one of the most common forms of flow-induced vibration in shell-and-tube exchangers for high velocity shellside gas flows in large exchangers. When a forcing frequency (such as the frequency of vortex shedding, turbulent buffeting, or any periodicity) coincides with the natural frequency of a fluid column in a heat exchanger, a coupling occurs. The kinetic energy in the flow stream is converted into acoustic pressure waves; this results in a possibility of standing wave vibration, also referred to as acoustic resonance or acoustic vibration, creating an intense, low-frequency, pure-tone noise. Particularly with a gas stream on the shell side, the sound pressure in a tube array may reach the level of 160–176 dB, the values up to 40 dB lower outside the heat exchanger shell [122]. Acoustic vibration could also increase shellside pressure drop through the resonant section and cause severe vibration and fatigue damage to the shell (or casing), connecting pipes, and floor. If the frequency of the standing wave coincides with the tube natural frequency, tube failure may occur.

Now we will briefly describe two additional mechanisms: vortex shedding and turbulent buffeting. It should be noted that these mechanisms could cause tube vibrations, but their influence on a tube bundle is less critical compared to the fluidelastic instabilities described earlier.

Vortex Shedding. A tube exposed to an incident crossflow above critical Reynolds numbers provokes an instability in the flow and a simultaneous shedding of discrete vorticities alternately from the sides of the tube. This phenomenon is referred to as *vortex shedding*. Alternate shedding of the vorticities produces harmonically varying lift and drag forces that may cause movement of the tube. When the tube oscillation frequency approaches the tube natural frequency within about ±20 percent, the tube starts vibrating at its natural frequency. This results in the vortex shedding frequency to shift to the tube's natural frequency (lock-in mechanism) and causes a large amplification of the lift force. The vortex shedding frequency is no more dependent upon the Reynolds number. The amplitude of vibration grows rapidly if the forcing frequency coincides with the natural frequency. This can result in large resonant amplitudes of the tube oscillation, particularly with liquid flows, and possible damage to tubes.

Vortex shedding occurs for Re numbers above 100 (the Re number is based on the upstream fluid velocity and tube outside diameter). In the region $10^5 < \text{Re} < 2 \times 10^6$, vortex shedding has a broad band of shedding frequencies. Consequently, the regular vortex shedding does not exist in this region.

The vortex shedding frequency f_v for a tubebank is calculated from the Strouhal number

$$\text{Sr} = \frac{f_v d_o}{V_c} \qquad (17.157)$$

where Sr depends on tube layouts as given in Fig. 17.61. The Strouhal number is nearly independent of the Reynolds number for Re > 1000.

The reference crossflow velocity V_c in gaps in a tube row is difficult to calculate since it is not based on the minimum free flow area. The local crossflow velocity in the bundle varies from span to span, from row to row within a span, and from tube to tube within a row [5]. In general, if the flow is not normal to the tube, the crossflow velocity in Eq. 17.157 is to be interpreted as the normal component (crossflow) of the free stream velocity [154]. Various methods may be used to evaluate reference crossflow velocity. In Table 17.36, a procedure is given for the determination of the reference crossflow velocity according to TEMA Standards [5]. The calculated velocity takes into account fluid bypass and leakage which are related to heat exchanger geometry. This method of calculation is valid for single-phase shellside fluid with single segmental baffles in TEMA E shells.

Turbulent Buffeting. Turbulent buffeting refers to unsteady forces developed on a body exposed to a highly turbulent flow. The oscillatory phenomenon in turbulent flow on the shell side (when the shellside fluid is gas) is characterized by fluctuating forces with a dominant frequency as follows [123, 154]:

$$f_{tb} = \frac{V_c d_o^2}{X_l X_t} \left[3.05 \left(1 - \frac{d_o}{X_t} \right)^2 + 0.28 \right] \qquad (17.158)$$

The correlation was originally proposed for tube-to-diameter ratios of 1.25 and higher. It should be noted that the turbulent buffeting due to the oncoming turbulent flow is important

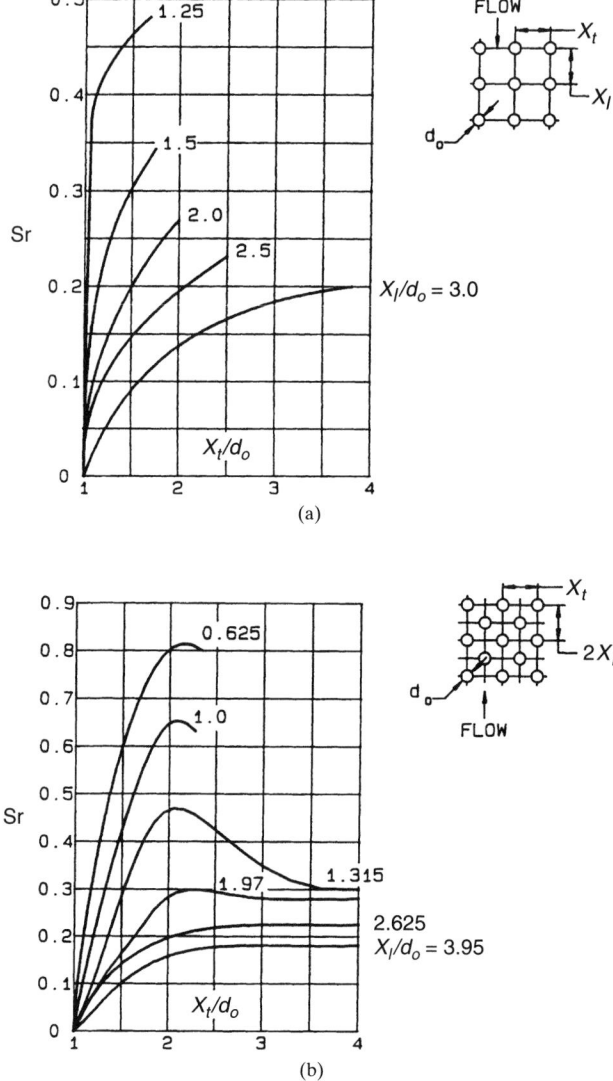

FIGURE 17.61 Vortex shedding Strouhal numbers for tube patterns:
(a) 90°, (b) 30°, 45°, 60° [5].

only for gases at high Reynolds numbers. The reference crossflow velocity in Eq. 17.158
should be calculated using the procedure presented in Table 17.36.

Tube Bundle Natural Frequency.   Elastic structures vibrate at different natural frequencies. The lowest (fundamental) natural frequency is the most important. If the vortex shedding or turbulent buffeting frequency is lower than the tube fundamental natural frequency,
it will not create the resonant condition and the tube vibration problem. Hence, the knowledge of the fundamental natural frequency is sufficient in most situations if f_n is found to be
higher than f_v or f_{tb}. Higher than the third harmonic is generally not important for flow-

TABLE 17.36 Reference Crossflow Velocity in Tube Bundle Gaps [5]*

<table>
<tr><td colspan="2" align="center">Reference crossflow velocity V_c</td></tr>
<tr><td colspan="2" align="center">$V_c = \dfrac{F_h}{M}\dfrac{\dot{m}_s}{a_x \rho_s}\left[\dfrac{m}{s}\right]$; $\dot{m}_s\left[\dfrac{kg}{s}\right]$; $\rho_s\left[\dfrac{kg}{m^3}\right]$; $a_x = C_a L_{bc} D_{otl}$ [m^2]</td></tr>
<tr>
<td>

$F_h = \left[1 + N_h\left(\dfrac{D_s}{p_t}\right)^{1/2}\right]^{-1}$

</td>
<td>

$N_h = f_1 C_7 + f_2 \Psi + f_3 E \qquad f_1 = \dfrac{(C_1 - 1)^{3/2}}{C_1^{1/2}} \qquad f_2 = \dfrac{C_2}{C_1^{3/2}}$

$f_3 = C_3 C_1^{1/2} \qquad C_1 = \dfrac{D_s}{D_{otl}} \qquad C_2 = \dfrac{d_i - d_o}{d_o} \qquad C_3 = \dfrac{D_s - D_{baff}}{D_s}$

$C_7 = C_4 \left(\dfrac{p_t}{p_t - d_o}\right)^{3/2} \qquad \Psi = C_5 C_8 \dfrac{D_s}{L_{bc}}\left(\dfrac{d_o}{p_t}\right)^2 \dfrac{p_t}{p_t - d_o}$

$E = C_6 \dfrac{p_t}{p_t - d_o}\dfrac{D_s}{L_{bc}}\left(1 - \dfrac{L_c}{D_s}\right)$

See below for C_4, C_5, and C_6.

</td>
</tr>
</table>

L_c/D_s	0.10	0.15	0.20	0.25	0.30	0.35	0.40	0.45	0.50
C_8	0.94	0.90	0.85	0.80	0.74	0.68	0.62	0.54	0.49

A linear interpolation C_8 vs. L_c/D_s is permitted

$$M = \left[1 - \dfrac{0.7 L_{bc}}{D_s}\left(M_w^{-0.6} - 1\right)\right]^{-1}$$

$$M_w = m C_1^{1/2} \qquad C_a = \left(\dfrac{p_t - d_o}{p_t}\right)$$

C_i	30°	60°	90°	45°
C_4	1.26	1.09	1.26	0.90
C_5	0.82	0.61	0.66	0.56
C_6	1.48	1.28	1.38	1.17
m	0.85	0.87	0.93	0.80

* In Ref. 5, U.S. Engineering units are used.

induced vibration in heat exchangers. For vortex shedding, the resonant condition can be avoided if the vortex shedding frequency is outside ±20 percent of the natural frequency of the tube.

Determination of natural frequency of an elastic structure can be performed analytically (for simple geometries) [127] and/or numerically (for complex structures) using finite element computer programs (such as NASTRAN, MARC, and ANSYS) or proprietary computer programs.

Straight Tube. A tube of a shell-and-tube heat exchanger with fluid flowing in it and a flow of another fluid around it is hardly a simple beam structure. Consequently, the natural frequency of the *i*th mode of a straight tube rigidly fixed at the ends in the tubesheets and supported at the intermediate baffles can be calculated using a semiempirical equation as follows:

$$f_{n,i} = \frac{\lambda_i^2}{2\pi}\left(\frac{EI}{M_{\text{eff}}L^4}\right)^{1/2} \tag{17.159}$$

where E represents modulus of elasticity, I is area moment of inertia, and M_{eff} is the effective mass of the tube per unit length defined by Eq. 17.156. Length L in Eq. 17.159 is the tube unsupported span length. The coefficient λ_i^2 is the so-called frequency constant which is a function of the mode number i, the number of spans N, and the boundary conditions. The frequency constant for the fundamental frequency $i = 1$ of an N-span beam with clamped ends, pinned intermediate supports, and variable spacing in the outermost spans is presented in Fig. 17.62 [127].

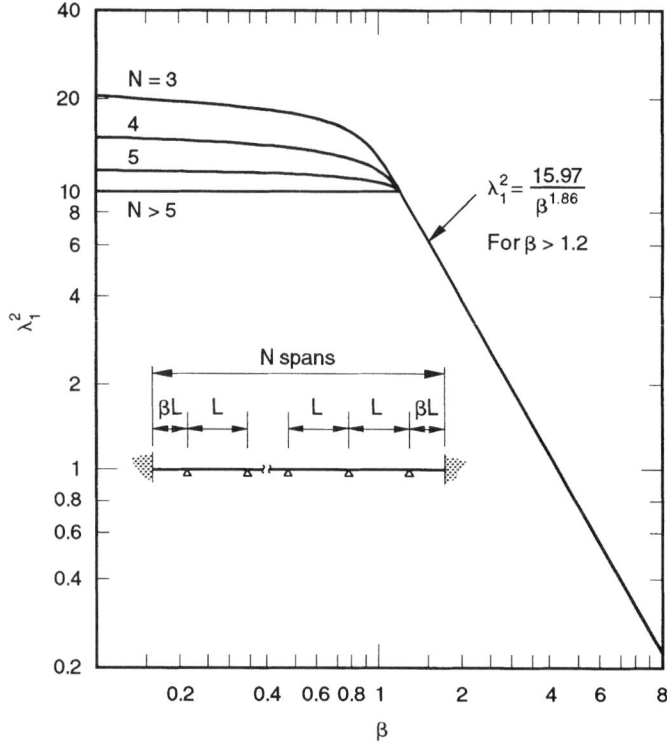

FIGURE 17.62   Frequency constant λ_1^2 for the fundamental frequency of an N span beam for Eq. 17.159 [127].

Various factors influence the tube natural frequency in a shell-and-tube heat exchanger as summarized in Table 17.37. In general, the natural frequency of an unsupported span is influenced primarily by the geometry, elastic properties, inertial properties, span shape, boundary conditions, and axial loading of the tube.

U Tube.   It is more difficult to predict correctly the natural frequency of U tubes than the natural frequency of a straight tube. The fundamental natural frequency can be calculated following the suggestion from TEMA standards [5]:

$$f_n = \frac{C_u}{R^2} \left(\frac{EI}{M_{\text{eff}}} \right)^{1/2} \tag{17.160}$$

where C_u represents the U-tube natural frequency constant, which depends on the span geometry, and R is the mean bend radius. The numerical values of C_u for four characteristic U-bend geometries are given in Fig. 17.63 [5].

Damping Characteristics.   Damping causes vibrations to decay in an elastic structure and depends on the vibration frequency, the material of the elastic structure, the geometry, and the physical properties of the surrounding fluid (in the case of a shell-and-tube heat exchanger, the surrounding fluid is the shell fluid).

The quantitative characteristic of damping is the logarithmic decrement δ_o. It is defined as $\delta_o = \ln (x_n/x_{n+1})$, where x_n and x_{n+1} are the successive midspan amplitudes of a lightly damped structure in free decay. The magnitude of the damping factor is within the range of 0.03 and 0.01 [122]. Statistical analysis of damping factor values compiled by Pettigrew et al. [128] reveals the data as follows. For a heat exchanger tubing in air, the average damping factor is

TABLE 17.37   Influence of Design Factors on the Tube Fundamental Natural Frequency (Modified from Ref. 155)

Influential factor	Variation trend		Comments
	Factor	Frequency	
Length of tube span (unsupported)	↑	↓	The most significant factor.
Tube outside diameter	↑	↑	
Tube wall thickness	↑	↑	A very weak dependence.
Modulus of elasticity	↑	↑	
Number of tube spans	↑	↓	The rate of decrease diminishes with a large number of tube spans.
Tube-to-baffle hole clearance	↑	↓	f_n increases only if there is a press fit and the clearance is very small.
Number of tubes in a bundle	↑	↓	
Baffle spacing	↑	↓	
Baffle thickness	↑	↑	A weak dependence only if the tube-to-baffle clearance is tight.
Tensile stress in tubes	↑	↑	Important in a fixed tubesheet exchanger.
Compressive stress in tubes	↑	↓	A slight decrease; under high compressive loads, high decrease.

0.069 with the standard deviation of 0.0145. For a heat exchanger tubing in water, the average damping factor is 0.0535 with the standard deviation of 0.0110. TEMA standards [5] suggest empirical correlations for δ_o that depend on the fluid thermophysical properties, the outside diameter of the tube, and the fundamental natural frequency and effective mass of the tube.

Acoustic Natural Frequencies.   The natural frequencies of transverse acoustic modes in a cylindrical shell can be calculated as follows [122]:

$$f_{a,i} = \frac{c_{\text{eff}}}{\pi} \frac{a_i}{D_s} \qquad i = 1, 2 \tag{17.161}$$

where
$$c_{\text{eff}} = \frac{c_o}{(1 + \sigma)^{1/2}} \qquad \text{with } c_o = \left(\frac{\gamma}{\rho \kappa_T}\right)^{1/2} \tag{17.162}$$

Here c_{eff} represents the effective speed of sound, c_o is the actual speed of sound in free space, γ is the specific heat ratio, and κ_T is the isothermal compressibility of the fluid. A fraction of shell volume occupied by tubes, *solidity* σ can be easily calculated for a given tube pattern. For example, $\sigma = 0.9069(d_o/p_t)^2$ for an equilateral triangular tube layout, and $\sigma = 0.7853(d_o/p_t)^2$ for a square layout. Coefficients a_i are the dimensionless sound frequency parameters associated with the fundamental diametrical acoustic mode of a cylindrical volume. For the fundamental mode $a_1 = 1.841$, and, for the second mode, $a_2 = 3.054$ [122].

According to Chenoweth [130], acoustic vibration is found more often in tube bundles with a staggered rather than inline layout. It is most common in bundles with the rotated square (45°) layout.

Prediction of Acoustic Resonance.   The procedure for prediction of the onset of acoustic resonance is as follows [122]:

1. Determine the first two natural frequencies of acoustic vibration using Eqs. 17.161 and 17.162. Note that failure to check the second mode may result in the onset of acoustic resonance.

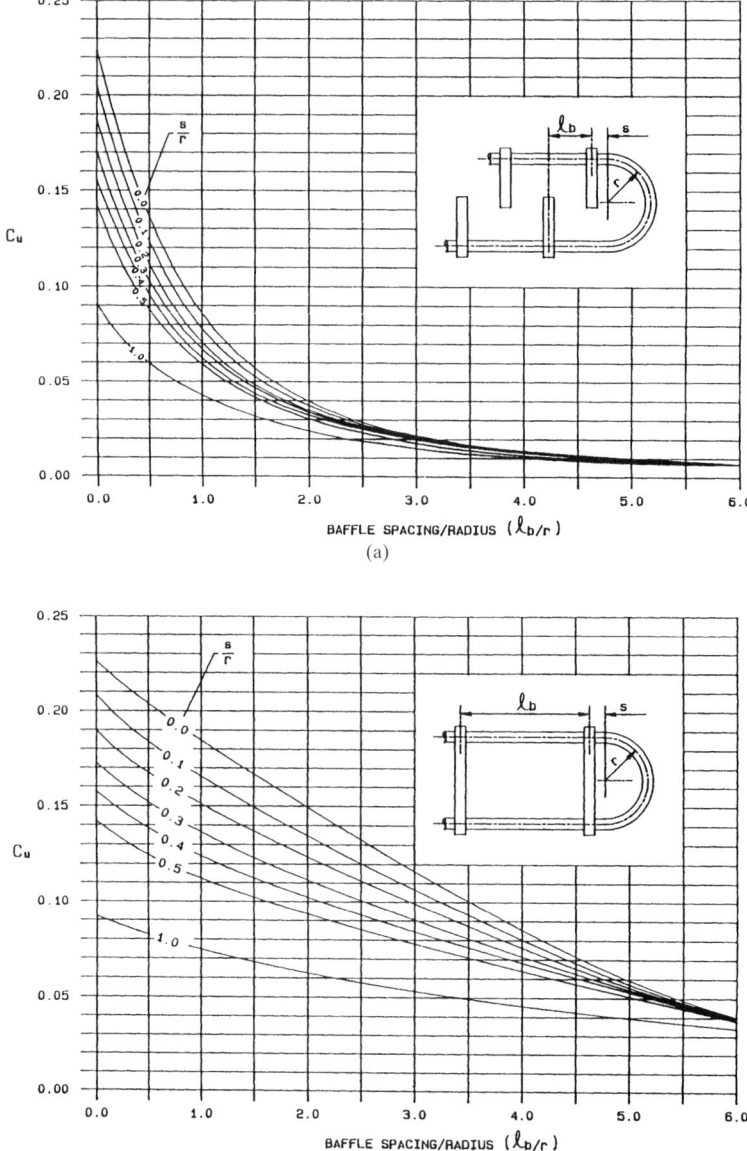

FIGURE 17.63 U-tube frequency constants for geometries shown in (a) and (b) [5].

2. Determine the vortex shedding frequency using Eq. 17.157 and turbulent buffeting frequency using Eq. 17.158. Also compute the natural frequency f_n of the tubes by using Eq. 17.159 or 17.160.

3. Determine the onset of the resonance margin as follows:

$$(1 - \alpha')(f_v \text{ or } f_{tb}) < f_{a,i} < (1 + \alpha'')(f_v \text{ or } f_{tb}) \tag{17.163}$$

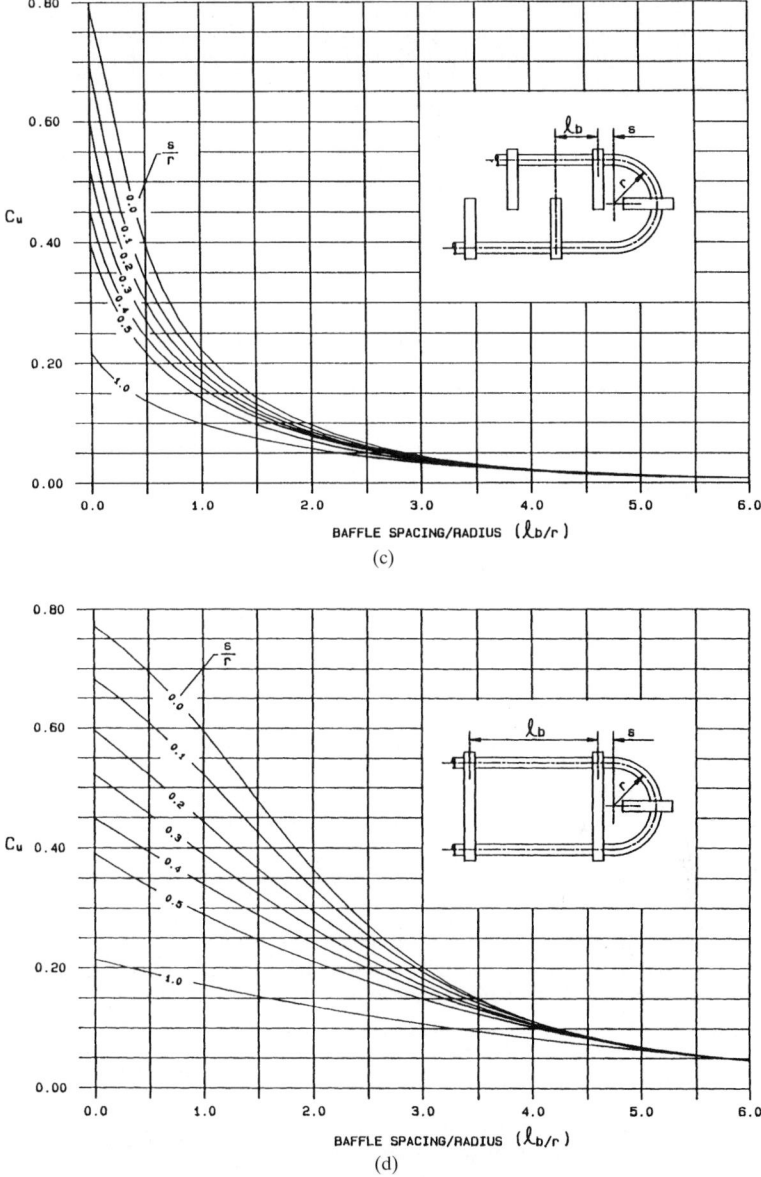

FIGURE 17.63 (*Continued*)   U-tube frequency constants for geometries shown in (*c*) and (*d*) [5].

where the coefficients $\alpha' = \alpha'' = 0.2$ as recommended by TEMA Standards [5] (i.e., f_v or f_{tb} within ±20 percent of $f_{a,i}$). Some measurements [129] suggest different values for this margin, $\alpha' = 0.19$ and $\alpha'' = 0.29$, and the maximum values being 0.4 and 0.48, respectively.

The acoustic frequency of an ith mode should be within the margin defined by Eq. 17.163 for resonance to occur. TEMA Standards [5] suggest somewhat different conditions. In addition to Eq. 17.163, TEMA recommends that the reference crossflow velocity (Table 17.36) must be above certain values involving an additional parameter that depends on the Reynolds and Strouhal numbers and the tube bundle geometry. For details, refer to Ref. 5.

Design Guidelines for Vibration Mitigation

Three major modes of tube failure deserve to be mentioned [122]:

- Fatigue due to repeated bending (i.e., if the stress level in the tube at the tubesheet joint is above the fatigue limit, a circumferential crack will grow about the tube periphery).
- Large-amplitude vibration, resulting in clashing of adjacent tubes at midspan, which will wear flats in neighboring tubes, leading to thinning of the tube walls with eventual splitting (collision damage).
- If there is clearance between a tube and its support, large amplitude tube motion can wear a groove in the tube at a support, in particular if the baffles are thin or harder than the tubes or there is a large baffle-to-tube clearance (baffle damage).

Heat exchanger vibrations can be reduced either by increasing the tube bundle natural frequency or reducing excitation mechanisms. Methods to accomplish that and eventually prevent vibration and potential damage can be summarized as follows [130]:

- Reducing the longest unsupported span length
- Reducing the shellside velocities by increasing tube pitch or using TEMA X- or J-shell styles
- Reducing nozzle velocities by adding annular distributors and/or a support plate at the centerline of the nozzles and/or vapor belts
- Changing the baffle type (multisegmental baffle, RODbaffle type bundle)
- Adding deresonating plates in the exchanger bundle to break the acoustic waves

FLOW MALDISTRIBUTION

A standard idealization of the basic heat exchanger theory is that the fluid flow rate is distributed uniformly through the exchanger on each side of the heat transfer surface. However, in practice, flow maldistribution is more common and can reduce the idealized performance significantly.

Flow maldistribution can be induced by (1) heat exchanger geometry (mechanical design features such as the basic geometry, manufacturing imperfections, and tolerances), and (2) heat exchanger operating conditions (such as viscosity or density-induced maldistribution, multiphase flow, and fouling phenomena). Geometry-induced flow maldistribution can be classified into (a) gross flow maldistribution, (b) passage-to-passage maldistribution, and (c) manifold-induced maldistribution. The most important flow maldistributions caused by operating conditions is the viscosity-induced maldistribution and associated flow instability.

Various problems related to flow maldistribution phenomena including flow instabilities are discussed extensively in the literature. Refer to Mueller and Chiou [131] for a review.

Geometry-Induced Flow Maldistribution

A class of maldistribution phenomena that are a consequence of geometric characteristics of fluid flow passages are called geometry-induced flow maldistributions. This type of maldistribution is closely related to heat exchanger construction and fabrication (such as header design and heat exchanger core fabrication, including brazing in compact heat exchangers). This maldistribution is inherent to a particular heat exchanger in question and cannot be influenced significantly by modifying operating conditions.

Gross Flow Maldistribution.   The major feature of the gross flow maldistribution is that the nonuniform flow happens at the macroscopic level (due to poor header design or blockage of some flow passages in manufacturing operation) and that it does not depend on the local heat transfer surface geometry. This class of maldistribution causes (1) a significant increase in heat exchanger pressure drop and (2) some reduction in heat transfer rate. In order to predict the magnitude of these effects, the nonuniformity is modeled in the literature as one-dimensional or two-dimensional. Some specific results are presented next.

One-Dimensional Flow Nonuniformity.   In this case, the gross flow maldistribution is restricted predominantly to one dimension across the free flow area. A method for predicting the performance of a heat exchanger with this type of nonuniformity is quite straightforward for heat exchangers with simple flow arrangements [100]. The key idea in quantifying the influence of the flow maldistribution on the effectiveness of a heat exchanger involves three interrelated idealizations:

1. The total heat transfer rate in a real heat exchanger with one-dimensional flow maldistribution is equal to the sum of heat transfer rates that would be exchanged in an arbitrary but previously defined number (N) of hypothetical, smaller units, having the same flow arrangement but without the maldistribution.

2. Each of the units defined by idealization 1 obeys the set of standard idealizations of the basic heat exchanger theory listed on p. 17.27.

3. The sum of the heat capacity rates for all smaller units is equal to the total heat capacity rates of the real maldistributed heat exchanger.

With these auxiliary assumptions, the temperature effectiveness of a heat exchanger can be calculated using the following equations after the maldistributed fluid stream is divided into N individual uniform fluid streams:

1. For counterflow and parallelflow arrangements:

$$P_{ms} = \frac{1}{C_{ms}} \sum_{n=1}^{N} C_{ms,n} P_{ms,n} \tag{17.164}$$

2. For crossflow exchanger (mixed-unmixed) with nonuniformity on the unmixed side only:

$$P_{ms} = \frac{1}{C_{ms}} \left[P_{ms,1} C_{ms,1} + \sum_{n=2}^{N} P_{ms,n} C_{ms,n} \prod_{k=1}^{n-1} \left(1 - \frac{P_{ms,k} C_{ms,k}}{C_s} \right) \right] \tag{17.165}$$

where P and C represent temperature effectiveness and heat capacity rate, respectively, and the subscript *ms* denotes maldistributed fluid stream side.

The method is proposed only for the counterflow, parallelflow and crossflow arrangement with one fluid unmixed throughout. Analytical expressions for temperature effectivenesses are provided in Table 17.6 for the uniform flow case and are used for individual N hypothetical units. The heat capacity rate of the maldistributed fluid in the *n*th subheat exchanger, as in a counterflow arrangement, is given by

$$C_{ms,n} = \left(\frac{V_n}{V_m} \right)_{ms} \left(\frac{A_{o,n}}{A_o} \right)_{ms} C_{ms} \tag{17.166}$$

where V_m represents the mean flow velocity, A_o is the minimum free flow area for the whole exchanger, and the subscript n represents the quantities for the *n*th exchanger. The heat capacity rate for other fluid in the *n*th heat exchanger can be calculated using the same equation, Eq. 17.166, but with the velocity ratio equal to 1 (due to uniform flow on that side) and other variables defined for that side of the heat exchanger.

The influence of gross flow maldistribution is shown in Fig. 17.64 for a balanced ($C^* = 1$) counterflow heat exchanger in terms of the performance (effectiveness) deterioration factor, $\Delta\varepsilon^*$. It can be seen that, for a particular value of V_{max}/V_m and given NTU, the greatest reduction in the heat exchanger effectiveness occurs when the velocity function is a two-step function over the flow area. The effect of flow maldistribution increases with NTU. Note that the reduction in the temperature effectiveness obtained using Eqs. 17.164 and 17.165 is valid regardless of whether the maldistributed fluid is the hot, cold, C_{max}, or C_{min} fluid.

Tubeside maldistribution in a counterflow shell-and-tube heat exchanger studied in [132] led to the following major conclusions:

- For $C_s/C_t = 0.1$, the performance loss is negligible for large flow nonuniformities for NTU < 2.
- For $C_s/C_t > 1$, a loss can be noticed but diminishes for NTU > 2.
- $C_s/C_t = 1$ is the worst case at large NTUs (NTUs based on the shell fluid heat capacity rate).

Fleming [133] and Chowdhury and Sarangi [134] have studied various models of flow maldistribution on the tube side of a counterflow shell-and-tube heat exchanger. It is concluded that high NTU heat exchangers are more susceptible to maldistribution effects. According to Mueller [135], the well-baffled 1-1 counterflow shell-and-tube heat exchanger (tube side nonuniform, shell side mixed) is affected the least by flow maldistribution. Shell-and-tube heat exchangers that do not have mixing of the uniform fluid (tube side nonuniform, shell side unmixed; or tube side uniform, shell side nonuniform in crossflow) are affected more by flow maldistribution. According to Ref. 136, the radial flow variations of the mismatched air side and gas side reduce the regenerator effectiveness significantly.

Two-Dimensional Flow Nonuniformity.   The two-dimensional flow maldistribution has been analyzed only for a crossflow exchanger. In a series of publications as summarized in Refs. 131 and 137, Chiou has studied the effects of flow maldistribution on an unmixed-

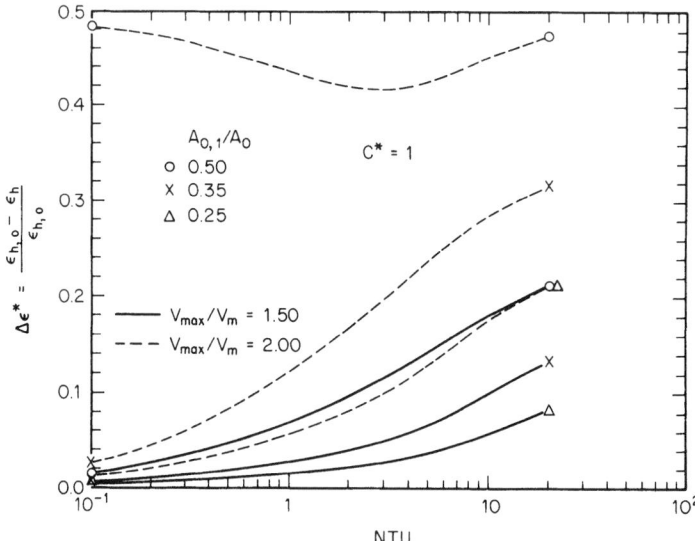

FIGURE 17.64   Reduction in heat exchanger effectiveness caused by gross flow maldistribution [100].

unmixed crossflow single-pass heat exchanger with flow maldistribution on one and both sides. If the flow maldistribution is present only on one side, the following general conclusions were obtained:

- For flow maldistribution on the C_{max} fluid side, the exchanger thermal performance deterioration factor $\Delta\varepsilon^*$ approaches a single value of 0.06 for all $C^* < 1$ if NTU approaches zero. The performance deterioration factor decreases as NTU increases. For a balanced heat exchanger ($C^* = 1$), the exchanger thermal performance deterioration factor continually increases with NTU.

- For flow maldistribution on the C_{min} fluid side, the thermal performance deterioration factor first increases and then decreases as NTU increases.

- If flow nonuniformities are present on both sides, the performance deterioration factor can be either larger or smaller than that for the case where flow nonuniformity is present only on one side, and there are no general guidelines about the expected trends.

A study of the influence of two-dimensional nonuniformities in inlet fluid temperatures [138] indicates that there is a smaller reduction in the exchanger effectiveness for the nonuniform inlet temperature than that for the nonuniform inlet flow. For various nonuniform flow models studied, the inlet nonuniform flow case showed a decrease in the effectiveness of up to 20 percent; whereas, for the nonuniform inlet temperature case, a decrease in the effectiveness of up to 12 percent occurred with even an increase in the effectiveness for some cases of the nonuniform inlet temperature.

Passage-to-Passage Flow Maldistribution. Compact heat exchangers are highly susceptible to passage-to-passage flow maldistribution. Neighboring passages are never geometrically identical because of manufacturing tolerances. It is especially difficult to control precisely the passage size when small dimensions are involved (e.g., a rotary regenerator with $D_h = 0.5$ mm or 0.020 in). Since differently sized and shaped passages exhibit different flow resistances and the flow seeks a path of least resistance, a nonuniform flow through the matrix results. This phenomena usually causes a slight reduction in pressure drop, while the reduction in heat transfer rate may be significant. The influence is of particular importance for continuous flow passages at low Re numbers.

A theoretical analysis for passage-to-passage flow maldistribution was conducted for the so-called plate-spacing and fin-spacing-type nonuniformities influenced by manufacturing tolerances. In the analysis, the actual nonuniform surface is idealized as containing an equal number of large and small passages relative to the nominal passage dimensions. The models include: (a) the two-passage model [139], (b) the three-passage model, and (c) the N-passage model [140]. Both triangular and rectangular passage cross sections were studied. The influence of the fin curvature was studied in Ref. 141.

In the N-passage model, there are N differently sized passages of the same basic shape, either rectangular or triangular. In Fig. 17.65a, a reduction in NTU for rectangular passages is shown when 50 percent of the flow passages are large ($c_2 > c_r$) and 50 percent of the passages are small ($c_1 < c_r$) compared to the nominal passages. The results are presented for the passages having a nominal aspect ratio α_r^* of 1, 0.5, 0.25, and 0.125 for the $\textcircled{H}$ and $\textcircled{T}$ boundary conditions and for a reference NTU of 5.0. Here, a percentage loss in NTU and the channel deviation parameter δ_c are defined as

$$\text{NTU}_{\text{cost}}^* = \left(1 - \frac{\text{NTU}_e}{\text{NTU}_r}\right) \times 100 \tag{17.167}$$

$$\delta_c = 1 - \frac{c_1}{c_r} \tag{17.168}$$

where NTU_e is the effective NTU when a two-passage model passage-to-passage nonuniformity is present, and NTU_r is the reference or nominal NTU. It can be seen from Fig.

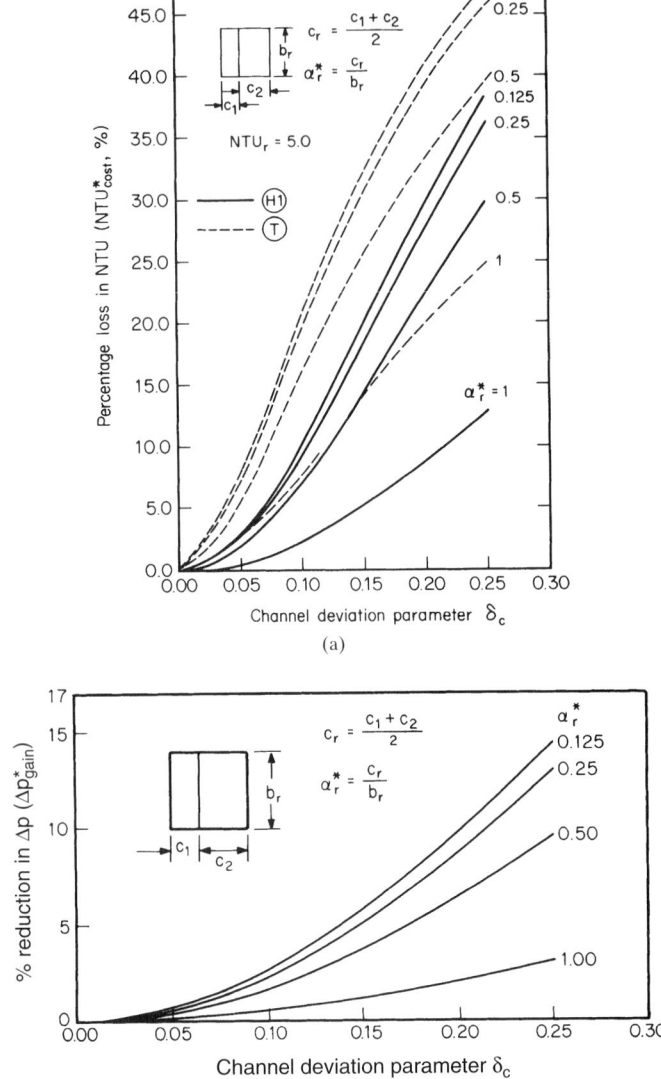

FIGURE 17.65 (*a*) Percentage loss in NTU as a function of δ_c, α_r^*, and thermal boundary conditions for two-passage nonuniformities in rectangular passages. (*b*) Percentage reduction in Δp as a function of δ_c and α_r^* for two-passage nonuniformities in rectangular passages.

17.65*a* that a 10 percent channel deviation (which is common for a highly compact surface) results in a 10 and 21 percent reduction in NTU_{H1} and NTU_T, respectively, for $\alpha_r^* = 0.125$ and $NTU_r = 5.0$. In contrast, a gain in the pressure drop due to the passage-to-passage nonuniformity is only 2.5 percent for $\delta_c = 0.10$ and $\alpha_r^* = 0.125$, as found from Fig. 17.65*b*. Here, Δp_{gain}^* (reduction in Δp) is defined as:

$$\Delta p^*_{\text{gain}} = \left(1 - \frac{\Delta p_{\text{actual}}}{\Delta p_{\text{nominal}}}\right) \times 100 \qquad (17.169)$$

The results of Figs. 17.65a and b are also applicable to an N-passage model in which there are N differently sized passages in a normal distribution about the nominal passage size. The channel deviation parameter needs to be modified for this case to

$$\delta_c = \left[\sum_{i=1}^{N} \chi_i \left(1 - \frac{c_i}{c_r}\right)^2\right]^{1/2} \qquad (17.170)$$

Here, χ_i is the fractional distribution of the ith shaped passage. For $N = 2$ and $\chi_i = 0.5$, Eq. 17.170 reduces to Eq. 17.168.

The following observations may be made from Fig. 17.65a and additional results presented in [140]: (1) The loss in NTU is more significant for the Ⓣ boundary condition than for the Ⓗ boundary condition. (2) The loss in NTU increases with higher nominal NTU. (3) The loss in NTU is much more significant compared to the gain in Δp at a given δ_c. (4) The deterioration in performance is the highest for the two-passage model compared to the N-passage model with $N > 2$ for the same value of c_{max}/c_r.

Results similar to those in Fig. 17.65a and b are summarized in Fig. 17.66 for the N-passage model of nonuniformity associated with equilateral triangular passages. In this case, the definition of the channel deviation parameter δ_c is modified to

$$\delta_c = \left[\sum_{i=1}^{N} \chi_i \left(1 - \frac{r_{h,i}}{r_{h,r}}\right)^2\right]^{1/2} \qquad (17.171)$$

Manifold Induced Flow Maldistribution.   Manifolds can be classified as two basic types: simple dividing flow and combining flow. When interconnected by lateral branches, these manifolds result into the parallel and reverse-flow systems, as shown in Fig. 17.67a and b; these were investigated by Bajura and Jones [142] and Datta and Majumdar [143]. A few general conclusions from these studies are as follows:

- To minimize flow maldistribution, one should limit to less than unity the ratio of flow area of lateral branches (exchanger core) to flow area of the inlet header (area of pipe before lateral branches).
- A reverse-flow manifold system provides more uniform flow distribution than a parallel-flow manifold system.
- In a parallel-flow manifold system, the maximum flow occurs through the last port and, in the reverse-flow manifold system, the first port.
- The flow area of a combining-flow header should be larger than that for the dividing-flow header for a more uniform flow distribution through the core in the absence of heat transfer within the core. If there is heat transfer in lateral branches (core), the flow areas should be adjusted first for the density change, and then the flow area of the combining header should be made larger than that calculated previously.
- Flow reversal is more likely to occur in parallel-flow systems that are subject to poor flow distribution.

Flow Maldistribution Induced by Operating Conditions

Operating conditions (temperature differences, number of phases present, etc.) inevitably influence thermophysical properties (viscosity, density, quality, onset of oscillations) of the flowing fluids, which, in turn, may cause various flow maldistributions, both steady and transient in nature.

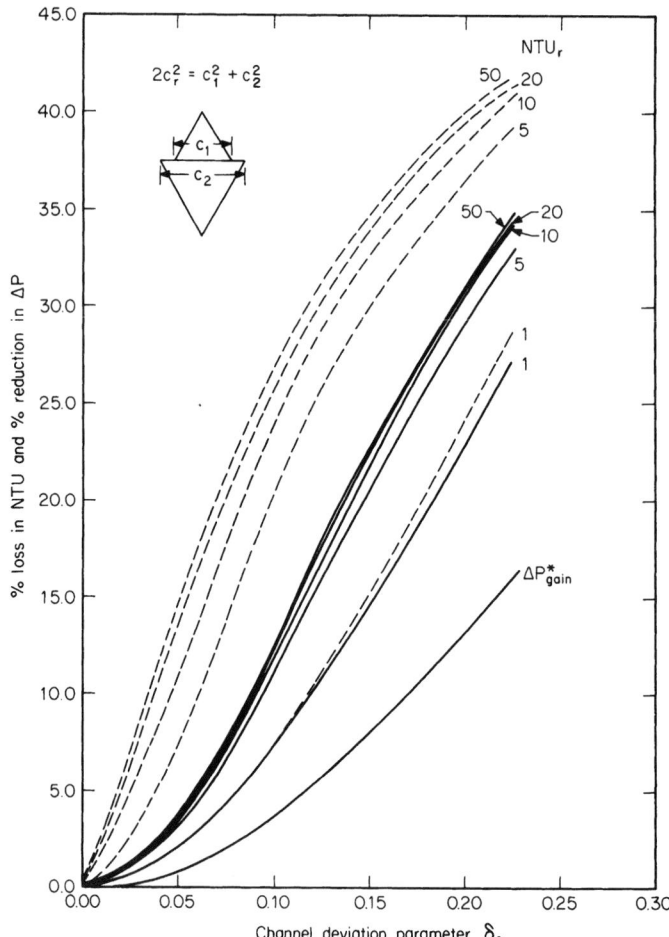

FIGURE 17.66 Percentage loss in NTU and percentage reduction in Δp as functions of δ_c for N-passage nonuniformities in equilateral triangular passages.

Viscosity-Induced Flow Maldistribution.   Viscosity-induced flow instability and maldistribution are results of large changes in fluid viscosity within the exchanger as a result of different heat transfer rates in different tubes (flow passages). The possibility for an onset of an instability phenomenon is present whenever one or more fluids are liquids and the viscous liquid is cooled. Flow maldistribution and flow instability are more likely in laminar flow ($\Delta p \propto \mu$) as compared to the turbulent flow ($\Delta p \propto \mu^{0.2}$). Mueller [144] has proposed a procedure for determining the pressure drop or mass flow rate (in a single-tube laminar flow cooler) above which the possibility of flow maldistribution that produces flow instability within a multitubular heat exchanger is eliminated. Putnam and Rohsenow [145] investigated the flow instability phenomenon that occurs in noninterconnected parallel passages of laminar flow heat exchangers.

If a viscous liquid stream is cooled, the viscosity of the liquid may either vary along the flow path influenced by the local bulk fluid temperature $\mu(T)$, or it may stay invariant, defined by the constant wall temperature $\mu_w(T_w)$. The total pressure drop between the inlet

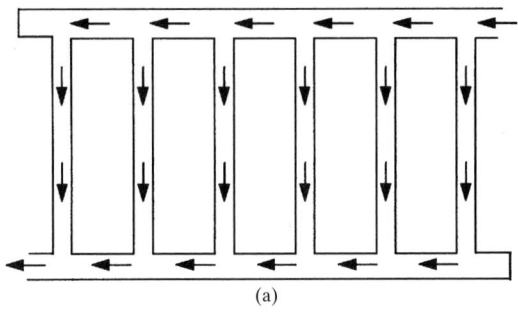

(a)

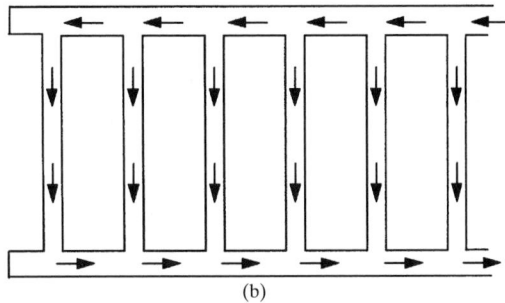

(b)

FIGURE 17.67   Manifold systems: (*a*) parallel flow, (*b*) reverse flow.

and outlet of the flow passage could be approximated as a sum of the two terms that are based on the two viscosity regions defined by an axial location x below which μ is dependent on T and beyond which μ is dependent only on the wall temperature. Assume that one can define an average viscosity μ_m such that, when used in the standard pressure drop equation, gives the true pressure drop for the tube section between the tube inlet and the location x. From x to the tube exit ($x = L$), the viscosity is μ_w. The pressure drop in the second zone should be calculated using viscosity μ_w. The total pressure drop is equal to the sum of the two above mentioned pressure drops. It can be shown [144] that the pressure drop calculated in such a manner behaves as shown in Fig. 17.68 as a function of the flow rate. The analysis is based on: (1) fully developed laminar flow in the tube, (2) the viscosity, which is the only fluid property that can vary along the flow path, (3) the only frictional pressure drop contribution significant, and (4) the wall temperature, which is constant and lower than the fluid inlet temperature. There will be no flow-maldistribution-induced instability in a multitubular cooler if the mass flow rate per tube, assuming uniform distribution $\dot{m}_m$ is greater than $\dot{m}_{\min}$ in Fig. 17.68.

Mueller [144] proposed the following procedure to determine the maximum pressure drop, $\Delta p_{\max}$, above which the flow-maldistribution-induced instability would not be possible. The case considered is for a viscous liquid of a known inlet temperature being cooled as it flows through the length of a tube of known constant temperature T_w.

1. From viscosity data, determine the slope m of curve ln (μ) versus $1/T$ where T is temperature on the absolute temperature scale.

2. With known m and the inlet (μ_i) and wall (μ_w) viscosities, determine the average viscosity (μ_m) using Fig. 17.69. This figure is based on the assumption that the fluid reaches the wall

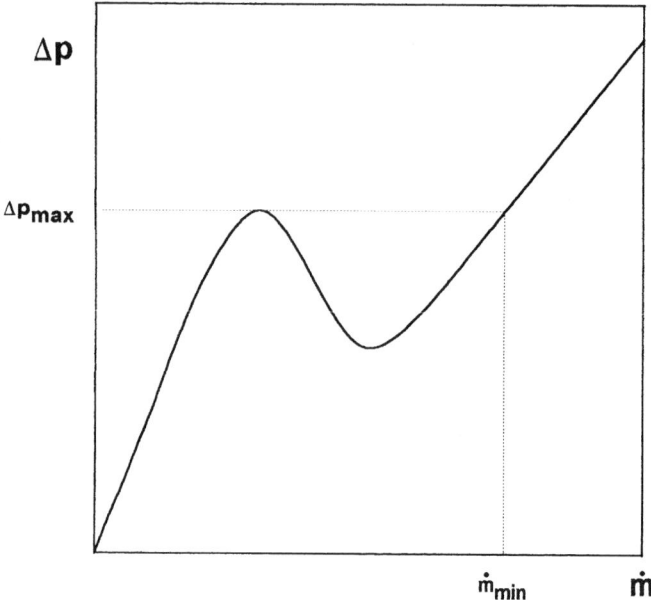

FIGURE 17.68   Pressure drop versus mass flow rate for a single flow passage in laminar liquid flow cooling [144].

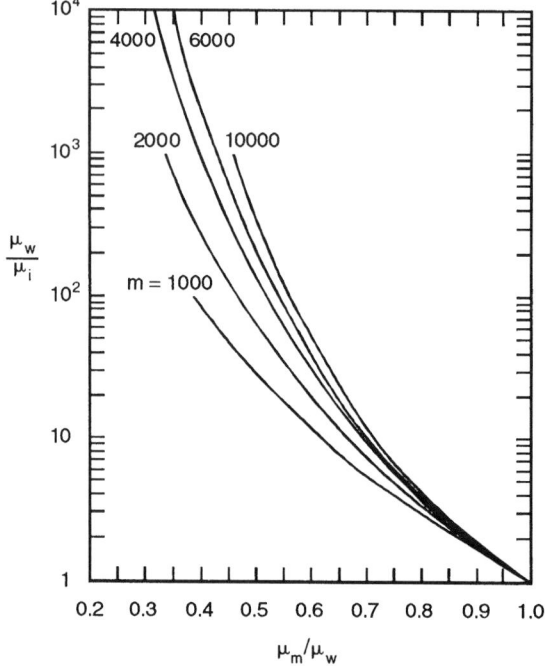

FIGURE 17.69   Viscosity ratio chart for various slopes m of $[\mu = Be^{m/T}]$ [144].

temperature within the tube. If the fluid exit temperature is still larger than the wall temperature, the average viscosity should be modified. The details are provided in Ref. 144.

3. With these viscosities, determine x from

$$x = \frac{L/2}{1 - (\mu_m/\mu_w)} \tag{17.172}$$

4. Calculate the mass flow rate from

$$\text{Gz} = \frac{kL}{\dot{m}c_p} \approx 0.4 \tag{17.173}$$

5. Calculate the maximum pressure drop from

$$\Delta p_{\max} = \frac{128}{\pi g_c \rho D_h^4} \dot{m}\mu_w \left[x + \frac{x^2}{L}\left(\frac{\mu_m}{\mu_w} - 1\right) \right] \tag{17.174}$$

If the pressure drop is found to be less than calculated by Eq. 17.174, the fluid flow length should be increased (either increasing the duct length or considering a multipass design) to eliminate the flow instability.

The preceding method can be used to calculate the pressure drop magnitudes for various tubes in a multitube unit and consequently to visualize the eventual flow maldistribution instability. For further details regarding the procedure, refer to Ref. 144.

Flow Maldistribution in Heat Exchangers with Phase Change. Two-phase flow maldistribution may be caused and/or influenced by phase separation, oscillating flows, variable pressure drops (density-wave instability), flow reversals, and other flow instabilities. For a review of pertinent literature, refer to Ref. 131.

Flow maldistribution problems in condensers may be severe and should not be underestimated [146]. Most problems with flow maldistribution in condensers, though, can be resolved through good venting and condensate drainage [2].

The flow maldistribution problem in evaporators is most severe when the major part of the pressure drop occurs in the two-phase heated region and cannot be prevented in some cases (reboilers and evaporators). However, in most heat exchangers with such problems, the economic penalty for maldistribution is small, and deterioration of the thermal performance is slightly affected [115]. The flow instabilities can be controlled by restricting the liquid circulation [2]. Falling film evaporators are particularly prone to flow maldistribution. This effect can be reduced through design modifications (the top tube plate must be exactly horizontal, addition of inserts into the top of each tube must be secured, etc.) [120].

Mitigation of Flow Maldistribution

Flow maldistribution in a heat exchanger may be reduced through modifications in the existing design or taken into account by incorporating its effect in the design methodology. Most gross flow maldistributions may result in minor performance reduction (smaller than 5 percent for NTU < 4 [131] in shell-and-tube heat exchangers). At high NTUs (NTU > 10), the performance loss may be substantially larger. The passage-to-passage maldistribution may result in a significant reduction in heat transfer performance, particularly for laminar flow exchangers.

Any action in mitigating flow maldistribution must be preceded by an identification of possible reasons that may cause the performance deterioration and/or may affect mechanical characteristics of the heat exchanger. The possible reasons that affect the performance are [131, 147]: (1) deterioration in the heat exchanger effectiveness and pressure drop characteristics, (2) fluid freezing, as in viscous flow coolers, (3) fluid deterioration, (4) enhanced fouling, and (5) mechanical and tube vibration problems (flow-induced vibrations as a consequence of flow instabilities, wear, fretting, erosion, corrosion, and mechanical failure).

No generalized recommendations can be made for mitigating negative consequences of flow maldistribution. Most of the problems must be solved by intelligent designs and on an individual basis. A few broad guidelines regarding various heat exchanger types follow.

In shell-and-tube heat exchangers, inlet axial nozzles on the shell side may induce gross flow maldistribution. Placing an impingement perforated baffle about halfway to the tubesheet will break up the inlet jet stream [131]. It is speculated also that a radial nozzle may eliminate jet impingement. The shell inlet and exit baffle spaces are regions prone to flow maldistribution. An appropriate design of the baffle geometry (for example the use of double segmental or disk-and-doughnut baffles) may reduce this maldistribution.

Flow maldistribution is often present in phase-change applications. A common method to reduce the flow maldistribution in condensers is to use a vent condenser or increase the number of tubeside passes [131]. To minimize the negative influence of flow maldistribution, one should reduce the pressure drop downstream of the vaporizer tube bundle and throttle the inlet stream to prevent oscillations. Also, for reboilers and vaporizers, the best solution is to use a vertical exchanger with the two-phase fluid to be vaporized entering on the shell side through annular distributor [147]. Rod-type baffles should be used whenever appropriate. Prevention of maldistribution in air-cooled condensers includes the following measures [115]: (1) selective throttling of the vapor flow to each tube row, (2) use of a downstream condenser to eliminate the effects of inert gas blanketing by having a definite stream flow through each tube row, and (3) matching the heat transfer characteristics of each tube row so as to produce uniform heat transfer rate through each tube row.

FOULING AND CORROSION

Fouling and corrosion, both operation-induced effects, should be considered for the design of a new heat exchanger as well as subsequent exchanger operation. Fouling represents an undesirable accumulation of deposits on heat transfer surface. Fouling is a consequence of various mass, momentum, and transfer phenomena involved with heat exchanger operation. The manifestations of these phenomena, though, are more or less similar. Fouling results in a reduction in thermal performance and an increase in pressure drop in a heat exchanger. Corrosion represents mechanical deterioration of construction materials of a heat exchanger under the aggressive influence of flowing fluid and the environment in contact with the heat exchanger material. In addition to corrosion, some other mechanically induced phenomena are important for heat exchanger design and operation, such as fretting (corrosion occurring at contact areas between metals under load subjected to vibration and slip) and fatigue (a tendency of a metal to fracture under cyclic stressing).

In order to understand the influence of fouling on compact heat exchanger performance, the following equations for h and Δp are derived from the equations presented earlier for fully developed gas flow in a circular or noncircular tube:

$$h = \begin{cases} \dfrac{\mathrm{Nu}\,k}{D_h} \text{ with Nu = constant} & \text{for laminar flow} \\[3mm] \dfrac{k}{D_h}\left[0.022\left(\dfrac{4L\dot{m}}{A\mu}\right)^{0.8}\mathrm{Pr}^{0.5}\right] & \text{for turbulent flow} \end{cases} \qquad (17.175)$$

and

$$\Delta p = \begin{cases} \dfrac{1}{D_h^3}\left[\dfrac{1}{2g_c}\dfrac{\mu}{\rho}\dfrac{16L^2}{A}\dot{m}(f\,\mathrm{Re})\right] & \text{for laminar flow} \\[3mm] \dfrac{1}{D_h^3}\left[\dfrac{0.046}{2g_c}\dfrac{\mu^{0.2}}{\rho}\dfrac{(4L)^{2.8}}{A^{1.8}}\dot{m}^{1.8}\right] & \text{for turbulent flow} \end{cases} \qquad (17.176)$$

For constant $\dot{m}$, L, A, and fluid properties, from Eqs. 17.175 and 17.176,

$$h \propto \frac{1}{D_h} \qquad \Delta p \propto \frac{1}{D_h^3} \qquad (17.177)$$

Since $A = \pi D_h L$, Δp is proportional to D_h^{-4} and $D_h^{-4.8}$ in laminar and turbulent flows, respectively. As fouling will reduce the flow area A_o and hence the passage D_h, it will increase h to some extent, but the pressure drop is increased more strongly. The thermal resistance of the fouling film will generally result in an overall reduction in heat transfer in spite of a slight increase in h.

The ratio of pressure drops of fouled (Δp_F) and clean exchanger (Δp_C) for constant mass flow rate is given by [151]:

$$\frac{\Delta p_F}{\Delta p_C} = \frac{f_F}{f_C}\left(\frac{D_{h,C}}{D_{h,F}}\right)\left(\frac{u_{m,F}}{u_{m,C}}\right)^2 = \frac{f_F}{f_C}\left(\frac{D_{h,C}}{D_{h,F}}\right)^5 \qquad (17.178)$$

If we consider that fouling does not affect friction factor (i.e., the friction factor under clean conditions f_C is equal to the friction factor under fouled conditions f_F) and the reduction in the tube inside diameter due to fouling is only 10 to 20 percent, the resultant pressure drop increase will be approximately 60 percent and 250 percent, respectively, according to Eq. 17.178, regardless of whether the fluid is liquid or gas (note that $h \propto 1/D_h$ and $\Delta p \propto 1/D_h^5$ for fully developed turbulent flow and constant mass flow rate). At the same time, the slight increase in h will not increase the overall heat transfer coefficient because of the additional thermal resistance of the fouling layers.

Fouling in liquids and two-phase flows has a significant detrimental effect on heat transfer with some increase in fluid pumping power. In contrast, fouling in gases reduces heat transfer somewhat (5–10 percent in general) but increases pressure drop and fluid pumping power significantly (up to several hundred percent). Thus, although the effect of fouling on the pressure drop is usually neglected with liquid flows, it can be significant for heat exchangers with gas flows.

Fouling

General Considerations.   The importance of fouling phenomena stems from the fact that the fouling deposits increase thermal resistance to heat flow. According to the basic theory, the heat transfer rate in the exchanger depends on the sum of thermal resistances between the two fluids, Eq. 17.5. Fouling on one or both fluid sides adds the thermal resistance $\mathbf{R}_s$ to the overall thermal resistance and, in turn, reduces the heat transfer rate (Eq. 17.4). Simultaneously, hydraulic resistance increases because of a decrease in the free flow area. Consequently, the pressure drops and the pumping powers increase (Eq. 17.63).

Fouling is an extremely complex phenomena characterized by a combined heat, mass, and momentum transfer under transient conditions. Fouling is affected by a large number of variables related to heat exchanger surfaces, operating conditions, and fluids. In spite of the complexity of the fouling process, a general practice is to include the effect of fouling on the exchanger thermal performance by an empirical fouling factor $r_s = 1/h_s$. The problem, though, is that this straightforward procedure will not (and cannot) reflect a real transient nature of the fouling process. Current practice is to use fouling factors from TEMA [5] or modified recent data by Chenoweth [148]. See Table 17.38. However, probably a better approach is to eliminate the fouling factors altogether in the design of an exchanger and thus avoid overdesign [149]. This is because overdesign reduces the flow velocity and promotes more fouling.

Types of Fouling Mechanisms.   The nature of fouling phenomena greatly depends on the fluids involved as well as on the various parameters that control the heat transfer phenomena and the fouling process itself. There are six types of liquid-side fouling mechanisms: (1) precipitation (or crystallization) fouling, (2) particulate fouling, (3) chemical reaction fouling, (4)

TABLE 17.38 Fouling Resistances of Various Liquid Streams (Adapted from Ref. 148)

Fluid	Fouling resistance $r_s \times 10^4$ (m^{2}K/W)	Comments
Liquid water streams		
Seawater	1.75–3.5	$T_{out,max} = 43°C$
Brackish water	3.5–5.3	$T_{out,max} = 43°C$
Treated cooling tower water	1.75–3.5	$T_{out,max} = 49°C$
Artificial spray pond	1.75–3.5	$T_{out,max} = 49°C$
Closed loop treated water	1.75	
River water	3.5–5.3	*Operating conditions for all water streams:*
Engine jacket water	1.75	For tubeside flow, the velocity for the
Distilled water or closed cycle		streams is at least 1.2 m/s for tubes of
condensate	0.9–1.75	nonferrous alloy and 1.8 m/s for ferrous
Treated boiler feedwater	0.9	alloys. For shellside fluid, the velocity is
Boiler blowdown water	3.5–5.3	at least 0.6 m/s. Heat transfer surface temperatures are below 71°C.
Industrial liquid streams		
No. 2 fuel oil	3.5	
No. 6 fuel oil	0.9	
Transformer oil, engine lube oil	1.75	
Refrigerants, hydraulic fluid, ammonia	1.75	
Industrial organic HT fluids	1.75–3.5	
Ammonia (oil bearing)	5.3	
Methanol, ethanol, ethylene glycol		
solutions	3.5	
Process liquid streams		
MEA and DEA solutions	3.5	
DEG and TEG solutions	3.5	
Stable side draw and bottom products	1.75–3.5	
Caustic solutions	3.5	
Crude oil refinery streams:		
temperature, °C		
120	3.5–7	Assumes that the crude oil is desalted
120 to 180	5.3–7	at approximately 120°C and the tubeside
180 to 230	7–9	velocity of the stream is 1.25 m/s or
>230	9–10.5	greater.
Petroleum streams		
Lean oil	3.5	
Rich oil	1.75–3.5	
Natural gasoline, liquefied		
petroleum gases	1.75–3.5	
Crude and vacuum unit gases and vapors		The values listed in this table are typical
Atmospheric tower overhead		values that reflect current trends to longer
vapors, naphthas	1.7	periods before cleaning. It is recognized
Vacuum overhead vapors	3.5	that fouling resistances are not known
Crude and vacuum liquids		with precision. Actual applications may
Gasoline	3.5	require substantially different values.
Naphtha, light distillates,		
kerosine, light gas oil	3.5–5.3	
Heavy gas oil	5.3–9	
Heavy fuel oil	5.3–12.3	
Vacuum tower bottoms	17.6	
Atmospheric tower bottoms	12.3	

TABLE 17.38 Fouling Resistances of Various Liquid Streams (Adapted from Ref. 148) (*Continued*)

Fluid	Fouling resistance $r_s \times 10^4$ (m²K/W)	Comments
Cracking and coking unit streams		
Overhead vapors, light liquid products	3.5	
Light cycle oil	3.5–5.3	
Heavy cycle oil, light coker gas oil	5.3–7	
Heavy coker gas oil	7–9	
Bottoms slurry oil	5.3	
Catalytic reforming, hydrocracking, and hydrodesulfurization streams		
Reformer charge, reformer effluent	2.6	Depending on charge characteristics and
Hydrocharger charge and effluent	3.5	storage history, charge fouling resistance
Recycle gas, liquid product over 50°C	1.75	may be many times larger.
Liquid product 30°C to 50°C (API)	3.5	
Light ends processing streams		
Overhead vapors, gases, liquid products	1.75	
Absorption oils, reboiler streams	3.5–5.3	
Alkylation trace acid streams	3.5	
Visbreaker		
Overhead vapor	5.3	
Visbreaker bottoms	17.5	
Naphtha hydrotreater		
Feed	5.3	
Effluent, naphthas	3.5	
Overhead vapor	2.6	
Catalytic hydrodesulfurizer		
Charge	7–9	
Effluent, HT separator overhead, liquid products	3.5	
Stripper charge	5.3	
HF alky unit		
Alkylate, depropanizer bottoms	5.3	
Main fractional overhead, and feed	5.3	
Other process streams	3.5	
Industrial gas or vapor streams		
Steam (non-oil-bearing)	9	The original data for fouling resistance are
Exhaust steam (oil-bearing)	2.6–3.5	given in U.S. Customary units with single-
Refrigerant (oil-bearing)	3.5	digit accuracy. The conversion into SI
Compressed air	1.75	units has as a consequence that the
Ammonia	1.75	apparent accuracy seems greater than the
Carbon dioxide	3.5	intent of the original data.
Coal flue gas	17.5	
Natural gas flue gas	9	
Chemical process streams		
Acid gas	3.5–5.3	
Solvent vapor	1.75	
Stable overhead products	1.75	
Natural gas processing streams		
Natural gas	1.75–3.5	
Overheat products	1.75–3.5	

corrosion fouling, (5) biological fouling, and (6) freezing (solidification) fouling. Only biological fouling does not occur in gas-side fouling, since there are no nutrients in the gas flows. In reality, more than one fouling mechanism is present in many applications, and the synergistic effect of these mechanisms makes the fouling even worse than predicted or expected.

In precipitation fouling, the dominant mechanism is the precipitation of dissolved substances on the heat transfer surface. The deposition of solids suspended in the fluid onto the heat transfer surface is a major phenomenon involved with particulate fouling. If the settling occurs due to gravity, the resulting particulate fouling is called *sedimentation fouling*. Chemical reaction fouling is a consequence of deposition of material produced by chemical reactions in which the heat transfer surface material is *not* a reactant. Corrosion of the heat transfer surface may produce products that foul the surface or promote the attachment of other foulants. Biological fouling results from the deposition, attachment, and growth of macro- or microorganisms to the heat transfer surface. Finally, freezing fouling is due to the freezing of a liquid or some of its constituents or the deposition of solids on a subcooled heat transfer surface as a consequence of liquid-solid or gas-solid phase change in a gas stream.

It is obvious that one cannot talk about a single, unified theory to model the fouling process. However, it is possible to extract a few parameter sets that would most probably control any fouling process. These are: (1) the physical and chemical properties of a fluid, (2) fluid velocity, (3) fluid and heat transfer surface temperatures, (4) heat transfer surface properties, and (5) the geometry of the fluid flow passage. For a given fluid-surface combination, the two most important design variables are the fluid flow velocity and heat transfer surface temperature. In general, higher-flow velocities may cause less foulant deposition and/or more pronounced deposit erosion, but, at the same time, it may accelerate the corrosion of the surface by removing the heat transfer surface material. Higher surface temperatures promote chemical reaction, corrosion, crystal formation (with inverse solubility salts), and polymerization, but they also reduce biofouling, prevent freezing, and precipitation of normal solubility salts. Consequently, it is frequently recommended that the surface temperature be maintained low.

Before considering any technique for minimizing fouling, the heat exchanger should be designed to minimize or eliminate fouling. For example, direct-contact heat exchangers are very convenient for heavily fouling liquids. In fluidized bed heat exchangers, the bed motion scours away the fouling deposit. Plate-and-frame heat exchangers can be easily disassembled for cleaning. Compact heat exchangers are not suitable for fouling service unless chemical cleaning or thermal baking is possible. When designing a shell-and-tube heat exchanger, the following are important in reducing or cleaning fouling. The heavy fluid should be kept on the tube side for cleanability. Horizontal heat exchangers are easier to clean than vertical ones. The geometric features of fluid flow passages should reduce to minimum stagnant and low-velocity shellside regions. On the shell side, it is easier to mechanically clean square or rotated square tube layouts with an increased tube pitch than the other types of tube layouts.

Single-Phase Liquid-Side Fouling.   Single-phase liquid-side fouling is most frequently caused by: (1) precipitation of minerals from the flowing liquid, (2) deposition of various particles, (3) biological fouling, and (4) corrosion fouling. Other fouling mechanisms are also present. More important, though, is the synergistic effect of more than one fouling mechanism present. The qualitative effects of some of the operating variables on these fouling mechanisms are shown in Table 17.39 [2].

The quantitative effect of fouling on heat transfer can be estimated by utilizing the concept of fouling resistance and calculating the overall heat transfer coefficient (Eq. 17.6) under both fouling and clean conditions. An additional parameter for determining this influence, used frequently in practice, is the so-called cleanliness factor. It is defined as a ratio of an overall heat transfer coefficient determined for fouling conditions and an overall heat transfer coefficient determined for clean (fouling-free) operating conditions. The effect of fouling on pressure drop can be determined by the reduced free flow area due to fouling and the change in the friction factor, if any, due to fouling.

TABLE 17.39 Influence of Operating Variables on Liquid-Side Fouling [2]

Operating variables	Precipitation	Freezing	Particulate	Chemical	Corrosion	Biological
Temperature	↑↓	↓	↑↓ ↔	↑↓	↑↓	↑↓ ↔
Velocity	↓ ↔	↑↓	↓	↓	↑↓ ↔	↑↓
Supersaturation	↑	↑	○	○	○	○
pH	↑	○	↑↓	○	↑↓	↑↓
Impurities	○	↓	○	○	○	○
Concentration	↑	↑	↑	○	○	○
Roughness	↑	↑	↑ ↔	○	↑ ↔	↑
Pressure	↔	↔	○	↑	↑	↑↓
Oxygen	↔	↔	○	↑	↑	↑↓

When the value of an operating variable is increased, it increases (↑), decreases (↓), or has no effect (↔) on the specific fouling mechanism listed. Circles (○) indicate that no influence of these variables has been reported in the literature.

Prevention and Reduction of Liquid-Side Fouling. Among the most frequently used techniques for reduction of liquid-side fouling is the online utilization of chemical inhibitors/additives. The list of additives includes: (1) dispersants to maintain particles in suspension, (2) various compounds to prevent polymerization and chemical reactions, (3) corrosion inhibitors or passivators to minimize corrosion, (4) chlorine and other biocide/germicides to prevent biofouling, and (5) softeners, acids, and poliphosphates to prevent crystallization. Finally, an efficient mechanical removal of particles can be performed by filtration. An extensive review of fouling control measures is provided in Ref. 150.

Heat transfer surface cleaning techniques can be applied either online or off-line. Online techniques (usually used for tubeside applications) include various mechanical techniques (flow-driven or power-driven rotating brushes, acoustic/mechanical vibration, chemical feeds, flow reversal, etc.). Off-line techniques include chemical cleaning, mechanical cleaning by circulating particulate slurry, and thermal baking to melt frost/ice deposits.

Single-Phase Gas-Side Fouling. Gas-side fouling may be caused by precipitation (scaling), particulate deposition, corrosion, chemical reaction, and freezing. Formation of hard scale from the gas flow occurs if the sufficiently low temperature of the heat transfer surface forces salt compounds to solidity. Acid vapors, high-temperature removal of oxide layer by molten ash, or salty air at low temperatures may promote corrosion fouling. An example of particulate deposition is accumulation of plant residues. An excess of various chemical substances such as sulfur, vanadium, and sodium initiates various chemical reaction fouling problems. Formation of frost and various cryodeposits are typical examples of freezing fouling on the gas side. An excellent overview of gas-side fouling of heat transfer surfaces is given by Marner [151]. Qualitative effects of some of the operating variables on gas-side fouling mechanisms is presented in Table 17.40 [2].

Control of fouling should be attempted first before any cleaning method is attempted. The fouling control procedure should be preceded by: (1) verification of the existence of fouling, (2) identification of the feature that dominates the foulant accumulation, and (3) characterization of the deposit.

Prevention and Reduction of Gas-Side Fouling. The standard techniques for reduction and/or prevention of fouling on the gas side are: (1) techniques for removal of potential residues from the gas, (2) additives for the gas side fluid, (3) surface cleaning techniques, and (4) adjusting design up-front to minimize fouling. Details regarding various techniques for gas-side fouling prevention, mitigation, and accommodation are given in Ref. 152.

Fouling Under Phase Change Conditions. Fouling is common on the water side in a boiler. Large heat fluxes are subsequently reduced, and elevated wall temperatures may cause tube

TABLE 17.40   Influence of Operating Variables on Gas-Side Fouling [2]

Operating variables	Particulate	Freezing	Chemical	Corrosion
Temperature	↑↓	↓	↑	↑↓ ↔
Velocity	↑↓ ↔	↓	↑↓ ↔	↑ ↔
Impurities	○	↓	○	○
Concentration	↑	↑	○	↑
Fuel-air ratio	↑	○	↑	○
Roughness	↑ ↔	○	○	↑ ↔
Oxigen	↔	↔	↑	○
Sulphur	○	○	↑	↑

When the value of an operating variable is increased, it increases (↑), decreases (↓), or has no effect (↔) on the specific fouling mechanism listed. Circles (○) indicate that no influence of these variables has been reported in the literature.

wall rupture. The most frequently used technique for preventing water-side fouling is the water treatment. Strict guidelines are developed for the quality of water [150]. Among the most difficult problems caused by fouling in boilers is the particulate fouling—especially deposition of iron oxide particles and various inorganic salts. In addition, corrosion fouling can be intensified by the presence of oxygen. Various factors that influence fouling in such conditions are: (1) local thermal conditions (heat flux magnitude), (2) concentration of suspended particles, (3) fluid characteristics (velocity, chemical properties), and (4) heat transfer surface characteristics.

The prevention of iron-oxide-induced corrosion can be accomplished by mechanical filtration of iron oxide, by the use of additives (iron oxide dispersants), and by adding inhibitors for corrosion. Standard procedures for reduction of corrosion in boilers include deaeration of the feedwater and the use of additives such as sodium sulfite.

It was reported [95] that fouling on low-finned tubes in reboilers may occur at a reduced rate compared to plain tubes. Also, the use of porous enhancements (porous coatings, high flux tube) demonstrates strong resistance to fouling.

Corrosion

Single-component corrosion types, important for heat exchanger design and operation, are as follows: (1) uniform attack corrosion, (2) galvanic corrosion, (3) pitting corrosion, (4) stress corrosion cracking, (5) erosion corrosion, (6) deposit corrosion, and (7) selective leaching [153].

Uniform corrosion is a form of corrosion caused by a chemical or electrochemical reaction between the metal and the fluid in contact with it over the entire exposed metal surface. It is usually easy to notice corroded areas attacked by uniform corrosion. This type of corrosion can be suppressed by applying adequate inhibitors, coatings, or cathodic protection. *Galvanic corrosion* is caused by an electric potential difference between two electrically dissimilar metals in the system in the presence of an electrolyte (such as water in a heat exchanger). It may occur at tube-to-tubesheet junctions as well as at the tube-to-baffle hole and baffle-to-shell contacts. Reduction of this type of electrochemical corrosion can be accomplished by selecting dissimilar materials to be as close as possible to each other on the galvanic series list for pairs of components in the system. In addition, insulation of dissimilar metals, application of coatings, addition of inhibitors, and installation of a third metal that is anodic to both metals in the galvanic contact may be used to minimize galvanic corrosion. *Pitting corrosion* is a form of localized autocatalytic corrosion due to pitting that results in holes in the metal. Pits caused by pitting corrosion are usually at places where the metal surface has surface deformities and scratches. This corrosion type is difficult to control. Materials that show pitting should not be used to build heat exchanger components. Adding inhibitors is not always efficient. *Stress corrosion* is a form of corrosion that involves cracks caused by simultaneous presence of the tensile stress and a corrosive medium. Cold working parts and U-bends in shell-and-tube heat exchangers are the locations where corrosion may take place in combination with an existing stress. The best prevention of stress-corrosion cracking is an appropriate selection of material, reduction of tensile stresses in the construction, elimination of the critical environmental components (for example demineralization or degasification), cathodic protection, and addition of inhibitors. *Erosion corrosion* is a form of surface corrosion due to the erosion of the heat transfer surface due to high-velocity fluid with or without particulates (e.g., fluid velocity greater than 2 m/s or 6 ft/sec for water flow over an aluminum surface) and the subsequent corrosion of the exposed surface. The erosion corrosion is more common at the inlet end of a heat exchanger flow passage. The selection of the correct material less prone to erosion and an adequate velocity range for a working fluid may reduce erosion and cavitation effects. For example, stainless steel 316 can sustain three times the water velocity flowing inside tubes compared to steel or cooper. Also, design modifications, coatings, and cathodic protection should be considered. *Deposit corrosion* (also called *crevice corrosion*) is a form of localized physical deterioration of a metal surface in shielded areas (i.e., in stagnant fluid flow regions) often caused by deposits of dirt and corrosion products. Stagnant areas (such as various gaps) may also be attacked by localized corrosion. Fouling and various deposits influence corrosion at shielded areas if the combination of fluid and heat exchanger surface material is inappropriate. The best prevention of this type of corrosion is a design in which the stagnation areas of the fluid flow and sharp corners are reduced to a minimum. Design should be adjusted for complete drainage, and, if possible, welding should be used instead of rolling in tubes in tubesheets. *Selective leaching* is a selective removal of one metal constituent from an alloy by corrosion. Additives to an alloy, such as arsenic or tin, may reduce the onset of the removal of a constituent from the alloy, thus solving the problem with selective leaching.

CONCLUDING REMARKS

The content presented in this chapter shows clearly the diversity and complexity of topics related to heat exchangers. Space limitation, however, has prevented the authors from thoroughly covering many equally important aspects of design and operation of heat exchangers (refer to Fig. 17.53). Let us briefly summarize some of the issues that should attract considerable attention of an engineer and/or researcher but are not discussed in this text.

The mechanical design of a heat exchanger is a very important consideration for trouble-free operation for the design life. Some of the important considerations are the desired structural strength and fatigue characteristics (based on the operating pressures, temperature, corrosiveness, and chemical reaction of fluids with material), proper selection of the materials and the method of bonding of various components, and problems during operation (such as transients, dynamic instability, freezing, and erosion). Also, the design and operational problems should be addressed for the flow distribution devices (headers, tanks, manifolds, nozzles, or inlet-outlet pipes), heat exchanger installation, maintenance (such as cleaning, repair, serviceability, and general inspection), shipping limitations, and so on. Heat exchangers must also comply with the applicable local, state, national, and/or international codes and standards. The details regarding these considerations can be found in Refs. 4, 5, and 154–156.

Manufacturing considerations are at least as important as the desired thermal and mechanical performance. These include the actual manufacturing of the components of a heat exchanger, the processing considerations (putting together assembly/exchanger), the manufacturing equipment (tools, furnaces, machines) and space, the stacking and bonding of exchangers (brazing, soldering, welding, or mechanical expansion), and leak-free mounting (joining) of headers, tanks, manifolds, pipes, and so on. A variety of references are scattered in the literature on this topic [157]. Basic information on brazing can be found in Ref. 158.

Process integration and system synthesis require a skillful manipulation of system components. For example, heat exchanger network synthesis requires the utilization of very specialized methods of analysis [111, 112]. The search for an efficient system operation requires a multidisciplinary approach that will inevitably involve simultaneous utilization of heat transfer theory and thermal and mechanical design skills as well as specific thermodynamic considerations and economic evaluation. The optimal design of a system cannot be achieved without careful thermo-economic considerations at both system and component (i.e., heat exchanger) levels.

The overall total lifecycle cost for a heat exchanger may be categorized as the capital and operating costs. The capital (total installed) cost includes the costs associated with design, materials, manufacturing (machinery, labor, and overhead), testing, shipping, installation, and depreciation. The operating cost consists of the costs associated with fluid pumping power, warranty, insurance, maintenance, repair, cleaning, lost production/downtime due to failure, and energy cost associated with the utility (steam, fuel, water) in conjunction with the exchanger in the network. Costing information is generally proprietary to industry, and very little information is published in the open literature [159].

Operation and exploitation of a heat exchanger, even in situations when the device is not operating in an unsteady mode such as rotary regenerators, should require considerations of transients and corresponding heat exchanger response. Erratic operation, startups and shutdowns, and/or requirements for optimal control of systems (in which heat exchangers represent important components) are some of the reasons why time-dependent behavior should be studied.

The variety of heat exchanger design types emphasized in the text often prevents unification of the analysis methods and design strategies. This text is primarily devoted to the most frequently used heat exchanger types—recuperators and regenerators. Those interested in the details of design and operation of agitated vessels, multifluid heat exchangers, micro heat exchanger applications, or cryogenic and/or various new heat exchangers design introduced by development of new energy sources and/or emerging technologies (such as solar collectors, high temperature applications, and bioengineering) should consult specialized literature.

NOMENCLATURE

Symbols used only once and/or symbols used only within the context of a specific topic are, as a rule, defined in the text, table, and/or figure.

Unless clearly specified, a regenerator in the nomenclature means either a rotary or a fixed-matrix regenerator.

Symbol, Definition, Units

A	Total heat transfer surface area (both primary and secondary, if any) on one side of a direct transfer type exchanger, total heat transfer surface area of all matrices of a regenerator, m^2, ft^2
A_{cross}	Cross-sectional area of the channel, m^2, ft^2
A_{ba}	Bypass area of one baffle, m^2, ft^2
A_c	Total heat transfer area (both primary and secondary, if any) on the cold side of an exchanger, m^2, ft^2
A_f	Fin or extended surface area on one side of the exchanger, m^2, ft^2
A_{fr}	Frontal or face area on one side of an exchanger, m^2, ft^2
A_h	Total heat transfer surface area (both primary and secondary, if any) on the hot side of an exchanger, m^2, ft^2

A_k	Total wall cross-sectional area for longitudinal conduction (subscripts c, h, and t denote cold side, hot side, and total (hot + cold) for a regenerator), m², ft²
A_{mb}	Minimum flow area at centerline of one baffle, m², ft²
A_o	Minimum free flow (or open) area on one side of an exchanger, m², ft²
A_p	Primary surface area on one side of an exchanger, m², ft²
A_{sb}	Shell-to-baffle leakage area, m², ft²
A_{tb}	Tube-to-baffle leakage area, m², ft²
A_w	Total wall area for heat conduction from the hot fluid to the cold fluid, or total wall area for transverse heat conduction (in the matrix wall thickness direction), m², ft²
A_{wg}	Gross window area, m², ft²
A_{wt}	Window area occupied by tubes, m², ft²
a_i	Sound frequency parameter, dimensionless
B_c	Baffle cut, percent of diameter, dimensionless
Bo	Boiling number (defined in Table 17.26), dimensionless
Bi	Biot number, Bi = $(h\delta_f/2k_f)$ dimensionless
b	Plate spacing, distance between two plates (fin height) in a plate-fin heat exchanger, m, ft
C	Flow stream heat capacity rate with a subscript c or h, $\dot{m}c_p$, W/°C, Btu/hr °F
C^*	Heat capacity rate ratio, C_{min}/C_{max}, dimensionless
C_m	Added mass coefficient (Eq. 17.156 and Fig. 17.60), dimensionless
C_{max}	Maximum of C_c and C_h, W/°C, Btu/hr °F
C_{min}	Minimum of C_c and C_h, W/°C, Btu/hr °F
C_{ms}	Heat capacity rate of the maldistributed stream, W/°C, Btu/hr °F
Co	Convection number (defined in Table 17.26), dimensionless
C_s	Heat capacity rate of the fluid stream that is not maldistributed, W/°C, Btu/hr °F
C_u	U-tube natural frequency constant (Eq. 17.160 and Fig. 17.63), dimensionless
C_r	Heat capacity rate of a regenerator, $M_w c_w \mathbf{N}$ or $M_w c_w/\mathcal{P}_t$, for the hot and cold side matrix heat capacity rates, $C_{r,h}$ and $C_{r,c}$, W/°C, Btu/hr °F
C_r^*	Total matrix heat capacity rate ratio, C_r/C_{min}, $C_{r,h}^* = C_{r,h}/C_h$, $C_{r,c}^* = C_{r,c}/C_c$, dimensionless
C	Coefficient (Eq. 17.155 and Table 17.35), dimensionless
c_{eff}	Effective speed of sound (defined by Eq. 17.162), m/s, ft/s
c_p	Specific heat of fluid at constant pressure, J/kg °C,[†] Btu/lbm °F
c_w	Specific heat of wall material, J/kg °C, Btu/lbm °F
c_o	Speed of sound in free space (defined by Eq. 17.162), m/s, ft/s
D	Diameter of a spherical drop, m, ft
D_{baff}	Baffle diameter, m, ft
D_{ctl}	Diameter, $D_{otl} - d_o$ (defined in Table 17.30), m, ft
D_h	Hydraulic diameter of flow passages, $4r_h$, $4A_o/P$, $4A_o L/A$, or $4\sigma/\alpha$, m, ft
D_{otl}	Diameter of the outer tube limit (defined in Table 17.30), m, ft

---

[†] J = joule = newton × meter = watt × second; newton = N = kg·m/s²; pascal = Pa = N/m².

D_s	Shell diameter, m, ft
D_w	Equivalent diameter in window, m, ft
d_e	Fin tip diameter of a disk (radial) fin, m, ft
d_i	Tube inside diameter, m, ft
d_o	Tube (or pin) outside diameter, m, ft
d_1	Tube hole diameter in baffle, m, ft
E	Modulus of elasticity, Pa, lbf/ft^2
Eu	N-row average Euler number, $\Delta p/(\rho V_m^2 N_r/2g_c)$ or $\rho \Delta p g_c/(N_r G^2/2)$, dimensionless
F	Log-mean temperature difference correction factor (defined by Eq. 17.17 and Table 17.4), dimensionless
F	Parameter, $[\rho_v/(\rho_l - \rho_v)]^{1/2} j_v/(d_i g \cos \theta)^{1/2}$, dimensionless
Fo	Liquid drop Fourier number, $4\alpha\tau/D^2$, dimensionless
Fr	Froude number, $G^2/(gD_h\rho_{\text{hom}}^2)$, dimensionless
Fr$_{\text{lo}}$	Froude number with all flow as liquid, $G^2/(gd_i\rho_l^2)$, dimensionless
f	Fanning friction factor, $\tau_w/(\rho V_m^2/2g_c)$, $\rho \Delta p g_c D_h/(2LG^2)$, dimensionless
f_a	Acoustic frequency, Hz, 1/s
f_n	Tube natural frequency, Hz, 1/s
f_{tb}	Turbulent buffeting frequency, Hz, 1/s
f_v	Vortex shedding frequency, Hz, 1/s
G	Mass velocity, based on the minimum free area, $\dot{m}/A_o$, based on the total flow rate for two-phase flow, kg/m^{2}s, lbm/hr ft^2
g	Acceleration due to gravity, m/s^2, ft/s^2
g_c	Proportionality constant in Newton's second law of motion, $g_c = 1$ and dimensionless in SI units, $g_c = 32.174$ lbm ft/lbf s^2
H	Velocity head or velocity pressure, Pa, lbf/ft^2 (psi)
Ⓗ	Thermal boundary condition referring to constant axial wall as well as peripheral heat flux with wall temperature
Ⓗ₁	Thermal boundary condition referring to constant axial wall heat flux with constant peripheral wall temperature
Ⓗ₂	Thermal boundary condition referring to constant axial wall heat flux with constant peripheral wall heat flux
h	Heat transfer coefficient, W/m^2 K, Btu/hr ft^2 °F
h_e	Heat transfer coefficient at the fin tip, W/m^2 K, Btu/hr ft^2 °F
$(hA)^*$	Convection conductance ratio (defined in Table 17.10b), dimensionless
I	Area moment of inertia, m^4, ft^4
i	Specific enthalpy on per unit mass basis, J/kg; Btu/lbm
i_{lv}	Specific enthalpy of phase change, J/kg, Btu/lbm
J_i	Correction factor for the heat transfer coefficient, $i = 1, 2, \ldots, 5$ (defined in Table 17.29), dimensionless
j	Fluid superficial velocity (volume flow rate of the respective phase divided by cross section area), m/s, ft/s
j	Colburn factor, St Pr$^{2/3}$, dimensionless
K	Pressure loss coefficient, $\Delta p/(\rho V^2/2g_c)$, dimensionless

$K(\infty)$	Incremental pressure drop number for fully developed flow, see Eq. 17.86 for definition, dimensionless
K_c	Contraction loss coefficient for flow at heat exchanger entrance (Eq. 17.65 and Fig. 17.35), dimensionless
K_e	Expansion loss coefficient for flow at heat exchanger exit (Eq. 17.65 and Fig. 17.35), dimensionless
k	Thermal conductivity, for fluid if no subscript, W/m K, Btu/hr ft °F
k_f	Thermal conductivity of the fin material, W/m K, Btu/hr ft °F
k_w	Thermal conductivity of the matrix (wall) material, W/m K, Btu/hr ft °F
L	Fluid flow (core) length on one side of an exchanger, span length for flow-induced vibration analysis, m, ft
L	Drop travel distance, m, ft
L_{bb}	Bundle bypass diameter gap (see Table 17.30), m, ft
L_{bc}	Central baffle spacing, m, ft
L_{bi}	Inlet baffle spacing, m, ft
L_{bo}	Outlet baffle spacing, m, ft
L_c	Distance from baffle cut to shell inside diameter, m, ft
L_{cp}	Distance of penetration, m, ft
L_f	Fin flow length on one side of a heat exchanger, $L_f \leq L$, m, ft
L_p	Bypass lane (defined in Table 17.30), m, ft
L_{pl}	Tube lane partition bypass width (defined in Table 17.30), m, ft
L_{pp}	Tube pitch parallel to the flow, the same as X_l, m, ft
L_{sb}	Diametral clearance, shell to baffle (defined in Table 17.30), m, ft
L_{ti}	Tube length between the tubesheet and baffle tangent to the outer tube row, m, ft
L_{tb}	Diametral clearance, tube to baffle, m, ft
L_1	Flow (core) length for fluid 1 of a two-fluid heat exchanger, m, ft
L_2	Flow (core) length for fluid 2 of a two-fluid heat exchanger, m, ft
L_3	No-flow height (stack height) of a two-fluid heat exchanger, m, ft
L_c	Height from baffle cut to shell inside diameter, m, ft
ℓ	Fin length for heat conduction from primary surface to either fin tip or midpoint between plates for symmetric heating, ℓ with this meaning used only in the fin analysis, m, ft
ℓ_b	Baffle spacing (defined in Fig. 17.63), m, ft
ℓ_{ef}	Effective flow length between major boundary layer disturbances, distance between interruptions, m, ft
ℓ_f	Fin height for individually finned tubes $= (d_e - d_o)/2$, offset strip fin length (see Fig. 17.47), m, ft
M_{eff}	Effective mass per unit tube length (defined by Eq. 17.156), kg/m, lbm/ft
m	Fin parameter $(2h/k_f\delta_f)^{1/2}$, 1/m, 1/ft
$\dot{m}$	Fluid mass flow rate, $\rho V_m A_o$, kg/s, lbm/hr
N	Number of zones or sections in a numerical analysis, number of hypothetical units of a maldistributed heat exchanger, dimensionless
N, N_r	Number of tube rows in the flow direction, dimensionless

N	Rotational speed for a rotary regenerator, rev/s, rpm
N_b	Number of baffles, dimensionless
N_c	Effective number of tube rows, $= N_{tcc} + N_{tcw}$, dimensionless
N_p	Number of separating plates in a plate-fin exchanger
N_p	Number of fluid 1 passages
N_{ss}	Number of pairs of sealing strips, dimensionless
N_{ss}^+	Parameter, N_{ss}/N_{tcc}, dimensionless
N_t	Total number of tubes in an exchanger
N_{tcc}	Effective number of tube rows in the crossflow zone (between baffle tips), dimensionless
N_{tcw}	Effective number of tube rows in the window zone, dimensionless
Nu	Nusselt number, hD_h/k, dimensionless
NTU	Number of heat transfer units, UA/C_{min}, it represents the total number of transfer units in a multipass unit, dimensionless
NTU_{cost}^*	Reduction in NTU (defined by Eq. 17.167), dimensionless
NTU_1	Number of heat transfer units based on fluid 1 heat capacity rate, UA/C_1; similarly, $\text{NTU}_2 = UA/C_2$, dimensionless
NTU_o	Modified number of transfer units for a regenerator (defined in Table 17.9), dimensionless
n, n_p	Number of exchanger passes
ntu_c	Number of heat transfer units based on the cold side $(\eta_o hA)_c/C_c$, dimensionless
ntu_h	Number of heat transfer units based on the hot side $(\eta_o hA)_h/C_h$, dimensionless
ntu_n	Number of heat transfer units based on the nominal side, dimensionless
P	Temperature effectiveness of the fluid 1 and 2 with subscripts 1 and 2 (defined in Table 17.4), dimensionless
$\mathbf{P}$	Wetted perimeter of exchanger passages on one side, $\mathbf{P} = A/L = A_{fr}\beta$, m, ft
$\mathscr{P}$	Fluid pumping power, $\dot{m}\Delta p/\rho$, W, hp
Pe	Péclet number, Re Pr, dimensionless
Pr	Prandtl number, $\mu c_p/k$, $V_m D_h/\alpha$, dimensionless
p	Fluid static pressure, Pa, lbf/ft^2 (psi)
p_f	Fin pitch, m, ft
p_t	Tube pitch, center-to-center distance between tubes, m, ft
Δp	Fluid static pressure drop on one side of a heat exchanger core, Pa, lbf/ft^2 (psi)
Δp_{bi}	Fluid static pressure drop associated with an ideal crossflow section between two baffles, Pa, lbf/ft^2 (psi)
Δp_c	Fluid static pressure drop associated with the tube bundle central section (crossflow zone), Pa, lbf/ft^2 (psi)
Δp_{gain}	Pressure drop reduction due to passage-to-passage nonuniformity, Pa, lbf/ft^2 (psi)
$\Delta p_{i\text{-}o}$	Fluid static pressure drop associated with inlet and outlet sections, Pa, lbf/ft^2 (psi)
Δp_s	Shellside pressure drop (defined in Table 17.31), Pa, lbf/ft^2 (psi)
Δp^*	$= \Delta p/(\rho V_m^2/2g_c)$, dimensionless
Δp_w	Fluid static pressure drop associated with an ideal window section, Pa, lbf/ft^2 (psi)

q	Total or local (whatever appropriate) heat transfer rate in an exchanger, or heat "duty," W, Btu/hr
q''	Heat flux, heat transfer rate per unit surface area, q/A, W/m², Btu/hr ft²
q_e	Heat transfer rate through the fin tip, W, Btu/hr
q_0	Heat transfer rate at the fin base, W, Btu/hr
q_w	Heat flux at the wall, W/m², Btu/hr ft²
R	Thermal resistance based on the surface area A; $\mathbf{R} = 1/UA =$ overall thermal resistance in a two-fluid exchanger, $\mathbf{R}_h = 1/(\eta_o hA)_h =$ hot side film resistance (between the fluid and the wall), $\mathbf{R}_c$ cold-side film resistance, $\mathbf{R}_s$ scale or fouling resistance, and $\mathbf{R}_w$ wall thermal resistance; definitions found after Eqs. 17.4 and 17.5, K/W, hr °F/Btu
R	Heat capacity rate ratio (defined in Table 17.4), dimensionless
R	Mean bend radius in Section on Flow-Induced Vibration, m, ft
R_i	Pressure drop correction factor; $i = \ell$ for baffle leakage effects, $i = b$ for bypass flow effects, $i = s$ for baffle spacing effects
Re	Reynolds number based on the hydraulic diameter, GD_h/μ, dimensionless
Re_d	Reynolds number based on the tube outside diameter, $V_m d_o/\nu$, dimensionless
Re_L	Reynolds number based on the plate width for condensation, $4\Gamma_l/\mu_l$; also Reynolds number defined in Table 17.25, dimensionless
r	Radial coordinate in the cylindrical coordinate system, m, ft
r	Tube bend radius (defined in Fig. 17.63), m, ft
r_h	Hydraulic radius, $A_o L/A$ or $D_h/4$, m, ft
Sr	Strouhal number, $f_v d_o/V_c$, dimensionless
St	Stanton number, h/Gc_p, $\mathrm{St}_o = U/Gc_p$, dimensionless
s	Plate width or tube perimeter, m, ft
T	Temperature, °C, °F, K, R
Ⓣ	Thermal boundary condition referring to constant wall temperature, both axially and peripherally
T_∞	Ambient fluid temperature, °C, °F
T_ℓ	Temperature of the fin tip, °C, °F
T_m	Fluid bulk mean temperature, °C, °F
T_0	Temperature of the fin base, °C, °F
T_w	Wall temperature, °C, °F
ΔT	Local temperature difference between two fluids, $T_h - T_c$, °C, °F
ΔT_{lm}	Log-mean temperature difference (Table 17.4 and Eq. 17.18), °C, °F
ΔT_m	True mean temperature difference, q/UA, °C, °F
U, U_m	Overall heat transfer coefficient (defined by Eqs. 17.4 and 17.6); the subscript m represents mean value when local U is variable, W/m²K, Btu/hr ft² °F
V	Fluid mean axial velocity, V_m occurs at the minimum free flow area in the exchanger unless specified, m/s, ft/s; a special function in Eq. 17.47, dimensionless
V	Exchanger total volume, m³, ft³
V_c	Reference crossflow velocity in gaps in a tube row (defined in Table 17.36), m/s, ft/s
v	Specific volume, $1/\rho$, m³/kg, ft³/lbm

We	Weber number, $G^2 D_h/(\sigma \rho_{hom})$ or $G^2 L/(\sigma \rho_l)$, dimensionless
X	Martinelli parameter $[(dp/dz)_l/(dp/dz)_v]^{1/2}$, dimensionless
X^*	Axial distance, x/L, dimensionless
X_ℓ	Longitudinal tube pitch, m, ft
X_t	Transverse tube pitch, m, ft
X_{tt}	Martinelli parameter for turbulent-turbulent flow, $[(1-x)/x]^{0.9}(\rho_v/\rho_l)^{0.5}(\mu_l/\mu_v)^{0.1}]$, dimensionless
x, y, z	Cartesian coordinates, m, ft
x	Quality (dryness fraction), dimensionless
x_n	Midspan amplitude, m, ft
x^+	Axial distance, x/D_h Re, dimensionless
x^*	Axial distance, x/D_h Re Pr, dimensionless

Greek Symbols

α	Ratio of total heat transfer area on one side of an exchanger to the total volume of an exchanger, $A/\mathbf{V}$, m²/m³, ft²/ft³
α	Fluid thermal diffusivity, $k/\rho c_p$, m²/s, ft²/s
α	Void fraction (defined by Eq. 17.115), dimensionless
α^*	Aspect ratio of rectangular ducts, a ratio of small to large side, dimensionless
α_f^*	Fin aspect ratio, $2\ell/\delta_f$, dimensionless
β	Heat transfer surface area density, a ratio of total heat transfer area on one side of a plate-fin exchanger to the volume between the plates on that side, m²/m³, ft²/ft³
β	Span length fraction at terminal ends (defined in Fig. 17.65), dimensionless
Γ_L	Amount of condensate produced per unit width of the surface, kg/s m, lbm/s ft
γ	Specific heat ratio, c_p/c_v, dimensionless
δ	Wall or primary surface (plate) thickness, m, ft
δ_c	Channel deviation parameter based on the passage width (defined by Eqs. 17.168, 17.170, and 17.171), dimensionless
δ_f	Fin thickness, at the root if not of constant cross section, m, ft
δ_o	Logarithmic decrement, $\ln(x_n/x_{n+1})$, dimensionless
ε	Heat exchanger effectiveness (defined in Table 17.4), it represents an overall exchanger effectiveness for a multipass unit, dimensionless
ε_p	Heat exchanger effectiveness per pass, dimensionless
ε_r	Regenerator effectiveness of a single matrix (defined in Table 17.9), dimensionless
ζ	Damping factor, $\delta_o/2\pi$, dimensionless
η_f	Fin efficiency (defined by Eq. 17.22), dimensionless
η_o	Overall surface efficiency of total heat transfer area on one side of the extended surface heat exchanger, see Eq. 17.24 for the definition, dimensionless
θ	Tube inclination angle, rad, deg
κ	Length effect correction factor, dimensionless
κ_T	Isothermal compressibility, 1/Pa, ft²/lbf
Λ	Reduced length for a regenerator (defined in Table 17.9), dimensionless
Λ_m	Mean reduced length (defined in Table 17.9), dimensionless

λ	Longitudinal wall conduction parameter based on the total conduction area, $\lambda = k_w A_{k,t}/C_{min}L$, $\lambda_c = k_w A_{k,c}/C_c L$, $\lambda_h = k_w A_{w,h}/C_h L$, dimensionless
λ_i	Frequency constant (defined by Eq. 17.159 and λ_1 in Fig. 17.62), $i = 1, 2, \ldots$, dimensionless
μ	Fluid dynamic viscosity, Pa·s, lbm/hr ft
ν	Fluid kinematic viscosity, m²/s, ft²/s
Π	Reduced period for a regenerator (defined in Table 17.9), dimensionless
Π_m	Harmonic mean reduced period (defined in Table 17.9), dimensionless
ρ	Fluid density, kg/m³, lbm/ft³
σ	Ratio of free flow area to frontal area, A_o/A_{fr}, also the volumetric porosity for regenerators $= r_h \alpha$, dimensionless
σ	Solidity of a tube bundle on the shell side, dimensionless
σ	Surface tension, N/m, lbf/ft
τ	Time variable, s
τ_w	Equivalent fluid shear stress at wall, Pa, lbf/ft² (psi)
$\phi(\ )$	Denotes a function of; a parameter defined by Eq. 17.36, dimensionless
φ	Friction multiplier of respective phases in two-phase flow (defined in Table 17.21 and Eq. 17.113a and b), dimensionless
ψ	$\Delta T_m/(T_{h,i} - T_{c,i})$, also a parameter in Fig. 17.31 and Eq. 17.41, dimensionless

Subscripts

a	Air side
a	Terminal point of the heat exchanger zone
ave	Average
b	Bulk
b	Terminal point of a heat exchanger zone
c	Cold fluid side
cf	Counterflow
conv	Single phase convection correlation
cp	Constant properties
crit	Critical heat flux (CHF) conditions
f	Fin
f	Two-phase friction
g	Gas side or gas phase
h	Hot fluid side
h	Hydrostatic effect
hom	Homogeneous
H	Constant axial wall heat flux boundary condition
i	Inlet to the exchanger
ideal	Ideal heat transfer conditions
l	Liquid phase
lo	Total two-phase mass flow rate in the channel flowing as liquid
loc	Local value

lm	Logarithmic mean
m	Mean or bulk mean
m	Momentum
max	Maximum
min	Minimum
NB	Nucleate boiling
n	Nominal or reference passage
o	Overall
o	Outlet
p	Pass
pf	Parallelflow
s	Scale or fouling when used as a subscript with the thermal resistance
s	Shell side
sat	Saturation
T	Constant wall temperature boundary condition
t	Tubeside
tb	Turbulent baffeting
tube	Tube side
v	Vapor phase
w	Wall or properties at the wall temperature
x	Local value at section x along the flow length
1	One section (inlet or outlet) of the exchanger
1	Reduced size passage side
2	Other section (outlet or inlet) of the exchanger
∞	Free stream

Superscripts

$-$ Mean value (in Table 17.23, denotes mean over the total length; in Table 17.24, denotes mean over the tube perimeter)

REFERENCES

1. R. K. Shah, "Heat Exchangers," in *Encyclopedia of Energy Technology and the Environment,* edited by A. Bisio and S. G. Boots, pp. 1651–1670, John Wiley & Sons, New York, 1994.

2. R. K. Shah and A. C. Mueller, "Heat Exchange," in *Ullmann's Encyclopedia of Industrial Chemistry,* Unit Operations II, Vol. B3, Chapter 2, pp. 2-1–2-108, VCH Publishers, Weinheim, Germany, 1989.

3. G. Walker, *Industrial Heat Exchangers—A Basic Guide,* 2d ed., Hemisphere, Washington, DC, 1990.

4. G. F. Hewitt, coordinating ed., *Hemisphere Handbook of Heat Exchanger Design,* Hemisphere, Washington, DC, 1989.

5. Tubular Exchanger Manufacturers Association, *Standards of TEMA,* 7th ed., New York, 1988.

6. K. K. Shankarnarayanan, Plate Heat Exchangers, *Proc. Symposium on Heat Exchangers,* Paper IT-3, Indira Gandhi Centre for Atomic Research, Kalpakkam 603102, India, February 1996.

7. R. K. Shah and A. C. Mueller, "Heat Exchanger Basic Thermal Design Methods," in *Handbook of Heat Transfer Applications,* 2d ed., W. M. Rohsenow, J. P. Hartnett, and E. N. Ganić (eds.), Chapter 4, Part 1, pp. 1–77, 1985.

8. W. Roetzel and B. Spang, "Verbessertes Diagramm zur Berechnung von Wärmeübertragern (Improved Chart for Heat Exchanger Design)," *Wärme-und Stofübertragung,* Vol. 25, pp. 259–264, 1990.

9. R. K. Shah, "Heat Exchanger Basic Design Methods," in *Low Reynolds Number Flow Heat Exchangers,* S. Kakaç, R. K. Shah, and A. E. Bergles (eds.), Hemisphere/McGraw-Hill, Washington, DC, 1982.

10. S. G. Kandlikar and R. K. Shah, "Asymptotic Effectiveness-NTU Formulas for Multipass Plate Heat Exchangers," *ASME J. Heat Transfer,* Vol. 111, pp. 314–321, 1989.

11. B. S. Bačlić, "ε-N_{tu} Analysis of Complicated Flow Arrangements," in *Compact Heat Exchangers—A Festschrift for A. L. London,* R. K. Shah, A. D. Kraus, and D. Metzger (eds.), pp. 31–90, Hemisphere, New York, 1990.

12. L. J. Huang and R. K. Shah, "Assessment of Calculation Methods for Efficiency of Straight Fins of Rectangular Profile," *Int. J. Heat Fluid Flow,* Vol. 13, pp. 282–293, 1992.

13. D. Q. Kern and A. D. Kraus, *Extended Surface Heat Transfer,* McGraw-Hill, New York, 1972.

14. R. K. Shah, "Temperature Effectiveness of Multiple Sandwich Rectangular Plate-Fin Surfaces," *ASME J. Heat Transfer,* Vol. 93C, pp. 471–473, 1971.

15. R. K. Shah, "Compact Heat Exchangers," in *Handbook of Heat Transfer Applications,* 2d ed., W. M. Rohsenow, J. P. Hartnett, and E. N. Ganić (eds.), Chapter 4, Part III, pp. 4-174–4-312, McGraw-Hill, New York, 1985.

16. R. K. Shah and D. P. Sekulić, "Nonuniform Overall Heat Transfer Coefficient in Conventional Heat Exchanger Design Theory Revisited," *ASME J. Heat Transfer,* Vol. 120, May 1998.

17. W. Roetzel and B. Spang, "Design of Heat Exchangers, Section Cb: Heat Transfer," *VDI Heat Atlas,* VDI-Verlag GmbH, Dusseldorf, 1993.

18. R. K. Shah, "Nonuniform Heat Transfer Coefficients for Heat Exchanger Thermal Design," in *Aerospace Heat Exchanger Technology 1993,* R. K. Shah and A. Hashemi (eds.), pp. 417–445, Elsevier Science, Amsterdam, 1993.

19. R. K. Shah and A. L. London, *Laminar Flow Forced Convection in Ducts,* supplement 1 to *Advances in Heat Transfer,* Academic Press, New York, 1978.

20. W. M. Kays and A. L. London, *Compact Heat Exchangers,* 3d ed., McGraw-Hill, New York, 1984.

21. W. Roetzel and B. Spang, "Thermal Calculation of Multipass Shell and Tube Heat Exchangers," *Chem. Eng. Res. Des.,* Vol. 67, pp. 115–120, 1989.

22. B. Spang, Y. Xuan, and W. Roetzel, "Thermal Performance of Split-Flow Heat Exchangers," *Int. J. Heat Mass Trans.,* Vol. 34, pp. 863–874, 1991.

23. Y. Xuan, B. Spang, and W. Roetzel, "Thermal Analysis of Shell and Tube Exchangers with Divided-Flow Pattern," *Int. J. Heat Mass Trans.,* Vol. 34, pp. 853–861, 1991.

24. B. Bačlić, F. E. Romie, and C. V. Herman, "The Galerkin Method for Two-pass Crossflow Heat Exchanger Problem," *Chem. Eng. Comm.,* Vol. 70, pp. 177–198, 1988.

25. R. K. Shah and A. Pignotti, "The Influence of a Finite Number of Baffles on the Shell-and-Tube Heat Exchanger Performance," *Heat Transfer Eng.,* Vol. 18, No. 1, pp. 82–94, 1997.

26. K. Gardner and J. Taborek, "Mean Temperature Difference: A Reappraisal," *AIChE J.,* Vol. 23, pp. 777–786, 1977.

27. P. G. Kroeger, "Performance Deterioration in High Effectiveness Heat Exchangers Due to Axial Heat Conduction Effects," *Advances in Cryogenics Engineering,* Vol. 12, pp. 363–372, Plenum, New York, 1967; condensed from a paper presented at the 1966 Cryogenic Engineering Conference, Boulder, Colorado.

28. R. K. Shah, "Thermal Design Theory for Regenerators," in *Heat Exchangers: Thermal-Hydraulic Fundamentals and Design,* S. Kakaç, A. E. Bergles, and F. Mayinger (eds.), pp. 721–763, Hemisphere/McGraw-Hill, Washington, DC, 1981.

29. H. Hausen, *Heat Transfer in Counterflow, Parallel Flow and Cross Flow,* 2d ed., McGraw-Hill, New York, 1983.

30. F. W. Schmidt and A. J. Willmott, *Thermal Energy Storage and Regeneration,* Chaps. 5–9, Hemisphere/McGraw-Hill, Washington, DC, 1981.

31. B. S. Bačlić, "The Application of the Galerkin Method to the Solution of the Symmetric and Balanced Counterflow Regenerator Problem," *ASME J. Heat Transfer,* Vol. 107, pp. 214–221, 1985.

32. R. K. Shah, "Counterflow Rotary Regenerator Thermal Design Procedures," *Heat Transfer Equipment Design,* R. K. Shah, E. C. Subbarao, and R. A. Mashelkar (eds.), pp. 267–296, Hemisphere Publishing Corp., Washington, DC, 1988.

33. G. Theoclitus and T. L. Eckrich, "Parallel Flow through the Rotary Heat Exchanger," *Proc. 3rd Int. Heat Transfer Conf.,* Vol. I, pp. 130–138, 1966.

34. G. D. Bahnke and C. P. Howard, "The Effect of Longitudinal Heat Conduction on Periodic-Flow Heat Exchanger Performance," *ASME J. Eng. Power,* Vol. 86A, pp. 105–120, 1964.

35. P. J. Heggs, L. S. Bansal, R. S. Bond, and V. Vazakas, "Thermal Regenerator Design Charts Including Intraconduction Effects," *Trans. Inst. Chem. Eng.,* Vol. 58, pp. 265–270, 1980.

36. R. K. Shah and T. Skiepko, "Influence of Leakage Distribution on the Thermal Performance of a Rotary Regenerator," in *Experimental Heat Transfer, Fluid Mechanics and Thermodynamics 1997,* M. Giot, F. X. Mayinger, and G. P. Celata (eds.), Edizioni ETS, Pisa, Italy, 1997.

37. I. E. Idelchik, *Handbook of Hydraulic Resistance,* 3d ed., CRC Press, Boca Raton, FL, 1994.

38. D. S. Miller, *Internal Flow Systems,* 2d ed., BHRA (Information Services), Cranfield, UK, 1990.

39. A. P. Colburn, "A Method of Correlating Forced Convection Heat Transfer Data and a Comparison with Fluid Friction," *Trans. AIChE,* Vol. 29, pp. 174–210, 1933; reprinted in *Int. J. Heat Mass Trans.,* Vol. 7, pp. 1359–1384, 1964.

40. W. M. Kays and A. L. London, "Heat Transfer and Flow Friction Characteristics of Some Compact Heat Exchanger Surfaces—Part I: Test System and Procedure," *Trans. ASME,* Vol. 72, pp. 1075–1085, 1950; also "Description of Test Equipment and Method of Analysis for Basic Heat Transfer and Flow Friction Tests of High Rating Heat Exchanger Surfaces," TR No. 2, Department of Mechanical Engineering, Stanford University, Stanford, 1948.

41. E. E. Wilson, "A Basis for Rational Design of Heat Transfer Apparatus," *Trans. ASME,* Vol. 37, pp. 47–82, 1915.

42. R. K. Shah, "Assessment of Modified Wilson Plot Techniques for Obtaining Heat Exchanger Design Data," *Heat Transfer 1990, Proc. of 9th Int. Heat Transfer Conf.,* Vol. 5, pp. 51–56, 1990.

43. D. E. Briggs and E. H. Young, "Modified Wilson Plot Techniques for Obtaining Heat Transfer Correlations for Shell-and-Tube Heat Exchangers," *Chem. Eng. Progr. Symp. Ser. No. 92,* Vol. 65, pp. 35–45, 1969.

44. R. K. Shah and M. S. Bhatti, "Laminar Convective Heat Transfer in Ducts," in *Handbook of Single-Phase Convective Heat Transfer,* Chapter 3, John Wiley, New York, 1987.

45. A. J. Ghajar and L. M. Tam, "Heat Transfer Measurements and Correlations in the Transition Region for a Circular Tube with Three Different Inlet Configurations," *Exp. Thermal and Fluid Sci.,* Vol. 8, pp. 79–90, 1994.

46. M. S. Bhatti and R. K. Shah, "Turbulent and Transition Convective Heat Transfer in Ducts," in *Handbook of Single-Phase Convective Heat Transfer,* Chapter 4, John Wiley, New York, 1987.

47. R. L. Webb, *Principles of Enhanced Heat Transfer,* John Wiley, New York, 1994.

48. R. K. Shah and M. S. Bhatti, "Assessment of Correlations for Single-Phase Heat Exchangers," in *Two-Phase Flow Heat Exchangers: Thermal Hydraulic Fundamentals and Design,* S. Kakaç, A. E. Bergles, and E. O. Fernandes (eds.), pp. 81–122, Kluwer Academic Publishers, Dordrecht, Netherlands, 1988.

49. A. Zukauskas, "Convective Heat Transfer in Cross Flow," *Handbook of Single-Phase Convective Heat Transfer,* S. Kakaç, R. K. Shah, and W. Aung (eds.), Chapter 6, John Wiley, New York, 1987.

50. R. M. Manglik and A. E. Bergles, "Heat Transfer and Pressure Drop Correlations for the Rectangular Offset-Strip-Fin Compact Heat Exchanger," *Exp. Thermal and Fluid Sci.,* Vol. 10, pp. 171–180, 1995.

51. T. A. Cowell, M. R. Heikal, and A. Achaichia, "Flow and Heat Transfer in Compact Louvered Fin Surfaces," *Exp. Thermal and Fluid Sci.,* Vol. 10, pp. 192–199, 1995.

52. A. M. Jacobi and R. K. Shah, "Heat Transfer Surface Enhancement through the Use of Longitudinal Vortices: A Review of Recent Progress," *Exp. Thermal Fluid Sci.,* Vol. 11, pp. 295–309, 1995.

53. M. Fiebig, "Vortex Generators for Compact Heat Exchangers," *J. Enhanced Heat Trans.,* Vol. 2, pp. 43–61, 1995.

54. T. J. Rabas and J. Taborek, "Survey of Turbulent Forced-Convection Heat Transfer and Pressure Drop Characteristics of Low-Finned Tube Banks in Cross Flow," *Heat Transfer Eng.,* Vol. 8, No. 2, pp. 49–62, 1987.

55. A. Ganguli and S. B. Yilmaz, "New Heat Transfer and Pressure Drop Correlations for Crossflow over Low-Finned Tube Banks," *AIChE Symp. Ser. 257,* Vol. 83, pp. 9–14, 1987.

56. H. C. Chai, "A Simple Pressure Drop Correlation Equation for Low Finned Tube Crossflow Heat Exchangers," *Int. Commun. Heat Mass Transfer,* Vol. 15, pp. 95–101, 1988.

57. H. Nakamura, A. Matsuura, J. Kiwaki, N. Matsuda, S. Hiraoka, and I. Yamada, "The Effect of Variable Viscosity on Laminar Flow and Heat Transfer in Rectangular Ducts," *J. Chem. Eng. Jpn.,* Vol. 12, No. 1, pp. 14–18, 1979.

58. W. Aung, "Mixed Convection in Internal Flow," in *Handbook of Single-Phase Convective Heat Transfer,* S. Kakaç, R. K. Shah, and W. Aung (eds.), Chapter 15, John Wiley, New York, 1987.

59. A. E. Bergles, "Experimental Verification of Analyses and Correlation of the Effects of Temperature-Dependent Fluid Properties," in *Low Reynolds Number Flow Heat Exchangers,* S. Kakaç, R. K. Shah, and A. E. Bergles (eds.), Hemisphere/McGraw-Hill, Washington, DC, pp. 473–486, 1983.

60. G. F. Hewitt and D. N. Roberts, "Studies of Two-Phase Flow Patterns by Simultaneous X-Ray and Flash Photography," AERE-M 2159, Her Majesty's Stationery Office, London, 1969.

61. Y. Taitel, D. Bornea, and A. E. Dukler, "Modeling Flow Pattern Transitions for Steady Upward Gas-Liquid in Vertical Tubes," *AIChE J.,* Vol. 26, pp. 345–354, 1980.

62. J. Weisman and S. Y. Kang, "Flow Pattern Transitions in Vertical and Upwardly Inclined Tubes," *Int. J. Multiphase Flow,* Vol. 7, pp. 271–291, 1981.

63. Y. Taitel and A. E. Dukler, "A Model for Predicting Flow Regime Transitions in Horizontal and Near Horizontal Gas-Liquid Flow," *AIChE J.,* Vol. 22, pp. 47–55, 1976.

64. D. Barnea, "Transition from Annular Flow and from Dispersed Bubble Flow—Unified Models for the Whole Range of Pipe Inclinations," *Int. J. Multiphase Flow,* Vol. 12, pp. 733–744, 1986.

65. D. Barnea, "A Unified Model for Predicting Flow-Pattern Transitions for the Whole Range of Pipe Inclinations," *Int. J. Multiphase Flow,* Vol. 13, pp. 1–12, 1987.

66. D. Barnea and Y. Taitel, "Interfacial and Structural Stability of Separated Flow," in *Annual Reviews in Multiphase Flow 1994,* G. Hetsroni (ed.), Vol. 20, Suppl., pp. 387–414, 1994.

67. N. Kattan, J. R. Thome, and D. Favrat, "Flow Boiling in Horizontal Tubes: Part 1—Development of a Diabatic Two-Phase Flow Pattern Map," *ASME J. Heat Transfer,* Vol. 120, pp. 140–147, 1998.

68. M. K. Jensen, "Boiling on the Shell Side of Horizontal Tube Bundles," in *Two-Phase Flow Heat Exchangers: Thermal-Hydraulic Fundamentals and Design,* S. Kakaç, A. E. Bergles, and E. O. Fernandes (eds.), pp. 707–746, Kluwer Academic Publishers, Dordrecht, Netherlands, 1988.

69. D. Chrisholm, *Two Phase Flow in Pipelines and Heat Exchangers,* Godwin, London, 1983.

70. I. D. R. Grant and D. Chrisholm, "Two-Phase Flow on the Shell-Side of a Segmentally Baffled Shell-and-Tube Heat Exchanger with Horizontal Two-Phase Flow," *ASME J. Heat Transfer,* Vol. 101, pp. 38–42, 1979.

71. G. F. Hewitt, "Gas-Liquid Flow," in *Handbook of Heat Exchanger Design,* G. F. Hewitt (ed.), Ch. 2.3.2, pp. 1–33, Begell House, New York, 1992.

72. L. Friedel, "Improved Friction Pressure Drop Correlations for Horizontal and Vertical Two-Phase Pipe Flow," *European Two-Phase Flow Group Meeting,* Ispra, Italy, Paper E.2, 1979.

73. D. Chisholm, "Pressure Gradients due to Friction during the Flow of Evaporating Two-Phase Mixtures in Smooth Tubes and Channels," *Int. J. Heat Mass Transfer,* Vol. 16, pp. 347–358, 1973.

74. D. Chisholm, "A Theoretical Basis for the Lockhart-Martinelli Correlation for Two-Phase Flow," *Int. J. Heat Mass Transfer,* Vol. 10, pp. 1767–1778, 1967.

75. J. G. Collier and J. R. Thome, *Convective Boiling and Condensation,* 3d ed., McGraw-Hill, New York, 1994.

76. V. P. Carey, *Liquid-Vapor Phase-Change Phenomena,* Taylor and Francis, Bristol, PA, 1992.

77. K. Ishihara, J. W. Palen, and J. Taborek, "Critical Review of Correlations for Predicting Two-Phase Flow Pressure Drop across Tube Banks," *Heat Transfer Eng.,* Vol. 1, No. 2, pp. 23–32, 1980.

78. P. J. Marto, "Heat Transfer in Condensation," in *Boilers, Evaporators, and Condensers,* S. Kakaç (ed.), pp. 525–570, Wiley, New York, 1991.

79. I. D. R. Grant and D. Chisholm, "Horizontal Two-Phase Flow Across Tube Banks," *Int. J. Heat Fluid Flow,* Vol. 2, pp. 97–100, 1980.

80. D. S. Scharge, J. T. Hsu, and M. K. Jensen, "Void Fractions and Two-Phase Friction Multipliers in a Horizontal Tube Bundle," *AIChE Symp. Ser. 257,* Vol. 83, pp. 1–8, 1987.

81. D. Butterworth, "Film Condensation of Pure Vapor," in *Handbook of Heat Exchanger Design,* G. F. Hewitt (ed.), Ch. 2.6.2, pp. 1–17, Begell House, New York, 1992.

82. S. Kakaç (ed.), *Boilers, Evaporators & Condensers,* Wiley, New York, 1991.

83. J. Van der Walt and D. G. Kröger, "Heat Transfer During Film Condensation of Saturated and Superheated Freon-12," *Prog. Heat Mass Transfer,* Vol. 6, pp. 75–98, 1972.

84. D. Butterworth, "Filmwise Condensation," in *Two-Phase Flow and Heat Transfer,* D. Butterworth and G. F. Hewitt (eds.), pp. 426–462, Oxford University Press, London, 1977.

85. J. W. Rose, "Fundamentals of Condensation Heat Transfer: Laminar Film Condensation," *JSME Int. J.,* Vol. 31, pp. 357–375, 1988.

86. T. Fujii, H. Honda, and K. Oda, "Condensation of Steam on a Horizontal Tube—The Influence of Oncoming Velocity and Thermal Conduction at the Tube Wall," *18th Natl. Heat Transfer Conf.,* San Diego, ASME/AIChE, pp. 35–43, August 6–8, 1979.

87. J. C. Chato, "Laminar Condensation inside Horizontal and Inclined Tubes," *ASHRAE J.,* Vol. 4, No. 2, pp. 52–60, 1962.

88. M. M. Shah, "A General Correlation for Heat Transfer During Film Condensation inside Pipes," *Int. J. Heat Mass Transfer,* Vol. 22, pp. 547–556, 1979.

89. M. K. Dobson and J. C. Chato, "Condensation in Smooth Horizontal Tubes," *ASME J. Heat Transfer,* Vol. 120, pp. 193–213, 1998.

90. V. Srinivasan and R. K. Shah, "Condensation in Compact Heat Exchangers," *J. Enhanced Heat Transfer,* Vol. 4, 1997.

91. S. Q. Zhou, R. K. Shah, and K. A. Tagavi, "Advances in Film Condensation including Surface Tension Effect in Extended Surface Passages," in *Fundamentals of Bubble and Droplet Dynamics: Phase Change and Two-Phase Flow,* E. Ulucakli (ed.), ASME HTD-Vol. 342, pp. 173–185, 1997.

92. V. P. Carey, "Two-Phase Flow in Small Scale Ribbed and Finned Passages for Compact Evaporators and Condensers," *Nucl. Eng. Design,* Vol. 141, pp. 249–260, 1993.

93. M. W. Wambsganss, R. K. Shah, G. P. Celata, and G. Zummo, "Vaporization in Compact Heat Exchangers," *4th World Conference on Experimental Heat Transfer, Fluid Mechanics and Thermodynamics,* Brussels, Belgium, June 2–6, 1997.

94. J. R. Thome, *Enhanced Boiling Heat Transfer,* Hemisphere, New York, 1990.

95. J. G. Collier, "Boiling within Vertical Tubes, Convective Boiling inside Horizontal Tubes, and Boiling outside Tubes and Tube Bundles," in *Handbook of Heat Exchanger Design,* G. F. Hewitt (ed.), Chapters 2.7.3–2.7.5, Begell House, New York, 1992.

96. S. G. Kandlikar, "A General Correlation for Saturated Two-Phase Flow Boiling Heat Transfer Inside Horizontal and Vertical Tubes," *ASME J. Heat Transfer,* Vol. 112, pp. 219–228, 1990.

97. Y. Katto and H. Ohno, "An Improved Version of the Generalized Correlation of Critical Heat Flux for the Forced Convective Boiling in Uniformly Heated Vertical Tubes," *Int. J. Heat Mass Transfer,* Vol. 27, pp. 1641–1648, 1984.

98. J. W. Palen and W. M. Small, "A New Way to Design Kettle and Internal Reboilers," *Hydrocarbon Process,* Vol. 43, No. 7, pp. 199–208, 1964.

99. R. K. Shah, "Multidisciplinary Approach to Heat Exchanger Design, in Industrial Heat Exchangers," J-M. Buchlin (ed.), Lecture Series No. 1991-04, von Kármán Institute for Fluid Dynamics, Belgium, 1991.

100. R. K. Shah, "Compact Heat Exchangers," in *Heat Exchangers: Thermal-Hydraulic Fundamentals and Design,* S. Kakaç, A. E. Bergles, and F. Mayinger (eds.), pp. 111–151, Hemisphere Publishing Corp., Washington, DC, 1981.

101. R. K. Shah, "Plate-Fin and Tube-Fin Heat Exchanger Design Procedures," in *Heat Transfer Equipment Design,* R. K. Shah, E. C. Subbarao, and R. A. Mashelkar (eds.), pp. 255–266, Hemisphere Publishing Corp., Washington, DC, 1988.

102. R. K. Shah and A. D. Giovannelli, "Heat Pipe Heat Exchanger Design Theory," in *Heat Transfer Equipment Design,* R. K. Shah, E. C. Subbarao, and R. A. Mashelkar (eds.), pp. 609–653, Hemisphere Publishing Corp., Washington, DC, 1988.

103. R. K. Shah and A. S. Wanniarachchi, "Plate Heat Exchanger Design Theory," in *Industrial Heat Exchangers,* J. M. Buchlin (ed.), Lecture Series N. 1991-04, von Kármán Institute for Fluid Dynamics, Belgium, 1991.

104. D. P. Sekulić and R. K. Shah, "Thermal Design Theory of Three-Fluid Heat Exchangers," *Advances in Heat Transfer,* Vol. 26, pp. 219–328, Academic Press, New York, 1995.

105. K. J. Bell, "Delaware Method for Shell-Side Design," in *Heat Transfer Equipment Design,* R. K. Shah, E. C. Subbarao, and R. A. Mashelkar (eds.), pp. 145–166, Hemisphere, New York, 1988.

106. J. Taborek, "Shell-and-Tube Heat Exchangers: Single-Phase Flow," in *Handbook of Heat Exchanger Design,* G. F. Hewitt (ed.), pp. 3.3.3-1–3.3.11-5, Begell House, New York, 1992.

107. G. Breber, "Computer Programs for Design of Heat Exchangers," in *Heat Transfer Equipment Design,* R. K. Shah, E. C. Subbarao, R. A. Mashelkar (eds.), pp. 167–177, Hemisphere, New York, 1988.

108. D. Butterworth, "Developments in the Computer Design of Heat Exchangers," in *Heat Transfer 1994, Proc. 10th Int. Heat Transfer Conf.,* Vol. 1, pp. 433–444, Brighton, UK, 1994.

109. K. J. Bell, "Approximate Sizing of Shell-and-Tube Heat Exchangers," in *Handbook of Heat Exchanger Design,* G. F. Hewitt (ed.), pp. 3.1.4-1–3.1.4-9, Begell House, New York, 1992.

110. D. P. Sekulić, "Second Law Quality of Energy Transformation in a Heat Exchanger," *ASME J. Heat Transfer,* Vol. 112, pp. 295–300, 1990.

111. A. Bejan, G. Tsatsaronis, and M. Moran, *Thermal Design and Optimization,* Wiley, New York, 1996.

112. B. Linnhoff, D. W. Townsend, D. Boland et al. (eds.), *A User Guide on Process Integration for the Efficient Use of Energy,* Pergamon Press, Oxford, 1982.

113. D. Butterworth, "Steam Power Plant and Process Condensers," in *Boilers, Evaporators & Condensers,* S. Kakaç (ed.), Ch. 11, pp. 571–633, Wiley, New York, 1991.

114. A. C. Mueller, "Condensers," in *Handbook of Heat Exchanger Design,* G. F. Hewitt (ed.), pp. 3.4.1-1–3.4.9-5, Hemisphere, New York, 1990.

115. G. H. Hewitt, G. L. Shires, and T. R. Bott, *Process Heat Transfer,* CRC Press, Boca Raton, FL, 1994.

116. F. Kreith and R. F. Boehm, *Direct Contact Heat Transfer,* Hemisphere, New York, 1988.

117. M. Cumo, "Numerical Methods for the Analysis of Flow and Heat Transfer in a Shell-and-Tube Heat Exchanger with Shell-Side Condensation," in *Two-Phase Flow Heat Exchangers: Thermal-Hydraulic Fundamentals and Design,* S. Kakaç, A. E. Bergles, and E. O. Fernandes (eds.), pp. 829–847, Kluwer, Dordrecht, Netherlands, 1988.

118. H. R. Jacobs, "Direct-Contact Condensers," in *Handbook of Heat Exchanger Design,* G. F. Hewitt (ed.), pp. 1–16, Hemisphere, New York, 1990.

119. R. A. Smith, *Vaporisers, Selection, Design, and Operation,* Longman, New York, 1986.

120. J. G. Collier, "Evaporators," in *Two-Phase Flow Heat Exchangers: Thermal-Hydraulic Fundamentals and Design,* S. Kakaç, A. E. Bergles, and E. O. Fernandes (eds.), pp. 683–705, Kluwer, Dordrecht, Netherlands, 1988.

121. J. G. Collier, "Nuclear Steam Generators and Waste Heat Boilers," in *Boilers, Evaporators & Condensers,* S. Kakaç (ed.), pp. 471–519, Wiley, New York, 1991.

122. R. D. Blevins, *Flow-Induced Vibration,* 2d ed., Van Nostrand, New York, 1990.

123. P. R. Owen, "Buffeting Excitation of Boiler Tube Vibration," *J. Mech. Eng. Sci.,* Vol. 7, pp. 431–439, 1965.

124. H. J. Connors, "Fluidelastic Vibration of Tube Arrays Excited by Cross Flow," in *Flow-Induced Vibration in Heat Exchangers,* D. D. Reiff (ed.), Proc. of a WAM Symposium, ASME, New York, pp. 42–56, 1970.

125. M. J. Pettigrew, J. H. Tromp, C. E. Taylor et al., "Vibration of Tube Bundles in Two-Phase Cross Flow: Part 1 Hydrodynamic Mass and Damping," in *Symposium on Flow-Induced Vibration and Noise,* M. P. Paidoussis, S. S. Chen, and M. D. Berstein (eds.), Vol. 2, ASME, New York, pp. 79–104, 1988.

126. T. M. Mulcahu, H. Halle, and M. W. Wambsganss, "Prediction of Tube Bundle Instabilities: Case Studies," *Argonne National Laboratory Report ANL-86-49,* 1986.

127. R. D. Blevins, *Formulas for Natural Frequency and Mode Shape,* Van Nostrand, New York, 1979.

128. M. J. Pettigrew, H. G. D. Goyder, Z. L. Qiao et al., "Damping of Multi-Span Heat Exchanger Tubes," in *Flow-Induced Vibration—1986,* S. S. Chen, J. C. Simons, and Y. S. Shin (eds.), PVP-104, ASME, New York, 1986.

129. R. D. Blevins and M. M. Bressler, "Acoustic Resonance in Heat Exchanger Tube Bundles—Part I: Physical Nature of the Phenomena, Part II: Prediction and Suppression of Resonance," *ASME J. Pressure Vessel Technology,* Vol. 109, pp. 275–288, 1987.

130. J. M. Chenoweth, "Flow-Induced Vibration," in *Handbook of Heat Exchanger Design,* G. F. Hewitt (ed.), p. 4.6.6-1, Hemisphere, New York, 1990.

131. A. C. Mueller and J. P. Chiou, "Review of Various Types of Flow Maldistribution in Heat Exchangers," *Heat Transfer Eng.,* Vol. 9, No. 2, pp. 36–50, 1988.

132. M. T. Cichelli and D. F. Boucher, "Design of Heat Exchanger Heads for Low Holdup," *AIChE, Chem. Eng. Prog.,* Vol. 52, No. 5, pp. 213–218, 1956.

133. R. B. Fleming, "The Effect of Flow Distribution in Parallel Channels of Counterflow Heat Exchangers," *Advances in Cryogenic Engineering,* pp. 352–363, 1966.

134. K. Chowdhury and S. Sarangi, "The Effect of Flow Maldistribution on Multipassage Heat Exchanger Performance," *Heat Transfer Eng.,* Vol. 6, No. 4, pp. 45–54, 1985.

135. A. C. Mueller, "An Inquiry of Selected Topics on Heat Exchanger Design," *AIChE Symp. Ser. 164,* Vol. 73, pp. 273–287, 1977.

136. J. A. Kutchey and H. L. Julien, "The Measured Influence of Flow Distribution on Regenerator Performance," *SAE Trans.,* Vol. 83, SAE Paper No. 740164, 1974.

137. J. P. Chiou, "The Advancement of Compact Heat Exchanger Theory Considering the Effects of Longitudinal Heat Conduction and Flow Nonuniformity," in *Compact Heat Exchangers: History, Technological Advancement and Mechanical Design Problems,* Book No. G00183, HTD-Vol. 10, pp. 101–121, ASME, New York, 1980.

138. J. P. Chiou, "The Effect of Nonuniformities of Inlet Temperatures of Both Fluids on the Thermal Performance of Crossflow Heat Exchanger," *Heat Transfer 1982, Proc. 7th Int. Heat Transfer Conf.,* Vol. 6, pp. 179–184, 1982.

139. A. L. London, "Laminar Flow Gas Turbine Regenerators—The Influence of Manufacturing Tolerances," *ASME J. Eng. Power,* Vol. 92A, pp. 45–56, 1970.

140. R. K. Shah and A. L. London, "Effects of Nonuniform Passages on Compact Heat Exchanger Performance," *ASME J. Eng. Power,* Vol. 102A, pp. 653–659, 1980.

141. J. R. Mondt, "Effects of Nonuniform Passages on Deepfold Heat Exchanger Performance," *ASME J. Eng. Power,* Vol. 99A, pp. 657–663, 1977; Vol. 102A, pp. 510–511, 1980.

142. R. A. Bajura and E. H. Jones Jr., "Flow Distribution Manifolds," *ASME J. Fluid Eng.,* Vol. 98, pp. 654–666, 1976.

143. A. B. Datta and A. K. Majumdar, "Flow Distribution in Parallel and Reverse Flow Manifolds," *Int. J. Heat Fluid Flow,* Vol. 2, pp. 253–262, 1980.

144. A. C. Mueller, "Criteria for Maldistribution in Viscous Flow Coolers," *Heat Transfer 1974, Proc. 5th Int. Heat Transfer Conf.,* Vol. 5, pp. 170–174, 1974.

145. G. R. Putnam and W. M. Rohsenow, "Viscosity Induced Nonuniform Flow in Laminar Flow Heat Exchangers," *Int. J. Heat Mass Transfer,* Vol. 28, pp. 1031–1038, 1985.

146. Z. H. Ayub, "Effect of Flow Maldistribution on Partial Condenser Performance," *Chemical Processing,* No. 8, pp. 30–34, 37, 1990.

147. J. B. Kitto and J. M. Robertson, "Effects of Maldistribution of Flow on Heat Transfer Equipment Performance," *Heat Transfer Eng.,* Vol. 10, No. 1, pp. 18–25, 1989.

148. J. M. Chenoweth, "Final Report of the HTRI/TEMA Joint Committee to Review the Fouling Section of the TEMA Standards," *Heat Transfer Eng.,* Vol. 11, No. 1, pp. 73–107, 1990.

149. J. W. Palen, "On the Road to Understanding Heat Exchangers: A Few Steps Along the Way," *Heat Transfer Eng.,* Vol. 17, No. 2, pp. 41–53, 1996.

150. J. G. Knudsen, "Fouling in Heat Exchangers," in *Handbook of Heat Exchanger Design,* G. F. Hewit (ed.), pp. 3.17.1-1–3.17.7-9, Hemisphere, New York, 1990.

151. W. J. Marner, "Progress in Gas-Side Fouling of Heat-Transfer Surfaces," *App. Mech. Rev.,* Vol. 43, No. 3, pp. 35–66, 1990; Vol. 49, No. 10, Part 2, pp. S161–S166, 1996.

152. W. J. Marner and J. W. Suitor, "Fouling with Convective Heat Transfer," in *Handbook of Single-Phase Convective Heat Transfer,* S. Kakaç, R. K. Shah, and W. Aung (eds.), Chapter 21, John Wiley, New York, 1987.

153. M. G. Fontana and N. D. Greene, *Corrosion Engineering,* McGraw-Hill, New York, 1978.

154. K. P. Singh and A. I. Soler, *Mechanical Design of Heat Exchangers and Pressure Vessel Components,* Arcturus Publishers, Cherry Hill, NJ, 1984.

155. E. A. D. Saunders, *Heat Exchangers: Selection, Design and Construction,* John Wiley, New York, 1989.

156. M. A. Taylor, *Plate-Fin Heat Exchangers: Guide to Their Specification and Use,* HTFS, Harwell Laboratory, Oxon, UK, amendment to 1st ed., 1989.

157. J. P. Gupta, *Fundamentals of Heat Exchanger and Pressure Vessel Technology,* Hemisphere, Washington, DC, 1986.

158. R. K. Shah, "Brazing of Compact Heat Exchangers," in *Compact Heat Exchangers—A Festschrift for A. L. London,* R. K. Shah, A. D. Kraus, and D. E. Metzger (eds.), pp. 491–529, Hemisphere, Washington, DC, 1990.

159. M. S. Peters and K. D. Timmerhaus, *Plant Design and Economics for Chemical Engineers,* 4th ed., McGraw-Hill, New York, 1991.

CHAPTER 18
HEAT TRANSFER IN MATERIALS PROCESSING

Raymond Viskanta
Purdue University

Theodore L. Bergman
The University of Connecticut

INTRODUCTION

Human civilization has been involved in thermal processing of materials since the Bronze Age. In the more recent past, the focus was on metals, alloys, and plastics. At present, different engineered materials, such as crystals, semiconductors, amorphous metals, ceramics, composites, and biomaterials are supporting rapid development of technology. An integral and critical component of materials processing and manufacturing is the management of heat addition/extraction, thermal flows, and change of phase, since temperature is critical in a wide range of manufacturing and materials processes. While this has been long recognized in the metals industry, the breadth of applications has grown dramatically. Temperature and time histories define the end product, since the state is determined by the properties. Temperature and heat-rejection limitations control the performance of a material. In many cases, it is a lack of an understanding of thermal phenomena in materials processing that hinders development and exploitation of new and existing materials and processes.

The array of processes where thermal engineering plays an important role not only includes the traditional heating and cooling of metals and alloys, glass, ceramics, and polymers for purposes of heat treatment, forming, casting, melting, and solidification, but also includes less traditional processes of coating, flame and plasma spraying, and plasma vapor and chemical vapor deposition. Heat and mass transfer and fluid flow issues underlie all of the above processes, and, recently, thermal engineering issues in materials processing have received increased attention as evidenced by recent symposia [1–5]. The types of problems encountered are very broad and range from the traditional conductive, convective, radiative, and phase-change heat transfer to less traditional ones such as heat transfer to sprays, laser, electron and ion beam heating, and materials processing in microgravity environments.

There are several important considerations that arise when dealing with materials processing. First, most of the processes are time-dependent, and the variation with time is generally of interest since the material undergoes a given thermal process to obtain the desired properties of the product. Second, most of the processes involve all modes of heat transfer, and conjugate conditions also arise because of the coupling between conduction in solids and convection and possibly radiation in the fluid. Third, the material properties are often strongly temperature-dependent, giving rise to strong nonlinearity in the energy equation. Fourth, the material properties may also depend on the shear rate and cause the material to be non-newtonian. Fifth, suitable thermal and mechanical treatments such as quenching or induction hardening produce extensive rearrangement of atoms in metals and alloys that involve nonisothermal phase transformations [6]. Sixth, the material undergoing the thermal processing may be moving, as in hot rolling or extrusion, or the thermal source itself may be moving, as in laser cutting or welding. Thus, the material properties and the processing itself affect the thermal transport and are, in turn, affected by the transport. This aspect often leads to considerable complexity in the mathematical modeling as well as the numerical simulation of the processes.

The focus of this chapter is on less traditional heat transfer processes that are not discussed in the previous seventeen chapters of this handbook but that are encountered in thermal processing of materials where heat and mass transfer are of critical importance, and improved and more detailed modeling is therefore needed. This is especially true when new materials are manufactured, when new properties are sought in traditional materials, when processes are designed for more economical manufacture, or when an increase in product quality and productivity is needed.

In preparing this chapter, the authors have sought to provide the reader with correlations that can be used to rapidly estimate heat transfer rates and/or temperature distributions. Because of the preceding issues, along with the facts that the desired accuracy of thermal control is very high in many operations (exceeding the accuracy inherent in application of some correlations) and an extreme range of geometries and materials exists, applicable correlations are not universally available. With the advent of commercially available computer packages, direct simulation of heat transfer processes is possible, making the need for correlations less critical than in the past. Therefore, we have also spent time discussing physical mechanisms and providing recommendations and guidelines for heat transfer analysis and/or simulation.

HEAT TRANSFER FUNDAMENTALS RELEVANT TO MATERIALS PROCESSING

Conduction Heat Transfer

Conduction occurs in all processing operations. When temperature-time diagrams are generated and used to relate thermal processing parameters to the overall properties of the processed materials, the inherent assumption is that conduction rates are sufficiently large to preclude the existence of significant temperature differences within the material (low Biot number).

The conduction analyses and general results outlined in Chap. 3 can be applied to many materials processing operations. In the discussion to follow, conduction heat transfer in several specific applicatons is discussed.

Conduction Heat Transfer in Beam-Irradiated Materials

Determination of the temperature distribution induced by laser, electron, or plasma beam sources is relevant in operations such as surface transformation hardening of metals, drilling, cutting, annealing, shaping, and micromachining. Descriptions of beam-generating devices as well as discussions of applications are available [7–15].

In the absence of phase change, the temperature distribution within the material is established by the size, power, and shape of the beam, along with the thermophysical properties of the material. Prediction of the internal temperature distribution in moving materials heated

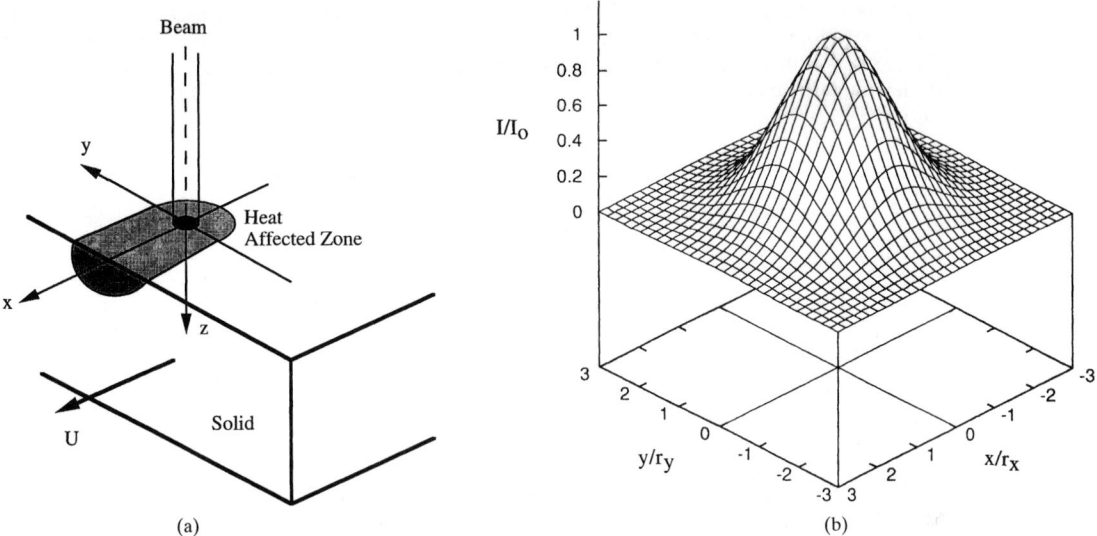

FIGURE 18.1 Schematic of (a) beam-surface interaction and the coordinate system and (b) the elliptical Gaussian beam intensity distribution.

by point sources has a long history [16, 17], and the results presented here are primarily concerned with exact solutions associated with continuous wave (CW) or pulsed Gaussian beams irradiating opaque or semitransparent moving material.

Consider the case where a Gaussian beam is directed normal to a moving solid as shown in Fig. 18.1. The solid absorbs the beam power, either on its surface or volumetrically. The Gaussian beam intensity distribution (after reflection) is [18]:

$$I = I_o \exp(-x^2/2r_x^2) \exp(-y^2/2r_y^2) \qquad (18.1)$$

where r_x and r_y are the axes of the beam ellipse and I_o is related to the total beam power P by [18]:

$$I_o = P(1 - \rho)/(2\pi r_x r_y) \qquad (18.2)$$

Here, ρ is the surface reflectivity. The energy absorbed in the material is described by:

$$q''' = [P(1 - \rho)/2\pi r_x r_y] \exp\left(-\frac{x^2}{2r_x^2} - \frac{y^2}{2r_y^2}\right) f(z) \qquad (18.3)$$

where $f(z)$ represents the effects of volumetric radiation absorption within the radiatively cold, nonemitting, nonscattering material. With the coordinate system fixed beneath the beam, the energy equation that describes the temperature distribution within the irradiated, constant property material moving at velocity U is

$$\nabla^2 T + q'''/k = \frac{1}{\alpha}\left(\frac{\partial T}{\partial t} + U\frac{\partial T}{\partial x}\right) \qquad (18.4)$$

In the absence of material removal, (1) the beam diameter, the size of the heat-affected zone, and the thermal radiation penetration depths are typically small relative to the material thickness, and (2) radiative heating is large compared to heat losses to the ambient or surroundings. Hence, an insulated semi-infinite material (initially at T_o) is usually considered.

Effect of Beam Shape on Temperatures in Stationary, Opaque Irradiated Materials.   Analytical solutions are available [18–23]. A primary quantity of interest for materials processing purposes is the maximum temperature rise, $\theta_{max} \equiv T_{max} - T_o$, and, at steady state, it is (in dimensionless terms):

$$\frac{\theta_{max} r_x k \beta^*}{P(1 - \rho)} = \frac{1}{\sqrt{2}\pi^{3/2}} K\left(\frac{\beta^{*2} - 1}{\beta^{*2}}\right)^{1/2} \tag{18.5}$$

for irradiation by a CW Gaussian beam [18]. Here, K is the complete elliptical integral of the first kind, and $\beta^* \equiv r_y/r_x$ is the ratio of the major axes of the elliptical cross section beam. For $\beta^* = 1$ (circular beam, $r = r_x = r_y$) the maximum temperature is

$$\frac{\theta_{max, \beta^* = 1} r k}{P(1 - \rho)} = \frac{1}{2\sqrt{2\pi}} \tag{18.6}$$

The influence of the Gaussian beam shape on the temperature distribution in a stationary semi-infinite target has been determined by numerical solution of Eq. 18.1, and the results are shown pictorially in Nissim et al. [18]. For $\beta^* = 1$, surface temperatures decrease with x and y, gradually approaching T_o. As the constant power beam is expanded in the y direction ($\beta^* > 1$) by, for example, sending a laser beam through a cylindrical lens, peak temperatures still occur at $x = y = 0$ but decrease relative to the $\beta^* = 1$ case due to lower incident flux at that location. Temperatures become more uniform at small and moderate y/r_x and increase at large y/r_x due to increased irradiation of these areas due to beam expansion. Temperatures and their gradients are reduced in the z direction (at $x = y = 0$) as β^* is increased, since the local irradiation is reduced as the beam is expanded. Moody and Hendel [23] calculate results where r_y and r_x are *simultaneously* increased and decreased, respectively, providing a constant irradiation flux and, in turn, higher values of $\theta/\theta_{max, \beta^* = 1}$. When the total beam power and irradiation area are held fixed, maximum surface temperatures vary with the beam shape as [23]:

$$\eta_{max} = \theta_{max}/\theta_{max, \beta^* = 1} = \frac{2}{\pi\beta^*} K\left(\frac{\beta^{*2} - 1}{\beta^{*2}}\right)^{1/2} \tag{18.7}$$

Effect of Material Motion on Temperatures in Opaque Irradiated Materials.   As the beam velocity is increased from zero, the surface temperatures are modified as shown in Fig. 18.2 [18]. At large Pe ($\equiv U r_x/\alpha$), surface temperatures are reduced with maximum values (for a particular velocity) remaining at $y = 0$ (due to symmetry) but migrating downstream from the center of beam irradiation as in Fig. 18.2b for the $\beta^* = 1$ case. Similar results are evident for

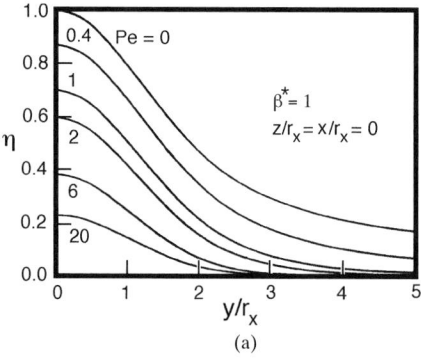

(a)

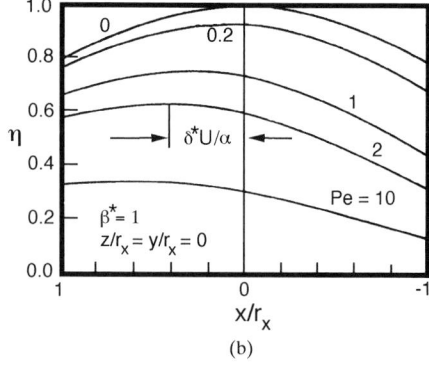

(b)

FIGURE 18.2   Steady-state temperature distributions in the x and y directions for an elliptical, CW, Gaussian irradiation of a moving, opaque material [18].

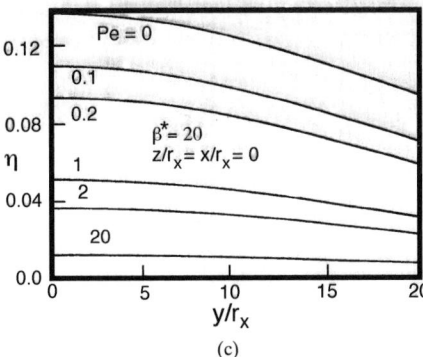

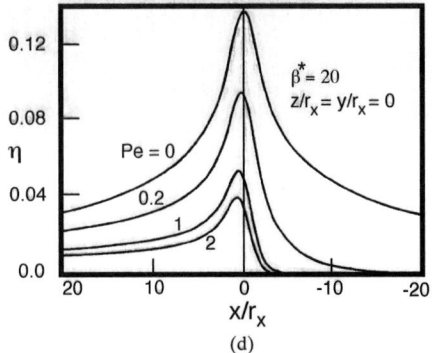

FIGURE 18.2 (*Continued*)   Steady-state temperature distributions in the x and y directions for an elliptical, CW, Gaussian irradiation of a moving, opaque material [18].

$\beta^* = 20$ in 18.2c and d, but maximum temperatures are reduced, exhibiting similar sensitivity to β^* as for the Pe $= 0$ case. The dimensionless distance between the point of maximum temperature and the center of beam irradiation ($x = 0$), $\delta^* U/\alpha$, has been correlated for a $\beta^{*-1} = 0$ (a Gaussian line source) and is [24]

$$\delta^* U/\alpha = 4.74\left(\frac{r_x^2 U^2}{8\alpha^2}\right)^{0.776} \tag{18.8}$$

Combinations of different beam shapes and material velocities in the ranges $1 < \beta^* < 40$, $0 < $ Pe $ < 10$ induce local surface temperatures (at $x = y = 0$), which are shown in Fig. 18.3. Surface temperatures beneath the beam at typical Pe values used in continuous processing (Pe < 0.2) are relatively insensitive to scanning speed. The predictions for elliptical beam heating of moving surfaces have been validated indirectly by comparing measured and predicted anneal zones of irradiated ion-implanted Si (arsenic at 6×10^{14} cm^{-2}) [18].

Irradiation of Volumetrically Absorbing Material with a CW Circular Gaussian Beam.
For the case of a Gaussian, pulsed beam, the solid's temperature distribution (for constant properties) may be estimated from Eq. 18.4, with Eq. 18.3 modified to describe the effects of pulsed beam irradiation. Specifically, Eq. 18.4 becomes [25, 26]:

$$\frac{1}{\alpha}\left(\frac{\partial T}{\partial t} + U\frac{\partial T}{\partial x}\right) = \nabla^2 T + \frac{I_o \kappa}{k}\, g(t)\, \exp^{-(x^2+y^2)/r^2 - \kappa z} \tag{18.9}$$

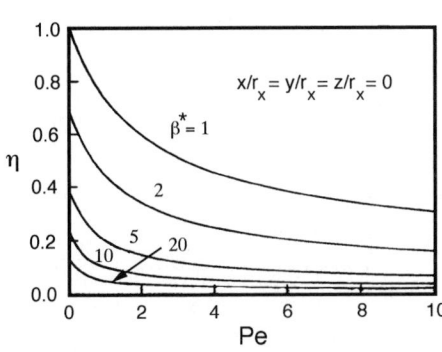

FIGURE 18.3   Schematic of the variation of η with scanning velocity and beam shape [18].

where κ is the extinction coefficient, $g(t)$ is a periodic pulsing function with period p, and r is the effective ($1/e$) beam radius.

When a circular CW beam is considered, the dimensionless extinction coefficient K ($= \kappa r$) and dimensionless time τ ($= Ut/r$) are used to present predicted results [26]. Exact solutions of Eq. 18.9 for various K and $\beta^* = 1$ have been found [26, 27]. Figure 18.4 includes predictions of the x-direction surface temperature distribution for a range of K and Pe ≈ 7. Note that as $K \to \infty$, the solutions asymptote to those of Fig. 18.2. As K becomes smaller, surface temperatures are reduced, and internal temperatures increase, as shown in Fig. 18.4b.

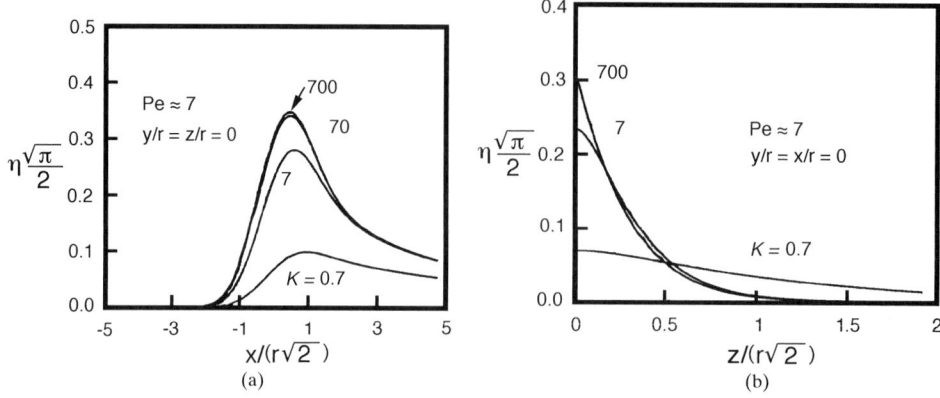

FIGURE 18.4   Steady-state temperature distributions in the x and z directions for a circular, CW, Gaussian irradiation of a moving, semitransparent material [26].

Irradiation of Opaque and Volumetrically Absorbing Material with a Pulsed Beam.
Pulsed sources can be used to tailor the material's internal temperature distribution. Pulsing is typically used to sharpen spatial temperature gradients. Solutions to Eq. 18.9 involving a single pulse for Pe = 0, $\beta^* = 1$ have been obtained by a number of researchers, and consideration of the general case of pulsed irradiation of a moving material with elliptical Gaussian beams is presented by Sanders [27].

Haba et al. [28] have computed the temperature distributions induced by pulsed irradiation of Mn-Zn ferrite with a copper vapor laser ($\lambda = 511$ and 578 nm) under the processing conditions of Table 18.1. This particular source is characterized by a top hat, circular intensity distribution.

TABLE 18.1   Processing Conditions of Haba et al. [28]

Peak laser power (kW)	250
Average laser power (W)	30–40
Pulse frequency (kHz)	5–10
Pulse width (ns)	10–15
Beam radius (mm)	0.1
Thermal diffusivity, α (m²/s)	1.9×10^{-5}
K^{-1}	∞
Pe	$0.5 \times 10^{-5} - 0.5$

For Pe = 0, $\beta^* = 1$ with a constant time-averaged power of 100 mW, the difference between maximum and minimum surface temperatures at $x = y = 0$ decreases as the pulse frequency increases, and eventually the thermal response converges to that of CW processing. At 6 kHz, a steady-state regime is reached after about 100–200 pulses. The normalized surface temperature distributions $\phi = ([T - T_o]/[T_{max} - T_o])$ are shown in Fig. 18.5 after the steady-state condition is reached. Since the beam intensity distribution is uniform for the CVL laser, the temperature directly under the beam is fairly flat. The sharpness of the transition at the beam edge is enhanced by decreasing the pulsing frequency. The temperature predictions of Fig. 18.5 were used to estimate microgroove dimensions in laser-machined Mn-Zn ferrite, and the predictions are in qualitative agreement with experimental results.

Figure 18.6 depicts the influence of a pulsating, circular, Gaussian source on the time variation of the material's surface temperature for Pe ≈ 7 and $K^{-1} \to \infty$ [26]. The dimensionless

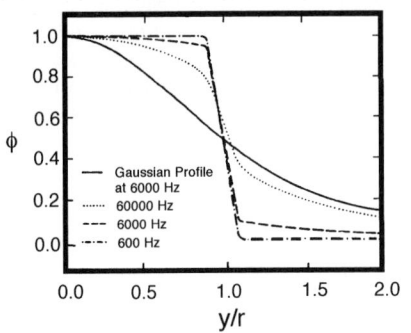

FIGURE 18.5 Computed time-averaged steady-state surface temperature distributions for $\beta^* = 1$, Pe $= K^{-1} = 0$ due to pulsed irradiation [28].

time of Fig. 18.6a is normalized by p ($= U_p/\sqrt{2}r$), so that the $p = 0.001$ case has 100 times more pulses than the $p = 0.1$ case.

The dimensionless pulse period is less than unity in many applications (the material travels less than r between pulses) resulting in relatively smooth temperature versus location behavior at any time. In contrast, for large Pe/p, local maxima and minima are noted, and surface temperatures are reduced, as shown in Fig. 18.6b. Surface temperatures are not reduced in proportion to Pe/p, since less time is available for lateral diffusion.

Effect of Temperature-Dependent Material Properties. Analytical solutions for cases of temperature-dependent thermal conductivity are available [22, 23]. In cases where the solid's thermophysical properties vary significantly with temperature, or when phase changes (solid-liquid or solid-vapor) occur, approximate analytical, integral, or numerical solutions are oftentimes used to estimate the material thermal response. In the context of the present discussion, the most common and useful approximation is to utilize transient one-dimensional semi-infinite solutions in which the beam impingement time is set equal to the dwell time of the moving solid beneath the beam. The consequences of this approximation have been addressed for the case of a top hat beam, $\beta^{*-1} = K^{-1} = 0$ material without phase change [29] and the ratios of maximum temperatures predicted by the steady-state 2D analysis. Transient 1D analyses have also been determined. Specifically, at Pe > 1, the diffusion in the x direction is negligible compared to advection, and the 1D analysis yields predictions of θ_{max} to within 10 percent of those associated with the 2D analysis.

The preceding simplified approach has been used in a number of studies involving more complex behavior, such as laser-induced *thermal runaway* of irradiated materials. For example, pulsed and CW CO_2 lasers are used to anneal ion-implanted Si for device fabrication. In this material, the extinction coefficient increases dramatically with temperature (κ increases by four orders of magnitude as T is increased from 300 to 1700 K [30]) resulting in potential thermal runaway during laser irradiation [31, 32]. The 1D form of Eq. 18.9 was solved for the case of Pe = 0, $\beta^* = 1$ pulsed ($p = 25$ ns) CO_2 irradiation of silicon to predict surface ($x = y = 0$) temperatures. For high heat fluxes, increasing temperatures lead to an increase of the rate of temperature rise, inducing thermal runaway and potential melting or vaporization of the solid phase.

Beam Penetration and Material Removal. Welding with electron, plasma, or laser beams can be modeled by considering the movement of a vertical cavity (along with the surrounding molten film) through the material to be joined (Fig. 18.7). The cavity depth-to-width ratio is usually about 10, and, to first approximation, the outer boundary of the molten liquid can be represented by a cylinder whose surface is at the melting temperature T_m.

Analytical solutions for the 2D temperature distribution around a cylinder moving at velocity U through an infinite plate have been derived [25], and solutions for moving *elliptical* cylinders also exist [33]. In either case, the local heat

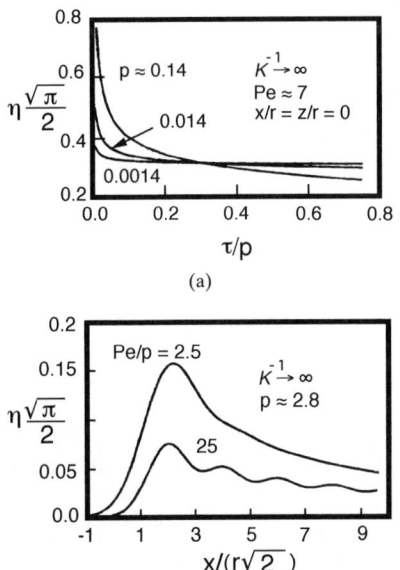

FIGURE 18.6 Effect of (*a*) the laser pulse period p on the time variation of the surface temperature, and (*b*) the material velocity on the surface temperature [26].

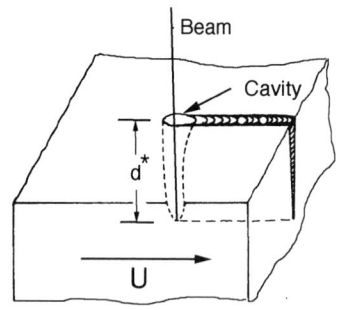

FIGURE 18.7 Schematic of beam welding or drilling.

flux is maximum at the front end of the keyhole, with more input power needed to sustain higher welding speeds.

Partial penetration of the beam through the workpiece (as shown in Fig. 18.7) is common. Predictions and experimental results showing the relationship between electron beam welding machine settings and penetration depths ($d*$) have been reviewed [34]. This led to the development of a correlation to relate $d*$ and independent parameters involved in beam welding

$$P/d*k\theta = 3.33 \mathrm{Pe}^{0.625} \qquad (18.10)$$

where k and Pe are evaluated at $(T_o + T'_m)/2$ where T'_m is the average of the liquidus and solidus temperatures for the particular alloy being welded, and $\theta = (T_{max} - T'_m)$. The characteristic length in Pe is the width of the fusion zone at the surface of the workpiece. Data scatter (for welding various aluminum and steel alloys) is of the order of ±40 percent for Pe < 1, improving to ±20 percent for Pe > 10.

The effect of 2D fluid flow between the beam and the solid has been considered [35–37]. When the circular beam penetrates the material, the size and shape of the elliptical region separating the liquid and solid phases cannot be specified beforehand but is determined by the balance between conduction and convective heat transfer rates, along with the latent energy release or absorption. Kim et al. [38] have solved the coupled differential equations of material fluid flow and heat transfer for the case when the keyhole surface temperature is the material's vapor temperature (T_v), the solid-liquid interface is at T_m, and the beam moves through a material initially at T_o. As the beam scanning speed is increased, the molten region becomes smaller and more elliptic with high heat transfer rates at the upstream edge. The required power to sustain welding is described by

$$P/k_\ell(T_v - T_m)d* = 4 + 15(\mathrm{Pe} \cdot \mathrm{Ste}_\ell^{1/2}) \qquad (18.11)$$

where $\mathrm{Ste}_\ell \equiv H/c_\ell(T_v - T_m)$ and Pe is based on the keyhole radius and liquid thermal diffusivity. For $\mathrm{Pe} \cdot \mathrm{Ste}_\ell^{1/2} > 0.1$, dimensionless penetration depths are correlated to within 10 percent of experimental data and predicted results based upon solution of the energy and Navier-Stokes equations.

Microscale Laser Processing.   Radiation and conduction heat transfer during laser processing of thin (microscale) semitransparent films has been reviewed [39]. During transient heating of semitransparent materials at the nanosecond scale (important in semiconductor processing), the thermal gradients across the heat-affected zones are accompanied by changes in the material complex refractive index (extinction coefficient). These complex refractive index variations, along with radiative wave interference effects, modify the energy absorption characteristics of the material and, in turn, the temperature distribution in the target. Recent studies have considered a wide array of processing scenarios, including pulsed laser evaporation of metals [40], laser sputtering of gold [41], and melting of polycrystalline silicon [42, 43]. In general, the studies have considered one-dimensional or two-dimensional heat transfer in the irradiated materials. Phase changes (vaporization or melting) have been accounted for. Finally, detailed predictions of microchannel shape evolution induced during laser machining of ablating materials using CW, pulsed, or Q-switched Gaussian sources has been achieved [44]. Modest [44] found that losses to the unablated material are virtually negligible for Q-switched operation, small for the regularly pulsed laser, and very substantial when the CW source is used. As a result, the microgroove walls are precisely shaped by the Q-switched source and are not as well defined when the CW laser is used.

Conduction Heat Transfer with Thermomechanical Effects

Elastic and/or plastic material deformation or induced internal stresses resulting from imposed thermal or mechanical loads can be important in applications where the structural integrity of the processed material is of concern. The evolution of internal stresses may, in turn, modify the material's thermal response by (1) inducing volumetric heating, or (2) modifying surface geometries (and, in turn, surface heat transfer rates) in operations such as rolling, forming, or pressing. Because of the breadth of applications and materials, the following discussion only highlights the thermomechanical response of several specific processes.

Elastic-Plastic Deformation During Flat Rolling of Metal Sheet.   Conduction heat transfer occurs in conjunction with pressing, rolling, and squeezing operations. Energy generation may occur in the compressed strip, and frictional heating will occur at the strip-roller interface. Contact resistance between the compressed solid and roller may significantly affect product temperature. Relative to elastic deformation, elastic-plastic deformation is much more complicated, since permanent material displacement and potentially high rates of internal energy generation can occur. Usually, considerable uncertainty is associated with the constitutive modeling of the solid material as well as the heat transfer rates at material boundaries where deforming forces are applied. Finite element models are typically used to handle the material deformation.

Flat rolling of a metal sheet, as seen in Fig. 18.8*a*, can be performed at room temperature (cold rolling) or at high temperature on a hot strip mill. As the strip enters the roll gap, it is

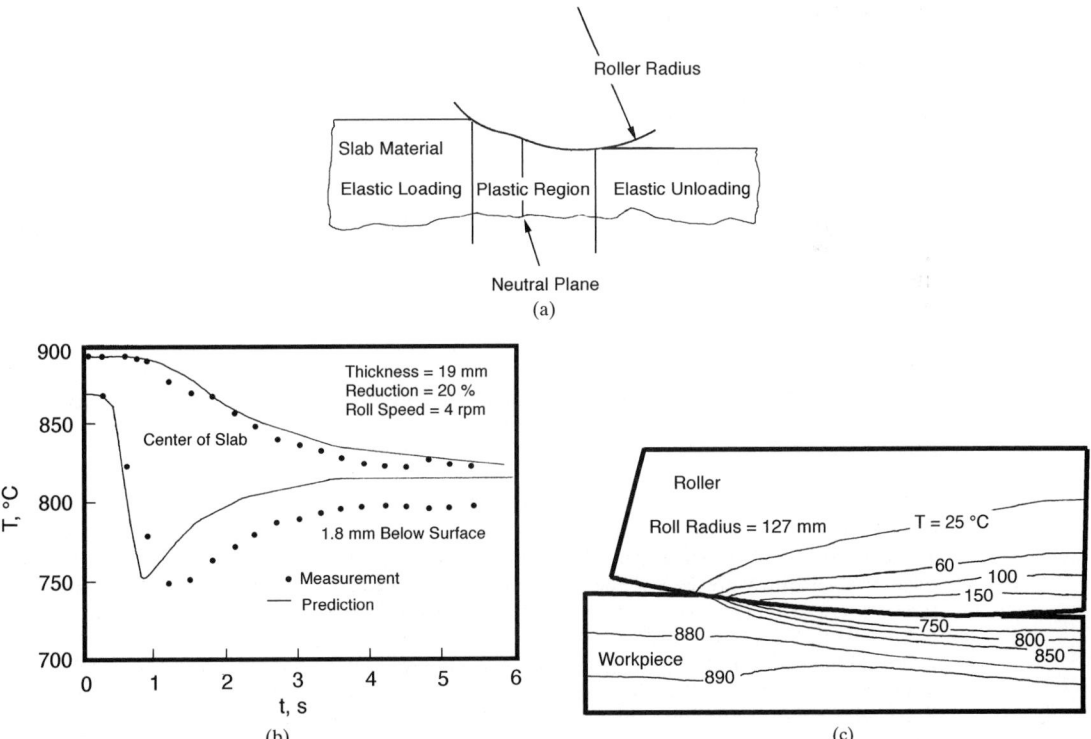

FIGURE 18.8   Thermoelastic and plastic effects during rolling of metal sheet showing (*a*) the physical system, (*b*) measured and predicted strip temperatures, and (*c*) predicted temperature distributions in the roll and strip [45].

elastically deformed. It is subsequently plastically deformed and elastically unloaded. Relative velocities between the roll and strip change through the bite, with identical roll and strip velocities occurring at the neutral plane of Fig. 18.8a. Frictional heating occurs at the roll-strip interface before and after the neutral plane.

Thermomechanical aspects of strip rolling have been reviewed [45, 46]. Investigation of cold rolling of metals shows that the heat transfer occurring between the roll and strip does not significantly influence the predictions of roll pressure, power requirements, or temperatures [47], but it plays a significant role in hot rolling. Limited experimental measurements of temperatures induced by rolling are available, and, with a parallel thermomechanical modeling effort, these data can be used to *infer* heat transfer conductances between the roll and sheet. Here, the conductance is defined as $C = q''/(T_s - T_r)$, where q'' is the heat flux at the interface, while T_s and T_r are the adjacent surface temperatures of the sheet and roll.

Conductance values have been estimated for a variety of processing conditions using the combined experimental/analytical approach [45, 48, 49]. Conductance values range from 2600 W/m²K to 30,000 W/m²K, depending on the operating conditions and material being processed [45]. It is emphasized that, since the thermal and mechanical effects are so closely coupled, inferred conductance values are highly sensitive to the thermomechanical constitutive model and the coefficients of sliding friction at the roll-strip interface. Nonetheless, estimates of the thermal response can be made via finite element modeling. A comparison between predicted and measured temperatures in a strip of low carbon steel is shown in Fig. 18.8b, and predicted roll and strip temperature distributions are shown in Fig. 18.8c [45]. A similar combined analytical/experimental/finite element modeling approach has been used to estimate the dynamic contact resistance between a high-temperature, mechanically deforming Pb-Sn sphere and a planar, highly polished steel surface with application to electronics assembly [50]. As the sphere softens upon approach of its melting temperature, contact conductances become extremely large, as expected.

Elastic Deformation Due to High-Intensity Localized Heating.   If plastic deformation is avoided, elastic deformation may still induce material cracking and/or failure. The thermomechanical response of an opaque material to high intensity localized heating has been considered. When strain rates are insufficiently high to induce internal heating (and deformations are small), the conduction solutions of the previous section may be combined with the classical equation of elastic stresses:

$$\nabla^2 \varphi = \beta \left[\frac{1 + \tilde{v}}{1 - \tilde{v}} \right] T \qquad (18.12)$$

to yield expressions for the thermomechanical response of the heated material. Here, β is the thermal expansion coefficient, $\tilde{v}$ is Poisson's ratio, and φ is the potential of the thermal-elastic shift ($\partial \varphi / \partial i = V_i$).

If heating of a material with a CW Gaussian source is considered, appropriate initial conditions are $\varphi|_{t=0} = (\partial \varphi / \partial t)|_{t=0} = 0$ with $\sigma_{xx}(r, 0, t) = \sigma_{rx}(r, 0, t) = 0$ on the surface of the material. In laser processing, high thermal stresses can be generated that, in the extreme case, can lead to fractures running along grain boundaries. The cracks can be observed even when maximum local temperatures are below the solid's melting temperature. Analytical solutions for the stress distributions and their time variation are available, with maximum stresses developing at $x = y = 0$ [51, 52].

Analytical solutions for thermoelastic stress distributions within moving material, irradiated with two-dimensional CW Gaussian beams ($\beta^{*-1} = 0$), have also been obtained [24]. For a material characterized by $k = 50.2$ W/mK, $\rho = 7880$ kg/m³, $c = 502$ J/kgK, $P/2r = 10^5$ W/m, $U = 4$ mm/s, $\beta = 10^{-5}$ K^{-1}, $\tilde{v} = 0.3$, and $\bar{\mu} = 10^5$ MPa (the material shear modulus), the dimensionless surface stress component varies with Pe as shown in Fig. 18.9. Here, Pe was varied by changing the beam radius, and the beam moves relative to the surface in the positive x direction. At large Pe, stresses are relatively uniform, while, at extremely small Pe, stress gradients

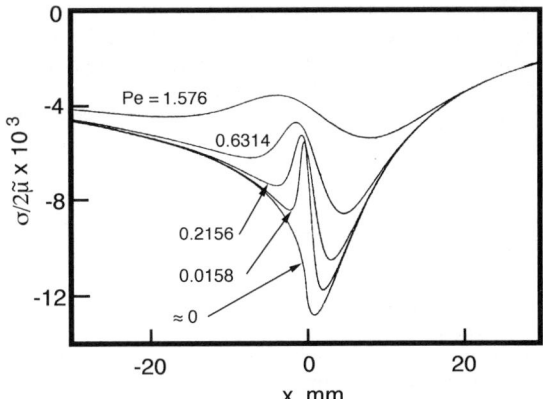

FIGURE 18.9 Dimensionless surface stress distributions within a moving solid material irradiated by a two-dimensional Gaussian beam [24].

are concentrated about the $x = 0$ location. At intermediate Pe, local stress gradients are concentrated just behind the beam center, due to the shift of maximum temperatures to that location.

Thermoelastic Instabilities During Planar Solidification of a Pure Material.   A wide array of materials processing operations involve solidification, and contact resistances between the solid and cooled surface may have a profound effect upon phase change rates and phenomena. During casting, for example, there is a thermal contact resistance at the mold-solid interface, since each surface is rough on the microscopic scale. If a bottom-chilled mold is considered, the contact pressure at the mold-solid interface will be initially determined by the hydrostatic pressure in the liquid, but, as solidification proceeds, temperature gradients within the solidified shell will induce thermoelastic distortion and influence the contact pressure locally. Local separation between the casting and the mold can occur, significantly prolonging solidification times relative to the perfect thermal contact case [53].

In response to the localized contact resistances at the mold-casting interface, the nominally planar solid-liquid interface can be affected, especially in the early stages of solidification. This coupled thermomechanical effect is a possible explanation for the long-wavelength perturbations sometimes observed during unidirectional solidification [54] and has been analyzed [55–57] using the idealization of elastic deformation of the solid phase (although plastic deformation is likely to occur in reality since temperatures are high).

Figure 18.10*a* shows the system geometry considered in Refs. 56 and 57. Note that the heat loss through the solid's bottom is sinusoidal with the local flux, prescribed as

$$q''(x) = q_o'' + q_1'' \cos{(mx)} \qquad (18.13)$$

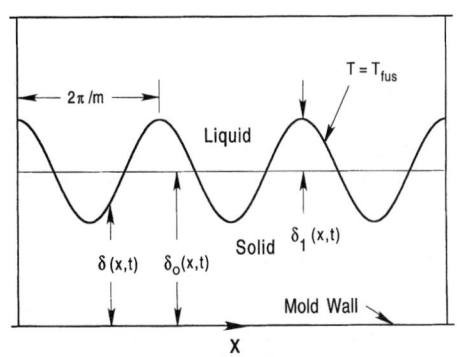

FIGURE 18.10*a* Thermoelastic effects during directional solidification of a pure material showing (*a*) the physical system.

where $q_o'' \gg q_1''$. Here, $\delta(x, t) = \delta_o(t) + \delta_1(x, t)$ is the location of the solid-liquid interface, and the liquid is initially at the fusion temperature T_{fus}. The two-dimensional heat diffusion equation, subject to Eq. 18.13 and

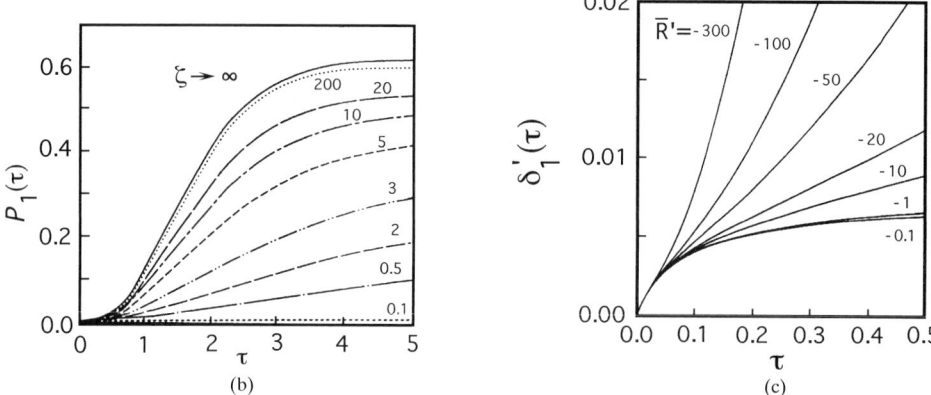

FIGURE 18.10 (*Continued*)   Thermoelastic effects during directional solidification of a pure material showing (*b*) perturbation in the metal-mold contact pressure versus time for different ζ, and (*c*) perturbation of the solidification front for different sensitivities of the contact resistance to the contact pressure [56, 57].

$$k \frac{\partial T(x, \delta, t)}{\partial y} = H\rho \frac{d\delta(x, t)}{dt} ; \qquad T(x, \delta, t) = T_{\text{fus}} \quad \text{at } \delta(x, t) \qquad (18.14)$$

was solved in conjunction with Eq. 18.12 to yield the predicted dynamic response in terms of the perturbation in contact pressure along the mold-solid interface, $P_1(t)$ [56], where

$$P(x, t) = P_o(t) + P_1(t) \cos (mx) = -\sigma_{yy0}(0, t) \qquad (18.15)$$

and $P_o(t) \gg P_1(t)$.

Figure 18.10*b* shows the time ($\tau = mq''_o t/\rho H$) variation of the interface pressure perturbation, and $\zeta = 2m\alpha\rho H/q''_o$. Note that $\zeta \to \infty$ as Ste$_s \to 0$. Figure 18.10*c* shows the dimensionless perturbation of the solid-liquid interface, $\delta'_1(t) = m\delta_1(\tau)$ versus $\tau = m^2 kT_f t/\rho H$ for various arbitrary sensitivities of the contact resistance to pressure $R' = R'(P_o)$, where P_o is the dimensionless unperturbed contact pressure $(1 - \tilde{v})P/\bar{\mu}\beta T_{\text{fus}}$. For high sensitivities, the solid-liquid interface shape becomes significantly perturbed and potentially unstable. Figure 18.10*c* is associated with $\zeta \to \infty$. At higher Ste$_s$, it is expected that the overall solidification process will become less sensitive to elastic stresses generated in the solid phase.

Single-Phase Convective Heat Transfer

Forced Convective Heat Transfer.   Convective heat transfer to/from a continuously moving surface has many important applications for metal, glass, paper, and textiles manufacturing processes. Examples of such processes are hot rolling, wire drawing, metal extrusion, continuous casting, glass fiber production, and paper production [58–60]. Knowledge of fluid flow and heat transfer is often necessary for determining the quality of the final products of these processes [61]. A number of different physical situations arise and can be characterized by the following two: (1) the ambient fluid is stagnant relative to the continuously moving surface, as in Fig. 18.11*a* and (2) the fluid is parallel, like Fig. 18.11*b*, or in counterflow, as in Fig. 18.11*c*, relative to the continuously moving sheet. In the latter case, two physical situations are encountered and require separate treatments: (1) the velocity of the moving surface U_s is greater than the free stream velocity of the fluid U_∞, $U_s > U_\infty$, and (2) $U_s < U_\infty$. In any case, the flow generated by the motion of a solid surface is of the boundary layer type, and convective transport characteristics can be predicted [62]. Other physical situations may arise and will be identified in this subsection.

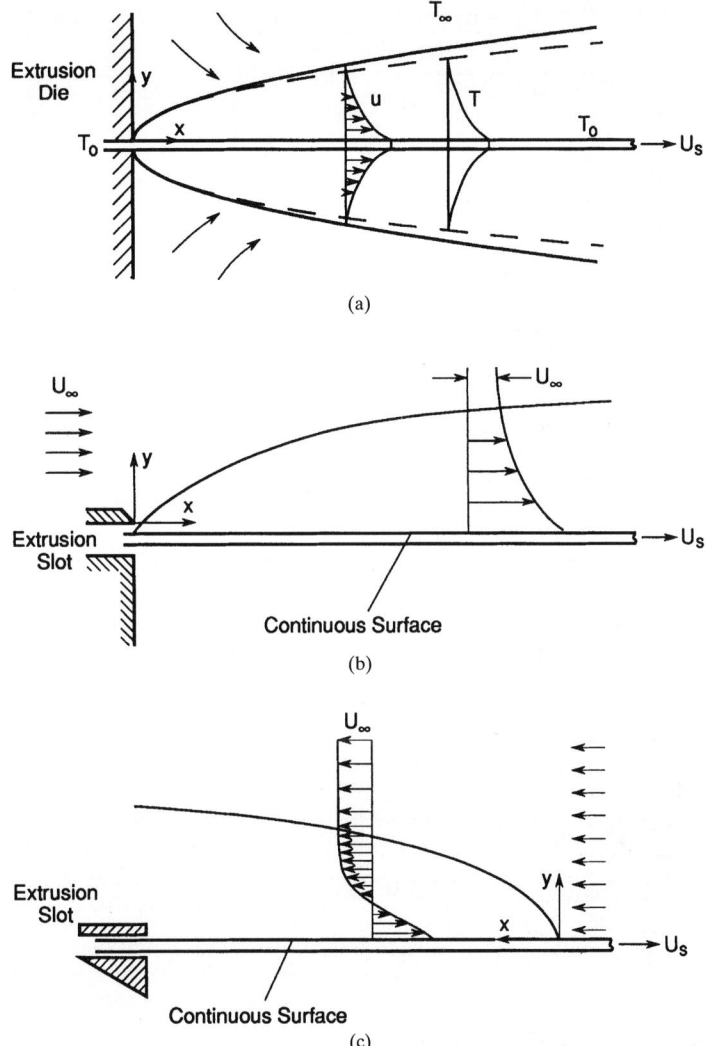

FIGURE 18.11   Sketch of the velocity and temperature profiles induced due to
(a) a moving isothermal surface at temperature T_o, (b) sketch of the boundary
layer on an isothermal moving surface and cocurrent, parallel stream, and (c)
sketch of the boundary layer on an isothermal moving surface and counter-current,
parallel stream.

The boundary layer, along a semi-infinite continuous sheet issuing from a slot and moving
in an otherwise quiescent medium, develops from the opening of the slot and along the direction of motion as depicted schematically in Fig. 18.11a. The heat transfer in such a boundary
layer is physically different from that of the classical forced convection along a stationary
semi-infinite plate. The heat transfer rate from a moving sheet is higher than that from a stationary plate due to the thinner boundary layers in the vicinity of the moving wall.

Laminar boundary layer flow and heat transfer from a moving plate to a quiescent fluid
under uniform wall temperature (UWT) and uniform heat flux (UHF) boundary conditions
have been studied. Similarity solutions of the incompressible boundary layer equations with

TABLE 18.2   Summary of Forced Convection Heat Transfer Correlations $[\mathrm{Nu}_x/\mathrm{Re}_x^{1/2} = G(\mathrm{Pr})]$ for a Moving Plate in a Quiescent Fluid

Correlation for $G(\mathrm{Pr})$	Remarks	Reference
$0.545\mathrm{Pr}^{1/2}$	Laminar, UWT, $\mathrm{Pr} \to 0$	Jacobi [70]
$0.807\mathrm{Pr}$	Laminar, UWT, $\mathrm{Pr} \to 0$	Jacobi [70]
$0.545\mathrm{Pr}/(1 + 0.456/\mathrm{Pr})^{1/2}$	Laminar, UWT, $0 < \mathrm{Pr} < \infty$	Jacobi [70]
$1.8865\mathrm{Pr}^{13/32} - 1.447\mathrm{Pr}^{1/3}$	Laminar, UWT, $0.1 < \mathrm{Pr} < 100$	Ramachandran et al. [71]
$0.5462\mathrm{Pr}^{1/2}/(0.4621 + 0.1395\mathrm{Pr}^{1/2} + \mathrm{Pr})^{1/2}$	Laminar, UWT, $0.01 < \mathrm{Pr} < 10^4$	Lin and Huang [63]
$2.8452\mathrm{Pr}^{13/32} - 2.0947\mathrm{Pr}^{1/3}$	Laminar, UHF, $0.1 < \mathrm{Pr} < 100$	Ramachandran et al. [71]

constant thermophysical properties have been obtained and empirical correlations developed. Probably, the most general correlation for local heat transfer at a surface of a continuously moving plate (sheet) in a quiescent fluid has been developed by Lin and Huang [63] and is given by

$$\mathrm{Nu}_x/\mathrm{Re}_x^{1/2} = 0.5462\mathrm{Pr}^{1/2}/(0.4621 + 0.1395\mathrm{Pr}^{1/2} + \mathrm{Pr})^{1/2} \qquad (18.16)$$

The maximum error of this correlation compared to numerical results is less than 1 percent for $0.01 \le \mathrm{Pr} < 10^4$. This and other available convective heat transfer results are summarized in Table 18.2. Comparison of the correlations for heat transfer over a stationary flat plate given in Chap. 6 with those in the table reveals that convective coefficients from a plate in motion are over 20 percent higher than those for a stationary plate with a fluid flowing over it. This is owing to the thinning of the hydrodynamic boundary layer. Consistent with the classical results, the heat transfer coefficients are higher for UHF than for UWT boundary conditions.

The problem of a stretching plate moving in a quiescent environment with linear [64–67] or power-law [65, 67] velocity profiles has been analyzed. Various temperature boundary conditions [66, 68] have been considered, including stretching of a surface subject to a power-law velocity, and temperature distributions for several different boundary conditions [67, 69] have been analyzed. Fluid friction and heat transfer characteristics have been predicted, including with suction and injection at a porous stretching wall with power-law velocity variation. For example, Ali [69] has reported similarity solutions of laminar boundary layer equations for a large combination of speed and temperature conditions by employing the most general power-law velocity and temperature distributions with various injection parameters to model flow and heat transfer over a continuously stretched surface.

Heat transfer associated with simultaneous fluid flow parallel to a cocurrently or countercurrently moving surface has been analyzed under the UWT boundary conditions when $U_\infty > U_s$ and when $U_\infty < U_s$ (Fig. 18.11c). Laminar [63] and turbulent [72] flow situations have been studied.

For laminar flow, Lin and Huang [63] have obtained similarity solutions over a wide range of Prandtl numbers $(0.01 < \mathrm{Pr} < 10^4)$ and summarized their numerical results in a form of an empirical correlation

$$\frac{\mathrm{Nu}}{(\omega\,\mathrm{Re}_\infty)^{1/2}} = (1 - \xi)^{1/2}\left\{\left[\frac{\mathrm{Nu}_B}{(\omega\,\mathrm{Re}_\infty)^{1/2}}\right]^n + \left[\frac{\xi}{(1 - \xi)}\frac{\mathrm{Nu}_s}{(\sigma'\,\mathrm{Re}_s)^{1/2}}\right]^n\right\}^{1/n} \qquad (18.17)$$

where $\sigma' = \mathrm{Pr}^2/(1 + \mathrm{Pr})$, $\omega = \mathrm{Pr}/(1 + \mathrm{Pr})^{1/3}$, $\xi = (1 + \omega\,\mathrm{Re}_s/\sigma\,\mathrm{Re}_s)^{-1}$. In Eq. 18.17, $\mathrm{Nu}_B/(\omega\,\mathrm{Re}_\infty)^{1/2}$ is the heat transfer parameter for the special case of the Blasius problem $(\gamma' = U_w/(U_w + U_\infty) = (1 + \mathrm{Re}_\infty/\mathrm{Re}_s)^{-1} = 0)$ and can be calculated from the correlation [73]

$$\mathrm{Nu}_B/\mathrm{Re}_\infty^{1/2} = 0.3386\mathrm{Pr}^{1/2}/(0.0526 + 0.1121\mathrm{Pr}^{1/2} + \mathrm{Pr})^{1/6} \qquad (18.18)$$

The maximum error in this correlation does not exceed 1.4 percent for $0.001 \leq \mathrm{Pr} \leq \infty$. The heat transfer parameter $\mathrm{Nu}_S/(\sigma' \, \mathrm{Re}_s)^{1/2}$ for the special case of the Sakiadis problem ($\gamma' = 1.0$) can be calculated from the empirical Eq. 18.16. The values of the exponent n in Eq. 18.17 have been determined [63] and range between 0.76 for $0.01 \leq \mathrm{Pr} \leq 0.1$ to 1.02 for $7 \leq \mathrm{Pr} < 10^4$. The corresponding maximum errors between the numerical results and the correlation are 5.7 and 1.4 percent, respectively.

Both similarity and integral approaches of solution of the incompressible, constant property boundary layer equations were employed to predict local and average fluid friction factors and heat transfer coefficients for turbulent flow [72]. For a limiting situation of stagnant fluid, the solutions were compared with published experimental data, and good agreement has been obtained. In contrast to results for laminar flow, it was found that skin friction and heat transfer are higher for $U_\infty > U_s$ than for $U_s > U_\infty$. Based on the integral method of solution local Nusselt number for cocurrent turbulent flow over a moving surface can be expressed

$$\mathrm{Nu}_x = 0.0198 \tilde{D}^{1/5} \overline{U}^{2/3} \, \mathrm{Re}_x^{4/5} \, \mathrm{Pr}^{2/3} \tag{18.19}$$

where

$$\tilde{D} = [2(U_s/U_i) + 7(U_\infty/U_i)] \tag{18.20}$$

and

$$\overline{U} = |U_s - U_\infty|/U_i \tag{18.21}$$

with $U_i = U_\infty$ if $U_\infty > U_s$ and $U_i = U_s$ if $U_s > U_\infty$. Equation 18.19 clearly reveals the dependence of the local heat transfer coefficient on the relative motion between the moving surface and the fluid. Skin friction and heat transfer predictions based on the similarity and integral methods of solution are in reasonably good agreement, indicating that integral solutions can be used for quick estimation of the skin friction coefficient and Stanton number. Experimental data are not available to validate the predictions for the range of conditions and parameters of interest. Stanton numbers based on the similarity solution (Fig. 18.12) are in very good agreement with the analytical results of Tsou et al. [74], and the integral solution yields Stanton numbers that are less than those obtained from the similarity solution.

The physical situation where the fluid stream parallel to the moving plate is in an opposite direction to the motion of the plate (Fig. 18.11c) is also encountered in materials processing. Using a similarity method, Klemp and Acrivos [75] found that a critical value of the moving surface to the free stream velocity ratio ($\lambda' = U_s/U_\infty$) was 0.3541. The inability to obtain similarity solutions of the boundary layer equations for laminar flow was attributed to the boundary layer separation from the moving plate. Similarity and integral solutions for fluid friction

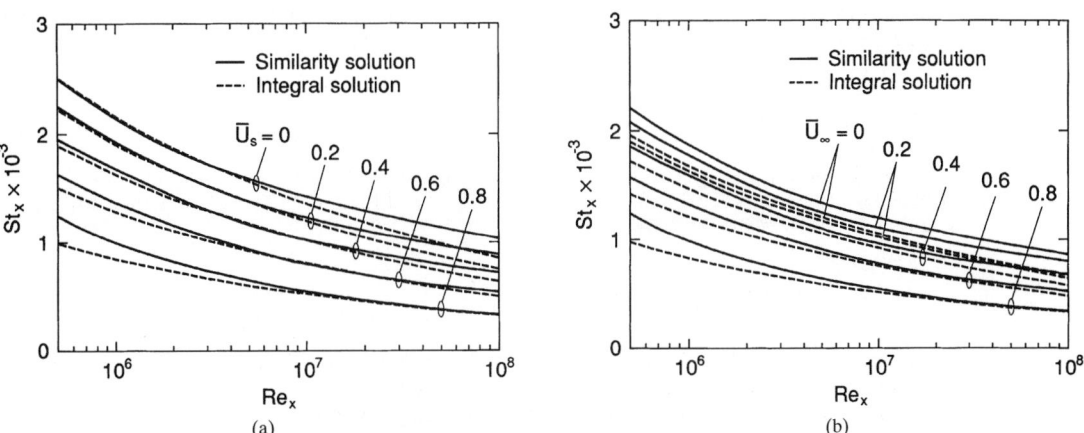

FIGURE 18.12   Local Stanton number results: (a) St for $U_\infty > U_w$ and $\mathrm{Pr} = 0.7$, and (b) St for $U_w > U_\infty$ and $\mathrm{Pr} = 0.7$ [72].

and heat transfer between a continuous, isothermal surface that is in relative motion to a parallel fluid stream that is in counterflow for the case of $U_s < U_\infty$ has been obtained [76]. The similarity solution for heat transfer was obtained for a range of velocity ratios λ' (< 0.3541) for which the boundary layer is attached to the moving surface. For values of $\lambda' < 0.3$, the integral method of solution may be useful, but the method predicts separation to occur for too high values of λ'.

Mixed Convective Heat Transfer to Moving Materials.   Buoyancy forces arising from the heating and cooling of the sheet of Fig. 18.11a modify the flow and thermal fields and thereby the heat transfer characteristics of the process. Simple empirical mixed convection correlations for local and average Nusselt numbers, based on the method of Churchill and Usagi [77], have been developed [71, 78] and are shown in Fig. 18.13.

For the UWT boundary condition, the local mixed convection Nusselt number for a horizontal, isothermal, continuous moving sheet can be expressed by the equation

$$\text{Nu}_x \, \text{Re}_x^{-1/2}/F_1(\text{Pr}) = \{1 \pm [F_2(\text{Pr})(\text{Gr}_x/\text{Re}_x^{5/2})^{1/5}/F_1(\text{Pr})]^n\}^{1/n} \tag{18.22}$$

where

$$F_1(\text{Pr}) = 1.8865\text{Pr}^{13/32} - 1.4447\text{Pr}^{1/3} \tag{18.23}$$

and

$$F_2(\text{Pr}) = (\text{Pr}/5)^{1/5} \, \text{Pr}^{1/2} \, [0.25 + 1.6\text{Pr}^{1/2}]^{-1} \tag{18.24}$$

The corresponding average mixed convection Nusselt number can be correlated as

$$\overline{\text{Nu}}_L \, \text{Re}_L^{-1/2}/2F_1(\text{Pr}) = \{1 \pm [5F_2(\text{Pr})(\text{Gr}_L/\text{Re}_L^{5/2})^{1/5}/6F_1(\text{Pr})]^n\}^{1/2} \tag{18.25}$$

For $0.7 \le \text{Pr} \le 100$, $n = 3$ provides a good correlation. The temperature difference for Gr is $\Delta T = T_o - T_\infty$.

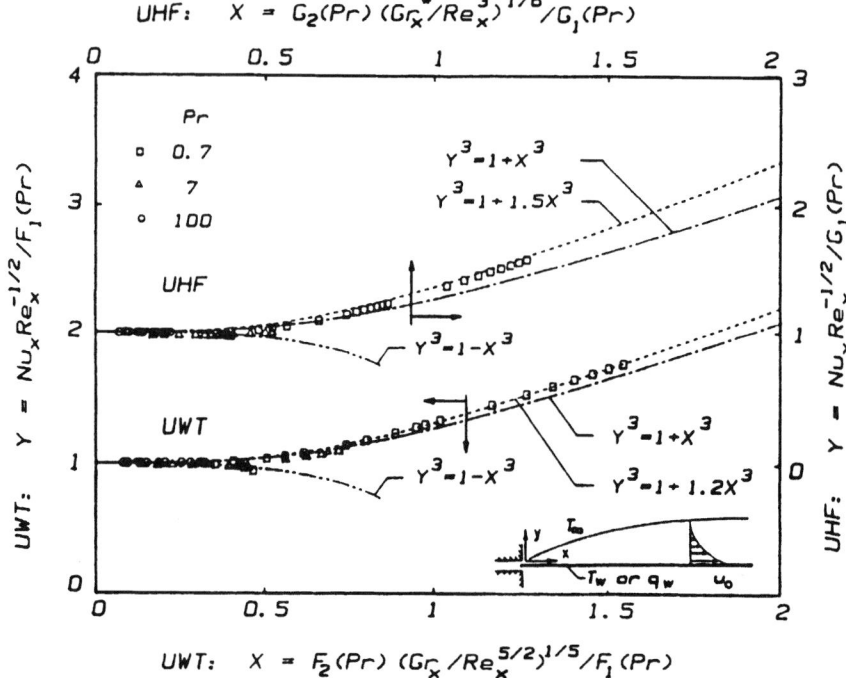

FIGURE 18.13   A comparison between the predicted and correlated local Nusselt numbers for the UWT and UHF cases [71], published with permission of ASME International.

For the UHF boundary condition, the local mixed convection Nusselt number for a horizontal, isoflux, continuous moving sheet can be expressed by the equation [71]:

$$\text{Nu}_x \, \text{Re}_x^{-1/2}/G_1(\text{Pr}) = \{1 \pm [G_2(\text{Pr})(\text{Gr}_x^*/\text{Re}_x^3)^{1/6}/G_1(\text{Pr})]^{-n}\}^{1/n} \tag{18.26}$$

where
$$G_1(\text{Pr}) = 2.8452\text{Pr}^{13/32} - 2.0947\text{Pr}^{1/3} \tag{18.27}$$

and
$$G_2(\text{Pr}) = (\text{Pr}/6)^{1/6} \, \text{Pr}^{1/2} \, [0.12 + 1.2\text{Pr}^{1/2}]^{-1} \tag{18.28}$$

The average Nusselt number can be written as

$$\overline{\text{Nu}}_L \, \text{Re}_L^{-1/2}/2G_1(\text{Pr}) = \{1 \pm [3G_2(\text{Pr})(\text{Gr}_L^*/\text{Re}_L^3)^{1/6}/4G_1(\text{Pr})]^n\}^{1/n} \tag{18.29}$$

Again, $n = 3$ also provides a good correlation of the results for the UHF case.

The correlations presented for the Prandtl number range of $0.7 \le \text{Pr} \le 100$ and for both buoyancy-assisted and buoyancy-opposed conditions agree well with the analytically predicted values for laminar flow [71]. No results for turbulent flow under mixed convection conditions have been identified in the published literature, but it is reasonable to speculate that buoyancy effects would probably be small under such conditions.

Laminar mixed convection adjacent to vertical and inclined moving sheets in a parallel freestream (Fig. 18.14) have been analyzed for both UWT and UHT boundary conditions [78]. The analysis covers the entire mixed convection regime, from pure forced convection to pure free convection. The numerical results are reported for different combinations of the velocities of the moving sheet U_s and the freestream U_∞ for which $U_s > U_\infty$ or $U_s < U_\infty$. Empirical correlations for local and average Nusselt numbers are reported for a wide range of the buoyancy parameter. The equations are, however, rather lengthy and are not included for the sake of brevity. It should be noted that the factor $(C \, \text{Pr})^m$ is missing in the denominator for the expressions of the local and average Nusselt numbers reported in the paper. In this factor, C is a correlation constant, and m is a constant defined by $U_\infty/(U_s + U_\infty)$ for $U_s > U_\infty$, and $U_s/(U_s + U_\infty)$ for $U_s < U_\infty$. The solutions presented by Ramachandran et al. [78] are not applicable when the strip velocity U_s is downward; that is, in the direction of the gravitational force.

Conjugate Heat Transfer to Moving Materials.   In the previous two subsections, the thickness of the material was assumed to be small, so that the controlling resistance was on the fluid side; therefore, the thermal coupling between heat transfer within the moving material and the convective flow and heat transfer in the fluid could be neglected. However, in many

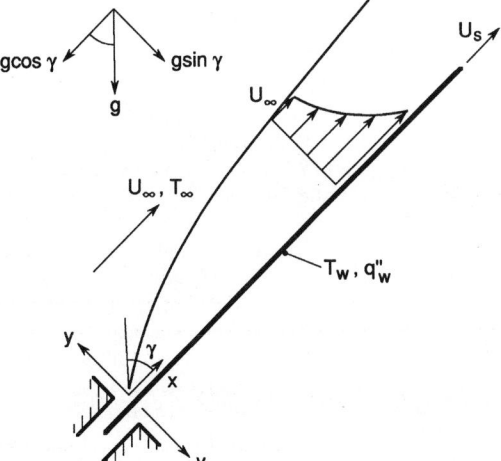

FIGURE 18.14 Schematic of the physical arrangement and coordinate systems (after Ramachandran et al. [78]).

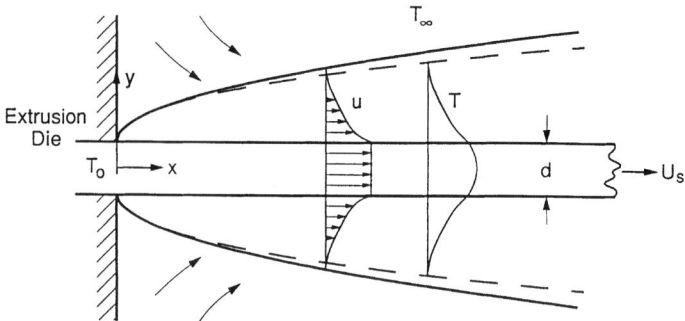

FIGURE 18.15   Schematic of velocity and temperature profiles in the conjugate heat transfer problem on a continuous moving flat plate.

processing problems, interest lies mainly in the temperature distribution in the material that is of finite thickness. Under these conditions, the heat transfer within the moving material is coupled with the convective flow and heat transfer in the fluid. This implies that the flow generated by the moving material is computed, along with the thermal field, in order to obtain the heat transfer rate in the material. The heat transfer coefficient is not assumed or calculated from existing correlations, as discussed in the preceding two subsections, but is obtained from governing equations for the fluid flow. The intimate coupling of heat transfer in the moving material and the external fluid results in a conjugate heat transfer problem. This is depicted schematically in Fig. 18.15 for a continuously moving flat plate in quiescent ambient fluid. Such problems have been studied, and recent accounts are available [62, 79].

A realistic model for heat transfer from/to moving material under thermal processing consists of at least two regions: (1) the moving solid material, and (2) the fluid in which the flow is induced by the surface motion and/or is forced externally. The temperatures and heat transfer in the two regions are coupled through thermal boundary conditions at the interface between the solid moving material and the fluid (Fig. 18.15). At the interface, the temperature of the solid T_s must equal the temperature of the fluid T_f such that

$$T_s = T_f \qquad \text{at } y = 0 \tag{18.30}$$

Also, the heat flux at the interface must be continuous and can be expressed as

$$k_s \left. \frac{\partial T_s}{\partial y} \right|_{y = d^-} - k_f \left. \frac{\partial T_f}{\partial y} \right|_{y = d^+} = q''_{\text{rad}} \tag{18.31}$$

In this equation, q''_{rad} is the net radiative heat flux at the moving material surface imposed by external sources such as radiant burners/heaters or electric resistance heaters. Both parabolic, boundary layer [80], and full, elliptic [61, 81] problem solutions have been reported. Because of the nature of the problem, the heat transfer results can't be given in terms of correlations. The interested reader is referred to Refs. 62 and 79 for citation of relevant references.

Impingement Heat Transfer to Gaseous Jets.   A single gas jet or arrays of such jets, impinging normal on a surface, may be used to achieve enhanced coefficients for convective heating, cooling, and drying. A disadvantage of impinging gas jets is that local heat transfer is highly nonuniform. This is owing to the complex fluid flow structure, which consists of the free jet, stagnation or impingement zone, and wall jet region. In the stagnation region, flow is influenced by the target surface and is rapidly decelerated and accelerated in the normal and transverse directions. In the wall jet zone, velocity profiles are characterized by zero velocity at both the impingement and free surfaces. Many materials processing applications such as

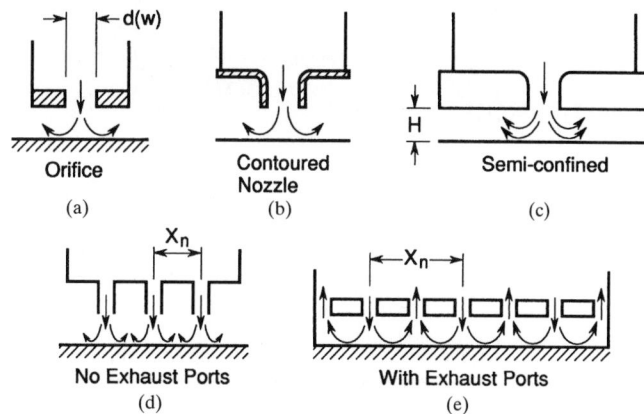

FIGURE 18.16 Flow geometries and arrangements for impinging gaseous jets.

annealing of metal sheets, tempering of glass, and drying of textiles and papers involve the need to cover large areas. Thus, impingement heat (mass) transfer schemes require use of an array of round or slot jets [82]. In addition to flow from each nozzle exhibiting free jet, stagnation, and wall jet regions, secondary stagnation zones result from the interaction of adjacent wall jets. In many such schemes, the jets are discharged into a restricted (confined) volume bounded by the target surface and the nozzle plate from which the jets originate (see Fig. 18.16). Both the local and the average rate of heat (mass) transfer depend strongly on the manner in which spent gas is vented from the system. Extensive reviews of available convective coefficient data for impinging gas jets have been reported [82–84], and results can be obtained from the references cited therein. Design needs are met by experiments, because turbulence models are not yet sufficiently reliable for simulating relevant turbulent flow parameters [85].

To account for the effect of mixing between the jet and ambient crossflow and to obtain meaningful results for different applications, the local convective heat transfer coefficient is defined by

$$h = q''/(T_w - T_{aw}) \tag{18.32}$$

where q'' is the convective heat flux and T_{aw} is the adiabatic wall temperature, which is defined through the effectiveness as

$$\eta = (T_{aw} - T_r)/(T_j^o - T_\infty) \tag{18.33}$$

In this expression, T_r and T_j^o are the recovery and total jet (nozzle) temperatures, respectively. The available results suggest that effectiveness depends on the nozzle geometry and axial nozzle-to-plate distance (H/d) and the radial displacement from the stagnation point (r/d), but it is independent of the nozzle exit Reynolds number [86, 87].

Single Jet.   Review of published data indicates that many different factors affect heat transfer between an isothermal turbulent jet and the impingement surface. As shown in Fig. 18.17, gas jets are typically discharged into a quiescent ambient from a round jet of diameter d or a slot (rectangular) nozzle of width w. The factors that influence local and average convective heat (mass) transfer include the following: nozzle (orifice) geometry, small-scale turbulence in jet, exit jet velocity profile, entrainment, nozzle-to-surface distance, confinement, angle of incidence, surface curvature, and external factors. For a single circular jet impingement on a flat plate, the local Nusselt number can be expressed in the general form

$$Nu_d/Pr^{1/3} = f(r/d,\ H/d,\ Re^{m(r/d,H/d)},\ Pr) \tag{18.34}$$

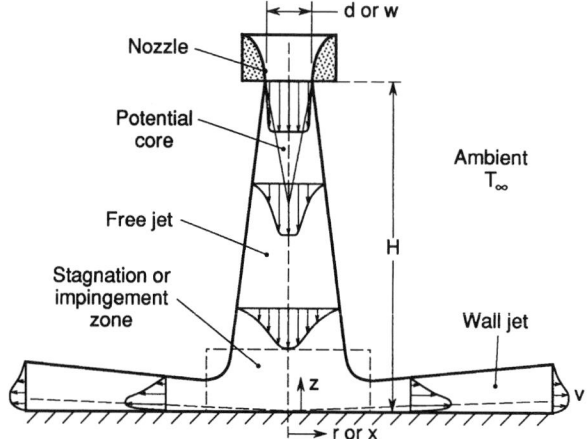

FIGURE 18.17   Schematic and coordinate system for an impinging jet discharging into a quiescent ambient.

Available empirical results show that the Reynolds number exponent m depends not only on r/d and H/d but also on the nozzle geometry [83]. Some experimental results illustrating the local Nusselt number dependence on the Reynolds and nozzle-to-surface spacing are given in Fig. 18.18. Numerous citations to extensive local heat transfer coefficient results showing the effects of Re and H/d can be found in Refs. 82–84.

The experimental database for single round and slot nozzles as well as arrays of round and slot nozzles has been assessed by Martin [82]. He has recommended the following empirical correlation for the average Nusselt (Sherwood) number for a single round nozzle:

$$\overline{Nu}_d = \overline{h}d/k = G(r/d, H/d)F_1(Re_d)\,Pr^{0.42} \tag{18.35}$$

where

$$F_1 = 2Re_d^{1/2}(1 + 0.005Re_d^{0.55})^{1/2} \tag{18.36}$$

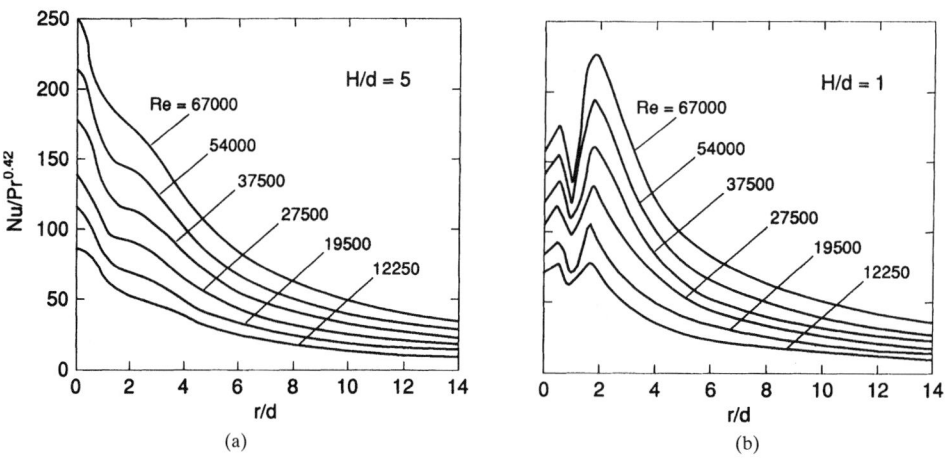

FIGURE 18.18   Effect of the Reynolds number on local heat transfer parameter $Nu/Pr^{0.42}$ for jet impingement heating with nozzle-to-surface parameter H/d: (a) $H/d = 5.0$, and (b) $H/d = 1.0$ (after Klammer and Schupe [88]).

and
$$G=\left(\frac{d}{r}\right)\frac{1-1.1(d/r)}{[1+0.1(H/d-6)(d/r)]} \qquad (18.37)$$

This correlation is valid for the following range of parameters: $2\times10^3 < \text{Re}_d < 4\times10^5$, $2\le H/d\le12$ and $2.5\le r/d\le7.5$. For $r/d<2.5$ results are available in graphical form. A similar correlation is available for single slot nozzles [82].

Arrays of Jets. The heating or cooling of large areas with impinging jets requires arrays; however, the flow and geometrical parameters have to be carefully selected to provide both a sufficiently high average convective coefficient and uniformity over the impingement surface. For arrays of round or slot nozzles, additional parameters describing, say, the round nozzle geometrical arrangement (in-line or staggered, confinement, and crossflow scheme) need to be specified [82, 84]. For arrays of nozzles, there is interference between adjacent jets prior to their impingement on the surface. The likelihood of such interference effects is enhanced when the jets are closely spaced and the separation distance between the jet nozzles (orifices) and the impingement surface is relatively large. There is also an interaction due to collision of surface flows associated with adjacent impinging jets. These collisions are expected to be of increased importance when the jets are closely spaced, the nozzle (orifice) impingement surface separation is small, and the jet velocity is large.

Convective heat transfer from a flat surface to a row of impinging, submerged air jets formed by square-edged orifices having a length/diameter ratio of unity has been measured [89]. Local Nusselt numbers were averaged over the spanwise direction, and averaged values were correlated by the equation

$$\frac{\overline{\text{Nu}}_d}{\text{Re}_d^{0.7}}=\frac{2.9\exp[-0.09(x/d)^{1.4}]}{22.8+(S/d)(H/d)^{1/2}} \qquad (18.38)$$

where x is the streamwise coordinate and S is jet center-to-center spacing. This correlation is appropriate for the range of parameters studied ($2\le H/d\le6$, $4\le S/d\le8$, $0\le x/d\le6$, $10{,}000\le \text{Re}_d\le40{,}000$). Equation 18.38 indicates that the spanwise-average Nusselt number has the maximum at the impingement line ($x/d=0$), decreasing steadily with increasing distance (x) from the impingement line. The decay is faster for $S/d=4$ than for $S/d=8$, since, for a given Re_d, the total mass flow rate for $S/d=4$ is twice that for $S/d=8$. For a given S/d, the Nusselt number decreases with increasing jet exit-to-impingement-plate distance.

The surface-average heat transfer coefficients are of importance, for example, when the target to be heated or cooled is being moved beneath an array of stationary jets. The speed at which the target should be moved can be evaluated with the aid of the surface-average heat transfer coefficient. The surface-average Nusselt number from the impingement line ($x/d=0$) to a particular streamwise (x/d) location is calculated from the equation

$$\overline{\overline{\text{Nu}}}=\frac{1}{(x/d)}\int_0^{x/d}\overline{\text{Nu}}\,d(x/d) \qquad (18.39)$$

Local convective coefficient measurements for impinging flows from arrays of nozzles show qualitatively similar results as from single nozzles [90, 91]. For practical engineering calculations, mean (area-averaged) convective transport coefficients are needed. For arrays of nozzles, the spatial arrangement of nozzles must be specified. The averaging must be carried out over those parts of the surface area attributed to one nozzle. For arrays of round nozzles, the surface area is different for in-line than staggered nozzles [82]. Empirical correlations for average Nusselt numbers for arrays of round and slot nozzles have been developed and are of the same form as Eq. 18.34, except that there is an additional correction factor on the right-hand side of the equation that accounts for nozzle area relative to the area over which the transport coefficients are being averaged. The equations are rather lengthy, and reference is made to Martin [82] for the relevant correlations.

The air, after impinging on a surface from two-dimensional arrays of circular jets, is constrained to exit in a single direction along the channel formed by the surface and the jet plate.

TABLE 18.3   Constants for Use in Correlation, Eq. 18.40 (from Florscheutz et al. [92])

	In-line pattern				Staggered pattern			
	C	n_x	n_y	n_z	C	n_x	n_y	n_z
A	1.18	−0.944	−0.642	0.169	1.87	−0.771	−0.999	−0.257
m	0.612	0.059	0.032	−0.022	0.571	0.028	0.092	0.039
B	0.437	−0.095	−0.219	0.275	1.03	−0.243	−0.307	0.059
n	0.092	−0.005	0.599	1.04	0.442	0.098	−0.003	0.304

The downstream jets are subjected to crossflow originating from the upstream jets. Average Nusselt number correlations appropriate for use in analyzing circular jet array impingement systems in which the flow is constrained to exit in a single direction along the channel formed by the jet plate and the impingement surface, have been developed by Florschuetz et al. [92]. The correlation adopted is of the form

$$\overline{Nu} = A \, Re_j^m \, \{1 - B[(H/d)(G_c/G_j)]^n\} \, Pr^{1/3} \tag{18.40}$$

where the coefficients A and B and the parameters m and n depend on geometric parameters. They can be expressed in the form of simple power functions such that A, m, B, and n can be represented as $C(x_n/d)^{n_x}(y_n/d)^{n_y}(z/d)^{n_z}$. In the above equation, G_c is the channel crossflow mass velocity based on the cross sectional area, and G_j is the jet mass velocity based on the jet hole area. Equation 18.40 was applied separately to the inline and staggered hole pattern data obtained by Florschuetz et al. [92]. The resulting best fit values for the coefficients are summarized in Table 18.3 for both in-line and staggered patterns.

Jet-induced crossflow has been found to have an important effect on impingement heat transfer [82, 92, 93]. In order to delineate its influence on average convective coefficients more clearly, Obot and Trabold have identified three crossflow schemes, referred to minimum, intermediate, and maximum, and correlated their experimental data. The best heat transfer performance was obtained with the minimum crossflow scheme. Intermediate and complete crossflow was associated with varying degrees of degradation. The average Nusselt numbers for air were represented by the equation

$$\overline{Nu} = A \, Re_j^m \, (H/d)^n A_f^r \tag{18.41}$$

where A is a regression coefficient and A_f is fraction of the open area (area of orifice/area attributed to one orifice). The regression coefficient A and the exponents m, n, and r depend on crossflow scheme and geometric conditions. The coefficients A, m, and r are summarized in Table 18.4. The exponent n is given graphically by Obot and Trabold [93] in a figure. The values of n depend on the open area A_f and range from about −0.1 at $A_f \approx 0.01$ to about −0.4 at $A_f \approx 0.035$, showing only a mild dependence on the exhaust scheme. Equation 18.41 is based on experimental data for Reynolds numbers from 1000 to 21,000 and jet-to-surface spacing between 2 to 16.

TABLE 18.4   Summary of Empirical Coefficients in Eq. 18.41

Flow scheme	A	m	r
Minimum	0.863	0.8	0.815
Intermediate	0.484	0.8	0.676
Maximum	0.328	0.8	0.595

From Obot and Trabold [93].

Impingement Heat Transfer to Liquid Jets.   Impinging liquid jets have been demonstrated to be an effective means of providing high heat transfer rates [94]. Such circular or slot (planar) jets are characterized by a free liquid surface. Propagation of a free surface jet is virtually unimpeded by an immiscible ambient fluid (air) of substantially lower density and viscosity, and jet momentum can be efficiently delivered to and redirected along a solid surface. Prediction of heat transfer to such jets requires analysis of the flow field and prediction of the free surface position, which is governed by a balance of pressure and surface tension forces acting on the surface. Other factors, such as stress relaxation when the fluid is discharged from a nozzle, drag against the ambient gas and free stream turbulence. Gravity can also affect flow in a free jet.

Liquid jet impingement cooling offers very high convective coefficients ($\sim$10 to 100 kW/m^2 K) and is relatively simple to implement using straight-tube, slot, or contoured nozzles. The liquid jet issuing from the nozzles can be aimed directly toward the desired target (heat load). Jet impingement heat transfer has received considerable research attention during the last three decades, and the research findings have been organized in comprehensive reviews [95, 96]. Since, in materials processing applications, the jets are expected to be turbulent, the results presented here are only for this flow regime.

Before discussing results, it is desirable to identify flow and heat transfer regimes, which differ substantially between planar and axisymmetric impinging jets. As illustrated in Fig. 18.19, a planar jet of width w divides at the stagnation line, and the inviscid flow downstream has half the thickness of the incoming jet and moves at the jet's velocity V_j. At the stagnation line, a laminar boundary layer will form, growing in the wall jet region. If the jet Reynolds number is sufficiently high, the boundary layer will ultimately experience transition to turbulent flow and transport. In the wall jet region, the turbulent boundary layer will grow in thickness but can remain thin relative to the overlaying liquid sheet if the Reynolds number is high. Outside the hydrodynamic boundary layer lies the inviscid flow region, wherein the effects of viscosity are negligible. Generally speaking, the hydrodynamic and thermal boundary layers are not likely to be of equal thickness and will therefore encompass the entire liquid layer in the wall jet region at different locations.

Stagnation Zone.   A jet issuing from a fully developed tube flow without a terminating nozzle will also be turbulent for $\mathrm{Re}_d > 4000$, where Re_d is based upon V_n. The manifolding and piping systems that supply liquid to nozzles are often turbulent, and, unless the nozzle has a very large contraction ratio, this turbulence will be carried into the jet formed. Stagnation

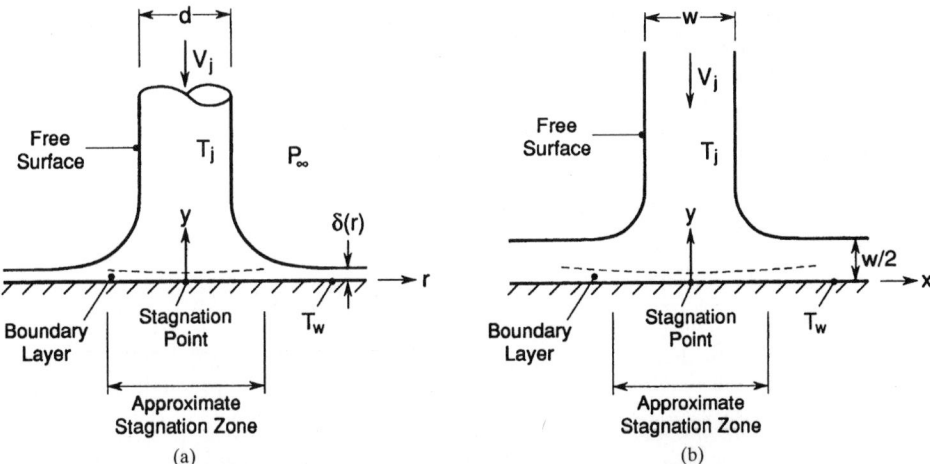

FIGURE 18.19   Impinging jet configurations for inviscid flow solutions: (*a*) axisymmetric; (*b*) planar.

zone heat transfer to turbulent liquid jets is affected by the velocity profile at the nozzle exit, free stream turbulence, and nozzle-to-target separation.

For fully turbulent jets issuing from a fully developed pipe-type nozzle at $H/d < 4$, Stevens and Webb [97] recommend the following stagnation zone Nusselt number correlation:

$$\mathrm{Nu}_{d,o} = 0.93\mathrm{Re}_d^{1/2}\,\mathrm{Pr}^{0.4} \tag{18.42}$$

where V_j is used in Re_d. This correlation is valid for $15{,}000 < \mathrm{Re}_d < 48{,}000$ and agrees within approximately 5 to 7 percent with the correlations of other investigators [98, 99] who employed pipe-type nozzles. For fully developed tube-nozzle at $H/d = 1$ and a dimensionless velocity gradient of approximately 3.6, Pan et al. [100] recommends the correlation

$$\mathrm{Nu}_d = 0.92\mathrm{Re}_d^{1/2}\,\mathrm{Pr}^{0.4} \tag{18.43}$$

for $16{,}600 < \mathrm{Re}_d < 43{,}700$ to an accuracy of about 5 percent. Note that these two correlations are for about the same Reynolds number range and are in very good agreement with each other.

For higher Reynolds numbers, other investigators [101, 102] report stronger dependence of stagnation Nusselt number on Reynolds number. Gabour and Lienhard [102] find, for $25{,}000 < \mathrm{Re} < 85{,}000$,

$$\mathrm{Nu}_{d,o} = 0.278\mathrm{Re}_d^{0.633}\,\mathrm{Pr}^{1/3} \tag{18.44}$$

which is reported accurate to within ± 3 percent of the data for cold water jets having $8.2 < \mathrm{Pr} < 9.1$ and tube diameters between 4.4 and 9.0 mm. Faggiani and Grassi [101] for $H/d = 5$ correlated their experimental data by representing the Reynolds number exponent as a function of the Reynolds number itself:

$$\mathrm{Nu}_{d,o} = \begin{cases} 1.10\mathrm{Re}_d^{0.473}\,\mathrm{Pr}^{0.4} & \mathrm{Re}_d < 76{,}900 \\ 0.229\mathrm{Re}_d^{0.615}\,\mathrm{Pr}^{0.4} & \mathrm{Re}_d > 76{,}900 \end{cases} \tag{18.45}$$

The stronger Nusselt number dependence on Reynolds number may result from an increasing influence of free stream turbulence; however, further evidence is needed to verify that conjecture.

Using a long parallel-plate nozzle to produce fully developed turbulent jets, Wolf et al. [103] correlated their stagnation zone Nusselt number data to an accuracy of 10 percent by the equation

$$\mathrm{Nu}_{w,o} = 0.116\mathrm{Re}_w^{0.71}\,\mathrm{Pr}^{0.4} \tag{18.46}$$

The correlation is based on Re_w from 17,000 to 79,000 and Pr between 2.8 and 5.0. Vader et al. [104] used a converging nozzle to produce uniform velocity profile water jets. These nozzles were intended to suppress but not to eliminate turbulence. The stagnation zone Nusselt numbers were correlated by an equation

$$\mathrm{Nu}_{w,o} = 0.28\mathrm{Re}_w^{0.58}\,\mathrm{Pr}^{0.4} \tag{18.47}$$

based on Re_w from 20,000 to 90,000 and Pr between 2.7 and 4.5. Note that the stagnation zone convective coefficients measured by Wolf et al. were about 69 percent higher than those measured by Vader et al. at $\mathrm{Re}_w = 50{,}000$.

Local Heat Transfer.   No general theory has been developed for local Nusselt numbers beyond the transition region, and local coefficient data are very sparse [95, 96]. Liu et al. [105] have divided the flow field into several regions and have developed expressions for the local Nusselt number. Limited comparisons between model predictions and experimental data at radial locations beyond the transition to turbulent flow have yielded good agreement. Correlation of experimental data of radial profiles of the local Nusselt number for turbulent, axisymmetric free-surface jets using the superposition of dual asymptote technique of Churchill and Usagi [77] has been less successful [98].

For turbulent planar jets issuing from convergent nozzles with a nearly uniform velocity profile, Vader et al. [104] recommend

$$\mathrm{Nu}_{x*} = hx/k = 0.89\mathrm{Re}_{x*}^{0.48}\,\mathrm{Pr}^{0.4} \tag{18.48}$$

where Re_{x*} is the local Reynolds number based on the free stream velocity $U(x)$ outside the boundary layer, $\mathrm{Re}_{x*} = U(x)x/\nu$, and Nu_{x*} is the local Nusselt number at this location. It applies to isoflux surfaces for $100 < \mathrm{Re}_{x*} < \mathrm{Re}_{x*,c}$, where $\mathrm{Re}_{x*,c}$ is the turbulent transition Reynolds number. The physical properties are evaluated at $(T_j + T_s)/2$. They recommend that the stagnation zone correlation be used for $\mathrm{Re}_{x*} < 100$. This correlation together with one developed by McMurray et al. [106] is illustrated in Fig. 18.20. Beyond transition to turbulence the correlation of McMurray et al. reduces to

$$\mathrm{Nu}_x = 0.037\mathrm{Re}_x\,\mathrm{Pr}^{1/3} \tag{18.49}$$

where $\mathrm{Re}_x = V_j x/\nu$. This correlation is supported by data for $6 \times 10^5 < \mathrm{Re}_x < 2.5 \times 10^6$.

For fully developed turbulent planar jets issuing from a parallel plate channel and impinging on an isoflux surface, Wolf et al. [103] correlated their local heat transfer coefficient data for water jets as

$$\mathrm{Nu}_w = \mathrm{Re}_w^{0.7}\,\mathrm{Pr}^{0.4}\,G(x/w) \tag{18.50}$$

where

$$G(x, w) = \begin{cases} 0.116 + (x/w)^2[0.00404(x/w)^2 - 0.00187(x/w) - 0.0199], & 0 < (x/w) < 1.6 \\ 0.111 - 0.00200(x/w) + 0.00193(x/w)^2, & 1.6 < (x/w) < 6.0 \end{cases} \tag{18.51}$$

The correlation applies to $1.7 \times 10^4 < \mathrm{Re}_w < 7.9 \times 10^4$ and is accurate to within 9.6 percent.

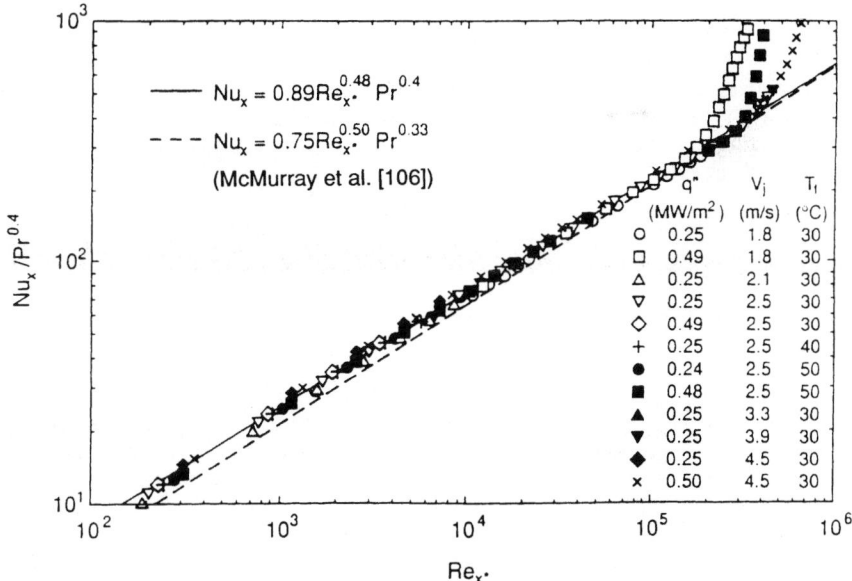

FIGURE 18.20 Correlation of single-phase convection data downstream of the stagnation line. Reprinted from D. T. Vader, F. P. Incropera, and R. Viskanta, "Local Convective Heat Transfer from a Heated Surface to an Impinging Planar Jet of Water," *International Journal of Heat and Mass Transfer*, 34, pp. 611–623, 1991, with kind permission from Elsevier Science Ltd, The Boulevard, Langford Lane, Kidlington, OX5 1GB, U.K.

Other Aspects of Jet Impingement Heat Transfer.   Current review of experimental studies and existing correlation for arrays of planar and axisymmetric water jets is available [95] and can't be repeated here because of space limitations. Suffice it to mention that the arrangement of the nozzles, jet inclination to target surface, surface roughness, jet splattering, and motion of the impingement surface need to be considered. For example, impingement of planar [107, 108] and axisymmetric [109] liquid jets on moving surfaces arises in thermal treatment of metals. In general, the results show that the use of heat transfer correlations for the stationary plate configuration to predict transport from moving surfaces may not be appropriate except for low surface velocities. In many cases, the surface velocity may exceed the jet impingement velocity. The results of Zumbrunnen [107] can be used to assess the importance of surface motion on convective heat transfer and on the applicability of empirical correlations to moving target surfaces.

Two-Phase Convective Heat Transfer

Boiling heat transfer, which can be used for accurate temperature control of, for example, moving steel strip, has been discussed in Chap. 15 of this handbook, and reference is made to it for fundamentals of different types of boiling. Only four specific types of boiling conditions that are encountered in materials processing and manufacturing are discussed here.

Boiling

Nucleate.   Jet impingement boiling is used in the production of metals under conditions where surface temperature and heat flux are typically very large, and acceptable cooling times are relatively short. Single and arrays of circular and slot jets have been reviewed, and the effects of parameters for free-surface and submerged jets, jet velocity, and subcooling have been discussed. The present discussion is restricted to free-surface (circular or planar) jets of water under fully developed and local nucleate boiling conditions because of their relevance to materials processing applications. Review of available fully developed nucleate boiling data by Wolf et al. [110] revealed that conditions are unaffected by parameters such as jet velocity, nozzle or heater dimensions, impingement angle, surface orientation, and possibly subcooling. However, the conditions depend strongly on the fluid used. For fully developed nucleate boiling of water, the data can be correlated by the empirical equation

$$q''_{\text{FNB}} = C\Delta T_{\text{sat}}^n \tag{18.52}$$

where q''_{FNB} and ΔT_{sat} have units of W/m^2 and °C, respectively. The coefficient C and the exponent n are summarized in Table 18.5. Despite significant differences in jet conditions and nozzle geometry (circular or planar), there is generally good agreement between their respective nucleate boiling characteristics.

TABLE 18.5   Fully Developed Nucleate Boiling Correlation [q''_{FNB} (W/m^2) = $C\Delta T_{\text{sat}}$ (°C)^n] for Water

Author	Jet type	C	n	Range of ΔT_{sat} (°C)
Ishigai et al. [111]*	Planar-free	42	3.2	26–47
Katto and Ishii [112]*	Planar-wall	130	3.0	21.33
Katto and Kunihiro [113]*	Circular-free	340	2.7	18–38
Monde [114]*	Circular-free	450	2.7	18–46
Miyasaka et al. [115]	Planar-free	79	3.0	26–90
Toda and Uchida [116]	Planar-wall	6100	1.42	16–68
Wolf et al. [117]	Planar-free	63.7	2.95	23–51

After Wolf et al. [110].
* Correlations have been obtained by graphical means and should be considered approximate.

The relationship between the critical heat flux and various system parameters depends on the specific flow conditions. Four different CHF regimes (referred to as V-, I-, L-, and HP-regimes) have been identified for free-surface jets [110]. In each regime, the critical heat flux depends on parameters such as the jet velocity at the nozzle exit (V_n), density ratio (ρ_ℓ/ρ_g), heater diameter (D), and has been shown to be markedly different in the different CHF regimes. To date, however, specific demarcations between the respective regimes have not been proposed. The following expression for the critical heat flux of a circular, free-surface jet:

$$\frac{q''_{CHF}}{\rho_g i_{fg} V_n} = 0.278 \left(\frac{\rho_\ell}{\rho_g}\right)^{0.533} \left(1 + \frac{\rho_g}{\rho_\ell}\right) \left[\frac{\sigma}{\rho_\ell V_n^2 (D - d)}\right]^{1/3} \left(1 + \frac{D}{d}\right)^{-1/3} \quad (18.53)$$

has been used as a foundation for the different correlations that have been proposed. Reference is made to Wolf et al. [110] for correlations for specific flow conditions, geometry, and so on, including a correction factor to account for subcooling. The critical heat flux data in the stagnation region for impinging planar jets of water have been correlated [115] by the equation

$$q''_{CHF} = (1 + 0.86 V_n^{0.38}) \left\{ 0.16 \rho_g i_{fg} \left[\frac{\sigma(\rho_\ell - \rho_g)}{\rho_g^2}\right]^{1/4} \right\} (1 + \varepsilon_{sub}) \quad (18.54)$$

where ε_{sub} is a correction factor for the effect of subcooling

$$\varepsilon_{sub} = 0.112 \left(\frac{\rho_\ell}{\rho_g}\right)^{0.8} \left(\frac{c_{pf} \Delta T_{sub}}{i_{fg}}\right) \quad (18.55)$$

In Eq. 18.54, V_n is in m/s, and the term in braces can be interpreted as a critical heat flux for pool boiling proposed by Kutateladze (see Chap. 15). The impingement data of Miyasaka et al. were restricted to a relatively narrow range of subcoolings ($85 \le \Delta T_{sub} \le 108°C$), where $\Delta T_{sub} = T_{sat} - T_f$. The dependence of q_{CHF} on the velocity revealed by Eq. 18.54 is not as strong as for water at atmospheric pressure in parallel flow on an electrically heated flat plate, but the strong linear dependence on subcooling is retained [118].

Transition. Information concerning transition boiling for impinging jets is sparse and is limited to the most fundamental quantities such as minimum heat flux and surface temperature [110]. In the transition boiling region, Fig. 18.21 shows that the heat flux increases with the jet velocity and subcooling [111], but $q'' \sim \Delta T_{sat}$ curve shifts toward higher heat flux and wall superheats with increased subcooling. The characteristics of the transition region between the maximum and minimum heat fluxes are seen to differ markedly as a function of subcooling. At low subcoolings ($\Delta T_{sub} = 5$ and 15°C), the heat flux decreases after the start of the quench and the onset of film boiling. It reaches a minimum at the onset of surface wetting and increases monotonically to the maximum heat flux. At higher subcoolings ($\Delta T_{sub} = 25°C$ and 35°C), the heat flux again declines at the start of the quench and reaches a minimum, but the subsequent increase is not monotonic. A nearly constant heat flux occurs, say, from $\Delta T_{sat} \approx 300°C$ to $\Delta T_{sat} \approx 500°C$ for $\Delta T_{sub} = 25°C$. At the largest subcooling, $\Delta T_{sub} = 55°C$, the data reveal that no film boiling occurred at the stagnation point, despite surface temperatures as high as 1000°C. The available results

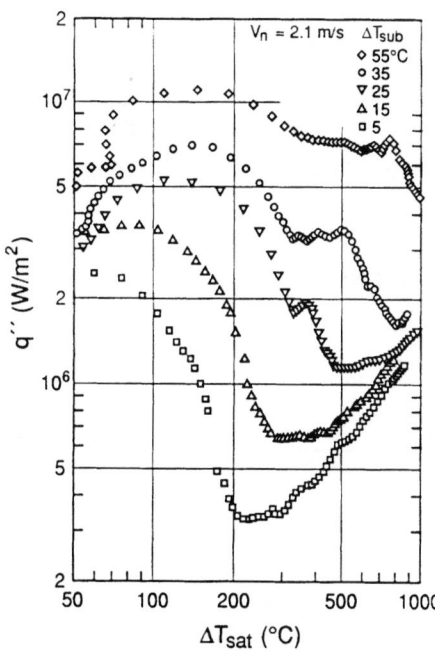

FIGURE 18.21*a* Boiling curve for a planar, free-surface jet of water showing the effects of (*a*) subcooling (from Ishigai et al. [111], used with permission).

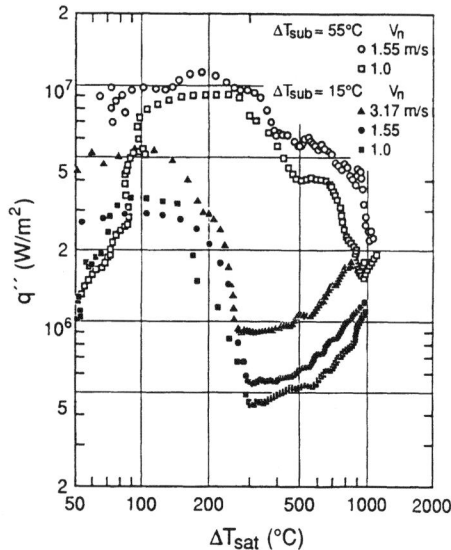

FIGURE 18.21b Boiling curve for a planar, free-surface jet of water showing the effects of (b) velocity (from Ishigai et al. [111], used with permission).

show that the jet velocity has little effect on the transition boiling regime, but the minimum and film boiling heat fluxes increase with increasing jet velocity [111]. The experimental results of Miyasaka et al. [115] clearly show that high jet impingement velocities and subcoolings can completely suppress film boiling, since the liquid jet reaches the hot surface and prevents formation of a stable vapor film. In the transition regime, violent nucleate boiling is evident, and boiling curves ($q \sim \Delta T_{sat}$) appear to be continuous extensions of the boiling curves for ordinary saturated pool boiling.

For jet velocities and subcoolings in the ranges of $0.65 \leq V_n \leq 3.5$ m/s and $5 \leq \Delta T_{sub} \leq 55°C$, Ishigai et al. [111] correlated the minimum heat flux for a planar jet by the expression

$$q''_{min} = 0.054 \times 10^6 V_n^{0.607}(1 + 0.527 \Delta T_{sub}) \qquad (18.56)$$

where q''_{min}, V_n, and ΔT_{sub} have units of W/m², m/s, and °C, respectively. The minimum temperature (T_{min}) consistently increased with increasing jet velocity and subcooling. For a free-surface circular jet, Ochi et al. [119] correlated transition boiling data at the stagnation point by

$$q''_{min} = 0.318 \times 10^6 (V_n/d)^{0.826}(1 + 0.383 \Delta T_{sub}) \qquad (18.57)$$

where q''_{min}, V_n, d, and ΔT_{sub} have units of W/m², m/s, mm, and °C, respectively. The correlation is based on data for velocities, subcoolings, and nozzle diameters in the ranges of $2 \leq V_n < 7$ m/s, $5 \leq \Delta T_{sub} \leq 45°C$, and $5 \leq d \leq 20$ mm. The minimum temperature was also shown to be a function of the nozzle diameter and jet velocity, with the effects of velocity being least significant at smaller nozzle diameters.

Wetting is a critical factor that controls cooling of a metal being quenched [120]. A relationship between the wetting (Leidenfrost) temperature and the cooling water (jet) temperature T_j for a circular jet of water impinging on a stainless steel plate has been established by Kokado et al. [121]. The measured wetting temperature at the stagnation point, which separates the wetted and nonwetted zones, is represented by an empirical equation

$$T_{wet} = 1150 - 8T_j \qquad (18.58)$$

where T_j is the jet temperature in °C. This relation is based on a rather narrow range of temperatures ($71 \leq T_j \leq 92°C$) and impingement velocities ranging from 2.0 to 2.5 m/s. At wall temperatures below T_{wet}, the surface is assumed to be in contact with the liquid (wetted), while, for wall temperatures in excess of T_{wet}, the surface is assumed to be insulated from the liquid due to vapor blanketing (nonwetted). Equation 18.58 does not imply that T_{wet} is independent of other jet parameters such as the jet velocity, nozzle dimensions, and fluid properties. Rather, it represents a relationship that is most likely unique to the conditions of the Kakado et al. experiments.

In a recent review, Klimenko and Snytin [122] have shown that, for a stationary surface, the Leidenfrost temperature depends strongly on the material and coolant property combinations; however, experiments involving extensive combinations of solids and coolants have not been performed, and minimum heat fluxes and Leidenfrost temperatures have not been measured when the surface is in motion. Experimental data [123, 124] show that the Leidenfrost temperature depends not only on the water subcooling but also on the fluid motion and the thermophysical properties through the thermal admittance, $a = \sqrt{k\rho c}$. The results demonstrate conclusively the importance of the material properties and fluid motion on the Leidenfrost temperature and the cooling rate during stable film boiling. The effects of surface condition (scale and surface roughness) and metal surface motion on the Leidenfrost temperature do not appear to have been investigated.

Film.   Experimental observations [110] reveal that there is no vapor film between the hot metal surface and an impinging jet of water. In the immediate impingement region, subcooled water is in direct contact with the surface, even if the surface temperature exceeds the Leidenfrost point. This condition is due to penetration of the vapor film by the liquid. Away from the jet impingement zone (i.e., parallel flow region), film boiling is observed. Film boiling heat transfer to water jets impinging on hot metal surfaces have been studied [121, 123, 125].

Analyses of film boiling on stationary surfaces are discussed in Chap. 15 of this handbook; therefore, only film boiling in a parallel-flow regime on moving surfaces that are prototypic of cooling applications in the materials processing industries is discussed here. The effect of surface motion on laminar [126] and turbulent [127] convection film boiling of water on a moving surface has been analyzed. Integral and similarity solutions of boundary layer equations have been obtained and were found to be in very good agreement with each other. The integral solution for turbulent flow parallel to a moving surface yields the following expression for the local Nusselt number:

$$\text{Nu}_x = h_x x/k_g = 0.0247 \tilde{\beta} \overline{U}_\ell^{0.6} \left(\frac{\mu_\ell}{\mu_g} \right) D^{0.2} \, \text{Re}_x^{0.8} \, \text{Pr}_\ell^{1/3} \tag{18.59}$$

where the subcooling parameter is $\tilde{\beta} = [\text{Pr}_g \, c_{p\ell}(T_s - T_\infty)/\text{Pr}_\ell \, c_{pg}(T_L - T_s)]$, the dimensionless relative liquid velocity is $\overline{U}_\ell = [|U_\infty - U_s|/U_\infty]$, $D = (2\overline{U}_s + n)$, and the dimensionless x-component of the interfacial velocity is $\overline{U}_s = U_s/U_\infty = 1/[1 + \tilde{\beta} \, \text{Pr}_\ell^{1/3}]$. The value of the exponent n of the velocity profile in vapor boundary layer was taken to be $n = 7$. The model predictions were compared with experimental data for forced film boiling along a stationary, highly superheated surface, and very good agreement has been achieved in zones of developed film boiling [128]. Since experiments cannot be performed in the laboratory corresponding to mill conditions, analysis provides the only means for obtaining the needed scaling relations. The extent to which plate motion can enhance and suppress heat transfer, respectively, downstream and upstream of an impinging jet has not been conclusively established [129].

Transient Boiling.   Studies of transient subcooled boiling using different test specimens have been carried out by a number of investigators, and a comprehensive discussion of this heat transfer mode as applied to heat treatment of metals is available [130]. Figure 18.22 illustrates the cooling curves that were determined using a 4-mm OD by 40-mm-long cylindrical gold specimen. The corresponding heat fluxes are shown in Fig. 18.23. As the liquid subcooling (water temperature) decreases (increases), the quench point is delayed, and the quenching (wetting) temperature is reduced. At water temperatures greater than 60°C (see Fig. 18.22), film boiling is relatively unstable; therefore, the boiling state shifts at once from film to transition and nucleate boiling heat transfer. Note from Fig. 18.22 that the maximum heat fluxes for $\Delta T_{\text{sub}} > 50$ K are very high, and for $\Delta T_{\text{sat}} > 50$ K, the effect of water subcooling (ΔT_{sub}) on the nucleate and transition film boiling is very large.

The convective coefficients during quenching of metals in water (at 20°C) increase with the thermal conductivity of the metal [132]. In addition, it has been shown that the convective coefficient of oxidized steel was significantly lower than that on unoxidized steel when quenched in water. It was found that the convective coefficient is highest for copper, followed by aluminum and then nickel. Müller and Jeschar [122] have shown that the convective coefficient depends not on the thermal conductivity of the metal alone but also on the thermal admittance of the metal $\sqrt{k\rho c}$.

Heat Transfer to Impacting Sprays.   Impacting sprays have broad industrial applications, including the metallurgy industry [124, 133–135]. The transport processes related to impacting sprays are very complex because they involve not only the dynamics of sprays but also single- and two-phase convective heat transfer. Nucleate and transition boiling as well as partial and stable film boiling can occur if the surface is above the saturation temperature of the

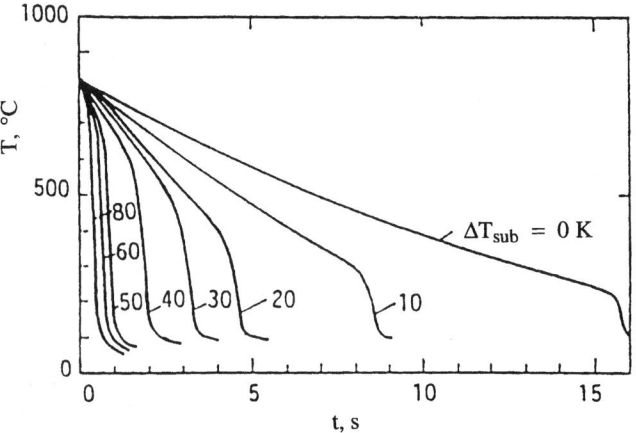

FIGURE 18.22 The experimental cooling curves of gold for a 4-mm-diameter, 40-mm-long vertical cylinder (from Tajima et al. [130]).

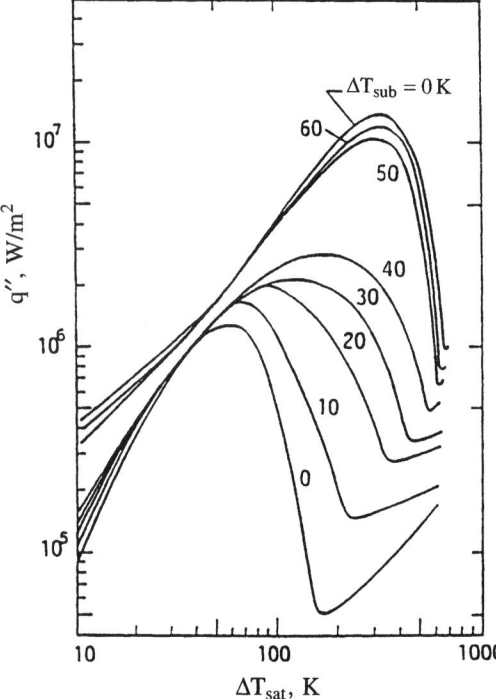

FIGURE 18.23 Subcooled transient pool boiling heat flux versus ΔT_{sat} for a 4-mm-diameter, 40-mm-long vertical cylinder (from Tajima et al. [130]).

coolant. Most frequently, the coolant fluid is water, but various synthetic coolants are also used [134]. In addition to the dynamics of the sprays, spray density, mean liquid droplet diameter, and the size distributions and the transformation of bulk liquid into sprays are important factors that influence the local heat transfer rate. A recent publication of the dynamics of the sprays is available [135], and this discussion is concerned with nucleate, transition, and film boiling heat transfer to sprays.

Typical heat transfer results to monodisperse sprays impacting on a heated surface are shown in Fig. 18.24. The liquid flow rate is varied over a wide range, while the droplet diameter is kept almost constant [136]. The heat flux versus surface temperature trends are similar to those of conventional boiling curves (see Chap. 15 of this handbook), and the heat fluxes are very high. The available experimental data [133, 134, 137–140] show that the volumetric spray flux $\bar{V}$ ($m^3/m^2 \cdot s$) is a dominant parameter affecting heat transfer. However, mean drop diameter and mean drop velocity and water temperature have been found to have an effect on heat transfer and transitions between regimes. Urbanovich et al. [141], for example, showed that heat transfer is not only a function of the volumetric spray flux but also of the pressure difference at the nozzle and the location within the spray field (Fig. 18.25).

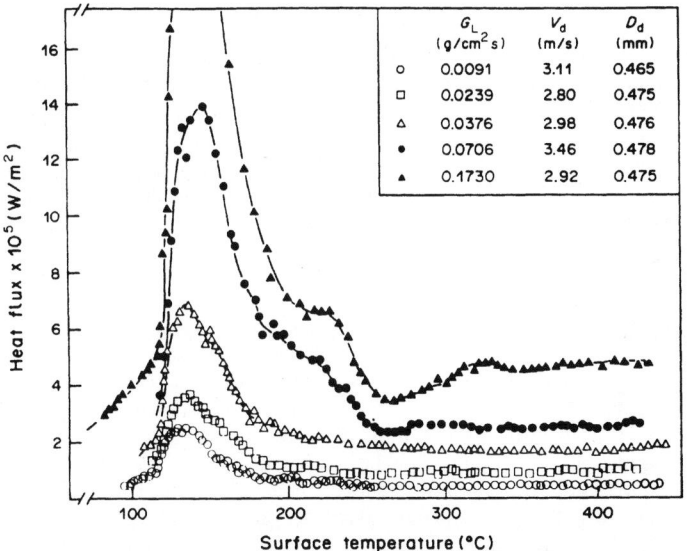

FIGURE 18.24 Heat transfer of horizontal impacting sprays at various liquid mass fluxes. Reprinted from K. J. Choi and S. C. Yao, "Heat Transfer Mechanisms of Horizontally Impacting Sprays," *International Journal of Heat and Mass Transfer,* 32, pp. 311–318, 1987, with kind permission from Elsevier, Langford Lane, Kidlington, OX5 1GB, U.K.

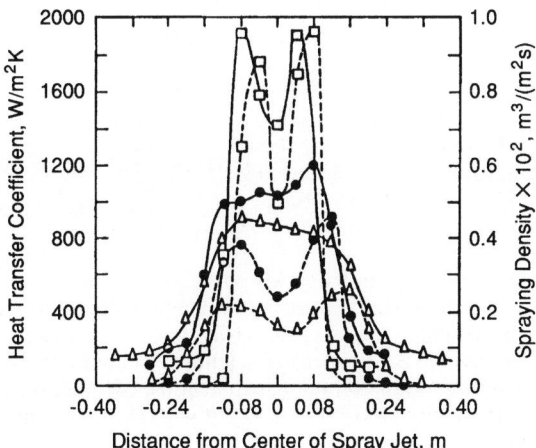

FIGURE 18.25 Variation of spraying density and heat-transfer coefficient along central axis of field of action of water jet on metal surface being cooled at water pressure before sprayer of 1.5 MPa (after Urbanovich et al. [141]).

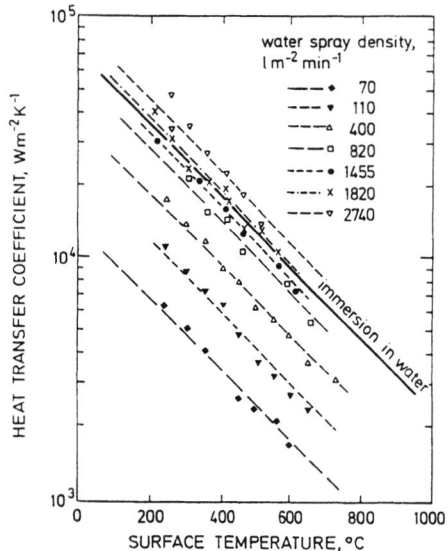

FIGURE 18.26 Dependence of the heat transfer coefficient on surface temperature during spray cooling at various spray intensities (from Bamberger and Prinz [131]).

A semiempirical analysis of heat transfer to impacting sprays has been developed by considering the major components of spray heat transfer to consist of (1) contact heat transfer to impacting droplets, (2) convective heat transfer to gas, and (3) thermal radiation heat transfer [142]. The model further assumes that the droplet interference is negligible (i.e., dilute sprays), and the three heat transfer components are independent of each other. The heat transfer data to a single impacting droplet have been correlated by the Weber number, surface temperature superheat, and thermophysical properties.

Some attempts to develop empirical heat transfer coefficient correlations have been made, and these are reviewed by Totten et al. [134]. For example, Bamberger and Prinz [131] have measured the heat transfer coefficient (including the radiative heat transfer contribution) during spray cooling of a copper billet with water (Fig. 18.26). An increase of the coefficient is revealed with an increase of the specific spray intensity and an identical dependence of the coefficient with surface temperature as that during immersion cooling with water. It should be emphasized that the correlations available [134] are not general but specific to the systems studied, owing to the geometrical arrangement, nozzle design and operating conditions, materials cooled, droplet distribution, and so on [132].

For practical spray heat transfer calculations, a summary of empirical and interpolation correlations for each regime of spray boiling curve, ranging from single-phase to film boiling, and of transition conditions between these regimes has been provided [139, 140]. As an illustration of how the hydrodynamic and other parameters affect heat transfer to sprays, the empirical equation for the critical heat flux q''_{CHF} is provided as an example [139]:

$$\frac{q''_{CHF}}{\rho_g i_{fg} \overline{V}} = 122.4 \left[1 + 0.0118 \left(\frac{\rho_g}{\rho_\ell} \right)^{1/4} Ja \right] \left(\frac{\sigma}{\rho_\ell \overline{V}^2 d_{32}} \right)^{0.198} \tag{18.60}$$

where q''_{CHF}, $\overline{V}$, and d_{32} are in W/m², m³/m²s, and m, respectively. In this equation, d_{32} is the Sauter mean diameter of the spray, but it can also be expressed in terms of the mass median diameter. Equation 18.60 correlated the experimental data to within a mean absolute error of 11.5 percent. The correlation suggests that q''_{CHF} of impacting sprays is significantly higher than that of pool boiling. Even compared with the q''_{CHF} for forced convection boiling, impacting sprays show a substantial advantage, because a higher q''_{CHF} can be easily obtained with a small amount of liquid flow rate. This is owing to the fact that CHF of impacting sprays is determined by quite different mechanisms than that of conventional pool or forced boiling. Caution should be exercised, however, in using the correlations suggested in the literature, because factors such as the nozzle characteristics, nozzle-to-target distance, surface orientation, thermal inertia of the material, and the surface roughness may be additional factors influencing CHF that are not accounted for.

Air-Mist Cooling.   The term *mist* refers to a two-phase jet containing air and very fine droplets of water moving at a high velocity. Mist is produced by air atomization of a water stream. Although no precise distinction exists, customarily, a relatively higher liquid flow rate is referred to as mist, and a lower one is considered fog [143]. The impingement of a gas/liquid mist on a surface is an attractive means of obtaining high heat transfer coefficients (e.g., in continuous steel casting and heat treatment [144]). In the mist jet, the water droplets are finer

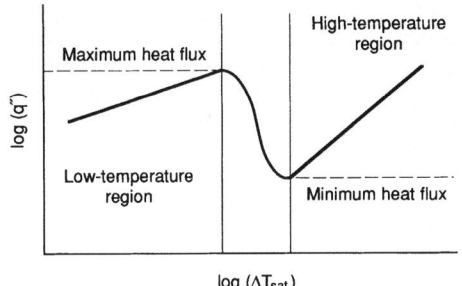

FIGURE 18.27 Schematic diagram of characteristics of mist-cooling heat transfer. From I. Tanasawa and N. Lior, *Heat and Mass Transfer in Materials Processing*, pp. 447–488, Taylor and Francis Group, New York. Reproduced with permission. All rights reserved.

and have larger velocities than in water spray, on account of which the mist jet provides uniform contact of water droplets on the impinging hot surface. This results in uniform cooling, which reduces the thermal shock and risk of thermal cracking of steel. In addition, mist cooling enhances the rate of heat extraction from the hot metal surface by increasing the Leidenfrost temperature as compared to water spray.

The mist-cooling heat transfer characteristics can be represented schematically, as shown in Fig. 18.27. They are similar to boiling heat transfer characteristics (see Chap. 15 of this handbook). The q_w'' versus ΔT_{sat} curve is divided into three temperature regions in the order of low to high temperature and has a maximum and minimum heat flux. The low-temperature region of the mist-cooling curve is relevant to electronic cooling and heat exchanger applications. A recent review is available [145] and will not be discussed here. The transition- and high-temperature regions are of interest to processing of steel and other metals [146] and will be discussed here.

Mist-cooling heat transfer characteristics have been reviewed by Nishio and Ohkubo [146], and an attempt has been made to correlate existing experimental data. The mist-cooling convective heat transfer coefficient is correlated in terms of relevant flow parameters such as droplet diameter d, droplet velocity V_d, and the volume flux of the droplets $\overline{V}$ by the empirical equation

$$h = Cd^m V_d^n \overline{V}^r \qquad (18.61)$$

The values of the exponents m, n, and r (based on experimental data and analysis) have been provided. Unfortunately, using a statistical approach, no universal correlation could be established between h and the mist-flow parameters. Therefore, theoretical models have been proposed to predict the coefficient between a parallel stream of impinging droplets and the hot surface in terms of droplet diameter, velocity, and volume flux [136]. The results have established that the spray-cooling convective coefficient is mainly controlled by the droplet volume flux. The droplet diameter and velocity have much weaker influence on heat transfer than the droplet volume flux. In the high-temperature region, the mist-cooling heat transfer coefficient h varies with the surface superheat ΔT_{sat} as $h \sim \Delta T_{sat}^{-1/3}$. Based on experimental data for silver, nickel, stainless steel, and fused quartz, Ohkubo and Nishio [147] correlated their experimental convective coefficient data at $\Delta T_{sat} = 500$ K by the following empirical equation:

$$h = 2.87 \times 10^4 (1 + 1.69\gamma^*) \overline{V}^{0.6} \qquad (18.62)$$

where h and $\overline{V}$ have units of W/m²K and m³/m²s, respectively. The thermal inertia (thermal admittance) ratio γ^* of the liquid coolant (water) to the solid plate is defined as $a_\ell/a_s = [(\rho ck)_\ell / (\rho ck)_s]^{0.5}$. Equation 18.62 correlates the experimental data with an uncertainty of ±30 percent.

The effects of parameters such as surface roughness and wettability and the thermal characteristics of the heat transfer surface layer on mist cooling in the high-temperature region have been investigated [147]. The available experimental data show that the convective coefficient is not a strong function of the air velocity for air velocities smaller than 20 m/s, but the heat transfer coefficient depends strongly on the surface superheat ΔT_{sat} (i.e., $h \sim \Delta T_{sat}^{-1/3}$). The surface roughness height and pattern have a weak effect on mist-cooling heat transfer. In mist cooling, surface wettability appears only in the superheat at the minimum heat flux and in the transition region, where the heat flux increases with surface wettability.

Solidification on a Moving Surface.   Solidification is an important process in materials processing and manufacturing, but it is beyond the scope of this discussion. Reference is made in the literature for recent accounts [148]. The present discussion is restricted to solidification of a melt on a moving prechilled or a continuously chilled substrate. This is a common method of coating surfaces or casting ribbons or strips by depositing the melt on a moving substrate [149, 150]. From the mathematical point of view, the problem of solidification of a melt on a moving substrate is very similar to the problem of convective heat transfer on continuously moving objects discussed previously. The major difference is that now there exist two phases: (1) the solid phase formed on the moving object, and (2) a liquid region in which the flow is induced by the motion of the object.

To avoid imperfections in the coating layer caused by rippling, backflow, and flow instabilities that occur during the impingement of the melt on the substrate [149], the coating arrangement illustrated schematically in Fig. 18.28 has been proposed. The melt is deposited in a direction parallel to the surface of the substrate. The position of the solid-liquid interface $y = \delta(x)$ for a pure substance has been predicted by Rezaian and Poulikakos [151] and for a binary alloy by Stevens and Poulikakos [152]. Assuming that the solidification takes place at a definite fusion temperature T_f, which is appropriate for a pure substance, the temperature at the solid-liquid interface $y = \delta$ must be continuous such that $T = T_s = T_f$. Furthermore, taking advantage of the boundary-layer approximations, the thermal energy balance at the interface $y = \delta$ can be written as

$$k_s \frac{\partial T_s}{\partial y}\bigg|_{y=\delta-0} - k \frac{\partial T}{\partial y}\bigg|_{y=\delta+0} = \rho_s U_s H \frac{d\delta}{dx} \qquad (18.63)$$

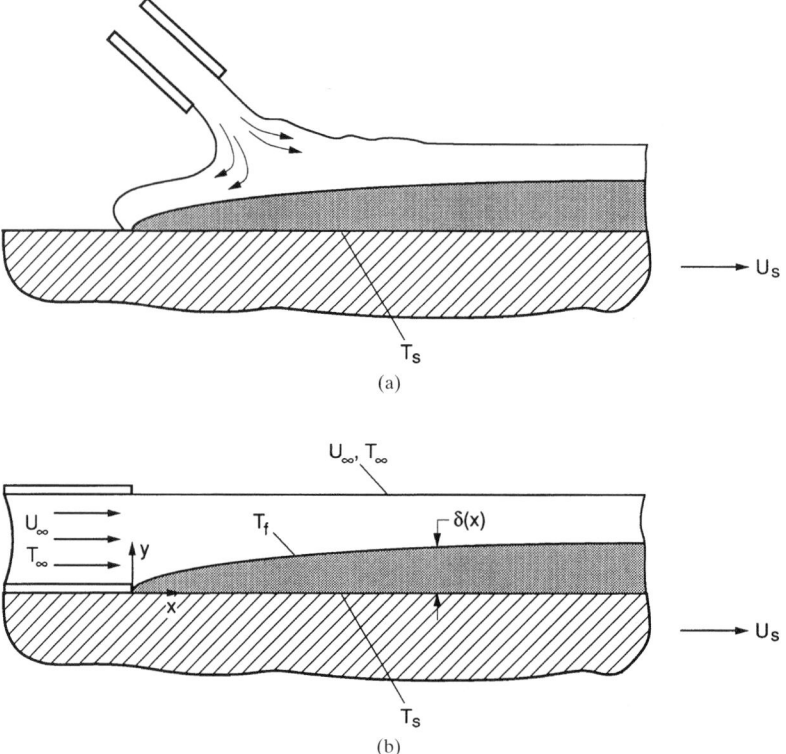

(a)

(b)

FIGURE 18.28   Schematic of the general coating processes (*a*) and the physical model and coordinate system (*b*).

By considering an infinitesimal pill-box with one side in the fluid and one in the solid, one can derive the interface condition

$$\rho \left[u \frac{d\delta}{dx} - v \right]_{y = \delta + 0} = \rho_s U_s \frac{d\delta}{dx} \qquad (18.64)$$

which insures continuity of transport across the interface. The solidification theories on moving substrates that have been developed [151–154] have not been validated using test data.

Radiation Heat Transfer

In many operations, radiation is the dominant heat transfer mode. The heated material's geometry may be simple (such as a sheet) or complex and/or discontinuous. It may be stationary (batch-processed) or moving. Radiant heating can be supplied directly via combustion or indirectly with electric or gas-heated elements. In general, the treated material and the surrounding gaseous environment are both radiatively participating. If the material is semitransparent to thermal radiation, it may be cold (negligible volumetric emission) or may be at a high enough temperature so that significant volumetric emission occurs. Processing with inert gases to prevent surface oxidation or combustion is common, so convection and conduction can occur in conjunction with radiative transfer, and these modes are coupled to the radiative exchange through heat transfer interactions at various solid surfaces. For other processes (such as chemical vapor deposition), specific gases are introduced at selected locations so that, even though the heat transfer may be radiatively dominated, understanding the advective transport of important chemical species is of equal or primary concern. A review of technologies in which heating is induced primarily by radiation is available [155].

Material temperatures induced by radiation heat transfer depend upon:

1. The spectral-directional absorption, reflection, and transmission characteristics of the load
2. The spectral-directional emission characteristics of the radiation source (or sink)
3. The spectral radiation characteristics of the medium separating the source and load
4. The geometrical configuration of the load relative to the source
5. The thermophysical properties of the load
6. Associated convection and/or conduction heat transfer processes

Items 1–4 determine the degree to which the radiation source and material load are thermally *coupled* and can be addressed with the heat transfer analysis methods outlined in Chap. 7 of this handbook. Items 5 and 6 may be quantified with an analysis, which takes into account the multimode heat transfer effects discussed elsewhere in this handbook. Because of the nonlinear nature of radiative heat transfer, few correlations exist that can be applied to relevant materials processing situations.

Source-Load Coupling for Opaque and Semitransparent Materials.   Because no materials exhibit true gray behavior, a primary issue in the design or operation of process hardware is radiative source-load coupling. The coupling "efficiency" may be estimated only if the spectral radiative properties of the source and load are known.

Figure 18.29 includes spectral properties for a paper product (i.e., the spectral, diffuse absorptivity of 62 g/cm^2 paper), along with normalized Planck blackbody distributions of sources at various temperatures [156]. In the absence of convection or conduction heat exchange between the source (s) and load (L), and assuming for the moment that the source and load are in an infinite parallel plate arrangement, an expression for the heat flux delivered to an opaque load can be derived using the analyses of Chap. 7:

$$q''_L = \int_{\lambda = 0}^{\infty} \frac{[e_{\lambda b,s}(T_s) - e_{\lambda b,L}(T_L)]}{\{[1/\varepsilon_{\lambda,s}(\lambda, T_s)] + [1/\alpha_{\lambda,L}(\lambda, T_L)] - 1\}} \, d\lambda \qquad (18.65)$$

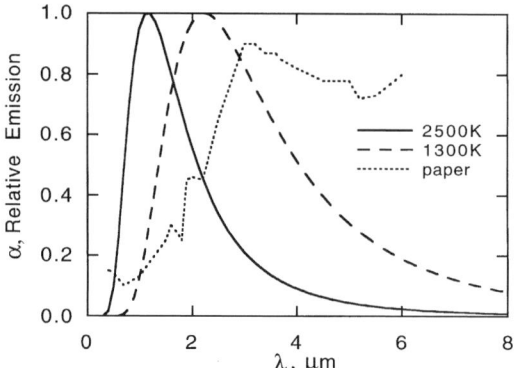

FIGURE 18.29 Normalized Planck distributions of gas (1300 K) and electric (2500 K) paper dryers, along with the spectral absorptivity of a typical wet paper product [156].

Inspection of Eq. 18.65 illustrates the general result that radiative energy can be transferred to the load only if the source emissivity and load absorptivity are both sufficiently large in spectral ranges characterized by significant source emission. Hence, the low temperature sources of Fig. 18.29 are more efficiently coupled to the paper product than the higher temperature (2150°C) source. Source-load coupling can be extremely inefficient, and, in cases where large spectral property variations occur, low-temperature sources can deliver more radiative energy to the load than their higher-temperature counterparts. In directly fired furnaces, very careful attention should be given to match the spectral characteristics of the flame to the absorptive characteristics of the load, since the spectral emission of combustion flames is highly wavelength-dependent.

In other cases (e.g. thermal processing of glass, silicon, plastics, or porous materials), the load may not be opaque, and the temperature distribution within it develops in response to volumetric absorption of the source radiation, along with possible scattering and/or emission. If volumetric emission from the material is negligible compared to volumetric absorption ("cold" material), analytical solutions for the temperature distribution within a one-dimensional sheet, cylinder, or sphere can be obtained. General features of the analyses, including recommendations for handling spectral effects, are discussed by Viskanta and Anderson [157]. See also Chap. 7.

An example of source-load coupling (nongray material properties) in a semitransparent solid is shown in Fig. 18.30 [158]. Here, a one-dimensional ($L = 305$ mm thick) slab of glass is irradiated uniformly on the $x/L = 0$ surface by a collimated Planckian source at $T_s = 2225$ K. Convective cooling occurs at $x/L = 0$ with $\bar{h} = 22.7$ W/m²K and $T_\infty = 300$ K, while adiabatic conditions are applied at $x/L = 1$. The results shown include the steady-state temperature distribution for an opaque solid (horizontal dashed line) along with temperature distributions (at various times) from an analysis that includes spectral variation of the absorptivity for the glass slab (volumetric emission was neglected). The "thermal trap effect" is shown and indicates that significant temperature differences may be induced within the semitransparent material by thermal irradiation.

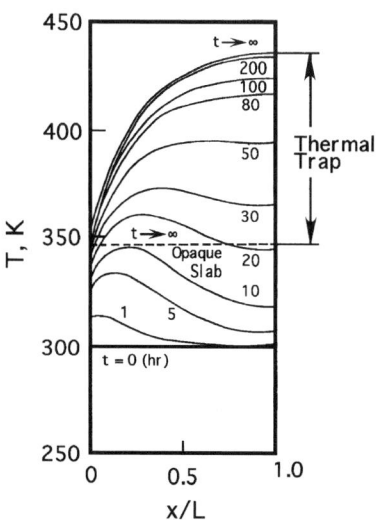

FIGURE 18.30 Temperature distributions in a plate of glass during irradiation from a Planckian source [158].

Load Properties. In strongly absorbing solids or liquids such as metals, conduction and advection are the only mechanisms by which heat is transferred throughout the interior of the material. However, in weakly absorbing semitransparent and partially transparent materials such as the dielectrics, internal radiation may also contribute to the transport of energy. Depending on the conditions, internal radiation may be negligible, of the same order of magnitude, or predominant over conduction and/or advection. If internal radiation is negligible, the material is opaque, and radiation heat transfer may be analyzed with a surface exchange analysis (Chap. 7).

Opaque Loads. Radiative properties of metals and opaque nonmetals (e.g., painted matter) can be predicted using classical electromagnetic theory, and details are available elsewhere [159]. However, the theory is only applicable to "clean" materials with optically smooth surfaces. When the surface roughness is significant ($\sigma_o/\lambda > 1$ where σ_o is the rms height of the surface roughness and λ is a characteristic wavelength), multiple reflections can occur between roughness elements, significantly increasing the surface absorptivity and emissivity in the case of metals. Since different finishing techniques for solids (e.g., lapping, grinding, or polishing) can lead to different surface roughness textures of the same σ_o, different values of ε, ρ, and α can exist for the same material characterized by the same value of σ_o.

In the case of metals processing, surface impurities such as thin layers deposited either by adsorption or chemical reaction (such as oxide layers) can increase the surface emissivity dramatically (Fig. 18.31). Because of the extreme sensitivity of the effective radiative properties of metals to minor surface roughness or contamination, it is recommended that measured radiative property values be used when possible.

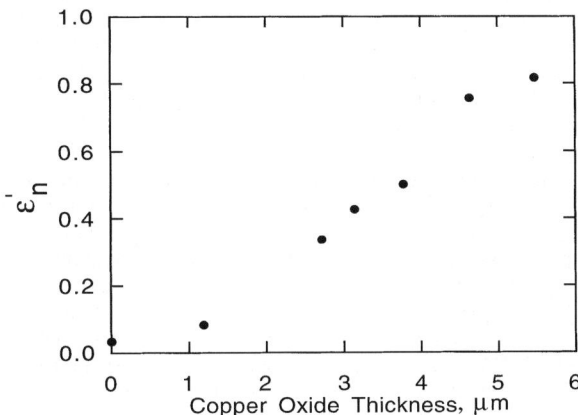

FIGURE 18.31 Effect of oxide coating on emissive properties of copper at 369 K [160].

Opaque nonmetals generally have high emissivities and can exhibit highly wavelength-dependent behavior [159]. Many nonmetals have behavior that deviates radically from that predicted by electromagnetic theory (Chap. 7), and available property measurements are less detailed than for metals. Again, recourse to use of measured property values is recommended.

Semitransparent Material. When the load is not opaque, radiative heat transfer occurs within the material in conjunction with conduction and/or advective transfer (if the load is moved with respect to a coordinate system). The thermal response of the load is, therefore, determined in part by volumetric radiation heat transfer, necessitating prediction or measurement of the relevant radiative properties.

If the material is homogeneous (no scattering), radiative properties can be predicted in the same manner as for gaseous media by postulating a radiative structure model, applying quantum mechanics to find the energy states of the postulated structure, and then calculating the spectra using transitional probabilities between the various energy states [161]. Only limited success with solids or liquids has been achieved due to dense atomic packing, presence of electrons, and so on.

In lieu of prediction, the spectral properties usually require measurement. For example, the absorption coefficient can be measured in several ways, the simplest and most common of which is to measure the transmittance of a sample of known thickness. Within spectral bands, the absorption coefficient is proportional to the natural logarithm of the spectral transmittance. Measured absorption spectra for various solid materials are shown in Fig. 18.32. Note that processing techniques can have an appreciable effect upon the value of κ_ν. For example, the strong band of the fused quartz spectrum at 2.7 µm is due to entrapped hydroxyl ions that come from water vapor incurred during the process [162, 163].

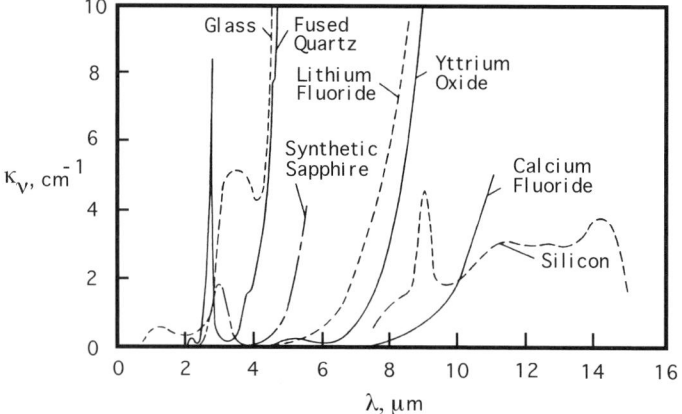

FIGURE 18.32 Absorption coefficient spectra of typical semitransparent solids [157].

Source Properties. For indirect heating, spectral emissivity property data is usually available from the manufacturers of the radiation sources. Directional effects may be tailored for specific applications by varying the shape and material of backside reflectors if tube or bulb sources are used. Indirect heating may also be accomplished using combustion if the flame is separated from the load by an impermeable panel (or tube) or porous ceramic plates. For panel heaters, directional effects are not as severe, since the panels or plates are usually of large area. Utilization of porous plates can be advantageous, since the products of combustion can be employed to augment the primary radiative heating effect with convective heating (or drying) of the load. As an example of different sources, schematics of gas-fired and electric-fired radiation paper dryers are shown in Fig. 18.33. Note that, for the electric fired dryer, the thermal radiation must be transmitted through various semitransparent materials (e.g., quartz and glass). In either case, moist air is removed between the source and load, in part to decrease absorption of radiation by water molecules and, in turn, increase the efficiency of the source-load coupling. A review of common types of radiation sources and reflectors, along with their spectral and directional characteristics, is available [164].

Intervening Medium Properties. The medium separating the source and load is typically a gas and (except in the cases of inert gases or dry air, which can be treated as nonparticipating) can affect source-load coupling. The influence of the separating medium can be espe-

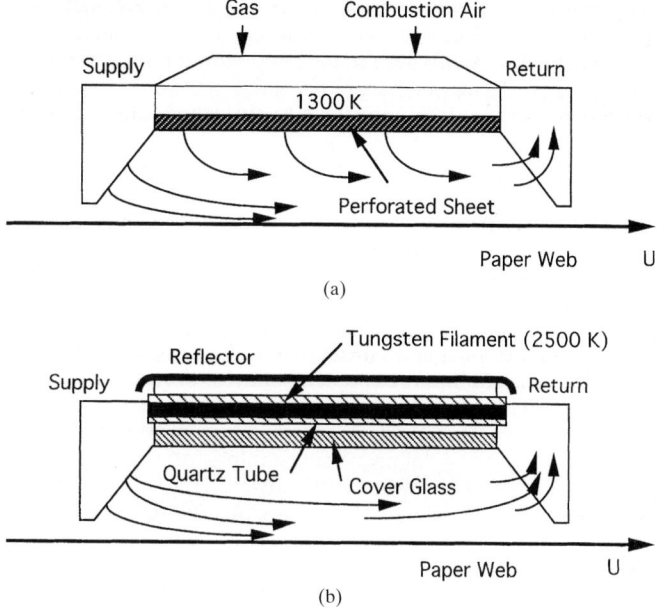

FIGURE 18.33 Radiation paper dryers using (*a*) gas firing and (*b*) electric sources [156].

cially severe in material removal operations such as drying or beam cutting or welding. For example, in laser-beam drilling of metal, the gas cloud formed in response to solid vaporization can either reduce *or increase* the energy transported to the solid at a particular instant in time [165]. Figure 18.34 shows the predicted overall thermal coupling coefficient (ratio of beam energy to energy delivered to the solid) for Gaussian irradiation of aluminum, along

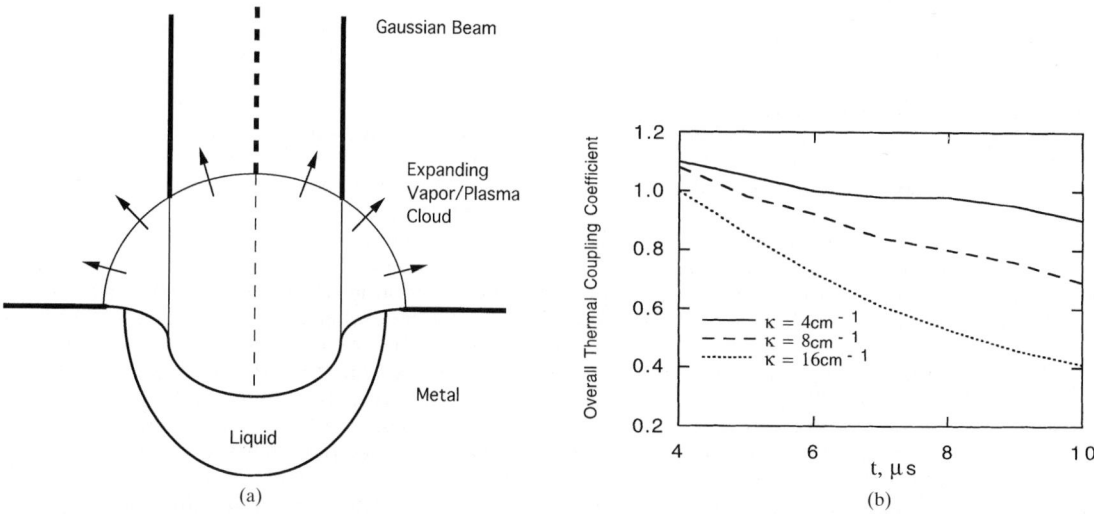

FIGURE 18.34 Schematic of (*a*) the physical system and (*b*) predicted overall thermal coupling coefficient for Gaussian irradiation of aluminum [166].

with a schematic of the physical system [166]. Note that the coupling efficiency is initially greater than unity. It decreases with time due to expansion of the vapor cloud and decreases as the extinction coefficient of the vapor cloud is increased.

Heat Transfer Analysis.   Surface-to-surface radiative heat transfer analyses range in complexity from solutions involving gray-diffuse surfaces [167] to more detailed analyses [159, 168, 169] discussed in Chap. 7. In the references, methods are outlined for handling specular surfaces, spectrally dependent properties, fully directionally dependent properties, and surfaces with varying temperature and/or varying incident radiation. When volumetric radiation exchange occurs in processed solids or processing gases, the radiative transfer equation (RTE, Chap. 7) describes these effects and must be solved to determine the radiation intensity distribution. In most real systems, solution of the RTE is performed in conjunction with solution of the energy and Navier-Stokes equations (Chap. 7), leading to a great deal of difficulty and, usually, cumbersome and unwieldy descriptions of the process.

To gain insight into the coupled behavior of real systems involving multimode effects, preoccupation with specular or spectral phenomena, or solution of the RTE may not be advisable or even feasible since (1) detailed surface properties are often unknown (or they change dramatically with the age of processing equipment as, for example, oxide layers are formed), and (2) computational hardware and software have not evolved to the point where predictions based upon the full set of equations are possible [170]. Fortunately, experience has shown that the predictions generated by simple radiation models (which may initially appear to be gross oversimplifications of reality) are often surprisingly robust in their ability to describe real behavior. This is because radiation transfer formulations, in general, are tolerant to the casual treatment of details, since local imperfections are averaged out in the integration process associated with evaluation of the radiation intensity along a long-of-sight in the medium. In the discussion to be presented in the next section, detailed radiation analyses will not be presented; rather, the coupling of the three heat transfer modes will be considered in order to gain insight into the thermal behavior of practical systems.

Complex Geometries.   In operations such as heat treating, surface radiation occurs in rather complex geometries. For example, parts can undergo surface-to-surface exchange with intervening solids obstructing the view of other surfaces. Alternatively, the material may be in motion (such as when carried by a conveyor belt), further complicating the analysis of radiative exchange between the part and the processing oven or furnace. Since analytical, tabular, and graphic values of F_{ij} have been generated for only a few relatively simple geometries, recourse to computational methods is often needed to evaluate shape factors for use in subsequent analysis of practical materials processing and manufacturing operations.

A thorough comparison of several computational methods for evaluating diffuse radiation view factors has been made [171]. Figure 18.35 shows several of the geometries considered. As evident, intervening surfaces are present in all of the cases, which range in complexity from the relatively simple six-sided shelf structure to the complicated multiple-sided truss.

Computations of the diffuse-view factors were made with several techniques, using available numerical algorithms. They included (1) a Monte Carlo method [159, 172], (2) a projected contour method [159, 173], (3) a hemi-cube method [173, 174], and (4) two double-area integration methods [173, 175]. Although the details of the different methods may be found in the references, it is relevant to note that both double-area integration methods rely upon discretization of the large surfaces into subelements. View factors between sub-

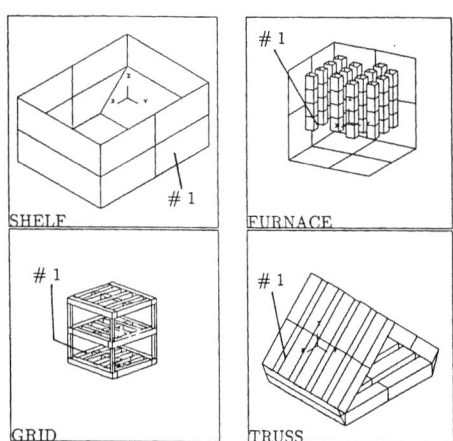

FIGURE 18.35 Geometries for testing diffuse view factor evaluation [171], published with permission from ASME International.

elements are determined and summed to find F_{ij}. In the simple method, obstruction by intervening solids is checked by connecting the centroids of the subelements with a line and then checking for interference between the connecting line and other solids. The refined double-integration method accounts for partial obstruction of the subelements with a more accurate "clipping method."

When dealing with realistic geometries, the accuracy of the computed view factors should be checked. For example, according to the enclosure rule (Chap. 7), the summation of view factors from an individual surface to the enclosure must equal unity in order to satisfy conservation of radiative energy. Sample predicted sums of the view factors from surface 1 (see Fig. 18.35) are shown in Table 18.6, along with the maximum error of $A_1 F_{1j}$ [171]. Monte Carlo-generated view factors are treated as the exact values, and only Monte Carlo-generated view factors whose estimated accuracy is better than 90 percent are used in the comparison exercise.

Several trends are evident upon inspection of Table 18.6. First, global conservation of energy is not guaranteed when numerical methods are employed to evaluate the diffuse view factors, and violation of the enclosure rule can be significant, especially for the more complex geometries (e.g., simple double area). Errors in predicted local view factors can be very large, again mainly for complex geometries. The Monte Carlo-generated view factors can also violate the enclosure rule (grid), and this difficulty is due to *ray leakage*, where individual emitted rays are not attributed to any surface.

Complex Geometries with Specular Reflection.   When processing metal parts, for example, specular reflection can be significant, and the directional dependence of the surface properties may impact the temperature of the material. Using a Monte Carlo approach, the effect of specular reflection has been considered for simple geometries, and specular behavior generally impacts the thermal evolution when open geometries are considered (e.g., Ref. 176).

With newer computational tools such as finite element and boundary element methods, conduction and convection in arbitrary three-dimensional geometries can be handled in a relatively straightforward manner. Some recent efforts have been made to link the FEM method (for ease in handling complex geometries) with alternative methods to evaluate view factors in order to predict thermal processing of complex shapes with specular surface radiative properties or specular reflection [177].

Figure 18.36 shows a finite element mesh representation of a connecting rod being heat-treated between two hot, black surfaces held at 1000 K [177]. The open (nonheater) sides are black and cold (0 K), while the connecting rod surfaces have emissivities of 0.1, and reflectivities (diffuse or specular) of 0.9. Steady-state connecting rod temperatures were predicted, assuming the rod's reflectivity is either purely diffuse or perfectly specular. As might be expected, most (97 percent) of the computational effort was spent evaluating diffuse and specular view factors.

Predicted steady-state temperatures for the rod of Fig. 18.36 are shown in Fig. 18.37, based upon specification of pure diffuse reflection, as in Fig. 18.37a, or pure specular reflection, as in Fig. 18.37b. As expected, the surfaces that face the heaters are at a high temperature, while those that are exposed to the cold openings are relatively cool. Local temperature differences of up to 100 K are evident when the diffuse and specular reflectivity predictions

TABLE 18.6   Sum of View Factors and Maximum Individual View Factor Errors [171]

Method	Shelf	Furnace	Grid	Truss
Simple double area	(1.021) 0.116	(1.147) 1.394	(1.361) 3.248	(1.590) 19.13
Refined double area	(1.001) 0.046	(0.976) 1.484	(1.000) 1.183	(1.002) 0.277
Hemi-cube	(1.000) 0.049	(1.001) 0.582	(0.998) 0.654	(0.919) 4.044
Projected area	(1.000) 0.048	(1.001) 0.490	(0.980) 0.788	(1.021) 2.115
Monte Carlo	(0.999)	(1.000)	(0.939)	(0.998)

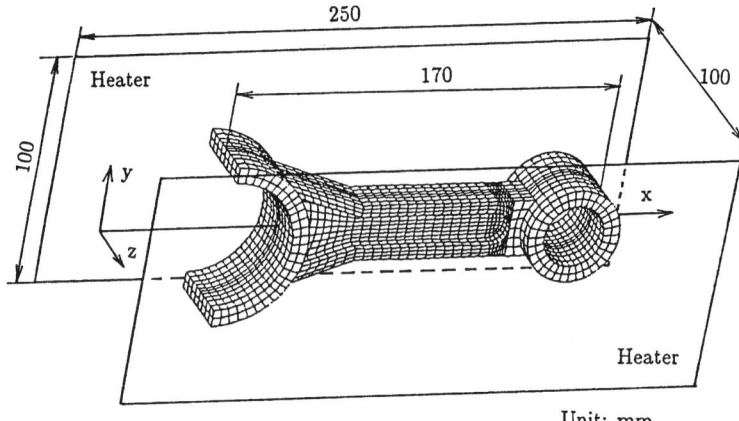

FIGURE 18.36   Finite element mesh representation of a connecting rod radiatively heated between hot surfaces [176], from S. Maruyama, *Numerical Heat Transfer,* 24, pp. 193–194, Taylor & Francis, Inc., Washington, DC. Reproduced with permission. All rights reserved.

are compared. The impact of specular reflection can be substantial for complex configurations, and the effects are most pronounced in holes and grooves that are heated from outside sources.

For multimode problems, it is sometimes advantageous to use a "dual grid" technique in order to minimize the computational expense associated with storing and evaluating view factors. A course mesh can be used for the radiation heat transfer, while finer meshes can be used for the conduction and/or convection heat transfer. This technique is discussed in detail, and associated computational error (which is small) is reported in Zhao [178].

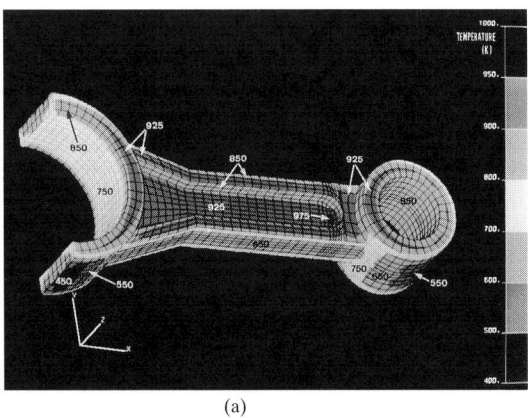
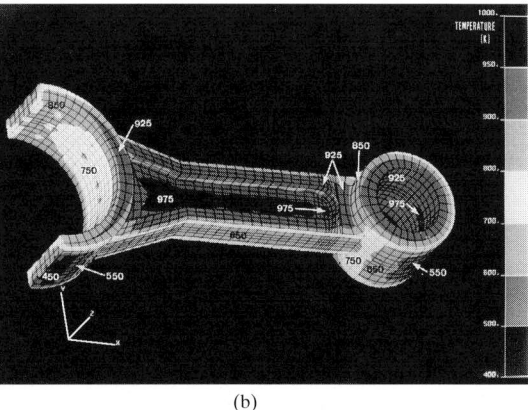

(a)   (b)

FIGURE 18.37   Predicted steady state temperatures assuming (*a*) purely diffuse part reflectivity and (*b*) purely specular part reflectivity [176] from S. Maruyama, *Numerical Heat Transfer,* 24, pp. 193–194, Taylor & Francis, Inc., Washington, DC. Reproduced with permission. All rights reserved.

SYSTEM-LEVEL THERMAL PHENOMENA

The discussion in the preceding subsections was concerned with more specific physical situations arising during thermal processing of materials. Clearly, a wide array of important manufacturing processes involve significant thermal effects, and an up-to-date discussion of some processes is available [179]. The interested reader is referred to several books in the area of manufacturing and materials processing that discuss important practical considerations without considering in detail the underlying thermal transport phenomena [180–182]. However, a few books have appeared recently that are directed at modeling transport phenomena in materials processing from the fundamental point of view [183–184]. Clearly, a large amount of work has been done on heat transfer and fluid flow underlying materials processing from a thermal system point of view. This section of the chapter can include only a few examples of important materials processing issues impacted by thermal phenomena.

Numerical analysis (solution of the integro-differential equations, describing conservation of mass and thermal and radiation energy, and the Navier-Stokes equations) of large-scale operations is complicated by several factors. These are: (1) the material's thermophysical properties may not be well known; (2) length and time scales can span orders of magnitude; (3) geometries can be complicated, invalidating applications of correlations developed for simple geometries and making radiation exchange analysis cumbersome; (4) the product may be discontinuous (making the analysis of moving product streams tedious); and (5) the material morphology may evolve in response to heat transfer phenomena. The preceding features are superposed upon the usual challenges of (for example) evaluating turbulent convective heat transfer or radiation heat transfer rates in systems where the radiation properties are highly wavelength-dependent. In the following discussion, several examples of system level thermal phenomena are discussed.

Heating of a Load Inside Industrial Furnaces

The high- and medium-temperature industrial furnaces used in materials processing can be roughly classified into two types according to the energy source [185]: (1) fossil fuel (mostly natural gas) and (2) electrical heating resistance or induction-type furnaces. The fossil fuel furnace can be either direct- or indirect-fired. Alternatively, they can be classified into batch and continuous furnaces. In the former, the load is stationary, and, in the latter, it moves through the furnace while being heated. In both types of furnaces, radiation, in general, is the principal mode of heat transfer, but convection may not be negligible in smaller furnaces and would have to be considered when predicting heat transfer to the load (working piece). There is a great variety of both directly and indirectly fired industrial furnaces tailored for different applications and materials to be processed [186]. In the directly fired furnaces, combustion of natural gas takes place in the chamber, and heat is transferred from the products in contact with the load. In the indirectly fired furnace, combustion of natural gas takes place in tubes, and the products of combustion heat the tube walls. The heat is subsequently transferred from the tubes to the load predominantly by thermal radiation. The indirectly fired furnaces are often referred to as *radiant-tube furnaces*.

Indirectly fired furnaces find applications in metallurgy, paint enameling, the pharmaceutical industry, and other situations where it is necessary to control the chemistry of the furnace atmosphere [186]. An important aspect of the heat treatment of metals, for example, is the effect of the furnace atmosphere on the stock being heated. In most cases, the need is to minimize or eliminate completely the undesirable effects of furnace gases, such as oxidation or decarbonization. Both directly [187, 188] and indirectly [189, 190], natural-gas-fired furnaces have been analyzed, but the details cannot be included here.

In this subsection, a few examples are presented that illustrate heating of opaque and moving continuous materials. The analysis is simplified by decoupling the combustion processes taking place in the chamber or in the radiant tubes, and focus is on the heat transfer in the oven (furnace) and to the moving load.

Indirect Fired Furnaces

Heat Transfer to Opaque and Moving Continuous Sheet—System Behavior.   A schematic diagram of an indirectly fired continuous furnace is shown in Fig. 18.38. Combustion of natural gas takes place inside the multiple radiant tubes located below the roof (crown) of the furnace. Radiant tubes can also be located at the ends or sides of the furnace, and the high-temperature products of combustion are kept isolated from the stock being heated (the load). The energy released due to combustion is transferred to the radiant-tube wall, and the heating of the stock material is accomplished predominantly via radiative heat transfer from the heated walls of the radiant tubes and from the surfaces of the enclosure. The refractory walls of the furnace interior, heated by the radiation emitted from the tubes, redirect the energy to the load, radiant tubes, and the refractory walls themselves. Furthermore, the furnace chamber (enclosure) may be filled with an inert radiatively nonparticipating atmosphere, such as argon or nitrogen, to prevent scaling or decarburization of the load during the heating process. The presence of gases in the chamber can result in convective heat transfer from gases to the load, radiant tubes, and refractory walls. The processes occurring are quite complex but have been analyzed using mathematical/numerical models [189–191].

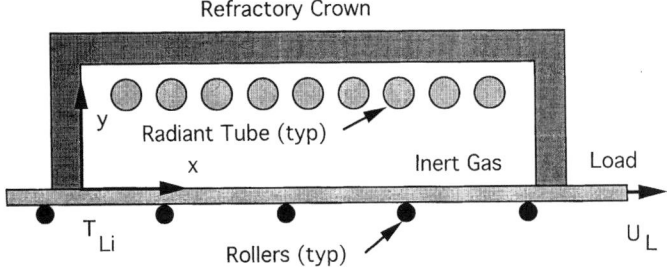

FIGURE 18.38   Schematic of an indirectly heated furnace.

The thermal system model for radiant-tube continuous furnace involves integration of the mathematical models of the furnace enclosure, the radiant tube, and the load. The furnace enclosure model calculates the heat transfer in the furnace, the furnace gas, and the refractory walls. The radiosity-based zonal method of analysis [159] is used to predict radiation heat exchange in the furnace enclosure. The radiant-tube model simulates the turbulent transport processes, the combustion of fuel and air, and the convective and radiative heat transfer from the combustion products to the tube wall in order to calculate the local radiant-tube wall and gas temperatures [192]. Integration of the furnace-enclosure model and the radiant-tube model is achieved using the radiosity method [159]. Only the load model is outlined here.

The transient, two-dimensional energy equation for the load (i.e., a plate, a sheet, or a slab) moving at a constant velocity U_L through the furnace can be expressed as

$$\rho_L c_L \frac{\partial T_L}{\partial t} + \rho_L c_L U_L \frac{\partial T_L}{\partial x} = \frac{\partial}{\partial x}\left(k_L \frac{\partial T_L}{\partial x}\right) + \frac{\partial}{\partial y}\left(k_L \frac{\partial T_L}{\partial y}\right) \tag{18.66}$$

Order of magnitude (scale) analysis reveals that, in this equation, the first term on the LHS is important only during the initial stages of the transient, and heat conduction in the x-direction can be neglected in comparison to the y-direction, owing to the large length-to-thickness ratio (typically greater than 100).

The boundary condition at the inner (top) surface of the load is exposed to the radiative and convective heat q''_{tot} such that

$$-k_L \left. \frac{\partial T_L}{\partial y} \right|_{y=0} = q''_{tot}(x) = q''_{rad}(x) + q''_{con}(x) \qquad (18.67)$$

where $q''_{tot}(x)$ is calculated from the furnace enclosure and furnace gas submodels. The adiabatic boundary condition is specified at the midplane of the load, as it is considered a plane of symmetry due to symmetric heating of the load from above and below. The presence of the rollers is neglected. The temperature of the load entering the furnace is known and serves as the boundary condition for the x-direction. Details of the numerical method of solution and results of simulations are available elsewhere [190, 191].

Some typical results for the total heat flux and temperature at the surface of the load are shown in Figs. 18.39 and 18.40, respectively. The results are for heating of steel and depict the effect of load velocity U_L for a constant load thickness. The steel plate moves from left to right through the furnace. The net heat flux to the load, plotted in Fig. 18.39 for various load velocities, reveals a highly nonuniform heating of the load at the lower load velocities. At the inlet to the furnace ($x = 0$), the load is relatively cold (see Fig. 18.40 near $x = 0$), and thus a large amount of heat is transferred to the load by radiation. However, at lower load velocities, the longer residence time of the load in the furnace causes the load surface temperature to approach the temperature of the radiant tubes. At the furnace exit, therefore, the net heat flux at the load surface decreases sharply, as evident from the figure for load velocities of 0.008 and 0.01 m/s. At higher velocities, the load surface temperatures are much lower than the radiant-tube temperatures, and the net heat flux at the load surface remains more or less uniform (Fig. 18.39). The effects of load and refractory wall emittances, load material, load throughput, and radiant-tube geometry have been studied, and results of numerical calculations are available [190, 191]. As the load emissivity decreases, more energy is reflected to the crown. When heat losses from the furnace to the ambient (exterior) are considered, the rising crown temperatures increase heat losses and, in turn, decrease the furnace efficiency (ratio of heat transfer to the load to heat from the tubes). Local temperatures decrease slightly. The effect of the refractory emissivity on load temperatures and furnace efficiency is small.

Heat Transfer to Opaque and Moving Continuous-Sheet Multimode and Conjugate Heat Transfer Effects. In the previous discussion, convective heat transfer coefficients were obtained from empirical correlations. Therefore, coupling between convection and radiation

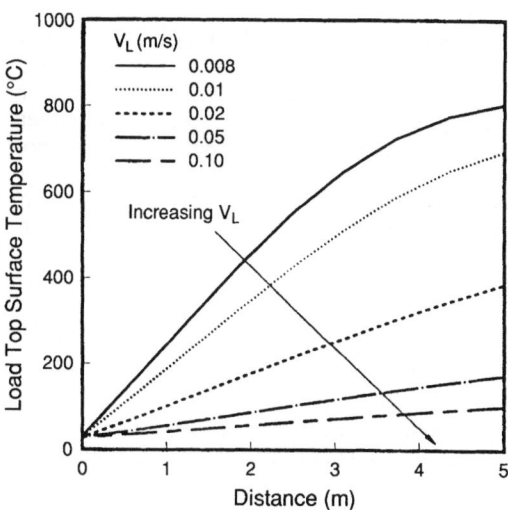

FIGURE 18.39   Variations of the net heat flux to the load with distance for varying load velocities [190].

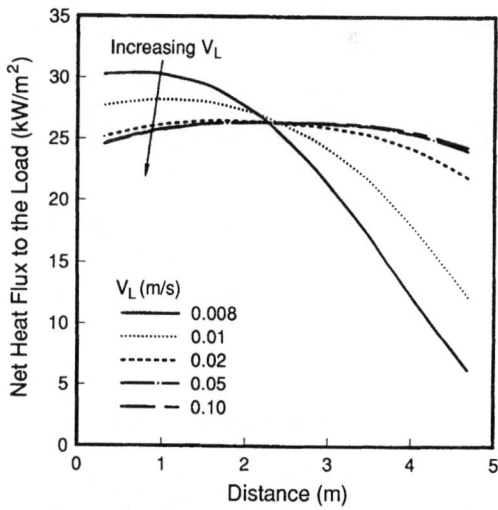

FIGURE 18.40   Variations of the load surface temperature with distance for varying load velocities [190].

is treated only approximately. In cases where convective heat transfer rates are comparable to (or exceed) radiation rates, or when chemical species distributions within the processing gas are of interest, more rigorous evaluation of convective transport may be justified.

Constant Temperature Sources. Zhao et al. [193] have analyzed the thermal response for a configuration similar to that of Fig. 18.38 for five constant-temperature (250°C) tubes and an adiabatic crown. Since the source temperature is fixed (piped superheated steam is assumed as the heating medium), overall load heating will change as processing conditions are modified. In their study, cold load is introduced into the 2-m-long, 0.75-m-high furnace at $T_{Li} = 20°C$ (for $U_L > 0$). As the sheet material travels, it is radiatively and convectively heated by the tube bank. Gas advection is induced by buoyancy and inertial forces, and air is allowed to enter and leave the system through 0.1-m-high openings at the inlet and exit of the furnace. The material thickness, density, and specific heat are 5 mm, 200 kg/m³, and 1000 kJ/kgK, respectively, and conduction within the load was ignored. The emissivity of the furnace walls was set to 0.1, $\varepsilon_L = 0.9$, and that of the tubes is unity.

System response was predicted using a gray-diffuse radiation analysis, together with the two-dimensional Navier-Stokes and energy equations, a k-ε turbulence model, and the ideal gas equation of state. At $U_L = 0$, convective and radiative heat fluxes at the load are balanced. As U_L increases, cool load is carried further into the furnace, as shown in Fig. 18.41a, and an overall counterclockwise gas circulation (not shown) is induced. The load temperature increases with x in Fig. 18.41a, leading to convective load cooling at $U_L = 0.01$ m/s in Fig. 18.41b. As load velocities are increased further, load exit temperatures are reduced, leading to the increased radiative and convective load heating shown in Fig. 18.41b. The need to account for convective heating depends strongly, therefore, on the conveyor speed used, with convection playing a less prominent role at higher load velocities due to the increased temperature difference between the source and the load, as well as entrainment of cool ambient air into the enclosure by the moving belt. Three-dimensional predictions (for an oven width of 1 m with insulated front and back sides) have also been obtained [178]. For the same operating parameters as for the two-dimensional furnace, radiative heat transfer to the load is decreased, and local convective heating (cooling) can be reversed as three-dimensional effects are accounted for.

Constant Power Sources. Predictions for laminar flow have been obtained for a two-dimensional (L long by H high) rectangular furnace with a uniformly heated crown (no tubes) and no openings at the oven entrance or exit [194]. Since the source heat flux is constant, only

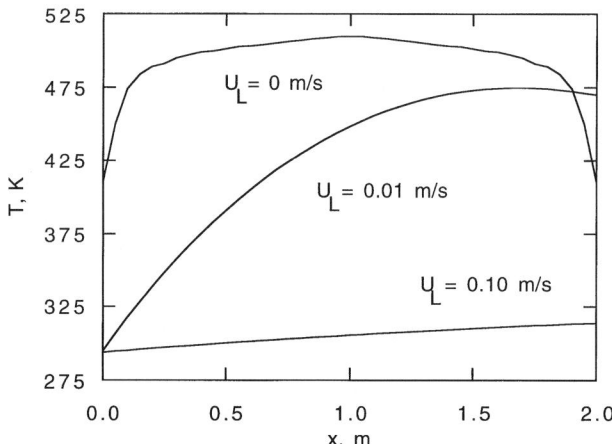

FIGURE 18.41a Multimode heating with an indirectly fired furnace. Shown is: (a) predicted sheet temperature distributions.

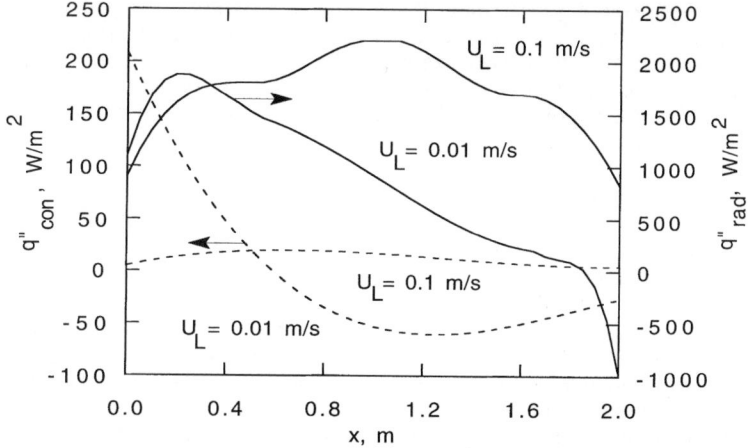

FIGURE 18.41b Multimode heating with an indirectly fired furnace. Shown is: (b) predicted radiative and convective sheet heat fluxes [193].

local temperatures (and, in turn, relative rates of radiation and convective heating) will change as processing conditions or material properties are varied. As the crown and load temperature distributions change, highly coupled multimode effects involving radiative heating and mixed gas convection, and conduction in the solids occurs.

If the nonparticipating gas convection is two-dimensional and laminar, and if the fluid is considered to be Boussinesq, appropriate nondimensionalization of the descriptive equations yields the following dimensionless parameters [194]:

Convection: $\mathrm{Re}_L = U_L H/\nu_g$; Pr; $\mathrm{Gr/Re}^2 = g\beta\Delta T H^2/\alpha_g U_L$; $A = L/H$ (18.68)

Radiation: $N_{r1} = \rho_\Lambda \chi_\Lambda \delta_\Lambda Y_\Lambda \Delta T/(H|q''_{so}|)$; $N_{r2} = k_g \Delta T/(H|q''_{so}|)$

$N_{r3} = U_\infty \Delta T/|q''_{so}|$; $N_{r4} = d_L k_L \Delta T/(H^2 |q''_{so}|)$; $\Gamma = T_H/T_{Li}$ (18.69)

where $T_H = T_{Li} + [|q''_{so}|/\sigma]^{1/4} = T_{Li} + \Delta T$ and U_∞ is an overall heat transfer coefficient used to quantify losses. The parameter $|q''_{so}|$ is the largest value of the source flux. In addition, the emissivities of the load and furnace walls ε_L and ε_w appear. The parameters in Eq. 18.68 are standard, while parameters N_{r1} through N_{r4} are the relative strengths of material advection to radiative heating, convection to radiation, ambient losses to radiation, and load conduction to radiation, respectively. The ratio N_{r1}/N_{r4} is the Peclet number.

Numerical predictions for base case conditions ($A = 10$, Re $= 500$, Gr $= 10,000$, Pr $= 0.7$, $\Gamma = 2$, $N_{r1} = 100$, $N_{r2} = 0.05$, $N_{r3} = N_{r4} = 0$, $\varepsilon_L = \varepsilon_w = 0.5$) were generated. The entire length of the furnace crown is heated, and losses to the ambient were neglected. As the belt speed is increased, as shown in Fig. 18.42a, convective load (source) heating (cooling) is enhanced (in contrast to the preceding constant source temperature case). As in the constant source temperature case, lower load temperatures $[\theta = (T - T_{Li})/(T_H - T_{Li})]$ result as the production rate is increased as required by conservation of energy. At small N_{r1}, as in Fig. 18.42b, higher load and crown temperatures are induced, and differences in these temperatures are relatively small, as shown in Fig. 18.10b. As $N_{r1} \to 0$, $t|\vec{r}_i| \to \infty$, while, as $N_{r1} \to \infty$, $T|\vec{r}_i| \to 0$, since convection heat transfer rates induced by high load velocity become large. As the load emissivity is reduced in Fig. 18.42c, more of the energy emitted by the crown is reflected, increasing its temperature. As the crown temperature rises, however, more energy is delivered to the load via convection, offsetting the reduced radiative heating of the load. If heat losses to the ambi-

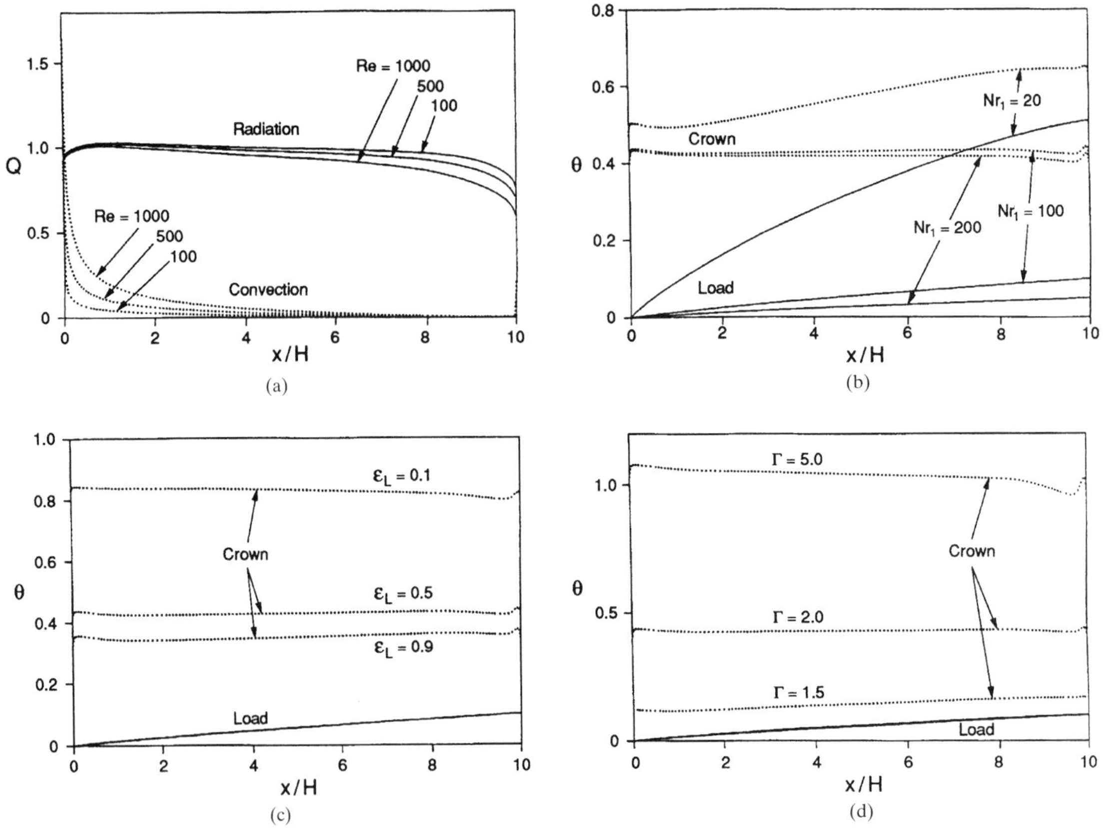

FIGURE 18.42   Predicted (a) effect of Re on radiative and convective load heating and sensitivity of load and crown temperatures to (b) N_{r1}, (c) ε_L, and (d) Γ [194].

ent were accounted for, lower load temperatures would be induced at lower load emissivities. As Γ is increased, the difference in crown-to-load temperatures increase, with dimensional load temperatures rising in proportion with Γ as shown in Fig. 18.42d.

Discontinuous Load.   Continuous thermal processing of discrete material occurs in operations such as painting, curing, and food processing. The processing of a general, discontinuous load has received some attention [195]. If the load of Fig. 18.38 is in discrete form, the analysis is complicated by the presence of moving boundaries and the need for continuous reevaluation of view factors as the load is carried through the furnace. Correlations for convective heat transfer coefficients at the load and furnace surfaces are unavailable, and heat transfer is inherently unsteady, necessitating rigorous analysis via solution of the Navier-Stokes, thermal energy and radiative energy equations.

A general processing furnace is shown in Fig. 18.43. Discrete, flat material (initially at 300 K) is carried (6 mm/s) through the oven, which is characterized by crown and floor temperatures of 600 K. The high-emissivity material and oven surfaces interact radiatively and convectively, while conduction occurs in the material. An exhaust is present at

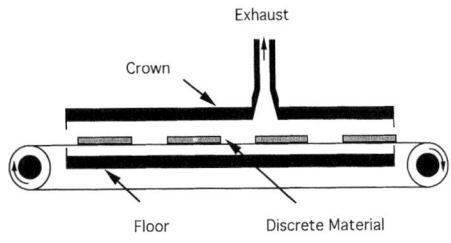

FIGURE 18.43   Schematic of thermal processing of moving, discrete materials. The belt is permeable.

approximately 2/3 of the oven length (720 mm). Other parameters are listed in Bergman et al. [195]. No benefits are associated with nondimensionalization of the model equations because of the number of length and time scales involved.

Figure 18.44*a* shows predictions of the volume-averaged material temperatures, as well as local air temperatures 1 mm above the top of the material slab, which were generated by solution of the transient, two-dimensional descriptive equations. Here, x is the location of the material's centerline, and an individual slab is entirely within the furnace over the range $0.12 \text{ m} \leq x \leq 0.72 \text{ m}$. The material is heated as it moves through the furnace, while buoyancy-induced mixing induces a high frequency fluctuation of the local air temperature. Figure 18.44*b* shows the predicted air temperature midway through the furnace. The low frequency response is due to the oven period p, which is defined by the slab and gap lengths normalized by the conveyor velocity ($p = 30$ s). Again, high frequency behavior is due to buoyancy-induced mixing. The high and low frequency behavior associated with processing of discrete material can both have a high impact on the thermal history of the heated product, and results of parametric simulations are available in Bergman et al. [195] and Son et al. [196].

Direct Fired Furnaces.   For direct fired furnaces, radiative heat transfer from the flame and combustion products as well as from the walls to the load is usually the dominant heat transfer mode. Convection from the combustion gases makes a much smaller contribution. The radiative transfer within the furnace is complicated by the nongray behavior of the combus-

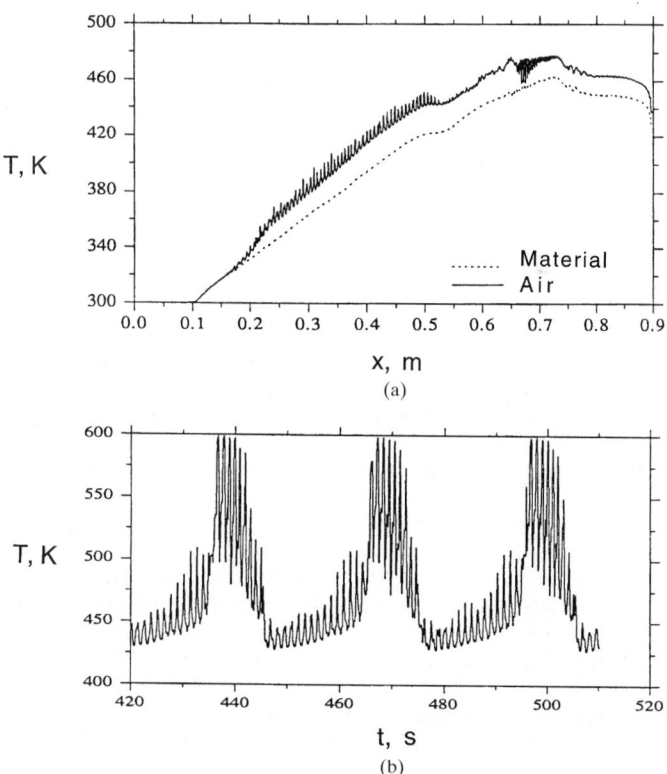

FIGURE 18.44   Predicted (*a*) volume-averaged material and air temperatures through the furnace and (*b*) local air temperature history. The furnace entrance is at $x = 60$ mm.

tion gases due to the presence of selectively absorbing and emitting species such as CO_2, H_2O, CO, CH_4, and possibly soot or dust particles, which are capable of absorbing, emitting, and scattering thermal energy. Because of the significant volumetric effects, few general results have been obtained.

Batch Heating.   Despite the complexity brought about by direct firing, identification of the relevant dimensionless parameters for direct fired furnaces has been attempted [197]. Figure 18.45 shows a typical furnace that might be used in a batch firing operation. Combustion occurs within the enclosure, and radiative exchange (and, to a lesser degree, convective transfer from the gas to the load) results in transient heating of the load, which is in sheet form. The descriptive equations for the system response (gas energy balance, load and wall heat conduction equations, the gas energy equation including absorption and emission described by Hottel's zonal method using the four-gray gas model in Chap. 7, and expressions for gas-to-surface and surface-to-surface exchange) were applied to the system and nondimensionalized, resulting in 17 dimensionless parameters [197]. The model was validated via comparison with limited experimental data [198].

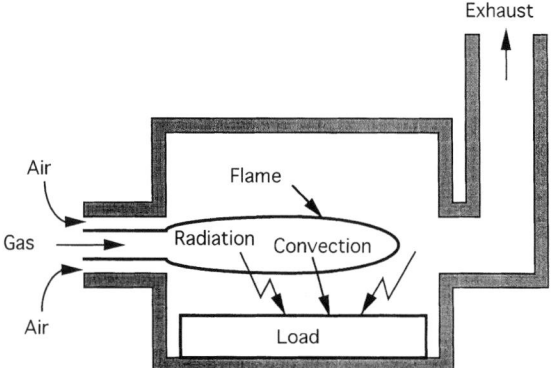

FIGURE 18.45   Schematic of direct fired batch processing of slab materials.

Because the dimensionless parameters identified were not all independent, only several primitive variables were subsequently varied in a sensitivity study to determine the dependence of the material's thermal response to variations in the load thickness such as: the refractory wall thickness, load and refractory thermal diffusivity ratio, the air/fuel mixture used in the combustion, the refractory and load emissivities, the furnace height, and the exposed area of the load.

The predicted effect of load emissivity, combustion space size, and refractory emissivity for a particular furnace and load is shown in Fig. 18.46 [197]. A furnace 5 m long by 1 m high by 1 m wide was loaded with a 0.15-m-thick sheet of iron, while the refractory walls were constructed of 0.5-m-thick red clay brick. The methane burners fired at a rate of 500 kW during operation. Additional process parameters and thermophysical properties are listed in [197].

The load temperature increases as its emissivity is raised (Fig. 18.46*a*), as expected. Decreases in the furnace volume (height) increase the load temperatures (Fig. 18.46*b*), since less refractory and gas is heated, and losses to the ambient are reduced. Changing the refractory emissivity (Fig. 18.46*c*) yields little impact on the furnace efficiency (defined as the time average of energy delivered to the load normalized by the time average of energy released by combustion), except at small refractory emissivity, when reradiation to the load occurs. The last finding is consistent with the results of field tests run on a 50-ton-per-hour rotary hearth furnace, which showed that application of high emissivity refractory coatings yielded no significant change in production rates or furnace efficiency [199].

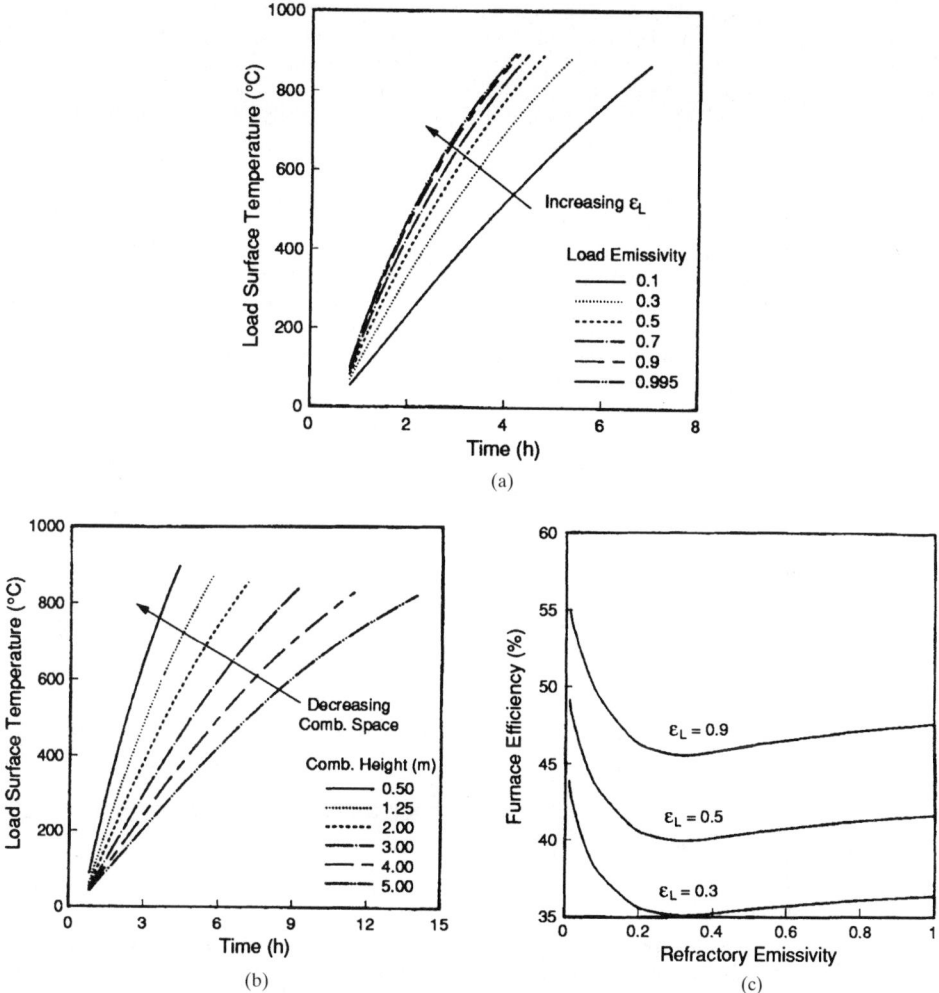

FIGURE 18.46 Predicted load temperature and its variation with (*a*) load emissivity and (*b*) combustion space size. Also shown is (*c*) the dependence of the furnace efficiency upon the refractory emissivity [197].

Quenching

The production of steel, aluminum, and other metal alloys having desirable mechanical and metallurgical properties requires accurate temperature control during processing. For example, the objective of quenching steel is to raise its hardness and improve its mechanical properties such as tensile impact and fatigue strengths, improve fear resistance, and so on. Although various quenchants (oils, polymers, water, and so on) are utilized, more often water has been used due to considerations such as safety, management, economy, and pollution. Applications include quenching of forgings, extrusions, and castings, as well as strip, plate, bar, and continuous castings and investment products. Three, but not all, techniques of quenching are illustrated in Fig. 18.47: (*a*) immersion, (*b*) spray, and (*c*) film. For immersion quenching the workpieces are dipped into baths of different liquids, and, for spray quenching, the workpieces are cooled by water sprays. For nonferrous metals, cooling water is introduced on the surface in the form of a film by way of a distribution system illustrated in Fig. 18.47.

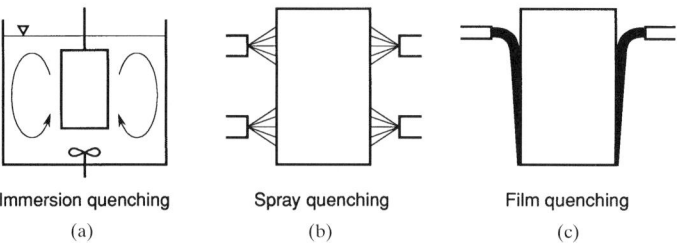

Immersion quenching   Spray quenching   Film quenching

(a)   (b)   (c)

FIGURE 18.47 Schematic diagrams of different quenching techniques: (*a*) immersion quenching, (*b*) spray quenching, and (*c*) film quenching. From I. Tanasawa and N. Lior, *Heat and Mass Transfer in Materials Processing*, pp. 535–547, Taylor and Francis Group, New York. Reproduced with permission. All rights reserved.

When water is employed in the quenching of continuously moving metals, liquid jet, spray, and mist impingement are possible choices of rapidly cooling a metal. For example, accelerated cooling is often a major objective in steel production, since it improves the mechanical properties of the final product by providing desired structural changes with respect to grain size and the ratios of ferrite, pearlite, and laminate. Optimum strength and toughness properties of hot-rolled steel, for example, can be achieved by refining the ferrite grain size and precipitation conditions through accelerated cooling approximately 15 K/s from the initial post-roll temperature. Controlled cooling rates of the order of 10^5 to 10^6 K/s or higher are needed to produce amorphous metals using rapid quenching solidification processes [200, 201].

Nucleate, transition, and film boiling heat transfer regimes are expected to occur during quenching of metal parts, moving hot rolled strip, and continuous castings at different locations and time. It is desirable to discuss the different heat transfer regimes occurring during the process before analyzing the temperature versus time history for a particular quenching application.

A typical temperature history for a metal as it is quenched in a stationary liquid bath is illustrated in Fig. 18.48. Four heat transfer regimes are revealed. Regime I depicts the temperature response from the moment of immersion to formation of a stable vapor film. During

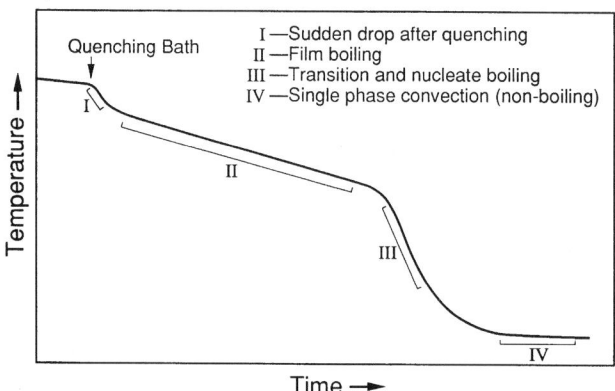

FIGURE 18.48 Schematic representation of a typical quenching curve illustrating surface temperature variation with time. From I. Tanasawa and N. Lior, *Heat and Mass Transfer in Materials Processing*, pp. 455–476, Taylor and Francis Group, New York. Reproduced with permission. All rights reserved.

regime II, stable film boiling occurs. Stable film break-up begins, goes through transition boiling, and ends in nucleate boiling in regime III. Regime IV denotes single phase heat transfer. Although regime III corresponds to the largest cooling rate, it is the least understood, even for conventional saturated boiling conditions, unrelated to metallurgical applications involving subcooled boiling with forced internal or external flow of the coolant (see Chap. 15 of this handbook). Reference is made to representative reviews on film and transition boiling [202, 203] and the mechanism of quenching nuclear reactor cores under loss-of-coolant accident conditions [120] for discussion of the processes relevant to the quenching of metals [134].

Quenching of Steel.   Film boiling is the exclusive heat transfer mode when quenching steel (for example, at temperatures of 700°C or higher) in water. Because of the formation of vapor film between the hot metal surface and cooling liquid (water), heat transfer from the surface to the coolant is impaired, resulting in a low convective heat transfer coefficient. The quenching of steel is commonly conducted at temperatures of 40°C or lower (i.e., with a subcooling of more than 60°C). The relation between the cooling water temperature and the surface hardness H_{RC} (Rockwell C scale) after the quenching of 0.45 percent carbon steel has been measured [130]. The results show that the hardness of steel is significantly affected by the cooling water temperature T_{cw}. When the coolant temperature exceeds about 40°C, the Rockwell C scale hardness H_{RC} decreases with an increase in T_{cw}. On the other hand, for $T_{cw} \leq 40°C$, the water temperature has little effect on H_{RC}.

Figure 18.49 shows the measured cooling curves (T vs. t) for a 6-mm-diameter and 80-mm-long steel (~0.45 percent carbon content) test specimen with thermocouples embedded at the center [130]. Each specimen was heated to 850°C inside an electrical furnace filled with nitrogen gas to prevent oxidation, and they were subsequently immersed in quiescent water. The coolant temperature was varied from 20 to 100°C to determine its effect on the cooling curve. The cross-sectional (Vickers) hardness values H_{mV} are indicated on the curves. Superimposed on the T vs. t (plotted on a semilog scale) are the continuous cooling transformation diagrams (CCT). It is noted that, when the cooling temperature exceeds 45°C, the temperature drop of

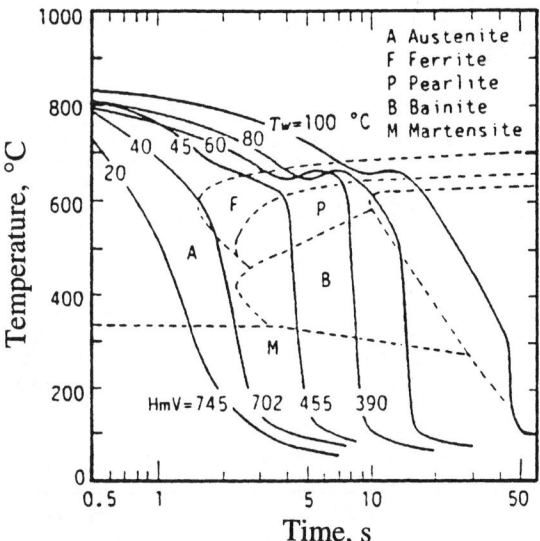

FIGURE 18.49 The measured cooling curves for water-quenched S45C (0.45 percent carbon) steel as a function of temperature (superimposed on a 2 TTT diagram) from Tajima et al. [130].

the specimen reveals a hesitation around 660°C, where the Ar' transformation occurs, and the effect of latent heat due to phase transformation is evident.

Transient temperature distribution during quenching of cylindrical steel (about 0.45 percent carbon) test specimens have been predicted [204] by solving numerically the transient, one-dimensional heat conduction equation

$$\rho c \, \frac{\partial T}{\partial t} = \frac{1}{r} \frac{\partial}{\partial r} \left(kr \, \frac{\partial T}{\partial r} \right) + q''' \tag{18.70}$$

This equation accounts for heat conduction in the radial direction only but accounts for heat generation due to phase transformations and physical property variation with temperature. When the temperature boundary condition on the surface of the cylinder was imposed in a form of a heat flux that was independent of the cylinder diameter, the predicted and measured centerline temperatures did not agree uniformly well. However, when the test specimen diameter was included as a parameter in the correlations for the surface heat flux in the film and nucleate boiling regimes as well as for the minimum and maximum heat fluxes, very good agreement has been obtained between the predicted and experimentally measured centerline temperatures of the test specimen (Fig. 18.50). From the results obtained, it was concluded that the cooling curves could not be accurately estimated without considering the curvature of the cylindrical test specimen in the expression for the heat flux (generated from the subcooled, transient boiling curve), which is used in the temperature boundary condition for Eq. 18.70.

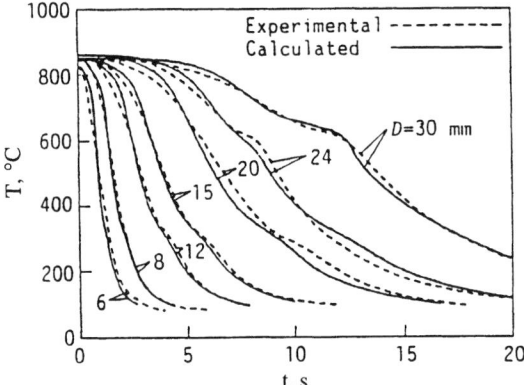

FIGURE 18.50 Comparison of measured and predicted cooling curves at the center of various diameter water-quenched 54°C (0.45 percent) carbon steel [130].

Strip Cooling with Liquid Jet Impingement. To achieve desired cooling rates and temperature control, several methods of cooling steel strip have been developed (Fig. 18.51). For example, an accelerated cooling method that exploits temperature-dependent structural changes with respect to grain size and ratios of ferrite, pearlite, and laminate is used to cool steel strip. The method is achieved by a variety of schemes, including sprays, planar jets, and round jets. The characteristics of these three types of cooling schemes were described and compared by Kohring [205]. The available results show that planar water jets are more efficient than spray nozzles in removing heat from a strip at 900°C and traveling at 10 m/s [93]. The cooling of a moving strip by an array of round jets is less efficient than with planar jets per unit water flow rate [206], but it provides greater flexibility and control.

Models have been developed to simulate the thermal behavior of a moving steel strip cooled by an array of planar [207] and round [206] jets. The strip moves on transport rollers,

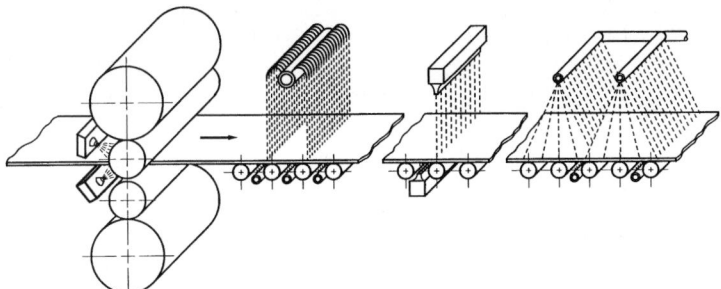

FIGURE 18.51 Schematic of cooling systems along a runout table for cooling strip steel. From I. Tanasawa and N. Lior, *Heat and Mass Transfer in Materials Processing*, pp. 535–547, Taylor and Francis Group, New York. Reproduced with permission. All rights reserved.

and an Eulerian coordinate system (the origin of which is fixed at a point immediately downstream of the last work rolls with $y = 0$ at the top of the surface and $z = 0$ halfway across the strip, respectively) is adopted. Assuming that the strip is infinitely long and the cooling system operates under steady-state conditions, the conservation of energy equation for the moving strip may be expressed as

$$\frac{\partial}{\partial x}(\rho_L U_L c_L T_L) = \frac{\partial}{\partial y}\left(k_L \frac{\partial T_L}{\partial y}\right) + \frac{\partial}{\partial z}\left(k_L \frac{\partial T_L}{\partial z}\right) \tag{18.71}$$

A comparison with Eq. 18.66 shows that, in writing, this equation, longitudinal heat conduction (x-direction), has been neglected in comparison to the lateral (y) and transverse (z) directions. The inlet and boundary conditions are:

$$T_L(0, y, z) = T_{Li};\ -k_L\left(\frac{\partial T_L}{\partial y}\right)_{\text{at surfaces}} = q''_y(x, y);\ -k_L\left(\frac{\partial T_L}{\partial z}\right)_{\text{at edges}} = q''_z(x, y) \tag{18.72}$$

Modeling of local heat transfer from the strip to quiescent air, single, and two-phase convection to water in the impingement cooling region, forced-film boiling region, and heat transfer between the strip and transport rollers, presents a challenge. The details concerning the specification of q''_y and q''_z in Eq. 17.72 and the numerical method of solution are given in the original publications [206, 207].

Figures 18.52 and 18.53 show that the temperature distribution is along the runout table in the upper and lower half of the strip, respectively. The results correspond to the following operating conditions: ultra low carbon steel (AISI 1005), a strip thickness of 3.556 mm, a strip speed of 10.57 m/s, an exit temperature from the finishing mill of 870°C, and a coiling temperature of 671°C. The results show that the difference between the calculated and measured coiling temperatures is very small, thus providing confidence in the model.

Starting from the exit of the finishing mill (cooling in the inlet quiescent zone by radiation to surroundings and convection to air), temperature decreases gradually, and the differences between the top and core temperatures are small (Fig. 18.52). However, the bottom surface temperature distribution (Fig. 18.53) is different due to the effect of heat transfer between the strip and transport rollers. Although this transfer is characterized by high local heat fluxes, the contact area is small, and the surface temperature recovers quickly after a sharp drop. Hence, the temperature difference between the bottom surface and the core remains small between neighboring rollers. The surface temperatures drop significantly in the jet impingement region (Fig. 18.52), where the local heat fluxes are the largest. Owing to heat conduction from the interior, the top surface temperature rises very sharply at the beginning of the film boiling region, since the cooling efficiency is significantly reduced relative to that in the impingement

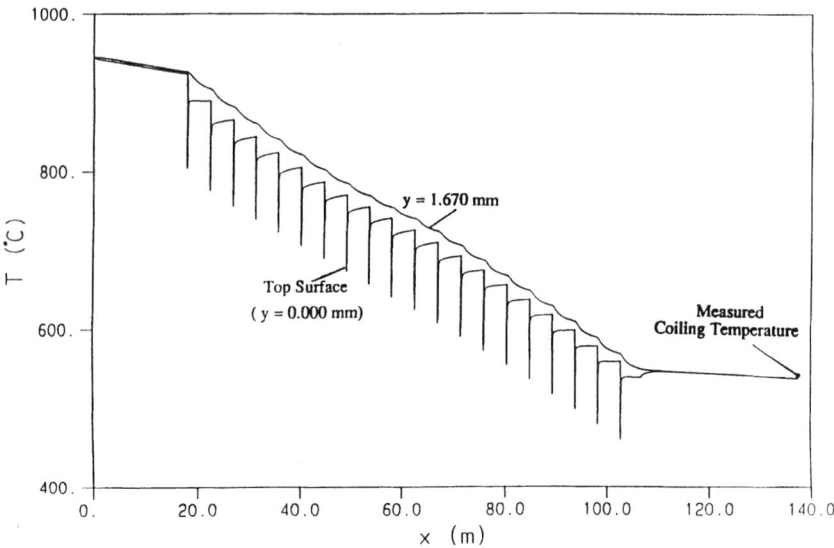

FIGURE 18.52   Top surface and core temperature distributions during quenching of a moving strip by planar water jets [207].

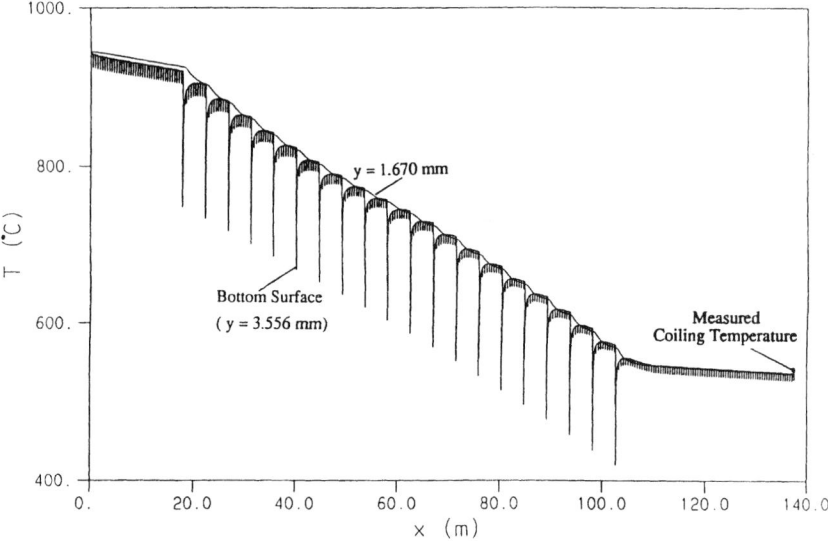

FIGURE 18.53   Bottom surface and core temperature distributions during quenching of a moving steel strip during quenching by an array of planar water jets [207].

region. The top surface temperature continues to rise along the entire film boiling region. The bottom surface temperature behaves in a similar manner, but, due to the cooling effect of the transport rollers, temperature recovery is somewhat smaller.

The model is capable of accurately simulating thermal behavior of steel strip cooled by an array of impinging water jets. The findings provide useful insights to a metallurgist for assessing the mechanical and metallurgical properties of steel, which are influenced by microstructures such as the grain size, dislocation density, precipitates, alloying elements in solution, and the second phase. Variation of these characteristics offers many possibilities for improved steel performance.

Processing of Several Advanced Materials

As evident from the previous section, process design and analysis is achieved by numerical modeling. Modeling is beneficial when highly coupled, multimode heat transfer exists along with nonlinear interactions between heat transfer rates and system morphology. In this section, several applications involving advanced materials are discussed.

Oxide Crystal Growth.   Large-scale modeling of oxide crystal growth has been reviewed [208]. Volumetric radiative exchange, conduction, and melt convection is expected to occur during the synthesis of materials such as yttrium aluminum garnet (YAG), gadolinium gallium garnet (GGG), and sapphire (Al_2O_3), all of which melt at high temperature and have semitransparent solid phases.

Brandon and Derby [209, 210] modeled Bridgman growth of oxide crystals in the system shown in Fig. 18.54a. Convective and surface radiative exchange between the ampoule and furnace walls drives system response, but this exchange process is straightforward and not of primary concern. In their formulation for the system model, the energy equations were applied in the oxide and ampoule materials, while buoyancy-induced melt convection was accounted for [210]. The melt and ampoule were considered to be opaque, while the crystal was treated as a gray absorbing and emitting medium. Heating and cooling was established by the thermal conditions shown in Fig. 18.54b, with $T_h = 2443$ K and $T_c = 2043$ K, bracketing the melting temperature $T_m = 2243$ K. The 150-mm-long ampoule with a 19.4-mm outer diameter and 3.2-mm wall thickness was pulled at a rate of 3.6 mm/h.

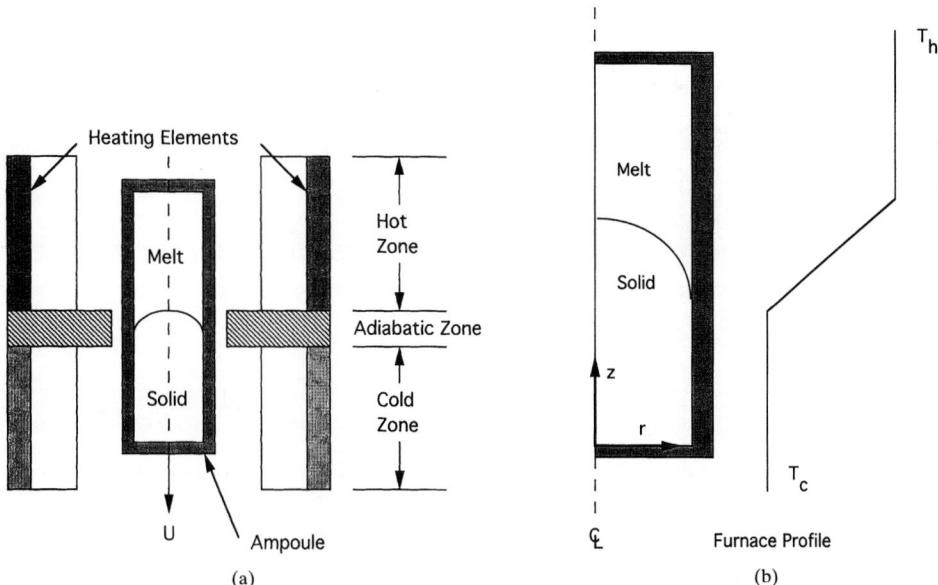

FIGURE 18.54   Model of a vertical Bridgman oxide crystal growth system showing (a) the system schematic and (b) the mathematical description. Reprinted from S. Brandon and J. J. Derby, "Heat Transfer in Vertical Bridgman Growth of Oxides: Effects of Conduction, Convection, and Internal Radiation," *Journal of Crystal Growth*, 121, pp. 473–494, 1992, with kind permission of Elsevier Science—NL, Sara Burgerhartstaat 25, 1055KV, The Netherlands.

Because of the high sensitivity of the crystal radiative properties to various dopants [211, 212] simulations were performed by parametrically varying the absorption coefficient of the solid. A gray-diffusive type analysis was used to estimate radiation heat transfer rates. As the solid becomes more transparent, the thermal resistance of the crystal is reduced, and the solid-liquid interface bulges upward, as shown in Fig. 18.54b. As the axial heat transfer is enhanced via the crystal's increased transparency, buoyancy forces are increased in the melt, leading to more vigorous convective flow. The balance between conductive-radiative cooling and convective heating of the solid-liquid interface induces a highly contorted boundary due to the high Prandtl number of the melt, along with ampoule wall conduction. From the practical perspective, large interface curvature increases the potential for the onset of instabilities, development of high stresses, and faceting of the crystal.

As mentioned previously, many materials-processing applications involve matter whose thermophysical properties are unknown. Because the influence of trace dopants upon the crystal's optical properties is uncertain, and since the simulations suggested that worst-case scenarios (in terms of extreme solid-liquid interface curvature) were associated with purely transparent crystals, large-scale modeling of Czochralski-grown oxide crystals used the assumption that the crystal is either perfectly transparent or entirely opaque to thermal radiation [213, 214]. For the transparent case, the surfaces of the crystal are treated as opaque surfaces, coincident with the assumptions that radiatively participating dopants may be introduced into the growth process from the surrounding vapor phase and subsequently deposited at solid-liquid or solid-vapor interfaces.

Figure 18.55 shows the predicted steady-state temperature and meridional flow streamlines for (a) no melt convection and stationary crystal, (b) no crystal rotation with buoyancy-induced melt flow, (c) modest rotation and mixed melt convection, and (d) high rotation rates with mixed melt convection [213]. The emissivity of the crystal-vapor interface of the system was specified to be 0.3, while the melt-crystal interface emissivity was set to 0.9 for the GGG crystal of refractive index 1.8. Additional property values and geometric details are listed elsewhere [215]. The crucible diameter is 200 mm, the crystal diameter is 100 mm, and thermocapillary convection was not included in the analysis.

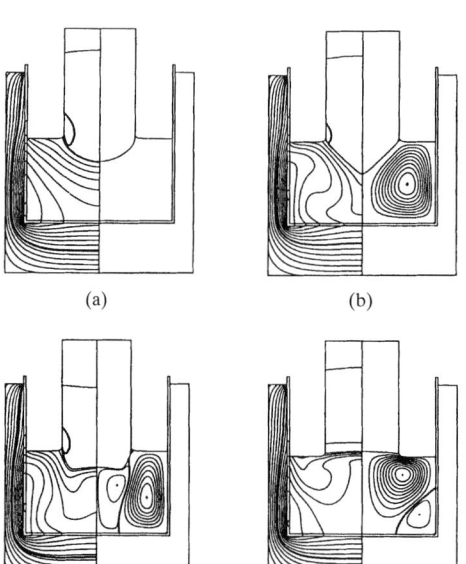

(a)   (b)

(c)   (d)

FIGURE 18.55 Temperatures (left) and streamlines (right) for GGG. Results are for (a) conduction and radiation only, (b) conduction, radiation, and buoyancy-induced melt convection, (c) conduction, radiation, buoyancy-induced convection, and modest crystal rotation, and (d) the same as (c), but with higher crystal rotation speed. Reprinted from Q. X. Xiao and J. J. Derby, "The Role of Internal Radiation and Melt Convection in Czochralski Oxide Growth: Deep Interfaces, Interface Inversion, and Spiraling," *Journal of Crystal Growth*, 128, pp. 188–194, with kind permission of Elsevier Science—NL, Sara Burgerhartstaat 25, 1055KV, The Netherlands.

The solid-liquid interface in Fig. 18.55a bulges into the melt because of the transparency of the crystal. Buoyancy-induced melt convection (Fig. 18.55b) sharpens the interface, but the interface still protrudes downward. As crystal rotation is applied, as in Fig. 18.55c, mixed convection is induced within the high Prandtl number melt. Melt convection, in conjunction with solid-phase conduction and radiation, flattens the interface to a shape similar to those observed experimentally upon crystal rotation. The interface is further flattened with increased rotation, as in Fig. 18.55d.

Although the radiation properties for the materials were not well known, the analysis is still valuable in that it clearly shows that volumetric radiative heat transfer may contribute to crystal degradation (large interface curvature). This example demonstrates that bracketing of actual system response can often be achieved in multimode materials processing problems with large-scale modeling, even without knowledge of particular thermophysical properties.

Rapid Thermal Processing of Silicon Wafers.   As mentioned previously, relevant length scales may span orders of magnitude, significantly complicating the thermal analysis. Rapid thermal processing (RTP) is a silicon processing technology used to perform thermal operations in integrated circuit fabrication such as annealing, oxidation, or chemical vapor deposition on a single wafer. RTP offers orders-of-magnitude faster throughput than conventional processing. The primary problem hindering broader application of RTP is control of thermal uniformity on the processed wafer and repeatability [216, 217].

A typical RTP factor (size 0 [1 m]) is shown in Fig. 18.56a. The thermal response of the silicon wafer is driven by radiation, with conduction and convection induced by radiative heating. The wafer's local radiative heating rate results from an interplay between (primarily) surface radiation effects and is determined by an imbalance between irradiation, emission, reflection, and, possibly, some transmission. Surfaces exhibit specular behavior, while the indirect heat sources (lamps) and associated windows have strongly wavelength-dependent absorption characteristics [217].

A desired temperature trajectory for the silicon wafer is shown in Fig. 18.56b [218]. Although temperatures vary depending upon the particular operation, it may be desirable to increase the wafer temperature from room temperature to as high as 1100°C at a rate of 20 to 50°C/s. To minimize thermal stresses or deposit reactive gases uniformly on the wafer (and, in turn, allow fabrication of even smaller submicron devices), thermal uniformity of (0 ± 2°C) is desired *at any time* during RTP.

Some efforts have been made to model RTP on a large scale [219]. Despite the preponderance of specular and spectral effects, diffuse and gray surfaces are typically assumed, since

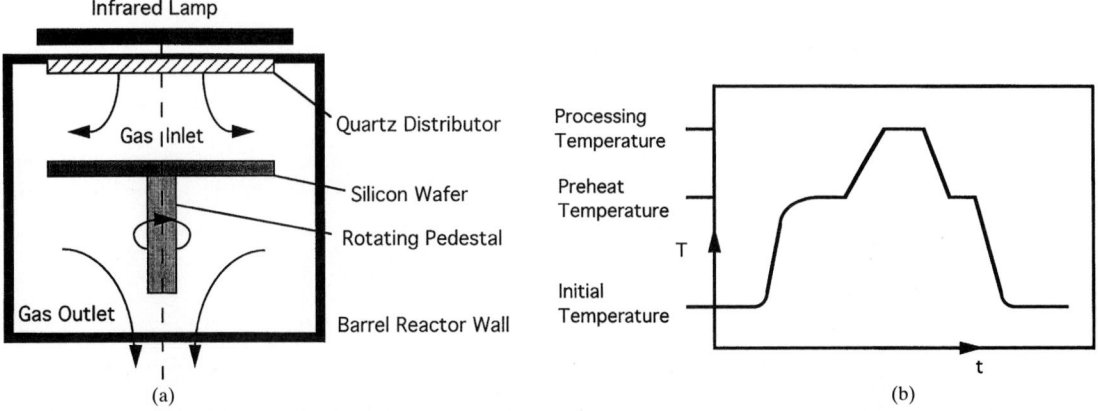

FIGURE 18.56   Shown are (*a*) typical RTP setup and (*b*) a typical wafer temperature trajectory.

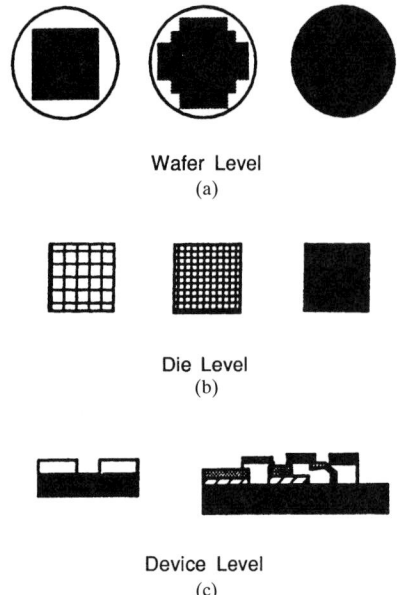

FIGURE 18.57 Patterns on the (*a*) wafer level, (*b*) die level, and (*c*) device level; published with permission from ASME International.

the gas flow and reaction rates as well as wafer conduction are coupled to the radiation solution. Complicating matters is the fact that the system's macroscopic thermal response is affected by microscale phenomena. The active (irradiated) wafer top is populated by submicron electronic devices. Top side patterns (square, diamond, and uniform coverage) are due to various, multilayered thin film structures ingrained upon the wafer. Patterns exist, therefore, at the wafer scale (0.25 m), die (chip) scale (100 μm), and device scale (1 μm), as shown in Fig. 18.57 [220].

Several studies have considered micro-macroscale surface radiative coupling as applied to RTP [221, 222]. The effective emissivity of the wafer depends on the thin film structure; therefore, the macroscale response is affected by the wafer level die pattern. Wong et al. [220] were concerned with the two-dimensional (r, θ) thermal response of the wafer induced by uniform, gray irradiation ($\varepsilon = 0.3$) from a 3000-K source. Simultaneous heat losses from the back side and edges of the wafer to 300 K surroundings induce a steady-state wafer temperature near 1200 K. The wafer (250 mm diameter and 500 μm thick) was populated with devices in square (Fig. 18.57*a*, left panel) and uniform (Fig. 18.57*a*, right panel] patterns. Using a one-dimensional thin-film radiative transfer analysis (solution of Maxwell's equations at device-level interfaces to account for coherent thin-film interference effects associated with the spectral distribution of the wafer emission) in the silicon dioxide region along with straightforward proportionalities (based on the assumption that the lateral device dimensions are large relative to the characteristic radiation wavelengths), the emissivity of the patterned wafer scale regions could be estimated.

For an oxide (device) thickness of 0.5 μm and a device population density of 0.8, predicted steady-state wafer temperature distributions are shown in Figs. 18.58*a* and *b*. The isotherms of

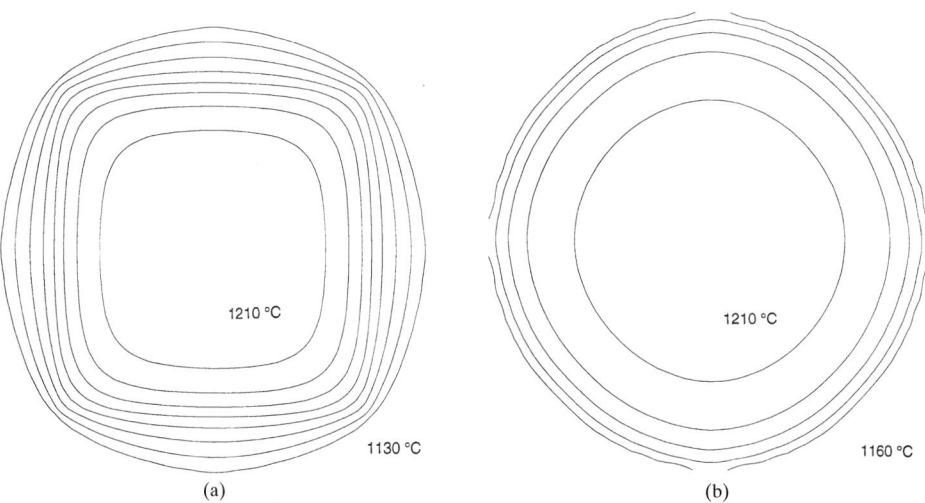

FIGURE 18.58 Temperature distributions (*a*) in the square region of Fig. 18.57 and (*b*) the same region, but with uniform wafer device coverage, from Wong et al. [219]; published with permission from ASME International.

Fig. 18.58*a* evolve in response to the square patterned region of Fig. 18.57*a*, while Fig. 18.58*b* shows the same region for the uniform coverage case. The square macroscale pattern clearly affects temperature uniformity (±88°C) relative to the uniform distribution case (±33°C). This example demonstrates that broad ranges of relevant length scales can be spanned with an approximate analysis and that micro-macroscale coupling can result in significant modification to system thermal response.

CONCLUDING REMARKS

The study of heat transfer related to materials processing is relatively new and rapidly developing. Some progress has been made during the last decade in understanding a few of the very many problems; however, much remains to be learned. The array of fundamental and practical design-related problems is extremely broad, and many could not even be mentioned in this very limited account. There are relatively few predictive equations useful for engineering design because of the inherent nonlinearities and conjugate heat transfer effects present in most real systems. Although there are some fundamental and generic studies that give insight into the thermal phenomena associated with processing of materials, many thermal phenomena related to specific materials processing and manufacturing technologies must be tackled on a system level, which requires simulation of multiple effects using computational approaches. It is with these limitations and opportunities in mind that this chapter was written. The authors hope that the work will be useful to those involved with thermal processing of materials. We thank our colleagues, who have provided many helpful suggestions.

NOMENCLATURE

Symbol, Definition, Units

a	thermal admittance, $J/(m^2 \cdot K \cdot s^{1/2})$
A	coefficient, defined as used
A	aspect ratio = L/H
A	area, m^2
B	coefficient, defined as used
c	specific heat, $J/kg \cdot K$
c_p	specific heat at constant pressure, $J/kg \cdot K$
C	contact conductance, $W/m^2 \cdot K$
C	coefficient, defined as used
d	nozzle or droplet diameter, m
$d*$	beam penetration thickness, m
d_L	load thickness, m
D	heater diameter, m
$e_{\lambda,b}$	blackbody spectral emitted flux, $W/m^2/\mu m$
$f(z)$	depth function for volumetric beam absorption, Eq. 18.3
F	functional relationship, defined as used
F_{ij}	radiation view factor between diffuse surfaces i and j
g	gravitational acceleration, m/s^2
$g(t)$	periodic pulsing function, Eq. 18.9

G	functional relationship, defined as used
G_c	channel crossflow mass velocity, kg/s·m²
G_j	jet mass velocity, kg/s·m²
Gr_l	Grashof number $= g\beta\Delta Tl^3/\nu^2$
Gr_l^*	modified Grashof number $= g\beta q''l^4/k\nu^2$
h	local heat transfer coefficient, W/m²K
$\overline{h}$	average heat transfer coefficient, W/m²K
H	latent heat of fusion, J/kg
H	nozzle-to-plate spacing, m
H	enclosure height, m
i_{fg}	latent heat of vaporization, J/kg
I	local beam intensity, W/m²
I_o	nominal beam intensity, W/m²
Ja	Jakob number $= c_{p,v}(T_s - T_{\mathrm{sat}})/i_{fg}$
k	thermal conductivity, W/m·K
K	complete elliptical integral of the first kind
K	optical thickness $= \kappa r$
l	characteristic length, m
L	length or thickness, m
m	geometrical wavelength, m^{-1}
m	constant, defined as used
n	constant, defined as used
N_r	dimensionless parameter, Eq. 18.69
Nu_l	local Nusselt number $= h/k$, defined as used
$\overline{\mathrm{Nu}_l}$	spanwise or steamwise average Nusselt number $= \overline{h}l/k$
$\overline{\overline{\mathrm{Nu}}}$	surface average Nusselt number $= \overline{h}l/k$
p	dimensionless pulsing period $= U_p/(\sqrt{2}r)$
p	period, s
P	total beam power, W
P	pressure, Pa
$\overline{P}_o$	dimensionless contact pressure $= (1 - \tilde{v})P/\overline{\mu}\beta T_{\mathrm{fus}}$
Pe	Peclet number $= \mathrm{Re} \cdot \mathrm{Pr}$, defined as used
Pr	Prandtl number $= \nu/\alpha$
$\lvert q''_{so}\rvert$	maximum source heat flux, W/m²
q''	heat flux, W/m²
q'''	volumetric source term, W/m³
Q	dimensionless heat flux $= q''/\lvert q''_{so}\rvert$
r	effective Gaussian beam radius, m
r	radial displacement from the jet stagnation point, m
r	coefficient, defined as used
r_x	axis of an ellipse, m
r_y	axis of an ellipse, m

$\overline{R}'$	dimensionless sensitivity of contact resistance to pressure $= mk\partial R_c/\partial \overline{P}_o$
R_c	contact resistance, K/W
Re_l	local Reynolds number $= ul/\nu$, defined as used
Re_j	jet Reynolds number based upon jet mass velocity $= V_j d/\nu$
Re_∞	local Reynolds number based upon free stream velocity $= U_\infty x/\nu$
Re_s	local Reynolds number based upon U_s
S	jet center-to-center spacing, m
St	Stanton number $= h/\rho c_p U_i$
Ste_s	solid-phase Stefan number $= H/c_s(T_{\mathrm{fus}} - T)$
t	time, s
T	temperature, K
T_f	fluid temperature, K
T_{fus}	fusion temperature, K
T_H	excess radiation temperature, K
T_j	jet temperature, K
T_m	melting temperature, K
T_o	initial temperature, K
T_s	solid temperature, K
T_v	vapor temperature, K
u	characteristic velocity, m/s
U	imposed workpiece or free stream velocity, m/s
U_∞	overall heat transfer coefficient, W/m²·K
V_i	velocity component in direction i, m/s
V_j	jet velocity at the point of impingement, m/s
V_n	jet velocity at the nozzle exit, m/s
$\overline{V}$	volumetric spray flux, m³/m²·s
w	nozzle width, m
x	Cartesian coordinate
x_n	nozzle separation distance in the x direction, m
y	Cartesian coordinate
y_n	nozzle separation distance in the y direction, m
z	Cartesian coordinate

Greek Symbols

α	thermal diffusivity, m²/s
α	thermal radiation absorptivity
β	thermal expansion coefficient, K⁻¹
β^*	beam shape $= r_y/r_x$
γ	inclination angle, degrees or radians
Γ	temperature ratio $= T_H/T_{Li}$
δ	position of an interface, m
δ^*	distance from beam center, m

ε	emissivity
ε'_n	normal total emissivity
ζ	dimensionless pressure perturbation $= 2m\alpha\rho H/q''_o$
η	dimensionless excess temperature, Eq. 18.7
η	effectiveness, Eq. 18.33
θ	angular coordinate direction
θ	excess temperature, defined as used
κ	extinction coefficient, m^{-1}
λ	wavelength of thermal radiation, μm
$\bar{\mu}$	shear modulus, MPa
ν	kinematic viscosity, $m^2\,s^{-1}$
$\tilde{\nu}$	Poisson's ratio
ρ	density, kg/m^3
ρ	reflectivity
σ	Pa stress or surface tension, N/m
σ	Stefan-Boltzmann constant, W/m^2K^4
σ_o	RMS surface roughness, m
τ	dimensionless time, defined as used
ϕ	dimensionless material temperature $= (T - T_o)/(T_{max} - T_o)$
φ	potential of thermal-elastic shift, m^2/s

Subscripts

c	cold
con	convection
cw	cooling water
g	gas
i	inlet
ℓ	liquid
L	load
max	maximum
min	minimum
o	jet stagnation
rad	radiation
s	solid or sheet or source
sat	saturation
tot	total
v	vapor
w	wall or surface
λ	spectral quantity per wavelength interval
ν	spectral quantity per frequency interval
∞	free stream value
0	average value
1	perturbation amplitude

REFERENCES

1. K. Samanta, et al. (eds.), *Interdisciplinary Issues in Materials Processing and Manufacturing,* ASME, New York, 1987.

2. S. I. Guceri (ed.), *Proceedings of the First International Conference on Transport Phenomena in Processing,* Technomic Publishing Co., Inc., Lancaster, Pennsylvania, 1993.

3. I. Tanasawa and N. Lior (eds.), *Heat and Mass Transfer in Materials Processing,* Hemisphere, New York, 1992.

4. M. Charmchi, et al. (eds.), *Transport Phenomena in Materials Processing and Manufacturing,* ASME, New York, 1992.

5. R. L. Shah, et al. (eds.), *Thermomechanical Aspects of Manufacturing and Materials Processing,* Hemisphere, New York, 1992.

6. K. F. Wang, S. Chandrasekhar, and H. T. Y. Yang, "An Efficient 2D Finite Element Procedure for the Quenching Analysis with Phase Change," *J. Eng. Industry,* 115, pp. 124–138, 1993.

7. J. F. Ready, *Industrial Applications of Lasers,* Academic Press, New York, 1978.

8. J. Mazumder, "Laser Heat Treatment: The State of the Art," *J. Metals,* 35, pp. 18–26, 1983.

9. D. Bauerle, *Chemical Processing with Lasers,* Springer, Berlin, 1986.

10. M. von Allmen and A. Blatter, *Laser-Beam Interactions with Materials,* Springer, Berlin, 1995.

11. M. I. Boulos, "Thermal Plasma Processing," *IEEE Trans. Plasma Sci.,* 9(6), pp. 1078–1089, 1991.

12. W. Steen, *Laser Material Processing,* Springer-Verlag, London, 1991.

13. C. J. Cremers, "Plasmas in Manufacturing and Processing," in R. K. Shah et al. (eds.), *Thermomechanical Aspects of Manufacturing and Materials Processing,* pp. 97–116, Hemisphere, New York, 1992.

14. J. Mazumder, "Applications of Lasers in Materials Processing," in R. K. Shah et al. (eds.), *Thermomechanical Aspects of Manufacturing and Materials Processing,* Hemisphere, New York, pp. 305–318, 1992.

15. W. H. Giedt, "Thermal Analysis of the Electron Beam Welding Process," in R. K. Shah et al. (eds.), *Thermomechanical Aspects of Manufacturing and Materials Processing,* pp. 349–358, Hemisphere, New York, 1992.

16. D. Rosenthal, "The Theory of Moving Sources of Heat and Its Application to Metal Treatments," *Trans. ASME,* 68, pp. 849–865, 1946.

17. N. R. DesRuisseaux and R. D. Zerkle, "Temperature in Semi-Infinite and Cylindrical Bodies Subjected to Moving Heat Sources and Surface Cooling," *J. Heat Transfer,* 92, pp. 456–464, 1970.

18. Y. I. Nissim, A. Lietoila, R. B. Gold, and J. F. Gibbons, "Temperature Distributions Produced in Semiconductors by a Scanning Elliptical or Circular CW Laser Beam," *J. Appl. Phys.,* 51, pp. 274–279, 1980.

19. K. Brugger, "Exact Solutions for the Temperature Rise in a Laser-Heated Slab," *J. Appl. Phys.,* 43, pp. 577–583, 1972.

20. M. Lax, "Temperature Rise Induced by a Laser Beam," *J. Appl. Phys.,* 48, pp. 3919–3924, 1977.

21. H. E. Cline and T. R. Anthony, "Heat Treating and Melting Material with a Scanning Laser or Electron Beam," *J. Appl. Phys.,* 48, pp. 3895–3900, 1977.

22. M. L. Burgener and R. E. Reedy, "Temperature Distributions Produced in a Two-Layer Structure by a Scanning CW Laser or Electron Beam," *J. Appl. Phys.,* 53, pp. 4357–4363, 1982.

23. J. E. Moody and R. H. Hendel, "Temperature Profiles Induced by a Scanning CW Laser Beam," *J. Appl. Phys.,* 54, pp. 4364–4371, 1982.

24. I. C. Sheng and Y. Chen, "Thermoelastic Analysis for a Semi-Infinite Plane Subjected to a Moving Gaussian Heat Source," *J. Thermal Stresses,* 14, pp. 129–141, 1991.

25. H. S. Carslaw and J. C. Jaeger, *Conduction of Heat in Solids,* 2d ed., Oxford University Press, Oxford, 1959.

26. M. F. Modest and H. Abakians, "Heat Conduction in a Moving Semi-Infinite Solid Subjected to Pulsed Laser Irradiation," *J. Heat Transfer,* 108, pp. 597–601, 1986.

27. D. J. Sanders, "Temperature Distributions Produced by Scanning Gaussian Laser Beams," *Appl. Optics,* 23, pp. 30–35, 1984.

28. B. Haba, B. W. Hussey, and A. Gupta, "Temperature Distribution During Heating Using a High Repetition Rate Pulsed Laser," *J. Appl. Phys.,* 69, pp. 2871–2876, 1991.

29. R. Festa, O. Manca, and N. Vincenzo, "A Comparison between Models of Thermal Fields in Laser and Electron Beam Surface Processing," *Int. J. Heat Mass Transfer,* 31, pp. 99–106, 1988.

30. E. H. Sin, C. K. Ong, and H. S. Tan, "Temperature Dependence of Interband Optical Absorption of Silicon at 1152, 1064, 750, and 694 nm," *Physica Status Solidi,* 85, pp. 199–204, 1984.

31. M. Miyao, K. Ohyu, and T. Tokuyama, "Annealing of Phosphorus-Ion-Implanted Silicon Using a CO_2 Laser," *Appl. Phys. Lett.,* 35, pp. 227–229, 1979.

32. A. Bhattacharyya and B. G. Streetman, "Theoretical Considerations Regarding Pulsed CO_2 Laser Annealing of Silicon," *Solid State Comm.,* 36, pp. 671–675, 1980.

33. T. Miyazaki and W. H. Giedt, "Heat Transfer from an Elliptical Cylinder Moving Through an Infinite Plate Applied to Electron Beam Processing," *Int. J. Heat Mass Transfer,* 25, pp. 807–814, 1982.

34. W. H. Giedt and L. N. Tallerico, "Prediction of Electron Beam Depth of Penetration," *Welding Research,* pp. 299-s–305-s, December 1988.

35. J. Dowden, M. Davis, and P. Kapadia, "Some Aspects of the Fluid Dynamics of Laser Welding," *J. Fluid Mech.,* 126, pp. 123–146, 1983.

36. M. Davis, P. Kapadia, and J. Dowden, "Solution of a Stefan Problem in the Theory of Laser Welding by Method of Lines," *J. Comp. Phys.,* 60, pp. 534–548, 1985.

37. Y. F. Hsu and B. Rubinsky, "Two-Dimensional Heat Transfer Study on the Keyhole Plasma Arc Welding Process," *Int. J. Heat Mass Transfer,* 31, pp. 1409–1421, 1988.

38. C.-J. Kim, S. Kauh, S. T. Ro, and S. L. Joon, "Parametric Study of the Two-Dimensional Keyhole Model for High Power Density Welding Processes," *J. Heat Transfer,* 116, pp. 209–214, 1994.

39. C. P. Grigoropoulos, "Heat Transfer in Laser Processing of Thin Films," in C. L. Tien (ed.), *Annual Review of Heat Transfer,* 5, pp. 77–130, 1994.

40. J. R. Ho, C. P. Grigoropoulos, and J. A. C. Humphrey, "Computational Study of Heat Transfer and Gas Dynamics in the Pulsed Laser Evaporation of Metals," *J. Appl. Phys.,* 78, pp. 4696–4709, 1995.

41. T. D. Bennet and C. P. Grigoropoulos, "Near-Threshold Laser Sputtering of Gold," *J. Appl. Phys.,* 77, pp. 849–864, 1995.

42. X. Xu, C. P. Grigoropoulos, and R. E. Russo, "Transient Heating and Melting Transformations in Argon-Ion Laser Irradiation of Polysilicon Films," *J. Appl. Phys.,* 73, pp. 8088–8096, 1993.

43. X. Xu, C. P. Grigoropoulos, and R. E. Russo, "Heat Transfer in Excimer Laser Melting of Thin Polysilicon Layers," *J. Heat Transfer,* 117, pp. 708–715, 1995.

44. M. F. Modest, "Three-Dimensional, Transient Model for Laser Machining of Ablating/Decomposing Materials," *Int. J. Heat Mass Transfer,* 39, pp. 221–234, 1996.

45. J. G. Lenard and M. Pietrzyk, "Rolling Process Modelling," in P. Hartly, P. Pillinger, I. Struyes, and C. Struyes (eds.), *Numerical Modelling of Material Deformation Processes,* pp. 274–302, Springer-Verlag, London, 1992.

46. A. A. Tseng and P. F. Sun, "Modeling of Polymer Processing and Its Application to Calendaring," in R. K. Shah et al. (eds.), *Thermomechanical Aspects of Manufacturing and Materials Processing,* pp. 377–387, Hemisphere Publishing Company, New York, 1992.

47. M. Pietrzyk and J. G. Lenard, "Boundary Conditions in Hot/Cold Flat Rolling," in *Proceedings of COMPLAS '89,* Barcelona, 1989.

48. M. Pietrzyk and J. G. Lenard, "Experimental Substantiation of Modelling Heat Transfer in Hot Flat Rolling," in H. R. Jacobs (ed.), *Proceedings of the 1988 National Heat Transfer Conference,* HTD-96, 3, pp. 47–53, ASME, New York, 1988.

49. A. N. Karagiozis and J. G. Lenard, "Temperature Distribution in a Slab During Hot Rolling," *J. Eng. Mat. Techn.,* 110, pp. 17–21, 1988.

50. T. E. Voth, *Thermomechanical Compressive Pre-Melting Response of Solder Spheres,* Ph.D. Thesis, University of Texas at Austin, 1995.

51. M. D. Bryant, "Thermoelastic Solutions for Thermal Distributions Moving over Half Space Surfaces and Application to the Moving Heat Source," *J. Appl. Mech.,* 55, pp. 87–92, 1988.

52. F. Kostrubiec and M. Walczak, "Thermal Stresses in the Tungsten and Molybdenum Surface Layer Following Laser Treatment," *J. Mat. Sci. Let.,* 13, pp. 34–36, 1994.

53. K. Ho and R. D. Pehlke, Metal-Mold Interfacial Heat Transfer, *Metall. Trans. B,* vol. 16B, pp. 585–594, 1985.

54. N.-Y. Li and J. R. Barber, "Sinusoidal Perturbation Solutions for Planar Solidification," *Int. J. Heat Mass Transfer,* 32, pp. 935–941, 1989.

55. O. Richmond, L. G. Hector, and J. M. Fridy, "Growth Instability during Nonuniform Directional Solidification of Pure Metals," *J. Appl. Mech.,* 57, pp. 529–536, 1990.

56. F. Yigit, N.-Y. Li, and J. R. Barber, "Effect of Thermal Capacity on Thermoelastic Instability during the Solidification of Pure Metals," *J. Therm. Stresses,* 16, pp. 285–309, 1993.

57. N.-Y. Li and J. R. Barber, "Thermoelastic Instability in Planar Solidification," *Int. J. Mech. Sci.,* 33, pp. 945–959, 1991.

58. Z. Tadmor and I. Klein, *Engineering Principles of Plasticating Extrusion, Polymer Science and Engineering Series,* Van Nostrand Reinhold, New York, 1970.

59. E. A. Fisher, *Extrusion of Plastics,* Wiley, New York, 1976.

60. T. Altan, S. Oh, and H. Gegel, *Metal Forming Fundamentals and Applications,* American Society of Metals, Metals Park, Ohio, 1979.

61. M. V. Karwe and Y. Jaluria, "Numerical Simulation of Thermal Transport Associated with a Continuously Moving Flat Sheet in Rolling or Extrusion," in H. R. Jacobs (ed.), *Proceedings of the National Heat Transfer Conference,* HTD-Vol. 96, pp. 37–45, ASME, New York, 1988.

62. Y. Jaluria, "Transport from Continuously Moving Materials Undergoing Thermal Processing," in C. L. Tien (ed.), *Annual Review of Heat Transfer,* 4, pp. 187–245, Hemisphere, Washington, DC, 1992.

63. H-T. Lin and S-F. Huang, "Flow and Heat Transfer of Plane Surfaces Moving in Parallel and Reversely to the Free Stream," *Int. J. Heat Mass Transfer,* 37, pp. 333–336, 1994.

64. J. Vleggaar, "Laminar Boundary-Layer Behavior on Continuous Accelerating Surfaces," *Chem. Eng. Sci.,* 32, pp. 1517–1525, 1977.

65. W. H. H. Banks, "Similarity Solutions of the Boundary-Layer Equations for a Stretching Wall," *J. Mécan. Theor. Appli.,* 2, pp. 375–392, 1983.

66. L. G. Grubka and K. M. Bobba, "Heat Transfer Characteristics of a Continuous Stretching Surface with Variable Temperature," *J. Heat Transfer,* 107, pp. 248–250, 1985.

67. M. E. Ali, "Heat Transfer Characteristics of a Continuously Stretching Surface," *Wärme-und Stoffübertragung,* 29, pp. 227–234, 1994.

68. V. M. Soundalgekar and T. V. Ramamurthy, "Heat Transfer Past a Continuous Moving Plate with Variable Temperature," *Wärme -und Stoffübertragung,* 14, pp. 91–93, 1980.

69. M. E. Ali, "On Thermal Boundary Layer on a Power-Law Stretched Surface with Suction or Injection," *Int. J. Heat Fluid Flow,* 16, pp. 280–290, 1995.

70. A. M. Jacobi, "A Scale Analysis Approach to the Correlation of Continuous Moving Sheet (Backward Boundary Layer) Forced Convective Heat Transfer," *J. Heat Transfer,* 115, pp. 1058–1061, 1993.

71. N. Ramachandran, T. S. Chen, and B. F. Armaly, "Correlation for Laminar Mixed Convection in Boundary Layers Adjacent to Horizontal, Continuous Moving Sheets," *J. Heat Transfer,* 104, pp. 1036–1039, 1987.

72. J. Filipovic, R. Viskanta, and F. P. Incropera, "Analysis of Momentum and Heat Transfer for Turbulent Flow over a Moving Isothermal Surface," *J. Mat. Proces. Manuf. Sci.,* 1, pp. 157–168, 1992.

73. H-T. Lin, K-Y. Wu, and H-L. Hoh, "Mixed Convection from an Isothermal Horizontal Plate Moving in Parallel or Reversely to a Free Stream," *Int. J. Heat Mass Transfer,* 36, pp. 3547–3554, 1993.

74. F. K. Tsou, E. M. Sparrow, and R. J. Goldstein, "Flow and Heat Transfer in the Boundary Layer on a Continuous Moving Surface," *Int. J. Heat Mass Transfer,* 10, pp. 219–235, 1967.

75. J. B. Klemp and A. Acrivos, "A Method for Integrating the Boundary-Layer Equations Through a Region of Reverse Flow," *J. Fluid Mech.,* 53, pp. 177–191, 1972.

76. M. V. A. Bianchi and R. Viskanta, "Momentum and Heat Transfer on a Continuous Flat Surface Moving in a Parallel Counterflow Free Stream," *Wärme-und Stoffübertragung,* 29, pp. 89–94, 1993.

77. S. W. Churchill and R. Usagi, "A Standardized Procedure for the Production of Correlations in the Form of a Common Empirical Equation," *Ind. Eng. Chem. Fund.,* 13, pp. 39–46, 1974.

78. N. Ramachandran, T. S. Chen, and B. F. Armaly, "Mixed Convection from Vertical and Inclined Moving Sheets in a Parallel Freestream," *J. Thermophys. Heat Transfer,* 1, pp. 274–281, 1987.

79. R. Viskanta, "Heat Transfer from Continuously Moving Materials: An Overview of Selected Thermal Materials Processing Problems," in J. S. Lee, S. H. Chung, and K. H. Kim (eds.), *Transport Phenomena in Thermal Engineering,* 2, pp. 779–788, Begell House, New York, 1994.

80. M. V. Karwe and Y. Jaluria, "Thermal Transport from a Moving Surface," *J. Heat Transfer,* 108, pp. 728–733, 1986.

81. M. V. Karwe and Y. Jaluria, "Fluid Flow and Mixed Convection Transport from a Moving Plate in Rolling and Extrusion Processes," *J. Heat Transfer,* 110, pp. 655–661, 1988.

82. H. Martin, "Heat and Mass Transfer between Impinging Gas Jets and Solid Surfaces," in J. P. Hartnett and T. F. Irvine, Jr. (eds.), *Advances in Heat Transfer,* 13, pp. 1–60, Academic Press, New York, 1977.

83. K. Jambunathan, E. Lai, A. Moss, and B. L. Button, "A Review of Heat Transfer Data for Single Circular Jet Impingement," *Int. J. Heat Fluid Flow,* 13, pp. 106–115, 1992.

84. R. Viskanta, "Heat Transfer to Impinging Isothermal Gas and Flame Jets," *Exp. Thermal Fluid Sci.,* 6, pp. 111–134, 1993.

85. T. J. Craft, L. J. W. Graham, and E. B. Launder, "Impinging Jet Studies for Turbulence Model Assessment—II. An Examination of the Performance of Four Turbulence Models," *Int. J. Heat Mass Transfer,* 36, pp. 2685–2697, 1993.

86. R. J. Goldstein, K. A. Sobolik, and W. S. Seol, "Effect of Entrainment on the Heat Transfer to a Heated Circular Air Jet," *J. Heat Transfer,* 112, pp. 608–611, 1990.

87. J. W. Baughn, A. E. Hechanova, and X. Yan, "An Experimental Study of Entrainment Effects on the Heat Transfer from a Flat Surface to a Heated Circular Jet," *J. Heat Transfer,* 113, pp. 1023–1025, 1991.

88. H. Klammer und W. Schupe, "Praktische Auslegung von Vorwärmekammern in Industrieöfen und der Wirschaflichkeit," *Stahl und Eisen,* 103, pp. 531–537, 1983.

89. R. J. Goldstein and W. S. Seol, "Heat Transfer to a Row of Impinging Circular Air Jets Including the Effect of Entrainment," *Int. J. Heat Mass Transfer,* 34, pp. 2133–2147, 1991.

90. R. Gardon and J. C. Akfirat, "Heat Transfer Characteristics of Impinging Two-Dimensional Air Jets," *J. Heat Transfer,* 88, pp. 101–108, 1966.

91. A. M. Huber and R. Viskanta, "Effect of Jet-to-Jet Spacing on Convective Heat Transfer to Confined, Impinging Arrays of Axisymmetric Air Jets," *Int. J. Heat Mass Transfer,* 37, pp. 2859–2869, 1994.

92. L. W. Florschuetz, C. R. Truman, and D. E. Metzger, "Streamwise Flow and Heat Transfer Distributions for Jet Array Impingement with Crossflow," *J. Heat Transfer,* 103, pp. 337–342, 1981.

93. N. T. Obot and T. A. Trabold, "Impingement Heat Transfer Within Arrays of Circular Air Jets. Part I. Effects of Minimum, Intermediate and Complete Crossflow for Small and Large Spacings," *J. Heat Transfer,* 107, pp. 872–879, 1987.

94. R. Viskanta and F. P. Incropera, "Quenching with Liquid Jet Impingement," in I. Tanasawa and N. Lior (eds.), *Heat and Mass Transfer in Materials Processing,* pp. 455–476, Hemisphere, New York, 1992.

95. W. B. Webb and C.-F. Ma, "Single Phase Liquid Jet Impingement Heat Transfer," in J. P. Hartnett et al. (eds.), *Advances in Heat Transfer,* 26, pp. 105–217, Academic Press, San Diego, 1995.

96. J. H. Lienhard V, "Liquid Jet Impingement," in C. L. Tien (ed.), *Annual Review of Heat Transfer,* 5, pp. 199–269, Begell House, New York, 1995.

97. J. Stevens and B. W. Webb, "Measurements of Flow Structure in the Stagnation Zone of Impinging Free-Surface Jets," *Int. J. Heat Mass Transfer,* 36, pp. 4283–4286, 1993.

98. J. Stevens and B. W. Webb, "Local Heat Transfer Coefficients under an Axisymmetric Single-Phase Liquid Jet," *J. Heat Transfer,* 113, pp. 71–78, 1991.

99. P. Di Marco, W. Grassi, and A. Magrimi, "Unsubmerged Jet Impingement Heat Transfer at Low Liquid Speed," in G. F. Hewitt (ed.), *Proceedings of the 10th International Heat Transfer Conference,* 3, pp. 59–64, *IChemE,* Rugby, 1994.

100. Y. Pan, J. Stevens, and B. W. Webb, "Effect of Nozzle Configuration on Transport in the Stagnation Zone of Axisymmetric Impinging Free-Surface Liquid Jets. Part 2. Local Heat Transfer," *J. Heat Transfer,* 114, pp. 880–886, 1992.

101. S. Faggiani and W. Grassi, "Round Liquid Jet Impingement Heat Transfer: Local Nusselt Numbers in the Region with Non-Zero Pressure Gradient," in G. Hestroni (ed.) *Proceedings of the 9th International Heat Transfer Conference,* 4, pp. 197–202, Hemisphere, New York, 1990.

102. L. A. Gabour and J. H. Lienhard V, "Wall Roughness Effects of Stagnation Point Heat Transfer Beneath an Impinging Jet," *J. Heat Transfer,* 116, pp. 81–87, 1994.

103. D. H. Wolf, R. Viskanta, and F. P. Incropera, "Local Convective Heat Transfer from a Heated Surface to a Planar Jet of Water with a Nonuniform Velocity Profile," *J. Heat Transfer,* 112, pp. 899–905, 1990.

104. D. T. Vader, F. P. Incropera, and R. Viskanta, "Local Convective Heat Transfer from a Heated Surface to an Impinging, Planar Jet of Water," *Int. J. Heat Mass Transfer,* 34, pp. 611–623, 1991.

105. X. Liu, J. H. Lienhard V, and J. S. Lombara, "Convective Heat Transfer by Impingement of Circular Liquid Jets," *J. Heat Transfer,* 113, pp. 571–582, 1991.

106. D. C. McMurray, P. S. Meyers, and O. A. Uyehara, "Influence of Impinging Jet Variables on Local Heat Transfer Coefficients along a Flat Surface with Constant Heat Flux," in *Proceedings of the Third International Heat Transfer Conference,* 2, pp. 292–299, AIChE, New York, 1966.

107. D. A. Zumbrunnen, "Convective Heat and Mass Transfer in the Stagnation Region of a Laminar Planar Jet Impinging on a Moving Surface," *J. Heat Transfer,* 113, pp. 563–570, 1991.

108. D. A. Zumbrunnen, F. P. Incropera, and R. Viskanta, "A Laminar Boundary Layer Model of Heat Transfer Due to a Nonuniform Planar Jet Impinging on a Moving Surface," *Wärme und Stoffübertragung,* 27, pp. 311–319, 1992.

109. S. J. Chen and J. Kothari, "Temperature Distribution and Heat Transfer in a Moving Strip Cooled by a Water Jet," ASME Paper No. 88-WA/NE-4, ASME, New York, 1988.

110. D. H. Wolf, F. P. Incropera, and R. Viskanta, "Jet Impingement Boiling," in J. P. Hartnett et al. (eds.) *Advances in Heat Transfer,* 23, pp. 1–132, Academic Press, New York, 1993.

111. S. Ishigai, S. Nakanishi, and T. Ochi, "Boiling Heat Transfer for a Plane Water Jet Impinging on a Hot Surface," in J. T. Rogers et al. (eds.), *Heat Transfer 1978,* 1, pp. 445–450, Hemisphere, Washington, DC, 1978.

112. Y. Katto and K. Ishii, "Burnout in a High Heat Flux Boiling System with a Forced Supply of Liquid Through a Plane Jet," in *Heat Transfer 1978,* 1, pp. 435–440, Hemisphere, Washington, DC, 1978.

113. Y. Katto and M. Kunihiro, "Study of the Mechanism of Burn-out in Boiling System of High Burnout Heat Flux," *Bull. JSME,* 16, pp. 1357–1366, 1973.

114. M. Monde, "Burnout Heat Flux in Saturated Forced Convection Boiling with an Impingement Jet," *Heat Transfer—Japanese Research,* 9, pp. 31–41, 1980.

115. Y. Miyasaka, S. Inada, and Y. Owase, "Critical Heat Flux and Subcooled Nucleate Boiling in the Transition Region Between a Two-Dimensional Water Jet and a Heated Surface," *J. Chem. Eng. Japan,* 13, pp. 29–35, 1980.

116. S. Toda and H. Uchida, "Study of Liquid Film Cooling with Evaporation and Boiling," *Heat Transfer—Japanese Research,* 2(1), pp. 44–62, 1973.

117. D. H. Wolf, F. P. Incropera, and R. Viskanta, "Local Jet Impingement Boiling Heat Transfer," *Int. J. Heat Mass Transfer,* 39, pp. 1395–1406, 1996.

118. F. Gunther, "Photographic Study of Surface Boiling Heat Transfer to Water With Forced Convection," *Trans. Am. Soc. Mech. Engs.,* 73, pp. 115–123, 1951.

119. T. Ochi, S. Nakanishi, M. Kaji, and S. Ishigai, "Cooling of a Hot Plate with an Impinging Circular Water Jet," in *Multi-Phase Flow and Heat Transfer III. Part A: Fundamentals,* T. N. Veziroglu and A. E. Bergles (eds.), pp. 671–681, Elsevier, Amsterdam, 1984.

120. R. A. Nelson, "Mechanisms of Quenching Surfaces," in N. P. Cheremisinoff (ed.), *Handbook of Heat and Mass Transfer,* pp. 1103–1153, Gulf, Houston, 1986.

121. J. Kokado, N. Hatta, H. Takuda, J. Harada, and N. Yasuhira, "An Analysis of Film Boiling Phenomena of Subcooled Water Spreading Radially on a Hot Steel Plate," *Archiv für Eisenhüttenwesen,* 55, pp. 113–118, 1984.

122. V. V. Klimenko and S. Y. Snytin, "Film Boiling Crisis on a Submerged Heating Surface," *Exp. Thermal Fluid Sci.,* 3, pp. 467–479, 1990.

123. R. A. Owen and D. J. Pulling, "Wetting Delay: Film Boiling of Water Jets Impinging on Hot Flat Metal Surfaces," in T. N. Veziroglu (ed.), *Multiphase Transport,* 2, pp. 639–669, Hemisphere, Washington, DC, 1979.

124. R. Jeschar, R. Scholz, U. Reiners, and R. Maass, "Kühltechniken zur Thermischen Behandlung von Werkstoffen," *Stahl und Eisen,* 107, pp. 251–258, 1987.

125. M. Lamvik and B.-A. Iden, "Heat Transfer Coefficient by Water Jets on a Hot Surface," in U. Grigull et al. (eds.), *Heat Transfer 1982,* 1, pp. 369–375, Hemisphere, Washington, DC, 1982.

126. J. Filipovic, R. Viskanta, and F. P. Incropera, "Similarity Solution for Laminar Film Boiling over a Moving Isothermal Surface," *Int. J. Heat Mass Transfer,* 36, pp. 2957–2963, 1993.

127. J. Filipovic, R. Viskanta, and F. P. Incropera, "An Analysis of Subcooled Turbulent Film Boiling on a Moving Isothermal Surface," *Int. J. Heat Mass Transfer,* 37, pp. 2661–2673, 1994.

128. J. Filipovic, F. P. Incropera, and R. Viskanta, "Quenching Phenomena Associated with a Water Wall Jet: II. Comparison of Experimental and Theoretical Results for the Film Boiling Region," *Exp. Heat Transfer,* 8, pp. 119–130, 1995.

129. D. A. Zumbrunnen, R. Viskanta, and F. P. Incropera, "Analysis of Forced Convection Film Boiling on Moving Metallic Strips and Plates During Manufacture," *J. Heat Transfer,* 111, pp. 889–896, 1989.

130. M. Tajima, T. Maki, and K. Katayama, "Study of Heat Transfer Phenomena in Quenching of Steel (Effects of Boiling Heat Transfer on Cooling Curves and Water Temperature on Hardness of Steel)," *JSME Int. J., Series II,* 33, pp. 340–348, 1990.

131. M. Bamberger and B. Prinz, "Determination of Heat Transfer Coefficients during Water Cooling Metals," *Mat. Sci. Tech.,* 2, pp. 410–415, 1986.

132. H. R. Müller and R. Jeschar, "Heat Transfer during Water Spray Cooling of Nonferrous Metals," *Zeitschrift für Metallkunde,* 74, pp. 257–264, 1983.

133. J. K. Brimacombe, P. K. Agarwal, L. A. Baptista, S. Hobbins, and B. Prabhakar, "Spray Cooling in the Continuous Casting of Steel," in *Proceedings of the 63rd National Open Hearth and Basic Oxygen Steel Conference,* 63, pp. 235–252, The Iron and Steel Society of the American Institute of Mining, Metallurgical, and Petroleum Engineers, Washington, DC, 1980.

134. G. E. Totten, C. E. Bates, and N. A. Clinton, *Handbook of Quenchants and Quenching Technology,* American Society of Metals, Materials Park, Ohio, 1993.

135. S-C. Yao, "Dynamics and Heat Transfer of Impacting Sprays," in C. L. Tien (ed.), *Annual Review of Heat Transfer,* 5, pp. 351–383, CRC Press, Boca Raton, Florida, 1994.

136. K. J. Choi and S. C. Yao, "Heat Transfer Mechanisms of Horizontally Impacting Sprays," *Int. J. Heat Mass Transfer,* 30, pp. 311–318, 1987.

137. E. F. Bratuta and S. F. Kravtsov, "The Effect of Subcooling of Dispersed Liquid on Critical Heat Flux with Cooling of a Plane Surface," *Therm. Eng.,* 33, pp. 674–676, 1986.

138. U. Rainers, R. Jeschar, and R. Scholz, "Wärmeübertragung bei der Stranggusskühlung durch Spritzwasser," *Steel Research,* 60, pp. 442–450, 1989.

139. I. Mudawar and W. S. Valentine, "Determination of the Local Quench Curve for Spray-Cooled Metallic Surfaces," *J. Heat Treating,* 7, pp. 107–121, 1989.

140. W. P. Klinzing, J. C. Rozzi, and I. Mudawar, "Film and Transition Boiling Correlations for Quenching of Hot Surfaces with Water Sprays," *J. Heat Treating,* 9, pp. 91–103, 1992.

141. L. I. Urbanovich, V. A. Goryainov, V. V. Sevost'yanov, Y. G. Boev, V. M. Niskoviskikh, A. V. Grachev, A. V. Sevost'ynov, and V. S. Gru'ev, "Spray Cooling of High Temperature Metal Surfaces with High Water Pressures," *Steel in the USSR,* 11, pp. 184–186, 1981.

142. S. Deb and S. C. Yao, "An Efficient Model for Turbulent Sprays," *Int. J. Heat Mass Transfer,* 32, pp. 2099–2112, 1989.

143. T. Ito, Y. Tanaka, and Z-H. Liu, "Water Cooling of Hot Surfaces (Analysis of Fog Cooling in the Region Equivalent to Film Boiling)," *Trans. JSME,* Series 13, 15, pp. 805–813, 1989 (in Japanese).

144. S. C. Koria and I. Datta, "Studies on the Behavior of Mist Jet and Its Cooling Capacity of Hot Steel Surfaces," *Steel Research,* 63, pp. 19–26, 1992.

145. Y. Hayashi, "Mist Cooling for Air Heat Exchangers," in C.-L. Tien (ed.), *Annual Review of Heat Transfer,* 5, pp. 383–436, CRC Press, Boca Raton, Florida, 1994.

146. S. Nishio and H. Ohkubo, "Stability of Mist Cooling in Thermo-Mechanical Control Process of Steel," in *Heat and Mass Transfer in Materials Processing,* I. Tanasawa and N. Lior (eds.), pp. 477–488, Hemisphere, Washington, DC, 1992.

147. H. Ohkubo and S. Nishio, "Mist Cooling for Thermal Tempering of Glass," in P. J. Marto and I. Tanasawa (eds.), *Proceedings of the 1987 ASME-JSME Thermal Engineering Joint Conference,* 5, pp. 71–78, ASME, New York, 1987.

148. C. Beckermann and C. Y. Wang, "Multi-Phase/Scale Modeling of Transport Phenomena in Alloy Solidification," in C. L. Tien (ed.), *Annual Review of Heat Transfer,* Vol. VI, pp. 115–198, Begell House, New York, 1995.

149. S. Kavesh, "Principles of Fabrication," in R. L. Ashbrook (ed.), *Rapid Solidification Technologies Sourcebook,* pp. 73–106, American Society of Metals, Metals Park, Ohio, 1983.

150. J. E. Flinn, *Rapid Solidification Technology for Reduced Consumption of Strategic Materials,* pp. 29–32, Noyes, Park Ridge, New Jersey, 1985.

151. A. Rezaian and D. Poulikakos, "Heat and Fluid Flow Processes During Coating of a Moving Surface," *J. Thermophysics Heat Transfer,* 5, pp. 192–198, 1991.

152. R. Stevens and D. Poulikakos, "Freeze Coating of a Moving Substrate with a Binary Alloy," *Num. Heat Transfer,* A, 20, pp. 409–432, 1991.

153. F. B. Cheung, "The Thermal Boundary Layer on Continuous Moving Plate with Freezing," *J. Thermophysics Heat Transfer,* 1, pp. 335–342, 1987.

154. F. B. Cheung and S. W. Cha, "Finite Difference Analysis of Growth and Decay of Freeze Coat on a Continuous Moving Cylinder," *Num. Heat Transfer,* 12, pp. 41–56, 1987.

155. T. L. Bergman and R. Viskanta, "Radiation Heat Transfer in Manufacturing and Materials Processing," in *Radiative Transfer—I,* M. P. Mengüç (ed.), *Proceedings of the International Symposium on Radiative Heat Transfer,* pp. 13–39, Begell House, New York, 1996.

156. K. G. Hagen, "Using Infrared Radiation to Dry Coatings," *Tappi Journal,* 5, pp. 77–83, 1989.

157. R. Viskanta and E. E. Anderson, "Heat Transfer in Semitransparent Solids," in T. F. Irvine, Jr. and J. P. Hartnett (eds.), *Advances in Heat Transfer,* 11, pp. 317–441, Academic Press, 1975.

158. M. H. Cobble, "Irradiation into Transparent Solids and the Thermal Trap Effect," *J. Franklin Inst.* 278, pp. 383–393, 1964.

159. R. Siegel and J. R. Howell, *Thermal Radiation Heat Transfer,* 3d ed., Hemisphere, Washington, DC, 1992.

160. R. R. Brannon and R. J. Goldstein, "Emittance of Oxide Layers on a Metal Substrate," *J. Heat Transfer,* 92, pp. 257–263, 1970.

161. J. C. Phillips, "The Fundamental Optical Spectra of Solids," F. Seitz and D. Turnbull (eds.), in *Solid State Physics,* 18, pp. 55–164, Academic Press, New York, 1966.

162. O. J. Edwards, "Optical Transmittance of Fused Silica at Elevated Temperatures," *J. Amer. Opt. Soc.,* 56, p. 774, 1966.

163. E. C. Beder, C. D. Bass, and W. L. Shackleford, "Transmissivity and Absorption of Fused Quartz Between 0.22 µm and 3.5 µm from Room Temperature to 1500°C," *Appl. Optics,* 10, pp. 2263–2268, 1971.

164. A. S. Ginzburg, *Infra-Red Radiation in Food Processing,* CRC Press, Cleveland, 1969.

165. J. F. Ready, "Effects Due to Absorption of Laser Radiation," *J. Appl. Phys.,* 36, pp. 462–468, 1965.

166. A. Minardi and P. J. Bishop, "Computer Modeling of Thermal Coupling Resulting from Laser Irradiation of a Metal Target," in S. I. Guceri (ed.), *Proceedings of the First International Conference on Transport Phenomena in Processing,* Technomic Publishing Co., Inc. Lancaster, Pennsylvania, pp. 1371–1388, 1992.

167. F. P. Incropera and D. P. DeWitt, *Fundamentals of Heat and Mass Transfer,* 3d ed., John Wiley, New York, 1990.

168. M. Q. Brewster, *Thermal Radiative Transfer and Properties,* John Wiley, New York, 1992.

169. R. F. Modest, *Radiative Heat Transfer,* McGraw-Hill, New York, 1993.

170. J. R. Howell, K. S. Ball, and T. L. Bergman, "Fundamentals and Applications of Radiative Heat Transfer: Implications for RTP," in R. B. Fair and B. Lojek (eds.), *RTP '94 Proceedings of the 2nd International Rapid Thermal Processing Conference,* pp. 9–13, RTP '94, 1994.

171. A. F. Emery, O. Johansson, M. Lobo, and A. Abrous, "A Comparative Study of Methods for Computing the Diffuse Radiation Viewfactors for Complex Structures," *J. Heat Transfer,* 113, pp. 413–422, 1991.

172. R. C. Corlett, "A User's Manual for Radsim-PC," University of Washington, Seattle, Washington, 1988.

173. A. F. Emery, *VIEWC Users Manual,* University of Washington, Seattle, Washington, 1988.

174. M. F. Cohen, D. P. Greenberg, P. J. Brock, and D. S. Immel, "An Efficient Radiosity Approach for Realistic Image Synthesis," *IEEE CG&A,* 6, pp. 26–35, 1986.

175. A. Shapiro, "FACET—A Radiation Viewfactor Computer Code for Axisymmetric 2D Planar and 3D Geometries with Shadowing," University of California, Lawrence Livermore National Laboratory, UCID-19887, 1983.

176. J. S. Toor and R. Viskanta, "Effect of Direction Dependent Properties on Radiation Interchange," *J. of Spacecraft and Rockets,* 5, pp. 742–743, 1968.

177. S. Maruyama, "Radiation Heat Transfer Between Arbitrary Three-Dimensional Bodies with Specular and Diffuse Surfaces," *Num. Heat Transfer,* 24, pp. 181–196, 1993.

178. L. Zhao, *Computational Study of Radiation and Mixed Convection in Enclosures,* Ph.D. Thesis, University of New South Wales, 1996.

179. W-J. Yang, S. Mochizuki, and N. Nishiwaki, *Transport Phenomena in Manufacturing and Materials Processing,* Elsevier, Amsterdam, 1994.

180. L. E. Doyle, C. A. Keyser, J. L. Leach, G. F. Scharder, and M. B. Singer, *Manufacturing Processes and Materials for Engineers,* Prentice-Hall, Englewood Cliffs, 1987.

181. J. A. Schey, *Introduction to Manufacturing Processes,* 2d ed., McGraw-Hill, New York, 1987.

182. S. Kalpakjian, *Manufacturing Engineering and Technology,* Addison-Wesley, Reading, Massachusetts, 1989.

183. J. Szekely, J. W. Evans, and J. K. Brimacombe, *The Mathematical and Physical Modeling of Primary Metals Processing Operations,* John Wiley, New York, 1988.

184. A. Ghosh and A. K. Mallik, *Manufacturing Science,* Ellis Horwood (John Wiley), Chichester, U.K., 1986.

185. R. Viskanta, "Impact of Heat Transfer in Industrial Furnaces on Productivity," in J. A. Rezies (ed.), *Transport Phenomena in Heat and Mass Transfer,* 2, pp. 815–838, Elsevier, Amsterdam, 1992.

186. R. Pritchard, J. J. Guy, and N. E. Connor, *Handbook of Industrial Gas Utilization,* Van Nostrand Reinhold, New York, 1977.

187. K. S. Chapman, S. Ramadhyani, and R. Viskanta, "Modeling and Parametric Studies of Heat Transfer in a Directly-Fired Batch Reheating Furnace," *J. Heat Treating,* 8, pp. 137–146, 1991.

188. K. S. Chapman, S. Ramadhyani, and R. Viskanta, "Modeling and Parametric Studies of Heat Transfer in a Directly-Fired Continuous Reheating Furnace," *Metal. Trans.,* 22B, pp. 513–534, 1991.

189. H. Ramamurthy, S. Ramadhyani, and R. Viskanta, "Modeling of Heat Transfer in Indirectly Fired Batch Reheating Furnace," in J. R. Lloyd and Y. Kurosaki (eds.), *Proceedings of the ASME/JSME Thermal Engineering Joint Conference 1991,* 5, pp. 205–215, ASME/JSME, New York/Tokyo, 1991.

190. H. Ramamurthy, S. Ramadhyani, and R. Viskanta, "Modeling of Heat Transfer in Indirectly Fired Continuous Reheating Furnace," in P. J. Bishop et al. (eds.), *Transport Phenomena in Materials Processing—1990,* HTD-Vol. 146, pp. 37–46, ASME, New York, 1990.

191. H. Ramamurthy, S. Ramadhyani, and R. Viskanta, "A Thermal System Model for a Radiant-Tube Continuous Reheating Furnace," *J. Mat. Eng. Performance,* 4, pp. 519–531, 1995.

192. H. Ramamurthy, S. Ramadhyani, and R. Viskanta, "A Two-Dimensional Axisymmetric Model for Combusting, Reacting, and Radiating Flows in Radiant Tubes," *J. Inst. Energy,* 67, pp. 90–100, 1994.

193. L. Zhao, G. de Vahl Davis, and E. Leonardi, "Convection in an Externally Fired Furnace," in G. F. Hewitt (ed.), *Proceedings of the 10th International Heat Transfer Conference,* Warwickshire, Brighton, 2, pp. 183–188, IChemE, Rugby, 1994.

194. R. B. Mansour and R. Viskanta, "Radiative and Convective Heat Transfer in Furnaces for Materials Processing," in S. I. Guceri (ed.) *Proceedings of the First International Conference on Transport Phenomena in Processing,* pp. 693–713, Technomic Publishing Co., Inc. Lancaster, Pennsylvania, 1992.

195. T. L. Bergman, M. A. Eftychiou, and G. Y. Masada, "Thermal Processing of Discrete, Conveyorized Material," in J. C. Khanpara and P. Bishop (eds.), *Heat Transfer in Materials Processing,* HTD-Vol. 224, pp. 27–34, ASME, New York, 1992.

196. Y. S. Son, T. L. Bergman, and M. T. Hyun, "Simulation of Heat Transfer in a Reflow Soldering Oven with Air and Nitrogen Injection," *ASME J. Electronic Packaging,* 117, pp. 317–322, 1995.

197. K. S. Chapman, S. Ramadhyani, and R. Viskanta, "Modeling and Analysis of Heat Transfer in a Direct-Fired Batch Reheating Furnace," in R. K. Shah (ed.), *Heat Transfer Phenomena in Radiation, Combustion and Fires,* HTD-Vol. 106, pp. 265–274, ASME, New York, 1990.

198. P. F. Sullivan, S. Ramadhyani, and R. Viskanta, "A Validation Study of Heat Transfer in a Direct-Fired Batch Reheating Furnace," in P. Cho and J. Quintiere (eds.), *Heat and Mass Transfer in Fire and Combustion Systems,* ASME HTD-Vol. 223, pp. 45–53, ASME, New York, 1993.

199. C. L. DeBellis, "Evaluation of High Emittance Coatings in a Large Industrial Furnace," in B. Farouk et al. (eds.), *Heat Transfer in Fire and Combustion Systems,* 250, pp. 190–198, ASME, New York, 1993.

200. I. Ohnaka and M. Shimaoka, "Heat Transfer in In-Rotating Spinning Process," in I. Tanasawa and N. Lior (eds.), *Heat and Mass Transfer in Materials Processing,* pp. 315–329, Hemisphere, New York, 1992.

201. R. Akiyoshi, S. Nishio, and I. Tanasawa, "An Attempt to Produce Particles of Amorphous Materials Utilizing Steam Explosion," in I. Tanasawa and N. Lior (eds.), *Heat and Mass Transfer in Materials Processing,* pp. 330–343, Hemisphere, New York, 1992.

202. D. P. Jordan, "Film and Transition Boiling," in T. F. Irvine, Jr. and J. P. Hartnett (eds.), *Advances in Heat Transfer,* 5, pp. 55–128, Academic Press, New York, 1968.

203. E. K. Kalinin, I. I. Berlin, and V. V. Kostiouk, "Transition Boiling Heat Transfer," in J. P. Hartnett and T. F. Irvine, Jr. (eds.), *Advances in Heat Transfer,* 18, pp. 241–323, Academic Press, New York, 1987.

204. M. Tajima, T. Maki, and K. Katayama, "Study of Heat Transfer Phenomena in Quenching of Steel (An Analysis of Cooling Curves Accompanied with Phase Transformation)," *JSME Int. J. Series II,* 31, pp. 98–104, 1988.

205. F. C. Kohring, "Water Wall Water-Cooling Systems," *Iron and Steel Engineer,* 62(6), pp. 30–36, 1985.

206. J. Filipovic, R. Viskanta, F. P. Incropera, and T. A. Veslocki, "Cooling of a Steel Strip by an Array of Round Jets," *Steel Research,* 65, pp. 541–547, 1994.

207. J. Filipovic, R. Viskanta, F. P. Incropera, and T. A. Veslocki, "Thermal Behavior of Steel Strip Cooled by an Array of Planar Water Jets," *Steel Research,* 63, pp. 438–446, 1992.

208. J. J. Derby, S. Brandon, A. G. Salinger, and Q. Xiao, "Large-Scale Numerical Analysis of Materials Processing Systems: High-Temperature Crystal Growth and Molten Glass Flows," *Computer Methods in Applied Mechanics and Engineering,* 112, pp. 69–89, 1994.

209. S. Brandon and J. J. Derby, "Internal Radiative Transport in the Vertical Bridgman Growth of Semi-transparent Crystals," *J. Cryst. Growth,* 110, pp. 481–500, 1991.

210. S. Brandon and J. J. Derby, "Heat Transfer in Vertical Bridgman Growth of Oxides: Effects of Conduction, Convection and Internal Radiation," *J. Cryst. Growth,* 121, pp. 473–494, 1992.

211. B. Cockayne, M. Chesswas, and D. B. Gasson, "Faceting and Optical Perfection in Czochralski Crown Garnets and Ruby," *J. Mater. Sci.,* 4, pp. 450–456, 1969.

212. J. Kvapil, B. Kvapil, B. Manek, R. Perner, R. Autrata, and R. Schauer, "Czochralski Growth of YAB:Ce in a Reducing Protective Atmosphere," *J. Cryst. Growth,* 2, pp. 542–545, 1981.

213. Q. Xiao and J. J. Derby, "The Role of Radiation and Melt Convection in Czochralski Oxide Growth: Deep Interfaces, Interface Inversion and Spiraling," *J. Cryst. Growth,* 128, pp. 188–194, 1993.

214. Q. Xiao and J. J. Derby, "Heat Transfer and Interface Inversion During the Czochralski Growth of Yttrium Aluminum Garnet and Gadolinium Gallium Garnet," *J. Cryst. Growth,* 139, pp. 147–157, 1994.

215. J. J. Derby, L. J. Atherton, and P. M. Gresho, "An Integrated Process Model for the Growth of Oxide Crystals by the Czochralski Method," *J. Cryst. Growth,* 97, pp. 792–826, 1989.

216. F. Roozeboom and N. Parekh, "Rapid Thermal Processing Systems: A Review with Emphasis on Temperature Control," *J. Vac. Sci. Technol. B.,* 8, pp. 1249–1259, 1990.

217. F. Roozeboom, "Rapid Thermal Processing: Status, Problems and Options after the First 25 Years," in J. C. Gelpy et al. (eds.), *Rapid Thermal and Integrated Processing II,* 303, pp. 149–164, Materials Research Society Symposium Proceedings, MRS, Pittsburgh, 1993.

218. C. Schaper, M. Mehrdad, K. Saraswat, and T. Kailath, "Control of MMST RTP: Repeatability, Uniformity, and Integration for Flexible Manufacturing," *IEEE Trans. Semiconductor Manufact.,* 7, pp. 202–219, 1994.

219. K-H. Lee, T. P. Merchant, and K. F. Jensen, "Simulation of Rapid Thermal Processing Equipment and Processes," in J. C. Gelpy et al. (eds.), *Rapid Thermal and Integrated Processing II,* 303, pp. 197–209, Materials Research Society Symposium Proceedings, MRS, Pittsburgh, 1993.

220. P. Y. Wong, B. D. Heilman, and I. N. Miaoulis, "The Effect of Microscale and Macroscale Patterns on the Radiative Heating of Multilayer Thin-Film Structures," in T. A. Ameel and R. O. Warrington (eds.), *Microscale Heat Transfer,* HTD-Vol. 291, pp. 27–34, ASME, New York, 1994.

221. P. Y. Wong, C. K. Hess, and I. N. Miaoulis, "Microscale Radiation Effects in Multilayer Thin-Film Structures during Rapid Thermal Processing," in *Rapid Thermal and Integrated Processing II,* J. C. Gelpy et al., eds., 303, pp. 217–222, Materials Research Society Symposium Proceedings, MRS, Pittsburgh, 1993.

222. P. Vandenabeele, K. Maex, and R. De Keersmaecker, "Impact of Patterned Layers on Temperature Non-Uniformity during Rapid Thermal Processing for VLSI Structures," in N. W. Cheung, A. D. Marwick, and J. B. Roberto (eds.), *Rapid Thermal Annealing/Chemical Vapor Deposition and Integrated Processing,* Materials Research Society Symposium Proceedings, 146, pp. 149–160, MRS, Pittsburgh, 1989.

INDEX

ABOUT THE EDITORS

Warren M. Rohsenow is professor emeritus of mechanical engineering and director emeritus of the Heat Transfer Laboratory at MIT. For his outstanding work in heat transfer, Dr. Rohsenow is the recipient of the Max Jakob Memorial Award and holds membership in the National Academy of Engineering.

James P. Hartnett is distinguished professor of mechanical engineering and founding director of the Energy Resources Center at the University of Illinois in Chicago. For his contributions to heat transfer, Dr. Hartnett has received the ASME Memorial Award and the Luikov Medal of the International Center for Heat and Mass Transfer.

Young I. Cho is professor of mechanical engineering in the Department of Mechanical Engineering and Mechanics at Drexel University, Philadelphia, Pennsylvania. Dr. Cho was the recipient of the 1995 University Research Award at Drexel University.